T0315405

FIBER OPTICS

ILLUSTRATED DICTIONARY

Advanced and Emerging Communications Technologies Series

Series Editor-in-Chief: Saba Zamir

The Telecommunications Illustrated Dictionary, Second Edition,
  Julie K. Petersen

Handbook of Emerging Communications Technologies: The Next Decade,
  Rafael Osso

ADSL: Standards, Implementation, and Architecture, Charles K. Summers

Protocols for Secure Electronic Commerce, Mostafa Hashem Sherif

Protocols for Secure eCommerce, Second Edition, Mostafa Hashem Sherif

After the Y2K Fireworks: Business and Technology Strategies,
  Bhuvan Unhelkar

Web-Based Systems and Network Management, Kornel Terplan

Intranet Performance Management, Kornel Terplan

Multi-Domain Communication Management Systems, Alex Galis

Fiber Optics Illustrated Dictionary, Julie K. Petersen

FIBER OPTICS

ILLUSTRATED DICTIONARY

JULIE K. PETERSEN

CRC Press
Taylor & Francis Group
Boca Raton London New York

CRC Press is an imprint of the
Taylor & Francis Group, an **informa** business

CRC Press
Taylor & Francis Group
6000 Broken Sound Parkway NW, Suite 300
Boca Raton, FL 33487-2742

First issued in hardback 2018

ISBN-13: 978-0-8493-1349-3 (pbk)
ISBN-13: 978-1-138-45575-7 (hbk)

Library of Congress Cataloging-in-Publication Data

Catalog record is available from the Library of Congress

Visit the Taylor & Francis Web site at
http://www.taylorandfrancis.com

and the CRC Press Web site at
http://www.crcpress.com

Preface

About This Dictionary

The reader might assume that the process of writing or using a fiber optics dictionary is dry and uninteresting, but that really isn't the case. Fiber optics is a vibrant field, not just in terms of its growth and increasing sophistication, but also in terms of the people, places, and details that make up this challenging and rewarding industry.

Fiber optics isn't as specialized as many people assume, either. Fiber optics forms the heart of the telephone industry, the nervous system of the computer network industry, and the organs of many medical, dental, experimental, and satellite technologies. That's part of the reason why this dictionary is so big. The Internet, the phone system, and wireless satellite systems are joined at the hip, with fiber optics landlines often supplemented by satellite links and vice versa.

Fiber optics is attracting attention from many different sectors. In Spring 2001, over 35,000 people from a wide variety of backgrounds attended a major international fiber optics conference. In spite of the inevitable peaks and slowdowns in the commercialization of any new technology, interest from professionals is growing and there are now thousands of training and certification courses for people who want to design, install, operate, and maintain fiber optic systems.

The Quest for Communication by Light

The fiber optics industry is very recent; most of the significant developments have occurred in the last 60 years. The application of fiber to underlying telecommunications infrastructures became important in the 1980s and the use of fiber spread to consumer products and local area networks by the late 1990s.

The history of fiber optics is based upon the efforts of many multitalented, tireless inventors, who traded social interactions for the thrill of discovery. These pioneers were passionate in their search for a way to communicate with light. Alexander Graham Bell was more excited about his Photophone, a light-based telephone, than almost anything he ever invented, even though it was a commercial failure. Bell recognized that he didn't have all the pieces of the puzzle to make it a viable technology and chose to move on, but that doesn't mean the Photophone was a bad idea; it just happened to be about 80 years ahead of its time.

The earliest pioneers recognized certain potentially ground-breaking properties of optical materials but weren't quite sure what made them work and, hence, were unable to fully harness their power. For example, in the 1600s, Rasmus Bartholin thoroughly described the doubly refracting birefringent properties of Iceland spar, a type of transparent calcite, but wasn't able to work out the mathematics. Later, both Wollaston and Nicol recognized Iceland spar could be assembled into new forms of prisms with special properties for controlling light, but it took many generations before scientists like Thomas Young began to unravel the mathematics that made this material so uniquely useful and applied that knowledge to describing the wavelike properties of light. No sooner had scientists become comfortable with the idea of light behaving as waves when Max Planck set the stage, in 1900, for a particle theory of light and Einstein elaborated and applied the new ideas in quantum dynamics, leading to our current understanding of the photoelectric effect. With the coming of the transistor and solid-state electronics, it was just a matter of time before smaller, less expensive fiber-based components could be constructed.

While optical science was evolving, the fabrication of pure glass was advancing as well. Many optical technologies in communications originated in much the same way as tongue depressors and penlights – doctors and dentists began using them to peer down people's throats.

Scientists have long suspected that glass and light had capabilities far greater than anything yet imagined, but they weren't sure how to combine the two and still keep the signal within the lightguide. In terms of communications applications, this was a big road block.

The idea of "bending" light isn't new; Colladon and Tyndall demonstrated it in the mid-1800s by directing light inside an arc of water. But the experiment remained an impractical curiosity until glass rods were shown to refract light in the same way. Even so, the phenomenon of refraction needed to be better understood before glass rods could be turned into effective fiber optic filaments.

By the middle of the 20th century, a few innovative scientists began coating glass with other materials, following the lead of Nicol, who had bonded together two pieces of Iceland spar with Canada balsam to create the Nicol prism. The new prism took advantage of the lower refractive index of Canada balsam and the birefringent properties of Iceland spar to split a beam of light and direct one beam out the side while the other continued forward. When the characteristic of light to refract off lower refractive index materials was finally harnessed in the form of cladding, fiber optics became a practical reality. From that point on, the quest for ideal proportions, purer glass, and more powerful, controllable light sources spurred the industry onto the next level of evolution. In the 1950s, the development of lasers provided the essential energy source that finally launched the optical communications industry.

With fiber optics now widely deployed, has it become just another ubiquitous technology, like telephone poles and automobiles? Perhaps in some ways this is true – cables for local area networks can be readily purchased on the Internet and optical couplers cost only a few dollars. But that doesn't mean the industry has reached its limits or that the technology is no longer dynamically evolving. Fiber-based networks are still in their infancy and the exploitation of the properties of light is still young and full of promise. In addition, there are many areas of interest in which problems of installation and deployment are tackled in innovative ways. For example, the city of Houston has signed an agreement to use a high-tech robot to navigate the city's sewers to connect hundreds of premises to the fiber optic broadband networks to complete the "last mile" between the populace and the fiber backbone.

Fiber optics is also becoming important in the signage, lighting, and medical industries. Lightweight, inexpensive colored light-guides, side-emitting filaments, linelights, and pointlights all have exciting applications in architecture, interior design, industrial safety, marketing, fine arts, and crafts. Hobbyists are using fiber filaments to light scale models and train sets. Inventive developers have created fiber optic "fabric" in which the fiber optic filaments are bent to deliberately release light at the joints where the weft crosses over the warp. Doctors and dentists use fiber optics for imaging and surgery. Wherever light is needed, there's a possibility a fiber optic filament can provide it.

Purpose of the Dictionary

It is the aim of this book to fill a gap in the literature on fiber optics. There is only one significant fiber optics dictionary on the market and it was last published in 1998. Many advances have occurred since that time that deserve to be documented. There is also a need for a text priced within the range of university students and technicians taking their certification training. This dictionary can meet that need as well.

Audience for the Dictionary

The *Fiber Optics Illustrated Dictionary* is suitable for a wide variety of beginning professionals in fiber optics, as well as students and instructors. It will also be of interest to professionals in other fields who want to get a beginning to intermediate introduction to optical technologies. The book covers historical antecedents, network protocols for telephone and computer networks, satellite technologies, telephone terminology, basic physics concepts, and units of measure important in optics. It also explains many math and light-refracting concepts through a combination of words and pictures so that concepts that are hard to understand at first are explained in two ways.

This book does not attempt to duplicate the information in the FOLDOC online dictionary or the Federal Standards documents. These dictionaries are readily available and searchable on the Internet and are well documented in Martin Weik's dictionary. Instead, the *Fiber Optics Illustrated Dictionary* takes a current and comprehensive look at the fiber optics field and the various applications of fiber optics, rounds out the picture with some introductory physics and fusion splicing information, and presents it in a form that is illustrated, cross-referenced, and enhanced by historical biographies and URL addresses for major not-for-profit and educational sites on the Web.

I hope you enjoy using the book as much as I enjoyed preparing it (despite the long hours and endless search for accurate and often elusive information).

I am indebted to the hard work and enthusiasm of the professionals at CRC who helped bring it to fruition, including Jerry Papke, who contributed the original concept, Chris Andreasen and the proofreading staff, who labored over many pages, and Jamie Sigal, Nora Konopka, and the folks in the production and marketing departments who all answered questions and moved the project along. Thanks also to Dawn Snider for her excellent interpretation of the cover.

Julie K. Petersen

About the Author

Julie K. Petersen is a technology consultant, author, educator, and outdoor enthusiast, and readily admits to being a technophile. Her whole house is wired with computer and video links, both inside and out, and there's rarely a day when she isn't configuring some new piece of equipment to broadcast over a wireless transceiver. Since TRS-80 computing days, she's been tweaking and fixing her own equipment and talks about configuring a wearable computer to interface directly with GPS data on the Internet.

"The technology is already here; it's just a matter of putting all the pieces together. What you do is take a head-worn display that projects an image on your retina with a laser beam that is eye-safe; such systems already exist. Then you have a body-worn GPS sensor with an interface and wireless link to the Internet that goes through a geographical server. The server matches your GPS coordinates with Web sites that offer information on maps, restaurants, nearby movie theaters, libraries, schools, etc. You could have a profile online for your preferences, and the display would change as your location changes. The process would be transparent, like a third eye, similar to the navigational images a fighter pilot sees projected over the landscape on the jet's transparent canopy, except even more natural. I've named it the *G-Eye*™ for geographic eye or GPS eye.

The image projected on the user's retina by the *G-Eye* system would be tailor-made to the viewer's preferences. It doesn't have to be a one-way communication either. If the wearer were a professional on the job, like a newscaster or research scientist, he or she could be wearing sensors with fiber optic faceplates to sense body changes or changes in the surrounding environment, pressure, temperature, light levels, etc., that could be fed back to the computer network to act as a hands-free 'body interface' or a roving human sensing system. The possibilities are endless. Some people may see this as far-fetched, the idea of the human organism as a sort of sensory node on a distributed network, but young people readily understand and adapt to concepts such as this, especially if the new technology promotes or enhances social interactions, which this obviously could.

The *G-Eye* would have been impractical a few years ago. Compact diode lasers, sensors that could respond quickly, and high-bandwidth network links weren't yet sufficiently developed. The limited capacity of the network infrastructure would have made such a two-way system impractical, but the new broadband fiber optic networks have astonishing speed and capacity, enough to individually outstrip the current collective traffic on the Internet. It's feasible to imagine the entire human populace interconnected through a combination of wired and wireless optical links and satellites.

Now equip the *G-Eye* with an optional digital video cam and microphone and you have an integrated network and digital phone/videoconferencing system that travels with you, instead of a half-dozen different, unconnected, bulky systems. If there are interruptions in the network radio link, then you could carry a length of fiber optic cable that jacks into the nearest cafe or network vending machine for a clean wired link. Fiber optic cables are lighter and more robust than most people realize. If there's an emergency, the problem of clogged airwaves (familiar to traditional cell phone users) could be alleviated by people having these pocket cables integrated with their *G-Eye* systems as a backup to wireless connections. There might even be an all-optical solution in certain circumstances. Imagine free-air optical transceivers mounted on buildings (like small satellite receiving dishes) that people jack into with optical modems, somewhat like a two-way infrared remote control. That way, if you're sitting in a park or at a sidewalk cafe, you could aim at a transceiver to maintain connectivity. People have a tendency to think in terms of single solutions when often the best solution is a variety of options. Why put just forks in your cutlery drawer when there's room for knives and spoons as well?

Unfortunately, I haven't had time to build the *G-Eye* system. The time demands of writing a comprehensive dictionary on the subject of fiber optics, which changes even as it is documented, is considerable and my spare time is almost nonexistent, but I'm fascinated by the depth and breadth of applications people are developing for optical waveguides, faceplates, and sensors and I'm sure there are many more surprises in store."

The author lives in the Pacific Northwest and enjoys reading, music, film, strategy games, and interesting cuisine. She advocates the use of technology to enhance the quality of life and solve human problems and especially encourages scientists and engineers to apply technology in ways that help reduce rather than extend the work week.

How to Use the
Fiber Optics Illustrated Dictionary

General Format   There are two sections to this reference: (1) a main alphabetical body, with numeral entries following Z and (2) several appendices with various charts, an extended section on ATM, a quick lookup acronym dictionary, and a timeline of telecommunications inventions and technologies.

Entries   Dictionary entries follow a common format, with the term or phrase in **boldface**, followed by its abbreviation or acronym, if applicable. Pronunciation is included in cases where it may not be obvious. Alternate names (e.g., William Thompsom, a.k.a. Lord Kelvin) are cross-referenced. The body of the entry is included next, with multiple definitions numbered if there are several meanings for a term. Finally, where appropriate, there are cross-references, RFC listings, and URLs included at the end, in that order.

Abbreviations   In many cases, the term and its abbreviation are described together so the reader doesn't have to look up abbreviated references to understand a particular entry; for example, cathode-ray tube will often be followed by (CRT) and Federal Communications Commission by (FCC) so the words and their commonly used abbreviations become familiar to the reader.

Web Addresses   Web addresses based upon Uniform Resource Locators (URLs) are listed for nonprofit, not-for-profit, charitable, and educational institutions and, in a few rare instances, for commercial enterprises with particular relevance for telecommunications or with substantial educational content on their Web sites. For the most part, commercial URLs are not included. If the address isn't listed, it can often be guessed (http://www.companyname.com/) or otherwise easily located through Web search engines listed in Appendix D.

RFCs   Request for Comments (RFC) documents are an integral part of the Internet, and extremely important in terms of documenting the format and evolution of Internet protocols and technologies. For this reason, RFC references are listed with many of the Internet-related references and can be found in numerous RFC repositories online. There is also a partial list of significant or interesting RFC documents listed according to category in Appendix F.

Diagrams and charts   Illustrations are included as close to the related definition as was possible in the space provided. Extensive listings of the various ITU-T Series Recommendations are included in almost every chapter because they are the standards upon which most Internet technologies, telecommunications standards, and commercial products are built. Charts are usually included on the same page as the related definition or the one following.

term or phrase —
abbreviation or acronym —
pronunciation —

> **Bootstrap Protocol** BOOTP. (*pron.* boot-pee) An IP/UDP client/server means of storing and providing configuration information. BOOTP evolved in the ARPANET days to allow diskless client machines, and other machines which might not know their own Internet addresses, to discover the IP address, the address of a server host, and the name of

definition —

> a file to be loaded into memory and executed. This is accomplished in two phases: address determination and bootfile selection; and file transfer, typically with TFTP. This has since evolved into Dynamic Host Configuration Protocol (DHCP). See

cross-references —

> Address Resolution Protocol, Dynamic Host Configuration Protocol, Reverse Address Resolution

Request for Comments reference number —

> Protocol; RFC 951.

Web address (URL) —

> http://www.urlgoeshere.org/

Contents

Alphabetical Listings 1

Numerals 1043

Appendices 1049

   A. Fiber Optics Timeline 1050

   B. Asynchronous Transfer Mode (ATM) 1052

   C. ITU-T Series Recommendations................... 1055

   D. List of World Wide Web Search Engines 1056

   E. List of Internet Domain Name Extensions 1057

   F. Short List of Request for Comments (RFC) 1059

   G. National Associations 1062

   H. Dial Equivalents, Radio Alphabet, Morse Code,
Metric Prefixes/Values......................... 1066

   I. ASCII Character and Control Codes................ 1067

α 1. *symb.* alpha, the first letter in the Greek alphabet. 2. *symb.* angular acceleration. 3. *symb.* angle, in geometry. Along with β, often used in geometric diagrams to designate the angle of incidence or refraction or other angles associated with light paths.

a 1. *symb.* acceleration. See Acceleration 2. *symb.* anode. See anode. 3. *abbrev.* area. 4. *symb.* atomic mass. 5. *abbrev.* atto-. See atto-.

A 1. *symb.* acoustic velocity. See acoustic velocity. 2. *symb.* ampere. See ampere. 3. *symb.* gain. See gain.

A & A1 leads See A/A1.

A & B Numbers 1. Designations for two of the wireless communication service categories available through Inmarsat satellite relays. Inmarsat A & B services are commonly used for ship-to-shore communications. The Inmarsat *A & B Numbers* vary depending upon the ship and the selected satellite, but include voice, facsimile, and data lines. See Inmarsat for a chart of service categories. 2. In mobile radio systems in general, but especially cellular, the *A Number* is a designation for the originating call point, signaling towards the network, and the *B Number* is the destination or answering point.

A & B bit signaling In communications networks, control or status information about the communications line itself may be interspersed with the data content that is being transmitted through that line. This is a form of in-band signaling. A & B bit signaling is a technique of inserting signal state information into particular bits at intervals in the data transmission, thus robbing a certain number of bits from the total

transmission. For example, *A bits* are used in voice communications implemented over T1 superframe (SF) networks to indicate outbound call signaling, with *B bits* as mirrors to the A bits. Through bit robbing, the A and B signal bits are carried in each 6th and 12th frame, respectively, of each of the 24 T1 subchannels. The types of supervisory information contained in these signal bits is relevant to switched voice or switched data services, including ring, busy, off-hook, and on-hook states. In extended superframe (ESF), A, B, C, and D bits may be robbed from the 6th, 12th, 18th, and 24th frames.

In some telephony systems, tone signaling is converted to A & B bit signaling for interoperability.

There is a trade-off when bits are robbed. Since the available bits are not all used for data, the total throughput is less when measured over time. However, for less demanding voice communications, for example, the difference in speed and quality of the signal is not subjectively apparent to the listener. Diagnostic T1 channel decoders typically show the A and B bit signaling status, along with other alarm, frame loss, or error conditions.

Newer T1 systems based upon bipolar eight-zero substitution (B8ZS) don't use this bit-robbing technique. See B8ZS.

A battery 1. A low voltage battery historically used to provide current to filaments or cathode heaters in electron tubes, now commonly used for small electronic appliances such as cameras, calculators, pen lights, etc. See battery. 2. A historic nonrechargeable

FCC-Designated Communications Frequency Blocks					
Block	Blk. Size	Frequency	Paired Frequency	Notes	Date
A Block	30 MHz	1850-1865 MHz	1930-1945 MHz	MTA Broadband PCS	1994-1995
B Block	30 MHz	1870-1885 MHz	1950-1965 MHz	MTA Broadband PCS	1994-1995
C Block	30 MHz	1895-1910 MHz	1975-1990 MHz	BTA Broadband PCS	1995-1996
D Block	10 MHz	1865-1870 MHz	1945-1950 MHz	BTA Broadband PCS	1996-1997
E Block	10 MHz	1885-1890 MHz	1965-1970 MHz	BTA Broadband PCS	1996-1997
F Block	10 MHz	1890-1895 MHz	1970-1975 MHz	BTA Broadband PCS	1996-1997

wet cell called an *air cell*, with carbon electrodes providing an average power of 2.0 volts. See talk battery.

A Block A Federal Communications Commission (FCC) designation for a Personal Communications Services (PCS) nonwireline license granted to a telephone company serving a Major Trading Area (MTA) that grants permission for broadcasters to operate at certain FCC-specified frequencies. See band allocations. See FCC-Designated Frequency Blocks chart.

A cable A 50-pin data cable commonly used for SCSI peripheral connections. See A connector.

A carrier *alternate carrier*. A Federal Communications Commission (FCC) designated nonwireline competitive telephone cellular service carrier which is not the established local wireline carrier (B carrier). See B carrier.

A channel In a system with two or more audio channels (e.g., stereo), the designation for the left audio channel, usually connected to the left speaker or microphone. Audio cables are sometimes color-coded to aid recognition, with white conventionally used for the left channel and red for the right.

A connector An ANSI-standardized 50-pin electrical data connector for interconnecting SCSI devices such as hard drives, cartridge tape drives, etc. SCSI and SCSI-2 device connectors are physically different, to prevent interconnection, but are electrically compatible so that they can be daisy-chained to coexist on the same bus. For some devices, manufacturers provide *P connector* (68-pin) to *A connector* adaptors to enable newer peripherals to be used in older computers. There are also a few P connector-like devices made with 50-pins so that a device can be connected without an adapter. See P connector.

A Connector – 50 Pins

The "slot-style" A connector, also known as a Centronics connector, was popular for years for SCSI devices. Newer formats have mostly superseded it, though Centronics-style connectors are still common on older dot matrix printers with parallel connections.

A interface See air interface.

A law See A-law encoding.

A link See access link.

A minus, A- The negative polarity of a voltage source, for example, the negative terminal of an A battery, often color-coded as black.

ITU-T A Series Recommendations – Examples		
Recom.	Year	Description
A.1	2000	Work Methods for Study Groups of the ITU Telecommunication Standardization Sector
A.2	2000	Presentation of contributions relative to the study of questions assigned to the ITU-T
A.3	1996	Elaboration and presentation of texts and development of terminology and other means of expression for Recommendations of the ITU Telecommunication Standardization Sector
A.4	2000	Communication process between ITU-T and Forums and Consortia
A.5	2000	Generic procedures for including references to documents of other organizations in ITU-T Recommendations
A.6	2000	Cooperation and exchange of information between ITU-T and national and regional standards development organizations
A.7	2000	Focus Groups: Working methods and procedures
A.8	2000	Alternative approval process for new and revised Recommendations
A.9	2000	Provisional working procedures for the Special Study Group on IMT-2000 and Beyond
A.11	2000	Publication of ITU-T Recommendations and WTSA proceedings
A.12	2000	Identification and layout of ITU-T Recommendations
A.13	2000	Supplements to ITU-T Recommendations
A.23	1996	Collaboration with the International Organization for Standardization (ISO) and the International Electrotechnical Commission (IEC) on information technology. Annex A is the Guide to ITU-T and ISO/IEC JTC1 cooperation.
A.30	1993	Major degradation or disruption of service
		There are also supplements available on quality and interoperability.

A plus, A+ The positive polarity of a voltage source, for example, the positive terminal of an A battery, often color-coded as red.

A port In a Class A, dual-attachment (dual ring) Fiber Distributed Data Interface (FDDI) token-passing network, there are two physical ports, designated PHY A and PHY B. Each of these ports is connected to both the primary and the secondary ring, to act as a receiver for one and a transmitter for the other. Thus, the A port is a receiver for the primary ring and a transmitter for the secondary ring. The dual ring system provides fault tolerance for the network.

Port adaptors can be equipped with optical bypass switches to avoid segmentation, which might occur if there is a failure in the system and a station temporarily eliminated.

FDDI ports can be connected to either single mode or multimode fiber optic media, providing half duplex transmissions. LEDs are commonly used on port adaptors as status indicators. Optical bypass switches may in turn be attached to the port adaptors. See dual attachment station, Fiber Distributed Data Interface, M port, optical bypass, port adaptor.

A Series Recommendations A series of ITU-T recommended guidelines for administration, working methods, and communication of information by personnel and working groups. They are available for purchase from the ITU-T and many in the A Series are downloadable without charge from the Net. Since ITU-T specifications and recommendations are widely followed by vendors in the telecommunications industry, those wanting to maximize interoperability with other systems should be aware of the information disseminated by the ITU-T. A full list of general categories is listed in Appendix C and specific series topics are listed under individual entries in this dictionary, e.g., B Series Recommendations. See ITU-T A Series Recommendations chart.

A-1 time An atomic time scale established by the U.S. Naval Observatory. The origin is set at 1 January 1958 zero hours Universal Time with a second unit equal to 9,192,631,770 cycles of cesium at zero field. See atomic clock, Universal Time.

A-law encoding A pulse code modulation (PCM) coding and companding scheme used outside North America as the CEPT standard. A-law is commonly used for analog-to-digital conversion for encoding speech by sampling the audio waveforms and applying logarithmic quantization. This is important in digital telephone communications. Since speech sounds have a fairly broad dynamic range in terms of linear encoding, A-law encoding reduces the dynamic range to reduce signal distortion and increase coding efficiency. See E carrier, Mu-law encoding, pulse code modulation, quantization, sampling.

A-scope, R-scope A specialized radar tracking scope for indicating the range of objects detected, displaying all targets as illuminated vertical blips, scanning repeatedly from left to right. See B-scope.

A/A1 Conductor leads in key telephone systems to implement hold functions. When a line is placed off-hook, the A lead is shorted or bridged to the A1 lead to put the line on hold. A similar concept is the MB/MB1 bridge that puts the affected line into an unavailable busy state. The bridged states may be indicated by LEDs, depending upon the phone design. Line sensing products that sense A and A1 lead controls on key telephone lines are of interest to firms that make heavy use of telephone services and automation, such as telemarketers. Line sensors can detect current and line status and, if desired, activate a relay to allow dialers and other devices to be interfaced with key systems. They may also provide key phone control leads for telephone equipment not using key system units.

A/B port See A port.

A and B Ports in FDDI Dual Attachment Station (DAS)

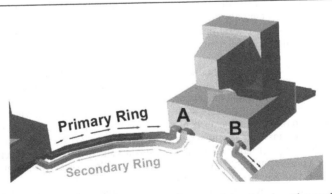

A & B or combined A/B port interface cards resemble Ethernet peripheral cards, with a small circuit board and ports facing the outside of a computer. However, unlike Ethernet, each has dual ports for accommodating the fault-tolerant dual-ring structure of FDDI networks and the connections are optical.

The ports may be single-mode (usually with ST- bayonet-mount ports) or multimode (with rectangular friction connectors). Note, once a port card is installed, it may emit laser radiation even if no cable is connected. Never peer into the port; the beam may be invisible and eye damage is possible.

A/B switch 1. A dial or switch with two settings for controlling sources of input and output to a circuit. Typically an A/B switch enables the user to mechanically complete a circuit between one of two inputs and one output or between one input and one of two outputs. A/B/C and A/B/C/D switches are also common. See A/B switchbox, switcher. 2. A setting on various appliances allowing a user to select between two operating modes, such as wireline or nonwireline, or between two optional frequencies. 3. In cellular communications, many new transceivers have an A/B switch that enables the user to select between a wireless or wireline connection when roaming.

A/B Switchbox – 25-pin Connections

A/B switchbox front and back. Passive switchboxes are commonly used to interconnect computers with various peripherals. For example, a serial cable leading to a modem could be plugged into the input connection, and A and B could each be connected to a different computer to share a modem.

Another configuration is to input a computer connection and attach A to a modem and B to a printer, so that a computer with one serial port can alternately use two peripherals.

A/B switchbox A very common, usually passive, connection-routing device selectable by a switch and providing receptacles or sockets for various connectors. Most inexpensive switchboxes provide passive, mechanical routing for low-voltage electrical circuits. More advanced switchboxes may provide automated switching or digital switching services. Mechanical switch settings are usually selected by a dial, a lever, or buttons. A/B switches are commonly used with video circuits and computer peripheral devices, though they are generic to almost any electrical device where line resources are shared.

In computing, A/B switchboxes help manage shared resources. They can be used to switch a serial communications line between a printer and a modem or facsimile machine, or between different printers, such as a laser printer and pen plotter. Serial boxes commonly have 25-pin D connectors, one for the input, which may be from the computer, and two for the output, which may be a printer and modem (or vice versa). A gender changer or converter (e.g., 9-pin) is sometimes needed to connect the selected cable. A/B/C and A/B/C/D switchboxes are also common. A *crossover* switchbox is similar to a straight switchbox, but provides multiple input and output combinations, and usually has four or more ports on the back for attaching the input and output connectors. *Switchers* are similar to switchboxes, and are fre-

quently used in live broadcasts and video editing to select among various video sources (cameras, VCRs) and computer-generated signals. Video switchers (sometimes called selectors) typically use RCA and BNC connectors to attach standard video cables. More recent video switchers may also have S-VHS ports. See switcher.

A/D 1. analog/digital. 2. analog to digital.

A/D conversion This term is used rather loosely to refer to both analog-to-digital and digital-to-analog conversion mechanisms (often because a transmission link performs both functions, one at each end). Technically, A/D conversion is the translation of analog to digital signals, often for transmission over data networks (e.g., voice transmitted over digital phone lines) or for sampling by computer applications such as speech or voice recognition software or music sequencing and editing software. The advantage of converting to digital format is that many types of processing can be applied to the data, including image or sound editing, sequencing, compression, encryption, error-correction, and more. Some common A/D conversion applications include:

- Analog sound capture through a microphone with the data being digitized for use over a digital mobile communications link or for use with a computer to capture music or voice as digital samples for later processing or playback.
- Analog image capture for transmission to a computer or videoconferencing unit for digital transmission over the Internet. See audiographics.
- The use of a computer modem at the receiving end of a traditional analog phone line for modulating analog telephone signals into digital serial transmissions for use by the computer processing the data.

See A-law, codec, sigma-delta modulation.

A/UX A 32-bit Unix operating system designed for use on Apple Macintosh computers in addition to or in place of the Apple operating system. A/UX is derived from AT&T's UNIX, BSD, with full POSIX compliance and System V Interface Definition (SVID) compliance. A/UX provides The X Windows System, sh, csh, and ksh. A/UX is sometimes also used to refer to the Amiga/UNIX OS.

A/V audio/visual. An abbreviation that has been used colloquially for a long time to refer to a wide variety of audio/visual media and devices, including film projectors, video tape players, laserdisc players, televisions, and just about any educational or entertainment broadcast or playback unit that provides both sound and images.

A/V switch A device that enables various audio/visual sources to be selected as needed. An A/V switch is particularly useful in situations where space or cost limits the available resources as when monitors, speakers, or other components are shared among multiple inputs. A/V switches are common in video editing studios and are now becoming common in home entertainment centers, that is, consumer audio/visual systems that allow selection between a variety of services or components such as satellite or cable television, DVD players, VCRs, etc.

AA See Automated Attendant.

AAA See authentication, authorization, and accounting.

AAAC all aluminum alloy cable. See ACSR.

AAAI See American Association for Artificial Intelligence.

AAAS See American Association for the Advancement of Science.

AABS See Automated Attendant Billing System.

AAC 1. See abbreviated address calling. 2. See Aeronautical Administrative Communications.

AAL ATM adaptation layer. See asynchronous transfer mode, and see the appendix for several pages of extended definitions and diagrams.

AAP 1. See Advanced Adaptive Protocol. 2. See applications access point.

AAPI See Audio Applications Programming Interface.

AAPT See American Association of Physics Teachers.

AAR See automatic alternate routing.

AARP See AppleTalk Address Resolution Protocol.

ab- A prefix commonly used with names of practical electrical units in the centimeter-gram-second (CGS) electromagnetic system, e.g., abampere.

ABAM An older Western Electric (now Lucent Technologies) cable designation for 22-AWG, 110-ohm, individually shielded, twisted pair cable that is typically used in central office trunk line, circuit line, T1, and T1 to E1 channel service installations. Newer adaptations of ABAM are often listed by vendors as ABAM/T1 cable. For T1, ABAM has a drive capability of up to about 200 meters and a loss of about 0.4 dB/30 meters. A higher gauge fuse cable is sometimes used in conjunction with ABAM for aerial installations. See category of performance for newer cable types. See fuse cable.

abampere, ab-ampere In the centimeter-gram-second (CGS) system, an absolute unit for current. Since the abampere is often too large for practical convenience, current is described instead in terms of amperes (one-tenth of an abampere). See ampere.

abandoned call See call abandons.

abandoned call cost An economic calculation to estimate the amount of revenue lost. Abandoned call cost estimates are primarily used by businesses whose customers order products or services through the telephone, or whose inquiries lead to sales later on. It's impossible to know how many of the calls would have generated revenue and how many would have been completed later, but business owners may benefit from rough estimates based on the number of abandoned calls times the percentage of anticipated sales resulting from those calls. See call abandons.

Abbe condenser A simple type of two-lens condenser invented by Ernst Abbe. It is used in photomicrography, where sufficient lighting is important. The condenser is located below the stage of a microscope so it can collect, direct, and spread light up onto the

object being examined and recorded. It aids visibility in high magnification environments.

Abbe number (*symb. – v*) A quantification of dispersion in an optical medium (sometimes called *Abbe constant* or *optical constant*). The numeric quantity is related to the index of refraction of a wavelength within a medium. It is a common dispersion index that is used along with a refractive index to describe the properties of commercial optical products and materials. The higher the Abbe number, the less variation there is in the index of refraction associated with differing wavelengths and the less the colors are dispersed. This is generally a desired property as there is reduced chromatic aberration. Plastics tend to have lower Abbe numbers than glass.

The Abbe number may be calculated by using Fraunhofer line index of refraction values and generally cluster between 20 and 70 in relation to index of refraction values of between 1.46 and 1.88. The following examples illustrate Abbe numbers common in the optical industry.

Medium	Approx. Number
SF11 glass	25.8
SF5 lead glass	32.3
BaF13 glass	45.0
nonachromatic lens	57.2
doped glass	57.6
borosilicate (BK7) glass	64.1
fused quartz	67.6
glass-ceramic	67.6
fused silica	67.7
synthetic fused silica	67.8

See Abbe condenser; Abbe, Ernst; ICO Prize.

Abbe refractometer A commercial instrument for measuring refractive indexes and mean dispersion in optical materials such as glass and translucent liquids and solids. It can also be used to measure purity, concentration, and dispersion in fluids. Depending on the manufacturer, Abbe refractometers range from analog to digital and from palm-sized to desktop models. They may be designed for white light sources or monochromatic light sources. See index of refraction, spectrometer.

Abbe, Ernst Karl (1840–1905) A German mathematician and physicist who began working at Zeiss Fabrications in 1866 and later became an owner. He developed a number of optical theories and invented a variety of optical condensers and metering instruments. See Abbe condenser, Abbe number.

abbreviated address calling AAC. In data network information routing, calling an address with fewer than the normal number of characters, usually from a table or file in which abbreviated address codes are stored. Similar in concept to speed dialing or abbreviated dialing on phone networks.

abbreviated dialing AD. 1. A feature of a phone which allows a short dialing sequence to replace a longer one. The abbreviated sequence can be programmed and associated with a longer number; then, when the shorter sequence is dialed, the system connects to the associated phone number. Also known as speed dialing. 2. A priority telephone service over special grade circuits, in which two or more subscribers can connect calls with fewer than usual dial tones.

ABC 1. arbitration bus controller. 2. See Atanasoff-Berry Computer. 3. automatic bass compensation. A circuit that increases the amplitude of bass notes to create more natural sound at low volumes. Used especially for playing back music recordings. 4. automatic bias control. See bias. 5. Automatic Bill Calling. A billing method for coin phone calls that is being superseded by calling card billing. 6. automatic brightness control. A circuit that senses ambient light levels and adjusts a display device automatically in order to optimize brightness levels for the viewer.

ABCD bits In network systems, a method for signaling using robbed bits, which provide in-band status information. The number of bits robbed depends upon the system. In Extended SuperFrame systems, four bits, designated ABCD, are utilized. See A & B bit signaling, Extended SuperFrame, robbed bits.

ABEC Alternate Billing Entity Codes. See Alternate Billing Services.

abend *ab*normal *end*. Abnormal or premature termination of a task or process, one that cannot be handled by available error recovery mechanisms. An undesired abend may cause the program or operating system to freeze or crash.

In workstation computers, abend problems with applications software are usually handled by the operating system so that the system itself does not crash, and there are usually mechanisms for killing individual processes that are locked or hung so that other processes are not affected. System-level abend problems on well-tuned networks are actually relatively rare. Some, not all, of the microcomputer single-tasking systems, and less robust task-switching or multitasking systems, experience abend problems that may require a system reboot. See abort.

aberration 1. Deviation from expected shape, behavior, or path. 2. Failure of an image to coincide point-by-point with its original, as in a television image or facsimile. 3. In optics, deviation of a viewed, transmitted, or projected image from its original, often due to limitations in optical components such as lenses, transmissions media, etc. Optical aberrations may include chromatic aberration, image distortion, curvature, astigmatism, and others. See astigmatism.

ABF air-blown fiber. See blown fiber.

ABIST See autonomous built-in self test.

ablation 1. Removal of a part. 2. The process of removing parts, such as small holes, grooves, or pits in order to encode information on a medium. Many computer storage media are recorded by ablating thin layers of plastic or metal, e.g., optical media such as compact discs.

ABM See asynchronous balanced mode.

ABME asynchronous balanced mode extended.

abnormal 1. Deviating from the normal, average, or expected. 2. A state, operation, or physical configuration that does not fit within expected, practical, or desirable norms.

abnormal propagation In broadcast transmissions, undesired influences from atmospheric or ionospheric changes that interfere with signal integrity. Terrestrial impediments, unplanned movement, and reflective interference may also cause the abnormal propagation of transmission signals. In fiber optic transmissions, scratches or breaks in the tiny fibers can cause the laser light beams to diverge from their expected paths, causing abnormal signals. In computer networks, on a larger scale, data files, mail messages, viruses, or other communications may abnormally propagate through a system in unexpected quantities or directions due to accidental or deliberate manipulation of headers and routing information.

abort 1. Stop prematurely or abruptly, cut off in miduse or transmission. 2. To terminate the transmitting or receiving of a message in progress. 3. To stop a software program or process in progress. An abend may be one type of abort, but *abort* more often signifies a situation in which a process is cleanly or voluntarily terminated without compromising system operating functions. 4. To terminate user access through a network or during a login, usually due to detection of unauthorized access or tampering.

abort sequence 1. A series of processes, functions, states, or steps leading to an abrupt end to the current function or transmission. Abort sequences may be safety mechanisms or a convenience to end a process that was initiated unintentionally (or which isn't behaving in the desired way). 2. At the algorithmic or network protocol level, a pattern of sequential data that signals that an abort should be initiated. Abort sequences may be specifically defined for certain systems. They may restore a previous state or abort in such a way that current work is minimally disturbed. Sometimes there are established applications or hardware procedures for initiating an abort sequence. It is important to design abort sequences so they cannot be accidentally initiated and so they are not initiated by data sequences that unintentionally resemble abort sequences.

Above 890 decision A 1959 decision of the Federal Communications Commission (FCC) granting permission for private construction and use of point-to-point microwave links. Thus, private companies, especially in remote locations, could utilize frequencies above 890 Mhertz for communications with oil rigs, power plants, gas pipelines, research stations, etc. The decision came about partly because of changes in technology, which made it less expensive and easier to use the higher frequency ranges for communications. This resulted in pressure to make these capabilities more widely available. Microwave Communications Inc. (MCI) was the first private commercial carrier service to take advantage of the Above 890 decision. See Telecommunications Act of 1996.

ABR 1. See available bit rate, cell rate. 2. See autobaud rate.

abrasion resistance A quality of a material to resist surface wear and tear during fabrication, installation, or use. Many rubbers, plastics, and metals are treated to increase their abrasion resistance. Network cables are often wrapped in a variety of gels, synthetic insulators, and metal sleeves to prevent abrasion, especially those used in harsh environments (e.g., deep sea installations). Neoprene and polyurethane are used for outdoor fiber optics cables, for example, while polyvinylchloride (PVC) is suitable for indoor cables.

Zirconia, a type of ceramic made from powder, is a strong, hard substance that is resistant to abrasion and other environmental degradation and thus is favored for fiber optic ferrules.

Abrasion resistance is quantitatively expressed in various ways, depending upon the industry and the type of material. See zirconia.

abs *abbrev.* absolute value. See absolute value.

ABS See Alternate Billing Services.

abscissa Conventionally the horizontal axis or X-axis in a Cartesian coordinate system.

Absent Subscriber Service, Vacation Service A service offered by local telephone carriers that retains the absent subscriber's phone number at a reduced rate so the subscriber will get the number back later, and that provides a standard recorded message to any people who call while the subscriber is away.

absolute 1. Relating to fundamental constants, phenomena, or other measurable, reliable, or stable parameters that can be used as a reference for additional measurement and observation. Viewed for its own characteristics rather than as it compares to others; authoritative. 2. Free from limitations; unrestricted; unconditional. 3. A defined "absolute" which is selected to be as close to an objective absolute as possible to provide a reference for measurement and calculations. See absolute potential, absolute refractive index.

absolute address In computer programming, the actual address in which a unit of data is stored (in contrast to a pointer to its storage location). 2. The binary address which directly designates a storage location.

absolute altitude Altitude described relative to the surface of the Earth, as distinguished from altitude measured relative to sea level.

absolute coding Machine level instructions that can be processed directly by a computer processor.

absolute delay The time interval between two synchronized transmission signals from the same or different sources.

absolute error 1. A means of expressing a deviation from a standard or expected value in terms of the same units as the units of the value. In statistical population distributions or other scatter distributions, this is a common way of indicating a deviation. 2. The absolute value, that is, the value without regard to sign, equal to the value of the error.

absolute gain In antennas, the gain (boost or increase) in a given direction and polarization when compared against an isotropic reference antenna, typically expressed in decibels. If a direction for the antenna is not specified, then radiant energy in all directions is assumed and gain is measured along a selected axis. See gain, isotropic antenna.

absolute luminance Light values (brightness) as measured on an objective scale as opposed to light values as perceived by human senses (which tend to perceptually vary according to contrast and proximity with other colors and light values).

Absolute and Relative Luminance

Luminance can be difficult to judge. Our eye-brain is influenced by the environment around the object we may be trying to assess. To most people, the circle on the right appears brighter than the one on the left, even though, on a scale designed for graphical paint programs, they both have luminance values of 90.

---

absolute position Position on an agreed-upon coordinate system, e.g., a system with a point of origin defined as the center of the mass of the Earth (geocentric).

absolute potential 1. The absolute capability of matter or a phenomenon to do work. There is currently no way to measure absolute potential energy in an entity, but potential energy can be observed or measured when factors change (relative potential). 2. In electricity, the absolute potential of a point infinitely distant from a point charge is defined as zero and then used as a reference potential. The absolute potential at a stipulated point is the work done against an electric field to move a unit charge from infinity to the stipulated point. Given a general point and point charges at specified positions, the absolute potential at the general point can be calculated along with the electric field intensity. See absolute, coulomb.

absolute power Power levels relative to a reference as expressed in quantitative units such as watts, volts, decibels, etc. A thermocouple power meter may display absolute power in terms of watts or decibels expressed in milliwatts (0 dBm = 1 mW).

The National Institute of Standards and Technology (NIST) Optical Technology Division and the NIST Electron and Optical Physics Division use a cryogenic radiometer for absolute power measurements in the detector calibration and spectral responsivity facility. In experiments at the Sandia National Labs, absolute power from X-rays is measured with time-resolved resistive bolometry with Sandia fiber optic-controlled noise-reduction technology.

absolute refractive index The absolute refractive index of a medium is the velocity of electromagnetic radiation in free space as it relates to the speed of radiation in the medium, usually specified for a given wavelength and temperature. It is a reference index against which the refractive index of other materials may be compared. Air has a low refractive index, similar to that of a vacuum and hence is useful as an "absolute" refractive index against which other materials may be assigned values. If a material has a refractive index of 3.1, for example, it indicates that light travels about 3 times faster through free space than it does for the specified material. In general, the longer the wavelength (e.g., red light), the less it refracts.

The "absolute" refractive index of a number of common optical materials is listed in the following chart.

Material	Absolute Refractive Index
diamond	2.417
ruby	1.760
flint glass	~1.74
quartz	1.544
crown glass	1.520
water	1.3333 (at 20°C)
ice	1.310

See index of refraction, Snell's law.

absolute scale In its generic sense, any reference or quantitative scale based on an agreed-upon fundamental or unvarying value. Many phenomena are adapted to a scale to help us understand their characteristics and provide an absolute reference from which to chart their relative attributes. Absolute scales are widely used by scientists in their research and descriptive statistics. A well-known example is the absolute temperature scale or Kelvin scale. See absolute zero, Kelvin scale.

absolute standard An assigned mass of one unit applied to a specified particle or object so that it can be used as a reference guideline.

absolute temperature Temperature measured or calculated with relation to an absolute scale such as the Kelvin scale. See absolute scale, Kelvin scale.

absolute unit The value of a quantitative measure such as amperes, decibels expressed in milliwatts, degrees Kelvin, geometric degrees, newtons, volts, watts, etc. In programming, absolute units are referenced to underlying physical quantities.

absolute URL On the Internet, a Uniform Resource Locator (URL) that describes a complete and direct path to a file, Web page, or other Uniform Resource. For example,

http://www.4-sightsmedia.com/stuff/page.html
is an absolute URL, whereas
../stuff/page.html
is a relative URL.

Absolute URLs are useful for upper-level files in an account with many cross-referenced files linked together. If within the *index.html* page at that address, for example, there are references to other pages on the same site, it is common to use relative URLs to name them. It saves time typing in long Web page addresses when coding in HTML, and it means that if the domain name changes from *4-sightmedia.com* to *newname.com* all the subreferences to other pages don't have to be changed as well, since they may be designated as *../Examples/file.html* rather than *http://www.4-sightmedia.com/Examples/file.html*. Even if the domain name stays the same, if all the files are moved up one level in the folder hierarchy or down one level, relative URLs don't necessarily have to be changed, but absolute URLs do. Thus, absolute URLs are best used for the top Uniform Resource in a linked hierarchy and are commonly used when a URL on another site is referenced, but they are not necessarily the best choice for subfiles or files in subordinate directories that may potentially need to be moved as a block.

absolute value A numerical notation and corresponding mathematical concept of the magnitude of a value without respect to its sign. Thus, the numeral -5 without respect to sign is written 5.

absolute vector A line or trajectory having both magnitude and direction with end points expressed as absolute coordinates. Absolute vectors are commonly used in graphical display systems.

absolute zero The lowest point in an absolute temperature scale system, zero degrees Kelvin; the low point at which there is thought to be no molecular activity and thus no heat energy, which can also be expressed as −273.15°C or −459.67°F. The Kelvin scale is named after William Thompson (Lord Kelvin).

absorbed dose The amount of radiant energy absorbed by a medium or object. This varies depending upon the type of radiation, distance, duration of exposure, and characteristics of the medium exposed to the radiation. Dosimetry systems (e.g., polymethylmethacrylate – PMMA) may be used for measuring absorbed dose in various materials. Absorbed dose may be measured by entrance and/or exit dosimetry or by absolute dosimetry (e.g., via calorimeter).

absorptance, absorption factor (*symb. – a*) A ratio of the radiant energy absorbed by a body relative to the radiation incident upon it. The absorbed electromagnetic or acoustic energy constitutes part or all of the transmitted radiation which combines with the reflected radiation to total unity (1). Absorptance is expressed as a percentage (based upon the energy absorbed) or assigned a value on a scale between 0 and 1. For example, acoustic damping materials may have an absorptance value of 0.78. Values may be expressed separately for different wavelengths (e.g., colors of the visible spectrum).

When the ratio of the absorbed radiation is related to the absorbed radiation by a theoretical black body at the same wavelength and temperature, it is called *monochromatic* absorptance. When absorptivity over a range of energies is being assessed, it is calculated as an *integrated* absorptance.

absorptiometer An instrument for measuring the optical absorbance of a substance. The instrument consists minimally of a source of electromagnetic radiation and a detector for measuring the amount of energy that passes through the sample substance. The material being measured is frequently liquid but may also be mineral (e.g., bone) or animal tissue. The instrument may measure a direct physical characteristic or may be used to assess the solubility of a substance (e.g., a gas).

absorption 1. The process by which particles penetrate and are subsumed by matter. 2. Penetration of a substance or wave into another substance. A sponge will absorb water and vegetation will absorb radio waves. 3. Dissipation, as of a wave, into another material as a result of its interaction with the other material. Sometimes this is desirable, as in sound-editing studios. See acoustics. 4. The process by which particles entering matter are reduced, or reduced in energy, as a result of interaction with that matter. 5. Reduction of energy as particles pass through or into another substance as a result of interaction with that substance. In radio wave frequencies, absorption tends to occur more readily at the highest frequencies, e.g., microwaves. Absorption can also be used to add information to a signal. See absorption modulation, scattering.

absorption band 1. The radiant energy of a range of electromagnetic waves or frequencies absorbed by a substance. The concept is useful in fiber optic cable fabrication. When Bragg gratings are incorporated into optical fibers to tune them to certain frequency ranges, the pattern is incorporated into the fiber with lasers corresponding to the absorption band of the doped fiber. 2. Depending upon the matter in which absorption occurs, a region of electromagnetic frequencies wherein the absorption coefficient reaches a relative maximum. See absorption coefficient.

absorption coefficient A measure of the fraction of electromagnetic energy (e.g., light) absorbed per unit distance in a medium (typically as a fraction per meter - /m). This may be used to express attenuation within a medium. The absorption coefficient + scattering coefficient = attenuation coefficient. See absorption index.

absorption current Current flowing into or out of a capacitor after its initial charge or discharge.

absorption factor See absorptance.

absorption fading Slow fading of transmission waves due to various absorption factors along the path. Complete fading or significant dissipation is known as absorption loss. Depending upon the transmission medium, degree of loss is sometimes expressed in decibels (dB) over distance.

absorption index A measure of the fraction of electromagnetic energy per unit distance at a given wavelength absorbed in a medium of a given refractive index. Thus, it is a more contextual measure than *absorption coefficient* that is useful for studying and describing transmission characteristics such as ionospheric absorption. See index of refraction.

absorption line In astronomy, a region of energy transition in atmospheric gases that results from the absorption of incident solar radiation. The width of the region is dependent upon a variety of factors including incident angle, proximity, time of day, motion, etc.

absorption loss The portion of a transmission that is lost due to interaction with another material through partial reflection or complete absorption into the material. This interaction may cause the conversion of energy into other forms, such as heat.

absorption modulation A means of modulating the amplitude of a wave, such as a radio carrier wave, by absorbing the carrier power using a variable-impedance device. See amplitude modulation.

absorption peak The maximum level at which a particular substance or entity can absorb electromagnetic or acoustic energy. When graphed, the absorption peak may be wide or narrow. For electromagnetic energy, the absorption peak is usually expressed in terms of wavelengths in micrometers (μm) or nanometers (nm).

In laser technologies, an absorption peak may be used as an absolute frequency reference to tune a system to facilitate long-term, stable operations and to reduce the need for recalibration.; the absorption peak for iodine is commonly used for this purpose. Variation in measured absorption peaks is used in a variety of disciplines to help distinguish one substance or entity from another.

In chemical analysis, a laser may be set to a sinusoidal modulation in order to pass in and out of a substance's absorption peak(s). The absorption characteristics of the probed substance may further be used to convert between frequency and amplitude modulation.

Absorption peak characteristics are used to assess optical fibers and select effective wavelengths for transmission. A single filament may have more than one absorption peak due to impurities. Generally, transmission frequencies are selected to work around these absorption peaks.

In photography, absorption peaks are specified for optical filters to provide the percentage transmittance level (usually between 10% and 85%) or highest wavelength transmitted for individual colors or types of light (e.g., fluorescent). The use of an appropriate filter aids in color compensation.

absorption wavemeter An instrument for measuring frequency or wavelength and sometimes the amplitude of the harmonics of that frequency by absorbing energy from the circuit being tested. When absorption is at its maximum, the wavemeter is tuned to the corresponding frequency of the circuit. This instrument is often used in conjunction with antenna systems.

absorptive medium A medium that tends to absorb radiant electromagnetic or acoustic energy rather than allowing the energy to reflect or pass through. Absorptive mediums are useful for acoustical damping and radiation shielding. An absorptive medium may help in mapping internal structures that can be sensed with sound, radar, or light to reveal tunnels, land mines, tumors, and internal organs that are more or less absorptive than the surrounding environment.

absorptivity See absorptance.

abstract syntax A means of specifying notational rules independently of the encoding used to represent the information. This is useful for defining and developing systems that may be implemented or expanded without foreknowledge of the final configuration of the system or by personnel other than those specifying the initial layers of the system. Abstract syntax is often used in open architectures and object-oriented environments. See Abstract Syntax Notation One.

Abstract Syntax Notation One ASN.1. A data definition notation system defined in 1988 as ISO X.208, superseding CCITT Recommendation X.409. ASN.1 provides flexibility and extensibility and supports the definition of a variety of basic and complex data types. ASN.1 grew out of a need for a way to relate abstract and transfer syntaxes that were emerging in the early 1980s, in a machine- and application-independent manner. Open Systems Interconnection (OSI) uses ASN.1 to specify abstract objects to facilitate the process of defining higher level layers without foreknowledge of specific lower layer objects that might later be incorporated into the system.

ABT See advanced broadcast television.

abuse numbers A database of phone numbers known to be inappropriate for outgoing calls (i.e., numbers not associated with typical business transactions). Some venders provide an option to track and highlight calls to specified abuse numbers so they can be readily identified on billing statements.

ABX See Advanced Branch Exchange.

AC 1. See Authentication Center. 2. See alternating current.

AC biasing In recording processes, a technique of adding a high frequency to aid in linearizing the recording head.

AC ripple Undesired modulation in an alternating current (AC) circuit. Filtering may be employed to reduce or eliminate ripple.

AC to DC converter A device for converting alternating current (AC) to direct current (DC). The current that comes from most wall sockets is AC current, but many devices including answering machines, feature phones, modems, etc. require DC current and will include a converter attached to the power cord or incorporated into the device.

It is unwise to interchange these power converters, as they have widely varying specifications. Most will list the voltage and amperage on the converter, and some will list the corresponding voltage and amperage on the device itself (usually on the underside). Installation of incorrect converter cords can damage sensitive electronic devices. If the device is NOT labeled, it is prudent to mark it as soon as you take it out of the box, with a felt pen or label, so that if the converter and the device get separated from one another, you can correctly match them again.

AC to DC Converters

The four AC to DC converters on the right convert alternating current from the main building power to specific amounts of direct current for powering sensitive electronic components. This power strip sensibly spaces and rotates the sockets 90° so the converters fit and don't cover up two or three sockets. Some converters have a regular plug, with the converter at a distance from the plug to provide even more leeway.

INPUT: 120VAC 60Hz 15W
OUTPUT: 12VDC 500mA

It is important to match the voltage and amperage settings listed on the converter to the specifications of the powered device. The diagram under the power specifications indicates the tip and ring polarity.

AC-powered phone Most small residential phones draw current from the phone line, but if the phone has extra features, such as electronic displays and speakerphones, or if it is a multiline business phone system, then dedicated alternating current (AC) from a wall socket is generally passed through a transformer to supply additional power to the phone. Battery systems also exist, typically for backup power or to hold stored settings in case the AC source fails. Private branch phone systems can consume a significant amount of power if many calls are being processed and may require power from both the phone switching cabinet (through the line) and from an AC power source serving the phone console.

AC13 A British private telephone signaling system. See SSAC13.

AC15 A British private telephone signaling system. See SSAC15.

AC/WPBX Advanced Cordless/Wireless Private Branch Exchange.

ACA 1. See American Communication Association. 2. See Australian Communications Authority. 3. See Automatic Circuit Assurance.

AcademNet A Russian academic/research network. http://www.academnet.magadan.ru/

Academic Computing Research Facility Network ACRFNET. A wide-area network connecting research facilities and laboratories across the U.S.

Academy of Motion Picture Arts and Sciences AMPAS. A professional, honorary organization composed of more than 6,000 professionals in the motion picture industry. It was founded as a nonprofit corporation in May 1927. Membership is by invitation of the Board of Governors to individuals with significant achievements. Life members are designated by a unanimous vote of the Board of Governors. AMPAS supports and advances the arts and science of motion pictures and recognizes outstanding contributions to the industry through various programs, especially through Academy Awards. With the increase in Internet content delivered in multimedia formats and with increased digital distribution of motion picture products (e.g., DVD), the film industry will likely have a strong influence on the future form and content of information transmitted through telecommunications technologies. Already, as of 2001, the computer games industry and the motion picture industry had begun to significantly overlap. http://www.oscar.org/academy/

ACAR aluminum conductor alloy-reinforced. See ACSR.

ACARD 1. Advisory Council for Applied Research and Development. U.K. advisory organization superseded in 1987 by ACOST. See ACOST. 2. Acquisition Card Program.

ACAT See Additional Cooperative Acceptance Testing.

ACATS See Advisory Committee on Advanced Television Service.

ACB 1. Annoyance Call Bureau. 2. Architecture Control Board. 3. ATM Cell Bus. 4. automatic callback.

Accelar routing switch A commercial switcher/router device from Bay Networks that makes switching decisions based upon Internet Protocol (IP) addresses embedded in the local area network (LAN) switch hardware, without proprietary protocols or appended bits. See IP switching.

accelerated aging, accelerated life test A design and diagnostic technique that involves subjecting a process, material, or mechanism to short-term conditions that simulate long-term use and environmental influences. Accelerated conditions simulate factors such as weather, movement, mechanical stress, chemical exposure, use, etc.

accelerating electrode A device in an electron tube, such as a cathode-ray tube, that increases the velocity of the electron beam.

acceleration (*symb.* – a) The expression of a change in velocity (speed in a particular direction) over time.

Acceleration is commonly expressed in meters per second per second. An international standard value for acceleration due to gravity on a free-falling object in a vacuum has been established as 9.807 meters per second per second.

acceleration voltage In a cathode-ray tube, the accelerating potential controlling the average velocity of electrons directed toward the imaging surface from an electron gun. The voltages are tuned in conjunction with the magnetic coil through which the electrons pass to create the *sweep* and image *frames* that help build the picture on the tube.

accelerator A system, process, chemical, organic substance, or device that acts on something to speed it up. Accelerators are used in many areas including, but not limited to, studies of elementary particles, chemical reactions, transmission circuits, and computer systems.

accelerator board, accelerator card A peripheral card designed to fit into a computer slot that increases the speed of the system, usually by increasing the CPU speed, or by taking over some of the more demanding of the CPU's functions, such as graphics manipulations. Games players love these.

accentuation 1. Intensification, emphasis. 2. In transmissions, the emphasis of a particular channel or frequency, often to the exclusion of others. Accentuation is found in the high frequencies in frequency-modulated (FM) transmitters.

Acceptable Use Policy AUP. A license or purchase agreement setting out limitations, restrictions, and acceptable uses which are binding to the purchaser or receiver. For example, a number of freely distributed network software programs stipulate that they may not be used or sold for commercial purposes.

acceptance angle, angle of acceptance 1. In microphone acoustics, a conical region at the front area of the microphone where the sound is effectively captured. 2. In fiber optic cable transmissions, an angle calculated with respect to the fiber's axis to be effective in "capturing" the incoming light rays and propagating them along the fiber when coupled into optical fiber bound modes. A laser beam entering the fiber at an angle that is greater than this conical acceptance angle is coupled into unbound modes. The acceptance angle is related to the diameter of the fiber conducting core and the cladding layer (the material that surrounds the fiber core). Acceptance angles vary, but for commercial plastic optical fiber, they are generally around 58°; for glass they may be similar to plastic or as high as 82°. Light guides made from quartz have smaller acceptance angles, which are dependent upon the fiber bundle length and the wavelengths being used, usually about half of a plastic fiber. See Brewster's angle, blaze angle, cladding, incidence angle, Littrow configuration. See acceptance cone.

acceptance cone A conical region within which signals are "captured" by a sensing device or optical transmission fiber with a circular cross-section. The shape of the cone is related to the acceptance angle around the axis of the active or inbound portion of the device or fiber. Depending upon the application, particles outside the acceptance cone and sometimes large-angle particles within the acceptance cone are excluded from processing or further transmission. See acceptance angle. See Acceptance Cone diagram.

acceptance pattern 1. In antennas, a diagramatic plot of off-axis power as it relates to on-axis power as a function of the position or angle of the antenna for a given plane (e.g., horizontal plane). 2. In fiber optics, a diagramatic plot of the total transmitted power as it relates to the launch angle of the transmission.

acceptance period A period, usually of a few weeks, during which a product or service is evaluated by the receiver as to its conformance to the agreed-upon specifications. It is more commonly a stipulation of custom installations than of off-the-shelf products. Acceptance differs from a warranty in that it applies mainly to initial configuration during the ramp-up or installation period, whereas a warranty may cover other factors and last several months or years after purchase and installation are complete.

acceptance test A test, which usually follows installation, that demonstrates that the product or services purchased conform to the agreed-upon specifications. An acceptance test may be contractually required by the purchaser before making final payments on the purchase.

accepted signal See call accepted signal.

acceptor An entity capable of receiving another entity, such as a compound, substance, or atomic particle. The providing entity may be called a donor. Donor-acceptor relationships are fundamental processes and may be used to polarize a molecule, for example, or be associated with certain photochemical processes. Donor-acceptor emissions are found in semiconductor technologies.

access 1. *n.* The point through which a circuit or communications device is entered, or the point at which the communications process is entered and initiated. 2. *v.* To gain entry into a circuit or communications device. Phones are generally accessed by dialing a number, although an *access code* may be required on a secure system. Dialing "9" first to obtain an outside line is a common access procedure. Account codes are sometimes used to assign billing to specific departments or individuals. Access codes may be used by installation or maintenance technicians to initiate services or procedures not available to the subscriber. Secured computer systems are accessed by *logging in* with or without a password. See access code.

access arm The positioning mechanism that supports a read/write head for reading from or writing to magnetic or optical storage media. On a computer hard drive, the access arm moves across the disk and positions the head directly to within thousandths of an inch of the area of magnetic particles to be read (or written). See seek time.

access attempt An attempt to gain entry to a system, facility, or process. In computer networks, successful and unsuccessful access attempts may be logged to provide information useful for improving network

efficiency or security. See authentication, firewall.

access carrier An interconnect agreement through which a telecommunications carrier can gain access to the services and network facilities of another carrier.

access charge 1. The charge made for access to a computer system or network. An access charge may be assessed on a periodic basis or per time or volume of use. Internet Service Providers typically charge flat monthly rates, although some will assess extra charges for storage, peak-time connects, access to chat areas, or special online services. 2. The Modified Final Judgment (MFJ) which broke up the Bell system included the rationale and stipulation that users should be able to choose a long-distance carrier, thus changing the way in which long-distance access charges were structured. Compensatory restructuring resulted in two categories of access charges: Customer Access Line Charges (CALCs), and Carrier Access Charges (CACs). The first applies to local phone loops and varies according to the subscriber (residential or business) and the characteristics of the service. The latter applies to service providers connecting to the local exchange circuits and varies according to factors such as distance. Adjustments and modifications in order to implement the many changes have subsequently occurred. See Telecommunications Act of 1996.

access code One or more characters that must be entered in order to obtain use authorization to a system such as a phone or network. Access codes are generally used for security, monitoring, and billing purposes. They can also be used by technicians to set up a system for use with specified features and, more recently, to program a telephone system. Some typical telephone access code implementations include (1) dialing codes to access an outside line or to dial a long-distance number (dialing "9" is common), (2) dialing an access code to bill the call to a particular line or department, (3) dialing a code to obtain authorized use on a privileged system.

access control 1. The policies, procedures, and system configurations controlling security or utilization of resources. Access control operates on many levels, including building access, system access, applications access, network access, device access, and computer operations access. See access code. 2. A physical or virtual control point, gateway, or other filter or security system that selectively allows data to pass through according to general or specific parameters, which may include priority level, data characteristics, sender, receiver, etc.

access control field An informational field in the header of a synchronous multimegabit data service (SMDS) cell which provides access to a shared bus, which in turn provides access to the SMDS network.

access control key A physical device, which may be electronic, mechanical, or both, for gaining or controlling access to a system, physical structure, or process. Common access control keys include traditional serrated house, auto, or office keys, more recent magnetic keys that resemble metal house keys, and

Acceptance Cone and Acceptance Angle in Fiber Optic Lightguides

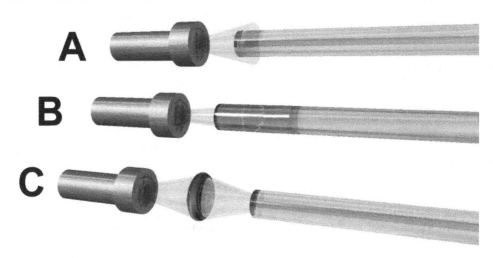

The acceptance cone *is a conical region within which radiant energy can enter a spherical or cylindrical conducting medium, whether it be an acoustical microphone or a fiber optic lightguide. Any radiant energy spreading beyond this region fails to be conducted within the waveguide. If the conducting core is narrow and the radiant energy source spreads at a wide angle, most of the light will be outside the cone of acceptance region (A). Narrowing the beam and/or increasing the size of the conducting core (B) can balance a smaller radiant beam with a broader acceptance cone for less loss at the endface of the optical fiber. Interposing a lens between the source of illumination and the fiber endface (C) is a common way of concentrating the beam to fall more closely within the acceptance cone so less of the light is lost.*

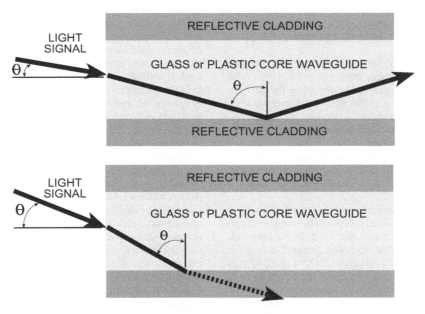

The acceptance angle *is a closely related concept, but more commonly refers to the reflective path of the light as it enters and continues along the wave guide, as opposed to the beam's initial path as it travels from the light source toward the lightguide. If a straight lightguide is aligned with a coherent beam, all or most of the light will travel through the waveguide, with a minimum of reflection and loss. However, this is rarely the case. There may be slight misalignments between the light source and the lightguide and the lightguide itself is often so slender that it bends (which is useful for installation and routing the light). Thus, much of the beam will reflect back and forth within the cladding, which has a slightly higher refractive index than the conducting core, but some of the light will exceed the angle at which the refractive index of the cladding is able to reflect the light beam back into the core to keep it within the waveguide. The light that is not lost through the cladding is said to be within the reflective angle of acceptance.*

card keys that resemble credit cards but contain access information in the ink or on magnetic strips or miniature circuit boards. Many hotels and workplaces now use electronic rather than metal serrated keys.

access control list ACL. A list, table, or database that provides a reference for various levels of security within a system. It can be as simple as a list of names in the hand of a doorway security guard, or as sophisticated as a tiered database of levels of security for different people and processes on a computer network.

On bulletin board systems (BBSs), many access control lists include a series of software flags for each user which can be toggled individually by the sysop to control user access to services such as chat, email, doors, downloads, etc. The "twit bit" is a flag that basically labels a user as a "loser," in BBS parlance. In other words, the twit is someone who has been flagged for limited access because s/he exhibits immature behavior and can't be trusted with access to any of the powerful features.

access control method A system for controlling access to systems, processes, or devices on a network. A variety of general guidelines and specifically defined systems for particular types of networks and protocols have been developed. Access control can be set up "by user," "by workstation," "by application," "by file," "by network," or a combination of these, which may be hierarchical. Examples of particular types of access control on specific types of networks include *carrier sense multiple access* (CSMA) on Ethernet systems and token passing schemes on IBM Token-Ring networks. See Media Access Control.

access coupler A connecting device used between physical network segments, such as fiber optic cable legs, to allow signals to be passed to the next leg. Access couplers are sometimes used in conjunction with relays and amplifiers, depending upon the type of signal and the distance being traversed. See SC-connector, ST- connector.

access delay In a packet-switched network, a performance measure for polling systems, calculated from the time of arrival of a data packet to the time it is retransmitted.

access group A group of accounts or individuals who have specific, defined levels and types of privileges within a system, which may be different from individual privileges and from other groups. For example, on a private branch exchange (PBX), a group of managers may be designated as having access to long-distance or outside lines, whereas a new employee may be assigned to a group with limited privileges until an evaluation period has passed. On a computer network, an access group may have certain read and write file privileges which differ from individual privileges and the privileges of other groups. Thus, they may be permitted to run only certain applications, look at certain directories, etc., according to settings established by the system administrator.

access line 1. The physical link between the subscriber box and the local telephone switching center. From the subscriber box to the telephones is considered inside wiring and may be installed by the subscriber or, for a fee, by the phone company. See local loop. 2. In BBSs, the line through which the caller accesses the BBS modem. There may be multiple lines, sometimes with different baud rate capabilities. Historically, BBSs have been accessed through phone lines but, increasingly, BBSs are interfacing with the Internet to provide online access through telnet. 3. In frame relay systems, a communications circuit that connects a frame relay device to a frame relay switch.

access link A link. A connection point in a network that enables access to resources on the other side of the link. In Signaling System 7 (SS7) it specifically refers to a dedicated signaling link that is not associated with any particular physical link.

access mechanism A device for moving and positioning an access arm, usually on random access read and/or write media.

access method Logical guidelines established by International Business Machines (IBM) in the 1960s for input and output access to computing resources, particularly those that are shared, as on local area networks (LANs). By consolidation of instructional sequences in common procedures, functions, and subroutines, the overall structure can be simplified.

access module AM. In general, an included or optional interface enabling transmission lines or peripherals to connect to a main unit such as a computer, microscope, milling machine, or telephony service unit. Depending upon the unit, it may include specific connectors, voltage or data selection switches, and sometimes signal conversion, compression or decompression electronics.

In microscopy, access modules can allow video recorders, still cameras, or projection devices to be attached. In computing, printers, modems, and other peripherals are commonly connected through peripheral cards, a type of access module. In wireless and wireline services, a mechanism to provide electrical or optical connections with various transmission cable ends (RJ-*x* jacks, BNC connectors, ST connectors, 25-pin D connectors, etc.) so they can be readily connected to service supply units.

Access Network Termination ANT. Typically a device that provides coupling to the user end of the communications line. Depending upon the type of transmissions service, such a device may provide signal splitting, framing and deframing, cell header processing, encryption/decryption services, and other termination/translation functions needed at the subscriber end of the communications circuit. Thus, a cable modem is an example of an ANT device as it provides access to a fiber optic line and termination/translation services for interfacing a personal computer. See access module.

access software provider Defined in the Telecommunications Act of 1996, and published by the Federal Communications Commission (FCC), under SUBTITLE A—Telecommunications Services as:
"... a provider of software (including client or server software), or enabling tools that do any one

or more of the following:

(A) filter, screen, allow, or disallow content;

(B) pick, choose, analyze, or digest content; or

(C) transmit, receive, display, forward, cache, search, subset, organize, reorganize, or translate content."

access tandem AT. A central office telephony switching system that provides distribution services for outgoing or incoming traffic between End Offices.

access tandem switches Specific types of switches that are used to connect End Offices to Interexchange Carrier (IXC) switches or to interconnect central office (CO) switches.

access time 1. The interval between a signal or instruction to access information or a device, and the time it takes to successfully retrieve that information, or interact with the device. Depending upon the system, the access time may or may not include the time it takes to *display* the requested information to the user. For example, in a database query on a computer system, the access time may be two seconds to search and retrieve a long list of names and addresses, but it may take twenty additional seconds to fully display all the listings, and the *access time* may not include the display time or may include the display time for the initial information, but not the time during which the software may be building additional viewable information below a scrolling window. Access time is described in terms of units appropriate to the average time and device involved. For example, access time for a process may be described in CPU *clock cycles*, for data access on storage media by *seek time* in *milliseconds* or *fetching* in *nanoseconds*. 2. In magnetic storage devices, the interval during which the access mechanism, once information is requested, moves across the medium to the desired location and successfully reads the data.

access unit AU. 1. In Token-Ring networks, a wiring concentrator that connects the end stations. The AU provides an interface between the Token-Ring router interface and the end stations. Also known as Media Access Unit (MAU). 2. In many X.400-based commercial software applications, the AU works in conjunction with mail servers to provide synchronization between post offices and other services, such as directories, address books, etc. 3. In building access or financial system access, access units are typically locks, card readers, or biometric scanning devices. Since these units are increasingly being monitored by remote terminals and even video cameras, fiber optics networks may be interfaced with access units to provide fast response and monitoring of access events. See access unit interface.

access unit interface The portion of an access unit that provides a bridge between the entity seeking access and the structure, process, or system to which the entity seeks access. The configuration and size of the interface depends upon the type of access system in place. A deadbolt is an example of a simple, traditional, mechanical access unit interface. Credit or ATM card-reading machines are common magnetic

access unit interfaces. Iris, retina, or hand scanners are more recent electro-optical access unit interfaces. See access unit.

access.bus A serial communications bus topology protocol developed jointly in the mid-1980s by Digital Equipment Corporation and Philips Semiconductors, for connecting peripheral devices such as mice, keyboards, card readers, scanners, etc. to computers through a four-wire serial bus. See Universal Serial Bus.

Accessible Information on Development Activities AIDA. Formerly known as IDAI, AIDA is a Development Gateway (DG) pilot project that encompasses tens of thousands of individual projects related to the International Development Markup Language (IDML). The DG provides support software and documentation to facilitate the conversion of data into IDML format. See International Development Markup Language.

accounting server A software application, sometimes operating from a dedicated, secured computer that monitors network usage, stores the information, and may assess charges for usage based on CPU time, real time, time of day, department, or some other measure appropriate to the type of use.

ACCS Automatic Calling Card Service.

accuracy 1. Degree of conformity to a stated or observed value considered to be optimal or correct. See calibration. 2. Precision. 3. Degree of freedom from error.

AC/DC, AC-DC 1. alternating current/direct current. See AC to DC converter. 2. An electrical appliance that can operate on alternating current (AC) or direct current (DC). 3. Euphemistically, a designation for something or someone operating in two modes or "swings both ways," especially in alt.*xxx* USENET discussion lists. See AC to DC converter.

AC/DC ringing A telephone ringing signal control mechanism that uses both AC and DC power sources. Ringing is usually provided through low voltage AC current transmitted through the line from a ringing generator to the subscriber's premises. The ringing generator may be powered by a -48V or -24V DC battery. Ringing cessation, amplification, special tones, or other phone options may be powered through DC sources associated with the line or with the telephone set. See ringing generator, ringing signal.

ACD See Automatic Call Distribution.

ACF Advanced Communication Function.

ACG See automatic call gapping.

achromatic 1. Uncolored; unmodulated; neutral; black and white; grayscale. 2. In the visual spectrum, lightwaves that are not dispersed or singled out according to a particular wavelength.

achromatic lens An optical lens designed to compensate for the chromatic aberrations that result from different wavelengths in the optical spectrum having different focal points, which can lead to blurring of an image. Multiple elements are used to construct achromatic lenses for precision applications such as astronomy and microscopy. The elements may be of different types of glass. Chemical coatings can further be used to improve lens quality by suppressing glare.

ACIA See asynchronous communications interface adapter.

ACK See acknowledge.

acknowledge, acknowledgment ACK. A message or signal from the receiver to the sender confirming receipt of data or accurate receipt of data. In handshaking, ACK sometimes also signifies that the receiver is ready for further data. ACK and NACK are commonly used on bidirectional communications systems, in which data can transmit in only one direction at a time. See negative acknowledge.

ACL 1. See access control list. 2. Applications Connectivity Link. 3. Association for Computational Linguistics.

ACLU See American Civil Liberties Union.

aclastic Having the property of not refracting light. Contrast with refraction.

ACM 1. See Association for Computing Machinery. 2. Automatic Call Manager. An administrative and operations system that handles inbound and outbound calls integrated with a database. Telemarketing, teleresearch, and collection agencies make use of these types of systems. 3. Address Complete Message. A call setup message in ATM networking that is returned to indicate that the address signals required for routing the call have been received by the called party. The ACM is sometimes sent in conjunction with other routing messages.

ACO 1. Additional Call Offering. 2. alarm cutoff. A switch that suppresses an audible alarm, while not affecting a corresponding visual alarm.

ACOnet Austrian Academic Computer Network. An ATM-based Austrian research network funded by the Austrian Ministry of Science, Transport, and Art. ACOnet interconnects about a dozen universities and provides international links to other countries through EBS-Vienna. http://www.aconet.at/

Acorn Tube

The acorn tube, developed for ultra-high frequency (UHF) applications in the late 1930s, is distinctive for its lack of a base and the leads protruding through a central glass girdle.

---

acorn tube A very small vacuum tube named for its

squat, rounded shape. It has electrodes leading directly through the glass on several sides. It was developed for use at extremely high frequencies (e.g., UHF) in the late 1930s. While most vacuum tubes have been superseded by transistors and other modern electronics, there are still high frequency applications where vacuum tubes are practical.

ACOST Advisory Council on Science and Technology. A U.K. organization which superseded ACARD in 1987, ACOST is a government advisory and coordinating body on policy and research.

acoustic Relating to sound phenomena, the science of sound, as well as biological structures and nonbiological apparatus for generating, conveying, controlling, or apprehending sound. In music, an acoustic instrument doesn't typically have an electronic pickup because it depends upon its physical structure and the surrounding medium to create and convey the desired sounds at the desired intensity. See acoustic wave.

acoustic converter Any device that is designed to interface with an audio sending and/or receiving circuit to provide amplification or conversion between analog and digital audio signals. The coupler is usually designed to exclude extraneous noise that could interfere with a signal and may be a self-contained unit or peripheral.

Acoustic couplers that resemble large suction cups were incorporated into early external modems to provide a way to interface telephone handsets with computers. The coupler was designed so that the outbound modem signal played into the mouthpiece microphone and the inbound signal played from the earpiece speaker into the modem. See acoustic modem.

acoustic delay line 1. In general, a transmission line designed to delay acoustic signal propagation. Why would you want to slow down a signal? Delay may reduce loss in systems where acoustic signals can be generated or travel faster than the transmissions system can handle them and "excess" data could be lost due to congestion or overrun. It may also be useful when sound is transmitted in conjunction with slower/higher bandwidth signals, such as video data in videoconferencing systems – a compensation system enables associated signals to be presented together even if they don't transmit together. 2. In historic computing machine design and programming, mercury acoustic delay lines were used for memory access timing purposes to create wait time before execution of subsequent instructions. An electrical pulse would be converted to a sound pulse as it traveled through a mercury tube and converted back to an electrical pulse at the end of the tube. If necessary, the signal would be sent more than once through the tube and sometimes multiple tubes were used to represent separate "registers." In historic calculators from the late 1960s, electromechanical delay line systems were used for register storage. They functioned by sending a series of pulses through a coiled wire calculated to provide a specific "tuned" rate sent to a transducer that converted the signal back into electrical energy. This was a more compact and efficient evolution of the bulky mercury acoustic delay lines in early computers. Delay lines have been superseded by integrated circuits in most applications.

acoustic echo canceller AEC. In many telephony devices, the microphone is only inches away from the speaker. Thus, sounds from the speaker can reach the microphone and interfere with the clarity of communications. Unpleasant screeching, echoes, or howling may occur. An acoustic echo canceller reduces or eliminates this unwanted interference. Echo cancellation is accomplished in a number of ways with one or more electronic filters placed in various positions in the signal path, depending upon the system. In some advanced systems, ambient background sounds are also filtered and 'cancelled' to improve the clarity of the audio signal.

acoustic feedback See feedback.

acoustic model In software applications, a means to apprehend and interpret sound input, such as speech, by breaking it down into smaller units and then using those together to build an aural representation or interpretation of input combined from these units into larger words and speech patterns.

Early attempts at speech recognition were hit-and-miss, and very person-specific, but new programs can transcribe speech into text with a useful degree of accuracy up to about 70 words per minute. Computerized speech used to be characterized by very flat, mechanized sounds but, with faster processors and better methods and sound samples, natural sounding voices can be generated. Many automated phone voice applications now use speech generation for messages, queries, and instructions. See phonemes, sampling, speech recognition.

acoustic modem A modulating/demodulating computer peripheral that converts digital signals created by a computer into audible tones that can be coupled with the transmitting end of a telephone handset or other audio transmissions device so they can be sent through an analog phone line. The device then converts the audible tones generated by the other end of the transmission back into digital signals for the receiving computer to interpret. The modem is usually attached to the computer by means of an RS-232 (EIA-232) interface, although some acoustic modems designed for the early Apple computers were connected through the joystick port.

Acoustic 300-baud modems were prevalent on personal computer systems in the late 1970s. These were gradually superseded by direct connect modems in the 1980s. In the 1990s, all-digital systems began to supersede analog phone lines.

Acoustic modems have many limitations. They tend to be bulky, as they need sufficient shielding around the transmitting and receiving electronics to prevent the tones from crossing over and interfering with one another. They are subject to interference from external noises. They work well only with old-style phone handsets – the newer, flatter ones don't provide sufficient shielding or contact with the couplers to transmit clean tones, and they do not generally employ any sophisticated data compression capabilities,

resulting in slow transmission speeds. See acoustic coupler, direct connect modem.

acoustic streaming Gradients of acoustic pressure in nematic liquid crystals giving rise to acousto-optic effects. This phenomenon has practical applications in light modulation and liquid crystal-based light valves. See acousto-optic, acousto-optic effect.

acoustic telegraph Messages conveyed by sound, such as bells ringing specific tones or sequences (still used in many European churches), drumbeats, horns, or shouts that are passed from one person to the next (still used on large sailing vessels, railroad lines, or other areas where other means of distance communication are not available).

In the early days of electrical telegraphs, a number of inventors were seeking ways to use tones to convey more information over a single line than was possible with the simple on/off system that was gaining widespread use. Experiments in trying to send tones led to the invention of the telephone, and so the technology leap-frogged over the acoustic telegraph. Basic telegraphic systems were used for a long time concurrently with the evolution of telephones.

acoustic transducer A mechanism that converts acoustic energy into other forms of energy or other forms of energy into acoustic energy. See acoustic wave, sonar.

acoustic velocity (*symb.* – a) The speed of sound (technically, *velocity* is the rate of motion in a direction). A measure of acoustic velocity is valuable in many communications fields and also facilitates the determination of other measures such as distance or the 'thickness' or permeability of a medium or particulate environment.

acoustic wave Longitudinal wave movement through an elastic medium (solid, liquid, gaseous, etc.). An acoustic wave or *sound wave* travels through a compressible medium through pressure and density changes along the direction of motion. The displacement of molecules within the acoustic medium produces high pressure (condensation) and low pressure (rarefaction) regions. The speed with which the wave travels through a medium depends upon the inertial and compressibility characteristics of the medium and may be calculated if these measures are known.

Sounds emanating from an acoustic wave may be audible or inaudible. Inaudible sounds may be subsonic (below human hearing) or supersonic (above human hearing). Differences in media result in various pitch and tonal qualities, some of which may be discerned by human hearing. Inaudible characteristics may sometimes be heard by other species or by acoustic measuring instruments. Since a vacuum is an absence of any medium and thus is not subject to physical perturbations, sound waves do not propagate in a vacuum.

A *surface acoustic wave* (SAW) is a wave that propagates along the surface of an elastic medium. Ripples along the surface of a pond provide an example of visible SAW motion but SAWs may be so small as to be undiscernible by the human eye (and may be particularly difficult to detect on rough surfaces).

Holographic interferometry techniques are being developed for the detection of SAWs.

Recent advancements in SAW science have resulted in a number of new electronic and chemical technologies. In microelectronics, Japanese researchers have been experimenting with SAWs as tiny linear motors using Rayleigh waves, SAW bandpass filters are used in digital radio modulation, and Sandia Labs has been turning SAWs into tiny chemical sensors for making handheld chemical detectors. SAW research has also resulted in some interesting discoveries related to X waves.

In fiber optic networks, an acoustic wave can be sent over a strand of fiber to produce a *notch* filter for controlling gain and tilt. See dynamic gain equalization processor, X wave.

acoustical Doppler effect When a moving object such as a train generates sound, there may be some interesting effects created by the compression and decompression of the sound waves relative to a stationary listener some distance from the train.

The engineer on the train hears the sound of the whistle at a constant pitch. The stationary listener hears the pitch increase as the train approaches and decrease as the train passes.

This effect is due to the compression of the sound waves relative to the stationary listener as the train approaches, which increases the frequency (pitch) of the sound wave. Conversely, as the train passes, the distance between sound waves becomes greater, resulting in a decrease in frequency (pitch).

When taking acoustic measurements of moving objects, this effect must be taken into consideration. Doppler effects, sometimes called *Doppler shift*, are important in remote sensing and guidance systems.

Acoustical Society of America ASA. A scientific society founded in 1928 after an initial meeting at the Bell Telephone Laboratories in New York. It began publication of its professional journal in 1929. ASA merged with three other societies in 1931 to form the American Institute of Physics. ASA has been involved in research, development, promotion, and standardization efforts in the field of acoustics. http://asa.aip.org/

acoustics 1. The art and science of sound production, transmission, and reception. 2. The sound-carrying capacity, in terms of quality, fidelity, and loudness, of an environment such as a concert hall or recording studio. See anechoic.

acoustics, engineering The art and science of sound control in electronic structures, including amplification, propagation, dampening, and the harnessing of sound to carry information, as in data and broadcast transmissions.

acousto-optic A-O. Incorporating acoustical and optical properties or providing conversion from acoustical to optical energy or vice versa. Acousto-optic devices are routinely used in acoustical, laser switching, and imaging applications.

Acousto-optic solid state components can be used to control laser beams for imaging, surveillance, and materials processing systems. Radio frequency control

signals are effective in accurately adjusting the angle of laser beams through acousto-optical deflection. In frequency-shifting applications, an acoustic wave can be used to diffract an incident optical wave to down- or upshift the laser frequency.

Piezoelectric transducers can be used with acousto-optic crystals to generate acoustic waves with a periodic variation in the refractive index. A portion of the incident beam (acoustic or optical) will be effective, providing a modulation mechanism. See acousto-optic modulation.

acousto-optic deflector In laser technology, a variable frequency, acousto-optic mechanism for controlling the angle of a coherent beam. Control signals deflect the beam within a certain range of angles and in certain specific directions, within the resolution parameters of the deflector. Acousto-optic X-Y scanners can be configured from two or more deflectors arranged to address both X and Y coordinate planes. Acousto-optic deflectors are used in spatial measuring, graphics, mirror tracking systems, and molecule manipulation applications. See Bessel beam.

acousto-optic effect The influence of an acoustic wave upon an optical phenomenon such that the optical energy is refracted. This effect can be quite precisely controlled and thus has practical applications for deflecting and modulating light beams. See acousto-optic deflector, acousto-optic modulation.

acousto-optic modulation A technique which can be used quite effectively for color control, dimming, and blanking in laser light beams. The beam is shone through an acousto-optic crystal. The modulation is applied with electrical impulses to the crystal to influence the intensity of the beam. Three beams can be used, red, green, and blue, as in a cathode-ray tube, to provide color modulation. This is known as polychromatic acousto-optic modulation. Other applications of acousto-optics can be used to tune filters. See acousto-optic, modulation.

acousto-optic switch In fiber optics, a switch that provides faster, more precise rerouting of signals traveling through an optical fiber network than is available through traditional mechanical switches. Experimental systems include acousto-optic switches that diffract light from individual incoming fibers to selected outgoing fibers.

acousto-optic antenna control An antenna control system based upon acousto-optic technologies. An innovative system with both transmit and receive capabilities was designed by Nabell A. Riza in the late 1980s to form the first all-optical phased array antenna controller. The system was extended to work with wideband signal processing applications while Riza worked with General Electric. Riza demonstrated how antenna arrays could be controled with nematic liquid crystal-based phase and time dealy control mechanisms which could, in turn, be applied to radar control systems. Riza was one of the first to suggest that optical micro-electromechanical systems (MEMS) could be used in the design of fiber optic switches for phased array systems. See multiplexed optical scanner technology.

ACP See activity concentration point.

acquisition 1. The gathering, receipt, and possession of data. 2. The process of orienting toward and acquiring data, that is, seeking a source; setting up the necessary protocols; aiming an aerial; scanning network inputs or broadcast frequencies; and receiving the transmission. 3. The process of gathering equipment, software, or businesses for a certain purpose.

acquisition and tracking A data detection or receiving system such as radar that seeks out a signal, locks in on it, and orients toward the source of the signal while receiving.

acquisition time The time required to seek out and lock on to the source of the desired signal. Commonly used in microwave transmissions such as radar and satellite communications.

ACR 1. abandon call and retry. 2. allowed cell rate. In ATM, an available bit rate (ABR) service parameter that describes the current allowable sending rate in cells per second. See cell rate. 3. attenuation to crosstalk ratio.

ACRFNET See Academic Computing Research Facility Network.

Acrobat An Adobe Systems commercial page layout software application used for creating documents containing text and graphics. The Acrobat Distiller takes PostScript code and interprets it into objects that can be displayed on a variety of platforms supporting the freely available Acrobat Reader program. The Acrobat Portable Document File (PDF) is popularly used to distribute documents on the Web. Since it retains font/image formating, the PDF files can be displayed or printed in professional layout formats. PDF is also becoming popular for forms handling, since the files can be designed to be selectively edited by a user. Thus, they are useful for electronic, paperless processing of government forms such as business license applications, for example.

In addition to software downloadable from the Net, a number of word processing and desktop publishing programs will export Acrobat-distilled PDF files and the resulting files may be several times smaller than the original, depending upon the contents. Distilled PDFs typically use a *.pdf* file name extension. See Portable Document Format.

acronym A word formed by taking the first letter or letters from each successive word in a phrase or compound term. An acronym is a special form of abbreviation that has wordlike cadence and sound properties and may, because of its ease of pronunciation and mnemonic qualities, eventually become part of the general vocabulary. Examples include scuba (self-contained underwater breathing apparatus), radar (radio detecting and ranging), and BASIC (Beginner's All Purpose Instruction Code).

ACS 1. See Advanced Communication System. 2. See Australian Computer Society. 3. automatic call sequencer. A simple form of automated phone call handler which hands off calls to available agents.

ACSE See Association Control Service Element.

ACSL See Advanced Continuous Simulation Language.

ACSnet The Australian Computer Society national computer network, which maintains links to national networks in other countries. There are close ties between the Australian government and the ACS, with the ACS providing information and critical debate to computing policies in government and, further, representing Australia in the International Federation for Information Processing (IFIP).

ACSR aluminum conductor steel-reinforced. Although aluminum is light and a good conductor, aluminum cables with steel cores tend to be bulkier and heavier than copper.

ACT 1. See Applied Computer Telephony. 2. See Authorization Code Table.

ACTA See America's Carriers Telecommunications Association.

ACTAS See Alliance of Computer-Based Telephony Application Suppliers.

actinism A property of radiant energy in the X-ray, ultraviolet, and visible parts of the spectrum to promote chemical changes.

actinometry The measurement of the radiation intensity emanating from a body. Actinometry is useful in assessing radiant energy from photochemical reactions and astronomical bodies.

ACTIUS See Association of Computer Telephone Integration Users and Suppliers.

activation fee, setup fee In many communications services, there is an activation or setup fee associated with starting a new account. This fee covers the service provider's administrative costs of installing the account and providing the new user with operating instructions, passwords, etc. Sometimes providers will waive activation fees in order to attract new subscribers.

active communications satellite A communications satellite that employs transponders (a type of repeater) or other means of amplifying and forwarding (relaying) a signal, usually with the frequencies shifted so the uplink and downlink transmissions do not interfere with one another. Unlike the larger passive satellites launched in the 1960s, newer active satellites can amplify a signal without the extra bulk needed in earlier systems. Virtually all current satellites are active.

active jamming The deliberate interposition of signals intended to disrupt communications such as radio or radar transmissions.

active line A communications channel that is currently being used. While no human communication may be taking place, if it is a data line, there may nevertheless be meaningful activity on the line, such as computer processes interacting with one another.

active lines In a television image, those lines that are visible to the viewer at any one time. Since a frame consists of many sweeps of the beam, only some of the possible lines may be seen by the viewer at any one time, but because the transition is so fast, the image is perceived as continuous. Those lines that are not active are *blanked*. See blanking, frame, scan line.

active matrix display Usually a liquid crystal display (LCD), active matrix is a means of brightening an electronic display by adding transistors to individual elements to maintain the image between successive scans or refreshes of the screen. Thus, the screen appears to refresh more quickly and gives a crisper, more contrasting appearance that aids in legibility. Color laptops frequently incorporate this technology and active matrix screens are gradually replacing passive matrix screens.

ActiveX Descended from Microsoft's Object Linking and Embedding (OLE) but intended to run over the Internet and to compete with Sun's Java, ActiveX facilitates adding animation, sound, and interactive element into Web documents. ActiveX components are similar to browser plugins, or Java applets. See ActiveX Controls, Java.

ActiveX Controls Microsoft ActiveX controls are interactive objects created individually by developers. They can be embedded in various Web-related applications. ActiveX controls can be programmed in a variety of languages, including Visual BASIC, Java, or C++. They can then be readily shared with other programmers. A number of commercial vendors of authoring and page layout display systems have incorporated ActiveX Controls into their software.

Advanced Communications Technology Satellite

ACTS, the Advanced Communications Technology Satellite, was a project of the late 1980s and 1990s in which the Space Shuttle was used to deploy a communications satellite with large reflective surfaces. Top left: ACTS relationship to the shuttle and the Earth, illustrated in 1987. Top right: ACTS in Earth orbit over the ocean after release by the Space Shuttle Discovery in the early 1990s. Bottom: Advanced Communications Technology Satellite (ACTS) parabolic antennas. [NASA/GRC image; NASA/JSC image detail; NASA/GRC image, August 1996.]

---

activity concentration point ACP. A place in a

network at which there is a high traffic load, in other words, a focal point for higher activity than is ordinarily found in other locations in the system, as in computer network servers.

activity reports Automated usage logs generated by computing devices that are usually accessible through a file or printout. Activity reports can provide information about times and types of use, errors, and sometimes transmitter/recipient information. Activity reports are commonly available on facsimile machines, high-end printers, and some electronic photocopiers. There are disadvantages to the glut of information and statistics that can quickly and easily be generated by electronic systems; there aren't sufficient hours in the day to evaluate all of it, nor is there sufficient storage space to keep hardcopy versions, and there is the danger of turning people into human robots, doing the drudge work of feeding information into computers or trying to evaluate the output all the time rather than using them to increase free time and quality of life.

ACTRIS See Association for Cooperation in Telecommunications Research in Switzerland.

ACTS 1. See Advanced Communications Technologies and Services. 2. Advanced Communications Technology Satellite. 3. See Association of Competitive Telecommunications Suppliers. 4. Automatic Coin Telephone Service. An automated system for handling payphone traffic, it directs the user on how much money to insert, handles calling card calls, provides diagnostic and tuning information to technicians, etc. In areas without ACTS service, calls are handled by TSPS operators.

Actual Measured Loss AML. A telephony system evaluation statistic used to express traffic management efficiency. It is sometimes compared to Expected Measured Loss (EML) to provide a measure of deviation from desired or mandated service efficiency levels.

actuate To initiate or activate into mechanical motion. Rotational motion, signal transmission, or mechanical button or switch positions are examples of devices that may be actuated.

actuator A mechanical or electromechanical positioning or regulating mechanism. Actuators have many uses, such as: (1) aiming antennas to remotely or automatically scan the arc of a satellite, (2) controlling active optics deformable telescopic mirrors, (3) controlling signal or chemical flow through regulating valves, or (4) providing access control in fiber alignment devices.

Actuators have become increasingly small and sophisticated and multiple actuators, ranging from a few to 500,000, may be used to control telescopic mirrors, for example.

Piezoelectric actuators are compact components typically created from multiple layers of ceramics. They are useful for beam deflection, sensors, valves, fiber-to-waveguide alignment devices, and optical and electrical switches.

In NASA/HR Textron/U.S. Armed Forces flight tests, fiber optics were found to sufficiently reliable and maintainable for use with smart actuator technologies. See AS-Interface.

actuator bus A transmission link for delivering actuator-related digital signals. See INTERBUS, PROFIBUS.

ACUTA See Association of College and University Telecommunications Administrators.

ADA 1. A high level, structured, data-typed programming language, somewhat like an extended Pascal, developed by and mandated within the Department of Defense, but not popular outside of this circle. It has been criticized by some programmers as being cumbersome and difficult to use. The language is named after Ada Lovelace, the technically astute daughter of Lord Byron. There have since been variations on ADA, including ADA++, which in turn has been superseded by ADA 95. See Lovelace, Ada. 2. Average Delay to Abandon. The average length of call duration for a caller held in a queue who hangs up before being connected with the callee.

ADACC Automatic Directory Assistance Call Completion.

adapter A device to connect one type of component, system or connector to other components, systems, or connectors to provide physical and electronic compatibility on each end of the connection. An adapter is used when the two connections do not naturally couple with one another. Related to adapters are connectors, which are most often small passive devices, simply passing information or current through, while adapters tend to be combined with active, signal-processing or enhancing components, or with gender changers, extenders, or splitters. See adaptor, adopter.

adapter card See peripheral card.

adaptive antenna array A series of antennas grouped and arranged so the combination of antennas provides enhanced reception or transmission over individual antennas. An antenna array can be configured to monitor signals or signal conditions or to use input from other sources, such as computers, and to adapt to them as appropriate. For example, in a directional antenna array, if the signal shifts due to movement on the part of the sending antenna, the array may be able to move or swivel to optimize the signal level (as in elliptical satellite orbit communications). For wireless communications, adaptive arrays with horizontal extensions are being installed on existing cell towers to direct beams towards mobile terminals in order to increase capacity.

Sector antennas may sometimes be housed in the same physical structure as adaptive array antennas to reduce the footprint of the assemblage, but care must be taken to arrange them so they do not cause distortion. See Butler antenna; antenna, smart.

adaptive channel allocation ACA. In systems where transmissions channels are dynamic for purposes of congestion control, resource allocation, or security, there may be mechanisms to change channel allocations between or during transmissions based upon scheduled, estimated, or computed loads. Commercial multiplexed wireless communications systems use ACA to maximize resources and respond to sub-

scriber demands. Many secured military communications systems use ACA for privacy and to prevent jamming.

adaptive communication A communications system that incorporates intelligence and feedback mechanisms to optimize signal or data transfer. In telephony, a cordless phone may automatically switch channels to find a better signal if the current one deteriorates. In the telephone switching system, a phone call may be routed through another trunk if congestion is detected. In computer network systems, the system may reroute packets if one of the hops in a journey changes or becomes unavailable.

adaptive differential pulse code modulation ADPCM. An ITU-T standard for voice digitization and compression in which sample rate speeds are related to the variation in the samples, thus using fewer bits than pulse code modulation (PCM), which is commonly used in digital voice coding, if the sample speeds are slow. An analog voice can be carried on an up-to-32 Kbps channel. ADPCM can be used over digital networks such as frame relay systems.

adaptive equalization A means of compensating for distortion or loss of detail in digital data or in the digital signals transmitted through a network link. Adaptive equalization can compensate for a number of types of distortion including delay, echo, and co-channel interference.

Trained adaptive equalization relies upon cooperative transmissions between the sender and the receiver to enable the receiver to make adjustments based upon the training interval. *Blind* adaptive equalization is based upon only the data received, without the benefit of training data.

Experiments have shown that adaptive equalization can improve high-speed data transmissions over multimode fiber at certain frequencies.

In the mid-1990s, Drewes et al. assessed adaptive linear equalization and decision feedback equalization for wireless ATM systems. Wireless data transmissions are subject to impulse prolongation from multipath propagation leading to intersymbol interference; one means of dealing with the problem is to use time division multiple access (TDMA) with adaptive equalization. Drewes et al. found that different algorithms were suitable for uplink and downlink transmissions, suggesting that least squares algorithms were best for radio broadcast channels that are different for each time slot.

In graphics and visualization applications, adaptive equalization is utilized to increase clarity of visual details without changing the color information by reducing overall dynamic range.

adaptive predictive coding APC. A means of converting analog signals to digital data through a predictive algorithm built upon linear functions of previous values for the sampled, quantized signals. The basics of speech encoding were developed in the 1930s, and predictive coding was applied to television broadcasts a couple of decades later. Improvements in the form of adaptive algorithms were developed by M. Shroeder et al. in the 1960s for speech transmission. The idea was to use preceding samples to continuously predict subsequent speech samples, to compare them, and transmit the residual prediction error. At the receiving end, the same prediction algorithm and information about the residual were used to reconstruct the samples. The result was more natural sounding speech.

The approach was not limited to speech encoding; it was generalizable to audio and visual signal processing tasks, such as sound or image compression. APC may be used with other types of predictive coding, as in still image compression in which different characteristics of an image are processed through different coding schemes for maximum compression.

APC is frequently used for speech and image digitization and is useful in environments where data compression lowers bandwidth or increases the speed of network transmissions. See Shroeder, Manfred.

adaptive routing A system of dynamic network routing that utilizes intelligence in addition to information in routing tables, to establish best routes, fastest routes, or alternate routes in the case of obstructions in the usual paths. Adaptive routing is intended to help optimize routing in a system that may change in its overall scope or topology or in its use characteristics at a given point in time. Optimization through adaptive routing is not always measured in terms of speed. In a network in which machines come and go (e.g., a large distributed network like the Internet), adaptive routing may be assessed in terms of reliable delivery of data in a constantly changing environment, for example. See hop-by-hop routing.

adaptor A person who physically or otherwise directly modifies a system, component, or instrument to perform a different function or to perform a function or set of functions in a different way. See adapter, adopter.

ADAS See Automated Directory Assistance Service.

ADB See Apple Desktop Bus.

ADC, A/DC 1. analog-to-digital converter. A technology that is widespread in telecommunications and increasingly important for wideband wireless low-voltage devices. 2. automated/automatic data collection.

ADC Telecommunications A global supplier of telecommunications quality assurance and testing products, founded in 1935.

ADCA 1. Aerospace Department Chairman's Association. 2. See Automatic Data Capture Association.

ADCCP Advanced Data Communication Control Procedures. A bit-oriented, ANSI-standard communications protocol related to High Level Data Link Control (HDLC).

ADCIS 1. Aged & Disability Care Information Service. 2. analog/digital CMOS ICs. 3. See Association for the Development of Computer-Based Instruction.

Adcock antenna A transmitting/receiving antenna with two or more vertical conductors arranged so that the pickup is minimized in the horizontal wires. Adcock antennas can be arranged in arrays to provide directional transmitting/receiving; one such array system resembles the configuration of the five dots on a throwing die.

ADCU Association of Data Communications Users.

add-on 1. More commonly known as three-way calling or add-on conference, a telephone subscriber feature that enables the connection of a third phone into an ongoing conversation. It is usually accomplished by putting the conversation on hold, calling the third party, and returning to the initial call with the third party linked into the call. 2. See applications processors, peripheral device.

add/drop multiplexer A/DM. In computer networks such as ATM networks, a system for sending a variety of types of data or data channels that are then "split out" as needed by individual computer terminals in a switch loop. A/DMs may also be used at add/drop points where local area networks connect to a long-haul network. The incoming data passes through flow control and add/drop control circuits before continuing along the transmissions path. A/DMs are common to ring architectures and tend to be asymmetric. Newer systems that incorporate some of the characteristics of digital cross-connects are capable of symmetrical behavior and may not require preassignment of ports, thus increasing flexibility and scalability.

In point-to-multipoint networks, A/DMs enable circuits to be added and dropped along the transmission path through a process of demultiplexing, cross-connecting, adding/dropping, and remultiplexing or through more recent architectures in which intermediate access points are added and dropped without de- and remultiplexing. In SONET byte-interleaved multiplexing, for example, lower rate signals may be associated directly with higher rate signals and added/dropped in one step.

For optical networks, A/DMs have become quite sophisticated. Some now utilize tiny micro-electromechanical system (MEMS) components and some are used in larger scale metropolitan trunking applications. The multiplexer may retain most of the optical data stream or may convert it into electrical signals. On fiber optic access and transport networks, a multiservice A/DM can provide interfaces for a number of network configurations (ATM, Fast Ethernet, TDM, etc.) to support heterogenous distributed networks. Optical A/DMs with passive thermal compensation have been devised to add and drop subsets of channels without converting between electrical and optical signals. Combinations of channels may be added or dropped while maintaining pass-through for priority channels.

A/DM concepts are not limited to physical media. Multiplexing is commonly used in programming where various types of data (image, sound, etc.) are bundled and transmitted together and then algorithmically "demultiplexed" at receiving terminals or processes, as needed. See interleaver, micro-electro-mechanical systems; multiplexing; switch, optical.

ADDACC See Automated Directory Assistance Call Completion.

additive increase rate factor AIRF. In ATM, an available bit rate (ABR) service parameter for controlling cell transmission rate increases is called the additive increase rate (AIR). AIR is signaled as the additive increase rate factor (AIRF) where AIRF equals AIR times the maximum number of cells permitted for each forward RM-cell (Nrm) divided by the peak cell rate (PCR).

Additional Cooperative Acceptance Testing ACAT. A method of telephone testing in which a technician at the central office works in cooperation with a carrier-provided technician at the carrier premises to test line integrity conditions such as noise, jitter, distortion, signal-to-noise ratios, and other typical transmission characteristics and possible sources of interference in a new installation.

address A locator, usually in the form of a number, of a position in memory or other storage medium, such as a hard drive or floppy diskette. A telephone number is a unique address on a phone system, used to establish a connection. An email address is a unique identifier used in the transmission, receipt, and storage of electronic messages over a network. There are directories on the Web that store the email addresses of specific individuals or companies on the Internet, or that can retrieve a name and address, given a specific email address. The individuals whose addresses are listed are not necessarily aware of the fact. See address, MAC; ego surfing; electronic mail.

address, Internet An Internet address, or Internet Protocol (IP) number, is a unique host name identifier on the Internet. IP addresses can be expressed as numbers, *255.0.0.0*, or as a full DNS name, *www.4-sightmedia.com*. A registration process is required to obtain a unique address on the Internet. See Domain Name Service, InterNIC.

address, MAC A Media Access Control (MAC) address is a device address on a network. See MAC address, Media Access Control.

address filtering Decision-making on a network as to which data packets will be permitted to continue. For example, a filter evaluates the source and destination Media Access Control (MAC) address and compares it against any specific restrictions or instructions that have been set up for the system. On a general level, address filtering can be used to keep out messages from unwanted sources, such as bulk commercial mail senders, and to reject messages to local destinations that may no longer exist, or that may be restricted. See firewall.

address resolution AR. On the Internet and local area networks (LANs) using ATM, the conversion of an Internet Protocol (IP) address or local address into its corresponding geographical/physical address. It may be done in stages, through a discovery process, with the layer address being sought first and other parts of the address, such as a Media Access Control (MAC) address, being resolved at a more local level. This hierarchical approach can streamline the amount of information that needs to be processed and carried initially and provides the flexibility to reorganize machines, switches, and routers at the local network level.

Address resolution is done by broadcasting from the sender to a number of nodes at the general destination and then responding to a specific destination, once information has been sent back from the appropriate end station to show where it is. See address, Address Resolution Protocol, MAC address, Media Access Control.

Address Resolution Protocol ARP. A protocol used to systematically, dynamically discover the low level physical network system that corresponds to an Internet Protocol (IP) address for a given host. ARP is used over physical networks that can handle broadcast packets (not all networks have a broadcast layer) to all the hosts, or the relevant hosts, on the system. By broadcasting to a general destination and then evaluating the responses by the local hosts, the specific address can be discovered and resolved without all the information about all possible destinations being stored at the originating system. See address, address resolution, MAC address, RFC 826.

address translation gateway ATG. A Cisco Systems DECnet routing software function for routing multiple, independent DECnet networks. ATG enables the user to establish address translation for selected network nodes.

addressee The intended recipient of a written message or data communication. See email.

addressing In computer programming and operations, a means of keeping track of stored information so it can be accessed in the future as needed.

ADF 1. See automatic direction finder. 2. automatic document feeder. A built-in or optional device on a printer, photocopy machine, facsimile machine, or scanner that holds a sheaf of paper, usually unattached single sheets, and feeds these pages individually through the machine. Some machines have a series of paper trays for different sizes or types of paper and can cycle through the trays as needed or automatically select the paper size.

ADIO, A/D I/O *abbrev.* analog/digital input/output.

adjacent Near; next to; directly before or after; beside. Having a shared border, contiguous with. If something is adjacent, then no other device or process of the same kind is between it and that to which it is adjacent. For physical devices, the adjacent entities may or may not be physically touching or connected by cables or other means.

adjacent channel In communications, a wavelength or stipulated channel bordering the signal in question. In AM radio communications, adjacent channels are relevant because a nearby signal may be very close to the desired signal and require fine tuning to get a good signal. A nearby signal may also overpower a weaker signal. In FM communications, adjacent channels are separated by *guard* channels to prevent interference. In fiber optics, adjacent wavelengths do not interfere with communications in the same way as wireless radio communications, but are still significant; it may be necessary to convert wavelengths such that they do not conflict with adjacent wavelengths or it may be necessary to separate adjacent wavelengths for adding or dropping them from a network at relevant connection points. See adjacent channel interference.

adjacent channel interference Due to demand, broadcast spectrums are subdivided into narrow bands to accommodate many channels. When broadcast channels are adjacent, the signal from one may interfere with those nearby. Most people have experienced this type of interference in AM car radios; as they move farther from the signal of the current selected station, adjacent stations (or stronger stations) may be heard over the desired station. For this reason, some of the better radios are equipped with adjacent channel selectivity circuitry which rejects the transmissions of adjacent channels to provide cleaner reception. See adjacent channel.

adjunct 1. Something that is additional to, or joined to, something else, but that is not essentially part of it. 2. Assistant, aide, associate. 3. A peripheral device that enhances a system, without being essential to its basic operation, such as a computer microphone, joystick (gamers would argue that this is essential), modem, telephone headset, etc.

adjunct service point In intelligent networks (INs), a point in an intelligent peripheral that processes logic interpreter service requests.

Adjunct System Application Interface ASAI. A set of AT&T technical specifications for the controlling of private branch exchange (PBX) telephone systems by computers.

ADM 1. adaptive-delta modulation. 2. add/drop multiplexer.

administrative domain AD. The group of network hosts, switches, and routers and their interconnections managed by a specified administrative authority, such as a system administrator on a small network or a network control center for a larger network.

Administrative Operating Company Number AOCN. In the telephone industry, AOCN providers handle a variety of national call routing and rating databases and services to telephone companies. AOCNs may also obtain NXX and other telephone codes on behalf of their clients.

Individual Operating Company Numbers (OCNs) are assigned to telephone companies to aid in this administration, as AOCNs serve multiple vendors. Vendors are required to select an AOCN.

NECA Services, Inc., a company evolving from the National Exchange Carrier Association, Inc., was established in 2000 to provide AOCN services to telecommunications vendors. CHR Solutions, Inc., is also authorized to provide AOCN services and may enter and update information in Traffic Routing Administration (TRA) databases. See Operating Company Number.

admittance (*symb.* Y or y) In an electrical circuit or material, a measure of the facility with which the current flows through the circuit or material. Admittance is rather whimsically expressed in *mho* units, which is *ohm* spelled backward, since ohms are used to express impedance, the reciprocal of admittance. Contrast with impedance.

ADN See Advanced Digital Network.

Adobe Systems Incorporated A California and Seattle-based company, Adobe is best known for Post-

Script, Acrobat, PageMaker, Premiere, and Illustrator, software products which are aimed at the large number of home and professional publishers, communications specialists, and graphics users. See Acrobat, PostScript.

Adonis A computer network of the Institute for Automated Systems in Moscow, Russia.

ADONIS Article Delivery Over Network Information Systems. A project of a group of well-known technical publishers for electronically publishing hundreds of scientific/technical journal articles. ADONIS data is made available on digital media and targeted for distribution to educational institutions and other relevant markets.

adaptor A person or body (e.g., corporate entity) that makes use of or takes on a particular concept, style of management, technology, or device. The term is often used in reference to those who are early to adapt a new or unproven (bleeding edge) technique or technology. See adapter, adaptor.

ADP automated data processing.

ADPCM See adaptive differential pulse code modulation.

ADQ See Average Delay in Queue.

ADR 1. achievable data rate. 2. aggregate data rate. 3. analog to digital recording. 4. ASTRA Digital Radio. Radio based on the ASTRA European satellite system.

ADS 1. advanced digital system. 2. See AudioGram Delivery Services. 3. automated data system.

ADSL See Asymmetric Digital Subscriber Line.

ADSL Forum An international association of ADSL professionals formed in 1994 to promote and disseminate information about asymmetric digital subscriber line (ADSL) services, fast communications over copper wires. The Forum provides technical and marketing information, including conferences and analysis of ADSL-related technology. http://www.adsl.com/

ADSP See AppleTalk Data Stream Protocol.

ADSTAR Automated Document Storage And Retrieval.

ADSU ATM Data Service Unit. A device for connecting to data interfaces (e.g., in ATM networks) to support networking through standard connectionless and connection-oriented adaptation layers.

ADTV See advanced-definition television.

ADU asynchronous data unit.

ADVANCE Project A project of the European Community Telework Forum (ECTF) to stimulate and coordinate leading global telework development throughout Europe, in conjunction with other organizations committed to this goal. The stimulation of new types of businesses, particularly small businesses and the support of existing businesses, are key goals of the project. See European Community Telework Forum, telework.

advance replacement warranty A type of warranty return/replacement service in which the replacement device or component is shipped prior to the returned item so the user can continue usage until the problem is corrected or the unit replaced. This service is valuable if the essential component's absence would reduce productivity. It's important to check billing policies on ARWs because some companies will bill a credit card until the return unit is received and then apply a credit, all of which may be prone to error and confusion if not monitored carefully.

Advanced Adaptive Protocol AAP. A network protocol designed to adapt dynamically to the available connection bandwidth, optimized to the application and device type.

Advanced Branch Exchange ABX. Not in common usage, but a phrase used to distinguish traditional voice-only telephone exchange branches from those providing newer integrated voice/data capabilities.

advanced broadcast television ABT. A general category encompassing audio/visual broadcast technologies that offer substantial quality and resolution improvements over traditional analog television services that prevailed until the 1990s. ABT systems typically offer better sound, higher resolution images, and interactive options. ABT services offered through cable are now usually digital, but analog/digital hybrid systems will continue to exist as the nature of wireless radio transmissions is analog. Because of entrenched commercial consumer television technologies, the implementation of ABT has lagged far behind its technological development. However, as the buying public has become better informed about ABT-related products through marketing and the Internet, the demand for advanced services and home entertainment systems has increased, particularly in North America, Western Europe, and Japan.

advanced common-view ACV. A time-referencing technique used to transfer frequencies and times of the various standards that contribute to Coordinated Universal Time.

Advanced Communications Technologies and Services ACTS. A European program for furthering communications technologies and infrastructures in the areas of multimedia, photonics, high-speed networking, mobile communications, and more. Over 200 projects have been part of the ACTS program providing valuable test and implementation information for European network development and deployment. See BLISS, BONAPARTE, BOURBON, BROADBANDLOOP, UPGRADE, WOTAN.
http://www.infowin.org/ACTS/

Advanced Continuous Simulation Language ACSL. The first widely successful commercial software language to facilitate the simulation or modeling of the behavior of continuous systems described by time-dependent, nonlinear, differential transfer functions. This generic simulations tool is useful in a number of fields, including aeronautics simulations, control system design, toxicology, heat and fluid movement analysis, and chemical process dynamics. ACSL components include graphic modeling, simulation, mathematical analysis, open application program interface, visualization, and others.
ACSL assets were acquired from MGA Software, Inc., in 1998 by Aegis Research Corporation with the intent of integrating ACSL with HLA Lab Works software.

Advanced Data Communications Control Procedures ADCCP. A bit-oriented, code-independent, data link communications control protocol (ANSI X3.66). ADCCP is an ANSI-standardized version of IBM's Synchronous Data Link Control (SDLC) Protocol and is related to ISO's High-Level Data Link Control (HDLC) Protocol Family and the CCITT X.25 link-level protocol. In 1996 it was released for public comment as a revision to ANSI X3.66:1979 by the Accredited Standards Committee X3. The revision included six subsections representing and specifying procedures, frame structures, classes, Exchange Identification (XID) command/response, and general purpose information field content and format for XID. X3.66 has also been adopted as a U.S. federal standard (FED-STD-1003; FIPS PUB 71).
A number of popular communications protocols, including ZModem, use 32-bit CRC error checking mechanisms based upon ADCCP. Department of Defense (DoD) interface standards for interoperability and performance for medium- and high-frequency radio systems use a 16-bit frame check sequence (FCS) as specified by FED-STD-1003.
The Link-Level Cluster Communications Protocol is a subset of ADCCP intended to facilitate the exchange of messages betweeen a master workstation and cluster workstations.

advanced-definition television ADTV, ADT, ATV. A general category of television technologies and related services that encompasses improved resolution and picture quality over traditional analog television up to the 1990s. See Advanced Television Systems Committee.

Advanced Digital Network ADN. A commercial leased-line 56 Kbps digital phone subscriber service.

Advanced Intelligent Network AIN. A telephone services architecture based around Signaling System 7 (SS7), and possible future versions of SS7, intended to integrate ISDN digital capabilities and cellular wireless services into a personal communications system (PCS). The AIN grew out of the Intelligent Network (IN) system initiated by Bell Communications Research (Bellcore) in 1984. It can dynamically process calls by evaluating 'trigger points' through the call handling process.
Currently a newer technology to AIN, called Information Network Architecture (INA), may coexist with AIN or eventually supersede it. See Information Network Architecture, Intelligent Network, Personal Communications System.

Advanced Metal Powder AMP. A durable metal powder technology suitable for use in high-capacity, very dense storage technologies such as backup tape cartridge media. AMP enables smaller particles to be used and can be coated with thinner coatings to create dense recording surfaces with higher magnetization levels than traditional media. AMP is used in Super DLTtape. See Advanced Thin-layered and High Metal Media.

Advanced Mobile Phone System AMPS. An analog cellular communications system utilizing frequency modulation (FM) transmissions, developed by Bell Laboratories in the 1970s based upon Bell mobile phone services with improved sound quality and features that were installed in the mid- to late-1960s. In 1972, a significant patent was awarded which described handoffs during travel between cells, setting the stage for future mobile phone services. AMPS was first implemented in 1978 in the U.S. and Korea. It uses the same bandwidth as a landline voice channel but is modulated onto a frequency-modulated (FM) carrier using frequency division multiple access (FDMA).

AMPS became the first standardized cellular phone service (1983) to use the 800 to 900 MHz frequency range, which is still the predominant type of cellular system in the world. NAMPS (Narrowband Analog Mobile Phone Service) is an interim enhancement to AMPS, which uses *frequency division* as a way of sectioning the bandwidth, a tradeoff that increases calling capacity but may also increase interference. AMPS is still a significant analog service but is slowly giving way to digital systems offering more features and better call security. See cellular phone, DAMPS, NAMPS. See AMPS, cellular phone, mobile phone, cell, cluster, roaming.

Advanced Network and Services ANS. A nonprofit organization founded jointly by the National Science Foundation, Michigan Education and Research Infrastructure Triad (MERIT), IBM, and MCI in September 1990 to develop a gigabit network to benefit American education and research. Initially ANS planned two independent networks running over the same system of physical lines. Various issues emerged as controversial, such as corporate access, cost of operations, and use of the MCI backbone topology, which was criticized as being insufficiently robust and lacking in redundancy.

Advanced Peer-to-Peer Networking APPN. A distributed networking system, now included in the Systems Network Architecture (SNA) developed by IBM. APPN workstations are dynamically defined to reduce the need for extensive changes when the network is reconfigured. APPN provides optimization of routing between devices, direct communication between users, direct remote station communication, and transparent sharing of applications over the network.

Advanced Peer-to-Peer Networking+ APPN+, APPN Plus. An enhanced IBM APPN which includes faster throughput, dynamic rerouting and congestion control, and other features to make it competitive with TCP/IP. See Advanced Peer-to-Peer Networking.

Advanced Radio Interferometry between Space and Earth ARISE. An advanced space communications/sensing project consisting of one or two 25-meter radio telescopes stationed in high Earth orbit (HEO). In conjunction with Earth-based telescopes, the ARISE will use very long baseline interferometry (VLBI) to obtain the highest resolution images of the most energetic astronomical phenomena. The data collected will aid scientists in studying the structure and evolution of the universe. The ARISE Web site is coordinated through NASA/JPL with information on the equipment, the science, and the potential

benefits of this type of cosmological research. See Very Large Array. http://arise.jpl.nasa.gov/

Advanced Satellite for Cosmology and Astrophysics ASCA. A cosmic X-ray astronomy mission in which the U.S. provided a scientific payload to the Japanese project, the fourth of its kind. ASCA (formerly called Astro-D) was launched in February 1993. It was the first such mission to use CCDs for X-ray astronomy. The technology is highly sensitive and especially useful for observing emission lines and absorption edges.

ASCA carried four large-area X-ray telescopes; two for use with a gas imaging spectrometer (GIS) and two with a solid-state imaging spectrometer (SIS). The observing program was available to participating Japanese and U.S. institutions and members of the European Space Agency. In July 2000, attitude control was lost and in March 2001, ASCA re-entered Earth's atmosphere.

Advanced SCSI Programming Interface ASPI. A SCSI host adapter-independent programming interface released by Adaptec in the late 1980s. ASPI permits multiple device drivers to share a disk controller by providing a consistent device driver interface. Typically developers have had the burden of supporting many different host adapters, writing several, sometimes dozens of individual device driver definitions and programs for their users. The user then either has to install and load them all or search through them at installation time, trying to locate the right device driver for the hardware peripheral, often a time-consuming, hit-or-miss process.

With ASPI, vendors can make their products ASPI-compatible, so software can talk to the hardware without many extra files or hit-or-miss installation effort on the part of users. While there are similar systems from other vendors, this is one of the more popular ones.

advanced telecommunications capability This acknowledgment of multimedia forms of communication is defined in the Telecommunications Act of 1996 and published by the Federal Communications Commission (FCC) as:

" ...without regard to any transmission media or technology, as high-speed, switched, broadband telecommunications capability that enables users to originate and receive high-quality voice, data, graphics, and video telecommunications using any technology."

Advanced Telecommunications Institute ATI. ATI, located at the Stevens Institute of Technology, promotes and supports the research of advanced telecommunications applications and services. http://www.ati.stevens-tech.edu/atihomepage/

advanced television, advanced TV ATV. A generic category for television broadcast technologies that supply better audio and/or video characteristics than are generally associated with the traditional NTSC system in North America. Various means of digital manipulation at the broadcasting or receiving ends can result in better picture viewing or sound without changing the underlying broadcast format, while others require a completely different way of sending and encoding a signal. High Definition Television (HDTV) is a type of advanced TV. See Advanced Television Systems Committee, ATSC Digital Television Standard.

Advanced Television Enhancement Forum ATVEF. A consumer electronics, broadcast, and cable networks industry alliance promoting the creation and distribution of enhanced television technologies at costs accessible to general consumers. See Advanced Television Forum, ATVEF Enhanced Content Specification. http://www.atvef.com/

Advanced Television Evaluation Laboratory ATEL. A world-class subjective evaluation facility, conformant to ITU-R Rec. 500. ATEL was one of three primary labs used to test advanced television systems that led to the Final Report of the Advisory Committee on Advanced Television Service (ACATS). ATEL is used by the Advanced Video Systems Group of Communications and Research Centre Canada (CRC) to conduct research and testing. The Advanced Video Systems Group conducts leading-edge research in video technologies and human visual perception as they relate to a wide variety of broadcast and multimedia applications. See Advanced Video Systems Group. http://www.crc.ca/

Advanced Television Forum ATVF. A nonprofit corporation that addresses global issues related to content and technology for enhanced TV technologies including commercial implementation of these technologies. http://www.atvf.org/

Advanced Television Systems Committee ATSC. An international committee establishing voluntary technical standards for advanced television systems. The ATSC has established *Recommended Practices* for the industry. The ATSC Technology Group on Distribution released the ATSC *Digital Television Standard* in September 1995 (Document A/53). The same year, the ATSC also published *Guide to the Use of the ATSC Digital Televison Standard* (A/54). The Digital Television Standard was, in large part, adopted by the Federal Communications Commission (FCC) in December 1996 and also adopted by Canada and some Asian and South American countries. The influential standards document was revised by the ATSC and released as A/53A in April 2001. It specifies the technical parameters of advanced TV systems, including input scanning formats, preprocessing, and compression parameters, the service multiplex, transport layer characteristics, and the transmission subsystem. The implementation of these standards may require licensing of patented technologies. For a summary overview of highlights of the standard, see ATSC Digital Television Standard. ATSC Standards documents can be downloaded from the Web. http://www.atsc.org/

Advanced Television Technology Center, Inc., Advanced Television Test Center ATTC. A private, nonprofit, corporate laboratory facility established in 1988 to test and recommend practical technology solutions for delivery and display of new U.S. terrestrial broadcast transmission systems. The ATTC, located in Alexandria, Virginia, was established as a result of Federal Communications Commission (FCC) Advisory Committee on Advanced Television Service (ACATS) research into advanced television (ATV) systems. It was determined that paper specifications for ATV would not be sufficient to fulfil the ACATS mandate and that a test facility was needed to evaluate various hardware configurations. The ATTC was colocated with the Cable Laboratories ATV facility.

The primary mandate of the Center is to facilitate the implementation of digital television technologies. The Center further supports education of engineers and other broadcast professionals through seminars and certification information. Articles and research reports are available online in Adobe PDF format. In 1990, Harris Corporation provided the radio frequency Test Bed used in testing the digital television systems. See Advisory Committee on Advanced Television Service, Association for Maximum Service Television, Harris Broadcast Communications. http://www.attc.org/

Advanced Thin-layered and High Metal Media ATOMM. A super-thin, super-smooth coating developed by Fujifilm that enables a magnetic layer over a nonmagnetic layer of titanium to be coated. This highly smooth surface improves read/write head-to-media contact and reduces spacing loss in order to support high-density recording and storage capacities. Using Advanced Metal Powder (AMP), the smaller, more thinly coated particles have a higher magnetization level. See Advanced Metal Powder.

Advanced Tracking and Data Relay Satellite System ATDRS. A NASA project to provide a shared communications service between the Earth and a geosynchronous orbit position. Among other things, the ATDRSS would facilitate launch and landing planning, testing, and execution. The system consists of relay satellites and two independent ground terminals. Planning studies for Phase B began in the early 1990s, and the satellites were expected to provide services until about 2012.

Advanced Very High Resolution Radiometer AVHRR. A broadband device for sensing passive radiation emitted from the Earth and its atmosphere. AVHRR technology is used on orbiting satellites, notably the National Oceanic and Atmospheric Administration's (NOAA's) Polar Orbiting Environmental Satellites (TIROS and NOAA-*x*) that have carried it as of 1987. The AVHRR provides global collection of data as the satellite orbits the Earth 14 times a day. Data formats include High-Resolution Picture Transmission (HRPT), Local Area Coverage (LAC), and Global Area Coverage (GAC). Data are both recorded and continually transmitted. The EROS Data Center (EDC) receives data from over the conterminous U.S. about six times a day and, since 1990, also receives global LAC and GAC data.

AVHRR data are suitable for many applications, including the research, mapping, and monitoring of vegetation (forests, grasslands, tundra), agriculture, and land cover. See Global Area Coverage.

Advanced Video Systems Group AVSG. A Canadian research group engaged in high technology multimedia and broadcasting studies, including broadcast television, high-definition television, and 3D-TV. The AVSG utilizes the Advanced Television Evaluation Laboratory for its research. More specifically, it studies video technologies as they relate to human perception. The Group has strong ties to the Video Quality Experts Group (VQEG), an international association of experts tasked with validating objective measures of picture quality for broadcasting. See Advanced Television Evaluation Laboratory.

Advisory Committee on Advanced Television Service ACATS. A committee of private sector individuals providing broad representation from the television broadcast industry reporting to the Federal Communications Commission (FCC) to recommend improvements to existing National Television Systems Committee (NTSC) television broadcast standards. The original North American NTSC standard was adopted by the FCC in 1941, with NTSC color standards adopted in 1953. Since then, there have been many improvements in technology, but sluggish commercial implementation and consumer adaptation of advancements hindered the commercial success of advanced technologies. ACATS was formed in response to this industry lag and to the fact that technologies in other nations appeared to be advancing ahead of U.S. standards.

Since its formation, ACATS has narrowed its focus and made recommendations on advanced television (ATV) service to the FCC. The Committee began to concentrate on advanced television technologies in 1987 and adopted/presented their Final Report in October/November 1995. Surprisingly, digital systems were not a significant focus of the Committee until 1990, when the convergence of computer technologies and broadcasting began to make a significant impression. In the *Final Report*, ACATS optimistically suggested ways in which commercial television could be brought into closer line with technological advances and provided advisement on ATV technical standards, based on theory and laboratory research.

Research was conducted primarily at the Advanced Television Test Center (ATTC), a private, nonprofit organization, Cable Laboratories, Inc. (CableLabs), a research and development consortium of American cable TV system operators, and the Advanced Television Evaluation Laboratory (ATEL), a facility of the Canadian Department of Communications. After narrowing many initial proposals, one EDTV system and five HDTV systems were laboratory tested from 1991 to 1992, resulting in the one analog system being eliminated from further consideration. It was then decided to combine the remaining digital systems into one "best system" rather than to continue the expensive process of developing and testing four separate systems which were, in many ways, converging. Thus, the *Digital HDTV Grand Alliance* was formed.

The *ACATS Technical Subgroup* continued to work

with the Grand Alliance and the Alliance was exhorted to retain a flexible approach and retain the public process aspect of development. The Alliance/ACATS system was tested and evaluated in the field in 1995, with the *Final Technical Report* based upon the results. At the time the *Report* was released, the broadcast system was the only one in the world to incorporate and support both scanning techniques (including progressive scanning formats). While the system was recommended for terrestrial ATV broadcasting, the Committee considered it to be sufficiently broad in its formulation to accommodate many computer media delivery technologies.

Thousands of public documents were generated during the course of the project, including a number of interim reports and recommendations. A report to the U.S. Congress was presented in 1989. A subgroup of the U.S. Government's Information Infrastructure Task Force endorsed the *Report*, along with the 1994 NIST/ARPA Workshop on Advanced Digital Video, and the Information Technology Industry Council. Much of the volunteer work and all out-of-pocket expenses were underwritten by Committee members. Laboratory work was funded by sponsors and grants. See Advanced Television Systems Committee, Digital HDTV Grand Alliance. http://www.atsc.org/

Advisory Committee on Public Interest Obligations of Digital Television Broadcasters PIAC. A committee established by U.S. Presidential Executive Order #13038 to study and advise on public interest responsibilities for those granted digital television licenses. Because airspace, that is, broadcast spectrum frequencies, is a limited and prized commodity legally belonging to the American people, those being granted licenses have a compensatory responsibility to serve the public interest, not to engage only in for-profit commercial enterprises. To fail to require commercial broadcasting companies to support public services and diverse subscriber communities through broadcasting would be like granting commercial industries unrestricted access to the resources of public parks without consideration for the needs and desires of the public itself.

NTIA is the Secretariat for the Advisory Committee. The Committee was comprised of members of the public, the broadcasting and computer industries, academics, and labor representatives.

The Committee's final report "Charting the Digital Broadcasting Future" was released in December 1998. Recommendations made by the Committee in the report include

- disclosure of public interest activities by broadcasters on a quarterly basis
- drafting of an updated voluntary Code of Conduct to reinforce public interest commitments
- adoption of a set of minimum public interest requirements for broadcasters in services for public benefit
- improvement of education through broadcasting
- balancing of the economic benefits of new multiplexing technologies with the choice of a

fee, contribution, or provision of multicasted channels for public interest purposes

- improvement of the quality of political discourse through free airtime before major elections and removal of prohibitions or bans on its sale to state and local political candidates
- cooperation with emergency communications specialists for effectively transmitting disaster warning information
- digital programming support for the disabled
- encouragement and furtherance of diversity in broadcasting
- exploration of alternative approaches inherent in the new television environment for serving public needs and interests

In addition to Committee recommendations, a number of dissenting opinions and alternate recommendations were submitted and provided in Section IV of the Advisory Committee's report.

The report itself was criticized by some as too lenient. In general, it recommends voluntary compliance and, in fact, provides for approximately two years of experimentation with new frequencies before the full mandate would take effect. From the point of view of detractors, the situation could be described as giving out experimental expense accounts and then saying two years later, "don't forget to make some voluntary charitable contributions with the money we gave you."

Some questions lingered after release of the report. Can commercial entities be relied upon to serve consistently the public interest without strong incentives and directives to do so? Will the full potential of new advanced television technologies be realized if subscribers are seen only as consumers and not as participants in the building of an information society? Thus, in October 1999, Vice President Al Gore wrote to the Chairman of the FCC tasking the FCC with taking

"... the next critical step: examining how broadcasters can fulfill their obligation to serve the public interest. Because of the critical importance of television to our nation, we believe that Americans should have the opportunity to participate in the process, we urge the Commission to institute a public proceeding to consider the public interest obligations of digital television broadcasters."

Thus, there are ongoing important issues faced by the government, the broadcast industry, and the public which remain relevant and subject to scrutiny and debate in an environment driven by a powerful broadcast industry that seeks voluntary self-regulation. See Alliance for Better Campaigns, Benton Foundation.

AE 1. acoustic emission. 2. Application Entity.

AEA 1. See American Electronics Association. 2. See American Engineering Association.

AEC See acoustic echo canceller.

AECS Plan Aeronautical Emergency Communications System Plan. A voluntary system of communication established and organized for the provision of emergency communications to the U.S. President and federal government representatives.

AECT See Association for Educational Communications and Technology.

AEEM Aerospace Engineering and Engineering Mechanics.

AEGIS Advanced Electronic Guidance and Instrumentation System.

Aegis System An advanced, automatic tracking and detection phased-array radar used by the U.S. Navy since 1973 to perform simultaneous searching, tracking, and missile guidance functions.

AEP See AppleTalk Echo Protocol.

aerial Conductive wires or structures used in transmissions. The term arose because most wires were originally suspended from poles, towers, or other aerial structures high enough to provide safety from interference and electrical hazards and to receive or send unimpeded signals. Sometimes aerials are distinguished as signal receivers, and antennas as signal senders. And sometimes the opposite distinction is made, so there isn't much consistency in usage. Since insect antennas can be considered as receiving units, it might make sense to call the receiver the *antenna*. Because of the lack of standardization of the terms, and because many of the same concepts of design and construction apply to both sending and receiving structures, this dictionary groups most of the information on aerials and antennas under the heading of antenna. See antenna.

aerial cable Transmission-receiving circuits strung through the air, typically supported by utility poles to keep them out of reach since many carry hazardous levels of current. Contrast with buried cable.

aerial distribution Aerial cabling configuration, with wires running through the air among buildings and poles. Various insulators and amplifiers or repeaters are used in many cable installations to protect signals from interference or to extend them over distance. Aerial distribution puts hazardous wires out of reach and is an alternative to underground or wall-based distribution. See distribution frame.

Aerial Experimental Association AEA. A research organization promoted by Mabel Gardiner Hubbard Bell, wife of A. Graham Bell, to support his strong interest in kites and aviation. It was established in 1907 by a small group of aviation enthusiasts.

aerial insert In cable runs that are predominantly covered, as in underground or building-based cables, a short segment installed overhead. Examples include a segment of cable from rooftop to rooftop or pole to pole in an otherwise covered system.

Aeronautical Administrative Communications AAC. A service of the aeronautical industry serving cockpit voice communications. Data connectivity that includes AAC is part of the Aeronautical Telecommunication Network (ATN).

aeronautical broadcasting Various government and commercial services providing information to the aeronautics industry, especially regarding meteorological conditions.

Aeronautical Mobile Satellite Service AMSS. A global mobile communications service implemented using Inmarsat geostationary satellites. Through a dedicated range of radio frequencies operating in discrete FDMA channels, the system provides information to aircraft worldwide (with some limitations near the Earth's north and south poles).

Three types of channels provide unidirectional dedicated communications and a fourth type of channel provides bidirectional communications. Channels are selected on the basis of the type of transmission (data or voice) and the length of the message. Information relevant to weather forecasting is also conveyed through AMSS. See Aeronautical Telecommunication Network.

Aeronautical Telecommunication Network ATN. A system of cooperative data networks that comprise a global aviation intercommunications structure which includes both fixed and mobile stations. It enables government air traffic control authorities and various aviation communications services with a variety of transmission types to interconnect. The system is being set up according to standards and guidelines developed by various prominent aviation and engineering organizations. It is based upon the Open Systems Interconnection (OSI) model. See Aeronautical Mobile Satellite Service.

Aerospace & Electronic Systems Society AESS. A society of the IEEE for members interested in the design, testing, and analysis of large, complex systems such as sensor systems for communications and navigation. The AESS sponsors individual chapters, conferences and panels and publishes AESS Transactions and the AESS Magazine. http://aess.gatech.edu/

Aerospace Industries Association of America, Inc. AIA. A trade association founded in 1919 that supports American manufacturers of commercial aircraft, engines, spacecraft, missiles, and related equipment. AIA represents its membership's needs and goals to the media, the public, other related organizations, and the U.S. Congress. http://www.aia-aerospace.org/

AES 1. Application Environment Standard, Application Environment Service. 2. atomic emission spectroscopy. 3. See Audio Engineering Society.

AESS See Aerospace & Electronic Systems Society.

AEW 1. aircraft early warning. 2. airborne early warning. Includes not only warnings of aircraft, but other airborne objects such as missiles and probes.

AF See audio frequency.

AFAST Advanced Flyaway Satellite Terminal. A family of commercial, modular, portable satellite terminals operating in the C-, Ku-, and X-band frequencies, from California Microwave, Inc. (CMI).

AFC 1. advanced fibre/fiber communications. 2. Australian Film Commission. 3. See automatic frequency control.

AFCEA See Armed Forces Communications and Electronics Association.

AFE 1. See analog front end. 2. antiferroelectric.

affiliate In the Telecommunications Act of 1996, published by the Federal Communications Commission (FCC), the term *affiliate* has a specific meaning as

follows:

"... a person that (directly or indirectly) owns or controls, is owned or controlled by, or is under common ownership or control with another person. For purposes of this paragraph, the term 'own' means to own an equity interest (or the equivalent thereof) of more than 10 percent."
See Federal Communications Commission, Telecommunications Act of 1996.

affine redundancy A phrase attributed to Michael Barnsley, who used it to describe the characteristics of fractals in terms of their self-similarity and their likelihood of looking more like parts of themselves, rather than parts of other things. See fractal.

affinity A relationship between structures or processes that are similar in function, form, location, or intention, particularly processes or queries aimed at acquiring the same resources or information. Data sharing in situations where processes execute in the same defined space where there are dynamic or predefined restrictions on routing is an example of an affinity relationship. See affinity routing.

affinity routing A network routing mechanism favored for applications where multiple users, large databases, or frequent update operations are prevalent. Affinity routing can be implemented by dedicating servers to a portion of the data frequently accessed and caching data to reduce disk seeks. It may incorporate selective partitioning. In some circumstances affinities may need to be eliminated to dynamically balance data workloads.

Forced affinity routing may also be called static distribution. See affinity.

affirmative In voice communications where signals are weak or noise is present, a synonym for "yes" which is intended to be clear and unambiguous.

AFI Authority and Format Identifier. In ATM, part of the network level address header.

AFIPS American Federation of Information Processing Societies. A national organization of data processing societies which organizes the National Computer Conference (NCC).

AFK An abbreviation for "away from keyboard," that indicates the particant in an online network chat is temporarily grabbing food, attending to the baby, or taking a short break.

AFM 1. Adobe Font Manager. 2. Adobe Font Metrics. 3. antiferromagnetism.

AFMR antiferromagnetic resonance.

AFNOR Association Française de Normalisation. The national standards organization of France. http://www.afnor.fr/

AFOSR See Air Force Office of Scientific Research.

AFP See AppleTalk Filing Protocol.

African Telecommunications Union ATU. Descended from the Pan-African Telecommunications Union (founded in 1977), the ATU was established in December 1999 by the 4th Exta Ordinary Session of Plenipotentiaries of the Pan African Telecommunications Union (PATU). The ATU seeks to make Africa an equal and active participant in the global information community by supporting and promoting the development of telecommunications policies, human resources, and technologies. http://www.atu-uat.org/

AFS See Andrew File System.

AFT 1. automatic fine tuning. 2. See automatic frequency control.

afterimage A visual image that may appear in pale outline or as a complementary color if an object is viewed for some time without moving, after the source of the image has changed or disappeared. The concept is important in designing display technologies. See persistence of vision.

AFTRA American Federation of Television and Radio Artists. A trade organization representing performers, founded in 1937. http://www.aftra.org/

AFV See audio-follow-video.

AGC 1. AudioGraphic Conferencing. ITU-T terminology related to transmissions protocols for multimedia. See audiographics. 2. See automatic gain control.

AGCOMNET A U.S. Department of Agriculture voice and data communications network.

aged packet In packet-switched networks, a data packet that has exceeded a prespecified parameter such as node visit count or elapsed time. Aged packets may be handled in a number of ways, depending upon their nature and the configuration of the network. They may be discarded, assigned a different priority, or returned to the originator.

Agency of Industrial Science and Technology AIST. A Japanese organization that is part of the Ministry of International Trade and Industry (MITI) that superintends research laboratories acknowledged for their technological innovation.

agent 1. Representative, broker, one who acts in place of or on the authority of another. 2. One who handles customer inquiries and procures services or products, often through other firms. Many long-distance providers are agents who procure services through other companies or through leased lines rather than by installing their own physical equipment. 3. On networks, a specialized software utility. Software agents are frequently used in client/server transactions to gather, organize, or exchange information according to security and priority levels usually established by the server. 4. On computers, in a general applications sense, agents are products (such as utilities or plugins) that do long, tedious or complex tasks, in conjunction with, and generally on behalf of, server software or user applications.

AGFNET Arbeitgemeinschaft der Grossforschungseinrichtungen. A German SNA-based computer network serving post-secondary institutions and research facilities.

aggregate bandwidth In a stream carrying more than one communication through some system of multiplexing, the aggregate bandwidth is the total combined bandwidth.

aggregation The bringing together or combining of physical, data, or radiant waves as in cables or transmissions. Aggregation typically refers to bringing together in terms of proximity, usually without a merging of information or electrical characteristics. However, some types of data are aggregated through an interleaving process, while still keeping individual portions true to their origins. Multiplexing is often used in conjunction with, or as a means of, aggregation. Agents sometimes aggregate, that is bundle, services for consumers. Cable companies sometimes aggregate certain types of stations into package deals for cable subscribers.

aggregate transmission The multiplexing of the transmissions of large numbers of users over a network backbone.

aggregator A service agent, broker, or liaison who coordinates negotiations on behalf of a block of subscribers, usually to get reduced rates. Billing is done by the service provider once the service has been established or facilitated by the aggregator.

Agility Communications A California-based company formed in 1998 to take advantage of commercial opportunities in dense wavelength optical networking. Agility is developing laser-based tuning for very high channel capacity communications based on Bragg reflectors. See Bragg reflector.

aging 1. *v.t.* A process of storing materials until their properties become essentially stable or reach a desired set of characteristics. 2. *v.i.* The characteristics of a material or process over time under a certain set of conditions. This may be an improvement, a deterioration, or simply a change.

agonic In magnetism, an imaginary line connecting all points on the Earth where the magnetic declination is zero. See declination, isogonic, magnetic equator.

AGP See Accelerated Graphics Port.

AGT 1. Alberta Government Telephones. 2. Audio-Graphics Terminal.

AGU 1. address-generation unit. 2. Automatic Ground Unit.

Historic Optics Book Illustration

A demonstration of the relationship between distance and light intensity, essentially, a historic photometer, as illustrated in the early 1600s by Peter Paul Rubens in de Aguilon's historic book on optics.

Aguilon, François de (1546–1617) A Belgian Jesuit who began a school for mathematics in Antwerp in

1611. Aguilon's most significant work was published as *Opticorum libri sex philosophis juxta ac mathematicis utiles* (*Six Books of Optics*), in 1613. The work is illustrated by the famous painter Peter Paul Rubens and includes images of binocular vision, stereography, and a historic photometer.

Ah ampere-hour.

AHT Average Handle Time. A call management phrase that describes the amount of time it takes, on average, to take a call, talk to the caller, and handle the caller's needs at the end of the call. For example, on a typical sales call, it may take a minute to connect with the desired person, fifteen minutes for the call, and twenty minutes after the call to log the caller's feedback and arrange to have a sales brochure sent to the caller.

AI 1. Airborne Interception. A radar-assisted fire control system used in military interceptor aircraft. 2. See artificial intelligence.

AIA 1. See Aerospace Industries Association of America, Inc. 2. American Institute of Architects. 3. Application Interface Adapter. A software utility which converts client function calls to standard SCSA messages.

AICE See Australian Institute of Computer Ethics.

AIEE American Institute of Electrical Engineers. It was consolidated with IRE to form the IEEE, an influential body of engineering professionals. See IEEE.

AIFF See Audio Interchange File Format.

AIIM See Association for Information and Image Management.

Aiken, Howard Hathaway (1900–1973) An American Harvard student and engineer who proposed development of a large-scale calculating machine, a historic forerunner of later electronic digital computers. The motivation for the machine was to create a system to solve cumbersome math equations, and the inspiration came from the writings of Charles Babbage and the Hollerith tabulating systems. Aiken was working on his doctorate when he conceived the idea and wrote a report. He subsequently received financial support in the 1940s from the President of International Business Machines (IBM), Thomas J. Watson, to build the Automatic Sequence Controlled Calculator, later renamed the Harvard Mark I. The success of the project led to the development of further computers in the series, including the Mark II, Mark III, and Mark IV computers, each building upon the experience of the previous system.

The success of the Mark I and the motivation provided by World War II spurred the development and financing of very large-scale computers. They were soon put into service by the U.S. Navy for calculating ballistics and other related equations, and Grace Hopper joined the computer project as a programmer. Aiken retired from Harvard in 1961. See Harvard Mark I to Harvard Mark IV.

AIM 1. amplitude intensity modulation. 2. See Ascend Inverse Multiplexing protocol. 3. See Association for Interactive Media. 4. ATM inverse multiplexer.

AIN See Advanced Intelligent Network.

AIOD See Automatic Identified Outward Dialing.

AIP ATM Interface Processor. A Cisco Systems commercial router network interface (ATM layers AAL3/4 and AAL5) for reducing performance bottlenecks at the User Network Interface (UNI).

AIR 1. additive increase rate. In ATM, a traffic flow control available bit rate (ABR) service parameter which controls cell transmission rate increases. See cell rate. 2. Airborne Imaging Radar. 3. All India Radio. 4. See Association of Independents in Radio.

air bridge In electronics, an aerially suspended interconnect, usually of metal.

air capacitor, air condenser A capacitor/condenser whose dielectric is air.

air cell A type of electrolytic wet cell once widely used in phone applications. Separate cells were connected to increase voltage. Polarization is reduced because oxygen from the air combines with hydrogen from the carbon electrode to form water. These historic cells had a useful life of about 1000 hours, and required ventilation. See dry cell, wet cell.

air column A channel of air, usually with certain size specifications or sound characteristics, within a piece of equipment, instrument, or chamber. Air column cables sometimes employ air as a dielectric, thus enabling a lighter, more flexible cable than one with a solid dielectric. See air-spaced coaxial cable.

air conditioning Running air through a system to alter its characteristics to make it suitable for people, equipment, or both. An air conditioner can affect temperature, humidity, and ion balances. Air conditioners are often used to cool work rooms in hot climates, and to cool equipment that generates heat but may be damaged by heat if the air temperature is not kept down. Many large supercomputing installations require cooling, and chip manufacturing plants condition the air to keep it free of dust, smoke, and other particles.

air core transformer A type of transformer designed to overcome some of the limitations of iron core transformers. At the higher frequencies used by broadcast communications, various problems such as the eddy effect and the skin effect will interfere with transmissions. Thus, air core coils and transformers, carefully tuned, can overcome some of these problems by eliminating the core.

air dielectric A component design configuration that uses air to provide a nonconducting medium in association with a conductor such as a cable or circuit. In cable manufacture, pressurized air around the conducting media can reduce interference. Components can be manufactured with air dielectric designs to be nonconducting for DC current for use in component crystal receivers. In general, air dielectric tuning capacitors have lower signal loss characteristics than solid dielectric-based tuning capacitors. See air dielectric cable, dielectric.

air dielectric cable A cable incorporating the nonconducting properties or air to promote higher velocity and lower attenuation than other types of cables. The air provides a margin between the conducting materials and the cable housings, reducing undesired interactions and limitations of capacitance, resistance, and inductance. Coaxial cables may use air dielectric properties through a pressurized fabrication around the conducting medium.

Air dielectric cables began to be generally available for communications applications in the mid-1980s. Standard lengths and connectors are commercially available. See foam dielectric cable.

Air Force Office of Scientific Research AFOSR. Descended from a small office of the Air Research and Development Command in 1951, AFOSR became the single manager for basic research within the U.S. Air Force in 1975. It provides the opportunity to direct leading edge research and technologies, through the Air Force Research Laboratory, to laboratories of the U.S. Department of Defense and U.S. industry. http://afosr.sciencewise.com/

air gap A region of air through which an electrical spark or magnetic current travels, as in spark gaps in gasoline engines.

air-incident recording AIR. A recording mechanism for magnetic media storage (tape, hard drives, etc.) that utilizes a recording layer over a substrate layer. The substrate helps protect the recording head from brief impacts with the surface, a system that works best in a sealed, stable environment. Contrast with substrate-incident recording.

air interface, airlink interface, A interface A radio frequency-translating interface for wireless communications.

In cellular communications, the air interface is the radio-frequency-based connection between a Mobile End System (M-ES) and a Mobile Data Base System (MDBS). If the user is traveling, the MDBS may change as the user moves from one cell to another. An air interface enables Cellular Digital Packet Data (CDPD) to be deployed over AMPS.

In local area wireless networks (LAWNs), the air interface is the radio frequency portion of a network that enables computers to exchange data without wires. In North America, LAWN air interfaces typically operate in unlicensed 900-MHz and 2.4-GHz frequency regions. In European HIPERLAN implementations, radio spectrum has been dedicated to wireless computer networks.

There are many ways to implement a mobile air interface and the International Telecommunications Union (ITU) has encouraged global standardization efforts for mobile phone technologies through its International Mobile Telecommunications 2000 project (IMT-2000). Several air interface proposals were part of this project (e.g., wideband CDMA). There are now two common air interface (CAI) standards for CDMA, cellular (TIA/EIA/IS-95A) and PCS (ANSI J-STD-008).

More recent air interface schemes, such as TDMA/TDD systems, can deliver capacity hundreds of times greater than older systems and as much as forty times greater than many 3G systems. Systems with rates up to 40 Mbps are commercially available.

Testing and troubleshooting of wireless networks presents a special set of problems. The airwaves are full of radio signals, all coexisting at various frequencies

and strengths. Nevertheless, vendors offer diagnostic instruments that measure field strength and, more recently, have begun to offer instruments that can decode the protocols (voice and data) used in the air interface, in order to fine-tune the system and fix or prevent potential problems. Devices for evaluating Quality of Service (QoS) are also available for auditing air interface transactions for a variety of wireless technologies.

See B interface, C interface, Cellular Digital Packet Data, D interface, E interface, Global System for Mobile, HIPERLAN, I interface, local area wireless network.

air time Time spent online, broadcasting, or engaged in two-way or multiple-connect wireless conversation. Service providers use accumulated air time as an accounting tool for scheduling, billing, and time management on shared systems.

air-blown fiber ABF. See blown fiber.

air-spaced coaxial cable A type of cable assembly design that incorporates air as a dielectric in order to minimize the loss of signal. Since there is no way to suspend the central core exactly in the middle of the column of air, air-spaced cables require spacers, usually of some type of plastic, inserted at intervals over the length of the cable, sufficiently far apart to let the air do its job (and to prevent moisture from entering), and sufficiently close together that a twist or bend in the cable doesn't allow the inner core to make contact with the next layer. See coaxial cable.

air-spaced doublet A type of focusing lens configuration that can outperform a number of other types of lens configurations (e.g., Petzval), but which are limited to monochromatic wavelengths and in terms of field of view. Air-spaced doublets are used in telescopes and multipurpose spotting binoculars.

aircraft earth station A mobile satellite transceiving station that, instead of being stationed on the ground, is installed on board an aircraft.

AIRF See additive increase rate factor.

airplane dial A type of rotary dial common on old radio systems that, when turned, moves a needle-like indicator back and forth in an arc, or straight line according to a marked gauge, similar to the gauges seen in airplane cockpits. Airplane dials are often used along with sliders on analog systems and with pushbuttons on analog/digital systems.

Airport Surveillance Radar ASR. Short-range radar coverage for airports and their immediate surroundings to facilitate the management of terminal area traffic and to provide the option of instrument approach assistance.

airtime, air time 1. The time during which a specific broadcast is active (airs). 2. Time allocated to a specific broadcast, whether or not it is used. 3. The time spent on a radio phone call. This information is frequently used in billing calls, as in cellular phone systems. Unlike wired systems where toll-free numbers or busy numbers are not billed, many wireless services bill for the amount of time the call is online, regardless of whether it is connected to a toll-free or local callee.

AIS 1. See alarm indication signal. 2. Automatic Intercept System. 3. See Association for Information Systems. 3. automated information system.

AIST See Agency of Industrial Science and Technology.

AISTel Associazione Italiana per lo Sviluppo delle Telcomunicazioni. Italian Association of Telecommunications Development. http://www.telecom-italia.org/

AIT 1. assembly, integration and testing. 2. Atomic International Time (more correctly known by TIA). See International Atomic Time. 3. See automatic identification technology.

AITS 1. Administrative Information Technology Services (University of Illinois) 2. Advanced Information Technology Services. 3. Associazione Italiana Tecnici del Suono. Italian Technical Association for Sound. 4. Australian Information Technology Society.

AIX Advanced Interactive Executive. An IBM implementation of Unix.

AJ anti-jam. A communications signal structured so that it is resistant to jamming or interference.

AJP *American Journal of Physics.*

aka also known as. 1. Alias, handle, nickname, pen name (nom de plume). 2. False or fraudulent name.

Al (*abbrev.*) aluminum.

AL Adaptation Layer. See ATM in appendix.

ALAP See AppleTalk Link Access Protocol.

alarm Warning signal, a signal indicating an error or hazardous situation. Alarm signals are generally designed with flashing lights or raucous noises to attract immediate attention. In electronic equipment, alarms are signaled by various messages, flashing elements, or sounds and may indicate the priority level and possible location or cause of the problem.

alarm indication signal, alarm indicating signal AIS. 1. In ATM networking, a signal indicating a failure. There are specific AISs in SONET circuits. Failure is declared if these conditions persist for a specified time period.

- A *line alarm indication signal* (L-AIS) is an error condition in which a defect pattern is detected in specific bits in five consecutive frames.
- A *STS-path alarm indication signal* is one in which specific consecutive bytes and the STS SPE contain all ones.
- A *VT-path alarm indication signal* is applicable to VTs in floating mode. The AIS alerts the downstream VT Path Terminating Entity (PTE) of an upstream failure. The defect is detected as all ones in specific bytes and three contiguous VT superframes.

2. Blue signal, blue alarm. A signal that overrides normal traffic during an alarm situation.

ALASCOM A commercial, regional communications service, consisting of satellite earth stations, fiber optic, and microwave links serving the state of Alaska.

Alaska Public Radio Network APRN. A local news network serving the unique needs of the State of Alaska, which has an unusual profile consisting of small, discrete, diverse ethnic populations spread over an enormous geographic region. Alaska further has a population that is unusually dependent upon radio broadcasting for news and social interaction due to its harsh and changing weather conditions and its scarcity of modern social/cultural amenities in geographically isolated communities. Thus, it faces technological and programming challenges beyond those of most other American states. APRN was founded in 1978. http://www.aprn.org/

albedo A ratio of the amount of electromagnetic radiation reflected by a body to the amount incident upon it. This reflectance may be described in the context of a portion of the spectrum (as the visible spectrum) or of the whole spectrum. The concept is used in telecommunications in relation to satellites and other celestial bodies. Albedo is complementary to absorptivity; it is often expressed as a percentage.

ALBO automatic line buildout. In data transmissions, a means of automatic cable equalization.

ALC 1. automatic level control. 2. automatic light control.

ALDC adaptive lossless data compression.

ALE 1. Application Logic Element. 2. Atlanta Linux Enthusiasts. 3. See automatic link establishment.

alert signal, alerting signal A transmission signal designed to gain the attention of an administrator or user. In computer networks, alert signals signify many things, such as imminent shutdown of a system, talk requests, new user logins, newly arrived email, etc. On telephone networks, alert signals are often used to indicate an incoming call.

Alexanderson alternator A high-frequency generator designed by E.F.W. Alexanderson that powered pioneer transatlantic communications. One of the historic uses of the Alexanderson alternator in broadcasting was at the Fessendon station which, in 1906, broadcast Christmas music to surprised and delighted listeners. Alexanderson received a patent for his generator in November 1911 (U.S. #1,008,577).

Alexanderson Alternator

A 200-kilowatt Alexanderson motor used for radio frequency alternation for the Radio Corporation of America (RCA) in New Jersey. [Scientific American Monthly, October 1920.]

Alexanderson antenna A vertically polarized wired antenna used for low frequency (LF) and very low frequency (VLF) transmitting and receiving that is not commonly used above amplitude modulation (AM) frequencies.

Alexanderson, Ernst F. W. (1878–1975) A pioneer developer of radio alternators in the early 1900s. GE had been contracted by Fessendon to develop a high frequency alternator for his pioneer radio station in 1904. Ernst Alexanderson was assigned to the project and achieved this significant engineering feat. He was involved in some of the early television development that was occurring in the 1920s and demonstrated a home television receiving unit. The Alexanderson alternator and Alexanderson antenna are named after him. See Alexanderson alternator.

Alexandre, Jean A French artist and inventor who was one of the earliest inventors of telegraph technology. Alexandre tried unsuccessfully to gain a direct audience with Napoleon to demonstrate what may have been dial-based electrical telegraph. Some have reported that his system was seen in 1801 or 1802, decades before the Wheatstone telegraph. He later went on to invent navigation and water filtration systems. See Salvà i Campillo, Francesc; telegraph history.

Alford, Andrew (1904–1992) A Russian-born American inventor of antennas for radio navigation and communication. In 1940, Alford co-authored "Ultrahigh-Frequency Loop Antennas" in *AIEE Transactions*. After working for many years in telegraphy and navigation firms, Alford joined the Harvard Radio Research Lab in the mid-1940s to devote more time to electronics design. He was the founder of the Alford Manufacturing Company which coinvented (with Kear & Kennedy) pioneer frequency modulation (FM) antennas. These led to systems that could simultaneously broadcast multiple FM programs from a single transmissions source.

Alford maintained a lifetime interest in antenna technologies. In his seventies, he continued to work and receive patents for his inventions, including a Doppler VOR ground station antenna for air navigation (U.S. #3,972,044 1974) and a two-frequency localizer guidance system and monitor (U.S. #3,866,228 1975 and #4,068,236 1978).

Alford antenna There were many antennas designed over the decades by Andrew Alford, most of which are called Alford antennas and many of which are still in use. His invention of the localizer antenna system won him a place in the National Inventors Hall of Fame. Junctions for an Alford FM antenna wind through the 87th floor of the Empire State Building. One of Alford's earlier designs is the horizontally polarized, omnidirectional *slot* antenna, introduced in 1946. It is commonly implemented as a long metal, tubular antenna with a long, narrow slot or series of slots. While not the most efficient antenna design for every use, it is easy to build and has some advantages over common dipole antennas. Thus, it is popular for amateur radio enthusiasts for weak signal communications and as television repeating units. The design

can be adapted to antenna beacons or used for fixed radio stations and satellite ground stations.

The Alford loop antenna is a rectangular loop antenna, with each of the corners slightly infolded toward the center to lower impedance at the nodes. It is used in navigation applications.

algebra A branch of mathematics in which generalizations and relationships are described and manipulated through numerals and other symbols using formal expression conventions. Algebraic concepts extend beyond the calculation of quantities to describe and manipulate transformations, functions, and dimensional spatial relationships.

Algebra is a fundamental tool that is used in almost every branch of science. It is especially useful where unknown information is to be extrapolated from known parameters such as performance characteristics as they relate to known physical parameters and laws (e.g., data rates in new cable fabrications) or for astronomical estimations (e.g., describing and measuring phenomena that are too distant, too transient, or too large to measure directly).

ALGOL *Algo*rithmic *L*anguage, *Al*gebraic *O*riented *L*anguage. A computer programming language developed in the 1950s by P. Naur, and others, for manipulating mathematical algorithms. C is said to be evolutionarily descended from Algol (with an intervening language called B).

algorithm A procedure consisting of a finite series of steps, defined to solve a problem or execute a task. The solution to the problem does not necessarily have to be known to create an algorithm to seek out a solution, or a path toward a solution. Logical/mathematical algorithms are widely used in the computing industry. The algorithm itself may not have a fixed number of steps, since an algorithm can be designed to be self-modifying, but the initial tasks, as set out by a programmer, for example, are finite. See brute force, heuristics.

ALI 1. See ATM line interface. 2. See automatic location identification.

alias *n.* 1. Pseudonym, assumed name, substitute or alternate name. 2. On operating system command lines, a short, easily remembered label for a longer, harder to remember label or command. Most systems will allow users to set up aliases at boot-up time, or in a file that can be reread while the system is running, to update the aliases. On Unix systems, a convenient alias is *ll* in place of *ls -la*. It's easier to type and displays more information in the subsequent directory listing, including permissions, file size, etc. 3. On Macintosh systems, there is a menu command to alias a filename. When selected, it causes an extra icon to appear, matching the original, under which the user can modify the name of the application, if desired, to better remember its function. This can be placed on the Desktop (or anywhere that's convenient), in place of the original icon which may be buried several folders deep or have an obscure name. When double-clicked, the alias then finds the original and launches it on behalf of the user. 4. Online, many users will assume an alias identity, known as a handle, or nickname, in order to present a friendlier, more interesting, or more obscure face to others. 5. In computer imagery, a visual artifact consisting of rough, staircased edges. This may result from low sampling, or from low resolution in the output device. See aliasing.

aliasing 1. In imaging, a visual artifact that causes rasterized images to take on a staircased effect when displayed or translated into resolutions that are too coarse to clearly resolve the image (usually those that are larger sizes than the original data). For example, an image of a circle 10 pixels high would be grossly distorted if displayed at 100 pixels in height unless smoothing (antialiasing) is applied. See antialiasing. 2. In audio, a frequency distortion that occurs in sampling when the sampling rate and the frequency interact in undesirable ways. A filter can sometimes reduce distortion.

align To bring into physical or conceptual association through similarities in spacing, orientation, function, or form, as in aligning fiber optic endfaces or data cables along a transmissions path, or aligning hypotheses as a result of experimental results to approach a new line of inquiry.

Alignment is a basic concept with many applications in optical communications.

- In fiber optics, physical alignment of fine filaments or filament bundles in relation to supporting or shielding structures is important for maintaining a light wave within the fiber and for providing insulation and protection. The axis alignment for individual fibers may also be important.
- The alignment of lenses for directing light waves within transmission paths may be crucial to the efficient functioning of a communications system.
- The alignment and deflection of light waves may be critical to logic operations in a circuit (on/off states) or may be used for add/drop multiplexing and routing capabilities.

Commercial software products aid fabricators in aligning and assessing fiber-to-fiber and fiber-to-laser assemblies in the production process. Hardware alignment systems facilitate manual or automatic alignment of optical technologies. For very fine adjustments in the nanometer range, piezoelectric control may be used.

Alignment in optical fiber component assemblies may occur after connection or during the attachment process. Alignment systems may be stand-alone or computer-controlled (commonly through PCI-based interface cards). With the increasingly small size and complexity of fiber components, automated alignment and clean room fabrication environments are increasing in importance. See aligned bundle.

aligned bundle A bundle of fibers or wires, in which the relative positions of each of the ends at one end are retained at the other end. In fiber optic transmissions, the bundling alignment is important to the quality of the transmissions and also influences the bend

radius and thus the physical requirements for installing the fiber bundle. See align.

alignment indicator 1. A diagnostic display (or sound) used in fiber optic sensors to aid in assessing fiber-to-fiber or fiber-to-laser alignments. See align. 2. An indicator used with a signal power sensor to align local wireless connections that have to pass over rivers, buildings, or irregular obstructions. Wireless local area networks may be used to connect terminals in separate buildings in situations where it is difficult to connect wires. An alignment indicator blinks, beeps, or provides a readout to aid the installer in adjusting transceivers to optimize the strength of the radio frequency (RF) signal. The sensing instrument and indicator may be combined with a telescopic sight and weatherproof housing. The wireless connection may be linked to a hybrid installation where wire or fiber are used in buildings.

alignment test In fiber optics fabrication, a test of the physical alignment or optical properties of two fiber filaments that are about to be fused. In simple manual fusion splicing, the alignment test may be based upon the physical characteristics of the fibers and the point at which they are joined. In local injection and detection (LID) systems, the actual light-guiding characteristics of the aligned fibers determine the positioning of the fibers for fusing. Light is injected into the wavepath and measured at the other end, prior to splicing, and the splice is performed at that point at which the light-guiding properties of the aligned fibers appears to be optimum. See fusion splice, local injection and detection.

ALIT See Automatic Line Insulation Testing.

all-dielectric cable A cable consisting of dielectric materials (insulating materials) that has no metal conductors as are found in most conventional cables.

all-wave antenna A multipurpose antenna designed to broadcast and/or receive a wide range of frequencies. All-wave antennas may include a number of different types of receiving structures on one basic supporting structure, and even better may be possible through careful antenna alignment (i.e., it may tilt or rotate manually, or electronically on servos).

All Call Paging A capability enabling a spoken message to be broadcast through a phone system, to all speakers and phones on that system. See hoot'n'holler.

all number calling Most people are now familiar with phone addresses consisting entirely of numbers, but in older phone systems in many regions of North America, a unique phone ID consisted of two letters, usually indicating the region or neighborhood, followed by five numbers. Thus, the number 525-1234 would have been called Larch 51234, Ladysmith 51234, LA 51234 or something to that effect. This was a more poetic and easy-to-remember system than the current all number system. All number calling was instituted to provide more numbers as human populations and the demand for phone lines increased. In most areas, all number calling was in place by the 1960s. Since numbers are difficult for many people to remember, companies will often request "gold numbers," numbers that correspond to letters, to spell out the name of the company or some aspect of its service.

all routes broadcast ARB. One of two types of route discovery frames that are common, namely, all routes broadcast (ARB) and single route broadcast (SRB). In frame-based networks such as Token-Ring networks, ARB is a common method of *source routing* in which a message is carried in an all routes broadcast (ARB) frame, and every possible route is traversed between the end stations. For efficiency, a spanning tree structure is typically used to organize the routing pattern. See all routes broadcast.

all routes explorer ARE. In ATM networks, a means of sending a transmission through all possible routes, which is useful for exploring paths for future transmissions. In source routing, an explorer frame is sent out to determine a path to a given destination. There are *all routes* and *spanning tree* explorer frames. See all routes broadcast.

all trunks busy ATB. A telephony trunk group condition wherein all the trunks in the group are busy. Statistical reports are generated indicating how often the condition occurs and the duration of ATB conditions. A tone indicator or recording may be provided to a caller indicating that all trunks in a specific routing group are unavailable. The tone sequence sounds like a fast busy signal.

ALLC See Association for Literary and Linguistic Computing.

Allan variance The computed half of a specified time average over the sum of the squares of the differences between successive readings of the frequency deviation sampled over the sampling period. Samples are adjacent in the sense that there is no "dead" time between successive samples. Allan variance is distinguished from classical variance (e.g., in time keeping) in that it converges to a finite value for most common types of noise.

Originally developed by David W. Allan for international time and metrology applications, Allan variance concepts have also been used for gene sequence analysis, residual noise analysis in a number of other types of systems (e.g., distinguishing noise from information), frequency stability measurments for oscillators (in the time domain), auditory-nerve spike train estimates beyond unity, diode laser spectroscopy water vapor measurement, and assessment of distance resolution in laser diode signals.

Allan variance can be displayed, along with other measures, on Stanford Research time interval counters. See relative intensity noise.

allcall Traditionally, a generalized signal transmission that might be intercepted by anyone with compatible equipment or signal processing algorithms. Since the advent of fax machines and email, users have mutated the meaning to mean a call or signal going out to all members of a distribution list. Since this causes confusion with the older radio-based term, the traditional meaning of allcall is now better described as anycall. See broadcast message, broadcast storm, anycall.

Allen, Paul G. (1953–) Paul Allen, Bill Gates' teenage Seattle high school friend and business partner, co-authored a number of early programming projects with Gates. Together they founded Traf-O-Data around 1972 and worked on commercial programming contracts. Allen discussed a number of ideas for creating and selling microcomputers with Gates, but Gates wasn't as interested in hardware as he was in software, and these ventures were not aggressively pursued. After graduation, Gates went to Harvard and Allen worked for Honeywell in Boston.

Allen learned of the Altair computer from the January 1975 issue of *Popular Electronics* magazine and discussed the article with Gates, conceiving the idea of developing a BASIC interpreter for the MITS Altair. Gates and Allen moved their business to Albuquerque to work in cooperation with MITS, and Allen became their VP of Software.

The most important alliance in Microsoft history was the contract to develop an operating system for International Business Machines (IBM), under controversial and competitive circumstances with Gary Kildall, the developer of the CP/M operating system. The text-based QDOS system, based upon a mid-1970s manual for Kildall's CP/M, was the Microsoft flagship to success. They purchased QDOS, developed by Tim Paterson, and developed it into PC-DOS for IBM and MS-DOS for Microsoft. Later Paul Allen left Microsoft to pursue other interests, including investments in a number of ventures, and in 1994 he founded the Paul Allen Group to monitor the performance of the various companies in which he has significant investments.

Charter Communications, cofounded by Allen through Vulcan Ventures, began upgrading its cable TV infrastructure in 2000 to also provide high-speed Internet access services. A Fortune 500 company, it now serves subscribers in 40 states through a concept Allen calls Wired World. In 2002, the company budgeted $3.5 billion to upgrade its coaxial and fiber broadband networks throughout the nation. Charter Communications also participates in the Cable in the Classroom program that provides cable connections and programming for schools.

Allen maintains regular contacts with the investment, computer, and entertainment communities. See Altair; Gates, William H.; Microsoft BASIC; Microsoft Incorporated.

Alliance of Computer-Based Telephony Application Suppliers ACTAS. A trade organization established to promote the distribution and development of computer-based telephone applications and standards. ACTA is associated with the Multimedia Telecommunications Association (MMTA).

Alliance for Better Campaigns A public interest group founded in 1998 to improve public participation in elections by promoting campaigns in which the greatest number of voters could be reached in the most engaging way. As would be expected, broadcast telecommunications media are central to many of the Alliance's aims. The Alliance supported a recommendation by the Gore Commission in 1998 to support voluntary provision of airtime for campaign messages, particularly in the month before voting. The Gore Commission further exhorted the Federal Communications Commission (FCC) to bring their jurisdiction into play if broadcasters did not voluntarily comply with the provision of air time for these campaign messages. See Advisory Committee on the Public Interest Obligations of Digital Broadcasters. http://www.bettercampaigns.org/

Alliance for Telecommunications Industry Solutions ATIS. An organization of industry professionals from North America and World Zone 1 Caribbean service providers. ATIS was initially the Exchange Carriers Standards Association (ECSA) in 1983, when it was created as part of the Bell System divestiture. It became ATIS in 1993. ATIS is concerned with a variety of issues ranging from telecommunications protocols and interconnection standards to general administrative operations of systems among competing carriers. ATIS has cooperated on many projects with the U.S. Federal Communications Commission (FCC). See Committee T1, Ordering and Billing Forum. http://www.atis.org/

alligator clip A long-nosed metal pressure clip with small teeth on the inner surface of the clip for grasping small objects or wires. Often two clips are mounted on a firm base to make them free-standing. They are commonly used in electronics to hold wires and various components, especially for soldering or gluing, or for establishing temporary electrical connections. They are also used to secure badges to clothing or baggage where a firm, temporary connection is needed.

allocate To apportion or earmark for a specific purpose. Resource allocation is an important aspect of computer and network operations. Memory, storage space, CPU time, and printers are queued and prioritized as part of the allocation process. Allocation is also essential to broadcasting and two-way radio communications, as there are only a limited number of frequencies available, and these must be carefully administrated to avoid interference and to maximize the number of regions in which they can be reused.

alloy A combination of a metal or metals with non-metals, or of metal with metal, made by the intimate fusing or amalgamation of the components. Alloys are intended to combine the better qualities of their constituents. For example, blending gold with a stronger metal may provide the greater malleability and beauty of gold with the durability of an alloyed metal.

ALM 1. airline miles. 2. AppWare loadable module. 3. automated loan machine. A type of commercial access point, similar to withdrawal ATMs, in which financial services in the form of quick loans can be negotiated through an automated teller machine.

almanac 1. Publication containing astronomical and meteorological data useful for navigation/positioning technology. 2. A file detailing satellite orbits and related atmosphere and time information.

alnico An iron alloy with *al*uminum, *ni*ckel, and *co*balt, sometimes with various combinations of cobalt, copper, and titanium added. It is commonly used to

make permanent magnets, used in many electronics components including speakers, motors, meters, etc.

Alouette-I, Alouette-A Canada's first research satellite, launched September 1962 to study radio communications in the northern reaches and the ionosphere. The project originated at the Defence Research Telecommunications Establishment. A Thor-Agena launch vehicle placed the satellite in orbit. Alouette somewhat resembled a fat metal pumpkin with slender antennas spiking out of the top and sides. The Alouette was followed by the International Satellites for Ionospheric Studies (ISIS) program in which Canada and the U.S. jointly developed several more satellites. Three years after Alouette-I, the Alouette-II was launched. Alouette-I operations were terminated September 1972, Alouette-II similarly lasted 10 years. The follow-up ISIS-I and ISIS-II satellites were in orbit for 20 years each. A huge number of scientific papers and many volumes of scientific sounding data were produced as a result of these long-lived projects. See ANIK.

ALOHA A method of radio wave transmission in which transmission can occur at any time. This means many transmissions may happen simultaneously and may cause interference, but sometimes it's a practical way to deal with unusual situations. The basic idea is to send out a signal, see if there's a response, and if there isn't, send again. Pure ALOHA and slotted ALOHA are variations. Pure ALOHA is very much a free-for-all and has been used for packet radio communications since the early 1970s. It has a low capacity rate, usually only about 18%. In slotted ALOHA, the transmissions are slotted according to time access, which may provide about double capacity of pure ALOHA. The name is derived from a failing satellite whose use was donated to researchers in the South Pacific. Since capacity outstripped demand, the loose ALOHA method fitted the circumstances.

ALOHANET An experimental frequency modulation (FM) transmission in which data frames are broadcast to a specific destination, developed by the University of Hawaii. See Aloha, packet radio.

alpha channel A portion of a data path, usually the first 8 bits in a 32-bit path, which is used with 24- and 32-bit graphics adapters to control colors. Popular paint programs like Adobe Photoshop allow the contents of alpha channels to be individually manipulated to create special effects.

alpha testing In-house testing of software or hardware. In software alpha testing, employees attempt to find and eradicate all the bugs, flow control, and user interface issues that can be determined by internal staff. See beta test.

Alphanet Telecom A new Internet protocol-based long-distance company based in Toronto, Canada. Phone, fax, and data transmissions will be jointly available as IP-based calling services leased through private carriers.

alphanumeric A set of characters comprising the upper and lower case letters of the English alphabet from A to Z, and the numerals 0 to 9. On some devices, lower case letters may not be included.

alphanumeric display A very common, usually inexpensive type of display on consumer appliances and electronics in which basic letters and numbers, and sometimes a few symbols, can be seen well enough to be understood for simple tasks. Alphanumeric displays are commonly based on liquid crystal diode (LCD) or light emitting diode (LED) technology. Alphanumeric displays are used in digital clock radios, microwaves, calculators, music components, handheld computers, and many other items.

ALPS See automatic loop protection switching.

ALT See automated loop test.

Altair The Altair was designed by Edward Roberts, William Yates, and Jim Bybee. The introductory price for the first three months was $395 for the kit, and $650 for a fully assembled unit. Programming was accomplished by means of small dip switches on the front of the computer; if the power was interrupted, the programmer had to start all over again and the available memory was infinitesimal by today's standards, only 256 bytes. It featured an 8-bit Intel 8008 central processing unit (CPU) and room for the addition of up to 15 peripheral cards. Later Altair-compatible buses incorporated Intel's upgrade to the 8008, the 8080, which was significantly faster.

Through marketing, a little luck, and the growing interest of electronics hobbyists, the Altair line was the first to capture successfully the hearts and imaginations of computer pioneers, and Micro Instrumentation and Telemetry Systems (MITS) sold more than 40,000 units by the time the company was sold in 1978.

Altair 8800 Hobbyist Computter

The Altair 8800 was available assembled or as a kit from MITS, a New Mexico-based company. It was introduced late in 1974 and was prominently featured in the January 1975 issue of Popular Electronics.

Unlike its commercially unsuccessful predecessors, the Altair became wildly popular with insightful hobbyists who grasped its potential and significance. The Altair bus, more commonly remembered as the S-100 bus, was quickly copied, and a number of clones, most notably the IMSAI 8080, began to appear. MITS set to work adding to its product line, creating a Motorola-based version, the Altair 680. The mass market computer had been born, and the industry quickly shifted into high gear, with far-reaching changes to society.

Paul Allen and Bill Gates, friends not long out of high school at the time of the release of the Altair, provided MITS with a BASIC interpreter just in time for it to be included with fully assembled versions of the machine, thus launching Microsoft Incorporated, the world's best-known software company. Steve Wozniak, inspired by the little kit computer, designed his own computer circuit board and, with Steve Jobs, formed Apple Computer, Inc., another of the world's most successful computer hardware/software companies. See Alto, Geniac, Intel MCS-4, Kenbak-1, IMSAI 8080, LINC, Mark-8, Micral, MITS, Scelbi-8H, Simon, Sol, SPHERE System, STPC 6800, TMS 1000.

Altair 680 A Motorola and American Micro-Systems, Inc. 6800 CPU-based computer from MITS, the same company which released the Intel 8008-based Altair a little less than a year earlier. The Altair 680 was featured in an article in the November 1975 issue of *Popular Electronics* as having a built-in TTY interface, a capacity of 72 program instructions, and room for up to five interface cards. The 680 was intended to appeal to hobbyists who liked the architecture of the MC 6800, and who were looking for a smaller, less expensive kit to build. The 680 was less than a third of the size of the Altair 8800 and much less expensive to build. See Altair.

MC6800-Based Altair 680

The November 1975 issue of Popular Electronics featured an article on building the Motorola MC6800-based Altair 680 by Edward Roberts and Paul Van Baalen. This was likely a strong factor in introducing the Motorola MC6800 CPU to hobbyist hardware designers.

Altair bus The original data bus that was developed by MITS for the Altair computer line. Later vendors changed the name to S-100 bus, and it became common in many different computers in the late 1970s and early 1980s.

Altair Users Group, Virtual There is a Virtual Altair Users Group on the Internet, comprised of hobbyists who still build, repair, and operate Altair computers. One of the participants, Tom Davidson, hosts an excellent Web site with schematics and circuit board images. http://hyperweb.com/altair

AltaVista One of the major World Wide Web search engines on the Internet, AltaVista draws from one of the larger Web database catalogs online. It was started by Louis Monier in Spring 1995 and was made a public search resource in December 1995. In June 1998 it was acquired by Compaq. A year later, a majority share was purchased by CMGI, Inc. http://www.altavista.com/

alternate access carriers A telephone service vendor other than the Local Exchange Carrier (LEC) can be authorized under competitive Federal Communications Commission (FCC) guidelines to provide alternate access.

Alternate Billing Services ABS. Telephony services that permit collect or bill-to-another-number services to callers. ABS is especially applicable to long-distance calls. Alternate Billing Entity Codes (ABEC) are an administrative tool used more specifically by Inter Exchange Carriers (IECs) to bill third parties for long-distance services. Some Billing Services providers provide code administration for ABEC.

alternate frequency A radio or optical frequency other than the "stock" or common frequency used on a system. Alternate frequencies are usually selected to prevent contention, interference, or to provide increased security. In some systems that provide stock and alternate frequencies, the two systems may not be mixed. Wireless systems are sometimes shipped with tables of suggestions for alternate frequency range groups.

Radio spectrum is a commodity that is carefully regulated by the Federal Communications Commission (FCC) and it is not easy to acquire additional frequencies. Thus, careful planning is needed to segment licensed frequencies for optimum use as primary or backup frequencies. The National Oceanic and Atmospheric Administration (NOAA) National Weather Service publishes a list of primary and alternate frequencies for amateur radio *Skywarn* storm spotting messages.

Shifting to an alternate frequency sometimes involves hardware adjustments to filter and oscillating components especially for radio frequency shifts from one band to another. However, with the increasing sophistication of digital control and communications systems, the capability to change frequencies in a communications system can be built into the system and may be accessible by a switch or dial.

Cable-based telephone services using time division multiplexing (TDM) are now capable of changing a group of lines to alternate frequencies while the lines are in service, to reduce or avoid noise interference. Newer cable modems for high-speed data communications have a feature called *frequency agility* that enables the system to identify sustained noise on the active frequency and switch to an alternate to provide a better connection.

Alternate Mark Inversion AMI. A line transmission code used for T1 and E1 lines in which successive *marks* alternate in polarity (negative and positive). This bipolar signal format is used on DS-1 lines, for example. A *mark* or a *1* is represented by alternating negative (minus) and positive (plus) voltages, with neutral representing zero. If two of the same signals occur in succession, a bipolar violation (BPV) occurs. The *ones density* requirement on lines using the AMI signal format are typically either B8ZS or HDB3. See B8ZS, bipolar signal, Coded Mark Inversion, HDB3.

Alternate Regulatory Framework ARF. A means of regulating local telephone companies intended to further competition within Local Access Transport Areas (LATAs). Since 1987, it has been called the New Regulatory Framework.

alternate route AR. An alternate data or telephone communications route selected when the initial choice is unavailable due to load or a break in the path. In telephony, sometimes called *second-choice route.*

alternate routing In both circuit switching and packet switching network systems, there are times when the initial attempt to trace and complete a transaction between a sender and the destination is unsuccessful. This situation can be due to high traffic, compromised intermediary switching links, destinations that are unavailable, etc.

In most circuit switching implementations, the transmission cannot go through until an end-to-end connection is set up, dedicating an established path to the call, so alternate routing to find another way to connect the requested call must take place before any data (or voice, in the case of a phone call) can be sent. In telephony, alternate routing usually involves locating a less busy trunk.

In contrast to circuit switching, packet switching does not require the establishment of an end-to-end connection before data can be sent; it can be sent regardless of whether it is known if the destination is reachable or available. Packet switching is used in dynamic environments where it is not known, or cannot be known, which routing nodes may be available, which route is most efficient, and whether the destination is online at any particular time. The packets are sent by various means, usually through hop-by-hop systems, and the packets from an individual message may be broken up and sent through different routes if a bottleneck or break occurs in the original path. At the destination, separated packets are reassembled, and there are usually several attempts to deliver the information before it is returned (in the case of most email) or abandoned (in the case of low-priority data). To facilitate alternate routing, packet system routers may have extensive routing tables listing a wide variety of connections within that region of the network. See router.

alternate use AU. The capability of a communications system to switch from one mode of service to another, e.g., between data and voice. See alternate voice/data.

alternate voice/data AVD. A transmission system that can be used for voice and data over one line, by alternating the services as needed, usually switched manually, as between voice through a telephone or data through a modem. Some modems are equipped with speakerphone capability to allow switching between voice and data, and further to detect the mode of an incoming transmission in order to switch to the correct mode automatically. More flexible and sophisticated systems are always being developed, and some success with newer, faster modems has been achieved to allow simultaneous voice/data communications. See simultaneous voice/data.

alternating current AC, ac. A very commonly used

form of electrical current with a periodically reversing charge-flow with an average value of zero. Unlike direct current (DC), alternating current (AC) varies continuously in its magnitude. For the supply of electricity to businesses and residences, it is set to reverse about 50 to 60 times per second, depending upon regional electrical codes.

Voltages in North America are supplied as 120 plus or minus about 10% for regular wall outlets, and to 220 for heavy duty outlets (for dryers, stoves, etc.). Voltages in Western Europe are set to 220.

Alternating current is typically used in commercial and residential power circuits leading to wall sockets, whereas direct current is typically used in battery-operated devices and sensitive electronic components. The large converters/transformers attached to the power cords of small components such as modems convert the AC power from the wall circuit to DC power compatible with the component. Given the greater sensitivity of electronic components, plugs now commonly have one wide leg and one narrow leg, to correspond with wider and narrower holes in newer wall or extension cord sockets. The wider and narrower pins correspond to the different characteristics of the wires to which they are connected, with one being a *hot* or live wire, and the other being a *neutral* or grounded wire.

Much of early communications technology was based on direct current (DC) as a power source. Telephones had *talking batteries* and *common batteries*. These batteries were large, leaky, wet cells, which were inconvenient if moved or exposed to fluctuating temperatures. Surprisingly, Thomas Edison was opposed to alternating current for the power supply for communications circuits, and hotly contested the concept with Nikola Tesla. More than fifty years after the invention of the telegraph, AC power for telegraph systems was still considered a novel idea, but the shortage of batteries, and their high cost, provoked French and Swiss engineers to experiment with AC generators, as described in the *Annales des Posts, Télégraphes et Téléphones* in September 1919. Eventually the advantages of AC power were better understood, and its use became common.

See B battery; direct current; ground; impedance; surge suppressor; talking battery; Tessla, Nikola.

alternator An electronic or electromagnetic device for producing alternating current (AC).

Alto A pioneering computing system developed at the Xerox PARC laboratories around 1973. The Alto was the inspiration for the graphical user interface incorporated into the Macintosh line of computers, and later into Microsoft windowing software. Some argue that the Alto was the first microcomputer, but that honor really belongs to the Kenbak-1 (1971), or perhaps the Simon (1949, 1950), and the commercially successful Altair (1974), since the Alto was never available to the general public in its original form, and its price tag was thousands of dollars. Nevertheless, many of the revolutionary graphical user interface ideas that filtered out to the commercial world were developed and implemented on the Alto. See

Altair; Kay, Alan; Kenbak-1; Macintosh computer; Microsoft Windows; Simon; Xerox PARC.

ALU 1. arithmetic and logic unit. An integral part of most computer processors' logic architecture for performing operations. 2. See average line utilization.

Alexandrite A nonmetallic crystalline material used in tunable solid-state lasers that operate in the near-infrared or ultraviolet regions. It has laser medical/cosmetic applications and is now used to write fiber Bragg gratings. It has also been described as a pumping mechanism for regenerative pulse amplification. Alexandrite is a variety of chrysoberyl first discovered in the Ural Mountains. In components manufacturing, it provides a broad tuning range with the capability to store and extract multijoule energy pulses. When cooled by air, additional tunability, at the top of the laser range, may be available. Alexandrite lasers operate around primary frequencies of about 790±60 nm, extending down to about 240 nm through second and third harmonics. Solid-state Alexandrite lasers operating in the ca. 250 nm range have excellent spatial coherence for precision fabrication applications. See fiber gratings.

aluminum A silvery, dull, malleable, light, inexpensive metallic element with good electrical conductivity and resistance to oxidation. Aluminum is somewhat brittle but is still commonly used in cables, antennas, reflectors, and other communications-related structures.

AM 1. See access module. 2. active messages. 3. active monitor. 4. See amplitude modulation.

AM broadcasting Transmission through amplitude modulation technologies on approved AM frequencies with the appropriate AM broadcasting license. In the United States, AM stations are spaced at 10 kHz intervals, ranging from 540 to 1700 kHz. See amplitude modulation, band allocations, broadcasting, FM broadcasting.

AM/VSB amplitude modulated vestigial sideband. See modulation, sideband.

amalgam *n.* Blend, composite, alloy, mixture.

amalgamate *v.t.* Unite, blend, consolidate, or merge. For example, amalgamating metals may help reduce the effects of chemical deterioration.

AMANDDA, AMANDA Automated Messaging and Directory Assistance.

amateur bands Frequency spectra set aside by regulatory authorities for the use of amateur radio operators. These are geographically subdivided, with some ranges designated for international use. Not all countries permit broadcasts by amateurs, licensed or unlicensed. In the U.S., the airwaves are legally owned by the American people and licensed in trust to qualified individuals and groups through the Federal Communications Commission (FCC).

amateur callsign A set of identification characters licensed to amateur broadcasters by a regulating agency such as the U.S. Federal Communications Commission (FCC). Callsigns in the U.S. indicate the country and region of the licensee.

Amateur Packet Radio Network AmprNet. A network of amateur packet radio hosts using TCP/IP network transmissions protocols with addresses assigned in the *.ampr.org* domain.

The sharing of packet radio communications among amateur radio buffs began with packet bulletin board systems (PBBSs) similar to the BBSs popular with computer hobbyists in the early and mid-1980s. The main differences between the two were that computer BBSs were primarily interconnected by land-based telephone lines and modems, whereas packet radio BBSs were interconnected by wireless radio frequency communications through terminal node controllers (TNCs) with a broadcast distance of about 20 miles or so. Relays were still necessary for long-distance packet communications.

When the Internet and TCP/IP became well established, many computer buffs shut down their BBSs and migrated to the Internet. Packet radio followed suit, forming the AmprNet to utilize low-cost global airwaves and simultaneous two-way communications.

Amateur Radio Emergency Service ARES. A public service organization of licensed Amateur Radio Operators of the American Radio Relay League (ARRL) who voluntarily provide emergency communications for public service events. ARES cooperates with state and local governments and the American Red Cross. http://www.ares.org/

Amateur Radio International Space Station

ARISS. An organization established to research and support the use of amateur radio in space; ARISS evolved out of the Space Shuttle Amateur Radio Experiment (SAREX). It serves as an educational outreach tool and experimental communications testbed. It also provides backup for emergency space communications and a medium for "off-duty" communication with friends and family members.

A Memorandum of Understanding was signed between ARISS and various national radio organizations in 1996 along with agreements with NASA and the Russian Energia. NASA liaises with the public through its Division of Education programs and Web site. Leadership and consultation are provided by the ARRL and AMSAT. ARISS designs, builds, and operates amateur radio equipment in cooperation with International Space Station programs. It established *ISS Ham* as a technical team to support hardware development, training, and operations while in orbit. While the initial communications of SAREX and ARISS were audio only, video is also an important aspect of radio communications and slow scan television (SSTV) is included in ARISS projections. See Space Shuttle Amateur Radio Experiment. http://ariss.gsfc.nasa.gov/

amateur radio operator, ham radio operator A radio broadcasting hobbyist permitted to transmit ra-

AMSAT-OSCAR Satellite Projects - Selected Overview

Phase II Satellites - developmental, low-orbit, operational, longer lifespan. See OSCAR.

Phase III Satellites - operational, high elliptical orbit, longer lifespan.

Phase IV Satellites - operational, high geostationary or drifting geostationary orbit, long lifespan.

Satellite	Launch	Notes
AMSAT-OSCAR 1	12 Dec. 1961	Phase-4A. 10 lb., beacon, 22-day orbit. Initiated by a U.S. west coast group. Nonrechargeable batteries. Elliptical orbit at 421 kilometers. Quarter-wave monopole antenna. Morse code telemetry. U.S. Air Force launched.
AMSAT-OSCAR 4	21 Dec. 1965	TRW Radio Club construction. Elliptical orbit at 34,000 kilometers (intended for circular orbit). No telemetry. Sleeve dipole and monopole antennas.
AMSAT-OSCAR 8	5 Mar. 1978	Phase-2D. Circular LEO at 910 kilometers. Several antennas. Battery failed June 1983.
AMSAT-OSCAR 10	16 Jun. 1983	NASA/NORAD #14129. Phase-3B. Similar to OSCAR I, with some improvements. Coatings provided better temperature control. On-board propulsion. High-altitude, elliptical, synchronous-transfer Molniya orbit at 35,449 kilometers.
AMSAT-OSCAR 13	15 Jun. 1988	NASA/NORAD #19216. Phase-3C. Linear analog transponder. Magnetorquer stabilization. Elliptical orbit at 38,000 kilometers, Molniya. Carried RUDAK-1, which failed.
AMSAT-OSCAR 16	22 Jan. 1990	NASA/NORAD #20439. PACSAT. Sun-synchronous near-polar LEO at 800 kilometers. Store-and-forward file server and AX.25 protocol. Digital repeater.

dio signals over specific frequencies. Most licenses that permit amateur transmissions require that the operator be licensed and fulfill certain requirements. In the U.S., also often called *ham radio* operator.

amateur satellite service A radio communication service using space stations or Earth-orbiting satellites for the purposes of the amateur radio-communications service. See amateur service, AMSAT, OSCAR. See AMSAT-OSCAR Satellite Projects Chart.

amateur service Radio communications services for the purpose of self-training, intercommunication, and technical investigations carried out by amateurs, that is, licensed or otherwise authorized persons interested in radio technique solely with personal, educational, or nonpecuniary aims. See American Radio Relay League.

amateur television AmTV, ATV. Black and white or color image broadcasts through amateur radio frequencies, with or without accompanying sound broadcast. Some amateur enthusiasts prefer to use ATV to mean fast scan TV over amateur bands, and SSTV for slow scan image transmission. With advancements in television technology, *advanced television* has begun to be identified with the ATV abbreviation, occasionally causing confusion. In the future it may be advisable to use AmTV to designate amateur television. See slow scan television.

amber A very light, transparent or semitransparent, warm golden substance from fossilization tree resin from pine trees that have been extinct for millions of years. Amber floats in water and occasionally washes up on the coasts of Europe after storms, intermingled with kelp and other natural debris. Sometimes insects can be found imbedded in the amber, preserved for centuries. Amber can be highly polished and has been used for jewelry for thousands of years.

The chief importance of amber to telecommunications is its static electrical properties, which can be observed by rubbing amber with a cloth (or on your hair) and using it to attract small fragments of tissue paper. In fact, the Greeks observed this property, and Plato recorded "... the wonderful attracting power of amber ..." in his *Timaeus* dialog. The Greek word for amber is elektron.

ambient *n.* Environment, atmosphere, mood, surroundings.

ambient light The light existing in an environment around and in addition to any deliberately established lights associated with a system. Ambient light may come from sunlight, reflective surfaces, phosphorescent materials, etc. and is typically composed of a variety of wavelengths.

Ambient light conditions affect mood and visibility and may be critically important to optical applications using precision instruments, those sensitive to light and those dependent upon specific types or levels of light for specialized applications (darkroom devices, microscopes, lasers, telescopes, etc.). The visibility and size of laser spots, for example, is affected by the amount and color of ambient light in addition to the distance traveled before the beam hits a reflecting surface.

ambient noise, room noise The general acoustic noise level of an environment, usually measured in decibels. The ambient noise in terminal rooms with printers or other equipment may be sufficient to cause hearing loss over time. Technicians who work long hours with high speed printers should wear ear protection.

ambient temperature The temperature in the environment around an object or system. Ambient temperatures may affect the durability, stability, and performance of many types of components, especially conducting materials that conduct not only transmission signals but also ambient temperature. For components that are especially sensitive to temperature extremes or fluctuations, the housings may be designed to control or mediate ambient temperatures.

AMDM ATM multiplexer/demultiplexer.

America Online AOL. A large, commercial Internet Services Provider (ISP) that provides access to the Internet, AOL-specific forums, news, email, and other features. AOL evolved from Quantum Computer Services, conceived by S. Case and J. Kimsey as a computer BBS providing online information and consumer services through modems. In 1989 Quantum was renamed to America Online and was launched with realtime chat, email, and special interest forums. Case became President of the company in 1990 and CEO in 1993. AOL became a publicly traded company in 1996. In 1998, it acquired Compuserve and ICQ, two well-known network services, as well as MapQuest, in 2000. In 2001, AOL completed a merger with Time Warner.

America's Carriers Telecommunications Association ACTA. A U.S.-based trade organization, representing commercial long-distance vendors (nondominant interexchange vendors). Of significance is the fact that ACTA has lobbied the Federal Communications Commission (FCC) to bar long-distance digital telephony over the Internet. The focus of ACTA is providing representation for its members to various legislative and regulatory bodies, and to further business activities of its members.

American Association for Artificial Intelligence AAAI. A nonprofit organization founded in 1979 to advance education in and scientific understanding of thought and intelligent behavior and their embodiment in machines. http://www.aaai.org/

American Association for the Advancement of Science AAAS. Descended from the Association of American Geologists and Naturalists, the AAAS was formed with a broader mission in 1848 to promote the development of science and engineering in the United States. http://www.aaas.org/

American Association of Physics Teachers AAPT. The AAPT supports professional and research physics and physics education through activities and publications, including the *American Journal of Physics*, *Physics Today*, and *The Physics Teacher*. Physics and engineering (applied physics) are at the heart of the understanding and development of all communications systems. http://www.aapt.org/

American Bell Telephone Company In 1875, the Bell Patent Association was formed by Alexander Graham Bell with investors willing to finance his telegraphy research. Two years later, in 1877, *The Bell Telephone Company* was formed by Bell, who included his associate, Thomas Watson. The company was formally incorporated in Massachusetts in 1878. Theodore N. Vail was hired as the general manager and had a long association with the company and its successors. In 1878, the Bell Telephone Company and the New England Telephone Company were consolidated into the National Bell Telephone Company. Then, in 1880, American Bell Telephone Company was incorporated. In 1881, American Bell purchased Western Electric Manufacturing Company and developed it into Western Electric Company, the equipment manufacturing arm of American Bell.

American Bell was the parent of the American Telephone and Telegraphy Company (AT&T). AT&T was established in New York as a subsidiary in 1885 for handling long-distance calls. These two were then merged into AT&T in 1899. See AT&T; Vail, Theodore N.; Western Electric Company.

American Civil Liberties Union ACLU. A prominent, nonprofit, nonpartisan, civil liberties organization founded in 1920 which now has more than a quarter million members. The ACLU monitors and protects freedom and takes action against violations of civil liberties wherever they may occur. The ACLU has a strong presence on the Web in light of the fact that many new freedom-related legislative actions have been taken as a result of the growth of the Internet. The ACLU publishes *ACLU Online* and the biweekly *Cyber-Liberties Update* electronic magazine. The ACLU deals with many telecommunications issues including Web censorship, online privacy, encryption, and more. http://www.aclu.org/

American Communication Association ACA. A not-for-profit association founded to promote academic and professional research, theory, criticism, and debate on human communications. ACA publishes *The American Communication Journal*, a professional, peer-reviewed, online publication. http://www.americancomm.org/

American Engineering Association AEA. A national nonprofit professional association supporting and promoting American leadership in engineering. http://www.aea.org/

American Electronics Association AeA. A Washington, D.C.-based professional association with offices in the U.S. and abroad, founded in 1943. AeA is dedicated to helping member companies excel in a global competitive market. AeANET is AeA's means of communicating industry news, surveys, and public policy issues to its membership.

In 2001, the AeA presented a public policy report to the 107th U.S. Congress asserting the importance of adapting to a new global Information Age. As its Public Policy priorities, the AeA listed expansion of science and math education, protection of privacy, simplification of Internet taxation, export controls, restoration of Presidential fast-track trade negotiating

authority, monitoring of China's conformance with World Trade Organization (WTO) agreements, and broadband deployment through forbearance in regulation and the promotion of competition. The report further lists statistics for the high-technology industry in the Quick Facts Appendix 3 section. http://www.aeanet.org/

American Institute of Electrical Engineers AIEE. Formed as a result of growing electrical development in the 1800s and the International Electrical Exhibition in 1884, to represent the profession and develop standards for the industry. Norvin Green, president of Western Union Telegraph Company, was the first president. Alexander Graham Bell and Thomas A. Edison were among the first six vice-presidents. AIEE was presented The Clark Collection in 1901 by Schuyler Skaats Wheeler. The Clark Collection was one of the world's great libraries of electrical technology. Andrew Carnegie further donated $1.5 million for AIEE premises. AIEE was merged with the Institute of Radio Engineers (IRE) in 1963 to form the IEEE. See IEEE, Institute of Radio Engineers.

American Library Association ALA. A governing body and support group for American librarians. The ALA provides member services, workshops, conferences, and administrative support. The organization has a long history of service to the public and its members. The ALA Code of Ethics goes back to a Suggested Code of Ethics proposed in 1930.

The author acknowledges the generous help received from many librarians in the creation of this dictionary. http://www.ala.org/

American Mathematical Society AMS. A large professional society dedicated to promoting mathematical research and education, founded in 1888. Headquartered in Providence, R.I., the AMS sponsors conferences, member services, online resources (e.g., MathSciNet) and a large number of mathematical publications. http://www.ams.org/

American Mobile Satellite Corporation AMSC. A commercial provider of seamless mobile communications services across North America under the SkyCell trademark. Hughes Communications is the largest shareholder, joined by AT&T Wireless, Singapore Telecom, and Mitel Corporation. A variety of services are marketed to government agencies, emergency organizations, and major corporations. AMSC is permitted to provide domestic mobile satellite services (MSS) in the upper L-band.

American Morse Code, Railroad Morse A system of dots and dashes used to represent characters for distance communications, quite possibly developed by Alfred Vail, while working with Samuel Morse. Due to the fact that American Morse includes some characters with internal spaces, which can be confusing to some, it is not often used, International Morse code is preferred. See Morse code.

American National Standards Institute ANSI. A significant U.S. private sector, nonprofit, standards-promoting body based in New York. ANSI was founded in 1918 by a group of engineering societies and government agencies. The ANSI Federation contributes the enhancement of global competitiveness of U.S. businesses by promoting the development and support of consensus standards and conformity assessment systems. Information on the many important ANSI standards is available online in the ANSI searchable database. http://www.ansi.org/

American Optical AO. A long-standing American optical firm known for its eyeglasses, lenses, and scientific instrument components.

AO began producing spectacles (eyeglasses) in 1833 after having originally been established as a jewelry shop. In 1838, Charles Spencer began marketing microscopes, setting up business as Spencer and Sons, in 1865. In 1869, American Optical Company was established by G. W. Wells. In 1843, William Beecher, AO's founder, produced steel eyeglasses on equipment of his own invention. Five years later, the product line was extended to gold frames. By 1898, AO was establishing industry standards for certain lenses. A research laboratory was established in 1909, one that was to attract a significant pool of talent, and AO was awarded a number of patents in the optics industry. In the early 1920s, the Spencer company introduced optical spectrometers, goniometers, and refractometers. In 1935, American Optical acquired the Spencer Lens Company, which operated as AO's Instrument Division as of 1945.

By the 1920s, the company had expanded from consumer eyeglasses into industrial safety products and expanded further into military optics in the early years of World War II.

Many renowned scientists in the optical community have worked at one time or another for American Optical. While AO didn't express much interest in fiber optics in the 1950s, W. Hicks, a recent AO employee who left to form another company, succeeded in fabricating a fiber filament, through fiber pulling, that could transmit light as a single-mode waveguide, in 1959. The potential of the single-mode waveguide was recognized by Elias Snitzer and described by him in a paper written in 1961.

During the 1960s and 1970s, many pioneering optical medical instrument components were produced by AO. In 1999 – 2000, American Optical was acquired by SOLA, an Australian lens company.

American Public Power Association APPA. A national American service organization representing local or publicly owned electric utility companies. http://www.appanet.org/

American Public Radio See Public Radio International.

American Radio Museum A diverse, well-selected collection of over a hundred years history of antique radio and electrical technologies, including a Tesla coil, Nipkow disc, Leyden jars, static generators, phonographs, and significant makes and models of crystal detectors and historic radios. Descended from the Bellingham Radio Museum, ARM was founded in the Pacific Northwest by Jonathan Winter. http://www.antique-radio.org/radio.html

American Radio Relay League ARRL. Founded in 1914 by Hiram Percy Maxim, with assistance from Clarence D. Tuska and a number of their colleagues, the ARRL is now a worldwide organization with almost 200,000 members, headquartered in the United States. Tuska was a youthful tinkerer and radio hobbyist when he met Maxim. The ARRL name is derived from the way in which amateur radio operators, constrained to certain power levels and frequencies, would cooperate by relaying messages from one person to another in order to send over greater distances or difficult terrain.

The ARRL cooperates with various radio groups and governing authorities such as the International Telecommunications Union (ITU) and the Federal Communications Commission (FCC). Its members have contributed to many of the technological milestones in communications history, including the pioneering of frequencies that were originally thought to be useless (and hence were assigned to amateurs). More recently, amateur radio enthusiasts have cooperated in satellite communications projects with AMSAT. The ARRL monthly publication *QST* has been available for more than 80 years. See AMSAT, International Amateur Radio Union. The ARRL's call letters are W1AW. http://www.arrl.org/

American Speaking Telephone Company A historic telephone company, based upon the Edison transmitter, established by Western Union in 1877 to compete with the Bell Company. With hundreds of thousands of miles of telegraph lines already installed throughout North America that could be adapted for telephone transmissions, Western Union was seen as a real threat to the Bell empire.

American Standard Code for Information Interchange ASCII. An important, alphanumeric 7-bit (128-character) communications standard widely used around the world for the transmission of textual messages. ASCII is a simple system, used on telegraph systems and computers. It doesn't support formatting attributes such as bold, italic or underline, and it is primarily useful for English and western European languages.

ASCII often functions as a lowest common denominator for textual communications since it is supported by most electronic mail, word processing, text editing, and desktop publishing programs, which may otherwise be incompatible. Differing formats are often resolved through ASCII translation and conversion. See ASCII for a chart showing the characters, control characters, and hex, decimal, and octal values for each. See ASCII and see Appendix for a chart, EBCDIC.

American Telephone and Telegraph Company AT&T. See AT&T for an explanation of the company's origins, history, and technologies.

American Voice Input/Output Society AVIOS. A not-for-profit organization dedicated to promoting and supporting speech applications research and technologies. Speech applications include voice recognition, speech recognition, and speech generation, all of which are now important input and output capabilities of computer systems and digital telephony networks. http://www.avios.com/

American Wire Gauge, Brown and Sharpe Wire Gauge AWG. A standardized wire diameter system, exclusive of covering, for nonferrous conductors such as copper and aluminum. With a range from 1 to 40, lower numbers denote thicker wires, higher ones thinner wires. Generally, for a specific material, the current-carrying capability increases as the diameter of the wire increases and the AWG number decreases. AWG 1 corresponds to a diameter of 7.35 mm with an amp rating of ca. 191, while AWG 40 corresponds to 0.799 mm and an amp rating of ca. 0.02. With finer wires manufactured and used for finely detailed electronics circuits, some charts extend the gauge sizes down to 0000 (11.68 mm).

Since heavier wires are usually more expensive, consumers tend to purchase the thinnest wire that will accomplish the task at hand. It's important to get wire that not only is adequate to carry the current desired, but that is strong enough to bend and stretch, especially around connectors, panels, punch-down blocks, etc. If the wire breaks at the connection point, it's not very useful. See Birmington Wire Gauge.

Ames Research Center ARC. A research organization dedicated to creating new knowledge and technologies within NASA's areas of interest. ARC was formed in 1939 by the U.S. National Advisory Committee on Aeronautics (NACA), which became part of National Aeronautics and Space Administration (NASA) in 1958.

AMHS See Automated Message Handling System.

AMI 1. See Alternate Mark Inversion.

Amiga Multimedia Personal Computer

The first of the Amiga line of computers, the Amiga 1000 was released in August 1985. It featured preemptive multitasking, built-in serial and parallel ports, a Motorola MC68000 CPU with coprocessor chips, two mouse/joystick ports, composite or RGB color graphics up to 640 × 400 pixels (more in overscan mode), and two-channel (16-voice) stereo sound.

Amiga computer A remarkable personal computer system for its time, the Amiga was developed by Jay Miner (hardware), Carl Sassenrath, R.J. Mical, et al. in the early 1980s. The original Amiga team members were part of the Hi-Toro company, a small develop-ment company whose members created the Lorraine, which was bought out by Commodore Business Machines in the fall of 1984 and became the Amiga.

The Amiga was well equipped for 1985 with full serial, parallel, and joystick ports, full-color graphics, the ability to run multiple screens simultaneously in different resolutions, NTSC video compatibility, built-in 4-channel (16-voice) stereo sound, fast graphics display with coprocessing chips, and a Motorola MC68000 CPU chip running at 7.15909 MHz with 32/16-bit internal/external addressing.

The Amiga had a fully preemptive multitasking operating system (working quite well in only 256 kilobytes of memory) which came with both a graphical user interface (GUI) and a text command line interface, both available for use at the same time. It helps to remember that in 1985 most personal computers lacked peripheral ports and employed single-tasking, monochrome graphics, and command line interfaces for prices ranging from $3,000 to $6,000. The Amiga 1000 offered everything built in, including monitor and sound, for under $2,000. The only other computer at the time significantly competitive with the Amiga was the Atari ST (the Apple IIgs never quite made the grade). Other Amiga models, including the 2000, 3000 and 4000, and updates to the OS were released over the next several years, followed by a new type of product from Commodore, the CD32.

The Amiga is historically significant not only for providing the first viable platform for desktop video, but for its many capabilities that have subsequently been incorporated into other systems (certain patented aspects of the Amiga have been used by prominent computer companies in today's mainstream products), showing the prescience and desirability of its design and features. Even a decade after its release, most personal computers lacked many of the Amiga's early capabilities, despite faster CPUs and other advances in technology.

In 1994 Commodore-Amiga folded due to problems in executive management and marketing. The Amiga product line was acquired by a German company, Escom, AG, (Amiga Technologies) and was later sold to Gateway, Inc., in 1997. Developers' conferences were reinstituted the same year. Gateway subsequently licensed use of the technology and trade identifiers to Amino Development Corporation, later known as the Amiga Corporation. Amiga conferences were still being held as of March 2001. See Amiga CD32; Commodore Business Machines; Mindset computer; Miner, Jay.

AMIS See Audio Messaging Interchange Specification.

AML 1. See Actual Measured Loss. 2. analog microwave link. 3. ARC Macro Language. A line-based interpreted programming language for the ArcInfo GIS, from ESRI. 4. Aurora Macro Language. An object-oriented, event-driven language for the Aurora Editor, a text editor from nText Research.

AMLCD active matrix liquid crystal display. See active matrix display.

ammeter, ampere meter An instrument for measuring the flow of electric current in alternating or direct current in ampere units. In communications circuits, where current may be very small (below one ampere), milliammeters and microammeters are used. When used as a measuring and diagnostic instrument, an ammeter is connected in series with a circuit to measure the current as it passes through. If the total current is above the range of the ammeter, or is such that it might cause damage to the sensitive instrumentation, part of it may be predirected through a shunt connected in parallel. See ampere, shunt.

Historic Ammeter or Ampere Meter

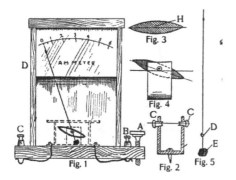

Historic drawings are often useful for describing basic mechanics and the forerunners to electronics as the essential components can be more easily visualized. The above diagram shows the basic structure and components of a historic ammeter, which is technologically descended from the galvanometer. [Popular Mechanics, May 1907.]

AMN See Abstract Machine Notation.

AMP See advanced metal powder.

ampacity The current-carrying capability, in amperes, of a circuit or cable. Typically ampacity is specified in product descriptions to indicate various types of cable assemblies, which may collectively consist of various combinations of wires and insulating materials.

AMPAS See Academy of Motion Picture Arts and Sciences.

ampere, amp (*symb. –* A) A unit of measurement of flow of electric current, named after A. Ampère. It is a practical meter-kilogram-second unit of electric current equivalent to a flow of one coulomb per second, or to the steady current produced by one volt when applied across a resistance of one ohm.

The international ampere was traditionally expressed as the steady current that will deposit silver at the rate of 0.001118 grams per second when flowing through a neutral silver nitrate solution.

The accepted scientific definition has since been replaced by a SI unit of electric current defined as a constant current that, in two straight parallel infinite conductors of negligible cross section placed one meter apart in a vacuum, would produce a force between conductors of 2×10^{-7} N/m. See volt, watt.

Ampère, André-Marie (1775–1836) A French physicist and mathematician who described and developed terminology for the nature of electricity. He also sought, in 1820, to formulate a combined theory of magnetism and electricity following some of the investigations of H. C. Ørsted.

In 1826, Ampère published an important paper, the "Memoir on the Mathematical Theory of Electrodynamic Phenomena, Uniquely Deduced from Experience" in which he described electrodynamic forces in mathematical terms. Many later experimenters built on Ampère's ideas, and his discoveries led to the development of magnet-moving coil instruments. The ampere unit of measure of electric current is named after him. See ampere, galvanometer.

André-Marie Ampère

André-Marie Ampère was inspired by the discoveries of Ørsted and worked with Arago to follow them up. Together they further investigated electrical and magnetic forces from which Ampère sought to formulate a unified theory to explain these phenomena.

Ampère's law In electromagnetism, the magnetic field associated with an electric current is proportional to the current. Ampère's law expresses this mathematical relationship and states that for a closed-loop path, the sum of the length elements times the magnetic field in the direction of the length element is equal to the permeability times the electric current within the loop. (Stating the rule in English is easier than calculating the integrals related to complex paths associated with irregular enclosed spaces.) Currents within these bounded spaces are positive or negative. Ampère's law has applications in assessing magnetic fields associated with conducting transmission wires and coils. Ampère's and Gauss's law together enable mathematical modeling of static magnetic fields. Ampère's law does not apply directly to a circuit with a charging capacitor. See Biot-Savart law, Gauss's law, Maxwell's equations.

Ampère's rule Based upon his discoveries in electromagnetism, André-Marie Ampère described a method for determining the direction in which a magnetic needle orients itself when in the vicinity of a current of electricity. See Biot-Savart law, left-hand rule, right-hand rule.

ampere-second A unit of electric charge flowing past a point in a current-carrying wire per second with a constant current of one ampere. Thus, amperes times seconds equals coulombs. See coulomb.

amplification See amplify.

amplifier A device or system that increases the magnitude or intensity of a phenomenon such as sound. This is accomplished in electronics through an increase in power, voltage, or current. Amplifying a signal doesn't necessarily make it louder, bigger, brighter, etc. than the original. The effect of amplification at the receiving end, or at a transfer point, may increase the signal that is received above its characteristics at the point it is received, but not necessarily above the original. Some systems are intended to increase the signal above the level of the original, as in public address systems and blow horns. Amplification systems seek to minimize the possible amplification or introduction of noise in the signal, while increasing the meaningful parts of the signal. See regenerative relay.

amplify To electrically, mechanically, optically, or conceptually enlarge; to increase the power or signal strength of; to make louder; to exaggerate. Amplification is a crucial process in many communications technologies that enables signals to be made effective, audible, or able to travel longer distances.

amplitude 1. The measure of the magnitude or extent of some property, movement, or phenomenon. 2. The magnitude of variation in some changing quantity from an established value such as zero, or from its extents. See amplitude modulation. 3. In a diagrammatic representation of a wave, the measure of the magnitude from the highest point in the waveform, to the lowest.

amplitude distortion Assuming a fundamental wave in a steady-state system, an undesirable condition in which the outgoing waveform differs from the incoming waveform sufficiently to affect the perception or informational content of the signal.

amplitude equalizer Corrective electronics, usually passive, designed to compensate for less than desirable amplitudes over a range of frequencies. Equalizers are used in audio recording and playback.

amplitude fading In an amplitude-modulated carrier wave, fading is the attenuation of the amplitude across frequencies, more or less uniformly. In passive laser communications links, atmospheric fluctuation is one factor contributing to amplitude fading and quadrature amplitude modulating (QAM) systems may be especially susceptible. Barbier et al. have described automatic gain control circuitry to help reduce fade.

In multimode, multichannel, optical fiber interferometers, assessment and reduction of amplitude fading are more complex. Kotov et al. have suggested that summing the signal magnitudes over various channels or selecting a channel with the largest amplitude

are methods that may significantly reduce output amplitude fading. See amplitude.

Amplitude Modulation

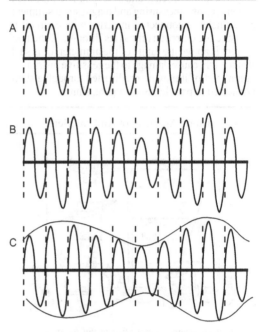

The top amplitude modulation (AM) diagram (A) shows an unmodulated 'carrier' signal. The middle (B) shows the signal modulated so that the amplitude varies through time. The bottom (C) shows the 'modulation envelope' which conveys useful information, such as magnitude of the modulation.

---

amplitude modulation AM. A very common means of adding information to a carrier wave. A basic radio wave carries no information. By varying or *modulating* the amplitude in a predetermined way, signals can be created which can be reconstructed as data, sound, or images at the receiving end of the transmission. This system was adopted in the early telegraph systems and is familiar in the form of AM radio broadcasts. AM radio typically requires about 10 kilohertz of bandwidth and is more subject to noise than frequency modulated (FM) radio. Designations of AM radio frequencies are under the jurisdiction of the Federal Communications Commission (FCC), and they have changed from time to time. In 1993, the FCC increased the upper limit of the AM band from 1605 kHz to 1705 kHz. Once frequency modulation (FM) was developed by Armstrong, it was thought that its superiority would overshadow amplitude modulation, but AM radio stations are still common decades later.

One of the simplest ways to modulate is to create intervals of current that are either on or off, as in Morse code telegraph communications and some types of binary computer signaling. Most computer modems use amplitude modulation and demodulation to convert from digital computer transmission signals to analog telephone transmission signals, and back again at the receiving modem.

Various types of amplitude modulation have been developed, and other nonamplitude modulation techniques exist, such as frequency modulation, in which the frequency of the signal, rather than its amplitude, is varied. See absorption modulation, amplitude shift keying, frequency modulation, modem, modulation, quadrature phase shift keying.

amplitude separation In television transmissions, the separation of the incoming signal into a video component and a synchronization signal component.

amplitude shift keying, intensity modulation, on/off keying ASK. A basic type of modulation that employs a constant-frequency signal, with two different signal levels used to represent binary values. In its simplest form, one state is represented by the lack of presence of the carrier, and the other by the presence of the carrier at a constant amplitude, hence on/off keying (OOK).

AmprNet See Amateur Packet Radio Network.

AMPS See Advanced Mobile Phone System.

AMR See anisotropic magneto resistance.

AMS 1. Account Management System. 2. See American Mathematical Society. 3. American Meteorological Society. 4. Attendant Management System. 5. automated management system.

AMSAT The Radio Amateur Satellite Corporation. A global organization of amateur radio operators who share an active interest in building, launching, and communicating amateur radio technology through noncommercial satellites. AMSAT was founded in 1969 as a result of the 1961 Project OSCAR satellite launchings. AMSAT was established as a not-for-profit educational organization to foster amateur participation in space research and communication.

Many early launchings have piggybacked as secondary payloads on weather satellites. More recently, AMSAT satellites have shared launch vehicles with other commercial and scientific craft.

In the mid-1990s, AMSAT became associated with ARISS amateur radio experiments aboard the International Space Station. Soon after, it became involved in the international *Phase 3D* satellite project, also called AMSAT-OSCAR 40 (AO-40). This project supports cameras, sensors, transmitters and receivers in several radio frequency bands, including S, K, U, V, L, and X bands. It is the fourth of AMSAT's high-altitude, global communications satellites, designed to replace earlier satellites that failed on launch or were no longer functional due to limited lifespans or technical failures. As of April 2001, efforts were underway to recover AO-40. In general, the satellite was in good condition, with some individual glitches that didn't impinge on overall health. However, since orbit 201, when it lost its solar lock, it was officially in a state of "hibernation" (unable to sense the Sun) with the magnetorque system off until solar lock could be reinstituted.

AMSAT now consists of a number of loosely affiliated organizations around the world, some bearing the AMSAT name with extensions, working together through cooperative rather than formal arrangements. AMSAT sponsors discussion lists and publishes a weekly online report on satellites, covering almost three dozen individual orbiting bodies, including space stations.

Many AMSAT enthusiasts are highly skilled technicians, and their knowledge and expertise have contributed to developing new technologies, in cooperation with a number of agencies, including the European Space Agency (ESA). See Amateur Radio International Space Station. See amateur satellite service and OSCAR for charts of the earlier satellite projects. http://www.amsat.org/

AMSC See American Mobile Satellite Corporation.

AMSC-1 A commercial satellite, operating in the L-band frequencies, owned by American Mobile Satellite Corporation. AMSC-1 provides voice, data, facsimile, paging, and other mobile communications services, particularly to commercial transport companies. Communication is through satellite phones or cellular/satellite hybrid phones.

AMSS 1. See Aeronautical Mobile Satellite Service. 2. Airborne Multi-Spectral Scanner. An aircraft-mounted scanning spectrometer for acquiring high-resolution imagery. 3. See Asian Mobile Satellite System.

AMTOR *am*ateur *t*eleprinting *o*ver *r*adio.

AMTS See Automated Maritime/Marine Telecommunications System.

analog 1. Relating to, similar to, linear, continuous with. 2. Circuits or devices in which the output or transmission varies as a *continuous function* of the input. Here are two examples commonly used to illustrate the distinction between analog and digital display and selection systems:

Time Piece Displays. Most analog watches have continuously sweeping minute and hour hands that move through a 360 degree arc through the action of internal rotating gears. Contrast this to a digital watch which stays on a one-minute or one-second setting until the next has been reached, and then 'flips over' the display to the next minute in discrete units.

Dials and Buttons. In older AM radios, the turning of an analog radio dial will move the station pointer in a continuous path through the various frequencies, and the transitions can be heard as the signals from various stations get stronger and weaker. In newer car radios, a push-button digital system is often used (sometimes in conjunction with an analog dial) to store the locations of preferred radio stations. Pushing the buttons 'jumps' to the desired stations without passing through the intervening frequencies.

Traditionally, phone conversations were processed as analog transmissions over copper wires. Gradually, digital switches and optical backbones were introduced, but the link to the customers' premises remained analog. When computers were first remotely accessed over analog phone lines through modems, it was necessary to convert the digital signals from the computer to analog signals through modulation. With the growing availability of mobile phones, ISDN, etc., end-to-end digital transmissions are possible and conversion from analog to digital is less often necessary. See digital, ISDN, modem.

anamorphic Capable of display in varying aspect ratios in the X and Y axes. Traditional television images are displayed at 4:3 aspect ratios and some movies are anamorphed (modified as to their aspect ratios) to fit television displays. Display systems may be anamorphic but, more often, the media being displayed have been put through an anamorphic process before being stored and distributed (e.g., DVD movies). Thus, these images are not so much *anamorphic* as they are *anamorphed*. It is entirely possible that image media may someday be truly anamorphic, stored in such a way that a computerized display system could process the incoming data and display at the desired aspect ratio in realtime. See anamorphing, letterboxing.

anamorphing Altering the aspect ratio of an image, optical beam, or other directional entity. The term is most often applied in two-dimensional situations where one dimension or the other is altered, rather than both. In imaging systems, this is typically in the X and Y axes. In fabrication processes, parts may be anamorphed in one or two out of three dimensions.

In laser optics, anamorphing prisms are used to reduce, enlarge, or correct a beam size or shape. Thus, an elliptical beam from a laser diode, for example, may be corrected in one dimension by small anamorphing lenses to create a circular beam.

anamux analog multiplexer.

ANC All Number Calling.

anchor 1. Something that serves to steady or hold, such as a guy wire or stake. 2. In hypertext programming, an element enabling links to related information. The anchor delimits the two ends of the hyperlink, designated with a tag as follows:

<A *link tag="location"*>*link text*</A>

anchor frame 1. In HTML coding, a frame (a defined section of the display) that contains at least one anchor tag (e.g., <A HREF ...>) pointing to addresses, data, or images to be associated with that frame. The TARGET tag can be used to specify that an anchor applies to a specific frame if there is more than one frame associated with a page. 2. In advanced television (ATV) technologies, a video frame used for prediction, most commonly an I-frame or P-frame. B-frames are not used as anchor frames. See ATSC Digital Television Standard.

Anchorage Accord An ATM Forum document comprising Foundation Specifications needed to assemble an ATM network infrastructure. This important suite established criteria for maintaining interoperability of ATM products and services. There were five dozen specifications listed in the Anchorage Accord, including intercarrier specifications, LAN emulations, interface requirements, physical layer specifications, traffic management specifications, and testing suites. The approval of the Accord was announced by the ATM Forum Technical Committee in April 1996. The step was an important one in forging working relations among theorists, specifications developers, and commercial implementers and was instrumental in furthering the acceptance and adoption of ATM as a networking technology. The Committee assured developers that specifications would be downwardly revised if interoperability problems were found in actual implementation. In August 1997, six additional specifications were announced, mainly to facilitate Internet connections and the transmission of converging multimedia data over ATM (e.g., voice or video over ATM).

The Accord documents can be downloaded free of charge from the ATM Forum specifications archive on the Web. See asynchronous transfer mode. http://www.atmforum.com/

ancillary charges Charges for optional or value-added services.

AND See Automatic Network Dialing.

Anderson bridge A device, usually employing a galvanometer, that measures reactance in order to determine capacitance or inductance by balancing against a frequency standard.

Andreessen, Mark Andreessen developed the first version of Mosaic, the precursor to the Netscape Navigator browser, in early 1993 while at the National Center for Supercomputing Applications. He was working with the Software Development Group developing for Unix. In 1994, he joined forces with Mark Bina, some of his colleagues at the University of Illinois, and developers from Silicon Graphics to form Mosaic Communications. They essentially rewrote the code, as the new Mosaic company didn't have the rights to market the version developed at the University. The company also had to change its name, so as not to infringe on the University rights to "Mosaic" as a tradename. The new company was called Netscape Communications and is now well-known for creating the Web browser known as Netscape Navigator.

Andrew File System AFS. A distributed file system named for Andrew Carnegie and Andrew Mellon. AFS grew out of a collaboration between Carnegie-Mellon University and International Business Machines (IBM).

anechoic 1. Not echoing or reflecting sound. 2. An environment without noise, or without significant noise. Sound recording rooms are designed to echo as little as possible, with thick, porous materials resembling foam egg crates absorbing the sound so it is prevented from reflecting back to the recording equipment. Speakerphones work better in anechoic environments. See acoustics.

angle Within the context of the central point in a circular reference, the displacement between two lines or surfaces originating or passing through the same reference point relative to one another, usually expressed in degrees or radians. If the two lines or surfaces are equivalent, the angle is considered to be zero. The number used to express the magnitude of the angle increases as the angle increases through a regular arc in a selected plane until they are again equivalent, thus a 360 arc in *degrees* or 2π arc in *radians*.

Basic Angle Designations

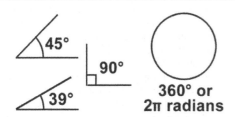

Angles are commonly designated with semicircles or squares (90°) are expressed in degrees or radians, based upon 360 degrees in a circle or 2π radians.

The concept of angle is intrinsic to almost every aspect of optics. Because light is said to travel in straight lines unless it is reflected or refracted in measurable, predictable ways in its interaction with common optical components, the geometry of angles enables the calculation of distances which, in turn, makes it poossible to model, design, fabricate, and use thousands of optical components.

angle of acceptance See acceptance angle.

angle of arrival The angle between the Earth's surface and the center of a radiant beam from the antenna to which it is radiating.

angle of beam The predominant range of direction of radiant energy from a directional transmitting antenna.

angle of deflection See angle of divergence.

angle of divergence In a cathode-ray tube (CRT), for example, the spread or divergence of an electron beam from an imaginary center position for that beam as it travels from the cathode to the coating on the inside surface of the front of the tube. A well-focused beam should spread as little as possible. Higher amplitudes tend to result in higher divergence. A perfectly straight beam has an angle of divergence that equals zero. See spreading loss.

angle of incidence The angle at which a radiant beam (or line) encounters an obstacle or theoretical reference, calculated in relation to the perpendicular (normal) from the surface of the obstacle. See incidence angle for a fuller explanation and diagrams.

angle of radiation The angle between the Earth's surface and the center of a radiant beam from the antenna from which it is radiating.

See normal, Brewster's angle, Snell's law.

angle brackets < > Symbols very commonly used in programming code as delimiters or arithmetic operators. These are best known as *greater than* (>) and *less than* (<) symbols. In HTML, the angle brackets delimit markup tags, e.g., <*p*> signifies a paragraph opening.

Angles – Degrees/Radians/Triangle

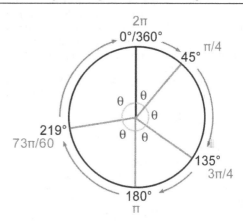

This diagram illustrates how angles are derived relative to the center of an imaginary circle and a beginning reference point. Shown clockwise from the top are equivalent measurements on two common geometric scales, degrees *and* radians.

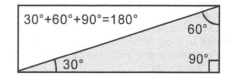

The Greek letter theta (θ) is often used to designate an angle and the Greek letters alpha (α) and beta (β) often refer to specific angles that are compared or mathematically summed (e.g., the combined angles of a triangle add up to 180°, which is a useful reference for calculations, especially when combined with Pythagorean concepts for right triangles [a triangle with a 90° angle]).

Angle of Incidence Example

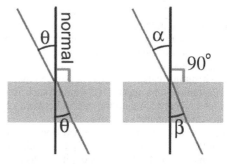

This simplified diagram of an electromagnetic incident wave encountering a dielectric with different refractive properties illustrates the angle of incidence (α) and the angle of refraction (β) as well as normal which is established, by definition, as 90° perpendicular (at right angles) to the plane upon which the incident wave makes contact with the intervening dielectric.

ångstrøm, angstrom (*symb.* – AAU, Å) A unit of measurement of length named after Anders J. Ångstrøm. Ångstrøm applied this unit to the measurement of wavelengths when mapping the Sun's spectrum. It is now also used to express atomic and molecular dimensions. It can be expressed as one tenbillionth of a meter, or one tenth of a nanometer, or 1×10^{-8} centimeters.

Ångstrøm, Anders J. (1814–1874) A Swedish scientist who researched the solar system and radiant waves. See angstrom.

angular misalignment loss In systems utilizing optical beams, a misalignment of fibers, mirrors, or connecting pieces resulting in the loss of beams that deviate from the desired path.

ANI See Automatic Number Identification.

ANIK The first domestic communications satellite, launched in 1972 by Telesat Canada, ANIK was fully operational by 1973. Circuits on the satellite were leased to Radio Corporation of America (RCA) until RCA had its own satellite. ANIK is actually a series of satellites, ANIKs C, D, and E were built in Canada's David Florida Laboratory (DFL) facility. The Canadian Broadcasting Corporation was the first television broadcast station in the world to use satellite broadcasting of their shows, utilizing ANIK in 1972. See Alouette-1, Canada Space Agency.

animate To bring to life, to give movement to, to move to action, to manipulate so as to simulate the effect of movement.

animation, cell One of multiple elements intended to create the illusion of movement through rapid sequential presentation of a series of cells. These are individual still frames that are similar to one another except in small details, drawn on cellophane or another transparent material, so that background images and other frames can be sub- or superimposed. Each cell is photographed once or twice, depending upon the speed of the movement, and the number of images needed. The human visual perception system functions in such a way that such a series of still frames presented at about 24 to 40 frames per second is perceived as movement. Humans are not able to resolve or distinguish each frame individually at those speeds. Film and computer animation models are based on this characteristic of perception. See frame, persistence of vision.

anisochronous In its simplest sense, something with varying (aniso) time intervals (chronous) such as the time interval between shooting stars or between keystrokes on a computer keyboard.

In signal transmissions, if the interval from one signal to the next does not necessarily equal other selected intervals in the transmission, it is considered to be anisochronous. In practical applications, where information may be sent in blocks, a reference block would contain a sequence of whole blocks within selected instants within the sequence, but would not necessarily map as whole block intervals to other selected sequences with intervals equal to the reference interval. Both telegraph and data transmission systems may have anisochronous characteristics.

ITU-T X Series Recommendation X.52 describes how to encode anisochronous signals into a synchronous user bearer. See asynchronous, isochronous.

anisotropic Exhibiting variance in a characteristic along a line, axis, plane, or other directional reference. A thick nonhomogenous liquid that has separated out into increasingly dense layers is anisotropic. The Earth's atmosphere is anisotropic in the sense that the gas mixture changes in relation to its distance from the Earth, becoming "thin" at higher altitudes. Crystals can be subcategorized as isotropic or anisotropic. This is an important consideration in optics, as a light beam passing through an anisotropic material will show different absorption characteristics depending upon its direction of travel. Anisotropic crystals may also emit different wavelengths (colors) of light depending upon the viewing angle.

Graded-index optical fiber in which the refractive properties change as you move outward to the edges is another example. In electromagnetic transmissions, it may refer to direction-dependent electrical or optical properties, e.g., polarized antennas. See dichroic, isotropic.

anisotropic magneto resistance AMR. A property of materials (e.g., alloys) exhibiting magneto resistance in a direction. In the manufacture of hard drive recording media, AMR is controlled and exploited through the use of very fine layers of recordable (magnetically alterable) materials. The use of AMR allows high capacity computer drives to store up to about 3 Gbytes per inch. By about 2003, AMR may be superseded by other technologies for very high capacity drives as research has uncovered other types of magneto resistance which are stronger at room temperature than AMR.

anneal To heat and subsequently cool to alter the properties of a substance (such as glass or wire), to make it stronger, less apt to crack or tear, or to fuse it with associated substances. Wires can be annealed to make them more durable.

announcement 1. The message that plays on an answering machine when the machine accepts an incoming call. 2. A message sent from a system administrator on a network to users, usually to let them know that the system may be shutting down temporarily for backups or maintenance. 3. A general message or page over a public address (PA) system.

annular ring A ring inserted around a hole as a support structure to hold a connection or wire, or to serve as an indicator. Small annular rings are used in printed circuit boards. Slightly larger annular rings are sometimes used on cables to indicate connection points.

annunciator An intercept device that indicates (with light or tone) the state of a circuit for information or diagnostic reasons. Information revealed by the annunciator may be as simple as the fact that the phone is ringing or more sophisticated, as in the state of a specified piece of equipment elsewhere on the line.

anode (*symb.* – P) 1. The positive terminal of an electrolytic cell. 2. The negative terminal of a current-providing cell or storage battery. 3. In a system of moving electrons, as in an electron tube, the direction to which the electrons flow or are attracted, originating from a cathode, and sometimes passing through a controlling grid. The anode is sometimes in the form of a thin plate of metal. See cathode.

Anode in an Electron Tube

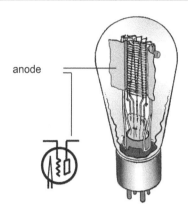

On the left is the symbol for a three-element electron tube. On the right is a tube drawn so the different elements can be seen behind the thin metal plate which is the anode, next to the grid (resembling a fine Venetian blind). The anode attracts the electrons emitted by the cathode (the filament, in this case).

Anonymous Call Rejection ACR. An optional telephone subscriber service that enables a blocked call (one that doesn't show up on a Caller ID system) to be rejected. A message is then played, advising the caller to disable call blocking and dial again so the recipient of the call can see who it is and pick up.

anonymous FTP A configuration of a File Transfer Protocol (FTP) data archive site that provides limited public access to users without the assignment of individual passwords. When you log into an FTP site, you will be prompted for a username. Type "anonymous" or "ftp" (in a text window, the command must be typed in lower case); you will then be prompted for a password, to which you respond with your full email address.

Assuming you have responded correctly to the prompts, and the system is set up for anonymous FTP, you will now have limited access to file directories, downloads, and perhaps uploads on the system. Many vendors are now using FTP sites to distribute demonstration versions of their software, and to dispense upgrades and technical support documents. A sample ftp login is illustrated under the entry for ftp. See Archie, ftp, File Transfer Protocol.

anonymous remailer An electronic mail transit point that deliberately obscures the identity and location of the poster to ensure his or her privacy. These remailers can provide protection to emailers from war-torn countries, for example, who are reporting information, or asking for assistance, and wish to protect their personal safety and anonymity. Anonymous remailers are occasionally used for illegal purposes, or to harass people on the Net, but generally, anonymous servers provide an important service. Refugees from political persecution have sometimes used them, and a number of celebrities on the Internet, wishing to safeguard their privacy, use anonymous remailers to post to public newsgroups.

ANS 1. See Advanced Network and Services. 2. answer.

ANSI See American National Standards Institute.

ANSII IISP ANSI Information Infrastructure Standards Panel.

Answer Back A signal (light or tone) that indicates the called party is ready to accept a call or transmission, or which acknowledges receipt of a transmission. See ACK, Answer Supervision.

Answer Back Supervision See Answer Supervision.

Answer Supervision A verification system that provides information between the local phone company and a long-distance service as to the successful connect status of a call. The signal is transmitted through the long-distance connection to make sure the call has been answered by the callee, and billing timing is initiated. In the past, long-distance calls were billed on an averaged wait-time-to-connect billing system without actual verification of the connection and, in fact, some small long-distance services still do it that way and initiate billing after a specified number of rings, before the called party answers.

ant A simple software agent sent out by a network node to probe the status (e.g., load status) of another node on the system. The ant returns to the sending node, which may be the same as the receiving node. See load-balancing system.

ANT See Access Network Termination.

antenna In its simplest form, a passive conductive device for transmitting and/or receiving signals, chiefly broadcast signals from radio, television, and radio phones. Most antennas for use with longer wavelengths are constructed from wires and metal cylinders or rods. Most antennas for use with very short wavelengths (microwaves) are designed as parabolic dishes.

A simple, vertical, one-quarter wavelength conducting wire can function as an antenna, if it is mounted where transmission waves can reach it and is connected at one end to a receiving device such as a radio. Most mobile whip antennas are of this kind, with maximum transceiving capabilities oriented along a horizontal plane, without much vertical capability. They are commonly seen on cars and trucks.

Antennas are mounted in many places, on TV sets, rooftops, mountaintops, in orbit, and on moving vehicles. They vary widely in shape, from thin rods, to branched, tree-like structures, to monuments like the Eiffel Tower in Paris, France.

Antennas can be designed to transmit selectively or in combination and include various grounding, directing, or reflecting components. Line-of-sight transmission antennas tend to be placed high, to reduce the number of obstructions, while receiving antennas tend to be focused in the direction of the desired transmission, to increase the signal and reduce interference from other signals.

An antenna generates two types of fields, electrostatic (along its length) and magnetic (associated with the antenna's current). They range from rabbit ears on older TVs, to high poles with guy wires in the yard of a CB radio enthusiast. Generally the higher and broader the antenna, the greater its range or scope, although there are exceptions to this general rule, based upon the shape and the frequencies involved. The Eiffel Tower was used by Lee de Forest as an antenna for sending a historic transatlantic radio broadcast. The orientation, length, and shape of an antenna will affect the type of frequency it can draw or transmit and its signal strength. A radio antenna, for example, is commonly designed so that its length is some multiple (e.g., double) or division (one half, or preferably at least one quarter) of the radio wave frequency.

Because radio waves vary in length and power, there is no one type of antenna that is best for all frequencies. The shape of an antenna must be optimized in relation to the length and characteristics of the waves it is transmitting or receiving. Some types of transmission, such as broadcasts from satellite cable stations or pulses from distance stars, must be captured with devices, such as parabolic antennas, that focus the waves. Due to their importance to telecommunications, this dictionary includes many listings under individual types of antennas. See also ground wave, Hertz antenna, ionospheric wave, isotropic antenna, J-pole antenna, Maxwell's equations, Marconi antenna, polarization, radio wave, satellite antennas, waveguide and the following antenna definitions.

antenna, extendible Inflatable and extendible antennas are particularly useful for applications that require light, collapsible equipment, e.g., space antennas deployed by rocket or shuttle. Keeping the equipment compact makes it easier to stow as payload and protects it from damage. Once it has been launched into space, however, an antenna needs to be extended to its full size to work effectively. Thus, different styles of antenna (from balls to umbrellas) have been developed to inflate and unfold once they are released or placed in position. Inflatable antennas also have potential for use in wildlife conservation, search and rescue, and military communications.

antenna, planar array A type of compact antenna array used in spread spectrum voice and data communications, military GPS applications and, when integrated with detectors, for certain imaging applications with millimeter/submillimeter wave receiver systems. Planar arrays are two-dimensional arrays (as opposed to linear arrays) used with a wide range of radio frequencies from about 800 MHz to over 27 GHz. Some of the advantages of planar antennas include their compact, more aesthetic design, compared to many grid parabolic antennas, and consistency of performance from one antenna to the next. For military GPS applications, they are used to reduce the chance of hostile interference through filtering. One disadvantage of planar antennas for precision imaging applications is their tendency to couple power into

surface waves. The HAARP antenna array is an example of a planar array used for ionospheric research. See HAARP.

Antenna Examples

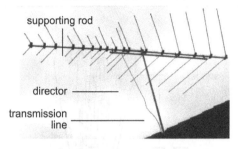

The roof-mounted antenna above is a type commonly used for television reception, mainly VHF frequencies.

This large parabolic tracking antenna aimed at the sky was used at the Kennedy Space Center Launch Complex to track space flight communications. The Apollo 7 Saturn IB space vehicle was being launched (far right) within view of the antenna, October 1968. [NASA/Johnson Space Center image.]

---

antenna, smart An antenna with computerized control that automates some of the functions of the antenna to improve its general efficiency or specific functions. In wireless communications, the demand for fast, effective services by an ever-growing population of users has spurred electronics engineers to develop advanced antennas that can sense and respond to situations more intelligently than basic electromechanical antennas that don't have the positioning, beam forming, or signal processing features of advanced smart antennas. Adaptive antennas enable transmission beams, and sometimes even the orientation of the antenna itself, to be tailored to current needs or capacities, a great boon to oversubscribed cellular systems or those that have a wide discrepancy between peak and low use times.

antenna effect In the case of improperly shielded loop antennas, or those in which the loop is incorrectly constructed or too closely spaced, the loss of the benefit of the structure may cause them to behave like simple whip antennas instead.

antenna gain An expression of the effectiveness or power of a signal from an antenna, usually selected at the point of its maximum radiation, when compared to a standard such as an isotropic antenna. Gain is commonly expressed in decibels. Gain is the greater power of transmission of a beam in a particular direction, as compared to a reference standard. See isotropic antenna.

Galileo High-Gain Parabolic Antenna

This 1989 artist's rendering of the Galileo spacecraft shows a high-gain parabolic antenna stowed between the two white flattish umbrella-shaped sun shields in the top half of the assemblage. Next to the small flat "sun screen" at the very top is a low-gain antenna, as is the vertical bar hanging down on the far left. The long bar protruding to the lower right supports magnetometers for sensing magnetic fields, and there are many other sensors built into the spacecraft. The antennas facilitate control of the vehicle and transmission of sensor readings back to Earth-based scientists. [NASA/JSC image originally painted by Ken Hodges.]

---

antenna impedance A ratio, at a specified point, of *voltage to current* such that impedance equals voltage divided by current. The impedance of any antenna will vary along its length according to a variety of factors. See resonant frequency.

antenna lobe, antenna pattern A 2D or 3D diagrammatic description of the direction angles and numbers of radiating patterns (or receiving patterns) of a specific type and configuration of antenna. The name is derived from the fact that waves tend to spread out in a more-or-less rounded or circular pattern, hence creating lobes in the diagram. Sometimes these are compared against a hypothetical isotropic antenna. The antenna pattern of a directional antenna and that of a general-direction antenna can be quite different.

antenna noise bridge A diagnostic device for determining the complex impedance of an antenna system. It is placed in series between the antenna feed line and its receiver.

antenna polarization A number of polarization structures and schemes to maximize the effectiveness or versatility of an antenna for different uses. This polarization may be linear or rotating circular. Once polarized, an antenna transmits and receives with the same polarization (unless, of course the antenna is reoriented between transmitting and receiving modes). Polarization is employed in ground-wave antennas but is less effective for ionospheric waves (sky waves). See polarization.

antenna stacking An arrangement of antennas in a vertical plane, one above the other, with a common transmission line, to improve gain and horizontal directivity.

antenna tuning The process of maximizing transmitting or receiving capabilities; if you're trying to do both with one antenna, sometimes the result is a compromise. This can be done through structure, by adjusting the sizes and positions of the various parts, and by orientation, by adjusting the angle and direction of the antenna. Even the degree of overlap of the tubing in dipole Yagi-Uda antennas can be important. Since antenna structures are tied to the length of the wavelengths concerned, structure is quite important. In directional antennas, such as parabolic antennas, computerized servos are often used to make small adjustments, and can be programmed to track a satellite in its orbit. See waveguide.

Anti-Terrorist Act and Effective Death Penalty of 1996 This Act deals specifically with the rights and lawful handling and legal counsel related to terrorists as well as setting forth the terms of restitution for victims. The Act amends the Justice for Victims of Terrorism Act and enables the Immigration and Naturalization Service and the Secretary of the Treasury to assess and designate security risks for the nation. See Anti-Terrorism Act of 2001.

Anti-Terrorism Act of 2001 ATA. Originally proposed as the Mobilization Against Terrorism Act (MATA), the second draft of ATA was put forth 19 Sept. 2001, in the wake of the terrorist attacks and destruction of the World Trade Center in New York City on 11 Sept. 2001.

This Act is a continuation in a long line of acts (including the 1996 Anti-Terrorist Act) seeking to balance the needs of law enforcement bodies with freedom and privacy concerns of American legal residents and citizens. In the 1990s, the pendulum swung to a great extent in favor of privacy advocates and software vendors who wanted to maintain a competitive advantage worldwide by providing strong encryption in software products. After the terrorist attacks, issues that were handled liberally were reviewed and again brought to the table, including wiretapping, electronic surveillance, and many other aspects directly related to telecommunications devices and the laws that govern their use.

The 2001 Act generated much discussion and controversy, as might be expected, based on the debates over the years. Civil liberties organizations expressed concern over the systematic and continued erosion of liberties and freedoms; the House Judiciary Committee proposed the Provide Appropriate Tools Required to Intercept and Obstruct Terrorism bill (Patriot bill) as an alternative to the Anti-Terrorism Act and the Senate proposed measures, as well.

One of the important arguments that came up with regard to the terms of the Act was the inclusion of sunset clauses (essentially, expiry dates) so that extraordinary measures implemented to cope with a crisis situation were not later used to harass ordinary citizens, as has occurred at various times in American history. See Security and Freedom Through Encryption Act.

antialias *v.t.* To compensate for a lack of resolution in incoming or displayed data compared to the source data. An image or other signal is said to be aliased when the viewing/display/detection resolution is less than that of the image/original resolution. In imaging, aliasing can create a moiré pattern or staircased "jaggy" look around edges. Antialiasing encompasses a number of strategies for correcting or compensating for the aliasing effect.

Antialiasing is a general concept that applies to a wide variety of detection and imaging technologies.

In television broadcast imaging (especially news shows), aliasing may be reduced by placing quartz or plastic optics in front of the image detector. The plate may be supplemented with digital image processing algorithms. Together they help reduce moiré that appears when the camera is aimed at "busy" clothing or surfaces such as houndstooth suits or fine-textured lattice fences.

In computer imaging, antialiasing is adding picture elements to create an illusion of gradual transitions between otherwise jagged or sharp transitions. Aliasing may occur at sharp tonal changes in a grayscale image, or at line boundaries in a monochrome image.

In low resolution raster images in grayscale or color, it is possible to use intermediary tones or colors between dark and light areas to reduce the effect of aliasing, providing the illusion that the shape or object is smooth.

In sound reproduction, the same principles can be applied to smooth out a rough sound transition due to low quality components, sound recording technologies, or digital sound sampled at low resolutions.

In radio astronomy, antialiasing filters may be used prior to digitizing signals from radio telescopes. In fluorescence detectors, aliasing may be applied before or after the signal is amplified (depending upon the instrument and resolution needs).

antilog, antilogarithm The number derived from a calculation in which the logarithm of a number has been supplied. Or expressed another way – the number from which a logarithm is derived. It is the inverse procedure of calculating a logarithm. Antilogs are handy for "collapsing" graphs or descriptive scales to put them within manageable spatial or numerical ranges. See logarithm for a fuller explanation.

antinode In a standing wave in an oscillating body, the point of maximum amplitude between the nodes on either side.

antireflection coating A plastic, liquid, film, or other coating applied to a surface to reduce its reflective qualities. In optics, coatings are usually applied to improve contrast and visibility, though selective screening of radiant energy is a common application, as well. The choice of coating depends upon the material to which the coating is to be applied and expected environmental influences (heat, humidity, chemicals, abrasion). There is sometimes a tradeoff in terms of efficiency and durability. There may also be a tradeoff in efficiency and ease of application of the coating (e.g., in terms of thickness or uniformity). Depending upon coating and application, coatings may be applied in a number of ways, including spraying, brushing, magnetic attraction, screen printing, pressure, fusion, gluing, or gravity bonding. Examples of objects/materials that are coated include safety glasses, gauge faces, the imaging surface of scanners, lenses, sensors, some types of resonating cavities, and certain types of windows. Semiconductor sensors (e.g., far-infrared detectors) may have antireflection coatings to increase transmittance from environmental sources or laser illumination sources. Some components combine highly reflective surfaces with areas treated with antireflection coating to selectively control the reflectance/transmittance of light over the extents of the surface.

Antireflective materials are generally selected for their high transmittance properties. Since electromagnetic radiation has different properties at different wavelengths, an antireflection coating will have different transmittance properties depending upon the source and composition of the incident radiation; the coating may be specifically targeted for a particular wavelength and thus acts as a filter, screening out wavelengths other than the one targeted to transmit (to not reflect). A broadband antireflection coating is one that reflects over a wide range of the spectrum.

For an antireflection coating to work well, it usually requires a "tight fit" with the surface that is coated. It may be bonded, fused, or held in place by gravity or friction. It may be applied to one side only or two or more surfaces. The thickness of the coating is based upon many factors and may need to be matched in depth to a multiple or fraction of a specific targeted wavelength (e.g., it may be half a wavelength thick). Since it is important to select the refractive index of the coating to balance the properties of the material to which it may be bonded (e.g., a glass lens), maintenance of the refractive properties through the interface between the layers is also important. In other words, it's not enough to calculate the refractive interactions between the coating and the material coated, it is also important to consider the refractive properties of the bonding agent if epoxy or something similar is applied. It is also important to bond or hold together the pieces with a minimum of intervening gaps, bubbles, or particles. The math gets especially complicated when the coating is more than one layer and computer modeling programs are often used to test multilayered coatings before fabrication. See reflection, refractive index, thin film.

antispoof 1. A mechanism for stopping or deterring unauthorized access to a premises or system by a person or program masquerading as someone/something else or otherwise misrepresenting its identity or authorization characteristics. Antispoof mechanisms are built into programs, firewalls, routers, and many other components in various types of wired or wireless data networks. See Trojan horse. 2. Spoofing is a mechanism for making a transmission appear to be active even if there is a time lapse during which data may not be sent. An example is when a transmission from a slow machine or a machine on an erratic connection is masquerading as active, to keep a link alive (from not timing out), even when no data is sent. This may be done in a number of ways, depending upon the application, including random signals, data padding, etc. An antispoof mechanism is one that detects this type of activity and disengages the activity or takes other appropriate action. See spoofing, facsimile. 3. Network spoofing is a situation where packets may be rerouted to a different destination for legitimate or illegitimate reasons. Antispoof mechanisms are designed to detect attempts at rerouting and ensure that the data reaches its original intended destination. See spoofing, network; spoofing, Web site.

Antique Telephone Collectors Association ATCA. The world's largest telephone collectors' organization, chartered in 1971. ATCA is a nonprofit corporation, based in the state of Kansas. It supports local and international telephone conferences and collectors' activities along with a telephone history site, telephone wiring diagrams, and other resources of interest in the development and fabrication of telephone equipment over the history of the technology. http://atcaonline.com/

Antique Wireless Association, Inc. AWA. Founded as a not-for-profit in 1952, the AWA supports research, preservation, and documentation of the history of wireless. It administrates the Antique Wireless Association Electronic Communications Museum in Bloomfield, NY and a virtual museum on the Web. http://www.antiquewireless.org/

antistatic A specialized tool or material that resists the buildup of static charges or which gradually dissipates a charge rather than sending out a quick discharge spark. There are antistatic wrist bracelets and antistatic mats for people who work on electronics, and antistatic packaging for the storage and shipping of sensitive electronic components. See static.

antivirus program A software program intended to detect and disable computer viruses, software programs designed to penetrate or vandalize a system without the consent or knowledge of the user. Some virus checkers run as background tasks and monitor any new files copied to the system. If a known virus or unusual program is detected, the user is alerted, and the software attempts to disable the intruder. It is almost always advisable to run good antivirus software, particularly if software is downloaded from bulletin boards, the Internet, or other public file archives. It is also a good idea to do so on any networked computer that shares file access with other computers. See virus.

anycall A generalized signal transmission that might be intercepted by anyone with compatible equipment or signal processing algorithms. An anycall broadcast does not assume that particular recipients will receive or respond to the message, yet is usually sent in the hopes that someone will receive the signal. Anycall broadcasts are useful for emergency calls for help.

Anycall signals are distinguished from allcall signals in that allcalls are directed to all users on a distribution list as opposed to anyone who is listening (or otherwise able to acquire the message). A car broadcasting a message through a megaphone while traveling through a city is an example of an anycall broadcast. The originators of the call don't know who is able to hear the message or how many recipients are reached by the message. In contrast, an email message posted to all members of a discussion list is an example of an allcall message.

In radio signaling, anycall has a more specific meaning in that the unspecified stations receiving the call follow a convention to stop scanning other frequencies in order to receive subsequent calls from the anycall frequency (which may be an emergency call) and will respond in pseudorandom fashion (in order to avoid a broadcast storm and signal contention). See allcall, broadcast message, broadcast storm.

anycast In IPv6, the proposed successor to IPv4, the primary protocol used for the Internet, anycast is related to communications between devices within a group, with the host device passing on some of the responsibility for routing updates to the closest member of a group.

anywhere fix The capability of a receiver to begin position calculations without an initial approximate location and approximate time, used in Global Positioning Systems (GPS).

AO 1. See acousto-optic (A-O). 2. active optics. 3. adaptive optics.

AOCN See Administrative Operating Company Number.

AOCS attitude and orbit control system. See telemetry.

AOL See America OnLine.

AOM acousto-optic modulator. See acousto-optic modulation.

AOR Atlantic Ocean Region. A longitudinal regional designation for geostationary satellites.

AOS 1. Alternate Operator Services. See Operator Service Providers. 2. Area of Service.

AOSS Auxiliary Operator Services System. A telephone operator system offering directory assistance, call processing, call detail recording, and similar services.

AOSSVR Auxiliary Operator Services System Voice Response. See AOSS.

AOTF acousto-optic tunable filter. A type of filter used in high-resolution spectromters. See acousto-optic.

AP 1. action potential. 2. aiming point. A target reference point for aiming an antenna or laser beam. 3. application program. 4. Applications Processor. An AT&T telephone add-on to provide more options. 5. array processor. 6. Associated Press. A commercial association with a long history of using long-distance communications services to gather and disseminate news.

Apache A freely distributable full-featured HTTP server for Unix systems, developed in the mid-1990s. It is the most prevalent server on the Internet and has been very influential in the growth of the World Wide Web system.

Apache is descended from a public domain HTTP daemon developed by Rob McCool in the mid-1970s at the National Center for Supercomputing Applications (NCSA), the same organization that spawned the Mosaic Web browser. After McCool left NCSA in 1994, a group of the HTTP daemon supporters began communicating through a discussion list organized by B. Behlendort and C. Skolnick, and the group started coordinating the development of their patches and enhancements. Thus, *NCSA httpd 1.3* became the base for *Apache 0.6.2*, which was released in April 1995. At that time, NCSA renewed development on the project and NCSA and the discussion members kept in touch.

In 1995, a new modular server architecture was developed by Robert Thau and incorporated into Apache 0.8.8, released August 1995, followed by Apache 1.0 in December 1995. In less than a year from the release of 1.0, Apache became the most widely used HTTP server on the Internet. See Apache Project, Apache Software Foundation.

Apache Project A global, volunteer, collaborative software development effort to create a commercial-quality, robust, full-featured, freely available implementation of a Web (HTTP) server. The principal participants in the project are known as the Apache Group or informally as the core. The Apache Group is now organized as the Apache Software Foundation. See Apache, Apache Software Foundation. http://www.apache.org/

Apache Software Foundation A not-for-profit corporation providing administrative, legal, and financial support for Apache open-source projects. Membership is open to those who have demonstrated a commitment to collaborative open-source software development. See Apache, Apache Project.

APAD See asynchronous packet assembler/disassembler.

APAN Asia-Pacific Advanced Network Consortium. This organization was established in 1997 to carry out research and development in advanced networking applications and services in the Asia-Pacific region.

APaRT See Automated Packet Recognition/Translation.

APC 1. adaptive-predictive coding. 2. advanced process control. 3. Aeronautical Passenger Communications 4. Association for Progressive Communications.

APCC The American Public Communications Council, affiliated with the North American Telecommunications Association (NATA).

APD avalanche photodiode. See photodiode.

APDU Application Protocol Data Unit.

aperiodic Occurring or recurring at irregular intervals. A repeating phenomenon or structure that does not have a regularly repeating nature. At the molecular level, a substance whose functions or structures are not regular or symmetric.

Human speech has an aperiodic nature that must be considered when designing compression algorithms, especially those that extract or compress the spaces between words in a predictive manner or those that apply regular algorithms to the irregular pitch and duration of uttered sounds.

Visual input of the natural world over time can be highly aperiodic (imagine the changing landscape as you drive down a highway). Our brains have adapted to recognizing certain shapes, sizes, and colors and assigning them meanings that we learn as we interact with the world, but it has been a significant challenge to develop image processing algorithms that can "recognize" aperiodic objects and events through vision detection systems.

Aperiodic phenomena are complex, with difficult to predict or calculate characteristics, especially at the detail level. As such, aperiodic transmissions with varying frequencies, pitches, transmission times, or other aperiodic properties are favored for security applications.

Aperiodic strip gratings are sometimes used to scatter electromagnetic energy.

Noise in optical or data transmissions is generally of an aperiodic nature.

aperiodic antenna In the positive sense, an antenna designed to maintain a relatively constant input impedance over a broad spectrum of frequencies. In another sense, a circuit or antenna structure that tends not to vibrate within the range of frequencies to which it is tuned.

aperiodic membrane In audio speakers, a resistive membrane and acoustical enclosure system coupled to a speaker for improving its mechanical performance. An aperiodic membrane system can help filter out harmonic distortions.

aperture 1. In the physical sense, an opening or hole, usually for controlling the admission of waves or particles, as in cameras, telescopes, and optical fibers. The size of the opening, and the speed with which it can be opened or closed, may be fixed or adjustable. In fiber optics, a variable attenuator can help control the amount of light transmitted between two coupled fibers. It may attenuate specific wavelengths or all the wavelengths passing through the fiber. This is particularly useful for instruments (e.g., spectrometers) that do not require the full intensity of the light signal that may be supplied by the illumination source (e.g., laser). An aperture may also function in a more virtual sense in that the light may be filtered by its line of travel rather than by passing through a hole. For example, fiber gratings can function as aperture filters to control the amount of light passing through a fiber by reflecting only the desired wavelengths in the destination direction. See acceptance cone, grating. 2. In a one-way antenna, the portion of the plane surface, perpendicular to the direction of maximum radiation, through which the major portion of the radiation passes. See aperture antenna.

aperture antenna An antenna characterized by a lens, horn, or reflector used as an aperture or directed region through which the majority of the radiant energy passes.

aperture distortion Aberrations in the focus, size, or shape of an image recorded through an aperture. Faults in an aperture, such as shape, orientation, perforations, jamming, speed of opening, etc., can cause undesirable effects. In a fiber grating "aperture" the spacing, precision, and composition of the grating must be carefully controlled in order not to introduce distortion.

aperture grill A focusing mechanism inside a cathode-ray tube (CRT), similar to a shadow mask, that helps target a beam on the inside coating of the monitor. An aperture grill consists of fine, aligned wires, and is said to have advantages over conventional shadow masks. See shadow mask.

aperture mask A thin grill or perforated sheet control mechanism that is commonly mounted inside an electron tube such as a color cathode-ray tube. The aperture mask is used to control more precisely and single out the electron beam, or portion of a beam, that passes through the mask to the inside surface of the display. See shadow mask.

aperture ratio In optics, especially photography, the ratio of the useful diameter of a lens to its focal length, the reciprocal of the *f*-number. In fiber optic grating "apertures," the relationships of the period, angle, and height of the grating facets to one another and to the incident wavelengths that pass through the grating. See aperture, *f*-stop, grating.

aperture stop See *f*-stop.

aperture tagging An older term for wavefront control or wavefront distortion correction. See micromachined membrane deformable mirror, wavefront control.

API See Application Program Interface.

APIC Advanced Programmable Interrupt Controller. Part of the Intel 440GX AGPset which provides input/output multiprocessor interrupt management.

apoapsis The point of greatest separation between two orbiting bodies. See apogee.

apogee The highest or most distant point, such as the apogee of Earth's orbit, that is, the point at which it is farthest from the Sun. The apogee of an orbiting artificial satellite is the point at which it is most distant from the Earth (which can be described in more than one way, but is usually from the center of Earth's gravitational field, or the center of an elliptical orbit). See apoapsis, geostationary, orbit. Contrast with perigee.

app See application.

APP See Ascend Password Protocol.

APPA 1. Alberta Professional Photographers Association. http://www.appa.ab.ca/ 2. See American Public Power Association. 3. Association of Higher Education Facilities Officers. http://www.appa.org/

apparent power In AC electrical power distribution, 1. the vector sum of the real power and the imaginary (*reactive*) power, 2. the square root of the sums of the squares of the effective power (the real and reactive power), 3. The root-mean-square (RMS) current times the root-mean-square voltage in the current.

The designations of real power, reactive power, and apparent power came about because alternating current (AC) is a more complicated phenomenon than direct current (DC) in terms of calculating power. In DC circuits, a fairly straightforward product of voltage times amperage provides a measure of power. However, in AC circuits, where sinusoidal periodic alternations of current and voltage are not necessarily in phase with one another, mathematical assessments of power have to take into consideration the alternating nature and phase differences of these waves in relation to one another.

Real power is derived by sampling the voltage in a large number of small time segments, then assessing the current in each and averaging the sum of the calculation. A wattmeter may be used to assess real power.

Reactive power, in an in-phase AC (or DC) circuit, will be zero (0) in which real power and apparent power are equal. However, there may be out-of-phase characteristics in the voltage or the waveform of the AC circuit and thus the power factor (PF) ratio may drop below one (1). Reactive power is the vector difference between apparent and real power.

In practical applications, if the apparent power increases, the power factor decreases and the circuit may adapt to satisfy the real power needs. See power, work.

APPC Advanced Program-to-Program Communications. An IBM set of operations and transactions to enable user-written programs to perform client-server network transactions.

APPC/PC An IBM application that implements advanced program-to-program communications (APPC) on a personal computer. See APPC.

append Add, affix, subjoin. It is very common in software programming to add the contents of a list, table, or file to the end of another file. Append is used most commonly to indicate additions to the *end* of a file; if the additions are in the middle of a file, or spread through various parts of the file, the term *merge* is generally used. See adjunct.

Apple IIGS computer A 65C816-based 2.8-MHz 16-bit addressing computer in the Apple II line, released in fall 1986 by Apple Computer, Inc. The Macintosh and PowerMac lines were more successful.

Apple Computer, Inc. A significant microcomputer hardware and software company located in Cupertino, California. Apple Computer was founded in 1976 by Steven P. Jobs and Stephen G. Wozniak, with Mike Markulla providing early business plan and financing support and Arthur Rock providing venture capital. Steve Jobs is known best for his marketing presence and administration tasks; Steve Wozniak is remembered for hardware design and computer-related technical tasks. Their initial product, leading up to the formation of Apple Computer, was a *blue box* designed to gain unauthorized access to long-distance lines, after which Wozniak developed a microcomputer circuit board, much like the original Altair kit, and this became the original Apple I computer. The Apple I was little more than a circuit board with neither case nor keyboard, yet the entrepreneurs sold about four dozen to excited hobbyists. They soon followed up with the Apple II at the West Coast Computer Faire in 1978.

Both Wozniak and Jobs had a strong commitment to providing computing services to education. Evidently the alliance of the young entrepreneurs was successful because Apple grew from those small beginnings to be one of the most significant microcomputer developers and retailers of the 1980s and 1990s, particularly with its Macintosh line, introduced in 1984 (following the less successful introduction of the Lisa a year before). Paired with the Laserwriter printer, the Macintosh launched a desktop publishing revolution. The subsequent PowerMac and G3 lines provided fast processors at lower prices than previous systems.

When sales flattened out and doom-sayers predicted the demise of the company, Apple responded by launching the *iMac,* a powerful, portable, individualist computer with an upbeat design and appeal similar to that of the Volkswagen Bug in the 1960s. The *iMac* evidently attracted more than loyal Macintosh users, with 16% purchased by new computer owners or those who had previously used other brands.

Apple Computer went public in 1980 and forged new directions, pioneering the graphical user interface developed at Xerox PARC, and incorporating the point-and-click style of interaction into the Lisa computer in 1983. The Lisa was ahead of its time and underappreciated. It did not sell well, probably due to the steep price tag. However, most of the characteristics of the Lisa showed up over the years in the Macintosh line, introduced in 1984, which eventually began to sell very well, after a slow start with the cute, but limited *Little Mac,* which had a small black and white screen and a single floppy drive.

Apple Computer continues to market computers and software, continually bringing out new desktop models and laptops, and continually updates its operating systems, e.g., OS X. See Jobs, Steven P.; Macintosh; Wozniak, Stephen.

Apple Desktop Bus ADB. A low-speed serial data bus to connect input devices to a Macintosh computer or other compatible hardware system. Input devices include graphics tablets, mice, keyboards, etc. ADB is a widely used, patented, Apple Computer, Inc., standard. Some versions of NeXT systems also conform to the ADB format so that Apple and NeXT keyboards and mice can be interchanged between Macintosh and NeXT computers. ADB devices typically communicate with the operating system through a low-level device handler. The ADB specification and licensing information is available through Apple Technical Publications.

AppleTalk A proprietary computer network proto-

col developed by Apple Computer, Inc., which functions independently of the layer on which it runs. Implementations vary, and include (1) LocalTalk and similar protocols (230 to 300 Kbps), commonly used among printers, Macintosh computers, and emulators; and (2) EtherTalk (10 Mbps), which provides broader multiplatform communications.

AppleTalk Address Resolution Protocol AARP. A protocol in the AppleTalk networking protocol stack that maps a data link address to correspond to a network address.

AppleTalk Control Protocol ATCP. A means for configuring, enabling, and disabling AppleTalk Protocol modules at both ends of a point-to-point link. ATCP uses the same basic packet exchange mechanism as the Link Control Protocol (LCP). See RFC 1378.

AppleTalk Data Stream Protocol ADSP. A connection-oriented protocol commonly used to establish a session for network data exchange between processes or applications. Established on DDP packet services, ADSP sets up a socket-based data exchange session that can transmit a continuous stream control on both sides of the session. ADSP is typically used by AppleTalk applications that establish a session for utilizing peer-to-peer services. For transmission of simple limited-data requests, see AppleTalk Transaction Protocol.

AppleTalk Echo Protocol ACTP. An AppleTalk transport layer network protocol in the AppleTalk protocol suite that enables a node to send a test packet to any other node through the Datagram Delivery Protocol (DDP) and receive an echoed copy of that packet, thus establishing the reachability of the tested node. It uses socket number 4.

AppleTalk Filing Protocol AFP. A client-server network file protocol that enables file sharing over an AppleTalk network. Thus, files stored on one computer on the network can be accessed remotely as though they were stored on a local storage device (e.g., hard drive). AFP provides the services for accessing an AppleTalk AppleShare server.

AFP file services can be implemented on other operating systems as well (e.g., Unix) to allow access to files on AppleTalk systems. AFP does not directly map to the Open Systems Interconnection (OSI) model, but it corresponds roughly to the high level Presentation and Application layers. See AppleTalk Session Protocol.

AppleTalk Link Access Protocol ALAP. An AppleTalk network protocol for communications over industry-standard hardware interfaces to other networks. Access may be through LocalTalk (LLAP) or EtherTalk (ELAP). ELAP handles interaction between standard Ethernet and AppleTalk proprietary protocols through an Address-Mapping Table (AMT) by encapsulating or enclosing data in protocol units of the data link layer.

AppleTalk Name-Binding Protocol ANBP, NBP. A protocol for translating entity names into numeric addresses that are used for locating resources on a computer network. Network endpoints have names of the form *Entity:Type@Zone*. Entities can be processes or applications. Since names are easier for people to remember than numeric addresses, it is common to have a mechanism like ANBP for translating people-friendly information into computer-friendly data for send-and-receive protocols to access network resources. ANBP is implemented through the *.MPP* driver.

AppleTalk Remote Access ARA. A mechanism to enable two or more computers, networked through AppleTalk, to share a serial device, usually a modem, on the remote system. In other words, if there is only one phone line and one modem, and four computers attached to the network, ARA can be set up so that any one of the people using the computers without a modem can access the modem through the other computer (one at a time) as though it were attached to the local machine.

AppleTalk Secure Data Stream Protocol A secure variant of AppleTalk Data Stream Protocol (ADSP) that establishes a network connection session after user authentication has been established using Authentication Manager.

AppleTalk Session Protocol ASP. A protocol that opens, maintains, and closes socket-based network connections. An ASP session establishes communications between an application or process and a server application. Sessions are asymmetric, initiated by the application or process, and responded to by the server. ASP is primarily used to provide services for Apple-Talk Filing Protocol (AFP). It is built on top of the AppleTalk Transaction Protocol (ATP). See Apple-Talk Filing Protocol, AppleTalk Transaction Protocol.

AppleTalk Transaction Protocol ATP. A basic, low-overhead protocol underlying network transactions, ATP is used to implement AppleTalk Session Protocol (ASP) servers. ATP is suitable for small data transactions. It has a simple request-response-done format that uses less overhead than the connection-oriented AppleTalk Data Stream Protocol (ADSP).

Applicability Statement AS. In the Internet Standards Process, an AS describes how and when Technical Specifications may be used in standardized or nonstandardized ways in the context of the Internet. An AS may not have a higher maturity level in a standards track than any Technical Specifications upon which it relies. A requirements document (comprehensive conformance specification) is the broadest form of Applicability Statement (e.g., Internet hosts). Technical Specifications are identified in an AS as to their relevance and interrelationships. Specific parameter ranges or subfunctions may be specified and guidelines for their implementation included in general or specific "domain of applicability" contexts. See Technical Specifications. For AS requirement levels, see RFC 2026.

application, applications program A catchall designation for computer software programs, especially high-level ones intended for endusers, such as databases, spreadsheets, word processors, graphics programs, telecommunications programs, programming tools, etc.

application-definable keys ADK. Keys that can be assigned to perform an application-specific function or to insert or display a menu, or symbol, or other feature for quick access, as desired by the user.

application framework The basic logical structure in an object-oriented development environment. When software is being designed, there is often a pre-existing set of assumptions within which the user interacts with the computer. For example, when a user sees something on a screen that looks like a button, he or she will expect something to happen when it is clicked, or double-clicked, depending upon the system, and the experience of the user.

These basic assumptions are cultural and experiential and are important in the design of software. If the software interface is obscure, or too radical to be understood, it may not be of practical use. A certain degree of consistency, immediacy, and familiarity are important factors.

By using an application framework, not only will the user be presented with a consistent set of stimuli and tools, but the programmer will have a context within which to create the software. The framework exists at several levels, at the user interface level, at the applications design level, and at the lower levels in which the parts, components, interactions, and processes are created.

In an object-oriented programming environment, it is easier to apply a framework, and to work within a framework, when shared objects, classes, and other programming primitives and structures are being used and reused. For this reason, most of the thinking about application frameworks has arisen in object-oriented programming environments, such as those utilizing Smalltalk, C++, and various graphical interface builders such as the NeXTStep Interface Builder or Apple Computer's MacApp. See application generator.

application generator AG. A software program that greatly facilitates the development of software applications code by providing a set of tools to describe the program, leaving the details to the software. It's a way of automating programming and taking out many of the drudge activities and details that are easy to mistype when coding in text with an editor.

This type of programming approach wasn't prevalent on desktop computers until Power Windows was released for the Amiga 1000 in 1986. It was one of the earlier microcomputer application generators, allowing the user to essentially draw the application as though using a paint program, placing buttons and icons, windows, and other structures where they were needed. Colors and logical relationships could then be dynamically adjusted with the mouse, and then presto! select *build* and it would automatically generate C, BASIC, or Assembler code. The code could then be edited and changed as needed.

With this type of programming environment, the programmer doesn't have to worry about counting pixels, about guessing what the interface will look like, or about writing reams of C code before even the smallest activity can take place on the screen. This is a very good idea. NeXTStep incorporated a very nice interface builder about two years later, which took aesthetics, utility, and logical linking to object-oriented structures several steps beyond Power Windows, and facilitated graphical creation of windows, menus, tables, buttons, and much more, providing a fast and easy way to create an interface and connecting structures that were consistent with the NeXT *application framework.* An hour with the interface builder could easily equal two days of coding by hand with a text editor. In the 1990s, other desktop systems began to come out with interface builders, an idea that has great practical value, especially as object-oriented environments became more prevalent. Some authoring systems also function as application generators, as do some programmable databases. If the software front-end that allows authoring and database configuration without programming also provides an option to save code that can be accessed and manipulated, usually with a text editor, and to link to operating system or program structures, then it is a form of application generator. See application framework.

application layer In layered hierarchy network systems, the layer that provides services to the applications programs (as in the Open Systems Interconnection [OSI]model) or the layer that runs the applications themselves, depending upon the network layer design. In OSI, the application layer ensures availability of parties, may provide authentication, checks the available resources, negotiates data, privacy, and error-checking parameters, and application-level protocols. See Open Systems Interconnection for a chart that describes layer relationships.

application program A very broad, generic term for almost any user-level computer program. That's not to say that system administrators don't use applications; they do. They just happen to be more technical applications aimed at a technical user level. Short, specific applications that do a single task or a small number of tasks are sometimes called utilities, such as a disk utility for formatting disks, or a conversion utility for changing a TIFF file to a BMP file, or a copy utility for duplicating disks. Application programs commonly used in telecommunications include Web browsers, FTP clients, chat clients, gaming clients (such as bridge, chess, Go, and casino applications), compression/decompression utilities, and file translation utilities.

Application Program Interface API. 1. An interapplication or intervendor interface that provides a somewhat standardized means of allowing programs to talk and work together. The Apache server and the Netscape server API are common interfaces used to implement network applications services. In a competitive environment, API conformance within consumer products is rarely perfect. By making subtle changes to a specification in an industry-leading product, vendors often slow down the competition (e.g., faster modems) and create a short business window during which they are the only company to support a particular product or specification (i.e., a temporary monopoly). However, a certain level of adherence to standards also has competitive advantages in that it is easier for third party vendors to support a leading product. 2. In the XOpen/Architectural Framework Technical Reference Model, the Application Program Interface is one of five basic elements and one of two interface types (the other being the External Environment Interface). The API is a specification for the data link between the Application Software and the Application Platform upon which all the services are provided, thus facilitating portability and interoperability among systems. The API includes the semantics, syntax, protocols, data structures, and other definitions necessary to ensure compatibility. See External Environment Interface.

Application Service Provider ASP. A vendor that distributes software functionality over data networks such as telephone networks, the Internet, or local/wide area networks. To get an idea of what this means, think of the different ways in which a person can get voicemail messages. The user can get a voicemail modem and set up the mailboxes on the computer and manage the messages that are received after a call comes through to the user's premises, or she or he can get a phone with built-in voicemail capabilities. On the other hand, an ASP, such as the local telephone carrier or a third-party applications phone services provider, can set up a phone line so that if it is busy or isn't answered within a certain number of rings, it will be redirected to a voicemail service associated with the carrier's equipment. That carrier is thus a voicemail ASP.

With the tremendous growth in digital telephony and data services, it is likely that markets for ASPs will grow, especially for services that are not easy for users to set up and manage themselves. See ASP Industry Consortium, Enhanced Service Provider.

Application Software Interface ASI. A means of working within a common application interface for provision of ISDN-related digital telephony services. See North American ISDN Users Forum.

Application Specific Fiber Platform ASFP. As implemented by Southampton Photonics (SPI), a platform comprising core technologies of specialty fiber design and manufacturing, fiber Bragg grating design and fabrication, and amplifier and laser technology. ASFP facilitates the volume manufacture of high-performance in-fiber components and subsystems.

application-specific integrated circuit ASIC. A computer chip or small, specialized circuit designed to enable or enhance a specific type of application. As examples, ASIC video cards have been designed to drive specialized monitors, ASIC modem cards provide functionality to specialized or enhanced modems, ASIC daughterboards sometimes provide hardware support to rendering and ray-tracing applications.

applications access point AAP. In general, an access point is a device or system that allows users to access a particular type of service. An applications access point is one that permits access to applications software. The access point is usually instituted for the management of resource sharing and/or for security reasons. Applications that may be shared through an access point typically include databases that are accessed or managed by multiple users or collaborative work applications in which changes and updates are relayed to the various participants.

applications processor A computerized system that can be integrated with a phone system to add functionality. Functions may include voicemail, Automated Attendant, Call Detail, networking (packet switching), and others. See peripheral device.

Applications Technology Satellite program ATS. A series (ATS-1, ATS-2, etc.) of satellite launchings carried out by the U.S. National Aeronautics and Space Administration (NASA) to test payloads and study space. Five of these craft in three configurations were manufactured by Hughes between 1966 and 1969. See Applications Technology Satellite Program chart.

Applied Computer Telephony ACT. A commercial product from Hewlett-Packard Company for integrating voice and data analysis technologies on HP systems. The system is used in conjunction with private branch phone exchanges to record, handle, and evaluate call-related transactions. See Hewlett-Packard.

APPN See Advanced Peer-to-Peer Networking.

approved circuit See protected distribution system.

approximate discrete Radon transform ADRT. A mathematical technique used in situations where substantial redundancy is expected or encountered. See discrete cosine transform, Fourier transform.

APR American Public Radio. See Public Radio International.

APS 1. See Advanced Photo System. 2. See Automatic Protection Switching.

APTS Association of Public Television Stations.

AQL acceptable quality level. An industry-established confidence level.

Aqua A new, aesthetically appealing, customizable user interface introduced by Apple Computer with Mac OS X for PowerPC- and Intel-based platforms. See Mac OS X.

Ar *symb.* argon. See argon.

ARA See AppleTalk Remote Access.

ARABSAT A communications satellite placed into orbit in the mid-1980s. The ARABSAT System was initiated in 1967 as a means of establishing a communications network for supplying cultural and social interaction for the League of Arab States. This effort was extended in 1976 with the founding of the Arab Satellite Communication Organization (ARABSAT).

In the early 1980s, the French Aerospatiale was commissioned to manufacture three satellites. ARABSAT 1A and 1B were launched in the mid-1980s. ARABSAT 1C was launched in February 1992 and ARABSAT 1A and 1B were turned off in 1992 and 1993. ARABSAT 1C is expected to be operational until 2002.

ARABSAT has control stations at Dirab, Saudi Arabia, and Tunis, Tunisia. Telephony and television services are provided, according to International Standards, to the Saudi Arabia and northern Africa geographic region.

Aragon, Dominique In 1820, Aragon described how an artificial magnet could be created by winding a coil around a piece of iron or steel that was carrying an electrical current. Soon after, electromagnets were developed.

ARAM audio RAM. A low-cost, low-grade integrated memory chip suitable for digital answering machines and other inexpensive consumer products.

aramid, aramid yarn A strong fibrous material commonly used to reinforce fiber optic cables, especially those that may be subject to rough treatment such as abrasion by rodents or small pellet shots.

Aramid yarn is favored for its low weight, flexibility, water resistance, low conductivity, and high strength properties. Aramid may be used in several parts of a cable assembly. Most often it is a reinforcing strength member, but it may also provide the material for the central element in a fiber bundle, and the ripcord used to open cable for attaching connectors. Dupont distributes a type of aramid yarn under the

Applications Technology Satellite Program		
Satellite	Launched	Notes
ATS-1	1966	Spin-stabilized synchronous altitude. Electronically despun antenna. Stationed over the Pacific Ocean. Successfully photographed Earth and provided a presidential communications link for recovery of Apollo 11.
ATS-2	1967	Gravity gradient stabilized. Insufficient thrust resulted in an elliptical orbit and it lost orbit after only 880 days.
ATS-3	1967	Synchronous orbit. Mechanical despun antenna, color camera that photographed tornados in 1968 and an eclipse of the Sun in 1970.
ATS-4	1968	Gravity gradient stabilization in synchronous altitude. Failed to reach intended orbit and lost orbit in 1968, two months after deployment.
ATS-5	1969	Synchronous orbit. Gravity gradient booms for stabilization didn't deploy correctly, but some of the experiments were successful. It was retired in 1984.

well-known tradename Kevlar. See Kevlar, swelling tape.

Aramid Yarn in Fiber Optic Cable

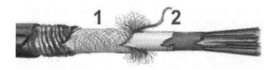

In a typical fiber optic cable assembly, the aramid yarn layer (1) provides strength and strain relief. It is often inserted between the outer strength member and an inner supporting structure, such as a water-resistant tube. A ripcord (2) may be provided to make it easier to peel back the layer in preparation for adding a connector.

ARB 1. Administrative Review Board. Established in 1996 by the U.S. Secretary of Labor within the Department of Labor. 2. Air Resources Board. 3. See all routes broadcast.

arc A very bright electrical discharge across a gap in a circuit. In fiber filament fusion splicers, carefully controlled, electrode-generated arcs are used to clean and heat the ends to be spliced so they will fuse into a continuous light-guiding path. See arc lamp; Aryton, Hertha Marks; fusion splice; Poulsen arc.

ARC See Ames Research Center.

arc converter A device used to convert direct current (DC) into undamped or continuous wave radio frequency (RF) signals. This technology was used in early radiotelegraphy. Many aspects of global radio communications in the early 1910s were based upon this technology. See Poulsen arc.

arc lamp An electrical lamp that exploits the tendency of electrons to jump a gap in a circuit, which can be harnessed to produce an intense light under certain circumstances. It was invented by Sir Humphrey Davy in the early 1800s and came into widespread commercial use in the late 1870s when it was incorporated into street lamps.

Arc lamps are important in optical 'etching.' Exposures from strong light sources can be used to chemically alter substrate materials through a template mask so that some of the materials are removed and some remain. In this manner, mercury arc lamps have been used in the semiconductor industry for circuit design through optical lithography. This is a technology that may someday be superseded by higher-precision electron-beam and laser etching techniques.

Archie Network archiving software developed by Peter Deutsch, Alan Emtage, and Bill Heelan. Named for the word *archive*, Archie is an Internet query tool that tracks the contents of anonymous ftp sites. It was introduced at McGill University, Canada, in 1990. Archie allows users to retrieve a list of FTP locations by submitting file search criteria to an Archie server. See Veronica.

Archimedes (ca. 287-212 BC) A Greek mathematician and inventor born in Sicily who made contributions to our understanding of volume and displacement, and who created the mathematical treatise "Measurement of the Circle" in which he described the calculation of the ratio of a circle's circumference to its radius. See Archimedes' principle.

Archimedes' principle A body immersed in fluid is buoyed up by a force equal to the weight of the fluid displaced. (This principle is humorously illustrated by actor Gary Oldman in a bathtub in the Cinecom Entertainment movie production of "Rosencrantz and Guildenstern Are Dead.")

architect One who designs a layout or topology, such as a building layout, circuit board architecture, network routing system, etc. The architect frequently is also the one who drafts the technical drawings associated with the layout and may or may not check electronics codes, building codes, or other regulations associated with the design.

architecture The design and layout of a process, system, or facility. The architecture involves the overall plan and topology, in addition to the relationships and interconnections between the individual parts. It may also include the *direction* of information paths or movement within the system. Good architectures usually try to incorporate, or at least balance, flexibility, robustness, efficiency, and scalability, whether it be the design of a building or of a microprocessor chip. See topology.

archival A format, medium, or protective system designed to facilitate preservation. Archival papers and plastic sleeves are acid-free, or free of plastics that may change the information or degrade rapidly. Archival data storage formats are nonvolatile (magnetic media such as video tapes, audio tapes, floppy diskettes, etc. are not very stable over time and may be damaged by proximity to magnets) and resistant to damage and degradation. In the data industry, archival file formats are as important as the materials on which they are stored, as the information is useless if it can no longer be read or deciphered.

archive A repository of records or files. A backup or duplicate of information made to preserve or prevent loss in compressed or uncompressed form. An archive generally contains information that needs to be kept over time, for one reason or another (legal, historical, etc.). Many archives are seldom or ever accessed. Computer data archives are becoming more prevalent and offer many search and retrieval advantages over traditional paper-based archives. See anonymous FTP, Archie, archival, FTP.

archiver A term for software tools that are designed to store files in such a way that they take a minimum of space and can be retrieved, reconstructed, and viewed at a later date. Software archivers often include compression algorithms and switches to allow an archived file to be scanned for header information without decompressing it. The degree of compression possible is very dependent upon the interaction between the type of compression algorithm and the type of data being compressed. Common software archivers include zip, lharc, Stuffit, and tar.

archiving The process of storing information, compressed or uncompressed, encoded or not encoded, such that it can be accessed and viewed at some future date, if needed. Archiving involves selecting a storage format, medium, and location and carrying out occasional or scheduled consolidation and organization of the objects or information. One of the big issues with archiving, besides space, is the development of efficient search and retrieval methods that make it possible to find a desired piece of information in a vast amount of data. See archival, archive, database, FTP.

ARCnet *Attached Resource Computer net*work. A popular pioneering local area network (LAN) developed by Datapoint Corporation in 1977 for use with thin coaxial cable. Incorporating a modified Token-Ring passing scheme, ARCnet provides high-speed baseband communications at 2.5 Mbps with either a bus or star topology. ARCnet became standardized as ANSI 878.1. Although not as widespread as it once was, ARCnet has been upgraded to include transmission over copper twisted pair wire and fiber optic cables.

Arcstar The brand name for the Nippon Telegraph and Telephone Corporation's (NTT) global services including NTT Worldwide Telecommunications Corporation, NTT Europe, ntta.com and Asian branches. These include managed bandwidth, Frame Relay, and Internet Protocol (IP) virtual private networks (VPNs). See Nippon Telegraph and Telephone Corporation.

ARD 1. advanced research and development. 2. See Automatic Ring Down.

ARIADNEt The ARIADNE network, an academic and research network operated out of Athens, Greece by NCSR Demokritos, a physics and sciences research association. http://www.ariadne-t.gr/

Ardire-Stratigakis-Hayduk algorithm ASH. A lossless compression algorithm, named after its creators at Western DataCom, developed between 1990 and 1993. It was intended for use over synchronous data communications with varying media characteristics. Unlike asynchronous transmissions protocols, framed data can contain a very large number of bits and does not have to be timed with start and stop bits. ASH provides a means to provide good compression ratios on various types of traffic in a multiuser network.

ASH incorporates interesting concepts from artificial intelligence. By using pattern-matching and predictive algorithms, data not yet transmitted and nonidentical strings can be processed and evaluated. As part of the compression methodology, ASH uses an Occurrence Optimized Codebook (OOC) for fast-cache access to commonly occurring tokens and strings. ASH safeguards against data expansion and latency. A patent has been sought for the ASH technology. See Lempel-Ziv-Welch.

ARDIS A commercial packet-switched nationwide wireless data communications service developed in the mid-1980s by Motorola and International Business Machines (IBM) and now owned by Motorola.

It was originally developed for field technicians and is appropriate for short messages and quick database lookups for a variety of applications. ARDIS is somewhat similar to CDPD except that it is a data-only service. It is used for wireless faxing and realtime messaging with any Internet address worldwide. ARDIS can be accessed through laptops, and personal data assistants. See RAM Mobile Data.

ARE See all routes explorer.

area code A three-digit code in a phone number that designates the region. See North American Numbering Plan (NANP). See the Appendix for a chart.

area code restriction A service for enabling the subscriber to deny telephone calls to specified area codes. It is not a blanket restriction as in some long-distance call-blocking services. The service is useful in offices and other environments where it is otherwise difficult to monitor phone use.

area network See local area network, metropolitan area network, wide area network.

Area of Service AOS. The geographical area supported by a vendor, carrier, or service provider.

Arena 1. The name of an HTML3 browser from the World Wide Web Consortium (W3C) designed as a proof-of-concept demonstration tool for HTML+ ideas preceding HTML3.

ARES See Amateur Radio Emergency Service.

ARF See Alternate Regulatory Framework.

argon (*symb. – Ar*) Argon is a colorless, odorless gas used in light bulbs. Argon plasma arc lamps can provide a continuous radiation source covering a broad spectrum from ultraviolet to infrared suitable for testing solar energy components for use in satellite communications. See argon laser.

argon laser A type of gas laser that primarily uses argon gas. This common type of laser can be used to produce green and blue light, which is useful for creating laser light show effects. It is similar to a krypton laser, except that argon produces a little more light. Argon is sometimes combined with krypton gas to produce an argon-krypton hybrid. Argon lasers are typically water-cooled.

ARI Automatic Room Identification. A telephony service used especially in the hotel/motel industry to identify call locations.

ARIB See Association of Radio Industries and Businesses.

Ariel 1 Historically, the first U.K. satellite project to study the ionosphere, launched in April 1962. The satellite's orbit decayed in 1976.

ARIES 1. Angle-Resolved Ion and Electron Spectroscopy. 2. The name of a commercial satellite service. See Constellation Communications, Inc. 3. See Australian Resource Information and Environment Satellite.

ARINC-429 Digital Information Transfer System DITS. DITS is a digital point-to-point hardware architecture and data specification for describing a digital bus for device communications for the aeronautics industry. The transmission hardware is based upon twisted-pair wires with one transmitter connected to one or more receivers. The data protocol is based upon 32-bit words consisting of a data portion and a label identifying the nature of the data. Development of the DITS specification and ARINC-compatible products aided in the commercialization of digital communications for aeronautics devices and facilitated the development of modular, off-the-shelf products. See ARINC-629.

ARINC-629 A commercial civilian communications protocol based upon a central twisted-pair bus architecture which can accommodate up to 120 peripheral devices or terminals at a 2-MHz bus rate. ARINC-629 evolved from ARINC-429, which was developed in the 1970s. ARINC products are manufactured by Excalibur Systems, Inc.

ARINC Incorporated An organization formally established in 1929 by Louis Caldwell and representatives of four major airlines as Aeronautical Radio, Inc. ARINC was established to be the dominant nongovernment single licensee and coordinator of radio communications. Upon its formation, the Federal Radio Commission (FRC) transferred to ARINC the responsibility for aeronautical ground radio stations, thus providing a within-the-industry means of administrating aeronautic rules and regulations established by the FRC.

ARINC continues to coordinate airline industry telecommunications activities, communications, and information processing systems. It publishes standards of relevance to avionics systems and analog and digital equipment.

ARISE 1. See Advanced Radio Interferometry between Space and Earth. 2. American Renaissance in Science Education. 3. Applications Reform in Secondary Education.

ARISS See Amateur Radio International Space Station.

Aristote Association A French association of organizations and firms engaged in the development of telecommunications and data transmissions technologies. The Secretariat is located at the École Polytechnique in Paris, France, with seminars and workshops held at the Arago and Poincaré amphitheaters. http://www.aristote.cea.fr/

ARISTOTELES Applications and Research Involving Space Technologies/Techniques Observing the Earth's Fields from Low Earth-Orbiting Satellites. A joint project of NASA and ESA carrying out scientific research data gathering on the Earth's gravity and magnetic fields. It is equipped with a Global Positioning Service (GPS) receiver, gradiometer, and scalar magnetometer.

Arkay CT-650 A commercial computer based upon a description of a paperclip computer, (a homebrew digital computer that could be constructed out of materials found around a home).

ARL 1. Air Resources Laboratory (NOAA). http://www.arl.noaa.gov/ 2. See Association of Research Libraries.

armature A portion of a magnetic circuit typically consisting of a conducting material, such as wire, wound around a core, which is moved within a strong magnetic field to create current. If the armature revolves, the wound material interacts with the magnetic lines of force, in a sense, cutting in and out, and the current generated by this interaction can be drawn out. The arrangement is used in generators and alternators, where the current may be drawn out by brushes. See coil.

Armed Forces Communications and Electronics Association AFCEA. An international nonprofit professional association for communications, electronics, and intelligence. Founded in 1946, the association has over 40,000 individual and corporate members in government, industry, and military sectors. AFCEA publishes SIGNAL Magazine, technical papers, and books. http://www.afcea.org/

ARM See American Radio Museum.

armor 1. Defensive or protective covering. 2. A type of heavy-duty waterproofing or other shielding used especially in underwater or underground installations. 3. Heavy shielding to protect facilities, equipment, and personnel from radiation or chemical contamination. 4. In computer technology, heavy security measures taken to keep users off a system, which may range from inaccessible facilities to extra software measures taken to discourage unauthorized intrusion.

armor-plated A physical or administrative high security system which the administrators deem to be virtually impenetrable. Highly resistant to access or tampering. Bullet-proof.

armored cable See aramid yarn; cable, armored.

armoring, ballistic A strong armor layer (e.g., coated steel) used to protect aerial components and cables from shotgun pellet intrusion. Ballistic armoring for fiber optic cables is generally about 0.25 mm thick. Sometimes dual armoring layers are used, in addition to aramid fibers.

The ballistic energy from pellets hitting a cable is dissipated by the steel armoring. Increased protection against larger pellets is offered by aramid fibers that cushion the fibers from the impact of a pellet.

ZETABON™ is an example of a steel-armored cable distributed by Dow Chemical Company. See aramid; cable, armored.

Armstrong, Edwin Howard (1890–1954) A gifted American researcher who invented the superheterodyne circuit and frequency modulation (FM). Armstrong was not only a good inventor but also a good theoretician; he was one of the few people who understood, on a physics level, many of the new inventions that were making possible the development of wireless communications.

In October 1914, Armstrong was awarded a patent for his wireless receiving system (U.S. #1,113,149). In the 1920s, he contested Lee de Forest in the invention of regeneration, but de Forest won the suit. After many years of painstaking research against conventional wisdom and the negative predictions of mathematicians and engineers, Armstrong proposed a method of wave modulation that varied the frequency rather than the amplitude of a wave. He was awarded a patent for FM radio signaling in December 1933 (U.S. #1,941,066).

Armstrong waged a long and tragic legal battle with

RCA over his patents to the FM technology which were posthumously upheld in his favor. During World War II, Armstrong furthered the art of radar transmission by suggesting the use of FM signals, rather than the short pulse radar bursts that were used at the time. His ideas are now commonly incorporated into frequency modulated (FM) radio, television, and radar transmissions. See frequency modulation, heterodyning.

AROS 1. Amateur Radio Observation Service. 2. Amiga Research OS. A project initiated in the mid-1990s to update, bug-fix, and enhance the Amiga Operating System. http://www.aros.org/

ARP See Address Resolution Protocol.

ARPA Advanced Research Projects Agency. See Defense Advanced Research Projects Agency.

ARPANET Advanced Research Projects Agency Network. The historical basis of the Internet, ARPANET was originally discussed by the ACM in 1967, presented to ARPA the next year and put into operation in 1969. The first widespread demonstration of ARPANET occurred at a Computer Communications conference in 1972, and a year later ARPANET traffic had grown to millions of packets of data transfer per day. By 1975, the ARPANET had been transferred to the Defense Communications Agency (later the Defense Information Systems Agency).

In 1982, ARPA endorsed TCP/IP as its protocol suite. In 1983, ARPANET split into MILNET and ARPANET (mandated to use TCP/IP), which together formed the Internet. Each was given a network number, and gateways were installed to provide packet forwarding between them. ARPANET was officially discontinued in 1990, largely due to the evolution of the Internet. See ARPA, BITNET, IANA, NSFNET, NEARNet, SPAN.

ARQ See automatic retransmit request.

array A type of data organization structure commonly used in programming. An array consists of an ordered list or matrix of information which can be visualized as 2D or 3D tables of information contained in cells which often have common characteristics, such as the size of the data cell (although the data in the cells may vary in length). Arrays form the structural basis for many types of databases, including tables and lists. Many software programs have built-in array-handling functions to automate common ways in which arrays are manipulated.

array antenna 1. An antenna with a number of directing, reflecting, or other elements arranged in a more-or-less regularly spaced, often symmetrical pattern. See antenna. 2. One antenna in an array of antennas that are organized and connected in such a way as to significantly boost power, range, and performance. These powerful antenna systems are used for picking up weak signals as in astronomy and military applications.

ARRL See American Radio Relay League.

ARRL Monitoring System A policing system of the American Radio Relay League to monitor and maintain the correct, licensed, and responsible use of amateur radio frequencies and procedures. For the most part, the amateur radio community successfully seeks to be self-policing, but there are times when unauthorized use reaches problematic levels or originates in countries that are not regulated by the U.S. Federal Communications Commission (FCC), at which time the ARRL will document the problems and call for assistance from appropriate domestic and international agencies. See American Radio Relay League.

ARRN Amateur Radio Repeater Network.

ARS See Automatic Route Selection.

ARSR See Air Route Surveillance Radar.

articulation Clear utterance or playback of sounds – the degree to which reproduced or transmitted sounds are clear enough to be understood by a listener. Mumbling is poor speech articulation. When people say "Testing, testing, ..." on a sound system, they are testing not only circuit functioning and volume but also the clarity of the sound and capability of conveying a message. Articulation does not have to be high fidelity to be understood. It depends in part upon the ability of the recipient to perceive and understand the message, and may not have to match perfectly the original. This is an important aspect of data communications as well. When conversations are converted from analog to digital, through a process called sampling and quantization, it is important to determine how much of the information is needed in order for the communication to be understood by the recipient. This information can be applied to compression and decompression systems for speeding up transmissions. See fidelity, intelligibility.

articulation index AI. A scale from 0 to 1 that provides an objective reference for the intelligibility of voice signals expressed. AI is quantified in terms of the percentage of speech units understood by a listener when the units are presented out of context. The index is a useful measure for testing and comparing voice transmission and speech compression/decompression technologies. See articulation.

artificial intelligence AI. Insights or behaviors attributed to an entity, usually a machine, that is not traditionally perceived by humans as having the capability to think in ways that involve problem-solving, insight, and other uniquely human characteristics. The field of artificial intelligence has spawned many useful approaches, languages, techniques, and programming algorithms. Expert systems, neural networks, robotics, vision systems, and natural language processing all have their origins in AI research. People interested in artificial intelligence come from a diverse range of backgrounds.

The origins of artificial intelligence, as they apply to computers, trace back to the 1950s, though Ada Lovelace proposed in the 1800s that "thinking machines" could be programmed to create music or art. Pioneer researchers include A. Turing, J. McCarthy, and N. Wiener. See expert systems; Intelligent Networks; Lovelace, Ada; robotics.

Artron *art*ificial neu*ron*. The familiar name for an electronically simulated neuron used in a maze-running robotic mouse in the early 1960s. See Melpar model, neural network.

AS 1. See Applicability Statement. 2. See autonomous system.

as is A term applied to products that are bought and sold with no implied or stated warranties. Condition may be guessed by inspecting and trying the equipment, but there is no way to know the completeness, remaining useful life, or technical functionality of the equipment. See fair, good.

AS&C 1. Advanced Systems & Concepts. A division of the U.S. Department of Defense associated with the Deputy Under Secretary of Defense (DUSD). ASCO, the DTRA Advanced Systems & Concepts Office, encourages seed projects for technologies developed to counter threats to national security. 2. Alarm Surveillance and Control.

AS-Interface, Actuator Sensor Interface, AS-i. An open network protocol for automating actuator and sensor control for industrial applications. It can operate as a controller or as a stand-alone automation support bus. It may also be used with higher level field buses to enable remote input/output operations. A single AS-i v2.0 master can support up to 124 actuators/outputs and 124 sensors/inputs. See actuator, INTERBUS, PROFIBUS.

ASA 1. See Acoustical Society of America. 2. Assistant Secretary of the Army.

ASARS See Advanced Synthetic Aperture Radar System.

ASC 1. advanced switching communications. 2. Aeronautical Systems Center. A U.S. facility located at Wright-Patterson AFB. Science and technology programs are carried out in the Wright and Armstrong Laboratories.

Ascend Inverse Multiplexing AIM. An in-band networking protocol from Ascend Communications that manages interconnections between two remote inverse multiplexers.

Ascend Password Protocol APP. A network User Datagram Protocol (UDP) used in servers to respond to password challenges from external authentication servers. APP Server utilities are available from Ascend for a wide variety of computer platforms.

ASAI See Adjunct Switch Application Interface.

ASAPI Advanced Speech API. An open, cross-platform speech applications programming interface developed by AT&T.

ASCA See Advanced Satellite for Cosmology and Astrophysics.

Ascend Password Protocol APP. A User Datagram Protocol (UDP) network security protocol defaulting to port 7001.

ascending node Intersection of a satellite's orbital plane with the Earth's equatorial plane.

ASCII (as-kee) The American Standard Code for Information Interchange was developed by the American National Standards Institute (ANSI). Also known as ASCII International Telegraph Alphabet 5, ASCII is the most widely used computer character set encoding scheme currently employing seven bits, thus making a total of 128 possible characters/symbols. ASCII is mainly suitable for English language communications. Since it is very limited in its letters and

symbols, many extensions to ASCII have been incorporated into key mappings on various computers to include symbols and western European characters. Sometimes called extended ASCII (even though the extensions aren't standardized), these 8-bit encodings provide 256 possible characters, but the higher 128 characters are not usually compatible across platforms. See EBCDIC, Unicode. See the Appendix for an ASCII chart.

ASCII editor A text editing tool that handles basic, simple characters standardized as ASCII text, which are cross-compatible and transferable over almost all seven-bit-compatible systems, which includes most computer networks. Since the ASCII standard does not support style attributes (bold, underline, oblique, etc.), it cannot be used for extensive formatting. Due to the limitations and simplicity of its character set, ASCII editors are very fast. They are also good for writing computer source code, which typically needs speed and compatibility more than style tags.

If you require style tags and indentation for text formatting, and want to transfer the documents across applications or platforms, the best supported format that includes them is Rich Text Format (RTF), also known as Interchange Format (developed by Microsoft and supported across its products). It's not 100% compatible across platforms, but it's pretty close and can be read and written by most word processors (with import and export menu options). Another good format for transferring more complex documents is generic Adobe PostScript, which includes not only text and style support, but image positioning, layout effects and more, or its cousin, Adobe Acrobat's Portable Document Format (PDF).

When designing Web pages with links to downloadable files, there should be more than one format available. If all Web site managers were to include these three: an ASCII version, an RTF version (which can be read into virtually any popular word processor), and a PostScript or PDF version, then the needs of low-end and high-end users would be well met, and at least one of the files would be accessible to virtually everyone using the Web. See appendix for an ASCII chart.

ASDSP application-specific digital signal processor. See digital signal processor.

ASE Application Service Element. An element of an application layer protocol in the Open Systems Interconnection (OSI) layered network model. It is combined with other elements to form the complete protocol. See Open Systems Interconnect.

ASFP See Application Specific Fiber Platform.

ASH See Ardire-Stratigakis-Hayduk.

Asia DAB Committee, Asia DAB ADC. A nonprofit association working in conjunction with the World DAB Forum to promote, support, and coordinate the implementation of digital radio broadcasting technologies in Asia. Digital Audio Broadcasting (DAB) promises to significantly change and improve the quality and richness of programming information available to consumers. See Digital Audio Broadcasting Forum. http://www.asiadab.org/

Asian Mobile Satellite System AMSS. A satellite-based mobile phone system of Asia-Pacific Mobile Telecommunications Satellite Pte. Ltd. The system was supplied by Hughes Space and Communications International, Inc. It operates in the L-band to provide a mobile infrastructure for communications in Asian-Pacific regions from Japan to Pakistan. Asia is a world leader in many aspects of wireless communications.

ASI 1. Advanced Study Institute. 2. artificial sensing instrument. 3. Astronomical Society of India.

ASIC See application-specific integrated circuit.

ASK See amplitude-shift keying, modulation.

Ask Jeeves A prominent and specialized freely accessible search engine on the Web, Ask Jeeves uses natural language processing to search information based upon sentence-like queries.

Finding information on the Internet can be a daunting prospect and many lexical search engines search on keywords or conditional statements and provide thousands or millions of hits on the queried topic. In contrast, Ask Jeeves will parse out English-like queries and provide a selective group of good hits from a variety of search sources based upon intelligent search algorithms, a process central to data mining. See data mining; search engine, Web. See appendix for a search engine chart.

ASL Adaptive Speed Leveling. A U.S. Robotics modem term for adjusting the speed of a serial transmission, depending upon line conditions, to optimize the transfer of data.

ASN 1. Abstract Syntax Notation. See ASN.1 2. See Autonomous System Number.

ASN.1 Abstract Syntax Notation 1. An ISO/ITU-T standard machine- and implementation-independent language defined in 1988 for the description of data, to facilitate the exchange of structured data among applications programs. ISO 8824, ITU TS X.208.

ASP 1. Abstract Service Primitive. In ATM networking, an implementation-independent description of user/provider interactions, as defined by the Open Systems Interconnection (OSI). 2. Adjunct Service Point. A network feature of peripherals designed to respond intelligently to processing requests. 3. administrative service provider. SCSA term. 4. analog signal processing. 5. See AppleTalk Session Protocol. 6. See Application Service Provider. 7. See ATM switch processor. 8. Attached Support Processor. 9. Association of Shareware Professionals.

ASP Industry Consortium An informational global advocacy group supporting Application Service Providers (ASPs), their delivery associates, and their customers, founded in May 1999. The group provides research information, education, and strategic guidance. The ASP Consortium currently serves hundreds of members in more than two dozen countries. It is managed by Virtual, Inc., a high-tech industry integrated management and marketing firm. http://www.aspconsortium.org/

aspect ratio The relationship of the proportions of the width to the height, usually of a rectangular form. A two-to-one aspect ratio, for example, is commonly written as 2:1. The aspect ratios of televisions and monitors are similar, but cinematic films, which are shown in theaters with panoramic screens, have a much greater width-to-height ratio. This is why letterboxed films have a dark strip on the top and bottom to preserve the full width of the image. Unletterboxed films have been modified to remove part of the picture from the sides. See anamorphic.

ASPI See Advanced SCSI Programming Interface.

ASQ 1. Administrative Science Quarterly. 2. American Society for Quality. ASQ sponsors ASQNet. http://www.asq.org/ http://www.asqnet.org/ 3. Applicant and Student Query. A system to automate the verification of student admissions information. 4. Application Status Query. A Web-based tool for allowing applicants to check their status, developed by DLA HROC. http://www.hroc.dla.mil/ 5. Automated Status Query.

ASR 1. Access Service Request. A request sent to a local exchange carrier (LEC) for access to the local circuit. 2. See Airport Surveillance Radar. 3. Automatic Send/Receive. A system that can send and receive messages unattended. 4. See Automatic Speech Recognition.

assembler A program that converts symbolic assembly language program code into machine instructions that can be directly executed by a computer CPU. On early microcomputers, in the 1970s, most serious programming was done with an assembler.

assembly language A low-level symbolic computer language which structurally and mnemonically fits somewhere between machine code and higher level languages such as C, BASIC, Java, and Perl. Languages like BASIC and Perl are typically run in interpreted mode (although compilers exist for almost everything, if you really want one). When compiled and assembled, C and assembly language are converted into machine language, which typically consists of the binary digits one and zero and is very difficult (for normal folks) to read and debug.

By coding in assembly language (which is difficult for most folks) and then using an assembler, a software utility to convert to machine code, the program can often be optimized to run faster and may be more difficult to reverse-engineer. See symbolic code.

assigned frequency See Federal Communications Commission.

assigned numbers A sequential numbering system administrated by IANA to organize and assist in the search and retrieval of Request for Comments (RFC) documents. See IANA, Request for Comments.

Association Control Service Element ACSE. An International Standards Organization (ISO) application layer service for establishing a connection, as part of the Open Systems Interconnection (OSI) model.

Association for Computing Machinery ACM. A well-known trade association serving more than 80,000 computing professionals in over 100 countries. Members participate in the exchange of ideas, discoveries and information in many areas of academia, government, and industry. The ACM was founded in 1947. http://www.acm.org/

Association for Cooperation in Telecommunications Research in Switzerland ACTRIS. A telecommunications research initiative for precompetitive research in telecommunication technologies, formed in 1995 by the Pro Telecom partners, an association of Swiss telecommunications companies. ACTRIS is centered in the Multimedia Communications Lab in Basel, Europe's largest experimental platform. It serves major Swiss telecom companies, as well as those in neighboring countries. The main objectives are to promote research and education in telecommunications.

Association for Education in Journalism and Mass Communication AEJMC. A professional association of post-secondary journalism and mass communications educators and administrators. http://www.aejmc.sc.edu/

Association for Educational Communications and Technology AECT. An organization committed to providing communication among professionals with a common interest in using technology for education. http://www.aect.org/

Association for Information and Image Management AIIM International. AIIM was established in 1943 as the National Microfilm Association. It supports users of document and content technologies with information about technologies and suppliers. AIIM seeks to provide practical, unbiased educational information to its global membership through conferences, articles, and member participation. http://www.aiim.org/

Association for Information Systems AIS. A professional organization serving as a global resource for academics specializing in information systems, founded in 1994. AIS publishes *The Journal of the AIS* and *The Communications of the Association for Information Systems*. http://www.aisnet.org/

Association for Interactive Media AIM. A Washington, D. C.-based nonprofit trade association dedicated to promoting consumer confidence and government support of interactive media products and related technologies. http://www.interactivehq.org/

Association for Literary and Linguistic Computing ALLC. An association for supporting the use of computer technologies for studying languages and literature, founded in 1973. ALLC publishes the *Literary and Linguistic Computing* journal through Oxford University Press. The association sponsors an annual convention in cooperation with the Association for Computers and the Humanities (ACH).

Association for Maximum Service Television MSTV. A professional association of local television stations established in 1956 to undertake studies to support development of Federal Communications Commission (FCC) television technical standards. In 1962, MSTV provided a voice for new television technologies in the All-Channel Receiver Act. In 1987, the association brought together broadcast organizations to petition the FCC to look into high-definition television (HDTV) and participated materially in the Advisory Committee on Advanced Television Services (ACATS). In 1996, the Consumer Electronics

Manufacturers Association (CEMA) and MSTV created the Model HDTV Station Project, licensed as WHD-TV, to educate broadcasters on the implementation of digital television services.

MSTV seeks to preserve and improve the technical quality of free, universal, community-based television service to the American public. MSTV provides its members with information on new technologies and policies, particularly digital television advancements and implementation information. See Advanced Television Test Center, Advisory Committee on Advanced Television Services, Digital Television Station Project, Inc. http://www.mstv.org/

Association for Women in Computing AWC. A not-for-profit professional association founded in 1978 to promote the advancement of women in computing. http://www.awc-hq.org/

Association of College and University Telecommunications Administrators ACUTA. An international nonprofit educational association serving colleges and universities and representing vendors serving the educational market. ACUTA sponsors the award for Institutional Excellence in Telecommunications and publishes news and a quarterly journal. http://www.acuta.org/

Association of Communications Technicians ACT. An organization to support technicians working in the field of communications. ACT is a member of the PCIA Federation of Councils, representing commercial and private mobile radio service communications industries.

Association of Competitive Telecommunications Suppliers ACTS. A Canadian-based association representing telecommunications equipment manufacturers and suppliers in order to support and encourage market competition.

Association of Computer Professionals ACP. A U.K.-based nonprofit organization providing courses that lead to recognized qualifications, founded in 1984. http://www.btinternet.com/~acp/

Association of Computer Support Specialists ACSS. A trade organization assisting and representing those who install, support, maintain, test, and repair computing systems. Members include vendors and manufacturers who provide customer support, as well as independent consultants and those who provide equipment and services to support these specialists. http://www.acss.org/

Association of Computer Telephone Integration Users and Suppliers ACTIUS. A trade organization in the United Kingdom which promotes awareness and acceptance of computer-telephone integration (CTI) technology through campaigns and educational programs.

Association of Independents in Radio AIR. A nonprofit organization representing and promoting the interests of a diverse membership of audio producers, audio artists, radio broadcast stations, and media arts centers. AIR sponsors online discussion groups, group health benefits, and the *AIRSPACE* journal. See Producers Advocacy Group. http://www.airmedia.org/

Association of Radio Industries and Businesses ARIB. A research and development organization, headquartered in Tokyo, Japan, that studies radio waves and developing radio systems and industries in telecommunications and broadcasting, in order to promote public welfare. A number of committees work under ARIB, including the Infrared Communications Systems Study Committee.

Association of Research Libraries ARL. A not-for-profit organization of the leading research libraries in North America. The ARL furthers and promotes the evolution of research libraries and scholarly communication. http://www.arl.org/

Association of Science-Technology Centers Incorporated ASTC. An organization of science centers and museums dedicated to fostering the public understanding of science through innovation and excellence. ASTC was founded in 1973 and has members worldwide in more than three dozen countries. http://www.astc.org/

Association of Wireless System Integrators AWSI. An organization to support integrators working in the field of wireless communications. ACT is a member of the PCIA Federation of Councils, internationally representing commercial and private mobile radio service communications industries.

ASSP 1. acoustics speech and signal processing. 2. application-specific standard product. An integrated circuit designed for a specific application.

ASSP Magazine A publication of the IEEE devoted to signal processing.

ASSTA See Australian Speech Science and Technology Association Incorporated.

Assured Link A telephone link meeting certain minimum transmission, loss (5.5 dB in the 300- to 3000-Hz bandwidth range), and service standards for a communications circuit for voice grade analog signals and sometimes one-way digital signals. See Basic Link.

AST See Automatic Scheduled Testing.

astatic galvanometer A device developed by William Thompson (Lord Kelvin) in 1858 to overcome the limitations of earlier instruments that were subject to interference from the Earth's magnetic field. Unlike previous galvanometers employing one needle, the astatic galvanometer uses two needles, each with a separate coil. The needles are oriented so that north and south poles effectively cancel out the Earth's magnetic interference. See galvanometer.

ASTC 1. See Association of Science-Technology Centers. 2. Australian Science and Technology Council. An association of professionals involved in technical and business communications. 3. Australian Society for Technical Communication Inc. A nonprofit society of professionals involved in communicating technical information.

ASTER Advanced Spaceborne Thermal Emission and Reflection Radiometer. A Jet Propulsion Laboratory (JPL) satellite imaging instrument project. Since 1998, ASTER has been obtaining moderate to coarse detail maps of Earth's temperature, emissivity, reflectance, and elevation characteristics. The

satellite that first carried ASTER, the EOS AM-1, is part of NASA's Earth Observing System (EOS). The ASTER instrument was subsequently launched in December 1999 aboard Terra to provide high spatial resolution instruments as one of five sensing systems. See Earth Observing System. See the NASA ASTER Web site for information on ASTER's progress. http://asterweb.jpl.nasa.gov/

astigmatism An aberration associated with lenses in which irregularities in the shape or general curvature of the lens cause more than one line of focus or in which a portion of the preferred line of focus is blurred.

Astigmatism is inherent in a number of laser beam-generating technologies such as laser diodes. Since the semiconductor-based emission facet of a laser diode is typically rectangular, it causes an astigmatic beam (a beam with an elliptical cross-section) that is usually corrected with one or more lenses.

ASTRAL Alliance for Strategic Token-Ring Advancement and Leadership. A vendor-supported organization formed in the mid-1990s to support migration to High Speed Token-Ring LAN technology. The group prepared a number of white papers and a draft standard for 100 Mbps Token-Ring transmissions. With the emergence of Ethernet as a widely-adapted technology, interest in Token-Ring networks was waning in the late-1990s. See High Speed Token-Ring.

Astro-D See Advanced Satellite for Cosmology and Astrophysics.

Astrolink A commercial global satellite communications service scheduled to come online in 2001. Astrolink International Limited is an independent Lockheed Martin venture. Lockheed Martin has been active in global communications frequency utilization conferences and on various ATM- and ITU-T-related technical and standardization committees and working groups.

Astrolink is targeted at providing multimedia applications over virtual private networks (VPN) with a focus on secure transmissions and connectivity between private and public networks.

The Astrolink system consists of nine geostationary satellites: five to provide global coverage, four to come online later. They will be operating over Ka-band frequencies with approximately 6-Gbps capacity per satellite using continuous beam uplinks and multifrequency TDMA.

astronomical unit AU. A unit of length defined as the distance from the Earth to the Sun, a measure that is generally given as about 149,579,000 to 149,599,000 kilometers. The variation in the unit comes about due to the variation in the Earth's movement in relation to the Sun and the measuring system and criteria used to establish the distance.

ASU application-specific unit.

asymmetric 1. Not symmetric, lopsided, irregularly proportioned, unbalanced, one-directional, having one side larger or longer than the other.

Asymmetric Digital Subscriber Line ADSL. A data communications service over traditional phone wires that supports mid-range speeds and the transmission of both voice and data over existing copper-pair wires. ADSL is becoming a popular cost-effective option for faster computer network access (e.g., Internet) in households and small businesses. It enables users to be online virtually all the time without interfering with the use of the phone for voice messages or having to dial the service provider each time access is desired. ADSL services are competitive with cable modem. "Asymmetric" refers to the discrepancy between the upstream and downstream transmission speeds. Most users download far more than they upload, so a compromise is established to balance cost and speed. Thus, download speeds are configured to be faster than upload speeds (e.g., 640 kbps versus 8 Mbps). Installation of ADSL service involves installing a network interface device (NID) that houses a splitter (a low-pass filter) at the subscriber premises. Separate lines run from the splitter to the phone and from the splitter to the ADSL remote transceiver (somewhat like a traditional modem) which, in turn, connects to the computer. Software to handle communications is installed on the computer. The combined telephone and ADSL signals travel from the loop to the subscriber NID where the signal is split. The telephone signal is then directed to the phone set and the ADSL signal is directed to the terminal unit (the remote transceiver) where it is processed and transmitted to a network interface card (NIC) installed in the computer. Much the same happens in the reverse direction. Phone and ADSL signals coming from the subscriber premises are sent to the central office where a splitter sends phone signals to the voice switching mechanisms and ADSL signals to the data network where an Internet Services Provider handles user requests such as Internet access, Web hosting, etc. Copper wire lines are not optimal for ADSL, as bridge taps and load coils on utility poles can interfere with ADSL signals. See Digital Subscriber Line for a fuller explanation. See cable modem, discrete multitone, G.lite, UAWG.

asymmetric transmission A transmission channel in which information flows more readily (faster) in one direction than the other, or moves primarily in one direction or the other at any one time, or in which a greater volume of information flows in one direction or the other. There are many instances in which information typically flows more in one direction than another, as in interactive TV, where most of the time the user is observing and not transmitting but may make an occasional request for a specific movie or file. The medium itself may not be inherently asymmetric. For example, a data upload over a modem is primarily one-way, but the line capacity is two-way, and the direction can be easily switched when uploading. The slower channel, or the one with a lower volume capacity, may be called the *back channel*.

asymmetrical compression In data compression techniques, some types of files can be compressed faster than they can be decompressed and some work the other way around. In designing compression algorithms, sometimes optimization in one direction or the other is preferred. In creating animation sequences, it is usually very important that they decompress and play quickly; otherwise the illusion of motion is lost. However, it is usually not a problem if the compression takes longer than the decompression because the computer can handle that while the user is working on other projects.

asymmetrical modem A modem designed to favor the transmission of the bulk of the data in one direction over the other. This is appropriate in situations where most of the communication is one-way, as in managing an archive site, where downloads typically outnumber uploads thousands-to-one. See Asymmetric Digital Subscriber Line.

asynchronous Not synchronous. A concept that applies across many areas of telecommunications, in which the timing of the information being received and transmitted is not predefined and may be unpredictable, as in many modem communications and interactive radio communications. This type of communication typically requires some means of indicating the starting and stopping points of the transmission. There are various schemes for handling this, from verbal cues ("Roger"), to start/stop bits, and various handshaking signals.

asynchronous balanced mode ABM. In an International Business Machines (IBM) Token-Ring network, a service in the logical link control (LLC) at the SNA data link control level that allows devices to send and respond to data link commands.

asynchronous communications interface adapter ACIA. A data formatting device that translates signals between the computer and a peripheral such as a modem.

asynchronous packet assembler/disassembler APAD. A mechanism to assemble a stream of bytes from an asynchronous source (e.g., a computer) into packets and transmit them to a network, and vice versa. In terms of serving a translation function, it can be thought of as loosely working like a traditional modem, which takes asynchronous digital signals from a computer and modulates and demodulates them for compatibility with an analog phone system. In the case of a PAD, however, the data is being packetized and sent over an X.25 network from one or more devices that are not directly X.25 compatible. The ITU-T has defined more than one standard for performing these translation functions to facilitate connections with X.25. Examples include:

X.3	basic packet data network assembly/disassembly
X.5	packet data network assembly/disassembly for facsimiles
X.29	packet data network assembly/disassembly control information and user data exchange procedures
X.39	packet data network assembly/disassembly control information and user data exchange procedures for facsimiles

See X Series Recommendations.

asynchronous transfer mode ATM. A high-speed, cell-based, connection-oriented, packet transmission protocol for handling data with varying burst and bit rates. ATM is a commercially significant protocol due to its flexibility and widespread use for Internet connectivity. ATM evolved from standardization efforts by the CCITT (now ITU-T) for Broadband ISDN (B-ISDN) in the mid-1980s. It was originally related to Synchronous Digital Hierarchy (SDH) standards.

ATM allows integration of LAN and WAN environments under a single protocol, with reduced encapsulation. It does not require a specific physical transport and thus can be integrated with current physical networks. It provides Virtual Connection (VC) switching and multiplexing for Broadband ISDN, to enable the uniform transmission of voice, data, video and other multimedia communications. See Anchorage Accord for information on acquiring ATM technical specifications. See the Appendix for details and diagrams on ATM and ATM adaptation layers.

AT 1. See access tandem. 2. advanced technology 3. AudioTex. A commercial telephony-based information service, offering announcements, messages, music, meeting schedules, etc.

AT, PC/AT Advanced Technology. The common name for a series of 80286-based personal computers introduced by International Business Machines (IBM) in the mid-1980s. This model was released about a year later than the Apple Lisa, at about the same time as the Apple Macintosh, and about a year before the Amiga 1000, Apple IIGS, and Atari ST computers. This is historically significant in the development of user interfaces, as most of the competing computers were evolving graphical user interfaces (GUIs) and included built-in serial ports and sound cards, while most of the AT systems were text-oriented (primarily MS-DOS), with sound and various interface cards optional. The IBM AT and licensed clones from other manufacturers were purchased primarily by business users, in part because the IBM name was well known in the business industry, and also because IBM had a decades-old tradition at the time of providing service and repair options to business owners. Two of the chief software products used on the AT were spreadsheets and word processors.

AT commands, Hayes Standard AT Commands A very simple control and reporting language built into Hayes Microcomputer Products, Inc. modems, and Hayes-command-compatible modems from other manufacturers. Originally modems were "dumb" devices; they had no significant memory or algorithms incorporated into the device to process commands or data from the computer. Hayes introduced "smart" modems in the early 1980s that could process a limited command set and enhance the utility of modems. This instruction set has since been incorporated into almost every make and model of computer modem, usually with enhancements by individual manufacturers.

The AT command set allows computer control of a modem and provides a way for the modem to report information back to the computer software. The AT stands for "attention" and is a way of alerting the modem that there is an instruction set following the "AT" which is to be acknowledged or executed. When you run a telecommunications program through your modem, the software is talking to the modem with AT commands along the path provided by the serial cable that typically connects the modem to the computer. If your software can be set to interactive mode, you can type the AT commands directly to your modem and see what happens. The AT commands are usually listed at the back of the manual that comes with a modem.

Many modem manufacturers have included supersets of the basic Hayes command set to provide control of proprietary or enhanced features specific to their products, so AT commands usually include most or all of the Hayes commands, and additional ones as well.

AT commands fall into a number of categories. There are commands for querying the status of the phone line, for querying the status of the modem, and for carrying out operations such as dialing, setting the transmission speed, setting the number of redials, setting the length of wait periods, etc.

Modems contain a number of registers in which information is stored, often in the form of a toggle (true or false) or integer setting. Thus, setting the register to *zero* signifies one thing, and setting it to *one* or another integer, when appropriate, signifies another. Thus, AT S0=0 sets the "S" register to zero. Since register S0 determines how many rings to AutoAnswer, setting it to zero effectively turns off AutoAnswer. AT S0=1 instructs the modem to AutoAnswer after it detects one ring. If you are running a computer bulletin board, or a friend is calling to send you a file over the phone line, AutoAnswer can be turned *on* (or you can type ATA "attention, answer" when you hear the phone ring). Remember to set AutoAnswer *off* when you are finished transmitting, or the next voice caller may get a nasty modem-blast in the ear. Some modems have enhancements that allow them to autodetect whether the incoming call is voice or data and to react accordingly so this doesn't happen.

```
AT S0=0 M1 DT 555-4321 W DT 123
```

attention; set autoanswer to zero rings

set speaker to be on (M1) during establishment of call (so you can hear dial tone and dialing) and off during connection (so you don't have to hear the modem sounds)

dial tone mode 555-4321

wait for dial tone

dial tone mode 123 (to dial an inner extension, for example).

AT commands can be combined. You needn't type AT in front of each individual instruction. For example, you might wish to initialize your modem, and dial out as a single string of commands.

AT&T American Telephone and Telegraph Company. A company established almost 150 years ago to create practical commercial applications from the early telegraph and telephone patents filed in the 1870s, primarily those of Alexander Graham Bell and Elisha Gray. Some of the patents became the property of the Bell System, and some served simply as competitive motivation to implement the new ideas and technologies. The American Telephone and Telegraph Company (AT&T) began as a long-distance subsidiary of the American Bell Telephone Company in 1885. In 1899 the two companies were again merged into one under the AT&T name. In the 1900s, AT&T was reorganized, becoming a holding company, the parent of the Bell companies and Western Electric. In the ensuing years, several additional reorganizations occurred, some voluntary, some mandated by U.S. justice authorities.

In the early 1900s, there was a period of substantial change in the phone industry, since the original Bell patents, protected for a term of 17 years, were expiring and independent companies were entering the phone market in substantial numbers. This situation resulted in independents collectively holding almost half of the phones until, by 1913, AT&T was again the majority holder, due to mergers and acquisitions, and was legally restrained from acquiring any more independents. AT&T was also mandated to permit independents to use the AT&T toll lines.

The Communications Act of 1934 further regulated the industry and established the Federal Communications Commission (FCC), which was given jurisdiction over the telephone and broadcast industries, a responsibility it still holds. In 1956, the U.S. government and AT&T entered into an agreement that AT&T would offer only phone-related services and not engage in common carrier communications such as computer network services. AT&T was further required to license Bell patents for royalties to interested applicants.

A number of antitrust suits ensued in the 1970s charging AT&T with monopolistic practices, and there were calls for divestiture resulting in divestiture proceedings in the 1980s. During this same period, AT&T was granted limited permission to engage in computer-related services.

While its political history was undergoing many ups and downs, the researchers in the Bell Laboratories provided an enormous amount of research and development in telephone technologies, beginning in the late 1800s and early 1900s. AT&T researchers developed the first two-wire telephone circuit, which is still in use today, the first practical transistor, and many other inventions that are in broad use. See Bell, Alexander; Bell Laboratories; Bell System; Carty, John J.; Kingsbury Commitment; Modified Final Judgment; Vail, Theodore.

AT&T TeleMedia Connection A Microsoft Windows-based videoconferencing product from AT&T Global Information Solutions providing audio/video, file transfer, and application-sharing utilities over ISDN through ITU-T H Series and Q Series Recom-

mendations standards and encoding techniques.

Atanasoff, John Vincent (1903–1995) An American physicist and inventor who developed a vacuum tube calculating device in the mid-1930s that foreshadowed the famous ENIAC computer that was operational in the post-World War II years. In 1939, with a small grant from Iowa State College, he pioneered the development of a binary logic computer called the ABC, or Atanasoff-Berry Computer, with assistance from Clifford Berry a recent graduate in electrical engineering. Unfortunately, the War brought the project to a halt. He left his academic position to become Chief of the Acoustics Division of the U.S. Naval Ordnance Laboratory where he worked on computing devices for the Navy along with atomic testing. In the 1950s, Atanasoff took a number of corporate positions and retired in 1961. See Atanasoff-Berry Computer; Berry, Clifford E.

Atanasoff-Berry Computer ABC. A pioneering binary, direct logic computer with a regenerative memory, designed and built by J. V. Atanasoff with assistance from Clifford E. Berry. After two years on the drawing boards, it was prototyped in 1939. It is significant not only for its historic place in the early history of computers, but also because it was designed with a separation between memory and data processing functions. The electricity needed to keep the memory refreshed, so the information wasn't lost, was provided by rotating drum capacitors.

Atanasoff had been working on the ideas that led up to the ABC since 1935 and related that the idea for the ABC came to him in a roadhouse in 1937 after he and his graduate students had developed a calculator for complex mathematics manipulation. Punch cards, which had been developed to store information for electromechanical devices in the late 1800s, were used to enter data into the ABC. In 1940, Atanasoff and Berry authored *Computing Machines for the Solution of Large Systems of Linear Algebraic Equations* with illustrations, in preparation for a patent application that was never completed due to circumstances associated with World War II.

Much of the information about Atanasoff's invention did not come to light until a long court battle in the 1970s between Sperry Rand and Honeywell. Unfortunately, after Atanasoff left for a position with the U.S. Navy, the computer was dismantled, without notifying the inventors. The Ames Laboratory is building a working replica of this historic invention. See ENIAC; Zuse, Konrad.

Atari Corporation A historically significant games and computer company established in 1972 by Nolan Bushnell. Atari shipped the first computer game to achieve wide commercial acceptance. "Pong" was a simple monochrome game with a ball and two paddles, a form of electronic table tennis that became wildly popular. Atari continued developing games but also subsequently introduced a number of microcomputers, including the Atari 800 and the Atari 520ST. The 520ST had a graphical interface and built-in MIDI and was competitive with the Amiga for the home market in the mid-1980s.

ATB See all trunks busy.

ATCA See Antique Telephone Collectors Association.

ATCP See AppleTalk Control Protocol.

ATCRBS air traffic control radar beacon system.

ATD 1. asynchronous time division. 2. Attention Dial. A modem command from the Hayes set that instructs a modem to dial the number following the command. Often a *T* or *P* will precede the number to indicate whether to dial as a *tone* or *pulse* signal. For example, *Attention Dial Tone* would be ATDT 555-1234. 3. advanced technology demonstration.

ATDRSS See Advanced Tracking and Data Relay Satellite System.

ATEL See Advanced Television Evaluation Laboratory.

ATG 1. See address translation gateway. 2. Art Technology Group Inc. An e-commerce platform development vendor.

Athena project, Project Athena A project of the Massachusetts Institute of Technology (MIT) Computer Science Lab begun in 1984. The goal was to take various incompatible computer systems, and develop a teaching network that could utilize the different resources of each in a consistent manner. The development of The X Window System originated from efforts to provide a graphical user interface (GUI) for Athena. See X Window System.

ATI 1. Accelerated Technology Incorporated. A commercial supplier of realtime operating system (RTOS) source code for embedded systems, based upon the Nucleus PLUS multitasking kernel. 2. Advanced Telecommunications, Inc. A commercial vendor of telecommunications-related system design, installation, and training. ATI is one of several firms associated with Applied Cellular Technology. 3. See Advanced Telecommunications Institute.

ATIS 1. Advanced Travelers Information System. A forum of SAE International, an engineering society for advancing transportation mobility. 2. See Alliance for Telecommunications Industry Solutions.

ATM 1. See asynchronous transfer mode. 2. See Automated Teller Machine.

ATM-PON asynchronous transfer mode passive optical networks. A type of optical distribution network, promoted as a means to implement large-scale, full-service subscriber telecommunications services. See fiber to the home.

ATM Adaptation Layer AAL. A layer in an ATM network. See asynchronous transfer mode in the Appendix for extended information and diagrams.

ATM cell The basic unit of information transmitted through an ATM network. An ATM cell has a fixed length of 53 bytes, consisting of a 44- or 48- byte payload (the information transmitted), and a 5-byte header (addressing information) with optional 4-byte adaptation layer information. Interpretation of signals from different types of media into a fixed length unit of data makes it possible to accommodate different types of transmissions over one type of network. See asynchronous transfer mode; see the Appendix for details and diagrams.

ATM cell rate In ATM networks, a concept that expresses the flow of basic units of transport used to convey data, signals, and priorities. Common cell rate concepts include leaky bucket and cell rate margin. See ATM Cell Rate Concepts Table.

ATM endpoint In an ATM network, the point at which a connection is initiated or terminated. See asynchronous transfer mode.

ATM endpoint address A location identifier functionally similar to a hardware address in an ATMARP environment, although it need not be tied to hardware. See asynchronous transfer mode, ATM endpoint.

ATM Forum, The An international nonprofit organization founded in 1991 to further the evolution and implementation of asynchronous transfer mode (ATM) technology as a global standard. The Forum provides educational information on ATM and specifications and recommendations to the ITU-T based on standards of interoperability between vendors, with consideration to the needs of the end-user community. The ATM Forum is a membership-by-fee group which includes a number of technical committees to discuss and report on specific issues such as signalling, traffic management, emulation, security, testing, and interfacing. See asynchronous transfer mode, UNI. http://www.atmforum.com/

ATM hardware address The individual IP station address. See asynchronous transfer mode, ATM endpoint address, Internet Protocol.

ATM line interface ALI. A device at the physical layer enabling connection to a variety of physical media allowing, for example, the accommodation of different line speeds.

ATM Link Enhancer ALE. A commercial error-correcting mechanism for satellite communications developed by COMSAT. The Header Error Control (HEC) specified for asynchronous transfer mode (ATM) is suitable for transmissions carried through low error rate media such as fiber optic cables. It becomes inadequate, however, in bursty transmissions environments such as wireless networks, particularly those that are satellite-based. To compensate for this limitation, COMSAT developed an ALE module which is inserted in the data paths before and after the satellite modems, to isolate ATM cells from burst errors. This module allows selective interleaving of ATM cells before they are transmitted through the satellite link, thus providing a lower bit error rate (BER) and an improved cell loss ratio (CLR). See asynchronous transfer mode, cell rate.

ATM models There are a variety of types and implementations of ATM networks, including Classical IP, LANE, IP Broadcast over ATM, and others. See asynchronous transfer mode, ATM Transition Model, Classical IP Model, Conventional Model, Integrated Model, Peer Model. See Appendix B for details and diagrams.

ATM slot A time indicator for the duration of one cell, usually described in microseconds. It will vary depending upon the cell-carrying medium. In ATM, one use of the term *slot* is to describe delay in switch performance. See asynchronous transfer mode, ATM cell.

A

ATM switch processor ASP. A modular component from Cisco Systems that provides cell relay, signaling, and management processing functions. It includes an imbedded 100-MHz MultiChannel Interface Processor (MIP) R4600 RISC processor, with ATM access to the switch fabric, to provide high call setup rates and low call setup latencies. It includes an Ethernet port and dual serial ports.

The ASP works in conjunction with a field-replaceable feature daughtercard which supports advanced ATM switch functions, including intelligent packet discard, dual leaky bucket traffic policing, and available bit rate (ABR) congestion control mechanisms. See asynchronous transfer mode.

ATM traffic descriptor A list of network traffic parameters, such as cell rates and burst sizes, and, optionally, a Best Efforts (BE) indicator, within an asynchronous transfer mode (ATM) virtual connection. This information is used to determine traffic characteristics and to allocate resources. See asynchronous transfer mode, BEC, cell rate, PCR, SCR.

ATM Transition Model A model lying between the Classical IP Models and the Peer and Integrated Models. See ATM models.

ATM Wireless Access Communication System AWACS. An ACTS project to support and influence emerging ATM wireless standards. The project considered various link level and system level simulations in order to research system concepts based on a 19-GHz air interface. It also addressed the feasibility of different modulation schemes and of directional antenna technology. See Advanced Communications Technologies and Services.

ATMARP ATM Address Resolution Protocol. ATMARP is the ATM Address Resolution Protocol (ARP) with extensions to support address resolution in a unicast server environment. ATMARP provides a means of resolving Internet Protocol (IP) addresses to ATM addresses. ATMARP use of public UNI addresses or ATM endpoint addresses is similar to Ethernet addressing; ATM addresses need not be tied to hardware. *InATMARP* (Inverse ATMARP) is used

ATM Cell Rate Concepts		
Abbreviation	Name	Notes
ACR	allowed cell rate	A traffic management parameter dynamically managed by congestion control mechanisms. ACR varies between the minimum cell rate (MCR) and the peak cell rate (PCR).
CCR	current cell rate	A traffic flow control concept that aids in the calculation of ER. The CCR may not be changed by the network elements (NEs). CCR is set by the source to the available cell rate (ACR) when generating a forward RM-cell.
CDF	cutoff decrease factor	Controls the decrease in the allowed cell rate (ACR) associated with the cell rate margin (CRM).
CIV	cell interarrival variation	Changes in arrival times of cells nearing the receiver. If the cells are carrying information that must be synchronized, as in constant bit rate (CBR) traffic, then latency and other delays that cause interarrival variation can interfere with the output.
GCRA	generic cell rate algorithm	A conformance enforcing algorithm which evaluates arriving cells. See leaky bucket.
ICR	initial cell rate	A traffic flow available bit rate (ABR) service parameter. The ICR is the rate at which the source should be sending the data.
MCR	minimum cell rate	Available bit rate (ABR) service traffic descriptor. The MCR is the transmission rate in cells per second at which the source may always send.
PCR	peak cell rate	The PCR is the transmission rate in cells per second that may never be exceeded, which characterizes the constant bit rate (CBR).
RDF	rate decrease factor	An available bit rate (ABR) flow control service parameter that controls the decrease in the transmission rate of cells when it is needed. See cell rate.
SCR	sustainable cell rate	The upper measure of a computed average rate of cell transmission over time.

on ATM networks supporting permanent virtual connections (PVCs). Inverse Address Resolution Protocol (InARP) supports dynamic address resolution enabling a protocol address corresponding to a given hardware address to be requested. See asynchronous transfer mode and the Appendix for a fuller explanation of ATM. See RFC 1293, RFC 1577.

atmosphere 1. Ambience, mood, feeling about a location or room. 2. A gaseous mass enveloping a celestial body. See atmosphere, Earth's.

atmosphere, Earth's The gaseous envelope surrounding the Earth which provides breathable air, moisture, weather variations, protection from the sun's radiation, especially the ultraviolet rays, and particles which deflect radiant energy that can be harnessed for telecommunications. Atmospheric pressure at sea level is approx. 14.7 pounds per square inch, with local weather variations, and decreases somewhat uniformly as altitude increases. Barometers are used to measure atmospheric weather, and barometric altimeters indicate altitude through changes in pressures in the atmosphere.

The atmosphere has been divided into three main regions. From the surface going away from the Earth, they are the troposphere, stratosphere, and ionosphere. See ionosphere.

ATN See Aeronautical Telecommunications Network.

atom A fundamental unit of energy or matter (depending upon how you look at it) that is the essential building block of molecules which, in turn, are fundamental building blocks of elements. An atom is chemically indivisible. However, from a physics point of view, atoms are described in terms of even smaller components, including protons, neutrons and their associated electrons. In the weird and wild world of quantum physics, there are even smaller units of energy called quarks and other atomic interactions yet to be fully understood.

atomic clock An instrument devised in the 1940s for precise timing and synchronization, it is now particularly important in the U.S. Global Positioning System (GPS) and many scientific research applications. An atomic clock uses the frequency associated with a quantum transition between two energy levels in an atom as its reference. It exploits the unique frequency characteristics of photons in a given transition. Atomic clock is actually a general category name for oscillators whose characteristics are based on quantum mechanical energy state transitions. Advanced atomic clocks can be accurate to within fractions of a second over hundreds of thousands of years. Space-based atomic clocks can be designed to be more accurate than Earth-based atomic clocks due to the lesser influence of the Earth's gravity.

Coordinated Universal Time (UTC) reporting centers make use of atomic clocks for establishing an international time reference. Atomic clocks can be used to validate satellite data for integrity and accuracy for use in navigational applications. Synchronization between transmitting and receiving telecommunications stations can be maintained by means of atomic clocks or by a set of less expensive timing devices which, in turn, can derive their timing from the more expensive atomic clocks.

In the early 1990s, atomic clocks were improved with the introduction of a Hewlett-Packard cesium-beam clock which was more rugged and more stable than previous models.

Europe's Geostationary Navigation Overlay System (EGNOS) is being designed to use cesium and rubidium atomic clocks to provide System Time (ST) for calculating precise navigation information for GPS-based land, marine, and air transportation systems. By using atomic clocks and GPS data from satellites, rather than traditional beacon-based navigational methods, accuracy can be established within seven meters.

Computers that are permanently or frequently connected to the Internet will sometimes be configured to poll the U.S. Naval Observatory atomic clock to synchronize their system clocks with the time on the Navy's clock. This synchronization is useful for time-sensitive file and database management and for time-critical e-commerce applications such as stock and auction transactions. Accurate time-stamp information is also useful for computer-based legal transactions and event tracking.

The National Institute of Standards and Technology (NIST) sponsors a Web site called "A Walk through Time" that features information and illustrations on time-keeping through the ages and the development of atomic clocks. See Coordinated Universal Time, Datum Corporation, Global Positioning System. http://physics.nist.gov/GenInt/Time/time.html

atomic laser A device or process that emits matter in the same general sense as an optical laser stimulates the emission of coherent pulses of light. Such a laser was developed out of research on Bose-Einstein condensates in the mid-1990s by W. Ketterle et al. at MIT. Rather than using mirrors to deflect light within an optical cavity, they used magnetic fields to deflect matter in a magnetic cavity, using sodium atoms, which are magnetically sensitive, as the "ammunition" for the atomic laser "gun." As energy is built up, a pulse of coherent matter manages to break through the magnetic barrier in much the same way as pulses of light break through the semitransparent mirrors of an optical cavity. The system was successfully demonstrated and described in 1997. By Spring 2002, the scientists had found a way to emit a continuous stream of atoms.

The potential for this technology in terms of designing complex crystal lattices, diffraction gratings, circuit boards and other fabrications important to the semiconductor industry may be very great. The system works in the environment of a vacuum, so it's not quite as easy to set up, but it is fascinating technology which will no doubt by harnessed in exciting ways.

For their discoveries, the inventors received a Nobel Prize in physics in 2001. See laser.

atomic number A number characteristic determined by experimentation, since the atom is far too small to be seen by any natural or microscopic means. This number is used to represent an element in a periodic table and describes electrons in relation to the protons in a neutral atom.

ATS 1. See Applications Technology Satellite program. 2. Automation Tooling Systems.

ATSC See Advanced Television Systems Committee.

ATSC Digital TV Standard - Examples

Doc.	Date	Description
A/80	Jul 99	Modulation and Coding Requirements for Digital TV (DTV) Applications over Satellite. Modulation and data coding for satellite communications are defined for a variety of programming types, including video, audio, data, multimedia, or others. It includes multiplexed bit streams such as MPEG-2.
A/64A	May 00	Transmission Measurement and Compliance for Digital Television describing test, monitoring, and measuring methods.
A/65A	May 00	The Program and System Information Protocol for Terrestrial Broadcast and Cable, providing a methodology for transporting digital television data and electronic program guide data with an amendment on Directed Channel Change (DCC) for program tailoring.
A/90	Jul 01	The Data Broadcast Standard defining protocols for data transmission compatible with digital multiplex bit streams according to ISO/IEC 13818-1 (MPEG-2 systems) standards. The standard encompasses both non-TV and TV programming, including Webcasting, streaming video, etc.
A/53B	Aug 01	The Digital Television Standard for advanced television (ATV) systems. The document specifies the parameters and video encoding input scanning format, along with preprocessing and compression parameters of the video encoding. It describes audio encoder signal formats, preprocessing, and compression, as well as the service multiplex and transport layer characteristics and specifications.

ATSC Digital Television Standard ATSC DTS. The

Advanced Television Systems Committee (ATSC) Technology Group on Distribution released the ATSC *Digital Television Standard* in September 1995 (Document A/53) along with Document A/54 which describes use of the DTS. The DTS is based upon the ISO/IEC MPEG-2 Video Standard, the Digital Audio Compression (AC-3) Standard, and the ISO/IEC MPEG-2 Systems Standard. It was part of an ongoing effort to upgrade consumer broadcasting programming and equipment to reflect improvements in technology.

The Digital Television Standard was, in large part, adopted by the Federal Communications Commission (FCC) in December 1996 as well as by Canada and some Asian and South American countries. This influential document was revised by the ATSC and released as A/53A in April 2001. It specifies the technical parameters of advanced TV systems. Examples of standards include those listed below. See Advanced Television Standards Committee.

ATSE Academy of Technological Sciences and Engineering. http://www.atse.org.au/

ATT See Automatic Toll Ticketing.

attachment Something connected to or with. In data communications, a note or file that is attached to the end of an existing file or other electronic communication. Commonly, binary files are sent with email messages as attachments because the message text part of many email systems cannot transcribe or transmit 8-bit binary code. The system will convert the binary attachment to a compatible mode (e.g., 7-bit text) and reconvert it back to binary at the destination. An email binary attachment allows you to send a picture, sound file, Adobe PostScript document, or other nontext transmission in conjunction with regular text.

attachment unit interface, autonomous unit interface AUI. Certain cables and connectors used to attach equipment to Ethernet transceivers. Commonly Ethernet connections are made via a printed circuit board installed in a slot in a computer with a BNC or RJ-45 connector protruding from the computer for making the connection to the Ethernet transceiver and cable. The ANSI/IEEE standard 802.3 (originally released as Document 1802.3-1991) defines an AUI physical layer interface called DB-15 or DIX.

attack time The time it takes for a signal or sound to go from its initiation to its full volume or power. On a violin, for example, it's the time interval from the moment the bow begins to move and a consistent note achieves its full volume and tone. The attack time on an electronic system is the time it takes from the initiation of a pulse, signal, or power-on action until the system reaches its intended activation threshold, output, or throughput level. See decay time.

ATTC See Advanced Television Test Center.

attempt An effort to initiate or establish a communications connection. In some systems that are billed on a flat rate or per-call basis, attempts are not billed. In other systems, such as those that bill by air time, the attempts are charged by the minute or second whether or not the call is connected.

attenuation The decrease between the power of the initial transmission and its power when received or measured at specified points, usually expressed as a ratio in decibels. Loss in power can result from distance, transmission lines, configurations, faults, and weather. See absorption, contrast with gain.

attitude and articulation control subsystem AACS. A spacecraft guidance system employed on the Cassini spacecraft mission to permit dynamic control of rotation and translation maneuvers. The AACS uses star and sun sensors to establish reference points for determining its position. The main engine and smaller engines are used for propulsive maneuvers. Sensors estimate attitude and rate of both the base body and the articulated platforms. A series of vectors, kinematically propagated in time, aids the system in determining motion of various bodies in relation to the base frame. The AACS works in conjunction with the command and data subsystem (CDS), which is the main processor on the craft. The CDS receives RF signals from Earth and sends information and control parameters to other systems, such as the AACS, accordingly. See Cassini.

atto- (*symb.* – a) Used as a prefix to represent a very minute quantity, one quintillionth of, 10^{-18}. See femto-.

ATU See African Telecommunications Union.

ATU-C ADSL transceiver unit-central office. A modem-like device installed at telephony central offices to process data communications received from the subscriber that are then forwarded through a data network to a subscriber-chosen data services provider. See Asymmetric Digital Subscriber Line.

ATU-R ADSL transceiver unit-remote. A modem-like device installed on the subscriber premises to process data communications. It typically interfaces between the computer and a network interface device splitter. See Asymmetric Digital Subscriber Line.

ATV 1. See advanced TV. 2. See amateur television.

ATVEF See Advanced Television Enhancement Forum.

ATVEF Enhanced Content Specification A foundation specification developed by the Advanced Television Enhancement Forum for the creation of HTML-enhanced television content. It focuses mainly on existing technologies rather than promoting new ones, laying out a means for providing and viewing broadcast programming on the World Wide Web or through specialized viewing software similar to Web browsers. It promotes the reliable transmission of a variety of types of broadcasts through Internet, cable, and land-based networks to computers, enhanced televisions, and dedicated enhanced TV consumer appliances. The specification is intended to encompass both one-way and two-way systems and both analog and digital systems. See Advanced Television Enhancement Forum, broadcast data trigger.

audible ringing tone An audible signal transmitted to the calling party to let the caller know that the called number is ringing. See busy signal.

audible sound Sound waves that are perceived by the ear/brain of a particular species. Audible sound ranges vary from species to species, with humans hearing generally between the ranges of 20 to 20,000 hertz. The upper ranges tend to drop off during the teenage years and decline gradually throughout a person's lifetime. Illnesses, very sudden loud noises, protracted loud noises, and sustained low level noises can have profound negative effects on a person's hearing. See audio, sound.

audio Pertaining to sounds, primarily those within range of human perception, from frequencies of about 20 to 20,000 hertz (the upper range especially tends to diminish as people get older). The comfortable hearing range varies in loudness from a few decibels to about 80 decibels. At volumes near and above 160 decibels, permanent hearing damage is almost certain. Sudden loud sounds, frequent exposure to loud sounds, or even long-term exposure to medium level sounds can damage the sensitive structures associated with hearing.

The types of sounds most commonly used for communication are speech and music. Most hearing is done with the ears, although some people augment their understanding of auditory information by reading lips or sensing physical vibrations through their fingers or bodies. Helen Keller was known for "listening" to symphonies through a sensitive sound board placed in the symphony hall under her chair. Many deaf or hard-of-hearing people use their fingertips pressed against the larynx of a speaker to aid in sensing auditory vibrations.

While humans can hear a broad range of frequencies, not all these frequencies are used in human speech. We can detect pitches up to about 18,000 to 20,000 hertz but don't utter sounds that high in conversation. Thus, telephone and other speech circuits typically are not designed to transmit the full hearing range of frequencies and will be optimized for the frequencies associated with the information being transmitted. See acoustic.

Audio/Visual Service Specific Convergence Sublayer AVSSCS. A multimedia convergence protocol for transmitting video over AAL5 using available bit rate (ABR) services. There is a particular focus on supporting MPEG over ATM as ATM has become a dominant networking medium and MPEG is a widely supported video format. ATM is capable of supporting simultaneous video and other data transmissions. See asynchronous transfer mode, MPOA.

Audio Applications Programming Interface AAPI. A library of functions designed to facilitate the design of audio applications. The functions can be called by compatible applications programs in order to interact with audio servers. Thus, the conversion, playback, or recording of audio can be accomplished without each aspect of the application being written from the ground up (i.e., without reinventing the wheel). Most operating systems now have AAPIs available for developers.

Audio Engineering Society AES. Since 1948, the AES has been promoting and fostering the development and advancement of audio technologies. The AES Standards Committee contributes information and technical expertise that supports national and international audio standards development. The Audio Engineering Society Historical Committee (AES HC) researches, collects, and preserves historical information and artifacts related to audio history. AES publishes the *Journal of the Audio Engineering Society* and various papers and conference proceedings. http://www.aes.org/

audio frequency AF. A spectrum of wavelengths that can be heard. For humans this is from about 30 hertz up to about 20 kilohertz, although the upper level declines to about 16 to 18 kilohertz by adulthood.

Audio Interchange File Format AIFF, Audio IFF. AIFF is a widely used audio file storage and exchange format descended from Interchange File Format (IFF). IFF is a flexible, multiplatform means of digitally encoding a variety of types of media-related information (not just sound). IFF was developed in the mid-1980s by Electronic Arts and Commodore-Amiga. In 1985, the format quickly became standard on the Amiga computer. Later variations of the concept and the file format were ported to other platforms to provide compatibility with Amiga files.

While TIFF, JPEG, and BMP have now superseded IFF for the exchange of image files, the Audio Interchange File Format concept has survived and migrated to other platforms and has been adapted by Apple Computer, Inc. as the standard audio file format for Macintosh systems.

AIFF facilitates the data storage and transmission of monaural (mono) and multichannel sound samples using a chunky format. On the Macintosh, it is stored in the data fork. AIFF is also supported on a number of professional workstations, including Avid Technology and Silicon Graphics (SGI), and has further been adopted as a standard audio format by the Open Media Format Interchange (OMFI) group. The original IFF format is documented in the Amiga ROM Kernel manual, Appendix H. Details of Apple Computer's AIFF are available from Apple Developer Technical Support.

Audio Messaging Interchange Specification AMIS. An analog telephony protocol that facilitates the exchange of voice mail messages among users on voice mail systems from different vendors. AMIS specifications were released in the early 1990s. Not all the features of commercial voice mail systems can be directed through AMIS. Depending upon the implementation, AMIS may not permit broadcasting to multiple users on another system, and there may not be a full complement of confirmation messages available from the other systems. A number of commercial products implementing AMIS systems or interfacing with AMIS systems have been developed by large vendors such as Toshiba and Lucent. Active Voice Corporation claims to be the first voice processing systems manufacturer to incorporate AMIS standards into its products. See Voice Profile for Internet Mail.

audio tape A type of magnetic storage medium used for audio recordings. Most audio tapes are small, so they can be used in portable tape decks or car stereos, with playing times ranging from 10 to 120 minutes. Common music tapes are 30, 45, or 60 minutes per side for a double-sided tape. Some audio tapes are designed as a continuous loop with the tape ends fused for continuous playing. Video tapes are sometimes used as high quality audio tapes.

Eight-track tape cartridges were introduced in the early 1960s and were popular for a few years. Cassette tapes were introduced soon after eight-tracks and eventually superseded them. By the 1980s, video tape-based audio recording approached CD sound quality.

Computers in the late 1970s and early 1980s used large magnetic tape spools and small audio tapes (e.g., cassette tapes) for recording data. Various types of magnetic tape systems (e.g., DAT) are still used for data backups. As CD players become less expensive and more prevalent they provide higher quality sound reproduction than most tape systems and a less volatile alternative to data tape backups.

audio-follow-video AFV. In many broadcast systems, audio and video are recorded and/or transmitted separately. In AFV, the audio signals are automatically routed together with their associated video signals

audio-on-demand AoD. Audio services provided to a user on request. AoD is one of the earliest services-on-demand (SoD) systems implemented in the telecommunications industry. In the days of operator-managed telephone services, imaginative service providers realized they could place a phone at the switchboard center near a radio or gramophone player and play music for the subscriber on request. It was an unsophisticated system, but the concept was timely, and the idea is now implemented with digitally automated technologies in the form of video-on-demand, and other custom request services. See services-on-demand, video-on-demand.

AudioGram Delivery Services ADS. A Nortel subscriber telephone service option that enables callers who get a busy signal or no answer to their ring to leave a message that will be delivered to the callee at a later scheduled time. Essentially it's a phone line answering machine service.

audiographics A multimedia network communications system suitable for distance learning, in which remote computer screens are shared as a conference and lecture interactive medium for dynamically sharing images, video, and text. Electronic Classroom, written by Robert Crago for the Macintosh, is an example of this type of application, designed to work over public switched telephone networks (PSTNs). Audiographics is sometimes called *telematics*. Some people like to make a distinction between audiographics, which is the transmission of still images and sound, and videoconferencing, the transmission of motion video and sound. With improvements in technology, the distinction is blurring. See whiteboarding, electronic; videoconferencing.

audiometer, sonometer An instrument for measuring hearing acuity invented by Alexander Graham Bell. Bell's use of the Audiometer to test hearing was reported in October 1884 in the *Deaf-Mute Journal*. At the April 1885 meeting of the National Academy of Sciences, Bell is reported to have demonstrated his Audiometer devised from two flat coils of insulated wire adjusted with graduated distances such that electrical current from an armature between magnetic poles was passed through one coil (interrupted by a rotating disk), while a phone was attached to the other coil. Thus, the current could be used to control the intensity of the sound, and the responses of the children being tested could be recorded and analyzed. About 10% of the students tested with this early instrument were found by Bell to be hard of hearing in their better ear.

While the Audiometer was initially used to test human hearing, many of the components developed for the audiometric industry (speakers, jacks, amplifiers, tone generators, transmission components, etc.) have since been adapted for telecommunications devices and testing systems (P.A. systems, telephony components, transmission line testing equipment, etc.).

Audiometers have become very sophisticated since their invention by Bell, and the term has become generic to a wide variety of audiometric instruments. Commercial audiometers now commonly include keyboards, internal digital storage for saving hundreds of audiograms, programmable functions, and serial interfaces for connecting to computers.

Audion & Electron Tube Controlling Grid

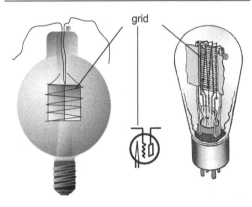

grid

The triode electron tube (left), developed from the Fleming valve by Lee de Forest, is one of the most significant inventions in electronics history. The third element, a controlling grid, added by de Forest to the two-element Fleming valve, enabled the flow of electrons, from the cathode to the anode, to be controlled.

Electron tubes are now more streamlined and sophisticated than the original Audion electron tube which resembled a lightbulb with a plate, filament, and grid inside and two wires running out the top.

Audion An extremely significant invention of the early 1900s, evolutionarily descended from simple *flame detectors*, that led to the three-element vacuum electron tube patented by American inventor Lee de Forest. The Audion was a tantalum lamp with a mostly evacuated glass globe sealed around a filament and plate. A simple wire bent in a zigzag pattern became a grid, providing control over the flow of the electrons from the filament to the plate in a way that had not been previously possible. Thus, electron tubes could be used to amplify signals, not just rectify them, as in the Fleming oscillation valve upon which de Forest's Audion was based.

This *triode* electron tube's control grid represented breakthrough technology which de Forest sold to AT&T at the bargain price of $50,000. It was used for decades throughout the electronics industry until it was superseded by the transistor for most consumer applications. Repeater devices based on the Audion enabled long-distance telephony.

Interestingly, like many inventions through history, the inventor himself didn't understand the detailed mathematics/physics behind *why* the Audion worked, creating problems in manufacturing. The only way to know if the tube was good was to test it, and the sensitivity varied from tube to tube. Edwin Armstrong was one of the few early scientists to grasp some of the physics associated with the Audion's functioning. He authored an article in *Electrical World* in December 1914, explaining the action of the Audion and how it could be more effective if more gases were removed from the bulb in manufacturing.

The term Audion was originally trademarked but has become generic for three-element tubes. See de Forest, Lee; Edison effect; electron tube; flame detector; Fleming oscillation valve.

Auditory Research Laboratory ARL. A lab at McGill University in Montreal, Canada specializing in the study of the perceptual organization of sound.

auger A tool designed for boring, or a bit that fits into a drill designed to make large bore holes, which can be used for wiring installations.

Augustine, Saint A philosopher who authored *De civitate Dei* (The city of God) in 428 AD. This important record of western knowledge includes historic observations of magnetic phenomena.

AUI See Attachment Unit Interface.

AUP See Acceptable Use Policy.

aural Heard or perceived through the ear; auditory. See acoustic, sound.

aurora Solar flare, a nuclear effect from the sun that can sometimes be seen by its influence on the Earth's upper atmosphere. The ionization that results causes the undulating light shows we know as the aurora borealis and aurora australis.

Aurora 1 A regional communications satellite in geostationary orbit over Alaska.

AUSEAnet The Australian-Asian network which supports multination VLSI project communications of the Assocation of Southeast Asian Nations (ASEAN) for sharing project information among participating countries.

Australian Communications Authority ACA. The governing body of Australia responsible for regulating telecommunications and radiocommunications, including the management of the radio frequency spectra and the National Numbering Plan. ACA also promotes self-regulation within the industry. The ACA was established under the Australian Communications Authority Act 1997 and is empowered under the Telecommunications Act 1997 and the Radiocommunications Act 1992. It falls within the Communications, Information Technology and the Arts portfolio. http://www.aca.gov.au/

Audion Basic Electronic Concepts

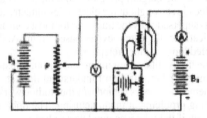

FIG. 2—CONNECTIONS FOR OBSERVING WING CURRENT AND GRID POTENTIAL

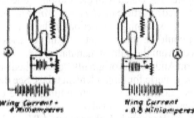

Wing Current = 4 Milliamperes   Wing Current = 0.8 Milliamperes

FIG. 3—CONNECTIONS GIVING TWO VALUES OF WING CURRENT

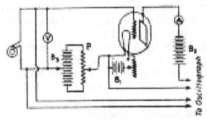

FIG. 5—CONNECTIONS TO OSCILLOGRAPH

Lee de Forest invented the Audion through trial and error and didn't fully understand why it worked; thus, it was difficult to commercially manufacture consistently reliable tubes. Edwin H. Armstrong's explanation of the workings of the Audion, published in Electrical World in 1914, provided the theoretical background needed to improve manufacturing consistency and hence practical applications of this important invention. See Audion.

Australian Computer Society see computer societies, national.

Australian Institute of Computer Ethics AICE. A national organization to provide research and education to the public and advice and expertise to leadership bodies in computer ethics in Australia, established in February 1998. AICE studies the social consequences of information technology, with its theoretical and practical ramifications.

Australian Resource Information and Environment Satellite ARIES. A low Earth orbit (LEO) sun-synchronous satellite project for deploying a hyperspectral sensor capable of identifying details of the Earth's surface that cannot be sensed by other types of instruments. Thus, resource, agricultural, and enviornmental products using reflected visible and infrared light gathered through about 96 spectral bands can be provided to commercial markets.

The CSIRO Division of Exploration and Mining has been coordinating the collaborative project. After years of planning and feasibility studies, ARIES received the go-ahead in 1995, and a timeline was established for it to become operational in 2001 (since delayed to 2002). A vertical tracking spatial resolution of 30×30 meters is possible on the system, which can be increased by viewing sideways or improved when resolving certain high-contrast objects. http://www.eoc.csiro.au/aries/

Australian Speech Science and Technology Association Incorporated ASSTA. A nonprofit scientific association seeking to advance research and understanding of speech science and its application in speech technologies for the benefit of Australia.

authenticate To establish the identity and authorization status of a user, device, process, or data seeking entry to a system or seeking to negotiate a transaction. Authentication is used at network access points, such as gateways and firewalls, in electronic transactions, such as purchases or contracts, and at password prompts to systems, servers, and applications. Data is often authenticated to see whether it has been altered during transmission, and email messages may be authenticated before being sent or received. See certificate, Challenge-Handshake Authentication Protocol, Clipper Chip, encryption, Pretty Good Privacy.

authentication, authorization, and accounting AAA. A network security approach for restricting, controlling, and recording remote access to network resources on an individual user and resource basis while maintaining an audit trail for system administration and billing.

authentication, basic access A basic authentication scheme for users accessing Internet services specified in the original HTTP protocol. This system is only minimally secure, as the username and password are transmitted as text that can potentially be captured and read. See authenticate; authentication, digest access; RFC 2617.

authentication, digest access A means to provide improved security for HTTP user authentication over *basic access authentication*. See authenticate, RFC 2069.

authenticator In packet networking, the end of the link that requires authentication, and specifies the authentication protocol to be used in the link establishment phase. See Challenge-Handshake Authentication Protocol.

authoring The process of using authoring systems to create computer software or multimedia presentations without a large expenditure in time in learning to program. The rationale for authoring is to free the creator from technicalities to concentrate on content and flow. Authoring involves developing a scenario, providing content, and putting them together in an interactive environment according to how the author wishes the user to interact with the software or presentation. Some authoring languages are very similar to English or BASIC. Others use graphical interfaces to allow the user to create the scenarios, and interrelate the program building blocks with a minimum of programming or text entry.

authoring system, authoring language A high-level computer programming language designed to be quickly and easily learned and used such that professionals (e.g., teachers) with expertise in their specific subject fields, but without programming experience, can develop computer software such as courseware and multimedia presentations.

authorization Permission to use a product or service or to gain entry to an area or structure. More and more, computerized means are being used to assign, track, and administrate access to secure areas or within corporate premises. Authorization can be monitored through video systems, magnetic cards, retina or fingerprint scanning, visual recognition of faces, passwords, and voice recognition.

authorization code A code that must be entered into a telecommunications system to gain access to the service or specific features of the service, or to generate statistical records of use. It is used for security or efficiency monitoring, frequently on touchtone phone systems to permit long-distance calls or gather departmental data. When used for security, the code is often typed before the desired number, although in cases of departmental billing or monitoring, it might be required after the number has been dialed.

Authorization Code Table ACT. A lookup table for determining whether a phone call is authorized on the list and should be permitted to ring through. If a number is not found on the list, it is called an *unmatched call* and may be rejected or forwarded to someone in authority.

authorized agent A person authorized to resell, release, or represent the product or services of a company in a somewhat cooperative, independent manner with state restrictions. It is common for large telephone carriers, both landline and wireless, to permit authorized agents to resell or repackage phone services such as long-distance services. In some cases, the agent is a software developer or Internet Services Provider (ISP) who works with the phone company to provide digital value-added services.

authorized user A person or entity authorized by the company or provider to use a service, system, or specific application or data file. This concept is important in communications security.

auto answer The capability of a telecommunications receiving device to automatically detect and respond to an incoming transmission. Facsimile machines, BBS modems, and answering machines are examples of devices with auto answer capabilities. Modems typically include the AutoAnswer command as part of their command set. Sophisticated telephone systems can be configured to automatically answer a voice call and, by using Caller ID, to display the person's file on a computer monitor. See Caller ID, Caller Name, bulletin board system, auto dial.

auto attendant, automated attendant An automated voice system, designed to provide a 24-hour a day substitute for an operator or receptionist, that answers incoming calls and plays a recorded message to the caller providing a number of touchtone options or selections from a touchtone-activated menu. Different systems can transfer calls to humans or voice-mail systems, perform transactions, provide information, initiate a faxback transmission, or initiate a fax tone for those with manual fax machines. The better systems allow you to go to submenus without waiting for the current recording to end and will give you an easy way to return to the main menu. Auto attendants are used by banks, mail order companies, information service companies, and others. See voice mail, Automatic Call Distribution.

Auto Busy Redial A surcharge phone service, multiline subscriber feature, or consumer phone feature in which the last number called can be redialed continuously until a connection is made. The system recognizes a busy signal, hangs up, and redials. There is a similar feature in most telecommunications software that is used to connect to BBS or Internet services which can cycle through a list of numbers, trying each one in turn, or which continuously attempts to connect with a specific number. The software can often further be configured to dial at specific intervals or for a specific duration of time. The Auto Busy Redial service is useful when combined with auto dial for voice communications. See auto dial.

auto dial, speed dial A phone feature in which a short code has been assigned to a longer number to allow the number to be dialed automatically with fewer keystrokes. See abbreviated dialing.

auto dial capability A software/hardware applications feature for dialing a phone number through a modem and setting it up for voice rather than data communication. It's very handy for dialing from a laptop, a cellular laptop link, or from a database on a desktop computer. Some phone solicitors use auto dial in conjunction with phone listings to maximize the number of call connects. Be aware that there are strict regulations governing the use of automated procedures for phone solicitations. See auto answer.

auto discovery, auto mount An automated process whereby a network server is alerted to a new device on the system and can gain sufficient information about its operating characteristics to bring it online and make it available to users. Device tables and databases are sometimes used to make this possible, and manufacturers are creating more devices that signal their presence and include electronically accessible information about the brand, model, capacity, and attributes.

auto start 1. The capability of an emergency power system to detect when electricity falls below a certain crucial level and start up standby generators to provide continuous service. 2. The capability of a computer system to restart or *reboot* after a power outage or power fluctuation sufficient to take down the system.

auto-negotiation In communications between two or more entities, the process of setting up parameters for intercommunication that are acceptable to both parties. Auto-negotiation commonly involves selecting the communications rate, though many other aspects may be auto-negotiated. For example, in the early days of modem communications, a user had to find out the communications parameters for the other party and match them on his or her system before attempting to make a connection. By the early 1980s, "autobaud" modems were being installed by bulletin board service (BBS) providers to automatically negotiate an acceptable connection speed (you still had to set some of the other parameters such as parity, stop bits, etc.), thus simplifying the process for busy or less technical users.

In traditional telephone modems, auto-negotiating a speed usually involves the autobaud modem (e.g., at the ISP) sending out a series of modem signal tones from the highest to the lowest available speeds until the user modem responds. By going from high to low, the fastest possible negotiation is chosen. In other words, if the ISP can provide connections at speeds up to 9600 bps and the user dials into the service at 2400 bps, the ISP's modem would start sending a tone signal for 9600 bps, if there is no response, the next tone might be 5600 bps, then 3300 bps, until it reaches 2400 bps at which point the user's modem should respond and the connection is negotiated. Sometimes the autobaud modem will go through the sequence a couple of times before abandoning the call, in case there was interference on the line that altered the signal.

Auto-negotiation is now an integral aspect of many communications networks. For example, ports on an Ethernetwork auto-negotiate during the linking phase of a connection (the same general idea as two modems negotiating a transmission speed for connecting over phone lines). The capability is now commonly built into hubs and network interface cards (NICs) as part of the 100BASE-TX and other standards (it is optional in the 100BASE-SX standard). Auto-negotiation pulses are transmitted in pairs, with a clock pulse and a data pulse and repeated at intervals. Since fiber standards did not originally include auto-negotiation in the sense of the 100BASE-TX standard, an emulation system for a link pulse was devised for fiber-based transmissions, with adjustments for the different timing needs of fiber loss versus twisted pair. See handshake.

autoanswer See auto answer.

autobaud The capability of a modem to detect the incoming baud rate and adjust its transmission speed and handshaking to match the rate in order to establish a connection. Useful in 24-hour a day, unattended services like BBSs and on systems that may be serving a variety of types of computers and modems.

autobaud rate Early modems had to be individually matched to the same baud rate in order to communicate successfully with one another, but since the mid-1980s, when 1200 baud transmissions were common, most modems have incorporated autobaud capabilities in which the called modem and the calling modem negotiate a common speed and then commence with user communications. Autobaud capabilities have been a great boon to bulletin board systems (BBSs) and Internet Service Providers (ISPs) as they must accommodate users calling in on a variety of types of computers and modems.

autodialing recorded message player ARMP. An automatic dialer which plays a recording to the person who answers the phone to keep him or her on the line until an agent can take the call. ADRMP systems are disliked by many callees who consider it intrusive to pick up the phone and be connected to a recorded message and asked to wait. Nonetheless, they are used by telephone solicitors and collection agencies to increase dialing and caller connect efficiency.

AUTODIN *Auto*matic *Di*gital *N*etwork. A global communications network of the U.S. Department of Defense.

Autoplex The first commercially significant semi-automatic telegraph key, jointly developed by Horace G. Martin and Walter Polk Phillips (developer of the Phillips telegraph code). The device was patented in 1902 by Martin. See 73, Phillips code.

Automated Attendant See Auto Attendant.

Automated Attendant Billing System AABS. In telephony, a system in which the caller dials collect and long-distance calls with the aid of an automated voice prompting system that seeks authorization from the called party, connects or rejects the call, and bills accordingly. Most telephone services in North America have become automated in this way with the use of speech recognition and synthesized operator-assist voices.

automated voice response system AVRS. A system designed to respond to voice commands without the intervention of a human operator. This type of system is often used over telephones by banks and mail order companies. It is sometimes used for security purposes and building access. It is applicable to visitor information systems installed in kiosks in amusement parks and other tourist attractions. See speech recognition, voice recognition.

Automated Directory Assistance Call Completion ADACC. A telephone directory call completion service made commercially available to telephone service carriers which automatically directs a call with information provided by the caller. In some areas it has superseded the familar Directory Assistance (DA) service. The service is usually billed to subscribers on an as-used basis. See Automated Directory Assistance Service.

Automated Directory Assistance Service ADAS. A commercial telephone service in which a speech recognition system is used to get information from a caller who has requested directory assistance. It requests the location and name and then either provides the phone number or compresses out pauses and patches the call to an operator who provides the phone number. If a caller is tentative or repetitive, the automated service can reprompt and then pass on only relevant information for the operator.

Automated Interchange of Technical Information MIL-STD-1840*x*. A data interchange standard.

automated loop test ALT. An automated system for testing traditional telephone service lines. Cables can be scanned for excessive noise, aberrations, water faults, or outright failure and traffic routed through a backup loop until the fault is corrected. Loop tests may be run on a regular basis (e.g., once a day), usually during low traffic times.

Automated Maritime/Marine Telecommunications System AMTS. The AMTS is a specialized network of public coast stations providing communications over inland waterways and ocean coastlines. Public coast stations (marine operators) are designated as common telephone carriers; they provide a means to connect marine radio transmissions to the public switched telephone network (PSTN).

In Spring 2000, Mobex Communications Inc. acquired Regionet Wireless, which had licenses for more than 3900 channels in the AMTS system serving the west and east coasts. This announcement came soon after the Mobex announcement of intent to acquire Waterway Communications System LLC, a holder of licenses for over 4000 channels serving the Gulf Coast and Ohio River barge industry. Thus, Mobex became a significant provider of marine mobile radio services in North America.

Automated Message Handling System AMHS. An interface between an automatic digital network and a local area network intended to facilitate the delivery of messages to a user's desktop station. AMHS was developed by NASA/JPL with funding from the U.S. Department of Defense and subsequently transferred to the Telos Corporation for servicing requests.

Automated Packet Recognition/Translation APaRT. A Cisco Systems technology that allows automatic network configuration and translation, e.g., Ethernet clients and a CDDI or FDDI server, so that workstations or switches do not have to be individually configured. APaRT recognizes and, if necessary, translates specific data link layer encapsulation packet types.

automatic alternate routing AAR. 1. A telephony service for placing calls that will be automatically routed through the most economical path available at the time. It is spelled with capitals when referring to a subscriber service specifically offered by a phone company. 2. A network service for obtaining a connection through an alternate route, if the primary route is unavailable, without user intervention.

Automatic Call Distribution ACD. A multiline phone capability or service that automatically manages and routes incoming calls to assigned lines. If there are no available lines at the moment the call is received, it is placed on hold, and may be configured

to play a recording such as "Your call is important to us, please stay on the line and your call will be answered in the order received." ACD systems can put the party on hold and play a recording, or they can be quite sophisticated, performing significant traffic direction and business transactions.

Mail-order companies, airlines, and other high phone-traffic businesses utilize ACD systems, although smaller companies are starting to use them as they become less expensive. An ACD system detects and answers incoming calls, searches a database for instructions on how to handle the call, responds to the call (as with a recording), and reroutes it appropriately as human operators become available. The routing itself can be programmed to the subscriber's needs, with a number of options available: *Uniform* distributes calls evenly, *Top-down* distributes the calls according to a list in the same order each time, so that calls go to the top of the list first, and work their way down, and *Specialty* distributes calls according to the callee who most appropriately can handle the call. ACDs can also be used to gather statistical data on the number of calls received, and how they are handled in order to fine-tune the system, and to respond to the business needs of the subscriber to improve call handling or change it as the need arises. See Centrex, private branch exchange.

automatic call gapping ACG. A scheme to help control system congestion at the call establishment stage in wireless intelligent networks (WINs). ACG introduces a minimum time gap between call attempts and permits a maximum number of attempts per defined unit of time if call volume is high. Implementations vary, but gaps of about a second are common. In the late 1990s, ACG was tested for performance improvements in a number of simulated and real networks, including A-link loads.

Prouskas et al. have more recently proposed a multi-agent system for controlling network load that may give better performance results for load control of network components than ACG.

automatic callback A subscriber option usually found on private branches in which a caller can key in a code or press a button for automatic callback if she or he has encountered a busy signal on an extension line. When the line is freed, the caller's phone and the callee's phone both ring so that the connection can be made.

automatic calling unit ACU. A device used to automatically dial numbers (a modem and the appropriate software can also do this) in order to save a human operator the time and inconvenience of dialing a lot of calls. This type of system is used by fundraisers, telemarketers, researchers, and others who make frequent calls to a predetermined list of numbers. On computer systems, ACU software is sometimes coupled with database directory programs or address books.

Automatic Circuit Assurance ACA. A circuit efficiency feature available to private branch exchange (PBX) telephone subscribers. ACA evaluates and reports on phone trunk usage or malfunctions. Trunks are typically evaluated on whether they are in use or locked, and whether holding times are long or short. ACA can also be used to chart unusual patterns to facilitate troubleshooting. When a potential problem is detected, an attendant is notified.

Automatic Data Capture Association ADCA. An Australian association of companies involved in manufacturing, distribution, and consulting for the data capture industry. ADCA promotes the development and maintenance of standards and education in data capture. Some specific data capture areas include radio communications, vision systems, optical character recognition, barcode interpretation, mobile pen computing, scanning, and others.
http://www.adca.com.au/

automatic dialer, autodialer A timesaving device enabling a user to program and store a short sequence of characters to represent a long number. When the short sequence is entered, the phone checks memory and, if the sequence is found, it automatically dials the corresponding long number. See speed dialing.

automatic direction finder ADF. An antenna that usually rotates or arcs back and forward and continuously monitors signals until it finds a strong one or one with specified desired characteristics. It then locks onto that signal or provides the direction or frequency information on some type of output device such as a monitor or dial.

Automatic Electric Company A historic telephone technology vendor supplying automatic switching systems based on Strowger technology, co-founded in 1901 by Almon B. Strowger, a mortician who reportedly wanted an automatic exchange because human operators were diverting business to his competition. This company was able to compete by installing working systems quickly, according to customer specifications, and was the largest company supported by telephone company independents. It was directed by Alexander E. Keith.

Originally most of the Automatic Electric systems were three-wire systems which used two wires plus the Earth as the third return-path conductor for the transmission. Later, they developed a two-wire system. In 1955, the company was merged into General Telephone and Electronics (GTE). See Strowger switch.

automatic exchange A central telephone switching office in which calls from subscribers are automatically routed to the callee through mechanical, electromechanical, or electronic switching. There are still a few operator-assisted exchanges around, mostly in remote locations or third-world countries, but automatic exchanges are found in most developed nations. The history of automated exchanges is interesting. Besides the economic motivation of not having to pay wages to operators, one of the early switching systems was designed by a mortician because he was apparently concerned that operators were channeling calls to his competition.

automatic exclusion Once a call has been answered, subsequent stations, nodes, or consoles are excluded from having access to the line.

automatic frequency control AFC. 1. Periodic sampling of a frequency modulated (FM) signal to focus the receiver on the approximate center of the transmission band. This came into widespread use in the 1930s. 2. A device that can seek a particular frequency or monitor the incoming frequency to keep the tuning accurate. AFC is common on FM receivers and other devices that must maintain operations within a very narrow range.

automatic gain control, automatic volume control AVC. A circuit designed to sense the level of incoming sounds and adjust their volume. It can serve two common purposes: to increase the dynamic range of the sound by making quiet sounds quieter and loud ones louder; or to condition the sound by making the volume more consistent (e.g., by quieting down the loudest sounds and strengthening the quietest sounds) when incoming signals are fluctuating more than is desired. Volume conditioning is widely incorporated into sound receivers with tuners, as the signal coming through an antenna can vary significantly due to varying broadcast characteristics and weather.

automatic hold A convenience in which the operator of a multiline telephone console or switchboard can switch between active call lines without having to push a hold button. This saves operator time and prevents caller frustration as the operator can't unintentionally disconnect the caller by forgetting to press the hold button.

automatic identification technology AIT. A general category of technological tools for facilitating and automating recognition of goods, processes, or individuals. There are software programs for automatically recognizing individuals by biometric data such as voice prints, fingerprints, facial features, retina scans, etc. Bar codes and radio frequency tags are used for recognition of goods in shipping and inventory management systems in many distribution and retail businesses. In March 2001, the U.S. Defense Logistics Agency published the *Defense Logistics Agency Automatic Identification Technology Implementation Plan* which details the integration of AIT into its business processes for collecting source data and asset information.

Used wisely, AIT has the potential to reduce repetitive work and manual label-reading and -processing. It can also be practical for surveillance and security applications, provided people's privacy rights are taken into consideration.

Automatic Identified Outward Dialing AIOD. A multiline phone option that records the extension number of the originating phone in order to facilitate billing. AIOD is especially common for the identification of long-distance calls. AIOD leads are terminal leads used to transmit this information to the phone carrier.

automatic level control See automatic gain control.

automatic light control A feature on many different types of cameras, in which the camera will adjust the settings to changes in lighting without manual metering or intervention by the user. This feature is particularly prevalent on small automatic cameras and many camcorders. Sometimes backlighting, high contrast, and other lighting situations can be preset with buttons so that a general ambience is made known to the camera, but the final settings are still automated. Professionals prefer automatic systems only if they also have a manual over-ride for tricky lighting situations.

Automatic Line Insulation Testing ALIT. Test equipment used by telephone companies to identify and test faulty outside cables, when not in use, by seeking below-threshold leakage resistance. Results can be communicated by various means to an attendant or test facility. ALIT testing is quick and is typically conducted during low traffic hours in order not to tie up the system.

automatic line hold See automatic hold.

automatic link establishment ALE. In radio communications and computer networking, the capability of a system to negotiate automatically a connection between two or more stations. Handshaking, identification, and authentication may be part of the process. In radio communications, ALE systems sometimes require a table or list of frequencies that are likely to result in a successful connection. In computer modem links, a record of communication parameters and modem speeds (if not autobaud) may be maintained to facilitate automatic connections. Automatic link establishment is now common and almost taken for granted, but until the mid-1990s, in computer communications especially, automatic connection was in no way guaranteed and, in many cases, not even possible.

automatic location identification, automatic location information ALI. A feature of enhanced 911 emergency systems that automatically provides information on the source of the call from a database.

automatic loop protection switching ALPS. A network flow protection mechanism that protects data flow by switching to a redundant or other backup system to continue service until corrections can be made to the faulty circuit. In a loop system, the backup system provides a different physical path for the data. This type of system is suitable for localized computer networks as opposed to public distributed networks. In telephony, there are similar systems for switching to a backup loop if the main loop is faulty.

Automatic Network Dialing AND. Also called Automatic Dialing (AD), AND is a means of using applications to dial a telephone number on a digital network or from various electronic peripheral devices on a network (e.g., a facsimile machine). For example, a database of phone numbers can be accessed by a computer-based dialing utility to dial the numbers in succession until a call is answered, to dial a single number repeatedly, or to dial specific numbers on a programmed schedule. The Telephony Application Programming Interface (TAPI) is one of the software programming tools that facilitates the development of applications to communicate over phone lines, using computers to handle the routine dialing tasks. AND is widely used by businesses that seek to automate their customer call-backs or telemarketing

calls. With Internet phone and Voice over IP becoming increasingly prevalent, AND is gaining in importance. See war dialer.

Automatic Number Identification ANI. 1. An identifier that provides the calling number to the callee, provided the callee has the service and equipment to display the information. It was historically distinguished from Caller ID by the number of rings within which the information was sent to the caller, but that distinction is disappearing. See Caller ID, Signaling System 7. 2. A multifrequency signaling parameter by which a long-distance carrier receives the caller's number from the local carrier for billing purposes.

automatic privacy A feature on some multiline systems that automatically locks out the ability of other people who pick up a phone to select the line already in use. These systems may also include a 'release' button that allows others to pick up the line and join the conversation.

Automatic Protection Switching APS. A network switching technique which varies from system to system. In some, it is a device that automatically switches from a primary to a secondary circuit if excessive error conditions are detected on the primary circuit. For Synchronous Optical Networks (SONETs), APS is defined in ANSI T1.105-1995. In SONET, APS carries the signaling bytes associated with establishing and releasing the protection of the optical facility.

Automatic Recall AR. A subscriber telephone service that allows a callee to dial automatically to the number that most recently tried to reach the callee. It allows the callees to reach the caller that they missed picking up or that they missed as a Call Waiting while they were already on the line. If the automatically dialed line is busy, the automatic dialing can continue for up to half an hour. If a connection is made, a ring alerts the user to the completed connection.

automatic recovery If there is a power outage or other problem that interrupts a phone system, bulletin board system, network, etc., automatic recovery is the capability of that system to power up to operating status and to recover as many of the original operating parameters and files as possible, as well as to recover or recreate the information that was contained in memory that is of importance to continued operations.

Automatic Redial A surcharge phone service that allows a caller to recall the most recent previously dialed number and dial it again by inputting a short code instead of rekeying the whole number. It is handy if the line was busy the first time it was called. Many business and consumer phones now have a redial button, thus decreasing demand for this service.

automatic rerouting The capability of a system to route a transmission through another leg, hop, or path when the original or expected path is not available. Dynamic routing in large systems often works this way. Large distributed systems where the physical and virtual pathways change constantly usually function with automatic rerouting. In some systems, such as Fiber Distributed Data Interface (FDDI), alternate routing is supplied in a dual ring system in which the port adaptor and various ports are quickly reconfigured to the secondary systems to prevent loss of data and connectivity.

automatic retransmit request, automatic repeat request ARQ. In its simplest form, as in CB radio, a verbal request for the sender to repeat a message that did not come through clearly or completely.

More standardized, automated ARQ systems exist, including one in which characters are sent in groups of a set length, and the sender waits for an *acknowledged* (ACK) or *not acknowledged* (NAK) signal (or no response) before retransmitting or continuing. In some systems, such as amateur radio communications, ARQ is called *mode A*.

In high-speed data transmission, error-detection fields are built into the data and used as check fields by the recipient. As in broadcast ARQ systems, an acknowledge (ACK) or not acknowledge (NAK) is transmitted to the sender and the sender responds accordingly.

Automatic Route Selection A phone service that automatically seeks and selects the desired circuit from available options (usually the least expensive carrier) for the path of an outgoing call. See Least Cost Routing.

Automatic Scheduled Testing AST. A form of telephone testing in which the telephone carrier provides test lines with associated responders for conducting loss and noise tests on a scheduled basis. Additional testing may be requested at an extra charge from the central provider. Test results are logged for each trunk line and provided to the carrier.

Automatic Sequence Controlled Calculator (ASCC). See Harvard Mark I.

automatic sounder A historic telegraph device that created audible clicks of the incoming transmission that could be heard and interpreted by the telegraph receiving operator. The term was also applied to a sounding device used for the teaching of telegraphic sending and receiving skills. When contact was made between the arm and the stop, the sounder's circuit was closed. See sounder.

Automatic Speech Recognition ASR. A term in telephony services for the capability to interpret a user's verbal response to prompts in order to facilitate call direction or other handling.

automatic switching system Various types of mechanical and electrical telephone switching systems became prevalent after the invention of the historic Strowger telephone switching system. These automatic systems enabled telephone circuitry to be controlled so that a call was connected by dialing a code rather than by asking a human operator to manually patch through the call. A number of large and small telephone switching manufacturers, including AT&T/Bell and the Lorimer brothers, created automatic switching equipment in the early 1900s, a trend that continued until the 1970s. By the 1980s, very few manual systems were in use except in rural or specialized situations, as they had been superseded by electronic switching systems.

Automated Teller/Transaction Machine ATM. Any automated walk-up or drive-up console system designed to carry out many of the transaction activities (usually financial) that have historically been handled by human bank tellers. Typical ATM functions include deposits, withdrawals, payments, transfers, and balance inquiries. ATMs are intended to provide services 24 hours a day or an option to those who prefer automated services.

ATMs are networked to a central system, if freestanding or off-site, or they may be directly linked to the local network when attached to the building with which it is associated.

automatic volume control AVC. A circuit in a radio receiver designed to prevent loud blasts from strong transmitting stations when the user moves the dial through the various stations. It can be disconcerting to tune through several weak stations and then hit a strong one that assaults your ear drums. AVC was designed in the 1930s to prevent these sudden gains and dramatic volume changes, and most systems now incorporate this feature.

automatic wakeup A timing device that creates an alarm or other alert to wake up a person. These can include a clock radio, alarm clock, bell, computer programmed sound file, or telephone signal.

Automation Tooling Systems ATS. A leading international provider of automation systems integration. ATS designs and manufactures factory automation systems, custom automation equipment, and high-volume precision components. The company has been developing custom automation solutions for fiber optics, optoelectronics, and photonics industries since 1994.

autonomous built-in self test ABIST. The capability of a system to automatically run built-in diagnostic routines.

autonomous switching A Cisco Systems router feature that enables the *ciscoBus* to switch packets independently, without interrupting the system processor, to provide faster packet processing.

Autonomous System Number ASN. A common administrative routing setup identifier, that is, routing through a collective numbered common domain. The ASN designates a system under common operations control, using common routing protocols, with the various routing tables dynamically maintained.

AUU, AUU bit ATM User-to-User bit. A payload type identifier (PTI) field bit used with AAL5 to indicate the end of a higher protocol packet (e.g., the IP packet). The PTI is defined for all AAL types with regard to AUU as shown in the chart.

signal condition	AUU bit	bit pattern
no congestion	AUU 0	000
no congestion	AUU 1	001
congestion	AUU 0	010
congestion	AUU 1	011

AUUG Australian Unix User Group.
AUXBC auxiliary broadcasting.

availability 1. The amount of time during which a telephone or network system is available for handling calls. It is expressed as the ratio of denied calls to attempted calls. See reliability. 2. In Global Positioning Service (GPS), the period of time during which a particular location, within *angle of elevation* parameters, has sufficient satellites to make a position fix.

Available Bit Rate ABR. In ATM networking, a layer service category related to flow control that may change subsequent to establishing the connection. ABR is related to Best Effort service. It is a class of service (CoS) defined by the ATM Forum that utilizes bandwidth on an availability basis for the transport of bursty data traffic. It is designed to approximate the traffic characteristics of existing local area network (LAN) protocols. In network implementations, there is generally an Available Bit Rate (ABR) command that facilitates the configuration of peak and minimum cell rates in kilobits per second. See cell rate, constant bit rate, variable bit rate.

available bit rate ABR.

avalanche diode A diode in which there is a breakdown region in the reverse bias that is triggered at a certain voltage. This makes them useful as voltage regulating components. See diode; zener diode; photodiode, avalanche.

avalanche noise In semiconductor junctions, a situation in which sufficient high-voltage energy is generated by some carriers such that others are physically impacted.

avatar 1. An embodiment in human form. 2. An electronic image or other embodiment of an individual that is computer-generated and holds some essence or presence of individuality or actuality beyond that of a photographic image or scan. This concept is prevalent in the imaginary world of computer gaming (and Internet chat areas), especially in virtual reality simulations.

AVD See alternate voice data.

average busy hour The hour in a day during which the most traffic is carried on a system. This information is important for configuring and tuning computer and telephone networks to handle traffic efficiently.

Average Delay in Queue ADQ. A measure of the average time a caller waits before a telephone call is processed or handled by an agent. It is important to keep this time as short as possible, to discourage the caller from hanging up or negatively perceiving the service.

average line utilization ALU. A telephony administrative statistic describing average bandwidth usage over a specified period of time. The information is useful in managing bandwidth in a multichannel system.

average speed of answer ASA. A telephony administrative statistic describing the average time it takes for an operator or automated system to answer a call, usually measured in seconds. ASA is used for system configuration, statistical, and staff training and management purposes.

AVHRR See advanced very high resolution radiometer.

aviation channels A set of broadcast frequencies set aside for aviation communications and aviation-related signaling/sensing purposes (e.g., radar).

AVIOS See American Voice Input/Output Society.

AVRS See automated voice response system.

AVSSCS See Audio/Visual Service Specific Convergence Sublayer.

AWA See Antique Wireless Association.

AWACS See ATM Wireless Access Communication System.

AWC See Association for Women in Computing.

AWG 1. See American Wire Gauge. 2. Association for Women Geoscientists. 3. Association of Washington Geographers.

awk An interpreted computer language common on Unix systems, developed by *A*ho, *W*einberger, and *K*ernighan (who is also an author of C). It has a C-like syntax. See Perl.

AX.25 A communications protocol designed for packet radio communications that operates at the link layer level. It is based on the ISO Open Systems Interconnection Reference Model (OSI-RM). Since its introduction, it has generally been superseded by NET/ROM, a more flexible means of transmission. See Open Systems Interconnection.

axial leads Leads on a component that are arranged to protrude along a linear axis in a common plane. In other words, they stick out the ends rather than out the sides.

axial ratio In elliptically polarized radiant energy, the ratio of the major axis to the minor axis.

axial ray See axis ray.

axicon An optical imaging element commercially manufactured as a conical lens (rotationally symmetric prism) with the capability of converting a beam of coherent light into a ring with a nondiffracting central region. Diffractive axicons, with very long focal lines, may be applied to 3D imaging technologies and have been demonstrated for generating Bessel beams for manipulating particles. Holographic axicons may be used as tools for generating volume intensity distributions.

Collimation of atomic beams was accomplished with axicons in the late 1980s and wider commercial distribution of axicons increased in the late 1990s and early 2000s. Commercial axicon components have conical polished surfaces, come in a number of standard types, and may be used in conjunction with focusing lenses. Axicon lens cone angles are typically about $160° \pm 20°$. Bessel beams may be generated through axicon lenses in the micrometer- and millimeter-wave ranges. See Bessel beam, laser.

axis 1. A reference, orientation, or vector in a coordinate system, typically depicted as a line when graphed. See normal. 2. A primary direction or line of motion. 3. An imaginary or implied line around which other elements appear to be oriented as, for example, the vertical axis of a tree trunk or the horizontal axis of a sea/skyscape. 4. A drawn line, usually straight, used as a reference in a graph or chart. 5. The longitudinal center or cross-sectional diameter of a wire or cable, commonly referenced for size.

axis of rotation A straight line around which a body or representation is symmetrically aligned or around which it rotates.

axis parabolic, axis paraboloidal See off-axis parabolic, on-axis parabolic.

axis ray, axial ray In fiber optic networks, a light beam or ray, from a laser source, that travels along a path in the waveguide that is coincident with the axis (longitudinal center) of symmetry of the light-guiding fiber. See fiber optic cable, waveguide.

AZERTY A designation for a computer keyboard or typewriter keyboard used in some European countries such as France. The letters represent the first six top left letters directly below the number/symbol keys. The layout is essentially an adaptation of a QWERTY keyboard with slight changes to accommodate some of the alphabetic differences in European languages (e.g., extra letters). See QWERTY.

azimuth 1. A geometric arc used in navigation and astronomy which is calculated, for example, between a fixed point on the horizon, and clockwise through to the center of a specified object. 2. A horizontal direction calculated from the angular distance between the direction of a fixed point, such as a navigational heading, and the direction of the object (boat, spacecraft, etc.). 3. A specific arc described in relation to a fixed point and a moving object or radiating transmission such as a rotating storage medium (drive, tape, etc.), or antenna. 4. The horizontal direction of a celestial point from a reference terrestrial point, expressed as an angular distance.

Ayrton, Hertha Marks (née Phoebe Sarah Marks 1854–1923) A British physicist, inventor, and author who investigated electricity, particularly electric arcs. She became the first female member of the Institution of Electrical Engineers (IEE).

Ayrton was the author of *The Electric Arc*, published by Van Nostrand in 1902, which became a standard textbook on the subject. In 1906 she was awarded the Hughes medal for her work on electric arcs and on

sand ripples. Ayrton was awarded a patent for the invention of an instrument that was used for dividing a line into any number of equal parts. E. Sharp published a memoir of Ayrton in 1926.

Hertha Marks Ayrton – Physicist, Inventor

Hertha Marks Ayrton was an intelligent and versatile observer, inventor, and author interested in the physics and applications of electricity. [ca. 1899 photograph courtesy of the Archives of the Institution of Electrical Engineers (IEE), London.]

---

azimuth-elevation mount A common type of antenna mount that facilitates two types of rotation, for adjusting both horizontal orientation (azimuth) and elevation (height). This type of mount is frequently used with parabolic antennas that work best when focused precisely toward highly directional beams. See parabolic antenna, microwave antenna, polar mount.

β *symb.* Greek letter *beta,* often used in geometric drawings to denote a specific angle. May also represent phase constant in electrical equations.

B 1. *symb.* magnetic flux. 2. *abbrev.* brightness, as a computer monitor or TV picture tube setting. 3. *symb.* byte, a unit of data commonly consisting of eight bits. See byte.

Historic Dry and Wet B Batteries

Older wet B batteries (right), with the individual battery cells connected in series, were cumbersome and awkward to handle. Historic wet cells were superseded by more convenient dry batteries (left). These examples are from the American Radio Museum collection. [Classic Concepts; used with permission.]

B battery A low-voltage source of direct current (DC) power, historically used to provide power to the plate, or anode, in electron tubes, or to relays in a communications circuit. B batteries ranged from about 22.5 to 130 volts, with 48 volts common in communications circuits. The early B batteries were wet cells, often consisting of a matrix of 1.5-volt cells combined. Later dry cells, with two leads, replaced the more cumbersome wet cells. Most had snap connectors, but some had 4-pin plugs. While B batteries are no longer common, there are still a few commercially available, ranging from 22.5 to 67.5 volts. Antique radio buffs will sometimes wire up a series of commercial 9-volt batteries to produce the functionality of an old-time B battery. See battery.

B Block A Federal Communications Commission (FCC) designation for a Personal Communications Services (PCS) license granted to a telephone company serving a Major Trading Area (MTA). It grants permission to operate at certain FCC-specified frequencies. In 1994/95, the FCC auctioned Broadband PCS A and B blocks to 18 winning bidders for total revenues exceeding $7 billion. See A Block for a chart of designated frequencies for Blocks A to F.

B Carrier A local wireline cellular telephone communications carrier. It is the designation for the local phone company, although the cellular system may be sold off after initial licensing by the local carrier. B Carriers are permitted to operate in four stipulated frequency ranges between 835 and 894 MHz. See A Carrier.

B channel 1. In a stereo system, the designation for the right audio channel, typically connected to the right speaker, often color-coded as red. 2. bearer channel. A channel in a circuit-switched ISDN connection with bidirectional data transmission capability. For a fuller description, see ISDN.

B interface An interface used in Cellular Digital Packet Data (CDPD) which is deployed over AMPS. The B interface connects the Mobile Data Intermediate System (MD-IS) to the Mobile Data Base System (MDBS). See A interface, C interface, Cellular Digital Packet Data, D, E, and I interface.

B minus, B- A negative terminal on a B battery. A negative polarity in a vacuum tube anode.

B plus, B+ A positive terminal on a B battery. A positive polarity in a vacuum tube anode or voltage source in an electronic transistor. See B battery.

B port In a Class A, dual-attachment (dual ring) Fiber Distributed Data Interface (FDDI) token-passing network, there are two physical ports, designated PHY A and PHY B. Each of these ports is connected to both the primary and the secondary ring to act as a receiver for one, and a transmitter for the other. Thus, the B port is a transmitter for the primary ring and a receiver for the secondary ring. The dual ring system is configured to provide fault tolerance for the network.

FDDI ports can be connected to either single mode or multimode fiber optic media, providing half duplex transmissions. LEDs are commonly used on port adaptors as status indicators. Optical bypass switches may in turn be attached to the port adaptors. Optical bypasses are provided to avoid segmentation which might occur if there is a failure in the system, and a station is temporarily eliminated. See dual attachment station, Fiber Distributed Data Interface, optical bypass, port adaptor. See A port for a diagram.

B Series Recommendations A series of ITU-T recommendations providing guidelines for the various means of expression of information, including definitions, symbols, and classification. These guidelines are available as publications from the ITU-T for purchase and some are downloadable without charge from the Net. Since ITU-T specifications and recommendations are widely followed by vendors in the telecommunications industry, those wanting to maximize interoperability with other systems need to be aware of the information disseminated by the ITU-T. A full list of general categories is listed in the Appendix and specific series topics are listed under individual entries in this dictionary, e.g., A Series Recommendations.

ITU-T B Series Recommendations		
Rec.	Date	Description
B.1	1988	Letter symbols for telecommunications
B.3	1988	Use of the international system of units (SI)
B.10	1988	Graphical symbols and rules for the preparation of documentation in telecommunications
B.11	1988	Legal time–use of the term UTC
B.12	1988	Use of the decibel and the neper in telecommunications
B.13	1988	Terms and definitions
B.14	1988	Terms and symbols for information quantities in telecommunications
B.15	1996	Nomenclature of the frequency and wavelength bands used in telecommunications
B.16	1988	Use of certain terms linked with physical quantities
B.17	1988	Adoption of the CCITT Specification and description language (SDL)
B.18	1993	Traffic intensity unit
B.19	1996	Abbreviations and initials used in telecommunications

B Series Standards A series of TIA/EIA documents related to cabling standards, many of which are directly relevant to fiber optic cable design and installation, as illustrated in the accompanying chart. The text of these documents is available for purchase from the TIA online. See TIA/EIA B Series chart.

B signal See Grade B signal.

B-911 A telephone emergency response system with a subset of the capabilities of a full 911 system. Most notably, it doesn't include Automatic Location Information (ALI).

B-CDMA Broadband Code Division Multiple Access. InterDigital Communication Corporation's commercial wireless local loop TrueLink product designed to provide enhanced broadband phone features through CDMA technology. InterDigital is collaborating with Siemens AG and Samsung Electronics Company, Ltd., in developing the proprietary B-CDMA technology. See CDMA.

B-DC, BDC broadband digital cross-connect. See broadband digital cross-connect system.

B-DCS, BDCS See broadband digital cross-connect system.

B-frame bidirectionally predictive-coded frame. In MPEG animations, a picture that has been encoded into a video frame according to information derived from both past or later frames in the sequence, using predicted motion compensation algorithms. This is a compression mechanism commonly used for storing large amounts of data on limited-space optical media. See I-frame, P-frame.

B-ICI B-ISDN (Broadband-ISDN) InterCarrier Interface. 1. A specification defined by The ATM Forum for the connecting interface between public ATM networks, for the support of user services across multiple public carriers. 2. An ITU-T standard for protocols and procedures for broadband switched virtual connections (SVCs) between public networks.

B-ICI SAAL Broadband Inter-Carrier Interface Signaling ATM Adaptation Layer. A signaling layer enabling the transfer of connection control signaling and ensuring reliable delivery of the protocol message. See asynchronous transfer mode, SAAL, AAL5.

B-ISDN Broadband ISDN. See ISDN for an introduction to ISDN concepts. B-ISDN was designed to meet some of the demands for increased speed and enhanced services on primary ISDN lines. It was geared to the needs of commercial users. It has since evolved into a strategy for delivery for many new telecommunications services including teleconferencing, remote banking, videoconferencing, interactive TV, audio, and text transmissions. Broadband ISDN is intended for services that require channel rates greater than single primary rate channels (i.e., voice at 64 kbps) and thus are offered over fiber optic-based telephone systems. B-ISDN services can be broadly organized as follows:

Category	Example activities
messaging, data	paging, electronic mail, data files (images, sound, formatted documents).
conversation	telephone, conferencing, audiographics, videotelephone, videoconferencing
interactive	distance education, services-on-demand, Web browsing, retrieval services such as news, stocks, etc.

The essential characteristics of B-ISDN services were approved in the I-series Recommendations by the

ITU-T in 1990. These developed into broader standards and specific recommendations for implementation, including network architecture, operations, and maintenance.

Recommendations to use ATM as the switching infrastructure for B-ISDN contributed to the formation of the international ATM Forum which promotes commercial implementation of ATM and related technologies.

Physical layer transmission for B-ISDN is accomplished through the Synchronous Optical Network (SONET) system. See I Series Recommendations.

B-LT See broadband line termination.

B-MAC, BMAC Broadcast Master Antenna Control. A device to control a communications antenna (e.g., microwave radio antenna). Traditionally it has been a self-contained unit, but computer software applications that emulate a controller unit are gaining popularity (with the traditional switches and dials being graphically displayed on the screen).

B-NT broadband network termination. See broadband line termination.

B-picture bidirectionally predictive-coded picture. In MPEG animations, a picture intended to become a frame that is encoded according to information derived from both past or later frames in the sequence, using predicted motion compensation algorithms. Once encoded, it is considered to be a B-frame. See MPEG encoder.

B-scope, E-Scope A radar screen displaying information on range (Y-axis) and bearing in rectangular coordinates. See A-scope, C-scope.

B-TE Broadband Terminal Equipment. An equipment category for broadband ISDN (B-ISDN) connecting devices, B-TE encompasses terminal adapters and terminals. See ISDN.

B8ZS binary/bipolar eight-zero substitution. A linecode substitution technique to guarantee density in network transmissions independent of the data stream. It is used on T1 and E1 network lines. The zeros can be replaced at the receiving end to restore the original signal.

Babbage, Charles (1791–1871) An English researcher who contributed a great deal to the theory and practice of computing and conceived his now-famous analytical engine by 1834. While Babbage's

B

TIA/EIA B Series Standards		
Document	Date	Committee/Description
TIA/EIA-568-B.3	1 Apr 02	TR-42
		Optical Fiber Cabling Components Standard (ANSI/TIA/EIA-568-B.3-2000)
		Specifies the component and transmission requirements for an optical fiber cabling system (e.g., cable, connectors).
TIA/EIA-568-B.2-2	1 Dec 01	TR-42
		Commercial Building Telecommunications Cabling Standard - Part 2: Balanced Twisted-Pair Cabling Components - Addendum 2 (ANSI/TIA/EIA-568-B.2-2-2001)
		Provides corrections to the 568-B.2 Standard.
TIA/EIA-568-B.1-1	1 Aug 01	TR-42
		Commercial Building Telecommunications Cabling Standard - Part 1: General Requirements - Addendum 1 - Minimum 4-Pair UTP and 4-Pair ScTP Patch Cable Bend Radius
		Applies to minimum 4-pair unshielded twisted-pair (UTP) and 4-pair screened twisted-pair (ScTP) patch cable bend radius.
TIA/EIA-568-B.1	1 Apr 01	TR-42
		Commercial Building Telecommunications Cabling Standard - Part 1: General Requirements
		This standard specifies a generic telecommunications cabling system for commercial buildings that will support a multi-product, multi-vendor environment.
TIA/EIA-568-B.2	1 Apr 01	TR-42
		Commercial Building Telecommunications Cabling Standard - Part 2: Balanced Twisted Pair Cabling Components
		This standard specifies cabling components, transmission, system models, and the measurement procedures needed for verification of balanced twisted pair cabling.
TIA/EIA-568-B.2-3	1 Mar 02	TR-42.7 *Commercial Building Telecommunications CablingStandard - Part 2: Balanced*
		Twisted-Pair Cabling - Addendum 3 - Additional Considerations for Insertion Loss and Return Loss Pass/Fail Determination (ANSI/TIA/EIA-568-B.2-3-2002).
		This addendum adds clause I.2.5 to TIA/EIA-568-B.2.

ideas for computers could not be easily built with technology available in the 1800s, the basic ideas were sound and have stood the test of time. Ada Lovelace collaborated with him in his work. There is a crater on the moon named after Charles Babbage. See Charles Babbage Institute.

babble Crosstalk from other communications circuits and the noise resulting from such crosstalk. This is typical of electrical circuits and is not a significant problem in fiber optic circuits except where electrical switches or loop sections are part of the system. The term generally implies a number of noise sources combined.

babble signal A deliberate transmission consisting of composite or otherwise confusing signals to obscure the intended transmission from unwanted listeners. A babble signal may be used as a jamming mechanism to deliberately interfere with other transmissions. See frequency hopping, jam signal.

BABT See British Approvals Board for Telecommunications.

BAC See binary asymmetric channel.

back bias 1. A technique for restoring the environment in a vacuum tube which may have been altered by external forces, by applying a voltage to the control grid. 2. A means of feeding a circuit back on itself before its point of origin or contact. One important application of this technique has been the creation of regenerative circuits in electron tubes, an important milestone in radio signal amplification. Regeneration was developed independently by E. Armstrong and L. de Forest and hotly contested in a patent suit. 3. In semiconductors, back bias is sometimes more commonly called *reverse bias*. It refers to an external voltage used to reduce the flow of current across a *p-njunction*, thus increasing the breadth of the depletion region.

back door A security hole that is accessible without going through the normal login/password procedure. A back door may be deliberately left by the developers or maintainers of a software application or operating system in order to gain entry later, sometimes much later. Back doors have legitimate uses for maintenance and configuration but are sometimes abused by disgruntled ex-employees or employees engaged in embezzlement or other illegal or unauthorized activities. See back porch.

back electromotive force, back EMF An electromotive force opposing the main flow of force in a circuit.

back end 1. A program that sends output to a particular device or front end. See client/server. 2. The final step in a transparent (to the user) task or process. 3. In networking, the manner in which a lower layer provides a service to the one above it. 4. In electronics, the final production stages of assembly and testing.

back end processor In computing, a chip or set of chips or separate computing unit that handles 'back end' tasks such as data storage and retrieval in order to free up the main CPU for processing tasks.

back haul See backhaul.

back lobe In a directional antenna, there is a main lobe and there may be additional lobes, one of which extends backward from the direction of the channeled signal, called a back lobe.

back porch 1. On a computer system, a file access point to the system or an application with limited privileges which may not be publicly announced or which may have a group password. In other words, there may be some files available to certain employees that may not be generally accessible by all employees. It's like a meeting place on a friendly neighborhood porch in a back yard where invited people are welcome to visit as long as they don't go inside the house and disturb the privacy of the home owners. This environment is somewhat like an unadvertised anonymous FTP environment in that users of the 'porch' do not have full privileges or access to all parts of the system. A back porch differs from a back door in that it is a circumscribed, known area, with limited privileges. A back door, on the other hand, may provide full privileges and is often not known to anyone but the person who programmed the software. A front porch would be a publicly visible, limited access area open to anyone. 2. In video broadcasts, the portion of a composite picture signal before the video signal which is between the edge of the horizontal synchronization pulse and the edge of the associated blanking pulse.

back projection A means of presenting information on a visual display system by illuminating, or otherwise activating, the display elements from behind. In its broadest sense, most TV and computer screens are back projection systems. However, a further distinction can be made that a projection system implies a larger display system, as would be used in a seminar, theater, or lecture hall, environments traditionally equipped with front projection systems (film projectors, slide projectors, etc.) that are separate devices from the actual display screen. In these environments, back projection screens are less common.

One of the main advantages of a back projection system is that the audience and various speakers can stand or sit directly in front of the display without obscuring the image projection with shadows. Back projection also tends to show up better in rooms where there is sufficient ambient light for people to take notes. The main disadvantage of such a system is that it usually requires specialized equipment for both the projection and the display screen, whereas films and slides can be shown on many types of surfaces, including a plain wall.

back reflection In a fiber lightguide, light that reflects back in the originating direction. Thus, it may interact with the original propagating signal. In most cases back reflection is undesirable and occurs where there are excessive bends, foreign particles, poorly fused joints, bad doping characteristics, or bad terminators. Fiber optic filament endfaces are commonly polished to fine tolerances and particular angles to control or eliminate back reflection. Sometimes a slightly concave endface can reduce back reflection better than a flat endface by reducing the fiber-to-air interface that exists at the coupling joint. Super polishing a concave endface can provide up to an additional -15 dB of back reflection for high-speed, high-bandwidth, systems such as broadband digital communications systems. An angled interface (e.g., 8°) is more difficult to "machine" and connect due to rotational alignment requirements, but provides even better contact and back reflection tolerances of up to -15 dB better than super polished endfaces (up to about -70 dB).

Once a fiber filament is polished, it is still important to ensure that the end is thoroughly cleaned without introducing scratches, otherwise back reflection can result from particle interference. Since a single-mode fiber core is only about 9 microns in diameter, even small particles can potentially obscure the core. Hand cleaning with an air blower and isopropyl alcohol or machine cleaning should be done just before coupling, especially if the fibers have been shipped or stored for any length of time. The endface should be checked with a magnifier (e.g., a microscope) before coupling, otherwise any stray particles could mar the surface when subjected to pressure in the joint. See attenuation, fusion splice. See acceptance angle, Littrow configuration.

back scatter See backscatter.

backboard A sturdy surface on which to mount electrical panel boxes, punchdown blocks, or other threading or wiring equipment that needs a firm backing and wouldn't be secure if mounted on plaster, wallboard, or some other brittle surface. Sometimes equipment is preinstalled and tested on a backboard, so it can be assembled lying down in a convenient position, e.g., off-premises, and then quickly mounted where desired.

backbone A primary ridge, connection link, or foundation, generally represented as longitudinal with branches. A telecommunications backbone is a major supporting transmission link from which smaller links, nodes, and drops are connected. Since the late 1990s, the number and scope of optical fiber-based communications backbones has been steadily increasing.

In 1999, RCN Corporation, a large regional Internet service provider (ISP), announced that it had selected a dense wavelength division multiplexing (DWDM) optical transport system for its east coast fiber backbone. The system was intended to support up to 40 wavelength paths transmitted over a single strand of fiber.

In spring 2000, Metromedia Fiber Network, Inc. (MFN) announced an acceleration strategy for deploying and extending their optical Internet infrastructure internationally throughout North America and Europe. The company's intention is to be the largest global provider of fiber-based infrastructure by 2004.

In late 2000, China Telecom began building China's largest capacity broadband network, projected to extend about 40,000 cable kilometers in a rapidly growing region that did not previously employ optical fiber. The system is being built upon Corning LEAF® fiber, an advanced non-zero dispersion-shifted fiber.

In August 2001, Cogent Communications, Inc. announced completion of the majority of an 80-Gbps bandwidth expansion to its 12,400-mile long-haul OC-192-based backbone that serves Internet Protocol communications to reach 45 of the 50 largest metropolitan service areas (MSAs).

Not all optical service providers are expanding, however. In late January 2002, it became known that Global Crossing was filing for the fourth largest corporate bankruptcy ever recorded in the U.S. Global Crossing had laid approx. 100,000 miles of fiber optic cables around the world, including submarine cables in the Atlantic and Pacific Oceans. Sale of the company as a whole or some of its assets were both put forward as Chapter 11 strategies for continuance of the company. See 6bone, Mbone.

backbone data circuit A main data communications circuit, usually of national distribution, from which there are secondary branches. The term was originally used to describe key USENET/email sites but is now used more generally. A backbone is sometimes defined in terms of the speed of communications and primary nature of the data, and it is sometimes considered the part of the circuit that customarily carries the heaviest traffic. A backbone can connect a mainframe with local area networks (LANs) or individual terminals or individual systems with peripherals such as modems, printers, video cameras, etc. Bridges, routers, and switches perform a variety of traffic control and direction functions within the system. More regional, medium-sized installations, as at universities and large corporations, may be called campus backbones.

Backbones can generally be categorized into three types: *distributed backbones*, utilizing multiple routers; *collapsed backbones*, with a configuration switching hub generally contained within a single building complex; and *hybrid backbones* which include collapsed backbones in individual building complexes interlinked with FDDI distributed backbones, for example. See campus backbone.

backbone radio circuit In packet radio communications, a packet-radio bulletin board system (PBBS) that provides automatic routing services for a number of users.

background communication Data communication that occurs while other user actions are taking place; it carries on in the background without intruding on other activities. For example, a user may be using a word processor while a file is uploading or downloading in the background. Single-tasking systems don't do this. Background communications are characteristic of multitasking systems and some task-switching systems, which will time-splice the processor between the two activities.

background noise Ambient noise, environmental noise, noise without significant meaning. If background noise levels are too high, they can interfere with communications. There are now digital systems, such as cellular phones, that can selectively screen out background noise and increase the clarity of a transmission from a noisy environment. This capability has both industrial and social communication advantages. The same types of algorithms are often used in videoconferencing and audio editing systems to enable users to condition the sound to filter out unwanted frequencies or noise.

background process, background task A computer program operating or waiting in the background, not in immediate sight or use of the user, often at a lower priority level, becoming active quickly when needed or brought to the foreground, or when other processes are idle. On data and phone systems, tasks such as system operations, archiving, cleanup of temporary files, print spooling, diagnostics, etc., are frequently run as background tasks and may function primarily on off-peak hours or when more CPU time is available.

backhaul In telephone and computer network communications, to send a signal beyond a destination and then back to the destination. For example, a phone call from Seattle to north San Francisco may be routed through Palo Alto and back to San Francisco. Backhauling happens for a number of reasons, including cost, availability, and traffic levels. Backhauling may also occur in companies with a number of branch offices. A call to one branch may, for various business reasons, be backhauled to another, in order to serve the caller's needs.

Backhauling on the Internet is quite common. For example, in some cases it may be cheaper or easier to Telnet to an ISP in a distant city with better rates and services, and then access ftp sites, chat channels, or other services by backhauling, perhaps even to the originating city, than to call out from a more limited local service.

backhaul broadcasting In cable broadcasting, to bring back a signal (haul) from a remote site (such as a big sports event or hot news tornado zone) to the local TV station or network head station for processing before being distributed to viewers.

backoff A retransmission delay which may occur when a transmission cannot get through to its destination, due to an interruption, collision, a medium already in use, etc. If a transmission fails, rather than trying again immediately, the sending or interim system may wait momentarily before retransmitting. The retransmission interval may be random or may be set within a certain range by backoff algorithms incorporated into the protocols being used. Backoff (one word) is the noun form; back off (two words) is the verb form.

backplane, backplane bus 1. In desktop computers, the physical connection between a data bus and power bus (both of which are usually on the motherboard) and an interface link or card (which are usually inserted into slots). See bus. 2. In phone exchanges, the high-speed line and power sources that connect individual components, often through circuit board slots. The speed and quantity of transmissions through the exchange are in large part determined by the capacity of the backplane. See bus.

backpressure, backpressure propagation In a network, the information that is being transferred is almost always accompanied by metadata describing the information content and, on large networks, about its progress from source to destination. In hop-by-hop routing, there is network communication about the location and subsequent routing as well. This overhead can sometimes add up if there is congestion on the network, and it may propagate upstream to form backpressure.

backscatter Backscatter is a phenomenon in which radiant energy is propagated in a reverse direction to the incident radiation, sometimes in a diffuse pattern. Backscatter usually happens when the radiant energy comes in contact with an object, particles, or various projections in an uneven terrain, or when it encounters outer boundaries or particles in a transmissions medium, as in fiber optic cables.

Sometimes backscattering is useful, and sometimes it is undesirable. In radar, the signals returned when radar waves hit a target and are reflected back to the sensing device are used to track the location and movement of the target. In directional antenna assemblies, backscattering of signals to the rear of the antenna may cause interference. See zone of silence.

backscatter, ionospheric In the E and F ionospheric regions (where many radio waves are bounced from the sender to the receiver) at the general angle at which the wave hits the ionized particles, some of the waves are propagated back in the direction from which they came. Backscatter may cause interference to the original signal or may result in the transmission being heard by receivers near the transmitting station (although the signal is generally weak). See E layer, F layer, ionosphere.

backup An alternate resource in case of failure or malfunction of the primary resource. The alternate may be identical (or as close as possible) to the original, as in data archives, or may be a substitute which is just sufficient for short-term functioning, as in a backup light source or power supply.

backup link A secondary link which may not typically carry traffic or may carry only overload traffic unless there is a failure in the primary link, in which case it becomes available for transmission until the fault is corrected. See alternate routing.

backup ring On Token-Ring networks, a second ring is often set up to provide a backup in case of failure of the first ring. Depending upon the setup, the system may switch automatically or may need to be switched manually. See Fiber Distributed Data Interface.

backup server A server system expressly designated to automate the handling of data protection tasks. The server can be configured to back up certain machines, directories, or files at predetermined times, or when processing overhead from other tasks is low. A backup server is usually configured with drivers for a number of backup devices, such as tape drives and magneto-optical disks, and may be secured against fire or public access to protect the backed up data.

backward channel A channel in which transmissions are flowing in the direction opposite to the flow of the majority of the data, usually the informational data. Some interactive systems are designed so that control signals and queries flow through the back channel, while the majority of the data flows through the forward channels, as in video-on-demand. Thus, the system can be optimized to accommodate faster data flow rates in the forward direction. Some simplified Internet access systems are designed this way, with a modem or other connect line set for faster data rates for downloading, and slower data rates for querying as, for example, for Web browsing.

backward compatibility The capability of a system to run legacy (older model) programs or to support older equipment. For example, 1.4 megabyte floppy drives are usually backwardly compatible with 770 kilobyte floppy diskettes; they can read, write, and format the older, lower capacity floppy diskettes. Similarly, a new version of a word processing program may be able to read and write data files created by an older version of the software.

Backward Explicit Congestion Notification BECN. In Frame Relay networking, a flow control technique that employs a bit set to notify an interface device that transmissions flowing in the other direction are congested and that congestion avoidance procedures should be initiated by the sending device for traffic moving in the direction opposite to that of the received frame.

backward indicator bit BIB. 1. In data networking, a signal bit or sequence of bits that is used to request retransmission when an error condition is detected. 2. A flow control status bit used in Signaling System 7 (SS7). In MTP Layer 2, a Message Signal Unit (MSU) indicator carried in bit 8 of the first octet in conjunction with the backward sequence number (BSN).

backward learning An information routing system based upon the assumption that network conditions in one direction will be symmetric with those in the opposite direction. Thus, a transmission moving efficiently through a path in one direction would assume this to be an available, efficient route in the other direction as well.

backwave In radiotelegraphy, an undesirable interference heard between code signals.

Bacon, Roger (ca. 1220–1292) An English philosopher, scientist, and a member of the Franciscan Order. In 1265, he completed an encyclopedic document of the knowledge of the time entitled *Opus majus*.

BACP See Bandwidth Allocation Control Protocol.

BADC binary asymmetric dependent channel. See binary asymmetric channel.

bad block In magnetic storage that is segmented into blocks, a section with write or read failures. Some operating systems will map out bad block sectors on a diskette or hard drive during formatting so they will not be addressed or used and will continue to format the remaining good parts of a disk. This is one of the reasons why the amount displayed for the usable portion of a disk can differ from the total storage capacity of the disk.

Bad Frame Indicator BFI. A means of signaling an error condition in a frame-based communications

medium such as a packet networking error condition alert or a cellular radio speech decoder frame error alert. In its simplest form, BFI uses binary logic to indicate an error-free frame (usually "1") or a bad frame (usually "0").

Baekeland, Leo (1863–1944) A Belgian-born American inventor of modest birth whose mother encouraged him to get a good education to improve his opportunities in life. He had an agile mind and emigrated to the United States to pursue his interests and professional connections. He is responsible for the invention of Bakelite, the first synthetic polymer, and Velox, a new type of photographic paper. See Bakelite, Bakelyzer.

baffle, heat A corrugated, latticed, or slitted structure that aids in controlling heat. In fiber optics a fiber routing tray for handling the positioning of multiple fiber cables can also serve as a heat baffle for channeling heated ambient air away from components or joints that might be adversely affected by heat. See heat sink.

baffle, light A device to selectively control the emission of light. When the device can be readily opened or closed or is frequently done so, it is more often called a shutter. When it is generally fixed, or is only infrequently opened or closed, it is usually called a baffle.

Baffles are common in scientific instruments that illuminate specimens or work stages. Such devices may have a baffle to prevent light from directly illuminating a sample and may optionally have a baffle to prevent stray light from disturbing nearby work areas. A baffle may be one of the components of an integrating sphere, which is a component installed in the entrance port of a monochromator. By rotating the sphere, the viewing angle can be controlled. A baffle may also be used to selectively cast a shadow against which a fiber optic light source can be directed for calibration purposes.

Since optical components can be impaired by dust and moisture, it is sometimes advisable to close or cover a baffle during storage or times of low use. Since a baffle has many small surfaces, it may be difficult to clean. Removing it and cleaning it in alcohol or water or vacuuming it, if it is difficult to remove, can help prevent contamination of nearby components. See stray light.

baffle, sound A device to direct sound and to prevent sound waves from interfering with one another. A baffle consists of a series of carefully spaced corrugations that provide a longer path within a limited amount of space. It can be constructed of wood, metal, or synthetics and works by lengthening the air path along the diaphragm through which the sound waves travel and by reducing interaction among them. Baffles are commonly used in speaker systems to improve the clarity of the sound.

BAFTA See British Academy of Film and Television Arts.

bag phone *slang* See transportable phone.

Bain, Alexander (1811–1877 [dates approximate; reports vary]) A Scottish chemist and clockmaker who developed an electrochemical paper tape recording system in the mid-1800s, suitable for telegraphic signals, at about the same time Samuel Morse was developing a somewhat similar system. The Bain system worked reasonably well except in situations with high noise on the line, which would create spurious marks on the tape.

Bain received a patent for his version of the telegraph in the 1840s which was contested by Morse but was sufficiently different to hold up in court. Morse subsequently bought out the Bain systems and converted them to his own. See Bain Chemical Telegraph.

Bain Chemical Telegraph A historic automatic printing telegraph based on chemical methods, patented in 1848 and 1849 (U.S. #5,957 & 6,328). If you have seen the output from a facsimile machine on thermal paper, you have the general idea of how it worked. Bain's system used paper that was coated with a chemical that was sensitive to electrical impulses on the receiving end of the transmission. When a message was received, the electrical impulses would initiate a chemical reaction that would change the color of the paper in the active areas, creating an image to match the one that had been transmitted, essentially a historic facsimile machine. Later enhancements of the general principles of the Bain machine led to very fast telegraphic systems. Seen Bain, Alexander; telegraph, history.

Baird, John Logie (1888–1946) Although historical research makes it clear that a number of people independently developed different aspects of television reception and display, in the late 1800s John Baird, a Scottish inventor, was one of the earliest successful experimenters. He was able to transmit a two-tone image of a face onto a small television screen in 1926 and by 1932 had developed a practical system for broadcasting images.

Baird used some of the principles of the Nipkow disc to develop his system. A light-sensitive camera was placed behind a perforated rotating disc, just as Nipkow had placed light-sensitive selenium behind a perforated rotating disc. The Baird system could only display a crude 30-line image at a frame rate a little less than half of that used now, but the 'proof of concept' technology launched an industry that is still going strong.

In the 1920s, in collaboration with Clarence W. Hansell, Baird patented the concept of using conducting rods or pipes to transmit images, a forerunner to fiber optic transmissions. However, it was Heinrich Lamm who successfully used optical fibers for image transmission. See Lamm, Heinrich; Nipkow, Paul.

Bakelite The development of Bakelite in 1907 revolutionized industrial production and heralded the "age of plastic." Inventor Leo Baekeland created this first synthetic polymer with a trademarked mixture of phenol, formaldehyde, and coloring agents. He was awarded a patent for Bakelite in December 1909 (U.S. #942,809).

This new material was hard and acid-, heat-, and water-resistant. It was quickly put to use in thousands of industrial products as a noncorrosive coating and chemical binder for composite materials. Bakelite

Frequency Range Designations					
ITU	Designation	Abbrev.	Frequency	Wavelength	Typical or Example Uses
2	extremely low	ELF	30-300 Hz	10 Mm-1 Mm	
3	ultra low	ULF	300-3000 Hz	1Mm-30 km	
4	very low	VLF	10-30 kHz	30 km-10 km	
5	low	LF	30-300 kHz	10 km-1 km	Facom distance measurement and navigation
6	medium	MF	300 kHz-3 MHz	1 km-100 m	AM radio
7	high	HF	3-30 MHz	100 m-10 m	CB radio
8	very high	VHF	30-300 MHz	10 m-1 m	TV channels, FM radio, land mobile radio (cellular), ISM, LAWN, amateur radio
9	ultra high	UHF	300 MHz-3 GHz	1 m-100 mm	TV channels, CB radio, land mobile radio (cellular), PCS, radar
10	super high	SHF	3-30 GHz	100 mm-10 mm	Satellite, amateur satellite, U-NII bands, radar
11	extremely high	EHF	30-300 GHz	10 mm-1 mm	Satellite
12	tremendously high	THF	300-3000 GHz		

Note: In the above frequency ranges, the lower limit is exclusive, the upper limit inclusive.

also provided new ways to create colorful, water-resistant, moldable household products, dials, small appliance casings, and even jewelry. Many early telephones and radios used Bakelite in their construction. See Baekeland, Leo.

Bakelyzer It looks like a B-movie adaptation of a Jules Vern diving bell on wheels, but it actually is a floor-standing iron pressure cooker devised by Leo Baekeland to mix simple organic chemicals into his versatile Bakelite synthetic resin. See Bakelite.

Bakken Library and Museum A museum located in Minneapolis, Minnesota, that houses a collection of about 11,000 books, journals, and manuscripts documenting the history of electricity and magnetism and their applications in life sciences and medicine. The collection focuses on 18th through 20th century works, including those of Franklin, Galvani, Volta, and other well-known pioneers. In 1969, the collection of historical electrical machines was added to the activities of the museum, including several Oudin and D'Arsonval coils and many electrostatic generators. http://bakkenmuseum.org/

balance To equalize, to counterbalance, to bring into harmony or equipoise, to offset in equal proportion, to arrange such that opposing elements cancel one another out or are of comparable weight, size, construction, value, strength, or importance. Balancing is commonly done in electrical circuits to equalize loads or to diagnose the location of breaks or interruptions in a line. Stereo volume is usually balanced to equalize the volume or perceptual evenness of the left and right channels.

balanced bridge A bridge circuit in which the measured output voltage is equal to zero. Bridge circuits are sometimes used diagnostically to seek out and measure unbalanced circuits in order to detect a break or anomaly in the wiring. See Wheatstone bridge.

balanced circuit A circuit in which the electrical properties are symmetric and equal with respect to ground. See balanced bridge.

balanced configuration A point-to-point High Level Data Link Control (HDLC) network configuration with two combined stations.

balanced line An electrical circuit consisting of two conductors with matched voltages at any corresponding point along the circuit, and which have opposite polarities with respect to ground. It is not uncommon to use more than one line to carry related transmissions or a split transmission, especially in newer multimedia applications. By matching voltages and setting opposite polarities, it is possible to reduce the incidence of crosstalk and interference, resulting in cleaner signals.

balanced modulation Modulation is a means of adding information to a carrier signal by varying its properties such as amplitude or frequency. In the early days of radio wave broadcasting, experimenters sought ways to manipulate or reduce the amount of bandwidth that was needed to carry the desired information. It was found in amplitude modulation (AM), using electron tubes, that the control grids of two tubes could be connected in parallel, and the

screen grids connected for push-pull operation such that the sidebands were singled out for transmission without the carrier. Double sideband modulation is another name for balanced modulation. In electronic music, balanced modulation refers to a way of combining electrical signals such that the voltage modulation need not always be positive and the output phase is valid for positive and negative signals. As in radio communications, only the sidebands from the original signal remain. See amplitude modulation, modulation, single sideband.

balcony A small ledge or platform for aerial jobs used by film crews, antenna technicians, or utility pole workers.

bale, bonfire A signal fire, one of the oldest optical networks, and one which could be used at night. In the 1400s in Scotland, a simple signal code, using one to four bales, was established by an act of Parliament.

ballast 1. A physical object that improves stability through its mass or can be jettisoned to reduce mass. Ballast is commonly used in boats and hot air balloons. 2. In an electrical circuit, a device that stabilizes a current or provides sufficient voltage to start up a mechanism (such as a fluorescent bulb) or transmission. Apparently about 50% of fluorescent lighting ballasts produced until the late 1970s contain hazardous PCBs of 50 ppm or higher in the potting material that surrounds the capacitor and should not be disposed of in landfills. They can be sent to authorized disposal centers.

balun balanced/unbalanced. A small, passive transforming device used to match impedance on unbalanced lines that are connected together, such as twisted-pair cable and coaxial cable, so the signal can pass through the differing types of lines. As with many interface devices, there may be some signal loss through the balun. See bazooka.

BAN 1. base area network. 2. basement area network. 3. Bay Area network. 3. See Billing Account Number. 4. See body area network.

band 1. The range of frequencies between two defined limits, usually expressed in hertz (Hz). See bandwidth. 2. A group of electronic tracks or channels. 3. A group of channels assigned to a particular broadcast spectrum, e.g., UHF (300 to 3,000 MHz). See chart of regulated band designations. 4. The range or scope of operations of an instrument. 5. An AT&T-designated WATS Service Area.

band allocations Frequency ranges for radio wave communications have to be shared, and devices communicating on similar frequencies can have devastating effects on one another. For this reason, the frequency spectrum is allocated and regulated in order to maximize use of the available spectrum, and also to designate waves suitable for different types of activities. In the U.S., this information is contained in the Federal Communications Commission (FCC) Table of Frequency Allocations and the U.S. Government Table of Frequency Allocations, which together comprise the National Table of Frequency Allocations. Other organizations such as the ITU have tables as well. The values in the tables change, and what is represented in the Frequency Allocations and Common Uses chart is a very generalized overview to provide a basic understanding.

When new frequencies are available, they may be allocated to amateur or specialized uses or auctioned for commercial use. When a user stops using an allocated frequency, it is reassigned. The available spectrum ranges have been established for various types of communications and some regions are unlicensed. Some of the more interesting unlicensed uses have been listed in the Frequency Band Allocations chart.

band center The computed arithmetic mean between the upper and lower frequency limits of a frequency band. This measure can be used to adjust modulation, to constrain it, or to provide the maximum possible amplitude range for an amplitude modulated (AM) signal.

band splitter A multiplexer that subdivides an available frequency into a number of smaller independent channels, using time division multiplexing (TDM) or frequency division multiplexing (FDM). See bandpass filter.

Frequency Band Allocations (Radio Waves) and Common Uses		
Name	Approx. Range	Examples of Applications
AM band	535-1605 kHz	Amplitude modulation, used commonly for radio broadcasts.
Videoconf.	around 24 MHz	Certain local videoconferencing systems.
Mobile	various	Frequencies around 48 MHz are used for consumer outdoor mobile intercom units.
Radar	10-200 MHz	Imaging radar applications.
Amateur	50-54 MHz	Amateur radio use. Frequencies allocated for amateur use are
	144-146 MHz	frequently changed as the FCC often puts a higher priority on commercial users. This is despite the fact that amateurs have contributed a great deal to radio communications technology.
FM band	88-108 MHz	Frequency modulation, used commonly for radio broadcasts and some low power FM transmitters (intercoms, bugs).
SAR	141 MHz	Synthetic Aperture Radar for environmental sensing and image processing.
Radar	300 MHz +	Approx. lower end of radar for remote sensing applications.
USDC	824-894 MHz	U.S. Digital Cellular FDMA and TDMA cellular phone services.
A-F block	1850-1910 MHz	Personal Communications Services (PCS) A to F block licenses
	1930-1975 MHz	granted to phone companies serving MTAs.
UPCS	1890-1930 MHz	Unlicensed Personal Communications Services (PCS).
S-band	2310-2360 MHz	Frequencies sensitive to terrain, making them unsuitable for some types of transmissions.
U-NII	5150-5350 MHz	Unlicensed National Information Infrastructure wireless communications, including PCS.
P-band	.22-.39 GHz	Experimental radar. SAR.
C-band	4-8 GHz	Microwave frequencies, more specifically 3.40 to 6.425 GHz. Satellite – larger antennas. VSATS. Incumbent telephony operations (2.0 GHz). Experimental radar. SAR.
L-band	1-2 GHz	Experimental radar. SAR.
X-band	8-12.5 GHz	Dedicated for use by the U.S. military for satellite communications. SLAR.
Ku-band	10.95-14.5 GHz	Now subdivided into fixed satellite service (FSS) at 11.7 to 12.2 GHz, and broadcasting satellite service (BSS) at 12.2 to 12.7 GHz. VSATs.
K-band	18.5-26.5 GHz	Satellite applications with smaller antennas, radar.
Ka-band	26-40 GHz	Satellite applications with smaller antennas, radar.
Q-band	36-46 GHz	Satellite, radar.
V-band	40-75 GHz	Radar band.
W-band	75-110 GHz	Radar band.

band, citizens See citizens band radio.

band-elimination filter BEF. A resonant circuit filter with a single, continuous attenuation band, in which the lower and higher cutoff frequencies are neither zero nor infinite.

band-stop filter, band-rejection filter A resonant circuit filter for locking out a specified range, or ranges, of transmissions according to their frequency ranges.

banded cable Two or more cables physically held in proximity to one another (aggregated) with metal or plastic straps or bands.

bandpass The range of frequencies that will pass through a system without excessive weakening (attenuation), expressed in hertz. See bandpass filter.

bandpass filter A device with a resonant circuit, often used in conjunction with frequency division techniques, that recognizes and selectively allows control of frequencies, letting through those that are desired. A band-reject filter is complementary in the sense that it recognizes and selectively screens out a range of frequencies in order to form a 'blackout' area within the full spectrum of available frequencies. See band splitter.

bandspread tuning A means of spreading a band of frequencies over a wider area in order to adjust the tuning more precisely. This is most commonly found in shortwave radios and more than one set of dials may be used, with differently spaced tickmarks on the tuning gauges to aid in adjusting the settings.

bandwidth 1. The extent of a range of frequencies between the minimum and maximum endpoints, typically measured in hertz (cycles per second). Technically, the term bandwidth is associated with analog systems. In recent years, it has been more loosely applied to mean data rates in digital systems and, hence, is sometimes expressed in bits per second (bps). 2. The range of the frequency required for the successful transmission of a signal. It may range from a few kHz for a slow-scan or sideband signal to 100 kHz for a frequency modulated (FM) signal. That is not to say that the bandwidth of a signal necessarily takes up the entire range of the band that may be designated for its use. See band spectrum table. 3. In a cathode-ray tube (CRT) device, the speed at which the electron gun can turn on and off. 4. The capacity to move information through a social, data, or physical system. 5. A numerical expression of the throughput of a system or network.

bandwidth allocation, bandwidth reservation In a network, the process of assessing and allocating resources according to flow, priorities, type, etc. It enables priority administration of the network traffic when congestion occurs.

Bandwidth Allocation Control Protocol BACP. In ISDN, a protocol providing mechanisms for controlling the addition and removal of channels from a multichannel link.

bandwidth augmentation 1. Adding additional frequencies or channels to an existing bandwidth range. 2. Replacing existing physical transmissions media with broader bandwidth media in a system where the data transmission is capable of broader bandwidth, but the physical media caused a bottleneck due to its inherent limitations (e.g., replacing copper wiring with fiber optic).

bandwidth compression Techniques for increasing the amount of data that can be transmitted within a given frequency range. The increased demand for broadband applications such as video has motivated technologists to find more efficient ways to use existing transmissions media, resulting in better compression schemes and better management of the direction of transmission in bidirectional systems. Compression can be very medium-specific. For example, in sending voice, blanks between words may be removed; in sending images, white pages may be compressed or eliminated; in sending complex multicolored images, lossy formats such as fractal compression or other lossy compressions such as JPEG may be used.

bang *colloq.* ! Exclamation point. 1. A common symbol used in many programming languages. For example, in C it represents a logical *not*. 2. Although its use is diminishing, it was at one time used in email addresses to designate a break between portions of an address, where an at sign (@) may now be used. Here is an example of a bang path:
{uunet,ucbvax}!galileo.berkeley.edu!username

bank A row or matrix, usually of similarly sized or configured components or data cells. Individual units in a bank are often interrelated, by shape, function, or electrical contact. In its simplest sense, a physical bank does not necessarily have connections between individual cells but may appear similar and be mounted in rows and columns. Banks may also be electrically related, either by induction, physical connections between the cells themselves, or by temporary electrical connections that occur when a bar drops down over a bank of cells, or a brush passes over the bank.

Many large-scale telecommunications devices and junctions are set up in banks. Punchdown blocks at switching centers are set up in banks, often on racks or panels. Memory banks can be physical rows of memory chips in a circuit board. Large Internet Access Providers (IAPs) may have banks of hundreds or thousands of modems, connected to phone wires. See bank switching.

bank switching A method of extending access to banks of components, such as memory chips, beyond the extent of any of the individual components, especially in situations where the operating system or microprocessor can address only a limited amount of memory at one time. By *paging* or *swapping* between banks, the virtual memory capacity is extended beyond the default physical memory or operating system (OS) or central processing unit (CPU) capacity. Bank switching is a tradeoff that may slow down memory access.

banner A clearly visible, often graphic representation heralding an advertisement or a new section in a printout or other text or image communication. Its purpose is to command attention and often (1) to demarcate the end of one communication or section and the beginning of the next one, especially in situations where many people are sending information through the same queue, such as a print queue or (2) to identify the type of communication or the person to whom it belongs.

bantam tube A squat electron tube with a normal-sized base that was once commonly used in small appliances or battery-operated mobile devices like portable radios. Modern transistors have since made most types of vacuum tubes obsolete. See acorn tube.

bar code Identification, information, and management code designed for optical scanning by a reading device. Black and white visual bar codes are familiar identifiers on consumer products. They assist the checkers in entering prices and adjusting inventory databases. Bar codes are frequently inserted on postal letters and packages that have been optically interpreted. In fact, the information encoded into video and audio discs is stored via bar codes.

Bantam port A connecting interface used on T1 systems for interfacing with receive and transmit (e.g., DS1 in/out) or external timing sources, typically with balanced 100- or 120-ohm termination. A mini-Bantam may be used on T1 daughtercards for supporting an external monitor. In circuit testing, the sleeve on the Bantam connection may be used to ground the circuit.

Port Configurations – Bantam

Unlike RJ-45 (left) or bayonet-mount ports (middle), which have distinctive shapes, a Bantam port is very simple and unassuming, just a small round hole (or dual holes for receive and transmit connections).

---

bar generator A device used to generate horizontal or vertical bars on an output device to determine and adjust linearity.

Baran, Paul (ca. 1926–) A Polish-born American engineer, Baran is acknowledged as a significant Internet pioneer. He conceived a Distributive Adaptive Message Block Network (a concept dubbed *packet switching* by Donald Davies) while working at RAND Corporation on U.S. federal communications infrastructure projects. The core concept was to create a decentralized system through which data could flow in any direction such that if a part of the system were lost, a portion of the system and the data would survive.

Baran worked briefly on the historic UNIVAC computer project before it was taken over by the RAND Corporation and, after jobs elsewhere, ended up at RAND. Baran's networking ideas were verbally presented at RAND Briefing B-265 in the summer of 1961. He subsequently authored a series of memoranda entitled "On Distributed Communications" which were released beginning in 1964. Around the

same time, Davies appears to have been independently developing distributed network concepts.

In the analog electronics world of the 1960s, Baran's idea was a far-fetched proposal, far ahead of its time (it was rejected as impossible by AT&T). As digital technology evolved, however, Baran's ideas became a practical possibility and formed the essential inspiration and format for distributed networks that evolved into the Internet.

A future-oriented thinker, Baran also predicted the e-commerce and online entertainment explosions of the year 2000, long before personal computers, the Internet, and the World Wide Web even existed. He prophetically described online comparison shopping from home through the use of product images and databases back in 1967.

Baran founded a number of commercial ventures and co-founded the Institute for the Future. He is a trustee of the IEEE History Center. In April 2001, Baran was awarded the Franklin Institute's Bower Award and Prize for Achievement in Science. See Davies, Donald; Internet. An oral history of Baran, transcribed in October 1999, is available online through the IEEE History Center.

Barbe, Jane The actor and singer whose voice has been heard by millions as the telephone voice that informs subscribers about changed numbers ("This number is no longer in service ..."), disconnects, and other situations. Barbe is heard on Bell and National Bureau of Standards systems and many national voicemail systems. She has provided the voice for Electronic Tele-Communications, Inc.'s Audichron time, temperature, and weather services since the mid-1960s. Fans of early television shows have seen her on the popular game show "I've Got a Secret" and "The Mike Douglas Show."

Barclay box relay A historic telegraphic relay which had better sound amplification and portability than conventional models.

Barclay insulator A type of early glass utility pole insulator invented by John C. Barclay. It is a type of spiral groove insulator and is numerically identified as CD 150. See insulator, utility pole.

bare metal, down to the bare metal The essentials of a machine or system. The low level systems functions. Programming "down to the bare metal" usually means programming in assembly or machine language or hand-wiring a prototype breadboard.

bare wire A wire without any kind of protective or insulating cover. The ends of insulated transmission wires are usually stripped of their covers to provide bare wire for a good electrical contact at a circuit junction.

Barge In A surcharge phone service or feature of a multiline subscriber service enabling someone (hopefully in authority) to barge into specified lines and interrupt a call in progress. It is a privilege that should not be used indiscriminately but may be important in emergency situations. See buttinsky.

barge out To abruptly leave a call in progress.

barium ferrite Barium, a silver-white, malleable substance which, in combination with iron, produces a substance that can be used in magnetic recording media. Methods of synthesizing barium ferrite nanoparticles through precipitation and spray pyrolysis are being studied.

Barkhausen, Heinrich Georg (1881–1956) A German physicist and educator, Barkhausen worked for Siemens & Halske and then accepted a professorship in communications engineering at the Dresden Technical Academy in 1911. He founded the first global college for weak-current engineering, work that was fundamental to the evolution and application of electron tube technology. Barkhausen discovered the Barkhausen effect in ferromagnetic materials and, in collaboration with K. Kurz, described Barkhausen-Kurz oscillations. See Barkhausen-Kurz tube.

Barkhausen effect A phenomenon described in 1919 by H. Barkhausen when he was studying magnetic and acoustic effects. He observed that a slow, continuous increase in the magnetic field applied to a ferromagnetic material would result in discontinuous increases in magnetization due to changes in elementary magnets in the increasing magnetic field. This suggested that magnetism was a phenomenon related to the larger domain of a ferromagnetic substance as opposed to being a discretely atomic function. The effect was sufficiently strong to be heard as clicking sounds when amplified through a speaker.

Barkhausen-Kurz oscillations In a vacuum tube, oscillation of electrons by means of electrodes and the grid through manipulation of the voltage across the grid and the plate such that the electrons flow back and forth between the filament and the plate. This phenomenon has practical applications for the generation of ultra-high frequency waves and aided in developing the principles of velocity modulation. See Barkhausen-Kurz tube, electron tube, klystron.

Barkhausen-Kurz tube, B-K tube A technology capable of generating microwaves that was developed around the same time as the magnetron, the Barkhausen-Kurz tube is a triode vacuum tube with the third element, the grid, operated at a high positive voltage with the plate at zero or negative voltage. This configuration produces an oscillating motion of electrons between the filament and plate. The tube dates back to the work of German scientists H. Barkhausen and K. Kurz who described the technology in *Physik Zeit* V.21, 1920. By the mid-1930s, there were a number of commercial vendors of B-K tubes, which were used in research and military sensing applications. See Barkhausen, Heinrich Georg; Kurz, Karl; magnetron.

barometer An instrument designed to measure atmospheric pressure. It is one of the tools commonly used to evaluate and predict weather patterns. Barometers are incorporated into a number of other instruments as well, most notably traditional altimeters. Newer altimeters sometimes incorporate Global Positioning Service (GPS) capabilities.

Barometers were important instruments in early studies of magnetism, particularly in Italy where members of the Accademia del Cimento used barometers in the 1660s to provide an airless environment to study whether the attractive properties of various substances were dependent upon air. Unfortunately, the difficulties of creating a vacuum and manipulating the materials within the small area hindered them from making any definite conclusions. See Toricelli, Evangelista.

barometric light A phenomenon observed at least as early as the 1600s, in which a glow or flash appears above the mercury in a barometric tube if it is moved quickly or shaken. The phenomenon is similar to that exhibited by neon light, although this was not known at the time. See barometer.

barrel distortion A type of visual aberration in which the outward corners of an image are contracted inward. This may happen on a convex or concave surface (depending upon whether it is backlit or frontlit) and is noticeable on older, more highly curved monitors and television screens. The opposite of barrel distortion is pincushion distortion. See keystoning.

barretter, barreter A device whose resistance changes in relation to temperature. The hot-wire barretter was devised in 1901 by R. Fessenden to improve the technology that was then being used to detect radio waves. The technology was incorporated into voltage-regulating devices consisting of a wire filament connected to the circuit in series contained within a gaseous envelope. In conjunction with a waveguide, a barretter can be used to measure electromagnetic power.

BARRNet Bay Area Regional Research Network. An association of university campuses and government research centers in the San Francisco area. See BBN Planet.

Bartholin, Rasmus (1625–1698) A Danish physicist and mathematician from the erudite medically inclined Bartholin family, Bartholin followed the example of his father Caspar and brother Thomas and studied in Italy from 1653 to 1656.

He is best known for his 1669 experiments with Iceland spar, a transparent form of calcium carbonate with interesting birefringent properties. Bartholin published his findings as Erasmi Bartholini in *Experimenta Crystalli Islandici Disdiaclastici* and followed them up with other scientific publications that included information about optics. His was an important contribution, inspiring a number of other prominent scientists to study and mathematically describe double refraction and its implications regarding our understanding of the nature of light.

Bartholin (sometimes transcribed as Erasmus Bartholinus) had an interest in astronomy as well and described the path of a comet he observed in 1665, which is richly illustrated in *Theatrum Cometum* published in 1666. See Iceland spar, Nicol prism.

Barton, Enos Melancthon (1840s–early 1900s) Barton co-founded Western Electric with telegraph/telephone pioneer Elisha Gray. He initially formed a partnership in 1868 with George Shawk. Shawk rejected an offer to partner with Gray, but Barton was interested in Gray's ideas and Gray & Barton was formed in 1869, evolving into the Western Electric Company in 1872. It is the spiritual forerunner of today's

Lucent Technologies. Graybar Electric Company, Inc., was spun off from Gray & Barton in 1925 in order to provide electrical distribution. See Gray & Barton; Graybar Electric Company, Inc.

base 1. Bottom; lower support portion; portion to which something is bonded; substrate. 2. In utility pole insulators, the base may be smooth or may have *drip points*, little knobs for the distribution of streaming moisture. 3. In layered semiconductor fabrication, the beginning or bottom layer of a component, which is often a supporting substrate. 4. In bipolar semiconductor components, a thin region of one type of semiconductor sandwiched between an emitter and a collector of another type to form a dynamic environment. In the base region, excitable (highly mobile) electrons act as minority carriers moving between the emitter and collector. See p-njunction. 5. In chemistry, a substance that gives up hydroxide ions in solution.

base film A substrate for holding magnetic particles as in audio and video tapes. The materials used for base film vary but generally have the characteristics of flexibility, resistance to wear, and affinity for holding the magnetic coatings that are applied to their surfaces.

Base Information Digital Distribution System BIDDS. A U.S. Air Force telephony communications distribution network installed in the 1980s with automated functions such as directory and operator assistance, reporting, record-keeping, and billing.

Base Information Transport System BITS. A U.S. Air Force military network system. Information about BITS is published on the Web as an aid to outside contractors planning and installing military information technology infrastructures.

base insulator A large support and insulating structure used on transmissions towers to insulate the tower from the ground.

base memory *jargon* The first block of memory, consisting of 640 kilobytes, in the older Intel-based desktop computers. Base memory is now more generally used to describe the typical or minimum memory with which a system is sold to the enduser. By early 2002, base memory configurations of 256 or 512 Mbytes wer common.

base rate 1. The basic rate without options or value-added services. 2. The basic charge per minute for measured service.

base station 1. A main transmitting and/or receiving station or central switching station, often one which serves as a junction between wireless and wireline communications paths, or between broadcast signals and cable subscribers. 2. In mobile communications, a fixed station within the transceiver system. See cellular phone. 3. In Global Positioning Systems (GPS), a receiver established in a known location to provide reference data for differentially correcting rover files. Baseline data can be correlated with position data from unknown locations collected by roving receivers to improve accuracy.

baseband 1. A simple type of transmission in which the signal is sent without altering it, as by modulation, and which does not require demodulation

through modems to alter the signal at its destination. A transmission that is not segmented by frequency division (multiplexing). This basic signal is centered on or near the zero frequency. See sideband. 2. A one-channel or one carrier-frequency data network such as Ethernet, that is alternately shared by the various peripherals, such as computers and printers, or allocated as requested. See narrowband, Token-Ring. Contrast with broadband.

baseband modem This phrase is an oxymoron, since a baseband signal is one that has not been modulated and thus doesn't require a modem to demodulate it. However, baseband modems do sometimes provide an interface device with some simple translation capabilities and may physically resemble standard modems, hence the name. Sometimes better termed a short haul modem, it is suitable for short distances.

baseband repeater A common main station repeating system used to retransmit a signal and, in some cases, drop out selected channel groups, e.g., voice channels, before retransmission. Over long distances with several legs, heterodyne repeating, which is less subject to loss or distortion from modulation and demodulation, may be used in conjunction with baseband repeating to exploit the better properties of each method. See baseband, heterodyne repeater.

baseboard raceway A cable conduit or pathway along wall baseboards. The raceway may run along the baseboards, or be built into them, so they won't be seen. Thus, wiring can be installed and hidden without tearing into the inner walls.

baseline 1. In coordinate systems, a scale, often the horizontal X axis, that establishes a reference structure on which related data can be depicted. 2. In radar, a line displayed to show the track of a scanning beam. 3. In typography, an imaginary line extending through a font (horizontal in Roman and Cyrillic fonts) for alignment. Desktop publishing software is not entirely standardized; some programs treat the baseline as the bottom edge of nondescending letters, and others treat it as the first unit beneath the bottom edge of these letters. 4. In a Global Positioning System (GPS), a pair of stations for which simultaneous data have been collected.

bash, bash shell, Bourne-again shell A popular, powerful, practical sh-compatible Unix command interpreter shell (command environment) released in the late 1980s by the Free Software Foundation. Bash is based on the Bourne shell, with some features from Korn (ksh) and C (csh) shells.

BASIC Beginner's All-purpose Symbolic Instruction Code. An English-like programming language designed at Dartmouth College in the early 1960s that evolved from the Dartmouth Simplified Code (Darsimco). BASIC was created to provide a programming environment that was faster and easier to learn than FORTRAN or lower level languages like assembler or machine code. BASIC was originally a compiled language that enabled students to write computer programs for the Dartmouth Time-Sharing System running on a General Electric (GE) mainframe computer.

The availability and sophistication of commercial software has increased dramatically since the early days of BASIC and more powerful programming languages have evolved (e.g., C/C++). Users no longer have to program a computer to use one, and the demand for fast, powerful software has necessitated the use of other languages. Thus, the original text-based BASIC has largely faded into computer history, survived mainly by object-based graphical versions (primarily Visual BASIC) that are useful for writing utilities or prototyping software interfaces and software flow (e.g., proof of concept demonstrations). With so much off-the-shelf software now available, the motivation for the general consumer to learn to program in BASIC has all but disappeared, though BASIC-like macro programming languages can still be found in most paint, database, and word processing programs for the purpose of automating common tasks. See VisualBASIC.

basic access authentication See authentication, basic access.

basic cable service The base service (lowest level) offered by a television cable company, consisting of a cable feed to the premises and broadcasting of a specific package of programs. The cable company's programming provisions, signals, and public, educational, and government access channels are government regulated under the Cable Act. Rates for basic cable services and equipment are regulated by franchising authorities that are, in turn, certified by the Federal Communications Commission (FCC). FCC rules are available through the International Transcription Service (ITS). See Cable Act of 1984.

Basic Call Model BCM. See Intelligent Networks Call Model.

Basic Control System BCS. An interrupt-driven computer satellite control system.

Basic Encoding Rules BER. Standardized rules for data encoding that provide support for the abstract syntax description language of Abstract Syntax Notation One (ASN.1). BER was developed in the 1980s, arising out of ISO X.409 and rewritten as ISO 8825 to provide a separation between ASN.1 and BER concepts. It was then introduced in the 1988 CCITT Recommendations as X.209 and became a tool for use in the development of open systems architectures. It became one of three major schemes that evolved in the early 1990s for encoding and is strong in the areas of extensibility and the ability to recognize encoding structure without knowledge of the originating ASN.1 type. See Abstract Syntax Notation One, Packed Encoding Rules, LightWeight Encoding Rules.

Basic Exchange Radio Telecommunications Service BERTS. A system developed in the 1980s to provide wireless services through radio signals to standard local telephone loops, especially to rural areas or for emergency services.

basic information unit BIU. In packet networking, a unit of data and control information consisting of a request/response header (RH) and a following request/response unit (RU).

Basic Link A voice grade circuit providing certain specified standard levels for services, transmission, and loss (8 dB in the 300 to 3000 Hz bandwidth). See Assured Link.

Basic Rate Interface BRI. There are two basic types of ISDN service available: BRI and PRI. BRI is an ISDN service consisting of two bidirectional 64 Kbps bearer channels (B channels) for voice and data and one delta channel (D channel) for signaling or packet networking at 16 Kbps or 64 Kbps. It requires two conductors through a U Loop, from the carrier to a terminator (NT1) at the customer premises. Except in the rare cases of extremely long phone lines with load coils, most existing phone lines can be used for BRI without significant changes to the actual wire. BRI is aimed at residential and small business users. See ISDN.

basic sequential access method BSAM. A basic means of accessing data stored external to a processor. Other common methods include *basic direct access*, and *basic partitioned access*.

basic telecommunications A Federal Communications Commission (FCC) general administrative category distinguished from *enhanced service telecommunications*. The concept applies to telecommunications that are facilitated by computer technologies without additional processing or protocol conversions. Basic phone service is one example. This service is regulated under Title II and mainly affects telephone service carriers rather than those offering enhanced computer data services. As such, basic telecommunications providers may be subject to fees and regulations that don't apply to enhanced services providers and are regulated in order not to stifle competition with enhanced services providers. Providers offering both basic and enhanced communications services have to maintain distinctions between basic and enhanced services when billing clients and marketing services to potential customers. See Federal Communications Commission.

Basic Trading Area BTA. An organizational designation for wireless telecommunications in which the United States is subdivided into almost 500 basic trading areas (BTAs) which are collectively grouped into Metropolitan Trading Areas (MTAs). The BTAs are used by the Federal Communications Commission (FCC) as a basis for assigning PCS wireless phone system licenses.

basket winding A technique of winding a wire coil, or other filamentary conducting material, such that the paths of the various turns of the winding do not touch except at junctions where they may cross. It is sometimes called lattice winding.

Basket winding is used in applications where a long length of wire, or a greater degree of surface area, needs to be organized into a small amount of space. Basket wound antennas and other devices can be quite aesthetic, resembling arabesque. See basket-wound tuners.

basket-wound tuners Historically, various types of basket winding with fine threadlike materials were commonly used in old radios to act as frequency tuners. The windings were of many shapes, cylindrical, circular, somewhat spherical. They varied in complexity from a dozen turns or so, to many hundreds of turns, in intricate patterns in many layers. By varying the shape, size, and the thickness of the wires, different frequencies could be selected. A radio often came with a selection of basket-wound tuners with electrical contacts on the base that could be plugged in as needed.

Basov, Nicolay Gennadiyevich (1922–2001) A Russian medical assistant, physicist, and engineer, Basov made important contributions to the field of quantum radiophysics and pioneered ammonia-based beams (early laser research). By 1957, Basov was designing and constructing optical quantum oscillators that had potential for gas and semiconductor-based oscillators. By the early 1960s, along with various collaborators, Basov was creating injection semiconductor lasers with gallium arsenide. Through investigations of short laser pulses, high-power single-pulse Nd-glass lasers were developed in the mid- to late 1960s. In 1964, along with Townes and Prokhorov, Basov was awarded a Nobel Prize. Under Basov's guidance in the early 1970s, an original chemical laser was developed and Basov et al. described infrared laser stimulation of chemical reactions. See laser.; laser history; Townes, C.

bass In audio, a low pitch; a deep audible tone.

bat switch See toggle switch.

batch An assortment of data or objects grouped to be processed during a single run of a program or process. See batch file, batch processing.

batch file A data file for grouping, storing, and facilitating the execution of complex sequences or frequently used computer commands. Batch files are a convenient way to store configuration parameters, frequently used groups of commands, a list of applications that are executed one after another, commands intended for deferred execution, scripts launched from Web pages, and startup commands for a computer system. ".BAT" is a familiar extension given to batch files on MS-DOS-compatible systems. Many systems provide job control languages (JCLs) or a variety of scripting languages for the quick creation of batch files. Perl is an excellent multiplatform programming tool for creating batch files use on and off the Web. Batch commands have traditionally been created with text editors, but graphical tools may be used. See batch, batch processing, JCL, Perl, Java.

batch processing Deferred or off-line processing of an assortment of data, programs, or objects handled during a single program or process run. Unless there is a fault condition, batch processing usually assumes once the job is initiated, it will run undisturbed and unattended. Email is often handled as a batch process, e.g., your Internet Services Provider may wait a specified period of time before posting a group of messages to your account rather than posting each one as it is received. Payroll accounts are often run as batch processes, as are many data collection programs, such as weather testing, astronomical observations, etc.

It is not uncommon for batch processes to run as background tasks, executing while users continue to use the system for other applications. Batch processes can be scheduled to run when network access is low, thus not putting a drain on system resources when many users are online. Batch processes can also be used to schedule transmissions, such as facsimiles, during hours when phone rates are low. See batch, batch file, realtime processing.

Batch Simple Mail Transfer Protocol BSMTP, bSMTP, Batch SMTP. A batch version of the Simple Mail Transfer Protocol developed in the early 1980s by E. Alan Crosswell to facilitate the reliable transmission of electronic mail messages over distributed computer networks independent of the transmission subsystem. Batch SMTP allows a series of commands to be bundled and sent to a remote machine for execution, rather than establishing a typical interactive SMTP session. The processing of special characters may also be supported through batch processing. This protocol was presented to the BITNET community in 1982 and installed on many BITNET mailer gateways. See Simple Mail Transfer Protocol. See RFC 821,

battery A group of two or more cells connected together in such a way that they produce a direct electric current (DC). While historians believe battery power may have been used for electroplating by the Parthians as early as the third century BC, the first significant records of modern battery experiments date from the work of C. A. Volta.

Battery-generated electricity was widely used in industrial applications and telegraph and telephone communications in the early 1900s. Edison was a strong proponent of DC current and received much opposition from Tesla and Westinghouse, who were advocating alternating current (AC).

Batteries are used widely in portable devices and as emergency or backup power for systems whose main power source is alternating current. The Sampling of the Evolution of Batteries chart describes a few common batteries and interesting technological adaptations. See B Battery, cell, storage cell, talking battery.

battery, rechargeable A direct current (DC) power source. A rechargeable battery is designed to readily have its power restored, usually through consumer-priced battery chargers or through an alternating current (AC) transformer attached between the wall socket and a battery-charging device. Rechargeable batteries are commonly used on palmtops, laptops, camcorders, etc. Most of them need to be fully discharged before being recharged, or a memory effect results in only a partial charge. Some can be trickle-charged when plugged into an outlet while being used. Larger rechargeable batteries for consumer electronics usually supply from two to five hours of charge.

battery, storage A type of battery that, once charged, will hold that charge for a practical amount of time without constant electrical refreshes from another source, such as alternating current (AC). Car batteries are a type of common storage battery that are recharged when the engine is running. They have a useful life of about three to five years if not completely discharged too often (by leaving lights on, for example). Storage batteries are often used as backup systems for alternating current (AC) systems. See battery, rechargeable.

battery backup A direct current (DC) backup system which kicks in if something happens to the primary power system. For example, many phones now have memory storage for names and numbers, or extra features like speakerphones or text display, that require more power than is provided by the current coming from the phone line. These phones may have a battery or AC power cord to help power the extra features which also functions as a backup battery to protect the contents of the electronic phonebook if the phone line power is interrupted by a power failure, or if the phone is disconnected and moved from one location to another.

Most computer systems have small batteries on the motherboard to protect the contents of certain types of chips that hold information such as configuration parameters. Lithium batteries are commonly used and should be replaced every 5 to 7 years or so.

Many microwaves, clock radios, and VCRs have backup batteries so that the time is not lost during a power outage. If your appliance flashes 12:00 after a power failure, it probably doesn't have a backup battery.

baud, baud rate A unit signifying a rate of transmission of data indicating the modulation rate, named after French engineer J.M.E. Baudot. The term is commonly used to describe modem data transfer rates (e.g., 9600 baud), although it originated from telegraph signaling speed in the 1920s. Note that the rate of transmission is not necessarily equal to the rate of acquisition of the data. Line interference, handshaking, error correction, and other factors can cause the actual rate of data received to be less than the raw transmission speed associated with the amount of data transmitted.

Baudot, J. M. Emile (1845–1903) A French engineer and inventor, Baudot made many contributions including a means, in the 1870s, to insert synchronization signals between baseband signals so time division multiplexing (TDM) could be used to combine signals into a bundle. He developed the Baudot code for telegraphic communications.

Baudot code, Murray code A data code used in asynchronous transmissions, named for its inventor J. M. Emile Baudot. It was widely incorporated into teletypewriter communications beginning around 1870. Baudot code was based on a marks and spaces character-representation scheme employing five equal-length bits to symbolize upper case letters. A simple method of reversing the polarity of the line was in use for about half a century before it was superseded by frequency shift keying (FSK modulation techniques). The character set was very limited and eventually standard codes such as EBCDIC and ASCII superseded Baudot code except for specialized communications, as for the hearing impaired.

See ASCII; EBCDIC; Baudot, J.M. Emile.

Bauschinger effect Straining a solid body beyond its yield strength in one direction decreases its yield strength in other directions.

bay 1. Harbor, indentation, arced enclosure. 2. An opening in a rack or panel into which modular components can easily be inserted. A patch bay has a series of regular openings designed to securely hold modular components while still providing easy access, ease of configuration, and swapping in and out as needed. See patch bay. 3. Part of an antenna array.

bayonet base A type of jack or jack-like base, as on a bulb, which has a small projection on one side and slips into a receptacle with a turn so the projection catches within a small trough and secures the inserted object.

bayonet nut connector, bayonet navy connector BNC. A quick-connect bayonet-locking connector, commonly used for coaxial cables for network or video transmissions. It is intended to provide a constant impedance. The outer shell has a small bayonet that inserts in a helical channel in the receptacle to aid in firmly securing and aligning the connector. This is good for securing connections that must not be interrupted or for securing cables against which there may be tension. See connector, F connector, RCA connector.

BNC Connectors

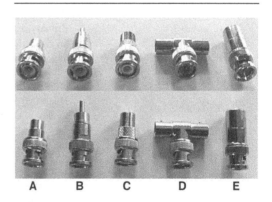

A   B   C   D   E

BNC connectors are widely used for coupling fiber optic and coaxial cables to switchers and patch panels, as they provide a secure connection that is not easily uncoupled by bumping or tugging. End and overhead views of a number of common BNC (bayonet mount) connectors are illustrated as follows: BNC to RCA plug (A), BNC to RCA jack (B), BNC to screw-type F connector (C), BNC to dual BNC male connectors (e.g., 'thin' Ethernet) (D), BNC terminator (E).

---

bazooka A device for isolating an outer conductor from other surfaces and connecting an unbalanced line to a balanced line. Bazookas are commonly used on the ends of coaxial cables that are to be connected to two-wire lines (e.g., copper twisted pair). See balun.

BB 1. See baseband. 2. See broadband.

BBC 1. Broadband Bearer Capability. A bearer class field that is part of an initial address message. See ISDN. 2. See British Broadcasting Corporation.

BBL 1. broadband loop. 2. BROADBANDLOOP Project. An ACTS project to define and test cost-effective broadband access network concepts facilitating the migration of fiber optic transmissions into local loops. The project demonstrates the upgrade of passive optical networks to high bandwidth capacity and the upgrade of wire-based networks to handle broadband services to the customer premises. The target group for the project is European residential subscribers and small- to medium-sized businesses. See BIDS, BONAPARTE, BOURBON, BROAD-BANDLOOP, UPGRADE, WOTAN.

BBN Bolt, Beranek, and Newman, Inc. A high-technology company in Cambridge, MA which developed, maintained, and operated the historically sig-

nificant ARPANET and later the Internet gateway, CSNET CIC, and NSFnet NNSC. See BBN Planet.

BBN Planet A subsidiary of Bolt, Beranek, and Newman, Inc., which operates a national Internet access network. See BBN.

BBS See bulletin board system.

BBT broadband technology.

Bc See Committed Burst Size.

BC 1. backward compatible 2. beam coupling. 3. binary code. 4. broadcast.

BCC 1. Bellcore Client Company. 2. See block check character.

BCD See binary coded decimal.

BCM See Basic Call Model.

BCOB Broadband Connection-Oriented Bearer. In ATM networks, information in the SETUP message that indicates the type of service requested by the calling user. The appropriate bearer class depends on the

Sampling of the Evolution of Batteries and Promising Battery Technologies

Type	Developer	Notes
Daniell battery	J. F. Daniell	A chemical battery used in early telegraph systems, ~1 volt.
Edison cell		A variable storage nickel hydrate (positive) and iron oxide (negative) cell with an electromotive force lower than that of a lead cell. This is a historic battery that was used in automobiles due to its ruggedness.
gravity/crowfoot cell		A voltaic wet cell for providing small currents at a constant emf.
Grove battery		Primary, 1.96 volts per cell. Zinc, platinum. Used by S. Morse.
A battery		Historically used as *talk batteries* in telephone installations and as low voltage batteries for electron tube filaments. Modern versions are now commonly used for cameras, calculators, and other small portable appliances.
B battery		Historical provider of low voltage power to the plates (anodes) in electron tubes and to communications relay circuits.
C battery		Introduced in the early 1920s, C batteries provided bias voltage to electron tubes for the control of the grid circuit and were often used in conjunction with B batteries to extend the life of the B battery. C batteries are now commonly used with small portable devices such as flashlights and portable boomboxes.
silicon solar	Bell Labs	Announced by Bell in 1954, there are now a number of variations on this technology from different developers. Kyocera introduced a multicrystal silicon solar battery in 1996 that has a conversion efficiency rate of 17.1%, considered good in the solar industry.
lithium nouveau		A battery chemistry based on lithium polymer which may provide longer life for power-hungry mobile phones, laptops, etc.
organic	McGinness et al.	Biological systems in the form of synthetic melanins that appear to have the form of an amorphous semiconductor "threshold switch" operating at threshold significantly lower than inorganic thin films. These also exhibit electroluminescence at the point of switching energy states, described in 1973.
electrochemical	Nogami et al.	A battery with its electrodes immersed in an organic solvent electrolyte solution with the electromotive force generating by doping/undoping processes associated with the electrodes, 1982.

specification of the network.

BCOB-A (Type A)	connection-oriented, constant bit rate, timing required
BCOB-C (Type C)	connection-oriented, variable bit rate, timing not required
BCOB-X (Type X)	transparent AAL, traffic type and timing requirements

UNI 3.0 and UNI 3.1 must support Type A and Type C or Type X as a substitute for the first two types for virtual connections (VCs). Internet Protocol over ATM signaling must permit Type C and Type X bearer capability in specified combinations. Type C and Type X both apply to multiprotocol connections.

BCP See Best Current Practice.

BCRS See Bell Canada Relay Service.

BCS 1. basic control system. 2. Batch Change Supplement. A development tracking system for documenting system features from proposal to finished product. 3. See beam control system. 4. See Boston Computer Society. 5. See British Computer Society.

BDF block data format, also referred to as data block format (DBF). A generic phrase for a block of digital data with a specific format with respect to the size and order of the data items within the block.

BDT See Telecommunications Development Bureau.

Be 1. See burst size, excess. 2. See Be, Inc.

Be, Inc. A computer software company founded by Jean-Louis Gassée, head of R&D at Apple Computer during the Apple II years. Be developed and released the BeOS (Be Operating System) in 1997. It is a fast, integrated-database, multiplatform OS aimed at the audio and graphics/video computer-using markets. In addition to BeOS, Be now provides integrated client-side software, development, and customization tools for Internet device and service providers and consumer electronics companies. In the early 2000s, Be was acquired by Palm. See BeOS.

BE 1. base embossed. A designation for glass and ceramic utility pole insulators with embossings on the lower edge, usually of the size, company, and/or patent date. 2. Bose-Einstein.

beacon 1. A signal, locator, or guidance beam or tone. 2. A transmitter that aids in monitoring radiant energy propagation.

beacon alert An alert frame in a Token-Ring or Fiber Distributed Data Interface (FDDI) device signaling a serious problem, such as a physical interruption of the signal or Media Access Unit (MAU). The frame includes information on the location of the break or the station that is down.

beam 1. *n.* A ray, shaft, or other directed energy or illumination, as an electron tube, radar, or light beam. 2. *v.* To direct or aim, as in a broadcast beam.

beam antenna An antenna that transmits and/or receives within a narrow, confined directional range.

beam control system BCS. A means for directing a beam. It may also have control capability for turning a beam on and off. A BCS is usually used for directing common electromagnetic beams such as microwave radio signals or a beam of light, but it

may also be used for high-precision control in atomic physics testing systems (e.g., study of neutrinos). The more sophisticated the technology, the more likely that beam control is automated with computer hardware and software. Logging of beam characteristics and activities over time may also be incorporated into a BCS.

beam divergence 1. As a beam travels through the air, various factors may cause it to spread out. This divergence may result in attenuation or dispersion of the signal strength of a transmission over distance. 2. The path of a beam may progressively move away from the axis of the original trajectory, resulting in divergence.

beam position monitor BPM. A mechanism for keeping track of the position of an emitted/transmitted beam of light (e.g., laser light) or other electromagnetic energy (e.g., radio waves). A beam position monitor is commonly used in systems where the direction of the beam can be controlled.

beam power tube An electron tube with a beam that is directed and concentrated in certain specific directions by a special electrode. Used, for example, in radio frequency (RF) transmitters.

beam splitter A device that produces two or more separate beams from one incident beam. Mirrors and prisms are commonly used to direct or split light beams. Coherent laser light is favored for most beam splitting applications. Once the beam is split, the split beams are usually of a lower intensity than the original beam, in proportion to how many times the beam has been split. Beam splitting is used in a number of industrial instruments and consumer devices and is also used for testing and diagnostic purposes.

In interferometers, a beam is split into two or more beams in order to compare the relationship of the beams when they are recombined. The interference patterns from the beams can provide information on influences from heat, vibrations, gases, etc.

bearer channel See B channel.

beat Short percussive tone, one instance of a repetitive sequence, a reaction from the impact of one object or process on another, the interaction of two different frequencies when certain portions of their cycles interact. See beat frequency, beat reception, zero beat.

beat frequency The frequency resulting when two different frequencies beat together on a nonlinear circuit. The beat frequency is equal to the difference between the two separate frequencies, typically expressed in cycles per second (hertz). When the beats are very close to the same frequencies, they can be set to generate an audible tone which may change when subjected to the influence of magnetic materials. See heterodyne.

beat frequency oscillator BFO. A low-current generator of beat frequencies in a nonlinear circuit. BFOs are used in a number of practical applications from metal detectors, which use search and reference oscillators, to older single sideband radio receivers.

beat reception The combining of two different frequencies, usually the external, incoming frequency, and an internally generated frequency which are then easier to amplify or otherwise condition as a single frequency than the incoming frequency would be by itself. See heterodyne. For contrast, see zero beat reception.

beating A wave phenomenon that occurs when two or more periodic waves of different frequencies combine to form a periodic amplitude pulsation. In audio, beating can quite often be heard and felt as an undulating pulse by those in listening range. See beat frequency, heterodyne.

Beaufort notation In meteorology, a code used for indicating the state of the weather.

Beaver Falls Glassworks A lesser-known historic utility pole glass insulator fabrication company founded by William Modes in Pennsylvania in 1869. See insulator, utility pole.

BEC 1. See Best Effort Capability. 2. Bose-Einstein condensation.

BECN See Backward Explicit Congestion Notification.

beehive insulator A type of early glass utility pole insulator, characterized by its beehive shape. See insulator, utility pole.

BEEP A generic application protocol framework for connection-oriented, asynchronous network interactions. In conjunction with other protocols, BEEP can provide reliability, privacy, and authentication options. Through transport mappings, BEEP specifies how messages are carried over the underlying transport mechanisms. Profiles may be defined for BEEP. For example, the RAW profile is a backwardly compatible, efficient, readily implemented profile for supporting legacy Syslog Protocol processing. The COOKED profile is intended for new implementations of Syslog Protocol handler, at the expense of more overhead. See RFC 3081, RFC 3195.

beeper Colloquial term for a portable device that alerts the user with an audible tone. Commonly incorporated into pagers, a beep signifies that there is a message awaiting the user, or some action to be taken as a result of the alert signal.

Beginning of Message BOM. In ATM networking, an indicator contained in the first cell of a segmented packet. The BOM segment is followed by Continuation of Message (COM) and End of Message (EOM) segments. A header is associated with the segments and they are passed to the Physical Layer for transmission. A two-bit Segment Type (ST) field identifies the type.

Beilby layer A microcrystalline or amorphous layer that is formed on the surface of metals by polishing.

Being There 1. A consumer-priced, Macintosh-based videoconferencing product from Intelligence at Large, which provides video, audio, whiteboard, and file sharing utilities over AppleTalk and TC/IP local area and wide area networks. 2. The title of a classic Peter Sellers movie.

bel *symb.* – B. A unit of relative power or strength of a signal, which is not commonly used because it is so large. It is used in conjunction with amplitude, usually by its tenth measure, the *decibel*. Named after

Alexander Graham Bell. See decibel.

Belden A major commercial manufacturer of communications media which has been responsible for influencing cable standards for many telecommunications systems.

bell 1. An audio device, often of resonant hollow metal, designed to emit sound when struck, vibrated by a column of air, or vibrated through electrical stimulation. 2. A digitally reproduced simulation of a physical bell, created either by sampling a physical bell and playing back the sound, or by analyzing the type of sound wave patterns produced by a physical bell and simulating them mathematically. A computer requires built-in electronics or a peripheral sound card in order to send sounds to a speaker, especially if it's good quality 16-bit stereo sound. Digital music synthesizers often have a wide array of bell sound patches from which to choose. 3. A phone bell that is activated by line current from the switching office to indicate that there is an incoming call on the line.

Alexander Graham Bell – Inventor, Educator

Alexander Graham Bell continued to think of himself as a teacher for the deaf long after he became famous and financially independent from sales of his technological inventions. He and Helen Keller were friends. Many of the members of his family were known as excellent orators. This portrait was photographed by Moffett Studio when Bell was approximately 69 years of age. [National Archives of Canada, Dept. of External Affairs collection.]

Bell, Alexander Graham (1847–1922) A Scottish-born American inventor, who was one of the original founders of the National Geographic Society. He emigrated to Canada with his family and subsequently found employment in Boston. He studied aviation, electricity, fresh water distillation, etc., but he is chiefly credited with the invention of the telephone. In fact, A. Meucci had tested animal membranes as vibrational devices many years earlier and a number of independent inventors in Europe and America were working on ways to transmit tones, and in some cases voice, over telegraph lines. Three significant contemporaries of Bell who achieved success in some of these technologies were Philip Reis, Elisha Gray, and Thomas Edison.

In March 1876, two years after working out the original concept, Bell and his assistant Watson reported having transmitted Bell's spoken voice over electrically charged wires, a story that caught the world's imagination and ushered in the telephone age. Bell's patent was filed just hours before a caveat to file was entered at the same office by Elisha Gray, for a similar harmonic telegraph. Bell's talking phone was not demonstrated publicly until some time later. One interesting historical note is that Bell's patent did not mention the use of a liquid medium, yet the rudimentary telephone that was later demonstrated did use a liquid medium, a structural detail that was mentioned in the Gray caveat.

Model of the First Telephone

A model reconstruction of what Bell has described as the first telephone invented by him in 1875, referred to as the "Gallows Telephone" due to its shape. [Library of Congress Detroit Publishing Company Collection. Published ca. 1920.]

There was significant critical interest in the invention. On 31 Aug. 1876, Anna Joy wrote to Bell, asking him "Was an experiment tried at your office for transmitting sound by Electricity? If so when was it tried? Who were the parties engaged? What was the result?" to which Bell gave an uncharacteristically cryptic reply:

"Apr. Sept. 1st, 1876. / The Manag. of the P. & A. T. Co. has forw. to me your note of the 31st ult. and I beg to state th. exp. were made with my app. for the tel. transm. of vol. s. under the direct. of Sir William Thompson at the office of the Pac. & At. Tel. Co. - but at what date I am unable at the pres. mom. to state. / Yrs. truly / A. Graham Bell"

In April 1877, Bell acknowledged a request from Boston professionals for a 'practical demonstration of my Electric Telephone' and made arrangements to publicly describe his invention three weeks later. Given all the hedging and the murmur of controversy over the veracity of the original claim, one wonders, in retrospect, if Bell's famous message to Watson was fabricated or exaggerated because the ambitious Bell

was anxious to get the patent before someone else, feeling he was on the verge of success.

In 1880, Bell devised a means to transmit phone signals through light, a forerunner to fiber optic communications that languished for almost 100 years before improvements in technology made the concept practical to implement.

Bell's Historic Telephone Invention

Alexander Graham Bell making a historic call by opening the New York to Chicago long-distance line on 18 October 1898. [National Archives of Canada image.]

A historic report in the Detroit News on 'Bell's First Telephone.' [Library of Congress Detroit Publishing Company Collection, ca. 1920. Photo by Underwood & Underwood, Inc.]

In 1882, Bell was granted United States citizenship though he continued to maintain a summer home in Nova Scotia, Canada.

Bell achieved enormous financial success and could have ceased working at a young age, but he continued to research aeronautic kites (he succeeded in getting a manned, motorized aircraft aloft in 1908), hydrofoils, and various technologies to aid the deaf, most notably the audiometer. For much of his life, he listed his occupation as teacher of the deaf.

While the invention of the telephone was not as revolutionary as the telegraph in technical terms, it was a highly significant, culture-changing evolutionary step that personalized distance communications and facilitated commerce in ways not previously possible. See audiometer; Bell System; Berliner, Emil; Edison, Thomas Alva; Gray, Elisha; Meucci, Antonia; Reis, Philip; Photophone; telephone history.

Bell asynchronous standards A series of full duplex standards developed by AT&T. These were widely supported by other manufacturers in the late 1970s and early-/mid-1980s. Other vendors and standards bodies began competing with the Bell standards, most notably Hayes, in the early 1980s. The V Series Recommendations by the ITU-T are now the dominant formats. Some Bell standards are shown in the Bell Serial Communications Standards chart.

Bell Atlantic A holding company created as a result of the AT&T divestiture in the mid-1980s. See Bell Operating Company.

Bell Canada, Bell Telephone Company of Canada The Canadian arm of the Bell system until the 1970s, when it became separated from the U.S. Bell system. Bell Canada was a member of the Stentor Consortium, along with BC Tel Ltd., SaskTel, and others. As companies merged and were bought out, the Stentor alliance dissolved and Bell became mainly focused on the provinces of Ontario and Quebec. Bell is the major telecommunications carrier and supplier of telecommunications equipment in Canada. Through mergers, BCT.TELUS Communications Inc. became the second largest telecommunications company in Canada.

In 1997, Bell Canada and TELUS Cable Holdings Inc. both applied to the CRTC for broadcast distribution licenses to conduct trials of broadcasting services, while distributing telecommunications services over the same digital networks. In 1999, both Bell and TELUS testified on issues of promoting electronic commerce by protecting personal information related to Bill C-54.

Bell Canada Relay Service BCRS. A 24-hour service that allows TTY users, who may be hearing impaired, to talk to one another or to a hearing person with the help of specially trained operators translating through teletypewriter terminals. The TTY equipment can signal up to 60 words per minute.

As an example of the service, the subscriber calls the BCRS operator and provides his or her name and number and the number of the person to be called. The operator requests billing information and then places the call. The operator then acts as a translator, conveying a text message by voice to the hearing callee, and a voice message by text to the hearing-impaired caller.

The call is kept confidential by the operator, and no record of the conversation is retained. BCRS services are billed at the same rate as normal phone charges.

Bell Communications Research Bellcore. An organization established as a result of the AT&T divestiture to provide a variety of central administration, training, standards, documentation, and quality services to the regional Bell companies who fund Bellcore and their subsidiaries. It is roughly equivalent to the Central Services portion of the pre-divestiture AT&T organization.

Bell Laboratories, Bell Telephone Laboratory, Bell Labs The research arm of the Bell system responsible

Bell Serial Communications Standards		
Standard	Speed	Notes
Bell 103	300	Asynchronous full duplex communications standard for transmitting at speeds up to 300 bps over publicly switched telephone networks (PSTNs). This standard was commonly used with computer modems in the late 1970s, but was superseded by Bell 212 in the early 1980s.
Bell 212	1200	An AT&T asynchronous full duplex communications standard for transmitting at speeds up to 1200 bps over publicly switched telephone networks (PSTNs). This standard was commonly used with computer modems in the early 1980s, but was superseded by Bell 201 in the mid-1980s.
Bell 201	2400	Asynchronous full duplex communications standard for transmitting at speeds up to 2400 bps over publicly switched telephone networks (PSTNs). This standard was commonly used with computer modems in the mid-1980s. Many other vendors began entering the modem manufacturing/standards industry at this time.
Bell 208	4800	Asynchronous full duplex communications standard for transmitting at speeds up to 4800 bps over publicly switched telephone networks (PSTNs). This standard did not particularly catch on in consumer markets. Many users leapfrogged from 2400 bps to 9600 bps as vendor participation and competition for faster speeds increased in the mid-1980s.

for many important discoveries and the development of thousands of telecommunications technologies and devices over the decades. The labs were established as a combined effort of the Western Electric Company and the AT&T engineering departments in 1907. It grew to be the largest industrial research organization in the U.S., and, in 1925, the engineering department of Western Electric was incorporated as Bell Laboratories, with the head office in New York City. In 1941, headquarters were moved to Murray Hill, New Jersey and larger plants were later established in Denver and Atlanta. Smaller field stations and satellite labs were regularly established over the years in many parts of the U.S. In 1934, AT&T's research division was merged into Bell Laboratories.

Bell Labs Museum An online resource sponsored by Lucent Technologies. You can visit the images and historical references at the Bell Labs Museum Web site. http://www.lucent.com/museum/

Bell Operating Company BOC. This is defined in the Telecommunications Act of 1996 and published by the Federal Communications Commission (FCC), as

"... any of the following companies: Bell Telephone Company of Nevada, Illinois Bell Telephone Company, Indiana Bell Telephone Company, Incorporated, Michigan Bell Telephone Company, New England Telephone and Telegraph Company, New Jersey Bell Telephone Company, New York Telephone Company, U S West Communications Company, South Central Bell Telephone Company, Southern Bell Telephone and Telegraph Company, Southwestern Bell Telephone Company, The Bell Telephone Company of Pennsylvania, The Chesapeake and Potomac Telephone Company, The Chesapeake and Potomac Telephone Company of Maryland, The Chesapeake and Potomac Telephone Company of Virginia, The Chesapeake and Potomac Telephone Company of West Virginia, The Diamond State Telephone Company, The Ohio Bell Telephone Company, The Pacific Telephone and Telegraph Company, or Wisconsin Telephone Company; and

(B)   includes any successor or assign of any such company that provides wireline telephone exchange service; but

(C)   does not include an affiliate of any such company, other than an affiliate described in subparagraph (A) or (B)."

See Federal Communications Commission, Telecommunications Act of 1996.

Bell speak *colloq.* A phrase to describe the substantial body of telephone jargon that grew up over the decades within the Bell system, particularly among technicians and scientific researchers.

Bell System The original holders of the Bell telephone patents formed by Bell, Sanders, and Hubbard in 1877, and incorporated in 1878, less than 15 years after the invention of the telephone. The company thrived and grew under the management of Theodore N. Vail. Since the term of exclusivity granted by a patent lasted only 17 years, the expiry of the Bell pat-

ents resulted in the founding of thousands of new independent phone companies. These gradually were merged and consolidated into the Bell System. In a 1984 court decision, divestiture of the American Telephone and Telegraph company (AT&T) removed the distinction between the Bell company and independent phone companies.

Bell Telephone Company of Canada Inc. Established in 1880, Bell Canada began by providing service to the larger centers in eastern Canada, most of which were interconnected within about 10 years. Bell Canada is under the jurisdiction of the Canadian Radio Television and Telecommunications Commission (CRTC).

Bell Telephonic Exchange The first telephone exchange in Ohio State.

Bellcore See Bell Communications Research.

Bellingham Antique Radio Museum See American Radio Museum.

BellSouth Corporation A large regional holding company created as a result of the AT&T divestiture in the mid-1980s. It is comprised of Southern Bell Telephone and South Central Bell Telephone Company and a number of other companies. BellSouth is cooperating with Nippon Telegraph and Telephone to provide large-scale integration of residential fiber multimedia telecommunications services. See Bell Operating Company, fiber to the home.

benchmark 1. A specified expression of performance based on agreed-upon test criteria. 2. A criterion expression, often numeric, against which other systems or processes are compared. Benchmarks are so system specific that it is hard to translate benchmark performance scores to real-life computing situations, and their validity is often hotly contested. See benchmark test.

benchmark test A criterion test for evaluating the performance of a system, often applied to the speed of processing. Although benchmark tests may be straightforward for simple electronic components, they are sometimes used to evaluate the system performance of complex systems, which is difficult to measure in objective units. For example, a computer with a 40-MHz CPU will perform more slowly on benchmark tests than a 200 MHz RISC chip CPU, yet a word processor running on one system may have the same apparent speed to the user as one running on a faster system due to many factors such as load on the system, user interaction, software optimization, address bus bottlenecks, amount and type of memory, etc. In a broad sense, benchmarks cannot be said to provide definitive performance measures, but they are nevertheless often established as a best-efforts way of comparing and contrasting systems with significantly different construction and characteristics. Even these are often considered "better than nothing" performance indicators. See Dhrystone, Rhealstone, Whetstone.

bend loss In cabling, attenuation caused by bends and twists in the wires or fibers. At each bend there is a tendency, especially in optical fibers, for the signal to want to continue to radiate in the same direction, resulting in slight losses through the cladding as the cable curves. See Bend Factors diagram.

bend radius In cabling and cabling enclosures, a description of the bend tolerance of a certain material at a certain radius, often under a certain pulling force. This measure is important for manufacturing, for selecting types and sizes of parts, and for installing pulleys, cables, and wires. See bend loss.

bend-insensitive fiber Fiber optic cable that is particularly resistant to losses when the fiber is bent. Bend-insensitive fibers were first being developed and described in the late 1980s.

There are various ways to reduce bend-induced loss in optical fibers. The materials, diameter and ratio of core to cladding, and the numerical aperture can all influence sensitivity to bend losses. For example, increasing the numerical aperture (e.g., NA = 0.16 for single-mode fiber) can confine the reflected light more tightly within the conducting core, producing a more bend-insensitive fiber.

Benedicks, Manson An American researcher who investigated the electromagnetic-influencing properties of germanium crystals in the early 1900s and found that they could be used to convert alternating current (AC) to direct current (DC).

Benjamin Franklin Institute of Global Education A resource and Web center that supports and promotes affordable access to education from global resources through distance education, founded in the mid-1990s by John Hibbs. http://www.bfranklin.edu/

bent pipe A description for a communications conduit, path, or transmissions medium that reflects an incoming signal at an angle, usually between 20 and 70 degrees, thus following a path that resembles a bent pipe. This is a very common configuration for Earth-satellite/satellite-Earth transmissions and for radio transmissions which are channeled by being bounced off the ionosphere.

Benton Foundation An organization established in 1948 that has promoted diverse and equitable public use of communications technologies for its social benefits since 1981. It is named after its founder, William Benton (1900–1973), a U.S. Senator, UNESCO Ambassador, and publisher of the Encyclopedia Britannica. The Benton Foundation provides news on communications policy and the social use of technologies; it supports a number of free online newsletters and discussion lists. The Foundation operates from an endowment along with additional funding from major communications industry vendors and philanthropical organizations. See Communications Policy Project.

BeOS An object-oriented, multitasking, fast, nonlegacy, microcomputer operating system developed by Be, Inc., under the leadership of Jean-Louis Gassée. Programmers claim it is a pleasure to program and that the environment is powerful and yet easier to learn and use than many others.

Bend Factors in Fiber Optic Lightguides

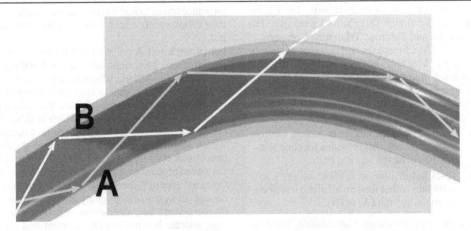

Light is guided through a fiber optic cable through a process called total internal reflection (TIR) *in which the outer* cladding, *which has a different refractive index from the inner conducting* core *(in this case, a multimode fiber), reflects the light beam back into the core. This process continues even if the fiber is bent to some extent, which is very useful for network cable installation and fiber optic probes for scientific or medical purposes. However, there are limits to how much a cable can be bent without bend-induced loss of signal.*

In the diagram above, the amount of loss in the bend depends upon the bend radius, the relationship of the core to the cladding in terms of size and refractive index, and the angle at which the beams encounter the cladding. If a light beam (A) passes the maximum point of bend, it may continue on through the lightguide. In contrast, a beam (B) traveling at a different angle such that it hits the cladding at the bend, beyond the critical angle at which it can be reflected back into the fiber, would be partly absorbed by the cladding and by any materials outside the cladding and would be lost as far as the fiber core is concerned. Commercial manufacturers design certain cables to minimize bend loss, by balancing cable parameters. Thus, bend-insensitive *cables often have a higher* numerical aperture.

Gassée is well known for his previous contributions to R&D at Apple Computer, when the Apple II line was being developed. BeOS is aimed at multimedia audio and visual applications.

BeOS was introduced to developers late in 1995 along with the Be computer. About a year later, Be, Inc. discontinued the hardware, to concentrate on software development, as their operating system software is able to run on several hardware platforms by various vendors. See Be, Inc.

BEP 1. See back end processor. 2. Bureau Economique de la Province de Namur. 3. Bureau of Engraving and Printing (Federal) 4. Business Enterprise Program (for minorities and handicapped workers).

BER 1. See Basic Encoding Rules. 2. See bit error rate.

Berkeley, Edmund Callis (1909–1988) An American educator and pioneer computer developer who worked on the Mark II construction project in 1942 and developed a lifelong interest in computer applications. Berkeley founded the Eastern Association for Computing Machinery in 1947, which became the respected Association for Computing Machinery (ACM) the following year.

In 1948, Berkeley started his own company, Berkeley Associates, to market his inventions. He authored many books foreshadowing computer and telecommunications history. In 1949, *Great Brains, Or, Machines That Think* was released with instructions on how to design computing devices. In 1950, Berkeley published *Computers and Automation,* a historic computing magazine. In 1956, he coauthored *Computers, Their Operation and Applications* with L. Wainwright. And, in 1959, he authored *Symbolic Logic and Intelligent Machines*.

Following up on the ideas in *Great Brains*, Berkeley described and constructed one of the first desktop microcomputers that became generally known through popular publications. The *Simon* (named after "Simple Simon") was made public in an electronics magazine in 1950 (as construction plans). Berkeley was also actively interested in the design and construction of small robots (quite prolific, in fact), which he marketed through Berkeley Enterprises, Inc. (originally Berkeley Associates).

Berkeley sought to bring computing concepts to hobbyists through the GENIAC computing device. Unfortunately, after disputes with his business partners, Berkeley lost the legal right to use the name, so he gave the name Brainiac to essentially the same technology and his former business partner used the Geniac name to market tube-shaped calculators.

The archival legacy of Edmund Berkeley from 1923 to 1988 has been donated to the Charles Babbage Institute by Berkeley Enterprises, Inc. and the Berkeley family. See Brainiac, Charles Babbage Institute, GENIAC, Simon.

Berkeley Internet Name Domain BIND. A popular implementation of the Internet domain name service (DNS) originally developed and distributed by the University of California in Berkeley. There have been numerous commercial implementations of BIND. As of 2001, BIND 4.9.8 (for older systems),

BIND 8.2.3, and BIND 9.1.2 were in widespread use on the Internet as free software. However, the purchase of support contracts aids in the continued development of the Internet Software Consortium's (ISC) versions of BIND.

Berkeley Standard Distribution BSD. A family of Unix-style operating systems originally released in 1974, adapted to the Digital VAX and PDP-11, and now widely ported to many systems. BSD was further developed by Bill Joy and others at the University of California in Berkeley, who released it in 1978. Joy subsequently wrote the well-known *vi editor*, and co-founded Sun Microsystems.

BSD flourished with the development of the ARPANET, the forerunner to the Internet, and the Computer Systems Research Group (CSRG) enhanced BSD with 32-bit addressing, virtual memory, and a fast file system supporting long filenames. They further introduced BSD Lite which was BSD without the licensed AT&T code, which could be freely distributed. The CSRG disbanded in 1992, and the community at large adopted BSD and developed FreeBSD. See FreeBSD, Unix, UNIX.

Berliner, Emil (later Emile) (1851–1929) A German-born American/Canadian inventor and musician who was keenly interested in acoustics, electricity, and physics as telephone technology began to emerge.

Berliner is best known for music technologies, but also made some significant contributions to historic telephone technology. In April 1877, he filed a caveat for a patent on a telephone transmitter and six months later is reported to have demonstrated several telephone devices at the Smithsonian Institution. In 1878, he received a patent for a transformer. Berliner joined the Bell Company and later founded Deutsche Grammophon and Gramophone Co., Ltd. F. Barraud's painting of a dog listening to a gramophone (trademarked "His Master's Voice") became the popular "Nipper and the Gramophone" trademark registered May 1900 by Berliner. RCA adopted the popular symbol, which is still recognized more than 100 years later. Berliner also founded the Esther Berliner fellowship to support women pursuing scientific research. See Gramophone.

Berners-Lee, Tim (1955–) A British physicist and programmer, Berners-Lee gained a spot in the history books with his Web project proposal introduced in March 1989, and his demonstration of World Wide Web software in the winter of 1989. The rapid acceptance and growth of the Web is a tribute to the viability of this concept. Prior to that, Berners-Lee developed Enquire, while at CERN in 1980, a hypertext system that no doubt formed the seed for his Web project. In 1994, he joined the Laboratory for Computer Science at MIT. He has won many awards of distinction for his work, including the 1998 MacArthur Fellowship. He is the coauthor, with Mark Fischetti, of *Weaving the Web*, a book about the origins and development of the World Wide Web. See World Wide Web.

Bernoulli, Daniel (1700–1782) Bernoulli was a Swiss mathematician born in the Netherlands, who pioneered the principles of fluid dynamics, now used broadly in aeronautics, electronics, and other fields. The Bernoulli principle is derived from his writings in *Hydrodynamica* which described basic properties of fluid pressure, density, and flow.

Bernoulli box An electronic, high-speed data mass storage and retrieval device based upon technology pioneered by Daniel Bernoulli. In general, the *Bernoulli effect* occurs as pressure acts upon a fluid medium in relation to the volume of the fluid. These forces will equal changes in kinetic energy associated with the fluid (whether liquid or gaseous). In general, as fluid flow slows, pressure increases, and vice versa. As applied to storage technology, increased rotational speed in a magnetic disk creates a cushion of air that controls the distance of the read/write head from the storage medium.

Daniel Bernoulli

One of the earlier removable computer storage media, called Bernoulli drives, were named after Daniel Bernoulli, a mathematician.

Bernoulli-Euler law In a homogenous bar, the curvature of its central fiber is proportional to the bending movement. It is a general concept, applicable to many fields including elastic theory and mechanical engineering.

Bernoulli's Theorem in a Field of Flow At every point in a steadily flowing fluid, the sum of the pressure head, the velocity head, and the height is constant.

Berry, Clifford Edward (1918–1963) An intelligent and mechanically gifted American who collaborated with J. V. Atanasoff in the development of one of the world's first digital computers. Atanasoff and Berry began working together on the project soon after Berry completed his Bachelor's degree in electrical engineering. The completion of a prototype computer led to a small grant to build a working version, in December 1939, while Berry continued his graduate studies. Berry then subsequently worked in corporate positions and applied for more than 40 patents in the

areas of spectrometry and electronics, 30 of which had been granted before his untimely death. See Atanasoff-Berry computer.

BERT bit error rate test/tester; block error rate test/tester. A diagnostic device that is used to test data integrity by transmitting a known pattern of bits and evaluating the subsequent bit error rate (BER), usually on a cable segment. See bit error rate.

BERTS See Basic Exchange Radio Telecommunications Service.

bespoken, bespoke Custom-made, made to order, made by engagement, requested item.

Bessel beam A nondiffracting optical beam, a recent technology that began showing some exciting practical results by the end of the 1990s. The potential for infinitely propagating beams for manipulating particles or transmitting information are exciting new fields of experimentation.

By 2001, U.K. researchers had demonstrated Bessel beam "tweezers" for manipulating, stacking, and aligning a variety of silica and biological structures. The linear momentum of light and its interaction with matter were exploited to trap particles. They also demonstrated the use of a laser beam as an optical particle guide, moving 1 µm spheres upward within the microscope slide medium within which the particles were held. In addition to applications in biology and medical imaging, this technology may have significant practical applications in optical fiber-based telecommunications.

Bessel beams can be generated with a glass element called an axicon. See acousto-optic deflector, axicon, laser.

Best Current Practice BCP. A process similar to the Internet Standards process, in that specifications are submitted to the IESG for review, but streamlined to provide industry leaders with a more flexible, and often quicker, consensual alternative to the Standards Track specifications for resolving individual policy and operations issues. See Internet Standards process.

Best Effort See available bit rate.

Best Effort Capability A capability offered on some ATM networks that tries to provide transmission but provides no guarantees of throughput. Might be used between two routers, for example. See ATM traffic descriptor, RFC 1633.

beta 1. In electronics, the current gain of a bipolar transistor in a grounded-emitter amplifier. 2. A version of software that is mostly complete, has been in-house tested, but requires wider input and trials from testers and users outside the company. See beta test. 3. (*symb.* – B) quartz. 4. (*symb.* – β) The second letter in the Greek alphabet, sometimes used to denote a specific angle in a geometric diagram.

BETA Business Equipment Trade Association. A representative for many of the large hardware manufacturers in the computing industry.

beta site A location or group of people designated to test and use a piece of nearly completed, internally tested software in working conditions more nearly like those in which the software will eventually be used.

beta testing The in-house process of testing and using a nearly complete software or hardware product to try to determine if there are still bugs or problems of usability, consistency, continuity, and ergonomics. Beta testing can take months or years, depending upon the state of readiness and the complexity of the software. Some developers use automated *monkeys*, programs that systematically climb through the software, to identify bugs or flow problems. It is a good process to use in conjunction with human testing.

In the author's experience, as much as 85% of software may be commercially introduced without sufficient beta testing. Because computing is confusing to the average user, users are hesitant to complain, thinking the fault is in their use of the product rather than the product itself, and sometimes this is true. But, upper managers often insist the product has to ship (whether it's ready or not) in order to generate revenues to stay in business even though, in most cases, it's bad economy. The cost of testing and correcting bugs before the product ships is almost always lower than the cost of repeated patches, upgrades, technical support, and loss of business due to consumer dissatisfaction when firefighting and corrections are done after the product ships.

If cars were sold with the same number of defects as many software products, consumer rights organizations would boycott the manufacturers. The author of this dictionary has contributed many hundreds of hours to beta testing and once found more than 300 bugs in two days of testing in a software product that the manufacturer insisted was "complete, ready to ship, and absolutely bug-free." There are responsible software houses that engage in extensive testing and quality control, and their efforts should be recognized and rewarded. For those that don't, caveat emptor or get your money back. See alpha testing, gamma testing, upgrade, user acceptance testing.

BETRS See Basic Exchange Telecommunications Radio Service.

beyond visual range BVR. Something that is outside human sight, or in some contexts, out of sight of human vision with binoculars. In a very general sense, it can mean something distant or obstructed. In telecommunications contexts it is more often used to indicate objects, communications means, or antennas that require 'line of sight' distances or unimpeded pathways to be effective. See line of sight.

bezel The rim or edge of a tool or piece of equipment, often angled or sloped. On a computer monitor, the housing edge around the cathode-ray tube.

BFA See Brocade Fabric Aware program.

BFI 1. See Bad Frame Indicator. 2. See British Film Institute. 3. Buckminster Fuller Institute.

BFO See beat frequency oscillator.

BFOC, bayonet FOC bayonet fiber optic connector. A quick-connect device without a screw thread.

BFT See binary file transfer.

BG, BGND background.

BGP 1. See Border Gateway Protocol. 2. See Byzantine Generals problem.

BHLI See Broadband High Layer Information.

bias 1. Expected or consistent deviation, inclination of outlook. 2. Deviation from expected value, systematic error. 3. In an electron tube, the fixed voltage that is applied between the control grid and the cathode.

bias distortion Inconsistencies or aberrations in the linearity of a signal. In finely tuned equipment, bias distortion is usually an undesirable property.

bias stabilization A means of controlling the bias in a circuit so that it does not fluctuate. Heat or signal variations can throw off bias, resulting in damage to components. See bias, bias distortion.

biasing To apply a small amount of positive or negative stimulus to a circuit, as in an electron tube, to shift it in one direction or the other.

BIB See backward indicator bit.

BIBO bounded input, bounded output. Input and/or output falling between specified values or other boundaries. A concept used in linear mathematics theory and calculations and in data networking analysis of traffic flow. While many models for network traffic are based on unlimited or scalable queues, for the purposes of modeling, testing, or design, it is often practical to establish bounds for input and output.

BICEP See bit-interleaved parity.

biconical antenna A balanced broadband antenna which resembles a bowtie in the sense that it has two metal cones mounted in the same axis, that meet at the narrow ends where the feed line is attached. The orientation of the assembly affects its polarity. It is suitable for transmissions in the VHF range.

BICSI A not-for-profit international telecommunications association headquartered in Tampa, Florida. There are regional offices in Australia, Brazil, and the U.K.

BICSI provides educational resources, technical publications, and support for cabling distribution design and installation. It was originally established in the early 1970s as the Building Industry Consulting Services, International, but is now formally known as *BICSI: A Telecommunications Association.* http://www.bicsi.org/

BID bridge identification code. See bridge.

BIDDS See Base Information Digital Distribution System.

BiDi bidirectional. 1. Capable of communicating in two directions either alternately or simultaneously. 2. Oriented or pointing in two directions, as directional antenna components with two main receiving or transmitting elements.

bidirectional reflectance distribution function BRDF. A function describing light reflectance from a surface at a given orientation from a source of illumination incident from a given direction. See Lambert's law.

bidirectional line-switched ring BLSR. A fault-tolerant topology for SONET that overcomes some of the problems associated with breaks in basic point-to-point ring topologies. In most ring topologies a secondary ring is in place, in case of a failure in the primary ring. In local area networks, this is a practical solution, but on long-haul phone networks, for example, it involves the costly installation of a secondary cable that is rarely used. In BLSR, if a failure occurs, the bidirectional portion of the link is engaged, and the traffic is routed in the opposite direction around the section that has failed. In fiber optic cables, this is a very fast transition and wouldn't be noticed by endusers under most conditions.

BIDS Broadband Infrastructures for Digital TV and Multimedia Services. An ACTS project to provide a comprehensive analysis of future broadband digital television and interactive multimedia services for European users. BIDS has established a database of information gathered from interactive digital TV trials and has analyzed a number of case studies. See BBL, BLISS, BONAPARTE, BOURBON, BROADBANDLOOP, BTI, UPGRADE, WOTAN.

bifurcated routing A routing technique that splits data traffic so that it continues through multiple routes (technically it would be two routes, as *bi*furcated means split into two branches).

BIG See broadband integrated gateway.

big-endian Stored or transmitted data in that the most significant bit or byte precedes the least significant bit or byte. Many file incompatibilities between computer systems, in which the file formats are otherwise almost identical, are due to platform conventions about whether the data is stored in big-endian or little-endian form.

BIGA Bus Interface Gate Array. Technology built into Cisco Catalyst systems to receive and transmit frames from packet-switching memory to its MAC local buffer memory external to the host processor.

BII base information infrastructure. The communications foundation for military establishments.

bilateral antenna An antenna whose maximum transmitting or receiving poles are diametrically opposite, that is 180° apart in a plane.

Bildshirmtext [*transl.* picture screen text] A German interactive videotext system from the German Bundespost. It is similar to the French Telecom Minitel service, except that the German Bundespost did not provide the terminal free. See Minitel.

billboard array antenna An antenna array that resembles a billboard in that it uses a large sheet metal reflector behind the stacked bipole arrays.

Billing Account Number BAN. An identifier that enables telephone carriers to bill individual customers or each of multiple accounts belonging to the same customer.

Billing Telephone Number, Billed Telephone Number BTN. In some situations, the telephone number billed may be one of several associated numbers but, for simplicity, all the calls are billed to one. This system is sometimes done with extension numbers. In other situations, the main number used may be different from the number to which the calls are billed, again, usually to simplify accounting or billing statements.

billion In North America and France, one thousand million ($10^9 - 1,000,000,000$). In the U.K. and parts of Europe, one million million ($10^{12} - 1,000,000,000,000$). (It used to be a huge number.)

BINAC Binary Automatic Computer. A joint project of J. Mauchly and J. P. Eckert who founded the Eckert-Mauchly Computer Corporation. The BINAC was developed under contract with the Northrop Aircraft Corporation, unveiled in 1949, and was historically significant not only as a successor to ENIAC but for its ability to store data programmed using C-10 code on magnetic tapes rather than on paper tape or punch cards. See ENIAC; Hopper, Grace Murray; UNIVAC.

binaries, binary files Files that have been compiled or assembled into machine-readable codes, usually 32-bit executable files, that are inscrutable to most human beings. Source code is higher level code (as in BASIC, C, or Perl) that can more easily be read and modified by a programmer. Binary files can be edited directly with a hexadecimal (base 16) editor. See attachment.

binary, base 2 A system of numeric concepts and numerals representing quantities in terms of ones and zeros with the smaller units on the right. Thus, two in the base 10 decimal system is written as "2." The same quantity expressed in binary is "10" or "00000010" with the one in the 'twos' position, second from the right. The columns from right to left are thought of as the "ones column," "twos column," "fours column," "eights column," etc. so that a digit in a specific column indicates the presence or absence of that amount. Thus, the following numeral in binary: 0011010 can be transposed to decimal by adding its values: $0+0+16+8+0+2+0 = 26$.

In electronics, binary values can be variously represented by pulses of unequal length, by amplitudes of specified magnitudes, by power on or off conditions, or by different tones.

Because most computers are two-state systems, the binary number system is used for programming and storage of data. Thus, zero and one can represent states such as *on* or *off*, *yes* or *no*, etc.

binary asymmetric channel BAC. A concept used in information theory related to Markov channels. As an example, images can be modeled as binary asymmetric Markov sources for transmission over communication channels. In mathematical descriptions of the capacity of physical or theoretic structures, the concept of BAC is useful for surface area calculations.

binary coded decimal BCD. 1. A system wherein each decimal digit is coded into a four-bit word. 2. A system wherein each octet within an ATM cell has each bit set to one of two allowable states, i.e., one or zero. 3. A system of coding high and low power transmissions. For example, BCD is used by the National Institute of Standards and Technology (NIST) to represent decimal numbers in order to disseminate time code information. At the start of each second of the 60-kHz broadcast, the carrier power is reduced 10 dB, putting the leading edge of each negative-going pulse on time. To create a binary zero (0), full power is restored 0.2 seconds later. Alternately, to create a binary one (1), full power is restored 0.5 seconds later. Position markers are signaled by restoring full power 0.8 seconds later.

binary file transfer BFT. Binary files are those that have been translated into a base 2 system to be more readily used by a computer. Binary files cannot be readily transported over 7-bit systems that typically use ASCII (or EBCDIC on some older systems) encoding unless they are encoded. Due to the encoding, binary files cannot be directly read by (most) humans or directly edited by most text editors, although a hexadecimal editor is sometimes used to make limited changes to binary files.

Binary file transfers are usually accomplished by (1) encoding the file into ASCII with a utility such as BinHex, (2) transmitting the file, and (3) re-encoding it at the destination. Most email clients now automatically convert binary file attachments. Multipurpose Internet Mail Extension (MIME) is one of the protocols used with mail clients to handle transparently the encoding and decoding of attached files.

binary phase-shift keying BPSK. A type of linear modulation in which the phase of a constant amplitude carrier signal represents two values through 180° shift reversals. When there is no phase change, a value of zero (0) is represented, while a phase change relative to the preceding wave period represents a value of one (1).

This basic modulation scheme is effective for amateur high-frequency radio transmissions and LowFER- and MedFER-band operations. BPSK is used in the SLOWBPSK program developed by Pawel Jalocha. The system was updated as PSK31 (phase-shift keying 31 baud) by Peter Martinez to function over a narrow 160-Hz phase-shift mode.

binary signaling Signaling based upon two states, whether it be digital or analog. Binary signaling is a common form of modulation with a variety of implementations, including arbitrary binary signaling, synchronous binary signaling, antipodal binary signaling (spread spectrum, Manchester), binary noncoherent signaling, binary orthogonal signaling (codewords placed at orthogonal axes). See binary phase shift keying.

binaural Related to two sound sources or two sound receiving sources, as human ears. Since humans are accustomed to using two sound sources to distinguish the quality and directionality of sound, monaural music tends to sound somewhat flat. Thus, stereo (binaural) sound systems have evolved to provide a more natural representation of sound.

BIND See Berkeley Internet Name Domain.

bind triangle In an International Business Machines (IBM) SNA implementation, a session setup message sequence.

binding post In electrical installations, a screw terminal with a corresponding nut around which U-shaped lugs or wrapped wires can be wound and secured with the nut. Sometimes there are two nuts, close together so the wire can be secured between the two screws. Binding posts tend to be used in temporary circuits, or in small installations. In medium- and large-scale telephone installations, mounting blocks and punchdown tools are much faster.

BinHex A very useful software archiving/translation tool that can be used to convert an 8-bit binary file into a 7-bit ASCII file through run-length encoding, so that it can be handled by 7-bit systems that may use different protocols but understand ASCII. Many email clients use BinHex internally to handle binary attachments to text messages. At the receiving end, the file must be converted back to its original form before it can be executed or otherwise used as originally intended. BinHex is a very widely used application on many platforms but is especially prevalent in Unix and Macintosh environments.

Binary Phase-Shift Keying

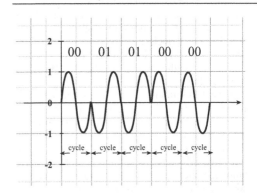

A wave can be altered in a variety of ways to represent information. By using two wave phases, with one of the phases shifted by half compared to the preceding reference wave, it is possible for the linear modulation of a wave to represent two values, zero (0) and one (1) which, in turn, can be transmitted in sequences of patterns to represent data values (e.g., ASCII characters). Note that the length of the period and the amplitude remain the same, only the phase is changed.

Biocomputing Office Protocol BOP. An Internet transaction protocol for transmitting command line and file data, somewhat analogous to SMTP-POP, but designed specifically to send command-line plus data input file block from the client to the server where it is analyzed and passed back to the client upon request. BOP was developed as an easier way for nontechnical professionals to access biocomputing resources. The server designed to implement BOP requests is called *bopper*, with *SeqPup* used as the initial client. BOP can be used with anonymous or password access and provides both deferred and interactive processing modes. See bopper.

biometric An objective measure or representation of a biological attribute, which may be a physical characteristic or the output of certain physical characteristics (e.g., handwriting). In technology applications, biometrics that are specific to an individual, such as fingerprints, iris or retina patterns, brain waves, and voice prints, are of interest for developing access, login, and authentication systems.

Fiber optic faceplates are being developed for use as biometric light-guiding surfaces in conjunction with detectors/sensors. This has good potential for 3D biometric imaging, remote biometric sensing, and larger sensing/scanning areas. See faceplate.

BIOS basic input/output system. A system in read only memory (ROM) on some Intel-based desktop computers, that supports the central processing unit (CPU) by supplying access to a variety of input/output devices, such as serial ports, joysticks, monitors, keyboards, etc. As these peripherals are basic to the functioning of the computer, they are frequently used and loaded from ROM into RAM for fast access as a system comes online during the powerup sequence.

Biot, Jean-Baptiste (1774–1862) A French professor of mathematics and astronomy, Biot collaborated with Arago in studying the refractive properties of gases, and discovered, in collaboration with Felix Savart, fundamental relationships in electromagnetic theory. The Royal Society awarded him the Rumford Medal for his work in the chemical polarization of light. A crater on the moon is named after Biot. See Biot-Savart law.

Biot-Savart law In electromagnetics, the magnetic field produced by a current flowing through a conductor may be described as a vector product inversely related to the distance of a point in the magnetic field to the source current. It is similar to Coulomb's law for electrical relationships. Using integrals, the computation may be applied to various arrangements of conducting sources by breaking the system down into smaller components. See Ampère's law, Coulomb's law, Gauss's law, right-hand rule.

BIP See bit interleaved parity.

biphase coding A networking bipolar coding scheme in which clocking information is carried in the synchronous data stream without separate clocking leads.

biphase-shift keying BPSK. A simple type of modulation scheme used in digital satellite transmissions. In BPSK, each phase of the carrier wave is shifted once with each complete cycle, with a shift indicating the change of the value (from one to zero or zero to one). See binary phase-shift keying.

bipolar 1. Having two mutually opposing or repelling forces, characteristics, or viewpoints. 2. Having two poles. 3. A circuit with both positive and negative polarity or alternating between positive and negative polarity. 4. In electronics, a structure prevalent in integrated circuits (ICs). 5. A device having both majority and minority carriers. 6. Having electromagnetic characteristics alternating between two poles. 7. A type of signaling in digital transmissions in which a binary value represents a signal amplitude of either polarity, and no value represents zero amplitude.

bipolar receiver A type of telephone receiver used extensively in the Bell System. It improved on earlier technology by using new magnetic alloys and employing a different acoustical system for the diaphragm. See ring-armature receiver.

bipolar signal A signal with two nonzero polarities; it can represent two states or three states in a binary coding scheme. See bipolar.

bipolar transistor A semiconductor commonly used in oscillators, switches, and amplifiers.

birdie 1. Twittering, squealing, or whistling noise, often high-pitched. Birdie is a descriptive term for auditory interference associated with electrical circuits. In older analog phone circuits, overloading sometimes caused crosstalk (conversations bleeding into one another) or, more commonly, birdies. In amateur radio systems, birdies may result from radio frequency (RF) leakage from nearby devices. In studios with both amateur radio and computer equipment close together, it may be necessary to power down everything to a skeleton system and gradually add them back in to locate sources of birdies. Monitors, cables, and unshielded devices are common culprits; even a computer keyboard can cause birdies. In circuit boards, birdies may result from improper grounding. Homemade or commercial radio frequency (RF) sniffers or more sophisticated spectrum analyzers can help locate sources of radio frequency leakage. Proper shielding and grounding, line filters, and toroids can help reduce birdies. If the birdies cannot be easily eliminated, it may be necessary to note the frequencies at which they occur and work around them. 2. A lightweight cable or wire installation accessory device. Once the conduit has been installed for a wiring/cabling installation, a birdie, attached to the wire by a long lead, can be blown with a compressed air tool so that it "flies" through the conduit, with the wire subsequently pulled through using the birdie as a lead. See pulling eye, snake.

birefringent A material with a molecular structure patterned the same along two axes but differently along the third. In a light-admitting material, this anisotropic structure will influence properties such as the index of refraction, which will vary depending upon the angle of incidence of any light that encounters the substance.

Observe a pebble in a bowl of water and one pebble will be visible, a little offset from its actual location. The offset is due to the *index of refraction* of the water compared to the adjacent air. If you look at a pebble through a translucent birefringent material, depending upon the angle, you may see two pebbles because birefringent materials are doubly refractive. Calcium carbonate ($CaCO_3$) is an example of a common natural substance with birefringent properties. See anisotropic, Iceland spar, index of refraction, refraction.

birefringent filter A mechanism for filtering wavelengths using intrinsic birefringent properties of materials used in the filtering component.

Birmington Wire Gauge A gauge standard for describing the diameter of iron wires (nonferrous wires are described with American Wire Gauge). The thinner the wire, the higher the number from 1 to 20, exclusive of the coating. See American Wire Gauge.

bis Second, update, revision, encore. In the V Series Recommendations of the ITU-T related to telecommunications, *bis* indicates a second version or update to a previously numbered standard. This was probably substituted for a revision number to prevent confusion between the series number and revision level. Similarly, *ter* designates three, or third.

BIS 1. See Bank for International Settlements. 2. border immediate system.

BISDN See B-ISDN.

BISSI Broadband Inter-Switching System Interface.

BIST built-in self-test. Testing capabilities included with a system. Many consumer laser printers have test modes that will create a printout detailing the operating parameters, settings, status, and problems, if there are any that can be reported on paper. Some of the newer printers also have LED screens or Ethernet links so test results can be reported on a built-in monitor or connected computer system.

Bisync (*pron.* bye-sink) Binary Synchronous Communication Protocol. A character-oriented serial network protocol that was developed in the 1960s, at a time when IBM dominated the network market. It is now mainly supported as a legacy protocol.

bisynchronous transmission A transmission that can flow in two directions on the same line or channel, usually at the same time. Traditional wireline telephones are bisynchronous, whereas some types of radios or intercoms transmit only in one direction, or in one direction at a time.

bit binary digit. A basic unit of digital information with two (bi-) states. Many schemes for signaling binary states have been developed: on/off (early telegraphs), high/low, one or zero (mark or space, data bits), black/white, dot/dash, etc.

bit-interleaved parity BICEP. In ATM networks, an error-monitoring method implemented at the physical (PHY) layer. The link overhead contains a check bit or word for the previous frame to flag errors.

bit-oriented Data communications that can encode control information in single-bit data units.

bit-oriented protocol BOP. In general, a network control protocol functioning at the data link layer. There are variations on bit-oriented protocols typically used for synchronous transmissions, including Synchronous Data Link Control (SDLC), Advanced Data Communication Control Procedures (ADCCP), and High Level Data Link Control (HDLC). See Transparent Bit-Oriented Protocol.

bit bucket *slang* A mythical container into which unwanted or unused code, email, chad from punch cards, or other computer information is discarded or lost. Information may also be deliberately discarded by sending it to the null device (dev/null) bit bucket. See chad, leaky bucket.

bit error A fault condition in which the value of an individual bit is changed by transmission or data interpretation errors.

bit error rate BER. A measure of transmission quality, usually expressed as a ratio of *error bits* to *total bits received*. A high bit error rate does not necessarily result in a faulty transmission. Error-detecting and correcting algorithms are incorporated into most current transmissions protocols. However, a high BER may result in slower transmissions, smaller packets, a higher percentage of retries, and perhaps even the necessity to connect several times to complete a file transfer, for example.

bit interleaved parity BIP. In ATM networking, a method used at the physical layer to monitor the error performance of the link. A *check bit* or *check word* is sent in the link overhead covering the previous block or frame. Bit errors in the payload will be detected and may be reported as maintenance information. Common implementations of error levels of BIP include BIP-8 (up to 8 errors), BIP-16, and BIP-24. In ATM trunking, when the BFrame is created on ingress, a BIP is generated and remains until the cell is extracted on egress of a switch. Cells may be dropped if a BFrame parity error is detected. Payload data may be BIP-checked separately and does not necessarily result in dropped cells.

In SONET implementations, distinctions are made between section (BIP-B1), line (BIP-B2), and path (BIP-B3) overhead, depending upon which part of the path is specified. Layers are hierarchical in SONET such that if section parity is correct, the layers beneath it should also be correct.

BIP is described more fully in Bellcore documents and in ANSI T1.105. See coding violation.

bit line BL. A concept in array-based memory technologies to designate the location of a specific bit, often in conjunction with a word line (WL). The word and bit information may be combined into a binary address to indicate the row and column of an array element, especially in a 2D array.

bit pipe 1. A generic descriptor for the physical or data transmission line of a digital circuit. A wider pipe is considered to have more capacity than a narrow pipe or pipeline. A bit pipe need not connect separate devices or systems; a pipe may be established between two processes on one device. 2. A telephone circuit used to transmit digital data packets.

bit robbing A process of commandeering bits in a transmission for something other than their usual purpose. Extra bits may be robbed to convey signaling information, especially if the signals are only occasionally needed. See robbed-bit signaling.

bit stuffing See zero bit insertion.

bitmap 1. A point-by-point digital encoding of graphics data for transmission, storage, or display. The displayed image does not necessarily reflect the format of the image file. For example, a vector file may be represented as a bitmap or pixmap image on a printer or computer monitor, or a pixmap file may be converted to a continuous tone image when printed to a dye sublimation printer. 2. A pixelated image. An image (picture or font) represented by discrete dots on a monochrome display, which is typically white, green, or amber (technically a multicolor image is called a *pixmap* rather than a bitmap). Grayscale images, with varying degrees of intensity on a monitor or varying sizes of monochromatic dots on a printed page, are often referred to as bitmap images, although they may be closer to pixmaps. A bit in computer data does not map directly to a point on the output device (in fact, several bits are usually needed to encode one image point). In the context of a bitmap, the term bit is used in its lay meaning to indicate a small amount, a section, or an individual point of the displayed image. See raster, pixels, vector.

BITNET, BITNET-NJE Because It's Time Network. An international, cooperative, academic network established in 1981 by Ira H. Fuchs (City University of New York) and Greydon Freeman (Yale University). It began as a cooperative project at the City University of New York, with Yale as the first outside connection through a leased telephone line. It ran originally on IBM mainframes and Digital VAXes communicating through EBCDIC formats. From there, it spread across the U.S. and became international when it was joined by the European Academic and Research Network (EARN) in 1982. A grant from IBM in 1984 helped establish support services for BITNET in the U.S.

BITNET promoted the noncommercial exchange of research and education information and was organized as a nonprofit corporation in 1987. By 1991, BITNET included almost 1500 organizations in 49 countries and for a while was the world's largest academic network. Participation declined thereafter due to the rapid growth of the Internet and BITNET's inherent interactivity limitations. In the late 1980s, it was merged with The Computer+Science Network (CSNET) to form the Corporation for Research and Educational Networking (CREN). In the end, CREN recommended to its members that BITNET dependency be terminated by December 31, 1996 in favor of other network systems, primarily the Internet. BITNET is based on an IBM communications protocol called Network Job Entry (NJE), which made it practical to connect mainframe computers through telephone circuits. It uses a *store-and-forward* system of transmitting information through nodes on the system. See Network Job Entry, RELAY, UUCP.

BITNIC BITNET Network Information Center. A support center for administering BITNET computer networks initially established with the aid of funding from IBM in 1984. After 1987, funding was member-based and volunteer-supported. See BITNET.

BITS See Base Information Transport System.

bits per second bps. A very common means of describing data transmission per second unit of time. A megabit per second, or Mbps, represents a million bits per second. Common consumer modems operate at data rates of about 9600 to 28,800 bits per second. T1, fiber lines, and other higher speed protocols and media can transmit at much higher rates.

BIU See basic information unit.

BL 1. bilateral. Having two sides. 2. See bit line.

black In politics, a designation for secret and/or classified information or activities. The designation has significant impact on telecommunications in a number of ways. Black operations may be used to tap into communications systems to eavesdrop on conversations or data transfers. Computing systems designated for black operations are typically equipped with special encryption systems and code-creating keyboards, wheels, algorithms and other means of encoding messages or data.

Fiber optic connections are favored over electrical connections for secured cable installations because it is harder to tap into fiber optic connections without detection. Wired connections can be tapped with sensitive instruments that sense emanations from the wire without necessarily touching the wire. A fiber optic transmission doesn't emit electromagnetic radiation in the same way as electrical connections. The only practical ways to tap into a fiber optic connection are (1) to bend the filament to cause the light to escape the reflective cladding layer (in which case the loss of power of the light beams could be detected beyond the point of the tap) or (2) to insert a clandestine tap segment in the link, which would involve temporary disconnection of the existing link (which may trigger an alarm), and would require higher technical expertise and much more precise components than a typical copper wire connection. See encryption.

black body A theoretical body which absorbs all incident light with no reflection and consequently appears black (without light) at all wavelengths.

Black Box A registered trademark of The Black Box Corporation of Pittsburgh, PA.

black box 1. *colloq.* A device whose internal workings are obscure or obscured. That is, the outside may not indicate what is inside, or how it works. 2. A device that is used by a lay person without technical knowledge of its construction or functioning. 3. A type of clandestine phone interface device used in a central office to gain unauthorized access to phone services by obscuring the fact that a long-distance call had been answered. See blue box, red box.

black box design A design model for inputs and outputs which function independently of the various ways the internal components might be configured. For example, a converter or transformer for matching two types of signals might be specified, with leeway given to a manufacturer as to the best way to implement and build the hardware itself.

black hole 1. A theorized invisible (thus perceived as dark) region in space with a small diameter in relation to its intense gravitational field. A black hole could perhaps result from the collapse of a massive star, in which the escape velocity equals the speed of light. 2. *colloq.* A fictional area into which things disappear when they can't be found, and those looking for them are sure they should be "right there." 3. In networks and computer systems in general, a point in the transmission link where data went in and apparently didn't come out. Black hole also refers to a metaphorical repository for lost data. Disappearing into a black hole may also be jargonistically described as disappearing into the ether, into the bitstream, or into the bit bucket.

black level A reference level on a display device corresponding to the lowest possible luminance setting, which typically appears as black (the absence of illumination), or nearly black, depending upon the characteristics of the display device.

black matrix tube A cathode-ray tube in which black fills the spaces between color phosphors on the inside front coating of the tube. The greater contrast between the lit phosphors and the small surrounding area results in a picture that appears to have crisper, brighter colors.

black recording In recording systems using amplitude modulation, black recording is the correlation between the maximum power of the transmission and the maximum density of the recording device. In recording systems employing frequency modulation, black recording is the correlation of the lowest frequency received and the maximum density of the recording medium. The phrase applies to various wired and wireless facsimile machines, printers, electronic photocopiers, etc. See black transmission.

black transmission, AM In an amplitude-modulated (AM) image transmission, black means that the greatest divergence in amplitude in the signal represents the black tones, and the narrowest divergence represents the lightest tones (or no tone at all). In white transmission, the opposite is true.

black transmission, FM In a frequency-modulated (FM) transmission, a black transmission means that the lowest frequency corresponds to black, and the highest frequency corresponds to white or no tone; in a white transmission the opposite relationship is used. Black transmission concepts in general can be applied to image scanners, facsimile machines, photocopiers, etc. See black recording

blackjack When Samuel Morse won a contract from the U.S. Congress in the 1800s to build a telegraph line from Washington, DC to Baltimore, MD, he initially tried to install the lines underground, alongside railroad tracks. There were problems with the line, however, and the wires were subsequently suspended from poles. To this day, millions of miles of communications cables are installed on utility poles. Since the poles were subject to weathering, they were coated with creosote, a preservative derived from coal tar that came to be called blackjack. It is still used to prevent dry rot and insect infestations.

Blake, Francis, Jr. (1850–1913) An American inventor, physicist, and photographer who became appointed to the U.S. Coast Survey in 1866, Blake was talented in mathematics and became skilled in telegraphy, astronomy, and hydrography. Upon leaving the Coast Survey in 1878, he pursued his creative ideas and invented a better telephone transmitter. He subsequently patented numerous other inventions. Emile Berliner was later to make practical improvements to some of the microphone technology developed by Blake. A number of Blake's papers are archived by the Massachusetts Historical Society, Boston. See Berliner, Emil; Blake telephone; Blake transmitter; telephone history.

Blake telephone A historic magneto telephone, known as the Blake Transmitter, which became the first standard telephone installed by the Bell Telephone Company of Canada. It incorporated a magneto generator, which was cranked to ring the central switching office, and a better quality transmitter invented by Francis Blake, Jr. A wet battery provided power to the system. The quality of the system was important to furthering Bell's success as a telephone company. See Blake transmitter.

Blake transmitter A pioneering telephone transmitter designed by Francis Blake, Jr., in 1878 so the diaphragm

could vary the strength of an already established current from a battery, rather than generating energy by means of electromagnetic induction, as in earlier models, thus producing a stronger sound. He received three related patents for carbon transmitting technologies in 1881.

blank 1. A transmissions gap, one in which no signal or data is coming through. 2. A space or nonprinted area on paper. 3. A spacer used to format a blank area in HTML documents for display on the Web. It may be created with (1) a blank image file, (2) a <PRE> (preformatted) tag, or (3) an   (nonbreaking space) tag. 4. An advancing key on a teletype, typewriter, typesetter, label-maker, or other device such that an unprinted area is established. The word "space" is often used interchangeably with blank.

blanking In a display device, such as a cathode-ray tube (CRT), the interval during which part or all of the display is suppressed. Blanking is used to suppress artifacts from the display sweep of an electron gun. The sweeping of a screen is often a repeated zigzag, but the part of the transmission intended to be seen is displayed on the straight sweep (usually the horizontal sweep), and the beam drops to the next line as it moves back to the other side, ready to sweep again (horizontal blanking). It is somewhat like the line feed and carriage return on a typewriter; as the carriage sets the typing position back and down to the beginning of the next line, it shouldn't make marks on the page. See blanking pulse.

blanking interval, blanking time The period during which a display is suppressed, usually to enable an electron gun to return to the next display position. See blanking, cathode-ray tube, frame, sweep.

blanking pulse A mechanism for suppressing a display, usually on a cathode-ray tube. It is sometimes accomplished by means of a positive or negative square wave. A series of pulses can be combined to create a blanking signal that is synchronized with the sweep.

blazed, blazing 1. Characterized by a somewhat periodic bright illumination, that is, sudden or undulating moments of greater light often associated with heat, such as a forest fire, torch, or firing line of muskets. 2. Illumination associated with sustained, intense heat, such as a well-stoked furnace or the desert sun. 3. Having been marked with a short slash, slit, or cut as an identifier, often as part of a group or series, as a line of blazed trees indicating a path. 4. A fabrication with a regular, periodic "sawtooth" structure that facilitates the filtering or concentration of energy through controlled diffraction, as in a blazed grating. See blazed grating.

blaze angle In a blazed grating, the incline of the individual "sawtooths" in relation to the mean grating surface. The angle is designed to reflect radiant energy of a desired wavelength in controlled ways to serve as a filtering mechanism. When "holographically" recorded and etched in an interference grating, the blaze angle can be established within certain tolerances but is not as precise as a machined grating due to the sinusoidal shape of the peaks and troughs in a photographically etched grating. Consistency of the blaze angles of individual facets is usually desirable. See Brewster's angle, incidence angle, Littrow configuration.

Blaze Angle in Fiber Grating

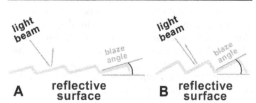

A reflective surface   **B** reflective surface

Blazing creates a sawtooth design in a grating that selectively reflects wavelengths, thus acting as a concentrator or filter. The blaze angle can be designed to control the angle of incidence of the reflected light beam, thus enabling the grating to be "tuned" to certain frequencies. In the asymmetric sawtooth configuration, the direction of the light source input is important for efficient exploitation of the angle of refraction.

---

blaze condition In a semiconductor component, a configuration in which a corrugated blazing structure is exhibiting diffractive control over incoming wavelengths such that they are selectively filtered or passed through the structure. This is, in part, a function of the shape of the blazed grating along with the layers associated with the grating. See blazed, blazed grating.

blazed grating A corrugated selectively reflective surface in which the grooves are regularly asymmetric. The shape of the pattern of grooves or corrugations is sometimes called a *sawtooth*. In this type of grating, the shape of the teeth and their orientation have higher significance than the grating period (although very fine teeth will effect diffraction efficiency). Blazed gratings can be mechanically or photographically fabricated but are generally mechanically ruled, since the sinusoidal characteristic of a photographic interference grating doesn't lend itself to blazing. The Hubble telescope utilizes a blazed grating. See blaze angle, échelle grating, grating, ruled grating.

blaze wavelength For a given blazed grating and its associated layered components, in a given index diffraction order, the wavelength for which the relationship between diffraction intensity and wavelength is at its maximum efficiency. See Maxwell's equations.

BLEU Belgo-Luxembourg Economic Union. A cooperative arrangement between Belgium and Luxembourg, established in 1921, to support currency bases and legal tender between the two countries.

blind transfer, cold transfer The transfer of a call without seeking the identity of, or receiving information about, the caller.

blind zone A zone where there are no transmission signals. A skip zone is one type of blind zone. See zone of silence.

BLISS Broadband Lightwave Sources and System. An ACTS project to bring to maturity key aspects of photonic networks and to demonstrate their practical implementation. Specific concerns include detailed studies of optical crosstalk and dynamic range. OEIC receiver chips are studied through PON and ATM ring solutions with verification and comparison to commercial systems. Some key components include photonic ICs, pulse sources, a variety of types of lasers, semiconductors, and other relevant devices and technologies. Trials were planned for a universal interface for different traffic types and physical media (copper, optical fiber, etc.) to provide ATM access. BLISS components are being used in other ACTS projects. See BROADBANDLOOP, UPGRADE, and WOTAN.

Bloch, Felix (1905–1983) A Swiss mathematician and physicist, Bloch was associated with a remarkable who's who list of the most renowned physicists of the early 20th century. He did graduate work under quantum theorist Werner Heisenberg in Leipzig, Germany, where his graduate thesis, published in 1928, made an important contribution to the knowledge of electron conduction. In Zürich, Switzerland, where Bloch worked for Wolfgang Pauli, he made his first unsuccessful attempts to formulate a theory of superconductivity but got a start in understanding different ways to approach the problem. He then traveled to the Lorentz Foundation, where he studied theories of electric conductivity, and then studied ferromagnetism at the University of Leipzig. Bloch worked with the highly regarded Niels Bohr as a Fellow of the Ørsted Foundation. While at the Institute in Copenhagen, he described theoretical relationships between classical and quantum theory.

Fermi encouraged Bloch to consider both theory and practice and, in 1934, Bloch moved thousands of miles to Stanford University where he taught theoretical physics and gave seminars with Robert Oppenheimer.

In 1936, Bloch sought to create a neutron source for the research of particle physics. Together with Luis Alvarez, he began work with U.S. Berkeley's cyclotron to determine the magnetic moment of the neutron. Thus, he resolved to build a cyclotron at Stanford, a project initiated in 1939. After a brief tenure at the Los Alamos Manhattan Project, Bloch worked on theoretical studies in radar and conducted microwave reflectivity research at the Harvard Radio Research Laboratory. Returning to Stanford in 1945, he had radio equipment built and cooperated with Purcell in dividing up the research. This led to progress in nuclear magnetic resonance (NMR).

Bloch and Purcell were awarded the 1952 Nobel Prize in physics. Bloch was one of the Stanford members who encouraged the Atomic Energy Commission to support construction of the Stanford Linear Accelerator Center (SLAC). In his later years he returned to his country of birth. See Hansen, William.

Bloch's theorem of superconductivity The lowest state of a quantum mechanical system, in the absence of a magnetic field, can carry no current.

block check character BCC. An error-checking technique developed for early teletypewriters in which a control character is appended to blocks for longitudinal checking and CRC. In packet networking, as a packet is assembled, the data is processed to create a BCC, which is then incorporated into the packet, checked at the receiving end, and acknowledged (ACK) or not acknowledged (NACK) if it does not match. The data can then be resent until the BCC matches or until the process is stopped. BCC is used in a variety of implementations, including most polled protocols.

block cypher A type of encryption algorithm that breaks plaintext data into segments (usually of a fixed size) and uses the same encryption key to transform each segment into a segment of ciphertext. See Blowfish, SKIPJACK.

block diagram A type of visual communications aid that uses simple shapes to symbolize objects, functions, relationships, conditions, and processes. A flow diagram or flow chart is a type of block diagram in which specific shapes have been standardized to have certain meanings within the context of the diagram. Rectangles, diamonds, and arrows are commonly used.

block transfer The process of moving data in a block, instead of in individual bits. Double buffering, in which a screen of information or block of data is built in the background and then instantly presented or displayed by transferring it from one area of memory to another, is a type of block transfer commonly used to reduce screen display delays. Various file transfer protocols make use of block transfer techniques, often reducing the size of the block if many errors are occurring.

blocking 1. Preventing entry/exit or transmission through. 2. Holding until time or space is available, as in a queue, or until the data, person, or object can be turned back. 3. A circumstance in which a call cannot be completed (the exchange may be overloaded or the line busy). See grade of service, call abandons. 4. Deliberate exclusion of certain parties from certain numbers (such as prevention of 900 calls, long-distance calls, etc.) 5. In business, an illegal practice preventing others from engaging in fair competition. 6. In vacuum tubes, creating very high negative grid bias to lower the plate current to zero.

blocking capacitor, blocking condenser A device in a circuit that blocks direct current (DC) while permitting alternating current (AC) to pass through.

blocking probability A performance measure describing the likelihood of data, or of a user, being rejected.

bloom On a cathode-ray tube (CRT) display device, the tendency of a phosphor excitation level to create a 'halo' effect of extra light that spreads out beyond the area being targeted. This tends to happen at higher intensity levels with lighter colors.

blooper 1. Goof, embarrassing error, bungle. 2. In transmissions through a regenerative relay, an unwanted signal created by the relay that is not part of the desired transmitted communication.

blower 1. *colloq.* A device to blow a current of air or gas, e.g., for installing air-blown fiber. See blown fiber. 2. A speaker or blowhorn.

Blowfish A 64-bit (8 bytes) encryption algorithm developed by Bruce Schneier, Blowfish has become the basis for a number of encryption schemes, including Kent Briggs' Puffer, Harvey Parisien's VGP, and Philip Zimmermann's PGP. Blowfish uses a variable-length key up to 448 bits in length. There may be restrictions on sales outside the U.S., due to Federal export restrictions. See Pretty Good Privacy.

blown fiber, air-blown fiber ABF. A fiber optic installation system designed by British Telecommunications PLC that enables faster, more flexible installation and reconfiguration of fiber optic cable systems by literally blowing the fiber lines into a grouped tube cable hose. Thus, existing or newly installed conduit can be fitted with fiber optic lines.

This system is often combined with point-to-point modular connectors that eliminate splicing. Since splicing is an exacting job in fiber optic installations, modular connectors are a great convenience.

blown fuse A fuse with a broken connection, due to some electrical abnormality on the circuit on which it is installed, which might have endangered other links in the system. Fuses typically cannot be reused and must be replaced with another with the appropriate voltage. A blown fuse will sometimes show a blackened area inside the glass. Circuit breakers have superseded fuses in many types of electrical wiring, but the phrase has remained and is often used to indicate a tripped circuit breaker. See circuit breaker, fuse.

BLS The U.S. Bureau of Labor Statistics. http://www.bls.gov/

BLSR See bidirectional line-switched ring.

BLU basic link unit. A generic term used in a variety of networks, refering to a basic transmission unit of control and data information.

Blue Book standard 1. A document published by Digital Equipment Corporation (DEC), Xerox Corporation, and Intel Corporation in 1980 to provide information on the Ethernet protocol standard, version 1. 2. The Blue Book or CD Plus standards are a subset of the Orange Book standards originally based upon the Red Book and Yellow Book digital audio and computer data optical recording formats. Blue Book is a special case of the Orange Book standards in which multisession data is recorded in two blocks with one session devoted to recording music and one devoted to recording digital data. Blue Book formats are supported by a number of major audio and computer data vendors. Blue Book discs can be played on audio CD players and on more recent CD-ROM drives. Besides their simplicity, the Blue Book standards served another purpose in preventing CD players from misreading the type of data on a disc (Orange Book) and trying to create audio from computer data and possibly even damaging the equipment (if you've heard a modem screech over a phone line, you get the general idea). See Orange.

blue box *colloq.* A small handheld device designed

to emit tones in the same frequencies as touchtone telephones, often used in the 1980s for connecting long-distance calls illegally through direct tones rather than dialing. Typically a connection was established through normal means, usually through a toll free 800 number and then the blue box was used to disconnect the remote ringing, without actually disengaging from the long-distance connection. A new connection could then be established within about a 10 second window, by punching in appropriate operator tones from a keypad on the blue box. (Some individuals have even learned to reproduce some of these tones by whistling, without having to use a blue box.) Newer systems can move these tones out of band or use more sophisticated monitoring and tracing of suspected connections to reduce the possibility of abuse.

Blue boxing probably originated in the very early 1960s, and the Bell System first apprehended a blue box user in 1961. The myth that blue boxing is done almost entirely by young college students is refuted by a report by AT&T that almost half of those caught stealing phone services with blue boxes are businessmen, many of them wealthy, along with a number of doctors and lawyers.

blue gun In a color cathode-ray tube (CRT) using a red-green-blue (RGB) system, the electron gun specifically aimed to excite the blue phosphors on the inside coated surface of the front of the tube. Sometimes a shadow mask is used to increase the precision of this process, so the green and red phosphors are not affected, resulting in a crisper color image. See shadow mask.

blue pages A convention in telephone directories in which government listings are printed on pages with a blue background to distinguish them from residential and business listings. Online directories of government information and email addresses are now sometimes called blue page listings.

blue wire A color designation used by IBM to indicate patch wires used to correct design or fabrication errors in situations where it is not practical to recreate the board with the corrections. See purple wire, red wire, yellow wire.

Bluetooth Project A combined effort of Ericsson, IBM, Intel, Nokia, and Toshiba, formed in 1998 as a special interest group, to develop a vision and path for a single, universal, low-cost wireless communications system that allows easy access from a wide variety of wireless consumer devices. See Service Discovery Protocol.

Bluetooth Service Discovery Protocol See Service Discovery Protocol.

BLV/BLI An operator call wherein the caller requests information about the busy status of a line or requests an interruption of a call on an Exchange Service.

BM 1. See benchmark. 2. See burst modem.

BMEWS Ballistic Missile Early Warning System. A U.S. Government long-range warning and tracking radar network designed to detect missile fire along the northern approaches.

BMP, .bmp A file extension standing for *bitmap*, often expressed as a three-character file name extension to maintain backward compatibility with operating systems that can't use a longer file extension. Technically, bitmap files are monochrome raster graphics files. While .bmp is used by some as a generic file extension name for any type of raster graphics file, bitmap also has a specific meaning for a standardized file format. See bitmap, raster.

BN 1. See background noise. 2. See border node. 3. See bridge number.

BNC 1. See bayonet nut connector. 2. See British National Corpus.

BNCC Base Network Control Center. The main administrative central facility for network operations within an organization or location. See Network Operations Center.

BO 1. body odor. See skunkworks. 2. See branch office.

board See printed circuit board.

Boardwatch A good prosumer-level print and Web publication dealing specifically with the telecommunications industry, particularly the Internet. http://www.boardwatch.com/

bobtail curtain antenna A phased-array, bidirectional, vertically-polarized wire antenna, intended for high-frequency transmitting and receiving.

BOC See Bell Operating Company.

body The main informational portion of a communication, sometimes sandwiched as a block between headers and trailers. Sometimes called the *payload*. In a picture file, the body is the portion that carries the object or raster information about the image. In a word processed document, the body is the portion that contains the informational text and accompanying illustrations. Contrast with header.

body, type In typography, the main portion of the shapes that constitute a character set (typestyle). The portion from which ascenders and descenders originate. Sometimes called x-height.

body area network BAN. A network based upon a body-worn communications network device. It is usually wireless, to enable mobility, but may also be wired if the user is stationary (usually seated at a computer, telephone, or games terminal). The trend in BANs is to incorporate them into clothing or body-worn harnesses to distribute the weight and decrease their visibility.

Boggs, David R. Along with Robert Metcalfe, the co-developer and co-builder of the first Ethernet systems in 1973 at Xerox PARC. Metcalfe and Boggs authored a frequently cited article "Ethernet: Distributed packet switching for local computer networks" in *Communications of the ACM* in July 1976. See Ethernet; Metcalfe, Robert.

Bohr's correspondence principle In an atomic system, the behavior of the electrons must increasingly approach that predicted by classical physics the higher the quantum number of the orbit.

bolometer A detection instrument for measuring the intensity of radiant energy through a thermal-sensitive resistor, a type of actinometer. Bolometers may be assembled in arrays.

Boltzmann constant (*symb.* – k) The ratio of the universal gas constant, R, to the Avogadro constant, Na. Named after Ludwig Boltzmann.

Boltzmann, Ludwig (1844–1906) An Austrian experimenter who built on the ideas of James Clerk-Maxwell, studying electromagnetism, thermodynamics, and statistical mechanics. Boltzmann demonstrated a number of Maxwell's predictions, confirming them, and published his results in 1875. The Ludwig Boltzmann Institute for Urban Ethology in Vienna is named after him.

BOM 1. BASIC operations monitor. A monitoring, debugging tool for BASIC programs. 2. beginning of medium. A phrase applied to the start of a tape or other serial storage media. 3. See Beginning of Message.

BONAPARTE Broadband Optical Network using ATM PON Access Facilities in Realistic Telecommunications Environments. An ACTS project to demonstrate the viability of broadband ATM PON as a cost-effective communications system and to demonstrate interoperability between ATM IBC island through the Pan European ATM Network. See BROADBANDLOOP, BOURBON.

bond *n.* 1. To join, adhere, or unite into a combined unit or system. Bond usually implies a semipermanent or permanent adherence, as opposed to wrapping a wire, which would not be considered a bond. A bond is often accomplished with a *bonding agent* such as glue, weld, or solder. See fusion splicing.

bond, electrical To form an electrical connection by joining two conductive surfaces, usually metal, to provide a low-resistance path for the circuit. In electronics, wires are often bonded to a small metallic pad on a circuit board. See bonding.

bonding 1. Joining two or more items with adhesive, weld or solder. In PC boards, there may be a bonding pad on the board or on a chip for the express purpose of providing sufficient space and electrical contact for a potential bond (usually solder). 2. An inverse multiplexing specification described by BONDING. See BONDING.

BONDING Bandwidth On Demand Interoperability Group. A set of protocols, known as the BONDING specification, developed by a consortium of data communications consultants and suppliers. BONDING arose from efforts to create a standardized inverse multiplexing protocol in order to improve interoperability among multiplexers from various vendors.

The BONDING specification describes a number of modes of interoperability for switched networks, so a sideband signal can be subdivided into multiple 56 Kbps or 64 Kbps channels, and recombined at the receiving end.

bong A tone transmitted through a phone line to indicate to the listener that additional information is required. The information is usually entered through a touchtone key pad or by speaking clearly.

Boole, George (1815–1864) An English-born mathematician, son of a maid and a shoemaker, Boole set up a school at the age of only 19. He taught himself mathematics and began publishing his ideas, introducing Invariant Theory. His 1854 publication "The Laws of Thought" introduced mathematical concepts applicable to computing operations and earned him the sobriquet of "father of symbolic logic." Boolean logic is named after him.

Boolean expression A type of expression often used in programming to control binary relational operations that may be executed or may express true or false. Boolean algebra in a broader sense in set theory involves the intersection and union of sets and elements of sets. It also provides a practical means for implementing logic in digital computers. Boolean algebra works readily on binary computing systems.

boom 1. Vertical spar, beam, pole, or suspended piping. 2. In video, a vertical bar, rod, or other support for microphones, cameras, or other equipment that need to be suspended over or near a source with a minimum of visual obstruction. 3. The horizontal supporting rods for many common antennas from which there may be secondary protruberances to increase transmission or reception.

boom pole A long pole with a spike on the end used by crews of *boomers* (telephone line installers) to guide a long telephone pole into a deep hole. A boom pole is sometimes called a pike pole.

boom truck roller See stringing roller.

boomer *colloq.* Telephone line installer. The name is derived from the boom poles installers used to hand guide telephone poles into their holes before machinery for this job became prevalent.

boot *abbrev., v.* To start, to power up, to get a machine going, to 'kick' something into operation. Derived from *bootstrap*, which is further derived from the phrase "pulling yourself up by your bootstraps." This term aptly describes how a computer has to launch its basic lower processes so it can recognize its own hardware and capabilities in order to further launch the higher level processes. See bootstrap.

boot ROM A read-only computer memory chip usually located on the motherboard as an essential part of a basic system. This chip provides the minimum necessary information for bringing the computer hardware online and may include diagnostic routines that test systems before bringing the whole system up. In simple terminals, the boot ROM may include all basic operating software needed or, as is the case on most self-contained desktop systems, it may include only the essentials and will seek a floppy diskette, hard drive, or other boot information for further instructions and parameters for launching the operating system, device drivers, and sometimes user applications.

On many Intel-based desktop computers, the information for accessing devices may be transferred to the BIOS during system startup. See BIOS.

BOOTP See Bootstrap Protocol.

bootstrap In a computing system, to bring up basic hardware and software systems in stages that are partially or wholly dependent upon the success of previous stages. For example, to bootstrap a computer from a power-off state, low-level hardware and software systems are brought online to the point where self tests can be performed and devices recognized.

These basic systems are then used to "pull the system up by its bootstraps" to the next level of operating system capabilities for processing input from the user, network configurations, and basic applications parameters.

Bringing a system online from a power-off state is called a *cold boot*. A *warm boot* is a *reset* from a power-on state during which the system typically re-reads the *boot ROM* and restores basic operating parameters without powering off the system or rerunning the low-level self-tests and device intitialization operations. Stable operating systems rarely crash or hang, but there are some microcomputer operating systems that do crash, and a cold boot is sometimes the only way to bring the system back into full operating mode. See device drivers.

Bootstrap Protocol BOOTP. An Internet Protocol/User Datagram Protocol (IP/UDP) client/server protocol for storing and providing configuration information for a network.

BOOTP evolved in the ARPANET days to enable diskless client machines and other machines that might not know their own Internet addresses to discover the IP address, the address of a server host, and the name of a file to be loaded into memory and executed. It is accomplished in two phases: address determination and bootfile selection; and file transfer, typically with TFTP.

BOOTP has since evolved into Dynamic Host Configuration Protocol (DHCP). See Address Resolution Protocol, Dynamic Host Configuration Protocol, Reverse Address Resolution Protocol, RFC 951.

BOP 1. beginning of packet. 2. See Biocomputing Office Protocol. 3. See bit-oriented protocol.

bopper A Biocomputing Office Protocol (BOP) server developed by Don Gilbert, based on *popper*, a Post Office Protocol server. Bopper provides biocomputing services to BOP-compatible clients. It was initially released for Solaris2 in June 1996.

Border Gateway Protocol BGP. An interdomain gateway routing protocol which is superseding Exterior Gateway Protocol (EGP). BGP is used on the Internet. See Exterior Gateway Protocol, RFC 1163, RFC 1267, RFC 1268.

Border Gateway Protocol Version 4 BGP4. A version of BGP which uses route aggregation to reduce the size of routing tables, and which supports Classless Inter-Domain Routing (CIDR).

border node In ATM networking, a logical node in a specified peer group that has at least one link that crosses the peer group boundary.

Border Node A means for establishing connections between networks of distinct topologies while limiting the flow of topology data across subnetwork boundaries. Thus, subnets with different NETIDs or defined clusters will have subnetwork boundaries. Border Node was defined originally as Peripheral Border Node (PBN) with a later release as Extended Border Node (EBN), described in 1997 by International Business Machines (IBM) for Advanced Peer-to-Peer Networks (APPNs).

EBN was developed to enable connectivity of multiple subnets. It facilitates interoperability, topology isolation, subnet partitioning, route calculation, optional security, optional exit access controls, and other functions.

Interior Border Node (IBN) and HPR border node (HBN) are subsets of EBN. IBN supports intermediate network routing, usually on the same APPN, but does not support APPN Interchange Node or SSE(CP) functions. HBN supports cross subnet path switching, ANR routing, and end-to-end routing.

BORSCHT An acronym used in the telephone industry to aid in remembering the components of a subscriber line interface (SLI).

B - battery (power source)
O - overvoltage protection
R - ringing
S - signaling and signaling detection
C - codec (analog/digital conversions)
H - hybrid (two-/four-wire conversions)
T - test access.

Boston Computer Society BCS. Formally, one of the largest computer user groups in the world, with a membership of over 25,000 at its peak, the Society served users of a variety of types of computer platforms. Jonathan Rotenberg, who was 13 at the time it was founded, ca. 1977, is credited with starting the organization. Despite a large and enthusiastic membership, the Society officially ceased to operate in September 1996.

bot, 'bot A term frequently used on the Internet to describe software robots that manage tasks on behalf of users and operators, especially in Internet Relay Chat (IRC) channels. Since IRC is an interactive social medium, these software programs have frequently been given personalities by their respective programmers and thus take on anthropomorphic characteristics not usually attributed to applications programs, hence the term bot instead of application. See avatar, robot.

bottleneck A point in a system, device, or transmission link that slows the rate of communication below the expected efficiency or below the capabilities of other links in the system. For example, a computer with a CPU capable of 64-bit processing may be impeded by a 32- or 16-bit data bus. As another example, you may have a fast serial card and an ISP with a T1 line, but your 9600 baud modem creates a bottleneck, limiting the upper speed of the transmission of data. Bottlenecks may be a constant limitation of a system or may be a limitation occurring only during times of peak traffic.

Bouguer, Pierre (1698–1758) A French mathematician, inventor, and author, Bouguer carried out measurements in astronomical photometry in the 1720s. Beginning in 1727, he was a multiple winner of the grand prize of the Académie Royale des Sciences. In 1748, he invented photometric and heliophotometric instruments.

Bouguer's significant 1729 essay on optics describes the relationahip between the absorption of radiant energy and the associated absorbing medium, now known as Bouguer's law. *Traité d'Optique sur la Gradation de la Lumière,* his treatise on photometry, was posthumously published as a first edition in 1760. In addition to his observations of absorption properties of radiant energy in atmospheric optics, it describes a number of types of photometers, including a method of goniophotometry. See Bouguer's law.

Bouguer's law A description of the relationship between an absorbing medium and the radiant energy absorbed in terms of the ratio of the transmitted and incident radiant energy intensity to the mass of the absorbing medium.

Bouguer studied illumination on two surfaces from light sources of the same kind. One was set at a fixed distance from the illuminated surface but had an absorptive material interposed between the light source and the illuminated surface, the other was set up the same way, but without the absorptive materials and with the light source set at varying distances from the illuminated surface. The intensity of the illumination on the first surface would vary depending upon the thickness of the intervening materials and the intensity of the illumination on the second would vary according to distance. By visually assessing the intensity of the two illuminated surfaces as the thickness or distance variables were altered and matched, Bouguer found that the relationships between the two could be perceived and quanitified. It's not a long stretch to realize that the intervening materials could also be swapped and the experiment performed again for a material of different composition (e.g., different translucency).

In contemporary applications of the concept, the relationship is usually calculated with respect to a specific wavelength with temperature and pressure held constant. It provides information on absorbancy characteristics or, seen another way, transparency. Bouguer's law is also known as Beer's law or, when the concepts of absorption in proportion to a concentrate and the thickness of the intervening materials are combined, it is called Beer-Lambert's law. See Lambert's law.

boule In fiber optics fabrication, a sooty, layered coating that builds up inside a supporting tube through a chemical deposition process. The boule is then further *sintered* to remove impurities and collapse the boule into a clear cylinder *preform*. The preform, which is typically composed of silica glass, can then be drawn out into a long fiber filament. See preform, vapor deposition.

bounce 1. To rebound, to come back, to deflect off of, to echo. The ionosphere is used to bounce radio signals over long distances. 2. In electronic transmissions, if data doesn't reach its intended destination and is routed back to the sender, it is said to have "bounced." This may happen when email is sent to an address that no longer exists, for example.

bounce, broadcast 1. In broadcast transmissions, if a signal hits a physical impediment, it may bounce, sometimes causing a zone in which there is interference in the transmission or no transmission at all. In other instances, the physical characteristics of the Earth and the ionosphere and the position of repeaters or satellites may be used to selectively bounce a signal in order to direct it. See ionosphere, Moon bounce. 2. In visual media such as television broadcast displays, bounce is an undesirable and unexpected variation in the brightness of the image.

BOURBON Broadband Urban Rural Based Open Networks. An ACTS Project building on a previous RACE project which studies issues of providing cost-effective, scalable access to ATM-based networking services in Europe and the broader Information Society. The project focuses both on users and technologies and involves the cooperation of Member States of the European Union. ATM and ISDN test beds are established in several countries. See BBL, BONAPARTE, BROADBANDLOOP, UPGRADE, WOTAN.

Bourseul, Charles (1829–1912) A Belgian-born French researcher who described, but apparently never followed up, a means of transmitting speech electrically through wires. His ideas were published in *L'Illustration de Paris* in 1854. See Meucci, Antonio; Gauthey, Dom; telephone history.

Bower-Barff process A process in which metal (iron or steel) is heated to red heat and then treated with superheated steam in order to reduce vulnerability to corrosion.

Boyle, Robert (1627–1691) A British physicist and chemist who developed pumps that could create near vacuums. Boyle subsequently observed that sound required a medium for its transmission. He also did numerous experiments on atmospheric pressure and discovered an important relationship between gas and pressure in 1662. In 1675, he published a treatise on electricity and observed that the attractive properties of amber did not require the presence of air. Boyle's law is named after him. See barometer; Boyle's law; Hauksbee, Francis.

Boyle's law, Marriotte's law At a constant temperature, the volume of a definite mass of gas is inversely proportional to the pressure such that the product of the volume (PV) is constant.

BP 1. bandpass. 2. base pointer. 3. beam position. 4. bypass.

BPAD Bisynchronous Packet Assembler/Disassembler. The BPAD Protocol is a transport protocol associated with X.25 networking.

BPDU See Bridge Protocol Data Unit.

BPI See bytes per inch.

BPON See Broadband Passive Optical Network.

BPM See beam position monitor.

BPS See bits per second.

BPSK See binary phase-shift keying.

BR 1. beacon receiver. 2. Bureau of Radiocommunications.

Bragg angle In the context of Bragg's law, the angle between the lattice plane and the incident X-ray beams (commonly expressed as theta – θ). See Bragg's law.

Bragg reflector A technology used in diode lasers that allows very fine control over the focus of the beam. A Bragg reflector is also called a *grating* *reflector*, due to the corrugated ridges used to direct the beam that change along their lengths. Bragg reflectors are being researched as a means of increasing throughput of data transmissions in existing cable installations. By finer focusing of the beams and multiplexing, capacity may be improved on fiber channels. See Agility Communications, Bragg grating, diffraction, quantum cascade laser.

Bragg grating A grate-like pattern that is "written" into a fiber during fabrication to modify the characteristics of the basic fiber filament to reflect wavelengths selectively. The performance of the grating may also be improved by straining the fiber medium at the time the grating is written.

Bragg gratings have provided significant advancements in waveguide control in fiber optics communications circuits. There are many types of grating design (and research continues) that enable the reduction of noise and delay, through filtering mechanisms that may be tunable and incorporated into optical waveguides. Gratings can facilitate channel filtering and gain equalization.

Research at the MIT NanoStructures Laboratory has resulted in new Bragg grating designs and fabrication techniques for lithographically "etching" the grating into the medium. For example, the lab has shown that a quarter-wave shift in the grating, to isolate a single wavelength channel in a multiwavelength system, can provide optical resonating functions, similar to that of a Fabry-Perot cavity. Thus, add/drop channel filtering capabilities can be built right into the fiber facilitating the development of all-optical transmission paths. Through electrical circuit modeling, multiple resonators can be cascaded to enable more complex functions. See add/drop multiplexer, Alexandrite, diffraction, fiber grating.

Bragg spectrometer A form of spectrometer useful in studying X-ray diffraction characteristics based upon the discoveries and observations of W.H. and W.L. Bragg in the early 1900s. X-rays are generated, filtered, and collimated (aligned into a fine beam) and aimed to strike a crystal surface at a specified angle. The rays reflected from the crystal are intercepted by a detector so that their characteristics may be studied and recorded. See Bragg's law.

Bragg's law, Bragg's relation A diffraction effect expressed mathematically as $nl = 2d \sin q$ by W. Lawrence Bragg in 1913 to describe the angles of incidence associated with X-ray reflections that occur when parallel rays encounter crystal structures (obstacles). Thus, the wavelength of an incident beam times a positive integer (sometimes expressed as an index $- m$) is equal to two times the distance between the atomic layers in the crystal $\sin q$. Depending upon which factors are known and substituted into the equation, diffraction angle, crystal plane separation, or the wavelength can be algebraically calculated.

The relational expression was based upon collaborative research with Lawrence Bragg's father, W.H. Bragg. The Braggs' observations were significant not only for their practical applications, but for providing evidence supporting theories about the periodic atomic structure of crystals. Bragg's law and the study of diffraction have since been applied to many other theoretical and practical fields of study beyond X-rays and crystals. See Bragg grating, Bragg reflector, Compton scattering.

Bragg, William Henry (1862–1942) A British physicist who studied X-rays and ionizing radiation and, in collaboration with his son, X-ray diffraction and its interaction with crystalline lattice structure. This latter research won the father/son team a Nobel Prize in physics, in 1915. See Bragg spectrometer.

Bragg, William Lawrence (1890–1871) An Australian-born British physicist who studied at Cambridge and became a lecturer there. In the early 1900s he collaborated with his father, W.H. Bragg, in the study of X-ray diffraction and crystal structures, an effort that jointly won them a Nobel Prize in physics in 1915. In 1915, they published *X-rays and Crystal Structure*.

In 1938, Lawrence Bragg became head of the Cavendish Laboratory at Cambridge. From 1953 to 1961, he served as director of the London Royal Institution.

braid A fibrous or filamentous, long, tubular intricately woven structure usually of plastic or fine metal that forms a covering over a conductive or insulating core in a layered cable.

Brainerd, Paul Brainerd founded Aldus Corporation in 1984, the year after the introduction of the Apple Lisa computer and the year before the release of the Apple LaserWriter printer. Aldus specialized in graphics applications, particularly for vector drawing and desktop publishing. Macintosh computers and Aldus software quickly became favorites with print industry service bureaus. The Aldus Corporation was one of the few developers that created some really good, quick, intuitive user interfaces. Good interface design is a rare talent in the software development industry. Aldus PageMaker and Aldus Freehand, developed by the Aldus Corporation, were acquired by Adobe Systems and Macromedia.

brainiac Probably originating from Edmund Berkeley's computing devices from the 1950s, this term refers to someone with good technical and/or mathematical intelligence of the kind that is not common. See Brainiac.

Brainiac Brain-Imitating, Almost-Automatic Computer. It is essentially the electromechanical GENIAC computer designed by Edmund C. Berkeley and Oliver Garfield, in the 1950s. There were disputes and a lawsuit between Berkeley and Garfield subsequent to which Garfield promoted his calculating technology under the name GENIAC, and Berkeley continued to promote computing devices under the name Brainiac. In writings on the Geniac/Brainiac technology, Berkeley described Brainiac computing experiments, in 1957 and 1958, and began discussing Brainiac's relationship to GENIACs and automatic computers, in 1958. In the late 1950s, Berkeley took steps towards exporting Brainiac abroad, as well. See Berkeley, Edmund; GENIAC; Simon.

branch 1. A junction point from which there is more

than one path along which to continue. 2. An instruction in a computer program which, when evaluated, can lead to a different destination for execution of the next step, depending upon the condition. 3. A substation, subsidiary office, or other facility which is a satellite of, or auxiliary to, the main operations.

branch circuit In a wiring installation, a separate circuit that, if damaged or tripped, doesn't affect the other branch circuits. This divides the power so the main circuit is not overloaded. On a circuit breaker panel, the branch circuit is a constellation of appliances and sockets wired to a particular breaker.

branch feeder In an electrical distribution system, a cable that connects the main cable and the subscriber distribution system, as between a phone switching center's main cable and a business distribution closet.

branch office BO. Subsidiary office (in the sense that a tree branch is subsidiary to a tree trunk) separate from the head office. There may be multiple branch offices. In large distributed computer networks, branch offices are often established to provide routing or switching services, customer services (including installation, maintenance, and repairs), and local marketing, billing, and tax procedures.

branching 1. Dividing, splitting into two or more paths or sections. 2. A hierarchical structure often used for database creation, search, and retrieval. 3. Branching electrical distribution systems for electrical installations and data networks.

branching filter 1. A device for separating or combining separate frequencies when used in conjunction with a guiding structure for the wave. 2. In computer networking, a software utility for selectively routing data into several paths or files based on specified characteristics.

Branly detector A device created in 1890 by Édouard Branly, consisting of a small, glass, metal-filled tube with a short wire inserted to make contact with the metal filings. When connected between a power source and a meter, current didn't pass through the glass unless a spark was discharged. The spark caused the filings to cohere and thus act as a conductor. This on/off quality of the Branly detector was very useful to the development of radio.

Branly, Édouard Eugène Désiré (1844–1940) A French inventor who devised the Branly detector in the late 1800s, a cohering device that contributed to the development of radio or, as it was then, wireless telegraphy. His technology was subsequently adapted by G. Marconi. Branly also investigated the transmission of nerve impulses. See Branly detector, coherer.

BRAS broadband remote access server.

Braun, Karl Ferdinand (1850–1918) A German researcher who discovered in the 1870s that certain minerals had a property of one-way conductivity of radiant energy; they could function, in a sense, as one-way gates. This discovery was an important early contribution to electronic circuitry that provided a transition from coherers to crystal detectors.

Braun invented the cathode-ray indicator tube or *Braun tube* in 1897, a significant device in the evolution of electronics. Braun's attitude towards science

was similar to Benjamin Franklin's. While Franklin was a shrewd and successful businessman, he also had a strong inclination to share knowledge that he felt would benefit humankind. Like Franklin, Braun published descriptions of his earlier discoveries rather than keeping them secret. However, given the importance of his discoveries, he subsequently patented his tuning transmitters in Britain, starting in 1899, technology that may have influenced Marconi's tuning patent of 1900.

A year after his invention of the CRT, Braun was hired to provide guidance on an underwater wireless telegraphy project that needed improvements. By rearranging the main components of the circuits and altering the coupling, Braun was able to greatly extend the range of the system. His employer and backers formed the Telebraun company which evolved into the well-known Telefunken.

Braun was awarded a Nobel Prize in Physics in 1909, along with G. Marconi, for his contributions to wireless telegraphy. See cathode-ray tube; crystal detector; Murgas, Josef.

BRCS Business and Residence Customer Service.

breadboard A board with numerous attachment points, often in a grid, that permits the prototyping of circuits. Breadboards often resemble a nest of colored worms, as they are frequently hand-wired with a lot of crisscrossing conductors with temporary attachments. Breadboards are handy for concept design, testing, teaching, temporary circuits, and convincing the boss that you have a good idea that will work. See proof of concept.

break Willful or inadvertent interrupting or stopping of a process, transmission, or broadcast. On computer terminals, a break can be sent in many instances with Ctrl-C or Esc, depending upon the software.

break in 1. *v.* Interrupt, or take control of, a circuit or process. This break may be from human or systems intervention or through an automated system. See Barge In, buttinsky. 2. *v.* Gain illegal entry to a system. See back door, hacking, Trojan horse.

Break key A specialized key included on some computer keyboards that permits a one-keystroke interruption of the current task, assuming the software supports, and correctly interprets, the input from the keystroke. Break keys are included since some of the common ways to interrupt tasks involve combination keystrokes, such as Ctrl-C, and hitting one key is easier, especially for less experienced computer users. See break.

break out box See breakout box.

breakdown potential, breakdown strength Dielectric strength, the maximum voltage that can be tolerated without breakdown.

breakdown voltage 1. The voltage at which an insulator or dielectric breaks, or at which ionization and conduction occurs in a gaseous environment. 2. The voltage that needs to be applied in a device to jump a gap (in air).

breaker 1. In electrical installations, a point in a circuit, usually a junction installed in series between the main electrical source and a branch circuit, in which excessive voltage is detected and the circuit tripped or broken in order to prevent overload, electrical fires or damage to appliances or the main panel. 2. In radio communications, anyone who drops in on a channel to communicate when others are already engaged in conversation. 3. In public online forums, a person who breaks into an ongoing conversation or thread with irrelevant or unkind and unwanted comments.

breaking strength 1. An industrial measure of the force needed to break a specific structure or material (or a combination of the two). Used in structural and safety design and selection and installation of appropriate wires and cables.

breakout An exit point for electrical conductors (wires, cables, etc.) along the length of the circuit, between the endpoints of the circuit. Breakouts can be used for additional installations or testing and diagnosis of the circuit. Breakouts are usually covered or capped in some way to prevent interference with the circuit and shock or fire hazards.

breakout box A diagnostic instrument used to tap into an existing circuit to evaluate its functioning, to see whether individual lines within a group are correctly connected, and, in some cases, to detect which signals are being transmitted. Breakout boxes often have indicators such as tones or light-emitting diodes (LEDs) to help assess the line and may include jumpers to temporarily switch connections. One useful application of a breakout box is to test a serial circuit, since computer equipment manufacturers don't consistently follow the RS-232 or RS-423 specifications. In the case of a temporary need for a null modem cable (e.g., for transferring information from one computer to another through telecommunications software), a breakout box can be used to cross the *transmit* and *receive* lines. See breakout.

Brewster's angle In the context of an electromagnetic wave encountering a dielectric material, there is a particular angle, called Brewster's angle, at which the polarization effect of the interaction is at a maximum, whether this is the polarization and reflection of unpolarized radiant energy (e.g., white light from the Sun) or the "cancellation" of energy through absorption, if it is already polarized parallel to the dielectric surface before contact.

Different materials refract light to different extents. Much of David Brewster's research centered on studying the different optical properties of crystals, including their reflective and refractive properties. The specific angle at which Brewsterian effects occur was found to be related to the *refractive index* of the dielectric material encountered by the propagating *incident wave* and the polarization and other characteristics of the incident radiant wave. Brewster's angle will be between $0° – 90°$ since no polarization occurs if the radiant energy hits the dielectric at an angle perpendicular to the material ($0°$ – straight on) or travels parallel to the dielectric's contact surface ($90°$ – thus not being reflected).

If the incident light is unpolarized when it encounters the dielectric, the reflected light will be polarized to different degrees, depending upon the angle, with Brewster's angle being the relationship where polarization is at its maximum. If the incident light is already polarized in a direction parallel to the dielectric surface, it will not be reflected.

Brewster's Angle Applied to Light

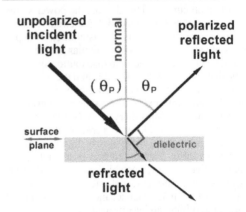

There is a particular angle at which unpolarized light, when encountering a dielectric surface, reflects as maximally polarized, called Brewster's angle, *expressed as* θ_P. *The angle of reflection, relative to the dielectric surface normal, is complementary to the angle of the refracted light, relative to normal. Conversely, if light that is already polarized (in a direction parallel to the surface plane) is incident to the surface, the polarizing effect of the dielectric cancels out the polarized incident light and no polarized light is reflected.*

Brewster's angle is consistent for incident light and dielectric materials with the same characteristics, but varies with the wavelength and index of refraction of the dielectric with which the incident light interacts.

Brewster's law With respect to electromagnetic energy encountering a dielectric material, the relationship between the angle of propagation of the reflected energy relative to the perpendicular of the surface (normal), and the refractive index of the dielectric material, can be described in an equation derived by Brewster, known as *Brewster's law*

$$n = \frac{\sin\theta_P}{\cos\theta_P} = \tan\theta_P$$

In this equation, θ_P (the angle at which the polarization effect for the reflected energy is at its maximum for a given material and wavelength) is *Brewster's angle*. The index of refraction of the dielectric material is expressed as *n*. Thus, the refractive index is equal to the tangent of the angle of the reflected energy (relative to surface normal). This information is useful for designing components that are intended to maximally polarize unpolarized radiant energy or to not reflect already-polarized energy (polarized parallel to the surface plane). The accompanying diagrams help clarify these relationships.

Examples of applications that exploit Brewster's law

and Brewster's angle include photographic filters, sunglasses, radio signal conditioners, fiber optic couplers, photographic films, and a wide variety of components for scientific instruments such as microscopes.

Brewster's law has broad and consistent applications in optics that are very useful but may sometimes be a hindrance. Novel ways to overcome some of its effects have been recently developed. In the 1990s, it was found that highly birefringent, layered, polymer film could be assembled in alternating layers to control light so that the net effect was to avoid polarization of the reflected light that is characterized by Brewster's angle. This led to the development of layered film components capable of reflecting light uniformly, which has many potential applications in communications, astronomy, and medical imaging. See acceptance angle, blaze angle, Brewster's angle, dichroic, incidence angle, Malus's law, Snell's law.

Brewster, David (1781–1868) Scottish-born British author, cleric, educator, and scientist who built telescopes as a child and entered university as a young adolescent. Brewster fulfilled his early promise and demonstrated talent in many fields of endeavor. He was admitted to the Royal Societies of Edinburgh and London and received many medals and honors during his lifetime.

He served as editor of the *Edinburgh Encyclopedia* for two decades and as president of the University of Edinburgh for almost a decade near the end of his life.

Much of Brewster's important experimental research was in the field of optics, with a focus on the structural components of the eye, as well as the interaction of light with the optical properties of crystals which, in turn, led to the formulation of *Brewster's law* to describe properties related to polarization. Around 1811-1813, he noted that there was a particular angle, *Brewster's angle*, at which the interaction of light with a mineral substance has a maximal effect related to polarization. Over the next two decades, he extended his studies into the spectral characteristic of glasses and gases and studied fluorescence in chlorophyll.

Brewster balanced his experimental observations with his talent for building things. He invented a compact kaleidoscope that was granted a patent in 1817, and also established the foundation for a new optical lens system that could transmit bright light suitable for lighthouses, in1835. See Brewster's angle.

BRI See Basic Rate Interface.

bridge 1. A link that provides a connection across a physical or conceptual gap. This link may or may not be intended to affect the quality or format of the objects or information crossing the gap. 2. In networks, a device commonly used to handle communications between separate local area networks (LANs) which may or may not use the same protocols. Thus, Token-Ring networks and Ethernet networks may be connected via a bridge. In Frame Relays, a bridge encapsulates LAN frames and feeds them to a Frame

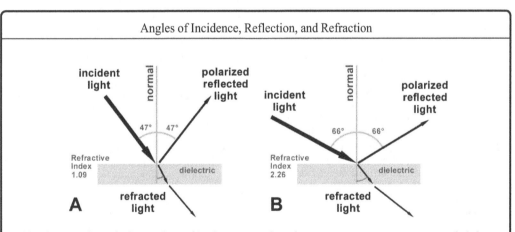

Angles of Incidence, Reflection, and Refraction

This diagram shows the basic relationships between incident electromagnetic energy interacting with dielectric materials with different properties. Imagine that the propagating wave is white light from the Sun traveling through air (which is itself a dielectric) to encounter a solid dielectric material. Materials with different refractive indexes (A and B) are shown for comparison. The collision of the incident light with the surface of the dielectric (a material that does not readily conduct energy) causes the direction of travel and polarity of the light to change. Depending upon the properties of the dielectric solids, some of the light will be reflected, some will be absorbed, and some will continue on down through the dielectric but at a different angle, called the angle of refraction.

In both (A) and (B), the angle of the incident light (the light hitting the surface) and the reflected light are the same relative to the surface normal (perpendicular to the surface) and there is a direct relationship between the index of refraction and the direction of travel of the scattered light. As the refractive index and the angle of the incident light increases, so does the angle of the reflected light, up to the point at which it is parallel to the surface (in which case it doesn't reflect).

Depending upon the properties of the reflective material, not all the light will be reflected. Some may be absorbed and some may be refracted through the material. The angle of refraction (light that passes through the material rather than reflecting off of it) is related to the angle of the incident light and its interaction with the refractive material.

Relay switch for subsequent transmission. It also receives frames from the network, strips the Frame Relay frame, and passes the LAN frame on to the end device.

Bridges in computer networks are associated with the Media Access Control (MAC) layer (or in OSI, the MAC sublayer). Bridges fall somewhere between repeaters and routers, although with increasing technological sophistication the distinction between bridges and routers is less clear. Bridges can be classified into general categories.

See brouter, extranet, Media Access Control address, repeater, router, spanning tree algorithm.

bridge, acoustic In acoustic instruments, the bridge elevates and spaces the strings and transfers vibrations to the body (soundbox) of the instrument. In electronic instruments, the bridge elevates and spaces the strings and transfers vibrations to the body of the instrument where they are, in turn, converted into electrical signals, usually by an energy-conversion device called a pickup.

Basic Categories of Network Bridges

Type of Bridge	Description
transparent	A general purpose bridge widely used on Ethernet networks that supports multiple bridges in a spanning tree configuration. A transparent bridge builds its own MAC address table based on source information from incoming traffic. Reference IEEE 802.1.
source routing	Specific to Token-Ring networks. Routing is determined at the source, rather than at the bridge and, hence, incoming frames contain routing information and an indicator as to whether it should be used. Reference IEEE 802.5.
source routing transparent	A less common hybrid configuration used in a small percentage of implementations.

bridge number BN. A ring network designator used in source routing. Together a segment number (SN) and bridge number (BN) comprise a route designator. When a destination is located on the ring itself, the bridge number is 0 (zero).

Bridge Protocol Data Unit BPDU. In ATM networking, a message type used by bridges to exchange management and control information. BPDU is a Media Access Control (MAC) management control protocol used to implement Spanning Tree Protocol (STP). It provides a mechanism for routing data traffic through a single conduit when more than one physical conduit exists (e.g., a backup loop for use if the root loop fails). The BPDU consists of flags, Hello time, and root, port, bridge, protocol, delay, and aging information.

bridger switching A technique for improving low return transmissions, as in cable networks, by sequentially turning on and off each leg of the distribution circuit. It is sometimes used in conjunction with high pass filters. While improvements in reliability can be attained in this way, it is at the cost of greater complexity and, hence, greater expense.

bridging clip A small piece of conducting apparatus used to connect nearby terminals, contacts, or other circuit elements that are close together, either for the purpose of changing a circuit (usually temporarily) or for testing it.

Bright, Charles Tilson (1832–1888) An English inventor and chief engineer for the Magnetic Telegraph Company. Bright was the first to undertake underground cable installation with gutta-percha as an insulating material. When Edison was installing the historic Washington-Baltimore line in the 1830s, problems with insulation and ground-breaking caused the line construction to be changed from underground to overhead, so Bright's success with an underground line was important. Later, Bright installed the first cables to be laid in deep water, first with a shallower line across the English Channel in 1851, and two years later a deepwater line between England and Ireland. Further lines around the world followed.

brightness The level of luminosity or amount of illumination emanating from a surface or display medium. Luminance is used to describe the lightness or brightness component of a television broadcast signal. Brightness across the visible spectrum is not equally perceived for different colors. See contrast.

Brillouin, Léon (1889–1969) A French-born American quantum physicist who studied band structures in crystalline solids. Brillouin is best known for describing Brillouin scattering, a quantum effect in acoustic modes in refractive materials.

In the 1920s, Brillouin made important discoveries in quantum dynamics, developing a means of approximating solutions to the Schrodinger equation. He accepted a position at the Sorbonne in the late 1920s. In the 1930s, he was associated with the Born Institute in Goettingen. In the 1940s, he took positions at American universities and authored *Wave Propagation in Periodic Structures*. In 1948, he became director of Electronic Education at IBM. Brillouin then joined Columbia University at a time when exciting evolutionary advancements in lasers were being made by scientists such as C. Townes. See Brillouin scattering.

The Laboratoire Léon Brillouin is named in his honor.

Brillouin scattering A frequency change/phase shift effect that occurs in scattered light from compressions/decompressions in an elastic, refractive material. Thus, photons are interacting with vibrational structures at the quantum level within the light-refracting medium as opposed to simply encountering incongruities within the structure (Rayleigh scattering). Brillouin scattering occurs when the light-refracting structure exhibits transverse and pseudo-transverse elastic compressibility (called *acoustic modes* due to the compression/decompression character of sound waves traveling through a medium). Brillouin effects are complex and subtle interactions that can be challenging to measure and that depend upon many factors, including density, refractive index, elasticity, temperature, humidity, vibrational intensity, the direction of the incident light, etc. Nevertheless, scientists are endeavoring to characterize and harness Brillouin effects for use in fiber optic and other technologies.

This effect was first studied by Léon Brillouin, ca. 1920. The mathematics of the scattering effect was not well understood until the study and development of lasers and fiber optics in the early 1970s gave it a wider context in which to be researched. This, in turn, furthered development of nonlinear fiber optics and the understanding of other optical phenomena and modulation technologies.

Brillouin scattering may impose limitations. For example, in optical fibers, increasing the power of the light in a cable causes it to reach a threshold at which point Brillouin effects impose physical limitations on light propagation. Acoustic noise arising from Brownian molecular motion can stimulate spontaneous Brillouin scattering causing back reflection.

Brillouin scattering is a useful tool for researching molecular excitation and Brillouin effects may be deliberately stimulated for use by testing and correction components. The effect can also be exploited in the development of fiber-based sensors. S. Burgett et al. have described the use of Brillouin optical time domain reflectometry to carry out strain measurments in precision-wound optical fibers and L. Thévenaz et al. have described strain and temperature sensing using embedded optical fibers. See acoustical Doppler effect, Bragg's law, Raman scattering, Rayleigh scattering, Stoke's shift.

British Academy of Film and Television Arts BAFTA. Established by Alexander Korda in 1946 as a club for the British film elite, the club evolved into the British Film Academy, founded in April 1947. It later merged with the Guild of Television Producers and Directors to form the Society of Film and Television Arts. BAFTA supports the Film Awards as well as Television and Children's Awards programs. http://www.bafta.org/

British Approvals Board for Telecommunications BABT. A U.K.-based telecommunications regulatory organization, the BABT provides third-party accredited regulatory and certification services. BABT Certification marks are commonly recognized in Europe. In April 2001, the BABT announced its affiliation with TÜV Süddeutschland AG. The merger resulted, in part, in the creation of BABT Product Service USA, which focuses on telecommunications conformity assessment.

BABT Product Service also has centers in several Asian countries. http://www.babt.co.uk/

British Broadcasting Corporation BBC. A television broadcast provider since the late 1920s, when it began its first experimental television transmissions, the BBC began widespread public broadcasting from London in 1936.

British Computer Society See computer societies, national.

British Film Institute BFI An independent Royal Charter body, founded in 1933, that promotes understanding of the moving image arts, including television and film. It fulfils the cultural and educational roles of the Film Council. http://www.bfi.org.uk/

British Indian Submarine Telegraph Company An underwater telegraph cable-laying company founded by John Pender, a British merchant, in 1869, not long after the laying of the first oceanic telegraph cable traversing the Atlantic Ocean. Through his involvement with the transatlantic cable project, Pender was inspired to propose a cable connection between Britain and India, where the British had established large centers of trade and colonization. In 1872, he also established the Eastern Telegraph Company. Within a decade, the company supported a majority of telegraph traffic to India and had expanded to include Australia, China, and Japan through overland and underwater cables. Eventually Pender achieved his ambitious dream when London, England and Bombay, India were linked in 1879.

British National Corpus BNC. A very large linguistic collection of written and spoken British English compiled from 1991 to 1994 under the management of an academic/industrial consortium led by Oxford University Press. In terms of technology, the collection is of interest in the fields of speech recognition and synthesis (telephony, computer text dictation and generation, etc.) and artificial intelligence (especially natural language synthesis and processing). BNC Online is hosted by the *British Library Initiatives for Access Programme*. The BNC is licensed material, but its associated SARA Client is freely available, as is limited demo access to the database.

British Standards Institution BSi, BSI. A U.K. standards body which provides input to various international standards associations, including ISO and ITU-T. It originated as a Committee within the Council of the Institute of Civil Engineers in 1901, appointed to consider standardization in the steel industry. The Committee of engineers and naval architects was joined by the Institute of Electrical Engineers in 1902. In 1918, it became the British Engineering Standards Association, granted a Royal Charter in 1929. The current name was adopted in 1931. BSI currently supports about 19,000 live standards. It owns the *Kitemark* quality mark. http://www.bsi-global.com/

British Telecom BT, British Telecommunications plc. Originally affiliated with the British Ministry of Defence and British Post Office, the telephone network and telephony research arms separated from the Post Office to become British Telecom. BT publishes the quarterly *BT Technology Journal* (technical papers) and *sphere* (BT technology) with copies available on the Web.

British Telecom Research Laboratories BTRL. The research lab of Britain's largest telecommunications provider, based in Marlesham Heath, Suffolk, England. British Telecom is involved in many international collaborative projects including research with the Massachusetts Institute of Technology in artificial intelligence natural language processing.

British thermal unit Btu. The unit of thermal energy required to raise the temperature of 1 pound of liquid water by 1 degree Fahrenheit at sea level. 1 Btu = 251.996 calories = 1054.8 Joules. Optical sensors have been developed to measure Btu's. While generally used to express the energy-producing or transferring capability of heating/cooling systems, it is also used to express the heat generated by large-scale computing systems to facilitate the design of building circulation systems associated with their installation.

brittle A quality of a physical substance lacking elasticity, one that is vulnerable to breakage. Substances may be brittle in one set of circumstances and not another, e.g., electrical components or connectors may be vulnerable below or above certain operating temperatures.

broadband A band of frequencies wide enough to be split into narrower bands, each capable of individual use for a variety of transmissions or for transmissions by a variety of users. Broadband transmission requires suitable hardware and cabling, capable of quickly transmitting and receiving a large amount of information. Fiber optics are often used. The entire breadth of the band is not necessarily used for transmissions, depending upon supply and demand. Also, there may be gaps between bands to prevent interference. Cable TV is a ubiquitous example of broadband transmissions where the band is split into all the different channels to which the recipients have subscribed. As in many broadcast media, broadcast technologies tend to be one-way, or mostly one-way, but with the increased demand for interactivity, more two-way communications over broadband are being developed. See baseband, telecomputer, wideband.

Broadband Connection-Oriented Bearer See BCOB.

broadband digital cross-connect system B-DCS. A digital cross-connect system that accepts a variety of optical signals and is used to terminate SONET and DS-3 signals. B-DCS accesses STS-1 signals and switches at this level and is appropriately used as a SONET hub for routing and other functions. B-DCS is commonly implemented with node termination via add/drop multiplexers (ADMs) and B-DC switches. Some newer optical systems that provide SONET-like capability can transmit without the B-DC switches. See wideband digital cross-connect system.

Broadband High Layer Information BHLI. An ATM information element that uniquely identifies an application (or session layer protocol of an application). BHLI is implemented in various ways, depending upon whether the codepoint is user-specific, vendor-specific, or ISO.

broadband integrated gateway BIG. A component of HFC (Hybrid Fiber Coax) networks that converts an ATM transmission into a signal that can be transmitted over the HFC. Working in conjunction with a connection management controller (CMC), the BIG strips information from ATM cells and orders and addresses them for further transmission. See connection management controller, HFC.

Broadband ISDN See B-ISDN.

Broadband Lightwave Sources and System See BLISS.

broadband line termination B-LT. Optical or electrical line termination (LT) on a broadband network that provides a physical and link between an access network and a local digital exchange. The termination unit will convert signals as needed and, in some cases, provide multiplexing functions for multiple attached devices.

Broadband Passive Optical Network BPON, Broadband PON. An optical communications network capable of providing high bandwidth services. Commercial implementations of consumer programming via BPON provide a range of up to about 20 kilometers on a single fiber serving a couple of dozen or more customers.

Broadband Telecommunications Architecture BTA. An architecture introduced by General Instrument for multimedia networking.

broadband terminal adapter BTA. A data communications device that interfaces a broadband ISDN (B-ISDN) connection to other terminal equipment that is not directly compatible with B-ISDN.

broadband transport manager BTM. In telephony, a transport mechanism for long-distance portions of a connection. In the late 1990s, Tellabs, Inc., planned field trials of a BTM for ATM networks.

BROADBANDLOOP BBL. An ACTS project for defining and testing a concept for a cost-effective broadband access network allowing fiber to be integrated into local loops as telecommunications bandwidth demand increases. The target user base for the project is residential subscribers and small- and medium-sized businesses. Field trials were set up for Denmark, Poland, and Portugal with some trials consisting of overlays to existing telephony and CATV networks. See BLISS, UPGRADE, WOTAN.

broadcast *v.* To transmit sound, images, or data over distance, in the context of more-or-less simultaneous receipt by a larger audience. Transmission can occur through a variety of media, over airwaves, satellite links, wire or fiber, or a combination of these. Reception can often be enhanced with antenna, cable, or satellite hookups. Radio, television, and Internet chat channels are common broadcast channels. Commercial and high power broadcasting is regulated. In the United States, the Federal Communications Commission (FCC) is the primary regulatory body and has jurisdiction over the allocation of broadcast frequencies. In Canada, the Canadian Radio Television and Telecommunications Commission (CRTC) handles many of the same functions.

In most of North America, very low power broadcasting is permitted without a license; otherwise it wouldn't be possible for people to use cordless phones, baby monitors, and wireless intercoms without being licensed. Generally these low power broadcasts are limited to a signal strength of 250 microvolts per meter, as measured three meters from the transmitter for FM transmissions, and 0.1 watts on a maximum three meter antenna for AM transmissions. This effectively limits the broadcast distance to 100 feet or so for FM and a couple of blocks for AM. Traditional broadcasts are typically in the range of 535 to 1605 kHz.

Commercial entertainment broadcasts are often financed by revenues from sponsors which are aired in the form of commercials. Since this revenue model has been successful for quite some time in the television and radio industries, it is not surprising that many broadcasters are turning to the same ideas in designing information to be viewed over the Web. In contrast to commercial stations, however, the Web is far less regulated and has many more participants, and it will be interesting to see how *Webcasting* evolves over the next several years. See television, radio, multicast, narrowcast, unicast.

Broadcast and Unknown Server BUS. In ATM networks, this server handles data sent by an LE client to the broadcast Media Access Control (MAC) address, all multicast traffic, and initial unicast frames which are sent by a LAN Emulation (LANE) client. It encompasses the functions that support establishment of a virtual circuit (VC) connection. See Media Access Control address.

broadcast data trigger Additional services are sometimes delivered with audio/video broadcast programming. Closed captioning or subtitles are examples of traditional broadcast services and others are becoming available as digital broadcasting over computer networks increases. Depending on the medium, certain standardized broadcast triggers have been defined and categorized, including transport type A and B triggers. Broadcast triggers are realtime data events associated with enhanced TV broadcasting delivered in a textual syntax based on the EIA-746A standard presented through the ISO-8859-1 character set (U.S. ASCII or Latin-1). By adding triggers to the data stream, the viewing box or software program receives a signal to interpret and present the additional information or services. In recent years, attributes have been added to the triggers to accommodate Internet broadcasts; these generally require two-way communications. The presence of two-way communications makes it possible to incorporate triggers as part of an on-demand interactive TV system.

The ATVEF has defined two modes of data transport that incorporate triggers. Transport A delivers triggers by the forward path and pulls data by the return path. Transport B delivers triggers and data by the forward path, but the return path is optional.

See ATVEF Enhanced Content Specification.

broadcast list 1. On computer networks, a list of users to whom broadcast messages are sent, usually by a system operator (sysop) or other privileged administrator. See broadcast message. 2. On fax machines, a list of recipients to whom the same fax will be sent.

broadcast medium In general, the physical substrate or underlying electromagnetic phenomenon that provides a conduit for electrical or other signals transmissions. Electromagnetic phenomena are the most common broadcast media, though a disturbance in a medium (metal, water, wood, etc.) that produces sound waves can also constitute a broadcast medium (albeit slower and less effective than electromagnetic media). In some networks, a physical layer capable of supporting broadcast messages.

broadcast message 1. A message sent to a selected group of users (or all users) on a computer or radio network. A common computer network broadcast message informs users that the system will shut down in 5 or 10 minutes. This message allows users to save work, close files, and finish up before being logged out. On networks, broadcast functions are usually available only to those with system privileges, as it is a capability that is easily abused. Schools are beginning to use broadcast phone or email messages to inform parents of registration, reporting, or meeting events with respect to their school-aged children. See allcall, anycall, broadcast list. 2. A message broadcast over a public broadcast medium, such as a news flash or Emergency Alert System (EAS) message. 3. A message broadcast over a paging or public address (PA) system.

broadcast over network In ATM networking, data transmissions to all addresses or functions on the system.

Broadcast Pioneers Library An education and research resource located in the Hornbake Library at the University of Maryland, College Park, founded in 1972. The collection includes correspondence, books, film, video, periodicals, historic photographs, scripts, and transcripts. More information is available online through the Pioneers' Web site. http://www.lib.umd.edu/UMCP/LAB/

broadcast standards Established in the late 1930s in the U.S., professional standards still exist as important guidelines for ethical business practices, safety standards, and standardized broadcast formats in the broadcast industry. See Canadian Broadcast Standards Council, Federal Communications Commission.

broadcast storm A broadcasting clamor that is excessively busy, frequent, or powerful that it overrides other communications. In radio broadcasts, a broadcast storms occur in times of emergency, when numerous operators simultaneously try to call for help or send messages to friends and relatives. In end-to-end systems, such as analog wireline telephone systems, broadcast storms as such don't occur (except, perhaps, in a different sense, on a party line) because excessive calling will result in a fast busy being sent to the caller, indicating that no trunks are available, rather than in many people talking at once. In data networks, however, a broadcast storm can occur as a fault condition in which some process goes wild and starts broadcasting to all workstations and disrupting user interactions and work. A storm may occasionally be caused on unsecured networks by a virus distributed by a vandal. See allcall, broadcast message.

Broadcast Wave Format See EBU Broadcast Wave Format.

broadcasting satellite service BSS. One of two divisions into which Ku-band satellite broadcast services have been split. BSS operates in the 12.2- to 12.7-GHz range. The other is fixed satellite service (FSS). See ANIK, Ku-band.

broadside array antenna A phased array (with harnesses) of antennas with the maximum radiation directed perpendicularly to the plane that holds the driven elements. This antenna arrangement can be configured as a billboard antenna by adding a reflecting sheet behind the array. See billboard antenna.

Broadway The internal development code name for The X Window System 11 Release 6.3 (X11R6.3) from The Open Group. See X Window System 11 Release 6.3.

Brocade Fabric Aware program BFA. A testing and configuration initiative for fostering end-to-end interoperability for storage area networks (SANs) in multivendor, heterogenous environments. Participating firms agree to specify, test, and support pretested SAN configurations with a mix of servers, switches, and storage subsystems. Brocade Fabric Aware qualifications may be presented after the completion of rigorous testing for interoperability within specified multivendor configurations. Brocade Communications Systems, Inc., has set up an interoperability laboratory testing environment to support the Fabric Aware program.

bronze An alloy consisting primarily of copper with tin and occasionally other elements added.

Brooks' law Adding manpower to a late software project makes it later. From Frederick P. Brooks, author of *The Mythical Man-Month*, a much-quoted provocative book about the engineering development culture.

brouter, b-router bridge router. Combination devices that function as links between different networks. The combination of a bridge and a router provides the physical and logical connections between networks, which may or may not have different protocols, and routing tables to facilitate the efficient transmission of information to the desired destination. A brouter typically performs its functions based on information in the data link layer (bridging) and the network layer (routing). See bridge, router.

Brown & Sharpe Wire Gauge See American Wire Gauge.

Brownian movement, Brownian motion Botanist R. Brown observed in early 1827 that pollen grains suspended in water were in a continual state of agitated motion. This motion has been widely observed for small particles suspended in fluids. It is said that the molecules of the suspension medium continually buffet the particles, resulting in the characteristic movement. Einstein later provided a mathematical explanation of Brownian motion. Theoretical models for queueing and aggregated connectionless network traffic, based on fractional Brownian motion,

have been proposed.

brownout 1. When power is partially, but not completely lost. Some companies use an industry-specific definition for a brownout, usually based on a relative or specific drop in voltage. Complete loss of power is called a blackout. 2. In cellular systems, a security precaution used by some companies to prevent fraudulent use. When brownout is in effect, there may be roaming areas in which a subscriber's system will not function.

browser 1. An object-oriented software development tool for inspecting a class hierarchy. 2. A software utility for displaying and traversing files and directories. 3. A software client for accessing the resources of the World Wide Web. See browser, Web.

browser, Web A historic milestone in software applications, designed to make it easy to access World Wide Web client/server resources stored on a variety of computers on the Internet. Prior to the development of Web browsers, there were many publicly available and valuable data repositories on the Internet, but access was through inscrutable line commands or uninspiring textual menus. The repositories themselves sometimes included images, but the images were seldom directly viewable over a remote link. They had to be copied and then loaded into a compatible application supporting the various formats in order to be useful on the local machine. Web browsers simplified this process by providing a 'front end' that transparently automated, integrated, and standardized access to the more sought-after types of information stored for distribution on the Internet. The Web browser client/server model was originally developed for NeXTStep by Time Berners-Lee, in 1989. As the concept spread through various systems and became associated with the Internet, another important change was taking place, the commercialization of the Internet from a research system to a system that could be used for commerce; this development got the attention of the general public, even those with no previous interest in computers.

A variety of Web browsers were quickly developed to meet the growing demand for Web access from different types of computers, resulting in phenomenal growth and interest over the next decade and beyond. Within a few years NCSA Mosaic, OmniWeb, Lynx (text browser), AWeb (Amiga Web), Netscape Navigator, and Internet Explorer had become popular browsers for 'surfing' the Web.

The Web gained in commercial prominence in the early 1990s, and browsers developed into practical tools for accessing, traversing, and displaying files on the Internet. A browser interprets standardized HTML tags that are used to describe the Web page, and displays the results on the user's system. Sun Microsystems Java applets are used by many developers to further enhance the capabilities of a browser with programming algorithms that aren't directly supported by HTML tag interpretation.

Browsers typically download HTML pages onto the local drive, so they can be more quickly redisplayed when the user moves back through previously viewed pages, and further facilitate the transfer of files through File Transfer Protocol (FTP) and other external utilities. Some browsers incorporate email functionality as well or will launch the email utility of choice when an email anchor is selected on a Web page. Most browsers support *plugins*, browser-compliant applications for processing multimedia file formats from a variety of vendors. Plugins are popular for enabling the user to play music, movies, streaming video news broadcasts, and other television-like forms of information and entertainment.

A Web browser's simple, accessible means of making available information on the Internet has resulted in an explosion of interest and participation, increasing from a handful of users in 1989, to more than 30 million in 1998. The number of data respositories available on the Internet has grown in conjunction with the increase in users.

One of the issues that becomes more important, as Internet commerce grows and data repositories include more and more personal information, is security. Security enhancements have been gradually added to Web browsers since the mid-1990s, but these are not impervious to skilled hackers. Web browsers were originally designed to be open and easy to use, and no one fully anticipated how quickly and aggressively banks, stores, and other institutions of commerce would set up their entire service line and customer access databases to be Web-accessible. As Web browsers become more powerful, they become more and more like operating systems that can be potentially accessed and controlled from remote sources in sophisticated ways. Already, by 1999, there were companies using their Web browsers to 'look' at information on the desktop of the individuals who installed the browsers on their computers. The companies didn't necessarily inform the user of this intrusion, or did so in small print or vague ways not understood by those not technically acquainted with computers.

If computers are to be secure in the future, it is important for the community at large to understand the potential for abuse and hold companies to high ethical standards in the matter of computer and individual privacy associated with Web browsers. Unfortunately, many people don't understand the technical or political issues and others don't consider the long-term consequences. While they may not be willing to give out name and address information on a moment-to-moment basis, they can often be persuaded to do so when offered the possibility of winning million dollar sweepstakes. Once their personal information gets into an unethical Web respository, it can be redistributed to millions of other computers within seconds; there's currently no way to undo this type of information theft, which may include social security number, credit card numbers, and more. Think twice before volunteering information to unfamiliar Web sites through your Web browser.

Another security-related browser issue is the use of cookies, identifiers within a browser that enable a Web site to recognize automatically a returning user without querying that person about his or her identity a second time. Users should take time to understand what cookies are, how they work, and whether they should be explicitly turned off in the browser to help safeguard privacy. Always read the privacy policy statements associated with each site before providing personal information and boycott sites that require more information than you feel they need to carry out a transaction.

Web browsers have opened up a world of communication, education, and opportunity for millions of people and will likely be an essential aspect of the Internet for a long time to come. Many Web browsers are freely available for download and Netscape Navigator is open source software that can be downloaded and modified by developers. See FTP, Java, HTML, HTTP, Internet, NCSA Mosaic, PDA macrobrowser, PDA microbrowser, SGML, World Wide Web.

browsing Searching or scanning through data for information or to get a general feel for the format or contents of a body of information. The information may take a variety of forms: text, files, directories, images, sounds, etc. See browser.

brush A conducting structure that provides an electrical connection between a motor and its power source.

brute force 1. A problem-solving method that involves trying every possible combination and permutation. This method is only practical for small problems of limited scope and is usually unwieldy for larger or more complex problems. Sometimes it is used in conjunction with other problem-solving methods such as heuristics. 2. A programming approach that involves reliance on a system's basic capabilities and processing power, rather than on efficient algorithms and elegance of design and concept. A brute force application generally does not run quickly on legacy systems.

brute force attack An attack on a security system using every possible combination, password, login name, or other entry data rather than using a targeted strategy. Brute force attack data are often generated automatically with computer software. This type of attack is usually easily detected and is often not very effective.

BS 1. See backscatter. 2. band signaling. 3. See base station. 4. See beam splitter.

BSAM See basic sequential access method.

BSCC BellSouth Cellular Corporation. A corporation serving about 10% of the U.S. wireless market, formed in 1991.

BSD See Berkeley Software Distribution.

BSE 1. back-scattered electrons. 2. Basic Service Element. 3. Basic Switching Element. In packet switched networking, a basic unit which may be combined with other BSEs to emulate a larger switching topology.

BSF bit scan forward. An assembly language bit manipulation in which a bit string is searched for a set or cleared bit, from low-order to high-order. See BSR.

BSFT See Byte Stream File Transfer.

BSI See British Standards Institution.

BSL See British Sign Language.

BSMS 1. billing and subscriber maintenance service. 2. Broadcast Short Message Service.

BSMTP See Batch Simple Mail Transfer Protocol.

BSP 1. Bell System Practice. Bell internal policies and procedures for creating instructional manuals for the servicing, support, and operation of phone equipment. 2. See byte-stream protocol.

BSR 1. bit scan rate. 2. bit scan reverse. An assembly language bit manipulation in which a bit string is searched for a set or cleared bit, from high-order to low-order. See BSF.

BSS 1. Base Station System. 2. See broadcasting satellite service. 2. Business Support System.

BSVC 1. Broadcast Switched Virtual Connections. 2. The name for an object-oriented, generic microprocessor simulation framework for building a virtual computer that evolved from a Motorola 68000 simulator supporting the 6850 UART.

BT 1. See British Telecom. 2. Burst Tolerance. In asynchronous transmissions mode (ATM) connections supporting variable bit rate (VBR) services, BT is the limit parameter of the GCRA. See cell rate.

BT cut crystal A type of crystal with vibratory qualities that makes it suitable for crystal radios.

BT Phonebase A service of British Telecom since the early 1990s that enables subscribers to make directory enquiries through a computer modem. The database is updated continuously and is more up-to-date than a yearly paper-based telephone directory. The call is billed at long-distance phone rates but is generally still less expensive than Directory Enquiry (U.K. Directory Assistance service). The typical connect speed is 2400 baud, though higher rates became available in some areas as of 1999. See TeleDirectory.

BTA 1. See Basic Trading Area. 2. See Broadband Telecommunications Architecture. 3. See broadband terminal adaptor.

BTag Beginning Tag. In ATM, a one-octet field of the CPCS_PDU used in conjunction with the Etag octet to form an association between the beginning of a message and end of a message.

BTBT band-to-band tunneling. Direct transfer of electrons from filled valence band (VB) states to empty states or recombination of electrons with holes in the valence band.

BTE 1. Boltzmann Transport Equation. 2. broadband terminal equipment.

BTI 1. British Telecom International. 2. Broadband Trial Integration. An ACTS project to demonstrate the role of Quality of Service (QoS) on Internet Protocol (IP) over ATM in order to develop optimization data for networks and to improve user perception of network services. The project involves development of an integrated IPv6 and switched ATM multicasting network with QoS support of user-controlled bandwidth and delay. The project is designed in three phases, the establishment of the technical platform, implementation of signaling and management of routers and switches, and the development of integrated

protocols. International connections will be through PVC-based ATM networks using UNI 4.0 SVCs for bandwidth management. See BID, BLISS, BONAPARTE, BOURBON, UPGRADE, WOTAN.

BTL Bell Telephone Laboratories. See Bell Laboratories.

BTM See broadband transport manager.

BTN See Billing Telephone Number.

BTRL 1. B Theory Research Labs. 2. Breward Teaching & Research Laboratories. 3. See British Telecom Research Laboratories.

BTS 1. Base Transceiver Station. In mobile communications, an end transmission point. 2. bit test and set.

Btu See British thermal unit.

BTU basic transmission unit.

BTU International A major supplier of thermal processing systems to the electronics industry, primarily semiconductor packaging and printed circuit boards.

bubble memory A type of nonvolatile memory; it doesn't have to be constantly electrically refreshed to retain the data. Bubble memory, as used in computers, consists of a thin layer of material that has magnetic properties. A magnetic field is used to manipulate a circular area such that the diameter becomes smaller, forming a bubble.

bucket truck See cherry picker.

buffer 1. A circuit or device designed to separate electrical circuits one from another. 2. A physical or electronic storage device designed to compensate for a difference in the rate of use or flow of objects or information. Generally a buffer is intended to increase speed of access and efficiency. In a computer, a buffer is often used as a storage area for frequently accessed information, so the software doesn't have to constantly access slower storage devices such as a hard drive, if sufficient fast access chip memory (e.g., RAM) is available. Cut and paste functions make use of a buffer. Data in a buffer tends to be temporary and volatile. See cache, frame buffer, RAM disk.

buffer, cable A layer of material to protect inner or outer components from abrasion, moisture, pressure, flexing, or tampering. In fiber optic cable assemblies, the buffer layer encircles the coating, cladding, and inner light-conducting core to provide protection from the elements. See swelling tape.

buffer box See Logical Storage Unit.

buffer condenser A condenser installed in an electronic circuit to provide protection to other components by reducing excessive voltages, especially surges.

buffer memory, buffer storage Electronic memory, usually RAM, used for information storage and retrieval, particularly for applications programs which make use of chunks of information that are frequently recalled. See buffer, cache.

bug A small, concealed listening device used in surveillance and espionage. Placing a covert bug in a room or on a phone line is almost always illegal. The term is also used in conjunction with small, hobbyist transceiver projects for electronics education, wireless intercoms, child monitors, and other legitimate uses. See bug, software/hardware; wire tapping.

bug, software/hardware A software or hardware error that adversely affects operations or user interaction. Grace Hopper is credited with relating the first story about a computer bug that was found by a technician, and for preserving the bug itself in a log book. This story has long been a part of hacker lore as the origin of the term "bug" in computer technology. The bug in the story apparently was moved to the Smithsonian Institution in the early 1990s (after an earlier unsuccessful attempt to have it accepted) but was not immediately exhibited. However, there are earlier anecdotes about bugs in industrial settings that indicate the term may go back decades, if not longer.

Removing bugs from software (debugging) is an art form, and not all programmers who are good at writing code are good at finding and correcting bugs. Unfortunately for developers, removing one bug often introduces one (or more) elsewhere. Unfortunately for consumers, some commercial software vendors release products knowing they are full of bugs, and there are no specific regulations prohibiting it. Because computer technology is technical, the user may not know whether a problem is from bugs or from incorrect use of the software.

Another unfortunate aspect to bugs is that companies often combine software enhancements with bug fixes and sell the new product as an upgrade with no guarantee that it is more robust than the previous version (sometimes it is less so). This situation is like buying a $15,000 car with a faulty engine, and having the manufacturer refuse to fix it and, instead, advise you to pay $5,000 to upgrade to next year's model. When you do, you find that the engine's been fixed, but the axles are defective, and the car has racing stripes that you didn't want in the first place. This situation in the software industry won't change until consumers stop buying substandard software and enhanced upgrades, and support instead the more responsible vendors who provide patches for bugs separately from releases of enhanced versions.

bug, telegraph A telegraph lever which, depending upon its position, can be used to send dots or dashes to partly automate transmission.

build 1. An increase in diameter of a line or object attributable to insulating materials. 2. In software development, the process of combining, compiling, or linking code so as to build an application.

bulb The sealed glass enclosure for an incandescent or fluorescent lamp. Bulbs provide protection for the gaseous environments and the delicate filaments that they enclose. See Edison, Thomas Alva; lamp.

bulk encryption Simultaneous encryption of a group or set of communications, such as multiple data messages or multiple channels on a broadcast medium.

bulk eraser, bulk degausser An electromagnetic device designed to save time by clearing the data from a large number of floppy disks at one time. By rearranging the particles on the physical disk, the electronic information is destroyed. It is handy for recycling the diskettes or for providing a measure of security with data that needs to be destroyed. It is wise to keep magnetic storage media away from computer monitors, which have magnets, or you may inadvertently erase or damage the data on them. A large-scale pirate software vendor, who was apprehended in Vancouver, B.C., is rumored to have had a bulk eraser in a storage cabinet wired to a button under the service counter to destroy evidence in the case of a police raid. See diskette.

bulk storage Media on which large amounts of electronic data can be stored. The amount of storage that constitutes large keeps increasing. In the mid-1970s, a tape holding 100 kilobytes was considered bulk storage! In the mid-1980s, a writable optical disk holding 600 MBytes was bulk storage. Now hard disks and tapes holding 4 GBytes or more are being bundled with consumer machines.

bulletin board system BBS. The forerunner to the Internet, BBS systems are typically individual computers set up for public or private modem access, by a number of users, on which there are shared files, mail, and chat services. The administrator is usually called the SysOp (System Operator). In the late 1970s and early 1980s, it was extremely rare for a BBS to be password-protected; there was open access to all. Unfortunately, persistent abuse has made this type of BBS almost extinct. In the mid-80s, there were still many BBSs running on TRS-80s, Color Computers (CoCo), Commodore 64s, Apple IIe's, and Amigas, with only 5 or 10 MBytes of hard drive storage for the entire system. BBSs have since become more sophisticated, offering credit card payment options, and increasingly are being linked to the Internet through telnet. See FidoNet.

bump contacts Small conductive lumps on electronic circuits that protrude to enhance electrical contact, such as those that allow chips to touch terminal pads.

bunch stranding A technique used to combine wires so they fit tightly together, with individual strands retaining the same directional relationship to one another to form a stranded wire. Stranded wire is useful in situations where flexibility is desired or when the electrical properties of the wires are influenced by proximity to others.

bunching An alternating convection-current effect in an electron stream caused by velocity modulation. Bunching is quantified as a parameter based on the relationship of the depth of velocity modulation to the absence of modulation. Used in electron tubes to generate ultrahigh and microwave frequencies. See klystron.

Bundesamt für Zulassungen in der Telekommunikation BZT. A German telecommunications approval authority established in the early 1980s.

bundled 1. Combined products or services, sometimes from a variety of manufacturers, offered at a combined price. Phone and cable companies often have bundled or packaged deals, such as regular telephone service and Caller ID-related services offered at a flat rate, or movie and educational channels combined. Software products are often bundled with computer systems. Operating systems are almost always bundled with computers, often along with various productivity applications, demo programs, and clipart libraries. 2. Individual wires or cables combined or interwoven to form a bundle for ease of handling and installation.

Bunsen cell A type of cell devised by R.W. von Bunsen that was an adaptation of an early wet cell in which the positive electrode was suspended in a bath of contained nitric acid (separated from the outer electrolyte solution) so that hydrogen would be oxidized and the cell depolarized. See dry cell; wet cell; von Bunsen, R.W.

buried cable An underground cable installation that cannot be altered without disturbing the soil, or accessing the cable under the soil through some entrance point. Buried cables are more aesthetic, as they do not clutter the landscape with utility poles and wires, but may be less easy to access to make changes or repairs. See Call Before Digging.

burn-in A diagnostic preliminary operation, sometimes at high temperatures, to test devices and circuits in order to identify those likely to fail. In electronics, many problems will show up early, in the first few weeks of operation, or under stress from heat and humidity. Sometimes called early failure period.

burn-in, monitor An undesirable ghost image on the coating inside a cathode-ray tube (CRT) monitor resulting from the persistent display of the same image or similar images. Monitor burn-in can be prevented by turning the monitor off when not in use, and using screen savers that darken the screen completely or, second-best, that move around small, low-contrast images. Many commercial products called screensavers do not save your screen. If a bright, still image covers most or all of the screen, it's not a screensaver, no matter what it is called, and persistent display of the image will cause burn-in.

burnout A condition in which a person's physical and psychological resources are severely depleted and stressed, and there are insufficient resources (e.g., time) for the body/soul to repair and renew itself. Overwork over an extended period with frequent stress-related deadlines usually leads to burnout. Burnout is prevalent in health professions where long hours and stressful shift work are common and in high tech startup companies where long hours and limited venture financing create deadline pressures and overwork.

Symptoms vary but may include lethargy, anxiety, shivering, vertigo, headaches, migraines, stomach upsets, insomnia, frequent colds, muscle aches, chronic inflammation, and apathy. Factors contributing to burnout can include overwork, lack of variety, low pay for the type of work done, lack of appreciation, poor time management, lack of rest and exercise, poor working conditions, long hours spent with a demanding public (such as sales or telesolicitation). See a health care professional if you suspect burnout and consider some lifestyle/priority changes.

burn rate The rate at which a seed or startup company uses its initial cash resources in the process of developing and marketing a new product or service with the goal of making the company financially viable. Burn rates for startup software development companies, for example, are minimally three to four million dollars per year and up (for a small company of about five to thirteen employees) for the first couple of years and can be much higher for telecommunications hardware development and manufacturing companies. Occasionally software products will come to market for much lower cash investments, but this is usually in situations where the developers have daytime jobs and invest in the startup in terms of "sweat equity" (time invested) rather than cash.

burn rate, employee The rate at which a company uses up its personnel resources in order to get the most productivity for the money paid in wages. Some companies take advantage of the fact that people can work extra hours and put in extra energy over the short haul (e.g., up to about two years). See burnout.

burst 1. Sudden increase in signal strength. See surge. 2. A color burst, or reference burst, is an oscillator phase reference in a color broadcast receiver. 3. In printing, the separation of continuous-feed or multipart pages into individual sheets, usually along a perforation. See burster. 4. In a Frame Relay network, a sporadic increase in a circuit where the total bandwidth is not continually in use. 5. In data communications, a sequence of more-or-less contiguous signals that are treated as a unit according to a predetermined set of criteria.

burst error, burst noise 1. A data burst sufficient to garble or interrupt data transmission. For example, scratches may result in a burst error on a laserdisc, audiodisc or magneto-optical storage disc. Some systems have software error-checking which will minimize the negative effects of a burst error, or which will make a best guess as to what the data should have been. 2. In audio transmissions, noise and interference that substantially exceed the ambient noise or the level of the desired transmission, e.g., sudden pops and clicks on an old vinyl recording.

burst mode A data transmissions mode in which the data is sent faster than usual, sometimes due to a device temporarily monopolizing the transmissions channel. See caching.

burst modem BM. A modem developed to send and receive information in bursts rather than as a continuous connection. Burst modems are typically used in applications where bandwidth is at a premium or for connections that may be expensive. Since many communications include blank spots or long pauses that eat up connection time, a burst modem can more efficiently handle traffic on packet data networks, multimedia services, and consumer communications networks (e.g., cell phones).

burst pressure The maximum pressure that a device or mechanism can tolerate before rupturing. This phrase is often used in reference to liquid or vapor conductors.

burst sequence In color video transmission, a mechanism for improving the stability of color synchronization by controlling the polarity of the color burst signals.

burst size, committed Bc, B$_c$. In Frame Relay networking, a cell traffic descriptor for the maximum amount of data (total number of bits) that the network will agree to transfer during a time interval Tc. It can be set individually for virtual circuits (VCs). See cell rate.

burst size, excess Be, B$_e$. In Frame Relay networking, a cell traffic descriptor for the maximum amount of uncommitted data, in excess of the committed burst size (Bc), that the network will endeavor to deliver during a time interval (committed rate time interval – Tc). This type of data is eligible for discard, if necessary, to provide quality of service (QoS). The excess burst size can be set individually for virtual circuits (VCs). Bursty traffic and congestion are two of the common traffic management challenges that are handled with a variety of procedures. See cell rate.

burst time interval A calculated interval related to signal bursts that is used to delineate a time period for assessing traffic in a network. In Frame Relay networks, for example, throughput is evaluated to gauge and adjust performance and user service levels. The committed rate time interval (Tc) is based on the ratio of committed burst size to committed (data) rate.

burst transmission A transmission in which the signal is intentionally sent in a group at significantly higher speeds than is usual, with as much informational content as possible. For example, in radio transmissions, the information may be sent in a burst at 50 or 100 times the normal speed and then played to the listener at normal listening speeds. Burst transmission is a technique for getting more information transmitted in less time.

burst transmission, isochronous In a data network, where many different devices may be operating at different speeds, burst transmission may be used to resolve some of those speed (data rate) differences in order to operate the network more efficiently.

burster In printing, a device that speeds and facilitates the separation of continuous feed or multipart pages into individual sheets, usually along a perforation. This type of equipment is sometimes incorporated into a multiple-capability machine (burster-trimmer-stacker) which also trims the pages to remove the tractor-feed strips and stacks the paper.

bursty data Data that come in spurts; it is often unpredictable. The nature of the data is important in the design of network traffic flow control procedures and protocols. See variable bit rate.

bursty information Information that alternates between intervals of low transmission and short bursts of high transmission incorporating a lot of data. Print jobs tend to be bursty, with long periods of idleness and short periods where it seems as if everyone on the network submits a job to the print queue at the same time (e.g., just before lunch break).

bus 1. An uninsulated conductor, such as a wire, bar, or printed metal patch on a circuit board, intended to provide an electrical contact point for adjoining conductors or devices. Commonly used in telephone and computer circuits. See backplane, edge connector, expansion slot. 2. A category of standards that facili-

tate the compatibility and interconnection of consumer electronics products. 3. One type of computer architecture in which a series of computer processing units (CPUs) are interconnected. 4. In its most basic sense, an uninsulated solid or hollow conducting bar or wire.

BUS See Broadcast and Unknown Server.

bus, data The data pathways internal to a computer that permit the transfer of data within the system and to peripherals associated with the system. Common buses include address and data buses which may or may not match the capacity of the CPU. Bottlenecks are possible at the bus if the information reaching the bus is greater than its physical or logical carrying capacity (e.g., a 16-bit address bus on a 32-bit CPU).

Bus Interface Gate Array BIGA. A Cisco Systems technology for allowing a Catalyst 5000 to receive and transmit frames from the packet-switch memory to the Media Access Control (MAC) local buffer memory, independently of the host processor.

bus mastering A capability of a peripheral device to take over functions and transfer data through a computer's system bus. This capability is incorporated into PCI video cards, for example, so the card can directly access system memory and provide other performance improvements. Bus mastering is a means of direct memory access (DMA) processing.

bus topology A network topology in which individual nodes are connected to a single communications line which is terminated at either end. Like a ring topology, this arrangement is fragile in that if one node or system goes down, it affects the entire network. See star topology, topology.

Bush, Vannevar (1890–1974) An American engineer and writer who devised the product *continuous integraph* or *product integraph* in the mid-1920s. This device was a semiautomatic machine for solving problems. He later invented the *differential analyzer*, an evolutionary descendent of the integraph, a mechanical apparatus for solving problems. Bush was President of the Carnegie Institution of Washington from 1939 to 1955.

bushing 1. A cylindrical lining in an opening/hole that aids in controlling the size of the hole, insulating it or providing a path, as for wires. See ferrule. 2. A cylindrical utility pole insulator with external ribs at one end. This type of insulator was typically used on high voltage leads.

busy 1. A system that is in use. For example, a printer may be busy processing data or handling a current job and thus is unable to process incoming data. Thus, it may buffer the data or send a busy signal to the sending station. 2. A telephone that is in use or off-hook and unable to receive calls, or a telephone trunk that is at capacity and can't finish processing the call. See busy signal, fast busy.

busy back A signal that communicates the call status back to the caller. See busy signal.

busy hour That hour during a specified period (day, week, etc.) in which the greatest volume of traffic is carried, as on a phone or computer network.

Busy hour traffic volume and characteristics provide

important data about the capacity, equipment, and switching needs of a network. The busy hour characteristics can be used to determine the amount of variation from low to peak usage times, the expected maximum requirements of the system, and other administrative and operational parameters. On phone networks, busy hour times have changed (become later in the day), probably as a result of the greater use of off-peak hours for facsimile transmissions and Internet usage.

busy lamp, busy light A lighted indicator on a device that shows it is currently active, or in use. The busy lamp is often paired with a second lamp that shows that the device is powered on. Thus, a printer may have two lights, a power-is-on indicator and a data indicator (this second lamp often flashes). Busy lamps are often included on modems to indicate whether data is being transmitted or received. These lights are very useful; a user can tell at a glance if a transmission is active.

On a phone console, busy lamps are useful indicator of which lines are currently in use, so the user can pick up on a different line and not cut in on someone else's conversation.

busy season In any service or product distribution, there are usually low periods and peak periods. In the airline industry, the peak periods are Thanksgiving, Christmas, etc. During these peak periods, there are greater demands on communications systems and the computer systems that support them. Thus, software and networks have to be designed to handle busy times without slowdowns or loss of business.

busy signal, busy tone A regularly recurring signal (beep) on a phone line, transmitted to a caller to indicate that the trunk is not available (fast beeps) or that the called party is on the line and cannot be connected with you unless he or she has a service such as Call Waiting. If so, and the party wishes to interrupt the call in progress to talk to the second caller, he or she can do so by typing in a code, and returning to the original call, if that person is still waiting. See Call Waiting, Audible Ringing Tone.

busy test In telephone networks, a diagnostic technique for determining the availability (or disability) of an expected service, such as a successful transmission through a newly installed line.

Butler antenna, Butler matrix antenna A passive array antenna in which the feed lines include combined junctions for more than one beam. Thus, multiple inputs and outputs are possible. Each beam-formed output from a Butler array can be fed to an individual receiver. Researchers have found that a Butler matrix can be configured with smaller numbers of components compared to many traditional antenna technologies. Butler antennas are now used in wireless telephony, including cell phones, direction-finding applications, and ionospheric research systems. See adaptive antenna array.

butt joint A wire connection or splice in which the ends of the conductors are butted up against one another and combined by soldering, welding, etc.

butt-in, butt-set See buttinsky.

butterfly capacitor A tuning device with good selectivity resembling a butterfly or bowtie, used for very high frequency (VHF) and ultrahigh frequency (UHF) transmissions.

butting The process of tightly aligning adjacent components, often along a planar surface. For example, linear and grid arrays of fiber optic tapers are butted within fine tolerances to produce larger imaging areas. The tradeoff is a slight loss of image resolution at the points where individual elements join and possible increased electromagnetic interference from the large number of individual charge coupled devices (CCDs) and their accompanying circuitry. There may also be overall geometric distortion, particularly in larger arrays, in which slight alignment errors accumulate over the extent of the surface. Nevertheless, the advantage of larger imaging areas and less edge distortion than typical glass lenses makes this a good solution for sensors, research, and medical imaaging.

buttinsky *slang* A type of telephone system that includes a transceiver worn on the hips that is used to "butt in" on conversations; it permits a technician to break in on or monitor a call. It is used in diagnostics and installation.

button caps Portable key sleeves designed to fit over individual buttons on programmable pushbutton phones or computer keyboards. These caps may be plain (for hand labeling), transparent, or preprinted, to indicate the newly programmed or temporary function of the key.

buttoned up Sealed, completely closed.

buzz *v.* 1. *colloq.* To catch the attention of someone at a distance or out of eyesight. To call for a quick chat on the phone. 2. To press a buzzer.

buzz test Diagnostic testing of the continuity of a circuit by placing a buzzer on one end and sending a current from the other end to see if the buzzer rings. Often used when the far end is out of eyesight, but not out of hearing. See buttinsky.

buzzer A signaling device, usually electromechanical, that makes a raucous noise when the circuit is completed, at an attention-getting volume and frequency. Buzzing noises can also be generated on computer systems. Buzzers have many uses; they can catch a person's attention, indicate a fault condition, or provide a tone that can be used as a diagnostic tool when tracing a circuit. See tone generator, tone probe.

buzzer leads The connecting posts or wires attached to a device that are intended for the connection of a buzzer.

BVR See beyond visual range.

BW See bandwidth.

BWF See EBU Broadcast Wave Format.

BX cable A type of cable used in electrical wiring consisting of insulated wires enclosed within a flexible metal tube.

B/Y signal In a color television signal, B/Y is one of the three primary signals (of RGB), providing blue (B) when combined with a luminance signal (Y).

bypass To shunt around the normal path; to provide an alternate path, often for temporary installation or diagnostic purposes.

Byron, Ada Augusta Lord Byron's daughter. See Lovelace, Ada Augusta.

byte A unit of data that is bigger than a bit (binary digit) and smaller than a word. The relationship of bits to bytes and bytes to words varies according to the system, but in most computing a byte is said to consist of eight bits. File sizes and storage on smaller media are usually displayed by the system in terms of bytes or kilobytes.

Many character sets encode each character within a byte of data. International character sets tend to use two-byte encoding schemes to accommodate the much larger number of letters and symbols. In Internet protocol documents, it is more common to use the term 'octet' (8 bits) instead of the term byte, presumably because the meaning of octet is more explicit. See kilobyte, megabyte.

Byte **magazine** One of the first popular small computer systems journals, founded in 1975, *Byte* magazine is still publishing in print and on the Web while many other computer magazines have come and gone. Robert Tinney's covers graced the publication for a decade and a half and Jerry Pournelle's columns have been a mainstay for decades. One of the most popular features of *Byte* magazine in the 1970s and 1980s was Steve Ciarcia's circuit cellar, an electronics column for hobbyists. At around the same time that Ciarcia left *Byte* for other activities, the tone of the magazine changed, it became more IBM-system-oriented, and "Small Systems Journal" was removed from the masthead.

byte protocol A list of special byte-sized binary patterns used for signaling or as masks. The byte protocol may include control instructions, delimiters, coordinates, or any type of information that can be encoded into a byte that is relevant to a particular application.

Byte Stream File Transfer BSFT. An Open Group Technical Standard means for transferring unstructured files among Open Group-compliant systems. BSFT is derived from File Transfer Protocol (FTP) but utilizes an ISO profile.

byte stuffing In its simplest sense, adding extra bytes to a data stream. Byte stuffing may be used to "pad" data to conform to certain file formats such as chunky file formats or those that must end in even-byte boundaries. Byte stuffing can also be used to mark time to keep a link live while waiting for continuation or acknowledgment. Byte stuffing can be used to represent characters that do not necessarily exist in the transmittable data set (e.g., symbols) or to distinguish between ambiguous characters/signals. Certain characters can be stuffed into the data stream to signal a specific character at the receiving point by mutual agreement, even if that character wasn't sent in the actual data. Byte stuffing can be used to distinguish control data (e.g., *esc*) from informational data ("Hi, Alice!"). Byte stuffing may signal something important or something different to come in the transmission scheme. It can be used in error correction schemes. Commonly byte stuffing is used to encode a signal that might be erroneously interpreted

as an end-of-transmission signal. For example, *esc-Y* may be sent over a network in place of an end-of-transmission (EOT) marker to distinguish it from an end-of-frame (EOF) marker.

In Point-to-Point Protocol (PPP), bit or byte stuffing (e.g., in V.42 modems) can reduce the effective data rate, making it necessary to calculate the extra overhead to determine whether or not to admit the stuffed data stream. In modem communications, the degree of byte stuffing can vary widely, up to almost double the original data.

byte timing circuit BTC. In the ITU-T X.21 Recommendation, a data timing circuit in interchange circuit B. For optimization, the transition to t=0 is required to occur within the same bit interval as the transition to c=ON. In call control character alignment, the Data Terminal Equipment (DTE) may be required to align call control characters transmitted to either SYN characters delivered to the DTE or to signals on the byte timing circuit. There are other timing circuits in X.21, including signal element timing (S) and DTE signal element timing (X).

byte-count protocol BCP. A common type of network protocol that utilizes a byte value in a field to designate the length of the payload (the message portion of a packet) rather than bit patterns or special characters. Thus, any pattern of bits can be transmitted without worrying that it might be misinterpreted. Ethernet (IEEE 802.3) is an example of a BCP.

byte-stream protocol BSP. Also called a *stream protocol*, a network protocol based upon the transmission of in-order byte-stream data, that is, bytes are written into and received out of a connection-oriented environment at the application level (the underlying transmission may be packet-based). Transmission Control Protocol (TCP) is an example of a byte-stream protocol, as is Internet Stream Protocol, V. 2 (ST2). In duplex transmission, two byte-streams are supported, one for each direction. Flow-control mechanisms may be incorporated to facilitate user control. See ST2+, Transmission Control Protocol.

bytecodes Byte-oriented machine instructions. In Java programs, the platform-independent codes that execute within a Java Virtual Machine (VM). Distinct bytecodes are used by the JavaVM for operations on various data types that are on the top of the operand stack. JavaVM bytecodes may also move operands between the frame and the operand stack. See Java.

bytes per inch BPI. In recording, a measure of the amount of data that can be stored on a medium. It is usually used to describe data density in media that are long and narrow, such as recording tapes. See superparamagnetic.

bytes per second Bps. Many people abbreviate this bps, which causes confusion because bits per second is conventionally abbreviated bps. Bytes per second (Bps) is a description of transfer or transmission rates in bytes over time. A byte is eight bits, or one octet. Data frame format sizes are sometimes described in terms of bytes or octets.

Byzantine agreement A core concept instrumental in the study and implementation of reliable, fault-tolerant processing sensing systems and networks. Byzantine agreement is agreement based upon consensus in a system processing data from multiple sources and in which the data may not always match. Agreement may be simultaneous (simultaneous Byzantine agreement – SBA) or eventual (eventual Byzantine agreement – EBA).

The name stems from the Byzantine Generals problem proposed by Pease et al. in 1980. Proofs and verifications of the concept and fault-tolerant circuits followed in the early 1990s, with implementations and improvements in efficiency emerging in the mid/late-1990s, along with authentication mechanisms for Byzantine agreement protocols. Early investigations focused on single faults, with mixed faults (hybrid faults) considered thereafter. By the late 1990s, algorithms to simulate Byzantine agreement were being released on the Internet so that the agreement process could be observed in progress and more widely implemented on distributed networks.

Mathematicians have approached the problem in a number of ways, since it is expedient to reach agreement quickly and efficiently. Bracha described expected rounds in a randomized Byzantine Generals protocol in 1987. Halpern et al. characterized EBA in 1990. Computational Logic, Inc. (CLI) verified and implemented the original Marshall et al. version of the algorithm. CLI developed a formal model of asynchronous communication based on the concept. Kesteloot suggested that executing a fault detection algorithm before running an EBA process may save on processing time. Garay and Moses have presented a polynomial-time protocol for coming to agreement. With regard to polynomial time algorithms, Kann has suggested the use of a consistent broadcast protocol for improving efficiency. See Byzantine Generals problem.

Byzantine fault A type of fault that is unanticipated or unexpected or that may result from intrusion into a network. When a network is invaded by a Trojan Horse or other type of program, it may exhibit transient behaviors or insert seemingly innocuous code that could be activated or triggered by various events or states within the system (e.g., a logic bomb). Byzantine faults can also occur in distributed networks where the number of nodes and the routing of information is dynamic and not necessarily predictable. Various means of detecting, assessing, and responding to Byzantine faults are being developed by vari-

ous organizations, especially those concerned with reliable transmissions over distributed networks, like the Internet, and those concerned about safeguarding computer security. See Byzantine agreement, Byzantine Generals problem.

Byzantine Generals problem (BGP) A paradigm for building consensus among distributed processes, proposed by Pease, Lamport, and Shostak in a SRI International Technical Report in 1980 and by Lamport, Shostak, and Pease in *ACM Transactions* in 1982. This problem statement has revealed important concepts in the understanding and implementation of fault-tolerant distributed networks and multisensor imaging and data collection systems. The task is to create a system that works in a reasonably correct and efficient manner in spite of faults or differences in information from multiple sources.

Simply stated, there are n number of generals, with one designated as a commander. Generals can intercommunicate. The task is to develop a communication protocol for the commander to send an order to the generals so that all generals are obeying the same order – if the commander is loyal, every loyal general obeys the order sent by that general. Since there may be faults in the system, any of the generals could be traitors (plotting a coup or holding erroneous information) and may send information inconsistent with that of the other generals.

The Byzantine Generals problem has many practical applications in telecommunications, particularly in parallel processing or distributed processing networks and in remote-sensing systems, where multiple sensors are providing data that may not exactly match and that must be resolved to make use of the information. In other words, in these contexts, multiple inputs that may differ may need to be agreed upon before they are processed (e.g., a remote-sensing satellite may have to resolve multiple inputs from separate infrared sensors or different types of sensors). The concept is useful in fault-tolerant networks for coordinating the operations of multiple independent processors. The Byzantine Generals problem does not specify whether a value has to be agreed upon simultaneously (*simultaneous Byzantine agreement*) or eventually (*eventual Byzantine agreement*). See Byzantine agreement.

BZT German telecommunications organization. See Bundesamt für Zulassungen in der Telekommunikation.

C 1. *symb.* capacitor or capacitance. 2. *symb.* Celsius or centigrade. 3. *symb.* the velocity of light.

C minus, C- *symb.* the negative terminal of a C battery. The connecting point at which the negative terminal of a grid bias voltage source is connected, as in a vacuum tube circuit. See C battery.

C plus, C+ *symb.* the positive terminal of a C battery. The connecting point at which the positive terminal of a grid bias voltage source is connected, as in a vacuum tube circuit. See C battery.

C++ A high-level programming language, a superset of C, developed primarily by Bjarne Stroustrup at Bell Laboratories. In endeavoring to be backwardly compatible with C, it is criticized by some as being unwieldy. Nevertheless, it is widely used in commercial software development and many programmers like it. See C language, object-oriented programming.

C battery A type of power cell first introduced in the early 1920s. Historically, C batteries supplied bias voltage to electron tubes that were used to control a grid circuit. C batteries are still commonly used in small flashlights, portable radios, and many small, portable, electronic appliances.

C Block A Federal Communications Commission (FCC) designation for a Personal Communications Services (PCS) license granted to a telecommunications company serving a Major Trading Area (MTA). This grants permission to operate at certain FCC-specified frequencies.

The organization of radio spectra for new technologies has not been a simple process. Problems arose with the original C and F Block license allocations assigned between December 1995 and May 1996; the auction rules established in 1994 by the FCC did not work well in practice. Almost 1000 C and F Block licenses were granted but, rather than creating a competitive market based upon many small businesses, the licenses ended up in the hands of a small number of large interests, some of which were over-ambitious and unsuccessful.

By 1997, the FCC Wireless Telecommunications Bureau had suspended payments on C Block licenses but later reinstated the obligation. Companies responded by declaring bankruptcy, tying up the assets in legal proceedings. As a result, the FCC cancelled certain spectrum allocations and scheduled the re-

auctioning of almost 500 licenses for the year 2000. After further delays, the auctions finally took place between December 2000 and January 2001, generating almost $17 billion in revenues. Once again, the intention of the C and F Block allocations was to encourage the development of innovative wireless communications and to increase the level of competition in the market. See A Block for a chart of frequencies. See Omnibus Budget Reconciliation Act.

C interface An interface used in Cellular Digital Packet Data (CDPD) that is deployed over AMPS. The C interface connects the Mobile Data Intermediate System (MD-IS) to the Intermediate System (IS). See A interface, B interface, Cellular Digital Packet Data, D interface, E interface, I interface.

C jack The USOC/Federal Communications Commission (FCC) code for a flush- or surface-mounted jack (as opposed to wall-mounted). The designation is typically a suffix added to RJ-11 (RJ-11C) or RJ-45 (RJ-45C), for example. See RJ for a fuller explanation and chart.

C language, 'C' A sophisticated, fast, widely used, medium-high-level programming language developed by Dennis Ritchie at Bell Laboratories in the early 1970s. 'C' was descended from B (the quotes are now usually omitted). C became widely distributed with Unix and was used for programming the Amiga computer in the mid-1980s. By the late 1980s, universities were teaching both C and Pascal as basic skill sets, and C has become widely used in the commercial software development industry. The chief advantages of C are power and flexibility. Its chief disadvantages are the logistics of keeping track of pointers and memory allocation and the many pages of code that are needed to accomplish basic tasks. The popular introductory book on C was written by Kernighan and Ritchie. See C++.

C lead In communications lines utilizing three wires, where one is positive and one is negative, the third wire or C lead may be used as a ground and can be manipulated to connect or release the circuit. In telephony, the control provided by the third wire is useful on trunk circuits.

C link, cross link In CCIS out-of-band telephone networks, a link to interconnect pairs of Signal Transfer Points (STPs).

C News A UUCP-based news-reading program developed from the earlier A News and B News programs. Along with InterNetNews (INN), C News superseded B News. The networks were changing, TCP/IP had been introduced along with Network News Transfer Protocol (NNTP). C News, written by Geoff Collyer and Henry Spencer of the University of Toronto, was released in fall 1987. C News was faster, more reliable, and supported a bigger article database than previous versions. A port of C News for AmigaOS was developed by Frank Edwards. See A News, B News, USENET.

C Series Recommendations A series of ITU-T recommendations that provides guidelines for general telecommunications statistics. These guidelines are available as publications from the ITU-T for purchase and some may be downloaded without charge from the Net. Note that the statistical yearbook was transferred to ITU-D. Since ITU-T specifications and recommendations are widely followed by vendors in the telecommunications industry, those wanting to maximize interoperability with other systems need to be aware of the information disseminated by the ITU-T. A full list of general categories is provided in Appendix C and specific series topics are listed under individual entries in this dictionary, e.g., A Series Recommendations.

ITU-T C Series Recommendations

C.1	1993	ITU statistical yearbook (deleted, activity transferred to ITU-D)
C.2	1996	Collection and dissemination of official service information
C.3	1993	Instructions for international telecommunication services

---

C-band A portion of the electromagnetic spectrum used extensively for the transmission of radio wave communications signals, especially those to and from satellites. The C-band extends from 4 to 8 GHz with satellite uplinks in the 5.925- to 6.425-GHz range, and downlinks in the 3.7- to 4.2-GHz range. Downlink and uplink frequencies are different in order to reduce interference between received and sent signals. C-band transmissions require relatively large receiving antennas, making them less popular for consumer services than Ku-band. See band allocations for chart. See C-band, optical.

C-band, optical In optical communications, an ITU-specified transmission band in the 1530 to 1565-nm range that is used for fiber optic transmissions and which recently is sometimes supplemented with L-band transmissions over the same cable (in multimode cables). See C-band.

C-Scope In radar, a screen that displays bearing and elevation information relative to the center of the region being scanned.

C-plane In ATM networking, as it applies to a Broadband-ISDN reference model, the C-plane is the control plane, a higher-level plane including all of the ATM layers, which bears control signaling information. It sits adjacent to the U-plane (user plane) and shares physical and ATM layers with the U-plane. The M-plane (management plane) enables the transfer of information between the C- and U-planes. In ATM networking, as it applies to Frame Relay bearer services, the U-plane parameters, such as throughput, maximum frame size, etc., are negotiated through the C-plane. See the Appendix for more detailed information on ATM.

C-stock Bell Telephone jargon for refurbished telephone equipment.

C/A Code, civilian code, S-code In the Global Positioning System (GPS), a Clear/Acquisition Code in which the carrier wave is modulated with a sequence of pseudo-random, binary, biphase signals to provide civilian locational information transmissions. This information is at a lower resolution level than the classified government GPS transmissions.

C/DTAC See Consumer/Disability Telecommunications Advisory Committee.

C64 See Commodore 64.

C7 The European analog to the North American Signaling System 7 (SS7), which is similar but not directly compatible. C7 is widely deployed for digital phone communications. See Signaling System 7.

CA See call appearance.

CAB See Canadian Association of Broadcasters.

Cabeo, Niccolo (1586–1650) An Italian scholar and experimenter who recorded a number of electrical observations and wrote the first Italian treatise on magnetism, *Philosophia magnetica*, in 1629.

cable Wire, fiber, or other conductive material in single or multiple (bundled) strands used for the transmission of light, heat, electricity, or data. Although the terms *wire* and *cable* are often used interchangeably, some technicians make a distinction based upon the bundling. If it is a single metal core, it's called a wire, if it is a combination of layers of two or more separately insulated wires, or if it is a fiber optic bundle, it's called a *cable*. A fiber optic cable may support a single fiber or may support a bundle ranging from 2 to over 400 fibers.

In telecommunications, the speed and quantity of data that can be conducted along a cable varies greatly with the materials that are used in its manufacture. Traditional phone lines are usually copper wires, while cable television broadcasting uses fiber optic cable with greater speed and bandwidth. The computer industry, with its demands for simultaneous transmission of data, sound, and video, has greatly increased interest in high-speed, high-bandwidth cable media. Wireless services such as cellular phone and individual satellite modem transceivers provide an option to physical cable links. See conductor, conduit, creel, fiber optic, swelling tape.

cable, armored A cable reinforced or wrapped in a strong, environment-resistant or vandalism-resistant covering, usually of wound metal. Armored cables are sometimes used to chain costly equipment, like computer terminals, to walls or work desks, to prevent theft. Armored cables are sometimes used on telephone book holders and handsets in public areas and penal institutions (to prevent the handset or the cord itself from being cut away and stolen).

External fiber optic cables may have an extra layer of armoring to prevent damage from lightning or to discourage rodents from chewing through the cable. See armoring, ballistic.

Examples of Basic Fiber Optic Cables

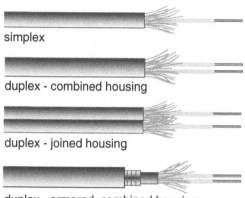

simplex

duplex - combined housing

duplex - joined housing

duplex - armored, combined housing

The top three cables are examples of basic, low-fiber-count, indoor fiber optic cables. The bottom is an internal/external armored fiber optic cable that is resistant to rodent and lightning damage.

cable, interconnect A cable intended for local connections such as fiber to the desk (FTTD), patch cables between systems, and point-to-point runs through buildings. Interconnect cables tend to be flexible, with a small diameter and bend radius, with adequate to good shielding. Interconnect cables may be single-mode or multimode and are usually compatible with standardized connectors.

Cable & Wireless (Marine) Ltd. The world's leading supplier of submarine telecommunications cable equipment and services in the 1990s, Cable & Wireless was restructured as part of Global Marine Ltd.

cable access, PEG By regulation, cable television broadcast providers must set aside and reserve channels for use by the public, educational institutions, and government entities (PEG). The cable company is limited from exercising editorial or content control over these public and government programming channels.

Cable Act of 1984 An Act of the U.S. Congress broadly deregulating the cable TV (CATV) industry. This significantly reduced the Federal Communications Commission's (FCC's) jurisdiction in this area. In 1992, the Act was partially repealed and further shaped by the Cable Reregulation Act of 1992.

Cable Act of 1992 Presented as the Cable Reregulation Act of 1992, it is now commonly called the Cable Act of 1992. This Act arose from the Cable Act of 1984 to provide a regulatory framework for steadily increasing cable services. Cable television

is among the more significant services affected by these regulations and the Act has recently been scrutinized for its relevance to interactive television services. See Telecommunications Act of 1996, basic cable service.

cable assembly A pre-assembled cable, ready for installation (e.g., fitted with jacks or other relevant attachments).

Armored Cable Assembly

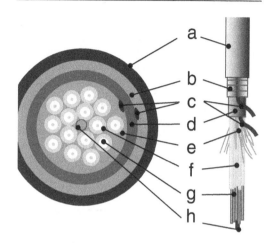

Armored cables have extra reinforcing in the cable housing to prevent damage. In this basic armored fiber optic bundle, the outer sleeve (a) provides protection against wind, solvents (e.g., water), and abrasion. It is often made of plastic (e.g., polyethylene). The next layer (b), between the sleeve and the inner jacket, is an armoring layer of materials that are difficult to cut, chew, or burn, such as steel tape. The armoring material also helps prevent the fiber from being inadvertently stretched during installation. Ripcords (c) may be provided directly under the armoring and the inner sleeve to aid in stripping the layer for splicing the cable to connectors or terminators. The inner sleeve (d) is a protective, often flame-retardant, layer to support the inner cable bundle. The inner bundle (e) includes torsion members, fillers, or other structures to support the numeric- or color-coded cladding layer (f) that keeps the lightwave within the fiber filaments (g). A central member (h) of solid or stranded plastic-coated steel, for example, may be included as a supporting structure. Gel-filled buffer tubes may surround individual fibers (not shown here).

cable bay An installation setup designed to accommodate many sets of cables in rows or a matrix, to facilitate easy access, identification, and maintenance.

cable connector An assembly for facilitating the coupling of cables to systems and components and to other connectors (e.g., adapters). Cable connectors come in many configurations and sizes in a variety of materials including plastic, ceramic, stainless steel, aluminum, resin, or rubber. Due to their flexibility in fabrication and relatively good water resistance, plastic connectors are common. Where electrical shielding is desired, metal may be used. A *cable assembly* is a cable with connectors already installed. In fiber optics, where preservation of the light beam is crucial to correct functioning, it is important to use well-designed and fabricated cable connectors. Splicing the fiber for attachment to a cable connector is also critical and special tools have been designed for this purpose. In multimode fibers, individual fibers may be color-coded or numbered to facilitate connections.

Labeling conventions are used in the fiber optics industry to identify certain configurations and coupling mechanisms (see Fiber Optics Connector Types chart). The abbreviations describing a cable connector are typically followed by a size. Thus, *SC/APC SM (3 mm)* would describe a standard connector with an angled physical contact supporting 3 mm single-mode fiber connections.

Good connections are critical to preserving the performance of fiber optic components. As new fiber optic connectors are developed, they may not couple directly with existing fiber optic instruments. Adapters may be available from the device manufacturer, for coupling with new cables. It is important to use adapters that preserve calibration accuracy if they are coupled to precision instruments.

cable core The inner, conductive portion of a sheathed or insulated cable. The fiber waveguide in a fiber optic cable.

cable coupler A hardware connector used to complete the circuit between similar or dissimilar cables with the same electrical or optical transmission characteristics. Depending upon the context, cable couplers may also be called *splice bushings* or *mating adapters*. A mating adapter enables the coupling of two cables with different connectors. See cable connector.

Cable Deregulation Act of 1992 In 1984, an Act of the U.S. Congress that broadly deregulated the cable TV (CATV) industry. The Act significantly reduced the Federal Communications Commission's (FCC's) jurisdiction in this area. In 1992, the Act was partially repealed and further shaped by the Cable Reregulation Act of 1992; rates were mandated to be lowered by an FCC-prescribed percentage in 1993 and again in 1994. It is now more commonly called the Cable Act of 1992. See Cable Act of 1992.

cable diameter An important indicator of the size and other properties of a cable. The diameter of a wire or fiber optic cable can dramatically affect its transmission properties, weight, flexibility, cost, and ease with which it can be interconnected with other components. It may also influence the distance over which it can carry a signal. See American Wire Gauge, Birmington Wire Gauge, fiber optic, multimode optical fiber, single-mode optical fiber.

cable drop The subscriber connection segment of a wired *cable access* installation. The cable drop is generally the section that originates at the *cable tap* on a utilities pole and ends at the subscriber's television or at a connector fed through the subscriber's wall to which the subscriber can hook the television.

Cable/Information Technology Convergence Forum CITCF. An organization representing the cable industry that was established to further communication between vendors and cable industry professionals.

cable loss An important property of a cable's transmission characteristics over distance. Loss of signal, or *attenuation*, is the gradual diminution of the signal to the point that it is no longer useful or can no longer be detected at the receiving end. Loss may be due to many factors that often occur together, including interference, the construction and materials used in the cable, the number of connections, the proximity of other conducting surfaces, the thickness, distance, weather conditions, etc.

Some types of transmissions can be carried over only a short distance. For example, external SCSI device cables are usually limited to lengths of about six feet and shorter cables are recommended. With Fibre

Fiber Optics Connectors		
Abbreviation/Type		**Notes**
ST	*straight tip, standard termination*	A metal or plastic housing with a bayonet connecting mount; a *ferrule* cylinder supports and aligns the fiber
FC	*fiber connector*	A threaded connecting mount, designed by NTT
SC	*standard/subscriber connector*	An inexpensive, plastic, molded connector commonly used in residential applications, designed by NTT; SC to SC and SC to ST duplex patch cables are common
LC	---	A newer, small-format coupler that takes up only about half the space of SC connectors when mounted in racks
The following abbreviations may follow the above connector types as suffixes (e.g., ST-PC):		
PC	*physical contact*	Promotes fiber to fiber contact in the coupling
AC	*angled contact*	Fiber end is polished at a slight angle, thus reducing movement at the coupled joint
APC	*angled physical contact*	
SM	*single mode*	
MM	*multimode*	

Channel technology, the distance for SCSI devices can be extended enough to use a separate maintenance room for aggregating the devices. For many types of optical fiber installation, the transmission distance may be only a couple of kilometers. See attenuation, multimode optical fiber, single-mode optical fiber.

cable map A diagrammatic record of the type and location of cables in a distribution system. For decades, hand-drawn cable diagrams were used. More recently, computer-generated diagrams and databases are sometimes used to keep track. To aid with local management, cables are frequently color-coded, marked, or bundled, and this information may or may not be redundantly recorded on the cable map.

Cable maps are particularly important in institutions and business complexes serving many rooms and buildings, and in submarines and ships that have substantial numbers of cables running through corridors and walls. In data networks, the physical cables between routers, switchers, and workstations are sometimes diagramed in the routing software, displayed as a color-coded schematic on-screen. These applications may have two types of maps included in them – a physical cable map, and a virtual connection map – one of which is laid over the other and each of which may be managed somewhat differently.

cable modem This phrase is used to describe both a device and a network service option that together enable computer access to high-speed transmissions via a broadcast cable network and subscriber service. Cable modem services are delivered through copper or fiber optic cables to the local drop (usually a *tap* on a utility pole ouside the premises) that leads to a connecting panel on the building. From there, the cable is connected to a specialized modem on the subscriber premises through a coaxial cable.

In many areas, the service is virtually 'live' all the time, although some ISPs will time out and reconnect specific Internet connections if the line is 'idle' for long stretches of time. Since the majority of consumer subscribers are assigned dynamic IP numbers, the ISP will time out the idle subscriber and assign the dynamic IP to another user, thus more efficiently utilizing resources. When the subscriber attempts to reaccess the service, a new IP number is assigned or, if the "lease" on the existing IP number is still available, the same number will be reassigned. Some cable modem users may have static IP numbers or may be using cable modem services to create a virtual local area network, in which case timeouts may not be used. Although many digital data service cables can handle two-way communications, most implementations tend to be asymmetric, giving more time to the downstream data (since most information is downloaded from source to user). Cable can deliver broadband, round-the-clock, fast access to data services such as the World Wide Web, without tying up phone lines. Cable modems enable users to download information about 20 times faster than ISDN modems, and about 80 times faster than 28.8K phone line modems. 360K cable modems began shipping in 1997; by 2000, the service was widespread in urban centers. The price

is competitive with ADSL and ISDN services. See Asymmetric Digital Subscriber line, ISDN.

cable plow A specialized plow designed to dig trenches specifically for the installation of underground cables on land or underwater. Modern ground-based systems can install numerous cables or cable conduits to depths of about six feet, depending upon the system, and at rates of up to several thousand linear feet per hour. Plows that create a trench by cutting are called open-cut plows, those that disturb the earth with vibrations are vibratory plows. Each has advantages but, in general, vibratory plow trenches are generallly easier to cover over again.

cable riser Vertical support for cables that are installed in walls and ceilings, in order to reach upper floors in multiple-story buildings.

cable run A conduit or other piping or pathing system that provides a means to thread cables or that constitutes the path of the cables. Cable conduit runs are used for a number of reasons. In some cases, they make it possible to add cable later, if the full cabling requirements are not known at the time the run is installed. They may provide extra insulation or fire protection, and they may be more aesthetic, enclosing a bundle of cables that might otherwise be distracting or unsightly.

cable tap In *cable access* installations, the tap is the physical connection, usually on a utilities pole, to which the subscriber line is attached.

cable television, community antenna system CATV. A television broadcast system that transmits licensed television programs and local programs to subscribers over a wired network, usually over fiber optic cable. Cable TV, that is, television broadcast delivered over wire, was established in Europe in the early 1930s, less than a decade after the viability of the television medium was first demonstrated. In North America, the distances between communities was much further, and cable TV was slower to develop. Satellite transmission options to cable television are becoming more widely available. See basic cable service.

Cable Television Relay Service Station CARS. A television relay station is a transceiving point between the original broadcaster and the subscriber. It may be a building facility, an unstaffed tower relay, a mobile relay, or other point at which the transmission is received, sometimes processed, and then retransmitted. For example, the broadcast may be sent over airwaves to the local relay station, which may be a local cable TV supplier. Once the various broadcasts are received, the local station subsequently sends the signals to the subscribers through physical cables. This way there is only one powerful antenna needed to serve the local area (otherwise each subscriber would need an antenna, the way it was before cable TV became available). It also gives the local station the capability of transmitting only those stations that the subscriber may desire or that fit the payment package arranged with the local station.

cable vault An enclosed area, often in a basement with extra fire-proofing, that encloses a large number of cables leading into a building or distribution frame. Cable vaults are usually used in situations where the type or number of cables pose extra fire or electrocution risks, and where it is desirable to restrict access to qualified personnel only.

cablehead The point at which a land cable and marine cable are joined. It may be indicated with a sign (as are seen on ocean beaches).

cableway A hole, slot, or other opening in a component unit or work surface that allows cables to be fed into the fixture from behind or beneath. Most computer desks now include slots and holes for cables. Some even put slots into the drawers, so a printer can be installed in a drawer and pulled out as needed.

Cabling Standards Update A quarterly newsletter describing high-speed network trends, technologies, standards activities, and economic indicators for a range of transmission media, including copper wire, glass and plastic optical fiber, coaxial cable, and wireless, published by Information Gatekeepers, Inc.

CAC 1. Customer Administration Center. A phone console used for maintenance and diagnosis of a multiline phone system. 2. See call admission control, connection admission control.

cache *v.* In the traditional sense, to cache something is to put it away or hide it in a secure place for later use. In computer terminology, the sense of putting it away for later retrieval is retained, but ease of access is also implied. Thus, to cache is to store information in an accessible location, as in RAM on a computer, so it can be retrieved quickly when needed. Many systems are specifically configured with cache memory, while others may use the hard drive as a cache location, which is not as fast but still may be effective in certain circumstances. Information is cached by an applications program for items that are often consulted or executed. This speeds up operations for priority activities. See cache, cache memory, RAM disk.

cache, cache memory A high-speed electronic memory buffer used in computing to increase apparent processing speed by more effectively managing resources. The cache storage is usually within a designated amount of random access memory (RAM) and thus is volatile, (although in its most generic sense, a hard drive would be a suitable cache device for a slow sequential storage medium like a tape drive). A hard drive controller card may itself include a cache.

The effectiveness of a cache depends upon a variety of factors, including the size of the cache, the ability of the software to utilize it, the types and variability of operations being done, the design of the caching logic, and the speed of the microprocessor. Since RAM access is typically faster than hard drive access, efficiency can be increased by storing frequently accessed information in the cache memory, where it can be written and retrieved more quickly than from disk. Information that is not found in the cache may then be added to the cache for future reference. In networking, a cache can be used to store frequently accessed information (often the locations or contents of data

files or applications) in order to serve it more quickly to users, as it is requested. FATs and hash tables may be stored in the cache to increase file access speed. BIOS device-controlling functions may be loaded from read only memory (ROM) into cache memory during a startup sequence. See cache, cache hit, cache miss.

cache controller In some computer architectures, a circuit that is specifically included to administer the storage, organization, and retrieval of cached information. This may be incorporated into a specialized chip.

cache hit A situation in which the data that was sought in a cache access was found and it is not necessary to access the slower storage medium (usually a hard drive), resulting in faster retrieval of the desired information. See cache, cache miss.

cache miss A situation in which the data that was sought in a cache access was not found and is consequently sought on the slower storage medium (usually a hard drive). It may subsequently be stored in the cache for future reference. See cache, cache hit.

caching Putting information in a storage area where it can quickly be retrieved when needed. It is a means of speeding up effective and perceived performance of a system. Disk caching and memory caching are two ways to speed up access to frequently used commands, device drivers, or frequently accessed data.

caching, data entry A means of speeding up data entry by retaining previously inserted information so that it can be reused or overwritten for subsequent entries. It's often more efficient to edit or retain the data in the field from the previous entry than to type it in from scratch.

CACM 1. California Association of Community Managers, Inc. 2. *Communications of the Association for Computing Machinery.* The communications journal of the ACM, one of the many ACM professional publications.

CAD 1. See computer-aided dispatch. 2. See computer-aided design/drafting.

cadence A rhythmic measure or beat. In telecommunications, many signals are identifiable by a pattern of tones and silences. Cadence has implications for telegraph, radio, and telephone communications where many audio signals are coded to particular rhythms (Morse Code, distinctive ringing, international variations in rings, and busy signals). People can learn to distinguish different types of data communications by pitch and cadence, as fax tones differ from data tones, and data tones vary further according to baud rate.

cadmium A bluish-white, malleable, ductile, metallic, noncorrosive element that is commonly used in protective coatings and platings.

CADS 1. code abuse/anomaly detection system. 2. computer abuse/anomaly detection system. Systems put in place to detect, log, and/or signal problems that may be due to tampering or other unauthorized use.

CAE 1. See Common Applications Environment. 2. computer-aided engineering.

CAFA computer-aided financial analysis.

cage antenna A multiwire antenna (imagine a ring of horizontal parallel wires somewhat constricted in the middle, resembling a cage) similar to a dipole antenna, configured to improve capacity and reduce loss.

CAI 1. See computer assisted instruction. 2. common air interface. An international interface standard defined to provide interoperability between wireless handsets and compatible networks.

CAL 1. CAN Application Layer. See CAN in Automation, Controller Area Network. 2. computer-aided learning, computer-assisted learning. See computer-assisted instruction.

CALC See customer access line charge.

calculator A device for facilitating fast, accurate mathematical computations. Early calculators (adding machines) could handle only simple arithmetic functions, while current ones include storage, automation, and programming capabilities for doing frequent or complex computations.

The calculator is the forerunner of the general purpose computer; in fact, the early computers were very large, very powerful calculators, and their histories run hand-in-hand. Then, as computer technology improved, memory and logic functions were scaled down and incorporated back into calculators.

In 1940, Remington Rand Inc. was advertising a "printing calculator" that was essentially a mechanical adding machine that used impact printing like a typewriter to record the tallies. Until the late 1960s, "advanced" calculators were too expensive for individuals and small businesses. By 1969, however, the Friden Division of Singer was advertising a desktop-sized version of an "electronic calculator ... that can remember up to 30 mathematical steps for you." Handheld calculators devised in the late 1960s became widespread in the early 1970s, costing about $200 for a very simple palm-sized arithmetical calculator. As prices came down, calculators superseded slide rules and abacuses for quick computations. Advanced calculators for under $15 are now commonplace.

The most celebrated early microcomputer was developed in 1974 by MITS, a company that was producing scientific calculators. With competition from bigger companies such as Texas Instruments, MITS needed a new source of revenue and developed the Altair. Since that time, calculators have been incorporated into many devices, such as cash registers and wrist watches, and even some computer keyboards. Current calculators include graphical displays, square root computations, multiple memory registers, and programming languages such as Forth. They are more powerful than computers from the early 19802. See abacus.

calendar routing An administrative method for directing inquiries according to the time of year, week, or day. Used especially in industries where inquiries are cyclic (travel industry), or where availability of personnel to assist callers is cyclic.

calibrate 1. To set, align, or mark a measuring or timing instrument according to an accepted standard. 2. To ascertain, record, or correct variations in a measuring or timing instrument with reference to another, or to an accepted standard.

California Education and Research Federation Network CERFnet. This research and education network was founded in the late 1980s by General Atomics, with aid from a National Science Foundation grant, and grew to be a national backbone by the early 1990s. CERFnet joined with other nets in 1991 to form the Commercial Internet Exchange (CIX). In 1996, the Teleport Communications Group Inc. (TCG), one of the largest competitive local telephone companies in the U.S., acquired CERFnet to provide Internet services to corporate and institutional clients. CERFnet is based on ATM and SONET architectures, with each Local Access and Transport Area (LATA) served by at least two backbone nodes.

call 1. *v.i.* To attempt to contact or to succeed in contacting another party or entity. A unit of virtual or human communication across some type of communications medium or at some distance. 2. *n.* A unit comprising a successful communication through some type of communications medium, or at some distance, between two or more parties or entities who are more or less simultaneously in contact, frequently with a 'give-and-take' character to the contact. Human participants in a call are generally called *parties* to the call. 3. *n.* In networking, a communications association between a user and a network entity or between two or more users across the network.

call abandons, abandoned calls Calls that are terminated by the originator before completion of the intended contact. For telephone calls, reasons for abandoning calls include fuzzy connections, wrong numbers, answering machines, being put on hold, obnoxious hold music (not all hold music is obnoxious, just some of it), even more obnoxious hold commercials, interruptions (children, doorbells), transfer to the incorrect person or department, etc.

Since any call connection in progress has impact on system capacity, abandoned calls have to be considered when structuring and managing a system. In commerce, if a high proportion of abandoned calls occur after a human operator has made verbal contact with the caller, it's important to determine and evaluate the reasons and take corrective measures to increase call completion. See abandoned call cost.

call accepted signal In telecommunications, a call control signal sent by the receiving data terminal equipment (DTE) to indicate acceptance of the incoming call.

call accounting system A system of recording the type and quantity of calls on a system. This information was originally recorded manually, and operators of public phone systems had elaborate card systems on which to record calls, particularly long-distance calls. Now accounting has been computerized, and the system can constantly monitor call volume, number of connects, number of abandons, peak hours, trunk allocation, and other statistics related to economics in general and call billing in particular. When used in private branch systems, it can further be used to track agent activities, length of calls, departmental use, etc. and integrated with revenue and customer databases to give an overall picture of the role of the phone calls within the company's business. See call card.

call admission control, connection admission control CAC. The set of actions taken by a network during a call setup or renegotiation to evaluate whether to accept or reject a connection or re-allocation request, based partly on the ability to supply Quality of Service (QoS). See crankback.

call announcement A feature in a telephone system in which an operator or other agent announces the call to the callee before connecting the call.

call appearance CA. 1. A telephony designation for equipment that provides easy handling of volume calls on a central console. Call appearance refers to the ability to see the calls that come in, usually through LCD indicators and/or LED displays for the various extensions, in order to manage and direct the calls and monitor which lines are in use. 2. A general reference to the format in which call information is displayed on a monitor or CED telephony display. In some systems, the user can custom configure Caller ID and other information.

call attempt Initiating a call that may or may not be completed. If a large number of call attempts are not completed, diagnostic and troubleshooting steps should be taken. Solutions may include training, additional lines, staff changes, or equipment changes or repairs. See call abandons, abandoned call cost.

call barring Prevention or elimination of all calls, or specific calls, associated with a specific phone. Usually implemented to prevent unauthorized use, or abuse. See call blocking.

Call Before Digging A safety sign to warn area workers that they must call for information on underground cables or hazards before digging.

call block A restriction put on a phone line to prevent connection of certain calls. More recently it has come to mean retaining anonymity from Caller ID by blocking the caller's identity from the receiver if the caller has keyed in a blocking code. Call blocking in this sense is free, whereas Caller ID costs money. This situation may seem backwards, like allowing a stranger through the door unseen, while the person opening the door has to pay money to see who it is and still may not get the information because that person is disguised (blocked). In terms of personal safety and security, it should have been set up the other way around, with the person answering the phone being provided the identity of the caller for free, and the caller having to pay to hide his or her identity. However, the system was probably set up with Caller ID as a subscription service because it generates more revenue for the phone company.

call card A manual call management and billing system in which the information about the caller, callee, distance, and duration of the call is recorded by the operator. See call accounting system.

call center A centralized telephone call facility handling a large number of calls. Call centers may be specialized for handling many incoming calls, such as those resulting from television marketing through toll free numbers, or many outgoing calls, as those originating from telemarketing or teleresearch firms. In these specialized environments, automatic call distributors, head sets, computerized dialing, and automated answering are commonly used.

call clearing The process by which a call connection is released and the call resources made available to other users. It is particularly important in end-to-end transmissions in which the line must be freed before it can be used again.

call control The entire process of detecting a call request, setting up the physical and logical connections, rerouting to available trunks if necessary, facilitating transmission, shutting down the call, and freeing the resources for other callers. Most of these functions are now computerized, although occasional operator assistance, directory assistance, or services for special needs users are still handled by human operators. Call control may go through more than one system, as when a call goes into or out of a private branch exchange (PBX) and through a public exchange, or where wireless and wireline services from different providers are used together to complete a call.

call control signal Any signal used in automatic connection and switching systems that controls the call sequence. In older systems, the control signals were transmitted by means of tones on the same line that was used for the voice transmissions. In newer Signaling System 7 (SS7) systems, the control signals and the voice transmissions are handled over separate channels. See Signaling System 7.

call data The statistical information associated with a call. This is used for monitoring, accounting, management, and planning, and these days is usually stored in a computer database, and sometimes organized and analyzed by computer software.

Call Detail Record CDR. A telephone record-keeping system, usually used for accounting and administrative purposes, that tracks and records details about incoming and outgoing calls such as the call duration, caller and/or callee, time of day, etc.

call diverter A subscriber surcharge service or phone peripheral device that intercepts an incoming call and forwards it to a phone operator or phone message, or to another number, as in Call Forwarding. Depending upon the service or device, the caller may or may not be aware that the call has been diverted.

call duration The period of time from actual connection of the call, until its termination. On phone lines and data networks, call duration information is used for statistical purposes for tuning the system, determining peak hours, and billing. It may also be used to detect and diagnose fault conditions.

call establishment The process of routing and connecting a phone call or data transmission path.

Call for Votes CFV. A formal process used as part of the sequence of events necessary to create a new public newsgroup on USENET.

Call Forward A surcharge or bundled phone service that permits the subscriber to automatically redirect an incoming call to another number. It is useful in cases where the callee is temporarily at another location, or where the callee wishes someone else to handle calls (such as an answering service). On consumer systems, the call forwarding is usually enabled by using a touchtone phone to dial a code (72# on a touchtone phone or 1172 on a rotary phone in N.A.) followed by the number to which the calls are being forwarded. It is disabled by dialing a code (73# or 1173 in N.A.). Some newer phone systems have an indicator light to show that the calls are being forwarded to prevent the subscriber from forgetting to deactivate Call Forward after returning to the original location. See Call Forward Busy, Call Forward No Answer.

Call Forward Busy Similar to Call Forward, except that calls are rerouted to a predetermined number only if the called number is busy; otherwise it rings through to the original number. See Call Forward.

Call Forward No Answer Similar to Call Forward Busy, except that calls are rerouted to a predetermined number only if not answered after a specified number of rings. See Call Forward Busy, Call Forward.

Call Forward Variable A combination of Call Forward Busy and Call Forward No Answer in which the call is rerouted to a predetermined number if a busy signal is encountered, or if there is no answer after a specified number of rings. See Call Forward No Answer, Call Forward Busy, Call Forward.

Call Girls One of the many colloquial names given to the early female telephone operators. Others include *Hello Girls*, *Central*, and *Voice with the Smile*. See telephone history.

call handoff In mobile phone systems based upon passing the transmission on to another transceiver while the call is taking place, as in cellular communications, the handoff is the point at which the call is transferred during the conversation. Mobile providers strive to create systems where the handoff is seamless and does not create delays, noise, or significant volume changes.

Call Hold A surcharge phone service or multiline subscriber feature in which the subscriber can put a call on hold, accept or place a second call, and then return to the original call. This service is similar to a hold button on a multiline phone, and the person on hold is not able to hear the second conversation.

call horn alert A mobile system set to beep a car horn to signal an incoming call, when the driver is away from the mobile handset or receiver.

call mix Telephone calls are of many kinds, as are logons on a computer. In a telephone system, the calls may be long or short; busy, abandoned, or completed; local or long-distance. On a computer system, the logons may result in downloads, modem access, running of applications, file maintenance, etc. The call mix is a statistical look at the types of usage that occur on a network.

call not accepted signal In telecommunications, a call control signal sent by the receiving data terminal equipment (DTE) to indicate rejection of the incoming call.

Call Park A subscriber service or console feature that allows a user to set the call so it can be answered on any other phone on the system. Call Park is useful in situations where the callees are moving around, and where they may be alerted to the presence of the call through a paging system. The parked call can then be retrieved from another line by dialing in a call unpark code.

Call Pickup A surcharge phone service or multiline subscriber feature that permits a subscriber to intercept a call to another prearranged number by typing in a code and then answering the other call. Suppose you and your housemate have separate lines, and your housemate has asked you to answer his or her calls; you can do so from your own phone. See Call Pickup Group.

Call Pickup Group CPUG. All the phones in a system through which Call Pickup is activated and that can intercept the calls of the others. See Call Pickup.

call processing A combination of computer and human operations in which the call is often set up and connected electronically and then handed off to a credit collector, researcher, telemarketer, technical supporter, or other agent, once the connection has been established. See call center.

call progress signal A telephone switching signal that indicates whether the call is generating a busy tone, a ringback tone, or an error. See ringback.

Call Record A data record of call details, which includes information such as date and time, call duration, call routing, stations used, time on hold, etc. This information may be used for billing and administration.

Call Rejection A subscriber surcharge or bundled telephone service that enables the callee to reject an unwanted call. There are two ways to put Call Rejection into effect. The subscriber can dial *60 (in N.A.), and follow instructions for entering an originating number to be rejected, or can activate Call Rejection immediately after hanging up from a call that is unwanted in the future. Call Rejection can be deactivated by dialing *80 (in N.A.).

call release time The duration during which a call is shut down and the line released for the next call.

call reorigination A handy feature in which calls can be initiated one time during a multicall session with a debit card, charge card, credit card, or calling card account. In other words, a series of calls can be made at the same time without having to re-enter codes or having to re-insert the card to make the subsequent calls. Between calls, a code is usually pressed, and the caller receives a signal to continue with the next call. This feature is particularly useful when having to make several calls at an airport to let people know your flight plans have been changed and you are catching a plane at a different departure gate.

call restrictor A physical or virtual call blocking mechanism that controls the type of outgoing calls that can be made on a line. Examples include blocking long distance calls from a phone near a public area, or blocking 900 calls from phones used by teenagers.

Call Return A subscriber surcharge option that allows the last caller, whether the call was answered or not, to be dialed back automatically. It can be handy for crisis centers and other emergency services.

call routing tree A diagrammatic representation of call routing configuration and logic. See call tree.

call screening The most familiar call screening is a receptionist who says the boss is in a meeting and can't be reached at the moment when the boss is actually watching the World Series with his or her feet up on the desk. More legitimate uses of call screening involve getting enough information from the caller to direct the call to the best person equipped to handle it. In automated systems, call screening is a setup that uses Caller ID, or some other identification tool, to monitor the origin of the call and to patch it through accordingly, or that uses a speech recognition system to direct the call.

call sequencer An automated system for evaluating incoming calls, queuing them if necessary, and assigning them to agents depending upon priority, availability, or caller characteristics.

call setup time In a circuit-switched network, such as most phone networks, the amount of time it takes to patch through the route from the caller to the destination in order to set up an end-to-end path for the communications. During the course of a call, the resources are dedicated to that communication and cannot be used by others. For a phone call, the call setup time includes the time it takes to dial and for the call to be switched through the system and the appropriate trunks to the destination. This time is usually not billed for land lines (wireless may be billed for air time) since it is not known during setup whether the call will be answered and how long it will last.

call shedding A situation in which automatic call handling systems are used to drop (shed) a phone connection if no sales agent is available to talk to the callee when he or she picks up the line. If all agents are busy when a callee answers an automatically dialed call, the callee may hear a recorded message and be put on hold. More commonly, however, the call is shed; the callee hears a click and a dial tone and has no opportunity to speak to a human agent. This practice is very annoying and illegal in many areas.

call sign See callsign.

call splitting A subscriber surcharge or private branch service in which a conference call participant can speak to any one of the other members of the conference privately, that is in nonconference mode. When a phone attendant is involved in the call, the attendant may relay the information privately to one of the called parties.

Call Stalker An AT&T commercial software package providing 911 emergency service agents with information about the caller, such as address and calling phone number.

call supervision A process for determining whether a telephone communication was actually answered, so billing is not activated unless a connection was made.

Call Trace A surcharge phone service or emergency service in which the tracing of the origin of the last call is provided and recorded in case it may be needed later for legal reasons. The results of the trace are not given to the customer under privacy laws but may be revealed later through proper legal channels.

Call Transfer A surcharge phone service, or capability of a multiline phone system, that allows a call to be transferred to any other phone on the system. Transfers are accomplished by typing in codes and the transfer number, or by keying a transfer button followed by the callee's line. Call transfer is commonly used in business, and the console often staffed by a full-time operator or receptionist. Callers are not tolerant of calls that are incorrectly transferred or accidentally terminated, and it's important that personnel responsible for transferring calls are well trained on the equipment and in business etiquette.

call tree A diagrammatic representation of call sequence information (usage) used for statistical analysis and planning. See call routing tree.

Call Waiting CW. A surcharge or bundled phone service that becomes active if a call comes in while the callee is already engaged in a call. Call Waiting signals the callee, either by an audio signal or blinking light, that there is another party trying to call, and provides the callee the option of ignoring, terminating, or holding the current call and then answering the second incoming call. This is useful for emergency calls or for ending a casual conversation to carry on with other calls.

Call Waiting can interfere with a transmission, or even cut off a call if the line is connected through a computer modem. Call Waiting can usually be temporarily disabled to avoid this problem, or the modem can be reconfigured to ignore this type of interruption. The first option is easier and preferable. Information on how to disable Call Waiting is listed at the front of most local phone directories.

Call Waiting for ISDN is specified within ITU-T Q.83 and Q.733 call completion services.

Call Waiting ID A surcharge or bundled phone service that combines Call Waiting and Caller ID capabilities. This enables the callee to determine the Caller ID (origination) of a call that is queued and waiting through Call Waiting services. The service requires a Caller ID-capable phone with a display to show the Caller ID data. See Call Waiting, Caller ID.

Callan, Nicholas J. (1799–1864) An Irish priest and educator who devised a historic induction coil in 1836. He also researched various aspects of electromagnets, condensors, and batteries. See induction coil.

callback facsimile A system in which you (1) dial a callback service, (2) key in your callback phone number, (3) identify the documents that are of interest (usually from a numerical list given by a voicemail system), (4) hang up, and (5) wait for a callback-enabled facsimile machine to automatically dial your fax machine and deliver the documents requested. A significant proportion of computer industry technical support and product information is now delivered this way. In the future, callback fax systems will likely be superseded by more flexible and economical email document delivery systems.

callback modem A modem that is set to receive a phone call through a network that acts as a callback request. A password may be required, and then a phone number to be dialed is provided to the system. The modem then is set by the computer to dial the number provided. Why do this instead of dialing directly? This system provides better security, so there is a record of numbers that have been connected to the network and data access. Sometimes toll charges are reduced; the toll is billed to the network number and handled by the business accounting office, rather than being billed to an employee.

Callender Rapid Phone Company One of the earliest automatic switching phone services, established in England in 1896 by musician and inventor Romaine Callender.

Callender, Romaine A Canadian music teacher/instrument maker and associate of A. Graham Bell, Callender founded the Callender Telephone Exchange Company in Ontario, Canada. Between 1892 and 1896, he submitted three series of patents for telephone switching inventions. He failed in trying to implement them in Ontario and subsequently traveled to New York to seek financing and open another firm. Traveling with him were two brothers, George William Lorimer and James Hoyt Lorimer, who assisted him in further experiments. The brothers finally succeeded in developing an automatic switching system in 1895. They later returned to Brantford, Ontario, and Callender sailed to England in 1896, where he formed the Callender Rapid Telephone Company. See Lorimer, George and James.

Callender switch A very rudimentary, early telephone switching system developed by Romaine Callender and the Lorimer brothers in the late 1800s. See Callender, Romaine; Lorimer switch.

Caller ID, Call Display A phone carrier 'added value' pay service that provides the call recipient with the phone number identity of the calling party. You may have to pay local and long-distance Caller ID charges separately. In North America, the Caller ID information is usually passed to the receiving phone between the first and second ring.

You need two things to take advantage of Caller ID: a subscription through the phone carrier to the Caller ID service and a phone or separate device with a Caller ID display. See call blocking, Class, ANI.

Caller Independent Voice Recognition An automated voice recognition system that can interpret voice input without being specifically tuned to a particular caller's voice. It is useful in phone applications that accept spoken numbers or commands for processing a call and in voice recognition word processing applications.

Caller Name A phone carrier added value pay service that takes an incoming Caller ID number (assuming the call is not blocked), looks it up in a directory listing database, and transmits the Caller ID number and its associated listing, if it exists, to the recipient's add-on Caller Name display or to a phone providing Caller Name display. This is not as flexible as a user-configured system where you can associate any name

or code you wish with a specific incoming number, but it is very useful for identifying a first-time caller or stranger (and it may be possible to use them together if you have compatible peripherals). See Caller ID, call blocking.

calling card A remote or off-premises phone service provided by common carriers to allow local and long-distance calls to be charged back to the subscriber's local phone number or other authorized billing number. There may or may not be surcharges associated with such a call. The name derives from a wallet card typically issued to the subscriber with instructions and digits to be dialed to gain access to the service. In many cases, you don't need the physical card to make the call, but automated phones are becoming prevalent in which the card is physically inserted in a slot or swiped through the phone to expedite the processing of the call.

Bell Canada claims a trademark over the Calling Card name, but the term is widely used in the generic sense, making it difficult to enforce the trademark.

calling jack In manual switchboard systems, the jack that is used by the operator to connect the call that came in through the *answering jack* to the circuit for the subscriber who will be receiving the call.

calling number display See Caller Name, Caller ID.

calling party, calling station A person or entity originating a call. See call.

Calling Party Number CPN. In telephony, a common channel signaling (CCS) parameter in the initial address message that identifies the calling number and is sent to the destination carrier.

calling sequence The sequence of numbers, letters, steps, and other information needed to connect a call through a traditional phone line or digital computer phone system. When calling through a modem, the calling sequence includes not just the number being dialed, but also the parameters for the line, the baud rate, whether it is pulse or tone, the speaker level, pauses, wait for tone to continue with extension numbers, etc. In computer software, the calling sequence may include linking to an address book or other database and saving statistic information gathered on the call.

CallPath A computer telephony integration (CTI) software product developed by IBM for integrating voice and data communications for telecommunications call centers, thus enabling them to function as more advanced *contact centers*. CallPath is open architecture software supporting multiple computer and telephony switch platforms. CallPath can be interfaced with Web-based applications and other in-house front-end applications. JTAPI implementations such as JavaTel can be run on top of Callpath.

In May 2001, Genesys Telecommunications Laboratories, a subsidiary of Alcatel, announced plans to purchase the Callpath assets with the intention of working jointly with IBM to deliver contact center solutions based on Genesys interaction management products and IBM DirectTalk and WebSphere platforms.

callsign, call sign In radio communications, a series

of identification characters assigned by local regulating authorities to every licensed radio operator or station. The callsign identifies the country, and sometimes also the region of the country. One of the most famous callsigns in radio history is 8XK which Frank Conrad used from his Pennsylvania garage, and which was later licensed as the history-making KDKA radio broadcast station. See KDKA.

calorie A unit of expended thermal energy – the amount required to raise the temperature of one gram of water by one degree centigrade (C) from 14.5 to 15.5°C. One calorie equals 4.186 joules. (Note: a dietary calorie is actually 1 kilocalorie – 1000 calories). See calorimeter, joule.

calorimeter An instrument for measuring energy expended as heat. For example, calorimeters can help assess the power of a laser by providing a reading on the amount of heat absorbed by the beam.

A calorimeter can be designed in a tower geometry that includes scintillating tiles. The application of energy in the active scintillator elements produces light, some of which may be re-emitted and shifted as to its wavelength with coiled wavelength-shifting fibers within the tile structure. The light can then be transmitted through an optical fiber to a phototube or photomultiplier tube where it is converted to electrical energy. See calorie, scintillator, wavelength-shifting.

CALS Continuous Acquisition and Life-Cycle Support (formerly Computer-aided Acquisition and Logistics Support). A Department of Defense (DoD) strategy for the creation, use, and exchange of weapons-related digital data.

CALSCH The calendaring and scheduling working group of the IETF. See iCalendar.

cam A compact digital or analog video camera intended for use in one location for an extended period, as opposed to a still or video camera that is carried around with the user to many locations (a few very tiny cams are intended to be body-worn or carried in a purse or briefcase). Cams are also distinguished by having few or no controls other than basic aiming and focusing capabilities. Cams are usually mounted on small stands or may be hidden within other real or simulated devices such as clock radios and smoke detectors. They may be wired or wireless and are often interfaced with a computer or a VCR for transmission or recording of the cam images.

Cams are becoming very popular for videoconferencing, baby monitoring, Internet security, remote monitoring (children at a day care center or wildlife that has been released after being fitted with a cam), and surveillance activities. In recent years the price of a small, high resolution color camera has dropped from $600 to less than $100. See camcorder.

cam, stump A small, high-quality video camera in a housing that resembles a bottle cap or tiny tree stump. These are favored for sports broadcasting applications as they can be easily mounted on helmets, stadium fences, sporting animals (hunting dogs, polo ponies, etc.). They are also suitable for use on remote sensing platforms and for surveillance applications.

CAM 1. carrier module 2. Call Accounting Manager 3. Call Applications Manager. A Tandem telephony software interface for linking computers with telephone switches. 4. See computer-aided manufacturing. 5. computer-assisted makeup, composition and makeup. A WYSIWIG terminal for previewing type composition and page layout. 6. See camcorder.

CAMA See Centralized Automatic Message Accounting.

camcorder A combination digital or analog video recorder and camera unit. Increasingly, consumer camcorders include playback, editing, and special effects capabilities. Newer digital camcorders can be used as both digital still-frame and motion recorders and can be interfaced directly with software for scanning, image processing, and Web applications. Camcorders may eventually supersede analog video cameras and still film cameras, since no film processing is required, and consumers frequently favor convenience over image quality (35mm film is about 16 times higher resolution than current consumer digital systems but digital quality is improving steadily). See cam.

Cameo Personal Video System A Macintosh-based commercial videoconferencing product from Compression Laboratories Inc. that supports audio, video, and file transfers. It works over Switched 56, ISDN, and Ethernet networks. Cameo uses a proprietary CLI PV2 compression scheme. See Connect 918, CU-SeeMe, MacMICA, IRIS, ShareView 3000, VISIT Video.

Campillo See Salvà i Campillo, Francesc.

campus A physical and geographic environment (primarily the grounds) associated with learning and/or research facilities, such as universities, hospitals, and some businesses.

campus backbone The primary network of wires/cables that interconnect a campus. See backbone circuit.

Campus Wide Information System CWIS. A system of interactive kiosks and public information sources that provides directories, product or course offerings, maps, calendars, and other general public services of interest to educational institutions, businesses, expositions, and shopping complexes.

CAN 1. Control Area Network. 2. See Controller Area Network.

CAN in Automation CiA. A nonprofit trade association founded in March 1992 to provide technical, product, and marketing information to promote and support Controller Area Network (CAN) technology. The association further develops and supports CAN-related higher layer protocols, including the CAN Application Layer (CAL), CAN Kingdom, CANopen, and internationally recognized standards. http://www.can-cia.de/

Canada Machine Telephone CTM. One of the earliest phone companies to use automatic switching, technology that was developed jointly by George and James Lorimer and Romaine Callender. The Lorimer brothers established CMT in Peterborough, Ontario, Canada in 1897, and there produced the first commercial Callender Exchange. The Lorimers continued to improve upon the technology until it bore little resemblance to the original Callender switching system. The company lost its technical expertise when James Hoyt Lorimer died, but his brothers George and Egbert continued to market the products in North America and Europe. Unfortunately, due to lack of reliability and long installation times, the company didn't thrive and was acquired by Bell in 1925. See Lorimer switch.

Canadarm A remote manipulator system designed and made in Canada for the U.S. space shuttle program. The National Museum of Science & Technology has constructed a full-size replica and produced an accompanying video for a traveling exhibit.

Canadian Amateur Radio Advisory Board CARAB. A nonprofit consulting group comprised of members of the Radio Amateurs of Canada (RAC) and the Radio Regulatory Branch of Industry Canada (IC). CARAB works as a communications liaison between RAC and IC. http://www.rac.ca/carab.htm

Canadian Association of Broadcasters, L'Association canadienne des radiodiffuseurs CAB/ACR. A trade organization founded in 1926 by 13 broadcast pioneers. The CAB supports over 500 radio, television, and specialty broadcast providers in Canada.

Canadian Broadcast Standards Council CBSC/ CCNR. An organization incorporated in 1990 to encourage high standards of broadcasting and professional conduct by private radio and television broadcasters. The CBSC keeps broadcasters informed about societal issues, administers codes of industry standards referred by the Canadian Association of Broadcasters (CAB), and provides information resources to the public. http://www.cbsc.ca/

Canadian Broadcasting Corporation CBC. The primary broadcasting organization of Canada, CBC is a public broadcasting service providing television and radio programming in both English and French. The CBC was initially established in 1936 to ensure Canadian content in broadcasting. CBC's first television broadcast took place in 1952, in Montreal. In 1966, it began color broadcasting, the first in Canada to do so. See ANIK, CKAC. http://www.cbc.ca/

Canadian Business Telecommunications Alliance CBTA. A national, nonprofit organization representing over 400 businesses and telecommunications users in Canada. The CBTA supports members and facilitates Canada's competitive participation in telecommunications markets through quality and innovation.

Canadian Datapac The world's first public data network which began operating in 1976.

Canadian Independent Telephone Association CITA. A national trade association supporting independent telephone service providers, founded in 1905. CITA is based in Toronto, Ontario and supports members in B.C., Ontario, and Quebec. CITA promotes the advancement and use of telephone services in communities served by its members and represents its membership in regulatory matters. http:www.cita.ca/

Canadian Information Processing Society CIPS. Founded in 1958 as the Computing and Data Processing Society of Canada, it became CIPS in 1968. CIPS defines and promotes information processing in Canada and supports the information technology (IT) profession.

Canadian Journal of Communication CJC. A scholarly professional journal that deals with many historical and sociopolitical aspects of communications in Canada and abroad.

Canadian National Museum of Science & Technology, Musée National Sciences & Technologie Canada's largest technological museum, located in southeast Ottawa, featuring permanent and special exhibits, traveling exhibits available for loan, school programs, workshops, lectures, publications, and more. http://www.nmstc.ca/

Canadian Radio Television and Telecommunications Commission CRTC. The Canadian regulatory commission, based in Ottawa, Ontario. This important organization is similar to the Federal Communications Commission (FCC) in the United States in that it allocates frequency spectrums and carries out other commercial and amateur radio and television broadcasting administrative functions.

Canadian Satellite Users Association CSUA. A trade association of broadcasters using Telesat facilities and suppliers of goods and services to CSUA voting members. THE CSUA sponsors an annual trade convention. See ANIK, Canadian Space Agency.

Canadian Space Agency CSA. One of the more ambitious of the CSA's various projects was the Communications Technology satellite (HERMES) project which was undertaken jointly with the U.S. Canada was to supply the satellite, and the U.S. the traveling wave tube amplifier. This high power, high frequency, communications satellite project got underway in 1971 and was intended to test direct-to-home broadcasting technology. HERMES was successfully launched in 1976 aboard a three-stage rocket. The satellite operated for almost twice its expected lifetime, almost four years.

Canada competes at the international level in spacecraft assembly, integration, and testing through its David Florida Laboratory (DFL), west of Ottawa, Ontario, established in 1972. Besides the HERMES satellite, the CANADARM and various ANIK satellites have been developed and manufactured at the DFL. See ANIK.

Canadian Standards Association CSA. A Canadian, independent, not-for-profit standards-setting body established in 1919. The CSA is a strong participant in international standards discussions and directions. It engages in a consensus approach to standards adoption and provides educational services, including publications, conferences, and seminars. The CSA operates a Certification & Testing Division and indicates that products or systems have passed a formal evaluation process at stated levels.

The CSA is recognized by the U.S. as a Nationally Recognized Testing Laboratory (NRTL), in order to eliminate the need for duplicate testing for products

C

marketed in both Canada and the U.S., and provides assistance to manufacturers marketing to the European Union.

CSA has an official mark recognized as indicating a product or system that meets certain industry standards. See Standards Council of Canada.

Canadian Telecommunications Consultants Association CTCA. A Canadian association of independent telecommunications consulting professionals. http://www.ctca.ca/

Canadian Wireless Telecommunications Association CWTA. A trade association representing the Canadian wireless telecommunications industry, including satellite, cellular, and other mobile communications services.

cancel Stop a process, function, or action. On a copying machine, to abort the current copy if it has not already gone through the machine and any additional copies that may have been requested.

In a computer application, to stop or abort the current operation or process. Control-C (two keys held down together), sometimes designated as ^C or *Ctrl-C* is a very common key code combination for aborting a process. It should be used with care as it may abort the user right out of the program. In many applications, a Cancel button is provided to close a dialog or window without carrying out any actions (when you change your mind), or to stop a process in progress. In some older systems, ^Y works in a manner similar to ^C. ^Z is somewhat related, and usually less dangerous; it may suspend the current process (rather than closing it down) and allow you to carry out other activities, so you can later return to the original process. With Unix system shell commands, a process can be resumed with *fg* (foreground) after having been suspended with ^Z.

On phone systems, many services are enabled and disabled, or canceled, by typing in two or three digit codes, sometimes followed by a # or * symbol. This applies to services such as Call Forwarding, Call Waiting, etc. It is advisable to cancel or disable Call Waiting before using a modem on a phone line in order not to be interrupted during a big data transfer. The codes for the subscriber's region for disabling various services are usually listed at the beginning of local phone directories.

candela (*abbrev.* – cd) A unit of luminous intensity, originally based upon the quantity of light generated by a single candle, it was later more precisely defined as the quantity of illumination emitted by a black body heated to the temperature at which platinum changes from a liquid to a solid state. See luminous intensity.

candlestick telephone A style of desk phone popular in the early 20th century that is characterized by a broad base with a slender, candle-like stem with a receiver on the top. The speaker was usually a separate unit, attached with a cord to the stem, which was hung on the stem when not in use. Some versions of the candlestick were adapted to hang on a wall, with a solid or accordian-style mounting bracket (e.g., the Western Electric accordian candlestick).

CAP 1. See carrierless amplitude and phase modulation. 2. See Cellular Array Processor. 3. See Competitive Access Provider.

Capabilities Exchange In Data Link Switching (DLSw), a Switch-to-Switch (SSP) control message that describes the characteristics of a sending Data Link Switching (DLSw) router to allow inter-router information exchange and to provide greater compatibility among different implementations. See Data Link Switching.

capacitance (*symb.* – C) The ratio between an electric charge and the resulting change in potential, or the time integral of the rate of flow of electric charge, divided by the related electric potential. Capacitance is measured in farads. See capacitor, capacity, Leyden jar.

capacitor An arrangement of conductors separated by dielectrics, which may be fixed or variable, designed to store electrical energy. Capacitors are used in a wide variety of electronic devices. See capacitance, capacity, condenser, Leyden jar.

Historic Capacitor

From Leyden jars to tiny solid state components, various means of storing electrical energy have been devised over the years. This historic capacitor from the American Radio Museum illustrates how much capacitors have changed, as most electronic capacitors now resemble stubby battery housings.

---

capacity 1. The maximum number of objects or occupants that can be contained on or in a system or environment under normal operating conditions (such as load, theater, or bridge capacity). 2. The maximum information-carrying capability of a communications system. The unit of capacity varies from system to system; on a network, it might be described generally in terms of number of users, or more specifically in terms of a calculation based upon speed, access, or load upon a CPU, or it may be based upon transfer rates for cells or frames.

Capasso, Frederico (1940s–) An Italian-born Bell Laboratories scientist who has made numerous contributions to electronics, particularly photonics. Capasso has contributed to bandgap engineering innovations in optoelectronics, semiconductor, and solid state electronics and, in 1994, co-invented the quantum cascade laser (QC laser). Capasso has developed components that function in ways not previously observed in nature and that are based on relative thickness and proximity, rather than chemical composition. See quantum cascade laser; Townes, Charles H.

Cap'n Crunch An infamous phone hacker (phreaker) from the 1970s and 1980s, John T. Draper (ca. 1943–) adopted this handle (techie nickname) and served a sentence for illegal (albeit creative) tampering with the phone system using technology and tones to make unpaid-for long-distance calls. His adventures and discoveries resulted in the phone company making some significant changes to their technology and plugging a number of security loopholes. Some of his exploits are described in Stephen Levy's book *Hackers* and in a 1971 article in *Esquire Magazine* entitled "Secrets of the Little Blue Box."

Legend has it that John Draper's monicker stems from a whistle he acquired from a cereal box of the same name, one which produced a 2600-Hz tone which could be processed by the phone trunk system as a hangup signal when blown into the telephone mouthpiece (a tip he received from a blind fellow). The line would stay connected, but the call would not be billed. This type of caller signaling is not possible on newer phone systems which use out of band signaling, because the voice conversation and the phone control signals are on different circuits.

Draper became associated with Steve Jobs and Steve Wozniak and wrote the first word processor for the Apple II computer, called TextWriter (which became EasyWriter).

In 1985, Draper wrote a series of Amiga computer technical tutorials, which he distributed free over the net, at a time when the Amiga was an underappreciated new entrant to the field of multimedia microcomputing. He now creates computer intrusion detection systems. See blue boxing.

Capstone chip A hardware security device that uses the same SKIPJACK cryptographic algorithm as the Clipper chip. It incorporates a Digital Signature Algorithm (DSA), a Secure Hashing Algorithm (SHA), a public key exchange, and various associated mathematical algorithms. It's a complex, powerful system, requiring almost 1 Gigabyte on an automated design system to set up the chip. The chips are being installed in various electronic devices for the U.S. Defense Messaging System. See Clipper chip, Pretty Good Privacy.

Capture Division Packet Access CDPA. A packet-oriented cellular communications network architecture designed to handle constant bit rate (CBR) and variable bandwidth multimedia telephony applications such as videoconferencing. Unlike some other protocols, CDPA is bandwidth-adaptable; it can support increased channel access for individual users for brief periods.

capture effect, captive effect In radio communications, signals often compete with one another if the frequencies are very similar or if two stations are coming in with approximately similar strength. In amplitude-modulated (AM) transmissions, the two

sound sources will be heard overlapping one another, and it's hard to make out what is being heard. In frequency-modulated transmissions (FM), the receiver will filter out the weaker signals, resulting in the capture of the weaker signal and the exclusive broadcasting of the stronger one. If the signals are equal in strength, the receiver may switch back and forth between the two, but it won't play them both simultaneously as in AM.

capture ratio The capability of a tuner to reject unwanted transmissions (other stations, interference) that are on the same frequency as those desired. The capture ratio is expressed in decibels, with a lower figure indicating better performance.

CAR computer-assisted retrieval.

car phone A cellular communications unit installed in a vehicle. While handheld, battery-operated systems are often called *car phones,* the phrase more properly distinguishes larger units that use power from the car's battery and connect to an antenna physically attached to the car (the center of the roof, or elsewhere). Generally they consist of two parts, a trunk or below-seat unit, and a handset. Car phones generally have higher power and better transmission than handheld cellular phones, although they lack the convenience of portability. See cellular phone, mobile phone, AMPS.

carbon dioxide laser, CO_2 laser A source of laser illumination based upon gaseous molecular action. An ammonia-based infrared laser was first described in a patent application in 1956 by R. Dicke. Carbon dioxide lasers were developed by A. Javan in the early 1960s.

From signaling and spectroscopy to welding, steel-cutting, etching, and delicate surgery, the carbon dioxide laser is suitable for hundreds of applications where power and precision are important. It has even been shown capable of halting infections and preventing extended tissue damage that could require amputation.

Early in 2002, Coherent Photonics Group announced that they had developed the first industrial Q-switched CO_2 laser. The laser emits high-energy, high-repetition, narrow pulses suitable for micromachining/drilling/PC board applications.

A basic CO_2 laser can be built in a lab with off-the-shelf parts. Note, CO_2 lasers generate high-energy beams that can cut through steel. Safety knowledge regarding assembly and use is essential. See helium-neon laser, laser history.

carborundum A substance with rectifying properties that was used in early radio wave crystal detectors. Unlike the popular galena, which required very delicate contact and tuning, carborundum could be clamped tight and sealed firmly within the detector unit, making it suitable for field work and rough handling. Much of the pioneer work on carborundum detectors was done by H. Dunwoody of the U.S. Army, who received a patent in 1906.

carcinotron An electron tube-based backward oscillator designed to generate extremely high frequency (EHF) signals. See magnetron.

card An electronic printed circuit board, especially one that is easily dropped into a *card slot* by a dealer or consumer. See printed circuit card, punch card.

card hopper In mechanisms that hold and feed punch cards, the holder in which the cards are stacked next to the feed mechanism for processing. See card stacker.

Card Issuer Identifier Code CIID. A calling card identification scheme. There are restrictions on which carriers can issue/use CIID cards.

card slot A slot-shaped data connector within an electronic system for the insertion of printed circuit board peripherals. Cards frequently consist of graphics controllers, drive controllers, serial and parallel ports, network connectors, and others. PCI is a common format for computer card slots. See edge connector, printed circuit board.

card stacker In mechanisms that hold and feed punch cards, the exit tray in which the cards are stacked after processing. There may be several of these, with a card sorter determining the destination stacker from holes in the cards. See card hopper.

Cardano, Girolamo (1501–1576) Also known as Hieronymus Cardanus, Cardano was an Italian mathematician, physicist, and physician to kings who authored many important historic publications, including *De subtilitate* (*On subtlety*, 1550), describing the accumulated knowledge about amber, with a definite statement that the properties of lodestone and amber differed in significant ways. He further described these differences. He also proposed the important mathematical notion of imaginary numbers and made a systematic study of probabilities. See amber, lodestone.

Girolamo Cardano – Magnetism Pioneer

A multitalented pioneer in mathematics and physics, Girolamo Cardano contributed important insights to our understanding of magnetism (e.g., properties of lodestone) and theoretical mathematics in the 1500s.

CardBus A 32-bit computer data bus designed for use with PCMCIA cards. The CardBus was designed to succeed the PC Card standard. See Personal Computer Memory Card Interface Association.

cardiode pattern A diagrammatic representation of the directional response of various transmitting and receiving devices: antennas, speakers, etc. It derives its name from the symmetrical, heart-shaped pattern that is typical. See antenna lobe.

Carnegie-Mellon University This U.S. educational institution is known for many contributions to telecommunications. One of the more familiar is the Andrew File System (AFS) used on computer networks. More recently, developers have created a working, campus-wide wireless data communications system which serves as a model for similar installations elsewhere and AFS is evolving into a powerful, distributed network protocol with a new name and some interesting new capabilities.

carrier 1. A wave of constant or known amplitude, frequency, and phase, which can be modulated by changing one of these characteristics. See carrier wave, carrier frequency, T1. 2. An entity that can carry an electrical charge through a solid. 3. An information-providing radiant energy from space. The four known categories of carriers are electromagnetic radiation, solid bodies, elementary cosmic rays, and gravitational waves.

carrier, communications A provider of communications circuits. *Common* (usually the local phone company) and *private* carriers are distinguished by degree of regulation and right to access of service by the public. The designation of *communications carrier* was intended to encompass companies with their own transmission facilities, as opposed to companies that lease or buy equipment or services for resale, but the general public often uses the phrase more loosely to include all long-distance companies.

carrier, GPS A GPS-related radio wave with at least one characteristic (such as frequency, phase, amplitude) that can be varied (modulated) from a known reference value. See Global Positioning Service.

Carrier Access Code CAC. See Access Code.

carrier band A range of adjacent frequencies that can be modulated to carry information, such as radio broadcast waves (without a carrier wave, multiple frequencies could not be transmitted without signal overlap and disruption). See band, carrier, carrier wave, modulation.

carrier bypass A phone service provider direct-connect link to the customer's lines, bypassing the local phone carrier. Some long-distance companies provide services through a carrier bypass in order to provide faster service or less expensive service by avoiding Carrier Common Line Charges. See Access Charge.

Carrier Common Line Charge CCLC. A charge paid by phone services providers to a primary carrier for using their switched network lines. Typically paid by long-distance providers. See Access Charge, carrier bypass.

carrier detect CD. A signal generated by a modem that operates over phone lines to indicate whether the phone carrier is present and the line can be dialed. Many modems have an LED to indicate the presence of the carrier signal. The command to the modem for carrier detect is typically &C1.

carrier frequency 1. The frequency of a carrier wave intended to be modulated by the wave containing the information. See carrier wave. 2. In the Global Positioning System (GPS), the frequency of the unmodulated fundamental output of a radio transistor. 3. The reciprocal of the period of a periodic carrier. See center frequency.

Carrier Identification Code CIC. A short code to identify uniquely a secondary phone service carrier for routing and billing. It was formerly three characters, but the Industry Carrier Compatibility Forum in 1988 and Bellcore in 1989 informed the Chief of the Common Carrier Bureau that four characters were needed to meet increasing demand. The numbers are issued by the North American Numbering Plan Administration (NANPA) to authorized entities. The expansion from three to four digits was termed the Plan of Record (POR). The implementation of this plan would not be trivial as it involved administrative changes and expenditures on the part of Local Exchange Carriers (LECs) and change-over expenses to anyone publishing materials with CICs (directories, letterheads, marketing materials, etc.). It also required procedures and priorities for conserving and reusing scarce CIC resources.

By the mid-1990s, Bellcore began assigning four-digit Feature Group D CICs. In 1998, this was further changed to a prefix (e.g., "10+10xxx") followed by the number, thus bypassing the subscription carrier (which would use the prefix code "1"). See Access Code, North American Numbering Plan.

carrier select keys Buttons included on a phone (usually a payphone) to provide the caller with a quick way to select a long-distance provider, thus not having to key in extra digits for access codes.

carrier selection Selection by a phone customer of a long-distance provider, usually done at the time of ordering the service, but it can be changed at any time. If you select a primary long-distance carrier, you will be able to access the service by dialing "1" plus the number. For alternate long-distance companies, you have to enter additional digits or access codes to complete a call. There are many long-distance companies, each offering better features and lower prices than the next. Evaluate these carefully before switching services, as there may be inconveniences, hidden charges, or limitations that are not apparent from the advertising literature and that may result in service that is limited and not necessarily cheaper in the long run. See Access Code, carrier bypass.

carrier sense The capability of a station to continuously monitor other stations to see if they are transmitting. See Carrier Sense Multiple Access.

Carrier Sense Multiple Access CSMA. A listen-and-send protocol used on local area networks (LANs). A system readying to transmit first probes the network to see if the line is clear; in other words, it ensures that another workstation is not transmitting. If the coast is clear, it sends the transmission. This protocol does not guarantee that collisions don't occur; it simply reduces the likelihood of an immediate collision. Various versions of CSMA exist to enhance its efficiency and provide greater collision detection and avoidance.

Carrier Sense Multiple Access with Collision Avoidance CSMA/CA. A version of Carrier Sense Multiple Access that is used in Ethernet systems in association with Media Access Control (MAC) protocols to integrate collision detection with time-division multiplexing (TDM). It aids in improving efficiency in CSMA systems. See Carrier Sense Multiple Access.

Carrier Sense Multiple Access with Collision Detection CSMA/CD. A version of CSMA with added traffic flow control capabilities to detect collisions, in order to increase the efficiency of flow of information on a local area network (LAN). CSMA/CD is not ideal for all implementations. In satellite communications, for example, the transmitting Earth stations cannot engage carrier sensing on the uplink due to its point-to-point nature. See Carrier Sense Multiple Access.

carrier shifting A technique of moving an entire modulated wave sequence in a positive or negative direction with respect to its midpoint, without changing the overall shape of the envelope. Carrier shifting is often used to manipulate mathematically a wave or to recreate a wave based upon only partial information (e.g., sideband). See phase shift keying.

carrier shifting fault An undesirable condition in transmitting a modulated carrier wave, in which the envelope, the range of amplitude-modulated signals above and below the midpoint of the waves, is unbalanced.

carrier signal A continuous radiant wave that can be modulated to add information to the wave. Carrier signals are modulated in a variety of ways; the two most familiar are amplitude modulation (AM) and frequency modulation (FM). One of the first researchers to search for a way to add information to a carrier wave was R. Fessenden, an American inventor who devised the hot-wire barretter and a high-frequency wave generator in 1901. Later, J. Carson studied the mathematical properties of carrier signals and proposed ways of carrying information by manipulating and recreating the signal at the receiving end, thus saving transmissions bandwidth. See Carson, John Renshaw; Fessenden, Reginald Aubrey; modulation; single sideband.

carrier synchronization In radio broadcasting, a carrier wave is used to carry a signal through a process called modulation, wherein information is added to the carrier wave. Various means of sending the modulated wave have been developed, some of which send only the sidebands, some of which send one side of the signal and recreate the other, etc. Consequently, at the receiving end, the receiver has to be designed so it can properly process the type of wave that is being received. In some cases, this situation involves the creation of a reference carrier that is synchronized with the received signal. See Carson, John R.; single sideband.

carrier to interference ratio CIR. A quantitative description of the effective transmission in relation to the (undesired) interference affecting that transmission. This ratio is of special interest to wireless communications engineers in designing transmissions components such as antennas and transceiving equipment. It is of increasing importance as vendors seek to increase the capacity of existing systems without degrading the signal to the point that the customers are unhappy with the service.

carrier wave A single-frequency wave that carries the transmission by being modulated by another wave containing the information. A carrier wave provides multiple channels and a means to reduce signal overlap through multiplexed broadcast waves. See carrier.

carrierless amplitude and phase modulation CAP modulation. A coding technique, based upon quadrature amplitude modulation, used in Digital Subscriber Line (DSL) transmissions. See discrete multitone, modulation, pulse amplitude modulation.

CARS See Cable Television Relay Service Station.

Carson, John Renshaw A mathematician and researcher at Bell Laboratories who contributed mathematics fundamentals related to modulation of communications waves that provided a way to recover the whole band from sideband transmissions. In 1915, he demonstrated that separate channels could be carried on each of the sidebands of a modulated carrier wave. In 1922, he provided a mathematical description of frequency modulation (FM) (and is somewhat infamous for having disagreed with E. Armstrong about the feasibility of FM transmissions). He was awarded the Franklin Institute *Cresson Medal* in 1939. See Armstrong, Edwin Howard; Carson's Rule.

Carson's Rule A method for calculating the minimum bandwidth of a frequency-modulated signal needed to transmit the desired communication. A larger number of subcarriers will necessitate a wider Carson's bandwidth. Named after John R. Carson.

Carterfone A commercial device developed in the 1960s for acoustically connecting two-way mobile radio communications to a telephone network system. Developed by Thomas Carter, who battled for the right to connect into the public phone network, the system became known through an important judgment by the Federal Communications Commission (FCC). Carter didn't sell many of the devices, but AT&T saw the precedence as threatening enough, in terms of its implications for other vendors, to obstruct its use. See Carterfone Decision.

Carterfone Decision A 1968 landmark judgment by the Federal Communications Commission (FCC) in which existing interstate telephone tariffs that prohibited subscribers from attaching their own phone equipment to existing phone lines was struck down. Carter Electronics had sought since 1966 to acoustically interconnect its private mobile radio systems to the national exchange network through a voice-activated system that started the radio transmitter. In pursuing its own right of access, Carter Electronics paved the way for other companies as well and, in a sense, foreshadowing the divestiture of AT&T.

As a result of the Carterphone Decision, manufacturers other than Western Electric, which had exclusive arrangements with AT&T, no longer were prevented from using the resources, and the interconnect industry was born. See Carterfone, Hush-a-Phone decision.

cartridge A common type of magnetic removable data storage that works somewhat like a floppy diskette but is physically larger in size and significantly higher in storage capacity. Cartridges commonly hold between 200 Mbytes and 1 Gbyte of data uncompressed. With the introduction of super-capacity disks, 3.5" floppies that can store more than 100 Mbytes of data, the discrepancy between low-capacity cartridges and high-capacity floppies is less than low-capacity floppy drives. Rewritable CDs are now beginning to compete with cartridges as backup storage devices due to the less volatile nature of the data.

Carty, John J. (1861–1932) Chief engineer of AT&T in the early 1900s after serving as the head of Western Electric's cable department. Carty developed the first two-wire telephone circuit and the phantom circuit, through which three conversations could be transmitted at one time over two pairs of wires. Western Electric purchased the rights to Lee de Forest's *Audion,* a three-electrode tube, in time for Carty to fulfill a promise he made in 1909 to provide a transcontinental telephone service to the U.S. west coast by 1914. See AT&T, phantom phone.

Carty, John J., Award for the Advancement of Science A triennial award for noteworthy and distinguished accomplishment in science. This award has been directed towards a different field every three years since 1932. It was established by AT&T.

CAS 1. Centralized Attendant Service. A centralized group of operators servicing systems which may have a number of branches within a region. 2. See channel associated signaling. 3. See Communications Applications Specification.

cascade To arrange or pattern into a series or succession of steps or stages, each dependent upon, or derived from the preceding, often in a falling or downward hierarchy. Computer menus, file systems, applications windows, and other graphical and logical structures are often developed with a cascade structure. Text editing applications sometimes have telescoping and cascading outline capabilities. Cascading principles and properties are now being studied with relation to quantum effects and harnessed for commercial applications. See quantum cascade laser.

CASE computer-aided software engineering, computer-assisted software engineering.

case sensitive 1. Computer software data or processes in which the case (lower or upper) is considered significant to the meaning of the text. For example, file names on Unix systems are case sensitive; that is, "MyFile.txt" is different from "myfile.txt". Case sensitive file names result in a greater range of descriptive naming possibilities. MS-DOS systems are case insensitive. AmigaOS is partly case-sensitive, forgiving about case when traversing directories in the shell, but allowing file names that can be distinguished from one another by case. 2. In word processing, search and replace routines can usually be configured to be case sensitive or case insensitive, depending upon your needs. 3. On the Web, URLs are case sensitive, but for the convenience of users many browsers will resolve the cases in a forgiving manner, to load a Web page even if the case is misspecified. This feature is not characteristic of all browsers, so it's usually better to type in the case correctly.

case sensitive password For access to secure computer systems, most password fields require an exact match, hence are case sensitive, in order to provide a greater variety of possible passwords and thus increased security. Case sensitivity also increases the total number of possible passwords, which is important if it is a multiuser environment with a limit on the number of characters in the password (e.g., eight characters).

Cassegrain antenna, Cassegrainian antenna A parabolic antenna arrangement in which the feed is located near the vertex of a concave surface of the main reflector, and a secondary reflector is located near the focal point and aligned to be within the focus of the main reflector. A beam is thus redirected from the feed unit through the secondary reflector to the main reflector to radiate a beam that is parallel to the axis of the main reflector. A Cassegrain feed is one type of arrangement; horn feed is another. Cassegrain antennas require more careful alignment and more parts than a horn feed antenna and thus tend to be used in higher end, more expensive applications. Due to the redirection of the reflection, they stay cooler than horn feed arrangements and are thus suitable for hotter climates. See antenna; horn feed; parabolic antenna; Ramsden, Jesse.

Cassegrain, Guillaume (dates unknown) Apparently a French founder and sculptor who developed a center mirror telescope even before Newton constructed a reflecting telescope. In 1672, Cassegrain proposed using a main mirror with a hole in the center, with a smaller hyperboloid mirror to reflect back the image through the hole to an eyepiece. This arrangement results in a compact design and minimal spherical and chromatic aberrations in the image. It is the same general configuration of some of the advanced telescopes of today. See Cassegrain antenna.

cassette tape A portable recording and playing medium consisting of a long narrow magnetic tape wound onto reels protected by a roughly rectangular plastic case.

Cassette tapes come in a variety of tape widths and are used for both sound and video.

Some audio tapes are wound onto the reel in a loop to enable continuous playing, but most are manually turned over or mechanically rewound. Cassettes are commonly used for consumer audio and backup and archiving of computer data. Very small cassette tapes are used in answering machines and small tape recorders designed for maximum portability (and, in some cases, minimum visibility). Cassette tapes for consumer audio almost completely replaced reel-to-reel tape, then began to be supplanted by audio CDs and other digital audio technologies. See CD, DAT, leader, reel-to-reel.

Cassini A spacecraft designed to travel through the solar system and send back information to be evaluated by scientists to teach us more about our planetary environment and the universe. The Cassini spacecraft has many tasks to perform on its way to the planet Saturn. Flyby targets include Venus, Earth, Jupiter, and various asteroids. Cassini's launch window was very small. It had to be launched between 6 Oct. 1997 and 4 Nov. 1997 to execute the various maneuvers and flybys needed to take it on its seven year journey to Saturn. The telecommunications and guidance systems associated with the Cassini mission are some of the most sophisticated to date and will teach us much about how far and how well we can transmit information to and from the corners of the Galaxy.

castellation An indented pattern or surface of a regular, repeated nature. For example, the battlements on castles are castellated. Castellated protruberances, or thin pads of conductive materials, are often incorporated into the edges of electronic circuit boards to provide contact points for electrical connections. Gratings are often castellated components.

cat, Cat *abbrev.* category. In cabling, *Category*, or *Cat* followed by a number denotes industry-specific cabling standards. See category of performance.

CAT 1. Call Accounting Terminal. An AT&T term for a microprocessor-equipped device that records call activity in order to provide automated accounting information. 2. See computer-aided teaching, computer assisted instruction. 3. See Council for Access Technologies.

catadioptric devices Devices that utilize both optical reflection and refraction to control the travel of light through the device, often to produce an image for viewing or transission. Depending upon the instrument, the catadioptric configuration may be intended to shorten the length of the barrel of a viewing device or process the light signal as it is being reflected and refracted. Catadioptric principles are used in a variety of astronomical telescopes, robotics imaging technologies, and some digital video cam applications.

Catarra See PDA marcrobrowser, SoftSource Corp.

category, wiring See category of performance.

category of performance Cabling and component standards that have been defined to promote and facilitate intercompatibility of products from different vendors. These standards are widely used in the phone and computer network industries, especially Cat 5. The categories of performance focus on the throughput of the transmissions rather than the specific materials used to construct individual cables. They are self certifying in the sense that the vendor is responsible for testing and maintaining quality and manufacturing standards to provide the performance categories detailed in the Categories chart.

cathode (*symb.* – K) 1. The negative terminal of an electrolytic cell. 2. The positive terminal of current-supplying primary cell. 3. In a moving electron system such as an electron tube, the electron-emitting portion, directed toward an anode, often a thin metal plate, usually passing through a controlling grid. The electron beam-emitting end of a cathode-ray tube (CRT). See cathode-ray tube, phototube, photomultiplier tube.

Cathode in Electron Tube

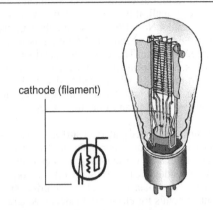

cathode (filament)

The cathode is one of the three essential elements of an electron tube, emitting the electrons that are attracted to the anode. In this tube, the filament acts as the cathode.

cathode ray An ionized region, composed of a stream of electrons influenced by an electric field, emanating from a cathode. See cathode-ray tube.

cathode-ray tube CRT. A display device consisting of a closed tube of glass with the air removed; it contains an electron-emitting gun at one end and a coated surface at the other. The cathode-ray electron beam emanates from the cathode and passes through a magnetic field that controls the beam. By sweeping across the coated inside surface of the glass, a *frame* is formed on a *raster* display and a *vector* is formed on a *vector* display, either of which can be seen through the glass from the outside.

Cathode-Ray Tube – Basic Parts

phosphor coating on inside of tube

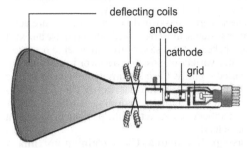

deflecting coils

anodes

cathode

grid

Cathode-ray tubes are essential components in many types of electronics devices. They are widely used as display devices for televisions, radar scopes, oscilloscopes, computer monitors, etc. This diagram of a historic electromagnetic-deflection cathode-ray tube provides an example of the basic, essential components.

The movement of the electron beam across the display surface excites the phosphors so that they selectively light up (fluoresce) and remain visible for a few moments. The sweep of the beam is very fast so that perceptually humans will 'see' the entire frame as one image rather than as a series of constantly refreshed lines.

Sometimes a grating called a *shadow mask* is inserted between the beam and the coating to further control and focus the beam to provide a crisper display.

Cathode-ray tubes can be *short persistence* or *long persistence*. Long persistence means the phosphors remain lit for a longer period of time, and the screen may not have to be refreshed as often to keep an image visible.

The refresh rate on most current monitors is 60 frames

Categories of Performance		
Category	Transmission Rate	Notes
Cat 1	Not used	
Cat 2	Not used	
Cat 3	Up to 16 megahertz	24-gauge wire. Typically used in voice communications and lower end data communications, such as Token-Ring and 10-Mbps Ethernet networks.
Cat 4	Up to 20 megahertz	Digital voice communications and data networks, e.g., Token-Ring.
Cat 5	Up to 100 megahertz	24-gauge wire with more stringent fabrication requirements than Cat 3 (e.g., better shielding). Typically used in higher end data communications and high-grade or digital voice applications, particularly high-bandwidth ones such as videoconferencing. Examples include FDDI, 100Base-T, 100-Mbps Token-Ring or Ethernet. See twisted pair for a diagram.

per second, a rate that is fast enough to appear stable and not flickering to the human eye. Color CRTs typically have three beams, red, green, and blue (RGB). The cathode-ray tube is fragile and large (regrettably) and is typically encased in a protective console. It is not advisable for laypersons to open the back of a CRT device, as there is a danger of electric shock from the stored charge. CRTs were being used as computer display devices by the early 1950s but did not become regularly associated with microcomputers until 1976. CRTs are commonly used for monitors (computers, video editing, scopes) and television screens. See Crookes tube; flat panel CRT; frame; Geissler tube; interlace; screen saver; shadow mask; Zworykin, Vladimir Kosma.

cathodic protection In many wiring installations, bare wire is used, so corrosion is a significant concern. One of the ways to prevent corrosion and buildup is by running a negative charge through the wire to repel negative ion materials such as chlorine.

CATNIP See Common Architecture for Next Generation Internet Protocol.

CATS See Consortium for Audiographics Teleconferencing Standards.

CATV 1. Cable Television. A system that delivers frequency-segmented television programming channels to subscribers through physical cables, usually 75-ohm coaxial cables. The full bandwidth of cable is not typically used, partly due to the extent of the subscriber's service and partly due to the insertion of non-program-carrying guard bands that act as separators to keep individual channel transmissions from interfering with one another. 2. Community Antenna Television. A large antenna, shared by a community, to intercept broadcast stations that are not accessible via small, individual antennas. Sometimes commercial communities (motels, resorts, condominiums) will make arrangements to install a powerful antenna and rebroadcast or channel the signals to individual units. In many cases, this will require a special license as there are laws protecting broadcasters from having their programs rebroadcast.

catwhisker A fine metal thread resembling the arched shape of a cat's whisker, used in early radio wave detecting crystal sets. The catwhisker contacted the crystal on one end and was secured to a metal conductive support on the other end. Some enclosed sets used a catwhisker that was fixed in place at the factory. Prior to the commercialization of crystal sets, it was known as a *feeler*. See crystal detector.

CAU See controlled access unit.

CAV See constant angular velocity.

cavity A depression, hole, indentation, or pit, which may be of any size. In various media, cavities of precise characteristics are created in the surface so they can later be used to deflect radiant waves or carefully focused laser light beams. The deflections pass into some kind of pickup mechanism (read mechanism) so the encoded information can be recreated and presented. Thus, cavities are at the heart of many optical recording technologies, CDs, for example.

cavity magnetron An early British innovation in radar

systems, developed in 1940, that enabled the use of extremely short waves (microwaves). Thus, it was possible to design more compact antennas and improve the quality of the information and images that could be resolved through radar systems. Smaller beam widths facilitated greater accuracy. In turn, the smaller, more accurate radar technologies greatly increased the number and types of applications that were practical, making it possible to mount radars on boats and planes. See magnetron.

CB radio See citizen's band radio.

CBC See Canadian Broadcasting Corporation.

CBDS 1. Common Basic Data Set. 2. See Connectionless Broadband Data Service.

CBEMA Computer and Business Equipment Manufacturers Association. See Information Technology Industry Council.

CBR An ATM traffic flow control concept. See constant bit rate.

CBS See Columbia Broadcasting System.

CBSC See Canadian Broadcast Standards Council.

CBT 1. See Canadian Business Telecommunications Alliance. 2. Computer-Based Training. See computer-assisted instruction.

CBTA See Canadian Business Telecommunications Alliance.

CBX Computerized Branch Exchange. A commercial private telecommunications system trademarked by the ROLM Corporation as their version of a private branch telephone exchange (PBX).

CBYD Call Before You Dig. The warning on signs by fiber optic cables to prevent contractors and other diggers from damaging underground installations.

CCB See Common Carrier Bureau.

CCC 1. clear channel capability. In communications, that portion of a data transmissions capacity that is available to users, the informational portion, above and beyond the various control and signaling transmissions associated with the functioning of the technology. 2. Communications Competition Coalition. A Canadian support and lobbying organization established to encourage Canadian telecommunications competition. 3. Center for Corporate Communications. http://www.communicationsmgt.org/ 4. Computer Communications Club. http://www.ccc.or.at/

CCD 1. See charge coupled device.

CCI See co-channel interference.

CCIA See Computer and Communications Industry Association.

CCIR 1. See Centre for Communication Interface Research. 2. See Comité Consultatif International des Radiocommunications.

CCIR video standard Similar to the EIA RS-170a standard for color video, CCIR is a dominant video format used in Europe just as NTSC is standardized in North America and Japan. CCIR supports a vertical resolution of 625 scanlines, 575 of which are typically displayed. It is an interlaced format with two scans of the screen combining to create a conceptual frame. The number of lines per field (two of which make up the frame) is 282.5 compared to 242.5 for RS-170a. The effective resolution when conversion

with raster-based displays is desired is 768 (compared to 640 for RS-170a). CCIR video is traditionally transmitted over 75-ohm well-shielded coaxial cables.
CCIRN Coordinating Committee for Intercontinental Research Networks. Established by the U.S. Federal Networking Council (FNC) and the European Réseaux Associées pour la Récherche Européenne (RARE). CCIRN promotes international cooperation and sponsors a number of working groups that meet in different parts of the world.
CCIS See Common Channel Interoffice Signaling.
CCITT Comité Consultatif Internationale de Télégraphique et Téléphonique. International Telegraph and Telephone Consultative Committee. An influential United Nations-sponsored international telecommunications standards committee based in Geneva, Switzerland. It changed its name to ITU in 1990. See

International Telecommunication Union.
CCITT Study Groups These subgroups, operating under the CCITT (now the ITU) study and make recommendations for specialized areas of telecommunications. See International Telecommunication Union.
CCNR (telephone) call completion on no reply (e.g., as in ISDN Q.733 call completion services).
CCP See Compression Control Protocol.
CCR See current cell rate.
CCS See Common Channel Signaling.
CCS/SS7 Common Channel Signal/Signaling System 7. See Signaling System 7.
CCSN Common Channel Signaling Network. See Common Channel Signaling.
CCT 1. Calling Card Table. See unmatched call. 2. See Consultative Committee Telecommunications.

Common CD Formats	
Format	Description
CD-Audio	Compact disc audio. A digital sound representation standard that is incorporated into CD-ROMs. Conversion from digital to analog for listening occurs in the computing hardware. Also known as *Redbook Audio*.
CD-I	Compact disc interactive. An interactive multimedia standard developed by Philips and Sony. CD-I players are designed to accept and play a variety of CD-encoded data and can typically be interconnected with a computer or TV playback system.
CD-Plus	Compact disc plus. A standard developed by Philips and Sony that enables audio CD players to play multimedia (graphics and sound) discs by skipping over the nonaudio segment that is stored on the first track.
CD-R, CD-ROM R	Compact disc recordable. A format for read/write CD-ROM systems. See compact disc for a fuller description.
CD-ROM	Compact disc read only memory. A standardized, widely used format for storing digital information on small flat optical platters that are read with laser technology and played on CD players. See compact disc for a fuller description.
CD-ROM	A computer peripheral that reads, and sometimes writes, digital information to a compact disc. Most consumer CD-ROM drives are read only, although read/write drives are now under $300 and may soon be a consumer item. A CD-ROM drive can be used to run applications, read text files, images (PhotoCD), and audio. Many CD-ROM drives come with software to play audio CDs through a speaker.
CD-ROM X	A compact disc read only memory extended architecture. A format developed by Microsoft that enables the interleaving of audio and video, rather than recording them on separate tracks. It requires a player that can understand the format. If played on a regular player, the audio will not be detected and played.
CD-RW	CD recordable/rewritable technologies that became prevalent in the late 1990s and which were almost immediately challenged by emerging DVD technologies.
CD-UDF	A standardized format for CD-recordable (CD-R) media that enables a variable packet-writing scheme to be used as an incremental approach to the recording of compact discs (CDs). It provides a means for easily recording files on a CD in much the same manner as on a floppy disk.
CD-V	Compact disc video. A standard for storing video images on compact discs that hasn't really caught on. It is being superseded by CD-XA which enables the interleaving of video and sound.
CD-WO	Compact disc write once. A format designed for mastering a CD. The CD is then used in-house or in limited quantities, or is sent to a duplication factory for mass production. Drives that are able to write a master CD were once out of the price range of small companies and consumers, but they have dropped to below $300 and can now be used by software developers, composers, and small record companies to produce masters or small production runs of specialized recordings.

CCTA See Central Computer and Telecommunications Agency.

CCTV Closed Circuit TV. See closed circuit broadcast.

CCU 1. camera control unit. 2. communications control unit.

CD 1. See carrier detect. 2. See compact disc. See Common CD Formats chart. 3. count down. A concept in broadcasting related to the signaling of the beginning of taping, editing, or live broadcasting.

CDA See Communications Decency Act.

CDCS Continuous Dynamic Channel Selection.

CDDI See Copper Distributed Data Interface.

CDE See Common Desktop Environment.

CDF See cutoff decrease factor.

CDLC See Cellular Data Link Control.

CDMA See code division multiple access.

CDMP Cellular Digital Messaging Protocol.

CDO See community dial office.

CDP 1. Cisco Discovery Protocol. 2. Customized Dial Plan.

CDPD See Cellular Digital Packet Data.

CDPD Forum, Inc. A not-for-profit organization established in 1994 to promote the development and acceptance of Cellular Digital Packet Data (CDPD). It supports vendors who develop and distribute CDPD products and services. http://www.cdpd.org/

CDR See Call Detail Record.

CDRH classifications A series of classifications to alert users of the dangers of incorrect use of laser-based components that emit external beams (more powerful lasers may be inside the assembly but must not emit to the exterior of the device). The laser safety classification categories are somewhat generalized since the wavelength of the light is an additional factor controlling the influence of the beam and is not specified for all categories.

CDT credit allocation. In packet-switched networks, such as OSI, CDT applies to transmission flow control.

CDTAC See Consumer/Disabilities Telecommunications Advisory Committee.

CDV 1. cell delay variation. 2. See Compressed Digital Video.

CDVT See cell delay variation tolerance.

CE Connection endpoint. 1. In ATM networking, a terminator at one end of a layer connection within a SAP. 2. circuit emulation.

CE Mark A sign that an object has been certified through the overseeing European regulatory body, the European Telecom Directive, and does not require further testing or approval within the individual participating countries. The mark provides identification of products that conform to certain specified safety, electromagnetic, and interoperability requirements. CE certification is required for all telecommunications terminal equipment (TTE) sold in the European Union. See Underwriters Laboratory, Inc.

CEBus Consumer Electronics Bus. A home automation standard managed by the CEBus Industry Council and accepted by the Electronics Industry Association (EIA). CEBus specifies a common format for connectionless peer-to-peer communications over standard electrical wiring. The CEBus HomePnP standard is a nonproprietary protocol based upon the IEA 600 open standard. In terms of functionality, it is similar in concept to the X-10 protocol in that it operates over 120-volt, 60-cycle home wiring.

CEBus is a two-channel specification, with one channel assigned to realtime control functions, the other to informational data. It uses a CSMA/CD protocol that includes various error detection and retry functions, end-to-end acknowledgment, and authentication. The Powerline Carrier uses spread spectrum technology to bypass electrical impediments in home wiring, spreading the signal over a range of frequencies rather than using a single frequency. See spread spectrum, X-10.

CEBus Industry Council CIC. A users group supporting the development and use of interoperable CEBus-based home network automation technologies. http://www.cebus.org/

CEDAR The Center of Excellence for Document Analysis and Recognition. An organization at the

---

CDRH Laser Safety Classification Categories

Class	Designation	Notes
Class I	EXEMPT	Visible, low-powered lasers considered safe for viewing.
Class II	CAUTION	Visible lasers in the 600 to 700 nm range at 1 mW or less. Do not stare directly at the beam or a reflection of the beam.
Class IIIa	DANGER	Visible lasers in the 600 to 700 nm range at 5 mW or less. Severe eye damage; avoid eye exposure to the beam or a reflection of the beam.
Class IIIb	DANGER	Visible lasers in the 600 to 700 nm range at 5 mW or greater and invisible lasers in the 700 to 900 nm range. Severe eye damage; avoid any exposure to the beam or reflection of the beam. Doesn't present fire hazard.
Class IV	DANGER	High power lasers exceeding characteristics of Class III lasers. They present a variety of dangers from eye and object damage to diffusion hazards, burns, and general fire hazards.

State University of New York at Buffalo that provides a number of interesting services including informational CD-ROMs.

CEI In ATM networking, a connection endpoint identifier.

Celestri A downsized version of the original M-star project, Celestri is a low Earth orbit (LEO), geostationary hybrid satellite system from Motorola. In May 1998, the Celestri expertise and technology was rolled into the Teledesic project, when Motorola Inc. bought in as a major partner. See Teledesic.

cell, ATM In asynchronous transfer mode networking, a unit of transmission consisting of a fixed-size frame comprising a header and a payload. See asynchronous transfer method, cell rate.

cell, battery Minimally, a receptacle containing an electrolyte and two electrodes arranged so the electricity can be generated from the cell by chemical actions. Development of modern cells stems from the experiments of C.A. Volta. Two or more cells can be combined to form a *battery*. See battery, storage cell.

cell, mobile phone In mobile communications, the basic geographic unit of a distributed broadcast system, within which a low-power transmitting station is located. Roughly hexagonal in shape, depending upon terrain. Its size varies with available channels, generally increasing as the radius of each cell decreases. Cells are further grouped into clusters. See cluster, cellular phone, mobile phone.

cell delay variation CDV. In ATM networking, a traffic flow buffering and scheduling concept. CDV parameters are associated with constant bit rate (CBR) and variable bit rate (VBR) services and relate quality of service (QoS) information by indicating the probability that a cell may arrive late. See cell rate.

cell delay variation tolerance CDVT. In ATM networks, a traffic flow control mechanism that allows cells to be queued during multiplexing to allow others that are being moved onto the same communications path to be inserted. Cells may also be queued to allow time for the system to insert control cells of one sort or another. See cell rate.

cell error ratio In ATM networks, the ratio of cells in a transmission that are errored to the total cells in the transmission, over a specified time interval, preferably as measured on an in-service circuit. See cell loss ratio, cell rate.

cell interarrival variation CIV. In ATM networks, a description of changes in arrival times of cells nearing the receiver. If the cells are carrying information in which the arrival of the cells at the same time is important to the synchronization of the final output, as in constant bit rate (CBR) traffic, then latency and other delays that cause interarrival variation can interfere with the output. For example, in videoconferencing, synchronization of images and sound might be affected by cell delays. See cell rate, cell delay variation tolerance, jitter.

cell loss priority field CLP. In ATM networks, a bit field contained in the header cell that indicates the cell discard eligibility of the cell. In congested situations, this cell may be expendable.

cell loss ratio CLR. In ATM networks, cell traffic is handled in many ways in order to maximize throughput, to synchronize arrival times where appropriate, and to minimize delays, latency, jitter, or loss. The cell loss ratio is a negotiated quality of service (QoS) parameter that depends upon the network traffic flow control setups. It is computed as a ratio of lost cells to the number of total cells transmitted, expressed as an order of magnitude. See cell error ratio, cell loss priority field, cell rate, leaky bucket.

cell misinsertion rate CMR. In ATM networking, a traffic flow evaluation parameter giving the ratio of cells that are received at the endpoint, that were not originally transmitted by the source, compared to the total number of cells correctly transmitted.

cell phone See cellular phone.

cell rate In ATM networks, a concept that expresses the flow of basic units of transport used to convey data, signals, and priorities. See the Cell Rate Concepts chart for further detail. See leaky bucket, cell rate margin.

cell rate margin CRM. In ATM networks, an expression of the difference between the effective bandwidth allocation for the transmission and the sustainable cell rate allocation in cells per second.

cell relay A type of fast packet switching network architecture using small fixed length packets that can be used for a variety of data types. The cell format is typically 53 octets comprised of 5 bytes of address information and 48 bytes of informational data. Cell relays can also provide quality of service (QoS) guarantees to a variety of services. See Frame Relay.

cell relay function In ATM networking, a basic service provided to ATM endstations. See cell relay.

cell relay service CRS. An ATM carrier service.

cell reversal In a battery, a reversal of the polarity of the terminal cells resulting from discharge.

cell site In cellular wireless communications systems, an individual transceiving unit. Multiple cell sites provide roaming capabilities. The cell site serves the local cell and slightly overlaps with adjoining cells to minimize dead spaces between transmissions when a subscriber passes from one cell region to another.

cell site controller Cellular radio operates with numerous cells, each associated with a transceiver. The cell site controller manages the various radio channels within that cell, allocating them when a user moves into range of the cell, and deallocating and reusing available frequencies as the user moves out of range again, or terminates the connection.

cell splitting A means of increasing the call capacity of a cellular system by splitting cells into smaller units.

cell switch router CSR. A network routing device that incorporates ATM cell switching in addition to conventional IP datagram forwarding, in order to provide improved service over traditional hop-by-hop datagram forwarding, especially with transmissions that pass through subnetwork boundaries. See RFC 2098.

cell switching In cellular mobile phone systems, the overall process of handling calls, monitoring signals as users move in and out of range of the transceivers in the various cells, and allocating and deallocating frequencies as needed to provide seamless service through a series of cells. Cells are designed to overlap somewhat so that there is no gap when switching from one to another and to compensate for the fact that the signal is weakest on the periphery of the transmitting area. The sophisticated moment-by-moment monitoring and orchestrating of this process is handled by monitoring systems and cell-switching software.

cell transfer In cellular mobile phone systems, the logistics of keeping an ongoing connection at acceptable volume and quality levels when switching the user from the transceiver in one cell to the transceiver of the cell that is being entered. This transfer involves allocating a frequency channel in the entered cell and deallocating and reassigning, if needed, the frequency channel of the exited cell.

cell transfer delay CTD. In ATM networking, the time elapsed between a cell exit event at the first point of measurement and the corresponding cell entry event at the second point of measurement for a particular connection. The cell transfer delay between the two points of measurement is the sum of the total inter-ATM node transmission delay and the total ATM node processing delay. See cell rate.

Cello A graphical Web browser created at the Cornell Legal Information Institute.

cellphone See cellular phone.

cells in flight CIF. In ATM networking, a descriptive phrase for a traffic service parameter, the available bit rate (ABR). CIF is a cell number limit negotiated between the receiving network and the source of the cells during the idle startup period, prior to the first RM-cell returns. See cell rate, Cells in Frames.

Cells in Frames CIF. The name given to a number of mechanisms for carrying ATM network traffic across a media segment and network interface card. CIF was developed by the Cells in Frames Alliance, a diverse group of professionals and commercial vendors. This group released the first CIF specification in 1996 for carrying ATM over Ethernet, Token-Ring,

Cell Rate Concepts in ATM Networks		
Abbreviation/Function Descriptor		Notes
ACR	allowed cell rate	A traffic management parameter dynamically managed by congestion control mechanisms. ACR varies between the minimum cell rate (MCR) and the peak cell rate (PCR).
CCR	current cell rate	Aids in the calculation of ER and may not be changed by the network elements (NEs). CCR is set by the source to the available cell rate (ACR) when generating a forward RM-cell.
CDF	cutoff decrease factor	Controls the decrease in the allowed cell rate (ACR) associated with the cell rate margin (CRM).
CIV	cell interarrival variation	Changes in arrival times of cells nearing the receiver. If the cells are carrying information that must be synchronized, as in constant bit rate (CBR) traffic, then latency and other delays that cause interarrival variation can interfere with the output.
GCRA	generic cell rate algorithm	A conformance enforcing algorithm that evaluates arriving cells. See leaky bucket.
ICR	initial cell rate	A traffic flow available bit rate (ABR) service parameter. The ICR is the rate at which the source should be sending the data.
MCR	minimum cell rate	Available bit rate (ABR) service traffic descriptor. The MCR is the transmission rate in cells per second at which the source may always send.
PCR	peak cell rate	The PCR is the transmission rate in cells per second that may never be exceeded, which characterizes the constant bit rate (CBR).
RDF	rate decrease factor	An available bit rate (ABR) flow control service parameter that controls the decrease in the transmission rate of cells when it is needed. See cell rate.
SCR	sustainable cell rate	The upper measure of a computed average rate of cell transmission over time.
UBR	unspecified bit rate	An unguaranteed service type in which the network makes a best efforts attempt to meet bandwidth requirements.
VBR	variable bit rate	The type of irregular traffic generated by most nonvoice media. Guaranteed sufficient bandwidth and QoS.

and 802.3 networks. As ATM is not tied to a particular physical layer, CIF has been defined as a pseudo-physical layer for carrying ATM traffic.

CIF provides a frame-oriented means of using ATM layer protocols transparently over a variety of existing local area network (LAN) framing protocols.

The Cells in Frames project has been funded in part by the National Science Foundation (NSF). More information is available through Cornell University. See cells in flight. http://cif.cornell.edu/

Cells in Frames Alliance An open membership organization that promotes the affordable deployment of ATM technologies and better use of existing local area network (LAN) infrastructures, while providing applications developers direct control over quality of service (QoS) networking issues.

Cellular Array Processor CAP. A parallel processing architecture, pioneered by Fujitsu, that has applications in a number of areas, including image processing, neural networks, cellular automata simulations, and high-speed, multi-user database searches. In the early 1980s, Mitsubishi Electric initiated a satellite 2D image processing system based upon CAP. CAP has been used on the Earth Resources Satellite Data Information System (ERSDIS) to process synthetic-aperture radar (SAR) data as well as on Japan Earth Resources Satellite-1 (JERS-1) images. Hughes Electronics has developed systolic CAP for large-scale parallel processing applications. The technology has come to be known more generically as array processing (AP) and Fujitsu's line of equipment has been using the AP- prefix since the mid-1980s.

Cellular Array Processor Program CAP Program. A project to develop a multi-computer peer network system to enable hundreds of users to simultaneously access a huge business register archived by the Australian Bureau of Statistics (ABS). Fujitsu Laboratories, pioneers of CAP technology, contributed a 12-node AP 3000 parallel computer to the project.

Cellular Data Link Control CDLC. An open data communications protocol suitable for wireless communications. In cellular phone systems, it provides a means to interconnect a data terminal and a cellular phone. The protocol includes various error detecting, correcting, and data interleaving features that make it suitable for wireless transmissions.

Cellular Digital Packet Data CDPD. An open standard originally developed and released by vendors in 1993, which was further defined by the CDPD Forum. Then it was passed on for maintenance and enhancement to the TIA in 1996. CDPD is suitable for packet data services for mobile communications, conceived as an extension to landline services for mobile users. The originators of CDPD wanted to develop a way to use existing cellular networks for wireless data, in other words, to overlay newer services on the existing infrastructure.

CDPD is a packet-based system, defined to operate over AMPS analog voice systems. Transmission speeds of 19.2 Kbps are possible over the traditional infrastructure. Internet Protocol (IP) is typically used with TCP. Packets are routed into and out of the CDPD network through an Intermediate System (IS), that acts as a relay. Specific routing functions and monitoring of mobility are handled by the Mobile Data Intermediate System (MD-IS). The MD-IS keeps track of locations through a Home Domain Directory (HDD) database.

CDPD is not intended to specify the various types of service that can be carried over the system but rather describes the architectural structure of the service and its integration with the existing infrastructure. Specific value-added services are up to individual vendors. See A interface, E interface, I interface.

Cellular Digital Packet Data Forum CDPDF. An association formed in the early 1990s to develop and promote a standard for cellular digital packet communications. It is now called the Wireless Data Forum. See Cellular Digital Packet Data, Wireless Data Forum.

cellular modem A modem integrated with cellular phone technology to provide ease of access to telecommunications services through mobile, wireless transmission.

Cellular modems are frequently used with laptop computers, often available in the form of PCMCIA cards, which are small, slender peripheral components that fit easily inside a portable device. Dialing and data transfer is controlled by the software. The communication may originate from a wireless system and hook into a landline system. Cellular modems are favored by traveling professionals, such as journalists, sales reps, and scientific researchers who may be relaying information to central facilities on a regular basis. See cellular phone.

cellular phone, cell phone An analog, digital, or hybrid mobile communications system, with hardware interfaces resembling traditional phone handsets, linked through a gridlike network of low-power wireless transmitters each servicing a geographic area, with a small amount of overlap with adjacent cells. As the user travels through these areas, or *cells*, the transmission is *handed off* to provide continuous service, freeing previous channels. Carrying on communications in various cells while on the move is called *roaming*. Power levels in each cell are optimized for demand and subscriber density.

The cell concept was introduced in the late 1970s to improve upon older single-transmitter mobile systems. The commercial cellular phone system in the U.S. was available by the early 1980s. More recently, digital cellular systems have been devised to increase capacity and call security (through encryption). It is estimated that there are now more than 150 million cellular phone users worldwide, and ambitious programs for launching communications satellites to handle seamless global cellular communications are in progress. See cell, cluster, cellular modem, mobile phone, PCS, AMPS, TDMA.

cellular phone security Cellular phone communications are not, for the most part, transmitted in a secure manner. In analog systems, the signal goes over airwaves that can be tapped by a radio scanner operating in the same frequencies. Although encryption is starting to be incorporated into digital cellular communications, it is not yet universal. Since cell phones are used for many sensitive business and law enforcement communications, security is of some concern to cell phone users. A number of security systems have been developed by cell phone providers, including cell phone attachment peripherals that provide fully digital voice encryption capabilities.

cellular priority access service CPAS. The capability of disaster response personnel to access wireless cellular communications services on a priority basis. This service has been incorporated as part of the U.S. national Cellular Priority Services (CPS) program as a result of experiences in disaster situations in which overloaded cellular networks hindered emergency response personnel communications. Incoming emergency calls can be cued for an available channel to facilitate call connections during times when cellular lines are congested. If voice channels are not available, feature codes can be used to contact the callee. After verification of the call request at a switching center, it can be queued on a priority basis. See Cellular Priority Services.

Cellular Priority Services (CPS) A U.S. national program to facilitate cellular communications among disaster response personnel such as National Security and Emergency Preparedness (NS/EP) users. The program was initiated as a result of cellular congestion during disaster situations that impeded emergency relief personnel from contacting one another. CPS is instrumental in the deployment of a uniform national program involving many different organizations and levels of government that participate in disaster relief and their associated communications needs.

CPS is involved in standards development, administrative processes, regulatory strategies, and competitive strategies for implementation of cellular and related personal communications systems. See cellular priority access service.

cellular radio Very similar to cellular phone and, in fact, a forerunner to cellular phone, in which a region is organized into cells, each with a transceiving unit that overlaps with the coverage of adjacent transceiving units. Cellular radio provides for reuse of frequencies and greater capacity than noncellular mobile radio services, and enables users to purchase cheaper equipment, since the power requirements are not as high. See cellular phone.

cellular security devices CSD. Add-on peripherals, or all-in-one cell phone sets, that incorporate various security means, primarily digital encryption, or that will transmit random noise to anyone attempting to monitor the communication. See cellular phone security, frequency hopping.

Cellular Telecommunications Industry Association CTIA. http://www.wow-com.com/

celluloid A durable, though flammable, plastic material composed mainly of cellulose nitrate and camphor. Celluloid has been used in many experimental technnologies, from the study of diffraction phenomena to the development of cinematic film.

In 1885, Hannibal Goodwin developed celluloid film, which became so widely used in the motion picture industry that the films were known for many years as *celluloids*. In spite of the greater resolution and durability of images captured on film, digital imagery, due to its flexibility and convenience, is superseding film in consumer markets. See digital video.

CELP code-excited linear predictive. An analog-to-digital voice encoding scheme suitable for sending voice conversations through digital networks such as local area networks (LANs) and the Internet. The technology was developed based upon research on acoustics and voice synthesis by Bell researchers M. Shroeder and J. Hall.

CELP is now widely used for digital encoding and speech synthesis. This makes long-distance carriers nervous, as existing phone lines are typically used for portions of the transmission, bypassing long-distance phone charges. See Schroeder, Manfred R.

Celsius scale A scale developed in the 1700s to describe temperature with the boiling point of water referenced as zero and freezing point as 100 degrees. A year later, the reference points were reversed by Christin, and the scale today continues to use zero as the freezing point of water and 100 degrees as the boiling point. Also called centigrade scale.

CEN/CENELEC Comité Européen de Normalisation (European Committee for Standardization)/Comité Européen de Normalisation Electrotechnique (European Committee for Electronic Standardization) CEN is one of three organizations responsible for overseeing voluntary compliance with standards in the European Union, while CENELEC develops technology and standards. CEN cooperates with ISO. See European Telecommunications Standards Institute. http://www.cenorm.be/

Center for Democracy & Technology CDT. A participatory organization promoting democratic values and constitutional liberties in a digital society. The CDT seeks practical solutions to protecting and promoting free expression and privacy in global communications technologies through education and consensus-building discussion and debate. The CDT has expertise in policy, law, and technology and tracks various issues, including information access and bandwidth regulations, cyber-terrorism, Congressional activities, digital authentication, and encryption. See American Civil Liberties Union, Electronic Frontier Foundation. http://www.cdt.org/

Center for Research on Information Technology and Organizations CRITO. A multidisciplinary research unit located at the University of California, Irvine. CRITO conducts theoretical and empirical research on the use, impact, and management of information technology (IT) in organizations. It is also home to the NSF Industry/University Cooperative Research Center, known as the CRITO Consortium, founded in 1998. http://www.crito.uci.edu/

centigrade scale See Celsius scale.

Central One of the many colloquial names given to the early female telephone operators. Others include Hello Girls, Voice with the Smile, and Call Girls. The term is now used to describe the central telephone office in a region. See telephone history.

Central Computer and Telecommuncations Agency CCTA. A U.K. government agency promoting good practices in information technology and telecommunications in the public sector.

Central IT Unit CITU. The Central Information Technology Unit was established in November 1995 in the United Kingdom to advise on government use of information technology. http://www.citu.gov.uk

central office CO. 1. Headquarters or main service- or administration-providing center. 2. In telephony, the switching station from which subscriber loops are established. In Europe, the term *public exchange* refers to the switching center.

The purpose of the office or central exchange is to provide a connecting point through which a subscriber can establish a connection to any other public subscriber on the circuit or to a trunk line leading to other central offices. The office secondarily provides power requirements, signaling and control devices, and subscriber line services. The lines are further equipped with protective devices, fuses, coils, etc., to guard against unusually high voltages.

central office battery The power source that provides direct current to the connected lines for phone conversations. Historically, this power was provided by a 24- or 48-volt *talking battery.* Later, the 48 volts was transformed from an alternating current at the central office. See talking battery.

central processing unit CPU. A circuit on a single chip that provides the basic, essential logic for performing general purpose computing computations and decisions. The most celebrated early computer-on-a-chip was the Intel 8008, released in the early 1970s, which became the basis for a line of early microcomputers and the inspiration for a whole new industry. CPUs may be set up in parallel, or may serve as the central processor for an individual system. Desktop computers are typically based around one CPU, while some workstations, especially those for scientific computations or high-end graphics, have multiple CPUs.

Circuitry within a chip can be organized in many different ways. The architecture of the chip affects other aspects of the system, such as memory management, bus addressing, and programming procedures, particularly machine language and assembly programming. The term *central processing unit* was coined somewhere around the late 1960s but did not become prevalent until the late 1970s. During the 1970s, the CPU was often called a main processing unit (MPU). See complex instruction set computing, reduced instruction set computing.

central wavelength In optical communications systems, the frequency at which the information-carrying signal is most effective or strongest for a particular physical structure. See cladding diameter.

Centralized Automatic Message Accounting CAMA. A billing and statistical system for recording calls on tape. It is sometimes also used to trace fraudulent use of phone services.

Centre for Communication Interface Research CCIR. A research center at the University of Edinburgh, founded in 1991. CCIR studies some interesting communications-related topics, including the design and implementation of telephony services, the development of multimedia and virtual reality interfaces, and the simulation of automated services. http://www.ccir.ed.ac.uk/

Centre National d'Études des Télécommunication CNET. The French organization that approves telecommunications products for the French market. It is now France Télécom R&D.

Centrex (from *Centr*al *ex*change) A commercial telephone service provided by local telephone exchanges in which the subscriber-specific switch is physically located either on the premises of the phone exchange (CO) or at the customer's premises (CU). Used primarily by businesses, as a lower cost alternative to a private branch exchange (PBX), Centrex systems have a number of extra calling features (Caller ID, Call Conferencing, etc.), with a wider selection of options than are available to residential subscribers. An on-premises Centrex system is similar to a PBX system in that it is located on the customer's premises, but a PBX is owned by the customer, whereas the Centrex system is leased from the phone company. Each choice has pros and cons in terms of upgrades and maintenance. A Centrex central office system can also be combined with a PBX system, but care should be taken not to order redundant options for the service, since many can be provided by either. A Centrex system (or PBX) can be combined with Automatic Call Distribution (ACD), to provide self-contained, sophisticated automated business telephone services, like those used by many mail order retailers.

Centronics A printer manufacturer well known for establishing a parallel data transmission standard for computer printers, especially dot matrix printers that were popular in the 1980s.

Centronics parallel data standard A data transmission standard established by Centronics and accepted de facto by much of the printing industry in the 1980s, particularly for cabling dot matrix and some daisy wheel printers to parallel ports on desktop computers. The cable is usually a flat or pinned D connector. This parallel standard is generally faster than similar serial cable attachments because data can be carried over eight wires at once rather than just one. Attachments vary, but most systems employ a pin connector on the computer side and a flat connector on the printer side. See A connector.

CEO Chief Executive Officer. Typically the highest member of the corporate hierarchy, in charge of overall business direction, goals, and strategies.

CEPT See Conférence Européenne des Administrations des Postes et des Télécommunications.

CEPT1, CEPT2, CEPT3 See E1, E2, E3.

CER See cell error ratio.

CERB Centralized Emergency Reporting Bureau. A Canadian reporting organization to safeguard the public.

Cerf, Vinton "Vince" G. (ca. 1930s–) Vinton Cerf is credited with some of the early ideas for network gateway architecture. He has held various engineering, programming, and teaching positions in businesses and educational institutions. In the early 1970s, he researched networking and developed TCP/IP protocols under a DARPA research grant and, in 1974, co-authored "A Protocol for Packet Network Internetworking" which describes Transmission Control Protocol (TCP). During this same period, he became a founding chairman of the International Network Working Group (INWG). In 1977, Cerf co-demonstrated a gateway system that could interconnect packet radio with the ARPANET. In 1978, he co-developed a plan to separate TCP's routing functions into a separate protocol called the Internet Protocol (IP). In 1997, he was awarded the National Medal of Technology. See Kahn, Bob.

CERFnet See California Education and Research Federation Network.

CERN Organisation Européenne pour la Recherche Nucléaire. The European Organization for Nuclear Research, located in Geneva, Switzerland, founded in 1954. http://www.cern.ch/

CERT See Computer Emergency Response Team.

certificate An authentication entity used in a variety of cryptographically ensured digital transmissions designed to safeguard privacy and authenticity in electronic messaging and transactions, especially contracts, payments, private information, etc. Certification is carried out through a technical evaluation and is administrated by a recognized, assigned certificate authority. While not a complete list of all definitions associated with the issuing and management of certificates, the Authentification Certificates chart provides a short list of some of the main aspects of digital certification. See encryption.

CEST Centre for the Exploitation of Science and Technology. U.K. industry-funded organization formed in 1988.

CEV controlled environmental vault. See cable vault.

CFB See Call Forward Busy.

CFCF A historic broadcast station, originating in Montreal, Canada, in November 1920. CFCF was originally radio station XWA, one of the world's first radio broadcast entities. Representing "Canada's First, Canada's Finest," CFCF became Quebec's first private television station in 1961 as well as a founding member of the CTV network. See CKAC, XWA.

CFDA See Call Forward Don't Answer.

CFF See critical fusion frequency.

CFGDA Call Forward Group Don't Answer.

CFP Channel Frame Processor.

CFR Confirmation to Receive. A notification in networks that a frame can be forwarded.

CFUC Call Forwarding UnConditional.

CFV See Call for Votes.

CFW See Call Forward.

CGA 1. Carrier Group Alarm. An out-of-frame alarm signal generated by a channel bank, which may be followed by trunk rerouting and error control. 2. See Color Graphics Adapter.

CGI See Common Gateway Interface.

CGM Computer Graphics Metafile. A standardized graphics interchange format.

CGSA Cellular Geographic Service Area. The physical area within which a cellular company provides services. CGSAs may include multiple counties and may even cross state lines.

CGSA Restriction A subscriber option to restrict calls outside a local Cellular Geographic Service Area.

chad The small punched out pieces or edge strips from encoded punch cards, tractor feed paper, or paper tape. Punch card chad were collected in *chad boxes* which had to be periodically emptied. Originally discarded, some bright marketing person started packaging them as *confetti* for parties and celebrations. The term appears to have originated around 1947 and the plural and singular are the same.

Chadwick, James (1891–1974) An English physicist who is credited with discovering the neutron in 1932 at Cambridge, England. Chadwick actively engaged in radiation experiments, some of which stemmed from the work of Ernest Rutherford.

chaff, window Materials such as metal strips or fine wires that are highly reflective to radar waves. Chaff may be strung or shot into the air for the purpose of scattering or deflecting electromagnetic waves.

chaining A common modular software execution technique in which a process is handed off to another program or launches other program modules on an as-needed basis. By having only the necessary components memory-resident, and by off-loading modules that are no longer needed, complex programs can be managed with limited resources.

Challenge-Handshake Authentication Protocol CHAP. An Internet standards-track protocol descended from a variety of semi-secure network implementations in the mid-1970s and 1980s, which provides a method for key authentication using Point-to-Point Protocol (PPP). It employs a three-way handshake upon link establishment to verify the identity of the peer and may repeat it at intervals. After link establishment, the authenticator *challenges* the peer, which responds with a hash value. If there is a match, authentication is acknowledged; if not, the connection should be terminated. The key is known only to the peer and the authenticator and is not transmitted. This protocol is suitable for small or medium connections with an established trust relationship. Large tables of keys are not practical, and the information must be available in plaintext form, with a secure central repository to store it. See RFC 1994.

challenge-response A means to query and respond in a system where an entity seeks access to a physical or network environment or to set of functions or resources.

There is more than one way to implement a challenge-response system. It may be incumbent upon the entity seeking resources to challenge the system to which it desires access, or the system with the access may challenge an approaching entity as to its intentions. In most cases it is easier to implement the system such that the approaching entity (which may be a person approaching a building or a computer function transmitting to a repository or restricted system or network) sends out a challenge to the repository of resources. In this way, it is not necessary to constantly poll or monitor the access point for approaching entities. Thus, the person or point seeking access issues the challenge and the secured system responds with access, access instructions, a request for more information (e.g., a password), or a denial. When it works the other way, the 'approaching' entity may not be known until it reaches a certain physical boundary such as the range of a motion detector or a gateway or access point in a computer system. In this case, the resource server may challenge the approaching

entity and request identification, more information, etc. before issuing the acceptance or denial of access. Challenge-response mechanisms are central to many types of intruder-detection systems and to digital certificate-based computer security systems. See certification, Challenge Handshake Authentication Protocol.

channel 1. In its most general sense, a path along which signals can be transmitted. 2. In radio broadcast, the electromagnetic frequency spectrum extending roughly from VLF to UHF (further above that is the microwave frequency). 3. A portion of the spectrum assigned for the use of a specific carrier, e.g., the FM band is divided into channels of a specified

Authentication Certificates – Summary of Basic Aspects	
Type	Description
certification authority	A screened, trusted, assigned authority tasked with evaluating applications for digital certificates and issuing, managing, and revoking the certificates as appropriate. The authority vouches for the binding between the data items within a certificate and for the association of the certificate to the entity using it.
certification authority workstation	A computer system designed to support the issuance of digital certificates and related documents and management aspects.
certificate revocation	The declaration, by a certificate authority (CA), that a previously valid digital certificate issued by that CA is no longer valid effective immediately or as of a specified date. The information may be held in a certificate revocation list (CRL).
certificate revocation list	An administrative tool used by digital certificate authorities (CA) to track and notify of certificates that have been revoked. The list may be used for public key certificates and/or attribute certificates and may apply to user and authority certificates. A certificate revocation tree is a hierarchical distribution mechanism for sending out information about revoked certificates, signed by the issuer of the tree.
certificate serial number	An integer value associated with and possibly carried with a digital certificate. The serial number is a unique number assigned by the authorized issuer.
certificate status responder	In the context of the Federal Public Key Infrastructure (FPKI), a trusted online server acting on behalf of a certificate authority (CA) to provide authenticated certificate status information as requested by certificate users.
certificate user	An entity such as a person, firm, or algorithm using and trusting the validity of certificate-related information (e.g., a public key value) provided in a digital certificate.
certificate validation	An act or process by which a certificate user establishes the trust relationship associated with a digital certificate. The process for ensuring the validity of a digital certificate, which may include verification of the certification path and the currentness of certificates in the path. Aspects that should be verified include expiry dates or possible revocation actions, the validity of the syntax and semantics, and the signature itself.

kilohertz range. 4. In a GPS receiver, the radio frequency, circuitry, and software needed to tune the signal from a satellite. 5. In audio, a sound path (e.g., stereo broadcast requires at least two sound paths). 6. In telephony, voice-grade transmission within specified frequencies and bandwidth. See circuit. 7. In a Frame Relay network, the user access channel across which the data travels. If a physical line is *unchannelized*, then the entire line is considered a channel. If the line is *channelized*, the channel is any one of a number of time slots. If the line is *fractional*, the channel is a grouping of consecutively or nonconsecutively assigned time slots.

channel aggregation Inverse multiplexing. Bonding multiple lines (e.g., phone lines) in such a way that transmission speeds are faster. This technology is sometimes incorporated into modems to speed upload and download times.

channel-associated signaling CAS. In association with channel banks, channel-associated signals are derived from analog electromagnetic signals and converted to digital signaling bits for transmission over digital lines.

In ATM networks, CAS is a form of circuit state signaling, in which the state for that specific circuit is indicated by one or more bits of signaling status, which are repetitively sent. In T1/E1 voice applications, CAS is used for in-band signaling information in the sense that it is carried in the data stream with each channel, rather than as a separate out-of-band transmission. However, it can be configured to use a specific channel (e.g., the D channel in ISDN) such that it is functionally similar to out-of-band signaling and is then considered to be an out-of-band technology. R2 signaling methods are commonly used in channelized voice networks but may be implemented in a number of ways. In general, CAS is set up on a network by entering global configuration settings into a router and specifying the controller to handle the specific signaling scheme. The signaling protocol and CAS are configured for a specified number of time slots (or may default to all channels). The framing characteristics and the line code are defined, and a clock source is specified. See facility-associated signaling, R2 signaling.

channel bank A network interface device that provides multiplexing and flow control functions on multiple transmission channels, such as voice and data, which are being brought into a single electrical data stream. A channel bank is not intended to provide switching functions. A channel bank is commonly used as an interface between multiple lines coming out of an analog system, such as a private branch exchange, which are then brought together and multiplexed onto a larger bandwidth digital transmissions medium such as a DS-1 stream associated with a T1 line. See multiplexing.

channel bonding A technique for increasing transmission speed by means of aggregating multiple lines. It is a way of getting a little more performance from existing lines without investing in higher end technologies, such as a T1 line. Channel bonding for mo-

dem communications is usually cost effective only for two or three lines; above that number, ISDN may be a better option.

channel hopping 1. Changing channels on a radio phone, usually to find one with clearer reception. 2. Changing channels constantly on TV, usually with a remote control, and often during commercials, to the consternation of other viewers who don't have control of the remote. 3. On IRC, switching chat channels and barging in, or listening in, on many conversations in a short period of time.

channel separation In broadcasting, the space interposed between designated adjacent channels. Since signals may have a tendency to interfere with one another, leaving a gap increases the likelihood of clean signals on either side of the gap. Channel separation is commonly used in radio transmissions, especially FM broadcasts, where the width of the separator may be greater than the width of the channel through which the FM transmission is passing.

Channel Service Unit CSU. A device typically integrated with a Data Service Unit as a CSU/DSU. It is often the first device in a digitally networked facility that routes information and may also protect equipment within the facility from electrical interference and damage. It may further regenerate a signal to maintain signal strength and integrity or convert the signal from one format to another. It is commonly used between incoming T1 or E1 lines and internal channel banks, at a central office offering end-to-end digital line services to subscribers, usually through leased lines. See Data Service Unit.

channel surfing Flipping through stations on a television looking for something worth watching, or avoiding the commercials. On a TV, it is usually done with a remote, from a comfortable position on the couch. Web surfing on the Internet is similar and may eventually evolve into channel surfing, as more and more companies are starting to broadcast multimedia on their sites. See channel hopping.

channelize To increase the capacity of a wide- or broadband transmission medium by subdividing it into smaller channels, often with a small gap between channels to reduce interference.

CHAP See Challenge Handshake Authentication Protocol.

Chappé, Claude (1763–1805) A French scientist and communications pioneer who collaborated with his brother Ignace. Chappé advanced communications in 1792 by adapting semaphores (visually encoded messages) to a pair of arms mounted on a system of towers. In 1793, this enabled a message to be transmitted 15 kilometers and a year later a message was sent from Paris to Rhine (a distance of about 150 miles) in only a few minutes, through coded positions representing characters and symbols.

There were disadvantages to this system, including visibility, the need for many towers, and the 24-hour monitoring that was necessary for adjusting the arms when a new message came through. Nonetheless, it was an improvement over previous systems and grew in France to over 500 signaling stations. The first formal *telegramme* is said to have been sent on 15 August 1794.

Chapuis, Robert J. (1919–) A French researcher and author who has enriched our understanding of telephone history. Chapuis has brought to light the inventions of individuals who might otherwise remain uncredited because they were not directly associated with the larger, more successful phone service companies. See Lorimer telephone.

character code The specific coding system, or protocol, used to formulate characters for transmission. Examples of character codes include Baudot code, Morse code, ASCII, UNICODE, and EBCDIC.

character set 1. All the symbolic character representations (letters, numbers, punctuation, diacritical marks, etc.) available to a system, such as a computer, printer, or typewriter. Character sets vary widely from device to device and culture to culture, and interpretation between different character representations is one of the big challenges in global computing. 2. In network programming, *character set* is associated with mapping tables that are used to convert octets to characters, related to *character encoding*. These are often identified by tokens. See ASCII, RFC 1345, RFC 1521.

characters per second In a telegraph key transmission in Morse code, the number of characters a human operator can send or read in a second. In a printer, the number of characters that are printed in a second. In a data transmission, the number of characters that are sent through a telephone line, through a modem, or through a network. These are sometimes expressed instead in terms of bits per second, since a character may take one byte in some systems and two bytes in others.

charge coupled device CCD. A sensing system that uses lines or arrays of light-sensitive photo diode elements. As light comes in contact with various CCDs, the intensity of the light is registered and translated into digital information. Resolution is related to the type, placement, and quantity of these elements. CCDs are used in many photographic and scanning devices, digital cameras, camcorders, computer scanners, robot vision systems, etc.

Charles Babbage Institute CBI. A leading archive and research center in the study of the history of information processing and its influence on society. The CBI, founded in 1979, operates out of the University of Minnesota. It is named after Charles Babbage, a pioneer researcher and inventor of early "thinking machines."

The Institute preserves historical documents and supports and promotes research in information processing history. Documents include oral histories, manuscripts, professional records, periodicals, obsolete manuals, photographs, and other reference materials. Collections of special interest include the Academic Computing Collection, the Amateur Computer Society Records, ACM publications, the Charles Babbage Collection, the James W. Birkenstock Collection of International Business Machines Corporation Records and British Tabulating Machine Company

Histories, EDUCOM records, Wallace J. Eckert Papers, the National Bureau of Standards Computer Literature Collection, Sperry Rand Corporation records, and many more.

The archives are open to the public for use on the premises. Photocopies may be delivered by mail or fax. Support is provided by the University Libraries, the Charles Babbage Foundation, and the Institute of Technology. CBI publishes the *CBI Newsletter* and awards the Adelle and Erwin Tomash Fellowship in the History of Information Processing. http://www.cbi.umn.edu/

chase trigger In digital recording techniques, there are some legacy analog time code technologies that cause problems when applied to a digital environment. Digital audio, for example, is recorded according to a system clock. If the frequencies vary, there may be audio artifacts that cause unwanted noise in the recording. A chase trigger is a means of synchronizing time code by starting the sound segment when a particular time trigger occurs. Once it is triggered, it follows its own clock speed irrespective of whether the underlying recording that initiated the trigger has remained steady. If the underlying signal is stable, it's not a problem, but in some recording environments stability is hard to guarantee. See house sync, reference clock, time code.

chat In online computer telecommunications, private or public message areas in which participants type messages to one another in a somewhat real-time manner. Internet Relay Chat (IRC) is the largest chat forum on the Internet, although some of the large service providers have their own subscriber chat channels. Chat lines can be set up to be keyword protected, to offer private or group conference conversations. Chats with celebrities are sometimes moderated to keep the conversation to a level in which the comments are not too rapid or overwhelming. Anyone can open a public chat channel on IRC. To create a new chat or join a current chat, you enter the IRC server and type #join mychatchannel (e.g., #join gardening) or select join from the menu, if using a menu-based IRC software client. It is not acceptable on IRC to make off-topic comments or to denigrate other participants or their viewpoints. See Internet Relay Chat, Netiquette, Netizen.

chatter 1. In circuits, a repetitive, undesirable, fast clicking or opening and closing of a circuit. Power fluctuations can sometimes cause chatter. Unchecked, it can lead to damage of equipment and interference with communications. 2. In servos, styluses, and other moving control mechanisms, quick, short oscillations in a direction other than the desired direction (often perpendicular to the desired direction) caused by friction, power fluctuations, improper calibration, or improper mounting (too loose or too tight).

cheapernet *jargon* Cheaper, maybe-not-as-fast, affordable networks, such as Ethernet running over thin coaxial cable.

check bit A bit, or a group of bits, used for a variety of error housekeeping functions. A single bit is often used for parity checking, whereas 7 or 16 check bits may be used for various cache functions. See checksum, parity.

checksum A computed value commonly used for assessing data integrity and detecting errors or anomalies. Checksums are used in file systems, encryption systems, and packet transmission protocols. In networks, checksums help determine, with a reasonable degree of confidence, whether a packet has arrived at its destination unchanged. See check bit.

cherry picker *colloq.* An industrial crane equipped with a one- or two-person bucket to raise workers to levels that cannot easily be reached by other means. These are used to access fruit trees, windows, utility poles, and other high places. See lineman.

Cherry Picker - Utility Pole Maintenance

An aerial bucket, popularly called a "cherry picker" facilitates the flexible and safe maintenance of primary power lines, transformers, secondary power lines, telephone, wires, and fiber optic data cables (in that order from top to bottom) that are typically strung on joint utility poles. See joint pole.

---

cherry picking *colloq.* Selecting only the calls most likely to bear fruit and assigning them appropriately. In other words, when a call comes in, over a phone or a modem, those callers who are in some way identified or prescreened to be the most likely to benefit the callee, usually by purchasing products or services or by investing in the company, are given priority.

Cherry picking is also applied to reader service inquiries. When magazine readers send in reader service cards, these inquiries are forwarded to the appropriate vendors. The vendor sheets sometimes include statistics gathered by the magazine, such as job titles and the number of boxes the inquirer checked. Thus, the vendor can select and respond first to those

most likely to produce revenue for the company.

Children's Internet Protection Act CIPA, ChIP Act. Introduced to the U.S. Congress by Senator John McCain in January 1999, the Act was never called to vote. However, it was later brought forward through an omnibus budget bill that was signed into law in December 2000 to come into effect in Spring 2001 as part of the Consolidated Appropriations Act. Also referred to as the Children's Online Protection Act, it requires that certain federally funded institutions have in place an Internet Safety Policy for protecting children from exposure to Internet content that may be inappropriate for children. The ChIP Act encompasses monitoring and filtering of objectionable content as well as monitoring of children while online. No sooner was the legislation introduced than it was challenged by civil liberties organizations as to its constitutionality.

chime A tone on a telephone, computer, or alarm system that sounds like a pleasant musical tone rather than a buzz or a beep. It's straightforward to generate many types of sounds and chimes on a computer system through sampling and storing sound files.

chimney effect The natural tendency of heat to rise. One type of cooling system that is based upon this effect consists of ventilation slots or holes in the top and bottom surfaces of a cabinet or component, facilitating natural air circulation.

China United Telecommunications Corporation China UNICOM. A state-owned corporation cosponsored by the Ministries of Electronic Industry, of Electric Power, and of Railways, founded in accordance with the "Company Law of the People's Republic of China." China UNICOM seeks to promote the reform of the telecommunication system and its potential resources, and to build a telecommunications network for furthering the industry of China. Telecom operations are under the control of the Ministry of Posts and Telecommunications. The corporation has international relationships with organizations such as GTE (U.S.), Nortel (Canada), and Singapore Telecom.

chip 1. In computers, most commonly a chip is a component integrated circuit attached to the circuitry by conducting legs that can be inserted into a PC board or onto another chip. The term is more commonly applied to the removable rather than the soldered components.

Chips come in many shapes and sizes, but most of those used in consumer appliances and desktop computers range up to about 2 or 3 in. in length. Sometimes two or more chips are closely coupled or laminated together. Larger, heat-producing chips may be coupled with a fan or require that a fan or heat sink be installed nearby with adequate air circulation. Chips can be general purpose, as with central processing units (CPUs), or very specialized, as with device driving chips optimized for a particular device. See semiconductor, silicon, large scale integration. 2. The transmission duration for a bit or single symbol of a PN (pseudorandom noise) code. 3. In spread spectrum wireless communications, a chip is a redundant data bit inserted at the origin of a transmission, which subsequently is removed when the transmission is deciphered at the receiving end.

Chip Protection Act CPA. Accepted into law in 1984, this act provided new intellectual property protection for the semiconductor industry which was experiencing tremendous growth at this time.

Chireix antenna, Chireix-Mesny antenna A bidirectionally resonant, cascading array, wire, or tubing antenna consisting of series-fed square loops with sides the length of a half-wave or quarter-wave. It is used for very high frequency ranges, typically VHF and higher.

chirp A short birdlike, often high-pitched, tone that may occur in various audio components if not configured correctly. It is usually an undesired quality requiring adjustments.

In laser light components, pulsing between different power levels or an instability causing variation around a central, desired wavelength can cause an optical chirp and accompanying chromatic dispersion. One of the skills learned by technicians is measuring laser chirp in modulated transmissions.

Laser chirp is usually undesired, but may have some advantages. It has been reported that positively and negatively chirped laser pulses may differ in their electron yields. Whenever there is an imbalance in phenomena generated by similar actions, the potential for generating logic signals exists, if ways can be found to harness the effects. It also means that one

type of chirp may be useful for different functions (or may be more disruptive) than another.

chopper A device for quickly opening and closing an electrical circuit or beam of light at regular intervals. A chopper may interrupt a signal to permit amplification, to demodulate a circuit, to interrupt a continuous stream of particles, or to send signals. The process is called chopping. See optical chopper.

Christie, Samuel Hunter (1784–1865) An English mathematician and educator who investigated the electrical properties of various metals and created a balanced bridge resistance-measuring circuit in 1833, which was used and described by Charles Wheatstone and became known as the *Wheatstone bridge.*

chroma A color as characterized by its hue and saturation, without reference to the relative brightness of the color. Black, white, and shades of gray are not considered to be chroma.

chroma key In cinematography, a technique for eliminating a specific region or background so that new imagery can be superimposed in the removed area. Everyone who has watched television shows, especially action shows, has seen chroma key effects. High speed chases on motorcycles and cars are often photographed from the front with the bike or car mounted on a frame that sways back and forth before a blue or green curtain (the chroma *key*). In other words, the moving scenery is simulated. This way the actors can be safely filmed, and their conversations recorded without extraneous noise.

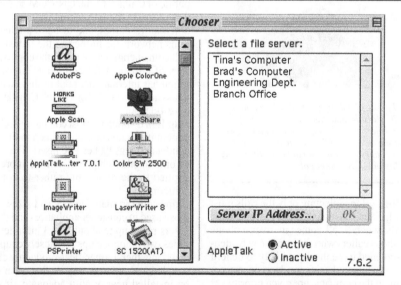

Chooser Utility for Viewing and Selecting Computer Network Resources

Each computer system on a network has some type of software utility similar to the above Chooser on the Macintosh that displays network resources and allows the user to choose those of interest. The icons on the left represent available network device drivers, including PostScript printing devices, scanner drivers, digital camera connections, etc. If AppleShare is selected, the system searches the network looking for other available systems and displays them in a list on the right. The user must then select a system and log on to that system with a valid password. Once logged onto the remote system, the user may select from a list of available mounted servers (usually hard drives) associated with that system that have file sharing attributes set to "on." The user can then send and receive data between systems.

Later, in the editing room, the foreground is super-imposed on a background of a road, desert, or foreign planet, and everything that was blue or green becomes the background. Chroma key effects are often cheesy looking, with a halo around the actors giving away the illusion, but with digital photography, it is becoming increasingly difficult to distinguish what is real and what isn't.

chrominance The color-carrying portion of a video signal, which for various reasons related to compatibility with black and white television sets, is carried separately from luminance. The chrominance comprises hue (color) and saturation (the intensity of the color).

chronograph 1. An instrument that graphically displays the time. 2. A diagrammatic representation of time intervals where some quantity or variable is expressed as a function of time.

chronoscope An instrument for measuring time in very precise increments that is used in scientific applications.

Chooser A software utility for selecting network resources as part of the Macintosh OS. See diagram.

CHU A specialized AM band radio station that broadcasts the time with brief tones each second, and the spoken time each minute in English and French. The time broadcast is Eastern Standard Time (EST), which is five hours after Coordinated Universal Time (UTC).

chuck A device to support and sometimes turn another, usually through pressure and friction. A lathe chuck is a hollow or geared area in which the metal or wood to be turned is held in place by a chuck. A drill chuck is a geared key mechanism for inserting or removing a drill bit held in place by pressure as the drill key tightens the aperture against the shaft of the bit.

A fiber optic splicing device may be equipped with V-grooves or chucks to hold the two pieces of fiber optic filament to be spliced. If it has chucks, they are typically either clamping chucks or vacuum chucks. With vacuum chucks, individual vacuum control to the chucks makes it possible to firmly secure one filament while aligning or rotating the other until the desired position is found. The chucks may also be cleared of particles with vacuum suction.

churn A term to describe customer subscription turnover in the cellular telephone and cable TV industries. Churn is of concern because the setup fees are often spread out over future monthly revenues and are not recouped unless the customer retains the service for some time. Churn is usually computed as some percentage formula based upon connects and disconnects.

CI 1. See congestion indicator. 2. Certified Integrator.

CIAC See Computer Incident Advisory Capability.

CIAJ Communications Industry Association of Japan. An organization committed to promoting the development of manufacturing of telecommunications equipment and related business activities.

CIC See Carrier Identification Code.

CIDR See Classless Inter-Domain Routing.

CIE The Commission Internationale de L'Eclairage (Internal Commission of Illumination) established in 1931, developed an international color model for light, with the aid of technology that enabled more precise measurement of wavelengths, especially those within the visible light spectrum. Maxwell's triangle was taken for the basic model and three primary colors selected and assigned to the system. The CIE chromaticity model is two dimensional but can be extended to three dimensions by reducing the value or amount of light, thus diminishing the brightness until it reaches black, the absence of light. See Maxwell's triangle.

CIF 1. See cells in flight. 2. See Cells in Frames. 3. See Common Intermediate Format. 4. cost, insurance, freight.

CIID See Card Issuer Identifier Code.

CIIG Canadian ISDN Interest Group.

CIK See cryptographic ignition key.

cinching An increase in pressure in which something is made tighter, becomes more difficult to undo, or becomes locked up. This effect is sometimes undesirable as on reel-to-reel systems, where the pull on a reel is greater than the speed at which it unwinds so the remaining tape, or other material, tugs and becomes very tightly packed. It can also happen on cable spooling or installation rollers and should be avoided as it might damage the cable components.

CIP 1. Carrier Identification Parameter. In ATM networks, CIP is a 3- or 4-digit carrier ID carried in the initial address message, used in establishing a connection. 2. Channel Interface Processor. A Cisco Systems channel attachment interface for their 7000 series routers which connects a host mainframe to a control unit.

CIPA See Children's Internet Protection Act.

CIPF See Canadian Information Processing Society.

CIR 1. See carrier to interference ratio. 2. See committed information rate.

circuit A physical or virtual collection of pathways, channels, or conductors interlinking given points or nodes in an orderly fashion to create communications or electrical links. Computer circuits include traces, wires, chips, resistors, capacitors, etc. A circuit can be open or closed.

circuit line The physical line carrying network traffic between user equipment and an IPX or IGX node. Most network services (voice, data, ATM, etc.) require that the circuit line be configured and 'activated' on the system to which it is attached before they can be used.

circuit line commands Commands typically used to set up and control a communications line for data, voice, and Frame Relay. Various operating states and parameters are established first to set up the general parameters for the line (as opposed to specific parameters for the voice or other communications service). The command set differs from manufacturer to manufacturer, but it commonly includes commands to activate the circuit line, down the circuit line, display or print the current configuration, and configure the parameters of the line.

circuit switching A type of end-to-end transmission system common in phone connections. In the process of setting up the connection, a number of resources are allocated to that specific call, most of which are tied up until the call is completed and the connection terminated. One advantage of this system is that it can guarantee a certain level of performance. A disadvantage is that the resources are tied up whether or not there is active communication. See message switching, packet switching.

circular antenna A horizontally polarized, half-wave dipole antenna formed into the shape of a circle except that the terminating ends do not touch to make a continuous loop.

circular magnetic wave A magnetic wave in which the lines of force describe a circular pattern.

circular polarization An electromagnetic wave whose lines of flux are oriented in a plane, usually horizontal or vertical, or where the "edge" of the field describes a circular shape. Circular polarization is used in antennas, where electricity serves to uniformly rotate the electromagnetic field through the antenna. It is possible to use one circularly polarized wave to communicate with another, or the circularly polarized wave can be manipulated to yield linearly polarized waves perpendicular to one another.

circular scanning Scanning in which the sweep of the sensor and/or the display monitor describes a full 360° arc, which can be pictured as a cone shape spreading out toward the direction of the region being scanned (e.g., some types of radar).

circulator 1. A process or device that moves something from hand to hand, or device to device. 2. In microwave transmissions, a multiterminal coupling device in which the transmission is passed down through adjacent terminals. 3. In radar transmissions, a device that alternates the signal between the transmitter and the receiver. 4. In data communications, a mechanism for allocating or transferring information or control among ports.

CISC See Complex Instruction Set Computing.

CISCC Collocation Interconnection Service Cross Connection.

Cisco IOS Cisco Internetwork Operating System. An OS incorporated as part of the CiscoFusion architecture to help the system administrator centralize, integrate, install, and manage internetworks.

Cisco Systems Inc. A significant vendor of routers, switchers, and related hardware and software for network systems. The author gained a greater understanding of the function and implementation of network routing systems through Cisco seminars.

CiscoFusion A Cisco Systems internetworking architecture that integrates scalable, stable, secure technologies with ATM, local area networks (LAN), and virtual local area networks (VLANs).

CiscoView A graphical device-management application that dynamically provides administrative, monitoring, and configuration information for Cisco internetwork devices.

CISE See Computer and Information Science and Engineering.

CISPR See International Special Committee on Radio Interference.

CITA See Canadian Independent Telephone Association.

CITEL Inter-American Telecommunications Commission.

citizens band radio, citizens radio service CB radio. Radio frequencies set aside for the use of relatively low power consumer radios and radio controllers (for model cars and planes). These have a limited range (up to about 10 or 15 miles for mobile units), although sunspot activity and local weather can sometimes provide some surprisingly long connections when broadcasting conditions are optimal. In the United States, CB radios are commonly used by truckers, travellers, and radio hobbyists. Communications over 150 miles are prohibited by the Federal Communications Commission (FCC). The frequencies originally allocated by the FCC were around 27 Mhz, but have been changed to around 463 to 470 MHz. Before computer bulletin board systems and the Internet, CB radio was a popular means of community interaction. Not all countries are free, and civilian use of radios is not permitted in some regions of the world. See OSCAR, AMSAT.

CITR Canadian Institute for Telecommunications Research.

CITRIS See Information Technology Research Center.

CITU See Central IT Unit.

City and Suburban Telegraph Company The first company in Cincinnati, OH to provide direct communication between homes and businesses, incorporated in 1873. In 1878, it contracted with the Bell Telephone Company of Boston, MA to provide Bell services in the Queen City area and, in 1882, contracted with American Bell to provide long-distance services. Its first payphone was installed in 1904 and mobile phones were introduced in 1946. In 1952, it became the first Bell company to provide 100% dial service. The company became Cincinnati Bell Telephone in 1971.

CIV See cell interarrival variation.

CIVDL See Collaboration for Interactive Visual Distance Learning.

CIX See Commercial Internet Exchange.

CJC See *Canadian Journal of Communication*.

CKAC The first Canadian television broadcasting station, which began experimenting with mechanical television transmitted over wires in 1926. A Baird disc camera and Jenkins scanning disc television receiver were early inventions that were tried around this time. Sound and images were transmitted separately so the sound could be played on a radio receiver. Shortwave bands were used for the images. Alphonse Ouimet, who later became the president of the CBC, was a technician for the first historic CKAC broadcast in 1931, a musical performance that was sent out to 20 viewers. See Canadian Broadcasting Corporation.

CL *symb.* left-hand circular or indirect polarization.

cladding 1. A coating, something that overlays, a protective covering, sheath. 2. A substance, such as metal, bonded to another to cover it by various means, such as pressure rolling or extruding. A process sometimes used in producing transmission cables. 3. A layer in a laminate (which may be planar or spherical) with a lower refractive index than its associative conductive materials, as in a fiber optic transmission cable, such that incident light is reflected away from the cladding rather than passing through. Glass and plastics are commonly used as fiber cladding materials but planar laminates may use other materials as well. Depending upon the materials and their refractive indexes, there are effective ratios between a cladding layer (as in a fiber bundle) and an inner conducting core as well as limits established by economics and whether the cable needs to be flexible. Different cladding arrangements have some interesting effects on light transmission in a fiber cable. For example, the materials in the cladding may be designed to reflect some wavelengths and absorb or transmit others. The cladding may also be designed so that it becomes thin at certain bend radiuses to emit rather than reflect light (thus radiating modes exceed guided modes). This may seem impractical, but it allows the fabrication of illuminated "fabric" sheets by weaving the fibers over a warp layer that deliberately bends the fiber filament at periodic intervals. See index of refraction, spilling, stray light, total internal reflectance, V number.

cladding alignment splicer CAS. A precision industrial tool for preparing a variety of types of cladded fibers (single-mode, multimode, dispersion-shifted, etc.) for assembly and installation. The device provides a tension and alignment mechanism (e.g., a groove) to facilitate precise splicing of cleaved fiber filaments. It may optionally have fiber end angle measurement, defect detection and pigtail continuity assessment capabilities. Passive alignment systems (on two axes) provide cladding alignment rather than core-to-core alignment. A CAS typically uses heat (fusion splicing) to join fiber filaments into a continuous waveguide. Automatic units can create a splice in about 10 to 20 seconds (not including a technician's assessment of the visual display of the splice and its accompanying data parameters).

A CAS typically includes a small built-in display or a connection to a computer display to provide an image of the assembly and alignment in the X and Y coordinates, usually magnified about 100 times. This enables a technician to visually inspect the assembly. Since there are many different splice modes, depending upon the type of fiber and components, the unit may be preprogrammed or programmable for quick setup. Depending upon the splicing capabilities, it may be possible to store and retrieve a log of splicing activities, including the selected splice modes, arc conditions, estimated efficiency (e.g., loss estimates), and optional comments. Stand-alone units may log from ca. 300 to 1000 splices. Units with computer connections could provide unlimited logging with removable mass storage, which may aid companies in monitoring quality control, production changes, staff training statistics, etc. See cleave, fusion splice.

cladding diameter In a cable that includes a cladding layer, such as a metal wire with a bonded coating, or a two-glass cladded fiber cable, the diameter that includes the cladding layer. In a perfectly round fiber, this can simply be measured; however, as the cable may be elliptical, the diameter may be calculated by taking the average of the smallest circular outer diameter and the largest circular inner diameter.

Cladding & Core Ratios in Optical Cables

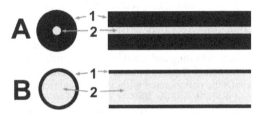

In single-mode fibers (A), the cladding thickness (1) is typically much larger than the diameter of the tiny filament that forms the core (2). In multimode fibers, the relationship is typically reversed. The core has a higher diameter in relation to the cladding thickness, providing room for multiple wavelengths to travel along the waveguide.

The ideal ratio depends upon many factors, including the cable materials, the wavelengths transmitted, and the degree of bend that might be expected of the cable during installation.

---

cladding glass A type of glass or other transparent material used in fiber optic cables that has a lower refractive index than the glass used in the inner core.

cladding mode In a transmission through a cladded conductor, a signal conducted through the outer cladding in addition to any signals that may be transmitted through the cladded core. See cladding beam.

clamping 1. Holding within an established operating, baseline, or midline range in a circuit, in order to maintain various processes or electrical charges at stable or safe levels. 2. In a cathode-ray tube (CRT), a process that establishes a level for the picture display at the beginning of each scan line within a frame.

clamping voltage An established level of voltage around, under, or over which an electrical device is permitted to operate. For example, clamping voltages can be used to establish a range within which a device operates, by setting them so that any fluctuations above or below that voltage will cause a system shutdown or other protective reactions.

Clark cell A type of historic low-volt energy cell using mercury and zinc amalgam in the cathodes and anodes.

Clark, David (ca. 1943–) David Clark has been a chairman and active participant in various Internet associations, including the IRTF and IAB. He has participated in numerous research efforts in high speed, very large networks, and network video applications, and various development efforts including the Swift operating system, Multics and Token-Ring local area networks (LANs).

Clark, James "Jim" H. (1944–) Formerly of SGI, in 1994 Clark co-founded Mosaic Communications Corporation with Mark Andreessen, which later become Netscape Communications, distributors of the most broadly used browser applications on the World Wide Web. See World Wide Web.

Clarke, Arthur C. (1917–) An English-born scientist and writer. With remarkable prescience, Clarke anticipated the age of satellites and long-distance communications. He was talking about it as early as 1942, while still in his twenties, and published an article about it called "Extra-Terrestrial Relays" in *Wireless World* in 1945. Clarke further wrote detailed descriptions of geostationary satellite orbits and satellite transmitting and receiving stations in the 1950s, years before the first Sputnik was launched. In the 1960s, he collaborated with Stanley Kubrick in the making of the movie *2001: A Space Odyssey* (1968), which has since become a classic. See satellite, Sputnik I.

Clark, George Howard (1881–1956) A Canadian-born American telegraph operator and significant collector of historical radio artifacts, Clark worked for the Boston and Maine Railroad. A scrapbook enthusiast as a child, Clark began collecting wireless radio materials in 1902. In 1903, he graduated from MIT in Electrical Engineering, specializing in radio work. He then worked for the Stone Telegraph and

Telephone Company in Boston. In 1915, while in the Navy, Clark helped devise a classification system for blueprints, documents, and general data. In 1918, he adopted the system for organizing his radio collection and coined the term "Radioana." In 1919, he joined the staff at the Marconi Telegraph Company of America and later the Radio Corporation of America (RCA). Clark devised the 'type number system' used by RCA for classifying equipment (e.g., vacuum tubes). From 1922 to 1934, he was in charge of exhibits of radio apparatus at shows and fairs on behalf of RCA. In 1928, he started a radio museum for RCA. The museum collection was turned over to the Rosenwald Museum in Chicago, IL and the Henry Ford Museum in Dearborn, MI. Clark's collection began to assume some importance when patent infringement cases came to court and the documents provided substantive evidence about the radio industry. At his death, the collection was given to MIT and, in 1959, turned over to the Smithsonian Institution. See Clark, George H. Radioana Collection.

Clark, George H. Radioana Collection An archive of historical radio memorabilia and documents collected primarily between 1900 and 1935 by radio engineer George H. Clark. It was transferred from the Massachusetts Institute of Technology to the Smithsonian's National Museum of American History in 1959. It is one of the most extensive collections of wireless radio history in the U.S., comprising more than 276 linear feet of shelf space at the

Amplifier Operations and Emissions Class Categories	
Amplifier Categories	**Description**
Class A amplifier	A single-ended circuit in which output current flows during the input cycle, as related to the grid bias and grid voltage. Provides good fidelity at low receiving levels.
Class AB amplifier	A circuit in which output current flows for more than half, but less than the full, duration of the input cycle. Better efficiency than a Class A amplifier but also has higher power requirements.
Class B amplifier	A circuit in which output current flows for half of the input cycle. More efficient than Class A or Class AB but has higher power requirements and can't be configured as a single-ended circuit.
Class C amplifier	A circuit in which output current flows for somewhat less than half of the input cycle. This provides high efficiency but also has higher power requirements.
Emissions Categories	**Description**
Class A0 emission	Incidental radiation emanating from an unmodulated carrier wave transmission.
Class A1 emission	A low-speed carrier wave (as those used for early telegraphy) unmodulated by an audio signal.
Class A2 emission	An amplitude-modulated carrier wave modulated by low audio signals to transmit simple tones or Morse code.
Class A3 emission	An amplitude-modulated carrier wave modulated by audio signals so intelligible conversation can be transmitted.

National Archives in Washington, D.C. The collection includes biographical information on the pioneers developing radio and photographs and documents on the growth and operations of radio companies, particularly National Electric Signaling Company and RCA. Interestingly, the collection is arranged according to a Navy filing system devised in part by Clark in 1915. See Clark, George Howard.

CLASS Custom Local Area Signaling Services. Telephone subscriber calling options including, but not limited to, Automatic Callback, Call Trace, Caller ID, Selective Call Rejection. In the past, when demand for these services was lower, they were billed individually, depending upon which ones were selected. More recently, phone companies have been offering monthly flat rate bundles on a variety of these caller options.

Class, facsimile For information on Class 1 and Class 2 facsimile standards and related concepts, see facsimile, formats.

Class, IP See IP Class.

class of service CoS. A general designation for an agreed or specified level of functioning or security, which varies from industry to industry. In telecommunications, network configuration and tuning and sometimes billing levels are established according to class of service parameters. See quality of service.

Classical IP A set of specifications for an asynchronous transfer mode (ATM) implementation model described in the early 1990s by the Internet Engineering Task Force (IETF) for local area internetworking. In Classical IP implementations, IP headers are processed at each router, creating latency and limiting throughput. Due to the increase in demand for multimedia capabilities, Classical IP is showing its age. One of the limitations of Classical IP is that direct ATM connectivity exists only between nodes with the same IP address prefix. See ATM models for a chart of some historic and new ATM models. See RFC 1577.

Classmark An electronic designation that identifies privileges and restrictions associated with a particular communications line or trunk. See class of service.

CLC 1. Carrier Liaison Committee. 2. Competitive Local Carrier.

cleaning arc A brief electrical spark generated by an electrode for the purpose of removing particulate matter from a surface. Cleaning arcs are used in fusion splicers for cleaning the ends of fiber optic filaments prior to joining since any undesired particles in the joint could interfere with light transmission. See fusion splicing.

clear 1. In computer monitor displays, to blank a screen, applications window, or terminal window. The *clear* command provides a clean slate, a visual working space without clutter, obsolete information, or distractions. 2. In programming, to set a storage location (a buffer, address, etc.) to a zero state, blank state (as with space characters), previous state, or default state. 3. In communications, a clear signal is one without noise or interference and of sufficient volume or intensity to be heard or seen distinctly.

clear channel 1. In telephone communications a transmission line that is used entirely for communication, and no control or other signaling bits are being transmitted. In other words, all the resources are available for the informational communication. 2. In radio communications, a station that is permitted to dominate a frequency and broadcast at a certain power level or up to a certain distance (e.g., 750 miles) during a specified time of day. A type of exclusive frequency arrangement.

clear to send CTS. A handshaking signal provided when the communication has been set up over a serial link, and the called modem is ready to receive information. See RS-232.

clearance In electrical installation, the shortest distance between separated live conductors, or between live conductors separated from physical structures, or between live conductors separated from associated grounds. See gap.

cleavage plane A planar direction in a material in which the molecular bonds that hold the substance together can be more readily cleaved. Many materials have a "grain," a general direction in which the fibers or lattice structure are aligned. Thus, most paper tears more readily in one direction than another and a number of materials used in electronic compoents will cleave more readily depending upon the orientation of the material with respect to the cleaver. For example, birefringent materials, those with a molecular structure that is the same in two planes and different in a third, will generally cleave more readily when nicked, broken, sawed, or sliced across the plane in which the bonds are weaker. Knowledge of the composition of materials and their cleavage plane is useful in materials science and materials and component fabrication. See cleave.

cleave To cut or break a cable or component so the broken surface meets certain needs, as in junctions, solder joints, or optical connections. With basic electricity-conducting wires, the angle or cleanliness of the break in a line is not usually critical, as wire can be readily wrapped or soldered in place without significant loss to the signal. However, with layered wires or certain electronic components, where a smooth or straight surface might be important to the electrical contact, or in fiber optic cable, where rough edges can degrade the light-carrying properties of the fiber, a precise cleave is critical.

Wire is usually cleaved with scissors, knives, or specialized cutters, depending upon the gauge of wire and the importance of a clean cleave. Fiber optic filaments are usually cleaved with mechanical blades, but some systems use ultrasonics for the very precise cleaves desired for fusion splicing. See cleavage plane, cleave angle, cleaver, crimp, fiber optic; fusion splice.

cleave angle The angle at which a surface is cut relative to a reference. Cleaving is often done prior to creating a glued or fused joint or adding a connector, but it may also be used to provide an unimpeded exit path for an optical signal (e.g., a fiber optic sign or lighting fixture).

In wires and fiber optic filaments, the angle is measured relative to the lengthwise conductive core and is usually described and handled in two planes. Thus, if the core is horizontal (in the X plane), the cleave angle is described in terms of its angle as it rotates through the X-Y plane. There's no reason why the angle in the Z plane couldn't also be described, but in most fabrication technologies for wire and cable splicing and attachment, the Z plane is kept at right angles to the X reference for simplicity (and because there's usually no added benefit in altering Z angles). Cleave angles vary depending upon the materials being cut, the cleaver, the purpose of the cleave, and any tension factors that may act upon a cable that is to be cleaved and spliced. A 90° cleave may be easier to cut and fuse and may provide a better unimpeded path for signals due to the smaller fused cross-section, but an angled cleave (e.g., 45°) may provide a stronger connection due to the larger fusing area and may withstand better forces against the fused joint. For end-emitting fiber filaments intended for signage, artworks, or lighting fixtures, an angled cleave can influence the shape and the amount of light that escapes from the endpoint. See cleave.

Cleave Angle Effects

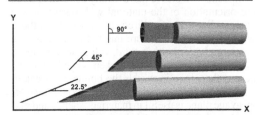

The angle of a cleave will influence the amount of surface area that is available for connecting or fusing a fiber optic filament. The smaller the angle, the greater the surface area. This has advantages and disadvantages. A larger surface area may provide added strength in certain directions and a stronger fuse, due to the larger surface area, but it also is more difficult to cut and match and creates a larger area in which contaminants or aberrations in a joint can interfere with light transmission.

A straight (90°) angle is used when the filament is being terminated, hand spliced, or spliced with a machine that is set to rotate the two ends independently prior to splicing. Angle cuts must be carefully matched if they are to be spliced, because even the smallest angle, if the ends are aligned perfectly, can cause a gap that interferes with light propagation.

---

cleaver, fiber optic A device specifically designed to provide the precise, clean cleaves required for fusing fiber optic filaments together, or for attaching connectors to a fiber filament. Tolerances are usually within 0.5°.

Traditionally, optical fibers to be coupled with other components were cleaved and then polished to provide the cleanest surface possible at the terminal ends. However, some manufacturers now promote products that produce a cleaner cleave and claim that polishing is not necessary. For complex bundled assemblies, polishing may still be necessary.

Cleave – Basic Types

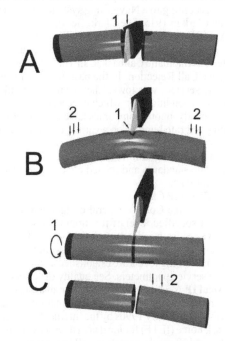

There are a number of ways to cleave a wire or cable, depending upon the materials. Wires and plastics can often be cut all the way through with a knife (A) or snippers that cut from both sides, especially if rough edges aren't a concern. Hard plastics and some metallic compounds often cleave better if nicked first and bent until they break at the point of the nick (B) (the effectiveness of snapping depends partly upon the direction of the grain). Glass may break more cleanly and evenly if nicked or scored (C) and then snapped. Nicking and snapping often provides the best cleave for fiber optic filaments.

Fiber optic cleavers come in many shapes and sizes. Some resemble microscope components, with viewing and testing capabilities, some have computer interfaces for magnifying or analyzing diffraction patterns in the filament that has been cleaved, some support polishing components, and some are compact units resembling staplers. Sizes range from handheld to table-top. Mid-sized cleavers for field work may have tripod mounts for securing in a mobile lab or an outdoor or in-plant location where there is no flat workspace for setting up the cleaver.

Since fiber scraps can get in the way and present a safety hazard (the shards are small and sharp) a tray for collecting the fiber end scraps is sometimes included with cleaving devices. It is important to shield the eyes with safety goggles and to remove any filament shards not discarded into trays when working with cleavers.

The cleaving blades may be specialized for the diameter range and type of fiber to be cut (glass or plastic). There may be different blade heights and rotations to accommodate different sizes. See cladding alignment splicer, crimp, interferometer.

Cleave Area and Strength

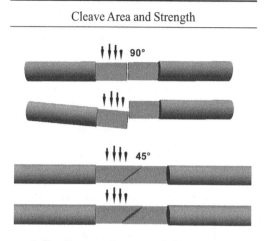

In fiber filaments, cleaves intended for splicing must be clean and precise to prevent interruption or back reflection of a light signal passing through the joint. The angle of the cleaves to be joined should match (in most cases) to facilitate fusion. Narrower angles (e.g., 45° from perpendicular in the orientation shown above) create larger surface areas for fusion and may resist breakage from forces in certain directions, but may also influence the lightguiding properties of the joint if the angles don't match rotational characteristics of the fiber (e.g., in polarized cables). Angle cleaves are generally more difficult to cut and splice, especially with hand tools.

CLEC See Competitive Local Exchange Carrier.

Clerk-Maxwell, James See Maxwell, James Clerk-.

click tones A signaling system common on phone systems, especially wireless phones, that alerts the user that the call is being processed.

clickstream *slang* A description of the flow of events and sites visited when a user navigates the Internet, particularly the Web, which is connected through clickable hyperlinks. Product vendors have an intense commercial interest in monitoring user behavior and maneuvering users to their sites.

cleavage client A system or application that serves the user but that may seek or require information or operating parameters through a host with a higher priority or greater capabilities. In the past, host and client have had almost opposite meanings for some computer administrators, but for consistency in this dictionary, and because the trend is in this direction, client is defined as the adjunct or subservient system or application. See host.

client application In a client/server computer software application, the client is typically the application used by the user to communicate to a source or destination through a related higher priority or more powerful (or just different) server program. The Net-

scape Communications Web browser is an example of a common client application that communicates to Web sites through a Web server, handles the traffic, and provides some measure of security.

client operating system On a network, the operating systems run on client machines, user terminals, and subsidiary machines. These do not have to be the same as the server operating system. A good server can handle a variety of client OSs and network between them seamlessly, using standard network protocols. For example, you may have a network that is configured with a Sun workstation and Sun operating system (SunOS, Solaris) as the main server, with a number of different client platforms connected to it, running different client operating systems and operating environments, such as Linux, Apple Computer's OS X, Be Inc.'s BeOS, or Microsoft Windows.

client/server model A computer processing method of improving efficiency, and sometimes security, by selectively distributing activities. In human enterprises, there is often a manager with an overall knowledge of the work to be done, security clearance, and the authority to designate tasks and respond to requests. In conjunction with the manager are workers with knowledge of specific tasks and needs, lower security clearance, and instructions to report their findings and to direct their questions and requests for resources to the manager. A client/server model on a computer system is similar to this. An ISP's Web server has the logic and security clearance to accept requests from many Web browsers, and to fetch the information and serve it back to the browsers which then format and display the information for the user. Most networks work on client/server models, where the server handles administrative details, file management, and security, and the client machines, usually terminals or desktop computers, handle input and output, local processing, and display.

CLIP calling line identification presentation (e.g., as in ISDN Q.81 and Q.731 number identification services).

clipboard In most operating systems, and in some software programs, an area of memory or a file on a hard drive designated to hold information (usually images or text, but may also be sound or video clips) that has been *cut* by the user for later retrieval. Most clipboards store only one clip at a time, with subsequent clips over-writing previous ones, so that only the most recent can be retrieved. Some clipboards can handle multiple clips, and some store the information on disk for later retrieval. On the Macintosh, for example, a user can save clips by copying or cutting them and storing them in the Scrapbook. Clips can be retrieved later by paging through the clips, selecting those desired, and copying and pasting them back into an application.

Clipper chip A microprocessor chip that provides security encryption features that can be incorporated into electronic devices. The Clipper chip has become the focal point for broad and heated debates over the privacy of global communications. The U.S. federal governing bodies had initiated plans to include the chip in consumer telecommunications products, to secure conversations from anyone but the government. The plans were initially announced in 1993 through the White House Escrowed Encryption Initiative. The system was designed by the National Security Agency (NSA). Three versions of the proposal, Clipper I, II, and III were promoted between 1993 and 1996.

Most people agree that there is a need for voluntary, widely available encryption options for government and private use, and vendors agree that standards provide a means for them to distribute products with intercompatibility. But on this issue, concern has been expressed about how the government is intending to implement and enforce encryption policies, and on their assertion that the system will work only if made mandatory. There has been a considerable outcry from vendors and the public, who questioned the robustness of the technology, and who are gravely concerned about too much power being in the hands of too few people.

In spite of the negative feedback, the Department of Commerce approved the Escrowed Encryption Standard (EES) as a voluntary Federal Information Processing Standard (FIPS) in 1994. One of the requirements would be that every Clipper chip would have its unique key registered with the federal government and held in split form by two federal escrow agents (NIST and the Treasury Department), creating accessibility for the federal government to wiretap secure communications. The debate over the chip, privacy, and law enforcement led in the fall of 1994, to the Encryption Standards and Procedures Act, which described federal policy governing the development and use of encryption technology for unclassified information. Back references were made to The Computer Security Act of 1987.

The public responded in many ways to the various proposals regarding the Clipper chip. Some sought to point out flaws in the process and design; others created free user-encryption programs that would defeat the Clipper system. One of the more significant challenges to the system was the X9 Accredited Standards Committee (ASC) announcement that it would develop a competing data security standard based upon triple-DES. The ASC sets data security standards for the U.S. financial industry.

The Clipper chip uses a nonpublic encryption algorithm called SKIPJACK which cannot be read off the chip and is designed so that it cannot be reverse engineered. According to the EES, when two devices negotiate a communication, they must both have security devices with Clipper chips and must agree on a *session key,* which may be a public key such as RSA or Diffie-Hellman. The message is then encrypted and sent with a law enforcement access field (LEAF), a serial number, authentication string, and a family key. When received, the LEAF is decrypted, the authentication string verified, and the message decrypted with the key. See Capstone chip, LEAF, Pretty Good Privacy, SKIPJACK.

clipping 1. In software applications, cutting out information, such as graphics, text, sound or video, usually for later retrieval or insertion elsewhere. See clipboard. 2. In graphics programs, the process of removing parts of an image, or of the display outside some designated boundary, usually the outer margin of a picture, or of an application's window. Information that is *clipped* may or may not be retained in memory. Often a program will retain the information, even if the user can't see it, so the user can quickly restore the information or scroll quickly through the image without recreating it or having to wait for the computer to reread it from disk. 3. In audio, a brief loss of sound, especially at the beginning or end of a transmission due to limitations of the technology (limited frequency range, direction flipping, ramp-up time). 4. In audio communications, especially phone calls over satellite links, the equipment may be operating part of the time in half duplex mode, transmitting in only one direction at a time, so gaps in the conversation may cause a switch in the other direction and clip part of the conversation.

CLIR calling line identification restriction (e.g., as in ISDN Q.81 or Q.731 number identification services).

CLLI See Common Location Language Identifier.

clock 1. A time-keeping and reporting device that uses various gravity (sand, weights) or oscillating mechanisms (radioactive decay, emissions, crystal vibrations) to track time. Quartz crystals have extremely consistent vibrations that are sometimes used to make very accurate clocks. 2. A device that provides regular signals for use as a timing reference. On a computer, instruction speeds are expressed in *clock cycles*.

clock bias The discrepancy between the time indicated on a clock and True Universal Time. See Coordinated Universal Time.

clock doubling A means of getting a little more performance out of a computer instead of having to replace the system. With constant demands for faster systems, balanced by the high cost of replacing a system that may be only a year old or less, some manufacturers have provided versions of the CPU chip or accelerator accessories that effectively double the speed of the CPU. This does NOT mean performance is doubled. The CPU is only one part of a system, and the bus rates, coprocessing chips, software design, operating system parameters, and other factors, will affect the actual performance increments to a great extent. In other words, the speedup is usually more on the order of 20% or so but, for graphics computations or resource-hungry software, that might be an important 20%. Sometimes it's worth it; it depends upon the cost of the doubler.

clock speed In computer systems, an expression of the speed of a central processing unit (CPU) or other processing chip, usually expressed in megahertz. Microcomputers in the 1970s ran at clock speeds ranging from about 1 to 4 MHz. In the 1990s, they ran at about 200 to 300 MHz. Current microcomputers in consumer price ranges run at over 700+ MHz.

Clock speed is *not* equivalent to system speed. Doubling the clock speed doesn't mean doubling the computing speed; sometimes the efficiency is just slightly more, and sometimes it is three or four times more. Determining the overall speed of a computing system is complex and requires evaluation of the general architecture of the system, the efficiency of the operating software, the amount of memory, the inclusion of coprocessing chips, and the type of application being run.

For example, the author's first 8-kilobyte RAM (yes, kilobyte, not megabyte), 1.8-MHz system ran telecommunications software and word processors very effectively at typing speeds of over 80 wmp. A feature-rich, well-written graphical word processor can run very efficiently on an 8-MByte RAM, 10-MHz system. The same software running on a 16-MByte 233-MHz system often is not *perceptually* faster because text entry, at its basic level, is not a computing intensive application.

In contrast with basic word processing, computing-intensive applications, however, can be dramatically affected by clock speed. A stock 1.8-MHz system is essentially incapable of doing 3D ray-tracing in a reasonable amount of time, whereas an older Amiga computer with a clock speed of only 7.16 MHz can render a complex 3D scene in 3 or 4 days, faster than many 25-MHz computers with different architectures. Amigas with 40-MHz accelerator cards can render the same scene in 3 or 4 hours, and dedicated graphics systems, running on parallel processing systems, or current Silicon Graphics Machines, for example, can accomplish the same feat in minutes or seconds.

Since computing speed is important to computer electronics designers, a number of measures have been established to provide information for comparing chips, systems, or architectures. These *benchmark* tests are not absolute measures of clock speed, but they provide some information that is helpful and they generate some pretty entertaining controversy. See benchmark, clock doubling, Dhrystone, Whetstone.

clone *n.* 1. Duplicate, exact copy, genetically identical individual. 2. A software program or device configured to masquerade as another device, either for diagnostic purposes, interim use, or fraud.

clone fraud A method of gaining entry to a system, or using a device, by simulating a user, serial number, or access code. Cellular phones are particularly susceptible to clone fraud, as it is not difficult to program a legitimate serial number into another cellular unit. See tumbling.

closed architecture A proprietary design that is supported and enhanced by peripherals that conform to its particular specifications, and that may not be manufactured by third party vendors, except perhaps by obtaining special permissions or paying royalties. Contrast with open architecture.

closed captioning CC. A broadcast technique for transmitting text, usually to be superimposed over a corresponding television image. CC is provided mainly as an aid for the hearing impaired, although in some cases it may also be used to provide subtitle translations. It is typically sent on the vertical blank-ing interval of the transmission, and a decoder may be required to interpret the signals.

closed circuit A broadcast circuit in which the sending and/or receiving components are limited to a certain frequency range or power level. Thus, a closed circuit radio system within a complex may be set to send and receive FM signals at 89 hertz. A radio station may have permission to broadcast at only 91.7 hertz on frequency modulated (FM) signals. In contrast, an open circuit is one that is not restricted to a narrow frequency range, as a CB radio, for example, which may be set to pick up signals broadcast over a variety of channels.

closed circuit broadcast, closed circuit TV A radio or television transmission that is broadcast to a small or restricted audience, often within a specific building complex or campus. Low power frequency modulated (FM) ranges are often used for this type of transmission because they are not as strictly regulated as higher power transmissions.

closet A room, cabinet, or case used for terminating blocks or patch panels for wiring configurations. The closet serves a variety of aesthetic, safety, organization, and security purposes.

cloud network Frame Relay network connections are now offered as a lower cost alternative for small businesses and educational institutions, and a cloud relay is one connectionless option of this type in which resources are shared, usually among four or five small subscribing organizations.

CLP cell loss priority. A one-bit ATM networking cell header toggle indicating the relative importance of the cell. CLP is important as there are various mechanisms in ATM for prioritizing cell traffic, or discarding cells in congested situations. See cell rate.

CLR See cell loss ratio.

CLTP See Connectionless Transport Protocol.

CLTS Connectionless Transport Service.

cluster 1. In cellular communications, a unit consisting of a group of adjacent *cells* within which channels are not reused. See cell, cellular phone, mobile phone. 2. A set of workstations or terminals in the same general physical or virtual networked grouping. These may share more than physical connectivity; they may also have shared devices that manage processing input and output, or specialized requests of the cluster. See cluster controller. 3. A combined unit of disk storage allocation, usually consisting of four or more sectors.

cluster controller A device controlling communications input/output for multiple connected devices.

clutter Wave reflections from obstructions such as terrain and buildings, which may show up as echoes or unidentifiable blips on a radar screen, thus interfering with scanning.

CMA See Communications Management Association.

CMC See connection management controller.

CMI See coded mark inversion.

CMIP See Common Management Information Protocol.

CMOL CMIP Over LLC (Logical Link Control). See Common Management Information Protocol.

CMOS Complementary Metal Oxide Semiconductor. A semiconductor chip that combines p-channel and n-channel MOS in a single substrate with push-pull circuits. Slow, but noise resistant, and good for battery-operated devices. CMOS RAM needs a small stream of constant power to preserve information stored in its memory, which is typically supplied by a lithium battery (available in photography and electronics stores). Default settings and sometimes video card and other peripheral parameters may be stored in CMOS RAM linked with a lithium battery on a computer's motherboard. See PRAM.

CMOT CMIP Over TCP. See Common Management Information Protocol.

CMR See cell misinsertion rate.

CMRS/PMRS commercial mobile radio service/private mobile radio service. The Federal Communications Commission (FCC) was directed, through the Omnibus Budget Reconciliation Act of 1993, to auction radio spectrum for CMRS. Unfortunately, the initial spectrum allocations in the C and F blocks, intended for small businesses, did not work out well in practical use. After numerous discussions and bankruptcy lawsuits, the FCC announced the cancellation of certain licenses and began to reauction portions of the C and F Block radio spectrum, beginning in 2000. See A Block for a chart of frequencies. See Omnibus Budget Reconciliation Act.

CMTS Cellular Mobile Telephone System.

CMYK A color model widely used in the paper printing industry. The initials signify cyan, magenta, yellow, and black, which are the four colors combined as tiny dots in *process color printing* jobs to simulate all hues and black. Black is included because the combination of the first three does not give a dark, rich, black pigment. Metallic colors cannot be produced within this color model, and extra runs through the press or spot application of metallics on a multicolor printer are necessary to accommodate metallic effects. Computer publishing software often seeks to simulate these colors on the monitor in order to provide WYSIWYG in the final printed result.

CN complementary network.

CNA 1. Centralized Network Administration. A means of consolidating network-related connections in a single location, usually a wiring closet or panel, rather than distributing them in various parts of the premises. 2. Cooperative Network Architecture.

CND 1. Calling Number Delivery. 2. Calling Number Display.

CNET Centre National d'Études des Télécommunication, now France Télécom R&D.

CNG A calling tone emitted by facsimile machines that lasts about half a second and repeats as many times as the software dictates, to signal its presence and to try to establish a negotiation with a receiving fax machine. Most machines default to about 45 seconds of tone sequence before they disconnect, if there is no successful connection. This time may not be enough for some systems or for a long-distance connection and some fax machines and fax modems have an option for extending it.

Most fax machines now automatically dial and emit the CNG. However, some of the older fax machines, or bargain basement varieties, still require a human operator to dial the number. The operator must then wait to hear a fax response and start the fax machine CNG by pressing a button. This method is a problem if the system that has been dialed has a sensing device to route incoming calls to a phone, modem, or fax machine depending upon the tone. If a human dials the line as a voice call, the switcher will route it to a phone, and then starting the calling fax's CNG does no good, as the phone has no way of routing the call back through the switcher to the fax machine. However, with increasing automation and decreasing cost of better fax machines, this problem is becoming less prevalent.

CNIS Calling Number Identification Services.

CNR 1. See Complex Node Representation. 2. customer not ready.

CNRI Corporation for National Research Initiatives.

CO 1. cash order. 2. See central office. 3. commanding officer.

co-channel interference CCI. A quantitative expression of inteference in a communications circuit when multiple channels are arranged in such a way that they may interfere with transmission on a neighboring or associated channel. This concept is especially important in wireless communications in which increasing numbers of subscribers are being accommodated within limited frequency allocations.

One way to reduce CCI is with guard bands, but the tradeoff is lost bandwidth. Another way of reducing CCI is with adaptive beam forming. See antenna, smart.

COAM See Customer Owned And Maintained.

coaxial cable A transmission cable consisting essentially of an inner conducting core surrounded by a conducting tube, each insulated and all wrapped together in an outer protective sheath. The inner core is a metallic conductor surrounded by a metal shield, that acts as a *Faraday cage,* with a dielectric material interposed between them. Typically, the signals are propagated in one direction along the conducting core.

Coaxial cable was an important development for the transmission of telegraph, telephone, and television signals as it was found to conduct radio frequency (RF) signals well. By the late 1940s, much of the eastern United States was interconnected with coaxial cable.

Coaxial cables are typically described in terms of their impedance; values from 50 to 95 ohms are common. The video industry makes extensive use of 75-ohm coaxial cables for interfacing cameras, frame synchronizers, and recording decks. In computer networking, 75-ohm cables are used for unbalanced E1 connections. Higher impedance 100- to 120-ohm twisted-pair wire is used for balanced E1 connections, and subrate cabling in trunk/circuit lines.

coaxial omniguide A lightguiding cable based upon layered film mirror technology developed at MIT in the late 1990s in a project led by Francis W. Davis.

By 2000, project participants had established Omniguide Communications to further develop and market a transmission cable based upon the technology that would reflect a wider range of wavelengths in a smaller space without changing the polarity or creating pulse distortion characteristic of traditional cables.

COBOL Common Business-Oriented Language. A verbose, high-level programming language once widely used for business applications and still taught in business schools, but which is slowly being replaced by other languages. See OO-COBOL.

CoBRA A commercial, portable, ISDN analyzer for installation, maintenance, and troubleshooting of ISDN Basic or Primary Rate networks, from Consultronics. Consultronics now markets the CoBRA-CQ as a portable local loop test set for ISDN, ADSL, G.lite and other formats.

COBRA A frequent misspelling of CORBA, Common Object Request Broker Architecture. See CORBA, Object Request Broker.

COBRAS Cosmic Background Radiation Anisotropy Satellite.

COCOT customer-owned coin-operated telephone. See payphone, private.

COD connection-oriented data.

code 1. A system of symbols, cyphers, characters, images, movements, sounds, or other meaningful marks or actions that serve to represent ideas and language in a way that is not commonly understood or recognized. Not all symbolic forms of communication are considered to be codes. For example, American Sign Language is not understood by many, but it is not considered a code in the sense that information on learning it is readily available in schools and libraries.

Social changes can alter the perception of whether something is a code. Before the development of the printing press and public education, text and reading were mainly restricted to the elite political leaders, and common people probably considered it as a sort of code. The use of coded information is common in wartime, or with politically or economically sensitive information. Some codes are exceedingly sophisticated and difficult to break. Until recently, most analog communications have not been coded to protect privacy, due to the difficulty of doing so. With recent digital technology, it has become much easier to code communications, and many software developers and equipment makers are adding encoding to their products. Many satellite communications, cell phone messages, and computer data communications are now encoded. 2. An abbreviated means of representing information in order to save time in its transcription or transmission, or to send it over limited transmissions devices, and sometimes also to shield it somewhat from prying eyes. Shorthand is a type of code intended to save time in taking oral dictation. Drumbeats or smoke signals are two types of codes designed to abbreviate information so that it is practical to transmit through these basic means.

Basic telecommunications codes have been in development since the 1600s. Schilling developed a needle telegraph code in 1832. Morse (Vail) code is a widely used alphabet coding system developed in the early 1830s. It is still often used in telegraph and radio communications, particularly in countries with limited access to computer equipment. See semaphore, Baudot code, Hollerith, Morse code. 3. Computer programming code is a system of linguistic and symbolic characters and syntax that serves to represent computer instructions so they can be run directly by the machine or compiled into machine-readable form.

Code designations in packet networking See Link Control Protocol codes.

Code Division Multiple Access CDMA. A digital, wireless communications service based upon spread-spectrum technology, which claims to provide about 10 times the capacity of analog. Access to the local exchange is wireless.

This technology was originally used in military satellites for its security features and resistance to jamming. Now more widely used in commercial applications, it provides access to many users at a time without the multiple user interference associated with other modulation techniques. The same frequencies in adjacent beams can be reused by assigning varying spreading codes to users. The method offers authentication of the source transmitter and is very secure against eavesdropping.

Frequency reuse logistics in AMPS and DAMPS systems are eliminated in CDMA by assigning codes to users so they can share carrier frequencies. The system capacity is not fixed but is influenced by the accumulated noise and interference associated with power levels and simultaneous users.

CDMA, supported by companies like Sprint and PrimeCO, is somewhat similar to TDMA, with somewhat less built-in support for private branch applications. B-CDMA is also in development as a proprietary technology by a group of vendors supporting InterDigital Communications. See B-CDMA, spread spectrum.

Technique	Description
DS-CDMA	Spread spectrum technology in which codes are used to modulate information bits such that each code is assigned to prevent the overlap of signals from user to user. The receiver regenerates the code and uses the information to demodulate the transmission.
FH-CDMA	A group of changing frequencies are modulated by the information bits in a two step process. First, the carrier frequency is modulated, and these modulated frequencies further modulate frequencies while still keeping them independent.

code rejects In packet networking, codes that are not used or not recognized are processed as code rejects.

codec en*code*/*dec*ode, *code*r/*dec*oder. A system to convert analog signals, such as video and voice, to digital signals for transmission, then back to analog at the destination. Codec mechanisms were originally installed on trunk lines, but as the cost of electronics dropped, they moved closer and closer to the home and office until now, with systems such as DSL, the codec is installed and performed right on the premises. Contrast with modem.

coded mark inversion CMI. In SONET and SDH networks, a two-level non-return-to-zero coding scheme. Binary values are coded in relation to a binary unit time interval (T). A one (1) is coded for a full time interval at one of two amplitude levels (low and high) such that the level alternates for successive ones. A zero (0) is coded by a positive transition from one to the other consecutive amplitude level at the midpoint of the time interval for half a binary unit time interval (T/2).

Code Red An intrusive program called a worm that used a unicode encoding technique to infect systems. A buffer overflow vulnerability in the indexing server was exploited to insert the worm onto a new system. Once a system had been infected, the worm used it to perform denial of service attacks on *www.whitehouse.gov* and, in some cases, defaced the server's home page. It spread by randomly generating IP addresses for new systems to infect. Windows NT and Windows 2000 systems using the Microsoft Internet Information Server (IIS) were vulnerable. In response, vendors such as Cisco Systems took steps to update systems to prevent this type of security breach and Microsoft issued a Security Bulletin MS01-033 with information on patches.

coding violation CV. In ATM networking, a coding violation results when bit interleaved parity errors are detected on an incoming signal. Each BIP error (typically up to 8, 16, or 24) increments a CV counter. In SONET, the section, line, and path errors are located in their associated overhead frames. Thus, in a BIP-8 system, up to $8 \times N$ coding violations may be associated with a frame. One or more coding violations in a second on a layer results in an errored second (ES) or a severely errored second (SES). See bit interleaved parity. See RFC 1595.

cohere To come together firmly, to be cohesive, to coalesce, to hold together, join, unite, merge. The term particularly applies to the action of small, discrete parts or granules.

coherent light Light in which the wave lengths are aligned or *in phase* to create a very straight, narrow beam, in contrast to light from lamps and flashlights that spreads out and quickly diminishes in intensity. Coherent light can be generated by lasers and by some light-emitting diodes (LEDs). Both lasers and LEDs are used as light sources for fiber optic cables.

coherer A device that causes particles to join, lump, or clump together when exposed to a nearby discharge of electricity or to a current running to the particles through a wire. As the particles are stimu-

lated to arrange themselves in a more coherent fashion, that is, to align themselves so that resistance is lowered, they collectively provide a better conducting surface. Many early coherers consisted of a glass tube corked at each end with filings sealed inside. The coherer was connected in series with a battery-driven electrical circuit.

Early experiments by O. Lodge in 1894, D. Hughes in 1878, and É. Branly in 1890 resulted in a cohering apparatus that could behave as an on/off switch by serving as a nonconductor, unless stimulated by a spark, and returning quickly to nonconducting status once the spark and the current had passed through. This useful device was adapted by Marconi for improvements in radio devices. See Branly detector.

Marconi & Castelli Coherers

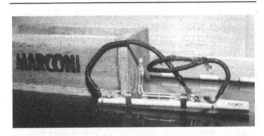

This historic Marconi coherer is only a couple of inches long, a delicate glass tubing supported by an ivory base. It is part of the American Radio Museum collection. Coherers were the forerunners to rectifying crystal detectors in crystal radio sets.

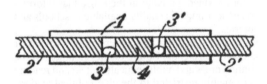

This diagram of a Castelli coherer shows a tube (1) within which are conductor plugs (2,2') separated by an iron plug (4) and two mercury pockets (3,3'). This coherer was used by Guglielmo Marconi in transatlantic experiments. [Scientific American, Oct. 4, 1902.]

coil In its simplest sense, a loop or number of continuous turns of wire or other material. The coil may have successive windings that are touching, or that may be spaced and stretched out like a spring. Coils are often used in wireless communications technologies where a long length of wire must fit in a small amount of space, where a broader conductive surface area is needed, or where the proximity of the wire loops changes its overall conductive properties.

In antennas, receptivity to electromagnetic waves is based in part on matching the length of a long wave; consequently, very long wires are needed for some applications.

There are many ways to wind and use coils. Tables

are published in electronics guides describing the length and diameter of cores, and the gauge and number of windings needed for the wire. Open coils with few turns are used as *load coils* in voice grade telephone wire installations. Wound coils, wrapped around a metal core, can be used to create an *armature*. Sending/receiving coils can be created with many windings over a core or a frame, utilizing the thickness of the wire, the shape of the coil, and other characteristics to control which frequencies are transmitted or received. Sometimes dual windings are used, that is, a smaller coil inside a larger one, with an insulating layer in between. A spark coil for a basic wireless transmitter can be constructed with an inner primary winding coil and an outer secondary winding coil encasing a soft iron core. Commercial induction coils, based upon the same structure as the simple spark coil, were used for decades to generate intermittent high voltage.

One unsettling historical fact is that X-ray coils were used in the early part of the century for sending wireless communication signals.

Load coils are commonly used on copper telephone wire installations to improve signals at voice grade levels, but they cause problems when data is sent at high speed through the wires, as in digital subscriber line (DSL) services; DSL transmissions are highly sensitive to noise and distance. See antenna, armature, basket winding, induction coil, load coil, winding, winding machine.

Armature Coil Windings

Two simple types of armature coils are shown here. On the left is a single coil, on the right, a double coil, wound in parallel. Armature coils can be quite large and intricate and are the basis of electric 'dynamos,' now more often called generators.

---

cold docking Hooking components into a base or desktop unit while one or, preferably, both units are powered off. This is done to prevent danger of electrical shock or damage to sensitive electronic components. See docking.

cold start Starting a system from a power off condition. In a computerized system, it also means there is no software online. From a cold start, many systems will run through physical and logical self-test sequences and bootstrap sequences to load device drivers or other software which may be needed to recognize and bring online the rest of the system, and

eventually the whole operating system.

Collaboration for Interactive Visual Distance Learning CIVDL. Videoconferencing technologies applied to distance education for engineering programs. The CIVDL is a member of the PUG Alliance.

Colladon, Jean-Daniel (1802–1893) A Swiss-born physicist and engineer, Colladon and his friend, Charles-François Sturm, traveled to Paris in 1825 to study mathematics and physics and to continue collaborating on scientific experiments. Both young men became assistants to J. Fourier. On their second attempt, they jointly won the prize offered by the Paris Academy for research on the compressibility of water. They accurately measured the speed of sound in water and provided important basic research as well as a chapter in the history of sonar. Colladon was instrumental in the conversion of city lighting to gas, in the 1840s and demonstrated water as a light guide. He developed a type of photometer, to aid him in measuring luminosity for his projects. Archival collections of correspondence and scientific papers are housed in the Geneva public and university libraries and the Swiss library in Bern. See Tyndall, John.

collapsed backbone A backbone is a main artery or trunk in a network system. A collapsed backbone is one in which the physical connections are incorporated into a centralized intelligent hub or *network center*, providing easier access and administration.

collate To assemble in the desired order. Many printing programs, word processors, desktop publishing programs, and photocopy machines now have settings that allow you to choose, for multiple printouts of a multipage document, whether it is to be printed sequentially or in page groups. Collation is the electronic substitute for lining up three card tables in a row with a pile of each page of a twenty-page document lying side-by-side, and having friends and co-workers walk down the line picking up one of each page. I'm sure most readers have done this at least once in their lifetimes. Collating settings and devices are great time savers.

collect call A call, usually on the telephone, in which the receiver pays for the call once it has been initiated. Most collect call systems require the prior approval of the person receiving the call before the call is permitted to continue. Person-to-person calls are generally more expensive than station-to-station calls. It is more difficult now to connect collect calls, as many people have answering machines to screen calls, and may not hear the operator requesting authorization.

collimate *v.* To make parallel, to cause to follow parallel trajectories. See coherent.

collimation 1. The process of making something travel parallel, with a minimum of divergence or convergence. 2. The process of making light waves travel parallel without diverging or converging. This process is useful in testing and aligning optical instruments and is essential for technologies that require a coherent beam that doesn't significantly lose power over distance due to spreading or scattering.

collision In data networks, there are commonly many

devices trying to send signals at the same time. If this happens at exactly the same time, collisions may occur. There are a number of mechanisms to manage collision-detection and traffic flow, including jam signals for preventing simultaneous transmissions. Typically, the jam signals will cause devices to back off and wait for a random period of time before trying again. The introduction of the random time factor reduces the chance of the same devices starting the transmission again at exactly the same time. Care must also be taken to ensure that not too many collisions occur. If there are many collisions, and devices are constantly backing off and trying again, then throughput may be compromised. Excessive collisions may mean that an additional router or bridge needs to be added to the system or that some devices need to be disengaged.

collision detection On data networks, the means by which the system detects that more than one device is attempting to transmit data at the same time. This detection may be done in a number of ways, with acknowledgments being one means of signaling a system that data has made it successfully through. If it hasn't, and no acknowledgment is received in a reasonable amount of time, then there may have been a collision and the system reacts accordingly. One type of mechanism triggered by collision detection is a jam signal, which alerts devices to back off until the jam is cleared. See collision, jam, jam signal.

collocation 1. Adjacent placement. 2. Physical placement of customer transport and/or multiplexing equipment within the carrier's premises.

collodion A viscous solution introduced into the processing of photographic prints in the 1850s.

color burst. See burst.

color carrier reference A continuous signal, related to a color burst signal, used for modulation and demodulation.

color code An identification system based upon colors or specified widths or patterns of color. Many industries color code their dials, wires, and components for quick recognition and selection. Electronic components such as resistors are often labeled as to their values with bars of colors in particular sequences.

Color Graphics Adapter CGA. A color graphics standard introduced by International Business Machines (IBM) in 1983 as their first color graphics controller card. Until then, IBM computers with native controllers displayed only in monochrome. CGA supported a display resolution of 320×200. It has since been superseded, first by EGA, and then by VGA, and now, almost entirely by SVGA.

color model A conceptual description of how colors are detected, perceived (usually by humans), or reproduced. Human color perception is an exquisitely sophisticated phenomenon, as is described insightfully and anecdotally in Oliver Sacks' *An Anthropologist on Mars*. Many, many color models exist, none of which is complete or generalizable to every situation. See CMYK, color space, Maxwell's triangle, Munsell's color model, RGB.

color monitor A monitor that uses color transistors or LEDs or is coated on the inside front of the tube with phosphors which when excited glow in particular colors (usually red, green, and blue), which combined can appear as any of millions of colors. Red, green, and blue are considered primary colors in light, because their combination in different intensities produces virtually any color. (Pigment systems define red, yellow, and blue as the primary colors.) Thus, most color systems in cathode-ray tubes employ three electron guns and are commonly known as RGB systems.

color space A model or scheme for objectifying the representation of color. Many color spaces exist, most of them devised to work with specific technologies. Color spaces for printing pigments assign numeric values to particular hues which are further coded so that the printer can mix the correct inks for use on the press.

color subcarrier A monochrome broadcast signal that is modulated with sideband information in order to convey color.

color television standards Different parts of the world have standardized on different formats and even different subformats, many of which are not intercompatible. The common ones for color television are NTSC, PAL, and SECAM.

colorimeter An optical instrument for measuring and comparing colors from different sources, often used to match or calibrate colors according to a color model or sample.

colorimetry A quantitative method of specifying colors through attributes such as wavelength (color), excitation purity (saturation), and luminance (intensity).

Colossus Mark I A code-breaking machine developed by Alan Turing and others, put into service in 1944 in Bletchley Park, England, to help decrypt messages from other nations, particularly Germany, transmitted during World War II. It was delivered under the leadership of Tom Flowers, representing the Telephone Research Establishment; Max Newman and Harry Hinsley played prominent roles. The existence of this machine was not publicly known until almost three decades later. See Manchester Mark I; Turing, Alan.

COLP connected line identification presentation (e.g., as in ISDN Q.81 and Q.731 number identification services).

COLR connected line identification restriction (e.g., as in ISDN Q.81 and Q.731 number identification services).

Columbia Broadcasting System CBS. This major U.S. network was granted its first commercial broadcast license in 1941 and not long after began to develop a color television system.

COM 1. See Component Object Model. 2. See continuation of message.

Com21, Inc. A publicly trading American-based global ISO 9001-registered supplier for the broadband access market, founded in 1992. The company provides ATM, DOCSIS, and EuroSIS products to cable service providers and operators for delivering high-speed Internet and telephony applications. Com21 has a research facility based in Ireland.

combination antenna An antenna designed to cover a range of frequencies, usually UHF, VHF, and FM, in a single unit. Combination antennas have a variety of elements including reflectors, Yagi-Uda arrays, and log-periodic components to accommodate a variety of signals with good gain. Since several signals are being received, the down-lead will usually require a splitter to feed the individual signals into the appropriate components, or in a combination component, into the appropriate input receptacles. See antenna, UHF antenna, VHF antenna.

COMETT Community Action Programme in Education and Training for Technology. An initiative of the European Union.

Comisión de Regulación de Telecomunicaciones CRT. The telecommunications regulatory commission of the Republic of Columbia. http://www.crt.gov.co/

Comisión Federal de Telecomunicaciones Cofetel. An administrative agency of the Secretary of Communications and Transportation of Mexico. http://www.cofetel.gob.mx/

Comisión Nacional de Comunicaciones The national communications commission of Argentina. http://www.cnc.gov.ar/

Comisión Nacional de Telecomunicaciones CONATEL. The national telecommunications commission of Honduras. http://www.conatel.hn/

Comité Consultatif International Télégraphique et Téléphonique CCITT. This important standards body is now known as the ITU. See ANSI, CCITT, International Telecommunication Union.

command buffer A portion of memory that stores recently executed commands or frequently executed commands, so that the command can quickly be fetched and re-executed if needed. A buffer is a type of simple memory cache used to speed up the overall performance of a system. See cache.

command line interface, command line interpreter CLI. The software interpreter that accepts text commands input by the user, attempts to fulfill the request by interpreting them into machine language, then responds with an answer, information, or error message. Most operating systems come standard with a command line interpreter; the Macintosh is a notable exception. On many computers, such as Amiga and Unix systems, new commands can be readily added to a *bin* directory and henceforth executed in the same manner as the default command set. See command line.

command path A location designator for directories on a system that hold system commands or commands that are to be activated from anywhere on the system without having to type the full path from the current directory. Most systems have a configuration file that allows common path names to be established at start-up time, and these generally stay active while the system is powered up. If path names are changed, it will be necessary to reread the path file to establish the new paths and, on some systems, you may have to reboot the machine (very inconvenient).

Commercial Cable Company A historic communications cable company founded in 1883 by John W. Mackay and James Gordon Bennett, Jr. The company laid some of the earliest cables between Ireland and the west coast of North America, and later to continental Europe as well. The company was hotly competitive with Western Union but needed land systems to be completely independent of Western Union. As a consequence, Mackay purchased a controlling share of Postal Telegraph Company.

Commercial Internet Exchange CIE. An alliance of CERFnet, UUNET, and PSI in 1991. Since that time, other services have formed agreements with CIX to allow unrestricted flow of traffic across networks in the CIX backbone. For a fee, service providers may access and send traffic across the network.

Commercial Internet Exchange Association CIX. A nonprofit trade association established to promote and support the use of the Internet for commercial activities. Its members consist of public data internetwork service providers supporting public data communications. CIX provides a forum for the exchange of ideas and information and encourages technical research and development. Membership is open to organizations offering TCP/IP or Open Systems Interconnect (OSI) public data internetworking services to the general public. http://www.cix.org/

Commercial Space Launch Act of 1984 A U.S. act of Congress that provided support for private satellite communications systems launching and operation. The regulation at present is light, mostly related to Federal Communications Commission (FCC) frequency assignments and the positions of satellite orbits, but this situation may change in the decades ahead as more and more satellites vie for space in Earth orbit. See Telecommunications Act of 1996.

committed burst size Bc. See burst size, committed.

committed information rate CIR. A service rate and traffic flow commitment level established for service in a Frame Relay network. That is, the CIR is a level that is agreed upon for data transmission rates. The user may use higher transmission rates, but the excess data will be marked as discard eligible (DE) in the case of network congestion. Since rates may vary, it is a computed average over a specific period of time. See cell rate.

committed rate measurement interval Tc. In networking, the nonperiodic time interval used to measure incoming data, during which the user can send only *committed burst size* committed amount of data and *excess burst size* excess amount of data. Generally, the duration of this *measurement interval* is proportional to traffic burstiness. See committed information rate, committed burst size.

Committee T1 An ANSI-accredited organization established in 1984 that develops and publishes U.S. network reliability standards and technical information of interest to network equipment developers, installation and maintenance personnel, and system administrators. The organization contributed to the ITU-T I-series recommendations for B-ISDN among others.

Documents related to safety, power, ISDN, SONET,

SS7, and wireless communications are available through Committee T1's sponsor, the Alliance for Telecommunications Industry Solutions (ATIS) in Washington, D.C. Committee T1 works in cooperation with organizations such as the Network Reliability Council. See Alliance for Telecommunications Industry Solutions. http://www.t1.org/

Committee T1 Technical Reports The Committee T1 provides a series of telecommunications technical documents available for a fee, and some that can be freely downloaded off the Internet in Adobe PostScript or Adobe Portable Document Format (which can be read with one of the many freely distributed Adobe PDF readers). Abstracts for *Approved ANSI T1 Standards* are also available. Since many of these are of direct interest to people developing, installing, and maintaining communications networks, a few are listed in the Committee T1 Technical Report Examples chart.

Commodore 64 computer C64. A low-cost 8-bit computer introduced by Commodore Business Machines in the early 1980s, aimed at the home and school markets. Listed at under $600 U.S., the C64 included a 6510 CPU with 64K RAM, a built-in sound generator, the Digital Research CP/M operating system, and game controllers and cartridge slot. It featured 320×200 pixel color graphics, was competitive with the Apple IIe (48K) and the Atari 800 (16K), and continued to be popular for a couple of years after the Amiga was introduced by Commodore in 1985. The C128 was an expanded version of the C64.

Commodore Amiga See Amiga computer.

Commodore Business Machines CBM. Formerly an office equipment company selling calculators and, later, the Commodore PET (Personal Electronic Transactor) computer, CBM is now best remembered for its introduction of the Amiga computer. In the mid-1980s, when Radio Shack had lost its enormous market share to IBM computers, Commodore acquired a computer named the Lorraine and launched it in the Fall of 1985 as the Amiga (despite protestations from

its developers that the operating system (OS) wasn't finished and that the hardware should have slots and more memory). Due to problems in management and marketing, CBM or Commodore-Amiga, as it came to be known, folded,with the Amiga assets bought out by a German company and later sold to Gateway, Inc. Licensing use was subsequently sold to Amino Development Corporation, now Amiga Corporation (though Gateway retained ownership of the patents). Commodore folded in 1994, but the Amiga didn't. Developers' conferences were reinstituted in 1997 and the Amiga2001 show was held in St. Louis in March 2001. See Amiga computer; Apple Computing; Miner, Jay.

Commodore PET Personal Electronic Transactor. One of the earliest commercially successful microcomputers, the PET was introduced early in 1977 by Commodore Business Machines. It was competitive with the Tandy Radio Shack TRS-80, which initially was also black and white with 4 kilobytes of RAM, but both computer systems were eventually overshadowed by Apple and IBM computers.

Common Applications Environment CAE. A set of standards intended to provide a framework for integrated systems, developed by the X/Open Company. See Single UNIX Specification.

Common Architecture for Next Generation Internet Protocol CATNIP. When IPv6, the successor to IPv4 for the Internet, was in the design stages, a number of proposed formats were submitted. CATNIP is one of three formats that were incorporated into the IPv6 specification by the Internet Engineering Task Force (IETF). See IPv6.

common battery In early telephone central offices, a 24- or 48-volt battery called a *talking battery* was used for supplying the power for a phone conversation. Later, starting around 1893, these were replaced by 48-volt *common batteries* at the central office which supplied the talking battery to each subscriber through the wireline, rather than each subscriber individually providing battery power. This practice

Committee T1 Technical Report Examples		
Number	Date	Title
TR-7	June 1986	3-DS0 Transport of ISDN Basic Access on a DS1 Facility
TR-13	Dec. 1991	A Methodology for Specifying Telecommunications Management Network Interface
TR-15	March 1992	Private ISDN Networking
TR-21	Sept. 1993	System and Service Objectives for Low-Power Wireless Access to Personal Communications
TR-36	May 1994	A Comparison of SONET and SDH
TR-45	Dec. 1995	Speech Packetization
TR-47	June 1996	Digital Subscriber Signaling System Number 1 (DSS1) – Codepoints for Integrated Services Digital (ISDN) Supplementary Services
TR-53	June 1997	Transmission Performance Guidelines for ATM Technology Intended for Integration into Networks Supporting Voiceband Services

made it possible for smaller home phones to be designed. See battery.

common bell A bell that rings when any of the designated lines on a phone system ring. It is often installed on main consoles, to allow an operator to intercept calls, or on night systems, so a single person can answer calls on several lines that would normally be answered individually.

common carrier A public communications service carrier, usually regulated and licensed by a government agency. A common carrier may not withhold service or discriminate against any public purchaser of the services.

Common Carrier Bureau CCB. A large department of the U.S. Federal Communications Commission (FCC) that recommends and implements regulatory policies for interstate telecommunications through enforcement, pricing, accounting, and program planning of network services and wireline services.

Common Channel Interoffice Signaling CCIS. An out-of-band telecommunications signaling system that encodes information and sends the signaling data over channels separate from the voice signals, using digital time-division multiplexing (TDM). This system is more efficient – full voice-grade paths are not needed for sending signaling information – and more secure than older signaling systems which used 2600 and 3700 Hz tones as supervisory signals.

Some of the key points in CCIS networks include Signal Transfer Points (STPs), tandem switches acting as routers, Signal Control Points (SCPs), data application servers, and Service Switching Points (SSPs) capable of switching tens of thousands of individual lines.

CCIS was introduced by AT&T in 1976. The system was significant in that it introduced a new out-of-band network, separate from the network carrying the voice conversations, for the telephone signaling transmissions. This type of system was inherently more secure than an in-band signaling system using tones that could potentially be introduced into the circuits by a user. The CCIT adopted CCIS as an international standard called Common Channel Signaling System 7 (CCS7 or more commonly now SS7). See Signaling System 7.

Common Channel Signaling CCS. CCS is a system that developed as local telephone carriers gradually linked up with regional systems, necessitating some common signaling standards for compatibility. Telephony required the transmission of two general categories of data, the informational content of a phone conversation and the supervisory/control signals associated with establishing, maintaining and disconnecting the calls.

As touch-tone technology developed and gradually superseded pulse dialing, and digital systems gradually emerged, the sophistication of the types of signaling that could be carried over phone lines increased. New services were devised that took advantage of digital signaling (e.g., Caller ID).

Originally both the signaling and the conversations were carried on the same channel. However, the blue-box antics of telephone phreakers in the 1970s revealed significant security weaknesses in this method, and out-band signaling took precedence, with content and supervisory data carried on separate channels. (In-band signaling still exists on many local branch systems but metropolitan and national networks tend now to be out-band.)

Standards for CCS include both in-band and out-band specifications. In early telephone signaling implementations, the signaling and call content had to be interleaved rather than overlapped, a situation that limited the types of information that could be transmitted about a call while it was in progress (think of the difference between a single-tasking operating system and a multitasking operating system on a computer to get the general idea). CCS permitted some of these limitations to be overcome, and it began to be more widely implemented in the early 1990s.

Thus, in ATM networks, CCS is a packet-based signaling architecture in which circuits share signaling channels in which the administrative and content signals may be transmitted at the same time (i.e., you can read data about a call on an appropriate device while the call is in progress). CCS channels may be cross connected.

CCS uses parameters that set up the network configuration, such as the switch type, debug level, data inversion mode, correspondence between maps and network interfaces and signaling instances, layer activation and timers, and data link flags.

In Transparent Common Channel Signaling (T-CCS), private branch exchanges can be interconnected with digital interfaces that use non-CCS protocols (e.g., a proprietary protocol) without the CCS signal having to be interpreted to process calls on the system. The proprietary signaling is preserved and transported transparently through the data network through a point-to-point connection. In other words, instead of routing the transmissions, a preconfigured route is used in conjunction with CCS frame forwarding to support transparency.

CCS has been defined for use with Signaling System 7 (SS7) telephony. CCS facilitates the establishment and take-down of calls, signal monitoring, internetwork transmissions, and special-case call handling (e.g., calling card connections). See Signal Transfer Point, Signaling System 6, Signaling System 7.

Common Channel Signaling Task Force One of a number of task forces of the Presidential National Security Telecommunications Advisory Committee (NSTAC) that looked into matters such as security of the public telephone network, in the early 1990s, and issued a Final Report in Jan. 1994. In May 2000, NSTAC issued a report on information technology (IT) telecommunications convergence issues for national security and emergency preparedness (NS/EP).

Common Desktop Environment CDE. An integrated graphical user interface for open systems featuring a standard interface for management of data and applications. CDE is an IETF platform Human Computer Interface (HCI) standard. See X Window System.

Common Gateway Interface CGI. A means of communicating instructions to a Web server through scripts or code, in order to enhance the utility of Web pages. HTML, a markup language used on the Web, was designed for formatting, not processing, data interactions. To extend the utility of HTML, the CGI can be used in conjunction with input to Web pages to process forms, messages, chat room interactions, database records, searches and more. Perl is one of the most flexible, powerful, and prevalent languages for implementing CGIs on the Web, especially for text processing, database searches, and forms parsing. Sun's Java tends to be used in situations where graphical menus, games, or images are desired. See ActiveX, Java, Perl.

Common Intermediate Format CIF. A subsection of the ITU-T H.261 standard that specifies various broadcast format parameters for ISDN videoconferencing. See ISDN. See Common Intermediate Format Types chart.

Common Location Language Identifier CLLI. A unique identifier system, developed by Bellcore, for certain regions and equipment. Thus, various exchanges, buildings, and facilities could be coded. A CLLI consists of four characters for the location, followed by two characters for the region, and five characters for the item.

Common Management Information Protocol CMIP. A standardized connection-oriented network management protocol based upon the Open Systems Interconnection (OSI) model. CMIP supports information exchange (as opposed to network functionality) between network management applications and management agents through managed objects. CMIP is part of the X.700 Recommendation of the ITU-T (also ISO/IEC 7498/4). CMIP was designed by industry and government participants to be the heir to the simpler Simple Network Management Protocol (SNMP). CMIP supports security features, including access controls, activity logging, and authorization. It works in conjunction with the Common Management Information Service (CMIS), which defines services for accessing information about network objects or devices.

A number of vendors have implemented CMIP. For example, Solstice CMIP has been developed to provide CMIP services on Sun Microsystems' Solaris 64-bit platform. In the early 1990s, AT&T and NCR released StarPRO CMIP compatible with BaseWorX UNIX-based systems. See Common Management Information Services.

Common Management Information Services CMIS. A standardized network services mechanism to enable peer processes to exchange information and instructions through a defined message command set. CMIS works in conjunction with Common Management Information Protocol (CMIP). The CMIS V.2 definitions and protocol were described in ITU-T X.710/711 Recommendations in 1991. CMIS was standardized in the mid-1990s as ISO/IEC 9595/2.

In 1997, S. Mazumdar of Bell Labs proposed a set of extensions that defined interfaces for providing CMIS-based services using the Object Management Group (OMG) object services such that CMIS-based objects could be made compatible with other managed objects in a native Common Object Request Broker Architecture (CORBA) environment.

Common Management Information Services and Protocol over TCP/IP CMOT. CMOT is an Internet Protocol information service mechanism in the context of ISO-standardized Common Management Information Services/Common Management Information Protocol (CMIS/CMIP) as it applies to a TCP/IP environment. CMOT was submitted as an RFC by Warrier and Besaw in April 1989 and updated October 1990 as a move toward international standards suitable for implementation over the evolving Internet.

CMOT provided a means for implementing the Draft International Standard version of CMIS/CMIP over Internet transport protocols in order to carry management information. See Common Management Information Protocol. See RFC 1189 (which obsoletes RFC 1095).

Common Intermediate Format (CIF) Types			
Format	Lines × Pixels	Defined within Standard	Notes
General		H.320	An umbrella encompassing the following CIF, FCIF, and QCIF standards, sometimes collectively called *p*64.
CIF	352 × 288 color	H.261	Suitable for large format videoconferencing. Requires two B channels to support both audio and video.
		H.221, H.230, H.242	Communications, control, and indication.
		H.711, G.722, G.728	Audio signals.
FCIF	352 × 288	---	
QCIF	176 × 144	H.276	Requires less bandwidth than CIF but also provides less resolution.

Common Part Convergence Sublayer CPCS. In ATM networking, a portion of the convergence sublayer of an ATM adaptation layer (AAL) that remains common to different types of traffic.

common part indicator CPI. In ATM networking, a 1-byte field used to interpret the remaining fields in the header and trailer.

Common Object Request Broker Architecture See CORBA.

Common Open Policy Service Protocol COPS. A simple, extensible client/server protocol model for supporting policy control over quality of service (QoS) network signaling protocols. COPS is a query and response protocol that enables policy information to be exchanged between a policy server and its clients (e.g., RSVP router). See RFC 2748.

Communications Act of 1934 A U.S. federal regulations act to organize and promote competitive communications technologies and services. This act established and described the responsibilities and jurisdiction of the Federal Communications Commission (FCC) which was descended from the Federal Radio Commission (FRC) formed from the Radio Act of 1927.

The Communications Act of 1934 was amended by the Omnibus Budget Reconciliation Act (OBRA) to preempt state jurisdiction in such a way that individual states were no longer regulating rates and entry by companies offering wireless services. It further organized wireless into two categories: commercial mobile radio services (CMRS), including cellular radio services and personal communications services (PCS), and private mobile radio services (PMRS), including public safety and government services.

Communications Act of 1996 See Telecommunications Act of 1996.

Communications Applications Specification CAS. A communications protocol developed in the late 1980s by Intel and Digital Communications Associates, Inc. (DCA) for use with computer peripherals to enable software to communicate with fax/modem interfaces. This protocol, along with Class 1, 2, and 3 fax standards, helped standardize computer facsimile communications, enabling software from different vendors to exchange data.

Communications Authority of Thailand CAT. A state initiative under the Ministry of Transport and Communications, established in February 1977. CAT is responsible for a national communications network linking to the global community. http://www.cat.or.th/

Communications Decency Act of 1996 A provision of the Telecommunications Reform Act that aroused extreme controversy and opposition by the Internet community as it made it a federal crime to send certain lewd, indecent, or other objectionable communications across networks. The Internet community rallied against it and, in a June 1997 milestone decision in the case of *Reno versus ACLU*, the act was declared an unconstitutional violation of individual rights to freedom of speech. See Telecommunications Act of 1996.

Communications Management Association CMA. Formerly the Telecommunication Managers Association, the CMA is a charitable business communications trade association based in the U.K. The CMA supports the role of managers in communications fields by providing and promoting educational activities and excellence in the use of communications technologies. http://www.thecma.com/

Communications Policy Project CPP. A nonpartisan initiative of the Benton Foundation to strengthen public interest and participation in the shaping of the National Information Infrastructure (NII). The Benton Foundation seeks to promote the use of communications for the greater social good and encourages democratic participation in policy debates and regulatory activities, especially those relating to open access to communications technologies and the promotion of diversity in services beyond the obvious commercial applications.

Since the mid-1990s, a portion of the Foundation's efforts has gone into educating the public about new digital environments and broadcasting media and the importance of the equitable allocation of radio spectra for positive social programming.

Among other things, the CPP advocates support for low-power television (LPTV) stations, as these provide diversity and a large proportion of social and educational content. Many LPTV stations broadcast local news and proramming for hobby, church, athletic, and community groups. LPTV stations often broadcast to remote or small communities that are not of commercial interest to large corporations because they don't have sufficient subscribers to generate a profit. There is a persistent danger that LPTV services can be crowded out by commercial interests with strong lobbies and economic bases if they are not actively protected and promoted by the public, the government, and communications agencies such as the Federal Communications Commission. See Benton Foundation, Community Broadcasters Association, National Public Radio, Public Radio International.

Communications Research Centre CRC. A major communications research agency of Industry Canada located at a secure site near Ottawa, Ontario adjacent to the Defence Research Establishment Ottawa (DREWO) and the Canadian Space Agency. The CRC engages in collaborative, innovative research in information technologies, communications, and broadcasting in support of Canadian knowledge-based economies. It further provides an independent voice for public policy development.

Communications Security Establishment CSE. A Canadian federal agency for providing information technology (IT) security solutions to the Canadian government. http://www.cse.dnd.ca/

Communicator III An IBM-licensed/Intel-based PC videoconferencing product with audio, video, whiteboard, and file transfer capabilities from EyeTel Communications, Inc. Communicator III works over Switched 56, ISDN, T1, Ethernet, and Token-Ring networks. It uses ITU-T H Series Recommendation standards and encoding.

Communique! A Sun SPARC-based videoconferencing program from InSoft that works over ISDN, FDDI, SMDS, Ethernet, ATM, and Frame Relay networks. It supports audio, visual, whiteboarding, file transfers, and a number of applications. CellB, JPEG, and Indeo standards and encoding are supported.

Community Broadcasters Association CBA. A U.S. national professional organization devoted to supporting and enhancing diversity and vitality in the community broadcasting field, with a special interest in Class A and low-power television technologies which are widely used in niche market and local community broadcasting. The CBA sponsors online news, workshops, and provides input into government policies. See Communications Policy Project, Community Broadcaster's Protection Act, low-power television, World Association of Community Radio Broadcasters.

Community Broadcaster's Protection Act A portion of the Omnibus Appropriations Bill signed into law by President Clinton in 1999 as a direct result of lobbying by the Community Broadcasters Association. The Act established a new class of television broadcasting in the U.S., making it possible for low-power television (LPTV) broadcasters to apply for permanent status. The CBA subsequently sponsored seminars to help educate broadcast companies and individuals on the implications and implementation of the terms of the Act and aided them in understanding Class A Compliance issues.

In January 2000, the Federal Communications Commission (FCC) adopted the Class A Notice of Proposed Rule Making (NPRM). Three months following, it released a report and order establishing the Class A Television Service, followed by a list of stations considered to be eligible for this Service. See Communications Policy Project, Community Broadcasters Association.

Community Broadcasting Association of Australia CBAA. The national representative body for community broadcasters in Australia. The CBAA provides representation, education, and support for licensed stations and licensee hopefuls including information about issues, ethics, intellectual property, fundraising, and Broadcasting Services Act requirements. The CBAA hosts the national community radio satellite service. See Community Broadcasting Foundation. http://www.cabb.org.au/

Community Broadcasting Foundation Ltd. CBF. An independent nonprofit funding body for community broadcasting in Australia, established in 1984. The CBF is supported by the Australian Dept. of Communications, Information Technology and the Arts (DCITA), and the Aboriginal and Torres Strait Islander Commission (ATSIC). It solicits funds and distributes grants for ethnic community broadcasting, print handicapped broadcasting, general community broadcasting, and policy development projects. See Community Broadcasting Association of Australia.

community dial office CDO. A type of central telephone switching office that is most often found in small rural communities. It is an unattended switching center that is serviced only as needed and maintained on an occasional basis by a traveling maintenance technician.

community radio A radio broadcast system that serves the cultural, ethnic, local news, special interest, or social needs of a community. Community radio stations are important because they are often the only venues for minority populations or isolated individuals to access programming matching their needs and interests. Many small groups are not served by large, for-profit broadcast corporations. Since most community radio stations are low-profit or no-profit ventures, they do not have the same lobbying power with Congress or the Federal Communications Commission as large, powerful broadcasting conglomerates. It is therefore up to listeners, foundations, educators, and related organizations to support the vital role played by community radio in safeguarding freedom of information and diversity.

The growth of the Internet and the capability of serving streaming audio to millions of listeners has broadened the reach of community radio broadcasting and the concept of community. While still not a profit venture in most cases, community radio stations can now broadcast to a wider spectrum of communities, based not just on geographical regions through low-powered transmitters, but to the entire world, through Web sites that can be accessed long distance without additional fees by all interested listeners with Internet access. See Community Broadcasters Assocation, National Public Radio, People's Communication Charter, World Association of Community Radio Broadcasters.

Community Radio Charter for Europe A set of priniciples and goals adopted by AMARC at the Pan-European Conference of Community Radio Broadcasters in September 1994. The Charter recognizes community radio broadcasting as a vital medium for fostering freedom of expression and information, cultural freedom and diversity, and local culture and traditions. It defines ideals and objectives to help radio stations achieve these goals. See World Association of Community Radio Stations.

compact disc A small, flat, circular, optical, digital random-access storage and retrieval medium. CDs are written and read with laser devices. CDs are used for audio recordings, audio/visual sound and graphics, and computer data and multimedia applications.

The CD format has been standardized to 120 mm (4.75") diameter. It consists of a thin layer of metallic film, etched with microscopic indentations called *pits* spiraling literally for miles around the recording surface. This structure is coated with a smooth plastic surface. The data is stored in a format that was developed by Sony and Philips and agreed upon by electronics vendors in 1981.

CD players first began to be marketed in Japan and Europe, and to a limited extent in Canada, in 1982. They did not begin to be distributed widely in the United States until 1983. By 1986, consumer players were inexpensive enough to promote an explosion of interest in audio CDs. See SPARS code.

compact disc interactive CD-I. A more recent version of CD formats with read-only players based around Motorola 68000 technology. It was developed by Sony and Philips and released in 1988. CD-I allows interactive multimedia use of compact discs. The CDs can be recorded with information in various forms, including computer data files, video images and still frames at more than one resolution, and audio in three formats.

Compact Disc Player – SCSI Connection

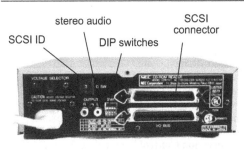

The back and front of a NEC external compact disc drive (CD) showing the various selectors, connectors, and components. Internal CD drives usually don't require disc caddies.

---

compact disc types and uses The two most common types of CDs are music CDs and multimedia computer application CDs. Music CDs are supplanting music on cassette tapes and vinyl records due to the greater clarity of the sound (no scratches or hiss) and greater stability of the medium (magnetic data, and the thin tapes themselves are somewhat fragile). CD-ROM discs hold about 680 MBytes of data, although actual informational content may be greater if the data has been compressed.

Typically, CDs are written once and read many times, although the data on PhotoCD discs may be extended in several sessions, with the new data being written to an unused section of the disc. A *multisession* CD player is needed to read discs that have been recorded in more than one session. See bar code, compact disc; digital video disc, laserdisc, PhotoCD.

compact disc video CD-Video. A variation on compact disc technology, announced in 1987, which delivered audio and video on one disc. The inner portion of the disc is the recorded music and the outer portion contains up to about five minutes of analog video and sound, similar to a small laserdisc. CD

players that support the video portion spin the disc faster than when playing the standard audio track on the inner portion of the disc.

compander A transmission device that *comp*resses and ex*pand*s a signal, usually to save transmission time. Modems that use compression techniques on-the-fly are companding devices and are typically installed at each end of a transmission line.

companding A combination and telescoped word derived from *compressing* and *expanding.* Companding is a process of compressing and expanding a signal and is used for a variety of purposes, including noise reduction, security, and increased transmission speed.

Compaq Computer Corporation A successful computer company established in 1982. Compaq shipped its first product a few months later, in January 1983, achieving phenomenal first-year sales. Compaq made the Fortune 500 list in 1986. It bought out Digital Equipment Corporation (DEC), one of the long-time, well-known companies in the computer industry, in 1998, and subsequently being bought out by Hewlett-Packard.

Competitive Access Provider CAP. A competitive local carrier that is permitted to compete with Local Exchange Carriers (LECs) and Inter Exchange Carriers (IXCs) to provide voice or data services. See Competitive Local Exchange Carrier.

Competitive Local Exchange Carrier CLEC. A competitive carrier that is permitted to compete with established local voice and data service providers, as a result of the deregulation in the Telecommunications Act of 1996. CLECs may build their own wirelines or lease existing lines for resale of services. CLECs include CAPs, IXCs, CATV service providers, and others. See Incumbent Local Exchange Carrier.

Competitive Telecommunications Association CAT. A Canadian-based association representing new entrants in the telecommunications service business, including interexchange carriers (IECs), competitive access providers (CAPs), and resellers.

complete document recognition CDR. A process that goes beyond object character recognition (OCR), in that it recognizes not only text and individual blocks or elements on a page, but the general layout and types of data. CDR software is quite sophisticated and can fairly reliably distinguish the difference between text and images, headlines and regular text, and columns and sidebars.

completed call In the telephone industry, completed call has a fairly specific meaning, describing a call that has reached and been answered by the callee, but it does not include the time that the callee actually spends on the conversation. In other words, the meaning of *completed call* concerns the establishment of the connection with the person being called and not the actual length of the communication.

complex instruction set computing CISC. A microprocessor architecture that accommodates complex machine language instructions in which a single operation may be comprised of many small instructions of different sizes, and which may take longer to execute than the same operation carried out on a *reduced instruction set computing* (RISC) chip.

CISC chips are more common on older architectures. A CISC processor command is translated into microcode, a series of smaller instructions, which are in turn queued and processed one at a time by a nanoprocessor. See reduced instruction set computing.

Complex Node Representation CNR. In ATM networks, a collection of node-related parameters that provide state information about a logical node. This information is useful in routing.

Complex Text Layout CTL. An IETF Human Computer Interface (HCI) platform standard.

Component Object Model COM. Microsoft's approach to object-oriented programming. The COM is a means for creating components that are reusable across a variety of applications, thus reducing programming time and increasing interoperability across applications. Microsoft's Object Linking and Embedding (OLE) provided a subset of the functionality now associated with COM. See Object Management Group. For a more complete discussion of the basic concepts associated with programming objects, see object-oriented programming.

Component Software Microsoft's description for object-oriented programming components associated with their Component Object Model. See Component Object Model, object-oriented programming.

composite Combined, bundled, aggregated, interleaved, entwined, mixed.

composite video A color composite video signal is one in which the luminance (brightness) and chrominance (color) are combined, with the chrominance modulated onto the luminance as a subcarrier. The signal may have to be separated by the receiver, depending upon the system. Videogame systems that plug into a TV set send out a composite signal, as opposed to an RGB signal that might be sent to a computer monitor.

compound modulation A successive modulation technique in which the modulated wave from one step becomes the modulating wave in the next step.

Comprehensive System Accounting CSA. Within the Open Source Software community, CSA is a set of C programs and shell scripts that help administrate individual online accounts. CSA facilitates accounting for users, jobs, daemons, and billing units and provides configuration parameters, accounts scheduling, and hooks to reporting applications. Thus, CSA makes it easier to monitor usage, frequency, and other access data for administrating and tuning networks and for billing purposes.

compress Condense, contract, shrink; reduce in size, transmission time, or byte count.

compression The act of reducing, shrinking, or shortening items or data in order to store or transmit the objects or information more easily. Data compression is based on the premise that most files or transmissions include white spaces, noninformational sections, or redundancies that can be removed without affecting or significantly degrading the meaning or

quality of the information when it is decompressed. Compression is sometimes also based on human perceptual characteristics or multiple means of representing the same data, some of which may be more space-conserving than others. See data compression, decompression, lossless compression, lossy compression, run length encoding.

compression algorithm The computer logic and code designed to automate the process of saving or transmitting data in less space or less time than if the data were stored or transmitted *raw* (unaltered). Compression algorithms are used on many types of data (video, still images, sound, text, etc.) and the degree of compression is often tied to the type of data and even the specific character of the particular data being compressed. A compressed file is not always smaller than the original.

Compression algorithms may be *lossless* (the information can be reconstructed to be the same, or to appear the same, as the original) or *lossy* (the information is reconstructed to be *essentially* the same as the original, or perceptually similar, but not identical).

Compression Control Protocol CCP. A protocol for negotiating data compression at both ends of an established Point-to-Point Protocol (PPP) link. CCP was introduced as a standards track protocol in the mid-1990s as a means to configure, enable, and disable data compression and to signal errors in the compression/decompression mechanism. CCP is similar to Link Control Protocol (LCP) except that CCP utilizes different timeouts, additional codes, and specific PPP Protocol field indicators and may utilize frame format modifications that may have been established with the link. See Link Control Protocol, Point-to-Point Protocol, RFC 1962.

CompTel Competitive Telecommunications Association. An association that includes WorldCom and a number of medium-sized communications carriers.

Compton scattering A form of photon scattering that results from stimulation by electromagnetic radiation. The scattering effect is small but important, as it occurs at a wavelength different from the incident radiation, scattering off of loosely bound "stationary" electrons. Thus, for light, a particle model is more effective at describing the effect than a wave model.

In Compton scattering, the scatter angles and energy levels in the scattered photons may be detected/calculated. At visible light ranges, the effect is very small, but becomes more apparent at the higher energy levels associated with X-rays or gamma rays. This is useful for telescopic radiation detectors. A Compton scattering telescope typically consists of a scintillating layer that Compton scatters gamma radiation. The scattered photons then encounter a second scintillating layer which absorbs them. Phototube detectors assessing the two levels can somewhat determine interaction between the two layers and the associated amount of energy deposited. Unfortunately, the Compton relationships don't include information about the angle of incidence of the incoming photons, so this must be determined or estimated by other means if the source of the radiant energy is sought. See Bragg's law, Rayleigh scattering, Thomson scattering.

Compton Scattering

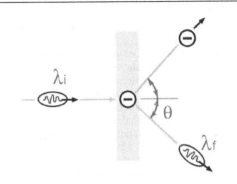

The Compton effect was studied in the early 1920s by A. Compton and may be expressed mathematically as

$$\lambda_f - \lambda_i = \lambda\Delta = (h/mc) * (1 - \cos\theta)$$

where λ_f is the scattered photon and λ_i is the incoming (incident) photon and h is Planck's constant. In terms of understanding the nature of light, this was important data confirming the hypothesis of a particle nature of light.

Compton, Arthur Holly (1892–1962) An American physicist and engineer who studied X-rays and developed theories and mathematical expressions describing their behavior, including reflection and polarization effects. He was also a pioneer in obtaining the spectra of X-rays by the use of ruled gratings, information that was invaluable in determining electronic charge. In 1941, he became chairman of the National Academy of Sciences.

Compton is best remembered for having observed and mathematically expressed Compton scattering, a subtle wavelength-shift effect that provided confirmatory evidence for a particle nature of light. In 1922, he published his observations in *Secondary Radiations Produced by X-rays*. In collaboration with A. Simon, Compton provided further confirmation of the effect by the *coincidence method,* describing how individual scattered photons and the electrons off of which they recoiled would appear simultaneously, an idea that required quantum rather than classical physics concepts to reconcile.

In 1927, Compton and C. Wilson were coawarded the Nobel Prize in physics. In 1991, NASA named its new space-based gamma radiation observatory after Compton. See Compton scattering.

CompuServe, CompuServ Historically, one of the earliest large-scale commercial computer service providers, CompuServe was initiated in 1969 as a time-share subsidiary of Golden United Life Insurance under the name CompuServ Network, Incorporated. In 1986, the service was purchased by H&R Block. By the late 1980s, CompuServe was also beginning to sell services to personal computer users and soon

after expanded into the European market. In 1989, CompuServe was one of the pioneering Internet mail relay carriers and, in the mid-1990s, extended its traditional BBS-style dial-up services to include Internet access.

CompuServe survived the online services shakeouts of the 1990s to become a large, commercial dialup Internet Services Provider (ISP). In 1997, it was bought out by America Online, Inc. (AOL) and positioned as an interactive service brand. In addition to public services, it provides CompuServe-specific services available only to members, including airline reservations, stock listings, chat services, etc.

computer A logic-processing device, which usually includes temporary or long-term storage and input and/or output devices for interaction with the user. It may or may not be programmable and may or may not be constructed with binary architecture (binary computers are prevalent). A computer doesn't have to be strictly electronic, and researchers have explored biological parts or processes for incorporation into computing devices. Quantum computers have been proposed, with science fiction possibilities, but none has yet been devised. However, individual quantum processes have been developed successfully and may someday be incorporated into computers.

The most common configuration for digital desktop computers consists of a central processing unit (CPU) for performing mathematical logical instructions, sometimes cooperating with coprocessing chips for graphics and sound; volatile storage, usually in the form of RAM; read/write semi-permanent storage, usually on magnetic or magneto-optical media; user-interaction input/output devices such as monitors, keyboards, mice, microphones, cameras, speakers, and joysticks; and program instructions in the form of operating systems and applications programs.

To enhance the usefulness of basic computers, printers, scanners, modems, and network interfaces have been developed, which communicate through a printed circuit board, or various SCSI, IDE, serial, parallel, USB, FireWire and networking ports. Many people make the mistake of assuming the software that runs the system, the operating system (OS), is the computer itself. While it is true that a particular OS is usually optimized for a particular platform, operating systems can be adapted to run on many systems. Early computers ran several operating systems (see TRS-80), and the trend is moving back in that direction. Linux, Be, Inc.'s BeOS, and Apple Computer's OS X, as examples, are designed to run on a number of hardware platforms, providing the user the freedom to choose his or her hardware/software combination.

Computer and Business Equipment Manufacturers Association CBEMA. See Information Technology Industry Council.

Computer and Communications Industry Association CCIA. A trade organization based in Virginia that represents data processing companies and common carrier service companies. The CCIA provides education and lobbying support to its members.

Computer and Information Science and Engineering CISE. A U.S. National Science Foundation Directorate that promotes basic research and education in the fields of computer information sciences and engineering. http://www.cise.nsf.gov/

Computer Control Company The company that originated the Series-16 minicomputers, washing-machine (and smaller) sized 16-bit-register computers that were bought out by Honeywell in the mid-1960s. See Honeywell Kitchen Computer.

Computer Emergency Response Team CERT. Established in the late 1980s by the Advanced Research Projects Agency (ARPA), based at Carnegie Mellon University. CERT provides assistance to computer operators wrestling with various network security and operations issues.

computer fraud Misrepresentation or theft accomplished on, with, or with regard to computers. The computer data may itself be the target of the fraudulent activities or a computer may be used as a tool to aid in noncomputer-related fraud (as in records theft, spying, or unauthorized access). Unsecured computer data, in the form of accounts, confidential business or investment information, personnel files, etc. is especially subject to tampering.

Computer Incident Advisory Capability CIAC. An archive and notices repository posted in conjunction with the Laurence Livermore National Labs Web site to inform the network community about security weaknesses and breaches that might compromise network systems, particularly the Internet. This posting is intended to provide technical assistance to help secure the Department of Energy (DoE) communications systems, but the posted bulletins are open to the Internet community. CIAC was founded in 1989 and is a founding member of the Forum of Incident Response and Security Teams (FIRST). CIAC provides training, education, technology watches, and trend, threat, and vulnerability data collection and analysis. CIAC publications that are transmitted to other parties are signed with a PGP encryption key.

Computer Science Telecommunications CST. A publication of the University of Missouri (Kansas City) School of Interdisciplinary Computing and Engineering.

Computer Security Institute CSI. A professional organization committed to supporting and educating information technology (IT), computer, and network security professionals, founded in 1974. http://www.gocsi.com/

computer societies, national A complete listing of the world's computer societies is outside the scope of this dictionary, but a sampling of some of the active and prominent societies that are accessible on the Web helps illustrate the types of organizations that exist and their general goals and priorities with relation to information technologies and computing science. See Computer Societies chart.

Computer Supported Telecommunications Applications CSTA. A computer telephony interface standard published by the European Computer Manufacturers Association (ECMA) in 1992. Work on ECMA

telephony standards was carried out by the ECMA Technical Committee TC32. CSTA was developed for integrating computers and telephone technology into a unified system, in a process described as Computer Telephone Integration (CTI). CSTA has been adopted as an ISO standard and is one of the most important international standards for computer telephony for the predictable future.

The standard describes information interchange among telecommunications and computer devices. It is sufficiently generic to encompass analog and digital private, public, and combination communications signaling systems, switches, and networks. Initial applications of the standard tend to focus on telephony/database integration and basic automation of services but, since it is a broad-based standard, new types of services and technologies will emerge as it becomes better understood and supported.

Computer Systems Policy Project A lobbying organization formed in 1991 to represent those who felt national networks were too oriented toward research and science and not enough toward everyday users. In actual fact, statistics show that a great majority of Internet use *is* devoted to conventional everyday user traffic, predominantly business and personal electronic mail and file transfers unrelated to research and science.

computer telephony integration CTI. Integration of computer database, dialing, and other features, with voice communications through a headset, handset, or other computer peripheral voice transmitting and receiving device. See Computer Supported Telecommunications Applications, computer telephony.

computer-aided dispatch CAD. A system in which the administration of services is aided by a computer. For example, emergency or law enforcement systems may track the location of vehicles and their direction of travel in order to dispatch calls in an efficient manner. Similarly, taxi and limousine services can be managed with the aid of a computer. Billing and mileage factors may also be stored by the system, and statistical measures tracked in order to enable a company to better manage its resources.

computer-aided learning, computer-assisted learning See computer-assisted instruction.

computer-aided manufacturing The process of using a computer to directly control manufacturing equipment, such as drilling machines, production lines, bottle cappers, saws, chisels, and any of the fabrications equipment which normally may have been driven by manual, electrical, or mechanical machines without the benefit of logic programming. One of the best applications of computer-aided manufacturing is integrating the production machines with parametric design software (usually used with CAD) so that objects that are generally the same, but perhaps different in specifics, can be manufactured with the same equipment, under the control of the computer. See parametric design.

computer-assisted instruction CAI. Instructional media and techniques used in conjunction with computer software, or entirely by computer software, in the form of tutorials, demonstrations, white papers, online mentors and instructors, educational software, tests, multimedia presentations, etc. CAI has been around for decades as a number of instructors were quick to grasp the significance of and opportunities provided by CAI, particularly for individual learning and distance learning. However, the resources and time to provide good educational programming and the prohibitive cost of systems up until recently greatly limited the practical application of these ideas.

Computer+Science Network CSNET. CSNET merged with BITNET in 1989 to form the Corporation for Research and Educational Networking (CREN); CSNET was discontinued in 1991.

computerTV See telecomputer.

Computists International CI. A professional association for information science, artificial intelligence, and computer science researchers. CI provides information on industry trends, leading edge technologies, research, and job opportunities. CI publishes a weekly *Computists' Communique* reporting on artificial intelligence, neural networks, genetic algorithms, machine learning, natural language processing, fuzzy logic, and computational linguistics. See artificial intelligence.

COMSAT Corporation Originally created by the U.S. Congress, COMSAT merged with Continental Telephone to form COMSAT Corporation, an international provider of satellite communications and networking services. COMSAT operates through the INTELSAT and Inmarsat systems and is currently the largest user of both systems.

COMSAT operates the COMSAT Laboratories for research, development, and technical consultation in pioneering satellite communications technologies. http://www.comsat.com/

concatenate To link, chain, or otherwise place adjacent objects or structures end-to-end. Thus, a female RCA plug can concatenate two RCA cables with standard male ends. A 2-port 25-pin data switcher can be used to selectively concatenate one or the other of two 25-pin cables (e.g., serial cables). A software "join" utility can concatenate two files, one after the other (as opposed to merging one file into another). See daisy chain.

concave Dished in, hollowed out, bowl-shaped (on the inner surface), or otherwise smoothly curving or arching inwards in two or three dimensions. Many optical components and networking structures have concave shapes, including resonating cavities, lenses, reflectors, parabolic antennas, etc. See parabola. Contrast with convex.

concentrator A point or device at which a number of elements are brought together either for simplicity of cabling and management, or to more efficiently provide a means to allocate shared resources. A star topology network is a type of concentrated configuration with a hub negotiating communications among the connected systems. A printer room with several kinds of printers available to the general office is another type of concentration point for general network services.

The concentrator itself may be a smart device, like a router, which processes the incoming information and sends the task or communication to the best destination or it may simply work on a first-come, first-serve basis.

A patch panel is a type of cable concentrator in the sense that it brings together the connecting points of a large number of individually cabled systems, usually to facilitate reconnections in different configurations. A modem pool is a concentrated assembly of individual modems, brought together in one facility, room, or closet, for easy access, configuration, and maintenance. See condenser.

concentrator, optical A device, such as a lens, that concentrates electromagnetic radiation into a higher energy level or smaller physical space. This is useful for increasing luminance, collimating beams, or channeling light into narrow openings such as the endfaces of fiber optic lightguides.

concentric In geometry relating to fiber optics, two or more broadly circular or rotating structures in the same general plane with a common center point or rotational axis. Thus, a bull's eye target comprises a series of differently sized concentric circles. A shaft within a rotating cylindrical tubing is concentric along the access of the shaft. Cable assemblies, with layers comprised of cylindrical fiber-conducting cores, cladding, tubing, aramid yarn, and outer armoring sheaths are concentric along the perpendicular plane to the axis of the length of the fiber core – in cross-section, the assembly will resemble a target.

Concentric structures are generally not concentric from every viewing angle. Just as a flat target has concentric circles only on the front and not on the side, the axis of reference is important. Concentricity may be associated with stepped or curved structures, as in Fresnel lenses or other types of curved, ridged components. However, they are only concentric in those planes in which the ridges share a common axis. In other words, they are concentric at the points where a knife, slicing down through all the layers, would pass through the shared center of all the reference components along the shared axis. Only an assembly of two or more perfect spheres of different sizes, with their radii sharing a common center, would be concentric in any given plane that passes through the center.

Technically the structures don't have to be circular to be concentric. A target could be comprised of different-sized squares rather than circles, as long as they share a common center point, but in practical applications concentricity is often associated with roughly circular or spherical components, especially those that rotate within one another.

concentricity error In fiber optic lightguides with two or more component layers (e.g., cladding surrounding the core), which may not be perfectly circular at any given point, and thus not perfectly concentric, the ratio of the cladding to the core, which has fine tolerances for optimum performance, may not be ideal. At this point, the reflection of the light beam off the cladding and through the core may be

Computer Societies – Sample List		
Organization	Abbreviation/Description	
Australian Computer Society	ACS	Per capita, one of the world's largest computer societies, established in 1966 in a merger of state-based computer societies. The ACS studies, supports, and promotes professional excellence in information technology. See ACSnet. http://www.acs.org.au/
British Computer Society	BCS	A professional society and registered charitable institution for supporting the field of information systems engineering, founded in 1957. The BCS is also a licensed engineering institution with accreditation-granting authority. http://www.bcs.org.uk/ http://www.infj.ulst.ac.uk/~bcs/
Computer Society of Bermuda	CSB	A nonprofit organization for fostering knowledge and applications in information technologies in Bermuda, founded in 1975 and incorporated in 1986. http://www.csb.bm/
Computer Society of India	CSI	Founded in 1965, the CSI is committed to promoting the interchange of information and the advancement of the theory and practice of computer technologies and professions. http://www.csi-india.org/
Computer Society of South Africa	CSSA	A not-for-profit professional corporation dedicated to the support and education of its members in various chapters throughout the region. http://www.cssa.org.za/
Computer Society of Sri Lanka	CSSL	Formed in 1977 to support and promote research and professionalism in the information technology field. http://www.ccom.lk/cssl/
Jamaica Computer Society	JCS	A communications and education body of Jamaica which seeks to promote the efficient and effective use of information technologies in Jamaica. http://www.jcs.org.jm/
Jordan Computer Society	JCS	A nonprofit organization for promoting the computer profession, founded in 1986. http://www.nzcs.org.nz/
Hong Kong Computer Society	HKCS	A nonprofit professional organization promoting education and applications in information technologies in Hong Kong, founded in 1970. http://www.csb.bm/
IEEE Computer Society	IEEE	A highly prominent American organization descended from the Subcommittee on Large-Scale Computing of the American Institute of Electrical Engineers (AIEE), founded in 1946. The AIEE and the Institute of Radio Engineers merged in 1963 to become the Institute of Electrical and Electronics Engineers (IEEE). The IEEE has members around the world. It promotes education and professionalism in a broad range of computing technologies and is responsible for the development of many telecommunications and computing standards. http://www.computer.org/ http://www.ieee.org/cshome.htm
Irish Computer Society	ICS	The ICS supports and promotes a broad range of information technologies through education and member services. It was founded in 1967. http://www.ics.ie/
Kuwait Computer Society	KCS	A Kuwaiti Public Welfare Institution, founded in 1982 as the Kuwaiti Society of Computers, renamed in 1990. The KCS promotes education and professionalism in computer sciences and information technologies. http://www.paaet.edu.kw/Info/HomePage/adel/kcs/kcs.htm
Lithuanian Computer Society	LIKS	An independent society of software users, amateurs, and professionals in informatics and computing science, officially registered in 1990. http://www.liks.lt/
Malaysian Computer Society	MCS	A society that promotes computer literacy in Malaysia, founded in 1998. http://www.geocities.com/Eureka/Concourse/2008/
Mauritius Computer Society	MCS	The MCS promotes personal and professional development and computer literacy in Mauritius. http://ncb.intnet.mu/mcs.htm
New Zealand Computer Society Inc.	NZCS	Promotes and fosters education, qualification, and professional development in information processing. http://www.nzcs.org.nz/
Norwegian Computer Society	NCS	An open, independent, self-financed society for promoting awareness and the advancement of information technologies for business and society. http://dataforeningen.no/ncs/
Singapore Computer Society	SCS	Singapore's largest information technology professional body, founded in 1967. The SCS promotes personal development and industry leadership in information technology. http://www.scs.org.sg/
Syrian Computer Society	SCS	A nonprofit national organization of scholars and engineers, based in Damascus, founded in 1989. The SCS furthers information technologies in Syria. http://www.scs-syria.com/

hindered by deviations from a perfect circle and core/ cladding ratio. See concentric, total internal reflection, critical angle.

concurrent Functioning, processing, or operating at the same time; parallel, in conjunction with; coexistent, simultaneous, synchronous.

concurrent programming Techniques and associated notation systems for parallel processing implementation. Distribution, synchronization, prioritizing, and signaling are important aspects of concurrent programming. For example, computer graphics special effects rendering is computing intensive, and farming out various objects to various processors or workstations, and then combining them in one frame when each is rendered, can greatly decrease the time it takes to create each image (called *render farms*). Not all types of operations benefit from concurrent programming. The overhead involved in setting up the distribution and coordination of the data must be smaller, in proportion to the effective processing that occurs, to make it worth processing in parallel.

concurrent site license In the software industry there are a number of common schemes for assigning software use rights. Exclusive operation on only one machine at a time is the most common, but it is also possible to get concurrent licenses that permit a specified maximum number of users to access the software at any one time from a networked server, or that permit up to a specified number of users (five is common) to install the software on individual workstations.

condenser An apparatus that concentrates or condenses a beam, ray, wave, or collection of particles. In the process of concentrating a substance, wave, or particles, the condenser may also secondarily store them, as in electrical energy. A device that focuses radiant energy, such as a lens for concentrating light or a parabolic antenna for concentrating satellite waves, can be considered a basic type of condenser. See concentrator; condenser, electrical.

condenser, electrical Condensers range widely in complexity and construction. They are used with a variety of types of electrical apparatus, for example, spark coils. Condensers employ a dielectric, that is a material that doesn't readily conduct direct current (DC). Dielectrics vary from paper to ceramic or glass, with the better insulators being used in higher voltage applications. A Leyden jar is one of the earliest condensers used for concentrating and storing electrical energy. In the Leyden jar, the glass acts as the dielectric. A variation on the same idea, using glass plates in a rack rather than a jar, were used for early wireless condensers. See capacitor, condenser, Leyden jar.

conditioning The processing of current to make it suitable for specific tasks. Some electrical appliances can tolerate variations in current or noise, while others are very sensitive to variations and noise, particularly small electronic components, requiring that the raw current that may come from a wall socket or other source first be conditioned to meet the needs of the device. See AC to DC converter.

conductance The ability, readiness, inclination, or disposition of a material or system to carry an electrical current, expressed in the practical unit *mho* (ohm spelled backward). The reciprocal of electrical resistance. See conductor.

conductivity method A pioneer experimental method of sending wireless communications using the ground as the conductor. It was demonstrated successfully at distances of over five miles by Preece, before Marconi demonstrated practical applications of wireless communications. Terminals of strong batteries were set up in series from a sending key, grounded at a distance of about fifty feet apart. A symmetric arrangement was set up at the receiving end, except that it used a telephone receiver or galvanometer. Other researchers experimented with this method, but little documentation of their efforts is available. See Preece, William; Steinheil, K.A.

conductor A material that readily carries an electrical current or heat. Some metals make especially good conductors (e.g., silver, copper, gold, aluminum) and are widely used in the manufacture of wire. Less conductive materials, such as rubber, used in specialized parts such as gaskets and seals, are sometimes impregnated with metal to increase their conductivity, while still retaining attributes that are difficult to achieve with metal alone. The term *conductor* originates from Desgauliers in the 1730s. Contrast with insulator.

conduit 1. In its most basic sense, a channel for directing physical objects or virtual data along its path. 2. A liquid conduit is a pathway often used for temperature regulation, dispersion of lubricants, or channeling of fluids from one area to another. See duct.

conduit, wiring 1. A tubular, hollow, physical pathway providing a channel for materials installed inside or directed through its core. Plastic, metal, and ceramic are common conduit materials. 2. A pipe that provides a protected pathway for wire, cable, or other conductive materials. Conduit is commonly used to run wires in a building and may also include insulation, color coding, and other attributes to protect or identify its contents. Conduit can be a good way to hedge against obsolescence, since it can be rethreaded more easily than cables that have been attached directly to the structure of a building inside its walls.

cone of silence See zone of silence.

Conference Européenne des Administrations des Postes et des Télécommunications (European Conference of Postal and Telecommunications Administrations) CEPT. An international standards body representing telecommunications providers in most nations other than Japan, Canada, U.S., and Mexico. It cooperates with CEN/CENELEC. See E1.

configuration Setup, organizational structure, architecture, topology, assemblage, physical and logical parts and interactions taken as a whole.

congestion indicator In ATM networking, a traffic flow control signal to reduce the allowed cell rate (ACR) in order to reduce the likelihood of increasing congestion. The information is contained in the RM cell. See cell rate, leaky bucket.

Congestion Manager CM. A network end-system module that enables applications to adapt to network congestion and enables a suite of multiple concurrent streams from a sender to a receiver with the same congestion properties to perform congestion avoidance and control. Congestion Manager was submitted as a Standards Track RFC by Balakrishnan and Seshan in June 2001.

The framework supplied by Congestion Manager integrates congestion management across all applications and transport protocols. It maintains parameters and exports an API with information about network characteristics and enables applications to pass information to the CM, to schedule data transmissions, and to share congestion information. Use of the CM requires explicit consent of the CM through the API.

CM may be elected for use on best-effort network systems that have well-behaved applications with their own independent per-byte or per-packet sequence number information and use the API to update the CM's internal state. See RFC 3124.

conical array antenna An antenna that can receive a range of VHF signals through a central rod with reflector elements extending out at right angles to the support rod and, at the other end, forward-oriented driven elements fanned out more or less from a single connection point on the rod.

conical monopole antenna A vertically polarized, broadband antenna shaped like a cone with the narrow end oriented towards the top. The frequency response is related to the size and angle of the cone.

conical scan In radar antennas, a circular scanning motion, often used on aircraft, that provides more complete information on the location and characteristics of the object of the scan. The conical scan can provide angular information.

Connect 918 A Macintosh and IBM-licensed PC-based videoconferencing product from Nuts Technologies that supports video, audio, whiteboarding, and screen sharing over analog, Switched 56, ISDN, and Ethernet networks. It uses ITU-T H Series and G Series Recommendations standards and encoding. See Cameo Personal Video System, CU-SeeMe, IRIS, MacMICA, ShareView 3000, VISIT Video.

Connected, Limited Device Configuration CLDC. A Sun Microsystems Java specification used by a number of major wireless telecommunications providers as a guideline for manufacturing and programming small Java-enabled communications devices. The audience for the CLDC specification is the Java Community Process (JCP) expert group and aforementioned developers. Participants include prominent network technology companies such as Nokia, Ericsson, Fujitsu, Sony, America Online, et al.

The "Limited Device" designation applies to relatively slow CPU, limited-resolution, limited-memory devices that are becoming prevalent as mass-produced hand-held assistants, including palm-top computers and schedulers, and advanced-feature cellular phones, etc. Thus, CLDC is a minimum-footprint specification for connected devices, devices that can interface with a computer or wireless network.

The full specification is downloadable free from Sun Microsystems' Web site.

connection 1. A system or circumstance of physically or logically joined entities, objects, or processes. An electromagnetic connection is one in which current from one system can pass into another, either through a splice, jack, plug, or other connector, or by a spark or induction system. 2. A *wireless connection* is one in which a signal is transferred without a visible physical connection. Wireless connections typically are based upon the transmission of sound or electromagnetic waves that originate in a transmitter (which commonly includes or is associated with a wave generator) and terminate in a receiver which may or may not convert the signals into visual or audio forms that may be directly perceived by humans.

connection management controller CMC. Works in conjunction with a broadband integrated gateway (BIG) to take data from incoming ATM cells for processing and routing. See broadband integrated gateway, HFC.

connection protocol The software protocol that negotiates a pathway for a transmissions connection session.

connection-oriented A type of communication in which the sender/receiver connection is established prior to transmission, as in a phone call. This may sound like a logical way to do things, but a substantial amount of network traffic does not follow this model. In sending an email message, for example, the message will be sent irrespective of whether the receiver is online at the time the message is sent. Then, if too much time elapses, or a certain number of attempts to deliver the message have failed, it will be returned to sender. Modem communications are connection-oriented. If there is no answering handshake at the other end of the transmission, no transfer of data takes place. Contrast with connectionless.

Connection-Oriented Transport Service COTS. A connection-oriented, end-to-end network communication service. COTS involves initializing the service, establishing a connection, transferring data, releasing the connection, and general cleanup associated with the release of the connection such that it may be reused, unbound, or closed. When a COTS session is initialized, it enables the COTS driver and associated application to be bound with a specific transport entity. Various buffer-handling or status utilities may be used during the connection. In the Open System Interconnect (OSI) model, connection-oriented network services are implemented by using the Connection-Oriented Network Protocol (CONP) and the Connection-Mode Network Service (CMNS).

connection-related function In ATM networking, a traffic management and policing function related to a network element (NE) where connection-specific functions are carried out.

connectionless A type of network transmissions architecture in which the data is sent without first establishing that the receiver is connected and available to receive transmissions. Large distributed computing environments frequently employ connectionless services. In contrast, phone calls and modem data transfers are connection-oriented end-to-end communications. Contrast with connection-oriented.

Connectionless Broadband Data Service CBDS. In ATM networking, a high-speed packet-based connectionless service similar to SMDS (Bellcore) that is defined by the European Telecommunications Standards Institute (ETSI). CBDS is appropriate for networks requiring high-volume, high-speed transmission rates and thus is favored for high-end graphics and publishing, videoconferencing and streaming, and scientific research applications. CBDS is favored over private lines or permanent virtual circuits (e.g., Frame Relay) for some inter-business communication needs.

Commercial implementations of SMDS or CBDS support multicasting and most major network protocols (TCP/IP, SNA, AppleTalk, etc.). CBDS can be integrated with existing Ethernet, Token-Ring or FDDI local area networks (LANs). At the present time, CBDS is more widely installed in Europe than North America. Refer to the ETSI ETS 300 series documents for specifications.

Connectionless Transport Protocol CTP. A protocol that provides a means to send to a recipient that may or may not be connected to the network at the time the data is transported. CLP allows end-to-end transmission addressing and error control, but does not guarantee delivery.

connectivity A property of mechanical and electronic systems that allows them to interconnect with other devices or systems for the purposes of transmitting or relaying information or signals. While connectivity generally refers to physical connectivity, the increasing importance of data in communications and hardware configuration has extended the term to software as well. When systems or devices can be readily interconnected in terms of hardware and software, they are said to be *compatible*. See connection.

connector 1. A device to join or combine two or more objects. When circuits are coupled with a connector, the two systems or objects are usually intended to communicate in the same way (otherwise the connector is usually referred to as an *adapter*). The connector may be incorporated into the device being connected or may be a separate item. See adaptor, gender changer, jack. 2. In a flow chart, a connector is a symbol that can be used to indicate a join in a flow, or the divergence of the flow into additional paths. See ST-connector.

Connon, John R. (1862–1931) A Canadian inventor and historian who devised various innovative mechanisms, including a type of dynamo, and who was granted a first patent for a cinematic camera in the late 1880s. Connon worked together with Rudolph Stirn, a German inventor, to develop panoramic technology. Connon's innovative camera could photograph a continuous image, without seams, while rotating 360 degrees (a similar camera was patented by M. Garella in England in 1857 but the author wasn't able to determine if the earlier camera was seamless). See panoramic camera.

Sampling of Standardized Connectors

A variety of common, standardized computer, phone, and video connectors.

1. 9- to 15-pin D-shaped computer data adapter
2. 25-pin D-shaped null modem data adapter
3. RJ-11 phone line splitter/joiner
4. and 5. stereo sound adapters
6. and 7. coax F and BNC video adapters
8. RCA video/audio splitter/joiner
9. RCA video/audio cross connector

A selection of video and data connectors, oriented to show the connecting pins.

1. 6-pin mini-DIN computer connector
2. RJ-45 10Base-T computer network connector
3. video BNC coaxial cable connector
4. audio or composite video RCA connector
5. Super-VHS (S-Video) video connector
6. SCSI-2 50-pin computer data connector
7. 25-pin D-shaped computer data connector
8. 50-pin flat SCSI data connector

Conrad, Frank (1874–1941) An American broadcaster who began as callsign 8XK in his Pennsylvania garage, which was later licensed as the history-making KDKA radio broadcast station. Conrad was also an avid inventor, with dozens of patents to his credit. His interests ranged from telegraphy to moving picture technologies. In 1919, he patented a radiotelegraph device (U.S. #1,314,789), followed by a wireless telephone (#1,528,047) in 1925. By the late 1920s, Conrad was designing various types of telephones, and in the mid-1930s his interests turned to television and motion picture electronics. The Frank Conrad Garage (where it all began) is an official preservation project of the Save America's Treasures effort. See KDKA.

consecutive Continued presentation of objects, data, or actions one after the other; successive, sequential, following. A distinction may be made between sequential and consecutive in that sequential implies that there are no gaps between succeeding steps, whereas consecutive implies there may be gaps or delays in successive presentations, depending upon the nature of the information of actions. For example, it would not be unusual to say a person was working on the computer on consecutive weekends, but the phrase "sequential weekends" would not normally be used, as there is a weekday gap between each weekend. Sequential events are always consecutive, but consecutive events are not always sequential. If this is confusing, think of the fact that a square is always a rectangle, but a rectangle is not always a square. See concurrent, parallel, sequential, serial.

Consent Decree, Dell A late 1990s judicial decree barring Dell Computer Corporation from telemarketing computer systems bundled with software that wasn't ready to ship. The proceeding was important because software was relatively new and established legal protections were still being worked out for new technologies. In this case, Dell promoted the Dell Software Suite as being bundled with Dimension computer systems. Consumers on the whole received the system without software and were not offered the opportunity to consent to the delay or cancel orders for a prompt refund. The Federal Communications Commission (FCC) charged that Dell violated the Mail Order Rule.

Since Dell is considered a major personal computer vendor, and vendors are focused on being first-to-market (which can be influenced by the readiness level of the software), this ruling can potentially affect many communications technology vendors, as they increasingly rely on software operating systems and utilities to promote their hardware products.

Consent Decree, MCI and British Telecom A 1994 decree in which British Telecom's acquisition of an interest in MCI (jointly called Concert Communications Corporation) were addressed as to the competitive effects of the merger. A modified decree was entered into in 1997, as a result of British Telecom's plans to acquire the remaining assets of MCI. When British Telecom sold its interest in MCI to WorldCom, and MCI sold Concert Communications Corporation to British Telecom, a motion to terminate the consent decree was tendered in 1998.

Consent Decree, Microsoft A 1995 decree with a 78-month duration involving Microsoft Corporation. Microsoft is a significant software vendor alleged by a number of competing vendors as engaging in monopolistic and unfair business practices. In this decree, Microsoft was enjoined not to enter into license agreements for operating system-related products with a duration exceeding one year and license agreements which would restrict OEMs from licensing, selling, or distributing non-Microsoft operating system software products. In later competition for market dominance of Web browser software, the antitrust division of the Justice Department alleged that Microsoft was not keeping to the terms of the previous agreements. The result was a long and complex anti-

trust investigation against Microsoft in the late 1990s continuing into the 2000s.

Consent Decree, Sprint A decree in 2000 between the Federal Communications Commission (FCC) and Sprint Communications Company, LP, regarding the illegal practice of *slamming* - switching consumers' long-distance telephone services without their consent. Sprint voluntarily notified the FCC of the practice, after a number of slamming enforcement actions taken earlier the same year by the FCC against other companies, and agreed to a voluntary monetary contribution in addition to returning consumers to their preselected carriers, terminating the agents involved, and implementing a stronger slamming prevention and detection program in the firm.

Slamming was not a new problem at the time. In an earlier consent decree in 1996, MCI agreed to voluntary contributions for slamming, indicating that the practice and the problems of its enforcement are longstanding.

Consent Decree of 1956 A historic agreement between the Justice Department and American Telephone and Telegraph (AT&T) to separate Northern Electric (later Northern Telecom) from Western Electric. In spite of the limited settlement, the Federal Communications Commission (FCC) received many complaints from manufacturers over the next two decades about AT&T refusing to buy or refusing to permit their subscribers to buy new improved telephone technologies. This situation lead to the Consent Decree of 1982.

Consent Decree of 1982 A landmark historic proceeding, this consent decree involved the divestiture of AT&T that took place in the mid-1980s under the direction of Judge Greene. It is now more commonly known as the Modified Final Judgment (MFJ), since it was a modification of the Consent Decree of 1956. See Modified Final Judgment for a fuller description.

console 1. Floor-standing cabinet, typically holding consumer broadcast receivers (radio, TV). 2. A primary operations physical unit that holds main electronic controls and monitors (such as lab equipment, medical monitors, industrial plant operations equipment, etc.).

console, computer operations A computer terminal for monitoring/controlling computer operations, printers, etc. On a secure network, the operating console is often password-protected to control access and may even be locked in a separate room to prevent access or physical theft. The main server sometimes serves also as the console, although on larger systems the server and the console may be separate systems.

console, telephone A primary multiline telephone unit used by an operator to answer and route calls (a replacement for the old physical cord-and-stereo-jack-style switchboards). These come in a wide variety of configurations. Some are programmable by entering letters, features, and numbers through the keypad, which may further be displayed on a small character display. See PBX.

consoleless operation Automated operations or routing, an option for companies whose needs are simple enough that they can function without a central unit, or without the expertise of an operator. See PBX.

Consortium for Audiographics Teleconferencing Standards CATS. A nonprofit organization, based in California, that promotes acceptance and development of audiographics teleconferencing standards. Audiographics teleconferencing in its ideal form is the simultaneous realtime use of images and sound, in a cooperative environment, by participants in different locations.

constant angular velocity CAV. A playback mode for magnetic and optical discs in which the disc rotates at a constant speed. A CAV disc generally requires more space on the disc to hold the same information as can be stored on a constant linear velocity (CLV) disc, but CAV format has the advantage of providing frames that can be viewed individually in 'freeze frame' mode as still images. See constant linear velocity.

constant bit rate CBR. In ATM networks, a cell rate traffic flow class of service (CoS) category that supports a constant or guaranteed rate of transport, and circuit emulation. Constant bit rates are important for types of communications that require synchronization of signals at the receiving end. For example, synchronization of sound and audio in a videoconferencing application is important, as unacceptable delays might occur if related cells are not arriving at the same time. See cell rate.

constellation 1. A group related by proximity and physical or conceptual connectivity (such as workstations, celestial bodies). 2. In GPS, the set of satellites used in a position calculation or all the satellites within communications range of a GPS receiver at a specific time. See Global Positioning Service.

Constellation Communications, Inc. A U.S.-based commercial provider of satellite communications services. Constellation is developing a low Earth orbit (LEO) system comprising 46 satellites called the ARIES satellite system. Eleven ARIES satellites will be placed in circular equatorial orbits at 2000 kilometers, and 35 will be divided into seven circular inclined orbits at the same altitude.

consult To seek advice, opinion, or information from reference materials, or from another person, presumably with expertise in the area of inquiry.

consultant Professional or other expert offering advice or information services, usually specialized.

Consultation Hold A surcharge phone service or multiline subscriber service that enables the operator to put an incoming call on hold while engaged in another call.

Consultative Committee Telecommunications CCT. A three-nation industry trade association that promotes trade expansion and the evolution of telecommunications equipment and services within NAFTA and North and South America. The CCT represents more than 50 industry telecommunications equipment and services suppliers, as well as regulatory and certification agencies. CCT liaises with CITEL and serves as industry advisor to the NAFTA Telecommunications Standards Subcommittee.

Consumer/Disability Telecommunications Advisory Committee C/DTAC. A Committee of business, academic, public, disability, and minority representatives established by the Federal Communications Commission (FCC) in November 2000 under the provisions of the Federal Advisory Committee Act. Three working groups focus on *Consumer Protection and Education*, *Access by People with Disabilities*, and *Availability and Affordability of Telecommunications Products and Services*.

C/DTAC first met in an open meeting in March 2001 to discuss the telecommunications needs of consumers and various underserved populations, with the meeting set up to be broadcast on the Internet. http://www.fcc.gov/cib/cdtac

contact A point in a circuit, usually at a junction, binding post, or terminal, where other parts of the circuit interconnect or are attached.

contention Competition for the same space or resources; disagreement over the allocation of resources with the implication that each of the disagreeing parties has some desire for or stake in the resources.

Contention can be a problem on systems where the demand for resources outstrips supply, especially if there is no mechanism for resolving contention, but contention is not always unexpected or undesirable. On networks, there is constant contention for resources, including computing power, routes, Internet access, printers, monitors, scanners, and modems. This is considered part of the normal operation of a network and is managed by a variety of strategies, including prioritization, overflow handling, buffering, polling, queuing, batch processing, packet rerouting, and timeouts.

In both linear and parallel processing systems, different resources may be assigned to different servers or groups of systems to prevent contention delays and facilitate arbitration of limited resources.

Contention mechanisms may be deliberately initiated in situations where there are multiple backup systems. For example, if a printer on a network fails, the remaining printers may compete for permission to complete the task and the original printer may stand down (or not participate at all if it has failed or been disconnected). See queue, queuing theory.

continuation of message COM. In ATM networks, a status indicator used in the asynchronous transfer mode (ATM) adaptation layer (ATM AL or AAL) to indicate that the cell is a continuation of a communication that has been segmented, that is, broken up and sent in different sections, sometimes over different pathways. See asynchronous transfer mode.

Continuous Redial A subscriber surcharge or bundled telephone service that enables the caller to redial a number that was found to be busy while making or receiving other calls. If a number is busy, the caller hangs up and dials *66 (in N.A.). The phone service will continue to try to connect with the busy number for up to 30 minutes. If the call connects, the caller is notified with a distinctive ring and can pick up the phone and take the call. Dialing *86 will terminate the continuous redial.

continuous wave A wave that is constant or unvarying in its major characteristics, such as amplitude.

continuous wave transmissions Any transmissions technology that employs continuous signals rather than pulses. Since most communications media rely on pulses or *modulation* to send meaningful information, continuous wave transmission is more specialized and generally used in signaling situations, such as security systems, in which an interruption of the continuous wave serves as an alert or system startup signal.

control field In many types of data communications that employ fields as information units, a control field is assigned to contain information about how to process related data.

control panel A console used to control operations of a system, vehicle, aircraft, or network. The control panel may consist of physical switches or dials or may be a text or graphical interface on a computer screen (often simulating switches and dials).

Control Panel In Apple computer operating systems, a collection of utilities, accessible through the Apple Menu, that provides access to many basic operating parameters, including sound, memory, monitor settings, network configuration settings, etc.

control segment In the Global Positioning System (GPS), the control segment is a global network of control and monitoring stations that ensure the accuracy of satellite positions and their clocks. This coordination is an important part of GPS, as the data is derived in part from the relationship of the satellites to one another and their signals. See atomic clock.

control terminal A workstation or personal computer configured to provide control of a network from any routing node. It acts as a command input and display console. Through remote access commands, it is possible to control a network from a node other than the one that is physically connected to the terminal (virtual terminal).

controlled access unit CAU. In the generic sense, a device in a link that selectively controls the entry and/or exit of people, things, or data to a system. A door with a lock is a simple example of a physical CAU. A revolving entry gate at a circus, with security personnel and ticket sellers, is another access unit in that people are funneled through a specific physical location. A firewall is an example of a computer network CAU in the sense that it controls access to or from the system behind the firewall. See Controlled Access Unit.

Controlled Access Unit When spelled with capitals, a CAU more specifically refers to an active wiring concentrator used in Token-Ring networks. The CAU transforms the logical ring topology into a star topology, for example, to facilitate installation in larger building environments. Two or more CAUs can be interconnected to produce a segmented network. Multiple CAUs intercommunicate through the main ring path. The CAU provides a means to attach units to the ring network as though it were a star topology, while still maintaining the basic ring configuration for the network traffic logic. However, it is more robust (more tolerant of a bad unit on the system) and more flexible (accommodating additional computers) than a basic ring network. See Lobe Attachment Module, Multistation Access Unit, Token-Ring network.

Controller Area Network CAN. A multimaster serial network bus originally developed for controller circuits in the automotive industry by Robert Bosch GmbH, Germany, in 1986. CAN is now being adapted for marine, medical, industrial, and other control and automation applications. The CAN bus is a high-speed (up to 1 Mbps), half duplex network communicating among microcontrollers. It is capable of interconnecting over 2000 devices and is often used in imbedded systems. It is particularly useful for short messages.

Data in a CAN system are transmitted and received using Message Frames. Standard CAN uses 11-bit identifiers, Extended CAN uses 29-bit identifiers. There are two ISO standards for CAN, specified for two different speeds. See CAN in Automation.

controller A software-supported computer hardware device that works in conjunction with the operating system, through the various system interfaces, handling input and output and control of that device. Thus, a disk controller provides functions to handle a hard drive. The most common desktop computer disk controllers follow SCSI and IDE standards. SCSI controllers are also commonly used for scanners, CD-ROM drives, and many types of cartridge drives. A serial controller handles serial communications into and out of the computer, usually to a printer or modem. RS-232 and RS-423 are two of the most common desktop serial interface standards.

When a computer first boots up, one of the processes that occurs is bringing the controller hardware and software online. The computer needs to locate the various devices and will often load a variety of software device drivers that support the hardware functions. See controller card.

Controller Area Network CAN. A multimaster-capable serial bus system that facilitates the networking of intelligent devices, including sensors and actuators. CAN is associated with the lowest layers of the ISO/OSI reference model.

controller card A computer circuit board card that connects through a slot (e.g., PCI), to provide an electrical and logical connection between a device and the main circuitry of the computer. There are a limited number of slots available for controller cards, usually more in a tower model, and extra power may be needed to handle the extra load. There may be jumpers or dip switches on the controller card to fine tune the settings, as on a graphics card or hard drive controller. The communications standard used by the controller card must fit that of the slot into which it is inserted. An EISA card cannot be put in a PCI slot, and vice versa. Software may need to be loaded onto the computer for the operating system to recognize the controller card functions. See controller.

CONUS Contiguous United States. A designation for the continental, contiguous U.S. consisting of 48 states. The term is used when referring to U.S. travel, transmission, or broadcast regions.

converter A very broad term for anything that changes the incoming signal to a different outgoing signal or that converts one type of physical connection to another. Converters are frequently used to convert between alternating current (AC) and direct current (DC) or to convert one cable type to another. A converter may be used to convert a frequency spectrum from one range to another, but this is pushing the boundaries of the term, as converter usually implies fairly simple conversions. More complex ones tend to have their own descriptive terms. When a converter is used simply to change one type of plug or jack into another without any electrical changes, it is more commonly called an *adapter*. When it simply links two components with no changes in the signals, it is called a *connector*.

convex Smoothly protruding and curving outwards, like the outside surface of a bowl. A continuous or mostly continuous surface that follows the outer surface of a spherical/elliptical shape. The term is generally applied to surfaces that do not constitute a full sphere. Thus, the outer surface of a ball is considered spherical, but if it were chopped in half or less than half, the smaller portion would be considered convex, whereas the larger portion, while still being primarily convex in geometry, tends to be semantically called spherical rather than convex.

If an elongated, rounded-end lozenge-shaped object like a vitamin pill were cut exactly in half between the two rounded ends, the portion that curves toward the midsection would be considered convex, but the portion where the curve straightens out to include the straight midsection of the pill would not. The outer surfaces of many lenses are convex as are the "back" surface of parabolic antennas. Contrast with concave.

Conway's law This saying has been variously restated (and probably improved over time). The idea is that there is congruency between the composition of the software team and the final design of the software (and in this version, an implied dig that a single programmer wouldn't ever finish the project), stated as, "If you assign *n* persons to write a compiler, you'll get an *n-1* pass compiler." Another version, not quite as apropos to computereze as the one just stated, but perhaps closer to the original, is "If you assign four groups to working on a compiler, you'll get a 4-pass compiler." It is attributed to Melvin Conway, an early Burroughs computer programmer.

Another Conway's law has been stated in Dilbertian fashion as follows, "There is always one person who knows what is going on. That person must be fired."

Cook, Gordon Author of the *Cook Report on Internet*, a newsletter devoted to issues concerning the commercialization and privatization of the Internet. Cook was formerly science editor at the John von Neumann National Supercomputer Center. Later he served for 18 months as a director for the U.S. Congress Office of Technology Assessment, assessing the National Research and Education Network (NREN)

Cook Report on Internet An independent, opinion-oriented by-subscription monthly online newsletter

focusing on Internet infrastructure policy and technology. The publication monitors telephony and computer convergence and new technologies being adapted by forward-thinking telecommunications providers.

Cooke, William Fothergill (1806–1879) A British researcher who collaborated with Charles Wheatstone in developing the telegraph. The two met in 1837 at a time when both were researching similar telegraph technologies. See telegraph history; Ronalds, Francis; Wheatstone, Charles.

cookie A token or other transaction acknowledgment or ID passed between transacting processes or programs to keep a record of an access or action. Cookies may be passed transparently between systems as part of normal operational protocols.

Data cookies are an integral part of Internet commerce, especially in the form of identifiers in Web link referrals and shopping cart purchases made online. Cookies can be passed from the browser to the shopping cart site to identify the visitor for later purchasing or statistical purposes and may or may not be acknowledged with information from the vendor being deposited on the buyer's system. The cookie may also track customers as they browse other sites (common on the Internet) and then return to finish their online shopping. Some people object to these types of cookies, which are automatically offered to the visited site by the browser and will disable this capability, and many people have objected to the *reverse cookie*, one that is deposited on the visiting browser's system, often without the knowledge of the user, as this opens a porthole for viruses, vandalism, and unfair trade practices. See Caller ID.

cookie monster An invasive software program, widely distributed in the mid-1980s, named for the popular children's television program character. The program would prompt the user with "Give me cookie ..." at increasingly shorter intervals, gradually taking up more and more CPU time, and if the user didn't type the word "cookie" it would eventually print so frequently, and steal so much CPU time, that it would make the terminal unusable. The original was rumored to have come from MIT. See virus.

Cooperative Research Action For Technology CRAFT. One of a number of programs of the European Union, CRAFT is a means to enable small- and medium-sized enterprises (SMEs) to engage third parties to carry out research on their behalf. The third parties may be commercial research organizations or research departments of academic organizations. In a CRAFT project, the SMEs are made up of at least three participants from at least two different EU states and the area of research must fall within the umbrella of the *RTD Framework Programme* which can include software development, for example.

Coordinated Universal Time, Temps Universal Coordonné UTC. An international astronomical time reference devised in 1970 by the ITU. UTC is related to the Greenwich meridian, that is 0 degrees longitude on the Earth's surface.

UTC uses a 24-hour clock, thus, 2:00 pm is 1400 hours.

Since UTC cannot exactly match Earth's slightly varying rotation, UTC was set to a UT1 reference with the Earth's position as of 0000 hours on 1 January 1958. Deviations are adjusted with leap seconds. Coordinated Universal Time is based upon the average period of the rotation around the Sun. UTC receives its frequency and time information from over 50 centers around the world and broadcasts it over a number of radio frequencies, with tones to indicate seconds and spoken words for upcoming minutes.

copper A malleable, metallic chemical element with high conductivity, which makes it invaluable in the manufacture of electrical wire and heating implements. It is the most widely used conductor in electrical work due to its properties, availability, and price. Gold and silver are also good conductors but not economically practical for most electrical work.

Copper Distributed Data Interface CDDI. An ANSI standard version of Fiber Distributed Data Interface (FDDI) that runs on twisted-pair copper wiring rather than on optical fiber.

copper twisted pair Copper is very commonly used for electrical wires, and twisting two copper wires together can improve its transmission properties. A pair of wires is intertwined in a helical pattern over distance in order to reduce capacitance. Over longer distances, the distributed capacitance can build up, and *load coils* are introduced, at intervals, to help balance capacitance and inductance. The use of load coils for telephone voice connections is common; however, they can cause problems when the same wires are used for data transmission. Copper pairs used for data transmission may be constructed differently from voice lines, with the insertion of a metal screen to differentiate the transmit and receive.

Compared to fiber optic cable, it is very easy to connect twisted pair. Cutting and splicing is relatively straightforward, whereas the cutting and splicing of fiber optic must be done with great care so as not to alter the alignment properties of the optic waveguides.

Sometimes the twisted pairs are further aggregated into *binders,* a group of 25 twisted pairs. This simplifies installation in multiconnection installations. Color coding is often used to keep track of the connections and binder bundles. See copper wire, load coils. See twisted-pair wire for a diagram.

copper wire The most commonly used transmissions medium for telephone calls and related telecommunications. Copper wires were first widely installed in the 1880s, superseding some of the earliest galvanized wires used for telegraph signals. Copper wire with an iron core was developed by Bell's Thomas B. Doolittle in 1883, and it became popular due to its combination of conductivity and durability.

Copper wire for phone communications was most commonly installed as a single wire, strung on utility poles, or as twisted pair. More recently, gel-filled multicables comprising up to almost 5000 twisted pairs have been used where many connections are required.

Many bare conducting wires have been strung without insulation, but insulation is often used to protect the wire from damage, interference, and corrosion. Rubber, gutta-percha, latex, plastic, and wound paper have all been used as wire insulators, as have air and jelly inside an outer core. See coaxial cable, copper twisted-pair, Copperweld, fiber optic.

Copperweld A trademark name for an early combination of copper wire with an iron core. It combined the flexibility and conductivity of copper with the durability of iron and increased the longevity of the wires.

coprocessor A computer processor that is not considered the main or central processing unit (CPU) but assists the CPU in handling heavier processing loads or specialized processing loads. CPUs are designed as general-purpose chips, and are not intended specifically for any one type of task. The Amiga computer, released in 1985, is an example of one of the first desktop computers to make extensive use of coprocessing chips to handle resource-hungry graphics and video operations in order to prevent these computations from slowing down the CPU. This arrangement is now more common.

The interaction of a CPU with support circuitry such as coprocessors is one of the reasons the raw speed of the CPU is not a perfect indicator of the performance of a system. Computers with coprocessing chips and average speed CPUs have often been shown to outperform faster CPUs if they don't have coprocessing support. Coprocessor chips are gradually becoming more common in desktop systems, with math coprocessors becoming prevalent in the 1980s, and graphics and sound coprocessors beginning to show up in many consumer systems in the 1990s.

COPS See Common Open Policy Service Protocol.

COPT Coin-Operated Pay Telephone.

copyright Certain legal safeguards conferred by government agencies. Copyright protections are granted to original works for a specified period of time, depending upon the type of work. Original drawings, musical compositions, software programs, and stories are copyrightable. Inventions are sometimes copyrightable, sometimes patentable, and sometimes both. It is important to include a copyright symbol © and a date with the name of the copyright holder, or the word *copyright* and the date, on each presentation of the original work or reproduction thereof. The C with the circle is recognized internationally by those countries cooperating in international copyright treaties, such as the Berne Convention.

A fee-based formal copyright process is available in most countries to provide a record of the type and date of the copyright materials. This record is a good source of evidence in a legal dispute, but the copyright registrar does not police the copyright; that is the responsibility of the copyright holder.

Many researchers, academics, and business employees are mistakenly under the impression that they automatically own the copyright for something they create. This is often not true, although the laws have swung slightly more in the direction of the creator in recent years. However, if it is work for hire or work paid for in the normal course of an employee's duties, the educational institution or the corporation *usually* owns the copyright to the work, unless there is a specific written agreement stating otherwise. If the employee, on the other hand, creates some original work outside of working hours and can *prove* it wasn't done under the direction of or using the resources of the employer, the employee may have a case for copyright ownership.

Copyright does not protect the owner if someone independently comes up with the same idea and has not copied the original idea. However, it can be difficult to prove the idea was conceived independently, if it is very similar to another, and that proof may be necessary in court if a legal proceeding is initiated. Most public libraries have excellent references on copyright requirements and registration guidelines, as do all U.S. government documents repositories. See intellectual property, patent, trademark.

CORBA Common Object Request Broker Architecture. In the current state of computing, there are many different vendors, many different computer platforms, and many different software applications, resulting in much duplication and incompatibility. Now that we have the Internet as a common ground for sharing development strategies, applications, and application-development tools, it is not necessary for these incompatibilities to exist, and neither is it necessary for a consumer to be forced to use any one particular computer platform.

CORBA is a strategy and a set of tools. It enables reusable programming objects to be used by many applications in a platform-independent manner. It is the combined effort of more than 500 vendors, engineers, and end users, organized as the Object Management Group (OMG). CORBA is a set of specifications for platform-independent, interoperable, distributed object-oriented applications. By using CORBA specifications, software vendors can create truly global software that can be distributed over the Internet and run on a multitude of systems.

CORBA is an infrastructure that provides general services, and request and response capabilities at a low level. The distribution of objects written in a variety of programming languages is supported. CORBA does not define the upper level architecture; this decision is left to individual developers. See Object Request Broker.

cord lamp In manual switchboards, or any type of cord panel where indicators are used, the lamp is a small bulb associated with a physical socket connection for a cord jack that lights up if the associated circuit is active. Cord lamps were used on old telephone switchboards to signal active switches. The modern equivalent is an indicator light on a multiline telephone console.

cordboard A historic switching panel, human-operated with long cloth-wound patch cords which plugged into jack receptacles on the desk level and interconnected, as needed, with jack receptacles on the wall corresponding to the local phone desk. See switchboard.

cordless switchboard Unlike cordless phones, which are used in wireless systems, a *cordless switchboard* is not a new wireless technology, but rather an older switching technology in which human operators used keys instead of patchcords to connect call circuits. While this was a great improvement over patchcords and was still used on many long-distance circuits until the 1970s, it was slow and expensive compared to all-electronic automatic switching systems.

cordless telephone A battery-powered wireless telephone handset with a short antenna and a separate charging element and AC adaptor, usually using very short-range radio signals. While a cellular phone is a type of cordless phone, the phrase *cordless telephone* is usually used to refer to very short-range phones used within buildings or circumscribed areas.

Cordless phones are predominantly analog, but more digital phones are being produced, resulting in more options for interfacing with a computer or providing secure or semisecure communications.

core 1. Center, inner, inmost. 2. A central strand or wire around which other conductive or protective layers, strands, or insulating materials may be wound. Usually the main conducting portion of a wire assembly. In most electrical installations, copper wire is used as a conductive core. In fiber optic cable, the core glass is usually surrounded by a layer of lower refractive cladding glass. See cladding. 3. A central bar, often of iron, around which a coil is wound to create an electromagnetic part. See armature, coil, electromagnet. 4. A small doughnut-shaped magnetic component used for computer storage, with polarity representing binary states. See core dump. 5. A central, removable strand around which other materials may be wound or braided in order to provide a brace for molding their shape.

Core, the 3D Core Graphics System. A baseline specification developed in the mid-1970s to encourage standards for device-independent graphics. This specification led to development of the Graphical Kernel System (GKS), an official standard for 2D graphics. For 3D graphics, GKS-3D and the Programmer's Hierarchical Interactive Graphics System (PHIGS) became official standards in the late 1980s.

core diameter A description of the thickness of an inner, usually conducting material, as of copper wire or the inner lightguiding core in a fiber optic cable as measured in cross-section through the center perpendicular to the axis of the length of the cable. See American Wire Gauge, Birmington Wire Gauge.

core dump A copy of the contents of core memory from a process error condition, usually consisting of undecipherable symbols and unprintables that can make a terminal or printer go crazy. On large systems, the output can be voluminous. Irate receivers of spam, unsolicited commercial email, have been known to retaliate by sending back large core dumps.

core to cladding offset See concentricity error.

core-to-core splicer An industrial fusion splicing device for joining fiber optic cables. A core-to-core splicer allows active alignment in three axes, compared to two-axis cladding alignment splicers. A core-to-core splicer provides alignment of single-mode, multimode, and active fiber cores and may include a built-in thermal sleeve shrinking mechanism. A splice log memory or computer link may be provided for keeping a record of splices and their statistics. Since this type of splicer may be portable, for field work, it may also include a pigtail port for facilitating cable termination and may have a built-in cleaver. See cladding alignment splicer, fusion splicer.

Cornell, Ezra (1807–1874) Cornell was talented in both business and mechanics and was associated closely with Samuel Morse and Hiram Sibley, founder of the company that became Western Union. Cornell contributed to early telegraph installations and helped Morse construct the historic Washington, D.C. to Baltimore, MD line that was funded through the U.S. Congress. Cornell designed early insulators from glass plates. Cornell remained a lifetime director of Western Union and was the chief founder of Cornell University.

corner reflector In its simplest form, two intersecting or joining flat reflective surfaces with sufficient angle between them (usually 20 to 160 degrees, and often 90 degrees on the reflective side) to allow reflectance of a beam. The corner may also be in three planes, shaped like the corner of a room where the walls join the floor or ceiling. Corner reflectors are common in radar applications. The materials vary, but mirrored glass and metals are often used.

corner reflector antenna A type of antenna which combines a primary radiating element in relation to two angled metallic surfaces, or rods arranged in a plane. Various styles of corner reflectors are used in UHF television reception and radar applications.

Corning Cable Systems A leading global manufacturer of copper and fiber optic communications products for voice, data, and video network applications. Corning Cable is wholly owned by Corning Incorporated, a publicly trading company that was originally established in 1851.

Corning Cable was involved in a number of important mergers in the early 2000s, including the acquisition of Siecor and Siemens' global cable and hardware businesses.

Corning Glass Works A historic glass company that employed many significant pioneers in glass technologies and their application. While working at Corning's Sullivan Park Research facility, scientists D. Keck, R. Maurer, and P. Schultz demonstrated a practical fiber optic waveguide that overcame the limitations of earlier attempts at reducing the losses associated with lightguiding over distances. For their work, they have received numerous awards, including the 2000 National Medal of Technology.

Corning Glass Works became Corning, Inc. In 1943, Dow Corning Corporation was formed and now produces a wide variety of silicon-based products.

Corning Museum of Glass From decorative art works to fiber optic components, glass has a long and colorful history that is exhibited in the extensive collections held in the Corning museum, located in Corn-

ing, New York. Scientific aspects of glass are documented and demonstrated in the Glass Innovation Center. http://www.cmog.org/

corona A halo, glow, or other luminous surrounding from various causes including refraction, particle movement, ionization, radiation, reflection. *St. Vitus' fire*, as reported by sailors, is likely a kind of corona effect. Voltages around power lines and antennas can sometimes ionize the surrounding air, resulting in a whitish-blue corona effect.

Corporation For Open Systems International COS. A nonprofit vendor-supported organization created in 1986. It was established to further the acceptance and use of data processing and data communications equipment, and to encourage multivendor product compatibility in these areas. COS is involved in various standards efforts, particularly those involved with test methods and certification requirements.

Corporation for Research and Educational Networking CREN. A nonprofit organization formed in 1989 when BITNET merged with the Computer+Science Network (CSNET) to promote and assist accessible, worldwide academic information exchange. In 1994, CREN announced a system to facilitate communications called Internet Resource Access (IRA), in essence an Internet-in-a-box system. In 1996, it recommended to its members that they discontinue support of BITNET in favor of other systems, primarily the Internet. Since the Internet has superseded BITNET as the dominant global communications network, CREN has increased its focus on seminars, educational materials, and software utilities that enable academicians and technology professionals to use and implement widespread, inexpensive network access for the purposes of research, communication (especially discussion lists), and distance education. CREN is a founding member of the Internet Society (ISOC). See BITNET, Internet Resource Access, LISTSERV. http://www.cren.net/

corresponding entities In ATM networking, peer entities with a lower layer connection between them, coordinated through protocol control information.

corrosion A wearing away, or alteration by chemical action, often leaving a residue such as rust or film as a byproduct of the corrosion. Many electrical wires and components are coated or bonded in order to prevent corrosion. Some elements resist corrosion, making them useful for applications in corrosive environments. See oxidation.

CoS See class of service.

COS 1. compatible for open systems. 2. See Corporation for Open Systems International.

COSETI Columbus Optical SETI. See SETI.

COSINE Cooperation for Open Systems Interconnection Networking in Europe. A program established by the European Commission to utilize Open Systems Interconnection (OSI) to interconnect various European research networks.

COSPAS/SARSAT A cooperative effort begun by the United States, the USSR, France, and Canada in the later 1970s. It supports satellite communications-related search and rescue operations which enables information such as the location of distressed aircraft or marine vessels to be communicated to rescue systems. COSPAS/SARSAT operates in conjunction with the emergency position-indicating radiobeacon (EPIRB) to support the Global Maritime Distress and Safety System (GMDSS).

COTS 1. commercial off the shelf. 2. See Connection-Oriented Transport Service.

Cotton-Mouton effect When light passes through a pure liquid in a direction perpendicular (normal) to an applied magnetic field, the liquid becomes doubly refracting. See Kerr effect.

coulomb A unit of electrical quantity named after Charles A. de Coulomb. A unit for the amount of electrical charge in the meter-kilogram-second (MKS) scale that passes through a circuit in one second at one ampere (unvarying) current. The international coulomb is the quantity of electricity that will deposit 0.0011180 grams of silver when passed through a neutral solution of silver nitrate in water. A coulomb can also be expressed as the quantity of electricity on the positive plate of a condenser of a capacity of one farad when the electromotive force is one volt. See ampere.

Coulomb, Charles Augustin de (1736–1806) A French physicist and engineer who experimented with applied mechanics and electromagnetism. In 1785, he demonstrated the laws of electromagnetic force between elements using artificial magnets with well-defined poles in which the associated phenomena could be more clearly observed. The Crater Coulomb on the moon and the coulomb unit of electric charge are named after him.

Coulomb's law A description of the magnitude of an electromagnetic charge. Two electromagnetic point charges will attract or repel one another with a force directly proportional to the product of their charges, and inversely proportional to the square of the distance between the two point charges. This phenomenon is more easily observed than many in the field of physics. Named after Charles de Coulomb.

Council for Access Technologies CAT. Originally founded in the early 1990s as the National ISDN Council, with the wide deployment of ISDN technologies, the group broadened its focus and became the Council for Access Technologies. CAT is an industry trade association promoting the implementation, standardization, and simplification of new telecommunications services offered over a variety of access technologies. CAT supports the exchange of technical information and information about compabitility and works with other organizations and end users to introduce new networking services. http://www.CATcouncil.org/

counter-rotating ring In ring network topologies, data typically travels in one direction along each path. In a counter-rotating ring, there are two signal paths, each one traveling in the direction opposite to the other. See Fiber Distributed Data Interchange, Token-Ring.

counterpoise 1. A state of balance, counterbalance, equilibrium. 2. A power or force acting in opposition such that the opposing forces are equivalent, or balanced. 3. A structure designed to balance the transmitting or conductive properties of a circuit.

counterpurpose Working towards one goal and paradoxically achieving the opposite result to what was intended. This term was coined by Doug West and has many applications in telecommunications technology. For example, when the Federal Communications Commission frees more radio frequency spectrum to ease a situation where demand exceeds supply, it often spurs the development of new technologies to take advantage of the newly available spectrum, with the result that demand exceeds supply by even more than was originally the case.

coupler In general, a connector to facilitate the joining of two elements; an elbow joint can couple two sections of pipe or a threaded (or bayonet) coupler can connect a camera to a microscope. In technology, the term often describes a mechanism for combining two or more cables or signals or for splitting a cable or signal into two or more paths. A coupler is a generic component with applications for mechanical, electrical, and optical media.

Optical couplers are used to branch optical power from a single fiber into multiple fibers or vice versa. A standard optical coupler handles single wavelengths with designated ratios. Broadband couplers may branch or combine optical power within a designated wavelength range in a constant ratio. For systems design or troubleshooting, couplers may be used to isolate circuits.

Optical couplers can be fabricated by aligning optical fiber cores closely enough for the signal to "jump" from one fiber to another. By twisting and heating the fibers, the conductive fiber cores will fuse to facilitate the transfer of light energy. The portions of the fiber that are tapered in the fabrication process will influence the reflective properties of the light beams and trap the *higher order modes* in a multimode fiber. These are known as *cladding modes,* as they are at the surface of the fiber where it is shielded by the cladding materials that keep light within the fiber waveguide. As the signal moves through the fused region, *lower order modes* remain in the original fiber. As the beams exit from the coupler and the tapered, fused region is left behind, cladding modes are converted back to *core modes.* The degree of taper, the proximity of the fibers, the length of the coupling region, and the wavelengths coming down the fiber are important parameters that hinder or facilitate signal resonance and coupling.

A 1×2 coupler, also known as a Y coupler or Y splitter, is a *directional coupler* that enables two paths to be joined into one or one path to be split into two. Y couplers/splitters are commonly used in fiber optic networks. A 2×2 coupler has eight possible individual paths through which the signal may travel for a bidirectional signal and four possible individual paths for a unidirectional signal. Combining the couplers increases the number of paths (and the complexity of the combinations and control mechanisms).

Couplers can be combined in tree-like branching structures to form multiple outputs from a single source. Note that loss is typically associated with splitting a signal. A sufficiently powerful source signal or signal amplification may be needed in a circuit with many branches.

A coupler may be connected to a fiber optic network by fusion splicing or through low-loss connectors attached to the coupler. *Pigtails* are protruding filaments intended for making connections. Commercial housings, some of which are rack mountable, are available for supporting multiple couplers. Depending upon the configuration, the coupler may require a termination adapter. See SC-connector, ST-connector.

coupler, star A type of coupling configuration in which a signal is split into multiple paths which may continue on through a circuit or be reflected back to multiple positions adjacent to and including the signal source. In optical star couplers, an input light beam is transmitted to all output ports.

coupling loss Loss in a conductive medium that is directly related to the physical properties of the coupling mechanism. Coupling losses can be due to poor materials, poor shielding, loose connections, rotational incongruities between the coupled ends, particles, heat expansion, humidity, etc.

Coupling loss is more prevalent and more significant in fiber optic cables than in most wired connections due to the precision needed to guide light along a link in a path.

The units for expressing coupling loss depend upon the type of conducting medium and conductive phenomenon. In other words, it will be expressed differently depending upon whether it is a wire or optical cable and whether the conducted signal is electricity, light, X-rays, or sound. It will also depend upon the relative magnitude of a typical loss and the type of signaling that is used. For example, coupling loss in a light transmission may be expressed in terms of loss of power (e.g., light intensity), whereas coupling loss in an acoustic medium may be expressed in terms of loss of intelligibility (e.g., the number of words correctly understood). See attenuation, coupler, fusion splice.

cover page The first page of a printout or facsimile that identifies information about the nature and date of the printout, and often the sender and intended recipient. On networks with shared printers, it helps identify the owner of the hardcopy. See banner.

coverage area In news and entertainment broadcasting and cellular communications, the geographic range of users/subscribers. Outside of the coverage area, signals will be weak or absent. This is not a problem for broadcast subscribers, who usually know they are either inside or outside the range of a certain channel, but for mobile communications users driving out of range in the middle of a conversation, it can be a problem. For this reason, some mobile communications will provide a signal that indicates that the limits of the range are nearby, and some handsets will have a light or message that indicates the user is outside the service range.

Fiber Optic Coupler – Duplex

Duplex fiber optic cables can be interconnected with a duplex multimode optical coupler for about $18.00. BellCore has published compatibility specifications that are followed by a number of manufacturers. Coupling ratios may vary, depending upon the application.

CP/M Control Program/Monitor, Control Program for Microcomputers. An operating system that came into widespread use in the late 1970s and early 1980s, written by programmer and professor Gary Kildall. It had a text-based, line-oriented interface and ran on the Intel 8080 and Z80 microprocessor families. Kildall formed Inter-Galactic Digital Research, later Digital Research, to market his computer software products. The forerunner of MS-DOS was derived from a CP/M manual by Tim Paterson and is basically the same in syntax and functionality. Digital Research continued to develop CP/M into CP/M86, and later DR-DOS. Kildall also created a multitasking version of the operating system and personal computer networking software long before they became common on personal computers. See Digital Research.

CPA 1. Charter Public Accountant. 2. See Chip Protection Act.

CPCS See Common Part Convergence Sublayer.

CPE See Customer Premises Equipment, Customer Provided Equipment.

CPI See common part indicator.

CPL commercial private line.

CPN See Calling Party Number.

cps See characters per second.

CPS See Cellular Priority Service.

CPU See central processing unit.

CQ In early radio transmissions, CQ was often used as a call to operators (all stations) as a way of getting general attention. There are some historians who believe CQ may also have been used as a distress call, and some have interpreted it as meaning "Come Quick" although this may have been attributed after the abbreviation had been around for a while. Baarslaq has written that the Marconi Company requested *CQD* to be established as a distress call (presumably *CQ Distress*) to distinguish it from a general CQ call in early 1904.

CQD A radio distress call which predated SOS. See CQ.

CR 1. See carriage return. 2. call reference. 3. *symb.* right-hand circular or direct polarization (ITU). 4. connection request. An Open Systems Interconnect (OSI) transport protocol data unit. 5. customer record.

cracker A specific subset or connotation of "hacker" for someone who gains unauthorized access to systems or applications. Since breaking into systems can sometimes be done through sheer persistence and brute force techniques, not all cracking is hacking in the sense of applying brilliant and elegant solutions. The term was derived from *safe cracker* since cracking a software password is somewhat analogous to cracking the combination on a safe. One of the most widespread activities of crackers is the deciphering of passwords for computer games so they can be distributed and played without paying fees to hard working game producers. See hacker.

CRAFT See Cooperative Research Action For Technology.

crankback In ATM networks, a mechanism that allows the release of a connection setup that has encountered an error condition (as in a failed Call Admission Control) to permit rerouting of the connection.

cramming An undesirable practice in which telecommunications services customers are billed for a number of enhanced features (e.g., Call Waiting) that were not requested or ordered.

crash System failure, lockup. On a computer, a crash usually implies a complete lockup of the operating system (this shouldn't happen and very rarely happens on good operating systems) and usually requires a reboot. If the operating system is robust and remains functional after an applications crash, the offending application can often be "killed" to remove it from the system and to prevent it from affecting other applications. It is generally done this way on Unix systems. Killing the application and processes associated with the application will also free any memory that may be inaccessible due to the crash.

There are many ways in which a system can crash. Less common causes are hardware failures or electrical anomalies. The most prevalent causes are software problems, especially applications memory management and operating system memory management. Endless loops and bad pointers can also cause crashes.

Cray, Seymour (1925–1996) For a whole generation of computer users, Seymour Cray's name was associated with some of the most powerful supercomputers in the world. Cray grew up with a strong interest in electronics and computing at a time when computers were mainly used by the military and not much was known about them. Cray founded Cray Research Inc. in 1972 and Cray Computer Corporation in 1989, companies known for high-end computers for over 20 years. The Cray 1 supercomputer was announced in 1975 and the Cray 2 in 1983.

Supercomputers attract less excitement now than they did a decade ago because the general processing speed and level of functionality of desktop computers have increased so dramatically that the distinction between high-end and low-end systems is less dramatic. Many supercomputing operations are now run on desktop computers internetworked with the Linux operating system. This operation is very cost-effective for research labs, educational institutions, and government agencies.

CRC 1. See Communications Research Centre. 2. See cyclic redundancy check.

CRC Press The publisher of this reference, CRC is one of the world's oldest and most well-respected technical publishers, based in Boca Raton, FL. CRC first published the *Handbook of Chemistry and Physics* in 1913, an enduring reference that has been reprinted more than 80 times as a definitive source for generations of science students and professionals. The company now has several imprints and ships books all over the world. http://www.crcpress.com/

credentials In computer security, documents or expressions of trust and confidence, particularly those that indicate the capability to perform a function or task. The Pretty Good Privacy data encryption system has an interesting aspect related to credentials in that parties can vouch for the good name and veracity of public key holders. See Pretty Good Privacy.

credit card phone A pay telephone equipped with a card slot instead of or in addition to a coin slot to read magnetized credit cards and calling cards. These are becoming increasingly common and are handier than phones where long credit card numbers and passwords have to be manually keyed.

Creed Telegraph System A system designed by Canadian inventor Frederick Creed with a typewriter-style tape-punching machine. It improved upon the speed and utility of existing manual punchers and began commercial distribution in 1908. His automated perforators could operate up to 150 words per minute. To this he added a translating and printing system, which eventually became a teleprinting transmitter/receiver system sold in England by 1927.

Creed, Frederick George (1871–1957) A Canadian inventor and telegraph operator who developed the Creed Telegraph System, an automated teleprinting transmitting/receiving system, in 1889. He then traveled to England to manufacture the Creed Printer. By 1898 he had demonstrated he could send telegraph messages at sixty words per minute, and his inventions were put into commercial use by 1913. Creed was a member of the Board of Directors for the International Telegraph & Telephone Company (ITT). See Creed Telegraph System, telegraph history.

creel 1. A woven basket for storing objects or critters (installation tools, caught fish, birds, picked apples). 2. A quantity, as might traditionally be held in a basket. The term is applied to fishing limits, for example. 3. A single support frame or multiple assembly for holding spools of cable on winding pegs. Winders may be attached to each peg or spools may be wound before hanging on the creel pegs, depending upon the application. The term comes from the fabric industry (e.g., woollen mills), where multiple spools of thread can be wound on a creel assembly and drawn off to other creels or mechanisms (warping creels or looms). Creels can support and organize multiple cables for production or storage. They are also used for supporting and spooling raw fiber in preparation for cladding.

Creighton, Edward (1820–1874) Creighton was experienced in building communications lines and roads and thus became the organizer of the western expansion of the telegraph system for Western Union, under Sibley and Cornell. Creighton surveyed the first transcontinental route across the wilderness in 1860, which followed the trails of the recently established Pony Express. Creighton worked in conjunction with the Overland Telegraph Company to carry out construction on the section west of Omaha.

In those days, building a line involved more than muscle and materials; it required bushwhacking through roadless, supplyless, lonely wilderness areas inhabited by wolves, coyotes, snakes, and bears. The vast tracts were without urban comforts of any kind. It also involved negotiating with native inhabitants and working out problems associated with roaming herds of buffalo who thought the telegraph poles were installed for their convenience as backscratchers. The line construction was originally estimated to take two years at a cost of over a million dollars. Under Creighton's supervision, the job was done in four months at a fraction of the original cost estimate. Fortunately, Creighton owned stock in Western Union and was able to benefit from this astonishing engineering feat. He used his gains to found Creighton University and to contribute to many civic projects.

CREN See Corporation for Research and Educational Networking.

CRF 1. See cell relay function. 2. See connection-related function.

crimp To squeeze so as to confine; to push in upon.

crimp tool, crimping tool A common handheld wiring installation tool that resembles a fat, snub-nosed set of pliers. It is designed to facilitate the cutting, stripping, and crimping of wires or fiber optic cable by providing rounded pressure points and leverage for exerting pressure. Crimp tools may have more than one gauge or setting for related tasks.

In fiber optic cable assembly, a crimp tool makes it easier to tighten and secure a crimping sleeve that slides over the jacketed fiber and its attached connector. It may also be used to tighten a strain relief boot that fits over the crimpling sleeve. Care must be taken to exert the correct amount of pressure when crimping, to avoid breaking a wire or fiber filament. See lapping film.

CRIS See customer record information system.

critical angle The angle at which a beam is reflected within a transmissions medium such as optical fiber. This angle can be very important to the strength of the signal over distance and the total distance the signal can travel. It may be modified by the angle of the beam, the thickness of the fiber, and various impurities (doped elements) that may have been introduced. See acceptance angle.

critical charge The amount of charge needed to initiate a process or to change the state of or value of data being stored or processed.

critical fusion frequency CFF. In a display device, the refresh rate frequency above which the individual scans are fused by human perception into a single frame or image. It is not a set number, as, for example, on a cathode-ray tube (CRT) display; it is related to the rate of persistence of the visible light from the excitation of the phosphors. As a rule of thumb, though, most systems show a nonflickering image at about 60 frames per second, or at about 85 Hz.

Critical Technologies Institute CTI. Organization established within RAND by an Act of U.S. Congress in 1992 and primarily sponsored by the White House Office of Science and Technology Policy. CTI works with the Federal Coordinating Council for Science, Engineering, and Technology to assure that technologies critical to national interests are identified and supported.

CRITO See Center for Research on Information Technology and Organizations.

CRM See cell rate margin.

CRMA See Cyclic Reservation Multiple Access Protocol.

Crookes Dark Space In a cathode-ray tube (CRT), as the gas pressure in the tube is gradually diminished, the glow surrounding the cathode detaches, leaving an area that is dark around the electrode, an area that may become quite large at low pressure levels. In a tube that has some air in it, this region can be more easily distinguished as being between the cathode glow on the inside and the negative discharge glow on the outside. Outside the negative glow is another region called the Faraday Dark Space. See Faraday Dark Space.

Crookes tube A simple experimental tube developed by William Crookes in the 1870s for studying electrical discharges. Essentially a variation on the Geissler tube, which was filled with various types of gases to observe their effects, the original Crookes tube was attached to a pump to evacuate the gases, and the two ends supporting the cathode and anode electrodes were sealed off to maximize the stability or controllability of the environment within the tube. (In fact, a complete vacuum was not achieved and, in some cases, the effect of rarified gases in the tubes was being studied.) The electrodes were connected to a source of electrical charge with the voltage controlled to form a crude ampmeter for detecting current. The Crookes tube also facilitated the discovery of X-rays (Röntgen rays). In some cases, a mineral substance was mounted within the glass tube to create fluorescing effects when voltage was applied across the tube electrodes. Like the Geissler tube, this illuminated effect made Crookes effects popular in English parlors at the time.

Later refinements of the Crookes tube made it possible to control cathode rays, an important step in the development of cathode-ray tubes, which are used in television, imaging, and display technologies. Crookes tubes are used in educational settings for studies in atomic structure for investigating electron flow and the influence of magnetic forces on this flow. See cathode-ray tube, Geissler tube.

Crookes Tube

The Crookes tube was an important experimental electron tube that was used by William Crookes beginning in 1870s, and later by other scientists, to research voltage levels and electromagnetic radiation. It also contributed to the development of pioneer display technologies such as cathode-ray tubes.

Crookes, William (1832–1919) An English physicist, chemist, and editor who developed scientific instruments for studying various phenomena and used them to make some important basic discoveries. Crookes founded the *Chemical News*, a weekly publication he edited for almost 50 years, though he is better known for investigating the effects obtained by passing electrical charges through various gases. One of his more notable achievements is the discovery of thallium in the 1860s. Further studies of thallium led to the invention of the Crookes tube in the late 1870s. He designed a vacuum-sealed glass tube for detecting charge that could be used to investigate the relationship between voltage and pressure. He later developed an instrument for detecting radioactive particles. Crookes tubes were subsequently used by Röntgen in his discoveries of X-rays. See Crookes Dark Space, Crookes tube.

cropping The process of truncating information, usually an image. Cropping refers to the removal of unneeded elements or data. For example, excess blue sky may be *cropped* from a photo during the development process or after the photo is printed. Cropping does not imply any change in size or proportions to the remaining portion of the image. In computer imagery, the cropping may involve removing the data only from the display area, the data regarding the 'hidden' or cropped information may still be in the computer's memory so it can be quickly restored, if needed. Cropping or *trimming* is often done by printers when cutting down print jobs that *bleed* outside print boundaries. See clipping, scaling.

cross assembler An assembly language programming and translation tool that enables assembly language symbolic coding to be written on one system to run on a different type of system. This is convenient because development machines often require more resources (more speed, memory, etc.) than the system on which the software will eventually run. There may not even be a machine available on which to run the software until the software is partly or mostly written, due to production schedules or cost. A cross assembler enables software development to get underway and continue somewhat independently of the hardware schedule.

cross connect A point in a circuit where a new or temporary connection is created by wiring between existing circuits or between facilities. Used variously for diagnosing problems, rerouting, or adding circuits.

Cross-Industry Working Team XIWT. An organization established to promote the understanding, development, and application of cross-industry National Information Infrastructure (NII) visions into practical technologies and applications, to facilitate communication between stakeholders in the public and private sectors. The XIWT Web site provides links to information, many reports, and white papers. http://www.xiwt.org/

crossbar switch In older mechanical telephone switching systems, a crossbar switch was similar to a relay, except that it was controlled by two external circuits and was used in more complex switching arrangements, such as those needed for long-distance connections. It was devised in the late 1930s, and AT&T developed a version based upon pioneer work by Swedish engineer Gotthilf Ansgarius Betulander. In 1938, the first crossbar central office went into service in Brooklyn, the same year the infamous "War of the Worlds" broadcast frightened credible residents of the area (who tied up phone lines in their panic). The crossbar switch eventually succeeded the widely used but troublesome panel switch in the 1950s and step-by-step switches in the mid-1970s. See Callender switch, Lorimer switch, panel switch, step-by-step switch.

crossover cable A cable in which a pair of wires are reversed at one end of the connection. This reversal is commonly done to convert a serial communications cable to a null modem cable. In this case, the transmit (Tx) and receive (Rx) wires are crossed, or switched over. In RS-232 specification cables these are lines two and three.

crosstalk A term for undesirable electrical interference, usually from nearby lines, in which the signal from one cable or component is close enough and strong enough to impinge on the signals in another cable or component. In telephone cables or switches, the crosstalk may be so excessive that a telephone conversation from another line can actually be heard. Crosstalk usually occurs in installations with inadequate spacing or shielding.

crosstalk, optical Fiber optic cables do not exhibit the same electrical leakage and crosstalk characteristic of tightly bundled or poorly shielded electrical cables; since the signal is a lightwave rather than an electrical signal, the term might not seem to apply. However, two types of problems related to crosstalk are characteristic of optical systems in the real world, where installation and structural considerations come into play:

- Some optical systems are hybrid, with electrical links or switches as part of the transmission system, in which case traditional crosstalk at the electrical interfaces may occur.
- In optical systems where many small fiber components are tightly bundled, creating numerous bends or connective joints, a number of types of light signal leakage are possible where there are breaks or weak points in the cladding, improper joints, or misalignments in the endfaces enabling light signals from one waveguide to interfere with another, resulting in optical crosstalk.

Optical crosstalk can be minimized by good coupling and cladding practices and the reduction of reflection around joints and bends. In terms of reducing electrical crosstalk at joints, many firms are developing and installing all-optical switching systems and linking components.

CRS See cell relay service.

CRT See cathode-ray tube.

CRTC See Canadian Radio Television and Telecommunications Commission.

cryptanalysis The research and analysis of cryptography, that is, message or data encryption. While cryptanalysis is generally the art and science of a broad range of cypher-related concepts, it is also more narrowly understood as the actual analysis and breaking of a cyphered message without foreknowledge of its content, structure, or any keys that might be needed to discern its contents.

cryptochannel A communications channel that is encrypted in some way to provide privacy and security to the conversants. When carried out through a computer network, a digitally encrypted mobile communications line, or a digital telephone line, many means can be used to hide the signal or the contents of the signal. These include key encryption, scrambling, frequency hopping, and others. Cryptochannels were not generally available to the public before digitally encrypted data communications were introduced to consumers in the late 1980s and 1990s; they were mainly used in government communications, particularly in the military. Now that encryption and secure channels are becoming available to almost everyone, it may change the way society communicates.

cryptographic ignition key CIK. A token for storing, transporting, and protecting cryptographic keys. A cryptographic module and a cryptographic ignition key may be used together to regenerate a key-encrypting key.

cryptographic key An input parameter used in key encryption security mechanisms to influence the transformation of information into secured data generated by a cryptographic algorithm. In general, cryp-tographic keys are intended to transform the data into a uniquely scrambled sequence that cannot be readily interpreted or transformed back into the original message without the appropriate key information. The correct corresponding key may be used to decrypt the message by an authorized recipient. This mechanism is important for securing computer data. While there are variations in the level of security depending upon how the key encryption is implemented, in general the longer the key, the more secure the data in terms of how technically difficult or time-consuming it would be to try to decrypt the message without the key.

Over the years, encryption keys have been getting longer and longer because key encryption aficionados keep finding ways to break the encryption algorithms, aided, in part, by more powerful computers and by multiple participants using their computer systems together to accomplish the task. Nevertheless, the cryptographic key model is one of the more secure, better understood, and accepted methods of securing data and is incorporated into various public/private key systems and digital certificate systems. See certification, encryption.

cryptography The process and study of concealing the contents of a message or transmission from all except the intended recipient. It is the primary means of security in telecommunications. The development of digital communications (ISDN, digital cellular, etc.) makes it easier to provide security, as typical unscrambled raw data or broadcast signals can be intercepted by unauthorized viewers. See certificate, Clipper Chip, cryptochannel, encryption, PGP.

Cryptoki A cryptographic token interface in a public key cryptographic system. It defines a cryptographic applications programming interface for developing devices to hold cryptographic information and perform cryptographic functions.

Cryptolope A type of electronic cryptographic container for the secure packaging of digital information, introduced in 1996 by International Business Machines (IBM). A Cryptolope is a public/private key encryption specification that provides a means to package and distribute control information and content in one package. The control information includes pricing, licensing, and conditions of usage. Also included are network addresses and usage data distribution instructions. Cryptolopes are implemented through Web browser plugins.

The Cryptolope package is organized in data layers, including a bill of materials (BOM) describing container contents; a clear text abstract of the contents, author, etc.; the encrypted contents; intellectual property rights, related copyrights, and usage rights, etc.

crystal 1. A substance characterized by a repeating internal structure occurring during the solidification of an element or mixture. The characteristic repeating structure is often manifested in the outward appearance. Many crystalline forms are transparent or nearly so. See piezoelectricity. 2. A piece of transparent, or semitransparent quartz, usually colorless. See quartz. 3. A crystalline material used in electronics for various purposes such as timing, rectification, and frequency evaluation. See crystal detector. 4. A wave-sensitive semiconductor used in electronics for applications such as radar detection.

crystal detector An elegantly simple, early form of radio device that superseded the coherer. A crystal detector took advantage of the rectifying properties of various natural and synthetic substances, commonly galena and carborundum. These materials have a property of allowing electrical alternating current impulses to pass through in one direction only. Thus, they can be used to convert AC frequencies to a direct current (DC) *half-wave.* AM radio signals are converted from radio frequencies to audio frequencies which are audible through headphones or speakers.

Historically, crystal detectors could be built on a very small scale and could be used without power sources or amplification, when carefully tuned and connected with high impedance headphones. In essence, they were the first portable radios and were popular for field and hobbyist uses.

The earliest sets used natural crystals, but later a number of synthetic crystals were developed, with various properties and degrees of sensitivity. Portability could be increased with sets that used crystals that could be tightly coupled with the catwhisker. Some of the more elaborate sets included tuning coils. Eventually crystal sets were superseded by vacuum tube radios, which provided amplification and a much higher degree of electronic manipulation and control Crystal radio sets are still sold as hobby kits from electronics suppliers, many of whom are on the Web. See catwhisker; coherer; Pickard, Greenleaf Whittier; piezoelectric.

crystal microphone An early type of microphone employing a piezoelectric crystal.

crystal pickup A particular type of stylus on an instrument such as a phonograph, created from a piezoelectric crystalline material that changes in shape and consequently generates an electrical impulse which is then interpreted by the electronics into sound.

crystal shutter A type of safety mechanism used in conjunction with crystal detectors to block excess radio frequency (RF) energy from reaching and possibly damaging the components.

CS communications satellite.

CSA 1. Callpath Services Architecture. 2. Canadian Space Agency. The CSA David Florida Laboratory is particularly known for its research into telecommunications technologies. 3. See Canadian Standards Association. 4. Center for the Study of Architecture/Archaeology. An organization devoted to the advancement of digital technologies suitable for the study of archaeology and architectural history. http://www.csanet.org/ 5. client-server architecture. 6. communication system architecture. 7. Communications Simulator and Analyser. A South African commercial product of C^2I^2 that functions as a generic interactive communications simulator and analyzer. The system supports many different protocols and media, including Ethernet and some of the newer

fiber-based systems. 8. Compliant Systems Architecture. A 1997 to 1999 EPSRC-funded project for designing, constructing, and evaluating a generic systems architecture compliant with individual persistent process applications. 9. See Comprehensive System Accounting. 10. computer system architecture.

Historic Crystal Radio Wave Detectors

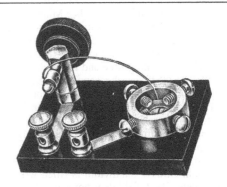

This diagram of a crystal detector, a historic radio receiver, clearly shows the catwhisker and mounting base for the crystal with which the catwhisker is normally in contact, with screws to hold the crystal in place. The mounting posts on the left are for connecting the wires for the headphones. The large knob provides fine adjustment (tuning) for the catwhisker.

This portable crystal detector, with its tuning coil receiver wound around a hollow core, is from the American Radio Museum collection.

CSC customer service center.
CSE See Communication Security Establishment.
CSI See Computer Security Institute.
CSMA See Carrier Sense Multiple Access.
CSNET See Computer+Science Network.
CSPP See Computer Systems Policy Project.
CSR 1. See cell switch router. 2. customer service record.
CST 1. See Computer Science Telecommunications 2. computer-supported telephony.
CSTA See Computer Supported Telephony Application.
CSU See Channel Service Unit.

CSU/DSU Channel Service Unit/Data Service Unit.

CSUA See Canadian Satellite Users Association.

CT 1. Call Type. 2. Cordless Telephone. 3. Conformance Test. A test to determine whether an implementation complies with the specifications of, and exhibits behaviors mandated by, a particular standard.

CT Innovation Alliance CTIA. A nonprofit trade organization of computer telephony developers and systems integrators, founded in March 1999. The Alliance was established by the Quebec Ministry of Industry and Commerce to position the province of Quebec as a center with global scope for computer telephony research and development. R&D areas include call accounting, voice and facsimile over Internet Protocol (IP), speech recognition, unified messaging, and others. http://www.ctinnovation.net/

CT3IP Channelized T3 Interface Processor. A Cisco Systems commercial fixed-configuration interface processor used with Cisco 7*xxx* series routers. The CT3IP provides 28 T1 channels for serial transmission of data, each with $n \times 56$ kbps or $n \times 64$ kbps bandwidth. Unused bandwidth is filled with idle channel data. The CT3IP does not support multiple T1 channel aggregation (bonding).

CTCA See Canadian Telecommunications Consultants Association.

CTD 1. See cell transfer delay. 2. Conditionally Toll Denied. 3. conductivity, temperature, and depth. Electronic devices that measure these variables are used in water sampling. 4. Continuity Tone Detector.

CTI 1. Call Technologies, Inc. 2. See Computer Telephony Integration. 3. See Critical Technologies Institute.

CTIA 1. See Cellular Telecommunications Industry Association. 2. Computer Technology Industry Association.

CTL See Complex Text Layout.

CTS 1. See clear to send. See RS-232. 2. Communication Transport System. 3. Conformance Testing Services.

CTSS Compatible Time-Sharing System. A developmental computer time-sharing system from the early 1960s.

CTTC coax to the curb. Coaxial cable installed into residential areas. See fiber to the home.

CTX See Centrex.

CU-SeeMe A Macintosh- and IBM-licensed PC-based videoconferencing program from Cornell University. It supports video, audio, and other utilities over Internet, with plans to make it Mbone-compatible. The encoding is proprietary. See Cameo Personal Video System, Connect 918, MacMICA, IRIS, ShareView 3000, VISIT Video.

cube 1. In geometry, a three-dimensional structure with six equally sized flat square planes all interconnected with each edge in full contact with an edge of each of the adjacent planes. Thus, the point at which the edges join forms a corner from which the cube may be referenced along three mutually perpendicular axes. A cubic structure may be real or imaginary and does not necessarily specify the size of the component planes, which may be infinite. 2. *colloq.* An

early model of NeXT computer that was shaped like a black cube, essentially a tower model, for easy access and upgrade. The shape of the NeXT later went to a more conventional thin slab, which could sit under the monitor.

cubic Embodying the geometric structure of a cube in one or more of its aspects. The simplest structure in a crystal is cubic, with atoms uniformly spaced along three mutually perpendicular axes. A regularly cubic crystalline structure is one that is isotropic, with the atoms spaced equidistant along each axis. In terms of the index of refraction of a dielectric material, most solids used in the optics industry are isotropic in any given plane (e.g., glass). However, it is possible to fabricate materials that are anisotropic (varying in density or composition) in order to control their refractive properties. See cube.

CUG closed (telephone) user group (e.g., as in ISDN Q.85 and Q.735 community of interest services).

cure *v.t.* To process so that the essential properties of a substance are changed, usually to improve them, as in curing a metal to give it strength or resilience, or curing a hide to preserve it.

curie A unit used for describing the strength of radioactivity, which is equal to 3.7×10^{10} disintegrations per second. It is named after Nobel scientists Pierre and Marie Curie, who did substantial pioneer work in radiation.

Curie point, Curie temperature A temperature at which peak levels of a dielectric constant occur in ferroelectric materials.

current (*symb. – I*) Movement of electrons through a conducting medium, usually expressed in amperes. Electricity moving through a wire or cable is current. See ampere.

current amplifier Any natural, mechanical, or electronic device that provides greater output of an electrical signal than the input signal. A public address system (PA) is a type of current amplifier, as are other microphone and speaker combinations.

current cell rate CCR. In ATM networking, a traffic flow control concept that aids in the calculation of ER and may not be changed by the network elements (NEs). CCR is set by the source to the available cell rate (ACR) when generating a forward RM-cell. See cell rate for a chart of related concepts.

curing oven A specialized heating device used in fiber optics cable assembly to cure bonding agents (e.g., epoxy) faster than by air drying, for example.

customer access line charge CALC. The charge for connecting a private branch phone exchange (PBX) to the central office exchange (Centrex).

Customer Owned And Maintained COAM. This designation is often used to describe customer-owned telephone devices, for example (a designation that became more important when the phone company no longer owned and controlled the phones within a customer's premises).

customer record information system CRIS. Also known in the general sense as CIS, customer information system, it is a computerized database for tracking customer contact, credit, and purchase information. In the last half decade, these systems have become so sophisticated, they can generate reports on customer buying frequency, purchase amounts, family relations, and much more, which concerns those wish to protect personal privacy. Stores that offer member cards for discounts are usually the ones tracking customer habits. Customers often have no input into how the information is used, with whom it is shared, how long it is kept on file, or which employees have access to the information.

cutoff decrease factor CDF. In ATM networking, CDF controls the decrease in the allowed cell rate (ACR) associated with the cell rate margin (CRM).

cutover That moment when a system is switched from one to another, as from an old system to a new one or from a primary system to a backup.

A cutover may occur if a system fails and has redundant resources in place to take over for the primary system. A ring-based network cuts over to a secondary ring, for example. A power system outage may cause a cutover to a backup generator.

It is usually desirable for the cutover to happen as quickly and uneventfully as possible, preferably so users on the system don't even notice the change or are only momentarily inconvenienced. See half tap.

cutter A mechanism for inscribing grooves in a recording medium such as a phonographic record. The mechanism is used to translate electrical impulses into physical patterns that can later be read and converted back into electrical pulses, usually auditory.

CW See Call Waiting.

CWIS See Campus Wide Information System.

CWSI CiscoWorks for Switched Internetworks. Integrated management control technology (for network topology, device configuration, traffic reporting, VLAN, ATM, and policy-based management) from Cisco Systems, Inc.

cyber- A prefix widely used with almost anything these days to indicate an electronic version of something. William Gibson is credited with popularizing the word "cyberspace" to describe an interconnected science fiction environment in *Neuromancer* in 1984. Cyber- and sometimes just cyb- have since been used in many contexts from computers to music, as in cybrarian, cyberceleb, cyberphile, cyberspace, Cyberspace, cybercast, cyberphant, cyberphobe, etc.

cybernetics A term introduced by American prodigy logician and mathematician, Norbert Wiener, who collaborated with Arturo Rosenblueth and a group of scientists from various disciplines in developing many fundamental concepts of artificial intelligence. He authored *Cybernetics: Or Control and Communication in the Animal and the Machine* in 1948 to discuss ideas about self-reproducing machines and self-organizing systems. Cybernetics refers generally to the field of control and communications theory, encompassing both human and nonhuman systems. Wiener further described feedback theory in mathematical terms and studied the flow of information from a statistical point of view. These disciplines have many practical applications in robotics.

In *Cybernetics*, Wiener poses some provocative (and revolutionary at the time) parallels between neuron states and electrical states of a binary computing device.

cyberspace 1. A term popularized by William Gibson in his popular science fiction/fantasy novel *Neuromancer* to describe a society in which the participants live in an abstracted reality, a "consensual hallucination experienced daily by billions...." See Cyberspace. 2. A content-rich virtual reality environment in which participants interact through a variety of sensory data input devices.

Cyberspace A conceptualization of the computing machinery and its associated culture as an intellectually and culturally habitable abstract space existing beyond the obvious physical role of tools and communications devices. It has been described as a meta-environment in which we can interact as part of a larger, perhaps not fully knowable, dynamic, digital organism. The Internet is seen as an important component or organizing feature of this environment. See cyberspace; Dyson, George.

Cyberspace Electronic Security Act of 1999 CESA. An act of the U.S. government that acknowledges fundamental changes in our society stemming from the development of the information superhighway and establishes a middle ground intended to enable law enforcement officials to carry out their duties while safeguarding individual freedoms and rights.

CESA describes the increasing demand for computer-provided information services and states that new risks arise as a result of their use. It contends that "Cryptography can meet these needs ... [as] an important tool in protecting the confidentiality of wire and electronic communications and stored data...." but that encryption can also "... facilitate and hide unlawful activity." The text of the Act expresses concerns that the normal tools of law enforcement for search and seizure are "... wholly insufficient when encryption is utilized to scramble the information in such a manner that law enforcement, acting pursuant to lawful authority, cannot decipher the evidence." With respect to the tools of law enforcement and their relationship to the pace of technology, the Act asserts that "Technology does not presently exist that allows law enforcement ... to decrypt such information...." In light of the strong encryption technologies that have been developed in the U.S. and abroad, there is evidence to support this last statement, especially given time constraints in particular instances such as terrorist activities.

In terms of individual freedoms and privacy, CESA states, "While means to aid investigators' and prosecutors' efforts to obtain plaintext are needed, the Act is not intended to make it unlawful for any person to use encryption in the United States for otherwise lawful purposes, regardless of the encryption algorithm selected, key length chosen, or implementation technique or medium used. Similarly this Act is not intended to require anyone to use third parties for storage of decryption keys, and this Act does not establish any regulatory regime for entities engaging in such an activity. Finally, this Act is not intended to affect export controls on cryptographic products."

These last statements are significant because they represent a departure from the tone and direction of previous proposals. They were included in large part because of criticism and debate between law enforcement supporters on the one hand and civil liberties supporters and vendors (who desire to compete in the global marketplace with strongly encrypted products) on the other.

It is an important Act and, due to space limitations, cannot be wholly included here, but the reader is encouraged to look up the CESA text and become familiar with its tenets and implications. See American Civil Liberties Union, Electronic Communications Privacy Act of 1986, Electronic Frontier Foundation, Electronic Privacy Information Center.

CyberStar A commercial global satellite communications system designed to provide broadband, interactive multimedia data transmissions. CyberStar is a venture of Loral Space & Communications. CyberStar operates on leased Ku-band transponders on the Telstar system.

CyberTelecom See Washington Internet Project.

cybrarian A compound word derived from *cyber[space]* and *librarian*. A cybrarian is a librarian/research professional setting up resources or conducting research and information retrieval online, especially on the Internet. Given the astounding volume of free information on the Net, and the difficulty of narrowing the search and finding relevant information, in essence locating the needle in the haystack, cybrarians provide valuable information, filtering, and organizational services. You can find almost any type of information on the Web, from people's names and addresses, to scientific abstracts and more, and this could easily be a full-time occupation.

cyclic fatigue Fatigue in a material from prolonged, repetitive, low-level vibration.

Glass and glass-like plastics suitable for fiber optic lightguides and ceramic ferrules in fiber optic connectors may become brittle and subject to breakage from overt stresses but are not considered highly vulnerable to cyclic fatigue in normal installations.

Tests for cyclic fatigue in optic fiber have been conducted using high-resolution micromechanical methods. Various types of matrix strains, fiber strains, and crack opening displacements (CODs) were detected by researchers in the mid-1990s. Rousseau et al. determined that the debonding of fibers "begins at the point of matrix cracking and rapidly increases."

So, while cyclic fatigue is not significant in fiber optics overall, slight existing cracks introduced from coupling stresses, installation strain, excessive bending, etc., could be exacerbated by the additional stress of cyclic fatigue. Also, the materials supporting fiber optics structures may be subject to cyclic fatigue, even if the inner core is reasonably resistant to repetitive forces, so it is still a factor to be considered in cable installation.

In cables in general, possible sources of cyclic fatigue include road traffic vibrations, wind on aerial cables, vibrations from construction machinery, etc.

cyclic memory A type of memory that can be accessed only when the process of memory access passes through that portion in its cycle that contains the information desired.

cyclic olefin polymer (COP) A synthetic material used to coat silver hollow glass lightguides suitable for the conduction of laser-emitted infrared radiation. It has good heat resistance, transparency and electrical characteristics. The material is marketed under a number of names by different companies, including ZEON's ZEONIX, Mitsui Chemicals' Apel, and others. In Summer 2001, Goodrich Corporation, a major developer of the material, sold its Electronic Materials Division to Sumitomo Bakelite Co., Ltd.

cyclic redundancy check CRC. A file integrity and data transmission error checking mechanism widely used in computing. The CRC is a calculate-and-compare system. A block of data, the total data content of a file, or a group of transferred data can be scanned to create a numeric sum total that provides a simple representation of its contents. This total is then compared with one that is computed in the same way after file compression, manipulation, or transfer. The computed values are evaluated. If they match, an error *may* have occurred, but with a low probability of likelihood. If they don't match, an error is likely, and the data transfer or manipulation may be repeated.

File transfer programs such as ZModem often included CRC methods to monitor data transfer. Some file compression formats, such as PNG, are divided into logical data chunks, with each chunk incorporating a CRC to provide a reference for data integrity so the file can be checked without opening the image in a viewer. See checksum, magic signature.

cyclic shift A system in which anything that comes out one end of the production line, register, stack, or other physical or logical conveyance goes back in at the beginning of the system in a continuous loop. Programmers sometimes use cyclic shifts to rearrange data.

cyclotron A device for devising nuclear manipulations that follow a helical path. There are many makes and models of cyclotrons, ranging also greatly in size. A common, basic configuration is an evacuated contained space in which charged atomic particles are guided and accelerated through a spiral path by various magnetic means. The centripetal path of the particles can be used to effect radiant emissions.

cypherpunk An individual advocating the prevention of tyranny through public access and widespread dissemination of electronic cyphers, encryption methods, and other digital security technologies, in order to ensure that their power and accessibility are not concentrated in the hands of only a few people or organizations. See Pretty Good Privacy; Zimmermann, Philip.

Czochralski technique A means of creating crystalline structures that are useful in semiconductor technologies. By passing the materials through a molten state, large single crystals can be grown. Drawing crystals from a melt is known as "crystal pulling."

δ, Δ *symb.* The third letter of the Greek alphabet, used in mathematics to symbolize *a difference in* or *change of/in*. For example, $a = \Delta v$ would be interpreted as acceleration equals *a change in* velocity.

d *symb.* deci-, a prefix used to denote one tenth, or 10^{-1}, as in *deci*meter, *deci*bel.

D 1. *symb.* dual polarization (ITU). 2. *symb.* electrostatic flux density. See flux.

D bank, digital bank, digital channel bank In data communications, a multiplexer combining analog signals from a number of low-rate data channels into a single high-rate digital signal at the transmitter or converting digital to analog at the receiving end.

D banks are commonly used in time-division multiplexing (TDM). In North American pulse-code modulation (PCM) systems, 24 channels are combined through TDM for transmission over a single line operating at DS1 rates. See D channel, time division multiplexing. See DS- for a fuller explanation and a chart.

D bit In an X.25 network, the D bit is a binary indicator at the beginning of a data packet, immediately following the Q bit, that signals successful packet delivery. Following the Q and D bits are a number of protocol fields including the piggyback and sequence fields used for flow control. The D bit indicates the meaning of the data in the piggyback field. If D is set to zero (0), the local data communications equipment (DCE) received the packet but it isn't indicated whether the remote data terminal equipment (DTE) received the packet. If D is set to one (1), it indicates that the remote data terminal equipment (DTE) has received the packet. See Q bit.

D Block A Federal Communications Commission (FCC) designation for a Personal Communications Services (PCS) license granted to a telephone company serving a Major Trading Area (MTA). This license grants permission to operate in a 10-Mhz block at certain FCC-specified frequencies. See A Block for a chart of frequencies.

D channel delta channel, data channel. A channel used for managing network connections. The D channel is the administrative channel for ISDN with Signaling System 7 (SS7). In ISDN, the D channel is a full-duplex channel that carries control and signaling information. It handles various call setup and teardown functions (e.g., establishment and handling of B channels) and signals subscriber service information such as Caller ID. The D (delta) channel controls the B (bearer) channel at 16,000 bps for Basic Rate Interface (BRI) or 64,000 bps for Primary Rate Interface (PRI) as shown in the ISDN Channel Functions chart. One D channel is typically associated with two or more B channels.

The D channel utilizes three layers and associated protocols for implementing communications. (Note that there are some differences in Layer 3 D channel implementations between North American and European versions of ISDN.)

D channel software monitors will generally show D channel statistics in terms of frames sent and received along with a number of diagnostic and error codes including overruns, timeouts, residue, buffer status, call attempts, etc. Buffering may be available for analyzing D channel characteristics remotely or at a later time. Filtering may be available for limiting services

ISDN Channel Functions		
Abbreviation	Service	Notes
BRI	Basic Rate Interface	16 kilobytes/second using DSSI to control the two B channels and/or the X.25 format user data.
PRI	Primary Rate Interface	64 kilobytes/second using DSSI to control all the B channels. In conjunction with NFAS, the D channel can also control B channels on multiple PRIs.

or for detecting and preventing unauthorized manipulation of D channel communications. Depending upon the product, the monitoring software may be linked into the system at the terminal (T) interface to monitor both forward and backward signal paths.

In Link Access Procedure (LAP), the recommended protocol for a circuit-switched network, the D channel carries the signaling information. See B channel, D bank, D channel stack, ISDN, protocol analyzer, Q Series Recommendations.

D channel stack In ISDN, a signaling stack that provides access to the resources of the D channel at the S/T reference point. The stack implements the three lower layers of the OSI reference model. Vendors such as Texas Instruments have developed D channel stack products as self-contained firmware designed to cooperate with the host-based Common ISDN API (CAPI). See D Channel.

D connector A standard cable connector that is housed in a shell that resembles the letter D. The elongated D-shape is a naturally keyed shape as the curved edges on one side provide a guide to the correct orientation for connecting the cable. D connectors, especially 9-, 15-, and 25-pin are commonly used to interconnect computers and peripheral devices such as printers, monitors, and modems. Most A/B switchboxes for computer data connections use DB-25 inputs and outputs. See DB-25.

D link In Signaling System 7 (SS7) networks, the D link is a diagonal link that connects a secondary Signal Transfer Point (STP) pair to a primary STP pair in a quad-link configuration. Thus, a local STP pair may be linked to a network gateway STP pair. Sometimes called B/D links since B links (bridge links) are a more generic designation for interconnecting STPs.

D region A region (as opposed to a layer) of the Earth's ionosphere that exists only in the daytime, starting from around 70 or 80 kilometers above the Earth's surface and extending up to and overlapping the E region, which is more clearly defined. The D region can have a significant impact on the propagation of radio waves, causing greater dissipation and attenuation when the region is active in the daytime. See ionospheric sublayers for a chart.

D Series Recommendations A series of recommended guidelines for general telecommunications tariff principles that may be purchased from the ITU-T. Since ITU-T specifications and recommendations are widely followed by vendors in the telecommunications industry, those wanting to maximize interoperability with other systems should be aware of the information disseminated by the ITU-T. A full list of general categories is in the Appendix and specific series topics are listed under individual entries in this dictionary, e.g., B Series Recommendations, C Series Recommendations, etc. See ITU-T D Series Recommendations chart.

D- conditioning As specified by the ITU for D-1 and D-2 transmission lines, a means to handle 9.6 Kbps operations by improving the signal-to-noise ratio and nonlinear (harmonic) distortion to within certain specified limits. For example, for D-1 conditioning, the S/N ratio parameter may be defined by a vendor as 28 decibels with the signal-to-modulation ratio parameters defined as 35 or 40 decibels, depending upon the order of the intermodulation distortion. D-conditioning varies with the type of service being offered over the D channel that is being conditioned. Not all conditioning is suitable for all types of services. For example, a D-conditioned line may not be optimal for voice communications.

D-1, D-2, D-3, etc. See digital video format for a chart of standardized formats.

D-scope, D-scan A type of C-scope radar display in which the target blips extend vertically to provide an estimate of distance. See C-scope.

D-VHS A digital version of VHS recording/playback technology developed by JVC. Since digital technologies are so easily transmitted to or from computer applications and over Internet links, formats like D-VHS may become important for Web-based communications. Digital variants are also important in video editing as no data is lost in the editing as in analog formats. Many digital video technologies are transitional. They use digital imaging but often store the information on tapes rather than disks or magnetic cards. They may also play older format analog tapes. See SVHS.

DAMPS, D-AMPS Digital Advanced Mobile Phone Service. Descended from AMPS, DAMPS was introduced as a 900 MHz frequency modulation (FM) second-generation mobile phone transmission technology in the early 1990s. DAMPS was capable of carrying three digital channels, compared to one analog AMPS channel, in a 30-kHz analog slot. Modulation is through a 2-bits-per-symbol nonbinary modem.

In DAMPS, bandwidth is allocated according to frequency division multiple access (FDMA) schemes. The two most prevalent means of dividing frequencies in DAMPS are time division multiple access (TDMA) and code division multiple access (CDMA). To increase capacity and security in mobile communications, in general, many cellular systems are being converted from AMPS to digital AMPS (DAMPS). Some legacy systems supporting AMPS, which is based upon frequency division multiple access (FDMA), have a dual mode for selecting multiple access (the formats are not directly compatible). See AMPS, North American Digital Cellular.

D'Arsonval, Jacques Arsène (1851–1940) A French physicist who proposed using the thermal properties of the ocean to generate energy. He was unable to achieve a net gain in power generation. Succeeding generations of researchers continued to pursue this idea with better success through experimentation.

D'Arsonval introduced the first reflecting, moving coil galvanometer in 1882, an improvement on previous arrangements. By means of a small, concave mirror mounted on the coil, the instrument could reflect a beam of light to a calibrated scale. It could measure the current and voltage of direct currents and was widely distributed in many forms. A number of electrical concepts and inventions are named after him. See galvanometer, D'Arsonval.

ITU-T D Series Recommendations	
Number	Description
D.000	Terms and definitions for the Series-D Recommendations
D.1	General principles for the lease of international (continental and intercontinental) private telecommunication circuits and networks
D.3	Principles for the lease of analog international circuits for private service
D.4	Special conditions for the lease of international (continental and intercontinental) sound- and television-program circuits for private service
D.5	Costs and value of services rendered as factors in the fixing of rates
D.7	Concept and implementation of "one-stop shopping" for international private leased telecommunication circuits
D.8	Special conditions for the lease of international end-to-end digital circuits for private service
D.9	Private leasing of transmitters or receivers
D.10	General tariff principles for international public data communication services
D.11	Special tariff principles for international packet-switched public data communication services by means of the virtual call facility
D.12	Measurement unit for charging by volume in the international packet-switched data communication service
D.13	Guiding principles to govern the apportionment of accounting rates in international packet-switched public data communication relations
D.15	General charging and accounting principles for nonvoice services provided by interworking between public data networks
D.20	Special tariff principles for the international circuit-switched public data communication services
D.21	Special tariff principles for short transaction transmissions on the international packet-switched public data networks using the fast select facility with restriction
D.30	Implementation of reverse charging on international public data communication services
D.35	General charging principles in the international public message handling services and associated applications
D.36	General accounting principles applicable to message handling services and associated applications
D.37	Accounting and settlement principles applicable to the provision of public directory services between interconnected Directory Management Domains
D.40	General tariff principles applicable to telegrams exchanged in the international public telegram service
D.41	Introduction of accounting rates by zones in the international public telegram service
D.42	Accounting in the international public telegram service
D.43	Partial and total refund of charges in the international public telegram service
D.45	Charging and accounting principles for the international telemessage service
D.50	Tariff and international accounting principles for the international teletex service
D.60	Guiding principles to govern the apportionment of accounting rates in intercontinental telex relations
D.61	Charging and accounting provisions relating to the measurement of the chargeable duration of a telex call
D.65	General charging and accounting principles in the international telex service for multi-address messages via store-and-forward units
D.67	Charging and accounting in the international telex service
D.70	General tariff principles for the international public facsimile service between public bureaus (bureaufax service)
D.71	General tariff principles for the public facsimile service between subscriber stations (telefax service)

D

	ITU-T D Series Recommendations, cont.
Number	Description
D.73	General tariff and international accounting principles for interworking between the international bureaufax and telefax services
D.79	Charging and accounting principles for the international videotex service
D.80	Accounting and refunds for phototelegrams
D.81	Accounting and refunds for private phototelegraph calls
D.83	Rates for phototelegrams and private phototelegraph calls
D.85	Charging for international phototelegraph calls to multiple destinations
D.90	Charging, billing, international accounting and settlement in the maritime mobile service
D.91	Transmission in encoded form of maritime telecommunications accounting information
D.93	Charging and accounting in the international land mobile telephone service (provided via cellular radio systems)
D.94	Charging, billing, and accounting principles for international aeronautical mobile service and international aeronautical mobile-satellite service
D.95	Charging, billing, accounting, and refunds in the data messaging land/maritime mobile-satellite service
D.96	Charging, billing, accounting, and settlement principles for Global Mobile Personal Communications by Satellite (GMPCS) for the international telephone service
D.98	Charging and accounting provisions relating to the transferred account telegraph and telematic services
D.100	Charging for international calls in manual or semi-automatic operating
D.103	Charging in automatic service for calls terminating on a recorded announcement stating the reason for the call not being completed
D.104	Charging for calls to subscriber's station connected either to the absent subscriber's service or to a device substituting a subscriber in his absence
D.105	Charging for calls from or to a public call office
D.106	Introduction of reduced rates during periods of light traffic in international telephone service
D.110	Charging and accounting for conference calls
D.115	Tariff principles and accounting for the International Freephone Service (IFS)
D.116	Charging and accounting principles relating to the home country direct telephone service
D.117	Charging and accounting principles for the international premium rate service (IPRS)
D.120	Charging and accounting principles for the international telecommunication charge card service
D.140	Accounting rate principles for international telephone services
D.150	New system for accounting in international telephony
D.151	Old system for accounting in international telephony
D.155	Guiding principles governing the apportionment of accounting rates in intercontinental telephone relations
D.160	Mode of application of the flat-rate price procedure set forth in Recommendation D.67 and Recommendation D.150 for remuneration of facilities made available to the administrations of other countries
D.170	Monthly telephone and telex accounts
D.171	Adjustments and refunds in the international telephone service
D.172	Accounting for calls circulated over international routes for which accounting rates have not been established
D.173	Defaulting subscribers
D.174	Conventional transmission of information necessary for billing and accounting regarding collect and credit card calls

ITU-T D Series Recommendations, cont.

Number	Description
D.176	Transmission in encoded form of telephone reversed charge billing and accounting information
D.177	Adjustment of charges and refunds in the international telex service
D.178	Monthly accounts for semi-automatic telephone calls (ordinary and urgent calls, with or without special facilities)
D.180	Occasional provision of circuits for international sound- and television-program transmissions
D.185	General tariff and accounting principles for international one-way point-to-multipoint satellite services
D.186	General tariff and accounting principles for international two-way multipoint telecommunication service via satellite
D.188	General charging and accounting principles applicable to an international videoconferencing service
D.190	Exchange of international traffic accounting data among administrations using electronic data interchange (EDI) techniques
D.192	Principles for charging and accounting of service telecommunications
D.193	Special tariff principles for privilege telecommunications
D.196	Clearing of international telecommunication balances of accounts
D.197	Notification of change of address(es) for accounting and settlement purposes
D.201	General principles regarding call-back practices
D.210	General charging and accounting principles for international telecommunication services provided over the Integrated Services Digital Network (ISDN)
D.211	International accounting for the use of the signal transfer point and/or signalling point for relay in Signalling System No. 7
D.212	Charging and accounting principles for the use of Signalling System No. 7
D.220	Charging and accounting principles to be applied to international circuit-mode demand bearer services provided over the Integrated Services Digital Network (ISDN)
D.224	Charging and accounting principles for ATM/B-ISDN
D.225	Charging and accounting principles to be applied to frame relay data transmission service
D.230	General charging and accounting principles for supplementary services associated with international telecommunication services provided over the Integrated Services Digital Network (ISDN)
D.231	Charging and accounting principles relating to the User-to-User Information (UUI) supplementary service
D.232	Specific tariff and accounting principles applicable to ISDN supplementary services
D.233	Charging and accounting principles to be applied to the reversed charge supplementary service
D.240	Charging and accounting principles for teleservices supported by the ISDN
D.250	General charging and accounting principles for nonvoice services provided by interworking between the ISDN and existing public data networks
D.251	General charging and accounting principles for the basic telephone service provided over the ISDN or by interconnection between the ISDN and the public switched telephone network
D.260	Charging and accounting capabilities to be applied on the ISDN
D.280	Principles for charging and billing, accounting and reimbursements for universal personal telecommunication
D.285	Guiding principles for charging and accounting for intelligent network supported services
D.286	Charging and accounting principles for the global virtual network service
D.300R	Determination of accounting rate shares in telephone relations among countries in Europe and the Mediterranean Basin

D

ITU-T D Series Recommendations, cont.	
Number	**Description**
D.301R	Determination of accounting rate shares and collection charges in telex relations among countries in Europe and the Mediterranean Basin
D.302R	Determination of the accounting rate shares and collection charges for the international public telegram service applicable to telegrams exchanged among countries in Europe and the Mediterranean Basin
D.303R	Determination of accounting rate shares and collection charges applicable by countries in Europe and the Mediterranean Basin to the occasional provision of circuits for sound- and television-program transmissions
D.305R	Remuneration for facilities used for the switched-transit handling of intercontinental telephone traffic in a country in Europe or the Mediterranean Basin
D.306R	Remuneration of public packet-switched data transmission networks among the countries of Europe and the Mediterranean Basin
D.307R	Remuneration of digital systems and channels used in telecommunication relations among the countries of Europe and the Mediterranean Basin
D.310R	Determination of rentals for the lease of international program (sound- and television-) circuits and associated control circuits for private service in relations among countries in Europe and the Mediterranean Basin
D.390R	Accounting system in the international automatic telephone service
D.400R	Accounting rates applicable to direct traffic relations in voice telephony among countries in Latin America and the Caribbean
D.401R	Accounting rates applicable to telex relations among countries in Latin America
D.500	Accounting rates applicable to telephone relations among countries in Asia and Oceania
D.500R	Accounting rates applicable to telephone relations among countries in Asia and Oceania
D.501R	Accounting rates applicable to telex relations among countries in Asia and Oceania
D.600R	Cost methodology for the TAF Group applicable to the international automatic telephone service
D.601R	Determination of accounting rate shares and collection charges in telex relations among countries in Africa
D.606R	Preferential rates in telecommunication relations among countries in Africa
D.Sup1	Cost and tariff study method
D.Sup2	Method for carrying out a cost price study by regional tariff groups
D.Sup3	Handbook on the methodology for determining costs and establishing national tariffs

D'Arsonval current A high-frequency, somewhat high-amperage, low-voltage current.

D'Arsonval galvanometer See galvanometer, D'Arsonval.

D'Arsonval movement A description used in contexts where a pointer associated with a dial moves to show a reading when stimulated by direct current.

D/A digital to analog. See D/A conversion.

D/A conversion digital to analog conversion. In the general sense, a process whereby any system or device converts a signal with discrete states (e.g., binary ones and zeroes) into a signal with theoretically infinite states (e.g., a radio frequency signal). In data transmissions, a process or device to convert discrete digital information to a continuous form for transmission over analog circuits, usually through one or more modulation processes. Thus, information from a computer can be converted by a computer modem to analog audio signals that are sent through a phone line, or digital signals from an Internet phone can be converted to analog pulses that can be heard over an analog headset. See A/D conversion, modem.

D/CLEC, D-CLEC Competitive local exchange carriers (CLECs) that specialize in data delivery services. D/CLEC services mainly arose as a competitive DSL-based option to expensive T1 services for data networking. As such, D/CLEC services have been of particular interest to small businesses.

D4 In T1 digital transmission lines, D4 is a type of channel bank. Channel banks carry out a variety of interface tasks, including time slot framing and detecting and transmitting signaling information. See SuperFrame.

DA 1. See desk accessory. 2. See destination address. 3. See Directory Agent. 4. See Directory Assistance. 5. discontinued availability.

DAA See data access arrangement.

DAB 1. See digital audio broadcasting. 2. dynamically allocatable bandwidth.

DACS See Digital Access and Cross-connect System.

daemon A computer process that lurks in the background to handle low priority or intermittent tasks, especially in Unix environments. Daemons carry out many tasks on computer networks, including low-level operating tasks, to automate some aspect of a system administrator's responsibility, and are transparent to most users. A daemon may be a continuous background process or intermittent, as needed. Daemons are useful as print spoolers, mail message managers, and general resource allocators, especially for client/server requests that are invoked irregularly.

DAF 1. See Data Administration Forum. 2. See Denver Advertising Federation. 3. See Destination Address Field.

Daguerre, Louis Jacques Mandé (1789–1851) A French artist and inventor who made significant improvements in photographic imagery technology in 1839. His early photos, called *daguerreotypes*, were impressed in silver plated onto copper. They have a very soft, low contrast quality to them, and the clarity of the image is affected by the angle at which the plate is held when viewed, due to the reflectivity of the metallic medium. They are fade-resistant, and many of the original daguerreotype images that still survive retain their images. See Talbot, Fox.

daisy chain *v.* To connect items individually, one to another in a series, usually through cable or connector hookups. Communication through a daisy chain of electronic units may be unidirectional or bidirectional. SCSI devices such as hard drives, scanners, cartridge drives, and CD-ROM drives are frequently daisy-chained to one another and to one controller on the logic board. When chaining SCSI devices, care must be taken to assure that each device has a unique ID number (usually from 0 to 7), and that the last member of the series (the one farthest from the SCSI controller) or *chain* is terminated, either with a physical connector attached to the outside, or by setting external or internal switches accordingly. Depending on the types of devices in the chain, it may be necessary for all devices to be turned on for other devices in the chain to function correctly, especially if the chain has been established for transferring electrical power as opposed to data signals. In other situations, devices may be individually turned on or off without interrupting the flow of data through the chain.

DAL See Dedicated Access Line.

Dall test A variation on the historic Foucault test for assessing the optical quality of lens components that includes a test lens placed in the path of the incident light beam. The test was described by Horace Dall in 1947. It is similar to the Foucault arrangement, but includes a lens placed between a point light source and the paraboloid surface under test to reveal any spherical aberrations opposite in sign to the surface at the radius of curvature. Thus, by determining the index of refraction of a lens that is flat on one side and convex on the other and setting the lens at a specific distance between the entry point of the light source and the reflecting mirror or lens under test, the two spherical components can be made to cancel one another out and reveal other characteristics of the reflecting component. This is a type of single reflection null test.

When a similar configuration is combined with at least two traversals of light through the lens, it is called a Ross test. The Ross test avoids some of the problems related to astigmatism. See Foucault test, Ronchi test, star test.

Dalton, Orv A prominent amateur radio enthusiast who contributed substantially to the design and construction of the first three OSCAR satellites. Callsign K6UEY. See OSCAR.

DAMA See Demand Assigned Multiple Access.

damped wave Radiant wave oscillations that gradually diminish in amplitude or that are being deliberately suppressed so that the amplitude diminishes.

damping The process of decreasing the amplitude of wave oscillations. The term is often used in reference to progressively suppressing sound waves (*sound damping*), though it can generally be used to indicate the suppression of a variety of types of energy, as electrical oscillations in a circuit.

Daniell battery A historic, fairly simple chemical battery, providing approximately 1.1 volts per cell which was suitable for providing constant, depolarizing current for early telegraphic systems, such as the Morse system. This battery tended to last longer than others in a closed circuit and was widely used during the first three decades of commercial telegraphy. The cells were usually housed in a box that was placed near a telegraphic station (often under a table or desk).

The Daniell battery was comprised of a copper electrode in a copper-sulfate solution on one side of a porous separator, and a zinc electrode in a diluted sulfuric acid or zinc-sulfate solution on the other side housed within glass or earthenware. This two-fluid battery was an important improvement over the voltaic pile. It is named after J.F. Daniell. See cell.

Daniell, John Frederic (1790–1845) An English chemist and colleague of Michael Faraday who applied his talents, in the mid-1830s, to the development of batteries that could last longer than those in use at the time. He invented a chemical battery known as the Daniell cell which was used in early telegraph systems. See Daniell battery.

DAP See Directory Access Protocol.

DAQ 1. data acquisition. 2. delivered audio quality.

DAR See digital audio radio.

dark conduction The property of a substance, such as a photosensitive material, to retain electrical conductance in darkness. It is usually a residual effect and tends to diminish over time until restimulated by light.

dark fiber, dry fiber Plain, unconnected fiber optic cable, not currently carrying a signal. Since fiber is often sold as the hardware portion of a subscriber service, this phrase was coined to indicate fiber that is sold just as fiber, with the purchaser doing the wiring of the components and transmitters. See dim fiber.

DARPA See Defense Advanced Research Projects Agency.

DARPANET A distributed network of the U.S. Defense Advanced Research Projects Agency, originating in 1969, from a desire on the part of the U.S. military to exchange information among different sites and to provide redundancy in the event of an attack. This project grew to become ARPANET by 1972. In 1983, ARPANET had grown so large that it was split into MILNET, specifically for U.S. military use, and NSFNET (National Science Foundation Network), which opened it up to researchers and scientists. See ARPANET, Internet, RFC 791.

DARS See Digital Audio Radio Service.

Darwin See Mac OS X.

DAS See Dial Access Switching.

DASD See direct access storage device.

DAT See digital audio tape.

data 1. Constituent basic elements of information that can be formally organized and combined to provide communication, most commonly through written means, though the term is not restricted to written communications. 2. Building blocks that can be manipulated and presented by electronic means, or which are used, interpreted, and organized by human perception and thinking.

data access arrangement DAA. A system for interfacing a communications device to the public switched telephone system. A modem interface is one example of a DAA that enables computers to transmit and receive data through the telephone system. A DAA may be customer supplied or carrier supplied and may require approval on the part of the common carrier providing the telephone line access.

Data Administration Forum DAF. A division of the British Columbia Advisory Council on Information Management, the DAF encourages corporate data administration and information management activities throughout the Canadian province and is active in the establishment of data administration standards.

data awareness Mechanisms for aggregating and switching network data traffic in the appropriate layer.

data base See database.

data bus See bus.

data carrier detect DCD. In telephony, DCD is a signal from the DCE to the DTE, indicating a valid signal between the DTE and DCE devices. It is typically used to set port status for a connection and to generate a signal indicating the loss of a connection. The DCE is commonly a modem or serially connected printer, and the DTE is the terminal or computer.

data circuit A circuit that uses transmission wires and components suitable for the fast transmission of digital information.

Data Communications Channel DCC. In SONET networking, a channel related to the OAM&P which includes security and performance information associated with facility and network elements (NEs). Both generic and vendor-specific information can be included. The DCC is incorporated into both the section and line overhead.

data communications equipment DCE. A category of devices specified by the Electronic Industries Association (EIA) that typically includes common serial communications peripherals, such as modems and printers. These in turn interface with data terminal equipment (DTE). In Frame Relay networking, DCE has a more specific meaning, as switching equipment that is separate from the various peripheral devices that are connected to a network or workstation. See data terminal equipment.

data compression The process of encoding data to store it in a smaller amount of space. Data compression is typically achieved with specialized software tools or with software built into data transmissions hardware. Data compression may be done in advance, if files are to be stored or transmitted later, or it may be done dynamically at the time it is needed, sometimes called *realtime* or *on-the-fly*, a capability that is built into some modems. There are many different general-purpose and specialized means of compressing data. Some data compression algorithms are paired with data decompression algorithms, for archiving search and retrieval and audio/video recording and playback.

Data compression tools do not always make the data smaller. For example, a very tiny icon file, when encoded with an image or general purpose data compression program, may actually be larger than the original by the time the header, decompression, or statistical information about the file is inserted by the compression program. Yet the same tool may quite effectively achieve as much as 60% compression on large images, so the selection of data compression technologies depends on finding the right tool for the job. To overcome this problem, three developers at Western DataCom have developed Ardire-Stratigakis-Hayduk (ASH), a compression scheme that incorporates some of the pattern-matching, and predictive concepts associated with artificial intelligence programming. This scheme attempts to broaden the scope of compression to handle many different types of data in the increasingly media-rich communications that are evolving. See Ardire-Stratigakis-Hayduk, Lempel-Ziv.

data compression approaches There are many practical approaches to data compression. One of the simplest is to remove redundant data, such as gaps, spaces, or repetitions. This approach is used in encoding voice conversations, which typically have many pauses. It is also useful for compressing graphics that have large areas of similar colors and text documents with repetition and blank spaces. Another means for compressing information is expressing it in a different way. For example, a bitmap image of a large letter *O* may require 15 kilobytes, whereas the mathematical definition for ellipses that can define the letter *O* may require only 5 kilobytes. A third way to compress data is to try to match the human perceptual recognition of the data image rather than the structural data characteristics of the original presentation. In other words, there are ways to display graphic images or to play sound files so that they look the same or sound the same to the general viewer/listener, even though the construction and dynamic range of the information may have been altered. Humans have a remarkable ability to conceptually add information or construct a view from a few clues. If you've ever watched a black and white TV show and "could have sworn" you had seen it in color, you've experienced one aspect of this phenomenon.

Data compression can be *lossy* or *lossless*, that is, it can retain most of the information in a file or all of the information in a file. A commonly used lossy image format in which most of the information is retained is JPEG, often for displaying Web graphics and videoconferencing images.

With the ever-growing volume of data unleashed by the capabilities of computer technology, and greater demands for perceptually rich multiple media, the demands for data compression to reduce file space, transmission time, and costs are very high. Some of the most promising recent data compression programs incorporate fractal and wavelet theories into their encoding techniques. See Ardire-Stratigakis-Hayduk, JPEG, Lempel-Ziv, PNG, wavelet theory.

data conversion The process of converting computer data stored in one format to another. The three most common reasons for converting data are achieving compatibility (upward, downward, and inter-application), saving space and/or saving time (compression/decompression), and needing to convert between digital and analog forms of information.

When software applications are upgraded, they often incorporate new features that are not available in the older versions. Data conversion may be necessary to store information in the older or newer file format, and some of the information may be lost in the conversion process.

Computer data conversions tend to happen within families of data. Graphics formats are frequently interchanged, audio formats are frequently interchanged, but there isn't much need to convert audio data into visual data, except for experimental applications. That is not to say computer data has to be rigidly defined; it doesn't. For example, the Interchange File Format (IFF) developed jointly in the mid-1980s by Electronic Arts and Commodore Business Machines is a broad specification for data definition that can be generically applied to text, sound, and graphics. Similarly, Adobe PostScript fonts, while following specific guidelines, are not just fonts, but rather are shapes that fit in the context of a larger picture, that of a page description language which is capable of describing many types of graphical elements besides fonts.

Many shareware and commercial data conversion utilities are available, especially for converting among the myriad graphics formats such as PNG, JPEG, Compuserve GIF (which now uses the PNG specification), BMP, ILBM, and TIFF. PNG, JPEG, and GIF are the most commonly used raster graphics formats on the Web, and TIFF is the most widely used graphics format in the publishing and document industry (faxes are also defined within the TIFF speci-

cation). ASCII is very widely used in text conversions, and Microsoft's Interchange Format (Rich Text Format or RTF) can be used for text conversions that retain formatting such as bold, indents, fonts, etc. For database information, dBASE formats are often used for converting between one program and another. See data compression, digital to analog conversion.

Data Country Code DCC. In networking, the DCC is a numeric code that specifies the country in which an address is registered for a public network. Each data country code is in Binary Coded Decimal (BCD) format, contained in two octets, in ISO 3166 format. See Data Network Identification Code.

data description In a data dictionary, a unit or group of information which may comprise one or more of the following: a definition of meaning and usage, attributes or characteristics, and category or classification information.

data dictionary 1. A reference set of data descriptions that can be machine-processed, and shared by a variety of applications. 2. In database management, a lookup reference of data descriptions with a format or relationship such that the database engine can efficiently save, extract, or scan information to/from the dictionary according to the needs of the database.

data element A basic unit of information defined generically, or for a specific application. For example, a data element in an employee database might consist of a name or job category. A data element may be further defined as including data items components or subcategories.

Data Encryption Standard DES. A cryptographic system consisting of an algorithm and a key comprised of a long series of numbers which are used together to transform data into information which appears unintelligible, and back into data by the person for whom the information is intended. DES was developed by the National Bureau of Standards (now the National Institute of Standards and Technology (NIST)) and is intended for public use and for government protection of certain federal unclassified data. See Clipper Chip, Pretty Good Privacy.

data entry The act of using a hardware interface to input data to a computing device. Data entry is commonly accomplished through a keyboard and mouse, but voice recognition systems, touch screens, and pen computers are broadening the choice of input devices. Typically, data entry is used to describe repetitive, discrete types of data, like database entries (names, addresses, order numbers, etc.), spreadsheet entries, etc. When the data is more fluid and conceptual and less repetitive, it is still, in its broadest sense, data entry but is more likely to be described in terms of the type of application being used, such as word processing.

data exchange interface DXI. A layer 2, frame-based interface installed between a packet-based router and a SMDS or ATM CSU/DSU. The DXI performs assembly and reassembly tasks on behalf of a router that may not have these capabilities. Since most routers now can handle these tasks, use of this particular type of interface is diminishing.

D

Data General Corporation DG. One of the better known computer companies in the 1970s, Data General was founded in 1968 to develop minicomputers and became a Fortune 500 firm a decade later. With almost half a million systems installed worldwide, Data General targeted high-performance computing environments, including scientific, technical, and industrial sites. By the late 1990s, DG was beginning to support server applications for Pentium II processors as well, a strong sign of the convergence of the workstation and personal computer markets. Data General became a division of EMC Corporation after an announced stock swap in 1999.

data grade circuit A distinction made to indicate the more stringent needs of computer data transmission, as compared to voice grade transmissions. Data is transmitted at different frequencies and is more precise and easily interrupted than a phone conversation. Phone conversations use a narrow frequency range and have a great tolerance for pauses, spaces, and extraneous noise, particularly since part of the processing equipment in a voice conversation is the human brain, which understands context and innuendo, as well as just the words. Data, on the other hand, requires a cleaner line, less interference and a greater frequency range and has low tolerance for pauses and spaces if they affect the integrity of the information that is being transmitted.

Voice grade circuits over phone lines are improved by load coils, a system of looping the wires that are strung along utility poles. Data grade circuits built in the same basic way are hindered by load coils, as they introduce noise at the higher frequencies used.

data line card DLC. In a digital telephone network, a link between the transmissions wire connecting the subscriber unit (e.g., modem) on one side and the digital switching matrix (DSM) on the other.

Data Link Connection Identifier DLCI. A means of assigning logical connections within a shared physical transmissions path. In Frame Relay networks, the DLCI is a unique 10-bit identification number (address) assigned to a virtual circuit (VC) endpoint that identifies the endpoint within a local access channel. It is a mechanism for keeping track of endpoint devices (e.g., routers) and is used for switching and multiplexing. When a network is configured, addresses are assigned and entry tables are created to map the DLCIs to one another for routing data over the network. If the Frame Relay is linked with a larger network, such as a global public network, the Frame Relay handles the routing of frames through the DLCIs.

A router may have multiple DLCIs. If there are multiple ports, identification numbers can be used again. If frame traffic is forwarded across the Frame Relay cloud (shared connectionless resources), the address ID number may change, as the mapping is handled by the Frame Relay switch (note that global IDs are optional Local Management Interface (LMI) extensions and their implementation is not widespread).

ISDN frames are similar to Frame Relay frames, and Frame Relay networks may be hooked into ISDN networks. Commercial products exist to interpret ISDN messages and route them through Frame Relay networks. In ISDN, the address is a unique 13-bit identifier. See Frame Relay.

Data Link Control DLC. A layer in the Open Systems Interconnection (OSI) model, DLC is responsible for a number of administrative and error-checking functions. In satellite communications, some special adaptations are needed at this level to accommodate the high bandwidth/delay characteristics of these transmissions.

data link layer DLL. In the Open Systems Interconnection (OSI) reference model, the layer that ensures transmission of data between adjacent network nodes. Bridges work at the data link layer. See Open Systems Interconnection.

Data Link Switching DLS, DLSw. Originally developed by International Business Machines (IBM), in 1993, DLS was submitted to the IETF as an informational Request for Comments (RFC). DLS defines a reliable means of transmitting SNA and NetBIOS TCP/IP traffic using IP encapsulation through multiprotocol router networks. See RFC 1795.

Data Link Switching Workgroup DLSW. The technical group that worked on the development of a new switching standard for integrating networks over TCP/IP. See Data Link Switching.

Data Link Switching Special Interest Group DLSw SIG. A vendor implementation group created in 1993 to address some of the issues raised in regard to RFC 1434 in which International Business Machines (IBM) provided preliminary information on Data Link Switching. This activity resulted in a new RFC being submitted to the IETF as RFC 1795 which obsoleted RFC 1434. See Data Link Switching.

data lump cable A data communications cable that includes integrated circuits (ICs) or other components underneath the cable shielding such that they produce a lump on the profile of the cable. These are popular for interconnecting small handheld devices to computers or computerized components for two reasons. The first is that an added battery source or conversion or translation software or hardware may be needed to make the handheld device operable with other devices. The second is that the shielding from the cable helps protect the components from short circuits from other objects that might be carried in a handbag, briefcase, or pocket.

A data lump cable may connect directly to a component or may interface with a cradle into which the component is coupled.

data mining The process of seeking out relevant information from a large storehouse of electronic data that may be on many different systems in many different formats. Data mining involves using intelligent strategies and algorithms to search for relevant materials based upon various parameters such as previous search history, data patterns, information correlations, preferences of individual users, keywords, and other triggers implemented to maximize the relevance of the information retrieved or flagged.

Data mining has become a topic of substantial interest

and development due to the vast amount of information that is flowing onto the Internet. Anyone who has used a search engine and received 300,000 *hits*, after narrowing a search a couple of times, can see the value in data mining algorithms that can carry out some of the work in advance. Data mining is of particular interest to researchers and marketing professionals and can be an important and legitimate way of searching, sifting, and sorting information for medical research, astronomy, inventing, investing, and information technology (IT) research.

Data mining on public networks has become the subject of controversy and political discussions due to the amount of personal information on the Web that is accessible to a wide audience and thus subject to misuse or abuse by individuals with questionable ethics or illegal intentions. Many children put personal information on the Web in the process of developing school projects. Much genealogical information is posted on the Web, and there is a great deal of information about people who don't even own or have access to computers, who thus cannot track or trace information that is being distributed about them by third parties. While individual pieces of information may not be harmful in themselves, data mining algorithms have been devised to develop profiles of individuals that indicate where they live, work, and shop, what they buy, how they spend their leisure time, and with whom they associate. These sophisticated profiles have never before in history been possible and there is currently very little legislation to protect individuals from their misuse by thieves, stalkers, or manipulative political or marketing bodies operating outside ethical constraints. See American Civil Liberties Union, Ask Jeeves, Electronic Frontier Foundation, cybrarian.

data multiplexing See multiplexing.

Data Network Identification Code DNIC. An ITU-T internationally specified system of network host identification that permits individual local networks, tied to public networks, to be located and recognized for internetwork communication, much as a country code and local phone number identifies a phone line. This data network identification scheme, somewhat analogous to a phone number, is used to locate hosts on interconnected public networks by means of X.75. The DNIC is the first four digits of a longer 14-digit international code. The first three digits are assigned by the ITU-T to specify a data country code (DCC) and the fourth digit is assigned by the national administration to specify the public data network within that country. Network Terminal Numbers (NTNs) are the responsibility of the administrators of the public network. See X Series Recommendations.

Data Numbering Plan Area DNPA. An ITU-T X.25-specified system of endpoint terminal identification implemented in the U.S. using the first three digits of a 10-digit network terminal number (NTN).

Data Over Cable Service Interface Specification DOCSIS MCNS/DOCSIS. An interoperable cable service delivery specification developed jointly by the Multimedia Cable Network System partners (Cox,

Comcast, TCI, Time Warner) along with CableLabs, MediaOne, and Rogers Cable. Numerous other vendors have contributed to the DOCSIS specifications process. The standards facilitate the provision of cable services through intercompatible hardware and business systems. Thus, DOCSIS provides an industry standard for cable Internet access and possible future services. The Motorola SURFboard cable modems installed by AT&T for their cable modem data services are DOCSIS-compliant, for example.

DOCSIS 1.0 supports the deployment of high-speed data services through various standards and protocols. An update incorporating Quality of Service (QoS) extensions for improved security and realtime delivery was released as DOCSIS 1.1. The DOCSIS descriptions also encompass existing ratified and de facto standards in the multimedia industry, including some of the ITU-T Series Recommendations, IEEE 802.*x*, MPEG-2 transport (downstream framing), and DES encryption schemes.

DOCSIS is an asymmetric specification, supporting data rates of 27 or 36 Mbps in the downstream direction and 320 Kbps to 10 Mbps in the upstream direction. Modulation is through QAM (downstream) and QPSK and QAM (upstream). Since USB and IEEE 1394 (Firewire) computer interfaces have become popular and are much faster than traditional serial ports, USB and Firewire 10Base-T subscriber interfaces are planned.

In DOCSIS 1.0, Baseline Privacy is a scheme for encrypting user data using the Cipher Block Chaining mode of DES with a 56-bit encryption key (it may be less for international locations). Regular key changes are specified for management purposes. RSA public and private key pairs are installed into cable modems in the manufacturing process. Data is encrypted only for transmission on the cable network and is not intended for high security in other environments. Authentication is not directly supported by DOCSIS 1.0 specifications. DOCSIS 1.1 extends Baseline Privacy with longer RSA keys (1024 bits) and the association of a digital certificate with each cable modem for authentication. The public and private key pair and digital certificate are installed in the cable modem during manufacture.

To aid implementors, Kinetic Strategies has prepared a research report, *DOCSIS Infrastructure Deployment Forecast: The North American Market for Standards-Based Cable Modem Products and Services 2000-20004* which is available for a fee. See Multimedia Cable Network System.

data over voice A means of including data on a transmissions line carrying voice signals by using frequency division multiplexing (FDM) to secure and utilize the remaining available bandwidth for the data signals. Thus, for example, a telephony device can be equipped with readouts for data about the call.

data processing A broad category of activities encompassing the manipulation of digital data, as in word processors, spreadsheets, paint programs, etc. In database systems, data processing has a more specific meaning, referring to the creation, access, retrieval, manipulation, and analysis of textual and financial data. Data processing is commonly used in payroll accounting, statistical analysis and reporting, customer profiling, and many other common tasks related to commerce and business management.

data protection A broad category of actions and systems that are designed to protect data. There are two general categories of data protection: keeping the data available and in its desired form (uncorrupted), and keeping the data safe from unauthorized use.

In the first category, data backups, archiving, mirroring, and other means are used to protect data from being lost or corrupted. This can occur at the local applications level, in the hardware, and at the overall systems level. Many file system directories are duplicated to provide access if corruption occurs in one. Some operating systems allow multiple versions of a file to be saved automatically, so that there is always a history of recent changes and a previous version that can be used if needed. Backup hardware in the form of tapes, cartridges, optical media, and redundant drives are used by many for scheduled or dynamic backups.

In the second category, passwords, digital encoding/ encryption, secure channels, proprietary formats, vaults, safe-deposit boxes, data certificates, digital signatures, etc., are all used to protect the data from unauthorized access, use, or abuse. See backup, backup file, encryption, mirroring, Pretty Good Privacy, RAID.

data rate A quantification of the input or transmission of computer data. Data rates are very situation specific. For example, in data entry jobs, the data rate may be the number of fields filled per minute or the number of customer orders entered per hour. In network communications, it may be the number of bits or packets transmitted per second. See baud rate.

data service unit DSU. A device used in ISDN systems to interconnect computers with digital phone services for end-to-end digital communications. It is similar to a modem in the sense that it fits between the computer and phone line service, but it differs in that it does not perform analog to digital and digital to analog conversions. The DSU is installed in the customer's premises and connects the synchronous communications system through a four-wire line (usually a leased line) to the local central office. The DSU is used in conjunction with a Channel Service Unit (CSU) which is installed at the central office.

data set ready DSR. A control signal commonly used in serial network communications and included in the pinout specifications for the ubiquitous RS-232 electrical connector. The DSR indicates whether the communications device is connected and ready to begin handshaking. For example, assume the user has dialed a BBS or Internet Access Provider (IAP), and the called modem has just picked up the line. The DSR senses the connection and provides a signal that lets the hardware/software know that it can continue to the next step of negotiating a connect speed and beginning the communications. See data terminal ready, RS-232.

data striping A means of distributing data across drives in an array. A fault tolerant means of providing data security that is incorporated into *redundant array of inexpensive disks* (RAID) systems.

data terminal equipment DTE. A communications data terminal hardware specification. See data communications equipment.

data terminal ready DTR. A control signal commonly used in serial communications and included in the pinout specifications for the ubiquitous RS-232 electrical connector. The DTR signals whether the communications device is connected and ready after it has successfully begun handshaking. For example, assume the user has connected with a BBS. The DSR verifies the connection, a connect speed is negotiated, handshaking begins, and the terminal is ready to continue communicating. The DTR signals this state of readiness. DTR serves an output for DTE devices and an input for DCE devices. See data set ready, RS-232.

data typing The process of specifying or determining the format of a variable, file, or block of data.

data unit DU. A generic term for any modular or connecting unit in a data path, although the abbreviation is most commonly applied to small, limited-function digital or digital/analog units as opposed to more complex systems. Examples include computer modems and passive switchers.

data warehousing A term used primarily by large corporations with very large databases until hard drives became bigger and less expensive and the Web made huge databases easily accessible through the Internet. Data warehousing is a system of information storage and retrieval for vast databases that are comprised of smaller storehouses of databases. Large databases present unique logistical and programming challenges. Many database storage and retrieval systems are limited in the number of records or files they can handle; in other words, many of them are not *scalable*. Consequently, various new strategies for data warehousing are being developed.

The Internet has made accessibility, through a local Internet Access Provider (IAP), so easy and inexpensive that companies are demanding increased access to databases not only online, but at branch offices in other states and countries. Thus, data warehousing through Web browsers is developing, even though the suitability of HTML to this task is somewhat limited. With one of the Web-friendly programming environments, such as Sun's Java, the job becomes easier, but the logistical demands of taking geographically divergent databases that may be in a variety of formats to meet local needs, and accessing them all as a conceptual unit over the Net, is an ongoing programming challenge that will probably continue to evolve for some time. Efforts to promote open systems and object-oriented programming strategies may contribute to streamlining the process of data warehousing. See CORBA, Open Systems Interconnect.

database Any collection of data organized in some form for storage or for storage and retrieval. A database can be as simple as a list of names or as complex as a relational, distributed, multisite archive of integrated images, ideas, text, facilities, actions, and processes.

A file system hierarchy on a computer storage device is a type of database, as is an employee file that includes pictures, birthdates, addresses, and social security numbers.

Database creation and management programs typically have text-based interfaces, graphics interfaces, or both. With text-based interfaces, information is organized into lines and fields and usually listed sequentially from top to bottom. In more flexible graphical databases, a *screen mask* or input template can be created almost as though using a paint program to draw the input screen. Lines, boxes, colors, and other visual elements can be used to make the database appealing and its functions and input actions apparent to the user. Fields are then assigned to the graphical elements, and the order of input is defined. More sophisticated databases include scripting or symbolic programming languages to allow automation of the database so error messages, prompts, help windows, and other applications elements can be presented when appropriate.

There are many ways to store data in a database: compressed or uncompressed, encrypted, encoded, or plain ASCII. The format of the data isn't usually what creates compatibility problems. More often the organization of the data, which can vary widely, is the hurdle that must be overcome when interchanging data among applications or systems. See data warehousing, expert system.

database reports Charts, graphs, lists, and other statistical reports that can be selected or computed from information in a database or more than one database. Reports are widely used for financial statements, business plans, demographics, research, etc.

database server A system, computer, or application specifically designed to provide database capabilities, security, and file access to multiple users on a system. There are two aspects to a database usually incorporated into a client/server model: the application that generates, searches, and retrieves the data and the data itself. Sometimes the data is on the server, and the application is on the individual user's machine. Sometimes it's the other way around, and sometimes all aspects of the database system are handled by the server. It depends on the sophistication of the system and the needs of the users. In high security situations, the server usually handles everything. In smaller networks, where security is less of an issue, the applications may be installed on individual machines to run faster, while the data is banked on the server, with password access, number of user restrictions, etc., centrally handled by the server software.

datagram This term is used in a general sense to mean a *unit of information* in a packet-switched network without regard to previous or following packets. Depending upon the network architecture, it may also have a defined format within that system. In layered architectures, the datagram may be associated with a specific layer or layers. A datagram may be encapsulated and subsequently decapsulated at the receiving end, for example when tunnelling through different systems. Internet Protocol (IP) datagram transmission over connectionless X.25-based public networks has been defined by a variety of organizations. Source and destination information are typically encoded in association with the datagram. See Point-to-Point Protocol, RFC 877, RFC 998.

DataSPAN Frame Relay Service A Nortel Frame Relay digital telephony service aimed at users with virtual private line networks (corporations, educational institutions, etc.) and high-speed interconnections. The DataSPAN service provides bandwidth on demand through space-saving multiple virtual circuits for each access port that connects to the user's equipment. DataSPAN services can transmit over permanent connections at DS-0, DS-1 (fractional or full) rates. It can support switched access from circuit-switched services at DS-0 rates. See Dialable Wideband Service.

date and time stamp A common function of computer applications and operating systems (OSs) that records when some event occurred. For example, files are usually date- and time-stamped as to the time of their creation or the time they were last updated (or both). Entries to databases are frequently date- and time-stamped, as are computerized physical premises access systems, electronic timecards, and many more. The only problem with date and time systems on computers is that not all computers take the time or date from a reliable source. Some have lithium-battery powered realtime clocks, but many do not. It may be up to the user to set the date and time manually, and a power outage can change the settings.

To improve time-stamping, many operating systems now have a utility that can automatically access a computer network and take the time from a reliable source on a network. This is useful for synchronizing with time-sensitive sites (auctions, stock exchanges) and for general file management.

Datapath Loop Extension, Datapath Extension DPX. In telephony networks, a commercial loop extension capability and card manufactured and licensed by Nortel. Datapath Services enable digital full-duplex synchronous and asynchronous fast transmission rates over standard twisted-pair wiring. The importance of Datapath Services is that they allow smaller, local exchange carriers (LECs) to provide services previously accessible only to bypass and interexchange carriers (IECs). Thus, high-bandwidth telephony applications such as multimedia services (e.g., videoconferencing) can be made more readily available to a broader range of subscribers.

The DPX may terminate either a channel bank or digital loop carrier, forming a link between the subscriber's connecting device and DS-0 channels on DS-1 network facilities on a digital public switched telephone network (PSTN), now also known as a public switched digital service (PSDS). The device that connects the user's computer to the wire loop and the DPX unit may be a modem or other similar data unit, a cluster control unit, or a terminal

D

interface unit (e.g., for ISDN connections). Depending on the wire gauge, the distance between the data unit and the DPX may be up to about 18,000 feet. In some situations, the DPX may replace a DLC in a channel bank.

Datum Corporation Manufacturers of time and frequency technologies used in computer networks, land-based wireline communications, and wireless satellite and cellular communications. Datum is a world leader in the manufacture of cesium atomic clocks, supplying Europe and most of the American Global Positioning System (GPS).

daughterboard, daughtercard *jargon* A printed circuit board that piggybacks onto a motherboard (which contains the main processing circuitry) in an electronic system. The daughterboard is frequently, though not necessarily, smaller than the motherboard and usually adds some specific type of functionality: more memory, acceleration to the main CPU, a device interface, etc. A fatherboard has further been described as a connection to a motherboard that provides a series of connectors, into which several daughterboards can be connected.

DAV See digital audio video.

DAVIC See Digital Audio-Video Council.

Davies, Donald (1924–2000) A researcher at the British National Physical laboratory, Davies developed concepts for distributed digital networks contemporaneously with Paul Baran in the mid-1960s. It is believed by Baran that each conceived the idea independently. This important new mode of data transmission became known as packet-switching and was incorporated into the ARPANET which evolved into the Internet. See Baran, Paul.

Davisson, Clinton Joseph (1881–1958) An American physicist, Bell Laboratories researcher, and winner of the Nobel Prize in physics in 1937 along with George Paget Thomson, Davisson studied electron diffraction and theories of electron optics. During World War II, he researched crystal physics and the theory of electronic devices.

Davy, Edward (1806–1885) An English physician who invented the electromagnetic repeater (electric renewer) and created one of the early copper wire-based telegraph systems in the mid-1830s. He demonstrated a needle telegraph in 1837 and received a telegraph patent in December 1837, half a year after Wheatstone and Cooke received their telegraphic patent. Later, after emigrating to Australia, he developed a process for refining copper. Thus, along with Morse, Cooke, and Wheatstone, he was one of the pioneers of telegraphic equipment and perhaps the first to develop a telegraphic relay. See telegraph history.

Davy, Humphry (1778–1829) An English scientist and educator who passed a current through potash in 1807, decomposing it and discovering a new element (potassium). Davy subsequently discovered more elements and clarified that some substances considered elements actually were not. He also proposed a theory of electrolysis. In the early 1800s, he observed the properties of carbon when connected to an electrical

source, and in 1808 he invented the arc lamp by connecting the terminals of a voltaic cell to a piece of charcoal, resulting in a brilliant light now known as an arc light, or electric arc. Michael Faraday became his laboratory assistant in 1813.

dB *abbrev.* See decibel.

DB 1. data bus. 2. See database.

DB-9 A common designation for a 9-pin D-shaped computer connector, used on many laptops and desktop computers, especially for serial connections through an RS-232 cable. DB-9 simply describes the physical connecting portion and does not define the electrical relationships of the pins to the wires in the cable to which the connector attaches. RS-232, on the other hand, defines specific pinouts and pathways for various types of signal and information data.

DB-15 A common designation for a 15-pin D-shaped computer connector most often used for monitor cables and Ethernet transceivers. DB-15 simply describes the physical connecting portion and does not define the electrical relationships of the pins to the wires in the cable to which the connector attaches.

DB-25 A designation for a 25-pin D connecter very widely used for computer data transfer, especially serial cables, and one end of many parallel and SCSI cables. DB-25 describes the physical connecting portion and does not define the electrical relationships of the pins to the wires in the cable connector. Many of the common, inexpensive A/B switchboxes are installed with DB-25 female connectors.

D-Style Connector – 25-Pin

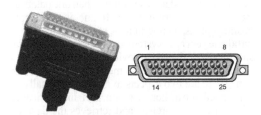

25-pin male D connector used primarily for serial and SCSI data communications.

DBD 1. database drivers. 2. See digital bearing discriminator.

DBMS See Database Management System.

DBS See direct broadcast satellite.

DBT 1. database thang. A tongue-in-cheek name for a Berkeley database data structure. 2. defect-based testing. 3. Deutsche Bundespost Telecom. 4. DBT. A versatile, extendible MATLAB Toolbox for radar signal processing suitable for signal intelligence and communications applications, developed by the Swedish Defence Research Establishment. DBT supports modeling of array antennas and simulation of many situations associated with these antennas.

DC 1. In telephone communications, Delayed Call. 2. See direct current. 3. disconnect conform. In Open Systems Interconnection (OSI), a transport protocol data unit. See Open Systems Interconnect.

DCA 1. Defense Communications Agency. A U.S. Government agency involved in military standards development. 2. Document Content Architecture. An International Business Machines (IBM) system of specifying a document series from draft to final document. 3. Dynamic Channels Allocation. A wireless concept that is used in DECT PCS services.

DCC 1. See data communications channel. 2. See Data Country Code. 3. digital command center. 4. See Digital Command Control Standards. 5. digital communications center. 6. See Digital Communications Conference. 7. See Distributed Computing & Communications Laboratory.

dcd The name of a freely distributable (under the GNU General Public License) Linux CD player, developed by David E. Smith, that allows CDs to be played on an appropriately configured computer. It provides a means to enter playlists, etc.

DCD 1. See Data Carrier Detect. 2. See Document Content Description.

DCE 1. See data communications equipment. 2. digital communications media. 3. See Distributed Computing Environment.

DCM 1. See digital switching matrix. 2. See dynamically controllable magnetic.

DCP 1. Department of Consumer Protection. 2. See Digital Communications Protocol.

DCS 1. digital communications system. 2. digital cross-connect system. 3. distributed computing system.

DCT 1. digital carrier trunk. 2. See discrete cosine transform.

DCTI See desktop computer telephony integration.

DDB 1. device-dependent bitmap. 2. digital databank. See data warehousing.

DDCMP See Digital Data Communications Message Protocol.

DDD See Direct Distance Dialing.

DDE See dynamic data exchange.

DDN See Defense Data Network.

DDOS distributed Denial of Service. See Denial of Service.

DDS 1. digital data service. 2. digital data storage. 3. distributed data system.

DE See Discard Eligibility.

de Coulomb, Charles A. See Coulomb, Charles A.

de facto standard A format, specification, or design, usually from a self-interested commercial source, that has become widespread. This often confers a large degree of industry control to the major stakeholders. Occasionally de facto standards are good, if a public-service body hasn't provided a standard, and if the standard brings down the cost of goods to make them more widely available to the general public. Sometimes de facto standards are bad, since the specification or products themselves may be of poor quality and may only have become widespread through aggressive advertising, consumer trust, or lack of understanding of the technology. By the time the consumer learning curve catches up with the technology, the standard may be too entrenched to change.

Very frequently, the cheapest product becomes the de facto standard, not because it is a good product, based on good design principles, but because it's more affordable. Sometimes the most convenient product becomes the de facto standard. If a product is easier to use, or more portable, it may outsell more flexible or powerful designs.

Sometimes the first product to hit the market becomes the de facto standard, and manufacturers are sometimes tripping over one another in the race to be the first. This scenario has occurred many times in the modem industry. The first to get the new, faster modem on the shelves often had a big say in specifying the format for communications and, consequently, a short-term de facto monopoly.

de Ferranti, Sebastian (1864–1930) An inventor who collaborated with Elihu Thomson and William Stanley in the development of the transformer. He was also responsible for developing the first high voltage alternating current (AC) distribution system at a time when direct current (DC) distribution systems were prevalent.

de Forest, Lee (1873–1961) A highly ambitious American inventor who harnessed the power of electrons by inventing the Audion vacuum tube, which was granted a patent in January 1907 (U.S. #841,387). This very significant technology has been used in electronics in many industries for decades, though eventually transistors superseded vacuum tubes, except for some specialized high frequency applications. Although he was loathe to acknowledge his predecessors, de Forest's invention stemmed from the work of T. Edison and J.A. Fleming. However, he is to be credited with the introduction of the electron tube *grid* unit, creating a *triode*, which was a significant advance over the design of the Fleming tube.

Lee de Forest's three-electrode vacuum tube made transcontinental communication possible, and the proliferation of vacuum tubes for radio wave detection caused the decline of crystal detector radio sets. In the 1920s, de Forest contested E. Armstrong for the invention of regeneration and won. (Lee was born *"De Forest,"* but later in life is said to have preferred *"de Forest."* His wireless company was spelled "DeForest.") See Audion, DeForest Wireless Telegraph Company, Edison effect, transistor.

de-encapsulation An important aspect of packet-switched networking in which *user data field* data in an encapsulated packet is extracted upon receipt.

dead band In guidance systems, a means of introducing hysteresis by preventing errors from being corrected until they have exceeded a certain specified magnitude, a safety precaution against the guided object reacting prematurely to interference or spurious signals.

Dead Link Check DLC. A Perl program to check HTTP links, with reporting capabilities, DLC was released in 1999 by Martial Michel. Since 'dead' links, those that no longer point to a valid address on the World Wide Web, can be a problem for users of a site, a utility that alerts the Webmaster to problems is a welcome tool. It can be used by itself or as an extension to Public Bookmark Generator.

dead spot A phrase often used in broadcast communications (radio, TV, cellular) to describe a region in which there are no signals, due to terrain or other obstructions. Cellular customers in particular are susceptible to dead spots, because they may be constantly moving between buildings, boulders, mountains, etc.

deadlock In competition for resources such as printers on a network, a route on the Internet, or space for a car on a road, deadlock occurs when there is more demand than there is supply and the contention is unresolved. Contention is commonplace on networks and there are many ways to handle the situation, but sometimes a system lacks a mechanism for resolving unexpected contention which may result in the system "freezing," looping (making repeated, unfulfilled requests), or losing the directive for resources (e.g., a document disappearing from a print queue). The term deadlock is usually reserved for contentious situations where all the parties involved in the competition for resources experience some type of protracted slowdown or lockup which is difficult or impossible to resolve except by extraordinary measures (e.g., a reboot). See contention.

debug *v.* To rid a system systematically of problems or *bugs*, especially in computer software. In software, bugs include syntax errors, looping errors, logical errors, and user interface design ergonomics problems. This process has become easier with the availability of debugging software and higher level programming languages, but it is still an arduous, exacting, painstaking activity. In the programming community, not all good programmers are good debuggers, and code is sometimes passed on to a programmer who has a particular mindset and talent for this exacting, detail-oriented work. Software databases for tracking and reporting bugs are becoming popular and automated programs to test software, in order to find problems and report bugs, may be included in the debugging arsenal.

decay Reduction through processes such as absorption, attenuation, erosion, or corrosion. Light or sound signals will decay over time and space due to interaction with the environment as energy is converted into other forms (e.g., heat). Physical structures, such as cables, will decay from abrasion and chemical interaction from heat, humidity, acidic or base solutions, or ultraviolet radiation from the Sun.

The term tends to be associated with conditions or phenomena that decline gradually, often in a somewhat smooth manner, as opposed to those that decline very quickly or in ways that would be graphed as spikes rather than a smooth curve. The measure of the speed of decay is somewhat related to the phenomenon itself. Light is a fast-moving phenomenon and thus decay patterns in light are much faster than those in a corroding cable.

In a fiber optic cable, dopants may be incorporated into the filament to influence the reflection of the light and wavelengths that are allowed to pass through the lightguide. The tradeoff is that the signal will decay more rapidly over any given distance as compared to an undoped cable with the same characteristics.

Since decay is related to distance, and transmission over distance is important in telecommunications, many methods to minimize decay and amplify signals have been devised, including electrical and optical amplifiers. See absorption, amplifier, attenuation.

DECCO Defense Commercial Communications Office (U.S.).

decibel dB. One tenth of a bel. A dimensionless ratio of two powers, which, in electricity related to telecommunications circuits, may be referenced to milliwatts. More familiarly in acoustics, a decibel may be calculated and expressed in *pascals* as a ratio of the sound pressure to a reference pressure.

Since that's a little difficult to understand without knowing the formulas and individual frames of references, it may be easier to understand decibels in terms of examples in acoustics. Sound volume is typically expressed in decibels, on a logarithmic scale (the bigger it gets, the proportionally louder it sounds) with lower numbers representing lower volumes. At the high end of the scale, around 175±25 dB, permanent hearing damage occurs. Even at sustained levels of 100 to 150 dB (and sometimes lower), loss may occur. Sudden contrasts from low to high volume sounds can be especially harmful to human hearing. Environmental noises tend to range from about 5 to 100 decibels, with low volume sounds emanating from appliances and traffic, and high volume sounds coming from horns, explosions, collisions, etc. See bel, neper.

decimal system Based on a base 10 numbering system, the most common one in human culture, probably due to the fact that we have 10 digits each on our hands and feet. Computers, on the other hand, are commonly based on binary systems, base 2, due to the fact that electrically powered computers are easy to design around systems that use two states: on or off, high or low, etc.

decimetric wave An ultra-high frequency (UHF) electromagnetic wave in the approx. 300- to 3000-MHz (3- to 30-GHz) range. Waves in this range are detected during solar flareups and those at the lower end of the scale are used in certain wireless voice and data communications. Wireless decimetric systems are used in military applications, aerospace, and communication systems of developing nations.

In fiber optic cables, amplitude modulation of laser light in the decimetric range may be used for testing embedded optical fibers (e.g., for strain tests) with resulting reflection from the fiber analyzed by a UHF-capable network analyzer.

DECNet Digital Equipment Corporation's proprietary Ethernet-based local area network (LAN). See Digital Equipment Corporation.

decollimate To cause to diverge from a parallel direction. For example, if a collimated laser light hits the endface of a fiber optic cable at the wrong angle or encounters dopants within the material, the light may be deflected from its original parallel path and be reflected in many directions. Lenses can be used to spread or concentrate a collimated beam. For example, a diffusing lens would spread a collimated beam, while a Fresnel lens could concentrate a beam by focusing it over an area that is smaller than the source beam. See collimate.

decompression The process of decoding, expanding, and otherwise reverse-engineering the information in a compressed file. Decompression is usually done to restore a file to its original state, or to restore the information to a form that is comprehensible to a viewer/listener/reader, but which may approximate, rather than duplicate the original state. Decompression may happen in advance of using the information or may be carried out as the data is being read, as in MPEG animation playback systems. See compression, data compression.

DECT See Digital European Cordless Telecommunications.

Dedicated Access Line, Dedicated Line DAL, DL. A private network connection between a business or individual and a phone carrier or network service provider. Calls through a dedicated line are automatically routed to the phone carrier or other service provider. Dedicated lines can be installed on various types of circuits, ranging from copper phone lines to fiber optic.

dedicated array processor DAP. A processor in a redundant array of inexpensive disks (RAID) system that specifies various array-specific tasks related to management of the multiple disks and the information and organization of the information on the disks. This information is particularly important if a problem has occurred with the data, and the disk array needs to be adjusted or rebuilt.

Dedicated Short-Range Communications DSRC. An international standards effort for short-range communications that is ambitious in the sense that it must contend with a heterogenous installed base, competing commercial interests, and varying levels of federal support. The standards effort has produced a number of drafts for two independent but related standards, including DSRC using microwave frequencies and a DSRC data link standard.

The long-term goal is to support an interoperable North American transportation system, and there is still discussion as to whether interoperability should be mandated or promoted. In the future, it is hoped that DSRC can provide a consistent framework for communications, including travel information systems, navigation aids, tracking and routing, and a wide variety of service-to-vehicle services that are possible through wireless data information services (maps, trip planners, emergency locators, etc.).

DSRC has applications for a wide range of transportation users, including casual travelers, professional travelers, trucking companies, emergency services, and transit authorities. The Transit Standards Consortium, for example, has investigated the use of a selected set of radio frequencies for DSRC applications for transit vehicles. See intelligent vehicle highway systems.

Deep Space Network DSN. A round-the-clock communications network that facilitates communication between Earth-based stations and space stations and interplanetary probes. Originally established in 1958, it is managed and operated by the Jet Propulsion Laboratory (JPL) with treaty agreements for foreign complexes. The three complexes that comprise the network are located in Goldstone, California; Robledo, Spain; and Canberra, Australia. See Goldstone Deep Space Communication Complex, NASA.

DeepView An exciting network collaborative development project that grew out of a demonstration of remote electron microscopy by Berkeley Lab researchers Parvin, O'Keefe et al. in August 1995. This historic achievement is important because electron microscopy is very sensitive to changes that can occur in a split second and thus is not tolerant of transmission delays while the viewer makes adjustments, waits for images to display, etc. The solution to this problem was to have a system onsite with the microscope that can monitor the environment and make adjustments as needed to take some of the burden off the remote viewer and reduce the time it takes to communicate commands and data offsite. However, this first successful experiment was a single-user solution. What if numerous users wanted to view and control the same microscope over the Web, for example in the same manner as many pan-and-tilt Web cams? Parvin, O'Keefe, and Taylor subsequently developed a system that came to be called DeepView, wherein multiple microscopes with different characteristics could be accessed by multiple remote participants. It is a scalable, distributed network concept for a common interface for accessing multiple microscopes that may change from time-to-time. The system also facilitates the exchange of information and data analysis. Combined with image processing that enables images to be improved, augmented, and combined, Deep View is a powerful system for coordinated microscope-based research.

Deep Space Network Parabolic Antenna

The 70-meter parabolic antenna for space communications located at the Canberra Deep Space Network site in Australia, Jan. 1990. [NASA/JPL image.]

---

default The initial setting, factory setting, reset setting, parameters, or configuration; the state or parameters that are in effect at the start of a program, available in a dialog box, or stored in a file. Sometimes the user can modify the default settings. Many customizable software applications or hardware devices have a button that the user can press to restore default settings in case erroneous or confusing settings have been subsequently established. In some systems, interrupting power to the system (e.g., removing a PRAM battery) can restore system defaults.

default carrier In telephone communications, the long-distance carrier assigned to handle calls for customers who have not specified an alternate carrier.

Defence Science and Technology Organisation DSTO. A division of the Australian Department of Defence, the DSTO has been investigating future technologies for use in defence applications, along with aiding in procurement, developing new capabilities, and enhancing existing capabilities since 1910. The DSTO has developed a trial high-frequency (HF) radio data network, as well as virtual reality simulations of naval and aeronautic systems. http://www.dsto.defence.gov.au

Defense Advanced Research Projects Agency DARPA. Formerly ARPA, an agency of the U.S. Department of Defense (DoD) that handles research and development in basic and applied research, and which takes on imaginative, high-risk/high-payoff projects which may result in dramatic technological advances on behalf of the DoD.

DARPA was established in 1958 to assure U.S. global technological advancement in coordination with, but independent of, the U.S. military research and development establishment. DARPA has, as part of its mandate, a design that is a deliberate counterpoint to traditional research ideas and approaches. It manages a budget of about $2 billion.

DARPA's technical staff is rotated every few years to encourage new perspectives and is drawn from industry, academia, and government laboratories.

Defense Data Network DDN. The packet switching network of the U.S. Department of Defense, established in 1982, which later became MILNET. DDN was separated from ARPANET (which evolved into the Internet). It includes classified and unclassified sections (DISNET*x* and MILNET).

defense information infrastructure DII. The facilities, networks, software and information that comprise both peacetime and wartime resources of the U.S. Department of Defense.

Defense Meteorological Satellite Program DMSP. A program initiated by the U.S. Department of Defense (DoD) in the mid-1960s, DMSP satellites survey environmental features such as pollution, clouds, water, snow, and fire, in the visual and infrared spectrums. This data is downloaded to ground stations and processed for U.S. military operations.

The primary sensor on the DMSP is the Operational Linescan System (OLS), which provides continuous visual and infrared spectra images of cloud cover over a sensing breadth of about 1,600 nautical miles. The data from the system aids military weather forecasters

in monitoring and predicting weather patterns, especially potentially dangerous weather systems such as thunderstorms and typhoons. It also provides information on electromagnetic fields that could influence long-range communications systems and ballistic-missile early warning radar systems based upon electromagnetic sensing systems.

Data are relayed from tracking stations in New Hampshire, Greenland, and Hawaii to the military meteorological center in Nebraska. The data can also be relayed directly from the satellites to appropriately-equipped transportable military tactical units. See Defense Support Program, Milstar.

Satellite Sensor Sensitivity Regions

The Defense Meteorological Satellite Program (DMSP) has been serving weather data collection needs for the U.S. military operations for several decades through 2.3-ton satellites operating in polar orbits at about 458 nautical miles from the Earth.

This diagram illustrates the two regions of sensitivity of the two onboard sensors of the Operational Linescan System on board the satellite.

Defense Satellite Communications System DSCS. A global military communications satellite network operating in super high frequency bands. DSCS is administrated by the DISA.

Defense Support Program DSP. A satellite data and defense support program operated by the U.S. Air Force Space Command as part of North America's early warning system. DSP satellites are intended to detect missile and space launches and nuclear detonations. See Defense Meteorological Satellite Program, Milstar.

Deflate Protocol DP. A means of compressing PPP encapsulated packets that was proposed in the mid-1990s. DP is based on the *deflate* compression format which was, in turn, based on Lempel-Ziv LZ77 compression. Deflate has a good compression ratio, is openly specified as zlib source code by Gailly and Adler, and is widely implemented in such popular compression programs as PKZIP and gzip. See Point-to-Point Protocol, RFC 1979.

deflect *vt.* To divert something from its original course, preferred direction, or current position. The term is most often used to describe the redirection of something that is moving, such as radiant energy or an object. Deflection is intrinsic to many semiconductor assembly processes and the functioning of communication systems.

Reflective shielding is used to deflect electromagnetic

radiation away from sensitive components. Many components are designed to deflect some wavelengths and allow others to pass, providing filtering capabilities.

Light travels in a straight line unless other forces act upon it to change its direction. Thus, the cladding in a fiber optic cable, which has a slightly lower refractive index than the conducting core, deflects/reflects the light back into the conducting core, enabling it to continue along the core and to follow paths that may curve. See reflect, refraction.

deflection modulation On a scanning cathode-ray tube (CRT), a system of deflecting the scan vertically from a base horizontal path. This makes the signal appear like a series of peaks above the baseline.

deformer A tool for deforming the shape of an object. It may be a physical tool that uses pressure or heat or some other mechanism for deforming an object, or it may be an algorithmic or virtual tool for deforming a digital or conceptual object.

degauss To demagnetize; to bulk erase; to provide electrical current-carrying coils to neutralize magnetism. Cathode-ray tube (CRT) monitors are equipped with magnets to influence the direction of the electron beams. Some monitors will have a degaussing button to reset the magnetic environment in the monitor. Named after Karl F. Gauss.

Delaney lamp detector A novel solution to creating a radio wave detector is the Delaney lamp detector, devised by U.S. Naval electrician Delaney. It was an electrolytic detector that consisted of an incandescent lamp bulb with the top broken off, the filament removed, and a 20% solution of nitric acid poured into the globe. This detector was found to respond well to signals originating nearby, without burning out from the oscillations of a strong signal coming from a nearby wireless station, a problem that was common to other types of electrolytic detectors. See detector.

delay Lag, hysteresis, gap, retardation of a signal or phenomenon, time lapse. Delay is a relative term in that it implies a lag as compared to other processes or compared to what was expected.

Delay is inherent in many types of telecommunications technologies, and systems design and operation must take into consideration a variety of types of delay. Transatlantic phone callers used to experience appreciable delays between the time one side of the conversation was heard and understood and the other side responded. This type of delay is now rare, or within acceptable levels. In the 1980s, the delay between sending and receiving email correspondence was typically about a week, now the delay is often only seconds or minutes.

Delay can be a serious problem in systems that use more than one line to transmit associated phenomena, as in many videoconferencing systems that transmit graphics and sound over separate communications lines. In packet switched networks in which packets are split, sent through different routes, and reassembled at the receiving end, portions of the transmission might delayed, affect the reconstitution of the message at the destination point.

In systems where synchronization of pulses is important, delay must be taken into consideration or may be deliberately introduced into faster aspects of the system to align the pulse timing of two or more components.

Delay is sometimes deliberately introduced in data transmission systems to prevent overly quick reactions to situations that might be nonmeaningful (as in guidance systems or brief power brownouts). Delay is inherent in store-and-forward systems that may collect email messages, or other types of network traffic, and dispatch them in batches or when CPU time is at a low ebb. See cell rate, hysteresis, jitter.

delay distortion In transmission technologies, distortion of the expected shape or characteristics of the signal as some aspects of it are delayed over other component aspects. Delay distortion tends to be cumulative over time and distance. In mathematical terms, it can be described as the different Fourier components of a signal traveling at different speeds relative to one another.

Delay distortion is more prevalent at certain wavelengths and with waves comprised of multiple frequencies. It may also be more significant at higher data rate speeds where processing of the signal at the destination must be handled more quickly and where there may not be as much time to respond to and correct distortion.

In fiber optic cables, delay distortion arises from light dispersing within the confines of the lightguide cable such that different wavelengths arive at the destination at slightly different times. See chromatic dispersion.

delay encoding A means of encoding within the bit period of a signal, a method that is more prevalent in radio signaling. A signal level transition can occur in the middle of the bit period or may be delayed to the end of the bit period. Binary data is represented by two signal levels representing zero (0) or one (1). If a zero is initially transmitted, the signal level does not change. However, if a second zero follows a previous zero, the signal changes at the end of the bit period. If it changes in the middle of the bit period, it represents a one. See time-delay modulation.

delimiter Any symbol, code, character, or other data used to signal a gap, break, boundary, or stopping/starting point. Delimiters are very important in both human and computer communications. People use spaces, commas, paragraphs, and other punctuation and symbology to serve as delimiters in written text. It greatly enhances the ease with which the information can be understood. Computer programs use symbols, punctuation, and spaces as common delimiters for arrays, lists, paths, and other types of information. Radio broadcast systems use spaces as delimiters between songs, or tones as delimiters before and after emergency announcements.

Dellinger fade-out In radio communications, sun spot activity may be associated with highly absorbing areas in the ionosphere which can impair short wave transmissions.

delta channel See D channel.

delta matched antenna See Y antenna.

delta modulation, difference modulation A common method for converting/encoding analog signals into digital signals (A/D conversion) for storage on digital media or transmission over data networks. This scheme was developed during the 1940s for telephony applications.

In delta modulation, a scanned value is compared to the previously scanned value, to see if it is greater or less than the previous value, sending a one (1) if it is greater, and zero (0) if it is less. (Actually the one and zero values can be reversed, just as long as the sender and receiver agree on which is assigned to the greater or lesser result.) Described another way, the sinusoidal analog wave is sampled, each sample is compared to the previous sample, and the modulated signal reflects the relationship by encoding it as a one or zero. Thus, the frequency of the sampling will influence the encoded signal in terms of how much information can be conveyed and how true it is to the original signal. This scheme is relatively simple and results in very fast encoding.

There are many different versions of delta modulation (e.g., linear and adaptive), and many are not compatible. Pulse code modulation (PCM) is the other common method for converting from analog to digital. Because more information is encoded, it's not as fast as delta modulation, but it can be more readily converted between different versions of PCM. Unlike delta modulation, PCM doesn't require an intermediary analog stage to convert between different versions, and thus it is more popular than delta modulation. See A/D conversion, delay encoding, sigma-delta modulation, pulse code modulation, sampling.

delta testing Final stage of testing a finished product on endusers who would normally be customers or users of the product. This testing is more commonly known as User Acceptance Testing. See alpha testing, beta testing, gamma testing, User Acceptance Testing.

Demand Assigned Multiple Access DAMA. In wireless communications, on-demand channel sharing. It is accomplished by assigning a call to a channel that is currently idle, or to an unused time slot, sometimes called Bandwidth On Demand. Thus, a single transponder can support hundreds or thousands of subscribers, though actual capacity depends upon the DAMA implementation. DAMA is common on mesh communications satellite networks.

Prototype DAMA systems were developed in the mid-1990s by the U.S. Defense Information Systems Agency (DISA). DAMA is a newer technology intended to increase channel capacity over methods common in the 1990s. It is typically used in single-hop satellite connections for a variety of point-to-point or multipoint voice, facsimile, or data services. DAMA assigns channels automatically, rather than manually or by establishing fixed channels. When a specific DAMA transmission is terminated, the channel is returned to a pool of available resources. DAMA is used in commercial satellite systems and on the U.S. Navy's UFO communications satellites (designed to replace lower capacity FLTSAT systems). Mini-DAMA is a type of U.S. Navy DAMA system that includes encryption and data transfer capabilities previously handled by devices outside the DAMA system.

demand circuits Data network segments whose costs are related to usage.

Demand Priority Protocol DPP. A client-server protocol mechanism that handles message exchange and sensing by multiple devices in IEEE 802.12 (100VG-AnyLAN) networks. The process is governed by the Media Access Control (MAC) states. The client requests access to the communications medium and the server evaluates the request and returns permission to the client, if the request is granted. Once permission is granted, the requester goes into a transmit frame mode and sends the data to the network.

demarcation point The point at which a telecommunications carrier's equipment or responsibility ends, and the subscriber's equipment or responsibility begins. This point may be at a junction box on the side of a building or a patch panel within a building. Residential lines tend to be demarcated outside, while multiline phone systems in businesses tend to be demarcated at a patch panel or terminal block inside the premises. The block itself is sometimes called a demarcation strip.

demodulation The process of taking a signal that has been manipulated to carry information, and extracting that information from a carrier wave or mathematically recreating that information from sidebands or other parts of the original transmission.

A crystal detector uses a crystal as a type of one-way valve to detect and demodulate radio waves. A modem is a device that demodulates an analog phone signal to convert it to digital information that can be understood by a computer. A broadcast receiving station takes modulated airwaves and demodulates them for local transmission through cable or airwaves. See modulation.

demon dialer, war dialer A function of a program, usually for telecommunications or telemarketing, that automatically calls a phone number repeatedly or calls down through a list of phone numbers, cycling back to the top until one answers.

DEMS 1. digital electronic management system. 2. digital electronic message service.

demultiplexing 1. In a multiplexed signal, a process for recovering signals combined within it, usually to restore distinct channels contained within the transmission. See carrier wave, band, channel. 2. In ATM networking, a function performed by a layer entity that identifies and separates Service Data Units (SDUs) into the individual connections of which they are comprised.

DENet Denmark's Ethernet network which interconnects academic institutions.

Denial of Service DoS. A network administration term for situations in which users are unable to access the system. DoS may be intentionally implemented by an administrator or may be the result of undesired network configuration problems, vandalism, or security breaches such as hacking or flooding. With the Internet being increasingly used for sensitive government and financial transactions (especially time-sensitive stock and auction transactions), DoS situations arising from deliberate tampering have become a significant security concern. Safeguards against unauthorized DoS are difficult to implement without also affecting legitimate users; sometimes security measures to prevent one type of abuse (e.g., excessive junk email) can leave a system more vulnerable to other types of abuse (e.g., DoS attacks), making it a challenge to configure a system for optimum use while keeping it secure. See Denial of Service attack.

Denial of Service attack DoS attack. A network service security breach in which access to legitimate users is blocked or significantly impeded by unauthorized users. Sometimes DoS attacks are launched to crash a system in order to take it offline or to reveal other security weaknesses through which an invader might further infiltrate the system.

Denial of Service attacks can be directed toward different aspects of a computer network. Filling up the available disk space, planting a program that greedily consumes system resources, or overwhelming the connection algorithms are various strategies that can cause DoS problems.

SYN-flooding is the transmission of a barrage of requests for a network connection to deliberately compromise normal use. When the server unsuccessfully tries to connect to an overwhelming number of unreachable return addresses (those that don't return a connection acknowledgment), a DoS condition results.

Denial of Service attacks have existed as long as there have been computer networks, but they achieved prominence in the media and in industry and government discussion groups in February 2000 when an ordered series of attacks was launched against some of the most prominent commerce sites on the World Wide Web. See Land attack, Trojan horse, virus, RFC 2505, RFC 2827.

dense wavelength division multiplexing DWDM. A transmission system capable of producing and simultaneously transmitting many wavelengths over a single fiber optic cable. On DWDM multiple channels can be carried on the same transmissions medium, in essence a virtual "fiber bundle" within a single conducting core. This substantially increases broadband capacity and is of particular interest to firms installing long-distance cables over difficult terrain such as mountains or ocean bottoms (submarine cables).

DWDM is speed and protocol independent, meaning it can carry a wide variety of transmission protocols, including SONET, Internet Protocol (IP), Ethernet, and SDH. Different types of communications can be carried at different speeds over the same physical network.

DWDM was commercially introduced in the mid-1990s and Sprint was one of the first high-profile companies to use it, implementing the technology for long-distance telephone services.

Due to cost and implementation, DWDM was originally favored for long-haul networks, but was beginning to be used in metropolitan area installations in 2000. DWDM can increase capacity over time division multiplexed (TDM) systems by a factor of four to eight times. Metro 1500 systems from Cisco, for example, can accommodate up to 32 wavelengths (channels) per strand of optical fiber. Gigabit Ethernet transmission speeds are readily implemented over DWDM optical networks.

By 2001, it was possible to transmit 1.60 Tbps over a distance of 2000 kilometers. By March 2002, Lucent Technologies had demonstrated a 64-channel differential phase shift keying (DPSK) system that could transmit over a long haul of 4000 km (2500 miles) with each channel transmitting at 40 Gbps through 100 km spans.

Testing and maintenance technologies for DWDM were also evolving in the late 1990s, with equipment capable of measuring wavelength, frequency, channel spacing, signal-to-noise ratio, and power being offered by companies such as Wavetek Wandel Goltermann in 2000. See lambda switching.

densitometer A photoelectric instrument for measuring the opacity or relative degree of light absorption of a material. Darker materials absorb more light and thus have a higher optical density. There are different types of densitometers, some that measure transmission and some that measure reflection. They are commonly used in the printing industry to monitor consistency and quality. See hygroscope.

density A general descriptive or comparative term describing the proximity of individual elements, hence the total number that can be fit within a specified area. Density is used to describe optical and magnetic storage capacities, screen and printer resolutions, compression efficiencies. As a general rule of thumb, especially for storage media, the denser the information capacity, the higher the cost.

Denver Telephone Dispatch Company One of the earliest phone companies in the United States, established by Frederick O. Vaille in early 1879.

Department of Health DOH. State, federal, and corporate health agencies have been devoting substantial resources to evaluating various distance education and physician support technologies and databases that can aid in health and safety. With telemedicine, for example, a surgeon in one region can oversee and direct an operation in a remote location. With shared databases, diagnosis and treatment research and information can be collectively used and managed. The applications of telecommunications to medical management are far-reaching, and we've currently only begun to implement the technologies.

depletion layer A barrier region in a semiconductor in which the mobile carrier charge density is not sufficient to neutralize the charge density of donors and acceptors. The donors contain impurities to facilitate electron activity, and the acceptors include trivalent impurities, forming "holes" to accept the electrons.

depolarization Demagnetization or the counteraction of polarization. In many transmission systems, the radiant wave signal may be horizontally or vertically polarized and may gradually lose that polarization as it passes through various media (moisture, particles, terrain obstructions, etc.). The wave may also be deliberately depolarized at the reception point. One of the important evolutionary steps in the history of the battery is the depolarizing characteristics of the Daniell battery.

Polarization and depolarization are significant in many technologies and are of particular interest to quantum physicists, antenna manufacturers, astronomers, and photographers. When light waves vibrate in more than one plane (e.g., sunlight), they are considered to be unpolarized. When the vibrations occur in only one plane, the light is considered to be polarized. This can happen through scatter, reflection, or refraction or can occur or be deliberately effected in a transmitter. Reflection off a highly reflective surface can yield unpolarized light. The angle of reflection off a less reflective surface will affect the amount of polarization to a different extent.

deregistration The process of dissociating two entities, which may be names, processes, objects, etc. Deregistration is a process that happens often in computer applications, particularly those associated with networks. The process of creating virtual links between applications, icons, and processes allows for keeping track or creating shortcuts. Often these virtual links are removed or reorganized, particularly in a dynamic network environment.

In public or pay systems, deregistration may be associated with billing or security, in that a subscriber, cellular handset, computer, or other entity is registered in order to assure service to the subscriber, or to provide privacy and authentication. In these systems, registration is often with a unique identifier, which must be deregistered before services can be discontinued, or before the identifier can be assigned to another.

deregulation A process of removing authority from a governing body in general, or of removing authority of the government body over specific jurisdictions. As examples, control over banking, telecommunications, or other large public or commercial systems have at various times been removed or reduced in order to stimulate competition and innovation. Sometimes deregulation has the intended positive effect. Sometimes deregulation causes a change in the system that attracts unscrupulous members of society, who seek new loopholes that can be used to take advantage of the system to the detriment of others.

DES 1. See Data Encryption Standard. 2. destination end station.

descramble Scrambling is mixing up a signal or digital data file so that it cannot be used without first being processed. Descrambling is recreating the original in lossless or lossy format so that it is human-readable or viewable. Scrambling provides security and protection from unauthorized viewing or copying and is a common way to commercially protect cable or satellite TV programs from being viewed by those who haven't paid for the service. In most regions, unauthorized descrambling devices are illegal. See encyrption.

Design Layout Record DLR. A database record of telephony design layout. A DLR may be searched by a circuit type with additional parameters supplied such as the U.S. state in which the circuit is located. A DLR search yields administrative and design information specific to the queried circuit.

Design Structure Matrix, Dependency Structure Matrix DSM. A system and project analysis tool for modeling information flow, as in the product development process. Parameters are charted in a matrix representation containing subsystems or activities and corresponding information exchange and dependency patterns. Research on DSM is ongoing at MIT.

Design System Language DSL. A predecessor to the PostScript page definition language, DSL originated in the mid-1970s at Evans & Sutherland Computer Corporation. Evans & Sutherland (E&S) became well known for their pioneering work in flight simulation and other 3D graphics software, and the Design System was one of the outcomes of a research project using an interpretive language to build complex graphics databases. The Design System Language was later put to use in CAD applications. See PostScript.

Designated Transit List DTL. In ATM networking, DTL is network routing policy data consisting of a list of nodes and any associated link identifiers, which specifies the path across a PNNI peer group.

desktop A broad term in the computer industry describing any common workspace found in a home or office (or garage) where the various user tools and appliances are contained. Desktop computers are those that can fit comfortably on a desk (as opposed to dishwasher-sized minicomputers or room-sized supercomputers). See desktop metaphor.

Desktop Management Interface DMI. A specification developed by the Desktop Management Task Force (DMTF) to allow easier, more consistent access to management database information and applications through Management Interfaces (MIs). The DMTF is a consortium of hardware and software vendors.

desktop metaphor A phrase to describe the conceptual rationale for the design and display of information on a computer output device, usually a monitor. The idea is to take objects and actions that are familiar to workers and home users and represent them in an easily recognizable form on the computer.

desktop publishing DTP. Using desktop computers to design, layout, and print files. It is now straightforward to create a file on a home computer and send it directly, via FTP, to a commercial printer. Many of the pioneer tools for the evolution of desktop publishing were developed at Xerox PARC, subsequently implemented and popularized on the Macintosh computer in the mid-1980s, and enhanced through the development of laser printers. DTP has had a resounding impact on the publishing industry and has put publishing into the hands of tiny companies and individual publishers.

Quark XPress, Adobe PageMaker, Adobe FrameMaker, and TeX are DTP software applications widely supported by users, service bureaus, and publishers. Ventura Publisher was popular in the 1980s. Quark is favored for ad layout, posters, and other large-format or complex color projects and, to a lesser extent for books, mainly those that have many illustrations and those that don't require a lot of extensive indexing and cross-referencing. Design professionals like Quark.

TeX is preferred for sophisticated technical symbols as are found in mathematics and physics treatises and textbooks and, although its interface is less intuitive, it has capabilities in symbol representation and formatting beyond most general purpose desktop publishing programs. Corel has been promoting its products as desktop publishing tools, and for illustrations it can work well; however, many service bureaus and professionals prefer some of the other products, at least at the present time. For creating and inserting PostScript *EPS* (encapsulated PostScript) illustrations into other desktop publishing programs, many people swear by Corel Draw! There are also a number of other publishing programs available and word processors that are sometimes stretched to perform basic desktop layout tasks.

desktop video Just as the Macintosh computer facilitated the development of a new mass market field called *desktop publishing*, the Amiga computer facilitated the development of a new mass market field called *desktop video*. Within two years of its debut in 1985, many video genlock and broadcast video products were introduced for the Amiga, the most significant being the NewTek Video Toaster, which is still used today by many local cable companies and some of the large broadcast networks. A desktop video-equipped computer enables the development, editing, and merging of computer images, video taped footage, and live broadcasts at a fraction of the cost of using older video equipment.

desktop computer telephony integration DCTI. The integration of computer processing and database search and retrieval functions, especially, with telephone communications equipment and procedures. For example, a database that is used to qualify and select potential or existing customers can be used to automatically dial on the behalf of sales agents. If the call is successfully connected, the next available agent can be notified and the information in the database can be displayed on the sales agent's terminal. This type of integration and automation is becoming increasingly prevalent. Videoconferencing is another example of DCTI.

desktop videoconferencing Video phone capabilities provided on a small desktop dedicated system or microcomputer equipped with a microphone, video camera, and fast network or phone access. Businesses are considering videoconferencing as an option to expensive travels, especially for meetings involving participants who are widely distributed geographically. Videoconferencing systems are affordable now, but the speed of the line greatly affects the refresh rate of the image. If the transmission rate is slow, the image will be blurry and slow to update, and it will look more like a series of still shots than natural movement. End-to-end fiber optics networks are well suited for the transmission of audio/video conferences.

despun antenna A platform-mounted rotating antenna used on satellites to orient a cone-shaped beam in a specified direction, usually a region on the Earth. There were early electronically despun antenna experiments on U.S. military satellites in the late 1960s which met only limited success. Later, mechanically despun antennas were also tried.

destination address DA. A flag, field, or other indicator commonly used to designate the receiving point for a data transmission, call, or physical correspondence. In Token-Ring, Ethernet, and Fiber Distributed Data Interface (FDDI) networks, the DA is a data field sent in the direction of the recipient that describes a unique Media Access Control (MAC) address. See address, Destination Address Field, domain name, Media Access Control.

Destination Address Field DAF. In frame-based networking, the DAF specifies the intended destination for the frame, which may be one or more stations (singlecast or multicast). In Ethernetworks, the DAF is usually implemented as a six-byte address, and, if the DAF contains all ones, it is considered to be an all-stations broadcast. In the IEEE 802.3 Ethernet standard, the Destination MAC Address follows the Start Frame Delimiter and precedes the Source MAC Address. See destination address.

destination end station DES. In ATM networking, the end or termination point of a transmission. It is used as a reference point for available bit rate (ABR) services. Since ATM has a number of traffic flow control mechanisms, which often depend on cells arriving at the destination at a certain time or in a certain manner, ABR and constant bit rate (CBR) distinctions are important with relation to the defined end station. See cell rate.

Destriau effect When exposed to an alternating electric field, certain phosphorescent inorganic materials suspended in a dielectric medium will luminesce, emit *electroluminescent light* rather than incandescent light. Certain semiconductors can be designed and excited to emit *carrier injection luminescence*.

detect Sense, become aware of, register a reading as a result of some influence, respond to.

detector 1. In radio electronics, a device to apprehend or detect radio waves. In early schemes, many materials were tried, including barretters, natural rectifying crystals, synthetic crystals, and electrolytic cups. The challenge with most detectors was solving the problem of amplifying the signal, or shifting the range of frequencies into the audible range. Current electronics have sophisticated and effective ways of achieving it, but in the late 1800s and early 1900s, ways of capturing and amplifying the waves were just being developed. 2. A substance or circuit that reacts when exposed to electromagnetic waves such as radio waves or light. The detector may also convert or rectify the electromagnetic oscillations so they can be incorporated into a circuit with practical applications. A crystal detector works as a type of one-way valve to rectify radio waves, without outside power sources. Later, vacuum tubes were designed that could accomplish this task in ways that allowed great selectivity over the frequencies detected (tuner circuits) and that could amplify the signal so the broadcast could be played through a speaker rather than through small headphones.

Fiber optics are now incorporated into many detectors. They are useful as detecting and imaging probes and as lightguides to draw a light signal from one part of the system to another. See crystal detector, Delaney lamp detector, electrolytic detector, electron tube, Lodge-Muirhead detector, magnetic detector, Massie Oscillaphone, optical detector, photodetector, Shoemaker detector, silicon detector.

Deutsch, Peter Along with Alan Emtage, one of the developers of the Archie system created originally at McGill University and a cofounder of "the Archie group," a group of volunteers dedicated to supporting and enhancing the Archie project. Archie is very widely used on the Internet.

Deutsche Telekom The German telephone services authority. DT includes a Technology Center, which engages in research, development, and demonstrations of innovative telecommunications systems. http://www.telekom.de/

device configuration management In networks, the configuration, tracking, and management of various devices, ports, and interface cards. These devices are increasingly software configurable, and the software used to manage them will often show actual images of the physical switches through a graphical user interface.

Device Control Protocol DCP. An assigned *well-known* port 93 on TCP/UDP systems.

device driver Computer software that controls, or provides an interface to, one of a variety of devices such as hard drives, printers, video cards, CD-ROM players. A device driver is a software interface through which the operating system interacts with various peripherals attached to the system. See display driver.

DFA See doped fiber amplifier.

DGEP See dynamic gain equalization processor.

DGPS Differential Global Positioning System. See Differential GPS, Global Positioning System.

DGPT Department General of Posts and Telecommunications, Viet Nam.

DGT Dirección General de Telecommunicaciones. Spanish telecommunications authority.

DHCP See Dynamic Host Configuration Protocol.

Dhrystone A relative performance test intended to evaluate system program execution other than floating point and input/output operations. As with many benchmarks, the information is useful only within a narrow viewpoint, as in controlled experiments. The Dhrystone was developed in the Ada programming language by R.P. Weicker of Siemens and is described in overview with other benchmarks in *IEEE Computer*, Dec. 1990. See benchmark, Rhealstone, Whetstone.

diagnostic programs Application programs used to run hardware test suites, to evaluate processor functioning, or both. Most computers will run a systems check of the hardware before booting up the higher level operating functions. Network servers now often have diagnostic tools with graphical user interfaces to show virtual and physical connections, traffic flow, routers, switches, and various configuration settings. Diagnostic software is available for many computers to self-check and indicate possible problems in processing rates or connections. Many diagnostic routines are intended to run on a regular schedule, to locate potential problems before they become serious. *Engine diagnostics* used to mean getting grandpa to stick his head under the hood of the car to listen to the engine. Now it means taking the car into a shop and getting it hooked up to a computerized engine evaluation system that displays various system parameters on a computer monitor with charts and graphs.

diagnostic techniques Various means of ascertaining and measuring the properties of a system or circuit in order to understand its characteristics, monitor its behavior, or detect any problems or anomalies. Diagnostic techniques form a part of troubleshooting a system and are often part of or combined with installation and testing. Some systems incorporate automated diagnostics that will check systems on startup, at random intervals, or at scheduled intervals. See ammeter, tap, voltmeter, Wheatstone bridge.

dial A circular, movable mechanism for entering a number or other code. While dials are used in many electromechanical devices, they are most commonly associated with rotary telephone sets. In a telephone system, turning the dial causes a pulse of a predetermined length to be sent along the line to indicate the desired destination address. This process allows the system to set up an end-to-end connection for a conversation or data communication. The dial was also historically known as a *finger wheel*.

The dial superseded the hook on old telephones and effectively obsoleted human operators for local calls within a few years. Dials were commonly used until the 1970s, when touchtone phones began to become prevalent in North America (many other countries still use dial phones). See Strowger switch.

Dial Access Switching DAS. A Cisco commercial network switching system that provides dial services (and dial backup services) for Frame Relay permanent virtual circuits (PVCs) in Cisco wide area networks (WANs). The PVCs are activated by ISDN calls to the switching network. DAS is comprised of the *DAS Server Shelf* hardware and *Dial-Up Frame Relay* software.

dial tone An audible signal that indicates that a phone line is active and ready to be dialed. The dial may be provided by the local phone carrier or by a private branch system. A different type of dial tone, called a *stutter dial tone,* has recently been introduced by phone companies offering voice messaging services to their subscribers, to indicate that messages are available to be retrieved.

dial-around service A competitive telephone service in which the user dials an access code prefix in order to hook into a third-party service by bypassing the normal local long-distance carrier. The 1010 (ten ten) long-distance service marketed through television advertising is an example.

Dialable Wideband Service DWS. A sophisticated digital telephony circuit-switched data service from Nortel that enables public telephone networks to offer a wide variety of digital services to their subscribers. DWS provides on-demand, trackable variable bandwidth connections which can be implemented over public telephone networks within the international number plan. See DataSPAN Frame Relay Service.

Dialed Number Identification Service DNIS. A service to display the number the caller dialed on an incoming phone call. This service is useful for regional numbers that are rerouted to a central administration or sales system. For example, an economic region may have several area codes, but there may be one sales office handling all the calls. The DNIS allows the person answering the call to see whether the caller dialed the local number, the toll-free number, or a long-distance number, giving some feedback as to which lines are being used.

DNIS does not provide the number of the calling party; a service like Caller ID is needed to provide this additional information.

dialing parity Defined in the Telecommunications Act of 1996 and published by the Federal Communications Commission (FCC), as

"... a person that is not an affiliate of a local exchange carrier is able to provide telecommunications services in such a manner that customers have the ability to route automatically, without the use of any access code, their telecommunications to the telecommunications services provider of the customer's designation from among 2 or more telecommunications services providers (including such local exchange carrier)."

This measure was designed to level the playing field and encourage fair practices. See Federal Communications Commission, Telecommunications Act of 1996.

dialup networking DUN. In the days before T1 and cable modem access to the Internet, most access to computer networks and bulletin board systems (BBSs) was through a modem and a phone line. Because modems and computer time were both expensive until the late 1980s, dialup networking was usually done on an as-needed basis, with the phone connection live only as long as was necessary to carry out the necessary data transfers (e.g., file uploads and downloads, reading mail, leaving messages). The earliest dialup networks were available only to corporations and government officials but, by the late 1970s, 300 baud modems and desktop computer-

based bulletin boards made dialup networking popular with hobbyists. Passwords were rare, courtesy was common, and various software programs were developed to facilitate the process and keep phone connections to a minimum. As modem technology improved, 1200 baud and 2400 baud modems were sold in the 1980s and, by the 1990s when Internet access was growing by leaps and bounds, 33 Kbps and 56 Kbps modems were generally available at reasonable prices. At this point many people added extra phone lines so they could stay online longer without tying up their voice lines, and service providers added extra modems to support the high demand.

With the availability of cable modem, T1, and other 24-hour access technologies in the late 1990s, dialup networking is becoming less prevalent, although it will probably still be available for a number of years. At some point, 24-hour fiber optic connections or satellite connections to every home and business might be standard, just as it is in some motels and educational institutions now.

DUN connecations are facilitated through scripts and files that store information for initiating a dialup connection. DUN files include phone numbers, modem settings, and other relevant parameters.

diaphragm A thin, flexible sheet or membrane, usually with a curved surface. In electronics, a diaphragm usually is designed to respond to vibrations and has an interface to a device that can interpret those vibrations into electrical signals. Or conversely, it may take electrical pulses and interpret them into vibrations. Diaphragms are common in microphones and speakers.

dichroic Literally, two-colored, though the term is often applied to phenomena with multiple-color interactions. A dichroic substance can cause an incident electromagnetic wave to pass some frequencies and reflect others – a very useful and often very beautiful characteristic. The iridescence associated with a hummingbird's feathers, when viewed from the right angle, gives some idea of the type of color effects that can be deliberately created with dichroic materials. Light propagated from a dichroic interaction has its associated electric field vector vibrating in a single direction. See absorption, anisotropic, dichroic filter, polarization.

dichroic filter/reflector There are a variety of reflectors that exploit dichroic phenomena to reflect some wavelengths and allow others to pass through. For example, a reflector may reflect only light in the visible range and filter out other frequencies by allowing them to continue on through the reflector. Dichroic filters differ from gel filters in that they reflect rather than absorb undesired wavelengths. This gives dichroic filters low susceptibility to heat degradation. Dichroic reflectors can be fabricated through vapor deposition processes in which fine layers of different materials are deposited within a thermal chamber. The layered composite may subsequently be stamped or cleaved into smaller components.

Depending upon their construction and the frequencies with which they will be interacting, they can serve as wavepass filters, bandpass filters that pass a range of wavelengths, or notch filters that reflect a range of wavelengths.

Dichroic filter arrays can be assembled by lining up multiple components on a flat or curved base substrate and selectively coating the surface with dichroic reflective materials. Thus, the surface could be patterned to selectively filter not only certain frequencies, but wavelengths coming from certain directions over a specified area.

Dicke, Robert H. (1916–1997) An inventor and educator, Dicke is known for his modeling of gravitational forces, his contributions to aerospace research, and his pioneer work in maser/laser technology.

In 1941, he joined MIT's Radiation Laboratory, becoming a Princeton professor a few years later. In May 1956, Dicke submitted a patent application for "Molecular Amplification and Generation Systems and Methods" describing ways to generate and am-

plify electromagnetic waves, such as those in the microwave frequencies, through the activity of resonant gases (e.g., ammonia). Dicke's patent anticipates Fabry-Perot interferometers and cavity-resonating gas lasers. Since an infrared light source is described in some of the drawings, it is a pioneer laser device (U.S. #2,851,652). Dicke's invention sought to establish more efficient ways to generate coherent microwaves or infrared waves and to improve the cavity-resonating (amplification) effect.

Dicke was a member of prominent astronomical and physics societies and won numerous awards for his work. See Javan, Ali; laser history; Townes, Charles.

Dicke radiometer A device developed by Robert H. Dicke to detect the very subtle radiation residual from the Big Bang, the theoretical cataclysmic expansion of our universe.

DID See Direct Inward Dialing.

dielectric Nonconducting material that provides an

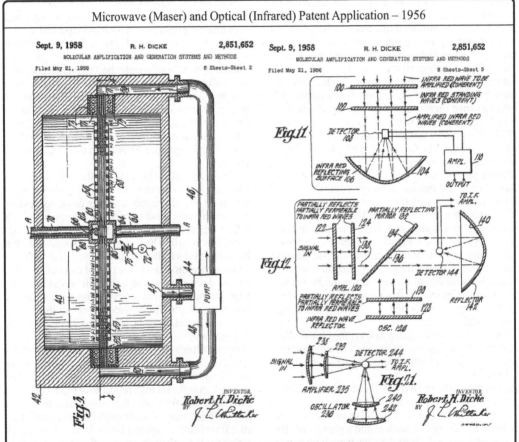

Microwave (Maser) and Optical (Infrared) Patent Application – 1956

These are excerpts from the diagrams that accompany Robert Dicke's May 1956 patent submission for "Molecular Amplification and Generation Systems and Methods" describing a means to generate and amplify electromagnetic waves, both in the microwave (maser) and infrared (laser) frequencies through resonant gases such as ammonia acting in an open cavity resonator. The patent is a milestone in the sense that it was rapidly followed by a number of interferometers and optical laser and gas resonating inventions. The Dicke patent was published in September 1958 and may have inspired more scientists than have given Dicke credit for the concepts described.

On the left is an example of a pumping system. On the right, a resonating cavity with coherent infrared radiation aimed into the reflective cavity through switching valves (forerunner to Q switches). [U.S. patent #2,851,652.]

insulating layer by impeding or resisting the passage of current. On an atomic level, a dielectric exposed to an electric field experiences slight changes that result in a *bound charge* at the surface of the material through polarization (as opposed to movement as occurs in conducting materials).

Dielectric materials are often used as cable shieldings or applied in layers between sheets of conducting materials in condensers. Common dielectrics used over the decades include paper, cloth, air, Bakelite, glass, ceramic, and certain synthetics. Glass and ceramic are the ones used in Leyden jars and utility pole insulators. See air dielectric, insulator.

dielectric breakdown The point above which which a dielectric (insulating) material will show conducting/sparking characteristics, usually expressed in volts.

dielectric constant DK (*symb. – K*). The dielectric constant is the degree to which a material can be polarized which determines the *bound charge* associated with the material. It describes how much an electric field is reduced within a dielectric material. The dielectric constant is related to the properties of a material such as composition, density, homogeneity, and temperature. The bound charge of a dielectric surface can be calculated by using Gauss's law.

The dielectric or *permitivity* property of a material influences the relative speed at which an electrical signal will propagate. The speed of the signal is roughly inversely proportional to the square root of the dielectric constant. Thus, low dielectric constants are associated with higher signal speeds and vice versa. As networking systems become faster and more sophisticated, the dielectric constants of fabrications materials becomes more important. See air dielectric, dielectric, Gauss's law.

dielectric feed In communications, a dielectric feed is a type of microwave lens that fits into the mouth of an antenna waveguide. It provides a wideband alternative to scalar feeds.

dielectric interaction Phenomena resulting from the presence or proximity of dielectric materials relating to the storage and discharge of energy that may cause interference or distortion to signal transmission systems. One solution to reducing undesired dielectric interaction is to use air dielectric cables in which energy-interacting materials are held away from conducting materials by a cushion of air.

differential cable A cable commonly implemented as a twisted-pair wireline with two wires transmitting the same data at the same time, except that one is transmitted as a positive (+) signal and one is transmitted as a negative (-) signal. It is sometimes also called a balanced cable. Since noise is typically introduced along the transmissions path, especially over longer distances, there will be slight changes in the signal at the receiving end. The "difference" is then taken between the two signals on the two wires in order to eliminate the portion associated with the noise. Twisted pair is used to cause the two separate lines to occupy, as nearly as possible, the same physical space along the transmissions path. This system enables longer cable lengths to be used (e.g., RS-422

serial specification) than in single-ended systems (e.g., RS-232 serial specification).

differential GPS DGPS. An implementation of the Global Positioning Service designed to improve local accuracy of the data. One or more high-end GPS receivers are placed at known locations where they receive GPS signals. These become reference stations, which estimate the variations of the satellite range measurements, forming corrections for GPS satellites within current view and then broadcasting the correction information to local users. See local differential GPS.

differential modulation A means of relative modulation based upon the detected state of the previous instant, rather than on an absolute predefined parameter. See delta modulation for an example of a commonly used, simple type of differential modulation.

differential phase shift keying DPSK. A means of relative modulation in which the previous state of the carrier signal phase is detected, and the subsequent state is based on the previous, rather than on an absolute predefined parameter. See phase shift keying.

differential polar relay A telegraphic transmissions relay in which the armature is polarized by contact with a permanent magnet and is operated by the difference in the strength of the currents. The direction of the currents constantly changes and can be controlled with a pole changer.

Differential Power Analysis DPA. A powerful analytical tool for extracting secret keys from cryptographic devices using statistical analysis and error correction techniques. Since a large percentage of sensitive computer communications is stored on media that may be carried about (portable computers, smart cards, etc.) or are transmitted over wireless networks, which can be intercepted by eavesdroppers, analytical tools for evaluating cryptographic integrity are important for developing and testing new technologies.

DPA can be used to attack a system to try to discern encoded information using hardware that is readily available. While many technologies are resistant to Simple Power Analysis (SPA), Differential Power Analysis can break many systems that are immune to SPA attacks. As is often the case, there may be a trade-off in time and power; a DPA attack may take longer than a SPA attack. DPA has been put into practical application by Paul Kocher and Cryptography Research.

differential pulse code modulation DPCM. A means of sampling a signal, subdividing it, and assigning values to the individual parts (quantization) in order to add this information to a carrier signal. This modulation can be done in a number of ways, and not all PCM transmissions are compatible. PCM is a very common means of converting analog to digital signals and is widely used in telecommunications. In differential PCM, a transmitted digital signal is used to represent the difference between consecutive analog signals. These differences are obtained by using a fixed quantization step size. See quantization, pulse code modulation.

differential quadrature phase shift keying
DQPSK. In general, quadrature phase shift keying
(QPSK) is a modulation scheme in which four signals are used, each shifted by 90 degrees, with each
phase representing two data bits per symbol, in order to carry twice as much information as binary
phase shift keying. DQPSK is a subclass in which the
difference between the current value of the phase and
the previous value of the phase are used instead of
the absolute value of the phase. See modulation,
quadrature phase shift keying.

differential Ziv-Lempel diffZL. A text compressor
combining Lempel-Ziv compression and arithmetic
coding with a form of vector quantization, described
in 1995 by Peter Fenwick. This compression scheme
is similar to Lempel and Ziv's original LZ77 scheme
but without explicit phrase lengths or coding for lit-
erals. It combines dictionary compression and vector quantization by using a standard scan to determine
the longest earlier phrase to match ensuing text to
create a reference phrase. A phrase includes a position code, sequence of zero symbols, and terminating nonzero symbol which is processed through arithmetic coders for displacement and data. Coding proceeds until an unexpected character is encoded; displacement coding accounts for most of the compressed output stream.
Performance of diffZL compares reasonably well
against LZB and LZ3VL. The unique characteristic
of diffZL is that it has no explicit phrase length or
literal encoding. The development of diffZL leads,
in part, to the suggestion that the limits of LZ77 compression may be about 3.0 bits/byte, a limit that has
very nearly been reached. See Lempel-Ziv.

Diffie-Hellman A fairly fast public key encryption
system described by Whitfield Diffie and Martin
Hellman in a 1976 IEEE issue of *Transactions on
Information Theory* entitled "New Directions in
Cryptography." This concept has since been incorporated into many encryption schemes, including
some Cellular Digital Packet Data (CDPD) systems
and the well-known Pretty Good Privacy (PGP)
program developed by Philip Zimmermann. While
the inventors patented the system, it came under dispute because of its public disclosure prior to the patent
application. See Hellman-Merkle, Pretty Good Privacy.

diffraction The reflection of wave or wavelike phenomena as they encounter an obstacle that is in their
line of travel can cause a complex interaction called
diffraction as the incoming waves are reflected away
from the obstacle and somewhat towards the incoming wave (depending upon the angle of incidence).
If the obstacle causing the diffraction is fixed and
ordered and the diffraction phenomenon is narrow
and homogenous, the diffraction pattern may be more
easily studied and exploited.
As an example, some of the sound waves emanating
from an audio speaker component may hit the interior of the speaker cabinet and reflect back across the
original sound waves, causing a complex pattern in
the sound waves where they interact. Some of the
waves hitting the cabinet interior will be absorbed and

some will be diffracted, depending upon the shape
and composition of the cabinet, the distance from the
speaker to the cabinet interior, and the reflectivity/
absorbancy of its construction materials.
Diffraction may have both positive and negative consequences. In acoustics, diffraction can sometimes
create a more complex or interesting sound, but it may
also cause unwanted interference, depending upon the
nature and magnitude of the diffraction.
Similar effects occur in the movement and interaction of electromagnetic phenomena. When light energy encounters an obstacle, a fringe pattern may result from the light diffracting from the obstacle as it
reflects back from the reflecting obstacle. This fringe
pattern may be seen with scientific instruments and
can provide information about the character of the
light and the obstacle. When viewed with optical instruments, diffraction patterns tend to appear elliptical.
In Fresnel lenses, diffraction through successively
angled prism-shaped projections serves to concentrate light. In rear projection systems, diffraction helps
display an image over a large surface.
Diffraction patterns in crystal structures have interesting properties due to the lattice arrangement of the
atoms. When hit with a stimulus (e.g., collimated X-
rays), some of the atoms in the upper surface will reflect back the beams, while some beams travel
through the upper surface to reflect from the next
level in the lattice, etc., creating a complex but predictable diffraction pattern that was first described
mathematically by W. Lawrence Bragg in the early
1900s.
In some cases, diffraction can be acoustically or optically controlled to produce complex effects or filtering mechanisms. See Bragg's law, diffusion, diffraction grating, dispersion, Fresnel lens, spot of
Arago, Wood anomaly.

diffraction grating A component designed to diffract
electromagnetic or acoustic phenomena as they impact a grating component, which is often a corrugated
structure. In optics, a diffraction grating can help
reveal some of the characteristics of light and can be
harnessed to provide some control over the direction
of diffracted light and the wavelengths affected by
its shape, its period (distance between corrugations),
and its materials. This is useful in filtering wavelengths and is important in the development of nar-
row-wavelength lasers and fiber optic filament
switching assemblies. See Bragg grating, diffraction,
dispersion, grating equation.

diffraction orders An ordered set of reflection patterns associated with an ideal diffraction grating that
diffracts incident radiant energy in discrete directions
at angles that can be calculated. See grating equation.

diffraction, Fraunhofer Diffraction in situations
where the factors that influence the complexity of the
diffraction patterns are somewhat controlled. For example, a light source should be a sufficient distance
from the reflecting obstacle and the beams monochromatic and parallel in order to create planar waves that
can be easily observed (somewhat like surface waves

hitting the side of a fixed dock). Hence, it is also known as *far-field diffraction*. Because laser beams can be collimated more readily than other wave sources, they are favored for this type of study.

Diffraction Types and Light Sources

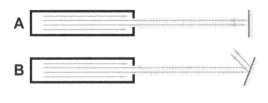

Several basic concepts in diffraction light sources are illustrated here. Imagine light beams generated by lasers that are aimed at obstacles set at distances that are far or near related to the breadth of the beam. A collimated laser beam can provide planar *light suitable for diffraction studies (A and B). When the parallel beams strike a planar obstacle sufficiently distant, they are reflected back in ways that are relatively straightforward to observe and predict with mathematical calculations. This is called far-field or Fraunhofer diffraction. (As with most things, it's actually a little more complicated. Since laser light is readily planar, you can place the obstruction closer and still get good results, but you will probably have to compensate by putting the viewing optics farther away.)*

Laser light can also be used to provide a point light source that creates a spherical *light beam when a lens is interposed to focus the beam on a small hole (C) through which it exits. When the spherical light beams encounter an obstacle, especially one close to the light source, the reflection pattern and its interaction with incoming lightwaves is more complex (D), as is the resulting diffraction pattern. This situation is called near-field or Fresnel diffraction.*

---

diffraction, Fresnel Diffraction that is more challenging to differentiate and calculate than basic Fraunhofer diffraction due to factors that complicate the interactions of the incoming waves and those that are reflected by an obstacle, such as nonparallel waves, or the close proximity of source and refracted waves is called Fresnel diffraction or *near-field diffraction*.

diffraction, selected area (SAD). The optical viewing of diffraction patterns in a selected area of a specimen, usually as viewed through an aperture for this purpose. This is a means to narrow the field of interest, for example, to view a single-crystal pattern. For even smaller views, such as a single particle, it may be necessary to use microdiffraction, with tradeoffs in precision due to optical limitations. See diffraction.

diffuser, Lambertian An ideal diffuser, with a spatially uniform lossless reflectance over a wide spectrum equal to unity (one). While no physical diffuser has perfect diffusing characteristics, there are some materials that have excellent diffusing properties and thus are useful as references for calculations, lab work, etc. Barium sulfate ($BaSO_4$) is a material with excellent diffusing properties useful as a reference standard for practical applications. Software modeling and rendering programs sometimes include Lambertian diffusers to study their influences on surrounding objects or to create certain lighting effects. See diffusion, isotropic antenna.

diffusing glass Glass that has been shaped, coated, laminated with plastic film, or sand blasted, to "break up" the incoming light in terms of its direction of travel such that the exiting light has a smoothly varying quality. Perceptually, diffused light looks "soft" and even rather than bright and narrowly focused.

Sometimes a second type of clear or colored glass is used as a laminating layer in the fabrication of diffusing glass. Thus, the exiting light emanates in many directions as it passes through the diffusing layer associated with the glass.

Small-diameter glass fiber filament arrays may be used in place of a sheet of glass, the filaments acting as prisms to diffuse light evenly.

Diffusing glass is useful for scientific instruments, lab experiments, projectors, interior lighting, windows (especially skylights, which are typically small), and other applications where a more uniform distribution of light is desired. Diffusing plastic is sometimes used in place of diffusing glass (e.g., in photo finishing) to reduce cost or weight. Light diffusion tends to increase scattering loss and glare. Scattering losses can be minimized by placing a reflector around the diffused light to reflect back up to about 90% of the light in the desired direction, as in a car headlight.

Materials can be tested for their diffusion characteristics with a spectrophotometer or modeled with ray-tracing computer algorithms. Either way has benefits. Physical testing gives a real-life measurement but may be cumbersome in terms of time and assessment, especially when evaluating unevenly diffusing materials over a wide surface area. Algorithmic testing can save time if many measurements are required or the surface area area is large or uneven, but is only as accurate as the theory and programming inherent in the software.

The fabrication of diffusing glass must be carried out such that laminated structures or coatings are tightly bonded to the glass with materials with matched (or otherwise appropriate) refractive indexes to minimize loss at laminating seams. Bonding materials must also be transparent, to let the light pass into the next layer of the laminate, and must be applied in a way that keeps out air bubbles or particles.

Since loss at joints and through materials with varying refractive indexes is characteristic of layered components, there have been suggestions for ways to incorporate the diffusing structures into the lens itself, whether this be a glass lens or a fiber optic lightguide.

Monolithic Lens/Diffusing Components	Diffusion Configurations

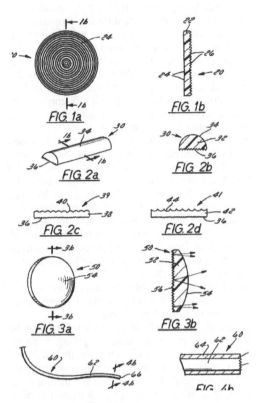

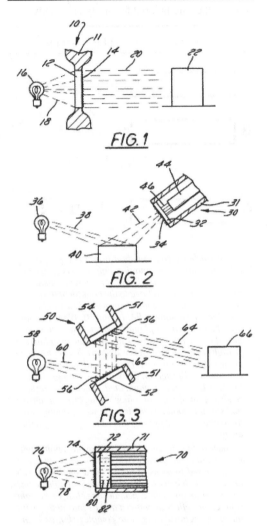

Some examples of monolithic diffusion lenses, that is, lenses that incorporate the diffusing microsurfaces in the lenses themselves rather than relying upon laminated coatings or plastics to provide a separate diffusing layer. The advantages of monolithic components include longer lifespan (coatings can peel away) and lower light losses from bubbles, particles, or differing defractive indexes. [Shie et al., U.S. patent #6,266,476, July 2001.]

Shie et al. have developed a number of designs for embodiments of monolithic diffusing elements that overcome some of problems of layered fabrications.

diffusion At the molecular level, a net transfer of mass due to random molecular motion caused by concentration gradients in the diffusing material. Molecules will move from regions of high concentration to low, a process that continues (unless interrupted) until the concentration reaches a state of equilibrium. As an example, sugar stirred into a hot cup of coffee diffuses through a process called osmosis until there is an even suspension of the sugar dissolved throughout the liquid. Certain ions are known to have higher diffusing properties than others due to their molecular motion. These properties can be exploited to create materials for facilitating diffusion in scientific experiments or commercial products.

An isotropic diffusing medium is one that is not considered dependent upon the direction of motion or

Some example configurations for incorporating a specially fabricated lens with a diffusing microstructure as it might be used in the path of a light beam to homogenize (diffuse) and propagate the beam in the desired direction(s). In this case the lens and diffusion surface are integrated, but the general concepts may apply to certain laminated structures, as well. [Shie et al., U.S. patent #6,259,562, July 2001.]

orientation of its motion and thus would diffuse evenly in all directions from the point of reference (see isotropic antenna). An anisotropic diffusing medium is one in which direction or orientation would influence its diffusion properties.

In the broader sense, diffusion refers to changes in the direction of travel of a constrained or point radiant energy source (e.g., a beam of light) such that it spreads over a wider area. The amount of diffusion that occurs in any given situation is dependent upon the angle of incidence of the radiant energy source, the diffusing materials, the power of the initial radiant energy, and ambient environmental conditions (e.g., humidity).

Diffusion is characteristic of other propagating sources besides radiant acoustic or electromagnetic energy. A drop of water hitting a flat surface straight on will diffuse in a more or less even pattern, as it spreads from the initial point of contact. See diffraction, diffusing glass, Fick's first law.

Laminated Diffusing Glass

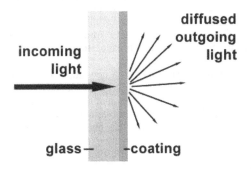

By coating a glass surface with chemicals or plastic film laminates, it is possible to alter the course of travel of a light beam to spread it over a wider area, thus providing more even illumination over a broader area. This is useful for microscope stages, for example, where even lighting aids in reducing contrasty point-light shadows, thus facilitating sample observation and imaging.

---

diffusion coefficient A mathematical means of expressing the transfer rate of diffusing atoms through molecular movement from the point at which they diffuse. This provides a way to evaluate or model the different diffusing characteristics of media and energy stimuli. See Fick's first law.

diffusion, semiconductor 1. Diffusion of dopant atoms in a semiconductor upon stimulation by heat. 2. The process by which dopants are introduced into a semiconductor medium to create p-n junctions, conduction channels, and other useful structures.

digest access authentication See authentication, digest access.

dig A Unix command providing information about computer network domain names.

digital A means of representing information in discrete units, rather than as a continuous stream. In communications technologies, information is typically represented in terms of binary units. There are many ways to represent information in a binary system: on/off, high/low, large/small, changed from previous state, loud/soft, fast/slow, lit/unlit, up/down, present/absent, etc. In digital computing, the binary units are usually ones and zeros. In electronic circuits, the units are often represented electrically by on/off, high/low, or change from previous state. Despite the simplicity of a binary system, it is powerful and flexible, and extremely sophisticated processes and

information can be achieved by combining, organizing, and variously encoding this digital information. Most communications systems prior to the 1970s were analog, but the trend is strongly towards converting analog signals into digital signals. Digital systems allow a far greater degree of control, security, compression, noise control, and modifiability compared to analog systems.

A simplistic explanation of the difference between digital and analog is often illustrated with watches. An analog watch has a hand that sweeps around in a 360° arc, showing hours and minutes and the positions in between. A digital watch has a readout that displays the time incrementally, usually in one second or one minute increments.

Another example is an analog dial on an AM radio that allows the tuner to be gradually adjusted through adjacent stations. As the dial moves, the signal volume increases and decreases, and there may even be periods where multiple radio signals overlap. This type of radio dial does not allow the listener to jump directly from a low frequency station to a high frequency station. With digital pushbuttons on AM and FM radios, the tuner can be set to jump to a specific frequency, and stations can be selected in any order, even if they are not in adjacent frequency ranges.

In an analog phone system, the phone equipment converts sound waves into electrical signals that are sent through the lines between the conversing parties, but it is also possible to encode the conversation as a digital signal and send it through a computer network or computerized phone system. Digital encoding allows the information to be compressed, modified, stored for later retrieval, or sent in conjunction with other data signals (such as a computer data file transfer) at the same time. See analog, ISDN, quantization, voice over ATM.

Digital and Analog Clock Examples

On the alarm clock on the left, a digital readout shows the time incremented in minutes. On the right, an analog display with a 'sweep hand' travels through the minutes and seconds in its arc in a continuous movement.

Digital 8 A video recording/playback system developed by Sony. It is a hybrid transitional format that records digital video signals onto Hi8mm tapes. See D-VHS, Hi8mm.

Digital Access and Cross-connect System DACS. A technology for reconfiguring a circuit, without manually changing the interconnections. It is similar to a multiplexer, except that changes can be made with software, rather than through physical rewiring.

Digital Advanced Mobile Phone Service See DAMPS.

digital audio broadcasting DAB. A transmission modulation technique that sends digital rather than analog audio signals. One of the more interesting developments in DAB is that impediments that cause interference to analog signal transmissions (e.g., mountains) are used as reflectors in digital signal transmissions to improve reception. In the process, the best regional transmitter is automatically selected to forward the signal to local listeners. Thus, a regular antenna, like the kind commonly mounted on a car, can be used to receive distortion-free programming with excellent sound quality.

DAB receivers have been commercially available since 1998, and as of the turn of the century there were more than 200 million DAB users worldwide, a number that is expected to grow substantially.

DAB signals are individually coded at the source, error protected, and time-interleaved in the channel coder. The services are then multiplexed in the Main Service Channel (MSC), combined with Multiplex Control and Service information, and transmitted through the Fast Information Channel (FIC). Transmission frames are formed in the Transmission Multiplexer. The DAB signal is shaped through Orthogonal Frequency Division Multiplexing (OFDM) and transposed to the selected radio frequency, where it is amplified and transmitted. At the receiving end, the digital signal is demodulated and channel decoded to process and error-correct the signal and then fed to the listener's radio device. Left and right audio channels are produced by processing through an audio decoder, data information is processed as is appropriate if textual or other enhanced features are included, and the final product presented to the user.

DAB is not restricted to audio broadcasting; the concept is applicable to a variety of types of programming, which means car radios, for example, could display images and textual information (song titles, artist names, lyrics, etc.) in conjunction with the radio programming. Given these characteristics, it's not surprising that there is much excitement in the radio industry about the potential of this new digital broadcasting technology. Within the next few years, it is likely to revolutionize the face and form of radio programming. Currently DAB is established as a standard for terrestrial broadcasting, but it has been developed with an eye toward future satellite broadcasting as well. See Digital Audio Broadcasting Forum.

Digital Audio Broadcasting Forum, DAB Forum, World DAB Forum A consortium dedicated to the commercial implementation of the DAB Standard through international coordination and cooperation among official governmental and regulatory authorities, manufacturers, network providers, and data broadcasters.

World DAB was initiated in Europe, but the technology developed by the consortium is being imple-

mented worldwide. In June 2000, the Asian DAB Committee was established during a conference in Singapore, and a cooperative Memorandum of Understanding was signed between the WorldDAB and the Singapore Broadcasting Authority. http://www.worlddab.org/ http://www.asiadab.org/

Digital Audio Broadcasting Standard, DAB Standard An ITU standard for digital broadcasting developed by the Eureka 147 Consortium, an international group of broadcasters, consumer electronics developers and scientists, and radio network operators. The work on the DAB Standard culminated in January 2000 and went onto the next stage, commercial implementation with the Technical and Commercial Committee of the Digital Audio Broadcasting Forum. See digital audio broadcasting. http://www.eurekadab.org/

digital audio radio DAR. A new audio broadcast technology that provides high quality sound over the airwaves and a wider selection of regional programming. It also integrates with various news, paging, and email services. There has been talk of putting DAR in the S-band, but a number of technical characteristics of DAR indicate this may not be the best choice. In 1995, the Federal Communications Commission (FCC) assigned a frequency spectrum for DAR use. See digital audio broadcasting.

Digital Audio Radio Service DARS. A high-quality-audio satellite-delivered radio programming service. Thus, listeners in remote areas who are without terrestrial broadcasting services or those who wish to receive higher-quality audio than might be available through a local analog-based station can receive high-quality radio transmissions.

Technologies change, so these numbers may change as capabilities improve, but during the mid-1990s, at least 12.5 MHz of radio frequency spectrum was needed to support a commercial DARS system. Since only 25 MHz of spectrum had been allocated to DARS in 1997 (2320 to 2345 MHz), when licenses were auctioned by the Wireless Telecommunications Bureau (WTB) of the Federal Communications Commission (FCC), only two licenses were made available to four applicants that filed by 1992. The auction bids for these licenses approached $90 million each and were finally awarded to Satellite CD Radio, Inc. and American Mobile Radio Corporation. The FCC stipulated that winning applicants must have a satellite operational within four years, with the system as a whole operational within six years.

digital audio tape DAT. A high-quality, high-capacity digital audio recording format suitable for high quality digital audio recordings and computer data storage. For audio recordings, the sound is sampled, quantized, and converted to a specified encoded format. The encoding includes error checking mechanisms and tracking information to facilitate searching for a particular location on the tape. DAT became popular in Europe in the early 1990s, but American vendors were so concerned about audio piracy on DATs that they effectively blocked the spread of the technology in the United States. DAT is now used to some extent for computer tape backup systems.

digital audio-visual DAV. Digitized audio/video data that typically bypasses a computer's main bus.

Digital Audio-Video Council DAVIC. A nonprofit association established in 1994 in Geneva, Switzerland, to promote global open interfaces and protocol specifications (DAVIC specifications) in audio-visual applications and services. There are over 200 member companies from more than 25 countries worldwide, representing manufacturing, service, research, and government agencies.

DAVIC concerns itself with the specification and development of tools rather than systems, with a focus on identifying and specifying components which are relocatable on a specific platform, and which are also cross-platform.

The DAVIC 1995 specification recommends SDH/SONET as the core network physical layer to which ATM cells, as standardized by various international bodies, can be mapped. Timing involves the use of a transmit clock derived from the network. Jitter is also managed with the network clock as the reference clock. There are five main entities within the specification, as shown in the chart.

Entity	Abbrev.	Notes
Content Provider System	CPS	
Service Provider System	SPS	
Service Consumer System	SCS	
CPS-SPS Delivery System	–	Connects CPS to SPS
SPS-SCS Delivery System	–	Connects SPS to SCS

Some technologies incorporated into DAVIC are the intellectual property of the contributors; they have agreed to make the technology available to anyone for free or for reasonable royalty fees. http://www.davic.org/

digital bank, digital channel bank See D bank.

digital bearing discriminator DBD. A digital RF-based system used in direction-finding applications, often with microwave frequencies. In conjunction with a radio-frequency receiver, it enables the direction of arrival of incoming signals to be estimated. A DBD can be configured by creating a circular array of open direction finding receivers, thus enabling a full 360-degree field of view and a high probability of intercept, without the need to rotate the antenna or to guess at the incoming direction in order to orient the antenna correctly. Each element in the array feeds individually into the network where the incoming data can be processed for spatial modeling.

digital camcorder A digital camera capable of capturing and storing information at a rate that is fast enough to create a series of digital frames which, when played back, show full motion video. See camcorder, digital video, dry camera.

digital cellular service A digital version of mobile cellular telephone communications in which the voice conversations are sampled, quantized, and encoded for transmission. This permits increased security, privacy, capacity, better handling of noise interference, and corrective processes when roaming across cells.

digital certificate See certificate.

Digital Command Control Standards, DCC Standards A set of National Model Railroad Association (NMRA) standards for packet-based digital command control. The DCC standards encompass the format of information that is sent via digital command stations to digital encoders using NMRA digital signals.

Digital Communications Conference DCC. An annual communications conference jointly held by the ARRL and TAPR radio associations. It is an international forum for beginning and expert amateur radio operations to discuss digital communications technologies and their applications.

digital cross-connect DXC. A centralized network component for aggregating and interconnecting a variety of digital signal links, ranging from a few to several thousand. For example, a multibank cross-connect running at 44 Mbps can carry almost 700 telephone voice channels per line. DXCs are used in wired, fiber optic, and digital radio systems. SONET ring-based optical interfaces for DXCs were introduced in the mid-1990s, by Lucent Technologies. DXCs typically have interface ports for remote access and monitoring. Smaller units are often rack mountable. Larger units may be floorstanding and may include cable management enclosures and prewired fuse panels. See add/drop multiplexer.

Digital Data Communications Message Protocol DDCMP. A station-to-station, byte-oriented, link-layer protocol developed by Digital Equipment Corporation (DEC) in the early 1970s that was used to develop DEC's network architecture as a processor-to-processor system (DECnet). DDCMP continued to evolve and had reached Phase IV by the mid-1980s. DDCMP functions over half- and full-duplex synchronous and asynchronous channels in point-to-point and multipoint modes. It provides management of the physical channel, message sequencing, and data integrity. DDCMP has been implemented on a wide variety of types of systems, including workstations, personal computers, and robots.

Digital Equipment Corporation DEC. A well-known computer hardware/software/services company which was established in the 1950s by Kenneth H. Olsen. DEC is perhaps best known for its PDP minicomputer series, the subsequent VAX series (VMS and UNIX operating systems), and the DEC Alpha. Many universities are equipped with VAX machines. In 1998, DEC was bought by Compaq, one of the leading makers of desktop computers. See Compaq.

Digital European Cordless Telecommunications DECT. Now called Digital Enhanced Cordless Telecommunications. An organization and wireless standard developed in Europe and adopted by the European Telecommunication Standard Institute (ETSI) in 1992. It was originally proposed as a unifying digital radio standard for European cordless phones. It has since been adopted by other countries, including Britain and some Asian countries. The DECT standard improves on previous technologies by supporting two-way calling as well as better mobility.

Open Systems Interconnection (OSI) principles have been incorporated into DECT in the sense that it consists of a physical layer, a data link layer, and a network layer.

DECT is implemented with transceiving base stations and mobile handsets. As it is optimized for capabilities different from those developed for cellular, it requires more cells to be used in a manner similar to cellular, due to the low power signals of DECT, but higher densities are then also possible.

DECT incorporates handover capabilities and Dynamic Channels Allocation (DCA) instead of fixed channels, with the hand unit scanning for the best signals.

Digital HDTV Grand Alliance A consortium of major U.S. and European entities with a stake in consumer electronics and broadcast television technologies. The Alliance was founded in May 1993 as a result of work by the Advisory Committee on Advanced Television Service (ACATS). In order to streamline the process of development and testing of advanced television (ATV) systems, the Alliance was tasked with taking the best of the best proposals evaluated and researched by ACATS and combining them into one superior technology, without making the specification too rigid or industry-centric. ACATS continued to advise the Alliance through the ACATS Technical Subgroup which was divided into six Expert Groups.

Based on the work of the Alliance and ACATS, an advanced system proposal was tendered in late 1993 and early 1994 which was subsequently approved for prototyping, construction, and laboratory and field testing in 1995. The results were communicated in the ACAT Final Technical Report. Requests from the Federal Communications Commission resulted in the addition of standard definition television (SDTV) multiple stream scanning formats being adopted for inclusion in the ATSC Digital Television Standard. See Advisory Committee on Advanced Television Service, ATSC Digital Television Standard.

Digital Loop Carrier DLC. Similar to a Local Loop Carrier, which provides a physical connection between subscribers and a main distribution switching frame, except that the DLC is committed to *digital* services over twisted-pair copper phone wires. The DLC is a system of switches and multiplexers which concentrates low-speed services prior to distribution through a local central switching office or controlled environment vault (CEV). By multiplexing signals up to a local terminal where it then splits to provide service to subscriber pairs, the cost of wiring can be reduced. DLC systems were developed in the early 1970s. See Next Generation Digital Loop Carrier.

Digital Micromirror Device DMD. A spatial light modulator semiconductor technology that has been incorporated into high-resolution Digital Light Processing (DLP) display devices. Tiny mirrors, only 16 microns square, can be organized in rows and columns to form the basis for a high-resolution display. DMD displays from Texas Instruments combine a CMOS SRAM with a movable micromirror mounted over each memory cell corresponding to a pixel on the display. The light is pulse-width modulated incident to the mirror by electrostatic forces controlled by the cell data. Filters or color wheels can be used to create color.

DMD was developed by Texas Instruments (TI) and first demonstrated by TI and Sony in the mid-1990s as a digital high-definition display system. DMD-based projection technologies are designed to be scalable and intended to provide images superior to cathode-ray tube (CRT) and liquid crystal diode (LCD). DMD technology also has applications in the digital color printing field.

Digital Millennium Copyright Act DMCA. A U.S. act enacted in 1998 through a comprehensive reform of U.S. copyright law to encompass changes brought about by evolutions in electronics that effect the creation of original digital works. The DMCA was also a step in the preparation for ratification of the international World Intellectual Property Organization (WIPO) treaties.

digital multiplexer A system for aggregating or interleaving two or more digital signals, so they can be carried over fewer transmission lines, and sometimes also to aid in synchronization of multimedia applications that may require more than one signal (e.g., audio and video for videoconferencing). The signal is frequently demultiplexed at the receiving end in order to separately handle the various component signals. See digital cross-connect.

Digital Network Architecture DNA. 1. An architecture that incorporates many aspects of the Open Systems Interconnection (OSI) model used by Digital Equipment Corporation (DEC) to develop applications. 2. A commercial network system from Network Development Corporation.

Digital Performance Archive DPA. A means for tracing the rapid developments in digital technologies as they pertain to performance arts activities. The archive is managed within the Arts and Humanities Research Board of the University of Salford, Manchester, U.K.

Digital Private Network Signaling System DPNSS. A standard for integrating private branch systems with E1 lines. DPNSS was originally developed as an open standard by British Telecom plc and U.K. PINX manufacturers, in the 1980s. In the early 1990s, DPNSS and ISDN internetworking was described, followed by DPNSS and Signaling System No. 1. Open documents are available from British Telecom.

Digital Research Inc. DR. Originally called Inter-Galactic Digital Research, Digital Research was founded by Gary Kildall and his wife at the time, Dorothy McEwen. Gary was the developer of CP/M (Control Program for Microcomputers), a popular text-based operating system for microcomputers. DR produced a line of good quality products starting with CP/M-80. GEM, the DR graphical operating system predated working versions of Microsoft Windows by several years, and DR-DOS was often described by reviewers and users as superior to MS-DOS.

DR's efforts were not limited to software. In 1984, the company released an expansion board for Intel 8088-based personal computers that allowed four terminals to be networked to a PC using standard RS-232. With Concurrent PC-DOS, it provided the user the ability to run up to four MS-DOS or CP/M-86 applications concurrently, along with the program running on each individual terminal.

Over the years, Digital Research introduced many basic desktop computing and networking tools that have become intrinsic to the industry. The company was purchased in the 1990s by the Novell Corporation, who subsequently transferred DR-DOS to Caldera who released it as OpenDOS. Unfortunately, Kildall, who pioneered so many fundamental contributions to the microcomputer industry, was found dead at the age of 52. See CP/M; Graphics Environment Manager; Kildall, Gary.

digital selective calling DSC. A synchronous transmissions system developed by the International Radio Consultative Committee (CCIR Recommendation 493). DSC is the basis for the Digital Selective Calling (DSC) communications service that provides automated access to coastal stations and marine craft. Four priority levels have been established, from routine to distress, with distress calls receiving priority handling and, in the U.K., routing to a Rescue Coordination Centre (RCC). Alarms may be associated with incoming distress calls received on marine craft. DSC calls include the caller identity, phasing signals, and error-checking signals. A dot pattern alerts scanning receivers of a call about to be received. Ships at sea are required to maintain continuous DSC watches at designated frequencies. The GMDSS Master Plan and the ITU List of Coast Stations list the DSC distress and safety call frequencies.

Digital Short-Range Radio DSRR. Initially perceived as an easy-to-license evolutionary heir to public domain Citizen's Band radio services, DSRR has since become a commodity for offering commercial services within the business community in North America and a venue for offering remote public services in Europe. It incorporates digital radio communications technologies for small low-range portable radio sets and has become an ETSI and TETRA standard for low-cost radio communications. DSSR operates in single- and double-frequency repeater modes in 933-935-MHz and 880-890-MHz bands. In North America and Australia, the double-frequency band is reserved for AMPS cellular services.

Enhanced Digital Short-Range Radio E-DSRR was introduced through the RACE MOEBIUS project, designed to make use of the INMARSAT HSD Satellite mode. Initially the system is being tested in remote European sites for applications such as the education of the children of itinerant travelers or personnel stationed in inaccessible areas. It is also being evaluated for its use in telemedicine applications.

digital signal hierarchy DS-. A North American time division multiplex (TDM) signal hierarchy, which is used in connection with data communications protocols. See DS-0 through DS-4, T1.

digital signal processor DSP. A specialized computer processor designed to work with digitized waveforms, often audio and video samples, in order to speed execution and provide more complex operations. Their computing power and flexibility allow them to be used for a wide variety of applications, such as the compression of voice and video signals, multimedia applications, medical imagery, combination phone/fax/modem devices, etc.

digital signal cross-connect panel DSX panel. A type of electrical cross-connect wiring bay or closet to facilitate the interconnection or patching of digital telecommunication facilities and equipment. This facilitates rearrangement, restoration, or monitoring of circuits. Bantam jacks are common in DSX panels. DSX panels connect a wide variety of equipment, depending upon the type of service (e.g., T1). Circuit connections may include channel banks, multiplexers, switches, repeater bays, and terminating connectors or circuits. See digital cross-connect, tap.

digital signature A type of digital identification associated with an individual or association that is sufficiently unique, secure, and resistant to forgery that it can be used for confidential and commerce-related online messages and transactions. A digital signature is essentially the electronic equivalent of a handwritten signature that can be traced to the person who created it. A digital signature was initially seen as a digitized version of this handwritten signature (somewhat like a rubber stamp signature), but it was quickly realized that a digitized version did not have the same verification characteristics (pressure, direction, speed, etc.) that were inherent in a handwritten signature, and other more abstract versions of the digital signature were developed that not only were more unique but were amenable to strong encryption techniques to ensure security. (The author feels there may still be some merit in developing algorithms that actually encode the pressure, speed, and other characteristics that can be measured in a person's signature through a special pressure-sensitive pad. There is more research that can be done in this area and specialized circumstances in which it could be applied, as in encoding signatures for local legal transactions to help prevent fraud.)

In a more technical sense, for the purpose of implementing software, a digital signature has been described as a value generated from an application via a cryptographic algorithm that embodies data integrity, message authentication, and/or signer authentication. A number of digital signature schemes are already in use for stock-related transactions, contracts, and general messaging. Digital signatures typically employ key encryption methods.

In Aug. 2001, the W3C described a Proposed Recommendation for XML digital signature processing rules and syntax to provide integrity, message authentication, and/or signer authentication services for creating independent, interoperable implementations. The XML Signature is a method of associating a key with referenced data. The XML namespace Uniform Resource Identifier (URI) and prefix for other sub-URIs for the W3C specification is `xmlns="http://www.w3.org/2000/09/xmldsig#"` See encryption, JEPI, Pretty Good Privacy, signature.

Digital Signature Standard DSS. A draft standard to permit the creation and transmission of a secure digital signature through a Digital Signature Algorithm (DSA) to provide authentication of documents and transactions. Web commerce is eagerly seeking means by which documents can be electronically secured in order to use them for trade, banking, stock transactions, contract negotiations, etc. and will probably quickly adopt this or another scheme when sufficient confidence in its efficacy is attained. See Electronic Certification.

Digital Steppingstones project A project of the Tomás Rivera Policy Institute (TRPI) that studies issues of access to technology and telecommunications networks and examines exemplary practical implementations in diverse environments, including libraries, schools, and community centers across the U.S. The Digital Steppingstones project focuses on underserved communities such as low-income and minority communities. An important aspect of this project is assessing access to the Internet and making recommendations for its broad and practical implementation in publicly accessible facilities.
http://www.trpi.org/dss/

Digital Subscriber Line DSL, *x*DSL. A data transmission service operating over existing copper public phone lines. Imagine turning on your computer, connecting to the Internet, and finding something interesting that you want to explain to a business colleague or friend. If you have only one phone line, and you're using a modem to change the computer's digital signals into analog signals that can be sent over the phone line, you would historically have to hang up the modem, wait for a dial tone, and *then* call your colleague or friend. Digital Subscriber Line is a family of two-way communications services that makes it possible for you to talk to your friend *without* hanging up the computer connection first. You can do both at the same time, which means you can talk through the phone while you navigate the Net together, discussing the things that you both can see. This is how it is done.

Phone services historically have been analog systems, and there are millions of miles of copper wires installed around the world to provide these services. With the development of computers, phone switching centers began, in the late 1980s, to convert to digital equipment and software. This enabled voice and data to be carried on one line at the same time, and instead of using a modem to change the computer signal to analog, and leaving the voice as an analog signal, it could be done the other way around. In other words, now the voice call is changed to digital and

the computer signal remains digital. This opens up a world of possibilities for faster transmission, better compression and security, and simultaneous data/ voice communications, without having to replace those millions of miles of copper wires.

That sounds very practical, yet relatively few people have switched to DSL services. One of the reasons is distance. While most subscribers are within the 12,000 feet or so in which DSL services can operate at their best speeds, about 20% of the population is not. Crosstalk and other types of interference are problematic as well and are still being resolved. For example, traditional phone lines have loading coils installed at intervals, to extend the signals on voice grade communications. Unfortunately, at higher digital data rates, these coils cause interference.

Perhaps more important is the way in which DSL services were deployed. Originally, subscribing to DSL involved having the phone company install a special voice/data splitter on the subscriber premises and, further, installing a special peripheral device in the subscriber's computer. This method was costly and not very practical, and most consumers are resistant to having proprietary peripheral cards installed in their computers. Most prefer the option of choosing a vendor and interface, and also of installing the hardware external to the computer, so things can be changed around as needed. For this reason, a number of commercial vendors have proposed several variations of DSL services, such as DSL Lite.

DSL was first developed by Bell Communications Research Inc. in 1987 to provide a means to deliver interactive TV and video-on-demand over copper wires. The name is somewhat confusing, since it is not the *line* that is installed, but rather the *interfaces* at each end of the line. The point of DSL was to create technology that would make use of existing lines. In fact, a DSL line typically consists of two telephone lines. Since the introduction of DSL, further variations have been adapted, as shown in the Digital Subscriber Line Services chart.

Digital Subscriber Line coding and variations
Since DSL is a multichannel service, it is necessary to split the available bandwidth to utilize it efficiently. This bandwidth splitting is typically done with echo cancellation (EC) or frequency division multiplexing (FDM).

There are two predominant schemes for subdividing available bandwidth into smaller units to individually evaluate their transmission suitability. This is useful over twisted-pair copper lines, which can vary widely in their characteristics. The two most common are discrete multitone (DMT) and carrierless amplitude and phase modulation (CAP), and others are being developed. Some of these modulation techniques have descended from the telegraph and radio broadcast industries, and some, such as wavelet encoding, are relatively new and still being explored. Each of these has various trade-offs in terms of availability, cost, speed, and susceptibility to interference, as shown in the Common Modulation Schemes chart.

Digital Subscriber Signaling System 1 DSS1. A telephone signaling system standardized as ITU-T Q.931 that is implemented over Layer 3 of the ISO communication model. It defines protocol for establishing, maintaining, and tearing down calls. DSS1 is commonly used in local loops providing Integrated Services Digital Network (ISDN) services for transmission over the D channel. Once the DSS1 signal reaches the local telephone switching office, it is usually transmitted to external nodes using Signaling System #7 (SS7). See D channel, Integrated Services Digital Network.

Digital Subtitle Encoder DSE. A PoliStream second-generation multimedia titling encoder that is capable of transmitting ideographic languages (e.g., Chinese) in varied font styles. It is a Digital Video Broadcasting (DVB) standards-compliant bitmap imaging system designed to transmit multilanguage subtitles. It accepts subtitle data from traditional controllers and generates realtime output of compressed bitmaps. The data may be sent through regular broadcast channels, satellite broadcast relays, or computer network links (e.g., Ethernet).

Digital Supervisory Audio Tone DSAT. A means of signaling using audio tones on cellular data networks. In AMPS cellular signaling, there are three designated supervisory audio tones (SATs). However, in NAMPS, there are instead seven subaudible digital vectors called DSATs. SATs and DSATs are used to verify the correct channel tuning after the channel has been assigned. The central office (CO) notifies as to the new voice channel and vector.

digital switching matrix DSM. A matrix format for digital switching that facilitates signal routing in high bandwidth applications. The device can be conceptualized as a grid of channels (e.g., 64 ¥ 64 data channels), which may be individually selected to input or output to any other selected channel, or the channels may be grouped into frames for combined processing. In commercial applications, the DSM may accept both digital and analog signals, but the signals will be converted to digital signals before being switched (routed). Pulse code modulation (PCM) is typically used for digitization in telephony applications. A commercial DSM for high-capacity telephonic switching is microprocessor controlled and may also include memory, time base, speech, and control components.

DSM is used in data switchers for realtime switching between synchronous video sources. It is also used in high-capacity telephone networks for routing calls to various application processing units. In a Datapath architecture digital telephone network, it is the switching unit between the data line card and the digital carrier module or trunk controller. See Datapath Loop Extension.

Digital Telephony and Communications Privacy Improvements Act Also known as the Digital Telephony Bill, this law was passed by the U.S. Congress in October 1994. The terms of the bill, drafted to be supportive of the efforts of law enforcement agencies, would require common telecommunications carriers to design networks in such a way that law enforcement agencies could access, in realtime, the contents of communications on their networks and transactional signaling. The Electronic Frontier Foundation came out strongly against the bill at the time, but after the events of 11 Sept. 2001, the same issues have been raised again and proposed as changes to the Security and Freedom through Encryption Act developed in the late 1990s.

When the Digital Telephony Bill was first introduced,

Varieties of Digital Subscriber Line Services			
Type	Abbreviation	Speed	Notes
asymmetric DSL	ADSL	6 Mbps +	Twisted-pair copper phone wires. The possible maximum rate of transmissions is inversely proportional to distance. Typically uses discrete multitone (DMT) line coding for data; frequency division multiplexing (FDM) or echo cancellation is used to subdivide the bandwidth.
high bit-rate DSL	HDSL	T1/E1 speeds	Symmetric. Longer distances can be supported through the use of repeaters. See high bit-rate Digital Subscriber Line.
single line DSL	SDSL		Still in development. Can be used over a single wire pair.
rate adaptive DSL	RADSL	up to 8.7 Mbps	Bandwidth can be tuned to subscriber needs. It works over longer transmission lines. Rate and speed adjust to the line length and quality.
very high rate DSL	VDSL	13-60 Mbps	Used in conjunction with FTTC or FTTB. Different downstream and upstream speeds. (Upstream speed is 1.5 to 2.3 Mbps.) Shorter maximum distance. Still in development.

wiretapping proponents suggested that electronic wiretapping was just an extension of current wiretapping practices as applied to electronic communications, but technologists and privacy advocates argue that the analogy is not tenable – the capabilities inherent in tapping into the National Information Infrastructure through electronics opens up far-reaching potentialities that are not safeguarded by current or proposed legislation in a way that is comparable to the safeguarding of traditional wiretaps.

As important as it is for law enforcement to find ways to keep up with crimes that are increasingly conducted through new electronic technologies, it is also important to examine the implications of expanding law enforcement tapping capabilities. How is electronic wiretapping different from traditional wiretapping? Here is a summary of just a few of the many complex issues involved:

Visibility. The establishment of an electronic wiretap can be remotely executed and is thus invisible, in contrast to traditional wiretaps where a telecommunications technician has to physically install and deinstall a tap. In a physical connection system, accountability is more readily determined and traced. Electronic wiretapping, on the other hand, is hard to monitor. A tap can potentially be engaged and disengaged through software that cannot be seen by anyone other than the user.

Duration of a Tap. There are laws and warrants determining the physical location and effective duration of traditional wiretaps. The physical removal of a tap or changing of a switch at a phone company ends the transaction when the court-assigned permit expires. Even with the same court order requirements in place, there are currently few safeguards consistently integrated into the network systems of Internet Services Providers or the Internet as a whole that can ensure that the process has been ended as required. Due to the nature of software and network communications, an electronic tap could be continued beyond the stated deadline with little chance of discovery. Some proponents feel that accountability needs to be ensured by agents outside of the system requesting the tap. ISPs might be a natural choice, but then the time and economic burden of compliance fall on vendors rather than on law enforcement agencies and vendors would, in essence, be entrusted with watching the police, a situation which has historically led to problems such as deal-making and, in some cases, corrupt alliances. Before electronic wiretapping can be fully endorsed and implemented, a new accountability technology needs to be built into global networks. Perhaps commercial vendors looking for new markets can help resolve these issues with innovative products.

Eavesdropping on Computer Users Outside the Purview of the Tap. A physical wiretap is limited in scope. You must install the tap at or near the premises being tapped. The tap may inadvertently pick up conversations of innocent individuals in the vicinity, but this is usually a local rather than a large-scale problem. With an electronic wiretap, the local physical limitation is gone. Theoretically, communications and activities of hundreds of millions of individuals could be monitored with the same mechanism used to monitor the alleged criminal, without any obvious sign that this is happening. This aspect must be addressed before broad-ranging powers are given to officials to tap a medium like the Internet with technology that is exceptionally powerful and generalized in its capabilities. See American Civil Liberties Association, Electronic Freedom Foundation, Security and Freedom through Encryption Act.

Digital Television Standard DTS. See ATSC Digital Television Standard.

Digital Television Station Project, Inc. DTSP. A collaborative project of the broadcast industry, including almost 300 broadcasters, computer manufacturers, content creators, and service providers. Based in Washington, D.C., the DTSP originated as the Model HDTV Station Project (WHD-TV), established by the Consumer Electronics Manufacturers Association (CEMA) and the Association for Maximum Service Television (MSTV) in 1996, to provide a hands-on educational facility for the implementation of digital television (DTV) technologies. DTSP is the January 2000 follow-up to this project, supported by the original sponsors and also by the National Association of Broadcasters (NAB). DTSP was established with an 18-month mandate to operate WHD-TV, a model digital television station, and provides continued opportunities to develop and test interactive television, data broadcasting, and overlay services such as closed captioning, lip syncing, etc. See Association for Maximum Service Television.

digital trunked radio system DTRS. A radio communications system in which communications channels are shared over trunks, as opposed to individual channels being shared by users as they become available. DTRS systems can be set up as individual or group channels (called talk groups) or combined as groups of groups, making it a useful system for government, emergency/safety, and industrial communications. Priority and encryption can be incorporated into digital trunked systems and they can be configured to intelligently manage traffic over the trunks to optimize efficiency of the system.

digital to analog conversion The conversion of data stored as discrete units, usually in ones and zeros on computer systems, to modulated analog wave patterns. A modem is a common device which performs digital to analog conversion when changing computer signals to modulated analog signals that can be carried electrically through a phone line connection. The process is reversed at the receiving end. See D/A conversion, modem.

Digital Versatile Disc DVD. Also frequently referred to as digital videodisc, when the content is primarily image-based, but the format is not limited to video images, and a large proportion of digital videodiscs also contain audio.

DVD is a vendor consortium-developed ISO-9660-

supporting optical disc format specification, similar to compact disc (CD), except that it is designed to store a much larger quantity of data. Standardization has not been a single process. The DVD Forum, a consortium of developers, and Philips and Sony, independent commercial developers, have provided somewhat different versions of DVD. The Philips and Sony technologies can be licensed on a royalty basis, while the DVD Forum specifications are shared technologies. The DVD format is gradually becoming standardized so that DVD consumer players and DVD-enabled computers can interchangeably use the discs. DVD is quickly growing in support and acceptance by consumers, especially for entertainment and computer data storage purposes. One of the more recent developments is the release of the DVD+RW format by the DVD+RW Alliance, which is not directly related to the DVD Forum.

DVD physical discs are the same diameter as the CDs popularly used for music (120 mm), but very slightly thicker, bringing the recorded surface of the disc a little closer to the laser pickup, permitting a higher resolution or areal density, i.e., smaller, more precise pits can be used to store the information, depending upon whether the disc is recorded single- or double-layered, and single- or double-sided. DVDs use higher density storage and different modulation and error-correction schemes than CDs. Further flexibility is possible through the use of dual lens apertures in the laser pickup to provide a dual CD/DVD player. While a DVD player can usually play CDs for backward compatibility, a CD player does not inherently have the capability to play DVDs.

DVD specifies more than the compression and playback format, it also provides functionality to build interactivity into the medium through menus, multiple languages, and other features. This functionality makes it attractive for educational software and games programming and allows movie makers to include extra features.

DVD designates MPEG-2 as the digital compression standard for video recorded on DVDs. MPEG-2 is a fast digital motion recording and playback specification. A DVD can be recorded on both sides for up to a total of about 18 Gbytes of data. This capability is very attractive to developers, as it means a two-hour MPEG-2 encoded movie can fit on one side. Sound is encoded in either MPEG audio or Dolby AC-3.

DVD can be played on a stand-alone system similar to a combination CD/laserdisc system; it works like a laserdisc player, but is small like a CD player. It can also be played on a computer through a DVD computer peripheral player.

DVD can provide many types of audio storage and playback but, because of its high storage capacity, vendors are particularly interested in offering movies on DVD rather than on cassette tapes or the larger-format videodisc formats. The DVD medium is more robust and convenient than tapes or large discs, provides better sound than most tape technoloogies, and can store motion pictures of up to 133 minutes in length, including subtitles. This capacity isn't sufficient to hold every type of movie, so longer films may be offered on two discs or are compressed with a number of innovative predictive coding methods.

It is probable that DVD, or something like it, will eventually supersede traditional video cassette tape movies. DVD is also a promising technology for electronic books (ebooks).

See Digital Versatile Disc player, MPEG. See land/groove recording and its associated cross-references for further details on how DVDs are recorded.

Digital Versatile Disc associations Much of the impetus for DVD development has come from industry alliances to promote and develop the technology. See individual entries for organizations of particular interest, which include the DVD Forum and the DVD+RW Alliance.

Digital Versatile Disc player, DVD player A consumer electronics component for decompressing and outputting the digital data from a Digital Versatile Disc (DVD). The player uses a laser light beam to scan and read the information coded into pits and grooves in the DVD medium. DVD players may play a single disc or may be capable of holding multiple discs (similar to CD disc changers and jukebox systems).

Most DVD players will also play audio CDs but the DVDs themselves are not directly CD-compatible. The DVD format uses a higher density of storage on the disc and different modulation and error correction schemes than CD. Consumer DVDs are designed in much the same way as laserdisc players and VCRs to play movies (multimedia) and audio recordings on a television set or through a monitor and separate audio components. DVD players are also available for personal computers and may be internal (similar to a CD-ROM player) or external (similar to an external CD-ROM or cartridge player). In the past, DVD players that worked on computers couldn't necessarily play the same DVDs as a separate consumer DVD player, but intercompatibility is now more common.

DVD players are also being designed as small portable entertainment devices (like small boom boxes or portable television sets). Movies are one of the most popular types of DVD products being sold and rented. Typically movies played back on DVD players with appropriate monitors support high quality component (Y/C) video at about 500 horizontal lines of resolution compared with only 210-225 horizontal lines of resolution in VHS. DVDs are mastered at CCIR601 4:2:2 ratios. DVD-Video supports multiple aspect ratios. When movies are played back, the consumer is usually given a choice of aspect ratios, selected from letterbox (4:3), pan and scan (4:3), and anamorphic (wide screen 16:9) formats. DVD players sold in Europe often support both PAL and NTSC while those sold in North America may only support NTSC (this is changing as higher definition formats are finally beginning to catch on).

Audio frequency response is generally in the range of 2 to 44,000 Hz with a signal-to-noise ratio of 110 decibels or more. The higher-end DVD players tend to have added features, such as progressive scan, various filters, digital video equalizers, surround sound, and slow-motion effects, that are not available on the less expensive models.

DVD+RW, one of the newer versions of Digital Versatile Disc technology, is promoted by the DVD+RW Alliance as a storage and digital video recording format that will be compatible with the majority of existing DVD players. It is expected that DVD+RW players/drives will be able to read CD-ROM, CD-RW, DVD-R, DVD-RW, CD-Audio and other similar discs. See Digital Versatile Disc, DVD Forum.

digital video DV. Technologies that enable the recording and playback of digitally encoded moving image information and sound fall into the category of digital video. Some of the big barriers to inexpensive digital video have been the large amount of data that is required to record even small segments of video and the wide bandwidth and processing that is required to quickly display high resolution color images. When still images from film frames using cell animation techniques are individually digitized and stored, each frame may require up to 24 Mbytes, if it is to approximate closely the image quality of 35mm film. Since each second of animation requires between 24 and 30 individual still frames, as much as 720 Mbytes may be needed to store a second of video. A full-length movie is usually ca. 7000 seconds or more, requiring more than 5,000,000 Mbytes of storage. That's a lot, and that's not including sound or data that might be added to provide search and retrieval markers interspersed with the images.

Digital video has been in development longer than most people realize. Much of the research in this area originated in the Research Center in Yokohama, Japan, in 1977. By 1979, Ampex, Bosch, and Sony were introducing technologies based on digital video concepts, and there was a move to introduce new standards to encompass emerging digital technologies.

In order to make digital video technology possible, a number of innovations and trade-offs have been implemented. Data compression and decompression techniques are used to store images in less space, but

Digital Versatile Disc (DVD) Formats	
Format	Description
DVD-Audio	A CD-like DVD with improvements in audio fidelity and higher capacity than traditional CDs.
DVD-R	Primarily a read format. It is a professional authoring format in which the DVD data, when ready to master, is recorded once to the DVD and then read as many times as desired.
DVD-RAM	A type of hard storage with random read-write access. It is a medium that can be written to, with an appropriate DVD drive, more than 100,000 times and functions much like a hard drive, except that the discs are easily swapped and stored (like cartridges). DVD-RAM discs can be used to store up to about 4.7 Gbytes of data.
DVD-RW	Similar to a DVD-RAM, in that it is a type of hard storage, but rather than random-access, it uses sequential-access, similar to streaming tape storage for up to about 4.7 Gbytes per side. It can be rewritten up to about 1000 times and thus is an appropriate choice for data that doesn't often change, such as backup data, or database information that is read more often than it is written.
DVD+RW	This rewritable DVD format is one of the more recent versions, promoted by the DVD+RW Alliance. Hewlett-Packard announced the release of its first DVD+RW drive in August 2001. DVD+RW is primarily intended for data storage and recording digital video onto 4.7-Gbyte discs. The Alliance plans to release software upgrades to support DVD+R as well. This format has been developed without the endorsement of the DVD Forum and is competitive to current formats.
DVD-ROM	A format for computer data storage drives that is generally available with SCSI and ATAPI interfaces with capacities ranging from about 2.x Gbytes to over 9 Gbytes. Blank rewritable cartridges are available in both single- and double-sided optical cartridges and recordable discs. DVD-ROMs may at some point supersede CD-ROMs.
DVD Video Text	Sometimes called DVD-Text, an optional means to store and access textual information related to a DVD. It can consist of consumer information for users of DVD players or DVD-ROM drives, text information for content providers or DVD authors, or textual supplements to a video or audio data stream to enhance the value of the main information on the DVD. A guidebook for DVD-Text is available from the DVD Forum.

the picture quality does not equal film, and fast processors and frame buffers are required to handle playback in realtime.

In spite of its limitations and technical requirements, developers are forging ahead with digital video products partly because digital video can be edited and manipulated in remarkable ways. Special effects that are impossible or difficult to achieve with analog film are possible with digital video. DV also has greatly increased possibilities for interactivity and access through the Internet. Of further importance is the fact that it doesn't have to go through a chemical photo-finishing process before it can be used. See animate, celluloid, D-1, Digital Versatile Disc, interactive video, MPEG, video-on-demand.

Digital Video Broadcasting DVB. A family of compatible television/media delivery standards including cable and satellite technologies. Digital Video Broadcasting standards are supported by the Digital Video Broadcasting Group and published by the European Telecommunications Standards (ETSI). In conjunction with the Centre for Electrotechnical Standards (CENELEC) and the European Broadcasting Union (EBU), ETSI has formed a joint technical committee to handle DVB standards. DVB is gaining acceptance in Europe as a digital video infrastructure and is gradually replacing the traditional analog environment. There are a number of subdivisions of DVB technologies.

DVB Technology Subdivisions

Subdivision/Description	
DVB-C	cable transmission standard
DVB-S	satellite transmission standard
DVB-T	Digital Terrestrial Broadcasting (DTTB) designed to be adapted to the needs of local frequency and geographical environments. It enables the development of single-frequency networks (SFNs). Configurable to support legacy 50- or 60-Hz systems.

See Digital Video Broadcasting Group.

Digital Video Broadcasting Group DVBG. A European trade consortium that provides support and specifications for traditional and emerging broadcast technologies, such as cable and satellites, and schemes for the protection of commercial programming. See Digital Video Broadcasting.
http://www.dvb.org/

digital video format A series of digital video standards introduced since the 1980s, including D-1, D-2, D-3, etc. See Digital Video Formats chart.

Digital Video Interactive DVI. A digital recording and playback chipset technology developed at the David Sarnoff Research Center. The technology was acquired by Intel Corporation which subsequently developed it into Indeo 2 and Indeo 3. It is now known

as Intel Video Interactive (IVI). See Intel Video Interactive.

digital videodisc See Digital Versatile Disc.

digital voice encoding A process of sampling and quantizing voice signals and storing them as digital data. Since voice encoding can require much memory, the information is usually compressed for storage and decompressed when replayed. Fractal and wavelet compression techniques are becoming popular choices. Much of the technology for digital voice encoding has come from the music industry. Research into the sampling and playback of synthesized music can be applied effectively in voice encoding and playback applications.

Digital voice encoding is used in speech and voice recognition systems and is used to send conversations over digital voice telephone channels, such as Internet telephone applications, ISDN, or digital PCS. To save memory and cut down on transmission time, interesting algorithms for removing pauses and spaces are used in conjunction with digital voice. Encryption to ensure privacy is also possible with digital voice communications.

Digital voice encoding can provide a data library of sounds, phonemes, words, or other units, for use in digital applications such as automated voice menu systems, speaking computer applications, digital talking books, answering machines, voice mail systems, and more. Digitally encoded voice tends to be more pleasant and natural sounding than mathematically generated voice and, hence, is favored for applications where instructions or responses are given to a human listener. See quantize, silence suppression, speech recognition, voice recognition.

Digital8 A digital video format developed by Sony. It records digital video data and stores it on Hi8/8mm tapes. The recording time is about 45 minutes due to the faster rolling speed of the tape compared to Hi8/8mm (which records about two hours). See Hi8mm.

Digital-S See digital video format (D-9).

digitize v. To convert from analog to digital, i.e., to take an analog signal and convert it into data that consists of discrete units, such as ones and zeros, usually by *sampling* the analog signal at discrete points and assigning a value to the data at that point. Since most broadcast media are transmitted as waves, they are analog systems of communication. However, in order to process the information or interface it with networks, it must be converted to digital format at some point in the transmission process. Thus, digitizing a phone signal makes it possible to add features such as compression, encryption, and voice recognition. Digitizing a video signal allows processing of the image: palette changes, overlays, image composites, split screen viewing, videoconferencing and more. A desktop scanner is a type of digitizer, as is a digital camera. See analog, digital, digitizer, pulse code modulation.

digitizer A device that converts a signal from analog to digital, usually by sampling the analog signal at discrete points over time and assigning a value to the measurement. Up to a point, the more frequent the

D

sampling, the better the encoding represents the original. Digitizers are commonly used in video and audio applications. The *sound patches* used in electronic music are digitized from analog sound samples.

Digitizers are not constrained to only two dimensions. A pen or robot arm can trace contours on a physical object and convert the spatial information into a 3D coordinate system for rendering, ray tracing, or CAD programs. One of the first widely distributed 2D digitizers for microcomputers (1986) was NewTek's DigiView for the Amiga. It is historically significant as the forerunner to the Video Toaster, which initiated the desktop video industry. See sampling.

DII See defense information infrastructure.

Dijkstra's Algorithm In ATM networking, an algorithm sometimes used in conjunction with link and nodal state topology information to calculate routes.

DIM See document image management.

dim fiber Fiber optic cable in which the carrier provides the means to carry the signal through the fiber but does not originate the signals at either end of the circuit. See dark fiber.

dimmed Lowered in illumination or visibility. Reduced in intensity. See ghosted.

DIMS 1. Document Image Management System. A hypermedia-based document management system jointly developed by HyperMedia and Aramco for the purpose of capturing the geological, geophysical, and reservoir engineering data at Aramco. 2. document and image management system. A system for automating and integrating the management of images of text and graphical documents that are typically scanned and stored as digital data.

DIN 1. Deutsches Institute für Normung. German institute for standardization, one of the major standards bodies in Europe. DIN is located in Berlin. The DIN specification is a widely used standard for computer connectors. 2. dual inline.

DINA 1. Danish Informatics Network in the Agricultural Sciences. 2. Distributed Intelligence Network Architecture. 3. Dynamic Intelligent Network Architecture. A research project topic at Carnegie-Mellon University.

dinosaur *slang* A large, obsolete, or aging system that requires excessive resources to keep in operation. Given the speed of technological obsolescence in computer technology, systems quickly become dinosaurs, but firms often hang on to them because the cost of installing and learning new technology may be higher than continuing to use old, slow, but tried-and-true systems. The term dinosaur applies especially to old room-sized supercomputers that are now less powerful than many desktop systems.

diode, Fleming tube Historically, an electron tube with only two electrodes, the cathode (electron-emitting) and the anode (electron-attracting) in which

Digital Video Formats	
Type	Description
D-1	A component digital video format approved by SMPTE in 1985 and introduced by Sony as the DVR-1000 in 1987. It is favored by digital effects videographers. D-1 records uncompressed 8-bit video. The cassette tapes are large, weighing up to six pounds.
D-2	A composite digital video format introduced by Ampex (1986) and Sony (1988) that enjoyed broad popularity until the late 1990s, superseding composite video tape recorders (VTRs) using 1″ tapes. D-2 (CCIR 601) records uncompressed video. The cassette tapes are large, weighing up to six pounds.
D-3	A composite digital video format developed by Matsushita and introduced by Panasonic as a competitor to Sony's D-2 format. D-3 records uncompressed video.
D-5	A component digital video format developed by Matsushita and introduced by Panasonic as a competitor to Sony's D-1 format. D-5 records uncompressed 10-bit video. D-5 has caught on in High Definition Television (HDTV). Some D-5 systems can play D-3 tapes. Panasonic has introduced a high definition 4:1 compression ratio version of D-5.
D-6	A high-definition digital video format developed by Toshiba/BTS and introduced by Philips. D-6 records up to 64 minutes per 19 mm cassette through a system of helical scan tracks with a track pitch of 22 micrometers. There is a longitudinal analog cue track, a control track, and a time code track.
D-7, DVCPRO	(DVCPRO is the more common name, D-7 is the SMPTE designation, but it is placed here for easier comparison with other D- formats.) A component digital video format introduced by Panasonic. D-7 records compressed 8-bit video at 34 mm/second. D-7 supports 16-bit, 48 kHz digital audio. Streaming video is supported through SDTI and IEEE-1394 (FireWire) standards.
D-9, Digital-S	A component digital video format introduced by JVC. D-9 records 3.3:1 ratio compressed 10-bit video. Promoted as similar in quality to Digital Betacam.

the electrons flowed freely and uncontrolled, which wasn't very useful. L. de Forest developed a Fleming tube into a *triode*, a three-electron tube, which enabled control of the electron flow.

In transistor electronics, a diode is a piece of semiconductor material, positive on one side and negative on the other, with a terminal at each end. Where the electrical positive and negative regions come together in the semiconductor, it is called a p-n (positive-negative) junction. Like a two-element electron tube, the electrons normally flow in one direction, serving as a rectifier, enabling conversion of energy from one form to another – an essential aspect of communications technology. An old crystal radio rectifier can change radio waves into audio waves that can be heard through earphones. A semiconductor rectifier can change alternating current (AC) to direct current (DC) for powering electronic equipment.

Forcing the flow of electrons to go opposite to the natural direction can be accomplished in some circumstances with sufficient voltage, resulting in reverse bias. This practice is sometimes useful in semiconductor technology for altering the information stored in a chip. See avalanche breakdown, erasable programmable read-only memory, Zener diode.

Fleming Valve – Two-Electron Tube

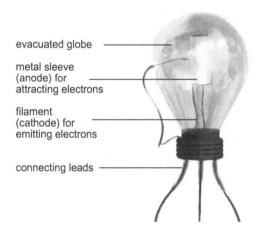

evacuated globe

metal sleeve (anode) for attracting electrons

filament (cathode) for emitting electrons

connecting leads

The Fleming valve was an important historical antecedent to three-element vacuum tubes but lacked the controlling grid that made it possible to control electron flow.

---

diode laser module DLM. See laser diode.

diode transistor logic DTL. A circuit board configuration wherein the logic is accomplished with diodes as opposed to resistors, reducing interaction between inputs and permitting many diodes to be used. However, depending on the design and application, there may be a trade-off in speed. DTL calculators during the 1960s can be found that include both transistors and integrated circuits and represent some of the early DTL devices. Sometimes referred to as diode logic (DL), the early implementations tended to be discrete-

component bipolar logic circuits (eventually superseded by integrated circuits). As integrated circuits (ICs) became more commonly used, DTL became an IC implementation of the basic diode logic. See transistor-transistor logic.

DIP 1. document image processing. 2. dual inline package. See DIP switch.

DIP switch dual inline package switch. A very small switch that is meant to be toggled to one side or the other (hence "dual"). It can be changed with a pencil or other pointed object. A few DIP switches are large enough to be toggled with the fingers. The early desktop microcomputers, such as the Kenbak-1 and the Altair, were programmed with DIP switches.

The smaller DIP switches are often found on SCSI devices, for setting SCSI ID numbers or for adjusting other settings. Graphics controller cards sometimes have DIP switches to adjust resolution or scan rate settings.

dipole antenna In its basic sense, an antenna with two poles mounted horizontally to produce one long rod that is a subdivision of the length of the wavelength it is designed to receive. Ideally the impedance (expressed in ohms) of the lead connecting the dipole antenna to the receiving equipment (radio, television, etc.) should match the impedance of the antenna at its strongest point (usually the center in a symmetrical antenna). The frequency response of a simple dipole antenna can be increased by placing the two rods proximate and parallel, and connecting them at both ends to form a *folded dipole antenna*.

DirecPC A commercial service from Hughes Network Systems allowing data communications through Very Small Aperture Terminal (VSAT) satellite systems with a personal computer. The satellite signal is received by a two-foot parabolic dish antenna that feeds the transmission to a peripheral card in a computer where the signal is demodulated, demultiplexed, decoded, and then sent to the software interface. The software can also orient the satellite dish.

An innovative version of this system for Internet use combines the satellite receiving system and a normal analog connection to an Internet Services Provider (ISP) as an upload/download hybrid system. This service allows the user to connect to an ISP through the normal phone line with a conventional modem, to interact through a Web browser. When files are requested, rather than downloading through the ISP phone connection, the files are transferred from the DirecPC network operations to the VSAT satellite, and then to the user's DirecPC dish, thus providing downloads at 400 kbps compared to about 39.6 kbps.

direct access storage device DASD. Quick access computer storage devices such as hard drives and memory chips.

direct broadcast satellite DBS. Originally this phrase was intended to describe a particular service transmitting in the 12.2 to 12.7 GHz range, a frequency approved by the World Administrative Radio Conference (WARC). However, since the availability of smaller, more convenient receiving dishes, the term is also used in a broader context, to describe any satellite that transmits a signal that can be picked up by individual home and business subscribers, without going through an intermediary station.

Since direct broadcasts are transmitted to small consumer dishes, and the size of the dish is related to the length of the radiant waves being received, the higher wavelengths are used, such as Ku-band frequencies. This permits a dish as small as two or four feet in diameter to be effective. Further, as DBS systems are transmitted to the small consumer dishes rather than to a station with a powerful receiving antenna, it was necessary for them to use high power transmissions, usually around 40 to 160 watts, far more than was being used for C-band communications in the 1980s when DBS began to develop. (Today satellites can deliver almost ten times that power.)

DBS to the home presents a number of the same moral challenges as widespread access to the Internet. DBS providers are concerned about illegal consumer copying and redistribution of programming, and consumers are concerned about program content. These issues are still being studied and resolved.

DBS system guidelines for Europe were originally established by segmenting the 11.7 to 12.5 GHz frequency spectrum into 40 channels, to be shared among the various European member nations.

U.S. DBS systems fall under the jurisdiction of the Federal Communications Commission, through guidelines established in the various Telecommunications Acts, and must conform to internationally established technical standards. See C-band, Ku-band, microwave.

direct connect modem A type of modem which became popular on desktop systems in the early 1980s, gradually superseding acoustic modems. The acoustic modem was used to connect data and phone lines by setting the handset of a telephone into two suction-cup style holders. Room noise could be a problem and the holders didn't always provide a good connection. In contrast, the direct connect modem was cabled electrically between the computer serial port and the modem, if it was an external modem, or between the computer and the internal modem card to the telephone line through the telephone jack that normally connected to a phone. This system offered faster data rates and was more convenient and effective in reducing noise and ensuring a good connection. See acoustic modem.

direct current DC, dc. A more-or-less constant electrical current flowing in one direction. DC current is supplied to many small appliances by batteries. Much of early communications history was based on direct current (DC) as a power source. Telephones had *talking batteries* and *common batteries*. The batteries tended to be large, leaky wet cells, that caused inconveniences if moved or subjected to temperature fluctuations. More than fifty years after the invention of the telegraph, AC power for telegraph systems was still considered a novel idea, but the shortage of batteries, and their cost, provoked French and Swiss engineers to experiment with AC generators, as described in the *Annales des Postes, Télégraphes et*

Téléphones in September 1919. Eventually the advantages of AC power were better understood, its use became common, and DC took its place as a current source for small, portable electric conveniences such as calculators, radios, wristwatches, laptops, cameras, etc. See AC to DC converter, alternating current; resistance; Tesla, Nicola.

Direct Distance Dialing DDD. A commercial name to indicate the capability of a network to connect a long-distance call without operator intervention.

Direct Inward Dialing DID. In the past, calls going through a central office (CO) to a private branch extension had to go through an attendant. With increasing automation, this routing is now rarely necessary. With DID, the called digits are passed through central office DID lines directly into the private branch exchange (PBX). DID lines do not offer a dial tone and hence cannot be used for direct outgoing calls.

Direct Inward System Access DISA. A telephone setup in which outside callers can dial into a telephone system, usually a private branch exchange (PBX), and have the use of the system's services as though they were on the premises using the system from the inside.

Direct Marketing Association, Inc. DMA. The largest trade association for users/vendors in the direct, database, and interactive marketing fields, founded in 1917. Due to the significant impact of the Internet on marketing technology and venues, the Association has also acquired two electronic commerce trade associations, the Association for Interactive Media (AIM) and the Internet Alliance (IA). Telecommunications are an integral aspect of marketing, and thus progress in electronics has shaped and been shaped by the activities of the DMA.

The DMA encourages and supports the growth and profitability of its membership and advocates adherence to high ethical standards. It provides leadership on behalf of the membership in government and public affairs. http://www.the-dma.org/

direct memory access DMA. A means to bypass the central processing unit (CPU) in a computer and interact directly with memory. This access is used to reduce processing time and increase speed.

direct outward dialing DOD. The capability of a private branch exchange (PBX) to dial calls outside the exchange without first dialing an access code (typically "9"), or going through an operator.

Direct Print Protocol DPP. A "thin" networking protocol for data transfer with a pair of DPP command sets, Direct Print Command, and File Transfer Command, for printing images over an IEEE 1394 serial bus, developed by HCL technologies. As the result of DPP interoperability events, a number of commercial firms have implemented DPP. There are other protocols that transmit data over IEEE 1394, including IP over 1394, SBP-2, and AV/C. Each of these has its own strengths and limitations in terms of the types of data that are efficiently handled. DPP has mainly been of interest to digital imaging and printing developers.

direct sequencing A spread spectrum frequency-changing broadcast technology. Spread spectrum broadcasts spread a transmission over a broader range of frequencies than is typically needed to contain the broadcast. A pseudorandom digital sequence directs a phase modulator to distribute the original RF transmission over a bandwidth that is proportional to the clock frequency of that sequence. The receiver must then be synchronized to the same pattern as the broadcast generator in order to remove the phase modulation and recreate the original signal.

Although it's not commonly done, it is possible to combine direct sequencing with frequency hopping. This combination would likely be used only in very high security transmissions, as the synchronization and receiving techniques are not trivial. See frequency hopping, spread spectrum.

direct set In ATM networking, a set of host interfaces that can establish direct communications at layer two for unicast.

direct view storage tube DVST. A cathode-ray display device introduced in the late 1960s to overcome the slow refresh and large storage buffer needs of early vector display monitors. By employing a slow-moving beam and a storage mesh, the DVST design substantially decreased the cost of display devices. See vector display.

directional antenna A radio antenna that is designed to concentrate its signal transmission or receiving strength, resulting in a stronger signal, but one which is not of equal magnitude in all directions. Commonly found in AM radio transmissions.

directory 1. A list, usually of names and associated information, often sorted or organized and displayed to enhance visual clarity. These may be the names of people, companies, institutions, files, or subdirectories, etc. 2. A table of organizational identifiers that provides addresses to individual items within the organizational path. This path is frequently hierarchical in structure.

directory, file On a computer file system, an organizational structure, comprising a file storage area, under which there can be further files or directories. A directory listing typically includes other information about the directory and its associated files or subdirectories, such as creation date, permissions, file type, and byte size. The display may be in text or graphical mode, and an icon resembling a file folder is often used to symbolize the directory.

Directory Agent DA. 1. A Novell Directory Services (NDS) database service that accepts advertisements for Service Location Protocol (SLP) devices from Service Agents, negotiates registration/deregistration, and answers queries. See Service Location Protocol, Service Agent, User Agent. 2. A database service for negotiating links between buyer and seller agents on a distributed network.

Directory Assistance DA. A telephone service in which the caller dials directory assistance to request the number of a person outside the local calling area, or within the local calling area if the number is not listed. The number may not be listed because the individual just moved in, just changed a phone number, or installed the phone just after the directory was published. Subscriber unlisted numbers may not be given out through Directory Assistance. In most areas, there is now about a $.75 charge associated with a Directory Assistance request, and there is usually a two-number limit on each request.

Directory Assistance used to be handled entirely by human operators, but there are now automated systems which will request the city and name of the person whose number is being sought and dispense the number, or hand over the call to the operator to complete the transaction, or to clarify the information provided by the caller. These automated systems combine speech recognition and speech synthesis to carry out their tasks.

directory caching A timesaving function of many operating systems that stores the local directory listing in memory so that each time the user accesses the directory information (for listing parts or all of the files), it will be displayed very quickly. For small directories there isn't much difference, but for very long directory listings it can be faster, especially if the information is being output to a shell or MS-DOS text window. Further, the information can be used to access more quickly the files in that directory. Since the directory information is read from memory rather than from disk, transfer time can be faster.

Directory Number Call Forwarding DNCF. An interim Service Provider Number Portability (SPNP) which is provided through existing available telephone services, such as call routing and call forwarding. The DNCF is set up so that calls are forwarded to a new number. While it is called Directory Number Call Forwarding, unlisted numbers can also be set up with the service.

Directory System Agent DSA. A directory application process in the Open Systems Interconnection (OSI) system. OSI utilizes a system of agents (helper applications) that can query a local database, communicate with other agents, or hand off requests to other agents when appropriate. The DSA specifically provides associated Directory User Agents with access to the directory information base (DIB).

directory tree The name given to a hierarchical directory file structure. The "tree" includes the current directory and subdirectories and files associated with the current directory both above it and below it in the hierarchy. Directory tree commands are used for file creation, deletion, renaming, protection, or for searching a series of files or all the files located in the tree.

Directory User Agent DUA. A directory application process in the Open Systems Interconnection (OSI) system. OSI utilizes a system of agents (helper applications) that can query a local database, communicate with other agents, or hand off requests to other agents when appropriate. The DSU specifically assists and represents the user in accessing a database through a Directory System Agent (DSA). See Directory System Agent.

dirty power Electrical power that is spiky, bursty, noisy, or otherwise unreliable. Dirty power can be dangerous to delicate electronic components. Laptops plugged in on ferries and trains should be used with power conditioners and good quality surge suppressors.

DISA 1. Data Interchange Standards Association. 2. Defense Information Systems Agency. 3. Direct Inward System Access. A subscriber option for external access to a private branch exchange (PBX), usually through a security code.

disable 1. *v.* To prevent from functioning. There are many ways to disable a process or mechanism, but the two most common are interrupting the power or the transmission. 2. *n.* A signal, tone, or command sent through a circuit to disable a device at the end of the transmission. Telephone companies have the capability of electronically enabling and disabling a number of phone services.

Disabilities Rights Office DRO. An office of the Federal Communications Commission (FCC) that helps to ensure that people with communications-related disabilities are given the same opportunities as others to access and use communications services. http://www.fcc.gov/cib/dro/

disaster recovery The procedures and resources for bringing a system back online after a disaster. Disaster recovery applies to everything from repairing or replacing facilities and equipment damaged by floods, hurricanes, or bombing to recovering data from a hard drive damaged by a lightning strike or a toddler poking paper clips into the circuit board. Data recovery is a fact of life. Whether or not you archive your data, your drive *will* fail at some point, or you *will* inadvertently overwrite an important file. Prevention is usually better but, if recovery of data becomes necessary, there are many shareware and commercial tools to assist with the task.

disc A round, flat data storage and retrieval medium. Usually these are *write once/read many* (WORM) devices, since optical media are not as easy to write, and rewrite, as floppy disks. Although in English disc and disk are used somewhat interchangeably, in computer parlance, disc tends to be used more often to designate optical storage media such as audio CDs, laserdiscs, etc., whereas disk is used more often as a contraction of *diskette* to designate an encased floppy diskette (flexible magnetic medium). See disk.

discard To throw away, be rid of, toss aside. This is usually said of something that is of no interest or further use (or anticipated future use) or that cannot be dealt with or used at a particular instance in time. The concept of discard is important in networks since there are many instances when too much data or bad data may create bottlenecks or other types of problems.

discard-eligible DE. In data-segmented networking systems, there are often packets or *frames* that are discarded for various reasons, including redundancy, congestion, timing out, low priority status, etc. Sometimes the decision to discard the packet is made on the part of the network's lower level functions, and sometimes the discard status is explicitly represented by a signal bit. In Frame Relay, there is a bit in the HDLC frame address field that can be set to mark the frame as expendable in case of congestion. See leaky bucket.

disclaimer A refutation of responsibility or liability. Software license agreements often have disclaimers of liability for any damage to computers on which the software is run. It's pretty unusual for software to damage hardware (yes, it is possible), but it doesn't happen under normal operating conditions. Nevertheless, attorneys prefer to have all bases covered.

disco tech *slang* A rather good pun for *disconnect technician*.

disconnect 1. To separate two discrete units or devices from one another, usually with the implication that the separation broke some type of electrical, inductive, or communications link between the two. 2. To terminate a communication, as in a phone call. 3. To break a circuit.

Discoverer XIV A historic satellite launched in Aug. 1960, it was the first satellite to be ejected from an orbiting space vehicle and recovered in midair. The satellite was launched into a north-south polar orbit in an orbit of 116 miles altitude at the perigee and 502 miles at the apogee. On the 17th pass, the Agena ejected Discoverer XIV from its nose and the re-entry vehicle fired retrorockets to slow the return trip. Once in Earth's atmosphere, it parachuted toward Earth and was recovered by a C-119 on the third pass over the parachute.

discrete Individual, separate, distinct.

discrete cosine transform DCT. A mathematical means of manipulating information by 'overlaying' cosines in order to analyze or use it from another point of view. DCT techniques are used in a number of digital compression schemes, including a lossy compression technique that provides a practical balance between a high degree of compression and relatively good perceptual clarity in the decompressed image (although very high compression ratios may create a blocky effect). An adaptive version of this technique is used in JPEG image compression. JPEG is one of the primary image representation formats used to display graphics on the World Wide Web. One of the disadvantages of DCT is a visual artifact called Gibb's effect that manifests as ghostly ripples running along distinct edges. See discrete wavelet multitone, Fourier transform, fractal transform, JPEG, MPEG, lossy compression, wavelet.

discrete line illumination, line illumination DLI. A light source with a narrow spectral width. Discrete line illuminators are useful in many technologies, including communications devices and telescopic and industrial quality assurance calibration systems. Some have connections for fiber optic cables for wavelength calibration of fiber-equipped spectrometers.

DLI sources include single-mode laser lights and cooled, very low pressure spectral calibration lamps. Spectral lamps are available commercially to provide a variety of defined wavelengths in materials such as neon, argon, xenon, and mercury.

discrete multitone DMT. A multicarrier technology used for transmitting multiple media over existing copper wires. It lends itself to a variety of data delivery services, including Digital Subscriber Line (DSL) and Hybrid Fiber-Coax (HFC). It has been selected by the American National Standards Institute (ANSI) as a standard line code for T1.413.

DMT uses discrete Fourier transforms for creating harmonics along the main lobes (and demodulates at the receiving end). Thus, DMT uses harmonics, or tones, as a means to divide the bandwidth into subchannels. The number of bits per tone is dependent upon the frequency of the tone since signal-to-noise ratios vary at different pitches. The utilization of these tones is somewhat like FM broadcasting; some tones are used for data and some are used as guard bands. A signaling tone is used in each data stream for timing. By dividing the available bandwidth into smaller units, portions of the available bandwidth can be individually tested to evaluate speed, availability, and suitability for transmission. This allows the optimization of transmissions over existing twisted-pair installations, which can vary widely in their characteristics from region to region. This scheme is currently used in ADSL installations and is being evaluated as a potential standard for VDSL. See carrierless amplitude and phase modulation, Digital Subscriber Services, discrete wavelet multitone.

discrete wavelet multitone DWMT. A multicarrier technology developed by Aware, Inc. for transmitting multiple media over existing copper wires, based on the same general principles as discrete multitone. DWMT is being promoted for use with Digital Subscriber Line (DSL) and Hybrid Fiber-Coax (HFC). DWMT divides the available bandwidth into smaller units to utilize suitable portions. It differs from DMT, which uses Fourier transforms, by using wavelet transforms for encoding subchannel bits. DWMT produces lower energy harmonics than DMT, making it easier to demodulate the encoded signal at the receiving end, where a forward fast wavelet transform (FWT) is used. It is less susceptible than other schemes to channel distortion and requires less overhead. See discrete multitone, wavelet.

discussion list A mechanism for people to intercommunicate and debate topics of interest over computer networks. In the early days, when most communications were facilitated by email transmissions among academic institutions and individual subscribers to a list, they were called *mailing lists*. However, now it is best to call them *discussion lists* to distinguish them from mailing lists that are used for bulk commercial or junk email messages containing product information or solicitations (similar to bulk mail delivered by the postal service). Discussion lists are ubiquitous and account for a significant proportion of communications over local networks. Because participation in discussion lists is so high, they are typically hierarchical and many are now moderated, to minimize abuse, commercial messages, and personal slurs.

The content of most discussion lists is text messages sent through email. In recent years, however, Web-based lists are growing in prevalence, and some lists now incorporate HTML tags and graphics. As bandwidth and transmission speeds increase, it is likely that video images and sound will become established in discussion lists. See mail distributor, USENET.

dish *colloq.* Common terminology for a parabolic satellite or terrestrial receiving or transmitting dish.

dish aperture The diameter of a parabolic communications receiving dish. In general, the larger the aperture, the broader the scope of signal apprehension (there are exceptions depending upon the nature of the signal and the shape and materials of the dish). The amount of aperture is also partly dependent upon the size and position of the feed apparatus, as it will block some of the signals.

dish focal point The distance, in a parabolic antenna, from the reflective surface to the focusing point of the signal. It is important to know this point in order to position the feed mechanisms efficiently. The focal point is dependent on the breadth and curvature of the dish, with flatter dishes generally having a focal point that is farther away from the reflective surface than more concave dishes of the same diameter.

disk, **diskette** A round, flat, flexible, encased medium, typically used for data storage and retrieval. The flexible medium inside the case is coated with magnetic particles that can be rearranged, thus providing read/write capabilities.

Common Modulation and Signal Subdivision Schemes		
Modulation or Subdivision Scheme	Abbreviation/Notes	
carrierless amplitude/phase modulation	CAP	Carrier signal is suppressed and reassembled at the receiving end. Single channel makes it more susceptible to interference.
quadrature amplitude modulation	QAM	Variations in signal amplitude are used to represent data.
discrete multitone	DMT	Frequencies are divided into discrete subchannels in order to optimize the throughput of each channel; faster, less susceptible to interference.
discrete wavelet multitone	DWMT	Provides some interesting means of implementing better performance and low-loss compression with less susceptibility to interference.

In the late 1970s and early 1980s, desktop computers were equipped with 8" disk drives, and the accompanying diskettes were flexible and highly subject to damage. By the mid-1980s, 3.5" disk drives were becoming popular, even though the diskettes were expensive at $6.00 each, because they had a hard shelled protective covering. Prices dropped until diskettes of double the capacity were only $.30 each. By the mid-1990s, superdisks were starting to be developed, but it was not until 1998 that they were widely announced. The superdisks are downwardly compatible with regular floppy disks but can hold over 100 Mbytes each. Diskettes are being superseded by USB and FireWire storge devices

Disk also is short for *disk drive,* a high capacity hard storage medium. See disc, diskette, hard drive.

disk controller A hardware peripheral circuit providing an interface between a computer's main circuitry and a floppy or hard disk drive. The most common formats for hard disk controllers are SCSI, IDE, USB, and FireWire. SCSI is predominantly used on Sun, Amiga, Macintosh, DEC, HP, NeXT, SGI, and server-level IBM-licensed desktop computers. IDE is predominantly used on IBM-licensed consumer-model desktops and some recent Macintosh computers. SCSI allows up to six devices to be chained to each controller port (the controller is considered the seventh device). IDE allows up to two devices, designated master and slave, to be chained to each controller port. SCSI disk controllers can be used to interface with hard drives, scanners, cartridge drives, CD-ROM drives, and redundant array of inexpensive disk (RAID) systems. IDE can interface with hard drives, cartridge drives, some CD-ROM drives, and some scanners. To enhance the limited capabilities of IDE, an Enhanced IDE specification has been developed. See FireWire, Universal Serial Bus.

disk mirroring See mirror, RAID.

disk operating system DOS. The low level operating functions of a computer that are used to read or write a floppy disk or hard disk and are engaged when the system boots up. The phrase has since taken on a more general meaning to encompass *all* the low level operating functions of a computer. In the late 1970s, there were several disk operating systems made available for the various microcomputers. For example, the Radio Shack TRS-80 Model I could be run with CP/M, TRS-DOS, LDOS and others, and you could select the one you wanted to run by reading it from tape (technically, a tape operating system), or from a floppy disk (and later from a hard drive). Gradually, however, as microcomputers came into the mainstream, people began to associate the operating software with the computer on which it came, and many do not realize that each computer hardware platform is capable of running a variety of operating systems. For modern computers, there are many choices, including Rhapsody, BeOS, Windows, MacOS, OpenStep, Linux with The X Windows System, etc. The user doesn't have to be tied to the operating system that comes bundled with the computer. In the future, if more developers adapt the various emerging international standards for open systems, particularly object-oriented technologies, applications may finally become operating system independent, and then users will be able to choose different computers the way they choose different cars, with the common object format being the "gas" that is standardized to power them all.

disk server A system dedicated to file storage, retrieval, and handling on a network with shared disk resources. It may also implement user access passwords, file locking, license restrictions, and other administrative tasks associated with dynamic file sharing on a multiuser network. On large computer systems, the disk server may manage many dozens or hundreds of disks and may require a room of its own with special fast cabling, such as Fibre Channel cabling. See RAID.

diskless computer A computer with RAM access only and no disk drives or which has no user-accessible hard disks. Typically, diskless PCs are simplified systems for inexperienced users or secured systems for preventing vandalism, theft, or the introduction of viruses. Diskless PCs are generally attached by a network cable to a secure console housing a storage medium such as a hard drive. Some, however, may allow the user to insert removable flash storage. Diskless systems are often found in kiosks, copying centers that offer access to computers by the hour, public shopping centers, libraries, amusement parks, and science and technology museum exhibits.

DISN See Defense Information Systems Network.

dispatch 1. *v.* To send out a verbal or written communication or package. 2. *n.* A missive, telegram, or other written or electronic communication, usually intended for someone out of normal hearing range, or intended to provide a record of the communication. 3. *n.* A communication intended for a group of recipients, such as field workers, taxi drivers, law enforcement officials, etc. Often sent verbally as a radio communication.

dispersion The spreading and gradual loss of signal strength that occur in electromagnetic or optical data transmissions over distance. Different media have different dispersion characteristics. Dispersion is usually, though not always, undesirable, and steps are taken to minimize it in most transmission technologies. There are a number of types of dispersion related to the wavelength composition, characteristics, and angles of reflectance of propagating light beams. In fiber optics, multimode cables are more subject to dispersion than single-mode cables, and hence have a shorter effective distance (this can be mitigated somewhat with graded index fiber).

In airwave broadcasts, undesirable dispersion, called scattering, can occur as a result of moisture, particles, and terrain. See diffraction, graded index, reflection, stepped index.

dispersion, chromatic In fiber waveguides, the discrepancy in the rate of propagation of light beams of varying wavelength. Since light is made up of a wide variety of wavelengths which have different properties in terms of speed, visibility, and ability to travel through "solid" matter, light pulses of differing wavelength regions will interact in specific ways with the conducting core and the reflecting cladding resulting in varying times of arrival and varying amounts of power (i.e., some wavelengths may be absorbed more than others), especially over longer cable runs. Graded index fiber that has been doped to customize its light-carrying property may be used to mitigate some of the effects of chromatic dispersion. See dispersion, modal.

dispersion, modal In fiber waveguides, the pulse spreading that occurs as the different beams of light, reflecting from the cladding back to the core, vary in how far they travel in a given period of time. In multimode fibers and those with larger core diameters, there is a greater prevalence of dispersion of the time of arrival of a light pulse due to the larger area through which the beams may reflect and disperse from one another. Graded index fiber, which has been doped to influence its light-carrying properties, may be used to mitigate the effects of model dispersion by matching the indexes of portions of the fiber to the beams that will be traveling through. See dispersion, chromatic.

dispersion, wave The separation of an electromagnetic wave into its component frequencies. For example, light rays passing through a prism will be dispersed into individual color frequencies. See diffraction, dispersion.

display *n.* A presentation of visual information, usually on a display medium such as a screen, wall, or computer display device such as a cathode-ray tube (CRT), plasma display, liquid crystal diode (LCD), or light-emitting diode (LED) surface. See cathode-ray tube, light-emitting diode, liquid crystal diode, multisync.

display driver The software that translates instructions from a computer processor to the correct operating instructions for an attached display device. A display driver may be written to control one specific type of display, a family of devices, or a wide variety of devices. Most computers have a set of generic display drivers compatible with the operating system. However, as there are many different types of products, with new ones all the time, many vendors will provide display driver software with their product. Sometimes the display driver will be incorporated into the peripheral card that interfaces with the display device; sometimes it is a two-way communications channel. Some display devices will send information to the driver about their capabilities and configuration parameters, so the driver can configure itself to optimally take advantage of the installed device.

display interface card A peripheral card that is interfaced between the computer processor and the physical display device in order for electrical signals and control data from the device driver to be transmitted between the computer and the display device. Sometimes the peripheral card itself comes installed with display driver software.

Some display interface cards have hardware settings to indicate the parameters of the installed display device. Thus, there may be a dial, jumpers, or dip switches that have to be set to configure the screen resolution, number of colors, refresh rate, and other parameters.

Display Support Protocol DSP. An assigned *well-known* port 33 on TCP/UDP systems.

Display Systems Protocol DSP. An assigned *well-known* port 246 on TCP/UDP systems.

distance learning, distant learning Receiving education through telecommunications media, including mail correspondence, email, Web sites, audio and video tapes, videoconferencing, etc. A number of educational institutions are providing course content, references, etc., on the Internet, especially on the Web, and with digital signatures they may also eventually provide testing, assignments, and critiques through electronic means. See audiographics, videoconferencing, whiteboarding.

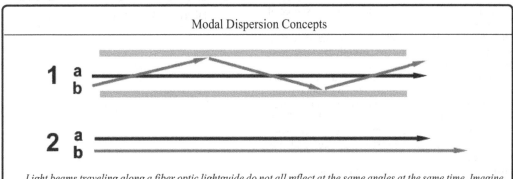

Modal Dispersion Concepts

Light beams traveling along a fiber optic lightguide do not all reflect at the same angles at the same time. Imagine two pulses sent through a fiber core at slightly different angles (1a and 1b) with 1a parallel to the reflective cladding and 1b at an angle such that it reflects back and forth in relation to the cladding and the conducting core but in the same general line of travel as the straight beam.

If you were to stretch out (calculate) the paths of two beams starting at the same time (2), b would be longer, and the discrepancy between the time of arrival *of a and b would increase as the diameter of the core and/or the length of the transmission link increased. Light travels very fast, so the effects are subtle, but may become significant over long haul multimode networks.*

distinctive dial tones This is a method by which dial tones are distinguished from one another by properties such as pitch, so that the caller can tell what type of call is being connected. This is particularly useful on private branch systems (PPBX) where internal and external calls require different sequences of numbers. For example, it is common to dial "9" to get an outside line on private systems, and the change in tone helps the caller know the call has been given outside access and will now accept an outside number.

distinctive ringing A subscriber option or feature of some phones that uses a different tone or ringing sequence to identify incoming calls as coming from inside or outside a branch system, or to signal another extension. Different phone carriers offer this option under a variety of names, such as Feature Ring and Ident-a-Call.

While distinctive ringing is common in businesses with private exchanges, there are now phones and peripherals for residential and small business users to distinguish calls on separate lines, or to identify another callee on a single line, with codes that make a different ringing sound or sequence. It's handy if you get many calls for teenagers.

distort To deform, contort, warp or pervert out of a normal sound, shape, or condition. Distortion is sometimes intentional, as in distorting the sound from an electric guitar or synthesizer to create some special effect. In most cases, however, distortion of a transmission is an undesirable fault condition.

distortion An undesirable change in the basic characteristics of a wave or data transmission sufficient to interfere with the information or its perception. Extraneous noise is not technically considered to be distortion. In visual images, distortion usually involves undesirable aberrations in the basic characteristics of the image, such as color, shapes, or lines that misform or obscure the original features, or of the entire image, in which case the outer contours may become squeezed or twisted. In sound distortion, the pitch, speed, or timbre of the sound may be altered, making it difficult to discern the content or source of the sound.

distribute To apportion, spread out, scatter, dole, deal out, or dispense; to give out, broadcast, or deliver to members in a group.

distributed backbone A backbone is a central artery or trunk in a network. A distributed backbone is one in which network segments are interconnected through hubs joined with backbone cables. Thus, there may be multiple segments or rings joined to one another through a backbone segment. See backbone, collapsed backbone.

distributed computing A system in which the computing processes are divided, parcelled out, or otherwise handled simultaneously, or in which computing services are apportioned, broadcast, or delivered to users. A local area network (LAN) or wide area network (WAN) are examples of distributed computing environments. Users of individual workstations can work independently of one another on tasks that are frequently carried out, yet they can share common files, applications, and devices distributed around the system. This system permits more open and efficient use of resources.

A render farm is another example of distributed computing in which a centralized system parcels out individual tasks related to constructing 3D models and images. These are assigned to various machines, and the completed processes or objects are reintegrated to create a completed rendering, thus potentially speeding up processing time. The Internet represents an example of distributed computing wherein routing and transmitting of messages occur through a cooperative network of many interconnected systems, and many common resources (Web sites, archives, applications, search utilities, online chats) are shared among users.

Distributed Computing & Communications Laboratory DCC. An experimental research lab at Columbia University that develops fundamental, novel networking technologies and exports them to academic and industrial organizations. Projects include mobile agents; programmable, active network systems; adaptive, self-managed network systems; and management technologies for Quality of Service (QoS). The DCC is under the umbrella of the Columbia Networking Research Center (CNRC).
http://www.cs.columbia.edu/dcc/

Distributed Computing Environment DCE. A middleware set of components from The Open Group that includes procedure, directory, file, and security services which may be bundled into a vendor's operating system or integrated into a system by a third-party developer to fulfil the above-listed functions.

Distributed Computing Environment DCE. A commercially developed set of services from Digital Equipment Corporation (DEC) that supports the development, maintenance, and use of distributed computer applications. It is based on the Open Software Foundation's DCE. (The Open Software Foundation is now The Open Group.)

Distributed-Feedback Laser Diode

DFB lasers have a layered, corrugated structure above the active lasing region. The spacing of the corrugations is related to the refractive index of the materials, allowing the corrugations to serve as a grating filter that selectively reflects the desired wavelength into the cavity from which it is emitted. The component is distributed in the sense that the light reflection and filtering process occurs across the length of the component.

distributed-feedback laser DFB. A compact semiconductor laser component used in lightwave transmission systems. DFB lasers are based upon Fabry-Perot (FP) lasers, but have an extra grating filter that provides a sharply concentrated wavelength output without the lower powered sidewaves that are characteristic of FPs. This makes DFBs suitable for tasks such as optical amplification and dense wavelength division multiplexing (DWDM) applications where finely concentrated signals allow closer proximity of neighboring signals without interference. L- and C-band transmission applications are common. DFBs are sold in various configurations in terms of power requirements, connector options, and the optical wavelength emitted. See Distributed-Feedback laser Fundamentals diagram. See Fabry-Perot laser.

distributed file system A file system that is distributed across more than one partition, more than one device, more than one workstation, or more than one system. Some operating systems can set up a distrib-

---

Distributed-Feedback Laser Fundamentals

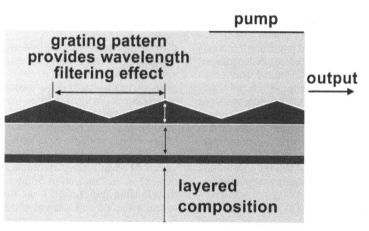

A distributed feedback (DFB) semiconductor laser is constructed out of a number of carefully organized active and passive layers, depending upon the intended size and function of the component. DFB lasers can be housed inside traditional computer chip packaging or may be fabricated of thin film layers for insertion into other assemblies. Silicon-based compounds are commonly used under the grating layer and as base substrates. Whatever shape or materials are selected, the basic concepts are the same – the organization of the layers and the grating period (distance from one corrugation to the next) serves as a reflective filtering component, causing some wavelengths to be reflected back and very specific tuned wavelengths to pass through to be output at the other end of the assembly.

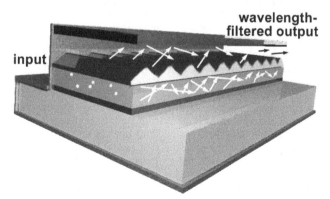

This is a highly simplified cross-section of the grating filter within a semiconductor component, illustrating the layered composition and corrugated structure of the grating. The grating period would be much finer than shown here and the number and thickness of the layers vary with the materials and fabrication techniques selected. Energy applied to one end of the component bounces along within the conductive layers and is selectively back-reflected and filtered as it passes across the length of the component (hence the name distributed-feedback *laser) leaving only a very narrow wavelength range by the time it exits the end as laser light. Altering the grating period allows the wavelength to be tuned to a specific frequency and the signal is very stable, qualities that are extremely useful for communications applications and the fabrication of scientific instruments.*

uted file system across partitions or across disks, in such a way that a directory or set of files appears to the user as though it were all in one place, thus making it transparent, appearing as a single logical unit, even though the physical underpinnings are different devices. See redundant array of inexpensive disks (RAID).

Distributed Management Task Force, Inc. DMTF. A trade organization promoting the development, unification, and adoption of management standards and initiatives for desktop, enterprise, and Internet environments. The organization seeks to enable integrated, cost-effective technologies to be used efficiently rather than in a crisis mode.
http://www.dmtf.org/

distributed network administration DNA. 1. A wiring and services distribution scheme that divides up the administration and actual physical connections into smaller units, which may be one per department, one per floor, or some other arrangement. This is convenient for systems where there are significant physical limits on the length of individual transmissions cables (as in some fiber optic installations), where operations or maintenance personnel are assigned to different sections of the building, or where different types of wiring are used for different departments or areas of the premises. See distribution, distribution frame. 2. A wireless services system in which administration is subdivided into particular regions or sections of the service (a local area, city, or state). This is a common administrative system for nationwide wireless systems that cover a large area of territory, but through which many users roam from region to region.

distributed programming platform DPP. A programming platform connected to a network such that the programming process is carried out transparently over a number of machines or CPUs to whatever extent is desired. Traditionally, programming has been done on discrete platforms or on connected platforms whereupon the programming process itself is not significantly different because of the connectivity. A DPP, on the other hand, is one in which the connectivity to other processors, in a distributed fashion, is exploited to take advantage of the increased power and redundancy supplied by the distribution. Thus, attributes such as parallel processing, background processing, fault-compensation, and other factors can aid in the programming. DPP also implies there is a means to evaluate the performance in terms of latency, reliability, and other functional trade-offs and priorities.

Distributed Queue Dual Bus DQDB. A connectionless packet-switched network protocol with a 53-octet cell and header/information structure somewhat like ATM, used for telecommunications services in Metropolitan Area Networks (MANs). DQDB is described in IEEE 802.6. It is descended from Queued Packet and Synchronous Exchange (QPSX). DQDB supports isochronous and nonisochronous communications.

DQDB employs the *slot* as the basic unit of data

transfer. A slot is further subdivided into a one-octet Access Control Field (ACF) and a 52-octet segment.

Distributed Single-Layer Embedded (Test Method) DSE. In ATM networking, an abstract method for testing a protocol layer, or sublayer, which is part of a multiprotocol Implementation Under Test (IUT).

distribution 1. A cabling term referring to the delivery of services through cables and/or wires. A distribution system is the combined media, connections, and topology that provide services through wires or cables, usually consisting of electricity, voice, or data network services. Many different types of distribution arrangements exist and are described throughout this reference. See distribution frame, horizontal distribution. 2. Apportioning, assigning, sending out, delivering of products, services, or computing processes.

distribution frame A centralized circuit management structure for creating, troubleshooting, and accessing a variety of incoming and/or outgoing lines, sometimes in the thousands if it is a commercial switching system. A distribution frame may be for supplying electrical power or may organize lines and connections for data communications. Frames are often built into closets, floors, or ceilings, depending upon available space and whether frequent access is required. The frame may include blocks, conduits, and other physical structures to facilitate cabling.

distribution panel A grid-like frame, usually of metal, with rows and columns of punched out holes through which cables can be threaded and mounted. These are often designed to fit standard 19-in. rack mounts.

distribution roller See stringing roller.

District Telephone Company Historically, one of the first entrepreneurial efforts to incorporate A. Graham Bell's telephone invention into a commercial product and service. George W. Coy, Herrick P. Frost, and Walter Lewis opened the District Telephone Company in New Haven, CT in 1878. Initially there were only eight conducting lines strung up wherever it was possible, and fewer than two dozen subscribers. The simple switchboard, designed and built by George Coy, was mounted on a table. District Telephone became Connecticut Telephone and then Southern New England Telephone, in 1882.

dithering A term that the imaging field has adapted from physical appliances in scientific labs where dithering devices are used to shake up items to minimize clogging, settling, or the effects of friction. In computer imaging, dithering is the blending and combining of pixels to simulate the effect of a shade of gray or a color that may not be available on the system.

divest To dispossess of authority, property, or jurisdiction; to take away, to deprive, to disinherit; to dismantle.

divestiture The act of breaking up, dispossessing, or otherwise dismantling an institution. In the U.S. legal system, divestiture is a process whereby the U.S. Justice Department oversees the alteration of a company's organizational and asset structure in order to deconcentrate power or enforce fair competitive restrictions. It is not a common procedure, but it occurs when a company's activities are monopolistic in nature or when a company has engaged in unethical or illegal practices that provide the company with an advantage. The process of divestiture thus is intended to re-establish safeguards to competitive access by new or smaller contenders. In the telecommunications industries, the three most high profile Justice Department actions against major vendors include the AT&T divestiture in the mid-1980s and the investigations into the competitive policies and practices of Microsoft Incorporated and Intel Corporation. At the time of writing, no conclusions have been drawn, and no divestiture proceedings have taken place, but divestiture is one of the possible steps the Justice Department may follow if any of the allegations of unfair practices are found to be true. For information on the AT&T divestiture, see Consent Decree of 1982, Judge Green, Kingsbury Commitment, Modified Final Judgment.

DL 1. See distant learning, distance learning. 2. distribution list. A list of parties receiving correspondence, items, or processes. Email and discussion forums are often handled through distribution lists.

DLC 1. Danish Lithosphere Centre. 2. See data line card. 3. See Data Link Control. 4. See Dead Link Check 5. See Digital Loop Carrier. 6. Disability Law Center. http://dlc-ma.org/

DLCI See Data Link Connection Identifier.

DLL 1. See data link layer. 2. See Dynamic Link Library.

DLM diode laser module. See distributed-feedback laser, laser diode.

DLP Digital Light Processing. See Digital Micromirror Device.

DLR 1. See Design Layout Record. 2. Deutschen Zentrum für Luft- und Raumfahrt. German Air and Space Center involved in planetary exploration and other space sciences. DLR supports educational programs, space forums, asteroid research, satellite research, robotics, and more. The DLR Asteroid Research Team is known as DART.

DLS 1. See Data Link Switching. 2. Distance Learning Section of the American Library Association & Association of College and Research Libraries. 3. Distributed Learning Subcommittee of the University of British Columbia within the Centre for Educational Technology. 4. Division of Laboratory Systems of the U.S. Centers for Disease Control and Prevention.

DLSw See Data Link Switching.

DLSW See Data Link Switching Workgroup.

DM See delta modulation.

DMA 1. Dayton Microcomputer Association. http://www.dma.org/ 2. See Direct Marketing Association, Inc. 3. See direct memory access. 4. See Document Management Alliance.

DMCA See Digital Millennium Copyright Act.

DMD 1. differential mode delay. 2. See Digital Micromirror Device.

DMI 1. See Desktop Management Interface. 2. Digital Multiplexed Interface. An AT&T interface that interconnects and multiplexes transmissions between T1 trunks and private branch exchanges.

DML Development Markup Language. See International Development Markup Language.

DMS 1. data management system. 2. decision management systems. 3. Digital Multiplex System. A series of programmable communications switches. Northern Telecom (NTI) provides one of the switches in common use in telecommunications, called the DMS-250 (there are other models as well). 4. diminishing manufacturing sources. 5. Division of Mathematical Sciences of the National Science Foundation. The DMS supports small-group projects aimed at exploring and developing the properties and applications of mathematical structures. http://www.nsf.gov/mps/divisions/dms/ 6. Dutch Meteor Society.

DMSP See Defense Meteorological Satellite Program.

DMT See discrete multitone.

DMTF 1. Desktop Management Task Force. 2. See Distributed Management Task Force, Inc.

DMUX digital multiplexer. See multiplexer.

DN See Directory Number.

DNA 1. See Digital Network Architecture. 2. See distributed network administration.

DNC 1. direct numerical control. Data connectivity between a computer processor and a computer-compatible numerical control (NC) device. 2. distributed networking computing. See distributed computing. 3. dynamic network control

DNCF See Directory Number Call Forwarding.

DNIC See Data Network Identification Code.

DNIS See Dialed Number Identification Service.

DNR Dynamic Network Reconfiguration. A feature on commercial International Business Machines (IBM) network that allows network address designations to be reorganized without closing down the system.

DNS See Domain Name System.

DNSSE See Domain Name System Security Extensions.

DOCSIS See Data Over Cable Service Interface Specifications.

document *colloq.* In computerese, a text file in a format specific to the word processor that created the file. The *.doc* file name extension is often used to identify this type of file. Document files in this sense of the word are not the same as standardized ASCII text files. Doc files provide formatting, font, and color information specific to the applications program that created the file. One of the best formats for exchanging files between applications that are not intercompatible is the Interchange format, also known as Rich Text Format (RTF). It holds more information than ASCII text files (font names and sizes, and text attributes such as bold and underline are retained), and it is widely supported on many platforms and by every major word processor and desktop publishing program. In fact, it would be great if more people would use RTF files on the Web for document downloads, instead of using word processor-specific document files that can't be loaded by many users.

document camera A camera specially adapted for photographing documents, usually for historical preservation, archiving, or replication. This type of camera is mounted and optimized for a suitable focal length and artificial lighting conditions. It may be combined with a bank of lights to illuminate the documents evenly. It is typically attached to a stand so that it can be readily raised or lowered like an enlarger.

More recently, document cameras have been equipped with electronics that enble the signal from the camera to be fed into a computer system. The camera may be a digital still camera, digital camcorder, or analog camcorder. In this way, documents can be digitized, stored directly on networks, or used in videoconferencing sessions.

Document Content Description DCD. A structural schema facility proposed in 1998 to the World Wide Web Consortium for specifying rules for the structure and content of XML documents. DCD utilizes a subset of the XML-Data Submission. DCD defines document constraints in XML syntax, using an RDF vocabulary, much the same as traditional CML DTDs. HTML coders will recognize the basic angle-bracket nested tag structure of a DCD. However, rather than the common paragraph, anchor, and image tags found in HTML, the contents of the DCD tags include element definition types and parameters. The DCD semantics are intended to be a superset of XML DTDs while maintaining conformance with RDF.

document database A digital repository of document information, usually consisting of document images which show documents as individual pages as they might be viewed in print form, or individually storing the elements of the documents (pictures, text, mathematical formulas, etc.), in such a way that the information can be manipulated, stored, and retrieved with word processors, drawing programs, etc. The design and programming of document databases is a great challenge, as it is not always known how the information might be used in the future. It is also becoming imperative to design systems that can handle vast amounts of information yet still search and retrieve the information with reasonable speed.

document image management DIM. Electronic storage, access, processing, and retrieval of documents stored in image format. Image format is a very common way to archive information that would otherwise require massive space. It means simply that text, graphics, and everything else is stored together as a picture, with no explicit differentiation of textual content. Microfiche is one of the systems designed to store image directory lists and image "snapshots" of newspapers, journals, documents, certificates, etc., in order to provide reasonably quick lookup and to save storage space.

document image processing DIP. The process of taking printed documents and converting them to a visual digital form to be manipulated and/or viewed electronically.

In the past, converting text information to image format somewhat limited what could be done with the images. It wasn't possible to apply search/retrieval,

and editing to the text itself, because it was not in a form amenable to database software. That is no longer true. While work still needs to be done on document recognition systems, there are now programs which can take information stored in image format and process the characters into text; some can even recognize which parts of the page are images, and which parts are text, in order to intelligently handle the content of the page. See document image management; zone, optical recognition.

Document Management Alliance DMA. A comprehensive standard for interoperability among electronic document management systems, sponsored by the Association for Information and Image Management. It is related to the ODMA, a standard for interoperability between desktop applications and electronic document management systems.

Document Object Module An interface model for allowing programs and scripts to dynamically access and process documents independent of language or platform.

Document Printing Application DPA. A standard for printing within an open, distributed network environment (ISO/IEC 10175) originally published in 1996. The Internet Printing Protocol (IPP) is a "lighter" version with some of the functionality of the comprehensive DPA standard.

documentation Text and/or images that, taken together, describe how a system works and how to use it. Good documentation goes beyond simply describing system features and functions; it describes them in such a way as to provide a context as to *when* and *why* they should be used. Documentation is often subdivided into tutorial and reference sections. This is generally a good idea, as tutorials are useful when you are first learning a system, and references are useful once you have grasped how to use a system. The worst computer documentation is a manual that simply repeats and describes the contents of the menus. The better ones tell you *what* the program does, and *how*, *why*, and *when* you should do it.

DoD See Department of Defense.

DOD See direct outward dialing.

DOH 1. See Department of Health. 2. See Document Object Module.

Dolbear, Amos Emerson (1837–1910) A Tufts College professor for almost three decades, beginning in 1874, Dolbear was an associate of Thomas Edison. Dolbear was awarded a patent for an induction-based wireless telegraph in 1886 (filed in 1882). He also developed an electrostatic version of the telephone, in the early 1880s, apparently improving upon the concepts of Philip Reis, but it lost out to the Bell phone in both utility and recognition.

In the 1890s, Dolbear described discoveries related to the effects of temperature on the properties of metals and their relationship to conductivity and stated "... that at absolute zero their electrical conductivity becomes infinite, or, as it is more generally stated, the electrical resistance of metals becomes zero.... so it seems altogether probable that the qualities and states of matter so familiar to us as solids, liquids and gases depend absolutely upon temperature and that at absolute zero there would be neither solid, nor liquid, nor gas, and that electrical and magnetic qualities would be at a maximum."

Dolbear wrote many magazine articles and books, including *The Art of Projecting: A Manual of Experimentation in Physics, Chemistry, and Natural History, with the Porte Lumière and Magic Lantern,* in 1883, and *Matter, Ether, and Motion,* in 1894. The Amos Emerson Dolbear Scholarship was established in 1947 to aid students of exceptional promise in electrical engineering and physics. The New England Wireless and Steam Museum, Inc. includes wireless artifacts from Benjamin Dolbear, Amos Dolbear's son.

Amos Dolbear – Physicist, Writer, Educator

Dolbear was an associate of Thomas Edison and a talented inventor in his own right. He described important aspects of conductivity and developed a number of telephone devices.

Dolby-NR Dolby Noise Reduction. A system developed by Ray Dolby to improve sound quality by reducing noise. Quality levels are designated with letters (Dolby SR is professional level quality). Dolby-NR is widely applied in the sound portion of the motion picture industry. See THX.

domain Sphere of influence or activity, province, dominion; the set of processes, items, or actions that constitute a sphere of influence or activity. There are two aspects to the concept of domain in telecommunications. The first is recognition and access to and from the domain. How does the domain provide a presence to other systems, and how do they recognize and acknowledge its presence? The second is what constitutes a domain, what is included within the organizational grouping that is called a domain? See domain identification and domain organization.

domain identification In most data network architectures, a domain is perceived from outside a system as a unique address or identifier describing the logical and sometimes physical access points to a system, since systems external to the domain frequently

know very little about the size and composition of elements within the domain.

Since computers work with binary digits, the domain on a computer network that allows it to be identified and signaled by other domains is usually an *address* that can be expressed as a numeral or alphanumeric series. This address will often have a name or symbol associated with it, to make it easier for humans to recognize and remember it. Information about the nature of the domain, such as its basic function or geography, are sometimes expressed in the name. For example, on the Internet the alphanumeric expression of a domain name typically indicates the type of domain (commercial, not-for-profit, educational, etc.) and the location (*.ca* [Canada], *.au* [Australia], etc.). This naming scheme is not completely consistent, but it is helpful in most circumstances. It also is not a guarantee of location, as a system may remotely dial into a domain in another country. See domain name, for further details on Internet naming, and domain name server, Domain Name System, firewall, host, Internet, server.

domain name A unique identifier, usually expressed in alphanumeric characters, to a computer domain. Local networks that are not connected to outside entities can setup this domain name in any way they choose or according to the domain naming parameters of the network software that is running on the system. On the Internet, however, which is a global distributed network, it is necessary to maintain an extensive database of domains in order that each one is uniquely identified. Thus, a domain name on the Internet must be registered with one of the assigned registration entities, the oldest of which is InterNIC. See domain name, Internet.

domain name, Internet A globally unique identifier for a domain that is continually or occasionally online on the Internet through an Internet Access Provider (IAP), which may provide additional services as an Internet Services Provider (ISP). This identifier is used by the system to locate a domain in order to send and receive files, email, messages, routing information, and other network traffic. A certain flexibility is inherent in this scheme in that levels below the IAP can be rearranged according to the needs of the local network.

The domain name is actually a name resolved to an Internet Protocol (IP) address, which is composed of numbers, but the name is what is familiar to most users, and it is automatically converted by the system. Once an Internet domain name has been assigned, such as *ourdomain.org*, subdomains can be locally assigned, such as

```
accounting.ourdomain.org
administration.ourdomain.org
sales.ourdomain.org
```

Similarly, usernames associated with email addresses can be expressed in a variety of ways:

```
max@sales.ourdomain.org
bighoncho@ourdomain.org
```

An email server at the local domain handles the processing of mail once it is received from the ISP, and it can route it according to the needs of the local system.

The domain name is expressed alphanumerically with dots between each of the levels or portions of the domain, with the assigned domain as the last two parts of the name. Since 1995, the number of registered domain names has risen from 100,000 to over 10 million.

Within the U.S., domain names are subdivided into various categories. Outside the U.S., the extension is usually a designator for the country. The North American Domain Name Extensions chart shows a sample of familiar domain name extensions. The appendix includes a more complete list of over 200 Internet domain name extensions from around the world. See Internet Corporation for Assigned Names and Numbers.

domain name registration A necessary administrative step for individuals and organizations who wish to have a unique domain name on the Internet. Unlike state registrations of company names, or companies in different industries that share a common name, there can be only *one* of a particular name on the global Internet. This fact has made domain names hot commodites, with companies buying and selling domain names the way logos are sometimes bought and sold. To register a domain name, you must have a site ready to come on line, and you will usually go through your ISP to establish the name for the site. In 1995 registration changed to a fee system; in 1998, it was about $100 for initial registration and $50 per year to retain the name. Now it's much less. See Domain Name System, InterNIC, IANA.

domain name server A computer with server capabilities that communicates with the Domain Name Service (DNS) on the Internet after having registered its own unique ID. The serving computer provides an Internet Protocol (IP) address to the DNS for a domain name that is not fully qualified (does not end in a dot). For individual users, this is usually handled by a serving computer at a local Internet Services Provider (ISP). For campus/commercial users, it may be handled by a local campus backbone server. The local *Start of Authority* is delegated to provide Domain Name Server services for the assigned domain. This information is essential to providing connections between individual systems or local area networks (LANs) and the Internet at large. See Domain Name System.

Domain Name System, Domain Name Service DNS. A domain name distributed database established in the early 1980s at the University of Wisconsin. DNS provides mapping between host names and Internet Protocol (IP) addresses. DNS evolved out of a need for a distributed system to handle a very large number of domain names. In older systems, host files were regularly distributed, until they became too large to be managed on most systems.

To become a node on the Internet, it is necessary to formally register a unique domain name. The extensions

familiar to Internet users as *.com*, *.edu*, *.net*, *.biz*, and *.info* are part of the domain naming scheme called *zones*. Each domain name is stored in a central repository on the Net, and addresses are resolved through this database.

Demand for domain names on the Internet has risen from a trickle to a flood. Businesses are realizing that the unique name requirement is different from traditional business naming schemes. For example, it is possible for two businesses in the same state to have the same business name if they are in different lines of business. On the Internet, however, there can be only one of each name in the world, assigned on a first-come, first-serve basis. This exclusivity has created an unprecedented demand for names, leaving second comers with little choice but to come up with a less desirable or memorable name or to change their signs, stationery, and other marketing materials, an expensive proposition.

To relieve some of the demand for domain names, new domain name extensions were introduced in 2001, theoretically increasing the availability of names to other comers. However, the demand problem wasn't fully resolved for two reasons. First, a new extension doesn't necessarily aid certain types of business (e.g., a book store isn't helped much by the fact that there are new *.aero* extensions available for aeronautics companies). Second, big corporations with large marketing budgets typically register multiple versions of their names, thus leaving smaller companies out of the running even when additional extensions are added. See domain, domain name server, InterNIC, RFC 830.

Domain Name System security extensions DNSSE. Specifications developed to improve the weak security aspects of the Domain Name System. As the Internet Domain Name System became increasingly important for the transmission of secure documents, such as sensitive government communications and financial information and services, it was felt that extensions for the use of cryptographic digital signatures should be developed. See RFC 2065.

domain organization In its general sense, the organizational structure of a digital network domain includes the operations, devices, and other elements under the general control of a processor, system, or network. The overall controlling and administrating entity may be called a *host* and may function as a server or have jurisdiction over a number of servers. The host maintains some type of access and management mechanism to a database or other record of other computers and devices on the system. This entity may be a single computer or software program, or a logical amalgamation of several computers or software programs. Similarly, security mechanisms are generally orchestrated by the host or other controlling member of the domain. The organization of the elements associated with the host and contained within the domain can vary substantially with the type of network and various devices that are included.

domestic arc A portion of an orbiting satellite's path or range that provides transmissions between the satellite and the country it is serving. There are a number of domestic satellites in use that specifically are launched to cover a particular country or territory, as in Alaska and India. Others serve domestic needs during a particular portion of their orbit. See satellite.

dominant carrier A designation for a long-distance telecommunications provider that dominates a particular region or market. In most cases, a dominant carrier is more stringently regulated in order to balance a monopolistic advantage with opportunities for competition.

dominant mode 1. The most effective or most prevalent mode of transmission or conduction within a material. For example, conduction (rather than convection or radiation) may be the dominant mode for heat transmission in a particular material. In fiber pulling systems in a furnace with a gaseous environment, thermal radiation has been found to be the dominant mode for heat transfer. 2. In a single-mode waveguide or one that can carry more than one wavelength at a time (e.g., multimode fiber) the mode that has the best propagation characteristics for that waveguide, which will depend upon the frequencies used and physical characteristics of the guide. Thus, the mode with the least loss or distortion of signal over the link is the dominant mode or the wavelength that is not cut off below the V parameter. See numerical aperture, V number.

DOMSAT Domestic Communications Satellite. A geosynchronous broadcast relay satellite. NOAA uses it, for example, in conjunction with the Geostationary Operational Environmental Satellite (GOES) system, for relaying meteorological data.

dongle A small hardware device or *security key* used for software or system security. Dongles were widely introduced on microcomputers to protect commercial products in the 1980s and were raucously opposed by hackers, technical users, and enough general users that their use has been almost abandoned in North America.

donor An n-type dopant, such as phosphorus, which is used in solar photovoltaic devices. The dopant puts an additional electron into an energy level near the conduction band to increase electrical conductivity. Doping is a common way of manipulating the properties of electromagnetic materials. See doping.

door A term often used on computer bulletin board systems (BBSs) to indicate a category of user access activity that is external to the BBS software itself. Thus, a separate software program, such as a game or quiz, external to the BBS software is launched when the user selects a specific door.

dopant An industrial chemical added in minute amounts to pure semiconductor materials, usually to improve the conducting properties of the materials. See donor, doping.

doped fiber amplifier DFA. A fiber optic cable that has been impregnated or doped with substances, usually rare earths, which alter its transmission properties. See doping.

doping A means of adding small amounts of materials with particular properties to another in order to

enhance it for the purpose for which it is being used. For example, in semiconductor manufacture, materials are doped to enhance or inhibit particular tendencies to give up electrons or form "holes." Optical fibers are often doped with rare earth elements to alter their transmission characteristics. See germanium.

Doppler shift A perceived or measured shift in frequency when the source of radiant energy moves relative to the position of the observer or receiver.

DoS See Denial of Service.

DOS See disk operating system.

double density DD. A term for the physical configuration of magnetic particles on a floppy diskette, and hence the maximum amount of digital data it can store. Three and one-half inch, double density diskettes hold approximately 720-880 kilobytes of data, depending on the platform. The diskettes themselves are interchangeable between systems, provided they are formatted for the operating system on which they are being used. Double density diskettes have given way to high density diskettes (1.4 Mbytes), SuperDisks (100+ Mbytes), higher capacity cartridge formats, and USB and FireWire devices.

double-pass autocollimation test A type of *null test* in the sense that it tests a mirrored optical assembly, such as a parabolic mirror, against itself. This is a very accurate test with the light path reflecting off the mirrored surface twice and is useful for assessing telescoping components. A Ronchi grating, with evenly spaced lines and facets, can be used as the reference image. Any spherical aberrations in a flat mirror under test will usually manifest as concave or convex out-of-focus images. if the mirror is parabolic, the bands will be curved and should be compared against an ideal reference. See Ronchi test.

double-superheterodyne DSH. A technology used in audio tuners as a conversion method for maximizing the selectivity of tuners and reducing distortion as much as possible in order to take full advantage of programming from broadcast sources that pack a lot of programming into limited frequencies and from shortwave sources. See superheterodyne.

DOV data over voice. Technology that allows data transmissions to be carried over traditional phone connections, usually copper twisted pair. See ISDN.

down converter A technique and device used in communications in which the incoming frequencies are shifted. There are two common reasons for doing this. By shifting incoming frequencies so they are different from the outgoing frequencies, it is possible to reduce interference between the two sets of signals. Some frequencies are shifted up when they are transmitted, in order to put them into a particular broadcast band or slot. A down converter is then needed at the receiving end to downshift the frequencies back to levels that can be used by the playing or viewing equipment that provides the information to the user.

downlink The common name for the *satellite to Earth* portion of a transmission, and the uplink is the Earth to satellite portion. The downlink is often frequency-shifted from the uplink in order to reduce interference between the two sets of signals. Uplink and downlink services may be carried by different providers and may be subject to different usage restrictions or billing arrangements

download To receive computer data from a source on another system, usually through a network or modem connection. Downloading is typically done with a communications program file transfer protocol (ZModem is commonly used), FTP client, or a Web browser. Common types of downloads include files from BBSs or Internet file archives and World Wide Web images and text.

When using browsers on the Web, the information that you view is typically downloaded or *cached* on your machine (because loading the source code of the Web page on your local computer allows the browser to redisplay the previously viewed pages more quickly.) This is usually convenient, but it has disadvantages as well and constitutes a security hole on your system. It may also fill up your hard drive. Make sure your cache is flushed (erased) when you are finished with the Web files. You can always save ones of interest in appropriate directories for later viewing while offline.

While downloading files from FTP sites or Bulletin Board Systems, make sure you don't accidentally overwrite an existing file of the same name. Not all download software will inform you of a duplicate filename. It is a good policy, at any time, to download into a separate directory or even a separate partition until you have run a virus checker on the downloaded files. All foreign files should be suspected of possible viruses until you have determined that they are problem- and virus-free. See FTP, upload.

downstream A designation for any of the systems, nodes, legs, or hops in a transmissions pathway that are subsequent to the current one. Thus, a printer is generally downstream from a computer, a radio listener is downstream from the radio broadcaster, and various workstations may be downstream from a server. See download. Contrast with upstream.

downtime The block of time during which a system is nonfunctional. Downtime on a computer may be caused by software or hard drive crashes, broken network connections, etc. Downtime on phone systems may be caused by power outages, overloaded lines, or breaks in the lines.

downwardly compatible Software or hardware designed to work in some way with older software or hardware, sometimes called *legacy* applications or equipment. Often the downward compatibility is only partial. For example, a software program may be able to export a file in the older format, but it may not include all the characteristics of the file when loaded into the earlier application. Similarly, a new computer may work with an older monitor, but that doesn't mean the monitor can support all the graphics modes that might be built into the graphics controller on the new computer. Downward compatibility is a way of safeguarding a financial investment and of maintaining a minimum level of continued data access and use of existing software. Contrast with upwardly compatible.

DP 1. See data processing. 2. See Deflate Protocol. 3. See demarcation point. 4. Dial Pulse. A standard Hayes modem command with the letter "P" used to designate a pulse dial setting for subsequent dialouts, thus ATP (ATtention, dial Pulse). See ATD.

DPA 1. Dearborn Protocol Adapter. A commercial Microsoft Windows PC board for interfacing host computers with automotive communications networks, from the Dearborn Group, Inc. 2. Defence/Defense Procurement Agency. 3. Demand Protocol Architecture. An architecture for dynamically loading protocol stacks as they are required. 4. digital port adapter. 5. See Differential Power Analysis. 6. See Digital Performance Archive. 7. Disabled Persons Assembly. Telecommunications technologies are an important tool for enabling disabled persons to lead fuller, more active lives as well as being a means for people with similar problems to intercommunicate and support one another through difficult challenges. http://www.dpa.org.nz/ 8. distributed processing architecture. 9. See Document Printing Application.

DPBX digital private branch exchange (PBX). Most private branches in North America are becoming digital, so the D is now commonly assumed when using PBX. See private branch exchange.

DPCM See differential pulse code modulation.

DPE distributed processing environment. See distributed computing.

DPLB Digital Private Line Billing.

DPNSS See Digital Private Network Signaling System.

DPO Direct Public Offering. See Initial Public Offering for description of this specific state-regulated subcategory for securities offerings.

DPP 1. See Demand Priority Protocol. 2. See Direct Print Protocol. 3. Director of Public Prosecutions (Australia). 4. Distributed Pipe Protocol. A client-server protocol that was implemented over TCP using remote method invocation (RMI) by Wang and Ouyang in 2000. 5. Distributed Processing Peripheral. 6. See distributed programming platform.

DPX See DataPath Loop Extension.

DQDB See Distributed Queue Dual Bus.

Dr. Dobb's Journal of Software Tools This journal has been a perennial favorite with programmers, providing technical information on a wide variety of platforms and programming languages since 1976. It originated as a newsletter in 1975 documenting Tiny BASIC. The name is somewhat a collapse of the originators' first names, Dennis and Bob.

Draft RFC A formal stage in the Request for Comments standards and information distribution process in which the proposal is submitted for evaluation and comment. On the Internet, this process is widely used to encourage open standards and professional and public participation. See Request for Comments.

drag line A wire, rope, or other line for threading wire and cable through narrow channels (pipes, conduits, walls, etc.). The drag line may be preinserted during building construction and left for later use. See birdie.

dragon A program running low-level secondary systems tasks, especially on Unix systems, which are generally transparent to the user. Monitors and statistical programs are often run in the background as dragons, and the results of their activities may be viewable by the system administrator or those with sufficient security clearance. See daemon.

DRAM 1. digital recorder, announce mode. 2. See dynamic RAM.

Draper, Henry (1837–1882) An American physician and inventor and the son of J. William Draper, Henry Draper was a pioneer in spectral analysis, carrying on research in astronomy and photography that was begun by his father. Henry made major contributions to spectroscopy and astronomical photography and created the first photographs of stellar spectral lines, in 1872. See Draper Catalogue, Henry.

Draper Catalogue, Henry The Henry Draper Catalogue is a spectral classification of stars established in 1890 by Edward Pickering, Anna Palmer Draper, Willamina Fleming, Annie Jump Cannon et al. in honor of Henry Draper. By 1915, more than a quarter million stars had been cataloged, and, in 1918, the first volume of the Catalogue was published. The Catalogue is still an important astronomical reference and has now been published online.

Draper, John (ca. 1943–) See Cap'n Crunch.

Draper, John William (1811–1882) An English-born American scientist, educator, and historian who developed some of the early photographic processes. Draper researched incandescent substances and his son, Henry Draper, carried on this line of research to become a pioneer in spectrum analysis.

Some of Draper's photographic research in the late 1830s apparently predated Daguerre's. He investigated photography of the very small and of the very distant (e.g., the Moon). His book *Human Physiology* (1856) contained the first-published micro-photographs. He is credited with creating the first American portrait, in 1840. Draper's ten-minute exposure was a big improvement over previous techniques that required many hours of exposure.

Draper was an associate of Samuel Morse and they cooperated on some projects. In 1862, he published *The History of the Intellectual Development of Europe*, in 1874, *The History of the Conflict Between Religion and Science* and, in 1878, *Scientific Memoirs*.

drift 1. Variation from a desired signal or current over time from factors other than line, load, environment, or warmup period. See calibration. 2. In radio technology, signal drift is not uncommon. For example, if you set an analog radio to a favorite station and gradually lose the setting as the tuning changes, this is drift.

DRiP See Duplicate Ring Protocol.

drop point A protrusion on an insulator or other object designed to channel moisture away from sensitive components or electrical currents. Drop points (sometimes called drip points) are typically on the lowest protruding edge where water is channeled and may have a variety of shapes and spacings. See insulator.

drive type The make, model, and configuration of a hard drive. A drive can conform to one of the common desktop standards, SCSI or IDE, for example, and, once it is formatted, it is important for the system to have a record of the drive type in order to keep track of sectors, blocks, partitions, and files, the various data configurations that can be set up dynamically on the drive during use. The drive type is really a combination of the cabling, data bus characteristics, physical properties of the drive, and magnetic data configuration which is superimposed on the magnetic recording surface.

Some types of drives can be used in combination with one another, and others cannot. For example, SCSI and IDE drives are not mixed on one data bus. A different controlling mechanism is used for each type. SCSI devices can chain up to six devices (the controller counts as the seventh device), while IDE drives can chain up to two, with one designated as a *master* and the other as a *slave*.

When a formatting software program is run with a new drive, or one that is being formatted for another computer system, it may query the drive for information about its characteristics and display that information on the screen. For example, it may show the brand, model, and size of the drive, and whether there are any existing partitions. Many drives now come preformatted, but it may still be necessary to set up partitions, if desired.

driver 1. In software, a program that includes code that can translate commands into instructions recognizable by a specific device, such as a facsimile modem, printer, scanner, hard drive, etc. Desktop publishing programs typically include a directory full of drivers for various printers which translates the print instructions from the software into the closest approximation possible by the printer through the print driver software. 2. In software event-processing, code that receives commands and distributes them appropriately for execution.

droid *colloq. abbrev.* android. 1. An anthropomorphic robot, generally more human than machine which may be a combination of biological and mechanical/electronic parts. 2. A company drone, someone who unquestioningly follows instructions and mechanically goes about the business of work (or living) without much enthusiasm or introspection, possibly due to apathy or unquestioned acceptance of authority. 3. A person hired as a human robot to do a mindless, repetitive, production-line job that offers few opportunities for variety or creative interaction. These are the kinds of positions that should be handled by machines, in order for people to have more leisure and creative time.

drop A short cable connection, often between a utility pole and a building, or between one panel and another, or a panel and other distribution entity, commonly used for supplying telephone, cable TV, or computer network services.

drop frame In television video broadcast recording and playback, North American television was designed to play at 30 frames per second. Then when color signals were introduced, the differences between black and white and color technology resulted in a compensatory adjustment of the frame rate down to ~29.97 frames per second, a situation that altered the time code and reduced its usefulness for station timing.

Drop frame mode, also known as *compensated mode,* is a technique in which the system skips ahead a very small amount at specified intervals, skipping over the first two bits in each minute. However, this is minutely too much of an adjustment, so each ten minutes, only a single bit is skipped. It's similar to the way in which we adjust our calendar to celestial events by introducing leap years (except that a day is added rather than skipped in a leap year) where needed, to better synchronize them.

Drop frame modes are important to broadcasters because programming is interspersed with commercial announcements, shorts, special programming, and other timing-related items, that need to be sequenced in a smooth, seamless way. See SMPTE time code.

drop loop In telephone wiring between the switching office and local subscribers, the circuit is called a *local loop* and the specific section of the circuit from the utility pole or other nearby junction point to the subscriber's home or office is the drop loop.

dropout, drop-out An undesirable low-level, irregular loss of information when transferring from one system, medium, or format to another. Dropout happens, for example, in video editing, when copying or editing tapes, especially with less robust formats and inexpensive equipment. Dropout can sometimes be seen as white dots appearing somewhat randomly on the screen. They are especially noticeable if the screen has large areas of dark or solid colors.

dropout, transmission A short interruption in a transmission, usually caused by a problem in the transmitting or receiving equipment. Different industries have different objective measures for the length of interruption that constitutes a dropout.

dropped call A call terminated without the express desire of the parties engaged in the call. In radio phone communications, dropped calls are not uncommon, as the signal can easily be interrupted by terrain, weather, or a stronger signal from another source or distance.

dry cable, raw cable, dark cable Conductive cable or wire with no added electronics and no signal passing through. The cable you find on spools in the hardware store is dry cable. Raw fiber optic cable is called dark cable.

dry cell A common, compact type of battery descended from the wet cell, but differing in that it employs nonliquid electrolytes in the form of paste or gel. Dry cells were invented by Gassner in 1888 and manufactured in the early 1900s. Since they do not

---

Digital Transmission Speed Categories

Signal Level/Description

DS-0	A 64,000 bps standard for transmitting digital data through pulse code modulation (PCM). A sampled signal is quantized and transmitted with bits that represent quantization levels being transmitted separately. A standard used in telephone systems.
DS-1	A frame format standard for transmitting data at 1.544 Mbps, developed in 1962. Used on T1 systems. It incorporates time division multiplexing (TDM) to combine 24 DS-0 signals, and adds a single framing bit. Signals are transmitted with bipolar (B8ZS) pulses or alternate mark inversion (AMI). In 1969, the standard was extended to SuperFrame to increase the signal-to-noise ratio, and later it was further modified to create Extended SuperFrame which is more robust. Europe uses a 32-channel 2.048 Mbps system which is somewhat similar but incorporates different synchronization and signaling formats. See Extended SuperFrame, SuperFrame.
DS-1C	This signal system was designated 1C because it fits somewhere between DS-1 and DS-2 in terms of its 3.152 Mbps signaling rates. Used on T1C systems. It was introduced by AT&T in 1975. DS-1 signaling bits are bit-interleaved into the information bits.
DS-2	A frame format developed for longer transmission lines and to accommodate AT&T's Picturephone technology (which was developed many years before the technology to use and support it became sufficiently widespread). Used on T2 systems. It combines four DS-1 signals or 96 DS-0 signals, employs two framing stages, and transmits at 6.312 Mbps. Europe uses a different ITU-defined system that operates at 8.448 Mbps (2.048 Mbps primary rate).
DS-3	A frame format developed for signaling over broad bandwidth signaling systems. Used on T3 systems. DS-3 uses Bipolar with Six Zero Substitution (B3ZS). The DS-3 signal combines 7 DS-2 or 672 DS-0 signals, is framed in two stages, and transmits at 44.736 Mbps. Through multiplexing, the asynchronous signals are transmitted over synchronous links. Europe uses a different ITU-defined system that operates at 34.368 Mbps (2.048 Mbps primary rate).

use liquid acids, they are easier to handle, and more portable than wet cells, and can be used in any orientation. They are commonly used in flashlights, small appliances, and many handheld devices. Many dry cells include toxic chemicals and heavy metals and should be recycled through local centers, not thrown in the trash. See Gassner, wet cell.

DS 1. Dansk Standardiseringsrad. The Danish Standards Institute, located in Hellerup. 2. digital system. 3. See Distributed Single Layer Test Method.

DS- A series of signal speeds for transmitting digital data through a variety of modulation and multiplexing schemes, designated DS-1 through DS-4, with higher numbers representing faster possible transmission speeds. This system is primarily used in North America and Japan. A similar system, which differs in data rates, encoding, and numbers of channels, the *E*-system, is used in Europe. The *DS*-system first was initially used by phone carriers for connecting main switching centers. Gradually, as the technology became less expensive, it began to be used in the backbones of larger private branch exchanges, and now it is used in telephone feeder plants, and local area network backbones as well. See Digital Transmission Speed Categories chart. See E-carrier.

DS Facility A categorization system for describing digital transmission capacity. See DS-, Digital Transmission Speed Categories chart.

DSA 1. data service adapter. 2. Digital Signature Algorithm. See Digital Signature Standard, Electronic Certification. 3. See Direct Selling Association. 4. See Directory System Agent.

DSAT 1. digital satellite. 2. digital satellite TV. 3. See Digital Supervisory Audio Tone.

DSC 1. See digital selective calling. 2. See Digital Subtitle Encoder. 3. Disability Statistics Center. 4. document supply center. 5. distributed statistical computing.

DSCS See Defense Satellite Communications System.

DSE 1. See Deep Sky Exploration. 2. See Distributed Single-Layer Embedded. 3. distributed software/systems engineering. See distributed programming platform. 4. See Dynamic Systems Estimation library.

DSH See double-superheterodyne.

DSL See Digital Subscriber Line.

DSLAM DSL access multiplexer. See add/drop multiplexer, Digital Subscriber Line.

DSM 1. See Design Structure Matrix/Dependency Structure Matrix. 2. See digital switching matrix.

DSP See Defense Support Program.

DSR See data set ready.

DSRC See Dedicated Short-Range Communications.

DSRR See Digital Short-Range Radio.

DSS 1. digital satellite system 2. See Digital Signature Standard. 3. direct satellite system. See direct broadcast satellite. 4. direct selling support. 5. direct station selector. A consumer broadcast system component that enhances channel selection for desired programs; the term is especially applied to satellite programming services selection.

Dstl An organization formed in July 2001 when the U.K. Defence Evaluation and Research Agency

(DERA) was split into a private company and the Dstl agency, which supports a number of sites supporting research in science and technology. Dstl traces some of its lineage back to organizations originating in the 15th century.

DSTO See Defence Science and Technology Organisation.

DSU See Digital Service Unit

DSU-CSU digital service unit-channel service unit. A digital connecting device, usually for linking a transmission line and a router. Some units combine a digital modem, router, and terminal server, a combination that is popular with Internet Services Providers. See digital service unit, channel service unit.

DSX panel See digital signal cross-connect panel.

DT See Deutsche Telekom.

DTE See Data Terminal Equipment, End Device.

DTL 1. database template library. 2. See Designated Transit List. 3. See diode transistor logic. 4. distance teaching and learning.

DTMF See dual tone multifrequency.

DTMX digital trunk manual (telephony) exchange

DTP See desktop publishing.

DTR 1. See Data Terminal Ready. 2. detailed transaction report. 3. discrete tone relations.

DTRS 1. digital tape recording system 2. digital telemetry recording system 3. Digital Trunked Radio System.

DTS Digital Television Standard. See ATSC Digital Television Standard.

DTSR 1. Dial Tone Speed Recording. 2. digital tape system recording 3. digital temporary storage recording.

DTT 1. digital tape transfer. 2. digital telecommunications/telephone trunk. 3. digital tie trunk. A digital telephony trunk line providing a direction connection between private branch exchanges (PBXs), in other words, tying them together. 4. digital trunk testing.

DTTB Digital Terrestrial Broadcasting. See Digital Video Broadcasting.

DTU 1. digital test unit. 2. digital trunk unit.

Du Fay, Charles François de Cisternay (1698–1739) A French soldier and scientist who discovered that electricity had two basic attracting and repelling properties, which could be demonstrated, for example, by rubbing amber with wool and rubbing glass with silk. He called these *resinous electricity* and *vitreous electricity*, making the distinction in a context that had eluded previous researchers.

Du Fay developed some of the ideas first investigated and described by S. Gray in England and made some important observations about the composition of the materials of the conducting medium. Perhaps most important of Du Fay's observations is that

"... an electrified body attracts all those that are not themselves electrified, and repels them as soon as they become electrified by ... the electrified body."

DUA See Directory User Agent.

dual attachment concentrator DAC. A connecting device used in double ring Fiber Distributed Data Interface networks which employ a token-passing scheme over a redundant ring network. Dual ports and attachment points are used in connection with the concentrator to reroute data if a problem arises. See Fiber Distributed Data Interface for more detailed information.

Dual Attachment Station DAS. A configuration of a Fiber Distributed Data Interface (FDDI) token-passing, dual attachment network. The dual attachments provide fault tolerance. They consist of a primary ring and a secondary ring, the first of which is usually used for data transmissions, and the second as a backup in case of problems. A Class A, dual attachment station (DAS) connects to both rings and a concentrator which, in turn, ensures the ring transmission is not interrupted. Failure in a ring causes a series of adaptations such that the ring wraps back on itself and temporarily eliminates the failed station. See A port, B port.

dual cable A two-cable configuration, usually in a local area network (LAN), often implemented to provide redundancy and fault tolerance, as in ring-based Fiber Distributed Data Interface (FDDI) systems. See Fiber Distributed Data Interface.

dual homing 1. A means of providing backup and fault tolerance on a network system, particularly characteristic of Fiber Distributed Data Interface (FDDI) networks. FDDI networks utilize stations, which can be eliminated through rerouting if a problem is found, without interrupting transmissions. Dual homing utilizes two concentrators, a primary and a secondary, with the secondary used as backup that is automatically activated if a problem occurs. See Dual Attachment Station, optical bypass. 2. In a Frame Relay network, a means of providing fault tolerance by using dual port connections in different locations.

dual mode There are many dual mode devices in telecommunications. Many modems are dual mode in order to support both vendor proprietary protocols and industry standard protocols. Dual mode monitors will sometimes support both NTSC and RGB signals. Some phones have dual pulse and tone dialing capabilities. Many cellular phones are now designed to support traditional analog signals and emerging technologies that use digital signals. Dual mode devices tend to come about when there are competing standards, or when technology is transitioning from one stage to the next. Some are autosensing, switching to the correct setting unaided, and some have to be explicitly set with a switch or software.

dual output fiber optic sensor A device with two sensors in one assembly. Thus, dual independent readings may be taken for redunancy or comparison. The sensors may be analog, digital, or hybrid.

dual tone multifrequency DTMF. Touchtone signaling on a phone system. The tones are actually a combination of two frequencies that are variously combined to provide unique codes for each key on the telephone, when pressed. This signal is sent through the line to indicate the desired number to be dialed or the desired selection from an automated phone menu system. The tones were chosen for frequencies that carry well on voice-grade lines.

Phone phreakers used to exploit these tones for dialing unauthorized long-distance numbers with small devices called blue boxes. With the increased use of out-of-band signaling systems, such as Signaling System 7, which send the signals separately from the conversational information, this practice is becoming less prevalent and will eventually be impossible. See touchtone phone.

Dublin Core A standard metadata scheme for describing document-like informational objects in order to facilitate data discovery on computer networks. Thus, core objects such as *Date*, *Creator*, and *Description* can be quickly and easily located and utilized for indexing and archiving purposes. The Dublin core is intended to supplement rather than supplant existing Web metadata searching and indexing methods and can be applied to physical as well as electronic objects. The Dublin Core Metadata Element Set (DCMES) became the first IETF metadata standard arising from this scheme. See Government Information Locator Service, International Development Markup Language.

Dublin Core Metadata Initiative DCMI. An open forum first held in Dublin, Ohio, in 1995 as a result of discussions in October 1994 at the 2nd International World Wide Web Conference. DCMI promotes the understanding and development of interoperable online metadata standards and specialized vocabularies intrinsic and related to the Dublin Core. Consensus is evaluated and administered by the Dublin Core Directorate. http://www.dublincore.org/

duct Protective pipe or tube through which lines or fluids are run. See conduit.

ductile A property of being malleable, a material that can be shaped, drawn out, flattened, or otherwise bent or manipulated without significant stress or breakage.

Duddell, William du Bois (1869–1942) An English experimenter and engineer who discovered that electric arcs created in a circuit with coils and condensers could generate very high-frequency audible tones in the low radio wave frequencies. Duddell fashioned a keyboard connection to control the oscillations, thus varying the pitch and creating the "Singing Arc," in 1899, arguably inventing the first electronic instrument and creating possibilities for wireless communications. The Duddell Medal and Prize was established in 1923 by the Council of the Physical Society. See Poulsen arc.

Dulbecco's observation The observation that credit for a scientific discovery seems to go to the one who gets the most publicity as opposed to the one who created the original invention or made the original observation. As stated by Nobel laureate Renato Dulbecco, "Credit generally goes to the most famous discoverer, not to the first." A closely related sentiment was expressed earlier by William Osler as "In science the credit goes to the man who convinces the world, not to the man to whom the idea first occurs."

To emphasize the importance of credit to a scientific career, Broad and Wade have stated, "It is difficult for a nonscientist to appreciate the overriding importance to the researcher of priority of discovery.... The desire to win credit, to gain the respect of one's peers, is a powerful motive for almost all scientists."

Priority for scientific discovery carries not only recognition and personal satisfaction, but has a highly significant bearing on future scientific opportunities and funding for future research.

dumb switch A switching device that channels signals through a desired pathway as needed without automation through digital intelligence. Dumb switches are commonly manually set passive devices. A dumb A/B switch can be used for two computers to share one printer or modem. Dumb switches are very common, particularly as A/B or A/B/C switches, because they are inexpensive and easy to set up and use. They make no logical or electrical evaluations or decisions about the incoming or outgoing data; they simply route it mechanically.

In its simplest form, a dumb switch does not alter or boost the electrical connection, though some may be equipped to amplify or condition a signal, without changing its informational content. Thus, switches may be electrically passive or active. In video, where a great deal of switching occurs, banks of both passive and active switches channel the desired video or audio feeds into the broadcast or recording channel. In contrast, on automated networks, "smart" switches may evaluate the incoming data and perform some rudimentary routing. Some switches are so smart, in fact, that the distinction between switches and routers at the high end is blurred. On layered networks, switches typically operate at the second layer.

Other than the inconvenience of threading cables behind desks and through walls, installing a dumb device switch is pretty straightforward. Ensure that the interconnected devices are compatible, and use gender benders and converters that are the right sizes and numbers of pins to hook everything together. Most dumb switches for computer applications have 25-pin female D connectors. Make sure all the systems are powered off before making any connections, and test new connections individually rather than all at once, so that a problem can be isolated and corrected right away. See A/B switchbox.

dumb terminal A minimally configured computer terminal, with no direct processing capabilities, that isn't very useful unless it is networked to a central processing system. Universities and libraries often use dumb terminals to provide cost-effective user access to the main system. The advantage is that the dumb terminals are inexpensive, easy to maintain, and not highly subject to abuse or vandalism. In their simplest form, they consist only of a touchscreen monitor or teletype interface; one step up is a keyboard or mouse and a monitor. Some not-quite-as-dumb terminals will include both keyboard and mouse, monitor, and sometimes a floppy drive or CD-ROM drive, but they still rely on a remote system for actual processing of data and commands. Hybrid terminals may include a processor for simpler tasks but still rely on the remote system for most of their computing power.

dumpster diving Searching for mechanical parts, discarded electronics, trade secrets, software and hardware manuals, access codes, login names, and passwords in large outside trash cans called dumpsters. Dumpster diving is practiced by corporate espionage agents and computer hackers. The practice has been around for a long time, but the term gained media prominence in the mid-1980s when a group of teenagers in San Diego was apprehended for computer piracy and revealed some of the ways by which it had come into possession of source code, computer passwords, phone numbers of BBSs and timeshare systems, and other confidential materials. Shredders are employed by many companies to protect sensitive documents from prying eyes and some will incinerate the documents rather than disposing of them in a dumpster. In the military, even greater precautions against theft are taken, with codebooks and other sensitive documents sometimes being printed in inks that will disappear if photocopied or exposed to moisture.

DUN See dialup networking.

Dunwoody, Henry Harrison Chase Dunwoody patented the use of carborundum in early radio wave crystal detectors in 1906 (U.S. patent #837,616). Carborundum (silicon carbide) was robust in the sense that it could be clamped down, and thus it was used in portable wireless telegraphy (radio) units. Tube radios, which could be readily amplified, eventually superseded crystal detectors.

duopoly A market situation in which two major sellers greatly influence the market, though they may not necessarily control it. In industries requiring licenses, a situation where exclusive operating licenses are issued to two businesses rather than one.

duplex Double, bidirectional.

duplex connection, duplex transmission Data transmissions in which a message can be sent in both directions along the same transmissions line or path. In many duplex systems, the messages can be alternately sent in one direction or the other, but not in both directions simultaneously, whereas in full duplex, the messages can be sent in both directions at the same time. Serial communications software often has half- and full-duplex settings.

duplex telegraphy A historic innovation in telegraphic communications in which two messages were sent in opposite directions, at the same time, over the same line by varying the strength of the current. This innovation was first put into practical use in the 1850s and was thereafter of great interest, since it could significantly improve the efficiency of telegraphic communications, in effect doubling the capacity of the line in the 1800s. Since two-way or duplex telegraphy had many commercial advantages, there was much interest in this idea and in putting it into commercial use.

J.W. Gintl, an Austrian telegraphic director and physicist, was one of the first to propose a practical means of duplex communications using two batteries. Two years later, Siemens & Halske were to patent a duplex telegraph, using only a single battery. In North America, Joseph B. Stearns refined the concept and Thomas Edison extended it, patenting a quadraplex telegraph (U.S. #480,567) that could transmit two signals in each direction by varying not only the strength of the current, but the direction as well (followed by patents for a number of refinements and variations).

Duplex or multiplex communications are now generally the norm for many types of communications, although wireless communications over the same frequency are still sometimes carried out in one-way or simplex mode, as are some simple homebrew communications systems. See Farmer, Moses; Frischen, Carl; Gintl, Julius; Stearns, Joseph; telegraph history.

duplexer A switching device that provides alternating transmitting and receiving through the same transmissions system (data line or antenna).

Duplicate Ring Protocol DRiP. A Cisco protocol for Virtual LAN (VLAN) switches and routers that identifies active Token-Ring VLANs. DRiP data is utilized for detecting duplicate configurations and for all-routes explorer filtering. DRiP has a Cisco HDLC protocol type value of 0x0102. See Token-Ring network.

DVB See Digital Video Broadcasting.

DVBG See Digital Video Broadcasting Group.

DVCPRO See D-7.

DVD See Digital Versatile Disc.

DVD Forum An international association of software and hardware developers and manufacturers of Digital Versatile Discs (DVD) technology. It was originally founded as the DVD Consortium in 1995. The original founding companies included Hitachi, Matsushita Electric Industrial, Mitsubishi Electric, Pioneer Electronic, Royal Philips Electronics, Sony, Thomson Multimedia, Time Warner, Toshiba, and Victor. There are now more than 200 member companies, with headquarters in Tokyo, Japan.

The Forum promotes acceptance of DVD-related products in the entertainment, IT, and consumer electronics industries. It defines DVD format specifications, publishes educational materials, creates DVD format books, and administers the DVD Verification Laboratories. Membership is open to organizations engaged in DVD research and development or manufacturing. See Digital Versatile Disc, Digital Versatile Disc player. http://www.dvdforum.org/

DVD+RW Alliance A voluntary alliance of industry-leading personal computer and optical storage developers and manufacturers, including Hewlett-Packard, MCC/Verbatim, Philips Electronics, Ricoh Company, Sony, Thomson multimedia, and Yamaha. The Alliance disseminates information relating to DVD+RW technologies through various events and publications. It has been working independently of the DVD Forum.

In spite of its quick acceptance and widespread popularity, the DVD format will likely be superseded by other optical formats, most probably blue laser technologies. Blue laser diodes have the potential to write optical media with more than five times as much data as current DVD formats. Thus, about 40 hours of broadcast recording or 2 hours of high quality cinematic entertainment could be recorded on a single side. See DVD Forum. http://www.dvdrw.com/

Dvorăk keyboard A type of keyboard layout designed by August Dvorăk and William Dealey after they studied the natural movement of fingers and of the hand over typewriter keys and researched ways in which to conform the key positions to the comfortable hand use, rather than conforming the hand to unnatural keyboard lettering layouts.

DVST See direct view storage tube.

DWDM See dense wavelength division multiplexing.

DWS See Dialable Wideband Service.

DXC See digital cross-connect.

DXI See data exchange interface.

dynamic bandwidth allocation The process of assigning bandwidth on demand or according to algorithms to maximize the efficiency of the system, rather than transmitting on particular frequencies or at particular times.

dynamic beam focusing In cathode-ray tubes (CRTs), the sweep of the beam from an electron gun across the inside surface of the screen that displays the image. This beam forms a series of motions which, if kept equal, are curved. To keep the distance equal across the sweep of the beams, earlier television screens and computer monitors were curved to match the length of the beams. Early flat screen monitors were rare and expensive. With more sophisticated hardware and software algorithms, manufacturers have devised ways of compensating the travel distance of the beam to adjust to the characteristics of a flat surface. One of these techniques is dynamic beam focusing, adjusting the beam focus as needed, depending on which part of the screen is illuminated, and its distance from the gun.

dynamic data exchange DDE. Any process in which data is transferred between systems or between applications without intermediary steps, such as saving the information and transmitting with a different application. See drag and drop, Object Linking and Embedding.

Dynamic Host Configuration Protocol DHCP. An expanded client/server configuration protocol descended from, and downwardly compatible with, Bootstrap Protocol (BOOTP). DHCP provides manual, automatic, and dynamic allocation of IP addresses and a complete set of TCP/IP configuration values. It utilizes ports 67 and 68 and retains BOOTP's *bootrequest* and *bootreply* packet formats. See RFC 1533, RFC 1541.

dynamic IP addressing When logging onto the Internet, or any system using the Internet Protocol (IP), it is necessary for a unique number to be assigned to the session to handle the flow of data to and from the user. Dynamic IP addressing is a scheme for automating the process of assigning an address when a user connects to an Internet Access Provider (IAP) or other network access point. As part of the user login, a unique number is assigned for that session. The number is typically freed when the user logs off, so the IAP can reassign it to the next user, if needed. Freeing the address is an important part of the process on large distributed networks like the Internet, where there may be millions of users online, with some of the large IAPs handling tens of thousands of simultaneous users.

dynamic gain equalization processor DGEP. An electronic processor for providing dynamic gain-flattening and tilt control in optical amplifiers. In commercial systems, DGEP provides gain equalization through patented all-fiber acousto-optical tunable filter (AOTF) technology. In single-mode optical fibers, acoustic waves can be applied directly to produce what Novera Optics calls *notch* filters for optical signal transmission. Notches may be shaped by controlling the frequency and amplitude of the acoustic wave. See acoustic wave.

Dynamic Link Library DLL. In software programming, a Microsoft product format for consolidating a number of frequently used routines, or routines that may not be available by default in an application, such as Visual BASIC. The DLL is an organizational programming tool that allows a 'library' of routines to be written once, bundled together, and thereafter linked into a program and called by the application program as needed.

dynamic RAM DRAM. Random access memory that requires a supply of current through the chip at all times in order to retain and refresh the stored information. When you turn a computer off, the data currently in RAM is lost. RAM is one of the most prevalent types of dynamic fast storage used in computers. Most systems these days require about 16 or 32 Mbytes of RAM for basic functioning. This amount is in stark contrast to desktop computers in the 1970s, which could run telecommunications programs, word processors, and spreadsheets in less that 8 kilobytes (not megabytes) of RAM, and systems in the mid-1980s, which could run music and graphics simultaneously in a fully multitasking environment in only 4 megabytes of RAM. See static RAM.

Dynamic Random Access Memory See dynamic RAM.

dynamic range A range of intensities, between the minimum and maximum extremes. It's a phrase that is often applied to concepts of light or sound. In imagery, the dynamic range of a scanner, for example, is the range of light levels, from the brightest highlight to the darkest shadow, that can be picked up and transmitted. In music, the dynamic range of a recorded symphony performance is the range from the softest note to the loudest, expressed in terms of decibels. Dynamic range is sometimes described more objectively in terms of the maximum and minimum levels of a parameter as measured by an instrument designed for that use. See gamut.

dynamic resource allocation In various types of communications, the administration, allocation, and dynamic reallocation of resources, such as frequencies, channels, processes, programs, and access to shared peripheral devices. Dynamic resource allocation usually entails intelligent algorithms for determining authorizations, priorities, and needs, and often includes sophisticated queuing, routing and multitasking capabilities.

dynamic routing In general, the creation and adjustment of communications paths on an as-needed or as-optimized basis, so paths will change to fit the needs of a situation as specified. In data networks, dynamic routing allows the system as a whole to stay online even if individual systems or routes change or are unavailable. This is accomplished through routers, which can communicate with other routers, usually those topologically nearby, and which may increase, or modify routing tables as needed.

Dynamic routing works well on large, changeable, packet-switched systems like the Internet. Routers can relay data around distressed or suddenly unavailable systems or trunks. On small systems, the overhead of dynamic routing may not be worth the loss of speed that the processing takes. Static routing may be used quite effectively on small systems with known, stable characteristics. See router, Routing Information Protocol.

dynamic sector repair A fault correction and prevention system built into hard drive systems, particularly multiple disk arrays such as RAID, that seeks faulty sectors on a disk, repairs the data if possible, and records bad sectors to prevent the system from trying to write to those sections in the future. See redundant array of inexpensive disks, SMART.

dynamic storage In computing, the allocation of temporary or permanent storage space in an intelligent manner, so unused space can be optimally used, and unneeded data is removed to allow the reuse of storage for other applications. It may also involve occasional reorganization of information if extra processing cycles are available. See garbage collection.

Dynamic Systems Estimation library DSE. An object-oriented noncommercial software library for studying multivariate time series analysis techniques and forecasting models. DSE runs in Splus and R. The library is suitable for applications such as studying the statistical implications of equivalence among different model representations, studying the forecasting properties of models, or studying small sample properties of estimators.

dynamically controllable magnetic DCM. Magnetic materials that can change permeability in real-time when stimulated by a magnetic field. DCM materials that have this property have been found in the VHF to microwave frequency ranges. Some communications antennas need to be transparent at some frequencies and reflective at others, and DCM materials are being tested for their effectiveness for this use.

dynamo The historical name for what is now termed a generator. A dynamo is a machine that converts mechanical energy into electrical energy (direct current). A friction-based bicycle light is an example of a simple dynamo. When the cyclist pedals, the wheel spins and rubs against the contact point for the generator. The generator takes this mechanical energy and converts it into light, so the cyclist can see and be seen at night. See alternator; generator; Siemens, Werner. 2. An energetic, dynamic individual.

dynamometer, electrodynamometer A sensitive current, voltage, and power detecting instrument similar to a D'Arsonval meter except that it uses a field coil, or coils, rather than a permanent magnet.

A dynamometer functions through a rotating coil controlled by the interaction between the magnetic fields of a moving coil and field coil(s). It can be used in conjunction with both direct current (DC) and alternating current (AC). See D'Arsonval galvanometer.

dynode In a photomultiplier tube, a component that enables the amplification of a signal through secondary emission by the stimulation of the release of additional electrons when it is struck by the photoconverted electron(s). It is made of reflective materials that will give up electrons and may be coated to improve its properties. Fabrication materials include BeCu (beryllium copper) and CsSb (cesium antimony).

A chain of several dynodes is placed in the path of the electrons emitted by a photocathode as they travel towards the anode in an evacuated electron tube. Typically about 10 or 12 dynodes are staggered in pairs at appropriate reflecting angles so that the electron path passes directly from one to the next in the path. The potential of each dynode is set relative to the potential of the next dynode in the chain, as each step has a multiplying effect on the energy as a whole (gain). As electrons strike the reflective material of the dynode, kinetic energy is transferred to the secondary electrons. The kinetic energy is determined by the voltage level of each dynode, with a relationship between the kinetic energy and the number of secondary electrons. Higher voltages (within operating ranges) result in higher numbers of secondary electrons, with a cumulative effect as the number of dynodes in the chain increases. Some tubes are suitable for use with a photomultiplier tube base that augments the capabilities of the tube and may provide external voltage stabilization capabilities for the last dynodes in the chain and adjustment control for individual dynodes (e.g., the focus dynode). Linear output from a connecting dynode (e.g., the 9th of 12 or 10th of 10 dynodes) may be processed through a scintillation component.

Thus, the effect of even a single photon can be studied with photomultiplier tubes, a capability that is especially useful in particle physics and low light inputs. See photomultiplier for further information and a diagram. See photomultiplier, scintillator.

Dyson, George (1953–) An American kayak enthusiast and technology historian, Dyson is the author of *Darwin Among the Machines: The Evolution of Global Intelligence*, a provocative book about the origins of computers and networking, and philosophical speculations about intercommunication and emerging digital intelligences.

In 2001, Dyson authored *Orion*, a book that reveals remarkable preliminary steps to travel to Mars that were carried out in secret by the U.S. government in the 1960s and later shut down.

Network Digital Intelligences

Dyson's philosophic/historic treatment of the emergence of intelligence in global distributed digital networks is even more applicable now than when it was released. As fiber optic networks with high-speed, high-bandwidth capabilities become more widely installed, the concept of the social and physical integration of networking increases in significance.

e 1. *symb.* basic unit of charge. Proton-associated charges are designated $+e$ and electron-associated charges are designated $-e$. See coulomb. 2. *symb.* energy as in $e = mc^2$. 3. *symb.* – natural logarithm with a value of about 2.71828. See logarithm. 4. *symb.* voltage, though V is commonly used. See volt.

E 1. See E notation. 2. *symb.* See exa-.

E & M signaling A signaling method communicated over two leads or wires, each one labeled E (ear) and M (mouth), one in each direction. Each lead may be grounded, open, or have signaling voltage applied. If station 1 wants to call station 2 (assuming 2 is onhook and not busy), voltage is applied to the M lead to provide a supervisory signal indicating a call, and the associated E lead at station 2 will be grounded. See A & B bit signaling.

E & M Signaling Examples

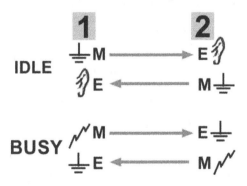

Two simple examples help illustrate E & M Signaling. If the circuits at both locations (1 and 2) are idle, M will be grounded and E open (in both directions). If both circuits are busy, voltage is applied at M to signal the busy condition and E will be grounded.

E Block A Federal Communications Commission (FCC) designation for a Personal Communications Services (PCS) license granted to a telephone company serving a Major Trading Area (MTA). It grants permission to operate at certain FCC-specified frequencies. See A Block for a chart of frequencies from A to F Block.

E interface, external interface A general interface so that Cellular Digital Packet Data (CDPD) deployed over AMPS interface with external networks. See A interface, I interface.

E notation Scientific notation for representing a large value. For expressing large values or entering them into a computer, E or e represents numbers as powers of base 10, which is useful if the times symbol could be mistaken for the letter "x" or when space is limited (e.g., on calculators). For example, 1,560,000,000 may be more succinctly written as a coefficient with a value ≥ 1 and <10 times a power of ten, yielding 1.56×10^9 or 1.56E+9.

E region A portion of the Earth's ionosphere, above the troposphere, that ranges from about 100 kilometers to about 130 kilometers above the surface of the Earth. This is also known as the Heaviside layer, or the Kennelly-Heaviside layer, and is used for deflection of short wave radio signals. See Kennelly-Heaviside layer. See ionospheric subregions for a chart.

E Series Recommendations A series of ITU-T-recommended guidelines for overall network operation, telephone services, service operations, and human factors. These guidelines are available as publications from the ITU-T for purchase over the Net. Since ITU-T specifications and recommendations are widely folowed by vendors in the telecommunications industry, those wanting to maximize interoperability with other systems need to be aware of the information disseminated by the ITU-T. A full list of general categories is listed in the Appendix, and specific series topics are listed under individual entries in this dictionary, e.g., D Series Recommendations.

E1, E-1, CEPT1 See E-carrier.

E-911 service Enhanced 911 service. A 911 emergency service with extra features, such as automatic number identification (ANI – the caller's phone number), the calling address (ALI – automatic location information), and selective routing (SR). Many emergency service personnel consider these 'essential' rather than 'enhanced' services, and the Federal Communications Commission (FCC) has been encouraged to support and adopt E-911 in its rule-making decisions. In 1999, the FCC issued a Report and Order requiring that all mobile phones capable of analog operation incorporate a special feature for reliably processing 911 emergency calls. Manufacturers were given until February 2000 to comply with the ruling. See 911 calling.

E-band A spectrum allocation for wireless PCS in certain regions such as Latin America (e.g., Brazil).

E-band, optical In optical communications, an ITU-specified transmission band in the 1280 – 1625-nm range.

The E-band spectrum has only recently become commercially viable, as fiber technology evolves. Vendors have begun releasing systems for incorporating E-band capabilities into coarse wavelength division multiplexing (CWDM) fiber optic systems carrying other wavelengths, thus increasing capacity. Suppliers of testing equipment for optical networks began releasing test equipment for portions of the E-band spectrum in early 2001. Nonzero-dispersion-shifted fibers tend to be used in these systems, a fiber type also suitable for C-band and L-band applications. See C-band, L-band.

E-carrier, European-carrier The European ITU-T-specified analog to the T1 high-speed communications system used in North America. They are the same in many general aspects, but they differ as to the overall hierarchy and details, such as speed of transmissions, number of channels, and the lack of bit-robbing. Since bit-robbing is not used for signaling purposes, a full 64 Kbps is possible (as opposed to the 56 Kbps in bit-robbed systems with similar characteristics). The analog to the superframe on T-carrier systems is the multiframe (16 frames) on E-carrier systems.

E1 transmits at 2.048 Mbps (compared to T1 1.544 Mbps) using time division multiplexing (TDM) and pulse code modulation (PCM) simultaneously on up to 30 64-Kbps digital channels. Two additional channels are used for signaling and framing. See J-carrier, T-carrier.

e-commerce, ecommerce electronic commerce. Many forms of electronic commerce have evolved over the last twenty years. Banks have been using computers longer than most businesses, and ATMs have been common since the early 1980s. But e-commerce capabilities have grown over the Internet, and now the term has a broader meaning for any type of electronically facilitated or direct electronic transaction carried out over a private or public network, in addition to traditional ATM-based electronic transactions.

E-commerce Experts Group ECOMEG. A technical working group of the WAP Forum that communicates commerce-related information to other groups. Since security is an important aspect of electronic commerce, ECOMEG provides the WAP Security Group (WSG) with the requirements for security features that are important to financial institutions, merchants, service providers, and others. See WAP Forum.

E-IDE Enhanced Integrated Drive Electronics. An enhanced computer peripheral connection format, descended from IDE. IDE hard drive controllers are very common on Intel-based desktop computers.

Enhanced IDE has a greater storage capacity than IDE and faster data transfer rates. See IDE.

E-Privacy Act The Encryption Protects the Rights of Individuals from Violation and Abuse in Cyberspace Act was introduced in May 1998 by Senators Ashcroft, Leahy, and Burns to promote privacy as it relates to the electronic age. The bill supports the widespread availability of strong encryption without back doors and protects the privacy of sensitive, confidential information such as health information and financial documents. The bill sought to support global competitiveness and to support the efforts of law enforcement agents through establishment of a National Electronic Technology Center (NET Center). See Encryption for the National Interest Act, Security and Freedom through Encryption Act.

E-TDMA See extended time division multiple access and time division multiple access.

e-zine electronic magazine. An electronic publication, usually provided over public networks, which retains many of the format, editorial features, and characteristics of a print magazine.

EA Equal Access. A moral and regulatory stipulation that all persons have equal access to telecommunications services.

EAC See European Astronaut Centre.

EACA See European Association of Communications Agencies.

EACEM See European Association of Consumer Electronics Manufacturers.

EADP See European Association of Directory and Database Publishers.

EAGLE Extended Area Global Positioning System (GPS) Location Enhancement. EAGLE is a commercial GPS system implemented by Differential Corrections Inc. (DCI) to provide services in North America. EAGLE employs a network of reference nodes or stations and a central processing hub to provide more precise GPS location information than is provided by an unenhanced GPS system functioning on civilian frequencies. EAGLE employs separate error estimates to generate local area corrections and frequency modulated (FM) subcarrier broadcasts to provide correction information to users.

Current EAGLE nodes are widely installed across North America, from Seattle and San Diego to Halifax and Miami, and link with Frame Relay networks. The frame relay system is used for the transmission of data to two network hubs which create separate estimates for the position and clock errors of each of the satellites being used. A grid of ionospheric corrections is also used. The composite corrections are then transmitted to all the FM stations in the network through a geostationary satellite sending to small dish antennas at the receiving facilities. The vector corrections are converted into local corrections to produce scalar corrections for each satellite. The resulting data stream is broadcast to the mobile user through a frequency modulated (FM) subcarrier.

EAM See electroabsorption modulator.

EANTC See European Advanced Networking Test Center.

Early Bird INTELSAT's historic commercial communications satellite, claimed to be the first commercial satellite, launched from Cape Canaveral in April 1965 to orbit over the Atlantic Ocean. By the end of the year, Early Bird was already providing more than 100 telephone circuit relays and dozens of hours of television programming. The INTELSAT projects evolved from the Early Bird system. See INTELSAT.

Early Packet Discard EPD. In ATM networking, a traffic flow control service guarantee technique used in situations where congestion occurs on ATM networks, usually in unspecified bit rate (UBR) services. Cells early in the packet set are discarded, perhaps right down to the final cell, which is not discarded, as it is needed as a signal for the receiving station that it is the end of the packet set. See cell rate.

early token release In a token-passing network, as a Token-Ring network, a means by which a station sends out a token without first checking to see if the receiving system has acknowledged the transmission. This can increase the efficiency of transmissions around the ring in some situations, as normally the system only sends one token at a time in one direction. See Token-Ring network.

EARN See European Academic and Research Network.

EARP See Ethernet Address Resolution Protocol.

EARSeL See European Association of Remote Sensing Laboratories.

Earth grounding Grounding an electrical circuit by placing a lead into the Earth. It works best when it is inserted a few feet into damp soil. Earth grounding was historically placed near outhouses, where the soil was usually damp and soft (before inside plumbing became prevalent). See ground.

Earth Observing System, Earth Observation Satellite EOS. A central project of NASA's Earth Science Enterprise (ESE) consisting of scientific research and data supporting a series of coordinated polar-orbiting and low-inclination satellites designed for long-term global observations and experimentation. See ASTER. http://eospso.gsfc.nasa.gov/

Earth Resources Technology Satellite ERTS-1. The historic first Earth remote-sensing satellite launched in 1968 from an Air Force base located in California. It was equipped with the controversial, but ultimately successful, Hughes Aircraft scanner. This program developed into the Landsat series in 1975, and two very similar satellites were launched in 1975 and 1978. See Landsat, scanner.

Earth station The portion of structures and transmission equipment associated with a satellite that are stationed on the Earth. They may include facilities, antennas, orientation systems, transceivers, etc., in a building or on a mobile unit.

earthing See Earth grounding, ground.

EAS 1. Earth and atmospheric sciences. 2. Eastern Analytical Symposium, Inc. http://www.eas.org/ 3. electronic acquisition system. 4. electronic articles surveillance 5. See Emergency Alert System. 6. engineering and applied sciences. 7. See Enterprise Application Server. 8. See European Astronomical Society. 9. experimental and applied sciences. 10. See Extended Area Service.

EASI See ETSI ATM Services Interoperability.

Eastern and Associated Telegraph Company An early submarine telegraph cable company, established as a merger of a number of smaller companies into the Eastern Telegraph Company, in 1872. Through a further amalgamation of global telegraph companies, it had become the largest by 1900 and served many major cable chains in Africa, India, and Australia. In 1929, it ended with the formation of Imperial and International Communications Limited.

EBA 1. Special Broadcasting Service Enterprise Agreement (1996) 2. See European Broadcasting Area. 4. See eventual Byzantine agreement.

EBCDIC (*pron.* eb-si-dik) Extended Binary-Coded Decimal Interchange Code. A family of 8-bit (256-character) encodings adapted by International Business Machines (IBM) from punch card codes, in preference to ASCII which is more widely used by the rest of the computing community. See American Standard Code for Information Interchange.

Ebone The leading European network backbone serving more than 60 networks in more than 20 countries, providing connections to major world networks and to the Internet. The Ebone connects directly to North American networks through the Ebone's trans-Atlantic fiber cable.

The Ebone initiative was launched as a not-for-profit organization financed independently of government funding and formally founded in 1991. The ISPs setting up the early Ebone formed the Ebone Holding Association. In 1993, Hermes Europe Railtel was founded, creating an independent European fiber-optic network. The two companies were subsequently combined as the Ebone under Global TeleSystems (GTS). In July 1999, GTS announced the acquisition of the remaining 25% of Ebone not already owned by GTS. The operating headquarters are located in London.

In April 2001, Global Knowledge, a major IT training company for professionals implementing and managing complex network systems, announced Ebone as their choice for providing European Internet training services. The system combines international private leased circuits (IPLCs), IP-based virtual private networks (VPNs), and hosted IP services. See backbone, EUnet, Mbone.

EBS 1. electronic broadcast service/system. 2. See Emergency Broadcast System.

EBU See European Broadcasting Union.

EBU Broadcast Wave Format, EBU BWF. An object-oriented audio file format developed by the European Broadcasting Union to facilitate interoperability among differing computer platforms. The EBU BWF defines the minimum necessary information required for audio broadcast applications. In 2000, the EBU included a provision for BWF files to include a SMPTE UMID and version 1 of the BWF specification was released. Technical descriptions are available through the EBU in Adobe PDF format. See European Broadcasting Union.

E

ITU-T E Series Recommendations

Definitions and Basic Descriptions

E.100	Definitions of terms used in international telephone operation
E.106	Description of an international emergency preference scheme (IEPS)
E.128	Leaflet to be distributed to foreign visitors
E.123	Notation for national and international telephone numbers

International Services

E.105	International telephone service
E.104	International telephone directory assistance service and public access
E.109	International billed number screening procedures for collect and third-party calling
E.110	Organization of the international telephone network
E.111	Extension of international telephone services
E.112	Arrangements to be made for controlling the telephone services between two countries
E.113	Validation procedures for the international telecommunications charge card service
E.116	International telecommunication charge card service
E.118	The international telecommunication charge card
E.120	Instructions for users of the international telephone service
E.122	Measures to reduce customer difficulties in the international telephone service
E.124	Discouragement of frivolous international calling to unassigned or vacant numbers answered by recorded announcements without charge
E.125	Inquiries among users of the international telephone service
E.150	Publication of a "list of international telephone routes"
E.152	International freephone service
E.153	Home country direct
E.154	International shared cost service
E.155	International premium rate service
E.163	Numbering plan for the international telephone service
E.171	International telephone routing plan
E.175	Models for international network planning

E.190	Principles and responsibilities for the management, assignment and reclamation of E-series international numbering resources
E.191.1	Criteria and procedures for the allocation of the ITU-T international network designator addresses
E.195	ITU-T international numbering resource administration
E.401	Statistics for the international telephone service (number of circuits in operation and volume of traffic)

International Network Management

E.410	International network management – general information
E.411	International network management – operational guidance
E.413	International network management – planning
E.414	International network management – organization
E.415	International network management guidance for common channel Signaling System No. 7

Supplementary Services

E.130	Choice of the most useful and desirable supplementary telephone services
E.131	Subscriber control procedures for supplementary telephone services
E.132	Standardization of elements of control procedures for supplementary telephone services
E.151	Telephone conference calls

Operator & Directory Assistance Services

E.104	International telephone directory assistance service and public access
E.121	Pictograms, symbols, and icons to assist users of the telephone service
E.127	Pages in the telephone directory intended for foreign visitors
E.115	Computerized directory assistance
E.114	Supply of lists of subscribers (directories and other means)
E.126	Harmonization of the general information pages of the telephone directories published by administrations
E.140	Operator-assisted telephone service
E.141	Instructions for operators on the operator-assisted international telephone service

Numbering Plans

E.160	Definitions relating to national and international numbering plans
E.163	Numbering plan for the international telephone service
E.164	The international public telecommunication numbering plan
E.166	Numbering plan interworking for the E.164 and X.121 numbering plans
E.168	Application of E.164 numbering plan for UPT
E.169	Application of Recommendation E.164 numbering plan for universal international freephone numbers for international freephone service
E.169.2	Application of Recommendation E.164 numbering plan for universal international premium rate numbers for the international premium rate service
E.169.3	Application of Recommendation E.164 numbering plan for universal international shared cost numbers for international shared cost service
E.195	ITU-T international numbering resource administration
E.190	Principles and responsibilities for the management, assignment, and reclamation of E-series international numbering resources
E.213	Telephone and ISDN numbering plan for land mobile stations in public land mobile networks (PLMN)
E.215	Telephone/ISDN numbering plan for the mobile-satellite services of Inmarsat

Routing Plans and Routing Data

E.148	Routing of traffic by automatic transit exchanges
E.149	Presentation of routing data
E.173	Routing plan for interconnection between public land mobile networks and fixed terminal networks
E.174	Routing principles and guidance for Universal Personal Telecommunications (UPT)
E.190	Principles and responsibilities for the management, assignment, and reclamation of E-series international numbering resources
E.350	Dynamic Routing Interworking
E.351	Routing of multimedia connections across TDM-, ATM-, and IP-based networks
E.352	Routing guidelines for efficient routing methods
E.353	Routing of calls when using international routing addresses

Billing, Accounting

E.230	Chargeable duration of calls
E.231	Charging in automatic service for calls terminating on a recorded announcement stating the reason for the call not being completed
E.232	Charging for calls to subscriber's station connected either to the absent subscriber's service or to a device substituting a subscriber in his absence
E.251	Old system for accounting in international telephony
E.252	Mode of application of the flat-rate price procedure set forth in Recommendations D.67 and D.150 for remuneration of facilities made available to the administrations of other countries
E.260	Basic technical problems concerning the measurement and recording of call durations
E.261	Devices for measuring and recording call durations
E.270	Monthly telephone and telex accounts
E.275	Exchange of international traffic accounting data between administrations using electronic data interchange (EDI) techniques
E.276	Transmission in encoded form of telephone reversed charge billing and accounting information
E.277	Conventional transmission of information necessary for the collection of charges and the accounting regarding collect and credit card calls
E.433	Billing integrity

Telephone Tones

E.180	Technical characteristics of tones for the telephone service
E.181	Custom

Signaling System 7 (SS7) and ISDN

E.145	International network management guidance for common channel Signaling System 7
E.167	ISDN network identification codes
E.172	ISDN routing plan
E.177	B-ISDN routing
E.191	B-ISDN addressing
E.184	Indications to users of ISDN terminals
E.213	Telephone and ISDN numbering plan for land mobile stations in public land mobile networks (PLMN)
E.215	Telephone/ISDN numbering plan for the mobile-satellite services of Inmarsat

E

E Series Recommendations, cont.

Signaling System 7 (SS7) and ISDN, cont.

E.330	User control of ISDN-supported services
E.331	Minimum user-terminal interface for a human user entering address information into an ISDN terminal
E.416	Network management principles and functions for B-ISDN traffic
E.671	Post-selection delay in PSTN/ISDN using Internet telephony for a portion of the connection
E.710	ISDN traffic modeling overview
E.716	User demand modeling in Broadband-ISDN
E.720	ISDN grade of service concept
E.730	ISDN dimensioning methods overview
E.735	Framework for traffic control and dimensioning in B-ISDN
E.736	Methods for cell level traffic control in B-ISDN
E.737	Dimensioning methods for B-ISDN

Facsimile Services

E.320	Speeding up the establishment and clearing of phototelegraph calls
E.323	Rules for phototelegraph communications set up over circuits normally used for telephone traffic
E.450	Facsimile quality of service on public networks – general aspects
E.451	Facsimile call cut-off performance
E.452	Facsimile modem speed reductions and transaction time
E.453	Facsimile image quality as corrupted by transmission-induced scan line errors
E.454	Transmission performance metrics based on Error Correction Mode (ECM) facsimile
E.456	Test transaction for facsimile transmission performance
E.457	Facsimile measurement methodologies
E.458	Figure of merit for facsimile transmission performance
E.459	Measurements and metrics for characterizing facsimile transmission performance using nonintrusive techniques
E.460	Measurements and metrics for monitoring the performance of V.34 Group 3 facsimile

Mobile, Satellite, and Marine Telephony

E.200	Operational provisions for the maritime mobile service

E.202	Network operational principles for future public mobile systems and services
E.201	Reference recommendation for mobile services
E.210	Ship station identification for VHF/UHF and maritime mobile-satellite services
E.211	Selection procedures for VHF/UHF maritime mobile services
E.212	The international identification plan for mobile terminals and mobile users
E.214	Structure of the land mobile global title for the signaling connection control part (SCCP)
E.216	Selection procedures for the INMARSAT mobile-satellite telephone and ISDN services
E.220	Interconnection of public land mobile networks (PLMN)

Data Services

E.300	Special uses of circuits normally employed for automatic telephone traffic
E.301	Impact of nonvoice applications on the telephone network
E.370	Service principles when public circuit-switched international telecommunication networks interwork with IP-based networks
E.417	Framework for the network management of IP-based networks

Network Management

E.170	Traffic routing
E.412	Network management controls
E.416	Network management principles and functions for B-ISDN traffic

Traffic Forecasting, Engineering, and Measurement

E.170	Traffic routing
E.490	Traffic measurement and evaluation – general survey
E.491	Traffic measurement by destination
E.492	Traffic reference period
E.500	Traffic intensity measurement principles
E.501	Estimation of traffic offered in the network
E.502	Traffic measurement requirements for digital telecommunication exchanges
E.503	Traffic measurement data analysis
E.504	Traffic measurement administration
E.506	Forecasting international traffic
E.507	Models for forecasting international traffic

Traffic Forecasting, Engineering, and
Measurement, cont.

E.508 Forecasting new telecommunication services

E.510 Determination of the number of circuits in manual operation

E.520 Number of circuits to be provided in automatic and/or semiautomatic operation, without overflow facilities

E.521 Calculation of the number of circuits in a group carrying overflow traffic

E.522 Number of circuits in a high-usage group

E.523 Standard traffic profiles for international traffic streams

E.524 Overflow approximations for non-random inputs

E.525 Designing networks to control grade of service

E.526 Dimensioning a circuit group with multi-slot bearer services and no overflow inputs

E.527 Dimensioning at a circuit group with multislot bearer services and overflow traffic

E.528 Dimensioning of digital circuit multiplication equipment (DCME) systems

E.600 Terms and definitions of traffic engineering

E.651 Reference connections for traffic engineering of IP access networks

E.700 Framework of the E.700-Series Recommendations

E.701 Reference connections for traffic engineering

E.710 ISDN traffic modeling overview

E.711 User demand modeling

E.712 User plane traffic modeling

E.713 Control plane traffic modeling

E.716 User demand modeling in Broadband-ISDN

E.731 Methods for dimensioning resources operating in circuit-switched mode

E.733 Methods for dimensioning resources in Signaling System No. 7 (SS7) networks

E.734 Methods for allocating and dimensioning Intelligent Network (IN) resources

E.743 Traffic measurements for SS7 dimensioning and planning

E.744 Traffic and congestion control requirements for SS7 and IN-structured networks

E.745 Cell level measurement requirements for the B-ISDN

E.750 Introduction to the E.750 series of Recommendations on traffic engineering aspects of networks supporting personal communications services

E.751 Reference connections for traffic engineering of land mobile networks

E.752 Reference connections for traffic engineering of maritime and aeronautical systems

E.755 Reference connections for UPT traffic performance and GOS

E.760 Terminal mobility traffic modelling

E.800 Terms and definitions related to quality of service and network performance including dependability

E.801 Framework for service quality agreement

E.810 Framework of the Recommendations on the serveability performance and service integrity for telecommunication services

E.820 Call models for serveability and service integrity performance

E.830 Models for the specification, evaluation, and allocation of serveability and service integrity

E.845 Connection accessibility objective for the international telephone service

E.846 Accessibility for 64 Kbps circuit-switched international end-to-end ISDN connection types

E.850 Connection retainability objective for the international telephone service

E.855 Connection integrity objective for the international telephone service

E.862 Dependability planning of telecommunication networks

E.880 Field data collection and evaluation on the performance of equipment, networks, and services

Traffic Performance, GoS, QoS

E.420 Checking the quality of the international telephone service – general considerations

E.421 Service quality observations on a statistical basis

E.422 Observations on international outgoing telephone calls for quality of service

E.423 Observations on traffic set up by operators

E.424 Test calls

E.425 Internal automatic observations

E.426 General guide to the percentage of effective attempts which should be observed for international telephone calls

E.427 Collection and statistical analysis of special quality of service observation data for measurements of customer difficulties in the international automatic service

E

E Series Recommendations, cont.

Traffic Performance, GoS, QoS, cont.

E.428	Connection retention
E.430	Quality of service framework
E.431	Service quality assessment for connection set-up and release delays
E.432	Connection quality
E.436	Customer affecting incidents and blocking defects per million
E.437	Comparative metrics for network performance management
E.438	Performance parameters and measurement methods to assess N-ISDN 64 Kbps circuit switched bearer service UDI in operation
E.439	Test call measurement to assess N-ISDN 64 Kbps circuit-switched bearer service UDI in operation
E.493	Grade of service (GoS) monitoring
E.505	Measurements of the performance of common channel signaling network
E.525	Designing networks to control grade of service
E.529	Network dimensioning using end-to-end grade of service (GoS) objectives
E.540	Overall grade of service of the international part of an international connection
E.541	Overall grade of service for international connections (subscriber-to-subscriber)
E.543	Grades of service in digital international telephone exchanges
E.550	Grade of service and new performance criteria under failure conditions in international telephone exchanges
E.720	ISDN grade of service concept
E.721	Network grade of service parameters and target values for circuit-switched services in the evolving ISDN
E.723	Grade of service parameters for Signaling System No. 7 networks
E.724	Grade of service (GoS) parameters and target GoS objectives for IN services
E.726	Network grade of service parameters and target values for B-ISDN
E.728	Grade of service parameters for B-ISDN signaling
E.770	Land mobile and fixed network interconnection traffic grade of service concept
E.771	Network grade of service parameters and target values for circuit-switched public land mobile services
E.773	Maritime and aeronautical mobile grade of service concept
E.774	Network grade of service parameters and target values for maritime and aeronautical mobile services
E.775	UPT grade of service concept
E.776	Network grade of service parameters for UPT

E.164 Codes

E.162	Capability for seven digit analysis of international E.164 numbers at time T
E.164.1	Criteria and procedures for the reservation, assignment and reclamation of E.164 country codes and associated Identification Codes (ICs)
E.164.2	E.164 numbering resources for trials
E.164.3	Principles, criteria, and procedures for the assignment and reclamation of E.164 country codes and associated identification codes for groups of countries
E.165	Timetable for coordinated implementation of the full capability of the numbering plan for the ISDN era (Recommendation E.164)
E.165.1	Use of escape code "0" within the E.164 numbering plan during the transition period to implementation of NPI mechanism
E.166	Numbering plan interworking for the E.164 and X.121 numbering plans
E.193	E.164 country code expansion

Human Factors and Interfaces

E.117	Terminal devices used in connection with the public telephone service (other than telephones)
E.133	Operating procedures for cardphones
E.134	Human factors aspects of public terminals: generic operating procedures
E.135	Human factors aspects of public telecommunication terminals for people with disabilities
E.136	Specification of a tactile identifier for use with telecommunication cards

Human Factors and Interfaces, cont.	
E.137	User instructions for payphones
E.161	Arrangement of digits, letters, and symbols on telephones and other devices that can be used for gaining access to a telephone network
E.333	Man-machine interface
E.434	Subscriber-to-subscriber measurement of the public switched telephone network
E.440	Customer satisfaction point
Supplements	
E.800SerSup1	Table of the Erlang formula
E.300SerSup1	List of possible supplementary telephone services that may be offered to subscribers
E.800SerSup2	Curves showing the relation between the traffic offered and the number of circuits required
E.300SerSup2	Various tones used in national networks
E.300SerSup3	North American precise audible tone plan
E.300SerSup4	Treatment of calls considered as terminating abnormally
E.800SerSup5	Teletraffic implications for international switching and operational procedures resulting from a failure of a transmission facility
E.300SerSup5	Modelling of an experimental test design for the determination of inexperienced user difficulties in setting up international calls using nationally available instructions or to compare different sets of instructions
E.300SerSup6	Preparation of information to customers travelling abroad
E.800SerSup7	Guide for evaluating and implementing alternate routing networks
E.300SerSup7	Description of INMARSAT existing and planned systems

EBU time code A European reference timing standard based upon 25 frames per second (fps) that is used in Europe, Australia, and nations supporting the PAL or SECAM video standards, endorsed by the European Broadcasting Union. Time code encoding is used to facilitate film audio/video synchronization and frame-accurate video editing. See SMPTE time code.

EC 1. end chain. 2. See exchange carrier. 3. European Community, European Common Market, European Union (EU). An organization of member European nations that have been developing, over a number of decades, a common currency, common passports, common network resources, and intercountry work, commerce, and decision-making alliances in order to promote trade, both within the EC and between the EC and other nations.

ECC 1. Electronic Commerce Canada Inc. A voluntary organization of public- and private-sector executives sharing information about electronic commerce. http://www.ecc.ca/ 2. Electronics Communications Committee. 3. elliptic curve cryptography. See elliptic curve. 4. Emergency Communications Center. 5. error correcting codes.

ECCA See European Cable Communications Association.

eccentric circle In a phonograph record, a blank, nonconcentric groove cut into the inner part of the platter to trip the automatic stylus pickup mechanism when the record has finished playing.

eccentricity 1. Deviation from normal or expected. 2. Deviation from a regular or expected path, as a straight line or a circle. 3. In orbits, the deviation from a circular path. 4. In conductive materials, as wires, the deviation at a particular point of the diameter of the conductor with the insulation, when measured in cross section.

ECCO Equatorial Constellation Communications Organization. A system of 11 commercial mini-satellites plus one spare in low Earth orbit (LEO), a concept descended from small-scale satellite systems developed by the Brazilian Space Agency. By launching a large number of small satellites into the same equatorial orbit, it is possible to arrange them so that there will be at least one within line-of-sight transmissions range at any one time.

ECCO is a joint venture of Constellation Communications, Inc. (CCI), formed in 1991, and Telebras. Telebras is involved in internationalization and general business administration of the system. The combination of the Constellation and ECO-8 programs resulted in the ECCO project. CCI includes well-known shareholders including Bell Atlantic, Raytheon, and Global Wireless, Inc.

In 1997, the Federal Communications Commission (FCC) authorized the first two phases of the project and, in 1998, Orbital Sciences Corporation was contracted to build and launch the systems. Commercial mobile and fixed-site voice, data facsimile, and positioning services were initially scheduled to come online in 1999 but then scheduled for Fall 2001.

The 12 satellites are designed to share a ring orbit

around the equator at 2000 kilometers using CDMA in the L-/S-band frequencies for down/uplink transmissions, and C- or Ku-band for feeders. The target service area is the equatorial belt between 23° north and south latitudes, comprising over 75 countries and a high proportion of the world's population. In spite of the large number of people who live in this region overall, many of them are in remote, small, rural communities with few or no wired communications services.

From the user's point of view, ECO-8 telephony services are functionally similar to cellular. The caller's desired number is transmitted through the satellite system to the local gateway where it is further relayed to a public phone circuit, wireless system, or to another ECCO satellite to link with another subscriber.

ECCO Working Group European Chapter on Combinatorial Optimization. This is a EURO Working Group that discusses recent and important issues in combinatorial optimization, founded in 1987.

échelle grating (*Fr.* – échelle meaning ladder) A high-resolution, stepped (blazed) optical grating structure used as a wavelength conditioner in certain spectrometers, telescopes, and optical communications components. Aluminum is commonly used as a coating on the grating to provide the reflective surface and there may be an additional coating to prevent corrosion of the aluminum.

An échelle grating has a broader groove than an echelette grating and thus fewer grooves over a specified length. It is used in near-Littrow, high reflection angles in high diffraction orders. In general, light is reflected from a collimated light source or reflecting mirror onto the blazed steps of the échelle grating where it selectively reflects off the angled surfaces of the grating to a detector or intermediary component such as a cross-dispersion grating. Since the orders overlap, practical embodiments of échelle gratings in optical systems may include additional elements to optically separate these orders (e.g., a prism as shown in the accompanying diagram). Échelle gratings have the advantage of being compact, while still providing relatively high resolution. They also affect a fairly wide spectrum of frequencies.

There have been a number of patents for fiber optic communication technologies incorporating gratings to support dense wavelength division multiplexing (DWDM). Bragg gratings have been cited in many of these. In July 2000, A. Sappey and G. Murphy submitted a patent application for a dense wavelength multiplexer/demultiplexer for propagating multiple optical channels with a select channel spacing as a single optical signal in the near-infrared range, using an échelle grating optically coupled to the collimating/focusing component.

A practical embodiment of the Sappey/Murphy invention would incorporate an échelle grating with a resolution of at least 20,000, with between about 50 to 300 grooves/mm, a blaze angle between 51 and 53°, and a free spectral range at least as large as the near-infrared frequency range to yield a multiplexed channel spacing of 0.4 nm or less with a separation

of at least 40 µ. See diagram. See blazed grating.

échellette grating Similar in structure and function to an échelle grating, except that the reflection angles may be lower and the grating period finer. Échellettes are associated with different dispersion characteristics compared to échelle gratings. The pattern of dispersed wavelengths tends to be slightly broader for an échellette grating, with less overlap of the orders than is found in échelle grating dispersion. For detection instruments, it is useful to be able to switch between échelle and échellette grating reflectors in order to view the light phenomena from a slightly different statistical point of view.

echo 1. Repetition of a sound (or other reflective phenomenon) due to reflection, with the echo gradually dying away through attenuation. Undesirable echo is sometimes experienced on phone lines and radio links where there is a delay or other technical problem. Deliberately induced echo may be used as a testing strategy and is intrinsic to reflected signal detection schemes such as sonar and radar. 2. Output to a command line or output window on a computer. Echo is a command used by many batch and other scripting languages to echo or print to a terminal, whether that terminal is a window on a computer screen, a teletype, or a printer.

ECHO European Commission Host Organisation. A noncommercial network host gateway to support and promote the use of network information services within the European Community (EC), established in 1980 by DG XIII/E. It is now managed by the Information Market Policy Actions (IMPACT) program. ECHO provides a means for businesses to understand the use and benefits of multilingual database information services. It provides a demonstration host, training, online access to directories of electronic information services, and a means to intercommunicate with associated individuals and organizations.

ECHO 1 A telecommunications and geodesic satellite launched into an orbit of 1600 km by the U.S. on August 12, 1960. ECHO 1 is historically important because it provided the first government satellite telephone links and television broadcasts on February 24, 1962. This satellite's orbit decayed in 1968. See ANIK for information on the first commercial television broadcast.

echo cancellation A technique for isolating and filtering unwanted echo signals which may accompany and interfere with the main analog transmission. Echo cancellation is often used on voice circuits, especially satellite transmissions, and may also be used in frame relay systems. In general, echo cancellation attempts to maintain a full-duplex circuit, although there are exceptions as in clear channel or ISDN calls. See echo suppressor, interference, noise.

echo check A diagnostic technique in which data is transmitted and then echoed back from the receiving end to the sender to check the completeness and integrity of the data.

ECHO satellites A series of satellites launched by the United States, beginning in 1960. The early

ECHO Project launched large highly reflective balloons capable of bouncing back radio signals. Without active relays, the communications signals were weak, but much was learned from these early experiments. The series included ECHO A-10 (never achieved orbit), ECHO 1, ECHO 2 (delayed from 1962 to 1964). Improvements in active relays superseded the ECHO Project. See ECHO 1, West Ford satellites.

echo suppression A means of reducing undesirable echoes, especially in satellite voice communications. Echo suppression differs from echo cancellation, in that echo suppression disables the reverse transmission while the person continues to talk, thus functioning more like a half-duplex line. See echo cancellation, interference, noise.

echo suppression disabler A means to coordinate *echo suppression*, the removal of undesirable echoes,

especially on satellite voice lines. Since echo suppressors limit the capability of the system to half-duplex transmission by suppressing the signal in the direction opposite to the sending signal, it is important to be able to disable the echo suppression to restore full-duplex operation. Echo suppression is typically disabled by sending a high-pitched signaling tone from an answering modem.

ECI 1. equipment catalog item. 2. End Chain Indicator. A boolean indicator signifying the end of a chain of data. In Systems Network Architecture (SNA) systems, it is contained in the sna.rh.eci field. 3. engineering change instruction. 4. external call interface. Generically, a hardware and/or software means to intercommunicate with a local or internal system and an outside or external system, as between a local network and the Internet or between a private branch exchange (PBX) and a public switched

---

Échelle Grating Incorporated into DWDM Multimode Optical Communications System

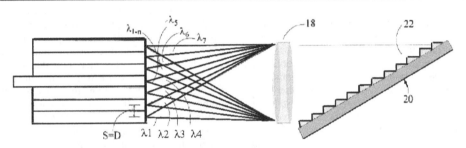

This diagram illustrates the basic components assembled with an échelle grating in a DWDM system. A pigtail harness (left) supports the structural output of multiple waveguides. Light passes through a collimating/focusing lens (18) that is optically coupled to the échelle grating (20). The grating subdivides the incident light beam containing multiple channels and angularly disperses the output wavelengths with good spatial and channel separation.

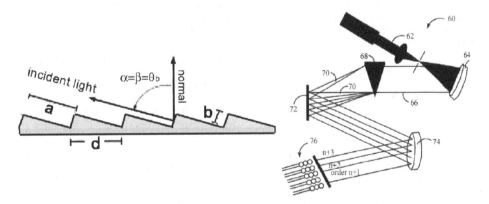

The schematic on the left is a closeup cross-section of the échelle grating, illustrating the incident light interacting with the blazed grating structure. The riser (a) and reflective step (b) combine to form a grating period. In this DWDM embodiment, the system would have a groove density of <300 grooves/mm and a blaze angle (θ_b) >45° to efficiently separate closely-spaced channels within a reasonbly compact area.

The schematic on the right illustrates the basic components of an échelle grating multiplexer/demultiplexer. Light from a single-mode optical fiber (62) is reflected to the prism (68) by the collimating/focusing mirror (64). The two-dimensionally dispersed wavelengths from the prism reflect off the échelle grating (72) and are directed to a concave collimating/focusing mirror (74) which then directs the beams toward an output fiber array (76). [Sappey and Murphy; U.S. patent #6,415,080, July 2002.]

telephone network (PSTN). 5. External Call Interface. On a CICS system, an application programming interface (API) that enables a non-CICS client program to communicate with a CICS program, calling it from a CICS server. Thus, the client can access server routines such as utilities and databases.

ECIS 1. European Committee for Interoperable Systems. 2. European Computer Industry Research Centre 3. European Conference on Information Systems. 4. European Council of International Schools.

Eckert, John Presper (1919–1995) An electronics inventor and collaborator with J. Mauchly on the historic ENIAC computer project. With Mauchly, he formed the Eckert-Mauchly Computer Corporation in 1946, which was acquired by Remington Rand Corporation in 1950. When Remington Rand merged with Sperry Corporation in 1955 to form Sperry Rand, Eckert became an executive with the company. The firm eventually merged with Burroughs Corporation to become Unisys. See BINAC, ENIAC.

ECL See emitter-coupled logic.

ECMA See European Computer Manufacturers Association.

ECMATC32 A project of the Standardizing Information and Communications Systems Technical Committee of the European Telecommunications Standards Institute jointly with ECMA, an international standards association, with a working agreement between the two bodies signed in September 2000. ECMATC32 working groups include TG14, for private integrated services and corporate network services and signaling, and TG17, for corporate open architecture and systems for IP-based services. Thus, the focus of the standardization efforts is private and corporate telecommunications systems.

ECN See explicit congestion notification.

ECOMEG See E-commerce Experts Group.

ECP 1. electronic commerce platform. 2. See Encryption Control Protocol. 3. See Enhanced Call Processing. 4. See Executive Cellular Processor.

ECPA See Electronic Communications Privacy Act of 1986.

ECSA See Exchange Carriers Standards Association.

ECT explicit call transfer (e.g., as in ISDN Q.82 or Q.732 call offering services).

ECTRA European Committee for Telecommunications Regulatory Affairs.

ECTUA See European Council of Telecommunication Users Associations.

ECTEL See European Telecommunications and Professional Electronics Industry.

ECTF 1. See Enterprise Computer Telephony Forum. 2. See European Community Telework Forum.

ECTUA See European Council of Telecommunications Users Association.

EDAC error detection and correction.

EDACS See Enhanced Digital Access Communications System.

EDDA See European Digital Dealers Association.

eddy current An electrical current induced by an alternating magnetic field, which can be found in good conductors such as iron and may contribute to signal loss in electrical circuits.

EDF erbium-doped fiber. See erbium doping.

EDFA See erbium-doped fiber amplifier.

edge connector A common type of thin, printed circuit board foil-imprinted extension used for making an electrical connection to a slot, usually inside a computer or in a peripheral card bay.

edge-emitting A component that emits acoustic or electromagnetic waves through the structural or functional side of the component. For example, in semiconductor laser diodes, which are fabricated in layers, the laser emission is from the side and thus each individual component must be stamped out before it can be tested. See vertical-cavity surface-emitting laser.

EDH See electronic document handling.

EDI See Electronic Data Interchange.

EDIFACT, UN/EDIFACT Electronic Data Interchange for Administration, Commerce and Transport. Arising out of work by ANSI and the United Nations, EDIFACT was developed during the 1970s and 1980s by the United Nations Working Party on Facilitation of International Trade Procedures to provide a standardized means to facilitate data exchange. EDIFACT provides internationally agreed-upon standards for platform-independent directories, syntax, and structures for character-format electronic documents.

The transmission of EDIFACT documents involves the creation of a flat file (e.g., through export from a compatible application such as a word processing program) that is passed to an EDIFACT translator where it is converted to an EDIFACT Message following the U.N. guidelines. The file, which includes routing and destination information, is sent over the network and, when it reaches the destination, is decoded according to the type of Message sent and stored or imported to a compatible application.

EDIFACT conventions are published by the United Nations in their Trade Data Interchange Directory (UNTDID). See Electronic Data Interchange, Trade Data Interchange.

EDIS See Emergency Digital Information System/Service.

Edison base The standard screw-in light bulb base common in North America.

Edison cell A historic variable-storage nickel hydrate (positive) and iron oxide (negative) battery cell with an electromotive force lower than that of a lead cell (about 1.2 volts). It was developed by Thomas Edison and became suitable for use in automobiles due to its ruggedness. See battery for a chart of other types of cells.

Edison effect A phenomenon that Thomas Edison observed in 1883 and patented in 1884. While working with electrical illumination, he sealed a metal wire into a bulb near the filament and noticed that electricity flowed across the gap between the hot filament and the metal wire, a discovery that became important to later electronic researchers in the development of broadcast technologies. See Audion.

Edison Electric Institute EEI. A professional association supporting U.S. shareholder-owned electric utility companies and their affiliates and associates worldwide. Members service about 70% of the U.S. electric customer base. The EEI advocates public policies and provides strategic business and marketing support. http://www.eei.org/

Edison General Electric A company formed by Thomas Edison in 1889. Throughout his prolific career, Edison sought practical, commercially viable applications of his ideas, and he actively marketed many of the products that were invented in his laboratory.

Thomas Alva Edison – Prolific Inventor

Thomas Alva Edison was a significant inventor of audio recording and playing technologies. [U.S. National Archives collection.]

Edison, Thomas Alva (1841–1931) An American inventor who, at the age of 15, learned telegraphy and soon became a very fast and competent telegrapher. In 1868, he invented a device to record votes, and later, in New York, invented a stock ticker. In 1876, he set up a pioneer industrial research lab in Menlo Park, New Jersey, where he turned out hundreds of inventions, including the phonograph, electric typewriter, and fluoroscope. He developed the ideas of others, as well, making practical improvements to Bell's early telephone devices.

Edison's inventive output was prodigious, and he pursued his interests almost to the complete exclusion of his business affairs, practical matters, and family. In all, he received more than 1000 patents.

Edison is probably best remembered for developing the incandescent lamp in 1879, following the invention of the arc lamp by Humphry Davy early in the century and a short-life incandescent bulb by Joseph Wilson Swan. His efforts were aided, in part, by the earlier research of M.G. Farmer, who supplied him with advice and a number of materials that might be useful in building a lamp. Almost immediately following his historic invention, Edison began an electrical utility company in New York City, in 1882, called the Pearl Street Central Station. This and other early utility companies provided direct current (DC) which has since been almost universally superseded by alternating current (AC), an approach promoted by Nikola Tesla and denounced by Edison. See Farmer, Moses; incandescent lamp; Tesla, Nikola.

editor A software program for manipulating information, particularly textual information. While graphics can be edited online, the tools to do so are not usually called editors, but rather paint or drawing programs or image processors. The term *editing* is used more in the context of line-oriented information carried out with word processors, desktop publishing programs, and text editors.

Programming editors are often optimized for the special formatting needs of programmers. For example, they may monitor parenthetical statements and alert the programmer if the parentheses or brackets are unbalanced, which would result in a syntax error in the program. They may provide different colors for different kinds of information, as for comments, variable names, or procedural labels. They may also provide the capability of telescoping or expanding the text (a feature also found in some word processors). Scriptwriting editors are often set up with templates that indicate the correct margins and line spacing required by the theater and motion picture industries.

Word processors are optimized for document creation and basic formatting of text. Extensive formatting of text and graphics shouldn't be done with a word processor. Yes, it's possible, and yes, individuals have created some great documents with word processors, but it's also possible to hammer with the backside of a hatchet. It just isn't very efficient or comfortable. Complex page layout should be done with a page layout program. In a business environment, the extra cost of the software is incidental compared to the extra hours, weeks, or months that have to be paid to someone to use a word processor for an unintended purpose. GNU-Emacs is one of the most powerful text editors in existence, developed by Richard Stallman in the mid-1980s. It is configurable, scriptable, full-featured, and freely distributable.

EDM electronic document management.

Edmund Industrial Optics A New Jersey-based designer and manufacturer of optical solutions for electronic imaging, biomedical and biometric applications, telecommunications, and semiconductor industries. Affiliated with Edmund Scientific, it is an international supplier of industrial optical components and devices, including coatings, lenses, illuminators, lasers, etc.

Edmund Scientific A significant supplier of scientific professional and hobbyist lab supplies and optical and robotics components. If you like cool techie toys and great science projects, this company has a wide variety of interesting products.

EDO RAM extended data-out random access memory. A type of faster random access memory (RAM) that began to become prevalent around 1997. Many Intel-based desktop architectures were designed with expansion slots for EDO RAM, as it was less expensive than previous types of RAM.

EDP See electronic data processing.
EDRS European Data Relay Satellite. See European Space Agency.
EDSAC Electronic Delay Storage Automatic Computer. A historic large-scale, stored-program, electronic, digital computing machine developed in 1949 at Cambridge University, England, under the leadership of Maurice Wilkes. See EDVAC, ENIAC.
EDTV See enhanced-definition TV.

EDVAC Vacuum-Tube Computer

Running the room-sized EDVAC required staff to operate and maintain hundreds of vacuum tubes and thousands of feet of wires. Modern handheld desktop calculators are now faster and more powerful. [U.S. Army photo.]

Educational Technology Branch ETB. A branch of the Lister Hill National Center for Biomedical Communications that conducts research and development in computer and multimedia technologies and disseminates the information to the U.S. National Library of Medicine, the world's largest biomedical library, which is located in Bethesda, MD.
http://lhncbc.nlm.nih.gov/
http://www.nlm.nih.gov/
EDUCAUSE A consolidation of CAUSE and Educom, established in July 1998, EDUCAUSE is an international, nonprofit association. EDUCAUSE administrates an extensive collection of higher education information technology (IT) resources for higher education institutions and firms supporting the higher education information technology fields.
EDUCAUSE supports and participates in advocacy projects, conferences, and seminars and other professional development activities. It is affiliated with various higher education and computing associations throughout North America. See CAUSE, Educom.
http://www.educause.edu/
Educom An association of colleges and universities dedicated to the support and evolution of educational computer network technologies. In July 1998, Educom was consolidated with CAUSE to form EDUCAUSE. See EDUCAUSE.

EEC European Economic Community. Now the European Union (EU). A European common market that has been in development for several decades. Gradually, European currency, European passports, inter-country networks, and greatly reduced border restrictions are being phased in. See European Union.
EEI 1. See Edison Electric Institute. 2. external engine interface. 3. See external environment interface. 4. See External Environment Interface.
EES Earth exploration satellite.
EEMA See European Electronic Messaging Association.
EEPROM electronically erasable programmable read-only memory. See erasable programmable read only memory.
EF&I engineer, furnish, and install.
EFCI See explicit forward congestion indicator.
EFF See Electronic Frontier Foundation.
effective competition A market regulation status level. Broadcast cable providers must meet certain criteria to claim *effective competition* status. See cable access.
effective radiated power ERP. The transmitting power of a broadcasting antenna. It is sometimes controlled by a directional antenna.
EFI&T engineer, furnish, install, and test.
EFS 1. See electronic filing system. 2. See Electronic Filing System. 3. electronic financial system. 4. See Electronic Form System. 5. See error-free seconds.
EFT See electronic funds transfer.
EFTA See European Free Trade Association.
EG See European Graphics Association.
EGA See Enhanced Graphics Adapter.
EGC 1. Economic Growth Center. A statistical and economic development archive at Yale University. 2. See Enhanced Group Call.
Eggebrecht, Lew Chief of Commodore Engineering at about the time Commodore-Amiga was sold to Amiga Technologies in Germany. See Amiga.
EGNOS European Geostationary Navigation Overlay Service. A European land and marine communications service which augments the U.S. Global Positioning System (GPS) and the Russian GLONASS system (similar to U.S. GPS), using the raw data to compute information and broadcast it through the GEO satellites. The EGNOS system combines ground-based and satellite segments to comprise the European implementation of the Global Navigation Satellite System (GNSS). See Global Navigation Satellite System, Wide-Area Augmentation Service.
ego surfing *slang* Searching for your own name on the Net, in the media, or in databases. There are many legitimate reasons for ego surfing, and everyone should probably do it once in a while to make sure names or net addresses are not fraudulently distributed or used by imposters to post to newsgroups. Such misuse could result in misunderstandings and embarrassment (and sometimes even litigation) directed at the legitimate owner of the name or address.
There are search engines on the Internet that allow

users to check newsgroups for postings under a particular name. Many Web directory services on the Web give out phone, address, and sometimes even personal information on individuals, drawn from phone books and less legitimate sources. You must alert these services if your personal statistics are being used indiscreetly or illegally and request that they stop. (There is usually information about you on the Net, whether or not you even use a computer.)

Ego surfing is routinely engaged in by those who enjoy studying family histories and are on the lookout for more information to add to their genealogy databases. Ego surfing also can help those who are publishing scientific, political, or other information with wider social implications to follow the dissemination, and sometimes the impact, of their communications, in order to engage in global dialog or to correct misrepresentations or misunderstandings.

EGP See Exterior Gateway Protocol.

egress 1. Exit, way out. 2. In frame relay networks, frames that are exiting *away* from the frame relay towards the destination. The opposite of ingress.

EIA See Electronic Industries Alliance.

EIA Interface Standards A collection of standards describing configurations, signals, and other communications parameters for various electronic connecting interfaces. These are often used in conjunction with ITU-T specifications for protocols and functions.

Probably most familiar of the Interface Standards are EIA-232-D and EIA/TIA-232-E, which are 1987 and 1991 updates to the decades-old RS-232 specification for serial transmissions between data terminal equipment (DTE) and data communications equipment (DCE). This standard has been widely implemented in desktop computers and other devices and is commonly used for communicating with modems, remote terminals, and printers.

Most systems support EIA-232-D and EIA/TIA-232-E through 25-pin D connectors, though minimally nine pins are needed to implement the specification, and 9-pin D connectors (EIA-574) are sometimes used. The EIA has also defined faster standards for serial communications, including EIA-422 (balanced signals), EIA-423 (unbalanced), EIA-485 (multipoint), and EIA-530 (EIA-422 with 25-pin D connector).

EIA standards additionally encompass wiring connectors and topology, including building wiring and network backbones.

EIA/TIA categories Standardized specifications for cable transmission speeds. See the EIA/TIA Transmissions chart for categories.

EICTA See European Information and Communications Technology Association.

EIG See Electronic Information Group.

eight hundred service See 800 service.

EIIA 1. Embedded Industrial Internet Appliance. A commercial development kit offered by Arcom Control Systems for adding TCP/IP connectivity to products. 2. European Information Industry Association.

EIMF European Interactive Media Federation. See European Multimedia Forum.

Albert Einstein – Physicist

Albert Einstein was a patent clerk in Switzerland for seven years before he was able to find the type of position he desired in an academic research environment.

Einstein, Albert (1879–1955) A German scientist who is considered one of the greatest in the history of physics, best known for his special theory of relativity. Einstein moved first to Switzerland, where he was educated as a teacher of mathematics and physics, and then later to the United States. Unable to find a teaching or research position, he worked in the Bern patent office for seven years. Within two years of starting as a junior clerk, Einstein had become a technical expert in the patent office. While there, he made

EIA/TIA Cable Transmission Categories		
Cat.	Transmission Speed Rating	Typical installations
Cat 1	No performance criteria specified.	
Cat 2	Rated to 1 MHz	Telephone wiring installations
Cat 3	Rated to 16 MHz	LAN, Ethernet, 10Base-T, UTP Token-Ring
Cat 4	Rated to 20 MHz	LAN, Token-Ring, 10Base-T, IEEE 802.5
Cat 5	Rated to 100 MHz	WAN, Fast Ethernet, 100Base-T, 10Base-T

good use of his free time by writing a large volume of theoretical physics articles.

By 1905, Einstein was writing about his ideas in electromagnetic energy and the photoelectric effect, building on and extending the work of M. Planck. In the same year, he wrote his paper on the special theory of relativity, reinterpreting classical physics, and supported Maxwell in hypothesizing that the speed of light was constant in all frames of reference. His concepts about the equivalence of mass and energy were tied in with the other writings at that time. This creative intellectual output was astonishing, and the later recognition and corroboration of his ideas catapulted him into the history books and out of the patent office into a series of posts at universities in Eastern Europe, Western Europe, and America. In 1940, he became a citizen of the United States.

In 1915, he published the definitive culmination of his writings on relativity, and in 1917 he proposed electromagnetic emission principles which led to the development of lasers. In 1924, he made further discoveries concerning the relationships between waves and matter.

In 1921, Einstein was awarded the Nobel Prize in physics for his studies of the photoelectric effect. A crater on the moon has been named after him.

EIPA 1. Electronic Information and Communication for Pedagogical Academies (Austria). 2. European Information Providers Association. 3. See European Institute of Public Administration.

EIR 1. electronic incident report. A report on anomalous or unauthorized activity on an electronic system. 2. Equipment Identity Register. A mobile services security database that aids in tracking lost or stolen communications devices.

EIRP 1. Effective Isotropic Radiated Power. 2. Equivalent Isotropic Radiated Power. An ideal radiator providing a de facto common reference for radiated power. See isotropic, isotropic antenna.

EIRPAC Eire (Telecom Éireann) packet network. An X.75 packet-switched national data network in the Irish Republic. A number of charging bands provide various degrees of EIRPAC connectivity to other nations. Band 1 provides domestic communications, Bands 2 and 3 service connections to a number of European countries, Band 4 supports destinations in North America, and Band 5 services communications with other countries. Local access is provided in major cities (e.g., Dublin) and dialup through the public switched telephone network (PSTN) is available to remote subscribers with modems approved by the Department of Commerce.

EIS 1. Epidemic Intelligence Service. A service of the U.S. Centers for Disease Control and Prevention. 2. See ESO Imaging Survey. 3. See Expanded Interconnection Service.

EISA See Extended Industry Standard Architecture.

EISCAT See European Incoherent Scatter Scientific Association.

EITA 1. enterprise information technology architecture. 2. See European Information Technology Association.

EIU 1. economic intelligence unit. 2. Ethernet interface unit. 3. external interface unit.

EKE 1. See electronic key exchange. 2. encrypted key exchange. See Encrypted Key Exchange Protocol, public key.

EKTS See Electronic Key Telephone Service.

Electric Telegraph to the Pacific Act A historic act of the U.S. Congress to solicit bids for construction of a government communications line connecting San Francisco to major centers. The line was to be open to the use of all U.S. citizens upon payment of the appropriate charges. In July 1862, the Pacific Railway Act was enacted to "... aid in the Construction of a Railroad and Telegraph Line from the Missouri River to the Pacific Ocean...." The Central Pacific Railroad Photography History Museum is a good source of photographs and information on railroad and communications history. http://cprr.org/Museum/

electrical wire Any conductive filament, usually metallic, that readily transmits current. Copper is very commonly used for telecommunications wires.

One of the earliest documented long wires was a thread that conducted electric virtues over 600 feet in 1729. The thread was strung during electrical experiments conducted by S. Gray and G. Wheler.

electricity A fundamental constituent of nature that is observable as positive and negative charges and currents through materials according to their conductivity. The discovery and harnessing of electricity is the basis of our current technological society. Electricity arises from various sources, including friction (static electricity), light (photoelectricity), heat, chemical activity, piezoelectricity (pressure, especially in crystalline substances), and mechanical energy (as from a generator). Ancient observers were aware of various electrical phenomena, but it was not until the last 200 years that we have begun to understand their various properties and the fundamental principles that underlie them.

electroabsorption EA. A phenomenon in which the application of an electric field or voltage to a substance causes a change in optical absorption that occurs very quickly. The change is sufficiently strong to be detected, which lends itself to practical applications. The time it takes to control and apply the electrical field is the chief limitation to the speed of the effect. When this phenomenon is associated with quantum wells, it is called the quantum-confined Stark effect.

Electroabsorption is exploited in the field of spectroscopy and has been used to develop modulators for demultiplexing and detection. Electroabsorption light modulators have been integrated with laser diodes. See electroabsorption modulator, self-electro-optic effect device, quantum well, Stark effect.

electroabsorption modulator EAM. Modulation of an optical beam by exploiting the electroabsorption properties of certain substances such that continuous-wave laser output (e.g., in dense wavelength division mutliplex optical networks) can be externally modulated to send digital information with rapid light pulses. In the late 1990s, the University of California, Santa

Barbara, demonstrated the first traveling wave electroabsorption modulator at 18 GHz. In March 2001, Alcatel announced the launch of a new EAM based upon the electroabsorptive properties of indium phosphide. The component was capable of operating at 40 Gbps with low power consumption.

A number of commercial vendors are now supplying SONET/OC-192 EAM drivers for driving EA-modulated lasers at speeds up to 12.5 Gbps. For example, in March 2002, Oki Semiconductor released an EAM module monolithically integrated with a DFB laser suitable for 10 Gbps over distances up to about 40 km and Mitsubishi and Alcatel launched new 10-Gbps EAM lasers.

electrode 1. An essential component of an electron tube. Any of the basic components, the electron-emitting cathode, controlling grid, or electron-attracting anode, are considered electrodes. Carbon arc electrodes are used for generating precise heat utensils for welding or fusing a variety of materials. They may be handheld or incorporated into precision cutting machines. The carbon burns away as the electrode is applied. Depending upon the application, compressed air may be used to blow away particles on the surfaces to be fused. Carbon arc welding requires safety shielding to protect eyes. Electrode-based devices require regular checking and cleaning and, depending upon the type and frequency of use, may require oxidized parts to be replaced. See fusion splicing. 2. A plate in a battery.

electroluminescence The direct conversion of electrical energy into light. This process is used in display technologies, with electroluminescent materials such as zinc sulfide doped with manganese. See electroluminescent display.

electroluminescent display EL display. A gridded display technology that incorporates an electroluminescent material sandwiched between outer panels. When exposed to high electrical fields, the inner material emits light. Like plasma panels, individual points are selectively lit through matrix addressing. EL displays are brighter than passive LCD displays but also require more power. See liquid crystal display, plasma display.

electrolysis The production of chemical changes by passing an electrical current through an electrolytic material. See electrolyte.

electrolyte A nonmetallic substance that, when chemically or electronically stimulated, becomes an ionized conductor. Electrolytic properties are widely employed in electronics. Electrolytes are used electrical cells, rectifiers, etc. Various acids were commonly used as electrolytes in early inventions.

electrolytic cell A power-conducting (as opposed to power-generating) cell comprised of a conducting liquid (the electrolyte) and two identical electrodes. Electrolytic cells are used in refining, reduction, and electroplating processes.

electrolytic detector, liquid detector A radio wave detecting device patented by R. Fessenden in 1903. It was discovered accidentally when Fessenden was seeking ways to improve on his hot-wire barretter. A broken filament led to the discovery that signals could be received better through the separate pieces of filament in an electrolytic solution than with a single piece of filament. From this, Fessenden combined nitric acid and a platinum wire into a rectifier that could detect both continuous and damped waves.

The electrolytic detector was an important milestone in radio history, as it provided a means to create a much more sensitive receiving instrument. Most electrolytic detectors required an outside power source, though some were manufactured with a built-in battery integral to the design. See barretter, Shoemaker detector.

electrolytic paper tape A type of paper tape used on some of the old telegraph systems in which a stylus passed an electrical signal onto the coated tape to produce an image of the message being transmitted. The image on early systems was often blue, though the amount of current on an electrolytic system can influence the color of the image.

electromagnet A device that has a significant magnetic field only when current is flowing through it. The strength of the wire is dependent on the size and type of materials used, the amount of current, and the number of coils. Electromagnets are used extensively in appliances, industrial hoists, telephones, public address speakers, and bulk erasers. You can make a simple electromagnet by wrapping a conductive wire around an iron nail and passing current through it (taking care not to touch any of the live wires). This in turn can be used to magnetize the end of a screwdriver by stroking the nail in one direction over the screwdriver. Quite handy for holding screws in place, but be wary of using magnetized screwdrivers near electronic components. See bulk eraser; Faraday, Michael; solenoid.

electromagnetic Embodying electric and magnetic properties. See electromagnet.

electromagnetic communications Communications that employ the propagation of transmission waves through space. Meaningful signals are sent in many ways, with light, radio waves, microwaves, etc., usually by modulating the transmission of the radiant energy in some way.

electromagnetic deflection The directing of the path of an electron beam by means of a magnetic field (often in the form of a coil).

electromagnetic field 1. A field of magnetic influence around a conductor produced by a current flowing through the conductor. See electromagnet. 2. Together, an electric field and its associated magnetic field. The magnetic field is perpendicular to the lines and direction of force. See right-hand rule.

electromagnetic interference EMI. Undesirable noise, degradation, overlap, or echo in an electromagnetic transmission.

electromagnetic pulse EMP. A large or fast-moving electromagnetic transmission that is quicker, or burstier than the immediately preceding and succeeding transmission. Sometimes a pulse is a natural phenomenon, such as lightning, or it can be a deliberate means of creating a signal or carrying information.

electromagnetic spectrum The range, or diagrammatic relationships, of the known types of electromagnetic radiation, organized by wavelength.

electromagnetic wave The radiant energy produced by an oscillating electric charge. Infrared, ultraviolet, gamma, visible wavelengths, and cosmic rays are some examples.

electromotive force emf. Descriptive of the pressure of the movement of electrons through a circuit, sometimes described as similar to the movement of liquid through a closed piping system. Indeed, for many decades the two kinds of movement (electrical and liquid) were assumed to be the same, and early scientists spoke of emf as an electrical fluid. The discovery of the relationship between magnetic and induced electrical forces helped scientists to understand emf. An external influence (battery, power supply, etc.) can cause an electron charge to flow through a conductive medium creating an electromotive force. See Faraday's laws, volt.

electron A minute, elementary particle of matter, carrying a negative electrical charge. Electrons are normally found surrounding a positively charged nucleus. The term is derived from the Greek word elektron (amber) but was first used with this specific meaning by G.J. Storey in 1891. See positron.

electron beam A stream of electrons traveling close together in the same trajectory, directed by a magnetic field. An important constituent of cathode-ray tube (CRT) technology widely used in TV and computer monitors.

electron microscope An optical-electronic instrument that provides magnification of minute structures by means of recording the movement of a focused beam of electrons. The results were originally displayed on fluorescent screens or photographic plates; computer monitors are also used. Early electron microscopes could enlarge images 100 times more than the finest optical microscopes of the time, but the images were limited to black and white still objects. New techniques were continually sought to increase the range of objects that could be imaged and the ways in which they could be represented. Computer enhancement and interpretation has opened a wide range of possibilities.

electron tube A device in which the movement of electrons is conducted within a sealed glass or metal container. While electron tubes were made of glass for many decades, some all-metal tubes came into use in the mid-1930s. The most common implementation of the electron tube is the *vacuum tube*, since the life of the electron-emitting materials could be extended by removing the air or encasing a controlled mixture of gases.

The most important evolutionary development in the history of the electron tube is the *Audion*, the commercial name for a *triode*, a three-element tube with a control grid invented by Lee de Forest. One of the most important adaptations of the vacuum tube is the *cathode-ray tube* still widely used in television sets and computer monitors. See anode, Audion, cathode, cathode-ray tube, vacuum tube.

A Variety of Electron Tubes

The most common types of electron tubes used in electronics for many decades were three-element (or more) vacuum tubes. Experimentation led to the development of many different types of tubes for different purposes, and numbering systems were set up to keep track of parts so consumers could replace broken or burnt out tubes. For the most part, semiconductor components have replaced electron tubes, except for some high frequency applications. This interesting assortment is from an exhibit at the American Radio Museum.

electronic bulletin board See bulletin board system.

Electronic Certification An electronic signature that serves the same purpose as a written signature on a physical document (usually a letter, contract, or administrative approval). Electronic Certification is accomplished through cryptography, typically key cryptography. See DSS.

Electronic Classroom A commercial Macintosh-based distance-learning audiographics multimedia videoconferencing tool written by Robert Crago, an Australian developer. Electronic Classroom provides images, QuickTime compressed video, and voice over public switched telephone networks (PSTN).

electronic commerce, ecommerce Financial and barter transactions conducted across data networks using electronic means of communications and agreements, including the exchange of documents, signatures, virtual money, etc.

In the earliest ecommerce implementations, people used bank machines to carry out simple deposit, withdrawal, and balance inquiry transactions. On the Internet, they simply communicated the terms of commercial transactions through email but, since then, some significant changes have occurred. There are now banking machines that can transact traveler's check purchases and loan negotiations and provide more complex information services. Similarly, with the dramatic growth of the World Wide Web, tens of thousands of companies and individuals have expressed the desire to conduct remote financial transactions, and sophisticated secured electronic commerce systems are being developed and promoted.

Shopping cart systems have become common on the Web, and online banking services are prevalent, along with new services such as PayPal and Billpoint that allow a user to manage funds within a virtual bank through email and Web browsers. By the mid-1990s, there were initiatives to standardize transaction mechanisms, efforts to promote private and secure ecommerce, and many individual ways to exchange money on the Internet. See certificate, digital signature, Electronic Data Interchange, electronic mall, encryption, JEPI, Pretty Good Privacy.

Electronic Commerce Service ECS. ECS is a set of electronic mail and verification services developed by the U.S. Postal Service (USPS) to offer secure electronic mail so that it becomes an electronic extension of the USPS physical mail system. USPS is cooperating with private firms to develop this technology.

Various aspects of the USPS plans include personal and professional certificate services through a Certification Authority (CA), time and date stamping (in essence an electronic postmark), certified email, return receipt, verification, and archiving.

It is important to consider that USPS email differs from personally forwarded email in some of its legal safeguards, and that USPS has a history of statutes and precedences which may make it attractive to business users.

As with all major milestones in U.S. postal history, the USPS released a commemorative stamp, in early 1996, to launch their electronic venture. The computer whose birthdate was commemorated was the ENIAC, which certainly deserves credit for its historical importance, but the system that should probably have been honored as the first large-scale computer is the Atanasoff-Berry Computer, which preceded the ENIAC.

Electronic Communications Privacy Act of 1986 ECPA. An Act of the U.S. Congress, adopted to address issues of privacy related to the growing prevalence and use of computers and related digital technologies, especially for communication. The ECPA was passed to address more specifically issues of privacy pertaining to electronic surveillance.

Prior to the ECPA, privacy and electronic surveillance issues were generally covered in Title III of the *Omnibus Crime Control and Safe Streets Act of 1968*. This Act was established a decade before personal computers began to be widely distributed and three decades before the use of the Web and inexpensive surveillance devices were widely available to consumers. Thus, it was felt that issues not clarified by Title III as to new and emerging technologies needed to be re-evaluated. The issues were and are complex.

The Federal Bureau of Investigation (FBI) and local law enforcement agencies were concerned about criminals using new technologies as a means to facilitate criminal communications and acts. Private citizens, on the other hand, were concerned that too much power in the hands of authorities could lead to invasion of the privacy of law-abiding citizens. Thus, the Act needed to be able to satisfy the needs of law enforcement officers without impinging on those who were not involved with criminal activities.

The ECPA was signed into law on October 21, 1986. Title III was thus extended to apply to both the transmission and storage of digital data, most notably electronic mail, and it extended the concept and definition of intercept to apply to electronic textual data (prior to this, most privacy laws pertained to spoken voice communications over phone lines, e.g., wiretapping). Controversy followed the lawmaking, with detractors saying that the final version differed from the proposed version and that it did not go far enough to protect civil liberties. See American Civil Liberties Union, Cyberspace Electronic Security Act of 1999, Electronic Freedom Foundation, Encrypted Communications Privacy Act.

Electronic Data Interchange EDI. A series of standards developed primarily for business communications, EDI is a scheme for network interchange of electronic messages and documents, often between different companies or government agencies. EDI software works in conjunction with applications software. Files are extracted from an application, converted into a standard EDI format, and passed on to the communications software for transmission. EDI is not secure in and of itself and may be combined with authentication and encryption schemes. Practical applications include the interchange of invoices, purchase orders, policy documents, RFQs, waybills, cost estimates, etc.

The primary international standard for formatting EDI messages is Electronic Data Interchange for Administration, Commerce, and Transport (EDIFACT). EDIFACT messages are included within an EDI envelope. The interchange IDs of sender and receiver must be agreed upon. Their addresses are described in the X.400 standard (ITU-T X.400/X.435). See X Series Recommendations.

electronic data processing EDP. A system for receiving, manipulating, translating, and storing data, sometimes in large amounts. For example, numbers may go in, and paychecks, employee statistical information, or sales demographics may come out.

Electronic Directory ED. An informational database based on a directory standard, such as X.500 or LDAP, intended to help integrate various directories on a network. Thus, access is improved over that of searching and querying various directories, and different formats and protocols are made transparent to the user. EDs are of interest especially to corporations, educational institutions and libraries. See X.500.

electronic filing system EFS. A generic phrase for various means of submitting forms and other administrative information or applications via computer networks (e.g., email) rather than through traditional paper submissions. In order to ensure the security and legality of these documents, which cannot be hand-signed by traditional means, the legal system has had to assess the laws of contracts and evidence and emerging digital certificate formats. While some of these issues were worked out in the late 1990s, and digital certificates have received recognition and endorsements from the U.S. government, the adjustment to new technologies is ongoing. XML and Adobe PDF are emerging as two popular formats for filling out and submitting forms and applications.

Electronic Filing System EFS. The U.S. Patent and Trademark Office (U.S.P.T.O.) system for submitting patent applications, biosequence listings, and pre-grant publications submissions. This system enables applications to be sent electronically through the Internet rather than through mail or courier systems on paper. To facilitate and standardize the process, the U.S.P.T.O. provides XML authoring tools, software, and a digital certificate for secure transmission. See Electronic Form System.

Electronic Form System EFS. An XML schema *definition* that provides an extensible specification for platform-independent electronic forms and surveys that can be transmitted through email or Web middleware.

Electronic Frontier Foundation EFF. A well-known, influential civil liberties association co-founded in 1990 by Mitch Kapor and John Perry Barlow, which acts as a forum and advocate for social and legal issues related to the information and electronic revolution. See American Civil Liberties Union, Electronic Privacy Information Center. http://www.eff.org/

electronic funds transfer EFT. A system through which financial transactions are enacted electronically over a network without the actual exchange of physical currency. Stock transactions have typically been carried out this way, especially in the last decade, and the exchange of actual stock certificates is not required.

Electronic Industries Alliance EIA. A national trade association representing the U.S. high technology industry. The related Government Electronics and Information Technology Association (GEIA) includes companies that manufacture or engage in research, development or systems integration to meet the needs of U.S. government agencies. http://www.eia.org/default.htm

Electronic Industries Foundation EIF. The philanthropic arm of Electronic Industries Alliance, established in 1976. EIF addresses social/educational issues in the electronics industry.

Electronic Information Group EIG. The EIG is associated with the Electronics Industries Alliance. http://www.eia.org/

electronic key exchange The process of exchanging public keys in a public key encryption scheme, so messages can be forwarded to the owner of the public key and decrypted with a corresponding private key. It is used for authentication of electronic communications. Public keys are sometimes uploaded to public databases on the Internet.

Electronic Key Telephone Service EKTS. An ISDN-1 standard for supplementary ISDN services. Key telephone systems enable multiline phone sets commonly used by businesses and institutions to directly select outside lines through the central office rather than through local switching equipment.. Commercial vendors of EKTS phone devices have built many features into these systems, including on-hook dialing, programmable function keys, indicator lamps and LEDs, bridging (similar to a conference call within a telephone number group), and more. Visual indicators typically show the line status

electronic leash Any mobile communications unit that ties the user to an employer or other entity to which the user has a responsibility or obligation to respond. It particularly applies to those devices that clip to a belt or pocket. Beepers and pagers are good examples, though short-range radios and cellular phones may be included. Electronic leashes are frequently used by professionals who are in the field or on call. Short-range radios are now also being marketed to families as child-monitors for family outings at parks or malls. (It's considered bad manners to keep a beeper or pager active while watching a live stage performance.)

electronic lock Any locking mechanism activated through software, image recognitions systems or other sensors. Electronic locks are being incorporated into laptops, mobile phones, and cars to deter unauthorized use. They are also incorporated into some door entry systems.

electronic mail, email The invention of email is attributed to Ray Tomlinson of BBN in the early 1970s. Email is the exchange of correspondence through computer networks. In its broadest sense, it includes text files, sound files, graphics, and data files. The exchange of electronic mail is one of the most heavily used capabilities of the Internet, enabling messages to be sent to almost anyone in the world with an Internet connection, often without surcharges (unless messages are unusually large), depending upon the Internet Services Provider (ISP).

Most email follows basic standardized formats in order to maintain platform independence on distributed networks like the Internet. Email headers contain not only information about the source and destination of the message, but they may include various amounts of information about the path of the message and the affiliations of the sender. The header may also include optional information about the format and display of the subsequent message.

On distributed network systems, it is not unusual for multiple emails from one source to one destination to take different routes or to be divided into pieces (packets) and sent through various routes, to be reassembled at the destination before being made available to the recipient.

In the mid-1980s, it was not unusual for an email message to take two or three days to travel from Boston to San Francisco, for example. Now it is unusual for it to take more than a few seconds. A message posted to a public newsgroup, such as USENET, can be sent from a home or office computer and subsequently downloaded from a local mail server in 15 or 20 seconds, sometimes even faster if the ISP is in a direct link with the backbone supplying the newsgroup and the newsgroup is unmoderated and does not hold the messages before posting them.

Until the late 1990s, most email was text-based, even though graphical email had been available for many years on specific platforms. With the increase in the use of Web browsers for reading and handling email files, however, senders began to send binary file attachments, HTML-format messages, and other graphics and sound formats, thus significantly increasing file sizes and transmission times for email.

By the late 1990s, the volume of unsolicited [commercial] email (colloquially termed spam or UCE) had reached such alarming levels that ISPs were forced to put blocks on email with certain characteristics or from certain sites. A remarkable proportion of messages were being sent from fake senders or even fake domain names. As much as 80% of a person's email may be spam, causing an annoyance and time factor in addition to the raw cost. The cost to the ISPs and their users who pay for disk space or connect time, etc. can be very high and is increasing daily as more people connect to the Internet and the multimedia capabilities of the technologies improve. Thus, there have been frequent calls to legislators to do something to stem the tide. On the one hand, advertisers want the freedom to contact potential customers; on the other, customers and ISPs rightfully don't want to shoulder the cost of unsolicited or undesired communications. See Forum for Responsible and Ethical Email.

electronic mail gateway In a network where there are links to other networks that may be using a variety of network mail protocols, a gateway system handles receipt, translation (if necessary), and delivery of the messages to the local system, the external system, or both. Mail filters and distribution agents may be incorporated at the gateway level to selectively process and deliver email. See electronic mail, Post Office Protocol.

electronic mall An online storefront, usually Web-based, that allows consumers to view, review, or request information about products and services. Although Web page layouts differ widely, electronic malls are nevertheless beginning to use some common conventions, such as shopping carts which are lists of products that are selected by pressing buttons on the Web page to queue for the final order entry and payment section. As no consistent electronic commerce money transfer schemes have emerged as yet, most of the electronic malls that are successfully getting orders are using credit card ordering.

Electronic Market A model for cross-border, remote electronic commerce in a broader context as defined by initial e-commerce studies of APEC and TELESA. The Electronic Market is a set of rules for international interconnection of electronic commerce

test-beds intended to verify the effectiveness of the concept and its implementation. See Integrated Next Generation Electronic Commerce Environment Project.

Electronic Media Forum EMF. A joint experimental project of the American Federation of Musicians, the International Conference of Symphony and Opera Musicians, and the Symphony, Opera, and Ballet Orchestra Managers' Media Committee. This forum tackles the issues of online performances, copyrights, and many other aspects related to the presentation and appreciation of the arts through emerging and converging multimedia technologies. http://www.electronicmediaforum.org/

Electronic Messaging Association EMA. An organization established to support the development and use of secure global electronic commerce services, including electronic mail, faxing, paging, and computing services. EMA contributes to education, public policy discussions, and connectivity and interoperability issues. http://www.ema.org/

Electronic Payments Forum EPF. An alliance of commercial, nonprofit, academic, government, and standards bodies, committed to furthering the development and implementation of electronic payment systems to promote global electronic commerce. The EPF is organized under CommerceNet, the Financial Services Technology Consortium (FSTC), and the Cross-Industry Working Team (XIWT) http://www.epf.net/

electronic phone Newer phone technology in which many of the mechanical parts common to traditional phones are replaced with circuit board logic design. This design results in components, such as cellular phones, that are smaller, lighter, and more portable.

Electronic Privacy Information Center EPIC. A public interest research center to focus public attention on emerging privacy and civil liberties issues, founded in 1994. EPIC is located in Washington, D.C., and works in association with Privacy International in the U.K. EPIC maintains policy archives with information on computer security, encryption, privacy, and freedom of information and speech. It also keeps an archive of previously classified government documents obtained under the Freedom of Information Act. It publishes a newsletter on civil liberties, tracks legislation, and recommends better texts on privacy and cyber-liberties. See American Civil Liberties Union, Electronic Frontier Foundation, Cyberspace Electronic Security Act of 1999. http://www.epic.org/

electronic receptionist A type of Auto Attendant system with extra logic and capabilities that allow the personalized notification of messages through a network, sometimes even with a graphic animation of a human receptionist delivering the message. It is likely that electronic receptionist virtual reality scenarios and videoconferencing will create some remarkable systems in the future. See Auto Attendant.

Electronic Rights for the 21st Century Act A large U.S. Senate bill (S 854) introduced in April 1999 by Patrick Leahy. It deals with privacy and encryption with several areas similar to the Security and Freedom through Encryption Act. No further action has been taken on this bill. See Security and Freedom through Encryption Act.

electronic ringing A digitally encoded sound file that is played instead of the conventional mechanical telephone ringer. Since it's digital, it can be made programmable, and thus any sound patch could technically be used as a ringing sound. Just as the error beep on most computer systems can be selected or recorded by the user, electronic ringers will probably be selectable in the future. Given the creative bent of many individuals, some likely ringers would include car horns, buzzers, dogs barking, voices saying "You've got a call." Now, if you consider the availability of Caller ID, it won't be long before someone invents a ringer that not only alerts the person about an incoming call but also speaks a message that says, "It's your girlfriend, better answer before she changes her mind!" Phones integrated into personal computers already exist, so it's just a matter of time before these are combined with electronic voice mail and customized ringing sounds.

electronic serial number ESN. In general, an encoded identifier for locating or authenticating a device. The degree of security of these numbers varies widely. In the cellular phone industry, the ESN is a 32-bit code that uniquely identifies a cell phone unit. It is standardized to include a unit number and information about the manufacturer. Also, there is reserved space for future or specialized use. In some cell phones, the ESN cannot be changed except by those who are handy with electronics and have a custom reprogramming cable and computer connection.

When using a cellular phone to make a call, the cell switching center uses the information to check the validity of the caller, and it sometimes checks a database of cell phones that are reported as stolen. If the number is valid, and there are no problems with the ownership of the unit or the currentness of the account, the call is put through.

The ESN is also used as an access code on the Web. Different cell phone manufacturers have various online services, including customer assistance, warranty information, etc., which can be accessed by users with legitimate ESNs.

Electronic Signatures in Global and National Commerce Act Passed by the U.S. Congress in June 2000 to facilitate global and interstate use and enforcement of electronic contracts through electronic signatures and records. An important aspect of the Act is related to the receiver's ability to access and deal with electronic documents and the related impact on electronic commerce. Thus, the issue of consent for the receipt of electronic documents, as opposed to traditional paper documents/signatures, on the part of the recipient is being widely discussed.

electronic switching system In old phone systems, call connections were established by human operators plugging phone jacks into jack panels. Later, mechanical switches and electromechanical devices automated the process. Still later, logically programmed solid-state components took over switching functions in conjunction with the telephone switching equipment to create electronic switching systems. See Electronic Switching System 1.

Electronic Switching System 1, Electronic Switching System 1A ESS1/ESS1A. An AT&T computer-controlled electronic telephone crosspoint switching system based upon the 1A processor. This is an End Office (Class 5) system that has resulted in a number of additional technologies, including the No. 10A Remote Switching System (RSS). The 1A processor handles maintenance and administrative tasks associated with the system switches and provides read-out data for operators and maintenance personnel. It is a digital switch that functions within an analog system that was first introduced as ESS1 in the 1960s and ESS1A in the 1970s when end-to-end digital systems were not yet common. It has generally been superseded by newer digital systems (e.g., ESS5). See 1A.

Electronic Technology Systems ETS. A European global testing and certification body, located near Berlin, Germany. ETS supports a number of related industries, including consumer electronics, telecommunications, and safety, and provides testing equipment for digital mobile markets.

electronic yellow pages EYP. An electronic database of business address listings. There are many EYP lookup services on the Web that allow the user to query for a business name or business-related keyword in order to locate a company. The company does not have to be connected to the Web to be listed, as many electronic yellow pages services get their listings from local communities or printed directories. See yellow pages, Yellow Pages.

electronics The art and science of investigating and harnessing electrons. Some lexicons make a distinction between electron movement through the air or through a vacuum (electronics) as distinct from electron movement through a wire or other similar conductor (electricity). While this distinction may be important in a theoretical sense, for the purpose of this general purpose telecommunications dictionary, the broadest sense of the term *electronics* has been used. The harnessing of electrons is fundamental to telecommunications technology. The term came into general use in the 1940s. See magnet, vacuum tube.

Electronics Technicians Association, International ETA. A not-for-profit international organization established in 1978 to support and promote the electronics service industry through education and certification programs. Headquarters are in Indiana. Some programs require that candidates first hold an Associate Certified Electronics Technician designation, and others do not. Programs include traditional and fiber optics installation certificates.

Electronics Telecommunications Research Institute ETRI. A research institute located in Taejon, South Korea, descended from the Korea Institute of Electronics Technology (KIET), established in 1976, and the Korea Telecommunication Research Institute (KTRI), established in 1977. In 1985, KIET and KTRI were merged into ETRI under control of the

Ministry of Science and Technology; then in 1992, it moved to the control of the Ministry of Science and Technology. In 1995, it was re-established as a judicial foundation through the Electrical Communications Act and three years later was restructured into four technology labs (and later into five labs to position it as a global leader in core information technologies). ETRI affiliation was moved to the Department of the Prime Minister in 1999. In 2000, the Korean national research institute was made an affiliate of ETRI. ETRI is focused on research in basic and advanced science/engineering and telecommunications technologies, computers, information technology, and semiconductors. Areas of particular interest include switching systems, mobile/wireless communications, and semiconductor technology in support of a national telecommunications system. ETRI publishes the quarterly *ETRI Journal*.

In 1994, the U.S. Air Force visited the Institute to learn about the research carried out there and to make initial contacts regarding possible future collaborative research programs. http://www.etri.re.kr/

electrophoretic display A display technology incorporating positively charged, suspended color particles sandwiched between two outer plates. The suspension medium consists of a contrasting color so the suspended particles are visible or not visible depending on their position within the suspension medium. The particles are moved to the front (visible) or back of the display (not visible) by applying negative and positive voltages through matrix addressing.

electroplating The process of using electrode position to cause a usually metallic substance to adhere strongly to another. Metals are often electroplated to prevent corrosion, to improve conductive properties, or to increase their value (as with gold or silver electroplating on jewelry or cutlery).

electroscope An instrument for detecting the presence of an electric charge, and whether it is positive or negative. It may also be designed to indicate the intensity of that charge.

electrostatic An electrical charge at rest, familiar to many people as *static* electricity. Static electricity associated with rubbing amber (*amber* in Greek is *elektron*) was known to the Greeks by at least 600 BC when Thales recorded that a 'fossilized vegetable rosin' (amber), when rubbed with silk, acquired the property to attract very light objects to itself.

electrostatic charge An electrostatic charge that is stored as in an insulator or capacitor. Subsequent discharge must be carefully controlled or prevented in electronic components, as they have the potential to cause harm. See condenser, Leyden jar.

electrostatic deflection The process of controlling the path of an electron beam by passing it between charged plates. This system is widely used in cathode-ray tubes (CRTs). See cathode-ray tube.

electrostatic focusing The process of using an electrical field to focus an electron beam.

electrostatic plate A type of printing plate incorporated into high speed laser printing processes that uses various metals, chemicals and photoconductors to create and transfer an image.

electrostatic printing A printing process in which the imaging is primarily done with electrical rather than chemical means; it is a dry replication process very common in photocopiers and laser printers. Typically, laser beams trace an electrostatic pattern that attracts very fine particles of powder, usually toner, that are subsequently fused onto the printing medium with heat.

electrostatic voltmeter A voltage-driven voltage measuring instrument that requires almost no power for its operation. The electrostatic voltmeter functions by using the attracting and repelling properties of magnetic bodies. It can be used on both alternating current (AC) and direct current (DC).

Elektro See Geostationary Operational Meteorological Satellite.

ELIU electrical line interface unit.

Land Mass-Distributed Orbits

The Ellipso project employs some unusual orbits, different from the common geostationary or regularly elliptic orbits used by other satellite service providers.

Ellipso A commercial satellite communications system based on groupings of low Earth orbit (LEO) *bent pipe* transponders, operating in a unique elliptical obit through CDMA technology.

The Ellipso orbits are adapted to the distribution of land masses and populations on the Earth. Two complementary and coordinated constellations of satellites comprise the Ellipso constellation. The Ellipso-Borealis Subconstellation covers northern latitudes through 10 satellites in elliptical orbits in two planes with apogees of 7846 kilometers operating in a three-hour orbital period. The Ellipso-Concordia Constellation provides coverage of southern latitudes through six quasicircular equatorial orbits at 8040 kilometers, and four complementary elliptical orbits to increase daytime capacity.

Ellipso is a service of U.S.-based Mobile Communications Holdings, Inc. Mobile's communications services are designed to enhance rather than displace existing phone and data services. Mobile Communications is targeting low-cost services especially to isolated and rural areas.

elliptic curve A geometric structure of interest to cryptographers because it can be incorporated into algorithms with calculations that can be performed with relative ease in one direction, but that are difficult or impractical to reverse thus foiling decryption efforts.

ELOT Hellenic Organization for Standardization, located in Athens, Greece. ELOT handles standardization, certification, and quality control in Greece. http://www.elot.gr/

ELSU Ethernet LAN Service Unit.

EMA See Electronic Messaging Association.

EMACS, GNU EMACS A powerful, flexible, configurable, network-savvy text editor written originally by Richard Stallman and promoted and supported by a large community of programmers. EMACS works on a wide variety of platforms, with GNU EMACS running principally on Unix systems. MicroEMACS is a popular derivative, and CygnusEd is a fast commercial editor with many EMACS-like characteristics. While EMACS has a bit of a learning curve, it has many useful capabilities not found in other editors. It can work directly as a mail client, can read and write files directly over an FTP link, and can be modified and extended with LISP. See text editor.

email See electronic mail.

EMBARC Electronic Mail Broadcast to a Roaming Computer. A North American paging service from Motorola based on the X.400 standard. EMBARC transmits 8-bit text and binary files containing up to 1500 characters. See ERMES, SkyTel, X Series Recommendations.

embedded Closely contained within a bounded environment. The human brain is embedded within a bony structure called the skull. A microphone and speaker are embedded within the handset of a telephone. Electronic components are tightly embedded within a small appliance such as a wristwatch, hearing aid, or handheld computer. Optical fibers are tightly embedded within the sheathing in a fiber optic transmission cable.

In general, in communications embedded devices are those in which electronics or transmission media are packed into a tightly constrained space. This arrangement was difficult in the days when electronic devices were based on vacuum tube technology. Transistors made it possible to greatly reduce the size of a device and the proximity of the components within that device. Large-scale integration of semiconductors took this trend one step further, enabling microminiaturization of logic-capable devices.

There are many reasons for embedding components: transmission speeds may be increased, power consumption may be lower, proprietary technologies may be hidden or obscured, or the footprint or size/weight of the device may be minimized. Embedded devices may also have disadvantages: fabrication may be more technical or costly, electromagnetic interference may be more prevalent, servicing may be more difficult, or the space for including enough hardware and software resources to meet the task at hand may be limited. See embedded devices.

embedded devices 1. A general category of components that are especially useful in mobile and space systems due to their low power consumption and compact size. 2. A quickly growing class of consumer devices in which the electronics are bundled into small housings to enable the production of products that are convenient and easy to power by battery sources. VLSI technology and circuit boards have made it increasingly practical to design small, sophisticated portable devices. Forty years ago the only embedded devices people regularly carried around were wristwatches and transistor radios. Now it is not uncommon to see a person carrying a watch, a cellular phone, a pager, a GPS device, an electronic scheduler, a garage door opener, and a calculator, all at the same time.

The programming of embedded devices is a particular challenge to programmers who have become accustomed to working on desktop computers with 128K of memory and 1600×1200-pixel screen displays. Embedded devices often have significant memory and display area constraints, equivalent in some ways to the microcomputers of the late 1970s. Older programmers are familiar with programming text for 8K 120×80-pixel text displays but probably didn't expect that skill to become useful again in the year 2001. To facilitate programming for embedded consumer devices, environment emulation programs for use on desktop machines have been developed. See Java.

embedded operations channel EOC. In a data transmission, a number of bits set aside for use for monitoring configuration, maintenance, or other performance data that are distinct from the user data transmitted in the normal channel but carried tightly within the data stream (in-band) as opposed to being transmitted separately (out-of-band).

EMC electromagnetic compatibility.

EMC and Radio Spectrum Matters Technical Committee ERM. A working committee of the European Telecommunications Standards Institute (ETSI) involved in developing standards for electromagnetic compatibility, radio spectrum, and a range of radio products (mobile, broadcast radio, vehicle transmitter systems, etc.). It's a broad mandate, beyond the scope of any single committee, so ERM works closely with the Operational Co-ordination Group and others. Significant progress was made in conjunction with the ETSI Mobile Standards Group Technical Committee in creating Universal Mobile Telecommunications System standards during 2000 and 2001. A Task Group was created focusing on air interface specifications, and a Guide was published to assist network operators in developing interoperable interface specifications. A Citizen's Band (CB) Guide was also published with findings regarding the impact of CB frequencies (27 MHz) on other radio technologies.

emergency access An alert system incorporated into some private branch exchange (PBX) systems that rings selected phones, or all phones, in the event of an emergency.

Emergency Alert System EAS. A new emergency system designed to replace the Emergency Broadcast System (EBS). The EAS is based upon digital technology, and all broadcast stations are required to have EAS decoders subsequent to January 1, 1997. The EAS provided for a one-year transition period during which old equipment had to be retained to ensure that new equipment could take over the functions effectively. Furthermore, television stations must be able to automatically send a visual message. Specific equipment for delivering the emergency messages has not been defined on a physical basis. Instead, the type of message functions that must be supported are described. The equipment allocated to these functions must be certified. Live monitoring, logging, and documentation requirements form part of the EAS. See Emergency Broadcast System, Global Maritime Distress and Safety System.

Emergency Alert System (EAS) activation signal A mandatory protocol of the EAS, consisting of a digital header, an attention signal, a message, and an end of message (EOM) code. The header contains information about the origin of the message, its nature, location, and time during which it is valid.

Emergency Broadcast System EBS. A warning system that has been replaced by the new digitally based Emergency Alert System (EAS). See Emergency Alert System, Emergency Digital Information System/Service.

emergency dialing A type of speed dialing in which a shortened sequence of numbers, or sometimes even one button, facilitates a fast connection with emergency telephone services such as the police, ambulance, or fire department.

Emergency Digital Information System/Service EDIS. An emergency and disaster information service established in California in 1990 for delivering official information to the news media and the public. As such it constitutes a Community Warning Systems (CWS) that supplements national disaster services such as the EAS. Due to the prevalence of earthquakes and floods in Californian urban areas, and the proximity of many high technology companies, it seems natural that California would be one of the early adopters of an electronic disaster information service. In the late 1990s, the California EDIS system was updated to include sound and graphics and to support satellite datacast capabilities. It is now managed by the Governor's Office of Emergency Services.

EDIS combines a 24-hour broadcast service, newswire, and a Website to provide authorized agencies with emergency or disaster-related text, graphics and sounds via computer networks. It distributes messages from the California OES Data Network, the National Weather Service, the Law Enforcement Telecommunications System, and a number of specialized science networks providing seismic and other disaster-related information. Transmitters are spaced around California, particularly in the Los Angeles and San Francisco Bay areas where population densities are higher, and broadcast on standard radio frequencies that can be accessed by amateur radio systems. EDIS information can be accessed by the general public and the news media through digital radio broadcasts, email, maps, and images optimized for computer transmission and display. EDIS is designed to be accessible even if public networks are 'gridlocked' due to excessive demands typical of disaster situations, through its satellite distribution network. EDIS provides a means to convey Emergency Alert System (EAS) messages as well as local information to the general public and emergency and disaster relief agencies. http://edis.oes.ca.gov/

emergency hold A capability in which an emergency professional (fire, police, crisis center, etc.) can keep a connected call live even if the caller hangs up or presses the switch hook. Since it is not unusual for people in emergency situations to behave erratically, or to hang up inadvertently and then try to reconnect, it gives the emergency personnel the opportunity to retain or regain contact with the caller.

Emergency News Network ENN. While this title is used generically to refer to a number of different state and national news services, it is often used specifically in connection with the Chicago-based *EmergencyNet News*, a service of the Emergency Response and Research Institute.

Emergency Number Service ENS. A provision of common carriers of land and mobile telephony services for quick access to emergency services. In North America, 911 is the familiar emergency number. In other countries, there are other systems. For example, in some locales, 999 is used to call for police assistance. See 911 calling, E-911, National Emergency Number Association.

emergency telephone A phone designed to provide service in spite of unusual circumstances, such as power outages. It may have backup batteries or a connection to emergency power in order to maintain operability. It may also have a direct wire or priority wireless connection to emergency services.

EMF, emf 1. See electromotive force. 2. See Electronic Media Forum. 3. See European Multimedia Forum.

EMI See electromagnetic interference.

emission 1. Outward radiation or conduction of waves. 2. Ejection of electrons, which could arise from a variety of causes such as impact or heat.

emission current A current produced in the plate of a cathode-ray tube (CRT) when the electrons beamed from the cathode pass into the plate.

emission velocity The initial velocity of electrons beamed from the cathode in a cathode-ray tube (CRT).

emissivity A measure of the ratio of flux emitted by a radiation source to the flux radiated by a black body with the same area and temperature.

emitter 1. A device that provides periodic pulses at regular intervals as is used, for example, on punch card machines. 2. An electrode in a transistor (minority carrier).

emitter-coupled logic ECL. A fast logic-circuit amplifier comprising a bipolar transistor pair with coupled emitters. It is one of three main logic families. ECL is faster than similar transistor-transistor logic (TTL) circuitry (with a tradeoff in power to keep the circuit active); the gate delay may be less than

1 ns, making it suitable for high-speed data circuits. See CMOS, transistor-transistor logic.

emoticon An apt name for an online grassroots evolution of tiny ASCII icons to represent moods, emotions, and expressions. Since networks before the Web were traditionally text-based, and since communications online can often be misunderstood without facial and voice expressions to express subtle distinctions, many users have adapted an iconography into text messages to provide a wider range of expression. Because of the difficulty of creating vertical drawings in a text-based message, most emoticons are designed to be read sideways, that is, with your head turned 90°, usually to the left.

Emoticons (Smileys)		
: -)	:)	two variants of the basic smiley, the most common emoticon
; -)	;)	a smile with a wink
: - D	: D	laughter, a big talker, delight
: - O	: o	surprise, astonishment, awe
: - /	: - (	smirks
: - \	: - |	frowns
: ~P		sticking a tongue out

EMP See electromagnetic pulse.

empty slot ring In a ring-based local area network (LAN) topology, such as Token-Ring, a free packet that travels around the ring through each node on the network, and is checked at each workstation for messages.

EMR See Exchange Message Record.

EMS See Expanded Memory Specification.

EMT electrical metal tubing.

EMTUG See European Manufacturing Technology Users Group.

EMU 1. Economic and Monetary Union. A long-term plan for the unification of Europe into a common market, resulting from the Maastricht Treaty of 1992. 2. European Monetary Union/Unit. A common market currency in effect as of 1 January 1999 based on measures taken by the Economic and Monetary Union.

emulate To copy, simulate, provide working functions of.

emulation, software To recreate the functioning of an application or operating system. Emulators abound in the computer world. There are Macintosh OS emulators for Amigas and NeXT computers, and Windows emulators for Macintosh and Amiga computers, etc. It is not strictly true that an emulator *always* runs more slowly than the original system. With fast hardware, one operating system can sometimes run an emulator faster than the OS may run on the machine for which it was originally designed.

En Banc An informal Federal Communications Commission (FCC) forum for the discussion of FCC rules and policies.

encapsulated PostScript EPS. A PostScript file format including *bounding box* information (explicit statements about the lower left and upper right extents of a rectangular area that includes the entire image contained in the file). The EPS bounding box information is commonly found in the header at the beginning of the program. Many programs cannot print a PostScript page without the EPS information, if it is an imported file, since the size and boundaries of the image may not be explicitly stated in the page description code. (It's like trying to find a frame for a picture without first knowing the size of the picture.) EPS is a widely supported vector format and is useful for desktop published documents, posters, typography, and other applications in which one wants to take advantage of the highest possible quality available on a particular output device. See PostScript.

Encapsulating Security Payload ESP. In TCP/IP networks, ESP is a transport layer protocol compatible with both IPv4 and IPv6 formats. It operates in both tunnel mode and transport mode. In secure network transmissions, the ESP may be used in conjunction with an authentication header mechanism. See link encryption, RFC 1827, RFC 2401.

encapsulation 1. Encasing or completely surrounding with a protective covering. When something is encapsulated, there's not usually much room between the covering and that which is being covered. Encapsulation is used for insulating, sealing, or separating, and it might make the contents water- or airtight. 2. In frame relay, a process by which an interface device inserts the protocol-specific end device frames into the frame relay frame. 3. A means of wrapping a bundle of network information in order to *tunnel* it through a system, to pass it through different protocols, without altering the wrapped data. It is subsequently de-encapsulated to retrieve the wrapped information. See Point-to-Point Tunneling Protocol, tunneling.

encoding The conversion of data from one form to another, sometimes to obscure the contents, sometimes to reduce the amount of space necessary to hold the information, and sometimes to change it to a compatible format. Compression and encryption are two means of data encoding.

Encrypted Communications Privacy Act ECPA. An Act introduced in March 1996 to address legal security levels for computer network transmissions and civil liberties by preventing unlimited access to computer records by the U.S. Government. This Act came about in part due to criticisms and concerns expressed after the enactment of the *Electronic Communications Privacy Act of 1986* (ECPA). One of the arguments for allowing the use of higher level encryption was to aid Americans in maintaining safe and secure business and personal transactions in a globally connected world. Another motivation for assigning this limiting measure to government came from U.S. industry assertions that the U.S. would have difficulty maintaining a competitive advantage in the global marketplace with respect to encryption technologies (with strong export restrictions on software, foreign nations were offering stronger encryption on competing products than the U.S. vendors were permitted to offer). See Electronic Communications Privacy Act of 1986, encryption.

Encrypted Key Exchange Protocol, EKE Protocol A password-based certificate-free protocol designed to increase security in digital networks. Introduced in the early 1990s by S. Bellovin and M. Merritt of AT&T Bell Laboratories (U.S. patent #5,241,599), it combines symmetric- and public-key cryptography to enable two parties, sharing only a relatively insecure password, to initiate a computationally secure cryptographic communication over an insecure network, such as a wireless telephony system. The system provides security against passive and active attacks and dictionary-style password-guessing attacks. Since its introduction, extensions and adaptations of this system have been suggested, including Japan's Dual-Workfactor Encrypted Key Exchange Protocol. See Kerberos authentication.

encryption The process of encoding data in a way that makes it difficult or impossible to decode. Encryption is widely used in times of conflict to hide the contents of sensitive communications. It is also used for encoding private messages or business transactions and documents. Encryption is often combined with authentication. Key encryption is a type of encoding which uses public and private keys to encode and retrieve messages. See Clipper Chip, key encryption, Pretty Good Privacy.

Encryption Control Protocol ECP. A means for configuring and enabling data encryption algorithms on both ends of a point-to-point (PPP) link. ECP uses the same basic packet exchange mechanism as the Link Control Protocol (LCP). See RFC 1968.

Encryption for the National Interest Act A House of Representatives bill (H.R. 2616) introduced in July 1999 by numerous representatives. It deals with privacy and encryption but is significantly different from the Security and Freedom through Encryption Act proposed in February 1999. It supports freedoms to use encryption for domestic purposes but does not explicitly support the production, transfer, or vending of encryption-related products. Encryption export restraints are discussed instead. The proposal further does not require key escrow or other access to support law enforcement efforts to decrypt communications related to criminal activities although it does delineate procedures to be followed for gaining access to plaintext of decryption information upon receipt of a court order. See E-Privacy Act, Security and Freedom through Encryption Act.

end delimiter An end of file (EOF), end of block (EOB), end of line (EOL), end of message (EOM), end of transmission (EOT), or other data signal that designates the final byte or section of a block of information.

end device 1. The terminal point, whether it be the source or the destination, of a data transmission. 2. A hardware or software terminal device, which may be a file server or host. 3. In a frame relay network, data is sent from the source to an interface in which it is encapsulated into a frame relay frame. At the destination end device, the frame relay frame is stripped off again (de-encapsulated). See Data Communications Equipment, Data Terminal Equipment.

End Office switches Hardware routing switches from which telecom subscriber Exchange Services are connected. These may be private branch trunks and their connections with Interexchange Carriers or Competitive Local Exchange Companies. Local End Office switches may be designated as Class 5 switches. Some systems use a combination of Class 4 (e.g., long-distance Tandem switches) and Class 5. The trend is for telephony and data services to be carried seamlessly over the same infrastructure.

Digital End Office switches are gradually superseding traditional analog switches, and various issues related to their function (e.g., trunk group echo) have resulted in changes and upgrades in End Office switches. Next Generation switches may be installed as alternatives to certain End Office switches. One of the most recent developments is the implementation of software-based Class 4 and 5 switches (Softswitches) that enable a telecommunications service provider with broadband access to offer reduced-cost services or End Office functions. See Exchange Service.

end to end digital path A communications link between two hosts capable of sending wholly digital data (without analog conversion along the way) over an arbitrary number of routers and subnets.

end to end path A communications link between two hosts capable of sending data over an arbitrary number of routers and subnets.

end to end service A network service that links one subscriber to another subscriber or another branch of the first subscriber. It typically consists of the local loops at each end of the transmission connected through an InterExchange Carrier (IXC).

end user 1. The service or product user at the end of the retail or service chain; in other words, it's not a user who is a wholesaler, broker, or agent. Some people use this phrase interchangeably with *user*, but it might be useful to maintain a distinction. 2. In telephone services, an end user is distinguished as a subscriber using loop start or ISDN. 3. In network communications, the computer users who are not the technical administrators of the system, e.g., students and faculty on a college campus.

endface A planar surface of a component, wire or cable. On a cable, it is the terminating surface more or less perpendicular to the length of the cable, which is attached to couplers, components, and other cables.

endface test In fiber optics fusing processes, inspection of the end surfaces of cleaved fiber optics filaments. The ends must meet certain specifications for smoothness and cut angle tolerances so the surface lets light pass readily and evenly rather than being distorted or reflected back into the originating fiber.

endoscope A medical exploratory instrument for remotely viewing inside-body structures. The endoscope was one of the first instruments to use glass fibers, in the early 1900s. In 1952, Harold H. Hopkins

E

applied for a grant from the Royal Society to develop glass fiber bundles into an instrument for endoscopic viewing. When the grant was approved, Narinder S. Kapany, a significant pioneer in fiber optics, was brought in as an assistant. Fiber-based endoscopy enables surgeons to view the interior of the body through smaller incisions, reducing the possibility of infection and speeding surgical recovery.

Energy Sciences network ESnet. The network of the U.S. Department of Energy chartered in 1986 as a result of combining MFEnet, which had operated since 1976, and HEPnet, a leased-line network established in 1980. In the early 1990s, ESnet began to offer international data services and implemented World Wide Web connectivity. In 1996, the ESnet base was relocated to Lawrence Berkeley National Laboratory. ESnet provides a high-end, advanced, reliable communications infrastructure to support the Department of Energy's scientific research and various projects. http://www.es.net/

Englebart, Douglas (1925–) An American innovator and engineer who served as a radar technician in World War II and subsequently joined the NASA Ames Laboratory. He later had the opportunity to pursue some of his pioneering ideas at the NLS laboratory at the Stanford Research Institute in the 1960s and at Xerox PARC in the 1970s. Englebart is credited with inventing the three-button mouse in the late 1960s, an accompanying keyset, and many of the graphical user interface ideas that are now prevalent on desktop computers. His concept of "windows" as a means of expressing a work area on a computer monitor was further enhanced by Alan Kay, the originator of Smalltalk, and others in the lab at the time. Englebart has many patents. See Kay, Alan.

enhanced-definition television, enhanced-definition TV EDTV. A general category of television technologies that forms a bridge between traditional NTSC technologies and advanced television (ATV) technologies. EDTV is a means to improve analog NTSC broadcasting without going to an entirely new system (e.g., digital technologies). Thus, enhanced signals within the existing broadcast channel or added signals provided through additional spectrum allocations can augment existing broadcast technologies. See advanced television.

Enhanced Call Processing ECP. A subscriber phone service with enhanced voice mailbox features that allow more than one user on the line to have an individual mailbox or for mailboxes to be individually configured to be listen-only, redirect, menu-directed, or leave-a-message. Commonly the ECP can be programmed through online menus, greatly enhancing applications that can be established on the system. ECP services can also be established through voice/fax modems by configuring the software on a computer. Thus, fax-back, voice mailbox, redirect, and other applications can be programmed to control the modem. The modem, in turn, handles the management of the phone line.

Enhanced Digital Access Communications System EDACS. A widely distributed commercial, trunked, public safety mobile radio communications system from Ericsson Inc./General Electric. EDACS operates in 800-MHz, 900-MHz, VHF, and UHF frequency bands and is utilized by private firms, public safety organizations (police departments, airports, and the U.S. Naval District in Washington, D.C.). The system is hierarchical, with talk groups organized into agencies, which are further divided into fleets and subfleets. Each radio frequency is assigned a Logical Channel Number (LCN) that is programmed into each radio on the EDACS system. Dual-format mobile radios from Ericsson/General Electric can be used on EDACS systems.

Enhanced Graphics Adapter EGA. A 16-color graphics display standard developed to supersede the Color Graphics Adapter (CGA) standard that was prevalent on Intel-based desktop computers in the mid-1980s. EGA could display up to 640×350 pixels. It has since been superseded first by Video Graphics Array (VGA) and then Super Video Graphics Array (SVGA) formats which offer higher resolutions. Part of the reason for improved graphics resolutions stemmed from competitive pressure from a large number of color graphics systems released in the 1980s by vendors supporting the Motorola $680xx$ chipset rather than the Intel chipset. The Apple IIGS, Commodore Amiga, Silicon Graphics workstations, and Atari ST all had high resolution graphics with thousands or millions of colors while most Intel/MS-DOS-based systems at the time were still displaying 16 colors (8 regular colors plus 8 half intensity). See Color Graphics Adapter, Video Graphics Array, Super Video Graphics Array.

Enhanced Group Call EGC. Broadcast messaging and routing services commonly used in safety and warning systems such as Inmarsat mobile satellite communications. For example, EGC facilities are part of the International SafetyNET maritime safety system. They are also incorporated into the FleetNET broadcast facility, which enables the simultaneous transmission of a single message from an authorized information provider to multiple Inmarsat-C Ship Earth Stations (SESs).

Typically, an EGC message is identified by an EGC network identification code (ENIC) that may be stored in the Ship Earth Stations system. The message to be sent includes an address specifying the geographic region for the destination. The message may be sent to individuals, specified groups, or all receivers within the specified network. Multiple recipients receive the message at about the same time. Commercial EGC receivers enable ship positions to be entered and updated manually and automatically. They may be combined with Global Positioning Service receivers in integrated units.

Enhanced IDE An enhanced computer peripheral device controller specification that provides faster transfer speeds and larger addressing space than the original IDE specification. See IDE devices and controllers.

Enhanced Parallel Port EPP. A parallel data transmissions specification intended to provide faster parallel speeds over common desktop computer parallel ports, without changing the basic hardware (cables, connectors, etc.) used to implement the connection. Parallel ports are commonly used with impact printers and some inkjet and bubble-jet printers. Ethernet, USB, and Firewire formats may eventually supersede EPP for desktop peripheral connections.

Enhanced Position Location Reporting System EPLRS. A digital, mobile, tactical radio system used by the U.S. military to provide an integrated network to support high-speed, automated warfighting systems and to enhance command-control capabilities. It also serves as a communications backbone for the military's Tactical Internet. ELPRS enables the collection of data from widely dispersed systems in battle areas, transmitting them back to command centers and combat force locations.

Each radio in the military network is assigned a unique identifier and unique time slots during which it can transmit to the Net Control Station (NCS) and other radio sets using time division multiple access (TDMA).

Automated functions enable EPLRSs to avoid jamming threats and to adjust to and overcome line-of-sight changes and limitations. The Raytheon EPLRS operates in wideband UHF frequencies using embedded cryptography for security at variable data rates up to 100 Kbps. In actual practice, U.S. military systems transmit at up to about 56 Kbps, but higher rates are expected with the integration of military and commercial systems in future upgrades. See Enhanced Trivial FTP.

Enhanced Serial Interface ESI. A freely distributed serial interface specification developed by Hayes Microcomputer Products. It is an extension to the COM Card specification for desktop computers.

Enhanced Serial Port ESP. A Hayes serial port specification introduced in the early 1990s that is intended to extend and replace the COM1/COM2 ports that are familiar to most users of Intel-based desktop computers. It provides two independent operations modes, to support existing standards and the ESP format. Depending on the version, defaults are set either with the Hayes Programmable Option Selection (POS) or with DIP switches.

Enhanced Service Provider ESP. A category of service provider recognized by the Federal Communications Commission (FCC) as subject to certain rules, regulations, and exemptions.

As digital phone services and data communications services became more prevalent in the 1980s with the increased use of computer modems and computer networks, the FCC needed to sort out how the regulations governing telephone services and interstate commerce should apply or be changed to apply to developing computer timeshare and digital service providers. Should enhanced phone service companies and computer bulletin board sysops be regulated as common carriers? This question led to a distinction between (1) basic telecommunications services, focusing on the physical transmissions capability and (2) enhanced services, encompassing everything beyond the basic physical transmissions capability. Thus, voicemail services, computer bulletin board systems, Internet access services, document transmissions services, etc. came within the category of enhanced services, and those who provided them were distinguished as Enhanced Service Providers (ESPs). In 1983, the FCC determined that ESPs would be exempt from access charges related to interexchange service, even though phone lines were being used to implement data communications services. While adjustments and changes have occurred since that time, including close scrutiny in December 1996 through the Access Charge Reform proceeding, the general concept to promote the competitive viability of ESPs continues to receive support. See ASP Industry Consortium, Telecommunications Act of 1996.

Enhanced Specialized Mobile Radio ESMR. A traditional nationwide wireless two-way radio dispatch service that has been expanded to offer services somewhat similar to cellular, but with more limited allocation of bandwidth in the 800 and 900 MHz regions. The technology began to catch on in the late 1990s, evolving from the well-established SMR radio networks. ESMR uses the same frequency band as SMR but uses newer digital technologies to increase capacity and the range of services over a wider geographic area. A single switching center may support 500 or more base transceiver stations (BTSs). To cost effectively support this many BTSs, they may be organized into small groupings (hubs) with each group connected, in turn, to intermediate stations. In this way, multiple circuits can share the resources of a fast digital transmissions medium (e.g., a T1 line).

The ESMR system is currently adding millions of new subscribers per year. More recent implementations of the technology go beyond traditional dispatch, offering digital two-way radio, cellular phone services, and packet-data services (within the general dispatch-related niche) within areas licensed to ESMR vendors. The technology is now being extended by commercial vendors (e.g., Motorola) to create virtual private networks for workgroups within the framework of the larger ESMR system. ESMR services are of particular interest to utilities, construction, sales, and transportation (bus, taxi, trucking) industries.

In Europe, ESMR is based on Terrestrial Trunked Radio (TETRA) standards (ETSI). See iDEN.

enhanced TV, enhanced Television A general category of broadcast technologies similar to traditional TV but characterized by additional services, which may include subtitles, closed captioning, on-demand viewing, and other enhancements arising from digital broadcasting and Internet communications. Two-way communications are necessary for many enhanced TV services and broadcast triggers are used to administrate the recognition and delivery of enhanced services.

The basic data categories supplied through enhanced TV are announcements, content, and broadcast data triggers. *Announcements* provide the user with options, services, annotations, or scheduling information through SDP/SAP, *content* is the broadcast programming (e.g., a movie), and *broadcast data triggers* help to administrate the enhanced TV services.

The ATVEF recommends that the PNG image format be used for Internet TV broadcasting whenever possible, with JPEG and GIF for services not yet available through PNG. See broadcast data trigger.

Enhanced Trivial FTP Enhanced Trivial File Transfer Protocol. An experimental implementation of Network Block Transfer Protocol (NETBLT) that uses the User Datagram Protocol (UDP) for its transport layer. It was designed as a means to improve data transfer throughput for the specialized needs of half duplex radio networks using the Internet Protocol (IP). Through ETFTP, transmission parameters can be customized for low-speed, long-delay radio links. See File Transfer Protocol, Simple File Transfer Protocol, Trivial File Transfer Protocol, RFC 998, RFC 1350, RFC 1986.

ENIAC Electrical Numerical Integrator and Calculator/Computer. A historically significant, post World War II, room-sized, vacuum tube, punch card computer dedicated in 1943, following the success of the Harvard Mark I. ENIAC was developed by John W. Mauchly and J. Presper Eckert at the University of Pennsylvania, under the guidance of John Brainerd. Adele Goldstine authored the technical guide "Manual for the ENIAC" in 1946, the year the computer was unveiled in Philadelphia. Hand wiring was necessary to configure ENIAC to handle different problems.

The ENIAC was derived in part from the Atanasoff-Berry Computer (ABC). At the very least, the ABC provided inspiration for the ENIAC, and it is possible that it also provided some design ideas. In the 1930s, Mauchly is reported to have visited with Atanasoff and left with notes on the ABC. See Atanasoff, John V.; Atanasoff-Berry Computer.

ENN See Emergency News Network.

ENOS Enterprise Network Operating System. An operating system with the power and reliability to effectively support a medium- or larger-sized business or institution.

ENS See Emergency Number Service.

Enterprise Application Server EAS. A temporary name for an Open Source project to create a multiplatform, open source Sun Microsystems Java application server suitable for enterprise firms, that maintains consistency with industry standards, particularly J2EE.

Enterprise Computer Telephony Forum ECTF. A nonprofit association established in 1995 to promote interoperability among computer telecommunications products and to provide a framework for Computer Telephony (CT) interoperability, including standards and education. http://www.ectf.org/

enterprise network, corporate network A network that broadly serves the needs of an enterprise (a larger business) and reaches most of the network-using members of the organization. Since larger networks often have different types of users, they may integrate a variety of technologies, such as a private branch

exchange (PBX), data/voice/videoconferencing, etc.

envelope 1. Boundary, encasement, encapsulated entity, bounding box, extent. 2. The globe around a vacuum tube or bulb. 3. In amplitude modulation, the extents of the frequencies of the modulated wave. 4. A means of describing the content and characteristics of a message.

Environmental Engineering Technical Committee EE. A working committee of the European Telecommunications Standards Institute (ETSI) engaged in defining the infrastructure and environmental aspects of all telecommunications equipment. Working groups of the EE include environmental conditions (EE 01) and power supplies (EE 02).

EE 01 involves the development of environmental classes and the classification environmental conditions and related equipment tests to verify that they are mutually compatible. Ecological aspects are not supported by this project, as it was decided that these are being addressed by the European Information and Technology Industry Association (EICTA). EE 02 involved amending and improving existing standards for power supplies and developing an ETSI Guide for power distribution for telecommunications networks.

ENVISAT A project of the European Space Agency, ENVISAT is an advanced polar orbit Earth imaging satellite system, intended to eventually supersede the ERS-1 and ERS-2 systems launched in the 1990s. It is designed to provide advanced imaging using a variety of radar and other sensing systems for scientific research and commercial purposes. Tests were successfully carried out in French Guyana in 2000, and ENVISAT is expected to provide imagery for a period of about five years from the time of launch. See European Polar Platform, European Remote-Sensing Satellite. http://envisat.esa.int/

EO 1. end office. 2. erasable optical.

EOB end of block. See end of file for a description that is conceptually similar to end of block markers.

EOC See embedded operations channel.

EOF See end of file.

EOM end of message. See end of file for a description that is conceptually similar to end of message markers.

EOS 1. Earth-observing satellite. 2. See Earth Observing System. 3. electro-optical system.

EOT 1. end of transmission. EOT signals enable a system to provide a variety of communications options, such as the initiation of an automated sequence (e.g., sounding an alarm if it has been an emergency transmission), signaling that the other end of the transmission may now transmit, shutting down the system if appropriate, turning on other equipment, etc. An EOT signal may be passive or active. A passive EOT might be signaled after 30 seconds of silence or 30 seconds without further transmissions, for example. An active EOT might be a specific sequence of bytes or frequencies that are recognized as meaning the end has been reached or is imminent. 2. end of tape. EOT flags or markers enable manufacturers to include capabilities related to the tape player or recorder that allow it to signal the end, automatically reverse to the beginning of the tape, or play the reverse side of the tape upon reaching the end of the side.

EOTC 1. Electro-Optics Technology Center. A department at Tufts University that researches and provides hands-on education in the areas in which optics and electronics are related, founded in 1984. 2. Executive Office of Transportation and Construction. 3. European Organisation for Testing and Certification. See European Organisation for Conformity Assessment.

Ephemeris A tabular prediction of the position of a celestial body or orbiting satellite. This may be calculated from Earth information or supplemented with information transmitted by the satellite itself, as in Global Positioning System (GPS) data.

EPIC 1. See Electronic Privacy Information Center. 2. European Project on Information Infrastructure Coordination Group.

EPIISG See European Project On Information Infrastructure Starter Group.

EPLRS See Enhanced Position Location Reporting System.

epoxy A synthetic resin useful for protecting or bonding materials such as fiber optic cable components. Epoxy has excellent adhesion properties for many types of materials, good water resistance, chemical resistance, and optical characteristics.

Epoxy has a tendency to set over time, even when stored in a closed container, rendering it useless for anything other than a doorstop. For this reason, epoxy is often mixed on an as-needed basis in the amount required for a specific task. It may be air cured for a few hours or heated to cure in a few minutes.

It is important to handle epoxy carefully as it can bond fingers and other body parts and generally get in the way if spilled or sprayed. It is best to have a separate set of tools for opening, handling, and mixing the epoxy so the glue doesn't damage fragile components. If the epoxy mix is sufficiently viscid, it can be applied with a disposable syringe as long as the needle bore isn't too fine. Be careful to wear safety goggles and never point the syringe at anything that is not intended to be bonded.

One common use of epoxy is as a bonding agent to attach fiber filaments within the ferrule of a terminating or coupling connector.

EPP 1. See Enhanced Parallel Port. 2. European Polar Platform.

EPPA 1. Employee Polygraph Protection Act (1988). 2. See European Public Paging Association.

EPRML extended partial response/maximum likelihood. See partial response/maximum likelihood.

EPROM See Erasable Programmable Read Only Memory.

EPS See encapsulated PostScript.

EPSCS Enhanced Private Switched Communications Service. A commercial switch-renting service from AT&T serving businesses.

Equal Access A requirement resulting from the AT&T divestiture Modified Final Judgment (MFJ) in the 1980s, that holds that each Bell Operating Company must provide network services access to competitive companies equal in type and quality to that used by the Bell Operating Companies themselves. See Feature Groups.

Equal Charge Rule A stipulation of the 1980s AT&T divestiture Modified Final Judgment requiring Bell Operating Companies to charge rates that don't vary as the volume of traffic varies. See Modified Final Judgment.

Equatorial Constellation Communications See ECCO.

erasable programmable read-only memory EPROM. A read-only computer memory chip that can be electronically erased and reprogrammed. An EPROM is like an Etch-a-Sketch® in that it's a general purpose device for imprinting information that can be changed or retained at the discretion of the programmer. The EPROM can be programmed for specific needs and doesn't require a constant current to keep the memory refreshed. In other words, it retains the information if the system is turned off. EPROMs are handy for technology that is frequently upgraded, such as computer peripherals (e.g., modems). The information most likely to change can be put in an EPROM, and the chip can be swapped out for an updated EPROM while retaining the rest of the circuitry.

erasable storage Any storage medium on which the information can be readily removed after it has been recorded. Most erasable storage media are magnetic in nature, as magnetic particles are amenable to being rearranged to remove the data encoding, or to overwrite new information. Semipermanent magnetic storage media include hard disks, floppy diskettes, data cartridges, and audio/video tapes.

Many types of memory chips are also erasable and are commonly used as temporary storage on computers. The most commonly used memory chips will lose the information when the power is discontinued.

A few optical storage media are also erasable, as some have been developed such that the layer in which the encoded pits are stored can be subsequently altered, but this property is not common to most optical media. See erasable, programmable read-only memory, superparamagnetic.

erbium A rare earth element commonly used in the doping of fiber optic cables to manipulate their transmission propagation characteristics, especially to enable cables to carry signals for longer distances. Erbium-doped fiber amplifiers (EDFA) have not had sufficient bandwidth to support wavelength division multiplexing (WDM), but recent research with erbium-doped amplifiers in modified silica and tellurite glass hosts indicates that broader bandwidth support is possible. See yttrium.

erbium-doped fiber amplifier EDFA. Erbium-doped amplifiers have become an important means of reducing signal loss on long fiber optic transmissions. Traditionally the attenuation of a light signal in a fiber optic cable over distance was handled by converting the signal to electromagnetic energy, amplifying it, and converting it back to an optical signal. The development of EDFAs provided a means of amplification without conversion, and they amplify far more efficiently than analogous electronic amplifiers. These fiber optic amplifiers came into use in the late 1980s and are now used to extend the range of fiber optic transmissions. See doping, simulated Raman scattering.

erbium doping A technique of using erbium, a rare earth element, to impregnate another material in order to alter its transmission characteristics. Erbium doping is a technique used in the manufacture of fiber optic communications cables amplifiers to minimize signal loss over distance. See doping.

ERC European Radiocommunications Committee. See European Radiocommunications Office.

ergonomics The study and application of human engineering, i.e., the design of systems and products that adapt them to the needs and comforts of their users, rather than the other way around. Ergonomic applications require a knowledge of human anatomy, movement, and orientation, as well as human perception and preferences — psychological and sociological. This information is then incorporated into design and manufacturing with a result that, more often than not, is an economic and social compromise. Nevertheless, ergonomic designs are to be encouraged. A number of interesting ergonomic adaptations can be seen in the design of chairs, computer keyboards, and phone head and hand sets.

Ericsson Telecommunications A major equipment supplier and research and development organization serving Canadian and world communications markets, especially in mobile phone industries.

erlang A unit of measure of telephone traffic. It has been variously interpreted in the telephone industry as equal to a full traffic path, to a specified number of calling seconds, or to a ratio of full traffic to no traffic. The term is based on mathematical analysis of the characteristics of telephone transmissions by the Danish telephone engineer A.K. Erlang. Erlang was analyzing traffic flow and congestion in the Copenhagen Telephone Company beginning in 1908, which led to changes in the design of telephone switches.

Erlang's theories have had practical applications in phone system design for many decades, but now they have to be re-evaluated in light of changing characteristics of phone calls since the rise of the Internet. Two-hour voice calls are rare; two-minute voice calls are common. But when computer users log onto the Internet, two-hour connect times are common, as are four-hour connect times. The theories used to develop trunk use and capacity algorithms may have to be reapplied to the new types of usage patterns. See queuing theory.

Erlang, Agner Krarup (1878–1929) A Danish mathematician, educator, and telephone engineer who studied the mathematical characteristics of telephone transmissions in the early 1900s and described his findings in a number of publications, including *The Theory of Probabilities and Telephone Conversations* in 1909. He described how random calls follow a Poisson pattern of distribution. This observation not only led to some practical design changes in telephone switching systems, it also was the beginning of the study of queuing theory, an area of research that has many implications for current research and applications in data network traffic. See erlang.

ERM See EMC and Radio Spectrum Matters Technical Committee.

ERMA Electronic Recording Method/Machine, Accounting. A historic banking system, first demonstrated in 1955. By a year later, the system had been enhanced with solid-state components and released as ERMA Mark II. In 1959, General Electric began delivering the system, and one was installed in a Bank of America location in California, considered to be the world's first electronic banking system.

ERMES European/Enhanced Radio Messaging System. A European wireless mobile communications paging protocol specified by the European Telecommunications Standards Institute (ETSI) in 1986. ERMES operates at 169.6 to 169.8 MHz at 6.25 Mbps. See EMBARC, SkyTel.

ERO See European Radiocommunications Office.

ERP effective radiated power.

error control In computing, there are many schemes, philosophies, and protocols for safeguarding the integrity of data. Error control encompasses several aspects of data handling: error detection and error correction, if appropriate or possible, that may be part of more extensive data recovery. One of the most rudimentary types of error control is detect-and-drop-if-bad. In other words, if cyclic redundancy checking (CRC), noise sensing, or some other error detection mechanism detects a problem, drop the transmission. While this method sounds harsh, it actually was the predominant strategy for file transfer protocols for many years.

Error control is related to every aspect of computing, not just file transfers. It involves user interaction with applications programs, file loads and saves, data protection while files are open, and information protection for cached data. The most common implementations of error control, however, are in network transmissions and dialup data transfers through modems. The arsenal of error control mechanisms is growing, and, more and more, error control schemes are a mix of software and hardware functions. Error control protocols now sometimes include sophisticated check, compare, and evaluate algorithms, and some incorporate artificial intelligence concepts.

There are a number of error-correcting protocols now widely used in data modems, including MNP4, HST, and V.42 (which includes MNP4 and Link Access Procedure (LAP-M)). See checksum, cyclic redundancy checking, Microcom Networking Protocols (for a chart), XModem, YModem, ZModem.

error free seconds EFS. A unit of measure of the quality of a transmitted signal expressed as a percentage of bit errors over a specified period of time. EFS is defined in the ITU-T O Series Recommendations (0.151). See bit error rate.

ERS See European Remote Sensing Satellite.

ERTS-1 See Earth Resources Technology Satellite.

ESA 1. emergency stand-alone. 2. See European Space Agency.

ESCA See International Speech Communication Association.

ESF 1. See European Science Foundation. 2. See Extended SuperFrame.

ESI 1. electro-scientific instrument. 2. electronic share-trading system. 3. Enhanced Serial Interface. 4. End System Identifier. In ATM networks, an identifier that distinguishes multiple nodes at the same level, in case the lower level peer group is partitioned. 5. environmental sensing instrument.

ESMR See Enhanced Specialized Mobile Radio.

ESMTP See Extended Simple Mail Transport Protocol.

ESN 1. See electronic serial number. 2. electronic switched network. 3. See emergency services number.

ESnet See Energy Sciences Network.

ESO Imaging Survey An astronomical imaging project of the European Southern Observatory, an intergovernmental organization headquartered in Germany. The project is an ongoing public imaging survey in support of very large telescope (VLT) projects. Data gathered from the original and PILOT surveys are in the public domain. http://www.eso.org/science/eis/

ESOC See European Space Operations Centre.

ESP 1. See Encapsulating Security Payload. 2. See Enhanced Serial Port. 3. See Enhanced Service Provider.

ESPA 1. Educational Software Publishers Association. http://www.uk.digiserve.com/espa/ 2. European Selective Paging Manufacturers Association. The ESPA protocol is a widespread paging standard that enables third party systems to connect with paging devices.

ESPAN Enhanced Switch Port Analyzer. An external network diagnostic and analysis tool which captures information that has been copied to a switched interface.

ESRIN The information systems center and main Earth observation center for the European Space Agency (ESA), located in Frascati, Italy. ESRIN processes data and images from ESA's satellites and combines them with data and observations from international satellites. ESRIN also provides information to the public regarding ESA projects. See European Space Agency.

essential service A regulatory distinction that provides special access to some types of telecommunications equipment or services and that provides more relaxed regulations in some aspects, and more stringent regulations in others. Essential services are classified differently in different nations, but tend to include some types of medical personnel, fire-fighting services, transportation administration, emergency broadcast channels or stations, etc.

ESTEC See European Space Research and Technology Centre.

ESTO See European Science and Technology Observatory.

ETA See Electronics Technicians Association.

ETACS Extended TACS (total access communications systems). The wireless transmission technology used in the United Kingdom and northern Europe, derived from U.S. AMPS systems. It is widely used for mobile phone services.

ETB 1. See Educational Technology Branch. 2. electronic term book. 3. electronic technical brief. 4. electronic test bed. There are many laboratories and educational institutions operating electronic test beds for the development and testing of electronic instruments and devices. One example is the U.S. Department of Defense (DoD) Ballistic Missile Defense Organization/Jet Propulsion Laboratory ETB where various characteristics of electronic devices, including radiation performance, are studied. 5. end transmissions block. A terminating marker, indicating the end of a transmission block. 6. Engineers Toolbox. A commercial engineering package providing platform-independent, Web/Java-based engineering analysis and reference modules.

ETF 1. Emerging Markets Telecommunications Fund, Inc. A long-term capital appreciation fund established in 1992. 2. European Teleconferencing Federation. An industry trade association for video- and teleconferencing. See audiographics, videoconferencing.

ETFTP Enhanced Trivial File Transfer Protocol. See Enhanced Trivial FTP.

Ethernet An important, widely implemented local area network (LAN) and metropolitan area network (MAN) network transmissions standard developed in 1973 by Dr. Robert M. Metcalfe and David Boggs and patented in 1975. Tat Lam designed the first transceivers for Ethernet, and Ron Crane provided hardware expertise for the eventual IEEE 802.3 standards. Crane and Metcalfe founded 3Com Corporation in 1979.

The early Ethernet ran at approximately 3 Mbps. Much of the early work was done by the Xerox research lab (PARC), and further development was undertaken by a multivendor consortium. Ethernet was formally specified as a production-quality standard called the DEC-Intel-Xerox (DIX) or "Blue Book standard," transmitting at speeds up to 10 Mbps. It was subsequently adopted for standardization for a wide variety of media by the Institute of Electrical and Electronics Engineers and designated IEEE 802.3 CSMA/CD in 1985.

Each Ethernet interface card requires an Organizationally Unique Identifier (OUI) which is assigned as a three-octet number for the IEEE. The organization further subdivides this locally into unique six-octet numbers known as a Media Access Control (MAC) address or Ethernet address. The IEEE organization handles identifier allocation by online registration forms or by phone at the IEEE Registration Authority.

Current Ethernet protocols can run over thick and thin coaxial cable, multimode fiber, and unshielded twisted-pair. Physical standards for running Ethernet include 100Base-TX, 10Base-5, 10Base-T and others.

E

It's easier to understand these physical standard designations if the three component parts are analyzed as follows: a "10" indicates a signaling speed of 10 Mbps, while the "base" stands for *baseband*, and the suffix describes the maximum run of an unrepeated cable segment (in hundreds of meters), if it is a number, or refers to fiber (F) or twisted-pair (T).

Ethernet is now a worldwide networking standard, having been adopted by the International Organization for Standardization (ISO) as ISO/IEC ANSI/ IEEE Std. 802.3 in 1992.

Ethernet Address Resolution Protocol EARP. The addresses of hosts within a particular protocol may not be compatible with the corresponding Ethernet address. That is, the lengths or values may differ. EARP deals with an incompatibility by allowing dynamic distribution of the information needed to build tables to translate an address from the foreign protocol's address space into a 48-bit Ethernet address. This can also be generalized to non-10Mbps Ethernet systems such as packet radio networks. See RFC 826.

Ethernet Digital Subscriber Line EDSL. A Digital Subscriber Service that uses copper wires running between subscribers and the central office as a shared communications medium. Crosstalk is a limitation, but, like Ethernet and unlike earlier xDSL technologies, EDSL has some ability to adapt to traffic interference. See Digital Subscriber Line. Contrast with Asymmetrical Digital Subscriber Line.

EtherTalk An IEEE 802.3 standard Ethernet protocol implemented for local area networks (LANs) on Macintosh and G3 computers by Apple Computer. See AppleTalk.

ETIS See European Telecommunication Informatics Services.

ETNO See European Public Telecommunications Network Operations Association.

ETO See European Telecommunications Office.

ETRI See Electronics Telecommunications Research Institute.

ETA See Electronics Technicians Association, International.

ETS 1. See Electronic Technology Systems. 2. See European Technology Services. 3. European Telecommunication Standard. See European Telecommunications Standards Institute.

ETSA See European Telecommunication Services Association.

ETSAG European Telecommunication Standards Awareness Group Advisory Committee.

ETSI See European Telecommunication Standards Institute.

ETSI ATM Services Interoperability EASI. A former European Telecommunications Standards Institute (ETSI) project, in conjunction with EURESCOM P813, that produced specifications for interoperability for asynchronous transfer mode (ATM) networks and ATM network services. The project was finalized in 1999 and closed in May 2000 with the remaining work being transferred to the Telecommunications Management Network Technical

Committee (TC TMN) and the Services and Protocols for Advanced Networks Technical Committee (TC SPAN).

Eudora Light, Eudora Pro A widely distributed commercial personal computer electronic messaging applications program compatible with the Macintosh operating system (OS) and Windows. Eudora Pro is a pay version, and Eudora Light is freely distributed by Qualcomm Enterprises.

EUnet European Unix network. A cooperative commercial backbone network in Europe, established without public funding in the early 1980s, at a time when most computer networks were focused on academic communication and research and development. EUnet evolved from the European Unix Users Group (EUUG) network connecting The Netherlands, U.K., and parts of Scandinavia. In March 1995, EUnet International was formed; the same year EUnet installed full Internet connectivity.

EUnet expanded and upgraded over the next several years to service 14 European nations. It was acquired by Qwest in March 1998 and joined with KPN as KPNQwest a year later. A large proportion of EUnet's commercial customers were nonprofit institutions. EUnet predated the Ebone and has had an on-again/ off-again relationship with the Ebone. See Ebone.

EUROBIT In 1999, EUROBIT was succeeded by European Information and Communications Technology Association (EICTA).

European Academic and Research Network EARN. A European networking system which joined with the BITNET system in 1982, making BITNET an international network. In 1994, EARN was merged with RARE to form TERENA. See BITNET, TERENA.

European Advanced Networking Test Center EANTC. An institution within the Technical University Berlin that has offered consulting, testing, and educational services for modern network technologies since the late 1880s. EANTC provides standard conformance and interoperability testing and performance measurement. EANTC engages in educational programs and collaborative projects in Asynchronous Transfer Mode (ATM) technologies. It further provides ATM group testing activities to promote multivendor interoperability of ATM systems.

European Association of Communications Agencies EACA. A trade organization that publishes ethical guidelines for responsible advertising and has issued a position statement regarding virtual advertising in electronic media. It promotes and supports self-regulation and the rights and responsibilities of its members. http://www.eaca.be/

European Association of Consumer Electronics Manufacturers EACEM. A trade organization established to promote and support production and distribution of consumer electronics products and services in the European Union. http://www.eacem.be/

European Association of Directory and Database Publishers EADP. A trade organization representing members in more than three dozen nations worldwide, EADP was founded in 1966. EADP engages in benchmark studies, compiles statistics related to the industry, promotes quality and excellence, and publishes a directory of its members and their products. http://www.eadp.be/

European Association of Information Services EUSIDIC. An independent communications vehicle for information industry professionals, founded in 1970. Members include senior personnel and managers from publishing, Internet, computer, and information firms, as well as government personnel throughout Europe. http://www.eema.org/

European Association of Remote Sensing Laboratories EARSeL. A scientific network of about 300 academic and commercial remote-sensing institutes, founded in 1977 through the European Space Agency (ESA), the Council of Europe, and the Europe Commission. EARSel promotes and stimulates education, research, and cooperation in remote-sensing and Earth observations. http://www.earsel.org/

European Astronaut Centre EAC. The base for astronauts from European Space Agency projects, established in 1990. EAC provides training and medical support to ESA astronauts and astronauts from partnering space agencies, especially those involved with the International Space Station. See European Space Agency.

European Broadcasting Area EBA. The European portion of the broadcasting region within the European Broadcasting Union, including the Euroradio and Eurovision services. (In addition to the EBA, the EBU broadcasts to North America and the Asia-Pacific region.) See European Broadcasting Union.

European Broadcasting Union EBU. The world's largest professional association of national broadcasters, the EBU has headquarters in Geneva, Switzerland and represents public service broadcasters through its office in Brussels, Belgium. It was founded in 1950 and, in 1993, merged with the union of eastern European broadcasters (OIRT).

The EBU represents and negotiates broadcasting rights; provides a range of technical, legal, and operational services; operates the Eurovision and Euroradio networks; and promotes and coordinates coproductions on behalf of members in Europe, North Africa, the Middle East, and elsewhere. The EBU also collaborates with other prominent associations, including the North American Broadcasters' Association, the Asia Pacific Broadcasting Union, the Arab States Broadcasting Union, and others.

In addition to member support and programming coordination, the EBU is involved in the research and development of new broadcast media, providing input into formats, standards, and digital broadcasting technologies. The more recent multimedia activities include a Digital Strategy Group (DSG), the On-Line Services Group (OLS), and the Multimedia Forum. See Euroradio, Eurovision. http://www.ebu.ch/

European Cable Communications Association ECCA. A trade organization promoting the interests of the European cable industry, headquartered in Brussels, Belgium. The ECCA fosters cooperation among operators in the industry and promotes member interests in Europe and abroad. The ECCA further works to assure fair access to various communications infrastructures, provides aid in the management of copyright issues, and encourages interoperability standards to facilitate compatibility of the various cable broadcasting and data services technologies. http://www.ecca.be/

European Commission EC. A significant European policy initiatives body working within the European Union (EU). The EC works in partnership with European institutions and governments of the member states of the EU. http://europa.eu.int/comm/

European Committee for Telecommunications Regulatory Affairs ECTRA. This CEPT-affiliated organization studies and develops telecommunications regulatory frameworks and policies for Europe in cooperation with the European Radiocommunications Committee (ERC), the European Commission, and the European Free Trade Association (EFTA). ECTRA was founded in 1990.

European Community Telework Forum ECTF. An organization formed in 1992 to further and coordinate European developments in telework through computer networks and communications venues and to provide an open forum for discussion of related issues. It is funded through a nonprofit European Economic Interest Group. See ADVANCE Project, TelePrompt Project, telework.

European Computer Manufacturers Association ECMA. An international industry association promoting the standardization of information and communication systems; ECMA was founded in 1961. http://www.ecma.ch/

European Council of Telecommunications Users Association ECTUA.

European Digital Dealers Association EDDA. A vendor organization comprised of resellers, service providers, and consultants of Digital Equipment Corporation (DEC) products and related third-party products.

European Electronic Messaging Association EEMA. An independent, international, nonprofit forum for electronic business, founded in 1987. EEMA encompasses hardware, software, and governmental manufacturers and service providers. EEMA maintains associations with a number of electronic messaging associations in other regions, including Japan, Australasia, the United States, and Russia. http://www.eema.org/

European Free Trade Association EFTA. An international organization entered into by Iceland, Liechtenstein, Norway, and Switzerland, headquartered in Geneva, Switzerland. EFTA manages and monitors trade relationships among member states, on the basis of the EFTA Convention, that evolved from the Stockholm convention. It further maintains trade relationships with countries not affiliated with the European Union and negotiates its position within the European Union as three of the member states are EU members as well. http://www.efta.int/

European Incoherent Scatter Scientific Association EISCAT. An international research and education organization based in Scandinavia that operates three geophysical research radar systems, an ionospheric heater, and a cynasonde for conducting high latitude upper atmosphere research. Funding is provided by research councils of the EISCAT Associates in Scandinavia, Finland, Japan, France, the U.K. (Rutherford Appleton Laboratory), and Germany. The UHF amd VHF EISCAT transmitter is located near Tromsø, Norway. The more recently constructed Svalbard Radar incoherent scatter radar facility is on Spitsbergen in northern Norway, and there is receiving equipment in Sodankylä in northern Finland. The three EISCAT incoherent scatter radar systems operate at 931 MHz, 224 MHz, and 500 MHz. http://www.eiscat.uit.no/

European Information and Communications Technology Association EICTA. A trade association that succeeded ECTEL and EUROBIT, founded in November 1999. EICTA represents Europe's information, communications, and technology (ICT) professionals from its headquarters in Brussels, Belgium. ICT is considered the fastest growing industry sector in Europe. It promotes and encourages technology convergence and faster, better information and communications systems.

European Information Technology Association EITA. An information technology trade organization with its secretariat in the U.K. EITA is initially focused on studying IT industries in the Baltic and Slovak regions and encourages participation from the countries in those areas.

European Institute of Public Administration EIPA. An educational institute providing public management training for public officials of European Institutions and Member States of the European Union, established in Maastricht in 1981. EIPA's main centers are in Barcelona and Luxembourg.

European Interactive Media Federation EIMF. See European Multimedia Forum.

European ISDN, Euro-ISDN A version of Integrated Services Digital Network (ISDN) implemented for the European networking system. It differs in a number of respects from North American systems. While the signals are compatible (e.g., transatlantic ISDN calls), the equipment is not.
Euro-ISDN is an evolution of a variety of European ISDN systems, which up to now have not been fully intercompatible. Euro-ISDN refers to ISDN facilities based upon harmonized European Standards that have been introduced to all Member States of the EU. See ISDN, European ISDN User Forum.

European ISDN User Forum EIUF. See ISDN associations.

European Manufacturing Technology Users Group EMTUG. A member support organization that provides resources and assistance in information and communications technology in the manufacturing industries.

European Multimedia Forum EMF. Descended from the European Interactive Media Federation (EIMF), EMF is the primary European trade organization promoting the competitive environment of emerging digital media industries worldwide. EMF facilitates communication and contacts within the digital communication and promotes and stimulates the trade of multimedia tools and services. http://www.emf.be/

European Organisation for Conformity Assessment, Organisation Européenne pour l'evaluation de la Conformité EOTC. An independent and non-profit body established by the European Commission, the European Free Trade Association (EFTA), and the European Standards Bodies, founded in April 1990. EOTC is dedicated to the achievement of a common market within the European Union and the promotion of worldwide commerce through the elimination of technical barriers to trade. It seeks to facilitate product conformance and market acceptance of conforming products. http://www.eotc.be/

European Polar Platform EPP, PPF. A modular satellite platform project of the European Space Agency (ESA) that will carry eleven ENVISAT-1 atmosphere and Earth observation instruments. The platform was designed to fit on the Ariane 5 launcher and, once launched, to use the ESA Data Relay Satellite system for data transmissions.
ESA participated in weather satellite projects in the early 1970s and has continued its involvement in environmental monitoring since that time. The PPF project was underway by the mid-1980s, with a planned launch date of 1995. By the mid-1980s, Britain had expressed doubts about the project, the launch date was deferred, and the design was modified and scaled back to be similar to the SPOT-4 system. The appointment of the British Aerospace (BA) as a major contractor brought Britain back into the picture. Further scaling back of the project occurred in the early 1990s. Many satellite projects run behind deadlines, and the PPF was no exception. A new launch date of 1999 was established that was later rescheduled to June 2001 (and then to October 2001).
The PPF consists of a mission-specific payload module and a general service module. The payload module includes the ENVISAT instruments and payload support equipment. The service module includes the main satellite support and control functions. The PPF follows a sun-synchronous orbit at about 800 kilometers, passing over the Earth's polar regions. The major contractor for the project is Matra Marconi Space (formerly BA). See ENVISAT.

European Project on Information Infrastructure Co-ordination Group EPIC. A project of the European Telecommunication Standards Institute (ETSI) carrying on from the work begun by the SRC6 and the European Project on Information Infrastructure Starter Group (EPIISG), which was closed in May 1996. Having identified important aspects of the European information infrastructure (EII) that would benefit from standards work, the role of EPIC was to establish, coordinate, and monitor a number of active standards projects. This was an ambitious goal involving coordination and cooperation with other standards bodies, including CEN, ISO, IEC, etc. As such, the project was closed in August 1997, and the work transferred to the Information, Communication, and Telecommunications Standards Board (ICTSB).

European Project On Information Infrastructure Starter Group EPIISG. A project initiated in 1995 to continue the work begun by the 6th Strategic Review Committee (SRC6) to identify areas suitable for standardization in terms of a European information infrastructure (EII) as part of the global information infrastructure (GII). The final report identified approximately 30 areas for which standards work would be appropriate. The project was concluded in May 1996 when it was superseded by the European Project on Information Infrastructure Co-ordination Group (EPIC). See European Project on Information Infrastructure Co-ordination Group.

European Public Paging Association EPPA. A trade association representing the paging industry in Europe and abroad, including major telecommunications companies and professionals in more than two dozen countries, founded in January 1994. The EPPA provides publications and seminars, encourages the development of and adherence to standards, and represents its members before various government bodies. http://www.eppa.net/

European Public Telecommunications Network Operations Association ETNO. A primary trade association, established in May 1992. ETNO promotes constructive dialogue among members and others involved in the development of the European information society, in compliance with European law. http://www.etno.belbone.be/

European Radiocommunications Committee ERC. See European Radiocommunications Office.

European Radiocommunications Office ERO. The permanent office of the European Radiocommunications Committee of the Conference Européenne des Administration des Postes et des Télécommunications (CEPT), located in Copenhagen, Denmark. It supports the work of the Committee and handles the radio regulatory administrations of CEPT member nations.

ERO was established in May 1991 as a result of a Memorandum of Understanding, which was replaced by the CEPT *Convention for the Establishment of the European Radiocommunications Office*. In January 2001, it was merged with the European Telecommunications Office (ETO). http://www.ero.dk/

European Remote-Sensing Satellite ERS. Sun-synchronous, quasi-circular orbit European satellite systems orbiting at an average distance of about 785 kilometers from the Earth. It takes about 100 minutes for an ERS satellite to orbit and the complete cycle repeats every 35 days.

The first European remote-sensing satellite, ERS-1, was launched by the European Space Agency (ESA) in July 1991 from French Guyana; it operated until October 2000. ERS-2 was launched in April 1995.

ERS systems are suitable for studying geological formations, soil and snow pack mapping, forestry, ocean characteristics, coastal boundaries, agriculture, floods, oil spills, and more. As specific examples of how ERS data are used, ERS-2 carries the Global Ozone Monitoring Experiment (GOME) and JPL researchers use ERS-2 radar data to monitor and track sea-ice dynamics, especially in the Antarctic region. Although it is a European satellite, a number of commercial firms distribute ERS data to the North American market. The ENVISAT system was intended to replace ERS systems in 1999 with the launch finally occurring in March 2002. See ERS Sensing Instruments chart. See ENVISAT, European Space Agency.

European Science and Technology Observatory ESTO. The monitoring arm of the Institute for Prospective Technological Studies (PROMPT), ESTO is a network of about three dozen European organizations that inform the IPTS of technological trends, events, and breakthroughs with socioeconomic or environmental significance. The IPTS then channels the information to science and technology policymakers. http://esto.jrc.es/

European Science Foundation ESF. A scientific research organization that supports and promotes

European Remote-Sensing Satellite (ERS) Sensing Instruments		
Instrument	Abbreviation/Description	
active microwave instrument	AMI	A synthetic-aperture radar (SAR) and a wind scatterometer. SAR image and wave modes have spatial resolutions of 30 meters and 10 meters, respectively. The wind scatterometer has a resolution of 50 kilometers. Data is transmitted in the C-band frequencies (5.3 GHz).
radar altimeter	RA	Nadir-viewing pulse radar with two measurement modes and a resolution of 10 cm (vertical). Data is transmitted in the K-band frequencies (13.8 GHz).
microwave radiometer	MWR	Nadir-viewing, passive radiometer that works in conjunction with the radar altimeter to improve accuracy.
along-track scanning radiometer	ATSR	Four-channel (spectral bands of 1.6, 3.7, 10.8, & 12 μm), infrared radiometer with a spatial resolution of one kilometer.

cooperation and collaboration in high quality scientific research in two dozen European countries, founded in 1974. The ESF seeks to bring together resources and scientists from different countries to work towards common goals and encourages interdisciplinary projects. The ESF covers the full range of scientific enquiry, so not all are telecommunications related, but much of the research in the physical and engineering sciences and the environmental and Earth sciences related to telecommunications in one way or another. http://www.esf.org/

European Space Agency ESA. The research, development, and administrative organization for space exploration for Europe, established in 1975. With headquarters in Paris, France, ESA is descended in part from the European Space Research Organisation (ESRO, 1962) and is roughly equivalent to the U.S. National Aeronautics and Space Administration (NASA). ESA is a multinational agency that directly employs about 40,000 people. It works in partnership with a number of space agencies, notably NASA, and uses data from or contributes to a large number of space projects, including the European Remote-Sensing Satellites (ERS), the Hubble Space Telescope, the Ulysses spacecraft (built in Europe), the Solar and Heliospheric Observatory (SOHO - built in Europe), the pending Galileo satellite navigation system, and many more. See European Space Research and Technology Centre. See ENVISAT, ESRIN, European Remote-Sensing Satellite. http://www.esa.int/

European Space Operations Centre ESOC. A mission control and tracking facility for most of the European Space Agency (ESA) space projects, located in Darmstadt, Germany. ESOC provides technical guidance on ground links and orbits. See ESRIN, European Space Agency.

European Space Research and Technology Centre ESTEC. The largest establishment of the European Space Agency, ESTEC houses the space, microgravity, and Earth sciences departments and provides technical expertise to the European scientific and industrial communities. ESTEC has laboratories and technical facilities for spacecraft components testing. The ENVISAT systems was assembled at ESTEC. See European Space Agency.

European Speech Communication Association ESCA. See International Speech Communication Association.

European Technology Services ETS. A commercial firm providing services to facilitate the import of telecommunications equipment into Australia under regulations imposed by the Australian Telecommunication Act of 1997 and the Radiocomms Act of 1992. It also assists companies in complying with the CE mark for products bound for the European Union.

European Telecommunications Office ETO. Established by CEPT/ECTRA to provide expertise for ECTRA members and input into the European Union's developing telecommunications policies on licensing and numbering, ETO was founded in 1994. In January 2001, ETO functions were transferred to the European Radiocommunications Office (ERO).

See European Radiocommunications Office. http://www.eto.dk/

European Telecommunication Informatics Services ETIS. A not-for-profit professional organization of major European public telecommunications operators established in 1991. It supports telecom operators, suppliers, and content providers and encourages cooperation among them. ETIS consists of a Council, Management Board, Central Office, and individual Working Groups. http://www.etis.org/

European Telecommunication Services Association ETSA. A trade association supporting members in more than a dozen nations who supply, install, and maintain telecommunications equipment and services. http://www.etsa.org/

European Telecommunication Standard ETS.

European Telecommunications and Professional Electronics Industry ECTEL. One of two main European communications trade organizations representing telecommunications equipment vendors and distributors. EUROBIT and ECTEL were succeeded by the European Information and Communications Technology Association (EICTA) in 1999.

European Telecommunications Standards Institute ETSI. The standards body in Europe that corresponds to the American National Standards Institute (ANSI) in the United States. As a member-driven, nonprofit standards body, ETSI's role is to promote and support communication, cooperation, and integration of technologies in the European Union (EU), Europe's common market. It enlists the cooperation of a broad range of network administrators, service providers, manufacturers, researchers, and users from more than 50 countries in and outside of Europe. ETSI was founded in 1988 and is currently headquartered in Sophia Antipolis, France.

ETSI has a number of subgroups, including the Radio Equipment and Systems (RES) 10 group, which is responsible for high speed wireless data standards. This is in the process of being reorganized into another body.

While ETSI is primarily focused on voluntary standards, some may be adopted by the European Union in their EU member Directives or Regulations. Key ETSI Standards include mobile radio communications (land, sea, and air), service provider access, ATM networking, xDSL, public safety standards, and many more. ETSI cooperates with the International Telecommunication Union (ITU). See CEPT, International Telecommunication Union. http://www.etsi.org/

European Union EU. A highly significant effort of the countries of Europe to merge into a more open, flexible common market. The effort has been ongoing for many decades, beginning at the end of World War I, and is a significant administrative challenge, given the deep traditions, varying cultures, and different languages spoken in the European nations. In recent years, the plan is falling into place. Common passports, economic units, and other important building blocks are being established that will make the EU a major player in the economics, business development, scientific research projects, and cultural development of our emerging global society. In January 2001, Sweden became the host country for the Presidency of the Council of the EU. In February, a new Treaty, amending the Treaty on the European Union and the treaties establishing the member communities of the EU, was signed as the Treaty of Nice.

It has been the habit of many North Americans to disregard the changes occurring in Europe; in fact, many North Americans barely understand the EU, but it is important both socially and economically to keep abreast of the development of this important market. The EU may someday be significantly larger than the North American market, as the EU's population is already greater than that of the U.S. and it is conducting ongoing negotiations with countries in eastern and central Europe. In May 2001, a summit was held in Russia to discuss the progress and future of the EU.

In telecommunications, the existence of the EU facilitates the establishment of inter-nation communications standards and shared backbones. Telecom professionals can now take their expertise to other countries to aid in the progress and competitive viability of telecommunications and information industries. It further allows specialist countries to market their expertise and to take charge of areas in which they have longstanding traditions. France and Switzerland have long been known for their expertise in developing and manufacturing finely calibrated technical instruments and systems. Belgium has become the administrative headquarters for many telecommunications and information technology associations. Germany, Italy, and Spain have long traditions in the invention of new technologies. Scandinavia is a leader in atmospheric, ionospheric, and other off-world research with their satellite transceivers in the remote polar regions. The Netherlands and Greece have key geographic positions at opposite ends of the EU, linking Europe with North American travelers through Amsterdam and with the Middle East and the African continent through the southeastern region. The U.K. is a leader in aerospace technology and is involved in many multinational remote-sensing projects. This rich source of expertise and resources, when administered toward common goals, will make the EU a significant player in the new world economies opened up by telecommunications technologies.

European Workshop in Open Systems EWOS. Working under the auspices of CEN/CENELEC, EWOS was formed in 1987. The organization has created publications and hosted seminars to promote and support Open Systems development. In the mid-1990s, the EWOS Expert Group worked on producing a technical guide on electronic commerce. EWOS has cooperated with a number of Open Systems groups and standards organizations in the development of OSI in Europe. See Open Systems Interconnection. http://www.ewos.be/

europium A soft, ductile, relatively expensive rare earth metal discovered in the late 1800s and separated in fairly pure form by Demarcay in 1901. Europium is used in conjunction with yttrium oxide to create the red phosphors in color cathode-ray tubes (CRTs) and is used as a doping material for plastics used in lasers. See doping, erbium, gadolinium, yttrium.

Euroradio A major world broadcasting service of the European Broadcasting Union, Euroradio has headquarters in Geneva, Switzerland. The three primary programming arms of Euroradio are sports and news; music; and popular music, drama, and feature presentations. Music programming has primarily focused on classical music and jazz but is expanding into contemporary music as well. See European Broadcasting Union, Eurovision.

EUROTELDEV European (Regional) Telecommunication Development.

Eurovision Network A large, international, permanent broadcasting network serving the European and Asian regions, operated by the European Broadcasting Union (EBU). The network provides international news, sports, and cultural programming. Eurovision includes over 50 gateways in Europe that provide full connectivity to domestic broadcasting networks through a mix of ground-based and satellite broadcast systems. The European Service utilizes the EUTELSAT W3 Ku-band satellite (a widebeam digital satellite), while the Asian Service utilizes the AsiaSat C-band satellite. Transatlantic services are provided through the INTELSAT Ku-band system. The digital services offered through INTELSAT were moved from ETSI to MPEG-2 in December 1999, thus increasing channel capacity without increasing bandwidth. See European Broadcasting Union, EUTELSAT.

EUSIDIC See European Association of Information Services.

EUTELSAT *Eu*ropean *Tel*ecommunications *Sat*ellite organization. The largest satellite operator in Europe, founded in 1977 and formally established in 1985, with 11 satellites in orbit, and a further six under construction. The EUTELSAT orbital test satellite (OTS) project, designed as a direct broadcast satellite test system, preceded the EUTELSAT F*x* series Ku-band systems. The EUTELSAT digital system is MPEG-2/DVB-compliant providing both 24 Mbps and 8/12 Mbps services. The system is monitored by the Eurovision Control Centre.

EUTELSAT provides hundreds of television and radio stations to subscribers in more than 40 European member countries equipped with DTH or cable television reception services. See Eurovision Network. http://www.eutelsat.de/

EUV extreme ultraviolet.

eV electron volt. See volt.

event In computer processing, a signal or other indicator that a device or process requires attention, is relinquishing resources, or otherwise needs to communicate its activities to a central processor or other control unit.

event driven An event-driven hardware device, process, application, or network communication topology is one that proceeds when triggered by an external event, such as a token, trigger, tickler, interrupt, or alert. Many telecommunications devices, appliances,

E

and input/output peripherals are event-driven or event initiators.

eventual Byzantine agreement EBA. In EBA, a number of coordinated processors (with some specified upper value for faulty processors) agree on a state or value among those considered to be reliable. Thus, a state of mutual agreement is negotiated among nonfaulty processors in a fault-tolerant system, but the processing does not have to be simultaneous. See Byzantine agreement and Byzantine Generals problem for a history and fuller explanation.

EWOS See European Workshop in Open Systems, Open Systems Interconnection.

EWP electronic white pages. An electronic database of personal, and sometimes business, phone and address listings. There are many EWP lookup services on the Web.

exa- E. A prefix for an SI unit quantity of 10^{18}, or 1,000,000,000,000,000,000. It's a gargantuan quantity. See zeta-, atto-

ExCa Exchangeable Card Architecture. An open socket architecture extension to PCMCIA 2.0 for use on Intel x86-based computers, introduced by Intel in the early 1990s. The software specification provides standardized socket, card, and client services. ExCA allows interfacing of PCMCIA devices with computers, particularly mobile computers, which are more likely to have PCMCIA slots. See Personal Computer Memory Card International Association.

exception 1. error or unusual occurrence, such as an abnormal signal, data falling outside a certain specified range or a deviation from normal program execution. Common exception conditions in programming include stack overflow and divide-by-zero errors. In software development, exception handlers can be included in the code to detect and manage error conditions and resume program execution. 2. In ATM, a connectivity advertisement in a PNNI complex node representation that represents something other than the default setting of the node representation.

excess burst size See burst size, excess.

excess noise, current noise Undesirable noise that results from current passing through semiconductor components.

exchange A central location for making connections, directing traffic, and redirecting traffic. A public telephone switching office or regional system is often called a telephone exchange.

exchange access Defined in the Telecommunications Act of 1996, and published by the Federal Communications Commission (FCC), as

"... the offering of access to telephone exchange services or facilities for the purpose of the origination or termination of telephone toll services."

See Federal Communications Commission, Telecommunications Act of 1996.

Exchange Access SMDS XA-SMDS, Exchange Access Switched Multimegabit Data Service. A connectionless, cell-switched, security-enabled data transport service for extending network features through standard interconnections with interexchange carriers (IXC). XA-SMDS is similar in structure to

ATM and is designed so that migration to ATM may be possible as ATM becomes more widely implemented. Multiple node local area networks (LANs) and wide area networks (WANs) can be interconnected without installing a dedicated path, at speeds ranging from 1.17 to 34 Mbps. XA-SMDS is a public level service, with a universal addressing plan, so various XA-SMDS networks can intercommunicate as desired.

exchange carrier EC. A telecommunications provider operating under specified territorial and operating parameters designated within the industry.

Exchange Carriers Association ECA. An organization established to support the interests and accounting administrative concerns of long-distance telephone companies.

Exchange Carriers Standards Association ECSA. More familiarly known as the Alliance for Telecommunications Industry Solutions (ATIS) since 1993, the Washington, D.C.-based ECSA was established in 1983 to develop and promote standards related to the needs of various telecommunications carriers. The ECSA works in conjunction with a number of committees, including the Carrier Liaison Committee (CLC), Information Industry Liaison Committee (IILC), and Telecommunications Industry Forum (TCIF). See Alliance for Telecommunications Industry Solutions for more information.

exchange line The connection between a telephone subscriber and the local telephone switching exchange. See local loop.

Exchange Message Record EMR. An industry standard for the exchange of sample, study, and billable data messages among local exchange carriers (LECs).

Exchange Service ES. Basic subscriber phone service with a unique local telephone number and access to the public switched telecommunications network. Includes residence and business services and private branch trunk line services. Private lines and Special Access services are not considered to be Exchange Services.

excitation The application of an external stimulus to a system resulting in a reaction or response. The application of a charge, potential, or electromagnetic influence.

excitation voltage The minimum or sufficient voltage required for a circuit to be functional.

exciton An excited state in a crystal substance with the characteristic of moving and recombining holes and electrons. See p-n junction, quantum.

execution In a software process, the carrying out of preprogrammed, realtime, or heuristic steps in order for the program to run through its instructions or logical structure. It may or may not be an interactive process.

execution time A measure of the time in steps, minutes, or machine cycles that a process, or a particular computer instruction, takes to be carried out.

Executive Cellular Processor ECP. In wireless Mobile Switching Centers (MSCs), the capabilities for intelligent call handling, mobility management, and system control and configuration. In Lucent

Technologies systems, the ECP, with the Operations and Management Platforms (OMPs), comprise the Access Manager.

EXFO Electro-Optical Engineering, Inc. Expertise in Fiber Optics. A publicly traded leading designer and manufacturer of global fiber optic test, measurement, and automation solutions for telecommunications. The products are aimed at handheld and modular instruments telecommunications markets and high-performance optical instrument component lab users and systems vendors. In August 2001, EXFO announced an agreement to acquire Avantas Networks Corporation, a pre-revenue company developing data communications and telephony testing systems.

exosphere A region beyond the Earth's surface at the edge of the atmospheric "envelope" surrounding the planet. See ionosphere.

Expanded Interconnection Service EIS. A collocation arrangement, in which the switch services for a private branch are located within the premises of the local telephone carrier.

expansion slots Peripheral slots in an expansion bay or a computer intended for the placement of controllers, cards, and other device interfaces, usually comprised of printed circuit boards, which are used to extend a system. VESA, EISA, ISA, MCA, and PCI are various common standards for the electrical and transmissions protocols used with slot peripherals for personal computers.

expert system An expert system is a type of information-handling approach which grew out of artificial intelligence research. Various types of expert systems exist for information creation, storage, and manipulation. An expert system is one that involves the manipulation and creation of information in a way that is rule based and evaluative, rather than search and query. The traditional means of providing information to computer users is through a database, which usually involves storing and retrieving the data on a keyword basis, but an expert system can take in a richer mix of inputs, or nontraditional inputs, including natural language queries, visual queries, or other contextual input. An expert system also incorporates the combined knowledge of many experts in that it is not just a collection of facts but may further include data relationships, means of analyzing and evaluating the data, and other pertinent evaluative characteristics. Expert systems grew out of efforts to mimic the ease and naturalness of human communications through machine interfaces, in order to enhance the usefulness of computers.

Because expert systems often handle different types of data, different types of input, and process the information in different ways from other types of information repositories, they sometimes require different programming languages than those commonly used for commercial applications. Cobol, Fortran, C, and BASIC are used for many programs used in business and educational settings. However, because expert systems often require a different programming approach, good text parsing languages like Perl, and good information parsing and rule-based languages like LISP and Prolog may be used.

Of the various types of products that have evolved from artificial intelligence research, expert systems are some of the most commercially successful results.

explicit congestion notification ECN. An IETF IP standard proposed in the late 1990s for detecting and managing end-to-end network transmissions and congestion. A *congestion experienced* (CE) bit in the header serves as a data congestion indicator in contrast to evaluating dropped packets from an overflowing router queue. It could be used to improve the delivery of delay-sensitive applications, such as broadcasts on the Web, or for improving security and intrusion detection. One might expect a trade-off in transmission efficiency from having to encode and detect the ECN; however, in testing it has been found that bulk and transactional data transmissions may be more efficient when packets are marked rather than dropped. See random early detection, RFC 2481, RFC 2884.

explicit forward congestion indicator EFCI. In ATM networking, a traffic flow control congestion, or impending congestion, indicator contained in the ATM cell header. The congestion signal is sent to the end destination to adjust accordingly. See cell rate, leaky bucket.

Explicit Rate ER. A network congestion feedback mode provided in available bit rate (ABR) service. Network rates that can be received are indicated within Resource Management cells. See cell rate.

Explorer See Microsoft Explorer.

Explorer I The first successful U.S. satellite launched on January 31, 1958. Its mission was scientific, and it included instruments to measure radiation in space. At first, it was thought that the instruments might be defective, as the readings were much higher than expected, but the measurements were later verified.

Explorer 8 The first NASA satellite launched by the United States. The Explorer 8 was launched on November 3, 1960, to study the ionosphere.

export To save information in a format that is not the native format of the application doing the saving. For example, a word processed document may be saved in ASCII to facilitate transfer over a 7-bit network. This procedure is often done to create a version of a file which is compatible with other applications or transport mechanisms. Exporting is usually done through a conversion filter, and there may also be filters for importing.

exposure Contact with radiant energy, bacterial or viral toxins, or chemicals. Sun exposure can cause fading, burning, melting, or other chemical reactions. Exposure to radiation from X-rays or laser light can can cause burns, deep cellular damage, chemical changes, or death to biological organisms at high doses. Exposure is a concern in industrial environments for both equipment and humans. It is also a consideration in medical environments, where exposure to viruses, bacteria, X-radiation, chemicals, and other contaminants or hazards may cause harm. See CDRH classification.

express circuit An interurban phone carrier circuit connected without multiplexing equipment.

eXpress Transfer Protocol XTP. A lightweight network protocol originally developed by Protocol Engines, Inc., in the late 1980s. In its early development it was sometimes also called Xpress Transfer Protocol. It is a reliable, realtime, *transfer* layer (combined network and protocol layer) protocol. It was designed to be implemented as a VLSI chip set. XTP is designed for parallel processing and the various functions such as address translation and flow/rate/error control can be executed in parallel. XTP utilizes control packet and information packet frame formats. In multicast mode, one-to-many transmissions can be supported. Protocol Engines, as a company, was no longer able to continue development after the early 1990s, but interest in XTP as a protocol continued from outside the company.

The protocol is used in the European RACE and DeTeBekom projects. The XTP collaborators are the University of Dresden, The University of Salzburg, and the University of Ottawa. A number of extensions were added to the XTP 3.6 standard to support Quality of Service (QoS). This version is known as XTPX (XTP eXtended). It might be best to abbreviate it as eXTP rather than XTP to distinguish it from the commercial multicasting protocol known as Xpress Transport Protocol. The author was unable to find clarification as to whether the Xpress Transport Protocols had its origins in the eXpress Transfer Protocol or whether the protocols were developed independently.

extended ASCII A colloquial designation for a variety of noncompatible 8-bit character code designations in which the first 128 characters conform to the ASCII standard, but the subsequent 128 characters (which mostly include symbols and accented letters) are variously assigned by different developers.

Extended-definition Television See Enhanced-definition TV.

Extended Digital Subscriber Line EDSL. A version of Digital Subscriber Line (DSL) services that supports 23 B channels and one 64-Kbps D channel transmitted over a single line. See Digital Subscriber Line, Primary Rate Interface.

extended graphics adapter EGA. A color graphics standard introduced by International Business Machines (IBM) in 1984, considered the successor to color graphics adapter (CGA). EGA was widely implemented by third party developers on Intel-based personal computers. EGA could display up to 640×350 in 16 colors. Actually, to say "16 colors" is stretching it a bit because, in fact, there were eight colors, plus eight half-intensity versions of those same colors, rather than 16 colors selected for their usefulness to a limited palette. Not long after, IBM introduced PGA, which had slightly better vertical resolution than EGA (640×400) but was otherwise not a significant evolution. See color graphics adapter.

extended graphics array XGA. A 1024×768-color graphics format used in liquid crystal display (LCD) data projectors.

Extended Industry Standard Architecture EISA. A 188-pin bus interface specification to succeed Industry Standard Architecture (ISA), which in turn succeeded the IBM PC/AT bus specifications. EISA supports 32-bit memory addressing, and 16- or 32-bit data transfers. EISA was designed to support 32-bit Intel 80386 and 80486 processors. The specification works with various system resources, including input/output ports, memory, and DMA channels.

On EISA boards, configuration is done with EISA Configuration Utility (ECU) software, rather than through hardware, using a CFG file supplied with the board. EISA boards, while faster, are somewhat physically compatible with legacy boards, preserving the old AT pin specifications on the upper 98 pins. The rest are used for the EISA bus signals. The slot into which an EISA card is inserted is assigned a unique address so that the system can recognize and initialize the interface.

EISA is widely supported by many manufacturers but is being gradually superseded by newer formats.

extended play A designation for a technology that plays beyond that generally expected to be the maximum limit. Historic phonograph cylinders played for two or three minutes, but some companies found a way to make them play for four minutes on standard equipment, thus creating extended play albums.

Extended SuperFrame ESF. A frame format for 1.544 Mbps communications (2.048 in Europe with 30 channels) evolved from DS-1 in 1962 and SuperFrame in 1969, widely used in T1 systems. ESF provides improved error correction and can be serviced without taking down the entire system. Twenty-four frames are combined to create one Extended SuperFrame. Six frames are used for frame synchronization, six for error tracking, and twelve for Facility Data Link (FDL). Signaling is accomplished through robbed bits in frames 6, 12, 18, and 24, except in transparent mode, in which the 24th channel is used in order to provide Clear Channel Signaling (CCS). Facility Link Data (FDL) is used to transmit to telephone monitoring stations.

extended time division multiple access E-TDMA. A type of digital transmission scheme favored by cellular providers over older analog-based systems. See time division multiple access.

Extensible Markup Language XML. XML is a markup meta language that allows more flexibility and complexity of presentation than HyperText Markup Language (HTML) and is not limited to Web publishing. Like HTML, it is based upon the Standard Generalized Markup Language (SGML - ISO 8879). Some have promoted it as the successor to HTML, but the industry has not yet formed a consensus on this possibility. XML has been recommended for use with the International Development Markup Language and is integral to the WebDAV Distributed Authoring Protocol.

XML MIME entities are of four types: document entities, external DTD subsets, external parsed entities, and external parameter entities.

In Jan. 2001, Murata et al. submitted a Standards

Track RFC to standardize five new XML media types to facilitate the exchange of XML network entities. XML MIME entities contain information to be parsed and processed by the receiving XML system and may include system-level commands. The proposed new media types were intended to overcome problems inherent in trying to adapt SGML text/sgml and application/sgml media types for use with XML. The new media types follow the conventions of the IETF media types tree, for consistency, and include

Media Type	Notes
text/xml	Preferred over application/xml for unprocessed, readable documents (as with plaintext). Use of the parameter is optional but recommended.
application/xml	Use of the parameter is optional, but recommended, to distinguish between the recommended utf-8 or utf-16 character sets; otherwise the default is us-ascii.
text/xml-external-parsed-entity	
application/xml-external-parsed-entity	
application-xml-dtd	

See IDML Initiative. HTML, SGML, World Wide Web Consortium. See RFC 1874 for SGML media types. See RFC 3023.

extension, file name A suffix, often preceded by a period, as a subsection delimiter on systems that require it, and as a visual locator on systems that don't. It helps to indicate the file format, such as *.txt, .bmp, .tiff, .ilbm, .frame, .wrd,* etc. File extensions need not be restricted to three characters except on some systems with older types of file structures. In the mid-1980s most types of computer platforms (Atari, Macintosh, Amiga, Sun, Apollo, NeXT, SGI, etc.) did away with the mandatory period and three-character limitation, as did Intel-based machines running OS/2 and Microsoft Windows NT. Surprisingly, many consumer Intel-based desktop computers running Microsoft Windows retained the limitation until 1996 and even later on some of the machines running legacy software.

extension, phone An extra phone line that uses the same phone number as the originally installed phone with which it is associated.

Exterior Gateway Protocol An Internet protocol developed in the early 1980s for the exchange of routing information between autonomous systems. See RFC 827.

External Data Representation XDR. A standard representation for platform-independent data structures for remote procedure call systems, developed by Sun Microsystems. See RFC 1014.

External Environment In the X/Open Architectural Framework Technical Reference Mode, the External Environment is one of five basic elements. The EE comprises all the external entities with which the Application Platform (AP) exchanges data. The EE links to the AP through the External Environment Interface (EEI). Printers, scanners, other computers in a network, public switched telephone networks (PSTNs), human operators, etc. are all common elements of the EE. See External Environment Interface.

external environment interface EEI. Generically, a transmissions linking device between a local device (e.g., personal computer) and outside devices (keyboards, mice, printers, scanners, public phone services, surveillance cameras, packet radio transmissions, etc.) Serial, parallel, SCSI, IDE, USB, Apple-Talk Data Bus (ADB), Ethernet, and Firewire peripheral cards are examples of some common external environment interfaces on commercial computing systems.

External Environment Interface EEI. In the X/Open Architectural Framework Technical Reference Model, the External Environment Interface is one of five basic elements and one of two basic interface types. It provides the data link between the application platform and the external environment (EE). The EE is comprised of the various entities and systems with which the application platform exchanges data. Thus, the EEI may link to various devices such as printers, modems, scanners, keyboards, mice, monitors, etc. Open systems EEI standards facilitate interoperability among platforms interconnected through the EEI. See Application Program Interface.

external memory Any memory outside the direct access memory peripherals or chips in a system. Thus, information stored on punch cards or paper tape would be considered external memory, as would removable cartridges or tapes. Chip memory and internal hard drives would be considered internal memory.

extinction potential The lowest voltage level at which plat current in a plat will flow in a gas-filled electron tube.

EXTN Extension. An industry abbreviation designating the last four digits of a phone number. A 10-digit number is expressed with symbolic characters as: NPA-NXX-EXTN.

extraneous emission Any emission in addition to, or external to, the desired emission. Thus, emissions outside the case of a computer system or outside the sheath of an insulated wire are considered extraneous. Since these emissions can interfere with radio transmissions (radio broadcasts, intercoms, cordless phones, etc.), they are strictly regulated by the Federal Communications Commission (FCC), and electronic components must conform to stated emission requirements.

extranet A larger network, based upon Internet-working technology, for connecting local area networks (intranets) and other authorized users within a virtual 'closed loop' system. The extranet is seen as having the potential to significantly support and change electronic commerce and to take some of the traffic load off the Internet. Those promoting the concept suggest that it will enable customers to connect

E

more directly to vendors than is typical in the current Web-based system and that security will be more easily supported on the system. Extranets will potentially be used for monitoring transactions, accessing databases, requesting product information, etc. For example, bookstores might be connected directly to book distributors, to expedite orders, shipments, and account handling.

Marshall Industries, Federal Express, and Dell Computer are some of the early adopters of the extranet concept. Extranets are also seen as a way to give priority services to elite customers. For example, Charles Schwab & Company has developed SchwabLink Web, a redesigned online trading and research service available only to the company's top 5,000 investment manager customers. The system is Java-based, conforming to CORBA and InterORB Protocol standards.

The original use of the term has been attributed to Bob Metcalfe in 1996, though a number of electronic commerce work groups may have begun using the term around the same time. Steven Telleen was describing intranets in 1994, with a definition that is now more commonly expressed with the term extranet. Telleen described intranets as an "infrastructure based on Internet standards and technologies that supports sharing of content within a limited and well-defined group."

The concept of an extranet is not new, even if current implementations represent a significant evolution. In the earliest days of telegraph and telephone technology, it was not unusual for related businesses to have direct connections to one another without going through the local switchboard exchange, and later incarnations of this system used leased lines and sometimes hot lines for direct connections between building complexes or cooperating businesses.

extremely high frequency EHF. The frequency spectrum designated as 30 to 300 GHz, typically used for satellite communications. These very short wavelengths can be apprehended with small antenna assemblies. See band allocations for a chart.

extremely low frequency ELF. The frequency spectrum designated as 30 to 300 Hz. Waves in this range are extremely long and not of much practical use for communications with our present technologies (the same was formerly said of microwave frequencies, and they are now widely used). See band allocations for a chart.

extrinsic semiconductor A type of semiconductor that includes impurities that contribute to its electromagnetic properties, usually to enhance them. See doping.

extrusion 1. Forming by forcing through an opening, which may contribute to the shape of the extruded material. Extrusion is often accompanied by a heating or cooling process in order for the extruded material to retain the desired shape. 2. A means to produce or apply insulating materials to wire or cable by forcing plastic or other materials through an opening.

eye phone 1. A project by a Norwegian group, Media Lunde and Tollefsen, Ltd. (MediaLT), that designs products for the visually impaired. MediaLT plan to use video telephony as a communication channel to a sighted eye. 2. A body-worn helmet or laser-to-retina transmission system associated with electronics for communications. The user views images on the inside surface of the eyepiece, on a tiny screen next to the eye, or through a direct projection on the retina. A number of companies have produced workable systems of different types. This technology is of special interest to the virtual reality (VR) community, and research is fueled in part by demand from the games-playing community, with spinoffs for business markets and handicapped individuals. Potentially an eye phone can be integrated with a mobile communications link to the Internet (stock market day traders would probably love this kind of system). 3. A visual interface on a computer system equipped with an Internet phone that enables calls to be clicked and controlled through a computer mouse/pen-type interface. This is a less common use of the phrase and will probably fade away as Internet phone technology becomes more familiar and common, but it may continue to be applicable to phone interfaces that can be controlled by eye movements (e.g., by someone with quadriplegia).

eyelet A small, washer-like flat cylinder or short tube for threading or supporting various wires, cables, or other narrow parts.

EYP See electronic yellow pages.

f 1. *abbrev.* farad. See farad. 2. *symb.* femto-. See femto-. 3. *symb.* focal length (usually italicized). See focal length. 4. *symb.* frequency. See frequency.

F 1. *abbrev.* Fahrenheit. See Fahrenheit. 2. *abbrev.* fiber. 3. *symb.* filament. 4. *symb.* off with N as the corresponding symbol for on. 4. *symb.* 15 in the hexadecimal number system. The symbols used are 0 1 2 3 4 5 6 7 8 9 A B C D E F. Thus, "A" in hexadecimal represents "10" in the familiar decimal system, and "F3" represents "18" in decimal. Hexadecimal numerals are sometimes preceded with "X" or "0X" to indicate that the subsequent digits are represented in the hexadecimal system. For example, "15" in decimal may be represented as "0X0F." See the ASCII chart in the Appendix for a list of decimal, hexadecimal, and octal equivalents up to 127 decimal.

F Block Federal Communications Commission (FCC) designation for a Personal Communications Services (PCS) license granted to a telephone company serving a Major Trading Area (MTA). The license grants permission to operate at certain FCC-specified frequencies. See A Block for a chart of frequencies. See C Block for a history and a more detailed explanation.

F connector A small coupling connector used at the end of coaxial cable that is common in video editing, broadcast components, and local area network (LAN) cables. The F connector is recognizable by its center pin or plug opening for the center pin that is commonly seen on consumer television sets and on the cables run to the house by local television cable providers. 75-ohm F connector cables are commonly used for TV, VCR, satellite, and radio frequency (RF) device connections. See connector for a diagram.

F-ES The designation for a fixed, i.e., nonmobile, end system in a digital cellular network. The F-ES can both send and receive data and typically receives data from a mobile end system (M-ES).

F region A region of the Earth's ionosphere in which F1 and F2 regions tend to form. The F1 region is active in daytime. The F2 region is commonly used for the propagation of radio waves, due to its high ionization levels. See ionospheric subregions for a chart.

F link In Signaling System 7 (SS7), a fully associated transmission link directly connecting two signaling endpoints. The F link connects the host directly to a Service Switching Point (SSP) or a Service Control Point (SCP) without passing through intermediate Signal Transfer Points (STPs). For security reasons, F links are generally used for local applications rather than for links between networks.

F port, fabric port. On a Fibre Channel network, a fabric-attached loop or node that connects point-to-point to an N port. In commercial products, the F port may be self-discovering. See FL port.

F Series Recommendations A series of ITU-T recommendations that provides guidelines for nontelephone telecommunications services. These guidelines are available as publications from the ITU-T for purchase. Since ITU-T specifications and recommendations are widely followed by vendors in the telecommunications industry, those wanting to maximize interoperability with other systems need to be aware of the information disseminated by the ITU-T. A full list of general categories is listed in Appendix C, and specific series topics are listed under individual entries in this dictionary, e.g., A Series Recommendations. See ITU-T F Series Recommendations chart.

Ionospheric Regions

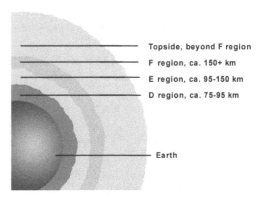

- Topside, beyond F region
- F region, ca. 150+ km
- E region, ca. 95-150 km
- D region, ca. 75-95 km
- Earth

An illustration of the general ionospheric regions enveloping Earth. These regions vary with temperature and distance. The F2 region is particularly important for reflecting radio waves over long distances.

F-number A numeric expression of the luminance associated with an aperture. Smaller F numbers are typically associated with higher luminance. Larger apertures are generally associated with higher F numbers up to the maximum luminance available. For lenses, the F number is the ratio of the focal length to the diameter of the aperture. On camera lenses, the F number associated with the maximum aperture is often inscribed on the lens casing. See candela, *f*-stop, luminance.

***f*-stop** An expression of the open region of an aperture that admits light in a fixed or adjustable lens assembly as in a camera or other imaging device. The scale is commonly used with optical lenses to indicate the amount of light entering a sensing or imaging device (based upon the diameter of the aperture).

Lower numbered *f*-stops indicate larger openings, hence a wider diameter and more room for light to enter. Typical 35mm camera *f*-stop settings range from 4 to 22.

There is a relationship between *f*-stop openings and the depth of field of the image in most lens assemblies (e.g., cameras). A wider aperture (lower *f*-stop) images over a broader area of the lens and thus is subject to greater curvature. A smaller aperture typically focuses through the central part of the lens, thus having less curvature over the extents of the image. The *f*-stop is often balanced with the shutter speed to further control the imaging factors. Some imaging devices (e.g., consumer cameras) can balance *f*-stops and shutter speeds automatically. See F-number, focal length.

ITU-T F Series Recommendations		
CATEGORY/Recommend.	**Description**	
TELEGRAPH SERVICE		
F.1–F.19	Operating methods for the international public telegram service	
F.20–F.29	The gentex network	
F.30–F.39	Message switching	
F.40–F.58	The international telemessage service	
F.59–F.89	The international telex service	
F.90–F.99	Statistics and publications on international telegraph services	
F.100–F.104	Scheduled and leased communication services	
F.105–F.109	Phototelegraph service	
MOBILE SERVICE		
F.110–F.159	Mobile services and multidestination satellite services	
TELEMATIC SERVICES		
F.160–F.199	Public facsimile service	
F.200–F.299	Teletex service	
F.300–F.349	Videotex service	
F.350–F.399	General provisions for telematic services	
MESSAGE HANDLING SERVICES	F.400–F.499	
DIRECTORY SERVICES	F.500–F.549	
DOCUMENT COMMUNICATION		
F.550–F.579	Document communication	
F.580–F.599	Programming communication interfaces	
DATA TRANSMISSION SERVICES	F.600–F.699	
AUDIOVISUAL SERVICES	F.700–F.799	
ISDN SERVICES	F.800–F.849	
UNIVERSAL PERSONAL TELECOMMUNICATION	F.850–F.899	
HUMAN FACTORS	F.900–F.999	

fab, fab plant A fabrication facility, a phrase often applied to a plant that produces computer chips or optical components.

fabric A generic term for describing the interlinked/ interwoven architecture or nature of a network's physical and logical interdependencies. A Fibre Channel fabric is a system with interconnecting Nx_ports capable of routing frames using only D-ID data in FC-2 signaling protocol fame headers.

The term is especially applicable to systems in which the speed of transmission across a network link, such as Fibre Channel, is near to, or exceeds, the speed of the processor in the machines that are being linked. Thus, the network as a whole begins to take on some of the characteristics of a machine organism, as opposed to discrete machines simply sharing information over much slower links. With intelligent routing algorithms, it becomes impossible in a larger network to fully know or predict the routes that data will take when being transmitted across the communications links. Taken as a whole, fabric-related networking is an interesting evolutionary development in digital technology and will change not just the speed of computing, but its very nature.

fabric/fibre loop port See FL port.

fabric port See F port.

Fabry-Perot interferometer A detection instrument with an optical cavity composed of highly reflective surfaces that can be variably separated. When illuminated with highly coherent light from a laser, the reflectors will bounce the light back and forth. If the reflecting waves resonate in phase, amplification of the wave, or *resonance*, occurs and the wave assumes enough power to "escape" through the reflective mirrors and stimulate a phototube that converts the light energy to electricity in order to measure and display its characteristics. The resonant process is sometimes called *constructive interference* due to the amplifying effect of the interaction between the waves.

To work effectively, the distance between the reflective surfaces is set in a mathematical relationship to the wavelength of the laser input (hence the variable separation of the reflective surfaces). An integral number of half wavelengths is related to the distance between the reflective surfaces. See Fabry-Perot laser.

Fabry-Perot laser F-P laser. A reasonable-cost semiconductor laser component with a constant index of refraction that lases simultaneously at different wavelengths; in other words, it emits a wider frequency spectrum than a distributed-feedback laser. Fabry-Perot laser diode amplifiers can be used to electro-optically modulate a signal and can be monolithically integrated with quantum well transistors. See distributed-feedback laser, Fabry-Perot interferometer.

face A geometric surface oriented toward the viewer, another component, the host system in general, or a radiant energy source. Thus, the endface of a fiber optic filament is the end where light enters and/or exits the fiber. The face of an imaging surface is the surface that is oriented toward the image and, generally, the source of illlumination (which may be reflected). See endface, faceplate.

face model A graphically modeled image of facial features, particularly eyes, lips, nose, and the general contours of the face. These images can be used in conjunction with computer applications for videophone, graphical answering services, simulations, police identification, artistic works, educational applications, and computer animations. Just as we now have synthesized speech for responding to user inquiries, we may someday have computer generated facial images responding to videophones and other interactive electronic devices, in effect, electronic receptionists. High speed data transmission technologies, such as fiber optic networks, increase the supply of and demand for anthropomorphic imaging technologies.

face time Time spent in a face-to-face encounter, as in social or business interchanges. A few years ago, one would never have thought to include this in a dictionary, but with so many human exchanges now being carried out remotely, by email, videoconferencing, etc., the distinction is becoming more significant. In fact, in the future, you might not know what a person looks like, even if you communicate with him or her audio/visually through a fast link, since people might design virtual environments and avatars to take their place in such encounters, projecting a personal image rather than an image of what they look like in real life and avoiding face time altogether.

faceplate At its most basic level, a protective plate, usually of plastic or metal, that fits over the front surface of a console or other device, which may have openings to accommodate various knobs or dials or sensors. The faceplate may be engraved, painted, or otherwise labelled to indicate settings. The term is also used to describe protective plates that have built-in connectors and sometimes some basic, compact electronics associated with those connections. Faceplates for supporting connections to wire or fiber-based data, voice, and video couplers may need to meet NEMA standards for electrical boxes.

In cable networks, specialized faceplates can be used to upgrade passive taps to addressable taps to facilitate connection and disconnection of subscriber links. Single- or multiple-gang faceplates canbe used in conjunction with fiber optic adapters (e.g., ST adapters).

In fiber fabrications testing, specialized microscopes may be equipped with a faceplate for inserting and steadying common fiber optic connectorsto facilitate inspection of the ends. See bezel; faceplate, fiber optic.

faceplate, fiber optic A planar surface comprised of an array of aligned fiber optic filaments that serves as an electromagnetic energy-guiding surface. Unlike traditional electrical faceplates that are mainly used as a protective surface, a fiber optic faceplate is a sophisticated component assembly that may replace or complement lens components for transmitting a signal from an input to an output surface. It may intensify the signal for imaging applications and may be customized for various thermal expansion or radiation-resistant tolerances. It can guide various electromagnetic phenomena, including X-ray emissions and visible light for use in X-ray crystallography cameras, dental X-ray technologies, and biometric sensing systems, when coupled with CCD and CMOS components. In a molecular microscope, a fiber optic faceplate can be used on an interfaced camera to compensate for interference from the monochromatic laser light source.

For some applications, a fiber optic faceplate may be used with or in place of a traditional energy-emitting component, such as a scintillator, providing higher resolution and lower scattering. It is capable of imaging areas (up to about a square foot) without the curvature distortion characteristic of lenses. The imaging area on a CCD array coupled with a fiber optic faceplate may be divided into quadrants to facilitate independent processing (e.g., timing) of sections of the imaging area.

When combined with electrical components, a fiber optic faceplate may be vulnerable to heat from nearby electrical discharges (e.g., sparks). An accumulation of tiny indentations in the surface would gradually compromise the utility of the plate. Some assemblies include components for monitoring electrical spikes or conditioning the power source to avoid this problem.

See faceplate, fiber optic taper, phototube, scintillator.

Fiber Optic Faceplate Bonded to Components

A basic fiber optic faceplate consists of tightly packed, coherently aligned optical fibers (left) bonded to electronic sensing technology (e.g., CCD imaging components). Attaching the faceplate to the image sensing components must be done with extreme precision to prevent signal loss or distortion – chemical gluing and oil coupling are two common bonding methods.

---

facet In transmission technologies, a planar geometric surface that reflects or refracts radiant energy. The term usually connotes more than one facet as in a multifaceted grating, lens, or resonating cavity. Thus, the facets of a gem may prismatically refract light,

the facets of a blazed grating may selectively reflect certain wavelengths, and the facets of a parabolic antenna may reflect radio waves toward a centtral feedhorn.

facilities An installation designed for a particular purpose. The term typically encompasses related buildings, equipment, and operations, though some people use it loosely to include the personnel as well. Facilities of significance in telecommunications include broadcast sending and receiving stations, computer terminal rooms, Internet link facilities (which may include tens of thousands of modems or terminal devices), and other wiring and access installations.

facilities-based carrier FBC. Phone carriers that use their own facilities and switching equipment to provide phone service, often long-distance service. Contrast this with those who lease or resell services from established carriers, although even facilities-based carriers enlist other carriers as needed.

facility-associated signaling FAS. In ISDN networks, a type of signaling in which the D channel is at the same primary rate interface (PRI) as an associated set of B channels. In contrast, in nonfacility-associated signaling (NFAS) the B channels are separate from the D channel PRI. Delivery of FAS is through a link access protocol. FAS over Internet Protocol is essentially the same as FAS over ISDN, but uses Internet Protocol (IP) as the transport mechanism. See channel-associated signaling.

facom A distance radio navigation system or measuring system. Facom is a means of analyzing local signals and received signals using the low frequency band for distances up to several thousand miles to determine distances. See band allocations.

FACOM fully automatic computer. A series of commercial large-scale computing systems first introduced in the mid-1970s by Fujitsu Limited, at about the same time the first microcomputers were being developed for commercial distribution. The FACOM M-190 is significant for being based upon large-scale integration (LSI) semiconductor circuitry that was new at the time. In 1981, Fujitsu released a large-scale general purpose system, FACOM M-380/382, the same year its first fiber optic communication systems were delivered.

facsimile device, fax device A device for sending an image transmission through wireless or wireline phone transmissions. The word 'facsimile' implies an exact copy, though on some of the cheaper fax machines, that's wishful thinking.

The two most common ways to transmit and receive facsimiles (faxes) are through dedicated fax machines and through fax modems. Often a fax generated on one type of system will be received on the other type. The most common form of fax machine is a dedicated system resembling a small printer that connects to a phone line. It sends faxes by scanning a piece of paper fed through the machine. Received faxes are printed in much the same way as they would be on a computer printer. Some faxes use continuous feed thermal paper, though more commonly now fax machines can work with sheet-fed plain paper. Many faxes optionally function as photocopiers and printers. See facsimile modem.

facsimile formats In order for facsimile (fax) machines to exchange data, a number of international standards have been defined for image encoding and transmission. In addition to this, dialup modem transmission protocols with compression and error correction for fax/modems, called the V Series Recommendations, are used for data transfer in conjunction with fax formats. Tag Image File Format (TIFF) is an important raster image encoding format widely used in facsimile transmissions, with TIFF-FX as an offshoot of TIFF developed specifically to facilitate faxing over computer networks.

Since a high proportion of facsimile transmissions are still basic black-and-white text documents, guidelines have been established for the encoding of minimal black-and-white images and text. These formats also have the advantages of small file sizes and fast transmission speeds. The Profile S and Profile F subsets of the TIFF specification support the transmission of basic black-and-white documents.

Basic Facsimile Formats

Class 1	EIA/TIA (EIA-578) standard for basic computer fax/modem interface.
Class 2	EIA/TIA standard for extended computer fax/modem interface which includes AT commands.

- -

Group 1	Single page transmission in six minutes. Common in the 1970s.
Group 2	Single page transmission in three minutes. Common in the late 1970s.
Group 3	(There is also a Group 3 bis, or Group 3 enhanced format.) 14,400 bps facsimile protocol. Two resolution modes include 103×98 dpi (standard), and 203×196 dpi (fine). Compression is supported. This is the most common protocol used with fax machines and fax modems, either Class 1 or Class 2. See V Series Recommendations V.27. Single page transmission in under 30 seconds.
Group 4	ISDN B facsimile protocol adopted in 1987 but not widespread in subsequent years.
	For metric equivalents and higher resolutions (e.g., 400×400), added in the early 1990s, reference the ITU-T T.30 recommendations.

Over the years, Group 3 has become the de facto standard for facsimile transmissions over standard phone lines, supporting two basic resolutions commonly called *standard* and *fine*. Three compression schemes are commonly used in Group 3 transmissions: Modified Huffman (MH), Modified READ (MR), and Modified Modified READ (MMR) (Group 3 and Group 4), with MH the most common.

Due to the limitations of data transmissions over phone lines with traditional modems, black-and-white imaging and bit- and stream-oriented transmissions predominate. However, as the Internet increases in importance and accessibility, byte- and file-oriented systems have been developed and adapted, allowing greater flexibility in the size and style of documents being transmitted. Email to fax gateways can channel network fax transmissions to appropriate servers and devices (e.g., a fax device can be assigned an email address) and other means of document transmissions have evolved alongside traditional facsimiles. Faxing over computer networks is of great interest to individuals and businesses, as it frequently saves long-distance phone charges. There are proposed and standardized formats for including fax information in existing email as opposed to distinct fax-over-Internet formats.

Another approach is to send original "facsimiles" such as word-processed files as email attachments. When viewed or printed, the files contain all the colors, text, images, symbols, and formatting found in the original files up to the capability of the display or printer upon which it is viewed, which is almost invariably higher quality than a traditional low-resolution, black-and-white fax transmission. Even if the file is printed on a fax machine that doubles as a printer, it will not be skewed or smudged (or blank) as are many scanned and manually transmitted fax documents.

Adobe Portable Document Format (PDF) has become an important vehicle for disseminating high-quality PostScript documents over the Internet, in essence, providing a high-quality "facsimile" that is the same as the original as opposed to being a lower-quality copy. With the wide availability of free PDF readers, PDF may supersede facsimile formats for many types of documents and forms (many state governments now use PDF for informational brochures and license application forms). For people who don't like to look at documents on computer screens or who require a physical printout on a fax machine, PDF files sent over the Internet (or over phone lines) could be fed to a dedicated reader on a PostScript-capable facsimile machine, resulting in perfect copies of the original (no smudging or blurring) at whatever resolution is available on the printer (most consumer laser printers now image between 400 and 1000 dpi).

Thus, the very concept of a fax transmission is changing, as more and more documents are created in electronic formats that can be transmitted directly, without first being scanned and encoded/decoded. In other words, legacy faxes, based on the current lossy scan-encode-transmit-decode scheme, may largely be superseded by future faxes based on lossless

F

compress-send-decompress schemes, with the compression needed only to speed transmissions, and thus not degrading the quality of the original. See facsimile mode, Huffman encoding, TIFF-FX, RFC 2301, RFC 2304, RFC 2305.

facsimile mode Facsimile machines have a number of operating modes, including various regular and fine resolutions, and can be manipulated to send in monochrome or grayscale, depending on the capabilities of the sender and receiver, and the software or hardware. Most fax machines and fax/modems send in Group 3 standard and fine modes, that is 203⊃×⊃98 pixels and 203⊃×⊃196 pixels. Other modes have been defined (Group 3 203⊃×⊃391 - superfine; Group 4 400⊃×⊃400 - standard), but they are not widely supported in consumer-priced products. A fax can be only as good as the weakest link in the transmission. If the sending fax sends in the lowest resolution, a higher resolution receiving fax doesn't improve the image. Conversely, if the sending fax uses fine resolution, but the receiving fax can only support standard resolution, the details will still be lost. Since the orientation of most faxes is *portrait*, and the orientation of most computer monitors is *landscape*, fax/modem software usually has zoom, pan, and rotate features to aid in viewing documents.

facsimile modem, fax modem A fax modem system consists of a fax-enabled data modem (one which works in two modes) hooked to a computer, sometimes combined with a scanner. Instead of creating a document, printing it, feeding it through the fax machine, receiving a printed page at another destination fax machine, and then perhaps even typing or rescanning (and OCR-ing) the printout back into a computer at the other end, the fax modem system sends a digital fax directly from the software application that created the document, to the receiving end. Or, if a scanner is used, the system sends the scanned file as the fax. If the receiving device is a fax modem system, rather than a fax machine, then the fax goes directly to the computer hard drive, and no paper is used in the transaction.

Optionally, a fax machine may connect to a local network so that users can be notified if a fax has been received or even select an option to view the fax on a computer monitor (thus providing hybrid fax machine/fax modem capabilities).

In business environments some people erroneously use fax machines when they should be doing direct data transfers. This is a common scenario: the main office of a corporation creates a new 80-page policy manual and wants to distribute it to all ten branch offices. The branch offices would like an electronic copy in order to customize it for their needs, or to easily make corrections as directed by the main office, etc. The typist types a copy, faxes 80 pages to each of the ten branch offices; 880 pages are generated in all, the original, and the 10 branch copies. Now the typists at each branch office retype the document into their word processors, thus duplicating the work already done.

Rather than always using a fax machine, there are better ways to distribute some types of documents. The first is a slight improvement. By using a fax/modem software program to send the document directly to each branch to another fax/modem program, no paper is printed until the documents are complete to each branch's satisfaction, and the completed customized documents can be OCR-scanned back into a word processing text file at the destination.

A better solution is to send the original file, in document format, through a modem or through the Internet, to each branch office, where the secretaries can load the received file directly into the word processor. This can be accomplished by putting the file on an FTP site, and notifying the branches that they can access the site and download the latest version of the file. If different word processors are being used, the original can be saved in Microsoft Interchange Format (also known as RTF or Rich Text Format), a widely supported format that can be read and saved by all major word processing programs.

The best solution to document exchange may be to have a secure centralized online document repository which can be accessed and modified dynamically by all branches through an Internet or private network connection. Fax machines are a great resource for sending short documents, but they are not the best solution for all document transfers, and the Internet provides distant branches with a way to dynamically produce and maintain documents without incurring long-distance charges. Integrated data, video, and voice services are increasingly offered by telecommunications carriers; facsimile communications are being superseded to some degree by email attachments, especially PDF files. See facsimile device, facsimile format, facsimile history, facsimile modes, Portable Document Format.

facsimile switch An external switching device that allows a single phone line to be used for more than one phone-related piece of equipment. Fax switches often can also handle telephone answering machines and computer modems. The fax switch is attached between the phone line plug and the various phone devices. When a call comes through, the device evaluates the tones and decides whether it's a voice call, a modem call, or a fax call, and routes the call to the appropriate device. Unfortunately, most fax switches can't detect when a manual fax machine is going to send a fax if the call originated as a voice call. If the person dials the phone manually and then wants to switch over to a fax call after the connection is established, many fax switches can't revert to data mode (newer ones may be switchable on receiving a particular code). In spite of that limitation, it's a great tool for homes, home offices, and small businesses that can't afford extra phone lines.

fade To diminish in strength, loudness, or visibility. In video or audio editing, fading is deliberately used to provide transitions that are perceptually pleasing. In data or broadcast transmissions, fade is usually an undesirable effect due to various factors such as distance, loss of signal, obstructions, interference, etc. Undesired fade can sometimes be reduced or eliminated by amplifiers, repeaters, robust wiring mediums, and good insulation.

fade margin Signal losses in satellite systems can occur from scattering, absorption, and various subtle types of interference. Consequently various fade margins are incorporated into the design of the systems, and they will vary depending on the degree of fade expected from various sources and on the length of the broadcast waves, with shorter waves generally being more subject to fade.

Fahrenheit scale A temperature scale that designates 32 degrees for the freezing point of water at normal pressure and 212 degrees for the boiling point of water at normal pressure, and other points relative to these. See centigrade scale, Celsius scale.

Fahrenheit, Daniel Gabriel (1686–1736) A Polish-born German scientist who established the widely used Fahrenheit scale. Zero degrees was designated as the temperature of a mix of ice, water, and salt, and 90 degrees was considered to be the temperature of the human body (in fact, it's closer to 98.6 degrees). See Rümer, Ole Christensen.

failsafe A designation that indicates that failure of a system is unlikely or impossible, or that backups are available if needed. In networking, few, if any, systems are completely failsafe, but there are steps that can be taken to prevent problems, such as the use of surge suppressors, backup power systems, redundant data storage or broadcast signals, etc. See fault tolerant, redundant array of inexpensive disks.

fake code See pseudocode.

fallback A contingency mode, plan, or operation. In communications, a designation for another speed or mode of operations if the current mode is not functioning as well as might be desired. Many modems may fall back (two words) to a slower speed if the connect negotiation doesn't work at higher speeds. Many communications programs may fall back to smaller packet sizes if there is a lot of noise, or other impingements on a data file transfer. In software, a fallback (one word) may be one in which the application or operating system goes to another mode or another program if some error condition or slowdown is detected. A network may go to a fallback route if the usual one is not available or not responding as expected.

falsing Spurious signals that accidentally are interpreted by a system as commands, or that are deliberately introduced to fool a system, usually for unauthorized purposes. In telephone systems, certain situations can be simulated by playing particular tone sequences, so the system is fooled into switching, transferring, or connecting long-distance or other types of calls. In transmissions control for satellites and other radio-controlled devices, environmental noise, falsely interpreted signals, etc. can have major consequences if the system thinks it's a command and acts upon it.

fan 1. Fan of science fiction. Since there are a large number of software developers who are science fiction fans, they have co-opted this term into many computer-related situations, video games, simulations, and virtual reality environments. 2. An active

cooling device (as opposed to passive devices such as heat sinks) often used to cool computers so that chips and other components are kept at optimum operating temperatures.

FAN See flexible access network.

fanfold See z-fold.

fanout A device that facilitates the separation of individual fibers in a fiber optic cable bundle, enabling them to be more easily handled, attached, configured, or repaired. After being fanned, individual strands may be channeled through a *furcation unit* for routing to its destination. See furcation unit. See fantail.

FANP See Flow Attribute Notification Protocol.

Fantail Fiber Optic Wiring Bundle

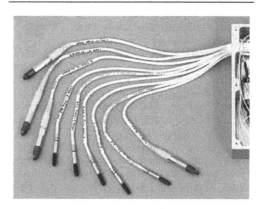

A fantail wiring bundle secured at the point where it connects to the electronic components, with connecting points for individual fibers fanned out for coupling with the appropriate connectors. Individual fiber pigtails typically have protective caps, in preparation for adding a connector, or will have a connector already attached to protect the precision-cut and polished terminal point. They may be color coded to facilitate the correct connections. [NASA/Langle photo, 1999.]

fantail A descriptive term for a wiring bundle that is secured somewhat near the point at which the wire connectors spread apart for attachment to a wiring rack, thus creating a shape that fans out from the bundled point. The strap that secures the bundle aids in holding the wires together if one or more of the wire connectors are disconnected, as they will hang a few inches below the rack for easy location and re-attachment, if desired. The fantail configuration also makes it easier to disconnect and reconnect the entire bundle of wires. See fanout.

FAQ See frequently asked questions.

far end crosstalk FEXT. When wires are packed together tightly, and signals are travelling through most or all the wires, the potential for interference from crosstalk increases. Far end crosstalk is a type of interference originating from multiple signals travelling in the same direction, typically through wire pairs, as in common copper twisted-pair installations. FEXT directly effects bit error rates (BERs), as it cannot be cancelled as easily or as effectively as near end crosstalk (NEXT). See near end crosstalk.

farad A unit of capacitance equal to one coulomb (of electricity) divided by (a potential of) one volt. Named after Michael Faraday.

faradaic Relating to an asymmetric alternating current (AC) produced by an induction coil.

faraday A measure of electrical charge transferred in the process of electrolysis per weight of an ion, or element, that is equal to about 96,500 international coulombs (or 96,490 absolute coulombs). Named after Michael Faraday.

Faraday cage A structure, usually mesh- and cage-like, to isolate a person, device, or electronic system from damage or interference from outside electrical sources. These may sometimes be seen in science museums where electrical devices, especially large Van de Graaff generators, are demonstrated. Named after Michael Faraday.

Faraday Dark Space In a cathode discharge tube, a region between the positive column and the negative glow that appears dark. Regions in the tube become easier to distinguish if the pressure is lowered in a tube that has some air in it (normally air is removed to extend tube life and effectiveness). Then it becomes possible to distinguish the Faraday Dark Space as a region just outside a pale negative discharge glow, which in turn terminates in Crookes Dark Space, which borders the outside glow of the cathode. Named after Michael Faraday. See Crookes Dark Space.

Faraday dynamo A historic electrical generator developed in 1832 by Michael Faraday.

Faraday effect A basic magneto-optical effect in which a plane of polarization of light in a magnetic field, traveling parallel to the lines of magnetic force, can be rotated to another plane by a transparent isotropic medium. Named after Michael Faraday, who described it in 1846, this effect was studied further three decades later by John Kerr. The distinction between the Faraday effect and the Kerr magneto-optical effect is that Faraday focused his attention on a beam that was transmitted through the magnetic material, while Kerr focused on a beam that was reflected off the magnetic material. The Faraday effect can be seen in a number of telecommunications technologies. In satellite communications, the plane of polarization of radio waves traveling through the ionosphere rotates about the direction of propagation, particularly at lower frequencies. See Kerr magneto-optical effect.

Faraday effect, acoustic The Faraday effect was described by Michael Faraday more than 150 years ago. It is a basic, fundamental effect that is exhibited in many different phenomena and is of continuing interest to scientists, both as to its properties and its practical applications. The acoustic Faraday effect is the acoustic analog of Faraday's magneto-optical effect. Typically, liquids do not propagate transverse waves, but L.D. Landau, in 1957, predicted that a quantum liquid phase of ^{3}He might exhibit transverse sound waves under specific conditions. Lee et al. have

observed rotation of the polarization of transverse sound waves in superfluid ^{3}He-B in a magnetic field, lending support to Landau's prediction. See Faraday effect.

Historic Faraday Electromagnet

This Faraday electromagnet, cobbled out of available materials, was wound partly from Faraday's wife's petticoat. [Classic Concepts illustration.]

Faraday-Stark effect A novel effect resulting from the combination of the Faraday magneto-optical rotation and quantum-defined Stark effect. This phenomenon enables an electrical field to be used to influence a Faraday or magneto-optic Kerr rotation. It is a photonic effect that was discovered in the mid-1990s which, along with linear polarizers, may have applications for future high-frequency modulation devices. The Faraday-Stark effect was described by Lee et al. in *Applied Physics* in 1996, and Faraday-Stark magneto-optoelectronic (MOE) devices were patented by Lee and Heiman (U.S. #5,640,021, 1997). See Faraday effect, Stark effect.

Faraday, Michael (1791–1867) An English physicist and chemist who was apprenticed to a bookbinder at the age of 13. He took time to read the books and to listen to local lectures by Humphry Davy, becoming his laboratory assistant in 1813. Faraday went on to conduct extensive experiments in electricity and magnetism. He passed electrical currents through solutions and observed their effects, adding new knowledge to the discoveries of A. Volta. Faraday demonstrated that the amount of an element deposited at an electrode is proportional to the current flowing through the solution. In 1831, he demonstrated that an electrical current can induce a current in a different circuit and made a historic entry in his journal linking electricity and magnetism. The following year he constructed a basic generator, calling it a *dynamo*. Faraday also studied the properties of metals and glass and developed new types of optics. He coined the terms *electrolyte, electrode,* and *ion.* Further important investigations of inductance in electrical circuits by other scientists grew out of Faraday's work. Many electrical effects have been named after him. See Davy, Humphry.

Faraday's laws Michael Faraday investigated the phenomena related to decomposition by galvanic current and made some important discoveries that have been investigated and variously stated by succeeding scientists. Generally, Faraday's laws are described as follows:

1. in electrolytic decomposition, the number of ions charged or discharged at an electrode is proportional to the current passed;
2. the amounts of different substances deposited or dissolved by the same quantity of electricity are proportional to their equivalent weights;
3. when passing a constant quantity of electricity through different electrolytes, the masses of the ions set free at the electrodes are directly proportional to the atomic weights of the ions divided by their valence.

Faraday called his discovery the "law of definite electrolytic action." It was opposed by Berzelius and those who adhered to Volta's theory of galvanism. Through subsequent experiments, Faraday's concepts have been refined and confirmed, and his discoveries are now known as Faraday's laws.

Farber, David J. Originally a computer consultant to the Rand Corporation in the late 1970s, Farber later became a cofounder of CSnet (Computer Science Network), NSFNet, and others. In 1995, he was awarded the SIGCOMM Award for lifelong contributions to his field. Farber has served on the boards of AT&T and several other industry telecom companies, as well as the Electronic Frontier Foundation and the Internet Society. In January 2000, the Federal Communications Commission (FCC) announced his appointment as Chief Technologist for the FCC. He is known for his online discussion list "Interesting People."

FARNet See Federation of American Research Networks.

Farnsworth, Philo Taylor (1906–1971) A precocious American musician and inventor who built an electric motor at about 12 years of age and described a television system to friends. He is reported to have shown a drawing of the idea to J. Tolman, a teacher, in 1922. Fortunately, Tolman later remembered the incident and produced the drawing, or the young Farnsworth might not have received credit for being one of the earliest inventors of television technology. Farnsworth kept working on the idea, submitted a patent application in January 1927, and successfully transmitted his first TV image in September 1927. The television patent was awarded in August 1930 (U.S. #1,773,980).

The versatile inventor also developed several types of amplifying systems, a system of pulse transmission, a projection system, a microscope, and a type of cold cathode-ray tube (CRT), securing hundreds of technology patents during his lifetime. It should be noted that biographers and Farnsworth himself credit his wife Elma "Pem" Gardener-Farnsworth as contributing significantly to the construction of his devices.

In September 1983, the U.S. Postal Service commemorated the achievements of Philo T. Farnsworth with a 20-cent stamp and first day cover and later issued a 33-cent portrait stamp from the Great American Inventors series. See television history. http://philotfarnsworth.com

---

Farnsworth Historic Imaging Tube

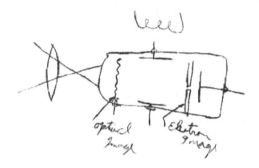

Farnsworth's teenage drawing of his concept of a historic television system emerged during patent disputes in the 1930s. [Philo T. Farnsworth, 1922.]

FAS 1. See facility-associated signaling. 2. See frame alignment signal.

fast busy A telephone busy signal that is distinctive in that it repeats at twice the rate of a regular busy. A regular busy signal indicates the caller's phone is unavailable (it's off-hook or in use), whereas a fast busy indicates that all trunk lines are busy, and the call cannot currently be routed to the destination.

Fast Ethernet A version of Ethernet enhanced to increase its 10 Mbps capacity up to 100 Mbps over copper or fiber, which brings it into the high speed networking range along with asynchronous transfer mode (ATM) and FDDI. This enhanced capability requires the upgrade of other devices such as hubs and network cards, partly because Ethernet hubs can be cascaded, whereas Fast Ethernet hubs are stacked. Fast Ethernet is an international open IEEE standard (802.3u, 1995) used in medium-scale networks such as campus backbones. See Fast Ethernet Alliance, Gigabit Ethernet.

Fast Ethernet Alliance An open trade association established to develop and promote Fast Ethernet technologies for existing voice-grade traditional copper twisted pair, founded in July 1993. A major goal of the Alliance was to standardize 100 Mbps Ethernet technology. By 1994, it had submitted 100Base-TX and 100Base-T4 wiring specifications for consideration to the IEEE for inclusion in the IEEE Fast Ethernet 100Base-T standard. The Alliance membership totaled more than four dozen telecommunications vendors including well-known companies such as NCR Microelectronics, National Semiconductor, Sun Microsystems. Fast Ethernet was adopted officially in June 1995 by the IEEE 802.3 group. See Fast Ethernet.

fast Fourier transform FFT. See Fourier transform, fast.

Fast SCSI A means of configuring SCSI to provide faster transmission speeds, up to 10 Mbps. There have been a number of variations on the SCSI standards. One of the more commonly implemented versions is SCSI-2. See Small Computer System Interface for a detailed description of SCSI technology.

FastIP, Fast Internet Protocol A 3Com commercial product in which just the first datagrams of the IP traffic are passed through the router and, if a direct path is found, subsequent ones may bypass the router using Next Hop Resolution Protocol (NHRP). It is embedded in local area network (LAN) adaptors and implemented in LAN switches rather than in Internet Protocol (IP) routers. See IP switching.

FAT 1. File Allocation Table. See FAT format. 2. final acceptance testing.

FAU See fixed access unit.

fault A defect, incorrectly functioning system, mistake, or accident. In fiber optic cables, faults may include crystallization during fabrication, bubbles, undesired particles (apart from deliberate fiber doping), misalignments, incorrect coupling or bonding, excessive bends, and dispersion problems.

fault isolation In electronic circuitry or software debugging, a troubleshooting strategy for isolating the location of a problem. In circuitry, it may involve shutting down parts of the system, wiring in shunts or bridges, or selectively stimulating particular areas. In software, it may involve setting breakpoints, printing debug messages, or tracing particular variables. See bridge, shunt, trace.

fault threshold The level at which a system's structure or function is considered to be compromised. This may be a certain number of lost packets, a specified electrical level, a certain number of physical flaws, or any other measures particular to a system that affect its integrity and functioning for a particular purpose.

fault tolerant A fault tolerant system is one that is designed so if a problem occurs, the entire system or important parts of the system will continue to function until the problem is corrected. Thus, system redundancy, backups, secondary routines or hardware paths, etc. can be incorporated to increase fault tolerance. Good computer operating systems are designed so that individual applications don't crash the system. The application itself may crash, or need to be "killed" (by killing the individual processes associated with the program), but the system can handle the crash without affecting other programs or the general operations and will clean up stray files, memory, etc. See Byzantine Generals problem, failsafe.

fax *colloq.* facsimile. See facsimile machine.

fax mode See facsimile mode.

FB See framing bit.

FBT See fused biconic taper.

FBus Frame Transport Bus.

FC 1. See feedback control. 2. frame control.

FC- connector A relatively common coupling part for fiber optic connections that physically resembles ST- connectors, but with a friction rather than a bayonet mount. FC- connectors are used to couple single-mode fiber connections. They are also available in hotmelt styles.

FC-*x* In the Fibre Channel hierarchical model, a series of services and protocols.

For more detailed information, see Fibre Channel, including the Fibre Channel Layers chart.

FC-*x*	Function	Notes
FC-0	physical	media, transport speeds, receivers, and transmitters are defined at this level
FC-1	encodings	transmission encodings and decodings
FC-2	signaling	a protocol for specifying mechanisms and rules for transfering data blocks, controlling data flow, and error detection mechanisms
FC-3	services	common services for N_Ports on a node
FC-4	protocols	upper level protocols in terms of channels and networks (e.g., SCSI and ATM) that map into the system

FCA See Fibre Channel Association.

FCIA See Fibre Channel Industry Association.

FCC See Federal Communications Commission.

FCC Glossary of Telecommunications Terms See Glossary of Telecommunications Terms, FCC.

FCLC See Fibre Channel Loop Community.

FCS 1. See Federation of Communications Services. 2. See Fibre Channel specifications. 3. See Frame Check Sequence. 4. Fraud Control System.

FCSI See Fibre Channel Systems Initiative.

FCW See *Federal Computer Week Magazine*.

FDD See floppy disk drive.

FDDI See Fiber Distributed Data Interface.

FDM See frequency division multiplexing.

FDMA See frequency division multiple access.

FDMS See Fiber Dispersion Measurement System.

feature code A number or character sequence used to activate a feature on a phone system, such as speed dialing, last number redial, etc. These are more common on multiline business phones than on residential phones.

feature connector A connector for coupling a peripheral card or device to another peripheral card, such as a video graphics adapter, so the second card can perform direct memory access (DMA) through the card's bus, without having to load the system bus. The feature connector is commonly used on VESA-compatible systems.

Feature Groups Designated groups representing various types of long-distance carrier switching arrangements that are part of the Bell Operating Companies (BOC) system.

Feature Group	Switching Arrangements
Group A	A subscriber line connection rather than a trunk connection to a local exchange carrier's network.
Group B	A trunk connection that uses an authorization code for billing. Used in areas where it is not practical to offer Feature Group D (Equal Access services), such as some older switching systems, and independent services.
Group C	The older long-distance services offered by local exchange carriers to AT&T before divestiture. Mutually exclusive with Feature Group D.
Group D	Equal Access services, facilities and signaling specifications, established since divestiture and implemented in the mid-1980s. Mutually exclusive with Feature Group C.

feature phone A phrase for phones that have extra features. Sometimes the features improve functionality (redial, speakerphone, channel, etc.), but sometimes they are just a marketing enticement and may not be very useful.

FEC 1. See Forwarding Equivalence Class. 2. See forward error correction.

FECN See Forward Explicit Congestion Notification.

Federal Communications Commission FCC. A significant U.S. federal regulatory organization originally created through the Communications Act of 1934, evolving from the formation of the Federal Radio Commission (FRC) in the Radio Act of 1927. The original mandate of the FCC was to regulate the broadcasting industry in the United States by the granting and administration of radio licenses. As such, the FCC administrated the allotment of frequencies, time slots, and callsigns. Since then, its jurisdiction has been broadened, reflecting the growth in telecommunications in general. The Commission is directly responsible to the U.S. Congress.

The FCC has a powerful role to play in the fair and equitable enactment and distribution of telecommunications resources in accordance with the Telecommunications Act of 1996. It is the responsibility of the FCC to see that the Act meets its goals of opening the telecommunications business to anyone, and of promoting fair competition in the industry.

The FCC now also oversees product emissions, ensuring that computing devices do not emit harmful radiation or unharmful radiation at levels that may nevertheless interfere with other radiant technologies such as radio waves.

The FCC overall organization consists of a number of commissioners, about nine offices (public affairs, plans and policy, general counsel, etc.) and six bureaus. See Primary Divisions chart. See Communications Act of 1934. http://www.fcc.gov/

Primary Divisions of the Federal Communications Commission	
FCC Bureau	Responsibilities
Common Carrier (CCB)	Enforcement, pricing, accounting, program planning, network services, and wireline services.
Wireless Telecommunications (WTB)	Domestic wireless communications, including paging, cell phone, PCS, and radio, excepting satellite communications. This bureau is further subdivided into Commercial Radio, Enforcement, Policy, Private Radio, Licensing, Customer Services, and Auctions divisions.
Mass Media (MMB)	Audio service, enforcement, policy and rules, video services, administration, and inspections.
Compliance & Information (CIB)	A national call center, and information resources, management, compliance, technology, and regional offices.
International	International planning and negotiations, satellite and radio communications, and general administration.
Cable Services (CSB)	Consumer protection and competition, engineering and technical services, policy and rules, public outreach, management.

Federal Communications Commission classes, FCC classes A series of designations or ratings applied by the FCC to electronics devices. These are primarily intended to help prevent interference from devices like computers that may affect electromagnetic broadcast waves such as radio and television signals. If you have tried to use a cordless phone near a computer and experienced interference, you are familiar with the type of problem excess emissions can create. Many commercial video devices, for example, are labeled 'For Commercial Use Only' to comply with FCC regulations.

Category	Notes
Class A	Computing devices rated for office use and that may not be used in the home.
Class B	Computing devices rated for home use.

Federal Computer Week Magazine FCW. A newspaper providing up-to-date news and product information to U.S. government computer technology users on sources and types of federal information technology (IT). It especially focuses on desktop, client-server, and enterprise computing and issues of volume procurement. FCW publishes online and print editions and maintains an online archive of past information.

Federal Information Processing Standard FIPS. A set of standards for document processing, search, and retrieval. Examples include FIPS PUB 180-1 (secure hash standard) and FIPS PUB 144 (digital communication performance parameters). A number of FIPS publications are based upon ANSI and CCIT standards.

Federal Networking Council FNC. The FNC reports to the Federal Coordinating Committee on Science Engineering and Technology and was chartered by the National Science and Technology Council's Committee on Computing, Information and Communications (CCIC). It provides a focal point and forum for networking collaboration among U.S. federal agencies with regard to education, research, intercommunications, and network operations. Since 1997, the various activities of the FNC have been carried out through the Large Scale Networking (LSN) group. http://www.fnc.gov/

Federal Standard Glossary of Telecommunications Terms See Glossary of Telecommunications Terms, Federal Standard.

Federal Technology Service FTS. A service of the U.S. General Services Administration (GSA) that provides information technology and network services to U.S. government agencies, including its mobile workforces. http://www.fts.gsa.gov/

Federal Telecommunications Standards Committee FTSC. A U.S. government agency that promotes the standardization of communications interfaces, including computer networks. The FTSC is chaired by the Chief of the Technology and Programs Division. Through the work of its technical subcommittees, it is the primary telecommunications standards mechanism supporting the National Communications System (NCS). The FTSC liaises with and evaluates the development of national and international standards and develops federal standards recommendations or the standards themselves in situations where existing standards are unavailable or unsuitable for U.S. government needs. The technical subcommittees have specialized expertise in various fields, including mobile, wireless, and multimedia telecommunications.

Federal Telecommunications System FTS. The intercommunications network used primarily by U.S. government civilian agencies. It includes interconnections to other agencies and to the public switched telephone network (PSTN).

Federation of American Research Networks FARNet. An organization comprised of commercial providers, some telephone providers, and mid-level NSFNet networks that meet to discuss issues related to these businesses and the Internet.

Federation of Communications Services FCS. A trade association representing the mobile communications industry in the British Isles. FCS promotes and encourages a healthy market environment for the communications industry in the U.K. and represents its members to the government and various telecommunications agencies. http://www.fcs.org.uk/

feed horn, feedhorn A basic signal-capturing component in satellite receiving antennas that is mounted at the focal point. It must either be rotated to correspond to the polarity of the incoming signal (horizontal or vertical) or be attached to a dual coupler. The focal length of the feed horn is dependent on the depth and diameter of the parabolic dish in which it is mounted. The feed horn is attached to a signal amplifier. See antenna, low noise amplifier, microwave antenna, parabolic antenna.

feedback *n.* 1. Information or phenomena that are reflected or translated and returned to the originating or transmitting source. 2. An opinion offered in response to some preceding event or information. 3. Returned information about data that has been received or passed through. In networks, there are many feedback mechanisms providing information data rates, congestion, traffic in the opposite direction, and the progress or success of a transmission.

feedback control FC. A means of controlling a system by sensing impulses or signals that are compared to a reference or desired value and responding accordingly. For example, when humans get cold and their body temperatures drop, the nervous system senses the difference and causes the body to shiver to help it generate sufficient heat to maintain life. In a telephone system, if a phone remains off-hook for more than a prescribed length of time, the system 'senses' the anomaly and responds with a beeping sound or message suggesting the subscriber hang up the phone. In Internet services using dynamic IP allocation, if a connected subscriber is inactive for a certain period of time, the service may disconnect the subscriber in order to make the IP number available to another subscriber. In robotics, feedback control is an important means to enable a robot to sense and navigate around its environment. Thus, feedback control systems are used throughout the telecommunications industry in a multitude of ways to start, stop, and maintain systems in order to facilitate efficient operations. See hysteresis device.

feedback signal 1. A signal that loops back around to its source. An undesirable audio or visual artifact can occur when the same signal that is being transmitted travels back through the original transmissions media. In sound systems, it commonly manifests as a piercing, shrieking sound, as when a microphone is located too near a speaker carrying signals from that microphone. If carefully controlled, audio feedback can sometimes be used to boost a weak signal. In visual systems, feedback often manifests as ghost images or wiggly distortions. 2. An intentional diagnostic looped back signal. In diagnostic systems, when a signal is transmitted and then compared with a reference when it returns (the returning signal is the feedback signal), it is possible to evaluate the similarities and differences between the two signals, or the information carried on those signals.

feeder cable 1. A primary cable, extending from a service provider or central switching location, to a distribution panel or end-user. In large installations, there may be a main feeder cable and branch feeder cables. 2. The cable that connects a primary distribution frame with intermediate distribution frames. 3. A main network backbone cable, which may have branch feeder cables leading to the main host computers. 4. A heavy duty, primary, or high bandwidth wire or cable intended to carry the main part of traffic from the transmission source to its primary dropoff points or hosts. Thus, fiber optic cables and 25-pair cables are common feeder cables.

FEFO first ended, first out. A priority queuing arrangement in which the first item processed, or the first process completed, is the first to be passed on, or further processed. Thus, processes that are finished are taken out of the queue in order to leave space or processing time for others. See FIFO, FILO, LIFO, LILO.

femto- (*symb.* – f) An SI unit prefix for 10^{-15}, a very, very small amount. In decimal, femto- is expressed as 0.000 000 000 000 001. See atto-.

FEP See Front End Processor.

FER Frame Error Rate.

Fermat, Pierre de (1601–1665) A French lawyer, linguist, and mathematician who made many contributions to our understanding of mathematics and optics, in spite of his recreational approach to mathematics, which meant that many of his discoveries initially went unpublished. Fermat's principle is named after him.

Fermat's principle When electromagnetic radiation travels by reflection off a surface from one point to another, it will take the path that can be traversed in the least amount of time.

Fermi level A value designated for electron energy at half the Fermi distribution function.

Fermi, Enrico (1901–1954) An Italian physicist who investigated atomic physics by systematically irradiating the elements, work derived in part from the investigations of James Chadwick.

ferric oxide A metallic compound commonly used to coat thin tapes or platters used in magnetic storage media. The ferric oxide molecules can be selectively rearranged by magnetic impulses in order to encode the desired information on the medium. There are other types of coatings available for applications such as sound or video recording; the differences in various coatings can affect the quality of the recording.

ferroelectric liquid crystal FLC. Crystals that are incorporated into spatial light modulators (SLMs) in optical computing technologies. They have the capacity for very fast bipolar switching. Surface-stabilized FLCs, created by suppressing the natural helical structure of FLCs, are used in a number of high-resolution color display technologies, including low power microdisplays. They are also suitable for use in optical shutters. The Ferroelectric Liquid Crystal Materials Research Center is located at the University of Colorado.

ferromagnetic Having the property of being very easily magnetized with high hysteresis, i.e., magnetism that changes readily with changes in the magnetizing force. See electromagnet.

ferrule A snug ring or cap encircling a tool, pipe, or wire; a short length of tubing or bushing (insulating liner) that helps to strengthen or secure a joint or coupling component. It is sometimes called a sleeve, though the term is usually applied to "hard" sleeves (as opposed to soft, flexible sleeves) made of sturdy materials. It may include a flange.

Ferrule Examples

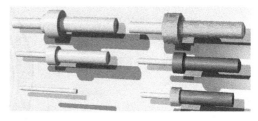

Ferrules are the most important single component in fiber optic connectors. They come in a variety of shapes, materials, and formats. They may be unflanged (bottom left) or flanged with plastic (left) or stainless steel (right), and they are commonly fabricated from zirconia (a type of corrosion and heat-resistant ceramic – right), though plastic ferrules (left) are now available. Ferrules similar to those shown above are commonly incorporated into standardized ST- and FC-connectors.

---

ferrule, fiber optic In fiber optics cable assemblies, a ferrule is the most important structural member. It surrounds, secures, and aligns the fiber filament and supports the surrounding housing. To meet the demands of optical alignment of fiber lightguides, ferrules must be exact and should not stress or overly bend the joint, otherwise interference from imperfect coupling could adversely affect the angle of the light beams or allow them to leak at the joints.

Ferrules are commonly made of zirconia (a ceramic made from a crystalline powder) or alumina. Zirconia is favored for its bending strength, resistance to corrosion, hardness, and heat resistance. It holds up well in the polishing process and over time once installed. More recently, plastic and opaque glass-ceramic ferrules have been improved to the point where they exhibit acceptable strength and durability for cost-effective alternatives to zirconia for certain applications in addition to which they may not require polishing.

Commercial ferrules for fiber filaments may come with pre-domed or pre-angled endfaces to facilitate termination. They may be semifinished blanks or whole finished standardized ferrules such as SC- and LC- styles. They are available with or without flanges. 3M has produced a line of connectors that use a V-groove rather than a ferrule for coupling optical fibers. Interferometers can be used to assess the characteristics of the ferrule-supported joint. See interferometer.

FES Fixed End System.

Fessenden, Reginald Aubrey (1866–1932) A prolific, Canadian-born, American inventor and radio pioneer who was one of the first to try to devise ways to carry information on top of a carrier wave. In the process of trying to achieve this, he developed a high-frequency generator in 1901 that could create radio wave, and a hot-wire barretter, which was developed into an electrolytic detector, for detecting radio waves. On Christmas Eve 1906, to the astonishment of those who heard the broadcast, Fessenden succeeded in transmitting voice and music, using an Alexanderson alternator, over public radio waves to the U.S. east coast. See barretter, carrier wave, electrolytic detector, radio history.

FEXT See far end crosstalk.

Feynman, Richard Phillips (1918–1988) A charismatic, individualistic American physicist who contributed greatly to our understanding of physics, especially in quantum electrodynamics (QED quod erat demonstrandum – that which has to be demonstrated), who developed Feynman diagrams and provided insights into the theory of computing.

FFT See Fourier transform, fast.

FGDC See Federal Geographic Data Committee.

fiber 1. A strand, filament, or other structure with long, slender threadlike qualities. 2. Colloquial for fiber optic (or optical fiber). See fiber optic.

fiber bundle Two or more fiber optic filaments held in close proximity, either with a supporting structure or sheathing. Combining fibers in a bundle enables more light signals to be delivered to the destination. It is not uncommon for hundreds of fiber filaments to be contained within a single bundle as the individual filaments are very small. Why use many fibers instead of a fatter fiber (*fiber rod*)? There are a

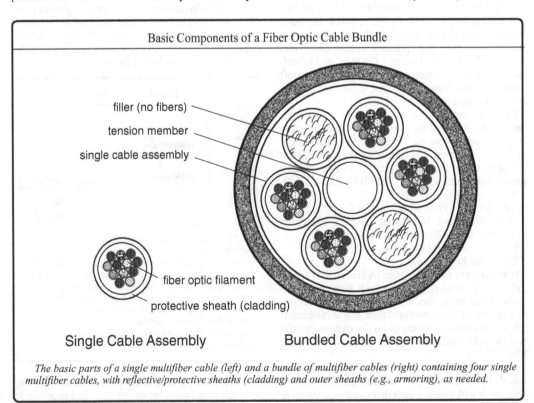

Basic Components of a Fiber Optic Cable Bundle

- filler (no fibers)
- tension member
- single cable assembly
- fiber optic filament
- protective sheath (cladding)

Single Cable Assembly　　**Bundled Cable Assembly**

The basic parts of a single multifiber cable (left) and a bundle of multifiber cables (right) containing four single multifiber cables, with reflective/protective sheaths (cladding) and outer sheaths (e.g., armoring), as needed.

number of practical advantages to bundling fibers, including flexibility and the capability of splitting off fibers along the path as needed as for communications "drops" (local service), signs, or ambient light fixtures.

The arrangement of fibers within a bundle is related to the purpose, length, size, weight, and philosophy of the fabricator. Often large numbers of fibers are randomly bundled with small gaps between fibers. The gaps may be useful in providing added flexibility to the cable, or may be filled with filler materials to provide structural cohesion to the bundle. Some bundles are loose along the running portion of the bundle but tightly aligned at the endfaces through a fusion joint process. This facilitates coupling and reduces the space needed for the coupler.

Bundles are sometimes carefully aligned for certain purposes, as in a single *lightline* or an arrayed faceplate of certain dimensions. Sometimes bundles are deliberately randomized in order to provide a randomly even light source at the point where the light exits the endfaces of the fibers in the bundle.

Sometimes fibers are bundled for ease of handling and installation. It is practical to bundle many fibers together and provide extra external insulation and armoring against the elements in bundles that are to be laid hundreds of feet underwater in deep oceans. Bundles are also practical in situations where the delivery of different wavelengths over the same cable is desired. See faceplate, lightline.

Fiber Channel See Fibre Channel Standard.

fiber creel A device for spooling fiber optic filaments to facilitate handling. J. W. Hicks was one of the first to spool optical fiber. See creel.

Fiber Dispersion Measurement System FDMS. A fiber Bragg grating measurement system that utilizes interference phenomena to evaluate grating transmission properties, developed by NASA. The system is quickly able to fully characterize fiber device phase, amplitude, transmission, and reflection from either direction.

Fiber Distributed Data Interface FDDI. An American National Standards Institute (ANSI X3T12, formerly X3T9.5) standard high-bandwidth 100 Mbps packet-switched protocol developed by the X3T9.5 committee.

FDDI Architecture Standard Documents		
Abbreviation	Item	Notes
MAC	Media Access Control	A network control mechanism for defining formats and methods. Like the PHY layer, the MAC layer is directly implemented in FDDI chips. The higher LLC sublayer provides data to the MAC.
PHY	Physical layer	An electronic signal encoding/decoding layer which mediates between the higher MAC layer and the lower PMD layer.
PMD	Physical Medium Dependent	The lowest sublayer, which specifies various physical media such as interface connectors, cables, power sources, photodetectors, etc.
SMT	Station Management	A node manager and bandwidth allocator. The SMT is further subdivided into connection management (CMT) which controls access, ring management (RMT) which provides diagnostic capabilities, and frame services.

FDDI Basic Port Types		
Port	Type	Characteristics
M port	Master port	Connects two concentrators and can communicate with DASs and SASs.
S port	Slave port	Connects single-attachment devices for interconnecting stations, or for connecting a station to a concentrator.
A port	Dual-attachment	Connected to the incoming primary ring, and outgoing secondary ring. See A port dictionary entry.
B port	Dual-attachment	Connected to the incoming secondary ring and the outgoing primary ring.

FDDI is packet-switched, based upon token-passing technology, with multiple frames travelling the ring at the same time, i.e., a dual-ring network with a primary and a secondary ring. On most configurations, the primary ring is used for data communications, and the secondary ring is used as a backup. A *concentrator* is typically used as an attachment point, with dual attachment stations (DASs) attached to both of the rings.

FDDI is typically used on local area networks (LANs) with fiber optic lines such as campus backbones. Two types of fiber optic cable are typically used: single-mode and multimode. FDDI supports synchronous and asynchronous transmissions, transmits over fiber optic cables to distances of about 200 kilometers, and can service about 1000 individual workstations, depending upon the topology. The specified wavelength is 1300 nm for data transmission. Short hops within the network may be handled by twisted-pair copper wires.

FDDI has a theoretical maximum speed of about 1/2 million packets per second; however, with padding and various other factors, achieved speed in commercial implementations is about one third of this.

There are four documents associated with the FDDI architecture standard, as shown in the FDDI Architecture Standard Documents chart. See A port, Copper Distributed Data Interface, Dual Attachment Station, dual homing, optical bypass.

Fiber Distributed Data Interface II FDDI-II. A specification based upon FDDI for high bandwidth rates of up to 100 Mbps but enhanced to handle circuit-switched (in addition to packet-switched) PCM data for ISDN or voice, in addition to data. FDDI-II supports both basic mode and hybrid mode transmissions, and in hybrid mode, it adds isochronous support to the asynchronous and synchronous services provided in basic FDDI. See Fiber Distributed Data Interface.

Fiber Distributed Data Interface frame format The frame format for FDDI frames includes a data frame and a token frame. The data frame contains packets from higher-level protocols en route to another node. The token frame contains three bytes and a preamble for setting the signal clock. The basic FDDI frame format is shown in the FDDI Data and Token Frame Formats chart.

Fiber Distributed Data Interface physical layer The FDDI physical layer implementation differs somewhat from other IEEE standards. It consists of a medium-independent portion (PHY) and a medium-dependent or medium-specific portion (PMD).

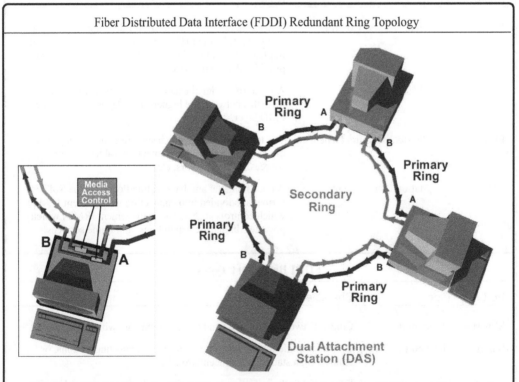

Fiber Distributed Data Interface (FDDI) Redundant Ring Topology

The dual-ring topology of this FDDI network provides fault-tolerant redundancy in the event that one of the workstations in the ring fails and is unable to pass through information intended for another workstation further along in the ring. A and B ports are provided on network interface cards (NICs), usually with standardized fiber optic connectors (e.g., ST- ports) for interconnecting the fiber optic cables with the other dual attachment stations (DASs) in the local network.

Fiber Distributed Data Interface ports There are four types of ports defined for an FDDI network as shown in the FDDI Basic Port Types chart.

fiber Bragg grating FBG. See fiber grating, Bragg grating.

fiber feed 1. Feeding a fiber optic filament from a spool for component assembly or installation. Mobile splicing labs often have spool creels for easy dispensing of the fragile fiber filaments. 2. In fusion splicing, the process of pushing together two ends of fiber filament at the correct speed and distance so they fuse evenly as the fibers melt away from one another during the heat process. See fusion splicing. 3. In fiber optic cable installation, the process of pushing a fiber cable along the path where it will be installed, whether that is in a wall, to a utility pole, or through an underground conduit.

fiber fuse effect An effect in germanium-doped fiber optic cables (which includes the majority of cables) whose discovery has been attributed to Russell et al. (The author was not able to procure primary documents but thought the anecdotal reports interesting enough to include here.) The effect is initiated with a heat stimulus applied to one part of a fiber optic cable, which causes the core to melt slightly to form a pit. The process then repeats without any further heat stimulus as the light propagates down along a length of cable for as long as the cable is consistent (not tapered, for example). This pockmarking destroys its integrity as a light-guiding vessel. It has been reported that if the laser input to the cable is reversed, the pits may disappear again. Germanium has some slight transducing capabilities that some say may account for the effect. See germanium.

fiber grating A structure within a fiber that may be established by doping and illuminating it with a pattern of light (e.g., ultraviolet) to "write" the grating into the fiber core. Gratings are important structures in fiber optic lasers and transmission lines that facilitate control of the optical processes within the fiber. There are a variety of types of gratings, including long-period gratings and Bragg gratings. Bragg gratings may have positive or negative index changes when stimulated with light. A variety of methods are available to increase the photosensitivity of a fiber grating, including hydrogen-loading or straining the fiber as it is being illuminated. Straining the fibers during grating imprinting does not appear to reduce fiber stability (Salik et al.). See Alexandrite, Bragg grating, doping, rare earth doping.

fiber handler An accessory used in conjunction with

Fiber Distributed Data Interface (FDDI) Data and Token Frame Formats

```
bits   64      8      8    (16 or 48) (16 or 48)   0      32     4      1
----------------------------/-----------/--------/--------------------
|   PA    | SD | FC |   DA    |   SA    | Info  | FCS | ED | FS |
----------------------------/-----------/--------/--------------------
```

PA	Preamble: frame synchronization with each station clock
SD	Starting delimiter
FC	Frame control
DA	Destination address
SA	Source address
Info	Information field: 0 to 4478 bytes
FCS	Frame check sequence
ED	Ending delimiter: end of frame
FS	Frame status

The above diagram shows the data frame format; the basic token frame format is as follows.

```
bits   16+     8      8      8
--------------------------
| Preamble | SD | FC | ED |
--------------------------
```

PA	Preamble: frame synchronization with each station clock
SD	Starting delimiter
FC	Frame control
ED	Ending delimiter: end of frame

a fusion splicing machine. The fiber handler may be fixed within the device, may be selectable from a number of options or may be swappable. The handler facilitates support of fibers, with different handles for different gauges or numbers of fibers from single fibers to linear fiber arrays to be made into ribbon cables.

Fiber in the Loop FITL. A fiber-based digital loop carrier system closely related to Fiber to the Curb (FTTC) and Fiber to the Home (FTTH). FITL provides main loop fiber connections to light-based/electrical multiplex/demultiplex and control terminal endpoints which connect, in turn, to subscriber lines (typicaly analog copper twisted pair). Recently, industry emphasis has shifted from FTTC and FTTH to hybrid fiber coax (HFC) installations to provide broadband video content to subscribers.

Fiber in the Loop A newsletter providing market factors in fiber installations and technologies around the world. Summarizes information from over 300 global sources; published monthly by Information Gatekeepers, Inc.

Fiber in the Loop, Integrated IFITL. A newer form of Fiber to the Curb implemented by Marconi Communications and offered by such vendors as BellSouth. It is promoted as a fiber optic version of ADSL services. Both ADSL and IFITL are provided by carriers for the delivery of Digital Subscriber Line (DSL) data services. Connection speeds are similar; the main difference is the media over which the data is transmitted.

IFITL offers the convenience of integrated voice, data, and video services through an RJ45 drop connection. Currently most IFITL subscribers are in the southeastern U.S., but the service is likely to spread to other areas. The loop length between an IFITL subscriber and the local Optical Network Unit (ONU) in newer Marconi MX services is relatively short, a few hundred feet, which should improve DSL services over those currently available through FITL services.

Fiber in the Loop Newsletter A biweekly publication (formerly *Fiber to the Home*) covering global developments in Fiber to the Home technology and its applications, published by Information Gatekeepers, Inc.

fiber laser A type of air-cooled laser based upon fiber (as opposed to diode lasers) that is smaller and requires less energy than many commercial gembased lasers but is also capable of higher power generation, to about 25 W. Fiber lasers were not initially favored for engraving applications, which require Q-switch capabilities, but are suitable for optical components testing, amplifier pumping, flexo-graphic printing, microsoldering, and microbending. However, developments in pulsed fiber lasers along with micro-electromechanical system (MEMS) technologies make it possible to use the modulator as a reflector with a switchable micro-mirror. This greatly extends the types of practical uses to which fiber lasers may be applied. Another advantage to fiber lasers is direct coupling, a feature that is often preferable to injecting the laser.

More recent fiber lasers based upon pulsed erbium design are suitable for freespace communications, fiber sensor technology, and rangefinding applications.

fiber loss Signal deterioration (attenuation, fade) in a fiber optic transmission. See fiber optic amplifier.

fiber optic, fiber optics, fiber optic cable, optical fiber A technology for guiding light through a fiber medium. The term *fibre optic* was coined in 1955 by Indian physicist N.S. Kapani, the inventor of cladding, high quality fibers, and many subsequent optical instruments and fiber-related technologies. Laser light is commonly used as the light source and may be modulated to carry information.

The end of an optical fiber, where light beams enter or exit the light-guiding medium, must be at a precise angle and polish if it is to be coupled with other components. There are different types of ends.

Fiber optic cables are commonly constructed of single or parallel, slender, transparent fibers of glass or plastic, encased in cladding materials that direct the light back into the core, and protective shielding materials to support the fiber and prevent damage. Light is transmitted through the length of the cable through internal reflection. The increasing capabilities of the medium to transmit signals over distance and the reflective nature of the transmission (which enable signals to be gently bent around curves) make this an excellent medium for data transmission and networking.

Fiber optic cables are generally categorized as single mode or multimode. Multimode cables provide an optical pathway that is wide enough that the signals refract at various angles as they travel along the cable. Eventually, however, they tend to run together, thus limiting cable lengths for this type of transmission to distances of less than one or two kilometers. Since dispersion in single-mode cables is not of the same nature as that of multimode cables, transmission distances are longer. Newer amplification and regeneration technologies are extending these distances.

There are different ways to generate the light that travels through the cable. Single-mode cables typically use lasers, whereas multimode cables typically use light-emitting diodes (LEDs). The wavelengths of light used vary with the installation.

The unenhanced capacity of fiber optic is about 2.5 Gbps, but technologies are being developed and implemented for boosting this. Fiber optic cables have far greater bandwidth, information-carrying capacity than traditional copper phone wires, and are suitable for transmitting high information content signals including full-motion video. See blown fiber, cladding, coaxial cable, copper wire, fiber optic taper, laser, multimode optical fiber, single-mode optical fiber.

fiber optic amplifier A device for increasing the power of a signal in a fiber optic lightguide. Ideally, a fiber optic amplifier will improve the signal to noise (S/N) ratio in a fiber transmissions medium and extend the distance of a transmission. Fiber optic cables most commonly use light from lasers or from light-emitting diodes (LEDs). The light source can be modulated by directly manipulating it on and off, or it can be indirectly manipulated by an outside controlling device. A lithium niobate modulator is an example of an external modulator that is used in the cable TV and digital network industries to extend a signal. Because it allows a stronger laser to be used, the signal can be sent farther. Pump lasers are a means to use semiconductor technology to amplify a laser signal. Doping with rare earth elements, such as erbium, alters the transmission properties of a cable, and the technique is commonly used to amplify signals. In general all-optical amplification systems are preferred, but some still use electrical amplification, which requires conversion of the light signal to electricity and back again after amplification. See doping, modulation.

fiber optic analyzer A diagnostic device for diverting and analyzing the coherent light signals transmitted through a fiber optic cable. In general, these instruments split or divert the light into its spectral components and compare the results with what is expected to be traveling through the cable. Discrepancies could indicate a problem. Traditionally, somewhat bulky systems have been used (about the size of a portable radio), but in 1999 T. Erdogan and a team of researchers at the Institute of Optics, University of Rochester, NY presented a solution based upon tiny grooves in the fiber that reflected light to a miniature mirror. Smaller than an eraser, the system lends itself to installation anywhere fiber optic communications cables are installed.

Fiber Optic Association, Inc. FOA. An international nonprofit society for fiber optic professionals incorporated in Massachusetts in 1995. FOA has been active in standards development, developing and administering a training curriculum and certification program, and promoting fiber optics technologies and their deployment. Members include engineers, installers, consultants, contractors, manufacturers, and sales personnel.
http://www.thefoa.org/foa.htm

fiber optic bundle An aggregated group of two or more fiber optic cables (which may have single or bundled multiple waveguides) plus tension members and protective sheaths as are needed to protect the inner cables. Filler cables may be included to support the structure to keep the inner cables in position. Optical cables are often bundled when supporting multiple transmission paths or different types of data types. There is a trade-off in weight and flexibility when cables are bundled, but there are also economic advantages and advantages in ease of installation and the amount of space required. The bend radius is usually greater in a single fiber, compared to a bundle, but a single-bundle cable and a multiple-bundle cable may have relatively the same bend radius. The tensile strength of a bundled cable may be higher. The number of cores within a tube can range from one to two dozen or more.

The manufacture and assembly of fiber optic bundles is a challenging profession. A bundle may contain thousands of optical fibers all of which must be positioned in close proximity within very fine tolerances. The terminating ends of the individual fibers must be exactly aligned for coupling with other components and must be highly polished to prevent interference and back reflection where the light beams enter or escape the fiber.

fiber optic cable A cable with at least one fiber optic filament, which may include a protective covering. Fiber optic cables for lighting and hobby applications are typically very simple assemblies of single or multiple fibers. Cables intended for industrial environments or high-speed data transmissions are fabricated and tested to higher standards to maximize transmission characteristics and minimize loss through the medium and the coupling points. End glow fiber optic cable is designed to maximize reflection along the waveguide so that the maximum amount of light travels to the end of the waveguide opposite from the incoming light source. Side glow cable emits light along the length of the cable and can be made to emit a variety of colors, making it practical as a substitute for neon. See InfoWatt.

Fiber Optic Cable

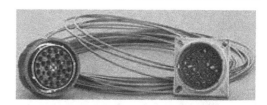

A fiber optic cable used in the NASA Fiber Optic Reliabiliy and Maintainability Program. NASA has made many contributions to fiber optics technology and uses fiber optics in its space shuttle programs and payload experiments. A novel Goddard Space Flight Center application was the development of a fiber optic solid state recorder to replace one of the reel-to-reel tape recorders in the Hubble Space Telescope. [NASA/GRC photo by Marvin Smith, 1995.]

Fiber Optic Data Bus FODB. A light, low-power, high-speed (up to 1 Gbps), IEEE 1393-compliant, fiber optic, ATM, ring-based data network developed in the late 1990s for use in future spacecraft. The system supports 1 master Controller Fiber Bus Interface Unit (CFBIU) and up to 127 slave Fiber Bus Interface Units (FBIUs). The system was developed by NASA/GSFC and the Department of Defense (DoD) to support realtime, onboard data handling for remote-sensing spacecraft. It is intended to be reliable, fault-tolerant and able to withstand the harsh environment of space.

Fiber Optic Data Transmission Experiment FODTE. A NASA/Langley Research Center experiment to assess how fiber optics equipment would be affected by a space environment.

NASA/Langley Fiber Optic Sensing Systems

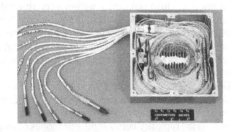

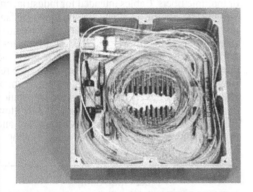

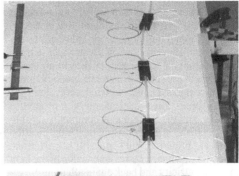

Fiber optics forms an important part of space science research. In 1999, a flight instrument for multiplexed reading of Bragg grating-based sensors was scheduled as part of the second Human Exploration and Development of Space Technology Demonstration to measure hydrogen leaks in the shuttle engine bay. The Integrated Vehicle Health Monitoring (IVHM) system is a network of optical fiber sensors for quickly determining the condition of space vehicles upon return from orbit, thus reducing costly between-flight inspections.

Shown above are fiber sensor packages for sensing gaseous hydrogen. [NASA/Langley photos, 1999.]

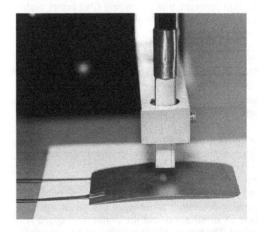

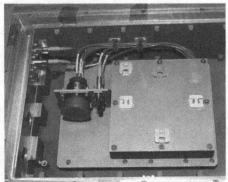

A number of types of fiber optic sensors have been developed by NASA research centers. The above photographs (1998) show open and closed examples. The bottom photo illustrates the fiber network box installed in a Radstone lid. [NASA/Langley photo, 1998.]

A fiber optic displacement sensor positioned above a high displacement Actuator (HDA) developed by NASA/Langley Research Center scientists. The HDA assembly may be used for high performance sensing applications, including auditory sensing, nondestructive, testing, and vibration sensing. [NASA/Langley Research Center photo, 1994.]

fiber optic field testing The process of assessing the optical power transmitted through a fiber optic path between designated points with equipment that can be readily carried and used by mobile technicians (installers, maintainers, etc.).

In its simplest form, the testing of a fiber optics transmission link involves projecting a light source into the lightguide and measuring the effective range and strength of the signal at another point. The light source and wavelength(s) are typically selected to match those that will be used on the system. Light-emitting diode or laser light compact units are commercially available to emit light at standardized frequencies in single or multiple modes. The strength of the light signal is set to correspond to the dynamic range of the signal through the transmission path. A power meter is then linked to the system to assess the power and loss statistics associated with the light at the other end of the transmission link from the light source. Adjustments, repairs, or replacements are made until expected or optimum readings are realized.

fiber optic flight control Aeronautic control systems based upon fiber optics cables rather than traditional wires. Optical fiber has a number of advantages in aeronautics, including light weight and immunity to lightning and radar interference. In the early 1990s, NASA successfully tested fiber-based engine speed and engine turbine exhaust temperature sensors on an F-15 aircraft at the Ames-Dryden Flight Research Facility through the NASA/Lewis Fiber Optic Control System Integration (FOCSI) program. The followup was multiple fiber systems on F-18 aircraft to monitor surface positions, air pressure, nose-wheel steering and other aspects.

fiber optic history See laser history.

fiber optic illuminator A device providing a suitable light source for transmission through a fiber optic waveguide filament or cable bundle. For communications networks, laser lights (with strong coherent beams) are preferred. For lighting or hobbiest applications where distance and precision are less important, incandescent quartz halogen lights are practical. A 75 W halogen lamp can illuminate about 200 side glow filaments and about ten times as many end glow filaments, depending upon the diameter and length of the fiber. Some light sources may be varied in intensity. Color or daylight filters may be optionally available.

For miniature hobby applications, battery-powered white LEDs may be used. Some commercial lighting and novelty illuminators have optional color wheels. See laser.

fiber optic laser sensor A device that uses a fiber optic laser for sensing a desired target object by projecting a beam towards the target and sensing light that is reflected back. Depending on the fabrication and the desired type of sensing, a FOLS device may have a coherent or diffused beam. The finer (more coherent) the beam, the more precise the targeting can be for small objects or those in proximity to others. The use of optical fiber enables FOLSs to be very small.

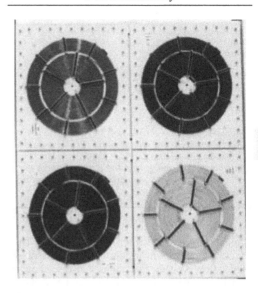

The FODTE incorporates 10 fiber optic cable samples with connectors, brackets, and nonmagnetic fasteners. Four optical fiber cables, configured in the form of a planar helix coil, are attached to thermally isolated mounting plates. The top photo shows the wound cables and assembly in the prelaunch stage, in 1984; the bottom is the same assembly in the post-launch flight stage, 1990. Later assessment revealed a number of patterns of discoloration as a result of space conditions. [NASA/LRC; NASA/JSC photo.]

fiber optic lever microphone FOLM. A microphone based upon a vibrating membrane that reflects light to enable intensity modulation of the light signal. FOLMS are useful for assessing high-frequency acoustic data in high-temperature settings (e.g., aeronautics).

fiber optic modem FOM. A fiber optic-based modulating/demodulating device for providing conversion between electronic and optical signals in a communications network. See modem.

Fiber Optic Woven Panel

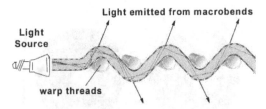

Supporting warp threads are woven across fiber optic filaments to provide structural support and sufficient tension against the fiber filaments to create a macrobend that enables light to escape through the cladding to provide a regularly distributed sheet of illumination.

fiber optic panel A planar surface (like fabric) consisting of woven, cladded fiber optic filaments. This is a unique concept in which the characteristic of cladded fiber is encouraged rather than discouraged. At each point in the woven fabric where a macrobend occurs, light is able to escape the cladding at regular points, creating an illuminated surface that can be used for a wide variety of applications. Lumitex, Inc. provides a commercial version of this concept. Lumitex panels consist of a number of layers of assembled and bonded fiber optic weave and may include a semitransparent diffusing layer to further control the spread of light. Where the fibers extend from the woven panel, they are collected together in a bundle and coupled with the illumination source.

fiber optic probe A probe used in medical or industrial applications that employs a bundle of fine, aligned optical fibers with light to transmit an image to a monitor, computer, or video recording device. Fiber optics probes are usually designed to be flexible so they can be threaded through small, curved spaces. This makes them valuable for medical examination of living bodies and industrial inspection of pipes and internal structures. A fiberscope is a similar tool, with a lens on one end, and a viewing eyepiece, or connection to a display monitor on the other end. See laser probe.

fiber optic rod A rigid fiber lightguide. Fiber optic rods come in a variety of materials and diameters and, for some applications, may be easier to install and more cost effective than flexible fiber filaments. They can be installed in locations and devices where the link is permanently coupled over a short distance (e.g., inside a panel or probe).

fiber optic sensing system FOSS. A system that incorporates fiber optics for transmitting the signal or data gathered by sensing devices (e.g., probes) to a display or processing system. Fiber optics sensing devices are used in industrial testing and measurement, contracting, network installation and maintenance, space science, and medical imaging. See illustrations on following page.

fiber optic splicing lab A fixed or mobile unit with

equipment to facilitate the joining of fiber optic transmission cables that require multiple joints to meet distance requirements for the lightpath (e.g., networking backbones). It may also have equipment for coupling the fiber filaments with terminating connectors. The assembly of fiber components requires certain environmental conditions, including a clean work space, protection from heat, moisture, and ultraviolet influences, effective racks, spindles and creels for inventory and spooling of filament lengths, good safety gear (e.g., goggles to protect from fiber shards or laser lights). It also requires cleaving, splicing, and testing equipment, disposal bins for waste, especially for sharp fiber optic shards. In mobile units, shock absorbers and stabilizing components are also important as cable splicing tiny optical filaments is a precision task. See fusion splicing.

Fiber Optic Probe and Probe Actuator

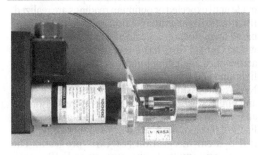

A small fiber optic probe and probe actuator used in space sciences. [NASA/GRC photo by Chris Lynch, 1998.]

---

fiber optic stub A short length of fiber optic filament that may be used as an intermediary component, as between a fiber taper and a CCD imaging component. Stubs have practical uses in situations where fusion splicing is not desired and stray light emissions should be kept to a minimum (e.g., potentially damaging laser lights). However, they introduce one more link in the signal chain which can produce loss of light or distortion of the signal at the coupling points and may not be practical for some applications. Stubs exhibit less loss, in general, than tapering components, which suffer loss through the regions where the taper occurs. In a sense, a stub can be seen as a 1:1 taper. Stubs come in a variety of polishes and end shapes (ball, tapered, cleaved, etc.) for different applications. New bonding technologies are reducing the need for fiber stubs in assemblies where the fiber is bonded to imaging or other components. See fiber optic taper.

fiber optic taper A fiber optic component designed to magnify or reduce image signals from its input surface to its output surface at various aperture sizes. It creates an effect similar to a glass magnifying lens except that the end surfaces of the fiber are flat, not curved (the curvature is in the light guide, thus eliminating the edge distortion typical of traditional magnifying lenses). The image enlargement or reduction is related to the diameters of the taper ends. Round to round, round to rectangular, and square to square tapers provide a variety of coupling options. Tapers are used in imaging applications in light sensors, fluoroscopy, video imaging, and medical, dental, and crystallographic radiography.

Enlargement/reduction ratios range up to about 1:5, above which loss through the tapered edges may be too high to be of any practical use. A fiber "taper" with a 1:1 ratio is called a fiber optic *stub*. See fiber optic stub, fused biconic taper.

Fiber Optic Taper Application

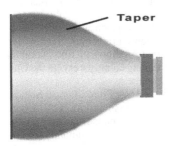

Taper

In an X-ray imaging system, a fiber optic taper may be bonded between a scintillator and active CCD imaging electronics, sometimes instead of multiple fiber optic components. The X-rays are converted to light energy that passes through the taper to the CCD components where they are further converted to electrical signals. Thermoelectric cooling of the CCD component can increase maximum possible exposure times.

Advancements in bonding technologies, which reduce loss and interference between fiber optic tapers or fiber optic faceplates and the imaging electronics, enable larger overall imaging areas and enlargement/reduction ratios to be realized. Fiber optic tapers and fiber optic faceplates may be combined (e.g., in X-ray imaging components). See faceplate, fiber optic.

fiber optic taper array A component comprised of multiple fiber optic tapers tightly aligned in a "tiled" pattern through pressure or heat. The array can then be bonded to other components such as CCD imaging electronics to create larger imaging areas than are possible with even the largest individual tapers. In a sense, a taper array is a specialized type of fiber optic faceplate.

Tapers come in a variety of sizes and shapes and the selection of a basic shape (or shapes) will influence the arrangment of the array and the effect of the enlargement or reduction.

Fiber Optics and Communications A newsletter summarizing information from over 300 global sources in the fiber optics industry. Published monthly by Information Gatekeepers, Inc.

Fiber Optics LAN Section FOLS. An industry consortium under the aegis of the Telecommunications Industry Association dedicated to promoting and educating about the technical advantages and economics of optical links in local area networks. http://www.fols.org/

Fiber Optics News An 8-page newsletter describing trends and projections for the fiber optics industry. Published weekly by Phillips Business Information, Inc.

Fiber Optics Sensors & Systems FOS2. A summary compilation from over 300 global sources in the fiber optics industry, published monthly by Information Gatekeepers, Inc.

Fiber Optics Weekly Update A review of over 300 global sources summarized for busy professionals. Published weekly by Information Gatekeepers, Inc.

fiber randomization In a fiber bundle, the arrangement of individual fibers with respect to one another such that a randomly even distribution of light occurs at branching junctions. Sometimes the fibers are arranged in specific rather than random patterns to facilitate a certain coupling pattern or ratio.

fiber ribbon A linear (flat) array of fiber filaments arranged in close parallel proximity and held together by jacket materials that may be color coded. They are similar to the ribbon cables used to link SCSI internal devices to a computer motherboard, though they usually contain fewer lines (typically up to 24 compared to wire ribbon cables that often have 50 or 60 lines). Creating wire ribbons is a fairly straightforward production task, whereas creating fiber ribbons is a significant technical challenge. It is difficult to mass splice multiple fibers that all meet necessary precision light-guiding characteristics. See fusion splicing, mass fusion splicer.

fiber ribbon splitter A tool designed to split fiber ribbon cables into smaller pieces, such as splitting a 24-ribbon cable into 6- or 12-ribbon sections for installation or for splicing in mass fusion splicers that only accommodate up to 6 or 12 splices at a time.

fiber rod A length of plastic or glass fiber for transmitting light, usually in the shape of a straight or slightly bent cylinder. A fiber rod is essentially a stiff fiber filament with a wider diameter than common thread-like fiber filaments used in longer communications lines. Quartz rods were used as light conductors in dentistry at least as early as the 1900s, and rods in a variety of glass and plastic materials are still practical for certain applications where a short length of secured nonflexible light-conducting material is appropriate.

Fiber to the Curb FTTC. Similar to Fiber to the Home except that the drop connects to a neighborhood Optical Network Unit (ONU - sometimes called a *pedestal*) serving a block of about a half dozen homes, rather than connecting directly to the network interface on the subscriber's premises. The link from the ONU to the subscriber's premises is typically a combination of copper twisted pair (voice/data) and coaxial cable (video).

FTTC has been implemented since about the mid-1990s. Initial installation costs are somewhat less than Fiber to the Home (FTTH) but maintenance and power consumption costs may be higher over time. IFITL is a variant of Fiber in the Loop (FITL) that implements Fiber to the Curb concepts to deliver voice, digital TV, and high-speed data as an integrated service rather than as separate services as is common now. See Fiber to the Home; Fiber in the Loop, Integrated.

Fiber to the Home FTTH. The system bringing fiber optic cables and telecommunications services to residential subscriber premises. In general, the link to the central office consists of a Host Digital Terminal (HDT), a Passive Optical Distribution (POD) network, and a link to the subscriber network interface (NI). Bellcore TA-NWT-00909 describes the generic requirements for fiber links in telephone local loops. FTTH is an ambitious undertaking, but BellSouth and the Nippon Telegraph and Telephone Corporation (NTT) announced in June 1998 that they were embarking on a joint project to make it a reality. The specifications being developed by the two telecommunications corporations will be disseminated through the international Full Service Access Network (FSAN). Trial regions were set up in the late 1990s, offering 120 channels of digital video, 70 channels of analog video and 31 channels of CD-quality digital audio. See Fiber in the Loop.

Taper Array Example

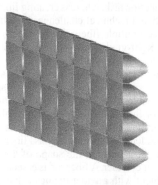

This diagram illustrates one possible taper array configuration, in which rectangular to circular tapers are coherently aligned and tightly bundled to create a specialized fiber optic faceplate capable of reducing an image (from left to right) when combined with electronic imaging components.

---

fiber, polarization maintaining A fiber waveguide with two or more segments that are fused at rotational angles such that they maintain the polarity of light traveling through the waveguide. See fusion splice.

FiberFax A commercial database published by KMI Corporation. FiberFax lists over 12,000 fiber optic installations and is searchable for major tops in fiber optics, including SONET, Frame Relay, FDDI, and FITL. See Fiberoptic Undersea Systems, KMI Corporation.

FiberGlobe Service A commercial market information database service from KMI Corporation that provides economic information on single- and multimode fiber optic cables, connectors, and transceivers

from 50 countries since 1993.

Fiberoptic Undersea Systems A commercial database tracking the links for 325 undersea fiber optics systems worldwide, published by KMI Corporation. See FiberFax, KMI Corporation.

Fiberoptics Market Intelligence® An analytic newsletter that describes developments and trends in fiber optics markets. The publication includes primary market information and articles on contract awards, mergers, standards development, research, and patents, published semi-monthly by KMI Corporation.

Fiberoptics NewsBriefs FNB. A weekly news service available on the Web without charge from KMI Corporation. It describes local, long-distance, and international submarine fiber optics markets and commercial releases of products serving those markets. http://www.kmicorp.com/fiberoptics_news_services/fiberoptics_news_briefs.htm

fibre Alternate spelling of fiber, common in most English-speaking countries except the United States. See fiber.

Fibre Channel See Fibre Channel Standard.

Fibre Channel Association FCA. An organization supporting Fibre Channel technology that merged with the Fibre Channel Community (FCC) as a working group in 1999. The FCA is based out of Austin, Texas. See Fibre Channel Industry Association.

Fibre Channel Group An association of graduate and undergraduate research and engineering students at the University of Minnesota supporting and learning about gigabit Fibre Channel-based networking technologies. The Group is supported by the Brocade Communications. The group provides useful fibre channel basics information on its Web site. http://gfs.lcse.umn.edu/fc/

Fibre Channel Industry Association FCIA. A not-for-profit, international trade organization supporting manufacturers, developers, vendors, and system integrators providing fibre channel technology and services. The FCIA was founded in 1999 as a result of a merger between the Fibre Channel Association (FCA) and the Fibre Channel Community (FCC). http://www.fibrechannel.com/

Fibre Channel Consortium FCC. An association in the University of New Hampshire (UNH) InterOperability Lab which provides support for product interoperability and educational demonstrations of Fibre Channel technology, founded in 1995 through a cooperative agreement of vendors. The Consortium tests Fibre Channel products in the lab on a scheduled basis. http://www.iol.unh.edu/index.html

Fibre Channel hub A device for facilitating the centralized connection of devices for sharing data in a Fibre Channel network. When used in a loop configuration, a Fibre Channel hub enables reconfiguration and automatic insertion and removal of additional or failed loop devices in order to facilitate the automatic administration of a Fibre Channel-based network. An arbitrated loop topology is a simplified means of using Fibre Channel technology without creating a Fibre Channel fabric configuration.

Fibre Channel Layers and Classes of Service

Layer	Functions
FC-4	Functional layer for interfacing with upper layer protocols and applications interfaces such as TCP/IP, ATM, SCISCI, HiPPI and others.
FC-3	Functional layer for common layer services needed to support applications, media, security, etc. It is still under development.
FC-2	Physical layer for signaling, framing, flow control, five service classes plus one intermixed class (see below).
FC-1	Physical layer for ordering of data and encoding, e.g., Escon 8B/10B encoding and coding (patented IBM technology). Control of media access ports and operators.
FC-0	Physical layer for rate-related media interfaces and transmissions such as 133 Mbps to 1.062 Gbps with room for higher future speeds. This is the level at which devices are physically linked with copper wire or optical cable.

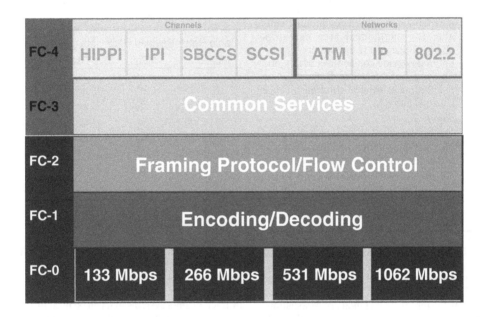

Fibre Channel FC-2 Layer Classes of Service

Class	Functions
Class 1	Switch configuration for a guaranteed, dedicated, connection-oriented circuit-switched channel is established. This class is suitable for high-speed multimedia and scientific data-handling applications.
Class 2	A guaranteed, connectionless, frame-switched circuit is established. This class is suitable for interactive, shared bandwidth, bursty transmissions.
Class 3	Similar to Class 2 but without the guarantee (unacknowledged). Suitable for high-throughput, multiple-destination messages (e.g., broadcast messages).
Class 4	Constant bit-rate (isochronous), guaranteed latency for digitized data.
Class 5	An optional combination where Class 2 and 3 connections may use unused Class 1 bandwidth, with priority to Class 1 frames.
Class 6	Guaranteed multicast service.

Fibre Channel Loop Community FCLC. A California-based organization supporting and promoting Fibre Channel technology with particular attention to mass storage in arbitrated loop topologies. The FCLC merged into the Fibre Channel Assocation as a working group in 1999.

Fibre Channel Topologies

point-to-point topology

Direct connections between two N ports wherein one is a server. This is a nonscalable topology by definition and arbitration is not required.

arbitrated loop topology

A topology in which multiple devices can share the media. It is a middle solution between the limited point-to-point topology and the more flexible but also more complex fabric topology, a compromise suitable for small local area networks (LANs). Because resources are shared, only one device con connect at any one time to a shared resource. Devices can be chained or arbitrated through a Fibre Channel hub.

fabric topology

The most powerful and potentially complex implementation of Fibre Channel, fabric topology encompasses more than one FC switch interconnected to other devices through one or more ports. F ports enable connections to other F ports or to N ports (node ports).

Fibre Channel specifications FCS. There are two general aspects of FC specifications, the ANSI Fibre Channel Standard (X3T11) and FCSI Fibre Channel Profiles developed by the Fibre Channel Association (now the Fibre Channel Industry Association) to assist implementors in understanding and developing the technology while maintaining interoperability. See Fibre Channel Standard, Fibre Channel Industry Association.

Fibre Channel Standard FCS. A high-speed, block-oriented, serial, fully bidirectional data transfer interface for interconnecting workstations, mainframes, display peripherals, and storage devices. Both electrical and optical media are supported by the standard. FCS has been standardized by the ANSI X3T11 committee.

The Fibre Channel Standard is intended to support both wire and fiber optic systems, from 133 Mbps to 1062 Mbps (and higher) at distances up to 10 kilometers (contrast this with a standard SCSI electrical cable transmission, for example, which has a practical distance of only a few feet). The actual maximum distance is dependent upon the medium and data rates. For example, single-mode fiber can transmit up to 10 kilometers, whereas shielded twisted-pair wire has a limit of 50 to 100 meters, depending upon whether the data rate is 25 or 12.5 MBps. Frame sizes may be up to 2,148 bytes, depending upon the size of the payload (the informational content). Development is underway to support higher speeds (e.g., 10 Gbps).

There are three general topologies (architectures) described for Fibre Channel networks.

FCS provides excellent opportunities for aggregating peripherals under desks, or in server rooms and secure areas, for standardizing a wide variety of computer peripherals, and for increasing architectural flexibility in the placement of equipment. Because it is a hot-swappable format, Fibre Channel devices can be added to or removed from a network without powering down the network.

FCS is a hierarchical, layered architecture, with five layers defined from highest to lowest.

Six data rate speeds have been defined as part of the Fibre Channel Standard.

Cable Rate	Payload Rate	Max. Distance	
Mbaud	MBps	Mini-Coax	9 µm Fiber
132.8125	12.5	35 m	10 km
265.6250	25	25 m	10 km
531.2500	50	15 m	10 km
1062.5000	100	10 m	10 km
2125.0000	200	–	–
4250.0000	400	–	–

See F port, Firewire, FL port, N port, storage area network. See Fibre Channel Layers chart and other entries prefaced by Fibre Channel.

Fibre Channel Systems Initiative FCSI. A group organized in 1993 to promote the interoperability, use, and distribution of Fibre Channel technologies. FCSI Profiles serve as guidelines for the implementation of fiber channel systems that can be used by component manufacturers and systems and service integrators of fiber communications technologies. See Fibre Channel Industry Association.

Fibreoptic Industry Association Limited FIA. A professional organization representing suppliers, educators, and installers in the fiber industry, inaugurated in February 1990. The FIA is managed from the U.K. by the FIA Secretariat with support from a management council of elected individuals taken from FIA members and other supporting individuals. The organization includes a number of semi-autonomous regional offices. http://www.fibreoptic.org.uk/

Fick's first law The mass movement of atoms (flux) from one point to another in a given time frame is equal to the negative diffusivity times the delta concentration (the mass per given volume) divided by the delta diffusion distance as expressed for a one-dimensional system by

$$F = -D(\delta C/\delta x)$$

For the condition $C(\lambda) = 0$, the equation can be

simplified to

$$F = D(C(0)/\lambda)$$

As applied to biological systems, the net rate of diffusion of a gas across a fluid membrance is proportional to (a) the difference in partial pressure, (b) to the area of the membrane, and inversely propertional to the membrane's thickness.

Fick's second law describes the time-variant diffusion of atoms in a material for a one-dimensional system (e.g., unidirectional diffusion from a planar surface) for a constant diffusion coefficient. This facilitates approximations for various factors related to diffusion. See diffusion.

FID Field Identifier, part of an ISDN Service Profile Identifier. See SPID.

FidoNet Established in 1984 by Tom Jennings with the second node belonging to John Madill, this networked bulletin board system (BBS) became a major communications tool for techie discussions, email, and file transfers as bulletin board operators all over the country started to establish Fido boards for their local users.

field In a scanning video broadcast display, a field is every other line of the full picture frame. Thus, it is all the odd numbered lines taken together, or all the even numbered lines taken together, in an interlaced image.

field, data A record-holding or record-entering entity in a database. The definition of field types facilitates program setup, management, and data manipulation by alerting the software as to the nature of the information being entered into a field. That is, a field may be given a data type (number, string, date, etc.), or it may be untyped, but either way this tells the system something about the data.

field mode In video image capture, a mode that captures only half of the scan lines in order to save an image in lower resolution, thus taking less storage space. See field, frame mode.

field winding A mechanism for energizing electromagnets in a generator. See winding.

Field, Cyrus West (1819–1892) An American industrialist who contributed significantly to the economic and political negotiations involved in the first transatlantic telegraph cable, originally completed in August 1858 and later successfully installed as a permanent cable. The concept was championed in part by Canadian inventor Frederic Gisborne, who appealed to the American business community for financing so he could establish eastern Canadian telegraphic installations through a combination of steamships and underwater cables. Maritime Canada was a sparsely populated wilderness at the time, with a great need for communications links and limited resources for their development.

As a youth, Field was apprenticed to a prominent and successful merchant before setting up his own paper manufacturing firm. He thus had an established network of contacts in the New York and Washington, D.C. areas that he could draw upon for support. As a successful businessman, he probably could have avoided hazardous and technologically difficult projects, but the telegraph cable idea evidently fired his imagination. After considering the feasibility of the project, with help from his brothers and other business associates, Field contacted influential people for technical assistance and support (e.g., Samuel Morse).

Cyrus Field – Transatlantic Cable Installer

Cyrus Field achieved fame for his contributions to the laying of the first transatlantic communications cable. His portrait was widely used in articles and song sheets commemorating the historic event. [Portrait from the Mathew B. Brady studio, ca. 1850s. Library of Congress American Memory collection.]

Over the next decade, Field, Taylor, Roberts, and Cooper made substantial investments in the venture, with the project's promotional lead, Cyrus Field, receiving most of the public acclaim for the ambitious project.

In time, the Canadian-American alliance interlinked maritime Canada and, after initial failure, linked Canada to the U.S. through the Gulf of St. Lawrence. These early trial-and-error cable installations no doubt provided valuable technical experience facilitating the development of longer, ocean-laid cables. In 1856, Field traveled to England on behalf of the transatlantic cable project and formed the Atlantic Telegraph Company. In December 1861, George Opdyke wrote to President Abraham Lincoln in support of the project. No doubt other supporters made similar appeals to prominent dignitaries. The Library of Congress and Cornell archives include examples of correspondence and letters of introduction from Cyrus Field to prominent persons, including the President. These were turbulent times in America, however, and the pursuit of a cable communication with the British Isles wasn't the easiest cause to champion in view of British-American relations. In spite of the technical and political difficulties, success was finally achieved after several failures and temporary successes, with a permanent cable installed in the mid-1860s. Instantaneous intercontinental communication was now possible, a revolution that forever changed humankind's commercial and social interactions.

After the laying of the transatlantic cable, Field became an advocate for the public acquisition of communications services. To garner support for this concept, Field penned an article called "Government Telegraphy" for the *North American Review* in March 1886 that states, in part:

"It appears to me that the time has arrived when the Government of the United States should purchase, in the interest of the people, all the telegraph lines in the country.... Communication by telegraph has become almost as common, and quite as necessary, as communication by letter.... As letters are sent to all parts of the United States for two cents, and papers and magazines at one cent for every four ounces, so, I think, a telegraph message should be sent to any part of the United States at the lowest price at which experience has shown it possible that the transmission can be effected...."

An illustrated history of the Atlantic Telegraph Company is available on the World Wide Web. See Gisborne, Frederic Newton; gutta-percha; transatlantic cable. http://www.atlantic-cable.com/

FIF See Fractal Image Format.

FIFO first in, first out. In programming, a means of processing data so the first item to be stored or placed on a stack is the first to be fetched, moved, or discarded. Imagine a narrow vertical tube for gerbils (or hamsters, if you prefer fluffy rodents); the first gerbil to squeeze in through the top is the first to slide out the bottom. In general terms of telecommunications, in a FIFO system the first person who calls is the first to be referred to an agent.

Fiber Optic Filament Spools

Spools of fiber optic filament are sold for a variety of lighting and hobby applications, including signs, lamps, and art pieces in sizes ranging from about .25 to 2.0 mm. Sheathed filaments that can be separated from the main cable as needed are also available.

---

filament 1. A fine metal conducting wire commonly used in tubes and bulbs. By passing a current through a filament in a specialized, enclosed environment, it becomes incandescent, giving off light. See cathode. 2. A fine cylinder of glass or plastic. In fiber optics technologies, glass or plastic filaments with good conducting properties (e.g., clear glass with a minimum of fabrication blemishes and impurities) are used as waveguides for the conduction of light signals. They are popular as novelty fiber optic lamps, signs, art pieces, and for communications technologies. See fiber optics.

file A collection of associated data stored so that a pointer to the information identifies and encompasses the contents of that file as an accessible, readable unit, even if stored in separate parts. This is one of the most common units of storage in a computer system – file hierarchies, file folders, file types, and file management are all computer structures and processes constructed to manage files.

file attachment Most email systems are text-oriented 7-bit messaging media. So how do you send someone an 8-bit binary file? To meet this need, many text email systems have the capability of sending binary files as file attachments to a message. Since binary files include symbols and characters which cannot be displayed in a plain text window, and since the symbols are not meaningful to humans, it is more practical to send the file (which may be sound, graphics, or a computer application) as an attachment, rather than as a postscript to the email text message. In most cases, all that is necessary is to specify the name of the file in the Attachment: text box or email message header, and the system will take care of the transfer of the information.

file cache An area of memory allocated by an operating system or computer applications program to temporarily store a file that may need to be accessed or modified frequently. Many database and spreadsheet programs use file caches to allow quick updates and redisplays of information, and the data may also be periodically stored on disk as a background task so as not to lose information and updates in the case of a software crash or power outage.

file extension A syntactic convention that aids in identifying computer data file types. There are many categories of computer files: text files, graphics files, sound files, and within these basic categories are many subcategories, such as JPEG, TIFF, etc. A convention of adding a period and a short suffix to identify the type of file, so it can be found at a glance, has become widespread, and some applications and systems will even enforce certain file extensions.

Since the mid-1980s, every significant microcomputer and workstation level operating system except MS-DOS has allowed file extensions of reasonable length (up to 16, 32, 64, or 256 characters for the whole file name, depending upon the system). MS-DOS restricted its users to only three characters and enforced the use of the period (dot) as the file extension symbol. Since there were so many DOS-based machines, users of other systems had to truncate file extensions (and the rest of the file name) when transferring files to other systems. This impractical three-character extension limit is still prevalent, even though most Windows-based systems now support longer filenames (it is still common to see HTML file extensions on the Web listed as ".htm" instead of ".html").

On most other systems, the dot (.) is not mandatory for specifying the extension. The user can save a file with no dots or with a dozen dots. However, since several early systems in the 1970s required a dot, users are used to this naming convention.

On many systems, a file extension also lists a version number, so backups and revision histories can be maintained at all times. For example, the following extensions may be automatically generated:

`mygreatimage.tiff.1` or `mygreatimage.tiff;1`

`mygreatimage.tiff.2` or `mygreatimage.tiff;2`

`mygreatimage.tiff.3` or `mygreatimage.tiff;3`

If the default for the revision level is three, then the next file to be saved under the same name will supersede the oldest file, in a first in, first out (FIFO) sequence, so that no more than three files with the same name are stored at any one time. This version number extension/revision system is very handy when something is saved accidentally, and the previous version needs to be retrieved.

file gap A blank inserted to indicate a stopping point, or a division between sets of information, especially on a sequential file recording system. On an audio tape or digital data tape, a file gap indicates the beginning or end of a song or file.

file server Generally, a system on a network that administrates the storage of and access to files, often through a client/server model, in which multiple users make requests to the file server through the client software. This system reduces redundant storage of files on individual systems and makes it easier and faster to update individual files. The server also handles file locks, so data files cannot be simultaneously updated and saved by multiple users. Usually, a dedicated file server is equipped with high storage capacity, and it may manage security levels for access to the files. Network File System (NFS) is a commonly used Unix file server system from Sun Microsystems that is implemented on many platforms.

file server, Frame Relay In a Frame Relay network, the file server is a device that provides connections with terminals, controls transmission flow, and provides end-to-end acknowledgment and error recovery.

File Service Protocol FSP. A file transfer protocol somewhat similar to File Transfer Protocol (FTP) originally developed for Unix by Wen-King Su. It is a low-load (nonforking), reasonably robust protocol that permits transfers to be resumed from the point at which they stopped if the server goes down temporarily or, if desired, allows for partial file transfers from a specified point.

FSP is somewhat like Anonymous FTP except that it doesn't require a username and password. For identification and logging purposes, the host domain of the user is recorded. It is also claimed that it is less prone to server attacks than systems with FTP servers.

FSP relies on datagrams rather than TCP sockets for its transmission connections and uses the same port for communications. See File Transfer Protocol.

file sharing Access by more than one user, sometimes at the same time, depending on the nature of the data, to files that may be on one system on a network or spread out over several workstations that are inter-accessible. On the Apple Macintosh, file sharing is easily set up via utilities in the Control Panels so that passwords can be assigned and files shared with designated users on the system. On a larger network, a particular machine or set of machines, usually with large hard disk storage capacity, may be dedicated to file serving and sharing activities. See file server.

File Transfer Protocol FTP. A user-level file sharing protocol established by the early 1970s on the ARPANET and now widely implemented on the Internet in the form of FTP archive sites. The concept of FTP sites was to provide a simple, consistent means of presenting and accessing file information on a variety of types of file archive sites, so the user could easily navigate the site and upload or download files unassisted. In other words, FTP sites have a consistent look and feel; once you've learned a few easy commands, you can log in, look around, and get what you need without having to worry about the individual characteristics of the system on which the files are stored.

Many FTP sites provide public access through a user login in which you type "anonymous" as the username, and your full email address as the password. If you have a Unix shell account with an FTP client, you simply type "ftp" (in all lowercase), followed by "help" when it activates, to learn its basic commands and capabilities. The inset shows an example of a simple anonymous FTP login.

In the example session shown, the user logs in as anonymous, supplies a legitimate email address as the password, and is dropped into a limited environment where basic directory traversing commands and file download commands can be used. The message "Guest login ok, access restrictions apply." is displayed. This session is very typical in that the user is prompted to disconnect if logging of his or her activities is objectionable and is notified that there are restrictions. The user's logon number is shown, in addition to the total number of people on the system. If the system is at capacity, the user may be asked to try again or may be provided with a message giving the addresses of mirror sites (sites with the same files in other locations).

For file transfers, the *get* command, followed by a filename, will initiate a file download. The commands *bye* or *quit* will end a session.

On Unix systems, you can type "man ftp" at a shell prompt to read the manual pages for FTP, which include a list of common commands. FTP file download capability is built into most Web browsers and works transparently with many Web file archives. Several variations of the File Transfer Protocol exist, and cutdown, easier-to-implement versions have also been developed, which are described on the

F

Internet in various RFCs. See File Service Protocol, Simple File Transfer Protocol, Trivial File Transfer Protocol, RFC 171, RFC 172, RFC 959.

file transfer protocol In its general sense, any program that facilitates the movement of files from one system to another, particularly through phone, null modem or other serial data links, or the Internet. There are many file transfer protocols, but two of the most popular implementations are ZModem, for telephone line transfers, and File Transfer Protocol (FTP) for Internet transfers. Other popular programs include Kermit, XModem, and YModem. More detailed information is included in this dictionary under individual listings for the various protocols.

film *n.* 1. A thin membrane, skin, or coating. 2. A thin, light- and/or chemical-sensitive material commonly used in the photographic industry. 3. The collective name for a sequential, related set of still frames, taken together, form a story or cohesive idea that is viewed by playing the frames through a projector. Also called movie.

filter *n.* 1. A porous material through which mixtures are screened in order to selectively prevent larger bits of the mixture from passing through. 2. A device or material through which particular waves, frequencies, or particles do not pass. A filter may be used in combination with another device, such as an amplifier, in order to filter out noise, while propagating the desired portion of a signal. Electrical and audio filters are common.

filter, file File filters are not necessarily exclusionary tools, as in some senses of the word "filter," but rather may be conversion utilities available in many application programs to input or output files in a format that is not native to the application. Thus, a TIFF file might be imported into a paint program with a proprietary format, through a filter, and may be exported through another filter to a JPEG file, for example, for use on the Web.

A filter may also be coordinated with a database to selectively provide access to higher priority messages or processes, while filtering out, or queuing those of lower priorities. Email filters are especially useful to those who get hundreds of messages a day, as often happens on email mailing lists. A good email client will let you set up filters that file the messages in separate folders to be selectively read later, so the user can more easily determine which messages to check first. Exclusionary file filters also exist. For example, an email file filter may exclude all messages received from *luser@hotmail.com* or relegate them unread to the bit bucket.

filter, network In network transmissions, there are physical filters and logical filters. Logical filters function on every level of the system from low-end operating functions, to high-end user applications. Logical filters employ algorithms to selectively block the continuation of certain information, such as extraneous packets, unrecognized characters, extra information not supported by the receiving protocol, unwanted email, messages from sites operating unlawfully, etc.

filtering Using physical or logical means to selectively permit access of only the desired information. Thus, unwanted information can be screened out, or a lower capacity system can be used to view or use part of the information according to its capabilities. For example, filtering out parts of a transmitted image makes it possible to display it on a system with low resolution, or a slow image display, a solution that may be preferable to no image at all. See compression, MPEG.

filtering agent, filtering client A software program that can be configured to selectively reject or keep information according to a set of parameters or keys. With the excess of information available through the Internet, filtering agents are increasing in importance. See data mining.

filtering traffic On a network, the selective acceptance or rejection of certain packets, messages, or processes according to a set of priorities and parameters. High and low usage times may also be factors in setting up filtering instructions. Traffic filtering is usually accomplished by combining a database with a list of priorities. See firewall.

fin waveguide A structure that can be used in conjunction with circular waveguides to increase the range of wavelengths that can be transmitted, by attaching a longitudinal metal fin.

Financial Services Technology Consortium FSTC. A not-for-profit consortium of banks, financial services institutions, technology companies, government agencies, research labs, and educational institutions. FSTC supports and promotes collaborative research and technical projects affecting the financial services industry in order to further the competitive health of the financial industry in the U.S. http://www.fstc.org/

finder A name used on several computer systems for applications that aid in locating information on a system, whether it be files, directories, or the specific content of files.

Finder On the Apple Macintosh, the graphical user interface and operating system processes through which the user interacts with the system. Multifinder allows more than one program to be executed at a time and is available on the more recent versions of MacOS. It is also a generic name for a file finding tool that comes with the operating system.

finesse In optical resonating cavities, the number of round trips a light pulse can make from one reflective surface to another and back before the signal disappears through scattering and attenuation.

Finger The name of an online information utility, based on the Finger Protocol, that allows the user to retrieve and display information about users of a system, or the owner of a particular account on the network, provided no firewalls exist to block the *finger* command (as a command it is spelled all in lowercase). Login and logout times may be displayed, or the length of time since the last login. If the user queried has particular dot files configured, such as *.plan (dot plan)*, additional information from this file will be displayed. Users often use the *.plan* file to list philosophies, home addresses, office hours, interests, or professional credentials. See firewall.

Finger Protocol A network information protocol which is an elective proposed Draft Standard of the IETF. See finger, RFC 1288.

Finkel, Raphael A. Finkel is probably best known as the first disseminator of the infamous Jargon File, distributed from Stanford University in 1975. He is also the author or co-author of handbooks and numerous articles on networks, data structures, and computer algorithms. See Jargon File, The.

FIPS See Federal Information Processing Standard.

firefighting Trying to fix something after the fact. Often used in a derogatory sense to indicate the frustration of trying to rescue a situation that would not have occurred if proper steps or prevention methods had been used in the first place. The term describes distressing, expensive catch-up or fix-up situations resulting from bad management decisions. For example, shipping a software product before it is fully tested and debugged can result in a great loss of confidence on the part of customers, and enormous extra firefighting expense to the company in terms of subsequent upgrades and tech support that would not have been required if the product had been properly completed before shipment.

firewall A physical screen created to prevent the spread of fire. It may be a wall of heavy, fire-resistance materials. See wiring vault.

firewall, network A computer network security configuration designed to limit completely or selectively access to a system. At one time, firewalls were usually implemented on a specific *gateway* machine, but hardware and software firewalls now are set up in a number of ways, using filters, proxies, and gateways at the circuit level. A network traffic firewall examines incoming packets and selectively lets them pass through, and it may also edit outgoing traffic in order to protect the identities of the senders, as in some government networks. Many local area network (LAN) firewalls are one-way, with unlimited access out of the LAN and selective access into the LAN. Systems with firewalls frequently log all activities through the point or points of entry, with or without notification. See packet filtering, proxy server.

FireWire FireWire technology, also known as i.LINK, is a real breakthrough for connecting a wide variety of peripheral devices to computer networks. It was developed in the mid-1990s by Apple Computer and quickly moved toward broader industry standardization. In 2001, the National Academy of Television Arts and Sciences (NATAS) recognized Apple's FireWire contribution to the multimedia industry by awarding the company the Primetime Emmy Engineering Award.

The FireWire serial communications data rate is fast, up to 200 Mbps when it was first released, with 400 Mbps supported soon after. FireWire cables are easy to connect and hot-swappable (they can be plugged and unplugged without rebooting the host computer). A FireWire bus can support up to 63 devices, compared to the seven devices commonly supported by most computer SCSI buses. The format is autoconfiguring, so it isn't necessary to set or keep track of peripheral ID numbers for the devices. FireWire devices don't need to be terminated by the user.

FireWire is a good standard for a world demanding high-bandwidth multimedia applications. It's much faster than USB and supports isochronous data transfers, providing guaranteed bandwidth for realtime audio/video data streams. Many vendors have developed PCI-format FireWire cards and software drivers for computers that don't have native support for FireWire. Cards will typically support between one and three FireWire devices.

With so many sought-after advantages, FireWire has caught on quickly on many computer platforms. It is likely that most operating systems will soon provide native support for the FireWire (IEEE-1394*x*) standard. Peripheral hardware vendors supporting FireWire include Canon, JVC, Kodak, and Sony. By August 2001, 60 GByte FireWire hard drives were available for less than $200.

In spring 2001, the 1394 Trade Association introduced an update to the FireWire standard, called 1394b. The previous speed of 400 Mbps was already ripping fast compared to previously common peripheral bus standards, but the new standard increased this data rate even further, up to 3.2 Gbps over glass optical fiber media. The updated format also uses a new bus arbitration scheme, called Bus Owner Supervisor Selector (BOSS). It enables pipelined unidirectional arbitration, in parallel with data transmission. By using bilingual mode, the 1394b specification is backwardly compatible with previous versions. The new features of 1394 are available in beta mode.

FIRMR Federal Information Resources Management Regulation.

firmware Programmed circuitry that is semipermanent. Software on a disk is easily changed and rewritten. Software on the circuitry of a microchip is not easily changed and rewritten. In between these are EPROMs, erasable, reprogrammable chips which can be changed with the right equipment, and which retain the information during a power-off.

FIRST See Forum of Incident Response and Security Teams.

first call date A record of the first time a subscriber line is used, sometimes used in billing or in settling disputes.

first in, first out See FIFO.

fish job *slang* Phrase to describe a difficult wiring installation in which the wiring has to be pulled and threaded through constricted or hard-to-reach spaces.

fish tape *slang* A smooth-surfaced, nonconductive (e.g., steel) metal tape that is threaded through tight areas, such as a wall or cable conduit. It is then attached to a cable so that it can be pulled more easily back through the wiring path. See pulling eye.

fishbone antenna An antenna named for its resemblance to the ribs of a fish because it includes a series of coplanar antenna elements arranged in pairs. The fishbone antenna is used in conjunction with a balanced transmission line.

Fisher, Yuval Author of *Fractal Image Compression*, which describes the current knowledge of fractal compression in down-to-Earth terms with C source code examples. See fractal transform.

FITL See Fiber in the Loop.

FIX Federal Internet Exchange.

fixed access unit FAU. A wireless telephony designation for a wireless phone unit that is not intended to be carried around, but rather to provide wireless communications within a limited region. Thus, local wireless phone service can be installed without going through a local phone provider, much like a fancy intercom unit, or it can be subscribed through an alternate vendor as a limited cellular or PCS service.

Fixed End System F-ES. A nonmobile data communications system through which a mobile subscriber accesses landline network services. F-ESs typically comprise modems installed into desktop computers. See Cellular Digital Packet Data, Mobile End System.

Fixed Radio Access FRA. Local telephone service based upon wireless radio technology transmitted to an antenna attached to the subscriber's premises. The concept did not catch on until it was marketed to consumers as plain old telephone service (POTS) with enhancements, as opposed to new wireless technology. The concept originally was more prevalent in the U.K. than the U.S., but it is beginning to influence vendors in the North American market. In typical installations, the digital signal from the transceiver is converted to analog and carried to various points on the premises through the existing copper wire, but the digital signals can also be fed directly through newly installed fiber optic cables, an option that would be of interest to individuals and businesses desiring wideband high-speed services. The local loop concept, in which a network of about 20 base stations could serve a community of 40,000 population, is less expensive to implement and serve than current cellular systems and thus could serve a niche somewhere between traditional telephone and mobile telephone subscribers.

fixed satellite service FSS. One of two divisions into which Ku-band satellite broadcast services have been split. FSS operates in the 11.7 to 12.2 GHz range. The other is *broadcasting satellite service* (BSS). See Ku-band.

Fixed Telecommunications Network Service FTNS. A category of licensed service recognized by communications authorities in Hong Kong. There are various types of services that constitute FTNS, including wireline, cable, wireless, and satellite. FTNS Operators are assigned tariffs on the basis of the specific types of services they provide, such as voice or data. Changes in the provision of local FTNS were made in 1995 when the Hong Kong Telephone Company's franchise expired, and the service was licensed instead to four companies.

Fixed Wireless Access FWA. In regions where the cost of installing wireline may be prohibitive, due to rough terrain or sparse population, or where regional growth outstrips wireline installation capacity, FWA provides a long-term or temporary alternative. It combines radio-based phone service, in the place of the local wireline loop, with common carrier phone service. See time division multiple access, code division multiple access.

Fizeau interferometer See interferometer, Fizeau.

FK foreign key. A designation in a key cryptography scheme. See key, encryption.

FL port On a Fibre Channel network, a switch that connects to a loop. Middle priority addresses are assigned to FL ports, giving them higher priority than N ports (node ports) and lower priority (in terms of loop control) than an end loop (NL port) endstation port. See F port.

flag *n.* 1. A device or signal used to attract attention or to indicate the state of a situation. In software programming and network operation, flags are frequently used to indicate the state of processes or variables, often under changing conditions.

FLAG Ltd. Fiberoptic Link Around the Globe. A commercial fiber services carrier with installations of more than 18,000 miles of fiber optic cable installed worldwide.

flame resistant, flame retardant A medium that is inherently resistant to catching fire or spreading flames or is treated or manufactured to increase these retardant properties. Flame resistant and retardant materials are used in many industries including construction, electrical installation, and clothing manufacture.

flammable A property of easily catching fire or continuing to burn readily.

flange A rim or rib on an object to add strength or to aid in alignment.

flash *v.* On a phone or intercom system, to send a signal through the line by pressing the switch button on the handset holder or the button designated as a *flash button*. The flash button is used on some local multiline systems to transfer a call and may be followed by the keying in of the number of the extension.

flash button A button designated on a phone or intercom system to send a signal that is the same as pressing the switch button on the handset holder. See flash, flash hook.

flash cut See hot cut.

flash hook See switch hook.

flash interference In television transmission and display, a flash is a very brief interference, sufficient to distort the picture information.

flash memory A type of nonvolatile, rewritable computer memory technology, developed by Intel, providing an alternative to large storage devices. Since flash memory is physically compact and doesn't lose its data when the device is not in use, it has been incorporated into PCMCIA cards for portable computing applications. Flash memory is also starting to be used in portable telephone devices and digital cameras. See memory, PCMCIA.

flash tube A bulb or tube used to create a bright, momentary burst of illumination through application of a high-voltage pulse. One-time flash bulbs were used in older cameras; electronically activated, reusable bulbs are now common.

flat connector Any of a large variety of electronic connectors that are basically flat, that is, wide and narrow. Within each subset of connector types (power cable connectors, data cable connectors, etc.) there are usually some that are called flat connectors due to their low profile. Flat connectors typically contain a single line of pins, pads, or holes for coupling with cables, backplanes, or components.

For portable devices, a flat connector is a small, low-resistance, narrow electronic connector. This type of connector is increasingly favored for low-power portable device slide-in connections for use in a variety of applications where quick and easy connections and disconnections are desired. Examples include small cards that slide in and out of data readers or small handheld devices that slide into cradles and docking bays.

flat panel Any of a number of types of display systems that are narrower and flatter than traditional CRT displays. These may be special flat panel CRTs, gas plasma displays, liquid crystal displays (LCDs), or light emitting diode displays (LEDs). Flat panel displays are especially favored on mobile systems, such as computer laptops.

flat panel CRT A type of cathode-ray tube (CRT) color display technology in which the electron beams are aimed parallel to the front of the display device, then deflected 90 degrees onto the viewing surface. This configuration permits the construction of a much flatter, smaller, more convenient display device. While this technology is still relatively new and expensive, the bulkiness of traditional CRTs makes the flat panel CRT commercially attractive. See faceplate, fiber optic.

flat plate antenna A commercial/industrial/military satellite communications focusing antenna based on microcircuit design. It is similar to a common parabolic antenna, except that it incorporates a series of concentric rings laid over a transparent sheet to create a lens that can be used to redirect signals.

flat rate service A very common subscriber billing technique. Flat rates usually arise in services where the overhead of keeping track of many different types and quantities of usage would cut into profits. Flat rate services are also attractive to many subscribers, as they know in advance what it will cost and don't have to watch the clock or keep track of usage. In computer network access and telephone services, flat rate billing is very common. Since users of these services vary dramatically in time of access, connect times, and types of services used while connected, it probably is more economical in the long run to assign average usage fees than to try to track and bill widely varying usage. Flat rates for businesses tend to exceed those for residential use by roughly a factor of three, depending on the type of service. Local phone calls in many areas in North America are billed on a flat rate. In Europe and some parts of North America, per-call charges are levied instead and long-distance services are usually billed on a per-call basis. The newer digital cellular technologies sometimes have a flat rate billing option.

flat top 1. Something with a flat surface on top, as a flat-roofed building or aircraft carrier. 2. The portion of an antenna that lies horizontal.

flat top antenna An antenna that has two or more parallel, horizontally strung wires.

flatbed scanner A type of desktop scanner that permits the object to be scanned to be placed directly on the scanning surface; the object lies flat and doesn't have to roll through a drum or other moving mechanism. This type of scanner is preferred for scanning books and other large or three-dimensional objects.

flavor A slang term for type or model. Programmers frequently refer to different flavors of software languages or operating systems to indicate that they are essentially alike, but different enough to create compatibility issues. The distinction is somewhat like a 'dialect,' in languages, or a 'model' in a type of car.

FLC See ferroelectric liquid crystals.

FLCD ferroelectric liquid crystal display. See ferroelectric liquid crystal.

FLEA memory flux logic element array memory. The whimsical acronym for a type of computer memory developed by RCA in the early 1960s. The FLEA was created photographically and was capable of storing 128 bits of information. Its processing speed was 100,000 items per second.

Fleming, John Ambrose (1849–1945) An English electrical engineer who investigated the Edison effect and experimented with improvements to wireless receivers in 1904. By modifying an electron bulb so that it incorporated two electrodes, and attaching it to a radio receiving system, the radio waves could be converted to direct current (DC). Unfortunately, this new *diode* was not a significant improvement over previous electron tubes, but it was important in the evolution towards more sophisticated tubes that came later. The most important of these was the *triode* in which L. de Forest took the two-element Fleming tube as the basis for the invention of the *Audion*, which included a controlling grid as a third element. See two-electrode vacuum tube.

Fleming oscillation valve An electron tube developed by J. A. Fleming, based on Edison's work with electric light bulbs. This diode tube was in essence a two element rectifier. While it did not achieve the practical utility of later tubes, it led to the development of the *triode* by Lee de Forest.

Fleming's rule See right-hand rule.

flicker A characteristic of display devices, such as cathode-ray tubes (CRTs), in which the scanning of the screen is visible to the human eye as a light-dark flashing flicker. Flicker can result from a number of causes, including the quality of the monitor, the mode of display (interlace or noninterlace), or the speed of the screen refresh as the electron beam sweeps the screen. Generally, slower sweeps will appear to flicker more, as do interlace screen modes.

Apparent flicker is eliminated on better multiscan monitors. Most individuals can comfortably watch displays that are refreshed at about 70 Hz to 80 Hz; above that level, the trade-off in cost and computing is not sufficient to justify the insignificant or nonexistent improvement.

While flickering on screens may be uncomfortable to watch, sometimes an interlaced mode has a practical purpose, as when an NTSC-compatible signal is being generated to output to video. See cathoderay tube, frame, interlace, multiscan.

flip-flop 1. Quick reversal of direction or opinion. 2. A circuit or logic state that can assume one or the other of two stable states (on/off, high/low, etc.). A trigger circuit or toggle.

floating point A mathematical representation system in which a number is expressed as a product of a bounded number (mantissa) and a *power of* scale factor (exponent) within a number base (e.g., base 10); hence, 123.45 can be expressed as $.12345 \times 10^3$. *Floating point* refers to the flexibility inherent in placing the decimal point by adjusting the exponent.

floating point unit FPU. In computers, floating point math coprocessing chips are often paired with central processing units (CPUs) to carry the processing load of the math calculations, which are usually cycleintensive, thus freeing the CPU for other tasks.

floating selection In graphical user interfaces, a selected text or image area that can be manipulated and moved separately from its background; thus it appears to *float* over the other elements on the screen. Floating a selected region is useful for cut and paste, drag and drop, and image processing applications.

flood 1. To inundate, overflow, or cover a broad area all at once. 2. In scanning and printing technologies, flood lamps are often used to process plates and provide illumination for the recording of images. 3. To inundate with data, often unintelligible, as an incendiary or retaliatory action. 4. The outpour of vast quantities of digitally generated information. See core dump, data mining.

flooding 1. Overflowing, inundating. 2. In networks, a technique of sending many identical packets through various routes so redundancy increases the chances of the data reaching its destination. 3. In networks, a deliberate act of vandalism in which data is directed toward a system, or an email address, to fill the hard drive space or tie up the processor, to render the system useless. Users caught flooding are usually denied further access to a system. See core dump, mail bombing.

FLOP floating point operation. Mathematical manipulation of a floating point number. FLOPS (Floating Point Operations per Second) is often used to describe and compare microprocessor speeds. See MFLOP.

floppy diskette, floppy disk A thin, compact, portable, flexible, read/write, random-access data storage medium originally encased in a soft protective case or, later, a hard protective case. Data is stored and modified by rearranging magnetic particles on the surface of the disk and, as such, the disk should be kept away from magnetic surfaces to reduce risk of loss. Generally, magnetic media are not reliable for long-term storage (see superparamagnetic).

floppy diskette drive FDD. A device for reading and writing data to a floppy diskette data storage medium

that became more prevalent in the early 1980s. There was usually one floppy drive built into a computer and sometimes a second external drive would be available in place of a more expensive hard drive. Prior to this, most consumer machines used tape drives to store data.

FLOPS floating point operations per second. A measure of the speed of mathematical computations. See FLOP.

flow In packet networking, a sequence of packets with the same source and destination addresses and other similar characteristics. The detection of a flow by various routing and switching mechanisms can trigger flow-based processing of that sequence of packets to improve efficiency. Flow detection may also include marking the flow with a label. For example, a new virtual circuit (VC) may be set up for a packet flow, thus removing or reducing the need for routing until the end of the flow is detected.

Flow processing is a means of handling high-speed data through systems not normally capable of very high rates of packet transmission. The bottleneck is the overhead in managing the IP datagrams associated with the packets. Cut Through Routing, developed by Ipsilon (a commercial switch vendor), enables significantly faster IP routing by detecting classes of IP flows and processing them accordingly. When processed through a VC, the transmission can be handled by a switch without individual routing. The signaling between IP switches is handled with the General Switch Management Protocol (GSMP) and the Ipsilon Flow Management Protocol (IFMP). See General Switch Management Protocol, Ipsilon Flow Management Protocol.

Flow Attribute Notification Protocol FANP. In packet-switched networks, a protocol for management of cut-through packet forwarding functions between neighbor nodes. FANP indicates mapping between a datalink connection and a packet flow to the neighbor node, and it helps nodes manage the mapping information. This allows the bypass of the usual Internet Protocol (IP) packet processing by allowing routers to forward incoming packets. See RFC 2129.

flow chart A somewhat standardized diagrammatic representation of processes, procedures, conditions, and directions of traffic or information flow. Flow charts employ geometric shapes, symbols, and connecting lines to indicate the relative importance and relationships of the concepts being illustrated.

Programmers are often required by managers to provide flow charts of their software designs. However, many argue that outlines and pseudocode are more useful in representing the relationships and flow within a software program than conventional flow charts because of the lack of correspondence between human interface actions and the looping and jumping structure of the code itself (in other words, the order and frequency with which the user interacts with the software rarely corresponds to the order and frequency of the algorithms and procedures that enable those functions to be executed). Another factor complicating the application of traditional management flowcharts to the programming process is the complexity of the decision-making points in a program and the inability to predict every instance and sequence of events that the user might take in a sophisticated program with many options (it would be like trying to chart the moves in a chess game before the game has begun).

Code usually changes many, many times before the full program is developed, even when the flow representation stays the same. Programming is a relatively young art and many of the algorithms are still being invented as the software is being developed and cannot be known in advance, before the problem is actually solved. Imagine trying to flow chart the high school course selections for a four-year-old child without knowing anything in advance about his talents, skills, and interests and then you can understand the difficulty of flow charting a software program before it is written. Thus, flow charts are best seen as tools for conveying to management and co-workers the general goals and structures of a project, but not as efficient tools for developing the individual components of the computer algorithms themselves, which must be worked out as they are encountered.

FLOW-MATIC See B-0.

Floyd-Steinberg A dithering algorithm, i.e., a means of creating a perceptual tone or range of tones by intermixing colors related to those tones to create the illusion of more colors. Dithering is one way of stretching a limited palette to make it appear as though there are colors that are not actually available. Dithering works best with dots of light or color that are very small, too small for the human eye to resolve. In the Floyd-Steinberg error diffusion algorithm, the error between the approximate output value of a pixel and the actual value of a pixel is sequentially diffused to its near neighbors. See dithering.

FLTSAT, FLTSATCOM The U.S. Naval fleet communications satellite system that is one of the primary U.S. satellite communications systems along with the Defense Satellite Communications System (DSCS) and the Air Force Satellite Communications (AFSATCOM). Together, the three systems comprise the Military Satellite Communications (MILSATCOM) system. Control of the systems is handled through the Air Force Satellite Control Network (AFSCN).

FLTSATCOM was developed in the early 1970s to provide communications for seagoing vessels, aircraft, and U.S. military ground crews worldwide. In addition to Naval and Air Force communications, FLTSAT was designed to provide fast communications between the U.S. President and Commanding Officers.

Transmissions are received by several hundred fixed and mobile user stations on sea, air, and ground terminals in addition to Communication Area Master Stations (CAMS) in the U.S., Guam, and Italy. Channels are allocated to the Navy, Air Force, and Command. The first units were launched in the late 1970s, with additional units added to the system in the 1980s. The geographical area covered by the near-geosyn-

chronous equatorial orbits ranges from about 70 degrees north to about 70 degress south.

In 1991, operational control of satellite programs was turned over to the 3D Satellite Control Squadron (3 SCS) which became the 3D Space Operations Squadron (3 SOPS). Then, in 1996, FLTSAT constellations were turned over from 3 SOPS to the Naval Satellite Operations Center in California.

The FLTSATCOM Laboratory has a computer simulation of the satellite system (minus the radio frequency capabilities) provided by The Aerospace Corporation that gives a graphical representation of the satellite and its operations.

When fully deployed, a FLTSAT unit resembles a trashcan with an umbrella on one end (the parabolic antenna) and rectangular, solar array 'elephant ears' protruding from the central body. The FLTSAT systems incorporate a number of antennas in different wave bands, including S-band, UHF, EHF, and SHF (Super High Frequency) radio frequencies and have an operational life of about five years. They weigh approximately one to two tons each and are about the size of a large motorhome, with the solar panel ears extending to over 40 feet. It is expected that the UHF Follow-On (UFO) system will eventually supersede the FLTSAT constellation.

fluorescent lamp A fluorescent bulb used for lamps typically consists of a long glass tube equipped with an electrode at each end, with specialized vapor and gases sealed inside the tube. When electricity passes through the tube, light waves are emitted, causing phosphors coated on the inside of the tube to glow. Manufactured since the late 1930s, the fluorescent lamp doesn't use a filament and provides more light than an incandescent lamp for the same amount of current. Since less current is required, the bulb emits less heat. Fluorescent lamps manufactured before 1978 may contain PCBs in the ballasts and should be disposed of according to guidelines for PCB disposal to avoid contaminating landfills and waterways.

flutter 1. A rapid, repetitive, agitated back-and-forth movement; any erratic vibration or oscillation. In most systems, flutter is an undesirable characteristic that interferes with the main signal. See drift, wow. 2. Undesirable phase distortion variations that may result from more than one frequency transmitting at the same time. 3. In radio terminology, also loosely called drift and wow.

flutter bridge A device to measure flutter (undesirable variations from a constant oscillation, movement, or signal). It is used for testing and diagnostic purposes for various playback devices that should be playing at a constant speed, such as phonographs, tape recorders, film projectors, or disc players.

flutter rate The speed at which an oscillating body moves back and forth, commonly expressed in units per second or minute.

flux 1. Stream, continued flow. 2. An expression of the rate of transfer across or through a unit area of a given surface, per unit of time. See watt. 3. A substance used to facilitate the fusing of materials, as the use of rosin in soldering or welding. 4. Magnetic lines

F

of force in a magnetic field taken as a group (*symb.* – B). When expressed in terms of density per unit area, it is called *flux density* (*symb.* – D).

fly-by *n.* A representation of movement from a point of view above the ground, commonly used in animations, especially video games with flight simulation. Fly-by animations give a wonderful sense of being inside the scene that is imaged. NASA has produced some wonderful fly-by animations of the surfaces of other planets such as Mars. Satellite geophysical data make it possible to create fly-bys of Earth's surface right down to individual buildings and streets. Virtual reality fly-by simulations are startlingly real, with participants ducking moving images so as not to be hit. See virtual reality.

fly-page See banner.

flyback retrace. In a cathode-ray tube (CRT), the movement of the electron beam tracing the image on the screen from the end of the trace to the beginning where it starts over on the next line. Flyback is usually associated with a blanking interval in which the beam is turned off so as not to interfere with the image already displayed. There is more than one type of flyback on a monitor. The flyback associated with scanning each line is similar to the line feed and carriage return on a typewriter, in that the scan finishes at the end of one line and flies back to the next line down (or two lines down in interlaced screens) and the beginning of the subsequent line, in a zigzag (sawtooth) pattern. The other type of flyback is when the full video frame is finished, the beam is at the bottom or last line of the screen and then flies back to the top or first line of the screen. (This example assumes a typical CRT in which the scanning is left to right and top to bottom.) See blanking, frame.

flying erase head A mechanism on prosumer and industrial level VCRs and camcorders that erases previously recorded video traces that might otherwise interfere with new information being recorded on top of the same section. This head is typically found on systems that support insert editing. Rainbows and other undesirable artifacts are thus avoided.

flytrap A firewall or other security system that logs unauthorized attempts at access to provide information that can help identify or apprehend the intruder.

flywheel 1. A wheel that works with other mechanisms to smooth out and reduce inconsistencies in the rotational speed of the equipment. 2. A wheel that is used with other mechanisms, whose purpose is to store kinetic energy. Flywheels are often coupled with power generators to continue the motion when the generator mechanism slows or is idle.

flywheel effect In a transmission that experiences fluctuations, the maintenance of a steadier, more consistent level of current, information, or oscillation by physical or logical means. Analogous to the function of a flywheel.

FM 1. fault management. 2. See frequency modulation.

FM broadcasting Transmission through frequency modulation technologies on approved FM frequencies with the appropriate FM broadcasting license. In the United States, FM stations are spaced at 0.2 kHz intervals, ranging from 88.1 to 107.9 kHz. Low power FM broadcast signals are used for mobile intercoms, indoor intercoms, monitors, and cordless phones. See broadcasting, FM broadcasting, frequency modulation.

FM transmitter In its basic form, an FM transmitting system includes a microphone, a circuit, and a frequency modulating (FM) transmitter. In more sophisticated forms, it includes the various commercial/industrial transmitters costing thousands of dollars for broadcasting from licensed radio news and entertainment and other FM communications stations. Building simple FM transmitters in the 88- to 108-MHz frequency range is a very popular hobbyist introduction to electronics. With current technology, it is possible to create very compact, working FM transmitters for under $30, to broadcast a few hundred feet or even up to two miles under good conditions. Before conducting hobbyist experiments with low power FM transmitters, it is important to learn the various Federal Communications Commission (FCC) restrictions on broadcasting, and to honor laws protecting the safety and privacy of individuals.

FMAS 1. Facility Maintenance and Administration System. 2. Fund Management Accounting System.

FMV See Fair Market Value.

FNB See *Fiberoptics NewsBriefs*.

FNC See Federal Networking Council.

FNEWS A fast full-screen news reader for UNIX, ALPHA-VMS, and VAX/VMS systems, similar to NEWSRDR and ANU-NEWS. News articles for groups are cached and dynamically loaded. Version 2.0, released in 1995, included access security for newsgroups and removed the limits on the size and number of newsgroups that could be loaded. FNEWS is a commercial shareware product.

Fnorb A CORBA 2.0 Object Request Broker (ORB) written in Python and a tiny bit of C by the Hector Project participants at the CRC for Distributed Systems Technology at the University of Queensland, Australia. Fnorb supports CORBA datatypes and full implementation of IIOP. It is freely distributable for noncommercial use. See CORBA, ORB.

FNR 1. Faculty Network Resources. 2. fixed network reconfiguration. Configuration of an existing network with static transmission lines to support upgraded services or a wider variety of services, often used as an interim solution instead of completely replacing a network.

FNS Fiber Network System/Service.

FO-2 A committee of the Telecommunications Industry Association (TIA) developing physical-layer test procedures and system design guidelines and specifications for distributors and users of fiber optic communications technologies. Since 2000, plenary meetings have been held jointly with FO-6, a TIA committee on Fiber Optics.

FOA 1. fiber optic amplifier. 2. See Fiber Optic Association, Inc. 3. First Office Application. Testing of systems within an office application once in-house testing is complete or nearly complete. Most of the problems in the system have been worked out and what is now sought is relevance and feedback from a real-world installation. In software development, it is known as *beta testing*.

FOC Firm Order Confirmation. A product or service agreement confirmation document.

foam dielectric cable A cable assembly that utilizes foam as a nonconducting medium around the conducting medium to reduce noise and increase transmission speed. Foam dielectric cables are sometimes substituted for air dielectric. They don't quite meet the performance characteristics of the best air dielectric cables, but they have advantages such as good moisture-resistant properties and no requirement to pressurize the cable housing. See air dielectric cable.

focal length (*symb. – f*) In a viewing or recording mechanism, the distance from the focal point on the surface being viewed or recorded to the center of a lens or surface of a mirror, as on a camera.

focus *n.* 1. In an optical viewing or recording mechanism, the point at which rays diverging or converging from a surface intersect in the mechanism (through a lens or on a mirror) to produce a clear, unblurred image of the surface. 2. In a projected image, the point on the projection surface in which the rays converge to produce a clear, unblurred image. 3. In a color cathode-ray tube (CRT), convergence of the electron beams on a precise point on the coated inner surface of the glass to provide a clear image on the front surface of the tube. 4. In human vision, the point at which the distance of the object being viewed, the angle of the individual parts of the eye, and the angle of two eyes are correlated so that the image appears clear and unblurred. 4. Center of attention or activity.

focus group A group organized to concentrate on or discuss a specific issue.

FOD fax-on-demand.

FODB See Fiber Optic Data Bus.

FODTE See Fiber Optic Data Transmission Experiment.

FOLDOC Free Online Dictionary of Computing. FOLDOC is a popular, searchable resource containing over 13,000 concise definitions related to computing topics. FOLDOC was established in 1985 and is edited and copyrighted (1993) by Denis Howe. The information is distributed for use under the terms of the GNU Free Documentation License. http://www.foldoc.org/

FOLM See fiber optic lever microphone.

FOLS See Fiber Optic LAN Section.

FOM See fiber optic modem.

footprint 1. An area or impression on a surface comprising a more or less contiguous region of contact with the bottom of some object or signal. 2. The desk space or floor space taken up by a piece of furniture or equipment, usually considered the area of actual contact, or the area of contact plus everything within its boundaries, and the small area surrounding it, which may be taken up by connectors or protruding knobs. 3. The terrain or surface of the Earth over which a transmission signal can be received. A trans-

mission footprint is a little less defined than a physical footprint, as a transmission tends gradually to decrease in intensity (this may be shown by contour lines on a map or chart), and there is often no definite cutoff point, unless specified as signals below a certain level. 4. An audit trail or traces left by a transaction or process which has concluded or aborted. 5. The resource requirements of a system. For example, the Amiga is said to have a *small system footprint* because it can adroitly handle preemptive multitasking, sound, and simultaneous animated graphics in a Megabyte of memory on a 25- or 40-MHz processor.

forecasting Predicting future events, usually based on an analysis and evaluation of past events. Forecasting is needed in all areas of telecommunications to choose technologies that are powerful and economical and that won't be quickly outdated. It is also used by system administrators to configure and tune systems to handle predicted needs and traffic loads. Businesses use forecasting to select local area network topologies and workstations, and by managers to organize employee loads and working schedules. See erlang, queuing theory, traffic management.

Foreign Agent A service enabling nodes to register at a remote location, providing a forwarding address to a home network in order for forwarded packets to be retransmitted to the remote location. Foreign agents are an important aspect of Mobile IP systems.

Foreign Exchange Service FX, FEX, FXS. A service that connects a subscriber's telephone to a remote exchange as though it were a local exchange. Commercial vendors provide a variety of multiplexing interface cards to telecommunications carriers to facilitate provision of subscriber Foreign Exchange Services.

Forrester, Jay Wright (1918–) A computer pioneer who investigated memory devices for computers in the 1940s and 1950s while working on the construction of the Whirlwind computer at the Massachusetts Institute of Technology (MIT). Forrester was at the forefront of transition technology from analog to digital systems and invented core memory with assistance from William N. Papian in 1951.

FORTH 1. Foundation of Research and Technology-Hellas. 2. *Fourth* Generation Language. An extensible, high-level programming language typically used in calculators, robotics, and video game devices.

FORTRAN *For*mula *Tran*slation. A high-level computer programming language that was commonly used in the 1980s for math-oriented applications, and from which BASIC has derived many of its syntactical characteristics. It grew partly out of conceptual ideas and examples of reusable code promoted by Grace Hopper, and further from the encouragement of John Backus that a language be developed that could express and solve problems in terms of mathematical formulae. With the advent of other languages such as BASIC, C, C++, Perl, and Java, the use of FORTRAN is declining.

forum Discussion group, private or public meeting, judicial assembly. Electronic forums are common on the Internet. See USENET.

forum, online A network virtual environment for discussions. Internet Relay Chat (IRC) channels, USENET newsgroups, discussion lists, and various meeting places on Web sites are examples of global forums where topics are ardently and enthusiastically debated. When forums include celebrities, they are usually moderated to keep the questions and comments to a manageable level.

Forum for Responsible and Ethical Email FREE. An organization for education and assistance to individuals and groups seeking to balance the need for freedom to send email and the need for respecting the personal privacy and economic rights of email recipients. FREE sponsors an informational Website, discussion groups, and provides up-to-date news on spam-related issues in the media and the legislature. http://www.spamfree.org/

Forum of Incident Response and Security Teams FIRST. A global coalition established in 1990 to foster the exchange of information and response coordination among computer security teams. Participants come from a variety of academic, commercial, and governmental organizations. Part of the motivation for establishing FIRST came from the 1988 events associated with the spread of the Internet Worm, which made it clear that economic damage and loss of productivity were very real threats to the global Internet and that coordinated efforts to respond to such incidents would be more efficient than the hit-or-miss response to the spread of the Internet Worm. FIRST fosters cooperation among technology and security experts and facilitates research and operational improvements to support secured networking. See Computer Incident Advisory Capability, Computer Emergency Response Team, WORM, virus. http://www.first.org/

forward error correction FEC. A means of ensuring a transmission in advance by duplicating information or otherwise improving the chances of its being received the first time. For example, characters or groups of characters may be sent two or more times (called *mode B* in amateur radio transmissions) according to a predetermined arrangement. Repeating characters, or groups of characters, in data transmissions gives a receiver an opportunity to compare the groups and, if any of the information doesn't match, request a retransmission. The basic idea is to minimize the back-and-forth nature of handshaking to speed a transmission while still giving information that may be used to check the integrity of the information being received.

Forward Explicit Congestion Notification FECN. In a Frame Relay network, a bit used to notify an interface device to initiate congestion-avoidance procedures in the direction of the received frame. See Backward Explicit Congestion Notification.

Forwarding Equivalence Class FEC. In Multiprotocol Label Switched (MPLS) networks, the FEC is a networking categorization scheme associated with packet-forwarding. A specific FEC includes destination address information and may include service information. FEC details are managed within a router's forwarding information base (FIB). See Multiprotocol Label Switching.

FOS2 See Fiber Optics Sensors & Systems.

FOSS 1. Facilities Operations Support Services. Services at the Stennis Space Center. 2. See fiber optics sensing system. 3. Future of Space Science.

FOT Fiber Optic Terminal. A connection point or device at which a fiber optic circuit connects to a copper wire circuit.

FOTS fiber optic transmission system.

Foucault test A type of optical null test (a test using the instrument itself) for determining the optical quality of a spherical surface relative to the center of curvature of the component with a point light source. See Dall test, Ronchi test, star test.

Foucault, Jean Bernard Léon (1819–1868) A French physicist best known for his studies of the speed of light and the rotation of the Earth through the use of pendulums, Léon Foucault also developed a gyroscope (1852) and a mechanical telegraph. In 1850 he was awareded the Copley Medal by the Royal Society of London for his work on the relationships between heat, magnetism, and mechanical energy. He also studied photographic processes and vision. See knife-edge focusing.

Fourier, Jean Baptiste Joseph (1768–1830) Fourier, a French mathematician and lecturer, discovered in the early 1800s that the superposition of sines and cosines on time-varying periodic functions could be used to represent other functions. He made practical use of these techniques in the study of heat conduction, work that was developed further by G. S. Ohm in the 1820s in his mathematical descriptions of conduction in circuits. Work on linear transformation mathematics that predated Fourier's publications was carried out by Karl. F. Gauss but went unpublished until after Fourier's descriptions. See Fourier transform.

Fourier analysis A means of representing physical or mathematical data by means of Fourier series or Fourier integrals.

Fourier transform A linear mathematical data manipulation and problem-solving tool widely used in optics, transmissions media (antennas), and more. The superposition of sines and cosines on time-varying functions can be used to represent other functions, in other words, to represent the data from another point of view. The result of such a transformation is to decompose a waveform into subsets of different frequencies, which together sum up to the original waveform. In this way, the frequency and amplitude can be separately and more easily studied.

A rudimentary application of Fourier series calculations were used to utry nderstand planet orbits in Greek times. Their development was in part hampered by the Greeks' mistaken assumption that the Earth was the center of the universe.

Fourier transforms differ from wavelet transforms in that they are not localized in space; however, they also share many common characteristics. Named after J.B.J. Fourier. See discrete cosine transform, wavelet transform.

Fourier transform, fast FFT. This optimized version of a Fourier transform was developed in 1965 by Tukey and Cooley. It substantially reduces the number of computations needed to do a transform, hence the name. FFT computations are used in many types of imaging applications (e.g., filtering a 3D image to display a 2D interpretation).

fox message A test sentence that includes all the letters of the English alphabet, commonly used to verify if all letters of the English alphabet are present and/or working correctly. Familiar to most as "THE QUICK BROWN FOX JUMPS OVER THE LAZY DOG" (which may then be repeated as all lowercase, as needed).

FPLMTS See Future Public Land Mobile Telecommunication System.

fps See frames per second.

FPU See floating point unit.

FRA See Fixed Radio Access.

fractal A term popularized by Benoit Mandelbrot to describe his geometric discoveries and descriptions of structures that can be described and reproduced as mathematical formulas and have the characteristics of self-similarity in increasingly fine degrees of detail.

Fractal concepts have since permeated almost every aspect of computing, especially computer image display, compression, and reconstruction. Fractal geometry provides a means to model surprisingly complex and natural-looking structures with simple mathematical formulas. See the *Fractal Geometry of Nature* by Benoit Mandelbrot.

Fractal Image Format FIF. A proprietary image compression format developed by Michael Barnsley and Alan Sloan, who together founded Iterated Systems, Inc. to exploit the technology. Very high rates of compression are possible. The technology is asymmetric – it takes a while to compress the information, but it decompresses relatively quickly. See fractal transform.

fractal transform, fractal compression A resolution-independent, lossy image compression technique providing a high degree of perceptual similarity with excellent compression results. Fractal compression works by storing image components in terms of mathematical algorithms, rather than as individual pixels of a particular location and color. The organization of the image is evaluated for its intrinsic characteristics of self-similarity, and those characteristics are coded so they can be reproduced by repetitions in increasingly fine detail, up to the resolution of the output device.

With their excellent image fidelity and high compression ratios, the trade-off in fractal compression is the time it takes to encode or decode and display the decompressed image. With faster processors, this is becoming less of a limitation. See lossy compression; discrete cosine transform; Fisher, Yuval; JPEG; Mandelbrot, Benoit; wavelet transform.

fractals, fractal images A term borrowed from fractal geometry to describe visual images that have recognizable visual and mathematical characteristics of

self-similar, repeating branches and curves resulting from the rendering of fractal formulas. Colored fractals can be beautiful, and they adorn many calendars, posters, and t-shirts. Many familiar fractal formulas have been given names, such as Julia Set, Mandelbrot Set, etc. See Mandelbrot, Benoit.

FRAD See Frame Relay access device.

fragmentation 1. State of being broken up, separated into units or groupings, having lost connections or cohesiveness, or physically or logically separating over time. 2. In hard drive storage, fragmentation is a gradual process of the available or used areas of a drive becoming smaller and more widely dispersed. When information is stored on a hard disk or other similar directory-based system, files are placed where there is room on the drive and sometimes spread over a number of areas on the drive. When a file is deleted, its directory entry is removed and the space it occupied becomes free for other files. However, over time (especially with a lot of disk activity), the free areas get smaller and farther apart, and files stored on the drive need an increasing number of sections and links to keep track. This fragmentation slows down the system. It is sometimes advisable to defragment or "defrag" a drive to optimize the tables and file data locations. Some operating systems have built-in utilities for rebuilding a drive or system. It is important that sufficient memory and swap space are available on a system before defragmenting a drive, and it is highly advisable to back up the data first.

frame 1. A bounded visual or logical unit or block of related information, sometimes delimited with visual or binary flags or markers. A frame is sometimes a natural unit, as in a cyclic event in which the information repeats in some general sense (though the content may vary), and sometimes it is an arbitrary unit, chosen for convenience or by convention. 2. A physical unit, border, containment area, skeleton (framework), or inclusive extent. 3. A full-screen perceivable image on a monitor or TV screen consisting of the sum of all the sweeps of an electron gun during a full cycle of oscillations across the screen. 4. A unit of information in data networks such as Frame Relay systems. 5. A contained group of information on an HTML layout, such as a Web page. 6. A housing or support structure for components or wiring. See distribution frame, rack.

frame, data In most networking architectures, a frame is a group of data bits of a fixed or variable size, often in a specified format. It is common for frames to be organized into two general types: those which carry signaling, addressing, or error detection/ correction information, and those which carry the contents of the communication itself (sometimes called payload), although even these are sometimes combined. The format and organization of the frames are defined by a data protocol, and there are many general purpose and specialized protocols in use, most not directly compatible with one another. Interprotocol frame traffic can be carried or tunneled through other protocols or can indirectly communicate through conversion agents or filters.

F

Frames are organized into larger units comprising a communication and then may be sent together to the destination, or they may be disassembled, sent along different paths, and reassembled at the destination. Frames may also be *encapsulated*, wrapped in an outer envelope, to carry them through a system that requires another format or to tunnel through a system without having the contents of the encapsulated package changed in any way. It is then de-encapsulated at the exit point or at the destination.

When frames carry different types of data, such as graphics in one and sound in another, they are sometimes sent simultaneously through separate wires or data paths and reassociated at the receiving end, as in videoconferencing. In these situations, synchronization or alignment of information is important and information for achieving this may be included. See Frame Relay, Frame Relay frame format, protocol.

frame, distribution A wiring connection physical supporting structure. See distribution frame.

frame, video In video displays that cyclically sweep the full screen to create an image, a frame is the extent of the sweep that is required to cover the full screen. In the NTSC system prevalent in North America, the sweep is ~29.97 frames per second and, on an interlaced screen, is further subdivided into two sets of fields (all odd lines or all even lines). The formats that are common in Europe (PAL, SECAM) display at 24 or 25 frames per second.

It is important to time the frame presentations at a broadcast station, so that news briefs, commercials, and regularly scheduled programming can be organized into precise time slots. NTSC displays are generally 525 scanlines, though not all the bottom scanlines may be visible on the screen. European standards are 625 lines. A frame is an important unit in video display not only for physical synchronization of the signals, but also because the rapid sequential presentation of still frames creates the illusion of movement, and the properties of this illusion must be taken into consideration if creating still-frame animation sequences. See station clock. See television signal for a chart of common formats.

frame alignment signal FAS. In frame-based transmissions, a sequence of bits intended to provide framing alignment information for synchronization purposes. In other words, it provides the necessary information, usually at the head of a sequence of frames, for the receiver to synchronize itself with the incoming signal. The signal may also include status, control, and error-related bits. The bits following the frame alignment signal are often allocated to more specific alignment or configuration tasks (e.g., channel setup), and thereafter there is usually information content.

frame buffer A storage area used for preconstructing digital images in order to facilitate the quick display of those images, especially if they are to be displayed one after another, as in a sequence of animation frames. The image in the frame buffer is not necessarily displayed all at once. For example, in video games, it is very common to store a wide, ver-

tically narrow *panoramic* landscape in a frame buffer and to display only a portion of the scene at any one time. Then, as the characters in the game move along the landscape, the display scrolls smoothly to right or left, without display artifacts such as flicker or jumping that may be caused by disk reads or off-screen reconstruction of the image. Frame buffers are commonly used for high-speed, high-resolution applications such as computer animations, arcade games, and video walls. See frame store.

Frame Check Sequence FCS. A mathematical algorithm that derives a value from a transmitted block of information and uses the value at the receiving end of the transmission to determine whether any transmission errors have occurred.

FCS is used in bit-oriented protocols such as SNA SDLC to determine if sent and received messages are the same. For example, in SDLC the two-byte (16-bit) FCS field includes a cyclic redundancy check (CRC) value used to assess the validity of the received bits.

frame grabber A computer hardware/software peripheral device designed to capture and digitize a frame, or series of frames, from a continuous signal, usually from an NTSC source (a frame-based video signal). The signal generally comes from live video, laserdisc, or prerecorded tape. It is sometimes called a video capture board or video digitizer. The faster a frame grabber can capture frames, the more true to the original signal a playback of the grabbed frames will appear. Generally speeds of about 24 frames per second are required for a video animation to appear natural to the viewer. See frame buffer, sampling.

frame merge 1. Over frame-based media, a stream merge. 2. In a Frame Relay network, frame merge can be used as a way to forward IP packets or portions of packets inside a frame, rather than on an individual cell basis, to improve the scalability of a network while avoiding problems with scaling virtual circuits (VCs). 3. Frame merging is useful in cases where data streams are coming from more than one source, but the software can only handle one input stream at a time. In these cases, there are utilities available to merge the data streams in various ways, depending upon the type of data that is being received and how it is intended to be processed. 4. The phrase *frame merge* is sometimes loosely used to refer to convergence of data and telephony services for companies that are seeking ways to integrate their business telephone services with their data services on the same permanent virtual circuit (VC). Data-telephony convergence over Frame Relay is a better way to describe this process.

frame mode In video image capture, a mode that captures a full frame of scan lines more or less simultaneously in contrast to most desk scanners, which capture a line or block of the image at a time. Full frame images preserve image integrity but also take more storage space than some modes. Most digital cameras are frame mode capture devices. See field mode.

frame rate, video The speed at which a series of images is presented or a screen of visual information is drawn, usually expressed in seconds. Due to *persistence of vision* in human perception, individual still images presented at about 20 frames per second or faster give the illusion of motion. At speeds of over 30 frames per second, no substantial improvement in the animation quality is perceived by most people.

Motion picture film is usually displayed on 35mm projectors at 24 frames per second. Home 8mm and Super-8mm projects are somewhat variable around 20 to 24 frames per second, since most have dials to speed up or slow down the film transport rate.

North America TV is broadcast at about 30 frames per second. (In actual fact, due to differences between black and white and color technology, the rate is closer to 29.97 frames per second.) On various European systems, such as PAL and SECAM, broadcast frame rates are 24 or 25 frames per second.

On computer systems, frame rates vary with the software that is creating the frames or with the software playing the frames. Smaller video windows can be played back faster than large ones, as they take less time to compose and require less computing power to display. Displays of 256 colors also refresh faster than 24-bit displays (~16.9 million colors), although refresh will vary with the system speed and type of graphics card used. On systems less well adapted to video, rates may vary from 20 to 30 frames per second.

Videoconferencing systems running over analog phone lines may refresh only at frame rates of 5 or 10 times per minute, as the voice-grade lines and modem create a bottleneck. On ISDN and other digital lines that run at faster rates, 20 or more frames per second may be possible, depending upon the type of system and the size of the image window. See drop frame, MIDI time code, SMPTE time code.

Frame Relay, Frame Relay network Frame Relay is a networking connection option often selected by smaller businesses as a cost-effective way to set up a reasonably fast and powerful wide area network (WAN) or local area network (LAN) that can connect with public networks. Frame Relay can be used across Integrated Services Digital Network (ISDN) and a number of other interfaces to interconnect multiple virtual LANs at lower rates than the cost of leased lines. Standardization efforts for the technology were initiated in the early 1980s and continued for a number of years.

In 1990 and 1991 vendors formed associations to facilitate development and deployment of the technology. Viable commercial implementations began emerging in the early 1990s.

Frame Relay is a connection-oriented, packet-switching protocol designed to provide virtual circuits (VCs) for interconnections within the same Frame Relay network. Virtual circuits may be permanent or switched (similar to Ethernet). Permanent virtual circuits (PVCs) are more prevalent, but switched virtual circuits (SVCs) are of increasing interest.

Frame Relay evolved from and is somewhat simplified over X.25. For example, Frame Relay is concerned with packet delivery without sequence and flow control, resulting in faster throughput and sometimes lower cost by trading off error correction at the network level. (Error correction can be implemented by intelligent user terminal equipment, depending upon needs.) Frame Relay has been shown to work in practical situations up to almost 50 Mbps.

Frame Relay operates at the physical (PHY) and data link layers of the Open Systems Interconnect (OSI) reference model. It is implemented as a Layer 2 protocol. The physical interface can interconnect multiple remote networks through Frame Relay switches. Frame Relay can transport a number of encapsulated transmission protocols, including the widely used TCP/IP.

Control signals must be provided to indicate the connection status of the link. In a Point-to-Point (PPP) system, Frame Relay framing is treated as a dedicated or switched-bit-synchronous link. See Asynchronous Transfer Mode, cell relay, Data Link Connection Identifier, framing bits, and additional entries prefaced by Frame Relay.

Frame Relay access channel A user access channel across which Frame Relay data travels. The access channel specifies the physical layer interface speed of date terminal equipment (DTE) and data communications equipment (DCE). An access channel may be categorized as unchannelized, channelized, or fractional. When the entire DS-3/T1/E1 is used at speeds of 45/1.536/1.984 Mbps, respectively, as a single channel, it is considered to be *unchannelized*. When DS-3/T1/E1 lines have one or more channels operating at aggregate speeds not exceeding those just listed for unchannelized transmissions, they are considered to be *channelized*, with the channel as any of N time slots in a given line. In T1/E1, consecutive or nonconsecutive time slots are grouped as Nx56,64 Kbps/Nx64 Kbps where $N = 1$ to 24 or 1 to 30 DSO time slots per channel, respectively, and are considered *fractional*.

Frame Relay access rate The data transmissions rate of the Frame Relay access channel. It is the maximum rate at which the user can insert data into the Frame Relay network.

Frame Relay access device FRAD. Another name for the switch, router, or other network device that assembles and disassembles Frame Relay frames as they are transported through a system. When data frames are sent over a Frame Relay network, they are packaged with various types of information, often at the beginning and end of the block of frames, and unpackaged again, often at the access point to the destination system to recover the structure and contents of the original communication.

Frame Relaying bearer service FRBS. A service providing bidirectional transfer of service data units (SDUs) from one reference point to another, retaining the order of frames. FRBS trades off some aspects of error processing (e.g., acknowledgments) for speed. A local label facilitates device identification over virtual connections.

Frame Relay cloud A Frame Relay network that is shared among a small number of participating subscribers in order to get the benefits of the technology at a lower cost. It is a suitable option for smaller businesses that do not have high networking demands but would like higher speeds than are available through dialup modem connections, for example. The system can handle a firm's voice and data communications needs. As data passes through the cloud, it is handled by switches, depending upon how the virtual circuit has been configured to accommodate each subscriber's networking needs.

Frame Relay devices The common devices that comprise a Frame Relay network include computers, terminals, and circuit-related equipment. They generally fall into two categories, data terminal equipment (DTE) and data circuit-terminating equipment (DCE). The DTEs (desktop computers, terminals, routers, etc.) are usually located on the subscriber's premises, while the DCEs (the various circuit-connecting and -switching devices) may be locally installed or may be managed by the Frame Relay service provider.

Frame Relay extensions A consortium of vendors seeking to enhance the basic capabilities of Frame Relay in order to meet the demands of the commercial marketplace extended the Frame Relay protocol with a Local Management Interface (LMI) specification in 1990. Organizations such as CCITT (now ITU-T) and ANSI developed versions of the LMI that are now generally adopted.

The LMI is a specification for information exchange between devices that is enhanced with capabilities such as global addressing, multicasting, and additional status messages. The *frame-relay lmi-type* interface configuration command provides a means to select the type of LMI interface, and the *Frame Relay keepalive* command enables LMI for serial lines.

LMI statistics can be displayed with the *show frame-relay lmi* EXEC command.

Frame Relay flow control Flow control, the management of movement of frames within and between networks, is not explicitly defined in the Frame Relay specification, and the ITU has defined general concepts and standards for handling flow and congestion. In practice, congestion can be prevented in Frame Relay networks by establishing committed information rates (CIRs) to each user, denying the connection if insufficient bandwidth is available, and by discarding frames above the CIR. Existing congestion can be signaled to the user in the form of backward explicit congestion notification (BECN) and forward explicit congestion notification (FECN).

Frame Relay Forum FRF. An international professional association of corporations, vendors, carriers, and consultants promoting the Frame Relay networking technology, and supplying commercial Frame Relay products and services, established in 1991. The Forum develops and promotes specifications to support the viability of Frame Relay and sponsors interoperability events for designers, manufacturers, and vendors to test their Frame Relay-based equipment. http://www.frforum.com/

Frame Relay Forum Implementation Agreements IA. A series of formal, approved agreements (standards) developed and/or supported by the Frame Relay Forum. The preceding chart gives a brief summary of IAs, organized somewhat functionally. The documents may be freely downloaded from the Net. See Frame Relay Forum chart.

Frame Relay frame format The format for a frame is based on Link Access Protocol D (LAP-D) for ISDN. Frames are also known as protocol data units

---

Frame Relay Frame Format

```
0          1          2          3          4          5          6          7          8
0123456789012345678901234567890123456789012345678901234567890123456789012345678901234567890
+-------------------/-----------------/---------------------------------------------------
+
|  flag   |      address    |   information   |         FCS        |  flag  |
+-------------------/-----------------/---------------------------------------------------
+
   1 octet    2 to 4 octets      variable         2 octets         1 octet

Header Structure

   +---------------------------------------------------------------+
   |         DLCI upper             |  C/R  |  0  |
   +---------------------------------------------------------------+
   |         DLCI lower    |  FECN  |  BECN  |  DE  |  1  |
   +---------------------------------------------------------------+


      DLCI    data link connection identifier
      C/R     command/response
      FECN    forward explicit congestion notification
      BECN    backward explicit congestion notification
      DE      discard eligibility
```

(PDUs). Frame Relay frames are similar to DXI and FUNI. Flags are used to indicate the beginning and end of a frame, which may be variable in length.

The format specified for a frame includes a 1-byte (8-bit) flag, followed by 2 to 4 header address bytes, followed by a variable number of information bytes, followed by a 2-byte CRC code (frame check sequence), followed by a 1-byte flag. There are a number of possible configurations of the address field; it may be two, three, or four bytes in length, as determined by the extended address (E/A) bit. Information for the Local Management Interface enhancements is stored (e.g., DLCI information) in the frame header.

Frame Relay installation Frame relay communications connection services are generally available for a monthly subscriber fee or per-data rate from a local commercial provider, depending upon the speed of transmission. One-time connect charges for installation and port configuration are common. Transmission speeds up to 56-64 Kbps are typical, although most vendors offer higher speeds for more money. In-house installations of Frame Relay networks are also available.

Frame Relay physical layer interface The specification for Frame Relay does not stipulate particular physical connectors or cables. In practice, however, unshielded twisted pair (UTP) is commonly used in ISDN implementations of Frame Relay.

Frame Relay service Frame relay network service consists of a combination of hardware, software, and transmission services. It provides multiple independent multiplexed data links to another destination or to several destinations through a process which is at least as transparent as a leased line and less expensive. See Frame Relay installation.

Frame Relay, voice over VoFR. Frame relay technology provides an opportunity to combine data and voice communications services over the same network. Analysis of typical voice communications indicates that much of it is unnecessary (background sounds, pauses, etc.) and can be screened out before transmission over data networks. This aspect offers possibilities for processing and compression to provide for efficient transfer of digitally encoded voice conversations. Initially, there was no uniform standard for carrying voice over Frame Relay and various schemes for its implementation had been developed. In July 2001, the Frame Relay Forum announced FRF.20, an IP Header Compression Implementation Agreement that defines packet encapsulation and compression negotiation to facilitate the transmission of voice over IP.

Frame Relay-capable interface device FRCID. A peripheral device that performs frame encapsulation within a Frame Relay. See bridge, encapsulation, router.

Frame Relay Implementors Forum An association of vendors supporting standards of interoperability for Frame Relay implementations. A common specification was first introduced in 1990 based on the standard proposed by the American National Standards Institute (ANSI). See Frame Relay Forum.

**Frame Relay Forum
Implementation Agreements**

Number	Date	Description
FRF.6	Mar. 1994	Service Customer Network Management Implementation
FRF.19	Mar. 2001	Operations, Administration and Maintenance Implementation
FRF.17	Jan. 2000	Privacy Implementation
FRF.12	Dec. 1997	Fragmentation Implementation
FRF.13	Aug. 1998	Service Level Definitions Implementation
FRF.14	Dec. 1998	Physical Layer interface Implementation
FRF.1.2	Apr. 2000	User-to-Network (UNI) Implementation
FRF.4.1	Jan. 2000	User-to-Network Interface (UNI) Implementation
FRF.2.1	July 1995	Network-to-Network Interface (NNI) Implementation
FRF.10.1	Sept. 1996	Network-to-Network SVC Implementation
FRF.18	Apr. 2000	Network-to-Network FR/ATM SVC Service Interworking Implementation
FRF.5	Dec. 1994	Frame Relay/ATM Network Interworking Implementation
FRF.8.1	Feb. 2000	Frame Relay/ATM PVC Service Interworking Implementation
FRF.7	Oct. 1994	PVC Multicast Service and Protocol Description
FRF.15	Aug. 1999	End-to-End Multilink Implementation
FRF.16	Aug. 1999	Multilink UNI/NNI Implementation
FRF.3.2	Apr. 2000	Multiprotocol Encapsulation Implementation
FRF.9	Jan. 1996	Data Compression Over Frame Relay Implementation
FRF.11.1	May 1997	Voice over Frame Relay Implementation
FRF.20	Jun. 2001	IP Header Compression Implementation

---

frame store A high-capacity digital video storage buffer. Frame stores are most commonly used in two categories of applications: (1) those that require image buffering to provide sufficient speed for continuous display (see frames per second), such as computer editing or display systems, and (2) those that require image buffering in order to create complex, composite, or multiple display systems (such as video walls). In the first instance, the device from which the frames are being displayed or the display software may not be fast enough to read and display at 30 or so fps. By using a frame store, sufficient frames can be buffered in fast access memory (or on a very fast drive) to provide quick display and the illusion of continuous motion. If the display software creates unwanted effects on the screen when loading the next frame, the transition can sometimes be smoothed with double-buffering or grabbing the next frame from the frame store rather than from a hard drive. In other words, the new image is preconstructed in memory while the current image is being displayed, and the buffered image can then be displayed instantly over the previous frame, rather than reading in and decompressing the frame and then displaying it line by line over the previous frame.

In the second type of application, a frame store can help compose a complex image, such as computer graphic effects for a movie, which may have been raytraced one frame at a time, but which, when combined with footage of the actors, needs to match the speed of the action. A frame store can also be used as a component of a video wall, say 20 monitors in a four by five grid, which shows 1/20th of the actual image on each monitor. Since this display takes some computing power to split up an image into 20 separate subimages, the image grid could be segmented and prestored, so all the monitors display the correct parts of the grid at the same time. See buffer, frame buffer, desktop video.

Frame Switching bearer service FSBS. In a sense, an enhanced version of FRBS, in that it includes the basic functions of FRBS, plus frame acknowledgments and other error and flow control services. It is more similar to X.25, from which Frame Relay was derived, than the basic, streamlined FRBS service and reintroduces some of the features of X.25.

frames per second A phrase describing display speed for TV broadcasts, video, and film animations. The two most important aspects that determine this speed are human perception and display technology. Through persistence of vision and expectation, humans perceive still frames displayed quickly one after the other as motion. It requires only about 15 to 30 frames per second (depending upon the amount of detail and speed of the action) for these images to appear to be continuous motion. Most animations are created with 24 to 30 frames per second. Since motion media can be displayed only at the fastest speed of the display medium (usually a cathode-ray tube), the technology also determines the display rate, with speeds of 15 to 60 fps being implemented, and about 20 to 30 fps most often used. See frame; frame rate, video; refresh.

framing bit FB. A noninformational bit that can be used for a variety of signals in frame-based networks. A framing bit can signal the frame beginning or end

and error conditions, and it can be used for synchronization, depending upon the quantity and pattern of frames. In its simplest implementation, the framing bit signals to the receiving equipment that a new frame is about to begin.

In North America, Superframe and Extended Superframe standards are used for implementing T1 network services. Each frame in a T1 basic Superframe (SF) includes eight information bits and one framing bit. In SF, there are two types of framing bits: terminal framing (Ft) and signaling framing (Fs) bits. In Extended Superframe (ESF), there are three types of framing bits: frame pattern sync (Fps), datalink (DL), and cyclic redundancy check (CRC-6). The datalink framing bits differ somewhat, depending upon whether ESF is implemented according to ANSI or AT&T standards. The ANSI format provides for the transmission of a Performance Report Message (PRM) that allows actual performance to be compared with established thresholds and an alert to be generated if anomalous conditions are detected. By evaluating the pattern of framing bits in a series of frames, synchronization can be established. In formats such as DS-1C, a framing bit can be stuffed to generate 26-bit information units allowing for synchronization and framing.

franchise A government granted right to offer community public right-of-way for exclusive commercial communications services, such as phone services or cable broadcast services. The franchise fees, or a portion of them, may be used by local government agencies, a portion of which may be allocated to local Designated Access Providers (DAPs) for facilities funding. Some of the earliest local phone companies may be partially exempt if they gained their exclusivity prior to regulation (grandfathering).

Franklin, Benjamin (1706–1790) An American businessman (printer), statesman, scientist, and philosopher who did numerous experiments in electricity and printing. He shared his discoveries openly and coined many of the terms now used to describe mechanics and electricity. He called vitreous electricity, demonstrated by rubbing glass with silk, *positive* electricity, and resinous electricity, demonstrated by rubbing amber with wool, *negative* electricity. He did experiments with lightning and stored electrical charges in a device called a Leyden jar, and he established that man-made electricity and atmospheric electricity had the same properties. These experiments were enthusiastically received and replicated throughout Europe, spurring much interest and development in the field of electricity.

Ben Franklin also developed some early document duplication techniques which he used on his own printing press to help him manage his voluminous records and correspondence.

Ben Franklin was a successful business owner at a relatively young age and always hoped to retire early to devote the rest of his life to scientific inquiry and his various hobbies, but the American Revolution and the overwhelming public demand for his diplomatic skills kept him occupied for long hours right up to the time of his death in his mid-80s. See electrostatic, Leyden jar.

Franklin Institute A significant organizer and promoter of activities related to general science, electrical education, professional development, and technological development in electronics. The Institute organized many key American and international exhibitions starting in the 1800s and is still well known for its educational activities and awards for excellence. The Institute was founded by Merrick and Keating in 1824 as The Franklin Institute of the State of Pennsylvania for the Promotion of the Mechanic Arts in honor of scientist and statesman Benjamin Franklin. Since 1926, the Institute has been publishing the *Journal of the Franklin Institute*. In 1933 it began construction of the Fels Planetarium and has administrated the Franklin Institute Science Museum since 1934. In the 1990s, multimedia theaters were added to provide enhanced educational presentation capabilities. See Benjamin Franklin Institute (not affiliated). http://www.fi.edu/

fraud *n.* Deceit, trickery; unauthorized access or use, especially under an assumed identify, such as a false username, or through the use of unauthorized equipment; misrepresentation of identity, products or services, especially for monetary gain.

Unfortunately, fraud is now rampant on the Internet. A criminal with fraudulent intent can contact millions of potential victims in seconds at minimal cost.

There are many different types of fraud on computer networks. Fraudulent vendors use names that sound or look like recognized businesses and create copycat Web sites selling substandard products (or no products at all). They use email with embedded Web page links to fool people into thinking they are verifying their passwords on financial sites when in fact a *Trojan horse* program is capturing their passwords or credit card numbers. Fraudsters also use fictitious email addresses, anonymous emailers, and elaborate mail routing to obscure their locations and identities. Naive users, especially teenagers, children, senior citizens, or adults from small communities, are being defrauded on a massive scale by Internet-based get rich/pyramid/multilevel marketing schemes, fake contests, promises of off-shore commission profits, black and gray market pharmaceuticals, videos, and CDs, identity and credit card number theft, and solicitations to patronize illegal pornographic sites. Unsolicited bulk email (spam) is one of the primary vehicles used by fraudulent companies to contact and negotiate with large numbers of potential victims to perpetrate these crimes.

Many of these schemes are so sophisticated that it may be difficult, even for a computer professional, to recognize the deceit. Many existing laws should be sufficient to protect victims from Internet fraud, but existing black market, theft, piracy, embezzlement, pornography, and false advertising laws (which cover the majority of Internet-related crime) are difficult to enforce unless resources for law enforcement and consumer education are increased. http://www1.ifccfbi.gov/

Fraunhofer region In an antenna, a region of the field from which the energy flow proceeds as though emanating from a point source near the antenna, also called far-field region. It is considered to be one of three basic regions without distinct transition boundaries that are identified as you move away from an antenna source. The Fraunhofer region is the one farthest from the source beginning at a point where the angular field distribution is considered to be independent of the distance from the antenna. See Fresnel region.

Fraunhofer spectrum The portion of the solar spectrum visible to humans, i.e., the portion where the spectral absorption lines can be clearly seen. The dark lines have come to be known as Fraunhofer lines.

Fraunhofer observed that the range of spectral lines varied depending upon which celestial body was the source of the light. This information was valuable in that it led to observations about light emanating from stars with different chemical compositions, allowing scientists to analyze the composition of bodies in our universe from a distance. The Fraunhofer spectrum can be taken as a baseline reference against which spectral shifts can be compared. See Wallaston, William.

Fraunhofer, Joseph von (1787–1826) A German physicist who was skilled at applied optics and lens design. As a scientist, he applied his knowledge to the study of the Sun, its spectra, and the diffraction of light. He systematically set about measuring the position of hundreds of solar spectral lines and classified the most prominent lines. He then developed a diffraction grating and a general grating equation for measuring the wavelengths of colors and of lines in the dark spectrum. Various aspects of spectral nomenclature are named after Fraunhofer.

FRBS Frame Relaying bearer service. See Frame Relay.

Free Software Foundation FSF. A Massachusetts-based association committed to the development, acceptance, and promotion of open, free software standards and applications to benefit the world at large. The freedom to copy and distribute software, and the freedom to modify, enhance, and improve software are encouraged by the FSF. Thus, the programming and user communities benefit by the availability of constantly improving software and standards, and programmers have a broad, ready base of software from which to learn and to improve their skills.

The FSF has developed the integrated GNU software system, which includes assemblers, compilers, and more. Donations to the FSF are tax deductible. http://www.fsf.org/

free-space optics FSO. A term for optical transmission technologies that do not require a physical waveguide such as a fiber optic cable. An infrared remote television control is an example of a free-space transmission device that projects data through line-of-sight "free space." Free-space optics is promoted as a means of completing "the last mile" which is the distance from major optical backbone transmission services that separates most homes and businesses in the U.S. from established light-based

F

communications. This is mainly due to the prohibitive cost of completing the last link between the transmission pipeline and millions of individual buildings. FSO links could be established between the backbone and local drops through building-mounted eye-safe infrared laser transmitters rapidly pulsing binary on/off signals.

FSO links based upon infrared lasers as the light source can transmit to about 1 km, depending upon weather, at transmission rates of up to more than a gigabit/second, with the potential for higher rates as diode technology evolves. In spite of the limitations of line-of-sight and the impact of weather (e.g., fog) on free-space transmissions, FSO systems have a big advantage in terms of speed of installation and cost of installation and maintenance over underground cable systems. They may be particularly useful for interconnecting local area networks in different parts of a building or industrial complex. The transmitters can be placed on windows or roofs, without the need to run cables over the parking lots or inside building walls and can transmit at speeds more than 60 times faster than wireless radio links.

FSO transmission links were pioneered in the early 1970s by groups such as Bell Laboratories, who researched lasers and light-emitting diodes (LEDs) for FSO communications. Interest in FSO links increased in the late 1990s, with companies such as Global Crossing and Lucent Technologies testing new systems. See line of sight, Photophone.

FreeBSD A Unix computer operating system descended from Berkeley Standard Distribution (BSD), that flourished with the development of the ARPANET, the forerunner to the Internet. The Computer Systems Research Group (CSRG) enhanced BSD with 32-bit addressing, virtual memory, and a fast file system supporting long filenames. They further introduced BSD Lite which was BSD without the licensed AT&T code and could be freely distributed. The CSRG disbanded in 1992, and the community at large adopted BSD and developed FreeBSD. In 1994, some of the CSRG briefly came together and further developed BSD 4.4 Lite. See Berkeley Standard Distribution, Unix, UNIX.

freeware A category of product, usually software, that may be distributed and acquired without cost. Freeware does not mean copyright free. A developer has the right to retain the copyright to an original work and still distribute that work or product free of charge, keeping the right to revoke freeware privileges. Freeware is *not* the same as public domain software, in which the owner has given up the copyright, and it is *not* the same as shareware for which there is a moral obligation to pay the stipulated fee. Freeware (public domain) and shareware are two common types of freely distributable software. See public domain, shareware.

freeze frame A mode of visual display in which only one screen-full, or cell, of an animated sequence is shown. On digital systems, it's easier to show a single frame of information; displaying a single frame on film or on a CAV laserdisc is pretty straightforward.

In analog systems, or those recorded with overlap of information or no firm transition from one "cell" (frame) to the next, it is more difficult, as VCR tapes and the freeze frame mode may be of limited duration and quality. Some analog/digital systems will take a digital shapshot of the frame to be displayed and display it as a digital image, usually with better results than trying to display the analog image. Many videoconferencing systems don't show actual real-time motion but rather snapshot a digitized freeze frame every few seconds, to provide the illusion of seeing what is going on at the other end without seeing actual movement. Such systems are sometimes distinguished as audiographics systems. As transmission media become faster, full-motion video will become standard.

frequency (*symb.* – f) The number of periodic occurrences or oscillations in a specified unit of time. Frequency designations are used to describe the varying periodic character of specific regions of the electromagnetic spectrum. In electricity, frequency is the number of times a current alternates in hertz (named after Heinrich Hertz). Radio signals are usually measured in kHz, or in MHz at high frequencies (above 30,000 kHz). Many older radio and electronics manuals will describe frequency in terms of cycles per second instead of hertz, as the unit name was widely used until the 1960s. See wavelength.

frequency bias An adjustment made to a signal, such as an audio signal, that biases it in one direction or another. In practical applications, it may be a constant adjustment to a signal frequency to prevent it from reaching zero. It may also be a high-frequency addition to a signal to bring up the low-frequency sounds to reduce distortion in regions that don't record well.

Too much bias can result in clipped or distorted signals, depending upon how the bias is applied. In audio recording, bias may be added to a signal to overcome flat regions in the magnetic recording medium. Playback equipment ignores the bias signal but provides a better original, provided the addition of the bias signal is carefully controlled so as not to introduce distortion. Different types of tape respond differently to bias. For example, common ferric oxide tape is referred to as *normal bias* tape. In computer sound programs, there may be utilities for adding or controlling bias in order to edit the characteristics or quality of the sound.

frequency departure The degree of variation of a carrier frequency or reference frequency from an expected, assigned value.

Frequency Division Multiple Access FDMA. One of the simplest techniques for increasing capacity over communications channels, since the radio frequency spectrum is not unlimited. FDMA is a way of dividing up the available spectrum according to frequencies. The communications station typically assigns a unique frequency or frequency sequence to each user currently engaged in communication, and it tracks these as needed to provide many simultaneous links. This technique is used in cellular phone and satellite transponder systems. See Code Division Multiple Access, Complex Scheme Multiple Access, Demand Access Multiple Access, Multiple Access, Time Division Multiple Access.

frequency division multiplexing FDM. A technique used to more efficiently utilize a fixed or limited amount of bandwidth by subdividing it into narrower channels. Typically, *guard bands* are inserted between communications bands to reduce interference. Multiplexing can be used to increase the number or types of transmissions within a fixed medium. For example, it may be used to simultaneously transmit voice and data.

George Ashley Campbell invented the electric-wave filter in 1915, a device used in FDM. FDM is still a widely used transmissions technique that is only now being superseded by other methods, such as time division multiplexing (TDM is prevalent in fiber optic communications systems). See single sideband, time division multiplexing.

frequency frogging See frogging.

frequency hopping In mobile communications systems, a spread spectrum technique in which frequencies are jumped during the course of a transmission. This hopping may be done for many reasons, such as to try to find a cleaner or more stable signal or to try to avoid detection (sometimes used in military zones).

Frequency hopping was invented by Hedy Lamarr (born Hedwig Eva Maria Kiesler) while trying to develop a secure guidance system for a torpedo, using radio signals that would not be detected and subsequently jammed. Her collaborator, George Antheil, suggested a way to synchronize the varying frequencies with paper tape, but the synchronization system was somewhat cumbersome, and it was not until the development of computer electronics that Lamarr's idea was fully implemented. Lamarr and Antheil received a patent for the technology in 1942 and it has since been extensively used in military and civilian communications systems. Unfortunately, neither Lamarr, better known for her film career, nor Antheil, received any of the compensation or credit due for the invention. See direct sequencing, multiple access, spread spectrum.

frequency modulation FM. A sine-wave modulation technique widely used in broadcasting that works by varying the frequency of a constant amplitude carrier signal with an information signal.

FM radio broadcast signals typically require about 200 kilohertz of bandwidth and are not as subject to noise and interference as amplitude modulation (AM).

Many scientists insisted that frequency modulation was not possible. Edwin Armstrong thought it was and devoted a decade of intense research to the problem, ultimately proving successful. FM radio stations began broadcasting in the early 1940s. In the United States, the Federal Communications Commission (FCC) approved FM stereo broadcasting in 1961. It has approved the range from 88 to 108 MHz for FM broadcasting.

In one type of telephony, a frequency-modulated carrier signal can be transmitted over wires. Frequency modulation can be used when digital data is routed through an analog system for part of the transmission.

FM is also commonly used for very short range communications for cordless phones, home and business intercoms, baby monitors, short-range television security systems, and burglar alarms. See amplitude modulation; Armstrong, Edwin; carrier; channel; modulation; Moonbounce.

frequency shift keying FSK. A modulation technique used in data transmissions, such as wireless communications, in which binary "1" and binary "0" (zero) are coded on separate frequencies. This scheme can also be adapted to regular phone lines by assigning binary "1" to a tone and binary "0" (zero) to a different tone. There are other keying schemes for carrying information such as *on/off keying* and *phase shift keying*. See frequency modulation, phase shift keying.

frequency swing In frequency modulation, the difference between the maximum and minimum values at a given frequency. In other words, the limits within which the oscillations range.

frequently asked question(s) FAQ. A query or question-and-answer list of questions that have been asked and answered many times, so often, in fact, that someone has taken the time to write up the question/answer and post it, usually in a public forum on the Internet. FAQs comprise an important part of the information base of the Internet, on private and public forums, chats, special interest groups (SIGs), USENET newsgroups, and Web data sites. All Internet users are strongly advised to read the FAQ *before* posting on any online forum or risk being soundly scolded or flamed by other users. See Netiquette, RTFM.

Fresnel equations Mathematical descriptions of electromagnetic wave behavior developed by Augustin Fresnel. For radiant energy incident on a dielectric medium, the equations describe the amplitude of the transmitted and reflected electric fields. In other words, when an incident wave encounters a dielectric (e.g., when sunlight travels through air and then encounters window glass or water) some of the energy is transmitted through the dielectric, some of the energy is reflected and the sum of the two energy values equals that of the incident wave.

Fresnel lens An efficient optical component with multiple faceted glass, acrylic, or materials designed to prismatically refract light to produce a brightly concentrated beam. A large light assembly designed from concentric rows of glass in a Fresnel configuration can propagate light for about two dozen miles, making it highly useful for lighthouses. The shape of the beam can be controlled to some extent by the positioning of the facets. By coloring some of the facets and rotating the lens or light source, the assembly can be made to flash.

Small Fresnel lenses can be used as substitutes for magnifying glasses for a number of applications and have the advantage of being very flat, compared to traditional curved optical lenses. The tradeoff is optical quality, with curved, smooth lenses providing a clearer image.

A coherent light source (e.g., laser light) can be shone through a Fresnel lens to see how the refractive surfaces of the lens alter the course of the light beam.

A Fresnel lens can concentrate light for use in projection systems and can be configured to provide parallel light arrays for large-scale projection walls by concentrating and "feeding" the light to lenticular lenses. Inventors have suggested that inflatable domed or cylindrical Fresnel lenses might be useful as concentrators for supplying space or terrestrial solar power systems.

Safety glasses (preferably welding glasses) should be used when installing or adjusting Fresnel lenses even if the light source is disabled, because sunlight can be concentrated sufficiently by the lens to melt metal, which means it can readily (and quickly) burn skin. See diffraction, nearfield diffraction, Ronchi grating.

Fresnel Equations – Reflectance/Transmission

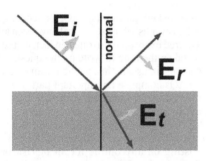

Fresnel mathematically described the relationship between incident electromagnetic radiation interacting with a dielectric substance and the resulting reflected and transmitted energy. He demonstrated that the sum of the energy associated with the reflected radiation (E_r) and the energy that continues to be transmitted through the dielectric (E_t) equal the energy associated with the original incident wave (E_i).

Fresnel region A region around an antenna between the physical equipment and the Fraunhofer (far-field) region. The transition between the Fresnel and Fraunhofer regions can be mathematically calculated if the length of the antenna and the wavelength are known. Fresnel approximations are used to describe diffraction patterns in the Fresnel region under certain wavelength, size, and distance conditions. See Fraunhofer region.

Fresnel, Augustin Jean (1788–1827) A French physicist and engineer who pioneered a transverse-wave theory of light as an explanation for the phenomenon of double refraction, developing Fresnel equations related to reflective and refractive processes. In 1819, Fresnel submitted his theories on diffraction to the Paris Academy. In 1822, he demon-

strated a faceted lens designed to refract light to concentrate a beam, a structure now widely used in lighthouses and scientific components. See Brewster, David; Fresnel lens.

friction feed A feed mechanism in a machine that relies on friction or pressure to feed the sheets (paper, card stock, thin metal plates, etc.) as in a printer, press, or photocopying machine. Friction feed devices are often made from rubbery materials to help adhere the medium to the feeder. See tractor feed.

friendly name A name that is easy to recognize and remember, used in place of cryptic or long names or codes. For example, a printer with a computer designation of LSL2345-b may be assigned a friendly name of *AdminLaser* in lists of available output devices. Domain names on the Internet have been given friendly names. The computer system doesn't require a familiar name like "coolsite.com" to locate a site; a binary address is more direct. But humans prefer language to numerals or binary addresses, and so domain names have been associated with data addresses to make it easier to use FTP or Telnet protocols, or to access a site through the Web. See alias.

fringe area A region just outside the major transmission area of a broadcast signal where the signal is degraded and inconsistent but generally present. Sometimes those receiving signals in fringe areas can improve the quality of service, up to a point, with better antennas.

fringing 1. An undesirable visual artifact, especially on cathode-ray tube (CRT) color displays, in which the electron beams are converging incorrectly so as to appear unfocused, with a fringe or edge of color slightly offset. 2. A visual artifact on an object displayed in a computer paint program, in which the color of the previous background of the object (e.g., a blue sky) shows up distinctly as a halo around the edges when the object is placed on another color (e.g., a red brick wall). A defringing option to blend the edge colors is available in many paint programs, to smooth the transition in a process called antialiasing. 3. An effect that occurs when coherent optical plane waves intersect one another and combine intensities to produce a set of spaced bright and dark regions. The fringing may be viewed or photographed depending upon the magnitude of the effect and the wavelength of the optical energy. The best results occur when the intersecting waves are of equal intensity, but this is often not the case in real life, where light tends to drop off as it continues along the line of travel. For deliberately creating fringe patterns, laser lights, with coherent monochromatic characteristics and a longer range are favored. See diffraction.

FRM focus-rotation mount. A pivoting antenna-focusing structure.

FRND Frame Relay network device.

frogging 1. An equalizing technique in which incoming high or low frequencies are inverted to become outgoing low or high frequencies. 2. Corruption of transmissions data in which incorrect data is inserted into or overwrites some of the expected data in a non-random way.

front end The portion of an application or device that interacts with or is accessible to the end user. On computers, shell command lines and graphical user interfaces are the most common types of front end. In audio/video equipment, the front end consists of the knobs and dials that are within easy reach of the user. The Web has become an interactive front end to the Internet. In commercial facilities, the front end is usually a store front or reception area that provides customer services, as opposed to storage or personnel-only work areas. In broadcast circuitry, the front end consists of the knobs and components which tune in the desired frequency.

front end system A system that acts as an intermediary gateway, filter, or console for a more powerful, but less user-friendly or accessible system. A desktop computer-based telecommunications server can serve as a front-end to a mainframe, sparing it for more computing-intensive tasks. An information kiosk with a simple touchscreen or touchpad input system that hooks into a more powerful network system is a type of front end for the general public. An automated teller machine (ATM) is a banking system front-end for the public.

FRSE Frame Relay switching equipment.

FRTE Frame Relay terminal equipment.

frustum The surface of a solid cone or pyramid that would be created if the top of the cone or pyramid were cut off parallel to its base. A concept of interest to mathematicians, programmers, and users of 3D modeling software.

FS federal standard.

FSAN See Full Services Access Network.

FSBS Frame Switching bearer service. See Frame Relay.

FSF A nonprofit educational association supporting GNU. See Free Software Foundation.

FSK See frequency shift keying.

FSO 1. Foreign Service Office/Officer. A foreign affairs diplomatic liaison. 2. See free-space optics.

FSP See File Service Protocol.

FSS See fixed satellite service.

FSTC See Financial Services Technology Consortium.

FTA See Federal Telecommunications Act.

FTIP Fiber Transport Inside Plant.

FTNS See Fixed Telecommunications Network Service.

ftp The command typed at an FTP site to access an archive based on File Transfer Protocol (FTP). It is usual to type the command in lower case; however, the name of protocol itself is usually written in upper case.

FTP See File Transfer Protocol.

FTP mail server A mail server that facilitates the retrieval of files from FTP archives by sending them to the user's email address. Since files on FTP sites can be text or binary, and some email addresses cannot directly accept binary files, the retrieved files may be sent as a binary *file attachment*. See file attachment, ftp, FTP.

FTR federal telecommunications recommendation.

FTS 1. file transfer support. 2. See Federal Technology Service. 3. Federal Telecommunications System. A government private telephone network. See FTS2000.

FTS2000, FTS2001 A nonmandatory program of intercity telecommunications services provided to federal agencies by the U.S. General Services Administration (GSA) through two networks (A & B) transmitting through fiber optic cable. Due to delays resulting from federal agencies failing to meet the 6 Dec. 2000 deadline, the project was changed from FTS2000, with Sprint and AT&T as major vendors, to FTS2001 with Worldom, Inc. (Jan. 1999) and Sprint (Dec. 1998) as major vendors. The change in vendors resulted in some controversy, with AT&T requesting that competition be reopened (Spring 2001), alleging material changes in the contract requirements, which had been relaxed to ease the process of agency compliance. In response, the GSA contended that data collection requirements were relaxed but service stipulations remained the same. The transition is ongoing and is expected to be fully in place by about 2005.

FTSC 1. Faculty Technology Support Center. 2. See Federal Telecommunications Standards Committee. 3. See FidoNet Technical Standards Committee. 4. Foreign Trade Service Corps.

FTTC Fiber to the Curb. Fiber cabling that reaches the drop near the home but does not include the drop onto or into the home. See Fiber to the Curb.

FTTH See Fiber to the Home, cable modem.

FTTHO Fiber to the Home Office.

FTTL See Fiber to the Loop.

FTTN Fiber to the Neighborhood.

FUBAR fouled up beyond all recognition. A phrase purportedly originating in military speak in World War II. Less polite versions of it fit the acronym as well. See foo.

FUD fear uncertainty doubt. A sales strategy attributed to Gene Amdahl, stated as "FUD is the fear, uncertainty, and doubt that IBM sales people instill in the minds of potential customers who might be considering [a competitor's] products." The marketing spin is that customers are safer with International Business Machines (IBM) products.

fudge *v.* To hedge, approximate, overstate, or talk around a subject so as to try to appear to know what you are talking about, to use "bafflegab"; to cobble together so it appears as though it might work, or so that it approximately works but may not be complete or robust.

fudge factor Tolerance factor, buffer, safety net. See fudge.

Fujitsu Limited A large Japanese commercial conglomerate originating in the 1920s. Fujitsu is known for a number of large-scale computing products, is a world leader in industrial robotics, and manufactures many consumer computer-related accessories (e.g., office-quality printers). See FACOM.

Fujitsu Laboratories, Ltd. A wholly owned subsidiary of Fujitsu Limited, founded in 1968. It is an international research laboratory which has recently

F

devoted significant resources to the development of fiber optic amplifiers and wavelength Division multiplexing (WDM) technologies for highspeed communication over longhaul optical networks.

Fujitsu Network Communications, Inc. A designer and manufacturer of fiber optic and broadband switching systems, and provider of telephone network management software which is marketed to ILCs, CLECs, VPNs, and cable TV companies.

full duplex A system that supports simultaneous transmission and receipt at both ends of the circuit. Full duplex operation requires a balance of hardware and software protocols to enable two-way transmissions. In some systems, full duplex operation creates a digital echo in which each unit of textual information is repeated. Some systems can technically support full duplex operation but are selectively operated in half duplex mode to improve the quality of the communication, as in some satellite voice systems and speakerphones. These systems have tones and sensors that coordinate the back-and-forth nature of the conversation so the transmission favors the direction in which the current user is transmitting. Systems with bandwidth limitations may work in full duplex mode for some operations (e.g., voice conversations) and then may switch to half duplex for more bandwidth-intensive operations (e.g., videoconferencing). See half duplex.

full scale The full functional range over which an instrument or device operates.

Full Services Access Network FSAN. A group of cooperating international telecommunications companies, including Bell Canada, BellSouth, BT, Deutsche Telecom, Dutch PTT, France Telecom, GTE, Korea Telecom, NTT, SBC, Swisscom, Telefonica, Telstra, and Telecom Italia. FSAN shares its documentation with relevant standards bodies. One of the groups associated with FSAN is the Optical Access Network (OAN). Nippon Telephone and Telegraph (NTT) and BellSouth are developing fully FSAN-compliant ATM-PON systems for 1999. See fiber to the home.

function key A configurable or special-purpose keyboard button. Function keys are often programmed as shortcuts to produce the same effect as typing several keys, or selecting an operation several menu items deep. Many computer keyboards have 10 or 12 function keys with a variety of uses, depending upon the currently active software. They may be located in a vertical line above the other keys in the keyboard, or they may be organized in two rows to the right or left of the keyboard. For touchtyping, the double row to the right or left of the keyboard is more practical and easier to use.

Many telephones have prelabeled or configurable keys for redial, speed dialing, and other optional functions.

functional specification The specification of an object or system in terms of how it will be used and/or what it specifically is designed to accomplish as an end goal. In its purest sense, a functional specification concentrates on user interaction or the task at hand (e.g., punching out keyboard key caps on a production line) and does not specify the parts, process, or equipment that may be needed to create the object or system, since there are usually many different ways in which the same end result can be achieved.

Thus, a functional specification for a computer would include descriptions of what a user might wish to do with the system and the means by which the user might interact with a system (visually, tactilely, auditorially, etc.) but would not specify the CPU, bus type, interface slot formats, specific model or styles of input/output devices, etc.

Sometimes it is wise to turn to functional specifications when technology becomes entrenched. A functional specification is one way to spur creative innovation. For example, a computer mouse is a ubiquitous means of interacting with a graphical user interface, but is it the best way? In a functional specification, one usually asks what does one want to accomplish with the computer. If the answer is to indicate a choice or selection in a certain context in a way that is natural and comfortable for the user, then perhaps a graphical pen, data glove, or eye-scanning headset might be suggested to fulfill the functional specification, rather than a mouse.

functional transparency The capability of a system to carry out its functions in such a way that the user doesn't have to see or worry about the inner workings or lower-level protocols and configuration of the system. The more natural and direct the interaction, the more transparent the system. In heterogenous distributed computer networks with many different types of computers and operating systems intercommunicating, functional transparency is a situation in which the user doesn't have to worry about the type of data or the network medium or protocols used.

In the earlier dialup modem days, users had to have enough technical knowledge to know how to configure terminal software to the correct baud and parity rates, etc. for each system to which they wanted to connect. Eventually autobaud modems added a certain degree of functional transparency by automatically negotiating and adjusting the data rate between the answering system and the calling system. With newer network protocols, such as email protocols on the Internet, once the initial configuration is installed, users are able to transmit a variety of types of data transparently over the system without worrying about compression protocols, file types, or whether the receiving system has the same parameters, operating system, or data rate capabilities as the sending system.

fundamental frequency 1. The lowest natural frequency in an oscillating system. 2. The reciprocal of the period of a wave. 3. The frequency most effective in a given situation (e.g., the one that transmits best over a particular waveguide).

fundamental group A group of trunks in which each local switching center is interconnected to one of a higher order.

furcation unit In fiber optics, a fiber installation and maintenance device that channels the individual strands of a fanned out bundle to their desintations. A furcation unit might be used, for example, as a connection between an outdoor bundled cable and an indoor multidevice hub.

fuse 1. *n.* A protective mechanism that reacts to break an electrical circuit when the current through the circuit exceeds a specific value. The mechanism may consist of a wire or chemical junction mounted in serial, which melts or breaks at a specified value, usually indicated with a number. Fuses are designed to break, if there is a problem, in order to protect more expensive electrical components from harm. The fuse was first patented by Thomas Edison in the early 1880s. Circuit breakers have replaced fuses in most new home electrical installations, as resetting the breaker is more convenient than replacing a fuse. See circuit breaker. 2. *v.* To join or blend together, usually by melting. To *fuse* implies a stronger or more consistent bond than does to *bond* (as with an adhesive), since there may be momentary heat or chemical alteration to create the bond.

fuse alarm A type of fuse connected with an audible device or a flashing light (or both) to indicate that the fuse is blown and must be replaced.

fuse block An insulated mounting structure for a fuse or bank of fuses. In smaller electronic devices, the block may secure a small clip that holds the fuse in place. In larger wiring installations, as in houses and offices, the fuse block may be a large metal electrical cabinet with several rows of fuse mountings. When further enclosed, it is usually called a fuse panel or fuse box.

fuse cable A section of cable spliced into electrically sensitive wiring installations (e.g., aerial cables prone to lightning strikes) that differs in specifications from the main wiring. A fuse cable is any cable that can defuse an electrical surge, but tends to be a higher gauge cable than is commonly used in electrical installations. Fuse cable lengths are usually kept as short as possible due to the added attenuation introduced by splicing in the different cables. In fiber optics installations, which are not sensitive to electrical charges in the same sense as wired communications, fuse cables are rarely needed except where the optical components are coupled to electrical devices.

fuse wire The wire inside the fuse housing that breaks when subjected to excessive current loads, breaking the circuit with the intention of protecting sensitive electronic parts. In the illustration of fuses, three different types of fuse wires can be seen, with straight wires on the three right fuses, a corrugated wire third from the left, and a flat zig-zag wire in the two left fuses.

fused biconic taper FBT. A process used for manufacturing couplers that are important in the fiber optics industry. FBT can meet the needs of wideband and ultra-wideband networks, a technology in which the coupling is crucial to providing a strong, reliable signal and can carry bidirectional signals. Multiple FBT couplers may be fused and/or cascaded in series, depending upon the network configuration. Because of the FBT, the light doesn't have to leave the optical fiber medium to pass through an optical interface, which is desirable for the prevention of signal loss. FBT couplers may also be used in conjunction with other types of couplers (e.g., Mach Zehnder dB). In dense wavelength networks, other technologies are being developed to solve coupling of fiber optic components, including planar waveguide components. These are not yet as common as FBT products. See fiber optic taper.

fused quartz, fused silica A glassy substance made from quartz crystals that is highly resistant to chemicals and heat. Quartz has remarkable noncorrosive and vibrational qualities that make it a valuable industrial material. See quartz.

fused semiconductor In semiconductor fabrication, the materials can be subjected to heat in such a way that they cool and recrystallize on a base crystal to form a tight electronic junction. See p-n junction, semiconductor.

fusion 1. A heat-induced liquid state. 2. A union of parts by applying heat or chemicals to liquify one or both of the parts (or a binding substance such as solder or weld) to form a permanent bond between them. 3. The union of atomic nuclei to form heavier nuclei, a process which generally requires enormous amounts of heat or pressure and can result in the release of a great amount of energy.

fusion sleeve A short strip of coating for supporting and protecting a fusion splice, as in a fiber optic cable with fused joints. The term tends to be applied to protective coverings made of soft (often shrinkable) materials. Sleeves from hard materials (e.g., metal) are more often called *ferrules* and long "sleeve" lengths are called *jackets*. The term *sheath* seems to be somewhat generically applied to both sleeves and jackets.

For fiber optic joints, fusion sleeves are commercially available in single or ribbon (mass fusion) styles. They may be preshrunk to reduce the amount of time it takes to install and shrink the sleeve the rest of the way to get a snug fit. Preshrunk sleeves also reduce the amount of time heat must be applied to secure the sleeve, reducing the possibility of heat-related damage to the joint and extending the life of the heating element. Commercial fusion splicers require about 1.5 to 3 minutes to process a sleeve heating cycle.

Sleeves may be transparent, for revealing the joint or colored for coding the joint (or the person who created the joint). Some companies will customize sleeves with code numbers, company logos, etc. on mass quantity purchases. See ferrule, fusion splice, splice guard.

fusion splice A thermally or chemically fused joint where two or more sections of cable have been combined to provide a continuous transmission path.

A soldered joint in spliced wires is a type of fusion splice, as is a thermal joint in a fiber optic cable. Even though a fusion-spliced joint has some mingling of molecules to provide a more solid connection than might be expected from oil or glue joints, it may still be fragile and require support and protection from knocking or bending. A sleeve and sometimes also a boot or extra support guard are typically placed over the joint for extra protection from strain and environmental damage. See fusion sleeve, fusion splicing, mechanical splice.

fusion splice viewer An instrument for imaging two or more fiber optic filaments that are secured for splicing. In the field this is sometimes done with simple tools such as magnifiers, but in mobile or fixed fusion splicing labs, viewing systems of various types are built into single and mass fusion splicing machines. Viewing may be through microscopes or LCD-based video imaging systems. Most fusion splicing machines enable viewing from two angles – the X and Y axes. The viewing system enables the technician to check the joints for debris or aberrations prior to splicing, facilitates alignnment, and allows the joint to be inspected after it has been fused.

In single-fiber splicing systems, it is reasonably straightforward to magnify the region of the splice to inspect the joint and make adjustments to alignment. Gauges are often incorporated into the display system to facilitate microadjustments. In mass fusion splicers, the problem of imaging the entire area over multiple filaments without making the viewing mechanisms prohibitively large sometimes results in tradeoffs in magnification and image clarity. Good lighting across the staging area can help compensate for reduced magnification, especially in portable splicing machines where size is a consideration.

fusion splicing A joining of two or more components (e.g., optical fibers) through a heat or chemical process that melts together the parts to be coupled. Fusion splicing is a common way to connect plastics, glass, and other fabrication materials for which the heat or chemical fusing process does not substantially alter the fused materials or interfere with their signal-transmitting characteristics. Since the molecules are intermingled, a fusion splice may be superior to other bonding methods (e.g., glue or oil).

In the fiber optics industry, protective sleeves are commercially available to cover splices which, depending upon the material fused, may be more fragile and subject to breakage or apt to pick up tiny particles if there are rough edges. Sleeves in different colors can be used to identify types of splices (or the individual who made the splice). A sleeve may have an extra strength component to support the splice and may be bonded to the splice or heat-shrunk to provide a tight fit, depending upon the application. Standard sleeve sizes range from about 30 to 50 mm.

While heat is commonly used to fuse optical fibers, chemical fusion through UV-cured resin compounds is possible. In the more common heat fusion process, the fibers melt away from each other as heat is applied by an electrode-generated arc, but are pushed together at the same time until the ends fuse. The heat must be applied in such a way that alignment is maintained and there are no combustion residues introduced within the splice. Some devices provide control over temperature and art time so that a fiber can be prefused prior to the final fuse. It is crucial that

the joint be nonreflecting so that the lightguide is continuous and undistorted from one fiber joint to the next.

Fusion splicing is still largely a hand-assembly and -inspection process, even with newer automatic fusion splicing devices. Fusion splicing equipment ranges from a pair of Kevlar-cutting scissors and a cable-stripping knife at the low end to sophisticated computerized inert-gas and vacuum-equipped automated systems at the high end.

Manual fusion splicing takes practice, especially when using manual or semiautomatic methods. Essentially, the process consists of the following steps:

- Strip the fiber filaments to an appropriate length.
- Clean the fibers without introducing scratches.
- Cleave the filament ends that are to be fused (cleanly and precisely within industry tolerances for angle and loss).
- Position the fiber ends within the fusion splicing mechanism. This is usually within a groove that aligns the fibers along two or three axes. The operator may have to align the Z axis in semiautomatic or manual machines.
- Clamp the fibers to maintain their aligned position.
- Close the mechanism and select a program for the splice (depending upon filament materials and diameter).
- Initiate the splice cycle (which takes a few seconds).
- Inspect the splice locally, if the fusion splicer has built-in capabilities for testing refraction patterns and tolerances, or remove the spliced cable and inspect it with other devices (e.g., an interferometer).

Fiber alignment in preparation for fusing may be manual, semiautomatic, or automatic. Most semiautomatic and automatic systems have some sort of computerized data display or ocular viewing system (e.g., similar to a microscope) to aid the technician in aligning fibers. Some systems use piezoelectric transducers to enable microalignment to more precise tolerances.

Multiple fused cables or filaments that are aligned in parallel proximity may not easily fit within sleeves. Fusion splice organizers with grooves to hold the cables and help keep out particles and moisture may be used instead.

Aligning two fibers in three planar directions (horizontal, vertical, and depth) may not be sufficient to maximize the light-guiding properties of the fibers when spliced. Through various fabrication or doping methods, the polarizing characteristics of the fibers may negatively influence the effectiveness of the waveguide if joined at different angles with respect to polarity. Thus, the rotational orientation of the fibers may have to be aligned with some of the more sophisticated fusion splicing machines to maintain the polarity through the length of the path.

Fusion splicing at its simplest involves splicing one fiber joint at a time. However, the demands of the marketplace for higher production rates for individual fibers or *fiber ribbons* with have resulted in machines called *mass fusion splicers* that can handle multiple joints.

For the assembly of complex cable arrays such as fiber optic patch panels, certified technical services are available for outsourcing the work. The fabrication of the fiber pigtails and fusion splicing may be carried out at the same location and documentation on the process and technical specifications may be provided as part of the service. Costs commonly range from $20 to $50 per splice. See cladding alignment splicer, core-to-core splicer, cleave, fusion sleeve, local injection and detection, mass fusion splicer.

Fiber Filament Fusion Splicing Basics

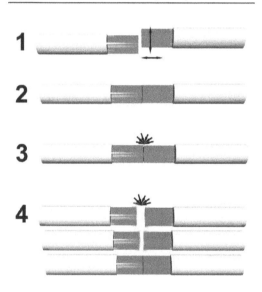

This simplified diagram illustrates the basic steps of heat splicing two sections of fiber filaments to form a single continuous lightguide. Fibers are cleaved and aligned in the Z direction in V-grooves or vacuum chucks (1) and then fine-aligned (2) in the X and Y directions until they are lined up core to core or cladding to cladding, depending upon the type of fiber and device used. A heating arc is applied across the region to be fused (3) which causes meltback as the filaments are warmed (4) which is compensated by moving the fibers closer together at a carefully controlled speed so they fuse without gaps or undue pressure.

---

fusion time The time it takes to complete a heat or chemically fused joint so it is stable enough, in terms of temperature and strength of the joint, to continue to the next step in the process or to be handled without compromising the bond. This may include a prefusing step that prepares the joint (e.g., by softening the ends) in preparation for fusing.

Times vary with the materials (glass, wire, or plastic), the type of splicer, and sometimes the ambient temperature and humidity. In general production work, where fast turnover is preferred, the process may take seconds or minutes, especially if splicing machines are used. With manual connections, such as fusing a connector to a fiber filament with epoxy and air-drying, the process can take as long as 18 hours. See fusion splice.

Future Public Land Mobile Telecommunication System FPLMTS. A standards effort initiated in the early 1990s to create a global mobile communications system that encompasses cordless and cellular technologies. It is intended to form a basis for integrated voice, paging, and data services at rates up to 9600 bps, and perhaps up to 20 Mbps, in both connection and connectionless modes, and for videoconferencing, global positioning services (GPS), and multimedia capabilities. See Global Systems for Mobile Communications.

future shock A state of human reaction to overload or too great a rate of change (as in changing technology), described in Alvin Toffler's 1970 book of the same name. In general, Toffler posits that humans, as biological/social organisms, have limitations as to how quickly they can adapt to change and contends that technological change could occur faster than they can adapt, resulting in an undesirable state of future shock.

In the 1980s and 1990s, people began to be more aware of the rate of change and obsolescence characteristic of computer technologies, but this idea was not generally considered in the late 1960s, when Toffler was researching and writing his book. Toffler's predictions were prescient, occurring half a decade before desktop computers became generally available, and are well worth keeping in mind, in terms of our ability to adapt to the computer revolution, to control its rate of change and evolution, and to assess and respond to how it affects our health and quality of life.

Fuzhou The capital of Fujian a Chinese province on the southeast coast. This region is home to global leading suppliers of ADSL modems and optical communication components, including Star Network Technology Co., Ltd.

fV *abbrev.* femtovolt (10^{-15} volt).

FVR flexible vocabulary recognition. A type of speech recognition in which a variety of words, not necessarily just those found in an associated database, can be processed. FVR is useful for wired and wireless telephony devices for dispensing instructions, schedules, answers to queries, and other commercial and industrial applications.

FVR is an advanced type of speech recognition. Because people's manners of speaking are varied, including dialects, accents, and slang, it is difficult to write algorithms to process natural speech spoken by many different people. Thus, programmers have developed a number of compromise solutions to simplify the interpretation process. One of these is to develop a limited vocabulary that is recognized by the software. FVR, on the other hand, overcomes this limitation in a number of ways, with larger databases and natural language processing algorithms.

It may still be necessary to train a flexible vocabulary recognition program to recognize the particular speech characteristics and pronunciation of an individual speaker.

FWA See Fixed Wireless Access.

FX 1. In the multimedia industry, an abbreviation for effects, as in SpecialFX. 2. See Foreign Exchange.

FXO Foreign Exchange Office. See Foreign Exchange.

FXS Foreign Exchange Station. See Foreign Exchange.

FYEO "For your eyes only;" another way of saying "private" or "confidential."

FYI An abbreviation for "for your information" that is commonly used on business memos, documents, email, and postings on the Internet.

FZA A data compression program developed by D. Carr (Gandalf Data Ltd.), that is derived from Lempel-Ziv (LZ), and favors high levels of compression over central processing unit (CPU) speed and memory. FZA is based upon packet-switched network techniques to compress information into a single frame or across multiple frames. See FZA+, Lempel-Ziv.

FZA+ An updated version of D. Carr's FZA data compression program. FZA+ was developed by A. Barbir.

γ Γ Gamma, the third letter of the Greek alphabet, used in mathematical diagrams and equations. Gamma is used to describe relativistic relations.

G 1. *symb.* conductance. See conductance. 2. *abbrev.* giga-. See giga-. 3. *abbrev.* grid (as in a vacuum tube).

g force (*symb. – g*) A unit of force of acceleration equal to that which would occur in a falling body acted upon by gravity at the Earth's surface, 9.81 meters per second per second. F forces influence the process of crystallization in an optical fiber during the pulling stage.

G Interface In Operation, Administration, Maintenance, and Provisioning (OAM&P), the G Interface is the user-to-computer interface of the Telecommunications Management Network (TMN). The G Interface is intended to promote consistency in the user interface and to reduce errors.

G Series Recommendations A set of ITU-T-recommended guidelines for transmission systems and media, digital systems and networks. These are available as publications from the ITU-T for purchase, and a few may be downloadable from the Net. Some of the related general categories and G Series Recommendations of particular interest are organized into general categories at the end of this chapter. A list of the general categories is included in the Appendix. See also I, Q, V, and X Series Recommendations.

G style handset The designation for older telephone handsets having a round design on the ear- and mouthpieces. They are similar to the newer, squared-off K style handsets, and both G and K style are heavier and more substantial than some of the newer cordless or cell phone handsets which are very flat and small. See K style handset.

G3Fax A facsimile machine that transmits through traditional phone networks using T.4 standards. For Internet facsimiles based on email formats, a G3Fax device is accessed through an IFax gateway that serves as a mail transfer agent (MTA) between the Internet and the traditional phone line-based fax machine. See IFax device, TIFF-FX, RFC 2305.

G-line A round, insulated wire used in microwave transmissions.

G-Line A Cornell University physics resource for X-ray beam-related research. The G-Line Laboratory is organized into three experimental stations, G-1 for the study of large molecule reactions, G-2 for general purposes, and G-3 for studying the growth of semiconductor films.

G-Scope, G-Scan A type of rectangular radar display in which a centralized blip is illuminated and becomes wider or narrower horizontally as the target moves nearer or farther away. Errors in aiming the radar are indicated by the horizontal and vertical placement of the blip (or *pip*).

G.lite An International Telecommunication Union (ITU) low-cost, splitterless alternative proposal to ADSL CPE. See Asymmetric Digital Subscriber Line, UAWG.

G/A ground to air communication.

G/G ground to ground communication.

G: Line In a network server configuration file, a global (network-wide) ban that is kept in memory until it is removed or expires. It is generally used to ban users who abuse their network privileges but may also be used to restrict access by specified domains or countries, or by servers with specified characteristics. While G: Lines usually default to a few hours, it is possible to configure them to last for months or years. Sometimes legitimate users are prevented from accessing a site because a particular domain or ISP has been "G: Lined." In this case it usually takes a concerted effort and some negotiations to regain or retain access. G: Line queries or protests should be directed, in most cases, to email address

abuse@domainname.com

(substituting the relevant domain name). See fraud.

GA go ahead. A common verbal and written communications convention that indicates that the communicator is finished and the listener is welcome to proceed (similar to "Roger" in radio communications). It is frequently used in question-and-answer style conferences on the Internet, particularly in moderated conferences where many people are queued to communicate, and each one has to wait for a turn.

GAB 1. Group Access Bridging, Group Audio Bridging. A telephony chat or party line service in which multiple users call the same number and are connected to one another at the same time. 2. See Group Asynchronous Browsing.

GAC See Global Area Coverage.

gadolinium A light, silvery, crystallizing, ferromagnetic rare earth metal isolated from yttrium in the late 1800s. Gadolinium has been used in the production of phosphors in color cathode-ray tubes (CRTs), and for making gadolinium yttrium garnets used in microwave technologies. See europium, yttrium.

gaff The spike that is attached to a utility pole climber's iron. See boomer.

gage Performance indicator. See gauge.

Gaia Global Astrometric Interferometer for Astrophysics. See interferometer.

gain Increase in power of a transmission, usually indicated in decibels (dB) when applied to audio gain. Gain is sometimes intentionally created by using various means to boost a signal. Unfortunately, doing so typically also increases noise and interference in analog systems. Gain is descriptive of an antenna's capability to increase its effective radiated signal, relative to a reference like an isotropic antenna or a center-fed, half-wave dipole antenna.

galena A bluish-gray lead sulphide mineral commonly used as a sensitive radio wave detector in crystal detectors in the early 1900s. Sometimes the galena was thinly coated with other materials to improve its properties.

gallium arsenide GaAs. A semiconductor substance used to produce electronic components, such as computer chips and solar panels (when combined with germanium). It is sometimes used in place of silicon for high speed devices. It withstands heat and radiation well, making it suitable for orbiting satellite applications.

If used alone, the bandgap of GaAs is too high for use as semiconductor lasers, so the material is impregnated (doped) with other chemical elements and, as such, is the main active material used to fabricate semiconductor diode lasers, emitting pulsed light in the infrared frequency spectrum.

GaAs lasers can be used at room temperature and cooler temperatures and, in general emit higher wavelengths at higher temperatures. GaAs lasers emit a wider, more diffused beam than gas lasers, but they are small, inexpensive, versatile, and rugged and suitable for many applications. See homojunction laser, indium gallium arsenide nitride, semiconductor laser.

Galvani, Luigi (1737–1798) An Italian physicist, educator and physician who experimented with minute levels of electricity in the leg nerves of frogs around 1786. Galvani's wife had observed twitching in a recently killed frog when a nerve was touched and pointed it out to her husband; Galvani followed up her observation with further experiments leading to a better understanding of nerve impulse transport in biological systems. His name has been applied to measuring instruments (galvanometer) and small levels of electricity on skin surfaces (galvanic skin response).

galvanic Related to or producing a direct electrical current through chemical rather than electrostatic means. See galvanometer.

galvanic cell A power cell that produces electricity through electrochemical rather than through electrostatic means. See battery.

Luigi Galvani – Physicist

Luigi Galvani did numerous experiments in nonbiological and biological electrical impulses. The galvanometer is named after him.

galvanometer Named after Luigi Galvani, an instrument for detecting low levels of electric current through the use of a magnetic needle or coil suspended within a magnetic field. The basic concept of a galvanometer is that very low levels of current, when channeled through a magnetic coil, can influence the magnetic field sufficiently to deflect the coil (or needle) in one direction or the other, depending upon the polarity of the current. The magnitude of the deflection (torque) is proportional to the amount of current. Thus, the torque can be read and measured when compared against a calibrated scale associated with the receiver, indicating the direction and strength of any current that may be present.

The historic development of the galvanometer followed from principles described by Ørstedt and Ampère and was furthered by Johann Schweigger, who presented a paper on the galvanometer multiplier in July 1820. Half a decade later, Ampère developed a working galvanometer, as did Nobili, who used two needles, each with its own separate coil, to reduce interference from the Earth's magnetic field. Nobili's innovation made it possible to position the device without worrying about the direction of the Earth's magnetic meridian, thus creating a *static* galvanometer. It also made it possible to use a longer needle, thus increasing the size of the gauge and the ease of getting an accurate reading. Nobili galvanometers designed by Ruhmkorff were popular in the latter 1800s. Most of the early galvanometers, known as tangent galvanometers, were constructed with a compass integrated into a surrounding electromagnetic coil. In 1858, William Thomson (Lord Kelvin) refined the technology and created the *astatic* galvanometer, the basis of more modern galvanometers. In 1880, D'Arsonval further improved upon the technology, by integrating a small coil with the metering needle, both housed within the field of a permanent magnet.

In telecommunications, galvanometers have many uses, including the testing and diagnosis of circuits. Thomson's mirror galvanometer, for example, was used for reading currents on long submarine communications cables. Following World War II, a dynamo transformer, a type of torque motor, based upon the moving-iron galvanometer, was patented by Mueller, an American inventor.

The versatile galvanometer is now also used in aeronautics, train, and power plant applications.

A number of scientific instrument museums include historic galvanometers in their collections, including the Robert A. Paselk Scientific Instrument Museum in Arcata, CA. Traditional galvanometers have been partly replaced by solid state current meters for industrial current-measuring use; however, mechanical galvanometers are still commonly used in schools to help teach the principles of electromagnetism. Commercial galvanometers vary in size and shape but most are about the size of a travel alarm clock and have a semicircular or vertical scale (similar to postal meters) for reading the measurements in amps, volts, or other relevant units. Electronic optical scanning galvanometers are becoming important in the scanning and printing industries. See galvanic; Galvani, Luigi; Schweigger, Johann.

Historic Galvanic Power Cell

A historic electrochemical power galvanic cell named after Italian physicist Luigi Galvani.

---

galvanometer, ballistic A type of galvanometer patented by the Leeds & Northrup Co. in the early 1900s intended to measure the total charge (as opposed to peak charge) or of a burst or short duration pulse of current. When the duration of the impulse to the coil is short, the coil gives a short twist proportional to the magnitude and duration of the current. Ballistic galvanometers are designed with a higher inertia coil. Some of the practical applications of ballistic galvanometers include the testing of various configurations in a prototype circuit board (e.g., combinations of capacitors) and the experimental determination of hysteresis curves for various substances or devices. See galvanometer.

galvanometer, D'Arsonval A historic, common coil-based galvanometer, consisting of a narrow, rectangular coil, suspended so that it can move between the poles of a permanent magnet to register a reading of direct current. This version of the galvanometer, developed by D'Arsonval in 1880, made the historic compass-based galvanometers obsolete. They were designed with compact, more reliable casings than the tangent galvanometers, since an external circular coil around the instrument was not used. Later adaptations of thermocouples or rectifiers to the D'Arsonval galvanometer permitted the conversion of alternating current (AC) to direct current (DC) in order to measure alternating current. See galvanometer.

galvanometer, mirror A device for measuring very low levels of current developed by Poggendorff in the 1820s and later refined by William Thomson (Lord Kelvin) in the 1850s that was a departure from the early compass needle style galvanometers. The *mirror* or *reflecting* galvanometer was used for measuring current in submarine communications cables in the late 1870s. There were two main systems in a mirror galvanometer, the galvanometer with a mirror that reflected a beam of light whose direction and extent was influenced by the polarity of a tiny amount of current, and a lamp at the receiving end with a calibrated scale that registered the light reflected from the galvanometer, shifting to the right or to the left depending upon the direction of polarity and amount of current. See galvanometer.

galvanometer, resonant A newer form of patented galvanometer technology that is being used in the printing industry for scanning images onto printing plates. This somewhat revolutionary technology uses computer electronics to control a raster-based laser scanner without the use of motors or rotating mirrors. Instead, the scanning mirror is associated with a torsion bar that is influenced by electromagnetic coils. Thus, in practical use very high scanning speeds are possible and there are no significant mechanical parts to wear out. See galvanometer.

gamma ferric oxide Ferric oxide is found as dark synthetic pigments and as natural red or black hematite. Gamma ferric oxide has been used since the 1940s to provide configurable magnetic recording surfaces, first on magnetic tapes, later on floppy diskettes.

gamma camera A device for detecting and recording gamma ray emissions. The basic components include a collimator, for preprocessing the gamma ray emissions entering at desired angles through gamma ray absorbing materials, a detector crystal (gamma ray scintillator), an array of photomultiplier tubes for converting the photonic energy into electrical energy while amplifying the electron emissions, output and logic circuits, and a processing unit (computer) for interpreting the data into meaningful information. Gamma cameras are used in radiopharmaceutical imaging for detecting radiation that has passed through bodily organs to reveal internal structures and anomalies.

G

gamma testing The testing of a product by out-of-house testers who fit the profile of endusers who would actually purchase or use the product being tested. It is the step before User Acceptance Testing (UAT) or delta testing and after beta testing (in-house testing). Gamma testers understand that the product is essentially finished but there may still be a few opportunities for user input into the final design (an important aspect of software design). The testers also understand that the product is considered to be bug- and defect-free but that a bug-free state is not 100% assured because people outside the company haven't used the product in a real life environment. See beta testing, User Acceptance Testing.

gamut 1. A range or series. 2. In imagery, the color gamut is the range of colors that can be perceived or reproduced, usually within the context of a particular situation or technology. For example, the gamut of an RGB cathode-ray tube (CRT) monitor is quite different from the gamut of printing inks or process color perceptual combinations of dots. The gamut range of videotaped images is narrower than that of film-based images. It is important to understand the differences in order to adapt the technologies to one another. See dynamic range. 3. The range of wavelengths within a spectrum that are usable, viewable, or of scientific interest.

GAN See global area network.

Gandalf FZA Compression Protocol One of several mechanisms proposed in the mid-1990s to enable compression of PPP encapsulated packets. Gandalf is based on FZA, which was developed by David Carr at Gandalf Data Ltd. and enhanced as FZA+ by Abbie Barbir. It is a derivative of LZ that optimizes at the expense of CPU cycles and memory in order to achieve high performance compression, available on a fee or royalty basis. See Compression Control Protocol, Deflate Protocol, Point-to-Point Protocol, RFC 1993.

gang To group or aggregate cables, components, objects, or picture elements. To mechanically or electrically combine components or devices so they can be controlled from one source.

Gang of Nine Nine vendor companies who formed a group in 1989 to promote and develop the industry standard architecture (ISA) which is commonly used for computer peripheral device connections on Intel-based microcomputers. Their work resulted in the 32-bit Extended ISA (EISA) standard.

gap Opening, space, small distance between objects or signals, small distance of a less dense material than the surrounding materials.

GAP See Generic Access Profile.

gap, electrical An opening in an electrical connection, which may close to allow a circuit connection or, more commonly, over which a spark will jump to provide a brief connection or discharge through the air, as the gap in a spark plug on an engine. An unplanned gap, through breakage, can interrupt an optical or electrical connection, thus interfering with a transmission.

gap, pickup The distance between a reading/recording head and the medium with which it interacts. The distance is often very precise, as in hard drive read/write heads and compact disc pickups, where a slight adjustment allows a much greater density of recorded pits in the disc.

gap, time An informational space or time gap inserted to indicate a stopping and/or starting point in recorded material or in a coded transmission.

gap loss The loss in signal strength that occurs when crossing a gap. Even very tiny gaps may substantially affect the quality of a transmission, and gaps are apt to occur at corners and junctions where transmission media change and in couplers.

garbage 1. In computing, meaningless information, spurious characters, nonsensical output. 2. A meaningless signal or electrical interference. See interference, noise.

garbage in/garbage out GIGO. *colloq.* A means of saying that you can't get something good out if you have put something bad into a system. In computer terms, it means that if you supply bad input, you're not going to get good output. In terms of hardware, a weak or poor signal at the origin is not going to result in a good signal at the receiving end. In programming terms, feeding the computer the wrong instructions or the wrong data is not going to result in the software performing in a correct or desired manner. In management terms, providing bad instructions and motivations to subordinates is not going to result in good work even if they are capable professionals.

garble A communication or signal in which some portion of the content has been changed to be undecipherable, unintelligible, out of sequence, or otherwise undesirably "scrambled." The change is usually in terms of the strength or content of the signal.

A phone conversation may sound garbled if the speed of transmission is inconsistent with the speed of the spoken message, resulting in words running together or pauses that are inappropriate. It may also sound garbled if there are power fluctuations causing the sound level to significantly fade in and out or if there are other signals overlaying the desired signal, making it unintelligible (e.g., party line noise, crosstalk). Thus, in voice and radio communications, the term tends to be associated with reduced intelligibility related to informational timing problems, power fluctuations, or multiple signals overlaying portions of one another.

In data communications, the term tends to be associated with portions of a communication arriving out of sequence or interspersed with nonsensical or scrambled data. It can also occur in encrypted messages if the decryption algorithms do not work correctly. A file or data communication may be garbled if packets are reassembled in the wrong order or have inappropriate portions merged into existing packets or between the appropriate packets. As in voice communications, garbled data can result from power fluctuations but it can also arise from programming bugs or programming deficiencies, such as insufficient or inappropriate error checking. Surge suppressors, power conditioners, robust and appropriate (for the task) error checking, good decryption algorithms, and data redundancy are all strategies for eliminating or mitigating problems associated with garbled.

Garden Valley Telephone Company GVTel. The oldest telephone cooperative in the United States, chartered in Minnesota in 1912 and still operating after more than 88 years.

Garfield, Oliver An author and engineer, in 1955 Garfield wrote a 63-page book *Simple Electronic Brains and How to Make Them* as a practical embodiment and complement to Edmund C. Berkeley's 1949 book *Giant Brains or Machines that Think.* The two men partnered for a while to produce computing mechanisms such as the Geniac, in the mid-1950s. Unfortunately, there were disputes and a lawsuit; Garfield and Berkeley ended up selling the computing devices as the Brainiac and Garfield got the name Geniac. The original Geniac (later the Brainiac) was an innovative computing device that could be configured to play simple games.

While the author of this text couldn't find hard evidence of wrongdoing, Garfield's business practices may not have been entirely legal. For a while Garfield used the Geniac name to promote his *GENIAC Pocket Calculator.* It was a repackaged Otis King calculator (a cylindrical slide rule) sold after the breakup with Berkeley through the Oliver Garfield Co., Inc. The Otis King calculators were serial numbered and shipped from England. Some of the Garfield calculators have all the original Otis King markings removed or, more likely, Garfield stole the design and cloned it without the identifying marks. It is unlikely he had permission from Otis King to distribute their calculators without attribution, so perhaps this sheds some insight as to why his dealings with Berkeley didn't work out. See Berkeley, Edmund; Geniac.

garnet A red, more or less transparent semiprecious mineral often used in jewelry and industrial applications. It somewhat resembles ruby but is generally a little less transparent and a deeper, muddier shade of red. Synthetic garnets can be manufactured from oxides of some of the rare earths. Lasers and some types of computer memory that use magnetic film are made from garnet. See gadolinium, laser.

GARP See Global Atmospheric Research Program.

gas laser A laser in which a gas or vapor is used as the active medium. The gas may be excited by a high-frequency oscillator or direct current. Ion lasers use noble gases such as argon. A pulse of voltage is used to ionize the gas and direct current (DC) can be used to maintain the state of ionization. Gas lasers were pioneered by A. Javan in 1960 and the argon laser was developed in the mid-1960s by William Bridges. See laser, laser history.

gas maser A maser in which a gas or vapor interacts with microwaves. Gas masers are found in scientific oscillators such as atomic clocks. See atomic clock; Dicke, Robert; laser history; maser.

gas plasma display See plasma display panel.

gas tube A type of electron tube that is not completely evacuated. Early electron tubes burned out quickly because of the gas inside the bulb. Later tubes were designed to last longer by being evacuated. In specialized applications, a trade-off is used and a small amount of gas is left in the tube to promote ionization to enhance current flow.

Gassner, Carl, Jr. A German-born inventor who first created a commercially practical dry cell, based on a sealed zinc container that, for the most part, superseded the wet cell. In November 1887, Gassner was granted U.S. Utility Patent #373,064 for his invention. Dry cells were welcomed because they didn't have the problems of leakage associated with wet cells, and they could be made smaller and more portable. They were an important innovation and are still in use today in the form of carbon-zinc batteries. See dry cell; Leclanché, Georges; Planté, Gaston.

gaston A random noise modulator sometimes used as an antijamming communications transmitting device.

gastroscope Developed around the 1930s as one of the historic fiber optic medical imaging technologies, the gastroscope makes it possible to peer inside the gastric (digestive) system. It is especially useful for locating and illuminating structures such as suspected tumors in the stomach.

GAT See Generic Addressing and Transport Protocol.

gate In an electrical circuit, in its broadest sense, a junction that selectively controls whether current gets through, when it gets through, or how much of it gets through.

gate, security In a software program, an input or process point that selectively lets users or other processes through, as a security gate, login gate, or system gateway for different types of information or protocols (network traffic, mail, etc.). In the more specific context of programming, a circuit that performs a basic logic operation and provides a single output from that operation. See firewall, gateway, proxy.

gate, telephone In telephone call distribution systems, a physical or virtual trunk through which calls are handled by a group of operators or agents according to some characteristics specified for each gate.

Gates, William Henry, III (1955–) Bill Gates is a prominent success story in the personal computer industry and has a major influence on the direction and type of software that is available to consumers around the world today. In the 1980s he went from entrepreneur to software magnate, earning billions of dollars by investing in technology and selling computer operating systems and basic business software tools. Gates began programming seriously in Seattle, Washington, while still in grade school, along with his friend and business partner, Paul Allen, who was two years older. Gates came from a privileged family with many solid business contacts, and Allen and Gates were both precocious entrepreneurs, successfully winning commercial programming contracts at a young age. Their first business partnership, based upon traffic analysis software, was called Traf-O-Data, founded around 1972.

The economic success of Microsoft began with the signing of the IBM OS/BASIC contract. Of particular importance is the fact that Gates reserved the right to market the results of the development (sold by IBM as PC-DOS) in competition with IBM, a contractual loophole that Gates had also achieved with MITS, who were dismayed because they thought they had bought the exclusive rights to Microsoft BASIC for the Altair. Gates' father was a lawyer and Gates had a long-standing interest in contracts and business deals, which probably accounts for the Microsoft-favorable outcome of the two pivotal contracts. Times have changed; today large corporations have teams of lawyers to develop and scrutinize contracts but, two decades ago, many deals were still established on trust and a handshake or on less extensive paperwork and scrutiny than is common now.

Since Gates and Allen didn't have an operating system to sell to IBM, they bought QDOS, developed by Tim Paterson. In one of computer history's greatest ironies, Tim Paterson acknowledged that he had created QDOS by using a mid-1970s version of the documentation for Gary Kildall's CP/M. Microsoft then hired Paterson to quickly develop it into PC-DOS. QDOS was syntactically and functionally similar to Kildall's CP/M. Microsoft's version of a PC operating system, sold in competition with the product they produced for IBM, became known as MS-DOS. In addition to operating systems, Microsoft began developing business tools, word processors, spreadsheets, etc., and buying up competitive products and companies including FoxPro, Stacker, Altamira Composer, Vxtreme, Vermeer (now known as Frontpage), Web Mapper, Softimage, Web TV, and others. Microsoft also now has stakes in several broadcast and cable companies.

In 1982, Kildall's Digital Research Inc. sued Microsoft Corporation and IBM for copyright infringement of the CP/M operating system. Digital Research won, but by this time Microsoft's momentum was so great, few people heard about the suit and Kildall's company lost its majority market share. The situation was still of consequence almost twenty years after QDOS was purchased, however, as Caldera, the new owners of Kildall's early and later operating systems technology, pursued legal avenues against Microsoft. In July 1996, Caldera, Inc. filed suit against Microsoft Corporation for damages and injunctive relief under U.S. antitrust laws for "illegal conduct calculated to prevent and destroy competition in the software industry."

Allen left Microsoft in 1983 for health reasons and to pursue other investments through his own company, but he maintained close personal ties with Bill Gates. He served on the Board of Directors of Microsoft for a time, announcing his intention to resign and become a strategy advisor as of November 2000. In the late 1990s, Gates was still head of the organization that had grown from a small group of half a dozen programmers and clerical staff to a campus in Redmond with tens of thousands of employees and assets in the billions. The controversy did not die down, however. In May 1998, the United States of America filed a complaint against Microsoft. Thus, Gates' actions as CEO came under severe scrutiny, a position

he resigned in January 2000. He continued to hold the position of Chairman, however.

Gates has spun off a number of other businesses, including a multimedia company and the Teledesic Internet in the Sky project that he is co-developing with Steve McCaw. He is clearly interested in expanding into the broadcast and telecommunications fields and heavily promotes the use of Microsoft's Web browser. Gates is one of the richest people in the world, and one of the most focused, aggressive, and successful business and marketing giants in the software industry. Allegations of antitrust violations are part of an ongoing investigation of Microsoft by the U.S. Justice Department on behalf of hundreds of software developers who don't feel they can compete against a company allegedly succeeding because of unfair business practices. Whatever the results of the investigation, part of Bill Gates' success is due to his ambition, tenacity, and all-encompassing competitive drive. Nonetheless, there is much controversy over his ethics, methods, and acquisition of products from other sources that he has marketed as his own.

gateway A transmission connection between networks that handles information flow and typically performs bandwidth and protocol adjustments and conversions, as needed. It may also combine (aggregate) network transmissions from several devices connected to a smaller number of network connections. Gateways are commonly used between dissimilar networks such as between local area networks (LANs) and the Internet (proxy servers), and between land services and satellite services, etc. When local nets are connected with external nets, the gateway may also perform security functions. Gateways are used on both wired and wireless networks. Wireless gateway networks use radio transceivers rather than physical connections for wire or fiber optic cables. Wireless networks also tend to have additional security features and encryption schemes to compensate for the fact that it's sometimes difficult to determine if someone is intercepting wireless transmissions.

Consumer-oriented gateways that are available as peripheral devices may include Web server software and other utilities to facilitate Internet connections and firewall, logging, or other security schemes to protect data transmissions or specific devices from unauthorized access.

A gateway differs from a proxy in that a gateway handles requests as though the requests originated from the gateway and not from the original client, thus serving as a server-side portal in a firewall. See firewall, proxy.

Gateway to Gateway Protocol GGP. A historic, experimental TCP/IP network transport layer routing protocol used to convey routing information through a distributed shortest path computation. The protocol was primarily used for gateway-related tasks such as routing datagrams. It is now considered obsolete and developers are cautioned not to implement it. See RFC 1009, RFC 1812.

gating 1. Selectively allowing certain waveforms, frequencies, or portions of waves to pass through a gate point. 2. Performing electronic switching by application of a certain current or waveform. 3. Using logic to control the passage of current or information through a system.

gauge An instrument or other indicator for measuring or testing. An indicator of the thickness or thinness of a substance or object, especially applied to wires and cabling. Position relative to another, as in the distance between the sides of a railroad track. See American Wire Gauge (Brown and Sharpe), Birmington Wire Gauge, Steel Wire Gauge.

gauss A centimeter-gram-second (CGS) unit of magnetic flux density. For example, if one line of magnetic force passes through one square centimeter, the field intensity is said to be one gauss. Named after Karl Friedrich Gauss. See lines of force, magnetic field.

Gauss – Mathematician & Astronomer

A portrait from a German postage stamp commemorates the achievements of Carl (Karl) Gauss who mathematically described important basic concepts in magnetism, surveying, and mathematics.

Gauss, Johann Karl Friedrich (1777–1855) A brilliant German mathematician and astronomer who devised the heliotrope, an instrument that could reflect sunlight over long distances, providing a means for making straight lines and calculations related to the Earth's surface. He investigated terrestrial magnetism in cooperation with W. Weber in 1831 and published some significant papers several years later. In 1833, he constructed an electric telegraph, just as similar devices began to be developed in the United States. The gauss unit for magnetic flux density is named after him.

Gauss's law As it applies to dielectrics (without getting into integrals), the total electric flux of a closed surface (a surface without holes) equals the charge enclosed divided by the permittivity. Gauss's law facilitates the assessment of a stationary enclosed charge associated with a surface by mapping the surface field outside the charge distribution. See Ampère's law, Coulomb's law, dielectric, flux, Gaussian surface, Maxwell's equations.

Gaussian curve, normal curve A Gaussian curve is often called a *bell curve* due to its shape, and a normal curve due to its experimentally observed or theoretical distribution (depending upon context). The concept and observations on larger statistical sets to confirm the concept were developed over a couple of centuries, with A. de Moivre noting in the 1730s that the binomial distributions described earlier by J. Bernoulli took on the characteristics of a continuous curve when developed using increasingly large sample populations. Around the late 1700s, J. Karl Gauss enlarged on the concept, finding many applications for its use. He thus popularized the bell curve that now bears his name.

When plotted, a normal curve provides a symmetric mathematical, statistical model and diagrammatic representation that resembles the outline of the curve of a bell. Its structure is based upon a combination of observed population distribution (in the statistical, not the human sense) and assumptions about its generalizability. In a diagrammatic representation of the curve, the majority of occurrences of a phenomenon fall within the center and highest point of the curve (the mean and mode). The changes in the direction of the curve and dispersion of frequencies are symmetric to either side of the mean and mode and are described in terms of *standard deviations*.

An interesting demonstration of the shape and distribution of elements in a Gaussian curve can be seen in a number of science museums. It is set up with vertical channels and ping pong balls (or ball bearings), in which the balls are dropped into the top of a device where they randomly roll down into a line of channels that can be viewed through a transparent panel. After many hundreds of balls have fallen down through the mechanism, the shape of the curve will emerge, with the majority of balls in the center, and fewer as you move left or right out to the edges (the tails of the distribution) due to the lower statistical probability of the balls moving farther and farther from the main center of the path.

Gaussian curve models are at the heart of many experiments and statistical theories. Relationships such as that between optical intensity and frequency will often fit Gaussian models, though they may be skewed in one direction or the other due to environmental effects.

Conformance to a normal "bell" curve is frequently an underlying assumption for the design and scoring of tests. Unfortunately, it is sometimes used as a justification for describing populations where the size of the sample being evaluated is not mathematically sufficient to justify assumptions about its relationship to the normal curve.

Gaussian minimum shift keying GMSK. A type of wave modulation from the frequency shift keying family used in wireless communications such as cellular phone systems and, notably, the Global System for Mobile Communication (GSM). GMSK is a compromise solution, as it is less efficient than some types of modulation, but it is popular for its cost benefit. In GMSK, a digital signal is modulated onto the analog

G

carrier frequency as the phase of the carrier is varied by the signal information. A Gaussian filter is applied before modulation. Most of the power is concentrated at 250 kHz and there is low incidence of channel interference. Commercial applications of GMSK for mobile telephony were described in the early 1980s and later put into practical use in GSM. In 1996, Thierry Turlette authored a frequently cited paper on GMSK "GMSK in a nutshell." In the late 1990s, more efficient algorithms for GMSK were suggested by researchers. See phase shift keying.

Gaussian noise Noise or other electromagnetic interference that conforms to a probability pattern consistent with expectations based on Gaussian statistics. Noise that is somewhat random across a range of frequencies is sometimes known as white noise or Gaussian noise.

Gaussian surface A theoretical surface providing a geometry analogous to a physical surface to facilitate modeling and calculations related to electric fields. See Gauss's law.

Gauthey, Dom A French monk who invented a tube system for channeling acoustics, in essence a "speaking tube" or historic manual telephone conduit (apparently derived from ship-based speaking tubes). It consisted basically of a tube with funnel shapes on the ends for speaking and listening. The interesting aspect of the idea is that it was suggested the tubes could be organized in relay, with human speakers (relay agents) at the junction of each length of tube. Thus, it would be theoretically possible to communicate a message more than 500 kilometers in less than an hour. He presented his invention to the l'Académie des sciences (Academy of Sciences) ca. 1782, transmitting sound over a distance of 760 meters. See telephone history.

Gb, Gbyte gigabyte. See giga-.

GB gigabit. See giga-.

GBCS See Global Business Communications Systems.

GBIC Gigabit Interface Converter. A small, sometimes swappable, fast gigabit network converter device that attaches to a port connection.

GBR ground-based radar. See radar.

GCA 1. See game control adapter. 2. ground-controlled approach. A radar-based aircraft landing system.

GCI Ground Control of Interception. A radar-based technique for directing aircraft in the interception of approaching craft.

GCMD See Global Change Master Directory.

GCRA See generic cell rate algorithm.

GCS 1. Government Communication Systems. 2. Great Canadian Scientists. A project to increase awareness of great scientists with Canadian affiliations and roots originating from a lack of print materials on the subject. Many of the profiled scientists have contributed to the telecommunications industry.

GCT See Greenwich Civil Time.

GDF group distribution frame. See distribution frame.

GDG See Global Development Gateway.

GDI See graphics device interface.

Geissler tube A type of sealed dual-electrode gas-filled tube that glows when current passes through it. The color of the glow varies with the types of gases used, and the wave characteristics can be influenced by the amount and type of current applied. Historic versions of the Geissler tube, named after the inventor J. Heinrich Geissler, resembled glass lanterns or candles mounted on a metal or wood base with a straight or spiral illumination up through the center of a second, protective glass bulb (sometimes filled with a colored liquid to enhance the effect of the primary illumination in the inner tube). Some of the smallest, earliest tubes lacked a stand and protective bulb but came in a variety of creative and charming shapes resembling short strings of hollow glass beads. Early Geissler-style tubes were used for signs and lamps, more recent variations are used as scientific research and calibration instruments. Over the decades, Geissler tubes have played a role in many pioneer aspects of electronics and telecommunications leading to the development of the light bulb, flash bulbs, a number of projection technologies, and the discovery of cathode rays.

In contemporary calibration tubes, the central region of the tube is usually narrower than the ends. The very narrow region in the center induces strong spectral lines for sample comparisons. The Geissler tube is useful for calibrating optical instruments and as a reference tool. It has also been used in conjunction with imaging technologies as a low-cost experimental rotating mirror photodetection device.

Educational versions of the Geissler tube often have more than one tube, each with different contents. These are useful for teaching about gases, induction, fluorescence, and other related concepts. For example, an energized Geissler-style tube can be used to illustrate magnetic field influences on charged particles. Geissler tube effects may be observed by the unaided eye or with scientific instruments such as spectroscopes. See cathode-ray tube, Crookes tube.

Geissler, Johann Heinrich Wilhelm (1814–1879) A German physicist and glassblower who established a workshop in Bonn, where he crafted scientific apparatus. One of his more significant inventions was a type of mercury air pump that is still in use today for evacuating the air from bulbs and various laboratory apparatus. In collaboration with Julius Plückers, Geissler did experiments to measure the density of water. He also created instruments for the measurement of vapor. Modern refinements of the Geissler tube, that he developed in the late 1850s, are still used for scientific calibration. See Geissler tube.

GEM See Graphics Environment Manager.

GEMINAX Global Enhanced Multiport integrated ADSL Transceiver. An integrated voice and data chipset provided by Infineon Technologies, GEMINAX was selected for the North American Litespan next generaltion network digital loop carrier multiservice broadband access platform. The chipset can be integrated into cost effective, very high density linecards for voice and DSL-based data network services. See Litespan.

GEMS See Global Environmental Monitoring System.

gender A designation widely used to cable connectors to indicate whether they are female "innies," or male "outies," that is, whether they have holes or extended pins. If it is a simple round or squarish connector like a switchboard or composite video connector, the female end is usually called a plug while the male end is usually called a jack.

Most common switchboxes for computer data connections have female connectors. Male and female connectors are both found on the backs of computers, and vary from platform to platform and sometimes even from brand to brand. This variation is inconvenient for users, who often have to pay extra money for gender changers, but cable and connector manufacturers don't seem to mind the extra business.

Gender Bending Adapters

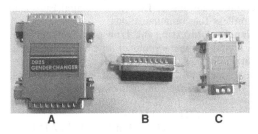

A   B   C

Gender benders are commonly used to make connections between cables with the same male or female contacts facing one another. Some common gender benders used with serial cables include (A) DB-25 regular; (B) DB-25 slimline; (C) DB-9 regular. The slimline models are convenient for portable devices, such as laptops, for connecting to modems or printers.

---

gender bender/changer/converter A small connecter designed to change the gender of a plug or receptacle in order to enable connection to another plug or receptacle. The male end is the one in which the contact points protrude from the connecter. The female end is the one in which the contacts are depressed. Thus, a simple gender bender has two female ends or two male ends. Switchboxes and adaptors sometimes also perform gender bending functions. Adapters/extenders (male on one side, female on the other) are sometimes erroneously called gender benders, because they superficially resemble them, but have different genders. Gender benders are usually passive devices that do not influence the signal passing through the device. See adapter, extender, switch box.

General Instrument Corporation GIC. A broadcast technology corporation which is significant because it was the first to suggest an all-digital broadcast system to the Advisory Committee on Advanced Television Service (ACATS) for consideration for recommendations to the Federal Communications Commission (FCC) and the U.S. Congress. Subsequent to the GIC proposal, the focus on digital systems greatly increased, and ACATS and other advisory bodies brought the thinking about television technologies into closer alignment with recent digital advancements.

In a separate but related matter, General Instrument Corporation sought, through the U.S. courts, to hold the FCC to the mandate to ensure "Competitive Availability of Navigation Devices" (e.g., cable set-top boxes) to maintain the right of competitors other than cable companies to produce and distribute these virtual navigation devices. However, as of June 2000, the court decision was to reject GIC's challenge to an FCC separation requirement for hybrid navigation devices. See Advisory Committee on Advanced Television Service.

General Magic A California-based commercial venture established in 1989 that focuses on providing innovative, cost-effective computer-Internet-telephony products for mobile applications. Products include Portico, a development name for integrated voice/data communications and information services, and DataRover integrated voice/data hardware devices

General Packet Radio Service GPRS. A wireless high-speed data communications standard operating over Global System for Mobile Communications (GSM) networks. GPRS is a packet-based air interface using existing circuit-switched GSM networks. GPRS offers enhancements to basic GSM. It provides functional connections with the Internet that were not possible with earlier systems. It can operate at speeds ranging from 9.6 Kbps to 115 Kbps (with plans for rates up to 384 Kbps) and offers live connections to mobile users such that users can be online continuously, without paying minute-by-minute fees other than when uploading or downloading information. The higher data rates possible with GPRS not only reduce connect time costs but enable broader-band services, such as videoconferencing, to be implemented on mobile devices. Full implementation of these higher data rates are somewhat dependent upon the evolution of the underlying GSM technologies but will likely be realized over the next two years or so. Many see GPRS as a bridge between GSM and 3G mobile services.

General Protection Fault GPF. A common fault condition encountered by users of Microsoft Windows software when applications memory conflicts occur. At the very least, applications should be closed up. Unfortunately, it may also be necessary to reboot the system.

General Radiotelephone Operator License GROL. A lifetime operator's license granted by the Federal Communications Commission (FCC) upon successful completion of competency requirements. The license is required for individuals who adjust, maintain, or service FCC-licensed marine, aviation, and international radiotelephone transmitters. It conveys the operating authority of the Marine Radiotelephone Operator License and requires the Written Element 1 and Written Element 3 exams to be passed.

G

Some related FCC radiotelephone authorizations include the First, Second, and Third Class Radiotelegraph Operator Certificates. These require exams, in some cases Morse Code competency, and are valid for five-year terms (renewable). A GROL-licensee may also be granted a Ship Radar Endorsement on qualifying and passing the Written Element 8 examination. This permits the repair, maintenance and internal adjustment of ship radar equipment. There are also licenses specific to the Global Maritime Distress and Safety System. See GMDSS Radio Operator License.

General Switch Management Protocol GSMP. An IETF standards-track packet network administration protocol to control a frame- or cell-based label switch. GSMP is an asymmetric protocol supporting master (controller) and slave (switch) interactions. GSMP provides a means to query and report connection, port, switch, and Quality of Service (QoS) information and statistics for MPLS label switch devices from or to a third party controller. Thus, GSMP can be used to query switches, to establish and release connections, to modify a multicast connection, and to manage switch ports. It can also notify the controller of asynchronous events. Multiple instantiations of a single controller can be used to control multiple switches and partitioning is supported for the control of a single switch by multiple controllers.

The capability to query and report switch-related configuration and statistical status is particularly valuable for remote switch operations and, as the protocol evolves, is desired for optical network configurations as well. Initially, GSMP was developed to work with static switch partitions; however, dynamic forwarding to multiple administrative domains might be feasible for future versions of GSMP.

GSMP does not make significant assumptions about the underlying hardware over which it is transmitted and thus can be implemented in a variety of types of networks. As a separate standards-track document, packet encapsulations for GSMP transport have been defined for ATM, Ethernet, and TCP networks:

- In *ATM* networking, GSMP packets are variable length and encapsulated directly in ATM AAL-5 with an LLC/SNAP header.
- For *Ethernet*, GSMP packets are transmitted after the Ethertype 0x880C identifier up to a maximum length of 1492 bytes.
- In *TCP/IP*, GSMP is transmitted after a prepended TLV header field of type 0x88-0C and an integer indicating the length of the following GSMP message. The message is processed after its successful receipt. TCP/IP encapsulation can also be transmitted with authentication.

With the evolution of GSMP towards an industry standard in the late 1990s, there has been interest from the commercial and developer communities in defining GSMP for other environments, including experimental switches and patented, proprietary switches. Thus, GSMP for FSR switches (VTT Information

Technology) and ForeRunner (Sprint) switches are of interest to developers. See flow, Ipsilon Flow Management Protocol, RFC 1987, RFC 2026.

Generalized Trunk Protocol GTP. A telephony protocol enabling complex Channel Associated Signaling (CAS) and Common Channel Signaling (CCS) protocols to be executed in peripheral devices in order to remove some of the processing load from the main processor.

generator In its basic sense, a machine that is designed to convert mechanical energy into electrical energy, or one that converts direct current (DC) into alternating current (AC). Dynamo.

Generic Access Profile GAP. A profile is defined by the Open Systems Interconnection (OSI) model as a combination of one or more base standards and association classes, necessary for performing a particular function. The GAP specifies well-defined compatibility levels for DECT products, as an extension of an ETSI-published Public Access Profile (PAP) incorporated into Digital European Cordless Telecommunications. (DECT). See Digital European Cordless Telecommunications.

Generic Addressing and Transport Protocol GAT. A digital telephony protocol for exchanging Application Protocol data units (APDUs) between service provision points. It may be used between a terminal and a network or within or among networks. The recommendations for GAT use with ISDN/B-ISDN is described in ITU-T Q.860.

generic cell rate algorithm GCRA. In ATM networking, an algorithm used to enforce a particular performance level with regard to cell traffic. The GCRA evaluates the cell to determine whether it conforms to the established cell traffic contract. In a network switch, the UPC function will typically incorporate an algorithm such as GCRA to enforce conformance with the specified parameters. See leaky bucket for a fuller explanation and example. See cell rate for related concepts.

generic flow control GFC. In ATM networks, a means of controlling traffic flow. A field in the ATM header can be used to designate flow control parameters. This field is evaluated en route, as appropriate, and is not included in the final delivered communication. See cell rate.

generic flow control field GFC. In ATM networking, traffic flow control is an essential aspect of moving cells from one place to another. In the ATM header, there are priority bits which can be set to inform the end-station that congestion control may be implemented by the switcher. See cell rate.

Generic Security Service Application GSSA. An IETF elective proposed general Internet standard. See RFC 2078.

GENIAC Genius Almost-Automatic Computer (in the tradition of the UNIVAC, ENIAC naming convention). The GENIAC was a personal computing device designed by Edmund C. Berkeley and possibly also Oliver Garfield in the mid-1950s. Berkeley was an insightful author, robotics entrepreneur, and the designer of what was likely the first desktop computer ever invented, the Simon. Garfield was an author, engineer, and business person in much the same vein as Berkeley. Unfortunately, the partnership ended up in disputes and Garfield ended up with the name Geniac while Berkeley continued marketing computing devices under the name Brainiac.

The GENIAC kit was originally sold through Berkeley Enterprises, Inc., documented by Oliver Garfield's *GENIACS: Simple Electric Brain Machines, and How to Make Them* bound with the *Manual for Geniac Electric Brain Construction kit No. 1*. Once assembled, it could play simple games and do basic computations. The kit sold for about $15.95 to $19.95 (about two days' wages). Logical functions were carried out electromechanically, through cascading rotary switches that could be configured for different computing experiments. For information on the GENIAC Calculator, see Garfield, Oliver. See Berkeley, Edmund C.; Brainiac; Simon.

geographic information system GIS. Any system in which terrain information is gathered, processed, and stored, usually for later retrieval for analysis, long-term comparisons, mapping, navigation, etc. An enormous amount of geographic information is gathered by orbiting satellites, geocoded, and stored on high storage-capacity computer systems, much of which is accessible through public computer networks. Land-based tracking of networks, utilities, and transportations systems is also carried out with GIS systems. See geoInterface, Landsat.

geographic interface This term has two meanings: (1) interfaces designed for users to access and manipulate geographic data, and (2) an evolutionary step in computer user interfaces which models the world in a simulated 3D environment as visual objects or functional constructs.

The second meaning represents a relatively new approach to computer user interfaces. The idea has been around for a while, but the resources to implement it have been cost prohibitive until recently. In the early 1970s, microcomputer user interfaces consisted of dipswitches for input and small blinking lights for output (see Kenbak-1 and Altair). They were replaced by simple monochrome, text-based interfaces by 1975 (see SPHERE System). In the early 1980s, developers began producing graphical user interfaces (based largely on 1970s research at Xerox PARC), and it was not long after that the first experimental geographic interfaces began to appear. See geoInterface, graphical user interface, object-oriented, Open Systems Interconnection.

geographic north The region on the Earth called the North Pole from which imaginary lines of longitude emanate to meet again on the other side at geographic south, as established by convention. Geographic north is the general direction in which the north-seeking needle on a compass points but it is not exactly the same as magnetic north, which changes as large geological formations change. See magnetic north.

Georgia Rural Telephone Museum Located in a former 18,000-ft cotton warehouse, built in 1911, the Georgia Rural Telephone Museum opened officially in 1995 in Leslie, Georgia. The building was established across the street from the Citizen's Telephone Company. It includes about 2,000 historic telephones and other examples of communications equipment. http://www.sowega.net/~museum/

Geostationary Meteorological Satellite GMS. A geostationary satellite system of the Japan Meteorological Association (JMA) and the National Space Development Agency (NASDA) that is affiliated with World Weather Watch (WWW). The GMS system senses the region over Asia and the western Pacific at 140 degrees E longitude. It is similar to the U.S. GOES system. The first GMS was launched from the U.S. in 1977, and new satellites have been put into orbit every four or five years since from the Tanegashima Space Center in Japan.

GMS-5, which has been operational since the mid-1990s, provides data on cloud patterns (including volcanic activity), water vapor distribution, and sea surface patterns to more than two dozen countries. The Japan Weather Association provides hourly GMS images online. GMS systems also serve as communications relays for forwarding information obtained from ships, aircraft, and communications buoys and provide assistance in emergency search and rescue operations. By the late 1990s, the GMS system was being designed to take on a broader role, serving not only as a meteorological sensing system and relay station, but also as an air traffic control support system. In February 2000, the GMS-4 system, launched in 1989, was taken out of service and moved out of the way into a higher region for its final orbit. The GMS-5 system, launched in 1995, is still operational and apparently in good working order, but a transition is being made to a new system of satellites, the Multi-functional Transport Satellite (MTSAT) series, beginning with MTSAT-1R and MTSAT-2. Despite its age, GMS-5 is expected to be able to operate until the MTSAT platforms come online around 2003. See Multifunctional Transport Satellite, National Space Development Agency.

Geostationary Operational Meteorological Satellite GOMS. Also known as Elektro, the GOMS system is a three-axis-stabilized imaging satellite that is part of the Russian Federation Planeta-C Meteorological Space System. The first of the series of satellites was launched in October 1994, with each platform expected to last at least three years. GOMS orbits over the Earth's equator at longitude 76°50' E. GOMS senses infrared images of the Earth's surface and cloud cover, provides continuous coverage of atmospheric patterns, detects potentially hazardous phenomena, measures sea surface temperatures, and monitors space-borne energy particles.

Aboard the satellite, a radiometric line scanner images the Earth, digitally encodes the information, and transmits it to a ground station for preprocessing and eventual dissemination as satellite products. The communications portion of GOMS includes satellite-to-ground transmissions, ground-station intercommunication relays, and transmission to users on an operational basis.

G

geostationary orbit A type of orbit that is timed with the movement of the body it is orbiting so the period is equal to the average rotational period of the orbited body. If, in addition, the orbit is circular, the satellite will appear to be not moving when viewed from the ground, hence the name.

In simpler terms, if you place a satellite in a circular orbit at about 35,900 to 42,164 km from the ground, it will appear to remain in the same place because of its synchronized relationship to the Earth's orbit. This type of orbit has some advantages for communications. The satellite is always available at the same location and not many high altitude satellites are needed to provide global coverage. The main disadvantage is that powerful sending and receiving stations are needed to send and receive signals to/from such a high orbit. Geostationary orbits were described by Arthur C. Clarke in the 1940s and 1950s in considerable detail and with remarkable foresight. Also called *geosynchronous orbit* and *fixed satellite* orbit.

Geostationary Orbit

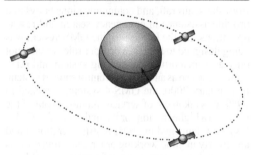

In a geostationary orbit, the movement of the satellite keeps pace with the movement of the Earth and is thus within the same general visual and communications region at all times, as seen from the Earth.

---

geosynchronous orbit See geostationary orbit.

GEOTAIL A Japanese research satellite launched in 1992 to study the structure and dynamics of the tail region of the Earth's magnetosphere. The orbit of the satellite was planned so it would cover the magnetotail over a wide range of distances. It contains instruments to measure the magnetic field, the electric field, plasma, energetic particles, and plasma waves. http://www.gtl.isas.ac.jp/

German Space Operations Center GSOC. A complex located near Munich, Germany, that was originally established to support the first German research satellite (AZUR), launched in 1969. GSOC is a facility of the German Space Missions Directorate (DLR) that prepares and executes national and international cooperative space flight projects through the main facility and a number of remote satellite ground stations. GSOC responsibilities include not only mission preparation, acquisition, rocket operations, and payload administration, but also the development of software support systems for mission support, data handling, and ground operations.

One of the interesting experiments undertaken by GSOC and its associates is the GPS Small Satellite Mission Equator-S (launched in 1997). Before this mission, Global Positioning System (GPS) receivers could be used only in near-Earth regions, well below the orbiting height of the GPS satellites themselves. This experiment successfully demonstrated that GPS receivers could be used at altitudes up to 34,000 kilometers. http://www.gsoc.dlr.de/gsoc.htm

germania A short name for germanium oxide. See germanium.

germanium Germanium is a "semi-metal" substance discovered in the 1880s. It is obtained primarily as a byproduct of zinc refining. There are reports that germanium may slightly convert transverse waves to longitudinal waves and vice versa. Materials with conversion capabilities are useful for semiconductor design.

The primary use for germanium is for fiber optic systems, and infrared optics account for another $13\pm2\%$ of the market. Germanium chloride ($GeCl_4$) is one of the chemicals that may be used in modified chemical vapor disposition (MCVD) processes for creating preform blanks for pulling optical fibers.

In pioneer vapor deposition fiber preform fabrication at Corning, germanium tetrachloride was used to form germanium oxide. See fiber fuse effect, vapor deposition.

GETS See Government Emergency Telecommunications Service.

GFP See global function plane.

GFC See generic flow control.

GGP See Gateway to Gateway Protocol.

ghost 1. A shadow image on a monitor with phosphor burn-in. When a phosphor-coated display device continually displays the same image (especially if it is a bright one) an undesirable pale image that does not disappear when the screen is refreshed may be burned into the display surface, thus interfering with the display of desired images. This ghost image indicates permanent damage to the monitor (unless the coating is replaced). See screen saver. 2. In audio communications, a quieter repeat or echo of a conversation. 3. In optics, a light-caused image in the dispersion plane resulting from periodic ruling errors in a diffraction grating (as opposed to scattering errors).

ghost, broadcast In broadcast images on a television screen, a slightly offset, pale copy of the desired image caused by secondary transmission of the original signal. Terrain can cause reflections of the direct signal that arrive at the receiver just slightly delayed and make the image appear slightly blurred or double.

ghost port In the early days of computing, software was in its development stages and was not always written to handle processes in a secure manner. Thus, when users left a system or terminated a remote connection in an unexpected way, the process itself did not necessarily terminate, allowing the next user into the currently open account or process. It would sometimes allow timeshare users, for example, to log in to the previous user's account through the ghost port (the unterminated session) and access the files.

Ghost ports are less common now, with standard operating systems and more security-savvy programmers developing the software, but hackers are still sometimes aware of security holes and programs and methods that create ghost port access to less robust systems. The phrase can also refer to an access port to a computer system that is visible to technical users but invisible to nontechnical users, in other words, one accessible through system commands or procedures that are inherent to the system (as opposed to being hacked into the system) but not generally known outside of techie circles. Thus, ghost port refers to a port that is left behind and is mistakenly assumed to be a new port or one that is visible or apparent only to a subset of computer users.

Ghostscript A PostScript graphics language that is part of the Free Software Foundation's GNU project. Ghostscript is almost fully compatible with Adobe's PostScript. Ghostscript is a great tool for viewing and printing PostScript files. Adobe PDF readers, freely available on the Internet, are now beginning to supersede the use of Ghostscript for viewing and printing PostScript files on some systems, but enhancements to Ghostscript for PDF compatibility have kept it alive on others.

giant magnetoresistance GMR. A resistance effect discovered in the late 1980s by P. Gruenberg in Germany and A. Fert in France. Large resistance changes were observed in materials comprised of alternating, very thin layers of metallic elements when exposed to high magnetic fields at low temperatures. Other scientists were excited by this discovery and began to study many different types of materials and configurations of layers to better understand the effect and to see whether it could be produced without subjecting the materials to very low temperatures. IBM researchers such as S. Parkin took a particular interest in this area of research. It was determined that very thin combinations of nonmagnetic metals between layers of magnetic metals could induce the nonmagnetic layer to change its orientation. Oscillations in the magnetic alignment were also detected, and it was noted that resistance was low or high depending upon whether the layers were in parallel or antiparallel arrangement.

GMR research has led to a new category of supersensitive hard disk drives and, since this is a general reproducible effect, no doubt many other practical applications will be developed. See Kerr magnetooptic effect.

.gif The conventional graphics file name extension used for CompuServe's proprietary Graphics Interchange Format (GIF) raster-format graphics files. See Graphics Interchange Format.

giga- (*abbrev.* – G when combined) (*pron.* jig-a) A prefix for 10^9 or 1,000,000,000 in the SI system. One billion. In computing, a giga is 2^{30} or 1,073,741,824 (a multiple of 1024). Giga- has long been used in supercomputing, mainframe, and scientific applications, but it was relatively unknown in lay language until the mid-1990s when gigabyte (Gbyte) hard drives dropped to consumer price ranges. It used to

be a lot of storage space. Ten megabytes used to be a lot of storage. In fact, whole community BBS systems used to run on five megabyte drives in the early 1980s. Now two gigabyte drives are considered to be average. See atto-.

gigabit GB. (*pron.* jig-a-bit) 1,073,741,824 (2^{30}) bits.

Gigabit Ethernet GbE. Ethernet networking capabilities capable of supporting half and full duplex transmissions at speeds of 1 Gbps (one billion bits per second). Fast Ethernet, the predecessor to Gigabit Ethernet, is a widely installed international open standard. In 1997, the IEEE approved the P802.3ab study group's proposed 1000Base-T standard for full duplex Gigabit Ethernet signaling over Category 5 networking systems. This approval led to the IEEE 802.3z Working Group ratifying a standard for Gigabit Ethernet, in June 1998, that included three physical layer specifications (one for shielded copper wire, two for optical fibers).

Gigabit Ethernet was developed because vendors and users wanted the benefits of a scalable high speed network to support existing Ethernet frame and protocol characteristics to enhance rather than obsolete existing systems. Gigabit Ethernet uses the same frame format, media access control, and flow control characteristics as slower Ethernets. Gigabit Ethernet provides a practical way to set up a backbone for interconnecting Ethernet and Fast Ethernet networks, and provides an Ethernet upgrade path as the technology becomes cheaper. The types of applications that require enhanced Ethernet implementations include scientific modeling, multimedia communications, data warehouse search and retrieval, and others. The most significant competitor to the faster versions of Ethernet is asynchronous transfer mode (ATM).

Gigabit Ethernet is implemented at the physical (PHY) and media access control (MAC) layers. It supports the same frame format and size, and carrier sense multiple access with collision detection (DSMA/CD) as Ethernet, and Fast Ethernet. Quality of service (QoS) is not inherent in Gigabit Ethernet, which is primarily a high speed connectivity mechanism, but is incorporated through other standards. RSVP is one way of providing quality through an open standard that can be incorporated into a Gigabit Ethernet system.

Objectives for link distances include multimode fiber optic links up to 550 meters, single-mode fiber optic links up to 3 kilometers, and copper-based links up to 25 meters, and Category 5 unshielded twisted pair (UTP) links up to 100 meters. A Gigabit Media Independent Interface (GMII) is also being studied. See asynchronous transfer mode, Ethernet, Fast Ethernet, Gigabit Ethernet Alliance.

Gigabit Ethernet Alliance GEA. A California-based multivendor open forum established in 1996 to promote the development and acceptance of Gigabit Ethernet technology and to actively support and accelerate the standards process. The GEA supports IEEE activities with regard to the development and ratification of Ethernet standards, particularly the

G

High-Speed Study Group, the IEEE 802.3 Working Group, and the IEEE 802.3z Gigabit Ethernet task force. The group provides technical resources for implementation and product interoperability. In June 1998, it ratified a standard for Gigabit Ethernet. See Gigabit Ethernet. http://www.gigabit-ethernet.org/

Gigabit Interface Connector GBIC. An interface, commonly sold as a hub-compatible peripheral card, used in Gigabit Ethernet uplinks. A hub may have slots for more than one GBIC. The GBIC enables an optical cable to be connected to an appropriate port by converting electrical impulses into laser light signals for transmission over a medium such as Fibre Channel. It is sometimes also called a Gigabit Interface Converter. See Gigabit Ethernet.

Standardized Cabling Specifications	
Specification	Description
1000BASE-CX	Wire-based transceivers or physical layer (PHY) devices for short-haul shielded copper cable connections up to 25 meters.
1000BASE-T	Wire-based transceivers of physical layer (PHY) devices for four-pair twisted-pair copper cable connections up to 100 meters.
1000BASE-SX	Optical transceivers or physical layer (PHY) devices for cabling through optical fibers at 770 to 860 nanometer wavelengths. Based upon Fibre Channel signaling for multimode fiber.
1000BASE-LX	Optical transceivers or physical layer (PHY) devices for cabling through optical fibers at 1270 to 1355 nanometers. Based upon Fibre Channel signaling for single-mode or multimode fiber.
1000BASE-LH	Long-haul multivendor specification.

Gigabit News Industry news on ATM/Gigabit networks along with analysis of competing technologies and market trends, published monthly by Information Gatekeepers, Inc.

gigabyte GByte, GB. (*pron.* jig-a-bite) 1,073,741,824 (10^9) bytes. Data rates are often described in gigabits per second (Gbps) or, for very fast rates, may be described as Gigabytes per seond (GBps). Gigabytes are also used to describe the storage capacities of a large number of tape and hard drive storage media.

Gigabyte System Network GSN. See Hippi-6400.

GIGAMO A gigabyte-class magneto-optical storage technology developed by Fujitsu Limited and Sony Corporation, announced in November 1998. It was the first widely available magnetic-induced super resolution (MSR) technology, providing 1.3 GBytes of storage on a 3.5-inch disc with a 5.92 MBytes per second transfer rate. GIGAMO retains the same cartridge size and disc diameter as ISO/IEC 15041 standards but has higher linear bit densities. The data storage capacity is about twice that of the widely adopted 640-MByte CD-ROM discs, and the technology is backwardly compatible, using the same write heads as earlier systems. See magnetic super resolution.

GIGO garbage in, garbage out. An abbreviation to describe a situation in which output cannot be better than its corresponding input, with the implication that it is the fault and responsibility of the developer or data entry person if the system gives back bad or incomplete information. See garbage in/garbage out.

GII See global information infrastructure.

gilbert A centimeter-gram-second (CGS) unit of magnetomotive force equal to 10 divided by 4p ampere-turn. It is named after William Gilbert.

Gilbert, William (1544–1603) An English physicist and physician who investigated electrostatic charges in various substances. He observed that magnetized iron lost its attractive power when heated to red heat and published *De magnete* (*On the magnet*) in 1600. He emphasized the distinctions between the magnetic effect of substances, such as lodestone, and the attractive properties of amber, a distinction previously promoted by J. Cardan in 1550 but at the time still not widely considered. In his treatise, he used the word *electrica* to describe attractive phenomena. Gilbert established that the Earth is a large magnet, thus explaining the general behavior of compass needles. The gilbert unit of magnetomotive force is named after him. See gilbert, versorium.

GILC See Global Internet Liberty Campaign.

Gilder's law Bandwidth capacity will roughly double in capacity every six or nine months or so. This law, in a sense, takes over from Moore's law, with chip processing capacities doubling about every 18 months to two years. With the emphasis away from single user systems towards network technologies, the same general idea is being applied to the growth and evolution of network transmission technologies. The term is attributed to Greg Papadopoulos at Sun Microsystems in honor of George Gilder, a business technology commentator and strong proponent of optical network technologies. See lambda switching, Moore's law.

Gill, Jonathan "Jock" (ca. 1946–) In 1992, Gill served as a Clinton/Gore campaign consultant on electronic publishing and email access and became the Director of Special Projects in the Office of Media Affairs from 1993 to 1995, during the Clinton Presidential Administration. He is probably best known for developing email access to White House documents and, in 1994, being the first manager of the U.S. President's Web site, allowing users to take a virtual tour of the White House and find information on contacting government agencies. Gill is the founder of Penfield Gill, Inc., a media communications and planning company.

GILS See Government Information Locator Service.

gimbal A mechanism or material that permits an attachment to be freely suspended or inclined in such a way that the suspended attachment remains level, or so the attachment can be inclined in any direction or several directions. Marine compasses and gyroscopes incorporate gimbal mechanisms.

gimp Extremely flexible wire or cable. Wire that can be easily threaded, woven, or spiraled. Gimp is wound up and attached to telephone handsets to allow a length of wire to tighten up like a spring when not in use, so the conversant doesn't have to interrupt the phone conversation to reach the refrigerator.

Gintl, Julius Wilhelm (1804–1883) An Austrian physicist and telegraph director who was one of the first to propose a practical means to transmit telegraphic communications in both directions at the same time. Up to this time, telegraphs were one way, thus tieing up the lines for returning messages until the current message was finished. Using two batteries and a compensation method, Gintl devised two-way communications in 1853. Two-way or duplex telegraphy was an important means of increasing line capacity that was later refined and extended by other inventors. See duplex telegraphy.

Ginzton, Edward Leonard (1915–1998) A Ukrainian-born American inventor and professor who emigrated to San Francisco in 1929, Ginzton headed up the Microwave Laboratory at Stanford. He developed further applications with the Klystron technology developed by the Varian brothers, the co-founders with Ginzton of Varian Associates, in 1948. In the late 1950s, he headed up the Stanford linear accelerator project. In 1959, he became the CEO of Varian Associates, remaining as Chairman until 1984 and serving on the Board until 1992.

Ginzton is remembered not only for his technical achievements with the Klystron tube and Stanford linear accelerator, but also for his pioneering management policies. Ginzton was instrumental in establishing employee incentives and benefits long before such practices were common or mandatory. His policies aided him in attracting top talent to the firm and provided a new model for employee relations in the emerging Silicon Valley community. He also chaired the National Academy of Sciences committee that advised on the Clean Air Act of 1971 and, with David Packard, supported the development of minority-owned businesses.

Ginzton has been recognized through numerous awards, including the IEEE 1969 Medal of Honor and induction into the 1995 Silicon Valley Engineering Council's Hall of Fame. When the Microwave Laboratory at Stanford University broadened in focus, around 1970, it began to be called the E.L. Ginzton Laboratory. It is an independent lab for engineering and physics research. See Klystron, magnetron, Silicon Valley, Varian Associates.

Giorgi System A system of measurement in which the units are meter, kilogram, second, and ampere (MKSA).

GIP See Global Internet Project.

GIS See geographic information system.

Gisborne, Frederic Newton (1824–1892) An English-born Canadian inventor who devised new ways of insulating communications cables against harsh environments. With the financial backing of Americans, most notably Cyrus Field, Gisborne was a strong motivational and administrative force in linking Europe and North America, in 1858, with the first successful transatlantic telegraph cable and the first successful permanent cable a few years later.

Leading up to this important historical achievement, Gisborne studied telegraphy in Quebec in the 1840s, excelling in his courses and subsequently accepting a head position with the Montreal Telegraph Company. He then was involved in the founding of the British North America Electric Telegraph Association (BNAETA) and, on its behalf, negotiated unsuccessfully with eastern Canadian governments to set up a Halifax-to-Quebec telegraph line. Thereafter, Gisborne took the position of superintendent of the Nova Scotian telegraph lines. He sought backing from the government for an underwater line from Halifax to Newfoundland, expressing an interest in solving the problems associated with undersea installation and maintenance. In 1851, the Nova Scotia Electric Telegraph Company was established to oversee existing lines and create new ones.

In 1852, Gisborne left his position with the Nova Scotia Electric Telegraph Company to establish the Newfoundland Electric Telegraph Company. His association with financier Cyrus Field began in 1854 when Gisborne contacted Cyrus' brother, Matthew D. Field, on behalf of the financially stressed Newfoundland telegraph project. The American collaborators purchased the assets of the Newfoundland Electric Telegraph Company and settled its debts, traveling the difficult wilderness journey to St. Johns, Newfoundland in spring 1854, to present the charter for a proposed New York, Newfoundland, and London Telegraph Company. The first undertaking of the expanded venture was to link Newfoundland with neighboring regions and with the U.S. Thus, submarine cables were installed between Cape Breton, Newfoundland, and Prince Edward Island, and telegraph lines were installed across the perilous Gulf of St. Lawrence (though not on the first try).

With the involvement of the U.S. collaborators, who envisioned oceanic telegraphy, Gisborne's ambitions increased. He became determined to establish a cable between Canada and Ireland. Despite months of harrowing travel and political arrangements further aggravated by financial difficulties, Gisborne continued to pursue his goal of establishing a transatlantic cable. Gisborne's contribution is sometimes overlooked, as he pulled out of the partnership near its successful completion, due to distrust and disagreements with his collaborators, but he was an important contributor, both to the inception and progress of the project. See Field, Cyrus West; Gooch, Daniel; gutta-percha; transatlantic cable.

GITS Government Information Technology Services.

Gladstone-Dale law The refractive index of a substance varies with a change in temperature or volume according to a formula in which the index of refraction (*n*) plus one, over the density (*r*), equals a constant (*k*).

glass A strong, brittle substance primarily composed of silica, that ranges from transparent to opaque, depending upon the quantity and composition of other materials contained within the material. Impurities can affect the color, opacity, polarizing and transmittance characteristics, index of refraction, strength, flexibility and other properties of glass. The glass in jars and windows typically contains about 75% silica combined with a number of oxides to enhance the fabrication properties of the glass.

Pure glass is generally preferred for fiber optics communications components (e.g., the conducting core), but impurities may be "doped" into the fiber to alter its refractive index or selective transmittance of certain wavelengths. There are special challenges associated with pulling out long filaments of pure glass. For example, the process itself may introduce bubbles or a crystalline structure within the glass, which would interfere with light transmission. The addition of oxides to enhance the fluidity of the glass is not practical because it introduces undesired impurities. It has been discovered that pulling fibers or fiber preforms at 0 gravity results in a fine, pure glass without the problems of crystallization associated with Earth-based fabrication. See pulling fiber, vapor deposition.

glass house *colloq.* A term to describe the large, glassed-in, controlled environments used to house and protect (and in some cases air-condition) large computer installations. These environments still exist, to some extent, in supercomputing systems, but technological advances have decreased the size and fragility of many computers, and glass houses are no longer needed for small- or medium-sized computing systems. Glass houses or clean houses are still used in chip manufacturing environments to provide a carefully regulated environment where temperature, humidity, and even tiny particles can affect the structure and functioning of certain delicate or microminiature components.

glass insulator A historic utility pole insulating safety device ranging in size from about 6 inches to about 18 inches that was very commonly used to support live wires on utility poles. See insulator, utility pole for a chart and more detailed information.

glitch 1. Unexpected, small, but annoying problem, usually causing a delay or minor informational error. This term is usually applied in instances where repetition of the problem is unlikely or infrequent. 2. Undesirable brief surge or interruption of electrical power.

Global Area Coverage GAC. One of the sample data set archives of the National Oceanic and Atmospheric Administration (NOAA) derived from the Advanced Very High Resolution Radiometer (AVHRR) sensors onboard TIROS and NOAA-*x* satellites. As the satellites orbit, they collect Local Area Coverage data, which is stored, combined, and compressed into GAC, available for download.

In order to compress the data from LAC to GAC levels, four pixels in a scanline are sampled and averaged and the fifth pixel is skipped, with the process repeating to the end of the scanline. The following two scanlines are then skipped, thus creating a pattern for each three scanlines on through the data file. GAC resolution is approximately 7.6 km.

Various agencies distribute data products based on GAC downloads. For example, the SeaWiFS GAC Level 1 consists of radiance data from combined north-to-south scan swaths with file sizes of about 19 MBytes. SeaWiFS GAC Level 2 is taken from Level 1A data that has been further processed, calibrated, and corrected. Unusual conditions may also be marked within the data. The National Science Foundation, the Office of Naval Research, and other organizations have provided support to a number of research and educational associations for GAC data. See Advanced Very High Resolution Radiometer.

global area network GAN. A network that is accessible to most or all nations in the world. The Internet is the closest thing we have to a GAN, although it is not yet ubiquitous or accessible by all nations or people.

Global Atmospheric Research Program GARP. A program in the 1970s to study atmospheric trends and patterns and to extend the range of daily weather forecasts over a longer period than was previously possible. The GARP Atlantic Tropical Experiment (GATE) was the first major experiment in the GARP, carried out in 1974. GATE is of interest, not only because it was international in scope, but because it involved intercommunication among a host of research ships, aircraft, and communications buoys. In the late 1970s and early 1980s, NOAA meteorological satellites were designed and launched to support the GARP. NOAA-7, for example, launched in 1981, had sensors for measuring the Earth's atmosphere, surface, and cloud patterns. The GOES system of sensing satellites also form part of the GARP. See GOES.

Global Atmosphere Watch GAW. A United Nations Commission for Atmospheric Sciences monitoring and assessment program that uses technology to track pollution, ozone, and other aspects of atmosphere composition around the world.

Global Business Communications Systems GBCS. An AT&T business, which was rolled in with the AT&T Laboratory restructuring of Bell Laboratories in 1995-1996, along with the Network Systems Group, AT&T Paradyne, Microelectronics, and Consumer Products. GBCS is moving into the area of multimedia and secure telecommunications services. Products include a Unix-based server that works on a private branch exchange (PBX) to provide videoconferencing capabilities over data networks such as Ethernet and IBM Token-Ring. The server software is called Multimedia Communication Exchange (MMCX) and was implemented first on Unix stations, with the intention of porting it to PC operating systems.

Global Change Master Directory GCMD. A NASA multidisciplinary database project originating from the Goddard Space Flight Center (GSFC) Global Change Data Center. The GCMD includes both national and international remote-sensing data sets contributed by more than 800 U.S. and worldwide organizations, including NOAA, NASA, DOE, NSF, EPA, USGS, educational institutions, and others. It is of interest to anyone interested in planet change, particularly climatic change. The data includes temporal and geographic information and some parts are interlinked with relevant external information. The GCMD relational database is now searchable online through the GCMD Web site. The data is of particular interest to climatologists, agriculturalists, hydrographers, mineralogists, and others.
http://gcmd.gsfc.nasa.gov/

Global Data Processing System GDPS. A system of preparing and disseminating cost-effective meteorological analyses and forecasts. The GDPS retrieves, assesses, decodes, sorts, and analyzes the data in preparing for making it available for distribution.
GDPS is one of three integrated core components of the World Weather Watch (WWW) system. It is administrated by the World Meteorological Organization. GDPS data are valuable for weather forecasting and meteorologically related agriculture, climatology, and aeronautics industries. See Global Observing System, Global Telecommunication System.
http://www.wmo.ch/web/www/DPS/gdps.html

Global Development Gateway GDG. A major World Bank initiative proposed through the GDG Principles discussion forum hosted by Bellanet. The GDG is sponsored by the World Bank to promote information exchange, access, and development.

global directory An internetwork computer database that stores various types of information related to the various networks. The information may be user login names and passwords, shared database resources, group member lists, device directories that can be accessed by more than one network, or pointers to various applications or documents common to the various networks. The Internet has a number of global directories of file databases, archives, etc. See Gopher, Archie, Veronica.

Global Environmental Monitoring System GEMS. Administrated by the United Nations Environment Programme (UNEP), GEMS was established as part of the Earthwatch program in 1975 to support and strengthen environmental monitoring in participating countries and to improve the collection and evaluation of environmental data. Computer-acquired and -processed information is an important part of this effort. See Global Resource Information Database.

global function plane GFP. An architecture within which the modular functionality for Intelligent Network (IN) services may be globally constructed. The GFP functions are described as service-independent building blocks (SIBs). The IN is viewed as a single entity within the GFP. GFP is defined in ITU-T Q.1201 which describes a generic IN GFP model, service-independent building blocks, and services and features offered through a global service logic.

Global Incident Analysis Center GIAC. A SANS Institute center that creates and disseminates reports of malicious activity on the Internet submitted by system administrators and network security professionals worldwide. GIAC maintains a large archive of security-related papers available for free download.
http://www.sans.org/giac.htm

global information infrastructure GII. A term used since the mid-1990s by international standards committees with regard to goals, standardization, and development of global interconnected telecommunications systems, including the technology, applications, and related services.

Global Information Infrastructure Commission GIIC. An independent, nongovernmental initiative inaugurated in July 1995 to promote leadership in the private sector and cooperation between private and public sectors in developing information services and networks. GIIC fosters economic growth, education, and quality of life through activities such as developing an accessible and diversified global information infrastructure. The GIIC operates as a project of the Center for Strategic and International Studies (CSIS). CSIS is a U.S.-based private organization founded in 1962 to conduct research in global public policy.
http://www.giic.org/ http://www.csis.org/

Global Internet Liberty Campaign. A human rights group that includes the American Civil Liberties Union (ACLU), the Electronic Freedom Foundation (EFF), and other member organizations.

Global Internet Project GIP. A private sector organization founded in 1996, consisting of senior level managers representing global software and telecommunications industries with high stakes in Internet development. As part of its activities, GIP encourages the education of world decision-makers in the potential evolution and uses of the Internet.
http://www.gip.org/

Global Land Information System GLIS. An interactive database system developed by the U.S. Geological Survey (USGS). It provides data that is valuable in the study of the Earth's land surfaces. Samples and information about GLIS products are available online. Topics within the GLIS database include climate, geology, hydrology, land cover, and others. There are aerial photographs, satellite images, and digital line graphs to serve a variety of needs. The aerial photographs, for example, include sources such as the National Aerial Photography Program, National High Altitude Photography, and various radar systems. http://www.earthexplorer.usgs.gov/

Global Maritime Distress and Safety System GMDSS. A maritime safety system which incorporates automated distress calls using Digital Selective Calling (DSC). In the late 1970s, maritime experts began to develop systems for updating safety and distress communications, resulting in the 1979 draft of the International Convention on Maritime Search and Rescue. GMDSS advocates a global search and rescue plan and a Global Maritime Distress and Safety

System (GMDSS) as a communications infrastructure for the overall plan. The system is based upon a combination of Earth-based and satellite-based radio services, emphasizing ship-to-shore marine signaling through relatively user-friendly consoles. GMDSS signaling is quickly superseding the decades-old Morse code-based system. In addition to the automation of distress signals, it calls for the shipboard downloading of maritime safety information as a preventive measure.

In 1996, the Telecommunications Act was written to encompass U.S. marine vessels, and ships were required to install GMDSS equipment by 1 Feb. 1999. All vessels subject to Chapter IV of the Safety of Life at Sea (SLOAS) convention must be fitted with GMDSS equipment (with stipulated exceptions) as must mobile offshore drilling units (MODUs).

Implementation of GMDSS has not been without problems. It has been criticized for false alarms and general reliability problems, and many nations have been slow to adopt the system. See COSPAS/SARSAT, NAVTEX.

global mobile personal communications services GMPCS. A phrase coined by the ITU-T to describe mobile communications through low Earth orbit (LEO) satellite systems but later broadened to include other modes of mobile communications (geostationary FSS, MSS, "Little LEOs" and wideband LEOs). GMPCS was the discussion theme of the World Telecommunication Policy Forum (WTPF), resulting in a set of principles and recommendations described in the "WTAC Report to the Secretary-General [of the ITU-T] on GMPCS," January 1996.

Global Navigation Satellite System GLONASS, GNSS. A satellite system deployed by the Russian Federation defense department, which has much in common with the American Global Positioning Service (GPS) in terms of satellite placement and the types of information transmitted. The 24 GLONASS system satellites are orbiting in three planes. Unlike the GPS system, GLONASS claims to plan to use the same levels of signals for civilian (CSA) and government use (SA), and civilian use is guaranteed for about the next decade. By the late 1990s, the project had been divided into two stages: (1) GNSS-1, the first generation Russian GLONASS and U.S. GPS system and (2) GNSS-2, the second generation system including civil access with improved positioning and services.

The MIT Lincoln Laboratory conducts research on the GLONASS system and reports progress and observations on the project on their Web site. http://vega.atc.ll.mit.edu/glonass/

Global Network Navigator GNN. A Web-based information service providing lists of and information about new services, sites, and related resources on the Internet.

Global Observing System GOS. A system for obtaining standardized observations of the Earth's atmosphere and ocean surfaces from ground, sea, air, and space-based observation platforms. GOS is one of three integrated core components of the World Weather Watch (WWW) system and is administrated by the World Meteorological Organization. GOS data are valuable in weather forecasting and meteorologically related agriculture, climatology, and aeronautics. Space data are available through the Environmental Observations Satellite (EOS) system comprised of five near-polar and five geostationary environmental observation satellites with a variety of imaging and sounding sensors. See Global Telecommunication System, Global Data Processing System. http://www.wmo.ch/web/www/OSY/GOS.html

Global One An international commercial joint venture of Sprint, Deutsche Telekom, and France Telecom.

Global Online Directory GOLD. A commercial product from VocalTec that works in conjunction with their Internet Phone software. Internet Phone lets you plug a microphone into your personal computer and use it as a phone transmitter to communicate with another person with Internet Phone capabilities. The computer speaker provides the equivalent of the phone receiver. Long-distance calls can be placed as though they were local calls through your ISP, without long-distance charges.

Internet Phone connections are full duplex, connecting through the TCP/IP transport protocol. In addition to the features of a conventional phone call, chat lines and other digital enhancements are available. GOLD is the global directory that stores information about Internet Phone users who can be contacted, just as the names of phone subscribers can be accessed through a phone directory.

Global Ozone Monitoring Experiment GOME. A satellite-borne ozone European Space Agency (ESA) research project launched in April 1995 on board the European Remote Sensing Satellite (ERS-2). GOME is a nadir-viewing passive spectrometer that senses atmospheric trace constituents by measuring solar radiation scattered by Earth's atmosphere. Cloud characteristics, aerosols, and surface reflection can also be measured. GOME is designed to sense in the visible and ultraviolet spectra from 240 to 790 nanometers. See GOME Data Processor. http://auc.dfd.dlr.de/GOME/

Global Positioning System GPS. A space- and ground-based 24-hour navigational system originally designed and used by the U.S. military (see Navy Navigation Satellite System), funded and maintained by the U.S. Department of Defense (DoD). It provides the means to monitor, update, and maintain orbiting satellite systems and to determine a location on or around the Earth through information from these systems.

GPS uses the known positions of satellites as reference points for discerning unknown positions on or above the Earth. There are now over 20 satellites in the system (some are spares), more-or-less evenly spaced, orbiting in 12-hour cycles at an altitude of about 10,898 miles (about 400 miles higher than the original NNSS). A system of sophisticated ground stations with antennas, coordinated by a master control station, administers, deploys, and maintains the satellites and updates them when needed to correct for clock-bias errors.

A variety of types of information can be computed from information from several satellites, including a location or position of a stationary or moving object, and coordination of time. This information can be incorporated into software applications in vehicle-mounted or handheld positioning receivers. From military operations, to recreational navigation on the ocean in a kayak, to airline navigation, GPS provides a wealth of data with which to determine latitude and longitude, altitude, and velocity. This information can further be combined with maps to record or suggest routes.

GPS satellites transmit timed binary pulses in addition to information constants about the current location of the satellite. The synchronized atomic clocks aboard the satellites permit the transmission of precise timing tags. The combination of the speed of transmitted electromagnetic waves and the atomic clocks installed in GPS satellites provides remarkably accurate timing pulses.

NAVSTAR Global Positioning System

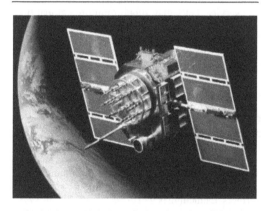

NAVSTAR Global Positioning System (GPS) satellite from the 24-unit constellation of planes and satellites that provides military and civilian navigation and positioning data. The nine-ton satellites orbit the Earth every 12 hours, emitting continuous positional signals. The signals can be used to calculate time, location, and velocity, depending upon the capabilities of the GPS receiver. The system is controlled and operated by the 50th Space Wing out of Schriever AFB, Colorado. [U.S. Air Force Space Command image.]

The GPS Master Control Station (MCS) is operated by the 50th Space Wing's 2nd Space Operations Squadron in Colorado. This squad is responsible for monitoring, controlling, and operating the GPS satellite constellation. The U.S. Air Force Space Command Space and Missile Systems Center in Los Angeles, California, is the executive agent for the Department of Defense (DoD) acquiring GPS satellites and equipment.

There are thousands of GPS users worldwide. Personal GPS devices can be purchased for as little as $180 to $450, and more sophisticated ones are used in all industries that rely on location information: airlines, shipping firms, ferries, military divisions, etc. For greater details on individual aspects of GPS, see differential GPS, EAGLE, Intelligent Vehicle Highway Systems, GPS Operational Constellation, GPS Navigation Message, local differential GPS, NAVSTAR, Precise Positioning Service, Standard Positioning Service, wide area differential GPS.

Global Resource Information Database GRID. A global network of environmental data centers cooperating to generate and disseminate key environmental geo-referenced and statistical data sets and information products. GRID centers are equipped to prepare, analyze, and disseminate environmental data that may be used as the basis for environmental assessments. GRID is associated with the Earthwatch program and the Environment Assessment Division of the United Nations Environment Programme (UNEP). It was established in 1985, evolving gradually out of the 1972 United Nations Conference on the Human Environment held in Sweden. Following the conference, UNEP was created to provide a focus for a Global Environment Monitoring System (GEMS) that led to GRID.

There are a number of GRID centers around the world, including in the U.S., South America, Europe, Russia, Nairobi (UNEP headquarters), Asia, and New Zealand. For example, the United Nations Environment Programme (UNEP) and 1987 World Commission on Environment and Development project in cooperation with the Government of Norway established a GRID environmental information center in Arendal, Norway. GRID-Arendal was opened as a nonprofit foundation in August 1989; it communicates and cooperates with other GRID centers.

Global Software Defined Network GSDN. A high-volume commercial virtual private network service from AT&T that utilizes AT&T's Worldwide Intelligent Network (WIN) to interconnect networks in the U.S. and other countries. GSDN selects an economical route for external calls and provides internal services, including order entry, tracking, file transfers, and teleconferencing services. GSDN is aimed at business networks.

Global Standards Collaboration GSC. A framework for the exchange of information regarding global standards development for interconnectivity and interoperability of systems and devices. Thus, the GSC has been bringing together senior officials from regional, national, and international standards bodies since 1988. Participating bodies include the International Telecommunication Union (ITU), the International Organization for Standardization (ISO), the European Telecommunications Standards Institute (ETSI), and others. Focal areas are discussed at large conferences. For example, global radio standardization was the key topic for the 2000 meeting in Sapporo, Japan. Other topics of current interest include number portability for mobile communications, universal personal telecommunications (UPT), and intelligent networks. http://www.gsc.etsi.org/

G

Global System for Mobile, Groupe Spéciale Mobile GSM. A digital cellular technology developed jointly by the telecommunications administrations of Europe. The Groupe Spéciale Mobile was founded in the early 1980s, and the Global System for Mobile (GSM) was first publicly announced in 1991 and has since been standardized in Europe and Japan.

GSM was the first fully digital system to provide mobile voice connections, data transfer services, paging, and facsimile at full duplex or half duplex rates up to 9600 bps. GSM operates in two frequency ranges: 890 to 915 MHz for signaling information and 935 to 960 MHz for information transmissions.

GSM is a set of standards specifying a digital mobile communications services infrastructure. It is based on a 900 MHz radio transmission technology and specifies related switching and signaling formats. An 1800 MHz Digital Cordless System (DCS) has also been added. Since mobile systems typically support roaming, and since the multicultural makeup of Europe provides a unique challenge in providing compatible services, interoperability has been emphasized in the GSM specifications.

GSM can be described in three categories: communications media, transceiving systems, and information systems, as shown in the GSM General Categories chart.

The GSM subscriber identity module (SIM), also known as a *smartcard,* is a security feature which handles encryption and authentication. It includes memory storage which can be used for dialing codes or other information related to the service. The SIM is also a means to download and display call-related information. See Future Public Land Mobile Telecommunication System, Personal Communications Network.

Global Telecommunication System GTS. A system of terrestrial and space-based data circuits for interconnecting meteorological telecommunications centers. GTS is one of three integrated core components of the World Weather Watch (WWW) system and is administrated by the World Meteorological Organization. GTS data are valuable in weather forecasting and meteorologically related agriculture, climatology, and aeronautics. The GTS provides rapid and reliable dissemination of observational meteorological data. The GTS is organized into a Main Telecommunication Network (MTN), Regional Meteorological Telecommunication Networks (RMTNs), and the National Meteorological Telecommunication Networks (NMTNs). The World Meteorological Centres (WMOs) are located in Russia, Australia, and Washington, D.C., and there are more than a dozen regional hubs in Asia, South America, Africa, Europe, and other locations. See Global Observing System, Global Data Processing System.

http://www.wmo.ch/web/www/TEM/gts.html

Global Title Translation GTT. A telephony and administration and routing function that enables added feature functionality in commercial systems such as Local Number Portability (LNP), calling card services, and mobile roaming support. GTT determines destination addresses in Signaling System 7 (SS7) and other relevant network systems. When a call is initiated, GTT determines the destination and may include additional information, depending on the feature service. GTT may support multiple global title addresses.

Global Transaction Network GTN. AT&T's extensive 800 service phone network, which was introduced in 1993. This service supports enhanced features, providing more flexible routing and numbering services which can be used, for example, by airline reservation systems.

Globalstar A system of 48 small *bent pipe* communications satellites orbiting at 1400 kilometers (LEO), for providing voice and data services (data files, paging, facsimile). Globalstar was established in 1991 as a joint venture of Loral Space & Communications, Ltd., QUALCOMM, Inc., and a number of corporate partners. Launching began in February 1998. In 1999, Globalstar launched four more systems, bringing the

Global System for Mobile (GSM) General Categories		
Category	Notes	
Media	GSM works over frequency-modulated (FM) signals using a combination of time division multiple access (TDMA) and frequency division multiple access (FDMA). Peak output power varies with the type of transmitter (mobile station class), ranging from 0.8 to 20 watts. Frequency hopping is used to reduce interference and multipath fading, and encryption increases security. The data rate is 270 Kbps.	
Transceiving	There is a base transceiver station (BTS) associated with each cell operating on fixed frequencies unique to its region. Honeycomb-like clusters are handled by base station controllers (BSC), which, in turn, are controlled (routed, switched, handed over) by Mobile Service Switching Centers (MSC).	
Information	There are databases associated with GSM that aid in the administration of subscriber information and those that aid in the administration of security and associated authentication mechanisms. There is also an equipment identity register (EIR), which keeps track of equipment types and configuration, and can block calls on stolen units.	

total satellites to 12. By 2000, the system was operational and the number of satellites increased to 48 over the next two years.

Globalstar services can be accessed with vehicle-mounted or handheld devices resembling cellular phones, and the system is integrated with cell phone services through dual-modem handsets. Remote users can access the system through Globalstar service providers, with fixed-position and wireline phones. Globalstar is intended to enhance rather than replace existing cellular and other phone services. Services are aimed at international business travelers, commercial vehicle operators, marine craft, field scientists, and others. The competitive aim is low cost for service and accessories.

GLONASS Global Navigation Satellite System. A Russian Federation Global Positioning System similar to the U.S. NAVSTAR system. GLONASS is managed by the Russian Space Forces. GLONASS provides all-weather positioning coordinates, velocity references, and time information from virtually any point on or near the globe.

More specifically, the GLONASS system aids in managing air and marine traffic, in supporting emergency and safety systems, geodesy, cartography, and ecological monitoring. Like the American GPS system, GLONASS is two-tiered. It provides *standard precision* navigation signals (SPs) with horizontal accuracy to about 63±6 meters and *high-precision* navigation signals (HPs) with authorization and specialized equipment.

The GLONASS system has launched more than four dozen satellites since the first was put into operation in October 1982. Early satellites had a lifespan of about two years, while later satellites lasted about four or five years. In general, the launches have been successful, although in 1987 and 1988 there were a number of failures. Since 1989, the orbiting platforms have included geodetic reference satellites. In February 1999, the Russian Federation made an open declaration of increasing international cooperation with regard to national satellite navigation technologies and international navigation systems. The constellation status is reported, within a few days, on the GLONASS Web site.

http://www.rssi.ru/SFCSIC/english.html

Glossary of Telecommunications Terms, FCC A short glossary of telecommunications terms related mainly to delivery of broadcast services to consumers provided by the Federal Communications Commission.

Glossary of Telecommunications Terms, Federal Standard Originally introduced in 1976 as MIL-STD-188-120, this archive has evolved through several revisions to become the Telecom Glossary 2000. It is provided through the U.S. National Communications System in print, CD-ROM, and Web formats, with increasing emphasis on electronic development and dissemination. The Glossary is mandated for use by all federal departments and agencies for the preparation of telecommunications documentation and is available from the National Technical Information Service.

Definitions from the previous standard were revised and updated by the *FTSC Subcommittee to Revise FED-STD-1037B*. The subsequent version, FED-STD-1037C (1996), was further updated to Telecom Glossary 2000, with many of the new definitions drawn from T1 Standards and Reports.

The Glossary includes standard definitions for telecommunications terms related to antennas, computers, transmissions media (e.g., fiber optics), networks, audio/video technologies, radio communications, etc. Sources include government publications and those of prominent telecommunications organizations including the ITU, ISO, and the American National Standards Institute.

The Glossary has been reviewed by the National Communications System Member Organizations, the Federal Telecommunication Standards Committee members, members of relevant industries and federal agencies, and by the general public. Telecom Glossary 2000 was discussed through members of the T1A1 Ad Hoc Glossary Group.

GMDSS See Global Maritime Distress and Safety System.

GMDSS Radio Maintainer License A license granted by the Federal Communications Commission (FCC) to maintain radiocommunications aboard ships that are equipped to comply with Global Maritime Distress and Safety System (GMDSS) regulations. It further confers the operating authority of the General Radiotelephone Operator License and the Marine Radio Operator Permit. A licensed maintainer is required by ships that conduct at-sea maintenance. The license requires FCC Element 1, Element 3, and Element 9 exams to be passed.

GMDSS Radio Operator License A license granted by the Federal Communications Commission (FCC) to individuals deemed capable of handling radiocommunications aboard ships that are equipped to comply with Global Maritime Distress and Safety System (GMDSS) regulations. The licensee may operate basic equipment and make antenna adjustments. The license further confers the operating authority of a Marine Radio Operator Permit. Applicants are required to pass the Element 1 and Element 7 written exams.

GMPCS See global mobile personal communications by satellite.

GMR See giant magnetoresistance.

GMS See Geostationary Meteorological Satellite.

GMSK See Gaussian minimum shift keying.

GNN See Global Network Navigator.

GNOME An open source user software environment and applications framework available to developers as part of the GNU project. GNOME is typically distributed with BSD and GNU/Linux distributions and is available for other platforms as well.

The goal of the GNOME project is to provide a user-friendly graphical desktop environment and a developer-friendly base of tools for creating GNOME-compatible software applications to augment or replace commercial operating systems and development environments. As the project has evolved, it is also being extended to include a set of commonly used office productivity tools.

One of the more interesting aspects of GNOME is that it can be internationalized in order to make it possible for developers around the globe to create applications in the language syntax that is most comfortable for them. One of the secondary goals of the GNOME project is to make GNOME developer tools and documentation available in every known world language (an ambitious goal but likely to result in a rich choice of options even if it is not literally achieved). GNOME also takes into consideration the maximizing of accessibility for people with disabilities, through the GNOME Accessibility Project. http://www.gnome.org/

GNSS See Global Navigation Satellite System.

GNU Acronym for "GNU's Not Unix!" A Unix workalike developed under the aegis of Richard Stallman of the Free Software Foundation (FSF). See Free Software Foundation, GNOME.

GNU as A GNU family of assemblers used to write software code for a variety of object file formats. The original GNU assembler for the Digital Equipment Corporation (DEC) VAX system was written by Dean Elsner. Many subsequent programmers and even some commercial vendors have enhanced and maintained the software. See Free Software Foundation.

GNU C compiler GCC. A C compiler supporting ANSI standard C, C++, and Objective C. The GNU C library includes ANSI C, Unix, and POSIX functions.

GNU Emacs A powerful, extensible, scriptable display editor distributed by Berkeley programmers with BSD, and by many other distributors and commercial vendors. Emacs is so powerful and so well liked by power editor users, many have half seriously referred to it as an operating system.

The first Emacs was written in 1975 by Richard Stallman. GNU Emacs, which was enhanced by Stallman with true LISP integrated into the editor, was introduced in the mid-1980s. GNU Emacs is widely available on Unix systems.

GNU graphics A set of graphics utilities for plotting scientific data, with support for GNU plot files on various systems and output devices, including PostScript, The X Window System, and Tektronix devices.

GNU's Bulletin A semiannual newsletter about various GNU projects, produced and distributed by the Free Software Foundation.

go local A command to instruct software to connect to a local connection (usually in the same room or vicinity), usually through a serial null modem interface.

GO-MVIP Global Organization for Multi-Vendor Integration Protocol. GO-MVIP is a nonprofit trade association which took over the development and promotion of MVIP in 1994, in order to assure its development and maintenance as a practical, robust common integration standard. GO-MVIP seeks to continue to develop and establish the design specifications for further versions of MVIP. See MVIP. http://www.mvip.org/

Satellite Sensing Systems

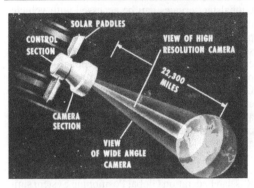

An artist's conception of an EROS meteorological satellite system, which was essentially the concept for the GOES satellites. This April 1961 drawing indicates the field of view and orientation of the proposed AEROS geostationary satellite. [NOAA In Space Collection image.]

An example of a visible image from the GOES 8 satellite on 16 July 2001 with a line drawing of North America and parts of Central and South America superimposed to aid in locating geographical features. [NOAA FSL image.]

GOES Geostationary Operational Environmental Satellite. A geosynchronous satellite system providing environmental monitoring data to various governmental organizations and the public. GOES systems began broadcasting, as we know it today, in 1974, having evolved out of early SMS weather satellite systems. Almost a dozen satellites were launched between the early 1968 magnetometer and infrared sensor-equipped satellites and the more sophisticated multisensor satellites of the late 1990s. With each launch, adjustments and improvements have been made to the systems. Vertical temperature and moisture sensors were added to GOES-4, and additional signal relay capabilities were added to later GOES systems. The original GOES satellites were spin controlled, meaning that they used the physics of spinning to maintain attitude control. Since the mid-1990s, however, with the launching of GOES-8, the configuration has changed to a three-axis-stabilized system. This configuration changed not only attitude control, but also the means by which the sensors were installed and operated. Doppler radar sensors were added to the more recent satellites and have become an indispensable aspect of creating and interpreting weather maps. The Pacific GOES has been used for operating the Pan-Pacific Educational and Cultural Experiments (PEACESAT) to enhance medical, cultural, and educational resources in the Pacific island nations.

GOES data is available to the Command and Data Acquisition Station, the National Weather Service, the Forecast Systems Laboratory (FSL) of the National Oceanic and Atmospheric Administration (NOAA), and many others. NOAA handles the operations of the satellites. GOES sensors provide continuous weather monitoring along with information that can be used to estimate rainfall during more violent storms (e.g., hurricanes) and snow accumulations and cover. Ice flows on large bodies of water can also be monitored with GOES data. Even the weather in space, in terms of various magnetic and energy particles, is monitored by GOES sensors.

The GOES Space Environment Monitor (SEM) mission provides a better understanding of space weather through the National Space Weather Program. Space weather forecasts aid not only in probe, shuttle, and satellite missions, but also in providing valuable cosmological information to physicists and astronomers. Satellite images with a variety of characteristics (visible, infrared, water vapor, etc.) from GOES sensors are available through the Forecast Systems Laboratory.

gold A malleable, metallic chemical element with high conductivity, which makes it useful for specialized electrical applications. Gold contacts are often found on sensitive electronics connectors in the computer and video industries. Copper and silver are also good conductors, with copper being the most widely used for electrical installations.

GOLD See Global Online Directory.

gold disk, gold disc 1. The master or final copy of a product (software, music CD, laserdisc, etc.) from which mass production replicas are made. 2. A special limited edition distribution. Collectors' edition. Gold disc music CDs sometimes are marketed as higher quality pressings with special inserts and special tracks that may not be included on a regular copy of the CD.

gold number, custom number, vanity number A phone number specifically selected so that it is easy to remember, particularly if the letters associated with the number spell out a word or other mnemonic. There is typically an extra fee associated with getting a gold number. Sometimes people get lucky, and their number just happens to be easy to remember or to spell something interesting.

Goldstone Deep Space Communication Complex One of the complexes of the NASA Deep Space Network (DSN) that provides radio communications for interplanetary spacecraft. The complex is further used for radio astronomy and radar observations of our solar system and the universe beyond. Other complexes in the Network are located in Madrid, Spain, and Canberra, Australia. See Deep Space Network.

Deep Space Communication Complex

Large beam waveguide parabolic antennas at the Goldstone Deep Space Communication Complex in the Mojave Desert in Goldstone, California. The 34-meter dishes are used for astronomical research and radio space communications. [NASA/JPL image, 1990.]

GOME See Global Ozone Monitoring Experiment.

GOME Data Processor GDP. A ground segment system designed to process data from the satellite-based Global Ozone Monitoring Experiment (GOME). Raw data is processed by the GDP into Level 1 *radiances/reflectances* and Level 2 *trace gas quantities* data that are made available to interested parties. A number of images and other data sets are also available, some of which may be downloaded by FTP. The GDP was jointly designed and developed by German, Dutch, and U.S. organizations and is administrated by the European Space Agency (ESA). Processing is handled at the DFD, the German Remote-Sensing Data Center. See Global Ozone Monitoring Experiment.

GOMS See Geostationary Operational Meteorological Satellite.

goniometer A detection instrument that can be set at a range of angles, sometimes in selected increments (or with optional heads), to assess the different patterns of light scattered from a sample. The results are useful in industrial analysis of materials such as polymers and particulate composites. Fiber optics enable the probing surface to be separate from the rest of the electronics. Goniometers may be built into spectrometers although some newer fiber optic spectrometers don't require traditional goniometers.

Gooch, Daniel (1816–1889) An English locomotive engineer and businessman who aided in the installation of the first successful permanent telegraph cable link between North America and the British Isles. At a relatively young age, Gooch was appointed the locomotive superintendent of the Great Western Railway (GWR). In 1840, Gooch contacted gifted engineering designer Isambard K. Brunel about designing an engine works for GWR (this association between the two talented engineers was to take an interesting turn more than a decade later). Using new technology and the results of experiments in atmospheric resistance, Gooch designed locomotives that could travel at faster speeds than previous models. He was also a supporter of wide gauge technology, despite the trend to smaller gauges. Gooch designed more than 300 locomotives in his career as an engineer. In the 1860s, Gooch resigned his position with GWR to put his efforts into telegraphic communications. He became Chairman of the Telegraph Construction and Maintenance Company and a director of the Anglo-American Company. In the meantime, Brunel had been designing the Great Eastern (originally the Leviathon). The Great Eastern was purchased by Gooch and his colleagues for cable laying and was instrumental in laying the first transatlantic communications cable, the feat that Gooch is best known for, despite his many other accomplishments. See Great Eastern.

good condition In many rating systems, a product with minimal abrasions from wear and tear, and mechanisms that are in good working order. Good condition does not describe the age of the product or its remaining useful life. Often intermediary between *fair condition* and *excellent* or *like new* condition.

goodput A generic measurement of network data successfully received, effective throughput; in contrast, discarded cells, or transmitted cells in a congested link, are called badput. See cell rate, throughput.

Goodwin, Hannibal (1822–1900) An American minister and inventor who created celluloid film in 1885 and received a patent for rollable film in 1887. For many years, motion picture films were known as *celluloids,* and individual animation frames used to create frame-by-frame animation are still known by the abbreviated form *cells.*

gopher The command for initiating a Gopher client on a text-based system is "gopher" (all lower case), and "xgopher" is a similar client command that works with The X Window System. See Gopher.

Gopher A document system developed by P. Lindner, M. McCahill, B. Alberti, F. Anklesaria, and D. Torrey at the University of Minnesota in the early 1990s to provide a local campus information server. The Gopher service quickly grew to become a worldwide resource. It is a client/server distributed document delivery system, i.e., a means of locating information on the Internet through a simple menu-like text interface (or graphical Gopher client) or of sending information through electronic mail. It is also possible to set bookmarks, Gopher information locations that are frequently used. Links to various Gopher servers together comprise a virtual community known as Gopherspace. The Gopher text menu interface is being superseded by graphical Web interfaces. See Veronica, RFC 1436.

Gopherspace The Gopher document system is composed of many widely distributed document repositories and Internet services in Cyberspace. Hence, the Gopher facilities online are called "Gopherspace" by many their users. See Gopher.

GORIZONT A Russian geostationary telecommunications satellite launched in 1996.

GoS grade of service. A phrase to describe service levels, which usually are individually defined on an industry basis. See class of service.

GOS See Global Observing System.

GOSIP Government Open Systems Interconnection Profile. A U.S. government version of the Open Systems Interconnection system which is required in many government data network installations.

Gosling, James Gosling is best known for his contributions to the Java programming language, developed at Sun Microsystems, Inc. He was associated with Bill Joy, Mike Sheradin, and Patrick Naughton on Project Stealth in 1991. Project Stealth's goal was to develop a distributed network in which the various electronic devices could intercommunicate. See Java; Joy, William.

Goubau 1958 Wave-Guiding Patent

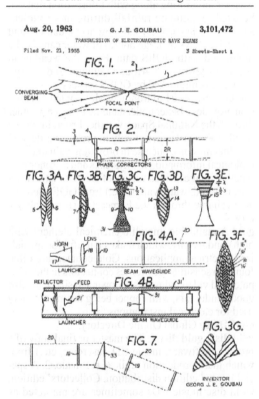

This selection of diagrams from Goubau's patent accompanies his description of a method for transmitting a substantially cylindrical wave beam by resetting the cross-sectional amplitude and phase distribution of the beam at intervals that are large compared to the beam radius. Goubau further describes a mechanism to phase-correct elements intercepting the beam to effect repetitions of the cross-sectional amplitude and phase distribution, as desired. The diagram illustrates some of the phase-correcting lenses for accomplishing his goals.

Goubau, Georg Johann Ernst A pioneer researcher into hollow optical waveguides utilizing lenses as a means to guide light. Goubau submitted a patent application for an invention to transmit electromagnetic wave beams, in November 1958 (U.S. #3,101,472). In 1961, he coauthored "On the guided propagation of electromagnetic wave beams" in *IEEE Transactions* and continued lens guide research with a number of scientists through the 1960s. See Dicke, Robert.

Gould, Gordon (1920–) An American engineer, Gould has been involved in maser/laser technology since the late 1950s. Gould studied optics at Yale University and later began a Ph.D. in physics at Columbia University, where Charles Townes was a professor. Townes had been conducting historic research on masers (microwave amplification) since the early 1950s. In 1957, Townes and Gould discussed the concept of using light amplification rather than microwave amplification for developing lasers.

Gould was an ambitious student who wanted to earn a patent for a historic invention. After discussions with Towne about patent application procedures, he left graduate school to work for TRG, worked on a prototype and filed a patent, in 1959. The patent was initially denied due to the earlier patent of lasers developed by Townes and Schawlow because Gould, at the time, could not produce substantive written or prototypic proof of lasers invented in 1958 But, after a subsequent legal battle, some of the claims were accepted and a patent was awarded in 1977 giving him rights to royalties for certain laser technologies.

In 1967 Gould accepted a professorship at the Polytechnic Institute of New York where he established and promoted laser research. Gould supports education in optics. See Dicke, Robert; laser; laser history.

Gouy, Louis-Georges (1854–1926) A French physicist and educator who described thermal motion in ions in support of molecular theories that applied to an electrical double layer associated with planar surfaces, now called the Gouy-Chapman theory. The theory, while flawed in some of its assumptions, applies reasonably well for certain aqueous solutions, but does not hold up as well for small distances and high surface potentials (e.g., edges).

Gouy balance An instrument for assessing the magnetization of a paramagnetic substance (one with unpaired electrons) as compared to the gravitational attraction of a counterweight, named after Louis-Georges Gouy. In other words, a device for determining magnetic susceptibility. Differences in apparent mass of a sample may be assessed in and out of a magnetic field. The number of unpaired electrons of a sample may be determined as the force with which a paramagnetic compound is attracted to the magnet is related to the number of unpaired electrons and can be compared with substanced with known parameters. A Gouy balance utilizes large magnets mounted stationary opposite a movable sample. Contemporary replacements for the Gouy balance do it the other way around, with movable magnets balanced on either side of a stationary sample.

Gouy phase shift A shift in the Gaussian mode that occurs when the beam passes through a focal point (sometimes called the *beam waist*), named after

Louis-Georges Gouy. The phase shift has been used as a means to split and statistically observe radiant energy characteristics and has been described as a mechanism for accelerating fast pulses. However, observation of the phenomenon was mainly of mathematical interest until the late 1990s, when systems were designed capable of transmitting in the terahertz (THz) frequencies–yielding beams that could be assessed for both amplitude and phase. This enables the Gouy phase shift to be more readily observed and interferometrically plotted. See tilt locking.

Government Emergency Telecommunications Service GETS. A service of the U.S. National Communications System for meeting national security and emergency preparedness (NS/EP) requirements for the use of public, defense, or federal telephone networks by authorized users. GETS provides emergency access and processing in local and long-distance telephone networks through a dialing plan and personal identification number (PIN). GETS services are provided through major long-distance networks, local networks, and leased networks through a universal access number and common telephone devices (desk phone, cell, phone, fax line, etc.). Once authenticated through the PIN, the caller receives special handling that may include priority and/or enhanced routing. See National Communications System. http://gets.ncs.gov/

Government Information Locator Service GILS. An ISO standard metadata scheme for describing government information resources with the intent of streamlining the management of information for efficient search and retrieval. GILS is an open, cost-effective, scalable data standard to facilitate the search for collections of information and specific information within a collection. GILS extensions allow specific categories of data to be managed and searched. GILS is based on the ISO 23950 search standard and can be approached in somewhat the same way as information searching may be approached in a reference library. See Dublin Core, International Development Markup Language. http://www.access.gpo.gov/su_docs/gils/index.html

GPF See general protection fault.

GPRS See General Packet Radio Service.

GPS See Global Positioning System.

GPS Control Segment A general overall category of the GPS system which comprises a main tracking station, in Schriever (formerly Falcon) Air Force Base, Colorado, and subsidiary tracking stations worldwide as part of the U.S. Department of Defense's Global Positioning System. The tracking stations take the signals from the satellites and incorporate them into orbital models that are further used to compute precise, individual, orbital data and clock corrections. Portions of this orbital *ephemeris* are sent via radio transmissions to GPS receivers. See GPS Space Segment, GPS User Segment.

GPS Navigation Data Satellites in the Global Positioning System (GPS) send out two microwave carrier signals, one of which provides navigation information in the form of a series of time-lagged data

G

frames sent over a specific time period. Subframes are also included for checking data integrity. The satellites are equipped with atomic clocks, and clock data parameters are sent and related to GPS time. Orbits are described by transmitting regularly updated ephemeris data. See Global Positioning System, Universal Coordinated Time.

GPS Operational Constellation The system of over 20 more-or-less evenly spaced, orbiting satellites (some of which are spares), equipped with atomic clocks, in the Global Positioning System (GPS) Space Segment. These satellites orbit the Earth twice a day at about 11,000 miles altitude, transmitting information used in the U.S. Department of Defense's Global Positioning System. The orbital planes are inclined at about 55 degrees in relation to Earth's equatorial plane. From any one point on Earth, it is generally possible to locate between five and eight satellites, four or five of which are typically used to compute location and timing information. The satellites transmit two microwave carrier signals; the L1 frequencies carry the navigation message (with data describing the orbit and clock parameters) and SPS code signals, and the L2 frequencies monitor ionospheric delay of PPS receivers. See Global Positioning System.

GPS receiver/display A fixed or mobile Global Positioning System (GPS) device that interprets GPS information and computes graphics, text, locations, maps, or other displays that provide the user information about position, time, and sometimes velocity. A graphical display of latitude and longitude is common. Receivers vary from room-sized systems to small handheld units from $180 up to hundreds of thousands of dollars. GPS consoles are used by surveyors, have been combined with map databases to provide car consoles, and have also been incorporated into 'smart cars' that can steer themselves. It is not unrealistic to predict that someday small GPS systems will be designed into wristwatch-style personal locators for travellers, sales representatives, hikers, et al. See Global Positioning System, Intelligent Vehicle Highway System.

GPS Space Segment A general overall category of the GPS system that consists of GPS satellites deployed and administered by the U.S. Department of Defense as part of its Global Positioning System. See GPS Control Segment, GPS User Segment.

GPS translator A Global Positioning System translating capability to support technologies that require precise positioning and trajectory tracking data. This type of data is of importance to missiles and rockets and other high-velocity acceleration devices. Translator GPS systems support midcourse corrections, safety operations, and downrange tracking. They are a cost-effective option to certain radar tracking systems and a higher-accuracy option to portable GPS receiver systems.

A vehicle or interceptor device installed with a GPS translator captures the GPS signals, translates them to an appropriate communications frequency, and transmits them to a ground-based station for further processing. The use of ground-based stations also makes it possible to store the data in larger storage media for subsequent playback and analysis, a capability that is impractical in most on-site GPS devices.

IEC is a commercial vendor of GPS translators that has been providing and developing the technology since the 1980s. It has supplied translator systems to both the U.S. Navy and the U.S. Air Force. IEC has recently introduced a family of digital GPS translators that can support encryption for greater security, and it is developing a smaller, more powerful Translated GPS Range System (TGRS), in cooperation with the Air Force.

GPS User Segment A general overall category of the GPS system that includes GPS receivers and users of the U.S. Department of Defense's Global Positioning System. See GPS Control Segment, GPS Space Segment.

Grade 1 to 5 twisted pair See twisted pair cable.

Grade B signal A radio-frequency broadcasting signal defined by the Federal Communications Commission (FCC). It is a measure of the strength of a television broadcasting station's signal at a specific location. The purpose of the rating is to define minimum acceptable standards of quality for viewing purposes in the context of defining whether a specific subscriber is served or unserved by the signal.

When the Satellite Home Viewer Improvement Act (SHVIA) was passed in 1999, it included a requirement that the FCC re-evaluate the Grade B signal standard to determine if it should be changed or replaced with something else for determining whether a household is unserved (unable to receive a signal of an acceptable level). Since it is impractical to go out and measure radio signal strengths in the vicinity of every subscriber in the U.S., the FCC created a computer model for satellite companies and television stations to predict whether a given household is served or unserved. This computer model went through several versions in order to include the effects of buildings, terrain, and land cover variations that could impede radio-frequency signals. If the predicted model is under dispute, a person may request a waiver from local TV stations serving the area. If the waiver is granted, the person becomes eligible to receive distant signals.

If the parties cannot agree whether a subscriber is served or unserved, the American Radio Relay League (ARRL) has been designated as an independent, neutral entity for arbitrating and designating the party, with input from the satellite provider and TV station, to conduct a field test of signal strength.

grade of service GoS. A service level indicator evaluated on an industry basis according to the type of service provided. In some industries a hierarchical category scale is applied to various levels or definitions of service. In telecommunications, grade of service is typically described in statistical terms related to the speed and probability of connecting, and the characteristics of the connection, etc. See class of service, quality of service.

graded index GI. A designation for a fiber waveguide

that has had its propagation characteristics for certain wavelengths altered by the deliberate inclusion of dopant materials in the fiber. Thus, undesirable effects such as dispersion can be mitigated in multimode fibers (especially those for longer distances) by selectively controlling the speeds at which the light beams travel within certain portions of the fiber. Beams that reflect in the lightguide at steeper angles normally would take longer to traverse the distance than beams that have a "straight shot" through the guide. However, by gradually changing the index of refraction of the lightguide out towards the outer diameters to enable faster propagation in this region, it is possible to compensate for the longer travel path. GI fiber operates in the O-band region around 1300 nm. See dispersion, light speed, stepped index.

gradient Gradual change in elevation, color, or texture along an axis. Gradual blend or transition. See gradient fill.

gradient fill A common feature of paint programs that allows the user to fill a defined area with graduated tones ranging from one specified color or shade of gray to another. The number of colors in the palette and the two end-tones selected will affect the smoothness and visual appeal of the transition, with more tones generally creating a more pleasing effect. Radial fills can be used to simulate 3D surface areas, as lighter areas appear as highlights.

Graham Act A U.S. 1921 act in which telephone companies were granted exemptions to the provisions of the Sherman Antitrust Act. It enabled AT&T, especially, to expand and exert further monopolistic control over the telephone networks. See the Kingsbury Commitment, Modified Final Judgment.

Gramme, Zénobe Théophile (1826–1901) A Belgian engineer who emigrated to France, Gramme developed a direct current (DC) generator, featuring a ring armature in 1869 and 1870. Together with Hippolyte Fontaine, Gramme opened a factory called Société des Machines Magneto-Electriques Gramme. In 1873, at the Vienna Exposition, it was noticed by a mechanic that an electrical connection from another generator could power the armature of the first generator, thus exhibiting the characteristics of a motor. It was an important historical advancement in industrial and transportation technologies.

grandfather clause A previously existing object, structure, statute, ownership right, or policy that may continue despite subsequent restrictions or regulations that would prevent its creation or continuance. A grandfather clause grants a type of pardon, special permission, or immunity. For example, military surplus purchased by a civilian in the 1960s may be regulated in the 1990s such that similar items may not be purchasable by current civilians (such as radiation bunkers). If the ownership is protected by a grandfather clause, when restrictions are imposed or reinstated, current civilian owners might not have to give up the property (but also may not be able to sell it, except perhaps back to the government).

Building codes are often subject to grandfather clauses. If you purchase a house built in 1920, it may not be subject to the same offset, materials, or safety regulations as current structures.

Voting and immigration laws have certain grandfather clauses. Immigrants to the country prior to a certain date do not require the same documents and eligibility requirements as later immigrants.

In telecommunications, phones and various electronic components built or installed before a certain date may not have to meet all current Federal Communications Commission (FCC) regulations.

grandfathered in Instated or installed before certain restrictions or regulations were put in place that would otherwise prevent creation, installation, or operation. See grandfather clause.

graphechon A special-purpose memory electron tube used in computer and radar applications. The graphechon can store an electrical charge pattern, similar to the functioning of an iconoscope, and recover the pattern at different scanning rates.

graphic equalizer A component providing a set of controls for adjusting the tonal qualities at several frequencies in an audio system, usually a music system. The equalizer is not a stand-alone component; it works in conjunction with other components such as receivers, phonographs, tape players, CD players, etc. It frequently has a series of vertical analog sliders for making individual adjustments.

Graphical Kernel System GKS. An official standard for 2D graphics in the mid-1980s, evolved from the Core. A 3D extension was subsequently developed, and GKS-3D became a standard in 1988. See Core, PHIGS.

graphical user interface GUI. A way of facilitating communication between a human and a device, usually a computing machine, by presenting the information in the form of visual metaphors. A graphical user interface works in conjunction with a variety of physical input devices, including speech recognition hardware, mice, keyboards, stylus pens, touchscreens, and joysticks. They provide a means to select and control the various visual elements, which commonly include menus, drag bars, buttons, icons, and window gadgets. Video games like Pong were early electronic adaptations of simple GUIs. Many of the earliest applied GUI ideas in general use today were developed at Xerox PARC and incorporated by Apple Computer into the Macintosh operating system.

graphics accelerator A chip or circuit board integrated into a computer system to relieve the CPU of some of the functions related to the processing and display of graphics. Graphics tend to be computing intensive, and sharing the load can significantly speed the display and refresh of images. Graphics accelerators are often sold as peripheral cards that can be plugged into a slot. See graphics coprocessor.

graphics controller, graphics display processor Specialized computer hardware to improve raster displays by taking some of the load from the CPU. Graphics controllers can speed up scan conversion, and the composition, display, and movement of graphics images and primitives. See cathode-ray tube, frame buffer.

graphics coprocessor A chip designed to speed computer graphics composition, display, or refresh by sharing the load with the system CPU. Coprocessors are sometimes designed for very specific tasks, such as updating a screen, or storing and displaying graphics primitives, hardware sprites, and the like. Unlike graphics accelerators, which are often sold to consumers as optional system-enhancing peripherals, graphics coprocessors are more commonly sold integrated into the system, often on the motherboard. See graphics accelerator.

Standardized Recording/Playback Formats

VHS
Super-VHS
Hi-8mm
CD+G
laserdisc

The variety of standardized recording and playback media for graphics and multimedia products is increasing. Some of the more common formats supporting both images and sound are shown here. It may be that DVD-related formats, which hold more information in less space, will eventually supersede most or all of the above technologies.

---

graphics device interface GDI. Physical and virtual connections between graphics hardware components and the computer CPU. Since graphics applications tend to be CPU intensive, it is very common for other graphics hardware (accelerator cards, frame buffers, blitters, etc.) to be incorporated into a system to facilitate the fast creation, display, and refresh of images on a variety of output devices.

graphics engine The part of a computer architecture supporting the graphics functions of the machine, particularly graphics composition, buffering, display, and fast refresh. Graphics engines are typically designed to handle many of the functions in hardware, so there is a minimum of on-the-spot software calculations. Enhanced graphics standards and graphics engines are being developed to support features such as realtime animation; hardware pan, zoom, compression/decompression; instant resolution-switching; and video signal support.

Graphics Environment Manager GEM. One of the first graphical user interfaces (GUIs) developed by Gary Kildall's Digital Research, the same company

that created the popular CP/M text-based operating system in the 1970s. GEM was first demonstrated publicly at the COMDEX computer industry trade show in 1983 and shipped a few months later. The interface greatly resembled the Macintosh interface that Apple had developed after observing development research at the Xerox PARC laboratories. GEM did not become widely distributed, with the exception of providing a front-end to Xerox's Ventura Publisher, a desktop publishing programming that was widely used for documentation page layout on Intel-based microcomputers in the later 1980s.

Graphics Interchange Format GIF. (GIF ought to be pronounced "gif" given that the G stands for "graphics," but its author apparently uses "jif.") GIF is a proprietary raster graphics format introduced by CompuServe, Inc. in 1987. It is an 8-bit graphics format developed with the patented Lempel-Ziv-Welch (LZW) compression, whose implementation requires a royalty agreement from Unisys Corporation. The level of compression varies with the type of image and number of colors, but 3 or 4 times compression ratios are common on a typical color image. Due to patent issues, CompuServe agreed in 1994 to secure a license agreement to distribute the LZW technology and issued the Graphics Interchange Format Developer Agreement to provide software developers permissions under CompuServe's software license agreement with Unisys.

GIF is particularly suitable for images that have a small number of distinct colors, as opposed to images that have a great variety of subtle color changes. It also handles line art, grayscale images (through color palette gray matching), and sharp color boundaries better than formats optimized for other characteristics. Because GIF is a 256-color format rather than a 24-bit color format (~1.6 million colors), 24-bit images will be dithered and adjusted and may not fully satisfy the needs of the user.

GIF will support transparency, which is sometimes desired in order for a background image to be displayed behind the GIF image or through parts of the GIF image as though there were holes. Transparency is often used by Web designers to produce special effects in Web pages, such as buttons with irregularly shaped edges.

GIF is one of the three most common graphics formats supported by World Wide Web browsers, the other two being PNG and JPEG. PNG is an open, nonproprietary format, developed to supersede GIF. In January 1995, CompuServe announced the GIF24 project for designing a replacement format for the original 8-bit GIF, and a month later officially announced that Portable Network Graphics (PNG) would be used as the basis for GIF24.

Support for GIF users is provided on the CompuServe Graphics Support Forum (GO GRAPHSUPPORT). See Lempel-Ziv-Welch, Portable Network Graphics.

graphics library GL. The writing of graphics routines for computers is time intensive and specialized. For that reason, many companies decide to purchase graphics libraries rather than to write their own. These library routines consist of a collection of commonly used graphics primitives and actions (lines, circles, dots, fills, patterns, etc.) that can be dynamically called from the graphics library or compiled and linked into the software executables as needed.

graphics mode A setting on dual-display mode systems (typically older IBM-compatibles) that permits the access and display of individually addressable pixels for the rendering of images and *graphics characters*. Some systems distinguish between text and graphics modes and will display only in one mode or the other. With faster processors, the trend is toward the more flexible graphics modes, with graphics characters. This frees the user from having to select a mode and switch between them.

Graticule Examples

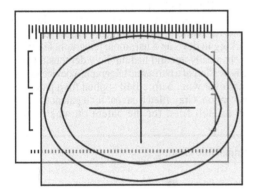

Graticules are overlay templates used for a wide variety of alignment and calibration purposes, especially in scientific viewing and measuring instruments, video titling, and the evaluation of video images. They are also used to display and analyze field positions in videotapes of team sports.

graticule graphical reticule 1. A diagnostic and measurement overlay screen used with cathode-ray tubes (CRTs). The screen is calibrated and placed on the front of the tube, with the tube image showing through so the relationship of the displayed image to the screen can be observed. A graticule may also be used with a spectrum analyzer, for analyzing bandwidth, with the analyzer calibrated to a specified number of graticule units. 2. A microscopic component incorporated into some eyepieces that provides an arrayed reference field for observation and count estimates of fibrous or particulate specimens (e.g., asbestos fibers). In this context, it is often called a reticule. It is useful in fabrication, contamination testing, and quality assurance. 3. A reference grid or outline image superimposed over a still or moving image to aid in analyzing or understanding the image or making sure it is within certain tolerances (e.g., the correct aspect ratio). This is especially useful for mapping, video element alignment (to avoid undesired clipping of titles or images), and video transfer to other media with the increasing variety of differ-

ent video formats (NTSC aspect ratios, widescreen, picture-in-picture, etc.). A graticule is the same basic concept as a reticule, but is more often used to assess recorded images, as in cartography or video, as opposed to real specimens as might be viewed with scientific devices, especially microscopes and telescopes. See graticule library, reticule.

graticule library A set of predesigned graticules for use as reference, calibration, or counting templates for superimposing over an image or text. A simple example of a graticule is a reticular grid on a piece of mylar that is placed over a picture to provide reference marks for hand copying the picture. More sophisticated computerized graticule libraries are stored templates in a variety of sizes, shapes, and configurations that can be digitally superimposed over video images or magnified viewing fields. For example, a graticule with reference marks for scale and for indicating the outlines of an antarctic icefield as it looked in 1970, could be superimposed over a satellite image of the icefield as it looks today. User-designed graticule templates could be included as part of a graticule library database for custom applications or future use. See graticule.

grating A series of narrow ordered slits or grooves specifically designed and oriented so they reflect electromagnetic waves in a spread or concentrated pattern in a desired direction. For some types of gratings, the period of the grating (the distance from one facet to the next) can be "tuned" to selectively admit or reflect optical wavelengths. Other types of gratings depend upon the shape of the corrugations to control light in a particular diffraction order. Thus, gratings are useful as filtering mechanisms. Antenna reflectors sometimes incorporate grating designs.

Gratings have existed for centuries for a wide variety of purposes. However, most grating components intended for use in scientific instruments have been developed since 1900. Diffraction gratings as they are used in modern components have been manufactured in a number of ways since the 1960s. They are now regularly incorporated into fiber optic filament and semiconductor components to act as lightwave filters. Gratings can be produced mechanically in a variety of materials or photographically in layered films. A photographic *interference grating* is a recording of a stationary interference fringe field (sometimes called a holographic grating).

Since gratings are often fabricated on thin films with fine tolerances and densities, it is necessary to match the light source with the structure. It may be necessary to lower the light intensity or increase the thickness of the grating and associated component layers in order to prevent the light (which is often laser light) from melting the grating. See blazed grating, Bragg grating, diffraction, échelle grating, Fresnel lens, interference grating, ruled grating.

grating arrow In scientific and engineering diagrams illustrating grating components with asymmetric corrugations, an indication of the orientation of the grating. For example, in a sawtooth-shaped grating (e.g., a blazed grating), the arrow would be oriented from the peak to the more distant trough (the one farthest from and connected to that peak) along the grating normal (typically illustrated along the horizontal [X] axis).

grating equation In the context of diffraction gratings, an equation that enables the ordered angles of reflected incident light to be calculated and modeled. Thus, $\sin\theta_m = \sin\theta + m\lambda/d$ with q representing the light source's incident angle upon a diffraction grating, at a wavelength of l and a grating period (the distance from one corrugation to the next) of d, q_m represents the diffraction angle such that the sin of the diffraction angle is equal to the sin of the angle of incidence plus the incident wavelength divided by the grating period. The structure of diffraction gratings and their effectiveness is in part due to the "tuning" of the angles to interact in desirable ways with specific wavelengths. See diffraction, grating, Wood anomaly.

grating, fiber optic A series of grooves in a fiber optic filament intended to control the propagation of light as it passes through the grating. For example, a grating can help control the direction of the lightwave, the wavelength(s) of the propagated light, or both. The grating is "etched" or "written" into the fiber and the fiber may be strained at the time the grating is written. Multiple overlapping gratings may be used to further condition the propagating lightwave. Long-period gratings may provide phase-matched coupling from transfering power from one optical mode to another within a fiber (e.g., between guiding and cladding modes). Permanent long-period gratings are typically laser written through an amplitude mask. Adjustable long-period gratings may be mechanically written through pressure that produces nonpermanent microbends.

Fiber gratings are used in sensors, scientific instruments, and fiber optic communication networks. See Bragg grating, grating.

gravity cell, crowfoot cell A type of voltaic wet cell suitable for providing small currents at a constant electromotive force. It derives its name from the way the lower and upper chemical solutions (e.g., copper sulphate over zinc sulphate) align themselves in relation to each other.

Gray, Elisha (1835–1901) A physicist and inventor who developed many early telegraph technologies at about the same time Alexander Graham Bell was working to develop a harmonic telegraph. Gray was mechanically apt and had publicly demonstrated an early version of a harmonic telegraph, a device to send tones over wire, before Bell applied for a patent for his version. Gray filed a caveat for a patent the same day as Bell filed for the patent on what is now

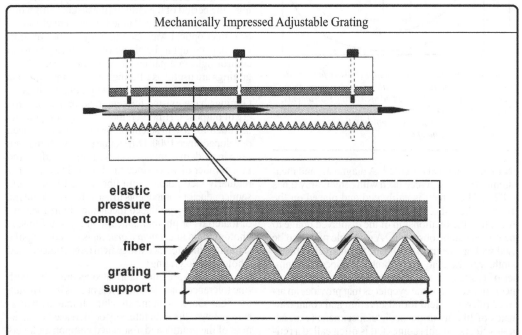

Mechanically Impressed Adjustable Grating

elastic pressure component

fiber

grating support

A long-period adjustable grating designed by L. Mollenauer of Lucent Technologies, Inc. provides an alternative to permanent gratings.

Instead of writing the grating into the structure of a single-mode fiber, as is traditional, the fiber is impressed into an external mechanical grating "form" to introduce microbends into the fiber that cause the propagating light to be conditioned for certain specified wavelengths. In other words, only selected wavelengths will escape the filtering process and continue propagating along the transmission path. By introducing a series of impressed gratings into the path, selectivity can be further controlled. With this type of configuration, dynamic changes in the wavelength selectivity and gain equalization are possible. This may be practical for many situations where small grating periods (the distance between corrugations) are not required.

considered to be the first telephone, thought of as a speaking telegraph at the time.

In 1867, Gray developed a new telegraph relay instrument. In collaboration with his partner, Enos M. Barton, Gray organized the Western Electric Manufacturing Company in 1869 and expanded by buying out the Ottawa, Illinois Western Union offices.

Late in 1873, Gray reports having noticed different vibratory properties in human tissue (an idea pioneered by Meucci) and described the placing of a galvanometer in the circuit with a microphone for transmitting human speech through wires. This observation resulted in a patent application that was not accepted until three years later as Gray had to substitute animal tissue for human to satisfy the Patent Office. Gray filed a patent similar to Bell's only hours after Bell, and his company was later purchased by Bell.

In the 1880s, telegraphs that would transmit handwriting were developed and Gray patented a *telautograph* which could lift the pen between letters permitting more natural characters to be transcribed, and sold the rights to a company founded with his name. See telephone history.

Gray, Stephen (1666–1736) An English experimenter who authored an article for *Philosophical Transactions* in 1720, describing various investigations of attractive properties and light-producing properties of various "electrics." He discovered that a substance electrified by friction could pass this property to another substance. He enlarged on the prior work of Gilbert, demonstrating that Gilbert's "non-electrics" could conduct electricity from one body to another and could be electrified if insulated with a conductor.

His association with the Royal Society of London indicates that he was likely familiar with the work and writings of F. Hauksbee, and he continued some of the interesting lines of inquiry first investigated by Hauksbee.

In the late 1720s, Gray began a fruitful collaboration with Granville Wheler, a member of the Royal Society. Gray and Wheler were to discover that substances could be roughly divided into additional substances that readily conducted "electric virtues" and those that did not. See inductance.

Gray & Barton A telecommunications company established by Elisha Gray and Enos Barton in 1869 when Elisha Gray bought out George Shawk's interest in the partnership. The physical facility was an electric shop abandoned by the Western Union Telegraph Company. In 1872, it became Western Electric Company, which supplied components to the Western Union Telegraph Company and later became an exclusive manufacturer for the Bell System. See Graybar Electric Company, Inc.; Western Electric Company.

gray market product One of several types of products that are not fully endorsed by the manufacturer. Examples include a product sold by an unauthorized distributor; a product that might be second rate in some way, which normally wouldn't be sold by reputable dealers; or a return item that is promoted as being new.

Gray Telephone Pay Station Company A company formed to commercialize the rotary payphones that were common from about 1930 to the 1960s.

Graybar Electric Company, Inc. A spinoff of the Western Electric Company in 1925, this company handled electrical distribution. The name derives from the original founders of the Gray & Barton company, founded by Elisha Gray and Enos Barton in 1869. In 1928-1929, the employees purchased the company from Western Electric Company, and it is still one of the largest employee-owned companies in the United States. Graybar Electric continues to do business after more than sixty years, globally supplying almost a million different electrical and telecommunications products. See Gray & Barton, Western Electric Company.

grayline The region in any particular place where the Sun is rising or setting. This information is of interest to those trying to determine a radio signal transmissions path using the characteristics of the Earth's ionosphere to help propagate the signal. There are software programs containing extensive databases of thousands of cities, designed to generate a grayline chart and display a world clock to assist amateur and professional radio operators in sending and receiving radio communications. Some grayline generators also include azimuth projection maps (Great Circle maps) which can help the user determine where radio signals may be arriving from different parts of the world, depending upon location and time of day. See ionosphere.

grayscale Visual information represented in shades of gray, i.e., with no color. In computing terms, many people confuse the terms *monochrome* and *grayscale*. Monochrome refers to one active color, whether it be white, black, green, or amber. Many older computer monitors were monochrome monitors. Grayscale refers to two or more (typically 16 to 64) shades of gray typically ranging between white and black. Grayscale monitors are less expensive than color and are very suitable for desktop publishing and other black and white and grayscale print-related applications.

grayscale monitor A monitor capable of displaying a variety of levels of light intensities (usually between 16 and 64), which are perceived as shades of gray.

grazing incidence An angle that is perpendicular or nearly perpendicular to a reference surface's *normal* such that an incident wave within the grazing incidence tolerances is reflected off the surface. For reflection at very low grazing angles that are almost parallel to the surface in an internal resonating cavity, a very highly reflective surface and short high-energy wavelengths (e.g., X-rays) are generally required. However, external resonators/reflectors have been used with wavelengths in the optical region. Assemblies with two adjacent dielectric materials with different indexes of refraction may exploit grazing angles to achieve total internal or external reflectance, depending upon the relationship of the conductive materials.

G

A grazing-incidence grating may be used in a mono-chromator.

A grazing incidence grating can be combined with a coated laser diode, laser cavity, and tuning component to provide sufficient signal dispersion for telecommunications frequency single-mode operation, provided the cavity length is balanced to the wavelength. See Littman-Metcalf configuration. Contrast with Littrow configuration.

great circle In geometry, an imaginary circle on the surface of a sphere that is defined as the intersection of the surface and a plane passing through the center of that sphere. See Great Circle.

Great Circle The name given to routes based upon Earth's spherical geometry, the Great Circle is a navigational concept to describe the shortest distance over the Earth's surface (assuming it was flat) between two specified points. Airlines use Great Circle routes to minimize travel distances, especially over long distances. The *polar aircraft route* from Vancouver, BC on the west coast of North America, which passes over Greenland and Iceland, to London or Amsterdam is roughly a Great Circle route. Great Circle geometry is also of interest to radio operators sending and receiving signals that are propagated through the Earth's ionosphere. See grayline, great circle.

Great Eastern A massive six-masted coal-powered paddlewheeler constructed in London between 1854 and 1858. The ship measured almost 700 feet in length and 120 feet in breadth. It was originally a passenger ship but was not financially successful in this role. Later it was purchased at auction by Daniel Gooch and his colleagues, Cyrus W. Field and Brassey, and outfitted for laying cable, a role that better suited the vessel. Large cable tanks were installed for storing and spooling the communications cable. In July 1865, the ship was used in an unsuccessful attempt to lay the transatlantic cable. A year later, the ship left Ireland with almost 3,000 nautical miles of cable in her hold, spooling more than half of it into the ocean over the next two weeks before docking in Newfoundland, Canada, and finally establishing the first permanent, successful telegraph cable to link the British Isles and North America. See Gooch, Daniel; transatlantic cable.

green 1. Young, inexperienced, naive, not hardened or aged. 2. Ecological, environmentally friendly, resource-conserving (as in power saver systems). Green products are represented as being kinder to the environment in terms of resource use, manufacture, materials (low in toxic materials or by-products), or operation than similar products by other manufacturers. See ISO 14000.

Green Book 1. The most technical of the three PostScript reference books produced by Adobe Systems, *PostScript Language Program Design*. See Blue Book, Red Book. 2. A standard Smalltalk reference *Smalltalk-80: Bits of History, Words of Advice* by Glen Krasner. See Kay, Alan. 3. A Compact Disc interactive (CD-I) standard that followed from the original Red and Yellow Book CD standards. Green Book standards, created by Sony and Philips, describe a hardware system and data recording format for multimedia disc recording that is suitable for entertainment programs.

green gun In a color cathode-ray tube (CRT) using a red-green-blue (RGB) system, the green gun is an electron gun specifically aimed to excite the green phosphors on the coated inside surface of the front of the tube. Sometimes a *shadow mask* is used to increase the precision of this process, so the red and blue phosphors are not affected, resulting in a crisper color image. See shadow mask.

green machine A physically robust computerized device or system designed and built to military specifications for field work.

Green Paper The popular name for a paper titled *A Proposal to Improve the Technical Management of Internet Names and Addresses* which was issued for comment by the U.S. Department of Commerce's National Telecommunications and Information Administration (NTIA) in 1998. See White Paper.

Green's theorem See Stoke's theorem.

green-gain control In a color cathode-ray tube (CRT), as in a television or computer monitor, a matrix resistor that can be varied to control the intensity of the green signal.

Greene, Harold H. (ca. 1924–2000) A German-born American Judge, Greene was a lawyer and prominent, highly respected member of the U.S. federal judicial system. He served the U.S. in military intelligence roles in World War II. In terms of telecommunications judgments, Greene is best known for the Modified Final Judgment (MFJ) and long divestiture proceedings leading to the breakup of AT&T in the 1980s. Judge Greene has also made important court decisions on Western Electric Company, Inc. (1991). See Modified Final Judgment.

Greenwich Civil Time GCT. See Coordinated Universal Time.

Greenwich Mean Time GMT. A time and geographic reference established by an agreement of 25 countries in 1884. Using astronomical instruments, a local time was established at the Greenwich Meridian in England, after which the world's regions referenced their time in relation to GMT. It is also known as Zulu time. GMT is normally expressed in 24-hour clock notation. The international standard for time for satellite communications and scientific research has since become Coordinated Universal Time, which is based on atomic rather than astronomical clocks. See atomic clock, Coordinated Universal Time.

Gregorian calendar See Julian calendar for the evolution of the Julian and Gregorian calendars.

grid A filtering structure, often used in vacuum tubes, that has the appearance of a grid or small set of blinds, and controls the flow of electrons from a cathode to an anode. Since this structure allows the flow of electrons to be manipulated, it may be called a *control grid*. A large part of electronics involves the harnessing of electrons through cathode rays, and grids are an essential control component in many devices.

GRID See Global Resource Information Database.

grid, reference A visual guide used in the background of drawing, painting, layout, and CAD programs to help align visual elements and objects. Sometimes a SNAP function will be available in conjunction with the grid, in which the objects will SNAP to designated positions on the grid, to align objects with mathematical precision that cannot be obtained visually. See graticule, reticule.

grid battery Within electron tubes, there is usually an electron flow element called a grid, which selectively controls the movement of electrons from the cathode to the anode. A grid battery supplies a bias voltage to the grid in the electron tube for this purpose. See electron tube, grid bias.

grid bias In an electron tube, a constant potential applied between the controlling grid and the anode to which the electrons are attracted. The grid bias, or *C bias,* is used to establish an operating point. A small cell may be used to supply voltage so the grid has a greater negative charge than other elements. See electron tube, grid battery.

grid cap At the top of some electron tubes there is a small cap that attaches to the controlling grid to act as a terminal. Sometimes a spring clip is incorporated into the grid cap to create the electrical connection.

grid modulation In a grid-controlled electron tube with a carrier signal, a voltage can be applied to modulate that signal in order to add information to the signal, as in a radio transmitter. See modulation.

grommet A ring-shaped insulator, usually made of plastic or rubber, used as a spacer inside air insulated cables or in panels which require wires to be strung through them, so the inserted wire doesn't touch the materials on the outside edge of the grommet.

grooming In network traffic management, the processing of signals by multiplexing and/or converting them to other wavelengths or modes. In SONET/SDH, the term is especially associated with combining lower-speed traffic (e.g., OC-12) with higher-speed traffic (e.g., OC-192) to reduce the need for extra switching operations. In SONET/WDM systems, it refers to the electronic multiplexing/demultiplexing of multiple low-rate tributaries associated with each wavelength in a multiple-wavelength transmission. The term is also more generically applied to the process of merging multiple low-bandwidth signals into a single time division multiplexed (TDM) signal in a variety of types of networks.

Given the variety and complexity of networks, grooming algorithms are not trivial to develop. The term is fairly recent, in this context, and the mathematics for generic and specialized grooming algorithms to handle high-speed multiple-channel transmissions is still largely in the development stage.

groove An indentation in a recording medium, often cylindrical or platter-shaped, that has minute variations which encode information. By creating both horizontal and vertical variations, it is possible to encode two tracks of information, as in a stereo vinyl record. Optical media usually store information in pits rather than grooves. See phonograph.

ground 1. The surface of the Earth. A large conducting body, such as the Earth, that provides a destination for electrical current. 2. A conductor that makes a connection with the Earth, through which power can drain. *v.* To put in or place in contact with a ground, such as the Earth or a conductor in contact with a ground. To make an electrical connection with a ground. To provide a path through which an electrical current will drain to the ground, such as a lightning rod on a house. A ground is usually established as a safety precaution to direct unwanted or unanticipated electrical charges away from areas where it might cause harm to structures or beings. On an ocean-going vessel, where the ground connection cannot be placed into the Earth or onto a structure connected with the Earth, a device can be grounded on the bed plate of an engine. 3. A voltage reference point in an electrical circuit. Although it may not actually be touching the ground, it is a reference point whose operation would not be changed if it were grounded to the Earth.

G

Utility Pole Grounding

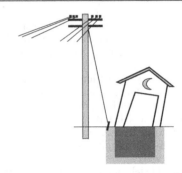

A long conducting spike pushed deep into damp soil makes a good ground for antennas and conducting structures. In the early days of telephones and telegraphs, grounding spikes were often driven into the moist ground near outhouses.

ground absorption A loss of transmission energy due to dissipation through the ground. Bounced airwave transmissions are particularly susceptible to ground absorption. Ground absorption may also be greater in regions of soft, uneven terrain.

ground button A button found on some electrical components and power strips to reset components that require a ground start after a power failure. See ground start.

ground clamp A device to connect a grounding conductor to a grounded object. Water pipes are commonly used as grounding objects, as are long metal spikes (often used near phone installations). At the turn of the century, when grounding pipes were not readily available, it was common for a telephone service ground wire to be pushed into the Earth in the damp ground in the vicinity of an outhouse.

ground junction In semiconductor fabrication, a ground junction can be formed by growing a crystal from a melt. See semiconductor.

ground lead A ground to which other conductors are attached so the ground lead can direct unwanted electrical current to the ground (usually the Earth). Heavy metal spikes are sometimes used as ground leads under external phone connection boxes.

ground noise Residual, usually low level, noise associated with a communications transmission in which no actual information is being transmitted, but which has low-level hisses or hums associated with the transmission devices or media. This type of noise also occurs in many analog audio recording technologies.

ground potential A reference potential associated with the Earth at a particular location. The ground potential at that position is considered the zero potential, and other potentials are referenced against it as a baseline.

ground return 1. A lead to the ground at the end of the circuit, for the return of a signal. 2. A type of circuit employing the ground as a return to complete the loop. The characteristics of ground circuits were discovered somewhat by accident by early experimenters, particularly when long telegraph cables began to be strung. It was discovered that it was possible to send signals along a single wire, as long as each end of the connection had contact with the ground, so the completion of the circuit happened through the ground rather than through a returning (second) wire.

ground scatter propagation A means of propagating radio waves through a series of hops between the Earth and the ionosphere, rather than following a great circle path. When the signal returns from the ionosphere to the Earth, contact with the terrain scatters it broadly in many directions. See ground wave, ionospheric wave.

ground start In telephony, it is necessary to take control of a line before it can be used. There are two common ways in which to do this, with a *ground start* or a *loop start*. The ground start is the type commonly found in business and other multiple line phones. When you pick up a phone, the plunger is released (off-hook) and the station detects a grounded circuit through the *ring* conductor. This is done so that transfer can be directed to the central office or main switching panel, if desired. See loop start.

ground state A reference descriptor, used to describe the lowest state or energy level of a system, as an atomic system.

ground station An Earth-based station (although the station may actually be located off the ground in a ground-based tower) used for sending, receiving, processing, or relaying communications signals. The term *ground station* typically describes services that are partly air-based or space-based, such as satellite communications systems. Traditional broadcasting stations are not usually called ground stations because they bounce signals from ground to ground or from ground to ionosphere to ground, without passing through a space transponder, relay, or other sky-based node. Ground stations may be primary senders/receivers or ground hubs, which relay information or strengthen signals. See M hop, satellite.

ground wave A transmitted radio wave that stays close to the ground. Radio waves travel in various directions from the point of transmission, some moving out through the ionosphere, others toward the ground. Ground waves are affected by the surface composition and topography of the surface over which they travel. Very rough or heavily vegetated terrain will interfere with the transmission of ground waves, while smoother surfaces, such as plains or calm waters, may permit transmission for hundreds of miles. See ionospheric wave, radio.

grounding strap A bracelet-like material or component usually worn on the wrist, that is common in the electronics assembly and repair industries. The strap prevents discharge from the hands that might damage static-sensitive components. If you must touch electronic components (as when adding memory to a computer), use a grounding strap or, at the very least, touch a ground such as the power supply, and then install the chips or boards without moving too much or shuffling your feet in the carpet. See ground.

Group 1 to 4 See facsimile formats.

group address A single address that is a logical name for a list of addresses. It may refer to multiple mailing lists, multiple devices, multiple users, or multiple receivers. A group address is used for management simplicity to provide a single reference point for a group of information.

Group Asynchronous Browsing GAB. A means to improve and evolve information search and retrieval data mining on the Internet through the World Wide Web and existing community-based browsing by groups and individual Web users. GAB was introduced by Wittenburg, Das, Hill, and Stead of Bellcore in the late 1990s. GAB includes a resource discovery utility called WebWatch, that draws the users' attention to changes in known documents.

group busy tone In telephone trunks, sometimes the system is at capacity and cannot route any additional calls until volume decreases. In this case, a group busy tone (a fast busy) may be sent out to those attempting to place a call. See fast busy.

group hunting In telephony, the process of searching for available lines in a designated group of trunks. See hunting.

group index, group refractive index (*symb. – N*) In fiber optics, a mathematical relationship between a given mode, the refractive index of the lightguide, and the velocity of light (in a vacuum). For a plane wave at a selected wavelength (λ), the group index equals the phase index (n) of λ minus λ ($\Delta n/\Delta \lambda$). See index of refraction.

group modulation The process of shifting or collectively modulating a set of signals that have already been individually modulated or multiplexed. It is often done to achieve a block frequency shift. Such frequency shifts may be done to bring a signal into a range that can be handled by the equipment or may reduce interference between incoming and outgoing signals in a repeater or relay circuit.

Grout, Jonathan, Jr. An inventor and entrepreneur who established one of the first marine reporting telegraphs in North America, which extended from Martha's Vineyard to Boston. It was an optical telegraph, a type that had been used in France and other European nations but, up to this time, was not definitely established in North America. The U.S. Patent Office recorded a patent issued to Grout in October 1800. Unfortunately, the original patent document was apparently lost in a fire in 1836. The service established by Grout, presumably based upon the patented system, operated during the early 1800s but went into bankruptcy by 1807. The old signal station still remains.

Grove battery A historic, dependable zinc/sulphuric acid and platinum/nitric acid primary cell normally used on closed circuits. The Grove battery provided 1.96 volts per cell. This type of battery was used by Morse in his early telegraph systems.

growler An electromagnetic circuit diagnostic and magnetizing/demagnetizing tool that emits a low growling sound when a short circuit is detected.

GRSU 1. Generic Remote Switch Unit. 2. Geographic Remote-Sensing Unit.

grunt *colloq.* A telephone pole and line installation crew member who works on the ground (as opposed to one who climbs the poles). Grunts are also commonly called groundmen or groundworkers.

GS trunk ground start trunk. See ground start, loop start.

GSA 1. General Services Administration. 2. See Global Standards Collaboration.

GSM See Global System for Mobile Communications.

GSMP See General Switch Management Protocol.

GSN Gigabyte System Network. See Hippi-6400.

GSO geosynchronous orbit. See geostational orbit.

GSOC See German Space Operations Center.

GSSAP See Generic Security Service Application.

GSTN general switched telephone network. A public switched telephone network (PSTN).

GTE Corporation Formerly General Telephone and Electronics Corporation, GTE is a major international telecommunications provider that originally supplied basic local telephone services but now includes long-distance, wireless, airline services, online directories, Web, and video services. It is building a national private coast-to-coast data network in the U.S. and wireless paging systems overseas. In 1997, GTE acquired BBN Corporation, an end-to-end Internet provider, and Genuity, Inc. to broaden its base of new telecommunications technologies.

GTN See Global Transaction Network.

GTP 1. general telemetry processor. 2. See Generalized Trunk Protocol. 3. Geophysical Turbulence Program. A scientific research program established through NCAR in the early 1960s. 4. Global Thinking Project. A Web-based educational environmental science project. http://www.gtp.org/ 5. GNOME Translation Project. A primarily volunteer effort to translate GNOME software applications and documentation to every known language. See GNOME. 6. Green Transport Plans. An employee travel plan service of the U.K. Department for Transport, Local Government, and the Regions.

GTT See Global Title Translation.

guard arm 1. A crossbar placed over wires, running in the same direction as the wires, to prevent contact with debris, people, animals, or other wires. 2. A wood or metal extension to prevent or selectively permit access (e.g., into a restricted facility). Horizontal guard arms commonly pivot to a vertical position to permit access and may be triggered by motion sensors or signals from automated password or key card systems.

guard band 1. A narrow broadcast bandwidth safe region interposed between communications channels in order to minimize interference between adjacent channels. Guard bands are particularly prevalent in frequency multiplexed systems, where maximum use of available bandwidth is achieved by dividing the available frequencies into smaller channels. 2. A safety zone in a circuit or chip, around the active portions of the circuit, to prevent electromagnetic interference with adjacent circuitry.

guard circle The smooth, ungrooved, inner portion of a phonograph record, or other revolving storage medium, protecting the stylus from moving into the center post and being damaged.

guard wire A wire near live wires, such as on utility poles, positioned so that if the live conducting wires break or fall, they will come in contact with the guard wire and be grounded, rather than causing danger.

guardian agent A pun on guardian angel in the sense that it is a software tool intended to protect innocent eyes from sites that some person in authority over the user deems unsuitable. With the vast and varied information on the Internet easily accessible through the World Wide Web, some parents and teachers are concerned about the type of Web browsing children might attempt. Software developers have responded to this concern by developing tools to lock out specific known sites or to flag sites with particular characteristics. There is no completely reliable way to screen out all such sites on the Web, especially since there is no consensus on what people consider objectionable. A child could quite innocently search for the word 'beaver' for a biology project. The resulting hit list wouldn't be restricted to Canada's national animal; it would also list every slang sense of the term found on commercial sites, home pages, poetry pages, sex sites, and more. There is a trade-off between user access to search engines and the potential harm from stumbling over an unintended site. While society has become more liberal in recent years, there are sites that are not suitable for children that may be screened out by guardian angel agent programs.

A new source of problems not yet fully addressed by guardian agent programs is junk email (spam). Irresponsible vendors are distributing massive quantities of email selling black market pharmaceuticals and fraudulent products at tempting prices, as well as distributing explicit photos that are obscene by almost any standards. With current Web browsers and graphical email clients, pornographic images automatically pop up in the email reading program without the reader having to click any links in the message.

Thus, large colored photographs of bestiality, underage girls engaged in sex with adults, or images of women naked in change rooms (who didn't know they were being photographed by hidden cameras) are designed to pop up instantly on the screen in the email client window or a separate window, before the reader has a chance to assess the message and delete it. These images can be seen by anyone in the same room with the computer. Children and many teens do not possess the emotional maturity to deal with this type of information and some children are frightened by it. There is no guarantee that there will be an adult in the room to explain the images or allay their fears. Many school-aged children have email addresses listed on personal Web pages or school Web pages which make them automatic targets for undiscriminating junk email robots. Email filters and guardian agent programs do not always screen these messages because the senders are careful to avoid using words or subject lines that are automatically filtered.

While concern focuses on ways of locking out particular types of sites and figuring out ways to stem the flow of obscene or illegal email, another serious danger on the Net is somewhat overlooked. Innocent adults and children will sometimes mistake the personality of a person on the Web as being the "real" person. This is understandable, but naive. It is easier to misrepresent oneself or to cover up hostile or dangerous intentions on the Net than in person. This type of danger is more difficult to detect with a software agent than obscene or tasteless Web sites.

guarding 1. The incorporation of points in a circuit where excess current or leakage are drawn off. 2. The process of maintaining a circuit in its busy state for an interval after it has been released, in order to assure a minimum period of time elapses before the actual disconnect occurs.

Guericke, Otto von (1602–1686) A German engineer, inventor, and statesman, von Guericke is credited as the inventor of the air pump, in the mid-1600s. The pump design was later improved by experiment-

ers such as R. Boyle and F. Hauksbee. Guericke also did experiments to study the rotation of the Earth that he described in 1672 in *Experimenta nova Magdeburgia de vacuo spatio*. In the course of his research, he created a spinning model to simulate the Earth. Because Guericke noticed that a feather was alternately attracted to and repelled by the spinning globe, the model has been credited by some as the first frictional generator, though von Guericke did not specifically design it to create friction for electrical experiments. See Hauksbee, Francis.

GUI See graphical user interface.

guided wave An electromagnetic wave whose path is controlled or directed by a structure or process acting as a conduit, channel, or waveguide. Physical waveguides can be quite sophisticated and are mathematically related to the wavelength of the wave being guided. Reflected waves are a type of guided wave in the sense that the reflection may be carefully organized to channel the wave in the desired direction. For example, a parabolic receiving antenna dish reflects transmission waves into a feedhorn mounted at a specific distance from the dish. See waveguide.

guru Sage, wizard, admired expert.

gutta-percha A latex substance present in a number of Malayan evergreen trees, including the *Palaquium oblongifolium*. The trees are like high-resin rubber trees that can be girdled or felled to tap the milky substance inside. The substance is processed and molded for shipment.

Gutta-percha came to the attention of Westerners in the 1840s through samples sent to London by Pacific region travelers. Gutta-percha was a tough, natural rubbery substance that came to be used extensively for insulation, general manufacture, a variety of adhesives, and even golf balls, until the late 1940s. For a while, almost any moldable rubber-like material was referred to as gutta-percha and many of these materials were used in the manufacture of bindings, frames, molded containers, and costume jewelry. When synthetic plastics became commercially wide-

---

Fiber Optic Coupling Module Applicable to Gyroscopic Applications

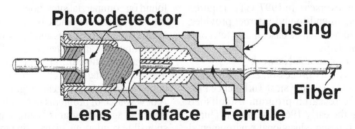

Goldner et al. have developed a modular system for interfacing a ferrule-terminated optical fiber to a photodetector. The fiber optic ferrule supports the fiber filament inside a protective, conductive housing. In this embodiment, the polished endface of the light-conducting filament faces a focusing lens supported against a cylindrical conductive material (e.g., metallic) that is soldered or epoxied into the housing. Assembled, the component comprises a modular photodetector housing with a fiber pigtail for attachment to a light source. The optional lens is suitble for applications such as a fiber optic gyroscope, which benefits from a light source with a small focusing area. [Diagram adapted from U.S. patent #6,422,765, submitted July 1999.]

spread, the use of gutta-percha declined. However, dental gutta-percha has had a useful life of more than 100 years, and it is still popular for root canal plugs and temporary fillings. Dental gutta-percha has recently been improved with the development of bacteria-resistant formulas.

Gutta-percha softens and stretches when warmed and hardens without becoming brittle when cooled. With the addition of stabilizers to protect it from oxidation, it can last longer in extreme environments than India rubber. This versatile material is historically important because it enabled the laying of communications cable in unfriendly environments, such as ground and deep sea installations. Gutta-percha was a significant factor in the success of the first transatlantic telegraph cable. See transatlantic cable.

guy wire A slender support line to brace and steady an apparatus that might sway or fall. Guy wires are often used in multiples, spaced around whatever they are supporting, and are frequently attached to narrow high structures such as transmission towers, aerials, masts, and poles.

Guy wires are usually thin, to minimize wind resistance. Consequently, they sometimes have small telltales, fine pieces of cloth or plastic attached to the wire, to discourage birds from flying into them or people from tripping over them. Guy wires are useful in for supporting structures that are exposed to surf or high winds.

GVNS global virtual network service (e.g., as in ISDN Q.85 community of interest services).

gyrofrequency The frequency at which charged particles naturally rotate under the influence of the Earth's magnetic field. The frequency varies with the type of particle.

gyroscope, gyro A device designed to maintain its axle in a constant vector while rotating. A gyroscope is typically designed to rotate through two axes that are perpendicular to the central structure and to each other. It is called a *gyrocompass* when it is oriented with the axle pointing northward.

Gyroscopes are becoming viable as an alternative to odometers for dead reckoning systems in mobile robots. This is partly because fiber optic gyroscopes (FOGs) are less subject to bias drift than traditional gyroscopes and are less prone to cumulative orientation errors.

gzip GNU zip, popular file compression program developed by Jean-Loup Gailly that incorporates Lempel-Ziv coding algorithms with 32-bit CRC. Gzip is widely used on Unix and IBM-compatible disk operating systems, especially for compressing/ decompressing files to save transmission times.

Gzip uses a deflate compression format derived from the freely distributable *zlib* source code distributed by Gailly and Mark Adler. Gzipped files may be unzipped by typing "gunzip [filename]" (the command must be typed all in lower case) at the command line, with relevant optional parameters. On a Unix system, see the "man" pages on gzip for more information on zipping, listing, and unzipping files. See compress, uuencode, zip, RFC 1952.

ITU-T G Series Recommendations		
General definitions and vocabulary		
G.100	1993	Definitions used in recommendations on general characteristics of international telephone connections and circuits
G.601	1988	Terminology for cables
G.701	1993	Vocabulary of digital transmission and multiplexing, and pulse code modulation (PCM)
G.780	1994	Vocabulary of terms for synchronous digital hierarchy (SDH) networks and equipment
G.810	1996	Definitions and terminology for synchronization networks
G.972	1997	Definition of terms relevant to optical fibre submarine cable systems
Transmission-related		
G.101	1996	The transmission plan
G.102	1988	Transmission performance objectives and recommendations
G.113	1996	Transmission impairments
G.114	1996	One-way transmission time
G.117	1996	Transmission aspects of unbalance about Earth
G.120	1988	Transmission characteristics of national networks
G.121	1993	Loudness ratings (LRs of national systems)
G.125	1988	Characteristics of national circuits on carrier systems
G.142	1988	Transmission characteristics of exchanges
G.171	1988	Transmission plan aspects of privately operated networks
G.172	1988	Transmission plan aspects of international conference calls
G.173	1993	Transmission planning aspects of the speech service in digital public land mobile networks
G.174	1994	Transmission performance objectives for terrestrial digital wireless systems using portable terminals to access the PSTN
G.175	1997	Transmission planning for private/ public network interconnection of voice traffic
G.221	1988	Overall recommendations relating to carrier-transmission systems
G.180	1993	Characteristics of N + M type direct transmission restoration systems for use on digital and analog sections, links, or equipment

G

ITU-T G Series Recommendations, cont.		

Transmission-related, cont.

G.181	1993	Characteristics of 1 + 1 type restoration systems for use on digital transmission links
G.671	1996	Transmission characteristics of passive optical components
G.712	1996	Transmission performance characteristics of pulse code modulation channels
G.773	1993	Protocol suites for Q-interfaces for management of transmission systems
G.801	1988	Digital transmission models
G.961	1993	Digital transmission system on metallic local lines for ISDN basic rate access

Echo-related

G.122	1993	Influence of national systems on stability and talker echo in international connections
G.126	1993	Listener echo in telephone networks
G.131	1996	Control of talker echo
G.164	1988	Echo suppressors
G.165	1993	Echo cancellers
G.167	1993	Acoustic echo controllers
G.168	1997	Digital network echo cancellers

Connection-related noise, distortion, jitter

G.103	1988	Hypothetical reference connections
G.105	1988	Hypothetical reference connection for crosstalk studies
G.111	1993	Loudness ratings (LRs in an international connection)
G.123	1988	Circuit noise in national networks
G.132	1988	Attenuation distortion
G.133	1988	Group-delay distortion
G.134	1988	Linear crosstalk
G.135	1988	Error on reconstituted frequency
G.141	1988	Attenuation distortion
G.222	1988	Noise objectives for design of carrier-transmission systems of 2500 km
G.223	1988	Assumptions for the calculation of noise on hypothetical reference circuits for telephony
G.226	1988	Noise on a real link
G.228	1988	Measurement of circuit noise in cable systems using a uniform-spectrum random noise loading
G.229	1988	Unwanted modulation and phase jitter
G.230	1988	Measuring methods for noise produced by modulating equipment and through-connection filters

G.441	1988	Permissible circuit noise on frequency-division multiplex radio-relay systems
G.442	1988	Radio-relay system design objectives for noise at the far end of a hypothetical reference circuit with reference to telegraphy transmission
G.823	1993	The control of jitter and wander within digital networks that are based on the 2048 kbps hierarchy
G.824	1993	The control of jitter and wander within digital networks that are based on the 1544 kbps hierarchy
G.825	1993	The control of jitter and wander within digital networks that are based on the synchronous digital hierarchy (SDH)

Software-related

G.191	1996	Software tools for speech and audio coding standardization

Optical submarine cable systems

G.971	1996	General features of optical fiber submarine cable systems
G.973	1996	Characteristics of repeaterless optical fiber submarine cable systems
G.974	1993	Characteristics of regenerative optical fiber submarine cable systems
G.975	1996	Forward error correction for submarine systems
G.976	1997	Test methods applicable to optical fiber submarine cable systems
G.977	2000	Characteristics of optically amplified optical submarine cable systems

Optical – various

G.911	1997	Parameters and calculation methodologies for reliability and availability of fibre optic systems
G.958	1994	Digital line systems based on the synchronous digital hierarchy for use on optical fibre cables
G.981	1994	PDH optical line systems for the local network
G.982	1996	Optical access networks to support services up to the ISDN primary rate or equivalent bit rates

Timing

G.811	1997	Timing characteristics of primary reference clocks
G.812	1988	Timing requirements at the outputs of slave clocks suitable for plesiochronous operation of international digital links

G.813	1996	Timing characteristics of SDH equipment slave clocks (SEC)
G.821	1996	Error performance of an international digital connection operating at a bit rate below the primary rate and forming part of an ISDN
G.822	1988	Controlled slip rate objectives on an international digital connection
G.826	1996	Error performance parameters and objectives for international, constant bit rate digital paths at or above the primary rate
G.827	1996	Availability parameters and objectives for path elements of international constant bit-rate digital paths at or above the primary rate

Connections, cable, and cabling

G.322	1988	General characteristics recommended for systems on symmetric pair cables
G.325	1988	General characteristics recommended for systems providing 12 telephone carrier circuits on a symmetric cable pair [12+12 systems]
G.421	1988	Methods of interconnection
G.422	1988	Interconnection at audio-frequencies cable pairs for analog transmission
G.612	1988	Characteristics of symmetric cable pairs designed for the transmission of systems with bit rates of the order of 6 to 34 Mbps
G.613	1988	Characteristics of symmetric cable pairs usable wholly for the transmission of digital systems with a bit rate of up to 2 Mbps
G.614	1988	Characteristics of symmetric pair star-quad cables designed earlier for analog transmission systems and being used now for digital system transmission at bit rates of 6 to 34 Mbps
G.621	1988	Characteristics of 0.7/2.9 mm coaxial cable pairs
G.622	1988	Characteristics of 1.2/4.4 mm coaxial cable pairs
G.623	1988	Characteristics of 2.6/9.5 mm coaxial cable pairs
G.631	1988	Types of submarine cable to be used for systems with line frequencies of less than ~45 MHz
G.650	1997	Definition and test methods for the relevant parameters of single-mode fibres

G.651	1993	Characteristics of a 50/125 μm multimode grades index optical fibre cable
G.652	1997	Characteristics of a single-mode optical fibre cable
G.653	1997	Characteristics of a dispersion-shifted single-mode optical fibre cable
G.654	1997	Characteristics of a cut-off shifted single-mode optical fibre cable
G.655	1996	Characteristics of a nonzero dispersion shifted single-mode optical fibre cable
G.661	1996	Definition and test methods for relevant generic parameters of optical fibre amplifiers
G.662	1995	Generic characteristics of optical fibre amplifier devices and sub-systems
G.663	1996	Application related aspects of optical fibre amplifier devices and subsystem
G.681	1996	Functional characteristics of interoffice and long-haul line systems using optical amplifiers, including optical multiplexing

Speech/Audio encoding and synthesis

G.115	1996	Mean active speech level for announcements and speech synthesis systems
G.192	1996	A common digital parallel interface for speech standardization activities
G.720	1995	Characterization of low-rate digital voice coder performance with non-voice signals
G.723.1	1996	Dual rate speech coder for multimedia communications transmitting at 5.3 and 6.3 kbps
G.724	1988	Characteristics of a 48-channel low bit rate encoding primary multiplex operating at 1544 kbps
G.725	1988	System aspects for the use of the 7-kHz audio codec within 64 kbps
G.728	1992	Coding of speech at 16 kbps using low-delay code excited linear prediction
G.729	1996	Coding of speech at 8 kbps using conjugate-structure algebraic-code-excited
G.764	1990	Voice packetization
G.802	1988	Interworking between networks based on different digital hierarchies and speech encoding laws

G

ITU-T G Series Recommendations, cont.		Modulation and multiplexing, cont.	

Modulation and multiplexing

G.711 1988	Pulse code modulation (PCM) of voice frequencies
G.726 1990	40, 32, 24, 16 kbps Adaptive Differential Pulse Code Modulation (ADPCM)
G.727 1990	5-, 4-, 3- and 2-bits sample embedded adaptive differential pulse code modulation (ADPCM)
G.731 1988	Primary PCM multiplex equipment for voice frequencies
G.732 1988	Characteristics of primary PCM multiplex equipment operating at 2048 kbps
G.733 1988	Characteristics of primary PCM multiplex equipment operating at 1544 kbps
G.734 1988	Characteristics of synchronous digital multiplex equipment operating at 1544 kbps
G.735 1988	Characteristics of primary PCM multiplex equipment operating at 2048 kbps and offering synchronous digital access at 384 kbps and/or 64 kbps
G.736 1993	Characteristics of a synchronous digital multiplex equipment operating at 2048 kbps
G.737 1988	Characteristics of an external access equipment operating at 2048 kbps offering synchronous digital access at 384 kbps and/or 64 kbps
G.738 1988	Characteristics of primary PCM multiplex equipment operating at 2048 kbps and offering synchronous digital access at 320 kbps and/or 64 kbps
G.739 1988	Characteristics of an external access equipment operating at 2048 kbps offering synchronous digital access at 320 kbps and/or 64 kbps
G.741 1988	General considerations on second-order multiplex equipment
G.742 1988	Second-order digital multiplex equipment operating at 8448 kbps and using positive justification
G.743 1988	Second-order digital multiplex equipment operating at 6312 kbps and using positive justification
G.744 1988	Second-order PCM multiplex equipment operating at 8448 kbps
G.745 1988	Second-order digital multiplex equipment operating at 8448 kbps and using positive/zero/negative justification

Modulation and multiplexing, cont.

G.746 1988	Characteristics of second-order PCM multiplex equipment operating at 6312 kbps
G.747 1988	Second-order digital multiplex equipment operating at 6312 kbps and multiplexing three tributaries at 2048 kbps
G.751 1988	Digital multiplex equipment operating at the third-order bit rate of 34,368 kbps and the fourth-order bit rate of 139,264 kbps and using positive justification
G.752 1988	Characteristics of digital multiplex equipment based on a second-order bit rate of 6312 kbps and using positive justification
G.753 1988	Third-order digital multiplex equipment operating at 34,368 kbps and using positive/zero/negative justification
G.754 1988	Fourth-order digital multiplex equipment operating at 139,264 kbps and using positive/zero/negative justification
G.755 1988	Digital multiplex equipment operating at 139,264 kbps and multiplexing three tributaries at 44,736 kbps
G.791 1988	General considerations on transmultiplexing equipment
G.792 1988	Characteristics common to all transmultiplexing equipment
G.793 1988	Characteristics of 60-channel transmultiplexing equipment
G.794 1988	Characteristics of 24-channel transmultiplexing equipment
G.797 1996	Characteristics of a flexible multiplexer in a plesiochronous digital hierarchy environment

Synchronous Digital Hierarchy (SDH)

G.774 1992	Synchronous Digital Hierarchy management information model for the network element view
G.774.01 1994	Synchronous Digital Hierarchy performance monitoring for the network element view
G.774.02 1994	Synchronous Digital Hierarchy configuration of the payload structure for the network element view
G.774.03 1994	Synchronous Digital Hierarchy management of multiplex-section protection for the network element view

Synchronous Digital Hierarchy (SDH), cont.

G.774.04	1995	Synchronous Digital Hierarchy management of the subnetwork connection protection for the network element view
G774.05	1995	Synchronous Digital Hierarchy management of connection supervision functionality (HCS/LCS for the network element view)
G.774.6	1997	Synchronous Digital Hierarchy
G.774.7	1996	Synchronous Digital Hierarchy management of lower order path trace and interface labeling for the network element view
G.774.8		Synchronous Digital Hierarchy management of radio-relay systems for the network element of view
G.775	1994	Loss of signal (LOS) and alarm indication signal (AIS) defect detection and clearance criteria
G.781	1994	Structure of recommendations on equipment for the Synchronous Digital Hierarchy
G.782	1994	Types and general characteristics of synchronous digital hierarchy equipment
G.783	1997	Characteristics of Synchronous Digital Hierarchy equipment functional blocks
G.784	1994	Synchronous digital hierarchy management
G.785	1996	Characteristics of a flexible multiplexer in a Synchronous Digital Hierarchy environment
G.803	1997	Architecture of transport networks based on the Synchronous Digital Hierarchy
G.831	1996	Management capabilities of transport networks based on the Synchronous Digital Hierarchy
G.832	1995	Transport of Synchronous Digital Hierarchy elements on PDH networks: Frame and multiplexing structures
G.841	1995	Types and characteristics of Synchronous Digital Hierarchy network protection architectures
G.842	1997	Interworking of Synchronous Digital Hierarchy network protection architectures
G.957	1995	Optical interfaces for equipment and systems relating to the Synchronous Digital Hierarchy

xDSL

G.991.1	1998	High-speed Digital Subscriber Line (HDSL) transceivers
G.991.2	2001	Single-Pair High-speed Digital Subscriber Line (SHDSL) transceivers
G.992.1	1999	Asymmetrical Digital Subscriber Line (ADSL) transceivers
G.992.2	1999	Splitterless Asymmetric Digital Subscriber Line (ADSL) transceivers
G.994.1	2001	Handshake procedures for Digital Subscriber Line (DSL) transceivers
G.995.1	2001	Overview of Digital Subscriber Line (DSL) recommendations
G.996.1	2001	Test procedures for Digital Subscriber Line (DSL) transceivers
G.997.1	1999	Physical layer management for Digital Subscriber Line (DSL) transceivers

General and miscellaneous

G.151	1988	General performance objectives applicable to all modern international circuits and national extension circuits
G.152	1988	Characteristics appropriate to long-distance circuits of a length not exceeding 2500 km
G.153	1988	Characteristics appropriate to international circuits more than 2500 km in length
G.162	1988	Characteristics of compandors for telephony
G.166	1988	Characteristics of syllabic compandors for telephony on high capacity long-distance systems
G.176	1997	Planning guidelines for the integration of ATM technology into networks supporting voiceband services
G.211	1988	Make-up of a carrier link
G.212	1988	Hypothetical reference circuits for analog systems
G.213	1988	Interconnection of systems in a main repeater station
G.214	1988	Line stability of cable systems
G.215	1988	Hypothetical reference circuit of 5000 km for analog systems
G.224	1988	Maximum permissible value for the absolute power level (power referred to one milliwatt of a signaling pulse)
G.225	1988	Recommendations relating to the accuracy of carrier frequencies
G.227	1988	Conventional telephone signal
G.231	1988	Arrangement of carrier equipment
G.232	1988	12-channel terminal equipment
G.233	1988	Recommendations concerning translating equipment

G

ITU-T G Series Recommendations, cont.

General and miscellaneous, cont.

G.241	1988	Pilots on groups, supergroups, etc.
G.242	1988	Through-connection of groups, supergroups, etc.
G.243	1988	Protection of pilots and additional measuring frequencies at points where there is a through-connection
G.411	1988	Use of radio-relay systems for international telephone circuits
G.431	1988	Hypothetical reference circuits for frequency-division multiplex (FDM) radio-relay systems
G.451	1988	Use of radio links in international telephone circuits
G.702	1988	Digital hierarchy bit rates
G.703	1991	Physical/electrical characteristics of hierarchical digital interfaces
G.704	1995	Synchronous frame structures used at 1544, 6312, 2048, 8488, and 44,736 kbps hierarchical levels
G.706	1991	Frame alignment and cyclic redundancy check (CGC) procedures relating to basic frame structures defined in Recommendation G.704
G.707	1996	Network node interface for the synchronous digital hierarchy
G.722	1988	7-kHz audio-coding within 64 kbps
G.761	1988	General characteristics of a 60-channel transcoder equipment
G.762	1988	General characteristics of a 48-channel transcoder equipment
G.765	1992	Packet circuit multiplication equipment
G.766	1996	Facsimile demodulation/remodulation for digital circuit multiplication equipment
G.772	1993	Protected monitoring points provided on digital transmission systems
G.795	1988	Characteristics of codecs for FDM assemblies
G.796	1992	Characteristics of a 64-kbps cross-connect equipment with 2048 kbps access ports
G.804	1993	ATM cell mapping into plesiochronous digital hierarchy (PDH)
G.805	1995	Generic functional architecture of transport networks

G.851.1	1996	Management of the transport network
G.852.1	1996	Management of the transport network
G.853.1	1996	Common elements of the information viewpoint for the management of a transport network
G.853.2	1996	Subnetwork connection management information viewpoint
G.854.1	1996	Management of the transport network
G.861	1996	Principles and guidelines for the integration of satellite and radio systems in SDH transport networks
G.901	1988	General considerations on digital sections and digital line systems
G.902	1995	Framework recommendation on functional access networks (AN) architecture and functions, access types, management, and service node aspects
G.921	1988	Digital sections based on the 2048 kbps hierarchy
G.931	1988	Digital line sections at 3152 kbps
G.941	1988	Digital line systems provided by FDM transmission bearers
G.950	1988	General considerations on digital line systems
G.951	1988	Digital line systems based on the 1544 kbps hierarchy on symmetric pair cables
G.952	1988	Digital line systems based on the 2048 bps hierarchy on symmetric pair cables
G.953	1988	Digital line systems based on the 1544 kbps hierarchy on coaxial pair cables
G.954	1988	Digital line systems based on the 2048 kbps hierarchy on coaxial pair cables
G.955	1996	Digital line systems based on the 1544 kbps and the 2048 kbps hierarchy on optical fibre cables
G.960	1993	Access digital section for ISDN basic rate access
G.962	1993	Access digital section for ISDN primary rate at 2048 kbps
G.963	1993	Access digital section for ISDN primary rate at 1544 kbps
G.964	1994	V-Interfaces at the digital local exchange (LE)
G.965	1995	V-Interfaces at the digital local exchange (LE)

h 1. *abbrev.* hecto-. See hecto-. 2. *abbrev.* horizontal. 3. *symb.* horizontal linear polarization (ITU).

H bend, H-plane bend A smooth transition in the orientation of the axis of an electromagnetic waveguide such that the axis remains parallel to the direction of the magnetic H-field polarization (transverse polarization). The H fields become distorted in a waveguide with an H bend. The degree of bend depends upon the frequency of the wave guided through the bend. The radius must exceed two wavelengths to prevent unwanted reflections.

H-bend rectangular elbows for radiowave communications typically have 90-degree bends with specific-sized openings through the flanges into the elbow. These are available to support a number of frequencies, including S-band and L-band (radar) communications.

H channel An ITU-T-defined transmission channel on packet-switched networks consisting of aggregated B channels (bearer channels), as are used on an ISDN system. See ISDN.

H drive In analog video, a periodic signal related to the horizontal component of a frame that is constructed with sequential, repeating line scans. The relationship between horizontal sync and vertical sync is such that the pulses can be combined on a single wire. Together they comprise a *composite* video signal. The H drive sends a short horizontal sync pulse during the horizontal blanking interval when the beam moves from right to left without tracing a line on the screen. A composite signal can be represented as Csync-red-green-blue and transmitted over four wires. Many computer monitors use a five-wire RGBHV system in which the H and V represent horizontal and vertical sync pulse components. See negative-going video.

H-plane bend See H bend.

H0 Channel In ATM networking, a 384-Kbps channel consisting of six contiguous DS-0s (64 Kbps) of a T1 transmission line.

hack *v.* 1. To quickly cobble together a program. 2. To create something quickly from available materials, a make-do solution, not necessarily elegant, although it could be, given limited resources. 3. To create a small, quick entertaining showpiece designed to illustrate a cool idea or interesting capability. See Schwabbie. 4. To seek to compromise or enter an area/process for which the person does not have authorization.

hack into To find a means of entrance other than the normal way, to compromise the security of a system by exploiting a weakness or lesser-known characteristic, to deliberately break into a computer system, network, or computer process without authorization.

hacker 1. A person who hacks into a system, i.e., gains entry by exploiting the hardware or software architecture through black boxes, stolen or guessed passwords, Trojan horses, design flaws, or back doors. Sometimes called *cracker* to signify someone using these techniques for illegal purposes such as cracking a password or serial number. See cracker. 2. A person who acquires a sophisticated, in-depth knowledge of a system and applies this knowledge to configuring or programming the system with a high level of expertise or complexity. An elite programmer, engineer, or technician. Two popular books on this subject are *Hackers* by Stephen Levy and *The Cuckoo's Egg* by Clifford Stoll.

Hacker's Dictionary, The An electronic and print dictionary that evolved from The Jargon File in the early 1980s. *The Hacker's Dictionary* was an expanded version of The Jargon File with added commentary, published by Harper and Row in 1983, edited by Guy Steele. The co-editors/contributors were Raphael Finkel, Don Woods, Mark Crispin, Richard M. Stallman, and Geoff Goodfellow. After nearly a decade in which it remained essentially unchanged, *The Hacker's Dictionary* was expanded beyond the artificial intelligence (AI) and hacker cultures to include terms from a broad variety of computers. The 1990s version, called *The New Hacker's Dictionary*, is maintained by Eric S. Raymond and Guy L. Steele, Jr. See Jargon File, The.

Hagelbarger, David W. (ca. 1921–) An American physict and engineer, Hagelbarger taught at Aeronautical Engineering until 1949, where he was also researching the use of analog computers for engineering applications. He then joined Bell Laboratories and became a colleague of Claude Shannon. Hagelbarger implemented some of the computer gaming ideas proposed by Claude Shannon in 1950, developing a penny-matching machine that was pitted against

Shannon's machine (long before computer-computer chess games became popular). Hagelbarger also co-authored articles on electronics with Shannon, worked on educational computer system concepts (e.g., CARDIAC), and developed data transmission error correction codes. In the early 1980s, he was co-developing experiments with remote computing terminals. See Hagelbarger code.

Hagelbarger code A form of burst error correction applicable to modems carrying communications data. Hagelbarger code is a *convolutional* or *recurrent* code in which up to 6 consecutive bit errors may be corrected, provided there are sufficient valid bits (at least 19 bits) prior to the error segment. It is named after Bell Laboratories researcher David W. Hagelbarger, who described the application of convolutional codes to burst correction in 1959, as well as developing circuitry to decode them. See Hamming code.

hairpin pickup coil A device with a one-turn coil, shaped like a hairpin, used for transferring ultra high frequency (UHF) energy.

hairpinning The routing of information or data through a switch in a main facility or network host and sending it out again through another switch or routing device.

HAL-9000 No computer-related dictionary would be complete without mention of the intelligent computer in the science fiction movie classic *2001: A Space Odyssey*. HAL stood for "*H*euristically Programming *A*lgorithmic Computer" and apparently the one-letter shift that spells out "IBM" was not intentional, or so say the makers of the film. If not, it's a strong enough coincidence to create an apocryphal legend.

HALE See High Altitude Long Endurance.

half duplex In a circuit, one-directional transmissions. Often half duplex circuits can transmit in either direction, but not simultaneously. Many systems which technically have bidirectional capabilities are operated in half duplex mode to reduce interference and echoes. Modems, satellite voice lines, some cellular radios, and speakerphones are often used in half duplex mode.

half-life A property of radioactive decay used as a quantitative measure, of interest to many different branches of science. Radioactive decay happens at widely differing rates for different materials, so *half-life* is not a fixed measure, but one based on our knowledge of the properties of the materials being described. The first half-life of a substance is the interval during which half the radioactive material is left unchanged. The second half-life is the next interval, during which half of the *remaining* radioactive material is unchanged, and so on.

These half life measurements are used by many scientists including astronomers, nuclear physicists, archaeologists, and geologists.

half tap A bridge that is placed across conductors without disturbing the normal functioning of the conductors.

half tap, network In data network communications, a duplicate path established between nodes or systems. A half tap provides redundancy where new

cable is being run, as in circuits where fiber optic is replacing copper, but where it's not desirable to disrupt the existing network until the new cabling is functional.

half tap, telephone In telephone communications, a duplicate service installed on the subscriber side of the demarcation point (usually on the customer premises). This may be done in instances where there is a problem with the original circuits, or where a new system is being installed and the old one is left in place until the new one has been tested and is known to be functional.

half wave antenna An antenna designed so that its electrical length is equal to half of the wavelength of the signal being received or transmitted.

halide glass A halogen-based glass that is becoming of commercial interest because it may be used as a host glass doped with rare-earth elements for use in fiber-based optical amplifiers and reflective gratings. Halide glasses are transparent in the visible spectrum and, with fluoride, luminesce in a region appropriate for telecommunications transmissions. Photochromic glasses may contain silver halide crystalline formations. Silver halide-coated glass or films are used as image recording media (e.g., for holographic images). ZBLAN is an important halide gas formed through gas-film fabrication techniques that has excellent properties for ultra-low-loss optical fibers. See silica, ZBLAN.

Hall, Robert N. (1919–) An American scientist who worked at General Electric from 1942 to 1946 and 1948 to 1987. He has been inducted into the National Inventors Hall of Fame for his invention of magnetron technology, a subject he studied during World War II. He has received dozens of patents, including patent #2,994,018 for his development of an asymmetrically conductive magnetron. Magnetron technology has been used in a wide variety of applications, including radar antijamming devices and microwave ovens.

Hall observed interesting semiconductor electrical properties that led him to discover alloyed p-n junctions. In 1950, he wrote "P-N Junctions Prepared by Impurity Diffusion" in *Physical Review*. By 1955, he had grown silicon-based crystals for use in transistors. In the 1970s, he turned his interests to solar energy research.

Hall is also a laser pioneer. He was group leader of a research team at the GE Research and Development Center that succeeded in creating semiconductor injection diode lasers, in 1962. This was to become an important light source for fiber optic communications, optical storage media (e.g., CDs), and laser printers. Hall was recognized for his work with lasers with the Marconi International Fellowship Award in 1989. See Dicke, Robert; Kao, Charles; Javan, Ali; Karbowiak, Antoni.

Hall constant A description of the relationship between current-carrying conductors and magnetic fields. The Hall constant = (transverse electric field) / (magnetic field strength) × (current density).

Hall effect If you take a current-carrying semiconductor with a magnetic field perpendicular to the direction of the semiconductor's current, a voltage is created that lies perpendicular to both the current and the magnetic field of flux. It has practical applications in generators and modulators.

Hallwachs, Wilhelm (1859–1922) A German physicist who developed a type of refractometer and who confirmed some of the pioneer photoelectric work of Heinrich Hertz. In 1888, Hallwach described his discovery of the photoemissive properties of certain substances when exposed to light by using an electroscope. He demonstrated that photoelectric cells could be used in cameras, a big boost to the evolution of television, which was just being developed at that time.

Hallwachs effect In a vacuum, a negatively charged body discharges when exposed to ultraviolet light. The effect is named after Wilhelm Hallwachs.

ham operator *colloq.* Amateur radio operator. A hobbyist radio operator engaging in noncommercial radio communications. Ham operators are primarily involved in personal, public service, and training communications over approved radio frequencies. Hams have also had a long history of voluntarily aiding in search and rescue, emergency, and disaster relief communications to augment government or commercial communications or in situations where no other support is provided. The frequencies in use for amateur communications are fairly standardized throughout the globe and hams have long communicated across international borders. In the U.S., ham communications and the issuance of ham radio licenses are administrated by the Federal Communications Commission (FCC). Ham operators come in all ages, shapes, sizes, and colors and represent a wide spectrum of abilities, professions, and technical expertise. See American Relay Radio League.

Hamming, Richard "Dick" Wesley (1915–1998) An American mathematician and software engineer, Hamming is best remembered for developing error correction codes for computing systems in the late 1940s. In 1945, he was working at Los Alamos in the computing department, executing calculations for the Manhattan Project.

Hamming later codeveloped L2, based upon L1 (Bell 1) developed by V. Wolontis and D. Leagus in 1956. In the 1960s, he authored "One Man's View of Computer Science" in the *Journal of the ACM* in which he describes his views on applied mathematical ideas and practical hands-on computer programming activities.

Hamming accepted a teaching position at the Naval Postgraduate School after leaving Bell in 1976. He lectured at the Naval School and became Professor Emeritus until his retirement in 1997. Hamming is the author of *Digital Filters*, a book on filtering applications in communications and broadcast technologies that is still being reprinted. The Hamming medal is awarded by the IEEE.

Quite a number of Hamming quotations have been passed down through his colleagues and students, including "It is better to do the right problem the wrong way than to do the wrong problem the right way."

Hamming code A linear forward error detection/correction code system named after R. W. Hamming of Bell Laboratories. Hamming developed the system in the late 1940s and described the system in *The Bell System Technical Journal* in 1950 in the context of fault-resistant large-scale computing systems. He referred to it as a *redundant systematic* code in which a certain ratio of bits was used for error detection and correction. Hamming acknowledged the tradeoff between redundancy and the accuracy and speed of the transmission.

Hamming code is a block parity mechanism that can detect single- and double-bit errors in data transmissions and correct single-bit errors per each message-bit codeword. The codes lend themselves to matrix representation. Block parity involves using more than one parity bit, each based upon a different combination of bits. The Hamming rule for determining the number of parity or error checking bits is related to the number of transmitted bits. See error correction, Hagelbarger code.

Hammond, Fred (1912–1999) A Canadian engineer, collector, and curator, Hammond was a co-builder of Hammond Manufacturing Company, in 1927, one of the largest historic electrical/electronic equipment manufacturers in Canada. The company started as Oliver S. Hammond's (Hammond's father) basement shop during World War I and evolved into O.S. Hammond & Son, including Fred Hammond and his brothers. In 1986, Hammond Manufacturing became a public company, trading on the Toronto Stock Exchange (TSE).

Hammond began building radios in the early 1920s and earned his first amateur radio license in 1929. He founded the Southern Ontario Chapter of the Quarter Century Wireless Association and helped build it into the largest local chapter. He was honored by the Canadian Amateur Radio Hall of Fame in 1996 and has received many other awards of appreciation and recognition over the years. Hammond will probably be best remembered for founding the Hammond Museum of Radio, a center that demonstrates and shares his love for radio technology for future generations. Hammond held amateur radio license VE3HC, inherited from his father. See Hammond Museum of Radio.

Hammond, John Hays, Jr. (1888–1965) An American engineer and inventor who developed radio control (RC) systems for vessels in the 1910s. As a schoolboy, Hammond was already experimenting with circuits and sensors. His social circle included many of the great inventors of the time, including Thomas Edison and Nikola Tesla, a fact that likely provided inspiration and encouragement for his talents. While at University, he met Alexander Graham Bell and studied radiodynamics and emerging telephony technologies. Starting as a patent clerk, Hammond familiarized himself with the patent process and, in a few years, amassed more than 100 patents. He studied many aspects of radio technology, including

H

frequency modulation (FM), radio tuning, telephony, guidance systems, and much more. Many of his patents were later purchased by the U.S. military for use in radio-controlled guidance systems. See frequency hopping.

Hammond Museum of Radio Named after its originator, Fred Hammond, the museum began as a personal collection in the early days of radio and grew steadily to the point where it was moved to a new, larger facility at the Hammond Manufacturing Company's South Transformer Plant in Guelph, Ontario, Canada, in September 1999. The collection includes hundreds of historic radio receivers and transmitters and represents many of the important developments in wireless technology. Many of the systems are still in working order or have been restored to working order. Of special interest is the *Collins Collection*, likely the largest operational exhibit of Collins Radio equipment in the world. The museum also hosts special exhibits to commemorate the discoveries and designs of a variety of radio pioneers. http://www.hammondmuseumofradio.org/

HAN See home area network. See fiber to the home, home ATM network.

hand off See handoff.

handle A pseudonym, a nickname, often very creative, humorous, or obscure. A handle indicates your personality, your interests, or helps preserve anonymity. Handles are frequently used on the Internet in various email messages or postings to public news forums or chat groups.

handoff, handover 1. The process of passing on a message or transmission to the next leg in a route that takes more than one type of communications medium or more than one transmitting region. A *make-before-break* handover is one in which the transfer to the new leg is carried out in such a way that the user does not perceive a break in communications. 2. The process of a communication being passed through various 'hands,' usually because the user is mobile, as from one zone to another, one station to another, one transmitter to another, or one frequency to another. 3. In cellular communications, the transfer of the call from one cell to the next as the subscriber moves through the various cells. Handoffs often involve frequency shifts. 4. The process of passing a caller to another agent, as from a receptionist to a sales representative or technical support person.

handset A human interface communications transceiver unit, most often associated with telephones. It's the part we pick up and hold to our ears and mouths in order to listen and speak on the phone. Handsets come in a variety of shapes, some of which have names in the telephone industry. The older round handsets familiar on rotary phones are G style, whereas the newer square ones more common on mobile phones and phones with the buttons on the handset are K style handsets.

handsfree A communications unit that does not require the user to hold it in order to be able to communicate with the caller. Headsets and speakerphones are examples of handsfree units in the telephone in-

dustry. Some phone systems permit handsfree menu selection or dialing through voice recognition. Car-mounted cell phones are becoming more prevalent, so the driver can have both hands on the wheel and concentrate on driving, rather than holding the cellular handset. For computer input devices, a voice recognition system can be used along with a headset to create a handsfree unit.

handsfree telephone Any telephone appliance that provides handsfree operation for some or most of its operations, such as a voice operated phone or computer (e.g., for spoken dialing), a speakerphone, a headset, etc. See handsfree.

handshake Communication between two systems to manage synchronization of the transmitted and received signals, often established with ACK or NACK signals, tones, keywords, or header packets. Handshaking is an essential component of most communications systems and is often incorporated into the transmission protocol itself. Handshaking can be done between people, between machines, or both. The most familiar form of handshake is the verbal "Roger" used on one-way-at-a-time radio links. It signals the other party that it's his or her turn to speak. In verbal communications, this "Roger" handshake is sometimes accompanied by electrical signals that set the half-duplex communications direction to favor the person who is currently talking.

There are also textual handshakes. In public chats on the Internet, where dialogs are typed rather than spoken, "GA" (Go Ahead) serves the same purpose as "Roger" on a radio link.

In modem communications, handshakes are used to acknowledge a signal, to coordinate baud rates, and to orchestrate the transmission, receipt, and data, so the signals don't override or clobber one another. On networks, a handshake can negotiate links between computers, printers, scanners, and other peripherals that might not always be online or might be shared (or transmit at different data rates from the main network). See auto-negotiation.

handwriting recognition A software application, often coupled with a scanning device or a stylus that resembles a pen, that recognizes and may also interpret written script. It may further translate the digitized handwritten text into displayable typewriter-style text, depending upon the application. Pen computing uses this type of technology and is of use to those who don't know how to type or don't want to. Since handwriting is widely variable, most systems must be trained to recognize an individual's writing and, even then, the results may not be perfect. Nevertheless, in the shipping industry, scientific field work, and other areas, handwriting recognition is useful as a form of user interface, and the technology will eventually improve to the point where anyone's handwriting can be recognized and interpreted by a computer. Fiber optic faceplates, that may be used in place of traditional optical scanning lenses, may add new dimensions to handwriting recognition as a security or input mechanism and may increase the active region over which the writing may be scanned.

Some handwriting recognition applications are controversial. The use of electronic signature pads for the acceptance of courier packages means that a company has your signature, in electronic form, in a database. It is not possible for an employer to monitor the activities of every employee at every moment and thus it is extremely important to secure, limit access to, and purge this database. But, since there has been no verbal or written agreement between the user and the courier as to how the handwritten signature may be used, there is little legal protection for the person who has innocently given their signature to receive their package. This opens the door to many types of personal intrusion and illegal activity, including identity theft, fraud, and blackmail. See Personal Digital Assistant.

hang up *v.* To disconnect from a transmission (two words when it is a verb). On modems, ATH is the Hayes-compatible command for hanging up. On phones, a hangup (one word when it is a noun) occurs when the button is pressed for at least a specific amount of time. In some areas, the callee may not be able to hang up this way if the caller is still on the line. It doesn't work the other way though; if the caller hangs up and the callee is still on the line, the transmission is disconnected. Many Internet Service Providers (ISPs) will automatically hang up (terminate) a computer connection if there is no activity after a certain amount of time, such as 10 minutes.

Hansell, Clarence Weston (1898–1967) An American research engineer and television pioneer in New York state, Hansell worked for a year for General Electric, then the Radio Corporation of America (RCA), and later for the U.S. government as a scientific investigator with the Technical Industrial Intelligence Committee in Germany, during World War II.

In the 1920s, Hansell worked with Scottish inventor John L. Baird on the development of a mechanical television system based upon the idea of using an array of transparent rods (essentially a fiber optic system) to transmit broadcast and facsimile images.

In 1925, Hansell founded the RCA Radio Transmission Laboratory at Rocky Point, N.Y. at which the world's largest radio transmitting station was developed. Due to the relationship between radio transmissions and the Earth's atmosphere, he also became interested in ionization effects and climatology.

Hansell was a member of many prominent science institutes and engineering societies. A collection of his papers from 1928 to 1967 are housed in the State University of New York (Stony Brook). See Baird, John.

Hansen, Holger Møller A Danish scientist who investigated the transmission of images through bundles of parallel glass fibers in the late 1940s and early 1950s. Hansen applied for a patent for cladded glass or fiber imaging in 1951, but was denied the patent due to the prior work of Hansell and Baird on television technologies in the 1920s. Without a patent, it was difficult to get funding for commercialization of the invention.

Hansen, William Webster (1909–1949) An American physicist and educator with a pioneering interest in the use of high-frequency radio waves (microwaves) in particle acceleration research. Hansen joined the staff at Stanford University in 1934. He was an associate of Martin Packard, working on a team with renowned physicist Felix Bloch.

In the 1930s, Hansen began his association with Russell and Sigurd Varian, working in the basement of the Stanford physics building. When the inventive Russell Varian sketched out an idea for a Rumbatron Oscillator or Amplifier, in July 1937, it was Hansen who had provided the basic rumbatron concept and the calculations to support the viability of Varian's idea, leading to the invention of the Klystron tube.

Hansen subsequently did important work in microwave theory and passed on the knowledge through courses at Stanford while the Varians developed practical applications of the Klystron technology for radar and communications. Following World War II, Hansen returned to his research interests and pioneer work in disk-loaded accelerators. He demonstrated a linear accelerator in 1947, sponsored by the Office of Naval Research. It led to the later creation of the Stanford Linear Accelerator Center (SLAC) and many discoveries in basic particle physics and X-ray spectroscopy. The basement lab had evolved into the Microwave Laboratory, which eventually became the Ginzton Laboratory and the Hansen Experimental Physics Laboratory. See Ginzton, Edward; Klystron; Mark accelerators.

hard copy An image or document that is readable by looking directly at the medium on which it is transcribed, as on a piece of paper, cardboard, stone, or parchment. A soft copy must be accessed with some type of technology in order to be viewed, manipulated, or displayed. Soft copies commonly exist on hard drives, floppy diskettes, tapes, CDs, and other magnetic or optical media.

hard disk drive, fixed disk drive HDD. A data storage device most often associated with desktop computing systems although it is also useful for storage in computerized milling machines and other industrial automation products. In the early days of computing, program code and data were stored on paper tapes, punch cards, and magnetic tape spools and cassettes. All of these early devices, as they were implemented at the time, were essentially linear/serial devices with limitations in speed and flexibility.

In the 1950s, IBM engineers claimed leadership in data storage with the invention of a high-speed random-access device. The release of the 305 Random Access Method of Accounting and Control (RAMAC), in 1956, made it possible to store five megabytes on 50 24-inch discs. The machine weighed more than a ton. In the early 1970s, sealed disks known as Winchester disks were introduced and the name became generic for disk drives for several years. Floppy disk drives were an inexpensive alternative to the new, expensive hard disk drives sold during the 1970s and the 1980s, but by the late 1980s, hard disk drives became the dominant storage medium and floppy drives were used mainly for program distribution and swapping small files among computer systems. In 1986, the American Society of Mechanical Engineers declared IBM's contribution as an International Historic Mechanical Engineering Landmark. The following year redundant arrays of independent disks (RAID) technology was patented.

The hard drive as we know it is based upon a rotating circular platter with a read/write head that never travels very far from any specific location on the platter, thus providing not only random access, but also high-speed access, compared to previous methods. One or more magnetic platters are permanently contained within a fixed housing (as opposed to cartridges or other portable storage devices), hence the name hard or fixed disk. When a drive is formatted, the magnetic particles are aligned to a specific pattern and, from that point, data is written by influencing the particles and read by detecting the state of the particles on the magnetic surface. Compared to floppy diskettes, the hard drive can hold far more information and is safe from dust and fingerprints. IBM's claim that hard drives would revolutionize computer storage was correct, as hard drives quickly superseded tape drives for most realtime applications. (Tape was retained for backup purposes due to its lower cost and was later reintroduced as a random-access removable medium, popular in the 1990s.)

Hard drives were originally expensive washing machine-sized devices purchased by institutions for mainframe computers, but by the early 1980s, hard drives were smaller and more accessible and available for the desktop market as well. A five megabyte hard drive in a breadbox-sized housing could be purchased for under $1,000 and soon smaller, higher-storage drives were available for a few hundred dollars. Another consequence of smaller drives was portability. By the early 1990s it was possible to equip portable computers with high-capacity drives and removable hard drives were built into some models. Hard drives have been developed in a number of formats, too numerous to list here. However, the most common hard drive controller/hard drive formats on desktop systems during the 1980s and 1990s were

1. Small Computer System Interface (SCSI) – a robust format that could be daisy-chained (usually up to seven devices) to include several drives or could be used in RAID systems. SCSI was installed in most Motorola-based machines including Macintosh, Amiga, and many workstation computers.

2. Integrated Drive Electronics (IDE) – a more limited master/slave format introduced to bring the price down on drives for popular Intel-based machines serving price-conscious consumer markets. These drives became prevalent on IBM/IBM-licensed computers sold to homes and small businesses in the 1990s. Due to limitations in capacity and expandability, a new enhanced IDE format was introduced as EIDE. EIDE was essentially a move to give IDE the capabilities familiar to SCSI users.

Improvements in hard drive capacity have been strongly tied to the ability of the read/write head to read and write data of finer precision and higher densities. Thin films were introduced in the late 1970s, along with the run-length-limited (RLL) data-encoding scheme. In the early 1990s, IBM introduced magnetoresistive head technologies based on discoveries in the late 1980s of high magnetic field effects on crystals. With more sensitive sensors came higher-capacity hard drives, with significant breakthroughs resulting from the development of giant magnetoresistive (GMR) head technologies.

SCSI and IDE had a relatively long reign, considering the pace of computer technology, but Universal Serial Bus (USB) and FireWire drives were making inroads by 2001. The USB data transmission standard was developed by a consortium of companies (Compaq, DEC, IBM, Intel, Microsoft, NEC, Northern Telecom) in the mid-1990s. USB hard drives have several advantages over previous formats, including small size, hot-swapping capability, portability, and high capacity, all at a reasonable price. FireWire was developed by Apple Computer and the IEEE 1394 Working Group. Its principal advantages are speed and ease of use, which make it suitable for hard drive and other demanding transmission technologies (e.g., video). Given their many benefits, USB and FireWire may quickly supersede both IDE and SCSI formats on consumer machines, although updated SCSI formats, including Ultra SCSI and Wide Ultra SCSI may continue to serve workstation and service provider environments where very high transmission speeds are desired. See disk controller, redundant array of inexpensive disks, superparamagnetic.

hard sectored A storage medium, usually magnetic, in which the various boundaries or sectors are physically designated with holes, pits, ridges, or other markers to indicate their extents. Hard sectored media are becoming less common than soft sectored as they are less transportable between different systems.

hard transfer A term for an electronic monetary transaction involving the actual exchange of funds between individuals or banking institutions. A hard transfer often follows a soft transfer. A paper check is a type of soft transfer. It is a monetary transaction that is not actually finalized until the money is withdrawn from the bank. Similarly, online there are many monetary transactions that are soft transferred and later hard transferred from the actual bank or other financial institution.

hard tube A type of electron tube that has a high vacuum environment within the sealed glass bulb.

hard wired See hardwired.

hardware The physical circuits and devices associated with systems, especially computerized systems, that are fixed or hard wired and unlikely to be altered by the user. Contrast with software (although the distinction is not actually cut and dry), which is selected and swapped out by the user, modified, or overwritten. See firmware, software.

hardware flow control A capability built into most of the high speed serial card modem combinations that helps to handle data flow control. Use of hardware flow control may also require the use of a hardware flow control cable.

hardware interrupt On computing systems, a call to the software to interrupt the current process in order that it may temporarily listen to or interact with a hardware device interfaced with the system. See interrupt, IRQ.

hardwired 1. A circuit that is intended as permanent or is not expected to change in the near future, and thus it is wired in such a way as to make it efficient to produce or easy to use, rather than making it amenable to change. Contrast this arrangement to patch bays and breadboards, which are intended to prototype temporary circuits and are easy to change. Programs or pathways built into computer motherboards are typically hardwired, whereas the various user-added peripherals, especially those that fit into slots or chips designed to be swapped out when better technology is developed, are considered to be modular or configurable and not hardwired. 2. People who are *hardwired* are said to be set in their ways, not amenable to change or open to new ideas. 3. An idea or system that is *hardwired* is one that is entrenched, difficult to change for various reasons, including politics, economics, or complexity.

Harmon, Leon D. (1922–) A Bell Telephone Laboratories researcher who initiated a project to simulate the functions of biological nerve cells by means of simple transistors. These could be closely associated with one another in arrays and were applied in a simulation of mammalian eye nerves. His work was featured in industry journals in the late 1950s and in the film *Thinking Machines* in 1960. Harmon chaired workshops sponsored by the National Science Foundation (NSF) in the mid-1970s.

Harmon was a pioneer cyberneticist interested in machine simulation of aspects of human perception. Long before most others, Harmon could see the practical applications of recognition technologies. In this vein, he coauthored a number of articles on character recognition, human face recognition, and automation of these processes by intelligent systems in the early 1970s. He also collaborated in a number of projects with Ken Knowlton, and together they did experiments in scanning images and reconstituting them with computer algorithms, thus creating some of the first examples of computer graphics and image processing as they relate to human perception. See Knowlton, Kenneth; neural network; Shroeder, Manfred R.

harness A securing system of straps, combination connectors, or other means used to consolidate multiple cables so they can be handled more easily as a unit.

Harris Broadcast Communications, Harris Corporation An international, publicly traded commercial provider of advanced broadcast technologies, Harris is one of the pioneer developers and providers of digital broadcast technologies. It contributed the radio frequency Test Bed for the Advanced Television Test Center in 1990, was first to market with a number of digital television exciter and transmissions products, the first to transmit a commercial HDTV signal, and the first to establish an operating digital television air-chain. Harris also broadcast the first major live high-definition television (HDTV) sports event in 1997. In 1999, Harris and the CPB/WGBH National Center for Accessible Media demonstrated digital closed caption and descriptive narration technologies at the National Association of Broadcasters conference.

In April 2001, Harris announced a business arrangement with Dotcast, Inc., to provide digital content through a revolutionary network that takes advantage of advances in broadcast technology. It provides popular computer services available on the Internet, through new digital broadcast technologies in a one-to-many relationship rather than a one-to-one relationship as is typical of Web browsers.

Harris Corporation is also a provider of computer security products, distributing network analyzing and scanning software to clients such as the Canadian Public Works and Government Services. Harris has contributed to the development of the European Digital Video Broadcast (DVB) standard. See Advanced Television Test Center; Association for Maximum Service Television, Inc.; KLAS-TV.

Harrison, John (1693–1776) A British clockmaker who devised a means, in the 1770s, to create a portable chronometer to aid in marine navigation by determining longitude, even when being bumped around by heavy seas. Several countries have acknowledged Harrison's contribution with commemorative postage stamps.

Harvard Mark I A historically significant, large-scale, automatic computer constructed by Howard H. Aiken and IBM engineers in the early 1940s. The concept was proposed by Aiken as he was finishing his graduate work at Harvard University in 1939. In his report, Aiken envisioned a calculating machine that embodied some of the concepts of Charles Babbage, one that could handle cumbersome mathematical equations too lengthy or time-consuming for humans. Aiken's concept led to support from International Business Machines (IBM) to build the machine at the IBM labs in Endicott. Although most often remembered as the *Mark I,* it was also known at the time as the *IBM Automatic Sequence Control Calculator.*

The Mark I was a 35-ton electromechanical behemoth that had a number of characteristics to distinguish it from basic calculating machines, making it a true historic computer. It used magnetically operated switches to handle the logic patterns, included central processing units and multiple storage registers, and could run (and rerun) instructions stored on prepunched paper tape. For the realization of Aiken's goals, it was capable of working out mathematical equations to 23 significant digits.

By the spring of 1944, the machine had been moved from the IBM labs to Harvard University and began to be known as the Mark I. At Harvard it was put into service by the U.S. Navy for military calculations in the aftermath of World War II. Three programmers were involved on the project in the 1940s; the best remembered is Grace Hopper, who joined the project in 1944, after Richard Bloch and Robert Campbell. See Aiken, Howard; Hollerith, Herman; Hopper, Grace.

Harvard Mark II The second in the series of large-scale computing machines developed under the direction of Howard Aiken, the Mark II was designed to replace some of the mechanical elements of the Mark I with electronics. It also took advantage of some of the improvements in electronics technologies that had occurred since the early 1940s. World War II and the need for fast, complex computations provided motivation for funding and building more advanced computers after the success of the Mark I computer. The Mark II was completed in 1947 and Aiken put his attention to the development of the Mark III.

Harvard Mark III Third in the line of large-scale computing machines developed under the direction of Howard Aiken, the Mark III was delivered in 1951 to the U.S. Naval Surface Weapons Center. It improved upon earlier Mark computers and on many competitors by incorporating drum memory with separate drums for instructions and data.

Harvard Mark IV Fourth in the line of large-scale computing machines developed under the direction of Howard Aiken, the Mark IV was the last in the series, with Aiken working on the project until 1952.

Hau, Lene Vestergaard (ca. 1959–) A Danish physicist who graduated from the University of Århus, Hau joined the Department of Physics at Harvard University where she has achieved the remarkable accomplishment of stopping light – without losing its energy.

In 1998, Hau's research group succeeded in slowing light to a speed of only 17 m/s by optically inducing quantum interference in a Bose-Einstein condensate. By January 2001, it was announced that this line of inquiry had led to the spectacular feat of stopping light emitted from a laser and releasing it again at full speed and intensity, in a sense creating a low-loss (or no-loss) atomic optical "capacitor."

Fiber optics has been one of the breakthrough technologies in communications media, but Hau's group has gone beyond this concept, creating a dynamic system of atoms and photons with optical properties with a nonlinear refractive index orders of magnitude greater than an optical fiber or any other transmission medium. These developments will have far-reaching consequences for new scientific research and applied technologies. It may be possible to create single-photon optical switches, entirely new classes of computer components, and light-based storage devices unlike any of their predecessors.

Hauksbee the Elder, Francis (ca. 1666–1713) An English artisan and experimenter who built on the work of R. Boyle and associated with Isaac Newton. Hauksbee did studies in static electricity and created a pump, in the early 1700s, that apparently improved upon earlier designs and prevailed for the next century

and a half. The availability of pump technology was important not only for commercial pumping of oil and water, but because the ability to create a good vacuum was invaluable to scientific exploration and the study of sound transmission, magnetism, and electricity. A Hauksbee air pump, ca. 1708, is listed in the King George III Collection at the London Science Museum, and there is a ca. 1720 Hauksbee pump in the Museum of the History of Science in Oxford.

When he was joined by James Hodgson, in 1702, Hauksbee was already employed by the Royal Society to demonstrate experiments. By 1705, he was an instrument supplier and became a fellow of the Royal Society, the same year he reported on his experiments with producing light in a mostly evacuated mercury vessel. At about the time he started giving science lectures with James Hodgson. Hauksbee's research led to further experiments, and the observation that lampblack particles would move up and down very rapidly and make an audible sound when a glass tube that had been rubbed was held above the particles. Following this observation, he devised a rotating wheel to allow the glass to be rubbed at a great rate, in essence inventing a friction generator.

In 1709, Hauksbee described his discoveries in his self-published *Physico-Mechanical Experiments on Various Subjects*. The book was republished in Italian in 1716 and in English in 1719 by J. Senex in a larger edition with several new experiments. In 1754, a French edition was released. Hauksbee also described experiments with capillary action in *Philosophical Transactions,* in 1712.

Hauksbee did not have an extensive formal education and was not highly literate, but his mechanical aptitude and talent for experimentation were exceptional and earned him the respect of his peers. See barometer; Boyle, Robert; Gray, Stephen; Guericke, Otto von; Hodgson, James. [Source for birth/death dates: Jeanette (Jan) Shermer, descendant.]

Hauksbee the Younger, Francis (ca. 1687–1763) An English instrument maker and scientist, the nephew of Francis Hauksbee, the Elder, listed here mainly to distinguish him from his uncle of the same name with whom he is often confused (Hauksbee the Younger was also a member of the Royal Society). Hauksbee carried on the tradition of experimentation and scientific inquiry of his uncle and set up an outlet for the distribution of scientific devices. The Charles Townshend Papers list a 1757 communication about a Francis Hauksbee having developed medicine for the treatment of venereal disease.

Hayes Microcomputer Products Inc. One of the early entrants to the modem market, Hayes set many of the industry's de facto standards for serial communications through modems. Hayes modem control commands are still widely used. See AT commands.

Hayes Standard AT commands See AT commands.

HBA See host bus adapter.

HBS See Home Base Station.

HCI 1. See Host Command Interface. 2. human computer interface. 3. See Human Computer Interface standards.

HCP hard clad plastic. A material used in fiber optic cable construction.

HCS hard clad silica. A material used in fiber optic cable construction.

HD See half duplex.

HDB3 High-Density Bipolar Three. A signaling scheme used in high-speed digital networks, especially phone networks. HDB3 is based upon Alternate Mark Inversion (AMI) and uses positive and negative pulse states. If four or more zeroes are sequentially transmitted, HDB3 inserts a violation code, an enhancement on basic AMI transmission. The insertion of the violation bits facilitates the reconstruction of the signal at the receiving end. HDB3 and other enhanced signaling schemes have been superseding AMI. See Alternate Mark Inversion, B8ZS.

HDCM See high-resolution direct core monitoring.

HDD See hard disk drive.

HDLC See High Level Data Link Control.

HDSL See high bit-rate Digital Subscriber Line.

HDT Host Digital Terminal.

HDTV See High Definition Television.

head A device for reading, writing, or removing data from a volatile storage medium (usually magnetic). VCRs, hard drives, floppy drives, and tape recorders all have heads that touch, or nearly touch, the surface of the storage medium in order to transmit the information to the logic circuits or mechanisms that decode the information into human-meaningful form or to write to the storage medium.

head thrashing If read and/or write heads on storage mechanisms encounter hardware or software problems, especially bad sectors, the mechanism may start to oscillate rapidly, sometimes uncontrollably. This can lead to damage to the head or the data if not terminated in time.

header 1. Identifying text printed in a block at the head of a file or document. Header information frequently includes file format, version, date of creation, author, and typographic information. Header files are common to word processing, desktop publishing, and EDI applications. 2. A commonly used system routine contained in a separate file and referenced during program compilation and linking. System windowing routines and graphics routines are frequently linked in from header files. A header provides modularity and a write-once-use-many solution to many programming tasks. 3. In ATM, the protocol control information located at the beginning of a protocol data unit.

header area In an EDI file, the area that contains the header information for the document. See EDI, header.

Header Error Control HEC. In ATM transmissions, an error detection mechanism contained in a byte at the end of the 53-byte ATM header. It corrects single bit errors and is efficient over transmissions media with low bit error rates (BERs) such as fiber optic cable. In ATM carried over wireless transmissions, the signal is not as clean as a fiber optic signal, and the BER rate can be substantially higher. Satellite transmissions tend to be especially bursty, a situation not handled well with a single bit error mechanism. Some satellite service providers have compensated for this by developing a variety of solutions, including interleaving of cells to isolate the data from burst errors. See ATM Link Enhancer.

headset A radio or telephone transceiver unit worn on the head or wrapped around the ear (sometimes referred to more specifically as an earset). Headsets are typically used by professionals who sit and take many calls, including receptionists, console attendants, telemarketers, and reservation takers and by those on the move, including truckers, warehousers, and ground staff. Headsets are also becoming a consumer item for use with cellular phones (so drivers can keep both hands on the wheel) and other hands-free applications.

heap memory A type of local memory storage that is dynamically allocated while a program is running. Heap memory is usually of more concern to applications programmers than to users, but there are some applications in which heap memory needs to be set prior to running the software in order to provide enough working room for memory-intensive applications. On some systems, heap memory is limited to a maximum of 64 kilobytes.

heat sink A structure for dissipating or radiating heat away from a heat-generating device such as a motor or semiconductor. Heat sinks often resemble open coils, flat fence rails, or other repeated, spaced elements, usually of metal, that are configured to increase their surface area, and thus their radiating capacity. Sometimes the component is called a cooling fin.

Some CPUs require surprisingly large heat sinks, especially accelerator chips intended to provide faster performance. See dissipate, baffle.

Heathkit EC-1 Educational Analog Computer A historic hobbyist computer, introduced in 1959 or 1960 as a Heathkit, a Daystrom product line that was very popular with computer hobbyists in the late 1970s and early 1980s (until it became cheaper to buy a system than to build one from a kit). The EC-1 (Educational Computer-1) was one of the earliest low-cost desktop computers, selling for $199, and one of the last of the analog computers. It had a steel chassis supporting rows of knobs and status lights, looking much like the Altair digital computer that came out a decade and a half later, and it could do basic calculations. It was marketed mainly to educational institutions teaching applied physics and mathematics. See Altair, Arkay CT-650, GENIAC, Kenbak-1, Simon.

Heaviside layer See Kennelly-Heaviside layer.

Heaviside, Oliver (1850–1925) An English physicist who started as a telegrapher. He later became interested in electricity and began publishing on that subject in 1872. He made thorough studies of Maxwell's equations and then set about simplifying them to two equations expressed in two variables.

Along with J. J. Thompson, Heaviside theorized about the electromagnetic reactions and mass of electrically charged particles in motion. See Kennelly-Heaviside layer, Maxwell's equations.

HEC See Header Error Control.

hecto- (*symb.* – h) A Système Internationale (SI) unit prefix for 100 or 10^2.

Heisenberg uncertainty principle Proposed by W. Heisenberg in 1927, the uncertainty principle has since become a fundamental principle of physics. (Heisenberg formulated a model of the structure of an atom in the 1930s which has held up well over time.) In studying movement of electronics, Heisenberg proposed mathematically that it is not possible to determine precisely both the *position* and the *velocity* of a material particle at the same time. The uncertainty increases as the size of the particle decreases.

Many researchers have generalized this principle and restated it in various broader contexts, but most commonly it is brought up when describing the results of quantum experimental results. It is said that these are determined in part by the point of view and methods of the researcher. For example, if light is studied as a particle phenomenon, it appears to behave as a particle phenomenon. If it is studied as a wave phenomenon, it appears to behave as a wave phenomenon, at least as far as the observers and measuring instruments are concerned. In other words, attempts to pin down precisely the location of an electron obscures its energy level, and vice versa, thus challenging the absolute nature of the world suggested by classical physics.

Heisenberg, Werner (1901–1976) A German physicist responsible for deriving a theory of atomic structure and proposing the uncertainty principle in 1927, which has since become widely associated with his name.

Heisenberg built on the work of previous physicists and mathematicians, including Hermann Weyl. For his contributions, he was awarded a Nobel prize in 1932. In the 1940s, he acted as the director to the Kaiser Wilhelm Institute for Physics. Near the end of the second World War, Heisenberg was captured by American troops and taken to Britain. When he returned to Germany, he helped found the facility that became the Max Planck Institute for Physics. During his later years, he was working to formulate a unified field theory of elementary particles.

helical antenna, helical beam antenna An antenna designed with a helical (spiral) conductor wound in a circular or polygonal shape. The axis of the helix is usually mounted parallel to the ground. The circumference size of the helix in relation to one wavelength affects the angle of radiation.

heliochrome [sun color] An older word for a color photograph, that is, one photographed in color as compared to one photographed in grayscale and tinted by hand using oil pigments. Color photography was not widespread until the 1960s.

heliograph [sun writing] A visual signaling system employing light signals, which was established around 1865 by H. C. Mance. The heliograph took advantage of the production of glass mirrors in the 1840s to increase the distance over which sunlight could be reflected. It used adjustable mirrors mounted on tripods and could convey messages in Morse code in daylight up to about 100 miles.

In the United States, leaf shutter versions of the heliograph were developed to interrupt the light signals instead of directing the angle of the mirror as was done with the earlier British heliographs.

Most visual signaling systems were superseded by wire telegraphy, but the heliograph survived for several decades, probably because it used Morse, which was then becoming widely accepted, and because it required no external power source.

Since heliograph signals and microwave transmissions share some of the same line-of-sight characteristics, heliographs were resurrected to research the placement of microwave relay stations, and the heliograph is still sometimes used for military communications in regional conflicts where other means of communication are scarce.

heliography [sun recording] A type of early photographic process, also called *sun drawing,* which was pioneered by French inventor, Joseph Nicephore Niepce, in 1816. Originally Niepce used a camera similar to the camera obscura to imprint temporarily an image of light onto paper coated with silver chloride. It was several years before he developed the process to the point where the image could be permanently preserved. See Daguerre, Louis; photography.

helionics The science of the conversion of solar energy to electrical energy.

heliotrope [sun turning] An early surveying instrument that employed the sun's rays to triangulate from mountain prominences. This instrument was developed and used for the highly successful engineering feat of surveying India in the 1800s. It may also have been used for signaling. It was later adapted as a heliograph by H. C. Mance in Britain and used for many decades for daylight signaling of military communications up to 100 miles.

helium-neon laser, He-Ne laser A type of low power atomic gas laser, now commonly used in light shows and monitors. This laser produces warm color tones in the red-orange range but gas lasers with different gas mixes may also produce radiant energy in the blue, infrared, and ultraviolet ranges. Until semiconductor laser diodes became prevalent, this was the most common commercial laser.

The helium-neon laser was described in 1959 and developed and demonstrated in December 1960 by A. Javan, not long after Townes and Schawlow had developed an optically pumped laser. The gas laser worked on a different principle from the Townes laser, converting electricity to a pure, continuous light beam by passing an electrical current through two inert gases. The light is then amplified by reflecting it between two mirrors at either ends of the laser device.

The device was immediately tried for telecommunications. The day after the laser was first successfully tested, Javan's lab workers successfully used it to transmit a telephone conversation by converting voice vibrations to light pulses that were detected across the room by a sensor. In essence, the researchers had realized the dream of A. Graham Bell when he first invented his Photophone. Despite Bell's great excitement about his invention and his vision of its potential, he never saw widespread commercial use of the device due to the difficulty of harnessing a consistent, coherent, sufficiently powerful synthetic source of light. Javan's invention, with the contributions of W. Bennett and D. Heriott, made the missing piece of technology available, a technology that since has been applied to many technologies in addition to optical communications. See carbon dioxide laser, laser history, Photophone.

helium recovery In fiber optics preform fabrication and fiber pulling, helium is commonly used in many aspects of the manufacture. Contaminants are introduced into the helium at different steps along the way and one-time use of the helium is wasteful. For this reason, manufacturers such as Praxair have developed draw stations that enable the helium to be recovered, cleaned, and reused at points where lower purity helium is effective in the fiber fabrication process. See vapor deposition.

helix, helical shape A spiral, continuous coil. Many types of springs employ a helix shape. In radio transmissions, a horizontal- and vertical-polarized wave combined as *circular polarization*, is transmitted in a helical fashion so that it can be picked up by both horizontal- and vertical-polarized antennas.

Hellman-Merkle A trapdoor knapsack cryptography system principally designed by Ralph Merkle, with input from Martin Hellman, who was a collaborator with Whitfield Diffie on another cryptography system. The Hellman-Merkle scheme was found to be breakable and was reported as such in 1982. See Diffie-Hellman.

Hello Protocol In Open Shortest Path First Protocol (OSPF), a mechanism for establishing and maintaining neighbor relationships. It may also be used for dynamic neighbor discovery on broadcast networks. The Hello Protocol elects the designated router on networks with at least two attached routers.

Helmholtz, Hermann von (1821–1894) A German physicist who expressed relationships between fundamental phenomena, such as heat and light, by treating them as manifestations of a single force, a concept we now associate with *energy*. Helmholtz further sought to generalize the concepts put forth by James Joule. Helmholtz encouraged the work of Heinrich Hertz, who became one of the true pioneers in the discovery of the physical existence and properties of radiant energy (radio waves).

The Hermann Helmholtz Association of German Research Centers (HGF) continues Hermann Helmholtz's tradition by supporting research on scientific, technological, and biomedical topics through a consortium of centers. See Hertz, Heinrich. http://www.helmholtz.de/

henry A unit of inductance in a circuit (self-inductance or mutual inductance of two circuits) such that the electromotive force of one volt is produced when the inducing current varies at the rate of one ampere per second. Named after Joseph Henry.

Henry Ford Museum & Greenfield Village The largest indoor/outdoor museum complex in the U.S. providing authentic historical artifacts and educational activities. Among other fields, it features exhibits on the history of transportation and communication, including Thomas Edison's Menlo Park Laboratory. The Henry Ford Museum is located in Dearborn, MI. http://www.hfmgv.org/

Henry magnet Joseph Henry experimented with electromagnets in the 1820s and 1830s, creating a number of different configurations. The Henry magnet is a type of experimental magnet, powered by a quantity battery, that he called a "quantity magnet." The horseshoe-shaped magnet was wrapped with several layers of insulated wire connected in parallel. Henry also experimented with a magnet powered by an intensity magnet that could power devices at a distance. Henry based his magnets on his studies of electromagnetism and the work of previous magnet researchers. As an improvement to earlier attempts, Henry's powerful quantity magnet could transform enough electrical energy into mechanical energy to be useful. Thus, Henry himself experimented with the practicability of the electromagnet for powering rudimentary telegraphic systems and he communicated his ideas to many eminent inventors who built upon his ideas and put them into commercial use. See Henry, Joseph.

Henry, Joseph (1797–1878) A gifted American physicist who began experimenting with magnetism in 1927, Henry produced a high-power industrial electromagnet in 1931. He incorporated his various discoveries into many practical devices, including telegraphs, relays, and electromagnetic motors.

Joseph Henry actively encouraged and assisted other researchers, in addition to carrying on experiments himself, a fact not always publicly acknowledged by the many inventors who benefited from his generosity. Morse, Wheatstone, and many other inventors were assisted by the information communicated on both sides of the Atlantic by Joseph Henry.

Joseph Henry advocated a science and research focus for the great Smithson endowment that eventually became the Smithsonian Institution, and he was appointed its first Secretary in 1878. He also co-founded the U.S. National Academy of Sciences. The henry, a unit of inductance, is named after him (1891). See Henry magnet.

HEP high energy physics.

HEPIC High Energy Physics Information Center. An informational link among worldwide HEP resources to assist researchers in locating sources and resources. One of the services is a global search facility for searching across HEP Web servers that is updated about once per month. HEPIC is funded by the U.S. Department of Energy Division of High Energy Physics. It is supported by the Fermilab Computing Division. http://www.hep.net/

hermetic seal A seal that is airtight and leaktight, and is sometimes used to preserve a gaseous environment, inside a sealed device, that is different from the gaseous composition outside. Hermetic seals are generally

H

intended to be permanent. Hermetic sealing is used in a variety of industries, including electronics and electrical installation.

Herrold, Charles D. "Doc" (1875–1948) An American inventor and educator, Herrold is one of the first experimenters to transmit voice over distance. He did this through Station FN in the Garden City Bank Building, beginning in 1909. He became known by amateurs for his station SJN broadcasts. When Lee de Forest's transmitter failed, Herrold provided music and news to the 1915 World's Fair about 50 miles away. He had other interests as well and was awarded a patent for the Arc Phone in 1915. In 1922, he was broadcasting at 833 kHz through station KQW from San Jose, California. See KDKA, radio history.

hertz Hz. A unit of frequency expressed as one cycle per second, named after H. R. Hertz.

Hertz antenna An antenna system that uses distributed capacitance to determine its resonant frequency, which, in turn, is influenced by the physical length of the antenna. This antenna is used in applications where ground reflection is not a necessary factor for its functioning. Unlike Marconi antennas, a Hertz antenna is not dependent on the ground or the body of a vehicle as a resonant conductor. This type of antenna is common for television and frequency modulated (FM) broadcasts. See antenna, Marconi antenna.

Hertz, Heinrich Rudolf (1857–1894) A German physicist who demonstrated important properties of electromagnetic radiation, discoveries that later experimenters applied to facilitate the transmission of radiant energy. By 1887, the physical existence of radio waves had been established. He also contributed a streamlined reformulation of Maxwell's equations that was widely accepted. The Hertz antenna and hertz unit of frequency are named after him. See Helmholtz, Hermann.

Hertzian waves Electromagnetic waves in the range from about 10 kHz to 30,000 GHz. James Clerk-Maxwell had proposed that rapidly vibrating electric currents would emit waves, and Hertz experimentally confirmed this proposition. The waves are named after him. See radio wave.

Hertzstark, Curt (1902–1988) A Viennese inventor who devised handheld calculator technology that was patented in the late 1930s, but wasn't produced until the mid-1940s. Its accuracy apparently rivaled many modern devices. See Zuse, Konrad.

hetero- Prefix for different, other, not usual. It is often used to describe a mix or variety.

heterodyne *v.t.* To produce a *beat* between two frequencies, which could be of various kinds: audio, optical, or radio. In radio heterodyning, an electrical beat can be selectively created and controlled by heterodyning a received signal current with a steady introduced current. The frequency thus formed can then be further processed by amplification, as in repeater stations or filtering. See beat, beat frequency, heterodyne repeater.

heterodyne repeater A frequency repeating system commonly used in the propagation of radio signals that uses heterodyning to create an intermediate frequency through demodulation, which is amplified, modulated, and retransmitted over the next leg. This technique is less subject to distortion and loss through modulation/demodulation than baseband repeating. Heterodyne repeating may be used in conjunction with baseband repeating if the signal is traveling through several legs and some channels need to be dropped off at the baseband stations. See Armstrong, Edwin H.; baseband repeater.

heterojunction In semiconductors, a common dynamic junction, usually of p-n type, in which the materials on either side of the junction are substantially different. See homojunction, p-n junction.

heterojunction laser A semiconductor laser component that has been designed to reduce losses resulting from light diffraction within the optical reflective cavity by engineering the p-n junction width and the index of refraction of the cavity. By replacing some of the materials in the junction so that the junction is not homogenous, the index of refraction can be reduced to retain more light within the reflective optical cavity, reducing losses. This is more efficient than the traditional homojunction laser diode.

It is also possible to retain the materials in the junction but replace the materials in the p-n regions, but this may be so effective in confining the light energy within the cavity that it may damage the component. There is a happy medium between laser efficiency in terms of producing photons, and confinement inefficiency to allow the light to escape from the cavity before it damages the component itself. Larger optical cavities with different layers are one strategy for increasing efficiency without destroying the component from within. See homojunction laser, semiconductor laser.

heuristic problem-solving An exploratory problem-solving strategy that employs successive trial and evaluation of the results in such a way that the results can be used in the subsequent trials to "home in" on a solution. Heuristics are often used in artificial intelligence programs where the result is not known in advance, and where brute force methods are inappropriate due to the large number of possible choices and outcomes. Chess playing programs, for example, use a combination of heuristics to handle novel situations and databases of known moves and strategies. Heuristics are common in robotics, where a robot may have to interact with an unknown or unpredictable environment. See algorithmic problem-solving, brute force problem-solving, neural network.

Hewlett, William R. (1913–2001) An American inventor and business tycoon, Bill Hewlett was a founder of Hewlett-Packard along with David Packard. An instrumentation engineer, Bill Hewlett invented an audio oscillator which was used by Disney Studios in the production of *Fantasia*. He was a past president and director of the Institute of Radio Engineers (now the IEEE), an honorary trustee of the California Academy of Sciences, and held many other professional and civic leadership positions. A prominent Silicon Valley pioneer and philanthropist, Hewlett was known personally by many of the prominent personalities in the emerging small-scale computer revolution.

Hewlett-Packard Company HP. In 1938, Dave Packard and Bill Hewlett, both graduates of Stanford University, began working out of a garage in Palo Alto, California. Their first product, an audio oscillator resulting from Bill Hewlett's research of negative feedback, was designed to test sound equipment. It was a new type of design, utilizing an incandescent bulb to provide variable resistance. This product was followed by a harmonic wave analyzer. Their first big client was Walt Disney Studios. Disney ordered eight oscillators for the production of the movie *Fantasia*.

From these beginnings, Hewlett and Packard formed a partnership on New Year's Day in 1939; a toss of a coin decided the company name. HP was officially incorporated in 1947.

Since then Hewlett-Packard has become a well-known supplier of calculators, computers, software, printers, and other accessories to the computing industry. The firm is known for good quality products (their calculators have been known to survive the "drop-kick" test) and a corporate culture that seems to produce happier employees than many other companies in the industry.

In 1996, David Packard, successful businessman and philanthropist, died at the age of 83, followed by William R. Hewlett, in 2001.

HF, hf 1. hands free. 2. See high fidelity. 3. See high frequency.

HFC See Hybrid Fiber Coax.

Hi band A video standard developed by the Sony Corporation in 1989 to support 500 lines of resolution on a TV screen. This resolution falls between NTSC and High Definition Television (HDTV). See Hi8mm.

Hi-Band IR An infrared-remote technology used for remotely controlling a variety of consumer devices such as car stereo systems.

Hi-OVIS Highly Interactive Optical Visual Information System. A Japanese cable television delivery system employing different types of cable for different parts of the service network.

Hi8mm A high-quality, analog, tape-based compact video system developed by Sony, Hi8mm became especially popular in the late 1990s. Hi8mm has many benefits, including small cassettes, only 2.5 × 3.6 in. for two hours of recording/playback at regular speeds. The long play time gives Hi8mm an advantage over S-VHSc (compact S-VHS). The small cassette format enables video camcorders and playback decks to be compact as well. The sound quality is excellent, compared to older technologies, and features stereo sound, though not all Hi8mm camcorders have two microphones. The video resolution of 400 lines and signal-to-noise ratio of 40 to 50 dB are equal to S-VHS and far superior to 240-line 35 to 45 dB VHS formats.

For analog recording, Hi8mm is likely to remain a viable, favorite format for a while longer, but digital camcorders are catching on quickly and will probably supersede Hi8 as it offers 500+ lines of resolution, CD-quality sound, and digital nonlinear editing capabilities without conversion.

HiFD A 3.5-in. floppy diskette system developed jointly by Sony and Fujifilm, introduced in 1997. HiFDs are double-sided, with a magnetic storage capacity of up to 200 megabytes.

hibernation A resting state, one of low energy usage and activity. A term often applied to the sleep mode on portable computers, which powers down during times of low activity to extend battery life. Hibernation has also long been applied to software applications that lie dormant waiting for some event causing them to become active, or to run at a higher priority level. The event that rouses the program can be many things, including the time of day, input from an interface device such as the mouse or keyboard, activity of users, or data from another program. See sleep mode.

hickey A spot, halo, or other imperfection in the ink or toner of a printout caused by undesired extraneous particles, dried ink, etc.

Hicks, John Wilbur "Will" An American physicist and inventor, Hicks was one of the many optical researchers who worked at one time at American Optical. After leaving the company to cofound Mosaic Fabrications, Inc., with P. Kiritsy, in the late 1950s, he was approached by John Johnson about the potential of fiber optic filaments for lengthening a light path through fiber optics rather than through longer tubes. Hicks succeeded in pulling optical fibers, in making fiber-based vacuum assemblies and, in collaboration with B. Gardner, in developing fiber optic faceplates (arrays of aligned bundles of fibers) which were all significant innovations at the time. He maintained close ties with American Optical (AO). In 1961, he worked with Elias Snitzer at AO to demonstrate the dielectric waveguide properties of optical fibers so small, they would carry light in a single wavelength mode.

Hicks' interest in fiber optics hasn't waned. Many of his visions of fiber have come true and he continues to promote an optimistic viewpoint of widespread data-over-fiber availability. He has been involved with a number of companies and has continued to tender proposals for fiber-based technologies through American Micro-Optical, Inc. See Johnson, John; Kawakami, Shojiro; Kao, Charles K.; Snitzer, Elias.

hierarchy A group of items, people, or processes ordered according to some type of structure or rank, usually top-down or bottom-up. A hierarchy may be nested. Hierarchies are generally designed to facilitate the location of some item within the hierarchy or to simplify the understanding of its contents. People in an organization, and files or discussion groups on a computer or network system are often assigned positions within hierarchies.

High Altitude Long Endurance HALE. Pilotless platform crafts intended to float above commercial aircraft at about 20,000 meters, which have been proposed as two-way communications links capable of carrying phased antenna arrays. The HALE systems are experimental and a number of systems from helium to jet-engine propulsion have been proposed. The cost relative to traditional satellites makes HALE transceivers attractive to developers. Actual deployment of these systems has to take normal aircraft safety and traffic patterns into consideration.

high ASCII There really isn't such a thing, since ASCII defines the lower 128 characters (0 to 127), but "high ASCII" and "extended ASCII" are often used to describe characters above decimal values of 127, which are different on each system. See extended ASCII.

high bandwidth *slang* Descriptive of a person with high level, wide-ranging intellectual abilities. The author first heard this phrase in the early 1980s among hard-core programming friends in computer users groups and suspects it originated spontaneously in a number of places and subsequently spread through networks such as Fidonet. Other computer-related terms that have been applied metaphorically to human intelligence include *high baud rate*, *high clock speed*, and *multitasking*.

high bit-rate Digital Subscriber Line HDSL. A digital transmissions technology that can transmit DS-1 or E-1 transmissions for longer distances over the traditional, unshielded twisted-pair wire that is widely installed in telephone circuitry. HDSL is typically used in digital loop carrier systems, private branch networks, and cellular antenna systems. Unlike earlier technologies, HDSL transmits over multiple lines without repeaters and uses techniques to pack more information in less bandwidth. It provides rates up to 1.544 Mbps over DS-1 or 2.048 Mbps over E1 in the 80 to 240 kHz bandwidth range. See Digital Subscriber Line, DS-.

High Definition Television HDTV. In 1987, the Federal Communications Commission (FCC) acknowledged that the NTSC television was out of date and formed an Advisory Committee on Advanced Television Service (ACATS) to recommend a revised television standard for the U.S. Most of the improved systems proposed to the FCC were analog or hybrid analog/digital. The FCC stated a preference for a simulcast HDTV system and identified four digital systems. These were extensively analyzed, leading to an ACATS recommendation that digital HDTV should be adopted. In 1993 an alliance was formed in cooperation with ACATS to create an HDTV system, releasing specifications in 1994.

high Earth orbit HEO. An orbiting region around the Earth into which certain types of communications satellites are launched. There are advantages and disadvantages to high orbits. The main advantage is that it takes fewer satellites to provide global coverage. Disadvantages include the higher cost of launching, the higher amplification needed for signals to travel the greater distances, and the effects of radiation. The lifespans of high-orbit satellites tend to be around twelve to fifteen years. Most high-orbit satellites travel at about 20,000 to 40,000 kilometers outside earth. High Earth orbit satellites are typically used for geostationary satellites such as the U.S. Global

Positioning System (GPS). See low Earth orbit, Global Positioning System, medium Earth orbit.

high fidelity A playback system that reproduces the original so well that it is indistinguishable, or almost indistinguishable, from the original source. This quality is often accomplished through fast transmission and wide bandwidths. In audio systems, high fidelity is frequently abbreviated *hi-fi*.

high frequency A signal frequency defined as the range from 3 to 30 MHz.

High Level Data Link Control HDLC. An ITU-T standard bit-oriented data link layer communications protocol originally developed by ISO for managing synchronous serial transmissions over a link connection. In HDLC there are separate bit patterns for control and data representation.

high level language A computer programming language at the user level, designed to be as close to a natural language as possible and generalizable to a variety of platforms. FORTRAN, BASIC, and COBOL are probably the best-known and most widely used high level languages. High level languages are often interpreted, but may be compiled for the specific platforms on which they will be running. As languages become closer to machine level, they also tend to be more symbolic and, thus, more difficult to read and write. They also increase in platform-dependency. For contrast, see assembly language, low level language, machine language.

high pass filter A filter designed so it doesn't pass waves below a specified cutoff frequency (greater than zero), and the transmission band extends upward indefinitely from that cutoff point. See low pass filter.

High Performance Computing Act An act of the U.S. Congress which was passed in 1991 to facilitate and promote the development and evolution of interconnected computer networks serving educational institutions, research laboratories, and industry.

High Sierra standard A compact disc standard introduced in 1986 by the High Sierra Group (named after the hotel and casino at which the group met). It was subsequently adopted by ECMA (ECMA-119) and ISO (ISO 9660) and released with slight revisions. See ISO 9660.

High Speed Data Unit HSDU. A physical data communications device to provide a high-speed data channel together with control signals from, for example, a serial connection. HSDU may be software configurable and may operate in synchronous or asynchronous modes or both.

High Speed Internet Access A summary of information published about high speed Internetworking, published monthly by Information Gatekeepers, Inc.

high speed networking This is a relative phrase and will change as the technology advances but, in the mid-1990s, high speed networking was generally considered to be around or over 100 Mbps. Examples of high speed network technologies include asynchronous transfer mode (ATM), Fast Ethernet, and Fiber Distributed Data Interface (FDDI).

High Speed Technology HST. A U.S. Robotics proprietary, high-speed, full duplex signaling and error control transmission protocol. U.S. Robotics manufactures modems complying with this protocol and some dual-standard modems that are both HST and V.32 bis capable. See Microcom Networking Protocol, modem, V Series Recommendations.

High Speed Token-Ring HSTR. An enhanced commercial Token-Ring technology developed by International Business Machines (IBM). HSTR can run Token-Ring and Ethernet on one medium, support source routing through data packet headers, and was based on existing standards, including Fibre Channel. Aimed at a market similar to Ethernet, HSTR appears to be primarily supported by those upgrading legacy Token-Ring systems. See Token-Ring.

high usage groups In the telephone industry, high usage groups are trunks between main switching offices that are established as priority routes to handle the majority of transmissions. High usage trunk groups are intended to hand off overflow traffic to alternate trunks. See erlang.

high-threshold logic HTL. A concept used in physics and electronics especially with regard to computer circuit designs. Voltage thresholds are voltage levels at which more or less voltage may have no further effect or at which more or less voltage (at the high or low thresholds, respectively) could damage a system or render it inoperable. They are also points at which state changes may occur due to unconditioned or noisy signals. In transistor-based devices, the threshold voltage is related to performance levels and is often used to control the characteristics and operation of a system. Conventionally, on portable devices, lowered power is associated with degraded performance and power leakage. However, the use of high-threshold transistors is seen as one means to extend battery life on low-power devices, such as cellular phones, while in standby power mode without adversely affecting performance.

High-Density Bipolar 3 See HDB3.

High-performance Network Forum HNF. Developers and promoters of the HIPPI GSN standards. European High-performance Network Forum (EHUG) works in cooperation with HNF and coordinates an annual network technical symposium. http://www.hnf.org/

high-resolution direct core monitoring HDCM. A commercial software imaging process for fiber optic fusion splicing systems such as the Micro-Core™ Alignment Fusion Splicer. It facilitates core alignment prior to splicing and splice loss estimates after splicing.

high/low tariff A charge that is selectively made according to a type or level of service. For example, some Internet Service Providers (ISPs) charge different rates for connection time depending upon the speed at which the subscriber's modem connects. A bulletin board service (BBS) may quote two levels of service depending upon whether or not the user wants to access the games sections. In phone services, two prices may be quoted for a subscriber call, one cost per minute over high density trunks and a different cost per minute over low density trunks.

HIPERLAN high-performance radio local area network. A European standard for local area networks (LANs) that are interconnected through radio frequency transmissions rather than through wires. Unlike North American radio-based LANs, that are largely based on sharing unlicensed 900-MHz and 2.4-GHz spectrums, HIPERLAN was allocated dedicated spectrums in the 5.15- to 5.30- and 17.1- to 17.3-GHz frequency regions by CEPT, in 1992. A draft standard was developed by ETSI and presented as a functional specification in 1995. The availability of dedicated bandwidth increases the probability of reliable service with less chance of outside interference. See air interface, local area wireless network.

HIPPI High Performance Parallel Interface. HIPPI is a point-to-point high speed data transfer technology created at Los Alamos in the 1980s. HIPPI operates over twisted-pair copper cable for distances up to 25 to 50 meters (longer with cascaded switches) and distances up to 300 meters, or 10 kilometers over multimode or single-mode fiber cables.

HIPPI was originally developed for supercomputing applications but is starting to be adapted to other environments with the dramatic drop in price of the technology, particularly the switches. Transmission speeds include 800 Mbps and 1.6 Gbps (simplex or duplex). HIPPI can be employed with SONET over distances and over satellite transmission links. HIPPI is an ANSI standard with a series of documents spelling out the standard and its various switching and encapsulation characteristics. See HIPPI-6400.

HIPPI-6400, SuperHIPPI High Performance Parallel Interface. Officially it is now known as Gigabyte System Network (GSN). A very high bandwidth, low latency network transmission technology which offers gigabytes-per-second transfer rates, much faster than the capacity of Gigabit Ethernet, ATM, or Fibre Channel. Based on the HIPPI standard, but with enhancements in error correction and lower latency rates, HIPPI-6400 uses a fixed-length cell of 32 bytes. Transfer rates that are as fast as, or faster than, the internal workings of an individual computer on a network will change computing in a significant way. With this development, the network is no longer a bottleneck and individual computers attached to the network can theoretically function as individual parts of the same organism, that is, as a massively parallel computing system. Fast transfer rates with high bandwidth also make it easy to support a diverse variety of protocols, providing flexibility. SuperHIPPI can support applications like uncompressed digital movies and HDTV signals. See HIPPI, Scalable Coherent Interface.

Hirschowitz, Basil Isaac (1925–) A South African physician and educator who traveled first to the U.K., then to the U.S. where he became a professor of medicine. Hirschowitz visited Hopkins and Kapany in London at the time they were conducting pioneer research on the use of fiber optics for medical viewing instruments, in 1954. Along with C. Wilbur Peters, he undertook a project to develop a flexible fiber optic endoscope, hiring Larry Curtiss to assist in the work. The resulting instrument aligned the optical fibers by embedding them in a glass matrix with a low refractive index to facilitate total internal reflection. The instrument was first used with a patient in 1957. The Hirschowitz endoscope is now in the Smithsonian Institution collection. See Kapany, N; Hopkins, H.

Historic Speedwell Located in New Jersey, this museum features telegraph history through the daily life of the Vail family and their association with Samuel Morse. Morse had a close association with the Vail family who provided space, materials, and expertise to assist him in fabricating the inventions for which he is known. The Vail family is also known for the great success of AT&T. See Vail, Alfred; Vail, Theodore.

hits 1. A quantitative expression of the number of system or application accesses of a specified type within a specified time period. 2. The number of attempts of illegal entry to a system within a specified time. This information is used to gauge security needs and adjust procedures, if necessary. 3. The number of successful accesses of a database or a Web site.

HNF See High-performance Network Forum.

HNS See Hughes Network Systems.

HOBIS Hotel Billing Information System.

Hockham, Gearge A. Along with Charles K. Kao, Hockham authored "Dielectric-Fiber Surface Waveguides for Optical Frequencies" in *Proc. IEEE* (1966). Together they promoted the concept of removing impurities from optical transmission glass to reduce loss and increase transmission, an idea that made fiber links a practical reality.

Hodgson, James (1672–1755) A British mathematician and lecturer, Hodgson was evangelizing scientific achievements at an important time in history when many significant historical figures, including Isaac Newton, Defoe, Hauksbee, and Boyle, were influencing the future of science and trade. Hodgson had been an assistant to the Astronomer Royal Flamsteed since 1696, but left that position in late 1702 to become a lecturer.

Hodgson was versed in natural philosophy, experimental science, and mathematics, and joined Francis Hauksbee the Elder at the Royal Society during its leadership by Newton. In 1703, he became a fellow of the Society.

As a lecturer, Hodgson advertised seminars in astronomy and philosophy promoting the discoveries of members of the Royal Society and demonstrated experiments. Hodgson began to write technology books by 1704 and apparently also tutored in the sciences. Hauksbee, a demonstrator for the Society, had an interest in these lecturers and the two began to work together, thus creating a forum for practical embodiments of Newton's theories and some of Hawksbee's own inventions. In 1706, Hodgson published *The theory of navigation demonstrated*. In 1709 he became a mathematics master at the Royal Mathematical-School in Christ-Hospital. In 1736, he published *The Doctrine of Fluxtions Founded on Sir Isaac Newton's Method*, one of the early textbooks on calculus.

Hodgson was an important impetus in bringing experimental science (which was still a relatively recent concept) and mathematics to the common man, primarily merchants, navigators, and tradespeople who recognized the value of new technologies to their professions. He was also one of the first to voice the basic principles of sonar, suggesting that sound could be used to estimate the distances of ships at sea and of land objects. See Hauksbee the Elder, Francis.

Hoff, Marcian Edward (Ted), Jr. (1937–) An American engineer who, in response to a request from a Japanese company for a calculator chip, designed the Intel 4004, in 1971. This highly significant invention was the first commercially successful microprocessor chip, and it launched the microcomputer industry.

Hoff became a manager in the Applications Research division of Intel Corporation in 1968 and joined the 4004 development project. In 1983, he became Vice President of Atari, Inc., which was well known at the time for home computer systems and, in 1984, was awarded the IEEE Centennial Medal. In 1990, he was appointed Chief Technologist at FTI Teklicon, Inc.

hold 1. To pause, to cause to remain in a particular position or situation. 2. While attempting a computer login, the user may be queued or put *on hold* until fewer users are on the system. This hold usually manifests as a pause after typing the username or after typing the username and password. On popular archive sites, there may be a hold period or increased lag time while accessing a system. Sometimes it's better to log on later during off-peak hours, or to find a less busy *mirror site*, if one exists. 3. On phone systems where the receptionist is busy with other calls, or the automated system is queuing the caller for the next available operator, the user is usually put on hold, sometimes for extraordinarily unreasonable lengths of time (especially if it's a long-distance call). On systems that don't have music or a recording informing you that you are in the queue, it is sometimes difficult to know if you are still on hold or have been cut off. On multiline systems, a call can often be put on hold by one individual and picked up by another (line hold). On some systems, the hold call can be continued only on the main console or by the person originally putting the call on hold (exclusive hold).

Hold Recall An optional telephone feature that alerts you to the fact that someone is on hold. On most multiline systems and some residential phones, the line on hold will be identified by a flashing LED or, on more sophisticated systems, with beeps or voice messages.

hold time 1. In circuits, the time interval after the clocking of a trigger circuit during which data must remain unaltered. 2. In welding or soldering, the time during which the welded object must be held relatively steady in order for the weld or solder to harden. 3. In telephony, the length of time a caller is kept waiting, and waiting ....

holding beam A diffused electron stream used to regenerate charges applied to the surface of a storage tube.

holding coil In a communications circuit, an additional coil in a relay for each direction of a transmission that can be opened or closed independently of the main circuit to enable a single circuit to accommodate alternate two-way communications.

holding gun In an electron storage tube, the source of the electrons that make up the holding stream.

holding time In telecommunications, the entire duration of a call from the time the connection is requested until it is completed and disconnected. The actual time during which the connection is established is only a portion of the holding time, although it's usually the time that is billed. Holding time is important, for example, in tracking sales calls. How long does the sales representative or telemarketer actually stay on the line trying to connect a call and handle the entire transaction, as opposed to the amount of time actually spent with the customer? If there is a large discrepancy, another method may need to be tried.

Holding time is also important in computer or telephone circuit planning and management because the time spent connecting and queuing the caller may affect capacity and efficiency as much as that portion of the call during which the communication takes place.

holding trunk In telephone communications, a queue wherein a call is held until availability is established or an alternate route is found.

holiday factor A concept in service and retail industries (including Internet services) that accommodates changes in rates of service use or product purchase during holidays. The transportation, retail, and telecommunications industries tend to have patterns of higher usage during holidays and need to factor in extra staff, lines, products, etc. to handle the demand. Holidays may also create decreased demand, in which case shutdown or reduction of unneeded systems may result in cost savings.

Hollerith card Historically, a sturdy, rectangular, piece of card stock organized into rows and columns that were individually labeled and further organized into sections. A circular hole was punched in the card to select the coded data in that region of the card. The punched holes could later be read and decoded to reassemble the original information. This concept was born out of the cards used to store loom patterns on Jacquard looms. In the 1920s, IBM, the evolutionary heir to the Hollerith Tabulating Company, introduced a patented rectangular hole that made it possible to encode almost twice as much information as the original round hole cards.

Hollerith cards were used to store historic computing machine code in the days before tape, diskette, and hard drive storage. The punched card was, in effect, a binary storage system with the unpunched or punched locations representing off or on and corresponding to no or yes responses to a predefined parameter.

As the storage capacity of these cards was quite limited, many cards were needed to store a body of information. Punch cards, in general, can be fully punched and read in again with mechanical or optical devices, or partially punched to provide a record without the mess and waste of *chad* and read again with sensing devices. In the case of Hollerith cards, the punched cards were read by an electrical reader that completed a circuit and activated a relay when a pin passed through a portion of the card with a hole. With the use of *gang punches* with selectable templates, up to 40,000 cards a day could be punched. The computer cards commonly used in library books, until bar codes became prevalent in the 1990s, are descended from Hollerith cards. See Hollerith Electrical Tabulating Machine, Jacquard loom, zero punch.

Hollerith code A 12-level code designed by Herman Hollerith in 1889 for use on Hollerith cards which, in turn, were used with the Hollerith Electrical Tabulating Machine. The card is sectioned and labeled and the code was implemented in the form of holes punched at specified intervals in designated rows and columns, with each column corresponding to an alphabetic or numeric character, in order to form a semipermanent record that could be read and interpreted in the future. This code was widely used in early computing days to store program instructions and data. Over time, variations were developed and EAI and ANSI standards were established to govern the configuration of the cards, holes, and data encoding.

Hollerith code survived much longer than the Hollerith Machine, which was superseded by electronic computers. Hollerith code continued to be used in microfilm records, e.g., for storing microfilm image data and is still covered in Information Technology (IT) courses. For the processing of textual information, it has been superseded by other systems, including EBCDIC and ASCII. See Hollerith Electrical Tabulating Machine.

Hollerith Electrical Tabulating Machine A mechanical-electrical data processing mechanism for quickly and accurately recording and classifying individual data from a variety of sources. It was developed by Herman Hollerith in the early 1880s.

Hollerith's machine came to prominence when it was selected (over a system developed by A. Graham Bell) to tally the results of the 11th U.S. Census, in 1890. Before Hollerith's invention, it had taken almost a decade to process the results of a U.S. census. With the use of the Tabulating Machine by the Population Division of the Census Office, thousands of entries could be tabulated in a single day, thus greatly expediting the recording of over 60 million cards, one for each person included in the census. The 1890 census was processed in only a year.

The Census Office continued to use the Hollerith Machine for further statistical analysis of specialized groups within the population. For example, in 1901, Bell was using the Hollerith machine to compile and process information from the census relating to the blind and deaf, with assistance from his wife on selecting and organizing the format and priorities related to the data (an important aspect of setting up the punch configuration and tabulation). After the census, other organizations with statistical tabulating needs looked into the use of the machine, including accounting and transportation firms.

There were two main aspects to the Hollerith Machine: the adjustable punching mechanism for creating the data cards and the electrical pin-based reading mechanism for processing the cards, activating relays, and incrementing a counter. The letters and numbers organized on the punch mechanism's keyboard could be altered to suit the type of data that were being entered.

The Hollerith Machine received much publicity as a result of the 1890 census and caught the attention of other institutions, especially statistical and auditing companies and departments. Variations and improvements were developed, with the tabulating and sorting functions sometimes implemented separately and sometimes housed in one unit.

The Hollerith mechanical tabulating mechanism was to be an important impetus in the development of advanced calculators, the forerunners to early computers. See Harvard Mark I; Hollerith, Herman; International Business Machines.

Hollerith Tabulating Data Card

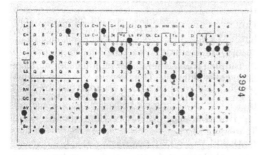

A sample of a punched Hollerith card in which the card has been organized in sections and labeled with alphabetic or numeric markers. [Railroad Gazette, April 1895.]

Hollerith, Herman (1860–1929) An American engineer who devised a historic tabulating machine, in 1884, which he further developed with the concept of punched cards as an information storage medium. In 1889 he received a patent for his machine to compile statistics (U.S. #395,782). Hollerith was subsequently contracted to carry out the storing and reading of U.S. census information, beginning with the 1890 U.S. census. Following the success of this invention, he formed the Tabulating Machine Company, in 1896, to market the technology. In 1911, it was merged with other companies to become the Computer Tabulating Recording Company, the historic beginnings of International Business Machines (IBM). See Hollerith Electrical Tabulating Machine.

hologram A type of imaging using lasers, based on the recording of an optical interference pattern produced by the interaction of two or more waves from the same source. The effect of viewing a holographic image is a sort of 3D, when moving the head around and focusing on different parts of the image. In most cases, it is not fully a 3D effect, as the image is usually recorded on some flat (with transparent depth) medium such as glass or transparent plastic (although the technology itself is not limited to this form of presentation). The image is typically more ethereal than a photograph, since it is viewed through "layers" of the transparent medium. Projected holographic images hold great promise for 3D virtual reality, and scientists have been working on holographic memory modules for computers that potentially can store enormous amounts of information in small three-dimensional components. It has been suggested that human memory may share some functional similarities with holograms.

Hollerith Tabulating System

The Hollerith Electric Tabulating System showing the punch on a movable arm poised over the card holder and a lettered and numbered keyboard hole alignments for the various data that were to be entered and stored on punched cards. [The Manufacturer and Builder, V. 22, 1890.]

A Hollerith reader, the sorting and tabulating mechanism that was used to process the individual cards into statistical data once they had been punched with information relevant to the application. [Railroad Gazette, April 1895.]

---

home In computer interfaces that distinguish regions

of the screen for different functions, one corner is usually designated as the *home* position, that is, a position to where a pointing device may return, or where a cursor might start again after the region has been cleared or a new window opened.

On English language-based systems, for example, the home position in a text window is usually the top left corner. If the text were being typed in another language that tracks from right to left, such as Hebrew, the home position would be the top right of the window or screen, or bottom right if it is bottom to top reading. In PostScript page layout programming, the home position is generally considered the bottom left corner, as it is with a number of printer graphics languages. Some systems have a *home* key as a one-step shortcut to position the cursor in the home position.

home ATM network HAN. A broadband home network providing connectivity with a variety of services and devices (computers, television, appliances, etc.). See ATM Forum, fiber to the home, RBB.

home page A World Wide Web concept referring to the first, primary, or main page of a set of hypertext linked pages on a particular host, or belonging to a particular individual or organization. Web pages are not inherently hierarchical, since any page can link to any other, but humans tend to grasp concepts more easily when information is organized in a top-down or bottom-up manner, and home pages reflect our preference for this type of organizational structure. A home page serves as a jumping off point, table of contents, or general site information map to help navigate the rest of the links. Commercial sites tend to have home pages that showcase product information and entice the user to explore the rest of the site. Personal sites often show family relations, professional credentials, and personal interests. Educational home pages usually provide information on course offerings, faculty, and facilities. See HTML, hypertext, World Wide Web.

home run A centralized wiring topology, like a star topology, in which cables to individual units or consoles all lead back to the central switching system. Most private branch systems, almost all key systems, and many of the smaller computer network systems have this type of cabling arrangement. In home run wiring, which has little or no redundancy, a severed line will cut off the end station from all other stations. See topology.

Homebrew Computer Club A historic, influential, electronic thinktank and tinkerer's organization, founded in California in 1975. Steve Wozniak is one of the most famous of the Homebrew members. The Altair was demonstrated at a Homebrew meeting, with homebrewers jumping on the opportunity to write applications for this early computer.

homeostasis A state in which there is a tendency toward balance, stability, or equilibrium. A state in which there is no change.

homing 1. Zeroing in on an intended destination or target. 2. Approaching an intended destination by holding some parameter of navigation constant (with the exception of altitude). 3. In guidance systems,

transmitting, receiving, and evaluating signals in order to locate a target. Bats and missiles use homing systems. 4. In telecommunications, homing is the selection of a route through which a call can be set to the next switching center, especially in toll systems, which may pass through several specified switching stations.

homodyne reception See zero beat reception.

homojunction A mechanism within a semiconductor in which the layers comprising either side of the junction are substantially formed of one material. For dynamic processes to occur within traditional semiconductors (those dependent upon materials rather than the composition of the layers), however, there needs to be some difference between the two layers associated with the junction such as differences in conductivity or doping. Doping, the impregnation of impurities with particular properties, creates a dynamic imbalance between the layers to facilitate energy transfer and release of photons, for example and it can reduce the contact resistance between the semiconductor and a metal contact, creating an n–n+ junction. See heterojunction.

homojunction laser A basic semiconductor laser with a single semiconductor material for the p-n junction doped to produce the dynamic energy transfer effects of the junction. The index of refraction is controlled by the type and amount of doping. This produces an effective laser, but some of the light is lost due to diffraction, so it is not of the highest efficiency. See heterojunction laser, semiconductor laser.

Honeywell Kitchen Computer Honeywell, in a blaze of optimism that is unique in entrepreneurial history, introduced one of their Series 16 computers as a home computer for the kitchen through the Neiman Marcus catalog for $10,000, in 1969. It could store recipes, was programmable, and included kitchen accessories. To put the price in perspective, it cost roughly the same as four new cars or a small starter home, or five years' gross wages for a person earning minimum wage. Even now, almost half a century later, there are few computers in kitchen environments, and it would be incredibly difficult to sell one for the modern equivalent of $80,000. As you might expect, the Kitchen Computer flopped.

The Series-16 computers, from which the Kitchen Computer was born, were originally designed by Computer Control Company and bought out by Honeywell. They were released at about the same time as the Digital Equipment Corporation (DEC) PDP-8 minicomputers. The Series-16 computers included the DDP-116 (1965), the DDP-516 (1966), and the DDP-316 (1969). The DDP-316 was essentially a physically reduced-size version of the DDP-516. By 1960s standards, the 316 line would probably be considered minicomputers, as they had 16-bit registers, with core memory and a memory read/write time of between 1 and 2 microseconds, which was pretty good compared to the low-cost microcomputers that came out a decade later. The main unit of the smaller models was about the size of a deep 17″ CRT monitor. Like the later Altair, the units had lots of switches

H

and status lights on the front panel. The Kitchen Computer was technically known as the H-316 Pedestal Model. Series-16 rack- and table-top models were available as well.

Despite its spectacular lack of success in the kitchen market, the Series-16 technology was not completely lost, as it eventually made its way into components used in the ARPANET (as interface message processors for connecting hosts), a more suitable market for cutting-edge, pioneer computing technologies. See Altair, PCP series computers.

hook switch, switch hook The hook switch was originally designed not just to terminate a connection so the next call could come through, but served also to disconnect from a battery source so it wouldn't be quickly used up, and later an electrical source.

Modern telephones draw current from the line and don't require a separate battery to operate the basic calling and receiving functions, but the hook switch, the hook on the side of an old traditional box phone or the buttons (plungers) on top of a traditional rotary desk phone, are still used for disconnecting a call, and sometimes for generating a tone (if they are held down briefly, which doesn't cause immediate disconnection). See hooking signal.

hook switch dialing On older wall box phones and rotary pulse phones, it was possible to dial a number by depressing the hook carefully for each number you wanted to dial. (Depressing the hook switch for too long would disconnect the line.) This is even possible on some of the older pay phones. See hook switch.

hookflash A signal-sending mechanism whereby the hook on an old-style phone or button plunger on a newer phone is quickly depressed to signal the initiation of a service or operation.

hoot'n'holler, holler down, shout down, squawk box A dedicated, four-wire, open phone circuit connecting speakers or speakerphones at each end of the connection round the clock. It's like a 24-hour public address system using phone lines with full duplex, two-way communication. Other phones on the system can be picked in order to listen to the conversations ongoing on the speaker system.

Hoot'n'holler systems are useful in industrial yards, institutions, and fast-paced financial floors where numbers of free-moving individuals look to centralized sources of information or engage in communal dialog at different locations.

hop *n*. 1. The extent of an individual transmission path between two nodes (with no intermediate nodes). 2. In radio, the extent of a transmission from Earth to ionosphere and back. 3. In frame relay, the extent of an individual trunk line transmission path between two switches. 4. In an IBM Token-Ring network, the extent of an individual transmission path between two bridges. 5. In cellular communications, where the user may be traveling through several transmission zones during the course of a call, a hop is a change in the radio frequency channel.

hop by hop/hop-by-hop routing In contrast to a system that predetermines a route before sending a transmission, hop-by-hop routing creates a route along the transmission path, a step at a time, by using routing information at switchers along the way. There are advantages to both. A predetermined route may be an efficient one, designed to speed the transmission through faster links or perhaps by choosing the shortest path. This is common on small or local systems. On the other hand, on a large system like the Internet, there may be millions of possible routes, too many to store in the routing tables at the source of the transmission. In this case, hop-by-hop routing is a scalable technique that makes use of the best information at each station to progressively build a path for the data. It has been suggested that ATM implementations of hop-by-hop datagram forwarding on the Internet are no longer adequate to handle traffic volume and improvements have been suggested. See cell switch router, RFC 2098.

hop channel In cellular communications, a radio frequency (RF) channel that is available to continue transmissions for a user with a call in progress who is moving through zones. Available channels are needed to continue uninterrupted transmission while the user is on the move. See cellular, hop, mobile communications.

hop count The sum of the number of hops that make up a route between its source and destination, or between a specified segment of the route. In radio communications, the number of times the wave bounces from the Earth to the ionosphere and back.

In networking, the hop count is the number of segments between individual nodes or routers, a number that is recorded in Internet Protocol (IP) packets on packet-switched data networks. In cellular, the number of times a radio frequency change occurred during the course of a call. Hop counts are one means to gauge the efficiency of a system and to configure or tune it for better performance.

hop off To exit one type of system and complete the route on another. For example, you may initiate a facsimile transmission on the Internet, that then *hops off* to a phone line and a dedicated facsimile machine. Or, you may make a voice call from a telephone that is routed through a voice translation program and interfaces with the Internet and becomes an email message at the destination. In this case, the *hop off* is from the phone system to the Internet, or, conversely, you can consider it a *hop on* to the Internet, if you are considering the Internet as the main portion of the transmission route.

Hopkins, Harold Horace (1918–1994) A British mathematician and physicist with an early interest in optics. After World War II, Hopkins took a position with the Imperial College in London. Hopkins applied for and received a grant from the Royal Society to develop glass fiber bundles for use in an endoscope. He hired N. Kapany, who became a significant pioneer in the field, to work on the project in 1952 and communicated his ideas to colleagues, including F. Zemicke. The idea spread and the race to publish resulted in A. van Heel describing the topic in June 1953.

Hopkins observed that a fiber could act as a filter as

well as a lightguide and continued researching aspects of fiber optics, lenses, and the improvement of these systems. In the 1960s, he took a position as a professor of optics at Reading University. See endoscope; Kapany, Narinder.

Grace Hopper – Programming Pioneer

Grace Murray Hopper was an American mathematician, physicist, lecturer, and one of the first computer programmers in the days when programming involved rearranging the wires within a vacuum-tube computer system.

Hopper, Grace Murray (née Grace Brewster Murray, 1906–1992) An American mathematician, physicist, and educator, Hopper was an originator of software compilers and developer of the COBOL programming language. She earned a Ph.D. from Yale in 1934 and spent many years as a lecturer, research scientist, and programmer for various organizations, including the U.S. Naval Reserve.

Hopper became involved in many of the important computer development projects at the end of World War II. In 1944, she joined Howard Aiken's Harvard Mark I project as its third programmer and later worked on the Harvard Mark II.

Hopper is perhaps best known for relating a story in which a technician found a bug inside a Harvard Mark II and solved a problem by removing it. She apparently glued the bug into the computer logbook and, in the 1970s, announced that she would be contributing it to the National Museum of American History. Thus, the term computer bug was popularized.

In the late 1940s, Hopper left the academic world to join the Eckert-Mauchley Computer Corporation where she had the opportunity to put her software theories to practical use. In the early 1950s, when new ideas about programming and reusing existing code began to evolve, she made what was probably her biggest contribution to the field. She became a cham-

pion of ideas that led to high-level languages, compiled software, and more efficient coding methods, even though many professionals claimed at the time that such things were impossible (they couldn't see past the physical wiring of computers to a time when electronics would be used to channel signals automatically). In spite of detractors, Hopper contributed significantly in the transition from paper tape and punch cards to coding languages such as C-10. She further proposed that computers could be programmed in English, an idea that was ridiculed, and developed a compiler for business use with an English-like syntax for nontechnical programmers that evolved into COBOL.

Hopper's contributions were too significant to pass unnoticed, in spite of a tendency to downplay the accomplishments of women at the time. In 1969, she was awarded the Data Processing Management Association Computer Science Man-of-the-Year Award. In 1973, she became the first woman recognized as a Distinguished Fellow of the British Computer Society and, in September 1991, was awarded the National Medal of Technology by the U.S. President. See A-0, B-0, bug, Harvard Mark I.

hops See hop count.

horizontal blanking interval, horizontal blanking time The period during which a display is suppressed on a cathode-ray tube (CRT) to allow the electron gun to return from the right side of the screen to the next display position down and on the left side of the screen (assuming left to right and top to bottom scanning as is present on most standardized frame-based video display systems). See blanking, cathode-ray tube, frame, sweep.

horizontal cross connect The interconnection between a horizontal distribution system and a telecommunications central wiring location such as an equipment or patch panel closet or bay.

horizontal distribution frame The equipment and structural elements that facilitate the interconnection of interfacility cabling configurations, as between subscribers and substations and central offices. The frame technically does not include the wiring but directs and contains it. Horizontal distribution frames are usually built into flooring or crawl spaces, hence the name. See distribution frame.

horizontal link, inside link In ATM, a link between two logical nodes belonging to the same peer group.

horizontal resolution A quantification of the amount of information that is contained on a single horizontal line of a rasterized output device such as a monitor or printer. On raster monitors, horizontal resolution is expressed in terms of pixels, usually about 800 to 1024. On black and white laser printers, horizontal resolution on consumer machines ranges from 300 to 1000 dots per inch (dpi), and on prosumer and industrial printers from 1000 to 2700 dpi. Thus, the total would be the number of inches times the dpi. A resolution of about 600 dpi or greater is needed to show clean lines and curves, without staircased artifacts, for common printed documents. A resolution of about 72 pixels per inch or higher on a grayscale

H

or color display is preferred for video displays. Colors and shades of gray can be used to antialias a display to give it a higher perceptual resolution.

horizontal scan rate A measure of the scan speed of electron beam display devices, usually described in hertz (Hz), as in cathode-ray tubes (CRTs) that sweep repetitively from left to right and top to bottom.

The horizontal scan rate describes how many horizontal scan lines per unit of time (usually seconds) can be displayed. At a particular scan rate, the number of lines that can be displayed decreases proportionally as the refresh rate increases. Multiscan computer monitors permit a variety of scan rates and resolutions, most ranging from about 40 to 75 Hz. See cathode-ray tube.

horizontal segment In wiring distribution systems, the wiring route from individual NAM or IO locations to the riser closets through ceilings or floors, usually up to a maximum of about 250 feet.

horn alert An electronic connection for sounding a horn or loud buzzer to signal an incoming transmission during times when the user might be some distance from the communications device. Horn alerts are used for after-hours phone calls or doorbells, for cellular phones in cars, and for a variety of security systems.

horsepower hp. A unit of power designated as equal to raising 33,000 pounds one foot in one minute, which can also be expressed as an English gravitational unit of raising 550 pounds one foot in one second. In the U.S., a unit of power equal to 746 watts. See watt.

host 1. One upon whom others depend for shelter or sustenance. 2. The main organizer and holder of an event. 3. It's a little difficult to define host as it relates to computer systems because different groups of computer personnel have given *host* and *client* opposite meanings in the past. For consistency with the English meaning of the word and popular usage, this dictionary defines host as a main server or controlling system, and the client as a subservient system in terms of priority or capabilities. See client.

host bus adapter HBA. A computer storage or network transmissions connectivity device. HBAs can be built into the motherboard or they may be available as optional peripheral cards. PCI Local Bus-compatible HBAs are prevalent, but there are also HBAs for other formats, including FC-AL (Fibre Channel) and CompactFlash. The HBA handles low-level hardware controller interaction, including data I/O to the controller registers and data transmissions.

HBAs are of interest for storage area networks (SANs), data warehouses, RAID systems, signal processing, video editing systems, and other high-capacity/high-throughput applications. HBAs for other formats such as CompactFlash can be used to insert very small CompactFlash or hard drive devices (e.g., matchbox-sized 320-MByte hard drive) directly onto a peripheral card.

Depending upon the type of HBA, there is typically a connection point on the card for attaching a fiber optic or copper duplex cable. If fiber optic connection is provided, there may be a gigabit interface converter (GBIC) as well. The HBA may have LED status lights to indicate power and port activity, similar to the status lights on a modem or network hub. As of Summer 2001, speeds up to 2 gigabits/second were possible, and most PCI-based HBA devices supported 32- or 64-bit addressing. There has been a trend toward adding larger buffers to increase performance. Although PCI is a widespread standard for peripheral cards, not all HBA PCI cards support all operating systems; there are different flavors of HBAs. Multiple HBA cards may be inserted into a system, up to a vendor-specified maximum, depending upon power and configuration.

HBAs are shipped with individual IEEE standard unique address identifiers. For Fibre Channel connectivity, a World Wide Name (WWN) is derived from the given IEEE address identifier to handle arbitrated loop activity. A software configuration utility may be used to establish a relationship between the HBA physical device and the logical HBA number assigned.

host carrier In telecommunications, the main carrier through which billing is channeled. In systems where a call goes through various networks or providers, carriers may have arrangements with the host carrier to bill through them to save paperwork and other administrative costs.

host computer A computer in a network providing primary operations and applications that are run through clients or remote terminals at other locations. A network may have more than one host, and some hosts may be specialized for modem access, email distribution, printing, and other tasks. The term host is related more to function than raw hardware capabilities but, due to resource sharing economics, the host frequently has greater capabilities (more memory, storage, peripherals, etc.) than the clients accessing it.

host site 1. A repository or other archive site accessed by remote users through client programs such as Telnet, FTP, Web browsers, and others. The host site is the one on which the administrative tasks and storage are carried out. 2. A computer bulletin board system, which typically hosts email, chats, games, and file uploads and downloads.

hot 1. Connected; live; ungrounded current-carrying conductor. A term frequently applied to electrical wires. 2. A hot chip is one that either runs at a high temperature and requires cooling, or one that has a fault that causes it to emit more heat than is normal and is likely to fail soon. See heat sink. 3. Stolen. 4. Topical, popular, desired by a large following. 5. Titillating, arousing. See hot chat.

hot cathode, thermionic cathode A hot cathode is one that produces a stream of electrons (a cathode ray) by means of thermionic emission. Thus, heat (thermal energy) that is associated with the cathode provides the energy boost needed to liberate the beam of electrons that comprises the cathode ray. Electrical current is used to provide the heat and replace the electrons that stream away from the cathode.

New types of hot cathodes have been developed for use with electronic devices, such as the waveguide tubes used in micro- and millimeter-wave communications. Barium-dispenser thermionic cathodes have been of particular interest to researchers. These new cathodes are suitable for environments where small size and low power requirements are advantages, as in Earth-orbiting satellites and deep space probes. NASA has a facility for cathode research and development and the evaluation of new materials, called the Thermionic and Non-Thermionic Cathode Research and Development Test Facility.

Thermionic cathodes are used in electronically pulsed injectors for high energy physics research on superconductors. They have also been considered for use as electron-beam scrubbing devices, that is, the beam could be used to clean hard-to-clean surfaces for specialized industrial applications.

Thermionic cathodes are also of interest in consumer applications. For example, Philips Display Components, the largest manufacturer of CRTs for computer monitors and television sets, has teams of materials scientists working on new generation versions and applications of cathodes.

For electron beams that are significantly brighter than conventional thermionic cathodes, thermal field emission (TFE) cathodes are being developed. These high-resolution emitters can be used in electron beam testing equipment and low-accelerating scanning electron microscopes (SEMs). See cathode-ray tube.

hot cut, flash cut The transition from one circuit to another while the system is in operation, hopefully without disruption to components or current users. Hot cuts are used when switching from an old wiring system to a new one, or when switching around physical routing paths. On individual computer systems, components are sometimes hot swapped, although it is *never* recommended. Never hot cut a component that is being accessed. It is especially inadvisable to hot cut most types of drives (floppy, CD-ROM, hard drive, etc.). (RAID systems are an exception.) Keyboards and mice are not usually damaged by hot cuts, but make it a habit to power off a system before making hardware configuration changes. See half tap, hot swap, redundant array of inexpensive disks.

hot docking Inserting a component into a docking bay (as in laptop docks or video bays) while the system is powered on. This is generally inadvisable. Whenever possible, power off all components before connecting electrical circuits. Some newer components are being developed for use in hot docking environments, such as high-capacity disk storage systems and consumer storage devices such as USB peripheral devices that include hard drives, card readers, and graphics tablets. Hot docking is very convenient for the user as it is not necessary to close applications and power down a system to attach or swap a peripheral device.

hot key combination 1. A combination of keys that when pressed simultaneously will perform a specific function or engage a memory resident program, such as a printer utility. It is handy for background processes that are frequently needed but would be distracting if running in the foreground along with other current process. 2. A combination of keys pressed simultaneously to perform a specific operating system function. For example, on an Amiga, *Amiga-Amiga-Ctrl* reboots the machine. On an IBM-compatible running MS-DOS, *Ctrl-Alt-Del* performs a similar function. 3. A combination of keys to access text style attributes and search and replace functions in older word processing programs developed before graphical user interfaces became common.

hot line A private, dedicated phone connection, sometimes indicated by the color of the phone. On a land line, when you pick up the line, it either connects automatically or does so quickly through the touch of a button or speed dialing. On a wireless service, the system may be configured so the phone can connect only with a specific number. Hot lines are used as emergency phones in buildings, on roadways, in brokerage firms, and by important personnel in government or military positions.

Fiber optic connections are particularly suited to secured hotlines used for sensitive emergency or government communications due to their speed, wide bandwidth, and relative immunity to electrical taps.

hot line service Phone service that expedites an automatic connection through a dedicated private phone. See hot line.

hot links In computer software applications, virtual links that form a connection between information in one document (such as text or images) and another, even if their native formats differ. For example, in a desktop-published document, there may be a hot link to text in a word processor and another to an image in a graphics program. Depending upon the system and the software, changing text in the word processor or in the graphics program may immediately effect a change in the corresponding desktop published document, or may effect a change when the page is refreshed or when *update links* is selected from a menu. As systems become more capable (multitasking, faster CPUs, more memory), hot links are more prevalent and updates happen more automatically. See drag and drop.

hot list In computing, a list of frequently used applications programs, directories, or Internet newsgroups, Web sites, or archives. A hot list is usually displayed as a text list or pull-down menu from which the user can quickly select the desired destination. See bookmark.

hot spot, hotspot 1. A location on a touch sensitive device that alerts the software to respond in some fashion to user input. 2. A screen location that responds when a cursor is moved into the region, or if the cursor is positioned and a mouse or key clicked to activate the hot spot. 3. A bottleneck or area of congestion in a network, component, or software routine. 4. An area of a circuit in which some component is generating more heat than would normally be expected and that may signify a potential problem. 5. A region of a document or image that includes an embedded link so some further action happens if the region is selected or activated. Hyperpage applications use hot spots for various links; graphics programs sometimes use hot spots to activate palettes or specialized drawing menus.

hot standby A backup or background system or program that is operating, but idle and available to take over if failure of the regular system occurs.

Hot Standby Router Protocol HSRP. A protocol that provides resiliency, fault-tolerance, and transparent network topology support for network routers. Standby routers inherit the lead position if the lead router in a group fails.

hot swap The process of connecting or disconnecting an electric circuit, component, or peripheral while the system is powered up. Hot swapping is done to minimize disruption to users of a system. It is highly inadvisable in most circumstances. Some systems are designed to handle hot swaps (some types of video components or redundant hard drive systems), but be sure you know what you are doing before attempting it. See FireWire, hot cut.

HotJava An adjunct to Java, the widespread, object-oriented, cross-platform programming language from Sun Microsystems that continues to grow in popularity for use on the Web. HotJava is a Java-enabled Web browser with support for JDK and SSL that is installed on the local computer system and enables Web sites with Java applications to run from a desktop system. Java support enhances a browser's capabilities. See Applets, Java.

Hotline Virtual Private Line Service A commercial Nynex subscriber service that uses public lines specially programmed and configured to operate as though they were private dedicated lines, with the connection activated when picking up the handset. See hot line.

House, Royal E. An American inventor who developed one of the first practical direct paper tape printing telegraphic receivers, patented in 1846. House continued to improve upon the original design and patented the improved version in 1852. See House telegraph.

House telegraph The House telegraph (U.S. #4,464) was a relatively complex printing telegraph. Two people were required to operate it, as one had to turn a crank to run the mechanism while the other operated the telegraph. It had a wooden base with the circuitry mounted on the top and a piano-keyboard-like series of keys underneath a hinged flap. The message was printed on a strip of paper similar to the stock ticker machines that evolved out of printing telegraphs.

The House telegraph is said to have been capable of transmitting up to 40 words per minute and was in common use in the U.S. in the latter half of the 1800s. The House telegraph formed part of the inspiration for the subsequent Phelps Combination Printer telegraph, designed by George Phelps to improve upon the House and Hughes telegraph systems. See House, Royal E.; Hughes telegraph; Phelps Combination Printer; telegraph, printing.

housing A protective enclosure commonly used to insulate, protect, or manage wires or electrical connections. Many housings are shaped like boxes, with one side open to provide access. Splice enclosures are a particular type of housing used to connect fiber optic cables between the head end and the node.

howl An irritating, unwanted wailing or screeching sound from acoustic or electric feedback that may occur, for example, when a speaker and microphone from the same transmission are placed too close together. Noise and echo canceling equipment can prevent or reduce howling.

howler, howler tone In telephone communications, a unit that creates a loud sound to signal that a phone has been left off-hook. For example, on some public exchanges, a recording will play first if a phone is left off-hook, "If you'd like to make a call, please hang up and try again ...," followed by a series of raucous beeps that can be heard up to about 15 feet from the phone.

HP See Hewlett-Packard.

HP 9830 Historically, one of the earliest desktop computers, coming out a few months after the Kenbak-1 and at almost the same time as the Intel SIM4, in 1972. The HP 9830 was the first desktop computer to really look like modern desktops, with a typewriter-style keyboard, numeric keypad, function keys, and status lights. It can't really be considered a personal computer, as it listed at just under $6,000 (more than the price of a car, in those days) and was primarily marketed to institutions and the scientific community. See Altair, Kenbak-1, Sim4.

HPA See high power amplifier.

HRPT high resolution picture transmission. A specialized image communication for very high resolution images such as those transmitted by satellites.

HSCI High-Speed Communications Interface. A single-port interface from Cisco Systems that provides full duplex synchronous serial communications.

HSCS high speed circuit switched.

HSDA high speed data access.

HSDU See High Speed Data Unit.

HSRP See Hot Standby Router Protocol.

HST See High Speed Technology.

HSV hue, saturation, value. In color imaging, a color model that allows settings to be adjusted along these three properties. Hue is the color, saturation is the amount or richness of the color, and value is the lightness or darkness. HSV systems for adjusting palettes on computer desktops, applications, and graphics programs are common.

HTL See high-threshold logic.

HTTP See Hypertext Transfer Protocol.

HTTPS See Secure Hypertext Transfer Protocol.

hub Focal point, center of attachment or activity.

hub, network 1. A connecting point on a network to centralize wiring and connection management. A hub may be passive or active and is often used in systems with star topologies. 2. A connection box on video or audio systems that permits centralization of cables and easy reconfiguration of devices. Often used in connection with switchers and, in many cases, the switcher itself may double as a hub. See bridge, router, switcher.

hub site The location of a hub, which may vary from a small box on a desk or rack to an entire closet or room, depending upon the size of the system. The hub site allows easy cabling and administrative access to a variety of connections. Hubs are often located at main wiring or logical junctions and may connect to external systems.

hue A color of the visible spectrum. Hue does not include white, black, or shades of gray, which are the presence of all colors (white) or absence of color (black) in various intensities (grays). Most people are familiar with hues as the colors of the rainbow: red, orange, yellow, green, blue, indigo, violet. In software, hue may also be called *tone* or *tint*. See intensity, saturation.

Huffman encoding Huffman is a fast, variable-length, tree-oriented encoding scheme developed in the early 1950s by David Huffman. It optimizes on the basis of more frequently occurring characters in order to achieve compression in fewer bits. Since it produces a coding table that can be reused for additional image encodings, it is efficient for certain types of multi-image applications, though not especially efficient for very short messages (due to header overhead). Data corruption on a small scale can significantly affect the content of a message when decoded, so Huffman is best used with robust transmissions protocols with built-in error-checking. JPEG image compression is based, in part, on Huffman encoding. Modified Huffman (MH) is widely used in facsimile transmissions.

Hughes, David E. (1831–1900) An English-born American music teacher who developed one of the first printing telegraphs, in essence the telegram and later teletype machines. The telegraph, with improvements by George Phelps, formed the basis of the American Telegraph Company, in competition with Western Union. In 1858, Hughes returned to Europe to demonstrate and promote the system and remained there as a resident. Hughes also invented the carbon microphone, in 1877, an important contribution to telephony. See House, Royal; Hughes telegraph; Morse, Samuel F.B.; Phelps, George; telegram.

Hughes, David R. (ca. 1929–) A West Point graduate and retired U.S. Army Colonel, Hughes is an acknowledged pioneer in internetworking and educational applications in distance learning. He is credited with teaching the first online college credit courses (1983). Hughes designed and supported the Big Sky Telegraph network and the Montana state METNET.

Hughes Network Systems A company (actually a group of companies under the Hughes umbrella) which has been involved in satellite communications since the early launches and has developed a number of associated innovative technologies. One such product is DirecPC, which allows a satellite feed to connect with a personal computer for data communications. See Applications Technology Network Program, DirecPC.

Hughes telegraph The Hughes telegraph began as an idea for transcribing music and ended up as a printing telegraph that used a tone to synchronize the mechanism between the transmitting and receiving printers. It was developed by David E. Hughes at about the same time the House printing telegraph was being marketed commercially and was patented in 1856 (U.S. #14,917).

Power to the Hughes telegraph was provided by a weight-driven clock system, similar to a grandfather clock, thus making it possible for a single operator to use the system, as opposed to two operators for the competing House telegraph. The essential designs of the House and Hughes systems were the same, a wood cabinet equipped with a piano-style alphabetic keyboard and the various mechanisms mounted on top of the cabinet. They differed mainly in detail and in the way they were powered.

The mechanism was not perfect and needed some refinement to be commercially successful, but there was a demand for telegraph machines at the time, and the rights were purchased by a newly forming company called the American Telegraph Company. ATC turned to George Phelps to improve the system, a move that made the Hughes telegraph moderately successful in North America and highly successful in Europe after Hughes traveled there to demonstrate the system, beginning in 1858. The Hughes printing telegraph enjoyed a long working life of more than a century in some European locations. In North America, the House and Hughes telegraphs led to improved designs that were highly successful early teletype machines. See House telegraph; Hughes, David E.; Phelps Combination Printer.

Hull, Albert Wallace (1880–1966) An American physicist who made important contributions to X-ray crystallography and who developed a number of types of electron tubes, including the magnetron, a tube capable of generating microwave frequencies. Hull published a description of magnetron technology in 1921 in the *AIEE Journal*, V.40. The magnetron became important in the development of radar and satellite communications systems. See magnetron.

Human Computer Interface standards HCI. A series of protocol platform standards from the IETF, including but not limited to Common Desktop Environment (CDE), Complex Text Layout (CTL), Motif, etc.

Hunnings, Henry (1843–1886) A British clergyman inventor who developed a carbon granule-based hearing aid in 1878, which replaced the electric hearing aid developed by A. Graham Bell in 1876. Hunnings also applied the carbon granule technology to improve upon microphones invented by Thomas Edison. Hunnings' improvements led to sturdy devices with better sensitivity. Similar carbon granule technology was also developed by Francis Blake, Jr., to improve telephone transmissions, and both Blake and Hunnings telephones were prevalent at the time.

hunt, hunting A process through which a call is routed by seeking the best path or first available path or device.

hunt group In telephone systems, a series of lines set up so calls can be assigned to the next available line in a group if the first accessed line is busy. There are different ways to organize hunt groups, from a straight sequential hunt to random hunts.

Hush-a-Phone decision A landmark communications case in 1956 which challenged the AT&T monopoly of phone line access. The company marketing Hush-a-Phone Company and Harry C. Tuttle wanted to attach a mechanical device to a phone set in order to screen out background noise. AT&T argued against it, supported by the Federal Communications Commission (FCC), but the decision was later overturned in the court of appeals, the main arguments including the mechanical rather than electrical nature of the device and the fact that it in no way harmed the phone equipment. A Hush-a-Phone is on exhibit at the Museum of Independent Telephony in Abilene, Kansas. See Carterfone decision.

Huygen's integral, Huygen's principle Wavefronts can be mathematically decomposed (integrated) into a series of point sources, with each seen as the origin of an expanding, spherical wavelet that can be represented as a free space Green's function. At any particular moment, the wavefront's shape comprises the envelope that includes the secondary wavelets. This is a useful though not complete representation of what actually occurs, as it does not fully account for diffraction interactions or differences in wavelength (e.g., between light and radio waves). See Huygen-Fresnel principle, Snell's law.

Huygen-Fresnel principle Fresnel and other scientists studied Huygen's principle, filled in the mathematics, and made some adaptations to the concept to account more other aspects, such as interference. The Huygen-Fresnel principle states that at a given instant, every unobstructed point of a wavefront serves as a source of expanding spherical secondary wavelets with the same frequency as the primary wave and that the amplitude of the optical field at points beyond the envelope (that was specified by Huygen) comprises the superposition of all the wavelets as to their amplitudes and relative phases.

This led to a more complete though still not fully developed description of wavefront behavior. Kirchhoff enlarged on the concept further by taking into consideration obliquity and his approach is still widely used.

From the Huygen concept, the Fraunhofer approximation, which is important in the study of optics and ultrasound, can be derived. See diffraction, Fourier transform, Huygen's integral, interference, wavelet.

HW-16 A three-band radio transmitter and receiver housed together in one case, by Heathkit. The transmitter was crystal-controlled and the device was electron-tube based, with the exception of one transistor component. It was popular with novice amateur radio operators in the late 1960s and early 1970s. The unit could be upgraded as the radio operator learned more and received an operator's license. Sockets on the front panel enabled crystals to be inserted. The HW-16 was most effective over the 40- and 80-meter

bands and can still be used over those frequencies.

Hybrid Fiber Coax HFC. A transmission system combining fiber optic cable with coaxial cable that can handle simultaneous analog and digital signals. It is less expensive than a full fiber or switched digital video installation but still provides greater bandwidth than traditional technologies built entirely on copper wire or coaxial. Network technologies such as ATM, SONET, and frame relay can be transmitted over HFC. Discrete wavelet multitone (DWMT) is being proposed as a suitable modulation scheme for existing HFC installations. See discrete wavelet multitone, Hybrid Fiber Coax architecture, SONET.

Hybrid Fiber Coax architecture A hybrid fiber coax technology for carrying video or telephone services, or a combination of both. For video, the bandwidth is typically divided into channels which can further be subdivided into phone lines. It is primarily a downstream technology, which serves broadcast TV very well, but it may not be as flexible for interactive TV and phone services. The downstream nature is not inherent in the cable, but rather in the transmission and amplification technology. Typically, optical fiber runs from the central office to a node servicing an area neighborhood. From that point, the signal can be converted to be carried via coaxial cables to individual subscribers. At the subscriber point, a device splits the video and telephone signals so they can be directed to the appropriate lines or devices within the premises.

This hybrid system balances some of the speed and bandwidth of a full fiber-based system, with some of the economic advantages of coaxial servicing individual neighborhoods. One disadvantage is that there is not an unlimited amount of bandwidth available for phone lines, and phone service must be planned and adjusted as needed. HFC technologies can put cable companies in a position to compete with telephone providers, which may create a shift in future market share. See Hybrid Fiber Coax.

Hyde, J. Franklin (1903–1999) An American chemist and inventor educated at Syracuse University, Franklin worked as a post-doctoral organic chemist at Harvard, then went to work at Corning Glass Works, in 1930. He remained with Corning Glass Works and then Dow Corning Corporation (which was founded as a result of his work) until 1975.

Hyde developed a means for converting silicon compounds to silicone and conducted extensive research on silicone rubbers. He is a significant pioneer in glass manufacture as the inventor of fused silica glass (patent #2,272,342). This invention is important to many areas of optics, including semiconductor technologies and optical lenses. It also formed the historic basis for practical fiber optic waveguides made from low-loss pure glass. Hyde was inducted into the National Inventor's Hall of Fame in 2000 for his development of silica glass. In 1992, the J. Franklin Hyde Scholarship in Science Education award was established in his honor. See Kao, Charles K.

hydraw See octopus.

hydroelectric Electrical power derived from the energy

provided by rapidly moving water or channeled water under pressure. Large dam projects are usually linked to a desire for hydroelectric power for residential, urban, and industrial purposes.

hydrolysis A process of chemical decomposition that occurs in the presence of moisture such that new compounds result from the reaction. Hydrolysis is of concern in maintaining insulating materials in underwater cable installations. Hydroxyl ions (OH-) may result from the ionization of water. See hydroxyl group.

hydrometer An instrument that measures, by displacement, the specific gravities of liquids. Used, for example, to measure the electrolytes in batteries.

hydroxyl group A group of atoms that occurs in organic molecules, a type of *functional* group. Hydroxyl comprises a polar group with a single oxygen atom (O) bound to a single hydrogen atom (H). The presence of a functional group alters a molecule's chemical properties. Hydrocarbons containing a hydroxyl functional group are known as alcohols. A metal joined with a hydroxyl groups is called a *metal hydroxide*.

Acidic concentrations have a low hydroxyl ion (OH-) to hydrogen ion (H) ratio. Base concentrations have a high hydroxyl ion to hydrogen ratio. If they are in balance, the concentration is neutral (pH 7).

Hydroxyl occurs naturally through hydrolysis and ozone photolysis reactions. Hydroxyl radicals are important in hydrocarbon reactions in the troposphere during the day. Hydroxyl groups also occur naturally in "spectral tuning" proteins in the eyes of creatures that sense color.

In the optoelectronics industries, the hydroxyl ions present in water molecules can interfere with optical transmission by absorbing some wavelengths. The term High-OH stands for "high hydroxyl content" and refers to transmission systems that are most efficient in the ultraviolet and visible spectra. Similarly, low-OH is most efficient in the near-infrared and visible spectra. Low-OH, clear, fused quartz is used in UV lamps, thermocouples, lightguides, fiber optic probes, and a variety of types of semiconductors. Low-OH synthetic fused quartz is used in components such as semiconductors and UV-illuminator sleeve tubings. See hydrolysis.

hygroscope An instrument for measuring the amount of moisture in a material. Handheld paper hygroscopes are commonly used in the printing industry to monitor paper moisture balance and the relative humidity of the air in order to adjust printing materials and processes for quality control. See densitometer.

hygroscopic 1. A material with a tendency to absorb and retain moisture. 2. A material that is able to absorb and retain moisture.

hyperlink, hypertext link A logical link between meaningful data organized within a random access database or markup language. Hyperlinks can be hierarchical or flat. They can be one-directional or bidirectional. Although hypertext links are most familiar to users in the form of virtual cards in a computer card catalog or as browser-accessible links on the

World Wide Web, a hyperlink in its broadest sense also applies to interconnected visual image links, where the user clicks on an icon or a part of a picture rather than on a word or block of text.

Hyperlinks on the Web have opened up global Internet interactions and cross references to immense, shared information storehouses. There are a number of popular games that are navigated through text or visual links. See browser, hypertext transfer protocol, World Wide Web.

hypertext A means of accessing information through referential links. This idea has been around for a long time and has had various implementations, with Bush developing a microfilm system and suggesting associative indexing in the 1940s. In the 1960s, D. Engelbart developed the On Line System (NLS) for storing research papers and other information in a hypertext-like manner. A number of other hypertext systems have been developed by various researchers, but the implementation of the concept on computer networks did not become commonly understood and recognized until the distribution of HyperText on the Macintosh computer in the late 1980s.

The most significant implementation of hypertext, which serves as a simple front-end to the Internet in the form of Web pages, is the Hypertext Markup Language. Hypertext tags can be imbedded in Web pages to allow them to connect to any other public page on the Internet. See Hypertext Markup Language.

Hypertext Markup Language HTML. A simple markup language for creating platform-independent hypertext documents for display and distribution over a computer network. HTML is a generic semantics implementation of Standard Generalized Markup Language (SGML – ISO 8879:1986).

HTML has a simple tag-based syntax that can be readily learned, for basic page display, by individuals without much prior programming skills. It can be readily configured to link to graphics and documents in other locations on the Web. Thus, HTML has become widely used on the Web to represent hypermedia, documentation with inline graphics, database query results, news, stock reports, course outlines, storefronts, and discussion lists.

The formal definition of HTML syntax is described in the HTML Document Type Definition (DTD).

HTML was designed by Tim Berners-Lee at CERN and has been in use since 1990 by the World Wide Web global information initiative. The Document Type Definition (DTD) was written by Dan Connolly in 1992. In 1993 a number of contributors provided enhancements, and the incorporation of NCSA Mosaic software allowed the inclusion of inline graphics. Dave Raggett derived forms material from the HTML+ specification.

In 1994, the HTML Specification was rewritten by Dan Connolly and Karen Olson and edited by the HTML Working Group, with updates by Eric Schieler, Mike Knezovich, and Eric Sink from Spyglass, Inc. Finally, the entire draft was restructured by Roy Fielding. The development and use of Web browsers began to spread.

H

Since then, the number of users of the Web, interacting through HTML, has climbed to more than 40 million, and many millions have authored personal, institutional, and commercial Web pages using HTML.

HTML has undergone a number of updates and revisions since its initial introduction. HTML pages are often enhanced with applets created with Sun's Java programming language and may also be enhanced with ActiveX objects. See browser, hypertext, RFC 2070 (Internationalization).

Hypertext Transfer Protocol HTTP. An application-level, generic, stateless, object-oriented network protocol intended for quick-access, distributed, collaborative, hypermedia systems. HTTP uses typed data representation, allowing system-independent data transfer. HTTP use is widespread, as it has been an intrinsic part of the World Wide Web initiative since 1990 and has been widely incorporated into Web servers and clients. It also provides a generic means of communication between user agents and proxies/gateways, and Internet protocols for email, search engines, and file servers.

HTTP communications on the Internet are typically over TCP/IP connections, with a default port of TCP 80. A *message* is the basic unit of HTTP communication, which uses a request/response paradigm for serving information. Once a connection is established between the client and the server, the client sends a request, and the server responds with control and error information and, if the request is successful, the requested content.

The syntax of the HTTP URL is as follows:

`http://<host>:<port>/<path>?<searchpart>`

There are other transfer protocols as well, including File Transfer Protocol (FTP), for the transmission of files over the Web. When used with a Web browser, it has a similar syntax to an HTTP URL, except that the prefix ftp: is used rather than http:. See File Transfer Protocol, MIME, Secure HTTP.

hysteresis 1. The diminution or retardation of effects upon a body from a force, when the force acting upon the body changes. For example, in a body that is magnetized by a changing magnetizing force (e.g., an electromagnet with a varying current), hysteresis is the amount by which the magnetic values of the body lag (due to friction or viscosity, etc.) behind those of the magnetizing force. 2. The difference in response of a system to a varying force or signal. 3. The difference in the ability of a system or device to respond and change according to a sudden force upon it. To give a simplified example, stomping on a car accelerator or brake does not result in an instantaneous change to a new speed. Hysteresis is the delay effect between the stomping action and the response of the vehicle to the action. Sports car drivers experience less hysteresis than motorhome drivers.

hysteresis curve A diagrammatic representation of a magnetizing force and its related magnetic flux.

In a hysteresis curve for magnetic materials that are subjected to a magnetic influence, then separated from the influence, then magnetized and separated again, it can be seen that materials retain some of their original magnetism after removal of the magnetic influence. This property can be shown to vary among substances by means of a hysteresis curve diagram. Thus, materials with a narrow curve are suitable for the cores of electromagnets in industrial applications; those with wide curves can retain their magnetic properties and are used accordingly.

hysteresis device A device or circuit intended to mediate a situation in which power levels, or other important operating aspects, are fluctuating outside of normal parameters. An example of such a device would be an emergency system that switches to reserve generators or battery power when voltages drop or which draw off extra power if voltages spike. In this case, a delay mechanism (a hysteresis device) may be deliberately introduced in order to prevent constant fluctuation or *fluttering*, so that the system switches to a reserve system only after a sustained or significant change in power levels occurs. Without the hysteresis device, the system might otherwise be constantly switching back and forth between main and reserve systems, a situation that would be impractical and perhaps even dangerous.

Hysteresis devices/circuits are also important in mobile communications based upon cell transmissions. In these systems, a person using a mobile phone, for example, may be traveling along a path that passes several transceiving stations. As the user moves from one transceiving region to another, the signal will change. Buildings and bridges will also affect the strength of the signal reaching the mobile unit. Mobile systems are designed to assess the incoming signals and select the best one, a situation that may change from moment to moment, especially in a moving vehicle. Whenever there is a switch from one transceiver to another, there is a slight interruption during which the system adjusts its settings. If the user is in a locality where there are several signals of similar strength, the unit could try to constantly switch from one to another, causing interruptions and inefficient use of the resources. A hysteresis circuit helps to ensure that excessive adjustments are not made.

Hz See hertz.

I 1. *symb.* current. 2. *symb.* incident (as in incident light). 3. A symbol commonly used to designate the "on" position on a rocker switch, with **O** commonly used for "off." 3. *abbrev.* intensity. The I is usually indicated on or near an analog dial on a computer monitor or TV screen, to allow the user to increase or decrease the amount of illumination of the display.

I interface, Inter-Service Provider interface An interface between two Cellular Digital Packet Data (CDPD) networks deployed over AMPS. See A interface, E interface.

I & R *abbrev.* installation and repair.

I Series Recommendations A set of ITU-T recommendations that provides guidelines for ISDN. These are available as publications from the ITU-T for purchase and a few may be downloadable without charge from the Net. Some of the related general categories and specific I category recommendations that give a sense of the breadth and scope of the topics are listed here. Since ITU-T specifications and recommendations are widely followed by vendors in the telecommunications industry, those wanting to maximize interoperability with other systems need to be aware of the information disseminated by the ITU-T. See also similar listings under Q, V, and X Series Recommendations that describe other aspects of telecommunications. See ITU-T I Series Recommendations chart following.

I signal One stream in a split signal in certain modulation systems. The transmission may be split into two streams: one is the *in-phase* or *I signal*; the other is the *quadrature-phase* or *Q signal*. In various data transmission schemes, it is common to split a signal and to alter the characteristics of one or both of the data streams so that they can be transmitted together without interfering with one another or creating excessive crosstalk. The signals may then be recombined or synchronized at the receiving end.

Streams may also be split according to their different transmissions needs, as in speech, which can be sent on voice grade lines, and graphics, which require better and wider transmissions media.

When prototyping hardware transmission devices (e.g., GPS receivers), it is sometimes expedient to model the behavior of the transmission in a computer simulation before building the hardware. Thus I/Q signals and other signal schemes are simulated according to their known mathematical properties. See quadrature amplitude modulation.

I-frame intra-coded frame. In MPEG animations, a picture that has been encoded into a video frame without reference to past or later frames, using predicted motion compensation algorithms. See B-frame, I-picture.

I-picture intra-coded picture. In MPEG animations, a picture that is to be encoded into a video frame without reference to past or later frames to prevent reference image errors. Once it is encoded, it is considered to be an I-frame. See MPEG encoder.

I-TV See interactive television.

I-way *slang.* An expression for the growing global telecommunications network, derived from a shortening of the phrase "Information Super Highway."

I/O input/output. Generally used in the context of computers as meaning input from users, applications, or processes and output to devices, applications, or processes. See input, input device, output, output device.

I/O bound input/output bound. A processor subjected to a processing load in excess of what it was designed to handle, or which causes processes and response time to be uncomfortably slow for the user, is said to be I/O bound. There are a number of ways to reduce the incidence of I/O congestion: more efficient algorithms; co-processing chips for computing intensive operations such as graphics, sound, or device management to ease the load on the central processing unit (CPU); faster CPUs; reconfiguration or reorganization of peripheral devices; distributed processing over a network, etc.

I/O device input/output device. A piece of computer hardware physically interfaced with a system, and electrically and logically configured to engage in two-way communication with the operating system and relevant applications. Many computing devices are primarily input or output devices. Joysticks and mice are primarily input devices; speakers and printers are primarily output devices. (Few devices are strictly one or the other, since signal processing, device status, and handshaking signals often are returned by the device to the system to improve the efficiency of their use.) Most monitors are output devices, but touchscreen

ITU-T I Series Recommendations

Recom.	Date	Description
I.112	1993	Vocabulary of terms for ISDNs
I.113	1997	Vocabulary of terms for broadband aspects of ISDN
I.114	1993	Vocabulary of terms for universal personal telecommunication
I.230	1988	Definition of bearer service categories
I.240	1988	Definition of teleservices
I.250	1988	Definition of supplementary services
I.371.1	1997	Traffic control and congestion control in B-ISDN: conformance definitions for ABT and ABR
I.311	1996	B-ISDN general network aspects
I.313	1997	B-ISDN network requirements
I.324	1991	ISDN network architecture
I.220	1988	Common dynamic description of basic telecommunication services
I.241.8	1995	Teleaction stage one service description
I.210	1993	Principles of telecommunication services supported by an ISDN and the means to describe them
I.310	1993	ISDN – network functional principles
I.319	1992	Principles of intelligent network architecture (also Q.1201)
I.330	1988	ISDN numbering and addressing principles
I.334	1988	Principles relating ISDN numbers/ subaddresses to the OSI reference model network layer addresses
I.410	1988	General aspects and principles relating to recommendations on ISDN user-network interfaces
I.510	1993	Definitions and general principles for ISDN interworking
I.601	1988	General maintenance principles of ISDN subscriber access and subscriber installation
I.610	1995	B-ISDN operation and maintenance principles and functions
I.620	1996	Frame Relay operation and maintenance principles and functions
I.120	1993	Integrated services digital networks (ISDNs)
I.121	1993	Broadband aspects of ISDN
I.122	1993	Framework for frame mode bearer services
I.130	1988	Method for the characterization of telecommunication services supported by an ISDN and network capabilities of an ISDN
I.140	1993	Attribute technique for the characterization of telecommunication services supported by an ISDN and network capabilities of an ISDN
I.141	1988	ISDN network charging capabilities attributes
I.150	1995	B-ISDN asynchronous transfer mode functional characteristics
I.200	1988	Guidance to the I.200-series of recommendations
I.211	1993	B-ISDN service aspects
I.221	1993	Common specific characteristics of services
I.470	1988	Relationship of terminal functions to ISDN
I.500	1993	General structure of the ISDN interworking recommendations
I.501	1993	Service interworking
I.511	1988	ISDN-to-ISDN layer 1 internetwork interface
I.515	1993	Parameter exchange for ISDN interworking
I.520	1993	General arrangements for network interworking between ISDNs
I.525	1996	Interworking between networks operating at bit rates less than 64 kbps with 64 kbps-based ISDN and B-ISDN
I.530	1993	Network interworking between an ISDN and a public switched telephone network (PSTN)
I.555	1993	Frame Relaying bearer service interworking
I.570		Public/private ISDN interworking
I.571		Connection of VSAT-based private networks to the public ISDN
I.580	1995	General arrangements for interworking between B-ISDN and 64 kbps-based ISDN
I.581	1997	General arrangements for B-ISDN interworking

ISDN BRI and PRI

Recom.	Date	Description
I.420	1988	Basic user-network interface
I.430	1995	Basic user-network interface – layer 1 specification
I.421	1988	Primary rate user-network interface
I.431	1993	Primary rate user-network interface – layer 1 specification

Protocols

Recom.	Date	Description
I.241.8	1995	Teleaction stage one service description
I.320	1993	ISDN protocol reference model
I.321	1993	B-ISDN protocol reference model and its application

ISDN – various

I.325	1993	Reference configurations for ISDN connection types
I.327	1993	B-ISDN functional architecture
I.328	1992	Intelligent Network – Service plane architecture (also Q.1202)
I.329	1992	Intelligent Network – Global functional plane architecture (also Q.1203)
I.331	1997	International public telecommunication numbering plan
I.333	1993	Terminal selection in ISDN
I.340	1988	ISDN connection types
I.350	1993	General aspects of quality of service and network performance in digital networks, including ISDNs
I.351	1997	Relationships among ISDN performance recommendations
I.352	1993	Network performance objectives for connection processing delays in an ISDN
I.353	1996	Reference events for defining ISDN and B-ISDN performance parameters
I.354	1993	Network performance objectives for packet mode communication in an ISDN
I.355	1995	ISDN 64 kbps connection type availability performance
I.357	1996	B-ISDN semi-permanent connection availability
I.364	1995	Support of the broadband connectionless data bearer service by the B-ISDN
I.370	1991	Congestion management for ISDN Frame Relaying bearer service
I.371	1996	Traffic control and congestion control in B-ISDN
I.372	1993	Frame Relaying bearer service network-to-network interface requirements
I.373	1993	Network capabilities to support Universal Personal Telecommunication (UPT)
I.374	1993	Framework recommendation on "Network capabilities to support multimedia services"
I.376	1995	ISDN network capabilities for the support of the teleaction service
I.411	1993	ISDN user-network interfaces – references configurations
I.412	1988	ISDN user-network interfaces – interface structures and access capabilities
I.413	1993	B-ISDN user-network interface
I.414	1997	Overview of recommendations on layer 1 for ISDN and B-ISDN customer accesses

B-ISDN – Physical Layer Specification

I.432	1993	B-ISDN user-network interface – physical layer specification
I.432.1	1996	B-ISDN user-network interface – physical layer specification: general characteristics
I.432.2	1996	B-ISDN user-network interface – physical layer specification: 155,520 kbps and 622,080 kbps operation
I.432.3	1996	B-ISDN user-network interface – physical layer specification: 1544 kbps and 2048 kbps operation
I.432.4	1996	B-ISDN user-network interface – physical layer specification: 51,840 kbps operation
I.432.5	1997	B-ISDN user-network interface – physical layer specification: 25,600 kbps operation

Multiplexing

I.460	1988	Multiplexing, rate adaption, and support of existing interfaces
I.464	1991	Multiplexing, rate adaption, and support of existing interfaces for restricted 64 kbit/s transfer capability

ATM-related

I.326	1995	Functional architecture of transport networks based on ATM
I.356	1996	B-ISDN ATM layer cell transfer performance
I.361	1995	B-ISDN ATM layer specification
I.363	1993	B-ISDN ATM adaptation layer (AAL) specification
I.363.1	1996	B-ISDN ATM adaptation: type 1 AAL
I.363.3	1996	B-ISDN ATM adaptation layer specification: type 3/4 AAL
I.363.5	1996	B-ISDN ATM adaptation layer specification: type 5 AAL
I.365.1	1993	Frame Relaying service specific convergence sublayer (FR-SSCS)
I.365.2	1995	B-ISDN ATM adaptation layer sublayers: service-specific coordination function to provide the connection-oriented network service
I.365.3	1995	B-ISDN ATM adaptation layer sublayers: service-specific coordination function to provide the connection-oriented transport service
I.365.4	1996	B-ISDN ATM adaptation layer sublayers: service-specific convergence sublayer for HDLC applications
I.731	1996	Types and general characteristics of ATM equipment
I.732	1996	Functional characteristics of ATM equipment
I.751	1996	Asynchronous transfer mode management of network element view

monitors are used for both input and output. Keyboards are typically input devices, except for those that have small LED displays to send configuration, status, or numeric keypad calculator information to the user. See input device, output device.

I/P input.

i.LINK See FireWire.

IA 1. See implementation agreement. 2. See intelligent agent.

IAB See Internet Architecture Board.

IAC 1. See Industry Advisory Council. 2. See Infor-

mation Access Company. 3. See Information Analysis Center. 4. See Institute for Advanced Commerce. 5. See interactive asynchronous communications. 6. See interapplication communications. 7. See Internet Access Coalition.

IAD See Integrated Access Device.

IAHC See Internet International Ad Hoc Committee.

IAM 1. incoming address message. 2. See initial address message. 3. intermediate access memory.

IANA See Internet Assigned Numbers Authority.

Sampling of IBM Desktop Computers

Type	Abbrev.	Description
IBM Portable Computer	IBM 5100	A larger desktop-sized computer introduced to businesses and educational institutions in September 1975. The Portable Computer came in a number of configurations, with varying amounts of memory up to 64K (a lot of memory in those days) at a cost ranging from just under $9000 to almost $20,000 (nearly the price of a house).
IBM Computing System	IBM 5110	A small-scale "affordable" computer that was transitional between high-priced desktops and mainframes, and later systems known as personal computers or microcomputers. Announced in January 1978, the 5110 was aimed at a wide portion of the business market, the market successfully penetrated by Tandy/Radio Shack Computers and the later line of IBM Personal Computers. It was available in configurations of up to 64K of memory.
IBM Personal Computer	IBM 5150	The first relatively low-cost personal computer introduced by IBM to realize significant sales to general consumers. The IBM PC was launched in 1981 to compete mainly with Tandy Radio Shack computers making big inroads in both hobbyist and business markets. Due to its better reputation for service and its licensing agreements with third parties, IBM eventually succeeded in taking the majority business market away from Tandy. Tandy did some things right: they opened a chain of computer centers to support the machines and to provide customer service.
IBM PCjr		An Intel 8088-based microcomputer, introduced in the early 1980s. The PCjr was intended as a low-cost home alternative to the IBM Personal Computer XT by IBM.
IBM Personal Computer	XT	Extended Technology. An Intel 8088-based microcomputer, introduced in 1983. The processing speeds of the various models of XTs ranged from 4.77 to 10 MHz (turbo XTs), with 16-bit data buses. A clock/calendar chip was not standard. Microsoft BASIC was included in ROM, and the computer could use cassettes for program reads and writes. DOS 2.1 was optional, but was needed in order to read and write floppy disk drives.
IBM Personal Computer	AT	Advanced Technology. An Intel 80286-based 16-bit microcomputer, introduced in the fall of 1984 by IBM as an updated alternative to the IBM XT. The processing speed of the AT was 6 MHz, with 256 kilobytes of memory. It came configured with a 1.2-MByte floppy drive, but the 20-MByte hard disk, graphics adapter, and monitor were optional. A clock/calendar chip was built in.

IAP See Internet Access Provider.

IAPP See Inter-Access Point Protocol.

IARL See International Amateur Radio League.

IARU See International Amateur Radio Union.

IBC See Integrated Broadband Communications.

IBM See International Business Machines.

IBM clone See IBM-compatible.

IBM smaller scale computing systems A series of desktop computers has been marketed by IBM for business, educational, and home markets since the 1970s. The first models were compact but expensive, costing nearly as much as a house, and accessible only to corporations or institutions with larger budgets. However, with the success of the MITS Altair personal computer, in 1975, and the introduction of low-cost desktop computers by other companies, it became clear that the market for computers was changing and IBM's ballpark price and promotional campaigns had to be adjusted to compete with startup companies developing new systems.

The first small-scale computer system intended by IBM to significantly exploit the new competitive market was the IBM 5110, announced in 1978. This lower cost successor to the IBM 5100 was unsuccessful in capturing popular attention, however, as it was overshadowed by the Tandy/Radio Shack TRS-80 and Apple computers and, to some extent, the Commodore PET.

Thus, in the early 1980s, IBM was scrambling to capture business and home markets, as demonstrated by their release of the IBM Personal Computer and the IBM PCjr. By the mid-1980s, they had successfully recaptured a large portion of the business market, however; the home market was still showing a preference for Apple, Atari, and Commodore-Amiga computers, while the educational market was largely based on Apple computers.

In the graphics industry, professionals were overwhelmingly selecting Apple Macintoshes over IBM computers due to the better graphics hardware and software available for desktop publishing and prepress. At the time, IBM computers were difficult to network and tended to be equipped with low-resolution monitors and text-based operating systems. It was not until about 1994 that IBM PCs made significant inroads in the publishing service bureau markets and, even then, companies adding IBM computers tended to hang on to Macintosh computers for large-scale printing jobs such as posters, billboards, etc., where reliable output and good printer drivers were important. Ironically, the reason IBM and IBM-licensed third party computers became better suited to the needs of the graphics sector was because of the market force for better games machines. Computer games require substantial computing resources, including more memory, faster processors, and better graphics and sound. With the exception of better sound, these are the same capabilities needed by the graphics industry, which they finally got in a roundabout way.

Over time, in business markets and, eventually, home markets, IBM computers prevailed. Part of the reason for the success of IBM-based computers in the mid-1980s was IBM's decision to license the technology to third-party manufacturers. Thus, IBM "clones" and IBM "compatibles" became prevalent in the mid-1980s, but IBM computers were preferred until about 1987, when it became clear to consumers that the quality of some of the clones was superior to IBM systems and, in many cases, less expensive. Once again, IBM had to adjust their marketing and manufacturing to compete with a market that was rapidly changing and evolving.

By the late 1980s, desktop computers were beginning to supersede mainframes for many types of computing applications in spite of the insistence of diehard mainframe reps that mainframes were here to stay. Since the mainstay of IBM up to this time had been their medium- and large-scale computers, it was important for the company to note and adjust to the increasingly important role played by microcomputers. By the mid-1990s, $1000 desktop computers were more powerful than many of the main- and miniframes sold to institutions a decade earlier for tens of thousands or millions of dollars. This remarkable trend for less expensive computers to have more and more powerful capabilities continues to this day. In chronological order, the Sampling of IBM Desktop Computers chart includes a brief list of early IBM computing systems. See IBM-compatible.

IBM Token-Ring See Token-Ring network.

IBM-compatible A de facto marketing term used by various companies to promote a desktop computer incorporating licensed Intel-based International Business Machines (IBM) technology to the extent that most, or virtually all software compatible with IBM personal computers would run on the third-party IBM-compatible machines.

IBN Institut Belge de Normalisation. A Belgian standards body of the Minister of Economic Affairs, located in Brussels. It is also involved in certification and accreditation activities. IBN is associated with the Comité Européen de Normalisation (CEN) and ISO. http://www.ibn.be/

IBS 1. See intelligent battery system. 2. See Intelsat Business Service.

IC 1. See integrated circuit. 2. See intercom. 3. interexchange carrier. See Inter Exchange Carrier. 4. intermediate cross-connect.

iCalendar, iCal An Internet calendaring and scheduling core object specification submitted as a Standards Track document by Dawson and Stenerson in 1998. The iCalendar spec is intended to provide a foundation for developing and deploying interoperable calendaring and scheduling services over the Internet. Since a number of different proprietary products from commercial vendors were beginning to be extended for use over the Net, a need was seen for defining a common format for the exchange of calendar and schedule information. Group or personal information managers may exchange information through the MIME content type defined in the specification.

As a result of interest in this most basic and common

I

type of application for Internet use, the Internet Engineering Task Force (IETF) initiated a calendaring and scheduling working group (CALSCH). Other protocols with a direct relationship to iCalendar interoperability have been defined, most of them arising from discussions of the CALSCH. CALSCH not only described and submitted specification drafts, but also administered interoperability testing. The work of CALSCH has also come to the attention of working groups developing separate but somewhat related protocols and formats, including the IPTEL working group.

The iCalendar spec is based upon an earlier vCalendar specification and has been further described in UML by Michael Arick as to its components, properties, and parameters related to the properties. In 1999, Mahoney and Taler submitted a draft of the Implementors' Guide to Internet Calendaring to aid in understanding the iCalendar effort and the relationships of the different protocols to facilitate the creation of conformant applications.

Progress towards a final embodiment dragged somewhat and the specification gained in complexity over time. Market-sensitive vendors began to be wary not only of the complexity but also of the time it was taking for the effort to solidify. CALSCH members eventually acknowledged that it might be best to simplify the project and denote some areas of iCalendar implementation as optional rather than mandatory, a move that sparked some renewed interest. See RFC 2445.

iCAL A Web-based commercial software calendar/scheduling utility available in personal and professional editions. Demo versions may be downloaded free from the Web.

Ical An X-based calendar/scheduling program developed by Sanjay Ghemawat. Version 2.0 was released in 1995. C++ Source code may be downloaded from the Web and through FTP.

ICAL See Internet Community at Large (natural history collections project).

ICALEP International Conference on Accelerator and Large Experimental Physics.

ICANN See Internet Corporation for Assigned Names and Numbers.

ICAPI 1. International Call Control API. 2. See Interface Control Application Programming Interface.

ICCB See Internet Configuration Control Board.

ICCC See Internet Channel Commerce Connectivity Protocol.

ICCF 1. See Interexchange Carrier Compatibility Forum. 2. International Civic Communication Forum. A nongovernmental organization (NGO) in the Ukraine, somewhat analogous to a nonprofit organization, that is assisting in providing guidelines for the establishment of further NGOs as democratic institutions. 3. See International Correspondence Chess Federation.

ICEA See Insulated Cable Engineers Association, Inc.

ICELAN 2000 A commercial software automation control system developed by IEC. It is a graphical network management and control product to provide

support and control over LONWorks nodes and applications. Based on Peak Components, this Windows-based software enables LONWorks users to install, maintain, schedule, and configure LONWorks networks. See LONWorks.

Birefringent Iceland Spar Mineral

Raw, transparent calcspar, called Iceland spar, prior to cleaving for use as refracting lenses.

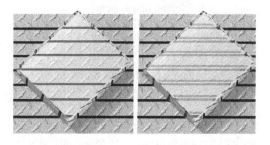

A rendered illustration of the birefingent refractive properties of Iceland spar (right), an anisotropic material, as compared to other common translucent isotropic materials such as glass (left). The directional crystal-like structure of calcite causes light to be refracted in two directions, resulting in a double image when viewed through the appropriate plane.

Iceland spar, calcspar (*symb.* – $CaCO_3$) A common, semihard, soluble, somewhat brittle, crystal-like mineral of the calcite group. Calcite is one of three mineral phases of calcium carbonate. It is the primary constitutent of limestone. The transparent form, known as Iceland spar or calcspar, was predominantly found in Iceland until sources began to dwindle and Mexico became a predominant supplier.

Depending upon impurities, calcspar may be pinkish, amber, or bluish and semitransparent. Impurities such as manganese enhance the mineral's ability to fluoresce under ultraviolet light.

Iceland spar resembles crystals when cleaved, as the faces within the material are rhomboid with blunted corners. In crystalline form, the structures are spiked. Iceland spar was widely used in optical instruments for almost 200 years.

Calcspar has the interesting property of doubling an image seen through certain planes. Combining two pieces will yield four images, depending upon the orientation of the pieces and the viewer.

In 1669, Danish physicist Rasmus Bartholin received a piece of calcspar from Iceland and studied its birefringent (doubly refractive) properties, publishing his findings in 1670. In 1690, Dutch physicist Christiaan Huygens enlarged upon the study of birefringence in a published treatise and described light emanations in terms of spherical wavelets. These observations and later ones by W. Wollaston and L. Malus were significant in the understanding of polarization and the wave nature of light.

In 1828, Nicol bonded two pieces of Iceland spar together with Canada balsam, which has a slightly different refractive index from the spar, thus developing a polarizing prism component. This became a mainstay of polarimeters and microscopes for almost 100 years. Calcspar is now used less frequently in specialized optical instruments, usually for polarizing light in the near-infrared and visible spectra. See birefringent, Nicol prism, refraction, Ulexite.

ICI See Interexchange Carrier Interface.

ICIA See Information and Communications Industry Association, Ltd.

ICM See Integrated Call Management.

ICMP See Internet Control Message Protocol.

ICO See International Commission for Optics in Appendix G.

ICO Galileo Galilei Award An international award given annually for outstanding contributions to the field of optics achieved under relatively unfavorable circumstances. In addition to the award, which was established in 1993 and has been awarded since 1994, the Italian Society of Optics and Photonics (Società de Ottica e Fotonica) donates a silver medal to the recipient. See ICO Prize.

ICO Global Communications A Craig McCaw company which at one time was planning a merger with Teledesic, LLC, the "Internet in the Sky" project underway by Gates and McCaw. This proposal was discontinued in late 2001. See New ICO.

ICO Prize An international award given annually to an individual who has made a noteworthy contribution to the field of optics that has been submitted for publication prior to the nominee reaching the age of 40. The award was established in 1982 and is administered by the ICO Prize Committee. In addition to a cash award, the Carl Zeiss foundation donates an Ernst Abbe medal to the ICO Prize winner. See Abbe, Ernst; ICO Galileo Galilei Award.

iCOMP Intel Comparative Microprocessor Performance index. A simplified means of evaluating and expressing relative microprocessor power, introduced by Intel in 1992. Intel, as a major vendor of microprocessing chips, sought a straightforward way to convey processor information to purchasers. The iCOMP is an index rather than a benchmark in a technical sense, as it narrowly describes instruction execution speed (not clock speed). Benchmarks involve sophisticated and careful evaluation of many performance factors, whereas an index is a basic indicator, in this case, a compilation based on four industry-standard benchmarks, without taking into consideration other aspects of the system architecture, including video display, device addressing, etc.

iCOMP is expressed on a comparative scale, which uses the instruction speed of the 25 MHz 486SX processor as a baseline, assigning it a value of 100, with subsequent processors rated relative to this.

icon 1. Pictorial representation, symbolic image, emblem. 2. In telecommunications documents and applications, a symbolic image, usually small and abbreviated, representing an object, program, state, or task. Visually similar iconic representations are sometimes used to show different aspects or states of the same thing, such as a *ghosted* icon to show something is in use or an iconized version of an application symbol to show something is loaded and available. Icons are used extensively in documentation and graphical user interfaces (GUIs) to represent concepts or contents. Some are specific to an application or platform, but some are common enough to be recognized across a variety of systems, e.g., folder icons to represent directories.

ICONET A satellite communications network being put into place by ICO Global Communications. See New ICO.

ICONTEC Instituto Colombiano de Normas Técnicas. A Columbian technical standards body.

ICTA 1. Idaho Cable Telecommunications Association. 2. Indiana Cable Telecommunications Association. 3. See Independent Cable & Telecommunications Association. 4. See International Center for Technology Assessment. 5. International Christian Technologists' Association. 6. See International Commission on Technology and Accessibility. 7. International Conference on Technology and Aging.

ID 1. identification, identifier. 2. See input device. 3. See integrated dispatch. 4. intermediate device.

IDA 1. See integrated data access. 2. integrated digital access. Digital systems intended to facilitate access to networks and/or application or information sources. The phrase is usually used with reference to systems where a number of features have been integrated into one easily used unit, such as a portable scheduler with a built-in wireless modem. 3. intelligent drive array. See RAID.

IDAI See Accessible Information on Development Activities.

IDCMA Independent Data Communications Manufacturers Association. An independent trade organization representing the interests of independent communications manufacturers. The IDCMA has spoken out on some of the FCC-related rulings regarding new technologies that may or may not be considered as customer premises equipment.

IDE See integrated development environment.

IDE devices and controllers Integrated Drive Electronics. A control mechanism and format for computer hard disk drive devices developed in 1986 by Compaq and Western Digital. IDE provides data transfer rates of about 1 to 3 Mbytes per second, depending upon other system factors, including the data bus. On common Intel-based microcomputers, the IDE uses an interrupt interface to the operating system. IDE has been highly competitive with the SCSI standard, another very common drive format. To get the production costs down, and because many Intel-based computers in the early 1980s did not come standard with controllers for extra peripherals, the IDE controller mechanism was incorporated into the drive. Each controller can handle two drives, a "master" and a "slave" (compared with seven, including controller, for SCSI).

IDE is more limited than SCSI (fewer devices can be chained, smaller addressable space, IRQs necessary, not compatible with RAID systems, etc.), but it is also less expensive and has become widely established. In order to overcome some of its limitations, a number of enhanced IDE formats now exist.

Most workstations and Motorola-based desktop computers (Suns, SGIs, Amigas, most Apple Macintoshes, NeXTs, and others) include SCSI controllers on the basic machine, making it unnecessary to purchase a separate drive controller to add SCSI peripheral devices to the computer. Some of the newer Macintosh and PowerMac computers support both SCSI and IDE. Most Intel-based desktop computers come with IDE controllers on the basic machine and SCSI controllers can be purchased as options. See FireWire, hard disk drive, SCSI.

IDEA See International Data Encryption Algorithm.

iDEN integrated Digital Enhanced Network. Digital phone technology developed and marketed by Motorola for workgroups. The phones may be used like two-way radios over a cellular network, thus overcoming the distance limitations of conventional portable two-way radios while also offering other services such as phone, messaging, and data transmissions. Fax and Internet access capabilities are also provided on data-ready iDEN units. The system is based on the concept of multiple workgroups communicating within a private virtual network that is part of a larger common infrastructure. See Enhanced Specialized Mobile Radio.

IDEN integrated digital electronic network.

identifier ID 1. In database management, a keyword used to locate information, or a category of information. 2. In programming, a variable name, extension, prefix, suffix, or other device to provide a means to easily recognize an element, or distinguish it from others.

IDF intermediate distribution frame. See distribution frame.

IDL See Interface Design Language.

IDLC See Integrated Digital Loop Carrier.

idle In a state of readiness, but not currently activated. Idle is often used as a power-saving measure, and may be a state in which only minimal power is used by the system until full power is needed, as in laptops that power-down the monitor and hard drives when they are not in active use.

idle channel code A repeated signal that identifies a channel that is available, but not currently in active use. See idle.

idle channel noise Noise in a communications channel that can be heard or occurs when no transmissions are active. For example, low level hums can often be heard in phone lines when no one is talking, but are not noticed when talking continues.

idle line cutoff In computer networks, it is not uncommon for Internet Services Providers (ISPs) or network administrators to set the system to log off any clients (machines or applications) that are inactive for longer than a specified period of time (e.g., 15 minutes). This frees up abandoned terminals or modem lines that are no longer in use.

idle signal 1. In networking, a channel which is open and ready, and which may be sending an idle signal, but through which no active or significant transmissions are occurring. 2. Any signal in a circuit intended to signify that no significant transmission is currently in progress. An administrative tool to allow potential users, operators, or operating software to detect available lines and put them into use, or to compile and record usage statistics for further evaluation and tuning of a system. See idle channel code.

IDML See International Development Markup Language.

IDML Initiative A collaborative initiative to improve global information exchange using XML in an international context through use of a standardized International Development Markup Language (IDML). http://www.idmlinitiative.org/

IDN See Integrated Digital Network.

IDSCP See Initial Defense Communications Satellite Program.

IDTV See Improved Definition Television.

IDU See Interface Data Unit.

IEC 1. See Inter Exchange Carrier. 2. See International Electrotechnical Commission. 3. See International Engineering Consortium.

IEEE Institute of Electrical and Electronic Engineers, Inc. The world's largest electrical, electronics, and computer engineering/computer science technical professional society, founded in 1963 from a merger of the American Institute of Electrical Engineers (AIEE) and the Institute of Radio Engineers (IRE). IEEE is a respected and influential organization that serves about a quarter of a million professionals and students in almost 200 countries. IEEE's activities are broad-reaching, including standards-setting, publications, conferences, historical preservation and study, and much more. See American Institute of Electrical Engineers, IEEE Standards Association, Institute of Radio Engineers, Organizationally Unique Identifier. http://www.ieee.org/

IEEE Canada Institute of Electrical and Electronic Engineers of Canada. The Canadian arm of the well-known IEEE, organized across the country into groups based on geographic regions. http://www.ieee.ca/

IEEE History Center The historical archive of the IEEE, including about 300 artifacts and a number of oral histories. The IEEE includes among its early members some of the pioneer inventors in the telecommunications field, including Thomas Edison and Nikola Tesla. It works in cooperation with the IEEE library in which IEEE publications are stored. See IEEE

IEEE Standards Association IEEE-SA. An international organization serving individual and corporate members with a portfolio of standards programs. IEEE-SA focuses on full consensus standards processes as well as innovative policies for standards development. It is affiliated with the IEEE and is empowered to formulate and promote international engineering standards to further globally beneficial applications of technology. Membership in IEEE-SA is not necessarily a requirement to participate in a standards working group. See IEEE.

IEEE 802.11 Standard for wireless local area networks (LANs) adopted in June 1997.

IEN See Internet Experimental Note.

IES, IESNA See Illuminating Engineering Society of North America.

IETF See Internet Engineering Task Force.

IF intermediate frequency.

IFax device An Internet-interfaced device capable of sending and/or receiving Internet facsimiles through existing Internet mail mechanisms as defined in RFC 822 and RFC 1123. In general, IFax formats must be MIME compliant.

IFax devices can also be used as gateways between the Internet and G3Fax (traditional) phone-based facsimile machines, with the IFax configured to handle the connection and dialup and any authentication necessary to prevent undue cost or unauthorized use. An IFax device can serve as a mail transfer agent (MTA) for one or more G3Fax devices. In general, Simple Mail Transfer Protocol (SMTP) should be used for such applications, although dedicated servers may use POP or IMAP. The IFax specification was developed by the IETF Fax Working Group and described as a Standards Track comment in 1998. See facsimile formats, G3Fax, TIFF-FX, RFC 2305.

IFCM See independent flow control message.

IFD See image file directory.

IFIP See International Federation for Information Processing.

IFRB See International Frequency Registration Board.

IFTL See Fiber in the Loop, Integrated.

IFWP See International Forum on the White Paper.

IGC intelligent graphics controller.

IGMP See Internet Group Multicast Protocol.

ignition Lighting, kindling, applying a spark so as to inflame or provide sufficient heat or current to set off a chain of events.

IGP See Interior Gateway Protocol.

IGRP See Interior Gateway Routing Protocol.

IGT Ispettorato Generale delle Telcomunicazioni. General Inspectorate of Telecommunications in Italy.

IGY International Geophysical Year.

IIA 1. See Information Industry Association. 2. See Irish Internet Association.

IICA See International Intellectual Capital Codes Association.

IIIA 1. See Integrated Internet Information Architecture. 2. See International Internet Industrial Association. 3. See Internet Information Infrastructure Architecture.

IIR Interactive Information Response.

IIOP Internet Inter-ORB Protocol. A wire-level communications protocol. See CORBA, Object Request Broker.

IISP 1. See Information Infrastructure Standards Panel. 2. Interim Interswitch Signaling Protocol. A basic call routing scheme which does not automatically handle link failures; routing tables established by the network administrator are used instead.

IITC Information Infrastructure Task Force.

IJCAI International Joint Conferences on Artificial Intelligence. An international biennial forum (in odd-numbered years) held since 1969.
http://ijcai.org/

ILD injection laser diode. See laser diode.

ILEC See Incumbent Local Exchange Carrier.

ILLIAC I A historic large-scale computer introduced in 1952 by the University of Illinois. It consisted of vacuum-tube technology and performed 11,000 arithmetical operations per second. See ENIAC, MANIAC.

ILLIAC II The successor to the ILLIAC I, the ILLIAC II was introduced in 1963. It was based upon transistor and diode technology and could perform up to 500,000 operations per second.

ILLIAC III The ILLIAC III was introduced in 1966. It was designed to process nonarithmetical data, and so was a departure from ILLIAC II, a special purpose machine.

ILLIAC IV Based on the new semiconductor technology, the ILLIAC IV was introduced in the early 1970s. It was logically designed after the Westinghouse Electric Corporation's SOLOMON computers developed in the early 1960s. The ILLIAC IV consisted of a battery of 64 processors which could execute from 100 million to 200 million instructions per second. It was significant not only for its speed, but also for the ability of its multiple processors to perform simultaneous computations. The services of the ILLIAC IV were made available to other institutions through high-speed phone line timesharing.

illuminator A radiant energy source that provides light which may be used directly or channeled through fiber optic filaments to another location. Illuminators commonly light buildings, microscopic stages, projectors, lighthouses, signal systems, calibration and aiming systems, and, when modulated, provide communications signals that can be sent over long distances through a lightguiding channel. Common sources of illumination for fiber optic systems are lasers and light-emitting diodes (LEDs).

Illuminating Engineering Society of North America IES, IESNA. A leading technical authority on illumination. For almost a century, the IESNA has been providing expertise on lighting practices through programs, publications, and services. Members include engineers, educators, scientists, manufacturers, and utility services personnel.
http://www.iesna.org/

ILMI See Interim Link Management Interface.

IMA Interactive Multimedia Association of Malaysia. Information about this standards-setting organization is available on the Web.
http://www1.jaring.my/cornerstone/ima/about.htm

IMAC See Isochronous Media Access Control.

image antenna A hypothetical antenna, used for mathematical modeling, defined as a mirror-image of an above-ground antenna, located below the ground symmetric to the surface, at the same distance as the actual antenna is above the surface.

image dissector A vacuum tube-based image scanning mechanism about the size and shape of a long flashlight, developed by television pioneer Philo T. Farnsworth in the early 1920s. It was a type of photomultiplier component enabling the transmission of straight line images by sweeping the image past an aperture at about thirty times per second. When Farnsworth applied for a patent for an electronic television system, in January 1927 (U.S. #1,773,980), the design included the "image dissector tube."

By the 1930s, Farnsworth had improved the technology so that it could transmit 300 lines per frame, leapfrogging over his competitor, John Logie Baird, in the U.K. At this point, Farnsworth began demonstrating and promoting his device in Europe. Gaumon British licensed the technology, with Baird in charge of incorporating the tube into a new television system. Baird took an unexpected path and hybridized Farnsworth's electronic system with a mechanical system and produced a 700-line image by 1935.

The technology caught the attention of other scientists. As RCA was working to simplify and improve upon photomultipliers, J. Pierce and W. Schockley (coinventor of the transistor) at Bell Laboratories were working on the concept as well, in the late 1930s. R. Winans joined Pierce in the effort and they published their results in the early 1940s. The mechanisms developed at that time are similar to those in use today.

The invention and its evolutionary descendents was an important component in image display/reading products and some military guidance systems for several decades until solidstate components began replacing vacuum tubes. Working as a television pickup it made it possible to broadcast film programs for television broadcasting. It differed from Zworykin's Iconoscope in light sensitivity and storage capability and thus was not as well-suited as the Iconoscope for broadcasting live performances.

Over the years, image dissectors became more compact than Farnsworth's original tube, measuring about 2-in. diameter by the 1940s and half that by the 1960s, while retaining the same general structure. By the 1950s, the basic flashlight-shaped image dissector had been integrated with a bulbous camera tube for commercial television sets.

Vacuum-tube-based image dissectors have been incorporated into optical readers, electronically scanned spectrometers, industrial defect detectors, and electronic astronomical star trackers. Modern versions have been used to image synchrotron radiation emissions in conjunction with phase-locked radio frequency signals, similar to the functioning of a stroboscope. The image can be scanned and viewed on an oscilloscope.

Image dissectors have applications in current optical systems. Goldstein et al. have developed an acousto-optical laser-scanning confocal microscope incorpor-

---

Fiber Optic Illumination Source and Beam Conditioning

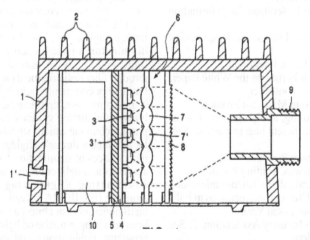

This example of a patented illumination source for fiber optic systems illustrates many of the basic concepts and components related to fiber light sources. The initial light source is provided by an array of light-emitting diodes (3, 3') from which the light beams propagate outwards (like flashlight beams). The light beams encounter an array of lenses matched to the LEDs. The lenses align the beams so they travel in congruent rather than varying paths (collimation). The collimated beams then pass through a transparent Fresnel lens (8), with facets angled relative to the center to amplify and "concentrate" the beam (like a lighthouse assembly). This conditioned beam propagates toward the connector for the light pipe (9) where a fiber or fiber bundle can be coupled to the light source. [Maas et al., U.S. patent #6,402,347, awarded June 2002.]

ating an image dissector tube for measuring head motion inside MRI and PET medical imaging scanners and image dissectors are used for certain photonic space-based star sensors.

See Baird, John; Farnsworth, Philo; iconoscope; photomultiplier; television history; Zworykin, Vladimir.

Farnsworth Dissecting Target Tube

Farnsworth continued to make adaptations to the dissecting tube technology he first conceived in 1922. In July 1930, he described a more sensitive dissector target tube, with amplification by secondary emissions through a "target" or auxiliary electrode, in his patent application for an electron discharge apparatus for the electrical scanning and transmission of television images.

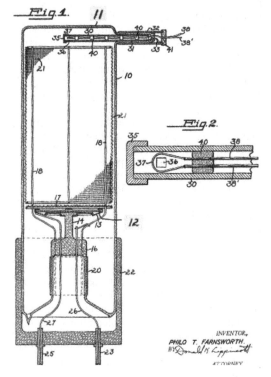

Excerpts adapted from the patent show the dissector tube (left) and a detailed view of the target component (right) housed at the top of the dissector tube. The image to be transmitted is focused through a window (11) just in front of the target and falls upon a photosensitive surface on the face of a front-silvered mirror (12). [U.S. patent #1,941,344.]

---

image file directory IFD. A data file structure providing location information for image data in the form of image information and data pointers. In a TIFF raster image file, for example, the IFD is an ordered sequence of tagged fields that begins on a word boundary somewhere after the header data. As more than one raster image may be in a TIFF file (as in a multi-page document or document with different versions of the same image), there can be more than one

IFD. Multiple IFDs can be organized as tree structures or as linked lists. See TIFF.

image intensifier In the optical spectrum, a device that increases the luminance of an image. The technology was first pioneered by French researchers in the 1930s and significantly developed two decades later by U.S. Army Corps engineer John Johnson under the direction of Robert Wiseman.

Image intensification in the optical spectrum is accomplished by means of a photocathode in a photomultiplier electron tube that amplifies the signal and turns it back into an image with increased luminance. Outside the optical spectrum (e.g., X-rays), the radiant energy is first passed through a scintillator, that converts the high-energy rays to frequencies in the optical spectrum, from which a photocathode can sense the signals. For imaging over a larger area than can be sensed by a single photocathode, a fiber optic faceplate may be used as a sensing source leading to an array of photomultipliers. Fiber optic filaments are sometimes also used to take the image out of the device and feed it into a computer or remote viewing device. The contrast or sharpness of the image may be processed with computer algorithms in conjunction with image intensification, prior to reconstruction of the intensified image.

The range of frequencies and the particular part of the spectrum that are intensified depend upon the system and the application. The system may be configured to intensify only certain parts of an image within stated bounds.

Image Intensifier – Basic Parts

The three basic components of an image intensifier include a photocathode (1) which emits electrons in proportion to the light falling upon the cathode, a microchannel plate (2), which is a finely fused array of glass channels coated with a resistive layer, and a phosphor screen (3). The phosphor screen typically emits light into a light-guiding component such as a fiber optic taper (4) which may be coupled to a charge coupled device (5).

The microchannel plate (MCP) provides a cascading amplifying effect to the electrons provided to it by the photocathode. The phosphor screen converts the amplified electron signal back into photons.

The fiber optic taper may be substituted with a fiber optic faceplate (a fiber optic array similar in form to the microchannel plate) and may further include a 180° twist in the fiber filaments to invert the image.

Image intensifiers are typically used with weak light

signals such as those from distant objects (stars or military targets) or night vision devices. An image intensifying system may be configured with shutoff circuits to prevent eye damage that could result from an image display that is too bright (as in night vision goggles). Commercial products are typically classified according to image resolution and degree of intensification that is possible.

Image intensifiers are also used outside the optical spectrum. They are particularly useful for intensifying X-rays, since there is a hazard involved with using strong X-rays and the lowest dosages possible should be used. Intensification enables the technician to get more information out of a weak signal. See Johnson, John.

Image Intensifier – Night Vision Goggles

Night vision goggles are an example of image intensifying technology. Fiber optic components are used in 2nd-, 3rd-, and 4th-generation image intensifier tubes that are used as step plates (2nd generation) and faceplates (3rd and 4th). Fiber optic inverters form the output component of the intensifier and may "flip over" the image inverted by an imaging lens. This is accomplished by twisting the axis of the fibers so the output at the end of the filaments is rotated 180° from the input light.

Officer M. Lewis of the U.S. Army positions night vision goggles on a flight helmet for a helicopter mission. [DefenseLink News Photo, April 1997.]

---

image inverter A software algorithm or physical mechanism for "flipping" an image right to left or top to bottom, or both. Many optical lenses will invert an image in the process of focusing or directing light through their structures and the image may need to be returned to its original orientation to be viewed. Many of the newer display devices, including wall-hung thin-screen displays and TFT displays for vehicles, are designed to invert an image so the display can be hung from the ceiling rather than being set on a shelf or mounted from the floor. An array of fiber optic filaments with a 180° twist may be used to invert an image that has been previously inverted with lenses or imaging plates.

imagesetter A professional-level graphics and type imaging machine, an imagesetter is similar to a high quality computer printer. Imagesetters are used in service bureaus, and traditional and digital printing houses, to create the image or the color separations used to control the ink distribution on the press. Typical resolution on these industrial quality machines is 1200 dpi to 2700 dpi (compared to 300 to 800 for most consumer machines) and they print on paper or film, or both.

While the distinction between consumer printers and imagesetters is blurring, with consumer printers now able to print up to 1200 dpi, there are still technical differences between commercial and consumer machines which are important to design, desktop publishing, and printing professionals.

Imagesetters do more than just print at higher resolutions; they also include more sophisticated and precise algorithms for halftone screens, may include higher quality fonts, may be able to print on special papers and even directly on aluminum, asbestos-based, or other more robust printing plate media. In addition, the distribution of the imaging materials on the printing medium is typically more precise and even. Further, a professional quality imagesetter has better alignment for subsequent printouts.

When printing color separations, especially for four- or five-color process printing, the consistency of the printing from one separation to the next is extremely important to the outcome of the final color printout, especially at resolutions of 175 lines or higher used in calendars, posters, and art prints.

In modern digital presses, the trend is to eliminate the separate imagesetter and incorporate the technology into the press itself. In the past, a computer file or traditionally photographed image was taken to a paper or metal plate through an imagesetter and, from there, the physical plate was attached to the press in order to create the printing job. Now it is possible to put a file on a floppy disk or cartridge and have the digital image sent directly to the press without the intermediary steps. It is even possible for a four-, five-, or six-color print job to be printed in one press run, rather than sending each color through the press in a separate pass, and aligning the plates each time. This new technology is revolutionizing the printing industry and eliminating a lot of intermediary steps and jobs in the process.

Image Inversion

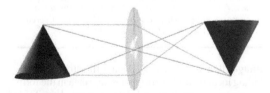

One of the more common ways to invert an image is to use a doubly convex lens, which directs the light to opposite sides from the incoming (incident) light. More recently, fiber optic arrays with 180° bends have been used to invert incoming light, sometimes to a remote viewing location.

IMAP See Internet Messaging Access Protocol.

IMASS Intelligent Multiple Access Spectrum Sharing.

Immediate Ringing A telephone or private branch system option in which there is no delay between the time of the reception of a call and the ringing of the telephone itself. Favored by those who want to provide quick responses to calls, such as emergency and crisis lines and certain businesses.

Immunity from Suit A legal agreement in which a license holder agrees not to sue the provider of a product or service. Microsoft and certain other large vendors are alleged to be asking for immunity from suit from some of their clients. In the author's opinion, purchasers should avoid signing any licenses that sanction neglect or mismanufacture on the part of the provider and should never sign anything that conflicts with constitutional rights or is coercive. Read license agreements carefully, *especially the small print*, and question and renegotiate anything that gives cause for concern.

IMPATT impact avalanche and transit time.

IMPDU See Initial MAC Protocol Data Unit.

impedance (*symb. – Z*) The total opposition, measured in ohms, offered by a circuit to the flow of alternating current (AC) at a given frequency. The ratio, in ohms, of the potential difference across a circuit to the current moving through that circuit. Design and insulating materials can substantially affect the level of impedance in a data cable, with low impedance cables generally costing more, but providing less noise and interference, and sometimes longer transmission distances. See admittance.

impedance bridge A device for measuring in ohms the impedance (combined resistance and reactance) of a portion of a circuit.

impedance compensator An electrical line which affects another circuit in such a way that the combination provides a desired consistent level across a specified frequency range. A compensator is used to minimize fluctuations and distortion.

impedance triangle A diagrammatic model for describing an impedance relationship. Imagine a right triangle with the sides respectively representing resistance and reactance, which change proportional to one another, and the hypotenuse representing impedance as related to the amount of the resistance and reactance combined.

implementation agreement IA. A generally agreed-upon means of describing a technology so that it can be put into production and/or commercial use. In order to support interoperability with global networks, many trade organization have adopted IAs as a means to standardize interfaces, protocols, and other network-related architectures and equipment in order to be able to produce commercially viable products. These IAs often become de facto standards or are integrated into the standards-development process of major standards-ratifying organizations. See Frame Relay Implementation Agreements.

import 1. Bring in from another source, region, or country. 3. Bring in from a non-native file format, protocol, or transmissions source.

import, file In software applications, to import is to bring in data from another program, file, or transmissions source, usually in a non-native file format. This is usually done through an applications filter or through drag-and-drop capabilities. In drag-and-drop imports, the program will either maintain links to the original imported file or convert the format to one consistent with the program into which it is imported.

import filter Many word processing, desktop publishing, and graphics programs have import filters, plugins, or modules which allow a number of file formats to be brought into an application and then saved in the native format of the application, or exported in the original format or a new one. This provides better compatibility between programs developed by different vendors. See export; import, file.

import script 1. A script which controls the assembly of a document by selectively importing information as specified. Often used in spreadsheets and databases. 2. A *very* handy feature of database software in which you can set up a form letter, and then have the script selectively build dozens or hundreds of personalized letters in a few minutes by automatically drawing in names, addresses, and variables from a database to merge with the form letter. Bulk mail companies often use import scripts to personalize letters, contest offerings, and envelopes. 3. In programming, an import script can set up documents or source code by selectively merging modules such as header files, modular routines, Unix "man" pages, etc.

Improved Definition Television IDTV. A picture broadcast and display system that provides better picture quality than conventional NTSC standards by incorporating field store techniques in the receiving circuitry. For example, the signal can be de-interlaced prior to display to reduce flicker. The originating signal is not changed.

Improved Mobile Telephone Service IMTS. Early mobile phone services were set up on systems based on large antenna transceivers with limited coverage and public operator-assisted broadcast services. The system had little flexibility or privacy, but it served as a forerunner for IMTS, in which the subscriber could place the calls directly; this in turn developed into current cellular systems where a larger number of smaller, automated transceiver systems allowed broader geographic coverage.

IMPS See Infinite Monkey Protocol Suite.

impulse 1. A nonrepetitive pulse so short as to be mathematically insignificant. 2. A very short nonrepetitive pulse which may not seem significant by itself, but which may impede transmission of the affected line or signal. Data transmissions are more sensitive to impulse interference than voice communications. 3. The uncontrolled desire to run out and get the latest techie toy, even though you don't really need it. Cell phones, faster computers, and scanners often fall into this category.

IMSAI 8080 An early 8080A-based microcomputer that used the MITS-developed Altair bus (S-100 bus); it was, in a sense, the first microcomputer *clone*. The 8080A was an enhanced version of the 8008 used on the first Altair. The IMSAI was introduced in 1975 by IMSAI Manufacturing with ads that compared it competitively against the Altair. See Altair.

IMSI See International Mobile Subscriber Identity.

IMTC The International Multimedia Teleconferencing Consortium. Their Web site includes information. http://www.imtc.org/imtc

IMTS See Improved Mobile Telephone Service.

IMUX See Inverse Multiplexer.

IMW See Intelligent Music Workstation.

in-band A transmissions scheme in which control and data signals are sent together over the same set of wires, or over the same frequencies, sometimes more or less simultaneously and sometimes interspersed with one another.

in-band signaling A type of signaling which is incorporated together with the data being transmitted. This is found, for example, in systems which encode signaling codes along with voice transmissions on the same wires (commonly copper twisted pair). In-band signaling has advantages and disadvantages. It doesn't require a separate set of wires to send control signals and thus is less expensive, but it does require more sophisticated handling of data and signals and has a higher potential for slowdown, errors, interference, or fraud.

In-band phone systems are at greater risk for security breaches and unauthorized use of services, because users can send in-band signals over the voice line and control certain telephone functions with illegal control devices such as blue boxes.

Newer out-of-band phone systems, based for example on Signaling System 7 (SS7), make unauthorized use through control signals on the transmissions line impossible, and these types of networks are increasing in prevalence as older equipment is replaced by newer networks. See ISDN, Signaling System 7.

in-line device A hardware device, commonly a peripheral which can be interposed between two other devices without interfering with the operation of the other devices, or intended to interface between two other devices to perform its function (and may or may not change the functioning of the other devices). Daisy-chainable devices are a type of in-line device, though not all in-line devices can be daisy-chained. See daisy chain.

INA See Information Network Architecture.

incandescent lamp A common type of illuminating bulb developed by Thomas Edison, originally consisting of a carbonized filament in a glass globe from which the air had been removed. However, the carbon tended to blacken the inside of the bulb and other solutions were sought, with tungsten coming into general use because of its high melting point. Experimentation with the internal environment of the bulb also resulted in the discovery that various gases could alter the glow or extend bulb life.

INCC Internal Network Control Center

incidence angle, angle of incidence The angle between the line of travel of radiant energy and the normal (perpendicular) of a reference surface (usually in the path of the emission). When the angle of incidence

is equal to the normal of a reflective surface, the radiant energy will be reflected back in the same direction from which it came, resulting in a Littrow condition. When the angle of incidence of radiant energy encountering an obstacle diverges from the normal of that object, interference patterns may result.

In optics, when reflective surfaces are built into a component, as in fiber optics cladding or a diffraction grating, the direction of travel of light energy can be controlled to propagate, concentrate, or filter desired wavelengths. See blaze angle, cladding, diffraction, grating, normal.

incipient failure A failure from degradation of a process or equipment in its early stages.

inclination 1. The angle of a surface or vector in relation to an associated horizontal. 2. A deviation of a surface or vector from horizontal or vertical.

incoherent scattering 1. A behavior of light in some circumstances whereby the phase of the light is random and unpredictable, as in LEDs. 2. A disordered scattering of transmission waves, such as radio, when they encounter a surface and are deflected.

incoming address message IAM. See initial address message.

increment 1. *n.* A small change in value. 2. *v.* To add to an existing quantity, as in a software programming loop. Incrementing an integer counter in a procedure is a very common way to keep track of quantities, operations, timing, and events. Although technically a negative value can be incremented, in programming this is usually called *decremented*.

incremental sensitivity A measure of the least amount of change that can be detected by a specific instrument or process.

incremental service delivery ISD. The delivery of a service in stages, as the user develops a need for more or different services. Many industries are "bootstrapped" this way to allow users to become accustomed to a technology at a low cost or at a beginner's expertise level while providing a means to "move up" when there is a need for next-level or enhanced services.

Incumbent Local Exchange Carrier, Independent Local Exchange Carrier ILEC. Sometimes called dominant carriers, ILECs comprise the RBOCs, independent phone companies, GTE, and others. See Competitive Local Exchange Carrier.

Incumbent Local Exchange Carrier duties The Federal Communications Commission (FCC) stipulates a number of duties, in addition to the Local Exchange Carrier duties, in the Telecommunications Act of 1996 as shown in the Incumbent Local Exchange Carriers chart.

Independent Cable & Telecommunications Association ICTA. A national, Washington, D.C.-based independent trade organization supporting private and alternate cable and telecommunications systems providers. ICTA members provide video programming and other services to residents of multiple dwelling units, primarily through shared tenant services (STSs), although a trend toward geographic clusters has been recently seen. Customers served by these services include apartment, condominium, and co-op dwellers, as well as motels, college campuses, and prisons.

independent flow control message IFCM. In Switch-to-Switch Protocol (SSP), the IFCM is transmitted as a 16-byte information message header of type 0x21 separate from the control message header.

index An organizational tool that provides a key to other types of information, or a larger body of information, stored elsewhere. Indexing is an extremely important aspect of database design, search, and retrieval. It provides a hook or jumping off point, a brief means of indicating the subsequent location of a hierarchy or list. An index in its broadest sense can point to records, further indexes, keywords, locations,

---

Incumbent Local Exchange Carriers (LECs) – FCC-Defined Duties

As defined in the Telecommunications Act of 1996:

"In addition to the duties contained in subsection (b), each incumbent local exchange carrier has the following duties:

(1)   DUTY TO NEGOTIATE – The duty to negotiate in good faith in accordance with section 252 the particular terms and conditions of agreements to fulfill the duties described in paragraphs (1) through (5) of subsection (b) and this subsection. The requesting telecommunications carrier also has the duty to negotiate in good faith the terms and conditions of such agreements.

(2)   INTERCONNECTION – The duty to provide, for the facilities and equipment of any requesting telecommunications carrier, interconnection with the local exchange carrier's network—

    (A)   for the transmission and routing of telephone exchange service and exchange access;

    (B)   at any technically feasible point within the carrier's network;

    (C)   that is at least equal in quality to that provided by the local exchange carrier to itself or to any subsidiary, affiliate, or any other party to which the carrier provides interconnection; and

    (D)   on rates, terms, and conditions that are just, reasonable, and nondiscriminatory, in accordance with the terms and conditions of the agreement and the requirements of this section and section 252.

(3)   UNBUNDLED ACCESS – The duty to provide, to any requesting telecommunications carrier for the provision of a telecommunications service, nondiscriminatory access to network elements on an unbundled basis at any technically feasible point on rates, terms, and conditions that are just, reasonable, and nondiscriminatory in accordance with the terms and conditions of the agreement and the requirements of this section and section 252. An incumbent local exchange carrier shall provide such unbundled network elements in a manner that allows requesting carriers to combine such elements in order to provide such telecommunications service.

(4)   RESALE – The duty

    (A)   to offer for resale at wholesale rates any telecommunications service that the carrier provides at retail to subscribers who are not telecommunications carriers; and

    (B)   not to prohibit, and not to impose unreasonable or discriminatory conditions or limitations on, the resale of such telecommunications service, except that a State commission may, consistent with regulations prescribed by the Commission under this section, prohibit a reseller that obtains at wholesale rates a telecommunications service that is available at retail only to a category of subscribers from offering such service to a different category of subscribers.

(5)   NOTICE OF CHANGES – The duty to provide reasonable public notice of changes in the information necessary for the transmission and routing of services using that local exchange carrier's facilities or networks, as well as of any other changes that would affect the interoperability of those facilities and networks.

(6)   COLLOCATION – The duty to provide, on rates, terms, and conditions that are just, reasonable, and nondiscriminatory, for physical collocation of equipment necessary for interconnection or access to unbundled network elements at the premises of the local exchange carrier, except that the carrier may provide for virtual collocation if the local exchange carrier demonstrates to the State commission that physical collocation is not practical for technical reasons or because of space limitations."

sequences, arrays, and much more. An index can comprise numbers, symbols, or lexical mnemonics, depending upon the context of the application. Some indexes are seen by the user and set manually; others are transparent to the user and set by the software. Databases, mass storage directory structures, and file hierarchies are typically indexed in one way or another for quick storage and retrieval.

An index is intended as a shortlist of what is contained in the database. It is a means of describing in brief what related information is held where in order to enhance the speed with which the related information can be found. The efficiency of an indexing system depends upon the the type of index used, the quantity of information being indexed, the overall structure of the database, and the types of information sought and retrieved from the system. If a system involves a small amount of data and a complex indexing system, then it is not likely to be efficient. If, on the other hand, a large amount of data can be relatively objectively categorized, based on objective information or good guesses as to what types of information will be sought, then an indexed structure is one way to store and utilize the information. See database.

index counter A very common form of feedback that allows a user or technician to monitor usage, or elapsed time or distance. Index counters and their electronic counterparts are found on tape drives, VCRs, microwaves, cars (odometers), photocopiers, and almost any appliance in which the location of information or tracking of usage for billing purposes is desired. Counters that give a rough estimate of the number of users who have visited a Web site, or at least the number of accesses to a particular page.

index of refraction, refractive index (*symb.* – n, *abbrev.* – RI) A numeric description of the refractive ("ray-bending") properties of a material, which depends upon its composition and density in relation to the composition and density of the immediate surrounding material(s).

Depending upon the reference scale, the index of refraction of a vacuum is given as 1.0 and other materials, which are naturally denser than a vacuum (which by definition has no matter), have higher numbers as given by $n = c/v$ where c is the speed of light in a vacuum and v is the speed of light in the medium. Air, which only very slightly refracts light, has a refractive index of 1.00029, very near that of a vacuum. The refractive effect can be observed with common materials. For example, a broom handle placed partly in water (RI 1.333) will appear to bend from the point at which it passes into the water. If the same handle is placed in a more viscous (denser) substance, such as benzene, which has a higher refractive index, it appears to bend even more and will be associated with a higher refractive index than water. Faceted materials like gems that glitter when moved about in light tend to have higher refractive indexes than dull materials such as plastic. Crown glass, which is used in many optical components, has a RI of 1.52 while diamond has an even higher index of 2.42.

That's the basic explanation. Actually, there are a couple of other important factors that are built into assumptions about the refractive index reference values. Light is not a homogenous phenomenon – it stretches from infrared to ultraviolet, with different properties at different wavelengths in the invisible and visible spectrums. Longer wavelengths tend to pass through matter more readily and shorter wavelengths tend to scatter more readily, so refractive index charts generally also specify the wavelength at which materials are referenced (e.g., 589 nm). Temperature is a factor, as well. Since the index of refraction is related to the density of a material and density changes when temperature changes, the temperature at which the index is referenced needs to be specified. Changes in density related to temperature are due to differences in molecular interaction. For example, the molecules in warmed air become more active and expand overall, thus becoming less dense and rising over colder air, the phenomenon that enables hot air balloons to fly. Thus, temperature influences density, which in turn influences refractive index. Thus, RI charts will generally specify the temperature at which materials exhibit a certain refractive index (e.g., 20°C).

(Since this is a reference on optics, the above explanation focuses on light as the incident beam, but remember that refractive concepts can be somewhat generalized to electromagnetic phenomena outside the optical spectrum.)

Index of Refraction - Basic Concept

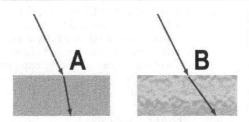

In this simple example, light traveling through air encounters matter with a different composition from the air. There is an interaction between the light and the molecules of matter such that the direction and speed of travel (velocity) of the light are affected. If it were a perfect mirror, all the light would be reflected, but assume it has some conductive (dielectric) properties. In this case, some of the light continues to travel through the material, but, the general path of the light is "bent" or refracted. In other words, it continues in a slightly different direction from the incident light, causing a refractive effect that continues until the light exits the material (or encounters structural changes or impurities in the material through which it is traveling). The degree of refraction can be expressed with a value relative to a baseline of no refraction = 1.0.

Different materials refract light in different ways. Materials with homogenous structures may have fairly simple, straightforward light-refracting properties. Crystalline or crystal-like materials such as Iceland spar and Ulexite refract light in unusual ways.

The index of refraction is one of the most important concepts in optics. The capability of refracting light in a particular direction by selecting a material with certain refractive properties and shaping it to exploit those properties is integral to the fabrication and implementation of fiber optics devices and all instruments that use lenses. The cladding in a fiber optic cable has a slightly higher refractive index than the light-conducting core, which accounts for its ability to refract light back into the core within certain angles to achieve total internal reflectance (TIR). Traditional lenses in telescopes, microscopes, and laser assemblies exploit the refractive properties of optical components to concentrate or diffuse light.

For practical purposes, when designing components, the index of refraction is related to the angle at which light diffracts through a medium relative to the "normal" (perpendicular) to the surface at which the incident radiation encounters the material. In a flat, planar surface such as calm water or window glass, this is a fairly straightforward calculation. For more complex surfaces (e.g., rough water, curved lenses), the tangent of a specified region around the point of contact of the incident light facilitates the geometric calculations for determining *surface normal*. See diffraction, diffusion, Iceland spar, Snell's law, Ulexite.

indicator light A light signaling a transmission, fault condition, readiness, or other state that requires attention. Indicator lights are common on appliances, modems, surge suppressors, hard drives, etc.

indirect addressing A common method in computer programming for creating a cross reference to additional related data. Since much of computer data storage cannot be determined in advance, indirect addressing makes it possible to use small segments of memory, or noncontiguous memory, hard drive space, etc. by creating pointers, directories, and other links to the main body of information.

indirect light Light that is not self-emitted, but rather is reflected from another source. For example, the Moon does not generate light on its own, but reflects light from the Sun.

indium gallium arsenide nitride InGaAsN. A semiconductor alloy developed in the early 1990s in Japan that can be produced through a metal-organic chemical vapor deposition (MOCVD) process. A small amount of nitrogen added to gallium arsenide gives the alloy some remarkable electrical and optical properties, dramatically altering its bandgap structure. The material holds potential for fabricating photovoltaic power sources for lasers and satellite communications. It may enable the development of solar power cells significantly more efficient that current silicon solar cells.

At Sandia labs, the material is produced in a chemical reactor in which indium, gallium, arsenic, and nitrogen are heated within the chamber, causing a chemical decomposition and crystal-forming process that results in the alloy. This brings the bandgap into more practical ranges than another popular semiconductor material – gallium arsenide. See gallium arsenide.

indium phosphide An alloy used in the production of semiconductor lasers, favored because its bandgap properties are suitable for producing laser light that can be used in short-haul fiber optics transmission lines.

INDIX See International Network for Development Information Exchange.

Indo-European Telegraph Company The company that successfully installed a wire communications circuit all the way from London to Calcutta, in 1884. The feat was largely inspired by the successful laying of the transatlantic telegraph cable two decades earlier.

induced Produced by the influence of an electric current or a magnetic field, usually by proximity.

inductance (*symb.* – L) The property of a material (generally in a circuit) to tend to resist change in the flow of electromagnetic current, resulting in changing lines of force. This tendency occurs where there is a flow of alternating current (AC) moving through a conductive material. The term is used specifically with reference to alternating current, as direct current (DC) does not exhibit the same alternate changes. Higher inductance values are generally associated with higher impedance values. Inductance is typically expressed in henries (H) or microhenries (μH).

A basic inductor can be created by winding a conducting wire, such as copper, into a coil. An understanding of the properties of inductance was a very important step in the development of induction coils. These could be devised to generate high-voltage charges, and thus a source of electricity.

Phone lines have been optimized over the decades for the cost-effective delivery of basic voice communications and were not designed for the specific electrical needs of data communications. Electrical surges, EMI interference, balun noise, and other problems can interfere with data delivery and slow down transfer speeds or even cause the line to be dropped. Thus, high-inductance noise filters are marketed to dialup modem users to process the electrical signals and improve data transmissions.

Heat-resistant, low-inductance power cables are increasingly available for telecommunications applications. These may be constructed in various ways, but PVC-insulated stranded, annealed, tinned copper wires are common. See induction, induction coil, resistance.

inductance analyzer A device for testing inductors and coils in telecommunications devices, switching power supplies, filter circuits, and similar products. Inductance analyzers typically test in a wide range of programmable frequencies and may include additional impedance, capacitance, and DC resistance testing capabilities. Automatic calibration and internal memory for storing and recalling settings and test setups may also be included (calibration kits may also be sold separately). Newer models may support several impedance settings in order to compare test results with those from older test sets and may include a serial or other standard interface for downloading settings to a computer system for storage and analysis.

The understanding of inductance and the use of test equipment for troubleshooting is an important part of courses for telecommunications technicians and generally involves about 30 hours of class instruction. See crosstalk, induction.

inductance bridge A diagnostic circuit configuration instrument that enables comparison of an unknown with a known inductance, similar to the concept used in a Wheatstone bridge. See Wheatstone bridge.

induction, electromagnetic An electric charge or magnetic field in a material resulting from the influence of a proximate electric current or magnetic field. In a circuit, induction may be deliberately created or may result from undesired influences from nearby circuits or electromagnetic components.

In computer circuits, especially those tightly integrated on a single chip, induction can be an important hindrance or limitation that must be taken into consideration in the chip's design and fabrication. Since it is difficult to build hundreds (or thousands) of prototype chips in order to test or measure the induction properties of a particular design, computer modeling programs have been developed to simulate the induction characteristics of prototype circuits. Sequence's Columbus-RF, for example, is a patent-pending technology for modeling resistance, capacitance, and inductance of chip circuitry.

On a larger scale, in structured cable systems, data cables that are very near to voice cables may generate undesired noise in the voice lines, partly through induction. See crosstalk, far end crosstalk, inductance, induction coil, near end crosstalk.

induction, logical A reasoning process in which general principles or overall concepts are derived by discerning patterns or relationships among individual or particular observations. The modeling of induction through heuristic problem-solving algorithms is of interest in robotics, artificial intelligence, and intelligent discovery and search procedures in advanced database systems. Inductive reasoning has applications for data mining on the Internet and might specifically be applied to discerning patterns in user inquiries that could be used to anticipate future queries or general needs. It can also be used to try to anticipate the needs of handicapped telecommunications product users based on overall observations of their patterns of use.

induction coil A historic electrical device that played an important role in early electronics inventions. It was a significant provider of high voltage current for many decades, and led to the creation of transformers for converting between alternating current (AC) and direct current (DC). It also led to various induction-based frequency converters. Today, induction coils are still used to offset capacitance in long communications wirelines.

A basic induction coil was created in 1836 and described the next year in *The Annals of Electricity* by Nicholas J. Callan. It consisted of a horseshoe-shaped bar of iron, wound with many feet of thick copper wire, and hundreds of feet of thin iron wire. By interrupting the primary circuit with a contact breaker, Callan could induce a charge sufficient to power an arc light. A year after Callan published his findings, an American, Charles Grafton Page, created an induction coil. See loading coil.

induction field In a transmitting antenna, a region associated with the antenna in which changing electromagnetic lines of force are active as current flows through the device. In long wireline installations, a field that is deliberately generated in order to offset capacitance. See induction coil, loading coil.

induction frequency converter A mechanically powered induction device connected to a source of fixed frequency current that utilizes secondary circuits to deliver a frequency proportionate in speed to the magnetic field. In its most general sense, frequency conversion has become a very important part of communications technology. The conversion of frequencies allows signals to be carried over a variety of media with different transmission characteristics, and further enables signals to be shifted so that incoming and outgoing signals are less likely to interfere with one another.

inductive connection, inductive pickup An electromagnetic connection between two devices or objects without direct electrical contact. The communication between the devices occurs from an electromagnetic influence through proximity to the changing electromagnetic lines of force. Some types of diagnostic tools use inductance to monitor or observe circuits without physically contacting the line. A number of surveillance devices also use this method for bugging a line, in order to avoid detection.

Regulations to protect privacy prohibit the unauthorized monitoring of communications through inductive surveillance devices. Fiber optic transmissions are immune to inductive pickup as the transmission of signals through light does not have the same characteristics as electricity of extending beyond the medium through which it is traveling. See wiretap.

inductive coupling The transfer of energy between two circuits that are close together, but not directly electrically connected. Thus, the interaction of the electromagnetic lines of force associated with the interaction of the circuits causes the transfer. The transfer may also occur due to self-inductance of each of the circuits (direct coupling). The transfer of energy may be desirable or undesirable. Unshielded or minimally shielded conducting wires that are too close together may create unwanted noise and interference through inductance.

inductive post A conducting bolt, screw, or post associated with a waveguide that provides inductive susceptance to allow tuning of the waveguide. It is usually mounted across the waveguide, parallel to the E field. See E field, waveguide.

inductive tuning In electronic devices such as radio tuners, a means of adjusting the tuning by moving a core in and out of a coil within which it is contained. The core is not in direct contact with the coil, but reacts to the changes in the electromagnetic field associated with the coil by inductance.

inductor A passive component that provides power

or power-related electromagnetic energy or energy control, traditionally called a coil. An inductor is able to "store" energy and to resist electromagnetic changes associated with the flow of current. Inductors come in a wide variety of types and are used in many aspects of telecommunications circuitry. Variations in the core material, the shape of the coil, and the number of windings will influence an inductor's properties. Tables listing tolerance codes have been standardized for inductors. For example, a tolerance code of K signifies ±10% tolerance.

Inductors are categorized in a number of ways: by their specific function (e.g., suppression), by their types of cores, or by their general size (e.g., miniature inductors for microelectronics). Inductor types include suppression, VHF, ring core, air core, laminated core, and many more. Suppression conductors are typically ferrite wound with enamelled or tinned copper wire all sealed with resin and/or a plastic sheath. The ferrite substance is chosen for having good magnetic properties, commonly manganese or nickel and zinc. Air core conductors have the winding built over a nonconductive core, commonly ceramic. See inductance, induction.

inductor, axial This inductor style is built with a central core and concentric leads on either end of the core. The core may be constructed from a variety of conductive materials. Axial conductors are used in power and radio frequency (RF) applications.

inductor, air core An inductor with a core with no magnetic properties (e.g., ceramic) that is used as the base or support for the conductive winding. Air cores are used in situations where low loss and low inductance are desired, as in high-frequency applications.

Industrial, Scientific, and Medical ISM. A set of segments of electromagnetic frequencies which do not require licensing by the Federal Communications Commission (FCC), excluding telecommunications applications. Typical ISM applications include particle acceleration, vibration generation, heating, ultrasound equipment, microwave ovens, humidifiers, etc. Class A ISM refers to industrial environments and Class B ISM is intended for domestic environments. In Canada, ISM is addressed by ICES 001.

ISM equipment operators are required by the FCC to "take appropriate measures to correct" interference to radio services unless those services are operating in the ISM frequency band. The conduction and field-strength limits for interference are dependent upon the equipment generating the radio frequencies and the specific frequency bands being used. Commonly used frequencies are in the 902- to 928-MHz and 2.4- to 2.4835-GHz ranges, but there are others.

An important growth area in ISM unlicensed frequencies is spread spectrum and frequency-hopping technology for local area wireless networks (LAWNs) for data communications. By the late 1990s, vendors had developed scalable high-speed ISM-frequency wireless networks that they felt would comply with all the FCC ISM regulations. With important approvals by the FCC beginning about 1999, the systems began to be put into commercial production.

There is also a high degree of correspondence between ISM and designated amateur radio bands, which sometimes causes problems for ham radio operators, especially given the recent increase in wireless consumer communications devices. These issues are being evaluated and debated by the ARRL and the FCC as the demand for radio frequency resources increases.

Industry Advisory Council IAC. A national trade organization representing information technology (IT) professionals who provide products and services to government agencies in the U.S. The IAC also serves as a liaison between the IT industry and the Federation of Government Information Processing Councils (FGIPC). http://www.iaconline.org/

Industry Canada A Canadian federal agency responsible for the protection of intellectual property and the allocation of licenses for use of radio frequencies. Formerly the Department of Communications. See Canadian Radio Television and Telecommunications Commission.

Industry Canada Emergency Telecommunications Branch A service department of Industry Canada that provides support for crisis situations. It administrates and liaises with a wide variety of communications and disaster departments and external organizations, including emergency broadcasting, priority access, emergency response coordination, and more. Several emergency and planning committees and working groups further the aims of the ETB. Working groups include the United Nations Working Group on Emergency Telecommunications (WGET), the Long-Distance Emergency Telecommunications Working Group, and the Wireless E-911 Working Group. http://spectrum.ic.gc.ca/urgent/

Industry Circuit Topography Act ICTA. A Canadian Act intended to protect integrated circuit topographies. See Semiconductor Chip Protection Act.

Industry Standard Architecture ISA. Formerly, a very common input/output bus architecture on International Business Machines and licensed third party computers developed originally on the IBM XT models, and carried through to later models. Originally it was an 8-bit architecture, but was upgraded to 16-bit. The expansion slots inside a computer have to follow a standard format so various manufacturers can create compatible peripheral cards. ISA was one of the common types of slots found in personal computers until the mid-1990s when it was superseded by *Peripheral Component Interconnect* (PCI), *Video Electronic Standard Association* (VESA), *Extended Industry Standard Architecture* (EISA) and others. See Extended Industry Standard Architecture, Peripheral Component Interconnect.

INETPhone A data telephone service connected and handled through the Internet, thus substituting the Internet for the long-distance segment of a phone call in a way that is transparent to the users. See RFC 1789.

Infineon Technologies A publicly traded leading global provider of integrated circuits for advanced communications sytems. See GEMINAX.

Infobahn *colloq.* The Information (Super) Highway, based on the German word *bahn*. The Information Superhighway is also colloquially called the I-way. See Information Super Highway.

Information Access Company IAC. A commercial electronic information vendor (purchased by the Thomson Corporation from Ziff Communications for almost half a billion dollars). IAC is one of the firms being watched by intellectual property rights advocates and writers to assess whether firms that distribute electronic information, especially over public networks, are tracking and compensating contributors fairly.

Information Analysis Center IAC. An interagency intelligence center located in the U.S. Embassy in Mexico City to assist the U.S. Ambassador to Mexico to collect and process intelligence for use by U.S. and cooperating Mexican law enforcement personnel. The information is collected and stored in electronic databases.

Information and Communications Industry Association, Ltd. ICIA. A U.K. trade association supporting information and service providers and operators, and hardware and software developers, especially those involved in electronic publishing and data distribution. Founded in 1978, ICIA evolved from the Videotex Industries Association. http://www.icia.co.uk/

Information and Software Industry Association ISIA. A trade association promoting the recognition, profitability, and standards of the information and software industry in Hong Kong, founded in July 1999. ISIA promotes awareness and use of information technologies, and represents and safeguards the interests of its members. It further promotes cooperation between Hong Kong and mainland China. http://www.isia.org.hk/

information content provider This is defined in the Telecommunications Act of 1996, and published by the Federal Communications Commission (FCC), as
"... any person or entity that is responsible, in whole or in part, for the creation or development of information provided through the Internet or any other interactive computer service."

Information Industry Association IIA. A U.S. trade association supporting businesses that develop and globally deliver innovative information products and services, founded in 1968. The IIA represented its members to government and inputs to the Federal Communications Commission (FCC) and the Federal Trade Commission (FTC). In 1999, IIA merged with the Software Publishers Association to form the Software & Information Industry Association. See Software & Information Industry Association.

Information Industry Association, Australian AIIA. A trade organization representing and promoting the information industry in support of the Australian economy. AIIA represents its members in government policy, promotes the value and applications of Australian information technologies, and provides educational and networking support to its members. http://www.aiia.com.au/

Information Infrastructure Standards Panel IISP. A national voluntary standards support panel established to facilitate the development of standards important to the Global Information Infrastructure (GII) and the U.S. National Information Infrastructure (NII). IISP promotes cross-sector efforts to identify, highlight, and resolve major standards issues, a mission that was approved in November 1997. http://www.ansi.org/public/iisp

Information Network Architecture INA. In the mid-1980s, Bell Communications Research began building its Intelligent Network (IN) to provide a broader range of telephone services and support for data transmission over traditional phone lines. From this grew Advanced Intelligent Networks (AIN), and then Information Network Architecture (INA) with its improved broadband support. There is some discussion as to whether INA will succeed or coexist with AIN, as AIN will meet the needs of many users for some time, considering the lag that exists between the time a new technology is introduced and when it is generally adopted by consumers.

Information Security Exploratory Committee ISEC. A committee tasked with the study and support of private sector information security. The ISEC was hosted by the Information Technology Industry Council.

information service This is defined in the Telecommunications Act of 1996 and published by the Federal Communications Commission (FCC), as meaning
"... the offering of a capability for generating, acquiring, storing, transforming, processing, retrieving, utilizing, or making available information via telecommunications, and includes electronic publishing, but does not include any use of any such capability for the management, control, or operation of a telecommunications system or the management of a telecommunications service."
See Federal Communications Commission, Telecommunications Act of 1996.

Information Service Industry Association of China CISA. A trade organization representing the Chinese information industry. CISA aids its members in creating and maximizing their competitive business strategies. CISA promotes the application of information technology (IT) in business, government, education, and individual market settings. http://www.cisanet.org.tw/

Information Services Association, Japan JISA. A nonprofit trade organization representing Japan's information technology (IT) services industry. Founded in 1984, JISA evolved from two organizations, the Japan Software Industry Association and the Japan Information Processing Center Association. JISA cooperates with the World Information Technology and Services Alliance (WITSA) and the Asian-Oceanian Computing Industry Organization (ASOCIO). http://www.jisa.or.jp/

Information Superhighway A catchphrase promoted by U.S. government representatives, particularly Al Gore of the Clinton administration, and the press, for the domestic and global communications infrastructure. See National Information Infrastructure.

Information Systems Auditability and Control Association ISACA. A global not-for-profit trade association of more than 17,000 information system (IS) professionals providing education, training, and certification support. http://www.isaca.org/

Information System Security Association ISSA. A not-for-profit international trade organization supporting the interests of information security (IS) professionals. ISSA supports communication among members, educational activities, and information security publications. ISSA is a founding contributor to the International Information Systems Security

Information Technology Research Centers – Brief Selection		
Title	Description	Abbreviation
Information Technology Research Center		ITRC
	An interdisciplinary research unit at the University of Arkansas devoted to advancing the state of research and practice in the development and use of information technology (IT). http://itrc.uark.edu/	
Information Technology Research Center		ITRC
	A nonprofit organization dedicated to the development, evaluation, and application of advanced technologies to enhance scientific research and education in information technology (IT). The Center is international in scope, operating out of Missoula, Montana. An example of an ITRC project is the TRIO ThinkQuest Project, U.S. Department of Education-funded project for sponsoring an online educational Web site development contest. http://www.itresearchcenter.org/	
Information Technology Research Center		ITRC
	A funded project in which the Innovative Computing Laboratory at the University of Tennessee Knoxville campus will be transformed into an IT research center with a broader scope, under the direction of Dr. J. Dongarra.	
Advanced Information Technology Development Research Center		
	A Japanese center to support research to improve road traffic systems, traffic information provision systems, and home information systems, as well as research on the application of geographic information systems (GIS) and other information technologies to design and construction.	
Center for Information Technology Research in the Interest of Society		CITRIS
	A California Institute for Science and Innovation, endorsed in December 2000. CITRIS is located at the University of California, Berkeley and is dedicated to promoting scientific advances in information technology fields critical to the California economy.	
German National Research Center for Information Technology		GMD
	A cutting-edge national and international research facility devoted to applications-oriented research in the information and media industries with a view to developing new products. http://www.gmd.de/	
Information Technology Research Centre		ITRC
	An Ontario-based Canadian information technology (IT) research center.	
Research Center for Communications and Information Technology		ReCCIT
	A Thai institute of King Mongkut's Institute of Technology Ladkrabang (KMITL), ReCCIT supports a number of cooperating labs in telecommunications, information technology, signal processing, and signal transmissions. Research projects are wide-ranging, including mobile and satellite communications, information science, multimedia, virtual modeling, circuit design, signal processing, and more. http://www.kmitl.ac.th/~reccit/	
Research Center Information Technology		RCIT
	Since 1990, RCIT has been developing large-scale, dynamic information systems for public institutions and companies in Europe. RCIT is also the European Telework Development (ETC) national coordinator.	

I

Certification Consortium (ISC²), offering comprehensive certification for information security professionals. http://www.issa.org/

Information Technology Association of America ITAA. A trade association representing the U.S. information technology (IT) industry. The ITAA responds to developments in governmental and international IT policy, promotes the interests of its members, and participates with other organizations in developing Internet policies. http://www.itaa.org/

Information Technology Association of Canada ITAC. A trade organization supporting Canadian information technology providers, ITAC identifies and focuses on issues affecting the IT industry and advocates initiatives promoting its growth and development. http://www.itac.ca/

Information Technology Industry Council ITI, ITIC. Formerly CBEMA, the Computer and Business Equipment Manufacturers Association, ITIC is a trade organization representing leading U.S. providers of information technology (IT) products and services. It includes well-known vendors such as 3Com, Amazon.com, Apple, Computer, Inc., Cisco Systems, Inc., Hewlett-Packard Company, IBM Corporation, and many more well-known firms. ITIC produces an industry Data Book with statistical information on computers and telecommunications equipment and services. See Information Security Exploratory Committee. http://www.itic.org/

Information Technology Research Center ITRC. There are many research centers operating under this name (or slight variations of the name), so only examples are listed in the Information Technology Research Centers chart, but since most of them are directly concerned with advancements in telecommunications technologies, this selection gives an overview of some of the Information Technology (IT) centers accessible on the Web, along with their goals, and their geographic distribution.

information theory The pioneer studies in queuing theory were developed and described by A.K. Erlang, a Danish engineer, in the early 1900s. Information theory, an evolutionary cousin to queuing theory, is a field of inquiry and mathematical modeling that was largely developed and disseminated by Claude E. Shannon while working at Bell Laboratories in 1948. Shannon took a theoretical, mathematical look at in-

formation, in terms of not only its content and structure, but also its source and purpose. Thus, signals and their frequencies, bandwidths, physical components, and electromagnetic characteristics were set in the broader framework of information and its human sources. This provided a broader view of communications and groundwork for more specific measures and descriptors of content and capacity that have real world usefulness. Information theory can be used to develop more objective system evaluation tools, compression techniques, and practical applications such as voice-over IP systems. See erlang, queuing theory.

InfoWatt A new electrical conductor technology that includes a fiber optics transmission cable in its core. This new conductor was intended to solve the problem of getting generated power more efficiently to consumers to alleviate existing distribution bottlenecks and, since fiber optics are not affected by electrical current in the same way as other electrical wires, a fiber optic conductor can be bundled into the cable in an electrically neutral core, saving space and opening up new opportunities for communications network delivery. The cable consists of a fiber optic core, a surrounding layer of thermoplastic composite strength members, and an outer wrapping of conductive aluminum. The structural components are nonconducting thermoplastic composite materials. Testing is underway and initial deployment is expected around 2003. InfoWatt was developed by W. Brandt Goldsworthy & Associates, Inc.

infrared Electromagnetic radiation with longer wavelengths which, in terms of frequencies, fall between the red part of the visible spectrum and radio waves. Although it cannot be seen by humans, infrared radiation is of commercial importance in remote sensing systems, remote control devices, video game consoles, and fiber optic transmissions. It is also being exploited for local area wireless networks (LAWNs).

Infrared serial data link standards are being adapted by a number of manufacturers. Infrared technology can be used to detect differences in heat and, consequently, movement of bodies emitting heat. Infrared detectors are used in many industries including electronics, construction (structural fault detection, heating, and insulation testing), and medical imaging. Infrared film is used in specialized photographic ap-

IrDA Network Protocol Layers	
Layer/Protocol	Notes
IrLMP	A mandatory link management protocol which manages resources and services and higher-level protocols which are made available to other devices. IrLMP sets up and maintains multiple connections.
IrLAP layer	Link establishment, maintenance, and termination. Similar to the half-duplex link control (HDLC) protocol.
physical layer	Provides point-to-point connections and communications between devices with cordless/wireless serial infrared half-duplex links.

plications. See Infrared Serial Data Link, snooper-scope, ultraviolet.

Infrared Communication Systems Study Committee ICSC. A research committee of the Association of Radio Industries and Businesses (ARIB), studying and promoting awareness and use of infrared communications systems. Centered in Tokyo, Japan.

Infrared Data Association IrDA. An organization established in 1993 to support and promote software and hardware standards for cordless/wireless infrared communications links. IrDA is headquartered in California. Infrared can be used with remote controls to control various consumer electronics devices and can also be used for data transmission between devices such as laptops, desktop computers, and peripherals. See Infrared Data Association Protocol.

Infrared Data Association Protocol IrDA Protocol. A multilayered networking structure from IrDA for defining hardware and software needs for infrared network communications. The IrDA protocol stack covers physical transfer of information, guidelines for link access, and link management. The layers are briefly described in the IrDA Network Protocol Layers chart.

Infrared Link Access Protocol IrLAP. A serial link access protocol from IrDA which provides three types of connectionless services and six types of connection-oriented services with four types of service primitives. IrLAP provides discovery, address conflict, and unit data services over connectionless services and connect, sniffing, data, status, reset, and disconnect services or connection-oriented services. IrLAP is primary-secondary or primary-multiple station oriented.

The IrLAP layer is intended to facilitate interconnection of computers and peripherals over a directed half-duplex medium provided through the physical layer. IrLAP stations can be operated in Normal Response Mode (NRM) or Normal Disconnect Mode (NDM), which correspond to connection state and contention state. IrLAP data and control are frame-oriented, with a frame including an address, a control field for determining frame content, and an optional information field.

infrastructure The structural underpinning or base that supports the other physical/conceptual layers and components associated with a system.

InGaAsN See indium gallium arsenide nitride.

INGECEP See Integrated Next Generation Electronic Commerce Environment Project.

ingress 1. Entrance, point of entry, way in, opening, doorway. 2. In Frame Relay networks, frames that are entering toward the Frame Relay from an access device. The opposite of egress.

initial address message IAM. In Signaling System 7 (SS7) networks, a signaling message sent in the forward direction that initiates seizure of a circuit, and provides address and routing information for the connection of the requested call. See Signaling System 7.

Initial Defense Communications Satellite Program IDCSP. A project of the U.S. military, IDCSP first launched three satellites in 1967. They included X-band transponders in the 26-MHz bandwidth, and supported experimental terminals for evaluating images, voices, digital data, and teletype channels using a variety of modulation schemes. The IDCSP was designed to shut down after five years of useful life.

Initial MAC Protocol Data Unit IMPDU. In packet-switched networking, the IMPDU encodes Media Access Control (MAC) Service Data Unit information. A number of MAC Protocol Data Units (PDUs) are derived from the segmentation of the IMPDU. See Media Access Control, Protocol Data Unit.

initial program load IPL. The bootstrapping of a system in that the operating system is loaded up first to make it possible to bring the other hardware and software systems online (monitors, disk drives, interface applications, etc.). Some systems provide a means to "soft boot" a machine (reloading the initialization software and OS without turning the power off and on again).

Many systems will now allow the user to set the device from which the computer will boot, especially if the computer has several possible boot devices such as hard drives, CD-ROM drives, and floppy diskette drives. At the present time, most systems have minimal startup programs stored in ROM chips and then default to boot the rest of the initialization of the OS from hard drives. They will typically seek operating system software on other devices if it is not found on the default drive and may be set to boot from a CD-ROM or diskette first (rather than the hard drive) if a disc/diskette is present.

In a more specific sense, this same bootstrapping process occurs with many computer subsystems. For example, there may be components or peripherals associated with a computer system that store initial parameters in ROM chips or on other storage devices that make it possible to bring the rest of the device's capabilities online.

Initial Public Offering IPO. A Securities Commission government-regulated mechanism for a company to offer a variety of types of shares (usually common and preferred stock) to the general public. There are a number of categories of public offerings, both state and federal, with levels of restrictions and guidelines depending upon the amount of investment sought. Telecommunications and biotech are two of the hot areas of recent years, and some high-profile stock offerings have been carried out in the technology industry, one of the most visible being Netscape Communications, developers of Web browsers/servers and other applications.

injection laser diode ILD. See laser diode.

inkjet printing An inexpensive color printing process in which inks from a series of ink "wells" are fired through a tiny opening called a nozzle. The firing is accomplished through heating the ink chambers to a high temperature so a vapor bubble is formed, which rapidly ejects the ink through the end of the nozzle onto the printing medium, where it cools and adheres. See dye sublimation printing, thermal wax printing.

Inmarsat *In*ternational *Mari*time *Sat*ellite Organization. Originally an international cooperative agency established in 1979, Inmarsat was then slated for privatization for 1 January 1999. It launched in 1992 and has provided global mobile satellite communications services (voice, data, facsimile), especially maritime services, since 1993. Inmarsat now serves over 80 member countries.

Inmarsat has a system of four geostationary satellites orbiting at 35,786 km using frequency division multiple access (FDMA). It provides transportation communications and Internet connect services. Five more are scheduled to be launched by the end of the century. Twelve medium Earth orbit (MEO) satellites are also planned.

Customers purchase services from a variety of packages depending upon whether they need phone, facsimile, Internet, emergency services, telemedicine, etc. The ICONET satellite system is a spin-off of Inmarsat communications services, originally known as Project 21. See Inmarsat Service Categories chart. See ICO Global Communications.
http://www.inmarsat.org/

INN 1. See InterNet News. 2. InterNode Network.

INP See Interim Number Portability.

InPerson A consumer-priced SGI-based videoconferencing system supporting video, audio, whiteboarding, and file transfers over analog phone lines and Ethernet networks. Video encoding is accomplished through HDCC compression developed in-house at Silicon Graphics with several audio compression formats.

input Information, in the form of a communication or signal, provided to a person, system, or circuit. Computer software input mechanisms include graphical user interfaces, shell windows, buttons, icons, dialog boxes, etc. Computer hardware input mechanisms include keyboards, mice, trackballs, touchscreens, joysticks, video cameras, and microphones. The input device on a telephone is relatively simple: a small speakerphone or diaphragm (microphone) in the telephone handset.

input device IDev (ID is sometimes used but may be confused with identification). An interface device for receiving and transmitting information from an input source (frequently human) to a processing system or remote location, usually a computing machine or electromechanical device. The input sensor and the transmissions unit are often housed together (e.g., telephone). There are a great variety of input devices including keyboards, mice, joysticks, light pens, touch screens, microphones (especially with speech recognition systems), infrared sensors, video cams, etc. The invention of the mouse, one of the most common computer input devices, is attributed to Doug Engelbart in the 1960s. Many of the input devices in common use today were pioneered by Ivan Sutherland in the early 1960s. See individual input devices.

INSPEC The world's largest English-language bibliographic database in physics, computing, and electronics. INSPEC evolved from *Science Abstracts*, which was first published in January 1898. The database regularly catalogs the contents of over 4000 technology journals, in addition to conference proceedings and other relevant literature. It currently holds more than 6 million records. See Institution of Electrical Engineers.

Institute for Advanced Commerce IAC. An IBM forum for studying fundamental aspects and trends in business. Through academic partnerships and conferences, the Institute tracks business and market characteristics with the goal of creating long-term corporate solutions. An important focus of advanced commerce is the application of computer and communications technologies, collectively known as e-commerce.

Institute for Telecommunication Sciences ITS. The applied research division of the U.S. National Telecommunications and Information Administration (NTIA). The ITS develops, tests, evaluates, and promotes advanced communications networks and domestic standards through its Boulder, Colorado, facility.

Institute of Radio Engineers IRE. The IRE was a historic professional organization formed as a result of the merger of the Society of Wireless Telegraph Engineers (SWTE) and The Wireless Institute in 1912, in order to establish and promote an international orientation for the consolidated organization. It served as a standards body, in cooperation with the U.S. federal government, and a professional support group for its members and the radio community at large. See American Institute of Electrical Engineers, IEEE.

Institute of Electrical and Electronic Engineers, Inc. See IEEE.

Institution of Electrical Engineers IEE. A U.K.-based professional engineering society founded in 1871 that now has almost 140,000 members worldwide. The IEE supports and promotes advancements in electrical, electronic, and manufacturing sciences and engineering and provides publications, historical archives, research databases, exhibitions, and educational activities for its membership and, in some cases, for the general public. IEE produces the INSPEC engineering science database. See INSPEC.
http://www.iee.org.uk/

Instrumentation, Systems, and Automation Society ISA. A standards-setting, international, nonprofit, professional society supporting instrumentation and systems engineers in more than 100 countries. ISA provides a number of publications and awards relating to the fields of instrumentation and automation, as well as certification resources. http://www.isa.org/

Insulated Cable Engineers Association, Inc. ICEA. A not-for-profit professional trade association dedicated to developing cable standards for the various control, power, and telecommunications industries, founded in 1925. ICEA generates documents of interest to cable designers, manufacturers, and vendors. http://www.icea.net/

insulated wire Conductive wire that has been coated, sealed, rubberized, clad, sheathed, or otherwise covered or processed to protect it from electrical leakage and external electromagnetic interference or corrosion. It may also be internally insulated if the wire is bundled with other wires or fabricated in layers that could interfere with one another if not separated with nonconductive materials.

insulation A material or particulate environment composed of atoms that do not readily give up their electrons. This inertial property can be exploited to create industrial materials resistant to the flow of current and exchange of heat between environments with disparate temperatures. Examples of common insulating materials include rubber, glass, and porcelain, but other substances can be insulators because insulation is somewhat contextual. The Earth's atmosphere is an insulator, shielding the planet from ultraviolet radiation, for example. When a storm occurs and electrical charges accumulate around clouds, they may overcome the air's insulating properties and manifest as lightning.

Historically, insulation was crucial to the successful installation of underwater telegraph cables, beginning in the 1800s. Gutta-percha, a rubberlike substance with excellent industrial properties for the time, made it possible to lay cables in corrosive salt environments, where attempts with other materials had failed. Insulation also made it possible to install underground telegraph and telephone wires and wires that could be used in harsh wilderness environments. In the 1930s, AT&T introduced a wire with improved insulation for telephone transmissions.

There are many primary and secondary ways in which insulation is used in telecommunications, including

- shielding conductive materials from heat or electrical interference,
- providing protection from external physical damage (erosion, corrosion, abrasion, tampering),
- providing protection and spacing among or between proximate or layered electromagnetic influences,
- providing a surface upon which marks or colors can be imprinted to aid in installation and maintenance, and
- providing protection to humans handling current-carrying wires.

There are some differences between insulating wires and fiber optic bundles. Wired telecommunications typically carry one signal per wire and wire is somewhat resistant to breakage if it is bent (a 180°+ bend can often be straightened out again without breaking a wire). While several wires may be bundled together (e.g., transatlantic telegraph cables were a collection of bundled wires), wire assemblages typically don't have the high number of strands found in fiber optic cables, and tiny fiber optic strands can break or easily become separated from the assemblage at junction points. Wires and optical fibers are also subject to different forms of environmental damage, resulting in different choices for the types and thicknesses of materials used to protect them.

A small gouge or scratch may not significantly alter the overall current-carrying characteristics of a wire but can significantly impair a tiny fiber from transmitting a consistent optical signal. Insulation has to be designed to accommodate these differences and stripping tools must be suitable for removing optical fiber insulating sleeves. Further, installers must be aware of electrical shock hazards when working around current-carrying wires, which is not a problem with optical fibers.

Industrial insulation is used for purposes other than covering wires. It may also be used to regulate the air temperature in facilities where material temperatures or operating temperatures are important, as in supercomputing applications or fabrication plants, where the room or chamber environments are important. Insulation is further used in atomic research facilities, as in supercooled environments, for studying specialized computing functions (e.g., quantum computing). See dielectric.

insulator See insulation.

Utility Pole Insulators

Insulators commonly sed to shed moisture and support conductive wires on utility poles. They were once constructed of glass (the early ones handmade), but now ceramic insulators are generally used and old glass insulators are collectibles.

---

insulator, utility pole Historically, the fact that glass would make a good insulator was suggested by E. Cornell, who assisted Samuel Morse in installing the historic 1843 Washington, D.C.-to-Baltimore telegraph line. He originally proposed glass plates and later described a more knob-like design, a larger version of which eventually became standard and widely used on utility poles until the 1970s.

Utility pole glass insulators are thick, threaded, mug- or thermos-sized objects, in clear glass or a variety of colors, most often blue or green. A number of hand-blown insulators were created in the late 1880s. The oldest commercial mass-produced ones, originating some time in the early 1850s, lacked threads but were colored. Molding processes for creating insulators were patented in the 1870s. The Oakman beehive insulator was favored by Western Union for telegraph poles.

Western Union used many thousands of Brookfield and Hemingray insulators over the years. The move to standardize insulators occurred around 1910; clear insulators were not produced until the 1930s. Ceramic insulators were introduced around 1908 by Locke Insulator, in order to undercut the cost of glass insulators.

Insulators were developed in many shapes and sizes, in a rainbow of gem-like hues. They provide a legacy of poetically descriptive category names such as slashtops, bat ears, eggs, beehives, and teapots.

Well-known glass insulator manufacturers, like Hemingray, shut down by the mid-1960s. Historic glass and ceramic insulators are found occasionally in secondhand stores and antique auctions, and older or more interesting ones are favored by collectors and sometimes sell for hundreds of dollars.

INTEGRAL International Gamma Ray Astrophysics Laboratory. A medium-size scientific mission selected in June 1993 by the European Space Agency (ESA) for the Horizon 2000 program. The ESA-led orbiting observatory mission is being carried out with contributions by NASA and the Russian Federation. INTEGRAL is involved in imaging and spectroscopy of celestial gamma-ray sources. Observations will be telecommunicated to ground-stations and made available to the global scientific community.

Integrated Access Device IAD. A data communications device that provides data and voice services, usually to small- and medium-sized businesses. IADs have generally been used to provide circuit-switched services, but as of January 2001, IADs supporting migration to packet-switched IP services were being offered commercially by Cisco Systems, Inc.

Integrated Broadband Communications IBC. A European Community-wide system of communications capable of supporting a wide range of service providers that was emerging in the mid-1980s and whose development was formally supported in a decision of the European Community (EC) in December 1978. It was felt by the European Council that telecommunications systems would benefit the EC's international competitiveness in general and the telecommunications sector in particular. It was also stated that a system that united rather than regionalized communications would be preferable and that common specifications were necessary but not sufficient to bring this about. The Single European Act was expressed to provide a good political and legal base for developing a European-wide scientific and technological strategy and industrial competitiveness in telecommunications.

One of the important contributors to the development of the IBC is the Research and Development in Advanced Communications in Europe (RACE) program. RACE was involved in overall IBC development and more specifically, the development of the Mobile Broadband System (MBS) being integrated with the IBC. In 1995, at the end of its specified term, RACE evolved into Advanced Communications Technologies and Services (ACTS) to represent the third phase of IBC implementation. See Mobile Broadband System, Research in Advanced Communications in Europe.

Now that many of the initial steps in establishing IBC have been taken, it is expected that more Europeans will stay at home to work, study, and socialize over computer networks, etc., thus increasing the importance of and demands on IBC. It is also expected that individuals with limited mobility can benefit from IBC and that educational, government, and health care services will be an important aspect of IBC.

integrated circuit IC. A single electronic component that incorporates what would normally require many traditional electrical circuits. This enables complex, sophisticated capabilities to be bundled into tiny packages and also often increases the speed of interactions and processing. A computer central processing (CPU) chip is one particular type of integrated circuit; a combination of circuits and chips included on a single card, like a peripheral card, is also an extension of the concept of an IC. Very large scale integration (VLSI) technology is the combination and interaction of many circuits in a combined package. In Canada, the Integrated Circuit Topography Act (1990) exists to protect registered integrated circuit designs as a form of intellectual property. Various U.S. and foreign copyright and patent laws also protect and publicly disseminate information on unique ICs.

Pioneer Integrated Ciruit

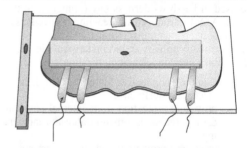

In 1958, Jack Kilby constructed a historic integrated circuit (IC) using germanium as the semiconductor, at about the same time that R. Noyce was working on the IC concept at Fairchild Semiconductor. Kilby's circuit was primitive by today's standards, mounted on a transparent synthetic base with four wire leads protruding from one side, but the invention was one of the most important in electronics history, following the milestone inventions of the triode in the early 1900s and the transistor in the 1940s.

The concept for making resistors, capacitors, and other common hardware on circuit boards out of silicon was new in the 1950s. Engineers from the old school evolving out of decades of experience with vacuum tubes and discrete components didn't immediately conceive the idea of using the new semiconductor technology for modeling all (or many) aspects of a circuit design. The earliest ICs included only a single transistor. Now, astonishingly, more than 100 million transistors can be packed into an IC.

Credit for the introduction of ICs, in 1959, has long been attributed to Robert N. Noyce, a Dane who was working at Fairchild Semiconductor and who helped form the Intel Corporation in America. Noyce was awarded a U.S. patent in 1961. However, it appears Jack St. Clair Kilby, a Texan, is the original "Father of the IC." Kilby apparently introduced the concept in September 1958, and Texas Instruments Incorporated (TI) applied for a patent on Kilby's concept a few months later. It has been reported that Kilby's patent was still being assessed as the Noyce patent was granted. In recent years, Kilby's contribution has been acknowledged and lauded and Texas Instruments has named a research center in his honor. Historians generally consider the two inventors to have independently developed their ideas at about the same time.

In engineering circles, the abbreviation IC is often used as a pun to refer to both integrated circuit and "I see." See semiconductor, very large scale integration. See Kilby, Jack; Noyce, Robert.

integrated data access IDA. A phrase usually applied to database access through shared resources or automated lookup tools that facilitate information-finding. A number of Web-accessible government archives are said to be IDAs. IDA also applies to a number of commercial database products that have various database discovery, search, and retrieval functions built into the product so that it can be readily deployed by an institution to set up information delivery services without a lot of time spent on in-house programming.

Integrated Digital Loop Carrier IDLC. A system designed to integrate Digital Loop Carrier (DLC) systems with existing digital switches as in a SONET network system. A basic installation consists of intelligent remote digital terminals (RDTs) and digital switch elements known as integrated digital terminals (IDTs), interconnected by a digital line. See Digital Loop Carrier.

Integrated Digital Network IDN. A digital network in which both the switching and the transmission are digital. Traditionally, communications switching has been analog, even if the data transmission was digital, necessitating modulation and conversion that limited transmission integrity and speeds. Gradually digital switches began to replace analog and digital phone; data services for general consumers began to become widely available in the late 1990s.

A computer with a modem is an example of a hybrid digital/analog transmission system. A computer generates digital data that is sent to the modem for conversion to analog signals for transmission through traditional copper phone lines. At the destination, it is remodulated back to digital data and interpreted by a computer. While the transmission was in effect, the phone line would be tied up to preclude voice transmissions. With the evolution and installation of digital circuitry throughout the phone system, a gradual transition to digital services such as ISDN and ATM is enabling enhanced communications services for business and residential customers as end-to-end digital transmissions systems are gradually supplanting analog or hybrid systems. It is now possible to transmit data and digital voice services over the same sub-scriber line. Terminal devices rather than traditional modems are used to interconnect subscriber premises to digital services over public networks through both copper and fiber optic media. See ISDN, Signaling System 7.

Integrated Dispatch ID. In general, computer-enhanced dispatch administration and/or messaging services integrated with traditional radio dispatch communications. This is of particular interest to emergency services call centers, where accountability and response times are important, and also to companies that have sophisticated dispatch tracking needs. See Enhanced Specialized Mobile Radio.

integrated injection logic IIL. A form of bipolar logic, reduced power circuit intended to provide greater efficiency over TTL chips.

Integrate IS-IS A proprietary routing protocol using one set of routing updates, developed by Digital Equipment Incorporated (DEC). DEC's version is based on the Open Systems Interconnection (OSI) routing protocol called IS-IS. The DEC implementation provides support for a number of other open and proprietary protocols by encapsulating them into Internet Protocol (IP).

Integrated Internet Information Architecture IIIA. An effort by Weider, Mitra, Sollins, et al. to develop protocol specifications and enhancements for some of the widely used Internet information systems. Based on the concepts that one solution will not fit all users and that users need a way to transition to other systems as their needs mature or change, the developers have focused on creating object-oriented informational and functional models for an Internet information architecture.

integrated messaging, unified messaging A term to describe the combination and consolidation of messaging services such as voice, video, facsimile, email, etc. through a networked computer system.

With a computer phone set, a scanner, and a printer attached to a microcomputer, it is possible to have all the capabilities of these various technologies integrated into one system. In fact, setting up the system this way provides *more* capabilities than these services have individually, since the computer software can be configured to monitor the calls, store accounting information, transfer data among the various systems, and use files directly, as in directly faxing a document from the word processor, without printing it and sending it through a dedicated facsimile machine. When a facsimile is received, it can be processed to turn it into text and images, or document and PostScript-format files can be sent directly, without any scanning or translation.

By attaching an Internet phone set to the computer, the computer can check the time at the desired destination, dial the call automatically from a database of names, connect the call, signal an alert when it is connected, keep track of the duration of the connection, alert you while making the call if you are attending to other business, and log the call, if desired, for future reference or statistical or business tracking.

By using an integrated voice, file, email service, you

I

can speak into the headset or a microphone and record a mail message, send it the same as normal email, which means the recipient can access it whenever he or she is online, and listen to it played on the destination computer as a sound file. This message can easily be combined with text files with binary files as attachments. The NeXTStep operating system has had this flexible type of voice/email/file capability built into its email system since the late 1980s, and Smalltalk object-oriented systems had it even sooner, so it is by no means a new concept. Unfortunately, it is not yet implemented on many commonly used platforms.

integrated model A network traffic routing solution supporting an exchange of routing information between ATM routing and higher level routing. This provides timely external routing information within the ATM routing and provides transit of external routing information through the ATM routing between external routing domains.

Integrated Next Generation Electronic Commerce Environment Project INGECEP. A trio of experimental projects to test online business/financial environments. INGECEP was proposed by the Telecom Services Association of Japan (TELESEA) to the APEC Telecommunications Working Group as an international interconnection of electronic commerce test-beds. In Japan project funding is provided by the Ministry of Posts and Telecommunications. INGECEP is promoted by the member companies of the Cyber Business Association (CBA) as follows:

- a debit-based commerce system introduced as a pilot project in 1995. It is associated with multimedia information (online malls, educational institutions, government sites, museums, etc.) provided by regional SMEs in cooperation with the Telecom Services Association of Japan (TELESA) on the backbone network using TCP/IP over ATM.
- a secure electronic credit-based commerce system utilizing Japanese cryptologic technology, HTTP, and MOSS carried out at the Japan Electronic Messaging Association (JEMA)
- an electronic money system launched to promote content-based business.

The first INGECEP trials between Japan and Singapore were conducted in July 1998. Increased interest after this trial necessitated guidelines for interconnecting multiple cross-border economies.

This is an important electronic commerce globalization effort. It defines, specifies, and tests cross-border remote technologies in the context of consumer trust, privacy, and language differences while also taking into consideration consumer protection, currency differences, import/export regulations, and money transfer transactions. After initial testing, a new approach labeled the Electronic Market system was employed to increase the scope of the project around the Asia-Pacific Region, including South Korea, Malaysia, and the United States.

Integrated Services Digital Network See ISDN, Signaling System 7.

integrated service A type of service more recently being designed and deployed on the Internet in addition to best effort services traditionally provided. Integrated services support special traffic handling based upon bandwidth, network latency, and other requirements not usually handled with best effort services. Examples include guaranteed service and controlled load service.

Intel Corporation One of the best known of the chip manufacturers serving the desktop computer market, rivalled mainly by Motorola. Intel's chips are widely installed in microcomputers worldwide.

Intel evolved out of an earlier company founded by Gordon Moore and Robert Noyce, who had worked together at the Schockley lab in Palo Alto. They founded a division of Fairchild Camera to exploit semiconductor technology, called Fairchild Semiconductor. They later cofounded Intel Corporation, which continues today as one of the world's leading chip design and manufacture firms.

The Intel 4-bit 108 kilohertz 4004 microprocessor became an important historical impetus in the design of desktop computers, with its successor, the 8008, becoming the world's first commercially significant programmable central processing unit (CPU). The 4004 was developed by Marcian (Ted) Hoff, introduced in November 1971. Three other chips accompanied the 4004, offered as the MCS-4 chip family. The Scelbi computer, first promoted in 1974, and the Altair, which came out as a kit a few months later, incorporated the successor to the MCS-4 family, the MCS-8, based around the 200-kHz 8008 (the 8008 was an enhanced version of the 4040) 8-bit microprocessor.

The 4004 was incorporated into many automated systems, including light controls, appliances, calculators, musical instruments, etc.

Gary Kildall developed a programming language for the early Intel processors called PL/M. The 8080 was incorporated into the Altair 8800, as it was in some of the S-100 bus (Altair bus) computers that became competitive with the historic Altair. Since then, the most significant evolution in Intel desktop computer chips is the Pentium series, introduced in the early 1990s.

The Intel Overview table is not comprehensive, but it provides an encapsulated look at some of the highlights in Intel chip development for microcomputer CPUs since the mid-1970s. See Hoff, Marcian; integrated circuit; International Business Machines; Kildall, Gary; Moore, Gordon; Motorola; Noyce, Robert.

Intel Video Interactive IVI. Intel purchased the Digital Video Interactive (DVI) chipset technology and developed it into Indeo 2 and Indeo 3, now known as IVI.

IVI has a number of interesting features, including transparency (e.g., for background overlays), scaling, and the use of an interframe codec for compression, based on relatively new wavelet compression, encoding the images into frequency bands so the image data can be represented at different resolution levels. Data can be password-embedded for protection. Key frames can be incorporated as reference points for random access. Brightness and contrast settings can be adjusted to adapt to the characteristics of the playback system.

intelligent agent A software application preconfigured or trained to handle tasks dynamically, or that has been trained to recognize certain characteristics of the input, which might be a person's voice, handwriting, or other specialized type of input that may vary from user to user. An intelligent email agent may be configured to screen out "spam," unsolicited commercial messages, to sort messages into folders according to sender or priority, or to forward messages to another address if the user is traveling or reading mail at another location.

The difference between a custom agent and an intelligent agent is that the custom agent is explicitly configured by the user, whereas the intelligent agent configures itself on the basis of monitoring the user's habits and interaction history. The agent then establishes actions and parameters based on intelligent analysis of the user's actions and preferences. In other words, a custom agent would require that the user explicitly instruct the email client to put all messages with "make money fast" in the subject line into a spam bucket, a file that contains unsolicited email.

An intelligent agent would notice that 15 messages in a row with "make money fast" in the subject line were moved to the other file area, and would subsequently do the transfer automatically on behalf of the user, perhaps prompting the first time it makes this decision in order to confirm that it is carrying out user preferences. See artificial intelligence, expert system.

intelligent answering A telephony industry marketing phrase for telephony-computer applications that pop up an information box on a computer screen based on the number that has been called or answered. The pop-up box provides information about the caller/callee contained within publicly available databases or in-house client lists or contact databases.

Intelligent I/O An open standard designed to provide a device-independent device driver architecture. Applied to redundant array of inexpensive disks (RAID) systems, Intelligent I/O provided faster drive access.

intelligent load balancing In computer telephony integration (CTI) applications, a mechanism for balancing call volume in centers that handle many calls or that forward calls to subsidiary call centers. Load balancing is based on statistical models for evaluating queues, call durations, call priorities, and the number of agents available to handle the calls. The intent, of course, is to streamline the service so that calls are handled quickly and efficiently, and distributed well over the types and numbers of agents available.

Overview of Some Common Intel Desktop Computer Central Processing Units						
Processor	Data Int. Bus	Data Ext. Bus	Address Bus	Clock Speed	Year Introd.	Notes
4004	4/8	4	12	1 MHz	1971	Separate program and data memory; 46 instructions
4040					1972	Enhanced 4004 with 14 additional instructions, and more space for programming and stack
8008	8	8	14	2 MHz	1972	Similar to 4040
8080	8	8	16	2 MHz	1974	Seven 8-bit registers, some of which could be combined into 16-bit register pairs; 256 I/O ports
8085					1976	An update to the 8080
8086	8	8	20	5 MHz	1978	Based upon the 8080 and 8085; 8-bit 64K I/O
80286	16	16	24	8 MHz	1982	
80386DX	32	32	32	16 MHz	1985	
80386SX	32	16	24	16 MHz	1988	
80486DX	32	32	32	25 MHz	1989	On-board cache, pipelines, integrated floating point unit
80486SX	32	32	32	20 MHz	1991	
Pentium	32	64	32	66 MHz	1993	Separate caches; superscalar
Pentium Pro		64	32	133-200 MHz	1995	CISC-RISC
Pentium II		64	32	233-333 MHz	1997	MMX
Pentium III				500 MHz	1999	MMX 2
Pentium IV				1400 MHz	2000	

Intelligent Music Workstation IMW. A five-year-long project which resulted in the 1994 release of a musical software/hardware environment in which commercial products can be integrated as modules. Developed at the Laboratory for Musical Informatics of the Department of Information Sciences of the University of Milan, Italy, funded by the Italian National Research Council.

Intelligent Network IN. See Advanced Intelligent Network.

Intelligent Networks Call Model INCM. Gaddis et al. described a Call Model for multipoint communications in switched networks in the early 1990s. The model provided dynamic multipoint, multiconnection communication channels (calls) for network clients. Protocols were defined for clients to create, manage, and manipulate telecommunications calls. The Model provided basic interconnection services for local and wide area networks. At about the same time, Hill and Ishizaki described a Call Model for distributed multimedia communications intended to encompass a number of types of media rather than being restricted to a specific type of data communication (e.g., videoconferencing).

In current practice, the INCM is a significant telecommunications Call Model central to advanced intelligent networks (AINs) that are typically implemented over SS7 networks. In general, this Call Model is a representation of service switching point (SSP) call-processing functions for establishing, maintaining, and taking down a call. The Call Model incorporates Points in Call (PICs), Trigger Detection Points (TDPs), and the triggers themselves. INCM is also sometimes called the Basic Call Model. See Universal Call Model.

intelligent routing 1. In data networks, an automated, dynamic, self-configuring routing system that takes most of the workload of configuration and maintenance from the human operator and handles it through software algorithms. These days, most routers and switchers are designed to handle routing intelligently and the distinction between routing and intelligent routing may gradually disappear. 2. In telephony call servicing, a marketing phrase to describe the automatic routing of a call to an appropriate operator or sales rep based on information and criteria contained in a list or more complex database. As an example, if a call comes in from ABC Copy Machines, from which a company leases equipment, a scenario can be set up to route ABC Copy Machines' calls to the equipment department or the print room, depending upon who usually talks to that vendor. Similarly a new caller, with a number that is not yet recognized by the system, might be routed to the information desk or to a new client sales rep.

intelligent transportation systems ITS. Transportation systems that incorporate new computer technologies, such as Global Positioning System (GPS), to improve efficiency. See Intelligent Vehicle Highway Systems.

intelligent vehicle highway systems IVHS. Advanced navigational systems which incorporate computer technologies such as Global Positioning System (GPS) and navigational databases. IVHS vehicles include sensors and compasses to interface with the computer control mechanisms and incorporate dead reckoning, maps, and GPS data to control direction and sometimes velocity. IVH systems can be configured for optimum efficiency and safety and could apply extremely well to specially designed mass transit pods or automated commuter systems. Even regular traffic could benefit from IVH systems. See guidance system.

intelligent workstation IW. A computer system with advanced features or knowledge bases suitable for business or scientific applications beyond that which a home user might need, for example, but which combines these enhanced features with accessible interface design so that the user need not be a computer expert to take advantage of their features.

Intelligent Workstation Architecture IWA. A framework for a computing system with advanced functions or applications such as expert knowledge bases, decision-making algorithms, intelligent search and retrieval functions, and other features that offer advanced computing wrapped up within an interactive, streamlined user interface design. Knowledge bases for complex data sets (scientific, medical, financial, etc.) that can be accessed and used by computer users with normal computer operating skills, but without computer technical-expert skills are good candidates for development within an IWA framework.

intelligibility In communications, the degree to which a message can be understood by sound and context. While articulation refers to the specific ability to make out a communication, intelligibility is the ability to make out sentences and phrases based not only on articulation, but also on context and inference. Thus, a poorly articulated transmission might still be decipherable in context, especially when enough information is given to figure out the nature of the communication. Intelligibility does not require perfect articulation or good fidelity. If a listener hears "Rog...ov....out" at the end of a CB radio conversation with a lot of noise on the line, it is still intelligible as "Roger, over and out" to an experienced radio operator. See articulation, fidelity.

International Federation for Information Processing IFIP. An international, nongovernmental, nonprofit organization comprised of organizations in the field of information processing. IFIP was established in 1960 under UNESCO as a result of discussions at the World Computer Congress, Paris, 1959. IFIP supports and promotes the research and development of information technologies for the benefit of all people. It hosts a number of Technical Committees to help fulfil these aims. http://www.ifip.or.at/

INTELSAT *In*ternational *Tele*communications *Sat*ellites. The largest commercial not-for-profit satellite communications services provider, founded in 1964. INTELSAT is a cooperative of more than 140 member nations and has 20 communications satellites in geostationary orbit, with further launches planned. INTELSAT operates as a wholesaler, with subscribers, many of them major broadcasting and telephone companies, paying for services according to their type and duration.

INTELSAT lays claim to having launched the world's first commercial communications satellite in 1965 (Early Bird) and the first global communications system in 1969. In 1980, they launched INTELSAT V, the first to use dual-polarization transmissions equipment. INTELSAT VI was a subsequent series of five satellites built by Hughes Aircraft Company.

In 1995, INTELSAT began providing global Internet access services through its satellite system. See Early Bird. http://www.intelsat.int/

Intelsat Business Service IBS. A commercial telecommunications service based on the INTELSAT satellite communications capabilities. IBS provides almost 10,000 communications channels for a wide variety of services, including voice, facsimile, data, videoconferencing, and telex.

Inter Exchange Carrier IEC, IXC. A telephony service provider permitted to provide long-distance services between Local Access and Transport Areas (LATAs), but not within a LATA region. It is also often written as Interexchange Carrier. The category is important as IECs are bound by a number of regulations to support their provision of services while still safeguarding competitive opportunities for other telecommunications providers who do not fit the definition for IECs. See Local Exchange Carrier.

inter- Prefix for between, usually between external and internal systems.

Inter-Access Point Protocol IAPP. A specification developed by Lucent Technologies, Aironet Wireless Communications, and Digital Ocean, IAPP is a means for different vendors to communicate with one another through roaming wireless mobile communications. IAPP describes a backbone-based handover process for mobile stations when implemented in conjunction with the IEEE 802.11 standard.

interactive 1. Reciprocal communication, that is, with a back-and-forth, or query-and-answer character. 2. Software which responds to the individual's input, usually in realtime or near realtime, as in multimedia applications. Video games are highly interactive, whereas archive searches over the Internet may be extremely slow (sophisticated searches can take days). Depending upon the circumstances, programs with slow interactivity may be better processed as batch files. Contrast with batch processing.

interactive asynchronous communications IAC. A means of interactively communicating over an asynchronous network connection that allows control of and communication with devices such as a computer modem over a serial connection. Typically the transmission line (e.g., serial line) will be initialized to set up communications parameters before carrying out interactive communications. IAC is useful in situations where the status and operating parameters of a device are broadcast back to the user.

Interactive Media Alliance, The TIMA. A nonprofit professional organization supporting various levels of technical and artistic expertise. TIMA fosters the exchange of ideas and knowledge regarding interactive media and promotes the advancement of the technology. TIMA is affiliated with the Technology Association of Georgia. http://www.tima.org/

interactive television Interactive TV, I-TV, ITV. TV broadcasting configured to provide a two-way dialog between the user and the broadcaster, enabled by computerization and two-way transmission circuits. Interactive TV has been implemented in a number of ways since the late 1970s, from educational programming to interactive music concerts and on-demand video, but the potential of this technology has only been hinted at so far.

One of the earliest interactive TV networks was the QUBE system from Warner Communications, which was first tested in Columbus, Ohio. Time Warner developed subsequent versions of this technology. Depending upon how it is implemented, interactive TV has been of more interest to educators than traditional passive-interactive TV for distance and self-directed education. See QUBE.

Interactive Television Association See Association for Interactive Media.

interactive video services IVS. Interactive video, in its broadest sense, is public or private image and sound broadcasting through public or private networks that is available upon request by the user. Due to the convergence of broadcast and computer technologies, it is now feasible to provide partial- and full-service interactive video services through a number of transmissions media: twisted copper pair, coaxial cable, fiber optic cable, and wireless. However, with the exception of fiber optic cable, the use of existing technologies, which were designed for other services, means that none of them are ideally configured for IVS, and vendors are hurrying to find ways to deploy services ahead of their competitors. Thus, a variety of technologies are emerging, in spite of the fact that the marketability of these services is not yet fully proven.

Interactive video services potentially include games, movies, and specialized channeling, such as stock quotations and industry-specific news. Some of these have been tried with varying success in different industries and regions, and some companies are devising ways to offer them over the Internet.

interactive voice response IVR. Systems that respond to voice commands or voice characteristics and may also prompt the user for further information or clarification. Phone systems that can recognize and respond to simple spoken commands are becoming more common, and software programs can interpret spoken commands and prompt users through synthesized speech.

Interagency Management Council for Federal Communications IMC. A representative body for telecommunications executives at key U.S. federal agencies, including the Department of Commerce, the Department of Defense, the Department of Education, the Department of Justice, NASA, the U.S. Postal Service, and others. It was established to provide a forum for participation in the planning and administration of the General Services Administration's long-distance telecommunications services provided through the FTS2000 program. Since then, it has become a focal point for the development and administration of federal technology programs.

interapplication communications IAC. A transparent means of intercommunication between computer software applications (e.g., between a word processor and drawing program). A similar concept was dubbed Compound Data Architecture (CDA) by Digital Equipment Corporation (DEC). The concept is now frequently implemented in layered architectures on multitasking systems, but in the early 1990s, surprisingly, it was not prevalent on desktop systems. Once a user has experienced the ease of moving around data and images among different types of applications or among applications from different vendors, it's hard to go back.

INTERBUS An open systems frame-based, data bus interface device standard and protocol for high-performance, distributed networks for manufacturing and process control. INTERBUS standards enable devices from different manufacturers to exchange information through standardized profiles for robotic controllers, peripheral drives, data encoders, industrial valves, etc. It is a bit-oriented, synchronous protocol that is used with sensors and actuators. INTERBUS is implemented on ring-based, token-passing networks and utilizes a single multipair cable to interconnect all devices, regardless of type or level of complexity. See actuator, PROFIBUS.

Intercarrier Interface ICI. One of the two interface ports of XA-SMDS systems which is used to specify how the carrier switch sends and receives data from an Interexchange Carrier's (IXC's) SMDS network. The other interface is the Subscriber Network Interface (SNI). See Exchange Access SMDS.

Intercast An Intel term for technology that allows a consumer to interface the TV set with a computer hooked up to the Internet, to receive "push technology" Webcasts or Netcasts, that is, digital broadcasts of information and entertainment transmitted over the Web rather than through television broadcast airwaves or television cable services. The digital information from the Web is displayed in the blanking spaces of the TV signals, so the TV can still receive normal TV broadcasts in addition to displaying Intercasts. See Webcast.

Intercept Service A service in which a call to a changed or disconnected number is routed to a recording or, if a recording is not available, to an Intercept operator. In the case of the latter, the caller will be verbally asked for the destination number and the operator will attempt to complete the call.

Interchange A commercial Internet connection service from Ziff-Davis, similar to some of the other large Internet Service provisions, but with a slightly more technological slant.

Interchange Carrier IC. A common telecommunications carrier that provides inter- or intra-LATA services through local public exchanges according to

definitions and regulatory guidelines established by the Federal Communications Commission (FCC) and the Telecommunications Act. See Inter Exchange Carrier.

Interchange Format See Rich Text Format.

intercom *abbrev.* intercommunication, intercommunicator. A set of at least two devices, minimally a transmitter (with a microphone) and receiver (with a speaker) or two transceivers, over which remote communication can take place in at least one direction. Most intercoms are audio only, but audio/visual intercoms are becoming more prevalent as the technology becomes more readily available.

Intercoms can generally be categorized as wired or wireless. Wired intercoms sometimes use existing wiring (e.g., doorbell wires in the walls of houses). Wireless intercoms use broadcast frequencies sent through the air or sometimes through building wiring using the AC sockets as an interface to the wiring for better transmission.

Many wired and wireless intercom speakers are wall mounted, like the PA systems in schools or hospitals, and the transmitter may be attached to a handheld microphone or operated through a telephone handset. Baby monitors are a type of wireless mobile intercom, in which one unit is placed near the baby and the other is placed near the parents or babysitter or attached to their clothing so they can move around. Intercoms are often incorporated into phone systems, so that the handset or speakerphone is the transmitter and the receiver is a speakerphone on another console (or on several consoles in broadcast mode).

The distinction between wireless intercoms and wireless radios is not a hard and fast one; there is overlap in capabilities between sophisticated intercoms and simple radio systems; the main difference is in ease of use. Intercoms and basic two-way short-distance radios tend to be unlicensed push-button devices, whereas wireless radios tend to be licensed devices, some requiring a higher level of expertise to operate. In this sense, a computer videoconferencing system can be called an intercom system. Once the software application is installed and launched, the user need only sit in front of a microphone and small camera in order to communicate with the person at the other end of the connection. No sophisticated skills are needed and even the push-button aspect of the communication has been eliminated.

Since videoconferencing systems aren't subject to the same distance restrictions as low-power wireless devices, it's possible that small flat-screen monitors with built-in speakers will eventually replace traditional intercoms. Parents will be able to readily see what their kids are doing in daycare or at school. Friends can keep in touch without making long-distance calls. Business associates can discuss important projects or interact in meetings from home or a remote office (or from the road with a wireless modem). When high-speed communications become available to a majority of users, Internet intercoms may well become one of the most prevalent telecommunications technologies. See public address system, videoconferencing.

Interdepartment Radio Advisory Council IRAC. An assemblage of committees, subcommittees, and working groups providing expertise and notification to the International Telecommunication Union (ITU) regarding the allocation and management of radio frequency spectra. IRAC develops procedures, processes requests, and assists in assigning frequencies to U.S. Government radio stations. IRAC includes the Frequency Assignment Subcommittee (FAS), the Spectrum Planning Subcommittee (SPS), the Technical Subcommittee (TSC), the Radio Conference Subcommittee (RCS), and the International Notification Group.

Interexchange Carrier Compatibility Forum ICCF. An organization that developed an expansion plan for telephony Carrier Identification Codes (CICs) when they became scarce in the late 1980s. The ICCF also served as a liaison in standardization efforts for fiber interconnectivity in the mid-1980s.

interface *n.* A hardware connection, or logical connection or translation point between two or more devices or transmissions media. Interfaces are an intrinsic part of interconnected computers, peripherals, and networks. Almost every aspect of data and electrical connections in the telecommunications industry uses a different format or version of a format, and the interface is the point at which all these different hardware and software junctions come together. Common electronic interfaces include docking bays, cradles, cable connectors, peripheral card connectors, card slots, and chip sockets.

interface, human-machine *n.* The point of contact or translation between humans and machines.

- A *hardware interface* is a device or system that translates human movement, speech, or sensory output into impulses (usually electromagnetic or mechanical, though chemical interfaces also exist) that the machine or computer device can interpret and compute into data and instructions or, conversely, that translates machine signals into sensory output or information meaningful to humans.

  Human-machine interfaces come in many varieties, including digital, analog, mechanical, chemical, or a mixture of these. Examples of hardware input interfaces include microphones, keyboards, joysticks, temperature sensors, serial connectors, video cams, data gloves, and pressure pads. Hardware output interfaces include monitors, speakers, pulsing lights, thermostat controllers, infrared device controllers, and more.

- A *software interface* is a system of algorithmic procedures/functions to meaningfully communicate information and options to humans and/or to interpret human communication and sensory into machine instructions.

  Common software interface conventions include the use of textual queries and responses, graphical pointers, folders, menus, and other culturally meaningful icons that indicate the state of the device, availability of services, current point in a process, etc.

Human-machine interfaces evolve through a system of trial-and-error combined with the sometimes idiosyncratic preferences of the people who design the applications or market the devices to the general public. At the present time, software interfaces are often developed intuitively by computer programmers with little or no input from users even though they are intended to satisfy the needs of a broad spectrum of people, rather than the more individual needs and preferences of the programmer, in order to produce products with commercial viability.

Unfortunately, interfaces become entrenched even if they are no longer appropriate or practical. Early versions of a technology are often designed to overcome pioneering design limitations. As the limitations are overcome, the increasingly inefficient interface may be retained because users have become accustomed to it or because it is expensive to change production lines. The QWERTY typewriter keyboard layout is a good example. It was designed for historic manual typewriters and laid out with the letters organized so they would *slow down* the typist to help prevent jamming that occurred with old-style mechanical keys. When electric typewriters and computer keyboards were developed without the jamming problems, the QWERTY layout was retained even though the original reason for the layout became irrelevant.

The concept of the interface is an important one as it influences how comfortably and efficiently humans can utilize a technology. Ease of use and interface design are essential to the success and proliferation of many telecommunications products. Interface design also reveals priorities; sometimes humans are expected to adapt to the limitations of a technology rather than the technology being designed to serve the needs of humans.

Interface Control Application Programming Interface ICAPI. A telecommunications call control library that facilitates network interface access for T1 robbed-bit signaling systems or T1/E1 CAS. It fits between the operating device drivers and the application. There are similar call libraries for ISDN and ANAPI. An ICAPI protocol uses bit transitions and in-band signaling to establish calls and transmit call information. Events and channel states may be logged.

Interface Data Unit IDU. In ATM networking, interface control information transferred to and from the upper layer in one interaction across the layer. A service data unit (SDU) may be passed across ATM Adaptation Layer 5 (AAL5) as an IDU corresponding to one protocol data unit (DPDU) in a one-to-one correspondence or, depending upon the type of service (e.g., message or streaming mode), the SDU may also be passed across AAL5 in more than one IDU.

Interface Device In Frame Relay networks, the Interface Device provides a link between an end device and the network through encapsulation. See encapsulation, Frame Relay Capable Interface Device.

interference Extraneous, unwanted signals that hinder transmission or perception of the desired signal. Types of interference include noise, static, pops,

I

crackles, echo, babble, chatter, crosstalk, cosmic noise, and background noise. See individual entries in this dictionary for details.

interference grating A grating created through a photoresist laser etching process by exploiting the interaction between the intersection of collimated light beams of the same wavelength. It is sometimes called a holographic grating due to the three-dimensional tapered effect between high and low points in the grat-

ing. The sinusoidal cross-section makes it difficult to impose a blazed pattern on an interference grating, putting some limits on the efficiency of this type. Interference gratings came into practical use in the late 1960s. They are favored over ruled gratings for a number of precision applications due to greatly reduced incidence of stray light, particularly in gratings with fine grating periods (reduced distance between adjacent facets). Photographically etched

Interferometer Examples		
Type	Abbreviation	Description
Fabry-Perot	FPI	A high-resolution interferometer utilizing multiple reflections from two proximate reflective surfaces. A Fabry-Perot interferometer with Michelson-type mechanical motion was designed by the National Bureau of Standards (NBS) in the early 1900s to make the first precision measurements of wavelengths published by the NBS.
		In cooperation with the University of Pittsburgh, MIT uses this type of interferometer with a cooled gallium-arsenide photocathode photomultiplier as a detector, with a computer as a controller, for night sky observations. There are FPIs installed in the Antarctic, Norway, and many other locations to remotely sense upper atmosphere wind and temperature conditions. Doppler concepts are incorporated into the sensor readings for meteorological observations.
		Fabry-Perot filters have also been suggested by Wickham et al. for use as sub-band tuning mechanisms (in conjunction with Bragg reflection gratings) for optically channelizing radio frequency (RF) signals for spectral analysis of the very high frequencies now used for communications signals. Bragg reflection gratings and FP filters have further been described by Bao et al. as a mechanism for determining wavelengths of transmitted or reflected light through a calibrated wavelength reference, thus creating a reference system.
Fizeau	FI	A basic type of image plane interferometer useful for noncontact testing of surface characteristics and for telescope design. The Fizeau interferometer produces a direct image from the source and does not incorporate the same degree of beam diversion characteristics as the other interferometers mentioned here. The FI is used in many astrophysical applications. More recent digital phase-shifting FIs have been developed by CSIRO for in-house precision metrology of optical components.
Mach-Zehnder	MZI	A historic interferometer descended from the Twyman-Green interferometer. With improvements, the MZI is still common as a calibration and diagnostic instrument. It is most often used in aerodynamics, thermal transfer, and plasma physics, but is also being studied in fiber optics research.
		The MZI is favored for many educational applications, as a basic model can be built by students in a rectangular or parallelogram configuration. Depending upon the alignment of the reflecting surfaces, interference fringes may or may not be produced and, by controlling the length of the optical path, phase shifts can be introduced in a controlled manner. The extent of the shift provides a means to monitor relative changes in the optical path, thus providing useful measurement information. The beam phasing characteristics of an MZI can also be modeled in computer software.
		The MZI is both simple and sophisticated, depending upon how it is implemented. It is not only a good student project, but also has been proposed for use as an atomic interferometer.
		MZIs are useful in the fabrication and testing of components and various sensors used in telecommunications. They have also been incorporated into integrated circuits for converting optical wavelengths and have been developed into directional coupler switches.
		... cont.

gratings are often easier to impose on nonplanar shapes than machined ruled gratings. On the other hand, they also tend to require higher intensity light sources than ruled gratings.

By placing photoresistive material in an interferometer and controlling the angle at which a light beam hits and rebounds from a reflective surface, it is possible to have a single light source etch a standing wave interference pattern, with the angle related to the distance between grooves. If two planar beams of the same wavelength and intensity (at the point of intersection) are aimed from slightly different directions onto a planar photoresistive surface, they will etch out a regularly spaced, grooved *classical* grating. Other types of waves and other imaging surfaces exist, some of which are useful for compensating for the characteristics of other components, but the abovementioned are common in the fabrication of semiconductor gratings. See diffraction, grating, interferometer, photoresist, ruled grating.

interference guard band See guard band.

interferometer Because of its wavelike properties, a beam of coherent light can be split and realigned in such a way that factors that interfere with one of the beams can be detected when compared to an unimpeded reference beam or, more simply, the split beams can be compared to detect subtle changes to one or more of the beams. Thus, a device that detects and displays interference between two or more light wave trains and, optionally, compares wavelengths against reference displacements is called an interferometer. Because light technologies can be very precise, compared to mechanical devices, interferometers can be used for very fine detection and calibration. The interference pattern information derived from an interferometer is useful in measurement or calibration, for example, to determine angular positions in satellite tracking. A series of horizontal or vertical measurements at precise distances along a path can yield data that can be processed to yield planar information (e.g., a height profile in an optical fiber).

There are a variety of types of interferometers, ranging in complexity from simple lab-built student models to more sophisticated instruments that incorporate integrated circuit (IC) concepts. There are also different modes for which an interferometer can be designed. The optical path length of a test and reference beam can be changed in their relationship in a series of phase shifts. In an optical fiber, the result is a lateral shift in the interferometric fringe pattern to measure the dimensions of a surface. Another approach is to scan down through a fiber to produce an interference signal along a series of points on a surface. With digital processing, the data can be assembled to generate a surface height profile.

Interferometers have been proposed as instruments to determine electrical states in silicon-based integrated circuits (ICs) and as diagnostic instruments for optical computer networks. One interesting application is the light-in-flight speckle interferometer

Interferometer Examples, cont.		
Type	Abbreviation	Description
Michelson	MI	A basic interferometer developed by A.A. Michelson to conduct the Michelson-Morley experiment in the 1880s. The scientists used the interferometer to determine whether a theoretical medium called the *aether* existed and could be detected. Michelson received a Nobel Prize in 1907 for his discoveries in optical science.
		In a Michelson interferometer, a monochromatic point light beam is split in two by a partially reflective material, such that one beam continues in the original direction, and the other is reflected (usually 90°) from the original course. The beams are recombined with the resulting interference patterns displayed on a screen. These are derived from the wavelike characteristics of light and can be analyzed to determine vibrational or thermal effects, which are useful in fabrication and diagnostics. The image in a Michelson interferometer is not viewed directly as in a Fizeau interferometer but, in viewing instruments such as telescopes, can be reconstructed.
		By calibrating the MI with a known reference, the wavelengths of other unknown materials (e.g., gases around astronomical bodies) can be studied through known interference characteristics. MIs are also used to develop standards in atomic lengths.
Twyman-Green	TGI	A basic type of interferometer used as an optical fabrication and diagnostic tool, developed by Frank Twyman and Arthur Green in the early 1900s. The modern TGI is based upon a monochromatic point light source at the focal point of a lens. By revealing patterns of optical interference resulting from unequal light paths, the TGI can be used to assess optical surface characteristics such as the flatness of a surface, performance of a component (e.g., a prism), or deviation from a reference shape.

presented by Swedish engineers for evaluating 3D shapes by using ultrashort laser light pulses. On a software level, algorithms for phase-stepping interferometry have been developed by Chinese scientists.

Interferometers are important in many aspects of astronomy. Not only are telescopes based upon interferometric principles, but Fizeau interferometers (FIs) are included as payloads on space missions, selected for their accuracy and capability of sensing over a wide field. The Global Astrometric Interferometer for Astrophysics (GAIA) mission, for example, includes three stacked, mechanically connected FIs designed to observe about 50 million stars.

Historic interferometers were based upon noncoherent, mixed-wavelength light, while interferometers developed since the invention of lasers typically use coherent, nearly monochromatic light. The Interferometer Examples chart includes a short list of some common, representative interferometers used in remote sensing, and component fabrication and testing applications relevant to telecommunications.

Interferometers may be further optimized for a particular task such as assessing the *cleave* (terminal cut) of an optical fiber by forming an interference pattern between the fiber surface and an optical reference. A CCTV camera or other display device may be used to enlarge and display the cleave so the angle and evenness of the cut can be seen. A divergence of angle can be counted as a specific number of fringes in a fiber of a specified diameter. The shape of the fringe pattern can indicate the evenness of the surface. The inteferometer may be customized to be self-calibrating for this task and accurate within a tolerance of one fringe, enabling a cut to be accepted or rejected (and possibly recut), before connecting to other components. Some interferometers can interpret the visual information into digital data for further processing, thus enabling lists or graphs of component characteristics such as angle of cut, smoothness, radius of curvature, etc., to be generated. See beam splitter, cleave, coherent light, ferrule, spectrometer.

interferogram A visual measurement diagram derived from output from an interferometer. A series of interferograms may be digitally processed to generate a surface height profile of the medium observed (e.g., optical fiber) See interferometer.

Interim Local Management Interface ILMI. A means of providing an ATM device with status and configuration information about virtual connections, and the registered ATM prefixes, addresses, services, and capabilities available at its ATM Interfaces through the Simple Network Management Protocol (SNMP) and an ATM Interface Management Information Base (MIB). ILMI is an open protocol that was developed as an interim solution by the ATM Forum in the mid-1990s to enable the exchange of UNI management information through direct encapsulation over ATM adaptation layer 5 (AAL5). However, the interim designation was dropped. ILMI is not universally implemented and meta-signaling may be used to serve this purpose on some systems.

Interim Number Portability INP. The use of various telephone subscriber services, such as call forwarding, call routing, and call addressing, to allow a call to be redirected to another location, usually on a temporary basis.

interior In ATM networking, an item such as a link, address, or node inside a PNNI routing domain.

Interior Gateway Protocol IGP. A family of network routing protocols for exchanging information with other routers and switches on the same system. When changes occur in the organization of the network, these changes are communicated to the routers, so the routing table databases may be revised accordingly.

Interior Gateway Routing Protocol IGRP. A Cisco Systems proprietary multipath routing protocol developed in the mid-1980s for routing within autonomous systems. Since then, the protocol has been further developed and many users have replaced Routing Information Protocol (RIP) with IGRP to run on large, heterogenous networks, like the Internet. IGRP was intended to run in a variety of network environments and enhanced IGRP has been developed to support TCP/IP, IPX, and AppleTalk.

IGRP-enabled routers send some or all of their routing tables to neighboring routers at regular intervals, a process that also enables distances among nodes to be calculated as the information propagates out through the network.

interlace A system used in frame-based video image display to display images in two-frame passes, with one pass imaging the odd lines and the next the even lines (or vice-versa), in an alternating pattern. Thus, in NTSC, for example, an interlaced screen is imaged in two fields of 262.5 lines (to make up the full 525 scan lines), each field taking 1/60 of a second. Some flicker can be seen on an interlaced display, so non-interlaced monitors, including multisync monitors, have become prevalent on computer systems. Generally, the faster the refresh, the more stable the image. See cathode-ray tube, field, frame, interleave, multisync, scan, scanning rate.

interleave *v.t.* 1. To arrange in alternating layers, rows/columns, or time slices. 2. In concurrent programming, a logical means to execute sequences in order to analyze the correctness of concurrent programs. 3. In networking, to transmit pulses through a single path through time-division from more than one source. 4. In graphics file storage and display, a means of arranging the image data so that all odd lines of the image and all even lines of the image are stored or displayed as a group. 5. In magnetic and magneto-optical data storage, a means and pattern of storing information on a disk so that the physical characteristics of the read/write sequence are accommodated without the drive head needing to "backtrack" to find the next section of data. 6. In multimedia applications, a means of slicing up the recording space so that different media (sound, graphics, etc.) are laid down in strips or sections on the tape or disc. 7. A data transmission error-correcting technique in which code symbols are arranged in an interleaved pattern before transmission and reassembled upon receipt.

interleaved video A video display in which a frame is constructed and displayed by alternately scanning all even lines and then all odd lines. This system of display is commonly seen on televisions screens and on some NTSC-compatible computer screens. A certain amount of flicker is usually noticeable on interleaved displays. See interlace.

interleaver In fiber optics systems, a multiplexing component that can increase channel density. Filters may be used to receive channels or route channel groups. See add/drop multiplexer.

intermediate frequency IF. In heterodyne receivers, the beat frequency created as a result of the difference between a locally generated signal and the incoming radio signal. See beat frequency.

Intermediate Signaling Network Identification In Signaling System 7 (SS7), a capability that allows an application process in the originating network to specify intermediate signaling networks for noncircuit-associated signaling messages, and/or to notify an application process in the destination network about intermediate signaling networks.

intermittent errors Fault conditions that happen occasionally, sometimes without apparent pattern, or occur from specific causes that happen seldom or irregularly. Difficult to anticipate and diagnose, intermittent problems are often not alleviated until a program has been run hundreds of times or a computer or phone network has negotiated thousands of calls.

intermodulation distortion A type of audio distortion that occurs when multiple tones interfere with one another in a way that is not harmonically related to the original tones.

internal modem A computer modem installed inside a larger system that is utilizing the modem. Internal modems are usually powered by the system in which they are housed and usually take the form of small PC boards or very small PCMCIA cards. Sometimes referred to tongue-in-cheek as "infernal modems," internal modems can be finicky to install in systems with several peripherals that require IRQs.

Internal modems are convenient in that they are out of sight and mind, and don't take up extra space – a real plus on laptop computers. They have disadvantages as well, as they are often machine- or platform-specific and often can't be reinstalled in a new computer of a different type, as can most external modems. External modems are easier to swap among systems, can be shared by a number of users through a switcher, and usually have status lights that are handy diagnostic tools. Since most internal modems install in a slot that faces the back of the system, they often don't provide status lights. In general, people prefer internal modems on small mobile devices and external modems on desktop systems or systems with shared resources.

International Ad Hoc Committee IAHC. See Internet International Ad Hoc Committee.

International Alphabet No. 2 An older alphabetic coding system of equal-duration pulses of negative and positive volts (called marks and spaces) in groups of five, to represent character signals. The beginning and end of each character was signaled by a start signal and a stop signal. The use of five elements in two possible polarities results in 2^5 or 32 character encodings. Even for a basic alphabet, this was somewhat limited, and schemes to double the number by allowing a code to represent one of two characters were devised.

Something similar happened later with computer character codes. International Alphabets evolved into ASCII and became widely implemented on computers, but there were only 128 characters, insufficient for accents or math symbols. Many developers added 128 codes for a total of 256 characters and called it "extended ASCII." Technically, the extra codes weren't standard ASCII and were not consistent across platforms. Another limitation was that they couldn't be used together, the user had to select ASCII or "extended ASCII" banks. In the mid-1980s, the Amiga and Mac removed this limitation, enabling individual addressing of letters in both. See Appendix for an ASCII chart. See ASCII, Unicode.

International Amateur Radio Union IARU. A regulatory agency and proponent of world amateur radio activity established in France in 1925. Amateur radio organizations throughout the world interact with a high degree of cooperation and communications. The IARU is essentially to global amateur radio communications what the American Relay Radio League (ARRL) is to American amateur radio. The IARU is organized into three regional organizations that parallel administrative divisions of the International Telecommunication Union (ITU). See American Radio Relay League. http://www.iaru.org/

International Atomic Time, Temps Atomique International TIA. An atomic time scale based on the coordinated efforts of more than 200 atomic clocks from more than 50 centers from around the world, which are maintained in France by the Bureau International des Pods et Mesures. Unlike the Coordinated Universal Time (UTC), which is adjusted occasionally in leap seconds to maintain some coordination with the Earth's axis rotation, TIA is not adjusted, but remains consistent with atomic time scales. Otherwise, TIA and UTC are very similar. See atomic clock, Coordinated Universal Time.

International Business Machines IBM. In the late 1800s, Herman Hollerith, an American engineer, evolved the concept of punched cards as a storage medium and applied it to the development of a tabulating machine, an early computer that could be used to store and process information in categories. This resulted in Hollerith cards, Hollerith code, and a machine which could tabulate the vast amount of census data gathered at regular intervals in the United States. The tabulating machine dramatically improved the efficiency of storing and analyzing census data, and Hollerith formed a company called the Tabulating Machine Company. This later merged with several other companies to form the Computer-Tabulating-Recording Company, which sold a wide range of industrial products.

Thomas J. Watson, Sr. left NCR to join the company

as general manager in 1914, and remained with the company for over four decades, eventually passing on the position to his son, Thomas J. Watson, Jr.

On Valentine's Day, in 1924, the name of the company changed to International Business Machines Corporation. IBM became an enormously influential company in the business and computing market, and funded or partially funded the research and development of several historic room-sized computing machines. IBM's research laboratory has contributed a great legacy of original and fundamental scientific discoveries of interest both inside and outside the computing industry. IBM inventions are awarded more than 1000 patents per year; in other words, IBM develops as many unique inventions in a single year as the best individual inventors of the 1800s developed in their entire lifetimes.

In 1975, IBM released its first microcomputer, the IBM 5100; it was not a commercial success, and it was not until 5 years later that the first of the long IBM PC line was introduced to the public. This time sales were good, particularly in the business market, and IBM and IBM-licensed personal computer technology became the most common platform for desktop computing. See Hollerith, Herman; IBM Personal Computer, Jacquard loom.

International Center for Technology Assessment ICTA. A nonprofit, bipartisan organization dedicated to helping government officials and the public in understanding technology and how it affects human society and the environment. ICTA explores and communicates the social, economic, ethical, political, and environmental impacts related to the manufacture, distribution, and application of technologies. ICTA also uses legal petitions, comments, and litigation to fight against harmful deployment of technology. http://www.icta.org/

International Commission on Technology and Accessibility ICTA. A commission to explore developments in technology that may assist people with disabilities and to promote and disseminate their understanding and use. ICTA was founded in the 1960s in Sweden in conjunction with Rehabilitation International (RI) and the Swedish Handicapped Institute. In 1969 ICTA and RI developed and adopted the International Symbol of Access (ISA). http://www.ictaglobal.org/

International Commission for Optics ICO. An international organization affiliated with the International Union of Pure and Applied Physics the supports and promotes the dissemination of knowledge in optics, founded in 1947. http://ico-optics.org/

International Correspondence Chess Federation ICCF. This is one of the more interesting historic bodies using telecommunications to enhance gaming communications. The ICCF has evolved from the *Internationaler Fernschachbund*, founded in 1928, which was succeeded by the International Correspondence Chess Association, in 1945, to become the ICCF in 1951. Correspondence chess hasn't only been conducted through postal mail services. ICCF members and chess players in general have always been technology conscious and have enjoyed their matches through the use of homing pigeons, telegrams, trains, planes, computer modems, and high-speed connections to the Internet.

International Data Encryption Algorithm IDEA. A European-designed, 128-bit, single-key encryption algorithm used for data security. It has been incorporated into Pretty Good Privacy (PGP) partly because it doesn't have the same U.S. export restrictions as other encryption algorithms. Use of IDEA is license-free for noncommercial use. See encryption, Pretty Good Privacy.

International Development Markup Language IDML. An Internet protocol and associated set of standards to facilitate development in a global context. In 1998, the Development Markup Language (DML) was seen as a means to support the markup of information on computer networks that describe developmental activities and mandatory data elements described by relevant standards. It was intended to be consistent with other metadata schemes and capable of multilingual markup. In February 1999, DML was renamed IDML. The IDML Working Group was formed to develop recommendations and a process for electing an IDML Advisory Group. It is recommended that XML be used for Site Description files even though other formats may be used. Extended IDML is a superset of the Core Activity Schema with additional audit trail and informational items. IDML was established as a pilot standards-track candidate in early 2001. See Extensible Markup Language, IDML Initiative.

International Electrotechnical Commission IEC. An international standards-development and recommending body, founded in 1906 as a result of a 1904 resolution at the Electrical Congress. The IEC publishes standards for electrical, electronic, magnetic, and related technologies and promotes cooperation among member countries. IEC standards form the core of the World Trade Organization's Agreement on Technical Barriers to Trade (TBT). Hundreds of technical committees and working groups carry out the mission of the IEC. Technical committee papers are submitted to a full-member National Committee members' vote in preparation for approval as international standards. http://www.iec.ch/

International Engineering Consortium IEC. A nonprofit professional organization supporting engineering research and education sponsored by universities and engineering societies, founded in 1944. The focus of the organization has broadened from a national to international purview and from electronics to information engineering. IEC sponsors courses, conferences, virtual exhibits, and a number of publications. http://www.iec.org/

International Federation for Information Processing IFIP. A nonprofit, nongovernmental, information processing research organization. IFIP was established in 1960 under the auspices of UNESCO after the first World Computer Congress of 1959. A number of technical committees provide expertise on technological matters. http://www.ifip.or.at/

International Forum on the White Paper IFWP. An international series of workshops, founded in June 1998, intended to bring together professionals and experts in law to respond to White Paper recommendations for assigned numbers on the Internet.

The process of open discussion and the goal of self-regulation were key aspects of these proceedings. The Internet community believed that it was possible for those in the industry to produce a viable system for assigning Internet addresses without government takeover of the process and used the Internet itself as an important venue for meetings, opinions, and sometimes heated debates over how addresses would be allocated and assigned.

As a result of initial discussions, a California non-profit public benefit corporation called the Internet Corporation for Assigned Names and Numbers (ICANN) was tendered as a draft recommendation as a means to coordinate the administration of Internet domain names and Internet Protocol (IP) addresses. The draft proposal, jointly presented by Network Solutions, Inc. and the Internet Assigned Numbers Authority (IANA), was discussed in September 1998 by IFWP with regard to how it fell within a model of common principles and structure specified by the *U.S. Department of Commerce Statement of Policy on the Management of Internet Names and Addresses*. Three proposals were presented to the Deptartment of Commerce in October 1998 by different groups, and testimony was presented on transferring the Domain Name System to the private sector. ICANN's bylaws and its first public meeting were held in November 1998. See Internet Assigned Numbers Authority, Internet Corporation for Assigned Names and Numbers, InterNIC.

International Frequency Regulation Board IFRB. An agency established by the International Telegraph Union in 1868 to manage the broadcast frequency spectrum. In 1912, the IFRB's *Table of Frequency Allocations* became mandatory. The frequency allocation table specified frequency bands for specific uses in order to minimize interference among stations. See Federal Communications Commission, International Telegraph Union.

International Information Systems Security Certification Consortium ISC². A nonprofit corporation established in 1989 to develop certification programs for security professionals working in the informations services field. http://www.isc2.org/

International Intellectual Capital Codes Association IICCA. A not-for-profit association tasked with defining a comprehensive lexicon of skills of interest to industries and users for employment/employee matching through the iCAP Catalog and the Intellectual Capital Inventory (iCAP) system. IICCA is responsible for the development, maintenance, and uniformity of the iCAP Catalog. See iCAP.

International Internet Association A fee-based Internet service that provides access to more than 20,000 databases from around the world.

International Internet Industrial Association IIIA. A professional association of Internet Service Providers (ISPs), Web developers, software developers and others directly influencing or being influenced by the development of the Internet. The IIIA is concerned with issues such as the addition of a larger available base of international Top Level Domains (TLDs). http://www.iiia.org/   http://www.iatld.org/

International Mobile Subscriber Identity IMSI. An ITU-T identification number assigned by a wireless carrier to a mobile station to uniquely identify the station locally and internationally.

International Network for Development Information Exchange INDIX. A coalition for organizations involved in development in information exchange. INDIX developed the CEFDA standards for data exchange and participated in the development of the International Development Markup Language (IDML). See IDML Initiative. http://www.indix.org/

International Organization for Standardization (International Standards Organization) ISO. An important international standards-setting body which has produced many of the specifications and documents used by telecommunications professionals. ISO is familiar to many through its ISO-9000 series of quality assurance specifications. ISO-9000 standards can be summarized as "Say what you do, then do what you say, and get it certified, if necessary." http://www.iso.ch/

International Radio Consultative Committee CCIR. A standards and regulatory-recommending body founded in 1927, descending from the International Radiotelegraph Conference in 1906, in connection with the International Telegraph Union. This organization was formed in response to public broadcasts over radio waves in the early 1920s. See International Telecommunication Union (ITU-R).

International Radiotelegraphic Convention One of the early international gatherings, resulting from the growth of telegraphy, resulted in a multinational consent agreement regarding Protocol and Service Regulations that was documented in November 1906. The convention was to be entered into force in July 1908 by Great Britain, Germany, Austria, Hungary, Denmark, and a number of other European nations, Japan, Argentine Republic, Brazil, Chili, Uruguay, Russia, Turkey, Persia (now Iran), the U.S., and Mexico. The convention defined various types of telegraphic establishments common at the time, including coast and ship stations, and delineated operating parameters to ensure cooperation in the use of telegraphic transmissions and designated frequencies, responses to distress signals, telegraphic charges, and other telegraphic matters of international importance. See Radio Communication Laws of the United States.

International Society for Measurement and Control ISMC. Formerly the Instrumentation Society of America (ISA), ISMC is a nonprofit professional organization supporting manufacturers and engineers involved in the theory, design, manufacture, and use of measurement and control instruments and computer systems. See Instrumentation, Systems, and Automation Society.
http:/www.isa.org http://www.isaca.com/

International Special Committee on Radio Interference CISPR. An international committee with members from a broad spectrum of the radio communications/engineering industry who work to promote international agreement on aspects of radio interference to facilitate international trade. CISPR is composed of a number of international organizations plus each National Committee of the International Electrotechnical Commission (IEC).

Through conferences and subcommittees, CISPR promotes and produces information guidelines, statistical methods, and standards related to the protection of radio reception from interference from consumer goods and industrial equipment, electrical supply systems, and broadcasting equipment. CISPR establishes limits and requirements for immunity to interference and takes into consideration safety regulations as they affect interference suppression of electrical equipment. Of particular interest to telecommunications are CISPR publications on electromagnetic compatibility (EMC) and emission standards. http://www.iec.ch/

International Speech Communication Association ISCA. A nonprofit organization promoting international speech communication, science, and technology originally founded in 1988 as the European Speech Communication Association, ISCA was established in 1999. It is now an independent, self-supporting organization. ISCA's interests include research in synthetic speech development and processing. http://www.isca-speech.org/

International Switching Center ISC. A gateway exchange whose function is to switch telecommunications traffic between national and international countries.

International Telecommunication Regulations ITR. A set of international regulations intended to supplement the International Telecommunication Convention while also recognizing the individual rights of nations to regulate their telecommunications sectors. The ITR framework seeks to promote the efficiency, harmony, and evolution of global telecommunications through established general principles for international telecommunication transport media and services offered to the public. Draft proposals of the ITR were presented at the world Administrative Telegraph and Telephone Conference in Melbourne, Australia, in 1988 per a resolution of the Plenipotentiary Conference of the International Telecommunication Union (ITU).

International Telecommunication Union ITU. A significant, influential, global United Nations standards agency descended from the International Telegraph Union. The ITU, headquartered in Geneva, Switzerland, provides extensive publications, promotes communication, sponsors international meetings and conferences, disseminates news, and develops standards and regulations. The ITU oversees a number of subgroups, called sectors (see chart).

The ITU is involved in a number of important communications venues for discussion and the dissemination of findings. Examples include:

- The publication of the ITU *Operational Bulletin* every 2 weeks to report on the administrative and operational information exchanged among administrators, service providers, and recognized operating agencies (ROAs); country codes and other statistical indicators are published in conjunction with the bulletin as annexes.
- The organization of the World Radiocommunication Conference (WRC) to administer international agreements in wireless telecommunications technologies.
- The organization of the World Telecommunications Standardization Assembly to determine needs and priorities related to standards development and dissemination.
- The convening of a Plenipotentiary Conference every 4 years (1998, 2002, etc.), an important meeting ground for member states where decisions on direction and policy are made and previous actions reviewed and debated. The 1998 conference was characterized by calls for greater involvement of the private sector in ITU activities. When the council is not convened, administration and oversight of the ITU is handled by the ITU Council.

See Telecommunication Standardization Bureau. For a brief description of ITU-T history, see International Telegraph Union. For series lists and individual recommendations, see Appendix C and alphabetized lists under letter designations. http://www.itu.int/

International Telegraph Union ITU. An old and influential organizing and standards-recommending body formed in 1865 when the telecommunications industry was beginning to boom. The ITU was created in response to the need for cooperation and formal agreements related to the installation and use of multinational telegraph systems. Twenty participating countries signed the first International Telegraph Convention.

After the invention of the telephone, the Telegraph Union drew up recommendations for legislation governing international telephony. Radio communications began to develop, so the Telegraph Union convened a preliminary radio conference in 1903 leading to the Radio Regulations and founding of the International Radio Consultative Committee (CCIR). In 1934, the name was broadened to International Telecommunication Union. It became an agency of the United Nations in October 1947, and the headquarters was transferred from Berne to Geneva in 1948.

The Union later became known as the CCITT, as there were a number of CCIs set up for different areas of communication in the 1920s; the CCIT and the CCIF were amalgamated in 1956.

In 1992, an important conference took place in which the organization was evaluated with the aim of updating it to align with the complex, changing environment of current and future technologies. The organization has recently been renamed International Telecommunication Union (ITU) because the fundamental objectives of the original organization remain essentially the same today as they were over 100 years ago, and the convergence of the many media and communications technologies through digital transmission has united many formerly separate areas. (Source: ITU-T Web site history.)

In Canada, communication with the ITU is accomplished through the Canadian National Organization for the ITU (CNO/ITU-T) and the Steering Committee on Telecommunications of the CSA (CSA/SCOT). See International Telecommunication Union; Morse, Samuel B.F.

ITU-T Subgroups (Sectors)

Abbrev.	Sector	Notes

ITU-R *Radiocommunication Sector*

Descended from the International Radio Consultative Committee (CCIR), this is the arm of the ITU responsible for researching technical and related regulatory issues. It regulates ground- and space-based radio frequency telecommunications.

ITU-T *Telecom. Standardization Sector*

Founded in March 1993, ITU-T replaces the International Telegraph and Telephone Consultative Committee (CCITT). The ITU-T endeavors to ensure efficient and on-time production of high quality standards covering all fields of telecommunications with the exception of radio, which is handled by ITU-R. The work of the ITU-T is handled by numerous study groups and is documented in tens of thousands of papers. Presently more than 2500 standards recommendations are in force that form a framework for global communications.

ITU-D *Telecom. Development Sector*

Facilitates global telecommunication development by providing organizing and coordinating expertise and assistance. The ITU-D works through conferences, study groups, the Telecommunication Development Advisory Group, and the Telecommunication Development Bureau (BDT).

International Traffic in Arms Regulations ITAR. Rules issued by the U.S. State Department, under the authority of the Arms Export Control Act, to control the export/import of defense-related articles and services, including information security systems such as cryptographic systems and TEMPEST suppression technology.

International World Wide Web Conference Committee See World Wide Web Conference Committee.

internet When spelled with a lowercase "i," generically refers to an interconnection of two or more data networks. While individual networks may be connected in any number of ways, it is common to interconnect them through the Open Systems Interconnect (OSI) model as it ensures a good level of compatibility with existing technologies and supports interoperability among a variety of types of systems. See Internet.

Internet A global communications community of more than 60,000 cooperating networks, evolving in the early 1980s out of ARPANET, now known as the *Internet* or colloquially as the *Net*.

The Internet consists of a distributed network of tens of millions of computers linked together through small and large communications services providers. By early 1995, the Internet had more than 4 million hosts and the term was officially defined by the Federal Networking Council.

The Net is defined in the Telecommunications Act of 1996 and published by the Federal Communications Commission (FCC) as

"... the international computer network of both Federal and non-Federal interoperable packet switched data networks."

The evolution of the Net has been influenced by a broad base of technical and lay interests and an equally wide range of commercial and public interests. The vocal promoters of the Net as a universal access communications medium to serve the public good have been joined by commercial interests seeking a way to use the Net to further private and public business interests. In the early days, the Net had a high proportion of users in technical and scientific fields and focused on cooperative communication and research. Since the mid-1990s, an overwhelming influx of commercial vendors has changed the character of the Net, but there has also been a large growth in cooperative nonprofit and community organizations.

In 1993, the United Nations and the U.S. White House came online, thus changing the ways to access and think about politics. Global doors have opened up to people doing genealogical studies and people are rediscovering friends they haven't seen since elementary school. The phone network is undergoing substantial changes due to competition from long-distance email and chat resources that are available on the Internet without long-distance phone costs.

The impact of the Internet on communications venues and global culture is highly significant and will likely exceed the changes brought about by the industrial revolution. The information glut and impact on personal privacy will be far-reaching as well. Speculations about the emergence of the Net as a form of digital intelligence may not be farfetched, and, with the cooperative communication possible among scientists and interested lay persons, research will move forward at an unprecedented rate. See ARPANET, Telecommunications Act of 1996, RFC 1958.

Internet 2 A consortium of more than 100 academic and nonacademic organizations working to develop a vision and implementation plan for the next generation of the Internet on a content and integration level. For information on the technical successor to the current Internet protocols and physical structures, see IPv6. http://www.internet2.edu/

Internet Access Coalition IAC. A lobbying organization that supports universal Internet access and monitors and comments on industry trends in Internet services provision, long-distance access, and other logistical matters that affect the ability of the public to access and utilize Internet services. The IAC sometimes works in cooperation with other organizations such as the Information Technology Association of America (ITAA).

Internet Access Provider IAP. A vendor who provides a connection to the Internet in the form of Frame Relay, ISDN, a dialup modem, or other physical or virtual connection, and who may or may not provide additional services, such as email, shell accounts, web hosting, etc. Providers with full services available, rather than just an access port to the Internet, are generally called Internet Services Providers (ISPs). See Internet Services Provider.

Internet Activities Board IAB. Established in 1983 to replace the Internet Configuration Control Board, the IAB subsequently came under the umbrella of the Internet Architecture Board. See Internet Architecture Board.

Internet Architecture Board IAB. Formerly the Internet Activities Board (and before that, the Internet Configuration Control Board), the IAB is a coordinating and policy-setting board for the Internet Engineering Task Force (IETF) and the Internet Research Task Force (IRTF). All three bodies were combined under the aegis of the Internet Society (ISOC) in the early 1990s and the IAB is now the technical advisor to the Internet Society. See RFC 1358 for a charter of the IAB, and RFC 1160 for a description of its organization and role. See Internet Engineering Task Force (IETF), Request for Comments. http://www.isi.edu/iab/

Internet Assigned Numbers Authority IANA. An organization which, since the early 1980s, has exercised authority over DNS operations, Internet Protocol (IP) number assignment, Root Name Servers, Request for Comments (RFC) documents, and protocol port number assignments. IANA is the central coordinator for the assignment of unique numbers for Internet protocols and serves as a clearinghouse for this purpose.

IANA also provides registration through a central repository for MIME types, that is, data object types identified by a short ASCII string which can be used to provide rich content types in conjunction with electronic mail.

Jon Postel has almost single-handedly spearheaded this effort, an enormous contribution by an Internet pioneer involved since the days of the ARPANET. IANA is chartered by the Internet Society (ISOC) and located at the Information Sciences Institute (ISI) of the University of Southern California. See domain name, naming authority, name resolution. http://www.iana.org/iana/

Internet Channel Commerce Connectivity Protocol ICCC. A channel-based connectivity protocol designed to facilitate electronic commerce, initiated by 3Com. ICCC is intended as a scalable, securable, channel-based electronic commerce infrastructure, based upon Extensible Markup Language (XML). Access to ICCC is through popular Internet browsers so that WWW-related Internet commerce applications can be built upon ICCC.

The concept is of an application-layer protocol that passes over the wire, to promote interoperability, with transactions accomplished through the widely established HTTP. "Shopping cart" programs on the Web illustrate the general idea of what ICCC is intended to accomplish through a standardized protocol model using existing standardized formats and a server-to-server transfer (rather than through the browser). See Open Buying on the Internet, Open Financial Exchange, Open Trading Protocol.

Internet Community at Large ICAL. A project funded by the National Science Foundation Division of Environmental Biology's Research Collections in Systematics & Ecology Program through the Museum of Paleontology at Berkeley to facilitate development of improved and new modes of communication among museums, donors, and research scientists, utilizing the World Wide Web. The main purpose of the Web project is to help reduce the number of orphaned or underutilized natural history collections. Additional support for ICAL-Entomology was provided by the National Science Foundation through the Bishop Museum.

Internet Configuration Control Board ICCB. A regulatory board established by the U.S. DARPA in the late 1970s to facilitate the creation of gateways between hosts and the network. The ICCB was replaced by the Internet Activities Board in 1983. See ARPANET, DARPANET.

Internet Control Message Protocol ICMP. A significant protocol in that it is an IETF-required standard on the Internet for reporting and error messages in Internet Protocol (IP) datagram routing. While not a reliability guarantee, ICMP can provide feedback regarding problems in datagram processing and delivery. ICMP messages are contained in the basic IP header. Examples include information on whether the destination is reachable, echo or redirect situations exist, time has been exceeded, or a problem exists with a parameter.

Currently the Net is run over IPv4, and migration to IPv6 is planned. ICMP for IPv6 is based on the same definition with some changes and is known as ICMPv6. See Classes and Format of ICMPv6 Messages chart. See IP, RFC 792, RFC 1788.

Internet Control Message Protocol for IPv6 ICMPv6. ICMPv6 is a required and integral part of IPv6 that must be fully implemented at every node. It is used for diagnostics and error reporting. ICMPv6 messages are preceded by an IPv6 header and zero or more extension headers, identified by a Next Header value of 58 in the header immediately preceding. ICMPv6 messages are organized into two classes, as shown in the Classes and Format of ICMPv6 Messages chart. See RFC 792.

Internet Corporation for Assigned Names and Numbers ICANN. A not-for-profit organization established as a result of studies and recommendations reported in the White Paper issued by the National Telecommunications and Information Administration (NTIA) of the U.S. Department of Commerce in June 1998. This was a significant step in privatizing management of the Internet domain name system. Amid discussions with other organizations such as the World Intellectual Property Organization (WIPO), ICANN developed policy and procedural guidelines for management of the Internet domain name system worldwide. One of the significant outcomes was the April 1999 announcement of a testbed for a Shared Registry System to be administered by five companies rather than one, with Network Solutions, Inc. to continue to maintain the registry database in order to ensure a centralized repository for unique domain names. See White Paper. http://www.icann.org/

Internet Engineering Steering Group IESG. The executive governing body of the Internet Engineering Task Force (IETF) and technical overseer for the Internet standards process, including final approval. The IESG is a member of the Internet Society (ISOC) and works within ISOC rules and procedures. See Internet Architecture Board.
http://www.ietf.org/iesg.html

Internet Engineering Task Force IETF. The IETF is governed by the Internet Engineering Steering Group (IESG). It is a large, international open community of network researchers and designers dedicated to the positive evolution of Internet architecture and operations.

The IETF is the primary Internet protocol development and standardization body.

The IETF has worked long and hard on IP Version 6 with the intention that it supersede IPv4. In 1997, the IETF made some significant changes to support more dynamic addressing schemes. Several draft standards for IPv6 were submitted in December 1998 followed by many proposed standards in 1998 and 1999. These are moving slowly through the standardization and implementation process. In the meantime, some adjustments have been made to lengthen the life of Version 4, which has address space limitations, so that it can continue to be a viable networking solution until vendors begin to implement and support IPv6 into the new century. See Internet Architecture Board, Request for Comments.
http://www.ietf.org/home.html

Internet Experimental Note IEN. A document system containing information on Internet specifications and implementations. The IEN is administered by the Network Information Center (NIC).

Internet Fax The terminology and goals for the development of Internet Fax systems and guidelines for the Internet Fax working group were submitted as an Informational RFC by L. Masinter in March 1999. Internet Fax is described as a document transmission mechanism between various devices and roles which may be differently configured. Several general categories of roles were defined as network scanner, network printer, fax onramp gateway, and fax offramp gateway. The common modes for Internet Fax were described as store and forward, session, and realtime. To support the concept of deploying facsimile services over the Internet, Klyne and McIntyre submitted a Standards Track RFC in March 1999 describing

---

Classes and Format of ICMPv6 Messages

Type of Message	Identification of Type	Message Type Number
Error message	Zero in the high-order bit of the message "Type" field	0 to 127
Informational message		128 to 255

ICMPv6 messages have the following format:

```
 0                   1                   2                   3
 0 1 2 3 4 5 6 7 8 9 0 1 2 3 4 5 6 7 8 9 0 1 2 3 4 5 6 7 8 9 0 1
+-+-+-+-+-+-+-+-+-+-+-+-+-+-+-+-+-+-+-+-+-+-+-+-+-+-+-+-+-+-+-+-+
|     Type      |     Code      |          Checksum             |
+-+-+-+-+-+-+-+-+-+-+-+-+-+-+-+-+-+-+-+-+-+-+-+-+-+-+-+-+-+-+-+-+
|                                                               |
+                         Message Body                          +
|                                                               |
```

Type	The type of message. Its value determines the format of the remaining data.
Code	Depends upon the message type. Used to create an additional level of message granularity.
Checksum	Used to detect data corruption in the ICMPv6 message and parts of the IPv6 header.

content feature schema for Internet Fax as a profile of a media feature registration mechanism for performing capability identification between extended Internet Fax systems. This was updated in August 2000. See RFC 2542, RFC 2879 (which obsoletes RFC 2531).

Internet Free Expression Alliance IFEA. An organization promoting a liberal computer communications environment in order to facilitate and safeguard the Internet as a free, open, and diverse forum for the exchange of discussions and data.
http://www.ifea.net

Internet Group Multicast Protocol IGMP. An IETF-recommended session-layer protocol for network transmissions to multiple sites. IGMP is a dynamic protocol that provides a means for end systems to request inclusion in a multicast group or video/data stream conference (group broadcast). A host can request membership into one or more groups at the same time. It can transmit datagrams to a group without necessarily having requested membership to that group. See RFC 1112, RFC 2236.

Internet Information Infrastructure Architecture IIIA. A framework and support system for information about Internet resources. Uniform resource names, characteristics, and locators are functional categorizations within this architecture, intended to facilitate the location of the desired information resources. See RFC 1737.

Internet International Ad Hoc Committee IAHC. The IAHC was a coalition of members of the Internet community cooperating to develop recommendations for the expansion of the Internet Domain Name System (DNS). It published a number of guidelines between 1996 and May 1997, made its Final Report in February 1997, and was dissolved in May 1997.
http://www.iahc.org/

Internet Message Protocol IMP. This is one of the historic Internet protocols, submitted by J. Postel in March 1979. It describes a means for transmitting messages between message processing modules over interconnected networks. Message processing modules are processes in host computers located in different networks that comprise a framework for internetwork message delivery. IMP was developed in the context of ARPA work in interconnecting networks and was tendered by Postel as a more general internal mechanism underlying a variety of user-interface programs, thus providing a messaging system suitable for heterogenous distributed networks like the Internet that was to evolve out of the ARPANET.

IMP was intended to support an environment in which processes run in hosts interconnected by gateways, with each network having many different hosts. The gateways are assumed to have minimal knowledge of which hosts are within their associated networks. IMP is implemented within a Message Processing Module (MPM). MPMs exchange messages by establishing full duplex communications and sending messages in a recognizable fixed format. The user creates a message with the chosen User Interface Program (UIP) with commands or an editor and then sends the message through a data structure shared with the MPM. The MPM discovers the unprocessed data, examines it, and determines the outgoing link in the route to an internal or external destination.

The MPM communicates through a reliable procedure using a transport level protocol such as TCP.

Internet Messaging Access Protocol IMAP. An electronic mail protocol descended from Interactive Mail Access Protocol and used for electronic mail servers. It provides a means to access electronic mail and news messages archived on a mail server as a dedicated or shared resource. Thus, an email client can access and read the mail messages on the remote server as though they were on a local storage medium. It is useful in situations where the reader is more concerned about reading the messages than downloading them to the local machine, especially if the messages are located at more than one site or on more than one account. IMAP is somewhat competitive with Post Office Protocol (POP), but both are useful depending upon the situation. POP is more appropriate for providing access to messages that will be regularly downloaded to a local machine and then deleted from the mail server archive. See MIME, Post Office Protocol, RFC 1730, RFC 2060.

Internet Network Information Center See InterNIC.

InterNet News INN. An NNTP/UUCP USENET newsreading system developed by Rich Salz for Unix systems with socket interfaces. This fast news program was first released in 1992. Later, in 1995, David Barr released a number of unofficial updates. Thereafter, maintenance of INN was taken over by the Internet Software Consortium (ISC). See C News, USENET.

Internet Official Protocol Standards The title of a Request for Comments (RFC) document released from time to time to inform the Internet community of the state of standardization of protocols used in the Internet that are determined by the Internet Architecture Board (IAB).

The memo itself is an Internet Standard and makes somewhat obsolete previous versions of the document. The document describes the standardization process, the Request for Comments documents, terms, and other important concepts and procedures related to standards used to create, maintain, update, use, and understand the Internet. Updates to the documents are released about once a year and are usually issued with round numbers (RFC 2900, RFC 2800, RFC 2700, etc.) to facilitate memorization and location. Due to the high volume of RFCs and standards that have been developed related to the Internet, the list includes only official protocol standards RFCs and does not constitute a complete index. The list is now determined by the Internet Engineering Task Force (IETF).

Internet Open Trading Protocol IOTP. An interoperable framework for electronic commerce over the Internet optimized for transactions between nonacquainted parties. IOTP is independent of the payment system and can encapsulate and support secure channel card payment, GeldKarte, Mondex, and others.

IOTP V. 1.0 was described in an Informational RFC submitted by D. Burdett in April 2000, and was supported by a description of digital signatures for IOTP presented by Davidson and Kawatsura.

In April 2001, W. Hans et al. submitted an Internet Draft for a *Payment Application Programming Interface* (API) for IOTP. It proposed a common interface for communication between the IOTP application core and the payment modules, increasing interoperability among these modules and providing a "plugin" mechanism for application cores.

In May 2001, D. Eastlake of Motorola submitted an Internet Draft to update RFC 2801 to document and correct errors detected since the submission of the original specification. See RFC 2801, RFC 2802, RFC 2803.

Internet Phone A commercial software/hardware system from VocalTec Ltd. that allows a computer user to place a telephone voice call through the Internet very much the same way that a call is placed through traditional telephone systems. The primary difference is that the voice conversation is converted to digital data and channeled through the user's Internet Services Provider (ISP) to the network, rather than through traditional telephone switching offices. The applications software works in conjunction with GOLD, the Global Online Directory that stores information about Internet Phone users who can be contacted online, just as the names of traditional phone subscribers can be accessed through a phone directory. See Global Online Directory.

Internet Policy Registration Authority IPRA. A top-level digital security certification authority (CA) in the Internet certification hierarchy. IPRA is X.509-compliant.

Internet Printing Protocol IPP. An application level protocol to facilitate remote printing over distributed networks based upon Internet technologies. The IPP model and semantics were described in an Experimental RFC by deBry and others in April 1999. It is a simplified model, including abstract objects, attributes, and operations that are independent of transport and encoding methods. Essentially the model is based upon a printer and a job object, with a job optionally representing multiple documents. Users can query printer capabilities, submit jobs, get status information on jobs, and cancel jobs. Security and internationalization aspects are also described in the specification.

The documents related to IPP include:

- Design Goals for an Internet Printing Protocol
- Rationale for the Structure and Model and Protocol for the Internet Printing Protocol
- Internet Printing Protocol/1.0: Model and Semantics
- Internet Printing Protocol/1.0: Encoding and Transport
- Mapping between LPD and IPP Protocols
- Internet Printing Protocol/1.0: Implementor's [sic] Guide

The IPP implementor's guide was submitted as an Informational RFC by Hastings and Manros in July 1999 to aid implementors in understanding and applying the information in the suite of documents related to IPP semantics, encoding, etc. The guide aids implementors in designing client and/or IPP object implementations and provides an order in which requests can be processed, in addition to error checking. See RFC 2567, RFC 2568, RFC 2566, RFC 2565, RFC 2569, RFC 2639.

Internet Protocol IP. A very significant protocol in that it is an IETF-required standard on the Internet along with the Internet Control Message Protocol (ICMP). There are other related IETF protocols, which are recommended or elective. IP is very widely used in TCP/IP implementations.

Internet Protocol provides addressing, segmentation and reassembly, and transport functions in conjunction with a number of associated protocols. Logical IP addresses are used to identify hosts by means of network and node addresses. A number of categories of networks are supported as IP Classes.

RFC 768 describes Internet Protocol. RFC 1602 is recommended for its description of the Internet standards process, and RFC 2200 is a useful standards track document for Internet Official Protocol Standards that further describes the standardization process. See IPv6 for more informaton and charts. See IP Class, IPv6, RFC 950, RFC 919, RFC 922, RFC 2200.

Internet Protocol Consortium IPC. The IPC administers the InterOperability Lab at the University of New Hampshire for testing protocols of importance to intercommunication on the Internet. There are three testing services and 17 consortiums currently supporting this effort. Of current interest is a test lab for testing IPv6 implementations, the version of Internet Protocol being phased in to coexist with and probably eventually supersede IPv4.

Internet Protocol (IP) Mobility Support A set of media-independent protocol enhancements submitted as a Standards Track RFC by C. Perkins, in October 1996, and updated/extended as a PPP IPCP option by Solomon and Glass in February 1998.

IP Mobility Support enables transparent routing of IP datagrams to mobile nodes on the Internet. A mobile node is identified by a home address, regardless of where it happens to be connected to the Internet at any particular time. While away from the original home address, the mobile node is considered to be associated with a "care of" address in much the same way as individuals who are traveling may use a care of address for postal mail delivery. The mobile node makes its care of address known by registering it with a home agent which, in turn, sends datagrams destined for the mobile node through a tunneling process. At the end of the tunnel, the datagrams are delivered to the mobile node. See RFC 2002, RFC 2290.

Internet Protocol Suite IPS. A standardized set of protocols based on a layered model that enables Internet systems to intercommunicate. Minimally, a host must implement at least one protocol from each layer, including the application, transport, Internet, and link layers. These layers form the basic architecture for managing the hardware and software functions that enable communication over the Internet using Internet Protocol (IP). In general, layered models are organized to illustrate more abstract applications-related concepts at the top and more basic hardware-related functionality and transport media at the bottom.

NASA has been conducting research to improve the efficiency of the IPS for satellite-based networks through its Satellite Networks and Architectures Branch. Since radio communications with satellites involve relatively long delays, special problems are involved in implementing IP. See IPv6. See RFC 1122, RFC 1349, RFC 1958, RFC 2502, RFC 2600.

Internet Relay Chat IRC. A worldwide "realtime" 24-hour text-based communications chat link on the Internet developed in the late 1980s by Jarkko Oikarinen. IRC was inspired in concept by MUT and in format by BITNET Relay Chat. Development began in August 1988, with the first server established in Finland as tolsun.oulu.fi. IRC II was released in 1989 by Michael Sandrof. A number of other developers released versions or variations.

By 1989, IRC had more than 50,000 users. In 1993, Request for Comments (RFC) 1459 was published to provide a consistent reference and basis for the IRC and clients intended to conform to IRC guidelines. It was followed in 1994 by the Client-to-Client Protocol (CTCP) to support IRC client communications.

Many IRC servers and computer systems are configured to provide a means to communicate remotely with IRC. Generally it is best to connect to one geographically close to the ISP, but some servers are busier than others and it is sometimes a good idea to select a low-use server that is farther away. IRC is typically accessed through port 6667.

IRC is an important meeting ground for people around the world. The form of an IRC chat is somewhat like a group conversation on a teletype machine, except that the output to the screen is much faster than the transmission and output to a printing teletype. Many celebrities, in and out of the telecommunications industry, have been known to participate in IRC conversations and to draw large crowds of participants around the world. To join a chat (a communications channel dedicated to a specified topic), you must have access to a provider that provides a port to IRC, a basic understanding of how to sign onto a chat, and a willingness to learn a few simple commands.

A command set must be learned to access IRC with a text-based client (there may not be a point-and-click graphical client available for every operating system). From the text line, a conversation on IRC is joined by typing #join gardening (or a topic of interest other than gardening). It's a good idea to visit the help channel by typing #join irchelp to get a feel for the way things work. There are thousands of IRC channels, so most common topics already exist; if you are seeking an uncommon topic, it will be automatically created when the command #join myweirdtopic is typed. The channel automatically disappears shortly after the last person leaves, except in the case of registered channels, but comes back (is recreated) as soon as it is re-entered.

Most IRC channels are public forums, but private keyword-protected IRC channels can be created at any time. Courtesy is very important on IRC. If a participant is rude, crude, inflammatory, or off-topic, he or she will be summarily kicked off the channel by an operator. If there is no operator present, usually everyone else will leave. Observe courtesy and Netiquette on IRC, and don't talk unless it's something worth saying. The operators or "ops" are hard-working volunteers who strive to make the IRC an open and fair forum for all.

The IRC software is freely distributable through a GNU General Public License from the Free Software Foundation. Many Internet Services Providers provide IRC access. Communications are predominantly in English, but other languages are sometimes used. See Internet Relay Chat operator, RFC 1459, RFC 2810, RFC 2811, RFC 2812, RFC 2813.

Internet Relay Chat operator An individual designated with certain responsibilities and powers for a channel on Internet Relay Chat (IRC), a public communications forum on the Internet. In an active chat channel, some of the participants are designated with "@" symbols next to their online nicknames. These operators or "ops" have jurisdiction over their channels and may set the guidelines for interaction and remove those who do not follow the guidelines. Because they establish and maintain law and order on IRC, many people call them IRC cops. In general, IRC ops are hard-working volunteers who make reasonable decisions and have kept IRC a viable communications medium in spite of the many people reluctant to follow guidelines of good taste and common sense. See Internet Relay Chat.

Internet Relay Chat (Server) Protocol IRC Server Protocol. A protocol for describing how Internet Relay Chat (IRC) servers may connect together to form a network. The IRC protocol was first implemented in the late 1980s and grew to support a worldwide network of servers and clients in just a few years. IRC Protocol is text-based and enables a simple socket program to connect as a client. Over the years, various developers have created operating systems-specific text and graphical clients to interact in IRC chat sessions using IRC Protocol. In general, IRC Protocol has been implemented over TCP/IP, though there is no restriction as to this.

Each IRC server has a unique name up to 63 characters and maintains a global state database that gives a picture of the IRC network so that each server is known by other servers. A hostmask can be used to group servers according to name in order to exclude hosts outside of the list. Servers hold netwide unique identifying nicknames (up to 9 characters), usernames, and connecting host informaton for each client currently connected to the IRC system.

Internet Relay Chat Protocol was submitted as an Experimental RFC in May 1993 by Oikarinen and Reed and has been updated by numerous RFCs since that time. See Internet Relay Chat, RFC 1459, RFC 2810, RFC 2811, RFC 2812, RFC 2813.

Internet Research Steering Group IRSG. The IRSG manages the Internet Research Task Force (IRTF) in conjunction with the IRTF Chair. Membership in IRSG is primarily those in chairing positions in the various research groups. The IRTF Chair is appointed by the Internet Architecture Board.

Internet Research Task Force IRTF. An organization engaged in discussion and Internetworking research to further the evolution of the Internet, especially with respect to technologies, architecture, protocols, and applications. The IRTF works in consultation with the Internet Research Steering Group (IRSG) and with the Internet Engineering Task Force (IETF) and the Internet Architecture Board (IAB). See RFC 2014. http://www.irtf.org/

Internet Resource Access IRA. An OEM product distributed by CREN in cooperation with IBM, announced in August 1994. This system was intended to facilitate Internet access for global academic and research communications by institutions that had largely been connected by BITNET networks up until this time. IBM provided hardware and software as a foundation for the CREN system. IRA provided connectivity between RSCS on IBM systems and

I

Internet Relay Chat Command Examples	
Command	Description
/basics	Very basic introductory information about IRC; a good thing to read the first time you use the system. Also try out /help newuser.
/bye	Drops the user out of IRC; /quit, /exit, and /signoff do the same.
/clear	Clears the current window; reduces clutter.
/date	Displays the current date and time for the local server or a specified server. The /time command performs the same function as the /date command.
/join <channel>	Changes the location to the specified IRC channel. For example, /join #buglovers puts the user in the channel with other insectophiles.
/help <info>	Self-explanatory and the command to type if you're really stuck.
/info	Provides information about the origins of IRC, its creators, maintainers, slaves, and other perpetrators.
/list	Provides a very long list of thousands of channels, and information about the topics and number of participants, so use this command with caution. The * (wildcard) character may be used to specify the characteristics of the listing, as can a number of useful arguments: -public shows only public channels; -private shows only private channels; -topic shows only channels with a specified topic.
/msg <nickname>	Sends a single private message to the specified person. Use /query if longer private conversations are desired.
/menus	A simple scripting feature for creating custom user menus for an IRC session. This is great for creating mnemonic commands or shortcuts.
/newuser	Information about IRC commands and IRC etiquette.
/nick	Sets the user's nickname. If the nickname is taken, another must be selected, or the default used.
/news	Information about changes, updates, new commands, and other IRC-related functions. It's a good idea to check this once in a while.
/query <nickname>	Initiates a private conversation with a specified user. Anything you type now is seen only by that user. The query command with no arguments cancels query mode.
/set <variable>	Sets various status, logging, and message parameters.
/who	Lists users on IRC; with a * (wildcard), it shows the local channel. A number of arguments can restrict the listing, e.g., -operators lists only operators.
/whois <nickname>	Provides more detailed information about the user specified, and his/her "actual identity."

Internet TCP/IP on open systems. See BITNET, Corporation for Research and Educational Networking.

Internet Safety Policy ISP. A policy required to be in place in federally funded institutions that provide Internet access to vulnerable individuals, especially children. See Children's Internet Protection Act.

Internet Secretariat An organization providing administrative assistance to a variety of Internet governing bodies.

Internet Security Association and Key Management Protocol ISAKMP. An application-level network protocol submitted as a Standards Track RFC by Maughan et al. in November 1998. ISAKMP utilizes security concepts for establishing Security Associations (SAs) and cryptographic keys within the Internet environment. It defines procedures and packet formats for peer authentication, SA creation and management, key generation, and threat mitigation for establishing and maintaining secure communications.

ISAKMP is distinct from key exchange protocols, as there may be various key exchange protocols with different security properties. A common framework facilitates intercommunication through SA attribute formats for negotiations, modifications, and deletions at all layers of the network stack. ISAKMP has been assigned UDP port 500.

Cisco Systems provides a no-charge ISKMP software distribution based upon the IETF ISAKMP to support Internet Key Management through this protocol. See RFC 2408.

Internet Services Provider ISP. A commercial vendor providing access to the Internet and some or all of its services. These services may include email, newsgroup access, World Wide Web access, Internet Relay Chat (IRC), telnet to other sites, Unix shell accounts, and more. Some providers have flat-rate fees for unlimited access, while others provide unlimited access during off-peak hours, and limited or pay access during times of heavy use. Others charge by connect time. Many distinguish between commercial and personal users, with separate fee scales for each, usually with more mailboxes and longer connect times for business users.

The ISP's link to the Internet may be through a variety of connections, usually 56 kbps or higher, up to T1 or even T3 lines. However, when dialing up through a regular modem on a phone line, you will not be able to receive and transmit information faster than the slowest point in the link (e.g., the modem speed). There are several large, well-known providers, as well as thousands of small, local service providers. The level of service of many small providers equals or exceeds those of the large companies, so shop around. See Internet Access Provider, National Service Provider.

Internet Society ISOC. A significant nonprofit international professional organization dedicated to furthering global cooperation and coordination of the evolution of the Internet and its associated technologies. ISOC was founded in January 1992. It grew out of standards development activities of the IETF and Internet Activities Board (IAB) in the early 1990s and counts among its members many of the early Internet pioneers. Fund-raising to continue to support the standards process was one of the important initial mandates of the Society. The ISOC oversees and/or works with a number of other agencies, including the Internet Architecture Board (IAB) and the Internet Engineering Steering Group (IESG). It supports and promotes Internet-related public policy, education, standards, and participation in the Society. See RFC 1310, RFC 1602. http://www.isoc.org/

Internet Software Consortium ISC. A group dedicated to developing production-level high-quality reference implementations of Internet technologies suitable for use by large-scale network providers and operators. Subgoals include compliance to key standards, straightforward implementation, and high interoperability. The ISC was formed with financial assistance from UUNET Communications Services and later from the Internet Multicasting Service and various sponsors. http://www.isc.org

Internet Standards process The orderly evolution of the Internet is of concern to many networking professionals, so the Internet community at large has developed various procedures to facilitate this process.

Internet Standards Process – Levels of Maturity		
Standard	Abbr.	Description
Proposed	PS	Entry-level for standards-track specifications as accepted by the IESG. To become a PS, a specification must be technically complete, generally well understood, received by the Internet community, and have design and reliability issues resolved.
Draft	DS	A PS may be promoted to DS after at least two independent and interoperable implementations from different code bases have been developed and sufficient successful operational experience has been obtained. DS status indicates confidence that the specification is mature and will be useful.
Approved	IS	A DS may be promoted to IS and assigned an STD series number if significant implementation and operational maturity are achieved and the IS promises to be of significant benefit to the Internet community.

For a technology to become an official or required Internet standard, it must go through a formal discussion, evaluation, and testing process. A protocol must pass through several defined levels of maturity, including Proposed Standard, Draft Standard, and (Internet) Standard and is documented in Requests for Comments (RFCs) notifications to the Internet Community.

The Internet Engineering Task Force (IETF) must recommend advancement at each stage for the protocol to pass to the next level, and specified waiting periods are imposed. Intellectual property rights (e.g., patents) must be identified and noted in Standards Track documents. When a protocol has successfully gone through the successive levels of the Standards process, it is assigned an STD number. The process is somewhat recursive in that it is described within itself in RFC 1602.

If a standard becomes outdated, the IESG may elect to retire it and appropriate notifications will be posted. For a standard to be revised, it must go through the full standards process again. The old standard will usually be superseded and retired to *Historic* status. However, if the new standard is sufficiently different or more mature than the previous standard such that both implementations have current implementation value, the two may coexist.

Non-standards-track specifications are labeled as Experimental, Informational, or Historic.

The Best Current Practice (BCP) subseries is a structure similar to the Internet Standards process within which proposals from community leaders can be fielded within the Internet community to stimulate and enable the development of guidelines for consensual policies and operations. See Internet Engineering Task Force, RFC 1311, RFC 1602, RFC 2026.

Internet Transparency In essence, Internet Transparency is a philosophy and design goal that supports the capability of the Internet to send anything anywhere. It holds that packets should be able to carry any type of data to any desired destination without the user worrying about format, routing, interoperability or other underlying aspects of an interconnected homogenous network. In actual implementation, there have been fits and starts and the occasional backslide in holding to this philosophy, but there appears to be a general desire to continue to work toward this goal and to extend the capabilities and overcome the technical limitations that stand in the way of a fully transparent Internet system.

In February 2000, B. Carpenter submitted an Informational RFC discussing this issue that was used as input to an Internet Architecture Board (IAB) workshop held in July 1999. The RFC documents some of the sources of loss of transparency and, in particular, issues such as firewalls, IP address allocation, intranet models, etc. See RFC 2775.

Internetwork Packet Exchange Protocol IPX. A network layer protocol that provides addressing, routing, and packet-switching functions for Open Systems Interconnection (OSI) model systems. IPX works on a best-efforts basis to deliver packets without a guarantee of successful delivery or verification of such (these are handled by other protocols).

Internetworking Alliance See World Internet Alliance.

Internetworking Over NBMA ION. A working group jointly chartered with the Internet and Routing Area of IETF, ION is a merger of the IPATM and ROLC groups. It focuses on issues of internetworking network layer protocols over NBMA subnetwork technologies, including encapsulation, multicasting, address resolution, optimization, and others. See ATM, Frame Relay, SMDS, X.25, ISSLL, ITU, RFC 1932.

InterNIC Internet Network Information Center. This is a service mark of the U.S. Department of Commerce; the name was associated for a number of years with an authorized central registry for domain names and IP number addresses on the Internet, hosted by Network Solutions, Inc.

Network Solutions/VeriSign was established as an exclusive provider of domain name registry services for Top Level Domains (TLDs) in 1991. InterNIC was established in 1993 in cooperation with the National Science Foundation (NSF) to continue domain name registration. To be part of the Internet, you need a unique identifier for the network and the individual host from which information is being sent. The domain name is associated with an IP number to create a unique address on the Internet. (More than one domain name can be assigned to an IP number, depending upon the administrative policies and services offered by individual ISPs.) In order to manage this administrative task, InterNIC kept track of registrations and domain name-IP number correspondences in a central database archive.

There has been a yearly fee since the mid-1990s for the registration and maintenance of domain names and the monopolistic nature of InterNIC has come under continued dispute, with various stalled proposals for providing additional domain name extensions and competitive opportunities for other name registries. In 1998, VeriSign GRS was separated from Network Solutions, Inc. at the time Network Solutions Registry was handling domain name registrar services.

By the late 1990s, many proposals for additional domain name extensions had been tendered and other registrars were being approved for granting domain names. A central database still needed to be maintained to ensure uniqueness and an orderly process for registrations, which were now in the tens and hundreds of thousands per month (and rapidly increasing). The demand for names was largely due to the increase of users on the Internet and the commercialization of the Net, which resulted in a domain name becoming an important branding and location tool for vendors and other organizations. With competition for the provision of domain name registration services, the price dropped from $200 in the mid-1990s to $50 in the late 1990s to $15 (and sometimes less if the domain was bundled with other services) in 2000. By 1999, some commercial firms were reported to be

I

applying for as many as 1000 domain names per month (usually for individual products in their product lines) and speculators had registered tens of thousands of common words and good potential names in the hopes of reselling them later at a profit. A clamor arose over demand for domain names, with new registrants claiming that all the good names were taken. The technical difficulties in simply extending the total number of IP numbers to meet demand are discussed elsewhere in this reference.

The Network Solutions Registry was created and registry services agreements extended until 2003. The commonly known established name extensions include .com, .edu, .gov, .net, .info, .int, .usa, and .org (along with assigned extensions for individual countries). Other extensions were subsequently added in 2001, including .biz to satisfy commercial demand and .info for information-based services, with plans to implement .pro, .aero, .museum, and possibly others in the future.

VeriSign TRS now provides registry services for the Internet domains and the InterNIC name is specifically associated with the Department of Commerce Web site. See domain, domain name, domain name server, IP address, and the Appendix for a list of country code domain extensions. http://interNIC.net/

interprocess communication environment IPCE. The concept of interprocess communication (IPC) became important when people began interconnecting computers and developing protocols to allow them to share data communications.

With the spread of timeshare networks in the late 1960s and 1970s, the various hardware and software mechanisms to facilitate IPC began to develop. They expanded when remote "smart" terminals and peer-to-peer networks were invented. Computer bulletin board systems (BBSs) and local area networks (LANs) in the 1980s and the Internet in the 1990s are important extensions of the general concepts of IPC. They have increasingly enabled users to utilize resources on remote systems and even to share files and programming environments as though they were local resources, thus extending IPC from something transient and part-time in its earlier implementations into a system in which many of the computers and processes are in 24-hour communication with one another. This doesn't just speed up the sharing of information and resources; it also creates a higher order of environment, somewhat like a cooperative or symbiotic digital organism. Given the advantages of access to greater resources through a larger, more sophisticated interprocess communications system, it is likely that this trend will continue. With high-speed optical connections, the distinction between processing speeds and bus speeds (that provide intercommunication among systems) becomes less critical and individual machines in advanced IPCEs may begin to lose their distinction as individual systems and be seen more as specialized aspects of a larger computing environment.

interrogate 1. In lower level software, to query the availability or state of a device or process. 2. In higher level software applications, to query a data or information resource in a systematic manner. For example, an intelligent agent may query a number of search engines on behalf of a user to find suitable avenues for further inquiry, thus automating and streamlining the process for the user. 3. In human terms, to systematically query the availability of specific data or information or to systematically query answers to general or particular questions, as in a database or other information archive.

interrupt A hardware system computing resource that causes a suspension of a process, usually to perform another temporary function. On some desktop systems, interrupts were implemented as a means of handling device requests to the CPU and were thus assigned IRQ numbers. This method has a number of significant limitations in that interrupts are often limited and must be carefully assigned to conserve resources and prevent conflicts, and no two devices can use the same interrupt simultaneously.

On an IRQ-driven system with several peripherals, it was sometimes necessary to disable one device (e.g., an internal modem) in order to operate another device (e.g., a sound card). This means of managing system resources was not common to all computers, but a significant number of Intel-based consumer machines sold in the 1980s and early 1990s had this form of interrupt-handling.

To overcome the problem of interrupt-handling, a number of vendors developed a system called Plug and Play, which allowed dynamic allocation of interrupts and power-on swapping of devices or device controller cards, provided that they support the Plug and Play format. (Don't just assume a component is Plug and Play; verify it.) While this doesn't fully change the underlying concept, it is at least a solution that aids consumers in getting the best use of their machines. See IRQ and accompanying Interrupt Request Numbers chart. See Plug and Play.

INTERSPUTNIK The Russian word for satellite is *sputnik*. The INTERSPUTNIK International Organization of Space Communications system of satellites delivers a variety of programming and data services, including the Voice of America (VOA), which has formed business relationships with a number of independent Russian radio stations, and Direct Net Telecommunications, which provides international digital voice and data services. See INTELSAT.

Interstate Commerce Act of 1887 An act established to regulate the growing interstate railroad business, with the intent of ensuring fair and equitable dealings between transportation carriers and the public. Later, the jurisdiction of the Interstate Commerce Commission was broadened to include regulation of communications services, including telephone, telegraph, and cable. Telecommunications services were later split off into separate communications acts as they grew in prevalence and importance and the Interstate Commerce Commission is now mainly tasked with regulating railroad lines, express companies, and similar transportation carriers.

InterSwitch Message Protocol ISMP. A mechanism

for encapsulating and transmitting control messages exchanged between network switches to maintain a dynamic record of the network topology. See link state advertisement, Virtual LAN Link Switch Protocol.

intersymbol interference ISI. ISI is a form of temporal distortion found in many aspects of telecommunications transmissions where two or more symbols are being transmitted in the same channel and overlapping of waveforms can occur or in systems where distortion is associated with the transmission of sequentially adjacent symbols. It is particularly prevalent where a lot of information is packed into a tight physical space or a tight time frame. Thus, it tends to occur in wire and optical networks where high speeds, distance, and attenuation are factors, in high-density recording data, and in high-speed wireless communications (e.g., digitized voice communications). Rolloff and distortion at transmission band edges can increase ISI. In optical fiber transmissions, spreading of the optical beams over distance can result in increased dispersion and ISI.

Various statistical models for detecting this type of interference have been proposed and research on mitigating this type of interference was beginning to be more comprehensively documented in the mid-1990s. Precoding (often in conjunction with Trellis coding) can sometimes help reduce ISI. Sometimes ISI is deliberately introduced into a transmission to shape the signal and may be systematically removed at the receiving end to mitigate other types of transmission problems.

intra- A prefix for inside, within. An intranetwork is a network within a company, home, or other confined locality. In many business contexts, it implies an Internet-compatible internal network, with many of the same functions, such as a Web server, IRC server, email server, etc.

intracellular electrode A device created in 1949 by Ling and Gerard. It consisted of a tiny glass capillary tube with conducting salt, no more than a few tenths of a micron in size. When used in a microprobe, it was possible to measure electrical currents in individual biological neurons. See neural networks.

intranet An internal network, as in a company or institutional local area network (LAN). The term was coined in part to distringuish internal networks from inter-business networks (extranets). Actually, the first meaning ascribed to intranets was inter-business networks, but even the person who coined the term gradually abandoned it in favor of the de facto LAN connotation. See extranet.

intrapreneur A person within an organization, usually a large one, who manages, takes risks, proposes and promotes ideas, leads, and generally behaves as an entrepreneur *within* and on behalf of the organization. See entrepreneur.

intruder An entity attempting to gain access or gaining access to a restricted system or system resource without proper authorization. See hacker.

intrusion detection The process of determining whether an intruder is attempting to gain access or

has gained access to a system or resource without proper authorization. This can apply to physical environments, where various motion detectors and other technologies may detect the presence of an object or person that should not be present, as well as virtual environments, where an unauthorized process or anomalous process or unusual pattern of activities may indicate unauthorized activities. System monitors, intrusion detection algorithms, usage patterns, incorrect password limits, and electronic alarms are all mechanisms used in network intrusion detection.

inverse multiplexer A multiplexer is a device which takes a circuit, broadcast signal, or given amount of data bandwidth and breaks it up into smaller segments. An inverse multiplexer does the opposite: it takes a number of smaller segments and puts them together to create a larger entity.

An inverse multiplexer is often used in conjunction with computers for high bandwidth applications to coordinate the signals, as in videoconferencing systems that require more than one data line to operate. As an example, imagine an ISDN data network set up for videoconferencing. Videoconferencing requires fast transmission of high-bandwidth resources: video and sound. Some videoconferencing systems are designed to run over two or three separate ISDN lines. In this case, the inverse multiplexer takes the data from the three sources, coordinates the timing, and sends this information to the computer system, which then displays the images and plays the sound together.

inverter 1. A device or circuit which reverses the polarity of a signal (from positive to negative, or vice versa). 2. A device which changes AC to DC or vice versa. AC to DC inverters (often called converters) are very commonly used in digital electronics that draw AC power from a socket. 3. A device or operation that inverts a signal. If the incoming signal is high, the inverted, outgoing signal is low, and vice versa. It is sometimes called a NOT circuit.

Inward Operator Personnel who can assist other operators (e.g., TSPS operators) in making call connections. Normally an Inward Operator does not communicate directly with callers, though phone phreakers have been known to do so.

InWATS Inward Wide Area Telephone Services. A subscriber service to receive incoming calls and be billed for them, rather than having the caller billed, somewhat like an automated collect call. This service is provided by a variety of local and interexchange carriers. See OutWATS, WATS.

IOC See ISDN Ordering Code.

IOL InterOperability Lab. Research, development, and vendor verification of interoperability of wireless communications products at the University of New Hampshire.

ION See Internetworking Over NBMA.

ionization 1. The process of dissociating atoms or molecules into ions and/or electrons. See scintillation. 2. The process of rendering a gas to be conducting by causing some of the electrons to detach from its molecules. 3. The process of rendering a solution to be conducting by electrochemical means, assuming the solution is one that contains a compound that can be made conducting.

ionization current A current resulting when an applied electric field influences the movement of electrical charges within an ionized medium.

ionoscope A camera tube that incorporates an electron beam and a photoemitting screen where each cell in the screen's mosaic produces a charge. This charge, or electric current, is proportional to the variations of the light intensity in the image captured. The ionoscope produced the television image which was then transmitted to the kinescope for viewing in the days of live broadcasts. Sometimes known by the general use and older trademarked term *iconoscope*. See kinescope.

ionosphere 1. A series of layers of ionized gases enveloping the Earth, the most dense regions of which extend from about 60 to 500 km (this varies with temperature and time of day). 2. The portion of the Earth's outer atmosphere which possesses sufficient ions and electrons to affect the propagation of radio waves. In this region, the sun's ultraviolet rays ionize gases to produce free electrons; without these ionized particles, transmitted radio waves would continue out into space without bouncing back. The deflected path of a radio transmission is effected by the direction of the waves and the density of the ion layers it encounters. See ionosphere sublayers, radio waves.

ionosphere, celestial A region around a celestial body comparable in ionic properties with the Earth's ionosphere.

ionospheric sublayers/subregions The Earth's ionosphere has generally been classified into a number of named regions, each of which has properties that make it somewhat distinct from others. These regions are largely hypothetical models, as they may change with the time of day or other factors and don't really form distinct layers as might be implied by the following chart. Nevertheless, the distinctions are useful as a basis for study and for determining good times for propagating radio frequencies through the ionosphere, even though further refinement and changes are likely in understanding of the regions. See Ionospheric Subregions chart.

ionospheric wave Sky wave. A radio wave moving into earth's upper atmosphere. When sky waves are reflected back, at about 2 to 30 MHz frequency ranges, they are known as *short waves*. See ionosphere, ground wave, radio, short wave, skip distance.

IP See Internet Protocol.

IP address, Internet Protocol address On a packet network like the Internet, a number in each packet is used to identify individual senders and receivers. Under Internet Protocol version 4 (IPv4), this is a 32-bit number, theoretically able to accommodate several billion possible addresses, although the actual total is lower due to allocation of subtypes within the system.

To be associated with the Internet, a unique network address number must be assigned. Once a network address has been assigned to a server, additional computers physically attached to that server (as a subnet) can be individually assigned numbers by the local system administrator (certain number patterns are suggested by convention for subnets).

The IPv4 address is a two-part address identifying the network and the individual devices on that network. It is written as four decimal numbers separated by periods with each number representing a byte of the 4-byte Internet Protocol (IP) address.

The decimal numbers are in the range of 0 to 255 (all zeroes or all ones are reserved for administrative use). Periods are used to decimal references to different parts of the network as follows:

255.255.255.255

The left part of the address represents the network. Depending upon the value in the first byte, the network address may be 1, 2, or 3 bytes long (see IP Class for further detail). A mask enables the rest of the address to be interpreted to remove a subnetwork number, if applicable, to determine the host number.

Ionospheric Subregions		
Name	Approx. Location	Notes
D region		A daytime phenomenon and hence not characterized in the same way as some of the regions which exist also at night. Daytime ionospheric activity in this region can impair radio wave propagation.
E region	100 to 120 km	The region which is most distinct in its characteristics and most apt to be classified as a layer.
F1 and F2 regions	150 to 300 km	F2 is always present and commonly used for radio wave propagation, and has a higher electron density than F1, which is only active in the daytime. The F2 region varies in height, and may sometimes go as high as 400 km in the hottest part of the day.
G region	outer fringes of F	Suggested as a distinct layer by some, but its existence as a definable separate layer is debated.

There are *static IPs* and *dynamic IPs*: those more or less permanently assigned and those assigned on an as-needed basis, respectively. Many Internet Service Providers assign temporary dynamic IP numbers to their subscribers to extend limited IP resources to the greatest number of people. As the Internet has grown, it has become increasingly important to manage and reuse IP number resources.

The IP address is located through an email or domain name lookup. IP addresses can correspond to more than one DNS, although a DNS does not have to have an IP address. The IP system is divided into classes, assigned roughly according to the size of the network. See IP Class, Domain Name System, Internet Protocol, InterNIC.

IP Broadcast over ATM An IP multicast service in development by the IP over ATM Working Group for supporting Internet Protocol (IP) broadcast transmissions as a special case of multicast over asynchronous transfer mode networks. See RFC 2022, RFC 2226. See the Appendix for details and diagrams on ATM.

IP Class A network categorization system that facilitates the identification of networks connected to the Internet as each network requires a unique address ID in order to be recognized on the Net by other systems.

The system was originally organized into three general classes, with some special cases. Classes A, B, and C were designated for unicast addresses; later, Class D was designated for multiclass addresses, and Class E was set aside for future use. Certain bits and

Internet Protocol (IP) address ranges were assigned to these classes and there were many discussions as to how to assign and administer public and private network classes and addresses. An IPv4 address is a 32-bit number within which the IP Class address is identified.

The unprecedented demand for IP numbers for linking computers to the Internet resulted in the original scheme being quickly oversubscribed. Thus, the original class scheme has been modified since its inception and IPv6 has been designed to accommodate many more Internet addresses. The specific comments that follow pertain to IPv4, with additional newly developed classes described in general terms. In general, an IP Class address (IPv4) is organized as four 8-bit decimal numbers (octets) separated by periods:

Class.Class Bit.Network ID.Host ID

e.g., 255.255.255.255 (general format)
e.g., 192.168.0.42 (local network ID)

The amount of data designated for the Network ID and the Host ID varies, depending upon the value in the leftmost byte and may be 1, 2, or 3 bytes.

Zeroes are used to designate unknown addresses with all zeroes (0.0.0.0) representing the default route. Loopbacks are designated with 127 (e.g., 127.0.0.1) and broadcast packets are designated with 255 (in other words, each system on the local network will receive the message if 255 is used).

	Internet Protocol (IP) Classes		
Class	Range	H/O Bits	Notes
Class A	0 to 127	0	A network service category similar to a private line for constant bit-rate (CBR) services such as voice communications. Class A networks have a 1-byte network number and a 3-byte host number with 7 bits allocated to the network ID and 24 bits reserved for the host ID. Thus, Class A can support up to 128 networks, each with 16 million hosts.
Class B	128 to 191	10	Class B networks have a 2-byte network number and a 2-byte host number with 14 bits allocated to the network ID and 16 bits reserved for the host ID. Thus, Class B can support up to 16,383 networks, each with 65,535 hosts.
Class C	192 to 233	110	A network service category for connection-oriented data (COD) that is suitable for bursty applications but capable of functioning at higher data rates than some other services. Through multiplexing, Class C services can be used for administering shared services. Class C networks have a 1-byte network number and a 3-byte host number with 21 bits allocated to the network ID and 8 bits for the host ID. Thus, Class C can support up to 2,097,151 networks, each with 256 hosts.
Class D	224 to 239	1110	A network service category for special and multicast networks. Address assignments range from 224.0.0.0 to 255.255.255.0
Class E	240 to 255	1111	A network service category for experimental networks.

The class definitions have been expanded and adjusted according to changing needs on the Internet, although the first three classes retain the general format from larger to smaller networks. See Internet Protocol Classes chart. See address resolution, IP address.

IP echo host service A network service protocol for sending packet IP datagrams after exchanging IP source and destination addresses. See RFC 2075.

IP forwarding The process of receiving an Internet Protocol (IP) data packet, determining how it will be handled, and forwarding it internally or externally. For external forwarding, the interface for sending the packet is also determined and, if necessary, the media layer encapsulation is modified or replaced for compatibility.

IP Multicast over ATM MLIS Internet Protocol multicasting over Multicast Logical IP Subnetwork (MLIS) using ATM multicast routers. A model developed to work over the Mbone, an emerging multicasting internetwork. Designed for compatibility with multicast routing protocols such as RFC 1112 and RFC 1075. By the late 1990s, IP multicasting was becoming an important mechanism for the delivery of broadcast data over the Internet and thus multicast technologies must be both flexible and robust to handle the demand of thousands or millions of users "tuning" in to the same Internet broadcast "station." See enhanced TV.

IP over ATM Internet Protocol over ATM. Implementing ATM involves the coordinated work of many computer professionals and market suppliers of networking products and services. As ATM is a broadly defined format intended to handle a variety of media over a variety of types of systems, there is no one simple explanation for how IP over ATM is accomplished. A number of subnet types need to be supported, including SVC and PVC-based LANs and WANs. There are also a number of relevant peer models, and end-to-end data transmission models, including Classical IP, TUNIC and others.

See asynchronous transfer mode for general information. See the appendix for diagrams and information about layers. See Internet Protocol, RFC 1577, RFC 1755, RFC 1932.

IP over ATM Working Group Merged with the ROLC Working Group to form Internetworking Over NBMA (ION). See Internetworking Over NBMA.

IP Payload Compression Protocol IPComp. A lossless Internet Protocol (IP) compression scheme for reducing the size of IP datagrams, submitted as a Standards Track RFC by Schacham et al. in September 2001. The protocol increases overall performance for hosts with sufficient computational power communicating over slow/congested links.

Compression is applied before any fragmentation, encryption, or authentication processes. In addition, in IPv6, outbound datagrams must be compressed before the addition of a Hop-by-Hop Options header or a Routing Header, since this information must be examined en route. In IPComp, datagrams are individually compressed/decompressed, since they may arrive out of order (or not at all). Inbound processing must support both compressed and noncompressed IP datagrams and decompression is carried out only after security processing has been handled.

In IPv4, compression is applied starting at the first octet following the IP header, continuing to the last datagram octet. The IP header and options are not compressed. In IPv6, IPComp is an end-to-end-type payload and must not be applied to routing and fragmentation headers. In IPv6, compression is applied starting at the first IP Header Option field that does not carry information needed by nodes along the delivery path. The compressed payload size must be in whole octets.

A number of applications of IPComp have been described, including IPComp using LZS (RFC 2395), IPComp using ITU-T V.44 packet method (RFC 3051) and IPComp using DEFLATE (RFC 2394). See RFC 3173 (obsoletes RFC 2393).

IP Security IPsec. A security architecture developed in the mid-1990s to resolve some of the issues of conducting secure transactions on the Internet, particularly business-to-business and electronic commerce transactions. The architecture encompasses protocols, associations, and algorithms for security, authentication, and encryption.

IPsec works at the IP network layer (contrast with Secure Sockets Layer) to provide packet encryption from a choice of encryption algorithms ranging from public-key encryption to secure tunneling. Originally, IPSec worked with an MD5 hashing algorithm, but this was found to be vulnerable to "collision" attacks, and reinforcement for MD5 and algorithm independence was added in later drafts.

IPsec protocols are developed through an IETF working group. They may be optionally implemented into IPv4 but are mandatory for IPv6.

IP switching Technology intended to improve transmission speeds and provide consistent bandwidth for Internet Protocol (IP) switching. On a network, IP switching seeks to bring transmission speeds up to the capability of the underlying physical transport medium. It does so by reducing delay in IP routing processing and by making the data transfer mechanism more circuit- than packet-switched.

IP Telephony WG iptel group. A working group of the IETF focused on research and development related to the propagation and routing of information for Voice-over-IP (VoIP) protocols. The iptel group defined the Telephony Routing over IP (TRIP) protocol. See Telephony Routing over IP.

IPATM See Internetworking over NBMA.

IPCE See interprocess communication environment.

IPComp See IP Payload Compression Protocol.

IPng IP Next Generation. See IPv6.

IPngWG IPng Working Group. A chartered Internet Engineering Task Force (IETF) group developing the next generation Internet Protocol known as IPv6. Members of the Working Group come from various telecommunications industries, including suppliers of data network hardware, network software, and the telephone industry.

IPO See Initial Public Offering.

IPRA See Internet Policy Registration Authority.

IPS See Internet Protocol Suite.

IPSec See IP Security.

IPSec Working Group A division of the Internet Engineering Task Force (IETF) working on standards specifications for the IP Security protocol (IPSec).

Ipsilon Flow Management Protocol In ATM packet networking, a protocol for instructing an adjacent node to attach a layer label to a specified Internet Protocol (IP) packet *flow* to route it through an IP switch. The label facilitates more efficient handling of the flow by providing access to information about the flow without consulting each individual IP datagram. This enables the flow to be switched rather than routed.

IFMP comprises the Adjacency Protocol and the Redirection Protocol. IFMP messages are encapsulated within an Internet Protocol version 4 (IPv4) packet. The IP header signals the IFMP message in its protocol field. It is used in conjunction with the General Switch Management Protocol. See flow, General Switch Management Protocol.

Ipsilon IP switch A commercial switch from Ipsilon, which identifies a stream of Internet Protocol (IP) datagrams for the IP source and destination addresses, and determines if they form part of a longer series. The Ipsilon Flow Management Protocol (IFMP) and General Switch Management Protocol (GSMP) are used in conjunction with specialized hardware to map flow to an underlying network, switching direct IP datagram flows across virtual circuits (VCs). This scheme is most suitable for smaller networks. See IP switching.

IPTel working group The IP Telephony working group, within the IETF Transport Area, formed in the late 1990s. IPTel focuses on propagation of routing information for Voice-over-IP (VoIP) protocols. It is responsible for a syntactic framework for call processing and gateway attribute distribution protocols. The group has defined the Telephony Routing over IP (TRIP) protocol to handle calls that need to be routed between domains. See iCalendar, Telephony Routing over IP.

IPv4 Internet Protocol, Version 4. Developed in the early 1980s, IPv4 was the Internet Protocol for the 1990s, expected to be superseded sometime in the next decade by IPv6. IPv4 features 32-bit addressing, which is suitable for local area networks and widely used there, but no longer sufficient to support the exploding demands on the Internet. See IPv6, RFC 791.

IPv6 Internet Protocol, Version 6. The Internet is a large, complex cooperative network supporting dozens of operating systems and types of computer platforms, tied together with many different circuits, cables, switches, and routers. As can be expected in a system this diverse, a flexible, farsighted vision of its future is needed to ensure not only that the technology does not become entrenched and obsolete compared to new technologies that are released, but also that it continues to retain the flexibility to pro-

vide universal access, much as is guaranteed by law for North American telephone systems. As such, its evolution is of interest and concern to many, and designers and technical engineers have labored long hours to propose future deployments and to develop transition mechanisms to allow the Internet to remain a living upgradable technology.

IPv6 is a significant set of network specifications first recommended by the IPng Area Directors of the Internet Engineering Task Force (IETF) in 1994 and developed into a proposed standard later the same year. The core protocols became an IETF Proposed Standard in 1995.

IPv6 is sometimes called IP Next Generation (IPng). IPv6 was blended from a number of submitted proposals and designed as an evolutionary successor to IPv4, with expanded 128-bit addressing, autoconfiguration, and security features, greater support for extensions and options, traffic flow labeling capability, and simplified header formats.

See 6bone, CATNIP, ICMP, Internet Engineering Task Force, IPv4, X-Bone, TUBA, Simple Internet Transition, SIPP, RFC 1752, RFC 1883, RFC 1885.

IPv6 addresses 128-bit identifiers for interfaces, and sets of interfaces, with each interface belonging to a single node. In most cases, a single interface may be assigned multiple IPv6 addresses from the following types: Anycast, Multicast, or Unicast.

IPv6 extension headers Separate headers are provided in IPv6 for encoding optional Internet-layer information. This information may be placed between the header and the upper-layer header in a packet. These extension headers are identified by distinct Next Header values. In most cases (except for Hop-by-Hop headers), these extension headers are not examined or processed along the delivery path until the packet reaches the node identified in the Destination Address (DA) field of the header. Thus, extensions are processed in the order in which they appear in a packet.

Extension headers are integer multiples of 8 octets, with multioctet fields aligned on natural boundaries. Extension headers in original drafts of IPv6 include Hop-by-Hop, Type 0 Routing, Fragment, Destination, Authentication, and Encapsulating Security payload. If more than one is used in the same packet, a sequence must be followed, both in listing and processing the extension headers. Details can be seen in the extension headers chart. See IPv6 Extension Headers chart. See RFC 1826, RFC 1827.

IPv6 flow A sequence of packets uniquely identified by a source address combined with a nonzero flow label. The packets are sent between a specified source and destination in which the source specifies special handling by the intervening routers. This may be accomplished by resource reservation protocol (RSVP) or by information in the flow packets that may be specified by *extension headers*. There may be multiple flows at one time, in addition to traffic not associated with a flow, and there is no requirement for packets to belong to flows.

IPv6 flow label A 20-bit field in the IPv6 header.

Packets not belonging to a flow have a label of zero, otherwise the label is a combination of the source address and a nonzero label, assigned by the flow's source node. Flow labels are chosen uniformly and pseudo-randomly within the range of 1 to FFFFFF hexadecimal, so routers can use them as hashkeys.

IPv6 from IPv4 developments Some of the changes proposed for improving and updating IPv4 incorporated into the draft documents for IPv6 include:

- increased address sizes (from 32 to 128 bits) and addressable nodes
- simplified autoconfiguration of addresses
- increased scalability of multicast routing
- new addressing provided through anycast addressing
- simplification of header formats
- improved support for extensions and relaxed limits on length of options
- flow labeling of packets to provide special handling capabilities
- removal of enforcement of packet lifetime maximums
- increased support for security, authentication, data integrity, and confidentiality

IPv6 header format The header format of IPv6, described in the draft RFC document, is shown in the chart below.

IPv6 over Ethernet networks IPv6 packets are transmitted over Ethernet in the standard Ethernet frames. The IPv6 header is located in the data field, followed immediately by the payload and any padding octets necessary to meet the minimum required frame size. The default MTU size for IPv6 packets is 1500 octets, a size which may be reduced by a Router Advertisement or by manual configuration of nodes.

IPv6 over Token-Ring networks Frame sizes of IEEE 802.5 networks have variable maximums, depending upon the data signaling rate and the number of nodes on the network ring. Consequently, implementation over Token-Ring must incorporate

manual configuration or router advertisements to determine MTU sizes. In a transparent bridging environment, a default MTU of 1500 octets is recommended in the absence of other information to provide compatibility with common 802.5 defaults and Ethernet LANs. In a source route bridging environment, the MTU for the path to a neighbor can be found through a Media Access Control (MAC) level path discovery to access the largest frame (LF) subfield in the routing information field. IPv6 packets are transmitted in LLC/SNAP frames in the data field, along with the payload.

IPv6 security The IPv6 Draft specifies that certain security and authentication protocols and header formats be used in conjunction with IPv6. These are detailed separately as IP Authentication Header (RFC 1826), IP Encapsulating Security Payload (RFC 1827), and the Security Architecture for the Internet Protocol (RFC 1825).

IPv6 transition IPv6 is a very significant development effort intended to supplant IPv4, the circulatory system of the Internet. Commercial implementation of IPv6 began in the late 1990s. Manufacturers and software developers are, in a sense, overhauling the Net in order to support the updated standard.

As part of the transition process, the 6bone testbed project has been set up to provide testing of IPv6 and various transition mechanisms. This provides a virtual version of IPv6 that can run on existing IPv4 physical structures. Various mechanisms for providing IPv4/IPv6 interoperability are being developed, including the Simple Internet Transition (SIT) set of protocols. SIT provides a mechanism for upgrade intended not to obsolete IPv4, but rather to gradually phase in IPv6, protecting the connectivity and financial investment of the many IPv4 users.

IPX See Internetwork Packet Exchange.

IR See infrared.

IRAC 1. infrared array camera. 2. See Interdepartment Radio Advisory Council. 3. internal review and audit compliance.

IRC 1. integrated receiver decoder. A type of satellite

IPv6 Extension Headers	
Extension Header	Notes
Hop-by-Hop Option	Unlike other headers, requires examination at each node.
Jumbo Payload Option	Used for packets with payloads longer than 65,535 octets. May not be used in conjunction with a fragment header.
Routing Header (Type 0)	Lists one or more intermediate nodes through which the transmission must pass. Similar to the IPv4 Loose Source and Record Route.
Fragment Header	Used by a source to send a packet larger than would fit on the path MTU, as fragmentation in IPv6 is performed only by source nodes.
Destination Options	A header used to carry optional information which is only examined at the packet's destination node.
No Next Header	A value (59) in any IPv6 header or extension header which indicates that nothing follows the header.

receiving device which can be integrated with a multiplexer. This device is used in digital TV broadcasting, especially with MPEG-2 encoded information. 2. See International Record Carrier. 3. See Internet Relay Chat.

IrDA See Infrared Data Association.

IRE See Institute of Radio Engineers.

IREQ *interrupt req*uest. On interrupt-driven systems such as widely distributed Intel-based desktop microcomputers, the insertion and use of a PCMCIA card causes an interrupt request signal to be generated to notify the operating system to suspend the current operation and temporarily process the request from the hardware devices attached via the PCMCIA interface. See interrupt, IRQ.

Iridium A series of low Earth orbit (LEO) communications satellites sponsored by Motorola. Iridium satellites began operations in the late 1990s. They in-

corporate FDMA/TDMA techniques and provide truly global voice, data, facsimile, and GPS services. The name is based upon the original estimate that 77 satellites would be needed to blanket the Earth, matching the element Iridium in the periodic table. The number of satellites needed for global coverage has been reduced to 66 (which were operational by 2002), but the name has remained.

IRIS A Macintosh-based videoconferencing system from SAT which provides video capabilities over ISDN lines with JPEG-encoded graphics. See Cameo Personal Video System, Connect 918, CU-SeeMe, MacMICA.

Irish Internet Association IIA. A professional association supporting, educating, and representing those doing business via the Internet from Ireland, founded in 1997. http://www.iia.ie/

IrLAP See InfraRed Link Access Protocol.

I

Internet Protocol Version 6 (IPv6) Format

```
 0                   1                   2                   3
 0 1 2 3 4 5 6 7 8 9 0 1 2 3 4 5 6 7 8 9 0 1 2 3 4 5 6 7 8 9 0 1
+-+-+-+-+-+-+-+-+-+-+-+-+-+-+-+-+-+-+-+-+-+-+-+-+-+-+-+-+-+-+-+-+
|Version| Traffic Class |           Flow Label                  |
+-+-+-+-+-+-+-+-+-+-+-+-+-+-+-+-+-+-+-+-+-+-+-+-+-+-+-+-+-+-+-+-+
|         Payload Length        |  Next Header  |   Hop Limit   |
+-+-+-+-+-+-+-+-+-+-+-+-+-+-+-+-+-+-+-+-+-+-+-+-+-+-+-+-+-+-+-+-+
|                                                               |
+                                                               +
|                                                               |
+                     Source Address                            +
|                                                               |
+                                                               +
|                                                               |
+-+-+-+-+-+-+-+-+-+-+-+-+-+-+-+-+-+-+-+-+-+-+-+-+-+-+-+-+-+-+-+-+
|                                                               |
+                                                               +
|                                                               |
+                   Destination Address                         +
|                                                               |
+                                                               +
|                                                               |
+-+-+-+-+-+-+-+-+-+-+-+-+-+-+-+-+-+-+-+-+-+-+-+-+-+-+-+-+-+-+-+-+
```

Version	4-bit Internet Protocol (IP) version number = 6
Traffic Class	8-bit traffic class field
Flow Label	20-bit flow label
Payload Length	16-bit unsigned integer. Length of the IPv6 payload, i.e., the rest of the packet following this IPv6 header, in octets. Any extension headers present are considered part of the payload, i.e., included in the length count. If this field is zero, it indicates that the payload length is carried in a Jumbo Payload hop-by-hop option.
Next Header	8-bit selector. Identifies the type of header immediately following the IPv6 header; uses the same values as the IPv4 Protocol field
Hop Limit	8-bit unsigned integer. Decremented by 1 by each node that forwards the packet. The packet is discarded if Hop Limit is decremented to zero.
Source Address	128-bit address of the originator of the packet
Destination Address	128-bit address of the intended recipient of the packet (possibly not the end recipient, if a routing header is present)

IRQ *interrupt request.* A system of implementing computer processor interrupts that is not common to all computer architectures, but which is characteristic of a large number of Intel-based microcomputers. Many desktop computers can readily accommodate several peripheral devices by just plugging them in and installing a software device driver. However, since Intel interrupt-driven machines are prevalent and some of the most frequent hardware configuration problems encountered by users on these systems are related to IRQ assignments, this section provides extra detail to assist users in configuring their systems. If a system locks up, freezes, or fails to recognize a new device, or a device which was working before a new device is installed, it may be due to an *IRQ conflict.*

When using an application program and an interrupt occurs, a signal is sent by the computer operating system to the processor which tells it to pay attention to the signaling process and temporarily suspend the current process. The IRQ is a number assigned to a specific hardware interrupt. The types of devices for which the system requires hardware interrupts include hard drives, CD-ROM drives, mice, joysticks, keyboards, scanners, modems, floppy diskette controllers, sound cards, and others. IRQs are limited in number and some are reserved for specific tasks.

A peripheral device often comes with a controller card that fits into an expansion slot inside the computer. Sometimes there are small dip switches or jumpers on the controller card or on the device itself (or both), which are set at the factory to a preferred, default, or mandatory IRQ number.

On systems that use a manual IRQ system for hardware devices, it is necessary to assign the interrupts to a corresponding device and a good idea to keep a list of the assignments. On older ISA bus systems, almost the whole process had to be done by hand by the user. With later EISA and Micro Channel buses, there is software assistance for detecting and managing IRQ assignments and sometimes it is possible to set the IRQs through software, rather than changing dip switches or jumpers.

In earlier systems, interrupts could not be used by more than one device at a time, some were reserved, and only eight were available in total. To complicate matters, some devices had to be associated with a specific interrupt, reducing the number of possible interrupt combinations on a system with several devices. The IRQ may need to be changed in two places: on the computer system and on the controller card or device. To accommodate more devices, more recent machines added a second interrupt controller, increasing the total number of interrupts to 16 (though again, not all were available, as some were reserved or used for linking).

In general, lower IRQ numbers are higher priority than higher IRQ numbers when two are signaled at

Interrupt Request (IRQ) Numbers and Functions		
IRQ #	INT	Notes
0	08h	Reserved for system timer.
1	09h	Reserved for keyboard.
2	0Ah	Reserved for linking (chaining, cascading) upper eight interrupts through interrupt #9.
3	0Bh	Serial port COM2 and sometimes COM4.
4	0Ch	Serial port COM1 and sometimes COM3.
5	0Dh	Originally assigned to a hard disk controller on 8-bit systems, later 16-bit versions reserved this for a second parallel port (usually designated LPT2). May be available for use by a soundboard, parallel printer, or network interface card (NIC).
6	0Eh	Reserved for floppy diskette controller.
7	0Fh	Reserved for first parallel printer, usually designated LPT1, by some software applications programs (e.g., word processors), but not reserved by the operating system, and thus may be available.
8	70h	Reserved for realtime CMOS clock.
9	71h	Reserved. Used for connection between lower eight and upper eight interrupts. Chained to interrupt #2. In some systems, used for graphics controller.
10	72h	Available. Often used for video display cards.
11	73h	Available. May be used for a third IDE device.
12	74h	Available, although it may be used by a bus mouse (e.g., PS/2 mouse).
13	75h	Reserved for math coprocessor-related functions.
14	76h	Reserved for non-SCSI controllers. Typically used for IDE drives (typical IDE devices include CD-ROM drives, cartridge drives, and hard drives).
15	77h	Available. Sometimes used for SCSI controllers, or a second IDE controller.

the same time, except that IRQs 3 to 8 come after IRQ 15 in priority.

Some peripheral controllers come factory set to a specific interrupt and cannot be changed. Two such cards with the same IRQ requirement cannot be used in the computer at the same time. There are situations where users actually must physically swap out cards to switch between devices. It is wise to ask about IRQ settings when considering the purchase of "bargain-priced" peripherals.

Since the interrupt system created administration and configuration problems for users on machines with several devices, some vendors developed the Plug and Play system, which works in conjunction with Windows 95 to ease the burden of setting and tracking interrupts manually. While this doesn't change the architecture of the system and while not all vendors have followed Plug and Play standards, it nevertheless assists users in managing their systems. See Interrupt Request Numbers and Functions chart. See interrupt, Plug and Play.

IRR See Internet Routing Registry.

IRSG See Internet Research Steering Group.

IRTF See Internet Research Task Force.

IS-54, IS-136 See North American Digital Cellular.

ISA 1. See industry standard architecture. 2. See Instrumentation, Systems, and Automation Society. 3. Instrumentation Society of America. See International Society for Measurement & Control. 4. Integrated Service Adapter. 5. Interactive Services Association.

ISC 1. international switching center. 2. See Internet Software Consortium.

ISC² International Information Systems Security Certification Consortium. See International Information System Security Association.

ISCA See International Speech Communication Association.

ISD 1. See incremental service delivery. 2. Internet Standards document.

ISDN Integrated Services Digital Network. ISDN represents one of the important technologies developed in recent decades to further the transition of communications networks from analog to digital. ISDN is a set of standards for digital data transmission designed to work over existing copper wires and newer cabling media. It began to spread in the late 1980s, and was becoming more prevalent in 2001, in competition with combined telephone/cable modem services.

ISDN is a telephone network system defined by the ITU-T (formerly CCITT), which essentially uses the wires and switches of a traditional phone system, but through which service has been upgraded so that it can include end-to-end digital transmission to subscribers. Some systems include packets and frames, as well (see packet switching and Frame Relay). Nearly all voice switching offices in the U.S. have been converted to digital, but the link to subscribers remains predominantly analog, so it has taken some time to work out the logistics of supporting competing switching methods.

ISDN provides voice and data services over *bearer channels* (B channels) and signaling or X.25 packet networking over *delta channels* (D channels). B channels can also be aggregated (brought together) as *H channels*.

ISDN provides an option for those who want faster data transfer than is offered on traditional analog phone lines, but can't afford the higher cost of Frame Relay or T1 services. ISDN transmission is many times faster (up to about 128 Kbps) than transmission over standard phone services with a 28,800 bps modem. Since the ISDN line doesn't have to modulate the signal from digital to analog before transmission and then demodulate it back to digital, but rather passes the digital signal through, it's faster. It is also possible to use an ISDN line as though it were up to three lines, sending several different types of transmissions (facsimile, voice call, etc.) at the same time. A terminal adaptor (TA) is a device commonly used to adapt ISDN B and D channels to common terminal standards such as RS-232 or V.35. A terminal adaptor takes the place of a modem and is provided in much the same way – as a separate component or as an interface card that plugs into a slot.

A network termination (NT1) device is also commonly used in ISDN installations, usually paid for by subscribers and located at their premises.

Not all cities or countries offer ISDN, but its availability is increasing. Many subscriber surcharge services, such as Caller ID, are available through an ISDN line.

ISDN is available in most urban areas with a choice of two levels of service as shown in the ISDN Basic Service Types chart.

ISDN ANSI standards There are many important American National Standards (ANSI) of Committee T1 related to ISDN available from ANSI. They are summarized by ANSI in the form of abstracts on the Web. ANSI also distributes related ETSI standards documents. Here is a sampling of those available for download for a fee. See the ISDN ANSI Standards chart; it provides a good overview of the issues involved in ISDN/B-ISDN implementation.

ISDN associations There are a number of professional trade associations associated with ISDN technology. Some of the more prominent national and international associations are listed in the ISDN Associations chart. There are also many regional (e.g., state) ISDN groups. See ISDN.

ISDN bonding protocol A protocol which facilitates the use of two ISDN bearer channels (B channels) to transmit a single data stream. The bonding protocol provides dialing, synchronization, and aggregation services for setting up a second call. Both synchronous and asynchronous bonding are supported by various standard and proprietary protocols.

ISDN Caller Line Identification CLI. A feature in which the call address of the caller is sent to the receiving device through the delta channel (D channel). This provides a means for the host router to authenticate the call and to apply any parameters which might be relevant to that particular call.

I

ISDN interfaces When ISDN services are established, a number of links and connections are set up to provide a path for digital transmissions between the telephone switching office and the customer equipment. Each interface link in the path has been designated and commonly used equipment given names to aid in installation and clarity in intercommunicating between the customer and the installer. The ISDN interfaces diagrams (following the ISDN ANSI Standards chart) provide two common scenarios. Note that these diagrams have been simplified and that geographical and equipment variations occur. There are some differences between ISDN deployment in Europe and North America, and local geographic differences are not indicated in the diagrams.

ISDN Ordering Code IOC. A system intended to facilitate the installation of ISDN services by providing the service provider with information about the customer's equipment needed for setup and configuration and smooth operation through a standardized code associated with the model of the ISDN equipment. This code is listed by a participating ISDN equipment vendor in the user manual that comes with the equipment. Prior to the implementation of this system, it could take hours to set up a new ISDN service.

IOC is a National ISDN initiative promoted by local exchange carriers (LECs), the National ISDN Council, the North American ISDN Users Forum, and Telcordia. Telcordia administers the registry and assigns an equipment supplier with an IOC upon registration. See ISDN associations/National ISDN Registry of Customer Equipment and Ordering Codes.

ISDN ANSI Standards	
No./Year Revised	**Title**
T1.113-1995	Signaling System No.7, ISDN User Part
T1.236-2000	Signaling System 7 (SS7) – ISDN User Part Compatibility Testing
T1.604-1990 (R2000)	Minimal Set of Bearer Services for the ISDN Basic Rate Interface
T1.603-1990 (R2000)	Minimal Set of Bearer Services for the ISDN Primary Rate Interface
TR No. 7	3 DSO Transport of ISDN Basic Access on a DS1 Facility
TR No. 15	Private ISDN Networking
TR No. 47	Digital Subscriber Signaling System Number 1 (DSS1) – Codepoints for Integrated Services Digital Network (ISDN) Supplementary Services
TR No. 62	Digital Subscriber Signaling System Number 1 (DSS1) Codepoints for Integrated Service Digital Network (ISDN) Supplementary Services (Supersedes TR No. 47)
T1.219-1991 (R1998)	ISDN Management – Overview and Principles
T1.217-1991 (R1998)	ISDN Management – Primary Rate Physical Layer
T1.239-1994	ISDN Management – User-Network Interfaces Protocol Profile
T1.218-1999	ISDN Management – Data Link and Network Layers
T1.216-1998	ISDN Management – Basic Rate Physical Layer
T1.602-1996 (R2000)	ISDN Data-Link Layer Signaling Specification for Application at the User-Network Interface
T1.241-1994	ISDN Service-Profile Verification and Service-Profile Management ISDN Interface Management Services
T1.625-1993 (R1999)	ISDN Calling Line Identification Presentation and Restriction Supplementary Services
T1.620-1991 (R1997)	ISDN Circuit Mode Bearer Service Category Description
T1.619-1992 (R1999)	MultiLevel Precedence and Pre-Emption MLPP Service, ISDN Supplementary Service Description
T1.616-1992 (R1999)	ISDN Call Hold Supplementary Services
T1.613-1991 (R1997)	Digital Subscriber Signaling System No.1 DSS1 ISDN Call Waiting
T1.612-1992 (R1998)	ISDN Terminal Adaptation Using Statistical Multiplexing
T1.611-1991 (R1997)	Sigaling System Number 7 Supplementary Services for non-ISDN Subscribers
T1.610-1998	DSS1 Generic Procedures for the Control of ISDN Supplementary Services
T1.609-1999	Interworking between the ISDN UserNetwork Interface Protocol and the Signaling System No. 7 ISDN User Part
T1.607-1998	ISDN Layer 3 Signaling Specifications for Circuit Switched Bearer Service for Digital Subscriber Signaling System No. 1 DSS1
T1.605-1991 (R1999)	ISDN Basic Access Interface for S and T Reference Points and Layer 1 Specification

No./Year Revised	Title
T1.601-1999	ISDN Basic Access Interface for Use on Metallic Loops for Application at the Network Side of NT, Layer 1 Specification
T1.650-1995 (R2000)	ISDN Usage of the Cause Information Element in Digital Subscriber Signaling System Number 1 (DSS1)
T1.642-1995 (R2000)	ISDN Supplementary Service Call DeflectionT1.403.01-1999 Network and Customer Installation Interfaces – (ISDN) Primary Rate Layer 1 Electrical Interfaces Specification (includes revision of T1.408-1990 and partial revision of T1.403-1995)
T1.250-1996	OAM&P – Extension to Generic Network Model for Interfaces between Operations Systems and Network Elements to Support Configuration Management – Analog and Narrowband ISDN Customer Service Provisioning
T1.632-1993 (R1999)	ISDN Supplementary Service Normal Call Transfer
T1.643-1998	ISDN Explicit Call Transfer Supplementary Service
T1.653-1996 (R2000)	ISDN Call Park Supplementary Service
T1.647-1995 (R2000)	ISDN Conference Calling Supplementary Service

B-ISDN (Broadband-ISDN)

T1.657-1996 (R2000)	B-ISDN Interworking between Signaling System No. 7 B-ISDN User Part B-ISUP and Digital Subscriber Signaling System No. 2 (DSS2)
T1.658-1996 (R2000)	Extensions to the Signaling System No. 7 – B-ISDN User Part, Additional Traffic Parameters for Sustainable Cell Rate SCR and Quality of Service (QOS)
T1.663-1996 (R2000)	B-ISDN Network Call Correlation Identifier
T1.644-1995 (R2000)	B-ISDN Meta-Signaling Protocol
T1.635-1999	B-ISDN ATM Adaptation Layer Type 5 Common Part – Functions and Specification
T1.629-1999	B-ISDN ATM Adaptation Layer 3/4 Common – Part Functions and Specification
T1.662-1996 (R2000)	B-ISDN ATM End System Address for Calling and Called Party
T1.656-1996 (R2000)	B-ISDN Interworking between Signaling System No. 7 B-ISDN User Part B-ISUP and ISDN User Part (ISUP)
T1.665-1997	B-ISDN Overview of B-ISDN NNI Signaling Capability Set 2, Step 1
T1.664-1997	B-ISDN Point-to-Multipoint Call/Connection Control
T1.654-1996	B-ISDN Operations and Maintenance Principles and Functions
T1.646-1995	B-ISDN Physical Layer Specification for User-Network Interfaces Including DS1/ATM (Supersedes T1.624-1993)
T1.640-1996	B-ISDN Network Node Interfaces and Inter Network Interfaces Rates and Formats Specifications
T1.652-1996(R2001)	B-ISDN Signaling ATM Adaptation Layer – Layer Management for the SAAL at the NNI
T1.645-1995	B-ISDN Signaling ATM Adaptation Layer – Service-Specific Coordination Function for Support of Signaling at the Network Node Interface (SSCF at the NNI)
T1.638-1999	B-ISDN ATM Adaptation Layer – Service-Specific Coordination Function for Support of Signaling at the User-to-Network Interface (SSCF at the UNI)
T1.637-1999	B-ISDN ATM Adaptation Layer – Service-Specific Connection Oriented Protocol (SSCOP)
T1.636-1999	B-ISDN Signaling ATM Adaptation Layer (SAAL) – Overview Description
T1.630-1999	B-ISDN ATM Adaptation Layer for Constant Bit Rate Services Functionality and Specification
T1.627-1993 (R1999)	B-ISDN ATM Layer Functionality and Specification
T1.511-1997	B-ISDN ATM Layer Cell Transfer Performance
T1.624-1993	B-ISDN User-Network Interfaces Rates and Formats Specifications (Superseded by T1.646-1995)

I

ISDN ring signal Unlike analog lines, in which an *in-band* ring voltage signal is used to ring the subscriber phone to indicate an incoming call, ISDN uses an *out-of-band* signal, that is, a digital packet on a separate channel, in order to leave established connections undisturbed.

ISEC See Information Security Exploratory Committee.

ISI 1. Information Sciences Institute. Funded by DARPA, ISI carries out research, development, and technology transfer on RSVP. 2. See intersymbol interference.

ISL 1. Inter-Switch Link. A Cisco Systems proprietary protocol that maintains virtual LAN information as network data traffic moves between switches and routers. 2. ISDN Signaling Link.

ISMC See International Society for Measurement and Control.

ISMP See InterSwitch Message Protocol.

ISNI See Intermediate Signaling Network Identification.

ISO See International Organization for Standardization. (In North America, it is sometimes called the International Standards Organization.)

ISO 9000 series A series of quality standards developed in the 1980s for documenting a company's processes and procedures. The ISO 9000 guidelines

further track the implementation of what has been documented. Audit, certification, registration, and accreditation programs exist for ISO 9000 and are mandatory in some industries, particularly for manufacturers and parts suppliers. http://www.iso.ch/

ISO 9660 A prevalent CD-ROM logical file format standard introduced by the High Sierra Group in 1986. It was subsequently adopted by both ISO and ECMA. It is published as ECMA-119, "Volume and File Structure of CD-ROM for Information Interchange." This compact disc standard aims at portability and future evolution and is widely supported for multiplatform CD-ROM applications. See Yellow Book standard, CD-ROM.

ISO/OSI Reference Model A hierarchical model of communications systems that is the basis for a number of implementations of services and protocols intended to be open for communication with other systems. It is a model for implementing heterogenous computer networks that can intercommunicate through a common networking model. The model was developed in the late 1970s by the International Organization for Standardization (ISO).

The ISO Basic Reference Model for Open Systems Interconnection defines communications types, layers, and the interfaces among them. Among these layers,

---

ISDN Associations – Sampling
Group/Abbreviation/Description
Canadian ISDN Resource Centre CanISDN
Resources and information for residence installations of ISDN. CanISDN includes a Web-accessible database of available ISDN services. http://www.canisdn.net/
European ISDN User Forum EIUF
A user association promoting the effective use of a well-coordinated ETSI-standardized ISDN format throughout Europe. The group has held biannual conferences since 1990 in various locations.
National ISDN Council NIC
An industry trade association promoting the deployment of ISDN services formed in the early 1990s. With ISDN firmly established as a technology, the group changed its name to the Council for Access Technologies. The ISDN Council work is still archived for reference on the Web site. http://www.nationalisdncouncil.com/index.html
National ISDN Registry of Customer Equipment and Ordering Codes
A central registry to assist service providers in setting up ISDN lines so that their switching equipment is compatible with customers' equipment. This uses the ISDN Ordering Codes (IOC) system, one of the National ISDN initiatives promoted by a number of ISDN associations. Using an IOC listed in an equipment instruction manual provides a service provider with the information needed to configure a newly installed ISDN line.
North American ISDN Users' Forum NIUF
Founded in 1988, the NIUF is coordinated by the National Institute of Standards and Technology (NIST). Management was established three years later through a Cooperative Research and Development Agreement (CRADA) with industry.
NIUF promotes ISDN applications development, implementation, acceptance, and furtherance, and provides services and opportunities for users and implementors to communicate their needs and goals to one another. http://www.niuf.nist.gov/misc/niuf.html
Vendors' ISDN Association VIA
A trade organization supporting the interests of vendors offering ISDN and ISDN-related equipment and services. http://www.via-isdn.org/

the model prescribes how the various levels and entities should interact with one another. It does so independently of the specific programming languages, operating systems, or application interfaces selected to run in conjunction with the framework provided by the model.

Adoption of the ISO/OSI Model has not been as widespread as the adoption of Internet models and protocols and a comparison between the two reveals some of the reasons. Documentation for the ISO/OSI

ISDN Interfaces – Two Examples

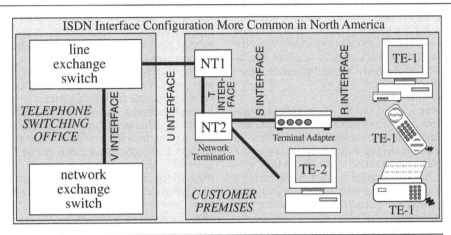

ISDN Interface Configuration More Common in North America

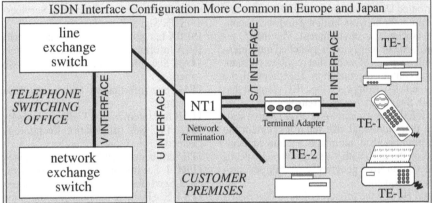

ISDN Interface Configuration More Common in Europe and Japan

The above diagrams are simplified illustrations of ISDN connections as they are commonly configured in North America and in Europe and Japan.

The V interface refers to the link between the public telephone network and the local line switch connecting to the customer's ISDN service. The U interface is the link between the telephone switching office's line exchange switch and the Network Termination (NT) device at the customer's premises.

In North America, where the customer typically owns the NT device, the NT converts the S interface into a T interface. This is the point at which North American and European and Japanese systems differ. In Europe and Japan, the NT device is typically owned by the telephone provider and includes NT2 functionality. It is capable of connecting multiple devices through the S/T interface.

The S interface or the S/T interface refers to the link between the NT device and either a Terminal Adapter (TA) or devices that have a TA built in (that are directly ISDN-compatible).

The NT device in either configuration can connect to devices that have a TA built in. Some of the newer facsimile machines and telecommunications devices may be directly ISDN-compatible and peripheral cards are available for computers that serve as TAs. Or, the TA can provide an interface to devices that are not directly ISDN-compatible, such as non-ISDN facsimile machines, telephones, computer modems, etc. In cases where a TA is used, the R interface refers to the link between the TA and the Terminal Equipment (TE-1). In cases where a TA is not used, the NT device connects directly to the TE-1.

In the European configuration, the Network Termination device (NT1) converts the 2-wire U interface link to a 4-wire S/T interface link and can handle multiple devices. The S interface and the T interface are electrically compatible and, in general, ISDN lines go through an NT1/2 or NT2 interface.

Due to incompatibilities between the two ISDN systems, some manufacturers are supplying equipment that supports both U interface and S/T interface connections. [Diagrams copyright 2001 Classic Concepts.]

was not as freely available as comments and standards related to the Internet. The model is more complex and systematic (as compared to the somewhat organic evolution of the Internet) and thus has been wrung through a longer, more arduous process of administrative hurdles. Nevertheless, the ISO/OSI is well known and many vendor systems intended for international markets implemented some or many of its architectural features. The basic model has been adapted in many aspects of computer networking and forms the basis of the European Community (now the EU) multination network integration effort. See International Organization for Standardization, Open Systems Interconnect.

ISOC See Internet Society.

isochronous 1. Uniform signals with embedded timing information, or which depend on an external timing mechanism. 2. In communications, a system in which the transmitter and the receiver use data clocks with the same nominal rate, although not truly synchronous. 3. In data transmission, a process using a specified number of unit intervals between any two significant instants. See asynchronous, constant bit rate, synchronous.

Isochronous Media Access Control IMAC. In Fiber Distributed Data Interconnect (FDDI) networks, a specification for network bridging and access control in an isochronous environment. The IMAC interacts with one or more circuit-switched multiplexers carrying a variety of media that require a constant bit rate and continued connection. Together with the HMUX, the IMAC comprises the HRC element that is FDDI-II. See Media Access Control.

isogonic 1. Exhibiting relative growth or relative scaling, such that individual size relations remain the same. 2. In magnetism, an imaginary line connecting points on the earth with the same magnetic declination (deviation from geographic north). See agonic, magnetic equator.

isothermal Having the property of even heat distribution across a more-or-less continuous area. This concept is especially important in heat-related fabrication. For example, mass fusion splicers that splice linear arrays of fiber optic filaments require an arc heat source that is as nearly isothermal as possible to ensure consistency of the joints across the array. Since isothermal properties from an arc are harder to maintain over a wider space, larger arrays may further require compensation in the distances between the fiber ends as the fusion process proceeds.

In addition to discontinuities in the arcing temperature across the array, fiber filaments on the inner part may hold heat more than outer fibers, but they are not highly heat conducting and thus can be placed relatively close together when mass fusion takes place. Thus, compensations to create an isothermal environment around the filament array are challenging but are gradually being overcome by commercial products. Materials more heat-conductive than fiber filaments (e.g., copper wire or traces in tiny semiconductor components) may require further compensa-

tions to offset the effect of heat that inner components may transmit to near neighbors more readily than to edge components.

isotropic 1. Exhibiting consistency in values or characteristics along axes in all directions, often from a central point or point of origin. This concept is used to describe certain physical objects, wave transmissions, and various scientific models. Isotropic crystals have a homogenous structure in the sense that light is scattered and absorbed according to the same general pattern within the structure of the material. See anisotropic. 2. In telecommunications, waves that radiate in all directions at the same time so the leading edge of an unimpeded wave emitted from a single point would form a spherical shape.

isotropic antenna An antenna which radiates in all directions at the same time. Since physical conditions will impede the transmission of waves in some directions, Earth antennas are not isotropic. The phrase is useful for distinguishing this type of antenna from those which direct or concentrate a beam and, also, as a theoretical model. An isotropic antenna is a useful reference point for describing signal variations such as antenna gain.

ISP 1. See Internet Safety Policy. 2. See Internet Services Provider. 3. Information Services Platform. 4. ISDN Signal Processor.

ISPBX Integrated Services Private Branch Exchange.

ISSA Information System Security Association.

ITAA See Information Technology Association of America.

ITAC See Information Technology Association of Canada.

ITAR See International Traffic in Arms Regulations.

ITI, ITIC See Information Technology Industry Council.

ITR See International Telecommunication Regulations.

ITRC 1. See Information Technology Research Center. 2. See Information Technology Research Centre.

ITS 1. See Institute for Telecommunication Sciences. 2. See Intelligent Transportation Systems.

ITU, ITU-T See International Telegraph Union, International Telecommunications Union.

ITU-T Recommendations A significant and extensive series of standards and publications to guide industry implementors and vendors in manufacturing and deploying telecommunications technologies.

ITV interactive TV. See interactive television.

IVDS Interactive Video Data Services. A service described by the Wireless Telecommunications Bureau (WTC) of the Federal Communications Commission (FCC) as a means of promoting innovative services and products for telecommunications.

IVI See Intel Video Interactive.

IVR See interactive voice response.

IVS 1. interactive voice service. See interactive voice response. 2. See interactive video service.

IW 1. interworking. 2. intraworking.

IWA See Intelligent Workstation Architecture.

IWS See intelligent workstation.

IXC interexchange carrier. See long-distance carrier.

.jpeg, .jpg File extension conventions commonly used on JPEG-format files, which are prevalent on the Internet. See JPEG, JPEG file format.

J *symb.* joule. See joule.

J box See junction box.

J Series Recommendations A series of ITU-T recommended guidelines for the transmitting television, sound program, and other multimedia signals. These guidelines are available from the ITU-T for purchase from the Net. Since ITU-T specifications and recommendations are widely followed by vendors in the telecommunications industry, those wanting to maximize interoperability with other systems should be aware of the information. A full list of general categories is listed in Appendix C and specific series topics are listed under individual entries in this dictionary, e.g., G Series Recommendations. See J Series Recommendations chart.

J- keys A historic series of telegraph keys labeled J-1, J-2, J-3, etc., up to J-51 (as far as is known) that were used by the U.S. Army Signal Corps and the U.S. Navy. Many are still used by amateur radio operators. L. Nutting has authored a guide to J-keys entitled *J-series Telegraph Keys of the U.S. Army Signal Corps*.

J-carrier A telecommunications system used in Japan, based on the T-carrier system previously standardized in North America. J-carrier is functionally similar to T-1, but some of the specifics differ (e.g., framing and line coding) so that the two are not directly compatible. It is a digitally multiplexed communications carrier system. See trunk carrier for a chart of transmission rates and number of voice channels. See E-carrier, T-carrier.

J-hook connector, J-hook A piece of J-shaped metal or tough plastic (like an unbarbed fish hook) used as an attachment device, often on the end of a cable or elastic fastener. A J-hook is open-ended, mainly intended for quick attachments/detachments for objects that aren't going to be turned upside-down. It is also useful for temporary attachments. When J-hooks are used for underwater tools and sensors, they may be attached to floats to mark their locations.

J-hooks often are used to support long lengths of aerial or piped cables to prevent them from touching the ground or bottom of the conduit while still enabling the cable to be easily moved, replaced, or pulled through the hooks.

A *J-hook tree* is a series of J-hooks on one supporting rod or stem. The hooks may be staggered or adjacent and may be on one side or several sides. This facilitates the stringing of several wires along a "virtual" or open conduit. In these applications, the J-hook may be coated with an insulating material or, if the J-hook connector is made of conductive materials, shielded wires may be used. See J-pole antenna. Contrast with snap-hook.

J-Phone Communications Co. Ltd. One of the largest mobile telephone operations in Japan, with a majority share held by the Vodafone Group in the U.K., J-Phone is the mobile telephony branch of Japan Telecom. J-Phone utilizes Packet Personal Digital Cellular (PPDC) technology, similar to the General Packet Radio Service (GPRS) used for GSM in Europe. Third-generation (3G) network equipment is supplied to J-Phone by Nokia, a prominent mobile equipment supplier.

J-pole antenna A simple, long single-mast antenna named for its vertical waveguide and shorter horizontal-to-vertical connecting J-like stub. It is an end-fed 1/2-wavelength antenna with the J portion of the antenna being a 1/4-wavelength voltage transformer.

The J-pole antenna may be used with short-range, low-power frequency-modulation (FM) transmitters and is of interest to amateur packet radio enthusiasts. It is practical for a range of frequencies and can be built by home hobbyists or purchased commercially. It is commonly constructed from copper pipe. There are a number of Web-based J-pole design pages that allow you to type in a desired operating frequency to calculate the dimensions for the J-pole antenna parts.

J-SAC *Journal on Selected Areas in Communications*, published by the IEEE. Each issue is devoted to a specific technical topic. Guest editors review articles for inclusion, with topics selected by the J-SAC Editorial Board. Topics range across many communications topics including ultra wideband radio, wireless communications, security, and more.

J-scope A type of circular radar screen displaying only the range of objects detected.

ITU-T J Series Recommendations

General Principles, Terms, and Definitions

J.1 Terms, definitions and acronyms applicable to the transmission of television and sound-program signals and of related data signals

J.2 Guidelines on the use of some ITU-T Recommendations in the J series

J.12 Types of sound-program circuits established over the international telephone network

J.13 Definitions for international sound-program circuits

J.51 General principles and user requirements for the digital transmission of high quality sound programs

J.64 Definitions of parameters for simplified automatic measurement of television insertion test signals

J.85 Digital television transmission over long distances – general principles

J.110 Basic principles for a worldwide common family of systems for the provision of interactive television services

Transmission

J.52 Digital transmission of high-quality sound-program signals using one, two or three 64-Kbps channels per mono signal (and up to six per stereo signal)

J.53 Sampling frequency to be used for the digital transmission of high-quality sound-program signals

J.54 Transmission of analog high-quality sound-program signals on mixed analog-and-digital circuits using 384-Kbps channels

J.55 Digital transmission of high-quality sound-program signals on distribution circuits using 480 Kbps (496 Kbps) per audio channel

J.56 Transmission of high-quality sound-program analog signals over mixed analog/digital circuits at 320 Kbps

J.57 Transmission of digital studio quality sound signals over H1 channels

J.61 Transmission performance of television circuits designed for use in international connections

J.68 Hypothetical reference chain for television transmissions over very long distances

J.73 Use of a 12-MHz system for the simultaneous transmission of telephony and television

J.74 Methods for measuring the transmission characteristics of translating equipment

J.75 Interconnection of systems for television transmission on coaxial pairs and on radio-relay links

J.77 Characteristics of the television signals transmitted over 18- and 60-MHz systems

J.80 Transmission of component-coded digital television signals for contribution-quality applications at bit rates near 140 Mbit/s

J.81 Transmission of component-coded digital television signals for contribution-quality applications at the third hierarchical level of ITU-T Recommendation G.702

J.88 Transmission of enhanced definition television signals over digital links

J.92 Recommended operating guidelines for point-to-point transmission of television programs

J.141 Performance indicators for data services delivered over digital cable television systems

J.142 Methods for the measurement of parameters in the transmission of digital cable television signals

J.144 Objective perceptual video quality measurement techniques for digital cable television in the presence of a full reference

J.145 Measurement and control of the quality of service for sound transmission over contribution and distribution networks

Human Factors, Interfaces

J.117 Home digital network interface specification

J.140 Subjective picture quality assessment for digital cable television systems

J.143 User requirements for objective perceptual video quality measurements in digital cable television

Distribution, Service Delivery

J.83 Digital multiprogram systems for television, sound and data services for cable distribution

J.84 Distribution of digital multiprogram signals for television, sound and data services through SMATV networks

Distribution, Service Delivery, cont.

J.86   Mixed analog and digital transmission of analog composite television signals over long distances

J.87   Use of hybrid cable television links for the secondary distribution of television into the user's premises

J.90   Electronic program guides for delivery by digital cable television and similar methods

J.93   Requirements for conditional access in the secondary distribution of digital television on cable television systems

J.113   Digital video broadcasting interaction channel through the PSTN/ISDN

J.119   RF remodulator interface for digital television

J.120   Distribution of sound and television programs over the IP network

J.150   Operational functionalities for the delivery of digital multiprogram television, sound, and data services through multichannel, multipoint distribution systems (MMDS)

J.151   RF remodulator interface for digital television

J.161   Audio codec requirements for the provision of bidirectional audio service over cable television networks using cable modems

J.162   Network call signaling protocol for the delivery of time-critical services over cable television networks using cable modems

J.163   Dynamic quality of service for the provision of real time services over cable television networks using cable modems

J.164   Event message requirements for the support of realtime services over cable television networks using cable modems

J.167   Media terminal adapter (MTA) device provisioning requirements for the delivery of real time services over cable television networks using cable modems

J.182   Parameter sets for analog interface specifications for the interconnection of set-top-boxes and presentation devices in the home

J.184   Digital broadband delivery system: out-of-band transport

J.166   IPCablecom Management Information Base (MIB) framework

J.168   IPCablecom Media Terminal Adapter (MTA) MIB Requirement

J.169   IPCablecom Network Call Signaling (NCS) MIB Requirements

J.180   User requirements for statistical multiplexing of several programs on a transmission channel

J.181   Digital program insertion cueing message for cable television systems

Testing, Measurement, Performance, Interference

J.11   Hypothetical reference circuits for sound-program transmissions

J.14   Relative levels and impedances on an international sound-program connection

J.15   Lining-up and monitoring an international sound-program connection

J.16   Measurement of weighted noise in sound-program circuits

J.17   Pre-emphasis used on sound-program circuits

J.18   Crosstalk in sound-program circuits set up on carrier systems

J.19   A conventional test signal simulating sound-program signals for measuring interference in other channels

J.21   Performance characteristics of 15-kHz-type sound-program circuits – Circuits for high quality monophonic and stereophonic transmissions

J.22   Performance characteristics of 10-kHz-type sound-program circuits

J.23   Performance characteristics of 7-kHz-type (narrow bandwidth) sound-program circuits

J.24   Modulation of signals carried by sound-program circuits by interfering signals from power supply sources

J.25   Estimation of transmission performance of sound-program circuits shorter or longer than the hypothetical reference circuit

J.26   Test signals to be used on international sound-program connections

J.27   Signals for the alignment of international sound-program connections

J.31   Characteristics of equipment and lines used for setting up 15-kHz type sound-program circuits

J.32   Characteristics of equipment and lines used for setting up 10-kHz type sound-program circuits

J.33   Characteristics of equipment and lines used for setting up 6.4-kHz type sound-program circuits

J.34   Characteristics of equipment used for setting up 7-kHz type sound-program circuits

J

ITU-T J Series Recommendations, cont.

Testing, Measurement, Performance, Interference, cont.

J.41 Characteristics of equipment for the coding of analog high quality sound program signals for transmission on 384-Kbps channels

J.42 Characteristics of equipment for the coding of analog medium quality sound-program signals for transmission on 384-Kbps channels

J.43 Characteristics of equipment for the coding of analog high quality sound program signals for transmission on 320-Kbps channels

J.44 Characteristics of equipment for the coding of analog medium quality sound-program signals for transmission on 320-Kbps channels

J.66 Transmission of one sound program associated with analog television signal by means of time division multiplex in the line synchronizing pulse

J.62 Single value of the signal-to-noise ratio for all television systems

J.63 Insertion of test signals in the field-blanking interval of monochrome and color television signals

J.65 Standard test signal for conventional loading of a television channel

J.67 Test signals and measurement techniques for transmission circuits carrying MAC/packet signals

J.94 Service information for digital broadcasting in cable television systems

J.101 Measurement methods and test procedures for teletext signals

Privacy, Security

J.91 Technical methods for ensuring privacy in long-distance international television transmission

J.95 Copy protection of intellectual property for content delivered on cable television systems

J.100 Tolerances for transmission time differences between the vision and sound components of a television signal

Interactive Systems

J.111 Network independent protocols for interactive systems

J.112 Transmission systems for interactive cable television services

J.114 Interaction channel using digital enhanced cordless telecommunications

J.115 Interaction channel using the global system for mobile communications

J.116 Interaction channel for local multipoint distribution systems

J.118 Access systems for interactive services on SMATV/MATV networks

J.200 Application environment for digital interactive television services

MPEG

J.82 Transport of MPEG-2 constant bit rate television signals in B-ISDN

J.89 Transport mechanism for component-coded digital television signals using MPEG-2 4:2:2 P@ML including all service elements for contribution and primary distribution

J.96 Technical method for ensuring privacy in long-distance international MPEG-2 television transmission conforming to ITU-T J.89

J.131 Transport of MPEG-2 signals in PDH networks

J.132 Transport of MPEG-2 signals in SDH networks

J.183 Time division multiplexing of multiple MPEG-2 transport streams over cable television systems

Supplements

J.Sup1 Example of linking options between annexes of ITU-T Recommendation J.112 and annexes of ITU-T Recommendation J.83

J.Sup2 Guidelines for the implementation of Annex A of Recommendation J.112, "Transmission systems for interactive cable television services" – example of Digital Video Broadcasting (DVB) interaction channel for cable television distribution

J.Sup3 Guidelines for the implementation of Recommendation J.111 "Network independent protocols" – example of Digital Video Broadcasting (DVB) systems for interactive services

J.Sup4 Terminology for new services in television and sound-program transmission

J.Sup5 Guidelines on the use of some ITU-T Recommendations in the J series

J.Sup12 Intelligibility of crosstalk between telephone and sound-program circuits

J-Sky A mobile Internet access service offered by J-Phone Communications.

J/LOCK A commercial Java-based cryptography library distributed by STI. It provides cryptography algorithms and interfaces for creating security systems and secured knowledge management systems. J/LOCK is based upon the Java Cryptography Architecture (JCA). It implements public key algorithms, message digest algorithms, and symmetric key algorithms and supports basic and commonly used security protocol data units and PKI profiles.

J2ME Java 2 Platform, Micro Edition. The Sun Microsystems Java programming language platform edition for consumer and embedded device space. J2ME has two configurations, the Connected Device Configuration (CDC) using the classic Java virtual machine, and the Connected Limited Device Configuration (CLDC), using the K Virtual Machine for severely constrained memory environments. See Java.

jabber Continuous (erratic or incessant) transmission of inappropriate, corrupted, or meaningless data (garbage), usually beyond the normal protocol interval. A network device broadcasting its availability redundantly is said to be jabbering. Jabber can lock up a system because it appears to other devices on the network to be busy.

Jabber may have limited diagnostic uses, but is mostly fault-related, from a variety of causes such as software bugs, excessive packet lengths, signal degradation, incompatible systems, hardware defects, or weather-related hardware aberrations or failures. Consequences range from bad data to system lockups. On Ethernetworks, the term is more specifically associated with the attempted transmission of packets that are longer than the maximum allowed Ethernet packet length. Jabber errors are sometimes called streaming errors. Network connecting components may include jabber circuitry to interrupt transmission until the fault is corrected.

Various components, such as fiber optic hubs, modems, "thin" Ether to "thick" Ether converters, and "thin" to fiber ST converters are equipped with LEDs to display error conditions such as jabber and data collisions. See garble, jabber switch.

Jabber A software development platform for creating collaborative applications and networks that has been widely adopted, partly because of support for realtime streaming of XML content, which is becoming popular on the Internet, and for its large-scale applicability. Jabber is distributed by Jabber, Inc.

jabber control, jabber lockup A protective mechanism to inhibit faulty data transmission caused by an overrun of data packets. See jabber, jabber switch.

jabber switch A fault-related circuit that may be incorporated into a fiber optic network media access unit (MAU). The device senses a jabber-related controller malfunction and "trips" in much the way a fuse will trip an electrical circuit. Jabber switches are often configured to reset automatically once the controller malfunction is corrected to minimize downtime. See jabber.

Jablochkoff, Paul (1847–1894). A Russian telegraph engineer who directed telegraph communications between Moscow and Kursk until 1875. En route to America, he stopped in France, where he joined the laboratory of Louis Bréguet (1804-1883), an electric clock and telegraph supplier to the French railways and navy. While working in the Parisian lab, Jablochkoff invented the Jablochkoff Candle arc lamp. His invention was awarded a U.S. patent in 1877 (#252,646). The English patent was taken by Robert Applegarth who developed a business from the candle. See Jablochkoff Candle.

Jablochkoff Candle The first widely distributed arc lamp, utilizing upright, parallel carbons separated by a narrow layer of plaster. The Candle was developed in 1876 by a Russian telegraph engineer, Paul Jablochkoff, and put into preliminary use in 1877. Compared to previous arc lamps, the Candle had a simpler design, making it amenable to fabrication and inexpensive distribution. The main disadvantage of the design was that the carbons had to be replaced if the lamp was switched off. Arc lamps for consumer and commercial lighting were gradually replaced by filament lamps between World War I and the 1940s. In the current semiconductor industry, arc lamps are used in optical etching of circuits. See arc lamp; Jablochkoff, Paul.

JACAL A symbolic mathematics system written by Aubrey Jaffer. JACAL runs in Scheme or Common LISP.

jack A contact junction between circuits consisting of a male or female receptacle at one end and usually a line at the other end. The jack portion typically has a protrusion and is colloquially known as the male end. Jacks come in a great variety of shapes and sizes. RJ-, RCA, and BNC jacks are commonly used in telephone, video, and networking installations.

jack in To attach oneself to an electronic system. To log on, or interact with a system through worn devices, such as data gloves, body suits, or implanted electronic components, especially those connected to the human nervous system. Often used colloquially to refer to interaction with virtual environments, where there is the illusion of "being" in the virtual space as opposed to maintaining a separation from it. See avatar, virtual reality.

jack panel, jack field A board or panel configured with several jacks or with plugs for inserting several jacks, for organizing or centralizing a number of related circuit connections. See cordboard, path panel, switchboard.

jacket A sheath that covers a cable to provide protection and sometimes outer insulation, as well as organization and identification. The jacket may bundle a group of related wires or fibers, and can enhance identification through colors, patterns, or materials, e.g., red and green are commonly used to identify tip and ring in phone wires.

jacking In underground wiring installations, a dig/install/haul process carried out together for efficiency. Jacking is the process of pushing a casing through a hole large enough to accommodate the casing as the

J

hole is dug just ahead of the installation of the casing, and the loosened soil is mucked back through the pipe. The trench created in the process is called a *jacking pit*.

Jacobson's algorithm An algorithm developed by Van Jacobson for computing smoothed roundtrip time (RTT) in network communications, based upon an ARMA model. It is used at the link level in layer-based systems. Karn's and Jacobson's algorithm are implemented for computing retransmission timeouts. See Jacobson, Van; Karn's algorithm.

Jacobson, Van An American computer scientist, Jacobson was group leader for the Lawrence Livermore Laboratory Network Research Group after which he became chief scientist for Cisco Systems and then for Packet Design. In the late 1990s, he was a collaborator in the Web Caching Project.

Jacobson is the author of many RFCs including RFC 1144 (1990), which describes compression schemes for supporting an increasing trend for linking the Internet up to individual desktops. He authored a number of articles in the early 1990s on congestion-avoidance in packet-switched networks. Jacobson is best known as one of the three principal creators of MBone multicast networking and for his popular Jacobson's algorithm for computing roundtrip information in networks. He was also an early proponent of the concept of "whiteboarding" which enables interaction between remote users in the other's computing space. See Jacobson's algorithm.

Jacquard, Joseph-Marie (1752–1834) An innovative French industrialist who devised a way of automating the storage and retrieval of loom patterns by punching the patterns into cards. This innovation met with strong objections from workers fearful of losing their livelihoods (which eventually happened) but is significant for becoming an important means to store computer programs until the early 1980s. See Babbage, Charles; Jacquard loom; punch cards.

Jacquard loom A type of automated loom that worked with pattern-encoded punch cards. This loom was devised by Joseph-Marie Jacquard in France in the late 1700s and early 1800s. The holes punched in the cards indicated whether or not a thread was to be woven into the pattern. By associating each card with a color, it was possible to quickly weave complex patterns, and the cards could be modified one by one and recombined, or stored and used in later projects. This caused a revolution in the textile industry. While the idea of encoding information with holes or slots had been used prior to the Jacquard loom, e.g., in music boxes, this large-scale industrial application was significant. It also served as a model for the storage of data and programs in large-scale computers a century and a half later. See punch cards.

JAD joint application design.

JAE See Java Application Environment.

jag 1. distortion caused by errors between the transmitter and the recording device, as in a facsimile. 2. distortion or staircasing (aka jaggies) which is an artifact of displaying an image in a resolution too low for the amount of information conveyed. See jaggies.

jaggies *colloq.* In image processing, a descriptive term for aberrations that often occur when the resolution of an image is too complex for the resolution of the device on which it is displayed or printed. The result is a "staircased," "aliased," or jagged appearance, especially along edges where there is a sharp contrast or transition from one color or tone to another.

Not much can be done with jaggies in monochrome images, but in gray scale or color images, jagged transitions can be smoothed with a technique called antialiasing. This involves selecting intermediate tones between the contrasting elements to visually create the illusion of a transition. Since human perception tends to want to blend such elements, it appears as a smoother line or shape. The technique of antialiasing is often used in the display of images on computer monitors because the resolution is relatively low on this medium, usually about 75 dpi, as contrasted with print, which is usually 300 to 2400 dpi. See antialias.

jam An accidental or deliberate fault condition which hinders or stops the subsequent flow of objects (e.g., printer paper jam) or data (radio, network, or radar signal jam).

On a data network, a jam may occur if more than one device senses idle time or a window, and each device tries to send a packet at the same time, thus causing collisions. A collision may occur unintentionally or be deliberately generated to test the collision-detection response in the network. When testing this on an IEEE 802.3 network, if a station is transmitting and detects a collision event, it should stop the transmission of data and transmit a 32-bit jam signal to indicate the collision.

JaM The name for one of the predecessors to the PostScript page definition language. JaM was the combined effort of John Warnock and Martin Newell at Xerox PARC in 1978, descended from work on the Design System Language originating at Evans and Sutherland in the 1970s. John Warnock credits John Gaffney with many of the essential ideas. JaM came to be used for various printing, graphics arts, and VLSI design applications. See PostScript.

JAM Jini Technology Access Module. See Jini.

jam 1. To fill up so as to compromise a system function, obstruct, stick a wrench into, deliberately impede, to deliberately insert nonsenical or garbled data.

jam signal A mechanism used in a data network to prevent redundant collisions. When a jam signal is received by more than one device trying to send at the same time, the devices typically wait for a random period and then try again, reducing the probability that they will simultaneously attempt to resend the data and cause further collisions. See babble signal, frequency hopping, jam.

jamming The blocking or obstructing of signals by physical or electronic interference. Jamming is done for reasons of political control or protests, vandalism, competitive obstruction, and national and personal security. Most types of jamming are restricted or prohibited.

JAMSAT Japanese affiliate of AMSAT. See AMSAT.

JAN Joint Army-Navy. The JAN specification is the forerunner of military specifications which have been superseded by the MIL designation. See MILNET.

JANET Joint Academic Network. A U.K. network established in 1984 which links universities and other academic and research facilities. JANET is funded by the Joint Information Systems Committee (JISC) and developed and managed by UKERNA (formerly by JNT Joint Network Team). The broadband aspect of JANET, capable of transporting video and audio simultaneously with the JANET data, is called SuperJANET (coined in 1989). See JNT.

Japan Amateur Radio League, Inc. JARL. Established in 1926 by a group of 37 radio communication enthusiasts dedicated to promoting the development and use of radio wave technology, JARL's first private license was granted a year later. Since then, the organization has grown and became involved in international activities. In 1985, a reciprocal operating agreement was signed between Japan and the United States. In 1986, JARL began launching a series of amateur radio satellites. JARL has the largest number of amateur radio stations of any country in the world. See JAS-1b, JAS-2. http://www.jarl.or.jp/

Japan Approvals Institute for Telecommunications Equipment JATE. A Japanese regulatory agency, established in 1984, which is roughly equivalent to the Federal Communications Commission (FCC) in the United States. JATE is authorized by the Minister of Posts and Telecommunications, under the provisions of the 1985 Telecommunications Business Law.

JATE grants technical conditions compliance approval and technical requirements compliance approval for public network telecommunications-related equipment. Once approval has been granted, the approval mark must be affixed to the approved equipment.

For non-Japanese manufacturers seeking JATE approval for their products, engaging a good technical translator may expedite the process.

Japan Electronic Industry Development Association JEIDA. A trade association promoting Japanese economic prosperity through the development of the electronics industry. http://www.jeida.or.jp/

Japan Institute of Office Automation JIOA. An organization accredited by the Japanese Minister of International Trade and Industry to build and promote an information society in Japan, while fostering interactive understanding between providers and users of information technologies. Founded in July 1981. http://www.jioa.or.jp/

Japan Electronic Messaging Association JEMA. The Japanese arm of The Open Group EMA Forum promoting electronic messaging since 1992. JEMA emphasizes overseas cooperation and promotes security research and global interconnections. JEMA is a member of the Asia-Oceania Electronic Messaging Association.
http://www.ema.org/
http://www.fmmc.or.jp/associations/jema/

Japan Personal Computer Software Association JPSA. A nonprofit trade association of software developers, publishers, distributors, dealers, and system integrators. JPSA carries out educational and market survey activities on behalf of its membership. http://www.jpsa.or.jp/

Japan Solderless Terminal JST. A line of commercial electronic cables developed by JST Mfg. Co., Ltd., since 1957, in Japan. JST products are used in many mobile devices, computing, photocopying, and robotics devices. The company makes many types of connectors, but is especially known for small 2- to 9-pin terminal cables, jumper, PC card, and crimp-style connectors.

Japan Telecom Co. Ltd. The third largest telecommunications carrier in Japan, with over 10 million users. Japan is known for its lead in the wireless market; its wireless subsidiary is J-Phone Corporation, a holding company majority-owned by Japan Telecom and J-Phone in the late 1990s. Other nations have a stake in Japan Telecom, including British Telecommunications P.L.C. See KDDI Corp., Nippon Telegraph and Telephone Corp. See J-Phone Communications Co. Ltd.

Japan, Communications Industry Association of CIAJ. A trade association committed to promoting healthy business activities and developmental efforts among telecommunications equipment manufacturers with the goal of improving the standards of living and quality of life in Japan and overseas. http://www.ciaj.or.jp/

Japan Computer Security Association JCSA. A trade association dedicated to increasing computer security awareness in Japan and assisting computer system developers, administrators, and users in developing secure computer programs, systems, and data. http://www.jcsa.or.jp/

Japan, Electronic Industries Association of EIAJ. A nonprofit trade association devoted to supporting and promoting the electronic industries of Japan. EIAJ was founded in 1948 and now has more than 500 members drawn from the consumer and industrial electronics sectors. EIAJ is involved in collaborative development projects, statistical forecasting, industry awareness, and professional development activities. http://www.jeita.or.jp/

JAR See Java Archive.

Jargon File, The The forerunner to *The Hacker's Dictionary*, the original *Jargon File* was an electronic dictionary started by Raphael Finkel in 1975 at Stanford University, and has since gone through many revisions with contributions from Mark Crispin, Guy L. Steele Jr., and the ARPANET communities. This well-known dictionary includes jargon, slang, and technospeak from the artificial intelligence (AI) and general computer communities, and is well-known to programmers worldwide. Don Woods later became the *Jargon File* contact, and in the early 1980s, Richard Stallman made contributions. The early 1980s also saw the first paper publications of The File and the transition to *The Hacker's Dictionary*. See *Hacker's Dictionary, The*.

J

JARL See Japan Amateur Radio League, Inc.

JAS-1b Japan's second amateur radio satellite, launched in 1990, 4 years after its first satellite was put into orbit. Six years later, in 1996, this was followed by JAS-2. See JAS-2.

JAS-2 Japan's third amateur radio satellite, launched in 1996 using an H-II launch vehicle.

JATE See Japan Approvals Institute for Telecommunications Equipment.

Java An object-oriented, platform-independent, threaded programming language that came into being largely because its two earliest contributors were not satisfied with C and C++, and wanted a way to develop programs with less effort and code. Thus, Bill Joy proposed an object environment based on C++ to Sun Microsystems engineers, and James Gosling, author of EMACS, developed a language called Oak. Communications between Patrick Naughton of Sun, Mike Sheridan, James Gosling, and Bill Joy resulted in the Green Project and collaborative work began.

Eventually, in 1995, Java was introduced by Sun Microsystems. Java requires significantly less code than C for many types of applications, is generally easier to learn, works well in conjunction with the Web, and has a good chance of becoming a widespread language of choice for software development. On the Web it is frequently used to supplement HTML as a means to interact with and convey information to Web users. HTML is a markup language, a type of language where a user can learn a few commands and install a basic Web page without too much technical knowledge. However, Web users with more sophisticated needs or the desire to interact with users in a more fluid and interactive manner than is possible with HTML generally select Perl and/or Java for implementing calculations, specialized interfaces, and more complex programming structures than are possible with HTML.

At first, Web users were slow to adopt Java. Those who were familiar with C and other powerful, fast development languages were uncomfortable with Java's limits and slow running times. However, Java has continued to evolve, the Web has continued to evolve (and now is accessed on faster systems and Internet links), and developers have begun to realize that there is an enormous middle ground of applications that don't have to run as fast as C to be useful and that can be implemented far faster and more readily with Java. Examples include basic menu selections, games based on strategy rather than speed, interactive database interfaces, more sophisticated interactive forms than are possible with HTML, statistical charts and graphs, and much more.

Java support from Sun includes the Java Development Kit (JDK), available for various Sun platforms, Windows NT, and Windows 95 Intel. Independent ports exist for other operating systems, including Linux, NeXTStep, and Amiga. Macintosh support is provided by Apple Computer's Macintosh Runtime for Java (MRJ), and Windows 3.1 support is provided by IBM.

Two applications environments are particularly relevant to personal communications devices.

- The *PersonalJava* application environment is designed to facilitate development of software for private network or Web-connected consumer devices that may be executing applets. This requires that a core set of software libraries be installed on the PersonalJava-enabled device. The PersonalJava AE comprises the Java Virtual Machine (JVM) and an optimized version of the Java class library. This environment is useful in situations where generalized applications or those that can't easily be predicted in advance will be used. The tradeoff is that the libraries require a certain amount of space.

- The *EmbeddedJava* application environment is designed to facilitate development of software for dedicated-function embedded devices which may be stand-alone or embedded. Only the class libraries needed to support a prespecified set of tasks are installed onto the EmbeddedJava-enabled device. This is useful in very specific dedicated applications where the purpose of the device is well defined and not likely to change in the near future. Space is saved and the resource "footprint" of the device is smaller (which often lowers the cost). The tradeoff is that future needs may not always be anticipated or met without changing the core set of library routines.

Java can be used in conjunction with the HotJava Web browser to allow Java programs to run on a desktop computer.

Java information and specifications are available through the Javasoft Web site. See J2ME, Java APIs, Java applet, Java Archive, JDBC, JOLT Project, JStamp. http://www.javasoft.com/

There is a good Java Frequently Asked Questions (FAQ) listing by Elliotte Rusty Harold on the Web. http://sunsite.unc.edu/javafaq/javafaq.html

Java APIs A number of important applications programming interfaces associated with Sun Microsystems' Java provide specifications and procedures for applications development.

Java applet An important component of Java object-oriented programming, an applet is a Java class used to extend Java. Applets can intercommunicate within the same virtual machine environment.

Applets are run within the circumscribed context of a Web browser, applet viewer, or other application that supports applets. This provides a measure of extensibility along with a certain amount of security, since the applet can normally only read and write files on the host machine through the application through which it is running. See Java.

Java Application Environment JAE. See Java.

Java Archive JAR. A powerful, Java open standard, platform-independent, compression file format for images and sound that brings together a set of files into one. In this way, Java applets and their associated components can be bundled and downloaded as a single file; however, it can also be used as a general-purpose compression/archiving tool, similar to ZIP. JAR files are very small, even smaller than PKZIP files, in many cases.

JAR has some support for data security. Individual portions of a JAR can be digitally signed and authenticated. JAR archives can be created with the JAR utility included with JDK, which functions in a manner similar to many common archive utilities. See PKZIP, RAR.

Java Community Process JCP. An open, internationally inclusive means of developing and revising Sun Microsystems' Java technology specifications and related support resource. More than 300 companies and individual participants are involved in this effort. Almost 100 Java technology specifications are being developed through the JCP, which was initiated in 1995. See Java.

Java Electronic Commerce Framework JECF. A securable, extensible framework for conducting electronic commerce, developed by Sun Microsystems. The initial component of JECF is JavaWallet, a client-side application distributed as part of the Java Commerce toolkit as a core component of the Java environment. Java Commerce APIs are used to implement basic services within the Java Commerce Client that can be used to develop online shopping malls and banking applications. See Java, JavaWallet.

Java name space A means of resolving names in a software program to Java runtime classes. In general, the system applies to classes, packages, and class members. Classes can be moved or removed from specified name spaces, which can be useful for security implementations. Classes themselves are part of a package. By handling things this way, rather than with global variables, name space conflicts are avoided in environments such as the Internet where the loading of dynamic, modular applications is prevalent.

A naming scheme was proposed, based upon Internet domain naming conventions, to provide unique package naming that included the name of the organization developing or providing the package. Thus, a unique package name might be:

    com.companyname.jdbc.coolapp

or, as some developers have chosen:

    companyname.javascript.coolObject

The names *java* and *sun* are reserved by Sun Microsystems.

There was some support for this concept and also some controversy. In general, developers have been following the guidelines for Java name space, but in some instances, the Java community has expressed a preference that core applications be placed within name space conventions with shorter, more generic names rather than the longer, company-linked names. See Java, Java telephony API.

Java Native Interface JNI. A Java native programming interface that ensures portability of Java applications across different platforms supporting Java. It is available with the standard Sun Microsystems Java Development Kit (JDK). Since there always seem to

J

Sampling of Java Applications Programming Interfaces	
Java DPI	**Description**
Java Media API	Java media applications programming interface.
Java Security API	The Java applications programming interface (API) for building authentication through digital signatures and other low- and high-level security features into Java programs. Support is provided for key and certificate management, and access control data. This provides a means for Java applets to be "signed" to ensure authenticity.
Java Speech API	JSAPI. The Java object-oriented open API for speech. Specifications for the development of speech recognition and synthesis applications. JSAPI supports speech dictation systems, employing very large vocabularies and grammar-based speech interactive dialog systems (command-and-control). The API provides three basic types of support: resource management, a set of classes and interfaces for a speech recognition system, and a set of classes and interfaces for speech synthesis. Related functions, speech coding and compression, are handled by the Java Media Framework and Codec support.
Java Telephony API	JTAPI. The Java telephone API designed to provide portability of telephony applications across applications and across different hardware platforms. JTAPI is a sanctioned specification extension to Java that is used in conjunction with toolkits (such as Lucent's Passageways and Sun's JavaTel), to serve as a guide for the creation of applications. JTAPI was jointly developed by Sun Microsystems, IBM, Intel Corporation, Lucent Technologies, Novell Corporation, and Nortel Corporation.

be a few platform-specific functions that people like to use, the JNI is intended to take advantage of functionality on a specific platform that is not within the Java Virtual Machine (JVM) environment. It enables native code (e.g., C++) to be integrated into Java applications. See Java, Java Virtual Machine.

Java telephony API JTAPI. Applications development tools based on the Sun Microsystems Java programming language that enable portable Java applications to set up, manage, redirect, and otherwise administer telephone calls handled through digital data networks. JTAPI was developed by IBM, Intel, Lucent, Nortel, Novell, and Sun Microsystems.

JTAPI implementations provide the interface between Java telephony applications and hardware or software telephony services. JTAPI provides a means to access telephony Call Control, Physical Device Control, Media Services, and Administrative Services.

JTAPI is an extensible, scalable specification appropriate for communications in first-party call control in consumer devices to third-party call control in distributed call centers. JTAPI development was begun in the mid-1990s by a consortium of computer and telecommunications companies who desired a portable, object-oriented means to integrate computers and telephony call control. JTAPI version 1.0 was released in October 1997 and version 1.3 was endorsed by the Enterprise Computer Telephony Forum in July 1999.

JTAPI is a java extension package comprising the classes, interfaces, and principles of operation in the javax.* name space (e.g., javax.swing).

JTAPI makes it possible to create applications that interact with and control telephone services. This is of interest to many developers, consumers, and business users of telephone services. The more obvious applications include call management, logging, dialing, and tracking software. Automated voicemail, facsimiles, and document distribution programs are also of interest. But there are also likely to be new and novel Internet telephony and personal digital assistant programs developed and designed to interface computer and telephone technologies in ways not previously possible.

Since JTAPI does not encompass every signaling protocol and since there is no way to anticipate every possible JTAPI application, some of the more innovative applications will require interfaces to extend and supplement the JTAPI specification.

JTAPI can run on top of existing telephony standards, including TAPI, TSAPI, Callpath (IBM), and SunXTL.

JavaTel is Sun Microsystems' JTAPI runtime environment for the Sun platform. See Java, JavaTel, javax name space, Telephony Application Programming Interface.

Java Virtual Machine JVM. Software routines for interpreting Java bytecodes into machine code. This interpretation/conversion process makes it possible to run Java applications on many different platforms. Each computer hardware architecture has a different way of interpreting programming instructions, based

on the central processing unit and its support systems. If you have a software program running indirectly within a virtual environment instead of directly on the host platform, a way to convert the program instructions to those expected by the host processor is needed. The JVM enables Java portability across many different systems. See Java.

Java XML, JXML An area of development and a mailing list devoted to Java and XML, particularly Java Class and Bean metadata expressed as CML documents, conversion of metadata to bytecodes, reversible conversion of Java Object Streams to XML documents, and other related issues.

JavaBeans A Sun Microsystems Java language object-oriented, platform-independent security model included in JDK. See Java.

Javan, Ali (1928–) An Iranian physicist of Azerbaijani descent, Javan has lived in the U.S. since 1949. In 1960, Javan invented a helium-neon gas laser, the first laser to emit a steady beam of light and the forerunner of electrical discharge pumped gas lasers. He was awarded The Franklin Institute's Ballantine Medal in 1962 for his achievement.

In 1975, Javan received the Fredric Ives Medal from the Optical Society of America. Javan founded Laser Science, Inc. in 1981, to develop and construct laser-based systems such as atomic clocks and optical communications systems for government agencies. The company was merged into Thermo Electron's Photonics Division in 1997. See Dicke, Robert; laser history; Patel, C; Townes, Charles.

JavaScript A cross-platform, scripted, open standard programming language familiar to most through the implementation incorporated into Netscape Web browsers. It is only superficially similar to Java, being slower and having a simpler syntax and limited functionality.

JavaServer Pages JSP. An industry collaboration project lead by Sun Microsystems to enable Web developers to develop and maintain dynamic Web pages for integration with existing business systems. JSP enables the development of platform-independent Web-based applications. It separates the user interface from the content generation so that changes in layout don't change the underlying content. JSP uses XML-style tags and scripts written in Java. Formatting tags (HTML or XML) are passed back to the response page. JSP is an extension of the Java Servlet technology, platform-independent Java server-side modules that fit into a Web server framework to extend the capabilities of the server with minimal overhead. JSP specifications are freely available to the development community so that Web servers and applications servers can be JSP-enabled. See Java.

JavaTel A platform-independent, scalable telephony applications toolkit based on the Java Telephony API, introduced by Sun Microsystems in 1997. JavaTel was designed to support computer telephony integration (CTI) by enabling the development of Java-based call center, voice response, Internet phone, and management applications. Thus, Java-based computer telephony applications can run on any Java-enabled device, rather than being constrained to CTI applications that run on SunXTL, Sun's proprietary implementation. JavaTel operates with the Solaris operating system and will run on top of Sun's earlier CTI implementation, the SunXTL system. See CallPath, Java Telephony API.

JBIG Joint Bi-Level Image Experts Group. A group formed after the JPEG group to concentrate on the task of lossless compression of bilevel, one-bit, monochrome images such as those commonly generated by printers, fax machines, etc. It is officially the ISO/IEC JTC1 SC29 Working Group 1 and is responsible for both JPEG and JBIG standards.

The JBIG format incorporates discrete levels of detail by successively doubling resolution. The image is divided into strips for processing, each with a horizontal bar and a specified height, with each strip coded and transmitted separately. The order and characteristics of individual strips can be specified by the user. The image can then be progressively decoded, one strip at a time, as received.

Once an image has been segmented according to strips and specified parameters, the resulting bilevel bitmaps are compressed with a Q-coder. Two contexts are defined by JBIG, the base layer, which is the lowest resolution, and the remaining differential layers. These provide contexts for optimization of the compression.

The JBIG format works well with the many common bilevel images that include text and line art. It is an accepted standard as ITU-T T.82. The JIBG2 standard, which represents work since the original JBIG specification, has been released as an International Standard (IS 14492). See JBIG, color; JPEG; MPEG.

JavaWallet A family of products developed in the Java programming language for enabling secured electronic commerce transactions. JavaWallet incorporates the Java Commerce Client, Commerce Java-Beans components, the Gateway Security Model, and Java Commerce Messages, which may be used independently of one another and may be bundled with other applications. JavaWallet may be used in Java-enabled browsers, as well. See Java.

JBIG Alliance Another name for the JPEG and JPEG committees officially known as the ISO/IEC JTC1 SC29 Working Group 1, sometimes abbreviated as ISO SC29/WG1. See JBIG.

JBIG, color; COLOR-JBIG A project to develop a JBIG-based, lossless, decompression system for document image processing for a variety of types of documents including bitonal, grayscale, and color. This is an interesting direction, since the original JBIG concept was to create, in a sense, a lossless, monochrome version of the JPEG file format. However, it is clear that a lossless color format has many applications (including commercial graphics, medical images, business documents, etc.). What remains to be seen is how a new format can improve upon the sophisticated and well-supported Tag Image File Format (TIFF) file format, which already supports lossless compression of monochrome, grayscale, and color images.

Since JBIG is an ISO standard, a color version of JBIG is of interest to the European community. Presumably the developers feel that there are capabilities and aspects of JBIG not already supported by TIFF. See JBIG, Tag Image File Format.

JBOD See just a bunch of disks.

JCL See Job Control Language.

JDBC Java database connectivity. This is a product from Sun Microsystems that facilitates the linking of Java programs to tabular databases. It provides connectivity to a number of standard database formats, including SQL, common spreadsheet formats, and flat files. See Java.

JDC 1. Japan Digital Cellular. See Personal Digital Cellular. 2. Java Developer Connection. The Sun Microsystems support forum and interactive message board for registered Java developers. (Nonregistered developers can read message, but not post to the message board.). 3. Journal of Design Communication.

JDS Uniphase Corporation A public company formed by the merging of JDS Fitel and Uniphase, JDS is a significant distributor of wavelength division multiplexing modules, monitors, and connectors for fiber optic cables.

In June 2002, the company announced a smaller semiconductor optical amplifier (SOA) for fiber-based communications links operating in optical C-band frequencies. The component has a unique integrated polarization-independent optical isolator meeting Multi-Source Agreement (MSA) standards.

JECF See Java Electronic Commerce Framework.

JECS See Job-by-Email Control System.

JEDEC Joint Electron Device Engineering Council. JEDEC was originally formed as the Joint Electron Tube Engineering Council (JETEC) in 1944. JEDEC is a standards-developing body of more than 300 member companies representing the electronics industry as part of the Electronic Industries Alliance (EIA). http://www.eia.org/jedec/

JEDI See Joint Electronic Document Interchange.

JEDIC Japan Electronic Data Interchange Council, Japan EDI Council. An interdisciplinary council of member organizations, including electronics organizations, manufacturers, software developers, and trade associations. It was formed in recognition of the importance of electronic data interchange (EDI) to Japan's consumer and industrial infrastructure and to encourage a common awareness and purpose similar to that fostered by the open EDI environments in Europe and North America. The JEDIC fosters educational, internationalization, and standardization efforts. http://www.ecom.jp/jedic/

JEEVES See Ask Jeeves.

JEEVES DNS Resolver A significant pioneer network domain name resolver (DNS) developed by Paul Mockapetris which was the precursor to the widely used Berkeley Internet Name Domain (BIND). See Berkeley Internet Name Domain; CHIVES DNS Resolver; Mockapetris, Paul.

JEIDA See Japan Electronic Industry Development Association.

JEMA See Japan Electronic Messaging Association.

JEPI See Joint Electronics Payment Initiative.

jerk A measure of the rate of change of acceleration, in other words, the first derivative of acceleration, similar to the relationship between speed and velocity and velocity and acceleration.

JET See Just-Enough-Time.

JFIF A minimal implementation of the JPEG family of image compression methods. This is often the implementation incorporated into Web browsers. See Joint Photographics Experts Group.

JHTML 1. An open source cross-platform HTML editor written in Java by Riyad Kalla. 2. A Mac-based plugin application for the Jedit text editor that enables it to generate HTML code. 3. See JavaServer Pages.

jiffy A unit of time equal to 1/60 of a second (North America), or 1/50 of a second elsewhere. Since the proliferation of computers, other definitions of a jiffy have been used, such as 1/100 second or a clock tick in the CPU. The term is most widely used in the film and video editing industries for editing timing purposes. See SMPTE Time Code.

jiffy box A box enclosure for mounting electronics for easy access. A jiffy box can be "opened in a jiffy" (quickly) because it generally slides or snaps open or is open-ended rather than being secured with screws. This is useful for prototype electronics that may change frequently or components that must be attached to others. The box may have ridges or bays for quick mounting of circuit boards.

Jini A Sun Microsystems network technology for providing a simple modular infrastructure for delivering platform-independent network services and for facilitating spontaneous interactions among programs using these services. Jini architecture, released in the late 1990s, typically consists of servers and clients registered with a lookup service. Upon registration, a client can specify the needed servers.

Some interesting applications have been developed with Jini technology, which can be adapted to applications that are computationally intensive and require the resources of a network of computers. For example, researchers in the Computer Graphics and Scientific Visualization Group in Italy used Jini to evaluate a novel distributed computing environment for scientific visualization (e.g., the modeling of fluid motion over wing structures).

In July 2001, GroupServe announced a Developer Web Site with expanded services for Jini technology, augmenting those offered by Sun. These include Jini-based email, database, and transaction interfaces. The services are accessed by proxies downloaded by the Jini Access Module (JAM). They run locally, remotely, or as shared resources. See Java, Jini Community.

Jini Community JiniCom. A community of Jini network services human and technical resources, JiniCom aids members in Jini development and hosts numerous community projects. It distributes the Jini Technology Core Platform Compatibility Kit (TCK) for testing Jini services for compliance to the Jini specification. To date, Jini standards distributed through the Community include the following, and more are in the draft standard stage (e.g., Internet Protocol interconnect standards):

Technology Core Platform Specification – Specifications for discovery and join protocols and formats, entry methods and templates, distributed leasing and events, transactions, and lookup services.

Helper Utilities and Services – Specifications for a set of standard helper utilities and services which extend the Jini Technology Core Platform. They encapsulate desirable behaviors in the form of reusable components to simplify the server/client development process.

JavaSpaces Service Specification – A distributed persistence and object exchange mechanism for code written in Java.

See Java, Jini. http://www.jini.org/

JIOA See Japan Institute of Office Automation.

JIPS JANET Internet Protocol (IP) Service. See JANET.

JIRO A development architecture for resource management distributed by Sun Microsystems as an extension to the Java platform. It provides an open, dynamic, extensible, scalable, network-centric management framework that can be integrated as a platform-independent system. It allows complex distributed environments, such as storage area networks (SANs), to be interconnected and managed. It does this through a standard management domain, including management services for logging, lookup, scheduling, events, security, and transactions through a standardized interface. The system locates and communicates with the services as FederatedBeans components (a cooperative concept based on the JavaBeans idea). See Java, Just-Enough-Time.

JIT See Just-In-Time.

JITT Just-In-Time-Training. A laptop-based intelligent feedback training project for astronauts and flight operators. The program was established in the mid-1990s by the NASA/Johnson Space Center.

jitter 1. Random or periodic signal amplitude or phase instability or degradation of relatively short duration. Jitter arises from various causes, including poor connections, overly long cables, incompatibilities between software and hardware, or weather. See wander. 2. Random or periodic temporal variations of short duration in a data stream. This is essentially a timing problem with relation to a clock source. With increased demand for wideband data services such as full broadcast video, reference clocks and related transmission signals become important aspects of data communications but can also provide one more source of signal interference. When expressed diagrammatically, jitter can be visualized as small timing differences between a reference clock representing the ideal signal and the jittered signal. At first glance, the two timing diagrams may look the same, but closer inspection reveals small deviations from the ideal clock backwards or forwards in time. On an oscilloscope, the jitter signal will be just slightly out of phase with the image and position of an ideal, expected signal. Networks may be tuned to tolerate a certain amount of low-level jitter, but persistent or high amplitude jitter should be investigated or corrected. See jitter, network; wander. 3. Unstable or erratic display on a television or computer monitor where the image deviates slightly but noticeably from the expected pattern in small jerky or wavy motions.

jitter tolerance Since jitter is a persistent possibility in high-performance network systems, especially those running through a number of interface devices from different vendors, many systems will specify a certain tolerance for jitter and will correct for jitter within certain parameters, when possible. The terms of the jitter tolerance depend upon the type of system, but may be specified in lost bits, timing disparities, or other characteristics. Since jitter can be transferred to an adjacent connecting device and thus can increase from one component to the next, the sum total of the jitter effects must also be below that which the system can tolerate.

jitter, network In networks, jitter refers to a number of problems arising from demultiplexing, incorrect physical connectors or regenerators, and latency times between consecutive transmission packets.

When data are serially transmitted, as is common in data networks, timing is a means to synchronize the data stream so the receiver can interpret, convert, or otherwise process the incoming information to make sense of the data and recover them for use on the local system or for conversion or forwarding to another system. The success or failure of this timing synchronization is partly dependent upon knowing the jitter characteristics of the transmission from previous experience or by dynamically analyzing the jitter in the incoming data stream and extracting useful information. This may be direct information such as the reference level and frequency of the signal or may be calculated to derive other information such as waveform characteristics and clock periods.

In SONET and other high-speed networks, timing is quite important and lack of synchronization can cause fluctuations in the data packets with respect to the reference clock cycle. This type of phase variation can be filtered with adjustment mechanisms. Jitter specifications for SONET network interfaces are described in ANSI T1.105.03-1994, and for computer networks in general in ANSI T1.102-1993. See jitter.

JitterTrack of Time Interval Error JTIE. A testing and diagnostic tool for measuring clock characteristics against a reference value with respect to short duration signal phase instabilities (jitter). The reference is measured over a specified time interval and evaluated for phase characteristics and anomalies. In networks, the TIE is typically measured in nanoseconds. See jitter, network.

JNT See Joint Network Team.

job In computer operations, a process submitted for later execution. The term was borrowed from factory terminology in the days when computers were large, slow, and very expensive to operate and maintain. Thus, demands for computing time exceeded resources, and it could take days or weeks for a job

(a computer program) to be processed, executed, and returned to the person who submitted the job (usually on punch cards or paper tape). In those days, jobs were commonly processed in queues, sometimes according to various priorities, and eventually returned to the user. Since paper media were frequently used to store the results, the finished jobs, along with the original program, were often sorted into cubbyholes in the same manner as postal mail.

Computing has changed. Systems are now fast and numerous and employ multitasking architectures. The term *job* is now mostly associated with background tasks and low-priority processes, or batch files that run in the background while the user continues to use the machine for other applications. Specialized applications and intensive scientific applications are still processed as jobs in the sense that they are submitted to an organizing authority (e.g., a server) and may take a long time to process, so the term is not outdated, but is much less frequently used.

Job Control Language JCL. A programming language for providing user instructions to a computer operating system, usually in the format of an interpreted scripting language. Although the phrase is now used generically, it was originally developed as a control language by IBM for the control of programs on older IBM batch-based computing systems.

Job-by-Email Control System JECS. A software application to facilitate communication between a remote computer (server) and a home or office computer. In other words, it enables a task to be emailed from one location to another, processed, and emailed back, rather than having it run realtime over a long-distance link. This is similar to batch processing in the days of timeshare systems, except on a larger scale over the Internet. It is an important concept and many jobs may be handled this way over large distributed networks in the future.

Jobs, Steven P. (1955–) An early entrant to the microcomputer industry, Steven Jobs began as an employee of Atari at the age of 17, hired to do video games development. Through the Home Brew Computer Club, he met Stephen Wozniak, an electronic hardware enthusiast, who was working as an engineer for Hewlett-Packard. Wozniak was designing telephone access devices and homebrew computer projects, and Jobs became interested in the business potential of these designs.

By 1976 Jobs had left Atari, and he and Wozniak together created a new company called Apple Computer. They were planning to sell a microcomputer in kit form, a project probably inspired by the Altair, a humble little history-making microcomputer first released as a kit in 1974. Both Jobs and Wozniak had a strong orientation and commitment to educational markets.

Despite his youth, Jobs displayed a futurist orientation, charismatic personality, and marketing flair. These traits have continued to keep him in the headlines for more than 20 years. Apple gained a foothold in the industry, and John Sculley was recruited to head the corporation. Under Sculley's leadership, Apple

J

became a billion dollar company and, as it grew, the two Steves receded into the background due to company growing pains, personal interests, and differences of opinion with the corporation, although not before becoming millionaires while in their twenties. Jobs left Apple Computer and founded NeXT, Inc. in 1985. This company designed some of the best computing hardware and software available in the 1980s. The elegantly simple hardware, robust operating system, stunning graphical user interface, straightforward built-in networking capabilities, Unix underpinnings, and various software utilities were as good as or better than many systems being sold a decade later. The NeXT hardware and operating system was aesthetic, well conceived, and reliable; business owners, frustrated with the limitations of current business computers, watched with a keen eye when the NeXT computer was released in 1987. Unfortunately, by not cultivating the early interest from the business community and targeting education almost exclusively, Jobs may have made one of his biggest mistakes.

The NeXT corporation was acquired by Apple Computer in 1996–1997. Interest in the NeXT in 1997 was due at least in part to its very good graphical user interface and integration with Internet services, which were now becoming important to consumers. By the mid-1990s, 8 years after its introduction, the consumer learning curve had improved and users began to appreciate the NeXT design and concept. Jobs' brash assertion in the 1980s that the NeXT was the computer for the '90s turned out to be more truth than bluster. Much of the NeXT philosophy is now incorporated into Apple computers.

A year before the NeXT was released, in 1986, Jobs purchased the computer division of Lucasfilm, Ltd., and incorporated it as an independent company called Pixar, cofounded with Edwin E. Catmull as vice president and CTO. Jobs has long been chairman and CEO of Pixar, a creative software, multimedia, motion picture company which made history with the Academy Award-winning "Toy Story," a computer-generated full-length motion picture distributed in 1995 by Walt Disney Pictures.

After a few years of quiet creative work, Steve Jobs' name again splashed across headlines in 1997 when Apple bought NeXT. Jobs was back as an executive at Apple, acting in an interim capacity, and speculation about whether he would again head Apple kept reporters on their toes. The management change and publicity created a flurry of activity at Apple, and stocks reacted accordingly. Jobs' return to the limelight showed that public interest in his activities hadn't declined after more than two decades. The revival of Apple Computer, at a time when analysts were predicting its demise, is in no small part due to Jobs' presence and creative inspiration.

Steven Jobs has a philosophical bent, as can be seen from his keynote speeches and interviews with major computing magazines, and it seems clear that his commitment to education and to harnessing the creative potential of computers for improving human lives is sincere. It is likely that he will never be far from the creative computing activities that will occur in the future, and will probably, in fact, be the inspirational force for many innovations yet to come. See Apple computer; Wozniak, Stephen.

JOFX Java Open Financial Exchange. A Java-based toolkit from Xenosys Corporation for developing Open Financial Exchange (OFX) applications and applets. JOFX is part of the LiveBusiness Foundation Classes for Java (LBFC), a set of Java frameworks, libraries, etc. for e-commerce. See Open Financial Exchange.

JOHNNIAC A historic large-scale computer built by Willis Ware, the JOHNNIAC was unveiled in 1954 by the Rand Corporation. Significantly, the first operator of the JOHNNIAC was Keith Uncapher, who became the first chair of the IEEE Computer Group, now the renowned IEEE Computer Society. See ILLIAC, MANIAC.

Johnson, John (ca. 1910s–) An American member of the U.S. Army who made significant contributions to the understanding of night vision image intensifying technology in the 1950s. Johnson's findings came to be called the Johnson Criteria and guided future developments in night vision for a variety of applications.

Johnson was also a pioneer in fiber optics, working in the Army Corps labs under Robert Wiseman who realized the potential of the technology for lengthening lightguides after hearing a lecture on the subject. Johnson approached American Optical, one of the leading lens firms at the time, about optical fibers, but was referred to W. Hicks, who had left the firm to form his own company. Johnson's concept and Hicks inventive skills turned out to be a good match, resulting in the development of fiber pulling, fiber-based vacuum assemblies, and fiber array faceplates. See Hicks, W.

Johnson, John Bertrand (1887–1970) A Swedish-born American physicist, Johnson developed the first sealed-cathode commercial cathode-ray tube (CRT) in 1922. He made important observations of thermal noise while working at Bell Laboratories in 1927 and described his observations in *Physical Review* (July 1928). The phenomenon came to be known as Johnson noise. Johnson was selected to receive the IEEE David Sarnoff award in 1970 for his contributions to electronics and communications. See Johnson noise.

Johnson, Reynold B. (ca. 1906–1998) An American research scientist and founding manager (1952) of the IBM Almaden Research Center. He later became president of Education Engineering Associates, Johnson pioneered the development of magnetic disk technology and computerized educational systems. A prolific inventor, Johnson received more than 90 patents in a range of communications systems, educational technologies, and magnetic storage devices. Johnson was elected to the National Academy of Engineering in 1981 for his contributions in engineering innovation and educational leadership. In 1986 he received the National Medal of Technology, followed two years later by the IEEE Computer Pioneer Award. The Reynold B. Johnson Information Storage Award was established in his honor in 1992. See RAMAC.

Johnson noise In electronics, heat-based agitation of electrons in conductors creates low frequency noise in the circuit. In communications circuits, the amount of noise is related to the receiver bandwidth and source temperature. Johnson noise is sometimes also called *thermal noise* and is characteristically emitted by all objects with temperatures above absolute zero. An understanding of Johnson noise is important to the design and production of antennas and to noise processing and filtering techniques in communications. See Johnson, John B.

joining In computing, the process of combining data files or streams. In the context of packet communications, the reassembly of packets that have been received disassembled. The term also refers to joining a conference, chat, or network community.

Joining is one of the most important and ubiquitous functions used in data transmissions. It is very common for data to be chopped up into pieces as it is routed, especially if there is a limit to file sizes in the sending or receiving systems. Join utilities are also used in connection with large files stored across more than one floppy or more than one hard drive partition, or computer. What is split apart usually needs to be rejoined when the data are accessed or moved later on.

In packet communications, individual packets from a larger data file are not necessarily transmitted through the same route in a distributed network. The concept of splitting the packets and sending them through many routes arose in the days when the U.S. military was looking for a means to safeguard data transmissions in the event of an offensive strike. It was proposed that if the data were traveling through different routes, it was less likely that the entire content would be lost. In subsequent computing applications, this was found to be a good model for many aspects of communications, including email, file transfer protocols, and much more, so joining the separate pieces of the communication at the destination became an important function of a system.

In file management, individual parts of a file are often stored where there is sufficient room and are not necessarily contiguous. *Pointers* are used as a virtual joining mechanism to tell the file retrieval algorithms where to look for the next "chunk" of data. Thus, the joining of a block of data that has been stored in separate sections often happens when an application is run and the program requires the entire contents of a graphics or text file.

In telephony and online communications, joining refers to entering a live communications venue such as a conference call, an Internet Relay Chat (IRC) talk session, or other community communication. Specific steps or commands for joining usually must be adhered to, especially in public discussions where standardized commands make it easier for people to join or leave.

Joining is an important function in computer workgroups. When users on several different computers are sharing or updating the same database, for example, it is important for the applications and operating system to keep track of who is joined into the workgroup application so that searching, retrieval, updating, and other common functions are handled so that one user doesn't wipe out the corrections or additions of another user.

In client/server applications, certain protocols and procedures can be put in place for a client to join a network system or specific process. This is a resource management tool to allocate resources on a more efficient or as-needed basis to conserve computing resources. Thus, many Internet Service Providers (ISPs) that provide 24-hour connections to the Internet, for example, may actually time out a user when a system is idle in order to allocate that user's Internet Protocol (IP) number to another user. When activity is detected from the first user, the system must again join the network and a new IP number will be dynamically allocated to allow the user to rejoin the Internet. This process is typically transparent to the user.

joint 1. Connection between two or more conductors. This may be a chemical bond, solder joint, or wires touching, clamped, or wound together. 2. A joining part, or space, between two sections, nodes, or articulations. 3. A junction where two or more structural members are combined.

Joint Bi-Level Image Experts Group See JBIG.

joint cache A shorter-term storage cache shared by a larger base of users or networks. Some significant joint cache proposals have been based on newly developed models for distributed network systems. Based on the premise that the Internet is a system as a whole accessed by joint users, a joint cache on the Internet is a mechanism for handling Internet traffic caching based on virtual rather than machine-specific or local-network-specific models.

As an example, a dynamic joint cache system was described by Dolgikh and Sikhov at the TERENA Networking Conference 2001, based on Zipf-like distribution. The model was based on research at the Samara Region Network for Science and Education. The authors proposed and tested an analytical model of a cache system that can be used to determine the scope, frequency of requests, and maximum efficiency levels for the most requested document in a cache system. While a simple example, this is an important basic concept that can be generalized to many Internet resources, including popular, high-traffic Web pages. This and similar efforts to conceive top-down structures based on the Internet as a whole, rather than bottom-up services based on individual networks or computers, are an important trend in the Internet development community, where the sum of the parts is seen as a larger resource that should be accessible to the greater Internet community. If the trend continues, computing applications that were impossible on smaller systems may emerge as possibilities on a global distributed network.

joint circuit Shared communication link.

Joint Electronic Document Interchange, UCL-JEDI JEDI. A project to survey, identify, and test a number of formats for electronic document interchange with an eye to standardizing the research community and facilitating the process therein. The project was initiated as the result of a call for proposals announced by UKERNA in 1994. The participants are studying popular interchange formats for word processing in academic and commercial environments. The project aims to identify format conversion methods and the relationships between *de facto* and internationally recognized standards.

Joint Electronics Payment Initiative JEPI. An idea initiated by CommerceNet and W3C in 1995, JEPI was aimed at developing and demonstrating payment-selection, negotiation, and purchasing electronic commerce scenarios in order to build a commerce mechanism practical for use in real world applications and which could be published as an open standard.

The initial inspiration for the technology came from Eastlake's Universal Payment Preamble and the W3C's PEP technology for HTTP transmissions payments. The Open Software Foundation and the Financial Service Technology Consortium were also involved in the early stages. The project was organized into four main groups including Browser Technology, Server Technology, Payments Systems, and Merchant Systems. Companies joining the effort were expected to commit to implementing the negotiation protocol in a product for testing.

By August 1996, the project had progressed to where Internet Draft documents were distributed in RFC format, in view of an eventual release of an IETF specification of the Universal Payment Preamble (UPP) and the development of an HTTP Extension Protocol (PEP) in the HTTP Working Group. By August 2001, the W3C working group was specifying and recommending the syntax and processing parameters for XML signatures. The project is ongoing. See e-commerce.

Joint Intelligence Virtual Network JIVN. A U.S. Government network for providing round-the-clock TS-SCI multimedia communications, including secured videoconferencing. This is one of the multimedia services supported on the Joint Worldwide Intelligence Communications System.

Joint Network Team JNT. An organization founded in March 1979 in the U.K. by recommendations of the Computer Board and Science Research Council (SRC) to study the networking requirements of the academic community and make proposals. The role was transferred to UKERNA April 1, 1994.

Joint Photographic Experts Group JPEG (*pron.* jay-peg). The Joint Photographic Experts Group was founded in 1986 to develop a standard for the compression of still, continuous-tone images. Soon after its formation, its goals were adopted jointly by the International Organization for Standardization (ISO) and the International Telegraph and Telephone Consultative Committee (CCITT), now the ITU-T. Research proposals for such an image compression scheme were solicited internationally, with a dead-

line of March 1987. By January 1988, the evaluators had narrowed down the suggestions and selected an Adaptive Discrete Cosine Transform method, culminating in a new standard described in ISO 10918-1 Recommendation T.8. Following the publication of the draft standard, work began on improving compression ratios further, and providing scalability. See JPEG file format.

joint pole, joint utility pole A shared telephone pole resource established in the early telegraphy and telephone days. When there were many small switchboards, rather than one large telephone provider, wires were everywhere, along with many poles to support them. It was not uncommon for individual wires to be running to each business and even to each neighboring business. Utility companies and subscribers and those with private lines quickly realized that the sharing of a telegraph/telephone pole had advantages, including cost, ease of maintenance, and aesthetics. Thus joint poles were designated for carrying transmission lines belonging to more than one entity. However, coordination of shared resources became an issue, so joint pole agreements and joint pole committees were established throughout the country in the early 1900s to manage joint poles. As electrical power superseded candles and oil lamps, electrical wires were also slung along telegraph poles, which came to be known as utility poles.

Utility poles are still with us and the demand for places to string communications lines has increased dramatically since the mid-1990s. Cable TV and fiber optic Internet access cables now share space with the telephone and electrical lines on utility poles. Deregulation has also complicated the administration of poles as to who is responsible for their installation, maintenance, and use. Joint pole committees are still important, perhaps more than ever. New communications companies, seeking to establish services at reasonable costs, have a vested interest in joint pole agreements and have shown interest in being involved in joint pole committees in their distribution areas.

Some relatively standard configurations for joint poles have developed over the years. In general, the poles are strung in a hierarchical arrangement from top to bottom, based upon the the electrical characteristics of the lines strung and the frequency with which they may need to be changed or serviced. Since ground wires aid in deflecting energy discharges from lightning and rarely need to be serviced, they are placed at the top. Moving down the pole, various primary and secondary transmission wires are connected. The more recent cable TV or fiber optics bundles running between poles are usually fat, well-shielded cables attached near the bottom, often running in bundles held together with short straps.

On poles with several types of transmission lines, certain distances are maintained between different types of wires to reduce electromagnetic interference. With fiber optic cables, which use light rather than electricity to transmit information, electrical interference isn't a problem except at points where the fibers are connected through electrical amplifiers or where the signal is converted to run on wires. However, optical fibers are dependent upon carefully designed splices and joints and, when bundled, don't have the same bend tolerances as small wires, so these factors need to be taken into consideration for fiber more than for wire. The design of insulators and connecting mechanisms on joint poles is also dependent upon the types of cables and power distribution levels carried on the pole. See joint trench, Joint User Service. See the joint pole illustrations and more detailed explanation on the following pages.

Joint Procurement Consortium JPC. A Bell consortium composed of a number of regional Bell holding companies including Ameritech, BellSouth, Pacific Bell, and SBC Communications, which reviews telecommunications product offerings and makes recommendations. In 1996, the JPC signed contracts with Alcatel for ADSL equipment for use over twisted copper pair networks as an alternative to fiber.

Joint Technical Committee JTC. The JTC is now called JTC 1. It is an International Standards Organisation/International Electrotechnical Commission (ISO/IEC) information technology standards body concerned with the specification, design, promotion, and development of systems used for the capture, representation, and processing of information. http://www.jtc1.org/

joint trench A means of aggregating cable installations so more than one department or company can share space within a single conduit or other wiring distribution system to save money and to limit the number of individual conduits installed in public areas. For utility services, guidelines and regulations require that other companies using a joint conduit must be contacted before any street upheaval or digging is undertaken. This is important in order to limit the disruption that inevitably occurs when major line changes or installations are made under or near public streets. See joint pole.

Joint User Service A tariffed, Federal Communications Commission (FCC) system for buying or otherwise sharing telecommunications services by mutual agreement. Local public utility service regulations have restrictions on how certain services may be shared and may require that all associated users be identified. See joint pole, joint trench.

Joint Worldwide Intelligence Communications System JWICS. A U.S.-secured global multimedia intelligence communications system. The system replaces the Defense Data Network's (DDN's) DSNET3 as the Sensitive Compartmented Information (SCI) portion of the Defense Information System Network (DISN) and is intended to facilitate the rapid exchange of audio/visual data. Data includes videoconferencing, graphics and scanned document files, Defense Intelligence Network (DIN) broadcasts, etc. JWICS was initially set up on a switched T1 backbone with T1 and, in some cases, slower connections, with a plan to transition to faster T3. Much of the communication is relayed via satellites. The system is designed to be installed at all major command sites in addition to the availability of portable versions such as the Mobile Integrated Communications System, which uses a self-contained JWICS system packaged into a set of transit cases. The lead contractor for the system is the Defense Intelligence Agency (DIA).

The goal of a secure system is not easily realized. In 1999 there were concerns about monitoring individual user activity on INTELINK as accessed through JWICS. It was realized that there were circumstances where unaudited use could be carried out if local access control (LAC) was not carefully implemented and contained. Until the configuration problems could be solved, it was recommended that government contractor access be limited to authorized individuals accessing the system through sites where U.S. Government or military personnel were available to oversee. See Joint Intelligence Virtual Network.

JOLT Project Java Open Language Toolkit Project. A collaborative effort to produce a freely distributable "clean-room" clone of Sun Microsystems' Java sufficiently compatible to pass Sun's Java validation suite. The participants instituted a plan to develop a development-quality Java compiler, an embeddable Java interpreter with a full class library, and documentation for all the JOLT components. The initial implementation was targeted for Linux/i386. It was also planned to embed JOLT into a freely distributable, full-featured Web browser. See Java.

Jones plug A multicontact polarized receptacle connector.

Josephson effect A quantum effect, which is not easy to explain, but as an example, imagine a nonsuperconducting material, such as a semiconductor or nonconductor, sandwiched between layers of superconducting material, so that the supercurrent tunnels through the nonsuperconductor and can variously be affected by magnetic fields. See Josephson junction.

Josephson junction A fast data technology sometimes used in place of silicon that provides a means to do very fast circuit switching. Josephson junctions can be connected together in series, provided their oscillating properties are matched. This is difficult, but has been achieved in devices called Josephson arrays. Josephson junctions have practical applications in many areas, but are of particular interest to researchers and engineers working with precision voltage metering, microwave electronics, and high-temperature superconductors. Named after British researcher Brian Josephson. See Josephson effect.

Josephson, Brian (1940–) A Welsh-born British physicist who received a Nobel Prize for physics in 1973 for his discovery of the Josephson effect. See Josephson effect, Josephson junction.

Joshi effect 1. In electronics, when alternating current is passed through a gas dielectric condenser and the gas is continuously irradiated with certain wavelengths of light, the associated fall or rise in the current is called the Joshi effect. 2. Similar to a theoretical model called the Prisoner's Dilemma (known to game theorists), Joshi is named for a Reed student from India who proposed a model for behavior on the

Joint Utility Pole Examples

A terminal (end) utility pole. Wires are aerially strung to only one side of the pole and a guy supports the other side. The lines on end poles may be terminated or may feed into a vertical conduit for connection to nearby underground wiring facilities.

A mid-block utility pole in which wires are supported and passed on to the next pole in the path and to the nearby buildings of local utility subscribers. Not all poles have transformers, but three are shown here, in cylindrical metal housings.

Joint Utility Poles

These two photographs illustrate somewhat different utility poles located several blocks apart. The one on the left is a terminating pole, connecting conducting wires on only one side of the pole with conduits for channeling lines to the ground. The pole on the right is a more common mid-block pole that primarily supports the lines and passes them along to the next pole (and to local utility subscribers) without ending the connections or directing them earthward. Both poles include a variety of wires, cables, insulators, and transforming devices.

In spite of their geographic and functional differences, these two poles clearly have many common structural characteristics, arising from various joint utility pole regulations and agreements. Most of the structural aspects have been worked out over the decades based upon electrical characteristics and safety and maintenance needs.

Recently, the role of joint utility poles has become more complex. Not only has the demand for phone services increased, but additional types of media, including copper-based high-speed data lines and fiber optic cables, have been added to the joint pole hierarchy, often severely stretching the space and weight limitations of joint poles, especially in densely populated urban areas. The problems of interference have increased as well. High-speed copper data lines, for example, have special needs that are different from voice lines. The loading coils added as induction structures to reduce attenuation (loss of signal) on voice lines may help facilitate voice transmissions over longer distances but they can also increase interference in high-speed data lines.

Utility Pole General Characteristics

Most utility poles are made of wood, usually cedar or fir, although some regions have poles made of metal or cement. Wood and metal poles are generally between 20 and 30 feet in height. Depending upon the type of pole and the terrain, the setting depth is usually about 4 or 5 feet into the ground. The accessible ground surface around the curb is termed the *grade*.

Joint Utility Pole Regions (Spaces) and Common Components

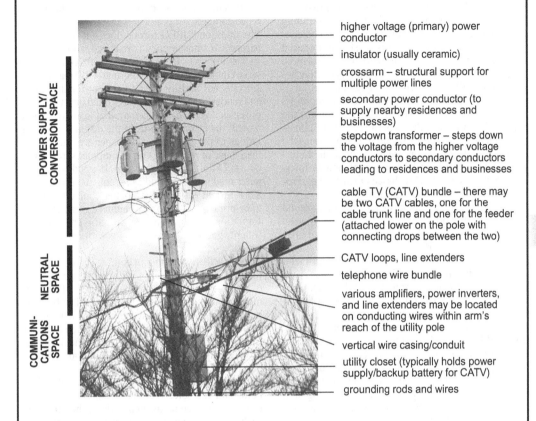

POWER SUPPLY/ CONVERSION SPACE

NEUTRAL SPACE

COMMUNI-CATIONS SPACE

higher voltage (primary) power conductor

insulator (usually ceramic)

crossarm – structural support for multiple power lines

secondary power conductor (to supply nearby residences and businesses)

stepdown transformer – steps down the voltage from the higher voltage conductors to secondary conductors leading to residences and businesses

cable TV (CATV) bundle – there may be two CATV cables, one for the cable trunk line and one for the feeder (attached lower on the pole with connecting drops between the two)

CATV loops, line extenders

telephone wire bundle

various amplifiers, power inverters, and line extenders may be located on conducting wires within arm's reach of the utility pole

vertical wire casing/conduit

utility closet (typically holds power supply/backup battery for CATV)

grounding rods and wires

There are often various cylindrical or box-shaped metal cabinets attached to a pole. These have a variety of functions. They may contain battery or other auxiliary power supplies (usually located higher up on the poles), may include tools or components, or may be cabinets associated with CATV installations (usually located lower on the pole than the CATV cable).

Mounting brackets must be designed and attached so they are strong enough to withstand the effects of wind, or the minor collision of a vehicle with the pole. In general, in urban areas, the pole should be about 5 feet behind the face of a curb and about 6 feet from the shoulder of the road, with differences related to local speed limits and odd intersection situations that may require other solutions. In general, utility poles are not permitted to be placed in traffic islands. Highways generally require greater offsets than city streets.

Hierarchical Organization of Components

While there are variations, in general, higher voltage conductors are located higher up on the pole, for safety and to minimize their effects on data-carrying lines. In most cases, conducting wires are required to be at least 12 feet from the

accessible ground (grade). In some cases, where alleys or driveways are involved, the requirement may be as high as 18 feet.

A static grounded wire may be at the top of the pole, to draw lightning away from the conducting wires below.

The power lines carrying high voltage between power supply substations are located near the tops of the poles and there may be one or more. Secondary power lines for carrying lower voltages to local subscribers are lower than the primary power supply conductors but higher than the communications media. Stepdown transformers between the primary and secondary power lines provide voltage conversion functions.

A multigrounded neutral (MGN), generally located below the primary power supply, is an uninsulated conductor for carrying unbalanced residual current. Various grounding wires and rods run down the side of the pole, as needed. There may also be terminating equipment on end poles, or conduits for connecting an end pole with underground wiring. A grounding rod at the base of the pole may be used to ground the MGN.

stock market that is particularly relevant in light of direct trading activities as they occur over the Internet (as opposed to traditional trading that was more often filtered through brokers). As in the Prisoner's Dilemma, people may well be better off not trading at all (not confessing), but given the social and economic dynamics, it's hard to resist. Joshi has proposed that with so many involved, the market becomes more fragmented and volatile with the result that it may become more difficult to achieve any gains.

JOTP Java Open Trading Protocol toolkit under development for client/server implementations of the Internet Open Trading Protocol (IOTP). JOTP clients are implemented as protocol commerce beans for the Java Wallet. JOTP servers are based upon the Enterprise JavaBean and Servlet component models. Interoperability is through XML-based Open Trading Protocol transaction transport mechanisms.

joule (*symb. – J*) An absolute meter-kilogram-second (MKS) unit of work or energy equal to 10^7 ergs. An SI unit of energy equal to 1 kg m^2 s^{-2}. Named after James Joule. See work.

Joule, James (1818–1889) An English physicist who studied the dynamics and efficiency of various types of engines. Joule demonstrated that when mechanical work is used in generating heat, the ratio of work to heat is a constant quantity. A joule, the absolute unit of work or energy, is named after him.

Joy, William (Bill) One of the codevelopers of Sun Microsystems' Java, attributed with the original idea for the programming language which eventually became Java. In the early 1990s, Joy met with the members of the Stealth Project to develop a language which could create short, powerful programs. See Gosling, James; Java.

joystick A hardware input device that receives and transmits signals containing directional information to a computing device. Commonly used for manipulation of computer software onscreen pointers and selectors. In its most common form, a joystick resembles an aircraft steering control, and may or may not include buttons. It is commonly used for video gaming applications. See mouse, potentiometer, trackball.

JPEG (*pron.* jay-peg) See JBIG Alliance, Joint Photographic Experts Group.

JPEG file format The JPEG image compression format was designed to be used with a wide variety of continuous tone images, without restrictions as to colors, resolution, content, etc. Depending on the software used to generate the file, the format provides the user with trade-off options between compression levels and the quality (lossiness) of the image, and is symmetric, with compression and decompression requiring about the same amount of time and processing power. Common file extension conventions to identify JPEG files are .jpeg and .jpg.

JPEG is not perfect for every type of image. Continuous tone images with many colors generally look good when rendered and compressed with JPEG, in spite of the substantial loss of information and reduction in file size. Crisply rendered images with few colors,

sharp boundaries, and thin lines tend to take on a fuzzy or speckled appearance when compressed into JPEG format, and should probably be processed with a different compression format more suitable for that type of image.

The traditional JPEG format does not support transparency. If transparency is required, another format such as Portable Network Graphics (PNG) or Compuserve Graphics Interchange Format (GIF) may be used. JPEG is not usually the best format for the storage and rendering of images to be printed on a traditional press, as it is a "lossy" format; that is, it does not retain all the information from the original. The resolutions of printed images on paper are much higher than those of renderings on a computer screen (1800 dpi vs. 75 dpi). An image that looks good on the computer may look fuzzy and inadequate on paper. The TIFF format is generally a better choice for images to be printed, as the format retains a great deal of information about the image, while still providing reasonably good compression ratios with common compression schemes.

JPEG is actually a family of compression formats and many implementations of it are quite minimal. For example, JFIF is a bare bones version of JPEG commonly found on the Web, and SPIFF has been formally defined to be upwardly compatible with JFIF. Variations on the JPEG format can often be identified by looking at the first few bytes in the file header. Work is ongoing on the newer JPEG2000 standard. See JBIG, MPEG, TIFF.

jukebox 1. A hardware appliance or peripheral that holds and can selectively or randomly access multiple data storage items, generally of the same type, such as audio CDs, tapes or records; computer diskettes; tapes; or cartridges. For audio applications, 20 to 100 items may be accessible, and access may be very quick, whereas for archival purposes, especially with high-capacity storage tapes, there may be 5 to 20 tapes, and access may be slow. 2. A software tool for selecting and playing digitized audio or multimedia files in the manner of a traditional phonograph jukebox from the 1950s. In fact, many software jukeboxes have user interface designs that mimic old-time jukeboxes. The jukebox analogy is now being extended to multimedia resources on the Web. Users can store their original audio or multimedia files on a Web site and users can access the jukebox through a Webpage interface as a shared resource.

JRG GII Joint Rapporteur Group global information infrastructure. A group of rapporteurs and experts from various ITU study groups brought together to further discuss and coordinate global standards-setting tasks.

JST See Japan Solderless Terminal.

JStamp A commercial battery-operated one-inch by two-inch circuit containing a realtime, native Java technology module developed by Systronix. The matchbox-sized module includes a 32-bit CPU, flash RAM and SRAM, and can handle realtime hardware interrupts. It's based on the aJile Systems Inc. chip design. The low-power-consumption module can operate up to 40 hours, depending upon the implementation. It is suitable for devices such as Personal Digital Assistants running Java applications, including graphics animations. This level of miniaturization is likely to spawn a whole host of carry-around applications that haven't existed in the desktop world due to lack of portability and high relative cost.

JTAG Joint Test Action Group. An international body, founded in 1985, JTAG seeks to develop electronics-related test methodologies and related standards. These standards are then recommended to various appropriate standards bodies, such as the IEEE.

JTAPI See Java Telephony API.

JTC See Joint Technical Committee.

JTIE See JitterTrack of Time Interval Error.

Judge Harold Greene See Greene, Harold.

Jughead Named in association with other Internet tools, including Veronica, Jughead provides a way for Gopher administrators to access and retrieve menu information from the various gopher servers on the Net. They stretched a little to put an acronym to this one, but ended up with "Jonzy's Universal Gopher Hierarchy Excavation and Display." See Archie, Veronica, gopher, Gopher.

Julian calendar A solar-based calendar in widespread use (some traditions use a lunar calendar) in the western world that arose historically from calendar reform instituted by Julius Caesar, after his conquest of Egypt. The Julian calendar is based on a 365-day year divided into 12 somewhat equal months with leap years used to reconcile the difference between the calendar divisions and actual celestial events (like adjusting a clock that runs slightly fast or slow in relation to clocks based on accepted time references). After some initial experimentation with leap years, they were standardized to occur every 4 years.

The actual solar year is slightly shorter than the Julian solar year, which is 365.25 days with the 4-year leap year convention. In the short run, this discrepancy doesn't bother most people but for theologists, farmers, and scientists (especially astronomers), the discrepancy can be important, especially since it adds up: after a century, the calendar is almost a day out of synch with the actual solar year. By the time the Italian Renaissance came about, when intellectual curiosity was high and systematic observations were being strongly established, the calendar was reformed, with input from a number of astronomers and theologians. In 1582, Pope Gregory XIII issued a papal bull establishing the Gregorian calendar, setting October 5, 1582, in the old calendar to October 15, 1582, in the new calendar, to synchronize the administrative calendar with the solar calendar. The day for using the leap year adjustment was changed as well, from February 25 to February 29. In essence, however, it is the same basic type of calendar as the Julian calendar. The Gregorian calendar coexisted with the Julian calendar for centuries, being first adopted by the Catholic communities. Gradually, however, nations switched over to the Gregorian calendar, though many of them not until the 1900s. Historians often have difficulty pinpointing historical dates related to the advancement of science in Europe because of the coexistence of the two calendar systems. In many cases, even when a historical record is dated, it might not be definitely known in which system it was recorded.

jumbo Large, oversized, bulky, etc. in relation to others of its type. Jumbo cables are typically those that are fatter overall or which carry larger numbers of optical filaments. Jumbo chips or LEDs are those that are largest in their category. Jumbo lenses are those that have a wider diameter or thickness than typical lenses.

jumbogroup In general, an aggregation of associated devices, transmission media, lines, or channels. In analog voice phone systems, a group within the hierarchy for multiplexing that has been established as a series of standardized increments.

The CCITT established a jumbo group as consisting of 6 mastergroups, each of which in turn comprises 10 supergroups or 600 voice channels. The ITU-T, successor to the CCITT, defines a mastergroup as comprising 5 supergroups, whereas in the U.S., commercial carriers follow a standard of a mastergroup comprising 10 supergroups. (Confused yet? The Voice Circuits Hierarchy diagram should help.)

More recently, a jumbogroup has been described as the highest FDM multiplexing level that contains 3600 voice channels as contained within 6 master groups. The important thing is to remember the jumbogroup's place in the telephone service hierarchy. That way, if capacities change and jumbogroup aggregations change, the relationship to other groups is still understood.

A jumbo group is occasionally called a hypergroup.

jump hunting In telephony, a means of searching for an available trunk or extension in nonconsecutive order by dialing the first number of the trunk to indicate which trunk to search. Also called nonconsecutive hunting.

jumper 1. A temporary or customized connection used to bypass or reroute a circuit, frequently in the form of a wire, often a short one; common on printed circuit boards. Jumpers are used to test circuits, correct errors, or make last minute changes in circuit board manufacture. They may also be used to set operating attributes on configurable peripherals such as hard drives (e.g., SCSI settings). Sometimes wire jumpers are equipped with alligator clip heads to facilitate quick connection. 2. A small, paired electrical protruberance on a printed circuit board, intended for configuring a circuit with settings that are infrequent. This type of jumper is common on hard drives, where there may be a row of seven or eight jumpers. The connections are "jumped" with very small U-shaped electrical tabs with plastic casings to set SCSI ID numbers or other configuration parameters. If the circuit is jumped, the tab is placed over both prongs in the circuit to complete the electrical connection. If not, the tab is removed or placed to one side on a single prong, so as to leave the circuit unconnected.

junction box, J box A wire or fiber optic encasement box for coupling cables. It is often made of metal or other moisture- or fire-retardant materials, and is commonly found in homes and businesses.

Junction boxes are frequently hidden in closets or attached to the sides of buildings near a utility/network drop. They are crucial to most wire and fiber transmission technologies. Every time a transmission link is interrupted, split, joined, amplified, or otherwise affected at a wiring junction, there is the possibility of damage, short circuits, noise, or loss of signal. The main functions of the junction box are

- to protect the wiring from moisture, heat, or other conditions that may damage the wires or fiber filaments or short an electrical circuit,
- to provide a means to connect sections of transmission lines so they can be combined to create a longer total line,
- to protect humans and animals from accidental electrical shock from contact with electrical wires, switchers, or amplifying components, and
- to provide a means to readily access, connect, or change the wiring or effect repairs at a particular point in a circuit.

The junction box is commonly placed at the junction or demarcation point between internal and external wiring. It may also be placed at a point in a line where other lines are split off from a main line or where sections of line are connected, or where the signal on a line needs to be amplified. In building wiring, the J box often also incorporates fuses or breaker switches to protect against overload and fire hazards. The simplest J boxes are those that house the wiring for a switch or other wire connection point in a wall or ceiling.

In fiber optics applications, the terminal and connecting boxes for the optical fiber splicing and connection points are also called J boxes.

Since fiber optic media are more difficult to handle and connect than most electrical wires, J boxes for fiber optics must be built to accommodate the differences between electrical current and laser light transmissions. They may have specialized termination components and usually lock the fibers in place more rigidly and precisely than wires in an electrical J box to accommodate fiber transmission and physical properties such as orientation and preferential bend radius. Blown fiber techniques were developed in part to reduce the number of junction boxes in a fiber optic transmission line.

In oceanic transmission cables, special techniques for installing and maintaining cables are necessary and the junction boxes must be built to withstand the rigors of salt water and cold temperatures. They must also be designed so that connections may be changed in an underwater environment without creating short circuits. Grounding and termination are accomplished differently from land-based installations.

Many of the present undersea junction installation and maintenance boats are descended from vessels that laid the first transatlantic cable in the 1800s and the mine-laying naval ships of World War I. Since marine junction boxes can't be hidden within walls, they must be designed so that they don't catch in fishing nets or interfere with nearby boating and diving activities.

In radio and computer electronics, J boxes are often constructed by hobbyists to add electronic switches or components to existing systems. The box supports and protects wiring connections and may include

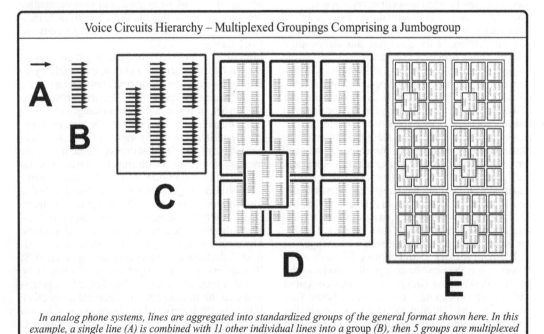

Voice Circuits Hierarchy – Multiplexed Groupings Comprising a Jumbogroup

In analog phone systems, lines are aggregated into standardized groups of the general format shown here. In this example, a single line (A) is combined with 11 other individual lines into a group (B), then 5 groups are multiplexed to form a supergroup *(C), 10 supergroups form a* mastergroup *(D), and 6 mastergroups form a* jumbogroup *(E).*

switches, buttons, status lights and connection terminals for data and power.

J boxes are distinguished from interface devices mainly in that they do not provide complex data conversion or modulation functions, but focus on simple wiring circuits as an extension to the main circuit. As in many aspects of electronics, the distinction is not always clear; there is overlap between interface devices and the more advanced J boxes (sometimes called smart J boxes). Advanced electronic J boxes have been designed for industrial installations to automatically evaluate the characteristics of an electrical line and log the information or report possible problems to an operator. These are essentially a J box and associated electronic devices housed in one unit, using the J box nomenclature for familiarity in marketing the products.

junctor A connective circuit extending between frames of a switching unit, terminating in a switching device on each frame, as in an internal network trunk.

JUNET Japan Unix Network. A noncommercial Japanese network dedicated to promoting communication among researchers in and outside Japan.

Junior Wireless Club Limited A pioneer amateur radio organization, formed in 1909. In April 1910, one of the young members, W.E.D. Stokes, Jr., made history when he represented the club before a U.S. Senate Commerce Subcommittee. As president of the club, he opposed a proposed bill to license wireless operators. Stokes also argued for a specified wavelength for amateur use to be allocated to amateur radio communications. In October 1911, the group changed its name to the Radio Club of America. See Radio Club of America.

Juniper Networks A software/hardware firm backed by prominent telecommunications firms 3Com, Lucent Technologies, and Northern Telecom. It has developed routing software in competition with Cisco Systems products and a number of channelized, long-haul, and multilink hardware interface products. See Junos routing code.

JUNOS routing code A router operating system (OS) announced by Juniper Networks, Inc. in 1998 that enables high-speed network forwarding across complex paths. It is competitive with the Cisco Internetwork Operating System (IOS) dominant on Internet-deployed routers. In 2001, an upgrade was released to support the use of multiprotocol label switching (MPLS) on virtual private networks (VPNs) over Internet Protocol (IP) networks.

just a bunch of disks JBOD. A disk data management system in which various organizational aspects of hard disk storage may span more than one disk. For example, some operating systems allow a user to transparently create large partitions (or file structures) that span more than one drive (as opposed to breaking up a drive into smaller segments) or even to create a file structure across more than one partition, thus making flexible use of the total virtual storage space available on a system.

As an example, imagine you have three 1 GByte drives and want to save a big animation file (e.g., a movie) that takes up 2.5 GBytes of drive space. Most consumer desktop systems aren't configured to do this; they will indicate that there isn't enough space, even if total disk space is 3 GBytes. In a JBOD system, the file management algorithms can manage the virtual space specified and split the file and file pointers appropriately over the various disk segment boundaries. The name is likely derived from RAID (redundant array of inexpensive disks) and has some aspects in common, in that the system is used with multiple disks, but it's not the same (in fact, it could be implemented across a RAID system). RAID is intended to provide redundancy on a system through mirroring or striping of data, whereas JBOD is intended to provide more flexible management of data across several disks. See redundant array of inexpensive disks.

Just-Enough-Time JET. A network transmission protocol proposed by Qiao and Yoo in 1999 for use with optical burst switching (OBS). It embodies the use of a delayed connection reservation (DR) and the capability of integrating delayed reservation with fiber optic delay line-based buffered burst multiplexers (BBMs). JET is considered by its developers to be more efficient for use with OBS than TAG-based OBS protocols.

In a given transmission path with a known number of hops, JET and OBS protocols can support multipath routing. Deflection routing can also be supported with the control packet selecting an alternate outbound link and setting the switch for the following data burst.

JET does not mandate buffer use or buffer size relative to the number of bursts or bits that can be stored or delayed simultaneously. Each burst will be buffered at the source during an offset interval and need not be buffered at intermediate nodes, if no blocking is encountered by the control packet. If a control packet is blocked, DR and BBM can be used to increase the effectiveness through buffer allocation and management. See optical burst switching.

Just-In-Time JIT. A term taken from the inventory and distribution field to describe computer processing systems that handle data translation or other tasks at runtime. For example, Just-in-Time compilers compile code on-the-fly instead of running a code interpeter (e.g., SmallTalk) in situations where this approach may be faster. See Java Virtual Machine.

JvNCnet John von Neumann Center network. A mid-level, northeastern U.S. regional network owned and operated by Global Enterprise Services, Inc. JvNCnet was the first T-1 research network and supported the networking needs of the U.S. National Science Foundation (NSF).

In 1989, the National Science Foundation announced a 3-year award to JvNC to establish Phase II, with input from academic and industrial institutions and representatives from the National Science Foundation and others. Phase II connected a number of prominent networks and organizations to JvNCnet, including Bell Laboratories, Rutgers University, Princeton University, Siemens Research, and JANET.

GES provides users with the hardware and telephone line necessary to connect to JvNCnet access points with a variety of transmission speed options.

JWICS See Joint Worldwide Intelligence Communications System.

JXML See Java XML.

k *abbrev.* kilo-. See kilo-.

K 1. *symb.* cathode (in pinout diagrams). 2. *abbrev.* Kelvin. 3. *symb.* 1024 (commonly used with quantities of data elements such as bits). Thus, 2^{10} or 2K = 2048 bits. Also a prefix as in Kbps (kilobits/sec). 3. *symb.* 1000 when used to indicate monetary quantities. For example, $50K typically indicates a salary of $50,000/year.

K plan, K-plan See keysheet.

K Series Recommendations A series of ITU-T guidelines for preventing interference in telecommunications systems (there are also interference-related topics in various other Series Recommendation documents). These guidelines can be purchased from the ITU-T. Since ITU-T specifications and recommendations are widely followed by vendors in the telecommunications industry, those wanting to maximize interoperability with other systems need to be aware of the information. A full list of general categories is listed in Appendix C and specific series topics are listed under individual entries in this dictionary, e.g., J Series Recommendations. See K Series Recommendations chart.

K-band A designated portion of the electromagnetic spectrum ranging from 10.9 to 36 GHz. The K-band range is commonly used for small antenna satellite transmissions. See band allocations, Ka-band, Ku-band.

K-carrier A four-wire broadband cable carrier system utilizing frequencies to about 60 kHz.

K-style handset The designation for the shape of newer telephone handsets which resemble older desk phone G style handsets, except that they have a more squared-off design on the ear- and mouthpieces. They are heavier and more substantial than some of the newer cordless or cell phone handsets, which tend to be flat and small. See G style handset.

K56flex modem A 56k data telecommunications modem technology developed by Rockwell Semiconductor Systems and Lucent Technologies to enable higher data throughput rates over standard analog telephone lines (POTS). The modem was competitive with U.S. Robotics' x2 technologies in the absence of an established 56k standard.

Higher data rates were achievable by looking at the structure of the phone lines and the prevalent patterns of modem usage. These modems were designed to do less conversion as analog phone lines were actually part of a predominantly digital system. In addition, the K56flex modems were optimized for downstream transmissions, with upstream being delivered at speeds up to 33.6 Kbps.

K56flex modems are backwardly compatible with the ITU V.34 standard to enable fallback when the server doesn't support K56flex. When the V.90 standard was approved by the International Telecommunication Union (ITU), Rockwell and Lucent announced that new modems would also be compatible with the ITU V.90 standard (formerly called V.Fast). Some K56flex modems could be upgraded to support V.90. See V.Fast, V Series Recommendations.

kA *abbrev.* kiloampere, 1000 amperes.

Ka-band The designated portion of the electromagnetic spectrum in the high microwave/millimeter range, approximately 18 to 22 GHz. The Ka-band is used primarily by small antenna satellite transmissions, and is intended to support future applications, for example, mobile voice. A 500-MHz allocation within this spectrum is earmarked for nongeostationary fixed satellite orbit services, and there are spectrums for local multipoint distribution services (LMDS), mobile satellite services, and geostationary satellite services. See band allocations for chart.

KA9Q Gopher server A Web-accessible Gopher server developed by Chris McNeil, with enhancements by McNeil and Peter Crawshaw. The Web version supports email, FTP, Gopher, NTP, Finger, and SLIP servers plus security through IP filtering. The name is based upon the underpinnings of Phil Karn's KA9Q NET/NOS.

KA9Q NOS TCP/IP Phil Karn's popular commercial TCP/IP software implementation for packet radio communications. The name of the software comes from his amateur radio callsign. It is available in Borland C++ and a 32-bit protected-mode version for DJGPP. It is popular for its compatibility with lower memory and CPU systems (not everyone wants to throw away perfectly good older computing systems). Amateur radio enthusiasts publish a number of amateur radio TCP/IP server gateways on the Web, which connect initially through telnet. Some of these are password-protected, and some can be accessed by anonymous login, somewhat similar to anonymous login sessions on FTP sites.

Kahle, Brewster Project leader for the Wide Area Information Server (WAIS) at Thinking Machines Corporation in Massachusetts (1989), and involved with the company since its founding in 1983. WAIS, Inc. was sold to AOL in 1995. Kahle designed the CPU of the Connection Machine Model 2 in the 1980s. He founded the Internet Archive and co-founded the Alexa Web information company in April 1996, and is active in promoting scholarly repositories of Internet history and documents that might otherwise have a short shelf life. See Wide Area Information Server.

Kahn, Robert "Bob" E. (1938–) A prominent American developer of network technologies, Kahn coauthored *A Protocol for Packet Network Internetworking (1974)* describing Transmission Control Protocol (TCP). In 1977, he co-demonstrated a gateway system that could interconnect packet radio with the ARPANET. In 1986, he founded the not-for-profit Corporation for National Research Initiatives to support and develop the National Information Infrastructure (NII).

Kahn has received many awards for his contributions to the development of the Internet, including the National Medal of Technology, in 1997. See Cerf, Vinton G.; packet radio.

Kalman, Rudolf Emil (1930–) A Hungarian-born American mathematician and engineer who studied at MIT and Columbia University in the 1950s. Since then he has held a number of professor and research positions at institutions such as IBM and Stanford and served as director of the Center for Mathematical System Theory at the University of Florida and chair of Mathematical System Theory in Zürich.

Kalman is considered a significant pioneer in the field of control theory and is particularly remembered for codeveloping linear filtering techniques with R. Bucy, in the early 1960s. He studied and modeled state-space concepts such as linear-quadratic control, minimality, and observability and applied them to systems analysis. In the 1970s, he was among those who introduced the use of algebra and geometry in linear and nonlinear control theory.

Kalman has received many engineering awards, including the IEEE Medal of Honor (1974). He was elected to the National Academy of Sciences in 1994.

Kalman filter A mathematical means of removing unwanted noise from a stream of data, developed by R. Kalman and R. Bucy. In 1960, when the development of computer algorithms was still largely in its infancy, Kalman proposed a recursive means of solving discrete-data linear filtering problems through a least-squares method. In "A New Approach to Linear Filtering and Prediction Problems," Kalman describes his objective as obtaining the specification of a linear dynamic system which accomplishes the prediction, separation, or detection of a random signal and points out some of the limitations of traditional solutions prior to the Kalman approach.

There is a discrete Kalman filter and an extended

K Series Recommendations	
Recom.	Description
K.1	Connection to earth of an audio-frequency telephone line in cable
K.2	Protection of repeater power-feeding systems against interference from neighbouring electricity lines
K.3	Interference caused by audio-frequency signals injected into a power distribution network
K.4	Disturbance to signaling
K.5	Joint use of poles for electricity distribution and for telecommunications
K.6	Precautions at crossings
K.7	Protection against acoustic shock
K.8	Separation in the soil between telecommunication cables and earthing system of power facilities
K.9	Protection of telecommunication staff and plant against a large earth potential due to a neighbouring electric traction line
K.10	Low frequency interference due to unbalance about earth of telecommunication equipment
K.11	Principles of protection against overvoltages and overcurrents
K.12	Characteristics of gas discharge tubes for the protection of telecommunications installations
K.13	Induced voltages in cables with plastic-insulated conductors
K.14	Provision of a metallic screen in plastic-sheathed cables
K.15	Protection of remote-feeding systems and line repeaters against lightning and interference from neighbouring electricity lines
K.16	Simplified calculation method for estimating the effect of magnetic induction from power lines on remote-fed repeaters in coaxial pair telecommunication systems
K.17	Tests on power-fed repeaters using solid-state devices in order to check the arrangements for protection from external interference
K.18	Calculation of voltage induced into telecommunication lines from radio station broadcasts and methods of reducing interference
K.19	Joint use of trenches and tunnels for telecommunication and power cables

K Series Recommendations, cont.

Recom.	Description
K.20	Resistibility of telecommunication equipment installed in a telecommunications center to overvoltages and overcurrents
K.21	Resistibility of telecommunication equipment installed in customer's premises to overvoltages and overcurrents
K.22	Overvoltage resistibility of equipment connected to an ISDN T/S bus
K.23	Types of induced noise and description of noise voltage parameters for ISDN basic user networks
K.24	Method for measuring radio-frequency induced noise on telecommunications pairs
K.25	Protection of optical fibre cables
K.26	Protection of telecommunication lines against harmful effects from electric power and electrified railway lines
K.27	Bonding configurations and earthing inside a telecommunication building
K.28	Characteristics of semi-conductor arrester assemblies for the protection of telecommunications installations
K.29	Coordinated protection schemes for telecommunication cables below ground
K.30	Positive temperature coefficient (PTc) thermistors
K.31	Bonding configurations and earthing of telecommunication installations inside a subscriber's building
K.32	Immunity requirements and test methods for electrostatic discharge to telecommunication equipment – generic EMC recommendation
K.33	Limits for people safety related to coupling into telecommunication system from AC electric power and AC electrified railway installations in fault conditions
K.34	Classification of electromagnetic environmental conditions for telecommunication equipment – basic EMC recommendation
K.35	Bonding configurations and earthing at remote electronic sites
K.36	Selection of protective devices
K.37	Low and high frequency EMC mitigation techniques for telecommunication installations and systems – basic EMC Recommendation
K.38	Radiated emission test procedure for physically large systems
K.39	Risk assessment of damages to telecommunication sites due to lightning discharges
K.40	Protection against LEMP in telecommunication centers
K.41	Resistibility of internal interfaces of telecommunication centers to surge overvoltages
K.42	Preparation of emission and immunity requirements for telecommunication equipment – general principles
K.43	Immunity requirements for telecommunication equipment
K.44	Resistibility of telecommunication equipment to overvoltages and overcurrents
K.45	Resistibility of access network equipment to overvoltages and overcurrents
K.46	Protection of telecommunication lines using metallic symmetric conductors against lightning-induced surges
K.47	Protection of telecommunication lines using metallic conductors against direct lightning discharges
K.48	EMC requirements for each telecommunication network equipment – product family recommendation
K.49	Test condition and performance criteria for voice terminal subject to disturbance from digital mobile phone
K.50	Safe limits of operating voltages and currents for telecommunication systems powered over the network
K.51	Safety criteria for telecommunication equipment
K.52	Guidance on complying with limits for human exposure to electromagnetic fields
K.53	Values of induced voltages on telecommunication installations to establish telecommunications, AC power, and railway operators responsibilities
K.54	Conducted immunity test method and level at fundamental power frequencies

K

Kalman filter. The discrete Kalman filter (DKF) attempts to estimate the state of a discrete-time-controlled process. The extended Kalman filter (EKF) handles nonlinear situations by linearizing (e.g., with partial derivatives) about the current mean and covariance.

The Kalman filter is now an important aspect of sonar and radar tracking, guidance, and navigation systems. It has also been used in satellite orbit calculations for various space missions. Recently, fiber optic gyroscopes (FOGs) in combination with Kalman filters, have been used in mobile robot systems. The Kalman filter fuses the FOG sensor data with the robot odometer to provide more accurate dead reckoning than is possible through traditional odometric systems in accessible price ranges.

Kangaroo Network A commercial hardware/software product from Spartacus/Fibronics designed to enable IBM mainframes to intercommunicate with other networks using TCP/IP.

Kangaroo Working Group A working group on telecommunications and the information society that works with the European Internet Foundation looking into issues associated with creating a level playing field in terms of Internet use and access in Europe, ensuring a balance between private industry and government. The Kangaroo Group has been actively involved in conferences since the mid-1980s and has actively debated Internet regulations and barriers to the use of cyberspace.

Kao, Charles K. (1933–) A Chinese engineer and significant pioneer in fiber optic communications, Kao studied in Britain and became head of the optical communication program at ITT's Standard Telecommunications Laboratories, Harlow, U.K., in December 1964. Along with G. Hockham, Kao championed the idea of single-mode optical fiber transmission systems and coauthored "Dielectric-Fiber Surface Waveguides for Optical Frequencies" in *Proc. IEEE* (1966). He followed this up with work on the structure of fiber communications and subsystems. Kao and Hockham were correct in proposing that loss in fiber signals was due to impurities in the transmission medium rather than inherent limitations in the glass itself.

During the period 1987–1996, Kao taught and became president of the Chinese University of Hong Kong. He became a Trustee for the S.K. Yee Medical Foundation in 1990 and is chairman of Transtech Services Ltd. He has received many honors from engineering societies. See Hyde, J. Franklin; Kapani, Narinder; Keck, Donald.

Kapani, Narinder Singh (ca. 1930–) A prolific Indian inventor, Kapani has been awarded more than 150 patents for various instruments and laser technologies. In 1954, in *Nature*, Kapani and Hopkins described a means to clad optical fibers to keep the light within the waveguide This was an important milestone in optical technologies, greatly extending the distance over which signals could travel through a fiber waveguide. The following year, Kapani developed a way to fabricate high-quality fiber

filaments and coined the term *fibre optics*. By 2000, Kapani had founded a series of companies, including K2 Optronics for concentrating on fiber optic communication devices. See Hansell, Clarence; Kao, Charles.

Kapor, Mitchell "Mitch" (1950–) The instigator of several historic high profile computer-related organizations, Mitch Kapor founded Lotus Development Corporation in 1982, and was the designer of the well-known Lotus 1-2-3 spreadsheet software. In 1990, he cofounded the Electronic Frontier Foundation (EFF), a nonprofit civil liberties organization. Kapor has chaired the Massachusetts Commission on Computer Technology and Law and served on the board of the Computer Science and Technology arm of the National Research Council, and the National Information Infrastructure Advisory Council.

Karbowiak, Antoni E. A researcher at Standard Telecommunications Laboratories in a group established by Alec Reeves in 1962. Karbowiak and his collaborators conducted pioneer work with optical fibers as potential transmission technologies at a time when most scientists considered the lossiness of (uncladded) fiber to be too great to be of any practical value. He became a collaborator with a young, talented Chinese engineer, Charles K. Kao, who is now considered a significant pioneer in fiber optics. In 1964, Karbowiak left STL to chair the electrical engineering department at the University of New South Wales. See Kao, Charles K.; Snitzer, Elias.

Karnaugh map A two-dimensional truth lookup table organized to facilitate combination and reduction of Boolean expressions. This is useful in digital logic circuit design. See Boolean expression.

Karn's algorithm A mathematical formula for improving network round-trip time estimations. It grew out of packet radio network algorithms but now is more widely applied. In layered network architectures, the algorithm helps the transport layer protocols distinguish among round-trip time samples. It is used in Transmission Control Protocol (TCP) implementations to separate various types of return transmissions and to establish whether or not to ignore retransmitted signals. It is also applied to backoff timers in Point-to-Point (PPP) tunneling networks. See ATM, Jacobson's algorithm, Point-to-Point Tunneling Protocol.

Kawakami, Shojiro A Japanese researcher who has contributed articles about many aspects of fiber optic technology, but who is chiefly known for his development of graded-index fiber and of a "photonic crystal" that acts like a prism in that it can separate optical signals with different wavelengths. The technology was developed by Shojiro (Tohoku University) and further developed in conjunction with Optoelectronics Laboratories, and NEC, in 1997. This invention has potential as part of add/drop multiplexer components in dense-wavelength transmission systems, as envisioned by NEC.

Kawakami was recognized as a Fellow of the IEEE Society for his contributions to fiber optics, in 1997. See Hicks, John; photonic crystal; Yablonovitch, Eli.

Kay, Alan (1940–) A precocious child and avid reader, Kay was inspired by the work of Seymour Papert at MIT in the 1960s. Kay was committed to the idea that computers should be easy, fun, and accessible, and began developing what was to become the Smalltalk object-oriented programming language. He became a group leader at Xerox PARC in the early 1970s, a period when tremendous innovation in microcomputer technology and user interfaces was stimulated at the lab.

Kazarinov, Rudolf F. A Russian physicist and engineer, Kazarinov originally carried out research at the IOFFE Physico-Technical Institute in St. Petersburg. He then came to the U.S. to conduct research at the Bell Laboratories Photonics Circuits Research Department. Kazarinov made significant pioneer theoretical contributions to semiconductor laser technologies beginning in the early 1960s. His contributions include the double-heterostructure laser, distributed-feedback (DFB) laser, and intersubband lasers. He also coauthored a number of patents, including hybrid lasers for optical communications, with Greg Blonder, Charles Henry, et al.

In 1998, Kazarinov was coawarded the prestigious Rank Prize in optoelectronics for his involvement in the 1994 development of the quantum-cascade (QC) laser, along with Federico Capasso. See Capasso, Federico; laser history.

Kbps kilobits per second; 1000 bits per second. It is sometimes written Kbits/s.

KBps kilobytes per second, 1000 bytes per second. It is sometimes written Kbytes/s.

KDC See key distribution center.

KDD 1. Knowledge Discovery in Databases. A branch of artificial intelligence applied to database query, search, and retrieval. 2. Kokusai Denshin Denwa Company, Ltd. A Japanese supplier of international telecom services, equipment, and facilities.

KDDI Corp. Japan's second-largest communications carrier, descended from Kokusai Denshin Denwa Kabushiki Kaisha (KDD), which was founded in 1953. In April 2001, KDDI announced Java support for its mobile phone services through its CLDC- and MIDP-conforming application interface called KDDI-P.

KDD R&D Laboratories, Inc. is the research and development division. It was founded when KDD was detached from the Nippon Telegraph and Telephone Public Corporation and moved into a new research facility for conducting research in international communications in 1960. It became independent of KDD in 1998 and was remerged, along with other firms, in 2000 to become KDDI's R&D division. The R&D lab has developed TDMA technology for satellite communications, submarine fiber optic cables, G3 facsimile coding technologies, magneto-optical discs, and data compression and transmission technologies. KDD Fiber Labs, Inc., a KDDI Group Corporation, develops fiber optics technologies, including WDM optical amplification and various types of light sources.

KDKA KDKA originated as amateur callsign 8XK,

operating from the garage of Frank Conrad. It is a historically significant Westinghouse Electric radio broadcasting station in Pittsburgh, Pennsylvania that used radio waves to report returns of the Harding-Cox Presidential race to the American public, on November 2, 1920. This was about 14 years after the earliest experimental broadcasts and a week after receiving its own official broadcasting license. By the following year, KDKA was making regular public broadcasts and radio broadcasting was booming, with more than 500 broadcasting stations sprouting up around the country. KDKA was still broadcasting under the same callsign more than 80 years later. See CFCF; Herrold, Doc; radio history.

KEA See Key Exchange Algorithm.

Kearney System KS. A parts numbering scheme developed for Western Electric telecommunications equipment, named after the town in New Jersey where the plant was located. The KS system has generally been superseded with vendor-specific and industry standard codes, although Kearney numbers are still found on some pieces of equipment.

Keck, Donald B. (1940–) Keck studied at Michigan State University, graduating in 1967 and went to work as a senior research scientist at Corning Glass Works in 1968, becoming involved in fiber-related projects with Robert Hall. He subsequently served as director of the Applied Physics division and VP of Optics and Photonics and then became VP and director of the Optical Physics Technology group.

Along with Maurer and Schultz, Keck was awarded the National Medal of Technology in 2000 for making low-loss optical transmission fiber a practical reality.

keepalive interval The period of time between keepalive messages. The amount of time depends upon the type of network and the type of activity taking place. For example, for a computer process, the interval might be measured in nanoseconds, whereas for a user activity, it might be measured in minutes. See keepalive message, keepalive signal.

keepalive message Messaging between network devices that indicates that a virtual circuit between the two is still active (alive). See keepalive interval, keepalive signal.

keepalive signal A network signal transmitted during times of idleness to keep the circuit from initiating a time-out sequence and terminating the connection due to lack of activity. See keepalive interval, keepalive message.

Kelvin balance, ampere balance A historical instrument for measuring the absolute value of an electrical current, named after its inventor, William Thompson (Lord Kelvin). It is essentially a galvanometer that measures the force produced by the magnetic field associated with the passage of current through a conductive medium.

In one of its historical fabrications, the instrument resembled a small reel-to-reel tape recorder, with two low, flat spools coiled with wires connected to one another in series positioned a few inches apart. A pivoting beam balance enabled a set of rings to move

K

freely between the coils. A finely incremented ruler-like gauge stretched the length of the instrument, in front of the coils, from the outer edge of one to the other. The whole thing was generally encased within a protective brass and glass enclosure. The instrument was sold with a set of weights. The current to be measured passed through the wire coils to create an attractive force referenced against a known weight. Two ampere balances were designated as legal standard instruments in 1894.

Kelvin effect When an electric current passes through a single homogeneous but unequally heated conductor, heat is absorbed or released. This effect is named after William Thompson (Lord Kelvin).

Kelvin scale A temperature scale proposed by William Thompson (Lord Kelvin), based on the efficiency of a reversible machine. Zero is designated as the temperature of the sink of the machine working efficiently, that is, complete conversion of heat into work, a situation possible only at absolute zero on a gas temperature scale. Zero degrees Kelvin (0 K) can also be expressed as -273.15 degrees Celsius ($-273.15°C$) or -459.67 degrees Fahrenheit ($-459.67°F$).

Kelvin, Lord William Thomson (1824–1907). A Scottish physicist and mathematician who made significant contributions to the field of thermodynamics, and applied his theories to the dynamics and age of the earth and the universe. He utilized the field concept to explain electromagnetism and its propagation. The concept of an all-pervasive "ether" was still prevalent at the time, so he explained a number of his observations within this context. He also developed the siphon recorder, a number of types of voltmeters, and an ampere balance, and was involved in laying the transatlantic telegraph cable. The Kelvin scale and Kelvin effect are named after him.

Kenbak-1 A discrete logic microcomputer designed by John V. Blankenbaker, introduced in 1971 as the Kenbak-1 Digital Computer. It featured 256 bytes of memory, three programming registers, and five addressing modes. The controlling switches were on the front panel of the machine. It was advertised in the September 1971 issue of *Scientific American*, 3 years prior to the introduction of the Altair, for only $750. One of the earliest microcomputers, the Kenbak-1 was apparently ahead of its time. Unfortunately, only 40 machines sold over the next 2 years, and the California-based Kenbak Corporation missed a significant business window by a narrow margin. One year after the company closed, the Altair computer kit caught the attention of hobbyist readers of *Popular Electronics* magazine and sold over 10,000 units. See Altair, Arkay CT-650, Heathkit EC-1, Intel MCS-4, Micral, Simon, Sphere System.

Kendall effect Distortion in a facsimile record, caused by faulty modulation of the sideband to carrier ratio of the signal.

Kennelly, Arthur Edwin (1861–1939) A British-born American mathematician and engineer who studied mathematical aspects of electrical circuitry. He also studied the properties of the Earth's atmosphere and its effects on radio waves and suggested that an ionized layer above the Earth could reflect radio waves, an idea soon after independently published by Oliver Heaviside.

Kennelly-Heaviside layer In 1902, A. Kennelly and O. Heaviside proposed, independently of each other, that an ionized layer surrounding the Earth could serve as a reflecting medium that would hold radiation within it. This led to the discovery of a number of regions surrounding Earth and utilization of the characteristics of some of these layers in long-distance wave transmission. It also led, in the 1920s, to confirming experiments in which radio signals were bounced off this reflecting layer. See Heaviside, Oliver; ionosphere; Kennelly, Arthur.

Kepler, Johannes (1571–1630) A physicist and astronomer from Swabia (now Austria) who studied the planets and endeavored to mathematically describe planetary motion. In 1604, he published *Astronomiae pars Optica* describing light travel, shadows, the functioning of the eye, and other concepts fundamental to modern optics. He also made important studies in optics related to better studying the planets. In 1611, he published *Dioptrice* which described the properties of lenses, including inversion and magnification, and a new form of telescope now known as the astronomical telescope. Kepler may have been the first to use the term "satellite" to describe orbiting moons.

Kerberos authentication An authentication system developed through the MIT Project Athena effort. Kerberos is a client/server security mechanism based upon symmetric key cryptography. Each user of the Kerberos system is assigned a nonsecret unique ID and selects a secret password. The secret password is provided to the Kerberos system and is not intended to be divulged by either party. The user then uses the password to request access from the system. The identity of the user is verified by generating a random number and presenting a problem that can likely be solved only by the authentic user, thus providing access to a message on the system. The symmetric nature of the system is in the use of the same encryption and decryption key. For security purposes, long, randomly selected strings work best with this system of cryptography; otherwise it may be vulnerable to password-guessing attacks.

Kermit Project A nonprofit, self-supporting project at Columbia University for the support of the Kermit Protocol and the development of Kermit-related technologies. The project also includes information on documentation, licensing, and technical support for users of the Columbia implementation of Kermit. See Kermit Protocol.
http://www.columbia.edu/kermit/

Kermit Protocol A packet-oriented, platform-independent file transfer protocol developed at Columbia University in 1981. Hundreds of Kermit implementations support the 7-bit and 8-bit transfer of text and binary files. They are commonly used over asynchronous, serially connected local area networks (LANs) and phone lines. Kermit is flexible and configurable.

Kermit is not the speediest protocol, as each packet is checked and acknowledged as it is transferred, but it is reliable, widespread, and well supported, especially in academic institutions; when all other protocols fail, it's often the one which will get the file transfer done. There are numerous terminal emulators based on Kermit, with VT52 and VT100 versions being common. Telnet, an important protocol for remotely connecting to a network host, has also been implemented with Kermit.

Kermit is a workhorse, but its use in its original form has declined. Most local area networks and the Internet now use other network connection and file transfer mechanisms such as ATM, Ethernet, and FTP, but traditional Kermit is still useful for phone links and small networks interconnected with basic serial connections.

Updated versions of Kermit may greatly extend its useful life. Internet Kermit Service is a file transfer service described by da Cruz and Altman at Columbia University based on a combination of the widely used Telnet Protocol and Kermit Protocol. It supports both anonymous and authenticated access. Kermit over Telnet enables the traversal of firewalls and a number of security options. By providing some advantages over File Transfer Protocol (FTP), this Kermit configuration is a practical option for distributed networks, including the Internet. The registered IANA port for Kermit connections is 1649.

Kermit is an open freely distributable protocol, so it can be used for software applications development, but Columbia's implementation of the Kermit protocol is copyrighted. See FTP, Kermit Project, XModem, YModem, ZModem. See RFC 2839, RFC 2840.

kernel 1. Line within a conductor along which the current-resulting magnetic intensity is zero. 2. Low level of an operating system at which processes and resources (such as memory and drivers) are created, allocated, and managed. Functions and operations at the kernel level form a bridge between hardware and software resources and are mostly or completely transparent to the user.

Kerr cell A device used to modulate light in conjunction with polarizers. The cell contains electrodes to direct the necessary electric field for inducing the Kerr electro-optical effect and the material being influenced by the combination of the field and a beam of polarized light. Photodetectors may be used in conjunction with the Kerr cell to determine if or when the effect occurs. Kerr cells have been used in the fabrication of high-speed optical shutters.

Kerr effect There are actually two Kerr effects named after John Kerr, so it is best to specify the one of relevance. For simplicity of reference, they are sometimes abbreviated as electro-optical Kerr effect (EKE) and magneto-optical Kerr effect (MOKE), especially in the field of microscopy. See scanning near-field magneto-optical microscope.

Kerr electro-optical effect A phenomenon discovered by John Kerr in 1875. It is an electro-optical effect in which certain substances become double refracting (birefringent) in the presence of strong electric fields or in which certain specific substances become double refracting in smaller electric fields. In other words, a single incident ray of light is refracted as two, with the two rays oscillating in mutually perpendicular planes. Isotropic liquids or gas, for example, show the Kerr effect and become optically anisotropic when subjected to a consistent electric field perpendicular to a beam of light.

To account for the effect, it is theorized that the application of the electro-optical energy causes a reorientation of a material's molecular structure. Since the effect is not universal across materials and levels of electromagnetic influence, it is often studied with certain parameters held within controlled limits, such as the constancy of the electric field and the wavelengths of light.

Since the Kerr effect can be induced through controlled conditions and occurs quickly, various researchers have suggested that it may have practical applications in troubleshooting optical transmissions or in increasing bandwidth in optical communications systems. See Kerr cell, Pockels effect.

Kerr magneto-optical effect The change in a light beam from plane polarized to elliptically polarized when it is reflected from the reflective surface of an electromagnet. The degree of rotation is directly proportional to the degree of magnetization of the reflective material. The transmissive aspects of this effect were first observed and described by Michael Faraday in the 1840s and researched further by John Kerr three decades later as to its reflective properties. Because of their relationship, the Faraday and Kerr effects are often described together.

This magneto-optical effect is useful for studying magnetic effects in superlattices and giant magnetoresistive (GMR) effects. GMR technology was discovered in the late 1980s and has since developed into a new, highly sensitive sensor design for disk drives. The effect can also be put into practical use for measuring current in power lines using a polarized laser to measure the degree of rotation. See Faraday effect, giant magneto-resistance.

Kerr, John (1824–1907) A Scottish-born physicist best known for discovering and describing the Kerr electro-optical effect. Kerr carried out research under the direction of William Thompson (Lord Kelvin). See Kerr electro-optical effect.

keV Abbreviation for kiloelectronvolt.

Kevlar The DuPont tradename for a strong, synthetic multipurpose material that is, ounce-per-ounce, about five times as strong as steel. In the 1960s, a new liquid crystal polymer fiber was invented by Stephanie Kwolek; Kevlar is the commercial embodiment of this fiber. Kwolek also invented Nomex, a fiber used in electrical insulation.

Kevlar is used in applications where strong, light, flexible materials are needed, such as bulletproof vests and protective sheathings for sensitive or electrically active materials. Kevlar tape and Kevlar strength members are used in fiber optic cables. A water-resistant sheath is often fitted over the Kevlar or Kevlar-impregnated inner layers for further

K

protection from the elements (typically PVC or polyethylene). Kevlar is sometimes mixed with building materials to increase strength and resilience in the event of industrial vibrations or earthquakes. Carbon steel or ceramic cutters are generally used to cut the Kevlar components that serve as strength members in fiber optic cables.

key n. 1. A small, physical security device, often made of metal, inserted into a matching lock receptacle to lock/unlock or activate/inactivate an object or structure. It's usually the shape that allows a lock to be opened or closed, but more recent data-compatible keys may have magnetic stripes rather than physical indentations. 2. In an image, the overall tone or value of the image, often used to adjust camera settings to balance the amount of light or to screen out certain colors or light intensities (e.g., chroma key). 3. A switch for opening or closing a circuit. 4. In a database, an organizational means to locate desired information without searching the entire content of the database. 5. On keyboards, keypads, phone pads, etc., a small, roughly cubic, raised, movable, input attachment intended to be depressed, usually by a finger, to make a selection. 6. The modern equivalent of the switch on an old phone.

key, telegraph A signaling device allowing the input of code, usually Morse code, and transmits it to the communications channel. The key superseded the

Key Encryption – Basic Concepts	
Concept	Description
key agreement	An encryption key establishment mechanism that is common to asymmetric cryptographic exchanges but may also be used in symmetric exchanges. In key agreement, a pair of entities, wishing to engage in a secured communication without prior arrangement, make use of public data (e.g., a public key) to negotiate a common key value unique to their communication (i.e., not known or used by other entities). When a Diffie-Hellman technique is used, key agreement is arranged without the need to transfer the key. See Diffie-Hellman.
key center	A trusted, centralized distribution point (e.g., separate server machine) capable of administering the use of key-encrypting "master" keys to encrypt and distribute session keys for secured communications.
key confirmation	The process of ensuring that participants in a key-secured communication are legitimate by determining whether they do indeed possess a shared symmetric key.
key distribution center	In symmetric digital cryptography, a key center that provides encryption/decryption keys to two or more entities that wish to engage in a secured communication through an agreed-upon key distribution protocol. These keys are often session-related.
key escrow	A security system component in which part or all of a cryptographic key is entrusted to a third party to hold "in escrow." The key "bank" or authority is responsible for storing and releasing the keys to a party involved in a communication, provided that party submits proper authorization. The authorized recipient can then use the key to decrypt a message.
	There has been considerable debate over the use of escrow authorities. On the one hand, some individuals feel no one should have any part of a communication other than the sending and receiving parties. On the other hand, some believe it is necessary to have a third party that can be served a warrant to hand over information critical to the maintenance of national security and the carrying out of law enforcement activities. It is also critical, for the system to work, for the authority to be highly reliable, accountable, and secure. See key generation, key recovery.
key establishment	The processes of key generation, storage, and distribution that together enable a secured key-related communication association to be established.
key exchange	The transmission or recording of a software key with another party, or swapping among two or more parties. See encryption, PGP, key generation.
key generation	The process of creating a software key for security uses. Once this has been done, it is expedient to keep track of information related to keys (location, password, etc.) so that key generation does not have to be done again. Portable devices for generating a key are sometimes used in conjunction with keyless security locks on building premises.

portfule of earlier systems. See telegraph history.

key encryption A personal or public identifier intended to establish the owner or recipient of a secure encoded message. Key-related negotiations may be symmetric or asymmetric and may be based upon public keys, private keys, or a combination of both. Key encryption may be based upon a long-term escrow system or upon short-term session-based communications.

A public key cryptographic scheme consists of a *public key* provided openly to anyone who wishes to send an encrypted message, and a *private key* used by the recipient to de-encrypt the received message. The Key Encryption chart provides a summary of some of the basic concepts. See certification, Clipper Chip, cryptography, encryption, Pretty Good Privacy.

Key Exchange/Encryption Algorithm KEA. An asymmetric key encryption algorithm similar to the Diffie-Hellman algorithm, that utilizes 1024-bit keys. KEA was originally developed by the National Security Agency (NSA) as a classified security mechanism whose status was changed in June 1998. See Clipper Chip, Diffie-Hellman, key encryption, SKIPJACK.

Key Contact A service of British Telecom that comprises an 11th phone number (nonmobile) in addition to the key numbers kept by a business subscriber in BT's Friends & Family Key Numbers service. See key numbers.

key illumination The lighting of a key on a keypad or keyboard to signal its status or facilitate its location. The keys may be illuminated to indicate that they

Key Encryption – Basic Concepts, cont.	
Concept	Description
key length	The number of symbols, usually expressed in bits or bytes, used in representing an encryption key. In general, the longer the key length, the greater the possible number of ways in which the data may be scrambled to ensure that it cannot be easily decrypted by an unauthorized party.
key lifetime/lifespan	The time span or expiry date associated with an established cryptographic key. The expiry period or lifespan may be determined in advance by an issuing authority, especially in session-related key assignments, or may be dynamically determined. The key lifetime is also determined in part by users. If a user loses or forgets or misplaces a key, its effective lifetime has ended even if the capability to use it still exists. Some key cryptography systems will explicitly include a parameter that determines the lifespan of a key (e.g., until the session is terminated). The capability to terminate a key is important in situations where key assignments must be reused in a dynamic resource-conscious system (e.g., session-related keys) or where the key users may cease to have authority (e.g., terminated employees).
key management protocol	A protocol developed to facilitate establishment of a key-administered transmission between entities wishing to secure the communication through key encryption.
key recovery	The process of determining the value of a cryptographic key that has been used to perform an encryption operation. Key recovery is a hotly debated political topic since law enforcement agencies have desired and at times secured the legal and technological capability to recover keys for decrypting secured communications. A key "escrow" system that included portions of keys (to facilitate key recovery) was at one time intended to be associated with software products exported out of the U.S. Key recovery by governmental agencies was repeatedly proposed and defeated during the 1990s. Key encapsulation is a means of storing information about a cryptographic key by encrypting it with another key so that only authorized recovery agents may decrypt and retrieve the stored key. See key escrow.
key space	The universe or space from which cryptographic key values may be taken. It is the total number of distinct transformations which may be supported by a cryptographic scheme, in other words, the realm of possible variations possible.
key update	The updating or derivation of a new key from an existing key. Also called rekey.

are active (or pending), as in a multiline phone system, or to enable them to be seen in low-light conditions (e.g., a burglar alarm keypad). Steady illumination or various flashing speeds and patterns may be used to indicate the line or device status.

key map A table that translates keyboard input values from one configuration to another, commonly used in computer software to transcribe the alphabets of a number of languages. This is useful for translation, alternate typing keyboard setups (e.g., Dvorak), graphics, and music applications.

key numbers Telecommunications numbers of particular interest, such as the phone numbers of frequently called family and close friends or business associates. British Telecom (BT) has a key numbers service that can be managed on the Web. The Web portal enables customers to set up and manage the Friends & Family Key Numbers list associated with their phone accounts. This service is without charge for residential and business customers for up to ten numbers and may include up to two mobile numbers. It is likely that this type of Web access to telecommunications services will increase, just as online banking is increasing, due to its 24-hour availability and update convenience; the user doesn't have to wait to call service representatives during business hours.

key performance indicator KPI. A statistic intended to indicate effectiveness in specified key aspects which are typically industry-specific. In the telecommunications industry KPIs may comprise the frequency or duration of calls, revenue per call, purchase trends, etc. KPIs are used to plan budgets, inventory, and investment, financing, and growth strategies.

Key Performance Indicator KPI. A commercial database to track mobile operators' key performance indicators such as minutes of use, churn, average revenue per user, and acquisition costs. The system was first released by EMC, a U.K. firm, in April 2000, and is incorporated into EMC World Cellular Information Services. The system computes average statistics for each KPI for regions and reporting periods.

key pulsing KP. A system for sending multifrequency signals from a pushbutton key telephone through a phone circuit to establish a connection. It is sometimes called key sending. In older, manually operated toll stations, pulsing was sometimes used by operators instead of dialing. Dial-operated pulse phones are on the decline, with touchtone phones replacing them.

key service unit, key system unit KSU. The internal electronics and logic that enable the selection of lines and other options in a key telephone system. This may be a small cabinet installed in a closet or some other area where the lines are not cluttering up the environment or causing an obstruction. See key telephone system.

key station Master station from which broadcasts originate.

key telephone system, key system KTS. A multiline telephone system in which individual phones have multiple keys or buttons that the user presses to select the line over which she or he wishes to communicate. In larger multiline key systems, there may be a main console through which the calls are channeled. This is not the same as a private branch system, in which a separate switching system is associated with the phones. In the key system, which is used in many small offices, the switching and selection of lines is done manually by the user. Some larger offices with private branch exchanges will use a hybrid system which also incorporates one or more key systems, sometimes in individual departments. New key systems commonly feature programmable function keys and LED status displays. See key service unit, private branch exchange.

keyboard A hardware peripheral interface device for detecting and transmitting user input to a computerized system through keys with assigned functional values. Descended from typewriter keyboards and typically arranged according to the historic "QWERTY" typewriter layout which, ironically, was designed to slow down typing in order to prevent key jamming on old manual typewriters.

A variety of keyboard layouts and shapes are available for various computer systems, some with better ergonomics than those which typically come with the system. See keypad; keyboard, touch-sensitive.

keyboard buffer Recent input is typically stored in temporary memory in order to prevent loss or corruption in the event that the system was not yet ready to respond at the time that the keys were pressed.

keyboarding Striking keys on a computer or other digital keyboard. This is distinct from typing in that typing is generally intended to immediately translate the keystroke into an image on a printing surface. Keyboarding, on the other hand, enables the keystroke to be stored and manipulated for a variety of purposes, including word processing, chatting, signaling, or printing at a later time. It is possible to "type" on a computer keyboard with a software program designed to immediately send the key to a printing device, but this is rarely used due to the greater convenience of editing the keyboard strokes before printing (or sending them electronically without printing).

keying A means of modulating a signal. This can be done in a number of ways, by varying the amplitude, frequency, or phase of a signal. See amplitude shift keying, frequency shift keying, Gaussian minimum shift keying, quadrature phase shift keying, phase shift keying.

keypad A key-based physical interface for various calculators, dedicated word processors, security systems, and computing devices. It is usually a compact group of functionally related keys, often consisting of numbers aligned in rows and columns to facilitate finger access. A larger grouping, as on computers or typewriters, is usually called a *keyboard*.

On a calculator, the keypad is generally configured as three or four columns by five or six rows, depending upon how many extra function keys (memory, clear, print, etc.) are integrated into the key pad layout. Many security entry devices and alarm systems are managed through a keypad with nine or twelve keys for entering numeric codes or alphanumeric passwords. The system may be wired or wireless and may trigger other mechanisms (e.g., a security camera) if used or if suspicious codes are entered. See keyboard, numeric keypad.

keysheet An administrative plan for phone extensions that tracks and illustrates the connections and features assigned to that phone. Keysheets are practical in institutional environments with many extensions, particularly if the extension phones have different capabilities and dialing privileges. A keysheet is even more important for keeping track if the phones are also individually programmable or if they are relocated on a regular basis.

Software exists for developing keysheet connection plans, diagrams, and overlays. Electronic forms are now also commonly used for individual members of departments to list the phones, faxes, and modems for inclusion in a keysheet database.

keystoning A visual aberration which occurs when an image is projected on a surface off-plane, that is, on a surface which is at an angle to the plane of the surface of the projecting lens. Thus, if a rectangular image from a film or slide projector, for example, were projected on a movable screen which was crooked, the image would be wide on one side and narrow on the other. See barrel distortion.

kHz *abbrev.* kilohertz, 1000 hertz. See hertz.

kiddie cam A video camera installed to monitor the activities of children, often over a network. These are common in household nurseries and increasingly common in daycare facilities. Some of these cams have been interfaced with the Internet to enable parents and caregivers to remotely monitor the activities of children. For the safety and privacy of children being monitored by Web-based kiddie cams, some kiddie cam services require a password to login to the remote viewing site.

Unfortunately, the kiddie cam moniker is also used by some sites to promote pornographic images of girls and boys who have been covertly photographed or who are too young to understand how their image is being used. This type of exploitation is generally illegal, but the sites promoting child pornography manage to stay online long enough for the images to be downloaded and shared among thousands or millions of Internet users.

KIF See Knowledge Interchange Format.

Kilby, Jack St. Clair (1923–) An American inventor and Texas Instruments Incorporated (TI) employee who was involved in designing the first integrated circuit chip, shortly after joining TI. The development catapulted the miniaturization and speed of electronics into a new level of evolution. Kilby's first IC was introduced in September 1958, shortly before Fairchild engineers developed historic ICs that were, for some time, considered to be the first (the patent was awarded to Noyce even as Kilby's application was still being assessed). Texas Instruments has named the Kilby Center, a silicon manufacturing research facility, in his honor.

Kilby is also responsible for providing miniaturized electronics that supported the portable calculator

market and off-Earth electronic devices that needed to be small and sparing on power consumption. In October 2000, Kilby's contributions were acknowledged with a Nobel Prize in physics jointly with two other scientists. See integrated circuit; Noyce, Robert; transistor.

Kildall, Gary (1942–1994) American educator and pioneer software developer. Kildall developed CP/M (Control Program/Monitor) over a number of years, beginning in 1973, with contributions from his students, when he was a professor of computer science at a California naval school. Kildall developed CP/M into a popular, widely used, text-oriented, 8-bit operating system in the late 1970s.

Kildall later founded InterGalactic Digital Research, which became Digital Research (DR), to market his software products. Digital Research developed GEM, an early graphical user operating system which predated functional versions of Windows. DR also created DR-DOS, which was competitive with MS-DOS, and claimed by many to be superior.

Kildall is also known for developing PL/M prior to CP/M, the first programming language for the historic Intel 4004 chip, and for co-authoring a floppy controller interface in 1973 with John Torode.

In the ensuing years, Gary Kildall lost one political battle after another with the rapidly expanding Microsoft, and Digital Research never flourished as one might expect for a company so often in the forefront of technology. Digital Research had a history of creating good products, but was overshadowed by its larger, more aggressive competitor. At one point DR won a lawsuit against Microsoft, but it may have been a case of too little too late; at that point Microsoft had so much momentum, it was unlikely Kildall and DR could regain their market share. Kildall is acknowledged as the original developer of many significant technologies for the microcomputer industry, but unfortunate circumstances cut short his life at 52. See CP/M; Gates, William.

kill 1. Remove or delete, as a word, line, or file. 2. Abruptly or prematurely terminate a process or broadcast.

kill file 1. An email or newsgroup filter that sends messages from particular people, or on particular topics, to the "bit bucket," that is, they are shuffled off to a file that never gets read, or is deleted unread. 2. A list of users banned or otherwise controlled from access to remote terminals or online chat services. See kill command.

kill command A software control command available to operators on various chat systems to disconnect a disruptive member from the site. Common reasons for killing a user include racial, cultural, or religious slurs, violation of chat rules, illegal activities, or excessive profanity. On Internet Relay Chat (IRC) there is a general set of guidelines and a code of etiquette to guide channel operators in the appropriate use of the IRC /kill command.

kill message 1. A textual message transmitted by an operator to a user in a computer-based "chat room" who is in the process of being removed from the chat area. This is an option on most systems for operators to inform the user as to why he or she is being removed. See kill command. 2. A software command sent to stop a process. This may be on a single-user system, a network, or a specialized system such as a transaction system based on digital data cards. A kill message can halt a process that has hung or gotten out of control, without taking down the operating system; it can stop suspect activities (e.g., possible hacking) on a network with remote terminals, or it can stop the use of a suspected stolen or lost ATM card. 3. An introductory message played over a telephone connection when the user has called a local or long-distance pay-per-call service. It may include consumer protection information or specifics about restrictions or potential call costs. Calls that may exceed charges of a specified amount may be required to include a kill message in the first few seconds of the call to allow the caller to hang up before charges accrue.

In spite of mandatory preambles and kill messages, there were complaints to and by the Federal Trade Commission (FTC) in the mid-1990s that telephone service vendors were abusing so-called toll-free 1-800 numbers in a variety of ways, including forwarding 1-800 calls to 1-900 numbers or otherwise manipulating the system to rack up charges on the callers' phone bills.

KILL message 1. A software message that causes a client/server connection to be closed by the related server for a variety of reasons, depending upon the system. For example, in Internet Relay Chat, it may be automatically invoked when duplicate nickname entries are detected; both entries are removed with the expectation that only a single nickname will reappear. This maintains global uniqueness. The KILL message may be available to operators but ideally should be handled by servers. See RFC 1459.

kilo- (*abbrev.* – k or K) Prefix for one thousand (1000), or 10^3. 10 kilograms = 10,000 grams when used for weights and measures. When used in the context of computer data, more commonly it is capitalized, as in Kbps (kilobits per second), and represents 1024. See k, K.

kilocharacter One thousand characters. See kilosegment.

kilosegment One thousand segments, with each segment consisting of up to 64 characters. It is used as a billing measure in some systems, such as X.25.

kilovolt-ampere kVA. A unit of apparent power. This is a general measure of power consumption for nonresistive devices such as certain types of lighting and computer components.

kilowatt kW. An SI unit of power required to do work at the rate of 1000 joules per second. See joule, kilowatt-hour, watt.

kilowatt-hour kW-hr. A unit of the energy used to perform work as measured over a 1-hour unit of time. One thousand watt-hours, or 3.6 million joules. This has practical applications as a description of the efficiency of different types of fuel, which can be expressed and compared in terms of kilowatt-hours.

kinescope 1. A cathode-ray tube (CRT) in which electrical signals, as from a television receiver, are displayed to a screen. 2. An early term for a motion picture, and probably the inspiration for the term cinemascope. In Britain, the term cinematograph was used to indicate a motion picture or motion picture camera.

kinetograph A device patented in 1889 by Thomas Edison for photographing motion picture sequences. See kinetoscope.

kinetoscope A device patented in 1893 by Thomas Edison for viewing a sequence of pictures, based upon the work of earlier experimenters going back as far as 1883. The loop of film images was illuminated from behind and viewed through a rapidly rotating shutter, thus creating a small motion picture film. See kinetograph.

King, Jan The young engineer who coordinated a number of significant amateur radio telecommunications satellite projects, starting with Australis-OSCAR 5 and continuing with the AMSAT satellites. King has written articles and technical reports on some of these activities, many of them for the QST journal. See AMSAT, OSCAR.

Kingsbury Commitment An important event on December 13, 1913 in which the U.S. Attorney General, James McReynolds, informed AT&T of violations of the Sherman Antitrust Act of 1890. AT&T voluntarily gave up controlling interest in the Western Union Telegraph Company, and agreed to stop buying up the independent telephone companies without first obtaining approval from the Interstate Commerce Commission (ICC). AT&T further agreed to provide independent phone companies with access to the long-distance network.

The Kingsbury Commitment derives its name from Nathan C. Kingsbury, the AT&T vice president who was appointed by Theodore Vail to correspond with the Attorney General. It is sometimes colloquially called the Kingsbury compromise. See Modified Final Judgment.

Kirchhoff, Gustav Robert (1824–1887) A German physicist who conducted pioneer work in spectroscopy and followed up on Ohm's work by providing further information and a more advanced theory of the flow of electricity through conductors. He also made adjustments to Huygen and Fresnel's description of the behavior of wavefronts by introducing the obliquity factor. See Huygen's integral, Kirchhoff's laws.

Kirchhoff's laws Laws for the flow of current first described in 1848 by G.R. Kirchhoff:

1. The current flowing to a given point (node) in a circuit is equal to the current flowing away from that point.
2. In any closed path in a circuit, the algebraic sum of the voltage drops equals the algebraic sum of the electromotive forces in that path.

KIS See Knowbot Information Service.

KISS Keep It Simple Stupid. A tongue-in-cheek, but all-too-relevant design and management philosophy.

Google is an excellent search engine on the Internet whose success is, in part, due to the relative absence of bells and whistles. The early Apple Macintosh computers held to this philosophy as well, developing a one-button mouse when many others were using mice with two or three buttons, and maintaining standards for the operating system that enabled new users to quickly figure out how to use it.

KISS method A system-independent architecture related to information systems modeling, which is described in terms of object-oriented concepts by its author, Gerald Kristen. KISS concepts are presented in a series of stages, and the model and presentation are sufficiently different from other works in the field of object-oriented (OO) programming that it has not excited a lot of interest in the OO programmers' community.

Kittyhawk A line of very small-sized (less than 2 × 3 in.) 20- and 40-MByte, 44-pin IDE hard drives distributed by Hewlett Packard for use in palmtop text and PDA computers.

KLAS-TV An early commercial adopter of all-digital advanced television (ATV) technology. KLAS-TV is a Las Vegas CBC affiliated owned by Landmark Communications, Inc., a privately held media company. Advanced television technologies were researched in the early 1990s by the Advisory Committee on Advanced Television Service (ACATS), with digital technologies coming in late in the process, but then becoming the central focus of ACATS evaluations. The ACATS Final Report was presented in 1995, paving the way for broadcast stations in North America to begin to implement higher quality, standardized digital television services. KLAS-TV was the first broadcaster in the Las Vegas region to offer end-to-end digital technology and transmissions, in April 2000. Prior to offering digital subscriber services, KLAS-TV delivered digital programming 2 years in a row to the National Association of Broadcasters (NAB) annual trade show at a Las Vegas convention center, establishing themselves as one of the pioneer commercial providers of all-digital broadcast programming.

System components for KLAS-TV equipment were developed and provided by Harris Broadcast Communications, who also provided the radio frequency Test Bed used in the ACATS evaluation of prototype digital broadcast systems. See Advisory Committee on Advanced Television Service, Harris Broadcast Communications.

Kleinrock, Leonard (1934–) In 1976, authored *Queueing Systems Volume II – Computer Applications*, a publication which helped the spread and acceptance of packet-switching technology.

Kleist, Ewald Christian von See von Kleist, Ewald Christian.

kludge, kluge Patchwork, improvised, or makeshift hardware or software, which can result from (1) time or material constraints, (2) sloppy workmanship, lack of foresight, (3) communication problems between decision-makers and implementors, or (4) staff changes or design changes during a project.

Kludge usually has a negative connotation, especially with software that tends to be sluggish from lack of structure and optimization, while well-conceived, but time-constrained projects are more often called "quick-and-dirty." Even well-begun projects can become kludgy after awhile, in which case engineers will generally advise, "Time for a ground-up rewrite!"

Klystron, klystron From the Greek "klyzo." A high-vacuum electron tube that uses electric fields to cause the "bunching" of electrons into a well-focused beam. The beam's kinetic energy is converted and amplified into ultra-high frequency radio waves (microwaves). Klystron was established as a trademark and as such is spelled with a capital letter, though the term is now also used generically for the historic line of Klystron tubes. Klystrons were used widely as oscillators and applied to radar transmitters until they were superseded for some applications by cavity magnetrons. Current commercial tubes are long-life, reliable, remanufacturable components.

Klystron Electron-Tube Inventors

Top: The Klystron and its inventors at Stanford in 1939. Clockwise from the left are Sigurd Varian, David Webster, William Hansen, John Woodyard (a graduate student), and Russell Varian. Small Klystrons were used for radar, navigation, and communications applications. [Copyright 1939 and 1951 Stanford News Service archives; used with permission.]

The Klystron evolved from pioneer versions of the magnetron which were developed in the 1920s. Russell H. Varian and Sigurd F. Varian respectively designed and constructed the first Klystron prototype at Stanford, beginning in 1937, in collaboration with William Hansen. The July 1937 notes of Russell Varian describe the "Rumbatron Oscillator or Amplifier" and input from William Hansen regarding a "spherical rumbatron with one core reaching to the center." The rumbatron moniker was based upon Hansen's previous work on cavity resonators. The Varians combined this with principles of velocity modulation to create the Klystron, publicly announced in 1939.

While Hansen went on to develop linear accelerators, the Varian brothers and other scientists, including their friend and associate, Edward L. Ginzton, used the technology to develop radar systems during World War II, cofounded Varian Associates in 1948.

The radio waves in historic Klystron tubes are drawn from a high-voltage electron beam in such a way that much of the energy is dissipated, resulting in low efficiency levels compared with succeeding technologies. However, klystron technology was never fully superseded, especially in broadcast applications, and some scientists felt the efficiency could be improved. The Lewis/Varian version of the klystron technology, developed in the mid-1980s, recovers the wasted energy by recycling the electron beam, effectively doubling the usable portion of the radio frequency and, consequently, reducing power consumption in UHF television transmitters. Commercial production of the new technology began in 1990 on the product now known as the CPI MDC klystron.

Commercial broadcasting klystrons come in a variety of configurations supporting frequency bands such as the C-band at different channel capacities (usually 6, 12, or 24). There may be separate tubes for image and audio amplification. Cooling with water is typical in klystron applications where the tube becomes hot. See bunching; Ginzton, Edward; cavity magnetron; magnetron; Varian, Sigurd and Russell; Varian Associates.

KMI Corporation A fiber optics and telecommunications market research and consulting firm founded in 1974. KMI is a subsidiary of PennWell and part of PennWell's Advanced Technology Division, with research headquarters located in Providence, RI. KMI publishes *Fiberoptics Market Intelligence*® as a semi-monthly newsletter and provides various commercial market studies, fiber optic systems wall maps, marketing workshops, and databases. See FiberFax, Fiberoptics NewsBriefs, Fiberoptic Undersea Systems, Undersea News Service.

KMID key material identifier. A term associated with Message Security Protocol.

KNET See Kangaroo Network.

knife-edge focusing Focus (wavefront) testing, as for scientific instruments such as telescopes. This is an aid to visually assessing a lens or mirror. Often a Ronchi screen or knife-edge bushing is placed as a mask at right angles to the light path before or behind the point of focus in relation to a lens or concave mirror to provide the knife edge. Focusing is then adjusted until the edge "cuts" the light beam exactly at the point of focus. If the lens/mirror is perfectly shaped, the lens aperture should darken evenly. If there are aberrations, the combination of light rays before or behind the knife edge will create lighter spots.

The technique, developed by Foucault (sometimes called *Foucault's knife-edge test* or *Foucault's test*), facilitates that testing of the surface quality of various lenses and reflecting elements. It is actually a special case of the Ronchi test, for components where the radius of curvature is double the focal length.

For more complex instruments, or those commercially fabricated, interferometric testing is now routine. However, for home-brew or less expensive components, the knife-edge test is still practical and can be set up with readily available materials.

A wavefront spatial filtering technique was suggested by O. von der Loehe (1988), developed from the Foucault test. It uses two orthogonal knife edges to measure the two components of a wavefront gradient. See pyramidic wavefront sensor, Ronchi grating.

Knife-Edge Focusing

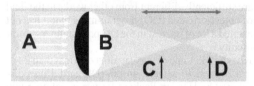

Knife-edge focusing is a fairly straight forward concept. In this example, incoming (incident) light that has been collimated (parallel beams) is aimed at a knife edge (B) used as a reference mask to determine the best focal point. A Ronchi grating or knife-edge bushing may be used as the mask. It is moved back and forth through the focal plane such that the patterns in the grating change as the light interacts with the pattern and will appear clear and even (rather than striped or patterned) at the sharp focus point between the before focus (C) and beyond focus (D) points. It takes a little practice and is a bit easier with a Ronchi bushing (rather than with a Ronchi screen).

Knife-Edge Testing – One Example

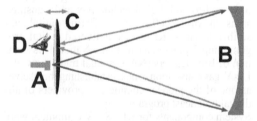

A knife-edge tester (Foucault tester) can be made fairly readily from a few materials. It is important that the support for the tester is stable and still. The light source (A) projects onto the surface being tested (in this case, a concave reflector – B) and is reflected back through the knife-edge Ronchi grating or bushing (C) to the eye of the viewer (D). The eye is positioned so that the returning light is blocked by the knife edge. Aberrations in the evenness of the surface will show up as bright spots, indicating a flaw.

Many commercial knife-edge testers magnify the image so that flaws are more readily apparent.

---

knife switch In old telegraph keys, a type of switch that could short key contacts in a series so the idle line was in a steady mark condition, with current flowing – also called a break switch. Opening the knife switch interrupted the current in all the sounder electromagnets on the line so that operators were made aware that a message was imminent.

knockout A raised or indented region of a receptacle which can be punched out or otherwise removed to provide access for wires, jacks, or other fittings. Common in general purpose electrical junction boxes.

Knowbot Information Service KIS. A uniform client/server means of interacting with, and displaying, information from a variety of remote directory services typically found on Unix systems, such as Finger, Whois, and others. A query to KIS uses white pages services to these types of systems and displays the results of the search in a consistent format. See Knowbots.

Knowbots In a Knowbot Information Service, programs that search and retrieve information from distributed databases as requested by the user. Knowbots may carry the information or may pass it among one another. See Knowbot Information Service.

knowledge base system, knowledge-based system, expert system A computerized system of storing the accumulated knowledge of humans in a system which accesses and manipulates the information using artificial intelligence programming strategies and rules to accomplish information delivery and problem-solving at a sophisticated level.

knowledge engineering Acquisition of knowledge from a human expert or experts and its incorporation into a computerized expert system.

Knowledge Interchange Format KIF. Computer language for the manipulation of knowledge data and interchange of knowledge among disparate programs. Intended not as a user interface, but as an internal representation for knowledge within programs or related sets of programs.

Knowledge Query and Manipulation Language KQML. A high level language that is part of the DARPA Knowledge Sharing Effort. It is a language and messaging protocol for exchanging information and knowledge as part of the larger project to create technology to facilitate development of large-scale, shareable databases. KQML can be used to support interoperability among intelligent agents in distribution applications. See Reticular Agent Definition Language.

Knowlton, Kenneth C. (1931–) An American pioneer in computer graphics and researcher at Bell Laboratories, Knowlton studied and experimented with many aspects of computer imagery, computer art, motion automation, and fast data storage, often in collaboration with L.D. Harmon.

In 1959, Knowlton co-authored a report for the U.S. Patent Office entitled *A Notation System for Transliterating Technical and Scientific Texts for Use in Data Processing Systems*. As his explorations turned to computer imagery, Knowlton grasped a concept often overlooked by artists using traditional "paint" tools in a computer environment: the computer can be used to generate types of art and images that cannot (or should not) be executed by human hands. In the course of his research, he and his collaborators produced a rich variety of mosaics, plots, grayscale

images, and even computer-based films.

Remember that these pioneers had to invent and write their own software systems for accomplishing their goals. In the 1960s, no commercial desktop paint and animation programs were available off-the-shelf (they weren't common until 1986). Many goals were attained by typing in raw numbers in low-level languages. To aid him in automating the animation process, Knowlton developed a motion language for creating short films. Together with filmmaker Stanley VanDerBeek, he developed a series of abstract animated films called *Poem Fields*. The tools invented to enable artistic visions to be derived from computer technology are applicable to many areas of scientific research and manufacturing including digitization, pattern recognition, artificial intelligence, robotics, and more. In 1979, Knowlton coauthored articles on visual perception and the use of sign language as a form of telephone communication, with Vivien Tartter.

Over the decades, Knowlton's interests in image processing and image generation remained strong and he has been awarded numerous patents for his work. As examples, in 1990, Knowlton and Wang Laboratories applied for a U.S. patent for what is essentially an ebook, an electronic book viewed through a video display (#5283864, 1994). Following this, in 1994, Knowlton and Wang applied for a patent for a document processing system including an optical scanner (#5517586, 1996). See Harmon, Leon; Shroeder, Manfred.

Knuth, Donald (1938–) Knuth's texts on data structures and algorithms are heavily used, and widely considered by programmers to be the "bible" of important basic programming structure information. Fundamental search and distribution trees and much more are in the Knuth texts. It would be difficult to develop sophisticated database software without them. Knuth is also known for authoring the powerful document system called TeX (*pron.* tek), which is one of the few that can handle complex mathematics-related text formatting.

KOLD-TV A pioneering digital broadcast station, KOLD 13 began using a networked digital video server in daily broadcasts in 1995. In October 1998, the Federal Communications Commission granted a number of commercial digital television station licenses, including Station KNSV-TV, Phoenix, and KPHO-TV, Phoenix.

Kompfner, Rudolf (1909–1977) An Austrian architect and engineer who worked in England as an apprentice architect in the mid-1930s where he invented a split-beam oscilloscope tube. Much of his physics background was self-taught, hands-on knowledge. Following a World War II interment, Kompfner was put to work on the development of a low-noise Klystron amplifier. By 1943, he had invented a traveling-wave tube after which he became a scientific officer and Distinguished Scientist. Following the war, he earned his doctorate at Oxford and went to work for Bell Laboratories, studying microwave tubes and continuing work on traveling-wave tubes.

K

He became one of the lead workers on the historic Echo satellite project.

Kompfner has received many engineering awards, including the Duddell Medal and the Stuart Balantine Medal of the Franklin Institute. In 1973, he received the IEEE Medal of Honor for his contributions to global communications technologies for his development of the traveling-wave tube. See Klystron, traveling-wave tube.

Konexx Modem Koupler A battery-powered commercial modem/modem adapter combination from Unlimited Systems. Konexx enables a modem to be hooked into various types of phone lines and cellular phone systems while traveling.

Kotel'nikov, Vladimir Aleksandrovich (1908–) A Russian contributor to fundamental research in communication theory, and theory and practical research in astrophysics. He is especially well known for the use of radio waves to locate and measure distances to planetary bodies and for contributions to early satellite communications. He is also remembered for publishing "On the carrying capacity of 'ether' and wire in electrical communication" (1933) that describes a theory of sampling and the representation of a continuous signal from discrete samples. His theories were later supported by the work of Claude Shannon. Kotel'nikov served as V.P. of the USSR Academy of Sciences and chaired the Scientific Council on radio astronomy. While at the Institute of Radioengineering and Electronics, he initiated research in the submillimeter wave band. See pulse code modulation; sampling theorem; Shannon, Claude.

KPI See Key Performance Indicator.

KQML See Knowledge Query and Manipulation Language.

Krum, Charles and Howard An American father and son team who worked together in the early 1900s to develop and patent a variety of telegraph transmitters and printing machines. One of their early successes was a printer created by interfacing a modified typewriter with a telegraph line, developed at the end of 1908. With a mechanical apparatus ready to use, it became necessary to develop some way to synchronize the pulses and the printing. For this, Howard Crumb applied for a start-stop patent in 1910.

krypton laser A type of gas laser which is primarily krypton that can be used to produce intense red light, or when used with certain optic enhancements, several colors. This is similar to an argon laser, except that it produces a little less light; sometimes argon and krypton are combined. Krypton lasers are typically water-cooled.

Kurz, Karl Information on this German experimenter is scarce (he seems to be overshadowed by Barkhausen), but he apparently collaborated with Heinrich Barkhausen in discovering Barkhausen-Kurz oscillations. See Barkhausen-Kurz tube.

KS See Kearney System.

KTH Kungliga Tekniska Hogskolan. The Royal Institute of Technology in Stockholm, Sweden. The institution provides education in optical communications within its photonics courses.

KTI Key Telephone Interface.

KTS 1. See key telephone system. 2. See Kill the Spams.

Ku-band A range of microwave broadcast frequencies from approximately 11 to 14.5 GHz which is further subdivided into fixed satellite service (FSS) and broadcasting satellite service (BSS). Ku-band is used primarily for data transmission, private networks, and news feeds. Satellites transmitting Ku-band signals tend to be powerful enough for the receiving dish to be small and convenient. Uplinks are in the 14- to 14.5-GHz range and downlinks in the 11.7- to 12.2-GHz range.

In November 2000, The Federal Communications Commission (FCC) extended permission to providers of nongeostationary satellite services to operate in certain segments of the Ku-band and issued rules and policies to govern their operations. It was felt that this would stimulate new competitive services such as high-speed Internet access, telephony services, and media broadcasts. It was further hoped that satellite transmission availability would increase services to rural areas. The FCC determined that Multichannel Video Distribution and Data Services (MVDDSs) could operate in the 12.2- to 12.7-GHz frequencies without interfering with incumbent Broadcast Satellite Services (BSS). See band allocations for a chart. See broadcasting satellite service, direct broadcast satellite, fixed satellite service.

KV Bell Telephone jargon for key telephones (K = key, V = voice). The term was derived from the Universal Service Ordering Code (USOC) commonly used until the time of the AT&T divestiture in the mid-1980s.

KVW Bell Telephone jargon for wall-mounted key telephones. See KV.

KWH See kilowatt-hour.

λ, Λ Lambda, the 11th letter of the Greek alphabet, used to symbolize wavelength, especially optical wavelength.

L *symb.* inductance. See inductance.

L CXR A backbone communications system based upon L carrier coaxial technology developed in the 1930s in the U.S., known generically by the public through warning signs as the *Transcontinental Cable* system. The system was intended for key government communications and civil defense, interconnected through facilities with underground repeaters, and maintained by Bell telecommunications workers, designed to transmit both telephone and television signals.

In the postwar years, the system was upgraded about every 10 years. The L-1 system of the 1940s gave way to an L-3 system by the 1950s and 1960s. (L-2 was used for a special installation between Washington, D.C. and Baltimore.) Later, through the use of repeaters every couple of miles, 20-tube L-4 cable centers were established approximately every 150 miles. By the 1970s, L-5 circuits had been established. The capacity of the system improved from a few MHz in the 1940s to 57 MHz in the 1970s with voice capacities increasing from a few hundred voice channels to more than 100,000.

Following the Cold War, the repeater stations, which resembled stone garden sheds over concrete vaults every few miles along the cable route, were sold. The old physical infrastructure was gradually replaced with fiber optic cables (with 20 times the capacity of the L-5 system) and satellite links. See L carrier.

L carrier An older, analog, frequency division multiplex (FDM), long-haul phone system that was common before digital services became prevalent (metropolitan areas using FDM more commonly used N carriers). U.S. Department of Defense L carrier communication systems in the post-World War II period stretched in two main east-west links from Washington, D.C., west to California and south to the tip of Florida, with many smaller trunk tributaries. See L CXR.

L multiplex LMX. Analog multiplexing circuits in L carrier coaxial transmissions media.

L Series Recommendations A series of ITU-T recommended guidelines for construction, installation, and protection of cables and other elements of outside plant telecommunications. These guidelines may be purchased from the ITU-T. Since ITU-T specifications and recommendations are widely followed by vendors in the telecommunications industry, those wanting to maximize interoperability with other systems need to be aware of the information disseminated by the ITU-T. A full list of general categories is listed in Appendix C and specific series topics are listed under individual entries in this dictionary, e.g., K Series Recommendations. See L Series Recommendations chart.

L system See L CXR.

L1 cache See level 1 cache.

L2 cache See level 2 cache.

L1 & L2 Designations for the two radio frequencies (1227.6 and 1575.42 MHz) broadcast by Global Positioning System (GPS) satellites. See Global Positioning System.

L2F 1. See Layer 2 Forwarding Protocol. 2. Legacy to the Future.

L2TP See Layer 2 Tunneling Protocol.

L2TPext See Layer 2 Tunneling Protocol extensions.

L-band A portion of the electromagnetic spectrum assigned for radio communications, ranging from 500 to 1500 MHz. Within this range, the frequencies between 950 and 1450 MHz are set aside for mobile communications. Global Positioning Systems (GPSs) use the L-band frequencies, as do some of the planet probe systems. See band allocations for chart.

L-band, optical In optical communications, an ITU-specified transmission band in the 1565- to 1605-nm frequency range. Until 2001, this spectrum was not widely supported.

The development of better efficiency, high-density erbium doping in the early 1990s facilitated the development of practical L-band systems. Benchtop erbium-doped fiber amplifiers (EDFA) may now be used with DWDM L-band networks (as they are for C-band), enabling optical amplification without having to convert optical signals to electrical and back again.

In January 2001, Lucent Technologies announced it had completed the first installation of a C+L-band optical network for NTT Communications (Japan). This increased capacity by opening up the previously

ITU-T L Series Recommendations

Recommendation/Description	Recommendation/Description
L.1 Construction, installation and protection of telecommunication cables in public networks	L.23 Fire extinction – classification and location of fire extinguishing installations and equipment on premises
L.2 Impregnation of wooden poles	L.24 Classification of outside plant waste
L.3 Armouring of cables	L.25 Optical fiber cable network maintenance
L.4 Aluminium cable sheaths	
L.5 Cable sheaths made of metals other than lead or aluminium	L.26 Optical fiber cables for aerial application
L.6 Methods of keeping cables under gas pressure	L.27 Method for estimating the concentration of hydrogen in optical fiber cables
L.7 Application of joint cathodic protection	L.28 External additional protection for marinized terrestrial cables
L.8 Corrosion caused by alternating current	L.29 As-laid report and maintenance/repair log for marinized terrestrial cable installation
L.9 Methods of terminating metallic cable conductors	
L.10 Optical fiber cables for duct, tunnel, aerial, and buried application	L.30 Markers on marinized terrestrial cables
L.11 Joint use of tunnels by pipelines and telecommunication cables, and the standardization of underground duct plans	L.31 Optical fiber attenuators
	L.32 Protection devices for through-cable penetrations of fire-sector partitions
L.12 Optical fiber joints	L.33 Periodic control of fire extinction devices in telecommunication buildings
L.13 Sheath joints and organizers of optical fiber cables in the outside plant	
	L.34 Installation of Optical Fibre Ground Wire (OPGW) cable
L.14 Measurement method to determine the tensile performance of optical fiber cables under load	L.35 Installation of optical fiber cables in the access network
	L.36 Single mode fiber optic connectors
L.15 Optical local distribution networks – factors to be considered for their construction	L.37 Fiber optic (non-wavelength selective) branching devices
L.16 Conductive plastic material (CPM) as protective covering for metal cable sheaths	L.38 Use of trenchless techniques for the construction of underground infrastructures for telecommunication cable installation
L.17 Implementation of connecting customers into the public switched telephone network (PSTN) via optical fibers	L.39 Investigation of the soil before using trenchless techniques
	L.40 Optical fiber outside plant maintenance support, monitoring and testing system
L.18 Sheath closures for terrestrial copper telecommunication cables	L.41 Maintenance wavelength on fibers carrying signals
L.19 Outside plant copper networks for ISDN services	L.44 Electric power supply for equipment installed as outside plant
L.20 Creation of a fire security code for telecommunication facilities	L.45 Minimizing the effect on the environment from the outside plant in telecommunication networks
L.21 Fire detection and alarm systems, detector and sounder devices	L.46 Protection of telecommunication cables and plant from biological attack
L.22 Fire protection	L.47 Access facilities using hybrid fiber/ copper networks

unused L-band frequency range. The improvement was made possible by Lucent's new DWDM L-band optical amplifier and dispersion-shifted fiber that is more suited to carrying multiple wavelengths in the L-band than in the C-band. By the end of 2001 a number of leading optical firms were supplying L-band components, such as dynamic gain equalizers that could be configured by the customer for L- or C-band ranges.

Optical circulators optimized for DWDM L-band applications typically come in two grades (A & B) and feature low insertion loss, high isolation and minimal distortion.

LAAS See Local Area Augmentation System.

label 1. A symbol, or group of symbols, often mnemonic, that identifies or describes an item, routine, record, file, application, or process. 2. In programming, a reference point, usually for a procedure, function, or subroutine. The label may have a specific meaning to the interpreter or compiler that processes the code instructions or it may be a convenience for the programmer for organizing code and simply be ignored when the code is processed, depending upon the environment. 3. In networks, a convenient name for a device to facilitate access. For example, a printer on a network known to the system as LZPTX5103, may be assigned a label such as *Building 3 Laser,* to make it easier for people to recognize. 4. In ATM and Frame Relay networks, a short, fixed-length identifier that facilitates packet/frame forwarding. See label switching.

label swapping In label switching networks, a router will commonly assign a new label to the received transmission in preparation for forwarding it to the next "hop" or leg in its journey toward the destination. See label switching.

label switching In ATM networks, a switching mechanism intended to take advantage of the flexibility and scalability of Internet Protocol (IP) routing. Label switching combines some of the aspects of Layer 2 switching with Layer 3 routing, so the distinction between switching and routing is not as clear as in other transmission schemes. A label-switching router can forward IP datagrams based upon a label associated with the packet and will usually assign a new label for the subsequent routing "hop" in the connection. See Multiprotocol Label Switching, optical-label switching, tag switching.

label-switched path A network route that is established based upon a label associated with a data packet in a network, with a new label typically assigned dynamically for each subsequent hop in the path (most routers do not spport static LSPs). The path must be established before test data or communications data can be effectively routed through the LSP. A signaling protocol is typically used to set up paths. See label switching.

labeled multiplexing In intcgratcd scrviccs digital networks (ISDN) and ATM networks, a routing mechanism in which multiplexing is carried out by concatenating blocks of channels with different identifiers in their labels.

labeling algorithm 1. A means of calculating the shortest path in network routing. 2. A means of inserting copy protection labels into data, such as compressed video data, so the information can be tagged, or otherwise identified, and can only be written or read under prescribed circumstances. This type of system is being developed to enable vendors to provide digital services to home consumers without fear that the products will be widely pirated and redistributed. It is being experimentally applied to the design of copy-protectable mass storage devices.

Laboratory for Computer Science LCS. An interdepartmental facility at the Massachusetts Institute of Technology (MIT) engaged in computer science research and engineering. LCS was founded in 1963 with support from the U.S. Defense Department as a result of the launching of the Sputnik satellite.

Members of the LCS have been involved in many important historical computer network developments including the ARPANET, Ethernet, Internet, and World Wide Web. They have also been involved in research and development of encryption technologies, electronic mail, and computer accounting software. Many prominent companies have been founded by LCS participants, included 3Com, Lotus Development, and RSA Data Security. LCS works cooperatively with relevant departments and with the MIT Artificial Intelligence Laboratory.

LAC Loop Assignment Center.

LACE See Low-power Atmospheric Compensation Experiment.

lacing cord A strong cord, sometimes coated or waxed, used to bundle wires strung along the same path.

ladar laser Doppler radar. See Doppler, laser, radar.

LADT See Local Area Data Transport.

lag To delay, linger, slacken, slow, be retarded, or tarry. Lag occurs in computer applications when the speed of the system is unable to match the speed of the interaction of the user. Lag is characteristic of dialup modem communications, where the speed of the data transmission doesn't match the speed of the computer processor. Lag occurs on data networks when congestion occurs; that is, the number of packets may exceed the ability of the system to handle and transport them. See cell rate, hysteresis, leaky bucket.

LAGEOS I The Laser Geodynamics Satellite, developed and launched by the Marshall Space Flight Center in May 1976. LAGEOS was one of the earlier remote-sensing satellites used for Earth sciences research.

Lakeside Programming Group A collaboration of programming friends, which included Bill Gates, Paul Allen, and Ric Weiland. The group created a payroll program in COBOL for a company in Portland, Oregon. They were informally named after the Lakeside private school attended by its members. At thc samc timc, thc group had a programming contract to build scheduling software for a school.

Eventually Gates and Allen formed Traf-O-Data in the early 1970s, to create a traffic analysis program. The partnership that led to the formation of the

Microsoft Corporation. See Gates, William; Allen, Paul; Microsoft Corporation.

LAI See Location Area Identity.

LAM 1. line adapter module. 2. See Lobe Attachment Module.

LAMA See Local Automatic Message Accounting.

lamb dip In laser technologies, a point in the absorption spectrum that dips (is lower in intensity) in relation to the overall absorption cross-section (which is generally Gaussian). When plotted, it graphically resembles a small crater (there may be small subsidiary peaks within the overall transmittance dip). The location of the lamb dip provides a reference for transition frequencies that are Doppler-shift-free, which is useful in certain laser spectroscopy applications.

A lamb dip may be observed in laser operation or may be generated by a sodium cell, for example, for tuning and actively stabilizing a laser (e.g., a dye laser). Passing a same-frequency probe beam through a laser in the direction opposite the laser pumping beam reveals the transitional lamb dip pattern.

lambda (*symb.* – λ) The 11th letter of the Greek alphabet, used as a symbol in a number of mathematical and logical contexts. Lambda symbolizes the null class, von Mangoldt's function, and wavelength.

Lambda-Connect Project A Lawrence Livermore ultracomputing development project funded by the Laboratory Directed Research and Development Program. In recent years, the performance gap between processing units and the links between the processors has been widening. This interesting project in parallel multiprocessing seeks to narrow the gap by overcoming the bottlenecks in transmitting data among processors via traditional electronic connections.

The LL team believes it can replace electron flow with photon flow (light pulses of different wavelengths). Speed improvements of up to 32 times are considered feasible. The use of optical interconnects also makes it practical to pack microprocessors in higher densities which, presumably, also improves data transport speeds and processing times.

Ultracomputing projects are of particular interest to strategic and scientific computing applications and may be used by the U.S. Department of Energy (DoE) Accelerated Strategic Computing Initiative for nuclear technologies simulations and testing and to the Department of Defense (DoD) for strategic planning and intelligence gathering and processing. See lambda switching, optical burst switching.

lambda switching Also called wavelength switching, this is a technique used in high-speed optical networks to switch individual optical wavelengths into different paths to route data through the network. It is more akin to circuit switching than packet switching and can be implemented as end-to-end connections. Fiber optic networks typically use lambdas (wavelengths) in the near infrared spectrum, translating into frequencies of around 100 terahertz (THz). In conjunction with multiplexing, which enables many separate wavelengths to be transmitted along a single fiber, lambda switching can be used to create virtual circuits. Qwest and other carriers began

to deploy lambda switching in their telecommunications networks ca. 2000 and 2001. See dense wavelength division multiplexing, Gilder's law, lambda, optical cross connect.

lambdasphere The environment within optical transmissions paths that channels many wavelengths. George Gilder is a speaker and well-known proponent of optically switched networks and is credited with coining the term *lambdasphere*. See Gilder's law, lambda switching.

lambert (*symb.* – L) A centimeter-gram-second (CGS) system unit of luminance, equal to the brightness of an ideal diffusing surface that radiates or reflects light at 1 lumen per square centimeter.

Lambert's law, Lambert's cosine law The reflection of radiant energy incident upon a small surface in a particular direction is proportional to the cosine of the angle between the reflected direction and the perpendicular (normal) of the surface.

This relationship provides a means to calculate how much of the incident light (or sound) is reflected, assuming the reflected light is constant in all directions (essentially an ideal diffuser) and the *angle of incidence* and associated *angle of reflectance* are small.

Given these parameters, the perceived intensity of the light is independent of the angle at which the reflecting surface is viewed, but will vary according to the angle of incidence at which the radiant energy encounters the diffusing surface, as governed by Lambert's law. Stated another way, the intensity of the incident light is relative to the angle of incidence. Body reflectance models follow Lambert's law.

Lambert's law was a prevailing relationship until the 1980s. Then, the advent of graphics and acoustics modeling software and semiconductor-based instruments (e.g., goniometers) caused it to be more closely scrutinized for its generalizability for wider angles and different surfaces.

In graphics, Lambert's law has been adjusted and generalized for smooth surfaces by L.B. Wolff, thus overcoming some of the "small angle" limitations of Lambert's law. In acoustics, Lambert's law's generalizability to diffusers was questioned, leading to the development and dissemination of diffusion coefficient data in much the same way absorption coefficient data has been available. See Bouguer's law, incidence angle.

Lambertian reflector A material that reflects light in many directions (essentially an ideal reflector), thus diffusing it according to Lambert's law. See Lambert's law.

lame A colloquial term often used in programming or electronics to derisively describe an uninspired or poor device, program, or solution to a problem. The term is based upon the concept of limping along on one leg and thus not embodying positive qualities such as speed, efficiency, or grace. A person who frequently comes up with weak or lame comments or solutions may be dubbed a "lamer."

lame delegation On the Internet, a situation in which one or more authoritative domain name servers (DNS) that convert Internet Protocol (IP) addresses into registered names are not responding correctly for the specified domain (which may have been called by a system higher in the hierarchy). Lame delegation occurs for a number of reasons:

- Secondary servers, which are expected to function as authoritative servers, may not respond correctly. If a secondary server is unspecified, or not working correctly, the situation is lame in the sense that the second "leg" isn't working properly. Since secondary name servers are not mandated, only recommended, this is not an uncommon occurrence; most people associate this meaning with the term.
- The name server may respond incorrectly from a cache rather than directly or may be configured incorrectly.
- There may be a lack of communication. If users change domain names, manage their own IP numbers, or switch Web hosts without coordinating with the name serving authority and the ISP, all sorts of lame delegation errors can occur.

Lame delegations can result in email delays, lookup slowdowns, and even a Web site "disappearing" from the Internet (not being accessible). If the primary and secondary servers are on the same physical machine and that machine goes down, there is no backup name server and a lame delegation would be in effect until the system came back online. See domain name server, lame.

laminate *n*. A structure composed of layers, often tightly sandwiched or bonded together. Laminated materials are often used in electronics, from early voltaic piles, which sandwiched moistened materials between layers of metal plates, to magnetic cores and semiconductors, in which layers of various materials are combined according to their electromagnetic properties. An individual layer in a laminated structure is called a *ply*. See semiconductor, thin film, voltaic pile.

Lamm, Heinrich (ca. 1908–) A German inventor who studied medicine at the University of Munich, Lamm was interested in creating an analogy to insect vision using a bundle of optical fibers. He acquired fibers from the Rodenstock Optical Works and painstakingly positioned the fibers to create a bundle that could transmit an image for a short distance. The practical limitations deterred Lamm and a similar concept had been patented already by Clarence W. Hansell who collaborated with television inventor John L. Baird, so Lamm ended up as a surgeon in America rather than a well-known inventor of fiber optics transmission systems. See Hansell, Clarence; Kapany, Narinder.

lamp An illuminating device that converts energy into light (usually visible light). In its basic form, a lamp consists of a light-producing source and a holder such as a wick in an oil-holding vessel, bulb in a handheld, battery-operated container (flashlight), fluorescent bulb in a fluorescent receptacle, light-emitting diode

in an electronic device, or filamented vacuum bulb in a desk or floor stand.

Lamps were used in some of the earliest communications technologies. By blocking the light from a lamp to signify dark or light, a binary communications code could be devised to send signals over distances. A shutter made the process easier. Shuttered lamps were used for many decades to send signals overland and among ships at sea.

Lamps are also used as sources for photographic lighting, especially as "flash bulbs" to provide supplemental illumination in low-light conditions. Not all lamps emit light in the visible frequency ranges. Infrared or ultraviolet light cannot be directly seen by humans, but the subjects illuminated may be seen with special equipment that translates the reflected light into visible frequencies or other forms of energy. See Edison, Thomas A.

LAMP See Large Advanced Mirror Program.

lampblack A sooty, dark carbon dust deposited by a smoking flame, as on the inside of a glass lamp globe. While its presence in lamps is usually undesirable, lampblack has commercial applications in the fabrication of some types of resistors.

LAN See local area network.

LAN adapter A hardware peripheral device or card that connects a computer to a local area network (LAN). Not all computers require LAN adapters; some come equipped with network cards and ports ready to attach to various connectors, commonly 10Base-T or 10Base-2. However, some require an intermediary device between the network card and the network cable in the form of a LAN adapter.

LAN aware Applications, systems, and devices that can recognize and appropriately respond to an interfaced connection to a local area network (LAN). This usually involves communicating with other devices on the net as security permits, locking and unlocking files as needed, querying appropriately, etc. Some operating systems are designed to be LAN aware, others run with third-party software.

LAN Channel Station LCS. 1. In frame-based networking, a channel protocol for LAN/mainframe intercommunications. In IBM mainframe-related LCS, host applications define a consecutive pair of subchannels for channel reads/writes through TCP/IP. LCS enables a local area network (LAN) MAC frame to be transported and provides a command interface for activating/deactivating and querying LAN interfaces. 2. A Bus-Tech commercial control unit emulation (also known as 8232) that provides a pass-through for exchanging data between a local area network/wide area network (LAN/WAN) and a mainframe through TCP/IP.

LAN Emulation See asynchronous transfer mode, LANE.

LAN Manager An early, commercial OS/2-based multiuser network operating system intended to run over TCP/IP or NetBEUI protocols. Microsoft and 3Com's LAN Manager came in two versions: for MS-DOS, Microsoft Windows, and IBM's OS/2, and for Unix/UNIX connections.

LAN Protocols A catchall phrase for a wide variety of protocols developed for local area computer networks, such as AppleTalk, Ethernet, TCP/IP, IPX, and others.

LAN segment In Frame Relay, a LAN linked to another LAN using the same protocol through a bridge. See bridge, hop, router.

LAN Server A commercial multiuser network operating system from IBM, based on OS/2 and NetBIOS. LAN Server supports a variety of client computer operating systems, including OS/2, Microsoft DOS and Windows, and Apple Macintosh. It has been superseded by the OS/2 server version.

LAN switch A local area network (LAN) switch, in its simplest sense, is a stand-alone box with connections, which simply directs traffic along one or more pathways. However, with improved technology, LAN switches are incorporating more and more intelligent processing capabilities, and some are almost indistinguishable from routers. Some are available as modular peripheral cards that fit into multiswitch card chassis.

LAN switches can help reduce or more efficiently handle congestion and can improve response time on networks. They have the capability to redirect the World Wide Web queries of users to local caches, and reduce Internet queries and can help balance network traffic among servers.

LANCE Local Area Network Controller for Ethernet.

Land, Edwin Herbert (1909–1991) An American inventor who studied briefly at Harvard University, Land is best known for his development of polarizers, in the form of polarizing film sheets, in 1929. It is less well known that Land was a prolific inventor who earned more than 500 patents in his lifetime, most of them directly related to optics (especially photography). He cofounded the Polaroid Corporation (originally the Land-Wheelwright Laboratories) with G. Wheelwright, in 1937, to develop and market his inventions and demonstrated "instant" photography to the Optical Society of America in 1947. He received a patent for this historic photographic technology in 1951 (U.S. #2,543,181).

In 1982, Land left the Polaroid Corporation but continued research on human visual perception.

The Edwin H. Land Medal was established by the Optical Society of America (OSA) and the Society for Imaging Science and Technology in his honor in 1992.

Land attack, land.c A malicious software program that became prevalent on the Internet in 1997, the Land attack was similar to an earlier SYN attack program. Land was programmed to attack systems using Internet Protocol (IP) communications by interfering with the stack. Certain routers were also vulnerable. A broad segment of operating systems from Apple, Cisco, Microsoft, and Sun and several BSD systems were found to be vulnerable. Some machines crashed while others hung or experienced slowdowns, depending upon how robust the operating system was and whether it could still function while one aspect of it was caught in an endless loop through spoofing

that tricked the system into calling itself. A more Microsoft-aggressive version called latierra.c was later released on hacker sites. Vendors posted patches to reduce or remove the vulnerability and updated subsequent releases of software to reflect these changes. See Denial of Service attack, Trojan horse, virus.

land/groove recording, land and groove recording A high-density recording technology for magneto-optical media. DVD disc surfaces are divided into annular zones, keeping the length of a sector and the recording density mostly constant through the disc. The zones are divided into two types of tracks: land tracks and groove tracks. These tracks, in turn, are divided into sectors. The hierarchical division into smaller units is similar to the storage and access formats used on hard disks, with adaptations to fit the nature of optical disks. Data are recorded on the land (the higher surface) and the groove (the indented surface), while address information is encoded in the pits. Permanent data that are readable but not writable by the user are kept separate from the user recording fields. See wobbled groove, wobbled land groove.

LANDA Local Area Network Dealers' Association. This is now called NetPros by LANDA. The organization promotes and supports excellence among Canadian resellers, consultants, and systems integrators in the information technology (IT) industries. http://www.netpros.ca/

landline, land-line, land line Communications circuits, especially telegraph and telephone, which travel through terrestrial wires and stations. Many mobile units interface with landlines, so that even if a call originates as a wireless call, it may be completed as a landline call to extend distance and free up wireless channels.

Landsat A series of satellites first launched through federal funding in the mid-1960s for remote sensing of the Earth from space.

The Landsat Earth sensing system launches were initiated in 1966 as a response to the announcement of plans for launching civilian Earth Resources Observations Satellites (EROS). As a result, NASA began to plan a satellite launch in order to secure information on Earth resources and provide for national security provisions in space.

This led to the launch of the Earth Resources Technology Satellite (ERTS-1) in 1972, with similar satellites launched in 1975 and 1978. The program was renamed Landsat in 1975. Landsats 4 and 5 were launched in the early 1980s, and jurisdiction was transferred to the U.S. National Oceanographic and Atmospheric Administration (NOAA).

In 1986, during the Reagan administration, jurisdiction was changed to a commercial company, EOSAT, and the primary users became large institutions that could afford expensive satellite data. Complete archiving of data was not always undertaken. EOSAT designed and built Landsat 6, which failed on launch.

In 1992, legislation was passed to return future Landsat missions to the public, and Landsat became part of NASA's Mission to Planet Earth program in 1994. Planning began for the Landsat 7 project.

Landsat satellites are in near-polar orbits, designed to be sun synchronous; that is, the satellites cross the equator at the same local sun time in each orbit. Thus, lighting conditions are kept uniform. The satellites are equipped with telemetry and remote sensing equipment, including cameras and multispectral scanners.

Data collected from early Landsat projects were stored in X-format, that is, band-interleaved by pixel pair (BIP-2). This format was superseded by EDIPS (EROS Digital Image Processing System) with a resolution of 3596 pixels x 2983 scanlines, in band-sequential (BSQ) or band-interleaved-by-line (BIL) formats.

LANE LAN Emulation. Local area network (LAN) emulation services and protocols running over asynchronous transfer mode networks. See asynchronous transfer mode and the appendix for greater detail.

language In computer programming, a means of representing instructions, procedures, functions, and data through symbols and syntax which can be interpreted into machine instructions to control the computer. Common high- and medium-level programming, scripting, and page description languages include Perl, Java, C, C++, PostScript, LISP, Pascal, BASIC, Cobol, and FORTRAN. A common markup language used on the Web is HTML. There are also job control languages, description languages, graphics languages, and low-level assembly and machine languages.

Language of Temporal Ordering Specification LOTOS. A language for the formal process and algebraic specification of computer network protocols for concurrent and distributed networks, described as ISO 8807-1990. LOTOS has been used internationally to specify many systems, especially by university groups. LOTOS is applicable to the specification of Open Systems Interconnect (OSI) model systems, for example.

Lankard, John R. In collaboration with Peter P. Sorokin, Lankard described tunable organic dye lasers in the *IBM Journal of Research and Development* in 1966 at about the same time dye laser technology was being developed by Schmidt et al. in Germany.

Lankard continued to study lasers over the subsequent decades, coauthoring articles on organic dye lasers in the 1960s, Q switches in the 1970s and laser applications in the 1980s and 1990s, such as laser etching and thin film packaging. See laser history; Sorokin, Peter P.

LANNET 1. large artificial neuron network. 2. A subsidiary company of Madge Networks, N.V., and leading supplier of next-generation Ethernet and ATM switching technologies for local area networks (LANs). LANNET was acquired by Lucent Technologies in 1998.

LANtastic A commercial peer-to-peer NetBIOS-based network operating system from Artisoft. It supports a variety of client computer operating systems, including Microsoft DOS and Windows, IBM's OS/2, Apple's Macintosh, and various Unix clients.

lap A device used for grinding piezoelectric crystals. Since the resonance frequencies of crystals are due in part to their size and shape, the lap provides a means to fine-tune the crystal. See detector, piezoelectric, quartz, Y cut.

LAP 1. link access procedure. 2. See Link Access Protocol.

laplaciometer An early analog calculator designed for complex mathematical work by J. Atanasoff and some of his graduate students in the 1930s. The laplaciometer was used to analyze the geometry of surfaces. These developments led to the design and creation of the historic Atanasoff-Berry Computer (ABC). See Atanasoff-Berry Computer.

Laplink A popular, practical commercial hardware/software networking utility introduced in 1986 for transferring files between computers, especially between laptop computers and office workstations. It is very common for mobile computer users to want to transfer the information from their laptops to their desk computers, and sometimes to transfer in the other direction as well (e.g., sales leads). Laptops typically have smaller hard drives and higher security risks than desktop computers, making it advisable to regularly move the data off the laptop. Transfers can be achieved in a number of ways, through a serial port, over a parallel connection, or through phone lines. In 1995 LapLink Host was added to the product line to provide technical support to remote workers.

Laplink was developed by Traveling Software, Inc., a Washington State company devoted to supporting the needs of mobile users, founded in 1982 by Mark Eppley.

lapping A technique for wrapping electrical tape, foil, or other ribbons around a central core so that the next edge overlaps the previous one, in order to create close contact and a good seal.

lapping film A finely abrasive polyster substrate sold in color-coded sheets according to grade. It is commercially available in a variety of minerals, micron grades, and backings. The film may be in resealable bags to preserve its properties and prevent contamination by moisture and solid particles. It is available in different shapes for manual or machine polishing applications. Successive grades allow finer and finer polishing, similar to the process of using finer grades of sandpaper to finish a wood product. Plain and adhesive backings are available. Aluminum oxide, silicon carbide, and cerium oxide lapping films are used for a variety of types of surfaces. Diamond lapping films are suitable for harder surfaces such as metals. Diamond lapping film is available in several styles: fine or course diamond particles can be bonded directly to film backing or the diamond particles can be encapsulated within a soft ceramic coating which is then bonded to the film.

Lapping film may be used to polish fiber filament ferrules, magnetic media, oxide and thin film disks, and precision metal components.

Lapping film can prepare optical fiber surfaces for bonding by flattening joint surfaces, polishing with progressively finer films, and removing excess bond (e.g., epoxy). Water may be used as a lubricant.

laptop computer A low-weight, battery-powered, or combination battery/AC-powered portable computer. Laptops fit easily on a lap, airline tray, train table, or other support surface in common moving conveyances. They range in weight from about 3 to 7 pounds. Some people distinguish notebook computers as mid-range between laptops and palmtops, at about 2 to 4 pounds. Larger portable computers are called "luggables" or transportables, and the smallest ones are called palmtops and programmable calculators. They weigh from a few ounces up to about 2 or 3 pounds.

A laptop battery usually lasts about 2 to 6 hours, and may be extended by using powersaver functions or turning off the monitor. Car lighter adapters may help power or recharge a laptop battery.

Laptops are equipped with flat, light, low power monitors typically incorporating LED, passive matrix, active matrix, or gas plasma components. The active matrix screens are brighter and easier to see in dim or bright lighting conditions than screens that depend upon optimal ambient lighting.

Many laptops are equipped with PCMCIA Type I or Type II card slots so the user can add lower power, compact peripherals such as extra memory cards, fax/modems, network ports, etc. USB and FireWire ports are also now common.

LArc A data archiving program descended from LZSS, developed by Lempel, Ziv, Storer, and Szymanski and further optimized and extended in the late 1980s by Okumura and Miki. Kazuhiko Miki reworked Okumura's version of LZSS with Pascal and Assembler to create the archiving tool called LArc, whose file handling was quick and compact. Huffman coding was later incorporated into the software by Yoshizaki to create an even faster and more popular version called LHarc. See LHarc.

LARC See Livermore Automatic Research Calculator.

Large Advanced Mirror Program LAMP. A research project involving the design and construction of a 4-meter-diameter segmented mirror with characteristics appropriate for deployment in space, completed in 1989. The segmented design enables large mirrors to be assembled. LAMP is the largest mirror designed for use in space, exceeding the Hubble astronomical mirror. See Large Optics Demonstration Experiment.

Large Optics Demonstration Experiment LODE. A project to research the control of large, high-power laser beams, completed in 1987. Since then, LODE data have been used in the development of space-based laser technologies and investigated for defense applications such as missile defense. See Large Advanced Mirror Program.

large scale integration LSI. A term describing the evolution in electronics from systems with many separate large-sized components to systems with a smaller number of integrated small-sized components. Since the distances between various parts of the logic and physical circuits are greatly reduced in LSI, the processing speeds are also much faster. LSI made possible smaller, more powerful electronic components such as calculators, computers, automated appliances, digital watches, clocks, timers, and much more. See very large scale integration.

large-core fiber A fiber cable with a core diameter that is broad in relation to most cables. A core of about 200 μm or greater can be considered large-core though the designation is somewhat dependent upon what is manufactured or commonly used at any particular time. For comparison, multimode fiber generally has a core diameter of about 56 μm.

LASE Laser Applications in Science Education. A workshop of the Optical Society of San Diego that featured a series of diode laser demonstrations activities.

laser *acronym* light amplification by stimulated emission of radiation. A device that stimulates photons to produce coherent, nonionizing radiation in the visible spectrum and infrared wavelength regions. While lenses and mirrors are commonly used to direct laser beams, the essential components of a laser are a lasing medium, a resonant optical cavity, and a pumping system (optical, mechanical, or electronic). Gas or semiconductors are used as the active lasing medium. The acronym was used by a number of students in the lab environment where C. Townes and his assistants worked on masers in 1957.

Unlike light from many other sources, such as incandescent bulbs, a laser beam remains very narrow and straight over long distances, a property called *coherence*. Many people have seen lasers in the form of business presentation pointers or rangefinders on guns to facilitate aiming. The pointer light usually appears as a small, round, red dot. Lasers make great cat toys, too.

Gems such as ruby and garnet (yttrium-aluminum-garnet) are commonly used in the production of lasers and many familiar laser devices project a red beam. Gallium arsenide is also used. Red lasers used in magneto-optical storage device read heads currently have wavelengths of about 650 nm, but blue lasers may become commercially available and will make it possible to record at higher areal densities. Blue lasers (ca. 410 nm) have a spot incident level almost 40% smaller than red lasers which would enable higher disc capacities and faster transfer rates.

Lasers are used for thousands of commercial, industrial, and medical applications. They function as high precision surgical cutting tools in medicine, as imaging components in millions of consumer printers, as read/write tools for audio/visual storage technologies, and as signaling tools in a variety of networking tasks, especially through fiber optic cables.

A very interesting new type of room-temperature quantum cascade laser technology originated in 1994. QC lasers offer greater control over frequency selection, and have many potential applications in remote sensing and industrial environments.

Basic Semiconductor Laser Component

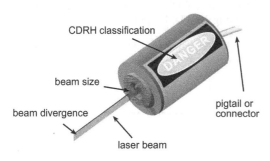

CDRH classification

beam size

beam divergence

pigtail or connector

laser beam

An active medium in a laser causes it to emit photons which can be collimated into a coherent beam of light. A laser module typically includes the lenses necessary to accomplish this task. A basic laser component, such as a laser diode module, may emit an elliptical beam that is usually corrected within the module housing to produce a circular beam.

The beam size *is the size of the beam as it exits the assembly (the diameter, in the case of a circular beam). The* beam divergence *is the angle of the beam as it spreads over distance. If it is an elliptical beam (as the beam from a laser diode), this is expressed with two values. The* pigtail *or standardized connector enables the laser component to be coupled with power sources or other components in a more complex assembly*

The direction in which the beam travels is the propagating axis. *If it fans back and forth (e.g., to produce a line pattern), the fanning angle is described in degrees.*

The CDRH classification label warns of laser danger and indicates the power emitted by the laser at a specified wavelength.

Most laser light is generated with longer wavelengths in the "hot" color regions (red and yellow), but the quest for an elusive "blue laser" began to bear fruit in the late 1990s. Blue lasers have the potential to revolutionize many aspects of laser storage, scientific, and medical devices because blue light has a very short wavelength and thus the beam from a blue laser is very tiny. This enables greater precision in cutting, shaping, writing, and reading mechanisms.

Laser light is a minimum loss communications tool when proper shielding (cladding) keeps the light beams within the fiber core. In very pure straight fiber optic cable, the loss over distances is low, and the transmission cannot be surveilled in the same way that electrical wires can be monitored through emissions that extend beyond the cable (a lightguide needs to be bent to read the signals, resulting in loss that can be detected beyond the point of intrusion). Laser light communications are not affected by electromagnetic interference (EMI) in the same way as many other means of transmitting information (there may be some possibility of EMI in long lines electromagnetically amplified at the splices, but optically am-

plified lines are being developed that may eliminate this problem).

One should never aim a laser pointer at a person or animal, as there is a possibility of harm, especially if the laser shines on a cornea or retina. While some types of light are safer than others, many are capable of causing permanent photochemical or thermal damage, even with very brief exposures. See argon laser, cladding, fiber optics, helium-neon laser, laser diode, laser history, overcoat-incident recording, quantum cascade laser, maser, YAG, yttrium.

laser class See CDRH classification.

Laser Communications Demonstration Experiment LCDE. A joint project of the National Space Development Agency of Japan (NASDA) and the Optical Space Communications Group (OSCG) of the Tokyo-based Communications Research Laboratory to develop the Communications Demonstration Equipment (LCDE) to be placed aboard the Japanese facility of the International Space Station. The project goal is to demonstrate 2.5-Gbps up- and downlink communications at 1.5 μ wavelength. The experiment is based upon erbium-doped fiber amplifiers and lasers.

Laser Communications Demonstration System LCDS. Established in the early 1990s by NASA through the Jet Propulsion Laboratory (JPL) to demonstrate improvements in technology and multidiscipline systems engineering related to space laser communications. See Optical Communications Demonstrator.

laser cutting A laser can be used as a high precision cutting tool for surgical procedures and industrial/commercial applications. Rubber stamps, wood blocks, and stencils cut with lasers have very fine, crisp, clean edges.

laser diode A type of semiconductor light-emitting diode (LED) that emits coherent light in response to the application of voltage.

Laser diode/lens combinations are used in a wide variety of applications, including alignment and measuring systems, interferometers, barcode readers, medical imaging systems, laser printers, and fiber optics communications systems.

Near-infrared diodes (NIDs) in the approx. 815±35 nm range are used in sensors and vision systems. Visible laser diodes (VLDs) operate around 662±27 nm, at wavelengths visible to humans, and are practical for use in low-voltage devices.

The basic components of a diode laser module (DLM) are a laser diode, drive circuit, and collimating lens, protected within a compact housing. The drive circuit controls the operating mode of the laser (e.g., continuous wave). It may also provide surge filtering, reverse-polarity protection, and other maintenance functions. Laser diode components with pumping mechanisms or temperature stabilization will typically be larger than basic DLMs. Cooling components may help extend diode lifetime. Some models permit user control of the focusing distance and consumer components may include a safety shutter to reduce the possibility of damage from laser light.

L

Laser diode component specifications include the diameter of the beam hole, the operating and output power, the output wavelength (in nanometers), and recommended operating and storing temperatures. Fiber-coupled laser diodes (FCLDs) combine small circuits with fiber pigtails for connection to fiber optic cables. They are available in continuous-wave and pulsed models, with single- or multimode connectors in a variety of wavelengths, usually from about 800 to 1550 nm. Prices per diode currently range from about $200 to $900 depending upon power genera-

tion properties. FCLDs are used in scientific research, medical testing and imaging applications, thermal printing, and optical measurement systems. Distributed-feedback laser diodes with wavelength concentration are useful for dense wavelength division multiplexing (DWDM) applications. See distributed-feedback laser diode, laser, light-emitting diode.

laser fax A combination device that incorporates the printing features of a laser printer with the scanning and transceiver capabilities of a facsimile machine. This is a handy tool for small offices where separate

---

Laser Diode Beam Shape, Correction, and Control

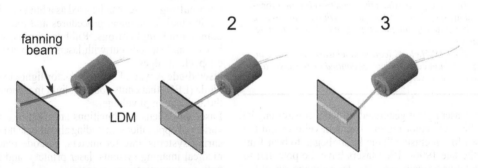

The emitting region (facet) of a semiconductor laser diode is typically rectangular due to the layered assembly. Thus, the beam is inherently elliptical in cross-section (astigmatic). One or more lenses may be used with the laser diode to collimate and focus the laser beam, correcting the astigmatism and generating a circular beam.

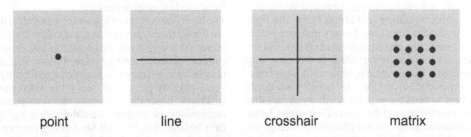

This highly simplified drawing of a laser diode module (LDM) illustrates how a laser beam can be rapidly fanned back and forth to produce a line when it hits a surface. Fanning in curves against a projection screen, fog, or smoke is the basic concept behind laser light shows and the technology is applicable to many other uses such as laser pointers, cutting and etching tools, sensor triggers, defense weaponry, and much more.

point   line   crosshair   matrix

Depending upon how the lenses and rotors are designed and organized, it is possible to control the shape of the laser beam. With the addition of fanning and multiple beams, many configurations are possible, as shown here.

---

high-capacity devices are not needed and space is at a premium. Ideally, a laser fax machine should be networkable so users can send and receive faxes without printing each one, and then select to print only the ones to be distributed and filed as hard copies. This saves paper and lineups at the fax machine.

laser history Laser history is a fundamental aspect of fiber optics history. Events leading up to the development of lasers and laser-based fiber optic transmissions include the invention of telescopes, television lenses, and the channeling of light through various media, including water (1840s), quartz rods (1880s), and bundles of fibers (1920s). In the late 1930s, scientists in Europe were on the verge of putting together the concepts fundamental to the laser, but the turbulent times and outbreak of the war shifted priorities and impinged on many areas of scientific research. It was not until the 1940s that solid theory and practical applications coalesced into modern laser technology.

Lasers were developed in part because scientists wanted a source of coherent light from which to carry out other types of research, such as the carrying of signals and "bending" of light. Traditional light sources would spread and dissipate too quickly for most experimental applications. In 1841, J.-D. Colladon attempted to direct the bright light from an arc lamp into a stream of water. J. Tyndall demonstrated a similar idea in 1870. In 1888, Roth and Reuss extended the concept to medical imaging, using glass rods to illuminate internal body structures. Einstein's theories in stimulated emission had, by 1917, set a theoretical foundation for laser technology.

J. Baird was experimenting with television technology in the early 1920s and received a patent for transmitting images using glass rods in 1926. C. Hansell extended this idea to bundled rods, work that was later referenced to deny H. Hansen a patent application. By the 1930s, H. Lamm had successfully transferredd an image using a fiber bundle.

Basic laser science was described by A. Prokhorov, a Russian physicist, in the late 1940s and early 1950s, and Prochorov continued to make important contributions to the field of masers and lasers over the next several decades. Another Soviet scientist, V.A. Fabrikant, described population inversion in the 1940s and coauthored a patent application submitted in 1951.

Experimentation with fiber bundles continued in the 1950s through the work of H. Møller Hansen, H. van Heel, H. Hopkins, and N. Kapany, but the practicality of light transmission without further developments in laser technology was limited.

In June 1952, Joseph Weber lectured on how Einsteinian coefficients could be used to amplify by stimulated electromagnetic emission. A practical embodiment of early lasers was invented by Arthur L. Schawlow, a Bell Laboratories scientist, and Charles H. Townes, a consultant to Bell, who first demonstrated and described microwave-based ammonia-beam lasers (masers) in 1954. The laser research was written up by Townes, Zeiger, and Gordon in *Physical*

Review V. 95. Building upon this research, Townes and Shawlow were applying the concepts to optically-pumped lasers by the late 1950s and Shawlow made later contributions to the development of laser spectroscopy. In 1956, R. Dicke applied for a patent on improved maser/laser resonating chambers (essentially the forerunner to Fabry-Perot interferometers and open-cavity lasers). Prochorov and Manenkov suggested the use of ruby in lasers in 1957.

By 1957, the idea of using longer wavelengths, perhaps in the optical region (lasers rather than masers) was beginning to occur to a number of researchers, including Townes, and perhaps also to Gordon Gould, an ambitious graduate student at Columbia who consulted Townes on patent issues before leaving school to accept a job at TRG. Dicke proposed the idea of infrared open cavity amplification in his patent application of 1956, so the fundamental idea of optical spectrum stimulated emission was not original to Townes nor Gould, but since Dicke's patent wasn't published until September 1958, it's difficult to know how many people were inspired by Dicke's work.

Townes continued his maser/laser research in collaboration with Schawlow at Bell Laboratories Dicke's patent was awarded and published in September 1958. When Townes and Schawlow circulated their laser design, in December 1958, Gould showed renewed interest in the idea, constructed a prototype laser in collaboration with TRG, and submitted a patent in 1959. The laser patent wars had begun. Gould's application was initially refused due to the Townes/Schawlow/Bell patent. TRG disputed the patent and lost because Gould was unable, at the time, to prove due diligence or produce substantial notes from the year 1958.

Later, the Gould patent claim was reopened when laser technology caught on and there was a strong potential for royalties. Gould was awarded the patent in 1977 for some of the claims in his 1959 application. Gould's conversations with Townes and his familiarity with the scientific papers circulated by Townes/Schawlow appear to have been strong factors stimulating his interest in laser technologies. How much Dicke's September 1958 patent contributed is still an open question. Most of Gould's engineering contributions were refinements of basic laser concepts rather than groundbreaking technologies, such as those developed in the early 1950s by Weber, Schawlow, Townes, and Dicke.

In 1958, G. Goubau applied for a patent for a means to control laser light with a lightguide and lenses, the beginning of refinements that would coalesce into modern fiber optics when lasers, lenses and fiber optic filaments reached a level where their capabilities could be built into a cohesive system for communications.

Also in the late 1950s, Theodore Maiman followed the line of research evolving from the optically-pumped laser and developed and patented the ruby laser. A practical embodiment of the gas laser, a different approach from optically pumped lasers, was invented by a Bell researcher, Ali Javan. Javan developed a helium-neon laser capable of emitting a pure, continuous beam by 1960.

At the atomic level, Russian physicist N. Basov conducted pioneer research in quantum radiophysics that was to become important in laser technology. Russian physicist and engineer R. Kazarinov lectured on homojunction semiconductor lasers in 1962 and co-authored a patent in March 1963, with Alferov, for a double heterostructure laser.

Dye lasers were described in 1966 by F. Schmidt et al. in Germany and P. Sorokin and J. Lankard in the U.S. By 1967, B. Soffer and B. McFarland had described how these new lasers could be tuned to different frequencies. While dye lasers were not as powerful as some of the other pioneer laser technologies, there were many potential uses for tunable lasers.

By the early 1970s, Kazarinov and R.A. Suris had described a distributed feedback double heterostructure laser. The Kazarinov/Alferov/Suris lasers are now important components in communications technologies and compact disc players.

For their contributions to laser technology, Prokhorov, Townes, and Basov were co-awarded a Nobel Prize in 1964. Chu, Cohen-Tannoudji, and Phillips were awarded a Nobel Prize in 1997 for their work in the 1980s on developing methods to cool and trap atoms with laser light. See laser, maser. See individual listings for many of the scientists mentioned above.

laser level A laser-based alignment tool that uses a straight, coherent laser light beam to sight along a plane. Laser levels in the $500 range are compact, usually water-resistant, self-leveling, and convenient for contractors, cable installers, and other personnel who require quick plumb lines and reference points. Laser levels have many advantages over traditional liquid-and-gravity-based (bubble vial) levels. They can instantly sight over longer distances (30 to 60 m, depending upon model), are pocket-sized, and may sight in several directions at once.

Less expensive laser levels in the $100 to $200 range resemble traditional levels and incorporate bubble vials. They are a transitional technology that fill a niche in the lower price range of laser levels. As the price of true laser levels decreases, these will probably become obsolete.

Low-end pocket-sized, single-beam laser levels in the under-$100 range are similar to laser pointers and are useful for spot-checking and may include a magnetic base for attachment to steel girders.

laser light pen A compact cylindrical laser assembly based upon a collimate diode laser (CDL). See

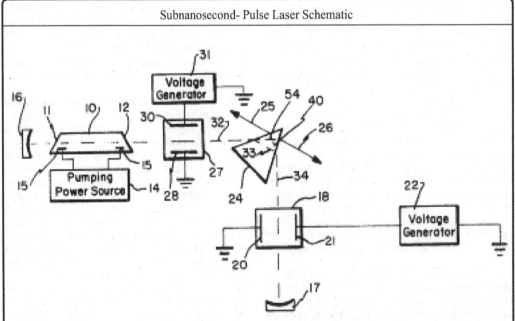

Subnanosecond- Pulse Laser Schematic

This schematic illustrates some of the basic components of a subnanosecond laser assembly, as described in 1971 by W. Simmons of TRW, Inc. Until this time, it had been difficult to fabricate lasers of sufficient power that could emit light pulses that were exceptionally fast, as there was a limiting relationship between the light pulse and the length of the laser cavity. This design overcomes that limitation by generating laser pulses until the optical laser cavity is mostly filled and then removing most of the radiation contained in the cavity. The remaining radiation is then amplified by laser action, exploiting the Fourier components of the pulse which match the Fabry-Perot resonances of the cavity. The light pulse is mode-locked in phase such that the pulse retains its shape during amplification (while reflected back and forth in the cavity). There is also the option of emitting a pulse train (a sequence of fast pulses). Subnanosecond-pulse capabilities are essential to many applications, including atomic research and telecommunications. [U.S. patent #3,701,956, October 1972.]

laser diode, laser pointer.

laser pointer A popular adaptation of laser light for use as a pointing mechanism. Remember the pointing sticks teachers and sales representatives used for lectures and visual presentations (and to wake up dozing students)? The laser pointer is the modern equivalent (with the exception that dozing students are no longer prodded). The beam is typically red or green and may project to about 150 m. Pointers come in a variety of shapes and sizes from small cylindrical models resembling AA batteries, to executive pen- and mouse-shaped models. While many laser pointers are sold as novelty or presentation items, they are also useful for installation, construction, and alignment of components, cables, and other communications-related equipment.

Most laser pointers project a pinpoint beam but some are designed to oscillate back and forth to project a line. The extent of the oscillation, called the *fan angle*, is usually about 30°. Some laser pointers include pulsed beams that can be used to create special effects, especially when combined with fog.

Most laser pointers are designed to emit laser light in ranges that are not harmful to eyes, but it is best to be cautious and never aim a laser pointer at anyone's face or at a reflective surface that might reflect the light beam into someone's face. See laser diode, laser level.

laser printer A printer that uses a computer-directed laser beam to render images. Laser printers typically use a specially treated drum that is influenced by the light of the laser. An impression is made on the drum by aiming very fine, high precision laser light beams at the coating, so the electrical charge is selectively altered. An electrostatic process then attracts the toner to the imaged areas (areas altered by the beam), and heat fuses the toner onto the printing medium, which is usually paper or card stock.

Most laser printers range from resolutions of 300 to 1200 dpi, although some can print at higher resolutions with special papers or plates capable of holding a very fine image. Many are enhanced with Adobe PostScript page description interpreters. Higher printing speeds and PostScript capabilities require additional memory.

Laser printing is considered a *dry* printing process, as opposed to offset printing on a press that uses wet inks. It is not advisable to use recharged toner cartridges in laser printers. In most consumer laser printers, the toner cartridge also includes part of the drum mechanism, which has a limited term of use. Even if the recharge toner is of good quality, it is still possible for the aging drum to stress the printer, perhaps even damaging it. The money saved on toner cartridges may be offset by the potential loss due to damage or reduced lifespan of the printer. See dot matrix printer, dye sublimation, inkjet printer, thermal wax printer.

laser probe A laser-based instrument used to illuminate and sense a phenomenon, process, or specimen.

laser range finder LRF. A device for determining distance through the use of a beam of coherent light and associated sensing systems to calculate the range. Gun sights, binoculars, camera autofocusing systems, and a number of security and surveillance systems use laser range finders. Depending upon the device, the range may be indicated symbolically with graphics or numbers. For example, the distance to an object may be displayed on the viewing screen in yards or meters, superimposed over the scene being viewed. Some laser frequencies can be harmful to the eyes; while many consumer devices use Laser Class 1 frequencies that are not considered dangerous, it is wise to be cautious.

laser show A laser show is an entertainment display created with colored laser lights crossing through the air, sometimes in a darkened room and sometimes falling on a display screen, domed theater surface, or fog/smoke medium. Some of the vector-based computer arcade games from the 1980s that successively drew colored lines around the screen give a general idea of a laser light show, except that it's three-dimensional and bigger.

Laser Fiber Optic Probe

A laser light is projected through optical fiber to illuminate and image a specimen mounted on a test stand at the NASA/Langley Research Center. [NASA/ Langley, 1993.]

---

laser-using communications equipment LUCE. Communications technologies based upon laser light beams emissions through air or through waveguides such as fiber optic cables. When transmitted through air, the light can be used as a traditional visible signal light or may trigger an electronic sensor. When transmitted through a waveguide, it functions much as traditional wired technologies but offers greater bandwidth capacity and lower susceptibility to electromagnetic interference.

laser, blue One of the more recent laser technologies, researched in the early 1980s and developed in the 1990s, blue lasers lagged behind others due to fabrication difficulties. Researchers had to find a way to generate more quantum "holes" in order to create the p-n junctions that were needed to stimulate photon emission in the blue wavelengths. Wide bandgap materials (e.g., zinc selenide – ZnSe) were needed to produce short wavelength blue light. Once some of the hurdles were overcome, blue lasers began to become a practical reality.

Shuji Nakamura is a significant blue laser inventor. He created high quality gallium nitride crystals with double the previous hole mobility, in 1991, and commercial blue LEDs in 1993 using a custom-built reactor and novel strategies. Meanwhile many well-known firms were advancing ZnSe-based blue and green laser technology, with practical implementations appearing in the mid-1990s.

Blue lasers provide a very precise, tiny laser beam with high potential for precision fabrication and high-density storage devices. Blue diode-based lasers emit light at approx. 460±30 nm, but were initially shorter-lived and more difficult to fabricate than conventional red lasers. However, as fabrication technology improves and the price drops, blue lasers can potentially increase the early 2000 storage capacity of red or infrared laser-based optical technologies by 4 to 10 times. By 2002, commercial blue/red DVD players for the Japanese market were being announced, with global distribution projected by 2004 or 2005.

laser, dye A type of tunable laser based upon injected organic dye molecules developed in the mid-1960s in Germany and the U.S. This laser has a broad emissions band and may be tuned by means of an adjustable diffraction grating incorporated into the laser resonating cavity.

laser, gallium arsenide GaAs laser. A type of laser commonly used in consumer CD and DVD players that emits light in the approx. 820 nm range. GaAs-based lasers commonly have a bandgap energy of about 1.45 electron volts.

laser, green A type of laser useful for interferometry, holography, and laser pointers; green lasers have wavelengths of around 527 nm and may send a beam 100 m or more. A compact diode-pumped green laser module may be small enough to fit in a hand.

LAT See Local Area Transport.

LATA See Local Access and Transport Area.

latency Delay or period of dormancy. The speed of acquisition or perception of a thought, object, or communication in relation to the desired speed of acquisition or perception.

Latency can result from the intrinsic properties of the communications medium or the communication itself. It can arise from the effects of the time it takes for information to transmit, or from the physical or logical pathways associated with the transmission. It can also result from crowding, congestion, misalignment, mistuning, unanticipated effects or traffic, and a large number of other possible factors. Response time is related to latency, with reduced latency usually desirable in this context. Every aspect of communications has to concern itself with latency.

In networking, there has been a lot of research and quantification of latency in order to design, evaluate, and tune computer architectures to carry out desired tasks. Here, latency is usually expressed in small units called milliseconds (e.g., latency in ISDN systems is ca. 10 msec). With slower communications pathways, such as slow modems over phone lines, latency may instead be expressed in seconds.

There are many ways to reduce latency: better algorithms, wider bandwidth, better transmissions media, more efficient hardware, different topologies, and new technologies.

Latency is sometimes intended in a dynamic system where connections change and the topology cannot always be anticipated. A queuing system is a means of using latency to good effect, as in email systems, which will hold messages until the intended recipients or routers along the communications path are available to receive them. Latency may also be used as a signaling system; in other words, delays are taken into consideration and used to convey information. See realtime.

Lewis Latimer – Inventor in Electricity

The son of a former slave, Latimer had great drafting skills and an inventive mind that earned him professional positions in some of the preeminent technology labs of the time. [Image ca. 1882, courtesy of the National Historical Society.]

Latimer, Lewis Howard (1848–1928) A skilled American drafter and inventor, Latimer worked in some of the most prestigious labs in America.

In 1865, Latimer received a Union Navy honorable discharge and became an office boy for a patent soliciting firm. Eager to do more, the ambitious young man self-studied mechanical drawing and convinced his employers to evaluate his drawings. For his efforts, he earned a promotion to drafting work and a dramatic increase in salary. He was later an assistant to Alexander Graham Bell and prepared drawings and descriptions for Bell's telephone patent. He subsequently received several patents of his own, beginning in 1874.

Latimer joined the American Electric Light Company and became a pioneer in the development of the light bulb. He received a patent for an improved electric lamp in 1881, and one for a process for manufacturing carbon filaments which he co-developed with Joseph V. Nichols in 1882. The filament patent was commercially successful, with carbon filaments replacing the short-lived bamboo paper filaments common at the time. In 1884, he became a member of Thomas A. Edison's research team. In 1890, he authored *Incandescent Electric Lighting: A Practical Description of the Edison System*, which became an engineering handbook. He was appointed as an expert witness on the Board of Patent Control of the company that evolved into General Electric (GE) and, in 1918, became a member of the Edison Pioneers.

LATNET A Latvian network service, concentrated mainly in Riga, where most of the scientific community is located. LATNET provides services to the RTD community and some businesses. It is operated by the Department of Computer Science at the University of Latvia in cooperation with the Riga Technical University. LATNET utilizes leased lines and radio links, operating with TCP/IP.

lattice model A flow control-related network access security model based upon the lattice format that arises from the ordering of finite security levels within a system. This is usually one of the first models discussed in courses related to security models because lattice-based access control is important for confidentiality and, to some extent, integrity. Lattice-based access control models were described by Sandhu in the early 1990s arising out of research in the 1970s (e.g., Denning, 1976). Sandhu's contributions were based upon work supported by a National Science Foundation grant and a National Security Agency (NSA) contract.

launch 1. To start, to set into operation. 2. To start, activate, or begin a computing process, operating system, or application. Programs are launched in a variety of ways, such as double-clicking on icons or typing the name of the program on a command line. Programs may also be launched automatically from preprogrammed script files, or transparently from within other programs.

launch, product In management, to begin a new program, project, or marketing plan, sometimes with a lot of fanfare in order to attract the attention of potential customers and the media. New software packages are often *launched* at industry trade shows.

Law Enforcement Access Field LEAF. In computer security, a section of classified data created in association with a Clipper chip or Capstone, and sent along with the encrypted message. The LEAF includes the session and unit keys concatenated with the sender's serial number and an authentication string. See Clipper chip.

LAWN See local area wireless network and wireless local area network.

laws of electric charges Stated simply: bodies with *unlike* charges will attract one another; bodies with *like* charges will repel one another; bodies with *no* charges will neither attract nor repel one another.

layer architecture Layer architectures are common in computer networks. Asynchronous transfer mode (ATM) is the most broadly implemented layer network architecture.

Defining a number of virtual and physical layers allows communications paths to be organized and administered so that many different developers and manufacturers can create processes and devices independently of one another, yet still apply them to the same system once standards and protocols for the various layers are established. Layers also provide a means to optimize the characteristics of the layer to the type of processes that occur within that layer. The layer architecture is usually described and diagramed horizontally, from bottom (physical or low-level layers) to top (virtual or user interface and applications layers) with variations depending upon the specific organization of the architecture. Layers typically communicate with adjacent layers directly above or below, or may pass through an intervening layer. The Common Layer Hierarchy chart shows a brief overview of some of the common types of layers.

Layer Two Forwarding Protocol L2F. A Cisco Systems Layer 2 application tunneling protocol in the TCP/IP suite introduced in 1996 and submitted to the IETF for consideration as a standard. The advantage of L2F was that it enabled virtual dialup connections with multiple protocols and unregistered IP addresses. L2F uses UDP port 1701. In 1997, Compuserve adapted L2F for establishing securable private networking services for dialup customers. In 1998, Valencia, Littlewood, and Kolar described the protocol as a Historic Request for Comments (RFC). See Layer Two Tunneling Protocol, RFC 2341.

Layer Two Tunneling Protocol L2TP. A securable network protocol that extends the Point-to-Point Protocol (PPP) model to enable it to tunnel over Internet Protocol (IP) which, in turn, enables virtual private networks (VPNs) to operate over public packet-switched networks such as the Internet. L2TP can project a PPP network connection to a location other than the point at which the transmission was physically received, enabling multilink operation across distinct physical Network Access Servers (NASs).

L2TP incorporates characteristics of Point-to-Point Tunneling Protocol (PPTP) and Cisco Systems' Layer Two Forwarding Protocol (L2F) to provide an extensible control environment for the dynamic setup, maintenance, and teardown of multiple Layer 2 tunnels established between logical endpoints in a transmission path. See RFC 2661.

Layer Two Tunneling Protocol extensions L2TPext. Extensions to the Layer 2 Tunneling Protocol including links, multicast, etc. The IETF has a working group responsible for the orderly development of extensions to L2TP, in addition to separating out the components of RFC 2661 for greater modularity. The IETF has submitted L2TP over Frame Relay and L2TP Security to the IESG as Proposed Standards. See Layer 2 Tunneling Protocol.

LB See leaky bucket.

LBA See Logical Block Address.

LBS See load-balancing system.

LC 1. lead channel. 2. local channel. 3. local company.

LCD 1. linear collider detector. 2. See liquid crystal display.

LCDE Laser Communications Demonstration Equipment; Laser Communications Demonstration Experiment. See Laser Communications Demonstration System.

LCDS See Laser Communications Demonstration System.

LCI International An American telecommunications system providing international voice and data services through owned and leased fiber optic networks. LCI is known as the first long-distance provider to bill both business and residential calls in 1-second increments, a service known under the Exact Billing service mark.

LCP See Link Control Protocol.

LCS 1. See Lan Channel Station. 2. See Laboratory for Computer Science.

LCU Lightweight Computer Unit.

LCV See line code violation.

LDAP See Lightweight Directory Access Protocol.

LDIP See Long Distance Internet Provider.

LDMC See Loop Data Maintenance Center.

LDMS See Local Multipoint Distribution Service.

LDU 1. local distribution utility. 2. See load distribution unit.

LE light-emitting.

lead-salt diode laser A type of laser that was experimental in the early 1980s and difficult to operate. Improvements in the theory and the increased availability of semiconductor technologies led to diode-based systems in the early 1990s that could be used to frequency modulate signals in a number of types of devices, including spectroscopes. Lead-salt diode lasers are now used in semiconductor processing, detection, and fiber optic communication applications. They can be tuned by controlling diode current at a constant cooled temperature. Lead-salt diodes have some performance and temperature limitations compared to quantum cascade lasers. See quantum cascade laser.

leader 1. The first segment, or part of a transmission or transmissions medium. 2. The first few centimeters on a magnetic tape (audio, video, etc.); the leader attaches and feeds the tape onto the spool. It is not intended for recording, and may be made of nonmagnetic material. 3. A packet, cell, segment, or other leading part of a data transmission, which contains information about data following, without including the data. It is a space or a signal to indicate impending information, rather than being a component of the information itself. See header.

leadership priority In an ATM network, an organizational function of a logical node assigning it priority which, in turn, enables it to be designated as the peer group leader (PGL).

LEAF See Law Enforcement Access Field.

Common Layer Hierarchy in Layered Computer Networks	
Layer	Notes
application layer	A high-level layer at which the user interacts with the network applications programs and utilities. Various types of text or graphical user interfaces may be implemented at this layer. The application may also include remote access mechanisms and information messaging and transfer services.
presentation layer	Data security and data representation during transfer.
session layer	A traffic directing layer that sets up a communication between applications, adjusts synchronization, if needed, and clears the communication when done.
transport layer	A generalized, network-independent means of interlayer communication between the high application-oriented layers and the lower level layers, supporting different types of connections.
network layer	Low-level network connection, routing, and flow-control functions.
link layer	Low-level connectionless or connection-oriented data transfer.
physical layer	Low-level electrical connections and interfaces between the computing platforms and the network cables and connections, and the link layer data transmissions.

leaf node A type of connecting point in a network, located at the end of a branch, so that only one connection is between the leaf node and the rest of the network.

leakage In electrical circuits, particularly those which are not well shielded, leakage of the electromagnetic radiation outside the boundaries of the physical medium can occur. This may interfere with other transmissions and devices.

The Federal Communications Commission (FCC) provides guidelines and regulations for shielding various radio and computer devices in order to minimize interference from leakage.

leaking memory See memory leak.

leaky bucket LB. A congestion-related conformance checking cell flow concept in ATM networking, an implementation of the Generic Cell Rate Algorithm (GCRA). Think of the bucket as a point in the network where cells may accumulate, depending upon varying rates of inflow and outflow. If cells are entering and leaving in equilibrium, that is, maintaining a sustainable cell rate (SCR), then the bucket will never be filled. If, however, inflow exceeds outflow, as the network experiences congestion, then the bucket may become full. There are various strategies for dealing with a full bucket, although prevention is advised. If, when the bucket becomes full, there are no further incoming cells, then it can be emptied in "bucket depth/SCR rate" amount of time. Bucket depth, the tolerance to cell bursting, can be set in relation to cell flow and retransmission timing. If, however, the incoming cells continue to accumulate, the bucket will overflow and must be handled in some manner, with cell discard as one of the options.

There is more than one way to implement a leaky bucket. Cisco Systems, Inc. suggests using *dual leaky buckets*, so a preconfigured queue depth threshold is set according to an agreed class of service (CoS) and quality of service (QoS). The first bucket can be configured to provide a service algorithm based on peak cell rate (PCR) and cell delay variation tolerance (CDVT) service parameters. The second bucket is based on sustainable cell rate (SCR) and maximum burst size (MBS). Nonconformant cells can be configured as cell discard, tag, or no change for each bucket. Dual discard thresholds can be supported to provide a delay mechanism for congestion cell discard rates. See cell rate.

leased line A line whose use is rented over a period of time from the entity that owns and manages the physical connection. Long-distance companies, specialized services, and businesses with direct private lines often lease lines from the local primary telephone carrier rather than installing their own.

Least Cost Routing LCR. A phone service that automatically seeks and selects the line through which to send a call with the least cost. See Automatic Route Selection.

LEC 1. See Local Exchange Carrier. 2. LAN Emulation Client. A LAN software client that keeps address translation and connection information for communication through an ATM network. See LANE, LECS. 3. Loop Electronic Coordinator.

least significant bit LSB. The LSB is the lowest order bit in a binary value. This is an important concept in computer data storage and programming that applies to the order in which data are organized, stored, or transmitted.

As an example, in the binary value 110 (as it is normally written with the larger values to the left), the zero (0) on the right, representing the number of ones, is the least significant bit (LSB) and the one (1) on the left, representing the number of fours (4) is the most significant bit .

One of the reasons data are not directly transportable among different systems is that some file formats or operating systems are standardized on LSB priorities and some are standardized on MSB priorities. In other words, some systems code/decode a binary value from smaller to larger (little endian) and some from larger to smaller (big endian).

least significant byte LSB. The least significant or lower-order value in a multibyte word. More often than not, a byte represents eight bits. See least significant bit.

Leclanché, Georges (1839–1882) A French engineer who invented a type of electrolytic battery cell later refined and used for signal bells and telegraph services. Leclanché studied in England and returned to France to continue his studies. In Belgium he was encouraged in his endeavors by Mourlon and established a small laboratory. After developing his dry cell, he opened a factory to produce batteries and electrical devices. Shortly before the death of his father, Leclanché returned to France and, with his own health in decline, made a tour of Europe, Egypt, and other countries (to collect Italian furniture). The Leclanché S.A. company of Switzerland, established in 1909, was a pioneer producer of portable lighting, paper capacitors, rechargeable batteries, and other battery-related products. It was associated with other companies in the late 1990s to become the Leclanché Group. See Gassner, Jr., Carl; Leclanché cell.

Leclanché cell A historic primary electrolytic cell developed by Georges Leclanché in the mid-1860s. This was an important time in battery history as technology – a transition from wet to dry cells and from cumbersome hard-to-move batteries to those that were encased and thus more practical and portable.

In its original form, the Leclanché battery was encased in a porous pot with a positive manganese dioxide and carbon electrode on the top and a zinc negative electrode on the bottom. The pot and zinc rod were immersed in a solution of [electrolytic] ammonium chloride. The electrolyte penetrated the porous container to reach the cathode. Moulin and Leclanché made commercial improvements to the battery and established a factory to create and distribute the invention. Later, Georges Leclanché's son, Max-Georges, made some changes to the commercial battery, replacing the porous container. The battery was used in many aspects of the emerging telecommunications industry and as a power source for bells and automotive lights. The TIS Standard for dry cells and batteries (TIS 96-2528 – 1985) refers to Leclanché-type cells.

LECS LAN Emulation Configuration Server. A LAN software server that maintains configuration information that enables network administrators to control which physical LANs are combined to form VLANs. See LEC.

LED See light-emitting diode.

left-hand circular polarization LHCP. A polarization orientation associated with antennas, for example, satellite antennas. Left-hand refers to a counterclockwise direction. The left or right sense of the polarization is dependant upon various factors, including the transmitter type and transmission frequency. Some systems can be switched from right- to left-hand and may benefit from this flexibility. Left or right orientations are also relevant to other types of polarization besides circular polarization, as in left- or right-hand *slant polarization*.

left-hand rule, Ampère's rule A handy memory aid, once widely used to determine an axis of rotation or direction of magnetic flow in a current. It originally came from Ampère's description of a person swimming in the same direction as the current (in a wire). When the swimmer looks left, it's the same direction that the north-seeking end of a compass will point if it is in the vicinity of the current-carrying wire. Since then, it was decided it was easier to use the left hand and actually look at the thumb and fingers, rather than imagining a swimmer. Extend the thumb and fingers of the left hand so that the fingers are held together and point straight in one direction, with the thumb at a right angle to the fingers, in an "L" shape. Now curl the fingers around a conductive wire, so that the thumb points in the direction of the current. The direction of the curled fingers is said to indicate the direction of the magnetic field associated with the current.

Some of the confusion associated with left- and right-hand rules stems from the fact that pioneer physicists did not originally know in which direction current was flowing in a circuit between negative and positive terminals. In fact, it was not always important to know, as long as the terms of reference were kept consistent in one direction or the other.

Sometimes a distinction is made between current in a motor and current in a generator. By this reasoning, using the left hand, the fingers will show the direction of the current for a conductor in the armature of a motor. Using the same hand relationship for the right hand will show the direction for a conductor in the armature of a generator. Since the universe appears to be right-handed in its general orientation, some physicists will assert that it makes sense to use the right-hand rule. See right-hand rule.

leg 1. A portion of a trip, broadcast, or transmission. 2. The transmission segment in a network between two physically distinct entities (such as a workstation, switch, router, node, etc.), between addressable entities, or a combination of the two.

legacy That which is inherited, or which remains from a predecessor.

legacy equipment/software Existing equipment, software, or operating procedures that are becoming dated but are still actively used are called legacy systems. For economic reasons, most legacy systems are maintained and enhanced, rather than scrapped in favor of new systems. Even when it's *not* economically practical, legacies are sometimes retained because managers are reluctant to let go of the emotional investment tied up in existing systems and procedures. A more practical reason for retaining legacy systems is that staff training costs time and money, and staff members may be reluctant to switch to a new system. Most computer operating systems are legacy systems, incorporating downward compatibility in order to work with older equipment.

Legacy to the Future L2F. A framework for providing an integrated modeling system of legacy and new code components to support simulations in the Los Alamos National Laboratory (LANL). L2F is implemented using JAVA and CORBA to facilitate communication among objects written in different languages. The system includes a setup server and an associated database, application servers, clients and optional security services. L2F has been cleared for open source distribution.

Lemelson, Jerome "Jerry" H. (1923–1997) A prolific American inventor and engineer, Lemelson received more than 500 patents by the time of his death, with more patents pending.

Lemelson began inventing gadgets in childhood and turned to technology in the mid-1950s. In the 1960s, he designed magnetic recording and beam switching and production line automation systems. In the 1970s, he began designing miniaturized devices such as portable audio/video systems and earned patents for optical circuits and display devices.

Sometimes the patents took many years to process. For example, a patent application for a videotape recording system for recording frame animation on tape media was submitted by Lemelson in 1962 and not granted until 1980.

Lemelson is credited with developing technology that led to the ubiquitous barcode reader and he sold many data processing patents to IBM.

In 1994, Lemelson and his wife founded the Lemelson Foundation to promote and support American invention and innovation.

Lempel-Ziv LZ. A pair of coding compression formats described in IEEE articles "Compression of Individual Sequences via Variable-Rate Coding," in 1977 (LZ77), and "A Universal Algorithm for Sequential Data Compression," in 1978 (LZ78). They are named after their developers Abraham Lempel and Jakob Ziv. LZ77 was presented by Ziv and Lempel as a dictionary-based scheme for lossless data (text) compression. LZ78 sends pairs of pointer and character data. These two important schemes are so universal and powerful that a significant number of data compressors have been based on Lempel-Ziv concepts.

LZ77 became the basis for LArc, LZARI, LHarc, and others. Lempel-Ziv schemes have been widely adapted and used in network and modem technologies, but the early versions were not developed for the demands of multimedia networks with widely ranging data characteristics (which scarcely existed at the time). Another limitation of the original Lempel-Ziv techniques was in performing efficient searches for previous matching strings. Enhancements and variations are addressing some of these limitations and adaptations to special circumstances. In 1994, Jung and Burleson proposed a parallel algorithm, architecture, and implementation for Lempel-Ziv compression exhibiting a scalable, regular structure suitable for VLSI array implementation. This has practical applications for data compression in portable digital data communications and wireless local area networks (LANs) where effective compression/decompression schemes can significantly improve throughput and reduce connection time costs. See Ardire-Stratigakis-Hayduk, differential Ziv-Lempel, LArc, LHarc.

Lempel-Ziv-Stac LZS. A data compression system developed by Stac Electronics, Inc., largely based upon the LZSS algorithm. See Lempel-Ziv-Storer-Szymanski.

Lempel-Ziv-Storer-Szymanski LZSS. A sliding dictionary compression scheme descended from LZ77 by Lempel and Ziv. LZSS was developed in the early 1980s by James Storer and Thomas Szymanski. Storer filed for a patent for data compression/decompression in 1987 (U.S. #4,876,541). LZSS differs from its predecessor mainly by using bit flags to identify the subsequent data as a literal or an offset, which can result in more compact data compression and faster decompression. A number of developers developed variations and optimized versions of LZSS, one of which evolved into LArc. See LArc.

Lempel-Ziv-Welch LZW. A dynamic dictionary, lossless data file compression scheme based on the Lempel-Ziv method, first described by Terry A. Welch in 1984 following the publication of the Lempel-Ziv format in the late 1970s. The developer community generated a substantial amount of debate over use of LZW, since it was distributed as an open standard for many years before the community at large was informed that LZW was being patented. Many programmers incorporated LZW into their software, thinking it was in the public domain. LZW has been incorporated into ARC and PKZIP. PKZIP is a particularly widespread archiving utility.

A patent for the technology is held by IBM (U.S. #4,814,746), with a similar one held and enforced by Unisys Corporation (U.S. #4,558,302). Some claim these two patents cover virtually the same technology, or that the Unisys patent is a subset of the IBM technology. A similar patent has also been held since 1989 by British Telecom.

The enforcement of the patent rights caused ripples of unease in the programming community, as several modem file protocols use LZW. Many graphics interchange formats incorporate LZW compression, including TIFF, which is widely used in desktop publishing programs, and the Compuserve 8-bit GIF format popular on the Web.

In telephone technology, LZW is used in Northern Telecom's Distributed Processing Peripheral (DPP) for the transmission of compressed data. At one point, Unisys issued a statement exempting freeware authors from paying license fees on the use of LZW in their programs in order to quiet the concerns of software developers who were distributing software without commercial gain. Later, Unisys asserted that *all* software developers would be subject to a minimum royalty ($.10) in order to protect the patent, with exceptions only for charitable institutions. See Ardire-Stratigakis-Hayduk, Lempel-Ziv, LZC.

lens A component with optical properties such that it lets light pass through while altering its path. A magnifying glass is a simple convex lens that spreads the light coming through from the object observed such that the object appears larger when it hits the human eye or an imaging apparatus. There are many different types of lenses. A prism is a type of lens that separates white light into its wavelength components so that the colors may be seen. A Fresnel lens diffracts light to concentrate or diffuse it.

Lenses may be combined in linear or planar arrrays to control light passage over a larger area than would be possible with one lens. It is usually not practical to make a single optical lens larger because the curvature causes image aberrations nearer the edges.

In free-space optics (FSOs), where diode lasers transmit signals through the air to connect buildings with one another or with a nearby fiber backbone, a lens can spread the light beam so that birds, tennis balls, or other obstacles that may briefly interrupt the transmission are likely to only pass through part of the beam rather than obstructing it completely. See faceplate; fiber optic; Fresnel lens; lens, lenticular; transfer lens.

lenticular Historically, lenticular meant having a lens shape, especially a lentil-shaped, double-convex lens as in a traditional magnifying glass. The term is now more often applied to lens configurations with arrays of closely aligned lens shapes designed to refract light at particular angles in succession as the viewer or the lens is rotated through a plane. It also no longer necessarily implies a double convex lens and some lens arrays with straight surfaces have been called lenticulars, though they are probably better referenced as gratings. See lenticular lens

lenticular image A specialized optical image imprinting technique in which a series of two or more related 2D images is interlaced and embedded under a ridged surface such that the net viewing effect is similar to a stereographic image or animation exhibiting depth or movement.

Early commercial lenticular images (sometimes called lenticulars) bonded an image that was printed on card stock behind the refracting lens. Now, with computerized image processing techniques, it is possible to take successive frames of an image, interlace the image at tolerances to match the lens density and imprint them on the back of the lens material itself such that the strips of the frame associated with the refracted light from a viewing surface are successively visible.

The source of the effect is the embedding of a series of frames behind the ridged surfaces of the base material to exploit the refractive properties of the individual lens facets, somewhat like a cross between a holographic image and a traditional cell animation. Thus, as the viewing angle changes, the frame associated with the ridges that can be seen from a particular angle becomes visible. Some have described it as an evolutionary descendent of stereograms. It has some advantages over stereograms in that many frames can be displayed in succession and no special viewing apparatus is required.

Lenticular images can be designed to move in horizontal or vertical directions. The number of possible frames in the series varies with the size of the lenticular surface, the differences between frames, and the orientation.

Lenticular images are created by taking a series of image frames, deciding on the orientation of the frames when viewed, digital image processing the frames to determine which parts must be imprinted on which surfaces of the lenticular lens, and imprinting them, usually through a lithographic process, serigraphic, or photographic process.

For many purposes, lenticulars have advantages over holograms. Lenticular lens bases may be more robust than holographic bases. They are typically easier to see in certain light conditions. The colors are photographic, as opposed to the subtle prismatic colors typical of holograms. They have advantages in size, as it is fairly straightforward to impress lenticular images as large as a small billboard.

Not surprisingly, lenticular images are popular as novelty items and point of purchase marketing displays. However, there are many scientific applications in which a lenticular image can help reveal the structures of an object or convey 3D information in an educational environment. For example, a lenticular image can help illustrate the 3D geometry of biological specimens or celestial bodies.

lenticular lens A flattish lens array designed such that, when viewed at particular angles, it reflects light in certain predetermined directions through successive, closely-aligned surfaces within the viewing scope of the user or imaging device. Each individual lenslet in the array is called a lenticule.

Commercial lenticular lenses are typically fabricated from coated plastic sheets (PVC, APET, styrene, etc.) that are flat on one side, ridged on the other. They are available in various shapes, sizes, and resolutions. A plastic rectangular sheet may have a resolution of about 10 to 75 lenticular ridges per inch (RPI) and a thickness of about 0.0625 in. Optimum viewing angles depend upon density and application but are generally around 35°. For offset printing processes, higher densities are available. As the density of the lenses and overall size increase so, generally, does the difficulty of laminating the material.

Many lenticular lens sheets are sold to the publishing industry for novelty products and signage. However, lenticular lenses composed of arrays of cylindrical lenslets are also sold as diffusion lenses for scientific purposes. See Fresnel lens, lenticular image.

LEO See low Earth orbit.

LES LAN Emulation Server. A local area network (LAN) software server which provides Media Access Control (MAC) address-to-ATM address resolution services for LAN Emulation (LANE). See LECS, LEC.

letterboxing A video display technique that preserves the aspect ratio of a wide screen cinematic production even when it is displayed on a different system such as a TV screen. Wide screen movies are often modified to fit a TV screen ratio, but then information on either side of the image is lost. In letterboxing, it may appear as if some of the image is gone, because there are larger black areas at the top and bottom of the image, but in fact, it is the letterboxed version that shows the *entire* image and retains the fidelity of the original picture. The difference can be quite dramatic. For example, in a number of scenes in the beautiful film *Baraka*, the nonletterboxed ver-

---

Lens for Concentrating Light Toward a Cable

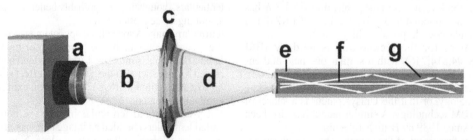

Depending upon the light source and the nature of the light emitted (e.g, degree of spread), a lens may be used to concentrate the light (b) emitted by a light-emitting diode (LED – a), for example. The lens focuses the light (d) to a point within the cladding diameter (e) of the lightguiding fiber optic cable. From there the light reflects within the core (f) and propagates along the waveguide (g).

sion shows only two people rather than the original three. See anamorphic, aspect ratio.

level 1 cache A small, fast static RAM buffer. On an Intel Pentium central processing unit (CPU) chip, a 16-kilobyte cache memory is incorporated into the chip.

level 2 cache An external, fast, static RAM buffer. On an Intel Pentium central processing unit (CPU) chip, a cache memory is incorporated into the processor in addition to the level 1 cache included with the CPU. On some of the Pentium chips, the level 2 cache is layered into the CPU (for faster access), and in some, it is a separate section, with a bus allowing it to communicate with the CPU. Level 2 caches may vary in size from 256 kilobytes to 1 MByte.

Leviton Manufacturing Company, Inc. One of the world's largest companies specializing in the design, development, and production of telecommunications wiring devices. The Leviton Voice & Data Division provides system solutions for network infrastructures with fiber and copper technologies. It produces connectors, cabling assemblies, panels, fiber optic test equipment, and other related products. The Voice & Data Division is located in Bothell, Washington.

Leyden Jar – Historic Capacitor

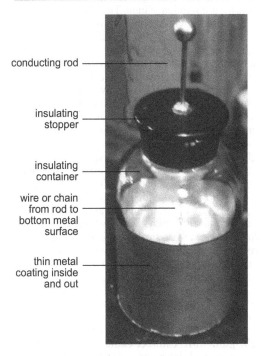

conducting rod

insulating stopper

insulating container

wire or chain from rod to bottom metal surface

thin metal coating inside and out

The Leyden jar was an important historic storage tank for electrical experimentation, acting as an electrical condenser/capacitor. [American Radio Museum collection.]

Leyden jar, Leiden jar A device that concentrates and stores electrical energy, thus serving as an electrical condenser; an early capacitor. The Leyden jar

was devised by E.G. von Kleist, a German experimenter, in 1745. It consisted of a nail in a bottle connected to a terminal of an electrical device, with the jar held in von Kleist's hand; he received an unpleasant shock from his experiment. A year later, Cunaeus and Peiter van Musschenbroek created a condenser consisting of a jar mostly coated inside and out with metal foil, with the inner coating in contact with a conducting rod that passed through the stopper (insulator).

The foil would typically cover about two-thirds or a little more of the surface of the jar, and the rod would be inserted through a stopper of cork or rubber. Sometimes a chain was attached to the bottom end of the metal rod. The jar was named after the town of Leyden (Leiden) in The Netherlands. It was subsequently discovered that a Leyden jar charge could be sent through wires over distance. Benjamin Franklin conducted numerous experiments with Leyden jars in his attic laboratory, and they remained prevalent for another 150 years.

LFACS See Loop Facility Assignment and Control System.

LFAP See Light-weight Flow Admission Protocol.

LGC Line Group Controller.

LGRS Local Government Radio Service.

LHA A lossless compression scheme developed by Yoshizaki. In terms of compression performance for text files, LHA was an improvement over the original LZ77 by Lempel and Ziv, through static Huffman coding, but not as efficient as the popular GZIP. It is also known as LZH due to shared .lzh/.lha file extensions. LHArc also uses .lzh file extensions, so there may be some confusion, but they are from the same family of compression utilities and somewhat downwardly compatible. LHA has been widely used in Japan and is a popular utility on the Amiga computer. See LHArc, LZHUF.

LHarc A fast, efficient dictionary-style compression/archiving utility descended from LZSS. LZSS was developed into LArc by Okumura and Miki in the late 1980s. Okumura then created LZARI, which incorporated adaptive arithmetic compression, from LZSS. Yoshizaki subsequently modified LZARI with adaptive Huffman coding instead of adaptive arithmetic coding to create LZHUF, which was then rewritten in assembler with an updated interface to create LHarc, which was fast and popular.

LHarc is a full archiving utility, allowing multiple files to be stored, listed, added to, or removed from a single archive. LHarc was especially popular on the Amiga, edging out earlier favorites due to its speed and versatility; there are versions for IBM-licensed PCs, Macs, Ataris, and Unix machines as well. See Lempel-Ziv, RAR.

LHCP See left-hand circular polarization.

LIC See light-guide interconnect cable.

Licklider, J.C.R. "Lick" (1915–1990) A computing pioneer who was instrumental in supporting a number of important early developments, including time-sharing and the ARPANET. He is best remembered for his inspiration and enthusiasm, and his ability to

L

get the funding and other resources necessary for various computer pioneers to build the stuff of dreams.

LIDB See Line Identification/Information Database.

LIFO last in, first out. A descriptive term for the order in which data are processed in a queue. For example, picture a stack of dinner plates in a plate well in a buffet; when the stack is refilled by the restaurant staff, the last plate on the stack (the one on top) is the first one removed by the next customer. Data can be handled in the same way. See FIFO, FILO, GIGO, LILO.

light Radiant energy visible to the human eye having wavelengths in the approximate range of 390 to 750 nm, that is, the transition to ultraviolet at one end of the spectrum and infrared at the other end. The phrase *white light* is used to describe light with a mixture of frequencies. The speed of unimpeded light is 3.00 x 10^{-8} meters per second, symbolized as *c* in mathematical calculations.

It was discovered that light could be broken up into its component wavelengths with prisms, and this aided researchers in understanding the nature of light and the colors associated with particular frequency ranges. The visible spectrum is specific to human perception; other mammals and insects have broader, shifted, or more specific perception of color ranges. A flower that to humans appears yellow may have other colors in the ultraviolet or infrared spectrum that are perceived by pollinating insects. Most dogs and cats are insensitive to colors as humans perceive them (Siamese cats reportedly can perceive color).

Light is the primary stimulus for sighted individuals sense to forms, other beings, and their orientation and movement in three-dimensional space. The interaction of the light waves hitting various objects, bouncing back through eyes and processed by a brain, constitutes the complex phenomenon called sight. Some creatures can see beyond the range of human-visible light. Dolphins use sonar (sound waves) to detect objects which may not be visible to humans, and thus can "see" inside some objects in a way not possible for humans without mechanical aids. See fiber optic, infrared, lamp, laser, light-emitting diode, spectrum, ultraviolet.

light guide Light-conducting material such as an optical fiber. The material provides a conduit or channel by which the light can be directed. The idea is similar to the concept of wave guide.

light path In general, the path traveled from one specified point to another by electromagnetic radiation in the optical freuquencies. In general, light travels in a straight line, but multiple reflected straight line segments can be combined to form a curved path. See lightpath.

light piping Bringing light into an area through fiber optic cables. Laser light will travel along a filament the size of a hair, and filaments can be bundled to provide more light. This is very handy for illuminating hard-to-reach places like small pipes or inside the human body for medical research or procedures.

light pulse The basic information-carrying transmis-sion medium in a fiber optic network. Coded electrical pulses from computer systems, existing phone systems, and other sources are used to stimulate a light source such as a light-emitting diode (LED) or an injection-laser diode (ILD) to generate the light pulses that are then funneled through a lens, corrected with prisms, if necessary, and transmitted through a cladded (shielded) fiber waveguide. The light propagates through the waveguide through a process called total internal reflection (TIF). The cladding layer around the fiber reflects beams that are within a certain range of angles back into the fiber core, preserving the signal over distance. At the receiving end, the light pulse is usually translated back into an electrical signal for digital decoding and other processing. Along the transmission path, the light pulse may be amplified or regenerated through a variety of electrical or optical means. See light-emitting diode.

light speed The speed of light in a vacuum is 186,292 miles/sec (299,800 kilometers/sec). When traveling through air, water, a fiber optic waveguide, or other dielectric (conductive) material, the matter will impede the speed of light and slow it down. In fiber optic cables, for example, the propagating light travels about 124,100 miles/sec.

In general, the higher the refractive index of a material through which the light travels, the more it impedes the propagation of light. This characteristic can be exploited in cable construction. The cladding around an optical fiber core is made from a material with a slightly lower refractive index than the core such that the light beams are reflected back into the core (within certain effective angles). Sometimes, as in multimode cables, the refractive index is graded to selectively control light speed based upon its distance from the core, in order to compensate for certain types of frequency dispersion. See dispersion, graded index, stepped index.

light-emitting diode LED. An inexpensive semiconductor p-n junction structure used in many electronic displays, particularly small ones. The LED lights up when a current is provided. LEDs are common in digital clocks, calculators, microwave readouts, electronic instrument displays, and much more. The LED typically resembles a small illuminated knob with a semiconductor within the knob (which is actually a lens), and leads coming out from the semiconductor/knob arrangement into the device circuitry.

LEDs are now also used to provide the light rays for certain types of fiber optic transmissions, especially in multimode cables (more precise and more expensive laser lights are used for single mode fiber).

light-guide interconnect cable LIC. Also called light-guide cable interconnect (LCI). A light-guide is a conduit for directing and containing a light-based transmission and an interconnect cable is one which specifically interconnects various equipment and devices. LICS are used in fiber optics transmission networks and typically have connection ends that can be more readily connected and disconnected for the purposes of installing and reconfiguration network physical topologies. See loose cable.

Light-weight Flow Admission Protocol LFAP. A protocol from Cabletron which allows an external Flow Admission Service (FAS) to manage flow admission at the network switch, allowing flexible FAS to be used by a vendor or user without unduly burdening the switch. See RFC 2124.

LIGHTCONNECT A leading global supplier of diffractive MEMS-based dynamic optical networking components such as gain equalizers.

lightline A multiple fiber filament assembly in which one or both cable ends is set in a row in a supporting frame to produce a line of light. Commonly one end of the assembly is round (to couple with the next link or the light source) and the other spreads the filaments side-by-side into a long line (essentially the shape of a window washing tool, round at one end [handle] and long and narrow at the other [wiper blade]). Applications include machine vision illumination, sensing, signage, artworks, and commercial lighting. For imaging or display applications, the individual fibers in a lightline can be calibrated to different intensities.

lightpath A point-to-point optical link that may pass transparently through intermediary nodes. Thus, it comprises a logical one-hop link, even if there is more than one physical hop.

LightSAR light synthetic aperture radar. See synthetic aperture radar.

lightwave communications Optical communications systems using high frequencies. This term helps distinguish optical communications from very short wave microwave communications. Fiber optic cables are used as the physical medium for transmission. This is distinct from *lightwave transmissions,* which involve transmission through air or space rather than through a cable as the physical medium. When homodyne or heterodyne detection schemes are used, they are called coherent optical communications systems.

lightwave transmission Optical communications systems based on transmitting a beam through air or space, without using a cable as a physical medium. This is a wide bandwidth, line-of-sight, short distance technology which is relatively inexpensive. It is suitable for building-to-building installations where it is impractical to string wires. This system is subject to loss and is somewhat dependent on weather, and thus specialized in its practical applications.

Lightweight Directory Access Protocol LDAP. A front-end client/server standard intended to provide a lightweight complement to Directory Access Protocol (DAP), LDAP is based somewhat on ITU-T X.500 and can access X.500 directories. It is a distributed, hierarchical protocol for accessing network entities and repositories and is more scalable for some implementations than, for example, Network Information Services (NIS). LDAP was developed in the early 1990s at the University of Michigan and submitted as a joint Standards Track RFC with the ISODE Consortium and Performance Systems International.

In LDAP, the protocol elements bypass some of the session/presentation overhead by going directly over Transmission Control Protocol (TCP) or the relevant transport protocol. Many of the protocol data elements are simply encoded as strings. A lightweight best error rate (BER) encoding is used for protocol elements. Extensions to the format such as authentication and server discovery are being discussed and developed. See RFC 1777.

LightWeight Encoding Rules LWER. LWER comprise one of three major encoding schemes used in open architecture development, developed in the early 1990s. LWER provide a means for creating encodings optimized for encoding and decoding CPU cycles and apply better to wide bandwidth communications between similar architectures than to slower mixed-system networks. See Basic Encoding Rules, Packed Encoding Rules.

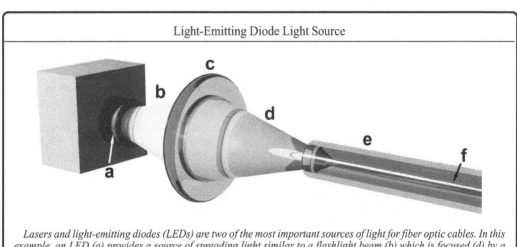

Light-Emitting Diode Light Source

Lasers and light-emitting diodes (LEDs) are two of the most important sources of light for fiber optic cables. In this example, an LED (a) provides a source of spreading light similar to a flashlight beam (b) which is focused (d) by a lens (c) towards the lightguide where it propagates by total internal reflection (TIR) by bouncing off the cladding (e), which has a lower refractive index than the conducting core (f). Depending upon the application, multiple LEDs may be used.

like new A subjective term describing a product that has been opened and stored or used, but is physically unmarred and functionally in good working order. The phrase is not intended to imply that the product will last as long as a brand new product in normal use, hence the phrase *like new* instead of *new*. See certified, refurbished, used.

LIM Link Interface Module.

LINC Laboratory Instruments Computer. One of the earliest small computers, developed at the Massachusetts Institute of Technology (MIT) in 1963, the LINC was the inspiration for Digital Equipment Corporation's (DEC's) PDP-8.

line code violation LCV. In E1/T1 networks, an indicator of bipolar violations (BPVs) or excessive zero (EXZ) errors. A line err second (LES) is a second time duration in which one or more LCVs are detected. A severely err second (SES) is one in which serious defects are detected such as 2048 LCVs or more for nonCRC signals.

line err second LES. See line code violation.

line of bearing LOB. In radio direction-finding systems, the direction and general position of a target or positional goal. A radio direction-finder is a device for tracking down the source of a radio frequency signal. Flashing lights or beeps are sometimes used on portable systems to indicate the direction or proximity of a target (e.g., wildlife with a radio collar). In mounted systems, a monitor may indicate a line of bearing relative to a selected reference point. Most direction-finding devices will have a certain margin of error, expressed in degrees, that can be used to calculate the region of error of a bearing over distance. Lines of bearing are sometimes graphically plotted by hand on a chart or map. A line of bearing is sometimes a calculated position based on a series of readings within a general range. In direction-finding, intersecting lines of bearing may indicate the position of a transmitter at or near the intersecting point.

Consumer systems for determining lines of bearing are more sophisticated compared to the old handheld beeping antennas prevalent for many years. It is now possible to use a Global Positioning System device with an automatically adjusting antenna to carry out direction-finding and transmitter tracking functions.

line of sight An unimpeded direct line of connection through "free space" as it relates to a particular transmission technology. Line of sight is a relative term. The line of sight for a flashlight beam is usually limited to 10 or 15 feet, as the light falls off quickly and stops if it encounters physical objects in its path. The line of sight for a coherent laser beam in the infrared spectrum (e.g., a VCR remote control) is a little longer than a flashlight beam and can pass through some types of objects. The line of sight for a handheld radio is a couple of miles because the radio waves can pass through walls and many other solid objects, but the signal is gradually scattered by very rough terrain, many buildings, and reflective surfaces. Line of sight is an important concept in radio communications and free-space optics as these are important options for linking people to fiber optic backbones

in regions where the fiber optic cables cannot be installed all the way to the premises. See free-space optics, Photophone.

line speed The maximum or actual speed of data transmission through a wire or cable. There are dramatic variances in line speeds depending upon the type of cable, the distance or character of the path, and the interface or modulation devices used to convert or send/receive the data.

Line speed for digital data transmissions is commonly expressed in terms of the number of bits or characters transmitted per unit of time (usually seconds). For comparison, a typical 64K ISDN connection to a page on the World Wide Web might download an average-sized image at 7500 cps. Many factors can reduce the line speed from its theoretical maximum, but ISDN and cable modem throughputs are significantly slower than 10M Ethernet, for example, which could download the same data at speeds of up to about 480,000 cps.

line trunk controller LTC. A telecommunications trunk controller that provides a means to use equipment contained within a central office. The LTC is capable of giving interfaces to outside ports; it is one of a number of peripheral modules that can provide trunk interfaces to a digital multiplex system (DMS). It may interface with multiple lines which, in turn, may be linked to a network by multiple speech links.

line utilization monitor LUM. A diagnostic and administrative utility providing statistical logs or displays of transmission line use.

line switching unit LSU. A generic phrase for a variety of devices that range from simple passive switchers to complex active voice or data transmission line switchers. A line switching unit can be used on a single computer or local area network (LAN) to switch between multiple peripheral devices such as printers and scanners or, in a more sophisticated architecture, between a disabled system and a standby system. In telephony, line switching units are typically floor-size cabinets with multiple lines and electronic switching circuits to manage those lines and their connections. LSUs may be manual or automatic. Automatic units are more likely to be used in remote locations or for backup or emergency systems triggered by alarms or fault conditions.

linear modulation A modulation scheme developed at Bristol University in 1991. The technology began to be used in the U.S. in the mid-1990s. The advantage of linear modulation is that it enables voice and data transmissions to be carried over narrow channels (e.g., 5 kHz), thus making it possible to increase capacity up to five times that of traditionally wider frequency-modulated (FM) channels. A pilot tone is inserted into the audio baseband, essentially splitting the band, enabling linear modulation to be achieved without distorting the signal.

The development of new modulation schemes that can support higher capacities and bit-rate densities has become particularly important in recent years. The demand for mobile services is growing, while the pool of available frequencies remains essentially the same. To complicate the picture further, the Federal Communications Commission (FCC) has issued a requirement that spectral efficiency for mobile devices be improved by the year 2005. Linear modulation is seen as one of the key technologies that may aid developers in meeting this requirement. See tone in band.

linear predictive coding LPC. A system for digitally encoding speech at low bit rates while retaining good clarity and recognition without extensive computing overhead. In LPC, speech signals are analyzed as to various common aspects, including resonance, intensity, pitch, etc. Formants (resonances or boosted frequencies) are filtered and stored in addition to the remaining signal, which is termed the residue. Portions of the speech are sampled at the rate of about 40 frames per second, resulting in manageable file sizes for storage and intelligible speech when the encoded data is reconstructed. In the encoding/decoding process, estimates are used and errors between a predicted signal (based on previous samples) and an actual signal are minimized for the series of sequential samples. LPC methods result in lower information bit rates than adaptive predictive coding (APC) methods. Since speech and music share many common aspects, LPC has been useful for computer music encoding for storing and extracting time-varying formant information.

LPC concepts have also been applied to the compression of imagery in that predictive techniques have been used to provide linear approximations of vector information which, in turn, can be used for lossless or lossy compression algorithms for geospatial imagery. See sampling.

linear programming Algorithmic symbols, procedures, and strategies in which a problem or statement can be expressed in a standard form within certain variable, sign, and coefficient constraints to find a solution to a problem. Both the program constraints and the problem itself are linear. While a limited form, many types of problems can be solved or expressed with this type of programming environment.

Variables within a linear programming problem can be seen as corresponding to decision factors in the problem to be solved. Linear programming is useful for solving or generating optional optimization solutions for production lines, transmission paths, or investment or management scenarios.

linear search & track processor LSTP. A system for processing radar search and track signal data. The Naval Research Laboratory Collaboration has been porting LSTP functionality to parallel computing platforms.

linear tape-open LTO. A tape storage technology developed jointly by Hewlett-Packard, IBM, and Seagate. LTO is an open format that incorporates linear multichannel, bidirectional tape formats with data processing enhancements such as error checking, data compression, etc. In spite of the many storage options available for data management that rely on optical or hard disk formats, there are still advantages to the use of tape, including high capacity and low cost. Hundreds of GBytes can be stored on a single tape cartridge in compressed format.

The LTO technology has been optimized into two open tape versions specified as *fast access* or *high capacity*. This decision was based on the observation that some storage needs are read-intensive (requiring speed of access) and some are write-intensive (requiring high capacity). The formats are respectively called Accelis and Ultrium. Licenses for third-party developers were made available as of April 1998. Commercial implementations vary, but transfer rates of 30 Mbps are available. IBM offers low voltage differential (LVD) and high voltage differential (HVD) versions for SCSI and Fibre Channel (FC). Quantum's Super DLTtape technology is somewhat competitive with LTO. See Super DLTtape.

line, communications In its most general sense, any path or transmissions link between two or more communicators, including any subscriber line, switches, routers, cables, etc. which might comprise the main transmissions pathway. The term *line* usually implies a physical connection, or series of physical connections including wire or fiber optic cable. Hybrid systems including wired and wireless connections are sometimes also called lines. Complete wireless connections are usually referred to in terms of *airways* rather than lines.

line, electric A circuit connection physical conductor consisting of shielded or unshielded wire/cable.

line conditioning Improvements and enhancements to a communications line to reduce interference and improve the quality of the signal. Some phone companies offer higher quality line service as an option, which may be important to those doing a lot of data communications over a phone line.

line finder An evolutionary improvement in step-by-step telephone switching systems that eliminated the need for a separate switch selection for each subscriber line. When the caller picked up the phone a relay would be used to find an available line-finder switch, hunting for the caller's terminal would be initiated, and the caller would be given a dial tone when the line was connected with the switching system.

Line Identification/Information Database LID. A national system of telecommunications information databases first deployed in the early 1990s. It is designed so that subscriber and carrier information can be readily accessed and cross-referenced. This information is used for information-querying, validation, and alternate billing administration. See Local Number Portability.

line impedance stabilization network LISN. A diagnostic instrument for measuring emissions, a LISN stabilizes the line impedance so that tests can be repeated against a reference at more than one point. This enables the device tested to be isolated from an external power source, for example. There are a variety of types of LISNs with low-pass or high-pass filters, depending upon their purpose.

LISNs have connectors for attaching the tested device and may also have a connection for additional diagnostic equipment or displays, such as a spectrum analyzer or electromagnetic interference (EMI) meter.

L

They are designed to test within specific frequency ranges (e.g., 10 kHz to 30 MHz).

LISNs are commonly used to see if devices are generating signals (radio frequency interference) that may affect their operation or nearby devices. They also can help determine whether the emissions are above certain maximum levels established as industry standards (e.g., ANSI) or in conformance with levels established by regulating bodies such as the Federal Communications Commission (FCC). Electromagnetic disturbance characteristics and standards are encompassed within CISPR 11 to CISPR 28. CISPR 16, in particular, specifies radio interference measuring apparatus and measurement methods pertinent to LISNs. Multiline ISNS are also known as V-Networks. See International Special Committee on Radio Interference.

line insulation test LIT. In telephony, a diagnostic test performed from the central office to determine line resistance and line voltages.

line light A linear source of illumination (long and narrow) as might be created with a fluorescent bulb or neon light. In most cases, the distribution of light is fairly even along the breadth of the light, though this is not necessarily so. A line light can also be created with an array of point lights, as from an array of light-emitting diodes or lasers and may be fed to a remote location through a fiber optic cable. See linelight.

line noise Electrical noise in a communications line which interferes with voice communications, or which causes spurious characters to show up in a data transmission. In older modems used over phone lines, line noise sometimes resulted in strange characters being displayed in the telecommunications software terminal window. Severe line noise can interrupt a data file transfer, or even cause the connection to be disconnected. With newer phone line services and newer, error-correcting modems, this problem is diminishing.

line of sight Many communication technologies require an unobstructed straight line of travel for the signals to reach the intended destination or be seen by the appropriate people, so this is an important concept that affects the design of many communications technologies. The term is generally used in the context of air- or spaceborne communications, though it is also applicable to sonar and may even be stretched to include nonvisible electromagnetic phenomena traveling through evenly particulate matter (e.g., X-ray probing in soft dirt).

Semaphore systems, for example, require that signaling arms or flags be visible to the receiver. Beamed light from a ship requires that no impediments block the way or bounce the beam in a direction away from the recipient (fog can rapidly diffuse a light beam). Very short wave radio signals, which may be readily reflected or absorbed by objects and even small particles, require a clear line of sight to project over distances. Line of sight thus refers to a straight, clear, direct, and generally unobstructed travel path. The line of sight for a particular technology is usually

expressed as an expected range or as an expected maximum.

Since some types of phenomena or wavelengths are more readily absorbed or reflected than others, line of sight is a context-sensitive designation. The line of sight for a flashlight beam or infrared remote inside a dark house may only be 15 ft depending upon where the visible or infrared light beam hits a wall, another person, or a piece of furniture. The line of sight for a 2.4 GHz home entertainment system transceiver might be about 150 ft, traveling through walls and people, but not having sufficient power to reach the entire neighborhood to interfere with all the other home entertainment transceivers (though it may cause strange images on the television next door). The line of sight for a radio modem or small handheld radio may be up to a couple of miles depending upon the hills and trees in the terrain and how densely the buildings are spaced. Thus, line of sight is dependent upon the type of phenomenon, its power, and its wavelength.

In order to extend line of sight, many communications technologies rely on repeater stations for propagating a signal past obstructions and over longer distances. Radio transmissions often use the Earth's ionosphere as a type of "repeater station" in the sense that the waves are directed at the ionosphere at a particular angle and bounced back toward the Earth to span longer distances than would be possible using a horizontal line of sight. GPS systems rely on multiple orbiting satellites that are spaced in relation to each other such that at least three are usually "visible" within the radio range of a receiver at any given time. The different coordinates obtained by the satellites within the line of sight are used to triangulate a location.

Mobile communications systems that rely on radio or light waves are subject to line of sight limitations. For this reason, many mobile data and voice services hook into landlines for a significant portion of the transit distance. The landlines have traditionally been copper wire phone lines, but increasingly broadband fiber optic land links are used.

line powered Any device that receives its power from the main system or transmissions medium to which it is attached. For example, most basic phones without extra features do not require a power supply or battery because they are powered by the current in the phone line. Some laptop peripheral devices, to minimize size and weight, derive their power from the laptop itself. See talk battery.

line printer A printing device that prints one line of characters, or a full line at a time. This phrase has been applied rather generically to most impact printers that print a line of characters by typing each character successively, but, technically, these are actually character printers. True line printers compose and "stamp" an entire line of text (they're quite fast and often very noisy), and compose the next line in the printer memory buffer while the current line is being printed. Line printers tend to be used in institutional and industrial environments where speed is more important than cost or low noise levels.

line status indicator On a telephone, modem, or other appliance that can connect to more than one line at a time or perform a variety of functions on one or more lines, there may be a character display or various lights to inform the user of the status of the line. On multiline telephones, a light usually shows which lines are in use so that the user can avoid barging in on a call inadvertently. On a modem, line status indicators may flash to show whether data are sent or received, and may indicate whether or not a carrier is present. On network ports, status indicators may indicate loss of frame (LOF) or loss of signal (LOS).

line switching See circuit switching.

Line Terminating Equipment LTE. In SONET networks, an element that originates and/or terminates a line signal. It can originate, access, or modify the line overhead, or terminate it, if needed. See SONET, Synchronous Transport Signal.

linear transponder A device commonly used in communications satellites and radio relay stations, which takes a small segment of frequencies, amplifies the signal strength across the range of frequencies, and retransmits them at a slightly different frequency range (by shifting or multiplying), so that the whole segment is adjusted up or down. This is often done to prevent the transmitted signals from interfering with the received signals. See store and forward repeaters.

lineman, line worker In the early days of the telegraph, when cable was being strung across continents, linemen were assigned to dig holes, cut down trees for poles, set the poles, climb them, and attach the wires, gradually working their way through wilderness, native encampments, and mountain ranges, until the coasts and settlements were interconnected. Once the lines were installed, they would test them, often with portable telegraph keys, and maintain the lines through inclement weather over hostile terrain.

The work was hazardous, no insurance or benefits were available, and linemen injured by electrocution, falls, and other hazards were dependent on the goodwill of their employers for assistance.

Maintaining the increasing number of wires and poles involved the dedication of many 24-hour crews and, until the mid-1960s in North America, much of the work was done by climbing the poles with belts and cleated boots (lineman's climbers), securing in at the point that needed repair, and doing the work manually with simple tools. Since that time, power tools, sophisticated testing equipment, and cranes with buckets (cherry pickers) for the line workers have increased in use to the point that it is uncommon to see a worker scale a pole in urban centers.

The line workers now are also responsible for digging, diagnosing, and installing underground transmissions lines, in addition to managing lines on utility poles.

lineman's climbers, line climbers A variety of pole climbing equipment used over the last century to allow installation and repair workers to scale utility poles in various types of weather. These range from climbing irons or spurs strapped on the legs, to cleats on the boots, used in conjunction with a heavy hip belt (sometimes called a "scare strap") that helped the line worker stay secured and oriented to the pole at a comfortable angle. In urban areas, line climbers have mostly been superseded by mechanized cranes, sometimes called "cherry pickers," although climbers are still needed in some circumstances, especially in areas of rough terrain. See lineman.

Magnetic Lines of Force

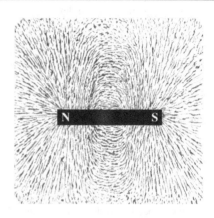

Iron filings sprinkled around a magnet on a light-colored surface or piece of glass will reveal particular types of patterns like those shown in this diagram, depending upon the shape of the magnet. These patterns will change each time the filings are sprinkled, since they are formed not because there are lines emanating from the magnet, but because the magnetic forces associated with the magnet cause the particles to interact with the magnet and one another in specific ways.

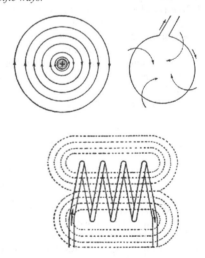

Lines of force are related to the shape and orientation of the objects with which they are associated, and the current flowing through those objects, in the case of electromagnets. Above are some examples of lines of force associated with different structures.

lines of force, lines of magnetic induction The sphere of magnetic influence of a magnet. For example, in a basic bar magnet, the lines of force can be conceptualized as radiating outward from the poles with no break between the north and south poles. Although diagrams of magnetic lines of force show them as discrete lines in one plane, the actual region of influence is three-dimensional. The *lines* themselves are a descriptive way of conveying the properties of the forces that cause them to crowd each other sideways. Areas where these lines converge, as at the poles, indicate areas of stronger magnetic force. The direction of the lines can be defined as the direction along the north pole if forces other than the magnetic force were hypothetically neutral.

Lines of force can be observed indirectly by sprinkling iron particles over a bar magnet and tapping the surface until they form patterns around the magnet. The lines are not fixed in one position; they will change if you sprinkle the filings again, but come about through their interaction, as each piece behaves as a tiny magnet, alternately attracting and repelling near neighbors. Collectively these lines are classed as flux. See flux, magnetic field.

link *n.* 1. In its broadest sense, a communications circuit or channel. 2. A specific leg in a circuit, as between two nodes, two networks, or two users. In ATM, it is more specifically defined as a logical link, that is, an entity with a specific topological relationship and transport capacity between two specified nodes in different subnetworks. See Ethernet, Frame Relay, asynchronous transfer mode, tunneling. 3. A communications medium over which nodes can communicate at the link layer. 4. A logical link between software entities, files, or processes, as between an icon and an executable program, or between an alias and a file directory. See alias.

link *v.* 1. To form a logical relationship between software entities, or software and hardware devices. 2. To interconnect hardware devices and/or cables. 3. In software development, while building executable files, to link the compiling source code into appropriate system resources, headers, or other system software specified.

link access protocol LAP. A generic category of protocols for establishing (setting up) data transmission connections across a wide variety of devices. As the name implies, LAP operates at a lower level called the link level, which is Level 2 in many common networking architectures. LAP is one of several protocols functioning at the link level to establish, control, and take down network links. In frame-based networks, balanced LAPs are more commonly used now than the earlier basic LAPs. However, there are many different implementations of LAPs for device connections including computers, modems, infrared remotes, etc. from different manufacturers.

Link Access Protocol Manager LAP Manager. A set of software utilities providing a standard interface between high-level protocols and link-access protocols for AppleTalk Link Access Protocol. LAP Manager enables a user to select among AppleTalk con-

nection files to specify a network to be used for the node's AppleTalk connection. When a connection is selected, the LAP Manager routes the communications through the selected link-access protocol and its associated hardware, thus acting as a switching mechanism. See Link Access Protocol, Link Control Protocol.

link aggregation token See aggregation token.

link attribute In ATM networks, a parameter used to assess a network link state, to determine whether it is a viable choice for carrying a given connection.

link connection In ATM networks, a connection which can be used to transmit information without the addition of any overhead.

link constraint In ATM networks, a restriction applied to the use of a specific link. In other words, restrictions as to whether it may be used as the path for a connection.

Link Control Protocol LCP. In order to support a variety of environments, Point-to-Point Protocol (PPP) provides a Link Control Protocol (LCP). When network link access has been initiated, the LCP agrees upon encapsulation format options, handles varying limits on packet sizes, detects common misconfiguration errors, and terminates a link.

The main phases for establishing, configuring, maintaining, and terminating a link are shown in the Link Control Procedures chart.

The Link Control Protocol packet is encapsulated within the information field of a Point-to-Point Protocol data link layer frame. In brief, the packet is configured as shown in the Link Control Protocol Packet Format diagram. See Compression Control Protocol, link access protocol, Point-to-Point Protocol, SNA Control Protocol, RFC 1171, RFC 1661.

Link Control Protocol codes LCP codes. These link-establishment packets are used to establish and configure network links. Link Control Packets are assigned as follows:

Code	Control code assignment
Code 1	Configure-Request
Code 2	Configure-ACK
Code 3	Configure-NAK
Code 4	Configure-Reject
Code 5	Terminate-Request
Code 6	Terminate-ACK
Code 7	Code-Reject
Code 8	Protocol-Reject
Code 9	Echo-Request
Code 10	Echo-Reply
Code 11	Discard-Request

See Link Control Protocol, Point-to-Point Protocol.

link encryption An internetwork security mechanism which, unlike an authentication header (AH), affords some protection from traffic analysis.

link MTU A unit describing the maximum packet size which can be conveyed in one piece over a communications link.

link state advertisement LSA. A mechanism for describing portions of the routing topology of a network within a network switch fabric. As implemented with Virtual LAN Link State Protocol, there are two basic types of link state advertisements:

switch link	lists all the functioning switch-related links and the cost associated with the use of each link
network link	lists all network link adjacent relationships

Each switch originates an initialization advertisement when it comes online (with no links), followed by one switch link advertisement that describes a complete list of all LSAs each time there is a change in the state of a neighbor.

Switches associated with multi-access links originate a network link advertisement that describes all the functioning fully-adjacent switches attached to the link. See Open Shortest Path First Protocol, Virtual LAN Link State Protocol. See RFC 2328, RFC 2642.

link state database LSD. A collection of link state advertisements for an entire network fabric that is maintained within the fabric-associated switches and used to calculate the system of best paths to all other associated switches.

link state routing An approach to routing on distributed networks that was pioneered on the historic ARPANET packet-switching network. McQuillan et al. described ARPANET routing in *IEEE Transactions on Communications* in May 1980. Within a few years, modifications to the basic concept increased the fault tolerance of the routing system and formed a basis for subsequent link state protocols such as Open Shortest Path First Protocol. See link state advertisement.

Linux A popular, well-supported, widely implemented, freely distributable, open source Unix-like operating system developed in the early 1990s by Linus B. Torvalds in Helsinki, Finland (he has since moved to California). Torvalds has been honored for his contributions to information and telecommunications technology.

Linux is mostly POSIX compliant, and features true 32- or 64-bit multitasking, virtual memory, TCP/IP drivers, shared libraries, protected mode execution, and more.

Linux supports the X Windows system, conforming to the X/Open standard. It also supports the common Internet protocols, including POP, IRC, NFS, Telnet, WAIS, Kerberos, and many more, as a client or a server. Popular Unix shells work with Linux.

Some excellent applications have been ported to Linux, with more being developed all the time. Linux has been ported to many platforms, including PowerPC, Amiga, Macintosh, Intel-based machines, Atari, and others. There are a number of commercially available versions of Linux, including Stac and Red Hat and it is a very popular choice for local area network (LAN) and Internet servers. It is reliable, powerful, and is being adopted by many large-scale research and development firms. The *Linux Journal* is a good trade magazine for Linux developers and users.

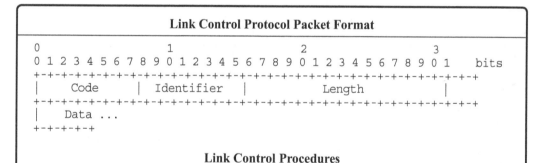

```
                  Link Control Protocol Packet Format

 0                   1                   2                   3
 0 1 2 3 4 5 6 7 8 9 0 1 2 3 4 5 6 7 8 9 0 1 2 3 4 5 6 7 8 9 0 1   bits
+-+-+-+-+-+-+-+-+-+-+-+-+-+-+-+-+-+-+-+-+-+-+-+-+-+-+-+-+-+-+-+-+
|     Code      |  Identifier   |             Length            |
+-+-+-+-+-+-+-+-+-+-+-+-+-+-+-+-+-+-+-+-+-+-+-+-+-+-+-+-+-+-+-+-+
|    Data ...
+-+-+-+-+
```

Link Control Procedures

Phase	Events	Notes
Phase 1	Establishment and configuration	Configuration packets are exchanged, a Configure-ACK packet sent and received, and the Open state entered.
Phase 2	Link quality determination	An optional phase in which the link is tested for quality.
Phase 3	Network layer protocol configuration	Upon completing optional link quality determination, network layer protocols can be configured and maintained or taken down.
Phase 4	Link termination	The link may be terminated at any time, usually at a user's request, but possibly also because of a physical event.

Linux International An international organization of developers, vendors, and users of the Linux Operating System and its associated programs. http://www.li.org/

Linux Terminal Server Project LTSP. An opensource project under the GNU Public License to create administration tools to facilitate the setup of Linux network diskless workstations. The motivation for the project was to provide a flexible, reliable, low-maintenance, cost-effective network setup for remote computing workstations.

liquid crystal display LCD. A low power display technology comprising a number of layers, one of which is made of liquid crystal, sandwiched together. The long crystalline molecules in the liquid crystal layer are used to deflect and polarize light. When exposed to an electric field, the crystals orient themselves in the same direction, no polarization occurs, and the light is absorbed; the display remains dark. Liquid crystal displays were originally developed by RCA's David Sarnoff Research Center and Westinghouse in 1963 and 1964. The first calculator to use LCD technology was introduced by Sharp in 1973. Simple LCDs include a set of alphanumeric characters and sometimes some symbols. More sophisticated LCDs incorporate raster display technologies, especially when used with computers. LCD panels are popular in phones, calculators, and other lowpower, mobile devices such as laptops. External light is needed for the user to see the display unless backlighting is added. An *active matrix* LCD includes a transistor at each display point to increase the speed at which the crystals can change state and add color. Tektronix further developed "plasma addressing," which incorporates some of the properties of gas plasma into liquid crystal displays.

liquid detector See electrolytic detector.

LIS 1. Link Interface Shelf. 2. Logical IP Subnetwork.

LISA Laser Interferometer Space Antenna.

LISN 1. See line impedance stabilization network. 2. See Low-Incidence Support Network.

LISP *List* Processing. A high level programming language introduced in 1958 by John McCarthy. It is used in many artificial intelligence applications, and as a macro scripting language in applications like AutoCAD from Autodesk Inc. and EMACS. Golden Common LISP is one of the more common implementations of LISP. LISP code is syntactically different from languages such as C, and has been retroactively called an acronym for *Lots of Insignificant Silly Parentheses* by subsequent generations of programming students due to its nested statements.

LISTSERV A significant discussion list software product introduced in 1986 that became prevalent on BITNET, Internet, and local area networks. As of spring 2001, L-Soft International, Inc.'s LISTSERV program was being used to manage over 170,000 local and public lists. LISTSERV is available in commercial and free LISTSERV Lite versions. See discussion list, Majordomo.

list server A computer file distribution system used for managing email, newsgroups, discussion lists, and other types of files received from one or many sources, and distributed to one or many subscribers to the service. Using servers to manage the traffic in a centralized manner can cut down on administrative overhead and provide a means to implement security and selective filtering as needed.

listed address LA. A location identifier, such as an email address on the Internet. The concept of the listed address has become important to Internet Services Providers (ISPs) in their management of "spam" (unsolicited bulk email messages). Checking to see if the source of spam is a listed address is one of the ways in which mail can be filtered or selectively processed.

LIT See line insulation test.

Litespan Alcatel's next generation network digital loop carrier multiservice (narrow/wide/broadband) access platform, the most widely deployed integrated digital loop carrier with DSL in North America, with a greater than 42% market share.

The Litespan family of products provides advanced SONET-based Next Generation Digital Loop Carrier (NGDLC) capabilities, providing ATM xDSL and TDM narrow/wideband subscriber services. Litespan 1540 is a flexible multiservice access platform enabling the delivery of multiple access services to a region from a single node. See GEMINAX.

lithium-tantalate A synthetic crystal first grown in the Bell Laboratories, lithium-tantalate was the first really practical alternative to natural or synthetic quartz for the development of a number of communications-related components such as filters. This substance has practical applications for high bandwidth transmissions. See quartz crystal, quartz crystal filter.

little-endian Stored or transmitted data in which the least significant bit or byte precedes the most significant bit or byte. Many file incompatibilities between computer systems, in which the file formats are otherwise almost identical, are due to platform conventions which store the data in big-endian or littleendian form.

Littman-Metcalf configuration A setup in which a resonating cavity external to a laser (often in the form of a diffraction grating) is used to reflect a selected portion of the incident light at low angles called grazing angles. This configuration may be used to produce a mode-locked, stable beam tuned to a desired frequency. See grazing incidence, grating.

Littrow configuration In an optical grating filter, a configuration in which light diffracted from the grating travels back along the same path as the incoming (incident) light. This occurs, for example, when a light beam of a specific wavelength hits a reflective planar surface at 90°. Imagine light traveling down the shaft of an imaginary invisible flagpole that is anchored in a flat reflective base. When the light hits the base, it is reflected back again along the direction of the shaft in the same path but in the opposite direction.

Awareness of Littrow effects is important to technologies where back reflection can interfere with signal transmission. For example, if a fiber light source hits the endface of a fiber filament conducting core (or any bonding materials that may have been used in the joint) at an angle that causes the light to be reflected back to the source, the signal will not propagate through the light pipe. Changing the shape of the endface, the method of fusing, or the angle of propagation may help mitigate back reflection. See grating. See acceptance angle, back reflection. Contrast with grazing incidence.

Littrow Configuration

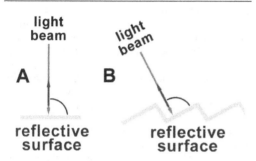

If the light reflected from a surface travels back along the same path through which it came (A), it is called a Littrow Configuration. *The absolute angle is not the key factor, but rather the angle of the incident light beam relative to the surface plane from which it is reflected, as illustrated when the surface is rotated to the same degree as the incident light beam (B).*

Livermore Automatic Research Calculator/Computer LArc. A supercomputer developed for the Lawrence Livermore National Laboratory (LLNL) in the mid-1950s by the Sperry-Rand Corporation. The LArc had multiple noninterleaved core memory boxes and data, instruction, and DMA access slots; the random access memory (RAM) was mechanical in nature.

LLC See Logical Link Control.

LLC encapsulation LLC encapsulation is a means to "envelope" a set of packets so that several protocols can be carried over the same virtual circuit (VC). Type 1 and Type 2 are defined for connectionless and connection services. See Logical Link Control.

LLC/SNAP LLC/SubNetwork Attachment Point. An encapsulation protocol used in Logical Link Control. In an ATM environment, this is the default packet format for Internet Protocol (IP) datagrams. See TU-LIP, TUNIC, RFC 1483.

LMI Local Management Interface. See Frame Relay extensions.

LMOS Loop Maintenance Operations System.

LMS 1. See Local Measured Service. 2. Local Message Switch. 3. Location and Monitoring Service.

LMSS Land Mobile Satellite Service.

LMST See lightweight multiband satellite communications terminal.

LMU Line Monitor Unit.

LNA See low noise amplifier.

LNP See Local Number Portability.

LNW Research Corporation A California company that sold microcomputers assembled and in kit form in the early 1980s. The LNW-80 was a TRS-80-compatible computer arguably better than both the Apple II and the TRS-80 in that it supported a faster CPU, more memory, and higher resolution color graphics. Unfortunately, the LNW computers didn't catch the attention of the public and never had the impact of the better-known brands.

load coil See loading coil.

load distribution unit LDU. A device for distributing electrical loads which may include fault detection circuits, alarms and control systems connected with the load. A load distribution unit can be used for disconnecting system loads to prevent damage to backup systems (e.g., batteries) during prolonged AC power outages or from low voltage exposure.

Load Number The Canadian counterpart to the U.S. Ringer Equivalence Number (REN) system. As there may be more than one phone device attached to a single line (modem, fax, answering machine, etc.), a system was established to determine and indicate the *ringer load,* that is, the electrical load on the phone line associated with a particular piece of equipment. In this way, the Load Numbers can be summed to show the total load, and make sure the line was not overloaded.

In the REN system, most lines can handle a load of up to 5.0. In Canada, the concept is the same, but the scale uses larger units; in other words, a standard phone might have a load of about 10 to 20 points, with the total load for a single line being about 100 points. Many electronic devices will show the Load Number somewhere on the main circuit board.

load-balancing system LBS. A system for dynamically balancing the execution of jobs to be processed, usually accomplished through software or a combination of software and firmware (as in routers). LBSs are particularly useful in environments where tasks are farmed out to various systems, as on a distributed computer network.

A load-balancing system is also used for dynamically balancing the transmission of traffic through a communications pathway, as in telephone circuits.

In general, load-balancing systems are used where prior knowledge of what resources may be available at any given time cannot be easily anticipated. LBSs are intended to facilitate and automate task or traffic delegation as transparently and seamlessly as possible.

Load balancing is becoming increasingly important as global voice and data networks serve greater numbers of users. Load balancing is distinct from load sharing in that balancing involves the efficient, well-distributed use of resources as opposed to the general availability of a resource to multiple users or processes.

Internet Service Providers (ISPs) use load balancers with Internet servers to efficiently manage response times and site access availability. Load balancers can be quite sophisticated, not only distributing user

L

requests across multiple servers, but also rerouting requests if changes occur in the available servers. Depending upon their sophistication, LBSs may automatically adjust to server changes without powering down the system and may support security features associated with the service.

In CORBA-based architectures, load balancing enables distribution of client processes among servers on a number of levels to balance network, operating system, and middleware-related processing loads. Othman, O'Ryan, and Schmidt have suggested a number of load balancing architectures to support CORBA systems. McArdle et al. have described two types of load-balancing strategies for CORBA-based Service Control Points (SCPs), including a novel ant-based algorithm and a distribution method based on mathematical minimization of expected communication flows.

SGI has a Network Load Balancing software product that collects Ethernet devices under a single IP address and balances the input and output loads across the Ethernets. Thus, the IP address of the load-balancing device is advertised (usually through its domain name) and individual IP addresses of the associated devices are not. The variety of approaches to load balancing is reflected in products from Foundry Networks' ServerIron line, which include Server Load Balancing (SLB), Global Server Load Balancing (GSLB), and Firewall Load Balancing (FLB). The Network Load Balancing component in Windows 2000 scales by distributing IP traffic across multiple cluster hosts.

Effective load balancing can lead to higher quality of service for busy multiuser systems or services, such as popular Web sites. Quick response times can make the difference between keeping or losing a potential customer. Fast network transfer speeds are not sufficient in themselves to solve all the aspects of quick access and adding more servers may not be economically feasible. Efficient delegation of tasks and traffic direction through good LBSs is one way to maximize the effectiveness of an existing system.

Load balancing is as much art as science. The system must anticipate and adapt to a changing environment and the analysis of the effectiveness of load-balancing algorithms is, in itself, a challenge. State aggregation and decomposition are two means of assessing dynamic load balancing. In 1997, H. Lin proposed a combination of these methods and introduced the concept of a correlation window for analyzing dynamic LBS policies. The Parallel Programming Laboratory at the University of Illinois conducts research in load balancing, particularly in object migration and seed load balancing, concepts of interest in parallel computing systems.

There are a variety of commercial LBSs for telephone trunks and Internet use along with several Internet-related open-source software such as Queue, Supersparrow (for wide area networks), and ANTS load-balancing systems are distributed under the terms of the GNU Public License. See ant, Supersparrow.

loading coil A small electromagnetic induction device which helps prevent attenuation of the signal on a wireline. Loading coils were developed in the early 1900s to improve long-distance transmissions in telephone lines and are still commonly used. By calculating the optimum size and spacing the loading coils carefully along a wireline, it was possible to extend a circuit by several times. With the advent of data communications, loading coils have become a mixed blessing. While they improve transmissions in voice grade lines, they tend to add noise and distortion in the higher frequencies used in data transmissions.

LOB See line of bearing.

Lobe Attachment Module LAM. In a Token-Ring network, an expansion device to extend the number of users that can be accommodated on a single segment of a Token-Ring network. Multiple LAMs (up to a specified number) can be connected to a Controlled Access Unit (CAU). It may also be used with a Multistation Access Unit (MSAU). See Controlled Access Unit, Multistation Access Unit, Token-Ring network.

LOC 1. line of communication. 2. See Loss of Cell.

Local Access and Transport Area LATA. The terms of the 1984 AT&T divestiture resulted in the creation of numerous geographically local telecommunications service areas, of which there are now more than 200 in the U.S. LATAs are determined, to some extent, by population densities. Originally, local Exchange Carriers were not permitted to connect calls across LATA boundaries, as that was the privilege of Inter Exchange Carriers (IXCs). Since that time, the rules have been modified to some extent. See Inter Exchange Carrier, Modified Final Judgment.

Local Area Augmentation System LAAS. A VHF-based Global Positioning System (GPS) augmentation system which functions along with the Wide Area Augmentation System (WAAS) to provide high-accuracy satellite-based navigation services for aviation, LAAS provides navigation and landing data where WAAS cannot be used. LAAS meets Category II/III aviation requirements in specific locations. Further details are available through the Federal Aviation Administration (FAA). See Wide Area Augmentation System. http://gps.faa.gov/

local area network LAN. A computer network within a specified geographical space, such as a building or region, or within an institutional entity such as a classroom or department. The network links computers and a variety of shared resources, typically files, application programs, and peripheral devices such as printers, fax machines, modems, and scanners. Connections between the computers are through wires, fiber optic cables, or wireless signals.

LANs are typically connected directly by telephone wire or coaxial cables, and thus are somewhat constrained in physical size and number of users, due to lower transmission speeds, network topology, signal reduction (see attenuation), and fixed-bandwidth limitations, than wide area networks (WANs).

There are a number of common ways in which LANs are connected, with various commercial, shareware, and freeware products to handle the software tasks associated with networking. One of the most popular server products is Apache, which is robust, freely distributable, and very widely installed. Commercial products include IBM Token-Ring, Microsoft NT, and Novell Netware.

As technology advances and becomes less expensive, direct cabling will probably decrease and wireless solutions become more common, thus reducing the distinction between LANs and WANs. See Token-Ring, virtual LAN.

local area data transport LADT. Data transmission services offered over existing local phone lines. In general, it is a low-speed, low-cost option practical for many casual users of data services. In some cases, the system has been enhanced to provide simultaneous voice and data (e.g., ISDN services). The growing availability of U.S. ISDN (USDN) services is increasing the likelihood that local ISDN-based LADTs may eventually provide transport among LADTs. The abbreviation also refers more specifically to AT&T's commercial LADT offerings, which include protocol conversion. See ISDN.

Local Area Signaling Services LASS. Commercial adjunct processor-based services used to create initial orders. A Bulk Calling Line Identification (BCLID) service provides private branch exchanges (PBXs) with information on calls from outside the PBX group.

Local Area Transport LAT. A proprietary communications protocol for terminal-to-host transmissions, developed as VAX systems by Digital Equipment Corporation (DEC).

local area wireless network LAWN. A local home, business, community organization, campus, or other phone or data network which uses wireless technology to provide the links. The communication is often carried out over radio frequencies (RF) in the frequency modulated (FM) or infrared range. Techniques common to LAWN include spread spectrum, originally developed for government operations, and narrowband. See Industrial Scientific Medical, spread spectrum.

Local Automatic Message Accounting LAMA. An automatic message accounting (AMA) system used in local telephone switching centers in conjunction with number identification information to collect billing data. It also automates the routing of long-distance calls through more than one local office.

local battery Equipment which draws power from a local source, rather than drawing current from the line to which it is attached. Most phones draw current from the phone line sufficient to operate the phone, but if they have extra features (LCD display, speakerphone, etc.), they may require additional power which comes from a battery or local wall socket. Laptop computers use local battery peripherals, such as PCMCIA modems, rather than those which require separate power, e.g., desktop modems, to maximize convenience and portability.

Local Bus Computer processors require a way to communicate with the many devices that make a computer useful: storage, printers, modems, input devices, scanners, etc. The local data bus is an interface that links the motherboard and various controller cards and other interface connectors that comonly communicate with external devices. Many different standards are defined for bus transmissions, and the bus speed does not always match the CPU speed, creating a processor bottleneck for some types of processes and activities.

A Local Bus is one of the newer, faster buses beginning to supersede other common buses, including EISA, VESA, and PCI.

local call A telephone service phrase referring to calls placed through the local exchange, billed on the subscriber's predetermined regular service plan. These are generally geographically close. There are three common types of billing systems for local calls: 1. unlimited calls for a flat rate monthly fee, 2. flat rate up to a certain number of calls, then a per-call charge beyond that, or 3. a per-charge call which may or may not be scaled according to the total number of calls for the month. The first option is widespread in Canada and the U.S., while many places in Europe use the third option. For contrast, see long-distance.

local differential GPS LDGPS. An implementation of the Global Positioning Service (GPS) designed to improve local accuracy of the data. A single GPS receiver is placed at a known location where it can receive GPS signals. It becomes a reference station which forms a scalar correction for GPS satellites within current view; broadcasting the correction information is provided to local users. Since there is degradation over distance, a series of "cells" would be needed to apply this system over a large geographical area. See differential GPS, wide area differential GPS.

Local Exchange Carrier LEC. A designation for a local telephone company, now more commonly distinguished as an Incumbent Local Exchange Carrier (LEC) or a Competitive Local Exchange Carrier (CLEC).

This is defined in the Telecommunications Act of 1996, and published by the Federal Communications Commission (FCC), as meaning:

"... any person that is engaged in the provision of telephone exchange service or exchange access. Such term does not include a person insofar as such person is engaged in the provision of a commercial mobile service under section 332(c), except to the extent that the Commission finds that such service should be included in the definition of such term."

See Federal Communications Commission, Telecommunications Act of 1996, United States Telephone Association.

Local Exchange Carrier duties The Federal Communications Commission (FCC) stipulates a number of duties in the Telecommunications Act of 1996 as shown in the Telecommunications Act chart.

local injection and detection LID. In fiber optics splicing and assembly, LID is a fusion splicing alignment process. Two fibers to be joined are lined up

L

end-to-end. Light is then injected into the conducting core of the fibers that are to be joined. One or both of the fibers are adjusted spatially in fine increments until there is optimum transmission of light through the dual-fiber assembly – somewhat like focusing a camera back and forth to determine the point at which the image is sharpest, except that light conduction rather than image focus is determined. The fibers can then be fused into a single lightguide which hopefully retains the good alignment of the unfused fibers. The splice is then inspected. See fusion splice.

local loop In telephone installations, a physical link through a wire pair connection between the subscriber, which may be an individual, a business, or a private branch system, and the switching office. The local loop once included the connection right to the subscriber's phone, but now the demarcation point is usually a patch panel or exterior connections box (although to-the-phone is still available for a fee).

Local Management Interface LMI. In Frame Relay networks, an extended specification for information exchange between devices. See Frame Relay Extensions for more detail.

Local Measured Service LMS. A telephone billing system in which subscribers pay according to the number of calls made or received (or both), rather than according to a flat monthly rate. Measured service is sometimes provided at a flat rate up to a specified number of calls, and then a per-call fee above that number (many banks set up checking charges this way as well). Generally, LMS is interesting to those who make very limited use of the phone or who have a line primarily for incoming calls. In some countries, all service is measured and even calls that are unanswered or that result in a busy signal may be billed in some areas.

local multipoint distribution service LMDS. A proposed terrestrial wireless communication service designed to send video over small cells. This would be competitive with urban cable TV services. The Federal Communications Commission (FCC) released a Notice of Proposed Rule Making on the LMDS proposal in December 1992. This led to various FCC proposals to segment the transmissions band, with the effect that the primary spectrum would be limited to the 27.5 to 29.5 GHz range.

Long-term advocates of satellite communications were concerned that this approach would limit future evolution and growth of satellite communications deployment.

Local Number Portability LNP. A telecommunications service to enable local telephone numbers to be retained even if a subscriber's carrier is changed. LNP was implemented in the mid-1990s and made it easier for phone services to compete. The management of LNP services is, in part, dependent upon the maintenance of regional databases containing the numbers and the vendors servicing those numbers. See Local Number Portability Administrator, Location Routing Number.

Local Number Portability Administrator A managing authority tasked with administering the Local Number Portability database for tracking and updating local numbers and the telecommunications carriers servicing those numbers. NeuStar, Inc. (formerly Lockheed-Martin Information Management Service) is currently the administrator for the Local Number Portability Administration Center (NPAC). See Local Number Portability.

local service ordering guidelines LSOG. Requirements published by individual communications carriers to instruct customers on how to pre-order or

---

Telecommunications Act – Stipulated Local Exchange Carrier Duties

"Each local exchange carrier has the following duties:

(1)	RESALE	The duty not to prohibit, and not to impose unreasonable or discriminatory conditions or limitations on, the resale of its telecommunications services.
(2)	NUMBER PORTABILITY	The duty to provide, to the extent technically feasible, number portability in accordance with requirements prescribed by the Commission.
(3)	DIALING PARITY	The duty to provide dialing parity to competing providers of telephone exchange service and telephone toll service, and the duty to permit all such providers to have non-discriminatory access to telephone numbers, operator services, directory assistance, and directory listing, with no unreasonable dialing delays.
(4)	ACCESS TO RIGHTS-OF-WAY	The duty to afford access to the poles, ducts, conduits, and rights-of-way of such carrier to competing providers of telecommunications services on rates, terms, and conditions that are consistent with section 224.
(5)	RECIPROCAL COMPENSATION	The duty to establish reciprocal compensation arrangements for the transport and termination of telecommunications."

order local connection products and services. LSOGs typically identify Order and Billing Forum (OBF) industry guidelines and carrier-specific requirements. After presenting the guidelines, the carrier will usually offer a variety of forms to the customer to sign up for services.

Gone are the days when you could simply ask a phone company to install a phone. Now you must read guidelines and fill out forms to identify various choices of equipment leasing or buying, various carriers for different services, optional "value" or "feature" packages, optimum proportion of data services, phone services, number of extensions, extra lines, etc. Many carriers now publish the forms online. This is a great convenience except that many of them offer the forms only in Microsoft Word (a proprietary commercial product) instead of using a format such as PDF, for which the vendor (Adobe) offers freely downloadable readers. Hopefully more vendors will use file formats with free readers for distributing Web forms to potential customers.

local service request LSR. The process by which a telecommunications customer requests local service. For example, a Competitive Local Exchange Carrier (CLEC) would carry out LSR procedures to order loops from Bell Telephone. Many carriers now provide the request forms on the Internet for download. They can then be mailed or, in some cases, emailed or otherwise submitted electronically. When a local service request is combined with other requests in order to set up an enhanced loop service, the procedure and forms are collectively known as an access service request (ASR). An administrative LSR may be required to identify a CLEC.

Local Services Provider LSP. A telecommunications provider that specializes in local services (e.g., local telephone services) rather than regional or national services (e.g., long-distance calling).

local signal transfer point LSTP. In mobile personal communications services, one or more local message routing translation and screening points. A local signal transfer point is associated with a LATA geographical area connecting to multiple service switching points (SSPs). One or more LSTPs may be connected to a regional signal transfer point. Mobile switching centers (MSCs) are connected to the signaling network (e.g., SS7) through a local signal transfer point.

LOCAL TV Act of 2000 An act which created a $1.25 billion federal loan guaranteed fund to assist residents in rural areas to receive urban and local TV broadcast signals through satellite transmissions. Thus, both profit and nonprofit organizations could access resources for offering local broadcast signals, with the government guaranteeing a portion of the loan. An amendment cleared the way for electric cooperatives and telephone systems to participate in the loan guarantee program.

LOCAL TV is an acronym for Launching Our Communities' Access to Local Television. The Act was drafted in part because the 1999 Satellite Home Viewer Improvement Act (SHVIA) did not include legislation to help rural Americans receive local broadcasting. See Rural Local Broadcast Signals Act, Satellite Home Viewer Improvement Act.

LocalTalk A proprietary local area network Apple-Talk-compatible protocol developed by Apple Computer and used on Macintosh computers and peripherals. LocalTalk is not a fast protocol, but all Macs come networkable right out of the box, with a simple serial cable, and something can be said for ease of use and convenience, particularly in school and work environments. See AppleTalk.

Location Area Identity LAI. A subscriber identity allocated and assigned on a location basis as part of the Temporary Mobile Subscriber Identity (TMSI) used in a Global System for Mobile Communications (GSM).

Location Routing Number LRN. A 10-digit number to uniquely identify a switch in a telephone circuit. In the 1990s, LRNs facilitated the implementation of competitive services such as Local Number Portability (LNP). Various regulatory agencies supported this AT&T/Lucent system and it became an industry standard that was subsequently adopted by the Federal Communications Commission (FCC), in 1997. See Local Number Portability.

LODE See Large Optics Demonstration Experiment.

lodestone, loadstone A natural magnetic material called magnetite, an oxide of iron. All magnetite can be readily magnetized. This material was used to create early compasses and was called magnes lapis, magnetic stone, or magic stone. It is called lodestone when it comes out of the ground already exhibiting magnetic properties. Lodestone is probably the same stone mentioned by Plato in *Timaeus* as "... the Heraclean stone," since the Heraclean stone was paired in the same sentence with the attractive properties of amber. See amber, magnet.

Lodge, Oliver Joseph (1851–1940) An English physicist who demonstrated that radio waves could carry a signal over distance, in 1894. Lodge is also known for his experiments with tuning in radio waves, ideas developed by succeeding scientists.

Lodge-Muirhead detector A simple, early type of self-restoring radio-frequency detector built by hobbyists and commercial manufacturers in the early 1900s. It employs a small steel revolving wheel, with the outer edge sharpened to a very fine edge, supported between slots on frame posts on either side of a rod passing through the center of the wheel. A small motor supplies the power to quickly turn the wheel. A hard rubber cylinder with a slot cut in the top sits directly under the wheel. Mercury is poured into the slot and makes contact with a binding post threaded in from the outside of the rubber. A thumbscrew is installed in the rubber to raise or lower the mercury level. When the wheel revolves, it makes contact with the mercury and the signal is translated through the mercury, by means of brushes, to the binding post which connects to the rest of the circuit, including a receiver. The battery power to the motor is controlled by means of a potentiometer. See detector.

LOF 1. See Loss of Frame. 2. loss of fix. The loss

of a desired target or calibration setting. 3. lowest operating frequency.

LOFAR See Low Frequency Array.

log-periodic antenna A periodic antenna is one in which the input impedance varies as the frequency varies. Log-periodic antennas have a variety of arrangements of active interconnected dipoles to provide broadband, high-gain capabilities. They are useful for very high frequency (VHF) signal ranges.

logarithm, log A mathematical means of expressing an exponent – the power to which a number is raised. For example, in the expression $x = 10^n$ the symbol n stands for the log of x. As an example, $\log(10) = 1$.

A number of phenomena have logarithmic qualities, such as sound volume, pH scales, and earthquake magnitude scales. These phenomena are not sensed on a linear scale, but rather on a logarithmic scale, in which the magnitude of the previous point on the scale is not linearly related (e.g., twice as much each time), but exponential. However, expressing the magnitude of these phenomena in increasingly huge numbers can be cumbersome, so an antilog may be applied to "compensate" for the logarithmic increase to create a scale with numbers that are smaller and easier to understand. Thus, a Richter scale scale for expressing earthquake magnitudes uses small manageable numbers, but it must be remembered that it represents a logarithmic phenomenon and thus the difference between 5 and 6 on the Richter scale is much greater than the difference between 4 and 5.

Logs are expressed in terms of a reference base and it is best to specify which one is used. Thus, $\log_2 3$ would read as "log, base 2, of three." A log may be classified as *natural* (commonly used in chemistry and physics) or *common*. Natural logs have a base of e (2.71828) and common logs have a base of 10. Log tables are published to aid in lookup of logs for exponents other than powers of 10, since these are time-consuming to calculate. Many calculators now have log keys to simplify the process. See antilog.

logic bomb A software program designed to penetrate a system, present a message (sometimes through graphics or sound), or damage memory or stored data, when some particular logical operation happens. A time bomb is a type of logic bomb which can, for example, wipe all the data off a hard drive when the bomb detects that it is April 1. Logic bombs are not always malicious, but they are seldom appreciated. They generally fall into the category of practical jokes, which are usually funnier to the perpetrator than to the intended object of the joke. See virus.

logical block In storage devices, such as hard drives and magneto-optical devices, the smallest addressable unit. Each block is associated with a unique number, usually starting with 0, and incrementing for each succeeding block. This allows the system to locate data, read and write to the device, partition the drive, etc. in an organized manner.

Logical Block Address LBA. A means for saving and retrieving information by accessing block addresses on a storage medium, rather than by using cylinder-head-sector addressing schemes. The blocks on the storage medium are addressed sequentially, usually starting with zero. SCSI peripherals use this addressing method. Some IDE drives are now beginning to use this method, but it may be necessary to request LBA mode explicitly.

logical drive A drive configured separately from the physical configuration of the drives. For example, a computer may have three drives, each with several partitions, but logically, the system may organize them into four logical drives with various partitions from different drives in such a way that the user "sees" four drives rather than three, of sizes set by software. Conversely, a system may have three drives, each with a couple of partitions, which are aggregated into one drive. Thus, the storage space appears to the user as one large drive. Not all operating systems can organize drives in this way. Some of the lower-end personal computers have limitations in the configuration of logical drives.

logical link A link between nodes or devices based upon an abstract rather than a physical topology. Thus, virtual LANs, logically direct connections, and other types of paths can be set up in association with the physical connections.

Logical Link Control LLC. The upper sublayer of the layer 2 Open Systems Interconnection (OSI) protocol. LLC provides data link level transmissions control. It is the default multiplexing layer for Internet Protocol (IP) over AAL5. It was developed by the IEEE 802.2 committee to provide a common access control standard for networking which is independent of packet transmission methods. It includes addressing and error checking capacities.

Logical Storage Unit LSU. A buffer unit which connects to PBX systems to store call information. Also known as a *buffer box* or *poll-safe*.

logical topology In a network, the connections and relationships between computers and various devices may not map in a one-to-one relationship with the physical topology. Thus, logical topologies, organized and managed in software, may be administered and diagramed separately or as an overlay to the physical connections. In large networks, logical topologies often require tracking and display programs to configure and troubleshoot the logical connections. From a user's point of view, many aspects of a logical topology can make computing more efficient and enjoyable.

As a simple example, there are workstation networks which allow hard drives to be mapped in such a way that they appear as one giant drive to the user, even though the data may physically be spread over a number of partitions and drives. Thus, the user doesn't have to worry about whether there is enough space, which drive to use, or on which drive or partition that old file was stored. The operating system takes care of all the housekeeping involved in managing the system. On a larger scale, the same can be said of whole systems. The user may be using an application program located on a machine in another room, another building, or another state, but it can be logically mapped in so that it appears to be running on the local machine. Commercial software programs with graphical user interfaces exist to help manage logical topologies and, in some cases, various management utilities come with the operating system.

logical unit number LUN. The LUN is an identification system used with SCSI devices which allows the computer and controller to distinguish and communicate with up to seven devices including the controller for each SCSI chain. Each device must have a unique ID so that the controller can administer more than one device when they are chained together.

logiciel Software, in French.

login script, logon script A login script is a file that includes commands or variable settings pertaining to the login and initial setup. The two most common functions of a login script are: 1. to set up the system to the specifications of a particular terminal, and 2. to set up the preferences for a particular user or set of users. In the first case, the login may include information about the characteristics of the terminal, such as hardware-specific keyboard mapping, graphics card settings, etc., as well as environmental variables, patches, and other initialization parameters. In the second case, the settings may include the user screen size and color preferences, preferred fonts, frequently used applications, permissions, and other characteristics of the user environment. See batch file, JCL, Perl.

LOLA The Language of Temporal Ordering Specification (LOTOS) Laboratory. See Language of Temporal Ordering Specification.

LON local operating network. See LONWORKS.

Long Distance Internet Provider LDIP. Cooperative services offered by companies that operate under restrictions set by the Modified Final Judgment (MFJ) who wish to provide Internet services.

long haul communication A call which extends beyond the local exchange area.

long key A key held down for longer than a prescribed period which signals an event separate from a short press of the same key. For example, a long key on a computer keyboard might cause the character associated with the key to repeat, or to initiate a key-related process.

long tone In telephony, a long key is one which signals for longer than other keys, in order to communicate through various automated phone menu systems. That is why many phone system menus instruct you to press the pound key (#) after typing in a number, so that a long key signal will be transmitted.

long-distance call, toll call A telephone service referring to any call outside the local service area. Long-distance calls are frequently completed and charged through a carrier other than the local service, and include extra digits in order to identify the desired destination. Long-distance calls are usually billed in a cooperative arrangement through the local carrier, although some services may be billed separately. Sometimes called a *trunk call*, though this phrase is less common.

The first recognized long-distance call is said to have been a one-way message from Alexander Graham Bell's father in Brantford, Ontario, Canada, to Bell in Paris, Ontario, in 1876. The first two-way long-distance call was between Bell and Watson in Cambridgeport and Boston, respectively.

The first transcontinental phone line went into service in 1915, connecting San Francisco and New York City, with Edison and Watson conducting the first conversation over this line. About a decade later, long-distance radiotelephone service was established across the Atlantic Ocean.

Some of the important inventions which made long-distance communications possible were Pupin's *loading coils*, de Forest's *triode*, Armstrong's *regenerator* circuit, and microwave antennas introduced in the early 1950s. Contrast with local call.

long-distance carrier IXC. Local long-distance providers which are competitive with incumbent local exchange carriers.

longitudinal redundancy check LRC. A data transmission error checking technique incorporating a block check on a group of data. An accumulated Block Check Character (BCC) is compared to the sending BCC; if they match, the block is considered to have been transmitted without errors. See cyclic redundancy check.

LONMark Certification A program to ensure interoperability among functional LONWORKS devices, developed by the LONMark Interoperability Association. See LONWORKS.

LONMark Interoperability Association A trade organization of companies supporting and promoting the LONWorks control automation system, established in 1994. The association aids in the integration of multivendor systems for building, home, transportation, and industrial environments. Permission to display the LONMark logo is awarded by the association to firms that successfully complete the LONMark Certification conformance tests. http://www.lonmark.org/

LONTalk A communications protocol associated with LONWORKS automation products that enables intelligent control to be associated with motors, fans, switches, sensors, valves, and other industrial, transportation, and residential/business automation applications.

The protocol is designed to be able to support more than 500 transactions per second and priority levels may be assigned. It follows the Open Systems Interconnection (OSI) reference model and is currently the only product of its kind to implement all seven layers of the model.

LONTalk is favored in part because of its reliability. Unlike some automation systems with one-way communications only or two-way communications without acknowledgments, LONTalk supports end-to-end acknowledgments and automatic retries. Third-party developers of LONTalk have further developed "heartbeat" techniques to enable individual nodes to "check in" with a main or management controller to let the controller know all systems are functional. Alarm conditions can be triggered if nodes do not respond as expected.

L

LONTalk is powerful. Each node supports Network Management Services (NMS) such that they can respond to LONTalk commands from any node supporting NMS functions. Hierarchical addressing through domain, subnet, and node addresses can be used to interact with the network at any level.

LONTalk can be implemented over a variety of types of media, including twisted pair, power line (similar to X–10), radio links, and coaxial or fiber optics cables.

The LONTalk protocol can be implemented into microprocessors under the EIA-709.1 control networking standard by companies such as Echelon Corporation. See LONWORKS, Neuron Chip.

LONWORKS An internationally installed, open network automation and control system for industrial and residential markets developed and trademarked by Echelon Corporation.

LON stands for local operating network. The LONWORKS system uses intelligent control nodes intercommunicating with a common protocol called LONTalk. Each node includes embedded protocol and control functions and a physical interface for coupling the node controller to the communications medium. Nodes may be a variety of drives, relays, and sensing devices and may be used for automation, production lines, security, and more. Local control nodes are the basic network devices for operational control and actuation. Supervisory nodes collect and log data from the local control nodes or coordinate their behavior. Routers provide connectivity and flow between LONWORKS network channels.

The LONWORKS protocol can be embedded into processors, from 8-bit microcontrollers to 32-bit microprocessors. The Neuron Chip is a low-cost, commercially available processor with LONTalk support built in.

In 1999, LONWORKS was approved as an open industry standard by the American National Standards Institute (ANSI/EIA 709.1-A-1999 Control Network Protocol Specification). The protocol is also approved by IEEE and other professional societies. Intelligent Technologies (IEC) is a significant third-party developer of LONWORKS-compatible products. See CEBus, ICELAN 2000, LONTalk, Neuron Chip, X-10.

Look Ahead See Query on Release.

Loomis, Mahlon (1826–1886) An American dentist and researcher who was intrigued by the fact that early telegraphs could be run with only one wire, with earth providing the conductor for the return circuit. He reasoned that if earth could act as one conductor, then perhaps air could act as another, especially since Benjamin Franklin's experiments had alerted scientists to the electricity in the air. In 1865 or 1866 Loomis devised an experiment in which he raised kites with equal lengths of fine copper wire and demonstrated that a signal could be transmitted from one to the other without direct physical contact. He received a U.S. patent for his improved wireless telegraphic system in 1872.

loop 1. A complete transmissions circuit, or electrical circuit. 2. In telephone systems, a loop comprises the wire transmission path that extends from the central office to the residential or business subscriber and back.

loop, communications hardware A circuit, conduit, or line which comprises a continuous path with starting and ending points meeting at the same geographical point. The start and end points may or may not be joined. A loop may or may not include nodes. Communications through the loop may be unidirectional or bidirectional. A loop need not be roughly circular, although it sometimes is; often a loop consists of two adjacent lines, one which sends, the other which receives. Some loops send and receive on the same line (especially if it's a wider bandwidth medium such as fiber), so the loop aspect is based more on the nature of the transmission than the configuration of the cable. See local loop, Fiber Distributed Data Interface, Token-Ring.

loop, programming In software, a programming loop is a series of instructions which will repeat until some event or condition occurs to cause the software to drop out of the loop, or to branch to a specified destination. An endless loop is one which, theoretically, goes on forever. In actual practice, an endless loop often indicates a fault condition and is usually externally terminated. See nesting, recursion.

loop antenna A type of radio direction-finding antenna with one or more complete continuous loops of wire, the ends of which connect to complete the circuit.

Loop Data Maintenance Center LDMC. See Loop Facility Assignment and Control System.

Loop Facility Assignment and Control System LFACS. A database inventory of records and assignments associated with outside telecommunications loop facilities, including connection points, terminals, cables, etc.

loop start In telephony, it is necessary to take control of a line before it can be used. There are two common ways to do this, with a *ground start* or a *loop start*. The loop start commonly used in residential and other single line phone lines. When a caller picks up, the plunger is released (off-hook) and the circuit sends a supervisory signal by bridging the two wires in the phone connection (traditionally called *tip* and *ring*) with direct current (DC). This is done so that the subscriber will get a dial tone and a circuit through which to connect the call. The central telephone switching office sends a signal to the phone the caller is trying to reach and rings the number until it goes off-hook when it is picked up by the callee. When the loop is detected, the ringing signal is no longer set. See ground start.

loop test, loopback test A procedure by which a circuit is connected in a loop in order to test for faults or differences in signal strength or data integrity after passing through the loop. This is commonly used in installation of new circuits or troubleshooting existing circuits. For example, when installing a modem, looping the circuit through the computer system before connecting it to the public phone system can help predetermine whether the basic circuits, software, etc. are correctly installed before adding the additional factors associated with the phone circuit.

loop testing system LTS. In telephony, a subscriber line diagnostic system for physical level testing of copper wire circuits.

loop-through wiring A telephone wiring configuration commonly found in residences and other circuits where economy tends to be a higher priority than high reliability or redundancy. The wire runs from the junction point where the telephone company's wire reaches the house, then travels from room to room in electrical parallel while being physically wired in series. Like lights in older, serially wired Christmas light strings, a break in the circuit will interrupt power to all subsequent phones in the circuit. The alternative is *home-run* wiring (a form of star topology) in which separate wires run from the phone company junction point to each device on the circuit so that a break in a circuit affects only the device on that circuit. Loop-through wiring is certainly adequate for a circuit with only one or two communications devices, but if the premises have a variety of fax, modem, and telephone devices, home-run wiring is preferable.

loose cable In the fiber optics network industry, jargon for general purpose outdoor cables (as compared to indoor interconnect cables, for example). Loose cables are used for aerial, underground, and outdoor conduit installations and sometimes for indoor use. Loose cables come in a variety of different types with varying fiber counts and degrees of shielding.

loopback test See loop test.

loran *lo*ng *ran*ge navigation. A system of distance navigation in which several radio transmitters (usually land-based) are used to send out pulsed signals from different directions in order to determine the geographic location of the craft using the loran system. Useful for air- and watercraft under some circumstances, but limited by the availability and distance of loran stations. See Global Positioning System.

LORG 1. Marketing jargon for large organization. 2. localized orbital/local origin. Methods that can be applied, for example, to shielding/chemical shift calculations.

Lorimer switch One of the first commercially promoted automatic telephone switches, patented by the Lorimer brothers in 1900 and put into service in 1905. While it had many improvements on its predecessor, the Callender switch, it probably owes some of the impetus for its development to this earlier invention. It was installed in a number of switching systems in Europe, but was never fully reliable. However, the technology was modular and could be extended, an important influence on future telephone switching systems. See panel switch, rotary switch, Strowger switch.

Lorimer telephone An early telephone design powered by a central battery system and dialed with a series of levers representing units, not unlike an old calculating machine or cash register. Setting levers configured a telephone number.

Lorimer, George William A Canadian employee of inventor Romaine Callender, George William worked as a telephone operator at the Callender Telephone Exchange Company. He and his brother, James Hoyt Lorimer, later accompanied Callender to New York where Callender was seeking financing to establish a new company after filing a series of patents on telephone switching technology that he was not able to implement in Brantford, Ontario.

In New York, the group succeeded in creating an automatic switching system, after which they returned to Brantford. Callender traveled to England to found the Callender Rapid Telephone Company, and the Lorimer brothers founded the Canadian machine Telephone in Peterborough in 1897. After the death of his brother, James, Egbert Lorimer joined George in marketing their technology. See Lorimer switch; Callender, Romain; Lorimer, James Hoyt.

Lorimer, James Hoyt The brother of George William Lorimer, James originally studied law, but became involved in telephone switching systems research with his brother and George's employer, Romaine Callender. Together the Lorimers founded the Canadian Machine Telephone company in 1897. James had a strong mechanical aptitude, and the brothers continued to improve on the Callender switching technology until it was patentable in 1900. James Hoyt met an untimely death after which no significant technological innovation occurred in the partnership, although the products continued to be marketed. See Callender, Romain; Lorimer, George William.

LOS 1. launch on schedule. 2. line of sight. 3. See loss of signal and Loss of Signal.

loss 1. A decrease in power of a transmission signal as it travels toward its destination, usually expressed in decibels (dB). Many factors contribute to loss, such as distance, type of signal, weather, signal modifications through switches and routers, equipment characteristics, etc. Loss through a circuit is cumulative. See amplifier, interference, noise. 2. In a network, a quantitative measure of a reduction in system resources or services arising from undesired factors such as faulty equipment or configuration, vandalism, or incorrect usage.

Loss of Cell LOC. In ATM networking, a performance monitoring function of the PHY (physical) layer in which a maintenance signal is transmitted in the overhead indicating that the receiving end has lost cell delineation.

Loss of Frame LOF. In ATM networking, a performance measure indicating whether frame delineation has been lost. The LOF is transmitted through the physical (PHY) overhead. On some systems, a LOF condition will be signaled on a port with a light-emitting diode (LED), or as a "yellow alarm."

loss of signal LOS. In a general sense, the sudden, undesired, or unexpected loss of a transmission or other signal such as a beep, alarm, light signal, code, speech, or data signal. LOS results from many causes: interference, a break in the circuit, a change in the surrounding environment (pressure, sun spots,

L

moisture, loss of light, etc.).

Loss in Optical Fibers

There are many sources of loss in fiber optic cables. Here are examples of some of the most common.

Imperfections, bubbles, impurities, and crystallization during fabrication can all lead to loss as the light beams are reflected by the impediments against the direction of the signal or out through the cladding (A1).

Coupling losses, in the form of misaligned joints (B1) or particles trapped in the joints (B2) can cause light to escape or reflect backwards.

Bend losses can be related to structures within the fiber or to the overall geometry of the fiber. Microbends in the form of pits (C1) or irregularities between the core and cladding ratios (C2) may contribute to signal loss. Macrobends from the cable being installed with too great a bend can cause light to hit the cladding at too high an angle to be reflected along the lightguide through total internal reflection (TIR), resulting in significant losses through the cladding (C3).

Loss of Signal LOS. In ATM networking, a performance measure indicating that the receiver is not getting the expected signal, or that there is simply no signal because nothing is currently connected. The LOF

is transmitted through the physical (PHY) overhead. On some systems, a LOS condition will be signaled on a port with a light-emitting diode (LED).

lossless compression A type of data compression technique which does not lose information contained in the image in the compression stage. Some compression algorithms average, sample, or remove image information in order to achieve a high degree of compression, e.g., JPEG. Others retain all the information, e.g., TIFF. See compression. Contrast with lossy compression.

lossy compression A type of data compression technique which selectively or randomly loses information contained in the image in the compression stage. These algorithms average, sample, or remove image information in order to achieve a high degree of compression, e.g., JPEG. New wavelet mathematics is providing some very interesting compression options which provide a high degree of compression with a surprising degree of fidelity to the original image when decompressed and displayed. Other techniques retain all the information, e.g., TIFF. See compression, discrete cosine transform, fractal transform, wavelet. Contrast with lossless compression.

LOTOS See Language of Temporal Ordering Specification.

loupe A compact handheld magnifying tool designed to comfortably enlarge small details. Loupes are typically used with one eye held close to the magnifying lens. They are useful for visual inspection of defects, cleaves, circuits, surfaces to be bonded or cleaned, and other structures that are just a little too small for normal viewing.

Because they are small, loupes are not high magnification devices; they typically magnify about 10x. Neverthless, they are useful as fiber optic filament endface inspection tools in the field.

Some loupes include mounts for attaching them to stabilizing equipment or combination lens devices.

It is important to remember that loupes will magnify a light beam and thus should never be held over a light source (e.g., laser light) that could harm the eye.

Lovelace, Ada Augusta (1815–1851 or 1852) Countess Ada Lovelace (nee Byron) was the daughter of the famed English poet Lord Byron. Ada Lovelace worked with the computer pioneer Charles Babbage, and is regarded as the first computer programmer for her description of how an analytical machine might compute Bernoulli numbers. She proposed the possibility of using computers to compose music or produce graphics.

A computer language (ADA) was developed by the U.S. Department of Defense and named in her honor. See ADA; Babbage, Charles.

low Earth orbit LEO. An orbiting region around the Earth into which certain types of communications satellites are launched. There are advantages and disadvantages to a low orbit. The main advantage is that it generally requires less power to transmit and receive at lower altitudes; the main disadvantage is that it requires a larger number of satellites to provide full global coverage. Other factors include lower radiation levels and lower launching costs. The lifespans of low-orbit satellites tend to be around 5 to 8 years. Most low-orbit satellites travel at about 500 to 2000 km outside Earth. A region called the Van Allen radiation belt at the outer regions of low Earth orbit is generally avoided between the LEO and medium Earth orbits (MEOs). LEO satellites are primarily used for cell phone and data communications. Communications designed for lower orbits require a larger number of satellites than those for higher orbits. This necessitates greater coordination to handle the larger number of systems and to deal with the shorter periods during which each satellite is within range. In contrast, high Earth orbit (HEO) systems can blanket the Earth with only three or four satellites. The trade-off is that higher-placed satellite transmitters require more power to beam the greater distances. See Ellipso, Global Positioning Service, Globalstar, high Earth orbit, Iridium, medium Earth orbit, Orbcomm, Teledesic.

Low Frequency Array LOFAR. An electronic array functioning as an electromagnetic imaging interferometer in the approximately 10- to 150-MHz frequency range. The goals of the LOFAR astronomical project include the study of solar and planetary radio emissions and imaging of the high-redshift emissions. It is hoped the LOFAR research may reveal new classes of physical phenomena in the process of investigating new regions of the electromagnetic spectrum. The work is being carried out by members of the LOFAR Consortium, which includes the U.S. Naval Research Laboratory, MIT, and the Netherlands Foundation for Research in Astronomy.

Ada Lovelace – Pioneer Software Designer

Ada Lovelace had an active interest in the sciences and the arts and speculated on the future capbilities of "thinking machines."

---

low level formatting In storage devices such as hard drives and cartridge drives, formatting is the process of arranging the magnetic media on the storage surface to conform to a recognized pattern so the operating system can further organize data on the drive (the next step is usually to high-level format [initialize] and partition the drive). Each operating system has its own file formats, the protocols that allow it to create directories, organize files and file pointers, and read and write information from and to the drive. Some operating systems are designed to recognize the file formats of other systems as well. For example, on Macintosh and NeXT systems, if a DOS/Windows disk is inserted in the drive, the Mac or NeXT OS will recognize the foreign drive and read and write data files to the drive (and perform minor conversions as necessary) in the format of the diskette, rather than the native operating system format. This provides the user with a lot of flexibility in terms of data transfer and conversion. This does not mean that executable files from other systems can be run on any platform, but rather that files can be moved about as needed. Most drives now come low-level formatted from the factory, but if you have serious data loss on a drive, sometimes as a last resort, you may need to reformat the drive to make it usable again.

low level language A computer control or programming language at the machine or assembly level at which individual registers, accumulators, and other aspects of the physical architecture can be directly or nearly directly controlled. Low level languages are rarely used these days except for writing simulators for various types of processors.

It is much more common now to use high-level programming languages to create source code, and then engage an intermediary program, called a compiler, to translate the high-level language into machine instructions. A certain amount of bit-twiddling can be accomplished in some of the medium- or high-level languages, but is needed only in limited circumstances. Contrast with high-level language.

low noise amplifier LNA. A component which amplifies and sometimes converts telecommunications signals, typically from satellite transmissions. In a satellite receiving station, the LNA takes signals from the feed horn, amplifies them, and then converts them or sends them to a separate low noise converter (LNC); from there they are transmitted to the receiver, usually inside a building. See feed horn, low noise converter, parabolic antenna, satellite antennas.

low noise amplifier probe LNA probe. A component that works in conjunction with a low noise amplifier to control the signal polarity, which can be set to either horizontal or vertical, in order to accommodate more channels on a single system. The LNA probe is typically built into the feed horn mechanism on parabolic antennas.

low noise block converter LNB. A component which converts amplified signals, usually to a lower frequency to send to a receiver. In telecommunications, it is commonly used with satellites and may be incorporated into the low noise amplifier (LNA). LNBs have a broader range than LNCs, as they are able to convert a range of frequencies (provided they have the same polarization) rather than just a single frequency, as in LNCs. See low noise amplifier, low noise converter, parabolic antenna, satellite antennas.

low noise converter LNC. A component which converts amplified signals, usually to a lower frequency to send to a receiver. In telecommunications, it is commonly used with satellites and may be incorporated into the low noise amplifier (LNA). LNCs work with specific frequencies. See low noise amplifier, low noise block converter, parabolic antenna, satellite antennas.

low pass filter A filter that passes transmissions below a specified cutoff frequency, with little or no loss or distortion, but effectively filters out higher frequencies. See high pass filter.

Low-power Atmospheric Compensation Experiment LACE. An experiment begun in the mid-1980s in which a spaceborne target with a single sensor was used to assess compensation schemes associated with laser beams traveling through the atmosphere from the ground. This information was needed to support laser defense system research and development. LACE was built by the U.S. Naval Research Laboratory (NRL).

LACE was originally a fairly simple sensor system carried on a shuttle. In 1986, LACE became a full satellite instead of a set of sensors on a host satellite. By 1987, the sensor arrays carried a total of 210 sensors capable of characterizing ground-based laser beams. In 1990, the LACE satellite was launched and successfully demonstrated that techniques for compensating for atmospheric distortion of laser beams originating from the ground were tenable.

low-power television LPTV. Television broadcast technology with limited power commonly used to serve a local region such as a rural community. Broadcast technologies are closely regulated by the Federal Communications Commission (FCC) and various spectra and licenses are granted to broadcast stations dependent upon operator qualifications, content, viewing audience, and the power and frequency of the communications. LPTV serves a large number of educational and social niche markets and small communities that may not be of interest to larger broadcasting agencies. See Communications Policy Project, Community Broadcasters Association, Federal Communications Commission.

lower sideband In electromagnetic signals, the lower frequency half of a wave. In modulated signals, especially amplitude-modulated radio carrier waves, the sidebands contain the informational content of the signal. See single sideband.

lower sideband suppressed carrier LSSC. A modulated carrier wave that has had part of the signal stripped away in order to save bandwidth. This lower sideband is rebuilt mathematically at the receiving end to recover the original signal information. See sideband.

LP 1. linear programming. 2. low power. 3. low pressure.

LPC See linear predictive coding.

LPRF 1. low-power radio frequency. 2. low pulse-repetition frequency.

LPTV See low-power television.

LRC See longitudinal redundancy check.

LRF See laser range finder.

LRN See Location Routing Number.

LRS line repeater station.

LSA See link state advertisement.

LSB 1. See least significant bit. 2. lower sideband. See sideband.

LSDU Link layer Service Data Unit.

LSI A term in the semiconductor industry describing capabilities aggregated onto a single chip. See large scale integration.

LSMA Large-Scale Multicast Applications.

LSN 1. See Large Scale Networking group. 2. local signal number.

LSP 1. See label-switched path. 2. See Local Services Provider.

LSR See local service request.

LSS loop switching system. See switching.

LSSC See lower sideband suppressed carrier.

LSSGR LATA Switching System General Requirements.

LSTP 1. See local signal transfer point. 2. linear search & track processor.

LSU 1. See line switching unit. 2. See Logical Storage Unit.

LTB Last Trunk Busy.

LTC See line trunk controller.

LTE See Line Terminating Equipment.

LTO See linear tape-open.

LTS See loop testing system.

LTSP See Linux Terminal Server Project.

LUCE See laser-using communications equipment.

Lucent Technologies A company created following the AT&T/Bell divestiture. Lucent was established with the Bell Laboratories research staff and a number of the electronics, network, and business communications groups, including Systems for Network Operators, Business Communications Systems, Microelectronics, and Consumer Products. The organization has become prominent as a developer of many new optical telephone network technologies. See AT&T; Barton, Enos.

Lucent Technologies Canada Inc. A wholly owned subsidiary of Lucent Technologies, based in Ontario, Lucent Technologies Canada Inc. formed as a result of the restructuring of AT&T after the divestiture. Lucent began in Canada as part of AT&T Canada Inc. in 1984.

LUF lowest usable frequency.

lug 1. A projecting attachment point, especially for electrical circuits. See terminal. 2. An attachment added to the end of a wire which provides an eye, or forked end, which allows the wire to be more easily attached to a bolt under a binding screw.

LUM See line utilization monitor.

lumberg The older term for a Talbot, now superseded by lumen seconds. See lumen seconds.

lumen (Latin – *light*) A standardized SI unit of light

(luminous) flux equal to the light produced by one candela intensity on a unit area of a flat surface of uniform distance from the light source. A lumen indicates photonic energy flow. A footcandle (a description of flux density) is one lumen per square foot (lux is now more commonly used). An indoor lamp might output 2,000 lumens. See candela, steradian.

lumen seconds, Talbot A standardized unit of luminous energy over time, equal to the illumination from a one lumen light source emitted for one second, usually expressed in millijoules for a specified wavelength. It is also known as a Talbot after W.H.F. Talbot, replacing the term lumberg. See photon.

lumen hour A measure of luminous energy over time, equal to the illumination from a one lumen light source emitted continuously over the course of an hour.

luminance (*symb.* – L) The luminous (light-emitting) flux reflected or transmitted, as measured from a particular direction, from a source such as a TV screen, light-emitting diode, or laser light source per unit area as measured in a specified direction.

A luminance unit is expressed as a candela per square meter (cd/m^2), sometimes called a *nit*. In fact, luminance measuring devices have been dubbed "nitmeters." Nitmeters may also have electronics for measuring lux. Luminance meters or "light meters" are frequently used for assessing ambient light or reflected light from spotlights used for photography. They measure an aspect of luminance called luminance flux or *lux*. Spectroradiometers can quickly measure low light levels. Luminance colorimeters can measure luminance and other light-related properties such as chromaticity and color temperature.

These terms luminance and nit have superseded older expressions of luminance such as footcandle, Lambert, or footLambert. In casual terms, luminance is often called brightness. See photometer.

luminosity A ratio of light flux to its corresponding radiant flux at a specific wavelength, expressed in lumens per watt.

luminous flux The visible energy (light) produced per unit of time, expressed in lumens. Luminous flux may be measured with an integrating sphere associated with a photometer. See lumens.

luminous intensity A measure of the quantity of luminous flux in a given direction at a frequency of 540×1012 Hz at a particular solid angle (1/683 W per steradian), expressed in *candelas*. Specialized photometers can be calibrated to detect luminous intensity.

At the U.K. National Physical Laboratory the candela has been measured with a reported uncertainty of 0.02%, using a cryogenic radiometer which equates the thermal effect of optical radiation with that of electrical power.

Luminous intensity standards have been developed for many technologies including public lighting, signage, and street lights. See candela, steradian.

LUN See logical unit number.

Luneberg lens A type of focusing lens used in antennas to increase gain for ultra-high frequency

(UHF) transmissions.

lux A combining word from *lu*minance and flux also referred to as illuminance. A basic metric unit for expressing illumination (a footcandle equals 10.76 lux). The illumination on a one-square-meter area on which the flux of one lumen is uniformly distributed. See flux, luminance.

luxmeter A type of light-measuring instrument that records intensities. Light meters are commonly incorporated into cameras to help to determine aperture and speed settings.

LTS lightwave transmission system. Transmission through a light-guiding medium such as fiber optics.

LVD low voltage disconnect. See load distribution unit.

LWER See LightWeight Encoding Rules.

Lynch, Daniel C. (ca. 1940-) Lynch organized the first TCP/IP Implementor's Workshop in 1986, which later developed into Interop in 1988, a large gathering of Internet, network, and other telecommunications professionals. Lynch is also known for his role in the ARPANET transition from NCP to Internet Protocol (IP). He is a cofounder of CyberCash, Inc. and has been a member of the Board of Directors since 1994.

Lynx A text-based Web browser developed at the University of Kansas in the early 1990s by M. Grobe, C. Rezac, L. Montulli, and many others. Lynx enables limited-resource devices such as portable or desktop text-based terminals to navigate the Web. It is also useful for fast searches of Web content for situations where viewing graphics is not desired. Lynx is descended from a client/server-based distributed computing hypertext browser. See Microsoft Explorer; Netscape.

LZ77 See Lempel-Ziv.

LZ78 See Lempel-Ziv.

LZARI Lempel-Ziv arithmetic. A lossless compression and archiving utility developed by Okumura in 1988, based originally on LZSS, but which incorporated adaptive and static algebraic compression to encode characters and position fields, respectively. Thus, it is a statistical compressor, rather than a dictionary compressor as was its predecessor. LZARI was not the fastest archiver for its time, but it had good compression performance. LZARI was later adapted into LHarc by Yoshizaki. See Lempel-Ziv, LHarc.

LZB Lempel-Ziv Bell. A lossless variable-length-code compression scheme developed in 1987 by Bell, based upon LZ77 concepts. In terms of compression performance for text files, LZB is a little better than LZH and its predecessors, but not as efficient as the popular GZIP. LZB has fairly small memory requirements for decompression making it suitable for devices with limited memory resources.

LZB 80 Linienzugbeeinflussung 80. A signaling system developed by the German Federal Railways. LZB systems are sold internationally for high-speed train systems control and safety.

LZC A dictionary-based lossless compression scheme developed in 1985 by Thomas et al. based upon LZW, which is patented. LZC incorporates a variable-size pointer scheme. It dynamically monitors progress and can flush and rebuild the dictionary to suit the circumstances. As examples, Unix compress and MacCompress use the LZC algorithm. LZC is also used in some schemes to increase disk space availability by dynamically compressing stored files. (Don't confuse this with the wavelet-related LZC developed by Thaubman and Zakhor.) See Lempel-Ziv-Welch.

LZC layered zero coding. A rate-scalable encoding scheme described by D. Thaubman and A. Zakhor in 1994. LZC takes advantage of a strong correlation among subband coefficients, resulting in good image compression performance. LZC and its descendants have been used in a number of applications related to scalable image compression and have been incorporated into embedded wavelet-based video coders. (Don't confuse this with the dictionary-based LZC developed by Thomas et al.) See wavelet.

LZFG Lempel-Ziv-Fiala-Green. A fast, lossless compression scheme developed in 1989 by Fiala and Green (U.S. patent #4906991), based upon Lempel-Ziv LZ77 and LZ78 concepts. LZFG is a sliding window scheme with data stored in a modified *trie* (Patricia tree) data structure. The position of the text in the trie is output. LZFG has some speed benefits over Lempel-Ziv-Jakobsson (LZJ).

LZH See LHA.

LZJ Lempel-Ziv-Jakobsson. A dictionary-based lossless compression scheme developed in 1985 by Jakobsson. It is based upon LZW containing point-

ers only, with the pointers able to point to anywhere in the previous character data to indicate a substring. See LZFG.

LZHUF Lempel-Ziv Huffman. The algorithm incorporated into the LHarc data archiving utility by Yoshizaki that replaces LZARI's adaptive arithmetic coding with adaptive Huffman coding to improve the speed of LZARI (LZARI already had good data compression). With additional work, LZHUF evolved into LHarc. See LHarc, LZARI.

LZMW A lossless compression scheme developed in 1984 by Miller and Wegman. While LZMW improved upon its predecessor LZ77 (Lempel and Ziv), there was so much interest in adapting LZ77 that LZMW was short-lived, superseded by LZH, LZB, and the efficient and popular GZIP.

LZP Lempel-Ziv prediction. A lossless dictionary and hash-based compression scheme developed in 1995 by Charles Bloom. Bloom designed the scheme to be fast, scalable, and retargetable. It is descended from LZ77 and shares some characteristics with LZNW and PPMCB. It is distributed for noncommercial use under a Public License. See Lempel-Ziv.

LZR A lossless compression utility developed in 1981 by Roden et al. that does not have the window limitations of the earlier LZ77 scheme. Interestingly, an LZR scheme has been suggested for the compression of repetitive DNA sequences. See Lempel-Ziv.

LZS See Lempel-Ziv-Stac.

LZSS See Lempel-Ziv-Storer-Szymanski.

LZW See Lempel-Ziv-Welsh.

m 1. *abbrev.* meter. See meter. 2. *abbrev.* milli-. See milli-.

M 1. *abbrev.* mega-. See mega-. 2. *symb.* mixed polarization (ITU).

M bit, Mark bit, More bit In X.25 network data transmissions, a signal bit used to indicate that additional packets in a sequence are to be expected. A bit set to 1 or "true" indicates further packets will be coming, whereas 0 or "false" signals that packets were (intentionally) not sent. This helps the receiving system to distinguish between packets not sent and lost packets. In Realtime Transport Protocol (RTP), the M bit can signal the transport of at least one complete media frame or the remaining fragment in a frame. For frames fragmented across multiple RTP packets, the M bit can signal frame boundaries. See D bit, Q bit.

M Hop

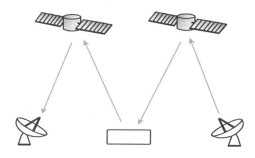

Hops between satellites can be repeated to pass through an intermediate station or hub (e.g., a ground station) en route to the final receiver. There may be a number of reasons for sending to an intermediate hub, including amplification, filtering of the broadcast channels, and redistribution to subscribers through Earth stations at more than one location.

M hop A type of pattern that results when communications transmissions are bounced from an Earth station to an airborne receiver, back to an Earth station or intermediary hub, up to an airborne receiver and back down to the final receiving station, thus resembling the letter "M." This is a common configuration in hub topology satellite communications. Newer satellites are being designed for intersatellite communication, so the signal goes from an Earth station to a satellite, to another satellite and then down to Earth again, thus forming a shape like three sides of a rectangle rather than the letter M, as shown in the *Basic hop/M hop* diagram.

M port In a Fiber Distributed Data Interface (FDDI) network, extra port on a concentrator for attaching other nodes in a branching tree topology.

M ports can be on both single attachment and dual attachment concentrators. The M port is an addition to the basic FDDI network. On a dual attachment station (DAS), a redundant link can be created by connecting the A and B ports on different concentrators on the M ports. On a single attachment concentrator, the M port may be connected to the S port. M ports are never connected to one another. The other end of the M port may be attached to a patch panel through a data grade cable. See A port.

M Ports in FDDI Ring-Based Network

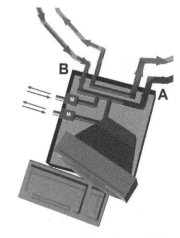

A and B ports are the main points of entry between FDDI cabling "rings" and dual attachment stations (DASs). However, it is also possible to configure M ports onto the ring for adding more stations.

ITU-T M Series Recommendations

Scope, Philosophy, General Principles

M.10	Scope and application of recommendations for maintenance of telecom. networks and services
M.15	Maintenance considerations for new systems
M.20	Maintenance philosophy for telecom. networks
M.21	Maintenance philosophy for telecom. services
M.32	Principles for using alarm info for maintenance of international transmission systems and equipment
M.560	International telephone circuits – principles, definitions, and relative transmission levels
M.1130	General definitions and general principles of operation/maintenance proced. to be used in satellite mobile systems
M.1140	Maritime mobile telecom. services via satellite
M.1301	General description and operational proced. for international SDH leased circuits
M.1535	Principles for maintenance info to be exchanged at customer contact point (MICC)
M.1537	Definition of maintenance info to be exchanged at customer contact point (MICC)
M.3100	Generic network info model
M.3600	Principles for management of ISDNs

Numbering

M.320	Numbering of channels in group
M.330	Numbering of groups within supergroup
M.340	Numbering of supergroups within mastergroup
M.350	Numbering of mastergroups within supermastergroup
M.380	Numbering in coaxial systems
M.390	Numbering in systems on symmetric pair cable
M.400	Numbering in radio-relay links or open-wire line systems
M.410	Numbering of digital blocks in transmission systems

ISDN

M.3600	Principles for management of ISDNs
M.3602	Application of maintenance principles to ISDN subscriber installations
M.3603	Application of maintenance principles to ISDN basic rate access
M.3604	Application of maintenance principles to ISDN primary rate access
M.3605	Application of maintenance principles to static multiplexed ISDN basic rate access
M.3610	Principles for applying TMN concept to the management of B-ISDN
M.3611	Test management of B-ISDN ATM layer using the TMN
M.3620	Principles for the use of ISDN test calls, systems, and responders
M.3621	Integrated management of ISDN customer access
M.3640	Management of the D-channel – data link layer and network layer
M.3641	Management info model for management of data link and network layer of ISDN D-channel
M.3650	Network performance measurements of ISDN calls
M.3660	ISDN interface management services

Monitoring, Maintenance, Performance, Service, Testing, Misc., etc.

M.34	Performance monitoring on international transmission systems and equipment
M.35	Principles concerning line-up and maintenance limits
M.50	Use of telecom. terms for maintenance
M.60	Maintenance terminology and definitions
M.70	Guiding principles on general maintenance organiz. for telephone-type international circuits
M.75	Technical service
M.80	Control stations
M.85	Fault report points
M.90	Sub-control stations
M.100	Service circuits
M.110	Circuit testing
M.120	Access points for maintenance
M.125	Digital loopback mechanisms
M.160	Stability of transmission
M.450	Bringing new international transmission system into service
M.460	Bringing international group, supergroup, etc. links into service
M.470	Setting up and lining up analog channels for international telecom. services
M.475	Setting up and lining up mixed analog/digital channels for international telecom. services
M.495	Transmission restoration and transmission route diversity: terminology and general principles
M.496	Functional organiz. for automatic transmission restoration

M.500	Routine maintenance measurements to be made on regulated line sections
M.510	Readjustment to nominal value of a regulated line section (on symmetric pair line, coaxial line or radio-relay link)
M.520	Routine maintenance on international group, supergroup, etc. links
M.525	Automatic maintenance procedures for international group, supergroup, etc. links
M.530	Readjustment to the nominal value of an international group, supergroup, etc. link
M.535	Special maintenance procedures for multiple destination, unidirectional (MU) group and supergroup links
M.540	Routine maintenance of carrier and pilot generating equipment
M.556	Setting up and initial testing of digital channels on international digital path or block
M.562	Types of circuit and circuit section
M.565	Access points for international telephone circuits
M.570	Constitution of circuit; preliminary exchange of info
M.580	Setting up and lining up international circuit for public telephony
M.585	Bringing international digital circuit into service
M.590	Setting up and lining up circuit fitted with a compandor
M.600	Organization of routine maintenance measurements on circuits
M.605	Routine maintenance schedule for international public telephony circuits
M.610	Periodicity of maintenance measurements on circuits
M.620	Methods for carrying out routine measurements on circuits
M.630	Maintenance of circuits using control chart methods
M.650	Routine line measurements to be made on line repeaters of audio-frequency sections or circuits
M.660	Periodical in-station tests of echo suppressors complying with Recommendations G.161 and G.164
M.665	Testing of echo cancellers
M.670	Maintenance of circuit fitted with compandor
M.675	Lining up and maintaining international demand assignment circuits (SPADE)
M.710	General maintenance organiz. for international automatic and semi-automatic telephone service
M.715	Fault report point (circuit)
M.716	Fault report point (network)
M.717	Testing point (transmission)
M.718	Testing point (line signaling)
M.719	Testing point (switching and interregister signaling)
M.720	Network analysis point
M.721	System availability info point
M.722	Network management point
M.723	Circuit control station
M.724	Circuit sub-control station
M.725	Restoration control point
M.726	Maintenance organization for wholly digital international automatic and semi-automatic telephone service
M.729	Maintenance organization for international public switched telephone circuits for data transmission
M.730	Maintenance methods
M.731	Subjective testing
M.732	Signaling and switching routine maintenance tests and measurements
M.733	Transmission routine maintenance measurements on automatic and semi-automatic telephone circuits
M.734	Exchange of info on incoming test facilities at international switching centers
M.760	Transfer link for common channel Signaling System 6
M.762	Maintenance of common channel Signaling System 6
M.800	Use of circuits for voice-frequency telegraphy
M.810	Setting up and lining up international voice-frequency telegraph link for public telegraph circuits (50-, 100-, and 200-baud modulation rates)
M.820	Periodicity of routine tests on international voice-frequency telegraph links
M.830	Routine measurements to be made on international voice-frequency telegraph links
M.850	International time division multiplex (TDM) telegraph systems
M.880	International phototelegraph transmission
M.900	Use of leased group and supergroup links for wide-spectrum signal transmission (data, facsimile, etc.)
M.910	Setting up and lining up international leased group link for wide-spectrum signal transmission
M.1010	Constitution and nomenclature of international leased circuits
M.1012	Circuit control station for leased and special circuits
M.1013	Sub-control station for leased and special circuits
M.1014	Transmission maintenance point (international line) (TMP-IL)
M.1015	Types of transmission on leased circuits
M.1016	Assessment of service availability performance of international leased circuits
M.1020	Character. of special quality international leased circuits with special bandwidth conditioning

M

ITU-T M Series Recommendations, cont.

Monitoring, Maintenance, Performance, Service, Testing, Misc., etc., cont.

M.1025 Characteristics of special quality international leased circuits with basic bandwidth conditioning

M.1030 Character. of ordinary quality international leased circuits forming part of private switched telephone networks

M.1040 Character. of ordinary quality international leased circuits

M.1045 Preliminary exchange of info for provision of international leased circuits and international data transmission systems

M.1050 Lining up international point-to-point leased circuit with analog presentation to the user

M.1055 Lining up international multiterminal leased circuit

M.1060 Maintenance of international leased circuits

M.1150 Maintenance aspects of maritime/land mobile telecom. store-and-forward services (packet mode) via satellite

M.1160 Maintenance aspects of aeronautical mobile telecom. service via satellite

M.1170 Maintenance aspects of mobile digital telecom. service via satellite

M.1230 Method to improve management of operations and maintenance processes in International Telephone Network

M.1235 Use of automatically generated test calls for assessment of network performance

M.1300 Maintenance of international data transmission systems operating in range 2.4 Kbps to 140 Mbps

M.1320 Numbering of channels in data transmission systems

M.1340 Performance objectives, allocat. and limits for international PDH leased circuits and supporting data transmission links and systems

M.1350 Setting up, lining up and characteristics of international data transmission systems operating in range 2.4 to 14.4 Kbps

M.1355 Maintenance of international data transmission systems operating in range 2.4 to 14.4 Kbps

M.1370 Bringing-into-service of international data transmission systems

M.1375 Maintenance of international data transmission systems

M.1380 Bringing-into-service of international leased circuits that are supported by international data transmission systems

M.1385 Maintenance of international leased circuits that are supported by international data transmission systems

M.1400 Designation for inter-operator networks

M.1510 Exchange of contact point info for the maintenance of international services and the international network

M.1520 Standardized info exchange between administrations

M.1530 Network maintenance info

M.1532 Network maintenance service performance agreement (MSPA)

M.1539 Management of grade of network maintenance services at maintenance service customer contact point (MSCC)

M.1540 Exchange of info for planned outages of transmission systems

M.1550 Escalation proced.

M.1560 Escalation proced. for international leased circuits

M.2100 Performance limits for bringing-into-service and maintenance of international PDH paths, sections and transmission systems

M.2101 Performance limits and objectives for bringing-into-service and maintenance of international SDH paths and multiplex sections

M.2101.1 Performance limits for bringing-into-service and maintenance of international SDH paths and multiplex sections

M.2102 Maintenance thresholds and procedures for recovery mechanisms (protection and restoration) of international SDH VC trails (paths) and multiplex sections

M.2110 Bringing-into-service of internat. PDH paths, sections and transmission systems, SDH paths, and multiplex sections

M.2120 PDH path, section and transmission system and SDH path and multiplex section fault detection and localization procedures

M.2130 Operational procedures for the maintenance of transport network

M.2140 Transport network event correlation

M.2201 Performance objectives, allocations and limits for bringing-into-service and maintenance of international ATM virtual path and virtual channel connections

M.3208.1 Leased circuit services

M.3208.2 Connection management of pre-provisioned service link connections to form leased circuit service

M.3208.3 Virtual private network

M.4010 Inter-Administration agreements on common channel Signaling System 6

M.4030 Transmission characteristics for setting up and lining up transfer link for common channel Signaling System 6 (analog version)

M.4100 Maintenance of common channel Signaling System 7

M.4110 Inter-Administration agreements on common channel Signaling System 7

Telecom. Management Network (TMN)

M.3000	Overview of TMN Recommendations
M.3010	Principles for Telecom. Management Network (TMN)
M.3013	Considerations for Telecom. Management Network (TMN)
M.3016	TMN security overview
M.3020	TMN Interface Specification Methodology
M.3101	Managed object conformance statements for generic network info model
M.3108.1	TMN management services for dedicated and reconfigurable circuits network: info model for management of leased circuit and reconfigurable services
M.3108.2	TMN management services for dedicated and reconfigurable circuits network: Info model for connection management of preprovisioned service link connections to form reconfigurable leased service
M.3108.3	TMN management services for dedicated and reconfigurable circuits network: Info model for management of virtual private network service
M.3120	CORBA generic network and NE level info model
M.3180	Catalog of TMN management info M.Imp3100
M.3100	TMN Implementors' Guide – defects and resolutions (M.3100 Series)
M.3200	TMN management services and telecom. managed areas: overview
M.3207.1	TMN management service: maintenance aspects of B-ISDN management
M.3210.1	TMN management services for IMT-2000 security management
M.3211.1	TMN management service: Fault and performance management of ISDN access
M.3300	TMN F interface requirements
M.3320	Management requirements framework for TMN X-Interface
M.3400	TMN Management Functions

Supplements

M.Sup1.1	Prefixes used in decimal system
M.Sup1.2	Transmission measure. conversion tables
M.Sup1.3	Normal (or Laplace-Gauss) distribution
M.Sup1.4	Methods of quality control
M.Sup1.5	Mathematical processing of measure. results of variations of overall loss of telephone circuits
M.Sup1.6	Statistical theory require.
M.Sup2.1	General observations concerning measuring instruments and measuring techniques
M.Sup2.10	Method for measuring frequency shift introduced by carrier channel
M.Sup2.11	Rapid verification test for echo control devices
M.Sup2.12	Automatic data acquisition and effective processing proced. for group and supergroup pilot levels
M.Sup2.13	Loop method for maintenance of 4-wire telephone-type leased circuits
M.Sup2.14	Automatic measuring device for carrier systems with large number of channels
M.Sup2.15	Detection of circuit faults
M.Sup2.16	Receiving relative levels at renters' premises for international leased circuits used for data transmission
M.Sup2.17	Results of investigation of service availability performance of international leased circuits made in 1982
M.Sup2.2	Measurements of loss
M.Sup2.3	Level measurements
M.Sup2.4	Measure. of crosstalk
M.Sup2.5	Measuring errors and differences due to impedance inaccuracies of instruments and apparatus. Use of decoupled measuring points
M.Sup2.6	Errors in indications given by level-measuring instruments due to interfering signals
M.Sup2.7	Measure. of group delay and group-delay distortion
M.Sup2.8	Measure. of sudden phase changes on circuits
M.Sup2.9	Vibration testing
M.Sup4.10	Transient analog circuit impairments and effect on data transmission
M.Sup4.1	Stability of overall loss and psophometric noise: results of routine maintenance measurements made on international network during first half of 1978
M.Sup4.2	Results and analysis of 10th series of tests of short breaks in transmission
M.Sup4.3	Character. of leased international telephone-type circuits
M.Sup4.5	Instructions for making future measurements of transmisison quality of complete connections for recording results of measurements
M.Sup4.8	Results and analysis of tests of impulsive noise
M.Sup4.9	Weighting of measurements relating to stability of circuits in international network according to size of circuit groups
M.Sup4.7	Instructions for making future measurements of transmission quality of international circuits and international centres and for recording results of measurements
M.Sup5.1	Requirements for transmission of television signals over long distances
M.Sup5.2	Setting-up and testing of international videoconference studios

M

M Series Recommendations A series of ITU-T recommended guidelines for TMN and network maintenance: international transmission systems, telephone circuits, telegraphy, facsimile and leased circuits. These guidelines are available for purchase from the ITU-T. Since ITU-T specifications and recommendations are widely followed by vendors in the telecommunications industry, those wanting to maximize interoperability with other systems need to be aware of information disseminated by the ITU-T. A full list of general categories is listed in Appendix C and specific series topics are listed under individual entries in this dictionary, e.g., L Series Recommendations. See M Series Recommendations chart.

M-ary A designation useful in assessing relative capacities in multilevel modulation schemes. M is used as the symbol for arithmetic equivalencies to represent the number of bits per symbol. Thus, in *M-ary signaling*, a symbol represents *n* bits, with *M* signal states, with $M = 2^n$. In turn, the number of symbols per second determines the *baud* rate. This can further be extrapolated to calculate capacity in relation to bandwidth.

M-Bone See multicast backbone.

M-quad, mini-quad A compact version of a quad antenna used for single or multiband radio communications. See quad antenna for a fuller explanation.

M-patch bay According to the Federal Technology

Recent and Widespread Macintosh Operating Systems	
Version	Description
Mac OS 7.6	A widespread version of the Macintosh operating system still used by millions of end-users for older Macs and supported by many developers well into the year 2000, even though OS 8.6 and its successors had been available for a few years and would eventually supersede it. One of the reasons for the longevity of 7.6 was its relative stability and networking capabilities. Another reason was that honest software buyers, with large installed bases of programs, could not easily move to OS 8.x and above without upgrading, a move that could cost hundreds, sometimes thousands of dollars per machine.
Mac OS 8.5	Code-named Allegra, and coexistent with the Rhapsody environment, this was a major Macintosh Operating System release (with upgrades and patches to bring it up to 8.6) that became well-established in the late 1990s. This release featured a more polished looking interface, better native graphics support, and increased support for Internetworking, as well as enhanced memory and file transfer capabilities. It was still being widely used in 2001 concurrently with Mac OS 9 and Mac OS X.
Mac OS 9.x	A new version of the Macintosh operating system announced in 1999 to supersede the OS 8.x line. With OS 9 came Sherlock 2, a Web-compliant searching tool, multiple-user environments (in the sense that more than one person could use the same Mac and keep preferences, file settings, etc. personalized and protected), voiceprint passwords, Internet file serving capabilities, Internet AppleScript automation tools, and more. In general, this release of the operating system brought the Macintosh operating system into the world of the Internet and made many aspects of the system customizable and more secure. Mac OS 9.1 was announced in 2001 to provide a transition to Mac OS X (which became prevalent by 2002).
Mac OS X	Macintosh Operating System Ten. Announced in May 1998, this new version of the Macintosh OS was scheduled to ship to developers in early 1999 with an original planned public release in Fall 1999 and an actual release in 2001. The system supports pre-emptive multitasking, advanced virtual memory, and memory protection, optimized for Apple's G3 PowerPC computers. OS X is built upon Darwin, the open source model core based on FreeBSD and Mach 3.0 technologies. Darwin is processor-independent and supports PowerPC- and Intel-based desktop computers. Thus, Open Source developers can create applications capable of running on a variety of hardware platforms. At the time OS X was released, the installed base of Macintosh platforms was over 25 million. Apple expected most existing Macintosh applications to run on OS X without alteration (though in practice this was not entirely so). Mac OS X.1 improved printing and networking capabilities as well as CD and DVD authoring capabilities. A new, aesthetically appealing user interface called Aqua was introduced with the Mac OS X system. Aqua included a new Finder and NeXT-like Dock to enable the desktop to be customized for quick access to commonly used applications. Aqua is built upon graphics technologies for the multimedia world, including OpenGL, Quartz, and QuickTime. See QuickTime.

Service (FTS), a patching system for monitoring and patching (interconnecting) digital data circuits at signaling rates from 1 to 3 Mbps.

m-s junction metal-to-semiconductor junction. A mechanism within a semiconductor acting as an electrical contact or as a barrier. See p-n junction.

Ma Bell *colloq.* A term to familiarly describe the Bell telephone system, and later AT&T. The Bell system was so ubiquitous and so well recognized that many came to refer exclusively to the corporation as Ma Bell.

MAC 1. See mandatory access control. 2. Media Access Control. In a layered network architecture, the lower half of the data-link layer that governs access to the available IEEE and ANSI LAN media. See Media Access Control for a fuller explanation. 3. See Message Authentication Code. 4. See multiplexed analog component.

MAC Access Arbitration MAA. In a layered network operating as a broadband fiber network (e.g., cable modem services), the Media Access Control (MAC) Access Arbitration is a sublayer which, along with associated sublayers, facilitates multiplexing and Quality of Service (QoS). See Media Access Control.

MAC address Media/Medium Access Control address. A network location identifier. See Media Access Control.

MACE Macintosh Audio Compression and Expansion. MACE is built into the Macintosh OS Sound Manager utility.

Mach line, Mach surface The division between regions of supersonic and subsonic flow.

Mach-Zehnder interferometer See interferometer.

machine dependent Software or peripherals designed to work on a specific system or architecture, and not readily usable on other systems (although sometimes modifications can be done). Low-level routines written to take advantage of a particular chip architecture or peripheral card are machine dependent. Most software executables are machine dependent, since they have usually been compiled from a higher level language down into low-level machine code for a specific system.

machine language A symbolic computing machine control language that functions at the lowest level possible on a system, in symbols readily understood by the computer, but inscrutable to most people. Machine language involves the most basic movement and processing of data, in terms that are specific to the computer architecture (usually binary). Thus, *move* and *add* instructions are used frequently in machine language programs. A move instruction transfers data between registers, and an add instruction performs a math operation (add, multiply, subtract, etc.).

Because machine language programs are cryptic, long, difficult to follow, difficult to read, and difficult to debug, assemblers were developed to codify and organize instruction sets so software could be written and debugged more quickly. Assembly language or assembler was a step up from machine language in that it used symbols to encode instructions

that would then be translated into machine language. Assembler was easier to code and debug than machine language, but the listings were long and many people had difficulty with the symbology and tedium. Later, higher level languages such as FORTRAN, BASIC, LISP, C, Modula, Perl, Java, etc. were added to make the task of programming easier still, and to adapt programming languages to specific types of tasks.

Some higher level languages are compiled into machine language executables that can be run directly thereafter, and others are interpreted into machine language at the time an instruction is processed. Compiled languages typically run many times faster than interpreted languages because the conversion to machine instructions happens only once, prior to writing the program. This machine language compiled executable is then stored as a file and used as needed. With faster processors, interpreted languages, which also have some advantages, such as direct feedback without compilation waiting times, have remained popular. Machine language coding is rare these days except for specialized coding.

Macintosh Plus – Apple Computing Platform

The Historic Macintosh Plus, released in the mid-1980s, had a portable all-in-one design, featuring a monochrome monitor, two serial ports, SCSI controller, graphical user interface, mouse, and networking capabilities through AppleTalk as standard features.

Macintosh, Mac A family of Motorola 68000-based personal computers developed and distributed by Apple Computer, Inc. The first of the Macintosh line was the Lisa computer, introduced in 1983. Its graphical operating system was described as radical by many. There were also vocal detractors who said "graphics interfaces will never be accepted in office environments." A graphical interface may not seem unusual now, but at the time, computers were almost exclusively text-based, and this new graphical user interface, which was accessed with a mouse, seemed extraordinary. Many business and computing professionals recoiled in suspicion. In retrospect, now that virtually all computers have graphical user interfaces, the harsh criticisms leveled at the Macintosh graphics now appear shortsighted.

The Macintosh graphical operating system was inspired by some of the brilliant research and development that was occurring in the 1970s and early 1980s at the Xerox PARC facility in California. The fledgling founders of Apple Computer were given a tour of the facility, and Steve Jobs came away energetically inspired by what he had seen, determined to add a new line of computers different from the Apple II line.

Unfortunately, while the Lisa was a good machine and most of its characteristics were later incorporated into the more familiar Macs, the initial price tag was high and it didn't sell well. The early Macintosh computers familiar to most people started with the Mac in 1984, followed by the Mac Plus and various subsequent Macs (there are now dozens of models). The 1984 release of the 128K 32-bit 4.7 MHz "little Mac" was accompanied by a significant product in 1985, the Apple Laserwriter. This new laser printer changed the way many people thought about computers. Since 10-pin dot matrix printers with extremely limited fonts were prevalent at the time, few had considered the potential of personal computers as publishing tools. With the Laserwriter and Adobe Systems PostScript fonts, publishers sat up and took notice and the desktop publishing industry was born, with very substantial repercussions to the traditional layout and printing industry that are still reverberating today. New all-digital presses are completely changing the way printed information is produced.

The Macintosh line became the preferred system in print publishing service bureaus in the late 1980s, and many of the high-end desktop publishing and graphics programs were available only for the Mac. It was not until the early 1990s that some of these important software programs were ported to Intel-based machines, and service bureaus began to use both platforms.

Almost 30 million Macintosh computers had been shipped since 1984, 79% of which were still in use, yet detractors continued to predict the demise of Apple and the Macintosh line (a prediction begun with the first introduction of the Mac line 14 years earlier). The first major change in the Macintosh line was the changeover to PowerMacs in the 1990s. In 1997, to the surprise of many, Steve Jobs took the position of interim CEO and later CEO and injected a sense of excitement into what had become an almost lackluster company. This change came about with the evolution from PowerMacs to G3s late in 1997. With the introduction of the iMacs in a variety of designer colors and styles, the Macintosh attracted a new generation of users. By 2000, Macintosh laptops had become competitive with the introduction of new models, including the iBooks and full-featured Titanium model. In July 2001, the G4 line was introduced, along with updates to the Mac OS X operating software. The portable models are particularly important to telecommunications since they are often integrated with handheld communications devices for business communications and favored for writing and reporting by journalists.

While the early Mac hardware was not inherently suited to video, it was popular for audio applications and many musicians adopted the Mac for the composition of electronic music. The Amiga was the preferred platform for video from the mid-1980s to the mid-1990s, until Commodore support for the Amiga dwindled, and then the desktop video market split onto Intel and Macintosh platforms. As the Macintosh hardware evolved, audio/visual capabilities were increasingly incorporated into the basic hardware, with high-quality stereo sound capabilities and some video capabilities built into A/V (audio/visual) models. This was important in maintaining the platform's viability in a market demanding greater graphics and sound capabilities for games, publishing, and multimedia/Web production.

MacMICA A Macintosh-based multipoint videoconferencing program from Group Technologies, which works over AppleTalk networks. See Cameo Personal Video System, Connect 918, CU-SeeMe, IRIS, Visit Video.

MacMiNT A text-based, Unix-like operating system ported from the Atari ST to the Macintosh. It can be used with freely distributable Unix utilities such as GCC, GDB, make, tcsh, perl, etc.

macro A programming routine or script that "bundles" or combines a number of steps, processes, operations, or other actions. Macros are typically used as time-savers for frequently performed functions. Scriptable macros are usually written in text editors, with simple BASIC-like commands. There are many different macro scripting languages, often developed to automate functions in a particular application such as a database or paint program.

Some macros are recordable; in other words, the user turns on a record feature in the software, then performs a number of operations which are to be used frequently in the same sequence, then turns record off and gives the macro a name. The sequence of events is stored and can be invoked later with a name or hot-key sequence. Batch and job control languages can be used to write macros. Perl is a good programming language for writing powerful macros for many types of applications, including Web site automation.

macrobend A significant bend in an optical fiber. Optical fibers are quite slender and somewhat flexible, but due to the nature of the transmission of light, they do not have the small-radius 180° (or more) bending capability of many types of wire and care must be taken not to snap or significantly stress the structure of the fibers because this will compromise their transmission capabilities. Macrobends can be found in installations where the fiber must go around corners to reach the locations desired or may occur in fiber splicing trays.

The degree of bend possible and considered significant depends upon the radius of the bend and the thickness (core diameter) and composition of the individual fiber(s), on the one hand, and the length of the waves in relation to the diameter of the fiber, on the other. When the radius of curvature of a fiber is large compared to the diameter of the fiber core, macrobend losses may occur. See macrobend loss, microbend.

macrobend loss Signal attenuation through absorption in an optical fiber transmission caused by curvature or macrobend in the fiber. The amount of loss is due to many factors, including distance, the diameter and composition of the fiber core, and the wavelength of the optical transmission. Slender fibers are generally more susceptible to bend losses than thicker fibers. Longer wavelengths tend to be more susceptible than shorter wavelengths. As the bend increases, the mathematical/physical relationship between the wavelengths "bounced" along within the fiber core is changed and a critical angle is reached at which the light waves exit the fiber and are lost. Macrobend loss may be reduced by increasing the core diameter of the fiber, but there is a tradeoff with fiber sensitivity, weight, cost, and the capability of the fiber to bend physically. See attenuation, loss, macrobend.

macroblock sampling A compromise technique used in video phone systems to provide a recognizable image in spite of slow transmission media. Standard telephone modems are too slow for full-screen, full-motion video images. By using an averaging system to sample the image, the image information can be sent more quickly. The smaller the blocks and the more frequent the sampling, the better the image fidelity, but the slower the processing. An alternative to macroblock sampling is wavelet video compression, which may provide better images through frame-by-frame compression.

MAE 1. See Metropolitan Area Ethernet. 2. See Merit Access Exchange.

MAE East The largest Metropolitan Area Exchange, located in Washington, D.C. The MAE is a ring system which provides Internet Service Providers (ISPs) with a relay point for exchanging packets with friendly systems through switched and shared Fiber Distributed Data Interface (FDDI) and switched Ethernet communications services. A MAE is, in a sense, a gargantuan wiring closet with thousands of lines of cables, switches, routers, and connections interconnecting many public and private network installations. The MAE system provides access and interconnections, but doesn't provide ISPs with political connections with the other services on the system. These have to be individually arranged by each ISP. See MAE West.

MAE West A Metropolitan Area Exchange established in 1988 in San Jose, California, providing switched and shared Fiber Distributed Data Interface (FDDI) and switched Ethernet communications services. Originally operated by Metropolitan Fiber Systems, MFS merged with Worldcom Communications in the mid-1990s to provide expanded nationwide services. MAE West interconnects with the Ames Internet Exchange (NASA) and well-known networks like CERFnet, BBN Planet, MCI, and others. See MAE East.

magic signature A file integrity mechanism built, for example, into PNG graphics files to detect file corruption. The signature is typically incorporated into the file header or the early part of the file image. Different levels of sophistication can be designed into magic signatures, with different byte lengths. See cyclic redundancy check.

Magic Wand In navigation and location applications, an electronic device proposed by Egenhofer and Kuhn for geographical information system applications. The Magic Wand would incorporate a Global Positioning System (GPS) receiver and a gyroscope to provide location and orientation information relative to the position and orientation of the wand. This information could be cross-referenced with a knowledge base including a digital terrain base enabling the user to find information on a topographic feature (e.g., a specific mountain). With a cell phone connection to the local area network or the Internet, the possibilities become vast, with the capability of user public information servers to fill in further details on features pointed to with the Wand.

Magic WAND Magic Wide Area Network Davis. A project to promote wide area networking in the University of California, Davis, campus community, especially to off-campus locations that serve its staff and students. The project was initiated in 1997.

Magic WAND Magic Wireless ATM Network Demonstrator Project. A joint European ACTS project to extend ATM network access transparently to mobile communications users. It is one of the larger ACTS projects. The project endeavors to specify, demonstrate, and promote standards for a wireless access system that retains the benefits of multimedia wired ATM networks. The project was described in the mid-1990s with trials and workshops conducted in 1998. Communication was aimed to be in the 5-GHz frequency range at transmission speeds of about 20 Mbps at ranges up to about 50 ms. Documents describing the project are available for public download. http://www.tik.ee.ethz.ch/~wand/

Magic WAND network A global community support group, initiated in 1996 in Japan by the mother of a disabled child to help others in similar situations through mutual support and communication. The network is also called Community-Based Rehabilitation.

Magic Wand Speaking Reader A Texas Instruments (TI) optical reader product released in the early 1980s. It incorporated text-to-speech capabilities for barcode-readable electronic books for children. As a scanner is passed over the barcode, the Magic Wand Speaking Reader strings stored allophones (speech units) together to generate a spoken message.

magnesium oxide A material suitable for insulating against water and heat when compressed around a conducting wire. When used in conjunction with copper wires, it is known as mineral-insulated copper-sheathed (MICS).

magnet 1. A body, person, or situation with attracting properties. It is called charisma when a person has "magnetic" qualities. 2. A body that produces an external magnetic field which can attract magnetic materials such as iron. Natural magnets are known as lodestone. Magnetic properties were described by

M

the Greeks at least as early as 60 B.C. Steel will hold magnetic charges for a long time; iron can be magnetized, but retains the magnetism to a lesser degree than steel, as do nickel and cobalt. Materials that magnetize readily, but lose the property quickly, are useful as cores for various electrical devices. Magnets are used in many industrial applications, generators, monitors, speakers, compasses, and many more. See electromagnet, gauss, lines of force, lodestone, magnetic field, solenoid.

magnetic bubble memory A form of magnetic storage in which the digital data are stored in "bubbles," small circular regions called *magnetic domains* on a thin film of magnetic material that is selectively polarized. Since the presence or absence of a bubble can be used as a binary toggle, ones and zeroes are easily represented. Magnetic bubble memory is nonvolatile through permanent magnets; the stored information is retained when the system is powered off.

Magnetic bubble memory was used in the early 1980s when other types of storage were expensive and unreliable, but was superseded by random access memory (RAM), floppy diskettes and hard drives by about the mid-1980s. Newer forms of bubble memory use circuits associated with crystals, an evolution of the older magnetic bubble memory that is not destructively read and rewritten. See magnetic core memory, random access memory.

magnetic circular dichroism MCD. Dichroism means having two colors. Circular dichroism (CD) is a phenomenon occurring when optically active matter absorbs circularly polarized light with a slight difference in left- and right-handed directions. In terms of polarization, it has the character of "ellipticity" (a phenomenon distinct from optical rotation) and occurs in certain asymmetric materials. Dichroic minerals may appear to be different colors depending upon the angle at which they are viewed. A spectropolarimeter can be used to measure CD. Nunes et al. have suggested the use of coherent laser-induced thermal grates for ultraviolet circular dichroism spectroscopy.

Magnetic circular dichroism is exploited in various spectrometry and X-ray technologies and is useful in exploring the properties and origins of magnetism.

magnetic core memory A type of random access memory developed at MIT in the late 1940s and early 1950s. Jay Forrester was head of the Whirlwind computer project, a system built for realtime control applications that spawned core memory technology when vacuum tube systems were still prevalent.

Magnetic core memory consists of an array of ferrite toroids (cores) with wires passing through toroids at each junction in the array, each of which represents a bit of memory. These toroid bits can be stimulated by current of sufficient intensity to alter the magnetic polarity of a specific point in the array, which can thus be made to toggle between two polarized states to represent binary ones and zeroes. The array configuration makes it possible to pinpoint a specific toroid by its X and Y coordinates and modify the polarity by running a portion of the current through a wire in each axis, thus not changing the positions surrounding the selected toroid but running sufficient current through the selected toroid to change its state. Like magnetic bubble memory, magnetic core memory is nonvolatile; it retains its data if the system is powered off.

By the 1970s, core memory was being gradually replaced by semiconductor technologies except for a number of specialized scientific applications. See magnetic bubble memory, random access memory.

magnetic detector A device designed to receive electromagnetic waves, pioneered by Rutherford and further developed by G. Marconi. The magnetic detector comprises a small induction coil with primary and secondary coil windings surrounding a glass tube with a soft iron wire, rather than the usual common soft iron core. The wire is connected to itself in a loop that runs outside the glass tube and coil windings and is set to move continuously through the tube. A magnet is installed near the secondary (outer) winding, adjacent to the circulating wire. This fairly sensitive detector was used in conjunction with early telephone receivers, but never came into wide use and had almost disappeared by the early 1900s. See detector.

magnetic disk A circular device coated on the surface with magnetic particles that can be rearranged to encode information. A number of computer storage devices, including floppy disks, cartridges, and hard drives, use magnetic disks. The disk shape is favored because it allows the disk to be rotated very rapidly under a read/write head. By controlling the position of the read/write head, fast access to any portion of the recording surface can be achieved, called random access.

Early magnetic floppy disks were vulnerable to damage because they had an opening for the read/write head where the disk might be inadvertently touched or scratched, and pliable coverings that could be bent, thus damaging the disk. With the commercial introduction of the 3.5-in. floppy diskette around 1983, these sources of trouble have been removed.

The chief disadvantage of magnetic disks is that they can be damaged by exposure to magnetic sources (be careful to keep them away from monitors and speakers), which gives them a somewhat limited shelf life. See gamma ferric oxide, superparamagnetic.

magnetic equator Similar to, but not coincident with, the Earth's geographic equator; also called the aclinic line as it is the point at which a dip needle is at zero (90°), between its two vertical positions. See agonic.

magnetic field The region of external influence associated with an electromagnetic body in which these forces can be detected or exhibit a measurable influence on magnetic materials or instruments. The intensity of a magnetic field is described in terms of the number of lines of force passing through a specified area, although the field is conceptualized as continuous. The influence of a magnetic field can be seen by holding a magnet near small magnetic objects. A magnetic field can be induced in certain materials by running current through them. See flux, gauss, lines of force.

magnetic field modulation MFM. A means of modulating a magnetic field very close to the recording surface of an optical disc such that the polarity of the magnetic field can be switched at very high frequency. This produces marks on the recording portion of the disc that are narrow and tall (crescents), enabling higher bit densities per unit area and, as such, the density is no longer limited by the wavelength of the laser reading the data after it has been recorded. The crescent-shaped aperture is formed between a front and rear (double) mask and crosstalk is reduced by this arrangement.

High-speed polarity changes enable the disc to be written in a single pass, thus significantly increasing write speeds. In earlier optical disc writing technologies, the magneto-optical coil had to be larger because of its distance from the recording surface. This limited the switching frequency due to magnetic inductance in the coil and hence the recorded marks on the disc could not overlap, limiting areal density.

With MFM write mechanisms, the bit density is limited not by the wavelength of the laser, but by the ability of the mechanism to resolve the individual marks in the recorded surface. Heat applied to the read-out layer which magnifies the data in the recorded layer provides higher resolution and supports increased disc capacity. By using a mask in conjunction with the temperature distribution in the recording layer, individual bits can be isolated even if the laser light covers more than one mark.

Newer laser technologies (e.g., blue lasers) will likely increase resolution even further. In combination with MFM, higher-resolution lasers will support very high capacity magnetic super resolution (MSR) discs. See overcoat-incident recording, surface-array recording.

magnetic induction The characteristic of certain permeable substances to become magnetized when placed near a magnetic source, without coming in direct contact with that source. Thus, a steel bar does not necessarily have to touch a magnet to be magnetically influenced by that magnet, or to influence other materials in the vicinity of the bar by induction. See lines of force, magnet, magnetic field, magnetic super resolution.

magnetic north The northerly direction in the Earth's magnetic field near, but not corresponding to, the north geographic pole, to which north direction-seeking poles of magnets are attracted. Thus, what is called magnetic north is actually Earth's south magnetic pole, located in northern Canada, since the north-labelled compass needle will orient itself toward the south pole of the Earth. See declination.

magnetic storage A medium designed so magnetic materials within it can be dynamically aligned and realigned to hold encoded data. Floppy diskettes, hard drives, and audio/data tapes are various forms of magnetic storage. Magnetic storage is inexpensive and very convenient in that it can be easily rewritten; however, it is subject to loss over time through superparamagnetic phenomena, and may be damaged by proximity to equipment with magnetic components, such as monitors. See bulk eraser, superparamagnetic.

magnetic stripe Typically a narrow strip on a portable medium such as a bank card or ID card, which is encoded with information of use to the cardholder. When inserted into a magnetic stripe reader, such as a cash machine, the information is used by the system as authorization for access and various transactions.

magnetic super resolution, magnetic-induced super resolution MSR. A technology for enabling very high storage capacities on magneto-optical media. MSR capacities can be up to ten times the bit density of earlier mechanisms. MSR is based on IRIS Thermal Eclipse Reading (IRISTER) technology, developed by Sony in 1991. Vendors such as Fujitsu introduced commercial versions in the mid-1990s, based on land groove recording techniques. They could hold 2 Gbytes on a single-sided disc and 4 Gbytes on a double-sided disc and be read by existing magneto-optical drives. See magnetic field modulation for a fuller explanation. See GIGAMO, land groove recording.

magnetic tape A narrow, very long magnetic sequential data encoding medium used for audio tapes and computer data backup tapes. In the late 1970s and into the early 1980s, magnetic tape storage was commonly used on microcomputers for storing and retrieving applications programs and data. Due to its slow, sequential nature, tape drives were soon superseded by 8-in. floppy diskette drives, except for backups or specialized applications.

magnetite A form of iron ore, readily magnetized, called *lodestone* when it is already magnetic as it comes out of a mine. See magnet, lodestone.

magneto An apparatus in which a magnet is put close to a wound helix or coil of conductive wire (usually copper) to stimulate a momentary electric current that changes direction when the magnet is withdrawn. Similarly, a magnetizable material may be associated with a coil and periodically stimulated by a magnet to produce the same effect. This is a simple magneto-electric machine. An alternator is a common application of a magneto that employs permanent magnets to generate ignition current for engines.

The discovery of magneto effects was an important step in telecommunications history. A. Graham Bell described its significance in a letter written in August, 1875:

"... the discovery of the Magneto – electric current generated by the vibration of the armature of our electro magnet in front of one of the poles – is the most important point yet reached....

... I feel sure that the future will discover means of utilizing currents obtained in this way – on actual telegraph lines.

I think some steps should be taken immediately towards obtaining a Caveat or Patent for the use of a Magneto-Electric Current.... specially as a means of transmitting simultaneously musical notes differing in intensity as well as pitch.

I can see clearly that the magneto-electric current will not only permit of the actual copying of spoken utterance, but of the simultaneous transmission of any number of musical notes (hence messages) without confusion.

The more I think of it the more I see that the method of making and breaking contact so many times per second – is only the <u>first stage</u> in the development of the idea.

When we can create a pulsatory action of the current which is the <u>exact equivalent</u> of the aerial impulses – we shall certainly obtain exactly similar results. <u>Any number of sounds</u> can travel through the same air without confusion – and any number should pass along the same wire.

It should even be possible for a number of spoken messages to traverse the same circuit simultaneously; for – an attentive ear can distinguish one voice from another – although a number are speaking together."

Historically, this letter shows the transitional period in Bell's thinking about telegraphy and the potential for telephony. Even though he mentions the "copying of spoken utterance," his main focus was sending a variety of coded tones over telegraph wires (a harmonic telegraph), not exclusively sending voice over phone lines. Since there were long debates over who invented the telephone and Bell retroactively claimed to have worked out telephone ideas a year before writing this letter, there are still questions as to whether Bell had actually sent voice transmissions over wires when he filed his famous harmonic telegraph "telephone" patent. Bell's first public demonstration of intelligible speech was not until 1876, with an instrument that worked in part due to the addition of a magneto-induction device. Clearly the magneto was recognized by Bell as an important discovery and was likely a key component in turning the harmonic telegraph into a voice-carrying telephone.

Bell's interest in magnetos was well-founded and shared by a number of inventors, some of whom had built magneto machines in the early 1870s. By the late 1870s, magnetos were being developed into small generators to create electricity for low-power appliances, including phone bells. The magneto enabled a subscriber to ring the operator by turning a crank on the phone to provide the necessary power. Thomas Edison shared Bell's interest in magnetos, patenting a number of dynamo-related technologies and, in 1880, a magneto signaling apparatus.

Currently, magnetos have potential for use in crank-powered light beacons or radio devices for backwoods communications and emergency rescues. If someone were lost in the woods with a magneto-based locator or handheld two-way magneto radio, the lost individual wouldn't be dependent upon batteries that wore out as long as he could turn a crank to generate additional current to power the device or to refresh rechargeable batteries. See electromagnet; Faraday effect; Gramme, Zénobe; induction; magneto phone; telephone history.

magneto-optical MO. Technologies in which light is used to detect or influence the magnetic characteristics of a medium.

In the context of communications technologies, magneto-optical refers to magnetic phenomena that can be controlled or read with optical technologies. Thus, a compact disc (CD) can be written by influencing the magnetic field associated with the materials within the disc and can subsequently be read by an optical laser pickup mechanism that senses the magnetic polarization of the encoded information.

Many important data technologies are based on magneto-optical phenomena, including removable high-capacity data storage drives such as the popular CD and DVD formats. Magneto-optical discs do not suffer from "bit rot" to the same degree as certain floppy diskette and recordable tape technologies which can experience data corruption over time, or from proximity to magnetic sources (e.g., monitors). Because a magneto-optical disc-reading head is optical, there is no direct physical contact with the recorded/recording medium and thus some of the problems of hard drive wear and tear are eliminated.

In general, vendors have sought to increase the numerical aperture (NA) of an optical lens to decrease the diameter of light incident or spot on the optic media in order to increase the track density and bit density of data on a disc. In other words, developers have worked hard to pack more information in less space in MO storage technologies. The ISO standard NA is 0.55 but companies like Maxoptix have achieved higher NAs (e.g., 0.8) with lower cost components that are readily available. See compact disc, Faraday effect, Kerr effect.

Magneto-Optical Filter Development Group MOF. A collaboration of astrophysical organizations dedicated to achieving high-time cadence synoptic magnetograms of the Sun. The group provides a public data archive of the synoptic data from the Magneto-Optical Filter installed in the Kanzelhöhe Solar Observatory.

magneto phone A historic phone mechanism based upon magnetic-electric effects for the generation of low power current. While some magneto phones were handheld with a button to disengage the instrument from the phone circuit, the more common ones were wall or desk mounted and employed a crank handle to generate electricity through a magneto to produce an alternating or "ringing" current. This sent a signal from the subscriber to a switchboard to ring a bell that summoned the operator. The operator would then crank the switchboard to ring the callee and patch together the connection. The power for the conversation itself was usually supplied by a "talk battery" connected to each of the phones.

Many of the old crank-handle magneto single-wire phone lines gave way to two-wire "common battery" systems in which the battery was located at the central switchboard. Yet, surprisingly, there were regions in the U.S. where magneto phones were in use right up until the 1980s! Magneto or battery power is no longer used in public phone systems except in some remote areas, because signal current to ring a bell or to notify the switching office that the phone is "off-hook" is now supplied through the phone line itself, and the connection is established automatically by dialing the desired number. See magneto.

Compact Magnetometer for Remote Sensing

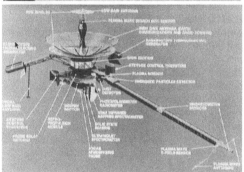

Top: The Lunar Portable Magnetometer mounted on the back of the Lunar Roving Vehicle during the Apollo 16 lunar exploration mission. Bottom: The Galileo spacecraft incorporated a large number of sensors, including magnetometers mounted on the long boom shown on the lower left in this artist's concept. [NASA/JSC images.]

magnetometer MAG. An instrument for detecting or measuring a magnetic field. A compass is a basic magnetometer. Magnetometers are used as navigation, measuring, and diagnostic instruments. They measure magnetic fields associated with the Earth, with various magnetic substances, and with electrical currents. A Barlington fluxgate magnetometer was installed in the early 1990s in Antarctica to measure magnetic field strength and direction. A magnetometer is one of the many sensors carried aboard the Galileo deep-space mission to the planet Jupiter, launched in 1989. It measures Jupiter's magnetic field, strong magnetosphere, and the characteristics and distortion in the fields caused by the interaction between Jupiter's magnetosphere and nearby bodies (satellites, asteroids, etc.).

Magnetic effects are fascinating to most people and a number of types of magnetometers can be built with simple materials (e.g., a plastic bottle and bar magnet), making this a good instrument to use as an educational tool. More sophisticated laser magneto-meters can detect subtle fluctuations in Earth's magnetic field and can be built by more ambitious hobbyists. Thus, a torsion balance, rare-earth magnets, a pair of coils, a laser source, and reflecting mirror, can be combined to create a sensitive Baker-Diverdi magnetometer.

magnetometer, induction IMAG. An instrument for detecting temporal variations in a geomagnetic field. An induction magnetometer may have several sensors with different characteristics incorporated into one housing. IMs can be used for a variety of purposes, including the study of electromagnetic characteristics of the ionosphere which, in turn, is useful in understanding transmission of radio waves. See magnetometer.

magnetomotive force Descriptive of the relationship of magnetic flux and reluctance through a magnetic circuit, somewhat analogous to electromotive force in an electrical circuit, although the magnetic circuit has a region of influence that differs from an electrical circuit in air. See magnet, reluctance.

magnetosphere A region within the solar wind flow in which there is an interaction between the solar wind and Earth's magnetic field. Fluctuations and interactions within the magnetosphere, especially as it relates to the ionosphere, are important to understand in terms of their influence on radio waves. The term is attributed to Thomas Gold, in 1959. The first image of the magnetosphere, taken by an ultraviolet camera, was captured during the Apollo 16 mission. Solar flares contribute significant energy to the magnetosphere and can substantially influence and disrupt radio communications. Strong flares may even disrupt power systems!

The magnetosphere also has its aesthetic aspects. When energy particles react such that electrons are directed into Earth's atmosphere, the result is the Norther Lights (Aurora Borealis) and Southern Lights (Aurora Australis) – eerie, waving emanations in the sky best seen at dusk.

NASA's Education Web site offers information on the history and characteristics of the magnetosphere. http://www.istp.gsfc.nasa.gov/Education/

magnetron In its simplest sense, a magnetron is a surface with magnets in or behind it designed to attract and trap charged particles. This basic implementation of a magnetron is useful in industrial environments where it may be combined with a vacuum chamber to coat surfaces such as metals and alloys in a process called magnetron *sputtering*.

In electron tube applications, a magnetron is a device that uses a magnetic field acting upon a diode vacuum tube to generate microwave frequency power. Magnetron tubes were especially important in the development of practical, long-range radar devices in the 1940s and later improvements on magnetron technology were incorporated into microwave ovens by Percy Spencer (Raytheon) in the mid-1940s. Traveling-wave tubes (TWTs) evolved from magnetron tubes in the mid-1940s.

Magnetrons are now commonly used to generate continuous-wave (CW) microwave-frequency signals.

M

With optional air- or water-cooled heads, different kilowatt output levels may be selected. See Barkhausen-Kurz tube, cathode-ray tube, cavity magnetron, Klystron, traveling-wave tube.

magnetron history Germany, Japan, and America all contributed to the development of the magnetron in the 1920s and 1930s. Its origins date back to the work of H. Barkhausen and K. Kurz in Germany, who described the shortest waves that could be produced by vacuum tubes, in 1920. The invention of the magnetron itself is attributed to Albert W. Hull, who described it publicly in the AIEE Journal, in 1921. August Zacek may have made similar discoveries in the early 1920s, as he ordered several special diodes which could have been used to study electron oscillations. In Japan, Kinjiro Okabe proposed a divided anode-type magnetron in 1928 that helped further the practical applications of the magnetron. In 1935, A. Arsenjewa-Heil and O. Heil described the concepts of velocity modulation and spatial bunching. In 1937, William Hansen and the Varian brothers designed and built a prototype Klystron tube, capable of generating microwaves; they announced their invention in 1939. In Japan, Kiyoshi Morita ordered magnetron prototype tube from JRC and there was close coordination between the Naval Research Institute and JRC in the mid-1930s for building magnetron tubes, directed in part by Shigeru Nakajima, resulting in a water-cooled, single-phase, 500-W oscillator, in 1939. In November 1939, John Turton Randall and Henry H.A. Boot announced the first cavity magnetron; within four years, it had become an important aspect of radar technology.

In the 1940s, Percy LeBaron Spencer noted the magnetron's ability to produce thermal energy in substances exposed to the microwaves, while researching magnetron radar applications. This led to the development of the microwave oven for which he applied for a patent in October 1945 (U.S. #2,495,429). Early microwave ovens were called "radar ranges."

As magnetrons became practical for industrial use after World War II, they were combined with generators for manufacturing applications such as surface coating (*sputtering*). Industrial use continues today in flat panel display manufacturing and other fabrication applications where frequency stability at a specific wavelength is important.

Magnetrons and solidstate electronics continued to be developed in the 1960s and 1970s, lowering the price and size of the components. Thus, microwave ovens became a common magnetron-based household product by the 1980s.

Magnetrons also have some novel uses in military defense. In the 1980s, the U.S. Air Force developed a tunable magnetron for imitating radar signals to draw enemy missiles from their intended targets.

Today, magnetron technology is important in satellite communications and industrial fabrication of optical components. See magnetron sputtering.

magnetron sputtering A physical vapor deposition process for coating metals, alloys, compounds, glass, and films using a coating chamber, a pump, and one or more magnetron generators to deposit fine layers of coating. The component being coated may be placed on a rotating table or a stationery jig. Fiber optic temperature sensors may be used as a process control component in sputtering chambers.

Since many optical instruments include coated components and thin films, sputtering is an important aspect of optics fabrication.

In the 1980s, Teer Coatings Ltd. developed an industrial coating system based upon two or more proximate magnetrons with opposing magnetic polarities. This configuration creates an active deposition zone within which the interacting magnetic fields can trap the ionizing electrons, resulting in superior coating efficiency over traditional single-magnetron systems. In the late 1990s, Makowiecki and Jankowski patented a sputtering process for producing thin boron-based films that have potential as ultrathin-bandpass filters and as low-radioactive elements in optical components.

MAHO See mobile assisted handoff.

mail bomb A vandalistic or retaliatory transmission sent through network email protocols with the intention of disabling an email address, the system upon which the address resides or, at the very least, to greatly inconvenience or annoy the recipient. Mail bombs take many forms, but the most common is a repeated message that eventually floods the recipient's email storage space or the storage space on the recipient's service provider's system, depending upon how it is partitioned.

Mail bombs are often sent to people who post absurd messages on public forums, or to originators of junk email (unsolicited email, especially of a commercial nature) to express the extreme displeasure of the recipients receiving the junk email. A mail bomb rarely solves the problem, however, since recipients often retaliate. See flame wars, spam.

mail distributor An agent, script, macro, or filter that takes incoming mail, evaluates the headers or other pertinent information, and distributes the mail accordingly. Thus, a single message might be forwarded to a number of users, different messages may be funneled to a single user, or groups may be set up to receive certain types of messages. The messages may include certain topics, which are keyed and processed, or may include priority or security information, which is handled accordingly. A mail reflector is the simplest type of mail distributor, which passes on mail with a minimum of evaluation and processing of contents (usually only the TO: header).

A mail distributor can be a big time saver when it is used to forward email to a mailing list or a discussion list. An address or database entry in the mail distributor can be used to expedite distribution to many recipients. This should not be used as a means to distribute junk email, more commonly known on the Internet as "spam," as there are regulations against this type of use, and users do not appreciate receiving it (many will boycott companies distributing commercial mail in this way). See discussion list.

mail filter A software utility or feature of an email

client, which automatically evaluates the sender, recipient, subject line, or content of a letter to sort it into designated categories. Mail sent to a specific domain name is often filtered by companies to individual employees' email accounts; junk email messages are often filtered out, and sometimes deleted unread. Some people filter personal and business mail into separate directories before reading the messages. Mail filters are a great convenience and worth the time it takes initially to set them up.

mail gateway Although there are standardized protocols for the distribution of email over networks, not all systems use the same protocols, and not all protocols are implemented in the same way. Thus, when mail passes from one system to another, if there is a mismatch, there needs to be a way to resolve the differences, or to tunnel or encapsulate the messages so they can reach the recipients. A mail gateway is a system in a computer network that handles mail channeling or the resolution of protocols.

mail list agent MLA. In SDNS Message Security Protocol (MSP), a mail list agent is one addressed by the message originator that represents a group of recipients. It provides message distribution services to the participants of that group on behalf of the message originator.

mail list key MLK. In SDNS Message Security Protocol (MSP), a mail list key is a token held by all the members of a mail list, or by a mail addressable group within the list.

mail reader A software program which permits email to be downloaded from a host system and read offline, so as not to incur connect charges or tie up a phone line. Most mail readers are actually mail readers and writers, and can be used to respond to the received messages or to compose new messages. They may also include filters to preorganize the mail before it is read, and a database interface which allows the messages to be organized and stored for later retrieval.

Some mail readers have been enhanced for use as online news readers as well, for following discussion threads on USENET and for posting to the various online lists. Posting is the same as sending email, except that the message will be publicly available and may be read by thousands or millions of readers. Pine, developed by the University of Washington, is one of the most popular mail readers. It is freely distributable, allows flexible processing of mail messages and files, and includes news reading and posting capabilities. See email, USENET.

mail reflector A mail node set up to pass messages on according to a predefined list. It does only the minimum processing needed to forward the mail to its intended recipients. For information on more sophisticated processing, see mail distributor.

mail server A software system which manages incoming and outgoing electronic mail on a network. Mail servers vary in complexity and features, but most will check the validity of an address; queue, deliver, and store messages (or return them if no valid address is found); forward mail, etc.

Due to overwhelming increases in the quantity of junk email on the Internet, some of the newer mail servers will check the validity of the sending address, and reject the mail if the sender does not appear to be legitimate. This may result in the loss of some real email messages: for example, if someone is about to change email addresses and close out an old account, he or she may send email letting you know the new address, then subsequently close the account before the message reaches its destination. The server may reject the legitimate message. However, some consider the trade-off worthwhile, in order to deflect the thousands, or sometimes tens of thousands, of junk mail messages that now flood the systems of most ISPs. See email, mail gateway, mail reader.

mailbox The part of an email client/server software system that comprises addresses and files which store electronic mail. Web browsers and dedicated mail processing software programs typically enable the management of multiple user mailboxes. Many Internet Service Providers will offer multiple mailboxes to their subscribers (e.g., six mailboxes per personal account). Many business accounts offer unlimited mailboxes.

mailto A URL designation for the Internet mailing address of an individual or service. In a Web page, a mailto can be used to make it easy for a person with a Web browser to send a message to the person or organization mentioned. The format to set up the hypertext link is:

```
Click to send email to<A HREF =
"mailto:stan@company.com"> Stan</A>.
```

In the above example, the name Stan will be highlighted in the Web browser to indicate that it can be clicked. When it is clicked, the browser will launch the user's email client, usually inserting the destination address automatically (stan@company.com), and enables a message to be written and dispatched to Stan without closing down the browser. It's very convenient but is falling into disuse because of junk email abuse – robot Web crawlers have been designed to quickly seek out mailto addresses on millions of sites, automatically adding them to junk emailing lists. See RFC 822, RFC 1738.

Maiman, Theodore Harold (1927–) An American physicist at Hughes Research laboratories who developed and patented a ruby-based laser. He described his pioneering laser research in *Nature* in 1960. In 1962, Maiman founded Korad Corporation to research, develop, and manufacture lasers. He eventually sold this firm to Union Carbide and formed Maiman Associates in 1968. Maiman is a member of numerous scientific professional organizations and has been awarded many prestigious prizes for his contributions to lasers. See laser.

main distribution frame MDF. A central wiring connection point in a larger more complex wiring system that includes more than one distribution frame. The main distribution frame is the one which connects the internal wiring with the external wiring. Within the premises, there may be secondary distribution frames in each department or each floor, depending upon the electrical needs and building configuration. See distribution frame.

main memory In a computer, there are sometimes a variety of types of memory, and there may be more than one memory bank. On some systems, where all the available memory is addressable by the system without significant restrictions, the concept of main memory is not important, as all memory is main memory. However, some systems make a distinction between system memory and expansion memory, and it may not be possible to address all the memory as one contiguous area. These systems treat the first memory, that which is addressed by default or used as first priority, as main memory. Extra memory for video display and other specialized uses is not considered main memory as it is not used as general-purpose storage by the system.

main station This is a loosely defined phrase because it is context-specific. The people within a system often have a tacit understanding of main station facilities and the definition varies widely, depending upon the industry. 1. In telephony, a user telephone set or terminal with a unique call designation used for originating calls and accepting calls on an exchange. If there are extension phones with the same phone number, one is usually designated as the main station. 2. In computing, an "intelligent" workstation (as opposed to a "dumb" terminal, for example) installed with a full set of capabilities or which functions as a primary or secondary server. A main station may also be a terminal with access to resources that may be otherwise restricted within the local facility, such as scanners, printers, modems, etc. 3. In broadcasting, a primary sending or receiving station, as opposed to a specialized or lower-resource relay station. A main station is more likely to be staffed or to have significant technological or broadcast power capabilities compared to other stations in the system. A main station is not always the largest or best-equipped station in a system; it may also be a clearinghouse or storage unit for a significant number of broadcasts or broadcast recordings, or it may be the highest-power station or one located on the highest prominence in the network.

mainframe The terms mainframe, miniframe, and workstation are all relative. The most powerful computers in the world are called supercomputers, and the less powerful computers that are above the consumer or workstation price range are called minis or miniframes. Mainframes fall between these two categories. In general, mainframes are typically the larger, more expensive, more powerful, faster systems with more storage capacity and the ability to handle many users on a network. Workstations and microcomputers are often used as smart terminals in conjunction with mainframes. Mainframes are used in larger educational institutions, large businesses, and scientific research facilities. Current consumer-priced desktop microcomputers are more powerful than the mainframes available 15 years ago.

mains A primary commercial alternating current

(AC) power supply. In North American domestic power grids, mains power is typically single-phase power routed through a breaker box to internal wiring. Triple-phase power is also available in some areas, usually carrying higher voltages. Mains electricity is used to power appliances, lights, heating units, industrial facilities, safety devices, etc.

Mains AC voltage for North America, South Korea, and parts of South America is 100 to 120V at 60 Hz. In Japan, it is 100V at 50 and 60 Hz. Mains AC voltage in Europe is around 220 to 250V, depending upon the country, with a target goal of 230V at 50 Hz by 2003 for unified Europe. One of the reasons for the many different types of electrical plugs in different countries is to prevent electrical shocks, fires, or damage to components resulting from the varying electrical properties of the mains power.

Increasingly, radio frequency (RF) signals are sent over mains wiring within buildings. This enables consumers to transmit audio, visual, and control (e.g., X10 components) signals through the wiring to other locations in the building without running dedicated wires or worrying about low-power wireless signal attenuation through walls and over distances. While this was not a problem in the past, the increasing use of mains for secondary signaling may become a concern as a source of interference.

In some cases, such as industrial applications, the power coming from the mains may not be exactly what is needed. The power may be conditioned to meet specific industrial or electronic needs. For use with sensitive electronic equipment, mains energy is often channeled through surge suppressors and transformers to filter and convert the energy. Many electronic devices require only 4 to 12V, far less than is coming from the mains supply and would be burned out without appropriate conversion.

Power does not always come out of the National Grid mains line; sometimes it is directed into the Grid. Alternate energy sources such as wind turbines sometimes generate more electricity than is needed and the surplus may be directed to the mains, depending upon power agreements and regulations in the area.

maintenance termination unit MTU. An electronic diagnostic device installed at a line termination unit (LTU) of a premises installed with telecommunications equipment, usually by the line provider. The MTU typically performs circuit tests for short circuits or open circuits that may be remotely monitored. This aids network operators in locating the source of a fault and determining whether the problem exists on the customer or connection side of the LTU. Multiple MTUs may be installed along a subscriber line to further pinpoint fault locations.

Major Trading Area MTA. A service area designation adopted in the early 1990s by the Federal Communications Commission (FCC) based upon an older Rand McNally classification of U.S. metropolitan regions. MTAs were identified by the FCC to administer and license wireless Personal Communications Services (PCS). There are over 50 MTAs in the U.S., built from contiguous Basic Trading Areas (BTAs) from almost 500 BTAs. Regional designations are somewhat important in administering wireless services, since the frequency ranges must be reused as efficiently as possible to provide service to as many areas and individuals as possible.

The FCC provides a Market Area cross reference that enables a user to search by MTA, BTA, or state and county to find corresponding markets. Thus, an MTA can be used to find corresponding BTAs or counties. For example, a search of the New York MTA yields a list of 20 BTAs, from Albany to Watertown, which are further linked to a list of counties within the BTA. http://www.fcc.gov/

Majordomo A widely used software program that automates the management of Internet discussion lists, developed and licensed by Great Circle Associates (GCA). It enables the remote administration of email list subscriptions, electronic mail messages, digests, and archives. Thousands of Majordomo lists are online, many with tens of thousands of subscribers. Lists may be open, private, or moderated. List management is handled through electronic mail (email). Some people call them mailing lists, but it is best to make a distinction between *discussion lists*, which are for the exchange of information and debate, and *mailing lists* which are used by marketing agents to email advertising messages.

The name is derived from head domestic or butler, the traditional manservant or master who handled visitors and the oversight of a house (domicile). Similarly, Majordomo does your bidding and handles your affairs, leaving you free to concentrate on communications with discussion list guests.

Majordomo source code is written almost entirely in Perl, making it possible for programmers to modify the source for their needs within the terms of the Majordomo License Agreement. Majordomo runs on a wide variety of Unix platforms and a Web interface is available as an add-on. Majordomo source code is available for free download within the terms of the licensing agreement. Version 1.94.5 was released in January 2000 and Jason Tibbits is developing Majordomo version 2. See LISTSERV. http://www.majordomo.com/

Make Busy A subscriber service or feature of a phone that causes the line to send a busy signal to an incoming call. This is like taking a line off the hook without removing the handset. While this might seem like an odd thing to do, it can actually save people money. If a business has more phone lines than agents ready to take calls (e.g., at lunch time), it may be better to send a busy signal than to let it ring indefinitely or to answer and put the caller on hold for a long time (which people don't appreciate). If the callers are dialing long distance, it can save them the toll charge, since the call is not answered.

Make Busy is also helpful to the receptionists on duty, since it can reduce the number of incoming calls to a manageable level. Depending upon the service, Make Busy can be assigned to one or more lines within a hunt group. See Ring Again.

malicious call A telephone call with annoying, abusive, obscene, or threatening intent. In many areas, malicious calls are unlawful, but it may be difficult to prove the malicious content of the call, especially since wiretapping laws generally prevent the caller from being taped without his or her permission.

In general, telemarketing calls are not considered malicious unless the caller is promoting illegal products or promoting them in a harassing or intentionally deceptive manner. Upon seeking help with malicious calls, you will generally be advised not to take the phone off the hook, but to answer it repeatedly and to immediately and quietly hang up, no matter how many times it takes. If the malicious caller utters what appear to be genuine threats, it is advisable to inform the phone company, the police, and other members of the household or office. If the malicious call appears to involve fraud, inform the phone company, the police, relevant consumer associations, and other members of the household. Some phone companies have call investigation centers that can assist the police with trapping and tracing persistently malicious calls, especially those of a threatening nature.

Malus's law When a beam of light that has already been once polarized by reflection hits a second surface at the polarizing angle, the intensity of the beam varies as the square of the cosine of the angle between the two surfaces. See Brewster's angle, Snell's law.

MAN See Metropolitan Area Network.

Management Information Base MIB. A set of data modules which contain the definition of a related set of managed object types. In SNMP management systems, it contains the logical names of informational resources on the network. In SONET network implementations, objects in the MIB are defined with a restricted subset of Abstract Syntax Notation One (ASN.1) as to their name, syntax, and encoding.

Management Information Services MIS. Corporate communications professionals whose job is to facilitate the acquisition, flow, use, storage, and retrieval of information within an establishment.

Mance, Henry Christopher (1840–1926) A British engineer who adapted the Indian heliotrope to a heliograph daytime signaling system using mirrors mounted on tripods. The angle of the mirror could convey line-of-sight dot and dash signals up to 100 miles. This system was used for military communications for several decades.

Manchester encoding An encoding scheme commonly used for baseband signaling in coaxial cable transmissions, especially 10Base-T network systems. There are variations to the encoding, but a typical differential Manchester employs a voltage transition in the middle of a bit period. A zero is represented with an additional transition at the beginning of a bit period. A one is represented with no transition at the beginning of a bit period.

There is a tradeoff between bandwidth and binary coding, as the coding consumes part of the bandwidth. In Manchester encoded transmissions, the amount of useful bandwidth is about twice the encoding signal. The Manchester encoding scheme is simple but useful, and can be used as one type of passband signal.

Manchester Mark I An early large-scale computing machine designed and built by Fred Williams, Tom Kilburn, and Max Newman in the late 1940s. It was significant in its ability to store programming information. The earlier prototype for this machine was colloquially called "Baby." The Manchester Mark I was composed of more than 1000 vacuum tubes. Input and output were communicated through switches, paper tape, and a teleprinter.

mandatory access control MAC. In network security, an access control service that enforces a security policy based upon a comparison of security labels and security clearances. Thus, access to resources can be controlled based upon the sensitivity of the desired information and the formal authorization of entities to access information of that level of sensitivity. An entity with access may not on its own enable other entities to access a resource made available to it.

Mandelbrot, Benoit (1924–) A Polish mathematician who emigrated first to France in 1936, and later to the United States. Mandelbrot extensively researched areas of complex geometry which have come to be known as fractal geometry. At least part of his thinking coincided with, or developed from, the work of G. Julia, who published important mathematical observations on the iteration of rational functions in the early 1900s. One family of fractal images called the Julia set is named after this predecessor.

Mandelbrot's early publications on fractals include "Les objets fractals, forme, hasard et dimension" (1975) and "The Fractal Geometry of Nature" (1982) which created an enormous stir, especially in North America, and fueled much of the fractal imagery since generated on computers.

Certain diffraction lenses have been found to exhibit interference patterns of a fractal nature.

mandrel In production fabrication, a cylindrical or tapered core or spool over which materials are pulled, slung, or wound. Mandrels may be used for temporary storage or for facilitating the dispensing of cable components (e.g., in a fusion splicer). See creel.

MANIAC A historic large-scale computer developed in the mid-1950s by the Los Alamos National Laboratory. The construction of its successor, the MANIAC II, inspired professors at Rice to initiate the Rice Computer Project. See Atanasoff-Berry Computer, ENIAC, Rice computer.

manipulation detection code MDC. Software algorithms for detecting whether data or processes have been changed or otherwise manipulated in the interim, over a set period of time, or during a dynamic session. Manipulation detection code is intrinsic to many types of workgroup applications in which two or more people may be using, changing, or interacting with the same application or database; the administration of the interactions must be carefully organized so they don't clobber each other's processes or data.

Manipulation detection code is also an important aspect of monitoring networks for intrusion or malicious tampering. The code may be something as simple as a log that keeps track of logins or file changes or failed password attempts or as sophisticated as specific sequences of data interposed into a system, program, or file that are assessed and possibly manipulated to detect tampering of various kinds or to send messages to the system administrator as needed.

MAP See Media Access Project, an important organization representing the public good.

Marconi antenna An antenna that requires the ground, or a large object to which it is mounted (such as a vehicle), to aid in resonance conduction. In other words, it is not a stand-alone antenna like a Hertz antenna. Marconi antennas are commonly used in amplitude modulation (AM) broadcasts. See antenna, Hertz antenna.

Marconi detector An adaptation of the Branly detector to which G. Marconi added a vibrating source to quickly set the coherer back to zero or nonconducting status. See detector.

Marconi, Guglielmo (1874–1937) An Italian who as a youth demonstrated wireless telegraphy to his mother in an attic laboratory in 1894, and experimented with radio waves in 1895. With further support and assistance from his mother, Annie Marconi, the 22-year old Marconi filed for a patent and demonstrated radio communications in London the following year, and received a British patent in 1897. Marconi traveled and lectured extensively, and kept in touch with other inventors in the field of radio communications.

Marconi's first communications were over very short distances, but in 1901 he showed that radio signals could be sent across the ocean between Canada and England, a distance of over 3000 km. He continued for many years to devise improvements in the technology, and to put them to practical application. In 1909 he was awarded a Nobel prize in physics along with K. Braun. Marconi began broadcasting from Marconi house in 1921 under the famous 2L0 callsign. See Braun, Karl Ferdinand; Murgas, Josef; Tesla, Nikola.

MARECS A European maritime satellite communications service established in the early 1980s; it is similar to the American MARISAT system.

Marine Radio Operator Permit MROP. A radiotelephone permit issued by the Federal Communications Commission (FCC) which is required for the operation of radiotelephone stations aboard certain Great Lakes vessels and for certain aviation and coastal radiotelephone stations. The MROP does not authorize the operation of AM, FM, or television broadcast stations. Issuance of the 5-year, renewable permit requires passing the Written Element 1 exam, which covers basic radio law and operating procedures. See General Radiotelephone operator license, Restricted Radiotelephone Operator Permit.

Marine Safety Office MSO. An office of the U.S. Coast Guard, located in Mobile, Alabama. The Coast Guard Inspection Department operates from a location near the main office. The MSO is responsible for protecting life, property, and the environment along the coastlines and navigable waterways of Mississippi, Alabama, and northwest Florida.

MARISAT Maritime Satellite. First launched in 1976, MARISAT was designed to provide mobile communications services to the U.S. Navy and other maritime clients. The European MARECS system is similar.

Maritime Identification Digits (MID). See Maritime Mobile Service Identity.

Guglielmo Marconi – Radio Pioneer

Guglielmo Marconi was a significant pioneer of radio technologies. [National Archives of Canada Marconi Company collection.]

A copy of a Western Union telegram from Marconi to Alexander Graham Bell, thanking him for his invitation to visit his summer home in Nova Scotia, December 19, 1899. In the end, Marconi declined this particular visit because Bell's location was not on the ocean, so it sounds as though Marconi was considering transatlantic communications at the time, a feat which he successfully achieved in 1901. [Library of Congress Alexander Graham Bell Family Papers Collection.]

Maritime Mobile Service Identity MMSI. An administrative identifier allocated and issued by various national maritime safety authorities to marine vessels. The MMSI is a unique, internationally standardized, nine-digit identification number, similar to a radio callsign or telephone number. It may be programmed by the vendor of the equipment or, in some cases, the operator. It is associated with a Digital Selective Calling (DSC) number (a system that en-

ables group or broadcast calling). DSC radios may be linked with Global Positioning System (GPS) receivers. MMSI numbers were developed for compatibility with the public telephone system and the number facilitates the routing of data and voice transmissions.

If more than one radio is installed in a vessel, it is to be programmed with the same MMSI. If a vessel carries an Emergency Position-Indicating Radio Beacon (EPIRB), it may also be assigned the same MMSI as other radios on the vessel. MMSI may be used for ship-to-ship and ship-to-shore communications. MMSIs are not exclusive to boats; INMARSAT satellite terminals also use MMSI numbers.

Three of the digits of the MMSI, called the Maritime Identification Digits (MIDs), indicate the country of location or registration, while six digits are used to uniquely identify the station. The number may be used for emergency identification or for more mundane matters such as call tracking and billing.

On the international level, the International Telecommunication Union (ITU) maintains a database of the MMSI of every vessel, called the Maritime Mobile Access and Retrieval System (MARS), which is accessible online. Individual nations also maintain databases. In the U.S., MMSIs are recorded by the Federal Communications Commission (FCC) and the Coast Guard. Some nations assign MMSI numbers free of charge (e.g., Industry Canada) and some charge an application fee.

In spring 2001, in the U.K., the Radiocommunications Agency introduced new procedures for the issuance of MMSI numbers to facilitate the use of portable VHF Digital Selective Calling (DSC) radio equipment. Thus, vessels that cannot be installed with fixed radios can realize some of the benefits of MMSI and the Global Maritime Distress and Safety System (GMDSS) through portable equipment. Portable equipment is allocated unique MMSI numbers, regardless of the numbers assigned to existing fixed equipment associated with a vessel. Portable MMSI numbers in the U.K. are issued a 2359 prefix exclusive to mobile equipment. Due to the unique nature of portable equipment, national radio licensing authorities will need to be notified if an MMSI-assigned device changes hands. It is likely that other nations will institute similar policies and procedures to accommodate the rising number of portable communications devices.

Mark accelerators A series of pioneering accelerators leading to the development of the world-renowned Stanford Linear Accelerator Center (SLAC), developed by William W. Hansen, who had earlier contributed to the invention of the Klystron tube (used in a variety of microwave communications and imaging technologies). The *Mark I* accelerator produced a 6 MeV electron beam. The *Mark II* was used for research in nuclear physics and the *Mark III* for a high-energy physics program. See Hansen, William.

Mark I See Harvard Mark I.

Mark-8 A pioneer Intel 8008-based personal computer kit. The Mark-8, a scaled-down hobbyist cousin of the PDP-8, was described in a June 1974 issue of *Radio Electronics* magazine by Jonathan Titus. However, it didn't achieve widespread commercial success. See Altair, Intel, Kenbak-1, Micral, Scelbi, SIM4, Sphere System.

mark-to-space transition, M-S transition In telegraphy, the momentary change when the system reverses polarity, or changes from a closed to an open circuit. At this point, a small amount of delay must be taken into consideration, which can be plotted on a timing wave. The reciprocal is the space-to-mark transition.

MARS Multicast Address Resolution Service. In ATM networking, a protocol used in IP multicasting.

MAS 1. See Multi-Agent Systems Laboratory. 2. See Multiple Address System.

maser microwave amplification by stimulated emission of radiation. A type of laser technology developed in the late 1950s. An internally-modulating maser consists essentially of a laser light source, a pair of reflectors, and a modulator between the source and one of the reflectors. See laser, laser history for a longer explanation.

mask A screen, stencil, or other object superimposed between a surface and light, pigments, or other media put on that surface so only the unmasked portions are seen or affected. When used with light, a mask is known as a *beam block*. See knife-edge focusing, Ronchi grating.

mask, data In computer programming, a mask is a set of data, flags, or bits used as a filter or operator to affect only those bits of data that correspond to the mask template, or which are not included in the mask template.

masquerade attack An attack on a system by an entity posing as another entity that has authorization to access that system or resource. See spoofing, Trojan Horse, virus.

mass fusion splicer MFS. A mechanism for splicing multiple pairs of fiber optic filaments, typically through the application of heat from an electrode arc. This is particularly challenging as filament splicing is a precision task and getting individual filaments to meet production standards can be difficult. In the case of mass fusion splicers, heat fusion from an arc source must be distributed evenly and quickly and in such a way that heat applied to one region of filaments does not cause side effects to those nearby. Depending upon how the heat is applied, the maximum number of fibers that can be handled at a time is also limited by the arc width and heat distribution technology. Thus, getting even heat distribution over a wider arc and delivering it to more than a dozen fibers is a daunting precision job. S. Morita et al. have developed a custom arcing mechanism and splicer capable of handling up to 24 fibers at a time.

MFSs are commonly used to increase production levels and to create multiple spliced cables for assembly as fiber ribbons. The number of fibers in a fiber ribbon continues to increase, with 24 fibers now common, yet most commercial MFSs splice up to about 12 fibers, which means multiple batches are needed to provide all the fibers necessary to assemble a wide ribbon cable. See fusion splicing.

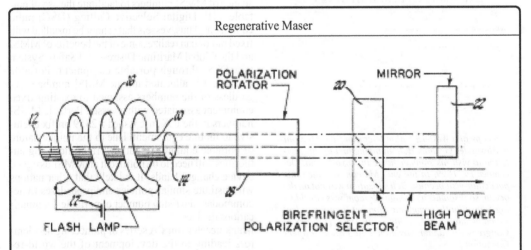

Regenerative Maser

POLARIZATION ROTATOR

MIRROR

FLASH LAMP

BIREFRINGENT POLARIZATION SELECTOR

HIGH POWER BEAM

The basic concepts of lasers were developed into practical devices in the 1960s and 1970s. This early solid-state maser design, based upon a ruby laser source, was developed by A. Vuylstere of General Motors Corporation in 1962–1963. A flashlamp serves as an excitation source for the pumped laser, with a pumping frequency of one per cycle.

The purpose of this design was to make masers more efficient by decreasing the pulse width of a laser beam while increasing power through the more concentrated beam. It works by varying the regenerative condition of the laser's reflective resonating chamber at specified times as a short-pulse transmission.

Many subsequent designs used the same general concept but further improved efficiency by modifying the interaction of the laser with the regenerative structures and processes within the laser cavity. By the late 1970s, with demand for laser technologies growing, designs utilizing repeated optical components in the form of chained amplifiers could provide higher-energy pulses. [U.S. patent #3,243,724, March 1966.]

Massachusetts Microprocessor Design Center
MMDC. An Intel microprocessor research and design center located in Shrewsbury, Massachusetts. In August 2001, over 200 Compaq microprocessor engineering and design employees joined Intel as part of an agreement for Compaq Computer Corporation to transfer key enterprise processor technology to Intel.

Massie Oscillaphone A simple type of loose contact electromagnetic wave detector long favored by amateur experimenters and educators. Two carbon blocks (battery carbon can be used) are set up adjacent to one another, about an inch apart, on a nonconducting base such as wood. The top surfaces of the carbon blocks are chiseled or filed so that they have a fine, thin edge. Holes are drilled through their surfaces, near the base, to provide room to insert a screw through each block, with the screwheads on the outside, to secure wires that connect with two binding posts. The top thin surfaces of the carbon blocks are wiped with a woolen cloth, and a light sewing needle or other similar conductor is laid across the top of the two blocks to create a contact between them.

When connected to a circuit including a battery power source, aerial, ground, and telephone receiver, an incoming radio wave will interact with the needle-carbon contact, causing the needle to adhere more closely to the blocks, lowering the resistance. This results in an increased flow of current which is translated into sound in the receiver. Further adjustments to the sensitivity of the needle can be made by placing a small magnet under the needle and adjusting its height. See detector.

mastergroup In analog voice phone systems, a hierarchy for multiplexing, organized as a series of standardized increments. See voice group for a chart. See jumbogroup for an illustration.

Matrix.Net A commercial service disseminating information on Internet hosts and providing Internet Performance Measurement products. Event Advisories on incidents and trends of particular importance to the Internet are provided on the Web site for free. Estimates and data on Internet hosts are based on raw data from the Network Wizards' global domain surveys.

MAU 1. See Media Access Unit. 2. See Multistation Access Unit.

Mauchly, John W. (1907–1980) An American physicist and engineer who collaborated with J. Eckert to build the historic ENIAC computer. See ENIAC.

Maurer, Robert D. (1924–) An American scientist and fiber optics pioneer who did graduate work at MIT and went to work for Corning Glass Works in 1952. As head of a research team that included D. Keck and P. Schultz, he succeeded in fabricating optical fibers that could carry far more information than existing copper wire links. Building on research from a number of groups in the early 1960s, the team developed a low-loss fiber waveguide that was practical for use as a telecommunications transmission technology.

Maurer was one of the first to dope silica with impurities (titanium oxide) to raise the refractive index

of the conducting core above that of the cladding so that the light beams would reflect off the cladding and stay within the conducting core.

In 1999, Maurer was coawarded the Draper Prize for his work in fiber optics engineering and, in 2000, Maurer, Keck, and Schultz were awarded the National Medal of Technology for their work at Corning. See Hyde, J. Franklin.

MAX See Media Access Exchange.

Maxim, Hiram Percy Founder of the historic Amateur Radio Relay League (ARRL), along with Clarence Tuska, a young fellow radio amateur, who became friends with Maxim after the older radio enthusiast had decided not to purchase radio equipment constructed by the precocious Tuska.

Maximum Transfer Unit MTU. In Internet Protocol (IP) networking, the largest size of IP datagram that may be transmitted through a specific data link connection. The MTU is not a fixed amount, but is a mutually-agreed value that can vary widely up to about 10 kilobytes. In a distributed network with a number of hops, a datagram may pass through nodes with different MTU sizes, necessitating queries and processes for handling the incoming data and its subsequent routing. If the relay or receiving MTU is smaller than the size of a transmitted packet, the packet must be subsected to segments smaller than the MTU and information about the process conveyed to the next link or recipient.

maximum usable frequency MUF. In a radio transmission signal path based upon propagating radio waves through Earth's ionosphere (e.g., through the Great Circle signal path from the eastern U.S. to Japan), the upper frequency level that may be usable. In general, higher frequencies are associated with lower refraction rates in the ionosphere. The MUF varies with terrain, region, and the influence of rays from the Sun.

Software programs can monitor and calculate the various factors that influence usable radio frequencies and will generate and display charts that can be used to aid radio operators and listeners. Since conditions constantly change, many software-generated maps are updated frequently, sometimes every few seconds or minutes, depending upon the system. Times associated with the maps are usually expressed in Zulu time.

maxwell An electromagnetic unit of magnetic flux – the flux-per-square centimeter equal to the magnetic induction of one gauss, or one magnetic line of force. It is named after J. Clerk-Maxwell.

Maxwell, James Clerk- (1831–1879) A precocious Scottish physicist who, building on the work of Faraday and Bernoulli and adding ideas of his own, contributed many important fundamental theories and equations related to electromagnetism and the nature of particles. He also made mathematical predictions about the composition of Saturn's rings that held up well over time.

Maxwell's equations A set of fundamental mathematical equations, originated by J. Clerk-Maxwell and further developed by Oliver Heaviside and

Heinrich Hertz, for expressing radiation and describing conditions at any point under the influence of varying electromagnetic fields. These concepts and equations are integral to many areas of science, and are of particular interest in understanding and developing transmissions media, antennas, and other basic building blocks in telecommunications. See Heaviside, Oliver; Hertz, Heinrich Rudolph; Maxwell, James Clerk-.

James Clerk-Maxwell – Mathematician

James Clerk-Maxwell is remembered for many of his mathematics and physics contributions related to fundamental laws and electromagnetism.

Maxwell's rule Every part of an electric circuit is acted upon by a force which tends to move it in a direction such as to enclose the maximum amount of magnetic flux.

Maxwell's theory of light In 1860, J. Clerk-Maxwell demonstrated that the propagation of light could be regarded as an electromagnetic phenomenon, the wave consisting of an advance of coupled electric and magnetic forces. If an electric field is varied periodically, a periodically varying magnetic field is obtained which, in turn, generates a varying electrical field and thus the disturbance is passed on in the form of a wave. Maxwell's theory predicted that the speed of light unimpeded was constant.

Maxwell's triangle An ordered representation of color relationships, in the shape of a triangle, developed in the late 1800s by physicist J. Clerk-Maxwell. His premise was that this model would contain all known colors. Red, green, and blue are identified as the three primary colors of light and are located in the three corners of the triangle. The colors progressively blend until, in the center, the combination of all the colors becomes white. A system of color notation was developed by laying a grid over the triangle. See color space, Munsell color model.

Mayer, Maria Goeppert (1906–1972) A Polish-born, American physicist who carried out fundamental research in models of the nucleus of atoms. For her independent work, she was awarded a Nobel Prize in physics, along with J. Jensen and E. Wigner.

MBone, mBone multicast backbone. See 6bone, backbone, multicast backbone, X-Bone.

MBS See Mobile Broadband System.

Mbus Message Bus. An open peer-to-peer coordinating infrastructure that provides integration for modular computerized systems design. The Mbus is especially applicable to "lightweight" distributed applications (e.g., limited-resource portable communications devices). It enables cooperation among modules serving a particular purpose while still supporting a variety of interoperable languages and communications standards. The Mbus framework is extensible and securable and can be implemented in a variety of multimedia and communications environments. It is intended to facilitate the design and assembly of complex systems out of simple components by providing a data channel through which application modules can find one another.

The Mbus is a local infrastructure providing transport layer functionality and addressing schemes including failure detection, session establishment and teardown, and component configuration. Mbus processes are message-based rather than object-based and are not programming language paradigm-specific. Components can be dynamically added, removed, or exchanged at runtime.

Mbus messaging may be unicast (to a specific address), broadcast (to all entities), or multicast (to qualified entities). The Mbus message itself includes a payload consisting of commands and their associated arguments/parameters that are processed by the messaged entities. The entities in the system periodically signal their presence to the Mbus group.

Message Bus profiles for local coordination and call control were submitted as Drafts to the IETF in February 2001 by Ott et al. See MBus, Multiparty Multimedia Session Control. http://www.mbus.org/

MCC Miscellaneous Common Carrier. See Radio Common Carrier.

McCahill, Mark P. Project leader in the development of the Gopher distributed networks query/search mechanism; Gopher was released by McCahill and Paul Lindner in 1991. It was one of the first accessible, nontechnical software applications that enabled teachers, researchers, and other professionals untrained in computer programming to access the storehouse of information that was accumulating on computer networks. Web search engines are, in a sense, the next generation version of Gopher for the Internet (text-based Gopher servers are still in use), providing point-and-click and graphical capabilities in addition to the basic query/search functions that aid in locating information online. See Gopher, Web browser.

McCarthy, John (1927–) A recognized pioneer in the field of artificial intelligence since 1955, McCarthy was one of the first to promote the basic concepts of computer timesharing in the late 1950s. McCarthy is also known as the originator of the LISP interpreted programming language that is used to automate computer-aided design processes and is popular in artificial intelligence research. See LISP.

McCaw Cellular Communications, Inc. A commercial communications services provider chaired by Craig McCaw, which was sold to AT&T in 1994 and renamed AT&T Wireless Services. McCaw is now collaborating with W. Gates et al. to develop the Teledesic satellite-based Internet system. See Teledesic.

MCF See Multimedia Communications Forum.

MCID malicious call identification (e.g., as in ISDN Q.81 and Q.731 number identification services).

MCNS See Multimedia Cable Network System.

MCNS/DOCSIS See Data Over Cable Service Interface Specification, Multimedia Cable Network System.

MCS-4 A significant early (1970s) chipset that inspired pioneer computer designers. See Intel.

MCVD modifed chemical vapor deposition. See magnetron sputtering, vapor deposition.

MD series A series of message-digest (MD) hash algorithms developed by Ronald L. Rivest that can be used, for example, to secure electronic mail communications.

The MD 128-bit algorithms have been incorporated into a number of significant data encryption systems including RSA Security Inc. cryptographic products. See Pretty Good Privacy, RSA Security Inc., RC6, RFC 1319, RFC 1320, RFC 1321.

MD Series Releases

Series	Description
MD2	128-bit one-way hash developed in the mid-1990s.
MD4	128-bit one-way hash that is faster than the previous MD4. MD4 was incorporated into P. Zimmermann's Pretty Good Privacy (PGP) 1.0.
MD5	128-bit one-way hash, an improved version of MD4. While reasonably secure, it has been suggested by Dobberlin that, if two files with the same MD5 hash were to be created (not an easy task), it might be easier to threaten the software. Leeming has suggested that a greater threat might be in finding two cryptographic keys with the same MD5 checksum, in which case the digital certificate for one could be used to access the other. PGP 2.0 and subsequent versions use this algorithm.

MD-IS See Mobile Data Intermediate System.

MDS-xxx A line of commercial digital switching products from Raytheon E-Systems.

MDT mobile data terminal.

MDX multidimensional extensions. In the context of the Microsoft Data Warehousing Framework, a syntax for querying multidimensional objects and data. MDX has a grammar similar to SQL.

mean opinion score MOS. A statistical quantification of reported subjective impressions. In other words, it is a value based upon people's perceived and stated impressions or preferences. MOSs are useful for assessing sensory impressions that are difficult to measure empirically, such as the quality or effectiveness of a perfume, massage, sound, or image. In audio communications, MOSs are used to determine whether a sound, such as speech, is pleasing, clear, or intelligible. In fact, the ITU-T has defined MOS more narrowly to focus on speech digitization and recreation and provides a rating scale from 1 to 5 for reporting the results. This helps programmers to tweak their Voice over IP (VoIP) software, for example, to balance file sizes and transmission speeds against sound quality. See P Series Recommendations.

mean time between failures MTBF. A performance indicator, the limit of the ratio of the operating time in a device to the number of failures as the number of failures approaches infinity. At the factory, test versions of a product are often subjected to extreme use to estimate in advance what MTBF rating might be under conditions of actual use.

mechanical splice A joining of two or more wropes, wires, or fiber optic filaments by twining, pressure, or proximity without the use of chemical or thermal bonding agents. Multifiber ropes may be spliced together by a number of different braiding or twining patterns to create a strong splice. Wires may be spliced by bending and folding back the ends, braiding, or spiraling the wires about one another. Both rope and wire pressure joints can be secured with tape or other mechanical coverings without twining, but the joint is typically fragile and may pull apart even with a small amount of pressure.

Fiber optic filaments may be joined by holding the carefully cleaved and aligned ends in close proximity and covering them with a dust and moisture proof splice joint that prevents rotation or strain to the joint. Unlike rope and wire, which may be braided and still be useful for most normal functions, fiber filaments cannot be braided to form a joint without eliminating the near-perfect end-to-end alignment that is necessary for maximizing light-carrying properties through the splice. Splice joint assemblies are usually intended for temporary fiber optic joints with fusion splicing preferred for permanent joints.

The choice of a mechanical or fusion splice depends upon the type of data that is transmitted, the amount of pulling and motion that is exerted on the splice joint, the frequency with which the configuration is changed, and the length of the cable run. Regular computer data and shorter cable runs are less subject to loss than high-end, broadband video signals or longer cable runs and may function well with mechanical splices. Patch panels that are frequently reconfigured are easier to change if mechanical splices are used. See fusion splicing.

Media Access Control, Medium Access Control MAC. Functions associated with the lower half of the data-link layer that governs access to the available IEEE and ANSI local area network (LAN) media (or medium, if there is only one). This layer supports multiple downstream and upstream channels. Devices such as network bridges are associated with the MAC layer (or sublayer in OSI).

Mechanical Fiber Splice

For a mechanical splice to be effective, the fiber filament ends must be precisely cleaved and aligned to form a continuous lightpath without gaps or particles that could cause back reflection (top). The joint must be firmly secured within a sturdy supporting structure (e.g., the Corning Cable Systems CamSplice™) to maintain the position of the filaments relative to one another (bottom).

Mechanical splicing is only recommended for joints that are well protected from strain and temporary joints such as patch panels that require frequent reconfiguration. For permanent splices, fusion splicing is generally preferred.

Media Access Control address, MAC address A MAC address is an important routing statistic widely used for managing data network transmissions. Route-related devices typically keep a list or table of MAC addresses which may be static or dynamic. Dynamic MAC address lists may be updated in a variety of ways depending on the stability of the configuration of a network and the size of a network.

The MAC address is used by a network bridge to determine whether a packet is to be forwarded. By copying an incoming source address to a MAC address table, the bridge builds up a port-related "picture" of device locations on the network. Since there are usually efficiency trade-offs on large networks, with machines being added and removed and MAC address tables becoming large and unwieldy, certain balancing mechanisms are built into bridge management. By limiting the lifespan of a MAC address, the system can be tuned to best serve the needs of the network. Thus, a *dynamic MAC address* that has not been used for a long time will be removed until a transmission is again received from that source. Low-use special purpose addresses (e.g., emergency systems) that must stay active can be assigned a *static MAC address* that isn't automatically deleted.

Media Access Exchange MAX. A system-level network access unit from Ascend Communications, into which peripheral cards can be inserted. A MAX can support multiple host ports or direct network connections, videoconferencing units, and remote LAN connections.

Media Access Project MAP. An important nonprofit, public interest telecommunications law firm that looks out for the First Amendment rights of individuals before the legal system and the Federal Communications Commission (FCC).

Over the years, broadcast agencies have been provided free use of the airwaves and, in return, have a legal responsibility to provide a portion of programming and resources for the public good. They are bound to uphold these obligations but may neglect them without citizen support groups like MAP, who take the time to lobby for the interests of the little guy. In recent years more free bandwidth has been broadly allocated to commercial broadcasters, particularly satellite broadcast frequencies.

It is important that citizens safeguard their rights, and that it be impressed upon the government, the FCC, and the broadcasters that these broader free permissions have inherent corresponding responsibilities. http://www.mediaaccess.org/

Media Access Unit MAU. In Token-Ring local area networks (LANs), a wiring concentrator that connects the end stations. The AU provides an interface between the Token-Ring router interface and the end stations. Also known as Access Unit (AU). See Multistation Access Unit.

Media Interface Connector MIC. An eight-pin modular RJ-45-8 plug. This resembles a common RJ-11 phone jack except that it is wider to accommodate connections for eight wires. This is the connector recommended for audio-visual applications by DAVIC specifications.

medium Earth orbit MEO. An orbiting region around the Earth into which certain types of communications satellites are launched, mid-way between low and high Earth orbits into which geostationary satellites are typically launched. The lifespans of medium-orbit satellites are about 10 to 12 years. Most medium-orbit satellites travel about 10,000 to 15,000 km outside Earth. A region called the Van Allen radiation belt between MEO and low Earth orbits (LEOs) is generally avoided. MEO satellites are primarily used for broadcast applications. See high Earth orbit, ICO, low Earth orbit, Teledesic.

Meet Me A commercial FTS2000 capability initiated by dialing an access number at a prearranged time, or as directed by an attendant, to establish a group conference call. Additional conferees can join a conference in progress with an Add On conference. It may be necessary to make arrangements for a Meet Me call several hours in advance, depending on how the system is administered. See FTS2000.

Meet-Point Trunk Telecommunication trunks configured for two-way traffic in jointly provided Switched Access Services (SAS), to interconnect End Offices and Tandems.

mega- (*abbrev. – M*) An SI unit prefix for 1 million, expressed as 10^6 or 1,000,000. To confuse matters, when used in conjunction with computer-related quantities, it often means 2^{20}, expressed as 1,048,576. The most common of these uses is in descriptions of computer storage capacity as megabytes (MBytes), in which 1 MByte is 1,048,576 bits. See kilo-.

Megaco Protocol A network media gateway control protocol used between elements of a physically decomposed multimedia gateway, that is, between a Media Gateway and a Media Gateway Controller. The Protocol was submitted as a Standards Track RFC by Cuervo et al. in November 2000. It provides a general framework suitable for gateways, multipoint control units, or interactive voice response units to interact. The protocol definition has common text with ITU-T Recommendation H.248.

A Media Gateway converts media from one type of network to a format compatible with another type of network. The gateway may be able to process and translate audio, video, or T.120 and can handle full duplex media translations, in addition to playing and executing media performances or conferences. Megaco Protocol connection model describes the logical entities/objects within the Media Gateway and can be controlled by the Media Gateway Controller. See RFC 3015.

megger An instrument for measuring values of very high resistance used, for example, for insulation resistance testing. See Wheatstone bridge.

Melissa See virus.

Melles Griot lasers A line of lasers from the Melles Griot optical company. The firm is a leading supplier of gas lasers and is well-known for helium-neon (He-Ne) lasers. Melles Griot also develops and distributes semiconductor laser technologies for research and commercial applications. In 1999, the firm announced acquisition of the Laser Power Microlaser Group, developers of blue, green, and near-infrared solid-stated diode-pumped lasers.

Melpar model An artificial neuron used at the Wright-Patterson Air Force Base in Ohio in the early 1960s to mimic human reasoning (or at least rodent reasoning). The Melpar model, familiarly called *Artron* by its inventors, was used as the "brains" of a maze-running bionic mouse, physically resembling the input mice used on today's computers. The bionic mouse brain comprised 10 Artrons, which was sufficient for a trial-and-error method of learning to run the maze. With a clean slate, the mouse took 45 minutes to complete the maze; eight tries later, it took only 35 seconds. See neural network.

meltback In fusion splicing of fiber and other materials, the receding away from the point where the fusion joint will occur. Machines that perform automated fusion splicing are designed to move the two ends to be joined closer together as they melt apart at a speed that is appropriate for fusing the two ends without excessive force or a gap. Different materials melt at different temperatures and will recede to different distances depending upon their diameter, composition, and environmental conditions.

meltback test In automated fusion splicing, a preliminary setup test performed to determine how quickly and how far a material melts back when heat is applied to the ends intended to be fused. This data is then used to manually adjust or machine calibrate the distance between the grooves or chucks holding the two materials. Adjusting fiber optic mass fusion splicers is especially challenging as the heat from the arc across a linear array of fiber filaments may not

be consistent, causing varying degrees of meltback. Compensations in the arc or in the movement of the chuck (or both) as fusion takes place is necessary to ensure precise splicing. See fusion splicing.

memory In a computing system, a storage area that is dynamically allocated and used by the operating system and various application programs. Most memory in desktop computers is random access memory (RAM), although some programs will also allocate hard drive storage as "virtual memory." Memory is one of the most basic elements of a computing system, along with the central processing unit (the CPU often also incorporates memory internally) and the input/output (I/O) bus.

Read only memory (ROM) is included in many computers to provide basic nonvolatile operating parameters to a system, particularly on startup. In the earliest microcomputers, a programming language was sometimes included in ROM. Random access memory (RAM) is dynamically allocated by the system and applications programs. RAM is further distinguished as static or dynamic RAM. Most desktop systems include about 8 to 64 Mbytes of RAM, and may be extended up to 64 or 256 Mbytes. RAM typically operates at about 60 to 80 ns, although this may change as newer, faster types of memory are developed.

Most types of computer chip memory are volatile, that is, the contents will disappear if the system is not constantly powered and refreshed. However, there are some types of chips that can retain information, such as erasable, programmable, read-only memory (EPROM) chips.

The price of memory fluctuates dramatically. In 1986, a megabyte of RAM was $600 U.S.; by the early 1990s this had dropped to $25, then increased again to $120. By early 1998, the price was down to $4 per megabyte and dropped to $1 in 2001.

Programmers tend to write code that fills available space. This results in applications that require more memory than many consumers have, setting off another round of buying. In 1978, the TRS-80 computer ran with 4 Kbytes of memory, and with 8 Kbytes it could do word processing and spreadsheet applications quite well. By the mid-1980s, the Amiga computer could multitask and run graphics programs concurrently with stereo sound quite comfortably in 4 Mbytes of RAM. Systems in the late 1990s rarely ran efficiently with less than 16 Mbytes, and most vendors recommended 32. Computers now commonly come installed with 256 MBytes or more.

MEMS See micro-electromechanical system.

Mensa Single Fiber Supertrunk System A commercial linear, synchronous, high bandwidth, point-to-point optical transceiver system designed specifically for supertrunk applications, distributed by Synchronous. It may be used for all analog or hybrid analog/digital signals at 200 MHz. It uses a 1550 nm DFB laser diode light source modulated by a Mach Zehnder. The receiver is based upon a high-response PIN photodetector.

The system provides dual-trunk performance over a

single optical fiber link. For international operation, Band V channels can be directly carried. External modulation enables the system to be used with standard installed cable and is said to eliminate laser chirp. The transmission link can be optically split or repeated.

mercury vapor lamp A lamp in which mercury vapor flows back and forth through a tube when made horizontal to complete the electric circuit and start the lamp. Ionized mercury vapor is then produced by the heat and current, creating light through the length of the tube. The light is very bright, with a greenish glow, and is generally used in industrial applications. See fluorescent lamp.

mesh topology A type of circular network backbone topology in which data can travel back along the backbone if a node becomes unavailable due to a disruption, such as line breakage or failure. The mesh nature of the topology stems from the appearance of the connections between a node and other nodes several nodes away. Some vendors and users prefer this over a ring topology as more than one route can lead to a particular location on the network. See topology.

message circuit noise On a correctly terminated circuit in a network, the background noise that exists when there is no test signal. The noise may arise from crosstalk, radio frequency interference, power line harmonics, or thermal noise. Noise is usually assessed by passing the noise through selected filters to determine its source and character.

Message Handling System MHS. On a network, MHS provides a means to store and forward messages among MHS users or applications. Unlike traditional telephone networks and the early two-way radio communications, most data networks do not need to establish an end-to-end connection before carrying out communications. Thus, the MHS provides a way to handle the messaging traffic under dynamic circumstances. See X.400 under X Series Recommendations.

Message Security Protocol MSP. A Secure Data Network System (SDNS) protocol for providing X.400 message security. With MSP, a message is given connectionless confidentiality and integrity, data origin authentication, and access control; nonrepudiation with proof of origin; nonrepudiation with proof of delivery.

MSP is a content protocol, in the application layer, and is implemented within originator and recipient MSP user agents. It is an end-to-end protocol that does not employ an intermediate message transfer system. MSP processing is carried out prior to submitting a message and after accepting delivery of a message.

An X.400 message comprises a content and an envelope. With MSP, a new message content type is defined with a security heading encapsulated around the protected content.

Three types of X.509 digital certificates are supported by MSP. The user's distinguished name and public cryptographic material are bound within an X.509 certificate which, in turn, is signed by a certification

authority (CA). The CA manages X.509 certificates and Certificate Revocation lists.

message switching A means of switching and multiplexing data packets by storing, queuing, and forwarding the message to the recipient. See circuit switching, packet switching.

Message Transfer System MTS. A general-purpose, application-independent, store-and-forward communications service within a Message Handling System (MTS). The MTS uses message transfer agents (MTAs) to relay messages. See Message Handling System.

message unit In packet networking, SNA, a basic unit of data processed by any layer.

meta-signaling A means to manage User Network Interface (UNI) vacant codes (VCs) signaling and associated broadcast channels incorporating a user part and a network part. Meta-signaling establishes point-to-point signaling VC and broadcast signaling VC (general broadcast and selective broadcast).

Metal Vapor Laser Variation – Schematic

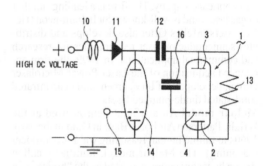

This schematic illustrates a variation on metal vapor lasers to make them more efficient without requiring increasingly complex or unwieldy tubes.

Voltage is transmitted through a choke coil (10), a charging diode (11), and a charging resistor (13) into the storage capacitor (12). The switching device (14) stimulates the capacitor to discharge its stored power into the circuit in the form of a preliminary charge between the cathode (2) and the grid (4) which provides variable impedance. This triggers the main discharge between the cathode and anode (3). The grid also serves as a discharge "buffer," improving the lifetime of the component. [U.S. patent #6,229,837, May 2001.]

---

metal vapor laser A type of atomic gas laser that efficiently emits light in the visible spectrum. A rare-gas metal vapor laser can also emit light in invisible ranges. Metal vapor lasers may be classified as neutral (e.g., gold, copper lasers) or ionized (helium-cadmium lasers). Traditionally they have been based upon a two-electrode structure to create a discharge pulse, though inventors such as A. Ozu have proposed variations that enable higher efficiency with a grid electrode placed near the electron-generating cathode to provide a preliminary discharge.

The Institute of Physics developed the first Bulgarian metal vapor laser (He-Cd) in 1970. The Bulgarian Academy of Sciences includes the Metal Vapour Laser Laboratory in Sofia, Bulgaria.

Metcalfe, Robert (1946–) An American engineer and journalist, Metcalfe is the acknowledged creator of Ethernet at Xerox PARC in 1973, along with David Boggs. In 1979 he founded the 3Com Corporation, and since 1990 has been involved with a number of publishing organizations. See Boggs, David; Ethernet.

METEOSAT Meteorology satellite.

meter A metric linear measure of length/distance equal to 3.28 feet or 39.37 inches. Europe, Canada, and many other regions are standardized on metric measures. The U.S. continues to use the British units of length (e.g., yard) except in scientific circles, where metric is used.

Metropolitan Area Network MAN. An urban network of high-speed hosts. See MAE East, MAE West, SMDS.

Metropolitan Fiber Systems MFS. A Competitive Access Provider (CAP) founded in the late 1980s. In the 1990s it established its own backbone, providing national network services. It was subsequently acquired by WorldCom.

Metropolitan Service Area MSA. An administrative designation used by many organizations providing commercial and public services in urban areas. The U.S. Government uses the designation for managing and analyzing data from the national census. A Metropolitan Service Area is a city with at least 50,000 residents or an urban area with at least 100,000 residents within the counties encompassing these areas. The Federal Communications Commission (FCC) has used this census designation to recognize over 300 MSAs for the purposes of assigning telecommunications licenses such as cellular telephone licenses. See Rural Service Area.

Meucci, Antonio An Italian-born Cuban inventor, chemist, stage designer, and engineer, Meucci made many pioneering discoveries in telecommunications concepts and devices, but his findings were not widely communicated to others, and hence not credited as to their impact on subsequent inventions such as telegraphs and telephones, which made telecommunications history.

Meucci developed rheostats, electroplating techniques, and experimented with passing electricity through the human body. While studying mild electrical charges, he discovered the "electrophonic" effect, which related nerve responses to specific applications of current through a wire. By the mid-1800s he had developed several devices for creating a vibrating electric current from spoken acoustical impulses. By using a copper strip and delicate animal membranes as diaphragms, he created one of the earliest telephone-like mechanisms. He emigrated from Cuba to the U.S. and applied for a caveat for his *teletrofono* which was granted in December 1871, 4 years before the patent of Alexander Graham Bell and the caveat of Elisha Gray. It is believed that Bell (and

Gray) may have had access to papers describing Meucci's invention and, due to greater resources and celebrity, overshadowed his inventions. See telephone history.

Meyer code A flag signaling code, employing left and right motions to create characters or syllables, and a forward motion to indicate ends or pauses. This code was in use until it was superseded in the First World War by International Morse code and American Morse code. See semaphore.

MFJ See Modified Final Judgment.

MFS See Metropolitan Fiber Systems.

MFSK See multiple frequency shift keying.

MH See Modified Huffman.

mho A practical unit of the measure of conductance, so named because it is *ohm* spelled backwards. See admittance, ohm.

MHS See Message Handling System.

MIB See Management Information Base.

Michelson interferometer See interferometer.

Micral The first fully assembled 8008-based microcomputer, the Micral featured 8-bit processing and 2 Kbytes of memory. It was designed in France by François Gernelle. The Micral sold for just under $2000, and, like its predecessor the Kenbak-1, was not commercially successful in the United States, an important market for microcomputers. It was introduced in May 1973 before the SPHERE, Scelbi-8H, Mark-8, and Altair computers. See Altair, Kenbak-1, Mark-8, MITS, Scelbi-8H, Simon, SPHERE.

micro-electromechanical system MEMS. Technology for integrating electromechanical functions into integrated circuits. MEMS-based actuators are used in single- and multimode fiber optic switches.

microbend In fiber optic cables, a small inconsistency, nick, or slight bend that might be introduced into the fiber during manufacture or installation or may occur after installation due to wear and tear. Microbends may introduce small changes in the optical transmissions path as the light "bounces" through the fiber, but are not likely to significantly degrade the signal. However, if the fiber is especially slim or the wavelengths especially long, microbend losses may occur. If there are a large number of microbends, the accumulated effect of the disruption over distance, especially in a cable with a small core radius, may be significant. See loss, macrobend, macrobend loss.

MicroCal Module A new type of integrated circuit (IC) designed to facilitate self-monitoring in virtually any type of wireless equipment, including base stations, mobile handsets, and subscriber units. The MicroCal Module scans the entire bandwidth, gathering data which is then fed back to a central maintenance center. The module, designed by Micronetics Wireless, was awarded a U.S. patent in 1996. Micronetics is working with a number of companies, including Nortel and Motorola, to integrate the module into wireless infrastructure equipment.

Microcom Networking Protocol MNP. A series of proprietary error control and data compression protocols designed for dialup modems, which are often used in conjunction with industry standard ITU-T-recommended error control mechanisms.

For example, MNP-4 works with modems that transmit at data rates up to 14,400 bps. MNP-4 is often implemented in conjunction with the V.42 error control protocol standard from the ITU-T. See Microcom Networking Protocols chart. See V.42.

microfiche A somewhat standardized optical archive system using thin transparent sheets of image-carrying plastic for storing scanned printed matter, especially newspapers, books, journals, etc. Since microfiche information is miniaturized to fit as much data on a sheet as possible, it is typically not human-readable without magnification. Microfiche machines backlight and magnify the data. Some photocopiers are designed to enlarge and print microfiche information, although the copies are often not very clear. Digital storage techniques are replacing microfiches and the quality is improving over earlier scans. Unfortunately, like microfiche archives, the originals are often destroyed for lack of storage space and funds. Microfiches are common in libraries, post-secondary institutions, and government archives but are being gradually superseded by scanned digital images.

micromachined membrane deformable mirror MMDM. A compact component for correcting aberrations in optical systems, such as distortion that may be caused by passing through an aperture or lens. In contrast to traditional glass lens correction devices, a MMDM is a thin membrane coated to give it a highly-reflective surface. Gold and aluminum coatings are common. The speed of response of MMDMs is faster and astigmatism is somewhat lower than other popular technologies (e.g., OKO mirrors).

By application of current through actuators incorporated into the membrane system, the shape of the membrane can be controlled through an attracting electrostatic force. The system is commonly set up with a flat spherical MMDM which is shaped as needed.

MMDMs have been found practical in the visual spectrum and have been studied for their effect at supercooled temperatures. See wavefront control.

microphone A device for apprehending sounds and transmitting them electrically or acoustically to a receiver or audience. A very simple microphone can be created by wrapping stiff paper into a funnel shape, attaching it to a string or wire, and stretching it to a receiver – another funnel on the other end. If the listener puts an ear near the receiving funnel while the speaker talks into the microphone funnel, the sound, while not loud, can be heard across a room. Add electronics to amplify the signal, and you have a basic microphone. Some microphones also include echo acoustics to make the sound of a voice more resonant. Many singers use this type of microphone to enhance their singing on recordings.

Microphones are widely used in camcorders, film cameras, tape recorders, and video recorders. Two microphones are needed for true stereo sound.

Microphones can be used as peripherals with computers for the creation of music and other sound samples, or for videoconferencing. See sampling, videoconferencing.

Microsoft BASIC A BASIC interpreter first released for the Altair computer in 1975. Paul Allen had seen the feature article on building the Altair in the January 1975 issue of *Popular Electronics*, so he and Gates talked about it in Harvard Square, and conceived the idea of writing a BASIC interpreter for the new kit-based machine. They contacted MITS, made a proposal, and set to work creating a BASIC that could fit into 4K of memory. The entrepreneurs had previous experience in looking at code for interpreters for various languages based on their business activities together through high school, and 8K BASICs

Microcom Networking Protocols	
Name	**Notes**
MNP-1	Asynchronous mode, half duplex transfer operation.
MNP-2	Simple error correction scheme, asynchronous mode, full duplex operation.
MNP-3	Error correction incorporated, synchronous mode.
MNP-4	Error correction incorporated, increased throughput. Often included with V.42 modems, along with MNP-5 data compression.
MNP-5	Simple data compression scheme. Often included with V.42 modems, along with MNP-4 error control.
MNP-6	Statistical duplexing and Universal Link Negotiation. Full duplex emulation.
MNP-7	Data compression scheme included.
MNP-8	MNP7 for modems which emulate duplex operation.
MNP-9	Data compression scheme included. Incorporates V.32 technology.
MNP-10	Dynamic fall-back and fall-forward adjusts modulation speed with link quality.

were available for the PDP-8. They developed the BASIC in a simulation environment, since it wasn't practical to write it on the Altair itself. Allen created a simulation environment for 8080 programming code and modified a symbolic debugger to understand the 8080 instructions. Gates laid out a design for the BASIC interpreter modeled on the BASIC he had encountered on a timesharing system at Dartmouth and began coding it, with assistance later from Allen. Monte Davidoff contributed some of the math routines, especially those for floating point operations. On the plane to Albuquerque to demonstrate the software, Allen created a bootstrap loader so the Altair would be able to read the data into memory, using a teletypewriter as an input mechanism. (Gates later streamlined the bootstrap loader.) On the first run at the demonstration at MITS, the BASIC didn't work. On the second try it did. This was a substantial achievement, given the short, hands-off development period and environment.

This first BASIC was later ported to many machines. Not long after the Altair kicked off the microcomputer industry, Microsoft BASIC Level II was bundled with the TRS-80 Model I in ROM in 1976, replacing Level I BASIC, and included with the Commodore PET. Microsoft also contributed some routines to the Integer BASIC designed by Wozniak for the Apple Computer, resulting in AppleSoft BASIC. Later, in 1984, Microsoft BASIC was incorporated into ROM on the IBM Personal Computer XT. In addition to the computer-specific 8-bit operating systems, BASIC was ported to run on the popular CP/M-80 operating system designed by Gary Kildall. At this point, Microsoft BASIC was still a text-based program.

Microsoft BASIC version 2.0, the first graphics-based BASIC for the Macintosh, was not announced until fall, 1984, a decade after the text version shipped. In 1985, Microsoft provided a windowing version on floppy diskettes for the Amiga 1000. Later Microsoft BASIC evolved further into Microsoft Visual BASIC, which differed chiefly in that graphically entered structures could be used to automatically generate code. See BASIC, Visual BASIC.

Microsoft Data Warehousing Framework A Microsoft commercial open, scalable architecture for creating, using, and managing integrated data warehousing applications.

Microsoft Data Warehousing Alliance DWA, MDWA. A trade association for those using and supporting Microsoft Data Warehousing Framework information technology (IT) applications and standards.

Microsoft Incorporated One of the earliest companies supporting the microcomputer market, Microsoft was founded by Paul Allen and Bill Gates in 1975 following their partnership as Traf-O-Data, which they formed around 1972. Although Gates and Allen had worked on programming projects together during high school in Seattle, they formalized Microsoft in 1975 in order to market a version of interpreted BASIC for the Altair computer. The trade name was registered in 1977.

Paul Allen learned of the Altair computer in a *Popular Electronics* article he saw in Harvard Square. Since the Altair so closely paralleled an earlier microcomputer hardware idea of his, he contacted Gates to let him know "someone else is doing it." They then talked in the Square about writing BASIC for the new machine. They contacted MITS, the makers of the Altair, and Allen, Gates, and Davidoff created a BASIC based on Gates' and Allen's experience with BASIC interpreters at Dartmouth. Six weeks later, Allen flew south and successfully demonstrated BASIC to MITS in New Mexico, setting the groundwork for their software development company.

The entrepreneurs moved their operations first to the Sundowner Motel across the street from MITS, and later to an eighth floor office in Albuquerque, New Mexico. Allen took a position as VP of Software at MITS, while keeping in regular contact with Microsoft and Gates. They hired high school friends to help out. Meanwhile, Gates began enhancing BASIC and porting it to new platforms that were introduced. When Gates briefly went back to Harvard, Ric Weiland and Marc McDonald formed the core at Microsoft. Marc McDonald designed and coded Stand-alone Disk BASIC, in consultation with Bill Gates. In the late 1970s, Microsoft BASIC was adapted to run on a popular text-oriented operating system called CP/M, developed in various versions by Gary Kildall of Digital Research between 1973 and 1976.

After three years in New Mexico, Microsoft relocated to Bellevue, Washington, near the co-owners' family members. At this location it was easier to recruit programmers as well. Microsoft now has a campus in Redmond, Washington, and has grown to be a large, financially successful enterprise.

While it has had occasional forays into hardware development, the primary focus of the company has been software, and a substantial portion of the revenues are derived from operating systems and business-related applications. Paul Allen left the company to invest in a number of other ventures, and formed the Paul Allen Group to oversee his various investments. Allegations of unfair business practices were leveled at Microsoft on numerous occasions during the late 1980s and the 1990s, and the company began to be scrutinized by the U.S. Justice Department. These proceedings are ongoing and are still not completely resolved. Bill Gates was the long-standing CEO until January 2000 when he resigned in lieu of Steve Ballmer who took the positions of president and CEO. Gates continued as chairman and chief software architect. See Allen, Paul; Altair; Gates, William; MITS; Traf-O-Data.

Microsoft Mobile Explorer MME. A multistandard microbrowser emulation environment optimized for low-resource devices such as mobile telephones capable of displaying HTML-like markup language pages. See browser, PDA microbrowser.

microwave A radio wave transmission frequency (1000+ MHz) generally used for radar and radio repeaters. Microwaves also provide the cooking power for microwave or *radar range* ovens. The generation of microwaves was initially achieved with magnetrons and Barkhausen-Kurz tubes in the early 1920s. Microwaves were sometimes called radio-optical waves in the early days of their development due to their position on the electromagnetic spectrum between light waves and conventional radio waves, and because some of their characteristics, such as propagation, were similar to light waves.

Microwave relay systems were in use as early as the 1930s by AT&T. The original magnetrons were developed into cavity magnetrons and traveling-wave tubes. Microwave generators and relays are now an important aspect of satellite communications. Microwaves are also finding increasing use in connecting local area wireless networks (LAWNs). While they are not used for the primary information-carrying aspects within the network, they are useful for interconnecting line-of-site separated LAWNs, or LANs, between buildings. Connection requires a license from the Federal Communications Commission (FCC). See magnetron, microwave antennas, short wave, traveling-wave tube.

Microwave Antenna Structures

Parabolic antennas constructed from mesh or solid materials are used for very short (microwave) radio waves. The curvature of the dish and the related placement of transmitting or receiving horns are important to the quality of the signal transmitted or received. [Classic Concepts photos; used with permission.]

---

microwave antenna Due to the very short wavelengths used in microwave transmissions, the physical arrangement of microwave antennas is quite different from those for UHF, VHF, and FM broadcasts. Microwave transmissions are directional for both up- and downlinks, quite different from the roughly isotropic, omnidirectional character of traditional television and radio broadcast waves. The common multibranched Yagi-Uda style antennas and fan dipole antennas are inexpensive and appropriate for VHF and UHF reception, but directional parabolic antennas are the norm for microwave signals.

The diameter of a parabolic dish antenna is a multiple of the length of the microwaves received and typical dishes range in size from about 2 to about 10 feet across, with the curvature of the dish determining the position of the feed horn which focuses the beams.

The first transcontinental microwave communication system began operations in 1951 through a system of relay stations between San Francisco and New York City. Within three years, there were more than 400 additional stations scattered across North America. See antenna, parabolic antenna, UHF antenna, VHF antenna.

microwave multi-point distribution system MMDS. MMDS is a system for distributing cable TV programming through microwave communications, more commonly known now as wireless cable. MMDS works in the frequency range of 2.50 to 2.686 GHz, and MMDS service providers are increasing in number. The signals are downlinked from the satellite to the local transceiver, and broadcast from there to subscribers within about a 50-mile radius, depending upon terrain. The subscriber receives the signal on a consumer-priced antenna mounted on or near the home, which is linked through a cable to a "black box" connected to (and sometimes sitting atop) the TV receiver. This box decompresses compressed digital signals and unscrambles signals intended to prevent unpaid/unauthorized viewing of the programs. The MMDS system is in the process of changing from analog to digital technology, opening up opportunities for digital multiplexing through highly linear radio frequency (RF) subsystems, thus providing more television channel choices for viewers.

microwave radar Radar systems employing microwaves have been extremely important in navigation, tracking, surveillance, guidance, and communications systems. Much of the early research in microwave radar was conducted at the Massachusetts Institute of Technology (MIT) radiation laboratory in the early 1940s.

Mid-Span Meet An interconnection point between two co-carriers. The Mid-Span Meet is the point up to which the carriers provide cabling and transmissions.

MIDI See Musical Instrument Digital Interface.

MIDI time code MTC. A standard developed to identify timing information associated with a stream of Musical Instrument Digital Interface (MIDI) data. See SMPTE time code.

MIDP See Mobile Information Device Profile.

Midwestern Higher Education Commission MHEC. MHEC was founded as an interstate agency in 1991 to promote resource sharing in higher education. As a subgroup, it includes a Telecommunications Committee that takes a regional approach to improving access, services, and costs of telecommunications services.

Milan Declaration on Communication and Human Rights This declaration was put forth in 1998, based in part on a number of global rights documents, including the Universal Declaration of Human Rights, the International Covenant on Civil and Political Rights, the American Convention on Human Rights, the European Convention for the Protection of Human Rights and Fundamental Freedoms, the Beijing Platform of Action, and other important acknowledgments of human rights.

The Declaration asserts the intrinsic relationship between freedom of opinion and expression and the technologies and venues available for their communication. It declares that "The Right to Communicate is a universal human right which serves and underpins all other human rights and which must be preserved and extended in the context of rapidly changing information and communication technologies, ..."

It affirms the need for equitable access to all communications media coupled with the mandate to preserve and sustain cultural rights and diversity. It underlines the importance of not reducing all information users to the category of consumers and affirms their role as communications producers and contributors. It calls for international recognition of community broadcasting as a vital contributor to human freedoms. See People's Communication Charter. The text of the Milan Declaration is available through the World Association of Community Broadcasters (AMARC). http://www.amarc.org/

Mill Street plant This historically significant power plant began providing three-phase alternating current (AC) in 1893. Partly due to the advocacy of T. Edison, most early power plants provided direct current, so the Mill Creek No. 1 hydroelectric plant was a precedent-setting installation, and many other similar AC power suppliers followed its example.

milli- (*abbrev.* – m) An SI unit prefix for 1 thousandth, 10^{-3} or 0.001. Thus, a milliamp is 1 thousandth of an ampere.

MILNET Military Net. The ARPANET was a historic computer network put into operation in 1969. In 1975, ARPANET was transferred to the Defense Communications Agency. Then, in 1983 it was split into MILNET for military usage, and ARPANET, which evolved into the Internet. MILNET is used for nonclassified U.S. military communications. See ARPANET, Internet.

Milstar A joint U.S. Army/Navy/Air Force satellite system for providing jam-resistant communications for wartime requirements for high-priority military users. It is a global constellation of 5-ton geostationary satellites orbiting at about 22,250 nautical miles. The first Milstar satellite was launched in February 1994, the second in November 1995. Six launchings are intended to support four satellites that are active at any one time. The satellites have operational lifetimes of about 10 years.

Milstar was designed to link ground, marine, and air command authorities. The satellites relay communications from terminal to terminal, anywhere on Earth. By transmitting from satellite to satellite, ground hops are reduced and security heightened. Milstar terminals provide a variety of data services, including voice, data, facsimile, or teletype communications.

MIME See Multipurpose Internet Mail Extension.

MIN See Mobile Identification Number.

Miner, Jay (~1930-1994) A gifted design engineer responsible for designing the hardware for the Atari 800 computer, the Amiga computer, and the Lynx color handheld game machine. In 1982, Miner joined Hi Toro to develop the Lorraine computer, which was subsequently sold as the Amiga by Commodore Business Machines. A proponent of open-mindedness and creativity, Miner included his dog's pawprint inside the case of the Amiga 1000. After the Amiga, he created the Atari Lynx, a fast color handheld game machine. Jay Miner was affectionately known as Padré, the Father of the Amiga, to the computing community. Following a serious illness and kidney transplant, Jay Miner devoted his remaining working life to developing medical devices, such as pacemakers, to aid society. Surprisingly, despite the fact that he understood that the creation of the Amiga was a remarkable achievement, Miner didn't anticipate the revolution in the video industry launched by his creation. In a computing industry where hardware architectures go out of date in a few months, the viability of the Amiga hardware for more than a decade, particularly for graphics and sound, is a tribute to its efficient and insightful design. See Amiga computer, Commodore Business Machines.

minifloppy A generic term for a number of floppy diskette technologies that store almost ten times as much data as a regular 3.5-in. floppy, but which are designed by some manufacturers to be downwardly compatible with 1.4-Mbyte drives. The price of storage on these high capacity floppies is substantially cheaper, and they may, in time, supersede current floppies.

minimal shift keying MSK. A type of modulation technique similar to quadrature phase shift keying (QPSK), except that the rectangular pulse in QPSK is a half-cycle sinusoidal pulse in MSK. See modulation, phase shift keying.

minimize button, iconize button Graphical user interfaces on several different operating systems include a small gadget on application or display windows which, when clicked, will shrink the window down to an icon. Thus, the program is available and can quickly be retrieved by double-clicking the minimized icon without shutting down the process and rerunning the program.

Ministère des Postes et Télécommunicaciones The telecommunications authority for the Democratic and Popular Republic of Algeria. Online communications are in French.

Ministerio de Comunicaciones The telecommunications planning and regulatory authority of the Republic of Columbia in South America. http://www.mincomunicaciones.gov.co/

Ministry of Information Technology and Telecommunications MITT. The Mauritius ministry that handles the formulation and implementation of

M

government policies in telecommunications and information technology. Telecommunications are governed by the Mauritius Telecommunications Act 1998. Online services are designed and published by the National Computer Board.
http://ncb.intnet.mu/mitt.htm

Ministry of Posts and Telecommunications MPT. The Japanese radio regulatory administration, MPT oversees radio communications, based upon the Radio Law of 1950. The MPT grants radio station and operator licenses, monitors and inspects stations and radio frequencies, and sets technical standards for radio equipment. http://mpt.go.jp/

Ministry of Telecommunications The telecommunications authority of the Lebanese Republic under the direction of the Minister of Post and Telecommunications. http://www.mpt.gov.lb/

Ministry of Telecommunications and E-Commerce The authority in Bermuda that oversees telecommunications, broadcasting, and frequency administration. http://www.mtec.bm/

Minitel A French Telecom service that provides free terminals for chat and electronic telephone directory videotext services. It is similar to the German Bundespost's interactive videotext system. See Minitel.

MIP See Multichannel Interface Processor.

MIPG See Multiple-Image Portable Graphics.

MIPS million instructions per second. A measure of processor speed used in system design and cross-system comparisons. MIPS describes the average number of machine instructions that a central processing unit (CPU) performs per unit of time of 1 second. This is a narrow definition of performance, as many other factors influence overall speed and efficiency. The Digital VAX-11/780 is defined as a baseline at 1 MIP. By the late 1990s, most consumer desktop models delivered about 3 to 10 MIPS and high-end minicomputers and mainframes ranged from 10 to 50 MIPS, with supercomputers comprising the top of whatever was state of the art at any particular time. See benchmark.

Mir A landmark "permanent" Earth-orbiting space station used for observation, experimentation, and scientific research about living and working in space. U.S. and Russian Mir missions began early in 1995, with the core module launched in February 1986. Mir consisted of a number of connected modules, docking components, solar screens, life support systems, and scientific instruments. The Mir capsule could hold two or three people fairly comfortably, and up to six for short periods of time. Travellers to and from Mir connected through the NASA space shuttle.

Amateur radio enthusiasts enjoyed regularly listening to Mir signals, and some have sent transmissions to the orbiting station.

A great deal was learned about the wonders and challenges of living in space from Mir. Information and photographs related to Mir missions can be seen through the Office of Space Flight Web site.

After orbiting the Earth for more than 15 years, the space station entered the Earth's atmosphere, in a controlled decline, on March 23, 2001.
http://www.nasa.gov/osf/mir

MIR See multimedia information retrieval.

mirror A highly reflective, usually polished surface that readily reflects light while absorbing very little of it. Water and glass have mirroring qualities, but some of the light is refracted or absorbed, making the image foggy or ghostly rather than crisp and detailed. Highly polished metal and silvered glass make excellent mirror surfaces. Dielectric mirrors are most effective within a narrow range of wavelengths and angles of incidence.

Mirrors were used for line-of-sight signaling long before electrical telecommunications methods were available. Hikers still regularly carry them for emergency signaling in the wilderness. Mirrors are also used in many types of computer devices, especially those which incorporate laser beams, such as laser printers. The mirror serves to direct the beam inside the mechanism onto the appropriate areas, such as a printing drum.

In fiber optics, it was discovered by MIT scientists that dielectric films would behave more like a metal if they were layered in a particular way, a capability that they subsequently applied to the development of improved lightguides that had the potential to support the transmission of a wide range of optical wavelengths, while retaining the polarity of the transmitted beams. The new technology, described in 2000, was also capable of reflecting light through small areas for increased miniaturization. See coaxial omniguide, heliograph.

mirroring A means of providing system backup security or redundant access by replicating data in different locations. The system can enable the user to access the mirror location if the original data storage location becomes oversubscribed or corrupted. Or, the system can be restored with information from the mirror. Redundancy is a very common property of computer systems. Some will mirror whole directory structures and files as a matter of course. Some hard drive systems are set up to constantly mirror information over several devices. While mirroring almost inevitably costs a little more in terms of memory or storage space and in processing time, it is usually worthwhile. See mirror site, RAID.

MIS See Management Information Services.

missile, fiber-guided A guided missile that flies in a high trajectory with a camera housed in its nose, which is connected to the launcher by a fine spooling fiber optic cable. The cable enables a remote human pilot to view the progress of the flight and guide it to targets within a 10-mile range using a joystick as the guidance controller. This weapon was initially released in the early 1990s by the U.S. Army as the FOG-M (Fiber Optic Guided Missile) but not widely fielded. See multiplexed optical scanner technology.

MITS Micro Instrumentation and Telemetry Systems. The historic creators of the Altair microcomputer, MITS, under the direction of Ed Roberts, originally sold radio transmitters (telemetry devices) for model planes. These products did quite well and got the company under way, but when the company moved into the area of calculator kits, there was a lot of competition from bigger names like Texas Instruments, and the Altair was in essence an effort to stave off bankruptcy. MITS developed the MITS 816 in 1972, and later the historic kit for the Altair 8800 in 1974. While the Altair is not the first microcomputer, it is to be credited as the first commercially successful microcomputer. In spite of the success of the Altair, the company was sold to Pertec, a manufacturer of peripherals. See Altair, Intel MCS-4, Kenbak-1, Mark-8, Micral, Scelbi.

MJ modular jack. Any jack designed to interconnect readily with various standardized receptacles in a circuit system. See RJ.

MLPP multilevel precedence and preemption (e.g., as in ISDN Q.85 and A.735 community of interest services).

MMC 1. minimum monthly charge. 2. See Mobile Multimedia Communication project. 3. See MultiMediaCard.

MMCA See MultiMediaCard Association.

MMCF See Multimedia Communications Forum.

MMCX See Multimedia Communication Exchange.

MMDC 1. See Massachusetts Microprocessor Design Center. 2. See Multi-Media and Digital Communications lab. 3. See Multi-Service, Multi-Carrier, Distributed Communications. 4. See multimedia desktop collaboration. 5. See Multimedia Development Center. 6. See multimodel data compression. 7. See Multiple Module Data Computer.

MMDM See micromachined membrane deformable mirrors.

MMDS See microwave multipoint distribution system.

MME 1. See Microsoft Mobile Explorer. 2. See Mobile Meteorological Equipment. 3. See Multimedia Message Entity.

MMF 1. See Mobile Management Forum. 2. See multimode optical fiber

MMI machine-to-machine interface. Since this can easily be confused with the abbreviation for man-machine interface, which was also traditionally MMI, it is preferable to use HMI for human-machine interface.

MMIC See Monolithic Microwave Integrated Circuit.

MMM See multimedia mail.

MMMS See Multimedia Mail Service.

MMMSec See Multimedia Mail Security.

MMS 1. marketing measurement system. 2. memory management system. 3. meteorological measurement system. 4. module management system. 5. multimedia survey.

MMSI 1. Manchester Museum of Science & Industry. 2. See Maritime Mobile Service Identity.

MMSP See modular multi-satellite preprocessor.

MMSS Maritime Mobile-Satellite Service.

MMTA See Multimedia Telecommunications Association.

MMU 1. Manned Maneuvering Unit. A human maneuvering unit used in untethered space walks

originating from the U.S. space shuttle missions. 2. memory management unit. Computer circuitry often built into central processing chips to handle administration of blocks of storage.

MMUSIC See Multiparty Multimedia Session Control.

MMX Multimedia Extension. Matrix Math Extension. See Pentium MMX.

MNLP See Mobile Network Location Protocol.

MNP See Microcom Networking Protocol.

MNRP Mobile Network Registration Protocol.

mobile assisted handoff MAHO. A process in which the handoff of a voice channel by a mobile station is assisted by the base station by providing information on the surrounding radio frequency (RF) signal environment.

Mobile Broadband System MBS. A wireless cellular network developed as one of the European RACE II Integrated Broadband Communications (IBC) projects. The purpose of the project was to develop third-generation, integrated mobile systems as part of a universal, cost-efficient, voice/data personal communications system.

MBS transparently transports Asynchronous Transfer Mode (ATM) cells over the air interface at 60 GHz at data rates up to 34 Mbps (with higher rates possible through multicarrier transmission). Two recommended sub-band frequencies for MBS are 62 to 63 GHz and 65 to 66 GHz.

MBS is supported over B-ISDN systems and differs from traditional cellular by its bursty nature and dynamically adjusting data transmission rates. A new channel structure and protocols have been developed in conjunction with the project to exploit the packet characteristics of B-ISDN connections. See Integrated Broadband System, Research into Advanced Communications in Europe.

Mobile Data Base Station MDBS. In CDPD mobile communications, a system which provides data packet relay functions between the Mobile End System (M-ES) and the Mobile Data Intermediate System (MD-IS). See Cellular Digital Packet Data.

Mobile Data Intermediate System MD-IS. In CDPD mobile communications, a system which provides routing and location management functions, utilizing a Home Domain Directory (HDD) database. The MD-IS communicates with the Mobile End System (M-ES) through the Mobile Data Base Station (MDBS). See Cellular Digital Packet Data.

Mobile End System M-ES. In CDPD mobile communications, the system through which the subscriber accesses wireless network services. M-ESs include modems installed in laptops, palmtops, personal digital assistants (PDAs), etc. See Cellular Digital Packet Data.

Mobile Identification Number MIN. Each wireless phone is assigned an identification number by the carrier. The MIN is not attached to the individual, as the phone may change hands or the individual may change locations.

Mobile Information Device Profile MIDP. A specification supported by a number of major wireless telecommunications service providers, MIDP is a set of Sun Java APIs that is part of the J2ME application runtime environment for mobile information devices, along with the Connected Limited Device Configuration (CLDC). MIDP was developed through the Java Community Process. It specifies aspects of storage, application life cycle, networking, and user interaction.

MIDP for PalmOS is an implementation of CLDC and MIDP optimized for PalmOS handheld platforms.

Mobile IP, Mobile Internet Protocol Mobile data networking through the Internet is coming into demand as the number of laptops and the availability of wireless modem services increases. Since the problems of maintaining contact with a network and network security are concerns on mobile systems, a set of extensions to Internet Protocol (IP) is being developed to handle the special needs of mobile users. Mobile IP uses a dual addressing scheme so that the communications node and the mobile unit can be tracked and administered. In simple terms, the location of the mobile system becomes a forwarding address to which packets are retransmitted. Security is incorporated to prevent an unauthorized person from intercepting the transmission. See Foreign Agent.

Mobile Management Forum MMF. A forum of the Open Group, announced in May 2000, as a means to pursue the objectives of the Open Group Wireless and Mobile Program. The MMF supports and promotes the deployment of interoperable wireless applications and devices into enterprise environments. http://www.opengroup.org/mobile/

Mobile Maritime Committee MMC. An informal committee of the U.S. Coast Guard that has existed for some time to express and address local maritime issues and problems. In February 2000, a more formal organization was established to promote actions to improve the safety, security, mobility, and environmental protection of the Mobile, Alabama, port. Members consist of port and waterway users and regulatory agencies. See Marine Safety Office.

Mobile Meteorological Equipment MME. Through the work of the Ad Hoc Group for Mobile Meteorological Equipment (AHG/MME), the Federal Directory of Mobile Meteorological Equipment and Capabilities was prepared to assist agencies in individual responsibilities and planning activities in response to requirements and emergencies. The Directory catalogs mobile meteorological equipment, software, and capabilities of U.S. federal departments and agencies to facilitate interagency cooperation. The Department of Defense (DoD) has been particularly active in the development and deployment of mobile systems.

Mobile Multimedia Communication project MMC project. A multidisciplinary research project coordinated within Delft University of Technology in the Netherlands. The MMC project was established to research solutions to wired and wireless Internet-style applications and bandwidth applications as they can be adapted to mobile networks such as cellular systems. The project ran from April 1996 to September 2000. Some of the work in the project has been trans-

ferred to the UbiCom (Ubiquitous Communication) program.

Mobile Network Location Protocol MNLP. In CDPD mobile communications, the MNLP provides a means to track the Mobile End System (M-ES), that is, the laptop modem, cellular phone, or other device that allows the user to link into the network, and to interlink the Home Mobile Data Intermediate System (MD-IS) and the Serving MD-IS. This works in conjunction with a Mobile Network Registration Protocol (MNRP) to verify the user's Network Entity Identifier (NEI), a security ID used to monitor and confine service to authorized users.

Mobile Network Registration Protocol MNRP. See Mobile Network Location Protocol.

mobile phone An audio broadcast system designed to provide mobile communications through hardware interfaces resembling traditional phone handsets. The earliest mobile phone systems were bulky, limited contrivances developed after the turn of the century and first demonstrated in 1919, but they were acknowledged as having an important place in future communications.

Historically similar to broadcast TV, a powerful transmitter was located to provide maximum range, up to perhaps 30 miles, for traveling subscribers. To increase the limited range and channel distribution of the single tower design, cellular networks were developed, which increased available bandwidth by providing many lower power transmitters, closely located to one another, over a wide geographic region. There are now a number of types of mobile phones, from short-range FM cordless phones with a range of a few hundred feet, to digital PCS and cellular systems with roaming capabilities that range from hundreds to thousands of miles. See cellular phone.

Mobile Solutions Partner Program MSPP. A Microsoft program initiated in August 2000 to promote development of mobile communications solutions based on .Net Mobile Web technology. MSPP supports vendors who are developing for the Microsoft Pocket PC operating system. The program was initiated in part as a competitive response to Palm's business lead in the mobile communications industry.

Mobile Subscriber Unit MSU. A main component of a mobile phone system consisting of a portable or transportable control unit and cellular radio transceiver. Convenience, size, transceiver power, and battery life are traded off in the various systems. Larger, more powerful units may be mounted to car batteries, and often split the telephone and the handset into separate units. Smaller handhelds frequently have less range and shorter battery life. See cellular phone, mobile phone.

Mobile Telephone Switching Office MTSO. A main component of a mobile phone service, which performs wireless relaying, switching, and administration tasks similar to those carried out by a wired telephone switching office, except that it must handle the specific technical needs of users who are moving and roaming (changing from one transceiving area to another) with signal monitoring and processing, handoffs, etc. In addition, the MTSO handles the link between the mobile services and connections to wireline services, as many mobile services are actually hybrid technologies, often taking calls from mobile users and connecting them with a wireline destination, and vice versa.

Mobilization Against Terrorism Act MATA. See Anti-Terrorism Act of 2001.

Mockapetris, Paul The developer of the JEEVES DNS Resolver, the first implementation of the Domain Name System, now incorporated into the Internet. Mockapetris is responsible for a number of significant Request for Comments documents related to the development of the Internet. His DNS Resolver spawned several subsequent implementations, the most significant being the Berkeley Internet Name Domain (BIND). In June 2001, Mockapetris was announced as Chairman of the Board for Nominum, Inc., a naming and address management solutions provider. See JEEVES DNS Resolver.

modal In applications programming, a type of user window, dialog, or other input or information display operation which does not suspend access to other processes. For example, suppose the user has selected a Quit function, and the software displays a dialog box that says, "Do you really want to quit? If so, the program will end without saving." Options to Quit or to Cancel will be presented. If the dialog allows the user to go back to the application without responding to the Quit/Cancel query, the operation is modal. If the user must reply before continuing with using the software, then it is not. While modal (multitasked) operations are preferred in many situations, in others, a response should be solicited before continuing, especially if it involves the possible loss of data.

mode In some older operating systems, a distinction was made between text mode and graphics mode, but most systems now work in graphics mode with text represented graphically. This system is more flexible.

modem modulator/demodulator. 1. A device which modulates and demodulates a signal. Digital data are typically modulated to be carried over analog transmission systems, and broadcast waves are modulated to add information to the carrier band. These are then demodulated again at the receiving end. 2. A computer hardware peripheral specifically designed to convert the digital signals generated by the computer into analog systems that can be carried across an analog transmissions medium such as twisted-pair copper wire, and demodulate them back into digital data at the receiving end. Many standards exist for the transmission of this type of data, and the sending and receiving modems must be able to negotiate a common format in order for the signals to be meaningfully received. Current modems commonly transmit at rates of 19,200 bps, 38,400 bps, and higher; most include facsimile transmission capabilities, and some include voice mail capabilities as well. They incorporate a number of error control, data compression, and modulation protocols in order to maximize speed of transmission over lines that people once claimed could never transmit data faster than 600 bps. See

error control protocol, data compression protocol, modulation protocol, serial port.

modem pool A set of modems usually servicing a network through which several users can dial out of the system, or through which a number of users can dial in, as to a BBS or Internet Services Provider (ISP). Most higher educational institutions have modem pools for users to access the system from home or classrooms, or through which they can dial out to community services or extra service providers. Often the modems in a pool will have different characteristics. For example, only a few lines may be high speed lines, due to higher cost, and the remainder may be a variety of slower, less expensive modems.

Some modem pools are extremely large. For example, one of the largest commercial Internet providers has over 100,000 modems in its pool.

A pool is a flexible way to maximize resources. A dozen modems can service a hundred workstations, provided the users do not need constant access to dialup resources. It is also easier for system administrators to carry out hardware maintenance and to maintain security when the modems are grouped and placed in a secure environment.

Modems

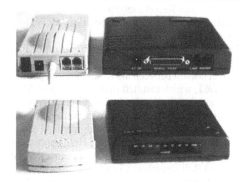

Two different computer modems: the Global Village on the left has a built-in 9-pin DIN connector, and operates at 33,600 bps; the SuperModem on the right has a standard 25-pin connection supporting V.34 standards. Each can be connected (daisy-chained) with a regular telephone set.

modem server A networked workstation application which manages the administrative and access tasks associated with a modem pool, or an intelligent modem hub which manages incoming and outgoing data from more than one user. In large modem pools, a system may be dedicated to assigning user requests for modems, for sending messages to the user (e.g., "All systems are currently busy, please make your request again in 15 minutes."), for evaluating which modems to allocate first (there may be different modems with different capabilities, such as access speed), and for assigning priorities and connect times, when appropriate.

modem standards This is one of the areas where *de*

facto vendor standards and industry standards (e.g., ITU-T V Series Recommendations) have continually leapfrogged one another, and engaged in an uneasy competitive race. The constant consumer demand for faster modems and the vendor desire to be the first to market with the next generation modem, have caused many vendors to develop their own standards ahead of the global cooperative standards process. For this reason, many modems are dual-standard modems, in order to support both the vendor and generally accepted industry standards. Some modems support either vendor or industry standards, which are often not compatible, and it is important to find out their status before purchasing.

In many cases, the early versions of modems supporting the faster speeds are the ones most likely to go out of date quickly. In the early days, many vendors followed Bell and Hayes standards, whereas in recent years, vendors have tended to go with the industry standards once the specifications are finalized and made available. The Hayes command set remains, although most vendors implement a superset of the original Hayes commands, which were quite simple and limited.

The Microcom Network Protocol standards for error control and data compression are widely supported modem standards. See Microcom Networking Protocol, V Series Recommendations.

modifed chemical vapor deposition MCVD. See vapor deposition.

Modified Final Judgment, Modification of Final Judgment MFJ. The name given to a historic 7-year antitrust lawsuit between the U.S. Justice Department and AT&T, which resulted in the breakup of AT&T. It is associated with Judge Harold Greene's decision regarding the 1983 to 1984 (clarification and revision) divestiture of AT&T. Under this judgment, AT&T was permitted to retain ownership of Bell Laboratories and AT&T Technologies (Western Electric), but the Regional Bell Operating Companies (RBOCs) were banned from manufacturing, and Local Access Transport Areas (LATAs) were created rather than retaining the existing local exchange boundaries.

Prior to the MFJ, charges were handled through Division of Revenues, but this was changed to an access charge tariff system. See AT&T; Greene, Harold; Kingsbury Commitment; Local Exchange Carrier; Willis Graham Act of 1921.

Modified Huffman MH. Huffman is a variable-length, tree-oriented data encoding scheme that optimizes on the basis of more frequently occurring characters in order to achieve compression in fewer bits. Modified Huffman is widely used in facsimile transmissions. See Huffman encoding.

modular Composed of separately organized entities, loosely or tightly coordinated or connected to create a larger whole. Modular programming is programming in which the larger application is composed of smaller associated elements such as blocks, objects, primitives, self-contained functions, etc. Object-oriented programming is a type of modular programming. A modular office is one in which the individual

components of the facilities can be changed around fairly easily; that is, desks, screens, phones, cables, etc. can be rearranged without undue effort. A modular phone system is one in which handsets or phone sets can be unplugged and moved or rearranged within a building or department. Modular software is software in which a number of separate or related utilities, tools, and functions can be used together in a number of ways. For example, there may be a variety of functions that do file conversions, image processing, filtering, special effects, etc. which can be used separately or in conjunction with a variety of programs. Some of the more flexible, stand-alone "plugins" exhibit these properties of modularity. For example, there may be a watercolor plugin which can be used independently to alter the contents of a graphics file, or may work as a plugin in the context of several programs such as an image processing program, a drawing program, etc.

modular multi-satellite preprocessor MMSP. A frame synchronizer designed to provide an interface between a host computer and synchronized mapper telemetry data. The MMSP takes the raw telemetry data, frame aligns and samples it, and transmits the information to the host computer, where it is further processed and the image information extracted from the data.

modulate To change gradually from one state to another. To tune or adjust. To vary the amplitude, frequency, or phase, typically to add information to a carrier wave. To change the velocity of electrons in an electron beam, as in a cathode-ray tube.

modulation A key element in the transmission of information. By changing or modulating an electrical pulse through a wire or other conducting medium, or an airborne electromagnetic wave, it is possible to convey information. Similarly, by manipulating its intensity and duration, light can be modulated to send information. Some of the simplest forms of modulation include turning a signal on or off, or varying it between high and low states.

For computer users, one of the most familiar modulating devices is the dialup modem, which takes a digital signal from the computer and modulates it to be carried over analog phone lines. At the receiving end, a modem then demodulates the signal, turning it back into digital signals that are transferred to the receiving computer.

There are many modulation techniques used throughout the telecommunications industry, some very simple, and some so sophisticated only computers can control them. The most common types of modulation are amplitude modulation (AM), frequency modulation (FM), and phase modulation (PM). Sometimes different modulation schemes are combined. Each scheme has its own unique characteristics.

Early detractors said frequency modulation was mathematically impossible, but Edwin Armstrong demonstrated, after 10 years of hard research applied to the problem, not only that it could be done, but also that it was a great thing. It has since been used in thousands of applications from radio programming to cordless phones and burglar alarm systems. Another important contribution to modulation was the work of John R. Carson, who demonstrated how a portion of a modulated signal could be transmitted, instead of the whole thing, and the original signal rebuilt at the receiving end, thus reducing bandwidth without loss of information. See amplitude modulation; Armstrong, Edwin; frequency modulation; phase shift keying; quadrature amplitude modulation; sigma-delta modulation; single sideband.

modulation protocol A data encoding technique used to convert digital data into analog signals. This determines the raw (uncompressed) speed at which the modem can transfer data. Current modems incorporate more than one protocol. See modem.

modulation, light A means of conveying information by manipulating a beam of light. The light can be directly influenced, by turning it on or off, or varying its intensity; it can be indirectly influenced by interposing shutters, gels, or other objects between the sender and the receiver. Light modulation is used in fiber optic transmissions, with lasers and light-emitting diodes used as common light sources.

moiré 1. In raster-oriented imagery, moiré is a visual artifact that appears as an undesirable, distracting secondary pattern which disturbs the intended appearance of the image. 2. In traditional printing on a press, especially process color printing, small dots are often interleaved to simulate the appearance of more colors. If the angles and patterns of these dots are not carefully controlled, a moiré pattern, resembling light through silk, may emerge. Better desktop publishing programs provide print settings to set the angle and type of halftone to match the technology on which the job is printed. 3. In video images, mixing high frequencies can create an undesirable, visible, low-frequency moiré.

moisture barrier A cover, sheet, bag, or other barrier, usually plastic, intended to retard or prevent moisture from coming in contact with building structures, wires, or electrical components. Moisture barriers are used to prevent rot, condensation, and electrical short circuits.

MOKE magneto-optic Kerr effect. See Kerr effect.

molding raceway A channel system incorporated into wood, plastic, or metal moldings to hold, protect, and direct interior wiring circuits. Molding raceways are of modular construction with a variety of fittings, so individual sections can be interconnected and holes can be punched where needed. Molding raceways are commonly used on baseboards and wainscots, where they blend naturally with the decor. See raceway.

Monolithic Microwave Integrated Circuit MMIC. Analog circuits incorporating a number of integrated functions operating at microwave frequencies. Many types of MMICs can be purchased for $5 or less per chip. MMICs provide support for wireless communications technologies, making it possible to design low-cost, high-bandwidth data transmissions links. MMICs can be used to convert between baseband and modulated microwave signals and, as such, can be used in conjunction with traditional integrated circuits (ICs) handling the baseband signals.

MMIC arrays developed by the NASA/Lewis Research Center and the Air Force Rome Laboratory were demonstrated in the mid-1990s in conjunction with NASA Advanced Communications Technology Satellite (ACTS) technology. These proof-of-concept MMIC arrays were in the K/Ka-band frequencies,

M

Radio Frequency Transmission Schemes		
Format	Abbrev.	Notes
ALOHA		A free-for-all style of transmission; any source transmits at any time, and continues to transmit if there is an acknowledgment. It is not a high-efficiency method, but there are circumstances where it is practical.
Code Division Multiple Access	CDMA	A hybrid scheme which incorporates time/frequency multiplexing to provide spread spectrum modulation. Thus, central channels can be handled without timing synchronization.
Frequency Division Multiple Access	FDMA	A traditional method of channel allocation in which bandwidth is subdivided into frequency bands, with guard bands providing a buffer between channels.
Packet Reservation Multiple Access	PRMA	A type of enhanced TDMA which incorporates aspects of S-ALOHA. Suitable for mobile transmissions.

indicating that high-density MMIC integration at 20 and 30 GHz was feasible.

Once MMICs became commercially available, they began to be of interest for many types of commercial and scientific applications. The Search for Extraterrestrial Intelligence (SETI) League, for example, considers the technology useful for constructing research devices for space communications and the possible interception of communications emanating from other regions of the Universe.

The Wireless Systems Innovation lab of NTT has developed uniplanar and multiplanar MMICs more compact and less expensive than earlier technologies that are suitable for radio, wireless area networks (WANs), and satellite transponders. With MMICs incorporated into beam-forming networks, it is envisaged that Earth station satellite terminals can be designed to be as small as cellular telephones.

monopole A slender self-supporting tower for attaching wireless antennas/aerials.

Moonbounce, Earth-Moon-Earth bounce EME. A means of using the Moon as a passive reflector for communications signals. Due to the great distances involved, very large antennas and strong signals are required, but given these in conjunction with the right weather conditions, Moonbounce transmissions have been demonstrated.

The first Moonbounced signal was recorded in January 1946 in New Jersey, where army engineers used a recently invented FM transmitter and receiver developed by E.H. Armstrong to send pulses to the Moon, which returned as a slight hum. This was a significant achievement as it not only showed the potential of FM broadcasts, but also demonstrated that radio waves could pass through the ionosphere and beyond.

Gordon E. Moore – Intel Founder

Gordon E. Moore, cofounder of Intel Corporation not only cofounded and headed up one of the most successful computer chip companies in the world, but is also well remembered for his predictions about semiconductor evolution now encapsulated as "Moore's law." [Photo copyright Intel Corporation; used according to Intel Press Room conditions.]

Moore, Gordon E. (1929–) An American chemist and business executive, Moore cofounded Fairchild Semiconductor, then Intel Corporation, in 1968, along with Robert Noyce, one of the inventors of integrated circuit technology (1959). In 1975, Moore became President and CEO until elected chairman and CEO in 1979. He retained the position of CEO until 1987 and became chairman emeritus in 1997. In May 2001, Moore retired from the board, having reached the age at which he had set mandatory retirement from the corporation.

Moore is a fellow of the IEEE Society and Chairman of the Board of Trustees of the California Institute of Technology. He was awarded the National Medal of Technology in 1990. See Intel Corporation; Moore's law; Noyce, Robert.

Moore's law Semiconductor chip technology will roughly double in capacity (and circuit density) about every year or so (later revised to every 2 years). This prediction, charted by Gordon E. Moore at a speech in 1965, turned out to be memorably prescient and has since been the basis for many industry planning decisions and forecasts. See Gilder's law; Moore, Gordon.

monochromator A device used in spectroscopic scanning devices. For slit to fiber transmissions, couplers/adapters may be incorporated into the device. For fiber to fiber transmissions, off-axis paraboloidal mirrors may be used to eliminate aberrations and to provide a point image from a collimated beam to support high-resolution applications. The device may include multiple gratings to permit selection of the desired spectral range. Monochromators are designed to support different optical wavelengths and more than one device may be needed, depending upon the wavelength selections desired. See off-axis paraboloidal.

Morse code A system of character encoding using dots and dashes, or long and short sounds or lights, that can be readily sent over distance over many types of transmission media due to its simplicity. International Morse code (continental Morse code) and American Morse code (railroad code) have been derived from this.

Morse code is flexible in that it can be sent with tones, clicks, dots and dashes, and lights, in a variety of media. In 1862, two Philadelphia inventors patented a signal light system using a shuttered oil lamp for sending Morse code which was intended to be mounted on the masthead of ships.

International Morse code developed from Austro-Germanic code, a variation on Morse code used in radio transmissions partly because American Morse code, while suitable for telegraph communications, was more difficult to interpret over radio waves. In 1851 it became the code of choice for transatlantic cable communications. Basic skill in Morse code has been a requirement of receiving amateur radio licenses for many decades. The code was apparently developed by Morse's collaborator, Alfred Vail, and is named for the inventor of the printing telegraph, Samuel F.B. Morse. See Morse code history.

Morse code history The original paper tape printing telegraph designed by Samuel Morse employed a system of numbers which were then correlated with words, according to a lookup reference. The lookup reference developed by Morse was very large and the system itself somewhat slow and cumbersome; it required the maintenance of a reference and the somewhat arbitrary assignment of nonmnemonic code number sequences to every word. A simpler, more direct system was needed.

Alfred Vail was from a family of fabricators and acted for years as assistant to and collaborator with Samuel Morse. Mechanically adept, he built many of the mechanical components designed by Morse. In the process of creating the mechanisms for the Morse printing telegraph key, Vail changed the orientation of the keying mechanism from horizontal to vertical, thus providing a more comfortable hand position. The change also resulted in a stylus which would lift up from the paper, leaving dots and dashes, rather than zigzag-shaped dips on the tape record that Morse's original mechanism produced.

Vail's assistant, Baxter, reported to Franklin Pope that Vail set to work simplifying Morse's unwieldy lookup code system. Vail apparently visited local printers to analyze typesetting cases to determine the frequencies of letter usage. Pope subsequently reported the story in 1888 in *The Century: Illustrated Monthly Magazine*. The code Vail developed evolved into American Morse code, and International Morse Code became a further streamlined variation. As Morse's assistant, Vail had agreed to turn over his inventions to the elder inventor. [Thanks to Karen Weiss and B. Neal McEwen for unearthing and reporting Vail's possible unacknowledged contribution to history.]

Morse sounder A type of early telegraph sounding instrument, which used audible clicks to broadcast the incoming message rather than a paper tape printout, which was slow. The sounder incorporated an electromagnet as a pole piece, mounted on a pivoting sounding lever with two stop positions. Releasing the magnet as it was energized produced the clicking sound. The duration of the clicks represented the coded dots and dashes of the Morse code system and were interpreted aurally by the receiving operator.

Typically the sounder was connected to the sending instrument with only one wire. The viability of the single wire circuit was observed by Steinheil in 1837 in Germany, and independently the following year by Morse in America. Both discovered that a second wire was not needed to complete the circuit if the two instruments were connected through the ground, using it as the return path for the circuit. This worked even over distance.

Morse, Samuel Finly Breese (1791–1872) An American artist and inventor in the 1800s chiefly known for the code that bears his name. He was a respected artist and one of the founders of the National Academy of the Arts of Design. In the 1820s he became increasingly interested in science and invented electromechanical telegraph devices, some of the first inventions to use electricity for communication.

With advice and assistance from J. Henry and L. Gale, Morse was able to construct a basic working design for the telegraph by 1837.

Samuel Morse demonstrated his invention to the presidential administration in 1838 and in 1843 won funding support from the U.S. Congress to construct a telegraph line between Baltimore and Washington, D.C. He sent his first public message over this line in May 1844, an event that launched a revolution in communications.

Morse became friends with the Vail family, who were talented fabricators and were able to assist him in constructing practical working models of his ideas. Many of Morse's inventions were built by Alfred Vail, Morse's assistant and collaborator. See Gale, Leonard D.; International Telegraph Union; telegraph; telegraph history; Vail, Alfred.

Mosaic, NCSA Mosaic NCSA Mosaic is one of the most significant landmark applications in the history of the Internet as it spurred the evolution of point-and-click visual interface that nontechnical computer users, even children, could understand and quickly learn to use. The simplified Internet access and support of images provided by Mosaic and its successors dramatically fueled the growth of the World Wide Web.

The first version of x-mosaic was programmed at The National Center for Supercomputing Applications (NCSA) by Marc Andreessen in 1992 with Mosaic 1.0 released in November 1993. Mosaic was subsequently enhanced and ported to support the X Window System, Apple Macintosh, and Microsoft Windows platforms.

In 1994, Andreessen and other NCSA personnel left to form Mosaic Communications Corporation based on the Mosaic concept. However, due to trademark ownership by the University of Illinois, the company and the software product were renamed Netscape Communications Corporation and Netscape Navigator, respectively, with the Board of Trustees of the University of Illinois retaining copyright ownership of NCSA Mosaic. The University of Illinois entered into an agreement for Spyglass, Inc. to negotiate commercial NCSA Mosaic licenses in 1994.

Other commercial Web browsers were in development in the mid-1990s, including OmniWeb by Lighthouse Design, Ltd., which was released in March 1995. In spite of the commercialization of browsers, the Software Development Group at NCSA continued development on the Mosaic project until Mosaic 3.0 was released in January 1997. See Gopher, Internet Explorer, Netscape Navigator, OmniWeb.

MOST 1. See Multidisciplinary Optical Switching Technology Center. 2. See multiplexed optical scanner technology.

motion pictures Any images which, when sequentially displayed, convey the appearance of motion, whether in real time or by presentation of a fast sequential series of still pictures, especially videos, film reels, and animated computer images. Traditional motion film pictures consist of a series of still images on a transparent medium played through a

M

projector, usually from 20 to 30 frames per second, with 24 or 30 being common, as these are the speeds at which human perception merges successive still frames into a cohesive impression of connected motion.

The development of motion picture photography owes some of its roots to a bet over a dispute as to whether a running horse lifted all four hoofs off the ground. Thomas Edison was one of the first to experiment with displaying a series of still frames in rapid succession in 1889. The first commercial motion picture, backed by the Canadian Pacific Railway, is attributed to Clifford Sutton in the early 1900s. See animation, celluloid, MPEG.

Motorola A significant computer chip designer and manufacturer and electronic appliances manufacturer since the 1960s. It is descended from the Galvin Manufacturing Company from the early 1930s. In 1974 it released the MC6800, the first in a long family of chips still being developed a quarter of a century later. One of the first microcomputers developed with the Motorola family of microprocessors was the Altair 680, released late in the fall of 1975. Since that time whole families of computers have been based on the subsequent MC68000 family of chipsets, including Macintosh, Atari, Amiga, Sun, Apollo, SGI, NeXT, and others. Motorola is also well known for products in the mobile data communications industry. The Motorola CPU Sample chart shows a brief summary of some of Motorola's best-known desktop computer microprocessors, prior to the collaboration with IBM to produce the PowerPC chips.

In 1998, Motorola teamed up with the McCaw/Gates Teledesic project to provide Celestri technology to the orbiting satellite network. See Altair 680. See Sampling of Evolution of Popular Motorola CPUs chart.

Mott insulator An interesting material discovered in the 1930s that appears to have the band structure of a conductor but acts, instead, as an insulator because of its energy gap characteristics. However, if the material is doped, it can be encouraged to behave as a high-temperature superconductors. Mott insulators have spurred closer scrutiny into band and energy gap structures to improve our understanding of their interactions and importance. When they are better understood, it will be easier to predict and harness the capabilities of poorly understood materials. The Mott insulator is named after Nobel laureate N. Mott.

Mountain Bell The familiar name for the Mountain States Telephone and Telegraph Company.

Mountain States Telephone and Telegraph Company An early telephone company, better known as Mountain Bell, which was formed in 1911 from the merger of the Tri-State and Colorado telephone companies, and the purchase of the Rocky Mountain telephone company.

mouse A hardware human interface device that receives hand and finger movements and transmits them to a computing device. They are then interpreted into actions by the operating system and applications software. The mouse is named for its basic shape, which typically consists of a palm-sized, rounded or squarish object, with one or more buttons under the fingers and a "tail," a cord that electrically connects the mouse with the computer. Mice come in various shapes and sizes: friction mice have a ball on the side that makes contact with a hard surface; optical mice

Sampling of Evolution of Popular Motorola Central Processing Units (CPUs)					
Processor	Introd.	Proc.	Data Bus	Addr. Bus	Notes
MC6800	1974		8	16	Used in Altair 680.
MC68000	1979	32	16	23	16 32-bit registers. Supervisor and user mode. CISC architecture.
MC68010		32	16	23	Virtual memory.
MC68020	1982	32	16/32	32	256-byte cache. Dynamic bus sizing.
MC68030	1987	32	16/32	32	Paged MMU on processor. 16-byte burst.
MC68040	1990	32	32	32	FPU, cached Harvard buses.
MC68060	1994	32	32	32	Superscalar pipelined. Power-saving.
The PowerPC family was created in collaboration with IBM and Apple Computer. This was a RISC-based line from the PPC 401 to 750 (G3), with speeds ranging from 20 to 500 MHz.					
PowerPC	1994	64	64	32	From 50 to 135 MHz
MPC750 (G3)	1998	64	64	32	300 MHz
MPC7400 (G4)	1999	64	64	32	Double-precision FPU, AltiVec instruction set. External L2 cache interface. 350 to 450 MHz.

require a grid or special pad. Laptop variations include finger pads and rollerballs, which are not strictly mice, but which employ the same basic movement and input concepts.

mouse history The invention of the computer mouse is attributed to Doug Engelbart and is variously reported as having been invented around 1959 to 1963. By the late 1960s, Engelbart was testing a three-button mouse in conjunction with a keyset that was used in the other hand. During the early 1980s, when the Apple Lisa was being developed (the first of the Macintosh line), there were discussions at Apple as to whether to use a two- or three-button mouse. The testing and rationale supplied by Larry Tesler indicated a one-button mouse was completely appropriate, and the Macintosh line still works very well with this device 15 years later. The majority of competing desktop computers use two-button mice.

MP See Multilink Protocol.

MP3 A popular abbreviation for MPEG-1 Layer 3 or MPEG-2 Layer 3. See MPEG.

MPEG Motion Pictures Experts Group. A series of international standards developed by a joint committee under the aegis of the International Organization for Standardization (ISO) and the International Electrotechnical Commission (IEC) to facilitate the development of digital video and audio formats and decoding schemes. Leonardo Chairiglione and Hiroshi Yasuda originated the MPEG development efforts in 1988. MPEG has received widespread acceptance for the playback of digital animations.

The development of media formats is somewhat dependent upon how the final compressed product will be used. In the case of MPEG series standards, consumer entertainment products (e.g., optical media) were taken into consideration, and thus the perceptual characteristics of the humans who would eventually be viewing or listening to the decoded MPEGs were factored into its development.

Video and audio technologies typically require a lot of bandwidth and file space, so a large part of the MPEG effort has concentrated on decompression schemes and fast playback algorithms. The compression itself is left up to the discretion of individual vendors.

MPEG is a family of standards numbered from 1 to 4, but the numbers do not necessarily indicate a progression. Some changes and development for specific needs occurred on the road from MPEG-1 to MPEG-4. What the different variants have in common is that they support the compression and playback of digital audio and video data (see chart).

MPEG is an asymmetric technology, based upon the use of MPEG encoders (to compress and store or transmit data) and MPEG decoders (to decompress and display data). It is asymmetric in the sense that the speed and processing power required for encoding is not necessarily the same as that needed to decode the MPEG data stream.

MPEG decoders are used to decompress and play back MPEG-based sound and videos either separately or together. The digital technology offers many appealing options for this process whether or not the MPEG video is combined with MPEG sound or with Dolby Digital sound. One of the reasons for the quick acceptance of digital videodiscs is their versatility. For example, on a DVD disc, there may be a video track, several sound tracks, and even data tracks for the display of textual commentary or subtitles. Thus, a favorite movie could be viewed several times, in its original form, in other languages, or in its original form with subtitles or other textual information. Many commercial entertainment DVDs include audio commentaries with the verbal thoughts and impressions of the directors and actors superimposed over the original movie (with the original soundtrack at a lower sound level in the background). This value-added means of packaging entertainment titles is one of the forces behind the quick acceptance of MPEG-based commercial products.

To be practical for the delivery of cinema-style education and entertainment programs, the playback display of MPEG data needs to be fast, so the technology is optimized with playback in mind.

Individual contributors hold a number of patents to various technologies which have been incorporated into MPEG. These contributors agreed in writing to provide the technology nonexclusively at fair and reasonable royalty rates.

There have been several enhancements to MPEG since its introduction as shown in the MPEG Versions chart. See animation, B-frame, I-frame, JPEG, MPEG decoder, MPEG encoder, P-frame.

MPEG decoder A mechanism for "unraveling" the data on an MPEG-encoded medium to make it accessible for playback. MPEG decoders are built into DVD players, Web browser plugin applications, and a number of other consumer products. The decoder sometimes has to handle more than just the fast decompression and display of frames. Many MPEG-encoded products come with several playback options, necessitating that the decoder be coupled with human interface algorithms to enable the user to select the MPEG options desired. On commercial DVDs, these algorithms are usually presented in much the same manner as application menus and buttons as are used on desktop computers. When the user selects an option, the information is used to configure and control the way in which the information is decoded for playback. See MPEG, MPEG encoder.

MPEG encoder A mechanism for creating MPEG-format compressed data styles for storage or transmission. While the format for MPEG compression has been standardized, the algorithms and hardware systems for creating MPEG-format data have been left up to the discretion of developers. Thus, MPEG-format files can be created in a number of ways on a variety of media and developers can create the MPEG files in a way that is appropriate for their application. MPEG decoders can even be incorporated into dedicated chips, for optimal portability or optimization for the task of encoding MPEG-format files.

In general, MPEG encoding seeks to minimize file sizes without significantly compromising picture

M

quality. This enables faster delivery of MPEG data over computer networks and makes it possible to fit more content on storage media. To accomplish these goals, a number of clever schemes for exploiting picture redundancy (where a frame is similar to a preceding frame) have been developed, such as *differential encoding*.

Differential encoding is when you analyze a series of image frames and make some assessments about their similarities and differences and then use this information to remove redundancies (and thus reduce file size). Since video sequences commonly include many frames in a row with almost the same picture information (e.g., a kite fluttering against a blue sky),

MPEG Versions	
Version	Description
MPEG-1	A relatively low-quality video standard initially developed for *progressive video*. Later adaptations also supported *interlaced video* but, in general, MPEG-1 has been superseded by other formats for applications that require higher quality video.
	Coding of Moving Pictures and Associated Audio for Digital Storage Media at up to about 1.5 Mbps. ISO/IEC 11172, standardized between 1993 and 1995.
	An optimized 1.5 Mbps bit stream for compressed video and audio, for compatibility with existing CD and DAT data rates. Non-interlaced color video is typically implemented at 352×240 (288 in Europe), which is relatively low resolution, as it derives from a CCIR-601 digital television standard. Replay speed is 30 frames per second (25 in Europe), fast enough for natural-looking motion. Sample precision is 8 bits.
MPEG-2	A higher quality format than MPEG-1 and supports both progressive and interlaced video and both two-channel and multichannel (surround) sound. It is a popular format that gained fast consumer acceptance and is suitable for a variety of digital videodisc (DVD), high definition television (HDTV), and other video applications. In consumer products, MPEG-2 sound compression technologies are not always used for the audio portion of an MPEG-2 disc; sometimes Dolby Digital is substituted (or included with the others). MPEG-2 supports a variety of bit-rates and picture resolution levels. Multiple audio and video streams can be multiplexed together for recording or transmission over a network.
	Generic Coding of Moving Pictures and Associated Audio ISO/IEC 13818, presented in draft form in 1993. ITU-T recommendation H.262.
	Similar in structure to MPEG-1, the documentation includes four parts in addition to the categories discussed in MPEG-1. MPEG-2 can address very low bit-rate applications with limited bandwidth needs, and support for surround sound multichannel applications. Video resolution is typically implemented at 720 to 550×480, somewhat similar to computer monitors, and a frame may be either interlaced or progressive formats.
MPEG-1+	MPEG-1 presented at MPEG-2 resolution. Frames are de-interlaced and compressed.
MPEG-3	Originally slated to support HDTV, but it was found that the MPEG-2 standards could be devised to support both DVD and HDTV and MPEG-3 was never developed. MP3 is sometimes mistaken for MPEG-3 but MP3 actually refers to either MPEG-1 or MPEG-2 Layer 3 formats.
	Merged into MPEG-2 when it was decided that MPEG-2 syntax could be scaled to support HDTV applications. Often confused with MPEG-1/2 Layer 3 (MP3).
MPEG-4	Developed to support the world of data communications over networks which are frequently shared or limited in their transmission rates. Thus, it is optimized for the world of computer networks, uses different compression techniques than the other MPEG formats, and is more appropriate for applications such as videoconferencing, whiteboarding, and streaming video rather than for high-quality optical media-based consumer entertainment applications.
	Very Low Bitrate Audio-Visual Coding. Launched in 1992 to develop new algorithms for providing support for a wider range of applications, and to improve efficiency. New applications include low-bitrate speech coding and interactive mobile communications.

there are opportunities to save time and storage/transmission space by encoding only those portions of the image that change from one frame or field to the next. As with many optimization schemes, differential coding comprises a tradeoff. If a series of frames must be reconstructed "on the fly," based upon a reference frame, with only the differences encoded, you cannot randomly jump to any frame in the sequence in the playback process and see the full picture without some fancy footwork and fast processing in the background. The other disadvantage to differential coding is that the playback algorithm must handle the reconstruction process and the error-checking accurately, or errors in decoding the reference frame, from which the others are derived, would be propagated through any frames that subsequently depend upon it. To help mitigate propagation errors that could occur from frame to frame in the playback of differentially encoded MPEGs, a scheme to insert the occasional full-frame picture has been developed. An intra-coded picture (I-picture) is a complete image that is inserted every few frames to stop any reference image errors from continuing for a large number of frames. How often these frames should be inserted is a matter of balancing file size and potential error levels. Since differential encoding is intended to reduce file sizes, inserting I-pictures too frequently would negate this advantage.

Sometimes images are constructed based upon the previous frames, but image processing can work the other way as well. A prediction-error picture (P-picture) is an image constructed from information taken from previous I- or P-pictures, while a bidirectionally encoded picture (B-picture) is one that uses later images to construct a previous image. Confused? Think of it this way: imagine watching a video of a car driving past a house. Most of the image stays the same from one frame to the next since the house isn't moving (assume a stationary camera photographing the scene). As the car moves past the house, however, some of the image changes. File space can be saved by encoding only the portions that change and constructing the successive frames from the information that is known rather than displaying every frame. Now here comes the tricky part. Suppose this scene opened with the car parked in front of the house before pulling out and driving by. There's no way to construct the portion of the house obscured by the car from previous frames because the information isn't there. With bidirectional predicted/interpolated encoding, the processor can look forward, to the frames of the house that show the portions that were obscured before the car drove off-camera, and insert that information into previous frames in the encoding process. Thus, differential encoding may be used to construct the scene based upon previous and later frames. Sound complicated? It can be, which is why MPEG is an asymmetric technology. Depending upon how it's done, a significant amount of processing may be required to encode a movie to fit onto a DVD, for example, and the equipment to accomplish the task may cost hundreds of thousands of dollars, compared to a couple of hundred dollars for the playback deck. See MPEG, MPEG decoder.

MP lambda switching MPlS (note MPLS in all uppercase is used as an abbreviation for multiprotocol label switching). multiprotocol lambda switching. See lambda switching.

MPLS See Multiprotocol Label Switching.

MPOA MultiProtocol Over ATM. A client/server protocol integration effort specified by a working group of the ATM Forum to provide direct connectivity across an ATM network between ATM hosts, legacy devices, and future network-layer protocols from different logical networks. This will enable the production of lower-latency, scalable ATM internetworks through a standardized virtual network with fewer router hops. See Anchorage Accord, LANE, IPv4, Next Hop Resolution Protocol.

MPOA Client An ATM term. A device which implements the client side of one or more of the MPOA client/server protocols, (i.e., a SCP client or an RDP client). An MPOA Client is either an Edge Device Functional Group (EDFG) or a Host Behavior Functional Group (HBFG).

MPOA over ATM sub-working group A group that seeks to solve some of the implementation problems associated with asynchronous transfer mode (ATM). It is integrating LAN Emulation (LANE), Next Hop Resolution Protocol (NHRP), Classical IP, multiprotocol encapsulation, and multicast address resolution in order to provide end-to-end internetworking ATM connectivity. MPOA is a packet-oriented protocol similar to LANE. The group provides courses, support, research, documents, and systems testing services.

MPOA Reference Model MultiProtocol Over ATM Reference Model. A specification approved by the ATM Forum in June 1997 for routing/switching over ATM networks. There are Internet MPOA resources (links, white papers, specifications, etc.) at the ATM Forum's site.
http://www.atmforum.com/atmforum/specs/approved.html

MPP See Multichannel Point-to-Point Protocol.

MRI See magnetic resonance imaging.

MRU maximum receive unit.

MS Mobile Station.

MS-CDEX A set of Microsoft DOS extensions for CD-ROM which allow MS-DOS to recognize the presence of a CD-ROM drive and access it accordingly.

MS-DOS Microsoft disk operating system. MS-DOS originated from a commercial text-oriented operating system developed from Tim Paterson's QDOS (which was based upon a CP/M manual according to Paterson), first released by Microsoft in 1981 to accompany IBM's Intel-based microcomputer system. This was a somewhat different move for IBM, as the company had often created its products in-house. But IBM was under pressure to release a successful microcomputer in order to avoid being locked out of the growing market; Radio Shack at one point had almost 80% market share with its TRS-80 line. IBM's move to look outside its own research and development resources provided a window of opportunity for emerging computer companies. They contracted to purchase an operating system from Microsoft and, in a remarkable turn of events, Microsoft managed to retain the rights to market the product themselves, in competition with IBM, who called their version PC-DOS.

Early versions of MS-DOS were intended for single-user applications, but by version 3.1 network functionality was added. At about this time, multitasking graphical operating systems were being released by several other vendors, and there was pressure on MS-DOS to provide features found on other systems.

MS-DOS became widespread through the 1980s, only slowly giving away to Microsoft's later graphics-based Windows products, which were developed in the mid-1980s as front-ends to MS-DOS. MS-DOS's text interface is now infrequently used. See MS-DOS history, operating system.

MS-DOS history The Microsoft disk operating system (MS-DOS) was originally Tim Paterson's QDOS. Paterson has reported that he gleaned ideas from a CP/M operating system manual published in the mid-1970s. CP/M was developed by Gary Kildall, a university professor and programmer, founder of Digital Research (originally Inter-Galactic Research). When IBM first approached Microsoft in the early 1980s to purchase the BASIC programming language for their personal computer, they apparently thought they were also obtaining the rights to CP/M. Bill Gates signed a nondisclosure agreement to work with IBM, and promised to supply BASIC. When IBM found out they hadn't purchased an operating system as well, Gates suggested they call Digital Research, developers of CP/M. Gates didn't want to lose the languages contract for lack of an operating system, and Microsoft didn't have an operating system that would meet IBM's needs. Up until this time, Microsoft had concentrated on languages, Digital Research had concentrated on operating systems, and Kildall had not expected the "gentleman's agreement" to change.

IBM was in a hurry. They paid a call to Digital Research (DR) at a time when Gary Kildall had another commitment. His wife/business partner was present to talk to the corporate giant, but was uncomfortable with signing the one-sided nondisclosure agreement presented by IBM without consultation with Kildall, especially since DR's attorney didn't like the terms of the agreement. IBM went back to Gates and, seeing an opportunity, Gates offered to supply an operating system as well as BASIC. IBM accepted. Kildall was probably surprised at not having the opportunity to meet further with IBM to negotiate terms, as was customary at the time.

Since Microsoft didn't have an operating system that would fulfil the contractual agreement with IBM, they went to a company called Seattle Computer Products (one wonders why they didn't first contact Kildall about licensing CP/M) and bought a simple operating system called QDOS (Quick and Dirty DOS) developed from CP/M by Tim Paterson. Microsoft

contracted Tim Paterson to make a few enhancements in collaboration with Robert O'Rear, and soon after delivered the product to IBM. It was released first as PC-DOS by IBM, and later as MS-DOS by Microsoft. There are some small differences between the two. This was the beginning of a long alliance and the launching of a tiny entrepreneurial effort into a major empire, with MS-DOS as the pivotal product. See Microsoft, MS-DOS, Digital Research.

MSA See Metropolitan Service Area.

MSAT Mobile Satellite. A commercial mobile satellite communications service developed jointly by AMSC in the United States and TMI in Canada.

MSAU See Multistation Access Unit.

MSB most significant bit. See least significant bit for explanation of MSB and LSB.

MSC Mobile Switching Center. A switch providing coordination and services between mobile network users and external networks.

MSK See minimal shift keying.

MSN multiple subscriber number (e.g., as in ISDN Q.81 and Q.731 number identification services)

MSP See Message Security Protocol.

MSR magnetic-induced super resolution.

MSS See multispectral scanner.

MSU 1. microwave sounding unit. 2. See mobile subscriber unit.

MTA 1. Macintosh Telephony Architecture. 2. See Major Trading Area. 3. Message Transfer Agent. 4. Metropolitan Trading Area. A Federal Communications Commission (FCC) designation for a region. This is an administrative process used to facilitate licensing for various communications services.

MTBF See mean time between failures.

MTI Moving Target Indication. In radar, the analysis of Doppler-shifted returning signals to select moving targets from various other nonsignificant or interfering objects (clutter).

MTM Maintenance Trunk Monitor.

MTS 1. member of technical staff. 2. Message Telecommunications Service. An AT&T designation for standard telephone service through direct dialing. 3. See Message Transfer System.

MTSO See Mobile Telephone Switching Office.

MTTR mean time to repair. A computed average of the amount of time it takes to bring a broken or faulty device back into service.

MTU 1. See maintenance termination unit. 2. Maximum Transfer Unit. 3. multiple tenant unit.

Mu-law encoding A pulse code modulation (PCM) coding and companding standard commonly used to compress audio signals. This is useful for speeding up data transmission and reducing storage space. Mu-law coding can improve the signal-to-noise (S/N) ratio without increasing the amount of data and thus is suitable for use in telecommunications applications. Mu-law is logarithmic in nature; it carries more information about the low-amplitude samples than high-amplitude ones and is based upon the known statistical characteristics of audio signals. Segments of digitized audio, called *samples*, are reduced to about half their original size. Zero code suppression is used.

Mu-law encoding compresses at almost the same ratio as A-law encoding.

Mu-law can encode speech in 8-bit format by sampling the audio waveforms at 8000 times per second. A-law/Mu-law pulse code modulation (PCM) is the most widely used scheme for encoding telephone conversations. A-law PCM is primarily used in European systems, while Mu-law PCM is primarily used in North American and Japanese systems. See A-law encoding, E carrier, pulse code modulation, quantization, U-law encoding.

MUD Multi-User Dungeon/Dimension. An addictive role-playing game prevalent on the Internet and most college campuses since MUDs were invented in the late 1970s. The first MUD has been attributed to R. Bartle and R. Trubshaw at the University of Essex.

mud ring A mounting bracket, for mounting a faceplate in a wall, for example. These can be used instead of receptacle boxes for terminating low-voltage cables. Mud rings should be mounted in dry areas away from alternating current (AC) wiring and should be positioned so that the faceplate mounted on top of the mud ring is straight.

MUF See maximum usable frequency.

Multi-Agent Systems Laboratory MAS Lab. A lab at the University of Massachusetts at Amherst involved in the research, analysis, and development of sophisticated artificial intelligence (AI) problem-solving and control architectures for single- and multiple-agent systems. The lab has developed realtime AI and control systems and techniques for the coordination of multiple agents.

Multi-Service, Multi-Carrier, Distributed Communications MMDC. Pioneering commercial technology from Tekmar Sistemi for radio communications over fiber applications. This is particularly applicable to in-building, short-range radio applications and is currently being used in the design and manufacture of a MMDC system for the Singapore Mass Rapid Transportation system for installation in trains, elevators, platforms, and stairways. The MMDC system will provide a platform for common commercial services, including FM/DAB Radio Rebroadcast, TV Rebroadcast, paging, and various digital communications services (AMPS/DAMPS, GSM, PCS, 3G,. etc.)

Multi-Vendor Integration Protocol MVIP. MVIP originated in 1989 from an initial consortium of three companies: Natural MicroSystems, Inc., Mitel, and GammaLink; four more helped the project to coalesce a year later. MVIP has since become one of the two common software/hardware bus standards in computer telephony (along with SCbus). The purpose of the standardization was to bring together the various telephone and computer technologies so they could readily interconnect and intercommunicate. MVIP provides an open, nonproprietary, uniform, yet flexible way of providing telephony components with computer equipment through open software development environments. In other words, phone-related technologies can be accessed and controlled through a desktop computer. The MVIP standard includes the capability to reconfigure "on the fly" to handle various call functions.

The original single-chassis standard designed for a synchronous environment was MVIP-90 and additional versions followed, including H-MVIP (high capacity MVIP), and MC-MVIP (multi-chassis MVIP). The Outline of MVIP Versions chart shows three MVIP formats.

MVIP supports Unix, OS/2, and Microsoft DOS or Windows. The work was officially taken over in 1994 by the GO-MVIP project. See Go-MVIP. http://www.mvip.org

MultiAudio A specification released by the Optical Storage Technology Association (OSTA) in July 2001. MultiAudio specifies a Table of Contents mechanism for retrieving, managing, and playing back audio files on popular optical formats such as CD or DVD media. Thus, direct reading of the Table of Contents, rather than lengthy disc initialization, is facilitated. It also facilitates the development of playlists and lists of various attributes of the files, such as genre, artists, album titles, etc.

The MultiAudio format is an extension of MP3 capability on consumer electronics devices and is backwardly compatible with current MP3 disc players. A standard scheme for storing the track name, year recorded, performer name, composer name, songwriter name, arranger name, album name, and genre is defined. Hopefully, future versions also include conductor and soloist (especially for classical music where guest conductors and soloists are common). The OSTA MultiAudio logo can be used through a royalty-free license by manufacturers demonstrating compliance with the specification through the self-certification requirements. The test disc is available through a onetime fee from OSTA. The logo will alert consumers to products featuring the enhanced playlist features. See MultiPlay, Optical Storage Technology Association.

multicarrier modulation MCM. A number of modulation techniques for multiplexed transmission of data by dividing the communications channel into smaller units and evaluating the units individually in terms of speed and suitability for transmission. MCM is used to implement Digital Subscriber Line (DSL) services over existing twisted-pair copper wires, which can vary widely in their characteristics.

MCM optimizes bandwidth usage for multiple media transmissions and reduces interference from impulsive and narrowband noise. See carrierless amplitude and phase modulation, orthogonal frequency division multiplex, discrete multitone, discrete wavelet multitone.

multicast An organizational arrangement and physical/data capability in which programming or data are distributed to more than one recipient from one or more sources. Broadcast television, in which one local TV station might broadcast to 50,000 local residents, is an example of a wireless, analog form of multicast communication.

There are a number of ways in which a multicast system can be configured. The one-to-many relationship may be a simple one, with one endstation transmitting to all the other stations receiving data, or it may be selective, in that static or dynamic groups are configured to receive the multicast transmissions. An example would be cable TV services. A single cable TV distributor may have 10,000 customers with premium services, 40,000 customers with basic services, and 20,000 customers with a combination of basic and premium selected services. Thus, multicast management groups are defined for who gets which cable stations.

In a dynamic computing network, multicasting is more subtle and complex, with the added capabilities that certain applications or users can be defined to receive certain services at specific times of the day through certain authorized privileges. A multicast group manager (MGM) is an application for handling the information needed to organize and administer the many and varied aspects of multicast network groups. More specifically, in the context of digital internetworking, multicast is a type of Internet Protocol (IP) address identifier for a set of interfaces. Frames sent from one endstation are received by one or more endstations as opposed to a singlecast or *unicast* topology in which frames sent from one endstation are received by only one station. In IPv6 *multicast addresses* supersede *broadcast addresses*. In ATM networking, the form of the multicast command is

```
atm multicast <address>
```

Xpress Transport Protocol is an example of a multicast protocol coexisting with other protocols (e.g., security mechanisms) that enables a number of different configurations of multicast groups. See anycast, unicast, IPv6 addressing, multicast backbone.

multicast backbone MBone, mBone. Technology that extends the Internet Protocol (IP) to support multicasting, developed by Steve Deering at Xerox PARC (who developed the IP multicast addressing system) with contributions by Steve Casner and Van Jacobson.

The MBone was adopted by the Internet Engineering Task Force (IETF) in 1992 and was first used to simulcast over in March 1992 at the IETF conference. It makes it possible to broadcast a specific transmission (game, surgical procedure, space mission, lecture, etc.) to many simulaneous users over the Internet, thus removing the need for dedicated lines.

MBone supports two-way transmissions of data between multiple network sites. Thus, with MBone, a single packet can have multiple destinations and pass through a number of routers before being split up to run through different paths. The packets reach their destinations at about the same time. Thus, it provides greater support for multimedia capabilities over the Internet. MBone systems are assigned Class D Internet Protocol addresses.

Multichannel Interface Processor MIP. A Cisco Systems interface router processor which provides up to two channelized T1 or E1 serial cable connections to a channel service unit (CSU).

Multichannel Point-to-Point Protocol MPP. A protocol from Ascend Communications similar to Point-to-Point protocol (PPP), but which supports multiple network channels in inverse multiplexed systems.

multichannel video programming distribution See MVPD.

multicore high-flex fiber A type of fiber optic cable bundle with multiple independent cores assembled such that flexibility is maximized and the bend radius comes close to that of wire. Such fibers can bend as much as 90° without significant loss of light transmitting properties, making them ideal for installation inside machinery, sensor assemblies, space vehicles, portable detection devices, and medical instruments.

Multidisciplinary Optical Switching Technology Center MOST. A multidisciplinary research center at the College of Engineering at UCSB, focusing on optical switching technology for advances in synchronous and asynchronous network communications (SONET, ATM). http://www.ece.ucsb.edu/MOST/

Outline of MVIP Versions	
Format	Notes
MVIP-90	Original single-chassis specification consisting of a multiplexed digital telephony bus with 512 × 64 Kbps capacity, distributed circuit-switching capability, and digital clocks architecture. MVIP-90 provides for 16 serial lines.
H-MVIP	High Density MVIP. A scalable, high-capacity, downwardly compatible superset of MVIP-90 adopted as a standard by the GO-MVIP technical committee in 1995. H-MVIP provides for 24 serial lines.
MC-MVIP	Multi-Chassis MVIP. A high-capacity version which provides several alternate physical layer connections with a common set of software interfaces for interconnecting single-chassis MVIP systems and telephony systems. MC-MVIP provides 1536 × 64 Kbps over copper twisted pair (MC1), FDDI-II for 1536 × 64 Kbps over copper or fiber (MC2), or SONET/SDH at 155 Mbps for 4800 × 64 Kbps over fiber.

multifeed dish, multifocus dish A type of parabolic satellite receiving dish which can be positioned to capture signals from more than one satellite at a time, with the signals reflected to a series of feedhorns.

Multifunctional Transport Satellite MTSAT. A Japanese system of satellites with updated transmission capabilities evolving out of the Geostations Meteorological Satellite (GMS) system. The Japan Meteorological Agency administers the operational plans for the satellite system, which will begin with MTSAT-1R and MTSAT-2. The system will make hourly observations for the northern hemisphere and observations each six hours for cloud-motion winds in the northern and southern hemispheres.

Newer sensing and transmission technologies are being introduced into the MTSAT series. In addition to previous capabilities of the GMS systems, MTSAT will quickly relay high-resolution digital images and will include a new infrared channel (IR4) for detecting low-level night clouds. The upgraded transfer capabilities necessitate the installation of new equipment and software. It is expected that the MTSAT systems will come online around 2003, superseding the still operational GMS-5. MTSAT data will be used by more than two dozen countries and by domestic Japanese services such as the Japan Weather Association (JWA).

multihop routing A ubiquitous approach to wired or wireless data transmissions that involves one or more relays or intermediate routing devices (nodes) in the transmission path between the sending and receiving devices. Multihop routing is commonly used to increase the transmissions range or to increase the flexibility of a system so that messages can be transmitted flexibly or to multiple destinations individually or simultaneously. The phrases *ad hoc* or *on demand* are often used in connection with multihop routing to indicate routing that is dynamically assigned to accommodate routing needs at hand.

Multihop routing is characteristic of distributed computer and relay radio networks. The intermediate device may dynamically assign the next hop or be involved in amplification of the signal, as in wireless communications. In analog networks or digital networks with analog relays, generally a price must be paid for the extra hops in terms of attenuation and sources of interference. In optical networks, the trend has been from electromechanical connectors and relays to all-optical relays to reduce attenuation in the various hops.

The length of a hop is related to the type of technology used. Hops in physical wireline networks may range from a few inches to a few feet, or with suitable cables, a few hundred feet. Hops in wireless networks are related to the power of the sending and receiving equipment (and associated regulations) and may range from a few feet to many hundreds of yards. High-frequency wireless propagation over distances of 2000 miles or more is usually accomplished by bouncing the signals multiple times between the Earth's ionosphere and the ground or a body of water. Another strategy is to aim the signal into the F region (a region above Earth that is active in daytime) and bounce the signal within the region such that it doesn't bounce up and down to the Earth, thus preserving some of the strength of the signal.

Depending upon the size or type of network, multihop routing architectures can be designed and configured in a number of ways. Distance-vector, hierarchical, and dynamic source routing are just a few examples. Single-hop topologies are sometimes combined with multihop topologies in a two-level architecture to take advantage of the relative stability of a single-hop signal and the flexibility and distance advantages of multihop transmissions. Some of the important issues related to the design and configuration of multihop networks include optimization, signal strength, and Quality of Service (QoS). See backscatter, ionospheric; M hop; Multihop project; propagation.

Multihop project A wireless network dynamic configuration project supported by the German Federal Ministry of Education, Science, Research, and Technology, initiated in May 2000. The project encompasses the development of wireless, channel-oriented, ad hoc multihop broadband transmissions with decentralized topologies as well as interoperability and WLAN issues.

multilevel coding A coding scheme or system, usually bandwidth-efficient encoding, for carrying more data through a channel in a specified period of time. This is one of many approaches to improving the efficiency of transmissions.

Multilink Protocol MP. Multilink is an Internet standards-track protocol which enables the splitting, recombining, and sequencing of datagrams across multiple logical data links. Originally designed as a software solution for implementing multiple simultaneous channels over ISDN, the concepts are generalizable to multiple PPP links between two systems. The purpose of multilink implementation is to coordinate multiple independent links (i.e., to aggregate a bundle) between a fixed pair of systems to create a virtual link with greater bandwidth than individual parts used alone. See RFC 1990.

Multilink Protocol Plus MP+. An extension to the PPP Multilink Protocol (MP) for managing multiple data links bundled by Multilink Protocol. MP+ adds an inband control channel and remote device management features to MP. Thus, an unconfigured system with the capability to answer incoming calls can be accessed and managed remotely prior to entering various protocol configuration parameters. See Multilink Protocol, RFC 1934.

multimedia A catchall term for sensory-rich communications media, that is, media that contain images, motion, sound, tactile feedback, etc. Multimedia educational and entertainment products are more common than multimedia business applications. Networks need greater bandwidth and faster processing speeds to handle multimedia traffic. See animation, frame buffer, MBone, virtual reality.

multimedia mail Electronic mail (email) that incorporates media capabilities such as audio and visual components whether through MIME or proprietary methods. These types of capabilities were pioneered in the 1970s and 1980s on research computers, implemented on a number of workstations in the early 1990s (NeXT, Sun, etc.), and began to appear on common consumer desktop systems in the late 1990s. Email created and viewed as HTML with graphics and embedded links is one example of the gradual integration of multimedia capabilities into day-to-day mail exchanges. The inclusion of multimedia files along with text-based mail messages is also becoming common, with .jpg, .tiff, and .pdf file formats commonly used for lossy, lossless, and PostScript file types, respectively.

MultiPlay A compatibility specification originally proposed by Oak Technology, Inc. in December 2000 and adopted and promoted by the Optical Storage Technology Association (OSTA). The MultiPlay specification was announced in December 2000, aimed at ensuring Compact Disc Recordable (CD-R) and Compact Disc ReWritable (CD-RW) disc compatibility in consumer CD and DVD players, along with personal computer CD and DVD players. The specification includes Red Book Audio, CD-Text, and VideoCD as the initial formats. The requirements for a consumer CD/DVD player to be considered MultiPlay-compatible include

- CD-Audio only players must play CD-Audio from CD-R and CD-RW discs,
- CD-Audio and CD-Text players must play CD-Audio and CD-Text from CD-R and CD-RW discs,
- DVD players capable only of CD-Audio (besides DVD Movie) must play CD-Audio from CD-R and CD-RW discs, and
- DVD players capable of CD-Audio and VideoCD must play CD-Audio and VideoCD from CD-R and CD-RW discs.

The MultiPlay logo can be licensed to manufacturers distributing compatible devices on a royalty-free basis. Compatibility is demonstrated through a self-certification process using test discs provided by OSTA on a onetime fee basis. See MultiAudio, Optical Storage Technology Association.

Multi-Media and Digital Communications lab MDCC lab. Located at Northern Oklahoma College, the MDCC lab is the largest Media 100 Inc. Finish/Intergraph installation in the world. Through Fibre Channel storage, video recording equipment, and multiple Intergraph stations, students can study and create multimedia productions.

Multimedia Cable Network System MCNS. A North American cable data communications industry specification developed by the Multimedia Cable Network System Holdings partners. The specification is known as Data Over Cable Service Interface Specification (DOCSIS), MCSN, or MCSN/DOCSIS. In 1997, VLSI Technology, Inc., announced a data security chip to support MCNS cable modem specifications. In 1998, Cisco Systems, Inc. announced integrated router/cable modem termination

equipment to support the MCNS. With this type of support for the specification, cable modem vendors began announcing cable modem devices. The system and associated services were especially attractive to residential and small business subscribers, and MCNS became the *de facto* standard. Some vendors continued to support proprietary schemes in addition to MCNS, such as 3Com, distributors of ATM-compatible modems, in the hope that niche markets would continue to be viable.

There were some objections to the development of DOCSIS, as it was felt it could not be deployed in Europe in support of a global standard. Nevertheless, some European vendors began selling MCNS/DOCSIS-compatible modems.

In the development phases, the MCNS approach was quite different from the other primary cable modem format, IEEE 802.14, in terms of how medium access control (MAC) protocol was integrated. While both formats traded off upstream speeds in favor of downstream, MCNS was more efficient in the downstream direction, while 802.14 broadband protocol was more efficient upstream. See Data Over Cable Service Interface Specification, Multimedia Cable Network System Holdings.

Multimedia Cable Network System Holdings MCNS Holdings. A North American cable industry group comprising Comcast Cable Communications, Cox Communications, Tele-Communications, Inc. (TCI), and Time Warner Cable. MCNS developed the MCNS Removable/Renewable Security System, in 1995, for securing voice and data communications and protecting from theft-of-service and denial-of-service attacks. It further developed the Data Over Cable Service Interface Specification (DOCSIS) in 1996. MCNS specifications have been widely adopted by cable modem manufacturers. See Data Over Cable Service Interface Specification, Multimedia Cable Network System.

MultiMediaCard MMC. A compact, removable, low-power-consumption, solid-state media card for storing and retrieving digital information that can be used with small, lower-power devices such as mobile communications and music appliances. MMC began as a joint project between SanDisk Corporation and Siemens AG/Infineon Technologies AG. It was first introduced in November 1997. Support for the concept has come from a variety of semiconductor suppliers, software vendors, and manufacturers of portable navigation, communication, and entertainment products.

MMCs are about the size of a postage stamp (32 × 24 × 1/4 mm), weighing less than 2 g. The connection is through a seven-pad serial interface. Read-only memory (ROM) is used for read operations and Flash technology is used for read/write applications. A single card can store, in its current form, up to about 64 Mbytes of data, a capacity similar to that of a hard drive of the early 1990s, but in a tiny, portable format. This is sufficient to contain the textual contents of about 100 books or the contents of a standard music CD. MMCs have potential not only for existing portable devices, but also for new hybrid devices such as cell phone/Internet appliances.

MultiMediaCard Association MMCA. A trade association founded in 1998 to support and promote the MultiMediaCard (MMC) standard and its global adoption. The MMCA develops and regulates open industry standards related to the MultiMediaCard technology. See MultiMediaCard. http://www.mmca.org/

Multimedia Communications Forum MMCF. A nonprofit international multimedia communications and software research and development organization founded in 1993. MMCF consists of telecommunications service providers, multimedia application and equipment developers, and multimedia users. MMCF promotes market acceptance of networking multimedia products and serves as a clearinghouse for multimedia-related specifications, standards, and recommendations. http://www.mmcf.org/

Multimedia Communication Exchange MMCX. A commercial phone/data server software developed by AT&T's Global Business Communications Systems (GBCS) for providing multimedia services for private phone branch exchanges. See Global Business Communications Systems.

Multimedia Communications Research Laboratory MCRL. A division of Bell Labs doing fundamental research in speech production and hearing, speech recognition, coding of audio, speech, and images, and echo cancellation. MCRL provides Lucent Technologies with innovative competitive products such as the electret microphone, capture and storage technologies, and media translation tools.

multimedia desktop collaboration MMDC. A software application enabling end users to significantly enhance desktop productivity through the utilization of multimedia communications technologies such as whiteboarding, realtime document collaboration, etc.

Multimedia Development Center MMDC. A center of the University of Nebraska associated with the Office of Information Technology Services.

MultiMedia Development Center MMDC. A service of Johns Hopkins University providing multimedia development tools (scanners, audio/visual tools, CD burning, etc.) and an introduction to advanced multimedia to staff and students of the Homewood Schools of Arts and Sciences and Engineering, by appointment.

multimedia information retrieval MIR. The access and retrieval or receipt of a transmission of multimedia data. MIR is commonly associated with databases on and off the Internet and is also now commonly carried out through email and certain videoconferencing products.

Multimedia Learning Lab MLL. Originally called the Multimedia and Visualization Laboratory (MVL), the facility was rededicated in January 2001 as the Multimedia Learning Lab. This is a teaching lab at the University of Arizona that facilitates study and creation of multimedia audio/visual technologies, including digitization, animation, and audio recording.

Multimedia Mail Service MMMS. A means to transmit media types not explicitly specified in the X.400 standard through the use of Externally Defined Bodyparts. It adds the concept of a global store for high-volume data to the Message Transfer System (MTS). Research, development, and manufacturing entities within Berlin, Munich, and other German centers collaborated in the development of the Multimedia Collaboration Services.

Multimedia Mail Security MMMSec. A German research project of the Open Distributed Multimedia Applications Group at the GMD Institute for Open Communication, funded by DeTeBerkom, Ltd. The goal of MMMSec is primarily to develop a security platform to support multimedia messaging and its integration within a multimessaging-system-enabled multimedia-mail user agent, in a user friendly manner. An X.420-Security-Bodypart was developed within the project to provide integrity, confidentiality, and digital signatures through encapsulation. See Multimedia Mail Service.

Multimedia Message Entity MME. In the context of a nonrealtime Multimedia Messaging Service (MMS), entities that may be located in a fixed network or Message Service between which different types of information may be transferred.

Multimedia Messaging Service MMS. A nonrealtime service in development by technical working groups for transferring different types of information between two Multimedia Message Entities (MME) in a store-and-forward fashion. MMS enables the development of unified applications for integrating the composition, delivery, and access of a variety of types of media communications (text, image, voice, etc.). See Multimedia Message Entity.

MultiMedia Telecommunications Association MMTA. A professional telephony/computer integration trade association descended from the North American Telephone Association. The MMTA is organized as an open public policy, market development, member support, and educational forum for telecommunications product and service developers and resellers. http://www.mmta.org/

Multimedia Virtual Laboratory MVL. The MVL is a research and development center of the Telecommunications Advancement Organization of Japan (TAO). This is a virtual facility with distant laboratories interconnected by high bandwidth networks to intercommunicate and share high-cost equipment.

Multimedia Visualization Laboratory MVL. See Multimedia Learning Lab.

multimeter, multirange meter A measuring instrument commonly used in the communications industry which has several scales for evaluating an electrical circuit, including voltage, resistance, and current.

multimodel data compression MMDC. A meta data software compression algorithm for managing the creation, invocation, and destruction of other software algorithms. MMDC is suitable for handling interleaved data streams of different types. It was developed by Ross N. Williams, implemented in Ada on a VAX/VMS system.

M

multimode optical fiber MMF. A multimode fiber optic transmissions cable, usually with a relatively thick core, acting as a waveguide to reflect and propagate light at many angles. A thicker core has advantages and disadvantages. More light signals can be sent at one time, but the signal transmission distance is shorter than single mode fiber, due to the interaction of the reflected signals within the core and gradual fading over distance. Thus, it is limited to about 2 km. Signals are usually transmitted through multimode cables with light-emitting diodes (LEDs) and received at the other end with a photodiode detector. This detector translates the signals back into electrical impulses. See single mode optical fiber.

Multiparty Multimedia Session Control MMUSIC. An architectural framework to support Internet teleconferencing. To date the MMUSIC Working Group has defined a number of IETF Draft documents to support this capability under the original umbrella of the Internet Multimedia Conferencing Architecture, including the Simple Conference Control Protocol, the Session Description Protocol, media alignment and messaging support, and others. See Mbus, RFC 2326, RFC 2327, RFC 2543, RFC 2974.

Multiparty Multimedia Session Control Working Group MMUSIC Working Group. An IETF Working Group chartered with the development of standards-track protocols to support Internet teleconferencing. MMUSIC's focus is on supporting the loosely controlled conferences common on the MBone. To support these functions, the MMUSIC Working Group has described the architectural framework for MMUSIC and interoperability scenarios and drafted protocols for various aspects of session initiation, control, and security.

multipath fading A signal loss characteristic of mobile communications in areas where buildings or uneven terrain bounce the transmission waves in many directions such that several "copies" reach the receiving station. The deflected signals may interfere with, or even partly cancel out, the primary signal. When a subscriber listens to an analog cell phone conversation or a motorist listens to an AM car radio, the sound volume may decrease. Digital transmission systems provide error-correction techniques to combat problems of fade, and spreading the spectrum over a broader bandwidth, if feasible, can sometimes help.

multiplayer gaming Many computer games are designed for one player to sit and play against the computer. Some of these games have a mode in which the user can elect to play against an opponent, usually on the same system, with the two players alternately using the input devices, or using two sets of joysticks, for example. These are usually known as single- or dual-player games.

Multiplayer gaming is implemented in a number of ways. There are games in which multiple players (usually three or more) can play on the same machine in much the same way as dual-player games, each one taking a turn, but multiplayer more often connotes a game in which multiple players can interact at the same time with the computer, with each other, or both. This opens up a whole new dimension in gaming, and multiplayer games are very popular due to the human element, and the higher sophistication that they often provide. Most multiplayer games can now be played over networks and, in fact, a large proportion of inexpensive Ethernet cards for home and small business computers are sold for playing fast-action multiplayer computer games. See MUD.

multiple access A means to increase communications capacity over a limited bandwidth medium. There are many ways to do this; some are specific to the type of technology, and whether it is analog or digital. In computer networks, multitasking and common address schemes and protocols allow multiuser access. Web pages on the Internet are a good example of multiple access technologies, since many thousands of people may be accessing the same site at virtually the same time. In cellular communications and transponders, there are at least three types of basic schemes that allow multiuser access to limited resources. See the Radio Frequency Transmission Schemes chart for some examples.

Multiple Address System MAS. In wireless communications, a designation for low-capacity not-for-profit, not-for-hire internal business networks with at least four remote stations.

In 1992, more than 50,000 applications were made to the Federal Communications Commission (FCC) for MAS facilities; these were to be resolved through a lottery system. In winter 1998, the FCC dismissed the MAS facilities in the 932- to 932.5- and 941- to 941.5-MHz frequency bands and refunded the filing fees. MAS Spectrum Auction No. 42 was held in November 2001. This auction was for 5104 licenses to cover 176 Economic Areas from paired frequencies in the ca. 928.85625- to 959.85625-MHz range. These licenses cover terrestrial point-to-multipoint and point-to-point fixed and mobile transmission excluding video entertainment.

multiple customer group access For many expensive technologies, there are options for different institutions or departments with lower access needs to share the technology. In Frame Relay systems, several groups can share the network through a Frame Relay cloud. In telephony, several firms with individual trunks may send and receive through a shared private branch system. For Internet service access, a group of users might share a limited pool of dynamically assigned Internet Protocol (IP) numbers or a small pool of fast modems, on a first-come, first-served basis.

Multiple Exchange Carrier Access Billing MECAB. A document prepared under the auspices of the Carrier Liaison Committee of the Alliance for Telecommunications Industry Solutions (ATIS), which establishes the methods for processing orders for access service provided by two or more local carriers, that is, a local exchange carrier (LEC) and a competitive local carrier (CLC).

multiple frequency shift keying MFSK. Frequency shift keying is a modulation technique used in data transmissions such as wireless communications in which binary "1" and binary "0" (zero) are coded on separate frequencies. This scheme can also be adapted to regular phone lines by assigning binary "1" to a tone and binary "0" (zero) to a different tone. In multiple frequency shift keying, greater use is made of different frequencies to represent information. These multiple frequencies may be transmitted one after the other, or simultaneously. See frequency modulation, frequency shift keying, phase shift keying.

multiple homing 1. In networking, taking an additional service line, which is often a leased line through a larger provider. This is sometimes done by Internet Service Providers (ISPs) as their customer bases increase. Thus, it becomes necessary to maintain and announce an additional set of routes through which addresses can be reached. 2. In telephony, adding phone connections so calls can be routed through more than one switching center to provide redundancy in case of problems with the initial service loop.

Multiple-Image Portable Graphics MNG. A format for using multiple subimages in a single frame, a multimedia graphics format originally developed as Portable Network Graphics (PNG) for animations. PNF was submitted as Draft 1 by Glenn Randers-Pehrson in June 1996. With Draft 12, in August 1996, PNF came to be known as MNG and, by December 1997, it was up to Draft 42 and considered essentially finished. See Portable Network Frame, Portable Network Graphics.

Multiple Module Data Computer MMDC. A Marine Product Group commercial instrument system computer that intercommunicates through a Controller Area Network (CAN). The CAN message center enables mariners to observe and control functions related to the operation and optimal performance of a boat. This is essentially a "smart boat" concept.

Multiple Station Message Waiting In telephony, a means of indicating on more than one telephone station that a message (e.g., voice mail) is stored and waiting to be accessed. This is usually indicated with a blinking light or brief text message.

Multiple Virtual Line MVL. A communications system that uses existing analog phone lines to provide higher speed digital communications and simultaneous voice. A mid-range option between cable and ISDN modems. An MVL switch must be installed at the switching center and subscriber point.

multiplex See multiplexing.

multiplexed analog component MAC. A means of television transmission used with direct broadcast satellite (DBS) programming. The various signal components – audio, luminance (brightness), and chrominance (color) – were transmitted independently. The advantage of this, particularly for DBS in European regions, is that the audio can be individually tailored to the language desired.

multiplexed optical scanner technology MOST. A low power-usage system for multidimensional scanning. It is a binary-switched scanner comprising flat-panel thin-film polarization switches. The technology has certain advantages over nematic liquid crystal

(NLC) scanners, including speed. MOST features large apertures, planar stacking of optical components, and storage for optimum beam settings. The use of remote fiber probes transmitting through a flexible fiber feed to the multiplexing device was proposed in conjunction with this technology.

MOST was developed by Nabeel A. Riza who proposed that optical wireless links had advantages over radio links for certain free-space applications such as guidance systems for military applications. In conjunction with a DARPA program, MOST is being developed as laser radar and laser communications systems, with other applications possible. See Photonic Information Processing Systems Laboratory; Photophone; Riza, Nabeel.

multiplexing Splitting the use of a circuit in one of a variety of ways so more than one signal can be carried at a time over a single transmission. This is an extremely important way of increasing channel efficiency and capacity in most telecommunications technologies.

Multiplexed phone circuitry began to appear in the late 1930s, and was widely established by the mid-1950s. A historically significant practical application of multiplexing was developed in 1953 by E.H. Armstrong and John Bose, in which they enabled a single FM station to simultaneously transmit more than one signal at a time. This made possible multiple simultaneous broadcasts, including the transmission of stereo broadcasts. Many different multiplexing schemes are included under individual headings in this dictionary. See add/drop multiplexer.

multiprocessing Using intercommunicating processors to solve computing tasks. The various processors may be housed inside one computer, or may be in separate systems, connected by a network. Multiprocessing is associated with parallel processors with distributed computer systems and often with high-end computing tasks such as scientific calculations, rendering, ray tracing, and computer animation. In a multiprocessing system, a centralized processor often assigns and distributes tasks, with the other processors feeding back information to the system or following instructions from this central administrator. The software development to optimize these kinds of systems still has room for development, and there are many opportunities to contribute to research and implementations in this area.

Not all tasks benefit from multiprocessing. In other words, it doesn't make everything faster. If the overhead involved in managing and coordinating the different processing tasks is greater than the speedup in the tasks themselves, then it's not a good candidate for multiprocessing.

One of the most exciting aspects of multiprocessing is that the speed of communication between the processors is an intrinsic part of the system. With higher and higher data rates, faster CPUs, and faster data buses, we may someday realize a time when the speed of the data transfer meets or exceeds the speed of its associated processors, in which case the various networked systems will function more like a single

M

organism than a set of separate computers.

Multiprotocol Label Switching MPLS. A connection-oriented routing system for switching network frames or packets. Network control and data transfer are handled separately and MPLS can support multiple network layer protocols. It is typically used in backbone networks. The IETF developed the scheme out of a number of switching technologies, most of them Internet Protocol-based, beginning in 1996. MPLS was intended to support scalable, efficient, high-speed forwarding, routing-controllable networks.

Since MPLS is a label-switching technology, the labels are important and are dependent upon the link layer used. MPLS transmission paths may be explicitly defined or controlled through information contained in the label. MPLS differs from conventional Internet Protocol (IP) forwarding in that the assignment of a particular packet to a particular stream is done only once, as the packet enters the network. This facilitates the use of explicit routing, without each packet carrying the explicit route. The assigned stream is encoded with a short fixed-length label that follows it to the next hop. Thus, at subsequent hops, rather than analyzing the packet each time to gather information from the network layer header, the label is used as an index to a table specifying the next hop, and a new label supplants the old label.

MPLS is particularly suited to implementation with Asynchronous Transfer Mode (ATM) and is of interest to IP-over-ATM. It has also been suggested for use with Frame Relay networks. MPLS provides explicit class of service (CoS) information, or information for inferring the CoS or precedence from the label. Routers that support MPLS are called label switched (or switching) routers (LSRs).

Concurrent with the development of MPLS was an Optical Label Switching (OLS) system proposed by Yoo in 1997. It shares a number of similarities with MPLS and also has some advantages. See Forwarding Equivalence Class, label switching, Optical Label Switching.

Multiprotocol Label Switching domain MPLS domain. A contiguous set of nodes that provide MPLS routing and forwarding operations.

Multiprotocol Label Switching edge node MPLS edge node. An MPLS node that connects an MPLS domain with a node outside the domain.

Multiprotocol over ATM MPOA. A multiprotocol virtual routing scheme for Asynchronous Transfer Mode (ATM) finalized by the ATM Forum in July 1997. Through bridging, devices in an Internet Protocol (IP) subnet can be interfaced with different physical ports on various MPOA edge devices while still appearing to be associated through the bridging. To legacy systems, the MPOA system appears as a traditional router.

Multipurpose Internet Mail Extension MIME. Since any type of binary file can be transported over the Internet using appropriate conversion routines, protocols, and compression algorithms, if needed, it is logical that text, graphics, and sound can all be transmitted through binary files. Yet, Internet Mail had provisions for only the transmission of ASCII documents. For this reason, developers sought to give email a richer set of capabilities, and released MIME specifications in 1992 as a general framework and format for the representation of many kinds of data types in Internet mail.

MIME enables a developer to write email software clients which can include pictures of the kids, sounds of the dog barking, and messages to loved ones all in the same message. This provides the user with a multimedia mail interface to the Internet. Since email accounts for at least 75% of the communications of regular consumers on the Net, enhanced email capabilities will probably be appreciated by many.

Content-rich email has been around on several workstation platforms since the late 1980s, but most of these applications were proprietary and could not intercommunicate. Text mail messages for the Internet were defined even earlier, in 1982 (RFC 833). Then, in 1992, a standards track protocol for a generalized Internet extensions format was submitted as a MIME Request for Comments (RFC) from the IETF Network Working Group.

MIME provides ways in which to include multiple objects in a single message, and to represent text in character sets other than just U.S.-ASCII. It supports multiple typefaces, images, and audio. While many email programs are MIME-compliant, not all of them support all the features of MIME, but it is expected that MIME implementations will increase, especially as MIME has been designed to be extensible; additional content types can be defined and supported. See RFC 1341.

multisession A feature of CD-ROM discs and drives which enables the CD-ROM to have information written to it in more than one session with access to everything on the disc. Since CDs were historically *write once/read many* (WORM) media, the early CD-ROM drives would not recognize additional information on the disc if it was written to the disk at a later date from the original information. This wasn't a problem if the consumer was buying a game disk with 600 Mbytes of information on the original disc.

Then Kodak developed a type of image storage format called a PhotoCD in which photographs are scanned, digitized, and saved on a CD-ROM. Since the user could bring the CD-ROM back to the processor to have another roll of film digitized and stored on the same disc (assuming there was still space remaining), it became important to create a CD-ROM format and a CD-ROM drive that could read the data from the multiple sessions. Most CD-ROM drives now are multisession.

Multi-Source Agreement MSA. An agreement signed by participating transceiver products companies, many of whom are from the small form-factor pluggable group. The initial goal of MSA in the early 2000s was to develop a common transceiver package for the 10 Gbps communications market and to efficiently provide the product and functional specifications to the industry. This resulted in industry standardization of the package size, pin configuration, and operating specifications of components for SONET/SDH and ATM optical network applications.

In spring 2002, major vendors, including Sumitomo Electric, Lucent Technologies, Fujitsu, Nortel, Agilent, Agere Systems, et al. announced their participation in the agreement in the development of internationally-compatible 10-Gbps optical laser diodes transmitters and receivers.

multispectral scanner MSS. A nonphotographic imaging system, commonly used on remote sensing satellites, that utilizes an oscillating mirror and fiber optic sensor array. As the mirror sweeps, it transmits brightness values a strip at a time. Multiple detectors are placed in an array in order to sense radiant energy in selected spectral bands.

The color seen on satellite image posters that can now be bought in poster and stationery stores is actually added by computers after the image is geocoded (the strips resolved into a cohesive image). In early days of colorizing satellite images, the color assignments were not as appealing as they are currently. The MSS technology was first developed for the Earth Resources Technology Satellite (ERTS-1) in 1968, a program which later developed into the LANDSAT satellites.

Multistation Access Unit MSAU, MAU. In a Token-Ring network, a passive concentrating device used in simpler or older Token-Ring installations to extend a Token-Ring setup by allowing a group of computers to be connected to the ring in a star topology configuration. The network still appears as a logical ring, however, with data passing through each computer. This makes the network a little more robust (a bad unit can be bypassed) and flexible than a basic Token-Ring. In general, these are being replaced by active Controlled Access Units. It is better abbreviated as MSAU to distinguish it from Media Access Unit (MAU). With extra ports, multiple MSAUs can be interconnected, provided the basic ring structure is retained. Each MSAU can accommodate Lobe Attachment Modules up to a specified number. See Controlled Access Unit, Lobe Attachment Module, Media Access Unit, Token-Ring network.

multisync A device that can handle transmissions at a variety of frequencies. This is a common feature of monitors enabling them to display at various resolutions when attached to different graphics display cards and computers. When installing a multisync monitor for the first time, it is very important to read the documentation with the graphics card or computer to determine the correct frequency settings for the monitor. If the settings are too high, you can damage the monitor. Most consumer color multisync monitors run at about 60 MHz.

multitasking Carrying out two or more tasks simultaneously. Many computer systems that claim to be multitasking are actually task switching. In other words, you can switch from one task to another, or one application to another, without closing the first task or application, but the background task freezes; in other words, it does not continue processing while the foreground task is active. If you can run a sound program in the background or format a floppy while doing something else, like using a paint program or a word processor in the foreground, without the sound stopping or jittering or without waiting for the floppy drive to finish, then you probably are working on a multitasking system. Most current workstations and many current microcomputers are fully multitasking. Historically, multitasking microcomputer operating systems have existed at least since the early 1980s, the most notable examples being some of systems that ran on the Tandy Color Computer (CoCo), Digital Research's multitasking version of CP/M, and the AmigaOS (1985), which had pre-emptive multitasking. In the late 1980s some of the other systems began to implement task switching and multitasking. In 1987, the NeXT computer was introduced with multitasking. Workstation-level computers such as Sun stations, SGIs, etc., have been multitasking for some time. By the early 1990s, most microcomputers had some degree of task switching or multitasking, and full multitasking on desktop systems is quickly becoming standard.

multithreaded Programming on microcomputers with threads became more common in the 1990s, although the practice has been around longer on mainframes and workstation-level computers. A system which is multithreaded can have a number of processes running independently of the others while still using common resources. Even though they function somewhat independently, they can also be programmed to communicate with one another at intervals or as certain events occur. See threads.

MultiUser Talk MUT. A simple multiuser chat program used on computer bulletin boards in the mid- to late-1980s, written by Jukka Pihl. It was superseded by more powerful, robust programs, most importantly, Internet Relay Chat (IRC). See Internet Relay Chat.

Munsell color model A three-dimensional, ordered representation of color relationships for pigments, developed in the early 1900s by an art instructor. For the purposes of illustration, imagine a sphere of no fixed diameter, in which white and black are the north and south poles and hues are represented around the sphere in terms of the basic colors of red, yellow, green, and purple and their complements, and are equally spaced around the perimeter, with the blended colors occurring as transitions at their junctions. But now, unlike a sphere, imagine chroma as a characteristic that radiates outward from the sphere's central axis, thus influencing the hues that it passes through. Although a complex model, it has the advantage of being somewhat scalable. As new pigments are invented, they can be incorporated into the model without redefining existing colors. See CIE, color space, Maxwell's triangle.

Murgas, Josef (1864–1929) A Slovakian-born priest, artist, and scientist, Murgas spent many years studying and working in Europe and then relocated to the U.S., using "Joseph" as his first name. Like Benjamin Franklin, Murgas was multitalented and tirelessly inquisitive and set up a laboratory in the attic for experiments. Over the years, he received numerous patents for his devices; he transferred these to a syndicate called the Universal Aether Telegraph Company. In 1904, Murgas patented a tone system of wireless telegraphy with a rotary spark for transmitting faster than the traditional Morse code system. He then built a high transmitting tower which was hailed in the local newspaper as the "World's First Telegraphy Tower." Many prominent citizens witnessed a test demonstration in September 1905 and a public demonstration in November 1905, after which Murgas traveled to New York to meet with Guglielmo Marconi and Reginald Fessenden. Unfortunately, gale force winds destroyed Murgas' transmitting tower and other bad luck befell him soon after. He was getting on in years and having financial difficulties, so he sold his important invention to Guglielmo Marconi, to prevent his discoveries from being lost to humanity.

Murgas' achievements have not gone entirely unrecognized. President Calvin Coolidge honored him by appointing him to the National Radio Commission. Liberty ship #2881 was named after him during World War II. A U.S. Senate Bill adopted 1 October 1985 urged the Citizen Stamp Advisory Committee of the U.S.P.S. to issued a commemorative stamp of Father Murgas to celebrate Slovak Heritage Month. See Marconi, Guglielmo.

Murphy's law Reportedly stated by Edward A. Murphy as "If there are two or more ways to do something, and one of those can result in catastrophe, then someone will do it." and also reported as "If there is any way to do it wrong, he will." However it was originally worded, Murphy's apropos observation was quoted a few days later, in a news conference, by Dr. John Stapp, a surgeon and research subject in studies of human tolerance to high-velocity ejections and gravity forces (Gs). The statement is often restated more simply as "If something can go wrong, it will." See Murphy, Edward.

Murphy, Edward A. An American engineer involved in human testing of some spectacular acceleration/deceleration experiments in the U.S. Air Force in the 1950s. Following his observations about the configuration of sensors, he is best known for a prescient observation about catastrophes now known as Murphy's law. See Murphy's law.

Murray loop test A type of diagnostic procedure which uses resistance through a bridge to locate an "open" in a length of circuit. It is similar to a Varley loop test, except that instead of adjustable dials, one arm is eliminated and a variable resistance arm connected in its place, and a third wire is not required. See Varley loop test, Wheatstone bridge.

Museum of Independent Telephony In Abilene, Kansas, the home of the United Telephone Company from 1898 to 1966, one of its former presidents, Carl A. Scupin, helped found the Dickinson County Historical Society and Museum. The Museum of Independent Telephony now shares premises with this museum.

Josef Murgas – Telegraphy Pioneer

Josef Murgas, pastor of the Sacred Heart Church. [Photo portrait courtesy of the Wyoming Historical and Geological Society, copyright expired by date.]

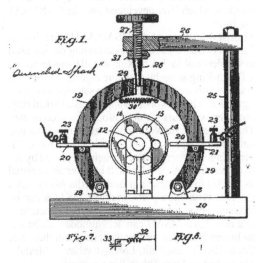

A "quenched spark" device from one of Joseph Murgas' patents from 1911. [U.S. patent diagram, public domain.]

Museum of Radio and Technology A nonprofit, volunteer-assisted antique radio technology museum, located in a converted elementary school in Huntington, West Virginia. It includes crystal radio sets, vacuum tube technologies, schematics, vintage books and magazines, and other educational resources and exhibits.

Museum of Television and Radio A nonprofit, New York-based preservation and education institution established in 1975 by William S. Paley. Its goal is to collect and preserve historic radio and television programs and make them available for public education and use. In 1991, it was moved to the William S. Paley building. It houses over 60,000 programs selected for their historic, artistic, and cultural value. The collection adds about 3000 programs per year. The museum works in conjunction with the *Museum of Television and Radio in L.A.*, California. Both institutions sponsor seminars and exhibitions.

Museum of Television and Radio in L.A. Also known as the West Coast Museum of Television and Radio, it was established in 1995 in Los Angeles. This museum is named after Leonard H. Goldenson, considered a pioneer of the broadcasting industry. It works in conjunction with the original *Museum of Television and Radio* in New York.

music on hold Background music on a phone line that is heard when a caller is put on hold. It can be set up to play music from a radio, cassette tape, or CD player, usually with a simple RCA jack near the phone panel. Recorded music is generally better, as most radio stations play advertising, which is not appreciated by most callers. Some people don't care for background music while on hold, but it's probably better than not knowing whether or not you've been cut off.

Musical Instrument Digital Interface MIDI. MIDI is a standard protocol for communication of sound information through a number of specified parameters. Functions provided on MIDI-capable instruments are assigned numeric values which can be digitally intercommunicated and remotely or locally controlled. MIDI capabilities are built into many musical editing and sequencing software programs.

MIDI-compatible instruments generally have DIN plugs for interconnecting the various MIDI devices, and usually include MIDI in, MIDI out, and sometimes MIDI through. A simple example of a MIDI setup would be a keyboard connected to the fast serial port or MIDI port of a computer with MIDI-compatible software. There may also be separate speakers, since most computer and keyboard speakers tend to be minimally useful for sound reproduction. MIDI allows the songs from the keyboard to be communicated to the computer and stored and edited. Conversely, compositions created on the computer, sometimes including custom sound patches, can be communicated back to the keyboard.

MIDI is not the only music protocol, but it's definitely the most widespread and best supported. It is built into numerous synthesizers, keyboards, drums, and software music editing and sequencing programs.

Musschenbroek, Pieter van See van Musschenbroek, Pieter.

MUT See MultiUser Talk.

mute A feature or device that allows sound to be turned off or lowered in volume. A mute button on a phone can cut out the sound to the mouthpiece so something can be said in the background without being transmitted over the connected call (handy if you have to call the dog or ask a co-worker a sensitive question). A mute on a stringed instrument dampens the vibrations of the bridge so that the sound of the instrument is much softer.

mutual capacitance Capacitance is the capability or inherent tendency of an object or substances to store electrical charge. This characteristic is related to the composition and structure of the object or substance, but also to the environment within which the entity exists. Influences from other objects will affect the capacitance of an object and increase its capacitance as they draw nearer to that object, hence the phrase *mutual capacitance*. As a somewhat imprecise example, but one that helps illustrate this idea, think of a satellite orbiting in space. As it draws nearer the Earth, the greater is its tendency to be pulled by the Earth's gravitational field. Similarly, with objects, the nearer they draw to another object, the greater is the tendency for capacitance, the capability to store a charge. Capacitance is also affected by the medium that exists between objects. Objects covered in plastic or a gel will have a different capacitance in relation to one another than objects separated by air.

Thus, capacitance may be expressed as it applies to a particular object but must also be taken into consideration as it applies to the atomic interactions between two (or more) objects. If the objects come into contact with one another, or are connected by a conducting substance, then the capability to store up an electrical charge with relation to one other changes in that the conducting surface effects a discharge or balancing of charges between the two objects.

In general, electrical conductors have higher capacitance than, say, a block of wood, but even people store up electrical charges. The human body acts as a capacitor when it picks up electrical charges through friction contact with carpets, for example. The next time you touch a conducting surface after storing up a charge, you may feel a spark as the excess electrical energy in your body is rapidly discharged. The extra charge is not always discharged quickly, however. If you hang around without generating a lot of friction with your feet nor having any contact with highly-conducting surfaces, you will still gradually discharge the stored up charge as it "leaks" more slowly into lesser-conducting materials such as the air, and wood or vinyl floors. See grounding, static electricity, triboelectricity.

MUX See multiplexer.

MVIP Multi-Vendor Integration Protocol. See Multi-Vendor Integration Protocol.

MVL 1. Man Vehicle Laboratory. A lab within the Center for Space Research (CSR) at the Massachusetts Institute of Technology, founded in 1962. 2. See Multimedia Virtual Laboratory. 3. Multimedia and Visualization Laboratory. See Multimedia Learning Laboratory. 4. See Multiple Virtual Line.

MVP multichannel video programming. Any of a variety of types of multiple-channel video programming genres, including television, network animations, Internet streaming video, or video programming delivered through traditional or fast data-rate phone lines. Due to the continued need to review, assess, and regulate video programming, the Federal Communications Commission prepared a report in 1999 on MVP that was released in January 2000. In its report, the FCC described a number of aspects of video programming, including the market share of

M

different means of delivering programming, and reiterated the belief that competition was the best way to promote broadbased access to video programming and competitive rates for the general public.

MVPD multichannel video programming distributor/distribution. A broadcast distributor of a number of video programming channels, usually provided through cable TV or satellite feeds. Broadband Internet is now considered to be a viable distribution channel. See MVP.

MVS Multiple Virtual Storage.

MXR 1. mixer. 2. multiplexer.

myriametric Myriametric waves are associated with the very low frequency (VLF) transmission range (3 to 30 KHz) and occur naturally in auroral phenomena. In the late 1980s, Horne described the plasma-based terrestrial generation of myriametric waves.

MZI Mach-Zehnder Interferometer.

n 1. *abbrev.* nano-. See nano-. 2. In mathematics, a symbol for a numerical quantity used to denote that any number can be substituted in place of the *n* or, quite commonly, *n* will represent any value within a stated number set such as any positive integer (whole numbers greater than zero). It is usually written in italics in lower or upper case. Scalable solutions to network configurations or database data sets are sometimes called *n* solutions to indicate that the system can be expanded to handle exceedingly large numbers or quantities far in excess of what might be needed in any given situation. See N definition #4. 3. *symb.* refractive index.

N 1. *abbrev.* noise, usually as it pertains to signal interference. Thus, S/N represents signal-to-noise ratio and C/N represents carrier-to-noise ratio. 2. *abbrev.* "north," on a magnet or compass. The north-seeking end of a compass needle points to a region near the Earth's North Pole called magnetic north. 3. *symb.* "on," with F as the corresponding symbol for "off." 4. *symb.* the last component, value, routine, link, or other virtual, conceptual, or physical aspect of a multipart system. In this context, it is usually written in italics. In a database, for example, *N* may be the largest conceivable number of records that might be expected to be needed or entered. In a distributed network, *N* may represent the largest number of hops that might be expected for a data transmission path. *N* is an important concept in both programming and physical topologies. The concept of *N* as a theoretical maximum enables planners to design and construct systems with the view of making the system efficient within the perceived constraints of *N*. See n definition #2. See N definition #1. 4. In a Central Office telephone code, any integer between 2 and 9. See NXX.

N connector A standardized medium-power, barrel-shaped cable/device coupler for interconnecting components in a system or devices on a network. N series connectors and adaptors come in a variety of shapes and configurations from simple connectors to extend the length of a cable to adaptors to change the "sex" of the cable between male and female to terminators for establishing the end link in a series of connections.

N connectors are selectively coated in nickel, silver, or synthetics over brass, phosphor bronze, or beryllium copper and typically support 50-ohm signals (though 70-ohm versions are available). They are widely used for connecting radio frequency (RF) coaxial components.

For Ethernet connections, 10Base-5 "thick" Ethernet cables with N connectors are becoming less common in favor of 10Base-2 "thin" Ethernet cables with BNC connectors or 10Base-T twisted-pair phone-style RJ-45 connectors. See F connector.

N port, node port. A port that enables an endstation device (peripheral, computer, etc.) to be connected as a node in a Fibre Channel network. The N port is connected to the switched fabric port or F port. The N port is assigned a higher address than FL and NL ports and thus would have lower loop control priority compared to these ports. A name server in the Fibre Channel fabric switch typically uses a discovery process to determine the fabric topology to obtain the addresses of other N ports for port logins.

The N port may be built into the connected device or may be contained on an interface board, such as a PCI-compatible board in a computer. High bandwidth devices, such as broadband storage devices, may have multiple Fibre Channel ports. See F port, FL port, NL port.

n region In a semiconductor, a region in which the conduction-electron density exceeds the hole density. The n materials interact with the p materials (the region with corresponding "holes") at the p-n junction in p-n semiconductors. See p region, p-n junction.

N Series Recommendations A series of ITU-T recommendations providing guidelines for the maintenance of international sound programming and television transmission circuits. These guidelines are available for purchase from the ITU-T. Since ITU-T specifications and recommendations are widely followed by vendors in the telecommunications industry, those wanting to maximize interoperability with other systems need to be aware of the information disseminated by the ITU-T. A full list of general categories is listed in Appendix C and specific series topics are listed under individual entries in this dictionary, e.g., K Series Recommendations. See N Series Recommendations chart.

N-1, N-minus-one The second-to-last or penultimate

node, device, link, router, hop, subroutine, loop, or other virtual or physical link or component. This concept is important in many respects. The second-to-last link in a programming loop, a physical or virtual network or other multipart system may have to be handled or configured differently, given that the next virtual or physical component in the system is the last. The penultimate item or data bit is often significant in programming routines. For example, in computer sorting algorithms, the number of iterations required in a particular function is frequently equal to $N-1$. The handling of software stacks, dynamic groupings of stored data, also is related to the concept of the penultimate item or data grouping in the stack. In routines where the program ends if it reaches the last or N data set, an awareness of reaching the penultimate

ITU-T N Series Recommendations

Recom.	Description
General, Definitions	
N.1	Definitions for application to international sound-program and television-sound transmission
N.54	Definition and duration of the line-up period and the preparatory period
N.2	Different types of sound-program circuit
N.3	Control circuits
N.4	Definition and duration of the line-up period and the preparatory period
N.5	Sound-program control, subcontrol and send reference stations
N.10	Limits for the lining-up of international sound-program links and connections
N.11	Essential transmission performance objectives for international sound-program centers (ISPCs)
N.51	Definitions for application to international television transmissions
N.81	Definition for application to international videoconference transmissions
Administration	
N.55	Organization, responsibilities and functions of control and subcontrol international television centers and control and subcontrol stations for international television connections, links, circuits and circuit sections
Testing, Monitoring, and Measurements	
N.12	Measurements to be made during the line-up period that precedes a sound-program transmission
N.13	Measurements to be made by the broadcasting organizations during the preparatory period
N.15	Maximum permissible power during an international sound-program transmission
N.16	Identification signal
N.17	Monitoring the transmission
N.18	Monitoring for charging purposes, releasing
N.21	Limits and procedures for the lining-up of a sound-program circuit
N.23	Maintenance measurements to be made on international sound-program circuits
N.52	Multiple destination television transmissions and coordination centers
N.60	Nominal amplitude of video signals at video interconnection points
N.61	Measurements to be made before the line-up period that precedes a television transmission
N.62	Tests to be made during the line-up period that precedes a television transmission
N.63	Test signals to be used by the broadcasting organizations during the preparatory period
N.64	Quality and impairment assessment
N.67	Monitoring television transmissions – use of the field blanking interval
N.73	Maintenance of permanent international television circuits, links and connections
N.86	Line-up and service commissioning of international videoconference systems operating at transmission bit rates of 1544 and 2048 Kbps
N.90	Maintenance of international videoconference systems operating at transmission bit rates of 1544 and 2048 Kbps
Supplements	
N.Sup5.1	Requirements for the transmission of television signals over long distances
N.Sup5.2	Setting-up and testing of international videoconference studios
N.Sup6.1	Effect on maintenance of the introduction of new components and of modern equipment design

set may be important to restarting, backtracking, or otherwise looping back and continuing program execution. See N definition #4.

n-ary A structure or element that may have *n* number of multiple applications or conditions or a specified limit that can be expressed as a positive integer greater than one (1) in some situations and greater than two (2) in others. The limit may be explicitly stated as ternary (three), quaternary (four), etc. If the number is large, the actual number may be used, as in 16-ary code in which 16 significant conditions exist. The term is often used as a synonym for "many" or "more than two" in general discussions where the *n* may be undesignated or unknown.

n-ary tree A data structure with multiple branching hierarchies with a maximum limit of *n* children in a node.

N-ISDN Narrowband ISDN, Narrowband Integrated Services Digital Network. Definitions vary but, in general, Narrowband ISDN is used to refer to earlier installations of ISDN providing lower data rates through copper wires as opposed to newer Broadband ISDN providing higher data rates through fiber optic cables. See ISDN for a fuller explanation.

N-scope A type of radar display in which the target appears as a pair of vertical blips coming from a horizontal time base. The direction of the target is inferred by the amplitudes of the related vertical blips. A target distance can also be determined by comparison to a pedestal signal along the base line.

NA 1. See naming authority. 2. network administrator/administration. 3. See Night Answer. 4. night attendant. 5. North America.

NAB 1. See National Alliance of Business and National Association of Broadcasters in Appendix G.

NABTS See North American Basic Teletext Specification.

NAC 1. Network Access Center. 2. See Network Access Control. 3. See Network Applications Consortium. 4. See Numbering Advisory Committee. 5. See null attachment concentrator.

NACIC See National Counterintelligence Information Center in Appendix G.

NACN See North American Cellular Network.

NADC See North American Digital Cellular.

NADF 1. See North American Directory Plan. 2. North Atlantic Directory Forum.

nadir In satellite imaging, a point on the ground centered vertically below a remote sensing platform.

name resolution A means of associating an assigned name with its origin, location, or other relevant characteristics. In a network where a name has been used as a mnemonic alias to allow easy recognition of an address, application, or process, there needs to be a mechanism to resolve the name into a form that can be easily recognized and subsequently located by the system. In other words, *mysite.com* has to be translated into a machine-readable address of the location of the host site for *mysite.com*. This is done through name resolution, usually through a lookup table or larger database. Sometimes the name itself will provide some information about its origin or date of es-

tablishment, e.g., the name is a set of alphanumeric characters assigned according to a system that can be understood by humans. See naming authority.

naming authority 1. A legislative or organizational body that assigns names, usually as unique identifiers. Various types of naming structures include hierarchical, flat, random, etc. There are many well-known naming authorities: the U.S. Library of Congress; R.R. Bowker (ISBN); IANA (Internet). On the Internet, the various registered domains may assign subauthorities and subnames for local machines. See IANA. 2. In a hierarchical document management system, a tree of entities which provides a unique identifier to each document. This task may be shared by subauthorities.

NAMPS Narrowband Analog Mobile Phone Service. An analog cellular communications technology which provides triple the capacity of an analog cellular voice channel by splitting the channel into 10 kHz bandwidth narrow bands. Narrowband standards were released by the Telecommunications Industry Association (TIA) in 1992 (IS-88, IS-89, IS-90). Digital mobile phone services are gaining ground on traditional analog systems. See AMPS, DAMPS, code division multiple access, time division multiple access.

NAMTS See Nippon Advanced Mobile Telephone System.

NANC See North American Numbering Council

nano- (*abbrev.* – n) A unit prefix for one billionth (North American system), or 10^{-9}, that is, .000 000 001. See nanometer.

NANOG See North American Network Operators Group.

nanometer (*abbrev.* – n) In the SI system, a length measure corresponding to one billionth (10^{-9}) of a meter. A nanometer is one thousandth of a micron. Nanometers are often used to describe the size of tiny particles (e.g., chemical substances) or electromagnetic wave phenomena (e.g., laser wavelengths) Light waves at 534 nm are in the visible spectrum (appearing yellow to human senses; wavelengths are also commonly expressed in microns). Smoke particles range from about 10 to 1000 nm. Electron microscopes image up to maximum ranges of about 5 to 0.5 nanometers.

With increasing miniaturization, the trend in the semiconductor industry is to use nanometers instead of microns, with a number of major vendors announcing an official changeover in spring 2002.

In synthetic materials fabrication, the region of 1 to 100 nanometers is of particular interest as it is less well understood and is subject to interesting interactions between molecular and macroscopic properties.

Distances traveled by certain phenomena such as seismic waves may be described in terms of nanometers per second (nm/sec). See nano-

NANP See North American Numbering Plan.

NANPA See North American Numbering Plan Administration.

NAP See Network Access Point.

NAPP See National Aerial Photography Program in Appendix G.

NAPT Native American Public Telecommunications, Inc.

narrative traffic A military term for transmitted secured or unsecured natural language communications. For example, in military space programs, the Special Communications Systems is a realtime, automated communications system on which narrative messages can be composed and edited. The Global Command and Control System (GCCS) is a secured system capable of handling narrative and data traffic. During times of emergency when traffic over networks is suppressed, narrative traffic may alternately be sent by other means such as courier.

narrowband 1. A term which varies in definition depending upon the industry and its bandwidth needs, and on the current state of technology. Narrowband usually represents the lower end of the available capacity or spectrum of a system. In some cases it is used to denote a single band within a multiplexed group of bands sent more-or-less simultaneously. In traditional telephony, it represents a sub-voice-grade line. In cellular communications, it represents one division of the broadcast spectrum consisting of a channel frequency (CF) of about 30 kHz, usually accomplished through frequency division duplexing (FDD). See AMPS, NAMPS.

Narrowband Analog Mobile Phone Service See NAMPS.

narrowband ISDN ISDN services at basic channel speeds up to 64 kbps, which is fine for voice and some data communications, but only adequate for applications like full-motion video, or video and sound. Efforts are being made to incorporate new standards into broadband ISDN (B-ISDN) that will remove the fixed channel structure limitation of narrowband ISDN.

narrowcasting A type of program delivery that targets specific people and often specific services to those people. If broadcasting is considered to be program delivery to a wide and sometimes scattered audience, from one to many, then narrowcasting can be seen as one to one or one to few. For example, electronic industries' personnel might subscribe to programming on circuit board fabrication. At an even more specific level is "pointcasting," that is, program services which target user-selected information, a type of electronic clipping service providing electronic information on specified topics of interest.

NARTE See National Association of Radio and Telecommunications Engineers in Appendix G.

NAS See network-attached storage.

NASA See National Aeronautics and Space Administration in Appendix G.

nasa7 A double-precision Systems Performance Evaluation Cooperative (SPEC) benchmark used in scientific and engineering applications. A benchmark tends to be a specific quantitative measure of a particular aspect of system functioning, and by itself conveys a limited picture of overall system performance. However, in the specific context for which it is intended, a benchmark can provide valuable infor-mation for design engineers, researchers, and manu-facturers. Nasa7 generates input data, performs one of seven floating point-intensive kernel routines, and compares the results against an expected reference measure. It is used to evaluate performance, memory, I/O operations, and networking factors. See benchmark, Rhealstone, Whetstone.

NASD Project Network-Attached Storage Devices Project. A project of the National Storage Industry Consortium (NSIC) to explore, validate, and document the technologies needed to enable the deployment of network-attached storage device systems and subsystems. The project was initiated because the trend towards large distributed networks is causing people to rethink the most efficient ways to implement storage capabilities, and more networks are using remote rather than local storage to handle large data repositories. Fibre Channel is seen as one of the technologies appropriate for fast-access remote storage. See National Storage Industry Consortium.

National See Appendix G for a long list of communications-related organizations prefaced with "National."

National Center for Supercomputing Applications NCSA. A research center at the University of Illinois, best known for the development of NCSA Mosaic, the historic Web information browser that preceded Netscape Navigator.

National Code Change A designated day in the U.K. when old telephone codes and numbers were changed to revamp the system. New codes and telephone numbers became available in August 1994 and the Code Change took place in April 1995. On the same day, the dialing code for international calls originating in the U.K. changed from 010 to 00.

National Electrical Code NEC. A code developed to safeguard public safety and property from hazards associated with the use of electricity. This includes wiring and electrical device construction, materials, installation, and maintenance and is adopted in many parts of the country as law for various building, equipment, and utility pole (below supply space) installations. The Code is developed by the American National Standards Institute (ANSI) and is published by the National Fire Protection Association (NFPA). See National Electrical Safety Code.

National Electrical Safety Code NESC. A code governing electrical facilities located in public rights-of-way to ensure the safety of the public and installation/maintenance professionals. It is published by the Institute of Electrical and Electronics Engineers (IEEE). See National Electrical Code.

National Geophysical Data Center NGDC. One of three data and information centers of the U.S. National Environmental Satellite, Data and Information Service (NESDIS).

National ID Card An identification concept proposed many times over the decades for a variety of reasons. In the U.S., there has been much opposition to the concept, with concerns about security breaches (people using the information in unethical ways to access private information in centralized databases), loss, replacement, forgeries, and more. In a sense, National ID Cards already exist for non-American legal residents (and a National ID Card doesn't solve the problem of illegal residents).

The commonly called Green Card (Immigration Visa for Resident Aliens/Permanent Residents) is a biometric national identification card issued over the course of many decades that includes a photo and a fingerprint (biometric identifier) tied to a federal database that already serves the purposes most people want a National ID Card to cover. To get a Green Card, you have to provide birth, background, educational, and other personal information to the U.S. federal government and be approved through a background check and interview, which is usually repeated approximately every 10 years. Asking every U.S. citizen to also carry a National ID Card in addition to the many driver's license/passport/birth certificate/social security documents already issued calls to question why yet another card should be issued and how it would be any different or better than the documents already routinely carried. Some people have even compared it somewhat radically to skin branding.

A National ID Card can be easily lost, as with any other card, and fall into the wrong hands. Wrongdoers often use forged or stolen documents; this might give them one more document to forge or steal and may not have any substantial security benefits in relation to the increased security risks associated with the loss of the card. Nevertheless, after the events of September 11, 2001, the issue of National ID cards has again been brought forward and will no doubt be debated for a long time. In the world of electronics, where the information on the ID card may be directly linked to a powerful central database, it must be carefully weighed whether a National ID Card program should be undertaken, especially given that, once the database exists, it may over time be commandeered for other purposes by future governments, or eventually be used by law enforcement agencies to track and profile individuals without their knowledge. Stranger things have happened in the past.

National Information Infrastructure NII. The name for the political, administrative, and physical underpinnings of an interconnected collection of public and commercial national narrowband and broadband data networks. One of the biggest stakeholders in the NII is the National Information Infrastructure Advisory Council (NIIAC), established in 1994 through a 1993 executive order. The NIIAC is responsible for advising the government on a national strategy for promoting development of the NII and the Global Information Infrastructure (GII).

The NII is a physically and regionally diverse system which is considered as a whole, mainly on the basis of interconnectivity. It includes small and large networks, wireless and wireline connections, public and private systems, and many sizes and types of organizations and individuals. The NII is also known by the catchphrase "Information Superhighway," although this describes the communications aspect of the NII and could be considered a subset.

National Public Broadcasting Archives NPBA. Housed at the University of Maryland in the Hornbake Library, the NPBA started as a cooperative project of several broadcasting and educational institutions. It was initiated by Donald R. McNeil, a Public Broadcasting System (PBS) board member, and officially dedicated in June 1990. NPBS provides an archival record of major documents and selected programming from U.S. noncommercial broadcasting history.
http://www.lib.umd.edu/UMCP/NPBA/

National Public Radio NPR. A major award-winning producer and distributor of public radio broadcast programs developed by independent producers. Based in Washington, D.C., NPR provides popular shows such as All Things Considered, Car Talk, and Morning Edition plus NPR hourly news on the Internet. It also hosts lively discussion groups online about issues raised on NPR programs. See Public Radio International. http://www.npr.org/

National Spatial Data Infrastructure NSDI. A U.S. Executive Order signed in 1994 under which federal agencies must document, and make accessible through the electronic Clearinghouse network, all new geospatial data collected or produced, either directly or indirectly, using the Federal Geographic Data Committee (FGDC) standard.

National Television System Committee See NTSC.

NATOA See National Association of Telecommunications Officers and Advisors in Appendix G.

natural antenna frequency An antenna's lowest natural resonance frequency when operated without external capacitance or inductance.

natural frequency The frequency at which an otherwise uninfluenced or unimpeded body will oscillate when stimulated to move. Knowledge of natural frequencies is important in structural engineering and scientific research.

The collapse of "Galloping Gertie," the Tacoma Narrows bridge is one of the more spectacular examples of how natural frequencies must be considered when building large or safety-oriented structures. The Tacoma Narrows bridge collapsed because of the interaction of a windstorm with the natural frequency of the bridge's movements. Without holes in the sidewalls to let the wind pass through, the accumulation of the bridge oscillations from the wind caused a resonance wave so great, it tore the bridge apart.

Natural frequencies aren't always harmful. They can be helpful tools for sensing devices. Many structures exhibit natural frequency vibrations that can be sensed with seismic instruments or light-based sensors. It has been proposed that fiber optic sensor arrays could be built into bridges and buildings to provide readouts of natural frequencies over time. In fiber optic sensors with micromechanical resonators (e.g., those made from metallic glass), a measured parameter (temperature, pressure, force, etc.) is expected to change the microresonator's natural frequency which may be both excited by and detected by light. This, in turn, provides data about the phenomenon or object sensed.

natural logarithm See logarithm.

natural magnet There are two types of permanent magnets. One is a substance that exhibits and retains magnetic properties without application of a current after it has been magnetized with another magnetic source. The second is a substance which exhibits magnetic properties as it comes out of the ground, without needing to be exposed to magnetic influences for it to become a magnet. The second type of permanent magnet is called a natural magnet. See lodestone.

natural wavelength The wavelength that corresponds to an antenna's natural frequency. Matching an antenna's resonant frequency to the characteristics of the wave being received (or transmitted) is an important aspect of antenna design.

NAVSTAR A series of Global Positioning System (GPS) satellites operated by the U.S. Department of Defense, whose navigational signals are available to civilian users. See Global Positioning System (GPS), GLONASS, Standard Positioning Service.

NAVTEX An international, automated weather and maritime navigational warning distribution system. NAVTEX sends warnings to ships as they move in and out of areas for which broadcast information is available that may be relevant to marine safety. See Global Maritime Distress and Safety System.

Navy Navigation Satellite System NNSS. A system of satellites moving in polar orbits about 700 miles above Earth, which preceded the Global Positioning System (GPS) used today. NNSS Doppler technology could compute group positions on or around the Earth to about 1-meter accuracy by means of multiple readings. The long time between transits over the same location (about 90 minutes), and the difficulty of determining instantaneous velocity led to the development of the GPS system. See Global Positioning System.

NBC See National Broadcasting Company in Appendix G.

NBFCP See NetBIOS Frames Control Protocol.

NBFM narrowband frequency modulation.

NBMA nonbroadcast multiple access.

NCACHE, DNS NCACHE Negative caching is a part of the DNS specification that deals with caching the nonexistence of a domain name or RRset, thus reducing message load and response time for negative answers. With the growth of network traffic and increased need for quick and frequent resolution of domain names, the importance of negative caching has grown to the point where it was suggested by Andrews and others that negative caching be routine rather than optional. See negative caching, RFC 1034, RFC 2308.

NCCS Network Control Center System.

NCIA native client interface architecture. An SNA applications-access architecture developed by Cisco Systems. NCIA encapsulates SNA traffic on a client computer, preserving the user interface from the native SNA system so that the end-user can work in a familiar environment and also have direct TCP/IP access.

NAVSTAR GPS Satellite

The 19th NAVSTAR satellite was launched in 1993, on board an Air Force Delta II craft. The NAVSTAR satellites are used for GPS data and geodesic research. [NASA/Marshall images.]

NCITS See National Committee for Information Technology Standardization.

NCO See National Coordination Office for Computing, Information, and Communications in Appendix G.

NCOP Network Code Of Practice.

NCSA 1. See National Center for Supercomputing Applications in Appendix G. 2. National Computer Security Association. Now the International Computer Security Association.

NCSA Mosaic A well-known Internet information browser and World Wide Web client developed at the National Center for Supercomputing Applications. Mosaic was the predecessor to Netscape Navigator distributed by Netscape Communications. Navigator was later declared open source software. See Mosaic for a fuller history; see Netscape Navigator.

NDIS See Network Driver Interface Specification.

NDSI See National Spatial Data Infrastructure.

NDT 1. network downtime. In telephony, the elapsed time from when network managers become aware of a problem until the moment at which the subscriber's service is restored. In computer networking, the time during which normal processes are unable to execute due to electrical outages, software crashes, network link disconnections, processing overload, malicious tampering, or unintentional bugs that interfere with processes outside of the application that has the bugs. 2. No Dial Tone.

NE See network element.

near end crosstalk NEXT. When wires are packed tightly together, and signals are traveling through most or all of the wires, especially in two directions, the signals originating at one end can exceed or interfere with weaker signals coming from the other end, resulting in crosstalk. With much higher speed transmissions media, such as gigabit Ethernet, which involve bidirectional signals in more complex systems of aggregated wires, this can be a severe impediment. One means of compensating for NEXT is to include a NEXT canceler, which detects and adjusts for noise in the circuit. See far end crosstalk.

near-infrared NIR. A region of the optical portion of the electromagnetic spectrum that has slightly shorter wavelengths than the visible spectrum perceived by humans as the color red. Certain snakes and insects have infrared-sensing receptors. Humans tend to sense infrared as thermal energy (heat). In general, near-infrared has frequencies ranging from about 700 to 3000 nm. Commonly-available commercial achromatic lenses are generally in the 700 to 1500 nm range. There are many filters designed to selectively include or exclude infrared radiation and some photographic filters selectively admit infrared radiation for specialized applications.

Infrared light is suitable for many types of data transfer and is commonly found in wireless remote controls and light-based wireless computer networks. It is widely used in remote sensing applications such as astronomy and in fiber optic communications systems. NIR-based spectroscopy is useful for materials analysis in a wide range of chemical and pharmaceutical applications. See infrared.

nearfield, near field In propagating electromagnetic or acoustic waves, the region near a source wave or reflected wave that is less than the length of one wavelength. Since there may be many wavelengths present, a specific frequency may be selected (which may be the strongest, the most central, or the most relevant to a particular task), or an average or other estimate of a range of wavelengths may be calculated to provide a nearfield estimate. If a range of wavelengths is being studied, the concept of nearfield is sometimes broadened to include the median or longest wavelength within the range and the interactions that occur between the wavelengths (e.g., diffraction). Thus, the nearfield is context-specific but, in general, inversely related to the wavelengths of interest.

Concepts of nearfield are important in observing or scanning radiant energy at very close ranges as there are special problems associated with collecting undistorted data from incomplete or interacting wavelengths. In laser technologies, measures such as beam width and beam separation or divergence from the intended path are also more difficult to calculate in the nearfield, but these calculations may be important in tasks such as aligning fiber optic light sources. See nearfield imaging, nearfield diffraction.

nearfield imaging Observing or recording an image at very close range. In electromagnetic imaging, it is a region in which the observing/recording device is so close to the source of propagating waves that the wavelengths may interfere with the process (e.g., by interacting with one another) or may be longer than the distance between the sample and the observing/recording device. This poses special problems in achieving undistorted viewing. Nearfield imaging is a challenge in many fields, including microscopy and spectroscopy.

Many of the same nearfield imaging issues apply to the monitoring of acoustic phenomena very near to the source of the sound waves.

There are a number of ways to tackle the challenge of nearfield imaging. Imaging at a range of frequencies and combining and processing the data may yield averages or patterns that provide information about the imaged sample. Hypercooling may reduce molecular movement, thus removing or reducing potential sources of interference. Using a fiber optic probe or taper to draw the image away from the surface is another strategy. See diffraction, Rayleigh scattering.

nearfield diffraction, Fresnel diffraction Diffraction in which electromagnetic waves incident upon an obstruction are diffracted in spherical waves originating from a point source. Since spherical waves overlapping any type of waves results in some complicated interactions, it can be challenging to calculate and mathematically model Fresnel diffraction. Fresnel diffraction must be taken into consideration in the design of antennas, imaging technologies (e.g., nearfield spectroscopes), and other devices in which there are radiant or reflective elements that are proximate and likely to have radiating energy that overlaps. For experimental purposes, a laser light source can be converted to a spherical light source by use of a lens and a small opening to spatially filter the light down to a point light source. See diffraction, Fresnel region.

NEC See National Electrical Code.

NECA See National Exchange Carrier Association.

neck The narrow portion of a cathode-ray tube (CRT) at the end where the electron beams are emitted from the cathode.

negative acknowledge, negative acknowledgment NAK A commonly used international communications control character which indicates that data was not received, or not received so that it could be understood. This is common to handshaking protocols, in which an acknowledgment is required before the sender can continue. See acknowledge.

negative bias In an electron tube, voltage applied to a control grid to make it hold more of a negative charge than the electron-emitting cathode. Manipulation of the control grid is what makes it possible to control the flow of electrons from the cathode to the anode, and thus to create different types of circuits and effects.

negative caching The storage of information about the nonexistence of an object, entity, or service. The availability of this information can reduce the time it takes to determine the components or configuration of a system or file. Negative caching statistics are commonly stored in tables or headers, depending on

N

the application. On computer networks, for example, storage of the nonexistence of certain servers or machines can lead to more efficient polling or routing of processes or network services. If certain printers or fax servers are offline for maintenance, negative caching information can be used to prevent overflowing queues. If certain hosts are offline, routing of packets can be renegotiated through other servers. If certain elements of an image file do not exist, negative caching can be used to prevent expensive production tasks on a printing press or milling machine holding up the line. See NCACHE.

negative feedback The propagation of an acoustic or electromagnetic signal in the general opposite direction of the source signal. This often results from backreflection of the outgoing signal or from the signal being circularly fed to the source through another source (e.g., a second microphone). In slower-moving phenomena (e.g., sound), the delay between the original source output and the feedback can magnify the effects of the phenomenon and cause echo.

Uncontrolled negative feedback may create undesired loss of gain, distortion, echoes, or other interference but it is not detrimental in all circumstances. In electronics, negative feedback in a circuit (sometimes called shunt feedback because the circuit is looped back upon itself) can help condition a signal. When configured as inverse or degenerative feedback, it may help reduce distortion, nonlinearity, and signal instability in an amplified signal. Optical negative feedback has been proposed as a means to reduce spectral linewidth and noise in frequency modulated (FM) systems.

negative glow A luminous glow which can be observed between an electron-emitting cathode and the Faraday dark space in a cold-cathode discharge tube. See Faraday dark space.

negative image An image in which the dark and the light values are reversed, or in which the complements of the colors are displayed instead of the normal colors; also called an inverse image. Photographic negatives contain a negative image. In desktop publishing, negative images are sometimes created so the printout can be processed some way in manufacturing. For example, an image printed on film for subsequent exposure to a printing plate might be printed in negative. Negative images are often used for posterization and other special effects. In monochrome television display systems, a negative image may arise from reversal of the polarity of the signals.

negative plate, negative terminal In a storage battery, the grid and any conductive material directly attached to the negative terminal, that is, the terminal that emits electrons when the circuit is active.

negative-going video In a four-field analog video sequence, a reference point in the subcarrier cycle (the other being positive-going) that is related to the start of an NTSC video frame. The positive-going and negative-going cycles alternate in sequence to form the video field.

A video frame is constructed by drawing scanlines in a sequential, repeating pattern, usually from left to right and top to bottom at a specified rate (e.g., 30 frames per second). When the electron beam is moving from right to left to begin the next line (or from bottom right to top left when it has finished the full screen) it is turned off or "blanked" so as not to interfere with the image on the screen. In an interlaced system, it takes two screen images to make up the full frame since only the odd or even lines are drawn with each half-frame. The process happens so fast that the human eye resolves two half-frames as a full image (although a bit of flicker may be noticed).

A horizontal drive (H drive) triggers a low-voltage negative-going pulse at the beginning of the horizontal blanking point (in standard systems this is at the right edge of the screen, the trailing edge of the scanline that has just been drawn) and ending where the edge of the sync for the next line begins.

Subcarrier/horizontal (SC/H) phase resolution is synchronization between the zero crossing point of the subcarrier sine wave and the negative-going (leading) edge of the horizontal sync in the first field and a specified line at the halfway point. This aids in resolving phase issues that are important in video tape recording. The timing provided by the leading edge of a negative-going pulse to the leading edge of the subsequent negative-going pulse can be used as a start-stop mechanism. Similarly, negative-going and positive-going edges of the same pulse can be useful for timing and integration of signals from multiple video components.

Some display monitors expect horizontal and vertical sync polarities to conform to certain specifications and, thus, some graphics output devices will include a hardware switch to output the signal with negative-going polarities. A *grab pulse* can be a positive-going or negative-going polarity output pulse for synching video to a component such as a frame grabber.

Negroponte, Nicolas (1943–) Outspoken author, philosopher, and educator, Negroponte is well known for his lectures and *Wired* magazine back-page editorials. He is the founder and director of the Massachusetts Institute of Technology's celebrated Media Laboratory, established in the late 1980s. Prior to that, he founded MIT's Architecture Machine Group, a think tank and research lab for discussing new approaches to human-computer interfaces.

neighbors A networking term used to describe nodes attached to the same link. See node.

nematic liquid crystal NLC. Nematics are rod-like organic molecules and nematic liquid crystals are mesomorphic (between liquid and solid) structures that show clumps of thread-like flaws when stressed/fractured, from which the name is derived.

Room temperature NLCs were developed by creating eutectic mixtures of MBBA combined with other compounds to broaden the effective temperature range, but their stability and dielectric properties were not suitable for commercial components. G.W. Gray of Hull University discovered cyanobiphenyl materials with room-temperature nematic phases with useful birefringent and dielectric properties. Nematic liquid crystals are anisotropic crystalline materials that exhibit some properties of solids, such as general orientation of molecules, and some properties of liquids, such has not having a specific positional order to the molecules, under certain circumstances (while in a thermotropic liquid crystal phase). There is an even more specific phase in which the molecules buddy up to each other but are not symmetric in reflection. This slightly skewed orientation is called a chiral nematic phase. Chiral nematic liquid crystals have a helical internal structure with interesting circular birefringent properties. Thus, circularly polarized light entering the material will travel at different speeds, depending upon the direction of the polarization and the wavelength, in relation to the orientation of the chiral NLC. Linearly polarized light will similarly be affected, causing the angle of polarization of the two beams of refracted light to progressively change as they move through the crystal. NLCs are used in optical scanners with birefringent thin films and large imaging surfaces. NLCs are the most common type of liquid crystal used in display devices. See multiplexed optical scanner technology, Schadt-Helfrich effect, smectic liquid crystal.

NENA See National Emergency Number Association.

neon gas (*symb.* – Ne) An inert gas with many industrial and commercial applications. When ionized, neon glows red. It was popularly used to illuminate signs in the 1940s and 1950s, and is still used for this purpose, along with other gases that emit other colors.

neon lamp A long glass illuminating tube with an electrode at each end and low-pressure neon gas inside, which may be angled into interesting shapes. When illuminated, it produces a red-orange light that can be seen in daylight and can penetrate fog better than most conventional types of lights. Neon has also been used in older tubes in the broadcasting industries, in simple oscillating circuits, and in commercial signs.

neper (*pron.* – nay-per, *symb.* – Np) A dimensionless mathematical unit for expressing relative measurements. It is used to express ratios that are useful in physics and electronics, as for voltage and current relationships.

The neper is similar to the decibel except that it is established upon a base of 2.718281828...; quantities expressed by nepers are based upon natural (Napierian) logarithms rather than base 10 logarithms. The neper is not a Système Internationale (SI) unit, but it is widely used and thus 1 Np has been assigned an SI equivalent value of 1. In terms of decibels, 1 Np = 8.686 dB.

The neper is named after the Scottish mathematician John Napier (Jhone Neper) who did historic research on logarithms. See decibel; Neper, Jhone.

Neper, Jhone (1550–1617) A Scottish mathematician, now more commonly known as John Napier, who did pioneer work in logarithms and published *Mirifici logarithmorum canonis descriptio* in Latin, in 1614. There were no computers in those days, so logarithmic tables had to be methodically calculated and inscribed by hand. Neper published a followup

document in 1617 describing a means of simplifying calculations using ivory numbering rods (Napier's bones) marked with numbers (a physical calculating device that facilitates computations just as a slide rule facilitates computations). Neper's logarithmic discoveries were important fundamentals for many future discoveries in mathematics, physics, and astronomy.

nephelometer An instrument for measuring or estimating light-scattering coefficients in fine particulate "clouds." Nephelometers are used to assess the properties of aerosols, dust, allergens, microbes, liquid suspensions, and other small groupings of fine particles by measuring light attenuation by scattering and absorption over distance.

Nephelometers incorporate many optical components, including a light trap to provide a dark reference against which scattered light may be assessed. They may also include concentrating or diffracting lenses, one or more photomultiplier tubes, and bandpass filters. If the light is to be separated into wavelengths, they may also include prismatic components such as dichroic filters. A Ronchi grating or "chopper" may be used for calibration. See dichroic, nephelometry, Ronchi grating.

nephelometry The science of light scattering or, stated another way, the study of "cloudiness" or of microscopic particles or surfaces that influence the passage and direction of travel of light. Nephelometry is useful in the study of weather, atmospheric pollution, drug solubility, immunology, allergens, electromagnetic wave propagation, dust, and microbes.

Lasers have become important components in nephelometers for measuring biosystem "clouds." Thus, laser/fiber optic nephelometers are useful instruments for counting fine particles and characterizinng suspended particles, tasks that are difficult by any other means. See nephelometer; Tyndall, John.

Nernst effect A potential difference (electromotive force) develops in a metal band or strip when heated. This is a transverse thermomagnetic effect, i.e., the force is perpendicular to the magnetic field. The effect is named after Walther Nernst (1864–1941) who described the third law of thermodynamics (the Nernst heat theorem) in 1905.

Nernst lamp A continuous source of near-infrared radiation developed by W. Nernst in the 1890s. It is useful in fields such as spectroscopy.

NESC See National Electrical Safety Code.

NESDIS National Environmental Satellite, Data and Information Service.

Net Citizen, Net Denizen See Netizen.

Net Police A generic term for the various individuals who moderate communications on the net for appropriateness, tact, good taste, honesty, and fair use. Although some resent the activities of the Net Police, for the most part, these folks are committed, caring, hard-working volunteers who want to see the broadest possible access to the Internet, and who encourage voluntary compliance with Netiquette in order to try to prevent government regulation of the Internet's open communications forums. See Netiquette.

NET/ROM A packet radio communications protocol which has largely superseded AX.25. It provides support for a wider variety of types of packets with automatic routing. See AX.25.

NetBIOS Frames Control Protocol NBFCP. Originally the NetBEUI protocol, NBFCP establishes NBF Protocol to run over Point-to-Point Protocol (PPP). NBFCP enables an end system to connect to a peer system or to the local area network (LAN) in which the peer is located. It is not suitable for interconnecting LANs. NBFCP defines a method for encapsulating multiprotocol datagrams, a link control protocol (LCP) for establishing and configuring the data link connection, and a family of network control protocols (NCPs) for establishing and configuring different network-layer protocols. See RFC 2097.

NETBLT NETwork BLock Transfer Protocol. This is a transport level networking protocol intended for the fast transfer of large quantities of data. It provides flow control and reliability characteristics, with maximum throughput over different types of networks. It runs over Internet Protocol (IP), but need not be limited to IP.

The protocol opens a connection between two clients, transfers data in large data aggregates called "buffers," and closes the connection. Each buffer is transferred as a sequence of packets. Enhanced Trivial File Transfer Protocol (ETFTP) is an implementation of NETBLT. See RFC 998.

Netcast Broadcasting through the Internet, for example, through streaming video. See Webcast.

Netiquette *N*ewsgroup *etiquette*, *N*etwork *etiquette*. An important, well-respected voluntary code of ethics and etiquette on the Internet. Many people have contributed to *Netiquette*, but it was mainly developed by Rachel Kadel at the Harvard Computer Society, and subsequently maintained by Cindy Alvarez. The whole point of having Netiquette is so that network citizens can enjoy maximum freedom by not abusing the rights and sensibilities of others, so that the Net will remain largely unregulated and unrestricted. This freedom depends upon the cooperation of everyone.

In the early days of BBSs, in the late 1970s, most systems were completely open and not password protected. Gradually the constant vandalism and lack of consideration for others caused passwords to be implemented. Eventually, by the mid-1980s, even this was not sufficient to curtail childish or destructive behavior and many of the system operators (sysops) gave up trying to maintain the systems.

Many of the same unfortunate patterns of abuse have damaged the USENET newsgroup system, which used to be a fantastic open forum for discussion, with many scientific and cultural leaders participating under their real names in the mid-1980s. Unfortunately, this system is now abused by bad language, inappropriate remarks, and get-rich-quick come-ons. Consequently, many groups have been forced to close up or go to moderated status, and most celebrities now use assumed names. If members of the Internet community realize that it is completely possible to voluntarily appreciate and respect the rights of others, the Internet can remain an open resource for all.

It's a good idea to read *Netiquette*. Its adherents encourage people to choose voluntary self-restraint and freedom over regulation. See emoticon, Frequently Asked Question, Netizen. Also, Arlene H. Rinaldi's "Net User Guidelines and Netiquette" in text format is available at many sites on the Internet, including: ftp://ftp.lib.berkeley.edu/pub/net.training/FAU/netiquette.txt

Netizen *Net* citizen or *Net* denizen. A responsible user of the Internet. Many founders and users of the Internet consider themselves members of a new type of global community that shares and promotes a vision of an open, freely accessible, self-governed communications venue in which participants voluntarily deport themselves with responsibility, integrity, charity, and tolerance toward the many diverse opinions expressed online. A Netizen is one who contributes to the positive evolution of the Net and respects online Netiquette. One could also more broadly say that anyone who uses the Net is a Netizen, although some people online have less polite terms for those who abuse their freedoms and those of others on the Net. See Netiquette.

netmask A symbolic representation of an Internet Protocol (IP) address that identifies which part is the host number and which part is the network number through a bitwise-AND operation. The result of this logical operation is the network number. Netmasks are specified for different classes of addresses, and are used in classless addressing as well. See name resolution.

NetRanger An intrusion detection utility from Cisco Systems now known as the Cisco Secure Intrusion Detection System. A service pack was issued to support the detection system sensor component to reduce the chance of the system being circumvented by an encoding vulnerability. See CodeRed, virus.

NETS See Normes Européenne de Télécommunications.

Netscape Communications Originally Mosaic communications, Netscape Communications was the original developer and distributor of Netscape Navigator, the best-known open-source browser on the Internet. The company was founded by Mark Andreessen and some very experienced business people from Silicon Graphics Corporation and McCaw Cellular Communications. It had one of the highest profile public offerings in the computer industry. See Andreessen, Mark.

Netscape Navigator The most broadly distributed and used Web browser on the World Wide Web, and the name of its related server software. Descended from Mosaic, the browser was developed by Netscape Communications and widely distributed as shareware until late 1997. At that point, Netscape made the decision, in 1998, to freely distribute the software as open source software and concentrate on marketing their server software. The first beta release was distributed in 1994.

network An interconnected or inter-related system, fabric, or structure. A logical, physical, or electrical grouping in which there is some electromagnetic or biological intercommunication between some or all of the parts. A broadcast network is a physical and communications association of directors, actors, production personnel, and technologies which together cooperate and are used to create and distribute programming to its viewers. A computer network is one in which computers are able to intercommunicate and share resources by means of wireless and/or wired connections and transmissions protocols. A cellular communications network is one in which a cooperative system of wireless communications protocols, geographically spaced transceivers, relay and controlling stations, and transceiving user devices are used to interconnect callers while moving within or among transceiving cells.

network-attached storage NAS. In general terms, a dedicated file storage device or system of associated devices on a network server. The NAS is intended to take the storage burden off the processing server so that a greater amount of storage and more specialized storage-related resources can be concentrated within the NAS. In more specific terms, NAS is the implementation of a storage access protocol over a network transmissions protocol (e.g., TCP/IP) such that the storage resources are concentrated and separate from the process server (and may even be in a remote location).

The advantage of NAS is that storage devices can be placed where it is convenient to install or maintain them and large storage closets can be established for high-end storage needs. The disadvantage is that the burden of the transmissions between remote users and the NAS is handled across the network, increasing traffic over what would occur if the storage devices were associated with individual workstations and decreasing access times, depending upon the distance and the relative load on the network.

NAS is often implemented within a storage area network (SAN), which is a broader concept, and may be designated as SAN/NAS. See NASD Project, National Storage Industry Consortium, storage area network.

network, broadcast *n.* A commercial or amateur radio or television broadcast station. A few examples of well-known broadcast networks include CBC (Canada), BBC (Britain), ABC, NBC, and PBS. Amateurs often run local or special-interest radio, television, or slow-scan television broadcasts. See ANIK.

network, computer *n.* 1. A system comprising nodes and their associated interconnected paths. 2. A system of interconnected communications lines, channels, or circuits. A small-scale computer network typically consists of a server, a number of computers, some printers, modems, and sometimes scanners, and facsimile machines. The highway system is a type of network, as is the very effective train system in Europe. See local area network, wide area network.

network access control NAC. Network policies, configurations, and administrative steps that control the data transmissions to a server, switching/routing component, network, workstation, or peripheral device. Password accounts, dedicated workstations and peripherals, gateways, firewalls, and employee passcards are examples of NAC components and procedures.

NAC policies and systems are of particular interest at institutions where sensitive information or expensive services must be handled efficiently or protected from tampering or misuse. NAC is important in business and educational local area networks and particularly important in classified government and military systems, especially on systems where links to the outside world through the Internet are desired without compromising internal security. In some cases biometrics (e.g., iris scans) may be used to control access to computer terminals or rooms. Monitors and logs may be implemented to oversee general use and to provide an audit trail to check back through unusual activities. See authentication, firewall, gateway.

Network Access Point NAP. A major backbone point which provides service to ISPs and is designated to exchange data with other NAPs. NAP was a development in the mid-1990s which arose from the change in the U.S. Internet from a single, dominant backbone to a shared backbone across four NAPs (California, Illinois, New Jersey, Washington, D.C.). See MAE East, MAE West, Metropolitan Area Ethernet, Public Exchange Point.

network address An identifier for a physical or logical component on a network. Components often have a fixed hardware address, but may also have one or more logical addresses. Logical addresses may change dynamically as the network is altered physically, or as the network software is tuned or protocols changed. Network addresses are typically associated with nodes and stations. See address resolution, domain name, Media Access Control.

network administrator 1. The human in charge of the installation, configuration, customization, security, and lower level operating functions of a computer network. On larger networks, these tasks may be divided among a number of professionals. See SysOp. 2. A software program that handles details of the job of a human network administrator. Activities automated with network administration software include monitoring, archiving, and system checks. See daemon, dragon.

Network Applications Consortium NAC. A trade organization seeking to support and promote generally accepted standards rather than a large number of fragmented proprietary standards for network applications in order to promote interoperability. http://www.netapps.org/

Network Control Protocols NCP. The Point-to-Point Protocol handles assignment and management of Internet Protocol (IP) addresses and other functions through a family of Network Control Protocols (NCPs) which manage the specific needs of their associated network-layer protocols. See Point-to-Point Protocol, RFC 1661.

network drive A drive accessible to multiple users on a computer network. On some network systems, users have to specify and access a particular drive to take advantage of the shared storage space. On other systems, the shared arrangement can be set up so that it is transparent to the user and, in fact, a volume may traverse several drives. Network drives are sometimes configured for data redundancy in case one drive or partition becomes corrupted. See redundant array of inexpensive disks.

Network Driver Interface Specification NDIS. A network protocol/driver interface jointly developed by Microsoft Corporation and 3Com Corporation. NDIS provides a standard interface layer that receives information from network transport stacks and network adapter card software drivers. The transport protocols are thus hardware-independent.

network element NE This is defined in the Telecommunications Act of 1996, and published by the Federal Communications Commission (FCC), as:

"... a facility or equipment used in the provision of a telecommunications service. Such term also includes features, functions, and capabilities that are provided by means of such facility or equipment, including subscriber numbers, databases, signaling systems, and information sufficient for billing and collection or used in the transmission, routing, or other provision of a telecommunications service."

See Federal Communications Commission, Telecommunications Act of 1996.

network fax server A workstation equipped with facsimile or fax/modem hardware and software so multiple users of the network can route a fax in and out through the server. This removes the necessity of having a fax system attached to each computer. The fax server can then also be located nearer its associated phone line. There are fax servers that can use Internet connections (T1, frame relay, etc.) rather than phone lines to send and receive messages.

network filter A transducer designed to separate transmission waves on the basis of frequency.

Network Information Service NIS. A client/server protocol developed by Sun Microsystems for distributing system configuration data among networked computers, formerly and informally known as Yellow Pages. NIS is licensed to other Unix vendors.

network interface NI. A junction or reference point in a network that supports or represents a change in the physical and/or logical structure of a link or, in some cases, represents a jurisdictional change (even if there is no physical change or data conversion). The point at which a phone service line connects to a subscriber's premises is a network interface and various aspects of this connection have been standardized. For example, the network interface electrical characteristics and interactions for an analog connection between a telecommunications carrier and the customer premises are described in ANSI T1.401-1993.

network interface card, network interface controller NIC. A PC board that provides a means to physi-

cally and logically connect a computer to a network. For microcomputers, typically these cards are equipped with BNC and/or RJ-45 sockets facing the outside of the computer and edge card connectors that fit into the expansion slots inside a computer. The cables resemble video cables, or fat phone cables, depending upon the type used.

Most systems require a physical *terminator* on the physical endpoints of the network (if the network isn't working, it may be because termination is missing or incorrectly installed). Separate software, not included with the computer operating system, may be required to use the specific card installed.

Many workstation-level computers come with network hardware and software built in, and Macintosh users are familiar with the built-in AppleTalk hardware and software. The trend is for microcomputers to use TCP/IP networking over Ethernet.

Network Integration Verification Test NIVT. A test designed to evaluate and improve three different Front End Processors (FEPs) handling routing and transport protocols in mixed high-performance routing systems with large Topology Databases to assess scalability. Information on NIVT and actual test reports are available through the IBM Web site.

network interface function NIF. A function associated with a specific interface link in a network. For example, there may be specific translation functions in the interface between a subscriber's computer and the Network Interface Device (NIC) supplied by a service provider, e.g., in ISDN networks.

network intrusion detection system NIDS. A means or set of procedures and/or programs designed to alert the system administrator or individual users about vandalistic attacks or unauthorized access to a computer network. There are many types of intrusion, including system flooding, virus insertion, account access, file access, and physical access. NIDS more often refers to the first four categories. (Physical access to the electronics inside a computer may be detected by video cameras, marks, fingerprints, or detection chemicals, but this type of intrusion is less prevalent than day-to-day attempts by unauthorized users to view, steal, or compromise data on a system.) An intrusion detection system is rarely just one application or device, but rather the coordinated implementation of a collection of policies, procedures, and tools to ensure system security.

In general, NIDS refers to systems that actively detect signs of intrusion (e.g., a utility that detects repeated entries of incorrect passwords) as opposed to passive systems that primarily deter intrusion (e.g., a password to access an account).

As networks and databases with sensitive information are increasingly connected to the Internet, they become more accessible and vulnerable to outside intruders. Electronic commerce sites on the Web are particularly vulnerable, as there are people who search for credit card numbers, bank accounts, personal identification, and other commodities that are recorded on computer systems. Two of the more important intrusion detection mechanisms include:

monitors Applications that display realtime or recent statistics and events. On older systems and mainframes, monitors commonly represent data as text displays arranged in columns. On some of the newer systems, graphical tables and graphs are also available. System monitors are often bundled with operating systems to allow sysops to view CPU usage, connection requests, numbers of users, entries and exits to the system, locations of machines being accessed, numbers or types of packets being transferred, URLs of users accessing Web servers, etc. Operators use a combination of live audio/visual monitoring and software utilities to generate an alert when anomalous patterns occur.

logs Records of activities. The data generated by monitors are often kept in running logs that can be archived indefinitely or stored for a period of hours or months, depending on the need. In addition to system monitors, software installation and system reconfiguration activities are often logged to record when changes were made, which files were added or deleted, and where the files were installed. Intruders are not always high-tech computer experts; often they are employees snooping or stealing data with little understanding of the electronic trail they leave when they engage in unauthorized access. Even if they do search for logs, with the intention of changing or deleting them, they may be unsuccessful if the system is configured to duplicate log entries on another computer or a protected directory.

One of the more difficult types of intrusion to detect is access by someone who has stolen a legitimate password. Since the intruder isn't breaking into the system, but logging on normally, the intrusion may go unnoticed for a long time. However, even this type of activity can sometimes be detected with a combination of monitors and logs. If CPU usage during the night or lunch hour is usually low, and the various monitors show John Doe's account becomes active during times when John isn't at his desk, there is reason to investigate the anomalous patterns. Intrusion detection systems are, in large part, dependent upon the ability of the system administrator and the capability of the system to determine normal usage patterns and variations from the norm.

The most difficult type of intrusion to detect is unauthorized activity by a system administrator or programmer. A small percentage of sysops take advantage of their privileged positions for personal gain. Since a high-level administrator or systems programmer has access to almost everything on a network, it's very difficult for others to detect tampering or unethical use of network resources. They are also expert at hiding a data trail when using one system to access another or when snooping on password-protected file systems. These types of intrusion often go unnoticed until funds disappear or until strange things happen after the individual leaves a project or firm. The primary ways to reduce system administrator intrusions are careful employee screening, built-in accountability policies and procedures, and good employee relations. For the most part, system administrators are intelligent, dedicated professionals, proud of their systems and concerned about maintaining good system security. Contrary to what might be expected, the majority of intrusions are probably not by system administrators, but by curious or mischievous programmers on the one hand and professionals trying to divert funds or information or save a buck on the other (statistics suggest that a surprising proportion of computer-related theft is by doctors and businesspeople as opposed to university students and teenagers).

Network intrusion detection is not a simple configuration that can be set up and used indefinitely. Computer security requires a responsive approach to a dynamically changing environment, like a farmer adapting to constantly changing weather and market conditions. A system administrator must monitor, fine-tune, and reconfigure on an ongoing basis to achieve network security. See cracker, firewall, hacker, virus, worm.

Network Job Entry NJE. A communications protocol developed by IBM that arose out of the widespread use of Remote Job Entry (RJE) protocol used for the remote submission of computer processing jobs. RJE could be used in conjunction with the mainframe Job Entry Subsystem (JES) to enable RJE-enabled workstations to submit jobs to a centralized mainframe system. Network Job Entry grew out of JES, extending JES and the functionality of RJE. NJE enables two JES subsystems on different host computers or in different local partitions to intercommunicate, thus supporting peer-to-peer communications of commands and job submissions on IBM host systems. NJE was developed at the time when mainframe prices were coming down and multiple mainframes began to be installed in separate locations. It enabled specialization of mainframe computing functions and efficient use of computers with different capabilities. While the protocol has been around for quite a while, it is less well known than RJE. BITNET was an important historical precursor to the Internet based on the NJE protocol. See BITNET.

Network Layer Packet NLP. In High Performance Routing on packet networks, a basic message unit that carries data over the path. See datagram.

Network Management Processor NMP. A network switch processor module used to control and monitor the switch.

Network Management Protocol NMP. A set of protocols developed by AT&T to control and exchange information with various network devices.

Network Management System NMS. An administrative service tool for subscriber networks that enables the system administrator to adjust performance characteristics and to set bit rates and user settings according to the levels of service available to subscribers.

Network News Transfer Protocol NNTP. A software application developed in the mid-1980s to provide a way to more quickly and efficiently query, retrieve, and reference information from newsgroups through NNTP servers. It also facilitates list management of newsgroup discussions. NNTP is a network news transport service. Newsgroups may be accessed through the Web from local clients using the NNTP URL scheme as follows:

```
nntp://<host>:<port>/<newsgroup-name>/
<article-number>
```

For global access to newsgroups, the news: scheme is preferable. See news, RFC 977, RFC 1738.

Network Operations Center NOC. A centralized around-the-clock facility for monitoring and maintaining a network, which may remotely service smaller centers such as POPs. NOCs typically provide a number of technical support and accounting services as well. Most large networks (computer, phone, broadcast) have a core staff dedicated to the physical and logistical tasks of keeping the system running, well-maintained, and current. One of the most prominent NOCs is the U.S. Air Force NOC, formerly the Air Force Network Control Center.

network prefixes Identifiers used to aggregate networks. Networks are divided into *classes* with the ability to serve up to a certain number of hosts. The prefix identifies the class, and hence the number of possible hosts.

Network Reliability and Interoperability Council NRIC. An advisory committee formed to provide recommendations to the Federal Communications Commission (FCC) and the telecommunications industry regarding network reliability and interoperability of public telecommunications networks. The original charter was filed in 1992. The most recent mandate was to operate until January 2002. The NRIC works through a number of focus groups and subteams. See Committee T1, New Wireline Access Technologies. http://www.nric.org/

Network Security Information Exchange NSIE. A forum for identifying issues of network security such as unauthorized or malicious entry or tampering that might affect national security and emergency preparedness telecommunications systems. Members exchange information on viruses, threats, incidents, and other attacks on public telecommunications networks.

network service point NSP. Cisco Systems technology that provides native SNA network service point support.

Network Solutions, Inc. NSI. In 1993, this company was awarded the contract for registering Internet domain names with the InterNIC by the National Science Foundation. NSI was acquired by Scientific Applications International Corporation (SAIC) in 1995. See InterNIC.

Network Terminal Number NTN. An identification number assigned to a terminal on a public network by the public network administrator. The ITU-T recommends that public voice/data and digital voice networks also assign NTNs. The NTN is a designation within the Data Network Identification Code (DNIC) for public networks interconnected with X.75. See Data Network Identification Code, X Series Recommendations.

Network Video NV. A freely distributable Sun SPARC-, DEC-, SGI-, HP-, or IBM RS6000-based videoconferencing system developed at Xerox PARC, which supports video, audio, and whiteboarding over Mbone networks.

Network Voice Protocol NVP. The historical forerunner of the Voice File Exchange Protocol proposed for the ARPANET in the mid-1980s, the NVP was submitted in November 1977 by Danny Cohen on behalf of a cooperative effort of the ARPA-NSC community.

We tend to think of voice carried over computer networks as a development of the late 1990s, but the idea has its roots much earlier and has been active on experimental systems since the 1970s and on some commercial systems since the mid-1980s. NVP was first implemented in December 1973, and was subsequently used for local and remote realtime voice communications over the ARPANET at some of the major research facilities in the U.S.

The development of secure, low-bandwidth, two-way, high-quality, realtime, digital voice communications was a major objective of the ARPA Network Secure Communications (NSC) project. It was, at the time, a high-priority military goal, intended to facilitate command and control (C^2) activities. By implementing the concept with digital technologies, encryption could be used to help protect the content of communications. NYP consists of a control protocol and a data protocol. See Voice File Exchange Protocol, RFC 741.

Neumann, John von See von Neumann, Janos.

neural computer A computing system theoretically designed to behave like the human brain in terms of performing logical, intelligent problem-solving and inferential "thinking" activities, which also may structurally mimic the interconnective structural topology of biological neurons in a centralized nervous system. A neural computer, like the human brain, configures itself through experiential learning, feedback, and internal reorganization over time. Neural computers are not entirely theoretical, except in their most ideal form. There have been many efforts and successes in the design of neural and bionic systems since the early 1960s, with worldwide efforts by major companies to design and implement practical neural computers on a small scale since the 1970s and on a large scale since the late 1980s and early 1990s. A neural computer is a specialized type of supercomputer, since "supercomputer" implies the state of the art in computing at any one time, and existing neural computers have demonstrated extremely fast processing and problem-solving speeds.

Neural net architectures tend to be highly parallel, with multiple registers, several layers, and a high level of interconnection between nodes. The concepts of neural computers date back to the 1940s, to the work of W. McCulloch, W. Pitts, A. Rosenblueth, and N. Wiener. See artificial intelligence; bionics; neural network; Wiener, Norbert.

neural network In a broad sense, a type of network organization that mimics the human nervous system, particularly the brain, in physical structure and connectivity or neural functioning as it relates to thinking, or both. Simulation of neural networks, and modeling of the complex reasoning, generalizations, and inferences characteristic of human thinking have long been of interest to programmers and scientists studying artificial intelligence. While the creation of androids, humanoid intelligent robots, is probably some time in the future, some interesting advances in programming have resulted from studies of neural network functioning. Software that has the ability to generalize and make choices, react, and further configure itself in response to feedback is being developed with practical applications in many areas, including robotics. Neural networks can aid machines and humans in unfamiliar environments.

Speculation about neural networks and "thinking machines" has been around at least since Ada Lovelace proposed, in the 1800s, that intelligent machines might someday produce art and poetry. In the late 1940s, Norbert Wiener, Arturo Rosenblueth, and their colleagues were discussing concepts related to "cybernetics," a term popularized by Wiener in *Cybernetics: or, Control and Communication in the Animal and the Machine*. In 1963, in *Electronics World*, Ken Gilmore described the work on bionic computers being carried out at Wright-Patterson Air Force Base in Ohio and modeling of individual neuronal circuits by companies like Bell Laboratories and the Ford Motor Company. In the 1950s and early 1960s there were already many experimental implementations of various aspects of neural networks, including electronic maze-running mice, pattern-recognizing machines, self-organizing machines, and simulations of human vision systems. See artificial intelligence; bionics; Harmon, L.D.; Melpar model; MIND; pattern matching; perceptrons; Sceptron; Wiener, Norbert.

neuroelectricity The very minute level electromagnetic fields generated by the activities of biological neurons. See neural network.

neuron In a biological system, cells specialized to code and conduct an electromagnetic impulse are called neurons. A network of interconnecting neurons is called a nervous system, and a network of interconnecting neurons with a main processing center is called a central nervous system, with the main processing center called the brain.

Neuron Chip A commercial microcontroller chip with the LONTalk automation control protocol embedded into the chip. This enables the chip to be used as a cost-effective controller for a wide variety of industrial and residential devices, including fans, switches, motors, motion sensors, valves, and more. The Chip includes three 8-bit inline central processing units, two of which are dedicated to LONTalk protocol processing, with the third dedicated to the node application program. The chip has built-in memory and 11 general-purpose input/output pins for interfacing with circuits.

Neuron Chips are programmed in Neuron C and can access the built-in LONTalk communications software and network management functions, as well as schedulers, and arithmetic/logic application runtime libraries. Neuron C is an object extension to ANSI C. LONTalk is an open standard for control automation networks. The Neuron Chip is available through Cypress Semiconductor and Toshiba Corporation. See LONWORKS, LONTalk.

neutral In stasis, in equilibrium, stable, balanced, normal, unaffected, neither positive nor negative, not tending to one side or the other, nor one state or another. Neither acid nor base.

neutrodyne In early radios, an amplifying circuit used in tuned receivers. Voltage was fed back by a capacitor to the circuit to neutralize it. See heterodyne, superheterodyne.

New England Museum of Wireless and Steam Located in Rhode Island, this museum preserves the original Massie station, the oldest surviving, originally equipped wireless station.

New Haven District Telephone Company A historic exchange that welcomed its first subscriber, Rev. John E. Todd, in 1878. By February 21, 1878, the company's first telephone directory included almost 50 subscribers, primarily physicians and businesses, listed according to professions. No numbers were assigned to the subscribers, as operators handled the calls and phone numbers did not come into use in this area until 20 years later.

New Ico Formerly ICO Global Communications (est. 1995), a London-based satellite communications service, spun off from the Inmarsat Project 21. Hughes Electronics has a large interest in the company, and Hughes Telecommunications and Space Company is building the satellites. Other ICO Global investors included COMSAT Corporation, Beijing Maritime, Singapore Telecom, Deutsche Telecom, and VSNL (India).

The original plan was to launch ten satellites plus two spares, into medium Earth orbits (MEO) at 10,000 km using *bent pipe* analog transponders. The satellites would be divided between two orbital planes, inclined 45° relative to the Earth's equator, orbiting once every 6 hours. Some innovations are planned; the solar wings carry gallium arsenide rather than silicon solar cells and the propulsion system is hydrazine-based. Thermal control is achieved in part with a sun nadir steering system, which orients the panels toward the sun, and the radiating surfaces away from the sun. C- and S-band capabilities will support 4500 simultaneous phone conversations.

Six ICONET satellites were scheduled to come online

in the initial stages to interface with 12 Earth stations. Start of service was scheduled for the year 2000 but has been adjusted to 2003 for the New ICO voice and packet-data services. New ICO is a McCaw-led acquisition that has resulted in an updated version of the ICO Global Communications project with some modifications to the satellites in production and the inclusion of third-generation (3G) wireless services in the new plans.

New Wireline Access Technologies NWAT. A Focus Group subteam of the Network Reliability Council established to examine reliability in key services deployed over the Public Switched Network (PSN). NWAT endeavored to identify, define, and clarify potential service reliability issues associated with new wireline technologies and to provide recommendations and potential solutions. The project ran from August to December 1995. Of particular concern were Hybrid Fiber/Coax (HFC) and Fiber-to-the-Curb (FTTC) access networks, which were evaluated against Digital Loop Carrier (DLC) and cable television (CATV) as benchmarks. Participants came from many key firms in the industry, including Bellcore, Cable Labs, Motorola, NYNEX, U.S. West, Time Warner, and others.

New York and Mississippi Valley Printing Telegraph Company An early American communications business organized by Hiram Sibley in 1851 which, as it expanded westward, came to be called Western Union, a name suggested by Sibley's associate, Ezra Cornell. Western Union subsequently installed the first transcontinental line in 1861.

newbie A telecommunications greenhorn; a new or inexperienced user. There's nothing wrong with being a newbie, but new users *must* read the introductory information, charters, and FAQs (Frequently Asked Questions) associated with the activity they wish to pursue. It's OK to ask questions on the Net, but the first question should always be "Where can I read the FAQ for this channel/newsgroup/discussion list?" Reading the FAQ will conserve bandwidth, save time, and can spare an individual a great deal of personal or professional embarrassment. See Netiquette.

news, Web access There are a number of ways in which programmers have implemented access to Internet newsgroups through Web interfaces. Traditionally, news has been read through Unix command line text interfaces, and many still read the various newsgroups this way. There are also dedicated newsreaders which run on individuals' machines.

When a browser is designed to support the display of newsfeeds, newsgroup articles can be accessed through the Web with two types of Uniform Resource Locators (URLs) as follows:

```
news:<newsgroup-name>
```

e.g., news:comp.sys.linux

```
news:<message-id>
```

News URLs are location-dependent. See NNTP, RFC 1036, RFC 1738.

newsgroup A private or public online forum, the largest of which is the USENET system. USENET has more than 35,000 newsgroups, covering every conceivable topic from *alt.religion* to *alt.bondage*. Most newsgroups function on a subscription basis; current software makes it reasonably easy to subscribe at the moment at which you would like to read the messages. Not everyone has access to the same USENET newsgroups; it depends partly on what topics your Internet Services Provider has downloaded for its subscribers. Postings on various newsgroup forums can range from one or two messages a day to several thousand a day. A *newsreader* software program can help sort out the topic *threads*.

In a text-based newsreader, the various newsgroups will be listed alphabetically; in graphical newsreaders, they may be hierarchically organized in menus. The following simple text-based excerpt shows the general format of newsgroup names.

```
alt.humor
alt.humor.best-of-usenet
alt.invest
comp.sys.mac.advocacy
comp.sys.next.software
comp.theory.info-retrieval
humanities.philosophy.objectivism
misc.business.marketing.moderated
misc.entrepreneurs.moderated
misc.legal
misc.legal.moderated
sci.astro
```

The above names have a hierarchical structure from general to more specific. The general topics listed above include *alt*ernate, *comp*uter, *humanities*, *misc*ellaneous, *rec*reation, and *sci*ence. Anyone can create a newsgroup, given sufficient community support and interest. Creation of a new USENET newsgroup requires a body of voters to ferry a proposal through a lengthy submission/acceptance process, which may take 4 to 7 months. This is necessary as a deterrent to frivolous group creation.

Some newsgroups are moderated. Unfortunately, due to inappropriate postings, open newsgroups are decreasing in number. This puts an unfair burden on moderators, who are generally volunteers, but at least it is a way to keep a forum alive. When you post to a moderated group, the posting is previewed for adherence to the topic, or content, or both. Some newsgroup moderators reserve the right to edit actual postings (although this is rare). Read the charter before you post if you don't wish to have your postings altered. If the message meets the requirements for the group, it is then posted by the moderator. This process can take from a few hours to a few days, sometimes even up to a week and a half.

If you are offended by the topic of a group, don't read the postings. Newsgroups have evolved with a very strong commitment to the tenets of free speech, and their participants vehemently guard their right to express and discuss their views online in the appropriate forum.

Netiquette has been developed to provide guidelines to the effective and courteous use of the USENET system. Read *Netiquette* and the charter for each group before posting; then enjoy; USENET is the closest individuals have ever come to having access to the sum total of human knowledge at any one time. It is a living, breathing "expert system" where you can seek answers and support on any topic, any time of day or night. See Call for Votes, *Netiquette,* USENET.

newsreader A software program for accessing, displaying, searching, and posting articles to public Internet newsgroups, particularly USENET. You need access to the Internet to read the postings on newsgroups. Newsgroups have certain customs and traditions, and you should read the newsgroup *Netiquette* before posting, as well as the *Frequently Asked Questions* (FAQ) document for that particular newsgroup. It is also wise to read the existing postings for several days before contributing, to understand the format and content of typical postings, and not to endlessly repeat a topic that may have been fully discussed. A good newsreader program will enable you to follow *thread*s, conversations on a particular topic. There are usually many discussion threads within any given newsgroup. Pine, a popular Unix-based email program developed at the University of Washington, can be used as a newsreader, as can a Web browser. Google Groups now provides searchable access to a huge historic newsgroups archive. See newsgroup.

newton A unit of force in the meter-kilogram-second (MKS) system of physical units of a size that will influence a body of a mass of 1 kg to accelerate 1 m per second per second. Named after Sir Isaac Newton.

Newton, Isaac (1642–1727) An English scientist and mathematician acknowledged as one of the greatest contributors of basic knowledge of our universe through his descriptions of the laws of motion and theory of gravitation (*Philosophiae Naturalis Principia Mathematica,* 1687) and the idea that earthly and celestial events might obey the same laws. He also studied the nature of light (*Optics,* 1692), described the nature of white light and its component colors, and laid much of the foundation for modern calculus.

Newton's rings A ring-shaped interference phenomenon that results from reflected radiant energy associated with transparent surfaces held very close together with a fine layer of air in between. The center of the rings is dark, with alternating dark and light bands with subtle colors emanating outward in concentric rings.

Robert Boyle was the first to describe and explain the phenomenon and Robert Hooke reported in *Micrographia* (1667) that he had observed the rings. Isaac Newton made use of the phenomenon to polish lenses. After carrying out experiments with prisms, Newton presented theories on the nature of color that were more accurate than Hooke's, which may be why the rings are called Newton's rings and not Hooke's rings, but perhaps they shouldn't be called Newton's rings either, because Newton actively discounted the significance of the rings, saying they were "not necessary for establishing the Properties of Light." Thomas Young, on the other hand, revisited Newton's observations and made some important mathematical explanations for the rings. In 1801, Young interpreted the interference as resulting from light interaction in the air between the reflecting/refracting surfaces, strong support for the wave nature of light.

You can generate the rings if you put a flat or convex transparent glass almost touching a convex piece of glass. Now carefully move one of the pieces closer and farther away from the other to vary the thickness of the layer of air. At the point where the air between the sphere and flat surface is of the same order as the wavelength of the light, the colored rings will appear through the glass.

Newton's rings are an interesting way to illustrate diffraction and are useful in polishing and checking lenses. Sometimes the phenomenon is undesirable, so there are various ways to reduce/eliminate the rings, including changing the distance between or orientation of the two proximate surfaces, using glassless carriers for darkroom work, or using a light dust of talcum powder on the glass lens. See diffraction, interference.

NEXT See near end crosstalk.

NeXT Unix-Based Workstation

The NeXT cube was promoted as the computer for the 1990s when it was released in 1988. Surprisingly, this marketing hype has held true, even though the NeXT is no longer manufactured. Many corporations are adopting Unix as their standard, as educational institutions have for years; graphical user interfaces are now ubiquitous, and Display PostScript still provides one of the best WYSIWYG solutions on any system. Many NeXTStep aspects are now in Mac OS X.

NeXT computer The NeXT computer was unveiled in early October 1988 by Steve Jobs' company NeXT, Inc. It included the first commercial erasable optical drive and incorporated VLSI technology. The programmable digital signal processor (DSP56001) came built in. The operating system was Unix-based, with a gorgeous graphical display interface incorporating Display PostScript. The fonts and graphics are all beautifully rendered in high resolution. The NeXT had some inspired input from Stanford, Carnegie-

Mellon, and the University of Michigan. The original NeXT was grayscale, with color added in later versions. The first NeXT was cube-shaped, with later hardware resembling more conventional desktop flat systems, known as NeXT stations.

The first NeXT was based on the Motorola 68030 processor with a built-in 68882 math coprocessor, and it came standard with 8 MBytes of RAM. At the time, most computers had 1 to 4 MBytes of RAM. The original price was $6500 and marketing efforts were aimed at higher education institutions, although business owners expressed early interest due to the networking capabilities of the system.

The first edition of CRC's Telecommunications Illustrated Dictionary was written on a NeXT computer, and even though the basic technology is over 10 years old, the computer hardware and operating system have stood the test of time in essentially their original form. The simple, stunningly aesthetic graphical user interface still beats most systems hands-down; the powerful object-oriented Unix-based operating system and shell connect seamlessly with the Internet, and the multitasking operating system allows dozens of processes to run happily at the same time. In over 3 years of 24-hour a day operations running multiple desktop publishing, Web browsing, and illustration programs at the same time, the author's machine didn't crash once. That's an enviable track record. After using a dozen different types of computers daily for over 20 years, the author has seen few systems that equal it (Sun systems provide similar performance).

The NeXT is an excellent networking computer, connecting easily to the Internet, other NeXT systems, and other types of computers through TCP/IP. It is also an excellent Internet portal, with a full complement of Unix tools, including Telnet, FTP, and others easily downloadable from the Net. OmniWeb, by Lighthouse Design, Ltd., is a powerful Internet graphical browser for NeXTStep that preceded many well-known graphical browsers.

In 1997, Apple Computing, Inc. bought out NeXT, Inc. and continued developing the operating system software under the development name of Rhapsody, now better known as Mac OS X. See Jobs, Steven.

Next Generation Digital Loop Carrier NGDLC. Developed in the 1980s as an evolution of Digital Loop Carrier systems, NGDLC is based on very large scale integration (VLSI) technology. ISDN was developed and promoted at about the same time that NGDLCs were implemented, so many were developed to accommodate ISDN. Whereas Digital Loop Carriers were designed to provide services over traditional copper phone lines, NGDLC was designed to work in conjunction with fiber optic cables or fiber/copper hybrid systems. See Digital Loop Carrier.

Next Generation Internet NGI. A U.S. federal multiagency research and development initiative established in 1997. NGI works with industry and academia to develop, test, and demonstrate advanced networking technologies and applications. The following federal networks are used as testbeds for the NGI initiative:

- NSF's very high performance Backbone Network Service (vBNS)
- NASA's Research and Education Network (NREN)
- DoD's Defense Research and Education Network (DREN)
- DoE's Energy Sciences network (ESnet) (proposed beginning in FY 1999)

NGI is coordinated by the NGI Implementation Team, coordinated by the Large Scale Networking Working Group of the Subcommittee on Computing, Information, and Communications (CIC) research and development of the U.S. White House National Science and Technology Council's Committee on Technology. http://www.ngi.gov/

Next Hop Resolution Protocol NHRP. An internetworking architecture, which runs in addition to routing protocols and provides the information that enables the elimination of multiple Internet Protocol (IP) hops when traversing a Next Hop Resolution Protocol network. Aims at resolving some of the latency and throughput limitations of Classical IP. See Next Hop Server, ROLC, NBMA.

Next Hop Server NHS. In an NHRP networking environment, the Next Hop Server locates an egress point near a given destination and resolves its ATM address, enabling the establishment of a direct ATM connection. See Next Hop Resolution Protocol.

NGDLC See Next Generation Digital Loop Carrier.

NGI See Next Generation Internet.

NGSO nongeostationary orbit.

NGSO FSS nongeostationary orbit fixed satellite service.

NHRP See Next Hop Resolution Protocol.

NICI See National Information and Communications Infrastructure in Appendix G.

nickel metal hydride NiMH. A rechargeable battery commonly used in portable devices. Hydride is a hydrogen compound.

nickel-cadmium cell NiCd, NiCad. A very common, sealed, rechargeable power cell that works well in low temperatures. The positive electrode is nickel and oxide, and the negative electrode is cadmium, with the plates immersed separately in a potassium-hydroxide electrolyte solution. NiCad batteries have been used in many small, portable telecommunications devices but have the disadvantage of a "memory effect," that is, they will not fully recharge unless first fully discharged, thus reducing the useful time of the battery.

nickname An easy-to-use, easy-to-remember substitute or secondary name. In most cases, it's a short name, generally one easy to remember because it is familiar or matches the personality or properties of the person or object for which it is designated. A nickname can also be a name given as a term of intimacy and affection between two people who are closely acquainted. Nicknames may also be names that are easy to type, to save time, as on public discussion areas of the Internet. See handle, NICname.

NICname, Nicname, Nickname On Internet Relay Chat (IRC), a name that can be set with /nick <putnamehere>. Only one person can have a specific nickname at any one time on IRC. See Internet Relay Chat, WhoIs.

Nicol, William (ca. 1768–1851) A Scottish educator and physicist who developed the Nicol prism from Iceland spar. There are few records about his early life, but he began publishing his research in 1828 and spent the latter part of his life studying crystals and fossils. He apparently developed his own lenses and invented new methods for grinding samples for microscopic inspection. Nicol's sister, who was about five years younger, married Edward Sang, a prominent mathematician and engineer.

Nicol prism A refractive component consisting of two blocks of Iceland spar (calcspar) cemented together along the diagonal plane with Canada balsam, a material derived from bark ooze that was a common bonding agent in optics for about 200 years. The Nicol prism was one of two common birefringent prisms, along with the Ahrens prism.

The unusually distinct birefringent properties of Iceland spar were described in 1670 by Rasmus Bartholin, but the mathematics of its properties were not worked out until almost 150 years later, by Thomas Young.

William Nicol put two equally shaped calcspar blocks together to produce the Nicol prism, described in 1828. The device was used in polarimeters as early as the 1840s. Historic polarimeters consisted of a sample tube mounted horizontally between an analyzing Nicol prism and a polarizing Nicol prism. Monochromatic light could be shone through the tube with the analyzing prism rotated to produce two light sources for comparison.

Nicol prisms were incorporated as substage polarizers into European scientific microscopes, beginning around the 1850s or 1860s. A rotating Nicol prism polarization analyzer could be mounted between the microscope nosepiece and the objective lens or over the eyepiece, depending upon the instrument. By selectively rotating the prism and the sample, the polarizing characteristics of the sample could be discerned. Nicole prisms continued to be common in microscopes as recently as the 1950s.

In the 1880s, William Thompson (Lord Kelvin) used a Nicol prism in his lectures and demonstrations on the polarization of light. Edwin Land credits his invention of polarizers at the age of 19 to a demonstration he saw as a schoolchild of a Nicol prism. His invention led to a patent and the establishment of the Polaroid Corporation.

Polarization is now typically accomplished with films and coatings. Polarizing beam splitters somewhat resemble Nicol prisms, having two blocks combined with a refractive coating layer, but the blocks are equilateral triangles and the resulting component is cubical. See Bartholin, Rasmus; Iceland spar; polarization, Wollaston prism.

NIDS See network intrusion detection system.

NIF See network interface function.

Nigerian Communications Commission NCC. The national regulatory authority for telecommunications in Nigeria, the NCC was established to facilitate, coordinate, and regulate private sector participation in the provision of telecommunications services. http://www.ncc.gov.ng/

NIMBUS A satellite program initiated in the early 1960s by the National Aeronautics and Space Administration (NASA), and now operated jointly with the Oceanic Atmospheric Administration (NOAA).

Nicol Birefringent Polarizing Prism

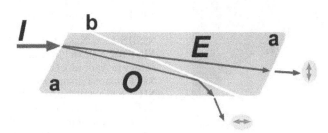

Iceland spar is a doubly-refractive (birefringent) mineral with interesting prismatic properties. Two calcspar blocks can be bonded along the diagonal with a material of slightly lower refraction index (b) to form a Nicol prism which makes it possible to more widely separate the two courses of light through the material and thus isolate the one of interest. In constructing the prism, the natural rhombus angle of ~72° can be ground down to 68° (a) for more effective results.

In this example, incident light (I) enters the material and is split into two beams, one slightly stronger than the other. The ordinary beam (O) is reflected off the seam due to its angle and the fact that it has a slightly lower refractive index than the bonding material (much as light reflects off the cladding in fiber optic filaments). The "extra-ordinary" beam (E) hits the seam almost straight on and passes essentially unimpeded across the bonded seam. Thus, the plane polarized light (E) exits in the same general direction as the incident light.

NIMBUS is used for research and development by atmospheric and Earth scientists.

NiMH See nickel metal hydride.

Nimrod Routing Architecture A scalable network routing architecture intended to support dynamic, heterogenous internetworks of arbitrary size, Nimrod was originally suggested by Noel Chiappa. It was thereafter refined by the IETF Nimrod Working Group and formally described in the mid-1990s.

Nimrod achieves scalability by representing and manipulating routing information at multiple levels of abstraction to accommodate expanding, diversifying networks. Nimrod is characterized by maps that represent internetwork connectivity and services, user route generation and selection, and user packet forwarding along established paths. It is applicable to routing within single and multiple routing domains in TCP/IP and OSI environments. See RFC 1992.

NIOD Network Inward/Outward Dialing.

Nipkow Disc – Historic Image Scanner

This Nipkow disc shows the spiral series of holes through which the light is beamed as the disc rotates. This historic example is from the American Radio Museum collection. [Classic Concepts photo.]

Nipkow, Paul Gottlieb A German experimenter who developed a rotating dial with a spiral arrangement of holes that he patented in 1884. It was an early electromechanical television system. This was later incorporated into television transmitting and receiving units. See Nipkow disc.

Nipkow disc A rotating disc with a sequential helical pattern of holes used by many early television experimenters to attempt the projection of television images. The perforated disc was rotated in front of the image to be transmitted in order to quantize the signal, in a primitive sense, by segmenting the image into lines. The photosensitive material selenium was placed behind the disc to register the dark and light areas of the image. Unfortunately, systems for amplifying the signal for transmission didn't exist at the time, and didn't become practical until other technological developments occurred. The disc is named after its inventor, Paul Nipkow.

Modern versions of the Nipkow disc are used in a variety of applications and typically are fabricated from plastic or glass, with thousands of pinholes or microlenses embossed into the substrate. They resemble translucent CDs and are used in optical scanners and confocal microscopy. In a confocal microscope capable of resolving very tiny images, the perforated disc rotates between a beam-splitter and a lens, providing expanded depth imaging through a form of optical sectioning. A rotation of the disc enables an XY section of the specimen to be acquired in realtime. Height can be evaluated through grabbing a frame and processing the image and Z data combined with it to provide a 3D topographical "map" of the microscopic specimen. See Baird, John Logie; television history.

Nippon Advanced Mobile Telephone System NAMTS. An early, first-generation, analog FM-based mobile phone system with digital processing, first introduced in Japan. Communication Services Limited (CSL) made the system publicly available in 1984.

Nippon Telephone and Telegraph Corporation NTT. The major Japanese telephone company and largest phone company in the world. In 1997, deregulation allowed NTT to begin operating internationally, and its first international subsidiary, NTT Worldwide Telecommunications Corporation, began servicing overseas corporate customers. See Arcstar; Japan Telecom Co. Ltd.; KDDI Corp.

NIS 1. Network Imaging Server. 2. See Network Information Service.

NISDN 1 See National ISDN-1 in Appendix G.

NIST See National Institute of Standards and Technology in Appendix G.

NITF National Image Transfer Format.

NIUF See North American ISDN Users Forum.

NJE See Network Job Entry.

NL port, end loop port In a Fibre Channel network, a port on an endstation that enables it to be connected to the Fibre Channel loop. The NL port is assigned the lowest addresses and thus has the highest priority in terms of obtaining control of a loop. See F port, FL port, N port.

NLA See Network Layer Address.

NLANR National Laboratory for Applied Network Research.

NLC See nematic liquid crystal.

NLPID Network Layer Protocol ID.

NMAA National Multimedia Association of American. http://www.nmaa.org/

Emily Noether developed mathematical group theories that are still widely used by physicists.

NMACS Network Monitor and Control System.

NMD See nonintrusive measurement device.

NMP 1. See Network Management Processor. 2. See Network Management Protocol.

NMR See nuclear magnetic resonance.

NMS Network Management System.

NNI 1. network node interface. 2. Nederlands Normalisatie-Instituut. A Netherlands standards organization established in 1959, located in Delft.

NNTP See Network News Transfer Protocol.

NOAA See National Oceanic and Atmospheric Administration.

noble gas A rare or inert gas. Examples include argon, krypton, neon, and helium. These gases are useful in illuminated signs and laser technologies. See argon, krypton, neon.

nodal clock In a network, a reference clock source for major timing functions associated with a node. A valid clock reference may be extracted from a variety of sources, including a Primary Linkage and Co-ordination Program (PLCP). The best nodal clock in a facility can be selected to provide a unified timing source for Building-Integrated Timing Supply (BITS) such that a group of network switches would appear as a node in the hierarchical network scheme.

node Junction, confluence, meeting point, terminal, intersection. A connection point in a network, which may consist of a router, switcher, dialup modem, computer, or other interconnecting device supporting the same protocol, or converting to the needed protocol. Together the nodes, equipment, and pathways constitute the network topology. See backbone, leaf node.

node port See N port.

Noether, Emily "Emmy" (1882–1935) A brilliant German-born mathematician, Emily Noether developed mathematical group theories which underlie many subsequent representations of modern physics. Einstein praised her contributions and offered to write her obituary. Noetherian Rings are named after her and grandmaster chess player Emanuel Lasker proved some Noetherian algebraic concepts (Lasker-Noether decomposition theorem).

noise Meaningless or otherwise unwanted sounds or signals interfering with the desired information or transmission in electromagnetic or acoustic communication systems. Noise can arise from bad shielding, wires too close together, overlapping transmissions, weather disturbances, irregular/reflective terrain, incorrect operation, deliberate human interference, random varying velocity, or faulty or incompatible hardware.

Noise in fiber optic networks is different from noise in wired networks. Electrical disturbances, voltage surges, and ground loops can have a significant impact on wired networks but may have little or no impact on fiber optic cables, especially end-to-end optical networks.

On the other hand, fiber optic cables can experience noise from the illumination source or the course of travel of the light through the cable. They may suffer *thermal noise* or noise from back reflection and are vulnerable to noise anywhere there is a junction with electrical components. Thermal noise may be especially problematic in systems using p-i-n photodetectors. Fiber optic networks may experience *beat noise* at the receiving end from amplification, depending upon the frequencies or from *relative intensity noise* from fluctuations in the emissions from the source illumination.

Another source of noise in optical networks is *modal noise*, which arises when the emitted light (e.g., laser light) travels through a multimode fiber in slightly different reflective paths, resulting in slightly varying distances for the total path or *phase delay*. The light reaching the output end may exhibit fluctuating interference patterns and may obscure whether the light pulse is on or off. Microbends and pits may increase modal noise.   *Modal partition noise* results from a fluctuation in intensity from the laser source affecting the longitudinal modes of a multimode transmission. Dispersion of the different modes can result in varying speeds along the travel path, causing fluctuations at the destination.

Intersymbol interference noise may result from uneven fluctuations of light pulses that overlap in the process of traveling toward the destination. Quantum noise resulting from the particle nature of light can contribute to *shot noise*. There is Poisson variance in the number of photons received in any given bit period such that the photonic energy fluctuates. This may increase as the optical power increases. See crosstalk, garbage, interference, loss (includes diagrams), intersymbol interference.

noise, modal See dispersion, noise.

noise canceling Techniques and technology to reduce or eliminate noise. This may be background noise or noise on the transmissions pathway. Noise cancellation can be through digital algorithms which analyze the information and screen out calculated noise (a feature now found on digital cellular phones), or may be through conditioning circuits in transmitters or receivers. Noise cancellation is sometimes achieved by adding noise, creating a "white" noise that may be less objectionable than bursty, intermittent noise.

noise filter An electrical circuit designed to detect or evaluate and exclude extraneous signals passing through a circuit. In digital circuits, fairly sophisticated analysis may take place. In analog circuits, simple exclusions of particular patterns or frequencies may be used.

non-facility-associated signaling NFAS. In ISDN networks, a type of signaling in which the D channel is at a separate primary rate interface (PRI) from an associated set of B channels. Multiple PRI lines can be supported through a single D channel using NFAS. In contrast, in facility-associated signaling, the D channel is at the same PRI as the associated B channels.

nonintrusive measurement device NMD. A device used to measure various parameters in analog voice transmissions over communications networks. Measured parameters include noise level, speech level, echo path loss, and echo path delay.

nonionizing That which does not cause ionization or change the ion environment around it. A number of transmissions media, including visible light rays and radio waves, do not cause ionization, but can be propagated by ionized particles with which they come in contact.

nonlinear distortion In an optical waveguide, the distortion that occurs over distance when more than one wavelength, with different transmission characteristics, or more than one pulse, which may reflect at different angles in the waveguide, travels at different speeds. Thus, a signal sent together doesn't always arrive at its destination in synch with its other components and the effect is cumulative over distance.

The gradual loss of synchronization can have deleterious effects not only at the destination point, but en route, as well, where crosstalk may result from nonlinear waveform interaction. Nonlinear distortion is influenced by the character of the original pulse, the breadth of the waveguide, the composition of the wavelengths (which may be prone to chromatic dispersion), the number of bends or obstacles (doping) in the lightguide, and the means of amplification. See chromatic dispersion, Raman scattering. See noise.

nonreturn to zero NRZ. A simple binary encoding scheme in which ones and zeros are represented by high and low voltages, and there is no return to a zero level between successive encoded bits, hence the name. Since transitions may or may not occur at each successive bit cell, the NRZ signal has spectral energy and, consequently, a direct current (DC) component that is a nonzero energy at DC. It is thus one

type of baseband signal. See Manchester encoding.

nonvolatile memory Circuits or components that retain their data, even if the electrical current is shut off. In computer circuitry, volatile memory is installed in greater quantities than nonvolatile memory. Nonvolatile memory is typically used for configuration settings (e.g., video parameters). See EPROM, read-only memory. Contrast with dynamic random access memory.

nonwireline carrier Also called an A Block carrier, for *a*lternate carrier; that is, a competitive phone services carrier that is not the established local phone company (usually a Bell carrier, hence B Block carrier).

NORC Network Operators Research Committee.

normal An imaginary line in a direction describing the perpendicular to another line or plane. In other words, for any given line or plane, there is a relationship that is perpendicular to that line or plane for any given point of reference.

The term "normal" is unfamiliar to many people and the word "perpendicular" is often substituted. The use of the word "normal" appears to date back to an Old English term for "rectangular" or "right." Since a right angle is 90° perpendicular to the reference angle by definition, this may account for the evolution of the term as it is now used.

For simplicity, picture a very slender flagpole anchored squarely in a flat concrete slab; assuming that any point in the center of the flagpole at a specified distance from the slab is equidistant from the slab in any direction (imagine invisible equal-length guy wires all the way around the pole), it is considered *normal* to the slab, even if the slab is removed from the ground and tilted in different directions – as long as the flagpole remains firmly anchored in the same position relative to the surface of the slab, it is considered *normal* to the plane of the slab. If you took away the flagpole and substituted an imaginary line, this line expresses the slab's *normal* geometry, even if the line were extended out the other side of the slab (poking into the ground like a pylon representing the negative direction of the normal vector).

While normal is usually visualized as a line, it should be remembered that the line could be imagined as being anywhere perpendicular to the reference surface, not just in the middle. For example, if you moved the flagpole from the center of the concrete slab to the edge, while still keeping it straight (i.e., maintaining the even lengths of the imaginary guywires), the flagpole is still considered normal to the plane of the slab. Thus, *for some shapes*, normal can be seen as a planar concept expressed as a line in the context of a specific given point of intersection with the reference plane. To understand the planar aspect, imagine walls in a house and how they are set at 90° angles relative to the plane of the floor. You can pick any point of intersection with the floor along the bottom of the baseboard of a wall and there will be an imaginary line normal to the floor concurrent with the plane of the wall. The relationship holds true for other selected points along the wall. The wall could be pivoted around one of those points, using it as an anchor, and any other point where the wall touches the floor in its new orientation will still be normal to the floor. However, the pivoting relationship between the wall and the floor only holds consistently for flat surfaces.

It is fairly straightforward to grasp the concept of normal for straight lines or flat planes, but what if the reference line or surface is curved? Imagine a toothpick sticking partway into a perfectly round orange. If the line of the toothpick is perpendicular to the tangent plane of the orange relative to where the toothpick intersects the curved surface of the orange, the toothpick is *normal* to the surface of the orange, whether it's the outside convex surface of the orange or the inside concave surface. In a perfect sphere, the relationship holds if the toothpick passes through the center and pokes out the other side. However, if the line of the extra long toothpick is offcenter, or the orange isn't perfectly round, normal cannot be assumed to be the same on both sides of the orange through a single line. Normal must be calculated relative to one point of intersection on a single reference surface if the calculations are to be generalizable to any surface. If you're having trouble visualizing the angle at which the toothpick must pierce the round orange to be normal, imagine equal-length guywires around the toothpick as you did with the flagpole. Now picture the stakes holding the guywires as spinning around the toothpick to mark a circle like a compass and slicing off a chunk of orange through the marks. If you place the round piece of orange on your tabletop, the toothpick will now be sticking straight up, perpendicular (normal) to the surface of the table. For a natural, asymmetric orange (or eggplant), normal will be related to the contours on the fruit's surface where the toothpick pierces the skin.

Determining normal for a point on a bumpy surface like the Earth's terrain or an optical diffraction grating is a little more complicated, since it will change every few inches or μm, but here is a way to visualize the relationship. Imagine flying a small aircraft (or a flight simulator) a few feet off the ground in the desert and maintaining that distance from the ground over a series of sand dunes. As you pull back, the nose of the aircraft pulls up. As you push forward, the nose of the aircraft pushes down. The relationship of the steering stick may not be perfectly perpendicular to the surface of the sand beneath it (due to lag), but it's close and it gives you a way to picture approximately where normal is for a given surface on a complex plane like desert terrain.

The concept of normal, when applied to bumpy surfaces, has a fractal nature in the sense that bumpy surfaces often appear more complex as they are more closely examined. Benoit Mandelbrot used the example of a coastline to describe this relationship in fractal geometry and he developed equations based upon self-similarity leading to the observation that the closer you look at a coastline, the longer it gets, due to the fact that slight indentations and protrusions can be more readily seen. Determining normal for a bumpy surface is somewhat like that. The closer a surface is scrutinized at the point where you want to determine the *normal* relationship, the more the quantity or shapes of the bumps and indentations might influence the angle of an imaginary steering stick or toothpick.

Thus, normal is dependent, in part, upon the scale and type of measuring apparatus (and geometry) used to determine the relationship of the point of intersection to the reference plane. Returning to the guywire analogy, imagine a toothpick placed normal to the surface of an irregular eggplant rather than a perfectly round orange; the angle of the toothpick would likely be different if the guywires were very close to the toothpick as opposed to some inches away from the toothpick.

The concept of normal, which is a way of conceiving of perpendicular relationships in 3D space in any orientation, is essential in many aspects of geometry, theoretical physics, robotics, industrial fabrication (especially telescopes and microscopes), mapping, geology, computer-aided design and drafting, and much more. It is frequently used to describe grating surfaces, layered semiconductor structures, and other industrial fabrications. The angle of incidence of a reflected wave is generally described using surface normal as a reference. Thus, normal is useful for describing light paths, reflectivity, and diffraction. Surprisingly, many technical references gloss over or don't mention the term at all. See incidence, normal wave.

normal distribution A theoretical construct based upon observations of the distribution of certain traits in sufficiently large populations. From these observations, it is possible to create a statistical representation of frequency distribution to form a bell-shaped curve with certain consistent mathematical properties across the curve. It is also called a Gaussian distribution after the observations of Karl Gauss. Thus, relationships such as the mode and the mean fall in the center of the curve and are equal and frequency tails off to either side through a number of symmetric standard deviations until it becomes zero.

Once it had been observed that many traits seem to follow a general statistical distribution pattern (many of which came out of studies or assumptions about human intelligence), the normal distribution or *normal curve* was then assumed to apply to many other frequency distributions, with this assumption then being built into tools designed to measure traits in a population (which has some self-fulfilling and circular aspects that may interfere with the development of objective measuring instruments). Thus, it is used as a basis for describing probabilities (though some might say possibilities). Many students ask professors if a class is being graded on a "bell curve." In fact, with small class sizes, there aren't sufficient numbers in the population set to justify the assumption of a normal curve but many instructors apply normal curve concepts to assigning grades anyway (or get nervous if exam results don't naturally follow characteristics of a normal curve).

In its most general sense, in a sufficiently large, natural population, a normal distribution describes a sample set in which there are many members with similar traits, then a gradually lessening number with slightly different traits diverging in either direction from the mean. For example, there may be a large number of men who are 5′10″ tall in the U.S., with diminishing numbers who are smaller or taller until a point is reached where there are no more people outside the sample set. When plotted in Cartesian coordinates, this trait maybe geometrically illustrated as a statistical curve resembling a symmetric bell shape, with the mean (average) and mode (most frequent) centered around those who are 5′10″.

In communications, the concept may be useful for some aspects of assessing, estimating, or predicting network demographics, peak time use, etc.

normal wave Energy in wave form (e.g., radio waves) that travels *normal* (perpendicular) to a reference plane or line. Since normal is a theoretical relationship (an imaginary line), the concept of a normal wave is not dependent upon whether the wave impacts the reference surface which may, in fact, be imaginary. In practical applications, a normal wave may fail to impact, reflect from, or travel through a reference surface (e.g., a receiving antenna). If the reference surface is reflective, and the wave reaches the surface, the wave would be reflected back in the direction from which it came, a situation called a Littrow condition. See normal.

Norman, Robert A researcher who published a clear statement of the laws governing magnetic attraction and repulsion in 1581. See magnet.

Normes Européenne de Télécommunications NET. An organization providing compliance testing for commercial telecommunications products for sale throughout the European Union to determine whether they conform to mandatory standards.

Nortel Northern Telecom Limited. A leading global digital network provider providing commercial data, voice, and video services. Nortel is a dominant public switching equipment supplier in Canada descended from Northern Electric. It is also known for manufacturing and distributing radar sets based on magnetron tube technology, particularly during the second world war. Nortel technology is leased by other companies. See Qwest.

North American area codes See the Appendix for a chart of area codes for Canada, the U.S., and U.S. territories.

North American Basic Teletext Specification NABTS. An Electronic Industry Association (EIA) and ITU standard that describes a means to modulate data onto a vertical blanking interval (VBI), the transition time when the electron beam in a video display travels from bottom right to top left with the electron beam turned off so as not to interfere with the image that is currently displayed. This is usually associated with an NTSC signal, as is standard in North American television broadcasting.

NABTS can be adapted to transmit Internet Protocol (IP) data so that broadcast companies can send vari-ous data services along with a television signal. RFC 2078 describes a one-way 36-byte packet structure that can be encoded into a single horizontal scanline of a television signal. Synchronization packets are located at the beginning, followed by address, index, and parameter information, followed by 26 bytes of user data, finally ending with forward error correction (FEC) data.

The full NABTS specification is described in EIA-516. The entire NABTS specification is not always implemented. See NTSC, RFC 2728.

North American Cellular Network NACN. A commercial provider of international cellular roaming services through their network backbone, serving over 7500 cities worldwide. Supported protocols are System Signaling 7 (SS7), X.24, GSM and IS41.

North American Digital Cellular NADC. A commercial digital mobile phone service launched in 1991, NADC was introduced as a second-generation system. IS-54 supports data rates of 48 kbps at a bandwidth of 30 kHz using digital phase shift keying (DPSK) modulation. IS-136 is based upon time division multiple access (TDMA). See DAMPS, time division multiple access.

North American Directory Plan NADP. An X.500-based client/server Directory System for providing global electronic directory and address book capability, distributed by ISOCOR.

North American ISDN Users Forum NIUF. See ISDN associations.

North American Network Operators Group NANOG. An association of Internet Service Providers which meets several times a year to discuss technical issues regarding the administration and operation of Internet-connected services.

North American Numbering Council NANC. A Federal Advisory Committee established and chartered with the U.S. Congress in 1995 by the Federal Communications Commission (FCC) to assist in adopting a model for administering the North American Numbering Plan (NANP). This identification scheme is used for many telecommunications networks around the world. NANC advises the FCC and other NANP member governments on general number issues and on issues of number portability (e.g., for mobile telephones). See North American Numbering Plan. http://www.fcc.gov/ccb/Nanc/

North American Numbering Plan NANP. A system of assigned codes and conventions introduced in 1947 for routing North American (World Number Zone 1) calls through the various telephone trunks of the public telephone network. In 1995, significant changes were made to the NANP, mainly due to increased demand for area codes, including changing the middle digits from 1 and 0 to 2 through 9. See Area Codes chart in the Appendix.

North American Numbering Plan Administration NANPA. A working group that develops and advises the North American Numbering Council (NANC) on processes for selecting a neutral NANP Administrator. It oversees a number of task forces and coordinates with them on issues related to cost recovery for the NANP administration.

North American Telephone Association NATA. Now known as the MultiMedia Telecommunications Association, this is an open public policy, market development, and educational forum for telecommunications products and services developers and resellers. http://www.mmta.org/

north geographic pole The point at which the imaginary lines of latitude converge at the north pole relative to the shape of the Earth and its alignment in its orbit around the Sun. The general direction in which north-seeking compass needles point is near "geographic north" in northern Canada. See north magnetic pole.

north magnetic pole A point in northern Canada, near the north geographic pole, to which the north-seeking tip of a compass points. What is called magnetic north could be seen as Earth's south pole, if the "north" end of a magnet were used to determine "north" on a compass or alternately as the north pole if a north-seeking (south) pole of a compass is used to determine the direction. If that sounds confusing, consider that the designation of north or south on a magnet is not an absolute measure but one assigned relative to the polarity of the Earth, for which we have already designated north and south geographic poles. Thus, "magnetic north" is the direction toward which a north-seeking (south) pole of a compass points. The north magnetic pole is not the same as the Earth's north geographic pole, despite its proximity, because the planet is a dynamic ecosystem whose magnetic properties change over time, whereas geographic north is a cartographically fixed point. See north geographic pole.

NOTIFY, DNS NOTIFY A mechanism for the prompt notification of network zone changes that was proposed as a Standards Track Comment in 1996. NOTIFY is a DNS opcode that enables a master server to advise slave servers of a change in data so that they may initiate a query to discover the new data. Traditionally many networks were configured to poll the server in order to discover any changes within the zone. This was a trade-off in terms of load on the system vs. the currentness of information. A NOTIFY transaction, on the other hand, establishes a means to initiate and expedite the update process when SOA RR changes occur (and, theoretically, other RR changes), thus reducing delay without imposing excess load on the system. NOTIFY uses a subset of the fields in the DNS Message Format. See RFC 1035, RFC 1996.

NOTIFY Set In distributed networks using DNS NOTIFY, the NOTIFY Set encompasses servers to be notified if changes to a zone have occurred that should be queried to enact updates. The set defaults to those listed in the NS RRset but, in some cases, additions or overrides may be possible to accommodate special circumstances or stealth servers that are not listed in the NS RRset. See NOTIFY.

NOV News Overview.

Novell One of the significant companies providing networking software (Novell Netware) to the business market. Novell is a public company established by a buyout of NDSI by Ray Noorda in 1983. In 1998, Novell began promoting Novell Directory Service (NDS) as a means to tie different networking platforms together. In the 2000s, Novell acquired Cambridge Technology Partners and SilverStream Software.

NOWT Netherlands Observatory for Science and Technology (Nederlands Observatorium van Wetenschap en Technologie).

Noyce, Robert N. (1927–1990) An American electronics engineer and significant pioneer of semiconductor technology, Noyce received the first American semiconductor patent (#2,981,877) and more than a dozen other patents. Early in his career, Noyce did research at the Philco Corporation. In 1956, he joined the Shockley Semiconductor Laboratory where he worked with transistors and soon met Gordon Moore, his longtime associate and business partner. Together they founded Fairchild Semiconductor and later, in 1968, Intel Corporation. Noyce was president of Intel until 1975 and then served as chairman of the board until 1979.

Noyce narrowly missed winning the Nobel Prize in physics. Although his patent was the first to be awarded for a pioneer integrated circuit (IC) invention, a patent application and verifiable invention by Jack Kilby at Texas Instruments was determined to predate Noyce's by just a few months. Kilby was awarded the Nobel Prize in 2000, a decade after Noyce's death. Nevertheless, Noyce made a meaningful contribution with his version of the new concept, as his design was commercially practical and his company, Intel, grew to be one of the foremost chip design and manufacturing firms in the world. The Robert N. Noyce award is presented annually to outstanding contributors by the IEEE society. See integrated circuit; Intel Corporation; Kilby, Jack; Moore, Gordon.

NPA 1. National Pricing Agreement. AT&T agreement. 2. See Numbering Plan Area. A three-digit area code. NPAs include special, reserved, and unassigned numbers.

NPR See National Public Radio.

NPSTC See National Public Safety Telecommunications Council.

NRC 1. See National Research Council. 2. See Network Reliability and Interoperability Council. 3. nonrecurring charge.

NREN See National Research and Education Network.

NRIC See Network Reliability and Interoperability Council.

NRSC National Radio Systems Committee.

NRZ See nonreturn to zero.

NSAI National Standards Authority of Ireland. A standards body for Ireland established in 1961, located in Dublin.

NSAP network service access point.

NSF See National Science Foundation.

NSFNET National Science Foundation Network. A network established by the Office of Advanced Scientific Computing through the National Science Foundation, which is used for the civilian computing operations of the U.S. Department of Defense. See National Science Foundation.

NSIE See Network Security Information Exchange.

NSInet NASA Science Internet, a network of the National Aeronautics and Space Administration.

NSP 1. See National Internet Services Provider. 2. Native Signal Processing.

NSSN National Standards Systems Network. http://www.nssn.org/

NSTAC National Security Telecommunications Advisory Committee.

NT Northern Telecom, Inc.

NTSC National Television System Committee. An organization formed by the Federal Communications Commission (FCC) which set black and white standards for the emerging television broadcast industry in 1941. By 1953, after the proposal and consideration of several television systems, the FCC adopted a 525-line color standard developed by Radio Corporation of America (RCA), which was downwardly compatible with previous black and white technologies. This is, in part, why luminance and chrominance information are carried separately. This system was accepted by the FCC and is widely used in North America and parts of South America.

In NTSC broadcasts, color, intensity, and synchronization information are combined into a signal and broadcast as 525 scan lines, in two fields of 262.5 lines each (Europe typically uses 625 lines). Only 480 lines are visible; the rest occur during the vertical retrace periods at the end of each field. NTSC is considered to run at 30 frames per second, although, in color television broadcasts, the actual playing rate is approximately 29.97 frames per second. See High Definition TV, PAL, SECAM.

NuBus NuBus is a simple Apple Computer 32-bit backplane card slot standard (ANSI/IEEE P1196) for the connection of peripherals to Apple Macintosh computers. The clock is derived from a 10-MHz reference. NuBus backplane space is limited to 74.55 × 11.90 mm (even though some models have larger slots). NuBus slots can support up to 13.9W of power per card, although more can be used if other slots are not filled.

nuclear magnetic resonance NMR. A technology used to reveal the inside of structures or biological organisms through a series of magnetic scanners or a magnetic field enveloping the body. It is used in addition to, and as an alternative to, X-rays in medical research and diagnostic imaging.

null Empty, having no value. A dummy value, character, symbol, or marker. Null values are sometimes used as delimiters to indicate the beginning and/or end of a value or data stream. Null characters sometimes are used as padding, to even out the size of blocks or to provide extra time for synchronization. Null is a very useful concept, regularly used in programming and network transmissions protocols.

null attachment concentrator NAC. In a Fiber Distributed Data Interface (FDDI) network, there are a number of types of node configurations, including single, dual, and null attachment concentrators. A null

attachment concentrator does not contain any A, B, or S ports but is configured with multiple M ports. The NAC may be used in a simple tree configuration. It does not support a secondary path for redundancy and thus cannot be inserted into a dual ring network. See Fiber Distributed Data Interface.

null modem A serial transmissions medium which functions in many ways as a modem, as it uses the same software, protocols, and serial transmissions media, except that there is no modem. In other words, instead of the signal going from a computer to a modem through a phone line to another modem and to the destination computer, the signal goes from the first computer through a serial cable with no modem connected to the second computer, and back again. The transmit (Tx) and receive (Rx) lines are swapped (usually lines 2 and 3). This provides fast local file transfer capabilities between machines.

null modem cable There are many ways to configure a modem cable, as long as the two computers talking to each other are talking the same language (transmissions protocol), but the most common configuration for a null modem cable is to take a standard RS-232c cable and cross (swap) the transmit and receive lines on one end, that is, lines 2 and 3. Or, rather than taking apart a cable, it is usually easier to get a null modem connector that swaps the lines. It looks very similar to an extender or other small coupler. See null modem.

number portability NP. A service which enables subscribers to retain a geographic or nongeographic telephone number when they change their location, their services provider, or their type of service.

This is defined with regard to switching services in the Telecommunications Act of 1996, and published by the Federal Communications Commission (FCC), as:

"... the ability of users of telecommunications services to retain, at the same location, existing telecommunications numbers without impairment of quality, reliability, or convenience when switching from one telecommunications carrier to another."

See Federal Communications Commission, Telecommunications Act of 1996.

Numbering Advisory Committee NAC. This name is used by a number of telecommunications advisory bodies worldwide including Australia, Hong Kong, Zaire, and others. In general, these committees provide forums for the exchange of views on numbering issues, including assignment, reassignment, storage, access, and the dissemination of public and governmental information for developing Number Plans. Recently, these committees have given increased attention to the allocation and availability of numbers for mobile telecommunications services.

Numbering Plan Area NPA. A three-digit geographic telephony area code. NPAs include special, reserved, and unassigned numbers. Within each NPA, there are 800 possible NXX Codes (also known as central office codes). NPAs are divided into two general categories:

Category	Description
Geographic NPA	The Numbering Plan Area code associated with a specific region.
Service Access Code	Nongeographic NPAs associated with specialized services that may be offered over multiple area codes, such as toll free numbers, 900 numbers, etc.

numeric keypad Any compact block of functionally related touchtone telephone, typewriter, calculator, or computer input keys. The most common type of numeric keypad is a group of about 10 to 18 numerical or function keys arranged in a block. These are usually physically organized to facilitate touch-typing or, in some cases, physically organized to slow down typing! On early touchtone phone systems, there was no point in entering numbers quickly as the switching on the network could not be accomplished as quickly as the numbers could be typed, so the digits were reversed to slow down digit entry.

The numeric keypad on a computer keyboard typically consists of 18 keys, with the numerals zero through nine, and symbols, usually consisting of period, plus, minus, asterisk (star), and enter keys. The remaining three keys differ widely on various computer platforms, but usually include symbols such as the tilde, slash, or pipe (vertical bar). See keymap, keypad.

numerical aperture NA. In a fiber lightguide, a quantitative description of the lightguiding capabilities of the "light pipe." If fibers were made of just one pure light-propagating material and were always straight and perfectly aligned with the incoming light source (also assuming there is no back reflection from the endface), the numerical aperture would probably be expressed in terms of the diameter of the fiber filament. However, fiber optic cables may be made from a number of materials (e.g., glass or plastic) with different refractive indexes and are not perfectly straight nor necessarily perfectly aligned with the source light. Cladding is laminated around the conducting core to reflect the light back into the filament and has a different refractive index from the conducting core. Mathematically expressing these different factors in relation to one another provides a numerical aperture rating that gives an idea of the light-guiding capacity of a fiber cable assembly. See acceptance angle.

Numéris, Numeris The name given to their line of data telecommunications services by France Télécom. Numéris is based upon Réseau Numérique à l'intégration de Service (RNIS) standards to provide its customers "... service that opens doors the world over for intelligent telephony, fast Internet access, and the efficient transfer of information." Data rates on analog lines are 33 Kbps (56 Kbps in some circumstances). Access to Numéris can be established through D ISDN (16 Kbps) or B ISDN (dual 64 Kbps) channels. Numéris Duet offers dual lines for simultaneous phoning and data communications (e.g., facsimile or Internet). Numéris Commerce offers D channel services for ecommerce applications such as bank card transactions.

nutating field In radar tracking, an oscillating feed from an antenna that produces an oscillating deflection of the radar beam.

Nutt, Emma N. Credited as the first female telephone operator. See operator, telephone, for additional information and history.

nuvistor A type of electron tube in a ceramic envelope with cylindrical, closely spaced electrodes.

nV *abbrev.* nanovolt.

NV See Network Video.

nW *abbrev.* nanowatt.

Nx64 A digital network channelized data transmissions system. As an example, a T1 line can be split into multiple Nx64k circuits for transmission across different ATM-based virtual circuits (VCs). When implemented through High-bit-rate Digital Subscriber Line (HDSL), it enables a network access provider to provide data services via two pairs of local loop connections without many of the transmission line hardware modifications required for some of the other data technologies. See Nx64 interface.

Nx64 interface A hardware network interface that enables Nx64-compliant connections at speeds between 64 and up to about 2048 Kbps, in multiples of 64 Kbps. This type of interface, in conjunction with Nx64 data formats, is commercially promoted to provide high bandwidth access to packet-switched backbone networks such as Frame Relay, X.21, and X.25. Thus, it is of interest, for example, to businesses with Frame Relay access to the Internet. A number of ITU-T V Series Recommendations relate to Nx64 standards (e.g., V.35). See Nx64.

NXX NXX is also known as Central Office Code, or CO Code. It is an industry abbreviation designating the three digits of a phone number preceding the last four (EXTN). A ten-digit number is expressed with symbolic characters as: NPA-NXX-EXTN. NXX is a reference to the exchange that services that specific area. N refers to any integer between 2 and 9 and X refers to any integer between 0 and 9. Each NXX Code contains 10,000 station numbers. See North American Numbering Plan, RNX.

NYNEX Corporation One of the Regional Bell Operating Companies (RBOCs) formed as a result of the mid-1980s AT&T divestiture, Nynex comprised several related companies, including the New York Telephone company (NY), the New England Telephone company (NE), and others (X) such as NYNEX Information Resources, NYNEX Mobile Communications, NYNEX Business Information Systems, and more.

In the early 1990s, NYNEX was developing interactive video network technology in cooperation with other vendors. In 1994, NYNEX representatives

N

made presentations through influential organizations such as the National Telecommunications Information Administration (NTIA) conference. In 1996, Bell Atlantic announced a merger with NYNEX, a plan that was carried out in 1997. The announced merger received Federal Communications Commission (FCC) approval to become the second-largest single telephone company, stretching from Maine to Virginia.

Nyquist, Harry (1889–1976) A Swedish-born physicist and engineer, Nyquist emigrated to the U.S. in 1907. Nyquist had an early interest in the transmission of pictures, resulting in the development of a historic facsimile system, AT&T's *telephotography* machine, in 1924. In the 1920s, Nyquist was also active in studying telegraph communications and providing theoretical observations related to transmission speeds and signal values. In 1928, he described principles for the digital sampling of analog signals in *Certain Topics in Telegraph Transmission Theory*. Unfortunately, theory came far ahead of practice; equipment at the time could not practically embody Nyquist's theories, but they are now the basis for digital sound sampling. The *Nyquist theorem* is named for this work; Claude Shannon who, like Nyquist, worked at the Bell Laboratories, cites Nyquist in his later development of information theory, in the 1940s. In 1927, Nyquist mathematically described Johnson noise, which is important in the understanding of interference in electronics. In the 1930s, he turned his interest to the study of amplifiers. Throughout his career he developed both theory and systems and is credited with more than 100 patents. See Nyquist theorem, sampling.

Nyquist frequency See Nyquist theorem.

Nyquist minimum The minimum bandwidth that can be used to represent a signal. This measure is used to limit the spectral width of a transmission signal in order to reduce the chance of interference and to maximize efficient use of the signal. See Nyquist theorem.

Nyquist theorem The Nyquist theorem is an important principle in telecommunications, where audio samples are used for synthesized voice, "music on hold," videoconferencing, Internet phone, and other multimedia and digital voice communications. Audio sampling is a process of taking digital slices of an analog signal in order to store and reconstruct the signal to preserve the original sound. In general, the slower the sampling rate, the coarser the recreated sound; the higher the sampling rate, the better the recreated sound. There are thresholds, however. Above certain thresholds there may be no perceptual improvement and artifacts and other technical interference will begin to negatively affect the quality of the recreated sound. Below certain thresholds, the sound

may not contain enough information to be intelligible. There are also thresholds in the relationship between the frequencies sampled and the sampling rate.

The Nyquist theorem is named for Harry Nyquist, who studied and described these important basic principles as they related to communications in the 1920s. The Nyquist theorem describes a sampling quality threshold relationship between the phase of the harmonically related sine and cosine functions over a specified time interval. In practical application then, an analog signal waveform sampling at equal time intervals requires a sampling rate of at least twice the highest frequency component in the analog signal to fully represent the characteristics of the original sound. Or stated another way, the highest frequency that can be accurately represented in a sampled signal is equal to one-half of the sampling rate. This translates to two samples per cycle and is called the *Nyquist frequency* or *Nyquist limit*. Thus, a full sample of 10 kHz of audio bandwidth would have to be captured at a rate of 20 kHz (or higher, within other thresholds) or information would be lost or introduce artifacts. In theory, the sampling rate can be infinitely high and the samples infinitely narrow. In practice, usually sound sampling rates of about 8 to 44 kHz are used.

If frequencies in the sample are higher than the Nyquist rate, an artifact known as aliasing will occur. Thus, in theory, if a sound sample possesses frequencies up to 10 kHz, but the sound is sampled at a rate at 18 kHz, then frequencies over 9 kHz should be filtered to prevent aliasing. In practice, signals above the Nyquist frequency may be filtered to reduce aliasing in the recreated sound sample. Theoretically, the filter would enable us to get back a pretty faithful representation of the original frequencies within the sample that are up to 9 kHz. In real life, filters have phase shift and slope characteristics that may interfere with a perfect re-creation.

Other limitations in sound sampling and re-creation are the digital storage and bus characteristics of the digital sampling system. In most systems, you need at least 8 bits to make decent-sounding voice, and 16 bits for a decent representation of music. Higher capacities are needed to accommodate higher dynamic ranges and more sophisticated sounds. Sound peripheral cards for computers and synthesizers with 16- or 32-bit sampling capacities did not become widely available until the late 1990s.

The Nyquist theorem has also been applied to video sampling, but it has been found that higher sampling frequencies may be needed for video compared to sound, with suggestions that four times the highest frequency may be necessary for a full re-creation of the original image. See Fourier transform; Nyquist, Harry; sampling; Shannon, Claude.

O A symbol used on many consumer electronics devices to indicate "off." On rocker switches, it indicates the side of the rocker which turns an appliance or component off. Its complement is "I" to designate "on."

O Series Recommendations A series of ITU-T recommended guidelines for specifications of measuring equipment that can be purchased from the ITU-T. Since ITU-T specifications and recommendations are widely followed by vendors in the telecommunications industry, those wanting to maximize interoperability with other systems should be aware of the information disseminated by the ITU-T. A full list of general categories is listed in Appendix C and specific series topics are listed under individual entries in this dictionary, e.g., K Series Recommendations. See O Series Recommendations chart.

O&M, O & M operations and maintenance.

O-band A transmission band specified by the ITU for optical transmissions in the 1260 – 1310-nm range.

O/P output.

O/R Originator/Recipient. A concept associated with the X.400 Message Handling System (MHS). The O/R address is used by the MTS for routing.

OAI See Open Application Interface.

OALC4 A family of relatively small-diameter fiber optic submarine cables developed by Alcatel. These cables are specifically intended for repeater-equipped systems. They can house up to 16 optical fibers within a welded steel tube. A gel substance protects the fibers from moisture and hydrogen effects. A steel wire vault surrounded by a seam-welded copper tube provides additional protection. High-density polyethylene provides abrasion resistance. Cables of the OALC4 family are suitable for use at sea depths of between 0 and 7000 to 8000 m, depending upon the ohms per kilometer rating.

OAM operations, administration, and maintenance. Various related management functions often associated with telephone and computer networks. In telephone networks, significant management and accounting tasks are associated with maintaining a dynamic environment in which subscribers all request different types and levels of service, and where the subscriber population is very mobile, thus changing their locations on a continual basis. Some systems have computer networks and entire facilities associated with just these aspects of the business. With mobile communications on the rise, these management tasks become even more intricate, and computer systems are used to facilitate the administrative tasks.

OAM Operations And Maintenance. Preventive maintenance information which, in an ATM B-ISDN environment, is included in the transmitted cells.

OAM&P operation, administration, maintenance, and provisioning.

OAM&P ANSI standards There are a number of important American National Standards (ANSI) of Committee T1 related to OAM&P, which are available from ANSI and described in the form of abstracts on the Web. See the ANSI Standards OAM&P chart for examples.

OAO orbiting astronomical observatory. Since 1966, a series of OAOs has been launched from Cape Canaveral to explore and measure astronomical phenomena that can more easily be seen from outside the Earth's atmospheric envelope.

OA See office automation.

OAS See Organization of American States.

OBI See Open Buying on the Internet.

object 1. A thing, article, entity, or unit of information. 2. An individually identifiable part, entity, or component. 3. In programming, an entity, often compartmentalized, that stores or receives data, e.g., a byte, block, register, segment, etc. 4. In the X Windows System, a software concept practically implemented as private data with private and public routines to operate on that data. 5. In typed objects, an entity that interacts as part of a defined operation.

object, programming In object-oriented programming, a reusable, modular, "wrapped up" collection of software characteristics, functions, and parameters at a basic level. For example, a button may be designed with certain visual and operational characteristics and stored for reuse in various applications, so that the code for the object isn't constantly reinvented. An object may consist of a collection of other objects to serve some related or higher function. See class.

Object Database Management Group ODMG. An independent standards organization now called the Object Data Management Group to reflect the broader

693

ITU-T O Series Recommendations

O.1 Scope and application of measurement equipment specifications covered in the O-series Recommendations

O.3 Climatic conditions and relevant tests for measuring equipment

O.6 1020-Hz reference test frequency

O.9 Measuring arrangements to assess the degree of unbalance about Earth

O.11 Maintenance access lines

O.22 CCITT automatic transmission measuring and signaling testing equipment ATME No. 2

O.25 Semiautomatic in-circuit echo suppressor testing system (ESTS)

O.27 In-station echo canceller test equipment

O.31 Automatic measuring equipment for sound-program circuits

O.32 Automatic measuring equipment for stereophonic pairs of sound-program circuits

O.33 Automatic equipment for rapidly measuring stereophonic pairs and monophonic sound-program circuits, links, and connections

O.41 Psophometer for use on telephone-type circuits

O.42 Equipment to measure nonlinear distortion using the 4-tone intermodulation method

O.51 Volume meters

O.61 Simple equipment to measure interruptions on telephone-type circuits

O.62 Sophisticated equipment to measure interruptions on telephone-type circuits

O.71 Impulsive noise measuring equipment for telephone-type circuits

O.72 Characteristics of an impulsive noise measuring instrument for wideband data transmissions

O.81 Group-delay measuring equipment for telephone-type circuits

O.82 Group-delay measuring equipment for the range of 5 to 600 kHz

O.91 Phase jitter measuring equipment for telephone-type circuits

O.95 Phase and amplitude hit counters for telephone-type circuits

O.111 Frequency shift measuring equipment for use on carrier channels

O.131 Quantizing distortion measuring equipment using a pseudo-random noise test signal

O.132 Quantizing distortion measuring equipment using a sinusoidal test signal

O.133 Equipment for measuring the performance of PCM encoders and decoders

O.150 General requirements for instrumentation for performance measurements on digital transmission equipment

O.151 Error performance measuring equipment operating at the primary rate and above

O.152 Error performance measuring equipment for bit rates of 64 Kbps and N x 64 Kbps

O.153 Basic parameters for the measurement of error performance at bit rates below the primary rate

O.161 In-service code violation monitors for digital systems

O.162 Equipment to perform in-service monitoring on 2048-, 8448-, 34,368- and 139,264-Kbps signals

O.163 Equipment to perform in-service monitoring on 1544-Kbps signals

O.171 Timing jitter and wander measuring equipment for digital systems based upon the plesiochronous digital hierarchy (PDH)

O.172 Jitter and wander measuring equipment for digital systems based upon the synchronous digital hierarchy (SDH)

O.181 Equipment to assess error performance on STM-N interfaces

O.191 Equipment to measure the cell transfer performance of ATM connections

Supplements

O.Sup3.1 Measuring instrument requirements – sinusoidal signal generators and level-measuring instruments

O.Sup3.2 Noise measuring instruments for telecommunication circuits

O.Sup3.3 Principal characteristics of volume indicators

O.Sup3.4 Consideration of interworking between different designs of apparatus for measuring quantizing distortion

O.Sup3.6 Crosstalk test device for carrier-transmission on coaxial systems

O.Sup3.7 A measuring signal (multitone test signal) for fast measurement of amplitude and phase for telephone type circuits

O.Sup3.8 Guidelines concerning the measurement of jitter

efforts of the organization to support Universal Object Storage Specifications (UOSS). See CORBA. http://www.odmg.org/

Object Definition Alliance ODA. A vendor association established by Oracle which aims to promote and develop new interactive TV and other multimedia services and networks that will operate over a variety of platforms. ODA seeks to establish associated technical standards for these products. Vendors include a number of high-profile financial institutions and retailers, and computer and media developers including Time-Warner, Apple Computer, Xerox, and Compaq. See video-on-demand.

object encapsulation A technique for combining related data and functions into an operational bundle, thus simplifying its use within a larger framework. The purpose is not to hide the intrinsic components of an encapsulated object, but to create a common superset of characteristics that work together and may be frequently used and reused. This technique is one type of modular approach to programming. See encapsulation, object-oriented programming.

ANSI Standards OAM&P Abstracts	
ANSI Standard	ANSI Document Title
T1.118-1992	G Interface Specification for Use with the Telecommunications Management Network
T1.204-1997	Lower Layer Protocols for Telecommunications Management Network Interfaces, Q3 and X Interfaces
T1.208-1997	Upper Layer Protocols for Telecommunications Management Network, Q3 and X Interfaces
T1.209a-1995	Supplement – Network Tones and Announcements
T1.214-1990	A Generic Network Model for Interfaces between Operations Systems and Network Elements
T1.215-1994	Fault Management Messages for Interfaces between Operations Systems and Network Elements
T1.221-1995	In-Service, Nonintrusive Measurement Device Voice Service Measurements
T1.224-1992	Protocols for Interfaces between Operations Systems in Different Jurisdictions
T1.226-1992	Management of Functions for Signaling System No. 7 Network Interconnections
T1.227-1995	Extension to Generic Network Model for Interface between Operations Systems across Jurisdictional Boundaries to Support Fault Management
T1.228-1995	Services to Interfaces between Operations Systems across Jurisdictional Boundaries to Support Fault Management (Trouble Administration)
T1.229-1992	Performance Management Functional Area Services for Interfaces between Operations Systems and Network Elements
T1.233-1993	Security Framework for Telecommunications Management Network (TMN) Interfaces
T1.240-1996	Generic Network Information Model for Interfaces between Operations Systems and Network Elements
T1.243-1995	Baseline Security Requirements for the Telecommunications Management Network
T1.240-1996	Generic Network Information Model for Interfaces between Operations Systems and Network Elements
T1.244-1995	Interface Standards for Personal Communications Services (withdrawn)
T1.246-1995	Operations Systems across Jurisdictional Boundaries to Support Configuration Management – Customer Account Record Exchange
T1.247-1995	Performance Management Functional Area Services and Information Model for Interfaces between Operations Systems and Network Elements
T1.250-1996	Extension to Generic Network Information Model for Interfaces between Operations Systems and Network Elements to Support Configuration Management – Analog and Narrowband ISDN Customer Service Provisioning
T1.252-1996	Security for the Telecommunications Management Network Directory

O

object inheritance A concept in object-oriented programming (OOP) which describes a hierarchical passing on of characteristics down through associated objects.

Object Linking and Embedding OLE. A software system developed by Microsoft Corporation which allows various applications programs that are OLE-compatible to share and exchange information. It is an interoperability system that lowers the distinction between various applications developed by different vendors so users can integrate the applications files and environments, and use them more as a suite of tools than as separate items. It further provides specification guidelines for the interface for accomplishing these tasks. OLE is a very good concept, in principle, and works well a lot of the time. Unfortunately, the various implementations are not yet perfect, as the OLE-compliant programs and OLE software programs installed on a system sometimes will clobber some of the other programs that don't support OLE, causing odd behaviors and situations where software has to be reinstalled, or OLE disabled temporarily. As OLE-capability must be incorporated into each software application by individual developers, there is some variation as to the completeness and dependability of these implementations.

When it works, OLE is a good for developing documents that take elements from a variety of text, image, sound, and other programs and combine them via links and drag and drop. Spreadsheet totals or statistics can be incorporated into stock offering documents, images can be incorporated into proposals, sounds can be incorporated into multimedia presentations, etc. without constantly opening and closing applications and converting various file formats with external utilities. OLE does more than just provide a way to insert information from one source into another; it further keeps a record of the links so that if source information in one document is updated, it will also be updated in subsequently linked documents.

OLE is used by various applications in Windows and Macintosh operating systems. See ActiveX.

Object Management Architecture OMA. An architectural framework developed by the Object Management Group (OMG) to lower the complexity and cost of developing new software applications. See CORBA, Object Management Group.

Object Management Architecture Board OMAB. A group established in 1996 by the Object Management Group (OMG) to oversee the OMG Technical Process, including the tracking and revision of technical specifications. See CORBA, Object Management Group.

Object Management Group OMG. A nonprofit organization of over 800 software developers, vendors, and end users whose aim is to establish the widespread use of CORBA through global standard specifications. Headquartered in Massachusetts and established by eight companies in 1989, OMG promotes the theory and practice of object technology for the development of distributed computing systems through a common architectural framework. OMG

seeks to establish industry guidelines and object management specifications to further the development of standardized object software, which it hopes will encourage a heterogenous computing environment across platforms and operating systems. See CORBA, Object Management Architecture, Unified Modeling Language. http://www.omg.org/

Object Request Broker ORB. The communications center of the Common Object Request Broker Architecture (CORBA) standard developed by the Object Management Group (ORG). It provides an infrastructure for program objects to intercommunicate, independent of the techniques used to implement them and the platform on which the software is running. Compliance with the ORB provides portability over many different systems. The ORB administers objects so an application need only request an object by name. There are now many commercial and freely distributable ORBs. See CORBA, Fnorb, Object Management Group, TAO. There is general information on CORBA at

`http://www.omg.org/`

There is a good list of ORB resources on the Web at `http://patriot.net/~tvalesky/freecorba.html`

Object Serialization Stream Protocol OSSP. A means to represent objects within a stream. Objects are grammatically represented and assigned a handle for use as a reference to the object. See byte-stream protocol, Java.

Object Services Management components of the Common Object Request Broker Architecture (CORBA) standard developed by the Object Management Group (OMG). A set of services for facilitating development productivity and consistency of implementation. The Object Services provide generic environments for objects to perform their functions, interfaces for the creation of objects, control of access to the objects, and administration of the location of objects. See CORBA, Object Management Group.

object-oriented programming OOP. A software development approach that follows a more natural and efficient evolution than many older reinvent-the-wheel approaches to programming. To understand the difference between non-object-oriented programming and object-oriented programming in a simplistic way, imagine a toy shop in which each elf is working in a separate little room, each with a separate set of tools, creating some kind of toy doll. At the end of the day, the creations are brought into a central room and it is discovered that some toys have been duplicated, none have interchangeable parts, and the end result is only a half dozen different toys. That's pretty much how traditional programming has been done, with an enormous amount of replication of effort. Every company writes the same sorting algorithms, there are hundreds of half-baked proprietary editors, and file search and retrieval methods are reinvented by thousands of programmers on a daily basis. It isn't very efficient. It isn't even very much fun.

Now picture a toy shop in which some general guidelines are set out for joints and limbs, and in which each toymaker has a magic replicator in which his or her components can be copied an indefinite number of times. Now imagine one of the toymakers is a mechanical wizard, and another is an artist, able to make beautiful embellishments. At the end of the day, instead of having a dozen toys, a limitless number of heads and feet, bodies and legs can be shared among all the toymakers. Not only that, but some particularly intricate mechanical parts and some wonderfully aesthetic ones can be used by all. Since guidelines were set out, the parts are interchangeable. The elves have created the basis for thousands of toys, rather than just a dozen. Assuming unlimited replication of individual parts, there's no limit to how often each component can be used. That's what object-oriented programming is, in the ideal sense. Once you create an *eye object* and give it certain parameters so that the color, shape, and various eye characteristics (contact lenses, eyelashes, ability to track a moving shape, etc.) can be individualized, you don't have to do it again; you can mix and match it with head, nose, and hair objects in thousands of different ways.

Similarly, in programming it is possible to create directory, menu, window, and button objects. Object-oriented concepts are not limited to physical attributes; the software can also incorporate more abstract user security objects, sort or fetch objects, and functions and behavioral characteristics associated with a type or class of objects.

Object-oriented programming is a modular approach that allows objects to be mixed and matched, or arranged in hierarchies, and customized to suit an individual application. Once created, they can be reused indefinitely. This can save development time and provides the basis for platform-independent software; it also gives a certain level of consistency to the interface, so users don't have a high learning curve for interacting with new applications programs. It further provides the programmer with a number of *levels* of interaction with an object. The developer can use the object in a transparent way, with the definition of the object encapsulated (bound together as an attribute or functional unit) by passing messages and parameters without worrying about how it was coded, or the programmer can take apart the object and use its individual components, or combine it with others to create a larger functional unit. This too is different from traditional programming. In many cases using someone else's non-object-oriented code involves a lot of study and adaptation to make it work in another setting, and it's rarely easy to mix and match parts of the code so that the characteristics can be inherited among the different parts. In contrast, program objects can be designed so that their characteristics and behaviors are known, so they can be immediately used without a long ramp-up period or restructuring. Object-oriented programming languages are still evolving. Smalltalk is one of the first object-oriented programming environments, developed 20 years ago, but many common languages currently used in commercial software development are not object-oriented; efforts to create object-oriented versions of traditional languages have not been fully satisfactory.

Nevertheless, the trend is toward object orientation, given its obvious advantages of portability and efficiency in many contexts.

For important and interesting information on taking the object-oriented model to global implementation and distribution, see CORBA and Object Management Architecture. See Open Systems Interconnect, Smalltalk.

OBRA See Omnibus Budget Reconciliation Act.

OBS See optical burst switching.

obscenity Obscenity, in its everyday sense, refers to actions or materials which are offensive, repellent, or vile. In a legal sense, it is more specific, as individual interpretations of what is offensive vary dramatically. Questions involving obscenity often conflict with individuals' rights and opinions regarding freedom of speech, and thus are important issues on the global Internet. See Communications Decency Act of 1996, Electronic Frontier Foundation.

OC operator centralization.

OC-*n* See optical carrier for definition and chart.

OCC See Other Common Carrier.

Occam's Razor A maxim well known to scientists, attributed to William of Occam in the 1300s, that it is vain to do with more what can be done with fewer (or less). It has been restated in many ways, in many contexts, but essentially, in science and in human spheres of activity, the idea is that the simplest explanation or one which doesn't require any additional hypotheses is usually the best, and often correct.

occlude To block, to obscure or limit from reaching the sight of the viewer. For example, a continuous stream of light can be pulsed by periodically occluding the beam. See chopper, knife-edge focusing, optical chopper, Ronchi grating.

OCIR overcoat-incident recording.

OCP operator control panel.

OCR 1. See optical character recognition. 2. Outgoing Call Restriction.

octal A base eight numbering system utilizing the numerals 0 through 7. See decimal, hexadecimal.

octathorp See octothorpe.

octet 1. A data unit widely used in digital networks. An octet consists of a sequence of eight data bits, sometimes called a byte (which is usually but not always eight bits and thus ambiguous). 2. In Internet Protocol (IP), octets are used as data units to describe an address or class designation as four octets (32 bits) separated by delimiters. See IP address, IP class. 3. In RFC descriptions of packets, an octet is a data unit for describing packet lengths.

octet rule In molecular physics and chemistry, an octet is a completed valance shell of eight electrons as is common to most elements. The octet rule is the manner in which atoms bond to one another as molecules so that valance shells fill to comprise eight electrons. There are exceptions, such as the common element hydrogen (H), which requires only two electrons to complete its electron shell.

octopus, hydra A visually descriptive name for a 25-pair cable common in multiple phone system installations. At the far end, the 25-pair wire is organized

O

into individual connectors (two, four, six, or eight wires) with phone cord connectors. An octopus is useful for stringing a single wire into a location where several phone connections are planned.

octothorpe, octathorp The # symbol, sometimes also called pound, hash, crosshatch, or number sign. It is used as an end signal (or "long" signal) on some touchtone phone menu systems. It represents a number sign in financial contexts, a suite number in postal addresses, and a *sharp* in music notation. See pound.

ODBC See Open Database Connectivity.

ODC See Open Development Consortium.

ODMG See Object Database Management Group.

ODP See open distributed processing.

ODU See optical channel data unit.

Odyssey A medium Earth orbit (MEO) satellite communications system intended to begin service in 1999 with 12 medium Earth orbit (MEO) satellites. Planned services included voice, data, facsimile, and Global Positioning Service (GPS). The project was discontinued in 1997 and TRW transferred technical expertise to the ICO Global Communications ICONET when it became a leading ICO shareholder later in 1997. TRW Inc. announced that it would turn back the license it had received for the Odyssey program to the U.S. Federal Communications Commission (FCC), in order to make the assigned frequencies available to other communications services. See ICO Global Communications.

oersted (*symb.* – Oe, Ø) A centimeter-gram-second (CGS) unit of magnetic intensity (field strength) equal to the intensity of a magnetic field in a vacuum in which a unit magnetic pole experiences a mechanical force of one dyne in the direction of the field. It can be expressed as $10^3/4pA$ m^{-1}. Named after Hans Christian Ørsted (sometimes transcribed as Oersted). See ampere.

Oersted (Ørsted), Hans Christian (1777–1851) A Danish physicist and educator who demonstrated the effects of current on a magnetic needle to a class of physics students around 1819. He reported on the magnetic effects of electric currents, information that brought together magnetism and electricity as never before, and resulted in a change in scientific thinking and the development of electric telegraphs in Europe and America. The oersted unit of magnetic intensity is named after him.

OFDM See orthogonal frequency division multiplex.

off-axis parabolic, off-axis paraboloidal OffAP. In optics, a geometric configuration used in reflectors in which the optical or symmetry axis of the reflector does not pass through the reflector but passes by nearby. The optical axis is parallel to the mechanical axis, in relation to the curved reflective surface, but generally falls outside the region of curvature of the reflective surface at a focal point that is accessible to the user. The parabolic surface is shaped to efficiently capture incoming rays within a desired field of view. For scientific applications, the reflectors may be coated with gold to enhance infrared reflectivity. On-axis and off-axis parabolic reflectors are used for broad spectral range illumination and light-collecting

applications, sometimes in place of optical lenses. Parabolic mirrors may also guide and focus beams, as in spectrometers.

OffAP mirrors can produce a point image from a collimated beam and eliminate aberrations in very fine diameter optical fibers in spectroscopic applications. Multiple OffAPs may be housed in an assembly.

There is a broad range of commercially available OffAPs for many applications and custom OffAPs may be ordered, but common focal lengths range from about 3 to 40 in. and off-axis distances range from about 1 to 20 in. for reflectors between 1 to 8 in. diameter. See monochromator.

off-hook On a phone set, the state of having the plunger or switch-hook in the active or "up" position so the circuit is connected. The term comes from the old wall phones on which the earpiece (receiver) was taken off a curved hook when in use (and when the battery power was engaged). When the phone is first taken off-hook, it alerts the switching exchange to the fact that the caller wants to use a line. The switching exchange returns a dial tone to the caller to indicate that the line is available for dialing. See on-hook.

off-peak hours Hours of low usage. In telephony, the hours between 11:00 P.M. and 7:00 A.M. are designated as off-peak in many areas, and calling rates are lower. The concept also applies to transportation systems, and fewer buses, trains, or subway cars may be in service during these hours.

off hours The times outside normal operating or working hours. Telephone and Internet services are often discounted in off hours or off-peak hours.

off the shelf Products and services that are ready to use without any customization. Products which can be readily purchased by anyone walking into a store or ordering from a catalog, and run with little or no configuration. Essentially the same as shrinkwrapped products.

office automation A catchall term for procedures and systems designed to streamline or increase the efficiency of business operations, often by installing technology that may or may not displace human workers. In some respects office automation has freed people from drudge work; it is no longer necessary to have rooms full of "human calculators" sitting and working out sums by hand, but technology has also introduced greater needs for training, storage of information, information retrieval, and other time-consuming activities that don't necessarily improve quality of life or shorten the work day.

Office of Science and Technology Policy OSTP. The science policy coordinating group for the Federal Government Executive Branch. The OSTP is led by presidentially appointed directors, organized into four divisions: environment, national security and international affairs, science, and technology. The OSTP provides expert advice to the President of the United States in matters of science and technology.
http://www.whitehouse.gov/WH/EOP/OSTP/html/OSTP_Info.html

Office of Telecommunications Along with the Office of Telecommunications Policy, this organization was rolled into the U.S. National Telecommunications and Information Administration in 1978 as a result of a reorganization.

Office of Telecommunications Policy OTP.

Office of the Director of Telecommunications Regulation ODTR. The National Regulatory Authority for telecommunications in Ireland, established June 1997 under the Telecommunications Miscellaneous Provision Act 1996. The ODTR administers the development of a liberal telecom market in accordance with the European Union and Irish law, allocates radio spectrum, and regulates broadcast transmissions and telecommunications equipment.

OFS Laboratories An international optical lab that has pioneered the production of optical fibers, connectors, cables, and attenuators. The Norcross, Georgia, plant began producing optical fiber in 1976 and is now the world's largest fiberoptic manufacturing facility. In addition to design and fabrication, OFS publishes a number of technical papers on fiber and photonic bandgap technologies.

In 2002, OFS introduced a fiber design that incorporates photonic crystal technology to "tune" the wavelengths passing through the fiber. See photonic crystal.

OFX See Open Financial Exchange.

OGT outgoing trunk.

ohm A practical unit in the meter-kilogram-second (MKS) system equal to the resistance of a circuit in which a potential difference of 1V produces a 1A current. Thus, if the values of two of these three are known, the third can be calculated. Named after Georg Simon Ohm. In 1908 the International Congress established the International ohm as the resistance offered to an unvarying current by a column of mercury at 0°C, 106.3 cm long, of a constant cross-sectional area of 1 square mm, and weighing 14.4521 g. In the U.S. in 1950, Congress defined the ohm as equal to one thousand million units (10^9) of resistance. See ampere, electromotive force, Ohm's law, resistance, volt.

Ohm, Georg Simon (1787–1854) A German physicist who, in 1820, investigated the conducting properties of various materials. He described the flow of electricity through a conductor and discovered the relationships among current, resistance, and electromotive force, information that greatly influenced subsequent theory and application in electricity. See Ohm's law.

Ohm's law In any specific direct current electrical circuit, the strength of the current is directly proportional to the potential difference in the circuit and inversely proportional to the resistance. Thus, current (in amperes) equals electromotive force (in volts) divided by resistance (in ohms), or $I = E/R$. See ampere, ohm, resistance, volt.

OHR See optical handwriting recognition.

OLE See Object Linking and Embedding.

OLIU Optical Line Interface Unit.

OLNS Originating Line Number Screening.

OLT optical line termination.

OM Operational Measurement

OMA See Object Management Architecture.

OMAT Operational Measurement and Analysis Tool.

OMG See Object Management Group.

OMSN optical multiservice node.

Omnibus Budget Reconciliation Act OBRA. OBRA is a 1993 U.S. Congress amendment to the Communications Act of 1932 which preempts state jurisdiction in such a way that individual states no longer regulate rates and entry by companies offering wireless services. The federally controlled spectrum was transferred to the Federal Communications Commission (FCC). It further organized wireless into two categories: commercial mobile radio services (CMRS), including cellular radio services and personal communications services (PCS), and private mobile radio services (PMRS), including public safety and government services. See Telecommunications Act of 1932.

omnidirectional Effective in all directions, radiating in all directions, or receiving from all directions. Functional in many directions without preference to any one. An omnidirectional antenna is one which is designed to send or receive signals in a maximum number of directions. A theoretical isotropic antenna is fully omnidirectional and often used as a reference for comparing antenna patterns or effectiveness. An omnidirectional speaker directs sound in all directions. Since this is structurally difficult to achieve, the speaker is usually a collection of speakers pointing in many directions, housed in one cabinet.

omnidirectional antenna An antenna designed to transceive signals through a wide range of directions. Since an antenna's capabilities are determined by shape and location, it is rarely completely omnidirectional, but broad omnidirectionality is achieved by maintaining equal field strength through the horizontal plane, and radiating in or out through the vertical plane. See isotropic, omnidirectional.

omnidirectional microphone A microphone designed to capture sound from all around its location. This is actually less common than directional microphones. Tape players, camcorders, digitizing sound sample microphones, phoneset microphones, and others have directional microphones to zero in on the crucial input, so they can screen out extraneous noises and conversations. Omnidirectional microphones can be said to capture sound "environments."

OmniWeb One of the earliest commercial Web Browsers, OmniWeb 1.0 was released in March 1995 for NeXTStep platforms by Lighthouse Design, Ltd. In spite of being a first release, it was a well-designed, full-featured browser, utilizing the Display PostScript and object-oriented capabilities of NeXTStep. It had flexible bookmark and other accessory capabilities not available in other popular browsers until about a year later. OmniWeb was subsequently ported to run on Macintosh Systems by The Omni Group and continues to be enhanced to take advantage of evolutionary changes in HTML and related Web languages.

OMSG optical multiservice gateway.

on-hook On a phone set, the state of having the plunger or switch-hook in the inactive, depressed, or

O

"down" position to interrupt the circuit, so it is not active while the phone is not being used. The term comes from the old wall phones on which the earpiece (receiver) was cradled on a curved hook when not in use (to conserve battery power). See off-hook.

on-line See online.

on-axis parabolic, on-axis paraboloidal OnAP. A geometric configuration used in reflectors in which the optical or symmetry axis of the reflector passes through the reflector. On-axis parabolic reflectors are used for collimating beams and off-axis paraboloidal reflectors are used for broad spectral range illumination and light-collecting applications. Paraboloidal reflectors are sometimes used in place of or in conjunction with optical lenses. Paraboloidal mirrors may be used to guide and focus beams and OnAPs have been used in experimental laser propulsion research. Higher-precision reflectors may be coated with gold. See off-axis parabolic.

on-ramp *colloq.* Access to a main communications link, such as highway, phone trunk, or networking service. The on-ramp is the link between the user's system and the main system. An Internet Access Provider (IAP) can be considered an Internet on-ramp.

on/off keying A type of modulation scheme similar to frequency shift keying (FSK), except that no signal is used for binary "0" (zero). See frequency shift keying, phase shift keying.

ONA See Open Network Architecture.

ONAC Operations Network Administration Center.

ONAL Off Network Access Line.

online 1. Having access to a system which is at least minimally functioning, and is largely automated. Interacting with a proceduralized system. A user is said to be *online* when he or she logs into a computer or a network, or accesses an automated phone system. 2. To bring a system *online* is to connect or power it up so that it is at least minimally functioning. 3. To bring an employee *online* refers to fitting the person into an organizational structure within an established system of priorities and procedures.

Online Public Access Catalog OPAC. An online service for providing access to bibliographic records. OPAC can be used to search records based upon author, title, subject, title keywords, and other search criteria. Many libraries worldwide offer access to OPAC through Telnet, Web interfaces, or internal Libsys systems. Similar to OPAC is the British Library Public Catalogue (BLPC); COPAC is a publicly accessible catalog from 20 of the largest research libraries in Ireland and the U.K. who are members of the Consortium of University Research Libraries (CURL).

ONT optical network terminal/termination.

ONU See optical network unit.

OOP See object-oriented programming.

OPAC 1. See Online Public Access Catalog 2. See outside plant access cabinet.

open 1. Unbounded, having no barriers or extents, unconcealed, exposed, uncovered, unobstructed. 2. An open circuit is one not currently connected, usually because no power is coming to it (as when it is turned off). A circuit breaker or blown fuse may create an open circuit. 3. An open transmission channel is one that is not currently in use or with channels available so that it may be used with appropriate facilities and authorization. An open channel may also imply one that is unsecured, where others can hear any communication that occurs.

open-space cutout In telephone wiring, a protective grounding mechanism, often used in conjunction with fuses and heat coils to guard against possible danger to people and equipment from large power fluctuations. If voltage is too high, the wire grounds by arcing across a small air gap between carbon blocks mounted on an insulator such as porcelain.

Open 56k Forum A consortium of telecommunications vendors promoting K56flex modem technology.

open air transmission A type of transmission which either depends on air for the propagation of the signals or which is commonly broadcast through the air. Radio, shortwave, and microwave transmissions are primarily open air systems.

open applications interface OAI. An interface built into a system and documented in such a way that third-party vendors can develop equipment and software applications that tie into that system.

Open Buying on the Internet OBI. An open, vendor-neutral, scalable, securable, business-to-business standards effort for the support of electronic commerce. The effort was initiated by a round-table dialogue by a number of Fortune 500 companies in October 1996. The founding participants included prominent firms such as American Express, BASF Corporation, Ford Motor Company, and others. The goal was the quick and effective implementation of interoperable Internet-based e-commerce solutions amenable to universal, high-speed access and paperless transactions. In June 1998, CommerceNet assumed management of the OBI Consortium to facilitate standardization efforts.

The OBI specification supports purchasing solutions for procuring high-volume, low-dollar, indirect products and services. OBI was first publicly demonstrated through the OBI Interoperability Showcase at the CommerceNet 99 conference. The first version of OBI was released in March 1997, with V2.1 released in November 1999.

Open Buying on the Internet Consortium OBI Consortium. An independent nonprofit organization dedicated to the development of open standards for business-to-business Internet commerce.

http://www.openbuy.org/

Open Collaboration Environment OCE. An environment created by Apple Computer which provides a means for third-party developers to create telecommunications applications that interface with the Macintosh operating system (MacOS). Thus, developers can create Internet phone, facsimile, network, and other telecommunications-related products for Macintosh owners.

Open Database Connectivity ODC. Microsoft's telephony software open application processing interface (API), part of a system that provides interoperability between Microsoft business-oriented database, spreadsheet, and word processing software, which is especially useful for digital telephony applications. The interface itself is independent of the application that provides the formatted data. In this way, call records and statistics can be stored and manipulated with popular software applications, providing computer-telephone integration and advanced call-recording capabilities.

Open Development Consortium ODC. An administrative concept introduced in November 2000 and formed early in 2001 to promote open standards for effective and collaborative exchange of development information. The consortium evolved out of ODC discussion list messages supporting informational concepts related to open development and open source software.

open distributed processing ODP. A framework for specifying systems, with an emphasis on distributed systems, defined in ISO/IEC 10746 and ITU-T X.900 as a four-part standardized reference model. It is network-independent and may be implemented with TCP/IP, OSI, NetBIOS, and others. See CORBA.

Open Financial Exchange OFX. A specification for the electronic exchange of financial data among businesses, consumers, and financial institutions over the Internet. The specification was jointly developed by CheckFree, Intuit, and Microsoft in 1997. The specification supports a wide variety of types of financial transactions including bill presentation and payment, banking, investment and stock tracking, pension account inquiries, and more.

In 2000, the OFX specification was made XML 1.0-compliant and certain tax form capabilities were added. A number of development toolkits are available for creating OFX-compliant applications, including JOFX (Java OFX). By late 2001, OFX was supported by more than 1400 payroll processing, brokerage, and banking firms. See JOFX.
http://www.ofx.net/

Open Group, The Formerly the Open Software Foundation, The Open Group is an organization which aids in the development and implementation of secure and reliable network infrastructures. The Open Brand is a registration mark (X) awarded by The Open Group to products which conform to the standard specifications. http://www.opengroup.org/

Open Network Architecture ONA. A system developed to encourage third-party vendors to supply public phone network products and services. Under the Federal Communication Commission's (FCC's) ONA, the telephone companies must provide the same service guarantees and levels to outside vendors' products that use the phone lines, as they use themselves. Network services must be stipulated as individual services in order to make them available to unaffiliated Internet Service Providers (ISPs). The Bell Operating Companies are required to comply with ONA.

open office An administrative and physical structure in which low walls or no walls are favored over high walls, movable walls are favored over fixed walls, and work stations are generally within view of more administrators and employees than in other office designs. Open office concepts are designed to promote flexibility and communication.

Open Shortest Path First Protocol OSPF. A TCP/IP distributed-computing dynamic routing protocol in the Interior Gateway Protocol (IGP) family of protocols, developed by the OSPF working group of the Internet Engineering Task Force (IETF). OSPF distributes routing information within a single autonomous system based upon link-state technology (a different approach from Bellman-Ford internet routing). OSPF includes explicit support for Classless Inter-Domain Routing (CIDR) and the tagging of externally-derived routing information. OSPF supports routing update authentication and IP multicast update sending/receiving. OSPF is a responsive protocol with low traffic overhead.

Routing is based upon destination information in IP packet headers without further encapsulation. OSPF detects and responds to topological changes, calculating new loop-free routes after a short period of convergence. See Hello Protocol, link state advertisement.

open skies *colloq.* Regulatory policies that are liberal enough to allow private use. Prior to 1972, the Federal Communications Commission (FCC) did not permit private American satellites to be launched for commercial communications. It then opened the doors on private domestic satellite launchings and operations, a move that created an opportunity for new competitive services to be established. MCI is one of the companies that got its start partly through recognizing and taking advantage of the opportunities presented by these openings.

Open Software Foundation OSF. This has now become the Open Group. See Open Group.
http://www.opengroup.org/

open system An open computer system is one which has few security barriers. Passwords may not be needed or individual users' directories may be open to all users. In many ways the Unix operating system and the Internet global network have been developed with an effort to keep them open and accessible; some people advocate that all systems should be that way.

Open Systems Interconnection OSI. An important layered architecture specification released as a standard by the International Organization for Standardization (ISO). OSI is designed to facilitate communications development between computer equipment and network software. Many vendors have opted to support this standard. Essentially, the communication is mapped onto seven layers as shown in the ISO/Open Systems Interconnection Reference Models chart. See ISO/OSI Reference Model for more information.

Open Systems Networking Initiative OSN. A trade organization promoting and supporting open network technologies such as high-capacity storage solutions for enterprise-level systems. The organization was founded by Cisco Systems, Quantum Corporation,

O

and others. The OSN architecture is intended to facilitate the interoperability and accessibility of data via industry-standard networks.

Open Trading Protocol OTP. See Internet Open Trading Protocol.

Open Video System OVS. A regulatory distinction for video services carried by Local Exchange Carriers (LECs), for example, who are not recognized as local cable service providers per Federal Communications Commission (FCC) guidelines and regulations. Thus, a local exchange carrier (LEC) could become

an OVS operator, but was also required to provide nondiscriminatory access on a portion of its channel capacity to unaffiliated program providers. By 2000, 25 OVS operators had been certified to serve 50 areas.

The mid-1990s was an important time during which technological advances made it possible for a wider range of types of providers to offer a broader range of telephony and cable programming options to their subscribers. In addition to this, the Telecommunications Act of 1996 opened the doors for telcos to offer

---

ISO/Open Systems Interconnection (OSI) Reference Model

From bottom to top, seven layers have been defined for OSI, ranging from physical and data layers through presentation and applications layers, thus following a low-level to high-level arrangement common in many computer architecture models. The selection of the layers was based on subdivisions chosen to distinguish well-defined functions and various levels of abstraction. In general, layers are related to services provided by the layers below. In actual implementations, the situation is a little more complex and many variations with sublayers or lesser-used layers exist.

Layer Function	Number	Purpose, Implementation
Application	Layer 7	User interfaces, applications programs, emulation, and other higher-level software implementations.
Presentation	Layer 6	Character sets, text handling. Data conversion into standard formats for transmission over a network or conversion from the transmitted format to something that user applications can understand. Encryption and other means of data security are also handled at this layer.
Session	Layer 5	Connection and session mechanisms for facilitating intercommunication between networks. File transfers, program sharing, data sharing, and basic traffic direction are managed at this layer to provide an orderly exchange of information (e.g., management of simultaneous requests from different users that might affect data integrity).
Transport	Layer 4	Process-to-process communications, addressing, and a number of end-to-end services. The data are packaged into packets in preparation for transmission. Identification is added so that disassembled packets that may travel different routes can be reassembled at the destination. Local network addresses may be defined at this layer. Error handling not already managed by lower levels may be handled here with respect to reliability and data integrity.
Network	Layer 3	Host-to-host communications and basic transfer units (packets), network addressing, forwarding, and routing. Addresses may be provided by the Network Layer to the Transport Layer.
Data Link	Layer 2	Activation, handshaking, interfacing, and other basic communications intended to initiate, regulate, and terminate a communications session. Common functions at this level include the management of wired connections (modems, Ethernet, etc.) or wireless transmissions, although the actual physical medium used is handled by the Physical Layer transparently to the Data Link Layer. Low-level error handling, data synchronization, and flow control are managed at this layer.
Physical	Layer 1	The transmission medium, electromagnetic properties, and other physical aspects associated with getting the signal from one place to another such as cables, devices, buses, signaling, etc. The physical layer is independent of the actual medium used and thus the layer model applies equally to electrical voltages in wired lines, radio waves in wireless transmissions, and laser light beams in fiber optic cables.

---

video through phone lines, to seek cable franchises, or to develop wireless cable systems. Together, these conditions resulted in blurred distinctions between cable and telephone companies since many services were now being offered over the same media. Thus, regulatory authorities were asked to clarify some of the issues and the FCC Cable Services Bureau became involved in this process.

On the other side of the equation, in 1996 the cable providers expressed concerns about telcos not being regulated in the same way as cable companies were by the Cable Act. Organizations such as the National Cable Telecommunications Association argued that the FCC exceeded its authority by preventing cable operators from switching to become Open Video Systems (OVS) providers, as defined by the FCC.

In January 2001, the FCC presented its seventh Annual Report on competition in the video markets. The document confirmed that cable TV was still the dominant video delivery technology, but with a declining market share of a couple of percentage points per year. Cable service rates increased slightly more than the rate of inflation in general over the study period. Interestingly, it was found that few telephone companies had sought OVS certification since 1996. See Telecommunications Act of 1996, Video Dial Tone.

operand A quantity or information which is being manipulated. For example, in mathematics, if you divide 200 by 10, then 200 is the operand. In computer algorithms, data, or the address of the data to be operated upon, are passed to an instruction in order for the instruction to act upon the data.

operating environment This has two meanings, depending upon the context. It is used in a limiting context to describe an operating system which isn't fully integrated or fully multitasking. It may be task-switching, or it may have a graphical user interface on top of a text-based operating system not fully implemented as multitasking. Many vendors have claimed multitasking operating systems which were not really so. In its second, broader context, it refers to the environment surrounding an operating system, that is, the system, the software that runs it, the peripherals and applications that support it, etc. An operating environment, in this broad sense, may encompass more than one operating system.

operating system OS. The most important software on a computer is that which lies between the user applications and the hardware. It's not possible to control the CPU, manage memory, access a disk drive, send images to a monitor, or transmit data through a network connection or modem without an operating system. The operating system handles interrupts, timing, the movement of data from one register to another, and all the nitty-gritty operations that are typically not seen or understood on a technical level by most users.

Microcomputer operating systems began to be developed in the 1970s. The early ones were text-based. One of the first widely used, popular operating systems was CP/M, designed by Gary Kildall. CP/M was the forerunner of QDOS, and hence, MS-DOS, and the syntax and commands are very similar. LDOS, TRS-DOS, UltraDOS, and other TRS-80 operating systems shared many common properties with CP/M. Kildall also later designed a multitasking operating system and a graphical user environment (GEM). In the early 1980s, Apple created proprietary operating systems for their Apple and Macintosh lines of computers, featuring the first widely distributed graphical operating system descended from pioneering work at Xerox PARC. This concept was so successful that it has since been adapted by virtually all subsequent vendors, including Atari, Commodore Amiga, Apollo, Sun Microsystems, Microsoft, SGI, The X Windows System, and NeXT. It's difficult to find a computer now that doesn't have a graphical user interface on top of, or in conjunction with a text operating system, or which is fundamentally a graphical operating system. The various versions of Windows are popular on consumer-level, Intel-based systems.

A number of multitasking systems were developed in the late 1970s and early 1980s, but the first widely distributed commercially successful preemptive multitasking operating system on a microcomputer was AmigaOS, introduced in 1985. Most of the workstation level computers had multitasking operating systems by the early or mid-1980s, and the other vendors began to follow this lead in the late 1980s, most notably OS/2 (Operating System 2), originally developed jointly by IBM and Microsoft. Many other operating systems released around this time were task-switching rather than fully multitasking. Microsoft Windows has since become widespread on personal computers, with Windows NT, originating out of the OS/2 collaboration, used on many server systems.

Unix is one of the most robust, earliest, and most important operating systems. A high proportion of institutional computing operations, much scientific research, and many Internet hosts run on Unix systems. Unix is freely distributable, powerful, flexible, dependable, well-supported, and runs on most computers. Linux is a popular implementation of Unix (as is BSD), available from a number of commercial and free distribution sources. Along with Apache, a freely distributable server software, Unix/Apache systems are used by thousands of Internet Services Providers to provide gateways to the Internet through the Web.

An operating system runs a computer, and computers are increasingly delegated control tasks beyond those humans can handle alone or in cooperation with one another. Navigational aids on aircraft are a good example. Fighter jets traveling at hundreds of miles per hour move too quickly for the human nervous system to react in time to control every aspect of the plane's behavior, so computerized systems handle many functions on the pilot's behalf. Extrapolate that type of control to appliances, houses, security systems, currency exchange, intelligent vehicle systems, and every aspect of human society that will someday be controlled by computers through a 24-hour Internet connection to all the other computers in the world. Given this broader outlook on our probable future,

O

the importance of carefully choosing a reliable operating system cannot be overemphasized. Buying the cheapest system or the most popular system, or passively allowing the choices to be made by profit-based corporations is, in the long run, a more far-reaching decision than selecting a President or other political leader. Why? Because a President who does a bad job can be voted out. An operating system that does a bad job cannot be disabled if it has been delegated important tasks such as running medical equipment, automated transportations systems, or environmental controls. Once a fighter plane has been designed with a dependency upon computer control, you can't just turn off the software in mid-flight. Similarly, if 5 or 10 years from now a family's financial transactions, medical prescriptions, education, and travel arrangements are handled by a computer operating system through the Internet, it may no longer be possible to turn off the computer at will. That bears thought and forethought and suggests that open source operating systems coexisting with commercial products can help maintain balance. See BSD, Linux, Mac OS X, Microsoft Windows, NeXTStep, OS/2, Solaris, SunOS, Unix, UNIX.

operational load 1. The full power requirement of a facility, which may be expressed in terms of averages or maximums. 2. The administrative and personnel requirements of an organization when in full operation.

operator An individual with system access privileges hierarchically higher or more powerful than general users. A telephone operator can manage calls and access services and control mechanisms that are not accessible by regular phone subscribers. A chat room operator can include or exclude specific users or set other restrictions or standards of use. A system operator has access to monitors, security devices, software programs, and other computer operations mechanisms not available to regular users.

operator, telephone An individual who handles the routing of calls from callers to callees and provides a variety of types of assistance such as directory assistance, long-distance procedures, billing options, etc. Many of these processes have now been automated, but in the early days, the telephone operator had complete responsibility for connecting, monitoring, and disconnecting calls through manual patchcords. He or she was also implicitly expected to provide emotional support, emergency information, local news and gossip, and business tips. In fact, the position of the phone operator included so much power (over connections to alternate business options) and information that a direct dial system was invented expressly to bypass the operator! Later switchboard systems required a higher level of concentration and expertise as many more lines would be serviced and the switchboards included lights to indicate connections and incoming calls, switches, cards for recording toll calls, and more.

While the names of the very first telephone operators have probably been lost to history, George Willard Coy is generally credited as the first telephone operator and Emma Nutt as the first female telephone operator. Nutt was instated following the relative failure of male telegraph operators to adapt to the task of courteously handling customer voice calls. Nutt apparently moved from a telegraph office to the phone exchange in Boston, hired in September 1878 by A. Graham Bell. She was paid a salary of $10 per month for a 54-hour work week (minus lunch). This was a year of great expansion, during which the New England Telephone Company was formed to sell licenses to telephone company operators in the New England area. Ms. Nutt apparently could remember every number in the directory.

Preferred Voice has created a synthesized voice speech attendant for delivery of information services named EMMA in honor of Ms. Nutt.

Records are spotty, but a small selection of notable first telephone operators for their regions prior to about 1913 is shown in the Selection of Pioneering "First" Telephone Operators chart. See Call Girls, Strowger switch, switchboard, telephone history.

Historic Telephone Switchboard & Operator

A telephone operator staffing the first switchboard in Idaho Springs, Colorado, wearing a headset. Manual switchboards still exist on islands, in remote areas, and in many undeveloped nationas. In North America, the main telephone system operates with high-speed digital switching circuits. [Denver Public Library Collection; copyright expired by date.]

---

operator-assisted call Any phone call in which the caller contacts the operator to handle some part of the transaction or connection, rather than direct dialing. There are usually surcharges associated with operator-assisted calls.

operator console, computer In computer networks, a console is a computer terminal which allows access to management and administrative functions that monitor and control the network. Common operations carried out on the console include user password assignments, new user account allocations, virtual configuration of devices and the network topology, server configuration, etc. Often the operator console is in a separate room for security reasons and, at the very least, is password protected to prevent unauthorized users, or well-meaning but uninformed users from accessing lower level system functions that could interrupt the functioning of the network.

operator console, phone In telephony, the main console of a multiline phone system. The operator console often is programmable and may include a hands-free earphone set.

Operator Service Provider OSP. Previously called Alternate Operator Services. A competitive provider of operator-assisted long-distance calls, especially third-party billing, collect, etc., which usually leases the services of existing phone networks. Large hotels sometimes provide AOS services to hotel guests for a premium. See splashing.

OPRE Operations Order Review.

Ops Abbreviation for operations, operators, operator services, and system operators. This is a particularly common abbreviation on Internet chat systems, where the operators (ops) or channel operators (chops, chanops) enforce accepted behavior on chat channels. See operator.

OPS off-premises station.

optical amplifier A type of device used as a cable repeater on fiber optic transmission lines, which functions without converting the optical signal. Early versions were based on semiconductor lasers. Most are now erbium-doped fiber amplifiers (EDFAs), that is, directly doped silica fibers, with the principal parameters defined by atomic composition.

optical attenuator A coupling component, usually of a standardized type (e.g., FC- connector) that attenuates the energy of the passing optical signal by a certain amount. OAs are compact, low-power components available in fixed, variable, and computer-controlled models. Multiple OAs may be combined into multichannel arrays to meet the needs of complex networks.

Thus, a fixed or variable amount of loss is intentionally induced in the received optical signal to adjust the signal to the desired level. This is useful for switching, gain tilt, modulation, and power control.

optical burst switching OBS. A mechanism proposed in 1999 by Chunming Qiao and Myungsik Yoo for managing bursty network traffic on the Internet. The research, supported in part by a National Science Foundation (NSF) grant, is intended to streamline network switching in anticipation of high-speed end-to-end optical networks of the future. The system combines aspects of optical circuit switching and packet/cell switching. While transition technologies are expected to combine optical networks with electrical switching, at some point there may be second generation all-optical networks that can be established as a layer beneath Internet Protocol (IP). OBS anticipates this evolution in technology.

Many broadband transmissions are inherently bursty (e.g., multimedia), as apparently are large numbers of self-similar traffic streams. Thus, a system more efficient for this situation than optical circuit switching was sought. The researchers have observed that

a packet can be sent along with the header, to reduce setup time and overhead, but that this approach has limitations as well. In OBS, a control packet is transmitted to set up a connection, followed by a data burst before the connection acknowledgment is received, essentially a one-way reservation system. In addition, data burst buffering at intermediate nodes is eliminated, reducing the wait for processing of control packets. Signaling is out-of-band. See Just-Enough-Time, Optical-Label Switching.

optical bypass In a Fiber Distributed Data Interface (FDDI) token-passing network, port adaptors can be equipped with optical bypass switches to avoid segmentation which might occur if there is a failure in the system and a station is temporarily eliminated. Normally, optical signals pass through the bypass switch uninterrupted, but if a station fails and is eliminated from the ring, the optical bypass can reroute the signal back onto a ring before the signal reaches the failed station, thus providing fault tolerance for the system.

The optical bypass switch is attached to the FDDI port adaptors between the attachment station and token-passing ring. See A port, dual attachment station.

optical carrier OC. 1. In general, a signal carried over optical media that is intended to provide control signals, information, or other signal data. Commonly lasers are used to provide carrier signals that may be given information content through electro-optical modulators (e.g., a quadrature amplitude modulator). See carrier, modulation. 2. A series of optical network transport levels defined in conjunction with SONET. See SONET Optical Carrier Transport Levels chart.

optical channel data unit ODU. A unit of data in an optical transmission network. In an ITUT G.709 optical transport network, for example, ODU overhead resides in columns 1 through 14 of three rows of the frame. It provides connection monitoring, path supervision, and client signal adaptation.

optical character recognition OCR. A software process, or combination of hardware scanning devices and software, which evaluates marks on a page and determines whether they have the predefined characteristics of text, symbols, and images, depending upon the software.

Most OCR programs use a combination of intelligent algorithms and character tables to process marks. These marks are typically text and character symbols, although some programs will also recognize mathematical and logical symbols if these are added to the program dictionary or the user dictionary. Some programs can automatically discern columns of text and images, and divide the page up appropriately, handling the regions separately. Once the document has been "recognized," the characters and symbols are converted to a common format, such as ASCII or one of the many flavors of extended ASCII, and stored as a file that can be further edited with a text editor, word processor, or desktop publishing program. In its strictest sense, OCR just recognizes the characters; software which handles images as well is

O

called *optical document recognition*. However, most OCR applications these days have some document processing and image recognition functions, so the phrase is used broadly here.

Most of the early work on pattern matching algorithms used in character recognition was done in the late 1950s and the early 1960s. By 1963, IBM had a system that could recognize Roman and Cyrillic characters at the rate of about 50 words per minute, the speed of a moderately competent typist, and faster than most typists can type information with many numerals. See optical document recognition, scanner.

Sample Optical Choppers

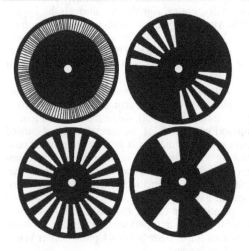

Optical choppers range from simple, regular patterns, to combinations of two or more blades (top right). High-precision choppers may have hundreds or thousands of chopping segments (upper left).

---

optical chopper A component for interrupting a beam of light, often at very fast speeds. Chopping can be accomplished with shutters, tuning forks, rotary discs/blades, or etched crystalline plates. Applications for choppers range from special lighting and art effects to highly-precise scientific applications and experiments. Scientific choppers typically have a data interface to a display component or computer.

Tuning fork choppers are intended to be used at a specific frequency and have small apertures at high frequencies, compared to rotary choppers. Despite the frequency and aperture limitations, tuning forks are robust and have many applications and can be used in a wide range of operating conditions.

Disc or rotary choppers resemble pinwheels in the sense that they have divided segments rotating around a central axis like a rotary fan. Most are mounted with ball bearings, but some are available with magnetic suspension.

Choppers can be fabricated as high precision instruments with highly accurate divisions in the chopping segments and carefully controlled rotational axes and speeds. The number of "slots" on a chopper may range from 2 to 2000+ and some choppers provide variable speeds for the rotations of the wheels. It is also possible to design chopper blades to chop more than one frequency at a time by having two sets of slots in the same wheel. It may be possible to optionally phase-lock the chopper to an internal clock or to a user-supplied external clock. A monochomator for dispersing light waves may be coupled directly to choppers with an appropriate coupling head.

Choppers may consist of coated gels or glass or one or more blades, depending upon the application. When there are two blades, one may be fixed while the other rotates. The alignment of the blades determines whether a beam is passed or blocked. The beam need not be completely blocked. With a general purpose scientific chopper with just two blades, it is possible to have chopping frequencies ranging from about 4 to 5000 Hz. With a gel chopper, different types or colors of coatings may selectively pass or chop certain frequencies.

Piezoelectric choppers are small enough to fit on a chip for use in integrated circuits. In conjunction with a collimated laser source, they may be used as compact optical sensors.

SONET Optical Carrier (OC-) Transport Levels				
Carrier	Rate (Mbps)	DS-3	DS-1	DS-0
OC-1	51.84	1	28	672
OC-3	155.52	3	84	2016
OC-9	466.56	9	252	6048
OC-12	622.08	12	336	8064
OC-18	933.12	18	504	12096
OC-24	1244.16	24	672	16128
OC-36	1866.24	36	1008	24192
OC-48	2488.32	48	1344	32256
OC-96	4976.64	96	2688	64512
OC-192	9953.28	192	5376	129024

Choppers are usually described in terms of their operational frequencies. Disc choppers with interchangeable blades and rotational speeds can support many applications. See knife-edge focusing, Ronchi grating.

Piezoelectric Semiconductor Chopper

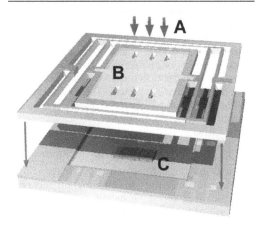

In the quest for ever smaller components, piezoelectric choppers have been developed, such as the one illustrated here, designed by H. Tohiyoshi.

Incident light (A) encounters a chopper plate (B) made from anisotropically etched Z-cut quartz, with chopper slits perpendicular (normal) to the surface of the component. Light passing through the slits activate photosensitive detectors (C).

---

optical combiner In general fiber optics terms, any system for aggregating optical cables in adjacent waveguides or for combining two or more optical signals for transmission through a single waveguide.

Optical cables are typically aggregated to facilitate installation and maintenance. Optical signals are combined for a variety of reasons. Transmission through a single rather than multiple waveguides can be cost-effective. Broadcast services available to all service subscribers may be combined with targeted services requested by individual users at appropriate points in the network. Combined signals may be useful for research purposes for studying optical efficiency, interactions, and various sources of loss.

As a fiber optic network component, the term more specifically refers to devices that combine or combine and regenerate combined optical signals with a minimum of loss and interference in the coupling/combining process. Combiners may be passive or actively integrated with switching mechanisms.

An arrayed waveguide grating (AWG) can combine (or separate) multiple wavelengths. An active self-switching combiner, as an example, takes signals from two or more ports and physically couples them so they are proximate, then phase-shifts the signals through optical amplifiers and brings them together in a combined signal. See multiplexing.

optical combiner, projection A projection surface or component that combines a projected image with other functions, images, or physical environments. For example, in military jets, the windshield may serve as a transparent viewing window as well as a projection surface for certain cockpit status displays. The same concept is applied to head-worn optical display systems, some of which have small transparent screens that enable the viewer to see ahead while at the same time viewing a small projected image (in some systems, the image may pass through the combiner to be projected on the viewer's retina).

Combiner projection technologies are useful anywhere information needs to be superimposed on a transparent surface so that the viewer can see both the projected image and the environment behind the projected image. Thus, they could be used with a video camera to project an image of the area behind a viewer so the viewer could see front and back at the same time or they could project status information over a video image or real-life environment. In terms of innovative networking possibilities, they could project a Web page coupled with a GPS reader that shows a map and local landmarks at the specific site at which a person is standing with a body-worn computer linked to the Internet.

Optical Communications Demonstrator OCD. A NASA/JPL laser-based extraterrestrial communication system capable of transmitting at speeds up to 250 Mbps and up to 1 Gbps with upgrades. A charge coupled device (CCD) array inputs to a two-axis reflector, which controls the orientation of a Q-switched diode-pumped laser communicating in the downlink direction. The ground station detects the light signal using silicon avalanche photodiodes. This is seen as a possible alternative to radio wave communications that are more apt to be distorted due to Doppler shift effects. See Laser Communications Demonstration System.

optical computer A type of processing hardware based on photons rather than electrons. This is a more experimental technology than is used with traditional computers, but it has possible advantages, particularly in speed and resistance to interference over current technologies, and may become more prevalent in the future.

optical connectors Connectors specially designed to couple fiber optic cable junctions so they interfere as little as possible with the path of the optical beams passing through the connectors. The connectors are usually used at points where the fibers connect with routing or switching circuitry, or with the optical interface to the system itself. Optical connections have to be well-engineered, as they must handle very precise beams and paths, and often must maintain the proximity and orientation of a bundle of optical fibers. Connectors for blown fiber installations are easier to install and maintain than a number of other types of fiber attachments. See blown fiber.

optical core The light-carrying central region of a fiber optic filament, commonly made from silica and germania.

optical coupler See coupler.

optical coupler module OCM. A component that provides functions in conjunction with coupled fiber optic cables such as branching/joining, multiplexing/demultiplexing, and filtering. OCMs are often modular and may be customized for operating wavelength and the size and number of fiber filament.

optical cross connect OXC. Optical connection components of telecommunications carrier networks that help operators manage the larger amounts of bandwidth enabled by newer capacity-increasing technologies. OXCs began to become generally available in the late 1990s.

OXCs can facilitate network administration by enabling a network to be more readily reconfigured or restored using optical switches. OXCs can handle data streams in the terabit ranges and work with a range of wavelengths and bit rates, making it easier to quickly effect rerouting. Currently OXC components are part of the optical portion of mixed wire and optical fiber networks, but as the optical market grows, demand for OXCs will likely increase. See lambda switching.

optical detector A substance or circuit which detects and converts electromagnetic waves in the form of optical waves. Solar panels contain substances that allow the conversion of light into electricity, which are widely used in the satellite industry to provide power for telemetry adjustment of orbits. See solar panel.

optical disc Any of a number of technologies used to store digital data which is subsequently accessed by light, usually a laser beam pickup. The information on an optical disc is commonly stored in a series of pits, that is, indentations in a metal disc coated with a plastic substrate. The placement, size, and proximity of these pits is defined in part for the specification for the type of optical technology. Newer technologies have been developed to increase the information capacity of a disc, as in digital videodiscs (DVDs) that are slightly thicker, to bring the disc closer to the laser pickup, with slightly smaller, tighter pits. Compact discs, videodiscs, PhotoCDs, and DVDs are popular forms of audio and visual storage media. Optical discs are also used for computer data storage, particularly for backups.

optical document recognition ODR. A software process or combination of hardware scanning devices and software, which evaluates the various elements on a page to determine whether they are distinguishable as text, images, symbols, or lines, and identifies them accordingly. The ODR in its simplest sense does not interpret the text into characters; it merely identifies page layout elements: images, columns, page numbers, text, etc. However, in commercial products, it is commonly combined with optical character recognition (OCR) capabilities in order to optically identify and interpret areas that are text into characters that can be edited with a word processor, text editor, or desktop publishing program. ODR is a somewhat more complex process than OCR; together they comprise powerful tools which are a part of image document management and processing.

ODR is particularly valuable for converting archived image information (e.g., microfiche) into documents that can be desktop published, or archived in databases for faster and more efficient search and retrieval. See image document management, optical character recognition, scanner.

optical eye pattern measurement procedure An optical fiber test procedure standardized within TIA/EIA-526, developed by the Optical Fiber Communications System subcommittee. It describes parameters for measuring the repetitive temporal characteristics of a two-level intensity-modulated optical waveform at an optical interface. Rise time, fall time, overshoot, and extinction ratio may be determined from the measured eye pattern. The waveform itself may be tested against a reference waveform mask to determine standards compliance.

optical fiber A flexible, light-conducting, filamentous plastic or glass medium used for optical signal transmissions. Optical fiber is typically from 2 to 125 μm thick, and is capable of carrying a variety of high-speed, wide bandwidth transmissions with relatively low loss, when correctly installed. Unlike wire, fiber is not subject to electromagnetic interference or most of the types of radiant eavesdropping techniques that can be used on wire. This has made it very popular for backbone, hazardous area, and multimedia installations. It also does not present the same potential fire hazard as wire electrical cables.

Two common types of optical fiber include step-index fibers and graded-index fibers. Step-index fibers have two layers: a lower refractive outer *cladding* layer and an inner core with a high refractive index. Graded-index fibers also are less refractive toward the outer edge, but rather than being two layers sandwiched together, as in the step-index fibers, the refractivity in the material overall decreases gradually in relation to its distance from the innermost point of the cylinder.

Fiber can be bundled without the electrical interference common to bundled wires; a group of fibers works together to provide greater capacity. Single-mode fiber transmissions, in which the signal can only follow one path through the filament, can travel greater distances without repeaters. See multimode optical fiber, single mode optical fiber.

optical fiber cable See fiber optic cable.

optical fiber ribbon A bundled fiber assembly in which individual fibers are laid side-by-side to form a flat ribbon or strip. This is convenient if they are to be installed in narrow areas such as inside walls or under carpeting. This way of arranging fibers is less common than bundling them into a cylindrical shape.

optical handwriting recognition OHR A specialized form of character recognition designed to separate joined shapes and recognize variations of particular letters, in addition to other generalized optical character recognition (OCR) functions. Most OHR systems have to be trained to recognize a particular style of handwriting, since there is so much variation in the letter forms in the way different people write.

optical insertion loss OIL. The coupling-associated

loss occurring at joints where optical transmission cables are connected into ports. The loss is due to a number of factors, including the quality of the cable assembly, the diameter and type of fiber (some are more subject to loss than others), the snugness of the fit, the amount of strain associated with the connection, and the ambient environment. In product literature, *low optical insertion loss* typically refers to the precision of the fit and resistence to strain in the joint between the connector and the port. There is sometimes a tradeoff between ease of use and reconnection and loss.

Optical Internetworking Forum OIF. An open trade organization fostering the global development and deployment of interoperable optical data switching and routing technologies. Membership includes service providers, equipment manufacturers, and end users. http://www.oiforum.com/

optical label switching See optical-label switching.

optical modulation depth OMD. An expression of the degree of modulation in an optical data transmission system. OMD is usually expressed in micrometers as a percentage per channel (e.g., 4% or 20%) at a specified frequency range. The OMD may have to meet certain standards (e.g., SONET).

Fiber optic CATV transmitting components may have controls for operator adjustment of OMD. Poor adjustment can result in reflections or clipping. See modulation.

optical network unit ONU. A device for connecting a user with a fiber optic network. The ONU creates a physical link and conversion services, if necessary, between the central office optical wavelength and the home or office optical wavelength, which may be different frequencies.

A number of ONUs are typically aggregated at a host digital terminal. Groups of host digital terminals are similarly aggregated at a control terminal, located at the central office of the local exchange.

In spring 2002, companies such as Salira began offering multiple-customer ONUs so that individual units for each customer were no longer necessary.

optical scanner See optical character recognition, scanner.

Optical Society of Japan OSJ. Founded in 1952 as a division of the Japan Society of Applied Physics (JSAP), it is now the biggest division of JSAP with almost 2000 members. OSJ publishes the *Japanese Journal of Optics* and the *Optical Review*. It provides educational support, technical working groups, and conferences for its members and sponsors a number of awards for excellence in the field of optics. http://annex.jsap.or.jp/OSJ/index-e.shtml

optical spectrum analyzer OSA. A desktop diagnostic instrument for measuring the power of a light signal at each of its emitted frequencies. It incorporates a diffraction grating to split out the wavelengths (in much the same way a prism splits white light into its constituent colors) and measures the power level of each wavelength. See reflectometer, tracer.

Optical Storage Technology Association OSTA. An international trade association promoting the use of writable optical storage technology, incorporated in 1992. OSTA develops technology "roadmaps," statistical illustrations of optical storage progress and trends, to define compatible product classes. It is not a standards-development organization per se, but it develops specifications and provides input to other organizations regarding practical implementations of optical storage standards. In July 2001, OSTA announced the development and approval of a new specification for organizing compressed audio files on optical disc, called MultiAudio. The intent is to ensure that discs with compressed audio files (e.g., MP3) are as straightforward and standard Red Book CDs. The format specifies a table of contents access mechanism for CD and DVD playlists and media playback. See MultiAudio, MultiPlay. http://www.osta.org/

optical time domain reflectometer OTDR. An instrument for reflecting light through a fiber optic waveguide to determine loss and integrity of the fiber light path. OTDR measurements may be uni- or bidirectional. Unidirectional measurements may be misleading if there are diameter differences from one fiber joint to the next. However, if the light will be traveling only in one direction and the OTDR is measured in the same direction, this may be sufficient. If the light is being alternately transmitted in two directions along the fiber optic path, bidirectional averaging (taking measurements from each end of the fiber path) will provide more complete data for assessing the lightguide. See bidirectional averaging.

optical transport network OTN. As defined in ITU G Series Recommendation G.709, OTN is an interface that builds upon previous SDH and SONET technologies to provide layered network services, but improves upon previous technologies with better optical channel management in the optical domain and forward error correction (FEC) to increase the distance over which data signals can be transmitted. G.709 standardizes the management of optical channels (wavelengths) without conversion to electrical through newer bubble and micro-electromechanical systems (MEMS) technologies. See optical transport unit.

optical transport unit OTU. In the ITU-T G.709 framing structure, a frame in which each row contains 16 forward error control (FEC) blocks of 16-byte interleaved codecs for a total of $4 \times 16 = 64$ FEC blocks. Currently three line and frame rates are defined.

Number	Line Rate kbps	Frame Rate kHz	Speed
OTU1	2,666,057.143	20.420	48.971 ms
OTU2	10,709,225.316	82.027	12.191 ms
OTU3	43,018,413.559	329.489	3.035 ms

In commercial implementations, OTUs are built into SONET optical transport platforms.

In "digital wrapper" implementations as introduced by Lucent Technologies, data is encapsulated in an optical transport unit frame similar to a SONET protocol frame so that any type of network traffic (ATM, IP, etc.) may be handled and forwarded over existing SONET networks. The OTU provides FEC information at the end of the frame to ensure the suitability of the optical signal for high-speed transmission.

optical-label switching OLS. A network switching system developed under the direction of S.J. Ben Yoo through a DARPA ITO-sponsored research project. It was first proposed by Yoo in 1997 and developed independently at around the same time as Multiprotocol Label Switching (MPLS).

OLS is a means of implementing packet-switching protocols over optical media. OLS is a scalable ultra-low-latency multiprotocol optical routing/switching system that shares some characteristics with MPLS. Underlying OLS is the packet-switching fabric comprising rapidly tunable wavelength conversion capabilities and scalable arrayed-waveguide grating. OLS is built upon this with a forward look to next generation all-optical Internetworks.

OLS routing uses an optical header with a label in the header to determine the packet forwarding. The data are held in the fiber path while the header is examined and may be replaced for the next leg in the transmission path.

OLS project goals as of 2000 included router connectivity in excess of 1024×1024 and aggregate switching bandwidth of almost a peta-bits-per-second. The multiprotocol optical routing is intended to be interoperable with circuit-, burst-, flow-, and packet-switched networks. See label switching, Multiprotocol Label Switching, optical burst switching.

OQPSK offset quadrature phase shift keying. See quadrature phase shift keying.

Orange Book A set of standards for Compact Disc readable (write once) optical media. The previous Red and Yellow Book standards established the basic standards for recording audio and computer data to a Compact Disc. The Orange Book extended these capabilities to enable multisession recordings to be created (though they are not recommended as masters). Red and Yellow Book data can be combined on one disc in whatever order is desired. See Blue Book.

ORB See Object Request Broker.

ORBCOMM Orbital Communications.

orbit The path described by a moving body in more-or-less stable balance with the gravity of the body being orbited so that it continues in that path for a significant period of time (usually at least a few hours or days, although orbits of artificial satellites can last for years in a stable orbit). The Earth is in orbit around the Sun, and the Moon is in orbit around the Earth. When the balance is lost and the orbit becomes smaller as the orbiting body is drawn inward, it is said that the orbit is *decaying*. Many types of orbits (orbits at different heights, with differently shaped paths) are used in telecommunications with artificial satellites.

Early communications satellite orbits tended to be circular or low and somewhat flatly elliptical (and tended to decay quickly). Communications were hindered by the necessity of locating the orbiting satellite and keeping it in range before it passed around to the other side of the Earth. Later satellites were put into higher, more stable orbits, which were often geostationary; that is, the orbit was synchronized with the movement of the Earth so that the location of the satellite was roughly above the same location at all times. The amount by which an orbit deviates from a circle is referred to as its *eccentricity*. See geostationary, satellite.

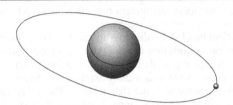

Elliptical Orbit

Satellite orbits tend to be circular or more flatly elliptical, and may be further controlled so that the satellite spends a greater part of the orbit over ocean or land masses, depending upon whether it is serving marine or terrestrial communications needs.

The orbits are often described in terms of their distance from the Earth as low, medium, and high Earth orbits (LEO, MEO, HEO). Geostationary orbits are a type of high Earth orbit in which the satellite's movement is paced such that it remains in the same position relative to the Earth.

Orbital Sciences Corporation Commercial developers of global satellite communications services. Orbital Sciences is a space and information systems company which designs, manufactures, and markets space-related infrastructures and products. See OrbLink.

OrbLink A global broadband commercial satellite communications network being developed by Orbital Sciences Corporation for deployment around 2002. The designers intend to use the newest technology to construct broadband services that can be offered at lower prices than existing services. OrbLink is based on seven medium Earth orbit (MEO) satellites orbiting at 9000 km in an equatorial orbit, transmitting in extremely high frequency radio bands. Services will include digital voice, data videoconferencing, computer networking, imaging, and other broadband applications.

Pending Federal Communications Commission (FCC) approval, intersatellite communications will be at 65.0 to 71.0 GHz at speeds up to 15 Gbps, supporting high-capacity intercontinental trunking. Two-way digital connections will be between 37.5 to 38.5 GHz and 47.7 and 48.7 GHz bands up to about 1.5 Mbps.

The seven satellites, plus one spare, are based on

Orbital's STARBus, a small, lightweight, geostationary technology, acquired through a purchase of CTA Incorporated's space system business. Each satellite will support 100 spot beams. As an economic note on the dynamics of new telecommunications technologies, the total cost of building and deploying the orbital network, according to Orbital Sciences, compares to the cost of installing two transatlantic fiber cables, which provide only about 10% of the trunking capacity of the proposed satellite system.

Orckit Communications A developer and manufacturer of high-speed local loop communications systems and participant in a number of network standards working groups. Orckit partners with Fujitsu Network Communications, Inc.

order entry The inputting, usually by voice or keyboard, of a customer request for a product or service. On the World Wide Web, Web *forms* are often available for customers to put in their own order entry. These Web applications are often in the form of *shopping carts* in which the customer browses various Web pages and enters each product desired into the shopping cart (order batch). The order is typically preprocessed by a CGI and then sent to the appropriate order fulfillment personnel for shipping and billing. Other automated systems allow customers to carry out a complete ordering transaction over a touchtone phone by using the keypad to enter codes and digits. See Automatic Call Distribution.

Organization of American States OAS. The OAS Web site provides information on the group. http://www.oas.org/

Organizationally Unique Identifier OUI. A globally unique 24-bit Ethernet address identifier for LANs and MANs managed and assigned by the IEEE Registration Authority online or by phone. The OUI is assigned as a three-octet field in the SubNetwork Attachment Point (SNAP) header, identifying an organization which further creates unique six octet numbers. Together they constitute a distinct Media Access Control (MAC) address or Ethernet address. See Ethernet, IEEE, Token-Ring.

originate/answer On a computer data modem, when the user wants to dial out to connect to another computer, a bulletin board, or an Internet access point, the commands for controlling the phone line and dialing the desired number come from the originating modem. "Originate mode" sets up a sequence of events which checks for a dial tone, dials, and handshakes with the receiving modem to establish the connection rate and protocol (or hangs up if the line is busy or is dropped). The receiving modem is set to "answer mode" so it detects an incoming call, answers it, and participates in the rate and protocol negotiation. Most of this is automatically handled through a terminal software program, but it may be necessary to set originate or answer through menu selections, or direct commands through the software to the terminal program.

originate restriction A security or specialized use restriction on a phone line which causes it to work only for incoming calls. Outgoing calls are blocked.

This restriction is sometimes set on phones adjacent to public areas to prevent people from monopolizing or misusing a phone line. Sometimes the originate restriction applies only to long-distance calls or calls outside a private branch exchange.

In some circumstances, the local phone company will partially disconnect a line by setting an originate restriction if the subscriber is behind in the payment of the phone bill. After paying the bill, it is usually necessary to request restoration of full service, as it is seldom done automatically.

originator Initiator, caller, inventor, introducer, founder. The person, entity, device, or station that first communicates a message or starts an action or process.

Orion A broadband data satellite service provider aiming at international common carriers and individual companies.

ortho-correction In satellite imaging, a correctional adjustment for distortion resulting from terrain.

orthogonal frequency division multiplex OFDM. A multicarrier modulation system which is similar to discrete multitone in that it utilizes Fourier transforms of data blocks. OFDM is suitable for Digital Subscriber Line (DSL) services. See Digital Subscriber Line, discrete multitone.

OS X See Mac OS X.

OS/2 Operating System/2. International Business Machines' 32-bit preemptive multitasking text and object-oriented graphical operating system targeted for Intel-based microcomputers in the late 1980s. It was originally developed for IBM by both IBM and Microsoft Corporation, and version 1.0 was released in 1987 to succeed MS-DOS. When an upgrade to OS/2 was well under way, Microsoft pulled out to concentrate on their own operating system in competition with IBM. Many of the same concepts that were part of OS/2 were incorporated into Microsoft Windows NT, which was marketed as a direct competitor to OS/2.

In 1991, IBM released OS/2 version 2.0. Version 2.1 added support for multimedia and Windows 3.1 applications. OS/2 had some commercial success in the early and mid-1990s, but by 1996, through aggressive advertising and bundling programs, Windows was better known and more widespread in North America. In spite of this, there are many strong supporters of OS/2, including a worldwide network of Team OS/2 Groups. It is still a popular choice in Western Europe and Canada. Team OS/2 information for OS/2 users is available at their Web site. See Team OS/2.

OS/2 SMP OS/2 Symmetric Multiprocessing. This version of IBM's OS/2 supports systems with multiple processors, making it suitable for Internet services, graphics, and corporate applications, particularly those which operate as various types of resource servers in a networked environment. See OS/2.

OS/2 Warp Operating System/2 Warp. By version 3 of IBM's OS/2 operating system, their operating system product was called OS/2 Warp. The Warp version added increased support for various peripheral

Overview of Early OSCAR Satellite Projects

Satellite	Launch Date	Technical Details	Notes
Phase I Satellites – Experimental, Low Orbit, Short Life Span			
OSCAR I	12 Dec. 1961	10 lb, beacon, 22-day orbit. Nonrechargeable batteries.	Initiated by a U.S. West Coast group. U.S. Air Force launched.
OSCAR II	2 Jun. 1962	Better coatings and temperature control.	Similar to OSCAR I, but incorporating improvements.
OSCAR *	Not launched	Phase-coherent keying.	Similar structurally to previous.
OSCAR III	9 Mar. 1965	First relay transponder. Solar backup.	Tracking and telemetry equipment. Approx. 3000 mi. range. 18-day transponder use.
OSCAR IV	21 Dec. 1965	High altitude, transponder. Solar, beacon, no telemetry.	Unplanned varying elliptical orbit. Two-way communication achieved. Link between Russia and U.S.
OSCAR 5	23 Jan. 1970	Controllable, magnetic attitude stabilization. No solar or transponder.	Seven analog telemetry channels. Australis-OSCAR 5 (AO-5). Built in Australia. First NASA-launched OSCAR.
Phase II Satellites – Developmental, Low Orbit, Operational, Longer Life Span			
OSCAR 6	15 Oct. 1972	Telemetry, command, transponder. Solar. Store-and-forward system.	Twenty-four telemetry channels. Two-way communications. Falsing and beacon problems. Life span exceeded 4 years. Educational materials printed.
OSCAR 7	15 Nov. 1974	Two transponders, linear frequency translation. Telemetry, radio teletype. Beacons. Up to 4500 miles low altitude.	Many countries contributed various technologies and parts. AMSAT-OSCAR 7 (AO-7). Relayed with OSCAR 6! Almost 7-year lifespan.
OSCAR 8	5 Mar. 1978	10-m antenna. Two transponders (Modes A & J) that could operate simultaneously.	ARRL operated. Cooperatively built by Project OSCAR, AMSAT and JAMSAT. Lasted 5 years.
Phase III Satellites – Operational, High Elliptical Orbit, Longer Life Span (see AMSAT)			

devices and reduced memory requirements. It was succeeded by OS/2 Warp Connect which offered networking through full TCP/IP capabilities. OS/2 Warp version 4 was aimed at corporate users. This version included increased networking features, speech-to-text speech recognition software, and built-in support for Sun Microsystems' Java. See International Business Machines, Java, OS/2.

OS/2 Warp Server IBM's Warp Connect system integrated with their local area network (LAN) server 4.0. This version of OS/2 was designed for handling file and device service sharing on networks. See OS/2.

OSCAR Orbiting Satellite Carrying Amateur Radio. A series of orbiting satellites originally developed in the homes and garages of a group of amateur radio enthusiasts. In 1962, the OSCAR Association was incorporated as Project OSCAR, Inc. The early OSCAR satellite projects began in 1961 and continue today in expanded and more sophisticated forms.

Early OSCARs used fairly simple beacon transmitters with nonrechargeable batteries, so they were only useful for a few weeks, but they showed what might be accomplished with relatively simple materials and a lot of cooperative effort. Solar cells and telemetry equipment were added to later versions in order to extend useful life and provide greater control over positioning. Relays were then added, with the aim of eventually providing two-way (bidirectional) communications.

OSCAR-AMSAT projects became increasingly sophisticated and, by the time the OSCAR 6, 7, and 8 were in orbit, telemedicine and search and rescue satellite communications were demonstrated to be feasible.

Pioneer OSCAR Satellite

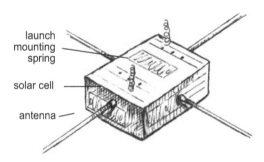

launch mounting spring

solar cell

antenna

The early OSCAR satellites were put together in a cooperative effort out of makeshift donated parts, yet were well-conceived, pioneer satellite technologies, increasing in sophistication with later projects. OSCAR III is shown here.

The deployment mechanisms of the early OSCARs were particularly interesting to scientists researching satellite installation. The building, launching, and especially the tenuous securement of domestic and international regulatory permissions to launch and operate were a great achievement for amateur enthusi-

asts, and benefits are still accruing from the hard work and voluntary contributions of radio amateurs. See Overview of OSCAR Projects chart. See AMSAT.

OSCE Organization for Security and Co-operation in Europe.

oscillation 1. Variation, fluctuation, continuing periodic reversal. Although oscillation in the general sense does not imply a regular oscillation, many waves, materials, and circuits studied or constructed by scientists exhibit fairly regular, predictable oscillating properties. See quartz. 2. The cyclic alternation of electrical properties in a circuit.

oscillator An electronic device designed to generate a low-current alternating current (AC) power at a particular frequency according to the values of certain constants in its circuits. In microcomputers, an oscillator can be used to provide a reference frequency for clocking. An oscillator is also useful for generating test signals. See oscilloscope, quartz.

oscilloscope A device designed to provide a visual representation of variations in electrical quantities as a function of time, displayed in the form of pulses or waves on a monitor. The size and form of the waves are traditionally tuned for optimum viewing with knobs, as on an old radio. Oscilloscopes are sometimes interfaced with computers to provide a means of directly adjusting and analyzing the oscilloscope signals through software. Oscilloscopes are useful for diagnosis and testing of electronic circuits.

OSF See Open Software Foundation (now the Open Group).

OSI See Open Systems Interconnection.

OSI Transport Protocol OSI TP. The ISO-recommended communications protocol used by X/Open.

OSN 1. operations system network. 2. See Open Systems Networking Initiative.

OSP See Operator Service Provider.

OSPF See Open Shortest Path First Protocol.

OSSP See Object Serialization Stream Protocol.

OST Office of Science and Technology. A U.K. government group founded in 1992 to coordinate science and technology issues across government departments.

OSTA See Optical Storage Technology Association.

OSTP See Office of Science and Technology Policy.

OT See Office of Telecommunications.

OTA Office of Technology Assessment (U.S.).

OTDR 1. See Office of the Director of Telecommunications Regulation. 2. See optical time domain reflectometer.

OTE The primary telecommunications carrier in Greece. OTE is government-owned.

OTGR Operations Technology Generic Requirements.

OTH over the horizon.

OTN See optical transport network.

OTOH An abbreviation for "on the other hand" commonly used in email and online public forums. See AFAIK, IMHO.

OTP 1. See Office of Telecommunications Policy. 2. See Open Trading Protocol.

OTU See optical transport unit.

OUI See Organizationally Unique Identifier.

out-band/out-of-band signaling Control signaling that is carried separate from the informational portion of a message. See Signaling System 7.

out-of-range alert In wireless communications, a beep or light that alerts the user that the handset is at the edge of its range and the user shouldn't move further from the source of the transmission.

outage Loss of power, service interruption. See blackout, brownout.

outlet 1. Exit, opening for egress, vent. 2. Plug receptacle in a circuit, usually for electricity or connectivity to data transmissions. 3. Source of goods, supplier.

outline font, algorithmic font, vector font A character set defined by mathematical algorithms that describe the shape of the letters with graphics primitives such as lines, arcs, ellipses, spline curves, etc. That way, when printed or displayed on a monitor, they will be drawn at the best possible resolution offered by the display system. Unlike bitmap fonts, which are hand drawn as raster images that cannot be significantly reduced or enlarged, vector fonts look good at sizes ranging from 4 points to 100 points and much larger. Outline fonts are resolution- and platform-independent, provided an interpreter is available on the system for the particular format that is being used. Since outline fonts are widely supported on many platforms, this is usually not a limitation.

output 1. That which results from, or comes out of, a process or system. 2. The combined signal and content information of a transmission. 3. The result of a computer process, e.g., the output of a word processing session might be a printed document, Web page, or a facsimile transmission.

output device A device that facilitates the communication or transmission of information, usually in another form or format. In most cases, an output device is a human interface in the sense that it facilitates the translation or movement of information between nonhuman-readable forms and human-readable forms, or between single-copy modes and multiple-distribution modes intended for a wider audience.

outside plant OSP. The various outside structures, devices, and cabling installations that together comprise a network. These may be installed above or below ground. Those supported by utility poles may be termed *aerial*. See joint pole for utility pole information and diagrams.

outside plant access cabinet OPAC. Solid, compact weather-resistant cabinets for housing remote-access network equipment and connections, usually in office and industrial park environments. The cabinets may be controlled for factors such as temperature and humidity in order to protect sensitive electronic components and are usually designed to deter vandalism. Besides leaving more space free inside the building, the outside access cabinet makes it possible for maintenance personnel to access the devices without entering the building premises.

outsourcing The process of assigning production or management tasks to an external consultant or organi-

zation. Outsourcing is practical when special expertise is needed, or the project is short and hiring new permanent staff would be impractical. Specialized design projects, advertising, documentation, and cyclic/seasonal projects are often outsourced. Network administration is often outsourced by small companies, whereas a company with a larger or more complex network would probably have an in-house system administrator.

Telephone answering services are a common form of outsourcing used by small businesses and home businesses. Utilizing an answering service is less expensive than hiring a receptionist – a good solution for small companies that don't receive a lot of incoming calls.

OutWATS Outward Wide Area Telephone Services. A WATS service for outgoing calls, which is available at bulk-use discounts. See InWATS, WATS.

OV2-5 A research satellite designed to measure solar and cosmic rays and magnetic influences. It was launched in September 1968 into a circular equatorial orbit at an altitude of 22,000 miles. A model of the OV2 was donated to the U.S. Air Force Museum by the Northrop Corporation.

overcoat-incident recording OCIR. A new technology for recording optical storage media that permits higher density data per unit area than previous methods such as substrated-incident recording. Traditionally, optical discs have used a substrate laid down over the recording surface to protect the data from abrasions, contamination, and oxidation. This limited the areal density of the recorded information.

OCIR technology was developed by Maxoptix (trademarked as OverCoat Incident Recording with patents pending). In OCIR, the recorded information is imprinted on top of the substrate (similar to hard disk media) and then covered with a protective layer of acrylic that is much thicker than hard disk and tape recording media, but thinner than standard optical recording substrates. Thus, the lens can be positioned closer to the recording surface, realizing a higher numerical aperture (NA) for recording at higher data densities. Maxoptix's goal was to realize 40 GBytes of recorded data at 30 Mbps transfer rates. See air-incident recording, substrate-incident recording, surface-array recording.

overflow 1. Traffic or data in excess of what is typically found on a system, or in excess of what the system is capable of handling. Some systems have additional or alternate circuits, lines, systems, or operators to handle overflow, while others may be slowed down in terms of speed of service, or may cease to function. 2. In telephone circuits, overflow traffic may be diverted to another trunk line. See erlang.

overflow, data In programming, an overflow occurs when an operation generates a result for which there is insufficient address or storage space.

overhead The portion of a task, data block, or operation that provides management information pertaining to the task, data, or operation, which is not part of its integral content. For example, the overhead in a graphics file may consist of a header containing size and palette information, which is not part of the image itself. The overhead in a parallel processing operation may be the time and processing it takes to handle the logistics of farming out the tasks and recombining the results of the processes. In networks, overhead exists in the form of protocol information, timing information, error data, security bits, routing, priority, and more. Given the amount of overhead in networks, it's a marvel that they can work so effectively.

overhead transparency, foil A transparent medium receptive to photocopy toner or various inks which is used in conjunction with a bright light and projector to project information on a large surface such as a screen or plain wall.

Overhead transparencies are often used for presentations, especially to illustrate lectures. Overhead transparency films come in a variety of compositions; some can be photocopied in black and white, some in color. Don't use regular transparency paper in a laser printer or photocopier, as the plastic may melt and destroy the internal mechanisms. Cardboard frames can be purchased to support the transparencies, which are somewhat flimsy and otherwise hard to hold and organize.

overlay 1. *n.* A keyboard template or sheath. See keyboard overlay. 2. *n.* A template, grid, pattern, image, or other reference information superimposed over a field of view. Overlays are used to measure, count, estimate, asses, evaluate, and embellish the information or environment over which they are used. See optical combiner, projected; graticule; reticle. 3. *v.* In programming, a technique in which a limited amount of storage is extended by reusing portions which are not immediately or subsequently required, or by initiating less commonly used routines only on demand. In telephone applications, overlays may be used to bring various tasks into memory as needed. Some versions of BASIC have commands (e.g., LSET) which allow a variable in RAM to be overwritten with a subsequent variable in order to prevent eventual slowdowns from *garbage collection,* that is, from the reorganization of storage to accommodate more information.

overlay area code A telephone area code assigned as a parallel code in an existing service area. These are commonly assigned to mobile services, like cellular and pager services, so that the area code is separate from the geographic code assigned to that region. These are not yet prevalent, but are expected to increase as mobile services are more widely distributed. See North American area codes for a chart of telephone and mobile service area codes.

overlay network A protocol or application-specific subnetwork, managed and configured independently of its underlying infrastructure, and interconnected by Internet Protocol (IP) encapsulation tunnels over production networks. Recent protocols are supported on overlay networks, including Mbone (multicast IP) and 6bone (IPv6).

overlay, video In video editing, it is common to *overlay* two video signals, or to overlay a computer signal over a video signal, or vice versa. Newscasts will often overlay a human weather forecaster over a computer-generated weather map. In cinema action shots, a stunt worker in a barrel may be overlaid on an image of the Niagara Falls. See chromakey.

override 1. To overlap, neutralize, take over, dominate. A stronger signal, such as an emergency signal, can override a regular transmission. A boss can override the decision of a subordinate; a priority transmission can override current transmissions. An operator can override a current phone conversation. Some private branch phone systems are configured so that someone in authority has the option of overriding other conversations, a power that should be used with discretion.

overrun To overwhelm, to swarm, to go above or beyond an edge or capacity, to overflow. A cost overrun happens when someone goes over budget or some other allotted quantity. A data overrun can happen when the receiving system isn't fast enough or smart enough to handle the incoming transmission. A printer overrun can happen if the print mechanism continues to function after the paper runs out (some facsimile machines still do this). Overruns often result in discard or loss of information. See cell rate, leaky bucket.

oversampling A process of redundant sampling used in some multiplexing schemes.

overscan, full scan An image output to a monitor that extends to the maximum outer extents of the cathode ray tube (CRT) or other scanning display device. Overscan on computers may be achieved by increasing the resolution of the display or by adjusting position and size controls associated with the display device. Overscan display modes are common in video applications, where the signal is not being optimized for the computer monitor, but for the video recording medium to which it is being output. Overscan may also be a screen option on some systems that are adapted for desktop video and usually adds about 10 to 30 pixels to each edge of the displayable resolution. (Thus, a 320 × 480-scanline interlaced image might become 360 × 525 scanlines in overscan mode, for example.)

Flat screen monitors are becoming more widely available, but in the past, cathode ray tubes had a significant curvature at the outer edges which would distort the image (like looking through a lens) at the outer edges. In order to minimize distortion, the image is usually not displayed to its fullest extent, but rather to the point on the front of the tube at which the curve begins. The edge of the monitor casing is usually designed by the manufacturer to fall approximately at the same point or slightly outside the point at which the overscan image falls. See cathode ray tube.

overshoot In general terms, a transmitted signal that travels to some point beyond the receiver (too far, too high, at too high a frequency, etc.) or a receiving system that over-responds to a stimulus signal. Overshoot may be a compensatory strategy, an undesirable condition, or a manageable condition (e.g., it may be damped or otherwise moderated). The

O

converse is *undershoot*.

Fiber optic sensor probes can help prevent undesired temperature or humidity overshoot in fabrication or controlled environment processes. At the same time, while fiber optic probes may be useful in preventing overshoot in fabrication processes they may, themselves, be subject to overshoot if they are held too near the object probed such that a resonant frequency amplifies the probe signal and gives an incorrect reading.

In electronic logical circuits, a derivative function may be used to assess the rate of change in a system to send an alarm or automatically adjust a component to halt, minimize, or reduce overshoot.

In more specific terms in optical communications, overshoot is an undesirable excess occurring as the result of a transition of a signal from one phase or signal level or type to one that is lower. Thus, excessive amplitude in a waveform beyond what is desired in a nonsinusoidal wave results in overshoot in fiber optic links. See hysteresis, optical eye pattern measurement procedure.

overtime period In a pay-per-time-connected service, the time that elapses after the paid-up period has been exceeded. When using a payphone, the time after the initial insertion of coins has run out is overtime, and the operator may request additional funds or terminate the call.

The same general idea applies to per-pay network access, time-sharing, or any other system in which a set amount of time is billed periodically, or is prepaid, with the option for the user to exceed the usage period as long as additional fees are paid, often at a higher rate.

overtone In wave phenomena that can be characterized as sinusoidal, an integral multiple of the wave frequency, a resonant *harmonic,* a combination vibration. Overtones may be numbered; the first overtone is twice the frequency of the fundamental reference frequency.

Musicians are familiar with audible overtones. For example, on a violin, high-pitched overtones may be generated by bowing a string at a certain resonant vibrating frequency such that the overtones can just be heard over the main note.

Light-based phenomena also exhibit overtones. In spectroscopy, absorption frequencies may exhibit multiple harmonic overtone bands. As these bands become more removed from the fundamental reference frequency, they progressively become more widely separated and lower in intensity. In near-infrared spectroscopy, the strongest absorption bands usually occur in the first and second overtones.

OVS See Open Video System.

OWT Operator Work Time.

OXC See optical cross connect.

oxidation The process of combining with oxygen, often resulting in a significant change in the material oxidized that may degrade or otherwise influence its integrity or usefulness for a particular purpose. Oxidation is a particular concern in external wiring installations or cables exposed to water or chemicals. See corrosion.

oxymoron A combination of contradictory, incongruous words. Puns sometimes have oxymoronic implications which may or may not be true. Satirical examples include: common sense, military intelligence, casual dress, friendly fire, and authentic reproduction.

p 1. *symb.* pico-. 2. *abbrev.* power.
P 1. *symb.* permeance. 2. *symb.* peta-. 3. *abbrev.* phosphorus.
p connector A power connector commonly used for attaching internal computer peripherals such as floppy drives, hard drives, CD-ROM drives, etc. Computer power connectors are largely standardized as 4-pin, keyed connectors. (In the late 1970s and early 1980s, they weren't always keyed.) It is preferable to call this a *power connector* to prevent ambiguity with 68-pin P connectors. See P connector.

Standardized Peripheral Power Connector

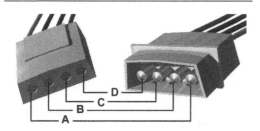

A typical power connector for computer peripherals such as hard disk drives standardized as (A) +12 volts – yellow, (B & C) ground – black, (D) +5 volts – red. Unlike earlier connectors, they are now typically keyed (notched on one side) to prevent incorrect connections.

P connector 1. An ANSI-standardized 68-pin electrical data connector commonly used for 8-and 16-bit data buses for computer peripherals such as SCSI drives, especially SCSI-3, as a P connector can support wide bus, high-density data transmissions. The Very High Density Cable Interconnect (VHDCI) connector is an *Alternative 4* P connector with the same pinouts as the 68-pin connector, but it enables multiple SCSI connectors to be connected to one backplate. See A connector. 2. See Polaroid connector.
p region In semiconductor component theory and engineering, a region in which the conduction-electron density characteristics result in positive "holes" that can be exploited for their dynamic interactive characteristics. The p region is related to n materials in the n region, where conduction-electron density

exceeds hole density. See n region, p-n junction.
P Series Recommendations A series of ITU-T recommended guidelines for telephone transmissions quality, installations, and local phone line networks. They are available for download from the ITU-T. Since ITU-T specifications and recommendations are widely followed by vendors in the telecommunications industry, those wanting to maximize interoperability with other systems should be aware of the information disseminated by the ITU-T. A full list of general categories is in the Appendix and specific series topics are listed under individual entries in this dictionary, e.g., M Series Recommendations. See P Series Recommendations chart.
P-47 A commercial cesium-doped yttrium silicate (Y_2SiO_5) powder with a fine grain for fabricating scintillating discs with fast delay times. In conjunction with an appropriate suspension liquid and binder, the powder can be used to coat a glass substrate with a uniform layer which is then baked for about 12 hours to produce a scintillating layer that can be combined with a variety of imaging components. See scintillator.
P-frame predictive-coded frame. In MPEG animations, a picture which has been encoded into a video frame according to information extrapolated from *past* frames in the sequence, using predicted motion compensation algorithms. See B-frame, I-frame, MPEG decoder.
P-picture predictive-coded picture. In MPEG animations, a picture that is to be encoded according to *past* frames in the sequence using predicted motion compensation algorithms. Once it is encoded, it is considered to be a P-frame. See MPEG decoder.
P-T pressure-temperature.
P/B peak to background (ratio).
P3P See Platform for Privacy.
pA *abbrev.* picoampere.
Pa 1. *abbrev.* pascal. 2. *abbrev.* protactinium.
PA 1. preliminary assessment. 2. See public address system.
PABX See Private Automatic Branch Exchange.
PACA 1. See Pacific and Asian Communication Association. 2. Picture Agency Council of America. A trade organization supporting stock image agencies in North America. http://www.stockindustry.org. 3. See Priority Access and Channel Assignment.

ITU-T P Series Recommendations

Telephone transmission quality, telephone installations, local line networks

P.10	Vocabulary of terms on telephone transmission quality and telephone sets
P.11	Effect of transmission impairments
P.16	Subjective effects of direct crosstalk; thresholds of audibility and intelligibility
P.30	Transmission performance of group audio terminals (GATs)
P.32	Evaluation of the efficiency of telephone booths and acoustic hoods
P.33	Subscriber telephone sets containing either loudspeaking receivers or microphones associated with amplifiers
P.35	Handset telephones
P.36	Efficiency of devices for preventing the occurrence of excessive acoustic pressure by telephone receivers
P.38	Transmission characteristics of operator telephone systems (OTS)
P.48	Specification for an intermediate reference system
P.50	Test signals
P.51	Artificial mouth
P.52	Volume meters
P.53	Psophometer for use on telephone-type circuits
P.54	Sound level meters (apparatus for the objective measurement of room noise)
P.55	Apparatus for the measurement of impulsive noise
P.56	Objective measurement of active speech level
P.57	Artificial ears
P.58	Head and torso simulator for telephonometry
P.59	Artificial conversational speech
P.61	Methods for the calibration of condenser microphones
P.62	Measurements on subscribers' telephone equipment
P.63	Methods for the evaluation of transmission quality on the basis of objective measurements
P.64	Determination of sensitivity/frequency characteristics of local telephone systems
P.65	Objective instrumentation for the determination of loudness ratings
P.66	Methods for evaluating the transmission performance of digital telephone sets
P.75	Standard conditioning method for handsets with carbon microphones

P.76	Determination of loudness ratings; fundamental principles
P.78	Subjective testing method for determination of loudness ratings in accordance with Recommendation P.76
P.79	Calculation of loudness ratings for telephone sets
P.82	Method for evaluation of service from the standpoint of speech transmission quality
P.84	Subjective listening test method for evaluating digital circuit multiplication and packetized voice systems
P.85	A method for subjective performance assessment of the quality of speech voice output devices
P.310	Transmission characteristics for telephone-band (300 to 3400 Hz) digital telephones
P.311	Transmission characteristics for wideband (150 to 7000 Hz) digital handset telephones
P.313	Transmission characteristics for cordless and mobile digital terminals
P.340	Transmission characteristics of hands-free telephones
P.341	Transmission characteristics for wideband (150 to 7000 Hz) digital hands-free telephony terminals
P.342	Transmission characteristics for telephone band (300 to 3400 Hz) digital loudspeaking and hands-free telephony terminals
P.350	Handset dimensions – formerly ITU-T P.35
P.360	Efficiency of devices for preventing the occurrence of excessive acoustic pressure by telephone receivers
P.370	Coupling hearing aids to telephone sets
P.501	Test signals for use in telephonometry
P.502	Objective test methods for speech communication systems using complex test signals
P.561	In-service nonintrusive measurement device – voice service measurements
P.562	Analysis and interpretation of INMD voice service measurements
P.581	Use of head and torso simulator (HATS) for hands-free terminal testing
P.800	Methods for subjective determination of transmission quality
P.810	Modulated noise reference unit (MNRU)
P.830	Subjective performance assessment of telephone-band and wideband digital codecs
P.831	Subjective performance evaluation of network echo cancellers

Telephone transmission quality, telephone installations, local line networks, cont.	
P.832	Subjective performance evaluation of hands-free terminals
P.833	Methodology for derivation of equipment impairment factors from subjective listening-only tests
P.861	Objective quality measurement of telephone-band (300 to 3400 Hz) speech codecs
P.862	Perceptual evaluation of speech quality (PESQ), an objective method for end-to-end speech quality assessment of narrowband telephone networks and speech codecs
P.910	Subjective video quality assessment methods for multimedia applications
P.911	Subjective audiovisual quality assessment methods for multimedia applications
P.920	Interactive test methods for audiovisual communications
P.930	Principles of a reference impairment system for video
P.931	Multimedia communications delay, synchronization and frame rate measurement

Supplements	
P.Sup1	Precautions to be taken for correct installation and maintenance of an IRS
P.Sup10	Considerations relating to transmission characteristics for analogue handset telephones
P.Sup14	Subjective performance assessment of digital processes using the modulated noise reference unit (MNRU)
P.Sup15	Wideband (7 kHz) modulated noise reference unit (MNRU) with noise shaping
P.Sup16	Guidelines for placement of microphones and loudspeakers in telephone conference rooms and for Group Audio Terminals (GATs)
P.Sup17	Direct loudness balance against the intermediate reference system (IRS) for the subjective determination of loudness ratings
P.Sup19	Information on some loudness loss related ratings
P.Sup20	Examples of measurements of handset receive-frequency responses: dependence on earcap leakage losses
P.Sup22	Transmission characteristics of wideband audio telephones
P.Sup23	ITU-T coded-speech database

Pacific and Asian Communication Association PACA. A nonprofit educational, literary, and scientific organization founded in March 1995. PACA supports and promotes the research, criticism, and application of artistic, humanistic, and social scientific principles of communication. PACA publishes the journal *Human Communication*, and sponsors various educational workshops and conferences. http://www.ukans.edu/~paca/

pack To compact characters or data together to conserve space, usually by removing spaces and any other unneeded characters. In the old 4-kilobyte computers from the 1970s that used BASIC as a programming language, "string packing" was a means to save precious memory.

Database entries and email messages often have a lot of empty space in them and so may be packed to reduce the storage size of files. Packing is a simple form of compression. See compression.

Packard, David (1912–1996) American businessman, philanthropist, and founder, along with William Hewlett, of the Hewlett-Packard computer company, one of the well-respected pioneering companies of the computing industry. The company had its humble beginnings in the Packard garage in Palo Alto, California, and has grown into a multinational company with over 100,000 employees. Packard also cofounded the American Electronics Association, and was a member of the President's Council of Advisors on Science and Technology for 4 years. Hewlett, William R.; Hewlett-Packard.

Packed Encoding Rules PER. Developed in the early 1990s, PER is one of three major encoding schemes used in open architectures development. Unlike BER, tags are ignored and length fields may be omitted. PER provides a means for creating more succinct encodings optimized for bits on the line and generally has lower bandwidth requirements than BER or LWER. See Basic Encoding Rules, LightWeight Encoding Rules.

packet 1. A generic term for a unit of data formed as a bundle with a certain specified organization, according to a protocol. Other designations for network units and bundles include *cell* and *frame*. Although packet formats vary, they most typically include a header, an information payload, and a trailer. The header may contain a number of pieces of information, including priority, source, destination, length of packet, etc. The payload is the message or information being sent, and may be split over a number of packets. The trailer may include flags, signals, and error detection or correction data. When a series of related packets is transmitted over a network, they may not all take the same route, and so disassembly, routing, and assembly procedures may be applied to transmitted packets; instructions to coordinate this process may or may not be included in some of the packets.

Sometimes packet-switched networks are connected to non-packet-switched networks, in which case tunneling takes place, or conversion through a packet assembler/disassembler, to accommodate the differences in formats.

P

packet assembler/disassembler PAD. In packet-based systems, information is converted into data units known as packets, and then transmitted. At the receiving end, these packets are apprehended and disassembled to turn them back into the information contained in the original content.

packet filtering The evaluation of packet structure or contents in order to selectively reject or accept passage of the packet through a network junction. See firewall.

packet radio Packet radio is a combination of computer equipment and radio transmissions used to exchange messages. Microcomputers and terminal node controllers (TNCs) are commonly used in packet radio systems. The computer is cabled to a radio transceiver at each end of the communication. Because computers have store and forward, or other types of scheduling capabilities, the operator doesn't have to be present when the message is sent or received. In radio, this is called *time-shifted* communications. The system could be configured to send at a time when interference is less likely to be encountered, or when a favorable time of day occurs at the sending or receiving end.

Packet radio transmission speeds are fast enough that various types of propagation can be used, including meteor-scatter. Due to the nature of packet transmission and its built-in error-correcting mechanisms, packet transmissions are reliable. Packet radio uses a number of protocols and favors the Open Systems Interconnection (OSI) reference model. Common protocols in use include NET/ROM, AX.25, TCP/IP, and ROSE.

packet reservation multiple access PRMA. An enhanced time division multiple access (TDMA), which incorporates aspects of S-ALOHA. It is suitable for mobile transmissions. See ALOHA, time division multiple access.

packet sniffer A diagnostic and snooping mechanism for examining the contents of network packets during transmission. See packet tracing.

packet switched radio See packet radio.

packet switching A computer communications technology developed in the early 1960s that bundles up information into discrete data *packets* which can be sent out in separate paths, like breaking up the cars on a train sending them on separate tracks, and putting them all back together again at the destination. In the 1960s, computing was becoming more accessible, generating greater interest in its use and spurring the manufacture of various types of systems. Practical packet-switched implementations began to appear in the 1970s, and separate server computers to handle various specialized purposes, such as accounting, opened the doors to the development of various types of distributed computing architectures.

The rise of ARPANET greatly influenced the development and acceptance of packet switching. With hosts springing up in distant locations and specialization and the variety of computing tasks increasing, packet switching was a practical way to facilitate intercomputer communications.

Gradually, layered architectures emerged, separating user functions and applications from lower level operating functions. This enabled information carried in packets to be communicated through many different types of systems, while still retaining the unique operating features and user interfaces of each system. Historically, telephone networks were built around circuit-switching. This meant that a dedicated path through the switching system had to be established (and was tied up) for the duration of the call. In a large global network where many institutions are online all the time, this is not a practical solution. A better way for large systems is to route information through whatever path is most practical at the time (since some systems may be inaccessible or offline without notice), to divide the packets up, if necessary, if routes change while the data are en route, and to resend any portions of the message that don't make it through. It works 24 hours a day, and will continue to try to send the data in a dynamically changing environment, even if intermediate hosts or the receiving party are temporarily offline. This essential flexibility is at the heart of packet-switching architectures and is incorporated into huge cooperative systems like the Internet. See circuit switching, Open Systems Interconnection, Systems Network Architecture, X.25.

packet switching network A communications network in which a channel is occupied only for the time during which the packet, a unit of data, is transmitted, a common distributed data network format. See Frame Relay.

packet tracing See packet sniffer.

pad connector Short for touchpad connector. A connector that enables a touchpad keyboard (with a flat surface rather than raised keys) or other flat input configuration to be attached to an electronic device (usually a computer or kiosk terminal).

PAG See Producers Advocacy Group.

page description language PDL. A means of providing commands to a system for the placement and formatting of page elements, such as text and graphics. Adobe PostScript is widely used, powerful page description language, and HTML is a very basic page description language extensively used to format information for viewing with a Web browser. Various printers include page description languages which are usually somewhere between PostScript and HTML in complexity.

pager 1. A general broadcasting loudspeaker connected to a phone or microphone, usually in a business, or educational or health care institution. See public address system. 2. A portable, wireless handheld device which can emit an audible, short verbal message or short alphanumeric message. These are often used by emergency workers, sales representatives, and business professionals. See paging.

paging Alerting a recipient that there is a message or item awaiting his or her attention. Public address systems can be used to page employees or clientele when packages are ready, when there is a phone call, or when lost children or items have been located or turned in. Pagers commonly known as *beepers* are portable wireless devices that will make an audible beeping sound to signal that a message or call is waiting, or that the user has to go to a certain location if paged. Portable wireless alphanumeric pagers can display a short message or telephone number to notify the user of a situation or phone message. Pagers are commonly used by professionals in the field, emergency workers, and industrial yard workers. See public address system, Short Message Service.

paging system PS. A system which allows a message to be broadcast broadly to anyone within range of the speaker, usually to attract the attention of a particular person or party, to give instructions, or to ask someone to pick up a message. Paging systems are common in hospitals, schools, and shopping malls. See public address system.

pair A pair of associated wires, often twisted together to facilitate electrical conductance and/or to reduce noise. Most phone networks are based upon decades-old circuits of twisted-pair copper wires. A pair of cables is used in a number of multimedia schemes, with one carrying sound and the other graphics, or one carrying sound and graphics data while the other carries timing information.

pair assignment The assigning of a specific current, transmission, or function to a twisted-pair wire. These are often designated with a code or color, in order to make interconnections quicker and less error-prone.

PAL 1. See phase alternate line. 2. See programmable array logic.

PALC plasma-addressed liquid crystal. See liquid crystal display, plasma display panel.

Paley, William S. An American experimenter and business tycoon who purchased and developed the Columbia Phonograph Broadcasting System (1927) into the Columbia Broadcasting System (CBS) in 1928. Under his leadership, the company grew and added new products and services to its line. In 1983 Paley retired from CBS, only to return 3 years later to work with Lawrence Tisch. In 1995 CBS was bought by Westinghouse.

In 1975, Paley established the Museum of Television and Radio in New York, an educational resource and archive of historical and culturally important broadcasts. The William S. Paley Foundation, Inc. has been established in his honor.

Palo Alto Research Center PARC. One of several Xerox research installations, PARC was founded in 1970 in the Stanford University Industrial Park. It is the site of many remarkable pioneer developments in the field of computers and telecommunications. The PARC was a hotbed in the 1970s for many original developments in object-oriented programming and computer interface design. Both Apple and Microsoft toured the facility in their early days and were inspired by their experiences there, particularly demonstrations of the Alto computer running Smalltalk applications. See Kay, Alan; Smalltalk.

PAM 1. payload assist module. A shuttle satellite deployment mechanism. The satellite in this context is considered the payload. 2. See port adapter module. 3. See pulse amplitude modulation.

panel switch A commercially successful electromechanical telephone switching system developed in the AT&T labs in 1921, based on Lorimer one-step selection concepts. It incorporated mechanical selectors to connect calls.

At the time the panel switch was introduced, independents were widely using the step-by-step switch developed a year earlier. The panel switch technology allowed customers to dial their own calls, albeit with a lot of noise in the early versions. The panel switch was widely used in the United States until the 1950s, when it was superseded by the crossbar switch, which had been developed in the late 1930s. See crossbar switch, Lorimer switch, rotary switch, step-by-step switch.

panoramic receiver A device used in radio communications which provides continuous monitoring of a specified band of frequencies. On a computer monitor, signals are displayed in graph form, with vertical blips moving horizontally along the X axis and amplitude graphed on the Y axis.

PAP 1. packet-level procedure. 2. See Public Access Profile.

paper tape An information storage medium. Paper designed to have specific areas of the tape encoded and punched or electrostatically recorded onto the tape, for subsequent reading by a paper tape reader or other interpretive device, such as a computer, stock ticker machine, player piano, or music box. This means of information encoding and storage was used to program early computers and had many characteristics in common with computer punch cards.

Early telegraph receivers used paper tape systems designed by inventors such as Bain and Morse. Later teletypewriter systems used tapes to save transmission time and money by being composed offline and sent only when complete. This also provided a way to correct significant errors before transmission, since a bad tape could always be repunched. Paper tapes have been superseded by tape drives, hard drives, floppy diskettes, magneto-optical discs, cartridges, and memory cards. See Bain, Alexander; Morse, Samuel F. B.

paper tape punch A device designed to receive or interpret coded information and translate it into physical locations on a paper tape and punch them accordingly.

paper tape reader A device which detects and translates the encoded holes in punched paper tape as the tape moves through the machine. The machine may be an interface to a display device, or may be self-contained. Older paper tape readers required that the holes be completely punched out and were usually read by optical means. Later machines could read semi-perforated or *chadless* tape, usually by means of physical sensors. See Hollerith code, paper tape punch, punch card.

PAR 1. Positive Acknowledgment Retransmit. 2. Precision Approach Radar.

parabola A plane curve that is frequently studied and described in various disciplines including physics, geometry, and art. Parabolic curves are observed in

P

the motion of objects and are used in the manufacture of reflectors and antennas. See parabolic antenna, parabolic reflector.

Parabolic Antenna Examples

feed horn
parabolic reflector
supporting struts

This roof-mounted parabolic antenna is about 8 feet across and uses a mesh parabola to reduce weight and wind resistance.

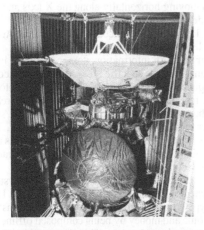

One of the bigger challenges in telecommunications is designing antennas and transmissions technologies that can communicate over vast distances in the inclement, radiation-high, temperature-fluctuating environments characteristic of space. The Cassini Saturn probe shown here in its testing phases, in 1996, is equipped on top with a parabolic radio antenna provided by the Italian Space Agency (ASI). [NASA/JPL image detail.]

---

parabolic antenna An antenna designed with a characteristic parabolic "dish" shape that captures a directional beam and focuses it, usually through a *feed horn*. This shape is especially appropriate for very short, directional transmission waves, such as microwaves, and the diameter of the antenna is designed to correspond with a multiple of the length of the wavelength being received. Parabolic antennas may be made from a variety of materials: solid metal, mesh metal, fiberglass. This style of antenna is commonly used for microwave satellite transmissions, though it is also used for some long-distance space applications, such as space probe communications. See antenna, feed horn, microwave antenna, low noise amplifier.

reflector An antenna, or other reflector, which utilizes the characteristics of the shape of a parabola to concentrate and direct reflecting beams. See parabola, parabolic antenna.

paradigm A clear or typical example, a standard, ideal, or archetype.

paradigm shift A fundamental, significant change in the way something is perceived or understood, particularly if it has been taken for granted, or assumed to be true for a long time, or by a majority of the population. In other words, the situation or thing itself has not changed, but our way of understanding it has.

A general paradigm shift occurred when humans, most of whom believed that the Earth was the center of the solar system and even the universe, acknowledged that the Earth revolves around the sun. The discovery that matter at the atomic level (quantum mechanics) did not behave according to accepted models of classical mechanics represented a paradigm shift in physics. Paradigm shifts often take a long time, sometimes decades or centuries (although transition periods are collapsing as education and television become widespread), and those who first propose new ideas and ways of looking at things are often pilloried or persecuted (even beaten to death or hanged) for their assertions. The suggestion that computers could be taught to be "intelligent," or to play games intelligently, was met with almost universal contempt in the 1960s and 1970s. In 1997, a computer beat a grandmaster chess player, an event that added credence to the argument that intelligent computers could be developed, and may someday surpass humans in specific or generalized intelligence, or develop machine intelligence of which humans are not capable.

parallel port An interface port on a computing system that permits the connection of parallel devices for the simultaneous transfer of data across multiple transmission wires. Most microcomputers are now standardized to 25-pin parallel D connectors, communicating with Centronics-compatible parallel protocols (although there are individual makers who use slight variations of the standard). Due to the increased speed of transmission over serial communications, parallel ports are commonly used for outputting to printers and other types of peripherals like cartridge drives. See serial port.

parallel processing Carrying out two or more tasks, more or less concurrently, usually with the intention of carrying out the processing at a faster speed, or otherwise more efficiently. See concurrent programming.

parameter A property which records, embodies, or determines a characteristic of an object or system. In communications, parameters affect many characteristics such as size, shape, speed, timing intervals, addresses, identities, etc.

parametric amplifier A type of low noise, radio-frequency amplifier which employs high-frequency alternating current (AC) for power. Used with microwave frequency electron beam devices.

parametric design The process of using general parameters, rather than individual measures, to automate computer-aided design and drafting (CAD) and computer-aided manufacturing (CAM).

USPTO Patent Resources

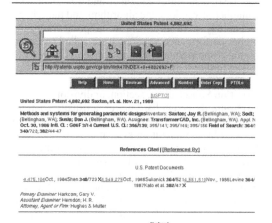

Abstract

A method for generating a parametric design on a computer without the use of a programming language. A drawing to create a master drawing from which other drawings of different dimensions can then be synthesized by modifying drawing. Instead of each dimension in the master drawing being given a fixed value, it is given a label. These labels run time by dimensions the user provides or dimensions which are calculated as described in a design plan. When run, the system uses simple language to prompt the user for each design value. It incorporates the response with entered into the design plan. If a response is unclear, a request is made for clarification. As a draftsman would do merges the design values with data from both the design plan and the master drawing to create a finished drawing parametric design, the user selects the controlling design plan to be used as a basis of the design. Acting on info

The U.S. Patent and Trademark Office is endeavoring to make patent and trademark information available online to the public. Patent abstracts, descriptions, and diagrams are currently available for recent patents and images are available for historic patents.

Parametric design incorporates a form of expert system and is particularly valuable in situations where many small variations on a basic design (bolts, boxes, modem covers, PC boards, telephone handsets, etc.) need to be designed and manufactured. In these cases, a computer program can be used to automate the design process, by providing guidelines, rather than single-part measurements, to turn out the many needed variations thousands of times faster than a CAD operator could draw each one by hand. One of the early patents for an applied parametric design computer program was awarded to Synthesis (OfficeCad), in Washington State, in the 1980s. It can be accessed online, along with other patents filed since the mid-1970s, on the U.S. Patent and Trademark Office's Web site. See CAD, expert system. http://patents.uspto.gov/

parametric equalizer A component device used in sound systems to selectively manipulate selected frequencies in order to adjust the sound, usually to suit the taste of the listener.

parasite An organism or process which feeds off another without providing a return. In technology, the term can refer to a process, or a mechanical or electrical device that monitors or uses transmissions clandestinely, or without the usual compensation to the provider of the transmission. Small wiretap devices are sometimes called parasites, especially if they draw their power from the line being tapped.

PARC, Xerox PARC See Palo Alto Research Center.

parity Equality, state of being the same, equivalent, matching.

parity bit A bit which is included in a transmission for error checking or status purposes. In telecommunications over a modem, most protocols allow the use of a parity bit appended to a data stream of a specified length, the parity bit set to zero or one, depending upon the preceding data. Parity values calculated and stored as the sent bits are checked against parity values calculated from the received bits. See parity checking.

parity checking A simple means of checking data integrity after a transmission by comparing the calculated value of the parity at the receiving end with the value calculated and stored at the sending end. Parity checking is very commonly used in file transfer through modems over phone lines.

First the transmitting and receiving ends negotiate a common protocol, for example, ZModem, then the parity setting is selected as odd or even (or none). Assume a parity setting of even for this example. Parity is calculated prior to sending, by tallying the ones or zeros in a group of bits (usually seven), and then assigning a parity value of *zero* if there is an even number of one bits and of *one* if there is an odd number of one bits, (or the converse, by looking at zero bits for odd parity). The sender transmits the data and its associated parity bit. The receiver calculates the parity of the received bits and checks to see if there is a match with the transmitted parity bit. If not, there is a problem.

The system is not foolproof; a match does not guarantee that the data were correctly transmitted, as the parity bit itself may have become altered along with the data, but mechanisms in most software evaluate the frequency of parity errors so that the user may be alerted and the transmission aborted, restarted from an earlier point, or resumed later, depending upon the protocol.

park drive In hard drives, "parking" the drive is a means to secure any moving mechanisms that may be damaged by being jiggled in transit. Some hard drives park automatically when not in use, and some use mechanisms to prevent damage if the unit is transported (e.g., drives in laptops). Older drives were often equipped with software-parking, and it was quite important to run the software command to park the drive before moving the system or removing the drive. This system is now uncommon. Mobile computers are equipped with self-parking drives.

park phone In telephony, parking is the process of putting a line through to a particular phone so that it can be picked up at another station, or to put a line on "soft hold" so the conversation can be continued from another phone.

park timeout In telephony, a time limit on a parked line after which it hangs up the line if the call is not resumed on another line (or the same line).

parking 1. In telephony, parking is the process of putting a line through to a particular phone so that it can be picked up at another station, or putting a line on "soft hold" so the conversation can be continued from another phone. 2. In computing, the process of securing a device so that any moving parts that might be damaged or that might damage other components are kept in check. For example, the read/write head of a hard disk drive could damage the head and the magnetic media if it skitters across the drive when it is moved. Parking the drive through software or built-in electronic sensors ensures that this type of damage doesn't occur. Depending upon the operating system and type of drive, a drive may have to be unparked or mounted after it has been parked. Older OSs and drives tend not to be auto-parking. It may be necessary to run a software utility to park the drive when powering down the system to remove the drive or to move the computer system. 3. In general consumer electronics, the process of tying down, bolting, or otherwise securing moving parts so they are not damaged or do not cause damage in transit. It is very common for laserdisc and compact disc players to have a transit bolt in the back to prevent the trays from moving around during shipment. Always take care to remove the transit bolt before use, save it by taping it to the manual or the underside of the player, and always reinsert the same bolt (the length may be important) before moving the equipment, especially if it is being shipped by a third party.

Parkinson's law C. Northcote Parkinson wrote in the 1950s that work expands to fill the time available for its completion. (For those who are perfectionists, and believe that if a job is worth doing, it's worth doing right, this is doubly true.)

partial response/maximum likelihood PRML. PRML is a means of digitally encoding analog data and reconstructing it. This concept can be applied to many aspects of technology but has been particularly useful in the development of improved data broadcasting and storage devices.

PRML-technology hard drives, for example, can provide high disc capacities and faster transfer rates than earlier technologies based on peak detection (the detection of voltage spikes resulting from magnetic flux reversals). Peak voltages become harder to detect as data are more densely packed (the peaks become difficult to distinguish from the noise). To overcome this problem, a new approach, based on digitally sampling the analog signal, was developed.

Vendors like Seagate have applied a PRML digital sampling and data reconstruction approach to the development of high-capacity drives that use partial response (PR) in magnetoresistive (MR) hard drive heads. These heads detect and sample an analog signal prior to Viterbi detection decoding. Together, PR and MR eliminate overhead in the electronic equalization (undershoot filtering) process, freeing up as much as an additional storage space. Maximum likelihood (ML) is used in the conversion of analog waveforms into digital data. Through Viterbi detection, all possible combinations of data are checked for the best match of least error with the incoming data. The assumption is that the least error pattern will most likely be correct and, in practice, it works quite well. Together PR and ML enable faster data transfer rates

through run length limited (RLL) coding and significant areal density increases over peak detection methods are possible.

As PRML technology caught on and was adopted by a number of vendors, the algorithms and underlying technology were improved to the point where the higher performance versions were called extended partial response/maximum likelihood (EPRML) to reflect further significant improvements over the earlier PRML drives. See Super DLTtape.

partition Subset, class, section, or division.

partition, drive On hard drives, a usually contiguous section of a disk individually initialized and handled by the operating system as a distinct unit. Some systems can format the individual partitions in a variety of formats, i.e., a 1-Gbyte hard drive with a NeXTStep 400-Mbyte volume on one partition, a 400-Mbyte Linux volume on another, and a 200-Mbyte Macintosh volume on a third, all recognized by the OS and readable/writable without any unusual technical expertise or demands upon the user.

On many microcomputer operating systems, disk volumes and files cannot cross partitions, but many Unix and workstation operating systems can handle volumes that cross partitions transparently to the user, e.g., two 500-Mbyte hard drives used together might appear to the user as a 1-Gbyte virtual drive. There are many schools of thought as to whether a hard drive needs to be partitioned. A few operating systems can only handle up to four partitions, each with up to 2 Gbytes of space and, consequently, a larger hard drive must be sectioned into smaller pieces in order to be handled by the operating system. Others don't have this limitation on the number of partitions, and can manage larger-sized partitions. In terms of disk management, in the case of problems, it may be easier to rebuild partitions or handle data recovery procedures if there are several partitions rather than just one. Redundant array drives are another way of handling error recovery. Often a small 200- to 500-Mbyte partition will be set aside as a "swap drive" and not used for other purposes. See RAID.

partition, memory In computer memory, a linked or contiguous section, separate from other sections, that is allocated for a specific purpose or process, such as video display or frame buffering.

party 1. One of the individuals in a transaction. A common legal term used to stipulate an individual or organizational entity. To be *party to* a transaction is to listen in or participate. In telecommunications, the transaction might be a telephone call, a conversation, or a computer communication.

party line In telephony, a line shared by two or more subscribers, so if one or more subscribers pick up the line and listen when someone else is engaged in the call, they can hear the conversation, and can't make further calls until the current conversation is disconnected. Party lines were very common on older shared phone circuits until the 1960s; they are now uncommon in North America. On ISDN lines and Frame Relay networks, a sort of party line system exists, but is rarely a hindrance to the user, unless too many subscribers are assigned to the line.

party line, following Following the *party line* is a phrase from politics that indicates acceptance and promotion of the administration's point of view. The administration might be a political party, a business entity, or other institution. It is sometimes used as a derogatory phrase for ambitious compliance, or for a person who doesn't think for him- or herself, but promotes the current popular point of view.

PAS 1. see profile alignment system. 2. See Priority Access Service.

pascal An SI unit of pressure equal to one newton per square meter.

Pascal A programming language descended from ALGOL, developed by Niklaus Wirth in 1970. Pascal became especially popular in the 1980s for teaching programming concepts and techniques. A structured, typed language, Pascal is somewhat similar to Modula II, and fits somewhere between C and higher level languages like BASIC and FORTRAN. It is less cryptic than C, but also less preferred by programmers in commercial development environments, yet is generally preferred over the less structured BASIC in educational environments. See Modula II, C.

Pascal, Blaise (1623–1662) A talented French inventor and mathematician, Pascal devised one of the earliest calculators, a "Pascaline," while still in his teens. It was a numerical base ten, movable dial, wheel calculator designed to assist his father in carrying out his duties as a tax collector. Pascal appears to have come up with the design independently, and probably was not aware of the earlier calculator developed by Schickard at about the time of Pascal's birth. Pascal also did research in fluid dynamics. See Schickard, Wilhelm.

pass through *v.* To move through a component device or leg of a network without significantly altering the characteristics of that which has just been passed through, or without being altered by that which is passed through. See passthrough device, tunneling.

passband 1. The range of transmission frequencies that can pass through a filter without a significant decrease in amplitude (attenuation). A passband filter allows selective screening out of irrelevant or undesired frequencies in order to create a device for a specific purpose, or to simplify its operation. 2. A signal that loses no spectral energy at direct currents (DC), unlike a baseband signal. A Manchester-encoded signal is one example of a passband signal.

passthrough device 1. A device chained between two other devices, which passes data through without changing them. For example, an external memory module might be attached to a computer, with an external hard drive attached to the memory module. The memory module passes through the hard drive signals in such a way that the hard drive works just as though it were directly attached to the computer. See daisy chain. 2. A device that provides access to and passes back the signals transmitted by another. Sometimes used as a diagnostic tool.

password A word or combination of characters which, when provided by a person or entity wishing to gain entry to a system or situation, is checked against certain characteristics, or a list of those who are authorized to have access. If a match is found, entry is permitted. Password protection systems are very common on computers and networks. It is very unwise to tape passwords to monitors or desks where anyone can see them. It is also unwise to use common words as passwords; a moderately long password with a combination of letters and symbols is safer. See anonymous FTP, back door, back porch.

patch *v.* To connect one circuit with another, usually through an intermediate line. For example, on old telephone switchboards, the operator would *patch* through a call by taking a jack connected at the other end to the main switchboard, and plugging it into the phone receptacle for the individual getting the call. A patch is a temporary connection, one subject to frequent change or used for diagnostic purposes.

patch, software *n.* In software, a patch is a piece of code that is inserted into the original code to override some of the original programming, or to add capabilities or data which weren't in the original code and perhaps should have been. A patch is distinguished from an upgrade in that it typically is intended to correct oversights or errors, whereas an upgrade is usually of greater scope, intended to enhance or extend the capabilities of the program. In many products, the two are combined.

patch, sound In electronic music, a sampled segment of sound stored digitally. The sound is measured and recorded, that is, "quantized," at rapid intervals in order to create a digital impression of the analog sound wave. For the most part, the more frequent the sampling, up to the limits of human perception, the more true to the original the sample tends to sound (the capabilities of the playback mechanism contribute as well). Sound patches can be generated by and used with many commercial sound synthesizers and computer synthesizer software. MIDI is a common protocol used in the music industry for communicating digitized sound between MIDI-compatible instruments and software programs. Speech and music sound patches are often used to enhance multimedia CD-ROM educational and entertainment products. More recently, messages composed from speech patches are becoming common on the Web. See quantize, sampling.

patch bay, patch board A hardware panel designed with multiple connecting ports such that the configuration of the patched in cables can be readily changed. In other words, it is set up so that temporary circuits, or those which are frequently changed, can easily be rewired. Dishwasher-sized patch bays are often equipped with wheels so they can be moved in or out of a work area, and usually have receptacles or terminals for easy insertion and removal of patch cords and/or wires. Patch boards are useful for prototyping, monitoring, and testing new circuit layouts. See patch panel.

patch cord A short length of wire or cable used to connect circuits. The connectors at either end vary, but are often RCA jacks or BNC connectors. Patch

P

cords are commonly used with patch bays, patch panels, and electronics components. Videographers and musicians often refer to video and audio connecting cables as patch cords, since video equipment connections are frequently reconfigured.

patch panel A hardware device, often wall-mounted, that facilitates the connection and reconfiguration of temporary circuits. Patch panels are often mounted in areas that are accessible to technicians but not to casual passersby, such as in maintenance closets. A patch panel may resemble a distribution frame, in that it has a grid of openings or connectors through which circuits can be routed. It commonly has mounted receptacles to match the types of jacks used in that particular circuit.

Fiber optic patch panels are typically made with greater precision and strain relief assemblies than CAT-5 Ethernet panels, for example, because the integrity of the coupling in fiber optic connections is crucial to its effectiveness.

Patch Panel - Fiber Optic

Patch panels facilitate the coupling and quick reorganization of wire or fiber optic connections. The panels are commonly installed on walls or in wiring closets where casual passersby can't knock or tamper with the connectors. This panel illustrates banks of common SC- (top) and ST- fiber optic connecting ports. Connectors may have locking mechanisms with alignment slots and knobs (e.g., bayonet mounts) to prevent accidental disconnection

Depending upon the type of panel and connectors, the panel may be passive, linking the connections straight across, or active, with electronics built in to the panel to influence the signals (e.g., providing amplification).

Patel, C. Kumar N. (1938–) An Indian-born American physicist with an interest in optics, especially molecular spectroscopy and laser systems. Patel studied in India and at Stanford University, then began his career at Bell Labs in 1961 where he became Executive Director of Research, Materials Science, Engineering and Academic Affairs, a position he held until 1993. He then served as Vice Chancellor of Research at UCLA until 1999 and as President & CEO of Pranalvtica.

Patel holds over 30 patents and has received many honors, including the OSA Lomb Medal and the National Medal of Science (1996) for the invention of

the first nitrogen CO_2 laser in 1964. He was also instrumental in the development of the Raman laser, a tunable laser demonstrated in 1969.

Patel has continued to actively pursue applications of lasers. In June 1997 he and his colleagues submitted a patent application for an optical bit rate converter suitable for time division multiplexed (TDM) multiaccess communications networks. See Javan, Ali.

patent A registration process formally established in the United States in April 1790 which provides a record of the ownership, development, and date and method of creation of unique products and processes. The first American patent was granted on 31 July 1790. By 1802, applications had increased to the point where a separate Patent Office was set up, and more rigorous scrutiny was established by 1836.

In the United States, the documents are processed and stored in a central government repository that is open to the public and intended to further technological progress by the encouragement of the dissemination of ideas. Japanese patents have been available over networks for some time now, and recent U.S. patents are now searchable on the Web through the U.S. Patent and Trademark Office site. The Clinton Administration announced on June 25, 1998 that over 20 million pages of patent and trademark information would be provided free to the public on the Internet by year's end. Content is supplied through the Commerce Department's large database of text and images. The collection will include the full text of 2 million patents dating from 1976, 800,000 trademarks and 300,000 pending registrations dating from the 1800s. Tiff images are available for historic patents. Patent applications must follow very specific format and content guidelines laid out by the patent office. Patent registration grants exclusive intellectual and certain commercialization protections to the inventor for a term of 17 years in the U.S. (international patents are similar). In cases of others coming up with the same idea simultaneously or previously, without knowledge that the idea has been patented, preference for the idea now goes to the inventor who first is granted the patent. This is a change from historical procedures in which an earlier inventor, if she or he had documents to prove the case, could have a patent from a later inventor overturned.

Many people incorrectly assume that the patent process exists to explicitly prevent others from infringing on patents, but it is the responsibility of the patent owners, not the patent office, to police the use and abuse of patented ideas. The patent does, however, define the nature and extent of the legal protection available to the inventor through the justice system. Granting of a patent does not include granting of a right to manufacture a product incorporating the idea, since other patents for other aspects of the invention may exist.

The most important aspect of the patent and the submission of patent applications is the *Claims* section, in which the inventor lays out, in point form, the characteristics which make the invention *unique* and *nonobvious*. Some or all of these claims may be accepted by the patent office, and the document is critiqued and rejected or returned to the applicant for revisions. Since uniqueness is often evaluated in a historical context in the *Prior Art* section, historical antecedents and current similar inventions must be described by the applicant thoroughly and succinctly. The invention must also be more than a half-baked idea, since the patent application must include a clear description of how to build or otherwise recreate the invention itself, without undue difficulty to a layperson or someone appropriately skilled in the area of specialization appropriate to a specialized invention. Hardware patents usually fall under the products category and software patents under the process category. Note that patents, copyrights, and other legal registration procedures may grant ownership to the *employer* of the inventor rather than the inventor, if the employee undertook the invention in the course of his or her normal work hours or duties.

One of the most famous patent clerks in history was Albert Einstein, who worked as a junior clerk in the Swiss Patent Office when unable to find work as a teacher or research scientist. While working there, he wrote some of his most startling, insightful treatises on relativity. See copyright, trademark.
http://www.uspto.gov/

Paterson, Tim Paterson developed a simple but historically important disk operating system for Seattle Computer Products in the late 1970s. The product was derived from Gary Kildall's CP/M operating system, which was the most successful and well known at the time, with over half a million copies distributed. Paterson created a basic operating system called QDOS (Quick and Dirty Operating System) which he has stated was derived in part from the program interface described in a CPM manual from the mid-1970s. Microsoft bought it, fixed it up a little, and provided it to IBM soon after. IBM released it initially as PC DOS 1.0. Meanwhile, Seattle Computer Products retained the rights to QDOS. Microsoft subsequently bought out all QDOS distribution rights for $50,000. The Microsoft financial empire essentially sprung from this transaction as the product was developed into MS-DOS and, eventually, after many facelifts and enhancements, evolved into Windows. See Digital Research; Kildall, Gary; Microsoft Corporation.

path A route, track, directional identifier, runway, conduit, or other end-to-end, hop-to-hop, or as-you-go means of delineating the track followed by a person, process, transmission, or data unit while traveling from one point, node, or endpoint to another. A file path is one which indicates the hierarchical organization and location of a specific file or grouping of files. A transmissions path is the specific or general direction of radiant energy travel.

path information unit PIU. In packet networking, a message unit consisting of a transmission header (TM) or a transmission header combined with a following basic information unit (BIU) or segment. See datagram.

Path Terminating Element See SONET path terminating element.

pattern matching, pattern recognition The process of comparing text, symbols, images, or other elements to determine whether they are the same, similar, or mathematically equal. The process of pattern matching is widely used in database search and analysis mechanisms, and its cousin, pattern recognition, is common to artificial intelligence applications including expert systems, robotics, and others.

Pattern recognition was in its infancy in the late 1950s and early 1960s, when computing systems were expensive, cumbersome, and programmed with punch cards. Nevertheless, early researchers at the time, sensing its potential, developed equipment and algorithms which could read a few handwritten letters, if they were plainly written. See Perceptrons.

pay phone, pay telephone See payphone.

payload The user information, and sometimes accounting and network administration information, carried in the upper layers in a layered architecture, within a cell, frame, packet, or other network data transmission unit. Separate from but associated with the payload, there is frequently signaling, header, error checking, and other data which relate more to the type and manner of transmission than to the information content from the user or process sending the transmission.

Payload Data Segment PDS. In communications satellites, the data services that are made available to authorized users. See Unified User Interface.

Payment Extension Protocol PEP. An HTTP payment extension protocol described by the JEPI project in August 1996 in conjunction with seven examples of the Universal Payment Preamble (UPP) that could be used over PEP. The purpose of the system was to develop a practical, automatable payment system for running over the widely distributed HTTP applications on the Web. PEP enables UPP to be embedded in HTTP to support Web client/server payment transactions. Examples of basic payment mechanisms include queries to determine what types of payment forms are supported, presentation of payment options, demand payment options, payment acceptance/rejection, and payment option queries. See e-commerce, JEPI, Universal Payment Preamble.

payphone, paystation phone Any self-contained public or private telephone unit that requires a per-call or per-minute fee, usually directly transacted with the phone. Although some human-operated stations exist most require payment by coin or stripe card.

The first pay telephones were attended by operators who collected the fees for the calls. One of the early coin box patents was issued in 1885, and William Gray installed a public coin phone in Connecticut in 1889 while employed by Pratt & Whitney.

payphone postpay Payphone calls paid after completion, usually with a calling card or credit card.

payphone, private Also known as COCOT, this is a customer-owned coin-operated phone, as might be found in a hotel lobby or tavern. COCOTs may provide only limited access to long-distance carriers.

PBX Private Branch Exchange. See Private Automatic Branch Exchange.

PC 1. personal computer. While many people use PC to refer specifically to IBM and third-party licensed hardware, in fact, PC correctly refers not only to Intel/IBM computers, but also to any personal computer or microcomputer priced in a consumer or small business price range. 2. printed circuit. See printed circuit board. 3. program counter. 4. protocol control.

PC card See PCMCIA card.

PC, IBM/Intel *colloq.* In general and marketing terms, PC is understood as a subset of personal computers, consisting of Intel-based IBM hardware, or third-party licensed hardware, running the IBM OS/2 software or, more commonly, running the Microsoft Windows graphical operating environment in conjunction with MS-DOS. For information on specific IBM desktop computers, see the listings under IBM Personal Computers.

PCA 1. point of closest approach. In a satellite communications system, a point on a segment of the orbit or ground track when the satellite is closest to a specific ground station. 2. Premises Cabling Association. 3. protective connecting arrangement. Commercial connecting rental agreement, required by AT&T/Bell prior to divestiture for telecommunications devices that were not AT&T/Bell, were connected to the AT&T/Bell system. See Carterfone decision.

PCB 1. power control box. 2. power control box. 3. process control block. 4. See printed circuit board. 5. protocol control block (in TCP and similar network protocols).

PCCA See Portable Computer and Communications Association.

PCF See photonic crystal fiber.

PCI 1. See Peripheral Connect Interface. 2. Protocol Control Information.

PCIA Personal Communications Industry Association. Formerly known as Telocator, PCIA is a national association representing the mobile communications industry.

PCL See printer control language.

PCM 1. See phase conjugation mirror. 2. See pulse code modulation.

PCMCIA Personal Computer Memory Card Interface Association. A professional association of electronics peripherals and semiconductor manufacturers and software engineers. See PCMCIA card, PCMCIA standards.

PCMCIA card, PC card A standardized computer peripheral card format, not much bigger than a fat wallet card, which is commonly used in portable computing applications. PCMCIA cards (since the mid-1990s called PC cards because it's easier to say) are microminiaturized devices with a thin edge connector, including memory cards, hard drive cards, fax/modem cards, network interface hookups, and more. They are used in radio phones, laptop and palmtop computers, digital cameras and camcorders, and various other portable electronic devices. The most common cards are called Type I or Type II (Type III is less common, and Type IV is vendor-specific). Most laptop peripherals use Type II cards, and it pays to have at least one Type II slot on a portable computer.

Hard drives and radio devices tend to use the thicker Type III cards.

PCMCIA standards A set of 8-bit bus standards which bears the same name as the organization which developed the standards, the Personal Computer Memory Card Interface Association. PCMCIA standards were developed and tested in the late 1980s and released for general use in 1991. While there is fairly good adherence to the standards, compatibility is not absolute. It's advisable to try cards before buying them, or to get them with a good return policy. The set of standards includes Type I, Type II, Type III, and Type IV. See PCMCIA card.

PCMIA Personal Computer Manufacturer Interface Adaptor.

PCN See Personal Communication Network.

PCO See Private Cable Operator.

PCP 1. See Predictor Compression Protocol. 2. Private Carrier Paging.

PCR 1. See peak cell rate. 2. See phase change rewritable. 3. problem/change report. 4. See processor configuration register. 5. See Program Clock Reference.

PCS 1. See Personal Communications Service. 2. personal communications software.

PD See phase drive.

PDA See PDA microbrowser, PDA macrobrowser, Personal Digital Assistant.

Web-Compatible PDA Macrobrowser

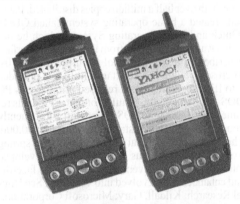

A full color, graphics-capable Palm personal digital assistant (PDA) connected via radio link to the Internet and installed with the SoftSource/Catarra HTML-compatible macrobrowser client, which provides the full-featured Web surfing of a desktop computer, with a pen to scroll, click, enlarge, or carry out other typical browser actions. This fully HTML-compatible combination of hardware and software has many advantages over more limited microbrowsers and WAP technologies, including the convenience of accessing standard Web pages and security features rather than requiring special software designed for WAP-based devices. WAP is appropriate for limited resource devices, but for full Web browsing, the SoftSource/Catarra server/client combination is currently the only product that provides unrestricted graphical Web-surfing on PDA mobile devices.

PDA macrobrowser A software application providing full Web compatibility on limited-resource portable devices through quick conversion and display algorithms transparent to users, Web page designers, and security providers. Macrobrowsers include display software technology installed on a Personal Digital Assistant (PDA) working in conjunction with a proxy server. Thus, unlike microbrowsers, information providers don't have to maintain two sets of Web pages or a separate type of digital certificate – a macrobrowser supports the existing Web infrastructure. Macrobrowsers began appearing in 2001 as PDAs with better memory and display technologies were released. They will likely co-exist with microbrowsers for a while, but may supersede them due to the improving power and resolution of handheld wireless devices and the relative ease of implementing Web pages and security features compatible with macrobrowser-enabled devices. See PDA microbrowser, SoftSource, Wireless Application Protocol.

PDA microbrowser A software application designed to provide limited Web access compatibility for constrained-resource devices. Microbrowsers began to appear on the market in 1999. In general, they implement limited-set, proprietary, or specialized adaptations of current Web browsing languages to run effectively on Personal Digital Assistant (PDA) devices with limited memory and display resolutions. The most common strategy for microbrowsers is to run a Web proxy gateway and to adapt Web pages to the limited-set languages compatible with these devices, through simplified/specialized markup languages. Microbrowsers are a reasonable way to support constrained-environment handheld computing devices, but put an extra burden of time and expense on Web page designers, programmers, and security providers, as two sets of Web pages and digital certificates must be maintained if microbrowsers and regular HTML browsers are to be supported on the same Web site. Microbrowsers will remain viable for a while for limited-resource devices but may eventually be superseded by PDA macrobrowsers for general purpose browsing, as handheld devices become more powerful and feature-rich. See CHTML, PDA macrobrowser, Personal Digital Assistant, WAP Forum.

PDC See Personal Digital Cellular.

PDF See Portal Document Format.

PDL See page description language.

PDP 1. See plasma display panel. 2. See power distribution panel.

PDP- A series of Digital Equipment Corporation (DEC) Programmed Data Processors (PDPs) that would today be called *minicomputers*. Right from the beginning, the PDP-x series filled a need for powerful, smaller-scale, lower-cost computers. The PDP-1 sold for only a tenth of the price of many computer behemoth mainframes. This made the PDP- line marketable to educational institutions and businesses that couldn't afford million-dollar computing systems and they became very popular with computer science students, with many user (and hacker) groups springing up around the machines.

DEC's first computer, the 18-bit PDP-1 was released in 1960 and a PDP-3 was built by a DEC client. The PDP-4 and PDP-5 followed in 1962 and 1963 and DEC released about one a year from then on. By the mid-1960s, DEC had launched a desktop model of the PDP-8. To be useful, it needed lots of peripherals, and its price was far beyond the range of personal computer owners, but for under $20,000, it was a transitional machine to the smaller scale mini- and, eventually, desktop computers of the late 1970s and beyond. By the early 1970s, the series was up to the PDP-16, but many purchasers were still using PDP-8, PDP-10, and PDP-11 machines into the 1980s. The PDP-x series was gradually superseded by DEC's VAX computers in the mid-1980s, but hobbyists still like to pick up the PDP-x computers at auctions and computer salvages. See VAX.

PDS See Payload Data Segment.

PDU See Protocol Data Unit.

PDUS Primary Data User Station. The combination of a ground station and a satellite image processing system.

peak cell rate PCR. In ATM networking, a traffic flow measure that describes the upper cell rate limit, which may not be exceeded by the sender. See cell rate.

peer entities In layer-oriented network models, entities within the same layer, usually diagramed and visualized as horizontally related.

peer model A networking model built with the assumption that internetwork layer addresses can be mapped onto ATM addresses, and vice versa, and reachability information between ATM routing and internetwork layer routing can be exchanged. See integrated model.

peering The voluntary exchange of routing announcements in order to effectively establish data paths among providers.

PEG Regulated Public, Educational or Government access. See cable access.

Pel See picture element.

Pender, John (ca. 1860–1896) A British merchant who succeeded in establishing ambitious historic telegraph cable links between Western Europe and the Far East and Australia. In 1856, John Pender became a director in the Atlantic Telegraph Company and thus was involved in the first transatlantic cable enterprise spearheaded by Cyrus W. Field. Pender's subsequent ventures indicate that he was inspired by the success of the Atlantic telegraph cable installation and its future economic impact.

Wanting to get in on the ground floor of the new industry, in 1864, Pender formed the Telegraph Construction and Maintenance Company (Telcon), foreseeing the future need for cable manufacture and maintenance. Not satisfied with this alone, however, he then founded the British Indian Submarine Telegraph Company, in 1869, with the goal of linking Britain and India. For building the local Falmouth link, Pender formed the Falmouth-Gibralter-Malta Telegraph Company but soon changed it to the more generic Eastern Telegraph Company as plans and

P

locations changed as the project advanced and the port location was changed to Porthcurnow (later Porthcurno – PK).

By 1879 Pender was not only a significant telecommunications magnate, but had realized the remarkable feat of building a telegraph cable link between London, England, and Bombay, India.

In 1882, Douro mail steamer and passenger ship sank off Cape Finisterre following a collision with a Spanish steamer. The John Pender telegraph ship, part of Eastern Telegraph's fleet, was nearby and took off a number of passengers, providing them with basic needs and enabling them to telegraph their loved ones with reassurances. See Field, Cyrus West; Porthcurno.

penetration Gaining access to a system, circuit, facility, or operation, usually for security reasons or unlawful access. Physical penetration of circuits or networks can be done through means of taps or black boxes. Logical penetration can be done through password-guessing, Trojan horses, viruses, and back doors. Bodily penetration can be done through overriding electronic security measures, entering as an impostor, or using insider privileged access in an unethical manner. See back door, Trojan horse.

penetration tap 1. Any means by which a conductor is accessed by piercing the outer layers of shielding and grounding and connecting to the current circuit, with the intention of not disrupting current transmissions. 2. A network connection technique which enables devices to be attached to the network cable without interrupting current network operation. The tap is carried out with a sharp tool which can pierce the outer and inner ground shielding of the network cable, such as a coaxial cable commonly used in Ethernet implementations.

penetration testing Testing a system for the integrity of its security. This is sometimes done by internal staff, contractors installing the security measures, or outside experts hired to try to penetrate the system. In the telephone and computing worlds, known "hackers" are sometimes hired to try to penetrate a system to try to identify security holes before the system is opened up to employees or the public, depending upon its nature. In 1998 it was found that cash cards, which were generally considered reasonably safe from decryption and unauthorized use, could be penetrated by measuring their electrical emanations and properties, a finding that calls into question the use of cash cards in place of traditional means of currency exchange.

Pentium An Intel Corporation 80586-based central processing unit (CPU), designed to succeed the 80486, introduced in 1993. Originally released at 66 MHz clock speed, other versions came out, including a 100 MHz version with a 16-bit cache and a 64-bit memory interface, and eight 32-bit general-purpose processing registers. The name is derived from the "5" in the processor line 80x86 due to a court ruling that a number cannot be trademarked.

Pentium II An Intel Corporation central processing unit (CPU) similar to the Pentium Pro. Unlike the Pentium Pro, which incorporates the level 2 (L2) cache into the chip with the CPU, the Pentium II operates with a cache inserted in a slot on the motherboard, thereby increasing the amount of time it takes for the two to communicate. It also incorporates MMX circuitry intended to improve graphics and multimedia-related operations.

Pentium MMX Pentium Multimedia Extension, Pentium Matrix Math Extension. The MMX is essentially a Pentium Pro chip enhanced with a number of new data types and floating point instructions that enhance computing-intensive operations such as graphics. Applications are becoming increasingly visual in nature, with more graphical user interfaces, image processing, rendering and raytracing, videoconferencing, realtime games, and virtual reality applications, so support for commonly executed graphics and math-intensive computing processes on the chip is intended to support these growing areas of interest. Also, by incorporating capabilities similar to those supplied by direct memory access (DSP), Intel can reduce its reliance on the DSP technologies of other vendors

The Pentium MMX incorporates what Intel calls Single Instruction, Multiple Data (SIMD) techniques to allow several processes to be carried out with a single instruction. See Pentium II, reduced instruction set computing.

Pentium Pro, P6 An Intel Corporation 80686 central processing unit (CPU) in the Pentium brand name line, introduced in 1995 as a successor to the Pentium processor. The Pentium Pro originally shipped as a 133-MHz CPU and shares a number of commonalities with the Pentium, including a 64-bit memory interface. It is a two-part chip in the sense that it has a CPU and a level 1 memory cache, plus a level 2 (L2) memory cache layered into the CPU rather than residing separately on the motherboard. It is a hybrid chip with an underlying RISC structure, but also includes a CISC-RISC translator for downward compatibility. The clock speed of the first version was 133-MHz, with other versions following.

People's Communication Charter A global movement by a number of international communications associations to demand the protection of the quality of communication services, their accessibility, affordability, and ease of use by the public in order to safeguard basic human rights. The Charter grew out of concerns about new and existing communication technologies conscripted around the world by self-interested governments or allocated preferentially to private parties for use as conduits for information delivery for propaganda or for-profit ventures, at the expense of communications supporting education, community needs, and civil rights. See Milan Declaration on Communication and Human Rights. http://www.waag.org/pcc

PEP 1. Packetized Ensemble Protocol. A high-speed, proprietary, full duplex transmission protocol from Telebit. It has error-correcting mechanisms and is said to handle line noise well; it is no longer in general use. 2. See Payment Extension Protocol. 3. See Public Exchange Point.

PER See Packed Encoding Rules.

Percent Local Usage PLU. A measure of telecommunications usage by time. PLU is a ratio of the local minutes to the sum of local and intraLATA long-distance minutes between exchange carriers, sent over Local Interconnection Trunks. Switched access and transiting calls are not included.

Perceptrons Self-organizing, pattern recognition systems built in the early 1960s at Cornell University. These systems were rudimentary, barely managing to recognize simple letters, yet studies and experiments of this kind led to the optical character recognition and handwriting recognition systems we now take for granted.

At the same time, at the Massachusetts Institute of Technology (MIT), researchers were developing pattern-matching systems for medical diagnosis, with a system designed to screen for cancer cells through a microscope. See neural networks, pattern matching.

perforator A tool to make a hole, to penetrate a substance, to punch an opening. Common three-hole punches are perforators. Electronic perforators are widely used to turn electronic signals into code pattern holes in punch cards and paper tapes. See chad, Hollerith card, punch card.

performance category See category of performance.

perigee The point in an orbit nearest the gravitational center of the body being orbited. See apogee, periapsis.

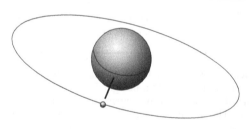

The perigee is the closest point of an orbiting object to the body being orbited.

period 1. Cycle, interval of time, portion of time encompassing a distinct culture (historical period). 2. Geologic time division that is part of an era and longer than an epoch. 3. The time interval between two consecutive orbits of a satellite through a specific point (usually the perigee) in the orbit. 4. In electronics, one interval of a regular, repeating event.

Peripheral Component Interconnect, Peripheral Connect Interface PCI. A very popular local bus standard developed by Intel in the early 1990s to support 32/64 bit data that was compatible with the new Pentium processors coming out at the time. It was designed with a newer chipset, to improve on the ISAs and VLBs that were then common, and to include bus mastering (use of the system bus). Since PCI's development, PCI slots have become common in Apple Macintosh and Intel-based IBM-licensed machines, along with upgraded versions of the VESA VL bus. The PCI Mezzanine Card (IEEE P1386.1) was designed to work with the PCI specification.

peripheral device 1. A piece of equipment which is not a main component of a system, but which, when connected to that system, enhances its functionality, speed, or storage capabilities. Peripheral devices generally cannot perform useful functions unless connected to the main system. Monitors, speakers, keyboards, scanners, video cameras, and printers are examples of peripheral devices. CD-ROM drives are an exception in that some are now designed to play audio CDs even if not connected to a computer. 2. In the telephone industry peripherals may also be called outboard processors, applications processors, or adjunct processors.

Perl Practical extraction and reporting language. A powerful, flexible, general-purpose, interpreted scripting language (originally spelled with a lower-case "p") developed by Larry Wall in 1986, and now extensively used for platform-independent scripting on multiple platforms on the Internet. The syntactical structure of Perl is quite remarkable (perhaps owing to Larry Wall's expertise as a linguist); useful, powerful routines can be written in a few lines or sometimes even in a few characters. An important tool for shell scripting, Common Gateway Interface (CGI) development, and much more. When combined with Penguin, it may be a serious contender with Java for object-oriented, Web-related interface design and Automation. The Perl Journal gives practical assistance to Perl programmers.

Permanent Number Portability PNP. A way for a telecommunications subscriber to transparently maintain an existing telephone number, even if changing to a different service provider in the same locality.

permanent virtual connection, permanent virtual circuit PVC. A logical communications channel (which may differ from the physical topology over which it is laid) established to stay the same for some time. In an ATM environment, there are two types of PVCs: permanent virtual path connections (PVPCs) and permanent virtual channel connections (PVCCs). PVCs provide manually configured connections between end systems. The addressing information, Virtual Path Identifier/Virtual Channel Identifier (VPI/VCI), must be put into both devices for connectivity. 2. In Frame Relay networks, a PVC is a logical link, with network management-defined endpoints and Class of Service (CoS). The link consists of an originating element address, data link control identifier, terminating element address, and termination data link control identifier. See RFC 1577.

permeability The porosity or penetrability of a substance. The degree to which liquids or gases can pass through a substance. Contrast with reluctance.

permeability, magnetic The property of a magnetizable material that determines the degree to which it will modify the magnetic flux in a region it occupies within a magnetic field. See magnetic field.

persistence 1. Perseverance, endurance, running the course, keeping on or with, tending to continue. (A quality essential to writing a reference of this magnitude, since documenting the telecommunications industry is like trying to gas up a car that's driving away at full speed.) 2. The tendency to continue a signal, echo, electrical charge, or data transmission after the actual communication has ceased or the message part has been received. 3. In a phosphor display system, the tendency of the phosphors to continue to fluoresce after the stimulus has stopped. This may be an undesired property, causing smear, or may be a desired property, enabling the image to remain viewable while the rest of the frame is being imaged.

persistence of vision A phrase that describes the way in which human visual perception "holds" an image for a brief moment, about a tenth of a second, even if the objects in the visual field have changed or moved. Thus, humans can only scan or perceive still images up to a speed of about 24 to 60 frames per second. Faster than that and they are no longer seen as still images, but as a series of moving or related images, especially if the forms in the images are closely related to the previous ones. Researchers Muensterberg and Wertheimer demonstrated in the early 1900s that this was a property of brain processing and perception more than a physical property of the retina. These characteristics of visual perception have greatly influenced the design and development of moving visual communications technologies. See frame, scan lines.

Personal Communication Network PCN. See Global System for Mobile Communications for the background and technology base for PCN. PCN was developed, starting in the late 1980s, as a modified form of GSM operating in the 1800-MHz frequency band (GSM is 900 MHz). It has smaller cell sizes, requires lower power, and is optimized to handle higher density traffic than GSM, but otherwise is essentially the same. The PCN standard was finalized in 1991. It is primarily used in the United Kingdom. See Global System for Mobile Communications.

Personal Communications Service. PCS. A low-power, higher frequency, standards-based, wireless mobile communications system, operating in the 1800- and 1900-MHz range, implemented in the mid-1990s. Most PCS systems are 100% digital. In contrast to cellular, which is limited to A and B carriers, PCS operates across six (A to F) carriers. In other words, cellular can be thought of as a subset of PCS in its broadest sense.

Three operational categories of PCS have been defined by the Federal Communication Commission (FCC) as shown in the PCS Categories chart.

In PCS, particular channels are assigned to specific cells, with provision for reuse. A channel is associated with one uplink and one downlink frequency. A specific number of channels is assigned to an operator's authorized frequency block. PCS service can be installed as a centralized or distributed architecture, and supports both *time* and *code division multiple access* (TDMA, CDMA). Designed to broaden market distribution of wireless services, the system may have more limited range than traditional cellular, but the cheaper connect times and handsets may be appealing to consumers. Industry watchers are predicting steady growth in mobile communications.

In Japan alone, there were more than 20 million Internet-capable PCS system subscribers by 2001. See AMPS, cellular phone, DAMPS, DCS, GSM, Personal HandyPhone Service.

PCS Categories

Category	Notes
narrowband PCS	PCS operating in limited bandwidth in the 900-MHz spectrum and not suited to high speed data communications, although low-bandwidth short text messages would work. Best suited to in-building and near outside-premises use, pagers, and cordless phones.
broadband PCS	PCS in the 1.9-GHz spectrum range for better quality voice communications and higher duplex-mode data communications.
unlicensed PCS	PCS in the 1910- to 1930-MHz range, suitable for in-house and in-company systems, and small independent service providers. Limited to low-power signals.

personal computer PC. A compact, relatively low-cost computer system designed for home, school, small business, and prosumer (high-end consumer) use. The first fully assembled, affordable PC with a keyboard and CRT monitor was probably the SPHERE computer released in 1975, but it didn't sell well. Subsequently, the Radio Shack TRS-80 series, followed closely by the Apple computers and the Commodore PET were all commercially successful. At the time of the introduction of personal computers in the mid- and late-1970s, the cost of a workstation-level computer was typically $40,000 and more, so the price tag of about $2000 to $6000 for a personal computer with useful peripherals (printer, modem, etc.) was revolutionary in terms of availability to individuals. In the early 1980s, when networks that could interconnect individual PCs began to proliferate and CPUs became more powerful, the distinction between personal computers and higher end systems began to blur – a progression that continues to this day, with personal computers of the 1990s being more powerful than minicomputers a decade earlier and laptop computers of the 2000s being more powerful than mid-range institutional computing systems of the late 1980s. The development of PC networks also opened up hybrid systems, with PCs sharing the computing power of mainframes and mainframes using PCs as I/O devices.

The term PC has been generically applied to systems used by individuals for personal, educational, and business purposes, and so does not fit the term "personal" in its strictest sense. Some people use PC to refer only to IBM-compatibles, which is not really a correct use of the term and has probably proliferated because "IBM-compatible" is such a mouthful. The distinction between a PC and a workstation is not as cut-and-dried as many people think. By the time you add a graphics card, sound card, CD-ROM drive, more memory, and network interface card to a personal computer, its cost is comparable to many off-the-shelf workstation-level computers. See Amiga, Atari, Intel, Macintosh, TRS-80, workstation.

Personal Digital Assistant

The Palm Personal Digital Assistant (PDA) provides handheld mobile computing through a color graphics display resolution better than early desktop computers. Full point-and-click Web browsing capabilities (right) are provided by the SoftSource/Catarra display client/proxy server programs communicating through a wireless radio link to the Internet.

Personal Digital Assistant PDA. A handheld computerized wireless device optimized for common time-scheduling and note-taking activities that many business and personal users particularly desire. These include calendars, account keepers, note-takers, calculators, alarm signals, modem connections, databases, etc. Some PDAs support handwriting recognition through a penlike interface, others have small text keypad input screens, and some have both. The more recent PDAs have color graphics displays and the capability of full Internet browsing without the HTML and security certificate restrictions of WAP-based limited-resource instruction sets.

PDAs were introduced in the late 1980s, with pen-recognition PDAs coming out in the early 1990s. Most PDAs work on batteries or AC power with a converter. Some work only with batteries. Battery life ranges from 2 to 5 hours on most systems, depending upon usage.

Apple ClockWorker is an interesting evolution in PDA technology. This little 300-MHz RISC chip with 30-MBytes of RAM and 70-Mbyte memory chip outruns many full-sized desktop computers. Even more surprising is that it is powered by a clockwork mechanism developed in the U.K. Twelve turns of the AppleKey are said to provide up to 3 hours of continuous use. The idea is not entirely new; analog wound watches have existed for decades, but this is an interesting adaptation to computer technology being tested in full-sized notebook computers. See PDA macrobrowser, PDA microbrowser, SoftSource, Wireless Application Protocol.

Personal Digital Cellular PDC. Formerly called Japan Digital Cellular, this is a time division multiple access (TDMA) digital cellular phone system used in Japan and, to a small extent, in the Asia-Pacific region. PDC services operate in the 800- and 1500-MHz radio frequency bands. It is an important standard due to the large number of subscribers (over 50 million) using PDC-based services. See Personal HandyPhone Service.

Personal HandyPhone Service PHP. A commercial 32 Kbps mobile data Personal Communications Service (PCS) popular in Japan. PHP was established in 1995 and began providing services to subscribers in 1997. In 1998, 64 Kbps services were introduced in some areas. The PHS network can be accessed by subscribers through various Personal Digital Assistants (PDAs) and notebook computers. The PHS network is separate from or totally independent of the public switched telephone network (PSTN).

Personal Identification Number PIN. A system of alphanumeric characters, usually numerals, which identifies a particular user or holder of an identification card. PINs are commonly used for credit cards, bank cards, ID cards, calling cards, and other forms of wallet-sized identification to access security doors, ATMs, phones, and vending machines.

PersonalJava applications environment See Java.

Personal Wireless Telecommunications PWT. An in-building wireless telecommunications transmission standard in North America (U.S., Canada, Puerto Rico) developed in the mid-1990s. It is similar to the Digital European Cordless Telecommunications (DECT) standard in Europe. It is intended for short distance, high-bit-rate, packet-based communications.

PWT uses unlicensed Personal Communications System (PCS) spectrum in the 1.9-GHz radio frequency band. Standards for the use of Frame Relay for mobile PWT-compliant devices (Project 4247) and for expanded PWT in the 1850 to 1910 and 1930 to 1990 MHz frequency bands were initiated within the TIA and EIA. Enhanced PWT uses licensed PCS spectrum.

peta- P. A prefix for an SI unit quantity of 10^{15}, or 1,000,000,000,000,000 – a really huge quantity. See exa-, femto-.

petticoat insulator A historic utility pole electrical line insulator that still has practical use. Many historians have suggested they were developed around 1910, but it was certainly much earlier, as glass or porcelain petticoat insulators were already listed as a requirement for outside wiring in the National Electrical Code of 1899. The earliest forms were single petticoats, with double-petticoats developed later.

P

The name refers to the outer underskirt-like shape of the insulator, which has flare for channeling moisture away from electrical wires, a shape practical for both glass and non-glass insulators. See insulator, utility pole.

PGP See Pretty Good Privacy.

PGP Inc. A company jointly established by Philip Zimmermann, the developer of Pretty Good Privacy, and Jonathan Seybold. See Pretty Good Privacy; Zimmermann, Philip.

PGP/MIME Pretty Good Privacy/Multipurpose Internet Mail Extensions. An IETF working group Internet messaging standard for the transmission of secure network communications. A variety of content types have been provided for MIME, and more continue to be added. Unlike S/MIME, PGP/MIME does not use public keys distributed through X.509 digital certificates. PGP can generate ASCII armor (required) or binary output for the encryption of data. The trend is for the signed portion of the message and the message body to be treated separately. PGP/MIME can support 128-bit encryption, although not all implementations will use the full 128 bits. See S/MIME, RFC 1847, RFC 1848, RFC 2015.

phantom circuit In telephony, a means of devising an additional circuit by utilizing resources from existing circuits on either side. Thus, three circuits can be configured to prevent crosstalk and used simultaneously with only four line conductors. The use of phantom circuits has, for the most part, been superseded by a variety of multiplexing techniques. See Carty, John J.

phantom group In telephony, a phantom circuit and the balanced circuits that flank it and from which it draws some of its circuitry.

phase alternate line PAL. A color television broadcast and display standard widely used in the United Kingdom and a number of European, South American, and Asian countries. The name originates from the fact that the color signal phase is inverted on alternate lines. The format was introduced in the early 1960s. It displays at 25 frames per second and can support up to 625 scan lines (not all are seen on the screen; some at the bottom may be obscured). It provides a better picture than the NTSC format prevalent in North America and is not compatible with NTSC or SECAM. PAL-M is a variation on PAL which supports 525 lines.

phase change rewritable PCR. A type of high-capacity optical storage technology, developed gradually over the period from the early 1980s to the mid-1990s. During the 1980s, Matsushita developed a number of PCR WORM drives, and released a read/write drive in 1991.

PCR enables multiple rewrites on the same cartridge. Using a pulsed laser diode at a higher power level, the recording surface of a disc can be changed between low reflectivity amorphous states to crystalline states, enabling data to be erased and written/rewritten. The data can be written in one pass rather than the two passes required for a number of magneto-optical technologies. Once the technology appeared commercially promising, Matsushita developed a combination PCR/CD drive, announced in 1994, and Toshiba led a development group to adapt phase change technology for creating rewritable Digital Versatile Discs (DVDs). At first, industry adoption and standardization efforts were not broadly supported.

phase conjugation A phenomenon discovered in the 1960s, phase conjugation is now a general concept used to describe a number of nonlinear optical phasing processes. Phase conjugation involves the precise reversing of the direction of the phase and propagation of a wave such that it travels back through the same path through which it originally arrived. Thus, optical phase conjugation is the precise reflection of a light beam back through its original path.

Phase conjugation has many applications. It can be used in the development of tracking systems, lensless imaging technologies (e.g., holograms), and defect detection systems. It can also be used to filter a signal or to regenerate a signal that has degraded en route, which would be a boon to many types of communication transmissions. NASA/JPL is using the concept to propose designs for very fine fiber optic-based probes for imaging in tightly confined spaces. See phase conjugation mirror.

phase conjugation mirror PCM. A reflecting mirror that may be used in conjunction with other mirrors in laser light beam directing systems, for example, but which is distinguished by its capability of precisely reversing the direction of a wave hitting the mirror. Contrast this with conventional mirrors, in which the direction of the reflected wave is related to the angle at which the wave hits the mirror. In addition, in a conventional mirror, only the sign of the wave vector component is changed, while in a PCM, the entire propagated beam reverses direction and the phase of the beam is conjugated or joined together. The phase conjugation process can be enhanced, depending upon the environment in which the process is carried out. Freon has potential as a stable medium. In the early 1990s, photorefractive polymers were developed in IBM laboratories. Since then, layered versions have increased their usefulness for industrial purposes. New polymer-based photorefractive compounds may replace crystals for some types of PCM applications as their technology improves and the cost dramatically drops. See phase conjugation, photorefraction.

phase drive PD. A type of optical data storage drive based upon phase-change recording such that the optical medium can be rewritten. See change rewritable.

phase jitter A particular type of undesirable aberration in which analog signals are abnormally shortened or lengthened. See jitter.

phase-shift keying PSK. A type of modulation scheme which distinguishes between a binary "1" (one) and a binary "0" (zero), by changing the phase of the transmitted signal 180° if the next input unit is a binary "0" (zero). If it is binary "1" (one), then a phase shift is not executed. See frequency modulation, frequency shift keying, on/off keying, quadrature phase-shift keying.

Phase-Shift Keying

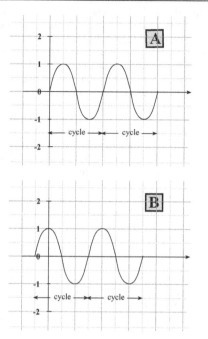

A wave period is one segment from the repeating sinusoidal cycles of the wave taken over time from a reference point on the wave. The period varies with the wave – longer wavelengths have longer periods. In A, the period of the wave begins at zero (0). In B, the wave has been shifted by a quarter of its period such that it is referenced from the highest point in the wave cycle rather than the point at which it crosses the X axis.

The length of the wave period hasn't changed, only the time point in the phase at which it is referenced, relative to the first wave. If the two different phases in the wave were plotted on top of one another, they would undulate with the same period length, shape, and amplitude – only the phase has been shifted.

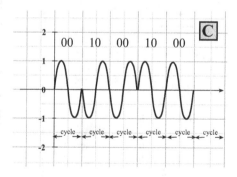

By creating a series of shifts in the waves, relative to the preceding wave, it is possible to use each individual wave to represent a binary value. Thus a half-period shift in a four-phase system changes a 2-bit binary value from 00 to 10 and from 10 to 00.

phase-locked loop PLL. A technology used in silicon-based integrated circuits (ICs), a PLL circuit controls an oscillator at a constant phase angle relative to a reference signal. The three basic aspects of a digital PLL are a controllable oscillator, a filter, and a phase detector/comparator combined within a closed-loop frequency feedback system. PLLs are useful for signal processing and synchronization applications such as controlling automatic phase adjustments in a signal. The signal can be referenced by the PLL in various ways; it can be based upon a carrier signal or linear or nonlinear baseband references.

PLL was traditionally analog, but there are now also digital versions and both are suitable for various types of applications. PLL has been around for several decades; it is commonly used to synch a reference broadcast signal to the horizontal oscillator of a television receiver, for example. Because it is a basic timing technology, it is found in components ranging from voltmeters and spectrometers to cell phones and space-based tracking and synchronization systems.

In communications devices, newer PLL circuits support products with higher data transfer rates, higher frequencies, and smaller footprints. Commercial dual phase-locked loop-based ICs are small, low-power-consumption components that can offer frequencies up to 2.5 GHz (in some cases, up to 4.8 GHz), making them suitable for radio transceivers for a variety of types of products, including cellular phones and PCS. PLL ICs can also be used as secondary circuits for providing intermediate frequency radio waves that are commonly used in cell phone receivers.

PLL circuits can be readily modeled in software for educational and design purposes. Java-based PLL modelers are available on the Web.

Phelps, George M. (1820–1895) An American machinist and inventor best known for his telegraphic key and printer inventions, although he also designed stock tickers (a type of specialized telegraph) and early telephone equipment. As a youth, Phelps was apprenticed as a machinist to his uncle, Jonas H. Phelps, to build scientific instruments. The Phelps and Gurley surveying instruments company evolved into Gurley Precision Instruments, which is still in business. George Phelps set up shop in 1850, in Troy, New York, and began designing and patenting a wide variety of precision electromechanical devices, including telegraph keys (e.g., a *camelback* key). He was known for elegance of design and superior workmanship. When approached about improving upon the popular but complex telegraphic instrument of R.E. House, Phelps joined with Jarius Dickerman to form Phelps and Dickerman and House's Printing Telegraph Instrument Manufacturer, located in Ferry Street in Troy. Thus, Phelps built House instruments for several years.

The American Telegraph Company purchased the Phelps and Dickerman holdings, retaining Phelps as a superintendent. After the American Civil War, American Telegraph was purchased by Western Union, again retaining Phelps for his knowledge and experience in the field. Western Union also acquired the patent rights to Phelps' printing telegraph. Phelps was assigned to work on a "harmonic telegraph," the forerunner to the telephone, a device first patented in the U.S. by A. Graham Bell.

Phelps was an associate of Thomas Edison and created some of the patent models for Edison's early inventions. Phelps became the superintendent of Western Union Telegraph in New York and remained as a staff inventor in his later career. He may also have been associated with the Field brothers, who were instrumental in laying the first successful transatlantic telegraph cable. See Phelps Combination Printer.

PHIGS Programmer's Hierarchical Interactive Graphics System. An official standard for 3D graphics from the late 1980s. The PHIGS+ extension added sophisticated rendering of realistic looking objects on raster displays. Simple PHIGS (SPHIGS) is a powerful, display-independent subset of PHIGS which incorporates some PHIGS+ features.

Phillips code A shorthand telegraphic code assembled/revised from existing systems by Walter Polk Phillips, published in 1879. Originally an American Telegraph messenger, Phillips became an accomplished press telegrapher (2731 wph) and his code was widely used for decades. See 73 in Numerals chapter.

phoneme A unit of speech, considered to be the smallest distinguishable unit, which may vary from language to language and among dialects of a particular language. Phonemes are of interest to programmers for speech recognition and speech generation applications. See speech recognition.

Photo CD Kodak Digital Science Photo CD System. An image storage and retrieval format developed by Kodak and introduced in 1992. PhotoCD is a means to store digitized still images in various resolutions on a compact disc so it can be read back from CD-ROM drives. It is used by many stock photo suppliers and graphic design professionals.

Conventional 35mm film shot with a traditional camera can be taken to photofinishers supporting PhotoCD and developed into both pictures and digital images. At the lab, the file is scanned with a high resolution drum scanner and saved onto PhotoCD discs. If there is room, additional pictures can be added to the disc later, and read back with a multisession CD-ROM XA drive and an appropriate software driver (including Apple QuickTime Photo CD extension, SGI's IRIX, Sun's Solaris, IBM's OS2/WARP, AmigaOS 3.1, IBM AIX, etc.).

A Photo CD disc can hold about 100 images, that is, about three or four rolls of film. The images are stored in Photo YCC color encoding, with multiple resolution levels. Pixel resolutions include: 2048×3072, 1024×1536, 512×768, 256×384, 128×192. The Photo CD Pro format also includes 4096×6144. See compact disc.

Photocopy Machine – Original Invention

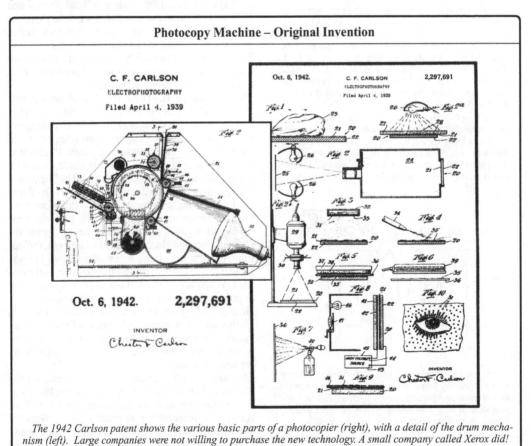

The 1942 Carlson patent shows the various basic parts of a photocopier (right), with a detail of the drum mechanism (left). Large companies were not willing to purchase the new technology. A small company called Xerox did!

photocopy A dry transfer replication process from an optically imaged source, sometimes also called a xerograph, after Xerox, the company that popularized the technology. C.F. Carlson was awarded a patent for a photocopy invention in 1942 and failed to sell it to some of the larger business-oriented companies. But a small company called Xerox took a chance on the technology. See the Carlson patent diagram.

photodetector PD. A component or biological system that responds to stimulation by light. Plants have photosensitive structures and mechanisms that enable them to detect sunlight and orient themselves towards it and certain natural and synthetic materials have photodetecting properties that can be incorporated into industrial device assemblies. Since light has a number of wave-like and particle-like properties and emits heat at different levels depending upon location and time of day, the definition of photodetector is somewhat broad, reflecting the capability of reacting to the presence of light without necessarily specifying what aspect of light is causing the reaction. In general, photodetectors are subclassified as thermal detectors and photon detectors.

Simple photodetector components may respond only to the presence (or absence) of light within certain parameters and some may be sensitive to light without discriminating its intensity or character. More sophisticated photodetectors may be "tuned" to detect specific wavelengths or regions of wavelengths and some are also sensitive to the magnitude of a light stimulus. Even at its most basic level, however, photodetection is an important capability at the heart of many systems. Photodetectors are widely used in imaging devices, security systems, robotic vision, and signaling and transmission systems.

In practical applications, the response of a photodetecting substance is often very weak and may require further processing to make it useful. Amplification of very subtle reactions to light has limits, due to noise that is introduced when a weak signal is amplified. Much of semiconductor technology is devoted to improving the signal-to-noise ratio of amplified signals. In addition, photodetectors are often environmentally sensitive. Light is ubiquitous and it is often challenging to detect only that light that is of interest. For example, a thermal-sensitive detector in a hot environment such as a desert, may need to be cooled in order to detect other sources of light (e.g., a signal light). An astronomical photodetector (for studying light from celestial bodies) works more effectively if placed in orbit around the Earth rather than in the observatory of a university in the middle of a large city, due to the interaction of ambient light sources.

Depending upon the type of detector, commercial photodetectors are typically described in terms of responsivity (the sensitivity and magnitude of their reaction to light), efficiency (how much signal is generated per photon stimulus), response time, signal-to-noise ratios and types of noise (e.g., Johnson noise), and the linearity of the response. Figures of merit may also be used.

Film photography is an example of directly harnessing the selective photosensitivity of certain chemicals by embedding them in a film substrate and briefly exposing them to light. The image captured in film can then be transferred to paper by yet another photosensitive process (with stray light excluded in a darkroom). Sometimes photodetection is only one step in a series of detection and conversion processes. For example, a scintillating device that converts electromagnetic energy outside the optical spectrum, such as X-rays, into optical wavelengths, may feed the signal to a photodetector. From there it may go to a photomultiplier that further converts the signal to electrical impulses. Thus, a photodetector assembly can indirectly detect wavelengths outside the optical spectrum.

A complex light impulse can be characterized by using a device in which multiple photodetectors are tuned to respond to different optical frequencies. The data derived from individual elements in the photodetector array can be signal processed to produce a complex overall statistical picture of the light-emitting characteristics of sample specimens or light-carrying transmissions media.

The creation of semiconductor photodetectors is as much art as science and much of the fabrication is at the molecular level, crossing boundaries in geology, quantum physics, chemistry, and biology. Structures for photodetectors can be grown in molecular beam epitaxy (MBE) systems on semi-dielectric substrates. Such components are being developed for new high-speed photodetectors, giving them properties for meeting the greater bandwidth and distance demands of microwave fiber optic links. See photoelectric cell, phototube, thermopile, traveling-wave tube.

photodiode A semiconductor photodetector component for converting light energy into electrical energy. See photodetector.

Sample Photodiodes

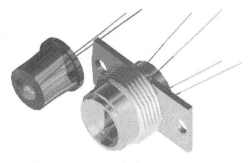

Photodiodes are semiconductor photodetecting components. They come in a wide variety of shapes, sizes, and levels of sensitivity to photonic energy. Illustrated here are common configurations for a gallium-arsenide diode (left) and an indium-gallium-arsenide diode.

---

photodiode, avalanche APD. A semiconductor component commonly made of silicon (Si) or indium-gallium-arsenide/indium-phosphide (InGaAs/InP). Silicon APDs are p-n junction solid-state detectors with high internal gain. They are reasonably immune to electric fields and sensitive enough to detect single photons at room temperature.

APDs are used for optical detection for a variety of applications including fiber optic communication receivers, fluorescence detectors, photon counters, time-of-flight ranging devices, and cryptography. Fiber optic receivers commonly use p-i-n photodiodes or APDs for detecting and converting an optical signal into an electrical signal.

New indium-gallium-arsenide/silicon (InGaAs/Si) APDs have been developed under a grant funded by AFRL/DARPA with separate absorption and multiplication (SAM) regions for use in near-infrared frequencies. These offer faster, more sensitive photodetection at wavelengths that were not previously practical. See avalanche diode, Zener diode.

photoelectric cell A type of electronic sensing device activated by light and widely used in security systems, automatic lighting systems (e.g., street lights), automatic doors, etc. A photoelectric cell can be made by coating cesium on one of the electrodes in a vacuum tube. This technology was used in early television cameras. See photodetector.

photography The art and science of registering light from objects in a scene and storing them in the form of an image. Later it became possible to produce multiples of these images by a number of means. Most photography involves capturing three-dimensional imagery in a two-dimensional format. Light is usually recorded from the visible spectrum, but there are cameras and films designed to record heat and infrared radiation which show images in a form different from the way humans perceive them, and electron microscopes record the movement of a beam of electrons.

Traditional photography was developed in the early 1800s by a number of inventors including Joseph Nicephone Niepce, a French inventor, who developed a process called heliography or sun drawing, on paper coated with silver chloride. Other pioneers included Daguerre (originator of the daguerreotype), Herschel, Talbot, and Archer. One of the earliest photos was captured with silver chloride by Thomas Wedgewood in 1802. More than 150 years passed before 3D photography, in the form of holographs, became practical. Newer digital cameras can immediately relay an image to a computer network so the image can be viewed almost instantly at great distances from the actual scene of the event. See Daguerre, Louis Jacques Mandé; heliography.

photometer An instrument for determining the intensity of transmitted or reflected light, sometimes called an *optical power meter*. A photometer is a type of radiometer and photometers that measure the intensity of frequencies beyond the human visual range are sometimes termed *radiometer/photometer* devices.

Photometers are used in scientific research, photography, and many aspects of experimental and commercial optics. Human visual senses are quite good at determining relative brightness, but photometric instruments are needed to make objective assessments of light intensity within and beyond the human visual range.

Photometers come in many shapes and sizes from simple photography or classroom models to high-end scientific research instruments. They may be used to measure power levels in laser beams, optical signals in modulated light beams, and solar radiation. Photometers are used to measure the intensity of traffic lights (which may dim over time) to make sure they are bright enough to be seen clearly by motorists. Goniophotometers are common in the lighting industry. Photometers aid in assessing light propagation through different types of waveguides in the design and development of optical network technologies.

The range of sensitivity of a photometer is dependent upon its price and intended application. The spectral range within which it is sensitive also varies, but commonly photometers measure visible and infrared frequencies. A basic classroom photometer may include several measurement scales with sensitivity to power levels ranging from about 20 microwatts to 20 milliwatts. Measurement scales may be linear or logarithmic. Some industrial photometers have optional, interchangeable sensor heads for different applications. Simultaneous measurements of more than one wavelength are possible with some scientific models. The reading from a photometer may be output to a built-in LCD display or may be transmitted to other devices such as oscilloscopes, recorders, or computer peripheral cards.

Photometers designed for microscopes may have an adjustable iris to enable the sample to be viewed while the light is measured. A housing for filters may also be included.

In astronomy, where light intensity provides information on the properties of celestial bodies, photometers are important research tools and may be integrated with spectrographs in telescopic systems. Sophisticated optical fiber-based photometers are now available for studying fast variable astronomical phenomena. Multiple fibers enable reference images to be assessed in conjunction with the phenomena being observed. Fiber optics may be used to link individual telescopes in a telescopic array.

The first known drawing of a photometer was by Peter Paul Rubens, who illustrated a book on optics by F. d'Aguilon, published in 1613. P. Bouguer described several simple photometers in a treatise published posthumously in 1760. This was an expansion of an earlier essay, published in 1729 and Bouguer is considered by many to be the inventor of the photometer. J.-D. Colladon developed a practical application of a photometer for his engineering projects in the mid-1800s. Prism-based spectrophotometers became available on the market after World War II but the technology remained relatively limited and expensive until the 1960s, when grating spectrophotometers became available. Since then advancements in electronics have made photometers increasingly small and powerful. By the 1990s, built-in filters, exchangeable sensing heads, LCD displays, and computer interfaces were readily available.

Fiber optics and lasers are now incorporated into a number of types of photometers. For example, in chemical photometry, a laser can be used as a light source for illuminating a sample to measure its photometric characteristics. When the coherent light hits the obstacle (sample), the light is scattered and may be detected by a fine fiber filament that directs the light that enters the fiber to a photomultiplier, where it is passed on to a processing system and display. See Aguilon, François de; Bouguer, Pierre; luminance; photopolarimeter; radiometer.

photomultiplier PM. A light-sensitive component that emits electrons in response to stimulus by photons (of sufficient energy levels). This is a very useful means to convert electromagnetic energy in the optical spectrum into electrical energy that can be used to activate and control other components.

photomultiplier tube PMT. Typically, an evacuated glass component containing a photocathode that emits electrons when subjected to photonic energy sufficient to trigger a photoelectric effect. The photocathode operates at a high negative voltage and the electrons emitted are accelerated towards a series (chain) of dynodes that are positioned along the electron path between the electron-emitting cathode and the electron-attracting anode. The dynodes generate additional electrons through secondary-emission multiplication.

PMTs can be configured with multiple anodes, arranged in linear (e.g., 1 × 16) or grid patterns (e.g., 8 × 8) for use with fiber faceplate scintillating applications, for example.

Photomultiplier tubes can respond to a wide range of wavelengths from ultraviolet to infrared, but responsivity and emission effectiveness are dependent, in part, upon the materials used. In general, PMTs are fast-response, low-noise components practical for a wide variety of applications, including laser technology, radiation measurement, spectroscopy, high energy physics research, and others.

Photomultipliers are sensitive enough to count photons at very low light levels (down to one photon) and thus are highly efficient at distinquishing signal from noise. Thermal noise can be reduced by cooling and ambient light and magnetic interference can be reduced with proper shielding.

Commercial photomultiplier tubes commonly have 14, 20, or 21 pins. The primary connections are to the 10, 12, or 14 dynodes, the anode, cathode, focus electrode, and shield.

Simplified Drawing of Basic Photomultiplier Components and Dynode Function

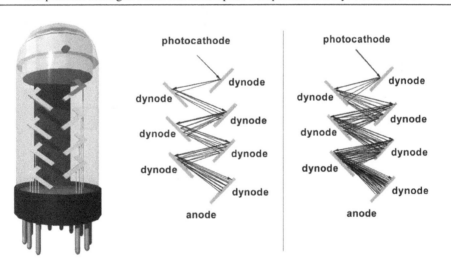

This is a highly simplified drawing of a basic photomultiplier tube used to convert and amplify a photonic signal. The photocathode at the top converts electromagnetic energy in the form of photons into electron emissions which are attracted to the anode at the base of the tube. As the electrons travel toward the anode, they encounter a series of dynodes in the middle of the tube, powered with voltages that are calibrated to one another to control the magnitude of electron emissions. As an electron from the cathode (or the preceding dynode) hits a dynode, it is reflected along with secondary emissions governed by the voltage applied to the dynode to the next dynode in the chain, causing a cumulative amplification of the signal. When the electrons reach the anode, the signal is processed by a small circuit within the base and output through the contacts coming from the bottom of the base to interface with other components (21-pin sockets are common). A magnetic shield that fits over the base can shield the electrical circuits from external interference. As illustrated in the line diagrams, the voltages applied to the reflective dynodes are related to the number of electrons emitted, with higher voltages (right) providing greater gain (within operating limits).

Thus, very weak signals, even as small as one photon, can be measured and manipulated with photomultipliers to facilitate research in particle physics and to fabricate sensors, and scientific and industrial quality assurance, testing, and sampling instruments.

A phototube is a simpler version of the photomultiplier tube (without dynodes). See dynode, photosensor.

photomultiplier tube base A mechanical and voltage distribution/dividing component for coupling with a photomultiplier tube. The tube base may optionally include a magnetic shield to protect it from Earth- and equipment-originating magnetic fields. The shield may also protect the coupled photomultiplier tube from ambient light and magnetic emissions.

The photomultiplier tube typically connects to the base through 14 or 21 pins. Outputs from the base, such as connections to the anode or a specific dynode, are typically through 50-ohm coaxial connections. Some versions include low-noise preamplifiers incorporated into the base for use with scintillation detectors.

Photomultiplier tube bases have also been designed for use with multiple photomultiplier tubes (e.g, in arrays). Voltages for the tubes in the assembly may

PhotoPhone - Bell and Tainter's Light-Based Communications Invention

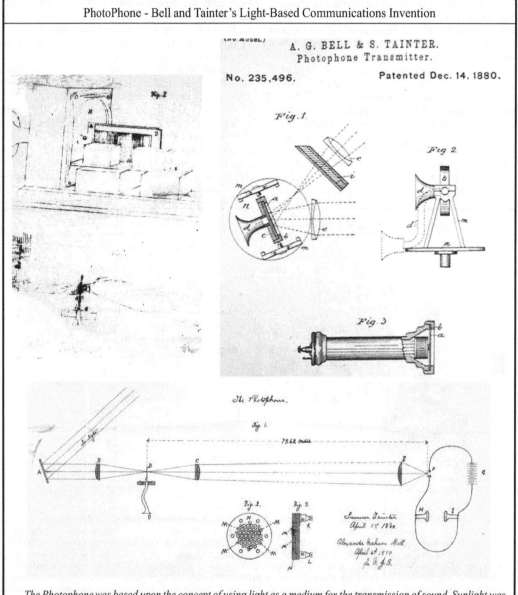

The Photophone was based upon the concept of using light as a medium for the transmission of sound. Sunlight was used to translate acoustic vibrations into light signals that were reflected to a receiver where they were converted to electrical signals through the use of light-sensitive selenium (the same material used for early television inventions). By substituting a parabolic surface, Bell found he could increase the intensity of the signals and was able to transmit signals over a distance of several hundred meters on a sunny day. Bell was very excited about the potential of wireless communications and took out four patents on the Photophone with assistance from Sumner Tainter. [Library of Congress American Memory Collection and U.S. Patent Office (upper right).]

be controlled individually, in groups, or unison. Cockcroft-Walton voltage multipliers have been suggested in place of resistive voltage dividers for PMTs that are densely packed, in order to minimize dissipated power. See dynode, photomultiplier tube.

photomultiplier tube chamber A housing for physically protecting, electromagnetically shielding, and cooling photomultiplier tubes. Depending upon the temperatures required, the housing may include single- or double-paned windows to prevent condensation or icing. A variety of materials are available for the windows, including Plexiglas®, Pyrex™, or fused silica. Fused silica is effective over a broader spectrum of wavelengths. See photomultiplier tube.

photonic crystal A photonic bandgap technology described and developed originally by E. Yablonovitch, developed further by Özbay at Ames Laboratory, Southampton Researchers, S. Kawakami and his collaborators in Japan, and a number of others.

These photonic crystals have periodic dielectric structures that exhibit large anisotropy, high dispersion, and photonic bandgap properties. The bandgap, which is similar in concept to gaps in semiconductor devices with a lattice-like structure and holes or "wells," makes it possible to selectively filter certain optical frequency ranges by means not available with conventional lenses or existing semiconductors.

Varying the refractive index of the component or introducing point defects within an otherwise perfect dielectric structure have the potential for localizing light, essentially trapping it selectively. The size of the holes could further be controlled to manipulate energy levels. Yablonovitch et al. have further described how 3D circuit designs could extend the technology into lower wavelengths. In 2002, Chen and Suzuki described an integrated fiber-photonic crystal system with a uniform bandgap and low insertion loss. This has potential for optical switches and routers. Also in 2002, OFS Laboratories introduced a new fiber design incorporating a photonic bandgap for tuning the transmission through the fiber.

There is much excitement surrounding photonic crystal technology. It has been suggested that highly-efficient light reflectors for fiber optics transmission sources (e.g., LEDs) and computers operating in the hundreds of terahertz computing range could be designed with the technology.

MIT has developed freely available software to model the dispersion relations in photonic crystals in order to visualize the band structures. It is available for download online as *MIT PHotonic-Bands*. See Kawakami, Sujiro; photonic crystal fiber.

photonic crystal fiber PCF. A type of microstructured optical fiber with low-index refractive materials fabricated within higher-index materials (e.g., silica). They may be categorized as low index (photonic bandgap) or high index guiding fibers that produce total internal reflection through a lower effective index.

PCFs were first demonstrated in the mid-1990s and have made it easier to harness the properties inherent in optical transmissions through novel fiber fabrica-

tion patterns. Dispersion properties, linearity, a broader range of numerical apertures, and other factors can be utilized and better controlled through PCFs, increasing the practical range of optical components and telecommunications devices that can be devised. See photonic crystal.

Photonic Information Processing Systems Laboratory PIPS. A research lab founded by N.A. Riza, a pioneering optical engineer, at the School of Optics and Center for Research and Education in Optics and Lasers (CREOL) at the University of Central Florida. The School of Optics offers interdisciplinary graduate programs in optics. See Riza, Nabeel. http://www.ucf.edu/

Photonics Components and Subsystems Newsletter PCSN. Global coverage of technology, applications and photonics markets. Published monthly by Information Gatekeepers, Inc.

Photophone A historic device that transmitted voice by means of light waves, invented by A. Graham Bell in 1880. Charles Sumner Tainter, an experienced scientific instrument-maker, had a significant hand in the practical embodiment of the idea.

Bell put great stock in the invention, filing for four patents for the device and its associated selenium cells. The concepts are still sound and the invention ahead of its time and worth mentioning in detail.

At least as early as 1878, Bell was developing the idea for the Photophone. He described the possibility of "hearing a shadow" by the action of interrupting a light beam incident upon selenium in a lecture delivered before the Royal Institute of Great Britain, in May 1878. In January 1879, he wrote a note describing how he had worked out the idea as

"... the art of causing electrical signals and audible sounds in distant places by the action of light. It has been discovered that certain substances such as silenium [sic] have their electrical resistance affected by light.

When a peice [sic] of silenium in a crystalline condition is placed upon the circuit with a telephone and voltaic battery a sound is audible from the telephone when a beam of light is allowed to fall upon the silenium.

When a galvonometer is substituted for the telephone the needle is deflected indicating the increase of current, when the light falls upon the silenium thus showing that the electrical resistance of the silenium is diminished under the action of light.

My invention consists in utilizing this property of silenium for the purpose of causing telegraphic signals from a galvenometer [sic] or audible sounds from a telephone in distant places without the necessity of a conducting wire between the transmitting and receiving stations....

The transmitting instrument consists of a powerful ~~series~~ source of light and of an apparatus for interrupting or varying its intensity.

The receiving instrument consists of a lens by means of which the distant light is focussed [sic] upon a peice of crystalline silenium, which is placed in circuit with a battery and either with a telephon[e] or galvenometer [sic]. When a galvenometer is used, signals can be made at the transmitting station by turning on or off the light. When the light is at its maximum the needle will swing to one side, and when at its minimum it will swing to the other.

The needle of the galvenometer can be caused to strike against suitable stops and thus open or close another circuit containing a Morse register or other telegraphic apparatus.

When the telelphone is employed at the receiving station audible sounds can be occasioned in the telephone by varying the intensity of the light at the distant station; for instance at the transmitting station let these be screens placed in front of the light each screen containing a multitude of small holes or narrow slits. It is evident that when the screens are placed so that the slits in the one coincide with the slits in the other, the light will pass freely through these slits to the receiving station. A very slight displacement, however, of one of the screens would suffice to shut off completely the whole light. I attach the movable screen to a metallic plate or disc arranged to be set in vibration by the voice of the speaker. "

A year later, in February 1880, Bell wrote an excited letter to his father announcing not only the birth of "Mabel's child," but also of his own "child" as he called his new invention.

"I have heard articulate speech produced by sunlight! I have heard a ray of the sun laugh and cough and sing! The dream of the past year has become a reality – the "Photophone" is an accomplished fact. I am not prepared at present to go into particulars and can only say that with Mr. Tainter's assistance I have succeeded in preparing crystalline selenium of so low a resistance and so sensitive to light that we have been ~~able~~ enabled to perceive variations of light as sounds in the telephone. In this way I have been able to hear a shadow, and I have even perceived by ear the passage of a cloud across the sun's disk.

Can imagination picture what the future of this invention is to be! I dream of so many important & wonderful applications that I cannot bring myself to make known my discovery – until I have demonstrated the practicability of some of these schemes. I want to utilize the invention before giving it to the public – I want to perfect it – so that we may talk by light to any visible distance – without any conducting wire between the speaker & the listener."

Two days after this letter was dated, Bell and Tainter had a sealed package (containing the Photophone and its description) deposited at the Smithsonian Institution in Washington, D.C. (There is also a receipt from the Smithsonian marked April 6, 1880 for a package from Bell and Tainter.) A copy of the packet was labeled "Copy of Sealed packet No. 1" in which was a written description of the invention in someone's handwriting other than Bell's or Tainter's, signed by Bell and authenticated by Tainter.

"The above description of the Photophone and the experiments made with it is correct[.] Sumner Tainter"

Bell was not an expert craftsman nor draftsman and Tainter, who collaborated with Bell to a significant degree at this time, assisted in constructing and describing the invention, in April 1880.

Bell continued a lively correspondence about the Photophone for the next few months, patenting aspects of the invention, communicating with the National Bell Telephone Co. about the embodiment of the device, the possibility of financing, and the fact that he had successfully transmitted sounds through light for a distance of several hundred feet.

On May 31, 1880, Bell wrote to his wife with continued enthusiasm that the "Selenium Cell" had been attached "to the interior of one of the parabolic reflectors and reflected an interrupted beam of sunlight on to it from the corner of the alley .... Result – a musical tone much louder and more distinct than any Photophonic effect yet observed." Thus, Bell was exploring ways to improve the technology and make it practical for use.

Bell wrote to his parents that he was planning to demonstrate the Photophone to a small group of friends on June 4 to have witnesses to his invention and intended to demonstrate it to the American Association for the Advancement of Science (AAAS) in August 1880. Later, in June, Thomas Watson, Bell's collaborator on the original wired telephone, wrote to Bell with some practical suggestions about using a diaphragm that could vary the amount of light passing through it as it was vibrating. Tainter continued with experiments on various components in the Photophone assembly.

In his address to the AAAS (September 1880), Bell acknowledges that the idea of using light to transmit signals was based upon fundamental concepts that were already extant and no doubt accessible to many inventors. He briefly credited David Brown of London as having developed the idea of using the undulating nature of a continuous beam of light as the source of the transmission of converted vibrations and of having created a "crude apparatus" for its embodiment. Also, there is a letter by J.F.W. printed in *Nature* in June 1878 that states "One is inclined to think that by exposing the selenium to light, the intensity of which is subject to rapid changes, sound may be produced in the phonoscope." This letter may have been seen by Bell who had clearly researched the scientific background of selenium. Bell further credits Sumner Tainter with a significant portion of Bell's described embodiment of the Photophone. This pulling together the strings of other people's discoveries is characteristic of many of Bell's inventions, including the original telephone. Bell was good at recognizing the pieces of an emerging puzzle, focusing them, formulating a practical embodiment, and publicizing his actions. As such, history has often credited Bell as the originator of ideas that were developed by earlier inventors, or by a number of people at about the same time.

In December 1880, Bell and Tainter received the first patent on the Photophone (U.S. #235,496). For a while the invention was known as a radiophone. In 1937, the packet delivered to the Smithsonian was opened, revealing the Photophone description and apparatus in sealed tin boxes.

Why didn't the Photophone catch on? One of the reasons was the quixotic nature of the Sun (and its absence at night) and the lack of an alternate source of illumination capable of transmitting coherent beams with sufficient power (practical room temperature lasers for providing a light source for signal transmissions were far in the future). Politics were a consideration, too. By May 1881, Theodore Vail of The American Bell Telephone Co., was bargaining for the rights to Bell's telephone and Photophone inventions "for a nominal price" in response to Bell's inquiries about obtaining financing for his laboratory. By December 1883, he was corresponding with Bell about the assignment of the "Photophonic Rec." which had been awarded U.S. patent #241,909.

In many ways, Bell himself was more of a theorist and idea man than an implementor, leaving the nitty gritty to others once he was sure a basic concept was sound, pursuing new ideas almost as soon as the current ones were off the ground. Two months after the invention of the Photophone, Bell was already talking about other interests and six months later, his correspondence about the invention was decreasing.

There were also practical hindrances to commercializing the Photophone. Even though wireless light-based communication on sunny days would have been better than no communication at all, it is frequently difficult for a marketing professional to sell the public or the executives of a corporation on the idea of a limited-use technology. Distance limits were a problem, as well. It was not until a decade after Marconi's demonstration of wireless telegraphy that anyone was able to extend the range of the Photophone to a practical distance of a few miles. Radio-based inventions promised better performance than light-based communications systems until lasers and cladded lightguides were developed a century later.

In an ironic twist of fate, Tainter recorded in his notes that the light transmission system worked with invisible light (infrared) and could pass through hard objects (the beginnings of radio technology). The inventors apparently didn't fully appreciate the significance of their observations, at first, and went on to develop a number of audio replay and recording technologies such as the Graphophone when they had actually been on verge of discovering radio waves beyond the visible and infrared regions of the electromagnetic spectrum.

Like the computer designs and concepts of Charles Babbage, the invention of the Photophone is one of the many examples of a sound technological idea that was envisioned far ahead of its time, where all the pieces of the puzzle took decades to fall into place.

In 1995, R. Hey, J. Robinson, and S. Stroud applied as a student group for a letters patent for an improvement on Bell's Photophone using instead a switched circuit to create variations in the intensity of light rays. The name Photophone has more recently been applied to a number of photonic apparatus, including a wireless phone with a built-in camera. It has also been used for a number of software applications, including an interactive communications package developed in the mid-1990s for integrating media and database information and relaying the information via radio links. See free-space optics; helium-neon laser; line of sight; selenium; Tainter, Charles.

Photoplastic SNOM Probes

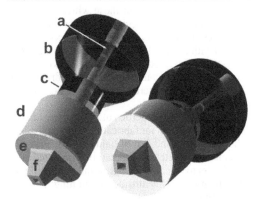

This hybrid fiber optic scanning microscopy probe design enables integrated-aperture tips of different shapes to be readily selected. It consists of an optical fiber (b) with a light-guiding fiber core (a) passing through a narrower etched optical fiber (c) into a ring-guiding structure with a photoplastic component (d). There is a metallic coating (e) on the end of the photoplastic component and an integrated scanning tip (f) of the desired shape and size with respect to the numerical aperture in which the coated tip of the fiber is held in place.

---

photoplastic SNOM probe An alternative design for fiber optic scanning probes from the IBM Zürich Research Lab. The hybrid probe includes a photoplastic component based upon SU-8 technology to hold interchangeable, reproducible integrated aperture tips. See scanning near-field optical microscopy.

photopolarimeter A device for measuring the polarization of light. It may be designed to be sensitive to multiple spectral bands. When combined with a photometer/radiometer, it is useful for applications in astronomy such as space probe missions to other planets. See photometer.

photorefraction A phenomenon in which illumination influences the electromagnetic characteristics of a material, thus altering its index of refraction. This effect can be exploited with dual laser beams to form advanced types of holographic images. In the 1990s, polymers with photorefractive capabilities were developed in the lab. These polymers have the potential to increase the industrial applications of photorefractive technologies and, when fabricated in layers, are useful for components such as phase conjugation mirrors. See phase conjugated mirrors.

photoresist A material that is chemically sensitive to the effects of light, such that light exposure strengthens or weakens the bonds between molecules. The material can be fabricated to be more sensitive to some wavelengths than others. Thus, light exposure followed by chemical processing enables a pattern to be created by dissolving away the portions that have been exposed to light (positive image) or those that have not been exposed to light (negative image). Hydroamber is an example of a photoresist product that can be masked, exposed to an arc lamp, and washed out with water to create a silkscreen stencil. For more precise images or those which are etched in three dimensions, especially for microminiature electronics components, photoresist material can be exposed with laser light. Thus, a series of ridges can be created (as opposed to the essentially two-dimensional dissolved or retained regions in Hydroamber). This is called an interference grating or, sometimes, a holographic grating.

The light-sensitive properties of photoresist materials make them useful for fabricating patterns for controlling the entry and exit of light through a layered component. The etched film can be combined with other layers to selectively control the input and output of lightwaves. See diffraction; grating; interference grating.

photosensor A component module comprising a photomultiplier tube (PMT) and a power supply. Photosensors are typically designed for fast response, compact size, and low power consumption. They are used as light detectors for electronics, fabrication quality assurance and testing, and security applications, and as specialized light detectors for radiation-detecting devices, scientific studies, and applications of luminescence and fluorescence.

phototube An evacuated glass electron tube for converting light into electric current through the photoelectric effect by emitting electrons in response to photons of sufficiently high energy. This enables light energy to be translated into electrical energy that can be interpreted by other electrical devices and signal processing systems.

A phototube is a basic cathode-anode arrangement without the amplification components that are included in photomultiplier tubes. Phototubes are used in many devices including photodetectors and absorption spectrometers. See faceplate, fiber optic; photomultiplier tube; photosensor; scintillator, traveling-wave tube.

photovoltaic A specialized semiconductor device which converts light into electrical energy. A basic photovoltaic cell is a slice of semiconductor material. These cells are combined into an integrated array called a *panel*, many of which are commonly called *solar panels*. A number of photovoltaic technologies such as monocrystalline silicon and polycrystalline silicon (wafers cut from silicon) are common. Photovoltaic panels are important for powering orbiting satellite transceivers and their various telemetry devices.

phreaker *jargon* A person who seeks to access telephone services by bypassing standard procedures or acquiring technical knowledge of the system. Phreaking involves unauthorized and often unconventional methods of accessing, controlling, or subverting the public phone system. Phreaking became prevalent in the mid-1970s, when in-band telephone signaling systems enabled control of phone switching systems through tones transmitted through the wires. A second level of phone phreakers, who simply wanted to use blue boxes to steal phone services and didn't care about the technical structure of the system, emerged in the late 1970s.

Due to the increasing technical similarity between personal computer systems and the evolving public phone system in the 1980s, there was a high degree of overlap between computer hackers and phreakers. Historically, intellectual curiosity and the possibility of entrepreneurial gain from selling illegal *blue boxes* motivated phone phreakers, but the allure of free long-distance phone services was hard to resist and phreakers often continued their activities far beyond learning about the system, repeating many of the same arguments used to justify software piracy.

Popularly, phone phreakers were assumed to be high school and college students but phone company studies revealed that about half of blue box users were business executives, doctors, and other working professionals.

The evolution of the phone system from in-band to out-of-band signaling greatly reduced the incidence of phone phreaking, as did the increase in the use of Internet email for long-distance communications. Phreaking has largely returned to being a subculture of technologically curious individuals, although theft of phone services still occurs. See blue box, cracker, hacker, in-band signaling.

PHS See Personal HandyPhone Service.

Physical Layer Convergence Procedure PLCP. A networking convergence procedure commonly used in Distributed Queue Dual Bus (DQDB).

physical layer signaling PLS. In layer-oriented network architectures, a way to transfer information from the physical interface components to the communications channel. In some standards, the PLS is treated as a specific signaling sublayer of the physical layer.

Pickard, Greenleaf Whittier A scientist who followed up on the observations of Karl Ferdinand Braun by testing hundreds of natural and synthetic substances for their conductivity. Braun had observed that some substances conducted radiant energy in one direction, inhibiting the backward return of a wave. Pickard researched these properties more closely, examining substances that exhibited this property, and to what degree they did so. In the course of his research, he found more than 200 substances which, in conjunction with a metal contact, could detect radio waves.

This led to the development of an early form of radio component called a crystal detector, which was first marketed in 1906 and patented by Pickard in 1908. See crystal detector.

pico- A prefix for an SI unit quantity of 10^{-12}, or 0.000 000 000 001 or one trillionth of – it's a very tiny quantity. See tera-, femto-.

picocell A limited-reach wireless services base station with low output, intended to serve a very limited area, such as a single building or industrial yard.

PICS See Platform for Internet Content Selection.

picture element PEL. The smallest individual unit that can be addressed and controlled on a computer video display.

Picturephone A pioneer videoconferencing system developed by AT&T Bell Laboratories. It has been decades since the development of the first prototype Picturephone system; it was not practical to introduce it to the public until 1970 and, even then, the technology, while appealing, was too expensive and cumbersome for consumer distribution. Since then the Picturephone concept has been refined to take advantage of advances in technology, and competing products have been developed to provide a number of practical videoconferencing options to businesses and individual consumers.

Picturephone Meeting Service PMS. An AT&T service that combined TV and voice transmissions to provide a means to carry out audiographics conferences. It was a little before its time and generally expensive, and there is now a lot of competition for videoconferencing technologies. See audiographics, Picturephone.

Pierce, John Robinson (1910–2002) An American physicist, writer, and musician who studied at Caltech, Pasadena, Pierce became a Bell Laboratories researcher and headed up the landmark development project that led to the invention of the transistor. He is also remembered for photomultiplier tube technology researched in conjunction with W. Schockley et al. and his work on pioneer satellite technologies, including the Echo and Telstar systems.

The multitalented Pierce also authored more than 30 science fiction articles and essays under the pen names of John Roberts and J.J. Coupling (which is probably a reference to atomic spin-orbit effects) and taught music at Stanford University until 1983. See image dissector.

piezoelectricity A form of electricity or electromagnetic polarity that arises from pressure, especially in crystalline substances. For example, compressing or twisting a quartz crystal causes its ends to assume opposite charges. The charge itself is not different from other charges; the name is descriptive of the manner in which it arises. A basic condenser can be constructed using a quartz crystal slice sandwiched between thin plates of metal foil. The vibrational interval of a quartz crystal is so constant that it is used for very fine clocks and to stabilize broadcast waves. See lap, quartz, Y-cut.

pigtail A short protrusion, usually wire, line, or rope, often intended for connecting to another component or transmission cable.

pike pole A long pole with a spike on the end to aid in erecting tall poles, masts, antennas, utility poles, and other tall narrow extensions which are frequently used to hold transmission wires. These are sometimes called boom poles, and their users boomers.

PIN See Personal Identification Number.

pincushion distortion A type of visual aberration in which the outer corners of an image are stretched outward and the centers between the corners curve in. The opposite of pincushion distortion is barrel distortion. See keystoning.

ping *packet internet groper*. In data communications, a software utility which employs an echo to detect the presence of another system and any delay which might be occurring in the connection. Often used as a diagnostic tool in conjunction with *traceroute*. Here is some sample output from ping, interrupted after three status outputs to the screen:

```
PING othercomputer: 56 data bytes
64 bytes from 192.42.172.20: icmp_seq=0. time=14. ms
64 bytes from 192.42.172.20: icmp_seq=1. time=2. ms
64 bytes from 192.42.172.20: icmp_seq=2. time=1. ms
^C
—othercomputer PING Statistics—
3 packets transmitted, 3 packets received,
0% packet loss
round-trip (ms) min/avg/max = 1/5/14
```

pink noise A slightly altered pure tone. This is sometimes used by "blue boxers" attempting to gain unauthorized use of phone lines for long-distance calls, to avoid detection up until the connection with the toll network. Once it reaches the network, however, only a pure 2600-Hz tone will be processed by the system. See blue box.

PINT Service Protocol PSTN/Internet Interworking Service Protocol. A digital telephony protocol to support the invocation of common telephone services from an Internet Protocol (IP) network. These services range from basic telephone or facsimile communications to content provision. PINT was submitted as a Standards Track RFC by Petrack et al. in June 2000. PINT is an enhancement to services offered through SIP 2.0 and SDP protocols. PINT services can be delivered over Intelligent Networks (INs), private branch exchanges (PBXs), cell phone networks, and ISDN networks.

PINT services are negotiated over what is presently two separate networks: the telephone network and the Internet. Calls are initiated on an Internet Protocol (IP) network, with the request passed onto the Global Switched Telephone Network (GSTN) for actual connection. The types of services that can be offered through PINT are defined as Milestone PINT services, and are restricted with an eye to the development and inclusion of further additional services. See RFC 2848.

pilot tone In linear modulation transmission schemes, a tone inserted into or above the frequency band to split or otherwise influence the signal with the intention of improving capacity or reducing noise (or both). A transparent pilot tone is one inaudible to human ears. See linear modulation, tone in band.

pipelining 1. In data networks, the transmission of multiple frames without checking for acknowledgment of individual frames at the time of receipt (this may be done later by a variety of means). 2. A technique

P

745

used in certain central processing units to aggregate processor instructions into a set of overlapping processing steps.

PIPS See Photonic Information Processing Systems Laboratory.

pits Minute indentations in the plastic medium of an optical recording disc which encode the information, and are read by laser pickup mechanisms.

PIU See path information unit.

pixel picture element. A unit representing the smallest resolvable area on a monitor or broadcast display. Typically used to describe individual picture elements in raster displays (e.g., computer monitors). The pixel is a useful unit in terms of storing raster images, creating image processing programs (e.g., computer "paint" programs), and describing the resolution of raster-based devices and image sizes (e.g., 1024×768 pixel display). See addressable graphics, pixmap, raster display.

pixel clock In video technology, this is a timing signal used to convert an incoming video signal into pixels, that is, into individual dots or display points. Pixel clocks are found in frame grabbers and can be output by some of the newer cameras. A frame grabber is a device that takes an analog video signal and converts it into a pixelized digital raster image (e.g., for freeze framing). Reference signals may be used to synchronize various aspects of video, especially when coming from different sources that may be at different points in the frame display cycle, that is, out of phase. The reference signal may be generated by the frame grabber or may be coordinated with an external reference source, such as the camera or a separate clock reference supplied to the frame grabber.

pixel jitter A type of undesirable interference that may occur in pixel synchronization systems such as pixel clocks. Jitter is variance in the timing of a clock that is expected to be even and reliable. Pixel jitter can cause a reference or signal being synchronized with a reference to occur slightly too soon or too late, degrading the signal quality. See pixel clock.

PKCS See Public-Key Cryptography Standards.

PKZIP A widely distributed, very popular, lossless data compression/archiving utility from PKWARE, Inc. that was first introduced in 1989. The program was initially released for DOS/Windows platforms, but was expanded to provide cross-platform support in 1997.

PKZIP has been evolving along with computer technologies. Newer versions can accommodate the increased storage capacities of bigger drives and the desire for users to secure their data for transmission over public networks. With version 4.5, the ZIP file's size is limited more by the capacity of the system than the capacity of the program archive. There is now also support for Public Key Infrastructure (PKI) digital signatures from recognized signature authorities. PKZIP files can be split to span several storage diskettes or cartridges or for reducing file size when sending file attachments through networks with file size limits. See Java Archive, RAR.

planar antenna array See antenna, planar array.

Planck's constant PC (*symb – h*) A formulaic description for a fundamental physical constant described by Max Planck in 1900 as a result of his study of blackbody emissions. Planck's constant satisfies a certain human need to see the world as ordered and held together with universal principles, yet leaves some room for dynamic (and as yet not fully understood) interactions.

Planck's observations in 1900 appeared to agree with experimental evidence, but only if one accepted that molecular vibration occurred in quantum units, that is, particular energy values that were in proportion to the frequency of the molecular vibration. This seemed surprising in light of wave theories that were accepted at the time.

Thus, light could be seen as consisting of photons with energy equal to the frequency of molecular vibration times a constant value, called Planck's constant, expressed as $h = 6.6262 \times 10^{-34} Js$. This inspired scientists such as Millikan to experiment with the photoelectric effect. Based upon Planck's model, Einstein suggested that light could have the same quantum nature and explained the photoelectric effect in terms of quantum mechanics, in 1905, for which he received a Nobel Prize. See quantum.

R.G. Planté – Physicist

A French postage stamp commemorating the contributions of Gaston Planté to storage cell technology. The original lead-acid cell was very simple, with only two plates, but later versions used multiple plates and were more effective than the original. [French Postal Service design by Mazelin, 1957.]

Planté, Raymond Gaston (1834–1889) A French physicist who was one of the first inventors of an electrolytic power storage cell (battery), historically called an *accumulateur*, in 1859. Planté and another scientist, the chemist Faure, gave seminars on *accumulateurs* in the early 1880s in Paris, inspiring further generations of engineers and theorists, including Henri Tudor, who created practical applications of storage technologies for the emerging automobile industry and mobile military installations, near the end of the 1800s. Planté's inventions were also important to the development of X-ray technology.

Planté battery Developed in 1859 by R. Gaston Planté, the Planté battery used a sheet of lead and a sheet of positively charged lead oxide separated by rubber, rolled into a tight cylinder, with two electrode connections pointing out of one end. The whole thing was then immersed in a dilute solution of sulfuric acid. This invention was the evolutionary forerunner to the lead-acid car batteries of today. A later version of Planté's device, with more elements, a housing, and terminals connected in parallel, was presented to the Academy of Sciences by César Despretz in 1860.

PLAR See Private Line Automatic Ringdown.

plasma display panel A type of display technology which consists of outer layers sandwiched around an inner layer of tiny neon bulbs imbedded in glass. The bulbs can be selectively lit by controlling voltages through matrix addressing. Unlike excited phosphors in CRTs, the bulbs remain lit until explicitly turned off, and thus do not need to be refreshed. Plasma panels are sturdy and flat and used in a variety of field applications. See liquid crystal display.

Plastic Optical Fiber Newsletter POF. A publication aimed at design engineers, installers, distributors, and end users of plastic optical fiber technologies, applications, and standards. Published 6 times per year by Information Gatekeepers, inc.

Platform for Internet Content Selection PICS. An effort of the World Wide Web Consortium since 1995, PICS seeks to create a means for voluntarily assessing and declaring the content of a Web site. This provides a way to implement filtering and searches. The first officially recommended PICS document was offered in 1996 as PICSRules1.1, and a number of Web browsers can now detect PICS ratings. PICS contains a language for defining Web page profiles and a number of rating systems have been developed, including ARC (Ararat Software), Net Shepherd CRC, SafeSurf Internet Rating System, and Voluntary Content Rating (VCR).

Platform for Privacy Preferences P3P. A development project of the World Wide Web Consortium (W3C) to provide a simple, automated way for Web users to better control use of their personal information on Web sites that they may visit and patronize. It is, in a sense, a checklist of questions related to a Web site's privacy policies that aids the user in assessing them. Thus, if a site is P3P-enabled, this information is available in a standardized, machine-readable format and can be automatically compared to the consumer's set of privacy preferences and discrepancies reported. A working draft of the first version of P3P was issued by the W3C in September 2001.

This type of development project has arisen due to the high volume of information, including personal demographics, gender, age, credit card numbers, etc. requested by online vendors. The problem is that an order form may *require* all or most of the fields to be filled out, thus forcing the consumer to volunteer information that makes him or her uncomfortable. The alternative is often a long, drawn-out search for a phone contact number (after the user has spent considerable time going through a shopping cart program and doesn't want to abandon the purchase). To better negotiate the volunteering of personal information and safeguarding of personal privacy, the P3P system is intended to arbitrate and streamline this process.

platform-independent Software or hardware which follows a standard, or is self-contained, so it can run without modification or significant modification, on a variety of computers. With faster processors, it is possible for more programs and languages to be designed as platform-independent, thus increasing the user's choice of systems. There exist standards designed to promote consistency in the design of software objects so that computer applications can be platform-independent as well. See CORBA, HTML, Java, Open Systems Interconnection, Perl, Unix.

PLC Power Line Carrier.

PLCP See Physical Layer Convergence Procedure.

PLL See phase-locked loop.

PL/M Programming Language/Microprocessor. The first programming language for the first commercially distributed microprocessor, the Intel 4004, designed by a consultant to Intel, Gary Kildall, who later founded Inter-Galactic Digital Research and developed CP/M. PL/M was developed to run on an IBM 360 computer to generate the code, which was burned into the ROM of the 4004. It was later modified by Kildall to support the successor to the 4004, the 8008.

PLM See Public Land Mobile. Some have used this to refer to Private Land Mobile, as well, but private mobile services are best distinguished by calling them PrivLM.

PLMR Public Land Mobile Radio.

PLS See physical layer signaling.

PLU See Percent Local Usage.

Plug and Play PnP. A format designed by a number of commercial vendors to overcome some of the problems associated with interrupt-driven computers based on IRQ assignment systems. Not all computer architectures use this means of handling interrupts, but many widely purchased Intel-based computers do. In the days when consumers had only one modem and one printer as peripherals, the IRQ-assignment design of these microcomputers was not seen as a particularly severe limitation. Unfortunately, as users added mice, sound cards, extra printers, scanner cards, network cards, and other device controllers to their systems, it became a significant problem to keep track of and set up IRQ numbers for interrupts so that there were no system conflicts, especially since some devices were designed to be associated with specific interrupts. As a result, some peripherals wouldn't work together on the same machine, and the user could run out of interrupts. At this point the architectural structure became a significant hindrance, and vendors developed Plug and Play to overcome some of the problems.

Plug and Play is a system that provides dynamic arbitration of system interrupts and sometimes also permits hot swapping of peripheral components or cards. To fully take advantage of Plug and Play, the user needs a Plug and Play-compatible operating system, a Plug and Play BIOS, and Plug and Play-compatible peripheral cards that don't have overlapping IRQ requirements.

Plutarch An ancient Greek philosopher who attempted to explain the static phenomena associated with lodestone, a natural magnet, and the attractive properties associated with the rubbing of amber. Although it took centuries for them to be understood, these early speculations led to an understanding of electricity. He also made an important observation about the differences between lodestone and amber, in that lodestone appeared to attract only certain substances, primarily iron, whereas amber attracted a multitude of objects, as long as they were very small and light.

PM performance monitoring.

PMA physical medium attachment. A device which connects physically with a network.

PMARS Police Mutual Aid Radio System.

PMDF A store-and-forward system from Innosoft International, Inc. for distributing electronic mail. PMDF is a mail transport agent (MTA) which directs messages to the appropriate network transport and ensures reliable delivery over that transport. PMDF is implemented for DEC Unix and VMS systems, and Sun Solaris and SunOS systems.

PMR 1. See poor man's routing. 2. private mobile radio.

PMS 1. See Pantone Matching System. 2. See Picturephone Meeting Service.

PMT 1. Photo Multiplier Tube. Light control technology commonly used in higher end drum scanners. See charged coupled device. 2. Photo Mechanical Transfer. A camera-ready layout tool used in the printing industry.

p-n junction, pn junction Within some types of semiconductors, a region of transition between p materials (*positive* holes) and n materials (*negative* electrons). See n region, p region.

PNG See Portable Network Graphics.

PNG Development Group A group initiated in January 1995 by the release of the first draft of the Portable Bitmap Format (PBF), which evolved into Portable Network Graphics (PNG). Thomas Boutell, Scott Elliott, Tom Lane, and many other early contributors developed an open source, freely distributable graphics compression format successor to Compuserve's proprietary Graphics Interchange Format (GIF). In an unprecedented development effort, the format went from first draft to final draft in only 2 months, a remarkable example to the standards-development community. See Portable Network Graphics.

PNM public network management.

PnP See Plug and Play.

PNP See Permanent Number Portability.

Pockels cell A component in laser assemblies that exploits the Pockels electro-optical effect to change the state of polarization of laser light. The effect can be further enhanced by combining the pockels cell with a polarizing component to control the opening or closing of a laser resonating cavity to provide switching capabilities. See Pockels effect, Q-switch.

Pockels effect A very small linear electro-optical effect, similar to the Kerr electro-optical effect, except that the change in the index is proportional to the applied electric field. Pockels-based cells require lower

voltages than similar Kerr cells. Optical shutters can be designed to exploit the Pockels and Kerr effects. See Kerr electro-optical effect, Pockels cell, Q-switch.

PODP Public Office Dialing Plan.

point of interconnection POI. The physical point of interconnection between various telecommunications networks. This may also represent a demarcation point for service and responsibility for connections.

Point of Presence POP. In data networks, a node consisting of a server, network connection, router, and one or more hosts, but not including a network operations center or network information center. A POP is sometimes serviced remotely from a Network Operations Center (NOC).

point of presence, telephony POP. In telephone networks, the central location of an Inter Exchange Carrier (IXC), the point at which the IXCs long-distance lines connect with the various local phone companies' lines. The point of connection for long-distance calls or for cellular calls.

point size A typographic unit of measure equal to ~.0139 in., approx. 1/72 in. (72.27). Point size is sometimes used to describe the sizes of graphic elements, and almost universally used to describe the height of a typestyle and distances between lines of type. It *may* include a small amount of space above and below the ascenders and descenders as it was originally based upon the physical size of a wooden or metal block on which the typeface was mounted, but does not include extra space added between the lines with *leading*.

It is not correct to refer to the size of a font on a computer monitor in terms of point size if you are talking about pixel resolution or size relative to the settings of the monitor. A 10-point font can vary from 8 scanlines to 20 scanlines (or more), depending upon the monitor settings and the degree of zoom. A 20-*scanline* font is not the same as a 20-*point* font. It is best to use point size to describe fixed media, such as the size of the font on a piece of paper once it has been printed. See ascender, descender, leading, pica, typeface.

point-to-multipoint A system in which a transmission originates from a single source and is transmitted to multiple destinations. Most broadcast media are point-to-multipoint.

point-to-point A system in which a transmission originates at a single source and is transmitted to a single destination. Many personal and business transactions, such as email addressed to a single recipient, are point to point communications. See Point-to-Point Protocol.

Point-to-Point Protocol PPP. A standard method for transporting multiprotocol datagrams over point-to-point links. PPP is intended for facilitating connections through a wide variety of hosts, bridges, and routers. PPP provides a method for encapsulating datagrams; a Link Control Protocol (LCP) for establishing, configuring, and testing the data-link connections; and a family of Network Control Protocols (NCPs) for establishing and configuring different network-layer protocols.

PPP is intended for simple, bidirectional, full duplex links transporting packets between peers. With PPP, different network-layer protocols can be simultaneously multiplexed over the same link.

PPP encapsulation requires framing to indicate the beginning and ending of the encapsulation. PPP consists of a 1- or 2-octet protocol field, followed by an information field of zero or more octets, followed by an arbitrary number of octets to pad up to the Maximum Receive Unit (MRU), which includes the Information field and Padding, as shown in the PPP Framing chart.

In order to establish communications, each end of the link must first send LCP packets to configure and test the data link. Once the link is established, peer authentication may be optionally carried out. PPP must then send NCP packets to select and configure one or more network-layer protocols. Once this has been done, datagrams may be transmitted. The link remains open until the LCP or NCP packets close it, or until some external event terminates the link. See Link Control Protocol, RFC 1171, RFC 1220 (Extensions for Bridging), RFC 1661.

Point-to-Point Tunneling Protocol PPTP. A networking protocol proposed as a draft in July 1997 by a group of commercial networking vendors including Ascend Communications, U.S. Robotics & 3Com, Copper Mountain Networks, Microsoft Corporation, and ECI Telematics. PPTP allows Point-to-Point Protocol (PPP) to be tunneled through an Internet Protocol (IP) network; that is, it provides a means of carrying PPP traffic so users can benefit from secure remote access over virtual private networks (VPNs). PPTP seeks to increase the flexibility of IP management.

Dial-in users of a common PNS can retain a single IP address. Users of other protocols such as AppleTalk could be tunneled through an IP-only system. In ISDN systems, where multilink PPP is typically used to aggregate B channels, a single PPTP Network Server (PNS) can handle the bundle, instead of grouping it at a single Network Access Server (NAS), thus making it possible to spread the bundle across multiple PPTP access concentrators (PACs).

PPTP specifies a client/server call-control and management protocol, allowing the server to control access for dial-in circuit-switched calls originated from a public switched telephone network (PSTN) or ISDN. It can also initiate outbound, circuit-switched connections. Flow- and congestion-controlled encapsulated datagram packet services are provided by an enhanced Generic Routing Encapsulation (GRE). PPTP provides for the decoupling of a number of Network Access Server functions in order to gain flexibility. See Point-to-Point Protocol.

pointcasting The broadcasting of specialized information or custom-selected information to a particular subscriber. Pointcasting is occurring over the Net, where it is possible to sign up for subscriptions for individualized information and have them electronically forwarded, somewhat like a traditional clipping service. See broadcasting, narrowcasting.

Poisson's bright spot See spot of Arago.

polar keying A transmission technique, commonly used in telegraphy, which employs two current states (as opposed to just turning the current on and off), i.e., current flowing in opposite directions, positive or negative voltages, to indicate *mark* and *space* signals. See telegraph, needle. The idea of using polarities led to later duplex and multiplex systems, and alternate polarities are now incorporated into digital transmissions systems.

polar mount A common type of antenna mount for parabolic antennas that is installed with the elevation mount pointed at the North Star (hence the name). Once installed, the hour axis, on which the dish swivels, is adjusted to correspond to the arc of the satellite orbit, and can be interfaced with a computer control device to simplify positioning. This type of mount is used with antennas that require orientation toward highly directional beams. See azimuth-elevation mount, parabolic antenna, microwave antenna.

polar relay A type of relay used in older communications switching systems such as telegraph systems, which uses a permanent magnet to center the armature, and a split magnetic circuit so the relay can be polarized to operate in one direction or the other. A vibrating circuit is sometimes added to a polar relay to increase its sensitivity and reduce its response time.

polarization *n.* 1. A movement or division in opposite directions or into opposite parts. 2. When radiation, especially light, vibrates perpendicular (normal) to the main electromagnetic beam (which may be straight or elliptical). In photography and sunglasses,

P

Point-to-Point Protocol Framing

```
+------------------+------------------+-----------------------+-
|    Protocol      |   Information    |   Padding up to MRU   |
|   8 or 16 bits   |    0 or more     |        ...            |
+------------------+------------------+-----------------------+-
```

A basic, unembellished frame as described in RFC 1171 looks like this:

```
+----------+----------+----------+----------+----------+----------+----------+
|   Flag   | Address  | Control  | Protocol | Informa- |   FCS    |   Flag   |
| 01111110 | 11111111 | 00000011 | 16 bits  |   tion   | 16 bits  | 01111110 |
+----------+----------+----------+----------+----------+----------+----------+
```

polarization is selectively controlled to influence the light reflections that pass through the lens. 3. In an atom, the slight displacement of the positive charge of a dielectric, when influenced by an electric field. 4. The orientation of molecules in a magnetic material, when aligned in the direction of magnetic lines of force. 5. In antenna transceivers, a way to process a transmission wave's characteristics and direction. Horizontal and vertical polarization provide ways to reduce certain types of interference and facilitate the directing of a beam. Vertical and horizontal polarization can be combined to create circular polarization. This creates a helical wave that can be received by either vertically or horizontally polarized antennas.

Polarization

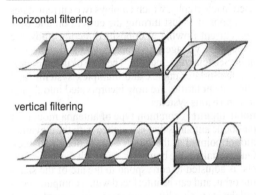

horizontal filtering

vertical filtering

An incoming wave can be separated into its horizontal or vertical polarized components by filtering. This has two advantages: it permits selection of incoming information and the transmission of more information in the same bandwidth.

Police Radio Service PRS. A mobile communications service for handling police communications such as dispatching, information requests, tactical instructions, and various administrative communications.

policy-based management In networking, an administrative adaptation to the management of networks that are too large to evaluate and maintain on a device-by-device basis. Policy-based management typically entails the development of a set of rules and guidelines for implementing network-wide decisions in areas such as security, quality of service (QoS), reliability, etc.

poll-safe See Logical Storage Unit.

polyvinyl chloride PVC. A weather-resistant polymerized vinyl compound, plastic or resin, widely used in piping and as insulation in low-voltage electrical installations. PVC is not suitable in high temperature environments, as the material will give out dangerous gases or will burn when subjected to heat.

poor man's routing A means of routing packets according to a route defined at the source, without using network layer routing algorithms. It can be a simple, appropriate means of accomplishing the trans-

mission on a small network with known characteristics, or on a larger network which is known to be stable and through which an efficient route has been precharted and considered to have a high likelihood of being available.

POP 1. See point of presence. 2. See Post Office Protocol. 3. Abbreviation for *pop*ulation. One population unit, in other words, one person. In wireless services, carriers are rated partly according to POPs.

Popow, Aleksandr Stepanowitsch (1859–1906) A Russian physicist and educator, who improved on the discoveries of Oliver Lodge and created a receiver with an antenna wire for better reception in 1895 or 1896. Around the same time, he sent a wireless message from an ocean vessel to his laboratory in St. Petersburg. See telegraph history.

populated In electronics, a printed circuit board (PC board) that has had components added to the traces, as opposed to a bare board.

port 1. Point of ingress or egress, or both. Data input/output point. 2. Entrance or exit to a network, firewall, or gateway; a transmissions interface. 3. Data input or output path to peripheral devices such as modems and printers.

port, access Access point to a virtual port, such as a chat room or a physical port, as a connecting point, switch, or card interface. Port management, security, protocol conversion, and traffic management are essential aspects of networks, especially mixed networks and those with shared resources. See port adapter.

port adapter In a Fiber Distributed Data Interface (FDDI) token-passing network there are two physical ports, designated PHY A and PHY B. Each of these ports is connected to the primary and secondary ring, to act as a receiver for one and transmitter for the other. The port adapters are typically configured as peripheral cards that fit into slots in an interface processor. In many systems, several slots are available, and it may be necessary to insert a blank card in open slots for ventilation, and to conform to emissions requirements for computer-related devices. Port adapters may be equipped with optical bypass switches to avoid segmentation which might occur if there is a failure in the system and a station is temporarily eliminated.

FDDI ports can be directly connected to either single mode or multimode fiber optic media, providing half duplex transmissions. LEDs are commonly used on port adapters as status indicators. Optical bypass switches may in turn be attached to the port adapters. See Fiber Distributed Data Interface, optical bypass.

port adapter module PAM. A Cisco Systems ATM module with ports that can be configured as redundant links for use by ATM routing protocols. Port status is indicated by LEDs, and each port can be configured to a variety of clocking options. Aggregate output traffic rates can be controlled to accommodate slow receivers of public network connections with peak rate tariffs.

port sharing A networked system in which two or more devices or two or more virtual connections

(VCs) share a single port. Common ports are printer and modem ports, and it is not uncommon for many computers to share a limited number of printers and modems. Port sharing is usually accomplished through a combination of software and hardware. In manual systems, the port may be shared by using an A/B switch device. For example, two computers may be sharing one modem and, since most modems cannot handle two ongoing transmissions, the switcher is set to one computer or the other. Multiple printers on medium or large systems may need more sophisticated management to handle the traffic, in which case an automatic switcher may handle the queues and access to the various printers or other ports. Not all ports connect to physical devices in the sense of a modem or printer. Some ports are virtual communications ports for chat lines, or other networked services, in which the handling of the port access is generally handled through software.

port-sharing device A device which selectively limits or administers shared access to a virtual or physical port. A switcher, from simple A/B switch to an intelligent logical switching device, is a common means with which to share resources on a networked system. See port sharing.

portability 1. An attribute of a hardware device that makes it easy to carry around. Pen computers, palmtops, and laptops are light and battery-powered to maximize their portability. 2. An attribute of software that makes it easy to move from one platform to another. Hardware-independence. Languages like HTML, Sun's Java, and Perl were designed to be portable. Unix software is portable to a wide variety of systems. 3. An attribute of a virtual address that allows the user to access a phone or computer network from a variety of locations. See mobile computing, cellular.

portable Easy to carry around, particularly devices which have desktop analogs as a desktop computer and a smaller portable computer, a desktop or wall phone, and a battery-powered, small portable phone. Portable is considered the category of devices that are easiest to use while mobile. Luggables and transportables are the next level up, as they are a little heavier and usually require more power consumption.

Portable Computer and Communications Association PCCA. A nonprofit organization established to provide a forum for the various industries to cooperate in the evolution of interoperable mobile computing and communications standards and implementations. http://www.outlook.com/pcca

Portable Document Format PDF. An Adobe Systems Inc. proprietary document encoding and display format used on many Web sites to distribute read-mostly document files. PDF files can be created and exported from many desktop publishing and layout programs, or with Adobe Systems Acrobat, which is designed for this purpose. Freely distributable PDF readers are widely available for download on the Internet. PDF files typically carry a *.pdf* file extension.

Portable Earth Terminal PET. A ground-based communications satellite terminal. The phrase is used generically and also refers to NASA units developed in the late 1970s.

Mobile Satellite Terminal

A portable Earth terminal for satellite communications. [NASA/GRC image by Brown.]

Portable Network Frame PNF. A format for using multiple subimages in a single frame; a multimedia graphics format intended as the foundation for the Multiple-image Network Graphics format (MNG) which is essentially Portable Network Graphics (PNG) for animations. PNF was submitted as Draft 1 by Glenn Randers-Pehrson in June 1996. With Draft 12, in August 1996, PNF came to be known as MNG. See Multiple-image Network Graphics, Portable Network Graphics.

Portable Network Graphics PNG. (*pron.* ping) A graphics standard developed during the first two months of 1995, with the Web in mind, which is recommended by the World Wide Web Consortium (W3C, WWWC). It received official "done" status in May, 1996. A standard for lossless, variable-transparency, cross-platform, "truecolor" graphics which includes, in the file, information about the authoring platform so the viewer can adjust the image accordingly.

PNG was developed as a result of an unexpected patent announcement by Unisys Corporation in the mid-1990s. Unisys claimed intellectual property ownership over the Lempel-Ziv-Welch compression algorithm, which was integral to many graphics compression schemes, the most visible of which was CompuServe, Inc.'s GIF format. GIF was well supported and widely used on the Web. Programmers were taken aback and reacted with concern, and soon there was a move to create an alternate format – not just a simple alternate, in fact, but a better format, which would provide an evolutionary improvement over GIF. At about the same time, CompuServe announced that they would begin development of a GIF successor, GIF24, followed by an official announcement a month later that PNG would be used as the basis for GIF24.

P

The Internet community of programmers, some of whose contributors included members of The World Wide Web Consortium (W3C) and Compuserve, began work on the development of PNG. The current consensus is that PNG is a good format and it is quickly being adopted by the Web community. The PNG specification was transcribed by Thomas Boutell and Tom Lane and released as RFC 2083 in 1997.

The features of PNG include an open software standard, good lossless compression ratios, 8-bit color palette support, 16-bit grayscale support, 24- and 48-bit truecolor support, alpha blending transparency for supporting different degrees of transparency, gamma correction, two-dimensional interlacing, text chunk support, multiple CRCs for error checking without viewing, security signature, full online references, and source code availability. PNG was not designed for animation support; it was intended for single images. MNG was subsequently developed to support animation. See Graphics Interchange Format, Lempel-Ziv-Welch, magic signature, PNG Development Group, Portable Network Frame, Virtual Reality Modeling Language, See RFC 2083.
http://www.cdrom.com/pub/png/png.html

Porthcurno The U.K. port location of historic submarine telegraph cables, established primarily by John Pender in the 1870s, Porthcurno (historically Porthcurnow, in southwest England) is now the port link for a new fiber optics cable, installed in 1997. It is also the home of the Centre for the History of International Telegraph Communications, including the Museum of Submarine Telegraphy images and artifacts documenting the region and telegraphic communications. The collection includes telegraph keys, sounders, relays, synchronizers, and more, some of which may be viewed online. The surrounding landscape includes historic cables under the sand and underground tunnels. http://www.porthcurno.org.uk/

portrait A descriptive word that refers to the direction of roughly rectangular objects, usually printouts, photographs, or monitors, which are oriented so the long side is vertical and the short side is horizontal. Contrast with landscape.

portrule A historic telegraphic device with a metal, toothed bar acting as symbolic contact points to define digits or code symbols. This device was later superseded by the simpler telegraph key.

POSIX 1003.0 Portable Operating System Interface UNIX. An open systems standards architectural framework, also known as IEEE 10030.0

POST See power on self test.

Post Roads Act A regulation of the telecommunications industry in the United States begun in 1866 with the Post Roads Act in which authority was granted to the Postmaster General to oversee rates for government telegrams and to assign rights of way through public lands. By 1934, after passing through some intermediate bodies, including the U.S. Department of Commerce, telecommunications became the primary responsibility of the Federal Communications Commission (FCC).

Post Telephone & Telegraph administration. Telecommunications operating bodies around the world that are individually controlled by their regional governments.

Postel, Jonathan B. (1943–1998) No dictionary about the Internet would be complete without a reference to Jonathan Postel. From the early ARPANET to the current Internet, he contributed three decades of low-key, passionate, dedicated service and volumes of fundamental information as a developer, advisor, protocol prototype implementor, and offical RFC documents editor. Like the underlying thread that runs through a tapestry, Postel held quietly to a vision, avoiding the fanfare and business opportunities that constantly presented themselves to pioneers of the computing industry. He chose instead to concentrate on the structure and orderly evolution of this most important communications medium, for the benefit of all. Over the years, Postel worked for a number of educational institutions and high technology companies including the Network Measurement Center at UCLA (ARPANET) and SRI International with Doug Engelbart. See IANA, Request for Comments.

postmaster The person responsible for configuring and maintaining a network mail server, often including administering users, setting up mailboxes, distribution lists, aliases, filters, etc. Many of the postmaster functions are actually handled by computer software such as dragons and mailer daemons.

PostScript, Adobe PostScript A powerful, high-level, device-independent page description language and document format widely used in desktop publishing and electronic document design. PostScript, from Adobe Systems, Inc., is used for bitmap and scalable images, scalable fonts, computer monitor display systems, and much more.

PostScript originated as the Design System language in 1976 at Evans & Sutherland Computer Corporation, a company renowned for its pioneering flight simulator programs. John Gaffney is credited by John Warnock as the inspiration behind many of PostScript's major design components. In 1978, when John Warnock joined Martin Newell at Xerox PARC, they reimplemented Design System as JaM (after their first names, John and Mark). At this time, the language was used for experimental applications in VLSI design, printing, and graphic arts, resulting in Xerox's printing protocol called Interpress.

Warnock joined forces with Charles (Chuck) Geschke in 1982 to form Adobe Systems Incorporated. He and Geschke further developed JaM in collaboration with Doug Brotz, Bill Paxton, and Ed Taft; PostScript was born of the effort.

One of the important developments that helped introduce the PostScript language and make consumers aware of its capabilities was the release of the Apple LaserWriter PostScript-capable printer in 1984. Although PostScript doesn't seem as remarkable now, in 1984 most people had 9-pin dot matrix printers and bitmap fonts that provided output that even the most undiscriminating viewer would admit was crude at best. Suddenly, with PostScript, a prosumer-priced laser printer could print legible, beautiful text down to 4 points in size in some of the finest fonts in the world. This launched the desktop publishing revolution, which is still having far-reaching impact on the publishing market, especially among small presses, self-publishing individuals, and genealogists.

PostScript fonts are some of the best computer fonts in the world. Adobe Systems maintains a large library for sale to consumers and service bureaus. PostScript fonts are scalable, in order to print out at the best resolution of the output device. Many people think that PostScript fonts and other scalable fonts are essentially the same, but most other scalable fonts are not integrated into a page description language, and so are not as flexible and powerful as PostScript fonts. Since PostScript is a programming language, fonts can be swirled, stretched, and individually rendered so that each letter differs from the previous in some essential way. The possibilities have even now not been fully exploited.

PostScript is commonly used to distribute documents on the Web, as is Adobe Acrobat format, a second-cousin to PostScript for displaying text and graphics. It is also possible to send high-quality PostScript documents through email, by sending the file as an email file attachment. This is a means by which people can send professional-looking text and graphics résumés, business documents, manuals, and much more over the Internet or be linked to a Web page for instant download. See vector fonts.

potentiometer 1. An instrument for measuring electromotive forces. 2. A device used to regulate a current by varying the resistances at either end. It can also perform the functions of a rheostat, which is more limited. Potentiometers are commonly incorporated into dials and computer input devices like joysticks. See rheostat.

POTS Plain Old Telephone Service. The basic analog phone service, which has been available from local phone companies and used in homes for years and years. No ISDN, no surcharge services such as Caller ID, Call Waiting, etc. See loop start.

potting To embed within an insulating or protective material or layer, usually for the purpose of reducing electrical interference or fire hazards. Potting is sometimes required in cases where higher voltage computer components may be interfaced with lower voltage phone lines.

Poulsen arc A device enclosed in a gas atmosphere with a strong magnetic field, which created an electric arc that could generate high frequency radio waves. It was found that larger versions of the Poulsen arc could generate even greater arcs.

The rights to market this technology were purchased in 1909 by C. Elwell, who formed the Federal Telegraph Company to design and build industrial arc transmitters for the newly developing broadcast industry. Eventually the technology was superseded by vacuum tube transmitters. See arc converter; Duddell, William; Poulsen, Valdemar.

Poulsen, Valdemar (1867–1942) A Danish scientist and inventor who built upon the work of William du Bois Duddell and devised a way to use an electric arc to generate continuous waves at high frequencies by placing the arc in a controlled atmosphere within a strong magnetic field. He collaborated with P.O. Pedersen in inventing a form of wireless telegraphy technology. In 1898, Poulsen recorded electronic waves on a thin conducting wire, a pioneer electromagnetic tape recorder that was called the *telegrafon* or *telegraphone*. He was awarded a patent for his magnetic recording device in November 1900 (U.S. #661,619) and the patent for the electric arc generator in 1902. His work with electric arcs provided a means to improve signal strength and stability in crystal-based radio sets.

One of Poulsen's exciting accomplishments that is uncredited in radio history is that he was broadcasting music from the town of Lyngby, Denmark, by spring 1909. He put a microphone near an early phonograph player on March 4, 1909 and transmitted to an amateur radio pioneer in Hellerup. See radio history.

power In the human sense, having the capability to exert political, physical, or other force over other objects, phenomena, or people. A physically powerful person has the ability to lift heavy objects. A politically powerful person has the ability to control the course of human events.

In the scientific sense, power has a much more specific meaning. Power is a measure of the rate at which *work* is done or, seen another way, the level at which *work* could be done. Thus, a more powerful current can generally drive more or larger electrical components, a more powerful engine can generally drive larger or heavier machinery (we say "generally" because there are other factors that may effect overall efficiency and linearity of the examples). The basic unit of work is the joule (J). Thus, power can be expressed in terms of watts and in its relationship to work (in Joules) over time. See power, electrical.

power, electrical Expressed for direct current (DC) in watts, the product of the electromotive force (in volts) and the current (in amperes). Thus, $P = E\,I$. In terms of resistance, according to Ohm's law, this can be expressed as $P = I^2\,R$.

Expressing power for alternating current (AC) is a little more complicated as another factor, the alternating phase of the current, must be taken into consideration. Alternating current alternates between positive and negative phases, as does the current, in a certain number of cycles per second (Hz). However, the phases of the two, while having the same periodicity, don't necessarily hit the same part of the sinusoidal phase at the same time. The separation of these phases is expressed in degress and there are designations for different aspects of the power (apparent, real, etc.) that must be considered in any calculations of AC power. See ampere, apparent power, Ohm's law, ohm, power, resistance, volt, watt.

power down To initiate or perform a sequence of operations in order to shut down a system. For example, a power down on a computer may involve closing files, asking the user to save data, logging out the user, etc. before actually terminating the power to the system. Power down sequences are designed to clean up systems, remove unwanted files, and prevent the accidental loss of data.

power hole digger A machine for digging deep narrow holes for utility poles, starting with the early telegraph lines. It was introduced in North America around 1915, although line workers using long spades still dug the holes in undeveloped regions for several decades after the introduction of the power hole digger. See pike pole.

Power Macintosh computer, PowerMac The successor to the original Macintosh line, the PowerMac is based upon IBM POWER RISC chip architecture. This popular line is gradually being succeeded by even faster Macintoshes built with G3 and G4 chips. The PowerMac series is not obsolete, however; a number of third-part developers such as XLR8 and Sonnet have developed reasonably priced PCI-format G3 and G4 accelerator cards that are compatible with most of the PowerMac machines, providing speeds ranging from 300- to 600- MHz and higher. See G3, Macintosh computer, POWER.

POWER, Power PC Performance Optimization With Enhanced RISC. A complex processor, one of the first superscalar processors, initially implemented by IBM with three integrated circuits (branch, integer, floating point). This technology was further developed as a microprocessor by IBM, Motorola, and Apple Computing in the early 1990s. The idea was to create a successor to the Motorola 68000 line and the Intel 80x86 line; the PowerPC was the result of this collaboration. The first version was the PowerPC 601, released in 1993, derived strongly from the IBM POWER specification. Since then, a series has been released, including the 603, 604, and G3, and the chips are incorporated most familiarly into the PowerMacs and Macintosh G3s.

power connector See p connector.

power hole digger A machine for digging deep narrow holes for utility poles, starting with the early telegraph lines. It was introduced in North America around 1915, although line workers using long spades still dug the holes in undeveloped regions for several decades after the introduction of the power hole digger. See pike pole.

power on self test POST. The process of checking internal systems prior to becoming fully operational that occurs in many electronic devices immediately after the system is powered on. For example, in computers, the system may check memory subsystems, configuration parameters, electrical voltages, the presence or absence of certain peripherals (e.g., monitors) before coming online in terms of initiating external data or operating systems that may be contained on a CD-ROM or hard drive. Many laser printers have self tests and may optionally display the results of the test and basic configuration parameters on a printed page. If the self tests fail, the system may power down, flash warning lights, or display a message on an LCD screen, depending upon how it is programmed to respond to problems and how early in the test sequence

the problem occurs. See power up sequence.

power save mode See sleep mode.

power up sequence The operational bootstrap and test sequence that a computer goes through when first powered on. This usually includes loading very low level routines, often from read-only memory (ROM), which then make it possible to load other routines and operating system capabilities from a hard disk, floppy, cartridge, or CD-ROM drive. It is very common for a computer to run through a hardware systems check in the power up sequence to test memory, sound, graphics, and other basic input/output devices. Device drivers and external device checks may also be performed, in addition to locating and interfacing with a network, if applicable.

If many devices are attached to a computer, you may have to power them up in the right order. If you turn on the computer before turning on external hard drives, CD-ROM drives, or video signal sources, the computer may not recognize the device or synchronize correctly with the signal. As a general rule, turn on peripherals before turning on the computer. Give a hard drive a moment to "spin up," that is, get the drive revolutions up to speed, before turning on the computer. Similarly, with a device such as a scanner or printer, which may also have test sequences, count to five before you turn on the computer. If a system is being powered up right after being shut down, it is important to wait 30 seconds or so before turning it back on. Some of the electronic components in a computer will retain current after the system is shut down, and a sudden surge of additional current may stress the circuitry. Give the current a few moments to drain off, then turn the system back on.

Newer peripheral bus technologies such as FireWire are more flexible in terms of monitoring a live electrical connection or data stream. They may be hot swappable and configurable on-the-fly, thus freeing the user from worrying about startup sequences.

PPDN Public Packet Data Network.

PPI 1. pixels per inch. See resolution. 2. See plan position indicator.

PPP See Point-to-Point Protocol.

PPS 1. packets per second. A means of quantifying network traffic by tallying the number of packets transmitted through a given point in a given amount of time. 2. Path Protection Switched. See SONET. 3. See Precise Positioning Service. 4. pulses per second.

PPSN Public Packet Switched Network.

PPTP See Point-to-Point Tunneling Protocol.

PRAM programmable random access memory. A chip that is sometimes used on computers to save semipermanent configuration settings such as monitor settings. The chip retains its information by being refreshed with power from a battery, usually a small lithium cell. This battery may have to be replaced every 6 years or so.

PRB Private Radio Bureau.

Precise Positioning Service PPS. One of the precise location data signals transmitted from Global Positioning System (GPS) satellites. This signal is for military and general government use, requires a specially equipped receiver, and is encrypted. Horizontal and vertical accuracy are about 22 to 28 m, respectively, and time accuracy is 100 nsec. See Global Positioning System, Standard Positioning Service (SPS).

Preece, William A British researcher who experimented in the late 1800s with conductivity methods for sending wireless communications. He was able to send a message a distance of 5 miles. See conductivity method.

Predictor Compression Protocol PCP. A method for transporting multiprotocol datagrams over PPP encapsulated links proposed in the mid-1990s. PCP is based on Predictor, a freely available high-speed compression algorithm implemented by Timo Raita in the mid-1980s. While not considered the fastest compression algorithm, it had the advantage of availability. See Point-to-Point Protocol, RFC 1978.

preform In fiber optics production, a rod that is created by a process of deposition and consolidation that is subsequently used as a *blank* or source rod for melting and drawing the medium into a thin fiber. The tube within which the preform is created through a vapor deposition process may be retained as part of the preform. See boule. See vapor deposition for a fuller explanation of the process.

prepaid phone card A credit card-like monetary storage medium which is charged or credited with a certain amount of prepaid phone access. This phone card can then be inserted in a card-compatible phone and will automatically allow access to calls up to the amount charged on the card. Copying machine cards are somewhat similar. Unfortunately, many phone cards are not rechargeable, which is an unfortunate waste of resources since a phone card can last many years, as do copy cards. This is mainly due to the way the accounting is done. The value of the card is not simply embedded in the card as it is on a copy card, rather it is handled through accounting software at the switching center. The best reason to get a phone card is to avoid inserting coins into a telephone, especially for a long-distance call during which you might be interrupted by an operator to add more change (which might not be handy). The denominations on a phone card vary from region to region, but amounts such as $4.95, $9.95, and $19.95 are common.

A variety of vendors offer phone cards; the service is not necessarily directly provided by the local phone company. This accounts for the different designs on the cards, the different ways in which they are promoted, and the different denominations that are available.

Presentation Time Stamp PTS. In MPEG-2 encoding, a timestamp that is encoded into the elementary packet stream. This is used for synchronization of different streams by comparing it against the System Time Clock (STC). A video decoder synchronizes the MPEG video data with the STC, with the assumption that the audio decoder follows suit. If the synchronization is within acceptable parameters, the decoded picture is displayed; otherwise, it is repeated, the STC is readjusted, or the next B or P frame is skipped over to maintain synchronization.

preset A setting configured in advance so it can quickly be accessed later. Thus, in video and audio editing, cuts and dubs are sometimes preset; in computer programs, times or online activities may be preset. Pushbutton radios can be preset to instantly tune to a desired frequency.

President's Task Force on Telecommunications Policy See Telecommunications Policy, President's Task Force on.

PREST Centre for Policy Research in Engineering, Science and Technology. U.K. group.

Pretty Good Privacy PGP. A powerful high-security encryption scheme developed in the early 1990s by Philip Zimmermann, based on the Blowfish encryption technology.

PGP provides privacy and authentication of transmitted messages. Only the person intended to see the message can read it. An intercepted message cannot be deciphered. Authentication provides assurance of the authenticity of the sender and that the message has not been changed. PGP is freely distributed to U.S. and Canadian citizens for noncommercial use by the Massachusetts Institute of Technology (MIT) in cooperation with Zimmermann and with RSA Data Security, Inc., which licenses patents to the public-key encryption technology incorporated into PGP. PGPfone has now been developed to provide for secure online digital phone conversations. In 1998, Network Associates acquired PGP technology. See Blowfish; International Data Encryption Algorithm; PGP Inc.; Zimmermann, Philip. There is a useful PGP-related informational FAQ on the Web. http://cryptography.org/getpgp.htm

preventive maintenance Regular inspection, testing, adjusting, and maintenance of equipment in order to prevent problems before they cause damage or affect service. Computers (especially floppy disk drives and the area around the fan) tend to accumulate dust every couple of years; the fuzz should be *gently* brushed out or vacuumed with a low-power, fine-nozzled vacuum, making sure the computer equipment is turned off and probably even unplugged. Don't use a metal nozzle for vacuuming and ground yourself first by touching the power supply casing to drain off accumulated static electricity. A good time to clean is when swapping or installing memory. Also gently check the various chips on the motherboard to ensure that they are firmly seated. Chips occasionally work their way up out of their seating or may loosen when a system is transported.

Monitors should be turned off when not in use, and should have a screen saver active when in use. Batteries in phones that have extra features (like LCD readouts) should be replaced regularly, *before* you lose all the phone numbers programmed into the system. See screen saver.

Price Cap Regulation A means by which local monopolistic phone companies are regulated so that rates remain the same for a specified period. Unlike Profit Cap Regulation, which did not carry large incentives to pare back staff or adopt more cost-effective, newer technologies, it was intended that Price Cap Regulation would provide incentives for innovation. Many companies changed from Profit Cap Regulation to Price Cap Regulation in the mid-1990s.

Primary Rate Interface PRI. One of two major categories of ISDN services, PRI caters to higher-end customers and businesses. See Basic Rate Interface, ISDN.

primitive A representation of a basic unit in computing. As examples, a graphics primitive may be a circle, line, or square; an audio primitive may be a phoneme or sound; a programming primitive may be a routine, procedure, or object. A primitive is some type of basic building block, usually one frequently used or reused. In networking, a primitive is an abstract representation across a layer service access point, where information is exchanged between a user and provider.

principal Main, central, overriding; the highest authority or administrator.

principle A general, or fundamental concept, statement, or truth. A basis for decision-making, actions, or operations.

print server, printer server A system that handles the logistics of requests to one or more printers on a network. Frequently there will be a variety of types of printers shared among users. These might include plotters, laser printers, dot matrix printers, high-speed page printers, and specialized color dye sublimation or thermal wax printers. The print server handles queuing; messages to users if a printer is not in service; alternate routing if the printers have been reorganized; scheduling, if some types of jobs (e.g., big ones) are to be run at night or after hours; and prioritizing, if some users have higher precedence. The print server can also be used to send messages to maintenance personnel if there is trouble with paper jams, empty trays, etc. Some printers have sufficient processing power to send status and error messages to the server, which in turn may be relayed to the user or the appropriate service center.

print spooler An application which manages and schedules a printing job to a printing device in such a way that the computer is not tied up, waiting for the print job to finish. For example, imagine sending a large plotting job to a plotter from a CAD program. If it is a single-tasking system and cannot handle the print job in the background, and if there isn't a large buffer in the plotter itself, it might take 10 to 40 minutes for the plot to finish, and the computer would be unusable for that period of time. In order to reduce wait time on this type of system, print spoolers were developed so that a plot could be printed to disk rather than to the plotting device, a process which might take 2 minutes instead of 20. The plot can then be spooled to the printer during a lunch break, after hours, or when the plotter is not tied up by another user. With the spread of multitasking systems and printers with large buffers, the use of spoolers is diminishing.

On larger networks, a print server may handle printing tasks as a type of "smart spooler." In other words, the print job might be sent to a file, or sent directly to the print server, and scheduled and spooled from there rather than from the originating machine. See spool.

printed circuit board PCB. A board upon which electronic circuits are mounted, with the circuit connections etched, foiled, or blasted onto the surface of the board, usually on the side opposite the majority of the components. The etched electrical pathways, called *traces,* provide flat, convenient electrical contacts without wires. This is a very practical, lightweight method of doing away with wires and enables mass production of PCBs.

The conventional wisdom is that printed circuit boards were first invented in the 1940s, but the American Radio Museum has a radio in its collection dating from the late 1920s which is designed with a copper, blasted circuit on the underside of the board as shown in the following photos. Thus, the founder of the museum has fascinating evidence that the technology was introduced almost 20 years before its previously acknowledged invention.

printer A device for transcribing information onto a medium which can be read directly or otherwise understood directly, or with a minimum of manipulation (as in mirror writing), by someone familiar with the communications medium (writing, illustrations, Morse code, seismographic charts, etc.). The printing medium is often paper, card stock, metal plates, or other portable media.

printer control character A character which has a specific control effect on the action of a printer. The effects include line feeds, page feeds, carriage returns, mode changes, font changes, page length control, and other features that might be specific to the printer. Control characters can be sent to a printer before sending the document, in order ot set up the printing parameters, or may be imbedded in the document itself to set typefaces, font sizes, text attributes, space, line feeds, etc.

Printer control characters are handled transparently by word processing programs, which send the appropriate characters through a printer driver without explicit programming by the user. In cases where a user is imbedding printer control characters in the document, a hex editor, or ASCII editor with hex capabilities, is often used because the control characters cannot always be entered from the keyboard, as they lie beyond the range of the alphabet. Printer control characters are often not displayable, or may display as unusual symbols.

printer control language A language designed to utilize the capabilities of a particular type of printer, or one conforming to a common standard such as Hewlett Packard Graphics Language (HPGL), widely used on plotters, for example.

printer driver A file or program providing information on the physical, operational, and control characteristics of a printer. This may include relevant control codes, available font shapes and sizes, paper feed controls, etc. Typically an operating system or applications program will interact with a printer through a printer driver stored as a computer file. Many printer drivers may be available, and if the relevant one is not online, often a substitute can be found from a maker with a similar printer. Many different laser printers will function with the same commands as Hewlett Packard or Apple laser printers, and many impact and inkjet printers will work with Hewlett Packard or Epson printer drivers.

Historic First Printed Circuit Board – 1928

Copper traces blasted onto the underside of a 1928 cabinet radio discovered by Jonathan Winter, curator and founder of the America Radio Museum (the wires were added later). This is currently the earliest known radio based upon a "printed" circuit board.

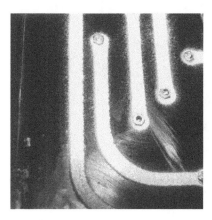

Detail close-up of copper blasted traces on the underside. Small bolts and nuts were used to provide the electrical connection to the vacuum tube components on the top side. This historic example of one of the first printed circuit boards in existence is part of the American Radio Museum collection.

---

printer emulation Any software which formats and outputs text to a device as though it were outputting to a printer. This way, for example, a word-processed program with all the text formatting, margins, and images can be sent through a facsimile transmission without first being printed and scanned. This saves paper and avoids problems such as slippage through the fax machine.

printing Printing is the production and reproduction of characters, symbols, and images on physical media for the purposes of preserving information or artistic works and/or to communicate or distribute them to others. Printing has long been associated with printing presses that enable many copies of a single master to be created.

There is evidence that movable type, in the form of clay discs, may have been used in Crete as early as 1500 BC, and type was used in southeast Asia by the 11th century. In the mid-1400s Johann Gutenberg introduced movable type to the west. Type was set by hand until the early 1800s, when the first typesetting machines began to appear.

Mass market microcomputer printers were introduced in the 1970s, and the home and prosumer desktop publishing markets became widespread in the mid- and late-1980s. Dry printing processes (e.g., photocopying with toner) became prevalent in the 1970s and have replaced a large proportion of wet printing processes (e.g., ink). A modern desktop laser printer costing $1000 is easier to use and more flexible than the letter presses of the 1960s that cost $30,000 or more.

Color printing processes are numerous, from traditional ink presses which use *spot color* while running the paper through the press several times, once for each color, to *process color*, which uses cyan, magenta, yellow, and black dots to simulate a wide range of colors. Computer printers use thermal wax, dye sublimation, inkjet, and colored ribbons to create printouts that rival commercial color photocopies and photographs, and may someday supersede them.

priority A level of access or usage which ranks higher than others. For example, on a computer system, operating system functions usually take precedence over user or network requests. On a server, such as a print server or network server, certain types of tasks may be handled first. In graphical user interfaces, a window which is clicked to the front may be assigned a higher priority, and greater proportion of processing time, than windows in the background. On most network systems, system administration functions take priority over user functions.

Priority Access and Channel Assignment PACA. A Supplementary Bearer Service (SBC) scheme for handling radio communications transmission access and channel assignment. PACA is of interest to emergency service administrators and providers because it can be used to queue priority calls (e.g., emergency calls) when all channels are busy. ANSI standard 664 describes PACA queuing for priority call handling. Enhanced Priority Access and Channel Assignment (PACA-E) was developed in the mid-1990s to support up to 15 priority levels to support emergency service Personal Communications System (PCS) users and is described in ANSI T1.706-1997. In 1998, the Technical Subcommittee T1P1 submitted a draft proposal to the American National Standards Institute for a Stage 2 Service Description for PCS-based PACA-E. This document defined and described call setup procedures for priority access and priority egress for PCS systems.

Typically, many more calls are made during storms, natural disasters, and times of local or national emergencies. This can overload land-based trunk lines and result in "fast busy" signals. Under these conditions, mobile networks also tend to become congested and PACA queuing has been studied as a means to handle busy channels in private and public wireless communications. PACA services can be accessed by the caller through a designated high-priority line or through dialing an assigned access code. In studies of the efficiency of PACA queuing, static queuing, and dynamic queuing conducted in the late 1990s, PACA compared favorably to the other methods.

Priority Access Service PAS. A program of the U.S. government, in association with the National Communications System (NCS), to provide a means for national security and emergency preparedness (NS/EP) telecommunications users to obtain priority access to available wireless channels when needed for initiating emergency calls. PAS was established because wireless airwaves are not unlimited and are frequently congested during emergency situations. Priority access provides some level of guarantee of completing an important national security or emergency-related call. See National Communications System, Wireless Priority Services. http://pas.ncs.gov/

prism A physical component for separating propagating light into its component parts or an algorithmic method for separating modeled light into its component parts. Prisms can be more broadly defined as separating electromagnetic energy into its component parts, but the term is largely used to describe prismatic effects in the optical spectrum.

Physical prisms come in many shapes, sizes and materials but are generally transparent to the wavelengths of interest. Thus, glass and crystals are commonly used for prisms for optical frequencies. What they have in common is the ability to refract light such that different frequencies are refracted in different directions.

A Selection of Common Commercially Available Prism Lenses

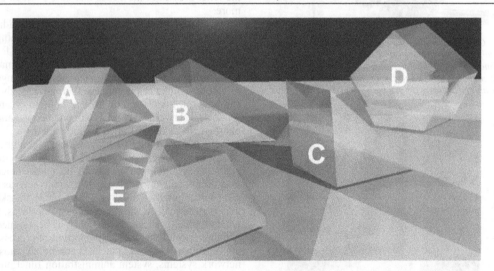

This illustration includes a number of common prismatic lenses for directing and separating light into its component parts. They are also useful for projecting, reverting and inverting images.

A – equilateral prism, *with three 60° angles, commonly called a* dispersing prism *since it disperses white light into its component colors.*

B – Littrow prism, *a type of dispersing prism, similar to the equilateral prism in that it can break white light into its components. Some versions will have a coating on the long surface (the plane of the hypotenuse) to divert the light beam such that the image is not inverted. See Littrow configuration.*

C – right angle prism, *a means of directing the light 90° from the incoming (incident) angle. This prism can be used to reverse and image left ot right or to invert the image, depending upon the orientation of the prism to the source and viewer.*

D – pentagonal prism, *similar to the right angle prism in that it can direct light 90° from the incident angle, but does so without inverting or reverting the image due to the light path being reflected twice within the prism at 45°.*

E – dove prism, *an interestingly-shaped prism with the top surface coated, this prism will rotate an image twice for one rotation of the prism relative to the viewer (in the longitudinal direction) Dove prisms come coated and uncoated.*

Assembled block prisms made from individual blocks of bonded glass or synthetics are not shown here, but they are a specialized type of prism designed to split light beams for selectively directing or polarizing light.

Traditional triangular prisms will refract light in the colors of the rainbow as well as infrared or ultraviolet light in frequencies invisible to humans (but visible to some insects, snakes, and birds). Some materials have natural refractive properties and others can be grown or manipulated to enhance their prismatic qualities.

A beamsplitter is a specific type of prism that separates and directs the separated paths of light so light of the desired properties is available for sending through optical components and undesired light is discarded through redirection, absorption or attenuation. Beamsplitters are commonly found in viewing scopes, interferometers, spectrophotometers, and many other instruments. See Iceland spar, Nicol prism, polarization, refraction.

Private Automatic Branch Exchange, Private Branch Exchange PABX, PBX. A private telephone exchange, usually located in a business or educational institution, which can handle switching and other functions automatically. In an automated exchange, an operator is not needed to handle outgoing calls; connections can be made by first dialing "9" to access an outside line. PABX was derived from Private Branch Exchange (PBX) which originally was attended by a switchboard operator. Since private exchanges are almost all automated, the terms PABX and PBX are now used interchangeably. See Centrex.

Private Cable Operator PCO. An independent cable operator providing video and telephony services to subscribers in niche markets such as multiple dwelling units. Many markets that are too small or specialized to be readily served by local exchange carriers (LECs) and thus alternative service providers have found viable ways to provide telecommunications services to these subscribers. These markets largely consist of apartment, motel, college, and housing cooperative dwellers. PCOs work in close cooperation with multiple dwelling unit (MDU) owners and managers. PCOs are supported, in part, by the Independent Cable & Telecommunications Association.

private carrier A privately owned, commercial public messenger service or telecommunications service provider that may or may not be in competition with a dominant commercial carrier or government-funded service.

private line In the early days of phone service, a private line was a line that went from one business or person to another, without necessarily going through a public phone exchange, or from one floor or room to another. It was not uncommon in the early part of the century for hundreds of wires to crisscross over a street between one building and another. As more people were connected through public wiring systems that could handle multiple connections, the meaning of the phrase changed and, until about the 1960s, a private telephone line came to mean one that was not a party line. That is, the phone line was dedicated to only one user, and there was no possibility of a neighbor on the same exchange listening to a conversation or tying up the connection so another subscriber couldn't dial a call. As private lines became the norm

and party lines began to disappear, "private line" began to take on a different meaning, similar to its early meaning, referring to a direct connection between two businesses, between a home and a business, or between different departments in a business or institutional complex.

Private Line Automatic Ringdown PLAR. A means of interconnecting two lines to form a "hotline" direct connection.

privileges 1. On a telephone system, particular functions and services available on particular consoles or to particular individuals in a company, sometimes through keying in an access code. 2. On a data network system, access to specific applications, processes, devices, or data files. More specifically, file privileges are a record of the actions an individual or group member can take on a file, typically *read, write, execute,* or *delete.*

PrivLM Private Land Mobile.

PRMA packet reservation multiple access.

PRML See Partial Response/Maximum Likelihood.

probe A detection or measuring device, often with a narrow tip, used to assess temperature, wind, humidity, current, voltage, amperage, polarity, or other properties of air or electrical circuits. A probe is often used in conjunction with an analog or digital readout displaying the results. A probe may be used as a diagnostic tool to troubleshoot or configure a network. On a computer system, a probe is a software process that seeks out specified information or detects certain actions or types of data. The results are typically reported to another program, which analyzes the information and acts accordingly. A probe can be used to locate a particular site, user, address, archive, or other type of information. See protocol analyzer.

process A software activity consisting of carrying out a set of predetermined or situation-influenced logical instructions, which may be low-level background processes integral to the operating system, router, or intelligent switcher, or higher-level processes related to the running of an applications program.

process switching On a network, packet processing at process level speeds, without the use of a route cache, as is used in fast switching. A Cisco Systems distinction.

processor configuration register PCR. processor configuration register. A computer processor-related programming register that typically contains a variety of bootstrap, break, data rate, and software configuration information. When spelled in capitals, it refers more specifically to the PCR in the Motorola 68000-family CPU.

Prodigy One of the earlier commercial online network services, established by IBM and Sears Roebuck. Prodigy used proprietary graphical user interface to provide access to the Internet. Many parents signed on to Prodigy to give their children educational access to online services. Like most of the largest commercial services, Prodigy earned revenues through ads on viewer pages. This practice is uncommon on the smaller, independent Internet Services Providers. Now that freely distributable Web browsers

P

are available, proprietary software is only used on some services (AOL). However, the ads remain, as many Web sites are subsidized by ad revenues.

Producers Advocacy Group PAG. An ad hoc collective of independent public radio producers formed in October 1995. PAG members are responsible for some of the best known and respected public radio programs in North America, including drama, documentaries, comedy, and popular music programs such as *The Thistle and the Shamrock*. They have also provided educational programs at home and abroad. PAG funding is from diverse sources including charitable trusts and endowments, the Corporation for Public Broadcasting, and corporate and individual donors. See Association of Independents in Radio.

PROFIBUS process field bus. An internationally standardized, open data bus standard suitable for a wide range of manufacturing and process automation applications. PROFIBUS facilitates communications between devices from different manufacturers. Originally a basic industrial communication protocol, PROFIBUS is now an extensive automation communications technology. It enables distributed digital controllers to be interconnected. Using PROFIBUS, a variety of automation and visualization systems can be jointly controlled and operated.

PROFIBUS International supports technical committees and working groups to maintain the open, vendor-independent nature of PROFIBUS technologies. See INTERBUS. http://www.profibus.org/

PROFInet An open, Ethernet-based automation implementation of PROFIBUS technology. See PROFIBUS.

profile alignment system PAS. In fusion splicing of fiber optic filaments, an automated, video-based alignment and measurement system. See fusion splicing.

profile alignment system, lens L-PAS. Similar to a profile alignment system with video-based alignment of fiber optic cladding and then lens effect associated with the fiber core.

Profile F, TIFF-F A black-and-white subset of the TIFF specification suitable for minimal facsimile document encoding but which has some extensions to the basic *Profile S* specification. See TIFF-FX.

Profile S, TIFF-S A basic black-and-white subset of the TIFF specification suitable for minimal facsimile document encoding. See TIFF-FX.

Profit Cap Regulation, Rate of Return Regulation Prior to the mid-1990s, the predominant means by which local monopolistic phone companies were regulated. Excess profits were required to be passed back to consumers as, for example, a rate reduction. By the mid-1990s, many companies changed to Price Cap Regulation. See Price Cap Regulation.

Program Clock Reference A synchronization reference clock used, for example, in MPEG decoding. The PCR can synchronize a Station Time Clock (STC). An MPEG-2 video decoder chip can be designed to include an internal counter useful in a Station Time Clock, which can in turn be accessed by other components through an interface.

program counter In general terms, a display or internal reference that keeps track of a location in a presentation (e.g., video or laserdisc program, TV broadcast, computer animation, etc.), sometimes to provide information to the viewer and sometimes as a reference point for searches or editing. In computer software execution, a program counter is a reference that monitors the location in a program that is currently being accessed. This is handy when debugging, testing a program, or tracing a logical path.

programmable Any device which can be controlled or altered through logical instructions without reconfiguring the physical connections.

programmable array logic PAL. In the design and manufacturing of computer circuits, a PAL is a circuit in which the OR array is predefined and cannot be changed, but the AND array is programmable. This was a simpler approach that followed the development of programmable array logic (PAL). See programmable logic.

programmable logic Circuit logic designed so that it can be reconfigured, as through linked flip-flops. In conjunction with memory circuits, programmable logic enables stored logic functions to be called upon and executed to configure the circuit. Programmable logic devices range from simple to complex (complex programmable logic devices [CPLDs]). The programmable logic array (PLA) is a more complex forerunner of programmable array logic (PAL). See programmable array logic.

programming language Instructions used by a programmer to control computer operations. Programming languages are roughly divided into low-, medium-, and high-level languages. Low-level languages are those which most directly translate into machine instructions and interact most directly with the hardware architecture of the system. Machine language and assembly language are considered low-level programming languages. Machine language programs are typically written in binary, with ones and zeros. Assembly language is similar to machine language, except that instructions are more symbolically represented, and routines can be written to pass control to different parts of the program. Machine language and assembly language are more difficult and time consuming for some to learn, and more difficult to trace and debug than higher-level languages.

Medium-level languages include those which are reasonably powerful, somewhat cryptic in their instruction sets and syntax, but comprehensible enough that some of the commands resemble written English. C is a common, somewhat medium-level language. It is powerful, but requires a good understanding of memory allocation, pointers, and arrays, and takes some time to learn and to apply. C is a compiled language, which means that the code is compiled down to machine code in advance, before the program is run. Higher-level languages like BASIC and various authoring systems were designed to be easy to learn and to use, with commands and syntax that are fairly close to written English. They often are run as interpreted languages (although compilers may exist), and so do not require knowledge of how to configure and run a compiler before they can be used. Interpreted languages are translated to machine code as the program is run, and thus will execute more slowly than a program which is precompiled. Interpreted languages tend to be more limited, but also more portable, than lower-level languages.

programming overlay, configuration overlay A covering made from membrane, plastic, cardboard or another material that provides information on a keyboard, keypad, or graphics tablet setup. Overlays were especially prevalent before graphical user interfaces, when complicated, difficult-to-remember control code combinations were used to run word processors and graphics programs.

progressive scanning A method of displaying broadcast video signals in which each frame is transmitted one after the other, rather than dividing the frame into sets of two fields interlaced. See interlaced.

Project Athena See Athena project.

Prokhorov, Aleksandr Mikhailovich (1916–2002) An Australian-born Russian physicist, Prokhorov (sometimes transliterated as Prochorov) studied propagation in radio waves and pioneered the study of coherent radiation, in 1947. In 1951, he published *Coherent Radiation of Electrons in the Synchotron Accelerator* for his Ph.D. thesis. Following this, Prokhorov researched a range of quantum electronics and radiospectroscopy topics. In the 1950s, he was appointed as head of the laboratory of oscillations at the Institute of Atomic Energy in Russia.

Together with Basov, Prochorov contributed important developments in microwave spectroscopy and ammonia-based molecular oscillators (the forerunner of masers and lasers). By 1957, Prochorov and Manenkov had suggested that ruby could be used in the construction of lasers. In 1958, Prohnorov suggested that a laser could be used to generate wavelengths in the infrared spectrum and that a open resonating cavity could be utilized (similar to an FP interferometer). For his contributions to fundamental principles in laser technology, he was coawarded a Nobel Prize for physics in 1964. See laser; laser history; Townes, Charles.

promiscuous mode In data networks, an open mode in which the network interface controller (NIC) passes all the frames which it receives, regardless of the destination address, to high-level layers in the network. This is usually only done in diagnostic situations, or by users gaining unauthorized access to information from the system. In normal operations, frames are evaluated and selectively passed along if the destination address maps to that device.

prompt *n.* 1. A mechanism for gaining the attention of the user to indicate that the system is ready for input or that input is required before continuing. 2. A prompt on a computer system may be in the form of a cursor, dialog box, flashing area, audible tone, or spoken message. 3. A prompt on an automated phone system may be a spoken question or suggestion to which the user can respond by typing in codes or, in some cases, by clearly speaking numbers or words.

PROMPT Institute for Prospective Technological Studies. Founded in 1989 as a result of reorganization of the Joint Research Centre to monitor and analyze new sciences and technologies. Monitoring is carried out by by the European Science and Technology Observatory (ESTO).

proof of concept A strategy for communicating an idea which is not readily accepted when communicated through verbal means alone. Proof of concept usually involves producing a prototype which is partially or mostly functioning, at least enough to show that the idea can work. Many new inventions or ideas are not believed until they are actually demonstrated. Edwin Armstrong spent years going against the stated assumption that frequency modulation was mathematically impossible, but eventually succeeded in his attempts. When it was shown that it *was* possible, resources were made available to develop and implement the technology. Proof of concept demonstrations are created to attract interest, support, or research and investment dollars.

propagate 1. Pass along, continue, extend. 2. To travel through a material or space, to cause to spread out over a greater area.

propagation The concept of propagation is intrinsic to every aspect of communication. When you send a message, sound, computer packet, modulated radio frequency signal, etc., you are propagating a phenomenon through space/time and any data along with any information coded into the propagation of the phenomenon. The study of propagation involves very fundamental research of transmissions media, such as radio waves, as well as higher-level research in ways to compose, encode, transmit, and route analog and digital signals. It also is dependent upon mathematical theories and applied engineering techniques for determining and implementing the most effective use of limited resources.

Radio wave propagation, for example, can be facilitated by a number of physical phenomena associated with the Earth, its atmosphere, and various solar influences. The ionosphere is often used to propagate radio waves, but there are other methods, including auroral, grayline (the sunrise and sunset zones), tropospheric scatter, and more.

Since it can be time-consuming and impractical to physically test alternative propagation models for radio or computer communications, programmers have developed modeling programs that enable users to try different scenarios for efficient propagation. If the software is sophisticated enough to develop the different options as well as running unattended, a great deal of data can be generated in a short amount of time, with the software even suggesting the best alternatives from those discovered. These can then be tested on real systems. Propagation models are often used to develop prediction models.

The U.S. Department of Commerce NTIA/Institute for Telecommunication Sciences makes software available online for modeling high-frequency radio wave propagation. See attenuation, backscatter, hop, ionosphere, routing.

propeller head *slang* A technical person who may lack social graces. Derived from beanies (caps) with propeller tops that were popularized by the Beanie and Cecil characters of the 1950s; the caps became especially popular in geek/hacker culture in the 1980s when prominent notables wore them to scifi cons.

prosumer A combination word derived from *professional* and *consumer,* intended to indicate a level of literacy or competency somewhere between a skilled layperson, and a technical professional. In other words, the reading level or skill level for the product or service requires more knowledge than possessed by an average layperson but not as much as might be required by a technician or design engineer in that industry. Computer technology has brought many new products into the marketplace that were previously used only by professionals with years or decades of experience. The printing, desktop publishing, and desktop video industries now include a large number of prosumer products.

PROTECT Act A bill introduced to the U.S. Senate by Senator John McCain and others in April 1999. It is strongly associated with the Safety and Freedom through Encryption (SAFE) Act and covers the use and export of encryption products. It seeks to guarantee the right to domestic use of encryption products and to ease export restrictions, but not to the degree suggested in the version of the SAFE Act debated at the time. In June 1999, it was approved as S 798 by the Senate Commerce Committee.

protected distribution system PDS. A transmission link with a structure that adequately ensures secure communications in terms of electromagnetic, acoustical, algorithmic, and physical safeguards for the communication of unencrypted but classified information. For example, a proprietary system that was electrically and or algorithmically different from all surrounding or interconnecting systems might be sufficiently secure to transmit sensitive information without encoding it prior to or during transmission. However, unencoded systems are difficult to fully secure and encryption is usually recommended except in very specific local, short-haul situations. In military/political jargon, these were traditionally called *approved circuits.* See encryption.

protocol 1. A code of etiquette and procedure. 2. An agreed-upon system of configuring a communications network and carrying out transmissions flow so that connected systems can "talk" with one another and exchange information. There are many different types and levels of protocols used in computer networking and telephony. Internet Protocol (IP) is one of the most widespread and important general network protocols and File Transfer Protocol (FTP) is one of the most prevalent of the specialized utility protocols.

protocol, networking In networking, a procedure for organizing and exchanging data transmissions that may include specific rules and/or formats. Protocols for various Internet transmission and application protocols are well documented in online RFCs (Requests for Comments), some of which are standards or draft standards. For technical information on protocols,

consult the RFCs. Many of the important protocols are briefly described in this dictionary under individual entries, but especially important on the Internet are Internet Protocol, Point-to-Point Protocol, and Transmission Control Protocol. See the Appendix for a list of RFCs related to Internet protocols.

protocol analyzer ProtAn. A diagnostic tool that can capture a segment of network data traffic to reveal its characteristics. A protocol analyzer is useful for educational programs, systems development, systems monitoring, and troubleshooting.

In the early days of computer bulletin board systems (BBSs), there were relatively few protocols; they were fairly simple in design, and problems with a network connection were often traceable to a hardware device or break in the physical link. Since then, networks have evolved, the number of protocols has burgeoned, and network capacity and speed are now further optimized through sophisticated software processes. A protocol analyzer has become an indispensable tool for network administrators and, given the learning curve on all these new technologies, more sophisticated expert system versions have been developed to help the administrator search and evaluate the data based on heuristics and common scenarios.

The information extracted by the ProtAn may be displayed as text, tables, graphics, or a combination of these. Hardware or software filters help the user to separate out data of particular interest. Much of the information extracted by the ProtAn is contained in the headers associated with packets, but general statistics and overall characteristics are also important. Wireless communications, using radio frequencies, differ somewhat from wired communications using conductive wires or fiber optics. There are ProtAns for both categories of media that are similar in concept but somewhat different in detail:

On wired systems, much of the protocol analysis is handled with software tools on a computer attached physically to the wired network. The analyzer may be used in conjunction with a decoding engine to display information about the captured data. Analyzers vary in design, but they can usually capture data from a live network or examine data previously captured and stored for later analysis. A ProtAn can reveal normal activities like traffic load and utilization (e.g., peak usage), server queries and lookup activity, domain name resolution, and the ASCII contents of a data stream. It can also help detect potential problems, including long pauses or duplicate or missing sequence numbers. A wide variety of packet types can be analyzed by freely distributable protocol analyzers, including AARP, ASCEND, BBOTP, FTP, LDAP, LDP, TCP, TELNET, TFTP, WLAN, and many more. Commercial ProtAns are also available.

In wireless communications, self-contained transceiver units with built-in software are more commonly used or may be designed as portable units with serial (or other) interfaces to transfer data to a desktop computer for further processing. Since wireless radio and telephony systems are increasingly prevalent, some commercial wireless protocol analyzers

(WPAs) can provide data analysis and evaluation of cordless phone signals, trunked radio data and voice channels, and digital private telephone lines.

A more general and encompassing tool for overall network performance testing and evaluation is a performance analyzer. This tool may incorporate a protocol analyzer, protocol simulator, and other query generation and evaluation tools or may be used in conjunction with separate modules handling the various functions.

In terms of network integrity and security, keep in mind that a protocol analyzer may be inserted into the communications circuit to evaluate its characteristics. Since some channels on a network may be involved in transmitting signaling information, the ProtAn could be used to manipulate signals on the network, thus interfering with normal operations. See probe.

Protocol Data Unit PDU. A data unit consisting of control information and user data which is exchanged between peer layers. In the Open Systems Interconnect (OSI) layered networking model, a PDU is associated with each of the seven layers (as well as a Service Data Unit (SDU)). The PDUs are used for peer-to-peer communications. At the lower layers of the model, the PDUs have been specified as Transport, Network, Data Link, and Physical, associated with segments, packets, frames, and bits, respectively. Services for the PDUs may be conveyed via Service Data Units (SDUs) in lower layers. Higher-layer PDUs are encapsulated in lower layer PDUs (a header is added). See asynchronous transfer mode, LLC, RFC 1042.

protocol simulator An implementation tool that enables developers and network service providers to test and verify protocols prior to setting them up on a network. A simulator typically allows the tester to set up data rates and basic operating parameters and to initiate and respond to queries and perform the other basic functions of a telecommunications network. It may be used in conjunction with a protocol analyzer to evaluate performance characteristics and perform fine-tuning prior to implementation on a client's system. A protocol simulator may be designed to simulate one specific protocol or related group of protocols or may be capable of simulating a variety of protocols, depending upon the product.

proxy An agent, deputy, or authority acting on behalf of another.

proxy, computer A software intermediary or agent that acts on behalf of clients and can act as a server or client. Proxy systems are frequently placed in points of a network where there are connections between LANs, or between a LAN and the Internet, or between a LAN and another external system such as an Internet Services Provider (ISP) or phone network. Proxies act as protocol managers and security administrators, and handle requests by servicing them or passing them through to other services. In cases where they are passed through, the proxy may interpret and modify a request before sending it on. Conceptually, a proxy server differs from a gateway in that requests are passed through as though from the original client, whereas a gateway handles the request as though it were originating from the gateway, thus forming a barrier between the server of the request and the client. A system can be configured as a security firewall to allow selective passing of messages through its portal. See firewall, gateway.

proxy server A software program that serves up requests on behalf of another process or machine. As such it can manage processes on a single machine, or broader server functions on a local or public network. In computer networks, a proxy is an intermediary server providing services between a client or clients and another server, usually a primary or secured server. Proxy servers are commonly used to improve network efficiency and security. By filtering requests, proxy servers can reduce the load on other servers and handle specific types of tasks or frequently requested tasks. If connected to the Internet, they can be used to manage common processing tasks or storage dedicated to warehousing frequently requested documents, files, or Web pages.

A proxy server need not necessarily be configured as a security mechanism. It may simply handle overflow requests for the primary server when it is busy without any concern about who is making the requests. However, a proxy server lends itself to screening requests and is often used to help secure communications, even if only in a minimal sense. Proxy servers can be configured to selectively control access to authorized users and, as such, can act as gateways or firewalls themselves, or they can work in conjunction with gateways or firewalls on one or both sides of the gateway or firewall link. See firewall; gateway; proxy, computer.

PRS 1. personal radio service. 2. See Police Radio Service.

PSAP See Public Safety Answering Point.

PSAP Pro A commercial E-911 database program developed by Public Safety Associates, Inc., in collaboration with MapInfo. PSAP Pro enables telecommunications providers to route E-911 calls to an appropriate Public Safety Answering Point (PSAP) for the dispatch of emergency personnel. The software includes data for the U.S., 10-digit emergency numbers, address information, geographical information (latitude and longitude), and jurisdictional boundaries for PSAPs. The software is based upon a product distributed earlier by Public Safety Associates and the National Emergency Number Association (NENA). See Pseudo-Automatic Number Identification, Public Safety Answering Point.

PSDN packet-switched data network. See packet switching.

PSDS Public Switched Digital Service.

PSE packet-switched exchange.

Pseudo-Automatic Number Identification P-ANI, Pseudo-ANI. A non-dialiable telephone number assigned to cellular transceiver sites or sectors to enable emergency call routing for mobile phone subscribers. Since mobile users are always on the move, there is no direct way to correlate the cell phone number with a house address or business street location.

By the mid-1990s, requests to the Federal Communications Commission (FCC) to facilitate improved wireless emergency location services resulted in some progress. In Texas, a system was developed in which virtual phone numbers were assigned to cell sites in fixed locations. These numbers are not dialed, but they enable a number and a location to be related to one another so that if a specific site is being used by a mobile user, it provides a ballpark idea of where the caller might be. FCC rules for mobile emergency location services began to be phased in beginning in 1996. The rule-making has changed several times to reflect improvements in technology since that time. There is more than one way to implement P-ANI for emergency services. P-ANIs may be statically assigned to cell sites or may be dynamically assigned when a call is received. The size of the region and the demand for phone numbers are factors that may influence the choice. In one implementation of Pseudo-ANI, a mobile user calls E-911 (Extended 911), the Mobile Switching Center (MSC) sends the P-ANI together with the WS-ANI to the appropriate emergency routers so that an Automatic Location Identification (ALI) host can correlate the P-ANI with a location. Feature Group D trunks may be used to transmit calls to E-911 routers. Once the emergency service has been contacted, the situation can be assessed, the caller can be contacted by a Public Safety Answering Point (PSAP) dispatcher, and the appropriate help dispatched to the caller's approximate location if the caller is unable to specify the exact location.

The P-ANI system may eventually become obsolete. Increasingly, Global Positioning System (GPS) receivers are being incorporated into cell phones, which means that GPS readings of the caller's location, within a couple of hundred feet or less, can potentially be transmitted to emergency services once E-911 services have been called. This would free up the nondialable numbers for use as dialable telephone numbers and enable more precise determination of the geographic location of a caller in distress. See Automatic Number Identification, E-911, Global Positioning System.

pseudocode 1. A software program or process listing written in somewhat plain language (not a programming language per se) which delineates the steps and algorithms in a process in order to outline or draft it so it can be transcribed into a programming language. Many programmers find pseudocode more practical and useful than flow charts for drafting software program flow. 2. P-code. A type of code which is compiled down to an intermediary stage without being specifically tied to one computer architecture. A number of high level languages can be compiled down to P-code, so they can execute faster (and sometimes to protect proprietary programming algorithms); they can then be distributed for a variety of computers. Each computer on which the P-code is run will need software to execute the program to do any further platform-specific translation of machine instructions which might be needed.

PSI 1. packet switching interface. 2. Policy Studies Institute. U.K. government consortium.

PSK See phase shift keying.

psophometer A diagnostic instrument that provides a visual reading of noise at various frequencies in audio circuits. It is a form of voltmeter incorporating a type of band-limiting filter called a weighting filter. Weighting filters aid in testing in such a way as to accommodate perception of sound at different frequencies and levels. Depending upon the model of instrument, various types of noise, including longitudinal noise and metallic noise, may be measured as to their presence, level, and relationship to loss in a transmission. Selective level meters that are used as voice band analyzers may also be suitable for use as psophometers.

The frequency ranges tested by this instrument are within human hearing ranges, with emphasis on frequencies used in speech (if measuring speech circuits) or music (with a broader range than speech circuits). Psophometers are used to test telephone, sound system, and broadcast networks. They may be integrated with other audio testing tools (e.g., distortion meter) and audio frequency generators. Psophometers for testing telephone circuits may be further specialized to accommodate local telephone standards (e.g., North American vs. European standards). The ITU-T has published recommendations for the use of psophometers on telephone-type circuits (O.41, P.53).

PSTN See Public Switched Telephone Network.

PSTN/Internet Interworking Service Protocol See PINT Service Protocol.

PSU Packet Switch Unit.

PSWAC See Public Safety Wireless Advisory Committee.

PSWN Public Safety Wireless Network.

PTE See SONET path terminating element.

PTFE polytetrafluoroethylene. A synthetic material useful for insulating wires in environments where they might be subjected to heat.

PTI Payload Type Identifier. An ATM cell header descriptor which indicates the type of payload in the cells, such as user or management. See cell rate.

PTN Public Telecommunications Network.

PTO public telecommunication operators.

Ptolemaeus, Claudius "Ptolemy" (ca. 93±6–ca. 164±14) A Greek philosopher and astronomer, Ptolemy was apparently one of the first to study and describe the interaction of light with matter. In book V of a Greek optics text attributed to Ptolemy (known to us through a partial Latin translation of an Arabic version called *Almagest*), Ptolemy cataloged and described angles of incidence and refraction. Arabic scholars such as Al-Haitham (Alhazen) also studied and described refractive optics, but the relationship between incident and refracted light was elusive and not clearly understood until almost 1500 years later, with the discoveries of W. Snell.

In *Mathematika Synthaxis* Ptolemy described a model of the universe and cataloged more than 1000 visible stars that could be seen from Alexandria.

Dioptrics is the science of light refraction and the

closely related *catoptrics*, is the science of light reflection, especially of reflective mirrors. The terms are mentioned at least as early as 1605 (Stevin) and may be descended from concepts described in Ptolemy's and Euclid's optics texts. See refraction, Snell's law.

PTS 1. Personal Telecommunications System. 2. See Presentation Time Stamp. 3. Public Telecommunications System.

PTT See Post Telephone & Telegraph administration.

Public Access Profile PAP. A profile is defined by the Open Systems Interconnection (OSI) model as a combination of one or more base standards and association classes, necessary for performing a particular function. The PAP is an ETSI-published profile incorporated into Digital European Cordless Telecommunications (DECT) as a test specification. See Digital European Cordless Telecommunications.

public address system PA, PA system. A system designed to receive and transmit amplified sound, especially voice or music, to a wide audience. It may be acoustical or electrical. An acoustical PA may amplify through a horn-shaped object that directs sound. An electrical PA is familiar to most people as the microphone, circuits, and speakers installed in most public schools and hospitals. See amplifier, intercom, loudspeaker, sound.

public key In public-key encryption schemes, this is the key given to the public to act as one of the tools to create a message addressed to the person to whom the key corresponds. The recipient then uses his or her private key to open and read the encrypted message. It is important to create a private key password that can be remembered for a long, long time; otherwise it isn't very useful to distribute it to the public because it will not be possible to decipher encrypted messages if the password is forgotten.

Public-Key Cryptography Standards A set of standards for public-key cryptography developed by RSA Laboratories in cooperation with members of an informal consortium that included Apple, Lotus, Microsoft, MIT, Sun, and others. The Open Systems Interconnect (OSI) Implementor's Workshop has cited PKCS as a means of implementing security-related aspects of the OSI standards.

PKCS are compatible with the ITU-T X.509 standard and support binary and ASCII data formats. They support algorithm-specific and algorithm-independent implementations and provide an algorithm-independent syntax for digital signatures, envelopes, and extended certificates, to facilitate interoperability. See RSA Security Inc.

public-key encryption An encryption scheme that often also incorporates an authentication scheme in which *public keys* are distributed for encryption of messages to the person owning the key, and *private keys* are established for decrypting messages. Sometimes the encrypted message is differentiated into two components: signature and message. It is possible in some systems to encrypt the signature to provide authentication without encrypting the message. Pretty Good Privacy is a public key encryption technology

that has been incorporated into various Internet applications, for example, in email through MIME. By the late 1990s, there was a trend to promote public key encryption as having the same legal and commercial force as handwritten signatures. See Blowfish, Diffie-Hellman, PGP/MIME, Pretty Good Privacy.

public-key infrastructure PKI. A general term for the foundation, architectural framework, and basic components that comprise a public key encryption system. In general, industries are in favor of the security provided by PKIs, but implementation has been slow, due to the extra steps and occasional confusion that arise from installing, using (and remembering to use) PKI-based systems. In spite of the difficulties, by the end of 2000 a number of firms were seeking solutions for the development of easier-to-use PKIs, especially for mobile business/commerce applications. See public key encryption, wearable public key infrastructure, wireless public key infrastructure.

Public Land Mobile PLM. A network administered for the purpose of providing publicly accessible, land-based mobile telecommunications services. It may be considered as intrinsic to or as an extension of the public switched telephone network (PSTN). A land mobile service is essentially a communications service between base stations and land mobile services, or among land mobile stations. Services thus provided include cellular telephony, PCS, and other governmental, industrial, and business mobile services.

Since PLM services are used in part for public safety and emerging Intelligent Transportation Systems (ITSs), there has been concern over the allocation of sufficient radio frequency spectrum to support the system as it grows and changes. It is anticipated that approximately 240 MHz of additional spectrum will be needed over the next decade to satisfy these needs.

Public Safety Answering Point PSAP. Services that support the administration of emergency services and the handling of emergency telephone calls. PSAP stations are typically centralized facilities with supervisors and a number of trained public safety communicators and dispatchers. PSAP responsibilities are somewhat tailored to their individual regions but, in general, their staff members monitor safety patrols and justice information systems, and process emergency service calls and nonemergency law enforcement-related calls.

PSAPs must satisfy a number of requirements in order to provide reliable, unified emergency services. For example, PSAPs must be able to handle calls from E-911 services from a variety of types of callers, including those using mobile phones. They must also be able to communicate with hearing-impaired callers. In many cases, upgrades to PSAP equipment have been necessary to accommodate the specialized needs of the growing mobile telephone community and the new regulatory requirements by the Federal Communications Commission (FCC). Some large phone service providers are now providing services to aid in the implementation of emergency regulations to help minimize the impact on and costs incurred by PSAPs. See Pseudo-Automatic Number Identification.

Public Safety Wireless Advisory Committee
PSWAC. A committee established in 1995 by the Federal Communications Commission (FCC) and the National Telecommunications and Information Administration (NTIA). The PSWAC provides advice on the various public safety agencies' wireless communications requirements through 2010. This includes identification of emerging technologies and recommendations as to their utility and role, and emphasizes the importance of semiconductor technologies.

Public Service Commission PSC. A state regulatory authority for communications. See Federal Communications Commission, Public Utility Commission.

Public Switched Telephone Network PSTN. The national telephone infrastructure consisting of the RBOCs, the IXCs, and LECs. The PSTN is intended to further the goals of universal access to telephone service described in the 1934 Communications Act.

Public Switched Telephone Network history
PSTNs have long been regulated in the U.S., beginning with the Post Roads Act in 1866. The Bell System and AT&T are an intrinsic part of the history of the PSTN; their dominance, research contributions, and many voluntary and required reorganizations have formed a colorful history in which the rewards of a competitive market, and the demands of independents to further fair competition have often been difficult to sort out and maintain.

AT&T was divested from Western Union in 1913 and mandated to provide access to independent carriers across its long-distance network. Further divestiture occurred in 1975, and a Modified Final Judgment, approved by Judge Harold Greene, came into effect on 1 January 1984. The Justice Department formed 22 Bell Operating Companies (BOC) from AT&T, which were organized into seven regional Bell holding companies. Until 1991, when the Supreme Court granted wider privileges, the BOCs were not permitted to provide electronic data services.

Service divisions that resulted from the restructuring fall into the following general categories, which are described more fully under individual headings in this dictionary: Local Access and Transport Areas (LATAs), Local Exchange Carriers (LECs), Independent Telephone Companies (ITCs), Interexchange Carriers (IXCs), and Other Common Carriers (OCC).

Public Utility Commission PUC. State regulatory commission with jurisdiction over phone companies. See Public Service Commission.

pull box In wiring installations, a box inserted into a long cabling run, especially at major junctions that allows easier access and working room for changing existing cables, removing them, or running new ones.

pull rig A piece of rigging equipment used to install or deinstall a line conductor. The pull ring facilitates this with a puller, a pulling rope, and a take-up reel for winding the line.

pulling eye A round open device incorporated into lines and other objects which are intended to be threaded through conduits or similar tight spaces. The eye is used to attach a line to thread the line. See birdie.

pulling fiber The process of creating long, slender fiber filaments for use in fiber optic components. There are a number of different ways to pull fiber, but they have certain characteristics in common. Most of them begin with a preform, created through a vapor deposition process, and take advantage of gravity to draw out the fibers.

The goal is to create fiber filaments that are consistent, without cracks, of the correct refractive index, and free of bubbles, crystallization effects, and undesired impurities (impurities are sometimes deliberately *doped* into the fiber to alter its properties).

Commonly a preform cylinder will be created, as in vapor deposition processes, prior to pulling the fiber into a long filament, usually through a gravity-based tower assembly.

In the simpler double crucible method, two concentric crucibles are mounted together, somewhat like a small water glass centered inside a larger one, both with holes poked in the bottom for the glass to flow through. The inner crucible holds the core glass, while the outer crucible holds the cladding glass. As heat is applied, the molten glass pours out through the hole in the bottom, due to gravity, and the core is imbedded within the cladding as the glass cools into a fiber filament. See doping, preform, vapor deposition, ZBLAN.

pulling glass *colloq.* Installing fiber optic cable, especially by pulling it through conduits and across aerial links with various implements designed for this purpose, as opposed to air-blowing. See blown fiber, pulling eye.

pulling strength This is an industry- and material-specific term but, in general communications terms, it refers to the maximum amount of pulling force that can be applied to a material or object (such as a communications cable) before its physical or data transmission properties are compromised.

Pulling strength varies from material to material; a fine electrical wire generally has a lower pulling strength than a medium-gage bundled wire of the same composition. Thus, pulling strength may be expressed numerically as the cross-section diameter of the conductor × the number of conductors × a breaking strength index agreed upon for the type of cable (e.g., 50 for copper wire). Fiber optic cable that includes a strength member has a higher pulling strength than copper twisted pair.

It is important to have a good intuition and established procedures regarding pulling strength when installing and deinstalling wire and optical cables, especially when pulling through conduits stretching around corners. Pulling strength must also be considered when handling cable storage drums and spool dispensers. For example, pulling from the bottom of a rotating drum, rather than the top, distributes the force in such a way that there is less stress on the pulled cable.

pulse 1. A rhythmic beating, throbbing, vibrating, or burst of electricity, sound, or light. 2. A briefly transmitted electromagnetic wave or modulation. 3. In telephony, a brief, timed signal sent out by a pulse-dialing phone to indicate the desired destination. 4. In radar, a brief burst of microwaves.

pulse amplitude modulation PAM. A common means of converting a continuous stream of information into a series of samples with assigned discrete values used in analog to digital conversions. PAM takes into consideration the fact that for many media, it is not necessary for the entire communication to be transmitted for it to be understood. For example, in a voice conversation, if you were to take samples of the speech at frequent intervals and transmit those to a human listener, enough of the information is retained that most people can still understand what was said, even if subtle parts of the original message are not included. The more frequent the samples, the closer to the original, up to the limits of human perception. For example, moving images played by a film movie projector are typically displayed at the rate of about 24 to 30 frames per second. Higher sampling rates (more frames per second) do not substantially improve the quality, since humans can't see the individual frames, and the whole is perceived as motion, rather than as a series of still images.

PAM is a baseband transmission multiplexing scheme used, for example, in Digital Subscriber Line (DSL) transmissions. PAM uses the amplitude of the modulated information to determine the amplitude modulation characteristics of the transmission. PAM's advantage is that it uses lower frequency bands, which are less subject to attenuation and crosstalk, two areas that have caused concern in DSL installations over existing copper wires. See carrierless amplitude and phase modulation, pulse code modulation.

pulse code modulation PCM. PCM is a means of sampling a signal and assigning values to the individual samples. As such it is a fundamental digitization technique that forms the foundation of many aspects of electronics, particularly voice-carrying transmissions and audio recording technologies. Because the signal at the end of the transmission carries the same data as the signal transmitted at the beginning of the transmission, the data portion is essentially a "lossless" process compared to analog signals in which the information contained within the signal can be affected by attenuation and other factors. (Attenuation can also occur in digital transmissions, but it compromises the strength of the signal more than the information content of the data itself; loss of data through aging or lossy compression/decompression results from implementation rather than from the essential characteristics of the digital transmissions technology.)

The digitization process inherent in PCM also enables error correction, compression, and encryption technologies to be applied to the data in ways that are not practical with analog transmissions.

PCM was one of the historic digitization schemes that did not come into use until sometime after the theory had been developed, but once it became practical, it became a significant aspect of communications that is still used today. A number of inventors were responsible for discovering, rediscovering, and applying PCM.

Pulse code modulation can be carried out in a number of ways but generally the process involves

- establishing a resolution at which the signal will be sampled,
- establishing cutoff levels within which the signal will be sampled,
- filtering the incoming analog signal,
- sampling the signal and assigning values to the samples,
- converting the quantized sample into a digital bitstream, and
- reconstructing the signal at the receiving end.

It is not uncommon for the digital signal to be converted back to analog at the receiving end, as on analog telephone voice lines or analog stereo equipment. However, more and more phone services and stereo systems have digital lines and components and it may not be long before these transmissions and playback systems are predominantly digital. Since PCM digitization can be applied in a number of ways, not all PCM transmissions are compatible, and different schemes are used in Europe and North America.

PCM is widely used in telecommunications. With A-law/Mu-law encoding, it is the most common encoding scheme for telephone conversations. Sampling rates may vary, but 6000 to 10,000 times per second is typical. Eight-bit sampling is sufficient for voice communications. In contrast, 16-bit sampling or higher is preferred for quality music sampling.

Pulse code modulation is sometimes used in multiple modulation schemes. For example, PCM subcarriers may frequency modulate another carrier. See differential pulse code modulation, pulse amplitude modulation, sampling. See individual listings for scientists Kotel'nikov, Reeves, Shannon. See the Pulse Code Modulation History chart for some of the highlights in the development of PCM and related technologies.

pulse compression In radar, a technique of using long pulses to increase the energy of the received signal while still retaining the resolution of short pulses.

pulse dialing A means of transmitting phone numbers through a phone line by converting the length of the signal that occurs with the rotation of the rotary dial into electrical pulses. Pulse dialing phones are becoming rare and pulsed signals will not work with online automated menu systems. The other common means of dialing is through tones, with each number assigned a particular tonal frequency. See tone dialing.

pulse dispersion The gradual spread of pulses as they travel over some transmissions medium.

pulse duration modulation PDM. A type of pulse modulation in which the length of time of the modulation is controlled to impart information. See pulse code modulation.

pulse generator A device for creating pulses of various specific amplitudes, shapes, repetition rates, and durations. Thus, a pulse generator can be used in a variety of pulse modulation schemes.

pulse inverter A wideband, low-distortion waveform-modifying device which enables an output wave to be the inverted form of the input wave.

P

pulse modulation A simple means of modulating a signal by applying pulses of current (often just on or off) without changing the frequency of the signal. See pulse code modulation, pulse width modulator.

pulse repetition frequency PRF. The number of pulses which occur in a unit of time, such as the number of pulses per second or minute. Pulse repetition frequency is important to many technologies, especially those that involve signaling or synchronization.

pulse width modulator PWM. An electronic circuit that can be configured to produce a pulse with desired characteristics (period, cycles, etc.). Thus it is essentially a component that regulates an electromagnetic pulse. There may be more than one modulator on a circuit board and each modulator may have different characteristics (e.g., 8-bits with 4 channels or 16-bits with 2 channels).

A PWM can be used for a variety of purposes, including device control (DC motors and servos), power bridge switching, light intensity control, and the modulation of transmission signals.

There are now all-digital PWMs, implemented as

Diode-Pumped Laser Optical Communications Receiving System

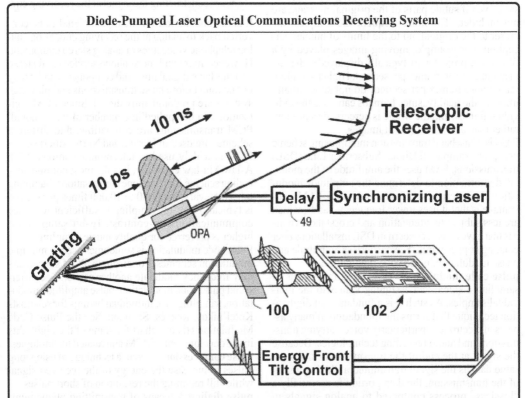

This diagram illustrates solid-state fiber lasers incorporated into an optical communications system and the close relationship between wired and wireless communications. It illustrates many fundamental optics concepts described in this reference. J-C. Diels et al. have developed a laser communication system that uses multiple lasers, lenses, diffraction gratings, and a modulator to disperse and modulate light pulses into coded words that are recombined, then sent and received by telescopic transceivers. A second laser generates reference pulses. When the references and main signal coincide, they are combined by a nonlinear crystal, with a detector recording the combined signal.

Fiber sources, pumped by diode lasers enable much shorter pulses to be generated which, in conjunction with faster modulation technologies, are well-suited for optical communications systems. Diode laser pumping is becoming more cost-effective and has a long lifetime compared to some pumping technologies.

In one embodiment of the system, a laser generates a fast pulse in the femtosecond range, along with a synchronization signal. The laser pulse is reflected off a dispersion grating. The pulses are converted by a modulator into a string of pulses representing coded words. The second grating recombines the coded words after which they are sent by a telescopic transmitter to a telescopic receiver. A reference laser receives the synchronizing signals from the main laser and generates reference pulses, the original and reference signals are combined by a nonlinear crystal and the output signal sent only if they coincide. The output is sensed and recorded by a detector. The diagram illustrates the optical detection system that processes the received signal. Parametric gain is used to select the appropriate channel for the incoming spectrum and to amplify the signal. The laser emits a synchronized train of pulses at a wavelength shorter than the incoming signal at the same repetition rate. The pulse from the grating serves as a pump for a parametric amplifying crystal. A Fourier transformation is performed on the amplified signal to regenerate the pattern of the originally transmitted pulse. Detection is then handled by a cross-correlator (100) and a detection array (102). [Image adapted from USPTO patent #6,421,154 filed October 2000.]

CMOS circuits, suitable for precision pulse width and pulse position modulation for computer peripherals and laser printers and copiers. See delta-sigma modulation.

pulsing Using discrete bursts of sound or electromagnetic energy to send signals through a communications medium. The strength or duration of the pulses may have relevance, depending upon the system. In older phone systems, pulses were used to indicate the number dialed. Beacons often use pulses to signal their presence and location. See pulse dialing.

pump, laser A device to create molecular action resulting in photons emitted as a laser beam. There are many types of lasers, and each may be used for different purposes, however, in general, the purpose of a laser pump is to stimulate a bound electron to go from one quantum level (usually from a *ground state*) to a higher quantum level, such that, when the excited electron decays, energy is emitted in the form of light. Since light is continuously pulsed or stream-emitted from the laser cavity, there needs to be a way to replenish the expended energy. The repetitious nature of the energy replenishment is what inspired the term "pumping." Pumping may be along the same axis as the resonating mechanism (longitudinal) or transverse to the axis of the active lasing medium.

In traditional lasers, the excitation is usually achieved by optical or chemical means. A ruby laser is optically pumped, while a helium-neon laser is chemically pumped. Electricity can also be used to stimulate a semiconductor to higher energy levels or to ionize gases. More recent laser diode components may use fiber amplifier or solid-state pumping.

Xenon flashlamps are common for pumping, or a laser diode or traditional laser may pump another laser. In a helium-neon (He-Ne) laser, helium pumps the electrons (excites them) into a metastable state of neon. A nitrogen, Nd:YLF, or Nd:YAG laser may be used to pump a dye laser. See laser. See Diode-Pumped Laser Communications diagram.

punch *v.* 1. To prod, poke, or perforate. 2. To perforate small regular holes in paper, card, fine metal, or other similar flat surfaces in order to create a semi-permanent record of a code. Early gramophone cylinders, player pianos, patterned looms, music boxes, and punch card reading computers employed this technique to encode and store information intended to be read back later. See Hollerith card. 3. To apply pressure to an enclosure in order to pop out a small section, to enable the threading of circuit lines. 4. To apply pressure with a punchdown tool to a wire and terminal block. See punch down, punchdown tool.

punch card Any sturdy paper or card stock in which parts of the card are punched out according to a code that can later be reread and decoded to produce the original meaning. Punched codes in various media have been around for centuries, with early ones incorporated into music boxes. The Jacquard loom incorporated punch cards to store loom patterns in the early 1800s, revolutionizing the textile industry and causing a substantial uprising among human weavers, whose jobs were obsoleted by the automation.

Punch cards can be used to store many sorts of codes and variations of the concept are used in music boxes, player piano rolls, older computing devices, etc. See Hollerith card, zone punch.

punch down *v.* To apply pressure, usually with a specialized punchdown tool, to wires looped around a terminal block to strip insulation from the end of the wire, and insert the conductive surface between the prongs of the terminal in order to make a solid, clean connection. [The verb form is two words, the noun form often combined into one word.] See punchdown block, punchdown tool.

punchdown block, terminal block, cross-connect block A multiterminal block designed to facilitate electrical cross connections, especially those which may change from time to time or where future expansion is planned. Very common in telephone installations where a multiline wire is threaded into a building and then, at the punchdown block, is split into pairs to wire several individual phones in different locations. See punchdown tool.

punchdown tool A handheld wire installation device that resembles a screwdriver with an extremely short, notched shaft. The handle may be spring-loaded. The ends are designed to fit particular sizes of terminals and used to quickly connect wires to a punchdown block. The wire is looped over the prongs of one of the terminals in the block, and the punchdown tool is pushed against them to spread the prongs and snap the wire securely into place. Some punchdown tools will punch and cut. See punchdown block.

Punchdown Tool

Telecommunications wires can often be attached to wiring blocks with screwdrivers and pliers, but the task becomes tedious if there are many wires to connect. A punchdown tool is specialized for attaching and trimming the wires more quickly and readily than general-purpose tools.

Pupin, Michael Idvorski (1858–1935) An Eastern European-born American inventor and educator. He studied low pressure vacuum tube discharges and X-rays, and invented an electrical resonator and inductive loading coils, paving the way for long-distance wired transmissions.

Pulse Code Modulation (PCM) Development – Highlights

- In the 1870s, J.E. Baudot inserted sychronization signals between baseband signals to multiplex the signal. This was the beginning of time division multiplexing (TDM).

- Edmund T. Whittaker developed a mathematical basis for sampling theory, in 1915, that is now applicable to pulse code modulation and time code modulation.

- Some sources cite that in 1926, P.M. Rainey at Western Electric was the first to patent PCM methodology. The author was unable to locate documents to confirm or deny this claim.

- In 1933, Russian scientist V.A. Kotel'nikov built on the mathematical basis described earlier by Whittaker and demonstrated how discrete samples could represent continuous functions. He later applied his theories to advances in radio astronomy and satellite communications.

- While working in Paris with ITT Laboratories, Englishman Alec H. Reeves described PCM as a means for transmitting voices. His invention was patented in France in 1937 (#833,929, #837,921) and subsequently patented in Britain and the U.S (#2,272,070 – 1942). Reeves was later to research optical communications at Standard Telecommunication Laboratories in the late 1950s and applied for further patents for PCM technologies in the early 1960s.

- In 1945, E.M. Deloraine developed a different method for digitizing speech, now known as delta modulation, which Reeves felt might have future applications when used in conjunction with PCM.

- In the U.S., Bell Laboratories explored the technology under Harold S. Black and developed a PCM system that was used by the U.S. Army Signal Corps. By 1947, Bell Telephone Laboratory researchers had an experimental 96-channel PCM system running between New Jersey and New York City and W.M. Goodall authored "Telephony by Pulse Code Modulation," in the *Bell System Tech Journal* the same month that Harold Black authored "Pulse Code Modulation" in the *Bell Lab Record*.

- In the late 1940s Claude Shannon, John R. Pierce, and Bernard M. Oliver developed information theory related to PCM, based on groundwork laid by Harry Nyquist and Hartley, that provided support for Kotel'nikov's theories. Oliver, Pierce, and Shannon authored "The Philosophy of PCM," in November 1948. In December 1948, Reisling's "Companding in PCM," was published in the *Bell Lab Record*.

- By 1951, the Nippon Telegraph and Telephone Public Corporation (NTT) labs were researching PCM. Dr. Kiyasu, who headed up electronic switching research for NTT in the 1960s, invented the reflected binary pattern scheme for encoding.

- Frank Gray, another Bell Laboratories engineer, developed a binary code system for pulse code communications described in 1953 and 1958 U.S. patents (#2632058, #2,632,058).

- In 1958, the Paris-based Laboratoire Central de Telecommunications of ITT included PCM in their research and development of electronic switching systems, work that was followed up in the London and U.S. labs of AT&T.

- In 1962, after almost a decade of development and testing, Bell Laboratories developed a working digital T1 carrier system that could transmit 24 separate voice communications over a single twisted-pair wire. The data for each conversation were encoded at 64,000 bps.

- By 1964, British Telecom had experimentally introduced PCM systems into junction cables.

- By 1965, T1 systems had also been designed and introduced for use in Italy. In Japan, a DEX-T1 pulse code modulation switch was being developed by NTT, with a prototype released in 1967. A.H. Reeves predicted that "... by the year 2000, pulse code modulation in some form will be the very backbone of the world's communication systems... for the year 2000 there is no doubt that optical PCM methods will have to be used...."

- In 1967, a PCM cable route was opened in London by British Telecom. The following year the London Post Office installed a PCM-based exchange, which opened in September 1968. This was the first to implement switching in digital form and demonstrated that digital switches could work within an electromechanical system.

- By the early 1970s, audio equipment and recording companies such as Denon and Nippon Columbia were using PCM techniques to create digital sound recordings. Through the 1970s, the technology improved and spread until, by the late 1970s, various international standards for PCM were adopted and commercial PCM-based recorders were available.

- By the early 1980s, consumer markets for audio CDs were being established.

Pure ALOHA See ALOHA.

Pure Voice See Qualcomm Code Excited Linear Predictive Coding.

purge *v.* To rid of, or remove. In computing, there are many instances where the operating system or applications software retains information, files, or backup files until explicitly instructed to remove them. For example, many workstations enable the number of backup versions of a file to be set automatically. Thus, each file may be saved under a slightly modified name or with the same name, but a different time stamp. *Purging* the directory deletes all the multiple copies of each file and retains only the most recent. Similarly, many desktop publishing and word processing programs save a history of edits and additions within the file, so that the information is available for recovery, if needed. This may result in *very* large files. These programs often will *purge* the file edit history if the user selects *Save As* instead of *Save*, resulting in smaller, cleaner files (but which are more difficult to recover if problems arise).

Many database programs retain information such as customer names and addresses even if the customer entry has been deleted. In this way, it can be recovered if the entry is needed in the future, but doesn't clutter up directories or slow down searches in the meantime. There may be a menu item called *Purge* which allows you to permanently remove these records, or in some programs, you can set the system to *purge* any deleted records that have remained inactive for longer than a specified amount of time.

Purkinje effect The human visual system is more sensitive to blue light in conditions of lower illumination and more sensitive to yellow light in conditions of higher illumination.

purple wire An IBM color designation for wires that have been incorporated during testing and debugging to accommodate errors (purple wires may actually be yellow). See blue wire, red wire, yellow wire.

push-pull circuit A circuit consisting of elements operating in a phase relationship rotated 180°. This makes it possible to cancel or filter certain elements and to amplify or magnify others. This circuit is used in oscillators, amplifiers, and transformers.

pushbutton dial A capability through a set of buttons, usually on a phone or computer keypad, for sending signals to a transmission system. The button "dial" generates a tone at a specific frequency, which is then transmitted and interpreted as a numbered location for the destination receiver. A few units have both rotary and pushbutton capabilities. See touchtone. Contrast with rotary dial.

put In file transfers over a network, e.g., through File Transfer Protocol (FTP), a line command or graphics "put" button for uploading a file. In other words, "putting" a file involves sending the file from the local system to a remote system. The analogous command for downloading/receiving a file is *get*.

putup A packaged wire or cable product, which may be wound on a spool, over a rack, or coiled in a box.

PVC 1. See permanent virtual connection. 2. See polyvinyl chloride.

PVCC Permanent Virtual Channel Connection.

PVN See private virtual network.

pW *abbrev.* picowatt. See pico-, watt.

PWB printed wire/wiring board.

PWM See pulse width modulator.

PWR An abbreviation for power sometimes seen in technical manuals or on components.

PWT 1. parking without ticket. A video surveillance system that enables parking management through computer software with video display capabilities that replaces traditional paper tickets. 2. See Personal Wireless Telecommunications. 3. project work team.

p*x*64 A videophone/videoconferencing standard within the ITU-T H.320 videoconferencing recommendations family. P*x*64, officially known as ITU-T H.261 is an interoperability standard for video data streams originally developed in the mid-1980s to support transmission of analog video services over Integrated Services Digital Network (ISDN) services. The name is derived from data rates based upon calculating p × 64 kbps with p = 1,2,...,30 (depending upon the available number of ISDN channels).

In 1990, the Video Codec for Audiovisual Services was completed and approved for Europe, with a modified version adapted in North America.

The encoding is hierarchically grouped, with video streams comprising Groups of Blocks (GoBs) of images or frames. Discrete cosine transform (DCT) algorithms are used to subdivide image parts into smaller building blocks.

The standards were originally specified for fixed-data-rate ISDN circuits, but were found adaptable to packet-switched circuits (e.g., the Internet), as well. Thus, in 1996, a standards-track RFC was submitted by Turletti and Huitema of Bellcore for an RTP Payload Format for H.261 video streams. In 1997, Zhu submitted a standards-track RFC for an RTP Payload Format for H.263 to enable packetization of H.263 video streams. This greatly increases the versatility of H.261, as packetization enables transmissions over distributed networks with standard User Datagram Protocol (UDP) datagrams. See RFC 2032, RFC 2190, RFC 2736.

pylon 1. A tower or other tall supporting structure for stringing wire over a wide span, buildings, people, or other structures. The pylon often serves to provide both structural support and safety. 2. A broadcast transmissions tower, often a tall one, for increasing line-of-sight distance transmissions by "towering" over terrain and buildings that might otherwise obstruct or reflect the signals.

pylon antenna A vertical standing antenna of slotted sheet-metal cylinders, sometimes combined in sections, one atop the other, to achieve greater height. The name comes from the cylindrical shape and orientation which resemble log pylons on a dock.

pyramidic wavefront sensor A wavefront sensor designed to continuously change gain and sampling, to better match system test performance with actual conditions. The device comprises a four-faced optical splitter (the active region is shaped like the upper surfaces of a pyramid) that is oriented with the tip

P

toward the radiant energy source. The four beams of light exiting from the pyramid are re-imaged with a display system to yield four images. The pyramid acts like a knife-edge test system, revealing optical aberrations carried by the light electromagnetic beam. The system has been tested in astronomical applications. See knife-edge focusing.

pyrheliometer An instrument for measuring infrared radiation which borders the visible spectrum on one side and radio waves on the other.

pyroelectric detector A detector that takes advantage of the characteristic of a pyroelectric material to build up a charge when exposed to pulsed or modulation radiant energy. Due to the quick response times of pyroelectric detectors, they are more useful for some types of applications and environments than other types of thermal detectors that have difficulty separating the impulse heat from ambient heat due to slower response times. See photodetector, pyroelectric, thermocouple.

pyroelectric A usually crystalline substance having polarized molecular properties in the absence of applied voltage. When stimulated by heat, the molecules become more active, expanding and altering the polarization of the material enough to build up an electrical charge on opposite surfaces of the material. Thus, a subtle electromagnetic impulse flows between the surfaces.

pyroelectricity Electromagnetic charges created through a change in temperature. Pyroelectricity refers to the means of generating the charge, not the nature of the charge itself, which is the same as others. Crystals have valuable oscillating characteristics and are commonly used in timing mechanisms and radio electronics and some have interesting pyroelectric activity when exposed to heat. See pyroelectric, pyroelectric detector.

pyromagnetic effect The combined effect of heat and magnetism in a material or circuit.

pyrometer An instrument for determining temperature, that is, a heat sensor for detecting and optionally quantifying thermal radiation. A pyrometer is generally used in situations hotter than those measured by a traditional mercury thermometer, especially where non-contact temperature is assessed. Temperature can be measured in a number of ways by electrical resistance, optical, or other radiant energy emissions. Thermal radiation may be assessed by measuring the brightness (intensity) of radiant energy within a narrow portion of the thermal spectrum. Another means is by evaluating data collected in two different wavelengths (usually red and green) to infer temperatures in the higher ranges (picture the way an electric stove element changes color as it heats). Infrared heat detectors (spectral radiation pyrometers)

have traditionally been combined with optical lenses and reflectors to make pyrometers. Initially, fiber optic cables were not seen as good infrared lightguides, but improvements in fiber optic technologies and control of spectral characteristics have made it practical to develop new fiber-based pyrometers. A fiber-based probe can be quite tiny and enables placement of the rest of the electronics at some distance from the probe, which is very useful for medical and industrial applications, especially those that involve insertion into constrained spaces or where there is high electrical activity that could interfere with electronic probes.

The temperature range to be sensed with a fiber-based probe and the length of the fiber are related, with lower temperatures having lower limits on the length of the fiber cable. The angle of acceptance of the fiber endfaces (or a lens associated with the fiber endfaces) determines the breadth of the sensing area and how closely the device must be held to the heat source for accurate readings. By combining fibers with lens focusing geometries, it is possible to design heat probes with very precise tolerances suitable for medical evaluation or industrial applications such as quality assurance and microfabrication. Unfocused fibers are suitable for general purpose applications where specificity of individual temperature regions is not critical. Where practical, backlighting the target object can provide a visual guide for aligning the probe.

Pythagorean Theorem

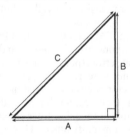

The sum of the squares of the sides (A, B) of a right angle triangle equals the square of the hypotenuse (C).

Pythagoras' theorem, Pythagorean theorem A mathematical rule that states that, in a right-angled triangle, the sum of the squares of the sides is equal to the square of the hypotenuse (the longest side). This theorem is widely used in mathematics for calculating distances and other measures. It is named for the Pythagoreans, a philosophical group connected with Pythagoras.

q 1. *symb.* quantum value. 2. *symb.* electrical quantity in coulombs. See coulomb.

Q 1. *abbrev.* quality. See Q factor. 2. *abbrev.* queue. See queue. 3. A merit indicator for a capacitor or inductor equal to the reactance divided by the resistance. 4. In a resonant circuit, an indicator of the sharpness or resolution of the resonance. It is calculated by taking the resonant frequency and dividing it by the resonant bandwidth. See Q-switch.

Q address A storage location for data, from which the information can be accessed and retrieved.

Q antenna A type of dipole antenna in which the feed line impedance is made to match the radiometer center impedance by the interposition of a vertical section, consisting of parallel bars between the two.

Q bit In an X.25 network, the Q bit is a binary indicator located at the beginning of a data packet, immediately preceding the D bit. The Q bit signals the existence of user data or qualified data in the form of control information. Protocols in higher layers can set this bit to one (1) to indicate control packets, otherwise a value of zero (0) indicates data packets. See D bit, M bit.

Q channel 1. In NTSC color television broadcasting, a frequency band in which green-magenta color information is transmitted. 2. In ISDN Basic Rate Interface (BRI) S/T interface implementations, an 800-bps maintenance channel. 3. In data transmissions, a channel associated with an I channel in modulated transmissions (e.g., phase-shift keying). See Q signal.

Q demodulation Demodulation of an incoming broadcast signal in a color television receiver to combine the chrominance signal and the color-burst oscillator signal in order to recover the Q signal.

Q factor (*symb.* – Q) quality factor. 1. In electronic circuits, a means of describing the desired characteristics of a system. The terms of the Q factor vary depending upon what is described (capacitance, inductance, etc.). In a digital circuit, for example, the Q factor may be used to characterize the signal-to-noise ratios of the two digital states. The concept has traditionally been associated with electrical voltages and wired communications, but can be generalized to optical communications. Generally, a higher number is used to indicate a more efficiently operating component. In lasers, for example, a maximal Q-factor may be associated with a filled laser resonance cavity, where the signal-to-noise ratio is high and switching may be triggered. The Q factor may be used as a figure of merit (e.g., for assessing bit error rates). See bit error rate. 2. See Q-factor.

Q multiplier A circuit used to enhance the selectivity of a component by feeding the signal back through the resonant network. This was used in early superheterodyne receivers, but various types of filters have, for the most part, superseded it.

Q output The reference output of an electronic flip-flop state, which may be one or zero.

Q Series Recommendations A set of ITU-T recommended guidelines for switching and signaling. These are available for purchase from the ITU-T. Some of the related general categories and specific Q category recommendations are included in charts on the following pages to give a sense of the breadth and scope of the topics listed here. A full list of general categories is listed in Appendix C and specific series topics are listed under individual entries in this dictionary. See also I, V, and X Series Recommendations.

Q signal 1. In various data transmission schemes, it is common to split a signal and to alter the characteristics of one or both of the two data streams so that they can be transmitted together without excessive interference or crosstalk. A Q signal or quadrature-phase signal is one of two common streams; the other is the in-phase signal or I signal, into which data are commonly split in various modulation systems. See quadrature amplitude modulation. See I signal. 2. A telegraph code shorthand signal consisting of two letters prefaced by a "Q" that is still well known to amateur radio operators. For example, QST is a general call preceding a message addressed to all members and amateurs. As such, it is also the name of the journal of the American Relay Radio League (ARRL). See QBF.

Q spoiling A technique used with lasers in which a more powerful burst or pulse is attained by inhibiting the action of the laser for a few moments, to allow an increase in the number of ions, and then Q switching to allow the extra burst of light to be emitted.

ITU-T Q Series Recommendations

General

Q.9   Vocabulary of switching and signaling terms

Q.1300   Telecommunication applications for switches and computers (TASC) – general overview

Q.1302   Telecommunication applications for switches and computers (TASC) – TASC functional services

Q.1303   Telecommunication applications for switches and computers (TASC) – TASC management: architecture, methodology and requirements

Q.1290   Glossary of terms used in the definition of intelligent networks

Q.1201/I.312 Principles of intelligent network architecture

Q.1202/I.328 Intelligent Network – service plane architecture

Q.1203/I.329 Intelligent Network – global functional plane architecture

Automatic and semiautomatic switching

Q.4   Automatic switching functions for use in national networks

Q.5   Advantages of semiautomatic service in the international telephone service

Q.6   Advantages of international automatic working

Signaling systems

Q.7   Signaling systems to be used for international automatic and semiautomatic telephone working

Q.8   Signaling systems to be used for international manual and automatic working on analogue leased circuits

Q.48   Demand assignment signaling systems

Q.50   Signaling between circuit multiplication equipment (CME) and international switching centers (ISC)

Q.55   Signaling between signal processing network equipment and international switching centers

Q.698   Interworking of signaling system No. 7 ISUP, TUP, and signaling system No. 6 using arrow diagrams

Q.700   Introduction to CCITT Signaling System No. 7

Q.701   Functional description of the message transfer part (MTP) of Signaling System No. 7

Q.721   Signaling System No. 7 functional description of the Signaling System No. 7 Telephone User Part (TUP)

In-band and out-band

Q.20   Comparative advantages of "in-band" and "out-band" systems

Q.21   Systems recommended for out-band signaling

Q.22   Frequencies to be used for in-band signaling

Q.25   Splitting arrangements and signal recognition times in "in-band" signaling systems

Q.25   Splitting arrangements and signal recognition times in "in-band" signaling systems

Phone features and signals

Q.23   Technical features of push-button telephone sets

Q.24   Multifrequency push-button signal reception

Q.27   Transmission of the answer signal

Q.28   Determination of the moment of the called subscriber's answer in the automatic service

Q.35   Technical characteristics of tones for the telephone service

Q.109   Transmission of the answer signal in international exchanges

Network access

Q.26   Direct access to the international network from the national network

Quality of transmissions; interference and noise

Q.29   Causes of noise and ways of reducing noise in telephone exchanges

Q.30   Improving the reliability of contacts in speech circuits

Q.31   Noise in a national 4-wire automatic exchange

Q.32   Reduction of the risk of instability by switching means

Q.33   Protection against effects of faulty transmission on groups of circuits

Q.44   Attenuation distortion

ISDN and B-ISDN

Q.71   ISDN circuit mode switched bearer services

Q.80   Introduction to Stage 2 service descriptions for supplementary services

Q.81   Number identification

Q.82   Call offering

Q.83   Call completion

Q.84   Multiparty

Q.85   Community of interest

Q.86   Charging

Q.87   Additional information transfer

Q.761   Functional description of the ISDN user part of Signaling System No. 7

Q.762   General function of messages and signals of the ISDN user part of Signaling System No. 7

Q.763   Formats and codes of the ISDN user part of Signaling System No. 7

Q.764   Signaling System No. 7 ISDN user part signaling procedures

Q.767   Application of the ISDN user part of CCITT Signaling System No. 7 for international ISDN interconnections

Q.768   Signaling interface between an international switching centre (ISC) and an ISDN satellite subnetwork

Q.850   Usage of cause and location in the digital subscriber signaling system no 1 and the Signaling System No. 7 ISDN user part

Q.860   ISDN and B-ISDN Generic Addressing and Transport (GAT) Protocol

Q.920   DSS1 – ISDN user-network interface data link layer, general aspects

Q.921   ISDN user-network interface, data link layer specification

Q.922   ISDN data link layer specification for frame mode bearer services

Q.923   Specification of a synchronization and coordination function for the provision of the OSI connection-mode network service in an ISDN environment

Q.930   DSS1 – ISDN user-network interface layer 3, general aspects

Q.931   DSS1 – ISDN user-network interface layer 3 specification for basic call control

Q.932   DSS1 – generic procedures for the control of ISDN supplementary services

Q.933   ISDN DSS1 – signaling specification for frame mode basic call control

Q.939   Typical DSS1 service indicator codings for ISDN telecommunications services

Q.940   ISDN user-network interface protocol for management, general aspects

Q.950   ISDN supplementary services protocols, structure, and general principles

Q.951   Number identification

Q.952   Stage 3 description for call offering supplementary services using DSS 1 – diversion supplementary services

Q.953   Call completion

Q.954   Multiparty

Q.955   Community of interest

Q.956   Charging

Q.967   Additional information transfer

Q.1901   Bearer-Independent Call Control Protocol

Q.1950   Bearer-Independent Call Control Protocol

Q.1970   Bearer-Independent Call Control IP Bearer Control Protocol

Q.1990   Bearer-Independent Call Control Tunneling Protocol

Q.2010   Broadband integrated services digital network overview – signaling Capability Set 1 (CS-1)

Q.2100   B-ISDN signaling ATM adaptation layer (SAAL) overview description

Q.2110   B-ISDN ATM adaptation layer – service specified connection oriented protocol (SSCOP)

Q.2111   Service-Specific Connection-Oriented Protocol in a multilink and connectionless environment (SSCOPMCE)

Q.2119   B-ISDN ATM adaptation layer – convergence function for SSCOP above the frame relay core service

Q.2120   B-ISDN meta-signaling protocol

Q.2130   B-ISDN signaling ATM adaptation layer – service-specific coordination function for support of signaling at the user network interface (SSFC At UNI)

Q.2140   B-ISDN ATM adaptation layer – service specific coordination function for signaling at the network node interface (SSCF at NNI)

Q.2144   B-ISDN Signaling ATM adaptation layer (SAAL) – layer management for the SAAL at the network node interface (NNI)

Q.2150   AAL2 signaling transport converter

Q.2210   Message transfer part level 3 functions and messages using the services of ITU-T Recommendation Q.2140

Q

ITU-T Q Series Recommendations, cont.

ISDN and B-ISDN, cont.

Q.2610	B-ISDN usage of cause and location in B-ISDN user part and DSS-2
Q.2650	B-ISDN – internetworking between Signaling System No. 7 B-ISDN User Part (B-ISUP) and digital subscriber Signaling System No. 2 (DSS-2)
Q.2660	B-ISDN – internetworking between signaling System No. 7 B-ISDN User Part (B-ISUP) and N-ISDN User Part (N-ISUP)
Q.2723	Extensions to SS7 B-ISDN User Part (B-ISUP)
Q.2725	B-ISDN User Part CS-2.1 See Q.2761 through Q.2764
Q.2726	B-ISDN User Part CS-2.1 See Q.2761 through Q.2764
Q.2727	B-ISDN User Part (B-ISUP) support of Frame Relay
Q.2730	B-ISDN – SS7 B-ISDN User Part (B-ISUP) supplementary services
Q.2735	Stage 3 description of Community of Interest supplementary services for B-ISDN using SS7
Q.2761	B-ISDN – functional description of the B-ISDN User Part (B-ISUP) of SS7
Q.2762	B-ISDN – general functions of messages and signals of the B-ISUP of SS7
Q.2763	B-ISDN – SS7 B-ISDN User Part (B-ISUP) – Formats and codes
Q.2764	B-ISDN – SS7 B-ISDN User Part (B-ISUP) – Basic call procedures
Q.2765	B-ISDN SS7 B-ISUP – Application Transport Mechanism (APM)
Q.2931	B-ISDN – Digital Subscriber Signaling System No. 2 (DSS-2) – User Network Interface (UNI) Layer 3 specification for basic call/connection control
Q.2934	B-ISDN DSS-2 switched virtual path capability
Q.2941	B-ISDN DSS-2 extensions
Q.2951	B-ISDN – Stage 3 description for number identification supplementary services using B-ISDN DSS-2 – basic call
Q.2955	B-ISDN DSS-2 – stage 3 description for community of interest supplementary services – basic call
Q.2957	B-ISDN DSS-2 – stage 3 description for additional information transfer supplementary services – basic call
Q.2959	B-ISDN DSS-2 – call priority
Q.2961	B-ISDN DSS-2 – support of additional parameters

Q.2962	B-ISDN DSS-2 – connection characteristics negotiation during call/connection establishment phase
Q.2963	B-ISDN DSS-2 – extensions for ATM
Q.2964	B-ISDN DSS-2 – basic look-ahead and other clauses
Q.2965	B-ISDN DSS-2 – Quality of Service (QoS) issues
Q.2971	B-ISDN – digital subscriber DSS-2 – user network interface Layer 3 specification for point-to-multipoint call/connection control
Q.2981	B-ISDN and B-PISN – call control protocol
Q.2982	B-ISDN DSS-2 – Q.2931-based separated call control protocol
Q.2983	B-ISDN DSS-2 – Q.2931-based bearer control protocol
Q.2984	B-ISDN and B-PISN – prenegotiation for multiconnection
Q.2991	Abstract test suite for the network integration testing for B-ISDN and B-ISDN/N-ISDN

Modeling, Intelligent Networks

Q.76	UPT functional modeling and information flow
Q.1200	Intelligent Networks (INs) recommendation structure
Q.1201	Principles of Intelligent Networks architecture
Q.1202	Intelligent Networks – service plane architecture
Q.1203	Intelligent Networks – global functional plane architecture
Q.1204	Intelligent Networks – distributed functional plan architecture
Q.1205	Intelligent Networks – physical plane architecture
Q.1208	General aspects of the Intelligent Networks application protocol
Q.1210	Q.12xx series structure for Intelligent Networks
Q.1211	Introduction to Intelligent Networks Capability Set 1 (CS-1)
Q.1213	Global functional plane for Intelligent Networks CS-1
Q.1214	Distributed functional plane for Intelligent Networks CS-1
Q.1215	Physical plane for Intelligent Networks CS-1
Q.1218	Interface recommendation for Intelligent Networks CS-1
Q.1219	Intelligent Networks user guide for CS-1
Q.1220	Q.122x series Intelligent Networks Capability Set 2 (CS-2) recommendation structure
Q.1221	Instruction to Intelligent Networks CS-2

Q.1222 Service plane for Intelligent Networks CS-2

Q.1223 Global functional plane for Intelligent Networks CS-2

Q.1224 Distributed functional plane for Intelligent Networks CS-2

Q.1225 Physical plane for Intelligent Networks CS-2

Q.1228 Interface recommendation for Intelligent Networks CS-2

Q.1229 Intelligent Networks user's guide for CS-2

Q.1231 Introduction to Intelligent Networks Capability Set 3 (CS-3)

Q.1236 Intelligent Networks CS-3 Management Information Model requirements and methodology

Q.1237 Extensions to Intelligent Network CS-3 in support of B-ISDN

Q.1238 Interface recommendation for Intelligent Network CS-3

Q.1241 Introduction to Intelligent Networks Capability Set 4 (CS-4)

Q.1244 Distributed functional plane for Intelligent Networks CS-4

Q.1248 Interface recommendation for Intelligent Network CS-3

Q.1290 Glossary of terms for Intelligent Networks (included in Q.9)

Q.1300 Telecom applications for switches and computers (TASC) – general overview

Q.1301 TASC – architecture

Q.1302 TASC – functional services

Q.1303 TASC – management, architecture, methodology, and requirements

Q.1400 Architecture framework for the development of signaling and organization, administration, and maintenance protocols using OSI concepts

Q.1521 Requirements on underlying networks and signaling protocols to support UPT

Q.1531 UPT security requirements for Service Set 1 (SS-1)

Q.1541 UPT Stage 2 for SS-1 on SC-1-1995: procedures for UPT functional modeling and information flows

Q.1542 UPT Stage 2 for SS-1 on CS-2 – procedures for universal personal telecommunication functional modeling and information flows

Q.1551 Application of Intelligent Network application protocols (INAP) CS-1 for UPT SS-1

Q.1600 Signaling System No. 7 interaction between ISIUP and INAP

Q.1601 Signaling System No. 7 interaction between N-ISDN and INAP CS-2

Q.1701 Framework for IMT-2000 networks

Q.1711 Functional model for IMT-2000 networks

Q.1721 Information flows for IMT-2000 CS-1

Q.1731 Radio technology independent requirements for IMT-2000 Layer 2 radio interface

Q.1751 Internetwork signaling requirements for IMT-2000 CS-1

Wireless communications

Q.14   Means to control the number of satellite links in an international telephone connection

Q.1000 Structure of the Q.1000 series recommendations for public land mobile networks

Q.1001 General aspects of public land mobile networks

Q.1002 Network functions for public land mobile networks

Q.1003 Location registration procedures for public land mobile networks

Q.1004 Location register restoration procedures for public land mobile networks

Q.1005 Handover procedures for public land mobile networks

Q.1032 Signaling requirements relating to routing of calls to mobile subscribers

Q.1051 Mobile application part for public land mobile networks

Q.1100 Interworking with Standard A INMARSAT system – structure of the Recommendations on the INMARSAT mobile satellite systems

Q.1101 General requirements for the interworking of the terrestrial telephone network and INMARSAT Standard A system

Q.1111 Interfaces between the INMARSAT standard B system and the international public switched telephone network/ISDN

Q.1112 Procedures for interworking between INMARSAT standard-B system and the international public switched telephone network/ISDN

Q.1151 Interfaces for interworking between the INMARSAT aeronautical mobile-satellite system and the international public switched telephone network/ISDN

Q.1152 Procedures for interworking between INMARSAT aeronautical mobile satellite system and the international public switched telephone network/ISDN

Q.1237 Extensions to Intelligent Network CS-3 in support of B-ISDN

Q

Q-band A microwave frequency spectrum ranging from 36 to 46 GHz, between the Ka-band and the V-band. Frequencies in this range tend to be used for radar and small aperture satellite transmissions. See band allocations for a chart of designated frequencies.

Q-factor A measure of frequency selectivity, or the "sharpness" of resonance in a resonant vibratory system which has one degree of mechanical or electrical freedom. See Q factor, Q-switch.

Q-switch In laser resonating cavities, a mechanism for opening and closing the cavity to allow or block the laser light to produce laser pulses rather than a continuous wave. This mechanism can be constructed with acousto-optic or electro-optic devices (e.g., Pockels cells). The term comes from the high signal-to-noise ratio (Q factor) that is characteristic of an amplified signal in a filled laser resonating cavity.

In acousto-optical switching, incident light from the laser is deflected when it comes in contact with acoustical waves (e.g., in a modulating crystal) and is scattered so that there isn't sufficient focused energy to exit the resonating cavity.

In electro-optical switching, incident light from the laser hits a polarizing "shutter" such that the light will pass or not pass. Thus, Pockels cells, in conjunction with a polarizer, act as a Q-switch to alternately allow the laser cavity to be filled or emptied in a process sometimes called *cavity dumping*. Pockels cells may also be configured/assembled with other components to allow a portion of the pulse to exit before the cavity is fully filled (amplified). See pockels effect, Q, Q-factor.

Q-Telecom A business unit of Info-Quest providing telecommunications services to Greece as one of four GSM suppliers in the country. In June 2002, it was announced that Q-Telecom would deploy Harris Corporation point-to-multipoint digital radio systems throughout Greece.

Q.SIG A global common channel signaling protocol (CCS), based upon the ISDN signaling protocol, used in the digital transmission of voice over digital networks such as ATM. In addition to the features in the ISDN signaling protocol, Q.SIG includes private branch exchange (PBX) features so a network of PBXs can interact as a distributed system. CCS systems are more prevalent in Europe than in the United States. See voice over ATM.

QA 1. quality assurance. 2. queued arbitrated. In DQDB, an information field segment used to transfer slots when they arrive through a nonisochronous transfer.

QAM See quadrature amplitude modulation.

QBE See query by example.

QBF, fox message QBF = "quick brown fox." The Q signal code to send a test sentence that includes all the letters of the English alphabet. A QBF message is commonly used to verify whether all letters available to a device or contained within a coding system are present and/or working correctly. It is familiar to most as "THE QUICK BROWN FOX JUMPS OVER THE LAZY DOG" (which is then repeated in lowercase, if needed). The idea is to convey the entire alphabet in the shortest sentence that is comfortably memorable as possible. In wireline devices, it is more often called a "fox message" since Q signals are associated more specifically with radio frequency communications. The phrase "fox message" is used more generically to test alphabetic communication signals and the physical integrity of typewriter or computer keyboard keys. See Q signal, Z code.

QC quality control.

QC laser See quantum cascade laser.

QCELP See Qualcomm Code Excited Linear Predictive Coding.

QCIF See Quarter Common Intermediate Format.

QCT Qualcomm CDMA Technologies. See Qualcomm Code Excited Linear Predictive Coding

QD See queuing delay.

QD-DOS, QDOS A historic microcomputer operating system (Quick and Dirty Operating System) developed by Tim Paterson, which was derived from a mid-1970s manual describing Gary Kildall's CP/M, and extremely similar in syntax and functionality. At that time, IBM was looking for an operating system for its line of microcomputers. IBM contacted Microsoft about contracting their (computer language) products, thinking they had also purchased the rights to CP/M. When they found that Microsoft didn't have an operating system, they went to visit Digital Research (originally Inter-Galactic Research), but the DR representative was reluctant to sign IBM's nondisclosure agreement on DR's behalf, especially when the attorney didn't like the terms of the contract. IBM went back to Microsoft and DR thought it would have a further opportunity to talk terms with IBM, especially since Microsoft didn't have an operating system that could meet IBM's needs at the time, as they had been concentrating their efforts on developing computer languages. Microsoft, however, promised one to IBM in a very short time period, and delivered on the contract by purchasing the code for QDOS from Seattle Computing, the company for which Paterson was working. They provided it to IBM who released it as PC-DOS. Microsoft subsequently purchased the distribution rights for QDOS for $50,000 and later released a slightly altered version of PC-DOS as MS-DOS (Microsoft Disk Operating System). Microsoft managed to stipulate contractually that they could retain the rights to sell the product they had developed for IBM, in competition with IBM. Thus, QDOS, derived from CP/M became IBM's product, rather than CP/M itself, and evolved into MS-DOS, and eventually Windows. See CP/M, Microsoft Corporation, MS-DOS, Digital Research.

QDU See quantizing distortion units.

QFA See Quick File Access.

QFC See Quantum Flow Control.

QFM See quadrature frequency modulation.

QIC quarter inch cartridge. See Quarter Inch Cartridge Drive Standards.

QICC 1. See Quad Integrated Communications Controller. 2. See Quad International Communications Corporation.

QIP A commercial Internet Protocol (IP) address management software product from Quadritek (now Lucent IPGSP) that facilitates the central management of network databases. In March 2000, Lucent announced that QIP had been ranked as the market share leader for standalone Internet Protocol (IP) address management products. The product is used by prominent companies such as MCI Worldcom's UUNET, Discover Financial Services, Ford Motor Company, and others. See Quadritek.

QJDP QIP/Windows 2000 Joint Developer Program. Lucent Technologies' initiative launched in December 1999 to foster development of design interfaces and requirements between QIP 5.0 IP address management software and Windows 2000. See QIP, Quadritek.

QL See query language.

QLLC See Qualified Logical Link Control.

QMS Queue Management System. See queue management.

QoR See Query on Release.

QoS See Quality of Service.

QPL Qualcomm PureVoice Library. See Qualcomm Code Excited Linear Predictive Coding.

QPSK 1. See quadrature phase shift keying. 2. See quaternary phase shift keying.

QR queuing requirements. See queuing theory.

QRP A designation for low-power amateur frequency radio transmissions. Low-power transmitters and receivers are an interesting subgroup of hobbyist radio, when used with respect for the privacy of individuals and within regulatory guidelines. Regulations for short distance, low-power transmissions are more lenient than for other types of broadcasts. QRP transmitters can be used for short-distance broadcasting, home security systems, door intercoms, climbing communicators, baby and child monitors, and other short-range projects.

QRP ARCI The QRP Amateur Radio Club International is a nonprofit organization dedicated to amateur design, construction, and use of QRP (low power) transmitters. See Amateur Radio Relay League, QRP. http://www.qrparci.org/

QSAM see quadrature sideband amplitude modulation.

QSDG See Quality of Service Development Group.

QTAM See Queued Telecommunications Access Method.

QTF See quartz tuning fork.

QTC QuickTime Conference. See QuickTime.

QTSS QuickTime Streaming Server. See QuickTime.

QTVR QuickTime Virtual Reality. See QuickTime.

QuAD Quorum Associate Distributor. See Quorum International.

quad- Prefix for four.

quad antenna A type of array antenna similar in principle to a Yagi-Uda antenna, except that it uses full-wavelength loops in the place of half-wavelength straight elements, thus providing greater gain over a similar Yagi-Uda antenna. As with many antennas, the feed is commonly 75-ohm coaxial cable. A two-element quad antenna is called a *quagi*.

Quad antennas come in a variety of configurations with the elements mounted on a boom or radiating out like the support threads of a spider web. In one common configuration, the prominent features include simple straight-pole elements mounted in a horizontal plane at right angles to one another with a diamond-shaped configuration on one pole mounted in the vertical plane. There are variations on quad antennas including cubical quads (which can resemble an open umbrella with the fabric missing) and hybrid quads.

Mini-quads – compact quad antennas – have some advantages that are appealing to amateur radio enthusiasts including reasonable cost, low wind loading, and small turning radius. Quad/mini-quad antennas in the 15±5 m frequencies are readily available.

Quad Integrated Communications Controller QICC (*pron.* qwik). A single-chip integrated CPU32+ microprocessor from Motorola designed for embedded telecommunications and internetworking applications. The MC68360 QICC is a next-generation MC68302 with four serial communications controllers, two serial management controllers, and one serial peripheral interface that operates at 4.5 MIPS at 24 MHz. QICC is useful for controller applications.

Quad International Communications Corporation QICC. A California-based international supplier of telecommunications products including Frame Relay.

quad wiring Wiring bundles consisting of four individually sheathed, untwisted wires brought together (aggregated) within a single cover. Quad wiring is often used for the internal wiring of two-line analog phones, with the lines inside generally color-coded green and red (tip and ring) for the first line, and black and yellow for the second line. This type of wiring is not recommended for data transmission installations. Quad fiber cables consist of four individual fiber cables bundled together within a single cover.

quadrature A state in which cyclic events are 90 degrees out of phase. In signal transmission quadrature, phasing is a common technique used to distinguish information in signals. It is also used to vary a signal so crosstalk between two closely associated transmissions is reduced.

quadrature amplitude modulation QAM. A modulation technique employing variations in signal amplitude. This modulation scheme is used in asymmetric digital subscriber line services, for example. It is a two-dimensional coding scheme that can be transmitted in a narrower spectrum, a combination of amplitude and phase-shift modulation. The QAM spectrum derives from the spectrum of the baseband signals as they apply to the quadrature channels.

QAM is similar to nonreturn-to-zero baseband transmission and multiphase phase shift keying (PSK), except that QAM does not have a constant envelope as in PSK.

QAM requires lower sampling frequencies and the spectral width can be optimized by keeping the baud rate lower, thus reducing the potential for crosstalk. See modulation.

Q

quadrature/quadriphase phase-shift keying QPSK. A type of phase shift keying modulation scheme in which four signals are used, each shifted by 90°, with each phase representing two data bits per symbol, in order to carry twice as much information as binary phase shift keying (BPSK), which can be seen as two independent binary phase shift key (BPSK) systems.

QPSK can be used to carry bit timing and can be filtered using raised cosine filters for out-of-band suppression. Even more sophisticated systems exist that employ differential encoding of symbol phases. Linear power amplifiers are used with the various QPSK schemes.

Staggered quadrature/quadriphase phase shift keying (SQPSK) is similar to QPSK except that the data channels are offset to shift the carrier 90°. The staggering facilitates recovery of I and Q channels. See frequency modulation, frequency shift keying, on/off keying, modulation, quadrature sideband amplitude modulation.

quadrature sideband amplitude modulation QSAM. A modulation encoding technique in which different signal amplitude states represent data.

Quadritek Systems Inc. A network products firm founded in 1993 to provide server-related solutions. QIP is Quadritek's (IP) network address management software product.

In March 1998, IBM and Quadritek announced a collaboration in which IBM would make changes to its Dynamic Host Configuration Protocol (DHCP) and Domain Name Server (DNS) to enable Quadritek and other vendors to use an open standard API to manage servers. Thus, Quadritek's QIP product could be used to manage multiple, distributed IBM DHCP/DNS servers. Quadritek is now Lucent Technologies IPSPG.

quadruplex circuit A circuit which is carrying two bidirectional transmissions simultaneously to make a total of four.

Qualcomm Code Excited Linear Predictive Coding QCELP. A proprietary algorithm from Qualcomm CDMA Technologies (QCT) that supports digital voice coding/decoding through code division multiple access (CDMA) methods. QCELP compression supports fixed and variable encoding.

In the mid-1990s, Qualcomm released the PureVoice vocoder (voice coder), based on 13-kilobit QCELP for use in cellular and Personal Communications System (PCS) products. The design goal of the PureVoice vocoder was to provide voice quality approaching that of wireline while still keeping the bit rate as low as possible to work within the capacity limitations of mobile communications devices. Since it is a software solution, it can be integrated into a wide variety of desktop and mobile devices.

In 1998, Qualcomm announced an agreement with Apple Computer to integrate the PureVoice QCELP-based audio codec technology into Apple's popular QuickTime multimedia software. The PureVoice technology is popular because of its relatively high quality and small file size, which is significantly smaller than audio files stored in .wav format, for example.

PureVoice was also adapted for use in email in the late 1990s, so users could send actual voice messages through computer networks such as the Internet. The excellent compression ratios inherent in the QCELP technology made this a practical application and enabled greetings to be communicated without traditional long-distance telephone charges. The PureVoice Player/Recorder and the PureVoice Converter are available online as licensed, freely distributable software from Qualcomm Incorporated.

Qualified Logical Link Control QLLC. A data link control protocol from IBM which works with the IBM SNA systems to allow them to operate over X.25 packet switched data networks.

quality 1. Meeting subjective and/or objective standards of excellence in operation, manufacture, aesthetics, or a combination of these. 2. In manufacturing, quality is more narrowly defined as conformance to high objective standards of appropriateness, functionality, and longevity within the context of related products. 3. In service industries, quality is generally determined by adherence to operating and ethical standards of the industry and degree of customer satisfaction. See quality assurance.

quality assurance Systematic actions which seek to assure satisfactory levels of manufacture, service, functionality, and longevity.

quality factor See Q factor.

quality of service QoS. This has a general meaning across many industries and somewhat more specific meanings in telecommunications networks. Quality of service is a performance descriptor and reference for the provision of services on a network. It includes parameters and values pertaining to data rates, acceptable delays, losses, errors, etc.

As part of the QoS requirements for an ATM network, four class of service (CoS) traffic types have been specified:

CoS	Characteristics
Class A	Connection-oriented, constant bit rate (CBR), with a strong timing relationship between source and destination. Constant bit rate video and PCM encoded voice are included in this category.
Class B	Connection-oriented; bit rate may vary, with a strong timing relationship between source and destination.
Class C	Connection-oriented; bit rate varies, no timing relationship between source and destination. TCP/IP and X.25 are included in this category.
Class D	Connectionless; bit rate varies, no timing relationship between source and destination. Connectionless packet data are included in this category.

There are many types of data, and how they are perceived in part determines how their quality is evaluated. Consequently, QoS requirements vary with the type of data. See cell rate, class of service.

Quality of Service Development Group QSDG. A Telecommunication Standardization Sector group of the International Telecommunications Union established in 1984 to help develop practical implementations of international telecommunication quality of service (QoS) standards. It is funded primarily by administrations and ROAs.

quantization A process in which a continuous range of values, such as an incoming analog signal, is subdivided into ranges, with a discrete value assigned to each subset. This is a means of converting analog data to digital data, and is used in musical sound sampling, modem communications, voice over data networks, radio wave modulation, and many other aspects of telecommunications.

Generally the frequency of the sampling influences the quality and fidelity of the outgoing quantized signal, within certain limits set by the capabilities of the equipment and the characteristics of the human perceptual system. Higher sampling rates tend to produce closer approximations to the original signal, but also require greater transmission speeds and bandwidth.

Quantization is used in a number of modulation schemes, including pulse code modulation (PCM), which is commonly used in voice communications. See modulation, patches, pulse code modulation, sampling, quantization error.

quantization, vector A vector version of scalar quantization, designed to reduce the volume of data files or the bit rates of data transfers. Vector quantization has practical applications for image and speech coding.

quantization error A number of aspects can introduce error into a quantized signal, including the amount of noise and interference accompanying the signal, the signal range or amplitude as it relates to the capabilities of the quantizing mechanism, the strength and complexity of the signal being quantized, and the mathematics used to carry out the conversion. Quantization error is sometimes assessed after a digital signal is reconverted to analog format, and the end signal is compared to the original, with the differences assessed subjectively (as in music systems) or evaluated with various measuring instruments.

quantize To convert a continuous range of values into discrete, nonoverlapping values or *steps*. This is an important means to convert analog to digital values.

quantizing distortion units QDU. A measure of the degree of degradation in a voice channel that occurs as a result of format and signal conversions (e.g., analog to digital to analog). This is described in the ITU-T G Series Recommendation G.113 (transmission impairments).

quantometer An instrument for the measurement of magnetic flux.

quantum (*plural* – quanta, *symb.* – q) A relatively recently discovered and investigated phenomenon related to the movement of electrons. Quantum theory

was first stated by physicist Max Planck in 1900. A quantum is a discrete quantity of electromagnetic energy (e.g., a *photon* of light energy), the smallest possible amount of energy at any given frequency v. Quantum phenomena are of great interest to physicists, and researchers are now investigating ways of enlisting quantum behaviors in the manufacture and use of various industrial products such as lasers and in operations associated with digital logic, with some surprising and provocative success. See Einstein, Albert; Planck, Max; quantum cascade laser.

quantum cascade laser QC laser. A new type of "NanoLaser" developed by Frederico Capasso and Jerome Faist at Bell Laboratories in 1994. The QC laser is a continuously tunable, single-mode, distributed-feedback device.

To understand how a QC laser works, imagine an electric current stimulating a number of electrons to cascade over a series of steppes (a terraced organization), squeezed through quantum wells in successive layers, dropping off energy in the form of photons (light pulses) as they contact and travel through each steppe. At each steppe, the electrons perform a quantum jump between well-defined energy levels. The photons emitted as a result of their activity reflect back and forth in an amplification process that stimulates other quantum jumps and emissions and results in a high output. This process can be exploited by creating a corrugated grating layer within a semiconductor which acts as a filtering device for specific wavelengths according to the grating period and, to some extent, the operating temperature.

QC lasers have a number of advantages over traditional diode lasers, including higher optical power and finer linewidth in terms of the specificity of the wavelengths emitted. They can operate over a wide selection of wavelengths in the mid-infrared range from 3.4 - 17 µm. QC lasers can be used in many applications, including medical diagnostics, radar heterodyne detectors, production process control, and remote sensing applications – particularly environmental monitoring in toxic environments due to gas-sensing capabilities.

The wavelength of the laser is determined by quantum confinement. Thus, it can be tuned selectively over a wide range of the infrared spectrum by varying the layer thicknesses and spacing of the different materials used in its manufacture. This differs from other technologies in that the output wavelength is not dependent upon the chemical composition of the semiconductors, but upon their thickness and positioning. These layers, created with a molecular beam epitaxy (MBE) materials-growth process, are sometimes only a few atoms thick. The QC laser also functions at higher temperatures than traditional diode lasers, making it practical for room temperature use. See Capasso, Frederico; distributed-feedback laser; Fabry-Perot laser; quantum well; vertical-cavity surface-emitting laser.

Quantum Corporation A prominent data storage device developer/distributor, founded in 1980. Quantum became well-known for computer hard drives in the

Q

late 1980s and the 1990s and, more recently, has introduced Super DLTtape technologies that provide fast transfer-rate, high-capacity storage on tape cartridges. See Super DLTtape.

Quantum Flow Control QFC. In ATM networks, a congestion avoidance scheme proposed for use on available bit rate (ABR) connections. For example, in a network in which VCI tunneling is implemented, the ATM device will send only after receiving explicit credit from a receiving ATM device at the other end of the connection. If tunneling is not used, buffer allocation and a credit manager must be included. If the buffer allocation is exceeded, noncomplying cells will be discarded.

quantum mechanics The study of atomic structure and behaviors using various measuring instruments and techniques. See Heisenberg uncertainty principle, quantum.

quantum noise When using a detector to investigate quantum characteristics in electromagnetic phenomena, there may be noise from random variations or fluctuations in the average rate of incidence of quantum interactions with the detector. These may be expressed in terms of photons.

quantum well QW. A quantum phenomenon associated with a structure fabricated from ultrathin alternating layers of wide bandgap (barrier) and narrow bandgap (well) materials. When an electron is caught in a well formed between the barriers, the probability of escaping the well is limited and the electron's energy level is affected. The quantization effects resulting from these events are related to the height and width of the fabricated barrier and can be derived through quantum mechanical calculations. Quantum well exhibiting structures can be constructed using crystal growth techniques (for use in quantum cascade lasers, for example). Quantum well components have unique properties that can be exploited for a variety of optical communications technologies.

Quantum well physics can be used to create modulators through the application of an electrical field perpendicular to the surface of the quantum well or alternately to its sides. Thus, the optical absorption of the QW is changed sufficiently to make it useful for signaling. This phenomenon can be exploited for semiconductor design to make small optical modulators with two-dimensional optical arrays.

A quantum well infrared photodetector (QWIP) is a multiple quantum well device based upon layered high-bandgap semiconductor fabrication. Bandgap discontinuity associated with the layers of differing materials creates quantized subbands in the potential wells. The phenomena associated with the layers is a photoexcitation of electrons between ground and first excited-state subbands. The carriers resulting from the photoexcitation are then able to escape from the potential quantum wells to generate a photocurrent. In the U.S. Naval Research Lab, quantum well physics has been used to develop a number of types of semiconductor lasers emitting mid-infrared light. Using "wavefunction engineering," engineers have designed complex, layered quantum well structures.

This has a number of possible applications, including laser radar (ladar).

Researchers at Imperial College demonstrated a new quantum well solar cell (QWSC) that may be more efficient than previous solar energy sources.

In 1997, scientists described an uncooled strained quantum well laser that could be used in SONET/SDH networks, especially short- and medium-haul transmissions.

See electroabsorption, quantum cascade laser, self-electro-optic effect device.

Quarter Common Intermediate Format QCIF, Quarter CIF. A standard for the transmission of video frames in the ITU-T H.261 standard. QCIF consists of 144 lines of luminance and 176 pixels per line (144×176 CIF format is optionally supported by H.261). This relatively low resolution creates an image that has a soft-focus, indefinite appearance, but has the advantage of using fewer system resources and less bandwidth. In fact, the standard was developed with the needs of circuit-switched networks in mind. For small windows, simple images, and small display devices, it has practical applications, and it is widely favored for videoconferencing, especially on ISDN networks. H.261 is usually implemented in conjunction with other related standards. See Common Intermediate Format.

Quarter Inch Cartridge Drive Standards QIC. An international association, established in 1987, to promote the acceptance and use of quarter-inch readable/writable data cartridge drives and media. These types of storage media are commonly used for computer backup, secondary storage, and temporary storage for files that need to be transported.

More than 100 QIC standards have been developed since 1988. QIC-40, QIC-80, QIC-3101, and QIC-3020 have been particularly prevalent in the tape cartridge field, although they are now being superseded by higher capacity formats. A complete list and fuller description of each standard are available on the QIC Web site. http://www2.qic.org/

quarter wave The distance, or elapsed time, in a conducting line or through a conducting space, which is 90° to a wave disturbance. This information in used in the design of antennas and in the quadrature transmission of signals, particularly in modulation schemes. See quadrature.

quartz A silicon dioxide mineral found or synthesized in crystal form and in crystalline masses, which is widely used in scientific research and telecommunications due to its oscillating qualities. Quartz is transparent, harder than glass, and varies in its oscillating frequencies depending upon its size and shape. Quartz crystal watches are extremely accurate, and quartz arc lamps are used for sterilization, due to the way ultraviolet light passes through the crystal. See piezoelectricity.

Quartz A 2D graphics engine from Apple Computer, based upon the standardized Portable Document Format (PDF). Quartz is incorporated into Mac OS X.

quartz (*symb.* – SiO_2) A mineral silicate of the silicon dioxide *quartz group* with a unique helical structure and piezelelectric properties that have many applications in electronics.

quartz crystal component A piece of quartz cut to a precise size for a specific purpose. Quartz has remarkable constancy in its vibratory qualities, making it suitable for extremely precise time devices. These vibratory qualities can be controlled by manipulating the shape and size of the crystal. Early radio sets were called "crystal detectors" as they used crystals (galena and carborundum were popular) to detect (rectify) and channel a radio wave. Quartz is commonly used in oscillators and filters. Quartz crystals are used to provide timing in watches and to stabilize broadcast waves. See quartz, quartz crystal filter.

quartz crystal filter The properties of quartz crystals make them useful for a variety of applications that require highly selective electrical circuitry, and hence they are used in the creation of various types of filters. Synthetic quartz crystals, developed in the 1950s, furthered the manufacture of quartz filters for use as electronic components. There were, in fact, few other materials that offered the advantages of natural or synthetic quartz until the development of lithium-tantalate crystals in the Bell Laboratories. See lithium-tantalate, quartz, quartz crystal.

quartz tuning fork QTF. A specialized instrument exploiting the piezoelectrical properties of quartz. In 1995, it was suggested by Karrai and Grober that QTFs could be used for measuring shear forces on scanning near-field optical microscope (SNOM) fiber tips. Subsequent research has led to the development of QTF mechanisms for non-optical distance stabilization in atomic force microscopes (AFMs) for the realization of apertureless SNOM. See scanning near-field optical microscope.

quaternary phase-shift keying QPSK. A modulation technique which is used to encode digital information to be transmitted over wire or fiber networks. It is a subset of phase shift keying (PSK), and is essentially a four-level version of phase modulation (PM). QPSK divides the bit stream into two streams, and sends them alternately to in-phase and out-of-phase modulators, where they are subsequently demodulated at the receiving end.

QUBE An Interactive TV information utility. Warner instituted the QUBE interactive educational TV network in the late 1970s. The first interactive television concert, broadcast live over the QUBE system in 1978, featured Todd Rundgren, pioneer multimedia recording artist.

quench To bring to a sudden halt, to cool rapidly, to quickly extinguish a flame, spark, or gas emission.

quench oscillator In some super-regenerator circuits, a type of ultrasonic oscillator which serves to quench, or rapidly reduce, the regeneration when it has almost increased to the point of oscillation.

quenched spark gap Early wireless transmitters used spark gaps in their spark transmitters, with several types of gaps: open gaps, rotary gaps, and quenched gaps, each with different strengths and weaknesses. Quenched gaps employed a racklike series of metal plates separated by thin layers of mica, resulting in a very small spark that is quickly quenched and does not tend to overheat as do open gaps. Due to improvements in technology and the need for regular cleaning to keep quenched gap transmitters working optimally, they were eventually superseded by continuous wave (CW) transmitters.

query 1. Request for data, in which the content of the data is the desired result. Common in database applications. 2. Request for data which provides information about the state (operating parameters, mode, security, etc.) or functioning (availability, readiness, status, responsiveness, etc.) of a system. Usually at a low operating level and generally transparent to the user.

query by example QBE. An idea introduced in the 1970s whereby a user interacts with a front end to a database by supplying examples of the type of information that the user wants to retrieve. Sometimes this is more practical than querying by keywords or algorithms. A number of popular database programs provide this capability.

query language A programming language intended to facilitate search and retrieval of information, usually from a database. Query languages are frequently in the form of interpreted scripting languages or graphical report generators, with commands that are similar to common English words, to make them easier to program by those without programming backgrounds.

Query on Release QoR. A telephone number portability mechanism suggested by Pacific Bell to trigger a database query, depending upon circumstances related to a subscriber's current carrier. In QoR, a telephone call setup signal is routed to the end office switch to which the dialed phone number was originally assigned (e.g., the NPA-NXX of the dialed number). If the dialed number has been transferred to another carrier's switch, the database is queried for routing information and the call completed to the new switch. The system is also called Look Ahead.

In 1997, the Federal Communications Commission affirmed its conclusion that this (interim) solution was not acceptable over the long term because it violated a statutory requirement for consumers to retain numbers without impairment of the quality, reliability, or convenience when switching carriers. The Commission felt that degradation in service in terms of postdial delays could compromise QoR forwarding and supported Location Routing Number (LRN) as statutorily acceptable. See Release to Pivot.

queue A stream of items or tasks waiting to be processed or executed, such as calls to an operating system, a network, or a phone system. Queues are used to maximize the use of existing resources, especially on shared systems. It's expensive to put a printer on every computer in a network and, since printing doesn't happen as often as data input/output, it's not efficient either. By allocating one printer to every few workstations, user print requests can be handled efficiently by the network, with simultaneous requests administered through a set of parameters. This also can improve resource choice. By sharing printers, it

Q

may be possible to offer a variety of types of printers and paper sizes, which is more practical and economical than trying to purchase several printers for each computer. See queuing.

queue administration Queues are widely used to manage resource-sharing on a network. Whether the resource is a printer or modem, applications program, data file, or gateway to the Internet or Web, computer systems create, manage, authorize, and prioritize access to these resources and services through queues which are usually transparent to the user.

On phone networks, queue administration may involve putting a caller on hold, checking to see if and when agents are ready to take the call, playing periodic messages to the caller, and assigning the call to the appropriate agent.

On computer networks, queue administration may involve logging in users as they sign on to the system, checking for the existence of devices when a resource request occurs (e.g., a printing job), determining if others are in the queue, and where to slot the new request (the size of the print job, or relative priority of the user requesting the job may be taken into consideration), and may even change the queuing arrangement dynamically if another printer comes online or a print request is canceled before the job is run.

Queued Telecommunications Access Method QTAM. An IBM communications control protocol which handles some applications processing tasks. QTAM is used in a number of telecommunications applications, including message switching, data processing, etc.

queuing delay QD. In its most general sense, a delay caused by queues or lineups within a system through which objects, information, or data are channeled. As a simple example, shopping in a downtown store with a queue at the cash register can result in a delay that causes a shopper to receive a parking ticket. When placing a phone call, a delay may be imposed on the caller by congested trunk lines or by the various switchers, routings, and routing priorities inherent in or configured into a system. Queuing delays can have many negative effects including customer frustration, loss of revenue, extra costs, and even death (in the case of emergency calls), so queuing theories and solutions are considered to be an important aspect of telecommunications.

The queuing delay in a transmission system can be assessed in terms of particular legs within a route or in terms of a transmission as a whole, from sender to recipient. This is often a dynamic process without a single catch-all solution. Queuing delay is often evaluated in conjunction with other types of delays, including processing, propagation, and transmission delays. Queuing refers to those aspects where data or objects "bump up" against one another, in the sense that (usually) similar objects congregate "behind" one another in order to pass through a gateway or other channel that cannot accommodate the queued entities simultaneously.

In cell-based transmissions, a queuing delay is a delay imposed on a cell due to the current inability of the cell to be passed on to the next element or function (because of congestion or errors). Depending upon the system and priorities, significant delays may have several results; the buffered cell data may be returned or destroyed. See queuing theory.

queuing theory Queuing, in its broadest sense, involves an understanding of mathematics, statistics, modeling, data flow, and human behavior as they relate to the ways in which machines may be configured, tuned, and operated so as to carry out worthwhile tasks and processes in an efficient and orderly manner. Researchers in queuing theory regularly come from fields such as probability mathematics, complex systems theory, and simulation research.

In the context of networking, queuing theory focuses on understanding, describing, and predicting patterns in transmission organization, priorities, delay, loss, and standards for quality of service (QoS).

One of the most important pioneers of queuing concepts was Danish telephone engineer A.K. Erlang, who studied and described telephone traffic in its mathematical context and practical applications in the early 1900s. Another significant contributor to the body of knowledge in queuing theory is Leonard Kleinrock, who was involved in the early development of the ARPANET and who authored *Information Flow in Large Communication Nets*, in 1961. He subsequently wrote *Communication Nets*, in 1964, which provides design and queuing theory for building packet networks, in spite of a common sentiment at the time that packet switching wouldn't work.

Queuing theory is an important aspect of performance evaluation and configuration in communications networks. Without a theoretical model for installing, configuring, and tuning a network, much time can be wasted in trying out the many different ways in which network traffic can be routed, especially in a heavily used system in which congestion and "bursty" traffic occur. Queuing applies to a broad range of environments, from individual circuit transmissions to global distributed networks.

A single queue system is one of the most basic models discussed in queuing theory. When all traffic is routed through a single channel on a first-come, first-served basis (e.g., a single cash register in a corner grocery store), the impact of extra traffic and overflow may be different from traffic management in multiple queue systems (a supermarket with multiple cash registers or multiple turnstyles at a sports stadium). The same concepts can be applied to telecommunications systems. A single phone routed through a dedicated line will be managed differently from a single line through a public phone system (wireline or wireless) where congestion might occur. The problems of queuing become more intricate when multiple users of multiple phones (or modems) are sharing network resources over a public network, especially in distributed networks where individual nodes may or may not be available at any particular point in time. See Erlang, Agner; queuing delay.

QUICC See Quad Integrated Communications Controller.

quick connection block A connection block for more quickly and easily making electrical connections without the necessity of learning to use specialized tools and wiring codes. Quick connection blocks are typically made of plastic or other nonconductive materials with metal terminal connecting points and may be preterminated for ease of installation. They are sold for a variety of uses, including electrical wiring, quick installation of multiple lights, computer network connections, and internal telephone line installation, especially for multiphone systems.

Quick connection blocks are especially appealing to small businesses installing their own phone systems. These blocks enable wires to be inserted with easily available screwdrivers, pliers, and wire cutters rather than punchdown tools. One common configuration for telephone quick connection blocks is a premade *66 block,* so-called because it supports up to 66 cross-connections.

Panasonic has a commercial Quick Connect Block, a premade 66 block, that connects to the standard building wiring and supports up to eight analog or digital phone stations for each block (depending upon the phone model).

Homaco, Inc. provides premade 66 blocks for telephone connections in 25-, 50-, and 100-pair sizes.

Comm-Omni International supplies cable termination and surge protection for cables for multiple-family residences and office buildings (e.g., private branch exchange terminations) with a quick connection block accessory with 50 6-pin interconnected rows.

Quick File Access QFA. A system for enabling faster access to data on tape-based storage systems such as computer cartridge drives. Prior to QFA and similar systems, tape drives had to read through blocks of data sequentially to find the desired data, resulting in slow read times, especially if the desired data were near the end of the tape. With tapes and data stored on tapes getting longer, this became impractical. Thus, a system of commands known as Quick File Access enables a block number to be read so that the tape can be fast-forwarded to the appropriate place in the tape to locate the data. Since the block information pointers cannot be easily manipulated on tape, they are usually held in a database. Unfortunately, if the database is compromised (or logging disabled), then data location can be a problem (especially if it is a backup tape to restore a crashed system upon which the database was stored). Nevertheless, QFA is much more convenient than traditional serial-access tapes.

quick-break fuse A type of fuse which breaks a circuit very quickly if a surge or other anomalous electrical condition occurs. Quick-break fuses are especially useful with electronics components, which are sensitive to electrical fluctuations and prone to damage.

QuickDraw A widely used proprietary computer drawing and display specification from Apple Computer Inc. QuickDraw can display screen images and processing PostScript files so they can be printed on nonPostScript-equipped printers.

quicksilver *colloq.* mercury.

QuickTime A proprietary cross-platform computer display, audio, and animation environment from Apple Computer Inc. that runs on Macintosh, Power-Mac, Windows 95, Windows NT 4.0, Windows ME, and Windows 2000. QuickTime is actually a suite of software applications for supporting picture display, multimedia authoring, and server support for streaming audio/video.

QuickTime allows some interesting applications to be developed and distributed, including frame-based animation, whiteboarding, video clips, teleconferencing applications, virtual reality environments, games, and more. The QuickTime format is widely supported on the Internet, with many Web-based multimedia applications distributed in QuickTime. Many digital cameras also support the QuickTime format.

Most recently, QuickTime has been enhanced to support streaming media in Internet browsers (HTTP, RTP, RTSP) to support more than 30 different audio and video file formats (AIFF, BMP, GIF, JPEG, MPEG-1, MP3, M3U, PICT, PNG, SGI, Targa, TIFF, VR, Wave, and more), and added modules for saving digital video (DV) camcorder formats for the development of digital video.

Most QuickTime software is freely distributed, including the QuickTime Player and PictureViewer. The QuickTime Streaming Server and Darwin Streaming Server are also freely distributed, with no streaming data license fees. See QuickTime chart.

quiet tuning In radio receivers, a tuning characteristic in which the signal is kept quiet, that is, not broadcast to the listener, except when the tuner is getting a clean, clear signal of a specified frequency on the incoming carrier wave. In other words, if it isn't a good signal, the receiver mutes the sound to save the listener from the distraction of weak or noisy stations.

Quorum International The marketing and distribution arm of Applied Electronics, a Hong Kong-based Original Equipment Manufacturer (OEM). Applied Electronics supplies many major computer and communications companies, including National Semi Conductor, IBM, Texas Instruments, and others. Products include security systems and pagers. In conjunction with MCI, the MCI-Q Program was developed to enable Quorum to purchase communications products and services at special rates. Through the Quorum Associate Distributor (QuAD), distributors receive discounted MCI Q-Connection rates and access to additional services (pagers, calling cards, etc.).

Quorum Teleconferencing Bridge A commercial, integrated voice and data device from AT&T to facilitate conference call set up and administration. The Quorum Teleconferencing Bridge connects and controls multipoint conference calls. It enables a local operator to set up, control, and monitor up to four simultaneous conference calls from participants in up to seven locations for each call (or a single meeting with up to 28 locations).

QWERTY A ubiquitous computer and typewriter keyboard configuration designation, named after the six lettered keys on the top left side. Although each computer keyboard has different symbol and function keys, most follow common QWERTY configurations. QWERTY was originally designed to slow down typing to prevent jamming on old manual typewriters (they jam easily).

Other keyboard layouts have since been proposed which consider ergonomics and physical properties, the most recognized being the keyboard designed by August Dvorak. The Dvorak keyboard was developed on the basis of studying finger motion and lettering combinations which were easier and more efficient to execute, and incorporating them into new keyboard character arrangements. A number of variations of this by other people have also been called Dvorak keyboards, even when they differ from that developed by A. Dvorak.

Unlike typewriters, it's easy to remap key positions on computer and alternate keyboards can be designed to put the letters anywhere the user desires. In spite of this, QWERTY keyboards remain prevalent, and manufacturers and teaching institutions are reluctant to change to other systems.

Qwest Communications A telecommunications company establishing fiber optic networks in over 100 U.S. and Mexican cities. Commercial services provided include dedicated business Internet access, Internet faxing, Internet phone (Q.talk), and video. In June 2000, Qwest Communications International Inc. completed a merger with U.S. West, Inc. announcing an $85 billion market capitalization with services in 14 states. In October 2001, Qwest and its principal shareholder, Anschutz Company, announced plans to purchase 14 million and 6 million shares, respectively, of the Netherlands company Koninklijke KPN N.V. The company further announced plans for acquiring Global TeleSystems, Inc., to embark on a significant European expansion.

QZ billing A telephone subscriber service in which the time and charges for an outgoing call (usually a toll call) can be obtained from the phone company. This is especially useful for business professionals and educational institutions that bill back to the calling department. QZ billing is being superseded by automated call accounting information.

QuickTime Applications	
Application	**Description**
QuickTime 3D	Apple Computer's 3D QuickTime cross-platform 3D rendering software.
QuickTime Conference	QTC. Designed on Apple Computer's QuickTime compression technology, QuickTime Conference supports videoconferencing in a window on the computer screen. Electronic whiteboarding is also supported, so participants can communicate and collaborate on shared drawing, text, or other projects. The software can be used to deliver Web events using QuickTime Live! software. See Simple Multicast Routing Protocol.
QuickTime Player	A software application for playing back audio and video files created by QuickTime Pro authoring software and other applications that export the QuickTime Pro authoring file format. QuickTime 5 has support for skip protection for streaming sources with uneven video delivery, media "skins" to enable the user to customize the look of the player, Internet TV channel display and audio control features, and support for plugins from Apple and third-party developers.
QuickTime Pro	A low-cost, commercial authoring system for creating QuickTime multimedia images and presentations. It supports many popular file formats that may be played on freely-downloadable QuickTime Player programs.
QT Streaming Server	An extension to QuickTime to deliver realtime multimedia over the Internet using the Real-Time Transport Protocol/Real-Time Streaming Protocol (RTP/RTSP). Thus, it can deliver video-on-demand and, when combined with broadcasting software, live streaming news, interviews, or entertainment through a network. QuickTime Streaming Server is available in Mac OS X Server and the Darwin Streaming Server is available through the Darwin open source project. The Streaming Server is available on a variety of platforms and the open source software makes it possible to port to additional platforms. The software provides Web-based server configuration, TCP-based broadcast support, skip protection, and other features.
QuickTime VR	An extension to QuickTime that adds cross-platform virtual reality capabilities through a movie-like presentation of images. The user can move through the scene, pan the surroundings, interact with objects, and much more. QuickTime Authoring Studio can be used to create virtual reality scenarios for display in QuickTime VR. QTVR Make Cubic enables users to build Cubic Virtual Reality software.

r *abbrev.* roentgen (røntgen). See roentgen.

R 1. *symb.* range. 2. *symb.* resistance. See resistance.

R interface In ISDN, a number of reference points have been specified as R, S, T, U, and V interfaces. To establish ISDN services, the telephone company and ISDN subscriber typically have to install a number of devices and links to create the all-digital circuit connection necessary to send and receive digital voice and data transmissions.

The R interface is the portion of the link between an ISDN terminal adapter (TA) on the customer premises, and customer non-ISDN station equipment (TE-2) such as phones, facsimile machines, computers using modems, etc. Some communications devices are being manufactured with ISDN terminal adapters built in, in which case, the R interface is not needed (or is considered to exist inside the component itself). See ISDN interfaces for a diagram.

R reference point In ISDN, the point in the digital communications path at which non-ISDN TE-1 devices connect to a Terminal Adapter (TA). See R interface.

R Series Recommendations A series of ITU-T recommended guidelines for radio systems, operations, and spectrum use. These guidelines are available for purchase from the ITU-T. Since ITU-T specifications and recommendations are widely followed by vendors in the telecommunications industry, those wanting to maximize interoperability with other systems need to be aware of the information disseminated by the ITU-T. A full list of general categories is listed in Appendix C and specific series topics are listed under individual entries in this dictionary, e.g., S Series Recommendations. See R Series Recommendations chart.

R-Y red-luminance. In the YUV video color model in which chrominance and luminance are separate, the color R (red) minus Y (luminance) equals V, which is used to calculate color differences. Associated with this is the color B (blue) minus Y (luminance) equals U. If both calculations were to be zero (0), it would indicate no color (chrominance). In video engineering, R-Y (red - luma) and B-Y (blue - luma) are multiplied by defined values to derive V and U, respectively. Color difference components with luma removed are represented as R´-Y´ and B´-Y´. Color component removal is a mathematical approach to decomposing a YUV video signal that is useful for a variety of reasons, including as a means to reduce bandwidth for transmission and for special effects. It is used in the PAL analog video standard and CCIR 601 digital video standard.

ITU-T R Series Recommendations	
Subseries	**Description**
BO	Broadcasting-satellite service (sound and television)
BR	Sound and television recording
BS	Broadcasting service (sound)
BT	Broadcasting service (television)
F	Fixed service
IS	Inter-service sharing and compatibility
M	Mobile, radiodetermination, amateur and related satellite services
P	Radiowave propagation
PI	Propagation in ionized media
RA	Radioastronomy
S	Fixed-satellite service
SA	Space applications and meteorology
SF	Frequency sharing between the fixed-satellite service and the fixed service
SM	Spectrum management
SNG	Satellite news gathering
TF	Time signals and frequency standards emissions
V	Vocabulary and related subjects

R/T 1. See realtime. 2. receive/transmit.

R/W read/write.

R&D See research and development.

R&E 1. See Research and Education. 2. research and engineering.

R&S research and statistics.

R1 signaling A type of signaling scheme commonly used in channelized voice networks in North America and Japan that is typically implemented through a T1 line. It enables supervisory and address signals to be transmitted between network switches using a single frequency for supervisory (line) information and multiple frequencies for address (register) information. T1 signaling is specified in ITU-T recommendations Q.310 to Q. 331. See channel-associated signaling, R2 signaling.

R2 signaling A type of signaling scheme commonly used in channelized voice networks outside North America and Japan. In association with E1 networks, it is similar to channelized T1. R2 implementations tend to be variants of ITU-T Q.400 to Q.490 recommendations; there is some flexibility in the channelized signaling parameters for setting the parameters for specific countries, an important capability in the multicultural European Union. Signal types are configured as line signals and interregister signals. See channel-associated signaling, R1 signaling.

RA 1. radar altimeter. 2. rate area. A region designated to receive products and services at described rates for the purposes of administration and billing. 3. See RealAudio. 4. Reliability Action Center.

RA number See return authorization number.

RA-EN See Radio Amateur Emergency Network.

RAC 1. See Radio Amateurs of Canada. 2. Radio Austria Communications. Based in Vienna. 3. See remote access concentrator.

RACE 1. random access computer equipment. 2. See Research in Advanced Communications in Europe.

RACES See Radio Amateur Civil Emergency Service.

raceway A duct or channel system designed to hold, protect, and direct interior wiring circuits. Raceways are typically plastic or metal modular construction, with a variety of fittings so individual sections can be interconnected and holes can be punched where needed. Raceways can be mounted on or in walls or floors. See molding raceway.

RACF 1. See Radio Access Control Function. 2. See resource access control facility.

rack, tray A support structure designed for the easy insertion, removal, and configuration of modular component systems. Racks are frequently equipped with rollers, although large ones may be attached to a wall for better support. They are generally assembled from rigid metal strips, interconnected to produce a strong open structure so the components can be quickly slid in and out of the individual bays from the front, and cabled to one another at the back. Racks are commonly used in the broadcast TV and video editing industries; they can also be found in telephone switching installations and on large computer networks with a variety of storage media. See distribution frame, rack unit.

rack mountable A component designed to specifications so it will fit easily and securely into a storage and support rack of a standard size for components from that industry. See rack, rack unit.

rack unit RU. A measure of distance based on physi-

cal rack structures that are common within various industries. Rack sizes have long been standardized in the power distribution, audio/visual, and portions of the retail sales industries.

Racks are especially common for supporting video processors, frame synchronizers, frame grabbers, and other video editing and broadcast components. In the video industry, the rack width for rack-mountable audio and video components is typically 19 in. and the vertical rack unit (RU) is 1.75 in. Thus, a 3.5-in.-high component would be described as having a height of 2RU. See rack.

racon See radar beacon.

RACS 1. remote access calibration/control/computing services/system.

RACT remote access computer/control terminal.

rad radiation absorbed dose. A quantification of radiation energy that describes how much radiation is delivered to 1 g of a substance by 100 ergs of energy. Radiation absorbed by body tissue is measured in roentgens.

RAD 1. rapid application development/design. 2. See Radiance. 3. random access device. 4. recorded announcement device. 5. remote antenna driver.

radar *ra*dio *d*etection *a*nd *r*anging. In its basic form, radar is a means of detecting distant or unseen objects by emitting radio frequency electromagnetic waves and measuring the reflected response. As such it can operate at night, during fog, and in situations where something is too distant to be seen by unaided eyes. Radar works on the principle that radio waves will deflect off of solid or sufficiently dense objects in a way that can be anticipated or controlled so the returning signal can be analyzed for the presence of the objects, their general shape and size, and their distance. Radar is a powerful and flexible technology that has become an intrinsic aspect of navigation, reconnaissance, and imaging systems.

Radar typically operates in ultra high frequencies (UHF) and microwave frequencies. See radar, history; sonar; Taylor, A.H.

radar, history Radar and sonar originated in much the same way in the late 1800s and the two technologies still share many terms and general principles. Johann Christian Doppler (1803–1853) made important contributions to both radar and sonar history by studying the way in which compressions arising from motion could alter sound frequencies, relative to the position of the viewer (or sensing instrument). It was discovered that this characteristic was also applicable to electromagnetic phenomena like radio waves and light, resulting in Doppler radar technologies about 100 years later.

The fact that electromagnetic waves could be reflected was first demonstrated by Heinrich R. Hertz (1857–1894) in 1886. In 1904, Christian Hülsmeyer (1881–1957) patented a "far-moving scope" which used reflected radio waves (early radar) for detecting marine vessels in order to prevent collisions at sea, a system that eventually superseded searchlights for detecting nearby vessels in the dark or fog. In 1906, William R. Blair described the use of electromagnetic oscillations to determine the incident energy returning from various reflecting surfaces. By 1916, interest in radio sensing devices was spreading rapidly. Dominik and Scherl had invented a radio echo sensing device in Europe and in 1917 Nicola Tesla wrote about radar concepts in the *Electrical Experimenter* in America.

The earliest commercial use of radio frequencies for bouncing signals was in the 1920s and 1930s, where it was used to determine the presence of marine vessels and aircraft to help prevent collisions. By the mid-1930s, when cathode-ray tube displays were becoming commercially viable, radar was used in both military and commercial navigation and documented in engineering textbooks. With the onslaught of World War II, interest in radar increased dramatically and substantial resources were devoted to its improvement and adaptation for military purposes. Since then, radar has been adapted to many industries and continues to be used for its original purpose, navigational safety, and for many diverse military and commercial applications. See radar.

radar beacon A device associated with a particular location or object that is capable of emitting a radar-range radio signal to indicate its geographic location.

> A *passive* radar beacon is one that sends out a radar signal from time to time that can be sensed by an appropriately tuned radar receiver within range. However, it does not have a receiver and cannot detect whether a radar signal has been aimed in its direction.

> An *active* radar beacon includes both a radar receiver and a radar transmitter. The receiver enables the beacon to sense a triggering pulse from a remote radar that causes the beacon to broadcast a returning signal or a series of returning signals. The information contained within the signal depends upon the system used, but can provide range, bearing, and, optionally, identification information.

Because radar technology tends to be expensive, radar beacons are more often installed in industrial shipping facilities and military applications than in consumer devices. Sometimes they are used in lighthouses.

Some less expensive radar beacons do exist, however, in the form of police speed detectors. The speed detector determines a vehicle's speed by sending out a radar pulse that essentially acts as a beacon to any suitably tuned automobile radar detection units within range.

radar detector A device designed to detect the presence of radar-range radio signals. These are used in military applications to detect or interpret radar-range communications. They are also sold for civilian use in the form of car-mountable devices to detect police speed-detection radar systems. The use of radar detectors is regulated and prohibited in some areas. See radar, radar beacon.

radar screen/scope A small display device, usually round or rectangular, which shows target signals as illuminated dots or blips. There may be grids and other alignment and location marks superimposed over the illuminated blips on the screen to aid in tracking and location. The term radar "scope" comes from the early days when cathode-ray tubes (CRTs) were being adapted for use as oscilloscopes and various types of radar screens. See radar.

radar systems Devices incorporating radio waves to detect the presence and characteristics of distant or otherwise unseen objects. Although radio echoes were observed in the 1920s and put into practical use in the 1930s, developments in radar guidance, detection, and identification systems did not flourish until the second World War. See cavity magnetron.

RADARSAT A Canadian satellite system launched in November 1995. The satellite supports a synthetic-aperture radar sensor that can deliver data in seven sizes based upon 25 beam positions. Images vary, depending upon the way the sensors are angled and the Earth viewed. Each beam position has a specific elevation angle and size. The two beam modes are single beam and multiple-beam ScanSAR. RADARSAT is part of the RADARSAT Network System.

Radar Scope

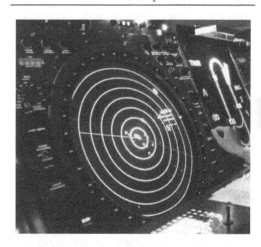

Much attention has been given to the development of radar technologies for navigation and military tracking purposes. This radar scope, on board a U.S. aircraft carrier, is used for aircraft approach control. Civilian and military air traffic control personnel ensure the safe and orderly flow of air traffic in commercial airports and on military vessels at sea. This photo was taken during NATO Implementation Force (IFOR) patrols of the waters of the Adriatic Sea.

The U.S. Air Force C-14B aircraft was the first to use ground-based radio beacons for a completely automated landing in August 1937. The system of five transmitting beacons used to accomplish this feat was developed by Carl J. Crane. [Detail of DoD photo by J. Hendricks, U.S. Navy.]

---

radial acceleration Acceleration in a circular trajectory, characteristic of a spinning solid or liquid substance. Radial acceleration is used in centrifugal separators to isolate particular particles or substances. The radial acceleration characteristics of various spiraling entities are of interest to astronomers. In optical media, radial acceleration is one of the characteristics measured to determine conformance with expected properties or standards, along with axial acceleration and radial runout.

Radiance RAD. A native file format for Radiance, a public domain Unix-compatible radiosity rendering engine. The file extension *.rad* may be used to distinguish RAD files.

radiant energy Transmitted electromagnetic energy such as heat, light, or radio waves. Radiant energy is typically measured in calories, ergs, or joules.

radio An appliance or other device designed for the transmission and/or receipt of radio wave communications. There are many types of radio technology: amplitude modulation (AM), frequency modulation (FM), shortwave, cellular, short-range (cordless phones, wireless intercoms), etc. With increased demand for wireless communications, harnessing and using radio waves efficiently has become extremely important in both scientific and commercial research. More details about radio communications can be found under individual listings in this dictionary. See crystal detector, detector.

Radio The publication name of a widely-distributed Soviet electronics journal which, in June 1957, announced the Soviet Union's plans to soon launch a satellite ("sputnik" in Russian), and provided details of the planned launch date, modulation techniques, and frequencies to be used. Sputnik I did in fact launch at the end of that year. See Sputnik I.

Radio Access Control Function RACF. In a PACA-E Personal Communications System (PCS), the RACF intercommunicates with a number of entities while handling the service logic, including the Radio Control Function (RCF) and the Call Control Function (CCF)/Service Switching Function (SSF). See Priority Access and Channel Assignment.

Radio Act of 1912 With increasing interest in radio broadcasting and demand on airwaves, the U.S. Congress passed an act which granted the U.S. Department of Commerce the authority to regulate amateur broadcasting in order to prevent interference with government stations and to increase maritime safety, largely due to the sinking of the *Titanic*. See *Titanic*.

Radio Act of 1927 As a response to the enormous rising demand for broadcast channels in the early part of the century, a conference was held to sort out the chaos. As of the Radio Act of 1912, the U.S. Department of Commerce took control of radio broadcasting. Zenith Radio Corporation applied for a license to operate at a frequency that was being used by other stations as well, and so was granted a license to broadcast at a different frequency. Zenith changed frequencies to one that had already been granted, instead of using the one that had been licensed. In the process of investigating the violation, it was found that the Department of Commerce didn't have sufficient jurisdiction to stop the actions of the broadcaster, and

one of the consequences was the creation of the Federal Radio Commission (FRC) in 1927. This was later to become the Federal Communications Commission (FCC) through the Communications Act of 1934. See Communications Act of 1934, Federal Communications Commission.

Radio Amateur Civil Emergency Service RACES. A public emergency service provided by reserve volunteer communications personnel to assist regular emergency services in times of extraordinary need. The organization was established in 1952. Planning guidance for RACES deployment is provided by the Federal Emergency Management Agency (FEMA). http://www.races.net/

Radio Amateur Emergency Network Raynet. Raynet was established as a result of a violent storm that pummeled the eastern coast of England in January 1953. Coastal towns and villages were ravaged and more than 300 drowned. Since telephone lines were destroyed as well, there was no reliable communication into or out of the area during the storm. The police authorities appealed to radio amateurs to help and the Home Office gave permission for amateur radios to coordinate rescue services, saving many lives. Thus RA-EN was born, and grew in capabilities in 1989 when legislation restriction was lifted to enable the organization to provide greater help. It is now known as Raynet.

Radio Amateur Satellite Corporation RASC, AMSAT-NA. This is the North American branch of the international amateur radio satellite community, a not-for-profit agency founded in 1969.
Over the decades, radio amateurs have made significant contributions to the evolution of satellite broadcasting technologies in spite of the budget constraints typical of amateur organizations. AMSAT-NA supports and promotes scientific research and development in radio communications satellites and space science in the public interest. The air waves, by legal right in the U.S., belong to the people, and radio amateurs are exercising not only their right to use this wonderful resource, but have furthered the art of wireless communications in countless ways. AMSAT has successfully launched more than two dozen noncommercial radio communications satellites into Earth orbit since the historic OSCAR satellites of the early 1960s. See AMSAT, OSCAR. http://www.amsat.org/

Radio Amateurs of Canada RAC. RAC provides liaison, coordinating functions, and policy decisions for the benefit of Canadian amateur radio organizations and individual amateur radio operators. http://www.rac.ca/

radio broadcasting Commercial radio broadcasting began in the early 1900s, arising out of the experimental broadcasts of inventor R. Fessenden in 1906. There were many amateur broadcasts between 1906 and 1920, including the regularly scheduled shows by Charles "Doc" Herrold, at the Herrold College of Wireless and Engineering in California, and the pre-KDKA broadcasts from the garage of F. Conrad in 1919. CFCF and KDKA are acknowledged as the first commercial stations, beginning in 1920.

Commercial broadcasting in Europe was underway by 1913, and the Eiffel Tower still stands as a historic reminder of the lofty ambitions of the broadcast pioneers. It was built for the Paris World's Fair in the 1800s and there have been several attempts to remove it since then, but its usefulness as a giant antenna is one of the reasons it was preserved. Lee de Forest participated in one of the first transcontinental broadcasts from the world's largest radio tower. The Radio Corporation of America (RCA), founded in 1920, is one of the best known and most influential of the early radio pioneers, and much of its history is related to the activities of David Sarnoff. Sarnoff was also instrumental in forming the National Broadcasting Corporation (NBC), in 1926. The following year the Columbia Broadcasting System (CBS) was formed (originally Columbia Phonograph Broadcasting until William S. Paley bought out the company in 1928). From 1921 to 1922 the number of commercial stations in the U.S. increased from five to over 500. In the early 1930s, record companies became nervous about competition from radio stations and began restricting the open broadcasting of audio recordings. From that point on, royalties and other means of enforcing payment for broadcasts were instituted.

By the late 1930s the wonderful music from bands and orchestras around the world could be heard through the magic of radio, and listeners who had never been to a theater to hear a live performance enjoyed the new form of entertainment. The advent of radio meant the eventual death of vaudeville, but some of the vaudevillian actors, perhaps best exemplified by George Burns and Gracie Allen, made a successful transition to radio, and eventually to TV programming.

By the early 1940s, frequency modulated (FM) broadcasting, made possible by the tireless efforts of inventor Edwin Armstrong, was beginning to catch on and, while it didn't supersede AM, it provided clean, clear transmissions that were favored by public broadcast and classical music stations. The invention of the transistor created a revolution in miniaturization and manufacturing. By the 1950s, radio had competition from TV broadcast stations, but unlike many technologies, it didn't lose its practicality and appeal. Radio stations in North America still outnumber TV stations, and radio sets continue to be in demand.

The next major milestone in radio broadcasting came with Sputnik and the exploration of space. In 1969, American astronauts sent sound and images from the Moon to Earth. Soon communications satellites were being launched into orbit in the 1970s and 1980s. This provided a means to develop mobile communications, and linked computers and radios as never before. Many of the pioneer communications efforts and new technologies were contributed by amateur radio enthusiasts, most notably through the OSCAR and AMSAT satellite programs.

With digital electronics, laptops, and cell phones, the importance of radio continued to grow, as wireless communications were integrated into increasingly mobile lifestyles. One of the significant recent events in radio broadcasting is the introduction of digital broadcasting, pioneered by Sweden in 1996. See AMSAT; ANIK; CKAC; Emergency Alert System; KDKA; OSCAR; Radio Corporation of America; Sarnoff, David; radio history.

radio broadcasting regulations Many different sets of guidelines and regulations have been developed to manage radio broadcasting. Some of these were intended to curtail unfair business practices, such as more powerful transmitting stations deliberately drowning out less powerful ones, and some were implemented to organize and coordinate the use of limited airspace, that is, the limited availability of broadcast frequencies. Others were put into effect in wartime to shut down broadcasting almost entirely, curbing the broadcast pirates, but also curbing responsible amateurs. In 1963 the Emergency Broadcasting System (EBS) was established, recently replaced in 1997 by the Emergency Alert System (EAS).

Several Radio Acts and later Telecommunications Acts have controlled American broadcasting over the decades. The jurisdiction has changed hands a number of times, from the U.S. Secretary of Commerce to the Federal Radio Commission (FRC) in 1927, to the Federal Communications Commission (FCC) in the mid-1930s. The FCC has retained its wide-ranging licensing and regulatory powers up to the present time. See Emergency Alert System, Federal Communications Commission, Radio Act of 1912.

radio button A physical button on a component, or iconic button in a software program, which permits selection of only one option from a group of mutually exclusive selections. Selecting any one option automatically deselects the previous option. The name derives from the action of pushbutton radio sets in which buttons can be pretuned to selected stations, and then pushed for the desired station, one at a time. Software radio buttons are often seen on input forms on Web pages.

Radio Club of America This historic amateur radio group is still operating after its birth in the early 1900s. It held its first meeting in 1909 and was founded as the Junior Wireless Club Limited in 1910 in New York City. The organization changed its name the following year to the Radio Club of America. The young members of the club successfully lobbied for the interests of amateur radio enthusiasts before the U.S. Senate Commerce Subcommittee. The first official meeting of the organization under its new name was in November 1911. As it grew, some of the great names in radio history became associated with the club, including Paul Godley, Edwin Armstrong, and David Sarnoff. It exists for the charitable, educational, and scientific furtherance of radio communications and provides scholarship funds for needy and worthy students of radio communications. See Junior Wireless Club Limited.

http://www.radio-club-of-america.org/

Radio Common Carrier RCC. Service providers of mobile telephone and paging services employing radio technology, as opposed to land line transmissions.

Radio Communication Laws of the United States
The sinking of the *Republic* and the *Titanic*, in addition to the International RadioTelegraphic Convention, were strong factors in the development of American regulations for radio communications. In 1910 and 1912, U.S. acts were approved resulting in the publication of the *Radio Communication Laws of the United States and the International Radiotelegraphic Convention*, by the U.S. Department of Commerce Bureau of Navigation Radio Service, in 1914. This document described the major international agreements and U.S. radio regulations and guidelines and spelled out the requirement for any steamer navigating U.S. waters carrying 50 or more persons to be
"... equipped with an efficient apparatus for radio communication ... capable of transmitting and receiving messages over a distance of at least one hundred miles, day or night. An auxiliary power supply, independent of the vessel's main electric power plant, must be provided which will enable the sending set for at least four hours to send messages over a distance of at least one hundred miles, day or night, and efficient communication between the operator in the radio room and the bridge shall be maintained at all times.
The radio equipment must be in charge of two or more persons skilled in the use of such apparatus, one or the other of whom shall be on duty at all times while the vessel is being navigated...."
The document further described licensing requirements for amateur and commercial operators, stipulated the designation of certain definite wavelengths as normal communications frequencies for a station, and standardized SOS (dot-dot-dot dash-dash-dash dot-dot-dot) in Morse Code as the official distress call, in addition to other basic tenets of radio communications. See International Radiotelegraphic Convention.

Radio Control Function RCF. In a PACA-E Personal Communications System (PCS), the point towards which Radio Terminal Function transmissions are directed and from which the RTF interacts with other network entities in the processing of a PACA-E (priority access) call. See Priority Access and Channel Assignment.

Radio Corporation of America RCA. An offshoot of General Electric founded in 1919 as a result of a merger with the Marconi Wireless Telegraph Company of America. In 1920, RCA made a significant agreement with WSA, AT&T, GE, Westinghouse, and others, to be the exclusive distributor of radio receiving sets and crystal detectors. In 1921, David Sarnoff joined the company as its general manager, and later moved up in the corporation, becoming vice president in 1926. Sarnoff was a colorful part of its history for many decades. See Armstrong, Edwin Howard; Sarnoff, David.

radio facsimile The transmission of the contents of pages including text and images by means of radio signals. Radio facsimiles were pioneered in the 1800s, and this early form of facsimile machine was in use at least as early as 1943. See facsimile.

radio frequency RF. Radiant electromagnetic waves that range from about 3 to 10 kHz at the lowest end to just about 300 GHz at the high end, a position that falls between the audio frequencies and the boundary of the visible spectrum where infrared is found. Radio frequencies are widely used for radio and television broadcasting, and for various types of wireless communication. The frequency range has been administratively subdivided into a number of categories so that limited airwaves can be assigned and licensed in an efficient way. In the U.S., this responsibility is managed by the Federal Communications Commission (FCC); in Canada it's managed by the Canadian Radio Television and Telecommunications Commission (CRTC). See band allocations for chart.

radio history Radio history is one of the most interesting histories in technology. As soon as humankind discovered that communications could be transmitted at the speed of light, they enthusiastically sought ways to achieve practical embodiments of the possibilities. The gradual discovery and harnessing of electromagnetic frequencies to carry meaningful communications changed society in fundamental ways and radio waves have since been developed for a bewildering variety of technologies, from interstellar telemetry to radar range ovens that almost instantly cook food.
The earliest radios were low-power crystal detectors that exploited the oscillating properties of natural and synthetic crystals to "capture" (resonate with) radio waves and transmit the signal to earphones worn by the listener. This wasn't a very easy way to share radio communications with a room full of people, so improvements for increasing the sound volume and the development of antennas to intercept the signals were pursued by radio inventors. The significant history of modern radio technology starts with Lee de Forest's invention of the *Audion*, a three-element electron tube derived from Fleming's two-element tube. The addition of the third element was highly important, as it enabled the flow of electrons to be controlled which, in turn, made it possible to build radio signal amplifiers for hundreds of different applications. De Forest claims he made voice broadcasts from New York around 1906. In 1916 he gained notoriety for broadcasting incorrect election results.
One of the uncredited pioneers of radio history was inventor Valdemar Poulsen. He is well known for his contributions to telegraphy and tape recording technologies, but most people are unaware that he broadcast music from the town of Lyngby, on March 4, 1909, by putting a microphone near a gramophone player. The radio transmission was received by 16-year-old Einar Dessau on his home-made amateur radio receiver in Hellerup
With the development of practical methods of radio broadcasting, creative minds quickly grasped the social, cultural, military, and commercial applications of radio communications. Experimental radio stations sprang up everywhere beginning around 1910 with many early broadcasts sent in Morse code. Consumer radio sets, called "talking machines," gradually

replaced hobbyist crystal radios.

Guglielmo Marconi was a young, ambitious inventor who became a prominent radio pioneer. He was the first to develop many radio technologies and the first to use or adapt many technologies developed by other all-but-forgotten inventors.

A true broadcasting pioneer, Doc Herrold began transmitting in 1909, the same year as Poulsen, and made history by airing music and news to the 1915 World's Fair. In 1916, he received callsign 6XF from the U.S. Department of Commerce. At about this time, Lenin began using radio to reach the public in Russia while Marconi set up broadcasting in Europe and North America. Wartime restrictions hampered broadcasting around this time but didn't stop the evolution of radio. In Montreal, Canada, Marconi's station XWA received an experimental license in 1918, a general license in 1919, and broadcast its first regularly scheduled musical concert in May 1920. XWA evolved into CFCF in November 1920. Station KDKA in Pittsburgh (under its original 8Xk callsign) broadcast election events in November 1920.

No doubt there are many uncredited radio pioneers as early callsigns were self-assigned, the technology was loosely regulated, and few listeners existed to credit the true pioneers. From the 1920s onward, however, regulated radio broadcasting flourished, making it easier to unravel and confirm its history.

Historic Presidential Radio Communication

President Herbert Hoover listening to an early electron tube-based radio set, around 1925. Hoover and Coolidge were two of the first presidents to exploit the informational and political potential of radio communications. [Library of Congress American Memory National Photo Company Collection.]

Calvin Coolidge was quick to exploit radio technology in the U.S. in the 1920s, for political communications, as was Vladimir Lenin in Russia. Radio is now an inextricable aspect of political campaigns and communications and is used by government agencies for coordination, national security, and emergency services.

The general public benefited from wireless letters and radio broadcasting.

The importance of the development of transistors at the Bell lab in the late 1940s cannot be overstated. Transistors not only made it possible to shrink building-sized computers down to the size of a large photocopier, but enabled portable radio technologies and mobile communications technologies to evolve and flourish from the 1950s onward. The semiconductor industry further enabled hardware engineers to combine many functions on a single chip, increasing processing speeds and reducing component sizes even further. Radio is still a highly significant communications technology, which is now being incorporated into telephones and other traditional wired devices. The evolution of the technology is ongoing. See Audion, CFCF; Conrad, Frank; crystal detector; Herrold, Doc; KDKA; Marconi, Guglielmo; Tesla, Nikola; telegraph history; XWA.

Radio Link Protocol RLP, RLP1. A protocol standardized by the Telecommunications Industry Association (TIA), in the mid-1990s, using SDI, a formal specification language. RLP is a data link layer circuit-mode protocol for connecting a Mobile Terminal (MT) with a Mobile Base Station (MBS) to provide a stationary digital radio interface through a public switched telephone network (PSTN) to another data-compatible telephony device. This makes it useful for cellular data transmissions applications.

Radio Technical Commission of Aeronautics RTCA, Inc. A private, not-for-profit corporation promoting consensus-based recommendations for communications, air traffic management, navigation, and surveillance issues. The RTCA was established in 1935. It now represents more than 270 trade, academic, and governmental organizations. The RTCA includes prominent names in the aviation industry, including the National Business Aviation Association, NASA, the National Air Traffic Controllers Association and various international airlines, and pilot associations. It makes recommendations to the Federal Aviation Administration (FAA) (also a member).
http://www.rtca.org/

Radio Terminal Function RTF. In a PACA-E Personal Communications System (PCS), the RTF is the point from which the PACA-E subscriber accesses the mobile communications network. Communications are directed from the RTF towards the Radio Control Function (RCF) through which it can interact with other network entities. See Priority Access and Channel Assignment.

radio wave An electromagnetic wave, commonly used to carry audio transmissions, in a frequency spectrum that ranges from 10 KHz to 200 GHz. Transmission waves such as radio waves are further classified into subcategories according to various properties; examples include ionospheric waves (sky waves), ground waves, short waves, and others. The characteristics of various transmissions media, chiefly the Earth's ionosphere, are exploited to aim and propagate these waves. Frequency divisions of radio waves according to wavelength (higher

R

frequencies have shorter wavelengths) have been designated as shown in the chart under band allocations. Sounds and other signals are converted to radiating waves for transmission, then converted back at the receiving end. See antenna, ionospheric wave, ground wave, radio, short wave.

Radio-Television News Directors Association RTNDA. An international nonprofit professional organization of network and local news executives in the broadcasting, cable, and multimedia industries. Established in 1946, the RTNDA promotes high standards of electronic journalism, the exchange of knowledge among members, public understanding of the profession, and journalistic freedom. It sponsors an annual international conference for professionals in the news industry. http://www.rtnda.org/

Radio-Television News Directors Foundation RTNDF. Affiliated with the Radio-Television News Directors Association, the foundation promotes excellence in electronic journalism through research and educational programs for news professionals and students of journalism. The work of the Foundation is supported by other foundations, corporations, and RTNDA members. Research results of interest are posted or summarized on the RTNDA Web site.

radiogram 1. A telegram sent through radiotelegraphy, also called "radiograph." See telegram. 2. Combined radio receiver and phonograph.

radiometer A device for measuring the intensity of electromagnetic radiation, which may include visible, UV, or infrared light and emanations beyond the optical spectrum. A photometer is a type of radiometer specialized for the measurement of light intensity. A spectroradiometer further enables color assessment by analyzing the spectral range or specific wavelength of sampled light in addition to measuring radiant intensity.

For scientific understanding and high precision applications, various research scientists and standards bodies have sought to derive absolute or reference measures for luminous intensity, luminous flux, and illuminance. A goniophotometric detector calibrated against the realized luminous intensity standard can be used to establish a unit of luminous flux. A similar instrument with a charge couple detector (CCD) may be used with standard intensity lamps to determinance luminance. See photometer.

radiometeorograph See radiosonde.

radiophone A device that transmits sound through radio waves. Although the term is less common, radiophones are everywhere; they have more individual names now due to their specialization (cordless phones, cell phones, etc.).

radiosonde, radiometeorograph A miniature, automatic radio transmitter usually sent aloft on an aircraft or meteorological balloon, to transmit back meteorological information, such as temperature, humidity, pressure, etc.

radiotelegraphy Transmission of telegraph signals through radio waves. The carrier wave was modulated to carry Morse code. The two main types were continuous-wave (CW), in which the carrier wave was

interrupted to form the coded symbols, and interrupted continuous-wave (ICW), in which the carrier was modulated at a fixed frequency.

radiotelephony The art and science of communicating through radio waves, often by means of various types of radiophones.

RADIR random access document indexing and retrieval.

RADIUS Protocol Remote Authentication Dial In User Service. A client/server network protocol for carrying authentication, authorization, and configuration information between a Network Access Server (NAS) (a client), desiring to authenticate its links, and a shared authentication server. RADIUS was submitted as a Standards Track RFC by Rigney et al. in June 2000.

A RADIUS client passes user information to designated RADIUS servers and acts upon the responses. The server receives connection requests and authenticates the user, returning all configuration information needed for clients to deliver services to the user. A RADIUS server may act as a proxy client to other servers.

Authentication transactions are handled as "shared secrets" that are not transmitted over the network. Any transmitted user passwords are encrypted and RADIUS can work with a variety of authentication schemes (e.g., CHAP). The officially assigned port number for RADIUS messages has been changed from 1645 to 1812. See RFC 2865, which obsoletes RFC 2138. See RFC 2868 for RADIUS support for compulsory tunneling and RFC 2869 for RADIUS extensions.

RADIUS Accounting Protocol An administrative client/server protocol used in conjunction with RADIUS Authentication and Authorization services to deliver accounting information from a Network Access Server (NAS) to a RADIUS accounting server. The RADIUS accounting server receives an accounting request associated with a database of users of a modem pool, for example. It returns an acknowledgment of the user request. The RADIUS Accounting server may act as a proxy client to other types of accounting servers, as well.

RADIUS Accounting was submitted as an Informational RFC by C. Rigney in June 2000. The officially assigned port number for RADIUS Accounting has been changed from 1646 to 1813. See RFC 2866, which obsoletes RFC 2139. See RFC 2867 for modifications for Tunnel Protocol support. See RADIUS Protocol.

RADL radio laboratory. 2. See Reticular Agent Definition Language.

radome A radar "dome," a housing around a radar antenna which protects it without interfering with the signals. Radomes are especially important in radar antennas exposed to the elements, as in an airplane.

RADSL See Rate Adaptive Digital Subscriber Line.

RAF Royal Air Force.

RAI remote alarm indication.

RAID See redundant array of inexpensive disks.

RAIN See redundant array of independent netports.

RAIS redundant array of inexpensive systems.

raised floor distribution A type of structure designed to accommodate a horizontal distribution frame for the attachment and management of wiring installations. It is typically designed so that the floor covering can be pulled aside or lifted to gain access for changes or additions. See distribution frame.

RAM See random access memory.

RAM disk An area of chip memory allocated and managed as though it were a disk drive. Unlike a disk drive, RAM is volatile; it requires a continuous source of power to retain its information and will lose the stored data if the system is turned off. A RAM disk is a means of disk caching that was popular when many systems had only floppy drives and no hard drives. It provided a fast way to access data without doing disk seeks or swapping out disks. RAM disks are now less prevalent.

RAM Mobile Data An open architecture, nationwide commercial data communications service offered by Ericsson, BellSouth, and RAM Broadcasting. It is similar to ARDIS, a packet data service offered by Motorola. Base stations are used for relaying messages to users or to other stations.

RAMAC Random Access Memory Accounting Machine. A historic large-scale computing machine announced by IBM, in 1956, for plant automation accounting and data processing applications. It included magnetic disk memory with a capacity of 5 million digits. Results of computations were recorded on punch cards. Over 1000 RAMACs were built before production was discontinued in 1961. See Johnson, Reynold.

Raman amplifier A mechanism based upon Raman scattering that provides gain in optical transmissions. This type of amplification is common to C- and L-band frequencies, but has also been suggested for use in the S-band. It is expected to increase power and lower costs in dense wavelength division multiplexed (DWDM) systems without the nonlinear waveform distortion characteristic of erbium-doped amplification mechanisms, especially in long-haul cables.

Raman scattering Low-level scattering associated with light deflecting off obstacles in its path. As a beam of light encounters an obstacle, there is interaction between the incident photons and the obstacle's molecules, resulting in a change in the direction of some of the light and a shift in its frequency, called Raman scattering. The scattered light is proportional to the intensity of the incident light.

Depending upon context, Raman scattering may be considered as noise or may provide useful information about the object from which it is scattered. When Raman scattering occurs in response to environmental conditions, or is an unexpected side-effect, it is called spontaneous Raman scattering (SpRS). When it is deliberately induced, or is an expected side-effect, as in an amplifier or spectroscope, it is called *stimulated Raman scattering* (StRS). When the wavelength emitted by a laser light source corresponds to the excited energetic vibrational level, it is called *resonance Raman scattering* (RRS).

Sometimes Raman scattering is deliberately induced and the result compared with spontaneous Raman scattering for research or test purposes. Raman instruments are considered far-field instruments due to the distance between the detector and the scattering point of the light, and their spatial resolution is generally limited to the approximately wavelength of the source light. For higher resolutions, scanning probe microscopes may be suitable. Some companies are now combining the capabilities of far-field and near-field instruments to enable more precise targeting in conjunction with higher spatial resolution.

In instruments designed to detect and assess Raman scattering, an optical tip may be used as a probe and/or an optical fiber bundle may deliver the Raman impulse to a monochromator and imaging detector. Undesirable Raman scattering may occur in optical transmission links carrying more than one wavelength due to interband interactions where power is transfered from shorter to longer wavelengths and may contribute to pulse distortion. Raman filters are available for some types of fiber optic probes. See Raman amplifier, Raman spectroscopy, Rayleigh scattering, scanning probe microscope.

Raman spectroscope An investigative instrument that induces light scattering from an illumination source aimed at a target sample. By assessing the scattered returning signal, characteristics related to the composition and density of the sample can be detected. This technology is becoming competitive with infrared (IR) analysis due to the reduced cost of laser diode light sources. Raman-detecting spectroscopes are available in a variety of wavelengths. The spectroscope may include a computer interface for the transmission of data for more extensive analysis. See Raman scattering.

RAMbus random access memory bus. See bus, random access memory.

RAMDAC random access memory digital-to-analog converter. A graphics adapter display circuit which converts the computer digital information for representing the screen image into analog signals that a cathode-ray tube (CRT) display monitor can use.

Ramsden, Jesse (1735–1800) An English instrument maker and engraver who invented micrometers, a pyrometer and a new type of eyeglasses and authored a description of a new "universal equatoreal." He is credited with furthering the technology of the Cassegrain telescope devised 100 years earlier by Guillaume Cassegrain. Ramsden was elected to the Royal Society in 1786 and received the Copley Medal in 1795.

RAND 1. random. 2. rural area network design.

random access memory RAM. A type of computer memory in which data in any part of memory can be accessed in any order, that is, it is not restricted to reading and writing data sequentially as in serial data, tapes, etc. RAM is a very fast access device almost universally incorporated into computing systems for use by applications for frequent operations or those that must be executed quickly.

In the mid-1970s, microcomputers typically had 4

Kbytes RAM and the price per kilobyte was about $100. Since then, the amount of memory installed in microcomputers has increased as prices have decreased. While there have been some interim fluctuations in prices, they have dropped dramatically as installed quantities have increased, as illustrated in this summary of the quantity/price changes over two decades.

Time Period	Typical Quantity	Approx. Price/Mbyte
mid 1970s	4 Kbytes	$100,000
late 1970s	8 Kbytes	$2000
early 1980s	128 Kbytes	$1000
1983-1984	256 Kbytes	$700
1985-1986	1 Mbyte	$700
late 1980s	4 to 8 Mbytes	$400
early 1990s	8 Mbytes	$250
mid 1990s	8 to 16 Mbytes	$200
early 1998	16 to 32 Mbytes	$4
late 1998	32 to 64 Mbytes	$1
late 2000	32 to 128 Mbytes	$0.50
late 2001	128–256 Mbytes	$0.10

random access storage RAS. A variety of types of memory and storage drives (e.g., hard disk drives) that enable data to be randomly accessed rather than sequentially accessed. These tend to be faster and more convenient than sequential access storage such as older tape drives. See random access memory.

random early detection (RED) An active router queue mechanism for detecting incipient congestion on a network through a number of congestion indicators. RED drops packets probabilistically rather than when the buffer overflows, with the probability increasing as the queue size increases. Signals from RED may indicate persistent congestion, information that is useful for network management.

RED can help control the average size of a queue on a system that experiences occasional transmissions bursts and thus reduces the chance of data loss. The maximum probability of a router packet being marked is set with the *maximum drop rate*. The random nature of RED reduces the tendency of synchronized processes to lock up when congestion is detected.

Dropped packets have traditionally been used as a congestion indicator, but with RED, packets may be dropped before a queue overflows and thus may not always be the best indicator of congestion for security purposes. RED also may not be the best mechanism for effectively handling a very large number of very tiny transmissions. See explicit congestion notification.

random number generator A device or algorithm intended to produce a truly random number or one that is at least difficult to predict. This is not as easy as it might seem, especially in computing devices that operate on rule- and clock-based principles. Their consistent modes of operation may cause them to generate the same random number, or the same pattern of random numbers, each time a random number generator is invoked. This is why most random number generators associated with computers are considered to generate pseudorandom numbers.

Sometimes the "randomness" of the number is at the whim of a software developer who has written or compiled a random number-generating routine. Many computer operating system-level or machine-level random number generators will fetch the same number the first time they are invoked and some will generate the same (or nearly the same) sequence of random numbers the first time they are invoked after a machine has been restarted or reset. Some random numbers are simply extracted from a list.

To overcome some of the problems of generating an unpredictable or "true" random number, programmers have tried several strategies based on looking at a changing timer or piece of data unconnected with the regular operation of a system. Various input operations on the part of the user, or processes associated with another application on the system, are sometimes sampled to obtain an unpredicted value. Scientists have investigated the use of radioactive decay and radio waves for generating a random value or seed. There are even semiconductor-based random number generating devices based on amplifying and sampling electronic noise.

In 1998, Matsumoto and Nishimura described a 623-dimensionally equidistributed uniform pseudorandom number generator that they called Mersenne Twister. This is a fast, efficient algorithm with a longer period than other generators. The source code is available online in C and Java.

DIEHARD is a suite of programs for testing random number generators developed by G. Gasram, with support from the National Science Foundation. Source code in C is available for several platforms.

With the growth of distributed networks like the Internet, it becomes possible to go outside the local system to find another system or timing device from which to fetch a number to use as a random value or a seed for generating a random number on the local system. There is even a site called www.random.org that provides a random number service for Web users.

Random numbers are not just intellectual oddities. They are used in a wide variety of practical applications, including fundamental research, gaming theory and applications, gambling systems, statistical studies, and software/hardware systems testing. They are especially important for generating encryption keys and cryptologic algorithms for securing data and digital voice communications. See encryption.

range 1. The extent, distance, or scope represented or traversed. 2. In a Global Positioning System (GPS), a fixed distance between two points, such as the distance between a GPS receiver and a satellite. 3. In mobile communications, the maximum distance of a transmission sufficiently clear to be useful.

rangefinder A device for determining distance without the use of a conventional ruler. Sonar is a type of rangefinding system, as are optical or acoustic distance sensors in robotic systems. Many automatic focus cameras have builtin infrared-based rangefinding systems in which a beam of invisible light is aimed at the object to be photographed and reflected back (up to a certain distance after which it is considered to be at infinity in terms of the lens capabilities of the camera). By calculating the time of flight of the beam from the camera, to the object, and back to the camera, the distance can be estimated accurately enough to focus the lens. Sometimes two modulated sub-beams of the same order of magnitude following one another are used so that a comparison can be made for accurate rangefinding.

Semiconductor lasers in invisible and visible light ranges are incorporated into many rangefinders. Invisible rangefinders are useful as surveillance devices, intelligent vehicle systems, robotics sensors, and level meters (e.g., grain bins). Visible light rangefinders are handy when the operator needs to know exactly where the beam is targeted. A fiber optic lightguide may be used to propagate the light from the receiving lens to a photodiode.

RAP See Route Access Protocol.

Rapid City IP switch See Accelar routing switch.

Rapid Transport Protocol RTP. An end-to-end, full duplex, high-speed, connection-oriented transport connection protocol. Flow control is adaptive rate-based (ARB) at the endpoints, and error recovery is handled via selective retransmission. RTP can be used to transport Systems Network Architecture (SNA) session traffic, for example.

RAR 1. An efficient data compression program developed by Eugene Roshal as shareware for DOS/Windows-based systems. Files are just a little smaller than those generated by PKZIP and can be made self-extracting. UnRAR utilities are also available for the Macintosh. See Java Archive, LHarc, PKZIP. 2. return address register, RA register. A programming register or data compartment (as might be found in a *stack*, for example) for storing and accessing return address information. Depending upon the architecture, other registers associated with the RA register typically contain variables, parameters, etc.

RARE See Réseaux Associés pour la Recherche Européenne.

rare earth Some rare earth elements are commonly used as doping agents which can aid in the propagation of signals when added to transmissions media, such as optical fiber, during manufacture. Rare earth doping is also being applied toward the design and manufacture of electrically pumped lasers that employ electronic circuitry. Erbium, Gadolinium, Europium, and Samarium are examples of rare earth elements.

rare earth doping A means of using small amounts of rare earth substances to alter the transmission-carrying capacity of a medium such as a fiber optic waveguide. Doping allows a signal to be amplified by the stimulation of the rare earth substances, thus increasing the transmissions capability and the distance that the signal can transmit. Since fiber optic cable is not a long-distance carrier in the same sense as other media, anything which increases the distance is a great boon to fiber cable manufacture. Transoceanic cable applications can particularly benefit from this technology. Erbium is one of the rare earths used in this process; Samarium is another, used to dope lasers. See doping.

RARP See Reverse Address Resolution Protocol.

RAS 1. See random access storage. 2. RAS. The nickname for Sun Microsystems raster-format files. The file extension .ras may be used to distinguish these files. See raster. 3. See remote access server. 4. Royal Astronomy Society. 5. Russian Academy of Sciences.

RASC 1. See Radio Amateur Satellite Corporation. 2. Royal Astronomy Society of Canada.

raser *acronym* radio amplification by stimulated emission of radiation.

raster A sequence of adjacent scanning lines on a cathode-ray tube (CRT) displayed quickly enough and closely enough together that they are perceived as a fairly uniform coverage of the display surface of the tube. The full coverage of the screen is called a frame. Most television broadcasts and computer monitor images start the raster at the top left corner, with each line sweeping horizontally left to right down the tube, and ending in the bottom right corner. There may be two sets of interleaved rasters displaying concurrently. Color raster systems typically employ three beams: red, green, and blue (RGB), the primary colors of light (the primary colors of pigment are red, yellow, and blue). Unlike a vector display, in which a straight line rendered at an angle appears reasonably straight (depending upon the resolution of the monitor), raster displays may have artifacts which cause the image to appear jagged or *staircased*. Anti-aliasing can perceptually decrease this effect. See antialiased, bitmap, interlaced, vector display, refresh.

raster fill The filling in of spaces between raster lines on a CRT to provide an image that appears brighter or sharper. See raster.

raster image processor RIP. A device to accelerate the process of data conversion, such as scan conversion on a monitor, or vector-to-raster conversion on a high-end printer. PostScript files, which are widely used in the printing industry, define vector-format files so that they can be output to a variety of types of devices at the highest quality possible for that device. Thus, service bureaus will take a PostScript file (as generated by a desktop publishing product such as PageMaker or FrameMaker) and RIP that file to their high-quality imaging device to create a raster-based paper printout or a paper, asbestos, or metal printing plate.

A printer device file is a file that includes information specific to the capabilities of a printer that is used in "rasterizing" an image. This can be used by a PostScript RIP, for example, to turn the vector-based instructions into objects and raster points that can be output to a printer. Adobe Acrobat distiller is a form of RIP that converts PostScript into high-quality PDF

R

files (popular on the Internet). These files can then be readily viewed on a variety of monitors and, if desired, printed to a variety of printers.

raster line A single line sweep (usually horizontal) of the electron beam on a cathode-ray tube. The time during which the image is rendered by exciting the phosphors on the inside front of the CRT. When the beam travels back to start the next raster line, it is suppressed in a process called blanking. See blanking, frame, raster.

rastering, rasterizing The process by which an image is converted to data, usually as a stream of bits. Rastering is a common process in document transfer, and is often accompanied by compression and decompression of the data in order to minimize transmission time.

RATCC radar air traffic control center.

rate The cost per object unit or unit of time of an equipment lease or service. Phone services are typically billed at a flat rate per month with individual surcharges for connect time for long-distance calls or cellular calls. Internet Service Providers (ISPs) typically charge a flat rate per month, although some add surcharges for popular services like email, file storage, and Web access.

Rate Adaptive Digital Subscriber Line RADSL. A means to optimize the throughput of data communications in a Digital Subscriber Line (DSL) service by adjusting the connection to compensate for variations in the line characteristics of the local loop. The desired bit rate may be initialized manually or automatically at startup or, if there is a way to monitor line conditions, may be adjusted as needed, based upon the performance characteristics of the line. This is in addition to the normal rate adaptation capabilities of the ADSL service. See chart under Digital Subscriber Line. See Asymmetric Digital Subscriber Line.

rate averaging An economic method for providing uniform, simpler pricing options for equipment or services which normally might vary widely in their costs of installation and operation to different groups of consumers. For example, phone companies have fairly uniform rates over a wide variety of terrains, services, and population densities. Postal services also employ rate averaging; in other words, a letter to the next town requires the same postage as a letter to the most distant part of the country from the sender's locality.

rate decrease factor RDF. In ATM networking, an available bit rate (ABR) flow control service parameter which controls the decrease in the transmission rate of cells when it is needed. See cell rate.

rate increase factor RIF. In ATM networking, an available bit rate (ABR) flow control service parameter which controls the increase in the transmission rate on receiving an RM-cell. See cell rate.

rate period In telephone service, a segment of time designated as a specific period in order to assign billing charges. Rate periods are determined by evaluation of phone call traffic volume, cultural customs, and time of day, and then usually established semipermanently so that subscribers become familiar with

peak and off-peak rate periods. Rate periods vary from country to country. In the U.S., for example, the least expensive rate period on weekdays is from 23:00 to 08:00, and cheaper rate periods are available on weekends. Companies often schedule fax transmissions to be sent out automatically after midnight to take advantage of the cheaper rate period. See rate period specific.

rate period specific When telephone calls which cross rate periods are billed at a higher or lower rate when the period changes, they are called *rate period specific*. International calls originating in the U.S. are usually not rate period specific, and the call is billed according to the rate period during which it was initiated. See rate period.

Rate Quote System A computerized telephone rate/quote system which can be accessed by TSPS operators.

rated voltage A designation of the voltage at which an electrical component is set to operate, or, if put in a variable voltage environment, the safest maximum voltage at which it can be used for extended periods without risk of hazard or component burnout.

RATP See Reliable Asynchronous Transfer Protocol.

rat's nest Mess; poor configuration; snarled, complicated arrangement of wires, machines, processes, or code statements.

RATS Radio Amateur Telecommunications Society. RATS broadcasts to a Java-enabled site on the Internet on a 145.790 MHz channel.

RaW read after write.

RAX See rural automatic exchange.

RAY See Rayshade.

RAYDAC Raytheon Digital Automatic Computer. A historic large-scale computer manufactured by Raytheon and named in the same tradition as the ENIAC. The RAYDAC began operations in 1953. See ENIAC, JOHNNIAC, UNIVAC.

Rayleigh disc An instrument for the fundamental measurement of particle velocity by means of acoustical radiometry.

Rayleigh expansion In the context of diffraction gratings, the consideration of electromagnetic variables in addition to what was known about the diffraction of light in the early 1900s. This improved our understanding of diffraction patterns within gratings and helped reveal the importance of wavelength interactions with grating surfaces.

Rayleigh's word came about as a result of observations about diffraction made by Wood a few years earlier that were anomalous in the context of existing theories about the diffraction of light from a grating surface. Rayleigh suggested that the diffraction pattern of the incident light as it propagated from the grating could be expressed as a mathematical series both within the region of modulation of the grating and outside the modulated region. This theory is still used to describe the outer region but has been found insufficient in some circumstances (e.g., resonance gratings) for explaining light propagation and reflection in the inner corrugations of the grating. See grating, Talbot effect, Wood-Rayleigh anomaly.

Rayleigh fading Fading, or loss of signal strength, as a result of interaction with the various objects or particles which are part of the environment of the transmission. This phenomenon is often found in mobile communications in which the interaction of the radio signals with the surrounding terrain causes signal fading. A number of techniques are being developed to reduce the incidence of fading. For example, in systems where long delays are acceptable, fade can be reduced by interleaving. Named after J.W. Strutt (Lord Rayleigh).

Rayleigh scattering Scattering of radiant energy by contact or interaction with minute suspended particles such as dust or moisture. Rayleigh scattering may result from fluctuations or impurities within a transmission medium that create small amounts of refraction compared to the transmission wavelength. (This refraction is typically not desired, though impurities are sometimes introduced to modify the transmission properties of a medium.) The phenomenon is named after J.W. Strutt (Lord Rayleigh).

In fiber optics, fluctuations in the composition and density of the glass or plastic waveguide may cause Rayleigh scattering. See attenuation, doping, Raman scattering.

Rayleigh, Lord (1842–1919) John William Strutt, an English physicist and mathematician who made fundamental mathematical contributions to the field of physics, including atomic physics, acoustics, and optics. In 1870, he published *On the Light from the Sky – Its Polarization and Colour*, which presented his ideas and calculations based on observations of the scattering of light and the relationship of the scattered radiation to wavelengths. In 1904 he was awarded a Nobel Prize for his discovery of argon. In 1907 he investigated anomalies and characteristics of resonance phenomena in the context of diffraction gratings. Rayleigh scattering is named after him. See Wood-Rayleigh anomaly.

RAYNET See Radio Amateur Emergency Network.

Rayshade A native scene description language used with the Rayshade 3D raytracing software for modeling light effects on 3D rendered surfaces/objects. Rayshade files may be distinguished by a .ray file extension.

RB 1. See radar beacon. 2. reverse battery.

RBBS remote bulletin board system. See bulletin board system.

RBOC See Regional Bell Operating Company.

RBS See robbed-bit signaling.

RC6 A block cipher encryption algorithm developed by RSA Security Inc. that was selected as a finalist when entered as a submission to the U.S. Federal Advanced Encryption Standard. RC6 was jointly developed by Ron Rivest (originator of the MD series message delivery algorithms), Matt Robshaw, Ray Signey, and Yiqun Lisa Yin of RSA Laboratories and was originally specified in 1998. It is the evolutionary descendent of the RC5 block cipher, which is based upon the concept of data-dependent rotations. See Rijndael.

RCA 1. remote control access. 2. See Radio Corpo-

ration of America. 3. root cause analysis.

RCA connectors A basic electrical connection format for cables carrying audio/visual and sometimes data signals. The RCA connector is a simple, peg-shaped jack or plug commonly used for single-signal cables (though they are sometimes also attached to the end of coaxial cables, with only one of the two coaxial conductors actively transmitting through the RCA end). RCA connectors are widely used in the audio and video broadcast and recording industries to interconnect components such as audio components, VCRs, DVD players, camcorders, switchers, and more. RCA connectors depend upon friction to stay in place, which means they are easy to plug and unplug, but should not be used for connections where a secure connection is vital or where there is tension against the connection (BNC connectors work better in these circumstances).

RCA Cables & Adapters

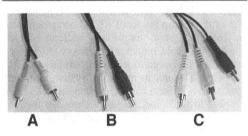

A   B   C

RCA-jack cables are frequently combined in one cable for convenience (the signals run on separate wires), with the jacks color-coded (yellow–video, red–audio mono/combined or audio right, white–audio left) to help consumers interconnect the correct plugs. Three common examples include A. video, audio–mono B. audio left, audio right C. audio left, video, audio right.

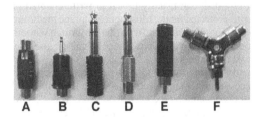

A   B   C   D   E   F

There are many adapters available to enable RCA cables and components to connect with other components (headphones, portable recorders, etc.). Some common adapters include A. gender bender or extender for adding an extra length of cable B. RCA to mini (ca. 1/16-in.) audio mono C. RCA to ca. 1/8-in. audio stereo D. RCA to ca. 1/8-in. audio mono E. RCA to audio mono female F. RCA splitter/joiner to split a single signal out to two receivers or to join two signals into one.

RCC 1. See Radio Common Carrier. 2. reduced complexity computing. 3. remote control center/circuit.

RCF 1. See Radio Control Function. 2. See Remote Call Forwarding. 3. remote control facility.

RCL runtime control library.

RCM remote carrier module.

RCP remote control panel.

RCS 1. radar cross-section. 2. See remote control system. 3. See revision control system.

RCU See remote concentration unit.

RD See routing domain.

RDB 1. receive data buffer. 2. remote database.

RDF 1. radio direction finding. In radar, a British term for a tracking system based upon locating the source of unidentified or foreign radio signals. 2. See rate decrease factor.

rdist remote file distribution program. A program to distribute and maintain file copies on multiple hosts on a network. See DHCP.

RDP 1. radar data processing. 2. rapid development program. 3. See Reliable Data Protocol.

RDT 1. recall dial tone. 2. remote digital terminal.

RE radio emergency.

REAC Reeves Electronic Analogue Computer. A historic large-scale analog computer series first introduced in 1948 by the Reeves Instrument Corporation. One of the selling points of REAC computers was the "patchbay system" for patching interconnections. The Aeronautical Computer Laboratory (ACL) was among the first organizations to purchase REAC computers. Documents about this system form part of the Marvin L. Stein Papers collection at the Charles Babbage Institute and are also included in the National Museum of American History Oral History Collection.

read A command commonly used in software application menus to provide the user with the ability to load data from permanent or semipermanent storage such as a floppy diskette, hard drive, cartridge drive, tape, RAM disk, or other medium. Files on a drive can be set with protections to be read only, or read/write, or write only so that they can't be read. Similarly a disk can usually be set to write-enabled or write-protected mode. Most optical storage media are read only, and cannot be rewritten or written without special, more expensive devices than are used by most consumers.

read only memory ROM. A nonvolatile, random-access data storage unit which is preconfigurable, and not changeable by the user by normal means. ROM chips are commonly used for kernel level operating instructions or other information for the low-level functioning of a system which needs to be quickly accessed and transparent to the user. See CPU, RAM, PROM, EPROM, kernel.

read-while-write RWW. A capability of some data storage devices to verify written data in realtime. This can be accomplished by executing code in one array while data are stored in another. In March 2000, Toshiba Corporation announced that they had developed the first 64-Mbyte read-while-write NOR flash memory device for use in various types of wireless, handheld consumer devices, and set-top boxes. See linear tape open.

Realtime Transport Protocol RTP. A packet-oriented data delivery services protocol for end-to-end services to support applications transmitting realtime data such as Interactive audio/video applications over unicast and multicast networks. RTP provides time stamping, rate control and source/payload identification as augmented by the Realtime Transport Control Protocol (RTCP) to enable the monitoring of data delivery in a scalable manner. RTP and RTCP are independent of the underlying network and transport layers and may be integrated into the processing of an application rather than implemented as a separate layer. RTP was submitted as a Standards Track RFC by Schulzrinne et al. in January 1996. See RFC 1889.

Real Time Markup Language RTML. A trademarked, proprietary set of extensions to the Hypertext Markup Language (HTML) for specifying the inclusion of streaming textual data on HTML-based Web pages using RTTP data as a source. Thus, live, streaming text data can be readily incorporated into a Web page. This is of interest to vendors of realtime services and information such as stock quotes, news, or racing results.

RTML is a product of Caplin Systems Ltd. based on Dynamic HTML (DHTML) combined with Java 1.1 or higher. Once it is installed, the user need not worry about programming any Java or other language. All that is required, once the streaming engine is installed, is to add a <SCRIPT> tag with optional attributes to the HTML pages. Newer browsers are required to view the streaming text messages. The software will alert the user if the browser is not compatible with the RTML feed.

Real Time Streaming Protocol RTSP. An extensible, application-level protocol for delivery control of realtime data. RTSP enables the on-demand delivery of multimedia streams delivered, for example, by service delivery protocols such as Rapid Transit Protocol. Sources of data may include stored or live streaming data feeds. The protocol supports multiple data delivery sessions. RTSP was submitted as a Standards Track RFC by Schulzrinne et al. in April 1998. See Rapid Transit Protocol. See RFC 2326.

real world A phrase to describe the application or testing of products or services in a real world environment, in other words, in the end population or facility for which the product or service is intended. Thus, a real world test of a new telephone might be to install it in a telemarketing firm. A real world test of a new computer might be to put it into a classroom or business. Real world testing is sadly lacking in the software industry. The pressure to get products to market tends to cause software companies to release a product after it has been tested in-house and through beta testers, but not by actual customers. This can have disastrous consequences as real world customers will always use the product in ways that cannot be anticipated by programmers and beta testers.

RealAudio A commercial on-demand, multiplatform, realtime audio player for multimedia-capable computers from Progressive Networks. The RealAudio format is widely supported on the Web and may be played through a Web browser plugin or launched separately (on multitasking systems). Earlier versions of the software (e.g., version 2.0) supported monophonic sound designed for the 14.4 and 28.8 kbps transmission speeds typical of dialup network connections at the time. Subsequent versions are being updated to support higher-bandwidth connections as well. RealAudio files typically use the .ra file extension.

realtime The term realtime is used somewhat differently by two groups of people: (1) marketing personnel and users and (2) technical systems designers and operators. First, a more general description of realtime.

Realtime is a description of computer processes that occur at a speed which corresponds with human perception of the speed of events in "real life," and in immediate response to requests. In other words, a ray tracing program that takes two hours to render and display a frame of an animation is not realtime, as there is a delay during which the viewer must wait for the image to be constructed and displayed. In contrast, a fast action video game, in which the motions are displayed at 20 or 30 frames per second so that they are perceived as natural motion, and in which the joystick, mouse, or other inputs from the user have immediate effect upon the game, is considered to be a realtime game. Realtime flight simulators are used to train pilots, and realtime rendering programs exist on some fast, high-end platforms.

In telephony, realtime processing involves handling calls as they are received. If callers are put on hold or experience delays in automated menu processing systems, the system is not providing realtime service. Realtime effects and processing, especially if they involve graphics, typically require fast, wide data buses, fast CPUs, and efficient mathematical algorithms for handling input, calculations, and display. In spite of the resources needed, humans seem to have a compelling interest in creating realtime scenarios and striving for real and fantasy simulations that mimic or outstrip the pace of life. This creates economic incentives for creating realtime simulations, especially in the entertainment industry, with audiences eager for these scenarios. Indeed, many of the advances in computer technology have been pioneered, fueled, and financed by the games industry. In a more technical systems implementation sense, realtime occurs in a computing system when computations are processed not only as expected, and with logical correctness, but also within certain predetermined or expected timing frames, and with a certain guaranteed minimum level of usefulness of the service. In this sense of the word, speed is not so much the issue, as is the appropriateness of the response time to the task at hand. Some realtime systems rely on sensor and other feedback mechanisms, and may be used not only in consumer computing operations, but also in industrial robotics or remote sensing applications. Realtime functionality is likely to be important in future space probes and the vehicles that deploy them, as well as in intelligent vehicle auto-navigation systems.

realtime capacity The capability of a system to handle calls, inputs, requests, or other stimuli as they are received. In configuring and tuning various types of networks, realtime capacity is one of the criteria many systems use as a reference point for smooth operations.

realtime diagnostics Tests which allow measuring, diagnostic, or display instruments to monitor and report events as they are occurring. Most electrical instruments work in realtime, reporting circuit status at the moment the instrument is applied to the circuit. This is not so easily done with sophisticated computer systems, where it is difficult to track everything happening on the system at any one time. More often software "monitors" (statistical display programs) for specific processes are used, which include the representation of statistics for load, CPU speed and processes, congestion, failed packet ratios, quality of service (QoS), etc.

Realtime Transport Protocol RTP. An IETF data format that provides higher video priorities to facilitate realtime multimedia transport over Internet Protocol (IP).

reassembly An important aspect of network communications in which an Internet Protocol (IP) datagram or other type of data unit, which has been split up at the source or en route and may have been transmitted in sections at different times and/or through different routes, is reassembled at the receiving end. The process of disassembly and reassembly allows packets to be transported through a large, dynamic network environment, like the Internet, which changes topologically in unpredictable ways. Reassembly and synchronization are also important in applications like videoconferencing, where more than one line may be used to transmit the various audio and video signals that make up the communication.

reboot To cause a system to return to its initial operating status, as it was at the beginning of a system startup, usually without turning off the power. This typically clears memory, closes all applications and files, sets initial test sequences and starting parameters for timing, sound, video, etc. and reinitializes devices. The term is derived from "boot," which comes from "bootstrapping."

If the power is turned off to reboot a system, it is called a *cold* boot. You should always count to 20 before flipping the power switch on again. Electronic components are sensitive to sudden power surges, and there is always some residual power in some of the chips that needs to drain off when electronics devices are turned off.

Most reboots are *warm* boots, in which the power to the system is not interrupted. Rebooting is seldom necessary in stable operating systems, which can operate 24 hours per day for years without crashing, hanging, or fragmenting memory. However, some operating systems do not handle error conditions or memory management well and may hang, freeze, or crash, in which case a reboot may be necessary in order to continue using the system.

receive-only device A device which can receive data but not send it. Technically, there are very few receive-

R

only devices in computer networking, since most devices employ *handshaking* to negotiate a transmission. For example, a computer printer may seem to be a receive-only device, but a printer has to be able to tell the computer when it is ready to receive, when it is busy printing and can't receive more data, and when it is available again for other jobs or other users on a network. This involves two-way communication. It may even signal the sender about its capabilities and configuration parameters. Most receive-only devices are passive devices or broadcasting devices such as simple PA speakers, buzzers, lights, etc.

receiver 1. A device for receiving signals, impulses, or data transmissions. 2. A device which captures, and sometimes converts electromagnetic waves or signals into a form meaningful to humans. Receivers are often combined with tuners to specify the frequency desired, and amplifiers to increase the power of the signal. See telephone receiver.

Recognized Private Operating Agency RPOA. An ITU-T designation for telephone companies providing internetworking services.

Recommended Standard 232 See RS-232 for an entry and accompanying chart of Recommended Standards.

RECON reconnaissance.

reconnaissance A preliminary or exploratory survey to gain information or data that can potentially be analyzed to yield information (it's often difficult to know in advance what data might later be useful or significant). Reconnaissance and surveillance are closely related activities. Reconnaissance is distinguished from surveillance in that surveillance is keeping watch over or observing someone, something, or some activities or phenomena. Reconnaissance is often used to support surveillance activities. Surveillance may also be used to obtain some of the data collected in reconnaissance activities.

Typing a few exploratory passwords into a computer system is a reconnaissance activity intended to see if it would be easy to breach a secure computer system. Using a computer to remotely view a scene captured by a video camera is an example of a surveillance activity that may also be part of a larger reconnaissance operation.

Recording Industry Association of America RIAA. A national trade organization representing the recording industry. The organization supports and promotes the protection of intellectual property rights and business prosperity of its members who collectively produce the vast majority of commercial sound recordings in the trade. The RIAA became prominent in the media when it opposed the alleged distribution of its members' copyright materials by unauthorized users through the Napster Web site.

In October 2000, the RIAA announced that it was going to develop a globally standardized system for identifying digital sound files in order for the copyright owner of the files to be able to track their use and to collect any royalties rightfully due for the materials. This was intended to enable the distribution of digital editions of the recordings quickly and easily over the Internet without jeopardizing the economic viability of the people creating and vending the music. http://www.riaa.com/

rectification 1. A condition in which current flowing through a material or circuit in one direction encounters greater resistance than current flowing through in the opposite direction. 2. The one-directional processing of an alternating current (AC).

rectifier 1. A material or circuit that offers greater resistance to an electrical current flowing in one direction than in its opposite direction. 2. A device for converting alternating current (AC) to direct current (DC). Rectifiers are commonly used on power transformers for electronics devices with power requirements different from the power coming directly from an electrical source. Vacuum tubes were used as rectifiers in early radios, with selenium rectifiers beginning to supersede them in the mid-1940s. See coherer, crystal detector, piezoelectric.

recursion 1. Returning, moving back upon. 2. A repetitive succession of elements or operations that affects the preceding elements or operations in a like (although not necessarily identical) manner according to a finite rule or formula. Recursive algorithms often generate data or images with the characteristic of "self similarity," with fractal display programs being a popular, visually appealing example. 3. See recursion.

red alarm In telephone transmissions systems, a critical failure alert signal which occurs if an incoming signal is lost or corrupted. This is implemented in various T3, T1, or SONET network systems.

Red Book 1. The original Compact Disc digital audio (CD-DA) specification, developed in the late 1970s and introduced in the early 1980s by Sony and Philips. Audio sectors, tracks, and channels are specified, along with other physical parameters. The format enabled up to 74 minutes of digital audio to be recorded at a sampling rate of 44.1 KHz, a rate that is sufficiently good to support the recording of classical music. The Red Book was followed, in the mid-1980s, by the Yellow Book, which specified CD-ROM parameters. See Yellow Book. 2. In telephony, books in the ITU-T (formerly CCITT) 1984 series of recommendations. 3. The Adobe PostScript Language Reference manual.

Red Box An environment for running Windows applications under Apple's Rhapsody on Intel-based computers and possibly on PowerPC-based computers, generally analogous to the Blue Box environment for running legacy Mac operating systems under Rhapsody. This would make it possible to have full compatibility with Intel-based operating systems such as Microsoft Windows on Apple hardware. See Blue Box, Yellow Box.

red signal An alert or failure signal or "stop" indicator used in many industries and in association with many different types of devices, networks, and gauges. See yellow signal for a fuller explanation or red and yellow signals.

red wire A color designation used by IBM to indicate wires used to establish a hardware patch to accommodate a code change. See blue wire, purple wire, yellow wire.

Reduced Instruction Set Computing RISC. A type of programming and system architecture which uses a set of simpler instructions performing single, discrete functions to carry out an operation than would be used in a comparable operation by a Complex Instruction Set Computing (CISC) design. Most of the newer computers tend to incorporate RISC architectures, although not all support circuitry enables the full capabilities of RISC architecture to be used.

Unlike CISC commands, RISC commands are the same size, which means that less time is required for subsequent processing of the instructions, because individual evaluation of the commands for size and conversion to microcode is not required. When RISC software is compiled, it is evaluated to determine which operations are not dependent on the operation or results of others, and slates them for simultaneous execution.

Due to the reduced instruction set and processing that takes place, the circuitry on RISC chips is simpler than on most CISC chips, resulting in a smaller physical size and, usually, lower heat output.

Not all chips are strictly RISC or CISC. For example, in the Intel line of processors, the Pentium chips are a transitional architecture that maintains some downward compatibility with the earlier CISC architectures, while still incorporating some of the advantages of RISC architectures. The chips tend to be larger and hotter than straight RISC chips, but meet a market demand through a transition period.

redundancy Replication, duplication, superfluity, repetition. Redundancy is important in computing because the loss of data, whether stored or in the process of transport, can have serious consequences to human safety, economics, or business transactions. See redundant array of inexpensive disks.

redundant array of independent netports RAIN. ZNYX Network technology that provides a scalable, modular architecture for delivering high-performance, high-availability, customized intranet connectivity through clusters of network servers. RAINswitch server switches were announced in 1997 to support RAIN implementation in Fast Ethernetworks.

redundant array of inexpensive disks RAID. A data storage, retrieval, and protection system using multiple disk storage devices, a system commonly used in networks. RAID consists of multiple hard drive storage devices linked together to provide data mirroring or data striping and parity-checking across disks in order to record the information redundantly. Duplication or data mirroring is primarily a function of software, whereas parity-checking requires a controller and is associated more closely with hardware. Many RAID systems are SCSI-based.

A basic low-end RAID center may consist of four drives, each with 2.1 Gbytes of storage, sometimes set up in a rack, talking through a centralized controller system, usually through a server.

Although a certain amount of storage is inevitably lost due to duplication of data, the big advantage of RAID systems is that they provide pretty good protection against data loss if a drive goes down. There is less protection if several drives go down but, since this happens rarely and since companies are reluctant to back up data sufficiently often on systems like tape

Redundant Array of Inexpensive Disks (RAID) – Level Specifications in Brief	
Level	Notes
Level 0	Striping, no redundancy or error correction. This can provide faster access, but does not protect data from loss.
Level 1	Disk mirroring. Complete redundancy. Provides data protection.
Level 2	Byte striping, dedicates at least one drive for parity information. Uses Thinking Machines, Inc.'s proprietary setup, which is not commonly used.
Level 3	Generally used instead of level 2. Block striping will improve performance if data are written in large blocks and simultaneous reads are used. Distributed parity information (originally required a dedicated parity disk, a stipulation removed in 1994). In other words, when appropriately implemented, better performance and data protection can be achieved.
Level 4	Similar to level 3, but larger data blocks are striped across disks; each drive is not necessarily involved in each access.
Level 5	Block striping, parity information distributed across drives. At least three drives are required for a minimum implementation. Each drive is not necessarily involved in each access. Parity information is also striped across disks. Provides data protection and, in many cases, will improve performance. This is a popular implementation of RAID.
Level 6	Not consistently specified or implemented.
Level 7	Similar to level 4, with larger data blocks striped across disks. Uses Storage Technology, Inc.'s proprietary caching mechanism and operating system.

drives, the RAID alternative works well in practice. Many RAID systems are "hot-swappable" which means that an individual drive can be pulled out and replaced while the system is online, thus not necessitating a system shutdown or inconveniencing current users on the system.

Specifications released in 1988 in the RAID paper proposed five levels. Since that time, changes and enhancements have occurred; the levels are not cut-and-dried since configuring various parameters, such as stripe size, creates overlapping characteristics between the different levels. Hybrid systems also exist. Generally, however, to provide an introductory understanding, the RAID levels can be summarized as shown in the RAID Levels chart.

In addition to redundancy and parity checking, a RAID system may have some intelligent monitoring incorporated into the system, which does periodic checks and analysis and reports anomalies to the controller. The controller can then signal a warning which allows the device administrator to check for potential problems, or swap out a drive before it fails. See dynamic sector repair, SMART.

re-engineer To step back from a system or process, take a new look at it, and redesign it, sometimes from the ground up, usually with the intention of making it more efficient and cost effective. Software often has to be re-engineered, as legacy systems tend to be slower and less efficient over time, due in part to the way they are upgraded and, in part, because of technological improvements and changes in hardware which are accommodated in a variety of ways. Market pressures also cause many software programs to be released before their time, in which case, they may be re-engineered before the next release. Work environments in companies that are growing or downsizing quickly often must be re-engineered as the ways of organizing facilities and staff appropriate to a small company are not necessarily appropriate to a large company.

reed relay switch A type of electronic telephone switch developed in the 1960s. Reed relay switches began to supersede crossbar switches, which were prevalent at the time, and some of the step-by-step switches still in use. Electronic switches opened up possibilities for many new types of caller services, such as Caller ID, Call Waiting, etc.

reel-to-reel tape Magnetic recording tape wound onto separate round reels which are usually about 4-in. to 8-in. in diameter. Although most tape is now distributed on cassettes rather than reels, reel-to-reel is still used in some professional recording studios, especially if eight or 16 tracks are required for sound mixing and dubbing. Gradually these reel-to-reel sound recorders are being superseded by digital recording media. See cassette tape.

Reeves, Alex H. A British engineer who is one of the significant pioneers of pulse code modulation (PCM) or, as he calls it, "coded step modulation." PCM is a fundamental quantization system used in audio recording and transmission technologies. Reeves conceived the idea while working at the International

Telephone and Telegraph Corporation (ITT) in Paris in 1937 and received several patents for the technology between 1937 and 1942. Reeves recognized the potential for PCM to counteract interference on transmitted speech communications. In 1965, Reeves authored *The Past, Present and Future of PCM*. See pulse code modulation.

reference clock A clock considered very accurate or stable, which is used as a reference for other clocks or processes, such as computer processes. Quartz crystal clocks are considered very accurate due to their vibrational properties and are often used in computers and watches. The speeds of various computer processes are described in *clock cycles*. Atomic clocks are used to establish Coordinated Universal Time and in satellite positioning systems which require accurate clock references.

In multimedia editing environments, a reference clock is used to provide "house sync," that is, long-term synchronization of various audio or other signals, which resolve to the reference clock rather than to the time code signal. See atomic clock, chase trigger, Coordinated Universal Time, quartz crystal, time code.

reference vector equalization RVE. A derivative of transparent tone in band (TTIB) modulation and a basic aspect of linear modulation systems, RVE utilizes equalization for linear amplitude and phase correction to improve transmissions. It is suitable for trunked narrowband mobile radio systems, for example. RVE enables a digital radio to equalize a transmission signal at the receiver increasing potential schemes for high bit-rate densities. See linear modulation, tone in band.

Referral Whois See RWhois.

reflect 1. To rebound from, to bounce off of, to impact and move away from as a result of the impact. 2. To provide a replicate image of, to mirror.

reflector/director elements On antennas, two or more protruberances from the main rod that are usually narrow and regularly spaced in an array. Reflector and director rods help improve gain and directivity of broadcast signals. See Yagi-Uda antenna.

reflectometer A device for measuring surface reflectance, that is *radiant flux*, the reflected radiant energy per unit of time. This is useful for designing and testing reflective mirrors, parabolic reflectors, and other components that need to meet certain regulatory and operational standards. The instrument may be passive, measuring ambient reflectance, or active, supplying a light source (e.g., laser) and measuring the resulting reflection. See interferometer.

reflectometer, optical time domain OTDR. A specialized reflectometer for locating and assessing possible sources of optical loss in longer fiber optic cables. The instrument emits a light pulse and converts and analyzes the reflections from the pulse. Depending upon the "echo" pattern, it may be possible to more closely identify the source of a problem.

refraction The change in the line of travel of an incident electromagnetic wave as it passes from one form of matter (or lack of matter, as in a vacuum) to another. At the point at which the incident radiant energy encounters and interacts with the changed environment having a different density and composition, the energy interacts with the surrounding molecules (or lack thereof) and its course is altered. See index of refraction for a full explanation. See birefringent; Ptolemaeus, Claudius.

refractive index See index of refraction.

refresh rate, scan rate The rate per unit of time at which information or an image is. The rate is usually dependent upon speeds that make it visible and intelligible to humans. This phrase is frequently applied to broadcast and computer display technologies, especially cathode-ray tube (CRT) displays in which the action of electron beams on the phosphors is very limited and must be reinitiated (refreshed) in order for the image to continue to be visible. Refresh is a general concept which applies to many different types of situations in computing, from individual phosphor refreshes to graphical user interface element refreshes.

The refresh rates of the phosphors on the inside coating at the front of a cathode-ray tube will affect the clarity and amount of flicker seen on the screen. Monochrome or gray scale monitors have longer persistence; that is, the image from the excited phosphors is visible longer, and thus do not need to be refreshed as often as color images.

Refresh of the entire CRT image is described as the number of times per second the frame is redrawn. Refresh rates slower than about 20 to 40 frames per second are perceived as flickering to the human eye, especially if the image involves fast action. Still or slow-moving images do not need to be refreshed as often.

The refresh rate of a computer program image is a combination of operating system and applications programming, and is not simply dependent on the hardware attributes of the system. In order to optimize speed on a computer display, the OS or programmer may choose to refresh only a section which has just been manipulated or changed. If the software does not keep track of what is transpiring on the screen or if several processes are active at once, it may seem that the display is slow to update or refresh a new window, gadget, or element drawn in a paint program.

In general, quicker refreshes are desired over slower ones, but the cost of more computing power and faster hardware puts some economic constraints on the refresh rates of various systems.

refurbished equipment Used equipment that has been serviced and tested by a technician to bring it back to original operating condition. If further work is done and substantial numbers of parts replaced or upgraded, it may also be referred to as remanufactured equipment.

Refurbished equipment, or *refurbs,* are usually cleaned up and made to appear new or nearly new. Refurbished items are typically sold at a discount of about 15 to 30% over the price of new ones.

regenerate To restore, bring back to original condition, recreate, duplicate. In electronics, signal regeneration is an important issue. Transmissions typically suffer from loss and interference over distances, and any means that can be used to maintain a signal or regenerate parts or all of a signal experiencing loss or change in some form is usually desired.

There are many physical and digital schemes for regenerating systems. In some digital systems, regeneration may involve putting a signal back into its original form at the receiving end. In a sense, single sideband transmissions are a type of regenerated transmission, since only a portion of the signal is sent. The opposite sideband and the carrier signal are mathematically constructed, and the original signal thus reconstructed at the receiving end. See regenerative repeater, relay, repeater, single sideband.

regenerative repeater A type of repeater used in communications that are characterized by uniformity of length and signal to correct the timing of the signal and retransmit the cleaned-up impulses. These are common in older teletype communications. See repeater.

Regional Bell Operating Company RBOC. One of a number of companies which were formed by the divestiture of AT&T, which originated from the original Bell Telephone Company through a long, colorful history of mergers and splits. In the mid-1980s Judge Harold Greene broke AT&T into seven RBOCs. For more detail, see divestiture, Consent Decree of 1982, Modified Final Judgment.

regional test bed RTB. A regionally allocated location, lab, or systems setup that enables designers, developers, or implementors to test their systems. Test beds can provide information such as whether the systems work, whether they work reliably, and whether they work compatibly with other systems. Test beds are important in almost every industry, since testing can be a complex activity involving specialized equipment and skills. Most designers and developers don't have the resources to set up individual test beds. Test beds are often funded by universities, consortiums, and government grants, especially when they are designed to test products that have a big chance of promoting the common good or being commercially successful and widespread.

register A repository for data, a storage area which may or may not also be used for data manipulation. There are many areas of telecommunications where registers are used. Many computer chip architectures have registers for holding information about to be moved or manipulated. Palette configurations for computer displays may be saved in color registers. Data modems have registers for setting various parameters, with Hayes AT command "S" registers common. Flags, configuration settings, etc. are stored in registers.

Registration Number As part of the Federal Communication Commission's (FCC's) jurisdiction over equipment which may emit radiant waves that interfere with other equipment, appliances, radios, etc. There is a process of submission, evaluation, and certification which warrants that the equipment has

R

passed FCC requirements. This Registration Number is not related to quality, suitability for a particular use, or other usability issues; it simply confirms that the equipment falls within acceptable emission standards.

Reis, Johann Philip (1834–1874) A German inventor who pioneered the transmission of tones and possibly also voice over wires. Reis accomplished this with various transmitters and other equipment that he developed and publicly demonstrated in Frankfurt to the Physical Society in 1861. No directly verifiable evidence indicates whether voice was transmitted at the 1861 demonstration, but Reis' subsequent work indicates that he recognized the potential for voice communications and concentrated many of his efforts in that direction, eventually developing a telephone design that was not unlike telephones actually put into production in the United States some years later.

Subsequent inventors, excited about the breakthrough, made improvements on Reis's early crude mechanisms, while Reis himself continued to study and improve the technology until his early death in 1874. See telephone history.

relative intensity noise RIN. The ratio of the mean square optical intensity noise to the square of the average optical power, frequently described in decibels (dB). A spectrum analyzer may be used to measure RIN. See Allan variance, Fourier transform.

relay *n.* 1. To resend signals, objects, or communications to another node after which they are further transmitted or transported. 2. An electromagnetic device in a circuit for providing automatic control, which is activated by varying electrical impulses. A relay is usually combined with switches to control when they open and close, and widely used to automate older telephone switching centers. Thus, it was important to design relays for durability, since they had to open and close circuits many millions of times. Because the relay is essentially a simple mechanism, it can be greatly varied by adjusting contact springs and windings, thus producing a large variety of types of relays. Multicontact relays were developed in order for numerous switching contacts to operate simultaneously. See crossbar switch.

Release to Pivot RTP. A number portability mechanism, similar to Query on Release (QoR), in which a telephone number that has been ported from a release switch returns the addressing information for routing a call (as opposed to the previous switch). If the number has been transferred, the information may be contained in the release switch or in an external database. See Query on Release.

reliability An expression of the dependability of a system under actual conditions of use. See availability, mean time between failures (MTBF).

Reliable Asynchronous Transfer Protocol RATP. A packet-based serial communications transfer protocol described by G. Finn in 1984. RATP is intended to facilitate reliable, easy-to-use communications between computers through public telephone circuits. Based on the ubiquitous RS-232 standard, RATP

enabled full-duplex, point-to-point communication more simply than some of the other protocols at the time as well as some capabilities not found in others. RATP is now considered a historic protocol. See RFC 916.

Reliable Data Protocol RDP. A data protocol for providing reliable packet-based data transport services such as remote loading and debugging. It is, in part, adapted from Transmission Control Protocol (TCP), and was submitted as an RFC by Velten et al. in 1984. RDP supports the bulk transfer of data for various monitoring and control applications, as needed, with a simpler set of functions than TCP. Version 2 was submitted by Partridge and Hinden in 1990 to address some problems discovered in testing in 1986 and 1987. It makes changes to the protocol header and corrects some minor errors. See RFC 793, RFC 908, RFC 1151.

Reliable SAP Update Protocol RSUP. A bandwidth-saving protocol developed by Cisco Systems for propagating services information. RSUP enables routers to reliably transmit standard Novell SAP packets only when a change in advertised services is detected by the routers. Network information can be transported in conjunction with, or independently of, the Enhanced IGRP routing function for IPX.

reluctance Opposition, or resistance in a magnetic circuit against the creation of magnetic flux. Similar to the concept of resistance in an electrical circuit. See resistance. Contrast with permeability.

REM 1. remote equipment module. 2. See ring error monitor.

remailer Any online electronic mail transit station which changes or prepends the header in such a way that the originating information is changed or obscured, or which intercepts mail and then forwards it on to its destination. Sometimes these remailers are LAN servers configured so that the header changes when incoming mail is served out to the local recipients. This is unfortunate in that the recipients cannot automatically reply to the sender and must manually type in the return email address in order to respond to their correspondents. This is not a recommended way of configuring a mail server and should only be done when a specific reason warrants it.

Remailers are sometimes used irresponsibly. Thousands of get-rich-quick and commercial products promoters use remailers to obscure the origin of their online postings because there is much legitimate opposition to unsolicited commercial messages on public forums and in private email.

Anonymous remailers are mail transit points which deliberately obscure the identity of the poster in order to ensure his or her privacy. See anonymous remailer, spam.

remapping On a computer system, remapping is moving data, often in the form of blocks, arrays, or tables, from one area of storage to another. Memory remapping, address remapping, file location remapping, and keyboard remapping are some common examples. Remapping is sometimes used to *double-buffer* computer graphics screens – building a screen in a background while the current screen is viewed by the user and then displaying it by remapping the entire image to the video display area. It can improve the likelihood of fast, clean transitions. See frame buffer.

remote access An important aspect of networking in which access to computing services, devices, and information can be gained through a remote device on the network, usually a computer terminal or phone line. On a phone line, remote access to an answering machine can enable a user to dial up the answering machine from a different phone, punch in some codes to see if there are messages available, retrieve those messages remotely, and even change the message on the answering machine through the phone line.

Remote access does not imply the level of operations that can be accomplished, only that the device can be accessed in some basic way. Remote access terminals vary greatly in their ability to interact with a server or other user functions. For example, on a basic text-oriented "dumb terminal" connected to the main computer with a serial line, the user may only be able to execute simple text commands and won't be able to display graphics or run sophisticated applications locally.

On the other end of the spectrum, some systems provide full access to remote applications, especially if they are connected with a fast transmissions protocol over fiber. In other words, there may be a graphical database program available on the server that the user can run on a smart terminal as though the terminal was the main computer. Not all operating systems can do this. The X Window System is designed to provide this type of capability in conjunction with various Unix systems. It has also been upgraded to provide similar services over the Internet. See X Window System 11.

remote access concentrator RAC. A network system for interconnecting numerous multiple remote telecommunications links to a local system such as a local area network (LAN). Businesses with wide area networks (WANs) and telecommunications providers make wide use of concentrators, as they aid in channeling and aggregating many types of equipment in order to serve broader needs such as large numbers of employees or telecommunications clients. High-performance concentrators may support several services, including access, switching, and routing.

Commercial RACs are created for many purposes such as interconnecting multiple ISDN B-channel lines to LANs through Ethernet or multiple computer modems to T1/E1 networks. Multiple RACS can sometimes be attached to one access switch. Depending upon the protocols supported by a RAC, it may be connected indirectly or directly to the public switched telephone network (POTS).

Many vendors promote a similar product called a remote access server (RAS), which is basically a lower-end version of a remote access concentrator (usually with fewer ports). Others refer to the operating software for a RAC as remote access server software. See remote access server.

remote access control facility RACF. A facility dedicated to the provision and management of remote access transmissions. An Internet Services Provider is a type of RACF that provides multiple subscribers with remote access to Internet services. A telephone switching office is another type of RACF, providing access to multiple remote telephones to its local subscribers.

Within a particular business, a remote access facility can be established as a management or security system to control, authorize, log, and secure data and physical resources (printers, modems, etc.) used by a variety of in-house or contract employees.

IBM provides the commercially trademarked Remote Access Control Facility system, first introduced for mainframes in 1976. The system is promoted to businesses for the management and securing of valuable corporate data. RACF supports OS/390 and z/OS software systems. See remote access.

remote access PBX A private branch telephone exchange which can be accessed from an outside line with appropriate authorization codes. Once logged on to the internal branch system, various features can be used such as voice mail messages, long-distance calls connected and billed through the PBX, etc.

remote access server The software and generally also the hardware on which the server software runs for managing transmissions between remote and local devices. Remote access to the Internet and off-site printers, security monitors, and other networkable devices can be facilitated by a dedicated server, especially in situations where numerous people will access the service through limited resources (e.g., a limited number of modems). See remote access concentrator.

Remote Authentication Dial In User Service See RADIUS.

remote batch processing A means of submitting a computing job remotely to a processing system and receiving it back as or when the job is processed. This is rarely done at the consumer level, but it is still common for high-end mathematical calculations, scientific research, and other intensive computing applications which may require large amounts of computing time or more sophisticated computing resources. In the earlier days of computing, remote batch processing, especially with punch cards, was the only type of service available, and it could take hours or days to receive the results of a simple calculation.

Remote Call Forwarding A service in which a phone number is located in the central office of one exchange and any calls made to that number are forwarded (essentially by internally making a second call) to a line in another exchange. This may be of value to businesses that want to maintain a local presence without the expense of a local office, so that customers can call a local number instead of long-distance.

remote concentration unit RCU. An off-site facility or device in which multiple devices/services of the same basic type are aggregated and handled together within one basic management device housing. Telecommunications services (such as Internet access or cable TV) and electrical power distribution are two examples of RCU applications.

remote control *n.* A device to allow control of another device without making direct physical contact. The control of the device may be through indirect physical means (through a remote controller and cable), a network (computer-controlled vending machines in another part of a building), or various wireless methods (infrared, FM, audible sound control, etc.). Remote control of computers on a network can be done through various telecommunications products, specialized remote applications and file serving software, or through operating systems which support this capability.

remote control access RCA. Access to restricted areas, rooms, or pieces of equipment through wired or wireless remote devices. Many environments and types of buildings control access through remote devices. Examples include remote garage door openers, key cards, parking barrier remotes, audio/video/computer components with security feature remotes and home automation remotes.

remote diagnostics Systems diagnostics which can be run from a remote location. It is common for higher-end routing and switching devices on a computer network to be controlled through software at a main administrative location. This software typically permits the running of test and diagnostic routines and may show graphical diagrams of problems or potential problems or bottlenecks. On phone systems, diagnostic checks can sometimes be carried out with devices that generate specific tones or signals, which can initiate processes at the other end of a phone line.

remote file access RFA. A capability for accessing electronic files from a system/terminal separate from the file server. RFA may be at some distance from the file repository. The Internet provides a transmission link to millions of file repositories around the globe that can be accessed and downloaded by various Web browsers and file transfer utilities. See File Transfer Protocol, ftp, Gopher.

Remote Imaging Protocol RIP. A protocol for facilitating the implementation of EGA-resolution color images and mouse control from a remote system. RIP was developed to overcome the limitations of text-based bulletin board system (BBS) communications (in the days before the World Wide Web). Vector-based commands are sent and rendered through a point-to-point network connection. The system has some limitations but is an interesting solution to text-based remote terminal connections through dialup links. See Remote Imaging Protocol script language.

Remote Imaging Protocol script language RIPscrip. A scripting tool from TeleGrafix Communications, Inc. for facilitating the development of graphical user interfaces for computer bulletin board systems (BBSs) supporting Remote Imaging Protocol (RIP). It is a 7-bit system to maintain compatibility with the many 7-bit communications systems common through dialup connections (e.g., X.25).

RIPscrip works in conjunction with RIPaint and RIPterm for developing graphical interface screens and protocol needs. See Remote Imaging Protocol.

remote job entry RJE. Generically, this is the entry of computer commands from a remote entry terminal, that is, a local terminal, for execution on a remote machine (e.g., a timeshared mainframe). This applies to individual job requests and to batch requests.

The term has changed meaning somewhat as computers have become more common and less expensive. Historically, remote job entry was a batch request submitted on computer punch cards. The cards were punched at a card-punching terminal by the programmer, then physically handed together as a "job" to a computer operator. The operator queued the job with others that were pending and then inserted the card bundle into the computer's card reader for execution. When the job was completed (or "crashed" due to errors), the operator would return the job cards to the programmer along with any relevant results (or error messages). These days, jobs are usually submitted electronically through software requests rather than physically through punch card stacks. See Remote Job Entry Protocol, remote programming.

Remote Job Entry Protocol RJEP. RJEP was first submitted as an RFC by Chuck Holland in the early 1970s. This generated quite a bit of interest and was corrected and developed further in RFC 407 in 1972. Remote job entry is a means by which a user can execute a computer processing job at a location (e.g., a mainframe) other than the local computer (e.g., a remote terminal). The user can thus submit a job to another system acting as a job server, which would process the job and deliver the results to the remote terminal through a TELNET connection. File Transfer Protocol (FTP) served as the file transfer mechanism. RJEP enables connections to be established individually or may leave the connection open for multiple submissions. See File Transfer Protocol, remote job entry, RFC 360, RFC 407.

Remote Mail Checking Protocol RMCP. This low-overhead mail checking protocol was submitted as an experimental RFC by Dorner and Resnick in 1992. It provides a client/server-based mail checking service. A program on the client's workstation uses RMCP to query a server to see if new email has been received for a specific user. It is suitable for use with remote mail servers, such as those implementing Post Office Protocol (POP). The protocol is based on the User Datagram Protocol (UDP) port 50. Some authentication in the initial communication between client and server, at the cost of complexity. See RFC 1339.

Remote Operations Service Element ROSE. An application layer service that provides the capability to perform interactive remote operations through a request/reply mode. ROSE is a generic information exchange technique which is not application-specific and not intended to define the operations it facilitates; this is left up to those implementing remote services. It is defined as ISO 9072-1, and as an X Series

Recommendation by the ITU (X.219). See Abstract Syntax Notation One, X Series Recommendations.

remote procedure call RPC. A means of making a request to a remote system so that it appears to the user as though the request is being fulfilled on the local machine. In other words, a user may open a word processing program and load in a file. The file may actually reside on a computer in another room or another city, but the user is unaware of any difference in using the file from the remote system or using a file from the local system as the RPC is transparent to the user. Another example would be the use of a terminal communications program which accesses a modem on another computer as though it were physically attached to the local machine. A number of conventions for making requests to a remote system, and fulfilling those requests, have been developed. The RPC standard is a system for defining the parameters of a remote communication.

Remote Procedure Call Model RPC Model. Based on concepts of remote procedure calls (RPCs) in which the caller and called procedure are typically on physically separate systems exchanging data through a communications link, the RPC Model operates as a control mechanism functioning through the caller's process and the server's process. An RPC interface provides a set of remotely callable operations provided by the server and provides a means to manage server resources made available to the remote caller.

As an example of one implementation, the caller sends a call message with parameters to the server process and holds for a reply. When a reply message is received, procedure results are extracted from the data and the caller resumes execution of local applications (or makes further calls). While the model does not exclude the possibility of concurrent processes on the part of the client and server, it is sometimes implemented with only one being active at a time.

remote programming The capability whereby a system can be programmed from a remote location, either through a data network or phone lines, usually after input of appropriate authorization codes. Remote programming enables a field worker or telecommuter to administer a system without being physically present. In computing, remote programming is often done by BBS operators who want to check and manage their systems when out of town. By dialing their own BBSs and logging in as the Sysop, they can validate new users, check mail, configure the bulletin board, and accomplish various maintenance tasks through a phone connection.

Remote programming is often implemented in high-end corporate and industrial software programs. The software is set up with security mechanisms so an authorized programmer working for the software vendor can dial into the customer's machine and do routine maintenance, software tune-ups, diagnostics, and configuration without having to travel to the customer's site, and during nonbusiness hours, if needed. This type of service is usually provided through a separate service contract for a specified period, or is billed hourly, as needed.

remote site A facilities or equipment location which is distant from the one presently occupied, or from which certain maintenance or administration tasks may be carried out. A sales representative with a laptop, a scientist with an intercom radio doing field research, and a computer terminal in an annex building are remote sites often directly or indirectly communicating with or through a main system at another location. Remotes sites may be fixed or mobile.

remote switching center RSC. RSC has three generally accepted meanings: (1) a secondary telecommunications switch facility located at some distance from a main facility (at a remote location), (2) a switching center that is designed to serve remote subscribers who are often hundreds of miles away, (3) a switching center in a central or remote location that is remotely managed and controlled through telecommunications links from another center intended for that purpose.

The first two meanings are traditional meanings, established before sophisticated software and digital links were available, and are still relevant. The third meaning is becoming more prevalent as telecommunications networks such as the Internet make it practical to configure, manage, log, and maintain switching services from a (remote) location offsite from the actual switching facility. Spelling the third as *remote-switching center* helps prevent ambiguity with traditional meanings.

Remote Telescope Markup Language RTML. Just as HTML is a markup language for representing information and handling user requests on the Web, RTML is a markup language for handling user requests for astronomical observations. It is a specialized request mechanism based upon XML.

RTML was initiated and has been adopted by the Hands-On Universe (HOU) Project as a proposed standard for users to send requests to observatory remotely controlled telescopes. RTML 2.0 was released by F. Hessman in May 2001. This version had changes and additions to RTML 1.1 and subsequently was refined by Denny and Downey. Denny then concurrently developed XRTML as an interim format; this came to be called RTML 2.0 while Hessman's fuller implementation, called RTML 2.0b, was renamed RTML 3.0. Objects were removed from RTML 2.0 to create RTML 2.1, which is specifically a request mechanism, with no support for the return of data. RTML 3.0, (released as 2.0b in July 2001) fully specifies a two-way communications protocol for submitting, updating, and acknowledging astronomical data requests, and for returning status information and data related to these requests. See Robotic Telescope Markup Language for the events leading up to RTML.

remote terminal A local computer terminal that enables remote access to services or accounts not available on the local terminal. Thus, a remote terminal in a branch office could be used to communicate with a centralized databank in the main office, for example. Remote terminals may be wired or wireless, or may be connected through a combination of wireline and wireless services (e.g., from the terminal to a wireless access concentrator to a satellite link to the main office).

Remote terminals are often classified as *dumb* or *smart* terminals. In general, dumb terminals are those which have limited capabilities and may have few applications that can be executed locally. Dumb terminals are used in situations where cost or security are concerns. Smart terminals are those that have significant functionality on their own, even if not connected remotely to another system.

Remote User Telnet Service A means by which a specific service of User Telnet may be accessed by opening a port connection (107), a mechanism suggested by M. Mulligan and submitted as an RFC in 1982 by Jonathan Postel. This enables remote access to another system for logging in and executing commands. It is now considered a historic service. See RFC 764, RFC 818.

remotely piloted vehicle RPV. A land, water, air, or space-based vehicle piloted from a remote location, usually through radio control signals (though other methods are theoretically possible). Military drones, children's toys, and robots may be controlled this way.

removable media Storage cartridges, drives, or diskettes which can be swapped out and replaced with another. This provides a less expensive, portable option to numerous fixed storage devices. In the mid-1990s various cartridge drives became very popular, as it was possible to store from 100 to 1000 Mbytes on a cartridge not much bigger than a floppy. The problem was that every cartridge drive had a different format and the formats were not intercompatible. More recently, super diskettes have been introduced, which use normal floppy-sized disks that can store 100 MBytes, but the drives are still downwardly compatible with 1.44 floppies, so that it's not necessary to have several devices attached to the computer. It is not clear, as of this writing, which of these technologies will prevail or whether another new one will leapfrog them before one or the other is firmly established.

REN See Ringer Equivalence Number.

REO removable erasable optical.

repeat dialing This is both a function of some phones and telephone/computer software programs, and a service of some phone companies in which a number found to be busy can be repeatedly dialed until the connection goes through, without the user dialing the number again. Repeat dialing is very commonly used in telecommunications software programs to dial up BBS numbers that are frequently busy.

repeater A device for receiving signals and retransmitting those signals in order to propagate or amplify the signal. Repeaters are commonly used in technologies with signal attenuation and fade. Repeaters are used in digital and in analog systems and are often spaced at intervals over paths that cover long-distances. In digital systems, it is possible to reconstruct the informational content of the signal; in analog systems, more often the signal is amplified, which means that accumulated noise and degradation are still limiting factors. Radio broadcast repeaters and microwave repeaters are examples of common implementations.

In networks, a number of devices assist in the conversion and propagation of signals (bridges, routers, etc.). A repeater is simpler than most of these devices, serving only to continue the signal and extend its range, or to clean it in its most basic sense, rather than to change the informational content of the data. See amplifier, bridge, doping, regenerative repeater.

Repeatered Submarine Fiber Optic Cable Systems RSFOCS. A market study on underwater fiber optic cable systems published spring 1999 by IGIC as an update to a previous study.

reperforator An instrument that translates received signals into a geometrically coded series of locations that are punched or otherwise impressed onto a paper tape. Early telegraph systems and most of the early computing devices used reperforators. These were then read with optical or tactile sensing devices to turn the code back into human-readable form. See chad.

reperforator/transmitter RT. A teletypewriter device which includes both a reperforator for punching received codes on paper tape, and a tape transmitting unit for sending the codes to a tape punching mechanism.

repetitive pattern suppression RPS. A means of data optimization which compresses digital communications by removing repetitive patterns and reproducing them at the receiving end.

replication A process commonly used on computer systems for security, redundancy, distributed access, or other backups. In large companies and on the Internet, whole file archives are often replicated or mirrored in order to provide access at a reasonable speed to a larger number of uses. Replication of data is a means to protect data in case of a serious problem with a system or storage device. RAID systems are a means of replicating data to preserve data in case of a fault. Replication in the form of regular backups is recommended for all important computer data that need to be preserved. In radio communications, transmissions are sometimes repeated to improve the chances of a message getting through. The replication may be of individual small units of transmission or may be repetition of a short message or signal. See ALOHA.

reprography Copying and replicating.

REQ request.

Request for Comments RFC. A significantly influential, formatted, open communications forum for technical experts which accepts, edits, numbers, publishes, and disseminates Internet-related documents including protocols, draft and official standards, notices, opinions, and research. Known as RFCs, these electronic documents form a body of more than 2200 contributions that provide a remarkable overview of the evolution of the Net, its structure, functioning, and philosophies.

There are a number of categories of RFCs. Some of them are tiered and cannot be submitted without passing through previous categories in a specified order, with specified waiting periods for comments and revisions from the RFC community.

Code	Category
(no code)	Unclassified
I	Informational
H	Historical
E	Experimental
PS	Proposed Standard
DS	Draft Standard
S	Standard

RFCs are not changed once they have been submitted, assigned a number, and distributed. Any changes regarding an RFC must be submitted as a new RFC. There are many excellent RFC repositories on the Internet, with good indexes and abstracts. Anyone can submit an RFC provided it is topical and follows the official format and procedures. This dictionary includes references to specific RFC numbers where the author felt the technical origin of the information would be of interest, and there is an overview of significant RFCs in the Appendix. For information on submitting RFCs, see RFC 1543.

Request for Discussion RFD. Similar to a Request for Comments (RFC) in that it is a means on the Internet to solicit and generate discussion on a specified topic. However, it focuses on intercommunication on a topic in a slightly less formal or definitive manner as opposed to specifying or defining a topic once certain conclusions or draft/final working models have been developed, as in RFCs. RFDs often precede RFCs but do not necessarily result in RFCs. See Request for Comments.

Request for Information RFI. A solicitation and notification of interest in receiving feedback or information on a specified topic, product, or process, without implying that the requester necessarily wishes to purchase or use that for which the information is solicited. See Request for Comments, Request for Discussion.

Request for Proposal RFP. A call for proposals for solving a problem or participating in a project. A Request for Proposal process is used by many institutions to initiate a project such as building a facility, setting up a new organization, or developing a new product or service. People who submit RFPs usually have a community or economic interest in shaping or otherwise participating in the project. A Request for Proposal may or may not be accompanied by a Request for Quote. See Request for Quote.

Request for Qute RFQ. A call for monetary quotations for a particular project or venture. The RFQ process is commonly used by institutions to assess various aspects of implementing a project and what that project will cost. Vendor input in the form of

RFQs not only provides information on the price, but may also give some idea of the timeline and type of materials involved. It also serves as a form of preliminary contract (and is sometimes the only contract) once a vendor has been selected to undertake a job. An example of a simple RFQ would be a faxed page explaining that the cost of purchasing and installing a new modem might be $200. An example of a more complex RFQ would be a 12-page document detailing the costs involved in setting up a video security system in a business office or factory that can be monitored and controlled (through pan and tilt controls) from a remote location through the Internet. A Request for Quote may accompany a Request for Proposal or may sometimes be combined with a Request for Proposal. See Request for Proposal.

reroute To make a temporary or permanent change in a data path. Rerouting frequently occurs in large, dynamic networks like the Internet. Systems where rerouting is common usually use a hop-by-hop method of routing in order to accommodate changes and to create new paths as needed. In electronics, rerouting of a circuit may be accomplished by a patch or shunt, a wire which bypasses the original path. On Fiber Distributed Data Interface (FDDI) networks using dual rings, rerouting to the second, backup ring is carried out if a problem is detected on the primary ring.

RES 1. regional earth station. 2. See Residential Enhanced Service.

Resale and Shared Use decision A decision of the Federal Communications Commission (FCC) to allow competition in value-added networks.

resampling The process of subsequent sampling of data, as image or sound data, and re-encoding it. Resampling usually occurs when the original sample was not of the resolution level or compression rate desired. Resampling may also occur in order to update or refresh information that may be changing, as in Internet videocam shots, videoconferencing, etc. See sampling.

research and development R&D. The study and associated development of theories, sciences, and technologies. In communications, much R&D focuses on creating marketable products. R&D often consumes a huge proportion of a startup business's budget. Many large corporations have R&D facilities to enable them to develop new products, including IBM, Lucent Technologies (Bell), and Xerox.

In many cases, R&D does not directly result in products that can be manufactured and distributed, but the process of R&D indirectly contributes in surprising ways or, sometimes, decades later, when manufacturing processes or needs change or catch up to the theories. As an example, Charles Babbage's historic analytical engine (a pioneer computer concept) research and design was apparently theoretically sound, but the technology to build his machines did not exist during his lifetime. As a more recent example, space research has resulted in many portable and wireless communications technologies that might not have developed otherwise.

Research and Education R&E. Those served by the National Science Foundation's (NSF) networking efforts.

Research in Advanced Communications in Europe RACE, RACE II. Also known as Research and Development in Advanced Communications Technologies in Europe and R&D in Advanced Communications Technologies in Europe, this program began with work by the RACE Management Committee that was tendered to the European Community Council and became the predecessor to the later Advanced Communications Technologies and Services (ACTS). RACE was organized in the mid-1980s to aid in the EC-wide establishment and unification of communications systems and providers. Following a definition stage, the RACE projects were carried out between 1987 and 1995 with financial contributions from the EC. Over the period of its tenure, more than 350 organizations participated in RACE projects and RACE collaborated with many important organizations, including ETSI, EFTA, EUREKA, ESPRIT, and others.

In the early and mid-1990s, RACE was involved in the research and development of telecommunications technologies leading to the development of the Integrated Broadband Communications (IBC) system and more specifically the Mobile Broadband System (MBS). With its evolution to ACTS, the third phase of IBC development continued the work beyond the original 5-year mandate of the RACE projects. See Integrated Broadband Communications, Mobile Broadband System.

Research Institute for Advanced Computer Science RIACS. A research facility located at the NASA/Ames Research Center in California. In conjunction with the NASA/Ames center, RIACs sponsors a Summer Student Research Program (SSRP) to provide students with the opportunity to team with research professionals in Moffet Field. Topics include automated reasoning, high-performance computing and networking, and applied research for NASA missions.

Research Laboratory of Electronics RLE. The first interdisciplinary research lab established at the Massachusetts Institute of Technology (MIT). The RLE, founded in 1946, is descended from the MIT Radiation Laboratory, which provided some significant discoveries during World War II. RLE brings together expertise from many quarters, including electrical engineering, physics, computer science, chemistry, aeronautics, linguistics, and others. The facility is primarily supported by the U.S. Department of Defense (DoD), the Department of Energy (DOE), and several national institutes and foundations (NASA, NSF).

Réseaux Associés pour la Recherche Européenne RARE. A European research network that was merged with EARN in 1994 to form TERENA. A number of technical reports were published by the RARE Working Group prior to the merger. See TERENA.

reseller A company which purchases a block of services (e.g., long-distance services) or numbers (e.g., cellular phone numbers) and resells them directly to customers. The reseller does not own and usually does not maintain the physical structures or services, but usually provides information and technical support to the resale customers. See agent, aggregator.

reset *v.* Return to the previous or default or factory operating settings. In computer operating systems, a reset command usually restores the system to the same configuration (or nearly so) as a fresh power-up, although some will restore memory, parameters, etc. without resetting some of the basic subsystems (e.g., sound). A software reset that doesn't involve a power-down is termed a "warm boot" and a complete power-down is called a "cold boot." If a software reset doesn't seem to correct problems, it's best to do a cold boot. But wait at least 15 seconds for the power to drain from the chips before powering up again.

In timing mechanisms or data counters, a reset usually restores the time or counting mechanism to a zero condition or industry default.

Residential Broadband RBB. A broad term for higher capacity/higher speed broadcast and networking services to residences. Larger providers such as NTT and AT&T are already engaged in upgrading cables and services to provider a greater range of programming choices to home subscribers. New switching technologies are an important technological and economic component of these upgraded services. See fiber to the home.

Residential Broadband working group RBWG. A working group of the ATM Forum Technical Committee which promotes a single set of global specifications to maximize interoperability of products from various vendors. The Residential Broadband group provides documents and recommendations regarding enhanced services to residential users of ATM-related technologies.

Residential Enhanced Service RES. A telephone subscriber service package offered by Nortel that includes enhancements such as increased security and features such as Call Redirect.

resinous electricity A historic term coined by Dufay to denote the type of electrostatic charge produced on sealing wax when rubbed with flannel (or amber rubbed with wool). Later, in the 1700s, Benjamin Franklin proposed *negative*, a term that superseded vitreous. See electrostatic, static electricity, vitreous electricity.

resistance 1. Opposition, counteracting force or retarding force against. 2. In electricity, opposition to the flow of current, usually expressed in ohms. There is great variability in the resistance of various materials. Those with low resistance, such as silver and copper, make good conductors. Resistance in a particular material may change with temperature, moisture, or the presence or absence of current. Resistance is defined as the reciprocal of conductance. See reluctance, resistor, ohm. Contrast with conductance.

resistor A component or system which provides resistance to electrical current. Some materials are naturally resistant, and this property may be further exploited by the way a circuit is configured (e.g., longer wires, more loops, etc.). Controlled current is useful in a number of circumstances and can be used to protect sensitive components or to provide operational control. Electronic resistors use standard color coding schemes to identify the degree of resistance they provide. Applying Ohm's law, it can be stated that the combined resistance of any two resistors connected in parallel can be expressed by dividing their product by their sum. See Ohm's law, resistance.

resolution Resolution is somewhat technology-specific, since it is often based not only on the size or discrete value of an individual unit, but also on the total area occupied by a block or line of units, and the units will vary depending upon the type of medium or technology described.

resonance 1. The enrichment of a sound by supplementary vibration. For example, the body of an acoustical stringed instrument is designed to increase the resonance of the string vibration by transmitting the sound through the bridge, the sound holes, and the body. 2. A greater amplitude vibration arising from smaller periodic vibrations with the same, or nearly the same, period as the natural vibration period of the system. This can arise in both electrical and mechanical systems. 3. The enhancement of an event by creating excitation within the system, as in particle reactions. See magnetic resonance imaging. 4. In a circuit, a balanced condition between inductive- and capacitive-reactance components.

resonance curve A diagrammatic representation of the relationship of various frequencies at or near resonance to a tuned circuit.

resonance Raman scattering See Raman scattering.

resonant frequency The frequency at which a maximum amplitude response occurs in a given object or system when acted upon by a constant amplitude sinusoidal force. In an antenna, for example, the transmitting current is greatest when the impedance level is lowest at a given frequency and power.

resonator 1. A device which increases and/or directs sound, as in a musical instrument, music box, or telegraphic sounder. In gramophones, a horn which amplifies and directs sound to the listener. 2. In telegraphic systems, a box which holds a sounder and directs the audible clicks to the ear of the operator. 3. In microwave communications, a hollow, metallic container in which microwaves are produced and amplified. 4. In crystal detectors, a piezoelectric crystal which oscillates when stimulated by radio waves.

resonator, optical A mechanism to reflect and amplify radiant energy in the optical frequencies through resonance patterns set up among the interacting reflected waves. An infrared resonating chamber was described in a 1956 patent application by R. Dicke. Optical resonators were described more fully by G. Fox and T. Li and by G. Boyd and J.P. Gordon, in 1961. Based upon a Fabry-Perot interferometer as a resonating mechanism for optical masers (pioneer lasers), Fox and Li described a resonating system in which the reflecting mirrors of the interferometers reflected the optical beam back and forth until, after many reflections, the relative field distribution was "synched" and the amplitude decayed exponentially. The difference between the oscillating maser and earlier masers was that the lowest-order mode could be made to dominate given the correct spacing, relative to frequency, for the resonance.

Resonance chambers or external Littman-Metcalf resonating cavities are now intrinsic to traditional lasers, laser diodes, and other electromagnetic devices that benefit from signal amplification.

resonator, unstable A resonating mechanism in which the relationship of the radiant energy and the chamber is such that the signal becomes unstable, that is, it progressively moves off the primary axis between the reflecting surfaces. This may be due to the character of the radiant energy source, fluctuating or incorrect power levels, the relationship of the reflecting mirrors to one another and/or to the beam, or flaws in the general assembly or components. It may also be a deliberate compromise between fabrication difficulty or expense and the level of signal needed to accomplish a particular task. (An unstable signal will eventually escape through the reflective surfaces or through a hole in the resonating cavity.)

In most cases, instability is an undesired quality. However, unstable resonation also has some particular characteristics that were described in the 1960s that may be exploited for certain purposes. By the early 1980s, unstable resonation was being incorporated into specialized commercial solid state lasers. Later, hybrid stable/unstable systems began to appear. The use of deformable mirrors in solid state hybrid lasers makes it possible to decrease diffraction influences on the focused beam.

More recently the diffraction properties of unstable lasers were found to be the source of surprising fractals as the resonating cavity spilled out around reflecting surfaces and formed aesthetically interesting interference patterns. See Q factor, Q-factor.

Resource Location Protocol RLP. A network protocol to assist in the location of a network resource in situations where other methods may not be practical. This was submitted as an RFC by M. Accetta in 1983. RLP uses a request/reply procedure to request an IP number or appropriate identifier. As an example, RLP can name a resource by an assigned protocol number and a variable-length protocol/resource-specific identifier. See RFC 887.

Resource Management RM. In an ATM system, cells contain information for managing bandwidth, buffers, and flow control aspects such as loads, traffic congestion, etc. These resource management (RM) cells are thus associated with the administration of the data transmissions. They are passed along the path through various switches to monitor and control congestion by adjusting the various cell rates (current, explicit, minimum) as needed. See cell rate.

response time 1. The time it takes to react appropriately to a given situation or signal. In business, response time to customer inquiries is often critical in making sales and getting repeat business. The response time to a phone call usually needs to be four rings or

R

less in a business environment, or eight rings or less in a residential environment, or the caller may give up and terminate the call before it is answered. In computer operations, response time of an input device, or the software, is important in terms of productivity and user satisfaction. 2. A command called *ping* can be used in network testing and management to determine a response time, or whether a host is even available to respond. See ping.

restart 1. Initiate again, begin again. Put back into service or operation, to power up again. 2. In computer operating systems, to reinstate operations without having to power down and power up the entire system. Many computers have a key sequence, menu selection, or restart button which will reinitialize the operating parameters without a full power up sequence. In some systems (e.g., Macintosh), the restart option will also do a clean shutdown of the system in order to ensure that files and applications are closed, and important processes finished so no data corruption occurs from a sudden shutdown. See reboot.

restore 1. To put back to its previous or original state, to renew, revive, to return to good operating order. 2. To return to the original position, location, or owner, to re-establish, to reinstitute service. 3. To get back data which have been erased or damaged, either by rebuilding or relinking pointers, file tables, and directory tables on the storage medium, or by accessing a backup archive of the data as they were last saved. 4. In programming, to return the value of a variable to a previous value, which may be a default or original value. 5. To recharge or refresh information in a memory circuit with continuous or periodic current. Lithium batteries are sometimes used for this purpose. 6. A gadget on the edge of the windows of most operating systems with graphical user interfaces (GUIs) which allows the window to be automatically sized to its original size without the user having to remember the setting or do it manually.

Restricted Numeric Exchange See RNX.

Restricted Radiotelephone Operator Permit RP. A fee-based permit granted without examination by the Federal Communications Commission (FCC) for the operation of most aeronautical and ground radiotelephone stations aboard pleasure craft (with some restrictions). The permit also covers the operation, repair, and maintenance of AM, FM, television, or international broadcast stations. See Marine Radio Operator Permit.

Reticular Agent Definition Language RADL. An object-oriented language distributed by Reticular Systems, Inc. to facilitate the development of intelligent software agents. RADL is an extension upon work by Shoham (AGENT-0) and Thomas (PLACA – Planning Communicating Agents). Reticular Systems also provides AgentBuilder, graphical tools for developing RADL programs. AgentBuilder can provide support for Knowledge Query and Manipulation Language (KQML), which can be used to share knowledge among multiple intelligent systems. AgentBuilder Enterprise supports realtime examina-

tion of remote agent operation, a Role Editor for defining agent roles, mobile agent support, CORBA/DCOM support, and an enhanced repository manager. See Knowledge Query and Manipulation Language.

Reticule Examples

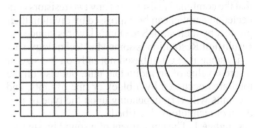

Reticules are grids, scales, or rules intended to facilitate testing, calibration, assessment, or counting, especially for microscope assemblies. They are typically imaged on transparent glass or plastic so they can be readily superimposed over an image and may include numeric references. In microscopes, they are usually made of glass and placed close to the image with which they are compared.

---

reticule, reticle An arrangement of rules, grid lines, dots, or wires serving as reference marks. Most often, reticules are printed on transparent materials so that they may be superimposed over a field of view but it is the nature of the marks and their use as a reference more than the fabrication materials that constitutes the reticule. It may also be imprinted on opaque materials. A reticule facilitates measuring, counting, and aligning an instrument, or viewed specimen.

While most reticules are placed parallel to the viewing plane and perpendicular to the viewing angle, a *tilt* reticule may aid in assessing angular deviations from a reference plane.

In microscopy, a reticule may be used for calibration and counting, or a micrometer may be used for calibration, followed by a reticule for counting. A reticule may be incorporated into an eyepiece (e.g., on the field stop of a microscope) or may be used in a work or staging area (e.g., over a microscope slide). While many reticules are preprinted, some are created dynamically as specimens are observed with only simple registration marks preprinted for alignment (usually on the outer perimeter). In telescopy, this type of reticule is handy as a reference for charting star movements. Star positions can be recorded on separate telescopes and separate reticules over time and assembled three-dimensionally to provide a map of the universe as seen from a number of viewing angles. This type of observation has led to a better understanding of the "shape" of space and the discovery of new celestial bodies.

Reticules are also used in autocollimators – devices that align a radiant energy source (e.g., a laser light source for a fiber optic lightguide) by projecting an image of a reticule that is reflected back to provide a basis for comparison. If there are differences between the original reticule image and the returned image that are outside of acceptable working parameters, the device makes adjustments as necessary or signals that there is a problem.

The name may derive from the word used to describe a woman's small mesh or ornately beaded drawstring purse. The term is used more often in associated with assessing real objects as viewed through scientific instruments as opposed to viewing video, photographic, or cartographic images, in which case the term *graticule* is more often used. See collimator, graticule.

RETMA Radio Electronics Television Manufacturers Association. See Electronic Industries Alliance.

retransmission consent In television cable broadcasting, consent is a local TV station's right to negotiate fees associated with program provision. It is common for a local station to purchase a variety of television programming from various distributors and broadcasters. They then resell these programs in various packages to local cable subscribers. See cable access, Cable Act.

retrofit *n.* To equip a device or system with new parts or capabilities that were not available, or perhaps not requested, at the time of initial purchase and installation. For example, it is very common for computer parts retailers to offer accelerator cards, storage device controllers, faster CPUs, and other enhancements to users trying to upgrade or extend the life of their systems.

retry Another attempt to perform the same operation if the previous attempt failed. This is an intrinsic part of most computing processes. If a process fails to read or write data from or to a storage location, it will retry a certain number of times before alerting the user that there is a problem. If a modem fails to connect, it may redial a specified number of times before signaling an error condition. If a network mail server sends email to a recipient who can't be found, it may try several times before bouncing the email back to the sender. For many operating systems, the retry parameters are set by programmers and are transparent to the user and not changeable. For individual applications programs, there may be user preferences for retries for various operations to be controlled.

return CR. A common designation for the carriage return key on a keyboard (sometimes called the *enter* key) or carriage return function in a program such as a word processing program. On most systems, the carriage return incorporates not just a *return* (which returns the cursor to the far left or right of the screen, depending upon the language), but also a *new line* (which drops the cursor down to the next line).

return authorization, return merchandise authorization RA, RMA. Permission from a manufacturer or vendor for a consumer or dealer to return a product, usually because it is defective, damaged, or does not meet the advertised specifications. Few vendors will process a return without prior authorization. Authorization is usually identified with a return authorization (RA) number. The RA is used for internal database tracking and inventory control.

return authorization number RAN. A number assigned to products returned to the vendor or manufacturer. The number documents that the return was authorized and provides a number for tracking and stock-handling purposes once the product has been received at the return point.

return loss A measure of the ratio of incoming to outgoing power, usually expressed in decibels, at a specified reference point. Return loss is a diagnostic means of evaluating various factors such as loss, quality, echo, etc.

return material authorization RMA. See return authorization.

REV reverse.

Reverse Address Resolution Protocol RARP. A client/server tool for reporting an Internet Protocol (IP) address to its client. The protocol is intended for workstations to dynamically determine protocol addresses when only the hardware address is known. The Ethernet address is mapped to the logical IP address. Used in TCP/IP. See RFC 903.

reverse bias See back bias.

reverse engineering A process of working backward from a finished product or process to determine the steps taken to construct it. Thus, a clock can be taken apart to see what makes it tick, and a software program can be taken apart to see what algorithms and conditional relationships were used in its construction. Software developers are generally nervous about having their software reverse engineered, because it is difficult to prove reverse engineering in cases of theft of intellectual property.

reversing error, reversal error A condition in which current divided through two circuits, such as through a component and a measuring instrument, will vary due to the deflection of the measuring instrument for the same current passing in both directions. See shunting error.

revision control system RCS. A system for managing, queuing, and logging revisions. This is important when designing a device or protocol, or creating a document (e.g., a software manual) and is especially relevant for quality assurance and projects with more than one contributor. There are both manual and computerized revision control systems. A simple example is a bulletin board posting to inform the people involved in a revision process.

Revision control is increasingly being built into software systems in factories and offices. There are automation systems that will keep track of changes in a manufacturing or packaging process to prevent errors and provide an audit trail. There are word processors that will log changes by individual editors with symbols or colors, as well as automatically date/time-stamp the changes, so that the process is documented and the changes can be traced backward, or the individual responsible for particular changes can be identified. Software designed for workgroups often has revision control features that prevent different contributors from changing a document at the same time and may notify members of a workgroup whenever new revisions occur.

RF 1. See radio frequency. 2. range finder. A device for determining distance (and sometimes also direction). See radar. 3. raster file. See raster. 4. rating factor.

RFA 1. See remote file access. 2. request for action.

RFC 1. See Request for Comments. 2. required for compliance.

RFD See Request for Discussion.

RFF raster file format. See raster.

RFI 1. Radio France International. 2. radio frequency interference. Electrical noise resulting from some wire or attachment acting as an antenna. 3. See Request for Information.

RFP See Request for Proposal.

RFQ See Request for Quote.

RFS 1. radio frequency shift. Unintentional drift or intentional shift (change) of a broadcasted radio frequency. This may be done for security purposes or to maximize the availability of bandwidth. See frequency hopping. 2. radio frequency system. 3. range finding system. 4. remote file sharing/system.

RG-58U A type of thin-wire cable used in 10Base-2 data communications cabling installations.

RGB red, green, blue. An abbreviation to describe the three primary colors of an additive (light-based) color model. This is the model used in most computer video display devices. The colors in the spectrum are created by systematically manipulating the amount of red, green, or blue that can be individually stimulated. In many computer software application palettes, the absence of any R, G, or B color is designated as 0 and the full intensity (of a transistor or cathode-ray tube gun) is designated as 255, with other colors created by mixing intermediate values. Thus, purple is created by mixing red and blue and white is created by displaying all three colors at full intensity.

RGP raster graphics processor. See raster.

RGS radio guidance system.

Rhealstone A type of computer processing system benchmark used in realtime multitasking systems. Run times for a set of operations (task switching, interrupt latency, etc.) are independently measured. See benchmark, Whetstone, Dhrystone.

rheostat A device with one fixed terminal and a movable contact, used to regulate a current by varying the resistances. Similar to a potentiometer, except that a potentiometer can connect to both ends of the resistance-varying element. See potentiometer.

Rhumbatron Oscillator An apparent misspelling widely disseminated on the Internet for Russell Varian's *Rumbatron Oscillator*, mentioned here to help the reader find the cross-reference. In Varian's original handwritten note of July 1937, he called it a *Rumbatron Oscillator* (at least in this instance). It's an important forerunner to the linear accelerator. See rumbatron, Klystron.

Riad computer A type of IBM System/360 series-compatible computer developed in Russia.

RI 1. radio interference. See attenuation, jitter, noise. 2. Rockwell International.

RIAA See Recording Industry Association of America.

RIACS See Research Institute for Advanced Computer Science.

RIAS See Research Institute for Advanced Studies.

RIB 1. See RenderMAN Interface Bytestream. 2. See routing information base.

ribbon cable A cable design in which the wires are encased so they are aligned side by side, in close proximity with insulating material separating the wires and holding the whole structure together like a long strip of ribbon. Typical ribbon cables carry between 9 and 60 wires. Ribbon cables are commonly used for parallel wiring connections, disk drives, and other data transfer applications, although some types of ribbon cables carry AC power. Ribbon cables are somewhat more fragile than other types of more heavily shielded cables and kinks in the cables can snap the conductors, if they are very fine. Wide ribbon cables can be difficult to attach to other types of connectors. Nevertheless, they are convenient to use in a number of places (inside computers and under floor coverings) because they are flat and flexible.

Flat "Ribbon" Cables

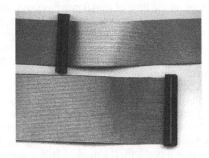

Examples of two ribbon cables: the top 34 pins, the bottom 50 pins. The flat flexible cables are commonly used inside computers where space is limited and narrow. Several devices, such as hard drives and CD-ROM drives, can be daisy-chained on one length of cable by inserting connectors that pierce the cables and make contact with the wires.

RIC regional information center.

Rice computer The Rice Computer Project was inspired by the MANIAC II at Los Alamos. Three professors, Zevi Salsburg, John Kilpatrick, and Larry Biedenham, initiated the project, which culminated in the development of a computer for research, and research into the development of computers. Joe Bighorse, as head technician, implemented most of the hardware design. The computer came online in 1959 and was fully functional by 1961. For almost a decade it was the primary computing machine on the Rice campus. Architecturally, the Rice 1 (R1) was descended from the Brookhaven computer and the MANIAC II. It was essentially a tagged-architecture computer, using 54-bit vacuum tubes plus two tag bits and seven error-correcting bits. It implemented indirect addressing capabilities and memory was stored in cathode-ray tubes (CRTs). After 1963, transistor-based logic was added.

Rich Text Format RTF. Also called Interchange Format. A document encoding file format developed by Microsoft that retains simple text formatting codes and basic text attributes (typestyle, size, **bold**, *italic*, underline, etc.). A widespread format that can be exported and imported between word processing, OCR, and desktop publishing programs. It is a good intermediary format to use when moving text with attributes from one system to another. Many people use ASCII to export/import text, and despair because the formatting is lost. You may wish to try RTF. Although there are different flavors of RTF, it usually works and can save hours of reformatting. Here is a very basic example of Rich Text Format showing the syntax for various formatting parameters.

```
{\rtf0\ansi
{\fonttbl\f0\fswiss Helvetica;}
\paperw9880
\paperh3440
\margl120
\margr120
\pard\tx20\tx7120\f0\b\i0\ulnone\fs20\fc0\cf0
Rich Text Format
\b0   RTF. Also called Interchange
Format. A document encoding file
format developed by Microsoft that
retains simple text formatting
codes, and basic text attributes
(typestyle, size,
\b bold
\b0 ,
\i italic
\i0 ,
\ul underline
\ulnone , etc.). A widespread format
that can be exported and imported
between most word processing, OCR,
and desktop publishing programs. It
is a good intermediary format to
use when moving text with attributes
from one system to another. Many
people use ASCII to export/import
text, and despair because the
formatting is lost. You may wish
to try RTF. Although there are
different flavors of RTF, it usually
works, and can save hours of
reformatting. Here is a very basic
example of Rich Text Format, showing
the syntax for various formatting
parameters.\
}
```

Rich Text Markup Language RTML. A proprietary programming language used by the Yahoo! Store to generate commerce-enabled Web sites. Yahoo! Store third-party developers provide services for automated Web site creation through their RTML tools. RTML is a combination HTML editor and commerce-specific components development tool for adding the types of features that businesses like to include in their Web sites to support product promotion and sales.

RIF See rate increase factor.

RIFF 1. raster image file format. See raster. 2. Resource Interchange File Format. A platform-independent multimedia specification developed in the early 1990s by a group of vendors including Microsoft Corporation.

RIG related interest group. See Birds of a Feather.

right-hand rule, Fleming's rule A handy memory aid widely used in mathematics and physics to determine an axis of rotation or direction of magnetic flow in a current. Extend the thumb and fingers of the right hand so that the fingers are held together and point straight in one direction, with the thumb at a right angle to the fingers, in an 'L' shape. Now curl the fingers around a conductive wire, so that the thumb points in the direction of the current. The direction of the curled fingers then indicates the direction of the magnetic field associated with the current. Using the same hand relationship, point the thumb in the direction of wire motion and the fingers will show the direction of the magnetic lines of force and the direction of the current, for a conductor in the armature of a *generator*. For the direction of current for a conductor in the armature in a *motor,* see left-hand rule.

Right-Hand Rule for Current Direction

The right-hand rule is a visual mnemonic device for remembering the relationship of the direction of the current in an electromagnetic field.

Rijndael A block cipher data securing system designed by Joan Daemen and Vincent Rijmen selected by the U.S. Government for its Advanced Encryption Standard (AES). Rijndael has variable block and key lengths and has been specified for keys of 128, 192, or 256 bits in length. It can be implemented on a variety of processors. Rijndael was based, in part, upon concepts developed by Daemen and Rijmen for the Square block cipher. See RC6, Square.

RIN See relative intensity noise.

ring 1. The sound made by a phone or other communications device to indicate an incoming call or imminent announcement (as on a PA system). The frequency of the tone and its cadence vary from country to country. In North America, phones typically ring once every 6 seconds. 2. Traditionally the red wire in a two-wire telephone circuit. The name originates from the configuration of a manual phone jack in an old telephone switchboard in which the large plug was divided into two sections, with an internal wire electrically connected to the *tip* of the plug and another wire to the *ring* around the plug partway up the jack, nearer the insulated cord. The ring is traditionally around –48 volts, with the negative charge used to help to prevent corrosion. See tip, tip and ring.

ring error monitor REM. In a Token-Ring network, "tokens" are passed around from machine to machine in a ring topology. A ring error monitor is a program that can be installed on machines in the ring to collect and signal ring-related errors in the form of soft errors and hard errors (MAC frames). Ring-related errors may include lost frames, tokens, or Frame Check Sequence (FCS) errors.

ring-armature receiver A telephone receiver which began to be widely installed in the 1950s following the development and widespread use of the bipolar receiver. It differed substantially from earlier receivers in its details, incorporating a lighter, more efficient dome-shaped diaphragm, driven piston-like by magnetic fields across the armature ring. See bipolar receiver.

ring-around-the-rosy A theoretical hazard situation on the phone system, in which a circular tandem connection exists, somewhat like an endless loop in a software routine.

ring, network A type of circular network topology. See Token-Ring.

Ring Again A telephone service that enables a caller to request notification if a busy number becomes available. When the number can be dialed, the caller's phone will ring and the call will be automatically placed on the caller's behalf, without the need to dial a second time, when the phone is picked up/activated. In general, only one Ring Again request can be set up at a time. See Make Busy.

ring topology A network topology in which each station in the network is connected in a closed loop so that no termination is required. In this architecture, data packets are passed around the loop through each intervening node until they reach the destination machine. Ring topologies are fragile in the sense that failure of one node or system affects the entire network. See star topology, topology.

ringback A usually undocumented, self-ringing telephone test number that can be used to verify a specific phone number. In some regions, when a two- or three-digit number, followed by the expected phone number, is dialed, a new dial tone is provided, which can then be used for ringing tests. Ringback functions aid service technicians in verifying the functioning of a newly installed or repaired system. They are also sometimes used for subscriber confirmation. In some regions, the ringback service connects the caller to a recording identifying the exchange in which the phone is located. See ringing tone.

ringdown See ringing current.

ringer, bell The mechanical or digital sound generator that indicates the presence of an incoming call. The interval and type of tones generated vary from phone to phone, and even more so from country to country. With the proliferation of digital devices, it is likely that ringers will eventually be configurable.

You'll be able to load in a sound patch and have the phone sound like anything desired: a cat, a parrot, a jet airplane, your spouse's voice, etc. In fact, paired with a Caller ID system, there is no reason why the ringer couldn't switch to a device that says out loud, "Hank is calling, want to talk to him?" (This definition was originally written in 1997. Since 2001, several wireless phone vendors have begun providing downloadable ringers to their mobile subscribers.)

Ringer Equivalence Number REN. When telephone equipment is purchased and several devices are placed on the same line, there is a need for a way to designate and organize the ringer load on the line so as not to exceed the available current. In the United States, a certification number has been developed for telephone products which indicates that they meet certain specified requirements and guidelines. This REN helps installers and consumers set up various pieces of equipment (phones, answering machines, fax machines) so that they will not interfere with one another when connected.

Various devices on the same line may have different REN values. If a certain maximum number of RENs is exceeded, the various devices cannot be guaranteed to work correctly and may not ring. For example, a typical residential REN of 5 requires about 3.6 V of ringing generator power, a power requirement that may be exceeded if the subscriber has a main phone, a fax machine, and four or more extension lines.

Ringing generators for regenerating the local signal are commercially available to boost current to ring more devices than may be serviced by the public telephone system. For example, a high impedance 25-Hz output signal can power an additional 5 to 7 devices, depending on their ringing current needs.

In some cases ring generators may be daisy-chained. For central office ringing supply systems and emergency environments where ringing reliability is important, redundant ringing generators may be used as a failsafe mechanism.

All-digital systems often supply telephone services to environments with a variety of digital and analog equipment. In these instances, an ISDN terminal adapter may provide analog ringing capabilities to local devices through hook status circuitry and a compact on-site ringing generator.

For data modems, the Ringer Equivalence Number is usually printed on the main chip in the center of the internal modem board, along with the Federal Communications Commission (FCC) registration number. It may also be listed on a label on the back of the device to assist the user in combining devices on a system. In most regions, the sum of the RENs on one line should not exceed 5.0. In Canada, the concept of Load Number is essentially the same as the REN. See ringing classification, ringing signal.

ringing The production of an audible signal at the receiving end by means of a mild AC or DC current to indicate signaling or the presence of an incoming call. In telephony, a ringing current is sent from the central office to the subscriber or from a local console or branch to a local phone device.

ringing cadence The rhythm (timing) associated with ringing signals on a telephone circuit. The ringing cadence varies depending upon the telephone company and the country but typically is about 2 to 4 seconds of ring interspersed with 2 to 4 seconds of quiet. In historic systems, the ringing cadence was often supplied through rather picturesque rotating drum mechanisms, some of which were filled with mercury to close the circuit when the contacts reached the bottom of the drum.

ringing classification A category established by the Federal Communications Commission (FCC) to describe ringer types on telephone devices. Class A ringers respond to signals between 16 and 33 Hz; Class B ringers respond to signals between 17 and 68 Hz. Telephone devices are typically labeled as to ringer types. See Ringer Equivalence Number.

ringing current, ringdown The current on a telephone system used to transmit ringing signals and ringing tones. This varies with the type of switching system and the distance over which it has to be carried. In traditional phone signaling systems, for example, the ringing signal is fed through the line as a 75 V current to generate a 20 Hz AC signal current for local calls, and 135 to 1000 Hz AC current for long distance. See talk battery.

ringing generator A low-power inverter system for generating ringing signals on telephone lines usually located at a central office or a local controlled environment vault. It is typically powered by a -48V central office or -24V DC private branch exchange power supply to generate a harmonic 20 Hz AC signal for local calls (higher current is used for long-distance calls). The generator produces a low-frequency alternating current (AC) that travels through the line to ring the alert bell in a subscriber telephone. It may also provide timing pulses. Multiple ringing generators may be used for larger telephone exchanges or to provide a variety of ringing frequencies. Newer units may be selectable for a variety of frequencies typically ranging from about 17 to 50 Hz.

There is no point in supplying ringing current to a line that is unavailable (off-hook), so a telephone system generally does not send a ringing signal until it has determined that the called line is available. If the callee's phone rings successfully, the ring is passed back to the caller so the caller can monitor the number of rings. The ringing is stopped when the call is answered or the caller abandons the call.

ringing key In a telephone switching system, a means by which the subscriber's telephone ringing is initiated by a key at the central office to indicate a call. Signaling current was sent as alternating current from a central office to the subscriber, a system that didn't work well for long-distance calls due to the loss of the signal over distance.

ringing load The sum total, within a given time period, of subscriber telephones that may be ringing at any one time. An estimate of ringing load is important when a centralized telephone system is being installed, in order to determine the number and type of ringing generators needed to provide the ringing

signals. See Ringer Equivalence Number, ringing generator.

ringing signal Any signal transmitted over a telephone to initiate ringing on the receiving phone to indicate that a call is being placed. Various schemes for sending this signal have been used over the years to make it possible to send the ringing signal over distances without the current interfering with actual call transmissions. Typically a ringing signal is transmitted through an AC waveform at 20 Hz for local calls in North America (25 Hz in Europe), with higher voltages for long-distance calls.

The ringing signal does not always activate the ringer on a telephone set. Modems use ringing signal detection circuits for autoanswer operations for computer bulletin board systems and phones for hearing impaired individuals or specialized environments may use the ringing current to activate a light rather than a bell. See ringing generator, ringing key.

ringing tone, ringback A tone generated in the caller's line to indicate a call is being routed to a receiving phone. The ring at the receiving end is initiated by a ringing signal sent from the central office or private branch switching system. This doesn't absolutely guarantee that the callee's phone is ringing. If the bell is defective or the routing of the call was in error (either through incorrect switching or because of dialing the wrong number), the callee may never hear the ring. See ringback, ringing signal.

ringing voltage The amount of voltage applied in an analog telephone switching system to cause the called phone or other phone device to ring, usually about 88V. There is a limit to the number of devices which can be rung when attached to a single line, too many, and there will be interference with the ringing circuitry. For more information, see Ringer Equivalence Number.

RIP 1. See Routing Information Protocol. 2. See raster image processor. 3. See Remote Imaging Protocol.

RISC See Reduced Instruction Set Computing.

Ritchie, Dennis M. (1941–) An American Bell Laboratories researcher who codeveloped Unix in collaboration with Ken Thompson. He is well known for his development of the C programming language, along with Richard Kernighan. See C, Unix.

Ritchie, Foster An inventor and early assistant of Elisha Gray who designed a writing telegraph in 1900 based on principles different from those originally patented by Gray a few years before. Known generally as *telautographs,* these devices could transcribe handwriting across short distances and were in use for several decades.

Riza, Nabell Agha (1962–) A Pakistan-born optical engineer and inventor, Riza is responsible for several important developments in optical technologies. Riza studied at Cambridge University, the Illinois Institute of Technology, and CalTech. He subsequently worked at General Electric's Liquid Crystal Display Laboratory and, in 1995, joined CREOL as head of the Photonic Information Processing Systems Laboratory. In January 2001, he launched Nuonics, Inc.;

R

the same year, he was awarded the ICO Prize for discoveries and developments in the field of optics, many of which have become fundamental technologies in optical engineering imaging and transmission technologies. See acousto-optic antenna control, multiplexed optical scanner technology.

RJ The Universal Service Ordering Code (USOC) for Federal Communications Commission (FCC) – *registered jacks* for connecting to a public network. USOC was developed in the 1970s by AT&T and the communications industry is widely standardized on this system. Each type of jack has a number of wiring configurations, depending upon the number of wires connected. Thus RJ-25 wiring uses the same jack as RJ-11 except that eight wires are connected instead of two. Wires are often connected in pairs.

To promote pair continuity in electrical receptacles with varying numbers of active wires, the wires are usually connected in pairs working out from the center. In the Sample of Common Wiring Jacks chart, C represents flush or surfaced mounted, W represents wall mounted, and X is complex line.

RJE See remote job entry.
RJEP See Remote Job Entry Protocol.
RL 1. radio locator. See radio beacon. 2. resistor logic. 3. return loss.
RLCM Remote Line Concentrating Module.
RLE See run length encoding.
RLP See Radio Link Protocol. 2. See Resource Location Protocol.

RM See Resource Management.
RMA, RA returned merchandise authorization. A coding system used to control and track returned merchandise. See return authorization.
RMF remote management facility.
rms See root mean square.
RMF read, modify, write.
rn A full-screen, configurable news reader developed by Larry Wall, the author of the Perl programming language, and released in 1984. Wayne Davison developed a superset of rn known as *trn*.
RNC radio network controller.
RNG See random number generator.
RNX Restricted Numeric Exchange. A local telephone exchange in which calls are restricted to the originating network. Suppose a number is 555-1111 when dialed from *outside* the local exchange. The first three digits are the NXX digits. From *inside* the restricted exchange, you might instead dial 105-1111, with 105 being the RNX digits. The call then rings through to the number without passing through circuits outside the local exchange. See NXX.
RO 1. receive only. 2. remote operation. 3. recovery operation.
ROA recognized operating agency. An agency tasked with operating a private or public telecommunications service such as the public land mobile network.
roaming Using wireless telecommunications services while moving around, either on foot or in a vehicle. The logistics of designing and managing

Sample of Common Wiring Jacks Described in Universal Service Ordering Code (USOC)		
Jack	Wiring	Notes
RJ-11		Can accommodate up to six wires, though typically only two or four are connected as one or two pairs. A very common type of single-line phone jack used for telephones, modems, and fax machines.
	RJ-11C/W	One pair of wires connected, as for a single line phone. Traditionally the green and red wires are connected as tip and ring for the first line. The connection is bridged.
	RJ-14C/W	Two pairs of wires connected, as for a two-line connection, e.g., line 1 might be a phone or answering machine and line 2 a modem or fax machine. Traditionally green and red are assigned as tip and ring for the first line, and black and yellow as tip and ring for the second line. The connections are bridged.
	RJ-25C/W	All three pairs configured. Thus, line 1 might be a phone, line 2 a modem, and line 3 a fax machine.
RJ-45		Can accommodate up to eight wires and is common for multiple line phones (up to four lines) and for data communications, especially Ethernet and Token-Ring.
	RJ-48C/X	Four wires are typically connected as two pairs to provide 1.54 Mbps digital data network services.
	RJ48S	Four wires are typically connected as two pairs to provide local digital data network services.
	RJ61X	Four pairs of wires connected to accommodate four phone devices. Three lines are bridged.
	10Base-T	Twisted Pair. Pairs two and three are connected.

roaming subscribers is quite complex and requires sophisticated software. See cellular phone. See Inter-Access Point Protocol.

robbed bit A bit commandeered in a transmission for something other than its usual purpose. This technique may be used to acquire extra bits for signaling information, especially if the signals are only occasionally needed. See robbed-bit signaling.

robbed-bit signaling RBS. In data communications, a means of taking one bit from a data path to provide signaling information. For example, in voice communications over in-band T1 systems, a bit may be robbed to indicate the hook condition of the line.

Roberts, H. Edward Founder of MITS, which developed and distributed the first commercially successful computer kit, the Altair 8800, that launched the microcomputer industry. Roberts codesigned the computer when the market for calculators began to slip. Roberts left MITS in 1977 and is now a physician. See Altair.

robot In its simplest form, a robot is a mechanical apparatus which automatically moves or senses according to a set program, or an adaptive program. In its most complex form, it is a sophisticated electromechanical logical device that can interact with its environment in ways which are ascribed to human intelligence. That is, it is adaptive, and responds in ways appropriate to its task or to its benefit. Robot arms are used in many production-line tasks, whereas many humanlike robots are portrayed in science fiction stories and films. A humanlike robot (in form and functioning) is called an android.

A robot is generally associated with hardware and physical forms, but with the development of software artificial intelligence techniques, online avatars that seem to have a humanlike presence on public forums and chat lines have also been called robots, or more commonly 'bots or bots. A bot can be online 24 hours a day, can monitor processes, log user activities and interaction, and more. Bots have been banned from some of the chat rooms due to badly programmed bots violating chat Netiquette. However, system operators sometimes have authorized bots running on Internet Relay Chat (IRC) to perform many useful functions. Bots are also associated with search engines. Just as IRC bots do housekeeping tasks on the chat channels, search bots do useful Web page search and retrieve jobs, like a crew of gofers ("go fer this, go fer that") working around the clock. See mailer daemon.

Robotic Telescope Markup Language RTML. The Berkeley Lab had acquired a 30-in. telescope for astronomical research on supernovas that was installed at the Leuschner Observatory. To automate the functions of the telescope, the scientists created a telescope control system to support scheduling, data-acquisition, database management, and image analysis, thus treating the telescope as a remotely controlled robotic device. Since other telescopes were being controlled by various, similar in-house systems, it seemed appropriate to come up with a simple, standardized, high-level language that could provide a user with access to a broad range of devices – the concept for the Robotic Telescope Markup Language. Initial discussions with telescope manufacturers indicated an interest in supporting RTML commands that users could access through the Web and the Hands-On Universe Project endorsed the concept and provided an initial plan for development. Thus, the first version was released as the Remote Telescope Markup Language. See Remote Telescope Markup Language.

ROC 1. rate of convergence. 2. record of comments. 3. Regional Operating Company. See Regional Bell Operating Company.

rocket camera A creative early 1900s invention for taking photographs at high altitude (close to 800 m) by attaching a camera to a stabilizing rod and equipping it with a small parachute.

Rocky Mountain Bell Telephone Company An early Bell telephone exchange, established in 1883, with financial assistance from American Bell Company. It rapidly acquired the Ogden Telephone Exchange Company, the Montana Telephone & Telegraphy Company, the Idaho Telephone & Telegraph Company, and the Park City exchange. This pattern of acquisitions continued for a number of years.

Rocky Mountain Telephone Company An early telephone exchange established in Salt Lake City in 1800.

ROD rewritable optical disc.

roentgen, røntgen (*symb.* – R) An international unit of X-radiation or gamma radiation which is equal to the amount of radiation which produces one electrostatic unit (*esu*) of charge in one cubic centimeter of dry air at 0°C and standard atmospheric pressure ionization of either sign.

Rogers receiver An early batteryless radio receiver first introduced in 1925 at the Canadian National Exhibition in Toronto.

ROH receiver (as on a telephone) off hook.

ROLC Routing Over Large Clouds. The ROLC Working Group has now merged with the IPoverATM Working Group to form Internetworking over NBMA (ION). See frame relay, Internetworking over NBMA, RFC 1735.

role-playing game RPG. This style of game is particularly popular on computer networks as it can be implemented with simple text interfaces and lends itself to including people who are anonymous in the sense that the players, located at different terminals, cannot see one another.

In general, game-playing is discouraged on computer terminals in schools and businesses unless it directly relates to course work, but role-playing games have been tolerated since many of them require players and playmasters to use certain programming commands and algorithms to set up the game and participate, thus justifying their educational value. Role-playing games can be addictive, however. More than one role player has failed a school term due to overindulgence.

Roll About A commercial, self-contained videoconferencing unit which includes the monitor, camera, microphone, and other components on a stand that can be moved from one office to another as needed.

ROM 1. See read only memory. 2. rough order of magnitude.

Rømer, Ole (Olaf) Christensen; Rümer, Ole Christensen (1644–1710) A Danish astronomer, physicist, and scientific instrument-maker who, in 1675, demonstrated the velocity of light as 11 minutes per astronomical unit (AU) based upon observations of the planet Jupiter and its moon Io. In the 1660s, he studied with Rasmus Bartholin at the Hafnia (Copenhagen) University and was entrusted with the editing of the great scientist Tycho Brahe's manuscripts. He was appointed in France by Louis XIV to tutor the Dauphin in astronomy, and also was appointed as the *Astronomer Royal* at the Danish court of Christian V.

Rømer invented a new type of thermometer and communicated some of his ideas to D. Fahrenheit in the early 1700s. In Denmark, he introduced a new system for numbers and weights that combined the concepts of weight and length. See Bartholin, Rasmus.

Ronalds, Francis (1788–1873) An English researcher who was a pioneer in modern telegraphy, Ronalds developed a frictional telegraph, using above-ground copper wire that he demonstrated ca. 1816, more than two decades before Wheatstone and Cooke in England and Samuel Morse in America developed the first widely known telegraph systems. Ronalds also experimented with other types of telegraph systems. Charles Wheatstone corresponded with Francis Ronalds around the time he developed a telegraph with W.F. Cooke. The British Navy failed to adapt Ronalds' telegraph system, choosing instead to continue their "tried and true" semaphore system of signaling. See Schilling, Pavel Lvovitch; Morrison, Charles; Wheatstone, Charles.

Ronchi grating A series of perfectly straight lines impressed into glass, acrylic, or other film surface. It is a specific type of diffraction grating in which the distance between the grating facets and the facets themselves are the same. When placed near the focal point of an incident beam of light, imperfections in the optical system will be revealed as deviations from the straight line Ronchi reference grating. This makes it suitable for testing optical components when used in conjunction with a diffuser for providing an even lighting source and, optionally, Ronchigrams, for comparison of the results to reference ideals. See diffraction grating, Ronchi test, Ronchigram, star test. See knife-edge focusing.

Ronchi test A way to test the optical quality of a traditional lens such as a telescopic mirror lens by using a Ronchi grating, with evenly spaced, straight lines as a reference against which aberrations in the optical system can be seen. In general, the tester is placed at the center of the point of curvature of a mirror, the reference light passes through a diffuser, then through the Ronchi grating, and reflects from the mirror at least once. The reflected light then passes through the grating to the eye of the observer or it may be projected or stimulate an imaging surface (e.g., a computer peripheral). The Ronchi test can be set up at home or in a scientific lab.

It is relatively straightforward to test flat optical reflectors with a Ronchi test, as aberrations will manifest as curved, turned in, or out-of-focus areas. If parabolic shapes are assessed, the resulting image (which may be curved) can be compared against an image of the ideal band pattern associated with the shape under test at a specified distance and angular orientation to the Ronchi grating. A Ronchi test may be followed up by a star test. See Dall test, double-pass autocollimation test, Ronchigram, star test.

Ronchigram Reference Images

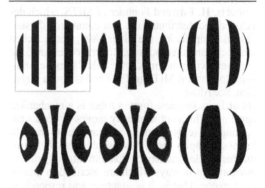

A Ronchigram is a reference image for comparison to light patterns from a reflective surface that have been reflected through a Ronchi grating. The top left image is a reference image for a regular mirror oriented flat in relation to the plane of the Ronchi grating. This is fairly easy to visualize and intuitively compare. The other images can indicate aberrant patterns for a supposedly regular mirrored surface or they can serve as references for other shapes, such as paraboloid mirrors, which are not as easy to intuitively check. For nonflat surfaces, the distance from the reflective surface to the Ronchi mirror becomes important if it is to be checked against a reference Ronchigram. A chart of Ronchigrams may also be accompanied by surface deformation diagrams for mirrors or lenses that correspond with particular Ronchigram shapes.

Ronchigram, Ronchi pattern A banded reference image that represents an expected pattern of light reflected through a Ronchi grating from a reflective optical surface such as a telescope mirror. It is possible to carry out a Ronchi test on a basic flat mirror without Ronchigrams, because aberrations will manifest as curved lines or fuzzy areas that can be readily seen, but when testing parabolic or other nonflat surfaces, there are more variables to control and the distance from the mirror, its orientation, and its shape may change the test results to manifest as a wide variety of patterns that can't always be intuitively assessed without a reference image. Computer software can be used to predict the Ronchi pattern, at a given distance, for a reflective surface of a given shape. See Foucault test, Ronchi grating, Ronchi test.

Røntgen, Wilhelm Konrad (1845–1923) A German physicist and educator who developed the vacuum tube (1895), the fluoroscope, and pioneer experimental studies in X-ray emissions in industrial radiography and medical radiology. He was the first recipient of the Nobel prize in physics. The roentgen unit of X-radiation is named after him. See roentgen, X-rays.

root mean square (*abbrev.* rms) The effective value of a quantity in a periodic circuit, measured through the duration of one period.

root name server On distributed computer networks, name servers provide a means to administer the transmission and routing of data among source and destination computers. Name servers are a type of address allocation and identification tool. Initially, the name servers handling root zones (indicated with a dot) also handled top-level domains (TLDs) such as the *.com* domains. However, due to the growth of the Internet and differences in administration of root name servers and TLD servers, it was proposed in the mid-1990s that a distinction be made between the two groups as the performance needs of root servers were felt to be greater than for many other types of servers. It was further proposed that requirements be formally laid down for root name servers so that vendors and implementors would have appropriate tools for handling the more stringent requirements of root name servers. See Domain Name Server, RFC 2010.

ROSAT Røntgen Satellite. A research satellite that has expanded knowledge of the universe and past events in this galaxy by aiding in discovering local hot X-ray plasma.

ROSE See Remote Operations Service Element.

Rosenblüth, Arturo A physiologist who collaborated with Norbert Wiener in investigations in artificial intelligence and self-organizing systems, much of which was documented in Wiener's book on cybernetics.

Ross test See Dall test.

rotary dial In telephony, a circular dial mechanism typically activated by placing a finger (or pencil end) in one of a series of punched out holes, and turning (dialing). The mechanism springs back to its original position after each selection. The alphanumeric selections are displayed under each associated hole. Turning the dial activates the telephone carrier's electrical loop for specified intervals that form a simple code to identify the number dialed.

Dials were not always circular. Some of the earliest dials consisted of levers, resembling the front of a small slot machine.

Rotary dials are normally associated with pulse dialing signals. Rotary phones are steadily being superseded by pushbutton phones. See pulse dialing, tone dialing, keypad, touchtone phone.

rotary hunt A means of routing selection for a call that has been placed in a multicircuit hung group. The system cycles through circuits, hunting for an idle one through which it then routes the call.

rotary switch A commercially successful electromechanical telephone switching system developed in the AT&T labs in the early 1900s, based on Lorimer one-step selection concepts and incorporating a permanently rotating motor. These were installed in Europe as a result of an International Telegraph and Tele-

phone buyout of Western Electric's International division in 1925. See Lorimer switch, panel switch, rotary switch.

ROTL Remote Office Test Line. A means of automatically testing telecommunications trunk lines from a remote location through a hardware switch. The remote testing may be conducted manually, through a control unit, or through automated Centralized Automatic Reporting on Trunks (CAROT).

Round Robin RR. A scheduling or resource-allocation system that "divvies up" tasks or resources more or less equally among all the participants in a generally cyclic pattern. Thus, a Round Robin system in a classroom with only one computer could be set up to allocate an hour to each student in a repeating rotation through the school year. In general, Round Robin systems deal with discrete events or resources that do not overlap (though there may be exceptions). A chain letter extended by each member in the chain and returned to the source when it is completed is another example.

Many sports use Round Robin systems (tennis, pool, bowling). They provide each player with the opportunity to play every other player.

Round Robin systems have been used by humans for centuries (maybe even tens of thousands of years) for scheduling work and play and for providing fair access to resources on a rotating basis.

The Round Robin system, as we now understand it, is at the heart of many scheduling algorithms that use the same general cyclic event-driven concept. Round Robin systems are used to allocate computers, modems, printers, and many other heavily accessed systems that are expected to be allocated on a more-or-less equal basis. They are especially useful when setting up paired Round Robin activities for large numbers of people or multiple Round Robins intended to take place concurrently. Other common Round Robin applications include

- token passing in a Token-Ring network. The token moves around in a circular topology to each machine in the ring,
- network routers such as the Cisco CSS 11800, that use Round Robin algorithms to handle data flow,
- telemarketing systems that assign incoming calls to available agents on a Round Robin basis, especially in situations where the agent is earning a commission and wants a fair share of the calls, and
- computer operating systems running multiple processes in which the computer moves on to the next process if the previous process has completed, but may preempt a process if it is not completed after a certain amount of time.

Weighted Round Robin schedules take into account the unequal capacities of different participants or processing systems. For example, in a telemarketing firm, if there is a crackerjack sales representative with seniority who always closes a sale, the Round Robin system of assigning incoming calls might be weighted to go to this sales representative more often than the other agents. Thus, the sales representative is happy and stays with the company and the company is happy because profits are maximized. Weighted systems tend to have the advantage of being more effective in systems with unequal members, but also have the disadvantage of being controversial when they involve people rather than computing processes. See queuing, Token-Ring network.

Round Robin, DNS A load-balancing resource-sharing system implemented on the Internet through recent version DNS/BIND 4.9 (Domain Name System with BIND) that arose in part because of heavy access to favored sites on the World Wide Web. This enables a cluster of servers to provide client requests to a very large number of users by allowing address records to be duplicated for a host, with different Internet Protocol (IP) addresses. The name server can then rotate in a Round Robin fashion through names with multiple address (A) records.

The DNS/BIND implementation is not robust in terms of handling server failures, as the Round Robin system is inherently intolerant of missing links in the pool of resources, but other programs can be used with DNS/BIND to handle possible server failures. See Berkeley Internet Name Domain, Domain Name System, Round Robin.

round trip delay RTD. A general phrase that refers to a number of aspects of network physics and behavior that contribute to delays in the full path back to the sender. It is a statistical measure useful in installing, configuring, and tuning networks. Some delays are evaluated in relation to what the elapsed time should theoretically be and some are evaluated relative to a test suite of alternate configurations. Sources of delay include the physical length of the path, the media and protocols used to send the information, data processing, compression and error-checking, data congestion, and even the speed of light itself.

route 1. *n.* Path taken by data or other transmissions. See traceroute. 2. *v.* To delineate a communications path. This may be fixed before the transmission or may be dynamic according to availability and load levels evaluated en route. See hop-by-hop.

Route Access Protocol RAP. An open distance-vector routing protocol for distributing routing information through Internet-connected systems ranging from large Internet Service Provider (ISP) systems to local area networks (LANs). The protocol was submitted as an RFC by R. Ullmann in 1993.

The Route Access Protocol is a generalized protocol, making no distinction between external and internal networks and is designed to accommodate both small and large systems. RAP is internally IPv7-compliant but is downwardly compatible with IPv4 networks. Distance-vector routing was selected for RAP to promote scalability. The protocol operates on TCP port 38 through a symmetric connection between RAP ports. Through peer discovery, it may also be used on User Datagram Protocol (UDP) port 38. RAP propagates routes opposite to the direction of the path of the datagrams using the associated routes. Source

restriction may be invoked to provide a certain measure of security and must be added to routes if security filters have been established in the Internet Protocol (IP) forwarding layer.

Once a connection is established, RAP peers need only send new or changed routing information. The routing information is purged by each system when the session ends. See RFC 1476.

route diversity An architecture providing a number of options in routing a transmission. This may be organized in a number of ways, depending upon the type of network, the load on the network, and the existence and quantity of redundant circuits. Route diversity can be as simple as an extra line to take the load if the primary line goes down or as complex as dynamic hop routing in a distributed computer network.

route flap, router flap A fault condition in which changes in routes propagating across a network (usually from losing one or more nodes) exceed the capacity of a router's processor and memory to cope with the change, and consequently impact its ability to route. Routers are generally selected to well exceed the number of routing paths expected to be needed to prevent this serious problem.

router 1. A device or mechanism for selecting a path. 2. In a simple network, an interface device which selects a path for the transmission packets. In layered networks, the router typically functions at layer two or layer three, depending upon the degree of automation and "intelligence" built into the router. There used to be somewhat of a distinction between routers and switches, but switches are becoming so sophisticated that the distinction is disappearing. Routers frequently include routing databases, in addition to algorithms to dynamically select routes. See bridge, switcher.

router, ATM In ATM networks, a router delivers and receives Internet Protocol (IP) packets to and from other systems, and relays IP packets among systems. Routers vary in sophistication, with some able to contribute significant network management functions, such as priority and load balancing. They can be protocol-dependent or protocol-independent. Also called an intermediate system. See LIS.

router, Frame Relay A Frame Relay-capable router has the ability to encapsulate local area network (LAN) frames in Frame Relay-format frames and feed them to a frame relay switch, as well as receiving frame relay frames, and stripping the frame relay frame to restore the information to its original form, passing it on to the end device. With improved technologies, the distinction between routers and bridges is lessening. Even switches now have many of the capabilities of routers. See bridge, gateway, switch.

routing Selecting or establishing a path. In telecommunications, the path may be used to transmit supervisory signals and data (either together, separately, or over separate paths). Routing may be static or dynamic. Dynamic routing is typically used in larger networks, especially those with individual systems in the path that may or may not be available at any given time. Static routing is suitable for small or local systems, or for those carrying priority or secured transmissions.

routing aggregation An administrative tool for organizing and optimizing the use and availability of routes to deal with the continually rising demand on networks. Users are encouraged to return unused addresses, and old addresses are assigned prefixes so multiple routes can be aggregated into one.

Routing Arbiter Database RADB. A routing database established by the Routing Arbiter project. One of several databases in the Internet Routing Registry.

Routing Arbiter Project A National Science Foundation-funded project given the task of coordinating routing for the new NSFNet architecture in cooperation with a number of educational and private business concerns. Route Servers were to be installed at connection points to reduce the need for peering. Due to NSFNet legacy database information, a number of large providers have shunned the Routing Arbiter and the project has changed to a service available through some of the Public Exchange Points. See peering.

routing code 1. In telephone communications, the area code. 2. In U.S. postal communications, the last four digits of the ZIP+4 code. 3. In networks, the data parsed by the router or switcher to establish a path to the intended destination.

routing computations, routing algorithms Mathematical schemes to compute efficient routes through a network. The algorithms may be straightforward if the topology and size of the network are known and are relatively stable. The situation is more complicated on the Internet, which is extensive, encompasses many different types of configurations, and which changes constantly as networks are added or changed.

routing domain RD. In ATM networking, a collection of systems that have been grouped topologically within one routing system.

routing information base RIB. A static or dynamic table of routing paths maintained within a router or a processor/memory system associated with a router. In distributed networks, in which systems may be added or removed dynamically, RIBs are generally updated dynamically as the information about the changes propagate through the network.

Routing Information Protocol RIP. A very common routing protocol from a family of protocols known as the Interior Gateway Protocols (IGPs). RIP evaluates the path between two points in terms of *hops* between the source and destination points. Each hop in the path is assigned a value which may be incremented by the router and entered into the routing table. The Internet Protocol (IP) address becomes the next hop. In order to prevent continuous routing loops, a limit is defined for a path's hop count. If the routing table entry exceeds the maximum hop count, the destination is considered unreachable. See RFC 1058, RFC 1723.

Route Servers Specialized servers from the Routing Arbiter project intended to hold clearinghouse routing information in a Routing Information Base at each network interconnection point in order to eliminate,

or at least reduce, the need for peering. These servers are not intended to actually transmit the traffic, but serve to handle the flow of information concerning pathways.

routing table, data network In data networks, a table detailing paths to specific Internet Protocol (IP) addresses. With the explosive growth of the Net, the number of paths, and hence the size of the tables, can become very large. Primary routers sometimes list more than 25,000 routes and the routers themselves must be designed to keep ahead of capacity. Discussions are ongoing as to the benefits and problems in various assignments of routing paths, with provider-based routing being favored due to greater ease of implementation, and geographic-based routing proposed because it has less of a tendency to concentrate power in the hands of a few large providers.

routing table, telephone network In telephone communications, routing tables serve to record and provide information for the processing of incoming calls. Thus, calls may not just go to a particular caller or workstation, they may be directed to automated voice services, voicemail, queued holds, recordings, etc. They may require the capability of stepping back through the route as well, depending upon the sophistication of the system and the selections available to the user.

Routing Table Maintenance Protocol An AppleTalk network routing protocol that is based upon Internet Protocol (IP) Routing Information Protocol. RTMP is a transport-layer protocol that manages routing information within an AppleTalk router and facilitates the exchange of routing information among AppleTalk routers to help keep the information current. Hop count is used as the routing metric for RTMP. The RTMP table includes information on the hop count to a destination network, the appropriate router port, the next hop router address, the status of the routing table entry, and the network cable range of the destination network. See Routing Information Protocol.

Routing Table Protocol RTP. A network communications routing protocol from Banyan System, Inc., VINES (Virtual Networking System). RTP is based upon Internet Protocol (IP) Routing Information Protocol. The protocol facilitates the distribution of network topology information among the various servers and enables servers and routers to identify others nearby. It works in conjunction with other protocols such as the Address Resolution Protocol (ARP). See Routing Information Protocol.

routing update Network configuration information provided by a router and, in some cases, costs associated with use of particular routes. Routing updates can be scheduled to be automatically sent out at specified intervals, and are typically broadcast if significant network configuration changes have been made.

row 1. Generally, about three or more elements more or less lined up in one directional plane. 2. In a 2D system, the grouping, between the left and right bounds, of more or less horizontally aligned elements arranged within a grid or tabular format. This meaning of the word is commonly used to describe screen

R

locations or positions within a spreadsheet or printed medium. See column.

Royal Society An independent national academy of science, founded in 1660 to support the British scientific community. The Society funds research, stimulates international communication among scientists, hosts conferences, produces reports and journals, and maintains one of the most remarkable historic archives of scientific endeavor in the world dating back to 1470 (which includes records and photos of many of the major advances in computing and telecommunications). Among its publications is the *Philosophical Transactions of the Royal Society* which was first printed in 1665. The society has recognized scientific achievement through awards since 1731. http://www.royalsoc.ac.uk/

RP See Restricted Radiotelephone Operator Permit.

RPC See remote procedure call.

RPE radio paging equipment. Systems that use radio waves to send a signal to a receiving unit. The signal may be a simple pulse to trigger a light, vibration, or sound, or it may contain information that can be printed out on a text-capable paging device.

RPG 1. Report Program Generator. A computer programming language for processing and displaying large data files. 2. See role-playing game.

RPM Remote Packet Module.

RPN reverse Polish notation. A mathematical notation system in which the values to be operated upon are entered first, followed by the operation to be performed. In other words, multiplying two numbers would be done like this:

2 [enter] **4** [enter] ***** (times)
which would display the result as 8

This system is used on HP calculators and others commonly used by the scientific community. Calculators marketed to the nonscientific community tend to use regular notation as follows:

2 [enter] ***** (times) **4** [enter] **=** (equals)
which would display the result as 8

Those who use one system often curse when they encounter the other, because it necessitates shifting mental gears to enter data and operations in the correct order.

RPOA See Recognized Private Operating Agency.

RPOP Remote Post Office Protocol. See Post Office Protocol.

RPS See repetitive pattern suppression.

RPT repeat.

RPV See remotely piloted vehicle.

RQS See Rate Quote System.

RR 1. radio regulation. 2. railroad 3. return rate. 4. See Round Robin. 5. rural route.

RRSF RACF remote sharing facility. See remote access control facility.

RS 1. radio satellite. 2. recommended standard. See RS-232 for an example using this prefix. 3. remote station. 4. reset. 5. See Royal Society.

RS-1 Along with RS-2, the first USSR amateur satellites, launched in October 1978.

RS-232 Recommended Standard 232. A decades-old single-ended standard for serial transmissions intro-

duced in the early 1960s. RS-232 is widely supported on desktop computers and other devices commonly used for communicating with modems, remote terminals, and printers. RS-232 specifies the electrical and physical characteristics of the connection. The most common implementation is RS-232-C (which is often transcribed as RS-232c for brevity) and many in the industry mean RS-232c when they say RS-232. The RS-232c specification defines a way to connect data terminal equipment (DTE) with data circuit-terminating equipment (DCE).

Most systems support RS-232c through a 25-pin D connector (DB-25), although fewer pins can be used to implement the specification, and 9-pin D connectors are sometimes used (or 25-pin connectors with some of the pins unconnected). A few systems specify more than 11 or 12 pins (pin 12 is not part of the spec. but some vendors assign a proprietary signal to the pin). The basic RS-232 pinouts more or less commonly used by manufacturers are as follows. The most important pins for establishing a basic connection are noted with asterisks and a minimal connection would require transmit, receive, and ground.

Pin	Abbrev.	Function
1	GND	earth ground
2*	TxD	transmit data
3*	RxD	receive data
4*	RTS	request to send (control signal)
5*	CTS	clear to send (control signal)
6	DSR	dataset ready (signals that the device is on)
7	GND	signal ground
8	DCD	data carrier detect (signal that carrier is on)
9	+12	DC voltage (Amiga)
10	–12	DC voltage (Amiga)
11	AUD	audio out (Amiga)
12		---
15		transmit clock
17		receive clock
18	AUD	audio in (Amiga)
20	DTR	data terminal ready
22	RI	ring indicator
24		auxiliary clock (provided by some vendors for local connections)

A null modem cable can be made from an RS-232 cable by swapping pins 2 and 3 on one end of the connection. This enables two locally connected computers to network through the cable with suitable communications programs running on each system. The initial RS-232 standard was superseded in 1987 by the standard defined by the Electronic Industries Alliance (EIA) as EIA-232-D, which was followed in 1991 by the EIA/TIA-232-E. RS-232c is similar to the ITU-T Recommendations for V.24 and V.28.

The single-ended handling of voltage in the RS-232 specification limits transmission distances to about 50 feet or so. For longer distances, it is preferable to use balanced pair voltage formats like RS-422, to transmit up to about 4000 feet. See RS-422. See EIA Interface Standards for a list of common standards. See Selected Overview of Recommended Standards for Communications chart for a summary of RS data/video communications standards.

RSA See Rural Service Area.

RSA An academic security research facility within the corporate structure of RSA Security Inc. The lab provides state-of-the-art expertise in cryptography and security technology to RSA and its customers. The lab personnel were active participants in the IEEE P1363 project specifying standards for public key cryptographic systems. See Public-Key Cryptography Standards.

RSA Security Inc. A prominent data security distributor providing products that aid companies in developing security/trust products and processes, especially for electronic commerce applications. The name is based upon the originators of the RSA cryptosystem, Rivest, Shamir, and Adleman. See MD series, RC6.

RSC 1. See remote switching center. 2. repair service center.

RSFOCS See Repeatered Submarine Fiber Optic Cable Systems.

RSS 1. remote switching system. 2. root sum square. A statistical calculation that is useful in assessing errors such as signal distortion.

RSU remote switching unit.

RSUP See Reliable SAP Update Protocol

RSVP Internet Reservation Protocol. An extensible, scalable protocol designed in the mid-1990s to provide efficient, robust ways to set up Internet-integrated service reservations, RSVP became an Internet standard in 1997. It has primarily been promoted by commercial interests, as it makes it possible to establish priority connections through reserved bandwidth, a feature of interest to large competitive business network users.

RSVP is appropriate for multicast applications, although it supports unicast as well. RSVP interfaces existing routing protocols rather than performing its own routing. The RSVP is used by a host to request a specific Quality of Service (QoS) from the network. RSVP attempts to make a resource reservation for the data stream at each node through which it passes. RSVP communicates with two local decision modules: admission control and policy control, to determine whether the node has sufficient resources to supply the QoS, and whether the user has administrative permission to make the reservation. One of the difficulties in implementing RSVP has been assessing fees for connections across more than one network. Some opponents of the system fear the establishment of "elite" Internet users based on economics rather than on quality of information or services offered. RSVP development has continued since 1995 as RSVP2. See STII.

RT 1. radio telephone. 2. See realtime. 3. See remote terminal. 4. reorder tone. 5. routing table. 6. See run time, runtime.

RTB See regional test bed.

RTC 1. realtime control. 2. runtime code. 3. runtime control.

RTCA See Radio Technical Commission of Aeronautics.

RTCP Real Time Conferencing Protocol.

RTD 1. realtime display. 2. See round trip delay.

RTDNA See Radio-Television News Directors Association.

RTDNF See Radio-Television News Directors Foundation.

RTE remote terminal emulation. See remote terminal.

RTF 1. See Radio Terminal Function. 2. See Rich Text Format.

RTFM Abbreviation for "read the freaking manual" used on public forums on the Net when a user asks a question that has been asked and answered hundreds of times and is clearly answered in the appropriate documentation, or FAQ. An exhortation for the user to look it up *before* using up people's valuable time asking again. See Frequently Asked Question.

RTL 1. Radio Television Luxemburg. News, sports, comedy, and community programming. 2. runtime license.

RTM 1. realtime/runtime monitor. 2. runtime manager.

RTML 1. See Real Time Markup Language. 2. See Remote Telescope Markup Language. 3. See Rich Text Markup Language. 4. See Robotic Telescope Markup Language. 5. See Runeberg Text Markup Language.

RTOP realtime operating system.

RTP 1. See Rapid Transport Protocol. 2. realtime protocol. 3. Realtime Transport Protocol. 4. See Routing Table Protocol.

RTTY realtime teletype. See teletypewriter.

RTS 1. realtime system. 2. Request to Send. Flow control, typically used in serial communications, which is an output for DTE devices and an input for DCE devices. See TxD, RxD, CTS, DSR, DCD, DTR, RS-232. 3. remote tracking station.

RTSP See Real Time Streaming Protocol.

RTU 1. remote telemetry unit. A system that enables a device, such as a communications satellite in orbit, to be controlled from a remote location, such as an Earth station or space shuttle. RTUs can typically be used to orient and move the remote system by activating and controlling various motors, gyros, and other positioning systems. The telemetry unit may or may not have feedback capabilities. If the RTU is one-directional, then coordinates or instructions are usually sent with the hope that they will work or with the understanding that another system (such as a shuttle or telescope) will monitor the effect of the settings. If the RTU is two-directional, various types of information will be relayed back to the RTU from the remote system and corrections made, if necessary. 2. remote terminal unit. A device to activate, control, or query a system in another location. Remote termi-

R

Selected Overview of Recommended Standards (RS) for Communications		
Designation	Type/Description	

RS-170   *Monochrome Video*: a standard for analog black and white (monochrome) video adopted in North America and Japan that was prevalent until color standards became dominant. The format evolved from historic broadcast specifications developed in the 1930s. RS-170 carries both timing and image information on a single signal. It formed the basis for broadcast television for many years. The format is still widely used in monochrome security cameras where a higher-resolution picture or lower installation cost are considerations. For television, RS-170 has been superseded by RS-170a color NTSC systems and is now gradually being superseded by RS-343 and digital video formats.

RS-170 broadcast standards specify 525 horizontal lines displayed as 2:1 ratio interlaced frames. These are alternate odd- and even-line half-frames that combine perceptually to create the full frame image, generating the effect of 60 frames per second. The range of the relative intensities of individual 'points' in the display is from no light (black) to full light ('white'), depending upon the voltage level. Not all of the lines are used for display; some aid in synchronization. Since there is no color signal (chrominance) to process and render, it is generally less expensive to manufacture monochrome systems. RS-170 signals are typically transmitted through 75-ohm well-shielded coaxial cables.

Europe uses a standard similar to RS-170 called CCIR. It differs in that it supports a higher vertical resolution (625 horizontal scan lines) and operates at the rate of 25 frames per second. See NTSC, RS-170a, RS-343.

RS-170a   *Analog Color Video*: an Electronic Industries Association (EIA) standard for analog color video adopted in North America and Japan in 1953. Consumer RS-170a-based televisions and video editing systems superseded RS-170 (monochrome) in the 1970s. RS-170a is also gradually superseding RS-170 standards for security cameras. Eventually, RS-170a will be replaced by digital video, but it is still prevalent in composite National Television System Committee (NTSC) systems.

The original RS-170 (monochrome) format specified 525 horizontal lines for display and synchronization purposes. The RS-170a standard was intended to support color while being downwardly compatible with the large installed base of monochrome systems expected to remain for some time after the color standard was introduced. There was also some consideration of video tape recorders that were becoming prevalent in the production and broadcast industries. RS-170a specified 1050 lines due to the four color fields needed for a frame through subcarrier repeats, but the effective resolution is compatible with RS-170, that is, 525 horizontal lines, 485 of which are displayed.

RS-170 RGB refers to video signals using the red-green-blue color model (e.g., many computer monitors) timed for compatibility with RS-170 specifications. See RS-170, RS-343, RS-343a.

RS-232   *Serial Communications*. A decades-old single-ended standard for serial transmissions introduced in the early 1960s. RS-232 is widely supported on desktop computers and other devices commonly used for communicating with modems, remote terminals, and printers. RS-232 specifies the electrical and physical characteristics of the connection. The most common implementation is RS-232-C (which is often transcribed as RS-232c for brevity) and many in the industry mean RS-232c when they say RS-232. The RS-232c specification defines a way to connect data terminal equipment (DTE) with data circuit-terminating equipment (DCE). See RS-232 for pinouts.

RS-330   *Monochrome Video*: an analog, monochrome composite video standard adopted in North America and Japan primarily for closed-circuit television systems. It supports a resolution of 525 horizontal lines at 60 frames per second. It is similar to RS-170 but does not require equalizing pulses.

RS-250c   *Serial Communications*: the most recent version of an analog color video standard for establishing acceptable performance in the transmission of broadcast-quality signals through various distances (short-, medium-, and long-haul) through a variety of wired and wireless links. The 250 series was first established in the 1950s. Unlike other video standards that specify resolutions, frame rates, and synchronization levels for displaying images, this one concerns itself with the delivery of a "clean" signal in terms of the signal-to-noise ratio, phase, gain, and other signal quality characteristics. Short-haul RS-250c requirements are easier to meet now that fiber optic transmission links are available.

RS-343/a *Monochrome Video*: an analog noninterlaced monochrome video standard used for nonbroadcast high-resolution cameras (625 to 1023 scan lines, not all of which are displayed), especially security cameras for monitoring and recording purposes. Many RS-343a cameras also output RS-170 for downward compatibility. Terminated 75-ohm coaxial cables are traditionally used for transmitting the signal but RS-343a is now also supported through RS-422 and fiber optic composite and digital RBG video links for wideband TV, medical video, military monitoring, and computer graphics applications. RS-343a is the monochrome version – which is confusing, since RS-170 and RS-170a are, respectively, monochrome and color. See RS-170, RS-170a.

RS-328 *Facsimile Communications*: a standard for facsimile transmissions introduced in 1966 to help improve interoperability among facsimile equipment from different manufactures which were, at that time, operating on a number of proprietary schemes. This came to be known as the Group 1 standard and other Group *x* formats have followed. Standardization has aided in implementing fax capabilities on other systems, including fax modems. See facsimile.

RS-366 *Parallel Dialing*: a parallel dialing standard for high bandwidth communications such as videoconferencing. RS-366 has been subdivided into Type I, II, and III. RS-366 has been implemented on a number of types of links, including RS-449, RS-530, V.25bis, and V.35 modem interfaces. It is compatible with RS-232 electrical specifications and is commonly transmitted through a DB-25 connector. It has further been implemented on fiber optic modems transmitting data at speeds up to 56 Kbps per channel and is also applicable to video transmissions over ATM networks (e.g., for switched virtual circuits).

There are commercial products to convert between RS-366 and Hayes AT commands. They resemble traditional modems in shape and size, with LEDs to signal various indicators. These converters enable a regular asynchronous modem to be used in conjunction with an RS-366 interface.

RS-422 *High-speed Serial*: a widely-used balanced/differential voltage twisted-pair standard for high-speed point-to-point serial transmissions. It is backwardly compatible with RS-232 but is faster, up to 100 Kbps, and can be transmitted over longer distances, up to about 4000 feet. RS-422 can be configured to support either software or hardware handshaking. The standard is not tied to any specific attachment device configuration, but is commonly implemented through cables with DB-9, DB-25, or 8-pin mini-DIN connectors. RS-422 is not inherently a multidrop standard, but 4-wire, half-duplex links can be constructed to provide some of the benefits associated with multidrop formats such as RS-485. See RS-485. See differential cable, RS-232.

RS-423 *High-speed Serial*: a balanced/differential voltage twisted-pair serial transmissions standard backwardly compatible with RS-232, but with multidrop capabilities and transmission distances of up to 4000 feet. Compared to RS0232 and RS-422, this standard has not been widely implemented.

RS-449 *High-Speed Serial*: a balanced/differential voltage twisted-pair standard for high-speed synchronous data transmissions. The signaling is associated with specific pin assignments for DB-9 and DB-37 connectors. This has been superseded by RS-530. See RS-232, RS-530.

RS-485 *High-Speed Serial*: a widely used balanced/differential voltage pair standard, downwardly compatible with RS-422, for high-speed serial transmissions, up to 100 Kbps. It has become popular in industrial and telecommunications applications for connecting multiple peripherals (printers, industrial fabricators, etc.) through multidrop transceivers/transmitters. Up to 64 devices may be connected, 32 for each multidrop line.

RS-485 supports drivers with higher voltage output ranges than RS-422. It can be used in half-duplex or full-duplex mode, but is commonly implemented as half-duplex. RS-485 may or may not need to be terminated, depending upon the configuration. As with RS-422, the format is not tied to any particular type of connector. See differential cable, RS-422, RS-232.

RS-530 *High-Speed Serial*: a balanced/differential voltage twisted-pair standard for high-speed serial transmissions that supports RS-422/RS-423 and the oldie-but-prevalent RS-232. It is specified for a DB-25 connector. There are commercial devices to convert between RS-530 and the older RS-232 serial format that is still supported on many computers. There are also cables to convert between RS-530 and V.35, X.21, and RS-449/RS-442 formats.

RS-530 can be supported over high-speed, point-to-point fiber optic modem links for distances of up to 30 km. Thus, a system with an RS-530 interface can connect through twisted-pair cable to a fiber optic modem, which then transmits to another fiber optic modem and twisted-pair at the other end of the fiber link. RS-530 supersedes RS-449.

R

nal units range from simple to sophisticated but, in general, refer to a remote terminal earmarked for a particular purpose, such as testing and diagnostics in a network system. A small-scale remote terminal is similar in concept to a television/VCR remote control but usually provides more control and better processing capabilities than typical remote controls. Older computers are often used as remote terminal units to control security systems and home automation systems (heat, lights, etc.).

RTV realtime video. Video that appears as natural movement, usually with at least 20 frames per second, and which further may be a live broadcast as opposed to playback from stored information. On the Internet, a technology called *streaming video* is becoming popular for airing live newscasts. In the past, most animations were developed, compressed, and stored in an Internet-accessible location for the user to download and play back on his or her local system. Streaming video enables the user to link into a live data stream and watch the action as it happens.

RU 1. In packet networking request unit, response unit, request/response unit. See basic information unit. 2. receive/receiving unit.

RUA Remote User Agent. A software agent that acts on behalf of a client making a request or supplying information from a remote terminal. See User Agent.

rubber bandwidth *jargon* A communications channel whose bandwidth can be dynamically altered, that is, it can be changed without terminating and reinitiating the transmission. This colorful phrase apparently originates from Ascend Communications, a supplier of networking-related products, to describe characteristics of an inverse multiplexing system.

Ruhmkorff coil An induction coil used to induce high voltages, first constructed in the mid-1800s by Eugene Ducretet, though the instrument is associated with Heinrich Ruhmkorff. This technology was developed into ignition coils. See coil, induction.

Ruhmkorff, Heinrich Daniel (1803–1877) A German physicist of the 1800s who constructed and distributed the induction spark coil, following pioneer work by Nicholas Callan and Charles Grafton Page and practical embodiments by Eugene Ducretet.

ruled grating A planar grating structure for controlling light passage in the optical frequencies. It is created mechanically by machining fine, parallel grooves into a surface supported by a substrated. Standard gratings are available for controlling specific wavelengths and other wavelengths can be requested as custom fabrications.

Ruled gratings are typically described in terms of the number of grooves per millimeter, the blaze angle, and the optimal wavelength to which it is "tuned." Ruled gratings tend to be favored for less precise applications and less dense grating patterns where stray light is less apt to interfere with overall efficiency and where efficiency in the optical spectrum is desired. A blazed grating is one in which the rules are slightly asymmetric. See blaze angle, interference grating.

rumbatron A term used by William Hansen for pioneer cavity resonating devices in the 1930s. The term

was subsequently adopted by Russell Varian, in collaboration with Hansen, to describe a Rumbatron Oscillator, the germinal idea for the Klystron tube. [The author was not able to find a firsthand reference as to whether Hansen spelled it rumbatron or rhumbatron. For consistency with Varian's Rumbatron Oscillator, it is spelled without the "h."] See Klystron.

run *v.* To initiate and execute a software program, or linked suite of programs which form an application.

run length encoding RLE. A lossless data compression technique that works well with data that include repeated sequences. The repeated sequences (white spaces in a document, a single background color in an image, etc.) are replaced with a code that indicates that what follows is a string of the same character of a particular length. If run length encoding is used on data with little or no redundancy, the encoded file may be *longer* than the original.

Rundgren, Todd (1948–) A multimedia recording artist who has managed to stay at the forefront of emerging interactive entertainment technologies, synthesizing the new capabilities in media into video and sound. Rundgren began programming microcomputers in the late 1970s, adapting Macintoshes, Amiga Video Toasters, and other systems to many new creative venues, producing new types of music albums, computer-generated rock videos, and interactive TV entertainment concerts. Since the mid-1990s, Rundgren has been president/CEO of Waking Dreams, which develops, licenses, and distributes products and services originating from creative and undervalued ideas.

run time The time during which a software routine or application executes. This can easily be confused with "runtime" which refers to CPU cycles and computer processing time. In contrast, run time is the overall length of time a job might take from submission to completion. Thus, a database query might take 5 minutes of *run time* or *execution time* to provide a result, yet require only 3 seconds of CPU *runtime* to execute. To confuse matters further, many people use these terms interchangeably and run time licenses are often spelled "runtime." See runtime.

run time license A onetime, per-use, or other distribution license granted by a software developer to allow a vendor to include the licensor's algorithms, usually for inclusion in an enduser product (often called a "customer application").

runtime In computer processing terms, this is the duration, that is, the CPU time (usually expressed in cycles), used to execute a routine or program. A process that takes 5 minutes for the user might require only 3 seconds of CPU runtime because computer time is often spent pondering or waiting for peripherals. Runtime is a reasonably objective value with comparative benefits for optimizing software algorithms and practical billing applications for shared access systems. See run time.

rural automatic exchange RAX. An automatic telephone exchange that didn't require a human operator to patch the connections, intended for rural communities. Siemens is credited with installing the first 40-line RAX in the U.K. in 1921. Radio-equipped, solar-powered RAX systems were introduced to rural areas in India in 1985.

Rural Local Broadcast Signals Act H.R. 3615. An act passed by the U.S. House of Representatives in April 2000 that extends loan guarantees to companies providing local television broadcasting through satellite transmissions. H.R. 3615 amends the Rural Electrification Act of 1936 to ensure access to TV broadcasting by multichannel video providers to all households in underserved areas that desire the service by December 31, 2006.

The Act was felt to be important because weather, emergency, and other crucial information was often disseminated through broadcast television, in addition to educational and entertainment programming. Other services, such as Internet access, may also be offered. Satellite access also serves as a competitive alternative for rural consumers. The Act will serve to improve access to more than 6 million satellite dish owners in rural areas. However, that still leaves the other 50% unserved and in need of support in having access to the same programming as urban subscribers. The Act was placed on the Senate calendar in May 2000. See Local TV Act of 2000.

Rural Local Television Signals Act U.S. regulations adopted in 1999 as part of the Satellite Home Viewer Improvement Act (SHVIA) for delivering local broadcast television signals to satellite television subscribers in unserved and underserved local television markets, a spectrum that would otherwise be allocated to commercial use. There was controversy regarding this issue and whether the terms of SHVIA sufficiently addressed the problem of rural access. These discussions lead to the Rural Local Broadcast Signals Act and the LOCAL TV Act of 2000. See Rural Local Broadcast Signals Act, LOCAL TV Act of 2000.

Rural Service Area RSA. An administrative designation used by many commercial and public service organizations, including telecommunications providers and public libraries. For telephone service, regions not defined as Metropolitan Service Areas (MSA) are in the category of Rural Service Area, which includes smaller cities (usually under about 50,000 inhabitants), towns, and rural regions. The Federal Communications Commission has recognized over 300 MSAs and over 400 RSAs in the U.S. and has further used this designation to license non-MSA cellular carriers.

Libraries use a similar categorization to define their public service regions as urban, suburban, and rural service areas. The rural service area is often defined as those areas outside the urban and suburban boundaries or may be defined in terms of its radius distance from the library facility, depending upon the geographical characteristics of the region. Since telecommunications services are typically contained within a fixed building setting, the designation of RSA is an important one because library services such as Internet access may need to be provided in some mobile form (similar to a Bookmobile service).

Providers serving RSAs have many unique problems with installing and maintaining profitable enterprises. Rural areas have smaller population-to-land ratios that necessitate more wires for fewer people and longer trips for service personnel. Rural areas often require a larger proportion of long-distance services compared to local services. To complicate matters further, the topography of rural areas may be rough and inaccessible, especially in mountainous regions. Even rodents pose a problem in rural areas, chewing through communications wires installed aboveground on rocky terrain. In terms of fast Internet access, ISDN and other services are disproportionately skewed toward Metropolitan Service Areas. Wireless communications might seem to be an ideal solution to many of these problems, but RSA providers often are smaller organizations that have difficulty competing with larger firms for Federal Communications Commission (FCC) licenses for wireless services.

rural telephone company This is defined in the Telecommunications Act of 1996 and published by the Federal Communications Commission (FCC) as:

"... a local exchange carrier operating entity to the extent that such entity—

(A) provides common carrier service to any local exchange carrier study area that does not include either—

    (i) any incorporated place of 10,000 inhabitants or more, or any part thereof, based on the most recently available population statistics of the Bureau of the Census; or

    (ii) any territory, incorporated or unincorporated, included in an urbanized area, as defined by the Bureau of the Census as of August 10, 1993;

(B) provides telephone exchange service, including exchange access, to fewer than 50,000 access lines;

(C) provides telephone exchange service to any local exchange carrier study area with fewer than 100,000 access lines; or

(D) has less than 15 percent of its access lines in communities of more than 50,000 on the date of enactment of the Telecommunications Act of 1996."

See Federal Communications Commission, Telecommunications Act of 1996.

Rural Utilities Service RUS. A U.S. Department of Agriculture agency that provides technical and funding support for rural utilities infrastructure projects involving electricity, water, and telecommunications. http://www.rurdev.usda.gov/rus/

RURL Relative Uniform Resource Locator, Relative URL. A compact representation of the location and method of access for a resource accessible over the Internet that is described relative to an absolute base URL. In contrast, an absolute base URL is one for which a specific location is established, such as *http://www.4-sights.com/* and remains the same no matter where the Web page holding the link, for example, is located. However, there are situations where an absolute URL is not the best solution.

Imagine establishing an extensive Web site at *www.4-sights.com* that has many document and graphics files hierarchically contained in subdirectories below the main directory. Assume the subdirectories are called *docs* and *pics* and that you are building a page in the *docs* directory (absolute URL - http://www.4-sights.com/docs/mydoc.html) that has many images linked from the *pics* directory. In terms of building the page in HTML, it is tedious to type *http://www.4-sights.com/pics/filename.jpg* each time a new image is added to the page with a link. Instead, a relative URL can be used and the image referenced as *../pics/filename.jpg*. In the context of the current location of the site, it means the same thing. The dot-dot-slash tells the system to go back up a directory and down into the *pics* subdirectory. Not only is this shorter and easier to type, but it saves a huge amount of time if the entire Web site must be moved en masse to another domain name or Web host (assuming you're not renaming directories or rearranging at the same time). With relative URLs, which reference the current location, the links will still point to the same locations in the directory structures as they did on the previous domain location. If the Webmaster didn't use relative URLs to build a site for which the domain name is likely to change, rap his or her knuckles.

Relative URLs were described in Standards Track RFC 1808 by Fielding in June 1995 and have since become an intrinsic part of HTML on the World Wide Web.

RUS See Rural Utilities Service.

Rutherford, Ernest (1871–1937) A New Zealand-born British physicist who contributed substantially to knowledge about atomic physics. Rutherford researched at the Cavendish lab studying ionizing gases and following up much of the work of the Curies and Philipp Lenard. He collaborated with Hans Geiger, developer of the Geiger counter, and influenced Paul Villard's studies of gamma rays.

RVA recorded voice announcement. A digital or analog recorded or synthesized voice announcement, as on an answering machine. The phone company uses RVAs to communicate with callers using touchtone menus and to alert the user to problems (such as an off-hook phone). See Barbe, Jane.

RW 1. read/write. 2. see real world. 2. remote workstation.

RWhois, rwhois Referral Whois. RWhois is a program for looking up information on the Internet. It is a primarily hierarchical, client/server distributed system for the discovery, retrieval, and maintenance of directory information on computer networks. RWhois facilitates deterministic routing of queries based upon tags, referring the user closer to the source of the information. The RWhois specification defines a directory architecture and a directory access protocol.

RWhois extends and enhances its predecessor, Whois (based upon WhoIs Protocol) in a hierarchical, scalable way in order to meet the increased demands on the Internet. The protocol and its architecture are structurally derived from the Domain Name System (DNS), and concepts from the X.500 Protocol and Simple Mail Transport Protocol (SMTP) have been incorporated into the specification. To use RWhois from a command line shell that supports the utility, type "rwhois" in lowercase.

Whois is useful for querying about IP numbers, NIC handles, and domain name registrants through domain registry services. For example, a VeriSign domain registry search for crcpress.com yields:

```
Registrant:
  CRC Press, Inc. (CRCPRESS-DOM)
    2000 Corporate Blvd., NW
    Boca Raton, FL 33431
    US
Domain Name: CRCPRESS.COM
Administrative Contact, Billing Contact:
    .

Record last updated on 14-Nov-2001.
Record expires on 19-Nov-2003.
Record created on 18-Nov-1993.
Database last updated on 28-Mar-2002
05:03:00 EST.
Domain servers in listed order:
NS1.DATARETURN.COM          216.46.236.253
NS2.DATARETURN.COM           63.251.95.25
```

The RWhois Operational Development Working Group is an IETF group chartered with coordinating, engineering, and operating the RWhois Protocol. See InterNIC, Whois, RFC 2167.

RWW See read-while-write.

Rx *abbrev.* receive. Often used in conjunction with Tx (transmit).

RxD receive data. Data channel, typically used in serial communications, which is an input for DTE devices and an output for DCE devices. See TxD, RTS, DSR, DTR, RS-232.

RZ return to zero. Return to a value of zero for a variable, file, or other entity. This is often used in binary signaling system contexts. See RZL.

RZ-code A variety of signaling codes that are able to return to a zero level to indicate a data value. RZ-AMI (alternate mark inversion), RZ-bipolar, and RZ-unipolar are examples. RZ codes are used in many transmission schemes for a variety of data encoding and timing synchronization applications. Some theorists make a distinction between RZ and AMI codes on the basis of frequency components, with reference to modulation, with RZ code streams having two components and AMI code streams having one. This distinction has practical implications in terms of filtering and resulting distortion in the RZ signal. See Alternate Mark Inversion, RZ.

RZL return to zero level. Return to a level of zero, which may be zero voltage, zero output (of radio waves, for example), or other varying phenomenon. Zero in this case, usually implies an absence of a phenomenon, but it may also be a reference point such as 0°F. RZL is often used in analog contexts. See RZ.

S 1. *abbrev.* sleeve or shield (ground shield) in electrical terminology. Cords, wires, and optical cables are protected by sleeves. The designation is sometimes used to describe plug configurations, for example, TRS refers to *tip/sleeve* or *tip/ring/sleeve* or *tip/ring/ground shield.* A common example of a TRS configuration is a balanced stereo plug on stereo headphones. 2. *symb.* sulfur. See sulfur.

S bus In ISDN networks the data transmission path interconnecting network terminating equipment (NT1 or NT2) to addressable devices. The S bus can supply up to 8 W normal power or up to 420 mW in restricted power situations. Voltages vary from around 31 volts for normal and from around 37 volts for restricted power. See S interface, SBus.

S interface In ISDN networks, a number of reference points have been specified as R, S, T, and U interfaces. To establish ISDN services, the telephone company typically has to install a number of devices to create the all-digital circuit connection necessary to send and receive digital voice and data transmissions.

The S interface is typically used in the U.S. to connect an NT1/NT2 network termination device at the customer's premises to terminal equipment (TE) or terminal adaptors (TA).

In parts of Europe, a combined S/T interface provides a four-wire electrical extension interface between network terminating equipment (NT1) and up to eight addressable devices. The use of four wires enables a pair of wires to be used for each direction to and from the NT*x* equipment. The attached devices are typically telephones or computer interfaces. See ISDN interfaces for a diagram.

S Series Recommendations A series of ITU-T recommended guidelines for telegraph services terminal equipment. The publications are available from the ITU-T for purchase from the Net. Since ITU-T specifications and recommendations are widely followed by vendors in the telecommunications industry, those wanting to maximize interoperability with other systems should consult the information disseminated by the ITU-T. A full list of general categories is listed in Appendix C and specific series topics are listed individually in this dictionary, e.g., R. Series Recommendations. See the S Series Recommendations chart.

S-band A radio wave broadcasting frequency spectrum ranging from about 1700 to 2360 MHz. S-band is typically used for radar, space communications, and some types of mobile services. It is terrain-sensitive. NASA uses S-band phase modulation to transmit and receive information between orbiting systems and the ground either directly or through relay satellites. NASA's orbiter communications *forward links* (uplinks) are modulated on a center carrier frequency of 2106.4 MHz for the primary system and 2041.9 MHz for the secondary system. Two frequencies are used to prevent interference if two orbiters are transmitting simultaneously. Similarly the *return links* (downlinks) are modulated on a center carrier frequency of 2287.5 MHz and 2217.5 MHz. The Department of Defense (DoD) forward links are at lower S-band frequencies than the NASA systems (the return link frequencies are the same as NASA) and are channeled through its own ground stations.

Radio waves are not used only for communication. They are also used for navigational tracking and the scientific determination of the characteristics of bodies in our solar system. For example, S-band Doppler effect experiments were conducted on Apollo 15 missions 14 to 17. These experiments enabled gravitational fields and other measurements to be calculated by observing the dynamic motion of the spacecraft through S-band radio waves transmitted between the craft and the Earth.

One of the limitations of S-band is that it is terrain-sensitive, which is why it has largely been used in space applications. However, if you are measuring terrain, this limitation becomes an asset. NASA has applied L-band and S-band frequency sensors to the remote sensing of geographical characteristics such as soil moisture and ocean salinity. The PALS (Passive/Active L/S-band Sensor) was first flown in July 1999 for these purposes. See band allocations for a chart. See S-band, optical.

S-band, optical For optical communications, an ITU-specified transmission band in the 1450- to 1530-nm range. The portion in the 1450- to 1490-nm range has been called the S+-band. The shorter optical wavelength bands are considered promising for increasing capacity in dense wavelength division multiplexed transmission systems carrying C- and L-band frequencies. However, novel means of amplification are required to fully realize the potential of S-band frequencies for this use as traditional erbium-doped amplifiers do not provide the needed gain. Bromage et al. at Lucent Technologies have suggested that Raman amplification could fill this need. See C-band, L-band.

S-band Linear Collider Test Facility A facility at DESY developed to serve as a test bed for the technical aspects of a large-scale 500GeV e+ e– collider to enable S-band technologies to be used more effectively with linear colliders for fundamental research in physics. See S-band.

S-band Single-Access Transmitter SSAT. A trans-

ITU-T S Series Recommendations	
Recom.	Description
S.1	International Telegraph Alphabet No. 2
S.2	Coding scheme using International Telegraph Alphabet No. 2 (ITA2) to allow the transmission of capital and small letters
S.3	Transmission characteristics of the local end with its termination (ITA2)
S.4	Special use of certain characters of the International Telegraph Alphabet No. 2
S.5	Standardization of page-printing start-stop equipment and cooperation between page-printing and tape-printing start-stop equipment (ITA2)
S.6	Characteristics of answerback units (ITA2)
S.7	Control of teleprinter motors
S.8	Intercontinental standardization of the modulation rate of start-stop apparatus and of the use of combination No. 4 in figure-shift
S.9	Switching equipment of start-stop apparatus
S.10	Transmission at reduced-character transfer rate over a standardized 50-baud telegraph channel
S.11	Use of start-stop reperforating equipment for perforated tape retransmission
S.12	Conditions that must be satisfied by synchronous systems operating in connection with standard 50-baud teleprinter circuits
S.13	Use on radio circuits of 7-unit synchronous systems giving error correction by automatic repetition
S.14	Suppression of unwanted reception in radiotelegraph multidestination teleprinter systems
S.15	Use of the telex network for data transmission at 50 bauds

Recom.	Description
S.16	Connection to the telex network of an automatic terminal using a V.24 DCE/DTE interface
S.17	Answer-back unit simulators
S.18	Conversion between International Telegraph Alphabet No. 2 and International Alphabet No. 5
S.19	Calling and answering in the telex network with automatic terminal equipment
S.20	Automatic clearing procedure for a telex terminal
S.21	Use of display screens in telex machines
S.22	"Conversation impossible" or prerecorded message in response to J/BELL signals from a telex terminal
S.23	Automatic request of the answerback of the terminal of the calling party, by the telex terminal of the called party or by the international network
S.30	Standardization of basic model page-printing machine using International Alphabet No. 5
S.31	Transmission characteristics for start-stop data terminal equipment using International Alphabet No. 5
S.32	Answer-back units for 200- and 300-baud start-stop machines in accordance with Recommendation S.30
S.33	Alphabets and presentation characteristics for the intex service
S.34	Intex terminals – Requirements to effect interworking with the international telex service
S.35	Answerback coding for the Intex service
S.36	INTEX and similar services – Terminal requirements to effect interworking between terminals operating at different speeds
S.140	Definitions of essential technical terms relating to apparatus for alphabetic telegraphy
Supplements	
S.Sup1	Minimal specifications for the bilingual (Arabic/Latin) teleprinter

mitter installed in the Hubble Space Telescope that sends data gathered by the Hubble to astronomers on Earth using S-band radio frequency signals. Hubble has two of these transmitters and two large communications dishes that direct the data transmissions to orbiting NASA satellites with single-access antennas, where they are collected and relayed to a ground station in New Mexico. From there the data are forwarded to the Hubble Space Telescope Science Institute in Baltimore.

When one of the SSATs failed in 1998, the other was rotated to shoulder some of the load of the failed transmitter. Since the Hubble was designed to be maintained and repaired during space missions, an S-band Single-Access Transmitter was one of the components stored in the Contingency ORU Protective Enclosure (COPE) in the Space Shuttle Discovery's cargo bay within the Orbital Replacement Unit Carrier. The Discovery crew was scheduled to replace the failed transmitter on Day 3 of Mission 3A. See S-band.

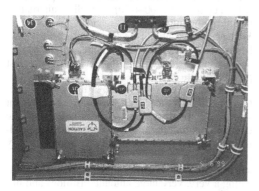

The S-band Single-Access Transmitter unit used in the space-based Hubble Space Telescope for transmitting data via orbiting satellites to data processing centers and astronomers on Earth. [NASA image.]

S-HTTP See Secure HTTP.

S-Video, Super Video A video transmissions standard in which information is carried in two separate signals: chrominance (color) and luminance (brightness). It is also known as S-VHS and as Y/C video with Y representing luminance and C representing chrominance (which in turn carries I and Q signals). S-Video provides a higher quality, sharper image than traditional *composite* video in which chrominance and luminance are transmitted as a combined signal. S-Video is commonly supported on newer monitors, camcorders, and other consumer and professional video devices. Beware of bargain basement S-Video cables; inadequate shielding can result in a type of interference called crosstalk and cancels out the benefits of S-Video. Most S-Video cables are 4-pin mini-DINs but some manufacturers have created custom cables/pinouts for S-Video (e.g., for game boxes). Note that it may not be enough to hook up an S-Video cable to get S-Video output/input; you may also have

to set a menu setting or flip a switch to direct the device to use the S-Video port as many devices default to a composite port. See I signal, Q signal.

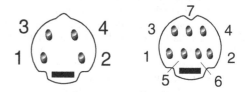

Two male mini-DIN S-Video connectors. On the left, the standard 4-pin cable compatible with a wide variety of camcorders, video monitors, VCRs, etc. On the right, a 7-pin mini-DIN connector for Macintosh computers (e.g., PowerMac AV) and some video capture cards with standard camera interfaces. Both cable ends are keyed to prevent incorrect insertion.

Pinouts – standard S-Video:

1	GND	*ground (Y)*
2	GND	*ground (C)*
3	Y	*luminance (light intensity)*
4	C	*chrominance (color)*

Pinouts – PowerMac AV S-Video (capture cards with female connections may assign pins 6 and 7 opposite to the Mac pinouts shown here):

1	A GND	*ground*
2	A GND	*ground*
3	Y	*luminance (light intensity)*
4	C	*chrominance (color)*
5	I²C clock	*composite video*
6	+12V	*no connection*
7	I²C data	*no connection*

S+-band See S-band, optical.

S/MAIL A commercial Windows-based product from RSA Data Security, Inc., released in August 1997. S/MAIL is a developer software kit for building S/MIME-enabled applications. It is a high-level toolkit providing a plugin engine for secure email messaging that enables S/MIME functionality to be integrated into a variety of application types, including EDI software, online service clients, and email clients. The S/MAIL toolkit includes core cryptographic components, message formatting, and a security user interface. See Multipurpose Internet Mail Extension, S/MIME.

S/MIME A commercial product from RSA Data Security, Inc. for providing interoperable, secure email. S/MIME facilitates the development of interoperable RSA-based security products for electronic messaging so that an S/MIME message can be composed and encrypted with one application and decrypted with another. It is based on standard MIME specifications integrated with the Public Key Cryptography Standards (PKCS). See Multipurpose Internet Mail Extension, S/MAIL.

S/N See signal-to-noise.

SA 1. See Service Agent. 2. source address. In networking, an address to identify the physical or virtual location of the system initiating a transmission. This is important for various reasons, depending on the device and the topology of the network, but is used for establishing point-to-point communications, for routing return information, establishing an efficient communications path, auditing secured communications, tracing a path, and many other functions.

SAA 1. Standards Association of Australia. See Standards Australia International, Ltd. 2. Supplemental Alert Adapter. A connection device for interfacing alerting devices to analog multiline phones. 3. See Systems Application Architecture.

Saco River Telegraph & Telephone Company SRTT. Maine's oldest independent phone company, established in 1889. The company has gone through many changes in its more than 100-year history. By fall 2001, it had published the Saco River Yellow Pages on its Web site and announced the coming availability of DSL services for its subscribers.

SAFE 1. Security and Freedom through Encryption.

SAFE Act See Security and Freedom through Encryption Act. See Clipper Chip, Pretty Good Privacy.

SAFENET The U.S. Department of Defense's (DoD's) military standard for a Survivable Adaptable Fiber Optic Embedded Network intended to provide mission-critical, networked communications. See Xpress Transport Protocol.

safety saddle A wide sling-shaped seat that fits around the buttocks of a worker suspended from a pole, line, building, crane, or other high prominence for the purposes of installing, testing, repairing, or maintaining equipment, especially telecommunications lines. The saddle is usually attached to a line or support hook by a hitch arrangement on the top of the saddle, above the user's belly, although those designed for use when climbing poles may wrap around to the other side of the pole to be hitched up as the lineworker climbs. Safety saddles are typically made of thick, tough, resilient materials (e.g., leather) as the worker's life depends upon their reliability.

sag, cable The characteristic downward caternary curve occurring in the center of horizontally hung cables due to the effect of gravity. The curvature is based upon a variety of factors, including the diameter of the cable, the "straightline" distance (the distance between two adjacent cable supporting structures such as utility poles), and the flexibility of the materials surrounding the conductive materials, if present. Wind, temperature, and humidity sometimes influence the degree of sag in lighter cables.

Cable sag estimates are important, since extra cable must be calculated and ordered to compensate for the effects of sag (the cable displacement), especially over long cable segments. Sag estimates must also be calculated for cables that are fragile and more likely to break and for fiber optic cables in which the angle of the light beam should be as straight as possible to prevent signal loss into the cladding. The sag estimate can help determine the maximum recommended distance between attachment points.

There are sites on the Web that enable users to calculate displacement cable sag errors by inputting cable tension, distance, weight, and gravity force data.

sag, voltage A short dip or decrease in voltage from a power source. In "mains" alternating current (AC) power sources, voltage sag can result from lightning storms, fallen trees, malfunctions, and other causes. Loss of power may result in a voltage sag while a backup power supply comes online or while a transformer feeds a load during the electrical fault. In public power distribution systems, those closer to the fault are usually the most affected; those farther from the fault may be buffered by intervening transformer stations.

Voltage sags may damage many types of sensitive electronic components, especially in their manufacture. As a consequence, the Semiconductor Industry Association (SIA) has developed voltage sag immunity standards for semiconductor manufacturing equipment. Certification according to the SEMI F47 standard indicates that equipment complies with certain voltage sag duration tolerances.

Similarly, the IEEE P1564 Task Force on Voltage Sag Indices met in 2001 to review a proposed five-step process for developing Sag Indices based upon a draft IEC document (61000-4-30), as well as alternative means of obtaining voltage sag indices.

Voltage sag susceptibility testing is a process for identifying weak links in a system through simulated production modes to identify problem areas and to test possible solution scenarios.

A power conditioner is a system installed in conjunction with the electrical distribution system to prevent sags. In electrically sensitive fabrications plants, the cost of a conditioner or other sag-prevention product may be less than the cost of loss or productivity or sag-caused damage.

sagan A tongue-in-cheek tribute to Carl Sagan, indicating a very, very large amount. "Billions and billions," as he would say with infectious enthusiasm in his popular TV series when referring to the many stars in the cosmos. Sagan's premature death was mourned by many amateur astronomers who got their first taste of the wonders of the galaxy and beyond through Sagan's show.

Sagan, Carl E. (1934–1996) An American astronomer, writer, educator, and inspirational host of the popular "Cosmos" television series on the U.S. Public Broadcast System (PBS). Sagan was the director of the Laboratory for Planetary Studies and the David Duncan Professor of Astronomy and Space Sciences at Cornell University. See sagan.

SAGE Semi-Automatic Ground Environment. A U.S. government security digital communications, detection, and craft control network.

SAIL See Stanford Artificial Intelligence Laboratory.

Salvà i Campillo, Francesc (1751–1828) One of the genuine pioneers of telegraphic technology, Salvà was a prodigy who received a degree in medicine at the age of 20 and went on to study communicable diseases, promoting the use of the smallpox vaccine. He was a prolific researcher in many fields and developed a type of underwater craft, a historic submarine, as well as an aeronautic balloon that was demonstrated in flight in 1784.

Salvà was already describing ideas for telegraphs by 1795, before the invention of the voltaic pile, and he designed an electrochemical telegraph signaling system around 1804 in Barcelona, Spain, more than three decades earlier than the Wheatstone/Cooke and Morse telegraphs. Each character was assigned to an electrical wire that produced gas bubbles in an acid bath at the receiving end when current was applied to the cable on the transmitting end. This system provided inspiration for Samuel Thomas von Sömmering a few years later. See Sömmering, Samuel Thomas; telegraph history.

SAM security accounts manager.

samarium Sm. A silver rare earth metallic element (AN 62) discovered spectroscopically in samarskite in the late 1800s. Samarium is used for doping calcium fluoride crystals for use in lasers. It is one of the rare earth metals used in carbon arc lights. See doping.

Samba An open source client/server system running on Linux systems, Samba facilitates peaceful coexistence between Unix and Windows platforms. Samba communicates with Windows clients transparently, enabling a Unix system to join a Windows-based "Network Neighborhood." In the other direction, Windows users can access file and print services and other resources on the Unix system. Communication is facilitated by the Common Internet File System (CIFS), the heir to Server Message Block (SMB) protocol. Samba has been ported to other operating systems as well, including VMS, NetWare, and AmigaOS. See Server Message Block protocol. Samba is freely downloadable from the Samba site. http://samba.org/

sampling Recording a signal by quantizing it at intervals in order to capture its basic properties, usually also accompanied by saving the samples in a digital or abbreviated form. It is a form of analog to digital conversion. In digital sound sampling, for example, the sound of a musical instrument or a voice can be digitally sampled a certain number of times per second in order to be able to play back the sound so that it retains and conveys the character of the original, although not necessarily all the information or format of the original.

Music synthesizers use sound samples to recreate the sounds of various traditional instruments. Computers use sound samples to alert or amuse users or to enhance video games or other applications. Telephone systems use sound samples to send voice mail announcements or instructions to users or to transmit voice conversations over digital systems.

In general, more frequent samples result in better fidelity to the original during playback. However, there are practical and perceptual limits. More frequent sampling makes higher demands on the equipment, requires more memory to store, and more bandwidth to transmit. Since humans can't distinguish the sample from the original above certain parameters, it is not practical to commit extra resources to recreating a signal above these limits. In commercial applications, a sampling rate of 44.1 kHz is used on audio CDs. Audio-only DVDs are higher, but some of this is related more to marketing and technical compatibilities than to increased perceptual enjoyment.

Some sounds, like the sound of a concerto played on a violin, are more complex than others (e.g., a doorbell), and require more frequent samples and a greater frequency range to retain the perceptual quality of the original. In some cases, different sampling techniques must be used at different pitch ranges in order to recreate the sound as it is heard by humans.

Sampling algorithms are often applied to sound transmissions, but they are also relevant to video transmissions or multimedia transmissions. In videoconferencing over slow transmission lines, the image is usually sampled rather than played in realtime, that is, a new still image is grabbed or digitized and transmitted every few seconds or every few minutes. Generally, as in sound sampling, more frequent samples provide greater fidelity to the original. In video sampling, rates of at least 20 to 30 frames per second are perceived by humans as natural motion. See animation, audiographics, pulse code modulation.

sampling rate The number of captures of an input, such as light or sound waves, per unit of time. Sampling rates for images are generally expressed in frames per second, with 24 or more appearing natural to the viewer. Sampling rates for audio are generally expressed in kilohertz (kHz); an instrument might be sampled at 50 kHz, that is 50,000 bits per second. Higher sampling rates generally require more sophisticated equipment, higher processing speeds, and faster transmission speeds, especially if the signals are sent over a network. See sampling, sampling theorem.

sampling rate, Nyquist A theoretical sampling frequency (in terms of rate not wavelength) at which the rate is the minimum separation of samples in a Fourier plane that enables a complete reconstruction of the original sampled data. The sampling process itself may introduce errors into the quantization, so the actual sampling rate needed to reconstruct a signal may be higher than the theoretical Nyquist rate. Depending upon the application, half the Nyquist rate may be referred to as the Nyquist frequency.

The mathematical and theoretical groundwork for defining the Nyquist rate is based on work by Nyquist in 1928 and Shannon in the late 1940s. The theorem has variously been called the Nyquist or Shannon or Nyquist-Shannon sampling theorem.

Sampling at rates below the Nyquist rate is called *undersampling* and results in less-than-perfect reconstruction of the original sampled data, but may be expedient in terms of resources. Humans are very good at perceptually filling in the blanks when presented with incomplete data. A picture of a familiar face can have most of the face removed with scissors or blotted out by a marking pen and still be recognizable to many. In the same sense, information can be left out of an image or sound sample and still be recognizable (if not optimal) for many purposes. Sampling beyond the Nyquist level is termed *over-*

sampling and is unnecessary in terms of reconstruction but may be useful in creating redundancy for error correction or transmission purposes.

The application of Nyquist rate theories is as much art as science. Since filters are introduced into quantization methods to create high and low cutoff values, some assumptions are made about the signal even before it is sampled. It cannot always be known in advance what these values should be and some adjustments may have to be made based on trial and error. The "complete" reconstruction of a signal is very hard to judge in advance as well. When choosing a sampling rate for audio, for example, and playing it back to a general audience, it might be acceptable to the audience and seem to them to be identical to the original analog signal. If, on the other hand, the same sample is played to a trained musician, such as a concert performer, that individual might notice significant differences between the original and the reconstruction. In fact, this situation happened in the production of early audio CDs, with many music lovers complaining that CD music sounded flat, a situation that has since been improved with experience, technological adjustments, and a better understanding of human auditory perception. See sampling, sampling rate, sampling theorem.

sampling theorem A theoretical basis for relating discrete representations, or samples, to continuous functions with the implication that the continuous function can be recreated from the discrete representations to a lesser or greater degree, depending upon the characteristics of the continuous function. In simpler terms, a sampling theorem is a theoretical context within which analog-to-digital and digital-to-analog mathematical conversion formulas can be developed.

Sampling theorems form a theoretical basis for developing formulas and practical applications such as analog sound digitization and reconstruction. These formulas are still evolving and it has been suggested that the Poisson Summation Formulas can be used to prove corresponding sampling theorems and, in turn, give rise to new uniform sampling formulas (Benedetto and Zimmermann, 1997). The Interpolation Identity has been proposed as a means to develop a new class of sampling theorems for obtaining efficient discrete-time (DT) systems and the efficient interpolation and reconstruction of samples (Eldar and Oppenheim, 2000).

Sampling theorems are often named after the scientists who developed them or who set the mathematical groundwork for subsequent sampling theories. There are sampling theorems named for the work of Whittaker in 1915 and for Shannon in the late 1940s. The theories of Whittaker and Shannon led to important advances in pulse code modulation (PCM). The Shannon sampling theorem is used in uniform sampling, for example, and is especially practical for certain types of transmissions, e.g., FM broadcasts. A variant on this is the Whittaker-Shannon-Kotel'nikov sampling theorem which may be used, for example, for sampling stochastic signals. A. Papoulis developed

a generalized expansion theorem for uniform sampling and M. Unser suggested an extension of this for nonband-limited functions. The Papoulis-Gerchberg theorem can be used to recover missing samples in finite-length records of band-limited data and has also been applied to wavelet subspaces (Xia et al., 1995). More recent applications of sampling theorems in wavelet technologies have resulted in some intriguing quantization and compression algorithms. See Kotel'nikov, Vladimir.

In practical applications, most samples are uniform (periodic) and Fourier transforms are often used in the context of sampling theorems. However, not all sampling environments are uniform; multichannel sets may be used or samples may vary according to time and traditional Fourier-based methods of handling the data may be impractical. Consequently, researchers have been developing general theories to encompass nonuniform sampling in shift-invariant spaces (e.g., Aldroubi et al.). See Fourier transform, pulse code modulation, sampling, sampling rate, wavelet theory.

SAN 1. See satellite access node. 2. See storage area network.

Sandbox A Sun Microsystems Java security block. Since Java applets are freely shared through public sites on the Internet, there is always reason to question whether they contain hidden viruses or other destructive or annoying capabilities. The Sandbox is a means of restricting doubtful applets to a confined area, that is, quarantining them so they cannot affect other data on the disk or other Sandboxes.

SANS Institute The System Administration, Networking, and Security Institute, founded in 1989, is a cooperative research and education organization serving security professionals and systems administrators worldwide. The Institute helps to promote research, incident awareness, and security certification programs related to global network security. http://www.sans.org/

SANZ Standards Association of New Zealand.

SAP 1. Scientific Advisory Panel. 2. See Second Audio Program. 3. Service Access Point. An interface point in a network, often associated with a specific layer. 4. See Service Advertising Protocol. 5. See Session Announcement Protocol.

SAPI Service Access Point Identifier.

SAR 1. See segmentation and reassembly. 2. synthetic aperture radar.

SAREX See Shuttle Amateur Radio Experiment.

Sarnoff, David (1891–1971) A Russian emigrant to America, Sarnoff was an ambitious, energetic radio operator and Marconi station manager at the Radio Corporation of America (RCA) widely reported to have intercepted the messages of the *Carpathia* when the *Titanic* struck an iceberg, and to have relayed the messages to relatives and friends of the passengers on the sinking ship. While it is likely that Sarnoff did play a part in relaying the messages, the RCA promotional information of this event included a doctored picture of Sarnoff at the console, which lends some doubt to the claim that Sarnoff handled the post single-handedly for 72 hours, as widely promoted. Sarnoff became the general manager of the Radio Corporation of American in 1921 and went on to play a large part in its history and development. He is also remembered for his association with the inventor E. Armstrong in the late 1920s which ended abruptly in 1935 with the removal of Armstrong's test equipment from the premises and the development of television. In 1926, Sarnoff was instrumental in founding the National Broadcasting Company (NBC).

The IEEE now honors Sarnoff's contributions with the David Sarnoff Award in Electronics. See Radio Corporation of America.

SART See search and rescue radar transponder.

SAS 1. single address space. 2. simple attachment scheme. 3. Survivable Adaptive Systems.

SASL See Simple Security and Authentication Layer.

SASMO Syrian Arab Organization for Standardization and Metrology.

SASO Saudi Arabian Standards Organization.

sat-, -sat A prefix/suffix often prepended/appended to satellite-related names and technologies.

SATCOM, SatCom Satellite Communications. SATCOM is both a colloquial abbreviation and the trade name of quite a number of satellite-related firms and organizations around the world.

- North American Treaty Organization Satellite Communications, known as NATO SATCOM, is a branch of the Communications Systems Division that supports the Satellite Communications Project, which in turn provides support to Major NATO Commands. See SATCOM Integrated Test Network.
- Satcom Resources is a commercial supplier of satellite communications systems.
- SatCom Systems, Inc. is a Federal Communications Commission-licensed supplier of U.S. market mobile satellite services through the MSAT-1 satellite system.
- SatCom Electronics, Inc. provides electronics products for wireless and broadband satellite communications, including DBS-TV satellite services.
- The British National Space Centre (BNSC) uses a variation on the name, S@TCOM for a program designed to help companies in the United Kingdom exploit satellite communications and navigations opportunities.

SATCOM Integrated Test Network SATIN. A central testbed for NATO SATCOM experiments in communications services and technologies in a SATCOM environment. SATIN can provide communications to other testbeds and can interconnect with other divisions and organizations through NC3A Satellite Experimental Terminal (SET) facilities. Examples of SATIN projects include ATM over SATCOM and Maritime over SATCOM.

satellite, artificial A manufactured object launched to orbit the Earth, Moon, or other celestial body. There are currently many communications and Global Positioning System (GPS) satellites in orbit around Earth which send and receive signals to and from ground stations and transportation vehicles (cars, trains, boats, planes, etc.). Satellites now provide the main means for wireless long-distance communications. The first artificial satellite was Sputnik I in 1957, followed by the first geostationary satellite in 1963. See global positioning systems, satellite antennas.

satellite, natural A celestial body in orbit around another.

satellite access node SAN. A terrestrial satellite link, usually consisting of an Earth station or Earth station hub.

satellite antennas Satellite antennas were originally launched into orbit for military monitoring and communications, space research, and cable TV broadcasting, but increasing numbers serve individual parabolic home receivers and data communications providers. GPS satellites orbit at about 18,000 km (11,000 miles) and broadcast satellites at about 36,000 km (22,300 miles) altitude.

In its basic form, a broadcast satellite system consists of a broadcasting station sending signals through an uplink dish aimed directly at a geostationary satellite antenna in synchronized orbit with the Earth. The signal subsequently is sent from the satellite to a downlink dish (parabolic antenna) attached variously to business complexes or rebroadcast stations, and subsequently directed to subscribers through coaxial cable. Some stations broadcast directly through scrambled signals to apartment blocks or individual households. The lead from the downlink dish feeds into the user's television or computer system. See antenna, C-band, feed horn, Global Positioning System, Ka-band, Ku-band, microwave antenna, parabolic antenna.

satellite broadcast frequencies The various ranges of frequencies over which satellite antenna transmissions take place. These are dependent on many factors, including the type of transmission, the type of satellite, and regulatory guidelines and restrictions. Broadcast stations typically operate in the C-band, with uplinks at about 6000 MHz and downlinks at about 4000 MHz to rebroadcast stations with powerful antennas. The frequency levels are tied to the size of the receiving dishes, with higher frequencies more difficult to accommodate technologically, but with the advantage of much smaller receiving dishes. Higher frequencies can broadcast to smaller receivers, making it possible for some frequencies to be broadcast directly to smaller consumer dishes. See C-band, Ka-band, Ku-band.

satellite closet A centralized wiring closet for interconnection of cables and equipment. In a number of satellite installations, the programming is beamed from the satellite to a central service provider with a satellite receiving dish and, from there, delivered by wire or cable to subscribers, necessitating local loop hookups. See distribution frame.

satellite communications A wide variety of radio, television, telephone, data, and other broadcast and two-way wireless communications provided by transmission via orbiting satellites to centralized distribution providers or individual subscriber satellite dishes.

The age of satellite communications began in the late 1950s, with the launch of Sputnik I, although it was described with remarkable insight by Arthur C. Clarke in the 1940s and 1950s in various articles and books.

The early satellites did not last long (from a few weeks to a few months) and power consumption and radiation problems had to be solved before widespread use became practical.

In less than four decades from their modest beginnings, satellite communications have developed rapidly and now hundreds of satellites of different designs orbit the Earth at various distances. Their lifespans now range from about 5 to 15 years, and most are powered by solar panels with battery backup.

It was not long after the first satellites were launched that they were used by commercial and amateur radio stations. The first television broadcast station to use satellites was the Canadian Broadcasting Corporation, transmitting through ANIK in 1972. Direct broadcast to consumers, rather than through intermediary stations, did not really become prevalent until the 1990s, when broadcasts of higher frequencies became practical and smaller, more convenient satellite dishes were manufactured.

Satellites are launched along with spacecraft or shuttle craft into elliptical or geostationary orbits from about 500 km to about 36,000 km above the Earth and are able either to passively transmit data back to Earth (these are becoming rare), or actively regenerate or otherwise amplify the signal and retransmit, usually at a different frequency to avoid interference of uplink and downlink signals. Satellites are general purpose with many transponders or specialized for data, voice, broadcast, etc. Broadcast satellites tend to be unidirectional, while data and voice satellites, as for mobile systems, are bidirectional. See AMSAT; Clarke, Arthur C; Global Positioning System; direct broadcast satellite, OSCAR, Syncom, Telstar, and the many listings under satellite services.

satellite communications control SCC. The Earth-based station facilities and equipment which control satellite transmissions, including signaling functions, access control, error correction, signal conditioning and noise reduction, etc.

satellite constellation A group of satellites in a cluster. A group of satellites is commonly used for Global Positioning System (GPS) applications, for example, where data from three or more related satellites are mathematically manipulated to yield precise positioning information of an Earth location.

Satellite Home Viewer Improvement Act SHVIA. An Act signed into law by President Clinton in November 1999. This significant Act modifies a number of existing communications and intellectual property acts, including the Satellite Home Viewer Act of 1988, the Communications Act, and the U.S. Copyright Act. It permits satellite broadcasting services companies to retransmit signals within their designated market area (DMA) without paying a royalty, beginning on January 1, 2002. To offset this privilege, satellite carriers must carry the signals of all full-power TV broadcast stations within that market, upon request.

The purpose of the Act is to promote competition among multichannel video programming distributors such as cable television and satellite suppliers, while increasing the programming range for subscribers. Satellite broadcasters would now be permitted to provide local broadcast TV signals to all subscribers within the licensed market (Designated Market Area [DMA]) as defined by Nielsen Media Research. This is known as local-into-local service and would initially include only major network affiliates. Unfortunately, as the service is optional, unprofitable markets such as rural areas would potentially be neglected by this Act without further legislative actions and incentives to promote the delivery of services in underserved areas.

The Act further permits satellite companies to provide distant network stations to eligible satellite subscribers. Some exemptions are contained in the Act, including an exemption for those subscribing to C-band services on or before October 31, 1999 and exemptions for certain recreational vehicles and trucks. Prior to SHVIA, consumers who terminated cable service had to wait 90 days before receiving satellite TV service. This is no longer mandatory.

The Act did not pass without subsequent opposition. A number of satellite providers jointly and individually filed suit against the FCC and the Copyright Office on the grounds that the local-to-local service provisions were unconstitutional. A number of public broadcasting organizations intervened in the interests of public television stations. Concerns over rural and other underserved populations resulted in other related Acts, including the LOCAL TV Act of 2000. See Grade B signal, LOCAL TV Act of 2000, National Rural Telecommunications Cooperative, Rural Local Television Signals Act.

satellite link A system of transmitters and receivers communicating with a satellite, usually through an active transponder which will amplify and shift the received communications to another frequency before retransmitting on the downlink. Uplinks and downlinks are often managed separately. For example, television broadcasts are primarily in one direction (though interactive TV applications are increasing), while phone and computer data communications are typically in two directions. See geostationary, satellite.

satellite scanner See scanner.

satellite scatter This has two opposite meanings, as scatter can be the undesirable diffusion and weakening of a signal or, conversely, a deliberate manipulation of the environment to enhance communications. In the latter, it was discovered that ionization of airborne particles could open up communications windows which otherwise were not available. Thus, burns from launched spacecraft or deliberate "seeding" of high regions with elements like barium could provide possibilities for detecting or sending transmissions through these temporary holes. There have also been experiments with heating the ionosphere with high-powered waves to form a type of "aurora" which can facilitate transmissions.

Due to their transient nature and the strength of the signals needed, these are not major sources of communications, but it's valuable to understand the nature of the various phenomena and receive occasional glimpses into frequencies emanating from space.

satellite services Profit and not-for-profit organizations which provide various types of satellite-based communications services. See American Mobile Satellite Corporation, AMSAT, ARIES, Astrolink, Constellation Communications, Inc., CyberStar, ECCO, Ellipso, ICO Global Communications, INMARSAT, Globalstar, OrbLink, Skynet, Spaceway, Teledesic.

Satellite Work Centers A telework organization similar to a branch office, placed in a residential or rural village area by a business entity, and made commercially viable by the implementation of new communications technologies. See ADVANCE Project, Shared Facility Centers, telework.

saturation To add or adjust such that no more can be absorbed, or contained. In color applications, saturation refers to color purity. Undithered colors on a computer monitor or printing inks made from primary pigments tend to be highly saturated.

SAW Surface acoustic wave. See acoustic wave.

Sb *symb.* antimony. See antimony.

SBE See Society of Broadcast Engineers.

SBus A Sun Microsystems data bus is used to support a standardized data format that can be transmitted over a wide variety of computer devices and services, including Fast and Gigabit Ethernet, SCSI, Token-Ring, ISDN, parallel connections, graphics adaptors, and frame buffers. Computers equipped with SBus slots can be extended with SBus-compliant peripheral cards in the same basic way that PCI peripheral cards are installed in Intel-based PCs and Macintosh systems.

Sbus cards may provide a data connection to another device or may provide conversion capabilities. For example, there are SBus cards that serve as PCMCIA adapters, providing one or more Type-II PCMCIA slots for inserting popular PC cards.

The SBus format is specified in IEEE 1496-1993.

SC- connector A standardized optical connector designed for CATV and data network hookups for use with hand or machine-polished fiber filaments. See ST- connector.

scalable Adjustable, able to increase or decrease in size, capacity, or other relevant characteristics, without significant degradation in quality of service or functioning.

Scalable fonts and images, usually defined as vectors, have the capability to adapt to lower and higher screen and printer resolutions, displaying at the best possible resolution for that particular device due to the internal algorithmic nature of the font definition. Scalable images are sometimes called *resolution-independent* images.

Scalable networks allow the system to accommodate to changing conditions. In static environments, scalability is not a critical factor, and nonscalable systems tend to be less expensive. In dynamic environments, such as the Internet or large WAN implementations, scalability can be a crucial factor, especially over time, contributing to the flexibility and usability of a system.

Many aspects of networks need to be scalable. The system software should be scalable to adapt to smaller and larger numbers of users, sometimes on a minute-to-minute basis. Physical storage mediums need to be scalable to accommodate less or more storage as needed. Routing protocols need to be scalable to accommodate changing topologies and numbers of workstations.

Scalable Coherent Interface SCI. A high bandwidth, scalable, media-independent network transmission technology developed in the late 1980s that operates up to about 1 Gbps. It is an ANSI/ISO/IEEE standard (1596–1992). SCI supports parallel distributed multiprocessing and cache-coherent interconnection and fits into the upper mid-range in throughput. It is faster than ATM, Fibre Channel, and Ethernet, but slower than HIPPI-6400, and does not have HIPPI-6400's retransmission capabilities.

Initial implementations of SCI tend to be high-end commercial/industrial and military supercomputing applications.

scaling *v.t.* 1. Sizing, adjusting to size. May be proportional scaling, or selective scaling in one or more axes. Scaling is a common operation in image processing. See cropping. 2. Adjusting to capacity, or number of members. In programming there are often scalable ways of designing algorithms. For example, an operating system may have a fixed number of windows which can be open at one time (e.g., maximum of 200) or it may be scalable, in which the maximum number of windows is limited only by the system resources, and becomes greater as greater resources are added (e.g., memory, storage space, CPU speed, etc.). Scalable systems are more flexible and less likely to go out of date, but are often more resource-intensive and sometimes more difficult to program. In network transmissions with a variety of protocols, scaling may occur to accommodate differences in bandwidth, data, or speed capacities of the various systems through which the transmission may travel.

scan converter A device for converting a video signal. With computers it is common to take the RGB signal that normally leads to the computer monitor and feed the signal through a scan converter so it can be recorded on a video tape.

scan line, scanning line On a display monitor, a narrow more-or-less continuous line illuminated by the movement of the electron beam across the inside surface of the tube. In television broadcasting and raster monitors, these are typically horizontal. On vector monitors, the scan line can be traced in any direction. See raster, vector.

scanner A device that samples objects, information,

S

or processes, and quantifies and records the results in some form that can be further processed or analyzed. Scanners, like digitizers, frequently sample analog information and convert it to digital, or convert a digital sample to a waveform for transmission. Typically the scanner does not process the information; its job is to *capture* the information, and often it can store that information in a variety of selectable formats. Once the information is captured, it is then sent "live" to a processing application or stored for later conversion or further processing. Scanners are used in a multitude of imaging and sensing applications, including those described in the Scanning Technologies chart.

scanning acoustic microscope SAM. A scanning microscope designed to assess the acoustical properties of materials at micrometer and nanometer ranges. There are different ways to approach this problem, but in general, a SAM uses focused beams to scan an object with some kind of detector to gauge the result. In general, SAMs have upper resolution limits that are partly determined by Rayleigh scattering.

Nearfield acoustic microscopes were developed to overcome some of the difficulties of imaging within the diffraction region close to a sample object. They may use electrons directed at the sample through a chopper, with the result picked up by a scanning electron detector, or they may use laser light aimed at the sample and detected and converted to electricity by photodiodes (scanning probe acoustic microscope).

scanning electron microscope SEM. A type of microscope designed to magnify at levels that are physically beyond the scope of traditional optical miscroscopes. SEMs are precision instruments that use a beam of electrons rather than a beam of light to image a tiny section of an object. SEMS evolved from transmission electron microscopes (TEMs), becoming high-end experimental instruments in the 1940s and commercially-practical scientific instruments in the mid-1960s.

SEMS work by accelerating electrons toward the sample through a system of condensing lenses and apertures such that a monochromatic beam is aimed at the surface of the sample through a magnetic lens. The beam is moved so that it scans in a pattern that covers the imaging area. The interactions between the electron beam and the sample are assessed and imaged with sensitive electron energy detectors. Magnifications are significantly higher than optical scopes, up to about 15,000x.

The interaction of the electron beam and the sample results in energy changes that can be detected and are generally interpreted so that higher energy levels show up as brighter image regions on the monitor, photograph, or other display medium.

scanning near-field optical microscopy SNOM. A more recent development in microscopy that enables spatial resolutions below the diffraction limit of light characteristic of far-field optical microscopy. Using a SNOM, the investigated sample is held at a distance smaller than the wavelength from the radiant source.

In the early 1990s, AT&T Bell scientists published near-field images that had been generated with a coated optical fiber probe, demonstrating that it was possible to resolve sub-wavelengths through small-aperture fiber optics.

SNOM capability opens up new areas of research in quantum-level nanostructures without spacial averaging.

Subsequent variations in the concept at UCSD, using silver particle suspensions illuminated by a focused laser instead of a fiber probe, have broadened its applications to include the study of magnetic thin film systems suitable for data storage. See scanning probe microscope.

SNOM Fiber Probe Simplified Example

A SNOM fiber probe is commonly made from a slender single-mode optical fiber with an aluminum coating. The fiber may be fabricated by fiber pulling or chemical etching methods.

Commercially, fiber probes are described in terms of diameter (e.g., 120 μm – d), the angle of the fiber tip (e.g., 12° – b) and its length (c), as well as the optical power for input/output, the wavelength for which the probe is optimized (e.g., 480-550 nm), and the optical efficiency for the specified numerical aperture (the uncoated tip – a). It may also be possible for the customer to specify the length of the covering that serves as a protective coating and handle (e). See probe, photoplastic for a variation on the basic probe.

---

scanning probe microscope SPM. A microscope designed to facilitate the viewing of surface objects and 3D biological systems by means that are not readily attainable with conventional scanning near-field techniques for probing or manipulating surfaces. SPMs are versatile instruments ranging from optical to scanning electron magnifications.

In the mid-1990s, E. Florin described a 3D scanning probe microscope that uses a 2-photon absorption process with fluorophores bound to a probe with the probe trapped by optical "tweezers" in continuous-wave mode. In September 2000, V. Kley submitted a patent application for a scanning probe microscope assembly and method for making confocal, spectrophotometric, near-field, and scanning probe measurements and images. In one embodiment of this invention, a fiber optic light guide transmits light from the source to the probe. In the context of a Mach-Zehnder interferometer, it is formed by two reference light sources (e.g., two lasers at different wavelengths) and two fiber optic guides. See scanning near-field optical microscopy.

scanning rate 1. The speed, in units per time period, at which a scanning device captures information. Many common technologies are expressed in terms of inches per second. Generally, for moving images, faster scan rates are associated with higher accuracy or fidelity in conveying motion. In still images, faster scan rates may compromise the resolution or fidelity of an image. 2. In cathode-ray tubes (CRTs), the scanning rate is the speed at which the electron beam sweeps the screen to refresh a full frame. This is usually about 30 frames per second. See cathode-ray tube, frame, interlace. 3. For technologies that use a beam for scanning (antennas, radar), the rate is more often expressed as a specified number of sweeps over a unit of time.

scanning rate, optical For optical scanning of images, such as those used with computer scanners and facsimile machines, scanning rate is often not as important as resolution. In other words, the fineness of the scan is described rather than the speed at which it is scanned, with resolutions of 1200 to 4800 dpi interpolated common on desktop scanners. See scanner.

scare-straps *colloq.* Telephone line worker safety belts.

scatter *v.* To separate, to distribute widely or randomly, to disperse, to diffuse in various directions.

scatter, transmission To diffuse or spread out in such as way as to lose the strength or directionality of a transmission signal. In most communication transmissions, this is undesirable. See spreading loss. 3. To enhance communications through exploiting delib-

erate, controlled scatter. See satellite scatter.

SCC 1. Specialized Common Carrier. A carrier competing with the dominant carrier in niche markets. 2. Standards Council of Canada.

SCCP See Signaling Connection Control Part.

SCE See Service Creation Environment.

SCEP See Service Creation Environment Point.

Sceptron spectral comparative pattern recognizer. An intelligent pattern-recognition system developed in the early 1960s by Robert D. Hawkins of Sperry Gyroscope. The Sceptron used quartz or glass fibers, a photocell, electrical current, and mechanical motion to recognize a spoken word and optionally print it on an illuminated display. When exposed to audio stimulation, the Sceptron "learns" the sound and can then recognize it. Other types of signals translated into audio frequencies can also be recognized. Sceptron was able to pick out a word from human speech.

The fiber optic array was the most unique aspect of Sceptron's design, loosely packing about 700 fibers into 1/4 cubic inch. When stimulated by mechanical excitation, the fibers, all of the same diameter but different lengths, would vibrate, each at its own natural frequency. A piezoelectric or electromechanical driver was provided to convert electrical signals into mechanical motion. A photocell then detected the motion of the fibers, registering the movement or lack

Scanning Technologies for a Variety of Devices	
Application	Notes
scanning antenna	Parts of a moving antenna which cause the directional scanning of the antenna beam.
facsimile scanner	A component of a facsimile machine which converts a sampled digital image to a waveform for transmission over phone lines. See ITU-T, facsimile, TIFF.
image processing	A computer input device which passes a beam over a 2D or 3D object or image, and converts it to digital data, usually a 2D raster image. See digitizer, optical character recognition, TWAIN.
scanning radio	A radio receiver that scans a range of transmission frequencies automatically, so the user can locate and listen to conversations occurring at the time of the scan. Often used to find emergency or cellular phone conversations.
remote sensing	In satellite remote sensing, a scanner employing an oscillating mirror that captures images in strips was first proposed in 1968 by Hughes Aircraft for use in orbiting satellites, and first deployed in 1972 in the first Earth Resources Technology Satellite. This technology revolutionized understanding and recording of Earth's topological and geological features. The program became Landsat in 1975.
robot vision	In robotics, a scanner samples visual input and processes the information in a way that provides data useful for tracking, navigating, or object sensing. Robot scanner interfaces vary widely, but often are small video cameras, light detection devices, or pattern sampling devices mounted on the robot itself. Robot scanners may be simple, to detect and record light or dark areas, or complex, to capture sophisticated patterns, further processed as faces or recognized as objects.
scanning software	A software program that searches data for new entries of a particular kind, such as newly uploaded files, email, user logins, etc., or which scans processes such as network load, states, CPU usage, etc.

S

of movement as a dark or light spot. Sceptrons were configured in pairs, one with a reference static mask, one as a memory mask, balanced together in a bridge circuit. See neural networks, pattern matching.

Schadt-Helfrich effect Also called the *twisted nematic effect*, this effect was studied and described by M. Schadt and W. Helfrich in "Voltage-Dependent Optical Activity of a Twisted Nematic Liquid Crystal," in 1971. It is a means of modulating light through a layered liquid crystal component.

Imagine a layer of nematic liquid crystals sandwiched between glass polarizing plates rotated 90° in relation to one another, with the crystals separated by thin spacers. A polymer or other orientation layer is introduced to more or less fix the alignment of the liquid crystals parallel to the plates to create a gradual twist down through the axis between the glass plates. Light is applied to one end of the structure, through the glass. The birefringent liquid crystals guide the linearly polarized incident light through the compo-

nent to the viewing surface on the opposite end. The light emerges and the end is lit. If voltage is applied to the component, it disturbs the orientation of the crystals and the polarized light is "canceled" and the viewing surface appears dark. Varying the voltage enables intermediary effects in the orientation of the crystals, resulting in gray scale modulation. See nematic liquid display, smectic liquid crystal.

Schäfer, F.P. Along with F. Schmidt, and J. Volze, Schäfer published "Organic dye solution laser," in Applied Physics Letters in fall 1966, an important historic landmark in dye lasers in particular and tunable laser technology in general. In 1968, Schäfer was honored for his scientific contributions with the Nernst Haber Bodenstein prize. Schäfer became a professor at the Max-Planck-Institut in Germany. Schäfer's interest in lasers never wained. Thirty years later he was still publishing numerous articles on laser technologies, including short-pulse lasers.

Schawlow, Arthur Leonard (1921–) An American/

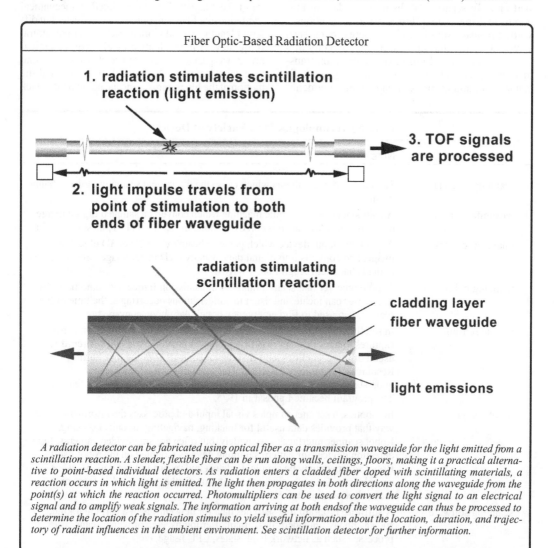

Fiber Optic-Based Radiation Detector

1. radiation stimulates scintillation reaction (light emission)

3. TOF signals are processed

2. light impulse travels from point of stimulation to both ends of fiber waveguide

radiation stimulating scintillation reaction

cladding layer
fiber waveguide

light emissions

A radiation detector can be fabricated using optical fiber as a transmission waveguide for the light emitted from a scintillation reaction. A slender, flexible fiber can be run along walls, ceilings, floors, making it a practical alternative to point-based individual detectors. As radiation enters a cladded fiber doped with scintillating materials, a reaction occurs in which light is emitted. The light then propagates in both directions along the waveguide from the point(s) at which the reaction occurred. Photomultipliers can be used to convert the light signal to an electrical signal and to amplify weak signals. The information arriving at both endsof the waveguide can thus be processed to determine the location of the radiation stimulus to yield useful information about the location, duration, and trajectory of radiant influences in the ambient environment. See scintillation detector for further information.

Canadian physicist and laser pioneer with a strong interest in spectroscopy, Schawlow studied at the University of Toronto and Columbia University. At Columbia he met Charles Townes, with whom became a collaborator in laser research. From 1951 to 1961, Schawlow worked at Bell Telephone Laboratories on superconductivity and nuclear resonance after which he took a position as a professor at Stanford University, retiring as Professor Emeritus in 1991. Schawlow has won many awards for his research, including the National Medal of Science and the Stuart Ballantine Medal. In 1981, he was co-awarded a Nobel Prize for his work in high-resolution electron spectroscopy that derived from the pioneer work. The Arthur Schawlow Medal of the Laser Institute of America is named in his honor. See laser history; Townes, Charles.

Scheduled Transfer ST. A media-independent upper-layer protocol that was originally developed as part of the HIPPI-6400 network transmission standard.

schematic A diagrammatic representation of an electrical circuit, floorplan, network, or other interconnected system. Electronics drawings have conventions and symbols for various types of components and connections.

Schilling von Cannstedt, Pavel Lvovitch (ca.1780–1836) A Russian diplomat posted to Germany, Schilling invented multi-needle and single-needle telegraph systems and a code to signal characters, numbers, and stop/finish/continue states. He collaborated with S.T. von Sömmering, who had developed an even earlier telegraph. Schilling's telegraphs have variously been reported as being demonstrated in 1820, 1832, and in between. Given Sömmering's inventions around 1809, it is plausible that Schilling's telegraph may have been demonstrated as early as 1920. See Salvà i Campillo; Sömmering, S.T.

Schmidt, F. Along with F.P. Schäfer and J. Volze, Schmidt published "Organic dye solution laser," in *Applied Physics Letters* in fall 1966, an important historic landmark in dye lasers in particular and tunable laser technology in general.

Schott Fiber Optics, Inc. SFO. A prominent international vendor of fiber optic lighting and imaging technologies that evolved from a fiber optic research project in 1954 (which originally became the fiber optic division of American Optical). Through the last half century, the company has developed glass core and coating technologies, winding and assembly processes, fiber drawing and faceplate fusing techniques. It has also designed and developed medical and scientific instruments using fiber components.

Schottky effect In a cathode-ray tube (CRT), a random charge variation that occurs in the emission of electrons in the strong electric field associated with the electron-emitting cathode. As an electric field is applied there is an electron discharge from the heated surface that reduces the amount of energy needed to stimulate electron emission.

Schottky diode A type of efficient rectifying component. Schottky diodes are incorporated into a number of electronic devices that require switching

or voltage control components. In TTL circuits they may be configured as clamp diodes. Schottkey diodes are found in optical beam control switching arrays and may be used as frequency doublers in RF traveling-wave tubes (TWTs). Synchronous rectification may now be done with field-effect transistors (FETs) rather than Schottky diodes.

Schultz, Peter C. (1942–) Schultz was educated at Rutgers University and went to work at Corning as a research scientist with a team that included D. Keck and R. Maurer. In the early 1970s, they developed a practical low-loss embodiment of fiber optics transmission lightguides, a technology that had eluded scientists since the early 1960s. Schultz is known for having codeveloped an outside vapor deposition process in 1972 that is still a standard in the industry. This led to the development of germania-doped fused silica.

Schultz joined SpecTran Corporation as VP of Technology in 1984, and later became president of technology at Schultz Galileo ElectroOptics, Inc. He became president of Heraeus Amersil in 1988. In 2000, along with Keck and Maurer, Schultz was awarded the National Medal of Technology.

Schroeder, Manfred R. (1926–) A German scientist and educator, Schroeder was an early researcher in chaotic dynamics and microwave waveguides who was hired by Bell Laboratories Research in 1954. At Bell he turned his attention to the study of speech and hearing and headed up the acoustics and speech research projects from 1958 to 1969. Since Bell was a telephone company, there were many practical applications for acoustics research including voice dialing, conferencing, speech recognition, text-to-speech applications, and more. In conjunction with B. Logan, Schroeder designed a speech compressor that was made available to the American Foundation for the Blind for the Talking Book program. In 1967, Schroeder was involved in the invention of linear predictive coding (LPC) and later in code-excited linear prediction (CELP), important contributions to the evolution of digital voice encoding and synthesized speech. While at Bell, Schroeder was named inventor/coinventor on 45 U.S. patents.

In 1969, Schroeder returned to Germany where he became a professor and lecturer in various aspects of physics, including number theory, chaos, fractals, and nonlinear dynamics. He continued to work at Bell Labs for 5 months of the year and to maintain a strong interest in experimental acoustics, particularly in the area of generating speech through phase changes, in collaboration with H.W. Strube. Schroeder's studies, especially in acoustics, have been honored with several medals and awards. See CELP; Harmon, Leon D.; Knowlton, Kenneth.

Schweigger, Johann Salomon Christoph (1779–1857) A German physicist who contributed substantially to the development of the galvanometer and is credited as the first inventor of a practical version of the galvanometer. Schweigger studied electromagnetism and observed in the early 1800s that current passing through a coil could increase the magnetic influence of a needle. Building on the work of Ørsted, Schweigger continued with his research on electromagnetism and various configurations of conductors and, by 1920, had developed an electromagnetic multiplier that could detect very small amounts of attraction or repulsion, thus developing a practical means for detecting and measuring galvanic current. See galvanometer.

scintillation A reaction that occurs when a charged particle with sufficient energy impacts a material which can be stimulated to emit energy in the optical spectrum as the electrons return to a nonexcited state. Exposing certain materials to ionizing radiation causes a scintillation reaction in which the materials fluoresce.

Scintillation may be stimulated intentionally or may occur naturally and may be desirable or undesirable, depending upon the circumstances. There is a branch of climatology that studies and predicts ionospheric scintillation patterns, which is of importance to satellite systems maintenance and radio wave propagation.

In practical applications in the fiber optics industry, scintillation is the process of converting electromagnetic energy outside the optical spectrum into light energy. See ionization, scintillator.

scintillation, ionospheric A reaction that occurs when electromagnetic energy passes through the Earth's ionosphere (such as a space probe communication, marine distress call, or orbiting satellite radio transmission) and encounters small irregularities in the plasma density along the path traveled, resulting in phase and intensity fluctuations. Some of the radiant energy is absorbed and reemitted in the optical spectrum.

Ionospheric scintillation is an important factor in radio signals that are reflected off or through the ionosphere. Scintillation can cause signal degradation, loss of lock conditions, fade, and cycle irregularities. Solar flares, which effect the ionosphere, can influence the degree of scintillation that might occur at particular times. Scintillation indexes have been developed to give a quick-glance idea of the level of scintillation that might be expected to occur in a particular equatorial sector at a selected frequency. This information can be combined with NOAA space environment data to generate scintillation graphs. For example, an intensity index of 0.5 or higher indicates moderate to high scintillation conditions.

scintillation detector A device that detects light emissions from a scintillation reaction, that is, one in which electromagnetic radiation is absorbed and reemitted in the optical spectrum. The detector then uses a photocathode to convert the light energy into electrical energy so that it can drive various measurement and display components to provide useful information. Fiber optics may be used to direct the light from the scintillator to the phototube (e.g., through a fiber optic array *faceplate*) or from the phototube to the processing components.

Since electromagnetic energy occurs in a very broad range of frequencies, materials may be chosen to create a scintillation reaction within specific frequency ranges (or to be more sensitive within those frequency ranges). For example, thallium-doped sodium iodide may be used in gamma ray detectors.

The energy level of the electromagnetic particle in part determines the magnitude of the ionization reaction and the intensity of the scintillation reaction, enabling the strength of the energy to be calculated. A discriminator may further be used to isolate pulses within a particular amplitude range.

Nishiura and Izumi have described an interesting type of fiber-based radiation detector. It is comprised of an optical fiber doped with scintillating materials which emit light energy when stimulated by radiation that is propagated through the length of the fiber to both ends. Detectors at each end of the fiber use differences in signal arrival times (*time of flight* - TOF) to calculate the location of the radiation exposure on the fiber path. Thus, the fiber line can be conveniently installed along a wall or ceiling and can be as long as the light is able to travel through the fiber core. In contrast to point-source detectors, it can detect radiation intensities along the length of the cable. The hardware interfaces with a software signal processing application to provide a visual display and numeric readouts of the radiant energy detected by the system.

Scintillation detectors are used for scientific and medical imaging, industrial safety devices (e.g., radiation detectors), and ionospheric interference predictors. See scintillation.

scintillator A component for converting energy from a high-energy charged particle such as X-radiation or gamma radiation into light energy. Thereafter optical fibers can optionally be used to direct the light (or a portion of the light) to a phototube or photomultiplier tube that further converts the signal to an electrical charge.

Scintillators are generally fabricated from materials with excellent light-transmitting characteristics, good temperature tolerance, and sufficient stability for accomplishing the task at hand. A high index of refraction is usually desirable and rapid decay of the excitation phase when a charged particle hits the scintillator generally permits a finer resolution for the processing of subsequent particles. Typically, the scintillator, and any associated light-sensitive components such as photomultiplier tubes, are shielded from extraneous light by dark housings and tape to seal the cracks.

There are many materials with scintillating properties, including liquids, gases, powders, and single crystals (e.g., yttrium-aluminum-garnet – YAG). Powder scintillators are typically mixed and bound to a glass substrate. Liquid scintillators may be a mixture of chemicals in a mineral oil suspension. Commercially available plastic scintillators are common. There may be a tradeoff between the response time and the longevity of scintillating materials. Some of the powdered materials may have quick response times, but will lose effectiveness over time with exposure to radiation. Single crystal scintillators have long life spans, but may have slower response times than some other scintillating materials.

The thickness of a scintillating layer depends upon the application. In electron microscopy, for example, it may vary from 0.5 - 1.0 mm.

Scintillation may be configured through the use of discrete crystals, bonded layers, or planar arrays consisting of strips, rectangles, trapezoids, etc.

Scintillators are used in electron microscopy, medical and dental imaging systems such as computer tomography (CT) and X-ray machines, and a number of experimental physics instruments. They may be bonded to fiber optic faceplates to jointly control and guide light emissions to electronic imaging components. See faceplate, fiber optic; P-47; phototube.

SCO An AT&T specialized network maintenance organization providing a single point of contact for resolving customer network faults.

SCO Unix The Santa Cruz Operation's adaptation of Unix. See Unix, UNIX.

scope 1. Generic term for any visual enlargement mechanism typically viewed through a narrow aperture, such as a microscope, telescope, or periscope. 2. *colloq.* A generic term for a bounded display device representing abstracted data that would otherwise be invisible to human eyes. Radar scopes enable translation of reflected radio signals into visual signals that are positionally displayed and oscilloscopes enable electrical signals to be displayed as waveform representations. See oscilloscope.

SCP 1. Satellite Communications Processor. 2. See Service Control Point. See Signaling System 7 (SS7). 3. See Session Control Protocol.

SCR 1. silicon-controlled rectifier. 2. See sustainable cell rate. See cell rate for chart. 3. System Clock Reference. A synchronization time reference used, for example, in MPEG decoding.

scrambler A device which rearranges or distorts a data communication or broadcast transmission to provide a measure of security. A descrambler is required at the end of the transmission to convert the information back into comprehensible form. Scrambling is a type of encryption, although the term *encryption* tends to be associated more with sophisticated data encryption schemes, and scrambling is associated more with the simple rearrangement of a few parameters, such as inverting audio frequencies. Black market descramblers are common, especially for television broadcast signals, since scrambling schemes are easier to decode than sophisticated key encryption algorithms. Scrambling and encryption are sometimes combined to maximize security. Scramblers are most commonly used to protect pay services like cable channels, but they are sometimes used to disrupt the transmissions of others, such as communications in war zones or traffic radar detectors.

Scribner, Charles Ezra (1858–1926) An American inventor and engineer, Scribner was chief engineer at Western Electric, after joining the firm at the age of 18. He had already patented a telegraph receiver at that age, and eventually held over 440 patents, more than any other single man in electrical history. One

of his most significant inventions was the multiple telephone switchboard. Scribner founded the Western Electric engineering department, which later evolved into Bell Telephone Laboratories.

SCRL See Signal Corps Laboratories.

scroll bar A graphical user interface (GUI) device which serves two purposes: to indicate additional information beyond what can be seen in the current window or box, and to allow the user to scroll up and down (or across) the contents of the box by clicking on and dragging the scroll bar or by clicking on arrow indicators at either end of the scroll bar. A well-designed scroll bar may have a third function – to give a proportional idea of how much information is being viewed or hidden. Displaying a scaling drag area to match the size proportional to the total of the information contained in the listing, helps the viewer to perceive whether a little or much is hidden from view.

Some scroll bars are designed so the scrolling speed accelerates the longer the listing and the longer the user holds down the scroll bar. This type of selectively accelerated scroll can be very handy and is perceived by most users as natural and comfortable.

SCSI See Small Computer System Interface.

SCSI connector An electrical/data connector supporting the data transmission requirements of the Small Computer System Interface (SCSI) standard. Traditionally, most SCSI devices (hard drives, cartridge drives, scanners, etc.) were connected with DB-25 or 50-pin flat connectors. However, more recent SCSI-2 devices use a connection with finer pins. See Small Computer System Interface.

Common SCSI Connectors

Traditionally, DB-25 (top left) and 50-pin flat connectors (top right) have been used to interconnect SCSI devices or to connect them to the host or peripheral card associated with the host. Since SCSI devices can be daisy-chained, it is important to terminate the last device in the chain. A common 25-pin SCSI terminator is shown (bottom).

SCT Secretaria de Comunicaciones y Transportes. A state agency given jurisdiction over communications for Mexico in the 1938 Law of General Means of Communications. It has played an important role in regulation and public services in Mexico's telecommunications history. In 1992, telecommunications were transferred to the Telecomunicaciones de Mexico.

SCTE 1. See serial clock transmit external. 2. See Society of Cable Telecommunications Engineers, Inc.

SCTP See Stream Control Transmission Protocol.

SDL See Specification and Design Language.

SDLC See Synchronous Data Link Control.

SDLLC SDLC Logical Link Control. A Cisco Systems IOS feature which can provide translation between Synchronous Data Link Control (SDLC) and IEEE 802.2 type 2.

SDH See Synchronous Digital Hierarchy.

SDN See Software Defined Network.

SDNS See Secure Data Network System.

SDP 1. See Service Data Point. 2. See Service Discovery Protocol. 3. See Session Data Protocol. 4. See Session Description Protocol. 5. See Simple Discovery Protocol.

SDSL single-line DSL. One of a number of optional configurations of Digital Subscriber Line telecommunications services. SDSL is intended to provide a basic DSL option delivering 1.544 Mbps data rates in both the upstream and downstream directions over existing copper twisted-pair wireline connections. See Digital Subscriber Line.

SDTI See Serial Data Transport Interface.

SDTP See Serial Data Transport Protocol.

Se *symb.* selenium. See selenium.

SEAL Simple and Efficient Adaptation Layer. See asynchronous transfer mode in the Appendix for information about ATM adaptation layers.

seal *n.* 1. A part designed to tightly close or close off a part, adjoining parts, or a container. Seals are sometimes used to make parts air- or watertight. 2. A security/tamper device designed to indicate whether the parts adjoining or underlying the seal have been opened or altered. It is common for technical components such as hard drives, internal parts to computers, and warranteed parts to be sealed by a plastic or metallic adhesive seal, or a dot of colored paint or resin. Breaking the seal may void a warranty.

search engine A software application designed to search and retrieve information from a database according to user-specified parameters or keywords. See WAIS.

search engine, Web A software application combined with a Web site to provide search and retrieval of a great variety of information, such as newsgroup postings, personal or business names and addresses, individual Web pages, or Web site topics. Many include advanced search features with logical operations for more specific or complex searches. Considering the millions of sites on the Web, Web search engines are indispensable. See Appendix C for a list of Web search engine sites.

SECAM sequential couleur avec memoire; sequential color with memory. A composite color television standard developed in France and used also in the French colonies and western regions of the former USSR. It supports up to 625 scanlines at 50 cycles per second at a frame rate of 25 per second at 4.42 MHz, with an inverted signal that makes it incompatible with PAL, the other common format in Europe, and NTSC, the format used in North America.

second 1. A brief unit of time defined as 1/60th of a minute in reference to a solar day. 2. A unit of time, designated in 1967 by the General Conference of Weights and Measures, as the duration of 9,192,631,770 periods of the radiation corresponding to the transition between the two hyperfine levels of the ground state of the cesium-133 atom. See atomic clock.

Second/Secondary Audio Program SAP. A secondary source of audio, usually a better quality audio or similar-quality audio provided in another language, that can be selectively chosen by a listener/viewer.

Section Terminating Equipment STE. In SONET networking, the STE may be a terminating network element or a regenerator between any two adjacent network elements (such as line repeaters or lightwave terminals). It can originate, access, and/or modify the section overhead, or terminate it, if needed. See SONET, Synchronous Transport Signal.

Secure Data Network System SDNS. A system which incorporates the SDNS Message Security Protocol (MSP), developed as a cooperative project between government and industry participants and sponsored by the U.S. National Security Agency (NSA). MSP was specified to be integrated into the X.400 Message Handling System (MHS) environment.

Secure HTTP, Secure Hypertext Transfer Protocol S-HTTP. A security-enhanced version of HTTP introduced in the mid-1990s. The original experimental version was designed to support multiple public-key algorithms, although a 1997 revision designates the Diffie-Hellman algorithm as the default. It is sometimes called HTTP/SSL. SSL was originally developed for transmitting private documents through the Internet via a browser (e.g., Netscape) and was later standardized as a transport layer security standard. See Hypertext Transfer Protocol, RFC 2616, RFC 2660, RFC 2716, RFC 2817.

Secure Multipurpose Internet Mail Extension See S/MIME.

Secure Public Networks Act SPNA. An act of the U.S. Senate (S. 909) introduced by Senators McCain and Kerrey, in 1997. The Act would make it lawful for U.S. persons to use encryption, regardless of algorithm, encryption key length, or implementation; however, it differs from the Security and Freedom through Encryption (SAFE) Act in that it would require U.S. persons to use government-approved third-party encryption key escrow agents to hold spare copies of every encryption key. This type of key recovery system has been brought forward numerous times in the last few years and has always met with strong debate from the business community and privacy advocates.

SPNA made it unlawful to use encryption to further

the commission of a criminal offense and made it generally unlawful to decrypt communications without authority. This effort was one of the few bills supported by Louis Freeh, Director of the FBI. Exports would be limited to encryption products with 56-bit keys or less. It established a relationship between digital signature certificates and key escrow systems. The act was approved by the Senate Commerce Committee in May 1997 but was not brought to a full Senate vote due to a number of other options that were introduced. See Encryption for the National Interest Act, Security and Freedom through Encryption Act.

Security and Freedom through Encryption Act SAFE. Important legislation proposed as alternative policy for protecting security and personal privacy while still promoting electronic commerce. It was a long process to work out the terms of SAFE, beginning in 1995. In February 1997, the SAFE Act was introduced with the same terms as H.R. 3011, proposed in 1996. The 1997 bill sought to ensure that Americans could use any type of encryption anywhere in the world and would be permitted to sell any type of encryption domestically. It further sought penalties against unlawful use of encryption and promoted the relaxation of export controls over encryption algorithms and encryption-related products. Through committees, the text of the original Goodlatte and Lofgren bill was significantly changed. Introduced into the U.S. House of Representatives as H.R. 695, in 1997, the bill amended title 18, U.S. Code to "... affirm the rights of United States persons to use and sell encryption and to relax export controls on encryption." As with other privacy-related bills, there was significant debate and the bill was competitive with others, including the Secure Public Networks Act.

In May 1997, the House Judiciary Committee unanimously approved SAFE, while agreeing to three amendments. In June, the House International Relations Subcommittee on International Economic Policy and Trade approved SAFE by a majority. Amendments and input from the House Intelligence Committee followed, with consideration of the Oxley-Manton amendment to criminalize the domestic use of strong encryption. This amendment drew much debate from citizens and law professionals.

By the 1999 106th Congress, there were two versions of the proposal. H.R. 850 dealt largely with the export of encryption products and had many implications for the Export Administration Act of 1979. Since it was first introduced, it broadened and became more lenient in terms of restrictions on personal and business use of encryption, essentially opening the door not only to using any encryption products but also in exporting them. It appears that the advocacy of privacy rights supporters and the competitive needs of U.S. exporters swung the pendulum in this direction. The Act subsequently went through many committees for markups and, in some cases, adoption.

Under the terms of the Act, it is lawful for persons specified in the Act to sell any encryption, regardless of the algorithm, in interstate commerce. Thus, there

would no longer be restrictions on the lengths of encryption keys, for example. It further states that the federal government and individual states may not require key escrow/key recovery information as was previously proposed in other bills.

In terms of export, a highly contentious issue, SAFE stipulates that the Secretary of Commerce (with concurrence from the Secretary of Defense) has exclusive authority to control computer-related exports, including information security products, except for military uses. The Secretary would carry out a one-time 15-day technical review of entities wanting to export encryption-related products, after which no export licenses would be required, except where national security might be involved (e.g., per the Trading with the Enemy Act). This is a significant change over previous policies that strictly limited encryption key lengths, for example. Pressure from the industry and competition from foreign bodies providing encrypted products and encryption algorithms were probably the basis for this change of heart. The Act expressly stated that nothing in the Act shall limit the authority of the President under the International Emergency Economic Powers Act, the Trading with the Enemy Act, or the Export Administration Act of 1979. Thus, there would continue to be prohibitions on the export of encryption products to countries "... that have been determined to repeatedly provide support for acts of international terrorism..."

To satisfy the concerns of law enforcement personnel, who increasingly worry that the ability of criminals to engage in surreptitious, global communications is outstripping the ability of law enforcement to monitor these communications, a stipulation in the Act stated that the Attorney General "... shall compile, and maintain in classified form, data on the instances in which encryption ... has interfered with, impeded, or obstructed the ability of the Department of Justice to enforce the criminal laws of the United States." SAFE prohibits the use of encryption for hiding messages related to criminal acts. See Anti-Terrorism Act of 2001, E-Privacy Act, Secure Public Networks Act.

Security Associations SAs. Elements for containing information for the execution of network security services such as Internet Protocol (IP) layer services (authentication, encapsulation, etc.), transport layer or applications layers services, or protection of negotiation-related network traffic. See Internet Security Association and Key Management Protocol.

Security Multiparts specification A specification for secure messages, separating the content data from the signature, and formatting them as multiple parts of a MIME communication. See S/MIME, RFC 1847.

Security Protocols for Sensor Networks SPINS. Based upon the premise that in the future, wireless networks will be comprising millions of small self-organizing sensors, Perrig, Szewczyk et al. have proposed a set of security protocols to explore security issues in sensor networks.

The SPINS work has arisen out of the SmartDust

program at U.S. Berkeley, in which prototype multi-hop networks of small, low-power, sensor devices are being built utilizing the TinyOS event-driven operating system. It was discovered that workstation models of security were impractical in a low-power, multi-sensor environment, including Diffie-Hellman algorithms and digital signatures. There simply is not enough resource space to implement the usual methods. RC5 was chosen as the cryptographic primitive, but other shared key algorithms could be used. The researchers have presented information on μTESLA (a tight version of the Timed, Efficient, Streaming, Loss-tolerant Authentication Protocol) to provide authenticated streaming broadcast; on SNEP (Secure Network Encryption Protocol) for data confidentiality, two-party data authentication, and data freshness; and on an authenticated routing protocol using SPINS building blocks. Results of the project indicate that it is feasible to add security to severely resource-constrained sensor networks through symmetric cryptography.

Seebeck, Thomas Johann (1770–1831) A German researcher who described thermoelectrical effects in 1823, after observing a connection between electricity and heat. This observation is now exploited in the fabrication of many types of semiconductor components, such as thermocouples, and is called the Seebeck effect. See thermocouple.

seek time A quantified description of the time it takes to locate specified information. In software, for example, this could be expressed as the average time in milliseconds or clock cycles it takes for a specific tool to locate queried information from a database of a given size. In hard storage devices, it could be expressed as the average time it takes for the read head to position itself on the track where the information lies. Industry definitions of seek times exist for specific types and sizes of devices.

segmentation and reassembly A common process in packet-based networking of dividing up the packets so they can be individually processed or routed, and reassembling them at the receiving end to recreate the original message or transmission.

seize To take control of a circuit or system so it cannot be used by others. Computer files are sometimes seized and locked so that the data cannot be inadvertently modified simultaneously by more than one user. This helps protect data integrity. Transmission circuits may be seized to prevent interference on the line or to preserve privacy.

Selective Sequence Control Computer SSEC. The successor to the Automatic Sequence Control Calculator, better known as the Harvard Mark I, instigated by IBM in 1948. By this time a number of different designers and manufacturers were getting involved in the development and marketing of large-scale computing machines and IBM was motivated by the competition. Machines built around this time represent a transitional evolution from calculator/tabulators to computing machines in common understanding.

selenium (*symb.* – Se). A photoconductive element (AN 34) first isolated in the early 1800s, selenium's light-sensitive properties were noted by British scientists in 1873, which subsequently led to many of the important pioneer experiments in television transmissions using selenium. The properties of selenium provided A. Graham Bell with the idea of transmitting sound via light which led to the invention of the Photophone.

As understanding of the capabilities of selenium evolved, it came to be used in television cameras. It is sensitive to heat and light in varying degrees, depending upon other factors, and has important rectification properties; it can be used to convert AC to DC power. In solid-state electronics, selenium is a p-type semiconductor. See Baird, John; Photophone.

Selenium Cells - Historic Forerunners

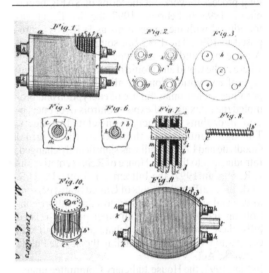

A selection of some of the selenium cells designed by Tainter and Bell in the process of inventing the Photophone, a means of transmitting sounds by light. Selenium was a light-sensitive material with rectifying properties that were exploited for sound and image communications. [U.S. patent #235,497, 1880.]

---

self-electro-optic effect device SEED. A quantum well-based photonic optical device used for photonic switching, developed in the AT&T Bell Laboratories in 1987. This multi-element device, when biased by an external voltage, creates an external field that shifts the wavelength of the onset of absorption which, in turn, causes the intensity of transmitted light to vary. Optical sensors can be used with a SEED system to detect the resulting light. This suggests the possibility of developing light-based switching mechanisms and optical logic. SEED latches were described in the late 1980s and the first SEED devices were created in the late 1980s and early 1990s. In December 1989, Bell Labs announced the development of a very high-capacity gallium-arsenide photonic integrated circuit capable of processing 2 kbits of optical information in parallel. Each element in the chip array is a Symmetric-SEED (S-SEED) which can function as a logic gate, memory cell, or switch. Illumination from a low-power light beam can be used with the S-SEED to cause it to switch in less than a billionth of a second. This has important implications for parallel processing, as complete arrays can be simultaneously accessed.

With the spread of optical communications technologies, scientists have been seeking ways to make the physical transmission path all-optical. SEED elements have possibilities as optical memory cells once the means to combine them more effectively into arrays as been worked out. This, in turn, could support the development of an all-optical switching mechanism (e.g., an optical ATM switch).

Within the U.S. Navy, J. Bechtel has described research on logic systems for solving Boolean equations based on a Symmetric-SEED (S-SEED) system. A single light source could potentially be split into a matrix of light beams, which are then modulated by a Symmetric-SEED-based spatial modulator and the individual members in the array reset according to Boolean inputs. The value of the remaining member, unaffected by inputs, would correspond to the Boolean solution with the result signaled by a matrix of equal intensity light beams onto a detector.

U.K. researchers, with help from a grant from the European Union, have developed a digital optical network for image processing based upon self-linearized SEED (SL-SEED) concepts. By exploiting the fact that SEED responsivity can be increased by application of voltage, a feedback loop can be established. The modulator photocurrents and detector influence one another until the photocurrents of both match. The photocurrent is proportional to the input light minus the reflected light. The detector photocurrent is proportional to the control light input. Control can then be "subtracted" from the signal. For image processing, an element must then interact with neighboring elements. See electroabsorption, quantum well, Stark effect.

SEM See scanning electron microscope.

semaphore A visual signaling system employing movable apparatus like arms or flags. Individual symbols, words, or instructions are made to correspond to distinct positions of the arms or flags. While electronic communications have superseded most semaphore systems, they are still sometimes preferred in situations where electronic messages might be overheard. See Chappé, Claude.

semaphore, programming 1. An access or exclusion indicator, such as a variable flag. Semaphores are useful for controlling file locks to preserve data integrity. In other words, they can be used to prevent multiple users from accessing a file simultaneously and changing data in a way that could disrupt the information or corrupt the data. 2. A low-level integer variable having only nonzero values; a primitive which can be used for synchronization in concurrent processing implementations.

semiconductor A material widely used in electronics due to its relative balance of electrical conducting and insulating properties (hence the name *semi*con-

ductor). Semiconductor materials are typically crystalline in structure, and their properties of enabling or impeding the flow of current are used in designing solid state electronic circuitry.

Materials commonly used to create semiconductor components include silicon, germanium, and gallium arsenide. Doping, the addition of other elements, may be part of the fabrication of semiconductors to further control and enhance their properties. Current flow in semiconductors is commonly controlled by electricity, but may also be controlled by the influence of light or magnetic fields. Semiconductors are important materials used in the manufacture of integrated circuits. See integrated circuit.

Semiconductor Industry Association SIA. The leading U.S. trade association for the microchip industry, established in 1977 by pioneers in the industry. The SIA promotes and supports the competitiveness of the U.S. semiconductor industry and represents its membership through input to government representatives. It also researches and reports on possible health hazards related to the semiconductor industry.

In 1982, the SIA formed Semiconductor Research Corporation (SRC) to plan, direct, and fund precompetitive silicon research projects at major academic institutions. In 1985, the organization submitted input to the U.S. government regarding the balancing of trade practices between Japan and the U.S. In 1987, SIA formed SEMATECH, a consortium of chip manufacturers dedicated to improving semiconductor manufacturing technologies. In 1997, the Focus Center Program was established to engage in long-term research to ensure the long-term viability of the industry.

Semaphoric Optical Signaling System

France had an extensive system of semaphore signaling before the telegraph was invented. Limited visibility in bad weather and the need for constant monitoring were two disadvantages of this system.

semiconductor laser A compact laser comprising layers of semiconductor and other components. A crystalline compound of gallium and arsenic (gallium-arsenide – GaAs) is used to make semiconductor lasers and dopants (deliberate impurities such as selenium) may be added to promote the movement of electrons through the material. With the application of current, the component lases with a slightly wider range of frequencies and a slightly wider beam spread than traditional lasers. See gallium arsenide; Kazarinov, Rudolf; laser diode.

semiconductor optical amplifier SOA. A solid state component for increasing gain in an optical transmission. This is important for achieving power and distance of sufficient magnitude for effective transmissions. See amplifier.

SEQUEL See SQL.

sequential A nonoverlapping succession or series, in chronological or data order, with no significant intervening time or data. See concurrent, consecutive, parallel, serial.

serial clock transmit external SCTE. A data stream common to serial cables used to connect computer modems. For example, on 25-pin serial communications connections, one of the pins may be assigned to carry the external clock signal from the data terminal equipment (DTE).

serial communication A means of transferring data one element at a time, often through a single wire or trace in a circuit. While it may not seem very fast or efficient, serial communication is easily implemented and very commonly used in computing systems. The RS–232 standard is the most common specification for the physical/pin connections for serial communications. See modem, parallel, RS–232.

Serial Data Transport Interface SDTI. An emerging packet data standard for the transport of audio, video, and data among various multimedia systems, including video servers, cameras, VCRs, editing systems, etc., especially in professional broadcast video environments. SDTI provides a network environment for video data exchange, without a lot of network overhead. It supports single-direction, point-to-point, compressed-data connections. SDTI has been well received due to the many advantages of digital systems, including the reduction of generation loss common to analog editing and transmission loss associated with analog broadcasting.

SDTI (SMPTE 305M) evolved from SDI (SMPTE 259M), which is used for transporting uncompressed audio/visual signals between digital broadcast and post-production devices. The two formats are mechanically and electrically compatible to support coexistence in the same facility.

SDTI, developed by the SMPTE PT20.04 Workgroup on Packetized Television Interconnections, extended this concept by enabling compressed video to be exchanged without the need for frequent compression/decompression processes that could slow or degrade the transmission. SDTI has been recommended by the EBU/SMPTE Task Force for Harmonized Standards for the Exchange of Programme Material as Bitstreams.

Serial Data Transport Protocol SDTP. A network data protocol which provides a means of transporting serial data streams over PPP links. SDTP arose out of the work of the TR30.1 ad hoc committee in the mid-1990s to provide a standard means for synchronous data compression. SDTP specifies a transport protocol and an associated control protocol (PPP-SDTP and PPP-SDCP) to be used in conjunction with PPP protocols. See Point-to-Point Protocol, RFC 1963.

serial interface card A printed circuit card which fits into a slot in a computer or other computerized device or piggybacks on a motherboard to provide standardized electrical connections for the synchronous serial transmission of digital data. The connection on the card is typically a 25-pin D connector. On consumer desktop computer systems, most serial interface cards support data rates up to about 28,800 or 38,400 bps. A serial interface card is a common way to connect remote computer terminals and data modems to a computer. See RS–232.

Serial Line Interface Protocol, Serial Line IP SLIP. Originating with an early 1980s 3COM UNET TCP/IP implementation, SLIP became a *de facto* standard encapsulation protocol for serial lines, used for point-to-point communications with TCP/IP. SLIP has now been superseded by Point-to-Point Protocol (PPP). See Point-to-Point Protocol, RFC 1055.

SERN See Software Engineering Research Network.

server A system which provides services to other computers connected to it through a network. A server may store and administer software applications, security measures, access to peripherals or external systems, etc. The server does not necessarily have to be an enhanced system, as servers can be specialized as print servers, mail servers, etc. (and several servers may be on a system), but servers performing the bulk of centralized or generalized tasks often have more memory, processing speed, and storage than other systems on the network.

The software is probably the most important aspect of a good server. Good network software is robust, configurable, and usually fully multitasking. There are many well-tuned network workstation options that are reliable and do not crash, except in the most unusual of circumstances. Shop around when selecting network server software; paying a few hundred or thousand extra dollars in terms of the initial cost can often be recouped in six months or less through savings on downtime, software reinstallation, and administrative costs that accrue on unreliable systems.

server agent In server/client systems, software that handles the major processing or protocols and serves a request from a client as a Web server, mail server, or FTP server.

Server Message Block protocol SMB. A client and/or server request-response network protocol for sharing resources such as files, ports, printers, and other useful services. SMB was defined by IBM in the mid-1980s and is prevalent on Windows-based systems. It works in conjunction with NetBIOS over TCP/IP. In conjunction with Samba, it can also be used with Linux systems. SMB is also known as Common Internet File System (CIRS). See Samba, SMB Project, RFC 1001, RFC 1002.

Service Advertising Protocol See Service Location Protocol.

Service Agent A network utility which, when queried, provides information about a network service (printer, modem, etc.) such as its URL.

Service Location Protocol SLP. An intelligent resource discovery and registration protocol developed in the mid-1990s. Described as a "quieter" alternative to Service Advertising Protocol (SAP), SLP includes extended attributes information to reduce network traffic queries. Thus, a printer may be described in terms of its capabilities (such as duplex printing, PostScript-capable, tabloid paper) and found transparently, without the user querying for its IP address. See Service Advertising Protocol, Service Agent, SLIP, RFC 2165.

service quality Standards of service established by businesses that include such things as service without outages, available lines without lag or busy signals, technical support availability, good data integrity, etc. This is not the same as quality of service (QoS), which has a more specific meaning.

Service-Specific Connection-Oriented Protocol SSCOP. A B-ISDN signaling ATM adaptation layer (SAAL) mechanism for managing the establishment, monitoring, and release of data exchanged between signaling peers. In the context of Q.2931, SSCOP provides error and flow control signaling services somewhat analogous to those provided by TCP for Internet Protocol (IP).

Service-Specific Convergence Sublayer SSCS. A component of the ATM adaptation layer (AAL) that coordinates protocols of the next higher layer with the requirements of the next lower layer, the Common Part Convergence Sublayer (CPCS). See asynchronous transfer mode.

services-on-demand SoD. Services provided to an audience on a request basis, rather than on a scheduled broadcast basis. The concept is not new; in fact, it has been available for media services for over 100 years, but new digital technologies are providing automated services, thus making available cost-effective SoD delivery options which were not previously possible. See audio-on-demand, video-on-demand.

Session Announcement Protocol SAP. A network protocol for sending announcements to users that is common to broadcast communications. It is considered distinct from broadcast content and data triggers. SAP version 2 was submitted as an Experimental RFC by Handley et al. in October 2000. A SAP announcer periodically multicasts an announcement packet to a known multicast address and port. The SAP recipient "listens in" on a SAP address and port for multicast scopes and thus learns of all the sessions being announced so that the sessions may be joined. See enhanced TV, RFC 2974.

Session Control Protocol SCP. A simple client/server network protocol to facilitate multiple conversations over a single TCP connection. With SCP, parties can establish (or reject) a virtual session over a single transport connection.

Session Description Protocol SDP. A network protocol used for announcements and other notifications for multimedia broadcasts. The protocol was described by the Multiparty Multimedia Session Control (MMUSIC) working group of the IETF and was submitted as a Standards Track RFC by Handley and Jacobson in April 1998. SDP provides session support for Internet multicast backbone (Mbone) services and more general realtime multimedia services. The protocol is not intended for the negotiation of media encodings.

SDP is designed to convey session directory information to recipients that can be used in conjunction with a variety of transport protocols, such as Session Announcement Protocol, Hypertext Transport Protocol, and others. See Session Initiation Protocol, RFC 2327

Session Initiation Protocol SIP. An application-layer signaling protocol for creating, changing, and terminating Internet-based telephony, conferencing, messaging, and events notification involving one or more participants.

SIP was originally developed with the IETF Multiparty Multimedia Session Control (MMUSIC) working group and continued, as of September 1999, by the IETF SIP working group. Internet Drafts were submitted by the SIP working group in May and October 2001.

SIP invitations provide a means to convey session descriptions so that compatible media types can be agreed upon by participants. Proxy servers are used to help route requests to users and to assist in firewall traversal. SIP runs on top of a number of different transport protocols. See Session Description Protocol. See RFC 2543, RFC 3050, RFC 3087.

Session Initiation Protocol Forum SIP Forum. A nonprofit association to promote and support Session Initiation Protocol, formed in June 2000. See Session Initiation Protocol. http://www.sipforum.org/

SET secure electronic transaction. A phrase used in electronic commerce to signify a transaction which is protected by various network security measures such as authentication and verification procedures, digital certificates and signatures, secure servers, etc.

set-top box *colloq.* A media device which sits on top of a TV set or within a home entertainment component cabinet to hook into the system in some way. Set-top boxes provide a variety of capabilities, including conversion of cable TV signals, provision of WebTV services, etc. Some set-top boxes are proprietary units offered through lease or purchase by a service provider or vendor.

SETI Search for ExtraTerrestrial Intelligence. An interesting, federally funded scientific project in which arrays of radiotelescopes are used to search for signs of intelligence in other parts of the universe. The movie *Contact* (Warner Home Video) provides an idea of a SETI-like project.

The rationale of SETI is that signals can be sent farther and faster using radiowaves than by sending spacecraft (SETI is also in favor of spacecraft missions and radio signals sent from spacecraft) and that it's worthwhile to send out signals in the hope that other life forms may intercept them or that humans may intercept the transmissions of other life forms. Unfortunately, to date, no signs of sentient communications have been detected, but SETI concepts have resulted in the discovery of interesting radiowave signals from distant celestial objects.

The Columbus Optical (COSETI) Observatory, a pioneering observatory located near Columbus, Ohio, conducts searches for extraterrestrial intelligence in the optical spectrum. http://www.seti.org/ http://www.coseti.org/

SF 1. single frequency. 2. See SuperFrame.

SF signaling See single-frequency signaling.

SFOCS See *Submarine Fiber Optics Communications Systems*.

SFTP See Simple File Transfer Protocol.

SGML Standard Generalized Markup Language. A markup standard adapted by the International Organization on Standardization (ISO) in 1986 which is not a language, but is designed for specifying the content and structure of a document or document language, with the assumption that the actual output

or display of the document may vary according to the output device. SGML allows the development of cross-platform applications and documents, and a document can be processed by an SGML compiler by referencing a document tag definition (DTD). HyperText Markup Language, widely used on the World Wide Web, is a descendant of SGML that incorporates some of its capabilities.

Yuri Rubinsky (1952–1996) was one of the pioneers who enthusiastically did much to promote the use of SGML through educational programs.

SGMP See Simple Gateway Monitoring Protocol.

SGRAM synchronous graphics random access memory. A type of memory optimized for use in memory-hungry graphics applications, particularly 3D rendering and ray tracing.

shadow mask A type of cathode-ray tube (CRT) color display technology which incorporates a thin, perforated metal plate mounted close to the front of the inside of the tube to create a mask through which red, green, or blue (RGB) phosphors can be selectively excited. See cathode-ray tube.

Service Environments and Systems

The following definitions are related to Service architectures that can be used in conjunction with Intelligent Networks (INs). See also Intelligent Network.

Service Access Code SAC.

A specific telephone prefix access code (e.g., 800) for a specific category of service such as toll free numbers. See Service Management System.

Service Control Point SCP.

A point that provides access to an Intelligent Network (IN) database, which is connected to a Service Management System (SMS), and which accesses Internet Protocol (IP) as needed. SCPs are a mechanism for providing advanced services by processing the format or content of transmitted information. Information contained in the SCP may be downloaded by phone service carriers.

Architecturally, the Service Control Point is somewhat self-contained, providing services to Intelligent Networks (e.g., those based upon Signaling System No. 7) that architecturally separate switch and service functions.

Service Switching Points (SSPs) can relay value-added service calls to the SCP which can appropriately complete the call. The ITU-T describes the SSP in Recommendation Q.1205.

Service Creation Environment SCE.

In the context of Service Control Points interfacing with Intelligent Networks (INs), the SCE enables new telecommunications services to be quickly designed and implemented without necessitating changes in the IN switching system.

Service Creation Environment Point SCEP.

A protocol used for defining and developing a service, for example, for implementing security features and validating services in networks (often in conjunction with LDAP). The SCEP creates service in conjunction with Service Independent Building-blocks (SIBs) and interfaces with the Service Management Point (SMP). The ITU-T describes the SSP in Recommendation Q.1205.

Service Data Point SDP.

A standard database designed for use with Intelligent Networks (INs). Contemporary phone services are complex compared to historic analog residential/business telephone lines. Users now have different services from different carriers and different variations of those services offered through bundles, specials, prepaid cards, and carrier offerings in specific regions. These services are stored, configured, and managed through software that must be associated with specific subscribers or paid calling card access. The SDP works in conjunction with Service Control Points (SCPs) to manage this enormous task.

In SDP, the customer information is separated from the logic that manages the services. This compartmentalization is important for the privacy of subscribers and also enables updates, maintenance, and new services to be easily integrated into the system without side effects in other areas. Fast processing and large data storage facilities are needed to make the system viable. The ITU-T describes the SDP in Recommendation Q.1205.

Shannon, Claude Elwood (1916–2001) A celebrated American theorist who contributed significantly to the study and understanding of information theory. The history of communications emphasizes the inventors, programmers, and hobbyists who have developed the mechanisms and operations of information systems, but few people at the time had taken a broad look at what information is, how it relates to the technology (e.g., channel capacity), and what the process of conveying information entails from a more abstract, theoretical, statistical, and broadly practical viewpoint.

Shannon, while working at Bell Laboratories, is credited with bringing together and clearly stating fundamental theories of information in 1948.

Shannon is often credited as wholly developing information theory, and his work may have been done independently of Erlang and Kotel'nikov, who developed similar theories many thousands of miles away, but it is derived at least in part from the work of Harry Nyquist, who is cited in Shannon's writings. Nyquist developed principles of communications rates and digital sampling in the 1920s.

Shannon has received many awards for his work, including the National Medal of Science (1966) and the John Fritz medal (1983). See Erlang, Agner; Hagelbarger, David; Kotel'nikov, Vladimir; Nyquist, Harry; sampling theorum.

Service Environments and Systems, cont.

Service Discovery Protocol SDP.

A wireless network resource discovery protocol from Bluetooth that enables applications/devices to seek out and find compatible devices with services that may be useful. Thus, using SDP, Bluetooth devices can connect wireless to nearby services such as printing, digital cameras, etc. SDP was specified by the Bluetooth SIG in 1999. SDP defines how a Bluetooth client application shell discovers available services and their defined characteristics, without prior knowledge of the services, as the device enters the accessible "region of access" of the services. SDP also makes it possible to detect when a service is no longer available.

In March 2001, RidgeRun, Inc., announced support for the Bluetooth technology by releasing an Open Source SDP enabling Bluetooth users to get broader access to on-demand wireless services.

Service Management Access Point SMAP.

In a network architecture, the SMAP interfaces with the Service Management Point and assists in managing user access and services. The ITU-T describes the SSP in Recommendation Q.1205. See Intelligent Network.

Service Management Point SMP.

In an Intelligent Network (IN) architecture (e.g., SS7), a centralized manager interfacing with a number of other elements such as the Service Creation Environment Point (SCEP), a Service Management Access Point (SMAP), and various Service Control Points (SCPs) and other elements to satisfy value-added service calls from the Service Control Point (SCP). The ITU-T describes SSP in Recommendation Q.1205. See Intelligent Network.

Service Management System SMS.

A centralized interactive computer system dedicated to coordinating network service-related information. The national 800 numbers are managed in a central database computer from IBM. This is one type of Service Control Point providing value-added services to Intelligent Networks (INs) known as SMS/800. See SMS/800.

Service Profile Identifier SPID.

When hooking up ISDN BRI services, the carrier provides the user with a SPID for each number installed, typically two. The SPID points to a memory location in the carrier's central office where ISDN parameters, including which services are enabled for a particular subscriber, are stored. As not all phone carriers have automatic SPID detection, some newer modems can determine what type of ISDN service is connected and configure the SPIDs accordingly. When connected to carriers with automatic SPID detection, they can configure themselves whether or not a computer is attached to the modem. Modems with these capabilities help compensate for some of the problems traditionally associated with the installation of ISDN services.

It is recommended that the subscriber keep a record of SPID numbers filed away somewhere, as it's easier to look up a lost SPID than to get it again from the phone carrier.

Service Switching and Control Point SSCP.

An entity in the physical plane that controls network resources and sessions, including the Service Control function, Service Data function, and Service Switching/Call Control functions.

Service Switching Point SSP.

An entity in the physical plane, the SSP is a point providing local access and an ISDN interface for a Signaling Transfer Point (STP), which, in turn, provides packet switching for message-based signaling protocols in an Intelligent Network (IN).

The SSP can be implemented as a central office switching system capable of communicating with a Service Control Point (SCP) in order to enable switched calls to be routed through a somewhat separate service environment that will then complete the call. The ITU-T describes the SSP in Recommendation Q.1205. See Intelligent Network, Service Control Point.

S

Shannon-Hartley Capacity Theorem A theorem that facilitates calculations for assessing an environment with additive white Gaussian noise and its implications for relative capacities in communications channels. In other words, capacity has been mathematically defined in relation to bandwidth, energy per bit, and noise power density, to yield useful information about total signal and noise power, and bandwidth efficiency (in bits per second per Hertz) in communication technologies.

Shared Facility Centers A telework organization similar to a branch office, but co-owned or partially community- or freelance professionals-funded, situated in a residential or rural village area, and made commercially viable by the implementation of new communications technologies. See ADVANCE Project, telework.

shared tenant services STS. A category of communications services applying to residents of multiple family dwelling units. In these types of residences, cable services, antennas and other communications facilities are often shared among a group of residents within a complex, section, or building. The distinction is important because, historically, many of these residents have been served by independent private carriers considered distinct from local exchange carriers (LECs) and thus are not bound by the same regulatory framework. With the advance of technology, private service provider services are becoming more like LEC services and thus are debated and evaluated in the context of changing technologies.

ShareView 3000 A Macintosh-based videoconferencing system from Creative Labs which supports audio, video, whiteboarding, application and document sharing, and file transfers over analog phone lines. An IBM-licensed PC-version called ShareVision PC3000 is also available. See Cameo Personal Video System, Connect 918, MacMICA, IRIS, VISIT Video.

sheath *n.* A close-fitting protective covering, usually tubular, often made of plastic. Sheaths can be used to bundle wires, to insulate, to protect from moisture or wear, or to provide identifying colors or symbols. They are commonly used on conducting wires and fiber cables. See conduit.

sheave The round, usually rotating track in a *stringing block* used to direct a line conductor that is being installed, removed, or temporarily redirected for maintenance or repair. See stringing roller.

shell, command shell A computer user interface input and display environment which translates user commands into operating system instructions.

Shenzhen Bordering Hong Kong, this is a major supplier of optical cross-connect, multiplexing, and single- and multimode optical fiber products. In addition to supplying components, the Chinese city puts major emphasis on linking its populace through fiber optic backbone network systems.

Sherman Antitrust Act An important 1890 U.S. act passed to prevent the establishment of monopolies that could hinder U.S. trade and competition based upon free enterprise.

SHF super high frequency. About 3 to 30 Ghz, used for satellite transmissions.

ship to shore telephone See marine telephone.

shock, electric A sudden, often hazardous, electrical stimulation to a living body which may greatly affect nerves and cause convulsive contractions through muscles, possibly endangering the heart muscle. It may also cause severe burning, confusion, and unconsciousness.

Light electric shocks are uncomfortable, but not always dangerous and are sometimes used as perimeter boundaries for livestock or secure areas. Light electric shocks are also used in animal experiments for studying the nervous system and are occasionally used in riot control and law enforcement.

It is unwise to open up or attempt to repair cathode-ray tubes (CRTs), which may store a considerable charge, without careful preparation and knowledge of safety procedures.

Electric shocks must be taken seriously and, if severe, may require contacting emergency services or the application of cardiopulmonary resuscitation (CPR). Never touch someone who is experiencing shock from an electrical source until the electricity is turned off or the source of the contact knocked away with a nonconducting material. Consult emergency first aid sources for information.

Shockley, William Bradford (1910–) An English-born American physicist who worked in the Bell Telephone Laboratories from 1936. He discovered the rectifying properties of impure germanium crystals at a time when vacuum tube rectifiers had replaced the old galena and carborundum crystal detectors. This led Shockley to explore the various impurities in germanium and he found electron drift toward the positive or negative pole under controlled conditions. When these solid-state rectifiers were combined, the transistor was born and vacuum tubes superseded. One of the most significant consequences of transistors at the time was miniaturization of communications devices and room-sized computing machines.

Shoemaker detector A type of electrolytic detector that incorporates a battery and, consequently, requires no outside power source. It consists of a glass tube with a platinum-sealed point, with a zinc strip rather than the platinum point coming in contact with a mild sulphuric acid solution. Shoemaker detectors were used commercially in wireless telephone receivers in the early 1900s. See electrolytic detector.

short circuit, short An unintended or harmful cross connection, of low resistance, of electrical circuits. Short circuits can occur from an excess of solder, incorrectly connected wires, conductive debris (such as a screw falling into a circuit box), worn-out insulation in bundled wire, water, or physical bumping of electric conductors, etc. The result is often a sudden flow of current in the wrong direction or of the wrong magnitude, which can potentially damage components. Some systems are configured to shut down or blow a fuse or breaker in the event of an excess of current or other abnormal electrical activity. See burst, spike.

short haul A short travel or installation distance. The actual length depends upon the situation or medium employed. A short haul for a SCSI cable is about 3 feet or less; above 6 feet, serious signal degradation occurs and special hardware is needed for distances over 12 feet (see Fibre Channel). A short haul for other media may be several yards or thousands of miles.

short haul modem A software/hardware combination used for short distance communications up to a couple of dozen miles, usually over a copper single-channel line. See baseband modem.

Short Message Service SMS. A global, wireless, low bandwidth, two-way service first distributed in Europe in the early 1990s and later in North America. SMS provides the capability of transmitting alphanumeric messages between mobile systems and external systems that support paging, email, and voice mail. The handsets used in these services can send or receive at any time, regardless of whether a data or voice call is in progress. SMS is appropriate for applications like stock quotes, paging, short fax and email messages, online quick banking, etc.

Short Message Service Center SMSC. A relay and administrative center for Short Message Service (SMS) which provides store and forward services. This is somewhat like an enhanced alphanumeric paging system with two-way service and guaranteed delivery. See Short Message Service.

Short Wavelength Fast Ethernet Standard TIA/EIA/ANSI-785. An industry standard for the transmission of data at 100 Mbps over 850-nm optical links approved in June 2001. This technology is also known as 100BASE-SX and provides a cost-effective upgrade path from 10 to 100 Mbps through the use of short wavelength opto-electronic devices. Overall this standard can lower the cost of implementing Fiber to the Desktop (FTTD) and Fiber to the Home (FTTH) services.

shortwave, shortwave Long-range radio transmission frequencies in approximately the 1.6 to 30 MHz range, above the commercial broadcast bands. Shortwave signals are easier to apprehend at night, due to lowered atmospheric noise and the fact that many shortwave broadcasters prefer to send in the evening hours. Coordinated Universal Time (UTC) is often used as the reference time for broadcasts. The Internet has sites that list broadcast times and frequencies for various shortwave stations around the world. See microwave, ionospheric wave, radio.

SHT Short Hold Time.

Shugart A historic disk development company. Shugart floppy disk drives were used on some of the earliest microcomputers.

shunt *n*. A switch, pipe, detour sign, or other diverting mechanism.

shunt, electrical In electrical circuits, a means to divert some or all of the current. A shunt is sometimes used to divert part of a current in order to prevent damage to sensitive measuring instruments. Temporary shunts are sometimes established with jumper wires or alligator clip connections.

shunt circuit, bypass circuit, detour circuit A circuit configuration through which a specific portion of the current is redirected or subdivided. Often used for diagnostic purposes, temporary arrangements, or circuits in which variable conditions are accommodated or where the original current can be more effectively used by dividing it. Shunts are sometimes incorporated into the internal workings of diagnostic instruments.

shunting error A condition in which current divided through two circuits, as through a component and a measuring instrument, will vary depending upon the frequency. See reversing error.

Amateur Radio Technology in Space

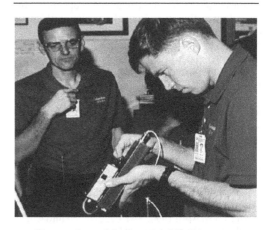

Two members of the Spacelab Life Sciences crew, McArthur and Searfoss, training with amateur radio equipment for Space Shuttle mission STS-58 in September 1993. [NASA/JSC image detail.]

Endeavour *Astronaut Linda M. Godwin uses the Shuttle Amateur Radio Experiment, as did several of the other STS-59 crew members, to communicate from space with ham radio operators and students on Earth. [NASA/JSC image, April 1994.]*

Shuttle Amateur Radio Experiment SAREX, SAREX-2, SAREX-II. A series of amateur radio experiments consisting of equipment and procedures carried as payload aboard a number of U.S. space shuttles. SAREX was designed to provide an opportunity for school and community groups to communicate with astronauts in space and for conducting communications experiments with ground-based amateur radio operators. SAREX was later integrated into the International Space Station project and has even been used for emergency messaging. Amateur radio has a long history of cooperation with U.S. aeronautical/space programs leading up to SAREX. In November, 1983, Astronaut Owen Garriott (W5LFL) took his amateur (ham) radio into orbit on the Space Shuttle *Columbia* STS-9 mission and communicated with Earth amateur radio stations. This tightened the relationship between ham operators and shuttle crewmembers and the experiments continued on future shuttle missions of *Columbia*, *Challenger*, *Atlantis*, *Discovery*, and *Endeavor*. Many astronauts are licensed amateur radio operators. In several cases, the entire crew of a space shuttle mission comprised ham operators (e.g., STS-74).

As examples of specific missions, a SAREX payload was carried aboard the Space Shuttle *Columbia* in 1993 and the Space Shuttle *Endeavour* in 1993 and 1994. The SAREX-2 amateur radio system was used to contact elementary and middle school children from space on October 22, 1993. SAREX-II was used in 1994 to contact school and Boy Scout groups and also to communicate with Russian cosmonauts aboard the MIR space station. Amateur radio communications were monitored in realtime by amateur radio stations via rebroadcasts and the telebridge system. SAREX-II was further used to communicate with the Star City training center outside Moscow, Russia, on 16 April 1994. SAREX evolved into Amateur Radio on the International Space Station program (ARISS).

Slow Scan Video Converter

The major unit in an early SAREX containing a slow scan video converter, control circuits, and power supplies. The small tape recorder enables recording of video image data on tape for retransmission through the SAREX equipment. [NASA/JSC image, June 1985.]

The American Relay Radio League (ARRL) chronicles the full history of the interesting SAREX and pioneer amateur radio experiments in their publications and on their Web sites. NASA provides instructional support materials for SAREX participants on their Division of Education Educational Services Web site. See Amateur Radio on the international Space Station, America Relay Radio League, AMSAT.

Si *symb.* silicon. See silicon.

SIA 1. Securities Industries Association. 2. See Semiconductor Industry Association.

Sibley, Hiram (1807–1888) Sibley founded the New York and Mississippi Valley Printing Telegraph Company which took on the name of Western Union, suggested by Ezra Cornell in the mid-1850s, when it began westward expansion. Sibley remained president during the expansion and Western Union installed the first transcontinental cable in 1861. After the failure of Western Union's first Atlantic cable, Sibley traveled to Russia to investigate the installation of a Siberian-Alaskan communications line, and the Russians offered to sell Alaska to Western Union. Sibley turned down the offer, but alerted the U.S. government to the opportunity. Along with his colleague, Ezra Cornell, Sibley helped to found Cornell University. See Western Union.

side circuit In telephone installations and other circuits where additional endstations are desired, but where resources are limited, a side circuit is a means to build an additional circuit using the resources of two adjacent circuits. See phantom circuit.

side lobe SL. In a directional antenna, any segment generating a lobe in the antenna pattern other than the primary/main lobe. Antenna patterns are often diagramed with a radiating region around the segment, a region of influence that appears lobe-shaped. In a Cartesian coordinate system, the lobe shows up as a pronounced hump or "shoulder." Side lobe peaks are often described in terms of their ratio to the main lobe peak and will vary, depending up the shape of the antenna, its electrical characteristics, and the frequencies transmitted.

Depending upon the configuration, an antenna may have several side lobes. In general, low side lobes are desired. Antennas that are likely to be physically adjacent to one another (toll booths, cell phones, even satellites, etc.) are often designed to deliberately suppress side lobes.

In radar sensing, side lobe signals can muddy the radar signal and cause confusion. Sometimes an omnidirectional antenna separate from the main antenna is used to send out a reference pulse that is lower than the signal from the main antenna, but stronger than the signal from its side lobes. This enables the transponder to make a few calculations and determine whether a signal came from the main antenna or the side lobe. It can then respond or not respond as appropriate.

Antenna pattern modeling programs are used for generating and assessing the configuration and influence of side lobes (including thermal emissions).

Side lobe interference in antennas can be a problem, but it is not restricted to larger physical structures; side lobe issues are also important to small-scale transmissions phenomena such as laser light paths. In optical network links, side lobes can result from various structures integrated into the optical fiber to enhance its transmission qualities. For example, Bragg gratings suffer undesired effects from side lobes in their spectral response. This can reduce the usable bandwidth.

side lobe dispersion An undesirable condition in which a directional antenna loses power due to dispersion of the signals or wave patterns out of the "sides" of the antenna (e.g., portions that are not part of the main lobe).

side-looking airborne radar SLAR. A self-illuminating (through microwaves) electronic image-creation system derived from a radar beam transmitted perpendicular to the ground track during acquisition from an aircraft. Thus, the signal hits the terrain at a rather flattened angle and the view of the terrain is vertical, revealing fine surface features useful in interpretation of the data. The imaging is provided in strips or mosaics, as is true for many satellite imaging systems. SLAR imagery is used by geologists, Earth resource scientists, cartographers, engineers, and others. SLAR encompasses real-aperture and synthetic-aperture radar (SAR). SLAR is not used for very precise topographic mapping, as the resolution is only up to about 30 m.

sideband The frequencies on either side of the main frequency or carrier band in a communications signal. These frequencies are within the *modulation envelope* of a transmission wave, but were originally not used because of problems with noise. Later, as technology improved and the demand for airspace continued to grow, sideband transmissions became more interesting, and it was found that one sideband could be transmitted, sometimes even without the carrier wave, and the original wave mathematically "rebuilt" at the receiving end. The advantages included lower power requirements for the transmission and a narrower wave overall, leaving more room for other transmissions.

sidetone In a telephone receiver, transmitting currents are directed into the receiver to make it possible for the speaker to hear his or her own voice (somewhat like an echo) as a form of feedback mechanism. This has to be carefully controlled so that it doesn't become excessive, and various anti-sidetone circuits are applied to minimize feedback and reduce transmission of acoustical noise.

In early telephones, the sidetone was loud enough to be distracting. In 1920, G.A. Campbell designed a circuit to reduce the excess current from the line and the local receiver, thus reducing sidetone and improving electrical efficiency.

Sidetones are also used in telegraph keying systems as well, to provide feedback to the person keying in the (Morse) code.

Sieve A protocol- and OS-independent, extensible mail filtering language proposed as a Standards Track RFC by T. Showalter, in Jan. 2001. Sieve is a language for filtering email messages at time of final delivery that can be implemented on either the mail server or a mail client. It is suitable for a variety of systems, including Internet Message Access Protocol (IMAP) servers.

Sieve uses IMAIL-compliant messages to enable a user to create filters for organizing incoming electronic mail (email). The language is intended to be powerful enough to be useful, while not being so powerful that it can break out of its operating environment or wreak havoc on a system through uncontrolled variable, loops, or programming bugs. Sieve also facilitates the use of graphical user interfaces (GUIs) for creating and manipulating email filters. See FLAMES, RFC 3028.

Siemens Telecom Networks A provider of telecommunications services and network equipment emphasizing robust, secure technologies to regional Bell operating companies and independent telephone and holding companies.

In March 2001, Siemens AG announced that it was cleared by the Federal Trade Commission to merge with Efficient Networks, Inc. (Texas-based suppliers of high-speed DSL networks). In December 2001, together with Cingular Wireless, the second largest mobile carrier in the U.S., Siemens announced that the two companies would be working to upgrade their services/systems to support 3G (third generation) mobile communications using Enhanced Data Rates for Global Evolution (EDGE) technology.

Siemens, Werner (1816–1892) An American inventor who, along with his brother, William, developed the dynamo, a device to convert mechanical energy into electrical energy without the use of permanent magnets. In the 1870s, he demonstrated that the velocity of electrical conductivity through a wire could approximately equal that of light.

SIF See SONET Interoperability Forum.

sigma-delta modulation SDM. A means of encoding the mathematical integral of a signal's amplitude rather than the amplitude of the signal itself by first integrating and then delta modulating the signal. It is sometimes called delta-sigma modulation, depending upon the sequence of the processes. It is a form of oversampling and noise-shaping signal conversion scheme.

The system came into practical use about two decades after delta modulation was developed in the 1940s, when VLSI technology began to emerge, making it possible to develop practical embodiments of the system. SDM helped overcome some of the limitations associated with delta modulation, a simple scheme for sampling an analog signal and encoding the samples based upon the previous state.

SDM is useful for quantizing and encoding audio signals for transmission over telephone circuits and is also used in image processing.

In sigma-delta modulation, an analog signal is quantized and the sum of the previous difference and the current signal compared rather than directly comparing the current and previous signal. This has greater spectrum independence than delta modulation and "conditions" the shape of the quantized signal in terms of its maximum possible range. SDM is also

S

known as pulse density modulation (PDM).

While the resolution of SDM is good, the filtering introduces latency and it is not as fast as might be desired for newer, high-speed communications technologies (e.g., multiplexed systems). Thus, research into improving the speed has yielded some derivative schemes. For example, reduced-sample-rate sigma-delta modulation (RSRSD) schemes require only a fraction of the speed necessary for sigma-delta encoding without trading off bandwidth. RSRSD may be applicable to A/D and D/A conversion for applications such as Extended Digital Subscriber Line (xDSL).

There are hobby kits available for demonstrating delta and sigma-delta modulation/demodulation processes. See A/D conversion, delta modulation, pulse width modulator.

Signal Corps Radio Laboratories SCRL. A facility established at Fort Monmouth in March 1918 as Camp Alfred Vail. The lab was active in classified research including the development of radio direction finding equipment and aviation radio communications. Following World War I, aviation communication research was moved to the Signal Corps Aircraft Radio Laboratory in Ohio. A few years later, a number of labs in Washington, D.C. and New York state were moved to the Fort Monmouth location and it was redesignated as the Signal Corps Laboratories. The lab was used for radio communications development and adjunct meteorological services. In the early 1930s, marine and aerial sensing were added and radar studies stepped up in the late 1930s.

Signal Transfer Point STP. A non-terminal node point which provides access to a database and packet switching for message-based signaling protocols for the Service Control Point (SCP) in an Intelligent Network (IN). STPs are widely used in Common Channel Signaling systems.

To provide redundancy in a telephone network, Service Control Points (SCPs) and STPs are usually paired up. Depending on the implementation, the STPs may be arranged hierarchically, with the lower layer serving the smaller or local region and the higher layer serving the larger region. The Intelligent Network is based around Signaling System 7. See Intelligent Network, Service Switching Point.

signal-to-noise S/N. A ratio frequently used in electronics and communications to quantify the proportion of a signal (or communication) that is desired and useful to the proportion of a signal that is undesired and distracting or destructive to the desired signal. It is preferable to have a high signal-to-noise ratio.

In electronics the signal can be an electrical or optical signal or a data stream and the noise can be various types of interference such as crosstalk, echo, spurious signals, etc.

When Internet Relay Chat (IRC) was first developed, it had a pretty good signal-to-noise ratio in terms of the content of conversations. As the Net has grown, this has changed to the point that public online discussions are often avoided in deference to private chats and moderated email discussions. See noise.

Signaling System 1 SS1. Historically, an international CCITT-standardized supervision single-tone signaling protocol to signal call requests between telephone switchboards. A 500-Hz signal tone at 20 Hz was used. Bell had a similar standard for manual ringdown signaling. These days Signaling System 1 is more often associated with newer digital signaling systems, i.e., DSS1. See Digital Subscriber Signaling 1.

Signaling System 2 SS2. Historically, a CCITT-standardized supervision dual-tone signaling protocol to handle pulse-dial selection. Dual 600/750 Hz signal tones were used. It was somewhat similar to historic Bell radiotelephone dial signaling systems.

Signaling System 3 SS3. Historically, a CCITT-standardized supervision single-tone signaling protocol used in one-way telephone circuits and not intended for use for multinational transit connections. This system was still in use until about the late 1970s.

Signaling System 4 SS4. Historically, an international CCITT-standardized supervision dual-tone signaling protocol for international and terminal transit. SS4 was the first to fully support global direct dialing signaling. SS4 is not entirely obsolete; it is sometimes used for signaling for Voice over Packet (VoP) devices.

Signaling System 5 SS5. Historically, an international CCITT-standardized supervision dual-tone signaling protocol used in T1 and E1 telephone trunks. Dual 2400/2600 Hz tones were used. SS5 is similar to Bell DDD trunks with supervision used in North America. SS5 is not entirely obsolete. Although it has generally been superseded by SS6 and SS7 for subscriber systems, testing and maintenance devices for central office equipment sometimes use SS5 signaling on digital trunk interfaces for testing gateways and SS5 is still used on analog international gateways.

Signaling System 6 SS6. Historically, a CCITT out-of-band signaling system developed in the 1960s, which is being superseded by Signaling System 7 in North America but is still popular in Europe. Both SS6 and SS7 owe their origins to Common Channel Interoffice Signaling (CCIS). SS6 was the first system to incorporate packet switching into public switched telephone networks (PSTNs). It supports a message-based protocol for requesting services, similar to the widespread X.25 standard. Signal units of 28 bits each were assembled into data blocks for transmission. See Signaling System 7.

Signaling System 7, Signaling System No. 7 SS7. SS7 is a common channel network signaling system, descended from Signaling System 6 and its precursor Common Channel Interoffice Signaling (CCIS). CCIS was an out-of-band that was inherently more secure than in-band multifrequency signaling systems prevalent at the time. It is sometimes referred to as Common Channel Signaling System 7 (CCS7).

SS7, introduced in the 1980s, is more flexible and powerful than earlier versions, making it possible to implement broadband digital services far in advance of basic voice circuits. One of the ways in which SS7 differs from SS6 is that it supports variable-length signal units (up to a defined maximum), while SS6 was constrained to fixed-length units.

Unlike earlier phone signaling systems, which operated through many semi-independent switching centers, SS7 brings the communications channels into a more integrated whole.

The international data rate for SS7 networks is 64 kbps, although faster data rates are being studied and implemented, such as 1.544 and 2.048 Mbps (international).

There are many factors influencing the adoption of SS7, including its flexibility and applicability to both wireline and wireless communications. But one of the first motivations for switching to SS7 was the development of phone services that could be dialed using a common area code, regardless of the geographical location of the subscriber (e.g., 800 numbers). Since local databases and switches could not be used to route calls that could be placed to any region, a central cross-reference registry was needed to associate the 800 number with a regional location and switching office. With SS7's message-passing capabilities, it became straightforward to associate a virtual area code with a geographical routing code. This opened the door to other types of virtual area code services, such as emergency 911, pay 900, as well as custom calling services (e.g., Caller ID) and enhanced services (e.g., number portability).

SS7 is now an important aspect of digital telecommunications services. Users have come to expect the many features that were difficult or impossible to implement over older analog tone-based signaling systems. With its out-of-band architecture, it is also inherently much more secure than historic in-band signaling systems.

SS7 is being gradually integrated into ATM/T1 and PCS/UPT networks. See Common Channel Interoffice Signaling.

Signaling System R1 SS R1. An international analog telephony signaling standard, equivalent to Bell's out-of-band 2600 Hz tone in DDD trunks in North America.

Signaling System R2 SS R2. An international analog telephony signaling standard, equivalent to Bell's out-of-band 3825 Hz tone inserted between voice channels in the carrier system.

signature 1. An identifying mark, usually a name, intended to relate a document or other transactional device to the individual associated with the document. A signature is often used as a means to acknowledge understanding or agreement to the terms of a transaction, e.g., a contract. See digital signature. 2. In a more general sense, a mark, style, method of doing things, or musical sequence that is identifiable as coming from a specific source or strongly associated with a certain person or group. The signing of a name in a person's handwriting, the creation of a painting with a distinctive and uniquely recognizable origin, or even a specific way of turning a phrase are all examples of signatures in the broader sense. On the Internet, individuals often come up with names, signa-

ture files, or icons to uniquely and quickly identify themselves to others in much the same way a corporation identifies itself with a trademark or logo. 2. In printing, a grouping of pages that is created in order to organize the pages for binding. Common sizes for signature groups are 8 or 16 pages. With many people now doing their own desktop publishing and submitting them to printers over the Internet for publication, it helps to understand some of the basic terms and procedures used to create documents in the printing industry. See fascicle, imposition.

SIIA See Software & Information Industry Association.

silence compression A technique used in voice over data network applications which involves removing the pauses and spaces that typically occur in many conversations. This reduces transmission time. Two common techniques typically used together include voice activity detection (VAD), which distinguishes speech from the surrounding background noise, and comfort noise generation (CNG), which creates a low type of static that gives humans a certain comfort level and trust that the line is still active and the call hasn't been cut off.

silent discard In packet networking, the discard of a packet without further processing. The system may log the event and may even store the contents of the discarded packet for later evaluation.

silica Silicon dioxide. See silicon.

silicon An abundant nonmetallic, tetravalent element (AN 14), widely used in semiconductor technology. Silicon comprises about 26% of the Earth's crust (by weight). Silica occurs in many common forms, including sand, quartz, flint, and opals. Silica is a main ingredient in glass manufacture. Highly pure silicon can be doped with a number of other elements for use in solid-state devices.

silicon detector An early type of radio wave detector similar in some aspects to electrolytic detectors. Silicon is used in place of the electrolyte, making contact with a platinum wire, and the thumbscrew contact with the silicon can be finely adjusted by filing the end of the thumbscrew to a fine point, using a spring with the thumbscrew to assure even pressure. The interaction of the thumbscrew and the silicon sets up a thermoelectric reaction which can be translated into audible waves in the receiver. See detector, electrolytic detector.

Silicon Graphics Incorporated SGI. A computer company known for innovative software and hardware workstation-level computers, especially those with good graphics and sound. SGI was founded in 1981 by James Clark, who later became affiliated with Netscape Communications Corporation.

Silicon Valley A region of California with a high density of high-technology companies, many of which pioneered computer technology. The economy, educational institutions, research labs, and climate were all factors that contributed to the growth of technology companies in Silicon Valley.

silicone rubber insulator A more recent type of insulator for installing conducting lines found on utility

S

poles. The top and bottom mounting surfaces are metal, while the layers of "skirts" down the body of the insulator are fabricated out of silicone rubber over a fibre-glass core. Silicone rubber is nonconductive, light, water-repellant, resistant to ozone and ultraviolet degradation, and easy to fabricate in a variety of shapes and sizes, making it an alternative for glass and ceramic insulators.

SIM4 Historically one of the early desktop computers, introduced over 2 years before the Altair, but several months after the Kenbak-1, in 1972, by the Intel Corporation, which was around the same time Hewlett Packard introduced the HP 9830. The single-board Intel computer was based upon a 4004 processor and was available in at least two models, the SIM4-01 and the SIM4-02. The SIM4-02 could be inserted into an Intel MCB4 chassis and programmed through a programmer card. See Altair, Kenbak-1, Micral.

SIMM single inline memory module.

Simon A historically remarkable computer project described in Edmund C. Berkeley's book *Giant Brains or Machines That Think* in 1949 and in *Radio Electronics* articles in the early 1950s. The name was based on Simple Simon. It was basically a desktop logic calculator that could be built for about $300 (about $4000 in today's money). In his book, Berkeley describes it as "... so simple and so small, in fact, that it could be built to fill up less space than a grocery store box, about 4 cubic feet."

The Simon was an electromechanical assembly for performing different calculating experiments, but it can probably be considered the first desktop computing kit considering the size of computer behemoths at the time. Simon was a papertape computer based on 129 relays and a stepping switch. In Berkeley's description, a two-hole tape reader was used to input numbers and operations and a four-hole tape reader was used to input instructions, but Berkeley points out that relays and other input modes apply just as well. Problems were entered in binary and answers were displayed on front panel lights (a design aspect used by many early microcomputers until the mid-1970s).

With assistance from William A. Porter, Robert A. Jensen, and Andrew Vall, Berkeley got a basic machine working. Considering that most people didn't know what a computer was in those days, it is amazing that Berkeley wrote about "machines that think" in November 1949 and published plans for actually building the Simon in 1950. Apparently more than 400 plans for the Simon were sold over the next decade.

Simon's little-cousin successor was the GENIAC, a computing "game machine" developed by Berkeley in the mid-1950s with documentation by his partner Oliver Garfield (until a dispute split the name from the technology). See Altair; Arkay CT-650; Berkeley, Edmund C.; GENIAC; Kenbak-1; Simplac.

Simplac A design for a transistor-based computer presented as a collaborative progress report documented by Edmund C. Berkeley through Berkeley Enterprises Laboratory in 1956. Milt Stoller had responsibility for the logical design of the machine. The machine was intended to have registers for three binary digits. The author is not sure whether this computer ever came to fruition. Berkeley had a lot of interests in robotics and artificial intelligence and was always beginning new projects. See Simon, GENIAC.

Simple Discovery Protocol SDP. An experimental minimal request/response multicast network recourse discovery protocol developed by Martin Hamilton. SDP payloads are application-dependent. SDP is not intended for bulk data transfers, due to the size of UDP packets.

Simple File Transfer Protocol SFTP. A simple file transfer protocol that fills the need for a specification that is easier to implement than File Transfer Protocol (FTP). It provides file transfer capabilities combined with user access control, listing of directories, traversing directories, file renaming, and file deleting. In other words, it incorporates the most common and necessary functions of FTP. See RFC 913.

Simple Gateway Monitoring Protocol SGMP. Developed in the mid-1980s and demonstrated in 1987, SGMP later evolved into Simple Network Management Protocol (SNMP).

Simple Internet Transition SIT. A set of Internet protocol mechanisms for hosts and routers designed to smooth the transition between IPv4 and IPv6, its successor. SIT eases the transition by supporting incremental upgrades of hosts through upgrading the DNS server with support of existing addresses. SIT employs a number of mechanisms to achieve interoperability and compatibility including:

- embedding of IPv4 addresses within IPv6 addresses
- encapsulation of IPv6 packets in IPv4 headers for transmission through IPv4 legacy routers
- dual IPv4/IPv6 protocol stacks model for hosts and routers
- header translation for IPv6 only routing topologies

simple line code SLC. A means of transmission through four-level baseband signaling that filters the baseband and restores it at the receiving end.

Simple Mail Transfer Protocol SMTP. A transmission subsystem-independent electronic mail protocol which establishes and negotiates communications between sender and receiver (or multiple receivers) across transport service environments. Transmissions may be direct, depending upon the transport service, or may pass through relay servers.

When a user mail request is generated, the sender-SMTP establishes a two-way transmission channel to the intermediate or ultimate destination-SMTP. SMTP commands are then sent between the two ends. Once a transmission channel is established, a lockstep negotiation of the transmission and identification of the recipient or recipients is carried out, and the mail data sent, with a terminating sequence to indicate the end. When successfully received, the recipient sends an OK reply. See electronic mail, email, RFC 821.

Simple Multicast Routing Protocol SMRP. A routing protocol from Apple Computing, Inc. which is used for AppleTalk network data from applications such as their QuickTime Conference, which in turn is used for videoconferencing, electronic whiteboarding, etc.

Simple Network Management Protocol SNMP. SNMP evolved from, but is not backwardly compatible with, the Simple Gateway Monitoring Protocol (SGMP). Essentially, SNMP communicates management information between network management stations and the agents in the network elements (NEs). SNMP was designed for TCP/IP-based network environments and manages nodes on the Internet. SNMP was originally designed as an interim solution with the intention that it follow generally along Open Systems Interconnection (OSI) guidelines. Over time, they were found more different than originally envisioned.

Along with MIB and SMI, SNMP has been designated by the IAM as a full Standard Protocol with "recommended" status. The SNMP Extensions working group was formed to evaluate and further develop the SNMP definition, with the mandate of retaining its simplicity. See RFC 1157.

Simple Raster Graphics Package SRGP. A low-level graphics package which incorporates features from a variety of graphics systems (such as GKS and PHIGS standards, The X Window System, Apple QuickDraw). SGRP typically functions as an intermediate layer between the applications program and the display device.

Simple Security and Authentication Layer A Network Working Group-proposed standard for providing a quick method of negotiating an authentication mechanism, even if the client has minimal knowledge of the system. See RFC 2222.

Simple Server Redundancy Protocol SSRP. A network protocol which provides resiliency for LANE services on ATM-based local area networks (LANs).

SIMULA object-oriented programming language designed by O. Dahl and K. Hygaard at the Norwegian Computing Centre between 1962 and 1967. It was intended for discrete event simulation, but gradually became a general-purpose programming language. SIMULA was one of the early languages incorporating object-oriented concepts. A number of versions of SIMULA have been developed over the years, with compilers for specific systems such as Control Data Corporation systems as well as IBM 360/370 and UNIVAC computers.

The Association of SIMULA Users (ASU), formally established in 1973, supports the development and use of the language and is one of the earlier computer users groups.

simultaneous voice/data SVD. A number of analog and digital techniques and standards which permit limited use of simultaneous voice and data through regular phone lines with computer voice/data modems. These might be considered medium level applications, since they do not support full realtime videoconferencing, but they allow whiteboarding and switching between voice and data as needed (alternate voice/data [AVD]). SVD is accomplished through multiplexing. In analog SVD, voice is multiplexed with data in digital SVD; data and digitally compressed voice are multiplexed into a digital data stream.

The ITU-T has established standards, draft standards, and specifications related to SVD. These are periodically reviewed and updated to reflect improvements in modem technology. V.61 has been specified for 14,400 bps standard for analog SVD, and V.70 for 28,800/33,600 bps for digital SVD.

simulator A software program, or software/hardware combination that models, reconstructs, or mimics an environment or situation, which may be real or imagined. Simulators are used in many areas of scientific research to enact scenarios; to test, confirm, or investigate hypotheses; to compare or contrast the effects of various changes to a system; or to monitor the evolution of a system. Simulators are also popular in the entertainment industry. Flight simulators have been developed into interactive, environmental video games with helmets, moving seats, and more, to provide a strong emotional/intellectual/tactile experience. Virtual reality simulators go a step farther, creating 3D effects which appear to inhabit the space around the user, sometimes so convincingly that the user will duck to get out of the way of a virtual image.

Sinclair ZX81 The successor to the ZX80, the ZX81 personal computer was introduced in spring 1982 and sold for under $200 (without monitor; it could be hooked up to a television set). It sported 8 kbytes of extended BASIC, math functions to 8 decimal places, a built-in printer interface, 1 kbyte of memory expandable to 16 kbytes, and a 32-column by 24-line display. It was also available as a kit for under $100.

sine wave A fundamental waveform present in almost all vibratory motion, which can be represented as a *sine curve* with periodic oscillations in which the amplitude of displacement at each point in the wave is proportional to the sine of the phase angle of its displacement. In telecommunications, the sine wave is important in many representations, but especially in alternating current (AC) circuitry and in representing sound. See oscilloscope.

sine galvanometer An early current-detecting instrument in which the coil is rotated until the reading needle again registers zero. This type is subject to interference from the Earth's magnetic field. See galvanometer.

SINGARS Single Channel Ground and Airborne Radio Systems. A tactical radio system. See Enhanced Trivial FTP.

single line repeater A mechanism for allowing two-way communication on a single line by permitting the transmission to be alternately broken in one direction in order to initiate or resume communication in the other direction. This is accomplished by an additional holding coil on each relay which can open

S

or close independent of whether the main circuit is open. See half-duplex.

single sideband Transmissions created by manipulating frequencies that are selected from one side of the *modulation envelope* of a transmission wave to recreate the original baseband transmission. Much of the credit for the development of single sideband technology, which is essential to frequency division multiplexing, belongs to John R. Carson, a mathematician with AT&T, and later Bell Laboratories, who mathematically demonstrated the relationship between the information in the sideband signals and the original baseband.

Sideband frequencies were not originally used because of problems with noise. Later, as technology improved and the demand for airspace grew, sideband transmissions became more interesting, and Carson demonstrated in 1915 that one sideband could be suppressed from the transmission and the other could even be transmitted without the carrier wave. Due to its predictable characteristics, the original baseband wave could then be mathematically rebuilt at the receiving end. In a sense, this was a type of "wave compression" accomplished by removing extraneous and redundant information. The significant advantages included lower power requirements for the transmission and a narrower wave overall (i.e., requiring less bandwidth), leaving more room for other transmissions. See frequency division multiplexing.

single sign-on SSO. A network security and management strategy to help reduce the number of passwords needed to access a variety of software and hardware resources on a network.

Single UNIX Specification Developed within the Common Applications Environment by the X/Open Company, the Single UNIX Specification is a collection of documents which includes interface definitions, interfaces, headers, commands, utilities, networking services, and X/Open Curses. This specification is distinct from the AT&T licensed source-code commercial product and is intended as a single stable UNIX specification for which portable applications can be built. It provides vendors a means to provide a "branded" product and assumes voluntary conformation to the specification. Basic components within the Specification are shown in the Single UNIX Specification chart. See Unix, UNIX.

single wire circuit A transmission path used in early telegraph lines and still used for telephone service in some rural areas. The single wire circuit relied on the conductive characteristics of the Earth to ground the circuit and complete the return path.

single-frequency signaling, SF signaling A telephony signaling system in which transmission is through a single designated frequency such as 2600 Hz in the U.S. and 2280 Hz in the U.K. SF signaling tends to be used in certain microwave transmissions and in two- or four-line wired networks. SF signaling is an in-band signaling scheme in that the signaling is transmitted in the same band or channel as the data or voice communications. When a phone system is on-hook, the designated signaling frequency is transmitted; when it is off-hook, the frequency is interrupted. Variations in signals can be produced within a single frequency by varying the level of the tones (as expressed in decibels - dB). In the U.S., a high level of -8 dB and a low level of -20 dB are standard as these are levels that can be readily recognized by the electronics in a phone receiver.

single-mode optical fiber A single mode fiber optic transmissions cable has a relatively thin core acting as a waveguide such that light is reflected and propagated at a consistent angle. A thinner core has advantages and disadvantages over multimode fiber. Signals cannot be sent at a multiplicity of angles in the tiny fiber core, but distortion is minimized and transmissions can reach longer distances. Thus, where multimode fiber in data network installations is limited to about 2 km, single-mode fiber can transmit to about 15 km. For other types of transmissions,

Single UNIX Specification – Components	
Components	Notes
XPG4 System Calls and Libraries	Internationalized, covering POSIX.1 and POSIX.2 callable interfaces, the ISO C library and Multibyte Support Extension addendum, the Single UNIX Specification extension including STREAMS, the Shared Memory calls, application internationalization interfaces, and other application interfaces.
XPG4 Commands and Utilities V2	Covering the POSIX.2 Shell and Utilities and a large number of additional commands and development utilities.
XPG4 Internationalized Terminal Interfaces	Including the new extensions to support color and multibyte characters.
XPG4 C Language	
XPG4 Sockets	See sockets.
XPG4 Transport Interfaces (XTI)	

longer distances are possible, sometimes up to 200 km.

Signals are usually transmitted through single-mode cables with laser diodes, in order to get the precise alignment needed for the fine filaments, and received at the other end with a photodiode detector. This detector translates the signals back into electrical impulses.

Single-mode fiber is divided into two general categories: non-dispersion-shifted fiber (NDSF) and dispersion-shifted fiber (DSF). DSF, in turn, is subcategorized as zero-dispersion-shifted fiber (ZDSF) and non-zero-dispersion-shifted fiber (NZDF). In DSFs, the core-cladding has been fabricated to shift optimal dispersion to higher frequencies. There are limits to how much of this type of compensation can be implemented, however, as shifting frequencies may interfere with other frequency "windows" traveling in the same lightguide, as in multimode fiber. NZDF is intended to overcome this type of interference by shifting the zero-dispersion point above the range of wavelengths that have been optically amplified. See multimode optical fiber.

sink 1. A device to drain energy from a system. Heat sinks are common on devices or components which run hot and need to be cooled for safety and to maintain operating temperatures. 2. A point where energy from a number of sources is directed, and then drained away. 3. A point in a communications system where information is directed.

sinter To cause to become a coherent mass without melting, through the application of heat. 1. In fiber optics, sintering of sooty deposits such that they form a clear substance is part of the process of creating *preforms* from which optical fibers may be drawn. See boule, preform, vapor deposition. 2. In the construction of multilayer electronic components, sintering is part of a *direct-write* process developed by Sandia researchers for precision printing of ceramic and metallic substrates with an ink-filled nozzle rather than traditional screening or etching processes. The direct-write process enables a high degree of precision and flexibility in the design of the components. The electronic inks are heated at low temperatures to evaporate fluids and the remaining dried metal or ceramic medium is fired to sinter the powders. Ink-written components have potential applications as conductors, voltage transformers, radio frequency filters, resistor networks, and other applications.

Sioussat, Helen Johnson (1902–1995) Sioussat was the Director of the Talks and Public Affairs Department of CBS radio from 1937 to 1958. Her extensive correspondence with many of the radio and television broadcast pioneers is historically significant and has been preserved in the Library of American Broadcasting at the University of Maryland Libraries. See Broadcast Pioneers Library.

SIPP Simple Internet Protocol Plus. One of three candidate protocol proposals eventually blended into IPv6 by the Internet Engineering Task Force (IETF).

SIR See substrate-incident recording.

SIS Standardiseringen i Sverige. The Swedish standards institute located in Stockholm.

SIT See Simple Internet Transition.

SITA See Société Internationale de Télécommunications Aéronautiques.

site license A legal arrangement granting specific use or distribution permission of a copyright product to a specified location, firm, or other entity. A site license is a common method for specifying and controlling software use and distribution within a firm, particularly if the firm wishes to install the software on a network for access by multiple users or on several user machines within the organization. Typically, software companies will offer site licenses with the first copy and installation of the product priced at one level, and discount subsequent installations. This is common in educational institutions. For example, the first copy might cost $1000 and permit installation on up to five machines, with subsequent installations, in groups of five, at $200 each. Network licenses typically specify how many users may simultaneously access the software, and the software itself may monitor and control access. Distribution of any sort, other than as specified by the license, in most cases is a criminal offense. See piracy.

Skanova A wholesale network provider within the Telia Group, Skanova operates the largest broadband network services network in Sweden.

skin 1. Outer protective layer. A skin is often used to isolate conductive materials and/or to provide insulation and, sometimes, identification through the use of colored or marked skins.

skin antenna An antenna used on aircraft, in which a region of the metal craft is delineated and isolated on its edges by insulating materials.

skin effect In electricity, a situation in which the current tends to pass through the outer portions, rather than through the core of a conductor. This is due to the magnetic field that arises in the wire which prevents penetration to the core of the wire. It may increase the effective resistance in long wires and interfere with transmissions in the high frequencies used in broadcast transmissions.

skinning Stripping an outer protective layer. This is commonly done with wires to expose the conductive material within in order to make a connection.

skip distance The distance traveled by a reflected radio wave from the transmitter to the point at which it reaches the Earth's surface or the receiving antenna. This distance is affected by the frequency of the wave, the angle at which it passes into the ionosphere, and various atmospheric characteristics and conditions. See ionospheric wave, radio.

skip selection In computer software applications, a selection that halts the current process, or lets it finish in the background, and allows the user to continue to the next menu or activity without waiting. In automated voice or tone systems, especially menu-driven touchtone phones, a key press that allows continuing to the next selection, menu, or local phone number without waiting for completion of the current message.

skip zone See zone of silence.

SKIPJACK The name of a symmetric encryption algorithm which is the basis of the Escrowed Encryption Standard (EES) incorporated into the Clipper chip. SKIPJACK can be used to encrypt a TELNET stream. It has also been described for use in conjunction with FTP Security Extensions and Key Exchange Algorithm (KEA) to provide for mutual authentication and the establishment of data encryption keys. See Clipper chip, RFC 2773, RFC 2951.

skunkworks *colloq.* A facility in which clandestine or time-pressured activities take place in an environment which is closed off to increase security. Government operations, sensitive research, and high-technology design often operate in environments that are without sunlight or adequate ventilation, and in which the participants are working long hours (without much free time for personal hygiene).

One of the most famous skunkworks was a Lockheed-Martin research "lab" established by Clarence "Kelly" Johnson in a small desert facility scraped together from salvaged materials. It was a tight, intense, jet aircraft research and engineering operation dubbed the Skonk Works after an Al Capp cartoon moonshine operation, while its official name became the Advanced Development Projects (ADP) division. Later the name was changed to The Skunk Works and the lab was moved to a location northeast of Los Angeles, California.

The term is more than an amusing historical name; it is also at the heart of some of the significant disputes about domain names on the Internet. Lockheed owns the service mark for The Skunk Works and initiated lawsuits in the mid-1990s against Network Solutions, Inc. for registering variations of the name on the behalf of parties other than Lockheed. It was decided at the time that a domain registrar supplies a service rather than a product and that NSI was thus not liable for contributory infringement against the mark.

sky maps Charts of the electromagnetic radiation in the radio frequencies emanating through space and around Earth. Much pioneer work in this area was conducted in the 1930s and 1940s by Grote Reber, an amateur radio operator, using a home-built 32-foot parabolic antenna. Cosmic frequencies can sometimes be detected when the ionosphere is temporarily affected by the burn of a spacecraft or deliberate seeding with elements such as barium.

The phrase is also generically used to describe images of the sky as humans see it looking up from Earth with or without telescopes. Some interesting utilities on the Web now enable users to enter a location and date/time to calculate and display a sky map that can be printed or downloaded. John Walker provides access to "Your Sky" free on his Fourmilab site. See radio astronomy. http://www.fourmilab.to/

sky wave See ionospheric wave.

SkyBridge A medium Earth orbit (MEO) satellite system from Alcatel providing commercial satellite-based networking solutions to service providers. SkyBridge was established in 1997. In March 2001, SkyBridge announced that it would also be implementing broadband communications services through geostationary satellites.

SkyCell Communications Limited An Indian cellular services provider incorporated in 1992. It is the first Indian cellular company to receive company-wide ISO 9001 certification.

SKYCELL An American Mobile Satellite Corporation (AMSC) satellite telephony communications service providing coverage for the North American continent and nearby islands, catering to mobile workforces, traveling executives, and government agencies. The company also offers a continent-wide regional dispatch service providing digital broadcast capabilities to up to 10,000 mobile users.

Skynet A U.S. domestic communications satellite service purchased in 1997 from AT&T by LORAL Space & Communications, Ltd. Skynet originated in the Echo satellite and Project Telstar efforts in the 1960s. Telstar 5 was launched 2 months after LORAL's acquisition and positioned at 97° west. Soon after, Satmex was merged into the firm, forming the LORAL Global Alliance. Orion Network Systems, Inc. was acquired in spring 1998. Telstar 6 was launched in March 1999 at 93° west; Telstar 7 was launched in September 1999 at 129° west. It has been called the most powerful communications satellite in Telstar history and has been joined by further satellites later named Telstar 10 and 12.

Skynet provides news, television broadcasting, distance learning, videoconferencing, and other data transmission services to about 85% of the populated regions of the world.

SL Symbol for left-hand slant polarization (ITU).

SL Mail A commercial SMTP and POP3 mail server daemon for Windows NT 4.0 from Seattle Lab, Inc.

slamming A reprehensible trade practice in which a long-distance supplier switches a person's long-distance service without his or her explicit informed consent. In the early 1990s, some companies did this by phoning potential subscribers and having them verify their name and address over the phone and then signing them up without actually asking for consent. Since that time, more stringent customer consent is required before a change in the service can be initiated, and the customer usually must initiate the request, or the company making the change must obtain written authorization or outside verification.

SLAR See side-looking airborne radar.

slave 1. A subsidiary structure, system, process, or device which takes direction or data from a master. Many computer peripherals are slave devices. 2. In programming, slave processes are sometimes used to gather and report information to a master controlling process. 3. In communications circuits, slave consoles, subsidiary switching centers, and substations are often used to supply low-density populations or workstations some distance from the main controller or switching center.

slave server In distributed networks using domain name systems, a slave is an authoritative server, identified in the name server's register, which retrieves zones using zone transfer. See stealth server.

sleeve A covering to protect cables, bars, and other

long narrow components. A sleeve helps keep out dirt, air, and moisture and may provide electromagnetic shielding. The term is used for shorter lengths of protective shielding. A longer length, that protects a span of wire or fiber optic cable, is more often called a *jacket*. A sleeve is used for identification (e.g., color coding) or to protect fragile sections such as areas where the jacket has been opened (e.g., for splicing cable).

Sleeves are often made of plastic, though flexible metallic sleeves may be used to provide additional strength or electromagnetic shielding. Plastic sleeves are sometimes designed so that they will shrink around the covered component when exposed briefly to heat to further ensure a tight seal. Some have a resin adhesive coating on the inside to ensure good contact. Transparent sleeves can facilitate inspection of a joint after the sleeve has been installed. See cable, fusion sleeve.

slide contact A small sliding ball or tab attached to a thin rod that acts as a contact mechanism on a tuning coil. Tuning coils were used in early radio sets to select a frequency. A radio might come with several tuning coils for selecting various frequencies, as desired. See tuning coil.

SLIP See Serial Line Interface Protocol.

SLM 1. See spatial light modulator. 2. System Load Module.

slot 1. In programming, a time or data "opening" into which other processes or data can be inserted. 2. A physical opening for connectors or wires/cables which is typically narrow and rectangular. The slots on the back of a computer allow external connection access to peripheral cards such as serial, graphics, or network interface cards (NICs). See slot types. 3. In building structures, an opening that may be built into a wall or floor in order to enable cables to be fed through the building.

slot types Most computers and switching stations have slots into which electronics peripheral cards can be inserted. In order for third-party suppliers to be able to develop options for consumers, a number of standards have been adopted for the shape and electrical configuration of these slots. Most of these slots are long, narrow-edge card configurations, with two to six slots in the typical desktop computer. Many computers will accommodate two different card formats. The software drivers for the cards inserted into these slots are sometimes supplied on diskettes, to be loaded on the system, and are sometimes supplied in hardware, on chips on the actual card. Some of the more common card slot types include PCI, ISA, ESA, ZORRO, and PCMCIA.

Slotted ALOHA See ALOHA.

slotting In setting up a network, the assignment of a circuit to available channel capacity.

slow scan television, slow scan TV SSTV. A type of black and white TV signal which can function within a narrow spectrum, similar to single-sideband transmissions for voice. SSTV has been used since the late 1950s by amateur television and radio operators to send series of images over radio frequencies. SSTV

can be viewed on a television set with a scan converter or on a computer monitor with the appropriate interface.

In the U.S., SSTV uses frequencies ranging from about 3.845 to 145.5 MHz to transmit a series of images which can be captured through a dedicated system or through a computer linkup. Interface circuits for setting this up are in the hobbyist price range. Hicolor mode can provide color images up to 320×240 in thousands of colors. Even higher resolution 640×480 24-bit images (millions of colors) can be transmitted, but they take 7 or 8 minutes, compared to low-resolution black and white images that take only 7 or 8 seconds.

Radio broadcasting is regulated throughout the world; those interested in SSTV technology will have to be licensed, usually for voice grade channels, by local regulatory authorities.

A related technology is *amateur TV* (ATV) which refers to *fast scan* amateur television. See amateur television.

SLP See Service Location Protocol.

SLR 1. send loudness rating. 2. single lens reflex.

Sm *symb.* samarium. See samarium.

Small Computer System Interface SCSI (*pron.* scuzzi). A standardized interface specification which provides a means for the central processing unit (CPU) and main circuitry on the motherboard to communicate with computer devices that are interfaced to the system. This requires standardization of electrical circuitry and data protocols because peripheral devices are manufactured by many different companies. One of the most common of these formats is SCSI, which is widely used to interconnect hard drives, scanners, cartridge drives, digitizers, CD-ROM drives, and more.

The SCSI standard is approved by the American National Standards Institute (ANSI), and several enhanced versions have appeared (variously called SCSI-2, extended SCSI, SCSI-3, wide-SCSI, etc.)

SCSI typically consists of a SCSI controller on a motherboard or a peripheral card, which is terminated and usually designated as zero or six, depending upon the system and one or more peripheral devices, set to SCSI ID number zero through five or one through six, depending upon which one is reserved for the motherboard, and terminated at the end of the last device. The devices can be hooked up end-to-end, that is, daisy-chained. Each SCSI controller can chain up to seven devices, with the motherboard or main controller counting as one. The cable for SCSI devices is either a 50-pin edge connector or a 25-pin D connector (or a hybrid cable with an edge connector at one end and pin connector at the other). SCSI-3 cables are wider.

Only one device can be assigned to each SCSI ID. Conflicts or lack of termination will cause failure to recognize a device or spurious errors. Many systems expect CD-ROM devices to be set to ID 3, although there is no inherent reason why ID 3 has to be assigned to only this type of device. Scanners often default to SCSI ID 4. The ID number will determine the priority setting for loading the device, thus boot disks are usually assigned a number closest to the number of the controller. In other words, if the controller on a motherboard is zero, then the boot hard drive should probably be set to one and a relatively low-use tape drive to five or six.

SCSI ID settings are sometimes on the outside of the device, with a thumb-turn switch or DIP switch, and sometimes on the inside, with DIP switches or jumpers.

Termination is accomplished either by placing a physical terminator in one of the cable connection slots, by setting DIP switches, or by setting jumpers inside the device. Automatic termination is available on some devices, which means that if the device senses that it is the last device in the chain, it will terminate automatically. These types of automatic terminators are sometimes specific to the slot. There will be two slots on the back of most SCSI devices so that they can be chained. Take care to follow instructions for which one to connect if the device is last in the chain and intended to terminate automatically.

Most SCSI devices can only work with cables up to about 6 feet in length, and 3 feet or shorter is generally recommended. Newer Fibre Channel Standard technologies can support longer connection lengths, allowing SCSI devices to be centralized in an operations room or wiring closet.

SCSI controllers are standard in many consumer and workstation computer systems, including Macintosh, Amiga, server-level IBM-licensed desktop computers, NeXT, Sun, SGI, some HP systems, and DEC. Most of these systems include an internal SCSI controller (for up to six hard drives and internal CD-ROM drives, etc.) and an external SCSI controller (for up to six scanner, printer, external CD-ROM, external hard drive devices, etc.). Thus, a total of 12 devices can easily be daisy-chained to these systems without any modifications to the operating system or the hardware, other than perhaps adding a software device driver and cabling. In the author's experience, SCSI is a good format. The inexpensive 8-year-old Motorola 68040-based computer used for the illustrations for this dictionary has two SCSI connectors (internal and external) with eight SCSI/SCSI-2 devices attached (scanner, tape drive, cartridge drive, six-disc CD-ROM drive, and four different kinds of hard drives). These are chained to the two controllers and worked together the first time they were connected without any compatibility problems.

SCSI drives are incorporated in mirroring and redundancy combination drive/backup systems such as *redundant array of inexpensive disks* (RAID) systems. These drives can be conveniently hot-swapped in and out if a drive fails and needs to be replaced, with the information rebuilt by the controller and software when the new drive is installed.

For consumer desktop Intel-based, IBM-licensed computers that come standard with IDE drives, a SCSI controller card can be added to the system to accommodate SCSI devices. However, on this type of system, it is important to determine whether appropriate device drivers are available for the peripheral, that there is no contention with the IDE drive, and also that any appropriate IRQ issues are settled.

Small Scale Experimental Machine Nicknamed "Baby," this historic computer was developed in the mid-1940s, based upon tube memory, a form of randomly accessible data stored in a Williams-Kilburn cathode-ray tube. Baby was a binary small-endian system that supported 32-bit words and a main random access storage capability of 32 words, which could be extended to 8192 words. It used several tubes for different functions, including a storage register that is still used in modern computers, the "accumulator," a couple of instruction tubes, and a tube for displaying the contents of the other tubes. A simple keyboard was used to set the bit sequences.

The project was undertaken by T. Kilburn and G. Tootill with equipment support from the Telecommunications Research Establishment (TRE). The system was initially used in 1948 for mathematical calculations that were laborious to execute by hand. See Williams-Kilburn tube, Manchester Mark I.

small vocabulary In speech recognition, it has been found that software can be designed to recognize a variety of voices, without special training of the system, if the total vocabulary of the recognition is kept small. These small vocabulary systems work well in specific environments such as stock buy/sell systems. While definitions of small vary, recent systems of this type typically recognize 200 or fewer words.

Smallhouse, Charles "Chuck" An amateur radio enthusiast (callsign WA6MGZ [now W7CS]) who contributed substantially to the first three OSCAR satellites' design and construction. See OSCAR.

Smalltalk An object-oriented computer exploration and development language developed through the Xerox Corporation in the 1970s. It was evaluated by four Xerox-selected companies in 1980, before being broadly distributed. By the mid-1980s, commercial versions of Smalltalk-80 were being released for a variety of platforms including IBM licensed personal computers and Apple II systems. Smalltalk has been favored by developers working in object-oriented programming environments and artificial intelligence applications. See Palo Alto Research Center.

SMAP See Service Management Access Point.

SMART Self-Monitoring, Analysis, and Reporting Technology. A preventive system implemented in data protection schemes such as RAID which uses predictive failure analysis to anticipate possible failures. Impending problems are communicated to the controller, which signals a warning so that faulty drives may be examined or replaced prior to any failure which might occur. See redundant array of inexpensive disks.

Smart Card A compact, thin card with embedded data. It may contain a microprocessor, memory, or both. It typically resembles a plastic credit card. Earlier Smart Cards incorporated a magnetic strip, but more recent cards may include a set of contacts embedded in the card.

There are various ways to categorize Smart Cards.

Some require contact with a sensor and some are read without direct physical contact. A contact sensor can consist of a reader with a slot or a sensing surface upon which the card is placed. Slot sensors generally require that the card be swiped across the surface to register the information. Basically the card is being scanned for information. Contactless cards typically incorporate tiny wireless transmitters and may include an antenna to increase the transmission range of the signal. There are also hybrid cards. For example, a wireless transmitter may be included for opening entranceways at a worksite while a second contact surface may be used to transmit more detailed information about the person holding the card when it is swiped through a card reader.

The common standard for Smart Cards is in the ISO 7816 series. A number of financial agencies have agreed upon a common specification for communications between Smart Cards and Smart Card readers, similar to the serial communication that occurs between computers and computer peripherals. This makes it possible to exchange virtually any type of data between readers and cards and increases the possibility of programmable universal cards for the convenience of users. The downside of universal cards is that, if they are stolen, a great deal of information may be in the wrong hands. However, various encryption and other security measures are being developed to help protect Smart Card users against theft.

The cards can also be categorized on the basis of the types of data contained on the card (information or algorithms) and whether they are reprogrammable (write once, read many (WORM) or rewritable). Since a Smart Card is somewhat like a tiny floppy disk or a very tiny computer, the range of uses to which it can be applied is exceptionally broad. It could not only facilitate telephone access for travelers, but could potentially keep track of where and when the calls were made. This information is valuable for corporations logging sales transactions, for example, and keeping records for the taxation department. A Smart Card could also help present and track prescriptions, employee purchases, medical histories, allergy shots, pet vaccination histories, automobile histories, and much more. Smart Card application interfaces have been introduced by a number of developers and a lot of interest in programming Smart Cards with Sun's Java language arose in the late 1990s. It is likely that Smart Cards will become a ubiquitous part of daily life, replacing many of the paper notepads and "dumb" cards now used by consumers and professionals.

Smart Card history The essential concept of the Smart Card was patented in 1974 by K. Arimura in Japan and R. Moreno in Europe. In the 1980s and 1990s, Smart Cards came into common use for many types of financial transactions, mobile communications devices, long-distance phone services, and authorized entry systems.

Smart Card Industry Association SCIA. A trade association supporting and promoting the development, utilization, and understanding of Smart Card technologies. http://www.scia.org/

SMASH Project A project dedicated to developing mass storage devices for multimedia applications for home use. This is intended to promote commercial offerings of video services to the home, with part of the goal of SMASH to provide a labeling algorithm system in the storage system to provide vendor copy protection. Thus, data on the storage device can be set so that it can only be stored or copied once.

SMASH seeks to develop realtime labeling methods for compressed video. Common schemes for this include spatial or discrete cosine transform (DCT). The SMASH Project also introduces two new realtime labeling techniques that can be used in conjunction with MPEG-1 or MPEG-2 format video information. See watermark.

SMAS Switched Maintenance Access System. A legacy telephony network system from Anritsu Company. The older systems (e.g., cross bar access systems) are gradually being updated or phased out.

SMATV Satellite Master Antenna Television. A satellite communications distribution system designed to send transmissions to hotels, motels, apartments, etc. Since these are sent mainly to commercial establishments, they are often used as marketing leaders or as pay-per-view revenue-generators.

SMB See Server Message Block protocol.

SMB Project A project for preserving the history of the Server Message Block (SMB) protocol while primary materials are still available. The first definition of NetBIOS was released by IBM in a Technical Reference in 1984. By the late 1980s, SMB File Sharing Protocol extensions were being published by Microsoft. http://samba.org/

SMDS See Switched Multi-Megabit Digital Service.

SME 1. Security Management Entity. 2. Small- and medium-sized enterprises.

smear 1. Descriptive term for a television signal display distortion in which the image is blurred and appears stretched in the horizontal direction. 2. Low-level frequency distortion in an audio signal. 3. In digital imagery, distortion of details resulting from sampling frequencies or compression algorithm compromises, so transitions which normally would be sharp and crisp in the original image exhibit blurring or smear.

smectic liquid crystal SLC. The molecules of nematic liquid crystals have a certain amount of orientation order but generally lack position order. At certain transitional temperatures, these materials may acquire a certain amount of positional order, called the smective phase. Thus, the material has some of the properties of liquids, but tends to form somewhat positional layers, resulting in a two-dimensional nematic liquid crystal. Depending upon the tilt and light-directing properties of the smectic liquid crystals, they may be subdivided into different types.

If the SLCs are encouraged to form a chiral orientation resulting in a helical orientation (as in the Schadt Helrich effect) they can be selectively used to modulate light to turn it on or off.

This form of liquid crystal has faster response time

and better contrast ratio than nematic crystal displays. See nematic liquid crystal, Schadt-Helfrich effect.

S/MIME Secure Multipurpose Internet Mail Extension. An IETF working group (inherited from the S/MIME Consortium) Internet messaging standard for the transmission of secure network communications. Unlike PGP/MIME, S/MIME public keys are distributed via X.509 digital certificates. S/MIME can support 128-bit encryption, although not all implementations will use the full 128 bits. See PGP/MIME.

SMPTE See Society of Motion Picture and Television Engineers.

SMPTE Registration Authority SMPTE RA. A format and specification authority for technologies related to the motion picture and television industries. For example, the SMPTE RA is approved by IEC and ISO for the registration of MPEG-related format identifiers. See Society of Motion Picture and Television Engineers.

SMPTE time code A standard developed by the Society of Motion Picture and Television Engineers which provides synchronization for information recorded on audio and visual video tapes. SMPTE time code digitally encodes hours, minutes, seconds, and frames.

SMPTE time code is recorded onto audio tracks and video tracks as follows: in audio as Longitudinal time code (LTC); in video as Vertical interleave time code (VITC).

A time code word consists of 80 bits (zero or one) per video frame, with 2400 bits per second corresponding to 30 frames per second for North American TV. In Europe, 2000 bits per second corresponds to the standard of 25 frames per second. See drop frame, Society of Motion Picture and Television Engineers.

SMR See Specialized Mobile Radio.

SMRP See Simple Multicast Routing Protocol.

SMS 1. See Service Management System. 2. See Short Message Service.

SMS/800 A centralized interactive computer system dedicated to coordinating network services related to toll-free 800/876/etc. numbers. These are managed in a central database-equipped mainframe computer from IBM that updates locally deployed databases. Access is through dialup, Internet, and dedicated connections with various security systems in place to restrict access to authorized users. The SMS/800 system supports a 24-hour-a-day, 7-day-a-week public service.

Within the SMS/800 system, a Responsible Organization (Resp Org) is an entity authorized to manage and administer a toll-free number customer using the SMS/800 system. Bell Operating Companies (BOCs) administer the SMS/800 system under the 800 Service Management System Functions Tariff.

The SMS/800 system tracks the availability of all toll-free numbers and permits Resp Orgs to access the database to search for available numbers and to change the status of existing numbers. The North American Numbering Plan (NANP) Administrator issues instructions to carriers for making toll-free numbers available or unavailable. Resp Orgs are assigned a logon ID code from a BOC and must meet certain certification requirements. Logon ID requests are processed by the SMS/800 Management Team in New Jersey.

Numbers are obtained by Resp Orgs from a common pool, with specific 800 number requests honored based upon availability on a first-come, first-served basis at the time the request is received in the SMS/800 Reservation Queue. Specific numbers may be reserved for a potential subscriber for up to 45 calendar days. If the period expires, the number is designated a spare. A Resp Org may reserve up to 2000 numbers, or up to 7.5% of the total available numbers, whichever is greater. Certain numbers are reserved for special purposes (e.g., hearing impaired) and are considered Closed. Resp Orgs are limited to a maximum of up to 3% of available numbers that are reserved at any given time. Resp Orgs must be willing and able to provide troubleshooting assistance and maintenance personnel. There is a customer record administration charge for each toll-free number assigned or reserved.

Hoarding of toll-free numbers by Resp Orgs is expressly forbidden within the Federal Communications Commission (FCC) Tariff guidelines. Acquiring more numbers than are intended for immediate use by a subscriber, or the sale of a toll-free number for an additional fee, contravenes the FCC's responsibility to promote the fair allocation and orderly use of toll-free numbers.

The SMS/800 Software Support organization publishes the *Service Management System (SMS)/800 Mechanized Generic Interface Specification* (SR-4592 - Feb. 2001). See 800. http://www.sms800.com

SMS/800 history The SMS/800 system originated in 1967, when AT&T introduced an inward Wide Area Telecommunications Service (INWATS) for business subscribers who wanted to purchase bulk calling to enable customers to reach them from a wide geographic region. In 1981, computerization made it practical to introduce a centralized database for managing national services such as 800 numbers and to assign parameters to specific numbers. This, in turn, made it possible to make the service more flexible and powerful. At about the same time, smaller companies and competing phone carriers were computerizing their customer databases.

When the Judge Greene divestiture proceedings led to the breakup of the Bell System in 1984, 800 services were required to be opened up to competitors. This necessitated the development of more sophisticated software to handle the management of databases from Bell Operating Companies (BOCs) and independent competing carriers.

Number portability soon became an issue, with so many competing phone carriers now offering 800 services. In 1991, the Federal Communications Commission (FCC) mandated that 800 numbers must be able to be moved among carriers according to the carrier selection of the subscriber.

By the mid-1990s, 800-number designations were

running out, due to the increasing demand. As a result, additional prefixes were released, to be assigned as the need arose. See SMS/800.

SMS/800 Management Team A team of administrators, consisting of a representative from each of the regional Bell Operating Companies (RBOCs), responsible for SMS/800 services. The team is headquartered in New Jersey.

SMS/800 Mechanized Generic Interface MGI. A means to interface with the SMS/800 centralized toll-free number database. The MGI facilitates the transfer of number and customer record administration between the SMS/800 and Responsible Organization (Resp Org) computer system over a network connection. It is a two-way interface delivered over a five-layer protocol model. The transport service is supported over the physical, packet, and link layers for error-free communication. The user program layer (UPL) supports specific applications messages.

Before active status on the SMS/800 system is granted, the MGI must be put through four test phases to confirm data communications integrity per specific field and laboratory testing requirements. A testing logon ID is assigned for the test period of about 4 months. Once access is authorized, an active login ID code is assigned. See SMS/800.

SMTP See Simple Mail Transfer Protocol.

Sn Symbol for tin (AN 50).

SNA See Systems Network Architecture.

SNA Control Protocol SNACP. A protocol which handles the configuration and enable/disable functions at the ends of a point-to-point link. Subdivided into two protocols that independently negotiate SNA with or without LLC 802.2. Similar to Link Control Protocol. See See RFC 2043.

SNACP See SNA Control Protocol.

snake A cabling aid consisting of a flexible, long, thin cord of metal or plastic used to feed wire and cable through conduit or through structures (ceilings, walls, attics, etc.) where space is tight, or access is limited. See birdie.

SNAP See SubNetwork Access Protocol.

snap-hook A loop-shaped connector with a normally closed hook that can be opened to add objects to the hook with the hook snapping closed automatically after insertion. Snap hooks may be locking or non-locking. Rock climbers are familiar with a number of types of snap hooks and similar hooks are used by workers who climb utility poles or towers to do installations or maintenance/repair. Snap hooks are also handy for slinging wire bundles and hanging up equipment that needs to be securely held. Contrast with J hook.

sneak currents Low-level undesired currents which seep into circuits and may, if continued long enough, cause damage. Sneak currents are those which do not cause immediate harm and are not sufficient to trigger safety mechanisms such as normal fuses and breakers. Sneak currents can result from causes such as worn sheaths and insulators, incorrect wiring, temporary contact due to settling, etc. See sneak fuse.

sneak fuse A special low-level current detection fuse

specifically designed to trigger if sneak currents are detected. See sneak currents.

Snell, Willebrord (1580–1626) A mathematician and astronomer from the Netherlands, who succeeded his father as a mathematics professor at the university of Leiden. Snell established a variety of methods for measuring the Earth, establishing some of the basic tools of geodesy. He further refined basic principles of light and predicted how light rays would act in an environment such as a glass rod.

In 1703, C. Huygens published *Dioptrica* in which he refers to Snell's observations about refraction made in 1621. For a time, R. Descartes was thought to be the originator of the mathematics of refraction, but it appears he got his information from Snell. The explanation of the phenomenon came to be called Snell's law. Snell's observations were important because refractive behavior is nonlinear and thus not easy to measure or mathematically describe unless the relationships are understood. See refraction, Snell's law.

Snell's law A description of the relationship between the angle of incidence and the angle of refraction in propagating radiant energy when it encounters a material of differing density. This relationship was verbally described by Willebrord Snell in 1621. In 1658, Fermat showed how Snell's law was generalizable to any propagating radiant energy traveling through any medium. See absolute refractive index; Brewster's angle; refraction; Snell, Willebrord.

Snell's Law of Refraction

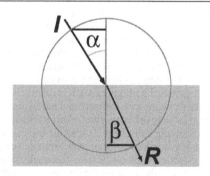

Snell described how propagating radiant energy (I) refracts when it encounters a medium of higher density (R). The energy is bent toward the surface normal of the medium such that the angle of incidence (α) and the angle of refraction (β) remain in constant proportion to one another in terms of the sines of the angles.

SNI See Subscriber Network Interface.

sniffer 1. *colloq.* nose. 2. A synthetic or electronic substitute for a nose, such as a chemical testing kit used for drug detection and identification by the U.S. Customs service. 3. A utility for peeling or ferreting out information, such as particular statistics from a log file. 4. A network traffic monitoring tool used for diagnostics and sometimes unauthorized snooping.

A sniffer is a useful system administration tool for monitoring and logging peak traffic times, network load, and possible problems. This information can help an administrator tune a system to operate efficiently. See packet sniffer, Sniffer. 5. A radio direction-finding tool for locating radio transmitters. See Ultra sniffer. 6. A detective or investigative journalist who sniffs out (investigates) information as desired by clients or editors.

Sniffer A registered trademark of Network General Corporation. The term "sniffer" has been used for decades in the search and rescue and law enforcement communities to refer to working dogs that are trained to sniff out fugitives, lost individuals, or chemical substances. The term has also historically been used in the radio community to refer to receiving devices that locate the source of radio signals. These direction finders help to trace signals emanating from sources such as wildlife radio collars or radio listening devices (bugs). With the advent of computer sniffers, Network General Corporation has trademarked the term in the context of software utilities and network analyzers and the firm endeavors to protect this registered trademark. See Sniffer Network Analyzer.

Sniffer Network Analyzer A commercial product for analyzing local area network (LAN) traffic, first introduced in 1988 by Network General Corporation. The software provides a log of traffic, a suite of alarms, and various statistics. The corporation specializes in fault and performance management solutions for enterprise networks.

Sniffer Technologies A commercial vendor of wireless monitors and network traffic analyzers and reporting products, including Sniffer Distributed, Sniffer Wireless, Sniffer Optical, and others.

sniffing *jargon* In computer networks, the process of looking at network data for testing, diagnosis, or unauthorized purposes. Because the Internet as a whole is a shared system, rather than a switched system (though switched portions exist), it is relatively easy to breach a system with software and hardware tools provided for network testing and administration. One type of scouting sniffer for clandestine use captures the first few hundred bytes of every remote session or file download/upload, so that the unauthorized snoop can scan the sessions for interesting information that might suggest a closer look.

Snitzer, Elias (ca. 1922–) An electrical engineer and professor emeritus at Rutgers University, Snitzer demonstrated with Hicks that tiny-diameter fiber cores could be used to transmit light as a single mode wavelength. He authored articles on dielectric waveguide modes in the *Journal of the Optical Society of America*, in 1961 and coauthored "Amplification in a Fiber Laser" in *Applied Optics*, in 1964. In 1990, he coauthored articles on rare earth doping of optical fibers.

Snitzer was elected to the National Academy of Engineering in 1979 and worked for many prominent firms in optical technologies, including Polaroid Corporation, American Optical Corporation, and Honeywell. In 1995, he became president of Photo Refrac-

tive Enterprises, Inc. Snitzer is responsible for many firsts in the field of optical waveguides and development of glass, erbium, and ruby lasers. See Kao, Charles K.

SNMP See Simple Network Management Protocol.

SNOM See scanning near-field optical microscopy.

snoop A testing and maintenance software tool for Solaris that enables the user to capture and inspect each packet in packet-based network traffic. It uses both the network packet filter and streams buffer modules for packet capture and displays them as received or logs them to a file. It is similar to tcpdump and related to etherfind (SunOS). The snoop utility can be used to tune network variables, to detect retransmissions, duplicate acknowledgments, and other aspects of the network configuration.

snooper, snooperscope, night scope A device designed to enhance night vision by sending out and intercepting an infrared beam. The incoming beam is interpreted into an image that shows objects not visible to the human eye.

snow An undesired aberration in a broadcast or display of a video image in which there are many randomly distributed speckles, often white. Snow can result from transmission problems, such as a weak or drifting signal, or from display device problems, as in a cathode-ray tube (CRT).

SNVT (*pron.* snivit) See standard network variable type.

SOA See semiconductor optical amplifier.

Société Internationale de Télécommunications Aéronautiques SITA. An international airline reservations and telegraphic transmissions service backbone network established in 1983.

Society for the History of Technology SHOT. An interdisciplinary organization dedicated to fostering and promoting an interest in the development of technology and how it affects society and culture, and to promoting scholarly study of related documents and artifacts. Members include individuals, professionals, museums, scientists, and librarians. The organization was formed in 1958 and is headquartered in the Department of History of Science, Medicine, and Technology at Johns Hopkins University.

Society of Broadcast Engineers, Inc. SBE. A nonprofit organization providing publications, workshops, a certification program, and liaison with important regulatory agencies such as the Federal Communications Commission (FCC) in order to promote education, standards, and professional competency in the broadcast engineering industry. The SBE was founded in 1963 and has since grown to over 100 chapters in the U.S. and abroad. It publishes the quarterly journal *The SBE SIGNAL*. http://www.sbe.org/

Society of Cable Telecommunications Engineers Inc. SCTE. A U.S. national nonprofit professional organization founded in 1969. The society includes over 13,000 members from around the world, representing a broad spectrum of cable professionals. The society provides education, certification, and standards development. http://www.scte.org/

Society of Motion Picture and Television Engineers SMPTE. An international organization originally founded in 1916 as the Society of Motion Picture Engineers. The T was added in 1950 to encompass the emergence of the television industry. The society includes over 8500 members in 72 countries, including engineers, technical directors, and production/post-production professionals dedicated to advancing the theory and application of motion-picture technologies. SMPTE contributes to standards development, encourages consensus-based recommended practices (RPs), and industry engineering guidelines (EGs).

In 1957, the society was awarded an Oscar for its contributions to the advancement of the motion picture industry. It has also received three Emmy awards for various recording and video systems and standards. SMPTE is best known for developing SMPTE time code methods, which are used for video editing. When video tape began to be widely used for recording and editing, a way was needed to synchronize edits, to locate specific places on the tape, and to dub sound to match the video sequences. SMPTE began in 1969 to develop a standard for digitally encoding time information in terms of hours, minutes, seconds, and frames onto audio or video tape. See MIDI time code, SMPTE time code. http://www.smpte.org/

socket 1. A means of providing unique identification to which or from which information is transmitted on a network. RFC 147 specifies a socket as a 32-bit number; even sockets identify receiving sockets, odd sockets identify sending sockets. Each socket is identified with a process running at a known host.

SOCKS An access and security technology designed to provide a framework for TCP and UDP client/server applications to conveniently, transparently, and securely utilize and traverse a network firewall. There have been a number of versions of SOCKS, with RFC 1928 representing version 5. Version 5 adds UDP and authentication capabilities, and extends addressing to accommodate the future needs of IPv6.

The protocol fits between the application layer and the transport layer and does not provide ICMP message-forwarding services. Traversing a firewall securely depends upon the various authentication and encapsulation methods selected and used in the negotiation between the SOCKS client and server. See firewall, gateway, proxy, RFC 1928.

sodium vapor lamp A lamp that glows a warm golden color, from the passage of electricity through metallic vapors in a cylinder encased in a glass tube. Sodium vapor lamps have been used as street and bridge lamps. See mercury vapor lamp.

soft copy A stored image, document, or file which is recorded on a medium which must be accessed with some type of technology in order to be viewed, manipulated, or displayed. Soft copies commonly exist on hard drives, floppy diskettes, tapes, CDs, and other magnetic or optical media.

soft transfer A term for an electronic monetary transaction which precedes the actual exchange of funds between individuals or banking institutions. A paper check is a type of soft transfer. It is a monetary transaction which is not finalized until the money is withdrawn from the bank. Similarly, online, many monetary transactions which are soft transferred and later "hard transferred" from the actual bank or other financial institution, such as a credit union.

SoftCard An early commercial product from Microsoft, from an idea suggested by cofounder Paul Allen. The SoftCard was an internal peripheral card equipped with a Z80 processor, which ran CP/M-80 from Digital Research. This card, when installed in an Apple II computer, allowed its user to install and run CP/M-compatible applications programs.

SoftSource Corporation/Catarra A Colorado and Bellingham-based firm that has developed a next-generation wireless mobile Personal Digital Assistant (PDA) macrobrowser that works transparently with the World Wide Web. This product may define the future of mobile wireless Web browsing. Recognizing the improved capabilities of newer PDA devices, the developers, Scott Sherman, Mike McMullen, Dan Suslo, and Steve Work, created an application engine and server technology compatible with HTML, that is not dependent upon specialized subsets (e.g., Wireless Application Protocol) or PDA languages characteristic of other microbrowsers. See PDA macrobrowser for an illustration. See PDA microbrowser, Wireless Application Protocol.

software Computer instructions stored on a medium which is reasonably portable and accessible by users. Actually, the distinction between hardware and software is much less clear than many people realize. It may seem reasonable to designate everything inside the computer as hardware and everything that holds information that can be inserted into external storage read/write devices as software, but that's not really the best distinction. Floppy disks and computer chips are tightly integrated hardware/software combinations whether external or internal, so the matter is really one of accessibility coupled with structure. Since computer instructions stored on disks are easily read, written, and moved, they are thought of as software. Since computer instructions on computer chips are not easily read or written and not easy for a lay person to access or move, they are considered part of the hardware.

The lowest level software functions are programmed into the computer chips themselves. At the hardware operations level, this software acts to start up the system, test it, bootstrap the device drivers to come online, and initiate the operating system to accept user input and output, and to otherwise communicate with the central processor. Some of these operating instructions may be read, in turn, from hard drives, CD-ROMs, cartridges, or other storage media. High level software interacts with the user through application programs.

Software is created with a variety of programming, editing, debugging, compiling, interpreting, and linking tools in a great assortment of languages, which are general purpose or optimized for specific types of programming. See programming.

Software & Information Industry Association SIIA. A trade organization representing companies that create, distribute, and facilitate the use of information in print and digital formats. Evolved from an organization originally founded in 1968, SIIA was established through a merger of the Software Publishers Association and the Information Industry Association, in 1999. SIIA promotes the information industry, represents its members in government policy and regulatory affairs, and provides a business development support network for top executives in the field. SIIA is based in Washington, D.C. http://www.siia.net/

Software Engineering Research Network SERN. An engineering and research joint venture of the Department of Computer Science and the Department of Electrical and Computer Engineering at the University of Calgary, Alberta. It is sponsored by the government of Alberta, University of Calgary, Motorola, Computing Devices, and Northern Telecom.

Software Publishers Association SPA. See Software & Information Industry Association.

solar cell In the 1940s, Bell Telephone Laboratories developed a storage cell from thin strips of silicon which had the characteristic of developing a charge in the presence of light. Since the silicon is not directly depleted in this process, solar cells are not subject to the limited life spans of traditional wet and dry cells. Solar cells have since been developed and refined in many ways and are used in many aspects

of electronics. See photovoltaic.

Solaris A popular 32-bit operating system from Sun Microsystems that is commercially distributed, as is their SunOS operating system and, more recently, OpenStep. Solaris is multiprocessing, multithreaded, and network-friendly (using NFS), based on an open systems architecture. Many large Internet Services Providers, university systems, and enterprise local area networks (LANs) run on Solaris. Solaris is available for various Sun SPARC, Intel-based, and Motorola-based systems.

solenoid A long, cylindrical, current-carrying coil with properties similar to a bar magnet, into which an iron bar will be drawn when current is applied to the coil. Solenoids are commonly used in circuit breakers which have replaced traditional fuses. See electromagnet.

Sömmering, Samuel Thomas von (1755–1830) A Prussian anatomist and inventor, Sömmering was one of the earliest inventors of telegraphic technology. Based on the work of F. Salvà, he developed an electrochemical wire telegraph which signaled letters and numbers through the application of current from a voltaic pile to a specific wire such that it created bubbles in an acid bath at the receiving end. Each wire corresponded to a character, with the wires and frame at each end looking very much like a threaded loom. This invention, which was demonstrated to the

ANSII Standards Related to SONET – Examples		
ANSI Standard	Title	Subtitle
T1.105-1995	Synchronous Optical Network	Basic Description including Multiplex Structure, Rates, and Formats
T1.105.01-1995	Synchronous Optical Network	Automatic Protection
T1.105.02-1995	Synchronous Optical Network	Payload Mappings
T1.105.03-1995	Synchronous Optical Network	Jitter at Network Interfaces
T1.105.03a-1995	Synchronous Optical Network	Jitter at Network Interfaces - DS1 Jitter
T1.105.04-1995	Synchronous Optical Network	Data Communication Channel Protocol and Architectures
T1.105.05-1994	Synchronous Optical Network	Tandem Connection Maintenance
T1.105.06-1996	Synchronous Optical Network	Physical Layer Specifications
T1.105.07-1996	Synchronous Optical Network	Sub-STS-1 Interface Rates and Formats
T1.105.09-1996	Synchronous Optical Network	Network Element Timing and Synchronization
T1.119-1994	Synchronous Optical Network	Operations, Administration, Maintenance, and Provisioning - Communications
T1.119.01-1995	Synchronous Optical Network	OAM&P Communications Protection Switching Fragment
SONET Multiplexing Techniques		
Technique	Type of interleaving	Notes
Interleaving	Interlaces individual bytes	Reduces overhead at receiving end
Single-stage interleaving	Direct byte interleaving	STS-N signal created directly
Two-stage interleaving	Direct byte interleaving	Accommodates European ITU-T rate

Academy of Science in Germany in 1809, predated Wheatstone and Morse technologies by three decades. Sömmering improved upon Salvà's design by extending the transmission range of the device and later collaborated with Schilling, who made some significant contributions to early electromagnetic telegraphy. See Salvà i Campillo, Francesc; Schilling, Pavel; Steinheil, Karl August.

sonar Sonar is currently considered to be the acronym for sound navigation and ranging (other phrases have been suggested). It is essentially a technology for generating and directing sound signals with the intent of analyzing the returning echoes to determine information about the size, shape, and relative distance of any objects encountered by the outgoing sound waves.

Sound requires a medium in which to travel. It is, in fact, a disturbance of a medium and, thus, sonar is used for probing elastic media such as water and, to a lesser extent, air (as in robotics).

SONET *Synchronous Optical Network.* SONET is a set of ANSI telecommunications standards which specify a modular family of rates and formats for synchronous optical networks.

SONET provides a standard operating environment for managing high bandwidth services, and incorporates multiplexing, service mappings, and standardized interfaces, so commercial vendors can develop interconnecting technologies.

SONET has been adopted by the ITU-T as the basis for the Synchronous Digital Hierarchy (SDH) transport system, and is a subset of this system. SONET is based on STS-1 which is suitable for T3, and SDH is based on STM-1, suitable for E4 transmissions.

Communication between nodes, to permit control, provisioning, administration, and security, is accomplished through the Synchronous Transport Signal (STS) transmitting at a line rate of 51.84 Mbps. The STS comprises payload information and signaling and protocol overhead. Since the two ends of a SONET transmission may vary in format and speed, data are converted to the STS format, transmitted, and, when received, converted into the appropriate user format. OAM&P is integrated into SONET. See detailed information in the following listings.

SONET ANSI standards A number of important American National Standards of Committee T1 related to SONET are available from ANSI and described in the form of abstracts on the Web. See ANSI Standards Related to SONET chart.

SONET frame The frame length is 8000 fps or 125 μsec. SONET uses Synchronous Transport Signal level 1 (STS-1) as its basic signal rate of 51.84 Mbps. SONET frames are organized in a row by column structure totaling 810 bytes. Transport overhead is contained in the first three columns and is subdivided to include section and line overhead. The remaining columns, from four to 90, are used for the Synchronous Payload Envelope (SPE).

The STS-*N* frame consists of frame-aligned, byte-interleaved *N* STS-1 signals.

The STS-*Nc* frame consists of concatenated STS-1

signals to form a multiplexed, switched signal that can be transported together. This is done to accommodate broadband services such as ISDN.

Following is an overview of some of the bit rate speeds for the Synchronous Transport Signal levels and how they compare to European equivalents. This chart only indicates bit rates; the frame formatting for each system differs even further.

U.S.	Europe	Bit rate
STS-1	—	51.84 Mbps
STS-3	STM-1	155.52 Mbps
STS-12	STM-4	622.08 Mbps
STS-24	STM-8	1244.16 Mbps
STS-48	STM-16	2488.32 Mbps
STS-192	STM-64	9953.28 Mbps

SONET Interoperability Forum SIF. An organization devoted to identifying and suggesting solutions to SONET interoperability issues, founded in 1994. Various workgroups work on topics such as remote login implementations and user requirements, interoperability with specific systems (e.g., TARP), architecture requirements, information models, access protocols and more.

The SIF findings and recommendations are published as SIF Approved Documents. Here is a small selection to give an idea of the scope of the documents. Most are available online; some can be obtained in paper format upon request.

Number	Focus	Title
SIF-007-1996	Graphical User Interface	*Design Principles for the Development of OAM Graphical User Interfaces*
SIF-009-1197	Remote Login	*NE-NE Remote Login Implementation Requirements Specification*
SIF-020-1998	Testing	*IS to IS Abstract Test Suite*
SIF-023-1998	Information Modeling	*Network View Model for Connection Management and Fault Management*
NSIF-031-1999	Architecture	*Architectures for an IP-Based DCN*
NSIF-038-2000	Security	*NSIF Requirements for a Centralized Security Server*

SONET multiplexing SONET signals can be multiplexed to make efficient use of network capacity.

There are a number of ways to accomplish this, as shown in the SONET Multiplexing Techniques chart. **SONET optical interface layers** SONET includes a hierarchy of interface layers. Each one builds on the previous; from high to low, they are path layer, line layer, section layer, and photonic layer. Individual layers communicate to peers on the same layer and to adjacent layers above and below.

SONET path overhead In SONET, path overhead is transported with the payload until the signal is demultiplexed at the receiving end. The path overhead supports four classes:

Class	Functions	Notes
Class A	*Payload independent functions* Required by all payload types	
Class B	*Mapping dependent functions* Required by some payload types	
Class C	*Application-specific functions*	
Class D	*Reserved for future use*	

SONET path terminating element PTE. The PTE is an element which multiplexes and demultiplexes the Synchronous Transport Signal (STS) payload and processes the path overhead as needed to originate or access it. If necessary, the PTE can also modify or terminate it. See SONET, Synchronous Transport Signal.

SONET timing In SONET networking, synchronization is accomplished by referencing a high accuracy clock and information from its slaves, so synchronization characters between equipment nodes are not used. Due to the high data rates carried by SONET, it is important to maintain clock accuracy. The three major timing modes supported are external timing based on a clock, generated free run/holdover timing from an internal clock, and OC-N signal line timing.

SOP standard operating procedure.

Sorokin, Peter P. (1931–) An American physicist who worked in collaboration with Mirek [Cevcik]

Stevenson, Sorokin developed the first uranium laser, in 1960, that would set the stage for future tunable continuous wave lasers. At about the same time, Schmidt et al. were developing similar technology in Germany. Sorokin continued to work with laser technology, developing a dye laser, demonstrated in 1966 at the IBM Laboratories with John Lankard. Dye lasers set the stage for tunable lasers.

In 1991, Sorokin was awarded the Arthur L. Schawlow Prize in laser science. See laser history.

sound Radiant mechanical energy produced by vibration, which requires a physical medium for its transmission (such as air), and is detected by hearing, accomplished through physical sound-detection, perception, and interpretation by the nervous system. Compared to light and heat, sound waves move very slowly. Human sound perception through hearing covers a frequency range from about 20 Hz to about 20 kHz, although lower, and sometimes higher frequencies are felt, though not heard, through vibrations in the body. Other creatures perceive broader, narrower, or more specific frequencies, and sound is a ubiquitous means of species communication. Enough is known about the nature of sound waves to record, reproduce, and modify them, and to propagate them over great distances. Humans can project unamplified voice through the air for a few dozen or hundred yards, depending upon atmospheric conditions. Whale songs will resonate for thousands of miles through water, although whale communication distances have been drastically reduced by interference from industrial shipping noise. See acoustics, sonar.

sound spectrograph An instrument for measuring the structure of speech and displaying it visually, developed in the early 1940s by Bell Laboratory researchers. This opened the door to more objective, quantitative measures of speech, information that is of interest not only to speech therapists, physicians, and educators, but also to developers of communications technologies.

sounder A sound amplification device incorporated into a communications receiver, usually a telegraph receiver, to make the code clicks audible to a human operator. Sounders were invented when it was noticed

SONET Interface Layers		
Layer	Peer	Notes
Path	Services and path overhead mapping	Transport services between path terminating equipment (PTE). Mapping signals to line layer format. Conversion between STS and OC signals.
Line	SPE and line mapping	Transport of path layer payload and overhead across physical medium. Synchronization and multiplexing.
Section	STS-N and section overhead mapping	Transport of STS-N across physical medium.
Photonic	Optical conversion	Transport of data across physical medium.

that telegraph operators had learned to interpret Morse code clicks and transcribe them manually faster than a paper tape could print the messages. A *mainline sounder* was an adaptation that allowed variable adjustments without a relay. See resonator.

SP stream protocol. See byte-stream protocol.

space division multiple access SDMA. One of two common optical multiplexing techniques which utilizes an angle diversity receiver, that is, multiple receiving elements receiving from different directional angles. See wavelength-division multiple access.

space-charge field In electronics, an electric field created outside the physical surface of a conductor or semiconductor.

space-to-mark transition, S-M transition In telegraphy, the momentary change when the system reverses polarity, or changes from an open to a closed circuit. At this point, a small amount of delay must be taken into consideration, which can be plotted on a timing wave. The reciprocal is the mark-to-space transition.

Spaceway A commercial constellation of geostationary communications satellites from Hughes Communications. Spaceway was formed from the merger of the Hughes Galaxy Satellite Services and the PanAmSat Corporation. Hughes Electronics is a subsidiary of General Motors Corporation. Spaceway is intended to be a global broadband communications system with service planned for 2003.

spade lug, spade tip A small, flat, notched (somewhat U-shaped) conductive connector attached to the end of a conductive wire in order to easily secure it to an electrical terminal by sliding the end around the mounting screw and applying pressure via a bolt, or by soldering. Spade lugs are still common inside small residential phone wire junction boxes, but in large installations, punchdown blocks and modular components are more prevalent. See lug.

spam *slang* A term widely used on the Internet to describe annoying, unsolicited, irrelevant, illegal, or worthless communications, usually in the form of email or public postings. It's generally said that the word originated as a tongue-in-cheek reference to a Hormel meat product called Spam, which is frequently pilloried and satirized in the media. Whether or not that is the case, the *spam* on the Internet, especially in the form of unsolicited email bulk promotion of get-rich-quick schemes and sex site promotion has become a big problem due to the intrusive way in which the spammers violate the space and privacy of recipients. Not as often acknowledged is the fact that spam causes substantial expense to ISPs, and general annoyance and expense to users who pay for email or extra storage space.

spamming *slang* Posting or emailing irrelevant, annoying, illegal, or unsolicited opinions or promotional materials, sometimes through anonymous mailers or with false return email addresses.

SPAN Switched Port Analyzer. A Cisco Systems network switch feature for extending the monitoring capabilities of existing network analyzers into a switched Ethernet environment. SPAN takes the traffic at one switched segment and mirrors it onto a predefined SPAN port. A network analyzer which is attached to the SPAN port can monitor traffic on other compatible switched ports.

spanning tree algorithm STA. A standard technique described in IEEE 802.1 which is incorporated into bridges in computer networks. For example, in Fiber Distributed Data Interface networks, it is incorporated into bridges that connect the primary and secondary rings. The spanning tree logic can prevent duplicate bridging and allows the backup ring hub to handle bridging if the primary ring hub fails. See Fiber Distributed Data Interface, Token-Ring.

Spanning Tree Protocol STR. A protocol based on IEEE 802.1d that provides resiliency through system and link redundancy that is especially suitable for virtual local area networks (VLANs).

spark A brief, bright, heat discharge, often from electrical or friction sources. Sparks are generated by spark plugs to fire up an engine or by matches or lighters to fire up combustible substances such as wood, cardboard, or lighter fluid. Unintentional sparks may be dangerous and may occur as a result of incorrect electrical connections (shorts, crossed wires), inadequate insulation, or contact with unintended conductive substances such as water.

spark coil A device incorporating an inductive magnetic core surrounded by helical windings of conductive materials used to generate a spark. The spark coil was typically used in conjunction with a condenser and vibrator (or interrupter) for telecommunications applications. Spark coils are still used to ignite internal combustion engines, but for many electronics applications, transformers began to replace spark coils in the early 1900s. See armature, coil, dynamo, induction, winding.

spark gap The distance across which an electrical spark jumps between electrodes. Adjusting this gap affects the behavior of the spark. If the gap is too large, the spark may not jump the gap.

SPARS code Society of Professional Audio Recording Studios. A three-letter code found in compact discs which indicates the analog or digital nature of a portion of the recording history. For example, ADD indicates that the original recording was analog, the mixing was digital, and the mastering stage was digital. See compact disc.

Spartan A family of satellites designed for remote sensing. In 1997 an unsuccessful Spartan mission occurred when astronauts failed to turn on the satellite before releasing it into space. Initial efforts to retrieve the satellite were unsuccessful and it was manually retrieved later.

In 1998, a 1 1/2-ton Spartan satellite was successfully used for capturing images of the Sun's corona from space. Before the Spartan was retrieved by the crew of the Space Shuttle *Discovery* in November 1998, it had already beamed hundreds of images to ground controllers. One of the interesting aspects of the Spartan 207 multipurpose satellite was a 132-pound Inflatable Antenna Experiment (IAE) that was deployed aboard the Space Shuttle *Endeavor* STS-77

877

mission. The IAE, resembling a flat, stiff parachute when inflated, was developed by JPL and L'Garde, Inc. as part of NASA's In-Space Technology Experiment Program.

Spartan Remote-Sensing Satellite

The Spartan 201 satellite, one in a series of remote-sensing meteorological satellites, as part of a pre-launch Crew Equipment Integration Test (CEIT) in the Kennedy Space Center. [NASA image.]

The Spartan 201 is held by the Space Shuttle Columbia's *Remote Manipulator System (RMS), December, 1997, over the Pacific Ocean. The RMS, designed in Canada, released the satellite into free flight. [NASA images.]*

---

spat 1. A unit of solid angle comprising the space about the vertex of an angle, equal to a sphere or 4π steradians. See steradian. 2. A unit of distance historically used by astronomers to represent 10^{12} meters (terameter), now rarely used.

spatial light modulator SLM. A computer-controlled optical or optical/electrical component that controls a beam of light through spatial dimensions through an array of optical modulators and associated detectors.

Spatial light modulation can drive 2D image displays as well as projecting 3D spatial images. The effect is more easily understood by envisioning 3D holographic images and then imagining that the image is created in realtime with light rather than being holographically recorded in a static medium. Spatial modulators can modulate laser or metal halide light beams to create 3D images based upon the human characteristic of persistence of vision and, in more

sophisticated systems, the basic motion concepts of frame animation.

Commercial SLMs are fairly recent and there are still many variations and experimental systems evolving. In general, however, SLMs work by sending a near-infrared or visible light pulse through various gratings and lenses to modulate the light, with the output pulse projected into an imaging system. Ferroelectric or nematic liquid crystals are often used as the retarding material with synthetic fused silica as the substrate. The phase of the lightwave may be modulated, with the intensity of the light kept constant, or the amplitude may be modulated, changing the beam intensity and the phase profile (which may be corrected with dual modulators connected in series, for example). Polarizing elements may be optional.

There are various ways to image the light beam. It can drive a 2D display like a computer/television display or it can be projected to a rotating 2D screen. Persistence of vision makes the successive images appear as 3D objects since one image is still briefly visible while the next is formed. Imaging successive frames at high speed and presenting the frames at a speed that makes it appear to the viewer as if there is natural motion (about 24+ frames/sec) is also possible, as frame rates in the order of over 2000 frames/sec (which could combine imaging frames and implied-motion frames as the image changes over a set of frames) can be achieved with this technology. Thus, many types of effects and images are possible and people can walk around the image and see it in three dimensions at the same time.

Spatial modulation technology has great possibilities for medical imaging, chemical analysis, the modeling of neural networks, optical computing, beam steering, spectroscopy, flat panel displays, entertainment, education, and volumetric computer-aided design.

SPEC See Standard Performance Evaluation Corporation.

Specialized Common Carrier decision A decision by the Federal Communications Commission (FCC) in 1971 to permit competition with AT&T in the provision of specialized voice and data services.

Specialized Mobile Radio SMR. A well-established analog, trunked two-way radio dispatch system favored by commercially dispatched passenger and cargo fleets, public safety, and local services (e.g., taxi). SMR enables a group of radio communications users to share a common channel through a central station. Allocated frequencies are in the 800- and 900-MHz ranges.

In North America, more than two million users subscribed to SMR services, a number that was gradually increasing until bandwidth bottlenecks began to impact the service. In response, the Federal Communications Commission (FCC) made available additional radio spectrum frequencies in the 900-MHz range, in the mid-1990s, and above 860-MHz in 1997, thus opening the door for SMR services to compete with niche areas of the cellular market.

With digital technology being adapted to SMR

services, considerable changes took place. First, Motorola developed a system called the Motorola Integrated Radio System (MIRS) and, more recently, Enhanced Specialized Mobile Radio (ESMR) is catching on, which makes ESMR competitive in some areas with cellular services. While SMR spectrum is more limited than cellular, it can be used over a longer range.

The upgrade to digital technology has greatly accelerated the increase in ESMR users in North America and Europe due to the improved quality of the sound and greater variety of services offered. See Enhanced Specialized Mobile Radio.

Specification and Design Language SDL. An ITU-T-defined language for the description and specification of the behavior of telecommunications systems. XDL is an extension of this language.

spectral spreading See spread spectrum clocking.

spectral transmission In a fiber optic cable, the ratio between the incoming light and the outgoing light at the other end of a fiber cable link. As light propagates through the fiber filament, it is reflected towards the cladding and back into the conducting core and may also be reflected in certain ways by the endfaces (points of entry and exit), and dopants or gratings within the fiber.

Whenever there are obstacles or microbends, some of the light is absorbed or lost through the cladding (if there are excessive bends or fractures in the cladding). It may even be stopped entirely or reflect backwards if the bend radius of the fiber exceeds the angle at which the light reflects in the desired general direction. Thus, depending upon the length of the cable, its components, the degree of bend, the wavelengths used, fiber impurities, fiber fluctuations in density, dopants (deliberate impurities), and grating factors, the outgoing light is typically less than the incoming light. Thus, incoming light versus outgoing light can be plotted with respect to wavelength for a cable with specified properties. This can be modeled with computer software or tested directly with measuring probes in installed or sample cables.

spectrograph An instrument for spreading light into its spectral components. By studying the brightness of the spectrum at each wavelength, it is possible to study the composition and characteristics of substances through their light-emitting properties and patterns. This is widely used by astronomers studying our solar system.

spectrometer An instrument that detects and records the spectral components of light. The main element is a dispersing component for separating light rays into their component wavelengths. Other typical components include a collimator to align the incoming light (which may beam directly to the collimator or come through a fiber optic probe), a prismatic grating to direct and condition the light reflected by the dispersing element, various concentrating lenses, and an optical detector. Once detected, the signal is sent to a display or recording device (e.g., charge coupled device) and may further be analyzed by a signal processor. Échelle spectrographs enable a larger amount

of the detector area to be used by creating higher dispersion over shorter wavelength ranges for each order. See échelle grating, interferometer.

spectrometer, mass An instrument for detecting and recording the differences in mass-charge ratios of ionized molecules or atoms that occur when a substance is stimulated by an electromagnetic force. This enables molecular properties to be separated/distinguished, which is useful for assessing the structural and chemical properties of substances.

spectrophotometer A versatile optical sensing and measurement instrument used in a variety of applications from chemical analysis to high energy physics research. The instrument may be used for evaluating the existing qualities of a substance as light energy passes through it, or may be used in conjunction with externally applied energy (e.g., radiation) to apply controlled influences to the sample under study. Thus, the spectrophotometer measures the characteristics of a substance through its absorption of light or the amount of light passing through the substance as compared to reference information.

A spectrophotometer includes a phototube or photomultiplier tube to convert photon energy into electrical energy so that the characteristics may be measured with electronic circuits and viewed with electric-based display components.

When the light emitted from a sample is dispersed with a monochromator, the resulting spectral lines can be analyzed and the results compared or combined with information from the photomultiplier tube.

When a monochromatic light source (e.g., laser light) strikes a substance and emits electromagnetic energy at a different frequency from the source energy, a phenomenon called Raman scattering occurs. This effect may be amplified with a photomultiplier tube so that the wavelength difference can be used to assess qualities of the sample.

Fiber optics are increasingly being incorporated into spectrophotometric and radiometric devices. Reflected or weak emissions may be detected with photomultiplier tubes and optical filters and the function of a monochromator may be handled by optical fibers. For general reference, the diameter of commercial side-on and smaller gauge head-on photomultiplier tubes ranges from about 10 to 38 mm, with effective areas of about 3 to 34 mm. Depending upon the type of tube and its intended use, spectral response is typically around 160 to 185 nm or 300 nm at the low end ranging up to about 650 nm or 850 to 900 nm at the high end. A few specialized near-infrared components have a spectral response between about 400 to 1200 nm. Larger gauge head-on tubes range from about 51 to 127 mm, with effective areas of about 46 to 460 mm. See diffraction, phototube.

spectroscopy A technique used by scientists to study the composition and/or characteristics of a substance based upon an analysis of its light-emitting properties. Spectroscopy is useful for chemical and crystalline structure analysis.

In electronics, spectroscopy enables semiconductor components to be tested and monitored. The property of photoreflectance can help determine electron mobility; by beaming laser light of a specific frequency into the semiconductor and influencing the dielectric properties, the optical response in terms of the reflectance coefficient may be determined.

In medicine, fiber optic probes may be used in conjunction with spectroscopes to assess the characteristics of healthy and diseased tissues to aid in medical diagnosis.

In astronomy, spectroscopy has enabled more detailed study of stars, beyond distance and brightness.

spectrum In general, a continuous sequence or range of some property or radiant energy. Many phenomena are described in terms of their characteristics or position within the spectrum of radiant energy, including electromagnetic, radio, sound, visible light, specific colors, etc. See band allocations, radio, sound, visible spectrum.

speech recognition The process of receiving, interpreting, and parsing spoken words. On computers, this is often accomplished with a microphone input device, an analog-to-digital peripheral card, and a software program that works independently or in conjunction with other programs such as word processors. It may also include *noise-canceling* features that help to separate the voice of the speaker from ambient room noise such as the hum of computers or low conversations in the distance.

Speech recognition is distinct from voice recognition in that voice recognition is the processing of the particular characteristics of a specific voice so that it can be recognized, such as in a security identification system. Voice recognition does not involve making sense of the content of the message, as does speech recognition. Speech recognition systems typically require a minimum sampling rate of about 3000 samples per second in order to reliably recognize words. Many systems sample at 8000 samples/sec.

Speech recognition can be used to dictate text, give commands, and receive information over a communications system in digital or altered form. Since speech recognition is a complex process, most current systems are specialized to recognize a specific limited vocabulary as spoken by a number of speakers in a specified language or a general (or specific) vocabulary as spoken by one particular speaker. More sophisticated systems can recognize and react to sentences and grammatical structures. Many speech recognition programs have training algorithms (*speaker adaptive* algorithms) included so the software can gradually adapt to the idiosyncrasies of a particular speaker's prounciation and mode of expression.

In the mid-1990s, speech recognition computer software began to be reliable and inexpensive enough to interest small businesses and individual consumers, and its use will probably spread through a variety of applications, perhaps adding another means of input to standard software such as word processors and electronic mail programs. See phonemes, voice recognition.

speech synthesis The reproduction of audible human communication, through the use of computers. There are many different ways to create synthesized voice. Sound samples of human voices uttering certain words, sounds, and syllables can be recorded as separate entities, stored digitally, and then combined and played back to create words and phrases. Other schemes, such as pure digital recreation of voice-like sounds are also available, but tend to have a distinctly mechanical quality to them.

The most famous synthesized voice in the world is probably that of Stephen Hawking, world acclaimed physicist who talks indirectly through words and phrases programmed into a computer keyboard installed on his wheelchair. When Mr. Hawking is finished composing the message, it is played to the listener through a speech synthesizer.

Synthesized voices are used in multimedia applications, on storybook CD-ROMs, in automated telephone mail order and banking systems, etc. See phoneme, speech recognition, voice recognition.

Speech Technology and Research Laboratory STAR. A division of SRI International which engages in world-class research in speech technology using engineers, linguists, and computer scientists. Technology developed in the STAR lab is fed to Nuance Communications for commercial development for telephony applications. Of particular interest to researchers is natural speech recognition (without the usual training to recognize a particular individual's voice) that can provide automated phone services or voice-based securities trading. Other areas of research include text-to-speech translation, visual information systems development, and digital encoding of audio signals.

speed dialing A means of keying in a shorter code to represent a longer one in order to speed up the dialing of long phone numbers. See abbreviated dialing.

SPF 1. shortest path first. 2. See stateful packet filtering.

SPHIGS Simple PHIGS. See PHIGS.

SPI 1. Security Parameters Index. 2. Service Provider Interface.

SPID See Service Profile Identifier.

SPIE – The International Society for Optical Engineering SPIE is a nonprofit, international professional society in Bellingham, Washington, dedicated to advancing research, engineering, and applications in optics, photonics, imaging, and electronics. SPIE produces educational publications, sponsors conferences and workshops, and now also provides Web resources in cooperation with the Institute of Physics (IOP). http://www.SPIE.org/ http://optics.org/

spilling In a fiber optic cable assembly, light that exceeds the critical angle within which total internal reflectance (TIR) is possible and thus passes through the cladding and is lost. Sometimes spilling is deliberately induced (e.g., by bending the fiber) to produce light effects at periodic intervals along the length of a fiber. In general, however, it is undesirable, resulting in losses that occur in the power of the light signal as it reflects through the fiber link. See cladding.

SPINS See Security Protocols for Sensor Networks.

SPIRITS A network architecture that supports

SPIRITS services originating in the Public Switched Telephone Network (PSTN) interacting with the Internet. In simple terms, it is an architecture to support popular telephone services over Internet phone connections, such as Internet-based Caller ID, Call Forwarding, etc. The SPIRITS architecture was submitted as an Informational RFC by Slutsman et al. in June 2001.

Implementation of SPIRITS services requires an Internet Protocol (IP) host installed with SPIRITS-supporting software and identification (e.g., PINs) for communicating with other SPIRITS servers. Once a host is SPIRITS-enabled, a user may connect to the Internet and register a service session and optionally specify the session duration.

The SPIRITS architecture consists of a number of service control, service switching, client, server, and gateway functions as well as a number of interfaces.

Interface	Notes
Interface A	A conduit for PINT requests/responses. Supports service session subscription, registration, and activation of a SPIRITS service.
Interface B	Notifies the subscriber of incoming calls and call information and submits a subscriber's choice of call disposition to the SPIRITS gateway.
Interface C	Client/gateway communication. The gateway may, in turn, communicate with the SPIRITS server or may act as a virtual server, terminating requests without relaying them.
Interface D	SCF to client communication, sending parameters associated with the applicable IN triggers. The SCF translates user requests into corresponding actions.
Interface E	PINT to SCF requests.

See PINT, RFC 2995, RFC 3136.

splashing When using competitive operator services for long-distance calls, if the caller places a call from San Francisco to Portland and the alternate service is based in Los Angeles, the call is said to be *splashed* if the billing is determined by the distance from Los Angeles to Portland. See Operator Service Provider.

splice *v.* 1. To unite or combine separate lines, usually by weaving together the individual strands. 2. In electrical splices, the joining of conducting wires to complete or extend a circuit. Care is usually taken to match the data lines so as not to cross one type of data channel with another, and bare wires are generally covered with an insulator such as a cap, electrical tape, or plastic shrink sleeve to prevent short circuits or shock. 3. In fiber optics, a joint where two ends of a fiber optic waveguide are mechanically joined or fused to facilitate the unimpeded travel of light across the splice joint. See ferrule, fusion splice, splice guard.

splice enclosure A component for protecting spliced cable where extra protection or safety requirements are needed. For example, a splice enclosure for electrical wires may provide extra protection against moisture, rodent chews, or electrical interference, as well as protecting from shock hazards.

For fiber optic cables, a splice enclosure can help align multiple cables and provide extra protection to spliced joints so that the waveguide is not interrupted or compromised. A splice enclosure is sometimes be called a splice organizer shelf or *fiber organizer shelf* and may be rack mountable. A *splice guard* may be used in place of a splice enclosure in some circumstances. See splice guard, splice tray.

splice guard An extra support component for a fused or mechanical joint that fits over a sleeve. It is usually shorter than the sleeve, in case the sleeve provides color or printed code information. For fiber optic cables, a splice guard is usually secured without heat or glue and provides extra protection against external forces that might compromise a joint, such as side or axial pull. A splice guard is usually applied and secured with a specialized crimp tool.

A splice guard is sometimes used instead of a *splice enclosure* in situations where only a few fibers need protection or where space is limited. Commercial products usually meet relevant Bellcore GR-326-CORE specifications (humidity, dust, and thermal aging, etc.). See splice enclosure.

splice tray A tray for protecting spliced cables, especially for temporary storage prior to installation. When the tray is used to organize multiple spliced cables for longer periods of time, a hardening gel may be used to support and secure the cables. See splice enclosure.

splitting A ubiquitous function in computing in which data or other transmitted information is broken up into streams, channels, chunks, or other units to facilitate processing or transmission or to conserve resources. Splitting is usually used in a context where the data will be reassembled or rejoined when it reaches its intended destination or is needed by some application. See joining for a detailed explanation.

SPNE signal processing network equipment.

spoofing 1. Deceiving, covering up the identity of, impersonating, or otherwise conveying an impression of being something else. 2. A means of gaining unauthorized access to a premises or system by deceit, impersonation, or other misrepresentation.

spoofing, facsimile Facsimile devices, especially the desktop models, have not traditionally been speedy machines and many are designed to time out if no data is received for a while (in order not to tie up phone lines). With the advent of faxing over faster packet-based network connections, facsimile delays can lead to time-out problems in the connection that would interfere with the function of the fax machine. Spoofing extends the tolerance for these delays so that the fax can finish with the machine disconnecting. By padding and other methods, delays and packet "jitter" can be smoothed over to spoof a realtime voice-line connection. A number of vendors have access server technologies that take into account the needs

S

of both store-and-forward and realtime facsimile protocols (ITU-T T.37, T.38).

spoofing, network In network traffic routing, a means of rerouting, or otherwise changing the destination of a transmission by mimicking the destination through responses, signals, or other identification. Spoofing has legitimate purposes, such as testing and modeling, or overcoming problems of latency in slower devices, but often the spoofing is intended to gain unauthorized access to a system or to impersonate a system and send substituted information. Reports of malicious spoofing on the Internet began to emerge in the mid-1990s.

Spoofing can be set up at the beginning of a session or can be inserted into a session after it has been established. The first method may be harder to detect, while the second may be a means to eavesdrop on information the source system thinks it is sending to the original destination, which is now the spoofed destination. In general, the spoofing is carried out by impersonating a trusted destination IP address. The spoof can sometimes be detected if the IP address is found to originate on the local system rather than on an external system.

Spoofing occurs for many reasons. It can enable a user to intercept programs or information or it can be used to send bogus responses to requests from the originating system. It can also compromise the ability of a system to carry out its normal functions. A denial of service attack is a situation in which a service (e.g., a Web server for a stock quote or online auction system) is made unavailable to normal users. By setting up a loop that ties up or floods the system, for example through a half-open connection that overflows, denial of service can be achieved. This doesn't usually compromise the information on the attacked system, but may harm it by denying legitimate users access. The host system may not be aware of the problem until customers begin to complain.

Since the mid- and late-1990s, various dynamic routing techniques and packet traffic monitors have aided in preventing and detecting spoofing. See Computer Emergency Response Team, Computer Incident Advisory Capability.

spoofing, Web site This is a subset of network spoofing that specifically spoofs or impersonates a Web site at another location. A vulnerability exists within browser software supporting HTML frames prevalent in the late 1990s. Subsequent versions of browsers have been updated to help prevent this type of spoofing (e.g., Netscape Communicator 4.51).

Web site spoofing is usually a malicious activity intended to mislead, inconvenience, or harm. A Web site can be spoofed to make a site look bad (e.g., maligning a commercial site) by inserting insulting, poorly designed, or shocking images or text into an otherwise normal-looking site. Spoofing can also be similar to the Trojan Horse concept wherein something desirable appears to be offered (but not given) to the user in order to gain entrance to a location or information repository. The user may think he or she is getting a free product sample, for example, when

the spoofed site is actually capturing names and mailing addresses (in fact this type of fraud can be carried out on a nonspoofed site, as well).

Not all Web site spoofs exploit browser vulnerabilities. Some of them rely upon human trust and error. One type of Web spoof impersonates the layout and functions of another Web site and may or may not spoof the original IP address or domain name (Web sites are sufficiently high in information content that an IP address or domain name change may not be noticed by the user).

For example, in summer 2000, someone spoofed the login page for the PayPal online financial system, which enables users to conduct online monetary transactions through user accounts. The Web page was designed to clandestinely capture names and passwords. The perpetrators then sent out bulk email that looked like a legitimate PayPal notification with a clickable Web address embedded in the email message. However, the embedded URL didn't link to *www.paypal.com*, it linked to *www.paypaI.com* with a capital "i" rather than a lowercase "l". People couldn't easily see the deliberate misspelling and anyone who linked to the site and tried to log on gave away his or her login name and password. This could then be used to access the real PayPal account to enable the perpetrators to transfer funds. Fortunately, the fraud was discovered quickly, before significant harm was done. This kind of spoofing has a good chance of landing the perpetrators in jail.

spool *v.* An acronym for "simultaneous peripheral operations on line," a means of running scanners, printers, plotters, video capture systems, or other peripherals while other computing tasks are ongoing. In earlier nonmultitasking computer systems the acronym was misleading, since spooling was a way of saving information in a file so it could run as a batch job later, rather than as a simultaneous process. Spooling as we know it now more accurately reflects the acronym. It is a technique for improving efficiency by accommodating the different operating speeds of a number of types of peripherals and processes by scheduling and optimizing the exchange of handshakes and data transmission. Spooling is especially popular for creating files that will be printed as a background task or sometime later (e.g., overnight, when printers aren't so busy, or when the computer isn't engaged in CPU-intensive functions).

On multitasking systems, it is not necessary to wait for a print, plot, or other peripheral job to finish to continue word processing or drawing, but it may slow current processes. Thus, the system may schedule the printing job to run while the user is away from the keyboard or engaged in computing tasks requiring less processing power. The resource management software routines can be designed to sense when resources are available and carry out the print job. On single-tasking or task-switching systems, however, with peripherals that don't support a *queuing* system, wait time can be a problem while a file is processed or printed. For example, plotters tend to print rather slowly. If the user must wait for 25 minutes while a plot is created, productivity is lost. By sending the plot to a software spooling process (or a hardware spooling buffer) and plotting it at a convenient time, the user can continue working uninterrupted.

spot of Arago In 1818, Augustin Fresnel, after whom the Fresnel lens is named, submitted a thesis on diffraction to a competition sponsored by the French Academy of Science. At the time, the wave theory of light was not scientifically accepted and Simeon Poisson endeavored to shoot down Fresnel's thesis on the basis that his theory would predict that a bright spot would appear in the shadow behind a circular obstruction. Dominque Arago, a member of the judging committee, took the time to try the experiment and verified Fresnel's prediction. It is sometimes called Poisson's bright spot, but it is more fitting that it be called the spot of Arago. See diffraction; Fresnel, Augustin.

spread spectrum A technique used with radio-frequency-generating systems to spread the emitted wavelengths over more than one frequency, in either predefined or random patterns (or a combination of these). In telecommunications, spread spectrum is generally divided into frequency hops, a system developed over 50 years ago by Hedy Lamarr, and direct sequencing, which utilizes the noise-like characteristics of pseudorandom sequences to control a phase modulator.

At the sending end, specific types of noise are incorporated into the signal, which is spread over a wider frequency range. At the receiving end, a lower bandwidth version of the original signal is recovered.

Spread spectrum is used for a number of reasons: it can reduce overall interference with other radio signals, can be more difficult to detect, intercept, or decode, and more difficult to jam, thus providing several types of security not possible with single-frequency transmissions. There are tradeoffs when using spread spectrum. More bandwidth than normal is required to transmit the signal.

Spread spectrum broadcasts are not only used in covert or private communications, but are now also used as a means to optimize increasingly limited broadcast space due to increased demand. In the United States, unlicensed spread spectrum transmitters are permitted to broadcast within specified frequency ranges. As an example, the 900-MHz cordless telephones and a number of wireless local area networks (LANs) operate in the unlicensed spread spectrum frequencies.

Pseudonoise, or direct sequence spread spectrum, is a technique used in local area wireless networks (LAWNs). Redundant data bits called *chips* are incorporated into the transmissions and the receiver must have knowledge of the spreading code to remove the added chips and decipher the incoming message. The insertion of chips provides a means to provide more frequencies within a given area, with a tradeoff in speed. The throughput of this type of system is about 2 to 8 MHz.

The Federal Communications Commission (FCC) regulates the amount of time that may be spent on any one channel. For example, in the ISM band of 902 to 928 MHz, the frequency hop on a particular channel must not exceed 0.4 seconds once every 20 seconds. Similarly, in the 2.4- to 2.484-GHz range, the interval is not more than once every 30 seconds. Recommendations are also being drafted by the IEEE. See frequency hopping, spread spectrum clocking.

spread spectrum clocking In electronics, many components emit radiation. The smaller and more tightly these components are bundled, the more likely there is to be undesirable electromagnetic interference (EMI) from neighboring parts. This is especially true in precision very large-scale integration (VLSI) electronic components generating high clock frequencies. Spread spectrum clocking is a means of spreading the spectral emissions generated by the clock signals to reduce the overall EMI. This is often done through frequency modulation (FM). See spread spectrum.

spreading loss In a beam of radiant energy, loss associated with the geometrical spread of the energy as distinct from absorption and scattering. Thus, the peak power level that emanates from the energy source is reduced as the energy propagates outward. It is also known as energy spreading loss (ESL).

To better understand this, imagine the light from a flashlight beam gradually spreading as it travels away from the source. When the beam hits the upheld palm of a friend standing several feet away, the strength or power of the light is lower than if the palm is in front of the light a few inches away. The rest of the light will spread around the palm, passing it by, even if no absorption or other type of loss occurs. Acoustic energy, like electromagnetic energy, exhibits spreading loss, because it radiates outward, which makes spreading loss an important consideration in sonar technologies as well.

When spreading loss occurs in energy forms traveling through the air, it may be termed free air loss. Spreading loss can vary as frequencies vary and as frequency compositions vary. For example, spreading loss in a coherent beam of laser light is far less than the spreading loss in a beam from a flashlight. Spreading loss becomes greater as the distance from the source increases (e.g., a radar echo traveling towards a target will exhibit spreading loss not only on the way to the target but also on the way back). It is especially important to account for spreading loss in very remote sensors, such as satellite imaging systems. Range-spreading loss associated with satellite imaging sensor data may be compensated for, based upon distance, incident angle, and known characteristics of the sensing medium. Some commercial satellite image products are routinely processed for range-spreading loss (e.g., synthetic-aperture radar products). In fiber optic transmissions, spreading loss is reduced or mitigated in a variety of ways, including reducing the number of components and installing active relays. Spreading loss in a number of media can also be compensated for by redirecting the spread energy back into the main beam. See absorption, attenuation, scatter.

S

Sprint A well-known telecommunications services company, Sprint was one of the early companies to make dialup access to nationwide ISPs affordable, in the early 1990s.

Sprint began as a small telephone company in Abilene, Kansas, founded by Cleyson L. Brown, in 1899. The Brown Telephone Company provided an alternative to the Bell system and grew and was reorganized as United Utilities. By the 1950s, it was the second-largest non-Bell company in the U.S. In 1972, the name was changed to United Telecommunications. By the late 1970s, the company began installing fiber optic links and soon after established UNINET, then the third largest commercial packet data network in the world. In the mid-1980s, Sprint became known for its long-distance services. Sprint established the first coast-to-coast, national fiber optic data and telephone system as Sprint International after which United Telecommunications became Sprint Corporation, in 1992.

In 1995, Sprint moved into the mobile market by acquiring wireless licenses for 29 major trading areas auctioned by the Federal Communications Commission (FCC). In 1998, Sprint introduced ION, simultaneous voice, video, and data services over existing connections.

There was an unsuccessful attempt to merge Sprint and WorldCom, a merger that did not receive U.S. government approval. Thus, Sprint focused its efforts upon building Sprint PCS into a major wireless services vendor.

SPS See Standard Positioning Service.

spud bar A tool somewhat resembling a cross between a pry bar and a long-handled, short-bladed spade. The handles may be D-shaped or ball-shaped and the blades may be replaceable. These are useful for breaking up tough soil clots for installing utility poles in areas where it is not practical or possible to bring in mechanized diggers (they're also handy for scraping asphalt lumps and old roofing or siding materials and for making holes in thinner patches of ice).

spur 1. A foot spike worn by line workers to improve contact and traction when manually climbing utility poles to carry out maintenance, testing, or repair. 2. A tributary or offshoot from a main line, as in a railroad track or communications spur line. Spur lines may run to the end of a cul de sac, a pier, or other terminal point where there is an obstruction or no further need for the line to continue.

Sputnik I The world's first artificial satellite, launched by the Russian Federation on 4 October 1957, studied the ionosphere and heralded the space age. It transmitted in a frequency range just above global frequencies for standard time signals, so that those listening in, which included Earth stations all over the world and a large number of amateur enthusiasts, would be able to monitor and report the status of the satellite. The project was announced in print by the Russian Federation several months before its launch so that results of communications from the satellite could be reported. See Radio.

Sputnik 2 The world's first successful launch and retrieval of a living creature, Laika, a dog, was carried out by the Russian Federation in Sputnik 2 on 3 November 1957. This craft also studied space radiation and used a slow-scan TV camera to relay images to the ground.

sputtering See magnetron sputtering.

SQL Structured Query Language. A widely used, structured data sublanguage coding system for querying database information (e.g., Oracle). Most professional database programs support SQL-format files or can import/export SQL instructions. SQL emerged from technical descriptions of relational databases documented by computing professionals such as E.F. Codd. Based upon this model, IBM developed SEQUEL, the forerunner to SQL. In 1979, Relational Software, Inc. introduced a commercial implementation of SQL that became integral to the Oracle database program. The company later became Oracle Corporation.

SQL has been supported and standardized by a number of prominent organizations, including ANSI and IS/IEC. Embedded SQL comprises SQL statements within a procedural programming language as well as certain extensions to standard SQL statements.

Square A data-securing block cipher system developed by Joan Daemen and Vincent Rijmen. Paulo Barreto and George Barwood developed a fast implementation of the algorithm for the Pentium II that is available on the Web. Square was used as the basis for the development of Rijndael, an important system used by the U.S. government. See Rijndael.

Sr *symb.* strontium. See strontium.

SR *symb.* left-hand slant polarization (ITU).

SRAPI Speech recognition API. A speech recognition and translation applications programming interface (API) developed by the SRAPI Committee, a Utah nonprofit corporation consisting of some well-known vendors of various audio and multimedia products. http://www.srapi.com/

SRGP See Simple Raster Graphics Package.

SS7 See Signaling System 7.

SSAC13 Signaling System Alternating Current No. 13. A British automatic 1VF signaling system for transmitting supervisory telephone signals between private branch exchanges (PBXs) in the 2280-Hz frequency. Using the two-wire section of a transmission path, signals may be sent in one direction or the other (but not both at the same time), although some implementations use four wires for two-way signaling.

SSAC15 Signaling System Alternating Current No. 15. A British private telephone line signaling system designed to operate in the 2280-Hz frequency. Each line has two transmitting lines and two receiving lines such that it is suitable for setting up a wide-area network (WAN) with two or more private branch exchanges (PBXs) and the appropriate wiring and interface cards. BTNR 181 is the British Telecom Signal Systems SSAC15 standards description.

SSB See single sideband.

SSCS See Service Specific Convergence Sublayer.

SSCOP See Service-Specific Connection-Oriented Protocol.

SSCP 1. See Service Switching and Control Point. 2. See System Services Control Point.

SSD 1. Secret Service Division. 2. Service Selection Dashboard (Cisco Systems). 3. shared secret data. Encryption information shared between negotiating systems for purposes of security. In mobile communications environments, SSD is divided into authentication procedures (SSD-A) and privacy/confidentiality (SSD-B) with SSD information shared between a user mobile handset and an Authentication Center (AC) and sometimes also a Visitor Location Register (VLR). SSD is documented in ANSI-41.

SSL See Secure Socket Layer.

SSP See Service Switching Point.

SSRP See Simple Server Redundancy Protocol.

SSTV See slow scan television.

ST 1. See Scheduled Transfer. 2. signaling terminal. 3. straight-tipped. 4. systems test.

ST Ports & ST- Connector – Basic Parts

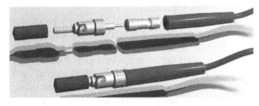

ST- ports (top) and their associated connectors are standardized fiber optic coupling components that are widely used.

The basic components in the ST- connector include (bottom image, left to right) a strain relief boot (to prevent excessive bending), sometimes a crimp sleeve, a 2.5 mm-ferrule connector, which may have a bayonet mount, and a dust cap. The necessity for a crimp sleeve depends upon the clamping or bonding method. The size of the hole to accommodate the fiber varies with the diameter of the fiber to be inserted.

ST- connector A common, standardized connector for coupling fiber optic cables. ST- connectors may vary in the details of their appearance but are essentially mechanically the same. They are fabricated in a variety of materials including metal (e.g., stainless steel), plastics, or ceramics.

ST2+ A connection-oriented, routable, multicast-capable Internet protocol for providing native ATM circuits for applications that require bandwidth guarantees and Quality of Service (QoS). ST2+ is an updated version of STII (ST2), specified in 1996, which has an Internet Protocol V.5 (IPv5) designation. See ST2+/UNI, RFC 1946.

ST2+/UNI Stream Protocol Version 2+. An evolutionary descendent of STII and a version of a previous ST2+ providing native ATM support for ST2+, this version of ST2+ was specified for UNI 3.1 in 1998. ST2+ is an ATM-based, connection-oriented Internet protocol for communication among ST2+ agents. The protocol is designed to extend UNI 3.1/4.0 signaling functions, to reduce certain signaling limitations, and to manage resources more efficiently in ATM and non-ATM networks. The protocol specifies the interaction between ST2+ and ATM on three planes that correspond to the ITU-T B-ISDN Protocol Reference Model:

Plane	Notes
user	Specifies encapsulation of the ST2+ DATA PDU into the AAL-5 PDU
management	Specifies the Null FlowSpect, Controlled-Load Service FlowSpec, and the Guaranteed Service FlowSpec mapping for UNI 3.1 traffic
control	Specifies encapsulation of the ST2+ SCMP PDU into the AAL-5 PDU, the relationship between ST2+ SCMP and PVC management (for ST2+ data), and the interaction between ST2+ SCMP and UNI 3.1 signaling

See byte-stream protocol, STII, RFC 2383.

STII, ST2 Stream Protocol Version 2. A connection-oriented Internet protocol which brings together SCMP (for signaling and control) and ST (streaming protocol). A connection must be set up between sender and receiver before datagrams are transmitted, and flow quality is accomplished through datagram scheduling by the routers. In contrast, a great deal of Internet traffic is connection-oriented, in that the receiving system doesn't have to be determined to be online before a transmission is sent. ST2 has an IPv5 designation. See ST2+, RFC 1819.

STA 1. See spanning tree algorithm. 2. Science and Technology Agency. Japan's research and development support, to plan and coordinate national science and technology. Founded in 1956.

stage A designated platform or work area that is sometimes housed in a special enclosure or room to provide the right environment for the work to be carried out. A microscope stage is the region where the sample is held in place for viewing. A fiber optic fusion splicing stage is the region where two filaments are aligned and spliced.

stand alone, stand-alone, standalone A system or device providing a self-contained service or function, independent of other major components. See turnkey system.

standard cell A fragile, special-purpose cell providing very small amounts of electrical current (1.019V) for short periods.

standard network variable type SNVT *(pron. snivit)* A common variables framework element

S

designed to support interoperability. The term is especially associated with the LONWorks control automation system, though it is applicable in a general sense to other types of network devices where predefined network variables have been associated with various units within the system (e.g., degrees, meters, volts, etc.).

Standard Performance Evaluation Corporation SPEC. A nonprofit organization supporing the establishment and maintenance of standardized, relevant benchmark computer performance evaluation tools that can be applied to various high-performance systems. http://www.specbench.org/

Standard Positioning Service SPS. One of the precise location data signals transmitted from Global Positioning System (GPS) satellites. This signal is available without charge for private and commercial use and is not encrypted. It provides information about the functioning of the satellite and its approximate location. Combined with information from three or four other satellites, the user can pinpoint a location with horizontal and vertical accuracies up to about 100 m and 340 nsec of time, depending upon the quality and accuracy of the receiving equipment. See Global Positioning System, Precise Positioning Service.

Standards Australia International, Ltd. SAI. An Australian standards organization with strong international activities focused on supporting business-to-business services based on knowledge sharing. The organization was established in 1922 as the Australian Commonwealth Engineering Standards Association and became the Standards Association of Australia (SAA) in 1988. Other changes were made, and the organization was incorporated as Standards Australia International in 1999.

SAI generates revenues through normal business activities and government-funded contributions to the national interest. It makes innovative use of the Web to develop and distribute intellectual property. Through a large number of technical committees, a staff of almost 300, and thousands of voluntary experts, SAI maintains over 6000 standards.

Standards Council of Canada SCC. The Standards Council of Canada is a federal Crown corporation responsible for promoting efficient and effective voluntary standards in Canada for the health, welfare, and economy of Canada. The SCC works in cooperation with, and manages, the National Standards System. The SCC coordinates input from the SCC to the International Organization for Standardization (ISO) and the International Electrotechnical Commission (IEC). See Canadian Standards Association, Telecommunications Standards Advisory Council of Canada. http://www.scc.ca/

standards reference model A communications framework for describing and classifying information processing standards by the Institute of Electrical and Electronics Engineers, Inc. (IEEE). Within the reference model, general sets of requirements are developed. The reference model divides standards into two categories:

Interface Type	Description

application program

Standards that influence application portability by describing and defining interoperability between applications software and the computer operating system. Consideration is given to systems, communications, information, and human-computer interactions.

platform external

Standards that influence system portability and interoperability, that is, the behavior of information processes which interact with their external environment. Data portability and user interfaces are important considerations.

Conformance to these standardization concepts provides a framework for vendors to develop and distribute products and services compatible with those of other vendors, and provides users with some assurance that products purchased from different vendors will work together.

standby processor A spare or secondary processor in a computer system that is ready to take over if there are problems with the first, or if extra computing power is needed on an irregular basis. The standby processor is usually a hot standby device, that is, it can come online without turning the system off or interrupting its functioning to a substantial degree. Sometimes the second processor is not idle, but is used for less intensive computing operations, while still remaining ready if it needs to come online as a substitute for the main processor. In some cases, the standby processor does low-level maintenance work on updating its databases and file structures, so that if the primary processor goes offline, the file information is known by the standby processor.

Standby processors are most commonly used in high-end systems that require a high degree of reliability. Examples would include medical or navigational applications where people's lives might be in danger if there were a processor failure.

standby time 1. A power-saving feature built into many consumer electronics. Camcorders, calculators, laptop computers, and various other devices that rely on limited battery power will often have timing mechanisms that monitor idle time, that is, time during which the device receives no input from the user. When the idle time expires, the device is powered down or put in a minimal power-consuming mode, in order to save battery life. 2. The amount of time a fully charged battery-powered unit can remain on before the battery runs out. This applies to many devices, including cell phones, cordless phones, short-range radios, laptop computers, etc. Standby time is often used as a marketing statistic to characterize a system and generally refers to idle time, rather than time during which the item is used. The use time (or talk time on a cell phone) is typically less if many features are used.

standing wave A phrase which is used to describe physical relationships and movements in real life as well as diagrammatic representations on display systems such as computers and oscilloscopes. Very large standing wave patterns are described by astronomers when talking about the movement, relationships, and symmetry of galaxies.

standing wave ratio SWR. A diagnostic measurement of a standing wave which is commonly used in transmitter and transmission line testing and maintenance. When cables are affected by moisture, wear, and loose connections, an impedance mismatch may occur which can be detected through the amplitude ratio of a standing wave. The transmitted wave is used as the signal source for measurement by an inline SWR meter.

Stanford Artificial Intelligence Laboratory SAIL. This research laboratory is the source of many pioneer ideas and developments in artificial intelligence and general concepts related to computers. SAIL contributed to the development of the LISP programming language. The SAIL facility also developed the SAIL programming language, an ALGOL-like language created in the early 1970s. The SAIL facility was closed down in 1990.

Stanley, William An inventor who created the first practical alternating current generator, after it was pioneered by Elihu Thomson. Thus, the first Stanley alternating current (AC) distribution system came into being in 1885, 3 years after Thomas Edison opened the first direct current (DC) power utility company. In conjunction with Elihu Thomson and Sebastian de Ferranti, Stanley also developed the transformer. See Alexanderson alternator.

STAR TAP The Science, Technology, and Research Transit Access Point. This is a persistent communications infrastructure project funded largely by the National Science Foundation to facilitate the operation and interconnection of advanced international networking systems. It supports applications, performance measurement, and technology evaluation, and facilitates the flow of network traffic to international collaborators. The project is managed at the University of Illinois Electronic Visualization Laboratory. See STAR TAP International Advisory Committee. http://startap.net/

STAR TAP International Advisory Committee A committee of international member groups connected to STAR TAP or interested in joining STAR TAP which oversees STAR TAP policies and operations. Members are affiliated with a wide variety of organizations including CERN, SURFnet, TransPAC, iCAIR, and Euro-Link. See STAR TAP.

star test In optics, a means of testing the alignment of reflecting surfaces at high powers by using illumination from a star as a reference. A star test typically uses the entire telescopic instrument rather than just the reflecting mirror or lens components. The pattern from a perfect optical test resembles a top view of a dinner plate with concentric rings in different widths and light values. If the lens is misaligned or aberrant, the pattern may appear eliptical, off-center, or blurred. Since a star test is done with the full assembly, it can also be used "on the fly" to make corrections. The star test is sometimes used as a followup to the Ronchi test. See Dall test, Ronchi test.

star topology A common type of network topology in which remote systems and nodes are connected point-to-point to a central system, and not to one another. Unlike some topologies with redundant connections, if the central system on a star network fails, the entire network is unable to intercommunicate. One advantage is that problems are easier to isolate, another is centralized administration and security. A star system provides the option of physically isolating the server from unauthorized access. Star topologies are used in many phone and data networks. See hub, topology.

Stark effect An effect observed and described by Johannes Stark, in 1913, following several years of research into Doppler effects in rays. He observed the application of an electric field could induce the splitting of spectral lines when he was studying the spectrum of hydrogen. Further investigation indicated that the spectra decomposed into several components, some of which were linearly polarized through the influence of a strong electric field and that this effect could be observed in substances other than hydrogen. Thus, at the atomic level, it appeared that the influence of an electrical field produced different results from that of a magnetic field. See Zeeman effect.

Stark, Johannes (1874–1957) A German researcher who studied the influence of electrical fields on spectral lines. Stark was awarded the Nobel Prize in physics in 1919 for his discovery of the Doppler effect in certain rays and of the distortion properties of an electrical field on spectral lines. See Stark effect.

start pulse In a start-stop teletypewriter system, a mechanism to release the receiving line relay and permit the receiving arm to move.

start-stop synchronization A method developed by Howard L. Krum for use with permutation code telegraph systems.

stateful packet filtering SPF. Packet filtering is a security and traffic management technique used in packet-based networks to selectively control network traffic based on information contained in packet headers. Many routers are equipped to support packet filtering. Dynamic packet filtering is a means of dynamically handling the traffic management after a stream of packets has passed through, to minimize the number of security gaps that might be left open by static packet filtering. Stateful packet filtering is a form of dynamic packet filtering that applies protocol-specific filtering rules and monitors state and context information associated with a network session. This may, in some cases, improve security (or appear to improve security), while still having some limitations and drawbacks associated with direct connections to internal hosts and no direct user authentication. See spoofing.

S

static *n.* 1. In radio transceivers or phone systems, static (also called *atmospheric*) is noise resulting from weather phenomena and related atmospheric electrical charges. Proper grounding can help reduce static interference. See interference, noise.

static electricity An electrical charge at rest, familiar to many people as *static* electricity. Static electricity associated with rubbing amber (*amber* in Greek is *elektron*) was known to the Greeks by at least 600 B.C. when Thales recorded that a "fossilized vegetable rosin" (amber), when rubbed with silk, acquired the property to attract very light objects to itself. See mutual capacitance.

static generator An early experimental device used to investigate static electricity and its effects. Static generators still are valuable as educational tools. See static, Van de Graaf generator.

static object A software term for an unchanging or uneditable object in an application or document. Images are often embedded as static objects in word processed documents, and can only be edited by locating the source document, changing it in the original context, and updating the static location (or reinserting it into the document as a static or dynamic object). Contrast with linked object.

static RAM SRAM. A type of fast electronic random access memory chip which is often used in computers in conjunction with dynamic RAM (DRAM) as they have different characteristics. Unlike DRAM chips, SRAMs do not need to be refreshed while in operation, thus providing fast access. As with many other types of memory chips, they require power to retain their information. See dynamic RAM.

static route A data transmissions path that is fixed and stored in a table or other form of database in a network router or high-end switcher (the distinction between routers and switchers is not as great as it used to be). Static routing is often faster than dynamic routing, but is not suitable for all types of installations. Static routing works well in small systems or those in which the routes are fixed and known, whereas dynamic routing is suitable for large, changing, distributed networks.

station A phone, computer, or other telecommunications device service office, console, or workstation.

station battery A battery used in early telephone switching stations which commonly provided 48V direct current (DC).

station clock A centralized timing clock which provides a local reference for broadcast or other telecommu-

---

Static Electricity Generator

discharging rods
conducting rods with brushes
collecting forks
revolving glass plates
drive wheel
Leyden jar

This excellent example of a static generating machine mounted on a base with Leyden jar condensers on either side is on exhibit in the American Radio Museum.

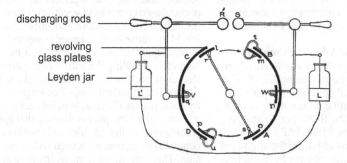

discharging rods
revolving glass plates
Leyden jar

This schematic diagram shows the basic components of a static generator which functions in essentially the same way as the one shown in the photos above. A Leyden jar is positioned at each side, with the revolving plates in the center. The discharge rods are shown across the top.

nications functions carried on at that station. Timing information may be actively transmitted from the station clock to other equipment. In broadcast stations in North America, the idiosyncrasies of black and white TV caused discrepancies in playback speeds of different media, even though they might use the same time-code basis. Thus, the playing time of a broadcast may be considered compensated when it has to be adjusted to match the station clock.

In networks, internal clocks and station clocks are used to provide timing information.

In very precise timing situations, such as in astronomical observatories, cesium-beam station clocks may be used for timing purposes.

STE 1. See Section Terminating Equipment. 2. Station Terminal Equipment. 3. Spanning Tree Explorer.

steady mark condition With multiple telegraph keys arranged in series, when not sending, an operator would close a switch to short the key contacts in order to leave the series unbroken. It puts the idle line into a *steady mark* condition.

stealth Secretive or furtive action or application. In computing, stealth products or processes intended to be active without detection, such as a viewing portal or recording system that allows people or processes to be monitored without their knowledge. Accounting, audit, or job performance applications are sometimes run in stealth mode. System operators often run stealth processes to monitor regular or unauthorized activities on computer networks or individual machines. Hackers use stealth measures to break into computer systems or to investigate processes or switching systems.

stealth server In distributed networks using domain name systems, a stealth server is similar to a slave server in that it is an authoritative server, but differs in that it is not identified in the name server's register and thus would be visible to other servers only if explicitly known by a static identifier. See slave server.

Stearns, Joseph B. (ca. 1830s–1895) An American inventor who designed and built Norumbega castle in 1886 near Penobscot Bay, Maine. He invented important aspects of duplex telegraphy that were patented in 1872 (U.S. #126,847 and #132,933). Stearns' system was initially used on the Franklin Telegraph Company lines and later on the Western Union lines. He sold the rights to duplex telegraphy to Western Union. Thomas Edison later extended the concept to create a quadruplex system in which two communications could be sent in two directions at the same time. There are some brief references that a J. Fisher may have introduced duplex telegraphy even earlier, but the author was unable to find any solid documentation to confirm this.

Documents such as Stearns' Western Union membership cards and the Stearns patents of 1872 to 1874 (and reissued patents of 1880 to 1882) are in the Western Union Telegraph Company Collection, 1848 to 1963. See Frischen, Carl; Gintl, Julius; duplex telegraphy.

Steinheil, Karl (Carl) August von (1801–1870) An astronomer and inventor (and associate of Gauss) who experimented with an Earth conductivity method to send wireless communications over distance in the 1830s. Steinheil was born in Alsace, but settled in Munich, Germany. With Gauss, he proposed the use of rail lines as a conduit for returning electrical signals. Steinheil developed a magnetic recording telegraph that marked high and low dots on a ribbon of paper (very similar in concept to Vail's code used on the Morse telegraph a few years later). In 1836, he contributed to the development of standards for measurement. In 1839, he pioneered electric clocks and reported some of his telegraphic discoveries in *Annals of Electricity, Magnetism, and Chemistry*. Ten years later he was involved in the organization of telegraphic communications in Austria. He is also known for inventing optical technologies. See conductivity method. See Sömmering, Samuel Thomas.

Steinmetz, Charles Proteus (1865–1923) A German-born American electrical engineer acknowledged for his genius in the investigations of lightning, alternating current phenomena, magnetism, and other discoveries which led to the development of safer power distribution systems and better motors, generators, and electrical appliances.

step bolt A bolt or large rivet or smaller bolt or rivet supporting a rung, intended to assist a climber in finding footing while climbing, as in a high-voltage line supporting tower. As towers age, the bolts tend to rust or even disappear altogether, causing risk to the climber, unless the bolts are maintained and replaced when needed. Tower maintenance workers are cautioned to place their gloved hands on the side supporting structures rather than on the step bolts, in case of a slip or fall.

step-by-step switch An early electromechanical automatic telephone switching system developed in 1920 and quickly favored by independents. It used rotating blades for setting switching connections. In competition with this, AT&T developed the first commercial panel switch in 1921. This type of switch was widely used until the mid-1970s, when crossbar switches superseded most panel switches and many step-by-step switches. Step-by-step switches must be modified to support touchtone dialing. See Callender switch, crossbar switch, Lorimer switch, panel switch.

stepdown transformer A power transformer that provides voltage conversion from a higher to a lower level. This type of system is commonly seen on utility poles carrying primary power to secondary power users and individual drop points. See diagram.

stepped index In a fiber cable, adjacent concentric layers with differing refractive indexes. The core has a slightly higher refractive index than the cladding, which causes the light to be reflected back into the core to continue along the waveguide. Stepped index is essentially an abrupt transition (though it need not be a large difference to be effective), in contrast to graded index fiber in which the refractive index decreases gradually as the distance from the center increases. See graded index.

Stepdown Transformers on Utility Poles

Stepdown transformers are common components of utility poles carrying primary power between substations (upper) and secondary power for residential/ small business subscribers (lower). Here the connections are shown between the higher-voltage conductors on the crossarms to the three transformers attached with brackets to the pole to the conducting line for the secondary power fed via connecting lines to the adjacent buildings. [Classic Concepts photo.]

---

steradian (*symb.* – sr) A derived SI unit of solid angle that is independent of scale. The name is related to *raidan*, which is an SI unit for plan angle. Both may be expressed in meter units.

There are about 12.5664 steradians in a sphere as derived from a conical shape radiating out from the center of the sphere with the area at the portion bounded by the cone at the surface of the sphere equal to the radius squared.

The steradian is useful in expressing light measurements. In conjunction with watts, the steradian is a standard unit of luminous intensity. See candela, luminous intensity, spat.

Stibitz Complex Number Calculator The first relay calculator, released in 1939, developed at Bell Laboratories.

Stibitz, George R. (1904–1995) An American mathematician and researcher at Bell Telephone Laboratories who developed relay arithmetic devices in the mid-1930s. Stibitz began to study binary circuits and the various applications of binary systems. This led to Stibitz's subsequent development, in collaboration with Samuel B. Williams, of an electromagnetic relay calculator, the Complex Number Calculator, which could manipulate complex numbers. By 1940, this machine had become the Bell Labs Model 1 and was incorporated into the telephone network in such a way that it could be remotely accessed, setting an early precedent for future network systems. Stibitz thereafter served as a consultant to the U.S. Office of Scientific Research and Development. Stibitz' continued interest in calculating machines and computers led to the awarding of a patent for a computing system that was a forerunner of modern digital computers, in February 1954 (U.S. #2,668,661).

In 1997, the American Computer Museum launched an annual award to honor living pioneers of the computer and information age and named it in honor of George R. Stibitz.

STL 1. Standard Telegraph Level. 2. Studio-to-Transmitter Link.

stock ticker A type of early stock reporting machine, somewhat resembling an anniversary clock with mechanical parts within a glass globe on a round base. This machine used a paper tape and telegraphic line hookup to provide fast reporting of stock activities with alphanumeric characters. In many ways, stock tickers led to asynchronous telecommunications.

Stoke's shift A loss of energy during an excited state of an electron where internal forces bring the excited electron to the first level singlet. In observations of fluorescence, the difference between the peak excitation wavelengths and the emission wavelengths.

The concept of the Stoke's shift is important in a number of fields related to telecommunications, including spectroscopy. In fiber optics, it applies to shifts that occur in doped in-fiber amplification technologies that help overcome signal loss through attenuation. See Stoke's theorem.

Stoke's theorem A theorem pertaining to vector calculus that describes a relationship between surface integrals over an open, spatially oriented three-dimensional surface and closed line integrals along the contour bounding the surface. It enables surface integrals to be mathematically reduced to line integrals and vice versa. This is useful for analyzing vector fields in electromagnetics. It is related to Green's theorem, which provides relationships between closed path line integrals and plane integrals for curves lying in a two-dimensional plane.

Stoll, Clifford (1950–) Author of *A Cuckoo's Egg* (1989), an account of computer espionage by foreign infiltrators as experienced by Stoll, who was determined to track down the source of a tiny, but puzzling accounting error, and found much more. This account of remote hacking was only the tip of the iceberg in terms of what was subsequently learned about computer penetration into unauthorized systems, computer theft, and fraud and thus has become a classic nonfiction techie thriller.

In his second book *Silicon Snake Oil*, Stoll takes a step back and looks at the pros and cons of digital technologies and how they have affected our world.

Stoner, Don (ca. 1932–1999) An American electronics experimenter and amateur radio enthusiast who, in 1957, suggested the amateur construction of a relay satellite capable of two-way communications. This idea was in advance of its time, preceding the widespread use of electronic transistors and the construction of government two-way communications satellites, and inspired the development of the OSCAR satellites. Stoner wrote about his ideas for amateur radio space research in *QST* in 1961.

In 1971, along with Pierre Goral, Stoner formed Stoner-Goral Communications (SGC) to sell radio equipment, 70% of which is not exported outside the U.S. Stoner retired in 1989 but continued to pursue the advancement of amateur radio until his death. He was editor for a time of *CQ* magazine and founder and head, for a time, of the National Amateur Radio Association. His callsign was W6TNS.

storage area network SAN. A somewhat catchall phrase for new systems for handling large amounts of data stored on a variety of types of media, accessed by a variety of remote users on a network. The model of a single hard drive and floppy drive on a dedicated workstation is impractical for heterogenous, high-capacity storage; high-demand storage needs on evolving local area and distributed public networks. A SAN is a general effort to efficiently organize, administer, and evolve storage solutions for heterogenous network environments. This more general concept is distinct from network access storage (NAS), which implies the consolidation of network storage resources. SAN is a broader concept, encompassing many different types of solutions for handling storage devices and data-need dynamics.

SAN is also more narrowly defined by commercial vendors as storage devices optimized for use on distributed networks. See network-attached storage.

storage cell, accumulator A secondary source of electricity, since it does not provide power immediately, but rather is charged up and then used. Car batteries are typically lead-acid storage cells that derive their power from a generator when the car is running, then store the power for later starting of the vehicle or operation of its electrical system when the motor is not running. See solar cell.

store-and-forward A technique for temporarily holding information until the conditions are right for transmitting the data to the receiver. This method is very common on data networks, where a router, local network, or individual machine may be offline or down. The data may be held indefinitely and transmitted when conditions are right (the right time, when traffic is lower, when the recipient is online, etc.), or may be bounced back to the sender after a certain interval or number of tries. It may even be abandoned, depending upon its nature and priority level.

store-and-forward repeaters Transmission devices that store and forward information when conditions are favorable. In radio receiving and transmitting stations, both Earth and satellite stations, the conditions for transmitting a received signal may not be optimal right away, due to weather, political unrest, high traffic levels, or the movement of a satellite out of transmissions range. The message is thus not sent until conditions improve or the satellite comes into a favorable position in orbit.

Storey, G.J. Scientist who first used the term *electron*, in 1891, to specifically describe an electric charge.

Storrer insulator A type of early utility pole insulator patented by L.W. Storrer in 1906, and first shipped by the Brookfield Glass Company in 1909. See in-

sulator, utility pole.

STP 1. shielded twisted pair. 2. See Signal Transfer Point. See Signaling System 7. 3. Spanning Tree Protocol.

STPC 6800 An early Motorola MC6800-based computer kit from Southwest Technical Products Corporation. It featured 2048 bytes of static memory, a serial interface, case, and cover in the fall of 1975, and sold for $450 without a monitor or keyboard. See Altair 680, SPHERE System.

strain insulator, link strain insulator A historic utility pole insulator that French inventor C. Priestley submitted for a patent in 1910. The Ohio Brass company acquired the rights and the patent was granted in 1912. It was in production, with variations, for about 30 years. This is also known as a *hog* or *pork liver* insulator due to the brown or tan coloring and the blobby dimpled saddle shape that allowed the conductive line to be fed through a channel connecting the dimples. The dimple channel configuration provided some protection in case the insulator broke, as it could still hold the cable. See saucer insulator, suspension insulator.

strain relief boot A protective strengthening sheath that fits over cables in sections where there may be pressure or pulling strain against the cable. Strain relief boots are often fitted over the point of coupling between a cable and cable jack. See ST- connector.

strand A single long thread of uninsulated wire. When two or more of these strands are combined or twisted around one another in the same bundle, it is called *stranded wire*. Wire is stranded for a number of reasons; it can make it more flexible and it may alter the electrical characteristics of the wire for some particular purpose.

StrataCom A commercial supplier of ATM-related telephone networking products, particularly Frame Relay switching systems and network management control software based on open, standards-based interfaces for integrating with other vendors' products.

stray light Light outside the effective wavelengths or wavelengths of observational interest. Undesired stray light may reach a viewing component or detector from a number of sources, including external reflected light, light scattered from imperfections in optical mechanisms, or flaws in grating corrugations. Interference gratings have been found to be much less susceptible to stray light effects than ruled gratings, especially when the grating corrugations are densely configured.

Stream Control Transmission Protocol SCTP. A reliable network transport protocol designed to transport public switched telephone network (PSTN) signaling messages over Internet Protocol (IP) networks. SCTP was submitted as a Standards Track RFC by Stewart et al. in October 2000.

SCTP operates over connectionless packet networks such as IP. It is modeled as a layer between the SCTP user application and a connectionless packet network service in the context of an association between two SCTP endpoints. See RFC 2960.

stream protocol See byte-stream protocol.

street price The price paid for a product after shopping around, as opposed to *suggested list* or *suggested retail,* the price the wholesaler or manufacturer has designated for the product, which is usually higher due to dealer discounts. The suggested list price is often imprinted on the product packaging and overlaid with a lower dealer price. The street price may equal the dealer's price or may be a commonly negotiated price; through discounts, volume, and competitive pricing, dealers often offer a street price much lower than the list price. The street price may also be higher than list price. For example, a ticket for a popular rock concert sold by street hawkers an hour before showtime may have a street price much higher than the original ticket price (the "scalped" price).

stringed insulators Electrical insulating objects mounted in multiples in order to increase the spacing distance to what is needed between a conducting line and its supporting structure. Higher voltage lines mounted overhead on utility poles or high voltage towers tend to require a longer string of insulators that can be directed away from the pole or cross arm. The insulators are commonly strung in single rows or in a "V" configuration. Depending upon their configuration and the number of strings, it may be difficult to access and maintain stringed wires from cherry picker (bucket) maintenance vehicles. Climbers manually scaling utility poles with strung insulators must also be careful not to move within the minimum air distance (MAD) arcing zone, which is especially tight in V configurations. The arcing zone is related to the length of the insulating string. See suspension insulator.

stringing The process of running wire or cable, especially as applied to stringing utility wires along outdoor utility poles.

stringing block See stringing roller.

stringing rope A narrow-gauge rope used to help string lines through narrow walls, conduits, or stringing rollers. It is also called a pilot rope or pulling rope. The end of the line to be installed will be attached to the stringing rope and then pulled through the narrow channel. For example,when threading a line up through the narrow vertical space in a wall from one floor to another, the stringing rope may be passed down from above, the line attached, and then the stringing rope pulled up again to feed or pilot the line through the wall or through a conduit pipe installed in the wall. If the stringing rope is used for aboveground utility line installation, the rope is used to pull conductive line through *stringing rollers,* over the track provided by the *stringing sheaves* that are the main part of the roller assembly. See stringing roller, stringing sock.

stringing roller A device with a rolling indented wheel attachment called a *sheave* that facilitates stringing wires and cables by enabling the pulled cable to run smoothly along the track supplied by the rotating roller. Stringing rollers also often have various types of hooks or eye sockets for mounting or attachment. Stringing rollers are commonly designed so that they can be mounted on utility pole crossarms or on vertical or horizontal insulators. Stringing rollers for outdoor use are typically made of materials that resist corrosion such as stainless steel and aluminum alloys. They may also have neoprene liners. When attached to crossarms or insulators on outdoor utility poles, a stringing roller may require an additional crossarm or insulator adaptor.

A *universal stringing roller* is a multipurpose roller designed with a number of fittings and adaptors for mounting it in various positions on various types of supports. This type of roller may also need an additional adapter when mounted on crossarms. A universal roller makes it possible to purchase mass quantities, even if they are used for different types of lines and mounting surfaces.

A *distribution roller* is a type of stringing roller intended to support multiple conductors. The roller is wider than a single stringing roller, with two or more side-by-side grooves to keep the conducting lines separated from one another as they pass over the roller. The distribution roller aids in managing multiple line support with a minimum of equipment and space.

A *boom truck roller* is a specialized type of stringing roller for temporary lifting and support of hot (electrically live) lines during maintenance, repair, or rearrangement. It can be mounted on a crossarm or insulated boom, as needed, and removed after the work is complete.

Stringing rollers are also called stringing blocks (as in block and tackle). They are generally sold with load ratings to support different types and gauges of wires and cables.

There are occupational regulations on the stringing of conductive lines. For example, under OSHA regulations, if a conductor or pulling line is pulled by an automated device, the lineworker is not permitted to be directly under the roller or on the crossarm, except as is necessary to guide a stringing sock through the stringing roller's sheave. See sheave.

stringing sock A device used with a stringing rope to enable it to attach to and pull multiple conductive lines through piloting by a single stringing line. It is sometimes also called a stringing board, depending upon its design.

stringing tool A tool designed to enable a single utility pole lineworker to transfer the conductive line from a stringing roller to a clamp (preferably one-handed) in such a way that the roller doesn't have to be removed. See stringing roller.

strip To remove the outer protective layers such as jackets, armoring, and sleeves, usually to reveal a conductive core. In electrical wiring, a protective plastic jacket that prevents shock and environmental interference or damage is typically stripped from ends to be wound or soldered together to provide a joint. In fiber optic cables, protective sheathing is stripped to add terminators or to provide an unimpeded end for cleaving, polishing, and fusing. Specialized stripping tools are available for either of these applications. With fiber optic cable it is especially important to avoid nicking or scratching the filament. Use a non-nicking tool intended for filament stripping and don't slide the cutting/stripping tool along the fiber filament unless it is specially designed to be used this way. Scratches could interfere with light transmission. After cutting the jacket, it should be gently tugged off the end of the fiber. If there is resistance against removal, cut closer to the end and remove the sleeve in sections (with as few cuts as possible). If there are several protective layers, they are often stripped one layer at a time, especially if they are made of different materials that require different types of cutters/strippers.

stripped insulator A conductive line insulator, as on a utility pole or tower, that has had the protrusions broken off, usually the outer skirt. This may happen as a result of aging, lightning strikes, or vandalism (e.g., target practice).

stripping, image assembly In traditional page layout and printing, the process of positioning page composites, in the form of negatives or positives, on a flat in preparation for creating the printing plate.

strobe *n.* 1. High speed intermittent illumination. 2. Older term for an electronic flash. 3. In asynchronous communication, input of parallel data to a register or counter. 4. A momentary intensified sweep of a beam on, for instance, a scope. 5. On a computer bus, strobe lines indicate when data are being transferred.

Stromberg-Carlson Telephone Manufacturing Company Founded in 1894 by Alfred Stromberg and Androv Carlson, who took advantage of the opportunity created by the expiration of the Bell patents to establish a competing phone company. Within 20 years, Stromberg-Carlson had become a leading independent telephone company. After acquisition by the Home Telephone Company, it moved to New York in 1904. During World War I, the company supplied communications equipment to the military.

The company subsequently introduced new materials into telephone construction, including Bakelite, a type of hard plastic. Around this time, the company was involved in consumer radio products and broadcasting, as well. It was FRC-licensed out of Rochester, New York, to operate station WHAM .

A number of innovators in telephone technology worked for the company in the 1930s and 1940s, notably Andrew W. Vincent, who left in 1946 to develop an improved dial telephone system.

As television technology evolved, the company provided consumer TV sets, many of which are now collector's items. In 1955, Stromberg-Carlson was merged into General Dynamics, a major defense contractor and, in 1984, was acquired by Comdial Corporation, which remains committed to telecommunications products.

strontium Sr. A soft element (AN 38) with a high refractive index and optical dispersion characteristics. It is used to fabricate glass for display devices.

Strowger, Almon B. (1830–1902) An American mortician and inventor who created the first commercially viable automatic telephone switching system, a step-by-step switch patented in 1889 and a dial-switch patented in 1891. This system enabled a subscriber to dial-connect a local call without going through a human operator. The first Strowger exchange was established in Indiana in 1892.

Strowger cofounded Automatic Electric in 1901, the largest telephone equipment manufacturer servicing Bell's competitors, the independent telephone companies. This was a successful fit, since Bell was creating its own switching technology, such as the panel switch, in competition with Strowger's technology; also, the Strowger switch was somewhat unmanageable in large installations, a limitation that was only a minor problem when the majority of Automatic Electric's customer base was small independent telephone companies. See Callender, Romaine; Lorimer, George and James; Strowger switch.

Strowger switch The first automatic telephone switch put into commercial service, in Indiana, patented in 1889 by Almon B. Strowger. Thus, direct dialing was born, and a human switchboard operator was no longer needed for connecting local calls. This also promoted a small revolution in phone design, since now dials were needed for callers to dial their own calls. The Strowger technology was further developed and put into service by the Automatic Electric company, cofounded by Strowger and directed by Alexander E. Keith. Surprisingly, the Bell system did not adopt the Strowger system until 20 years after its introduction. See Callender switch, Lorimer switch, panel switch, step-by-step switch.

Structure of Management Information A standard for object naming and describing mechanisms for the purpose of network management. See RFC 1155.

STS See shared tenant services.

STSK Scandinavian Committee for Satellite Telecommunications.

STU Secure Telephone Unit. A telephone designed to include cryptographic protection for voice, data, and fax transmissions.

STU-3 Secure Telephone Unit 3. A secure telephone unit used for government communications. See STU.

stump cam See cam, stump.

STUN serial tunnel.

Sturgeon, William Credited with producing the first electromagnet in 1823.

SU See subscriber unit.

Submarine Fiber Optics Communications Systems A newsletter published monthly by Information Gatekeepers Inc. (IGI) to provide market intelligence on new developments in underwater fiber optic technology, markets, and applications.

Submarine Telegraph Company A pioneer London-based underwater cable-laying firm known for some of the earliest marine telegraph cable installations. It was descended from the English Channel Submarine Telegraph Company founded by the brothers Brett in 1847. Using lead to weigh down the cable, an initial cable was installed between England and France in August 1850. However, cable capacitance was not fully understood at the time and the transmission was poor and of little use. In addition to this, the problem of boat anchors in the relatively shallow strait resulted in the cable being severed. The company reformed into the Submarine Telegraph Company and a new deadline for connecting France to the British Isles was established.

In September 1851, the company laid a functional underwater cable between Dover, England, and Calais, France. Over the next decade, it installed additional cables, totaling almost 900 miles of communications links. The success of these historic cable installations was based in part upon the use of gutta-percha as an insulating material for preventing salt water corrosion to sealed transmission lines. In 1863, Cyrus Field, the promoter for the first successful transatlantic cable, contacted the company regarding the duration of their cables, to which John W. Brett responded with details as to their duration, extent, and locations. See Field, Cyrus West.

Subnetwork Access Protocol SNAP. An evolution of the Logical Link Control (LLC) method, with backward compatibility with Ethernet, which facilitates communication of entities at a given network layer. SNAP was developed by IEEE to support multiple-standard, public and private Network Layer protocols. SNAP expanded 8-bit SAP space to 40-bit (5 byte) protocol ID, and uses the first 5 bytes in the LLC Protocol Data Unit (PDU). SNAP supports more upper-layer protocols than previous methods. It also allows Ethernet protocol type numbers to be used in IEEE 802 frames, to provide easy translation between Ethernet and IEEE 802 frames.

subscriber loop The circuit between the telephone company's central office and the subscriber station. In earlier times, the subscriber station extended all the way to the phone, but more recently this demarcation point has been changed to the service box outside or inside the premises to which the interior wiring usually attaches. It's still possible to get service right to the telephone; it just costs more.

Subscriber Network Interface SNI. One of the two interface ports of XA-SMDS systems used to connect an end user to the SMDS network. The other interface is the Intercarrier Interface (ICI). See Exchange Access SMDS.

subscriber unit SU. The device or system at the end of a circuit. This may be a phone, handset, or computer terminal.

substation A facility or piece of equipment that offers less in some way than a main station within a network of stations. In other words, it may have a smaller physical size, lower capacity, fewer units, a smaller staff, lower priority, etc.

In electrical facilities, a substation is a high-voltage electrical switching facility used to supply lines, circuits, and generators within a larger system. The substation may serve a local community, act as a relay station, and convert voltages from one level to another or from one type to another (e.g., AC to DC). Electrical substations are common in the industrial landscape. They usually have high fences, safety standoffs, and signs to warn the public of the dangers of injury or death from electrical shocks.

In multiple phone systems, a substation is a phone

Subscriber Loop – Telephone System

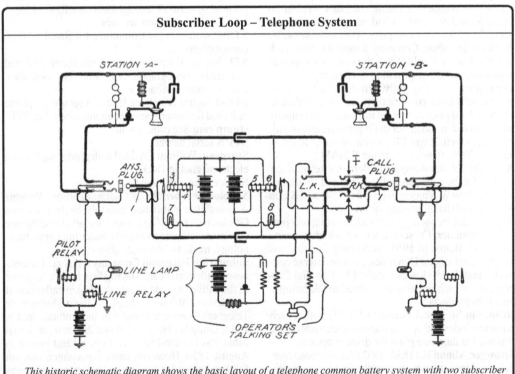

This historic schematic diagram shows the basic layout of a telephone common battery system with two subscriber lines in contact with one another through a manually operated cordboard switchboard.

console that is subordinate to the main console. It typically has more limited buttons, features, or callout capabilities than the main console. In cases where the substation has the same capabilities as the main console, the distinction is based more on where the call is initially directed (to the main console), with the substation only receiving the call when it has been redirected from the main console or another substation.

substrate-incident recording SIR. A recording mechanism used in standard magneto-optical recording in which a transparent substrate is laid down over the recording layer. The laser light aimed at the recorded layer when it is being read passes through the substrate. The substrate helps protect the recording layer from fingers, abrasions, oxidation, dust, etc. that would interfere with the recorded data. This substrate makes it possible for media like CDs to be picked up and moved around, in contrast to hard drives, in which the recording surface cannot be touched or moved without damaging the media. The substrate coating on optical discs has some disadvantages as well. Because the laser light must pass through an extra layer, there is a limit to the size, at the lower end of the scale, which can be used, thus limiting the resolution of the disc in terms of data density per unit area. Contrast with air-incident recording. See overcoat-incident recording.

sulfur A yellow, brittle element (AN 16) occurring naturally in and around thermally active phenomena such as hot springs and volcanoes. It is found in gypsum, barite, epsom salts, and iron pyrites. Sulfur has many uses. It is a component of explosive chemicals, sulfuric acid, and is used to vulcanize rubber.

SUMAC SuperHIPPI Media Access Controller. See HIPPI-6400.

Sumitomo Electric Industries, Ltd. SEI. A Japanese firm specializing in the manufacture and distribution of electrical and optical wires and cables, founded in 1897. In 2002, SEI announced at the Optical Fiber Cmmunication Conference that it had developed a record-breaking fiber in terms of low loss characteristics, with attenuation of only 0.151 dB/km at a wavelength of 1568 nm.

Sun Microsystems Computer Company SMCC. A California hardware and software manufacturer established in 1982, Sun's computer systems are commonly found in higher educational institutions, scientific research and medical imaging applications, and as servers for local area networks (LANs) in corporations, educational institutions, and Internet Services Provider (ISP) premises. Many of Sun's products are aimed at telecommunications applications for voice and data. The company's products cover a wide range from desktop systems to high-end research and supercomputing systems. The SunOS and Solaris operating systems are well known.

In 1996, Sun acquired Integrated Micro Products (IMP), including their fault-tolerant computer specifically targeted to the telecommunications industry, and Cray Research's high-end server system.

Sun's JavaSoft is the developer of the well-known Java object-oriented, platform-independent, general purpose programming language. JavaSoft collaborated with Lucent Technologies to develop a Java telephony application programming interface as part of a series of Java Media APIs to provide an open framework for Java applications development.

Sun Microsystems Inc. and Motorola Inc.'s Multimedia Group joined forces to develop products for cable operators to deliver high-speed data communications and Internet access to the home through Motorola's CyberSURFR™ cable modem.

Sun's XTL Teleservices for Solaris is a set of telephony software services and open application programming interfaces that extend Solaris LIVE, an integrated multimedia environment.

SunXTL A teleservices product delivery vehicle developed by Sun Microsystems, known as Sun XTL Teleservices Platform for Solaris. SunXTL provides Teleservices development support for applications intended to run on personal workstations. The types of teleservices which can be implemented with this technology include integrated voice mail, answering machine, automated dialing, faxing, etc. Because these are generated within the computing environment, they can be integrated with input and output from word processors, address books, databases, and spreadsheets.

SunXTL is a foundation library for telecommunications-related applications, which includes call control functions, data stream access methods, and data flow control.

SunXTL API A SunXTL Teleservices object-oriented applications programming interface which facilitates development of personal desktop applications with C++, including on-screen phone graphical user interfaces, remote workstation access, personal voice mail, etc. for telephony hardware peripherals.

SunXTL Call Objects The SunXTL API provides developers with C++ *XtlCall* objects to control various aspects of a telephone call, including querying the call state and the numbers associated with the call, the call's current status, and its data type or media class. It can also request a change in call state. The XtlCall objects also have callback methods for the asynchronous notification of state changes.

SunXTL Provider Configuration Database The SunXTL Teleservices configuration database is a repository for installed providers. The database provides information on each provider and how to invoke it, and lists its characteristics and capabilities. The database describes telephony resources such as available bandwidth, number of available lines, types of voice services available, etc. A graphical user interface (GUI) tool *xtltool* is provided for browsing and editing the Provider Configuration Database.

SunXTL Provider Interface A SunXTL Teleservices open interface providing third-party developers and Independent Hardware Vendors (IHVs) with a way to use the Provider library to ensure compatibility and compliance with basic system protocols. This message set can be extended with user-specific features. The provider interface fits between the server and/or

S

datastream multiplexer and the various drivers.

SunXTL Provider Library A SunXTL Teleservices library which works with the Provider Interface to keep the Provider information distinct from system services. The library provides interfaces to various data streams and services, including the Provider database and various server functions.

SunXTL Server The SunXTL Teleservices server provides administration, message passing, and security to networked personal workstations running SunXTL Teleservices applications.

SunXTL System Services The SunXTL Teleservices System Services provide an intermediary between the application view of a call object and the provider's implementation of the call. Interprocess message passing, object identification and creation, call ownership, and security are handled by the server.

Super DLTtape SDLT. A tape data format developed by the Quantum DLTtape Group that provides reliable high-capacity storage at good transfer rates. The format is competitive with the linear tape-open (LTO) format. Search and data rate speeds are similar, with LTO slightly faster and SDLT with slightly higher capacity (currently about 10% more).

SDLT is based upon magneto-resistive heads densely packed into clusters that are joined in an advanced thin-film medium, resulting in higher capacity and faster transfer rates than traditional magneto-resistive technologies. Advanced Metal Powder (AMP) materials enable small, smoothly coated particles with higher densities to be packed into the medium. In addition to the physical properties of the technology, SDLT uses partial response/maximum likelihood (PRML) data-handling techniques, further increasing capacity and performance.

Developers have mapped out four generations of SDLT technology with predicted capacities of more than a terabyte of data on a single cartridge:

Generation 1   SDLT drives backward-read-compatible with existing DLT products, released in March 2000 and fourth quarter 2001 supporting capacities of 110 and 160 Gbytes at transfer rates of 11 and 16 MBps. Ultra2 and Ultra 160 SCSI, LVD, and HVD interface support.

Generation 2   SDLT 640 planned for release in 2003, with backward-read capability, increased capacity, and Ultra320 SCSI and Fibre Channel interface support. Planned capacity of 3200 GBytes at 32 MBps.

Generation 3   SDLT 1280 planned for release in 2005, with backward-read capability and the intention to support emerging or prevalent interfaces at the time of release. Planned capacity of 640 GBytes at 50+ MBps.

Generation 4   SDLT 2400 planned for release in 2006, with backward-read capability and support for prevalent or emerging technologies at time of release. Planned capacity of 1.2 TBytes at 100+ MBps

Given the dramatic increases in hard drive capacity in 2001, backup technologies such as tape cartridges have become especially important. See Advanced Metal Powder, linear tape-open, partial response/maximum likelihood.

super server A high end server which consists of a number of computers networked together with communications links that are as fast, nearly as fast, or faster than the processing speed of any one individual computer, so the collection functions as a fast, integrated, distributed, unified entity. With very fast transmissions media and protocols like HIPPI and SuperHIPPI, the distinction between individual machines becomes less critical, and the processing algorithms for carrying out the tasks are more crucial to the concept of the system as an organism. A super server can also be a single machine with multiple CPUs, set up to function together to handle higher-end processing requests at faster speeds, or of greater complexity than might be achieved with a typical one-CPU system. A number of interesting distributed processing supercomputing applications have been configured at several U.S. research labs using Linux on personal computers communicating through fast network links.

Super Speed Calling A telephony subscriber option, which is essentially the same as Speed Calling in that it allows an abbreviated set of characters to be dialed to invoke a longer number. The distinction is more of a marketing distinction to describe enhanced systems where a name can be entered, which is easier to remember, rather than just a number (usually four digits or characters). See abbreviated dialing.

Super Video See S-Video.

super video graphics array SVGA. A graphics standard common on IBM and licensed third-party computers, supporting a variety of palettes and resolutions, including 800×600; 1024×768; 1280×1024; $1,600 \times 1,200$; 1024×768 (16 or 256 colors). See video graphics array.

supercomputing A term applied to high-end computing applications provided on the best hardware/software available at any particular state of the technology.

Supercomputers originated sometime in the 1950s, when the viability of computers as a commercial item became apparent. Some of the earliest supercomputers were designed by IBM and shipped in the early 1960s.

The supercomputers of 30 years ago had fewer capabilities and were slower than many handheld calculators of today, (they were also much larger in physical size). The definition of supercomputing is thus a relative one, since most desktop computer systems now are faster and more powerful than mainframes running multiuser networks in many universities and colleges 15 years ago. Supercomputers tend to be characterized by faster (or multiple) CPUs, wider data buses, faster network links, and larger, faster-access storage devices than those available as consumer products. They also may be run with more sophisticated distributed processing algorithms, although writing parallel applications is an art and much research and discovery is yet to be done.

Supercomputers tend to be used in scientific research and military applications.

superframe In its generic sense, superframe is used to describe a period in time during which a specified number of downstream and upstream frames are transmitted. Thus, the transmission time of a superframe will be related to the bit rate. The concept of the superframe is used in the context of frame timing and alignment.

SuperFrame standard SF. A 1969 improvement to the original 1962 DS-1 standard for a frame format for 1.544 Mbps transmissions (2.048 in Europe with 30 channels) which improves the signal-to-noise ratio and combines 12 frames into one SuperFrame. Frames 6 and 12 are used for robbed-bit signaling. This has since been superseded by Extended Super-Frame, which provides increased error detection and removes the need to take down an entire line for servicing. See Extended SuperFrame.

supergroup In analog voice phone systems, a hierarchy for multiplexing has been established as a series of standardized increments. See voice group for chart. See jumbogroup for a diagram.

superheterodyne receiver An early improvement in radio receivers designed to be more sensitive than radio frequency receivers of the time. The superheterodyne receiver incorporated a signal detector working in conjunction with a local oscillator to mix the signals, producing an intermediate frequency which was then amplified and passed on to a second detector, and from there to the earpiece. The superheterodyne circuit was invented by the "Father of FM," Edwin Howard Armstrong. See heterodyne.

SuperHIPPI See HIPPI-6400.

SuperHouse A trademark of BellSouth, to signify a house designed with information services resources built in (conduit, wiring, etc.) to support computing and Internet applications. This is often called a "smart house" and, in fact, SmartHouse has been trademarked by the National Society of Home Builders.

SuperJANET See JANET.

superparamagnetic Phenomena which contribute to magnetization and signal decay of magnetically recorded information over time, thus limiting the useful lifespan of magnetic recording media. The density of recording information is related to the superparamagnetic effects as well, resulting in a practical superparamagnetic limit. Studies for arranging magnetic data in particle array systems to study superparamagnetic effects to develop practical schemes and optimize recording density are being carried out at the IBM Research labs.

superpose 1. To place or lay over, with or without contact with that which is overlaid. 2. To overlay upon another such that all like parts of the overlay coincide with the overlaid.

superposition principle A principle which can be applied to networked electrical circuits to solve current values in individual branches of the network. See Kirchoff's laws.

Supersparrow A wide area Web server load-balancing system (LBS) distributed as open-source software under a GNU Public License. Supersparrow, based on Border Gateway Protocol (BGP), was developed by Simon Horman and initially released in December 2000.

Supersparrow aids in distributing Web access to available servers on the Internet. In many circumstances, users are given the option to select *mirror* sites that include the same content stored on servers that may be faster or geographically closer to the user. Mirror sites are an especially welcome option to high-traffic primary sites. However, users have traditionally had to select a static mirror site manually from a list of Web links. It's time-consuming to read the lists and stats and sometimes difficult to determine the most promising option. Often some of the links are unavailable or the addresses out of date. Supersparrow aims to effectively automate the process of utilizing mirror sites using BGP to determine an efficient data path (least-cost path). BGP was selected because it has information on efficient paths to points on the Internet and provides for failure recovery.

Unlike the Linux Virtual Server (LVS), Supersparrow does not require a single contact point for incoming traffic. Supersparrow is compatible with VA Linux servers and can accommodate connections to Apache Web servers. See load-balancing system.

superstation A television broadcast station whose signal reaches a very wide audience by being retransmitted beyond what would be possible by standard airwaves. The extended viewing area is often reached by a satellite transmission further extended through cable.

supertrunk A high-end data transmissions cable system which carries multiple high-bandwidth services such as several video channels.

surf *colloq.* A common term for riding a wave, a technology, a trend, or other force or medium. *Channel surfing* describes a television watcher who uses a remote control device to skim programs from channel to channel, particularly during commercials. This is done with the hope of finding better programming (or as thumb exercises for couch potatoes). *Surfing the Net* means to travel, in the virtual sense, through the myriad resources and sites on the Internet, especially through the Web, a graphical interface to the Internet. See browser, Internet, World Wide Web.

surface plasmon resonance SPR. A quantum electro-optical effect resulting from the interaction of light with a metal surface. At a specific resonant wavelength equal to the quantum energy of plasmons (electron packets), photon energy is transferred to the plasmons.

SPR is a form of total internal reflectance (TIR) occurring at the interface between materials with differing refractive indexes. If a thin conducting film is placed between two optical media, surface plasmons in the film will couple resonantly with matching light frequencies. The resonance wavelength can be calculated by measuring the light reflected by the metal surface. Under conditions of resonance reflection, incident light is absorbed rather than reflected. See reflectometer, total internal reflection.

surface-array recording SAR. A process for accessing both sides of an optical disc simultaneously through independent read-write heads on both sides of the disc. By using simultaneous access by the read-write heads, rather than one or the other at one time, companies like Maxoptix (which has trademarked Surface Array Recording) have realized almost double the transfer rate and disc capacity of previous products. Thus, unlike some previous optical media devices (laserdisc players), users need not flip over the media to read the second side. See air-incident recording, overcoat-incident recording, substrate-incident recording.

surge Large, sudden changes in a circuit current or voltage. See burst.

surge protector, surge suppressor A device that conditions or filters electrical current to reduce power fluctuations. It is placed where it can provide protection to subsequent devices or components in a system. Ground lines are sometimes used to drain off the surge. Surge protectors may be built into a building's electrical system to drain off surges from voltage spikes or lightning strikes.

Some limited protection from lightning storms may be possible with a consumer surge protector, but a direct hit will likely damage the protector and interconnected systems. Surge protectors are recommended for laptops that get plugged into circuits of a dubious nature (motel or ferryboat sockets, etc.).

suspension insulator A type of high-voltage utility pole insulator patented by F. Locke in 1905. The suspension insulator was supported out and away from the pole or crossarm with wires rather than mounted on the top of the pole or on a crossbar. Sometimes more than one suspension insulator was mounted on a single pole. Historic Locke insulators were rigid fixtures screwed to a porcelain disc. Locke used eye hooks in a later 1910 patent. Locke's ideas were refined and improved over the years and suspension insulators became an important means to increasing the level of power that could be transmitted across overhead utility poles.

There were problems, however; suspension insulators with ceramic or cement cap-and-pin designs were heavy and high maintenance; they tended to deteriorate quickly. Thus, later suspension insulators were simpler, all-metal designs (though still relatively heavy), consisting of a series of spaced insulating rings protected by larger arcing rings to protect from electrical surges (e.g., lightning strikes). See strain insulator.

sustainable cell rate SCR. In ATM networking, the upper measure of a computed average rate of cell transmission over time. See the Appendix for expanded explanations of ATM.

SUT 1. System Under Test. See ATM. 2. Service User Table. A telephone use authorization term.

SVC See switched virtual connection.

SVD simultaneous voice and data.

SVGA See Super Video Graphics Array.

SVHS, Super VHS, Superior Video Home System A video recording and playback standard introduced in 1987 by JVC for the high-end consumer (prosumer) markets. The format was intended to be less expensive than commercial systems but significantly better quality than the VHS format widespread at the time. SVHS is downwardly compatible with VHS. VHS tapes can be recorded on SVHS systems, but not the reverse, as the image signal differs (audio is handled in the same way as VHS). The SVHS tapes have holes in them to enable record/playback equipment to recognize the cassette as SVHS.

SVHS more nearly represents the resolution of television broadcast signals and thus retains the clarity of the image far better than VHS. With a higher initial resolution, SVHS is also more suitable for video editing as less detail is lost in copying.

At the time SVHS was introduced, consumer VHS decks were in the $199 to $600 range, whereas SVHS sold in the $500 to $2000 range, a little expensive for the average consumer, but attractive to those with professional aspirations and limited budgets. SVHS is a 1/2-in. tape format that was less expensive than the 3/4-in. formats that dominated the commercial industry at the time of its release and SVHS gradually found a niche, especially with videophiles using desktop video systems such as the NewTek/Amiga Video Toaster.

By 2000, the cost of consumer SVHS decks had dropped to below $200, with TBC-equipped versions selling for under $500. At this point the format became very popular for desktop video and multimedia editing for the Web. Hi-8mm/8mm camcorders that have recently become popular usually have S-Video output ports. See Hi-8mm, S-Video.

Swelling Tape in Fiber Optic Cable

Swelling tape may be incorporated into fiber optic cables, especially those installed outdoors in humid areas, particularly if freezing occurs that might damage sensitive fiber filaments. In this assembly, the components include coated optical fibers (1), swelling tape (2), water-resistant tubing (3), a supporting strain relief layer, usually of aramid yarn (4), one or more rip cords to facilitate the unpeeling of the cable for making connections (5), a reinforcing strength member (e.g., steel), and the outer sleeve (7).

SVHS-C, Super VHS-Compact A compact (small format) version of SVHS designed to be used in portable devices such as camcorders. The image quality is higher than VHS, but the tape recording times are shorter than a standard cassette. In general, SVHS-C began to be overshadowed by Sony-developed Hi-8mm formats offering the benefits of small size, high quality, and longer tape recording times. Nevertheless, JVC continues to support and improve upon the SVHS-C product line, introducing digital options such as time-lapse and still shots. Standard 45-minute VHS-C can be used in JVC SVHS-C units. See SVHS.

swelling tape A tape used in cable assemblies to prevent water interference or damage to the inner conductive materials. Swelling tape may be semiconductive or nonconductive and both may be used in a single cable assembly in different layers of the assembly (usually with intervening layers). It is used in a variety of types of cables, especially outdoor cables, including power and fiber optic cables (power cables may have more than one swelling tape layer).

switch *n.* 1. A mechanical, electrical, or optical device that breaks or completes a path in a circuit, or changes the path. In fiber optic networks, considerable effort has been expended on developing optical switches to reduce the energy conversion and loss that occurs in hybrid optical/electrical switch systems. Bell Labs is one of the research centers delving into this area, with some success in developing optical switches based upon micro-electromechanical systems (MEMS) technologies. See switch, optical; switcher. 2. An electronic circuit designed to carry out a logic operation. New switches have capabilities that were once found only in routers, and some can do switching at the application level (fourth layer). 3. In software, a means to direct a routine; a branch.

switch hook See hook switch.

Fiber Optic Communications Switch

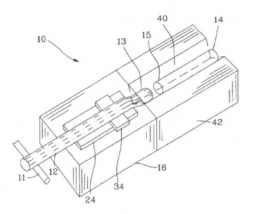

A patented fiber optic switching apparatus proposed by Rosete et al. in October 2000 to provide a direct means to switch signals on or off without lenses or stepping motors. An electrical charge applied to the electromagnet influences the switching member (24) and causes the ends of the connecting fibers (13 and 15) housed in the V-groove to come out of alignment, thus switching to an off state. Alternately, if two V-grooves are aligned one above the other, the fiber (12) can switch between to inputs (or outputs). [USPTO patent diagram.]

---

switch, optical A means to break, divert, or complete a transmission path through optical rather than electro-

mechanical components. Optical switches are often favored for optical networks to reduce the amount of energy conversion and loss that may occur in hybrid networks.

Bell Labs has developed a MEMS-based optical switch with response times below 70 μs using a mirror connected to an electrostatically-activated seesaw driven by a flat plate. Electrically stimulating the plate causes it to pull down and displace the mirror resulting in two states in which the optical signal can pass through or is deflected. By using a spectral grating to demultiplex the wavelengths and direct them to an array of mirrors, an optical add/drop multiplexer can be devised.

Historic Manual Communications Switch

This historic telephone switchboard cordboard shows the jacks and receptacles which were manually connected to complete a circuit between the caller and callee. Indicator lights helped operators keep track of calls in progress.

---

switchboard In its most general sense, any device into which a number of incoming and outgoing circuits are routed, where the routing of the individual circuit connections can be changed manually, mechanically, or electronically.

Human-operated telephone *cordboards* are probably the most picturesque of the various historic switching boards. The earliest ones required that a foot pedal be pumped to generate the power to ring the subscriber's phone. A manual telephone cordboard was often built into a wall, and was hand-connected with simple jacks and cables. In early telephone history, young men were hired to staff switchboards, as women were not permitted to work in most clerical positions, but the arrangement had problems. Some of the male employees were rude to callers, chewed tobacco while talking, and used excessive profanity. As a result, in the mid-1880s, women were hired, and eventually replaced men entirely until the late 1970s when a few male operators re-entered the field.

Most male and female switchboard operators are now being replaced by automated switchboard systems with voice recognition and touchtone menu dialing functions.

The large panels of switch connections in early cordboards and switchboards have been replaced by multiline phone consoles in many businesses, although some large phone installations or central switching services still have wall panels. Even here, human receptionists are becoming rare, with many small businesses adopting computer voicemail services instead. See Coy switchboard.

switchboard cable This has two meanings. It was originally a patch cable used in old manual cordboard telephone systems to patch two circuits together to create the end-to-end connection for a phone call. Now that manual switchboards have been replaced by automated switching systems, the switchboard cable is considered to be the one which connects a central office switchboard with an associated automated system, such as a computer.

Switched 56 The name of a 56-Kbps switched network voice/data service provided by some local telephone companies which allows calls among several points through one-pair or two-pair copper wires. Switched 56 can be used for voice, file transfers, Internet access, facsimiles, connections to other local area networks, and videoconferencing. This service is gradually being superseded by ISDN services. See DS-0.

Switched Multi-Megabit Digital Service SMDS. A high-speed wide area networking (WAN), connectionless, packet-switched cell relay transport service based on IEEE 802.6, offered by telephone service providers. It provides capabilities to interconnect LANs (Ethernet, Token-Ring, etc.) and WANs through public switched telephone networks (PSTNs).

SMDS can be integrated with transmission technologies such as ISDN, DS-x, and Frame Relay with associated bandwidths ranging from 56 or 64 kbps to 34 Mbps or more. It works with asynchronous, synchronous, and isochronous data and can be used over optical fiber. It can provide congestion control to protocols such as Frame Relay, which don't have congestion control as an intrinsic part of their specification or implementation. See cyclic reservation multiple access.

switched virtual connection, switched virtual circuit SVC. A generic term for a logical communications connection. In ATM systems, there are two types of SVCs: switched virtual path connection (SVPC) and switched virtual channel connection (SVCC). SVC provides on-demand connections between communicating end systems. Using signaling software, the Virtual Path Identifier/Virtual Channel Identifier (VPI/VCI) information will be dynamically allocated to the participating end systems.

Switched Voice Service SVS. The standard service offered with FTS2000. See FTS2000.

switcher An audio/video component that provides easy reconfiguration of several circuits. In a sense, the button on the receiver that enables selecting CD, phono, or tape is a switcher, although the meaning is more often ascribed to a separate component with a number of connectors, inputs, and outputs. See A/B switchbox, switch.

switching There are three types of switching commonly used in telecommunications networks: message switching, circuit switching, and packet switching. In message switching, the entire message is relayed, intact, through a variety of nodes or service points, from the sender to the recipient. Circuit switching is commonly used in end-to-end direct communications, as in telephone connections. Packet switching involves the segmentation, routing, and reassembly of communications so different parts of the messages may be transmitted at different times and through different routes. While this may not sound very efficient, it is an excellent way to manage data in a large, dynamic, distributed environment, and provides many possibilities for data sharing, representation, authentication, and filtering as desired. The Internet is built upon packet switching concepts. See circuit switching, packet switching, switch.

switching simulation The simulation on a computer, usually with graphical output, of the circuitry and operations of a switching system. Hayward and Bader did early computer simulations of telephone switching networks in 1955.

SWR See standing wave ratio.

SXGA super-extended graphics array.

symbol A character, icon, or other agreed-on abbreviated representation, useful in representing objects, quantities, languages, arithmetic and logic operations, rules, layout schemes, qualities, sounds, ideas, and others. Examples of symbols include street signs, logos, musical notes, electronics diagram elements, computer buttons and icons, words, punctuation marks, and arithmetic operators.

symbolic Representative of something else, or of some greater meaning, usually in an abbreviated form. See symbol.

symbolic code A computer code which represents programs in source language, with symbolic names and addresses, in contrast to machine code, which has hardware-specific names and addresses. Use of mnemonic symbols in higher level languages aids in programming and debugging, and symbolic names and addresses can be used to increase portability to other platforms. See assembly language.

symbolic debugger An essential software programming and debugging tool used to control and monitor applications under development. A debugger facilitates stepping through execution of the code, setting breakpoints and temporary branches, determining and changing the contents of variables, and viewing the design and functionality of software-in-progress. A symbolic debugger eases the task of finding code segments (many current software applications have tens of thousands and even hundreds of thousands of lines of code) by jumping to a symbol, such as a label name, to avoid searching through cryptic code.

symbolic language A computer programming language that uses mnemonic symbols, rather than machine code or actual hardware names and addresses, in order to make the process of creating the code more comfortable for humans and easier to read and debug. See symbolic code.

symbolic logic A written symbol language developed to express logical and mathematical concepts and arguments in a way that is more specific to itself and less ambiguous than natural human languages.

sync advancer A video component for resolving horizontal and burst phase signals for multiple camera signals with different characteristics due to cabling. In video, where signal pulse timing is important for combing signals, pulse discrepancies can arise from simple things like cables of uneven lengths, sometimes even resulting in delays that must be resolved before further processing the signals.

sync separator A video device for taking a composite video input and deriving the characteristics of the signal (such as composite sync, vertical sync, horizontal sync, back porch, field, ID) so that the signal can be processed to synchronize with other video components. For example, a synch separator can be used to provide loop-through viewing from a camera to a monitor while synching the camera feed to another component. The sync separator provides the pulses required by other components and may or may not be genlockable. Different models are available for positive-going or positive-going video (or both). See genlock, negative-going video.

synchronize 1. To cause to occur at the same instant in time. 2. To precisely match two waves or two functions. 3. To assess the characteristics of an input signal and process it to conform to characteristics expected by a receiving link/component. 4. To assess the characteristics of two or more input signals and process them to conform to timing, pulse, or other characteristics in order to form a single signal or multiple signals with compatible characteristics.

synchronous Signals with the same timing reference and the same frequency. See isochronous, asynchronous.

Synchronous Data Link Control SDLC. In IBM Systems Network Architecture (SNA) systems, SDLC is a bit-oriented, link-level protocol that provides a means of moving data between Network Addressable Units (NAUs). The Data Link Control layer lies between the higher layers and the Physical Control layer and communications links, and passes information through.

SDLC is a subset of the High-level Data Link Control (HDLC) protocol.

SDLC is packet-oriented, with each frame comprising a header, information, and trailer. It transparently provides flow control, multipoint addressing, error detection, and multimessage capabilities. See Systems Network Architecture.

Synchronous Digital Hierarchy SDH. A fiber optics transmission technology for efficient transport of digital signals. Different versions of SDH are defined; those used within North America and Japan and those used in Europe and elsewhere.

The North American version of this general technology is called SONET and is standardized by the American National Standards Institute (ANSI). SDH standards are coordinated by the International Telecommunications Union (ITU-T) and published primarily as G Series Recommendations documents. See G Series Recommendations, SONET.

Synchronous Optical Network See SONET.

Syncom-4 Communications Satellite

The Syncom IV-4 communications satellite as it is released from the payload bay of the Space Shuttle Discovery. *It was deployed from the Shuttle with a disc toss motion in August 1985. [NASA image]*

Syncom 3 The first geostationary satellite, designed for telecommunications use, launched on 19 August 1964 by the United States.

syntax 1. The rules of structure or grammar for a language, natural or computer. 2. In programming, the words and symbols valid within the structure and scope of a computer language.

syntax error An error in a programming statement which indicates an unrecognized word, symbol, or structure. Syntax errors are usually flagged and displayed by debuggers and compilers so that the error can be corrected. See syntax.

synthesized voice A mechanically or electronically generated speaking system. Synthesized voices are now used on phone systems, computers, and certain public address systems. The voice may be constructed from recordings of natural human voices, pieced together electronically from sound samples, or may be entirely synthetic. See speech synthesis.

Syslog Protocol A Berkeley Software Distribution (BSD) protocol that provides service options related to the propagation of event messages over a network; Syslog Protocol has now been ported to other operating systems as well. The protocol was submitted as an Information RFC by C. Lonvick in August 2001. Syslog messages may have varying content in varying format, since they originate on a number of types of platforms. Syslog services originate, are relayed, and are received in device, relay, or collector modes. A sender transmits a message without knowledge of whether the next leg is a relay or recipient. A collector may also be a relay and may selectively collect or relay multiple messages. A relay may alter the format of a message, but it is generally not recommended unless it is unable to discern the proper implementation of the format, in which case it should modify it. Messages are transported over User Datagram Protocol (UDP) using port 514 and are usually received over port 514. See RFC 3164.

sysop system operator. The sysop, or systems administrator, sometimes also called the *super user*, due to his or her higher access privileges and power over other accounts on a system. The system operator is a technical expert with high security privileges, managing a bulletin board system or computer network. On small systems, like BBSs, these tasks may be performed by one person. On medium-sized systems, sometimes assisting sysops are assigned intermediary privileges, between those of the super user and regular system users. Larger systems often split the installation, security, administration, file management, diagnostic, and tuning responsibilities of a network among a number of system operators, or may even have an entire facility devoted to the administration of the network.

system disk A disk that includes "boot" information, that is, low-level operating information, from which a computer system can be started. This may be a floppy drive, hard disk, CD-ROM, or other disk with system files somewhat transparent to the user. Without certain system files, a computer cannot configure itself to recognize peripherals, available memory, monitor types, etc. See bootstrap, operating system.

system integrator A commercial vendor offering a variety of network design and implementation services according to the configuration needs of various customers.

system reliability architecture SRA. Systems designed as fault tolerant and reliable, and which function even while undergoing maintenance checks and procedures. SRA implies systems which incorporate redundancy, the ability to *hot swap* components, fast recovery from power failures, and online upgrading of software.

System Services Control Point SSCP. A point in a network host system within the Virtual Telecommunications Access Method (VTAM) that initiates host applications, so that they can be associated with dependent logical units (LUs) and connections initiated and terminated. SNA systems use the SSCP to set up terminal sessions and, more recently, to link with sessions with other systems such as Unix, for example. In SNA, one SSCP in the domain handles this type of interoperability. It consists of software in the host processor for handling connections to the Network Control Program (NCP) which, in turn, manages data link protocols and routing functions.

The SSCP initiates sessions by issuing an Activate Physical Unit (ACTPU) command and subsequent Activate Logical Unit (ACTLU) commands. The SSCP communicates between SLUs and the host application. If the necessary resources are not available, an error command to the SLU is issued to prevent or abort establishment of a session.

Systems Network Architecture SNA. One of the first significant layered architectures, introduced in 1974 by IBM Corporation.

Layered architectures like SNA were developed when computers became smaller and less expensive, resulting in an increase in mass production and a greater variety of hardware configurations and operating systems. Thus, interconnectivity and specialization challenges were posed and new markets opened up. Layered architectures helped resolve these needs.

Systems Network Architecture (SNA) provides a cohesive way for users to communicate between systems for transmitting and receiving, by specifying the operating relationships of various components of the different systems. To achieve this, various communications peripherals, i.e., adapters, modems, data encryption devices, etc. are designed to be consistent with the implementation of the SNA specification.

SNA came out of the mainframe environment and thus was designed as a star topology host environment for supporting multiple terminals. However, more recent additions make it possible to support multihost, multidomain networks with peer-to-peer distributed computing topologies, as well.

SNA shares many common overall concepts with the Open Systems Interconnection (OSI) hierarchically layered network model, which evolved at about the same time. Both are seven-layer models, arranged from the physical and data link layers at the lower end to presentation and transaction (application) layers at the higher end. While the two systems are not directly compatible, there is now some SNA support for OSI protocols to improve interoperability.

In brief, the layers described for SNA from high level to low level include:

Layer	Functions
transaction services	application program functions for application intercommunication
presentation services	content of data messages is defined
data flow control	traffic flow and logical data grouping between two end users
transmission control	retransmission and error recovery, flow control, end-to-end acknowledgments
path control	data routing, parallel transmission over multiple links, segmentation and reassembly of data packets
data link control	transmissions, retransmissions across a link
physical	the physical media over which data are transported

See System Services Control Point, Token-Ring.

T 1. *symb.* tera- 2. *abbrev.* terminal. 3. *abbrev.* tip (as in tip and ring). 4. *abbrev.* trunk.

T connector A generic name for many types of cable connectors shaped roughly like the letter "T." This is a common shape for splitters, Ethernet cabling, certain types of terminators, and adaptors. The purpose of a T connector is similar to that of a Y connector: to join two inputs into one or to split one input into two. If it is a terminating connector, its function is to electrically indicate the end (terminating point) of a chainable data connection. See terminator.

T interface In ISDN, a number of reference points have been specified as R, S, T, and U interfaces. To establish ISDN services, the telephone company typically has to install a number of devices to create the all-digital circuit connection necessary to send and receive digital voice and data transmissions.

The T interface is the point between the phone company's switching device that serves the subscriber and the subscriber's building wiring. Thus, the T interface is the point between the central office switch and the user's phone connection. See ISDN interfaces for a diagram and relationships.

T Series Recommendations A series of ITU-T recommended guidelines for terminals for telematic services. These guidelines are available for purchase from the ITU-T. Since ITU-T specifications and recommendations are widely followed by vendors in the telecommunications industry, those wanting to maximize interoperability with other systems need to be aware of the information disseminated by the ITU-T. A full list of general categories is listed in Appendix C and specific series topics are listed under individual entries in this dictionary, e.g., S Series Recommendations. See T Series Recommendations chart.

T1, T-1 A communications system that can be carried over ordinary twin cable pairs, or fiber optic, that provides significant speed and bandwidth improvements over earlier technologies.

The T1 time division multiplexing (TDM) pulse code modulation (PCM) system is capable of carrying multiple simultaneous conversations (24 over twin cable pairs), and began to be incorporated into central office trunk switching technologies in the 1960s with more significant, widespread implementation to subscribers in 1982.

T1 digital communications represented a significant change over existing analog communications systems. Pulse code modulation (PCM) was pioneered in the 1940s and 1950s, but it was not until the early 1960s that some practical success was achieved. It was then implemented on T1 lines developed in the late 1950s by AT&T. Proponents wanted the improved transmissions technologies to be compatible with existing switching systems, potentially saving billions of dollars by using, rather than replacing, central office switching circuitry.

The capacity of T1 was originally stated as 1.544 Mbps (U.S., Canada, Japan), although European ITU-T standard implementations are faster, 2.048 Mbps, and upper limits tend to change as more efficient techniques are incorporated to improve the throughput of a system as a whole. It is a low loss transmissions system when delivered over fiber optic cable, but is subject to crosstalk in long metal wire installations.

While high-capacity 22-gauge is preferable for T1 transmissions, the system is not limited to this and can work with a number of paper or plastic insulated cable pairs, or staggered twist cable in use in telephone systems for decades. A single cable can handle up to almost 5000 channels.

Due to its cost, T1 is still primarily used in large installations, phone trunks, government installations, campus backbones, and medium and large enterprise networks, but installation and usage costs are still dropping and T1 may soon be accessible to small businesses, as well. See T-carrier.

T-Bone A freely distributable Java-based distributed network broadcasting system that permits interconnection of JavaBeans on different systems. It can be used for remote monitoring and control, push channels on the Internet, stock ticking feeds, and online discussions. See Java, JavaBeans.

T-carrier The generic term for T-1, T-2, etc. communications technologies standardized for high-speed data transmissions in North America. It served as the basis for the European E-carrier and the Japanese J-carrier systems, which are similar in concept, but somewhat different in implementation details. T-carrier systems are a generalized technology suitable for the transmission of voice, data, facsimile, multimedia audio and visual information over one system. See T1. See T- Carrier Transmission Systems chart.

T-CCS Transparent Common Channel Signaling. See Common Channel Signaling.

T/R transmit/receive.

TAB See tone above band.

table Collection of ordered data, often stored in arrays or printed in columnar form. Routing tables are an instrinsic aspect of many network linking devices.

TABS Telemetry Asynchronous Block Serial Protocol. A medium-fast AT&T network protocol.

Tabulating Machine Company Herman Hollerith, who developed Hollerith punch cards to store census data from the 1890 U.S. census, founded the Tabulating Machine Company in 1896 to market his products. Eventually, the company was sold out to the Computer-Tabulating-Recording company, which, in turn, became International Business Machines (IBM) in the 1920s.

TACD Telephone Area Code Directory. See North American Numbering Plan. See Appendix I.

TACS Total Access Control/Communications System. A first-generation, analog cellular FM-based radio system introduced in the United Kingdom by Cellnet and Vodaphone in the early 1980s. It is similar to the U.S. AMPS system. TACS used frequency shift keying (FSK) signaling with each TACS call assigned to a different frequency through frequency division multiple access (FDMA). There are now a number of variations of the original TACS system, including Extended TACS (ETACS) which used additional frequencies, International TACS (ITACS), International Extended TACS (IETACS), Narrowband TACS (NTACS), and Japan TACS (JTACS).

Tag Distribution Protocol TDB. A Cisco network protocol for building routing databases for the handling of tagged datagrams that are accessed by tag switches and tag edge routers. The tag bindings established by routers are communicated to neighboring routers through TDB. See tag switching.

Tag Image File Format See TIFF.

tag switching A Cisco proprietary data link layer/network layer routing/switching architecture intended to provide performance and scalability through existing network infrastructures.

Tag switching is a method of using *tag edge* routers and assigning software *tags* to each IP datagram in a sequence in order to identify router paths. In order to reduce the time needed for each router to send datagrams across the network, the tag edge routers append a special string of bits, called the *tag*, to the datagrams before they are transmitted across the backbone. The tag provides routing information to other routers (tag switches) so that they are freed from table lookups and processing. The tag does not stand for a specific path through the network but rather represents a general class of forwarding. Tag switching is similar to IP switching, except that nonstandard tag bits are appended. See IP switching, label switching, Tag Distribution Protocol.

TAI See International Atomic Time.

tail 1. A short, slender connecting length of fiber filament that may be sheathed. The tail is often associated with components intended to be connected in the field. Fiber tails are usually capped as particles and scratches can significantly impede performance. See pigtail. 2. A sheathed fiber bundle connecting an illumination source to a light fixture up to a practical limit of about 30 feet if no special added links or amplifiers are used.

tail circuit A final segment in a connection between a central switching location and the subscriber.

Tainter, Charles Sumner (1854–1940) An American scientific instrument-maker and inventor. Tainter, who was largely self-taught, was a significant collaborator with A. Graham Bell on the invention of the Photophone, a device for wirelessly conducting sound through light as a transmissions medium. He also worked with Bell on the development of the Graphophone, an improved phonograph system. Tainter took the Graphophone idea a step further, developing it into a dictation system. He began to suffer ill health and lost his lab to fire in 1897. He continued working on audio technologies, inventing a means of duplicating phonograph records. A number of Tainter's documents are housed in the Smithsonian National Museum of American History. See Photophone, selenium.

Type	Signal	Bandwidth	Typical cable	Notes
	DS-0	64.000 Kbps	---	
T1	DS-1	1.544 Mbps	19, 22, or 24 gauge	Originally developed for digital voice, also used for data communications. European E1 standard is similar.
T1C	DS-1C	3.152 Mbps	19, 22, or 24 gauge pairs	
T2	DS-2	6.312 Mbps	low capacitance	
T3	DS-3	44.736 Mbps	fiber optic or microwave	
T4	DS-4	274.760 Mbps	fiber optic or microwave	High-speed, high bandwidth applications.

T-Carrier Transmissions Systems Overview

ITU-T T Series Recommendations

General, Classifications

T.0   Classification of facsimile terminals for document transmission over the public networks

T.1   Standardization of phototelegraph apparatus

T.2   Standardization of Group 1 facsimile apparatus for document transmission

T.3   Standardization of Group 2 facsimile apparatus for document transmission

T.4   Standardization of Group 3 facsimile terminals for document transmission

T.6   Facsimile coding schemes and coding control functions for Group 4 facsimile apparatus

T.80   Common components for image compression and communication – basic principles

T.121   Generic application template

T.300   General principles of telematic interworking

T.400   Introduction to document architecture, transfer and manipulation

Standardized Test Charts, Images

T.20   Standardized test chart for facsimile transmissions

T.21   Standardized test charts for document facsimile transmissions

T.22   Standardized test charts for document facsimile transmissions

T.23   Standardized colour test chart for document facsimile transmissions

T.24   Standardized digitized image set

Multipoint Transmissions

T.122   Multipoint communication service – service definition

T.125   Multipoint communication service protocol specification

T.126   Multipoint still image and annotation protocol

T.127   Multipoint binary file transfer protocol

T.128   Multipoint application sharing

Security

T.36   Security capabilities for use with Group 3 facsimile terminals

Character Sets, Encodings, Imaging

T.50   International Reference Alphabet (IRA) (Formerly International Alphabet No. 5 or IA5) – Information tech. – 7-bit coded character set for information interchange

T.51   Latin-based coded character sets for telematic services

T.52   Non-Latin coded character sets for telematic services

T.53   Character coded control functions for telematic services

T.61   Character repertoire and coded character sets for the international teletex service

T.351   Imaging process of character information on facsimile apparatus

Terminal Characteristics

T.561   Terminal characteristics for mixed mode of operation MM

T.562   Terminal characteristics for teletex processable mode PM.1

T.563   Terminal characteristics for Group 4 facsimile apparatus

T.571   Terminal characteristics for the telematic file transfer within the teletex service

Procedures, Routing, Miscellaneous

T.10   Document facsimile transmissions on leased telephone-type circuits

T.10bis   Document facsimile transmissions in the general switched telephone network

T.11   Phototelegraph transmissions on telephone-type circuit

T.12   Range of phototelegraph transmissions on a telephone-type circuit

T.15   Phototelegraph transmission over combined radio and metallic circuits

T.30   Proced. for document facsimile transmission in the general switched telephone network

T.31   Asynchronous facsimile DCE control – Service Class 1

T.32   Asynchronous facsimile DCE control – Service Class 2

T.33   Facsimile routing utilizing the subaddress

T.35   Procedure for the allocation of ITU-T defined codes for nonstandard facilities

T.37   Proced. for the transfer of facsimile data via store-and-forward on the Internet

T.38   Proced. for real-time Group 3 facsimile communication over IP networks

T.39   Application profiles for simultaneous voice and facsimile terminals

T.42   Continuous-tone colour representation method for facsimile

T.43   Colour and grayscale image representations using lossless coding scheme for facsimile

T.44   Mixed raster content (MRC)

T.45   Run-length colour encoding

T.60   Terminal equipment for use in the teletex service

T

ITU-T T Series Recommendations, cont.	
Procedures, Routing, Miscellaneous, cont.	
T.62	Control procedure for teletex and Group 4 facsimile services
T.62bis	Control procedure for teletex and G4 facsimile services based on recommendations X.215 and X.225
T.63	Provisions for verification of teletex terminal compliance
T.64	Conformance testing proced. for the teletex recommendations
T.65	Applicability of telematic protocols and terminal characteristics to computerized communication terminals (CCTs)
T.66	Facsimile code points for use with recommendations V.8 and V.8 bis
T.70	Network-independent basic transport service for the telematic services
T.71	Link access protocol balanced (LAPB) extended for half-duplex physical level facility
T.85	Application profile for recommendation T.82 – progressive bi-level image compression (JBIG coding scheme) for facsimile apparatus
T.86	Information tech. – digital compression and coding of continuous-tone still images: registration of JPEG Profiles, SPIFF Profiles, SPIFF Tags, SPIFF Color Spaces, APPn Markers, SPIFF Compression types and Registration Authorities (REGAUT)
T.87	Information Tech. – lossless and near-lossless compression of continuous-tone still images – Baseline
T.88	Information tech. – coded representation of picture and audio information – lossy/lossless coding of bi-level images
T.89	Application profiles for recommendation T.88 – lossy/lossless coding of bi-level images (JBIG2) for facsimile
T.90	Characteristics and protocols for terminals for telematic services in ISDN
T.100	International information exchange for interactive videotex
T.101	International interworking for videotex services
T.102	Syntax-based videotex end-to-end protocols for the circuit mode ISDN
T.103	Syntax-based videotex end-to-end protocols for the packet mode ISDN
T.104	Packet mode access for syntax-based videotex via PSTN
T.105	Syntax-based videotex application layer protocol
T.106	Framework of videotex terminal protocols
T.107	Enhanced human-machine interface for videotex and other retrieval services (VEMMI)
T.120	Data protocols for multimedia conferencing
T.Imp120	Implementors' Guide for SG 16 recommendations
T.123	Network-specific data protocol stacks for multimedia conferencing
T.124	Generic conference control
T.134	Text chat application entity
T.135	User-to-reservation system transactions within T.120 conferences
T.136	Remote device control application protocol
T.137	Virtual meeting room management – services and protocol
T.140	Protocol for multimedia application text conversation
T.150	Telewriting terminal equipment
T.170	Framework of the T.170-Series recommendations
T.171	Protocols for interactive audiovisual services: coded representation of multimedia and hypermedia objects
T.172	MHEG-5 – support for base-level interactive applications
T.173	MHEG-3 script interchange representation
T.174	Application programming interface (API) for MHEG-1
T.175	Application programming interface (API) for MHEG-5
T.176	Application programming interface (API) for Digital Storage Media Command and Control (DSM-CC)
T.180	Homogenous access mechanism to communication services
T.190	Cooperative document handling (CDH) – framework and basic services
T.191	Cooperative document handling (CDH) – joint synchronous editing (point-to-point)
T.192	Cooperative Document Handling – complex services: Joint synchronous editing and joint document presentation/viewing
T.200	Programmable communication interface for terminal equipment connected to ISDN
T.330	Telematic access to interpersonal messaging system
T.390	Teletex requirements for interworking with the telex service
T.501	Document application profile MM for the interchange of formatted mixed mode documents
T.502	Document application profile PM-11 for the interchange of simple structure, character content documents in processable and formatted forms

Procedures, Routing, Miscellaneous, cont.

T.503   A document application profile for the interchange of Group 4 facsimile documents

T.504   Document application profile for videotex interworking

T.505   Document application profile PM-26 for the interchange of enhanced structure, mixed content documents in processable and formatted forms

T.506   Document application profile PM-36 for the interchange of extended document structures and mixed content documents in processable and formatted forms

T.510   General overview of the T.510-Series Recommendations

T.521   Communication application profile BT0 for document bulk transfer based on the session service

T.522   Communication application profile BT1 for document bulk transfer

T.523   Communication application profile DM-1 for videotex interworking

T.541   Operational application profile for videotex interworking

T.564   Gateway characteristics for videotex interworking

T.611   Programming Communication Interface (PCI) APPLI/COM for facsimile Group 3, facsimile Group 4, teletex, telex, email and file transfer services

Information Technology, IT ODA

T.81   Information tech. – digital compression and coding of continuous-tone still images – requirements and guidelines

T.82   Information tech. – coded representation of picture and audio information – progressive bi-level image compression

T.83   Information tech. – digital compression and coding of continuous-tone still images: compliance testing

T.84   Information tech. – digital compression and coding of continuous-tone still images: extensions

T.411   Information tech. – Open Document Architecture (ODA) and interchange format: introduction and general principles

T.412   Information tech. – Open Document Architecture (ODA) and interchange format: document structures

T.413   Information tech. – Open Document Architecture (ODA) and interchange format: abstract interface for the manipulation of ODA documents

T.414   Information tech. – Open Document Architecture (ODA) and interchange format: document profile

T.415   Information tech. – Open Document Architecture (ODA) and interchange format: open Document Interchange Format (ODIF)

T.416   Information tech. – Open Document Architecture (ODA) and interchange format: character content architectures

T.417   Information tech. – Open Document Architecture (ODA) and interchange format: raster graphics content architectures

T.418   Information tech. – Open Document Architecture (ODA) and interchange format: geometric graphics content architecture

T.419   Information tech. – Open Document Architecture (ODA) and interchange format: audio content architectures

T.421   Information tech. – Open Document Architecture (ODA) and interchange format: tabular structures and tabular layout

T.422   Information tech. – Open Document Architecture (ODA) and interchange format – identification of document fragments

T.424   Information tech. – Open Document Architecture (ODA) and interchange format: temporal relationships and non-linear structures

Document Transfer & Manipulation (DTAM)

T.431   Document Transfer And Manipulation (DTAM) – services and protocols – introduction and general principles

T.432   Document Transfer And Manipulation (DTAM) – services and protocols – service definition

T.433   Document Transfer And Manipulation (DTAM) – services and protocols – protocol specification

T.434   Binary file transfer format for the telematic services

T.435   Document Transfer And Manipulation (DTAM) – services and protocols – abstract service definition and proced. for confirmed document manipulation

T.436   Document Transfer And Manipulation (DTAM) – services and protocols – protocol specifications for confirmed document manipulation

T.441   Document Transfer And Manipulation (DTAM) – operational structure

T

Talbot effect In observing an image through a magnifying glass that appears through a diffraction grating, the image appears to be in focus whether the magnifier is held close to the image at the normal focusing distance or whether it is held farther away where the image would be expected to be blurred and at which the grating itself appears blurred. Talbot also observed alternating bands of complementary colors, depending upon the distance of the magnifying lens from the grating surface. As the magnifier was moved farther from the image surface, the sequence of observed colors repeated in a regular pattern and the diffraction grating would become alternately focused, unfocused, and focused again at particular distances. As it turns out, the Talbot effect is doubly periodic for the patterns across the grating and the viewing distance from the grating. In other words, the wave fields exhibit lateral and longitudinal periodicity and create a near-field "self-imaging" effect.

W. Henry F. Talbot demonstrated his observations to scientific societies and followed up the demonstrations with an article on optical science, published in 1836. As interesting and important as his observations were to optical science, they didn't receive much attention until Lord Rayleight provided mathematical interpretations of the Talbot effect (sometimes called Talbot distance), in 1881. This rediscovery and mathematical description of the effect made it possible to design and fabricate different types of diffraction gratings through photographic film processes. Surprisingly, very few scientists followed up on this research and it languished until recent advancements in optics highlighted its importance.

The Talbot effect is now of interest in theoretical and applied optics. It is exploited for testing and calibration purposes and the production of gratings for scientific instruments such as interferometers. Optical coupling in laser arrays may be based upon the Talbot effect. By placing a laser array in a Talbot cavity, the diffractive radiant energy may be laterally phase locked. This, in turn, has the potential to improve the efficiency and density of phased arrays and amount of light emitted from these components. See diffraction grating, Ronchi grating.

Talbot, William Henry Fox (1800–1877) An English inventor who developed a salt and silver nitrate photographic paper print process in the early 1930s. Photos up to this time had been printed on ceramic and silver-plated copper. Talbot was able to produce photographic silhouettes that he called *sciagraphs*, by embedding light-sensitive chemicals in the paper and exposing parts of it to light. He further realized that an image created on a glass plate could be replicated many times in the same manner, a variation on the concept of steel engravings (an early type of manual replicating process). He also made important observations in optics related to diffraction and polarization that he demonstrated and recorded.

The Fox Talbot Museum of Photography has a project underway to develop a searchable archive of the ca. 10,000 letters that make up Henry F. Talbot's correspondence. See Daguerre, L.J.M; Talbot effect.

talking battery In historic telephone central offices, a 24- or 48-V battery supplying power for a phone conversation. Around 1893, these were replaced by *common batteries* at the central office which, in turn, provided subscribers with *talking batteries*. See battery.

tandem Two, dual, pair. Acting together, in conjunction with, partnership.

Tandem Connection TC. Tandem connections, that is, paired or redundant connections, are used at the discretion of network carriers, and software with TC are mostly interoffice network applications rather than general public subscriber network applications. In SONET, the Tandem Connection layer is optional.

Tandem Connection Overhead TCO. In SONET, an optional overhead layer between the Line and Path layers as defined in ANSI T1.105. The layer deals with the reliable network transport of Path layer payload and its associated overhead.

TANE The Telephone Association of New England. A regional association providing information, education, and support to its membership. http://www.tane.org/

tangent galvanometer An early current-detecting instrument employing a card to record the degree of deflection. This type of galvanometer is subject to interference from the Earth's magnetic field. See galvanometer.

TAO Project TAO is part of the Satori project at Washington University, being developed by the Distributed Object Computing group, funded partly by the DARPA Quorum program. It is a high-performance, realtime Object Requester Broker (ORB) designed to provide end-to-end network quality of service (QoS) guarantees to applications by integrating CORBA middleware with operating system input/output subsystems, communications protocols, and network interfaces. TAO is freely distributable to researchers and developers. See CORBA.

tap A splitting/joining component in which each input (or output) is coupled with two or more outputs (or inputs), at least one of which is intended for "tapping in" to the system through a probe (network analyzer). Tap assemblies are generally designed to facilitate network monitoring without disrupting the normal functioning of the network.

tap, fiber optic Taps for coupling into and monitoring data or power in a signaling/communication system. In hybrid electrical optical networks, such as fiber-based cable TV/modem services, taps can be used to monitor energy distribution, network efficiency, quality of service, reliability, and bandwidth allocation. Some sources refer to the link between a cable TV loop and local neighborhood drops as a tap, but this is probably better described as a drop connector or distribution device, rather than a tap, unless it also includes monitoring components.

Taps are commercially available for single- and multimode fiber optic networks and can be purchased with commonly available connectors. Some models are polarization-maintaining and may optionally include isolating components. Designs for monitoring systems in which test points are distributed throughout the network to isolate specific parts of the link have been proposed. Some Ethernet switches will include an RF-45 port that enables a computer to be tapped into the network for systems analysis and maintenance. Most tap assemblies are palm-sized or a little larger but innovative small-format taps that combine the tap coupler within the body of a standard fiber optic connector are being pioneered by companies such as the FONS Corporation.

Taps may be used where local drops from a loop or backbone cable are connected for sharing data from a single transmission source, or may be inserted at other points in the link to monitor network transmissions (e.g., a fiber-based network running in full-duplex mode). Fiber optic transmission taps are designed to split or join the light signal with minimum loss.

Taps generally do not require electrical power unless extra monitoring or peripheral link electronics are included. Most taps with extra monitoring electronics will continue to couple the lightguide inline whether or not the monitoring electronics are active. A tap may be a permanent component in a network link or may be temporarily inserted for testing or maintenance assessments. The more demanding the application (e.g., Gigabit-speed computer networks) the more precise the coupling in the tap must be.

tap loss In a tap coupler, the loss of power between the input power and the output power related to the physical coupling and assembly. See tap.

tap port The port on a coupler that has the lower power of two unequal outputs.

tap test A means of locating an object or component by tapping around the general region of the object and feeling its location or inferring the location from changes in sound or other properties. A common tap test is locating a stud within a wall by tapping across the wall in a horizontal direction until a deeper, more solid, less hollow sound indicates the presence of something denser behind the area tapped.

When fiber optic sensors are embedded within components, it is sometimes difficult to locate the position of the sensor. By performing a tap test at the same time as watching sensor output, it may be possible to locate the sensor but may be difficult to determine its orientation.

TAPI See Telephony Application Programming Interface.

tariff Scheduled rate or charge between a carrier and its subscribers, usually published, and sometimes regulated by government agencies.

TAS Telecommunication Authority of Singapore.

taut sheath splicing The process of splicing a wire or fiber optic joint that does not have much give in it for maneuvering. This occurs in the field where installed cables are repaired and maintained after they have been secured to poles, towers, walls, floors, or within piped conduits such that there is little slack for working with the fiber. See fusion splicing.

TAXI Transparent Asynchronous Transmitter/Receiver Interface.

Taylor, A.H. A member of the U.S. Naval Research Laboratory who observed in 1922 that a radio echo from a steamer could potentially be used to locate a vessel. This observation was not put into practical use until some years later. See radar, sonar.

TBOP See Transparent Bit Oriented Protocol.

Tc See Committed Rate Measurement Interval.

TCAP See Transaction Capability Application Part.

TCI Telecommunications, Inc.

TCIF Telecommunications Industry Forum.

TCM trellis code modulation. See trellis coding.

TCO See Tandem Connection Overhead.

TCP/IP Transmission Control Protocol/Internet Protocol. An internetworking transmissions protocol combination developed in the 1970s on the ARPANET to enable the intercommunication of various types of computers across wide area networks (WANs). It was widely adapted by educational institutions by the 1980s and by corporations by the 1990s. Although it appeared for a time that Open Systems Interconnection might overtake TCP/IP, it has now become an international standard which has been implemented on most microcomputers since the mid-1990s.

TDC See time-to-digital converter.

TDM See time division multiplexing.

TDMA See time division multiple access.

TDRSS Tracking and Data Relay Satellite System.

TDS See Terrestrial Digital Service.

TDU 1. tape drive unit. 2. See Topology Database Update.

te wave transverse electric wave.

Team OS/2 A strong, independent, international support and advocacy group for IBM's OS/2 (Operating System/2). It provides education, demonstrations, resources, Web links, and other services for OS/2 users and the general community. See OS/2. http://www.teamos2.org/

Technical and Office Protocols TOP. A protocol development effort to support the needs of engineering and office environments. TOP was initiated by Boeing, and is now part of the Manufacturing Automation Protocol/Technical and Office Protocols (MAP/TOP) users' group. TOP was designed to conform to the Open Systems Interconnection (OSI) model.

Technical Reference Model TRM. A common framework and vocabulary for creating and communicating digital information services infrastructures, components, and their relationships in order to facilitate the development of systems that can intercommunicate. Various prominent bodies, including the American National Standards Institute (ANSI), the International Telecommunication Unions (ITU), the Institute of Electrical and Electronics Engineers (IEEE), the International Organization for Standardization (ISO), and others have made efforts to provide basic frameworks for the development of consistent, interoperable, multiplatform environments, thus providing flexibility in the choice of platform while still maintaining a means to exchange data. The TRM essentially consists of three main entities, the Application Programming entity, the Application

T

Platform entity, and the External Environment. Further, there are two interfaces, thus comprising five basic aspects. Providing a data link between the Application Programming and Application Platform entities is the Applications Programming Interface. Between the Application Platform entity and the External Environment is the External Environment Interface. See Applications Programming Interface, External Environment Interface.

Technical Specification TS. In the Internet Standards Process, a TS is any formal description of a convention, format, procedure, protocol, or service. The description may be complete in itself, or may contain references to other specifications. Certain conventions guide the general format of a TS, and a statement of scope and intent for use is required, but a TS does not specify its application within the context of the Internet; this is defined in an Applicability Statement. See Applicability Statement.

Technology Policy Working Group TPWG. A working group within the Committee on Applications and Technology founded in the mid-1990s. The TPWG addresses broadbased, overlapping technology issues related to interoperability and scalability of new telecommunications and information services. The group fosters partnership and cooperation between industry and government agencies.

tee connector/coupler See T connector.

TEI See Terminal Endpoint Identifier.

telautograph A historic telegraph machine invented in the late 1800s, which could transmit handwriting over short distances. The earliest models used a pen writing a continuous line, and did not leave breaks between letters or words. Subsequent improvements were made by E. Gray, F. Ritchie, and others, which allowed the pen to be lifted off the paper when desired. These devices were used for several decades. Modern versions of the telautograph, using electronics, are known as *telewriters*, and were superseded by facsimile machines. See facsimile machine.

Telco, TelCo Abbreviation for telephone company, a local or regional telephone carrier.

telebusiness The British counterpart of telemarketing, teleresearch, and telesales.

Telecom Developers A telephony industry trade show, the forerunner to the Computer Telephony Conference and Exposition, held regularly in the spring.

Telecom Information Exchange Services TIES. A service of the International Telecommunication Union (ITU) which provides member resources and access to the ITU Terminology database (TERMITE). See TERMITE.

Telecom Services Association of Japan TELESA. A nonprofit group of member companies providing Internet-related services. The group has established a consortium to conduct field trials of electronic commerce systems in cross-border contexts and is responsible for initiating the Integrated Next Generation of Electronic Commerce Environment Project (INGECEP) which is carried out in cooperation with APEC. See Integrated Next Generation of Electronic Commerce Environment Project.

Telecommunication Standardization Advisory Group TSAG. A division of the International Telecommunications Union which interprets global standardization concepts and goals into practical implementations.

telecommunications 1. Meaningful wired/cabled or wireless transmission and receipt of signals over distance. 2. Broadcast, telegraph, phone, and computer network communications, frequently with a give-and-take quality or by choice of the receiving party, carried through a variety of media, including wires, fibers, air, etc. 3. The term is sometimes used to indicate a broader scope of communications *telephony,* to include video, for example (although telephony's meaning is not quite as narrow as thought by some). 4. This is defined in the Telecommunications Act of 1996, and published by the Federal Communications Commission (FCC), as follows:

"... the transmission, between or among points specified by the user, of information of the user's choosing, without change in the form or content of the information as sent and received."

See Federal Communications Commission, Post Roads Act, Telecommunications Act of 1996.

Telecommunications Act of 1996 This is the first substantial overhaul of telecommunications regulations since 1934, signed into law by President Clinton on 8 February 1996. The goal and intent of the law is to enable open access to the communications business and to permit any business to compete with any other telecommunications business. The chief impact of this act is on phone and broadcast services. Regulatory responsibility is largely shifted away from state courts and regulatory agencies to the Federal Communications Commission (FCC), but much of the administrative workload remains with the state authorities.

Some of the changes include the lifting of some long-standing restrictions, with the Regional Bell Operating Companies (RBOCs) now permitted to provide interstate long distance services. Telephone companies can now provide cable television services and cable companies can now provide local telephone services.

The FCC and individual states are responsible for implementing the terms, and the FCC has published an implementation schedule for this important act regarding the various issues of interconnection, universal service, access, assignment of broadcast licenses, etc. See Above 890 decision, Commercial Space Launch Act of 1984, Communications Act of 1934, Federal Communications Commission.

Telecommunications and Customer Service Foundation TCSF. A Canadian-based association formed to promote excellence in customer service in the telecommunications industry.

Telecommunications and Information Infrastructure Assistance A U.S. Department of Commerce grant program established in 1994 to assist local government and nonprofit organizations in funding projects which contribute to the design and development of the national information infrastructure (NII).

Telecommunications and IP Harmonization Over Networks TIPHON. An ETSI working group resulting from the result of the shift from traditional analog telephony to digital telephony and the desire of vendors to offer voice services over digital networks such as the Internet. Project TIPHON addresses market demand and service compatibility and transmission over network boundaries. The ETSI-trademarked TIPHON group works jointly with the International Multimedia Teleconferencing Consortium (IMTC) to organize interoperability plugtests so vendors can test their Internet Protocol (IP) products with other developers' telephony systems. See Voice over IP.

Telecommunications Association, Latvia A Latvian trade organization fostering the development of telecommunications technologies and products, the liberalization of the telecommunications business sphere, and the adjustment of regulations in Latvia to further growth and development in the field. In February 2001, the association admitted the IP Telephony Association with the view of improving progress in the standardization and regulation of the Latvian communications sector.

telecommunications bonding backbone TBB. See telecommunications main grounding busbar.

telecommunications broker An entity (person or business) that assists in negotiating contracts for communications services on the part of a user, or who purchases specialized or bulk telecommunications services with the intent of reselling these services to consumers, sometimes at discount rates.

telecommunications carrier This is defined in the Telecommunications Act of 1996 and published by the Federal Communications Commission (FCC), as:

"... any provider of telecommunications services, except that such term does not include aggregators of telecommunications services (as defined in section 226). A telecommunications carrier shall be treated as a common carrier under this Act only to the extent that it is engaged in providing telecommunications services, except that the Commission shall determine whether the provision of fixed and mobile satellite service shall be treated as common carriage."

See Federal Communications Commission, Telecommunications Act of 1996, telecommunications carrier duties.

telecommunications carrier duties The Federal Communications Commission (FCC) stipulates a number of duties in the Telecommunications Act of 1996 as follows:

"Each telecommunications carrier has the duty—

(1) to interconnect directly or indirectly with the facilities and equipment of other telecommunications carriers; and

(2) not to install network features, functions, or capabilities that do not comply with the guidelines and standards established pursuant to section 255 or 256."

telecommunications closet A wiring panel or room or other centralized, secured or separated administrations center for equipment junctions and/or demarcation points. Larger systems may have a series of panels, punchdown blocks, racks, or other furnishings to secure and organize the wiring system. See telecommunications main grounding busbar, wiring closet.

Telecommunications Development Bureau BDT. An agency established as a result of the Plenipotentiary Conference in Nice, France in 1989 to set up technical assistance in developing countries for coordinating, standardizing, and regulating telecommunications in third-world countries. France is the location for many regulatory and standardization bodies. BDT activities began in 1990.

Telecommunications Electric Service Priority TESP. A governmental restoration initiative that promotes voluntary inclusion of telecommunications facilities considered critical to national security and emergency preparedness (NS/EP) in existing electric utility emergency priority restoration systems. TESP is administered by the U.S. Office of Priority Telecommunications (OPT) at the National Communications System (NCS) in cooperation with the U.S. Department of Energy (DoE), state governments, and utility services.

TESP promotes the voluntary modification of existing electric utility emergency priority power restoration systems to include telecommunications facilities that may be critical to NS/EP. See National Communications System.

telecommunications equipment This is defined in the Telecommunications Act of 1996, and published by the Federal Communications Commission (FCC), as:

"... equipment, other than customer premises equipment, used by a carrier to provide telecommunications services, and includes software integral to such equipment (including upgrades)."

See Federal Communications Commission, Telecommunications Act of 1996.

telecommunications facilities The integrated structures and equipment that enable telecommunications to be conducted and managed. This may include secure rooms for servers or patch bays, consoles, PBX systems, satellites, telephones, facsimile machines, modems, wires and cable, video cameras, radio transceivers, etc.

telecommunications grounding busbar TGB. See telecommunications main grounding busbar.

Telecommunications Industry Association TIA. TIA began in 1924 as a small group of communications suppliers. Later, it became a committee of the U.S. Independent Telephone Association (USTSA). This group split from the USTSA in 1979 to become a separate, affiliated association, and TIA was formed in 1988 through a merger of USTSA and the EIA Information and Telecommunications Technologies Group.

A national trade organization representing about 1000 member companies which provide communications and information technology products, materials, and services, TIA provides a forum for discussing industry information and issues, organizes industry trade conventions, and serves as a voice for manufacturers

T

and suppliers of communications products for matters of public policy and international commerce. TIA is accredited by the American National Standards Institute (ANSI) to develop standards for a variety of communication products. See TIA Fiber Optic Communication Standards chart. http://www.tiaonline.org/

Telecommunications Information Network Architecture TINA. A common architecture for building and managing communications services developed by the Telecommunications Information Network Architecture Consortium in the early 1990s. This architecture logically separates the physical infrastructure and the applications from the need to communicate directly with each another. Control and management functions are integrated and can be placed on the network independent of geography through a single Distributed Processing Environment (DPE). See Telecommunications Information Network Architecture Consortium.

Telecommunications Information Network Architecture Consortium TINA-C. An international association of over 40 telecommunications operators and manufacturers who first came together at a TINA Workshop in 1990 and formed the consortium to cooperatively define a common architecture (TINA)

to be promoted as a global standard for building and managing telecommunications services. This work draws heavily on the work of other organizations and standards bodies in order to take advantage of ongoing studies and developments, to expedite the progress of the TINA project, and to promote the harmonious cooperation of various groups with similar goals. http://www.tinac.com.

telecommunications lines Physical lines, usually metal wire or fiber optic cable, over which communications are transmitted, usually by electrical impulses or light. Contrast to wireless communications.

telecommunications main grounding busbar TMGB. An important component of a telecommunications electronic grounding system that extends the building grounding electrode system for a telecommunications network infrastructure. Typically one per building is housed in an accessible communications closet as a central attachment point for a telecommunications bonding backbone (TBB) with one or more telecommunications grounding busbars (TGBs).

A TMGB is designed to facilitate low-resistance contact between lugs and busbars. It should be directly bonded to the electrical service ground and to a TGB which, in turn, connects to a permanent metallic

TIA Fiber Optic Communication Standards

Document	Designation	Description
TIA/EIA-455-5B	FOTP-5	Humidity Test Procedure for Fiber Optic Components
TIA/EIA-455-78A	FOTP-78	Spectral Attenuation Cutback Measurement for Single-Mode Optical Fibers
TIA/EIA-455-87B	FOTP-87	Fiber Optic Cable Knot Test
TIA/EIA-455-98A	FOTP-98	Fiber Optic Cable External Freezing Test
TIA/EIA-455-157	FOTP-157	Measurement of Polarization Dependent Loss (PDL) of Single-Mode Fiber Optic Components
TIA/EIA-455-164A	FOTP-164	Single-Mode Fiber, Measurment of Mode Field Diameter by Far-Field Scanning
TIA/EIA-455-171A	FOTP-171A	Attenuation by Substitution Measurement for Short-Length Multimode Graded-Index and Single-Mode Optical Fiber Cable Assemblies
TIA/EIA-455-191A	FOTP-191	Measurement of Mode Field Diameter of Single-Mode Optical Fiber
TIA-526-2	OFSTP-2	Effective Transmitter Output Power Coupled into Single-Mode Fiber Optic Cable
TIA-559		Single-Mode Fiber Optic System Transmission Design
TIA/EIA-604		Fiber Optic Connector Intermateability Standards
TIA/EIA-620AA00		Blank Detail Specification for Single-Mode Fiber Optic Branching Devices for Outside Plant Applications
TIA/EIA-785	100 Mbps	Physical Layer Medium-Dependent Sublayer and 10 Mbps Auto-Negotiation on 850 nm Fiber Optics

Relevant Technical Committees

FO-2.1	Single Mode Optical Communication Systems	FO-6.2	Terminology, Definitions, & Symbology
FO-2.2	Digital Multimode Systems	FO-6.3	Interconnecting Devices and Passive Products
FO-2.3	Optoelectronic Sources, Detectors, and Devices	FO-6.6	Fibers and Materials
FO-2.6	Reliability of Fiber Optic Systems & Active Optical Components	FO-6.7	Optical Cables
		FO-6.9	Polarization-Maintaining Fibers, Connectors, and Components
FO-6.1	Fiber Optic Field Tooling & Instrumentation		

structural element of the building. The TMGB should never be bonded to a secondary electrical conduit or pipe, as this may result in different ground potentials between the TMGB and the communications equipment grounding. If multiple closets exist in the same building, the TGBs should be bonded to one another and to the TMGB through approved insulated wires to form a TBB.

The TMGB resembles a metal cribbage board in that it has pairs of holes punched (all the way) through a thin rectangular board and is supported by brackets and insulators with noncorrosive fasteners. The holes are drilled according to recognized size and spacing standards (e.g., NEMA). It is made of a conductive material such as copper, and may be plated with another metal such as nickel.

TMGBs are generally designed to conform to ANSI/TIA/EIA-607 specifications and BICSI recommendations and come in a variety of widths and lengths. Once installed, TMGBs should be labeled with warnings not to remove the structure or disconnect any of its components.

Telecommunications Management Network TMN. A global network management model for Network Elements (NE) and Operation System (OS) and the interconnections between them. Global standardization provides greater incentives for common interface development. Discussions of O&M aspects of intelligent transmission terminals began, and TMN was first formally defined in 1988, with the recommendation for M.3010 (Principles for TMN) published in 1989, in addition to others over the next three years. OSI Management, originating in ISO, was adopted as a framework for TMN to provide transaction-oriented capabilities for operations, administration, maintenance, and provisioning (OAM&P). Elements of a TMN interface consist of various definitions, models, and profiles, including architectural definition of TMN entities, OAM&P functionality, management application and information models, resource information models, communication protocols, conformance requirements, and profiles.

Telecommunications Policy, Office of OTP A government agency, established in 1970 as an Executive Office of the President during Richard M. Nixon's administration. The OTP evolved from the 1968 President's Task Force on Telecommunications Policy. Clay T. (Tom) Whitehead was the first OTP Director. Some of the staff were taken from the earlier Office of Telecommunications Management (OTM). The OTP was rolled into the U.S. National Telecommunications and Information Administration in 1978 resulting from reorganization.

Telecommunications Policy, President's Task Force on A significant milestone in telecommunications policy development which came about partly because of controversies regarding cable and long distance services. The Task Force was established in 1968 during the term of President Lyndon B. Johnson. Some of the important outcomes of Task Force research included the establishment of a government agency to deal with telecommunications policy and increased emphasis on competition and deregulation.

Telecommunications Policy Research Conference TPRC An annual forum, first convened in 1972 as the OTP conference, for public and private sector scholars and decision-makers to discuss recent empirical and theoretical research and the needs of the telecommunications industry. See Telecommunications Policy, Office of. http://www.tprc.org/

Telecommunications Reform Act An act by the U.S. government opening up local and long distance markets to competition. The act included a highly controversial provision called the Communications Decency Act (CDA) which was, after a great deal of discussion and input from the Internet community, declared unconstitutional. The Reform Act significantly altered regulations of the telecommunications industry.

Telecommunications Regulatory Email Grapevine TREG. An informal organization that carries on regular online discussions about real world issues associated with taking products and services through the various regulatory processes. This self-help group answers queries and shares experiences, archiving the information on the Web.

telecommunications relay service A 24-hour telephone service to assist hearing impaired individuals to intercommunicate and to communicate with hearing subscribers. This service may have a variety of telebraille, TTY, and voice options. In general, the subscriber calls a telephone agent, who intercepts the call, translates it, and relays it to the callee on behalf of the caller. The calls and call content are confidential.

Telecommunications Research Establishment TRE. A once-secret facility at Malvern where communications research such as radar research was conducted during World War II. The TRE was established in 1940, evolving from the Ministry of Aircraft Production Research Establishment (MAPRE). Some of the early research in digital storage devices was carried out here towards the end of the war. Many of the researchers at this facility made significant scientific achievements in a number of fields including radio astronomy. See Small Scale Experimental Machine.

telecommunications service This is defined in the Telecommunications Act of 1996 and published by the Federal Communications Commission (FCC), as

"... the offering of telecommunications for a fee directly to the public, or to such classes of users as to be effectively available directly to the public, regardless of the facilities used."

See Federal Communications Commission, Telecommunications Act of 1996.

Telecommunications Service Priority TSP. A program of the Federal Communications Commission (FCC) for identifying and prioritizing telecommunications services that support national security and/or emergency preparedness (NS/EP) missions. The TSP regulates, administers, and operates priority restoration and provisioning of qualified NS/EP telecommunications services to support emergency readiness and response to local, national, or international events or crises that might harm Americans or their property. http://tsp.ncs.gov/

T

Telecommunication Standardization Bureau TSB. The TSB provides support for the standardization sector of the International Telecommunication Union (ITU). As such, it helps to coordinate the work of the ITU-T, provides secretarial services, assists in disseminating information, and ensures the publication of various references resulting from the work of the ITU-T. See International Telecommunication Union.

Telecommunications Standards Advisory Council of Canada TSACC. A Canadian industry-government alliance formed in 1991 to develop strategies for Canadian and international standardization in information technology and telecommunications. Information on telecommunications technologies is provided on their Web site. http://www.tsacc.ic.gc.ca/

Telecommunications Technology Association TTA. Established by the Korean Ministry of Communication in 1988, beginning operations in 1992. http://www.tta.or.kr/

telecommuting Virtual commuting to the work site, that is, communicating through various telecommunications methods instead of physically traveling to the work site. A number of factors have contributed to the increasing desire for, and availability of, telecommuting jobs: increasing congestion in cities causing higher housing costs and less availability of housing; increased traffic congestion; more families with two working parents who don't want to leave children unattended; improved telecommunications services, with faster and better transmission, more hookup services through phone lines, and videoconferencing options.

Telecommuting is not for everyone; many people prefer to work under direction or to work in close physical proximity to co-workers, but many work better undisturbed and will use the time saved by not commuting to produce a higher-quality product. There are also increasing numbers of businesses willing to provide telework options so that they can recruit highly skilled workers from diverse regions. See telework, virtual office.

telecomputer, computerTV A TV broadcast system-computer integrated system that allows a user to control program selection or menu options for viewing, such as split screen for more than one show, digital effects, sound options, integration of TV and phone (e.g., on-screen Caller ID on the TV when the phone rings), email and Web access, shopping from home, etc. This is an example of the convergence of the computer and broadcast industries. Standards for ATM for the home are being promoted so that standardized commercial consumer systems can be developed which allow these many technologies to link and work together. See Broadband Residential, fiber to the home, Home Area Network, WebTV.

teleconference A telephone conference where three or more participants share in a conversation. Conference call buttons or codes are available on some local multiline systems, and operators can set up conference calls across public lines for participants who are distant from one another. See videoconference.

telecopier See facsimile machine.

teledensity A measure of the number of telephone lines per 100 POPs (individual people) used to assess service distribution, economic compromises, revenues, etc.

Teledesic, LLC A privately owned constellation of literally hundreds of satellites orbiting at 700 km (LEO) designed to provide switched broadband bidirectional network services, including Internet access, data, voice, videoconferencing, and interactive multimedia. It is designed to operate at up to 64 Mbps for downlink and up to 2 Mbps for uplink. The top transmissions speed is more than 2000 times faster than standard modems operating over wired phone lines. Connection is through small parabolic antennas. The Teledesic group approached the Federal Communications Commission (FCC) in 1994 for a 500-MHz frequency allocation within the Ka-band for this service.

In May 1998, Motorola Inc. joined the venture as the prime contractor, bringing in its Celestri technology, along with Boeing Company and Matra Marconi Space, a European satellite manufacturer. In August 2001, the company announced that either Lockheed Martin Commercial Space Systems (LMCSS) or Alenia Spazio (an Italian vendor) would be selected as the prime contractor to build the network as both had experience in building nongeostationary-orbit satellite systems. In November 2001, a talked about merger with another McCaw company, ICO Global Communications was discontinued. In February 2002, an agreement was signed with Alenia Spazio to contract the systems.

Teledesic LLC is a McCaw/Gates company scheduled to launch its satellites in the early 2000s and to be in service by 2005.

TeleDirectory A telephone directory service from British Telecom for personal computer users who have a frequent need for directory assistance services (e.g., five or more numbers per day) and wish to access the number online. Enquiries are billed on a per-number basis. See BT Phonebase.

Telefunken A German radio station founded in 1903 soon after Marconi's wireless demonstrations in London, England, excited the imaginations of radio experimenters and future broadcasters.

telegaming Gaming over a distance communications medium (telephone, computer network, postal service). Telegaming has been around for a long time. For centuries, people have played long distance chess and backgammon games by messenger and, more recently, by mail or phone. Currently it implies an unbroken connection, since that is now possible through computer networks and games like *chess* and *go* are routinely played on the Internet. Video arcade games are played on local networks, usually on an Ethernet link, although the term telegaming doesn't apply as well to an activity in which the participants can see or hear one another in the same or next room.

telegenic Having characteristics that appeal to television audiences, such as charisma, talent, humor, relevance (news), or other qualities favored by broadcast networks and viewing audiences.

telegram Originally *telegramme* (France, 1793). A printed record of a telegraphic communication. Early telegraph signals were transcribed on paper tape as wiggly lines; later, audible signals were interpreted by human operators and written down by hand; and, finally, devices that could interpret the signals into text and impress them on paper as telegrams were devised. For decades the telegram was delivered into the hands of the intended receiver or at least brought to the doorstep. Courier services and facsimile machines are superseding telegram services. See telegraph system, teletypewriter.

telegraph fire alarm That telegraph signals could be used to report fires through signal boxes was realized not long after the invention of the telegraph, and many of the larger communities installed this type of safety system by the early 1900s. The Boston Fire Alarm system was one of the first, following a published description of its feasibility by William F. Channing in 1845. Later, with the help of a telegraph engineer, Moses G. Farmer, Channing supervised the 1851 city funding and 1852 construction of the first fire alarm telegraph in the world. Originally based on manual crank boxes, painted black, the mechanisms were later changed to pull switches, and eventually dials. By 1881, the fire boxes were changed to red.

telegraph history The telegraph was a system of equipment and data encoding that enabled communication over distance, originally through drum beats, signal fires, and signal towers, and later by wires powered by high-intensity batteries. As with many technologies, the telegraph was invented in a number of places at about the same time, and many of the early models were never practically or commercially implemented.

In a sense, the technology has come full-circle. Many of the earliest modern telegraphs were optical telegraphs which gradually gave way to electrical systems and now, 200 years later, we are returning to the use of optics, channeled automatically through fiber rather than being transmitted by humans through air. Lesage had created a *frictional telegraph* as early as 1774, and A. Ampère and P. Barlow proposed early designs as well. J. Munro reports that someone identified only as C.M. described an electric telegraph in *Scots Magazine* in February 1753 that suggested a multi-wire system (similar to those later implemented by Campillo in Spain and Sömmering in Germany). If so, it is the earliest recorded reference to a modern telegraph system. Samuel T. von Sömmering created a 35-wire telegraph based upon electrochemical concepts which, in turn, was derived from the work of Francesc Salvà i Campillo in Spain, in 1795.

Optical telegraphs were developed by Claude and Ignace Chappé in the early 1790s and were probably built upon the ancient tradition of signal fires. The Chappé system used physically coded letters and symbols relayed through a system of towers by human "transceivers." The concept spread to other parts of Europe, including Denmark, where an optical postal telegraph was established in 1801.

One of the first practical commercial implementations of a nonoptical telegraph was in 1837 by C. Wheatstone and W. Cooke in England. The telegraph in America owed much of its design and development to Samuel Morse and Alfred Vail. Morse's original telegraph caveat (an intention to file a patent) described a mechanism with a horizontally moved key which made corresponding zigzag marks on a moving paper tape to represent numbers, which were then looked up to find the corresponding words in a reference dictionary prepared by Morse. Vail improved on the mechanics of the key, making it move up and down instead of side-to-side, thus forming dots and dashes with breaks in between on the paper. As this system was simpler and more direct than doing a dictionary lookup, it evolved into the system now known as Morse (Vail) code. Their telegraphic invention was demonstrated to the Presidency in 1838. Morse subsequently won funding from Congress to construct a telegraph long distance line, carried out the project with assistance from Ezra Cornell, and began to spread telegraphy throughout America in the mid-1800s. Both Wheatstone and Morse received advice and encouragement on the development of telegraphic instruments from Joseph Henry in the 1830s. Morse, unfortunately, didn't duly credit Henry's assistance.

Historic Telegraph Communications Technology

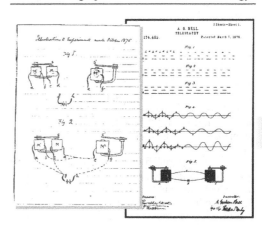

One of the early Bell telegraph patent documents. There were many inventors at the time independently making similar discoveries, and substantial competition to be the first to patent and commercialize the new communication technologies.

In its simplest form, the telegraph consists of a sender (a keying device), a receiver (with a sounder or printer), and a simple code for conveying characters. Early telegraph receiving machines used paper tapes to record messages (Morse's telegraph created a wiggly line), but operators began to recognize the slightly audible incoming clicks and could copy messages faster than a paper tape could print them, so machines were soon equipped with *sounders* and *resonators* to amplify and direct these clicks. Not surprisingly, many inventors sought ways to translate the signals into letters that could be recorded directly, as in a telegram or teletype-style printout. One of the first to succeed was David Hughes, a schoolteacher, in 1856. In America, messages were sent by shutting current on and off, while in Britain, Wheatstone introduced *polar keying,* a means of using polarity to convey signals. The concept of polarity is still used today in high-speed data transmissions.

In 1866, M. Loomis demonstrated that signals could be sent from one airborne kite to another, when each was strung with fine copper wire of the same length, without direct physical contact. This later lead to his 1872 U.S. patent for a wireless improved telegraphic system, although it was some time before his discoveries were put into practical use.

By the 1880s, scientific investigations and demonstrations had confirmed the viability of wired and wireless telegraphy. The end of the century then became a time of creative application of the concepts and evolutionary improvements in speed and practicality.

In 1895 and 1896, in Russia, A.S. Popow was conducting experiments with wireless telegraphy and succeeded in sending a shipboard message to his laboratory in St. Petersburg. Unfortunately, due to the secrecy surrounding Russian naval technology and inventions in general, Popow's discoveries were not communicated to the rest of the world, and he did not receive credit for his early experiments.

In the late 1800s, *telautographs* that could transcribe handwriting were created by several inventors such as E. Gray and F. Ritchie. While these were used for several decades, they didn't originally work over long transmission lines and were superseded by telewriters and, eventually, facsimile machines.

In 1886, Amos Dolbear, a Tufts University scientist and writer, was awarded a patent for a wireless telegraph based on induction.

In 1889, F.G. Creed invented a High Speed Automatic Printing Telegraph System. By 1898, his Creed Printer could transmit 60 words per minute and his technology was widely sold in many countries. He broadened his enterprise in 1923 by demonstrating marine wireless printed telegraphy, a system eventually used for marine safety.

Wireless telegraphy was of interest almost from the beginning of telegraphic history. In the early 1900s, V. Poulsen and P. Pedersen used an electric arc to generate high-frequency waves, setting the groundwork for wireless communications. Poulsen also developed the *telegrafon,* a historic electromagnetic tape recorder. Tape recorders were later used to develop dictation and telephone answering machines.

The telegraph had a revolutionary impact on communications, changing forever the concept of distance. It networked the predominantly rural early settlers of North America and spurred the installation of the first transatlantic cable, providing instant (by 1800s standards) communication with Europe. Prior to the oceanic cable, messages typically took 2 months or longer to travel in ships from one continent to the other. News, business, warfare, and family contacts were dramatically affected by the availability of fast long distance communications.

See Creed, Frederick George; heliotrope; Davy, Edward; Dolbear, Amos; Morse, Samuel F.B.; Popow, Aleksandr Stepanowitsch; Salvà i Campillo, Francesc; Sömmering, Samuel Thomas; Steinheil, Karl August; telegram; telegraph system; telephone; Wheatstone, Charles.

telegraph key A mechanical switch on early telegraph systems that enabled a circuit to be opened and closed in order to generate transmissions through a signal such as Morse code.

telegraph signals For telegraph signals through wires, two main methods were used: *polar* transmission, in which the polarity was changed to reverse the current; and *neutral,* or *open/close* transmission, in which open current (space) was interspersed with no current (mark).

telegraph system An apparatus for sending and/or receiving information over distance, coded in some fashion, usually in Morse code dots and dashes. A basic telegraph circuit consists of a key to translate finger or other mechanical pressure into signals, a relay sensitive to the very small current that may be coming through the wire, and a receiving device which can express the message by means of audible tones, paper tape code, or printed letters.

Telegraph systems have coexisted with, rather than been superseded by, telephone systems for a number of reasons, including the expense and time delays of setting up long distance toll calls to some areas, and the importance, in some situations, of creating a written record in the form of a telegram. With electronic telephony advancing and facsimile machines proliferating, the telegraph is becoming more historically interesting than practical. See telegraph history; telegraph, needle.

Telegraph Network System

A telegraph network map of the United States published ca. 1870 which shows the Pony Express Mail & Telegraph Route (in spite of the fact that the Pony Express was very shortlived).

---

telegraph, needle A type of five-needle telegraph devised by Charles Wheatstone and put into service in England in 1837. Faulty equipment lead to the gradual realization, by telegraph operators, that two needles were sufficient and, eventually, only one needle and one dial were used to efficiently convey messages. The needle telegraph also represents the development of *polar keying*, which employed positive and negative voltages for indicating *mark* and *space* signals. See polar keying.

telegraph, printing Early telegraph papertape and manually operated sounding systems did not satisfy the needs of inventors and users who wanted quick, automated written messages. Thus, the development of printing telegraphs was of interest to many. One of the first successful systems was developed by A. Vail in 1837, employing a type wheel. Later D. Hughes developed a practical working type wheel system in 1855, which became established in Europe, but didn't catch on well in America, where Morse systems were in use. Improvements to printing telegraphs continued and, in 1846, R.E. House developed a printer that printed telegraphically transmitted letters directly. Further improvements to House's system resulted in a patent in 1852. In 1905, Donald Murray published "Setting Type by Telegraph" in the *Journal of the Institute of Electrical Engineers* and went on to improve telegraphy in a number of ways. The necessity of noise-free transmissions and technical expertise to maintain the equipment prevented printing telegraphs from coming into widespread use until decades later. See teletypewriter.

Historic Automatic Telegraph System

An automatic telegraph sender. The wheels shown at the top represented characters that could be selected and placed in order to spell out a message. This example is from the American Radio Museum collection.

---

telegraphese A terse, abbreviated mode of messaging (or speaking) which has the character of a telegram. Since telegrams were often charged by the letter or by the word, a compact style of communication emerged in order to keep the cost as low as practical.

telegraphone, telegrafon This is not only a type of telegraph instrument, but more important, was an early electromagnetic tape recorder, designed in 1898 by Danish inventor Valdemar Poulsen. Poulsen succeeded in recording electronic waves on a thin wire

of steel, and improved on the technology enough to receive a U.S. patent in 1890. This developed into dictating machines sold through the American Telegraphone Company. See tape recorder.

TeleLink Project The full name is TeleLink Training For Europe Project. This is a European Community (EC), Euroform-funded project which seeks to promote and develop telework training opportunities and qualification guidelines. This includes qualification level certification (currently at the vocational level) for teleworkers and a system of TeleLink centers around Europe. See ADVANCE Project, telework.

telemarketing The promotion of products and services through telephone calls to individual premises. There are various regulations governing when telemarketers may call, whom they may call (e.g., calls to a person at his or her place of business must be stopped if the callee requests it), and what they must say to identify themselves and their affiliations. There are also restrictions on where they may obtain names, and how they must dial the call. Many scams have been perpetrated through telemarketing schemes, and it is important for the callee to get sufficient information to ascertain that the offering is legitimate. If you don't wish further calls from the source, you should request that your name be taken off their list. See war dialer

telemarketing broadcasts The promotion of products and services through mass market advertising usually providing a 1-800 or 1-900 number for the interested buyer to call. Automated systems for taking the caller's name and billing information through touchtone selections are becoming prevalent.

telemedicine Medical information and services and medical education provided over distance through telephone, radio, facsimile, videoconferencing, and the Internet. Information such as medical imaging results can readily be transferred as data, since much of it is digital in nature. Teaching and other communications among medical professionals and their patients are possible through newer technologies.

telemetry, telemetering The art and science of gathering information at one location, usually in terms of some quantity, and transmitting that information to another location for storage, analysis, or evaluation. Weather balloon data gathering and transmission through a radiosonde to a weather station for interpretation is one example of telemetry. The transmitting of information from space probes is another. Telemetry equipment is typically included on artificial satellites to aid in the control and orientation of the satellites.

Teletext A commercial computer service offered by NBC, which was discontinued in 1985. Many of these early computer services came and went, but they are coming back in updated forms now that there is a large user base drawn to the Web.

telephone A communications apparatus designed primarily to convey human voice communications. In its simplest form, a telephone consists of a transistor that converts sound into electrical impulses, and a receiver, which converts them back again into sound. Additional technology is used to amplify and direct the communication between these two basic devices. The design of the telephone set has gone through five overlapping phases in its development. See the Telephone Development Phases chart. See telephone history.

Historic Telephone

The earliest telephone was a simple device that looked more like a pinhole camera than current familiar desktop phones and mobile handset phones.

---

telephone amplifier A device to amplify sounds at the receiving end of a call. This can be incorporated into the handset, headset, or speakerphone, or may be an add-on to provide even more amplification for the hard of hearing. Most handset telephone amplifiers draw current from the phone line, but many speakerphones and add-on amplifiers require a separate power source. The amplifier is often adjustable through a dial or slider on the side of the phone.

telephone answering machine An electronic or mechanical device for answering calls and often for recording them digitally or on tape. Telephone answering machines based on reel-to-reel mechanisms have been available since the early 1960s, but small cassette and digital answering machines did not become common until the late 1970s and early 1980s.

Most households now have answering machines to respond to calls, take messages, or screen calls. Many of these will include information on the time and date of the call, and some will record the identity of the caller, if Caller ID is activated on the subscriber line. Computer voicemail applications can also be hooked to a phone line through a data/fax/voice modem to allow the software to function as a full-featured answering machine with multiple mailboxes.

telephone answering service 1. A service offered by commercial vendors in which a human operator or voice-automated system will answer the subscriber's phone line when it is call forwarded, or when the answering service number is called directly and forwards the message to the subscriber. This service is widely used by small businesses, freelancers, and real estate agents. Sometimes these services are combined with paging. 2. A service offered by local phone companies in which a human operator or voice-automated

system will take calls and forward messages to the subscriber, or through which the subscriber can use a touchtone phone to retrieve messages.

telephone central office See central office.

telephone circuit An electrical connection consisting minimally of a transmitter, receiver, amplifier, and connecting wires, and more commonly comprising a system of two-way audio and signaling connections between local exchanges and subscriber lines and telephones.

Telephone Company of Prince Edward Island A historic telephone exchange, incorporated in 1885, the year after the phone exchange was first established on the island.

telephone exchange Switching center for telephone circuits. See central office, private branch exchange.

Innovative Optical Telephone

An innovative optic telephone, based on the stimulation through a diaphragm of a flame from an acetylene burner. The impulses were then further transmitted optically through a light-sensitive selenium cell and reflector. The optic telephone was developed by Ernst Ruhmer, and was used for long distance communications. [Scientific American, November 1, 1902.]

---

telephone history The telephone was a significant evolutionary development, occurring a few decades after the invention of the telegraph. While the telegraph revolutionized telecommunications by making communications over great distances possible, the telephone personalized it, and many inventors were excited by the potential of sending tones, or even voice, over phone lines.

The use of tubes and strings to magnify sound and channel acoustic vibrations existed at least as early as the time of Robert Hooke, long before the development of modern telephones, but such devices, like the acoustic tubes demonstrated in 1682 by Dom Gauthey, were physically limited as to loudness and distance. It was not until electricity and magnetism were harnessed that amplified, long-distance modern telephony was possible.

In the early 1800s, German inventor Philip Reis observed that a magnetized iron bar could be made to emit sound. In America, Charles Page made a similar discovery, terming the sound "galvanic music." Subsequently, a number of inventors advanced telegraphic and microphonic technologies leading up to the invention of the telephone. Belgian inventor Charles Bourseul described his idea for transmitting tones in 1854, but wasn't able to implement a fully working version before Philip Reis and Alexander Graham Bell developed their own telephonic devices. Reis first demonstrated the transmission of tones through wire in Frankfurt in 1861. He reported in a letter that he could transmit words, but there is no direct way to verify the claim.

Around the time of Reis's death, an American physicist, Elisha Gray, was making numerous experiments in telegraphy and developed early concepts for harmonic telegraphy, the transmission of tones, and telephony.

In the mid-1800s, Italian-born Antonia Meucci was successfully experimenting with wires attached to animal membranes to transfer sound through current, but news of his significant discoveries did not become widely known outside Cuba. When he later emigrated to the U.S., he filed a caveat for a patent, in December 1871, for a *teletrofono*.

Bell's Telephone Demonstration

Here Alexander Graham Bell demonstrates his telephone invention. The inset shows one of his early sketches of the invention, from the famous Bell notebooks. Bell achieved great financial success from commercializing his discoveries.

The better-known precursors to the telephone in America and later variations appear to have been invented more-or-less independently by Elisha Gray and Alexander Graham Bell, but Bell filed his telephone patent (it was actually a precursor to the telephone, a harmonic telegraph) a few hours before Gray filed a caveat (intention to file within 3 months) in February 1876. The murky history of the invention of the telephone at this point stems in part from the fact that many innovations *were* being developed simultaneously and also because the inventors understood the great commercial potential of their devices. Hundreds of lawsuits were threatened and filed over

the next few decades, although some claims were more amicably settled. For example, in January 1877, Bell wrote to Gray rescinding any previous accusations he may have made that Gray copied from Bell's work. (In fact, both men may have copied from a third source, Antonia Meucci. It has been suggested, but not confirmed, that both Bell and Gray had access to Meucci's *teletrofono* documents when they were in the hands of Western Union.)

Emil Berliner was an inventor with a strong interest in music and the improvement of the quality of transmission of sound (which applied equally well to telephony). In April 1877, he filed a caveat for a patent for a telephone transmitter, three and a half months before Thomas Edison applied for a patent for a similar device.

In a 1911 lecture on the origins of the membrane telephone, Bell described how he worked out the idea in discussions with his father while on a family visit in Canada in the summer of 1874, 2 years before it was successfully implemented. Bell and Watson reported that Bell first spoke intelligibly over wires in March 1876. The transmission succeeded by use of a liquid medium, something not mentioned in Bell's patent. This voice capability was not publicly demonstrated until some time later, which seems odd given the magnitude of the reported achievement. Ironically, Bell had been discouraged by investors from trying to make a talking telegraph and was prodded to concentrate on a harmonic telegraph instead.

Gray had publicly demonstrated rudimentary telephone-related technology before the Bell patent was filed, and later successfully earned a number of telephone-related patents. He designed a telephone in the 1870s not unlike the second-generation switch-hook phones that employed separate ear and mouth pieces which came into use in later years.

The first commercial telephone exchange was established in Connecticut, U.S., and became operational in 1878. It was followed the same year by the second commercial exchange in Ontario, Canada.

The Bell patents formed the basis of the early Bell System in the United States, a company that has influenced the development of communications, and thus the course of history, in countless important ways. The Bell Telephone Company of Canada was incorporated in 1880.

By this time, telephone technology began to spread to other nations outside Europe and North America. The first telephone exchange was established in Japan in 1890 in the Tokyo/Yokohama region. In 1926, automatic step-by-step switches were introduced in Japan.

The most interesting evolutionary step in telephone technology, besides the growth of wireless communications, is probably the videophone, descended from early picture telephones such as the Picturephone. The Bell Labs were transmitting pictures in the late 1920s and demonstrated the early technology to the Institute of Radio Engineers in 1956, but it was not until 1964 that a practical experimental system was completed and the Picturephone was

T

exhibited cooperatively by Bell and the American Telephone and Telegraph Company (AT&T) at the New York World's Fair.

Currently many companies are scrambling to be the first to get a cheap, publicly accepted version of a picture telephone or as they are known now, audiographics systems, videophones, or videoconferencing systems. With the growth of the Internet and the drop in price of small video cameras, they began to be common computer peripherals in 2002.

Another significant change in telephony has been the sending of voice over computer networks by means of a specialized handset attached to a computer. This permits the connection of long distance calls world round without any long distance toll fees. The technology threatens to dramatically change the established economic structure of the telephone system, and it is difficult to predict whether the same revenue-generating model that has worked for about 100 years will be viable in the future, given the current rate of change. In fact, some of the long distance carriers, worried by this threat to their survival, have lobbied for this type of transmission to be blocked. See Bell, Alexander Graham; Berliner, Emil; Bourseul, Charles; Callender switch; Gray, Elisha; Meucci, Antonio; Photophone; Reis, Philip. See telegraph history which has a common ancestry and additional details.

telephone landline density A measure of the number of installed phone lines per 100 people.

telephone pickup Any of several devices for connecting into an ongoing telephone conversation, usually for monitoring purposes.

Telephone Pioneers of America TPA. A nonprofit organization founded in 1911, with chapters throughout the United States and Canada. Originally consisting of telephone pioneers with 25 years of service or more, with Theodore N. Vail as its first president, membership later opened up to a wider group, now numbering almost 100,000, as fewer pioneers remained from the original group.

TPA engages in a number of community-oriented activities, with a particular focus on education. A somewhat analogous organization serving non-Bell employees is the Independent Pioneers. http://www.telephone-pioneers.org/

telephone receiver The portion of a handset, headset, or speakerphone which converts electrical impulses into sound. On a handset, the receiver is the part that you hold up to your ear. Inside a basic traditional receiver is a magnet, with coils wound around the poles connected in series and a light, thin, vibrating diaphragm mounted very close to the magnet poles. When current passes through the coils, the diaphragm vibrates, producing sound by moving the air next to it. Early receivers used a bar magnet, which later was replaced by a horseshoe magnet. See telephone transmitter.

Telephone Relay Service A telephone service enabling handicapped individuals to communicate over telephones through third party interpreters. It is usually provided free of charge.

telephone repeater An amplification device employed on telephone circuits to rebuild and maintain signals across distances, which otherwise would be subject to loss.

telephone signaling Any device that indicates an incoming call, usually a bell, but may also be a light or moving indicator.

telephone switchboard A centralized distribution point for managing telephone calls. Early switchboards consisted of a human operator answering calls, and plugging a large physical jack into the receptacle of the person to whom the call was being patched. The first commercial switchboard in North America

Overview of Telephone Development Phases			
Type	Time period	Notes
original invention	late 1800s	Proof of concept, the first discernible, intelligible human voices can be heard over distances.
hand crank phones	late 1800s, early 1900s	Phones were large, to accommodate a battery, and had to be cranked to send a ringing current. Hand-crank phones were still in use in rural areas, including some of the San Juan Islands in the 1960s.
dial phones	early 1900s to 1980s	Common batteries and automatic switching systems made it possible to create smaller, line-powered phones and rotary dials so the subscriber could direct dial a local call, and later, long-distance calls.
touchtone phones	late 1970s to present	Phones that sent tones rather than pulses through the line, which were interpreted according to pitch. This made automated menu-controlled systems possible.
digital phones	early 1990s to present	Interface speakers or headset peripherals that attach directly to a computing device or desktop system to enable the user to talk into a digitizing program that samples the sound and transmits it over public data networks.

went into operation in Connecticut in 1878. Switches were mechanized in the mid-1900s, although it was not uncommon for human switchboard operators to staff manual switchboards in rural areas and private branches until the 1950s. Although mechanical switching stations still exist, updated switchboards function electronically.

telephone tag Colloquial phrase for two parties attempting to contact one another by phone, not reaching the other person, and leaving messages with an answering machine, operator, or voice mail system. Doing this back and forth a few times is telephone tag.

telephone transmitter The portion of a handset, headset, or speakerphone which converts sound into electrical impulses. On a traditional handset, the receiver circuit connects to the part that you hold next to your mouth. Inside the mouthpiece is a movable diaphragm with an attached carbon electrode, behind which another carbon electrode is fastened securely inside the housing. Between the electrodes are carbon granules (it's possible to build a simple phone transmitter using the core of a carbon pencil laid across two conducting surfaces connected to wires and a diaphragm). When a current is applied, resistance decreases, as a result of the carbon granules compressing more closely together. Thus the current increases and attracts the diaphragm more strongly. The diaphragm vibrates to produce an electrical impulse that corresponds to the movement of air caused by the speaker's voice. An induction coil may also be used to increase the voltage to compensate for signal loss through the transmissions medium. See Blake transmitter, coherer, telephone, telephone receiver.

telephone user interface TUI. The use of telephone equipment, usually a handset or headset or telephone line attached to a peripheral card, to interact with computer software. Instead of using a keyboard and mouse as the input devices, voice or touchtones over the handset or phone line are used to control the actions of the computer. For example, you may have a computer set up like an answering machine to answer calls, respond to callers, and log time, date, and caller messages. Then, from a remote location, you may call the line attached to the computer, and by speaking or pressing touchtone buttons, have the computer send back information about the calls or replay the calls.

telephony The science and practice of transmitting audio communications over distance, that is, over a greater distance than these communications could be transmitted without technological aid. The term has broadened from audio communications to encompass a wide variety of media, typically now including visual communications that accompany sound communications (as in audiographics and videoconferencing), although it is preferred that the more general term *telecommunication* be used for audio/visual transmissions. Most telephony occurs over wires, but wireless services transmitted by radio waves and satellite links are increasing.

Telephony, in its simplest sense, is not a high bandwidth application; each conversation requires only a narrow channel, but because of its continuous bidirectional nature, bandwidth needs increase as the number of simultaneous calls increases. Traditional telephony media, such as copper wires, are no longer strictly used for oral communications; they now service a large number of data transmission services such as Internet connectivity, facsimile transmission, and more. Due to increased demands for lines with greater speed and accuracy than are needed for simple voice transmissions, fiber and coaxial technologies are being used to upgrade data lines and, consequently, the phone lines. See HFC, telephone, telegraph.

Telephony Application Interface TAPI. A standardized telephone interface developed by Microsoft and Intel Corporation for the creation of a variety of

T

Semiautomatic Telephone Switching System

A schematic for a historic semiautomatic telephone switching system (it still required a human operator to turn a spring-loaded knob to send the dial pulses through the wire). [Scientific American, *October 11, 1902.*]

device-supporting software applications. TAPI can be used to create call control software for telephony devices for computerizing common functions and sought-after features. As an example, TAPI can be used to create a Caller ID-type function on a computer, with the computer answering a Caller ID-enabled phone line, assessing the Caller ID information, and perhaps relating it to a database of names or other information associated with the number and logging the call or notifying the user of the call.

Since not all phone systems are equipped with TAPI interfaces, there are now third parties, such as Ryan Technologies, that provide protocol conversion modules that enable TAPI-based applications to link through the module to the phone system. In 1999, Siemens extended the utility of TAPI by introducing a TAPI interface compatible with a national ISDN terminal. This, in conjunction with the Optiset phone, turns a computer into a powerful telephony terminal. See Telephony Services Application Programming Interface.

Telephony Routing over IP TRIP. A policy-driven inter-administrative domain protocol for routing voice-over-Internet calls, developed by the IETF IP Telephony (iptel) working group. TRIP is independent of the signaling protocol used. It uses Border Gateway Protocol (BGP-4) to distribute routing information between administrative domains. It thus enables digital telephony calls to be routed between digital network domains and supports the exchange of routing information between providers, thus building up a forwarding information base.

In August 2001, TRIP was published as an Internet Draft and, in September 2001, was submitted to the IESG for consideration as a proposed standard. TRIP may also be a part of future protocols for the propagation of routing information between gateways and their associated signaling servers (a process called gateway registration).

Telephony Services Application Programming Interface TSAPI. A set of guidelines developed by a group of developers, including Novell, Inc. and AT&T, for interconnecting corporate telephone systems into the data network server in medium and large business networks. The specification describes the physical link that can be used to implement software-based call control from a private branch exchange (PBX) switch, for example, so that control is handled from the originating point in the local area telephone network. Thus, TAPI's call tracking capabilities make it more powerful and suitable for enterprise environments than the more desktop-oriented Telephony Application Interface (TAPI). See Telephony Application Interface.

telephoto, telephotography Visual information conveyed through conventional photographs or digital photographs from data received remotely. Journalists, geographers, navigators, and others use telephotos to send or receive visual information from remote sources through wired or wireless communications, and to print them in various resolutions through photographic, laser, or other means. Satellite photos of the Earth's surface are extremely popular examples of telephotos. Many of the images now printed in national newspapers are telephotos sent through wireless modems by journalists using digital cameras and laptops.

Teleport Communications Group TCG. At one time, a national competitive local telecommunications provider with fiber optic SONET networks in over 50 large markets, acquired in early 1998 by AT&T.

teleprinter 1. Teletypewriter. 2. A Western Union trade name for printing telegraph terminals. See telex.

TelePrompt Project A European Community (EC) project funded by a consortium of academic and commercial groups designed to develop and further technology-based distance learning resources for European teleworkers. The term *teleworking* in Europe is roughly equivalent to the term *telecommuting* in North America.

teleran An aerial navigational guidance system employing information received through television waves and radar transmitted to aircraft by ground stations.

TELESA See Telecom Services Association of Japan.

telesales A British term for telemarketing.

Teletype A name trademarked by Teletype Corporation for a variety of teleprinting devices used in communications. See teletypewriter.

Teletype Corporation An early printing telegraph company, the Morkrum-Kleinschmidt Corporation, which was acquired in 1930 by the Bell System and renamed Teletype Corporation.

teletypesetter A machine for remotely controlling typesetting machines. When these were originally put into service, teletype machines relied on a five-unit code that was insufficient to transmit all the characters needed by a similar teletypesetting machine. Thus, a six-unit signal code was developed for teletypesetters to increase the size of the character set from 32 to 64.

teletypewriter TTY. A printing apparatus which, in its common form, resembles a typewriter on a pedestal with continuous feed or tractor feed paper so that it can print unattended. Sometimes it is used to send and receive signals over phone lines and for transmitting messages or computer data in text form.

The teletypewriter superseded key and sound telegraph systems because it could operate unattended, be read by individuals without knowledge of Morse code, and achieve transmission speeds of 60 to 100 words per minute. The earliest teletype-style printers and start-stop synchronization methods were developed by Charles and Howard Krum. See Baudot code; Krum, Charles and Howard; telegraph, printing; Teletype; telex.

teletypewriter code A five-unit code that employs elements of uniform length. Start and stop pulses are used to distinguish each character in the transmissions. See Baudot code.

teletypewriter exchange service Any commercial service which provides teletypewriter communications sending and receiving services through a switching exchange. Similar in concept to a long distance telephone exchange. TWX is one such service of the Bell System, established in 1931, subsequently owned by AT&T. See Telex.

Compact Television Camera

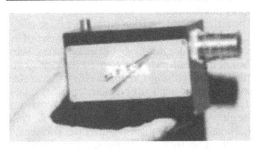

A tiny television camera, not much bigger than a human hand, designed in the days when television cameras were large and heavy. The technology was designed for the demanding task of space observation and very long-distance communications by the Marshall Space Flight Center. Similar cameras were quickly adapted for commercial telecommunications. [NASA/Marshall Space Flight Center image detail, date unknown.]

television TV. A system of sending and receiving broadcast moving images (even if the object in the transmission isn't moving), usually in conjunction with sound, although some closed-circuit television systems don't include sound circuitry. Television broadcasts can be transmitted through air or over cables, with cable TV (CATV) increasing in popularity. Air transmissions are captured with a television antenna designed for a portion of the broadcast spectrum (although three-in-one antennas exist for UHF, VHF, and FM signals). At the receiving end, a television set (tuner and monitor combined), or a VCR tuner and monitor are typically used to display the broadcast.

television broadcast band The various frequencies which are assigned and regulated for television broadcast transmissions. Due to the proliferation of programming and the increased availability of access through satellite transmissions, there is constant pressure to increase available frequencies and channels, and hundreds of programming channels are now available. See band allocations for a chart.

television camera A lens-equipped, optical-sensing pickup device designed to capture moving images and transmit or pass them on to receiving, editing, and broadcast equipment. The type of signal generated by the camera varies according to the receiving or editing equipment, and varies from country to country. Television cameras have traditionally been expensive, large, heavy, analog, high-resolution apparatuses. This is all changing, with small handheld digital and

Common Television Broadcast Formats		
Name	**Abbreviation**	**Notes**
National Televisions Systems Committee	NTSC	The North American standard since the 1950s. 525 vertical lines. NTSC uses negative video modulation and FM sound.
Phase Alternate Line	PAL	The predominant standard in the United Kingdom and parts of Western Europe since the early 1960s. 625 vertical lines. There are a number of variations of the PAL system, including PAL-B, PAL-H, PAL-M, etc. PAL uses negative video modulation and FM sound.
Sequential Color and Memory	SECAM	Developed in France and used in North Africa, Russia, and parts of Europe since the early 1960s. 625 vertical lines. There are a number of variations of the SECAM format, including SECAM-B, SECAM-H, etc.
High Definition Television	HDTV	Introduced in Japan and proposed as a global standard, but not readily adopted by American and other manufacturers, some of whom would prefer to enhance current standards rather than adopt a new one. 1125 vertical lines at 60 frames per second. HDTV is supported by some Internet push channels and can be viewed with an interface peripheral and a computer with a fast connection.
Multiplexed Analog Components	C-MAC	Developed in the U.K. and recommended by the EBU as a European standard.

T

analog personal cameras beginning to rival the quality of traditional TV cameras for only a fraction of the price. See NTSC, PAL, SECAM.

television history Television, perhaps more than any other of the major communications technologies, arose in fits and starts in the late 1800s with many geographically diverse announcements of success and few demonstrated working systems. One of the important discoveries in the history of television was the photoconductive characteristics of selenium, which responded to the amount of light hitting the surface. A French researcher, M. Senlacq, suggested in 1878 that selenium might be used to register the shapes of dark and light areas on documents. British researcher Shelford Bidwell was able to successfully transmit silhouettes by 1881, and the now famous German inventor, Paul Nipkow, after whom the Nipkow disc is named, patented an electromechanical television system in 1884. But the transmission of moving images and shades of gray in high enough resolutions to be practical eluded the early inventors. Although patents for television-related technologies began to appear in the late 1800s, it was not until the 1920s that television transmission and reception as we know it was demonstrated by inventors such as John L. Baird in the west and Kenjito Takayanagi in Asia. Baird's first significant success was in 1926, the same year Tekayanagi transmitted Japanese script with a cathode-ray tube.

In the U.S., a precocious 15-year-old, Philo T. Farnsworth, described an idea for a television to his schoolmates and reportedly showed a sketch to his teacher in 1922. In 1927 he succeeded in building a working model.

Experimental television stations sprang up in the late 1920s and, by the mid-1930s, regular public broadcasting began to develop. In Europe, television images were being transmitted by 1931.

Television sets were available by the late 1930s, but it took time before the technology became affordable for home use. By the late 1940s, there were at least 20 broadcast stations in North America, with hundreds of hopefuls clamoring for the limited licenses.

Black and white televisions came into widespread use in the 1950s in North America and color television was common about 15 years later. By the mid-1980s, melon-sized portable televisions became inexpensive and wrist-sized consumer TVs had been developed. Commercial sponsorship provides much of the funding for television in North America, thus controlling, to some extent, the type of programming which is available, influenced by majority consumer demand or perceived viewer preferences. In many other countries, television is funded and controlled by local governments.

The next major step in television broadcasting was the launching of communications satellites such as the Telstar 2 in 1962 which permitted intercontinental communication. Commercial application of satellite television broadcasting was pioneered by the Canadian Broadcasting Corporation through the ANIK satellite in 1972, followed in the late 1970s by

Turner and the Public Broadcasting System (PBS) in the U.S.

In North America, satellite television broadcasts can now be received by consumers on small parabolic dishes that are served by monthly subscription services, with hundreds of potential stations available. Television is widely used for mass-media entertainment, education, distance monitoring, and local security monitoring.

The influence of television on world culture is significant and substantial, with a preponderance of the programming originating in the United States. Thus, the role models depicted through television programming and advertising sponsors have a strong effect on viewers, and implicitly promote American values, styles of dress, and cultural priorities to all parts of the globe. See Baird, John Logie; Farnsworth, Philo T.; Nipkow, Paul Gottlieb; Nipkow disc; Takayanagi, Kenjito; television; television camera; Zworykin, Vladimir.

television relay A station designed to pass on a television broadcast signal to the next station so the signal is protected from loss. The relayed signal is not intended for reception by viewers until it reaches the destination station.

television signal The coding of images can be accomplished in a number of ways, and there are several standards, each of which is preferred in a different part of the world. Common formats related to the broadcast and display of moving image signals are shown in the Common Broadcast Formats chart.

telework Work at home or at satellite locations made possible through computer and telecommunications technologies. In 1988, Jack M. Nilles proposed a broad definition of telework as "... all work-related substitutions of telecommunications and related information technologies for travel," thus, employer/employee interactions across distance through new technologies. This term is more common in Europe and is roughly equivalent to the term *telecommuting* in North America. See ADVANCE Project, European Community Telework Forum, TelePrompt Project.

telex *tele*printer *ex*change. Generic term for a communications service developed near the end of the second world war that uses teletypewriters to transmit through wire lines and automatic exchanges to produce a written message at the destination. In Europe, this technology used audio frequencies over phone lines. See Baudot code, Telex, Western Union.

Telex A global message service established in the United States by Western Union in the early 1960s. This was competitive with AT&T's TWX service.

Telkes, Maria A physicist who did pioneer work in the development of solar energy in the early part of the 20th century. Solar energy has subsequently become an extremely important power source for orbiting communications satellites.

Telnet Protocol A widely supported 8-bit, byte-oriented network protocol for remote terminal access, originating from the days of the ARPANET. Telnet allows the user to log on to another system through a TCP/IP network, and perform file functions and other activities. Telnet is spelled in lowercase when used as a command to launch a remote utility that uses the Telnet Protocol. The form of the Telnet command is:

```
telnet [IP_address|host_name] [port]
```

(with the command entered in lower case). See RFC 318, RFC 854, RFC 855 to RFC 861 (various options).

TELSTAR 1 A historically significant low-altitude communications satellite that broadcast microwave transmissions and tracked satellites in the 1960s. This AT&T endeavor is claimed to be the first active communications satellite, launched 10 July 1962 by the United States, although some RCA engineers launched a transmissions satellite earlier. It is the first transponder-equipped satellite. Prior to this, satellites were passive transmitters, but the use of transponders for amplifying the signals was preferred from this time on, and some satellites now include as many as ten transponders. The TELSTAR had some early problems that were fixed in 1962; it ceased functioning in 1963. By 1964, two more TELSTAR satellites had been successfully launched and TELSTAR 3-D was launched in the mid-1980s.

Telstar Communications Satellite

The Telstar 3-D satellite being put into Earth orbit from the payload bay of the Space Shuttle Discovery in the mid-1980s. [NASA/JSC image detail.]

TEM wave transverse electromagnetic wave.

template 1. A pattern, guide, table, or mold used to provide the basic configuration, format, or design for creating a new version, or multiple versions of a project with few or no changes. A template is intended to save time by automating the creation of new versions. A word processing template can be used to set up documents which are reissued frequently with only minor changes (e.g., form letters).

Temporary Mobile Station Identifier TMSI. A dynamically assigned mobile station identifier (MSID).

TENET See Texas Educational Network.

tensile strength A descriptor for the greatest amount of longitudinal stress that can be borne by a particular material before it will rip apart. The units used to describe this property vary from industry to industry. It is an important factor in many manufacturing and industrial applications.

tension tester An industrial device that tests the tension parameters in a newly spliced fiber optic cable assembly. This is often sold as an option to a cladding alignment splicer. See cladding alignment splicer.

tera- T. A prefix for an SI unit quantity of 10^{12}, or 1,000,000,000,000. It's a trillion, a very large quantity, but considering there are now hard drives with terabytes of storage space, it's not as big as it used to be. It comes from the Greek root *terat* or *teras* meaning "monster." See peta-, pico-.

TERENA Trans-European Research and Education Networking Association. A European network evolving from the European Academic and Research Network (EARN) and the Réseaux Associés pour la Recherche Européenne (RARE). TERENA was established from the merger of these organizations in 1994 to promote and participate in the high quality international information infrastructures to benefit research and education. TERENA includes members from more than three dozen countries, as well as a number of high-profile computer developers/vendors, and the CERN and ACMWF international treaty organizations.

TERENA has been responsible for BITNET support in Europe including data collection and the distribution of nodes and routing tables. See BITNET.

terminal 1. An endpoint, extremity. 2. A conducting device, often a small metal post or receptacle, provided for facilitating a good electrical connection. 3. A device or system which provides remote access to a central computer. 4. An endpoint in a communications line, or one which can be, but is not necessarily, extended to other circuits.

Terminal Adapter TA. A device available in various configurations from a number of vendors, which provides protocol adaptation and interfacing with an ISDN line. A TA enables a variety of consumer electronic products such as computers, fax machines, etc. to connect to the ISDN service.

Terminal Endpoint Identifier TEI. An identifier for distinguishing between several different devices using the same ISDN transmission links. Values may be dynamically assigned to TEIs ranging from 0 to 126. Fixed TEIs are assigned values between 0 and 63. The value of 127 is reserved for TEI broadcast, which aids in carrying out management functions. An ISDN device must be assigned at least one unique TEI value, either by preassignment or dynamically, by the local exchange, as needed (sometimes called Auto TEI). The TEI Management Protocol is used to dynamically assign values in a request/response interaction. Dynamic allocation is usually used on point-to-multipoint links but may be used on point-to-point links

terminating office In a transmission such as a phone call or telegraph message, the terminating office is the switching center which is the final one that connects directly to the subscriber line or other receiver of the communications. In Internet dialup communications, the local ISP would be considered the terminating office.

TERMITE A terminology database which contains all terms appearing in printed glossaries of the International Telecommunication Union (ITU) since 1980, contributed by a variety of industry professionals, including technical editors and translators. English, French, Spanish, and Russian source terms are included. Access to the database is available through Telecom Information Exchange Services (TIES). ITU activities are now also being archived in this database. http://www.itu.int/search/wais/Termite/

terminator A physical device or setting to indicate the end of a data path in a connection in which more devices may be added. If a chain is not terminated, the end of the system is seen as an open port and the system either continues to try to send signals to (and to expect signals from) another device that isn't there or, in fiber optic networks, to experience undesirable back reflection and instability.

SCSI-format cartridge drives sometimes have autotermination built in to the device and the user may have to use a specified port on the device, if it is the last in the chain. Some devices terminate internally with a setting on a switch or a small internal or external dipswitch. Other devices, such as scanners and external hard drives, are terminated with an external terminator attached to one of the SCSI input/output connection mounts. Internal hard drives are often terminated with a set of resistors that can be removed or, in some cases, with jumpers.

Most SCSI chains can be terminated with standard 50- or 68-pin SCSI terminators, but there are exceptions, including a proprietary "black" SCSI terminator distributed by Apple Computer for some of their older computers and certain Apple laser printers that support font storage on attached hard drives.

On an Ether network using "thin" cables (10baseT), a terminator is required on each end of the data bus if the chain or "ring" is not closed. If "thick" Ether (10base2) connections to a plug-and-play hub in a star topology are used, separate termination is not required.

In a wired network, devices such as SCSI-format CD-ROM drives, hard drives, and scanners can be chained along the same data path, but a terminator is required on the end device (usually the one farthest from the motherboard) to prevent the system from seeking devices beyond the last one. Failure to terminate a SCSI chain can cause immediate or sporadic problems with data access on the chained devices. In a fiber optic network, the problem of back reflection can be even more serious than in wired networks. Many aspects of wired networks are designed to handle signals in two directions along one wire. Even when separate wires are used, the strength of the data signal is controlled such that damage to components is unlikely (possible, but not common). In fiber optic networks, an open, unterminated port may allow laser light to escape, which can be a danger to eyes. Improper closing of a fiber port can result in the laser light being reflected back down the waveguide in the wrong direction. This not only disrupts the path and amplitude of the light data signals, but also may interrupt or damage the laser source.

Passive terminators are most common, but advancements in networking and greater demands on the technology are giving rise to various types of active terminators. Longer cable runs and higher data rate performance may be accompanied by higher noise sensitivity that may be mitigated by active terminators with voltage regulation. Active SCSI terminators may autoselect between low-voltage differential (LVD) and single-ended (SE) modes. Because of the extra electronics, active terminators are sometimes a little larger than passive terminators commonly cost about 50% more than their passive counterparts.

Fiber optic terminators come in a variety of formats, including SC and FC and generally follow Telcordia standards. Some fiber terminators are doped and are typically sold to support specific wavelength ranges. Many of them resemble standard fiber connectors but they are designed to eliminated back reflection rather than to facilitate the attachment of another length of cable or device.

Terrestrial Digital Service TDS. A commercial private digital data transmission service offered by MCI to subscribers over local exchange carrier (LEC) T1, DS-3 systems. Thus, customers would have fast transmission links to MCI Services. There was some debate over this service option through the Federal Communications Commission (FCC).

TESC Technology Subcommittee.

tesla A meter-kilogram-second unit of magnetic flux density equivalent to one weber per square meter. Named after Nikola Tesla.

Tesla coil An air-core transformer for creating high-voltage discharges at very high frequencies.

Tesla, Nikola (1856–1943) An engineer and inventor born in Smiljan Lika (Austria-Hungary) who developed the alternating current induction motor, an essential part of alternating current distribution systems. Tesla began his research in Hungary, and then emigrated to the United States in 1884. He created a number of unique inventions and also improved upon those of others.

In America, Nikola Tesla and Thomas Edison came into regular contact with one another, not always with happy consequences, and an enmity grew between the two men. When it was proposed in 1912 that the Nobel prize be awarded jointly to Edison and Tesla, Tesla eschewed any association with Edison, and the prize went to a Swedish scientist instead.

Tesla's inventive mind turned power generated devices into interesting applications such as aircraft power systems and robotic submarines. In 1888, he was awarded a patent for an electromagnetic motor. Tesla earned more than 700 patents in his lifetime and produced many more unpatented ideas and inventions.

Tesla was somewhat temperamental and eccentric. One of his most practical contributions was the adaptation of alternating current into everyday applications. His colleague, George Westinghouse, further implemented many of Tesla's ideas. The tesla unit of magnetic flux is named after him.

TESP See Telecommunications Electric Service Priority.

test board A switching panel used for making temporary connections in conjunction with the panel or equipment being tested. By diverting some of the signals through the test panel, problems can sometimes be more easily isolated or identified. A test clip can also be used for the purpose of making quick temporary connections. See breadboard, shunt.

test jack/plug A connecting hole in a circuit or panel for inserting a corresponding plug (jack) and cable for making temporary connections for testing and maintenance. Test jacks and plugs come in many shapes and sizes depending upon what type of circuit is tested. A telephone test set is a portable telephone handset that has clips for temporary attachments to a conducting line or test jacks for temporary insertion into in a test jack plug frame.

test jack frame TJF. In private branch exchange (PBX) telephone systems, a frame for inserting test cables for maintenance and testing, usually mounted in or near the main PBX cabinet and terminating the cabinet connections.

test pattern Any pattern generated for a particular transmission medium that indicates the integrity of the various characteristics of its signal, which may include resolution, signal strength, stability, linearity, contrast, brightness, colors, sound range and quality, etc. 2. In video editing, a series of bands of specific colors. 3. In television broadcasting and television set calibration and diagnosis, a pattern (known to some as the Indian head pattern) which includes particular lines and line widths, ellipses, tonal gradations, and numerical values that allow the diagnostician to determine problems and make adjustments. This test pattern was frequently used in the 1950s and 1960s by local stations as a visual signal to viewers to indicate that there was no programming currently in progress, although this use has greatly declined due to the multitude of programming now available.

tetrode A four-element vacuum tube. The three-element tube, called a triode, was developed by Lee de Forest. This no doubt inspired experimenters to try other configurations. The four-element tube followed, consisting of a filament, plate, and two grids rather than one. The second grid, the tetrode or screen grid, was positioned between the first grid and the electron-attracting plate (anode).

Texas Educational Network TENET. A Texas education communications infrastructure dedicated to fostering educational innovation and excellence among educators and students. TENET developed through the collaboration of the Department of Information Resources, the Texas Education Agency, and the University of Texas. TENET provides various resources, including publications, discussion forums, and professional development seminars and facilities. http://www.tenet.edu/tenet-info/main.html

TFT See thin film transistor.

TFTP See Trivial File Transfer Protocol.

TFTP Multicast Option A protocol option for the Trivial File Transfer Protocol (TFTP) to enable multiple clients to concurrently receive the same file through multicast packets to increase network efficiency. TFTP Multicast Option was submitted as an Experimental RFC by A. Emberson in February 1997. See Trivial File Transfer Protocol, RFC 2090.

theremin An electronic musical instrument incorporating radio frequency oscillators in which two similar frequencies were combined to provide a lower, human-audible frequency. This was done by combining a reference frequency with a variable frequency. The theremin was played by interposing a hand to vary the capacitance between two projecting electrodes, thus controlling the pitch and volume. It was first constructed in 1920 and became popular in the late 1920s.

The process of mixing signals of slightly different frequencies is called heterodyning and was incorporated into many radios over the next couple of decades. A transistor version of the theremin still exists, and Mystery Science Theater 3000 fans are familiar with its eerie sounds.

It was named after its inventor Leon Theremin, who originally called his invention an "œtherphone." See heterodyning, Theremin, Leon.

Theremin, Leon (1896–1993) A Russian engineer and inventor who devised electronic musical instruments, most notably the "œtherphone" (theremin) while a student at the University of Petrograd. He traveled to America in 1927 to play a concert, and 2 years later licensed the Radio Corporation of America (RCA) to manufacture a "thereminvox." While in the U.S., Theremin also experimented with multimedia concerts, combining light shows and dance with the theremin music, later returning to do research at the University of Moscow. See theremin.

thermal noise Random noise arising from heat generated by the motion of charged particles. Thermal noise in electrical circuits is undesirable if it interferes with transmission.

thermal circuit breaker A breaker mechanism that trips when heat generated by excessive current expands the conductor. See circuit breaker.

thermion An electrically charged particle (a positive or negative ion) emitted from a heat source. See thermoelectron.

thermionic emission The emission of electrically charged particles under the influence of heat. Thermionic emissions are characteristic of *hot cathode*-ray tubes. Cathodes without thermionic emissions are called *cold cathodes*.

thermionic valve See vacuum tube.

thermistor An electrical resistor comprising a semiconductor with a high, nonlinear temperature coefficient. The resistance of the semiconductor varies sufficiently in relation to the temperature to make it useful in a number of applications. See thermostat.

thermocouple, thermal junction A device that measures temperature at the junction of a pair of joined wires employing dissimilar materials, with the difference in potential proportional to the temperature, determined by an instrument connected to the other ends of the wires.

Semiconductor thermocouple assemblies are fabricated by connecting two dissimilar metals and interconnecting the thermocouples in series. Materials with a high thermoelectric coefficient are typically used (e.g., antimony). When the junctions are dissimilar (e.g., different metals), the difference in heat absorption can be used to generate voltage, thus signaling the detection of heat. Treating one of the metals to increase its absorption through chemicals or coloring may magnify the effect (improve signal-to-noise ratio). See potentiometer; pyroelectric detector; Seebeck, Thomas; thermopile.

thermocouple wire A wire used with a thermocouple which is made of iron or particular alloys calibrated to the appropriate specifications.

thermodynamics The art and science of heat-related phenomena, their properties and relationships.

thermoelectron An electron (negative thermion) emitted from a heat source. See thermion.

thermography A printing process in which nondrying inks are treated to simulate a raised, engraved surface. After passing through the press, the ink is dusted with a compound which, after the excess is removed, is exposed to heat, causing it to fuse with the ink to form a raised surface.

thermopile A component with broadband absorption characteristics suitable for detecting radiation such as infrared light. Thermopile detectors are fabricated from multiple thermocouple devices connected in series. They may be made with wired junctions or film junctions, with film versions generally providing advantages of size, portability, and response times over larger wired assemblies. See photodetector, thermocouple.

thermoplastic A material with industrial significance because it can be heated and reshaped and rehardened by cooling. It has various uses including insulating and information recording. Contrast with thermoset.

thermoset A resin or plastic material which can be shaped and cured, but once this has been done, cannot be reshaped and cured again, as with thermoplastic. Contrast with thermoplastic.

thermostat 1. A sensing and regulating device triggered by temperature which is useful in turning machines on or off, for controlling fire safety devices such as alarms and sprinklers, and for regulating heating and cooling systems. 2. A device which regulates temperature, by measuring it and controlling heating equipment (or heating and cooling equipment) in order to maintain the temperature at the setting selected on the thermostat. This is usually accomplished by triggering the heating circuit when the temperature varies a certain amount below or above the desired setting. Thermostats that can be programmed for specific temperatures at scheduled times during the day are increasingly common. Temperature regulation (cooling) in large supercomputing implementations is important. See thermistor.

THF See tremendously high frequency.

thin film A very fine layer or combined layers used to enhance or change the properties of a material. Thin films are typically "grown" in chemical vapor deposition processes. There are many different types of film and their structure depends upon the chemicals used, the properties and combination of the component layers, temperature, and other fabrication parameters. Thin films can be grown with low dielectric constants, making them suitable for use in integrated circuits and as filters (e.g., DWM ONU filters).

Magnetic thin films have unusual megnetotransport properties useful for the development of sensing devices and magnetic recording technologies. Thin films can be grown at high temperatures for use in superconductivity research.

Thin films have thousands of applications as filters, dielectric mirror surfaces, barrier layers, polarizing layers, and more. Thin films can be deposited on fiber optic filaments to filter a signal or prevent backreflection. Antireflecting coatings for optical lenses in eyeglasses and imaging devices can be made with thin films. Calculators and wristwatches use silicon thin-film solar chargers.

Traditionally, thin film-treated wafers were placed between fiber endfaces to filter the light crossing from one fiber to the next. However, it has been suggested that better performance is possible by stacking between 20 and 150 layers of alternating high/low refractive-index films onto a substrate and using this thin film component in place of treated wafers. See sputtering. See Fiber Optic Probes diagram.

thin film transistor TFT. A technology used in display devices which creates a correspondence between a transistor and pixel on the screen so that pixels can be independently controlled. Used in color (RGB) active matrix LCD panels. This technology has been applied to portable display projectors and similar devices.

Thompson, Joseph John (1856–1940) An English experimenter who investigated electricity and X-rays. He was awarded a Nobel prize in physics in 1906 for gaseous conductivity of electricity.

Thompson, Ken (1943–) Principal developer, along with Dennis M. Ritchie, of the Unix operating system in 1969. It is quite a distinction considering its widespread use and utility. Unix has since evolved through extensive support by the programming community and exists in a variety of forms, although all bear similar features. Thompson also authored B, which was a predecessor to C. See Unix, UNIX.

Thompson, William See Lord Kelvin.

Thomson, Elihu (1853–1937) Inventor of one of the first alternating current (AC) generators, in 1878. At the time, the predominant form of power was direct current (DC). This was a significant achievement because it enabled the transmission of much higher voltages, necessary to cross some of the distances desired. Improvements to the concept were soon developed by William Stanley. Thomson also experimented, in 1892, with electric arcs. He collaborated with Sebastian de Ferranti and William Stanley in the development of the transformer.

thread 1. In piping, a helical indentation used to match and secure separate sections. 2. One of a number of continuing elements, themes, or trains of

Fiber Optic Probes and Thin Film Filter Technologies

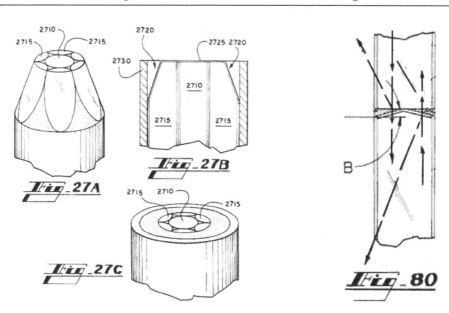

Figures 27a – 27c from a patent developed by M. Wach et al. show different views of a fiber optic light-scattering probe consisting of a tightly aligned multifiber ring surrounding a central fiber. Stepped index silica core/cladding fibers are suitable for this type of device. The central fiber, insulated by a light-blocking film or coating, may be used to deliver light while the surrounding ring fibers detect it or vice versa. Figure 80 illustrates the application of thin-film filtering to a complex contoured surface (in this case the cone-shaped endface of a fiber filament that is coupled with another fiber). Filters such as the one illustrated in Figure 80 can help reduce back reflection at coupled joints. Index-matching epoxy or gel can be used to fill in the gap between fibers.

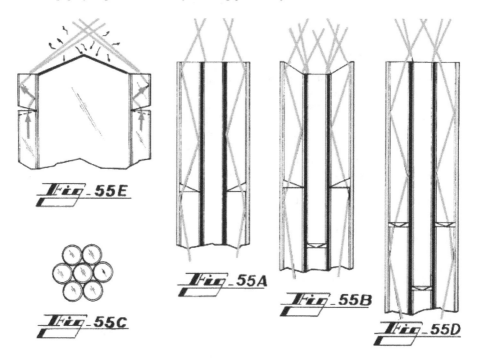

Figures 55a – 55e illustrate how different configurations are possible for manipulating light through adjoining segments of a multifiber probe. Various endface polishes, angles, and thin film filters make it possible to use the basic idea in a variety of applications. [Diagrams adapted from USPTO patent #6,416,234 , submitted August 2000.]

thought related to a common forerunner. 3. In programming, a flexible process organization mechanism by which individual processes can use common resources, but continue to operate unimpeded by other threads, if needed, in order to improve program efficiency or to increase simultaneous access to various system or applications resources. Common in object-oriented systems.

thread, discussion In online newsgroups, a topic of conversation characterized by the same (or similar) subject line, and theme and direction of discussion. Threads are a very convenient way to follow one line of thought through the myriad opinions discussed in the general context of a newsgroup. Good newsreading software will organize threads into groups and subgroups much the same way computer directories (folders) are organized on the computer operating system. Thus, the user can selectively open and read, or close and ignore, a thread.

three finger salute *slang* A descriptive phrase for rebooting the operating system (without powering down the system) with three designated keys held down simultaneously for MS-DOS/Intel-based IBM-licensed systems (Ctrl-Alt-Delete) and Amigas.

throughput Production; output; nonredundant information or items of relevance moving through a system. Throughput is used in industrial and computing industries to describe the efficiency of a system or end-result of a communication (how much information got through).

The measurement of throughput is quite specific to the system and information or objects being transferred, so there are few generalized standards for time intervals or total data against which to compare the throughput (end result). Nevertheless, relative measures of throughput, as compared to another manufacturer, another type of machine, or when processed in a different manner, can be very useful in tuning a production line system. Relative measures of data throughput in different parts of a network, or over different data protocols or operating systems, similarly can be used to improve the configuration and efficiency of a computer network.

TI-99/4 A Texas Instruments home computer introduced early in 1980. It featured 16 Kbytes RAM, sound capabilities, 16-color graphics on a 13-inch color monitor, extended TI BASIC, and cartridge-like solid state program modules for a list price of $1150 U.S.

TIA See Telecommunications Industry Association.

TIB See tone in band.

TIC See Token-Ring interface coupler.

ticket In telecommunications, a record of a transaction or paid toll, fare, or fee. The ticket indicates either that the transaction has been confirmed and it's OK to bill the client, or that the transaction and billing have both taken place (as in many credit card transactions). Tickets traditionally were on paper, but electronic tickets are becoming prevalent, with online transactions sometimes going directly through to the credit card company from the vendor without any slips or other paper confirmations.

tickler In computer applications, a program designed

to hibernate until a certain time or until certain events take place, and then become active to remind the user of something timely or important, such as appointments, anniversaries, events, etc. These applications have variously been called ticklers, reminders, and naggers.

tickler, electronic In electronics, a feedback or regeneration device consisting of two small coils connected in an electron tube, one to the anode (in series), the other to the grid-circuit.

tie *n.* Fastener, electrical strap, bundler. A strip, usually of plastic or Velcro™, to hold wires away from one another, bundle them together, or fix them in place, sometimes to a post or other secure structure.

tie line 1. In power systems, transmission lines that connect neighboring systems. 2. In telecommunications, a line for directly tying two telephone or telegraph connections together without going through a public switching center, often called a "dedicated" line. For example, the President might have a tie line connected directly to top advisors that doesn't go through any outside switches or connections, to ensure security, reliability, and speed for the connection. In the days before telephone switching centers and central offices connected local telephone subscribers, tie lines were common. For example, in the 1800s, a dozen telephone or telegraph tie lines might be wired out of a single office to connect it with other local businesses. The streets were often cluttered with hundreds of lines running between windows and buildings and the utility poles holding the lines were complexly wired and difficult to maintain.

Tie lines are still used for a variety of security and business-related purposes. They are especially useful in office complexes composed of several buildings that have a large volume of calls between the buildings. A tie line may be set up to connect directly simply by lifting the handset or pushing a button, as in a *hot line*. A *tie trunk* connects to telephone switching systems such as two private branch exchange (PBX) systems and may require dialing access code prefixes reserved for that purpose on a particular system or the system may be locally automated to interpret an extension number to be translated into a tie line call.

With the evolution of voice over digital data networks, the concept of tie lines has been adapted to computer networking. Access concentrators can be used to consolidate separate voice and data lines used for communications within a company into a single network and voice over network communications can then replace traditional telephone tie lines. Thus, existing Frame Relay links used for data, for example, could also be used for voice communications, removing the need for dedicated phone lines.

tie trunk A telephone (or telegraph) line directly connecting private branch exchanges (PBXs). See tie line.

tie line control system In power distribution, a system for administering the amount of electrical energy purchased by a subscriber (usually an industrial complex) from a utility company.

tie wrap A plastic, fabric, jute or other type of tie material used for holding together a bundle of cables for ease of placement or movement as a unit. Plastic tie wraps sometimes have a hook mechanism that catches and holds the tie when it is cinched tight. Velcro brand tie wraps are easy to unwrap and readjust, if needed.

TIES 1. Telecom/Information Equipment and Services. A government-to-government program which provides U.S. and Russian support for the expansion of international commerce in high technology. This is a subgroup under the U.S.-Russia Business Development Committee (BDC). 2. See Telecom Information Exchange Services.

TIFF Tag Image File Format. A very widely used platform- and application-independent, lossless, color, raster image file format that encodes the data as *strips* or *bands*. The TIFF format is used in faxes, image processing programs, scanned files, and many graphics creation programs. It is well supported by service bureaus and the printing and graphics design industries. Files are often identified by the *.TIF* or *.tiff* file extensions.

Creation of the format took into consideration the needs of the desktop-publishing industry and other related graphics applications, with the goal of making image information broadly interchangeable. TIFF was created to be extensible so that it may accommodate future needs.

TIFF has gone through a number of major revisions but, in general, fields are identified with unique tags so that various applications can elect to include or exclude particular fields depending upon their needs and capabilities. The core fields comprise Baseline TIFF. A TIFF file consists of three main parts: an image file header, a directory of fields, and the file data. Descriptions and definitions of baseline and extended fields are documented in the TIFF Technical Notes (TTN). An adaptation called TIFF-FX has been defined for facsimile applications. See facsimile, scanner, TIFF-FX, TWAIN.

TIFF-FX A subset of TIFF adapted to generating documents with minimal, lossless grayscale and color attributes for use as facsimile messages. The format uses some of the Baseline TIFF fields in addition to extensions pertinent to facsimile transmissions. Since the format can be used over both traditional and host-based transmissions media, TIFF-FX is suitable as a downwardly compatible, standardized facsimile format for data network communications. *Profile S* (TIFF-S) is a subset of TIFF, related to TIFF-FX, that defines a minimal black-and-white format to enable fast easy transmission of simple facsimile documents. *Profile F* (TIFF-F) is a slightly extended version of Profile S that is still restricted to black-and-white transmissions. TIFF-F was originally introduced by Joe Campbell and a group of fax experts; then, in 1998, with increased interest in Internet faxing connectivity, it was formally described by the IETF Internet Fax Working Group.

In essence, the image data to be faxed are compressed and inserted into a TIFF-FX file with the informa-

tional fields encoded with data specific to the image. The byte order is from least to most significant (an important detail, since the full TIFF specification can be set to either big- or little-endian and conversion may be necessary before transmitting as a TIFF-FX variant). While the full TIFF specification is somewhat flexible in terms of the ordering and structure of fields, TIFF-FX recommends that multiple *image file directories* (IFDs) be organized as a linked list. The MIME Content Type for these files is image/tiff. The Application parameter is TIFF-REG (optional). See IFax device, image file directory, TIFF, RFC 2301, RFC 2306.

TIIAP See Telecommunications and Information Infrastructure Assistance.

tiling 1. In printing, a technique for printing a large image on pieces of paper that are small, relative to the size of the image. Commonly used for billboards, banners, and wall-sized murals. Most computer printers have options for tiling, in order to print large images on letter sized paper. 2. In digital image display, a visual artifact common to heavily compressed images which causes a blocky, mosaic-like appearance to otherwise smooth lines and transitions. See DCT, JPEG.

tilt locking A method of frequency-locking a laser beam to an optical cavity by misaligning the laser with respect to the resonating cavity such that a non-resonant spatial mode is produced. An assessment of the interference between the carrier and the spatial mode yields a quantum noise-limited frequency discriminator. Tilt locking uses interference between the carrier field and a directly reflected phase reference signal (e.g., a non-resonant higher-order spatial mode). Thus, the encoding/decoding of spatial modes is optical rather than electro-optic.

Interference between the two spatial modes may be assessed by detecting the reflected beam on a two-element split photodiode such that each lobe of a transverse electromagnetic (TEM) mode is incident to a separate side of the photodiode. The error signal is derived by subtracting the photocurrents from each side of the photodiode.

Tilt locking may be useful for frequency stabilization, conversion, or interferometric gravitational wave detection.

TIMA See Interactive Media Alliance, The.

TIME Time Protocol. A network date/time protocol submitted as an RFC in May 1983 by Postel and Harrenstien. TIME provides a site-independent, machine-readable date and time. The Time service provides the time in seconds since midnight January 1, 1900. This is useful for systems that do not have a built-in date/time clock and for systems that need to be coordinated to preserve or aid data integrity or process administration. TIME can be accessed through port 37 over Transmission Control Protocol (TCP) or over the User Datagram Protocol (UDP). See RFC 868.

time code A system of encoding timing information on a recording medium, usually along with the information that is being stored. This technique is

T

commonly used with audio/visual recordings that will later be edited, dubbed, or otherwise manipulated or played within strict time constraints. Time code is typically stored as hours, minutes, seconds, and frames. Time code was developed in the late 1960s when analog recording tapes became prevalent, in many cases replacing film. The system was developed because video tape lacked the sprockets which previously had been used on film to synchronize sound and images. In the 1990s, another transition was made from analog video tapes to digital recording technologies, and the time code techniques used for analog video and audio encountered certain problems when applied to digital recording technologies. See chase trigger, MIDI time code, reference clock, SMPTE time code.

Time Division Multiple Access TDMA. A digital technology designed to overcome some limitations of analog cellular mobile communications. Time slot assignments allow several calls to occupy one bandwidth, thus increasing capacity for various wireless technologies. E-TDMA (Extended TDMA) provides even more time slots. TDMA is widely supported by AT&T Wireless Services. It is similar to Code Division Multiple Access (CDMA). See Demand Assigned Multiple Access.

There are a number of TDMA implementations, with three primary ones: European TDMA (GSM), Japanese TDMA (PHS/PDC), and North American TDMA (IS-136). See AMPS, DAMPS, cellular phone, time division multiplexing.

time division multiplexing TDM. A technique for combining a number of signals into a single signal by allocating a time slot in the combined signal with a multiplexer. At the receiving end, a demultiplexer is used to separate the interleaved signal back into its original signals. Some of the early developments of this technique were accomplished by J.M.E. Baudot in the 1870s. In current usage, TDM allows a variety of types of communications, audio and video, to be transmitted at the same time in one interleaved signal.

time signals From around the mid-1800s before time zones were established, to the present day, people have sought to devise ways to determine the time and synchronize their activities. The first time signals were drums or bells that were regularly sounded in local communities based upon the sun's position. Later, in the 1860s, the U.S. Naval Observatory used the telegraph to transmit time signals, and soon Western Union was sending standard time signals, a tradition they continued for a century. Telegraph time signals were similar to current Coordinated Universal Time signals, in that audible clicks were used coming up to the hour, just as tones now signal the upcoming minute. See Coordinated Universal Time, Greenwich Mean Time.

Time T An ITU-T designation for 2359 hours Coordinated Universal Time (UTC) on 31 December 1996.

time-delay modulation In optics, a form of phase modulation signal encoding the involves modulating a signal in the luminance channel with delays between pulses as the "carrier" signal.

time-space processing Also called temporo-spatio processing, this type of processing is used in applications and devices such as infrared motion detectors or radar systems which are using time and space as interrelated factors in their decision-making in terms of processing data or signaling an alarm condition. Time-space processing is also used in adaptive beam-forming antenna mechanisms.

time-to-digital converter TDC. An instrument for sampling a short time interval between two electrical signals, usually in pico- or nanosecond resolutions. TDCs are useful for measuring the leading edge of the time interval between transmitting and receiving pulses for tuning, maintaining, and troubleshooting electrical systems. They may also be used for detectors, imaging systems, laser rangefinders, and time-of-flight measurements. In clock correction systems, a TDC may be installed between a local clock and a processing system that derives optical data from the TDC.

There are tradeoffs in timing resolution and linearity between digital and analog interpolation methods used with TDC data.

Commercial TDCs typically include LEDs for displaying status and may support multiple independent channels (usually 4, 8, or 16). For longer time intervals, a counter and oscillator may be used.

timing Configuration of a system so successive repetitions are controlled for the desired interval (which may be desired to be variable), or so certain events begin and/or end at designated times or according to certain events. Timing is important in magnetic storage mechanisms, motors, signal amplitude sequences in electronics, and broadcast equipment configurations. Timing is also important on networks, where, for example, video and audio signals may be sent separately or on separate lines, but have to be coordinated at the end to provide services like videoconferencing. Constant oscillators are often used in conjunction with very precise timing devices. See atomic clock, quartz, SMPTE time code.

timing signal 1. A signal generated according to an accepted standard of time, usually for the purpose of providing a precise or objective baseline against which to measure events. 2. A signal generated by measuring some repetitive event which is then compared to some standard or clock. 3. A signal simultaneously recorded with data to provide a measure or standard against which the data can be analyzed. 4. A regularly emitted signal against which other time-related events can be synchronized.

TINA Telecommunications Information Networking Architecture. A networking telecommunications software architecture intended to be developed into a global standard. See Telecommunications Information Network Architecture Consortium.

TINA-C See Telecommunications Information Network Architecture Consortium.

tinned wire Wire that has been treated with tin to provide insulation and/or to facilitate soldering. Common on copper wire.

tinsel A fine, very long thread or strip of metal sometimes wound around electrical conductors between the insulator and the main core or wire. Because of its properties, tinsel is used in cables that need to be tightly wound or very flexible (such as phone handset cords).

tint Lighter or darker values of a particular color; hue. Tints are created by successive additions of white or black pigments, or by successively increasing or decreasing values of red, green, and blue (RGB) at the same time. Greater amounts of each color of RGB will produce lighter tints and lesser amounts will produce darker tints.

tip 1. The line or connection attached to the positive side of a circuit or battery. 2. In two-wire telephone wiring, the tip is traditionally the green wire attached to the positive side of the circuit at the central switching office. The name originates from the configuration of a manual phone jack in an old telephone switchboard in which the large plug was divided into two sections, with an internal wire electrically connected to the *tip* of the plug, and another wire to the *ring* around the plug partway up the jack nearer the insulated cord. See ring, tip and ring.

tip and ring Historically, the tip and ring designations derive from the configuration of a phone jack from a manual switchboard, called a cordboard. The *tip* was the positive circuit connected to the tip of the jack, and the *ring* was negative, located slightly away from the tip encircling the jack, sometimes called "sleeve." Later the tip and ring became standardized to correspond with the green and red color-coded wires traditionally used to install phone line services. Often telephone wire is composed of four wires with red, green, black, and yellow sleeves. Since dual lines have become more common in small businesses and in some homes, the black and yellow lines are used for tip and ring, respectively, for the second line. While these codes are standardized in North America, there are variations in other countries and in larger installations with multiple phone lines.

tip jack, pup jack One of the simplest connectors, a tip jack has a single, usually round, contact point plug that fits into a matching single-hole plug.

TIPHON See Telecommunications and IP Harmonization Over Networks.

TIRKS Trunk Inventory Record-Keeping System. A commercial product from Telcordia that aids in planning, inventorying, and assigning the telephony circuit order control and circuit provisioning of interoffice equipment and facility inventory. FEPS is a component of TIRKS that provides a range of automated software tools for planning and provisioning interoffice facilities and transmission equipment.

TIROS Television Infrared Observation Satellite. A historic series of global polar-orbiting meteorological satellites developed by GSFC, built by RCA, and managed by the Environmental Science Services Administration (ESSA). It was followed by the TIROS Operational System (TOS), then by Improved TIROS (ITOS), and subsequently the NOAA satellites. The TIROS spacecraft included low-resolution television and infrared cameras.

The TIROS craft 1 to 10 were launched between 1960 and 1965 into low Earth orbits. TIROS-N was launched in 1978. In general, they looked like cylindrical "mirror balls," studded as they were with solar cells, and were about the size of an oil barrel sliced in half. Spiny leglike antenna protruding from the flat end of the cylinder gave them an insect-like look.

TIROS systems provided the first meteorological data for weather forecasts that were received from space. Continuous coverage of the Earth's surface began in 1962 and proved the feasibility of space data for meteorological research and forecasting.

Remote-Sensing Satellite

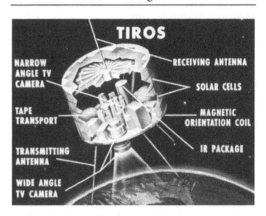

An artist's conception of the antennas, sensors, and solar cells in the TIROS 7 meteorological satellite, as envisioned in April 1961. [NOAA In Space Collection.]

TIROS N/NOAA Program A remote-sensing satellite program based on the TIROS satellite series that was initiated to improve upon the operational capabilities of the original TIROS numbered series from the 1960s and its successors in the 1970s.

In contrast to the earlier barrel-shaped TIROS systems, the TIROS-N satellites were longer and more rectangular, and about three times larger. The systems were three-axes stabilized and Earth-oriented.

Improvements in sensing technologies and solar power were incorporated into the later TIROS/N series which includes a number of successful and failed satellites beginning in 1978 and continuing into the late 1990s and present. ITROS-N satellites carried the Advanced Very High Resolution Radiometer (AVHRR) for day and night sea surface and cloudtop sensing. An atmospheric sounding system was also included, as well as a solar proton monitor to detect energetic particles from the Sun that might signal an upcoming solar storm.

Titanic, RMS The famous, ill-fated "unsinkable" ship that sank in 1912, with hundreds of lives lost, while crossing through ice fields north of Canada. The radio operator of the *Titanic* sent distress calls, but two of the closest ships didn't receive the communications, as 24-hour watch programs and radio regulations had not yet been established (this changed after the sinking of the *Titanic*). Fortunately, however, some of the sea-goers were saved by the *Carpathia*, who came to their rescue after receiving the radio distress call.

The sinking of the *Republic*, an earlier ocean-going ship, had a strong influence on legislation requiring wireless communications systems to be installed on ocean-going vessels. Due to the wireless distress calls, all but two of the hands on the *Republic* were saved. The sinking of the *Titanic* resulted in further legislation associated with keeping those communication lines open and monitored 24 hours a day. See the JASON project, MARECS, MARISAT.

TL See tie line.

TLD See top level domain.

TLF See trunk link frame.

TLP See transmission level point.

TLS transparent local area network (LAN) service.

TLS Protocol See Transport Layer Security Protocol.

TLV type, length, value. An encoding approach used in Basic Encoding Rules for the information content of elements.

TM 1. terminal multiplexer. 2. traffic management. A term associated with network transmission cell traffic flow, monitoring, and control. See cell rate.

TMA Telecommunication Managers Association. See Communications Management Association.

TMC traffic management center.

TMGB See telecommunications main grounding busbar.

TMN See Telecommunications Management Network.

TMS 1000 A one-chip 4-bit microcomputer introduced by Texas Instruments in 1972. Arguably the second microcomputer ever released with the Intel MCS-4 chip set, it never gained popularity.

TMSI See Temporary Mobile Station Identifier.

TOGAF See The Open Group Architectural Framework.

toggle 1. In general, a two-state process or switch. 2. Flip-flop. An electrical current that alternates in intensity at two distinct levels, or which has two states: on or off. 3. In software applications, a button or icon that has two states or positions. 4. In computer programming, a flag that is either set or not set, on or off. Toggles are very commonly used to keep track of software configuration settings.

toggle switch A switch that moves between two positions typically representing two states: on/off, high/low. (Three-state switches are sometimes also loosely referred to as toggle switches.) A traditional buildinng switch that turns a light on or off (in contrast to a dimmer switch) is a common type of toggle switch.

In aviation and audio equipment, a small narrow switch rounded on the end and tapered more finely at the point of attachment, roughly the shape of a baseball bat is sometimes called a bat switch. This type of switch is common in the aviation and audio industries where quick toggle adjustments are needed and many components are crowded for space in a small area.

token On a Token-Ring network, a status/priority information block used in coordination of traffic on the network. A token consists of a 24-bit frame that operates at the Media Access Control (MAC) level. It is continually passed around the ring in one direction and consists of a start delimiter (SD), an access control (AC) field, and an end delimiter (ED). Most of the information for controlling events is contained in the AC. It is further subdivided into a 3-bit priority field, a token bit (zero indicates it is a token), a monitor bit, and a 3-bit reservation field that lets the

Token-Ring Basic Frame Components		
Item	Abbreviation	Notes
Starting Delimiter	SD	Indicates the beginning of a frame.
Access Control	AC	Contains the Priority, Token, Monitor, and Reservation bits. In a frame, a workstation can only change the Reservation bits in the access control field. Only the active monitor can change the "M" bit, and only the workstation or device changes the Token bit.
Frame Control	FC	Indicates the type of frame: data frame or maintenance frame. A maintenance frame is used by the protocol to manage the ring.
Destination Address	DA	Physical or NIC address of the receiving workstation or device.
Source Address	SA	Physical or NIC address of the sending workstation or device.
Information		Layer control, routing control, and data.
Frame Check Sum	FCS	Error checking at the destination.
Ending Delimiter	ED	Indicates the end of a frame. If one of multiple frames or the last frame in a transmission, it's an I bit; if an error occurred, it's an E bit.
Frame Status	FS	Indicates if a frame has been recognized and copied by the destination station.

station get in the priority queue for future transmission of frames. A token is reissued after use by a station of the suitable priority, and continues on its way. See token-passing, Token-Ring frame, Token-Ring network.

Token-Ring interface coupler TIC. A Token-Ring local area network (LAN) port that is typically installed on a computer peripheral *network interface card* (NIC). The TIC facilitates the connection of the computer to the local network.

token latency In a token-passing scheme, the time it takes for the token to make it all the way around a token-passing local area network. See token-passing, Token-Ring.

token-passing A process on a Token-Ring network by which status/priority tokens are used as a mechanism to coordinate traffic around the unidirectional ring. A token is a 24-bit frame that operates at the media access control (MAC) level. It consists of a start delimiter (SD), an access control (AC) field, and an end delimiter (ED). The token-passing continues around the ring where each station checks the priority before adding frames to the traffic on the ring. Once a token has been used, assuming the proper priority, it is sent out again by the transmitting station to continue its journey around the ring through each station, carrying out the same sequence of events, according to priority levels.

Token-Ring frame A Token-Ring frame in the IEEE 802.5 specification consists of the components specified in the Token-Ring Basic Frame Components chart.

Token-Ring Interface Processor TRIP. A Cisco Systems high-speed interface processor with two or four Token-Ring ports, which can be independently set to speeds of 4 or 16 Mbps.

Token-Ring network TR network. A local area network developed by IBM in the mid-1980s, based upon a star or ring topology, that is, with nodes connected either directly to one central hub, or in a continuous loop not requiring termination. The token-passing is a scheme for data transmission between the stations which prevents collisions from different workstations sending messages at the same time. A workstation cannot transmit until it receives permission, that is, a *token* of the proper priority. The token is passed from station to station around the ring in one direction. At each station, the priority is checked before a frame is transmitted on the ring.

Token-Ring uses a source-routing system that moves information among stations based upon information in data packet headers, thus utilizing inexpensive bridges, a system different from Ethernet LANs. Speeds are up to about 16 Mbps, a limitation that has been addressed in High Speed Token-Ring, and throughput is about 60 or 70%, somewhat higher than Ethernet. Frames hold about 4000 bytes. Token-Ring LANs are sometimes combined with Ethernet LANs. They typically run over copper twisted-pair cables, although some are now implemented with fiber.

Since IBM's introduction, Token-Ring has been developed into a Media Access Control (MAC) level standard protocol by the IEEE (802.5).

A Fiber Distributed Data Interface (FDDI) network employs various token-passing concepts using dual rings to provide redundancy and fault tolerance. See Ethernet, Fiber Distributed Data Interface, High Speed Token-Ring, ring topology, topology.

TokenTalk An Apple Computer Macintosh-based implementation of a Token-Ring local area network. See AppleTalk, Token-Ring network.

toll call Any call to a location outside the local service area, so called because it is billed at a rate above and beyond the local subscriber service. A long distance call.

toll denial Denial of service outside the local service area. Toll denial may be part of a private exchange in order to limit calls to local calls, except as authorized. Toll denial may also be set up on an individual subscriber line (e.g., a subscriber who is behind in payments on long distance charges), or a line in a college dorm, or other location where potential toll abuse or fraud may occur. It may still be possible to make toll calls by going through an operator or entering authorization codes. See toll diversion, toll restriction.

toll saver A feature of answering machines and some computer software programs that lets you call into an answering machine from a long distance location and know whether there are any messages *before* the line is connected. The system is based upon the number of rings, usually four or two. With toll saver enabled, if there are no messages, the answering system will ring four times before answering. If there are messages, it will answer on two rings. That way, if the caller hears three rings, there are no messages, and he or she can hang up before the machine answers on the fourth ring and save the long distance charge.

toll terminal A phone system set up only for long distance calls (toll calls). Toll terminals expedite long-distance calls or compartmentalize billing or reporting on long-distance calls. A toll terminal may be in a secure location, accessible only to authorized callers.

tone above band TAB. A form of linear broadcast signaling used in conjunction with single sideband suppressed modulation. In TAB, corrections are applied to the received signal, as needed, to produce recognized pilot tones above the frequency of an information signal (as opposed to its center), to correct the accompanying information signal. Transparent tone above band (TTAB) is the same essential idea except that the tone is explicitly chosen to be outside the hearing range audible to humans. See tone in band.

tone dialing A system of audio tones called *dual tone multifrequency* (DTMF) used to generate distinct signals with which phone numbers and symbols can be transmitted. Frequencies are selected in such a way that dual tones are not harmonically related. Each tone is actually a combination of two tones, high and low, which are decoded when sent down the line to the switching office. The high tone is usually slightly louder than, or at least as loud as, the low tone. The tones range in frequency from about 697 to 1633 Hz. Two advantages of tone over pulse dialing are speed

T

and flexibility. As direct-dial long-distance services became available, numbers became longer, and it takes more time to dial a number on a pulse phone, partly because of the mechanical act of rotating the dial and partly because the dial has to return to the base position. Touchtone systems also provide more options. With a combination of digital processing and touchtone signals, automated menu systems can be accessed through the phone. These are now widely used by banks, mail order houses, and others. See pulse dialing, DTMF, touchtone dialing.

tone generator Any device or software application that generates tones. These may be at a particular frequency or may vary. Tones can be generated in the audible frequency range for humans, or higher ranges for electronic detection. Tone generators may be used as diagnostic devices on telephone networks. See buzzer.

tone in band TIB. A form of linear broadcasting modulation in which corrections are applied to the received signal, as needed, to produce recognized pilot tones at the center of a baseband information signal, thus making it possible to correct the accompanying information signal. Forward signal regeneration may be used to enhance TIB. Linear modulation techniques provide narrower channels than frequency modulation techniques, thus making it possible to increase the number of channels in a specified amount of space.

By the mid-1980s, transparent tone in band (TTIB), which uses a tone frequency inaudible to the user, was being developed for use in mobile radio networks, for example, to help reduce fading loss.

TTIB has become a standard in North America and Europe. See linear modulation, tone above band.

tone probe A network diagnostic device for testing voice and other audio networks by detecting and amplifying acoustical signals. The probe may be passive, detecting only existing signals, or may be active, sounding a tone that is read by a detector somewhere else on the system or by the probe generating the tone (which can then read the echo).

tone receiver unit TRU. The electronics in a telephone receiver which detect and interpret touchtone codes.

TOP See Technical Office Protocol.

top-down An organizing or processing hierarchy that distributes itself downward, usually in a branching pattern. Top-down often implies higher priority or more generalized functions or items at the top of the hierarchy. Thus, a top-down outline lists more important concepts first, a top-down personnel chart usually shows executive managers at the top, a top-down phone system starts with priority-listed agents, etc.

Top Level Aggregate TLA. An IPv6 prefix and a coveted commodity to the Internet community, the proposed assignment of Top Level Aggregates to privileged companies caused controversy. To quell the objections, TLAs were significantly increased and TLA requirements were removed from the IPv6 specification. Many Internet developers are concerned with preventing the development of a VIP

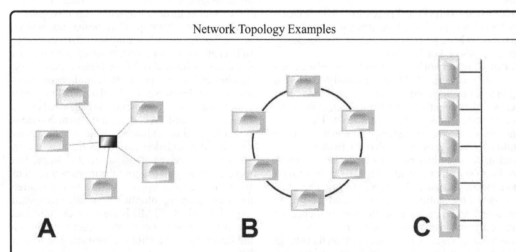

Network Topology Examples

A   B   C

These are three common topologies for interconnecting local area networks (LANs).

The star topology (A) is a popular way to interconnect computers, printers, and other peripherals with signal negotiations through a central hub (e.g., "thick" Ethernet). If a system on the network becomes nonfunctional, the network as a whole continues to intercommunicate.

The ring topology (B), as in Token-ring networks, is an older network topology that is still used in many LANs. In its simplest configuration, it is not fault-tolerant, however; if one system goes down, the network goes down, due to the missing link in the ring. In FDDI ring-based networks, there are dual rings to provide redundancy in case of a break in the link in the main ring.

The bus topology (C) is popular with daisy-chainable peripherals and with "thin" Ethernetworks that don't have too many systems attached. It is also used in certain backbone connections with LANs as the tributaries. The ends of the network are typically be terminated on bus topologies to inform the system that there are no more devices beyond a certain point.

system on the Net. See IPv6, Next Level Aggregate.

top level domain TLD. A hierarchical subset of the domain name system (DNS) which has been further subdivided into generic top level domains (gTLDs) and country code top level domains (ccTLDs). The distinction is administrative rather than functional; TLDs all theoretically have the same access and connectivity to the Internet. Within individual designations, some domain name extensions are open and some are restricted. For example, the popular *.com* designation used primarily for business entities is unrestricted, whereas the *.gov* designation is restricted for assignment to U.S. federal government agencies. Country code TLDs are open or restricted to varying degrees depending on the country maintaining the designation. Country code TLDs include *.ca* for Canada, *.uk* for the United Kingdom, *.au* for Australia, etc. By 1999, it was estimated that almost 100,000 domain names were being registered per month, with demand steadily increasing. See the Appendix for a list of domain name extensions.

topology, network topology A schematic representation and configuration of the geometric and electrical connections and relationships of a network and its various routing components. Depending upon the number of servers, terminals, routers, and switches, a variety of possible configurations are practical, including token rings, stars, and others. See backbone, mesh topology, node, star topology. See Network Topology Examples diagram.

Topology Database Update TDU. The refreshing of interconnections and routing information within a network system to reflect the current connections. The Topology Database may have upper limits as to the number of nodes that can be specified to describe the system. In many systems, there are commands for querying the Topology Database parameters and logs as well as commands for setting the basic operating parameters. For example, the frequency of updates may be logged to help a system administrator determine if configuration parameters are optimally set.

topology management In networks, the configuration, tracking, and management of connecting devices, particularly switches. The software used to manage them will often show graphical images of the various devices and their interconnections in the system.

TOPS See Traffic Operator Position System.

Torricelli, Evangelista (1608–1647) An Italian physicist and mathematician who invented the torricellian tube, now known as a mercury barometer. Barometers later became important in weather forecasting and in altitude-measuring instruments, particularly for aeronautics and ionospheric experimentation. The technology was also important in the evolutionary development of the vacuum tube, the basis for electronics for many decades until transistors and semiconductors were developed.

ToS See Type of Service.

total internal reflection In a light-guiding cable, the total reflection resulting when light rays guided through a waveguide (e.g., fiber optic filament) impact another material and are reflected back into the original waveguide. For example, in a fiber optic cable, the cladding has a slightly lower refractive index than the conducting core. When light impacts the cladding at angles that are not too steep (beyond the critical angle), the difference in the refractive index causes the light beam to reflect back into the light core.

Total internal reflection is not limited to fiber optic cables. It is characteristic of any optical component that has refractive quantum interactions at the interface between optical components. R.W. Woods was one of the first to document some unusual refractive properties in metallic surfaces, in 1902. See refraction; surface plasmon resonance; Woods, R.W.

Total User Cell TUC. In ATM networking, a recurring count of the transmitted Cell Loss Ratio (CLR), stored in the TUC field. The information is useful in assessing throughput. The CLR may be calculated using the Total User Cell (TUC) and the Total Received Cell (TRC) counts.

touchscreen A specialized computer monitor which is activated by contact with a finger or pointing instrument. The idea was that it was more natural for people to point than to use a mouse of keyboard. Unfortunately, holding an arm up for any length of time is uncomfortable, so touchscreens haven't become popular for extended or repetitive work. However, they are suitable for occasional input, as in kiosks and directory systems, and for some types of childhood education.

Townes, Charles Hard (1915–) An American physicist and professor, Townes joined Bell Laboratories in 1933 and worked there until 1947 after which he became a professor at Columbia University. Beginning in the early 1950s, Townes became a significant pioneer in maser/laser technology and collaborated on many projects with Arthur Schawlow. A number of his students and those who interacted with Townes in the Columbia lab also distinguished themselves in subsequent laser technologies.

In 1961, Townes became a professor and Provost at MIT and stepped down from Provost in 1966 to devote more time to research. In 1967, he became a professor at the University of California (Berkeley).

Townes has received many awards for his work, most significantly a Nobel Prize in physics coawarded in 1964. Since 1980 the Optical Society of America has annually awarded the Charles Hard Townes Award for outstanding work in quantum electronics. See Basov, A.; Capasso, F.; laser; laser history; Prokhorov, A.; Schawlow, A.

TP 1. test point 2. transaction processing. 3. transport protocol. 4. twisted pair.

TPA See Telephone Pioneers of America.

TPDU See Transport Protocol Data Unit.

TPEX twisted pair Ethernet transceiver.

TPI See tracks per inch.

TPOS See Training, Planning & Operational Support.

TPWG See Technology Policy Working Group.

TQM total quality management. A management philosophy and means of putting it into action to develop and maintain quality principles in commerce, service, and manufacturing. Quality assurance and ISO-900*x* certification are two kinds of tools in the TQM arsenal. See ISO 9000.

TR transmit/receive.

tracer A diagnostic tool for tracing a link through a circuit. In fiber optics, a tracer shines a visible light into the lightguide so a technician can check whether it is following the appropriate path and making it all the way through the lightguide (e.g., there are no breaks or excessive bends). An invisible beam (e.g., infrared) may also be used if an appropriate detector is at the destination point to detect the incoming light. A fault locater is a type of tracer that uses a powerful laser light source that can illuminate through the fiber jacket into the fiber to reveal anomalies such as a break in the lightguide. See optical spectrum analyzer, reflectometer.

traceroute A Unix shell utility written by Van Jacobson at Lawrence Berkeley Labs which seeks out and displays the path of a transmission packet as it travels from host to host, detailing the hops in the path. Traceroute sends an IP datagram to the destination host, then it iterates through each router, decrementing TTL, discarding the datagram, and sending back ICMP messages until the destination host is reached. Traceroute is extremely useful as a network diagnostic and optimizing tool and often used in conjunction with ping. See ping. The Traceroute Example inset below shows a sample of traceroute output showing the IP numbers, hops, and packets.

trackball A hardware peripheral device which receives tactile directional input and transmits it to a computing device. The information transmitted is similar to that of a mouse or stylus, and is often used in conjunction with graphical user interfaces (GUIs). Unlike a mouse or stylus, a trackball is generally fixed in place, with physically separate buttons. Trackballs are common in video arcade games and on laptop keyboards to increase portability. See joystick, mouse.

tracks per inch TPI. A measure of the density of a recording medium such as a phonograph record or formatted hard drive. The tracks may be physical (as in grooves) or virtual (as in a logical segmentation on a drive that varies according to the file standard and/or operating system).

TRACON Terminal Radar Approach Control. An airfield navigation radar system designed to aid in landing approaches and takeoffs. In commercial airports, the TRACON is typically housed in the air traffic control tower (ATCT). The Federal Aviation Administration (FAA) has almost 200 of these radar approach facilities in the U.S. A TRACON control room may accommodate a number of operators and typically has backup generators to ensure air traffic safety in the event of power outages. As cities grow, the need to support increased and more complex air traffic grows as well. To meet this need, the Federal Aviation Agency (FAA) has been working on projects to consolidate individual TRACONs into what is termed Large TRACONs.

tractor feeder A sprocketed paper alignment device used mostly on impact printers. It provides more precise control of positioning than most roller feeders by preventing slippage. The tractor feed resembles a pair of short regular series of inverted cleats which fit through corresponding holes in tractor feed paper. Tractor feed mechanisms are especially useful for multipage documents (invoices, checks, etc.) on dot matrix printers.

trade secret Information, data, process, or procedure which would lose its commercial value if revealed to outsiders. Nondisclosure agreements (NDAs), policy statements, and inservice training are mechanisms which restrict external communication of trade secrets. If trade secrets are not specifically identified by an employer, it may be more difficult to stop or prosecute an offender who has willfully or inadvertently revealed them. See patent, copyright, nondisclosure agreement, trademark.

trademark A legal designation for the right of an association to the use of a mark in trade. It can consist of a word, phrase, or symbol sufficiently unique

---

Traceroute Example

```
$ traceroute abiogenesis.com
traceroute to abiogenesis.com (207.173.142.184), 30 hops max, 38 byte packets

 1  chapman.nas.com (198.182.207.6)  322 ms  252 ms  251 ms
 2  orthanc.nas.com (198.182.207.1)  251 ms  247 ms  275 ms
 3  milo-s4.wa.com (204.57.232.1)  277 ms  257 ms  296 ms
 4  dilbert-fe4-0.wa.com (192.135.191.254)  530 ms  303 ms  273 ms
 5  dogbert-f4-0.nwnexus.net (206.63.0.254)  263 ms  282 ms  286 ms
 6  borderx2-hssi2-0.Seattle.mci.net (204.70.203.117)  306 ms  277 ms  373 ms
 7  core2-fddi-1.Seattle.mci.net (204.70.203.65)  288 ms  511 ms  *
 8  bordercore1.Denver.mci.net (166.48.92.1)  299 ms  323 ms  293 ms
 9  electric-lightwave.Denver.mci.net (166.48.93.254)  294 ms  351 ms  313 ms
10  H3-0.1.scrlib01.eli.net (207.0.56.162)  329 ms  394 ms  591 ms
11  gw2-CALWEB-DOM.eli.net (208.131.46.46)    730 ms  gw1-CALWEB-DOM.eli.net
    (208.131.46.30)  507 ms gw2-CALWEB-DOM.eli.net (208.131.46.46)  357 ms
12  abiogenesis.com (207.173.142.184)  362 ms  358 ms  350 ms
```

to be distinguished from others in the same general industry. Provided it is not already owned by another entity, a trademark becomes valid as soon as a company uses it in trade according to certain stipulations, provided the company continues using it. You cannot come up with a trademark idea, not use it, and claim it later if someone else uses it. Trademarks may be registered federally for a fee, although this is not required. The motivation for trademark registration is to provide prima facie evidence in the event of a legal dispute. Unregistered trademarks must be identified with a ™ symbol, and registered trademarks must be identified with an ® symbol. Policing of trademark violations is not a responsibility of the federal government or of the agency that registers a trademark; it must be done by the company seeking to protect the trademark. Trade names are similar to trademarks and are registered at the state level. There can be more than one company using the same trade name, if the line of business is sufficiently different to prevent confusion.

The Commerce Department's large database of patent text and images, 800,000 trademarks and 300,000 pending registrations dating from the 1800s, has been gradually uploaded to the Web since 1998. See copyright, patent. http://www.supot.gov/

Traf-O-Data The first business partnership of Paul Allen and William R. Gates, founded around 1972, growing out of the business experiences of the less formally organized Lakeside Programming Group. Allen and Gates worked together on a variety of software projects including an automobile traffic flow system. At about the same time, they prototyped an early microcomputer based upon the recently released Intel 8008, with the participation of a hardware designer, Paul Gilbert. They made some tentative attempts to sell this early machine, but it didn't work during an early demonstration.

At about the time Gates graduated from high school, Allen encouraged him to join him to start a company to build and sell computers based upon the Intel 8080 chip set, but Gates wasn't fired up about the idea of a hardware company, and his parents were encouraging him to continue his post-secondary education. The majority of Gates and Allen's subsequent efforts were software-related, most significantly, a BASIC interpreter for the Altair, and a disk operating system for IBM, which resulted in the founding of Microsoft Corporation. The subsequent success of the company was due to sales of software, primarily operating systems and business applications, and Apple Computer became the small business success story in computer hardware.

traffic A term often used on large communications systems to describe communications signals, data, cells, or packets which comprise the information and signaling associated with the transmissions. The term is sometimes also used to describe the overall flow and pattern of traffic in the context of the system it is on, as in traffic congestion, traffic flow, traffic monitoring, traffic engineering, traffic routing, etc. In telephony and various systems that are largely analog

based, the term may be used to describe the flow of information in the context of calls, as in call attempts, call connects, call volume, etc. In digital systems, it is often used in a more specific sense to indicate numbers of packets or cells. See leaky bucket.

traffic capacity A measure of the capability of a system to carry a certain maximum number of calls, cells, frames, or packets, depending on how capacity is measured or data carried on that particular system. The maximum capacity may not be a fixed number, as on systems where multiple channels may be aggregated to carry certain broad bandwidth types of information. On other systems, where one or two wires can carry one and only one communication, traffic capacity is more likely to be expressed as a set amount. *Traffic capacity* is used in systems design and marketing to provide buyers a general guideline as to the capability of a system. For example, a rural phone switching system in a small community may only require a capacity of 20,000 call seconds (usually expressed in increments of *hundred call seconds*) per hour during peak calling times, providing a measure of whether a system can meet or exceed those needs. See traffic concentration.

traffic concentration A measure of communications traffic that indicates how different peak traffic periods are compared to traffic on the whole. In telephony, the service day is usually broken down into hours for traffic monitoring to determine high and low traffic periods. Traffic concentration is typically expressed as a ratio of the traffic (number of calls) during the busiest hour to the total traffic during a 24-hour period. This may further be calculated over a period of weekdays, weekends, or 7-day weeks to provide statistical averages.

traffic engineering In more traditional communications systems, traffic engineering is the estimation and application of the type, quantity, and configuration of devices and equipment required to meet the needs of a predicted number of users of the system being designed. Experience, trial and error, probability theory, similar system comparison, and insight are all brought into play in designing a system to meet current and predicted future needs. This system works quite well with traditional telephone switching systems and local area networks.

As systems become more complex, however, and distributed digital data systems more prevalent, traffic engineering, once the initial equipment is set in place, becomes an even more esoteric process, with much of the configuration and selection carried out by computer software, not just through physical connections. Virtual networks are now laid over physical networks, resulting in several levels of traffic engineering. When these networks are interfaced with a global network like the Internet, then prediction of the number of potential users becomes more difficult and less important than incorporating flexibility and scalability into the system to handle unpredictable traffic loads and activities. At this level, traffic engineering becomes a collaborative activity between programmers and traffic engineers, with careful evaluation of

T

data and message priorities – scheduling and incorporating deliberate delays, for example, so some types of traffic can be set to transmit during off-peak hours. On data networks, many traffic decisions are now built into the software on servers, gateways, routers, and some of the high-end switchers.

traffic load The sum total of the traffic on a particular system or portion of a system, such as a specific trunk, leg, or hop, measured at a particular point in time, or during a particular specified range of time.

traffic monitor A mechanism for keeping track of traffic on a system. On data networks, some of the software included for system administrators allow the monitoring of various processes, and typically display them as sampled or realtime graphical charts. Thus, CPU usage, number of users, number of transmissions, level of traffic, etc. can be visually assessed, and statistics derived from computerized traffic and analysis can be stored and evaluated to make changes as needed. The results of the monitoring are sometimes directly incorporated into other software on the system which makes adjustments to priorities, number of users permitted online at one time, and other parameters that can be changed to increase or decrease the traffic to optimize use of the system.

Sometimes traffic monitoring is a very simple mechanism. One of the simplest and most familiar applications is on modems, where little blinking lights give some indication of how much data are traveling through the modem, and when. If a user is downloading a large file from a Web site and those lights stop blinking, it's possible that the connection has been dropped or there is a glitch in the system. In ATM systems, traffic monitoring is an essential aspect of preventing network congestions and bottlenecks, and some cells will be flagged as to their discharge eligibility. See cell rate, leaky bucket.

Traffic Operator Position System TOPS. A system to support telephony operators from a toll switch, TOPS was developed by Northern Electric (later Northern Telecom) in the early 1970s, a time when manual cord-based operator terminals were poised to be gradually replaced by automatic stored-program switches. The historic TOPS-1 utilized an Intel 8008 CPU; the first TOPS office was installed in 1975 in Alaska. As computer technologies improved, TOPS was upgraded to faster, more versatile processors.

traffic overflow A situation where demand exceeds capacity. When overflow occurs, the traffic is either routed to another trunk or leg, or it is rejected and a signal sent to the user in the form of a signal (like a fast busy) or a text message (as a broadcast on a computer system). In some data networks, particular packets may be discarded if there is traffic overflow, or they may be routed back to the sender.

traffic path The physical or virtual pathway taken from the sender to the receiver. This may be fixed, as in direct wire communications and smaller data networks, or it may be flexible, as in switched systems and many larger data networks. On very small systems, where the setup is known and doesn't often change, a fixed traffic path may be the fastest and

most practical implementation. In dynamic data networks, the communication may not be quite as fast, but the system has an advantage in that it can tolerate and adjust to unpredictable changes.

A path in a phone system usually passes from the subscriber to a *demarcation point*, usually a junction box between the internal and external wiring, through one or more *switching centers* which are connected with pathways called *trunks*, or sometimes through a wireless link during part or all of the transmission. In a digital computer network, the data may pass from the sender through various *servers*, *gateways*, *switches*, and *routers*; each section in the path is called a *hop* or a *leg*, with the various terminal points generically called *nodes*. A data network transmission may also be wireless for part or all of the path.

traffic policing In ATM networking, a mechanism which detects and controls cell traffic according to specified parameters. See cell rate.

traffic recorder A means to record traffic on a specified transmissions channel in order to monitor capacity, load, efficiency, etc. It may or may not be paired with software that helps analyze the traffic.

traffic reporting In networks, information on traffic flow gathered and organized by external analyzing devices or internal monitor agents. This information about packet volume, distribution, collisions, and errors may be reported in the form of tables, ASCII graphs and charts, or images, and is essential for configuration, tuning, and troubleshooting.

traffic shaping In ATM networking, a mechanism which shapes or modifies bursty traffic characteristics in order to create the desired traffic. See cell rate.

train *v.* To instruct, teach, indoctrinate, or drill. Training is an important component of any computing system that must recognize input beyond point and click or keyboard instructions. In pen computing, handwriting must be interpreted into commands, and it is usually necessary to train the system, by successive trials and feedback, to recognize an individual's style of writing. Some camcorders have eye-controlled systems which need to be trained to track the direction of focus of the user's eye. In voice recognition systems, software is trained to recognize an individual's mode of speaking. OCR systems can be trained to improve recognition of unfamiliar or unusual character sets. While software training systems are not perfect, they have evolved to the point where they do much useful work, and improvements in technology and software algorithms indicate that some day natural methods of computer input may supersede keyboard and mouse entry for many applications.

training An automatic feature of some hardware and software systems to evaluate the characteristics of incoming signals (timing, delay, etc.) or information (handwriting, voice, etc.) and improve performance or recognition through successive adjustments and feedback. See artificial intelligence, expert system, robotics.

Training, Planning & Operational Support TPOS. A support function within the Operations Division of the National Communications System (NCS). TPOS coordinates telecommunications operational planning among NCS elements, provides educational seminars, cooperates with other federal and regional organizations, and participates in the Regional Interagency Steering Committee (RISC) meetings. See National Communications System.

transaction 1. An agreement, or exchange of information or goods, between two or more parties or entities. 2. A business transaction, which may be subject to various recording requirements (contracts, taxation, audit records, etc.). 3. An entry into a database or spreadsheet.

transaction, network Any of a number of situations in which information or signals are exchanged, passed on, recorded, or evaluated. Examples of network transactions include protocol determinations and conversions, security-level evaluation and processing, routing, transmission between nodes, error processing, etc.

Transaction Capability Application Part TCAP. In the SONET specification, there are several chapters regarding the function, formats, and definitions for TCAP provided in the ANSI T1.114-1996 document.

Transaction Team A group with the Office of General Counsel that coordinates the Federal Communication Commission's (FCC's) review of applications for major transactions and changes. The Team was announced in January 2000 as a response to an unprecedented number of significant mergers in the telecommunications industry and was tasked with streamlining the review process while still safeguarding the public's interests.

transaction tracking A system of recording each instance of a transaction and, sometimes, the actual transaction or its outcome. In database or spreadsheet applications, for example, each transaction may be recorded as entered, in order to prevent loss of data due to a series of transactions not having been saved at regular intervals. This type of live recording of transactions frees the user from worrying about "saves" and ensures that, under most fault situations, no more than the most recent entry may be lost. See ticket.

transatlantic cable A communications link across the Atlantic Ocean incorporating a bundle of wires or, more recently optical fibers, within a tough, corrosion-resistant insulating protective sleeve. The cable is laid along the ocean bottom in regions and at depths where the likelihood of severing by boat anchors is less. Maintenance of oceanic cables is a complicated business, so much emphasis is put on building and installing them correctly in the first place.

The invention of the telegraphy in the 1800s provided a strong motivation for laying a cable to bridge the communication gap between North America and Europe. Since postal messages could take from several weeks to several months to traverse the Atlantic, there was a potential goldmine in the provision of cable services. Samuel Morse was one of the more prominent inventors who suggested the feasibility of a cable between North America and Europe.

Only three decades elapsed between the development of the first commercially practical telegraphs in England and America and the laying of the first working oceanic cable. The first attempts did not succeed, however, due to problems with installation, insulation, and capacitance over long distances, but improvements in insulation, and trial and error experience, resulted in eventual success.

The first working transatlantic cable was installed in August 1858, initiated by F.N. Gisborne's efforts to interlink maritime Canada, and financed and promoted by Cyrus W. Field. A more permanent installation was realized by Field and his associates in July 1866. An additional link was established between Ireland and Canada, in 1894. Cables to other continents and through other oceans followed as the technology improved. In the 1870s and 1880s, companies were formed to lay cables to South America and Africa.

With increasing demands for high-speed wideband communications links for data communications, transatlantic cables are more important than ever; the old wire cables are being replaced by fiber optic cables and new cables are being installed to link regions not previously interlinked. See British Indian Submarine Telegraph Company; Field, Cyrus West; Gisborne, Frederic Newton; gutta-percha; Pender, John; Submarine Telegraph Company.

Historic Transatlantic Cable Telephone Call

Prime Minister Richard Bennett (with cabinet members) speaking over a transatlantic telephone to George Perley at the British Empire Trade Fair at Buenos Aires, Argentina, in March 1931. [National Archives of Canada National Film Board image.]

Transcontinental Cable A generic euphemism for the U.S. Defense communications system stretching from Washington, D.C. to California and Florida during and after World War II. It was called this because of the labeling of the dig warning signs located in fields and near garden shed-like repeater stations positioned every few miles along the cable routes across the country. See L CXR.

transceiver Device which transmits and receives within a single unit, often sharing circuitry to reduce size and weight. See transmitter, receiver.

transcontinental telegraph The first American transcontinental telegraph line was initiated by Hiram Sibley, with encouragement from Ezra Cornell, in the mid-1800s. Sibley, the founder of what was to become Western Union, made a reasonable estimate that it would take 2 years and about $1 million to complete the project.

The telegraph line had a significant impact on the new Pony Express service, which had been in operation for less than 2 years when it shut down in October 1861, after inducements to stay in service at least until the telegraph line was completed.

Not surprisingly, problems other than weather plagued the construction of the line. Buffalo discovered that telegraph poles made good scratching posts, sometimes bringing down the poles in their enthusiasm. Native Americans sometimes made off with the wires because the lines stretched through their treaty lands; within days or hours, exquisitely woven copper wire bracelets would appear in local trading markets. Remarkably, despite the great distance, harsh conditions, and the small size of the work crew (only about 50 line workers), the transcontinental telegraph was completed in October 1861. It had taken only 4 months at a fraction of the projected cost, one of the most stunning achievements in western engineering.

transducer A general term for a device which converts one form of energy to another, a process used throughout communications. When sound waves from a telephone conversation come in contact with a telephone mouthpiece diaphragm, the diaphragm causes small polished carbon granules to cohere and the energy is converted to electrical impulses that are transmitted along the phone line. When the mechanical movements of a phonograph stylus are turned into electrical impulses, and then, at the speaker, converted again to audible sound waves, the signal has gone through (at least) two transducers.

transfer lens A lens that propagates light in a useful direction and may also concentrate or diffuse the light, as needed. The *effecitve plane* or *working plane* is that region over which the propagated light is optimized for the needed purpose, which is often uniformity and consistency of the light beam (usually perpendicular or nearly so to the plane of the light beam).

In the context of light transfer from an illuminator to a fiber optic lightguide, a lens designed to propagate the light from a laser or light-emitting diode into an optical fiber core with a minimum of feedback to the laser. If the laser light hits the fiber lightguide in such a way that it reflects back, it can cause jitter or fluctuating performance in the laser.

A hyperbolic lens surface, for example, can collimate laser light and further may transmit the light towards the fiber lightguide at angles that are efficiently related to the axis of the fiber and the angle most likely to reduce reflections from the end of the fiber. Thus, there are compromises in which the most efficient balance of factors is sought.

Toroidal, circular, or spherical lenses may be used as transfer lenses, depending upon the type of assembly, power of the laser light, and diameter of the lightguide. The precision of the application (e.g., high-speed telecommunications) dictates, in part, how sophisticated or efficient the transfer lens may need to be, as will the degree of curvature of the fiber filament endface.

A transfer lens may also couple photomultiplier tubes to CCD components in instruments such as spectrometers. Since traditional optical lenses for this purpose tend to be expensive and bulky, some instruments now use a fiber optic faceplate to provide a lighter, thinner coupling interface. Depending upon any size differences between the coupled parts, the fiber array may use fiber tapers rather than straight fiber filaments to guide all the light into a smaller or larger CCD components. See diffusion; face plate, fiber optic; Fresnel lens.

transformer An electrical device for changing the qualities of a current by mutual induction. Transformers did not come into wide use until the early 1900s, gradually superseding spark coils for providing power for communications and electronic components. They were similar to spark coils in that they had a core surrounded by conductive windings. However, the core used soft iron sheets rather than a bar. Like a spark coil, two sets of windings, one within the other, were commonly used. The core could be closed or open. Transformers required alternating current and direct current was still prevalent at that time, but the use of alternating current allowed the elimination of a vibrator, as the natural alternations of the current caused inductive discharge. This was more compact and practical than a spark coil.

Since many modern electronic appliances (modems, printers, answering machines, model trains) have electrical requirements different from that which comes out of a household socket (110 or 220 AC), the power cord may be equipped with a transformer which modifies the current to the needs of the device being powered (9 or 12 V is common). It is important to use the correct transformer; if the voltage is too high, it will likely blow the components.

Transient Mobile Unit A mobile communications term for a unit that communicates through a foreign base station.

transistor A small device developed in the late 1940s which provided a means to amplify signals with very low power consumption and very little heat. The name, derived from "trans-resistor," has been attributed to John R. Pierce who worked with Schockley at Bell Labs. The importance of the transistor to the development and evolution of electronics cannot be overstated. A new world of tiny components opened up, including portable radios, hearing aids, computers, satellites, and much more. See transistor history.

transistor history The invention of the transistor in 1947 is widely attributed to William Shockley, John Bardeen, and Walter H. Brattain of Bell Laboratories, although Ralph Brown is sometimes also mentioned. Starting in 1951, when they began to be commercially produced, transistors replaced large, power-hungry, cumbersome vacuum tubes, enabling electronics to be smaller and less expensive, and to run cooler and faster. (There are still some high-frequency applications where the use of vacuum tubes is practical.) The development of the transistor was foreshadowed by the 1926 patent application of Julius Edgar Lilenfeld, who had devised a way to control the flow of current in a solid conducting body by establishing a third potential between the two terminals. Later, in the 1970s, many types of transistors were succeeded by semiconductors. See de Forest, Lee; Kilby, Jack St. Clair; Pickard, Greenleaf.

transistor radio A radio developed in the 1950s, based on small semiconductor transistors instead of larger electron tubes. The smaller size and power consumption of transistors made it possible to design handheld portable radios, which became popular and widespread in the early 1960s. Portable radios were not new, since the early crystal detector sets required no outside power and could be carried around in a small case. However, practical, amplified, battery-driven portable radios did not become widespread until the development of small, low-cost transistor components.

Historic Transistor Patent

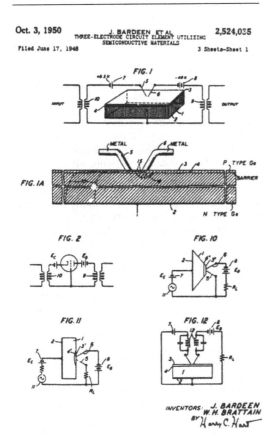

The landmark patent for the transistor, an invention that dramatically changed electronics. [U.S. patent document, public domain.]

---

transistor-transistor logic TTL. One of three main logic families, TTL logic circuit design is similar to diode transistor logic (DTL), but with multiple emitter transistors. TTL is based upon bipolar components, thus, when coupled, one can be used as the input component and one as the output component without any intervening resistors or other connecting components. This makes it faster than earlier technologies, though not as fast as a similar circuit designed with emitter-coupled logic (ECL). TTL is sometimes called multi-emitter transistor logic. See emitter-coupled logic.

transit In network communications, if a provider wants to send data to a destination but does not have the needed routing information, the provider may arrange temporary, permanent, fee or no-fee access (transit) to the destination through a second provider that has the necessary information or access. See peering.

transliterate To spell or represent the characters of one alphabet in the closest possible corresponding characters of another alphabet. Transliteration does not imply translation of the actual meaning of words composed of those characters. Western European languages transliterate reasonably well. It is harder to transliterate between Cyrillic and Roman characters, and is decidedly a challenge to transliterate between pictographic or symbolic alphabets, such as Asian languages and sequential, phonetic alphabets such as western European languages. The difficulties in transliteration on computing systems have led to many alternate keyboards, character mappings, and input systems. See Unicode.

transmission The sending of information through electrical signals. It is very common for information to be added to a signal through various forms of signal modulation, often with a carrier wave. See carrier wave, modulation.

Transmission Control Protocol TCP. A widely used Internet and local area network (LAN) connection-oriented, packet-switching transmission protocol descended from previous DARPA editions. Together with Internet Protocol, the TCIP/IP combination is a means for the transport of host-to-host information over layer-oriented network architectures. See Internet Protocol, RFC 793.

transmission header TH. In packet networking, a header that includes control information, and may be followed by a basic information unit (BIU) or segment. Used for network routing and flow.

transmission level point TLP. In installation, testing, and maintenance, a point in a conducting line at which the transmission level is measured according to whatever types of signals are sent through the conductor. In telephony, for example, the power level of a voice communication through an alternating current (AC) conductor may be measured in terms of decibels relative to a reference point.

transmission medium Any material through which transmission is facilitated either due to its inherent characteristics or through inherent characteristics enhanced by technology. Common transmission

T

media include air, light, wire, coaxial cable, fiber optics, etc. Broadcast transmissions primarily are sent through air, fiber, and coaxial cable. Computer transmissions are typically sent through copper wire or coaxial cable, although the use of fiber is increasing. Various media vary greatly in the amount of information (bandwidth) they can carry at any one time. See individual media for more detailed information.

transmitter 1. That which transmits or sends through some means such as chemical, optical, or electrical signals. 2. A device that sends out a signal, such as a transmitting antenna, telegraph instrument, or modem. A transmitter may also include various mechanisms to amplify, compress, modulate, or encode a signal. See telephone transmitter.

transparent A transparent technology is one in which the inner workings are not apparent to the user. For example, in computer operating systems with graphical user interfaces, the user sees the applications through point-and-click icons, text windows, and resizable gadgets and dialogs. The conversion of the information into operating instructions, the device drivers, queuing mechanisms, buffers, priority and security mechanisms, and binary arithmetic are essentially in the background, and thus invisible or *transparent* during typical interactions.

Transparent Bit-Oriented Protocol A protocol based on Bit Oriented Protocol (BOP) implemented on a number of Motorola communications products. TBOP is an access port type that accepts BOP frames and turns them into X.25 packets for transmission across a Frame Relay or X.25 network. It then depacketizes them for delivery to an end user device. Thus, BOP is used to transparently pass network information among devices with different protocols. TBOP will pass aborted and damaged frames as well. See Bit Oriented Protocol.

transparent tone in band TTIB. See tone in band.

transponder, radio In radio communications, a transceiver that transmits information automatically, on receipt of an appropriate interrogation signal.

transponder, satellite In satellite broadcasting, a device which receives and retransmits electromagnetic signals. Broadcast satellites employ this technology with multiple transponders. With compression, the capacity of a transponder can be significantly increased.

Transport A, Transport B See broadcast data trigger.

Transport Protocol Data Unit TPDU. In the Open Systems Interconnect (OSI) layered network model, the transport layer organizes data into a TPDU, that is, a packet that has had transport layer data added. Similarly, the session layer organizes data into Session Protocol Data Units (SPDUs). The TPDU contains the Transport Service Access Point (TSAP) address and the user's data (payload). See Protocol Data Unit.

Transport Layer Security Protocol TLS Protocol. A client/server-supporting protocol for securing network communications over the Internet. TLS is intended to deter eavesdropping, tampering, or forging of messages. It was submitted as a Standards Track RFC by Dierks and Allen in January 1999.

The TLS Protocol is primarily intended to provide privacy and data integrity between two communicating applications. It is composed of the TLS Record and TLS Handshake protocols. The TLS Record Protocol is layered over a reliable transport protocol and provides connection security in terms of privacy and reliability through encapsulation of higher level protocols. The TLS Handshake Protocol enables authentication of the client/server relationship and the negotiation of an encryption algorithm and cryptographic keys prior to data transmission. See RFC 2246.

Transport Service Data Unit TSDU. In the Open Systems Interconnect (OSI) model, an item of information passed by the network transport user to the transport provider. In the Transport Protocol Data Unit structure, all the data of the TSDU comprise the User Data. In the ISO-SP protocol, up to four session protocol data units together may comprise the TSDU.

transportable cellular phone A transportable phone consisting of a handset, antenna, and battery, usually bundled together in a carrying bag (sometimes known as "bag phones"). It's a little heavier than a self-contained, handheld cellular phone. This type of phone can operate on up to 3 W of power. It can be operated independently of a car battery and is typically used for field work that requires a phone with greater mobility and, sometimes, a larger antenna and longer life battery. It may include fax and modem features, and may be used in conjunction with a laptop. Journalists, scientists, and business people who need something a little more full-featured than a simple handset system use transportable systems.

trap *v.* To confine, narrow in on, circumscribe, or surround, especially with the implication that the object, person, or function cannot subsequently escape.

trap door A hidden device or software mechanism that causes a naive user to "fall" into an application, environment, or system, while assuming it is a legitimate user account. The user then tries to access normal applications, for example, leaving a trail of activities that can be logged, so that the trap door programmer can later search the log for usernames, passwords, or other information that can be used to penetrate an account. Since the user may begin to notice that the environment is not quite the same as normal, a trap door program may "inadvertently" crash or log out the user (or, on a network, say the system is going down and advise the user to log out) in order to escape detection. The next time the user logs on to the genuine account, everything is normal. See back door, Trojan horse, virus.

trapping In printing, a technique for assuring that adjacent inks meet, or slightly overlap so there won't be an undesirable, paper-colored gap between the inks if the registration of the press is slightly off. When a user creates desktop publishing page layouts intended for printing at high resolutions, settings must usually be adjusted to maximize trapping to ensure the quality of the final product. Typically, small, light-colored objects like fonts are *trapped* (slightly overlapped) over dark ones, so that the detail in the small objects is not lost. See choking.

traveling user TU. In a Secure Data Network System (SDNS), a traveling user is one who is visiting a Message Security Protocol-equipped (MSP-equipped) facility other than the usual one where the user reads and sends messages. In networks in general, a TU is someone who may interconnect or interact with a network from a variety of facilities while traveling. This type of network access is increasing as mobile systems become more prevalent.

traveling-wave tube TWT. A tube in which a stream of electrons interacts in a more-or-less synchronized manner with a directed electromagnetic wave so energy transfers from the stream to the wave. The basic components of a TWT include an electron gun for producing a high-density electron beam, a microwave circuit supporting a traveling electromagnetic wave through which the beam travels, and a collector to collect what is left of the electron beam that passed through the microwave slow-wave circuit. Amplification is attained through proximity of the electron beam to the traveling electromagnetic wave housed within a structure designed to propagate the wave.

TWTs evolved from magnetron microwave-generating tubes, emerging during the latter part of World War II. The TWT was designed by inventor Rudolf Kompfner, and later improved by Kompfner and John R. Pierce at Bell Telephone Laboratories. The Hughes Aircraft Company subsequently became a significant developer of military and commercial TWTs. The TWT was originally developed to support emerging radar technologies and is now also an important component in communications satellites and missile-seeking circuits. See cavity magnetron; Kompfner, Rudolf; magnetron; phototube.

TRE See Telecommunications Research Establishment.

tree structure A common structure in computer programming and file organization. A tree branches from a main trunk to its various branches, just as the roots successively branch into subdivisions as you move away from the main trunk. Data branching structures are found in various database data storage schemes, file directory structures, fractal images, and more. Physical branching structures are found in network topologies; for example, branches may come off a backbone and branch further in individual local area networks (LANs). Physical branching also occurs in phone circuits, with a main switching station supplying local private branches, which further subdivide to service individual lines within the local branch. Programmers use various types of tree structures including binary trees, B trees, B* trees, etc.

TREG See Telecommunications Regulatory Email Grapevine.

trellis coding A source coding technique used in a variety of contexts, from high-speed modems to MPEG decoding, to produce a sequence of bits from an incoming stream that conforms to certain desired characteristics.

TRL transistor-resistor logic. Also called resistor-transistor logic (RTL). Varying numbers and types of transistors and resistors combined in a circuit to comprise and control the logical operations of which the circuit is capable.

TRIBES Tri-Band Earth Station. A commercial satellite tracking station from California Microwave, Inc. (CMI), operating on C-, X-, and Ku-Band frequencies, aimed at government applications. TRIBES-Lite is a downsized version.

triboelectricity Electricity generated by friction, as in a bicycle wheel-mounted generator that becomes charged when the wheel is spun and rubs against the generator contact surface. Triboelectricity can also be generated by rubbing a balloon on a person's hair, enabling the balloon to magnetically be attached to walls or other surfaces until the electricity dissipates again. See mutual capacitance, static electricity.

tributary Secondary, or subsidiary peripheral, signal or process. Subsidiary peripherals receive their control data from servers or devices higher up in a hierarchical network. Subsidiaries may be aggregated to create a combination medium or signal. See tree structure.

triode An electron tube with three primary elements: an electron-emitting cathode, an electron-attracting anode, and a grid superimposed so that it can be used to control the flow of electrons. The invention of the triode by Lee de Forest, who named his commercial triode the *Audion*, is one of the most significant developments in the history of electronics. Prior to the addition of the third element, it was not possible to control the electron flow to any useful degree. See Audion.

TRIP 1. See Telephony Routing Over IP. 2. See Token-Ring Interface Processor.

Trivial File Transfer Protocol TFTP. A simplified, lock-step client/server version of File Transfer Protocol (FTP) for reading and writing files over networks. TFTP was originally developed to work with diskless workstations but has since been applied to other applications. TFTP is easily implemented over Internet Protocol (IP) and provides only the most basic file transfer features. It is encapsulated in UDP (standard over port 69) rather than TCP.

A multicast option suggests a way to use multicast packets to allow multiple clients to concurrently receive the same file. See Enhanced Trivial FTP, Simple File Transfer Protocol, RFC 1350, RFC 2090.

trn A versatile, full-screen news reading program developed by Wayne Davison as a superset of *rn*, written by Larry Wall. This incarnation adds threads through a hierarchical database.

Trojan horse A computer program that appears to be one thing, like a gift or incentive, but actually hides something more malicious or invasive, such as a virus or other program that can infect a host system and potentially change or damage it. The name comes from the Trojan War during which Troy was penetrated by the gift of a great wooden horse. Once inside the fortress, soldiers swarmed from the horse to do battle with the surprised recipients.

T

Thus, a Trojan horse program is a program disguised as something desirable or benign, a pirate program, computer game, popular utility, appealing Web site, etc., in order to attract users. If the Trojan horse is an executable masquerading as a game, for example, running it unleashes its ability to change or take over a system. Unlike the Battle of Troy, however, the victim may not realize his or her system has been invaded. The computer version of a Trojan horse can be much stealthier than its historical counterpart while surveilling or damaging a system.

Trojan horse programs are actually a particular type of masquerade program, by virtue of the fact that they offer something free or otherwise appealing.

One example of a Trojan horse program is a front-end that looks like a normal interface, but is really a visually or functionally identical layer on top of or instead of the interface that leads to something desired or appealing. It is designed to capture information to be used later for penetration of the system. For example, a programmer might create software that looks like a popular Web site or a network login prompt and then circulate information to entice people to use it. The user sits down at the terminal, registers for the service or types in an existing name and password, and gets an error message (from the Trojan horse) that the password was entered incorrectly, try again, or that displays something appealing and masks the fact that the username/password has been captured. The user sees a normal login prompt without being aware of the deception, types the password again, and logs in successfully or simply continues to browse the site if the password prompt was fake. The writer of the Trojan horse has grabbed a name and password, with a low chance of detection by the user, and may now be able to access the user's account by means that are difficult to detect as unauthorized access.

An Internet masquerade program was distributed in bulk email messages in summer 2001, when PayPal online banking services were becoming popular. The businesslike innocent-looking email message had an embedded link to a fake PayPal Web site called *paypaI* instead of *paypal*. The difference is that it used a capital "i" instead of an "l" (el), something few people would notice. When users tried to log on to the fake PayPal account with their usernames and passwords, the information was captured so that their real PayPal bank accounts could be accessed by the people who designed the masquerade. Fortunately, the ruse was discovered and reported quickly, but this example illustrates the potential for harming a large base of computer-naive Internet users. See back door, trap door, virus.

troposphere The lower layer of Earth's atmosphere, which contains clouds and most of the air, varying from a height of about 10,000 m at the poles to about 18,000 m at the equator. In radio transmissions, some frequencies can be bounced off the troposphere, that is, bent back to earth through super-refraction. See ionosphere, tropospheric scatter, tropospheric scatter transmission, tropospheric wave.

tropospheric scatter The dispersion and propagation of waves resulting from the varied and discontinuous physical properties of the troposphere. This can be predicted and controlled sufficiently to be useful in communications. See tropospheric scatter transmission.

tropospheric scatter transmission A method of electromagnetic wave propagation, employing frequency modulation, that exploits the irregular propagation properties of the troposphere. Tropospheric scatter transmission is a way to propagate, for example, microwave transmissions for thousands of miles, in segments up to about 500 miles per hop.

tropospheric wave The troposphere includes a diversity of moisture, heat, and other properties which can result in electromagnetic waves that undergo abrupt changes sufficient to create tropospheric waves distinguishable from the original tropospheric scatter transmission waves.

trouble unit A systems diagnostic or descriptive measure to indicate the expected performance of a circuit over a given period of time.

TRS See Telephone Relay Service.

TRU See tone receiver unit.

true bearing A bearing given with relation to geographic north (north as it is shown on a globe), rather

	Trunk Carrier Implementations for Voice/Data Lines					
Desig.	J-1 Mbps	J-1 Channel	T-1 Mbps	T-1 Channel	E-1 Mbps	E-1 Channel
-0	.064	1	64 Kbps	1	.064	1
-1	1.544	24	1.544	24	2.048	30
-1C	3.152	48	3.152	48		
-2	6.312	96	3.152	48	8.448	120
-3	32.064	480	6.312	96	34.368	480
-3C	97.728	1440	44.736	672		
-4	397.200	5760	274.176	4032	139.268	1920
-5					565.148	7680

than magnetic north (the north to which the north-seeking end of a compass needle points).

true north The Earth's geographic north. See magnetic north.

TrueType A system of widely used scalable vector-format fonts developed by Microsoft Incorporated. See vector fonts.

truncation 1. Cutting or chopping off at an end, sometimes abruptly. 2. The quick or abrupt termination of an operation or process. 3. The removal of characters from the end of a word or numeral, as in reducing a four-decimal numeral to two decimals without altering the numerals remaining (i.e., not rounding). Truncation applies to either the leading or trailing end, but more often than not, the part truncated tends to be the trailing end.

trunk A communications link between two switches or distribution points. A generic term that applies to many technologies, but is most often used in connection with telephone and network lines. Trunks can be set up to be bidirectional (left to right to left) or unidirectional (right to left or left to right). See path, route.

trunk carrier In the past, this has referred to any telephone voice network trunk connection. It is now also more specifically associated with the digitally multiplexed carrier systems used for digitized voice and data in North America, Europe, and Japan. The *T-carrier* system (for Trunk-carrier) in North America was the first of the *X*-carrier systems to be established, with Europe and Japan quickly adopting versions of it called the E-carrier and J-carrier systems, respectively. In general concept and format, these systems are similar. In actual implementation, some important differences preclude direct compatibility. The Trunk Carrier Implementations chart illustrates the similarities and differences in terms of data rates and number of voice/data channels available in the three systems. See E-carrier, J-carrier, T-carrier.

trunk exchange Hierarchically, a higher-level, specialized telephone exchange facility dedicated to interconnecting trunk lines (which, in turn, connect telephone services users and carriers). The concept of a telephone trunk appears to date back to sometime around the 1890s. In the earliest manually operated switchboard systems, two operators were used to set up a trunk relay in which one would answer the calls coming from outside the exchange and the second would then connect the call to the appropriate local subscriber. In some of the historic telephone exchange diagrams, the first operator was designated "A" with the second designated "B" to illustrate the steps involved. Outgoing calls from the local exchange would go to Operator B and be passed on to Operator A to be connected to a line outside the local area.

trunk group A group of trunks sharing essentially the same electrical characteristics and often the same physical characteristics connecting to the same switching endpoints or connections. Multiple trunks are common in areas where one trunk would not be sufficient to carry the traffic. If the main trunk is busy, traffic may be manually or automatically switched to the next one in the group, and so on. See trunk hunting.

trunk hunting A call management system that seeks an available communications trunk over which to route a call. Economy is achieved by hunting the most frequently used trunks first. See hunting.

trunk link frame TLF. In a crossbar telephone switching center there would be a frame supporting the links where subscriber telephone lines connected into the central office (CO) and a frame supporting the wires connecting to other switching offices. The lines connecting to other switching centers were terminated on the *trunk link frames*. The subscriber's line could thus be connected to another subscriber in the same local service area or to one of the trunk link circuits leading out of the service area on the trunk link frame connected to the appropriate remote office.

TS 1. See Technical Specification. 2. transmission scheme.

TSACC See Telecommunications Standards Advisory Council of Canada.

TSAG See Telecommunication Standardization Advisory Group.

TSAP See Transport Service Access Point.

TSAPI See Telephony Services Application Programming Interface.

TSAPI Service Provider TSP. A software driver that enables a TSAPI device to be adapted to a vendor's private branch exchange (PBX) system. The TSP may be used in conjunction with other computer telephony integration (CTI) products, but is the minimum essential software that links the systems. See Telephony Services Application Programming Interface.

TSB 1. Technical Service Bulletin. Bulletins issued by the Teleccommunications Industry Association. 2. See Telecommunication Standardization Bureau.

TSDU See Transport Service Data Unit.

TSO time share operation.

TSP 1. See Telecommunications Service Priority. 2. See TSAPI Service Provider.

TSR tag-switching router. See tag switching.

TSRM Telecommunication Standards Reference Manual.

TTA See Telecommunications Technology Association.

TTAB transparent tone above band. See tone above band.

TTC See Telecommunications Technology Committee.

TTIB transparent tone in band. See tone in band.

TTL See transistor-transistor logic.

TTS text-to-speech. A type of speech synthesizer.

TTY service Teletype service. Depending upon the area, this is a teletype service made available to those with hearing impairments, at a price commensurate with that paid by regular telephone subscribers. The program is supported through the combined fees paid by all subscribers in a region. Eligibility for the service is usually determined by local health services agencies.

TUANZ Telecommunications Users Association of New Zealand.

T

TUBA TCP and UDP with Bigger Address. One of three candidate protocol proposals eventually blended into IPv6 by the Internet Engineering Task Force (IETF). See IPv6.

TUC 1. total user calls. An accounting/administrative count of the total number of calls made per a specified period. 2. See Total User Cells.

TUG Telecommunication User Group.

TUI See telephone user interface.

tumbling A type of cell phone fraud that involves successively switching the electronic serial number for each call too quickly for the cell operator to detect the user.

tungsten A heavy metallic element with properties similar to chromium and molybdenum used in electrical installations, filaments, and contact points, and for hardening alloys.

tuning To adjust to resonate at a particular wavelength, as setting an instrument to a specific pitch or setting a radio antenna or tuner to receive a particular frequency of radiant energy. See tuning coil.

Tuning Coils

Historic radio tuning coils resemble large spools. They often came in sets, designed for different frequencies, and could plug into the circuit by means of two or more prongs.

Since tuning coils could be purchased in sets, they often included a base to keep the coils in order and protected from damage. These excellent examples are from the American Radio Museum collection.

---

tuning coil A winding coil specifically configured to pick up certain frequencies of radiant energy, particularly radio waves. Early tuning coils consisted of nothing more complicated than a coil of fine conductive wire wound around a wooden or rubber core, with the coil in circuit with a connecting pin or pins. Tuning coils for consumer sets tended to range from the size of a sewing spool to about the size of a human hand. Often they were stored like thread in banks or rows on little wooden shelves, so that the appropriate frequency coil could quickly be selected and inserted in a connector on the radio. The wire on each coil would be slightly different, to pull in a different frequency range, with different thicknesses and spacing between successive windings.

Sometimes a small sliding tab or knob attached to a bar would be placed along the edge of the winding, in order to make contact with a specific portion of the winding to provide further fine control. This was called a slide contact.

Other tuning coils used a type of intricate basket weaving in various patterns supported by a slender frame in such a way that a lot of wire could be wound into a small space and no cylindrical spool used. See basket winding, coil.

tunnel *n.* 1. A hollow tube, conduit, or passageway through an obstruction. 2. In software, an intermediary program that provides a temporary relay between connections without interpreting or otherwise changing the content of the communication. A tunnel often provides a temporary *portal* for passing data through a system such as a proxy, and ceases to exist when the ends of the connection are closed.

tunneling 1. Encapsulating a network transmission in an IP packet for secure transmission over a network. See virtual private network. 2. To temporarily reroute a network transmission packet in order to utilize routers which would not normally be able to route the transmission to the original destination due to not having the needed destination entry.

TUR Traffic Usage Recorder.

Turing, Alan Mathison (1912–1954) A British mathematician who traveled to Princeton in 1936, where he studied as a graduate student, and wrote *On Computable Numbers* ... He proposed some provocative ideas in ordinal logics and created a cipher machine while at Princeton, but Turing is best remembered for his description of a hypothetical device which could handle logical operations and manipulate symbols on infinite paper tape. This *Universal Turing Machine* is described as a *finite state* machine due to the finite set of instructions from which individual actions were derived. This provided the roots for thinking about computers in terms of algorithms and equations, and for positing devices that could be used for all possible tasks. Many of the conceptual roots of computer science, particularly general purpose machines and reusable code, were developed through his research. See Turing machine.

Turing machine A hypothetical model devised by Alan M. Turing, which could handle logical operations and manipulate symbols on infinite paper tape. This *Turing machine* is described as a *finite state* machine due to the finite set of instructions from which individual actions were derived at that point. Many extrapolations from and implementations of these basic concepts have contributed to the development of computing devices. See Turing test.

Turing test In 1950, Alan Turing published "Computing Machinery and Intelligence" in *Mind*, a philosophical journal. In it much of his thinking of the last few years was put into print, providing an effective inspiration to many future scientists and providing some of the concepts that developed into the field of artificial intelligence (AI). The Turing test was a provocative assertion that computers would, in the succeeding 50 years, be able to pass for a human under certain test conditions. Imaginative efforts to support this prediction have led to a wide variety of interesting programs, and prizes are offered for innovative software that can pass the Turing test.

turnaround time The time of a transaction, especially one that passes from one hand to others and back, as in sending something to another department or external service bureau. For example, the turnaround time for ordering a network connection may be two weeks by the time the order is tendered, the installer arrives and the system is up and running. 2. In network transmissions, the time it takes to send a transmission and receive an acknowledgment that the transmission was received, or can continue. 3. In detecting or *pinging* another system, the time it takes for the signal to reach the other system, and report back statistics on the connection. The phrase *turnaround time* is sometimes applied loosely to the time it takes for a human to send out a signal and receive an acknowledgment, and sometimes more precisely to the number of clock cycles or actual measured time it takes for the signal to leave the sending site, reach the receiving site, and report back to the sending site, used for system diagnostics and tuning. Or even more specifically, in handshaking or half-duplex applications, turnaround time is the interval that occurs when the system switches to communication in the other direction. In terms of half-duplex satellite phone conversations, this turnaround time is perceived as a blip, lag, or break in the line by the callers, and, if it is long, can distract from the conversation. See hysteresis.

turnkey system A self-contained system that can be purchased, installed, or relocated as a unit or as a package. Turnkey systems typically arise from two situations: 1. The technology is complex and the vendor combines options in such a way that the system meets a need but need not be configured or technically understood by the user. The user simply purchases it, turns it on, and uses it. For example, many video outlets bundled Video Toaster systems as turnkey solutions for desktop video applications. 2. Many options are available for configuring a system, and the vendor best knows how to combine and configure the individual components to meet the needs of the purchaser. Private branch telephone systems and multiline telephone systems with lots of options are often bundled and purchased this way.

turnstile antenna An antenna comprising two dipole antennas perpendicular to one another with axes intersecting at their midpoints.

turntable 1. A round, rotating platter, frequently with a central shaft or outer rim to hold objects in place, commonly used for playing audio and visual media, especially vinyl phonograph records. 2. A round, rotatable, platelike surface designed to give easy access to objects placed on it by turning. A Lazy Susan. Turn-

tables are used in loading platforms, cupboards, microwave ovens, and Chinese food restaurant tables.

TVM See time-varying media.

TVRO television receive-only

TWAIN An image-oriented communications protocol standard widely used in computer scanning devices and digital cameras (which are, in essence, portable scanners). Most high-end graphics programs can handle TWAIN-compliant devices, as can many optical character recognition systems. TWAIN is supported and promoted by a number of graphics industry vendors, including Eastman Kodak, Hewlett-Packard, UMAX, Adobe Systems, and others. See Tag Image File Format.

tweak freak A technical user who is exaggeratedly interested in the inner workings, whys, and wherefores of the tiniest technical details in a system. This characteristic is an asset when diagnostics and fine-tuning are needed, and a liability in general social situations.

twinning Systems configured in parallel to create redundancy, or alternate means of access or transmission. Twinning is sometimes done to install a new system while an old one is still used, or to provide a backup in the case of emergencies. In newer technologies, twinning can provide backups or transition systems until the old system is little used or phased out, such as twinning word processors with typewriters or physical facsimile machines with software facsimile programs.

twist length Also called *lay length*, this is the distance from one twist to the next in a twisted-pair conducting cable. For example, a cable with a lay length of 4 inches will correspond to three twists per foot or 3 TPF. In electrical-conducting wires, the twist length is kept as short as possible and helps to electromagnetically couple the cable while reducing crosstalk.
In light-conducting fiber optic cables, the twist serves a physical bundling purpose rather than an electrical coupling purpose. Helical twisting together of the tiny fiber optic filaments facilitates the installation of the cable and provides added strength. See twisted-pair cable, twists per foot.

twisted-pair cable Twin strands or cores of intertwined, insulated copper wire, a wire type used for many decades in the telephone industry. The twists are organized to help reduce interference in wires that are kept as close to inhabiting the same physical space as is possible. Thicker cables tend to provide cleaner transmissions, at a higher cost.
Unshielded twisted pair (UTP) is commonly sold as four pairs of 24-gauge wire. Shielded twisted pair (STP) is commonly used in applications where there may be interference from other nearby electrical sources.
Although the theoretic data transmission limit of twisted pair has been underestimated many times, and improved with new ideas and data protocol schemes, it is generally accepted that their practical capacity under normal operations is about 56 kbps. Twisted pair is now also commonly used in Ethernet LAN connections. Twisted-pair cable is sold commercially

T

in several grades of transmission performance. See category of performance. See twists per foot.

UTP and CAT-5 Cable Assemblies

Simplified images of four types of cables commonly used in telephony, video, and data cabling.

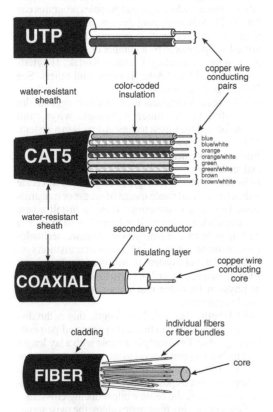

Top: A basic unshielded twisted-pair (UTP) cable; two-wire and four-wire UTP cables are especially common in telephone wiring (the copper pairs are entwined within the sheath). Shielded twisted pair (STP) looks similar, but has an insulating shield under the water-resistant sheath. The outer sheath is not necessary for all types of wiring, but is commonly used.

Middle: Category 5 unshielded twisted-pair (UTP) cable suitable for the higher demands of data cables. Pairs are typically identified by color schemes that include a solid color and a paired striped wire of the same color. For applications such as Ethernet, the wires are usually attached to an RJ-45 connector that resembles a fat phone cable connector.

Lower Middle: Coaxial cable, so-called because two conductors are housed within the covering, separated by an insulating layer to reduce crosstalk and to prevent short circuits that would occur if the two conducting layers touched. This is commonly used for video transmissions.

Bottom: Fiber optic cable based upon light-guiding filaments rather than electricity-conducting wire. This is commonly used for fast digital data communications.

The three top cables use copper for the central conducting core.

twists per foot TPF. A measure of the number of twists (helical intertwinings) per foot length in a pair of wires called twisted-pair wiring cable. The density of the twists per length affects the bulk and flexibility of the cable and the degree of electromagnetic emanations associated with the wire when it is used as an electrical conductor. Increased twisting generally increases the transmissions length. Uneven twists are generally used to reduce interference along the line from crosstalk and external sources.

There are industry standards for the number of twists per foot. For example, for Category 1 and Category 3 plenum and PVC for cables for regular applications, there are at least two TPF. Category 3 cables for telephony applications typically have 3 TPF with a gauge of 22 to 24 AWG. Cat 5 cable, suitable for the higher demands of fast data communications, may range from 8 to 36 TPF. The number of twists depends upon wire gauge, data transfer rates, and the number of twisted pairs included in a bundle. High-quality cables usually have evenly spaced twists that do not overlap. Different twist lengths in bundled cables can further aid in minimizing crosstalk. American Wire Gauge (AWG) #18 cable has 5 TPF. See twist length.

two-electrode vacuum tube, Fleming oscillation valve A historically important electron tube developed by John A. Fleming. The two-electrode tube consisted of a simple filament (cathode) and an electron-attracting metal sleeve (anode) that fit around the top of the filament, both of which were housed in an evacuated tube. With no controlling mechanism, it wasn't of much practical use, but it was a history-making invention nonetheless. Lee de Forest acquired a Fleming tube, experimented with it and created a three-element tube by adding a grid. This made it possible to control the flow of electrons from the cathode to the anode, an innovation that subsequently opened the door to the entire electronics industry. See Audion; de Forest, Lee; Fleming, John.

two tone keying A means of using two tones, one for *mark* and one for *space*, to modulate a telegraph signal so as to create two channels transmitting in the same direction.

two-phase coding A means of increasing data transmissions by splitting the signal into two orthogonal channels, one which is in phase and one which is a quadrature signal. They are then transmitted simultaneously with this 90 degree phase offset, each operating at half the data rate of the originating signal. Thus, a signal might be transmitted at the same data rate in half the bandwidth, but for the trade-off of converting it from a baseband signal to a passband signal when splitting it into two channels. The increased potential for noise, when two signals are transmitted simultaneously is also a consideration.

two-way trunk A network trunk which operates in - oth directions. In the case of telephone service trunk ines, it refers to one which can be seized from either nd of the connection, as opposed to one-way trunks, hich may be set up to send only or receive only.

WPD traveling-wave photodetector. See photodetector, traveling-wave tube.

TWS two-way simultaneous. A mode in which a router optimizes communications over a full-duplex serial line.

TWT See traveling-wave tube.

TWX *Teletype-Writer Exchange*. A Bell system printing telegraph service which could operate over the existing long distance network, established in the 1930s. In 1970, Western Union purchased the TWX service from AT&T and merged it into its own Telex service. See Western Union.

Twyman-Green interferometer See interferometer.

Tx, TX transmit.

TxD transmit data. A data channel, typically used in serial communications, which is an output for DTE devices and input for DCE devices. See DTR, DSR, RTS, RxD, RS-232.

Fleming Valve – Two-Electron Tube

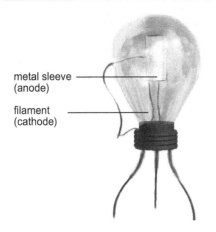

metal sleeve
(anode)

filament
(cathode)

The Fleming valve was a landmark invention that led to the evolution of three-element vacuum tubes. The Fleming valve was limited in practicality, however, in that it only included two elements, a filament (cathode) and a metal sleeve (anode) that fit over the filament to attract electrons. There was no way to control electron flow. It was not until Lee de Forest added a controlling grid to create a three-element tube that the Audion was born to create the electronic age.

TYMNET A historic, commercial, X.25-based data network access service descended from the ARPANET. TYMNET became available in 1974, not long after Telenet was offered, at dialup-line speeds of 300 baud! The service was offering faster data rates (1200-baud) within a few years. By the mid-1980s 2400-baud service was offered, with 9600-baud following in the late 1980s. This is pretty slow compared to current T1, ISDN, DSL, and cable modem standards, but 9600 was pretty ripping in the 1980s.

TYMNET grew and internationalized, with nodes assigned regionally, which may not seem impressive to current Internet users, but in those days, regional nodes made it possible, in some countries, to dial TYMNET as a local call from anywhere in the region. It was a pretty exciting development at the time and a foreshadowing of global network access.

A large number of companies eventually offered time-share or database information services over TYMNET, including Compuserve, Cybershare, the European Space Agency (ESA), Dow Jones, Dun and Bradstreet, Xerox Computer Services, and many more. TYMNET could be accessed in a number of ways, including through Datapac links. With the advent of the Internet, TYMNET has been all but forgotten, but it holds an important place in the growth of shared computing resources and access to computers by the general public.

Tyndall, John (1820–1893) A multitalented Irish physicist who rose far above his modest background, Tyndall studied the atmosphere, thermodynamics, electromagnetism, light guiding, and solar chemistry. He passed on his loves of science and mountaineering as an educator and popular lecturer.

In 1870, he demonstrated to the British Royal Society that light could transmit through an arcing stream of water, a concept pioneered by J.-D. Colladon in 1841. This idea of *total internal reflection,* in which the optical rays cannot escape the medium within which they are traveling, is fundamental to fiber optic networks.

Tyndall succeeded Faraday as director of the Royal Institution of London. Many scientific instruments are descended from his work, including fluorometers and UV and infrared spectrometers. See Colladon, Jean-Daniel.

Type 1/Type 2/Type *x* card A standardized, compact data card commonly used for modems, memory, and other plugin accessories for computers, digital cameras, and more. See PCMCIA card, PCMCIA standards.

Type 1/Type3 font The designations for the most commonly used PostScript-format ASCII or binary outline (vector) font format first released by Adobe Systems in the mid-1980s. The character shapes for symbols in the fonts are defined in PostScript so that they can be printed from the vector description to the best resolution possible on a PostScript-compatible output device. In simple terms, this means they look good if they are printed small and they still look good when they are printed billboard size, because the shapes are mathematically derived rather than scaled up, as with bitmapped raster images.

Type 1 and Type 3 fonts are not just character shapes, they are graphic expressions of programming algorithms, which means the characters can be swirled and manipulated and set across curved surfaces in a way that drawn fonts cannot. In theory, PostScript fonts could be animated. Type 3 fonts have some additional capabilities for special effects.

PostScript Type 1 fonts are largely responsible for the desktop publishing revolution that occurred in the mid-1980s when Macintosh computers and PostScript-capable LaserWriter printers were paired up with desktop publishing software. Before 1984, typesetting machines costs hundreds of thousands of dollars and the general public couldn't do much better than to use a special proportional typewriter. See PostScript.

T

Type of Service ToS. Type of Service is both a generic concept in networking and a specific parameter related to Internet Protocol communications. It describes and determines various networking parameters related to speed, security, format, etc. As with many types of network services, there may be trade-offs between speed, flexibility, and security, depending upon how ToS is specified at any particular time or on specific types of networks. ToS is now mandated by the Requirements for Internet Hosts specifications described in the various relevant Requests for Comments (RFC) documents. At the present time, support for ToS is somewhat uneven and there are known problems on some prevalent systems.

In Internet Protocol (IP), a byte in the IP header supports Type of Service information. The byte is divided into the precedence field, the TOS field, and a reserved bit, only one of which may be set at a time. It is recommended that routers be cognizant of the ToS value for a route in the routing table. If the routing protocol does not support ToS header settings, the ToS must be assigned the default value of zero (0).

Physical Link Security Type of Service was described as an RFC by D. Eastlake in May 1993. It was submitted as an experimental protocol providing a ToS to request maximum physical link security in addition to existing types of services in the Internet Protocol (IP) Suite. This ToS requests protection against surreptitious observation by agents labeled as outside the traffic.

When transporting SNA network traffic over TCP/IP networks, there are issues with preserving the SNA Class of Service (CoS) parameters. Cisco addressed this issue by developing a data-link switching enhancement (DLSw+) to improve response time and the effective use of bandwidth while supporting SNA Type of Service (ToS) over TCP/IP to ensure preservation of SNA CoS traffic parameters and aid networks in prioritizing SNA traffic.

Programmers have created tables and software utilities to streamline the processing of ToS settings. For example, *iptables* allows a table to be constructed with predetermined values that can be matched to the datagrams being processed. Thus, only those with matching patterns are processed. Masking utilities are available for accomplishing a similar task, but with more power and flexibility.

In NIKHEF ping code, TOS bits can be set to enable the user to set priorities for the ping query. Settings are between 0 and 255. See RFC 1455, RFC 1812.

u *symb.* A letter sometimes substituted for μ (micron) when special symbol sets are not available. See mu.

U interface In ISDN, a number of reference points have been specified as R, S, T, U, and V interfaces. To establish ISDN services, the telephone company typically has to install a number of wirelines and devices to create the all-digital circuit connection necessary to send and receive digital voice and data transmissions.

The U interface is a full-duplex link that works over a single pair (2-wire) cable. It interfaces a line terminating switch in the telephone switching office with a small interface device called a Network Termination device (NT1 or NT2) at the customer premises. In the U.S., the NT1 converts the 2-wire U interface into a 4-wire S/T interface which, in turn, can support multiple devices in a single bus loop configuration, such as a telephone, computer, or facsimile machine. Alternately, in parts of Europe, the NT2 interfaces with an S interface. See ISDN interfaces for a diagram.

U reference point A demarcation point in ISDN services installed in North America, where the local loop connects with the NT1 device. See U interface. See ISDN interfaces for a diagram.

U Series Recommendations A series of ITU-T recommended guidelines for telegraph switching. These guidelines are available for purchase from the ITU-T. Since ITU-T specifications and recommendations are widely followed by vendors in the telecommunications industry, those wanting to maximize interoperability with other systems need to be aware of the information disseminated by the ITU-T. A full list of general categories is in Appendix C and specific series topics are listed under individual entries in this dictionary, e.g., T Series Recommendations. See U Series Recommendations.

U-band Astronomical emanations ranging from 3200 to 3950 Å that are usually detected with U polarimetric filters from the U, B, V + R, I Johnson-Cousins broadband photometric system. U-band frequencies are emitted by some celestial bodies and help us gain a picture of our universe. U-band energy emanating from some galaxies has been detected and U-band flares are occasionally recorded.

U-band, optical In optical communications, an ultralong-wavelength ITU-specified transmission band in the 1625 to 1675-nm range. Distributed-feeback and continuous-wave laser diodes are available in U-band frequencies. U-band systems are not as common as C-band/L-band, but it has been suggested that U-band signals may be multiplexed with S-band signals in dense wavelength division multiplexed (DWDM) systems in much the same way.

ITU-T U Series Recommendations	
Recom.	Description
U.1 (03/93)	Signaling conditions to be applied in the international telex service
U.2 (11/88)	Standardization of dials and dial pulse generators for the international telex service
U.3 (11/88)	Arrangements in switching equipment to minimize the effects of false calling signals
U.4 (11/88)	Exchange of information regarding signals destined to be used over international circuits concerned with switched teleprinter networks
U.5 (11/88)	Requirements to be met by regenerative repeaters in international connections
U.6 (11/88)	Prevention of fraudulent transit traffic in the fully automatic international telex service
U.7 (03/93)	Numbering schemes for automatic switching networks

U-C user-central. In ADSL, the standardized interface between the twisted-pair local loop to the subscriber premises and the plain old telephone service (POTS) splitter on the network (usually the central office) side of the link. U-C$_2$ is a less-standard interface between the POTS splitter and the ATU-C (network ADSL Transmission Unit). While functionally

Sampling of UL Standards Related to Telecommunications		
UL No.	Descriptions of UL Numbers from Underwriters Laboratory	

UL 1409	*Low-Voltage Video Products Without Cathode-Ray-Tube Displays*: Antenna signal amplifiers, CATV adapters and digital converters, channel balancers and processors, distribution amplifiers, commercial TV cameras, disc players, electronic viewfinders, internal distribution amplifiers, laser disc players, modulators, picture tube degaussers, power packs, power supply-battery chargers, satellite receivers and dish controllers, single-channel converters, teletext and television decoders, television descramblers, tuner adapters and power supplies, UHF amplifiers, tuners, and converters, VHF amplifiers and tuners, video printers, video-production -processing, -receiving, and -recording equipment, and video tape recorders.	
UL 1418	*Cathode-Ray Tubes*: Bonded frame, laminated, prestressed picture tubes (CRT), rebuilt picture tubes, picture tubes for business equipment, dental, and medical equipment.	
UL 1410	*Television Receivers and High-Voltage Video Products*: Household and commercial television receivers and monitors, and health care facility television equipment.	
UL 1412	*Fusing Resistors and Temperature-Limited Resistors for Radio- and Television-Type Appliances*: For use in appliances that do not involve potentials greater than 2500V peak.	
UL 1414	*Across-the-Line, Antenna-Coupling, and Line-by-Pass Capacitors for Radio- and Television-Type Appliances*: For nominal 125- and 250-V, 50- to 60-Hz circuits, includes double protection capacitors rated 1.0 B5F maximum.	
UL 1419	*Professional Video and Audio Equipment*: Video tape recorders, audio/video editing equipment, audio/video receiving and processing equipment, signal transmission equipment, television cameras, video digitizers, video monitors, metering equipment, and similar equipment.	
UL 1492	*Audio-Video Products and Accessories*: Audio and video products intended for use on supply circuits. Audio products and accessories intended for household use and involved with the reproduction or processing of audio signals. Video products that are intended for household or commercial use, that receive signals in ways such as off the air, through a CATV/MATV cable system, from a video-recorded medium, and from image-producing units. Auxiliary products and accessories intended for use with audio or video products wherein the auxiliary and accessory products are separate and do not perform the desired function, but are used in addition to or as a supplement to products mentioned above. Cellular telephones and similar transceiving devices used on a vehicle, boat, or the like, where the telephone interconnects to the telephone network through a radio transmitter and receiver. Portable audio or video products of the types described above that are intended for use with a vehicular, marine, or any other battery circuit as the power supply means.	
UL 6500	*Audio/Video and Musical Instrument Apparatus for Household, Commercial, and Similar General Use*: This standard applies to the following apparatus that is to be connected to the supply mains, either directly or indirectly, intended for domestic and commercial and similar general indoor use and not subject to dripping or splashing: radio receiving apparatus for sound or vision; amplifiers; independent load transducers and source transducers; motor-driven apparatus which comprise one or more of the above-mentioned apparatus or can be used only in combination with one or more of them, such as radio-gramophones and tape recorders; other apparatus obviously provided to be used in combination with the above-mentioned apparatus, such as antenna amplifiers, supply apparatus and cable-connected remote control devices; battery eliminators; electronic musical instruments; electronic accessories such as rhythm generators, self-contained tone generators, music tuners and the like for use with electronic or nonelectronic musical instruments; video apparatus intended for entertainment purposes in health-care facility locations; cellular telephones and similar transceiving devices used on a vehicle, boat, or similar location where the telephone interconnects to the telephone network through a radio transmitter and receiver; portable audio or video apparatus intended for use with a vehicle, marine, or any other battery circuit as the power supply means.	
UL 1685	*Vertical-Tray Fire-Propagation and Smoke-Release Test for Electrical and Optical-Fiber Cables*: Limits for each fire test to make the tests equally acceptable for the purpose of quantifying the smoke. The cable manufacturer is to specify, for testing each "-LS" (limited-smoke) cable construction, either the UL vertical-tray flame exposure or the FT4/IEEE 1202 type of flame exposure. The same test need not be specified for all constructions. Cont...	

similar, the U-C is distinguished from the U-R interface due to the asymmetry of the link. See U-R.

U-DSL The U interface in a Digital Subscriber Loop (DSL) system.

U-law, μ-law A pulse code modulation (PCM) coding and companding data ITU-T standard used in audio systems on computer multimedia peripheral cards. This takes some of the load of specialized applications off the central processing unit (CPU). It is often implemented in addition to A-law companding and is suitable for compression of voice communications. Note, this is technically μ-law, but many keyboards don't support the Mu (μ) character and so it is alternately written as Mu-law or U-law. In fact, it's even sometimes written M-law since the Greek symbol for uppercase μ is *M*. See A-law, Mu-law.

U-plane In ATM networking, as it applies to Broadband-ISDN reference model, the U-plane is the user plane, a higher-level plane including all of the ATM layers, which bears user application information. It sits adjacent to the C-plane (control plane) and shares physical and ATM layers with the C-plane. The M-plane (management plane) enables the transfer of information between the C- and U-planes. In ATM networking as it applies to Frame Relay bearer services, the U-plane parameters, such as throughput, maximum frame size, etc., are negotiated through the C-plane. Synchronization and coordination between the U-plane and C-plane are described in ITU-T Recommendation Q.923. See the Appendix for more detailed information on ATM.

U-R user-remote. In ADSL, the standardized interface between the twisted-pair local loop to the subscriber premises and the plain old telephone service (POTS) splitter on the premises. U-R$_2$ is a less standardized interface between the POTS splitter and the ATU-R (premises ADSL Transmission Unit). While functionally similar, the U-R interface is distinguished from the U-C interface due to the asymmetry of the link. See U-C.

U.K. Education & Research Networking Association UKERNA. The trading name for the JNT Association, which has managed the development and operation of the Joint Academic Network (JANET), since 1994, under agreement with the Joint Information Systems Committee (JISC) of the U.K. Higher

Sampling of UL Standards Related to Telecommunications, cont.	
UL No.	**Descriptions of UL Numbers from Underwriters Laboratory**
UL 1577	*Optical Isolators*: Optically isolated switches and insulation systems, photocouplers.
UL 1651	*Optical Fiber Cable*: Single and multiple optical-fiber cables for control, signaling, and communications as described in Article 770 and other applicable parts of the National Electrical Code.
UL 1690	*Data-Processing Cable*: Electrical cables consisting of one or more current-carrying copper, aluminum, or copper-clad aluminum conductors with or without either or both (1) grounding conductor(s) and (2) one or more optical-fiber members, all under an overall jacket. These electrical and composite electrical/optical-fiber cables are intended for use under the raised floor of a computer room (optical and electrical functions associated in the case of a hybrid cable) in accordance with Article 645 and other applicable parts of the National Electrical Code.
UL 2024	*Optical Fiber Cable Raceway*: Covers the following types of optical fiber cable raceways and fittings designed for use with optical fiber cables in accordance with Article 770 of the National Electrical Code:
	Plenum. Evaluated for installation in ducts, plenums, or other spaces used for environmental air in accordance with the National Electrical Code as well as general purpose applications;
	Riser. Evaluated for installation in risers in accordance with the National Electrical Code as well as general purpose applications;
	General Use. Evaluated for general purpose applications only.
UL 1459	*Telephone Equipment*: Cordless telephones, key systems private branch exchange equipment, telephone answering devices, dialers, and telephone sets.
UL 1863	*Communication Circuit Accessories*: Telecommunications equipment such as jack and plug assemblies, quick connect assemblies, telephone wall plates, cross connect enclosures, network interfaces, and connector boxes.
UL 1950	*Practical Application Guidelines On-Line Service* (PAGOS): A reference service providing information for understanding and applying the requirements of UL Standards for Safety. Of interest is the UL 1950 Standard for Safety of Information Technology Equipment, Including Electrical Business Equipment.

U

and Further Education Funding Councils. Thus, UKERNA has responsibility for the education and research communities' networking programs in the U.K. It further researches and develops advanced electronic communications facilities. See JANET.

UA See User Agent.

UAC User Agent client. See User Agent.

UART See universal asynchronous receiver-transmitter.

UASs unavailable seconds. A measure of duration, in seconds, during which a service or entity is not available.

UAT See user acceptance testing, user application testing.

UAWG See Universal ADSL Working Group.

UBR See unspecified bit rate.

UCA See Utility Communications Architecture.

UCC See Uniform Commercial Code.

UCF See UNIX Computing Forum.

UCITA See Uniform Computer Information Transactions Act.

UCM 1. universal controller module. 2. See Universal Call Model.

Uda, Shintaro One of the designers of the Yagi-Uda antenna, a sensitive, directional antenna which worked in the higher frequency ranges and became the model for thousands of antennas that came later and are still in use. See Yagi-Uda antenna.

UDLC See Universal Digital Loop Carrier.

UDP See User Datagram Protocol.

UECT See Universal Encoding Conversion Technology.

UFO 1. See UHF Follow-On. 2. unidentified flying object.

UHF See ultra-high frequency.

UHF Follow-On, Ultra-High Frequency Follow-On UFO. A U.S. Naval, Air Force, and Command communications satellite constellation intended to supersede the aging FLTSAT and LEASAT satellite systems. The system is designed to provide interim Global Broadcast Service (GBS), EHF, and Ka-band transmissions. UFO provides more modern capabilities and more secure communications than the older satellite communications systems. Increased channel capacity is available with Demand Assigned Multiple Access (DAMA) technology. UFO is intended to provide global coverage of four significant geographic areas, including the U.S. and three major oceans. See FLTSAT.

UHTTP See Unidirectional Hypertext Transfer Protocol.

UI 1. Unix International. A consortium of computer software and hardware vendors promoting the development and implementation of Unix, and of related and other open software standards. See Unix, UNIX. 2. See user interface.

uk.telecom An online USENET newsgroup established in August 1991 to discuss topics related to telecommunications in the U.K.. Topics include services, prices, technical specifications, equipment functioning and options, ISDN, and the various telephone carriers providing services.

UKERNA See U.K. Education & Research Networking Association.

UL See Underwriters Laboratories Inc.

Ulex, Georg Ludwig (1811–1883) A German chemist, Ulex is remembered for having discovered Ulexite, the "TV rock," a unique mineral that projects light through its structure by internal reflection. Ulex found this rock in Chile in the mid-1800s. The mineral was named in his honor by renowned geologist/mineralogist James Dwight Dana (1813–1895), a correspondent of Darwin. See Ulexite.

Ulexite A fibrous substance of hydrated sodium calcium borate hydroxide. It is named for its discoverer, Georg Ludwig Ulex. Ulexite is a borate from the class of carbonates. The natural substance is found in the American southwest, South America and Kazakhstan. It ranges from translucent white to transparent and the chains of sodium, water, and hydroxide are linked in an interesting way.

Like Iceland spar, Ulexite is a somewhat brittle, complex mineral with unique optical properties; when it has a veined structure, it can channel light through its fibers. For example, if a 1-in. chunk of polished Ulexite is placed over an image, with the fibers perpendicular to the image plane, the image is channeled through the mineral and projected virtually undistorted upon the opposite (top) surface, much to the delight of the viewer. This capability has caused it to be dubbed the "TV rock."

Synthetic versions of Ulexite are fabricated in a variety of marbled colors to have aesthetic appeal while still retaining some of the optical qualities of the natural substance. Synthetic Ulexite may be coated on one side to maximize its light transmission properties. See Iceland spar.

ULP See upper layer protocol.

Ultra Sniffer A kit-based, handheld radio communications receiver designed to facilitate radio direction finding, especially for "fox hunts," during which hobbyists get together to try to find a hidden transmitter. This unit, from VK3TJN/VK3XAJ, is a little larger than a deck of cards. It is attached to a 2-m center beam with three elements attached in the center at 90° from the center beam. The unit is designed to overcome some of the limitations of other radio direction finders in terms of overcoming interference and selective tuning without increasing complexity (and knobs). See fox, sniffer.

ultra-high frequency UHF. A designation for a range within the radio frequency spectrum commonly used for broadcast communications, which ranges from 300 to 3000 MHz.

ultra-high frequency (UHF) antenna A category of antennas which are designed to take advantage of the particular characteristics of ultra high frequency (UHF) waves. Because of the wavelength differences between UHF and very high frequency (VHF) waves and the relationship of the rods on the antenna to the length of the wave, it is possible to make UHF antennas relatively small, with more branching elements compared to VHF antennas. However, as UHF television broadcast signals are generally weaker than those from VHF, there is a greater potential for loss, and they must be designed and installed with greater care to be effective. See antenna, combination antennas, VHF antennas.

ultraviolet Electromagnetic radiation with shorter wavelengths, between the violet part of the visible spectrum and X-rays. Although it cannot be seen by humans, ultraviolet radiation is of commercial significance because it can degrade many types of materials and pigments. Commercially, it is used in a variety of lamps, such as arc lamps, and can be used to remove data from erasable/programmable computer chips.

In astronomy, ultraviolet sensing devices use ultraviolet radiation focused through a spectrograph to study the characteristics of celestial objects. Telescopes and some satellites are equipped with this capability. The Hubble Space Telescope and the International Ultraviolet Explorer satellite enable study of objects using ultraviolet light near the visible spectrum. The Johns Hopkins Ultraviolet Telescope extends the range to the *far ultraviolet*, that is, the region further from the visible spectrum. See infrared.

umbrella antenna An antenna that resembles an umbrella in that the lines extend out and down from a central pole.

UML See Unified Modeling Language.

UMP1X The USOC code for telephony-related maintenance plan, tier 1, per line.

UMSP See Unified Memory Space Protocol.

UMTS See Universal Mobile Telecommunications Systems.

unattended call A situation that occurs when, for example, an automatic dialing system dials a line, then tries to pass the call to the first available human agent, but no agent is available. Consequently, the call is abandoned. This type of calling occurs in the telemarketing industry and the call is terminated in order not to irritate a potential customer. Unattended calls are also used by collection agencies and the system hangs up if no agent is available or if the call is answered by an answering machine, thus not impinging on the agent's time.

unattended systems Devices or systems which function without a human operator or without significant human attention except for installation and routine maintenance and upgrades. Unattended systems have become prevalent since the late 1970s, when computer automation became inexpensive enough to incorporate into a wide variety of components and machines. Computer bulletin board systems were some of the first information-rich systems to function 24 hours a day, and phone systems now are frequently automated with menu selections and voice mail options. Recently, faxback systems allow users to request product information or technical support in the form of fax documents. The system logs the phone call, gets the customer's document selections from a list of options, then dials back the user's fax machine and transmits the requested documents. Unattended systems generally function 24 hours per day at a sig-

nificant cost savings over human operators. Many businesses are willing to give up personalized service in favor of the economy offered by automated systems. See Auto Attendant.

unbalanced line A transmission line with two conductors (such as coax or a telephone circuit) with unequal voltages with respect to the ground. In phone circuits, this is generally an undesirable condition.

unblanking The portion of the sweep in a cathode-ray tube (CRT) where the beam is turned on, with pulses from the generator. See blanking.

unbundled Products or services which are sold separately. For example, a company may release a graphics card/monitor combination as a package deal, and later unbundle the items, that is, allow them to be sold separately in order to clear the products, get a higher return, or respond to market demand for one product over the other. Contrast with bundled.

Unbundled Network Elements UNE. Telephone network services that are sold or leased through competitive local exchange carriers (CLECs) as unbundled services from an incumbent local exchange carrier (ILEC). These physical and functional services, when broken down (unbundled) into discrete components, make it possible for them to be mixed and matched into a variety of new services that may be optionally resold or leased to endusers by a variety of providers. UNEs came about as a result of the competition-supporting provisions of the Telecommunications Act of 1996. UNEs include such aspects as local loops, switches, information (databased) services, etc.

The UNE model and Congressional decisions regarding UNE were controversial. In 1997, AT&T responded to a Circuit Court decision that defeated key provisions of the Federal Communications Commission's (FCC's) rules on UNEs. AT&T's position was that the rulings could open the door to competitive restructuring of existing services and monopolization of new services by competing providers, but at a higher cost, which would not result in the desired competitive benefits to users.

Unbundled Network Elements – Platform In telephony parlance, UNE-Platform (UNE-P) services are combinations of Unbundled Network Elements (UNEs) that provide finished (end-to-end) services to Competitive Local Exchange Carriers (CLECs) that are functionally equivalent to retail service offerings. UNE-P products are intended only for resale to endusers, not for the use of carriers themselves. Examples of UNE-P services include plain old telephone service (POTS), public access line (PAL), ISDN Pri and Bri, digital switched service (DSS), and Centrex services.

UNC See Universal Naming Convention.

underfill In semiconductor circuit assembly, material that seals components or fills holes or gaps, often around solder joints where chips attach to circuit boards. Underfill may insulate from possible electrical shorts or may provide structural support. It may also help prevent abrasians and corrosion. Underfill requires extra materials and time, and may need to

U

be cured, and thus is not always used, even when it might improve fabrication quality.

underfill, optical A situation in which the amount or intensity of light present does not fully fill an aperture or grating structure or meet operational minimums of a component. If a point light source is much narrower than an opening through which it is shining, the opening is said to be underfilled. Similarly, if a light source is smaller than the *acceptance cone* of a light-carrying component, it is said to be underfilled. In grating structures, which have different geometries depending upon the direction of the incident light, the structure may be optimally filled or overfilled in one plane and underfilled in others. In coupled optical fibers, particles or back reflection may result in loss of light at the junction and underfill of the succeeding link in the light path.

There is a relationship between the F number (aperture diameter) and an illumination target (grating, mirror, light pipe, etc.) such that a higher F number may underfill the aperture. Sometimes underfilling is desired and may improve resolution. At other times it may result in incomplete or ineffective functioning of a system. See acceptance cone, overshoot.

Undersea News Service An international submarine fiber optics news service that features selected press releases from major newswire services, compiled and published by KMI Corporation. See KMI Corporation.

undershoot See overshoot.

Underwriters Laboratories, Inc. UL. UL is a not-for-profit organization established in 1894. It provides conformity, safety, and quality assessment services and publications to a variety of organizations, including manufacturers. In addition, UL provides educational materials, input to international safety systems, and assistance to various regional authorities.

UL publishes a catalog of its standards and the standards themselves in print, on microfilm, CD-ROM, and diskettes. The UL also sponsors a UL Standards Electronic BBS (accessed directly, or through the Web). The majority of the UL published standards have been approved as American National Standards by the American National Standards Institute (ANSI). UL has a number of publications of interest to professionals in the communications industry, including WireTalk for the wire and cable industry.
http://www.ul.com/about/wtalk/index.html
UL provides ISO 9000 standards quality registration through its accredited RvA, Registrar Accreditation Board (RAB) and other international quality affiliations. It provides information on new international environmental management standards through ISO 14001.

The full UL catalog is available on the Web, but the Sampling of UL Standards chart shows some telecommunications-related UL Standards which may be of interest and which provide an idea of the types of use and safety issues concerned. http://www.ul.com/

Underwriters Laboratory Inc. assessment The UL provides a number of conformity assessment services for product certification. These include listing, classification, field engineering, and various types of safety and performance testing. The UL Conformity Assessment services chart shows some of the services relevant to telecommunications.

Underwriters Laboratory Inc. Mark UL Mark. UL provides a number of listing marks to indicate that products or systems have been evaluated by UL and conform to certain specifications. Those shown in the UL Listing Marks chart are relevant to telecommunications.

UNE See Unbundled Network Elements.

UNI User Network Interface, User-to-Network Interface. As specified by the ATM Forum, an ATM network

---

UL Conformity Assessment Services Related to Telecommunications

Service	Brief Description
Listing Service	A UL Listing Mark indicates that representative samples have been tested and evaluated according to nationally recognized safety standards.
Classification Service	A UL Classification mark indicates that products have been evaluated for certain properties under specified conditions.
Component Recognition	A service for factory-installed components in complete products.
Certificate Service	A service for completely installed systems.
Field Engineering Service	A service for installed products without UL Listing Marks or UL Classification Marks.
Testing Environ. Products	Evaluation of innovative environmentally friendly products.
LAN Cable Performance	Safety evaluations and evaluation of LAN cable according to industry performance specifications, including TIA/EIA standards.
Energy Efficiency	Electrical appliances are certified according to U.S. or Canadian standards for energy efficiency through the UL Energy Verification.
SDS Verification Testing	Verification of input/output products to Honeywell's Smart Distributed System (SDS) for compatibility of components to an industrial control communications network.

switch which interfaces user equipment to private or public ATM network equipment, or connects between Customer Premises Equipment (CPE) and public network equipment. See PMP, PCR, OCD, SCR.

Unicamp A group at the State University of Campinas, Brazil, formed in 1975 to deveop optical fibers for TELEBRAS (Telecommunicações do Brasil). The majority of optical fiber produced in Brazil is based upon research in optical communications, nonlinear optics, and other phenomena researched by Unicamp.

unicast A type of Internet Protocol (IP) address identifier for a set of interfaces. Unicast transmits a single Protocol Data Unit (PDU) to a single destination (unlike multicast, where it may go to multiple destinations). The format of the ATM subinterface unicast command is: atm smds-address <address>. See anycast, IPv6 addressing, multicast.

Unicode A character-encoding standard to support text-encoding in data files. Unicode, Inc. was originally a collaboration between Apple and Xerox, who produced the original specification. They were later joined by Adobe, Aldus, Borland, IBM, Microsoft, NeXT, Novell, Sun, and others. Unicode has been rolled in with an ISO specification as a subset of ISO 10646. Unicode is loosely based upon the widely supported ASCII standard, but in a greatly extended form to include major world languages not represented with Roman characters, including Cyrillic, Greek, Arabic, Hebrew, Japanese Kana, Chinese bopomofo, Korean hangul, and others. Symbols, punctuation marks, mathematical symbols, and technical symbols are also supported. Unicode uses a 16-bit character set, supporting over 65,000 characters. Each character is assigned a unique 16-bit value. No special modes or control or escape sequences are needed. Unicode comprises the first 65,536 code points of the ISO 10646 standard; the rest are reserved for future use. See Unicode Consortium.

Unicode Consortium A nonprofit association founded in 1991 to promote and support the acceptance and implementation of the Unicode character-encoding standard. The Consortium publishes a pamphlet on the Unicode specification. The Unicode Technical Committee, descended from Unicode, Inc., now functions as part of the Consortium to actively maintain the standard. See Unicode.

unidirectional Moving, responding, or transmitting in one direction, or in only one direction at a time.

Unidirectional Hypertext Transfer Protocol UHTP. A robust, unidirectional IP multicasting resource transfer protocol suitable for one-way broadcasting over the Internet or over the television vertical blanking interval. The protocol allows many viewers to simultaneously access the broadcast site. See broadcast data trigger.

unified memory architecture A system on which the video display drivers are integrated into the motherboard, and system random access memory (RAM) is used to buffer graphics displays, rather than having them as separate systems. Some systems use this very effectively, providing graphics coprocessor chips, and allowing greater video graphics memory and more control over memory for programmers, applications, and users. On other systems, this type of integration slows down the graphics rendering and overloads the CPU. This is not the fault of the concept, but rather a result of how it is implemented.

Unified Memory Space Protocol UMSP. A connection-oriented network protocol corresponding to the session and presentation layers of the Open Systems Interconnect (OSI) model. UMSP was submitted as an Experimental RFC by A. Bogdanov in December 2000.

UL Listing Marks Related to Telecommunications	
Listing Mark	Brief Description
UL Listing Mark	Commonly seen, and indicates that samples of the product conform to UL safety requirements according to UL published standards for safety.
C-UL Listing Mark	Canadian market products evaluated according to Canadian safety requirements.
Classification Mark	Products evaluated for specific properties under specified conditions. These usually consist of industrial and building materials and equipment.
C-UL Classification Mark	Classification Mark products intended for the Canadian market.
Recognized Component Mark	Specific to components used in products sold as complete units, and thus not usually seen from the outside. There is also a Canadian version.
International emc-Mark	Products which conform to electromagnetic compatibility requirements of Europe and/or U.S. and/or Japan and/or Australia.
Field Evaluated Product Mark	A product which is evaluated in the field rather than in a laboratory.
Facility Registration Mark	A facility which has passed UL quality assurance standards, specifically ISO 9000-series and ISO 14001 (environmental).

U

UMSP uses transport layer service for reliable delivery (with acknowledgment data). UMSP creates a network environment for organizing 128-bit address space distributed among Internet nodes. The protocol defines connections management algorithms and network primitive formats; it does not control local node memory. Connection parameters may be set in a number of ways and systems with high protection levels may be configured without restricting application functionality. See RFC 3018.

unified messaging See integrated messaging.

Unified Modeling Language UML. A widely used modeling language for specifying, constructing, visualizing, and documenting the artifacts of software systems. UML is intended to streamline and simplify the process of software design. It is a product of the Object Management Group. See Object Constraint Library, Object Management Group.

Unified User Interface UUI. In the Envisat satellite data communications services, the UUI is a single interface to User Service Facilities that enables users to access Envisat data services from any station or access node using a standard Web browser. The UUI interprets the browser commands to the service functions in the Envisat-1 Payload Data Segment (PDS).

Uniform Commercial Code UCC. An adopted code for conveying, clarifying, and permitting commercial activities within the provisions of the UCC Act as they apply to commerce within the 50 U.S. states and some of its territories. The Act sets forth the terms and conditions for commercial policies and activities described within the Act, concepts applicable to law, including actions, contracts, and remedies related to commercial endeavors and disputes. The UCC does not strictly dictate the terms of contracts and agreements between parties, but it helps to provide guidelines and default terms that provide a measure of consistency and security for those conducting commercial transactions. The Permanent Editorial Board for the Uniform Commercial Code provides oversight and permission for the distribution of UCC information. The Board publishes reports and drafts related

to UCC. See Uniform Computer Information Transactions Act.

Uniform Computer Information Transactions Act UCITA. This was formerly an Article of the Uniform Commercial Code. However, with the growth and prevalence of concerns specific to computing, it was felt that there was a need for a separate, related Act. This became especially true when a large amount of electronic commerce began to flow across the Internet. UCITA is a uniform commercial code for software licensing and other computer information-related transactions adopted by the National Conference of Commissioners on Uniform State Laws in July 1999. See Uniform Commercial Code. http://www.ucitaonline.com/

uniform line A line which has essentially identical electrical characteristics throughout the transmission path.

Uniform Resource Agent URA. An architecture for an agent system to provide Internet information access and management. Encapsulation of protocol-specific actions enables the addressing of high-level Internet activities. It is a structured mechanism for abstracting characteristics of desired information and distancing access processes from the client.

The URA system was submitted as an experimental RFC by Daigle et al. in October 1996. See RFC 2016.

Uniform Resource Characteristic URC. A data format for including meta-information, information outside the resources in question, for the identification and location of these resources on the Internet. See Uniform Resource Identifier. When URCs were proposed, in the mid-1990s, it was suggested that they be used in conjunction with Uniform Resource Names (URNs) instead of URLS, to remove location dependencies.

Uniform Resource Identifier URI. A means to identify resources on the Internet. Because of the size and structure of the Internet, these resources may exist in one or more locations concurrently or may at times not be available at all.

The syntax and encoding of the names and addresses of objects has been gradually developed since 1990,

---

Common Uniform Resource Locator (URL) Schemes

Scheme	Name	Type
ftp	File Transfer Protocol	Files and directories
http	Hypertext Transfer Protocol	Internet resources, Web pages
gopher	Gopher Protocol	File directories in Gopherspace
mailto	Electronic mail address	Internet electronic mail address
news	USENET news	Newsgroups and individual articles
nntp	USENET news using NNTP access	Alternate means of accessing news
telnet	Reference to interactive sessions	Interactive telnet remote logon sessions
wais	Wide Area Information Servers	WAIS databases, searches, documents
file	Host-specific file names	Accessible files from various hosts
prospero	Prospero Directory Service	Resources on the Prospero service

with URIs used to manage registered protocols or name spaces. A URI uses network protocols to express an address which maps onto an access algorithm. This is important because the Internet functions with many different protocols for the transmission and sharing of data. In most cases, the data can be converted to accommodate diverse formats. However, some types of information are impractical to convert, such as names and addresses of resources. By creating a type of object that can be labeled for recognition and retrieval and a name space in which these objects can reside, access and use of this information can be facilitated. A Uniform Resource Locator is an example of a URI.

One of the more interesting developments on the Internet has been the establishment of broadcasting channels over which video and audio radio and television programming can be viewed through Web browsers and various other specialized software programs. This has necessitated the definition and organization of URIs appropriate for digital broadcasts. See Uniform Resource Characteristic, Uniform Resource Locator, Uniform Resource Name, RFC 1630, RFC 1736, RFC 1737, RFC 2396, RFC 2838.

Uniform Resource Locator URL. A compact string representation for a resource available on the Internet. URLs have been in use since 1990 as Universal Resource Identifiers in WWW. A URL is a means to locate resources, by providing an abstract identification of its location. Generally, a URL follows this format:

    <scheme>:<scheme-specific-part>

Examples:

    http://www.abiogenesis.com/telecomdict

    ftp://www.peanut.org/

Scheme names consist of a sequence of lowercase characters from a to z, numerals 0 to 9 and the characters "+" (plus), "." (period), and "-" (hyphen). It is recommended that upper case be treated as lower case in resolving a URL.

A number of specific schemes for particular protocols are standardized or commonly used and there is a process for registering new ones. Common schemes (typed in lower case when used in a URL) are shown in the Common Schemes chart. See RFC 1630, RFC 1738, RFC 1808, RFC 2396.

Uniform Resource Name URN. Similar in concept to Uniform Resource Locators as a means to identify a resource or unit of information on the Internet, but intended to manage an object space of names expected to have a longer shelf life. URNs provide a globally unique means of identifying information about a resource or access to the resource itself. Functional specifications for URNs were proposed by Sollins and Masinter and presented as a Request for Comments in 1994. See Uniform Resource Characteristics, Uniform Resource Identifier, RFC 1737.

Uniform Resource Name Namespace for Object Identifiers URN Namespace for OIDs. On the Internet, an Object Identifier is a tree of nodes, syntactically described as a series of delimited digits. For example, the Internet OID is 1.3.6.1. The OID namespace specifies how an Object Identifier (ASN.1) is encoded as a Uniform Resource Identifier. The ISO/IEC Joint Technical Committee is the declared registrant of the namespace.

The scheme was originally submitted by M. Mealling as an Informational RFC, in November 2000, and updated in February 2001. See RFC 3061 which obsoletes RFC 3001.

Uniform Resource Name Namespace for Public Identifiers URN Namespace for PIDs. A namespace designed to enable Public Identifiers to be expressed in Uniform Resource Identifier (URI) syntax. Within XML, a public identifier is a simple string and, historically, public identifiers are not legal URIs in the context of the Web. The URN namespace enables public identifiers to be encoded in URNs in a reliable, comparable way through introduction of a formal public identifier namespace (publicid). The URN namespace scheme was submitted as an Informational RFC by Walsh et al. in August 2001. See RFC 3151.

uniformity The capability of a broadcast or other communications medium to deliver a steady and consistent signal within the desired range.

uninterruptible power supply UPS. A safety and steady-service device which protects equipment and data by guaranteeing a sufficient and steady source of electrical power in the event that other power sources are interrupted or lost.

UPS systems may take their current from an alternating power supply while the system is up, store charges from this source, and then switch to an alternate source, such as a direct current storage battery or separate alternate current generator in the event of power disruptions to the normal supply.

UPSs are used on computers, phones, lighting systems, and in emergency centers.

Power outages can create severe problems on computer systems, particularly to network servers, queues, backup file systems, and applications which are reading or writing to storage media at the time of a power outage. UPS systems can prevent loss of files in the process of being saved and prevent possible corruption to the medium on which they are being written.

unipolar Having only one pole, direction, or polarity.

unique user identifier UUI. An administrative function for uniquely recognizing or storing data on behalf of an individual user. This may be a name, number, symbol, token, or biometric equivalent, depending upon the system. A *username* is a type of UUI commonly used to enable access to restricted computer systems. UUIs are used in association with thousands of different types of secured systems and services. The New Zealand Customs Service, for example, uses a UUI system to register users and to administer their EDI clients using the Customs Computerized Entry Processing System.

unit vector In mathematics/geometry, a vector

U

indicating a direction only, having a length (magnitude) of one (1 – unity). |U| = 1. Thus, a vector may be converted to a unit vector by dividing each component of the vector by its magnitude. See vector.

United States Geological Survey USGS. The USGS carries out fundamental and applied research in geological surveying, cartographic data collection, storage, search, retrieval, and manipulation. It is responsible for assessing natural ecological events, energy, land, water, and mineral researches. The USGS conducts the National Mapping Program, and publishes thousands of reports and maps each year.

United States Telephone Association, United States Telecom Association USTA. An organization founded in the 1800s which promotes the well-being of the industry, and provides technical and standards assistance, discussion forums, and publications for its members. USTA represents more than 1200 local exchange carriers (LECs).

USTA provides representation before Congress and various regulatory bodies, training courses, technical bulletins, conferences, and media relations.

USTA arose from the National Telephone Association, established in 1897. This organizational strength provided a voice for independents in the dominant Bell marketplace. The National Telephone Association later became the United States Independent Telephone Association (USITA). The Kingsbury Commitment, an important step toward cooperation between Bell and the Independents, entered into in 1913, may have averted government takeover of the telephone industry arising from charges of the monopolistic control exerted by Bell at that time.

After the mid-1980s divestiture of AT&T, USITA became the United States Telephone Association (USTA) and it is now known as the United States Telecom Association to reflect the broader technology base of local exchange carriers (LECs). http://www.usta.org/

United Telephone Company A historic phone company founded in 1898 by Cleyson L. Brown in Abilene, Kansas, and later expanded to other communities. It operated there until 1966, and then moved to Shawnee Mission, Kansas, where it forms the local division of Sprint Corporation. See Museum of Independent Telephony.

UNIVAC Universal Automatic Computer. A historic, large, general-purpose electronic computing system in active use in the 1950s, descended from the ENIAC. It was designed and built in the mid-1940s by the Eckert-Mauchly Electronic Control Corporation, but taken over before its completion by Remington-Rand. UNIVAC was advertised as the UNIVAC File-Computer "electronic brain" by Remington Rand Univac, a Division of Sperry Rand Corporation. UNIVAC was the first significant commercial nonmilitary computing system, available for a little more than $1 million.

A mercury delay line, incorporating a long tube of mercury, was installed inside the computer housing as a memory device. There were a number of input/output modes, including magnetic tape, and various peripherals, such as printers. The clock speed of a UNIVAC wasn't much different from the personal computers first introduced in the mid-1970s and it took a great deal of care and expertise to get the system up and running and to maintain the vacuum tube-based hardware. It was programmed with X-1.

In spring, 1951, the U.S. Census Bureau acquired an 8-ton UNIVAC system. In 1952, the UNIVAC was used to (correctly) predict the presidential election returns but the results were not made public until after the election. Due to the media exposure, UNIVAC became so well known that the name became a generic term for large computing devices. There is an original UNIVAC in the Smithsonian museum. See ENIAC.

Universal ADSL Working Group UAWG. A commercial consortium formed to promote an easy-to-deploy, fast version of Digital Subscriber Line (xDSL) based on ANSI T1.413. Since traditional ADSL installations require a splitter to be wired to the subscribers' premises and a custom modem installed in their computer, there have been a number of initiatives to simplify the installation process and, hence, the cost, and to allow the subscribers a choice of modem hardware. See G.lite.

universal asynchronous receiver-transmitter UART. UART chips and UART circuitry perform a conversion function within a computer. When a computer software program generates data that travels from the computer to the serial card in a peripheral slot, or through a serial device to an external modem, the parallel data generated by the computer are converted by the UART into serial data that are then transmitted through the modem. The same process occurs in reverse at the receiving end. This is *not* the same process as is performed by a pair of connected modems, which modulate and demodulate a signal, convert it from digital to analog to transmit through the phone line and from analog to digital when received. The UART does its job before modulation/demodulation occurs in the modem. A UART chip may be in the computer, or in the modem itself.

Universal Call Model UCM. In telecommunications SS7 routing, an extension of the Basic Call Model associated with Intelligent Networks (INs). The UCM provides mediation between the Originating Call Model (OCM) and the Terminating Call Model (TCM). The UCM receives responses from an initial address message (IAM) and its associated calling line ID, whereupon the UCM activates seizure of the originating channel. It then puts out a request for analysis of the A- and B-number and route and, when the information is received, looks up the dialed number in a dialing base for routing. During the call, the UCM maintains routing and bearer circuit status. In systems using PRI ISDN, the UCM is bypassed. Also called line concentration module. See Intelligent Networks Call Model, Virtual Switch Controller.

Universal Digital Loop Carrier UDLC. Digital public switched network (PSN) carrier systems comprising a central office (CO) terminal near the switching system, a remote terminal at the customer's premises, and a digital transmission link connecting the two. Functional criteria for digital loop carrier systems are described in Bellcore TR-NWT-000057. Digital switching streamlined the system, enabling central office terminals to be integrated into the digital switch.

UDLCs were introduced in North America in the early 1970s. Connections to analog interfaces are through twisted-pair copper wires, as are those between the remote terminal and the network interfaces.

Universal Encoding Conversion Technology UECT. A Digital Equipment Corporation (DEC) proprietary software system for converting documents to and from Unicode. UECT has been incorporated into the AltaVista search engine, one of the significant search tools on the Web from Digital Equipment Corporation.

universal mailbox A centralized computer point of access for a variety of types of messages, including email, digitally encoded voice messages, facsimiles, etc., so the user can look at one listing to determine what to read and when to read it and to simplify the filing and cross-management of document databases. See integrated messaging.

Universal Mobile Telecommunications Systems UMTS. In conjunction with general packet radio service (GPRS), UMTS encompasses next-generation wireless telecommunications technologies. UMTS is a broadband, packet-based data transmissions technology base supported in Europe and also potentially in Japan and other Asian countries. It is not directly compatible with emerging American mobile standards.

UMTS technologies are characterized by high mobility and flexibility in terms of available data rates and are considered suitable for wireless Internet access with global roaming capabilities. Since wireless data technologies have lagged somewhat over the years from lack of support and interoperability, UMTS standards were developed to improve the situation and encourage market support for UMTS deployment.

UMTS standards were developed by two groups of prominent telecommunications vendors. Trials of UMTS were carried out by Nortel Networks and British Telecom (BT) in 1999. In 2000, ETSI finalized the first series of 3GPP specifications into the UMTS standard. The UMTS first series specifies a wide variety of services, including radio access, functions for applications development, multimedia messaging, and much more. The Release 99 first series enables developers to move ahead with the rollout of 3G services.

As a result of standardization, companies such as Nortel Networks have entered into agreements with developers to create 3G wireless dual-mode modems supporting the UMTS standard.

Universal Mobile Telecommunications Systems Initiatives UMTS Initiatives. Third-generation mobile telephony research and deployment efforts that are being carried out by a number of companies in a variety of regions, but particularly in Western Europe.

UMTS initiatives seek to support and foster the establishment of broadbased UMTS standard-based wireless communications systems:

- In May 2000, Lucent announced an initiative to create a Bell Laboratories Research & Development Centre to partner with Italian universities to enable startup businesses and researchers to test ideas and products related to UMTS integration an implementation.
- In November 2000, the German Bundestag (Parliament) approved UMTS funding for education and research to promote future development of UMTS technologies.
- In September 2001, Europolitan Vodafone announced costs for their UMTS initiative in Sweden and a collaboration agreement with the Vodaphone Group.

Universal Naming Convention UNC. 1. In general, a convention for logically mapping a name to a process or device such that its explicit file path or routing path is transparent to the user. The UNC can be used to set up a virtual network with various printers, storage devices, scanners, etc. linked in as resources regardless of where they may be located on the network. 2. In terms of file storage, a convention for identifying a shared file on a computer network without explicitly identifying its storage location to the user. This increases ease of access and transparency to users who shouldn't have to worry about the location of a file on a virtual storage system (which may be a device on another machine a few feet or a few thousand miles away) except as desired.

Universal Payment Preamble UPP. An electronic commerce payment mechanism developed by JEPI, based on work by Don Eastlake. The mechanism was described through an RFC document including several examples in August 1996 as to its relationship between the HTTP Payment Extension Protocol (PEP). UPP provides a uniform vocabulary for a uniform syntax and naming options common to payment systems. Common parameters could be specified within PEP-specified header fields and in payment-system-specific headers. See JEPI, Payment Extension Protocol.

universal payphone A payphone with a wide scope of payment options including coin, calling card, credit card, collect, etc.

Universal Serial Bus USB. An open serial data bus standard developed by a consortium of prominent computer products and telecommunications services providers in the mid-1990s. It allows peripherals to be attached to a computer through a single peripheral attached to the motherboard, with other devices chaining or attached in a star topology. Commercial USBs are designed to support many devices, sometimes up to 64 (the host computer is considered a device). A USB will sometimes also provide additional power to devices that might require it. One of the basic goals of USB development was ease of use. It was intended for personal computer users to be able to easily attach and detach external computer

U

peripherals without a lot of technical expertise. Thus, characteristics such as hot swapping and the capability to attach multiple devices were desirable design goals. Through a process of enumeration, peripherals are assigned unique addresses for managing runtime data transmissions and transactions.

The Human Interface Device (HID) Class is a USB Core-compliant aspect of USB intended for user input devices that require relatively slow data rates (1.5 MBytes/sec) compared to high-speed storage access, audio/visual, or networking transmissions (12 MBytes/sec). HIDs include keyboards, mice, joysticks, graphics tablets, etc. Traditionally the USB HID Class is connected to the computer through a wire, but interest in wireless versions is strong and implementations are being suggested.

The format is rapidly gaining popularity and many personal computers now come with USB ports built in. Older computers with PCI slots can be adapted for use with USB through peripheral cards. It is common for a peripheral card to have two USB ports.

With the success of USB 1.1, work continued on USB 2.0 to give it even better performance characteristics. It is estimated that higher data rates of more than 400 MBytes/sec may be possible without substantial hardware changes (with the exception of hubs) through the use of microframes. Transmission speeds for a variety of attached devices would be individually negotiated, providing backward compatibility and flexibility in device data rates. See FireWire.

Universal Service Order Code USOC. An identification system for tariff services and equipment introduced in the 1970s by AT&T, and later adopted by the Federal Communications Commission (FCC). Since divestiture, the code is even less universal than before, with individual Bell operating companies developing billing systems somewhat independently of one another.

Universal Time See Coordinated Universal Time.

Universal Transverse Mercator projection UTM. A map projection technique that preserves angular relationships and scale. UTMs are used in many planimetric and topographic maps. A UTM consists of a series of identical projections, each 6° of longitude oriented to a meridian, taken from around the world's mid-latitudes.

Universal Unique Identifier UUID. A unique identifier originating from the Network Computing System (NCS) and the Open Software Foundation (OSF) distributed computing environment. In February 1998, Leach and Salz defined the format of UUIDs guaranteed or extremely likely to be different from all UUIDs generated until the year 3400 A.D., depending upon the mechanism chosen for generating the UUID.

In a data communications equipment (DCE) transmissions cell, it is a broadly unique 128-bit identifier assigned to an object. The UUID is typically used in global contexts where it is a challenge to assign a guaranteed unique ID. Thus, a combination of data are combined to produce the UUID, which may include time stamps, random quantities (or seeds for

random quantities), and the hardware address of the originating network device, etc.

In the context of Universal Resource Identifier (URI) schemes on the Internet, a UUID enables network resources to be uniquely named without regard to location; they are thus not tied to a physical root namespace. These are also known as Globally Unique Identifiers (GUIDs). UUIDs are useful in that they need not be assigned and administered by a centralized authority (beyond node identifiers), as are domain names. They also have potential as transaction IDs, a property of particular interest to e-commerce transactions.

In Unix applications, the *Uuid class* provides a means for creating and converting *Uuid* objects to support network UUIDs.

Universal Wireless Communications UWC. A wireless communications collaborative program initiated by wireless operators and vendors in 1995. The program is built on the TIA IS-136 Time Division Multiple Access (TDMA) radio frequency standards, along with IS-41 Wireless Intelligent Network (WIN) standards.

Universal Wireless Communications Consortium UWCC. A Washington State LLC, established to support carriers and vendors of IS-136 TDMA/IS-41 WIN standards. The UWCC sponsors a number of working forums, including the Global TDMA Forum (GTF), the Global WIN Forum (GWF), and the Global Operators Forum (GOF). See Universal Wireless Communications. http://www.uwcc.org/

Unix A widespread, powerful operating system, originally developed in 1969 by Ken Thompson at AT&T Bell Laboratories. The trademarked version of Unix is spelled all in caps as UNIX, whereas Unix spelled in upper and lower case is used generically in the computer industry to refer to the many freely distributable flavors of Unix that have been implemented by different groups. UNIX has gone through a number of hands, from AT&T, to Novell Inc., to the X/Open Company Limited. See UNIX.

UNIX UNIX is a powerful, widespread, cross-platform, Internet-friendly, multitasking, multiuser operating system. When spelled in all capitals, UNIX is a registered trademark, licensed exclusively through the X/Open Company Limited. See Single UNIX Specification, Unix.

UNIX Computing Forum UCF. A comments and feedback forum through the Santa Cruz Operation, Inc. (SCO), which provides UNIX server operating systems and related products.

Unlicensed Personal Communications Services, Unlicensed PCS UPCS. A number of low-range communications systems can be used without broadcast licensing. These are commonly used for applications such as cordless phones, intercoms, monitors, etc. Some are incorporated into short-range wireless local area network (LAN) data and phone systems. Specific frequency ranges have been assigned to UPCS services by the Federal Communications System (FCC). UPCS are permitted within the 1890 to 1930 MHz frequency ranges and are further subdivided for use with asynchronous (1910 to 1920 MHz) and isochronous (1890 to 1910 and 1920 to 1930 MHz) communications. See band allocations.

unlisted phone number A service requiring a fee, in which a phone listing is not published in printed directories or available through directory assistance. Some carriers also provide unpublished service, which is excluded from printed directories, but may be listed with directory assistance, as a partial privacy measure. People pay to prevent their numbers from being listed for a variety of reasons: to avoid crank calls, undesired telephone solicitations, harassment from ex-spouses, etc. Some carriers make it possible for callers to leave a message for an unlisted number, which the caller may or may not return at his or her option. This is useful for emergency calls. See unpublished phone number.

unmatched call A call that does not have a corresponding match in a Service User Table (SUT). Call matching is a way of determining whether the call is authorized and should be permitted to ring through. If there is no match to the destination number in any of the relevant lookup tables, such as the Authorization Code Table (ACT) or Calling Card Table, the card will likely be rejected or may be redirected to someone in authority.

unpublished phone number A service, usually requiring a fee, in which a phone listing is not published in printed directories, but may or may not (depending upon the carrier) be available through directory assistance. Thus, if listed with directory assistance, it is a midway solution between a listed and an unlisted number. Some people choose unpublished phone numbers to avoid crank calls and undesired telephone solicitations. Some carriers will subscribers to exclude addresses from a published listing, without charging extra. For further privacy, see unlisted phone number.

unshielded Unprotected from emitting or receiving electromagnetic interference or broadcast signal interference. Most cables are shielded with plastic and/or metal foil, but since this increases the weight and cost of the cable, there are still circumstances where low shielded or unshielded cables are used. In video applications, well-shielded cables are recommended. Monitors should be shielded to protect users from radiation exposure, and computers shielded to prevent interference with nearby broadcast devices, such as radios. Improper or insufficient shielding may result in Federal Communications Commission (FCC) rejection in the manufacture of new products.

unshielded twisted-pair UTP. A very common type of cable consisting of one or more pairs of twisted copper wires bound together. UTP is frequently used for phone wire installations intended to carry faster data rates. See twisted-pair cable.

Unsolicited Commercial Electronic Mail Act of 2001. A bill proposed by Rep. Heather Wilson in February 2001 to protect Internet users and providers from unsolicited and unwanted electronic mail. Between 1995 and the present, the volume and frequency of unsolicited electronic mail, also known as "spam," rose dramatically, resulting in significant load to service providers along with lost productivity and undesired costs to users who had to filter out the messages and sift through them to locate legitimate mail. Because the cost to the sender of sending email is often insignificant, "junk" email is a far greater problem than junk postal mail. The cost of sending out a million junk postal solicitations is typically more than $30,000, providing an economic deterrent to excessive mailings. In contrast, the cost of sending out a million email messages may only cost pennies to the sender but may result in significant forwarding, storing, and filtering costs to service providers and recipients, especially as junk email file sizes increase due to added images and HTML-format tags.

Junk email is now widely used to promote fraudulent money-making schemes, off-shore sheltering of illegal gains, young teen pornography, black market pharmaceuticals, and gray market consumer items. Thus, many feel that stronger legislative constraints on unsolicited electronic mail should be put in place to protect recipients and providers from bearing the cost and inconvenience of these solicitations.

unspecified bit rate UBR. An unguaranteed ATM networking service type in which the network makes a best-effort attempt to meet the sender's bandwidth requirements. See available bit rate, cell rate.

unsupervised transfer, blind transfer A phone call transfer in which the recipient is not advised as to the identity of the caller. This is common on automated systems in which the caller can select an extension by way of the keypad on a touchtone phone.

unused A product which may have been opened, or taken home and returned, but which has not been used. It may have slight abrasions and, if sold, may carry a warranty that differs from a new warranty.

UPC Usage Parameter Control. A network mechanism for monitoring and controlling traffic and guaranteeing service for legitimate uses. See traffic policing, traffic shaping.

UPCS See Unlicensed Personal Communications Services, Unlicensed PCS.

UPGRADE An ACTS project intended to increase the capacity of the existing single fiber European Communications network to higher bit rates over existing hardware to serve the needs of future communications. Capacity will in part be increased with new modulators, switches, and semiconductor laser amplifiers. Various European networks (Deutsche Telekom, 1998 EXPO in Spain, etc.) are involved in testing the systems. Test results and components from the project will be used to update single fiber links to ca. 1300-nm wavelength. See BLISS, BROADBAND, and WOTAN.

uplink In broadcast communications, the uplink is the leg from an Earth station to a satellite. From the satellite back to the Earth is a downlink. The distinction is made partly because of the different technologies used in satellite and Earth stations, but also because uplink and downlink services can often be purchased separately.

upload, send, transfer To transmit a broadcast or transfer data from the current device to another one, usually at a different location or desk. Computer data are often uploaded from a personal computer to the Internet or to a mainframe. Information from a laptop may be uploaded to a desk computer. Telecommunications software, Web browsers, and FTP are common ways in which people upload files. Broadcasts may be uploaded to a satellite link. See upstream. Contrast with download.

Ground-Air Data Transfer

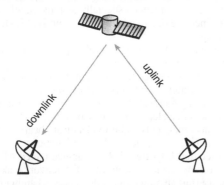

The uplink is the transmissions path from the Earth to the satellite, which is often at a different frequency from the downlink in order to reduce interference between incoming and outgoing signals.

UPP See Universal Payment Preamble.

upper layer protocol ULP. In hierarchical network models, a protocol that operates at a higher level of the model, which usually consists of application and user transactional functions. In the Open Systems Interconnection (OSI) reference model, ULP more specifically refers to protocols in any of the layers higher than the layer currently being referenced, although colloquially it often means the next higher layer.

upper memory area UMA. A section of memory on Intel-based IBM and licensed third-party computers commonly used to buffer video data which can be accessed and read by a video graphics display card.

upstream Generally, the transmission going in a direction away from the reference point. Thus, the stream of data from a personal computer to a mainframe would be considered upstream. Sometimes the designation implies from a smaller or less powerful system to a larger or more powerful system, so its use is not completely standardized. In cable networks, the transmission from the transmitting station to the cable television headend is the upstream direction. See upload. Contrast with downstream.

uptime An uninterrupted interval during which a system or process has been in active service. The active, functional time between failure or maintenance periods. Contrast with downtime.

upwardly compatible A device or program intended to work with later upgrades or revisions. Upwardly

compatible may also mean compatible with a larger or more complex version. For example, a handheld device bar code device may be designed to be upwardly compatible with a desktop computer. Upward compatibility in terms of later versions is much more difficult to achieve than downward compatibility, since future changes or improvements cannot always be anticipated. Contrast with downwardly compatible.

URA See Uniform Resource Agent.

URI See Uniform Resource Identifier, URL, URN, RFC 1630, RFC 1738, RFC 1808.

URL See Uniform Resource Locators. See RFC 1738.

URN Uniform Resource Name. See RFC 1737.

U.S. West One of the regional companies created when AT&T was divested in the mid-1980s, comprising Mountain Telephone, Pacific Northwest Bell, Northwestern Bell, and other related firms servicing the "Fourth Corner."

USB See Universal Serial Bus.

USDC U.S. Digital Cellular. A telephone standard which uses frequency division multiple access (FDMA) and time division multiple access (TDMA) techniques in the 824 to 894 MHz range.

USDLA United States Distance Learning Association. See distance learning.

USDN U.S. ISDN services. See ISDN.

used A term describing a product that has been opened and used, with no implications as to the quality, age, or remaining useful life of the product. Used equipment is generally represented as being in working condition, as far as is known. See certified, fair, like new, refurbished.

USENET Created in late 1979, shortly after the release of a Unix V7 which supported UUCP, USENET is an important communications medium best known for its more than 35,000 public newsgroups. USENET was developed by Tom Truscott and Jim Ellis at Duke University, and Steve Bellovin at the University of North Carolina. The first two-site installation was described in January 1980, at the Usenix conference and, after modifications by Steve Daniel and Tom Truscott, it became known as A News.

As soon as it caught on, A News volume began to steadily increase. In 1981 Mark Horton, from UC Berkeley, and Matt Glickman enhanced the software to better handle the increasing volume of information. This 1982 version was known as B News.

Two years later, administration of the software was taken over by Rick Adams from the Center for Seismic Studies. Moderated groups capability was added, in addition to compression, a new naming structure, and control messages. A rewrite by Geoff Collyer and Henry Spencer of the University of Toronto was released as C News in 1987.

In 1992, Rich Salz released InterNetNews (INN), a program optimized for NNTP hosts, but with support for UUCP. INN was designed for socket-oriented Unix hosts. Enhancements and bug fixes to INN were released by David Barr, beginning in 1995. Maintenance of INN was taken over by the Internet

Software Consortium.

UUCP gave way to TCP/IP, and TCP/IP's greater compatibility across platforms was a means to provide wider access to newsgroups. The Network News Transfer Protocol (NNTP) was also developed and, in 1986, a means to use this for news articles was released. For more information, see the "USENET Software: History and Sources FAQ" on the Internet. See FidoNet, newsgroup, RFC 822, RFC 1123, RFC 977, RFC 1036, RFC 1153.

user Sometimes called *end user*. Although often used to indicate a nontechnical consumer of a product or service, a user also generically refers to anyone interacting with that product or service, as opposed to developing or distributing it.

user acceptance testing, user application testing UAT. The testing of a product by actual users (those who fit the profile of potential buyers or users) in conditions similar to what the use environment would be in order to ensure acceptance of the product and sufficient design ergonomics and explanation (menus, manuals, etc.) for the operator to be able to use the product without significant intervention or assistance. UAT occurs when the product is considered to be finished and in good working order (bug-free). This is an extremely important aspect of product development, as many entrepreneurs have "surefire" ideas not readily appreciated or desired by users (e.g., potato-flavored ice; yes, someone actually tried it). This can also be referred to as delta testing or enduser testing. See beta testing, gamma testing.

user account An account assigned for a specific individual on a computer network or on a multiuser machine. A user account is a security system configured by the system administrator. The sophistication of the security can range from a simple name prompt at the time of login, to name and password logins at various levels of access, and different protections attached to directories, processes, and programs.

User Agent UA. A network service used by clients to find available services on behalf of the user. See Directory Agent, Service Agent, Service Location Protocol.

User Datagram Protocol UDP. An IETF-recommended protocol for the Internet which provides a datagram mode for Internet Protocol-based (IP-based) packet-switched network communications. UDP is primarily used with the Internet Name Server and Trivial File Transfer. The format of UDP header is shown in the User Datagram Protocol Header chart. For more details about UDP, see See RFC 768.

user event In programming, a type of input event which is signaled through an input device such as a mouse, joystick, keyboard, or touchscreen, and interpreted into a response by the operating system or applications program. User events typically include button, window, or menu selections/adjustments and movement of icons, windows, or objects. The most challenging types of user events tend to occur in fast action video games and realtime graphics input processing.

User Glossary Working Group UGWG. A group within the User Services Area of the Internet Engineering Task Force (IETF) which has created an Internet Users' Glossary. See RFC 1392.

user group, user's group, users' group An organization of users of a particular product or service. A support group. With the introduction of computers, society took a technological leap that was difficult for any one individual to understand or bridge. In order to facilitate the use and understanding of complex systems, programming languages, and technologies, many users' groups sprang to life, beginning in the mid-1950s, to provide mutual support and assistance in sharing information and meeting technological challenges.

U

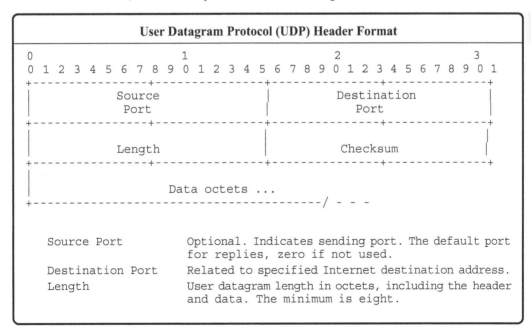

User Datagram Protocol (UDP) Header Format

```
                User Datagram Protocol (UDP) Header Format

 0                   1                   2                   3
 0 1 2 3 4 5 6 7 8 9 0 1 2 3 4 5 6 7 8 9 0 1 2 3 4 5 6 7 8 9 0 1
+--------------------------------+--------------------------------+
|            Source              |           Destination          |
|             Port               |              Port              |
+--------------------------------+--------------------------------+
|                                |                                |
|            Length              |            Checksum            |
+--------------------------------+--------------------------------+
|                                                                 |
|                     Data octets ...                             |
+-------------------------------------------/ - - -

    Source Port         Optional. Indicates sending port. The default port
                        for replies, zero if not used.

    Destination Port    Related to specified Internet destination address.

    Length              User datagram length in octets, including the header
                        and data. The minimum is eight.
```

The proliferation of users' groups is important not only for the support they provided to members, but also because this venue provided a forum for computer hobbyists, amateur radio groups, and many other amateur and professional enthusiasts to brainstorm ideas and contribute to the developing fields. The development of computer technology was no longer in the hands of large educational institutions and corporations. Individuals and small companies, particularly in the 1970s, had a window of opportunity during which they were able to make highly significant contributions to the development of the industry.

user interface UI. The communications link through which a person interacts with a machine. On computers, the UI, in its broadest sense, includes the various symbolic text, images, sound, and other sensory cues and gadgets presented to the user, with which the user interacts. This is commonly done through peripheral devices such as touchscreens, keyboards, mice, joysticks, microphones, data gloves, and others not yet invented. It is considered the highest layer of the computer system structure, with the machine instructions for physical operations comprising the lowest layer.

The user interface is the single most important aspect of computing and should never be undervalued. Computers were designed to serve the needs of people; if people are forced to adopt unhealthy or uncomfortable habits to interact with computers, or if computers take time away from people rather than freeing them from repetitive tasks or drudgery, then human needs are not adequately served by the technology. User interfaces and software applications should be designed with the goal that the purpose of the technology is to improve the quality of life.

The design of user interfaces is an art. It demands common sense, a knowledge of ergonomics, psychology, philosophy, electronics, aesthetics, and a large dose of sympathy for a broad range of users. As such, user interfaces have developed in fits and starts, with many software programs providing very poor support for users, forcing the user to conform to the idiosyncrasies of the machine (or the programmer who wrote the software), rather than the other way around. See user interface history.

user interface history The earliest telecommunications interfaces consisted of physical semaphores (smoke, flags, and arms) and telegraph keys sending coded messages that needed to be decoded and transcribed when received. The received message was usually presented as a long paper tape inscribed with wiggly lines, dots and dashes, or punched holes. Computers up until the 1950s used a similar model. This kind of user interface wasn't very friendly, so inventors, even in the earliest days of telecommunications technology, sought ways to encode the alphabet, so that letters could be directly sent and received (eventually resulting in teletypewriters) without the operators doing the translation. But the basic methods prevailed for decades, mainly because they could be used anywhere, with the simplest of equipment.

Telegraph key codes are still a requirement of attaining amateur radio licenses.

With the development of personal computers, user interfaces took a leap. The Altair microcomputer, sold originally as a hobby kit in 1984, had no monitor, mouse, or keyboard. It was programmed by means of flipping little dip switches; if you made a mistake, you had to start again. Yet within 2 years modern microcomputers, inspired partly by high-end systems with better resources than the Altair, came into being in the form of the TRS-80 and Apple computer, and keyboards and monitors became standard. Almost every change since then has been an evolutionary refinement or logical addition rather than a revolutionary change. Even the lifelike and startling three-dimensional virtual reality world represents, for the most part, an evolutionary development, albeit an exciting one.

Early computer user interfaces consisted primarily of monochrome screens displaying limited text, often with no lowercase letters, and large rectangular graphic blocks. While ingenious computing pioneers wrung astonishing surprises from this primitive technology, it was obvious that improvements were needed in order for a computer to be more fun, versatile, and consumer-friendly.

User-to-User Indicator UUI. In ATM network Adaptation Layer 2 (AAL2), a 5-bit indicator in a 3-octet-header CPS packet that follows the length indicator (LI) and precedes the header error control (HEC). Initially the PPT and UUI were separate fields, but were merged in the mid-1990s into one field, sometimes called the CPS-UUI field. The UUI field enables upper layers (users) to somewhat transparently convey parameters, for example.

UserID User Identification. A unique computer account designation used to gain access to a secure or monitored system. A UserID is frequently paired with a password for system access. Historically many systems accepted only eight characters for the UserID and, for backward compatibility, this limitation persists on many systems today. On networks using the most common mail systems, the UserID typically forms the first part of an email address.

USGS See United States Geological Survey.

USITA Formerly United States Independent Telephone Association. See United States Telephone Association.

USKA Union Schweizerischer Kurzwellen-Amateure Union (Union of Swiss Shortwave Amateurs). A member organization associated with the International Amateur Radio Union. http://www.uska.ch/

USOC See Universal Service Order Code.

USOP User Service Order Profile.

USP1X The USOC code for telephony-related maintenance plan, standard, level 1, per line/circuit, UNI.

USTA See United States Telephone Association.

UT1 A time reference based upon Earth's axis rotation. It is related to Coordinated Universal Time in that UTC was set to synchronize with UT1 at 0000 hours on January 1, 1958.

UTAM Since some of the frequencies used by incumbent carriers have now been designated for USDC services (1890 to 1930 MHz), companies are changing their operating equipment and software to operate instead in the 2.0 GHz microwave C-band. UTAM Inc. is an open industry resource for assisting in frequency relocation. See band allocations.

UTC See Coordinated Universal Time.

UTDR Universal Trunk Data Record.

Utility Communications Architecture UCA. A comprehensive suite of communications protocols based upon open systems for use by electric utilities providers/maintainers. UCA began at the Electric Power Research Institute (EPRI), in 1988. UCA is a flexible, scalable architecture that can control various types of devices ranging from small local devices to those in major installations and control centers.

UCA Version 2 has been developed to support the communication needs of Energy Management Systems, Supervisory Control and Data Acquisition Systems, Intelligent Electronic Devices, Remote Terminal Units, and others. The standardization effort is intended to help utilities providers to intercommunicate through a variety of physical media, to reduce installation, operations, and migration costs, and to achieve secure communications through standard mechanisms.

UCA documents are being assessed by an IEEE Standards Coordinating Committee for review as potential IEEE standards.

Utility Communications Architecture Forum UCA Forum. A group dedicated to promoting UCA and assisting individuals and companies in understanding, utilizing, and furthering this technology. http://www.ucaforum.org/

utility pole A sturdy tall pole installed in the ground (or *on* the ground in mountainous areas, supported by rocks and guy wires). The pole is used to support utility wires for power and telecommunications and may have crossbars and insulators. Utility poles are raised with the aid of long poles with spikes on the end called *pike poles*, or with industrial machines designed for the job. Most poles are made from logs, although some areas have metal poles; at one time in history, it was thought that metal poles would soon replace all the wooden poles, a prediction that didn't hold true. In some areas, especially avalanche areas, it was necessary to reinstall poles once or twice a year, a costly, time-consuming business, so various alternatives were tried, including laying the wire along the ground. Unfortunately, rodents like to chew through the wires, causing almost as much interruption to service as the avalanches. Transmissions in inclement regions are now often sent with microwave transceiving systems rather than with wires, a solution which requires less maintenance. See joint pole for more details and diagrams.

See Universal Transverse Mercator Projection.

UTM UUCP Map A logical map describing the interconnections between intercommunicating UUCP-capable systems. A UUCP Map Entry is issued when a host is registered on the UUCP system. The main routing information in UUCP-base networks is related to the UUCP Map and is contained in a *pathalias* database. Thus, for a message from *mysite.org* to be delivered to *yoursite* the path might be expressed as

`mysite.org  site1!site2!midsite3!yoursite!%s`

Because exclamation marks are traditionally used to separate the nodes, this is commonly called a "bang path" and veterans of the early email days can remember typing bang paths into their email *TO:* headers.

In the early 1980s, the UUCP Map was still small enough to be represented on a single page, with systems such as *ucbvax*, *menlo70*, *decvax*, and *chico* representing familiar interlinks to those who were using the system at the time. From this point, UUCP grew and spread to the point where static connections/routing maps were no longer practical in the way they were in the early UUCP days. By the 1990s, map updates generally came from domain registrations rather than from manually submitted registrations. By 2001, it was announced that UUCP Maps would likely be replaced with XML for future registrations. There are sites on the Web that enable the UUCP Map to be queried for a specific entry. For example, a search for *ucbvax* lists the Internet mail routing entry as *cs.purdue.edu!ucbvax!%s* and the UUNET mail routing entry as

`decwrl.dec.com!decvax!purdue!ucbvax!%s`

The UUCP Map can also provide information about site administrators. In Europe, UUCP Map entries are available through the backbone "netdir" service.

Joint Utility Pole With a Variety of Cables

A typical utility pole with crossarms bearing primary power lines, ceramic insulators, below which are transformers in cylindrical containers feeding power to secondary power lines, and local power drops to nearby residences and businesses. Below these are the telecommunications cables carrying voice and data services.

U

UTP See unshielded twisted-pair.

UTR Universal Tone Receiver.

UTS Universal Telephone Service.

UUCP UNIX-to-UNIX Copy. UUCP was a basic networking system developed in the mid-1970s by AT&T Bell Laboratories; it was distributed with UNIX in 1977. UUCP was quickly adopted by many educational and research institutions for disseminating mail. Two years later USENET, the global public news forum, was established using UUCP.

As UUCP spread and became an important medium for computer connectivity, it was ported to many other computer architectures. In 1997, UUCP mail routing was taken over by the UUCP Project. By the mid-1980s national networks in other countries were being set up with UUCP, establishing it as an important catalyst for intercommunications and the development of distributed networks. Due to its increasing importance, formats for the transmission of electronic mail through UUCP were then standardized for mixed computer environments along lines developed by ARPA. In 1987, UUNET was founded to provide commercial UUCP and USENET access. See BITNET, Unix, UNIX, USENET, UUCP Project, UUNET, RFC 822, RFC 920, RFC 976.

UUCP Project A project initiated in the early 1980s to enable the exchange of electronic mail among communicating sites using the UUCP store-and-forward transport system. The UUCP Mapping Project endeavored to create a single worldwide database of systems interconnected through UUCP, in addition to the determination of optimum data paths between systems. In 1997, the UUCP mail routing for UUNET site, created by Eric Ziegast, was turned over to the *UUCP Project* currently coordinated by Stan Barber. See UUCP.
http://www.uucp.org/

UUI 1. See Unified User Interface. 2. See unique user identifier. 3. See User-to-User Indicator.

UUID See Universal Unique Identifier.

UUNET A Unix-based network provider and backbone (long-haul network). UUNET provides Internet name serving, connectivity, MX forwarding, and news feeds. The formation of UUNET was probably due in part to the Acceptable Use Policy (AUP) focus on research and education enforced by the National Science Foundation's NSFNET. This stimulated commercial establishment of computer networks.

In the 1980s, UUCP over long-distance dialup lines was the primary means by which providers, institutions, and individuals received their messages. However, Internet connectivity has become ubiquitous and the situation has changed. In the 1990s, UUNET is the only remaining significant network that uses UUCP transport for USENET messages. UUNET Canada, Inc. is located in Toronto; UUNET Technologies Inc. is located in Virginia.

URN See Uniform Resource Name.

UV See ultraviolet.

UWB ultra wideband.

UWC See Universal Wireless Communications.

UWCC See Universal Wireless Communications Consortium.

UXT*xx* The USOC code for telephony-related surcharges for emergency reporting services. The *xx* designates the region. For example, UXTMN refers to UXT services for Minnesota.

v 1. *symb.* volt (may also be capitalized). See volt.

V 1. *symb.* vacuum tube. See Audion, electron tube, vacuum tube. 2. *symb.* voltmeter.

V & H Coordinates, V H Coordinates Vertical & Horizontal grid coordinates. Imaginary coordinate points on a virtual grid used to determine straight-line mileage between two specified points, with each exchange's location represented by a pair of V & H coordinates. This is used for various products and services charged on a distance or mileage basis, as are long distance calls.

The V & H system is based upon a 'flattened Earth' system from a Donald Elliptical Projection, developed by Jay Donald of AT&T in the mid-1950s. The basic idea is to create a triangular distance calculation over a flattened surface.

V & H Tape Vertical and Horizontal Coordinates Tape. A recorded tape provided primarily to assist with billing, it includes NXX types, major and minor V & H coordinates (latitude- and longitude-like regional designations), LATA Codes, and other information related to long distance accounting and service areas. It can be purchased from Bellcore.

V drive In analog video, a periodic signal related to the vertical component of a frame that is constructed with sequential, repeating line scans. In standard systems, the V drive sends a pulse so that the electron gun returns from the bottom right corner of the video frame to the top left corner (during the vertical blanking interval) in order in position for imaging the next frame (or half-frame in an interlaced system).

The horizontal and vertical sync are related so that pulses can be combined on a single wire, together comprising a *composite* video signal. A composite signal can be represented as Csync-red-green-blue and transmitted over four wires.

Many computer monitors use a five-wire RGBHV system in which the H and V represent horizontal and vertical sync pulse components. See H drive, negative-going video.

V interface In ISDN, a number of reference points have been specified as R, S, T, U, and V interfaces. To establish ISDN services, the telephone company typically has to install a number of devices to create the all-digital circuit connection necessary to send and receive digital voice and data transmissions.

The V interface is the reference point between the telephone switching office's exchange terminal switch and the line terminal switch. Thus, one side connects to the public telephone exchange and the other connects through the U interface to the subscriber's network termination (NT*x*) device. See ISDN interfaces for a diagram.

V number (*symb.* – v) In fiber optic lightguides, the *normalized frequency parameter* for describing the number of modes in a given fiber optic waveguide. The transmission of different guided or radiant modes in a fiber is based upon the materials used and their respective refractive indexes, the core diameter, and the relationship of the core to the reflective cladding that surrounds it.

Mathematically, the V number can be expressed as 2π (which is the number of radians in a full circle) times the radius of the fiber core (in microns) which yields the circumference of the core. This is divided by the wavelength (in microns) times the numerical aperture of the fiber (square root $n^2_1 - n^2_2$) where the numerical aperture (NA) is derived from the core ($n1$) and cladding ($n2$) relationship. Symbolically, the V number may be expressed as $V = (2\pi a/1)NA$.

Smaller V numbers tend to be associated with single-mode fiber transmissions, while larger V numbers are associated with multimode fibers that can transmit more than one wavelength. Within multimode fibers, a larger V number is associated with a larger numerical aperture up to practical tolerances.

The V number provides information that is useful in determining relationships between core and cladding relative to (1) the wavelengths that are intended for use with a particular cable and (2) the beam width of the illumination source for the lightguide. See cladding, dominant mode, index of refraction, numerical aperture.

V Series Recommendations A set of ITU-T recommended guidelines for interconnecting networks and network devices. These are widely implemented in computer modems. The V Series specifications are available for purchase from the ITU-T, and a few may be downloadable from the Net. Some of the related general categories and specific V category recommendations of particular interest are listed here. See also I, Q, and X Series Recommendations.

ITU-T V Series Recommendations	
Number	Description
V.21	A 1984 dialup modem standard supporting data rates up to 300 bps. Most modems in the late 1970s and early 1980s were acoustical couplers, with direct connect modems just beginning to catch on in the mid-1980s.
V.22	A 1988 dialup modem standard supporting data rates up to 1200 bps with fallback to 600 bps. It is interesting to note that in the 1980s, many technical people insisted it was impossible to support speeds faster than 1200 bps over standard phone lines, and that this was probably the fastest speed at which modems would ever transmit data.
V.22 bis	A 1988 update to the V.22 standard supporting data rates up to 2400 bps through frequency division techniques, with link negotiation fallback to 1200 bps, and fallback to V.22.
V.23	Dialup modems supporting data rates up to 1200 bps with fallback to 600 bps and a 75-bps *back channel* or *reverse channel*.
V.26	A 1984 standard for modem supporting data rates up to 2400 bps on four-wire leased telephone lines.
V.26 bis	A 1984 update to V.26 supporting 2400 and 1200 bps over public phone dialup lines.
V.26 ter	A 1988 update to V.26 and V.26 bis that supports both echo cancellation (public phone lines), and point-to-point two-wire leased phone lines.
V.27	A 1988 standard for modem data rates up to 4800 bps with manual equalizer for use with leased telephone lines.
V.27 bis	A 1984 update to V.27 that supports data rates of 4800 and 2400 bps with automatic equalizer for use with leased telephone lines.
V.27 ter	A 1984 update to V.27 and V.27 bis that supports data rates of 4800 and 2400 bps over public phone dialup lines.
V.29	A 1988 standard for modem data rates of 9600 bps for four-wire leased telephone lines.
V.32	A standard for public dialup and two-wire leased line modems with rates up to 9600 bps with fallback to 48,800 bps. This standard was approved in 1984. When modems incorporating V.32 were first introduced in the late 1980s, it was not unusual for a single modem to cost between $600 and $2,000. Modems of this speed can now be found for $10, as they have been superseded by much faster data rates of 33,600 bps. Many of the V.32 modems were dual-standard modems that supported V.32 in addition to various proprietary protocols from individual manufacturers. See V.32 bis.
V.32 bis	A 1991 standard for public dialup and two-wire leased line modems that supports rates up to 14,400 bps with fallback to 12,000, 9600, 7200, and 4800 bps. Like V.32, a number of these modems are dual-standard modems with proprietary protocols included, and they are generally all backward compatible with V.32. Many of them support MNP-2 to MNP-4 error control and MNP-5 data compression, in addition to V.42/V.42 bis error control and data compression.
V.32 ter	An update to V.32 bis designed by AT&T that supports data rates up to 19,200, and fallback to 16,800 and V.32 bis. This format is freely distributable and has become a de facto standard. The corresponding ITU-T approved format for 19,200 is V.34.
V.33	A 1988 standard for modem data rates of 14,400 bps for four-wire leased telephone lines.
V.34	A 1996 standard for dialup modems and fax/modems with rates up to 33,600 bps in full duplex, and up to 28,800 bps in half duplex for facsimile transmissions. (These are sometimes also called VFast, or VFC, but those are actually slightly different interim protocols developed by vendors who were anxious, in the mid-1990s, to get products to market while the work on the V.34 standard was being completed. Some modems became V.34/VFast hybrids.) V.34 modems are designed to adapt to the line by probing the connection and adjusting according to quality and capacity. They interact with the telephone circuit through handshaking. Optional control data can be sent through an auxiliary signaling channel. V.34 modems are not as subject to noise as earlier modems, due to the implementation of multidimensional trellis coding.
V.34 bis	A 1996 standard that supports the higher data rates of 56,000 bps and 31,200 bps.

	Digital Modems, Digital/Analog Hybrids, Wideband, & Parallel Transmission Modems
V.19	A 1984 standard for modems for parallel data transmission over telephone signaling frequencies.
V.36	A 1988 standard for wideband synchronous transmission modems using 60 to 108 kilohertz group band circuits.
V.37	A 1988 standard for wideband synchronous transmission modems for signaling rates higher than 72,000 bps, using 60 to 108 kilohertz group band circuits.
V.38	A 1996 standard for wideband data circuit-terminating equipment for rates of 48,000, 56,000, and 64,000 bps for use on digital point-to-point leased circuits.
V.70	A 1986 standard for digital simultaneous voice and data (DSVD) modems in which a data transmission and a digitally encoded voice transmission can be sent at the same time over a single dialup phone line. DSVD are typically downward compatible with standard dialup modems based on more recent high-speed technologies (e.g., V.34). This type of technology lends itself well to teleconferencing applications, and a number of vendors have incorporated V.70 to this end.
V.75	A 1996 standard for digital simultaneous voice/data (DSVD) transmission terminal control procedures. See V.70.
V.76	A 1996 standard for multiplexing using V.42 LAPM-based procedures. V.76 multiplexing has been incorporated into V.70 systems, although it is not limited to V.70.
V.80	A standard for the application interface for communications through data terminal equipment (DTE) of a synchronous H.324 bit stream, such as video, for asynchronous transmission over public switched telephone networks. V.80 works in conjunction with a number of other related technologies and standards. H.324 is an ITU-T approved standard that provides the foundation for combining video, voice, and data communications on a single analog phone line using a 28,800 bps data rate connection. This opens the door to a variety of practical, reasonably priced stand-alone and computer-based videoconferencing products that display up to 15 video frames per second. H.263 and G.723 are video and voice compression schemes used in H.324.
V.90	A 1998 standard for digital to/from digital or digital to/from analog connections, supporting download data rates up to 56,000 bps with upload data rates up to 33,600. During development it was known as V.pcm and some vendors call it PCM. This standard reflects the gradual conversion of data communications to digital format. This standard has its own Web site at http://www.v90.com/

	Half-Duplex Modem ITU-T-Recommended Transmission Standards
V.17	A two-wire scheme for facsimile machines and fax/modems used in conjunction with extended Group 3 facsimile standards for image transfer at rates of 12,000 bps and 14,400 bps.
V.27	A modulation scheme for dialup facsimile machines and fax/modems used in conjunction with Group 3 facsimile standards for image transfer at rates of 2400 bps and 4800 bps.
V.29	A modulation scheme for dialup facsimile machines and fax/modems used in conjunction with Group 3 facsimile standards for image transfer at rates of 7200 bps and 9600 bps.

	ISDN- and Digital Communications-Related
V.100	A 1984 standard for interconnection between public data networks (PDNs) and public switched telephone networks (PSTN).
V.110	A 1996 standard for ISDN support of data terminal equipment (DTE) with V-series interfaces. Listed also as I series 463.
V.120	A 1996 standard for ISDN support of data terminal equipment (DTE) with V-series interfaces with provision for statistical multiplexing. Listed also as I series 465.
V.130	A 1995 standard for an ISDN terminal adaptor framework.
V.230	A 1988 specification for a general data communication interface layer 1.

V

The chart on the following pages includes a number of V Series Recommendations of particular interest. Note that while a larger number often denotes a more recent standard, there are exceptions, and the categories are intermingled resulting in the number schemes series being interleaved.

V5 A telephony standard adopted by the European Telecommunications Standard Institute (ETSI) for use in digital local exchanges in Europe, Australia, South America, and the Far East. Exchanges around the world are gradually being upgraded to V5.

V-band A frequency band commonly used in radar applications. See the band allocations listing for a chart and details of frequencies.

V.Fast A vendor-developed format which was brought out while the ITU-T standards bodies were developing the V.34 protocol. V.Fast is similar to V.34, but not the same, and consequently some V.Fast modems are not compatible with V.34, while others are hybrid and support both. See V Series Recommendations listing under V.34.

VAC See vehicle access control.

vaccine A whimsical name for a virus protection software program that signals the user if anomalies or known viruses are present, and allows the virus to be disabled or deleted. As quickly as virus detectors and vaccines are written and disseminated, virus creators come up with new ways to create mischief or outright destruction on people's computer systems. With the Internet foreshadowing a not-too-distant day when every computer is linked to the net, perhaps 24 hours per day, more opportunities for vandalism exist, and education about virus detection and protection needs to be made known to the computing public.

vacuum An enclosed space in which the gases are at pressures below atmospheric pressure. Since many vacuums are actually near-vacuums, categories of vacuums from low to ultrahigh, depending upon the pressure, have been described. The discovery and creation of vacuums have aided scientists in making important discoveries and provide environments in which various phenomena take place, or fail to take place, due to the absence of gases. For example, a filament in a bulb can burn much longer in a vacuum.

ITU-T V Series Recommendations, cont.	
Error Control and Data Compression Protocols	
V.41	A 1972 code-independent error control standard.
V.42	A 1996 error control standard that greatly enhances the functioning of modems over standard phone lines. Phone lines tend to be noisy, slow, and somewhat unreliable for high speed data communications. In the past, if there were errors on the line, modems would react by exhibiting line noise, losing the connection, or aborting the current operation. With new error control capabilities built in, some of these problems are overcome through filtering and selective retransmission.
	V.42 includes two error control protocols for dialup modems. These are link access procedures for modems (LAP-M) and Microcom Networking Protocol (MNP-4), such that connections with a variety of modems are supported.
	V.42 uses a filtering process somewhat like the error correction schemes incorporated into a number of file transfer protocols, such as XModem and ZModem.
V.42 bis	A 1990 standard for data circuit terminating equipment (DCE).
Miscellaneous Modem/Transmission/Network-Related Standards and Protocols	
V.7	A 1988 guide to terms concerning data communications over telephone networks.
V.8	A 1994 guide to procedures for starting sessions of data transmissions and setting up connections parameters over general switched telephone networks.
V.8 bis	A 1996 guide to procedures for the identification and selection of common modes of operation between data circuit-terminating equipment (DCE) and between data terminal equipment (DTE) over general switched telephone networks, and on leased point-to-point telephone-type circuits.
V.18	Interoperability guidelines for communications services for the hearing impaired.
V.24	A 1996 set of definitions for interchange circuits between data terminal equipment (DTE) and data circuit-terminating equipment (DCE) that specifies the characteristics of interfaces, including pinout circuitry. This is similar to the RS-232 specifications.
V.25	A 1996 standard for dialup automatic answering equipment and automatic calling equipment.
V.25 bis	A 1996 command set designed for synchronous communications through serial ports and connections.
V.54	A modem diagnostics standard implemented in high speed modems.

vacuum column In some magnetic tape drivers, there may be a vacuum mechanism to control the tape loop. This provides a lower air pressure 'suction' next to the tape and the drive mechanism.

vacuum gauge An instrument for measuring the degree of vacuum in an enclosed space. There are a number of types of vacuum gauges, including manometers, ionization gauges, and thermal conductivity gauges.

Evacuated Electron Tubes

Vacuum tubes fulfilled thousands of roles in electronics for over fifty years, before the invention of the transistor and modern semiconductor technologies. Even now they are appropriate for certain high-frequency applications. The most significant step in the evolution of vacuum tubes was the triode, which included a controlling grid to harness electron energy. These excellent historic examples are from the American Radio Museum collection.

---

vacuum tube A ubiquitous, essential, and versatile electron tube that was common until about the mid-1960s, after which it was superseded by various early electronic transistors, and later more sophisticated semiconductor technologies.

The vacuum tube in its basic form consists of an electron-emitting filament (cathode) and a metal plate (anode) to which the electrons are attracted. Various types of control grids might be interposed between the cathode and the anode (or in other locations), with the whole thing sealed in a glass vacuum tube to control the internal environment and to prolong the life of the filament. The first vacuum tubes adapted for radio broadcasting equipment were developed in the early 1900s.

While most small vacuum tubes have passed out of use, cathode ray tubes (CRTs) are still widely used in television sets and desktop computer monitors. There are still situations in which vacuum tubes can be a better solution than the solid-state components now commonly used in electronics. For high power-level, high-frequency applications, vacuum tubes are sometimes more efficient and still merit consideration. See Audion, cathode ray tube, electron tube.

vacuum tube amplifier An important development stemming from Lee de Forest's invention of the Audion, developed by Harry DeForest. The vacuum tube amplifier was incorporated into telephone repeating units, which extended communications distances.

VAD See voice activity detection.

Vail, Alfred (1807–1859) An American scientist and inventor, and associate of Samuel Morse, from whom Morse adapted a number of ideas related to the building of telegraph systems. Vail was very mechanically apt and continued over the years to make technical improvements to the technology. In 1837 Vail made an agreement with Morse to turn over the rights for his inventions in return for a share of the commercial rights. In 1848, due to the mounting workload, and the lack of sufficient appreciation and compensation, Vail decided to terminate his relationship with Morse. Morse code may be one of Vail's biggest contributions. When Morse was designing a system for coding messages, he initially created a complex numeral-letter relationship that required a time-consuming dictionary lookup to decode the words. Vail apparently came up with the simpler system, after altering the mechanics of a telegraph instrument so it moved vertically rather than horizontally, to allow the instrument to leave spaces (thus yielding dots and dashes) when recording a message. See Morse code, telegraph history.

Vail, Theodore N. (1845–1920) Theodore Vail was from the same Vail family that had a close association with Samuel Morse at the time of his invention of the telegraph, and Vail learned telegraph code from the elder Morse. Vail became a telegraph operator, and later devoted most of his life to the management and promotion of universal telephone services.

Vail became company general manager of the first Bell company in 1878 in Boston. He was the first president of the Telephone Pioneers of America and a cofounder of Junior Achievement Inc. (1919). He was instrumental in continually expanding the company's offices and services, and directed the building of the first transcontinental telephone line, completed in 1914 and officially opened in 1915.

After what has been regarded as a brilliant career, where he maintained his commitment to quality and service, Vail resigned in 1887 due to his disgust at the narrowly commercial vision of the Bell financiers. In 1907, he was induced to return to AT&T as president, at a time when Bell was rapidly buying independents. Bell also purchased Western Union stock, and installed Vail as president of what had once been Bell's chief rival. With the death of J.P. Morgan in 1913, Vail voluntarily took steps to reduce AT&T's monopolistic buyouts and control of the long distance networks, in order that reorganization could be done from the inside rather than being imposed by regulatory authorities. Vail retired as president in 1919. See Kingsbury agreement.

valance band In semiconductors, a region representing the energy states of bound electrons, those that are not free to move readily within the material, as opposed to a *conduction band* representing the energy states of electrons that may move more readily to provide energy change/conduction.

In a diode semiconductor component, the valance and conduction bands are separated by a gap or "electron holes" to provide a p-njunction for directional movement of electrons within a material. The application of current can cause a shift in energy levels such that energy is released as the material returns to a state of rest. This energy, in the form of photons, may be directed outside the component to create semiconductor lasers, for example. See light-emitting diode, p-n junction, semiconductor laser.

value In imagery, the relation of a color to white or black, or levels of gray. Sometimes called lightness. In color, the intensity of the color.

value-added Services offered, usually for a fee, in addition to standard product or subscriber services. Options.

VAN value added network.

Van Allen radiation belt A region of space surrounding Earth in which there is high-intensity particle radiation which is sufficiently destructive that it is avoided for communications satellite orbits. This region is roughly between the low Earth orbit (LEO), starting at about 1000 kilometers, and medium Earth orbit (MEO). It is named after James A. Van Allen.

vanadate (*symb.* – VO$_4$) A rare earth crystalline substance grown by various means, sometimes two or three bars at a time. Vanadate may be doped with a variety of materials such as neodymium (Nd), which is a component of dydimium (neodymium is also used in electromagnet read/write components in magnetic storage disks).

Vandate is grown in labs, often using the Czochralski technique. Once grown into glasslike rods in a variety of colors, it can be annealed, cut into shapes that include beam-splitting cubes or birefringent wedges. It may then be polished, and incorporated into fiber optic components such as diode lasers.

Nd:YVO$_4$ (neodymium-doped yttrium-vanadate) has a short absorption depth compared to Nd:YAG (neodymium-doped yttrium-aluminum-garnet). Highly birefringent, transparent YVO$_4$ has been developed for use in passive fiber optic components such as interleavers, isolators, compensators, and circulators. See Czochralski technique, doping, YAG

Van de Graaff generator A device for creating electrostatic effects by charging insulated electrodes to high energy potentials. Van de Graaff generators are popular exhibits in science museums. They typically are configured as melon-sized metallic spheres atop a central pole, and visitors can place their hands on the globe and watch their hair stand on end. It is named after Robert J. Van de Graaff, an American physicist of the early 1900s.

vapor deposition A process used in the fiber fabrication industry for creating a preform cylinder (typically silica glass) from which fibers are pulled. There

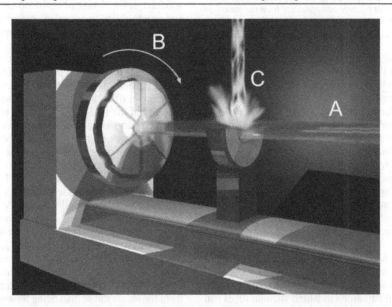

Vapor Deposition Preforms – Modified Chemical Vapor Deposition (MCVD)

In modified chemical vapor deposition (MCVD), oxygen is bubbled through chemical solutions chosen to influence the physical and optical properties of the fabricated material. The resulting vapors are drawn into a tube (A) that is attached to a lathe (B). A torch (C) is drawn along the tube as it rotates.

Inside the tube the vapors undergo chemical reactions and fuse to form glass on the inside of the tube. Rotating the preform tube helps consolidate the preform blank with an even consistency. The preform blank is fabricated with the correct core-to-cladding ratio and subsequently put into a high-temperature furnace as part of a fiber drawing assembly in which the molten medium is dropped from the top of the assembly so that gravity can draw down the glass to form a long, slender fiber filament.

are a number of different processes for vapor deposition based upon chemical vapor reactions that occur in the presence of heat, including modified chemical vapor deposition (MCVD), outside vapor deposition (OVD), plasma chemical vapor deposition (PCVD), and vapor axial deposition. The types and combinations of vapors vary, but argon or argon and helium are commonly used as a carrier gas in the deposition process and helium is used to sweep out impurities in the preform consolidation of the vapors. See diagram. See boule, helium recovery, preform.

van Heel, Abraham C.S. (1899–1966) A physicist from the Netherlands, van Heel had the opportunity to spend 1925 at the Institute of Optics in Paris, working with Fabry, after whom the Fabry-Perot interferometer is named. He also studied at Leiden University and took a position at Delft University in applied physics where he led the optics group from 1947 to 1966.

In 1948, van Heel became cofounder of the International Commission for Optics (ICO). In the 1950s he did significant research into the loss of light in plain optical filaments. His observations led to the landmark development of cladding, an outer layer that could contain light within a lightguide through internal reflection. In 1954 van Heel published his findings in *Nature* magazine and the fiber optics industry developed dramatically from that point on. See Hopkins, Harold.

van Musschenbroek, Peiter (1692–1761) A Dutch educator and experimenter who created the Leyden jar condenser (named after the region) in 1745, independently of other inventors, including E.G. von Kleist. It was a simple but practical capacitor, and effective enough to harm a person not careful in its handling. Van Musschenbroek also described a way to carry out experiments with the jar so that a person would not be mortally shocked. Many subsequent experimenters, including Benjamin Franklin, devised variations on the basic Leyden jar.

vaporware A derogatory term for software that has been announced prior to completion, or prior to distribution. The negative aspects of the term stem from two sources, truth in advertising and the fact that many announced software products never hit the shelves. Because the software industry is new and confusing to a great portion of the public, unsupported product announcements have not been as strongly condemned and regulated as in other industries, a situation which may change as the buying public becomes more familiar with the new technologies and tired of unsubstantiated promises.

variable bandwidth Bandwidth that can be tailored, usually on an on-demand basis, to the capacity needs of the current transmissions. The ability to adjust capacity allows the system to more efficiently allocate resources and provides a mechanism for setting up accounting systems that bill on an as-used basis.

variable bit rate VBR. A data transmission commonly represented by irregular groups of bits or cell payloads, followed by unused bits or payloads. VBR traffic is generated by most media other than voice.

In an ATM environment, a VBR service can be realtime or non-realtime, and is guaranteed sufficient bandwidth and quality of service (QoS). See cell rate for a chart.

variometer An instrument for measuring magnetic declinations, particularly of the Earth. See declination.

Varian, Russell H. (1899–1959) An American inventor, Varian worked in the physics laboratories at Stanford and, in 1948, cofounded Varian Associates with his brother Sigurd and Edward L. Ginzton. He served as the President and a Board member until his death. He was awarded numerous patents related to thermionic tubes, magnetic resonance, and various radar technologies. Varian was a member of several prominent engineering organizations and a Director of the West Coast Electronic Manufacturer's Association. He was a Sierra Club Committee member and acquired land to further conservation efforts connected with the Sierra Club. He and his wife Dorothy set the groundwork for Castle Rock to become a state park in 1968.

Varian had an early introduction to television technology through Philo Farnsworth and became well-known for the co-development, with Sigurd Varian and William Hansen, of the Klystron tube technology that is used to this day in broadcast TV. This important invention in the late 1930s enabled the generation of ultra-high frequency radio waves (microwaves) and supported the evolution of radar and communications technologies. See Klystron.

Varian, Sigurd F. (1901–1961) An American engineer and commercial pilot, Sigurd shared in many collaborative inventions and business ventures with his brother Russell. He helped cofound Varian Associates and invented a number of types of practical devices, including pumps, filters, heaters, and a precision high-speed drill press. His interest in navigation was a good match for his co-invention of the Klystron, an electron tube to generate ultra-high frequency radio waves (microwaves) that could be used to improve navigation through radio direction-finding technologies. See Klystron.

Varian Associates, Inc. A pioneering microwave technologies development firm, founded by the Varian brothers (Russell and Sigurd) and Edward Ginzton in 1948 in San Carlos, California. Dorothy Varian, Russell's wife, was also active in the founding, development, and operations of the company, and served as the treasurer until 1951.

Varian Associates built upon the Klystron tube invented by the Varian brothers and William Hansen and developed further by Ginzton, Marvin Chodorow, and others. Varian Associates and the Sperry Gyroscope Company's lab contributed to important developments in radar sensing, navigation, and microwave communications technologies over the following decades. Varian was also instrumental in creating some of the components important in chemical spectroscopy and medical imaging. Varian, Inc. was spun off from Varian Associates in 1999 to focus on life sciences and health care products.

V

Varian Associates was the first company to move into the Stanford Industrial Park, in 1953, the heart of the emerging Silicon Valley. There were a number of divisions within the firm, including the Microwave Power Tube Division and the Electron Devices Group, which separated from Varian Associates to become Communications and Power Industries (CPI) of Palo Alto, California. The Microwave Power Tube Division of CPI continues the tradition of developing and improving klystron technology. See Ginzton, Edward; Klystron; Varian, Russell and Sigurd.

Varley loop test A type of diagnostic procedure which uses resistance through a bridge to locate a fault in a length of circuit. Variable resistance is connected in series with the resistance of the broken or defective line. It is similar to the Murray loop test, but with a third wire, and more commonly used. See Murray loop test; Wheatstone bridge.

Varley, Cromwell Fleetwood A British researcher and technician who investigated ionization, and was involved in early Atlantic telegraph cable installation. He was hired by Western Union to evaluate the telegraph system in the U.S. Varley standardized many of the lines, systems, and diagnostic techniques. The Varley loop test is named after him. See Wheatstone, Charles.

VAX *V*irtual *A*ccess *E*xtension. A series of minicomputers from Digital Equipment Corporation (DEC) that was developed in the mid-1970s as the successor to the DEC PDP-*x* series. The VAX computer system was released in 1977 with 32-bit computer capabilities compared to the 18- or 16-bit systems characteristic of the PDP- line. The VAX 11/780 was popular in educational institutions around the world, often replacing or being added to the still-popular PDP-11 machines. Both dumb and smart remote terminals (ADS, Gigi, Tektronix, Ramtek, and others) could be readily interfaced with the VAX to support one or two hundred users at a time. The VAX 11/750 followed, along with several other models, including the MicroVAX.

The early VAX machines were generally running VMS as the operating systems, but Unix also became a popular VAX option over time. See PDP-.

VBE VESA BIOS extensions. A high-resolution VESA BIOS standard that can be implemented in hardware or software to provide control of video graphics displays on a computer, typically an Intel-based International Business Machines (IBM) or third-party licensed computer.

vBNS very high speed Backbone Network Service. A research network established in 1995 by the National Science Foundation.

VBR See variable bit rate.

VBX Visual BASIC Extension.

VC virtual connection, virtual circuit. A generic term for a logical communications medium established on request. A connection typically includes a concatenation of channels forming an end-to-end path. A circuit refers to transmission in both directions. Three types of VCs include permanent (PVC), smart/soft permanent (SPVC), and switched (SVC). In an ATM

environment, data to be transmitted by a VC is segmented into 53 octet quantities called cells. This consists of 5 octets of header, and 48 octets of data.

VCC virtual channel connection. A generic term to describe a logical connection. In an ATM environment, a virtual channel refers to the unidirectional transport of ATM cells associated by a common unique identifier value. Virtual circuits (VCs) can be combined to form a virtual channel.

VCI Virtual Channel Identifier. A value in the header of each ATM connection cell which identifies that connection. See VC, VCC, virtual channel.

VCL Visual Component Library. A library used for applications development for Borland Delphi products.

VCSEL See vertical-cavity surface-emitting laser.

vector Any quantity having both magnitude and direction.

vector display A cathode ray tube (CRT) vector display is one in which the sweep of a beam follows a vector (line or stroke) and illuminates (and refreshes) the *specific* part of the display that is needed to render the desired colors and shapes. In other words, the beam doesn't follow the sawtooth scan characteristic of raster displays. This is in contrast to a television screen, in which a beam constantly sweeps the screen to form a *frame* in which the beam travels across the entire display area on a constant, repetitive basis.

In a vector monitor, a vector generator takes the coordinates supplied by the processor and converts them into analog voltages that are used to control the direction of the beam as it excites the phosphors coated on the inside of the CRT. Vector graphics appear very 'crisp' and clean, but refreshing large areas or the entire screen is generally slow and impractical for many applications.

Vector monitors were prevalent during the 1960s and '70s, but have largely been replaced by mass market raster monitors. Tempest was an early video game that employed vector graphics, whereas Space Invaders used raster graphics.

vector font A set of textual characters or symbols defined by vector algorithms (usually lines, spline curves, and arcs) rather than by relative positioning of dots within a grid (raster or bitmap font). A vector font looks smooth and appealing at almost any size, except very tiny sizes, and displays at the highest resolution available to the output device on which it is being displayed or printed. In general, the higher the resolution, the smoother the lines and more attractive the overall look of the font. Contrast with raster font.

vector quantization See quantization, vector.

vehicle access control VAC. Control of various devices related to transport vehicles. Examples include garage doors, access gates, and other types of barriers that may be encountered and opened with a device in the vehicle, such as a radio-controlled remote control. Newer VACs often include security features and frequency programming for customizing the frequency emitted by the devices (this is handy in storage facilities where many doors are close together). The frequencies available for these types of devices vary, depending upon the country.

Vendor's ISDN Association See ISDN associations.

verifying punch A punch card perforator that also verifies the punches to each card to see that the perforations are correct, automatically replacing those that are defective.

Veronica Named tongue-in-cheek for a comic-book character reference to a related tool that queries FTP sites named Archie. What Veronica does for Gopher information is similar to what Archie does for FTP sites. Veronica is an Internet keyword query tool that searches Gopher sites, sometimes called *Gopherspace*, and typically displays the results of the search in the format of a Gopher menu. Users are linked transparently to the source, and may not see the location of the source unless they explicitly ask. Veronica was introduced in 1992 by S. Foster and F. Barrie of the University of Nevada (Reno).

Computer programmers love to come up with acronyms, and Veronica is no exception. It's a stretch, but it has been said that Veronica stands for Very Easy Rodent-Oriented Netwide Index to Computerized Archives. See Archie, Anarchie, Gopher, Jughead, WAIS.

Versatile Interface Processor VIP. A Cisco Systems commercial router interface card that provides multilayer switching.

Version Fast Class VFC. A vendor-developed interim format for modem-based serial communications, which was commercially implemented while the ITU-T was working out the V.Fast standard. See V.Fast, V Series Recommendations.

versorium A device for detecting electrical properties of various materials, invented by William Gilbert in the late 1500s. The versorium resembles a compass needle in that it is a horizontal movable needle balanced on a small support stem, but differs in that the needle itself is not made of magnetic material, like lodestone, but rather of wood or a nonmagnetic metal. Gilbert used this sensitive device for evaluating attractive properties of different materials when altered by rubbing. Descendants of this instrument are now called *electroscopes*. See Gilbert, William.

vertical-cavity surface-emitting laser VCSEL. (*pron.* vixel) In general, traditional lasers emit light from the edges of the structure, and the laser cavity runs along the horizontal length of the device. VCSELs differ in several important ways. The light is emitted from the surface rather than from the edges,

VCSEL Implementation of an Optical Neural Network

An example of an optical neural network implemented with a VCSEL array that is spaced apart from a mask that is, in turn, spaced apart from an array of optical detectors with the same configuration of rows/columns as the laser array. This implementation was proposed by R. Webb of British Telecom, in a June 1998 patent application for a telecommunications switch (U.S. patent # 6,307,854).

the laser cavity is vertical rather than horizontal, and the beam is narrower. Otherwise, the general operating principles are similar to regular lasers. VCSELs were first reported in the mid-1960s, by Melngailis, who coauthored many articles related to optics, especially infrared technologies, between 1965 and 1969 and has coinvented semiconductor-based optical wave guide devices since that time.

The commercial implementations of VCSELs are fairly recent, however. It was not until the late 1970s, in Japan, that wavelengths suitable for telecommunications applications began to be generated with VCSEL technology. Room temperature VCSELs began to be developed by the mid-1980s and practical laser devices based upon VCSELs were beginning to emerge in the early 1990s.

Surface-Emitting Semiconductor Laser

A cross-section of a VCSEL-based surface-emitting semiconductor laser showing the layered structure. This configuration was designed by N. Ueki of Fuji Xerox Company and submitted Aug. 1998 for a U.S. patent (#6,320,893). The device features a stable transverse mode, reduced threshold current, and higher output than previous semiconductor laser embodiments.

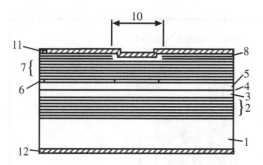

Layer (1) is the substrate supporting the lower reflective layer (2). The reflective layer bounces light back up into the central quantum well region through an undoped spacer layer (3). Between the lower and upper reflective layers is the quantum well active layer (4) composed of an undoped quantum well layer and a barrier layer. Above this is another spacer layer and the optical confinement layer (6).

The thickness from the bottom of the lower spacer layer (3) to the top of the upper spacer layer (5) is an integral multiple of the oscillation wavelength divided by the refractive index. This physical/mathematical relationship fosters a standing wave between the reflective surfaces such that the antinode of maximum light intensity occurs at the quantum well active layer.

In VCSELs, light is emitted from the surface through the sandwiching of reflective mirrors around the laser cavity, which is usually only a few wavelengths long. Metallic mirrors were not found to be very effective but distributed Bragg reflectors are effective in reflecting a high percentage of the reflected light to be guided back toward the laser cavity. VCSELs may be either optically or electrically pumped. The positioning and quantity of quantum wells can be used to control the optical properties associated with VCSEL devices. In current configurations, VCSELs can be "tuned" to between about 400 – 1600 nm through the selection, combination, and thickness of the layers and doping of the quantum wells.

One big advantage of surface-emitting laser components is that the assembly can be tested before it is stamped out of the fabrication materials, in contrast to edge-emitting lasers that must be stamped out before the emissions can be tested. This represents a significant time/materials/cost savings. They are also more efficient in terms of the amount of light that is emitted in relation to the power drawn from the circuit. In addition to this, the light beam is narrower and rounder than that from regular edge-emitting lasers. This is significant in terms of the suitability of VCSELs for fiber optic networks. If you get more light for the same amount of power and the beam fits into a smaller space, it becomes eminently suitable for use with tiny optical fiber waveguides.

Because they are surface-emitting, whole new classes of circuits can be constructed with VCSELs. They can be mounted on printed circuit boards or other suitable substrates, or can be mounted directly on other semiconductor chips (e.g., for optical detection systems), depending upon their function. They can also be organized into arrays.

Commercial and experimental VCSELs have been developed for a number of uses. There are now commercial VCSEL-based laser diodes in the 780 and 850 nm ranges. VCSEL multimode fiber optic transceivers are available for Fast Ethernet, ATM, FDDI, Fibre Channel, and Gigabit Ethernet networks, providing transmissions up to about 2000 m. VCSEL array optical 'neural' networks have been proposed as well. Edge-emitting lasers still have some important advantages, however. Currently, their maximum optical power is higher than that of VCSELs, and the wavelengths emitted are more suitable for optical fiber transmissions. VCSELs currently are suitable for very short distances, but some development needs to occur before they can be used in place of edge-emitting lasers for longer transmission lines. See laser, quantum cascade laser.

Vertical Service Code VSC. In telephony, subscriber-dialed codes that access value-added features and services from various telecommunications carriers. Examples include Call Forwarding, Call Trace, Automatic Callback, and others. These services are invoked by typing in a number or 'star code' (a number prefixed by an asterisk). The Industry Numbering Committee has developed guidelines for the assignment of VSCs. Care must be taken, when requesting VSCs for new services, that the number does not conflict with codes established in other regions within the North American Numbering Plan (NANP).

Examples of some of the more commonly available VSC assignments are listed in the VSC Assignment Examples chart. The North American Numbering Plan Administration (NANPA) publishes a more complete list. See chart. http://www.nanpa.com/

vertigo A perception by an individual that the environment is moving around the individual, or the individual is moving in relation to the environment when there is no physical motion or a misperceived physical motion. For example, watching a big-screen movie with realistic action may cause a viewer to feel as though he is moving, when in fact he is sitting still in a theater seat. The brain's misinterpretation of motion as it relates to body position causes a dizzy-like feeling and may cause or exacerbate motion sickness. It can even make a person feel like he or she is falling. Illness, nervous tension, or migraines can make vertigo more acute. (Vertigo is often confused with the dizziness that occurs when blood supply to the brain is reduced.) Sensitive individuals may experience vertigo after long periods of watching text scroll by on a computer.

Since a fairly significant number of people experience vertigo at one time or another, developers of multimedia products involving moving platforms, headsets, and other motion-controlling or body-worn devices need to take this into consideration when designing and testing their systems. Games developers have discovered that a certain number of people become 'seasick' from playing fast action video games. See burnout, carpal tunnel syndrome.

Very High Density Cable Interconnect VHDCI. A type of compact computer peripheral connection cable that enables multiple 68-pin SCSI connectors to be connected to a computer backplate. See P connector.

very high frequency VHF. Electromagnetic waves in the approximate range of 50 to 300 MHz, part of which is allocated for amateur use (50 to 54 MHz and 144 to 146 MHz) with some regional variations.

Very Large Array VLA. A system of 27 movable astronomical telescopes in a Y-shaped configuration 50 miles west of Socorro, New Mexico. It comprises a giant virtual telescope in that the signals can be interferometrically combined to provide a bigger picture of the cosmos covering the viewable extents of the separate telescopes. Thus, the VLA has an effective

Vertical Service Code (VSC) Assignment Examples		
Number	Name	Notes
*51	Who Called?	Provides the subscriber with the directory number and date/time of unanswered calls. This is similar in function to Caller ID devices that log incoming calls and date-/timestamp them.
*57	Customer Originated Trace	Provides the recipient of a harassing or threatening call with the ability to request a trace of the last call received.
*60	Selective Call Rejection	Enables the subscriber to reject incoming calls from a selection of calling parties.
*61, *81	Distinctive Ringing/Call Wait. Activate/Deact.	Enables the subscriber to have incoming calls from a select list of calling parties identified using a ringing sound that is distinctive and recognizable.
*66, *86	Automatic Callback, Activate/Deactivate	Enables a subscriber to automatically have a call placed to the last number dialed when that number becomes available (no longer busy).
*70	Cancel Call Waiting	Enables the subscriber to cancel call waiting. There are several different types of Call Waiting services, including regular, deluxe, and selective Call Waiting.
*72, *73	Call Forwarding, Activate/Deactivate	Enables the subscriber to control the redirection of calls to another number/ station. This is handy if you are expecting an important call and are planning to be at another location.
*77, *87	Anonymous Call Reject., Activate/Deactivate	Enables the subscriber to set up the rejection of calls from callers who are using anonymous dialing (those that don't show up on Caller ID services).

V

diameter of about 27 km. This type of antenna array was popularized in the movie *Contact* based upon the book by Carl Sagan. See Very Large Array, Expanded. http://www.aoc.nrao.edu/

Very Large Array, Expanded EVLA. Based upon the infrastructure of the Very Large Array (VLA), the EVLA will be an updated virtual telescope of movable radio-based telescopes with far greater sensitivity and angular resolution than the VLA, operating at frequencies from about 1 – 50 GHz. It will enable more accurate spectroscopic assessments of the cosmos at different wavelengths. The distance of the new stations being added to the current array may be as far away as 250 km. The communications links between existing and new antenna dishes are being updated with fiber optic links as part of the Phase I expansion. See Very Large Array.

very large scale integration VLSI. In the semiconductor industry, VLSIs are integrated circuits (ICs) that combine hundreds of thousands of logic and/or memory elements into one very small chip. This type of circuitry revolutionized the cost and manufacture of computers. VLSI has enabled the manufacture of palm-sized computers more powerful than room-sized computers from a few decades ago, which were dependent on vacuum tubes and wires for their circuitry.

Very Small Aperture Terminal VSAT. Very small commercial terminals for two-way satellite transmissions in the United States, and one-way communications in countries with restrictions. VSATs are generally organized in a star topology, with the Earth station acting as a central node in the network. This Earth station operates with a large satellite dish and a commercial quality transceiver.

In some VSAT implementations, the signal from the transmitting Earth station to the satellite is amplified and redirected to a hub Earth station. Since all trans-

missions pass through this hub, two hops are needed for intercommunication between satellites. (This results in a bounce pattern known as an M hop.) Some newer configurations, modulation techniques, and amplification systems are included onboard the satellite, so that an interim hop to a hub is not required.

VSAT systems are appropriate for centralized business and institutional networks. Commercial VSATs typically communicate in C- and Ku-band frequencies.

very short reach VSR. Connections spanning a short distance with the implication that it has fairly high bandwidth characteristics or needs. Thus, the term is being widely used in the fiber optics networking industry to describe short distances that require high-quality, fast connections without overengineering the components or making the cost prohibitive. Due to the demand for this type of technology, the Optical Internetworking Forum has defined a set of four implementation agreements for very short reach (VSR) intra-office interfaces using the OC-192 format and data rate. Short reaches remove the pressure to transmit data serially; parallel cables are practical for VSR connections. This can be implemented with multiple wavelengths over a single fiber or with multiple fibers (or both).

Cisco Systems has been a leading company working with vendors to create functional, standardized OIF-approved SONET/SDH interfaces optimized for VSR interconnections for routes, switches, and dense wavelength division multiplexing (DWDM). See Very Short Reach Interfaces chart.

VESA See Video Electronics Standards Association.
VF access voice frequency access.
VFast See V.Fast, V Series Recommendations.
VFC See Version Fast Class.
VGA See Video Graphics Array.
vgrep visual grep. A variant of a very powerful, useful Unix command. See grep.

Very Short Reach Interfaces			
Spec.	Gbps	Meters	Notes
VSR-1	1.25	ca. 300	Similar to Gigabit Ethernet multimode fiber (MMF) technology. A full duplex link based upon twelve parallel 850-nm bidirectional VCSEL channels running over two 12-fiber ribbons. One fiber carries CRC error correction, one carries parity information, and the remaining 10 are for data. The lasers and photodetectors can be built in arrays.
VSR-2		ca. 600	Based upon ITU G.691, a serial 1310-nm single-mode fiber interface. Compatible with existing central office (CO) fiber links. Based upon uncooled Fabry-Perot lasers (rather than VCSELs) or those with similar optical characteristics.
VSR-3	2.5	ca. 300	A full duplex bidirectional link based upon twelve 850-nm VCSELs and multimode ribbon fiber technology. Four fibers transmit and four are unused. The lasers and photodetectors can be built in arrays.
VSR-4	10	ca. 85/300	A full duplex serial link over two multimode fibers based upon 850-nm VCSELs.

VHDCI Very High Density Cable Interconnect Standard. A standardized computer interface specification. See P connector.

VHF See very high frequency.

VHF antennas A category of antennas designed to take advantage of the particular characteristics of very high frequency (VHF) waves. Because of the wavelength differences between VHF and ultra high frequency (UHF) waves, and the relationship of the rods on the antenna to the length of the wave, VHF antennas tend to be larger and more varied in their shapes than UHF antennas, and can be installed with less precision and still be relatively effective. They are not as broad, however, as a single UHF antenna can cover the entire UHF band, but a VHF antenna is usually optimized for a particular range or specific stations. See antenna, combination antenna, fan dipole antenna, UHF antenna.

VHS Video Home Systems. A widely used video format developed by JVC that is compatible with millions of home user systems. VHS and Beta formats were released at about the same time. Beta was acknowledged as being superior, but was also a bit more expensive, so VHS won the marketing wars. It is slowly being superseded by S-VHS, 8mm, and Hi-8mm formats, in addition to a number of digital formats, including DVD. S-VHS systems are downwardly compatible with VHS tapes, that is, you can play a VHS tape in an S-VHS system (but not the other way around).

VIA See Virtual Interface Architecture.

Vibroplex Trade name of a type of semi-automatic telegraph key introduced in the later 1800s, more commonly called a *bug key*. This particular type of bug key was patented in 1904 by Horace G. Martin. By using a vibrating point for automatically generating dots and dashes, it relieved telegraph operators from physical and mental strain.

video capture board See frame grabber.

video chipset A logic circuit in a computing device that handles the processing, and sometimes acceleration, for video graphics display. Various means of configuring this circuit, and integrating it with the system, control the speed, resolution, and palette set of the display.

Video Dial Tone VDT. On-demand consumer video programming services provided through the existing public phone copper wire infrastructure. This includes services such as movies-on-demand, videoconferencing, interactive programming, financial services, and interactive shopping. Until recently, the concept was feasible, but the communications rates through telephone lines were just too slow. With faster data-rate technologies such as DSL, phone lines are now seen as a viable delivery infrastructure for these types of services and may compete with cable services for the growing market. See Open Video Systems.

Video Electronics Standards Association VESA. An industry standards body established in 1990 that develops various peripheral standards for Intel-based microcomputers. This organization is responsible for defining Super VGA (SVGA) and the VESA local bus

for peripheral device interfaces with personal computers. See super video graphics array, VESA VL.

video floppy A 2-in. digital image storage floppy released in the mid-1980s to hold video images of 360 lines of resolution. Various methods are now used to store digitized images, including flash memory, 3.5-in. floppies, and various proprietary cards.

Video Graphics Array VGA. A graphics standard common on Intel-based International Business Machines (IBM) and licensed third party computers, supporting 640 × 480 (16 colors) and 320 × 200 (256 colors). It has been superseded by super video graphics array (SVGA). See super video graphics array.

video switcher A generic phrase for a wide variety of types of passive routing boxes for video signals. Home systems sometimes have simple switchers to select between a VCR and a laserdisc player. Professional switchers may have banks of connectors, sliders, and settings and may also provide other functions such as frame synchronization and amplification.

Passive Audio/Video Switcher

This simple consumer switcher provides switching for composite video and audio inputs and outputs through standard RCA receptacles. It can switch between four different devices.

---

video tape A magnetic recording medium resembling common audio tape designed to store both images and sound. The most common formats for video tape are VHS, S-VHS, S-VHSc (compact S-VHS), Beta, 8mm, and Hi-8mm. S-VHS and Hi-8mm are sufficiently good for many professional applications, although higher quality formats are preferred for commercial broadcast quality tapes. Many tapes can store audio in two places, intermixed with the images or on a separate track along the side of the tape, for high fidelity recordings that come close to the quality of CD. Recording times range from 20 minutes to several hours depending upon the type of tape and the quality settings.

All television programming used to be live. The performances were saved only in the minds of those who watched them. Then, in the mid-1950s, taped broadcasts became practical and the live aspect of television changed forever. Broadcasts could now be archived, played as reruns, broadcast during convenient times for a specific timezone, or sent overseas to other markets. Station managers could re-air programs without providing royalties to the actors, thus reducing costs (actors weren't happy about this). A whole

V

series could be shot in a period of weeks and then aired over a period of months, freeing up the actors and production staff to work on the next project.

VHS-taped entertainment has been widely available through video rental stores since the early 1990s.

In the 1980s, less expensive camcorders (camera/recording combinations) were utilized by consumers to tape special events, weddings, birthdays, graduations, and amateur movies. By the mid-1990s digital camcorders began to appear and, by the late 1990s, the price dropped to the point where they became consumer items, with film use declining.

video tape recorder, video cassette recorder VTR, VCR. A recording and playback device specifically designed to record simultaneous motion images and sound. The input is usually through video patch cords from microphones and cameras for live recording, or from camcorders and CDs, phonographs, and tapes for re-recording or editing. VCRs have been common consumer items since the mid-1980s; prior to that, they were generally used in the television and video editing industries.

One of the earliest VTR patent applications was submitted in 1927. One of the early 'portable' video tape recorders, based upon new transistor technology, looked just like a large reel to reel audio recorder. It was introduced by Ampex in 1963. This desktop model weighed about the same as a fairly large TV set, but was nevertheless only one-twentieth the size of previous floor-standing models. It used a single-head helical scanning mechanism and could record 64 minutes of programming on standard 8" reels. VCRs are much smaller now, and use convenient cassettes rather than reels. They have been improved to support high fidelity sound and high resolution images.

video-on-demand VoD. A commercial interactive video system in which the user can request a specific video to be played at a particular time, unlike traditional TV programming where the station determines which programs are to be broadcast, and when. A number of these programs have been tried in various regions with mixed success. It's difficult to institute a pay service in competition with hundreds of 'free' channels on TV, which are primarily financed by advertising sponsors. The most successful video-on-demand systems appear to be those installed in motels and hotels that cater to business people attending professional conferences. Thus, one could say tongue-in-cheek that success depends in part on what the market will bare. See audio-on-demand, services-on-demand.

videoconferencing The transmission of coordinated motion images and sound through computer networks. This is an exciting area with many systems vying for the front row seat. Technologies to transmit speech and images have been around since AT&T's Picturephone system, which was developed many years before it was introduced to the public in the 1970s. However, full motion video as found in videoconferencing, or still frame video and sound as found in audiographics, didn't reach practical speeds and consumer price ranges until the mid-1990s. Even

then, they were mostly of interest to educational institutions and corporations. By 1997, however, consumer systems were beginning to be practical, especially with the proliferation of tiny monochrome and color video cameras similar to those found in security systems. See audiographics.

vidicon A television with a photoconducting pickup sensor.

VidModem A patented signal-processing technology from Objective Communications Inc. that can accommodate simultaneous two-way video, voice, and data over standard copper wires. VidModem uses FM signals and compression to transmit a 24 MHz FM signal through the 20 MHz bandwidth that is supported on phone lines.

Vines A commercial virtual network based upon Unix system V, from Banyan Systems.

VIP See virtual IP.

Virtual Interface Architecture VIA. An association of vendors who seek to describe and promote a generic systems-area network in order to facilitate the development of software for various X86- and RISC-based computers and their interconnections. VIA was established in 1996 as a small vendor consortium, and has grown to over 50 companies. See Scheduled Transfer.

virtual IP, virtual Internet Protocol VIP. A function that enables the creation of logically separated switched IP workgroups across the switch ports of a Cisco switch running Virtual Networking Services (VNS) software.

virtual LAN virtual local area network. A local area network in which the internal mapping is organized other than on the geography (physical relationship) of the stations. This allows the system to be segmented into manageable groups. Network software is used to administrate bandwidth and load, and to maintain a correspondence between the virtual LAN and the physical LAN. Newer versions of software will even allow configuration and connections to be established through software with graphical user interfaces that display the equipment itself as graphic images, with lines to indicate the various interconnections. See local area network.

Virtual LAN Link State Protocol VLSP. A protocol submitted as an RFC by Cabletron in 1999. It is based upon the OSPF link-state routing protocol described in RFC 2328.

VLSP provides interswitch communication between switches running SecreFast VLAN as part of the InterSwitch Message Protocol (ISMP). Its function is to dynamically determine and maintain a fully-connected mesh topology map of the network switch fabric based upon best path trees. Identical link switch databases are maintained by each switch and call-originating switches use this topology database to determine routing paths for call connections. Switch states are distributed through the switch fabric by flooding. VLSP supports equal-cost multipath routing and provides fast updates of topological changes. See InterSwitch Message Protocol, link switch advertisement. See RFC 2642.

virtual office A company or department loosely connected physically, but is communications-linked through various business telecommunications options such as cellular phones, videoconferencing systems, satellite modems, the Internet, etc. Some of the participants may be working at home or traveling. Some corporations mistakenly consider the 'virtual office' to be a new concept, but publishers and their associated writers have successfully employed this business model for decades. See telecommuting, telework.

virtual path connection VPC. A path connection established on ATM networks along with an associated quality of service (QoS) category that defines traffic performance parameters.

virtual private network VPN. A secure encrypted connection across a public network that enables organizations to utilize a public network as a virtual, private communications tool. Through a process called tunneling, the packet is encapsulated and transmitted. A VPN is a cost-saving measure for businesses that don't want the expense of setting up an internally funded secure network, and yet desire interconnectivity between remote branches and departments accessible through the relatively inexpensive services of an ISP. VPNs provide a cost-effective alternative to laying cables, leasing lines, or subscribing to frame relay services. The disadvantage to VPNs over public networks is the response time.

virtual reality VR. A phrase to describe electronically generated environments that interact with human senses to provide the illusion of the "real world" or to provide a fantasy world experience that cannot be achieved in the real world. Sensory headsets, goggles, helmets, implants, gloves, shoes, body suits, computers, monitors, chambers, and a whole host of visual/tactile/auditory two- and three-dimensional inputs are used to create virtual reality worlds. See Virtual Reality Modeling Language.

Virtual Reality Modeling Language VRML (*pron.* ver-mul). VRML was originally released in 1994 by Tony Parisi and Mark Pesce. Initially dubbed Virtual Reality markup language by Dave Raggett, VRML is a file-format standard, built in part from the Open Inventor File Format, which was made freely distributable later in 1994 by Silicon Graphics, Inc.

VRML provides a means for creating 3D multimedia and shareable virtual environments. Its inventors describe it as a 3D Web browser. It can be used in geographical, architectural, and industrial modeling; simulations; education; and games.

In August 1996, when the version 2.0 specification of VRML was released, JPEG and PNG were specified as the two image formats required for conformance with the specification.

VRML plug-ins are available for a number of browsers. The files tend to be very large, but there are unique opportunities, too, like taking a virtual ride on Mars Pathfinder, for example, an experience that's worth the download time. VRML 97 was approved in January 1997 as International Standard ISO/IEC 14772-1. See Joint Photographics Group Experts, Portable Network Graphics, virtual reality, VRML.

Virtual Switch Controller VSC. A type of soft switch characteristic of telephony networks that separate the switching layer from the call control layer, enabling a higher degree of interoperability among various protocols and controller systems. The VSC directs calls across multiservice packet infrastructures when combined with industry-standard protocols and interfaces rather than proprietary systems. VSCs can be implemented over newer signal systems such as SS7. Some implementations are also called Media Gateway Controllers (MGC). See Virtual Switch Controller chart on previous page.

Virtual Tributary VT. In SONET networking, a sub-STS-1 signal designed for switching and transporting data. A VT Group (VTG) is defined as 12 columns, which can be formed by interleaved multiplexing, and a group may contain only one type of VT. VTs operate in two modes: locked and floating. The VT types are as follows:

VT Signal	Rate	Digital Signal	Rate
VT-1.5	1.728 Mbps	DS-1	1.544 Mbps
VT-2	2.304 Mbps	CEPT-1	2.048 Mbps
VT-3	3.456 Mbps	DS-1C	3.152 Mbps
VT-6	6.912 Mbps	DS-2	6.312 Mbps

Virtual Trunking Protocol VTP. A virtual local area network (VLAN) autoconfiguration protocol from Cisco Systems. VTP is a Layer 2 messaging protocol that enables centralized VLAN switch configuration changes (additions, deletions, name changes, etc.), with the changes communicated automatically to all other switches in the network.

virus A virus is computer code that is designed to be functionally similar to a biological virus in the sense that it uses its host to spread itself through a system. Just as a biological virus exploits the characteristics of its host to survive and replicate, a computer virus is designed to exploit the characteristics of a computer system and the activities of its users to survive and replicate. There are a number of subcategories of viruses, which may be organized according to how they are spread or how they affect a system. One major category is a computer worm, a virus that doesn't specifically require a host file (although it still needs a host) in order to replicate.

A virus insidiously takes advantage of normal computer functions to spread itself, just as biological viruses take advantage of normal biological functions. A virus may be associated with or attached to a process, file, or pattern of user activities (e.g., file management functions).

John von Neumann was one of the first theoreticians to delve at depth into the analogies between computing and self-reproducing biological systems, formulating a basis for biological analogies for computer viruses. Fred Cohen carried on this theoretical

V

985

tradition in the early-1980s with research and a number of academic publications specifically related to computer viruses. Cohen is credited with the coining of the term in relation to computer algorithms. Cohen's formal definition of a virus has been disputed at length, as it specifically refers to a program that modifies other programs "by modifying them ... to include a (possibly evolved) copy of itself." While this definition describes some aspects of viruses, it is somewhat more narrow and noninclusive than more recent definitions of computer viruses.

The creation and spread of viruses have some legitimate research and testing applications. Unfortunately, there are malicious uses as well and these are the ones that make the news. Since a virus is often intrusively or surreptitiously introduced into a system, the spread occurs without operator knowledge and consent and may not be immediately detected by users. Although the person inserting a virus may or may not intend explicit harm to the infected system, the recipient is almost always inconvenienced and suffers from a sense of invasion of both security and privacy. Users may also incur losses of time and money unanticipated by the person introducing the virus, especially now that systems are interconnected through the Internet and viruses can quickly spread to large segments of the population. Sometimes a virus contains bugs (programming errors) that cause damage unanticipated by the programmer.

On personal computers in the 1970s and early 1980s, viruses were frequently spread through the sharing of files on floppy disks. Now viruses are most often

A Sampling of Representative Computer Viruses	
Name	Description
Adore Worm	A network worm exploiting vulnerabilities in Linux machines, the Adore Worm uses random Class B subnet hosts as a pathway for downloading a portion of itself from an Asian Web server. The worm is stored on the local infected machine and executed through *start.sh*. It then moves and replaces */bin/ps* and */sbin/klogd* with files that allow entrance to the system and begins transmitting sensitive data to a number of email addresses, subsequently removing itself from the system (and restoring original files) through a cron daemon (timer program) in order to reduce the chance of detection. Linux vendors became aware of the worm and took steps to reduce the vulnerability of Linux software.
APost	An example of a typical email attachment virus targeted at the Microsoft Windows OS, detected September 2001. APost is an uncompressed executable file which, when run, displays an *Urgent! Open* dialog box. When the user clicks the dialog button, an error message is displayed, APost checks for a *README.EXE* file, creates one if absent (on drive root directories, including network drives), then adds a subkey to the user's autostartup key with a path for the APost file. It thus starts each time Windows is loaded and then connects to Microsoft Outlook, grabs the mail server login/password information and replicates itself to the email addresses listed in the address book. Once APost has invaded another machine, the original email host is deleted to obscure its origin. Windows-host email viruses are not uncommon; the Sircam, Mawanella, Magistr, and VBSWG.X all spread through the execution of email-attached files. It is best never to run email executables from questionable sources. (There are even a small number of email viruses that don't require an executable to be run in order to invade a host machine, so beware of unsolicited email.)
AutoStart 9805	The first widespread, significant malicious worm to infect Macintosh computers, detected in 1998, AutoStart 9805 caused unexplained disk activity at regular intervals. This worm is limited to PowerPCs/compatibles running MacOS with active QuickTime applications (with QuickTime CD-ROM AutoPlay enabled). The worm "eats" its way through data, with different variants of the worm targeting different types of files, replacing the data with garbage so that it is not recoverable. Fortunately, the specificity of the worm prevents widespread damage, but the nonrecoverability of data may be disastrous for those infected.
ExcelMacro/ Laroux	Considered to be the first real Microsoft Excel macro widely distributed as a virus, Laroux was discovered in 1996. Once a system is infected with this Visual Basic for Applications (VBA) program, Laroux becomes active when Excel is run and it will infect workbooks as they are accessed or newly created. While not as malicious as some viruses, this one is common and causes mischief and inconvenience by replicating itself.

picked up by uninformed users through network file downloads, which may include email with file attachments, but which most often is through downloads of public domain and shareware programs from bulletin board systems and the Internet. Reputable download sites will usually check uploads for viruses before making them publicly available, but not all site administrators have the time or in-depth knowledge to check every file. More recently, viruses have been spread through bulk mailings of email attachments to users who know how to send and receive email but otherwise have little knowledge of computer technology.

In keeping with the general tone of humor in the computer community, many viruses have names. In fact, sometimes the name itself is the inspiration for the development of a virus or simply a joke about a proposed virus, e.g., the Paul Revere virus, for example, warns you of an impending attack to your system, once if by LAN, twice if by C/:, etc.

There are thousands of viruses and dozens that are particularly virulent or significant in various ways. There isn't room to list all of them, but here are a few examples of viruses and specialized viruses called 'worms,' to give a basic understanding of their scope and common means by which they are spread.

Virus attacks deserve serious attention. In an information-based culture, economic and social damage from a virus can be as significant as theft or destruction of important documents in a file cabinet. In fact, they have a greater potential for harm since the Internet doesn't just reach into one file cabinet, but into

A Sampling of Representative Computer Viruses, cont.	
Name	Description
Internet Worm	A self-reproducing program released onto the Internet in 1988, the Internet Worm affected about 10% of the hosts on the Internet by exploiting a Sendmail weakness. This is one of the more famous worms for a number of reasons. It spread very rapidly, caused more harm than was anticipated by the programmer, R.T. Morris, Jr., and was one of the first to gain broad media attention. The law and network administrators are not tolerant of potentially destructive activities on the Net, and this instance and its originator were not treated lightly.
ILOVEYOU	A "love letter" Visual BASIC script (VBScript) worm that spreads through email in the manner of a chain letter. The worm uses Microsoft Outlook to spread itself through email; it further overwrites VBScript and may spread using a mIRC client. When executed, the program copies itself to a Windows OS system directory and to the Windows directory, adding itself to the system registry so that it becomes active if the system is restarted. It adds keys to the registry and replaces the Internet Explorer home page with a link to an executable file. If downloaded, the file is added to the registry as well. The executable portion of the code is downloaded from the Internet and functions as a Trojan Horse to attempt to steal a password and modifies the system so that it becomes active each time Windows is started. The Trojan registers a new window class, creates a hidden window and quietly remains resident. After a startup and a certain timer status, it emails them to a specified address, presumably the author's. The program sets itself up to replicate over an IRC channel if the user joins IRC and uses Microsoft Outlook to mass mail itself once to any people listed in the address book with the subject of ILOVEYOU.
Melissa	A rapidly spreading virus that became global within a short time and was discovered in spring 1999. Melissa proliferates through email. When infected, Melissa inserts comments into user documents based upon the television series "The Simpsons." Melissa is also capable of sending out information from a user's computer, information that might be sensitive in nature. The virus was apparently initially propagated through the alt.sex discussion group in Trojan Horse manner, that is, it was contained in a file that was purported to have the passwords for X-rated Web sites. Users who downloaded the document and opened it in Microsoft Word made it possible for a macro to execute, sending the file called LIST.DOC to people listed in the user's email address book. Other Word-format documents can be infected as well, contributing to the quick spread of the virus, and may slow down a system if large email attachments are mailed through a Word email client without the direct knowledge of the sender with the infected system. There has been at least one variant of Melissa discovered.

millions, spread throughout the globe and further can disrupt not just the files but the information exchange mechanisms themselves. Viruses can have life or death consequences if inserted into transportation or health service computer systems. Damage from viruses varies, depending upon the nature of the virus and how it is spread. Common problems include corrupted files, interruptions by messages or questions requiring an answer, slowed transmission or processing times, and deleted or filled up storage.

Because of their potential to disrupt work and financial transactions, many organizations have been established to deal with the issue of viruses (in addition to those focused on general computer security). Here is a small sampling.

- In 1990, the Japanese Ministry of International Trade and Industry took steps to prevent the spread of viruses in Japan and initiated the management of Computer Virus Incident Reports through the Information-Technology Promotion Agency Security Center (IPA/ISEC).
- The European Institute for Computer AntiVirus Research brings together the resources of universities, industry, security professionals, government, and the media to unite against efforts at writing and distributing malicious computer software. http://www.eicar.org/
- The U.S. Department of Energy (DOE) supports the Computer Incident Advisory Capability (CIAC) and an information site for reports of virus-related hoaxes. http:hoaxbusters.ciac.org/

There are about two dozen significant virus-specific discussion groups on the Internet. Of particular interest are the moderated USENET group comp.virus and email discussion list Virus-L. See back door, back porch, logic bomb, Trojan horse, worm, WildList Organization International, and entries prepended by virus.

virus, hoax A report of a computer virus that doesn't actually exist. Virus hoaxes are unfortunately somewhat common. They can cause mental anguish, work stoppage, and loss of productivity without the perpetrator actually writing and distributing a virus. The Good Times virus warning, initiated in 1994, is an example of a virus hoax that was still circulating several years later.

virus, wild A virus that is extant in the general computer community. A wild virus is one that has the potential to do harm and infect systems because it has been "let loose" to replicate indiscriminately or according to a pattern established by the person programming the virus. Many viruses are developed for academic research or testing purposes and are not distributed into the computing community outside the laboratory or research environment. See virus.

Virus Bulletin A commercial, international U.K.-based publication providing information and assistance with virus prevention, recognition, and han-

dling. The Bulletin provides news information on security-related conferences as well as hosting an annual conference.

Virus Catalog, Computer (CVC) This is considered one of the better technical sources of information on computer viruses for a variety of platforms, published by the Virus Test Center in Hamburg, Germany.

virus checker, antivirus software A computer program that searches and, in some cases, disables viruses on a computer system or network. Virus checkers are a means to detect the undesired contamination or spread of computer viruses before they reduce productivity or harm a system. Computer viruses have been around almost as long as cost-effective computers, but it was the late-1980s before virus checkers became widely available and it took a few years of experimentation for them to become really effective. The virus checker itself should be obtained from a reputable source, since a virus checking program is a good place to hide a virus. Most virus checkers will bring up a dialog box or other alarm warning if a potential virus is detected. They may provide options enabling the user to deal with the virus in a number of ways. It should be remembered that a virus checker is only one aspect of computer security and that security from a network standpoint requires many more sophisticated tools and techniques than a standalone computer to remain secure and optimally functional. In general, virus checkers seek out anomalous patterns or disk access activities. They also typically include a database of viruses with known characteristics that are explicitly sought and identified. A computer running a virus checker is sometimes said to be 'inoculated' against viruses. Since inoculations are generally effective, but not perfect, it's a reasonable analogy. Checkers can be initiated at computer startup and, now that most computers are multitasking, will often run as background tasks. It is usually a good idea to disable virus checkers when system maintenance, new installations, and reconfiguration options are run so that the virus checker doesn't interfere with the process or display an alarm when the maintenance is being carried out. It's also important to remember to restart the virus checker.

Many institutional computers are installed with virus checkers as a matter of policy. These have become quite sophisticated and can detect and sometimes disable many different types of viruses. However, each time a new application, a new processor, or a new mode of operations is introduced into a computer system, it has the potential to host a new type of virus that exploits the evolving technology. Like biological viruses, there will never be a 'perfect' virus checker that can anticipate or detect every type of virus.

Virus checkers also have disadvantages. They can slow the system or interfere with system extensions, and it takes diligence and time on the part of the user to install, configure, and manage the virus checker itself. In most cases, the management involved in using virus checkers is considered good insurance against the massive damage that can occur from a virus that wipes out or corrupts important or sensitive information.

In the absence of virus checkers, file backups and redundancy of live files are two strategies for reducing loss from virus attacks. Self-contained systems also help but are almost impractical in the trend toward a global computing environment. To help prevent the infection and spread of viruses, back up your data, investigate the available virus checkers, download only from reputable sites, and never open email or execute email attachments from questionable sources. See virus.

visible spectrum The region of light waves perceived by humans as color, ranging from approximately 380 to 700 nanometers, or 3800 to 7000 angstroms. Technology cannot reproduce all of the colors of the visible spectrum, but humans cannot always distinguish between very closely related colors either. For practical purposes, the approximately 17 million colors displayed on better quality computer monitors and the approximately 10 million colors that can be printed with pigments on a press are sufficient for most personal and commercial needs. Outside the visible spectrum are the infrared and ultraviolet wavelengths.

VisiCalc Visible Calculator. A historic early computer spreadsheet program, introduced in 1979, which was developed by Dan Bricklin and Bob Frankston for the Apple computer.

VISIT Video A Macintosh- and IBM-licensed PC-based videoconferencing system from Northern Telecom Inc. that supports video, whiteboarding, and file transfers over Switched 56 or ISDN. An extra transmissions line is needed for audio. See Cameo Personal Video System, Connect 918, MacMICA, IRIS, ShareView 3000.

Visual BASIC A basic Microsoft Windows programming application development product with a graphical user programming interface from Microsoft Corporation. Suitable for prototyping, although extensive use of dynamic linked libraries (DLL) may be needed for extensive applications development.

visual ringer A small lamp on a phone console, usually a light-emitting diode (LED), which lights up when the phone rings. This is convenient in a noisy environment or for those who are hearing impaired. It is also common on multiline phones, to indicate which of the multiple lines are currently ringing.

vitreous electricity A term coined by Dufay to denote the type of electrostatic charge produced on glass when rubbed with silk. Benjamin Franklin later proposed *positive*, a term that superseded vitreous. See electrostatic, resinous electricity.

VLAN See virtual LAN.

VLS Protocol The Virtual LAN Link State protocol developed by Cabletron Systems, Inc. is part of the InterSwitch Message Protocol (ISMP) providing interswitch communication between network switches running Cabletron's SecureFast Virtual LAN (SFVLAN). VLSP is used to determine and maintain a fully connected mesh topology map of the switch fabric. Within the switches are identical databases describing the topology maintained. Switches from which calls originate use this topology database to determine a routing path over which a call can be connected. VLSP supports equal-cost multipath routing and recalculates routes quickly with a minimum of routing protocol traffic. See RFC 2642.

VLSI See very large scale integration.

VLSP See Virtual LAN Link State Protocol.

VMI V Series Modem Interface. A standard software front end and software layer that provides an entry and exit point to modem functions implemented through a variety of modem standards, for DSP Software Engineering modem products.

VNS virtual network system, virtual network service.

vocoder *voice coder*. A late 1930s invention that provided a means for analyzing the pitch and energy content of speech waves. This technology led to the development of a device designed to transmit speech over distance without the waveform. The transmission was expressed at the receiving end with a synthetic speaking machine. This general concept has evolved into linear predictive encoders.

VoATM See Voice over ATM.

VoFR voice over frame relay. See Frame Relay, voice over.

voice activity detection VAD. A capability of digital voice communications systems to distinguish between information, such as speech, and the silences in between the speech elements. Typically, voice conversations consist of only about 40% talking, with the rest being pauses, silence, or low-level background noise. By transmitting information only when it is meaningful and filtering out the silent moments, it is possible to create a significant savings in the amount of data that needs to be transmitted. See silence suppression.

voice-activated system A system such as a computer, phone, door, etc. that responds to the sound of a voice, which might be any voice or a specific voice. Voice-activated systems are calibrated to separate out the patterns and frequencies common to human voices from general background noise or other sounds. Technically, there's no reason why it couldn't also be configured to recognize the bark of a dog. This should not be confused with a speech-recognition system that recognizes actual words, not just a general or particular voice. Sometimes the two are combined.

Voice File Interchange Protocol VFIP. A voice file interchange format proposed in 1986 when the ARPANET was still predominant. VFIP was designed to facilitate the interchange of speech files among different computer systems. The specification defined a header for describing voice data that includes a DTMF mask, information about duration and recording rate, and the encoding format. There was no requirement for the header to be explicitly attached to the file containing the speech data. See Network Voice Protocol, Voice Profile for Internet Mail, RFC 978.

voice grade channel A transmission circuit sufficiently fast (usually up to about 56 Kbps) and suitable for transmitting clear voice conversations within frequencies between 300 and 3300 Hz. It is not

V

typically suitable for other faster, higher bandwidth uses. Voice is a relatively low bandwidth application and does fine over copper wires, but others such as data transfer and video images require more.

voice group In analog voice phone systems, a hierarchy for multiplexing has been established as a series of standardized increments. See jumbogroup for a diagram.

Group Name	Composition	Number of Voice Channels
group	1 group	12 voice channels
supergroup	5 groups	60 voice channels
mastergroup	10 supergroups	600 voice channels
jumbogroup	6 mastergroups	3600 voice channels

voice over ATM VoATM. A growing area of interest, voice over ATM involves the digital transmission of voice conversations (which traditionally have been carried over analog phone lines) over asynchronous transfer mode (ATM) networks. Typically this involves taking a synchronous voice signal, segmenting it into cells, each with its own header, and interleaving the cells into the network with cells from other sources, eventually delivering the cell packets to their destination where they are converted back into a synchronous data stream.

Since various queuing delays on the network will affect the transmission of the cells, the receiving buffer must have timing capabilities to organize the arriving cells so as to not leave gaps in the synchronous output signal. Delays of greater than 50 milliseconds of the conversation roundtrip must be avoided in order to prevent echo on the line. ATM networks make use of echo cancellers to reduce echo delay problems. Delays of greater than 250 milliseconds must also be avoided, as they result in perceptual discomfort on the part of the participants in the conversation.

In order to maximize bandwidth over a public network, in which thousands of phone conversations coexist, compression techniques are commonly used to reduce transmission time and resources. See echo canceller, jitter, silence suppression, voice activity detection.

voice over frame relay VOFR. See frame relay, voice over.

voice over IP VoIP. Voice over IP involves digitizing conversations and other human vocalizations so they can be transmitted over data networks. This usually involves compression of the sound, as voice applications tend to be somewhat bandwidth intensive (though not as much as music and other types of sounds). Commercial VoIP offerings usually include familiar phone services like Caller ID, and newer ones like follow-me services that allow forwarding to cell phones or pagers. Some systems are designed to use the public switched telephone network as a fallback if there are problems with transmission over the data network.

Voice over IP Forum VoIPF. A group within the International Multimedia Teleconferencing Consortium (IMTC) that promotes and recommends voice over Internet Protocol (IP) technologies. See voice over IP.

voice over networks There are now a variety of ways in which wide bandwidth data networks can be used to send telephone voice calls. The call can be initiated through a regular phone line that connects to a private or public network, or through a computer voice system hooked directly to a network. Thus, Internet Service Providers (ISPs) are emerging as collaborators and competitors with traditional copper line phone service carriers.

voice over packet VoP. Devices capable of encoding voice signals for transmission over digital packet data networks. These devices typically support ITU-T G Series Recommendations (e.g., G.729a) and may also have echo cancellers, fax and modem support, and voice-band signaling support features. VoP devices are useful for packetized cable telephone, Voice over IP (VoIP), and DSL access.

Voice Profile for Internet Mail VPIM. An experimental profile submitted by G. Vaudreuil in February 1996 to define a digital computer-based voice messaging mechanism. Now known as *VPIM, version 1*, it was based upon the Audio Message Interchange Specification (AMIS), which facilitates message interchange among voice mail messaging on systems from different vendors. VPIM differs from AMIS in that it is a digital specification based on common internetworking protocols. The VPIM took into consideration common limitations of voice messaging platforms as they were implemented at the time. In September 1998, *VPIM, version 2* was submitted as a Standards Track protocol by G. Vaudreuil and G. Parsons. It represents contributions by the VPIM Work Group of the Electronic Messaging Association (EMA). Significant changes to the original experimental specification were made based upon demonstrations at two EMA conferences in the mid-1990s.

VPIM is a profile for using MIME and ESMTP protocols for digital voice messaging services. It specifies a restricted set of Internet multimedia messaging protocols for the provision of a minimum common set of features for internetworking among different voice processing servers. See Audio Messaging Interchange Specification, Network Voice Protocol, Voice File Interchange Protocol, RFC 1911 (VPIM, version 1), RFC 2421 (VPIM, version 2).

Voice with the Smile One of the many colloquial names given to the early female telephone operators. Others include Hello Girls, Central, and Call Girls.

voicemail A type of data communication in which a voice message is digitally recorded, usually through a small microphone interfaced with a computer, and sent through an email or voicemail client as an attachment or message. In order to hear the message, the receiver must have the capability to replay the message on the destination system. This is usually done either directly through the voicemail client or, if sent as an email attachment, it can be played with a separate player utility that is compatible with the type of sound file in which the message is stored.

voicemail, electronic A system for intercepting an incoming phone call, playing a prerecorded digital message, and recording a message left by the caller. Many voice mail systems support multiple messages, multiple mailboxes, and menu hierarchies accessed through touchtones entered from the caller's phone keypad, and may also allow a facsimile message to be transmitted manually, since many voice modems and voicemail systems support data and facsimile communications as well.

Voice mail systems are not used just as fancy answering machines; they are also employed in faxback systems, technical support systems, and for providing product information and purchase options to callers. Because electronic voice mail applications are digital, they can be programmed to provide a wide variety of services, according to the needs and imagination of the programmer and user.

volt (Symbol *v* or *e* for voltage) A SI unit of electrical potential. When a difference of electrical potential occurs between materials or portions of materials where there is a pathway between them, electrons seek a direction of flow which balances that potential. A volt is a unit of electromotive force (EMF) equal to that needed to produce a one ampere current through a one ohm resistance. In any given circuit, voltage, current, and resistance are related, so any one of those values can be computed if the other two are known. The unit is named after Alessandro Volta. See ampere, ohm, Ohm's law, resistance.

Volta, Conte Alessandro Giuseppe Antonio Anastasi (1745–1827) A physicist who pursued many of the ideas proposed by Luigi Galvani by studying the varying electrical properties of different materials. He questioned Galvani's explanation of 'animal electricity' and proposed that the reaction of the muscle to stimulation of a nerve was due to unequal temperatures, and set up more rigorous experiments to determine what was happening. He showed how electricity could be generated by chemical action, which became known as *galvanic electricity*. This was the forerunner of the electrolytic cell. In 1800 he described his invention of the voltaic pile.

Volta devised a condensing electroscope to respond to very sensitive charges and, with it, was able to demonstrate contact charges (though some were actually chemical interactions). The volt, a unit of electromotive force, is named after him. See Faraday, Michael; volt; voltaic pile.

voltaic pile Alessandro Volta developed a system of layers of metal plates and paper or briny cloth that exhibited a difference in potential between the top and bottom, which could be varied with the materials used and the number and organization of the layers. Volta attributed this difference to 'contact' electricity, though we now know that chemical factors play a role. Volta later modified the pile design to create what he called a *crown of cups*. The metal plates were placed in separate cups containing liquid, some distance apart. The plates were the poles or electrodes. Each cup is now known as a *voltaic cell* and a pair is known as a *voltaic battery*. See capacitor, thin film.

Voltaic Piles – Historic Capacitors

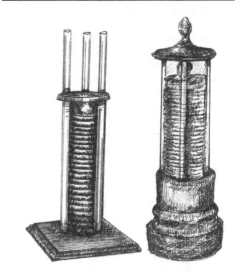

Two historic voltaic piles show the alternating layers of materials piled within supporting rods.

---

voltmeter, voltameter A galvanometer or other instrument such as an ammeter, connected in series with a resistor, calibrated to indicate electric pressure from electromotive force, or voltage differences in potential at different points of an electrical circuit. The voltmeter is connected in parallel across the circuit being tested and must have a higher resistance than that of the circuit being measured. In the past, sometimes also called a coulomb-meter or coulometer. See volt.

von Bunsen, Robert Wilhelm A German chemist in the 1800s who did numerous experiments with wet cells and made improvements on the early inventions leading to modern batteries.

von Guericke, Otto (1600s) An early experimenter who devised a machine that amplified and demonstrated the properties of negative and positive electromagnetic forces. Von Guericke used a spinning large sphere, molded out of sulphur, to investigate theories related to the spinning and magnetism of the Earth. He noted also that holding certain substances up to the sphere would produce a spark. He discovered basic principles of air pumps and demonstrated characteristics of vacuums with his Magdeburg hemispheres in 1663. A university in Magdeburg, Germany is named after him.

von Kleist, Ewald Christian (1715–1759) A German physicist who discovered in 1745 that an electrical charge could be held in a glass vial with a nail or piece of brass wire inserted. A similar jar was developed independently by P. van Musschenbroek, known as the Leyden jar. See Leyden jar.

von Neumann, Janos (Johann "John") (1903–1957) A Hungarian-born American mathematician who developed novel and influential theories in advanced mathematics, game theory, and quantum mechanics. He was fluent in several languages, gregarious, and considered brilliant even by the elite scientists he worked with on the Manhattan nuclear bomb project.

Von Neumann produced a body of work in the mid-1900s that significantly influenced the design and evolution of computing machinery, including practical implementation ideas, conditional control, self-modifying code, program storage, and much more. Von Neumann collaborated with Mauchly and Eckert on the EDVAC computer project. See EDVAC.

John von Neumann – Mathematician

Many of von Neumann's contributions to game theory have applications and consequences for practical applications outside the realm of pure mathematics. These can be applied in the design and operation of 'thinking' machines.

---

von Neumann machine A classification of computing systems, based upon the work of John von Neumann, that includes single-instruction, single-data computation, which requires the repeated access and fetching of instructions and data.

VoP See voice over packet.

VOR VHF omnidirectional range.

VP virtual path. A generic term to describe a logical connection consisting of combined *virtual channels*. In an ATM environment, it refers to the unidirectional transport of ATM cells belonging to virtual channels with the same endpoints, associated by a common identifier value. Related abbreviations include VPCI (Virtual Path Connection Identifier), VPI (Virtual Path identifier), VPL (Virtual Path Link), and VPT (Virtual Path Terminator). See virtual path connection, virtual private network.

VPC See virtual path connection.

VPN See virtual private network.

VQ vector quantization.

VRAM Video RAM. Memory chips designed to enhance graphics display. Bisynchronous input/output. Related is SVRAM, Synchronous VRAM which reads only in or out at one time. See WRAM.

VRML Virtual Reality Modeling Language. A programming language for developing 3D interactive image environments. This is a popular goal of video gaming and simulation developers. VRML allows you to create a sequence of images that can be presented in a World Wide Web environment in combination with a VRML client/browser. There are standalone and Web browser-compatible VRML clients available from several vendors. See Virtual Reality Modeling Language for historical background.

VRML Review Board VRB. Originally the VRML Architecture Group (VAG), founded in 1995, the VRML Review Board participates in and oversees Virtual Reality Modeling Language development, documentation, and formal specifications. http://vag.vrml.org/

VSAT See Very Small Aperture Terminal.

VSC 1. See Vertical Service Code. 2. See Virtual Switch Controller.

VSR See Very Short Reach.

VSX Verification Suite for X/open. See X Windows System.

VT-100 A data terminal, and terminal emulator, originally developed by Digital Equipment Corporation (DEC), which has become an industry standard and is widely used in telecommunications. The VT-100 emulation setting in telecommunications programs works with almost any remote system and is probably the one to select if you get strange characters or formatting in your software. Web browsers are quickly overtaking VT-100 as a front-end to online sessions, but VT-100 is still a valuable standby when connecting to remote systems in text mode. For the most part, it has been superseded by VT-220 and other newer standards, but it is still a good fallback if compatibility is a problem.

VTAM Virtual Telecommunications Access Method. A data communications access method used in International Business Machines' (IBM's) Systems Network Architecture (SNA). See Systems Network Architecture.

VTP See Virtual Trunking Protocol.

VTS Vehicular Technology Society.

Vulcan Street plant The historic location of the world's first hydroelectric central power supplier. It provided direct current, as did many of the early power stations. The use of direct current vs. alternating current was a subject of hot debate in the early days of power stations, with Thomas Edison supporting direct current against strong opposition by Nikola Tesla and Westinghouse, who felt alternating current was a better choice. Alternating current first gained a foothold in Europe, where the high cost of batteries spurred inventors to look for other solutions. See Mill Creek plant.

W 1. *abbrev.* wait. In the Hayes modem command set, W is a dial string modifier that directs the sequence of events to wait for the dial tone before continuing with any further operations. 2. *abbrev.* watt. See watt. 3. The USOC FCC code for wall mount jack. 4. *abbrev.* white (as in B & W TV). 5. *abbrev.* wideband, which is now often used as a prefix with other telecommunications names and abbreviations. 6. *abbrev.* wide as in wide area network (WAN). 7. *symb.* work. 8. *abbrev.* World and Wide and Web (and other 'W' words) simultaneously when used with a number modifier, for example, W3. This shorthand format is a reference to mathematical notation.

W-CDMA *abbrev.* wideband Code Division Multiple Access. See Code Division Multiple Access.

W-DCS See wideband digital cross-connect system.

W2XBS The Radio Corporation of America's (RCA's) first television broadcasting station, located in New York city. W2XBS was established in 1928 and gave the popular cartoon character "Felix the Cat" the exposure that made him a 'star.' In February 1940, it became the first station to provide television coverage of ice hockey, basketball, and a number of other athletic events. In March 1940, W2XBS broadcast the first opera presentation on TV.

W3 See World Wide Web.

W3C See World Wide Web Consortium.

WAAS See Wide Area Augmentation System.

WABI See Windows Application Binary Interface.

WABIserver See Windows Application Binary Interface server.

WACK *w*ait (with) *ack*nowledgment. A signal sent by a transmissions receiving station indicating that there needs to be a wait or delay before transmitting a positive acknowledgment (ACK). For example, the receiver may acknowledge receipt of the last block of transmitted data but not be ready to receive more data, a common situation with network printing devices or data relays or routers.

WACS See White Alice Communications System.

wafer A fine thin disk, usually cut from a larger piece of the substance. Many materials are cut into wafers including semiconductor materials, quartz crystals, and synthetic gems used in optical systems. Silicon is one of the most common materials used in semiconductors. Many electronic chips are layers of wa-

fers and photovoltaic panels are arrays of wafers. Since many wafers used in electronics and other industries are extremely thin, production methods are very specialized. A traditional metal saw is not appropriate, especially since the part cut away and lost by the saw blade, the *kerf*, would be larger than the width of the wafer itself. See thin film

WAG Wireless Application Group. See WAP Forum.

WAIS See Wide Area Information Server.

Wait on Busy U.K. term for a Call Waiting type of optional subscriber service in which a caller encountering a busy signal can wait on the line until the call in progress is over and be automatically connected.

walkaway An individual who leaves a process without completing it. This is an important concept in programming and many aspects of networks. Operators, software developers, and hardware manufacturers must anticipate possible walkaways and decide when to turn off a system or time out an abandoned process.

- On ATMs, a user may walk off and forget to take the bank card. Some machines will beep loudly to alert the user before she or he is out of earshot. Some are programmed to recall and hold the card if it is not removed in a specified amount of time.

- On computing systems (especially shared network terminals) processes are often timed out and the user may be logged off after a certain period of inactivity. This is particularly important if there are limited logins available or limited IP numbers assigned to Internet connections that must be shared among many users.

- In telephone networks, a user might walk away without hanging up the line. In this case, the phone signaling system may detect the inactivity and send the message "If you'd like to make a call, please hang up ..." If this fails to get attention, it may be followed by a loud intermittent raucous buzzing sound.

- On information kiosks or computerized library card catalog systems, users often leave the information system in various states that are confusing to a new user accessing the terminal. The software may be programmed to return to the main menu after a period of inactivity.

wait state In computer programming and processing, a time during which the processor waits. This may be explicitly established or may be dependent upon other events. Wait states are introduced for many reasons, for timing, synchronization, to reduce power demands, etc. They are especially prevalent in systems where there is a disparity between the processing or data transfer speeds of various interconnected components. The wait state can help manage these discrepancies by 'slowing down' the faster components relative to others.

Wake-on-LAN An IBM system for enabling centralized network systems to 'wake up' or power on remote workstations. This is useful for remotely updating software, logging files, and uploading or downloading messages or other files.

walk time Propagation delay in a Token-Ring network. Walk time plus service time combines to form scan time, the mean interval between the arrival of tokens at any given station.

walk-through A process of being presented or going through the basic physical, visual, or functional aspects of a product or process. Thus, a network ISP might walk a user through installation and setup of a modem. (Use a hyphen when it's a noun, and a space when it's a verb.)

walking code 1. Computer code that "walks through" a region of memory, code, or file data unit by unit. Depending upon the function, data may be altered as it travels through the code. Examples include tree walking code for hierarchical data structures or stalk walking code for searching through a stack. 2. An intelligence professional, code expert, or messenger who memorizes information to be imparted at another time or location or who speaks in a code or natural language that can only be understood by a few people. 3. In computer simulations, code that enables the user to walk through a modeled environment such as planetary terrain. 4. In robotics, software algorithms or heuristic procedures that enable a robot to navigate terrain in a manner somewhat like legged creatures.

wall outlet A wall-based electrical or fiber connection point for various devices designed for easy connections, often positioned above the baseboard or at shoulder height. Some devices are semipermanently wired into wall outlet (e.g., wall phone), but many wall outlets are designed for the modular, temporary installation of electrical cords, transmission cords (e.g., extension phone), and other connections.

wall telephone A telephone designed to mount on a wall at a comfortable height for the user. Wall phones were common until the 1960s when they began to be superseded by smaller, portable, plugin "extension" phones. Wall phones are still used in premises where theft might be a problem (restaurants, pay phone booths, etc.).

Wall, Larry Software author of *rn*, a popular newsreading system, and *Perl*, a significant interpreted scripting language widely used on the Internet. Larry Wall has also authored several bestselling programming books, most notably books on the Perl programming language. See Perl.

Wollaston prism A unidirectional light-refracting component with a high spectral range. A Wallaston prism is constructed by combining two similar triangular prisms with their optical axes perpendicular to one another. Epoxy is commonly used as a bonding agent. In use, it is oriented with the optical axes of the two bonded prisms perpendicular to the incident light beam.

A Wollaston prism can be made from a number of types of transparent crystal-like materials (e.g., quartz), but the use of Iceland spar helps maximize beam deviation. Iceland spar is also used for Nicol prisms.

A Wallaston prism can be used to split a laser beam into two beams. The exiting beams will be oppositely polarized from one another, with the angle of deviation related to the wavelength of the beam. The two sections of the prism may be the same size or the wedge angle may be adjusted to further control beam divergence. See Iceland spar, Nicol prism.

Wollaston Prism

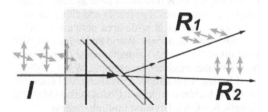

In a Wollaston prism, the incident light beam (I) is refracted into two oppositely polarized beams (R_1, R_2) when it encounters the bond between the two prisms.

The primary difference between the Wallaston and Nicol prisms is that the Nicol prism laterally diverts one of the beams through the first block while the other beam travels through the second block, whereas the Wallaston prism refracts the beams in almost symmetric directions through the second block.

Wollaston, William Hyde (1766–1828) A multitalented English chemist and physiologist who established a lab at the Royal Society, in 1793, and improved upon the voltaic pile (the forerunner of storage capacitors), in 1813. Wollaston made many contributions to the development of optical instruments and their application for studying optical properties in various materials.

In 1802 he developed a refractometer to measure the refractive indexes of various materials and, in the same year, discussed his verification of the unique refractive properties of Iceland Spar to the Royal Society. In 1809, he devised a crystal goniometer to study crystal angles that was more effective than traditional contact goniometers.

Sometime around the late 1700s, Wollaston observed thin, dark lines in the solar spectrum when light was passed through a thin slit and a dispersing prism. J. Fraunhofer later made the connection between these lines and the spectral properties of materials and they are now called Fraunhofer lines or the Fraunhofer spectrum.

Wollaston also had an interest in crystallography and devised instruments to help him study and infer their atomic structure. See Wollaston prism.

WAN See Wide Area Network.

WANMC See Wide Area Network Management Center.

wander Timing deviation or drift. In networking, especially high speed networks in which synchronization is important, wander and jitter can contribute to signal degradation. Physical factors such as connectors, regenerators, or temperature variants can contribute to wander due to propagation delay. Over longer distances, this effect can become magnified, with the pulse position gradually shifting.

In SONET networks, wander has a more specific meaning; it consists of a phase variation tracked and passed on by a phase-locked loop. This is managed by tracking the incoming signal and passing it through a filter, to extract timing data.

Wang Global A firm that formed an alliance with Microsoft to provide local area network (LAN) services. Wang Global was merged into Getronics NV, a Netherlands information technology (IT) services firm, in 1999.

WAP Wireless Application Protocol. See WAP Forum, Wireless Application Protocol.

WAP Forum, WAP Application Protocol Forum Ltd. An industry trade association representing members from all sectors of the mobile communications industry supporting the Wireless Access Protocol (WAP). The Forum is focused on developing and promoting interoperable, securable wireless network technologies and standards and in supporting their deployment and utilization. WAP pursues these goals through the establishment of Charters that are handled by *Specification Working Groups* (SWGs) and *Expert*

WAP Forum Groups Summary

WAP Subgroup	Description

WAP Architecture – chartered Aug. 2000 by the Specification Working Group
The development of wireless architecture to provide a framework for specifications built around Wireless Application Protocol (WAP). This is an ongoing project requiring consistency, completeness, and conformance to WAP specifications. Open to all members.

Wireless Application Group (WAG) – chartered Jan. 2001 by the Specification Working Group
Responsible for specifying WAP application technologies, including application enablement features such as programming interfaces, content formats, and user agents (UAs) to enable data services on handheld devices. WAG will be active as long as specifications and their maintenance are required and is chartered to cooperate with various standards bodies and specification groups. Although one of the more recent Working Groups, much of the effort within the WAP Working Groups has concentrated on this group. Open to representatives of member companies.

Wireless Protocols Group (WPG) – chartered Oct. 2001 by the Specification Working Group
Responsible for defining protocols related to unreliable and reliable data transfer between network entities, provisioning of WAP client devices, and supporting protocols and protocol services. Open to all members.

Wireless Interoperability Group (WIG) – chartered Sept. 2001by the Specification Working Group
Ensure the continued development of WAP certification, the continued review of WAP conformance and specifications, and the development and validation of test suites for conformance. Open to individuals employed by member companies.

Wireless Telephony Applications (WTA) – chartered Feb. 2001 by the Specification Working Group
Responsible for defining a framework for making simple, secure, extensible mobile network services accessible through WAP services. It further assists the WAE group in defining a WAE architecture that enables a single WAP application-level architectural model. Open to member companies.

Wireless Security Group (WSG) – chartered Feb. 2001 by the Specification Working Group
Responsible for specifying WAP security protocols and services as well as interaction with entities that serve security purposes. Open to member companies.

Multimedia Expert Group (MMEG) – chartered Mar. 2001 by the Technical Working Group
A coordination and information-sharing group responsible for defining requirements, use cases, and/or recommendations focusing on mobile multimedia applications and services and for positioning WAP as the preferred platform for developing wireless multimedia services. Subgroups concentrate on specific technologies such as Smart Cards (SCEG), telematics, wireless developers, and others. Open to member company representatives and multimedia communications experts.

W

Working Groups (EWGs). Groups are represented in the WAP Forum Groups Summary chart.

The WAP Forum has also initiated a developer registration and content verification program and a WAP certification program for manufacturers to apply for certification testing for compliance to the WAP protocol suite. See Wireless Access Protocol. http://www.wapforum.org/

WAP Security Toolkit WST. A suite of security mechanisms to support security aspects of the Wireless Applications Protocol developed by the WAP Forum for wireless portable devices. See WAP Forum, Wireless Transport Layer Security.

WAR See Wireless Application Reader.

war dialer An automated dialing system that sequentially dials a new number for each succeeding call, sometimes taken from a computer database. War dialers are used by those dialing to a large number of phone-access BBS systems, by collection agencies, telemarketers, and teleresearchers. There are restrictions in some areas on the use of war dialers for commercial solicitations.

war room A strategy and decision-making facility, often related to critical big business or government activities. The war room may be a closed, secure environment with no equipment other than perhaps tables and chairs, or it may include sophisticated electronics for monitoring and communications.

The U.S. Department of Defense opened a new war room in the Pentagon in 1969. It was equipped with a variety of high technology systems, including closed-circuit television, teletype communications, and various data processing systems and has since been updated to reflect the evolution in technology. War rooms are also common to specific command centers and are often interlinked.

For a lighthearted introduction to war rooms, the Hawk Films/Kubrick/Peter Sellers movie *Dr. Strangelove* is a classic parody of war room activities. See skunkworks.

warble tone A tone resembling a bird's warble in that it fluctuates in tone periodically, sometimes quickly. Because people tend to notice fluctuating tones more readily than steady sounds, warble tones are often used as signal tones, as on public address systems. A warble tone is usually one of the options included with multitone generators, which also feature sirens, steady tones, and timed pulses for various public or employee alert needs. Warble tones are used diagnostically in conjunction with an integrating detector device to measure crosstalk in a transmissions line.

WARC See World Administrative Radio Conference.

warm boot Restarting a system without powering it off. On a computer, this is often accomplished by selecting a 'restart' menu item or using a designated combination of keyboard keys. On a computer, a warm boot typically resets most of the operating system and basic hardware configuration parameters, but may not go through the entire repertoire of hardware systems test sequences that may be stored in ROM. See reboot.

warm image assessment WIA. Using video imaging to assess a component during heat treatment. For example, in fiber filament splicing, an arc is sent across the splice area to melt together the fiber ends and the WIA system locates the core and processes information as the fusion arc causes the light-conducting core to glow. See fusion splicing, profile alignment system.

warranty A promise on the part of a manufacturer, retailer, or service provider that the goods or services provided will meet stated terms of manufacturing quality, use, or lifespan. Most warranties are limited to manufacturing defects, and refunds, replacements, or damages up to the original price or replacement price of the product. Terms of warranties in the computer industry tend to be about three months, although longer warranties on equipment are now beginning to be honored, ranging from one to five years. Few warranties will cover abuse, loss, or damage from natural disasters.

Washington Internet Project A communications project with a portal on the Web providing news, views, discussion groups, an index of regulatory proceedings, and other communications venues for the discussion of significant Internet issues related to federal and state governments. It is also known as CyberTelecom. http://www.cybertelecom.org/

WASI Wide Area Service Identifier.

watch, watchpoint A means of monitoring program functioning, usually within a software debugger, while the application is being executed. Watch commands are often used in conjunction with trace and break commands and watchpoints are set much the same way as tracepoints.

water bore A device that uses a highly pressurized jet of water to bore holes intended for the insertion of underground conduit and cabling.

WATM Wireless ATM. A number of initiatives are under way to provide better support for wireless services over asynchronous transfer mode (ATM) networks. One of the proposals is for a Radio Access Layer (RAL) by the Olivetti Research Laboratory, which has developed a prototype wireless ATM local area network (LAN). See Research in Advanced Communications in Europe.

WATS Wide Area Telephony Service, Wide Area Telecommunications Service. A discounted long-distance service. This service originated from AT&T, but the name became generic and is now broadly used. WATS services can be incoming, outgoing, or both; WATS lines can be installed with incoming and outgoing services handled over the same line (although this may limit the service). As with many discount services, the savings are dependent upon the pattern of usage. If WATS lines are used more often and longer, the savings may be negligible or nullified.

Watson, Thomas A. (1854–1934) A multitalented machinist and famed assistant to Alexander Graham Bell, Watson filed for a telephone patent for a two-bell ringer in 1878. He also designed the Watson board, a very early and not entirely practical, telephone switchboard. By 1901 he was operating the largest shipyard in the nation, but was replaced as head of the company and left the firm to pursue geology. With encouragement by Bell, he subsequently studied voice (Bell was a superior orator) and became an actor and playwright.

Thomas Watson – Telephone Pioneer

Thomas A. Watson, approximately 1914 or 1915. [Library of Congress American Memory Collection.]

Watson, Thomas J., Sr. (1874–1956) Watson joined the Computing-Tabulating-Recording Company in 1914 and became President the following year. In 1924, the Computing-Tabulating-Recording Company became International Business Machines (IBM), a significant firm in computing history. After four decades at the helm, he passed the position on to his son, Thomas J. Watson, Jr. (1914–) who also "inherited" a number of coveted board positions in the business community. See Hollerith, Herman; International Business Machines.

Watson, Thomas J., Jr. (1914–1993) The son of business magnet, Thomas J. Watson, Sr., Watson was granted a degree at Brown University. According to his biography, he was not very academically inclined and got his degree with help from scholarship donations by his father. Watson, Jr., became a sales representative for IBM and then took time away from the company to serve in the U.S. Army Air Corps during World War II. He rejoined IBM in 1946 and became Vice President and a member of the Board. In 1956, he succeeded his father as Chief Executive Officer (CEO). Following a heart illness, Watson stepped down from the position of CEO in 1971, but remained as Chairman of the Executive Committee until he retired in 1985.

Watson, Research Center Thomas J. The Thomas J. Watson Research Center is headquarters for the IBM Research Division, with eight laboratories worldwide. Historically, the lab is descended from the Watson Scientific Computing Laboratory that was opened in February 1945 by Thomas J. Watson, Sr.

and Nicolas Murray Butler, President of Columbia University. The IBM/Columbia facility was expanded in 1953 and ceased operations in 1970. A new lab, also named after the longtime President of IBM, was established by IBM in 1961. This center focuses on physical, mathematical, and computer sciences; technology; semiconductors; and information services.

Watson, William (1715–1789) An English experimenter who demonstrated in 1746 that electrical current could be sent through a wire about 3 kilometers long, using the Earth as a return conductor, a technique later applied in many technologies including early two-way telegraph systems.

watt W. An absolute meter-kilogram-second (MKS) unit to describe electrical power that is equal to the amount of work done at the rate of one absolute joule per second. Described a different way, a watt is the electrical power expended when 1 ampere of direct current (DC) passes through a resistance of 1 ohm. For large units of power, the kilowatt (1000 watts) is typically used. See Ohm's law; power, electrical; Watt, James.

Watt, James (1736–1819) A Scottish inventor who pioneered the steam engine, after whom the watt is named.

WATTC See World Administrative Telegraph and Telephone Conference.

wattmeter An instrument for measuring electrical power in watts. The wattmeter is similar to a dynamometer, which measures force or power in that it employs a moving coil and a field coil; however, the windings on the coils differ from the dynamometer. See dynamometer.

wave A periodically oscillating or undulating process, or physical or electromagnetic phenomenon. The length of a wave is related to its frequency. High-energy waves such as X-rays are shorter than lower-energy waves such as light waves.

WAVE A commercial product from MPR Teltech Ltd. which permits the simultaneous realtime connection of up to eight different sites through ATM switches for broadcast TV quality videoconferencing.

wave audio See waveform audio.

wave division multiplexing WDM. A means of using separate channels grouped around distinct wavelengths to increase the capacity of a fiber optic transmission system. Proposed methods for multiple wave division are known as dense wave division multiplexing (DWDM). See wavelength division multiplexing.

wave filter A device such as a transducer that separates waves on the basis of frequency. Some loss occurs during this process, depending upon the method used and the characteristics of the wave. See wave trap.

wave length See wavelength.

wave packet A short burst or pulse of waves.

wave trap A device usually placed between the receiver and the incoming waves that excludes unwanted waves, especially undesired frequencies or interference waves. Like the receiver itself, a wave trap is often tunable to optimize control over incoming waves. See wave filter.

W

waveform The shape, or spatial characteristics, of an electromagnetic wave. 2. A graphical representation of the spatial characteristics of an electromagnetic wave, as on a scope or 2D or 3D coordinate system illustration or modem. Waveforms are typically graphed according to amplitude across time. Waveforms with certain recognizable shapes, when graphed, have been given names to distinguish them from one another.

waveform audio A digital representation of sound waves, often created by sampling through a pulse code modulation (PCM) technique. Waveform editing on microcomputers became well supported in the mid-1980s. The early Macintosh supported audio waveforms, and many pioneer computer musicians used Macintosh music sequencing and editing software to create electronic music compositions. The Atari ST was released with a basic midi device built in and the Amiga in 1985 came with built-in multiple channel 8-bit stereo sound. By the late 1980s, 16-bit third-party sound cards were available for most of these computers. Todd Rundgren, musician, composer, and multimedia designer, began using Macintoshes and Amigas to create digital sound videos and CDs in the mid and late 1980s.

There are many file formats for storing sound waves, and sound files can be played from Web pages on the Internet with browsers that support sound (assuming the computer has a sound card; most types of computers came with built-in sound cards by 1986). The .wav extension is commonly used to designate audio files of a particular format on the Internet.

waveform editor An applications program on a computer, or specialized electronic device, which allows the user to display, evaluate, and alter the characteristics of a wave. A computer display often uses a graphics system to represent the wave, as in traditional oscilloscopes. The dials on the simulated computer scope are often represented as buttons on the screen or may be input from a joystick or specialized peripheral.

Waveform editors are used to alter the characteristics of music patches, voice, or speech files. Adjustments can alter the volume, tone, harmonics, echo, and other characteristics, and when converted to digital form on the computer, the adjusted files can be stored and replayed later, or cut and pasted to create songs or speeches. Digital sound editing with a waveform editor can be combined with digital video sequences to synchronize the sound and video and to "put words into people's mouths." See waveform audio.

waveform monitor An oscilloscope or oscilloscope-like computer applications program that surveys an input signal and displays its characteristics on a screen. Typically there are dials or graphical user interface gadgets and buttons to adjust the displayed wave.

waveform control, waveform distortion compensation A type of image "sharpening" mechanism in which certain types of distortion (e.g., image sharpness or focus), which may be introduced as the wave passes through an aperture, are corrected on the other side. It was historically called *aperture tagging*.

The type and means of correction depends upon the application, the character of the wave, and the precision needed. In conjunction with a wavefront detector, correction may be applied by phase modulators, data processing algorithms, or micro-electromechanical systems (MEMs). Wavefront control to correct aperture distortion is applicable to a number of digital imaging, laser, telescopic, and scintillating systems.

At supercooled temperatures, micro-machined membrane deformable mirrors (MMDMs) have been studied as potential wavefront correctors.

waveguide A device for confining and channeling the propagation of electromagnetic waves, often through a hollow round tube, hollow rectangular tube, coaxial cable, or fiber optic cable. The interior environment of the waveguide will vary with the type of wave being channeled, since it must allow sufficient room relative to the characteristics of the wave so as not to change or diminish the signal. Thus, waveguides are more practical for high frequency waves such as microwaves.

waveguide dispersion The process by which an electromagnetic wave becomes distorted as it passes through a waveguide. Since the dimensions and shape of a waveguide interact with the phase and velocity characteristics of a wave, the waveguide's geometric properties may cause dispersion of the guided signal.

waveguide laser A gas laser that incorporates a tube as a waveguide to channel the direction of the laser beam.

waveguide lens A device used with microwaves, in which the waveguide elements act as lenses to produce the required wave phase changes through refraction.

waveguide phase shifter A device that takes the phase of the incoming waves and adjusts them in terms of their output current or voltage.

waveguide propagation A type of long-range communication that makes use of the atmospheric waveguiding channels that arise between the ionospheric D region and the Earth's surface. See ionospheric sublayers.

waveguide scattering Scattering of an electromagnetic wave that occurs due to the geometric characteristics of the waveguide structure, as in an antenna or fiber optic cable, not due to the materials of which the waveguide is constructed.

WaveLAN A commercial IEEE 802.11-compliant wireless local area network (LAN) system. WaveLAN was originally introduced in North America by AT&T in 1990 and 1995 and in Australia and New Zealand by Lucent Technologies in 1998. The system has been installed in more than 50 countries overall.

Without extension antennas, the system has a range of up to about 400 meters. Early versions of WaveLAN used 915 MHz, while newer WaveLANs use 2.4 GHz direct sequence spread spectrum (DSSS) technology and are designed to co-exist with standardized 10-MBps DSSS technologies.

wavelength The distance, when measured from any point on a wave, to the corresponding point in the phase of the next wave in a related type or series of waves, such as sound or light waves. A wave's length is sometimes described in units of distance and sometimes in terms of the time it takes, from the phase of a wave to the corresponding phase of the next related wave, to travel through the same point.

Waves can differ dramatically in length, as can be seen, for example, from the various categories of designated radio bands, which range from a single millimeter (EHF band) to tens of kilometers (VLF band) when expressed in distance, or from 30 to 10 kHz when expressed in frequency or cycles per second. See band.

Wavelength-Agile Optical Transport and Access Network See WOTAN.

wavelength-shifting WLS. The property of a

Basic Wave Concepts

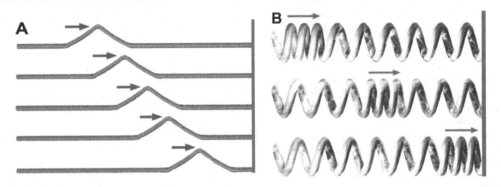

Wave motion in familiar objects can illustrate some of the basic concepts related to waves in general, including those that can't be seen, such as acoustic and electromagnetic waves. Imagine a string attached to a fixed point (diagram A) that is given a shake to set it in motion. The kinetic energy from the pulse is transmitted to the string and moves towards the fixed point in the direction away from which the wave was initiated. The string vibrates in only one plane and is known as a traveling wave. Thus, a traveling wave moves perpendicular (normal) to the longitudinal axis of the medium, in this case, the string. If there were no friction or obstacles to impede its progress, the pulse would continue indefinitely. However, even if the wall were infinitely far away, the wave would eventually die out due to attenuation from the string's interaction with air, which serves to gradually dissipate the energy.

Waves can move in other ways than the simple pulse in the string on the left. In the metal spring (diagram B) a pulse initiated by pulling longitudinally on the spring results in a different type of motion. The kinetic energy alternately compresses and expands the elastic segments of the spring, pulsing the energy longitudinally along the axis of the spring. Longitudinal wave compression in a medium is characteristic of sound waves. Without a medium in which the molecules can alternately move closer together and farther away, there can be no sound. That's why there is no sound in a vacuum.

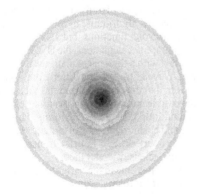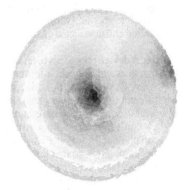

Complex waves can also be combinations of basic wave motions. For example, the ripples at the surface of a pool of water are similar to the string above in that they move horizontally away from the source of the pulse, but they move in concentric circles outward rather than in one direction. Water waves also embody aspects of longitudinal waves, since the water alternately compresses and decompresses as the wave moves through it. If the waves hit an obstacle (e.g., a cliffside or the side of a boat), some of the waves will be reflected back across the incoming waves, causing interference patterns (right) that are similar to the patterns of light observed with interferometers.

component that changes the wavelength of applied energy to re-emit it at another wavelength. For example, a scintillator can convert X-ray energy into light energy which, in turn, can be converted to electrical impulses to make practical use of the information. Fiber is sometimes used in a scintillating component to convert heat energy into light energy at a specific wavelength (e.g., blue light in a calorimeter) and may also be used to channel the light energy out of the wavelength-shifting component to other components. See scintillator.

wavelength division multiple access WDMA. One of two common optical multiplexing techniques in which each transmitter transmits at different wavelengths within a narrow spectrum, and the receiver extracts the desired wavelengths with a bandpass optical filter. See space-division multiple access.

wavelength division multiplexing WDM. A means of multiplexing different wavelengths through the same strand of fiber to greatly increase the capacity of data transmission over fiber optic cables. Optical signals at different frequencies do not interfere with one another. The technology permits a substantial amount of data to travel over even one strand, and when the strands are bundled, it permits transmission in the terabits per second range, ample for high bandwidth applications like video, data, and simultaneous voice.

WDM has become increasingly important to fiber optic communications, with improvements in both speed and distance making it practical for national backbones. In October 2001, Fujitsu announced having successfully tranmitted WDM signals over a distance of 7400 km at 2.4 terabits/sec. Advances in repeater configuration, exploiting Raman amplification technology, were also expected to increase power and reduce overall costs. See add/drop multiplexing, frequency division multiplexing.

wavelength selector A mechanism for selecting a single channel from many channels in a multiplexed fiber optic transmission. Typically a coupler affixed to a fiber Bragg grating is used, though research indicates that adjacent long-period gratings may help reduce loss for selected and unselected channels. See fiber grating.

wavelength shifter A device or process that takes an incoming series of waves and shifts their frequencies so the outgoing waves are related to the incoming waves, but in a different range. It is very common in satellite communications for the incoming signals to be shifted so they don't interfere with subsequent outgoing signals. In photocells, wavelengths may be shifted by means of compounds so that the length of the outgoing waves is related but greater.

wavelet analysis Wavelet analysis involves looking at time and frequency. A prototype function (*analyzing wavelet* or *mother wavelet*) is used as a starting point for dilations and translations by creating a high-frequency reference and a low-frequency reference, which are analyzed, in turn, for aspects of time and frequency. See wavelet theory.

wavelet filter compression An analysis low/high-pass filter technique used in wavelet compression by selective quantization in which images are decomposed into frequency bands. Wavelet encoding combined with vector quantization (VQ) has been shown to be a good means of compressing image data, and several schemes for accomplishing this have been developed. See wavelet.

wavelet packets Calculated linear combinations of wavelets that retain many of the properties of the parent wavelets from which they are derived.

wavelet theory A set of mathematical concepts and techniques related to the representation and manipulation of oscillating wave forms *according to scale*. Wavelet analysis provides a means to use approximating functions contained within finite domains and are well-suited to representing data with sharp discontinuities.

Wavelet theories and algorithms are being applied to audio and image compression with some practical and interesting results. When wavelet concepts are used in image compression, they share some characteristics with discrete cosine transform (DCT), although the functions used are more complex than cosines. Wavelet compression is sometimes used in conjunction with other methods, such as vector quantization, to provide low-loss, high-compression ratios. They have also been used in turbulence studies, human vision, radar systems, astronomy research, and fractal imaging. Wavelets are used in conjunction with, and sometimes instead of, Fourier transform methods, depending upon the application. Unlike Fourier transforms, wavelet transforms are not limited to sines and cosines, and can comprise an infinite set. See discrete cosine transform, Fourier transform, wavelet transform.

wavelet transform, discrete wavelet transform DWT. A linear mathematical technique that is a subset of wavelet packet transform. DWT is used to generate a data structure with segments of various lengths. In a sense, a transform is a means of rotating a function so it can be visualized and analyzed with a different set of tools, those tools being not only mathematical algorithms, in this case localized frequency wavelet functions, but also conceptual models. To simplify calculations, the DWT is factored into a product with a few sparse matrixes using self-similarity. Wavelet transforms differ from Fourier transforms in that they are localized in space, and an infinite possible number of basis functions can be applied to them, unlike Fourier transforms which use only sines and cosines, but they also share some common basic properties. See Fourier transform, wavelet, wavelet packets, wavelet theory.

wavelet types Due to the variety of possible types of wavelet transforms, some wavelets have been grouped into families, on the basis of *vanishing moments*, and subclasses, on the basis of the number of coefficients and level of iteration. Many more await development and discovery. The Daubechies wavelet family has a fractal structure. Others are Symmlet, Coiflet, and the simpler Haar family, often used to introduce wavelets in educational contexts.

wax master An original physical master intended as a prototype for making one or more copies, usually with a more durable material. Since wax is easily shaped, stretched, changed, and otherwise manipulated, it is a good medium for creating prototypes of production parts. Wax is used to design jewelry, sculpture, certain types of audio recordings, prototype components, and more. Once sufficient copies of the original are made, the wax is sometimes reused for other projects.

way station An intermediate office in a communications line. An intermediate phone in a way circuit, one which is not the main console.

way wire, way circuit A party line circuit that connects a number of subsidiary stations to a main switching or relay station. See party line.

WBC wideband channel.

WBEM See Web-Based Enterprise Management.

WCAV Web Clipping Application Viewer. See Web Clipping application.

WCP 1. See Web Clipping proxy. 2. See Wireless Certificate Profile. 3. wireless communications platform/product. 4. See wireless communications protocol.

WDM See wave division multiplexing.

WDMA See wavelength-division multiple access.

WDP 1. See Wireless Datagram Protocol. 2. Work/Workforce Development Program.

wearable public key infrastructure WPKI. A public key infrastructure and accompanying applications described by Muller and Smart in June 2000 for use with the Bristol University Cyberjacket. It is a low-computational-overhead system for providing authentic records of meetings and conversations and a password system where the user doesn't have to remember passwords to facilitate wearable computing.

Web *colloq.* See World Wide Web.

Web address See Uniform Resource Locator.

Web browser A display and hypertext client used as a front-end to Web-related services on the Internet. For a fuller explanation see browser, Web.

Web Clipping application WCA. A software application technique used by Palm Computing to provide the capability to utilize a small Web site or related type of database on a wireless Palm PDA device. These have also been called Palm Query applications (PQAs). A WCA is a form of application partitioning in the sense that part of the application is on the device and part of the information is presumably being accessed from a remote proxy server that connects, for example, to a remote Web site on the Internet. This approach is intended to reduce download times and to provide a solution for Web access with appliances with limited resources (as with wireless handheld devices).

A Web clipping application uses a subset of HTML (hence the term *clipping*) to create basic forms, graphics, and hyperlinks, which is then compiled with the Web Clipping/Query Application Builder (WCAB or QAB). The application runs inside the Web Clipping Application Viewer WCAV. Web Clipping can be used to implement a number of types of micro-

browsers. See microbrowser, Web Clipping proxy.

Web Clipping proxy WCP. Web clipping proxy servers provide an environment for accessing Web-type information via Web Clipping applications. The application connects with the WCP through a wireless connection, makes a request that is handled through the Internet by the WCP and sent as a compressed product back to the wireless device and Web Clipping app. It may also be used for simulating and testing Palm OS Web Clipping applications. See Web Clipping application.

Web hosting A service in which Internet Service Providers (ISPs) enable a business or individual to store Web pages on the ISP's computer system, instead of on the user's computer. Since an ISP's machines are typically connected to the Internet all the time and since a certain amount of Web server setup is needed to make a Web site function properly, it is very practical to have the ISP manage the administration of the Web server. The design, management, and updates of the individual Web page are left up to the user or can be handled by the ISP for a fee.

An ISP can also arrange for the registration of a *domain name* for the Web site, with the user paying the fee to the ISP, who passes it on to the registration authority, or with the user paying the registration authority directly and the ISP handling the setup of the necessary computer configuration. See domain name.

Web master See Webmaster.

Web search engine The World Wide Web is an enormous repository of information and it changes all the time, so there's no practical way to find a particular page without a little help. Enterprising programmers quickly realized that tools were needed to make it easier to locate information on the Web. As a result, they developed 'search engines,' which are application programs in Web format designed to facilitate the location and retrieval of information according to user-specified parameters or categories. Most search engines provide a text window for the user to type in a keyword, after which the user clicks a "Search" button, resulting in the display of a listing of relevant Web pages to which the user can jump by clicking on the highlighted hypertext link.

It is not unusual for a search keyword to result in hundreds, thousands, or even millions of 'hits,' that is, pages that include the specified keyword. It's obviously not practical to try to visit several thousand sites, so most search engines allow the user to narrow the search by adding more keywords and providing operators such as *AND* and *OR* to focus the search. This can reduce the resulting *hits* to a more manageable number.

Search engines also provide means for businesses and individuals to get their Web pages listed so others can find them. There is usually a button at the top or bottom of the page for this purpose.

There are thousands of search engines on the Web. Many sites have their own local search facilities, but there are also a dozen or so very prominent Web search tools that are commonly used. The applications listed in the appendix all perform general searches

W

of the Web, with the exception of DejaNews (now part of Google Groups), which searches newsgroups. Many also have 'specialties' to set them apart from the others. Note that these search engines catalog millions of pages and they are not necessarily up to date. The best search engines seem to have a lag of about three weeks to three months from the time a site is submitted until it is added to the database. See the appendix for a list of major search engines.

Web server A client/server model system on a multi-user network that serves requests for HTML-based Web pages that are part of the World Wide Web. A Web browser is a type of client software that communicates with the server through HyperText Transfer Protocol (HTTP), and displays the information received from the server in the form of a Web page containing a variety of text, graphics, and sound. Most Internet Services Providers have Web servers. See HyperText Transfer Protocol, server, World Wide Web.

Web Service Provider WSP. A commercial provider of computer network access to the World Wide Web or to a local Web-based network. Most, although not all, Internet Service Providers (ISPs) provide Web access and often include it in their monthly subscription cost. They may or may not provide a diskette with a Web client, called a browser. If not, Web browsers are widely available for download, or through friends, as many of them are freely distributable or shareware. Some service providers use proprietary servers and the user can only use the browser provided for this service; others allow users to select the browsers of their choice. See Web browser, Web server.

Web site On the World Wide Web, which is a mechanism for viewing many portions of the Internet, there are businesses and individuals who have organized data files in such a way as to provide a virtual community, storefront, library, educational resource, or other form of informational/educational/entertainment source for the public or authorized users. When a Web user accesses the site with a computer program called a *browser*, she or he is presented with various communications media (text, graphics, sound, etc.) available on the site.

A Web site is comprised of a series of related files, managed by a client/server Web system that serves the Web files, most of which are called Web *pages*, to processes that request them over the network. These files typically consist of text and graphics organized into hypertext relational links through a markup language called HTML. There may also be sound, forms, Sun Microsystems Java applets, and various scripts that provide additional functionality to the basic HTML layouts.

Web sites are as varied as the people who design them. To mention just a few, there are commercial sites promoting products and services; scheduling sites providing listings of television, radio, and other broadcast venues; weather sites enabling lookup of weather conditions almost anywhere in the world; educational sites through which *distance education* is becoming more and more available; personal sites journalizing the day-to-day activities of individuals; genealogical sites detailing family histories; and chat sites in which opinions are readily offered and debated. See World Wide Web.

Web streaming A capability of a Web browser through addition of a plugin, or which is incorporated directly into more recent browsers, that allows audio or video playback as the data is transmitted to the browser. In older browsers, if the user clicked on a sound or video file, the browser downloaded the sound or video to the local computer from where it was played after the download. More recently, through Web streaming utilities, the audio or video is played as it is received in 'realtime,' albeit sometimes at a lower quality level and with variations due to the connection speed to the Internet.

A number of commercial plugins and applications are available to take advantage of Web streaming, ranging from $20 to over $3000. A freely distributable version, called RealVideo, is available from Progressive Networks.

Web TV See WebTV.

Web-Based Enterprise Management WBEM. A distributed management system based on Web technology. WBEM uses a midlevel manager approach, employing HTTP and Web browsers, to provide access to management data and reporting mechanisms. There have been a number of efforts to standardize distributed management, with mixed results so far. WBEM is supported by HMMS, an open standard of the Desktop Management Task Force, and the HyperMedia Management Schema (HMMS) under the aegis of the Internet Engineering Task Force (IETF).

WBEM provides a means to extend the system through HMMS and utilizes HyperMedia management Protocol (HMMP) to link HMMS to run over HTTP.

Weblock *colloq.* The slowdown that may occur when trying to access extremely popular sites on the World Wide Web. The Web version of a traffic gridlock, where everything temporarily slows to a halt.

Webcast, Netcast A broadcast using the World Wide Web as the communications medium. While traditional Web browsing involves clicking on desired pages, text, and images, and thus 'pulling' the information to the user, Webcasting involves a method of "pushing" the information at the user in the sense that the user's screen is continuously updated without the user having to click the screen or manually request an update (like watching television). Since early browsers did not inherently function this way, it was usually necessary for the consumer to first download additional software to his or her computer. Subsequent browsers, such as Netscape Netcaster, directly incorporate the push capabilities and can be used in conjunction with Channel Finder, a directory of Webcasting channels.

Webcasting is a significant development. It means the advent of digital TV through a computer, with access to a multitude of "channels" on the Internet. When this catches on, TV will probably never be the same, particularly since very specialized information, such as stock and finance channels, art and music channels, etc. can be made available through the Web without the major sponsorship that is necessary to participate on the big broadcast television networks. HDTV-quality push reception can be accessed with a TV set interfaced with a computer peripheral. See Intercast.

WebCrawler One of the significant commercial search engines on the Internet, known for its quick simplicity and very cute little spider mascot. Web Crawler was one of the earlier search engines on the Net. http://www.webcrawler.com/

Webhelp Inc. Enhanced 411 A telephone Directory Assistance service that provides enhanced information services, including movie information, driving instructions, etc. The service is answered by the local carrier, compressed into a digital audio file, routed through a message gateway to an offshore agent who answers the information request. The caller can elect to receive the return information in a number of ways, including hanging up and waiting for a callback or Web-enabled phone message. The company's goal is to make the world of Web information accessible through the telephone.

Webmaster A professional who is responsible for a World Wide Web site. A Webmaster's duties typically include coding, installation, and maintenance of the site and *may* also include page layout, content decisions, and the production of graphics and text. Many individuals have responded to the demand for Webmasters by learning a little bit of layout and HTML and hiring themselves out as professionals, with mixed results. Proper creation, presentation, and maintenance of a good site, especially a commercially viable site, involve a large number of professional skills including marketing knowhow; writing, editing and proofreading; graphics design and production; Web statistics monitoring, analysis, and reporting. Choose your Webmaster carefully. In fact, you may do best to team up a user-interface and CGI-savvy programmer with a market-savvy artist and writer. HTML skills alone are *not* sufficient to create good database, point-of-purchase, or shopping cart software, and these may be essential to a business presence on the Web.

WebNFS Web Network File System. A proposed specification for a client/server protocol as an extension to recent versions of the Network File System (NFS). NFS is an RPC-based, platform/transport independent protocol and WebNFS is intended to provide semantic extensions to NFS. It helps make file handles faster and easier to obtain, and may improve transit of firewalls and system scalability. WebNFS clients assume the availability of a WebNFS server registered on port 2049. See Network File System, RFC 1094, RFC 2054, RFC 2055.

Website Meta Language WML. An extensible, freely distributed offline HTML-generation toolkit for use by Web designers creating Web sites on Unix systems. WML is written in ANSI C and Perl 5. It is distributed under the GNU General Public License

(GPLv2) and may be used free of charge in educational and commercial environments.

Webspace Derived from *cyberspace*, Webspace refers to the Web environment, the sum total of the hardware, software, people, transmissions, and interactions that comprises the World Wide Web community.

WebTV A simple consumer-oriented Web access device that uses a TV as the display. It was purchased in August 1997 from the California-based WebTV company by Microsoft Corporation. Sony Electronics, Philips Consumer Electronics, and others have products based on the WebTV technology.

Webzine An electronic publication available on the Web, either free or through subscription.

WECO See Western Electric Company.

Wehnelt interrupter A type of early electronic apparatus used in laboratories to provide interruptions ranging from 100 to 1000 per second. The positive side of a circuit was connected to a platinum electrode conducting through a well-insulated primary winding coil. This, in turn, was connected to a lead plate in a dilute solution of sulphuric acid. A tube for circulating water was included to provide cooling.

WELL, The The Whole Earth 'Lectronic Link. This Web site is somewhat like a bohemian meeting ground for intellectuals and artists. Established in 1985 by Stewart Brand and Larry Brilliant, the WELL's discussion community patrons include many artists, educators, writers, and high-level technical professionals. http://www.well.com/

Werts, Alain (ca. 1940s–) A French engineer working for Thomson CSF who proposed the use of silica-based fiber optic technologies in *L'Onde Electronique*, in 1966, but was unable to find funding to pursue the idea. Charles Kao's ideas were more widely disseminated and developed and Kao became known as a major pioneer in the field. See Kao, Charles.

West Coast Computer Faire A historic meeting place for computer hobbyists, many of them pioneers in the microcomputer field. Established in 1978 by Jim Warren, it was then the largest computer show in the world.

West Ford satellites A set of satellites launched by the United States beginning in 1961. The pioneering West Ford project launched surfaces into orbit that contained millions of slender copper dipoles to provide a reflective "blanket" around Earth. The first launch attempt was unsuccessful due to a failed ejection, but the second launch, in 1963, demonstrated that communication could be achieved, at least with high-powered ground stations. At that time, concerns were expressed over side effects from certain types of satellites and space debris, and since the feasibility of active relays was successfully demonstrated, the project was discontinued. See ECHO satellites.

WESTAR The name of a family of Western Union communications satellites, WESTAR I was the first U.S. domestic communications satellite, launched in 1972. A WESTAR satellite was launched with the Challenger space shuttle mission in February 1984.

W

Unfortunately, the Payload Assist Modules (PAMs) didn't function correctly; the WESTAR was launched into a lower orbit than had been planned, and it had to be retrieved by the shuttle Discovery some months later.

Western Electric Company WECO. Originally established as the Gray & Barton company in 1869 by Elisha Gray and Enos Barton. In 1872 it became known as Western Electric Company and, in thirty years, grew to be one of the world's largest manufacturers. The Graybar Electric Company, Inc. was spun off from WECO in 1925 and is still doing business as an employee-owned distributor of telecommunications products.

In 1881, Western Electric Manufacturing Company became Western Electric and acquired exclusive rights to manufacture and provide Bell Equipment.

In 1915, Western Electric took over Western Electric Company of Illinois and, in 1925, Western Electric Research laboratories were consolidated with part of AT&T's engineering department to form Bell Telephone Laboratories, Inc.

See Gray, Elisha; Gray & Barton; telephone history.

Western Union, Western Union Telegraph Company Western Union was originally organized as the New York and Mississippi Valley Printing Telegraph Company in 1851 by Hiram Sibley, and named Western Union Telegraph Company in 1956 by his business associate Ezra Cornell (cofounder of Cornell University), when it merged with Cornell's New York & Western Union Telegraph Company. Cornell was the largest shareholder for over fifteen years.

Western Union installed the first North American transcontinental line in 1861. Sibley traveled to Russia and was offered an opportunity to purchase Alaska on behalf of Western Union, but turned it down, passing on the information and the historic opportunity to the U.S. Government.

In 1868, Western Union hired C.F. Varley to evaluate the U.S. telegraph system with the result that he introduced many standards for coordinating the various lines and systems.

Western Union acquired the rights to new technology, called duplex telegraphy, in the early 1870s. This enabled two messages to be sent concurrently, one in each direction, over the same wire.

In 1873, Western Union entered the international market by a takeover of the International Ocean Telegraph Company.

In 1877, Western Union went into competition with the Bell system, acquiring patents or licensing rights from Edison, Gray, and others, and forming the American Speaking Telephone Company. As Western Union had thousands of miles of telegraph cable strung across the country, this posed a very real challenge to the Bell system, and they reacted with a patent infringement lawsuit that was upheld two years later in the Supreme Court.

Western Union had a short alliance with the Bell system from 1908, when AT&T gained control of the company, to 1913, when it voluntarily sold it off again to forestall government breakup of the company.

In 1943, the Postal Telegraph Company, founded in 1881, was merged into Western Union.

In 1950, Samuel Morse's original telegraph instrument was presented by Western Union to the Smithsonian Institution. Other versions of the Morse telegraph are replicas.

In 1970, Western Union acquired TWX (from AT&T) and merged it with its own Telex system. Western Union International was acquired by MCI International in 1982.

In the mid-1980s, the world was changing rapidly. Communication through the Internet was catching on, overnight couriers were in heavy competition with the telegraph industry, and divestiture changed the competitive atmosphere of the phone industry. As a result of this and other changes, the 138-year reign of the Western Union Telegraph Company ended in 1989.

Western Union Telegraph Company Collection This extensive historical collection was presented by the Western Union Telegraph Company to the National Museum of American History as a gift in 1971. It consists primarily of manuscripts, telegraphs, and photographs previously housed in the Western Union Museum. A tremendous amount of telegraph history is contained in these archives. See Western Union.

Western Union Telegraph Museum A historic repository of Western Union documents and photographs established in 1912 by in-house engineer H.W. Drake. By 1930, the collection had its own room at Western Union and historical instruments were being added until the late 1960s, when the materials were warehoused. In 1971, the collection was transferred to the National Museum of American History, then called the National Museum of History and Technology. See Western Union.

Westinghouse, George (1846–1914) An American inventor who pioneered many aspects of alternating current (AC) power, generators, and the railway air brake (1868). He also made practical applications from ideas derived from or shared with Nikola Tesla, and had more than one disagreement with Thomas Edison. His light and power system was used in the 1893 Chicago World's Fair, the same year he won the historic contract to develop alternating hydroelectric current from the Niagara Falls.

WestNet One of several National Science Foundation funded regional TCP/IP networks, WestNet is located in Salt Lake City and serves the states of Arizona, Colorado, New Mexico, Utah, and Wyoming.

wet cell A basic electricity-providing apparatus employing a positive and negative terminal, each separately in contact with a liquid electrolyte medium (battery acid is an electrolyte). Wet cells are inconvenient in that the electrolyte tends to evaporate and may spill, so other types of cells have been devised, chief among these, the dry cell, available since the early 1900s. A type of wet cell called an air cell became widely used in phone applications. See air cell, Bunsen cell, dry cell.

wetting Adhering a uniform, smooth film of solder to a surface, usually a base metal.

wetting agent 1. A substance which, when applied to a surface, prevents it from repelling wetting liquids to enhance the adhesion of liquids to the surface. 2. An agent that facilitates the smooth, even spread of liquid over a surface.

WFS See Woodstock File Server.

WFWG Windows For Workgroups. See Windows.

WGDTB Working Group on Digital Television Broadcasting.

WGIH Working Group on Information Highway. A working group of the TSACC.

WGS World Geodetic System. A very extensive geologic data set, updated every decade or so to take advantage of improved data that can be acquired with improved technologies.

Wheatstone, Charles (1802–1875) An English physicist who researched acoustics and invented the concertina. He also experimented with electromagnetic and solar clocks, and developed a speaking machine based on earlier work by W. Kempelen. Like Samuel Morse, he received encouragement and assistance from Joseph Henry, and invented a telegraph that predates the one designed by Morse. In collaboration with W.F. Cooke, Wheatstone's telegraph was operating in 1837 and he may have received information and assistance from an earlier telegraph inventor, Francis Ronalds. In 1840, Wheatstone proposed an underwater cable between England and France. The Wheatstone bridge that bears his name was developed by Samuel Christie and described by Wheatstone in 1843. See Cooke, William Fothergill; polar keying; Ronalds, Francis; telegraph, needle.

Wheatstone bridge, resistance bridge A device employing a galvanometer and a group of interconnected resistors for measuring resistance against a comparative standard. This tool can be used for determining faults in a length of wire, so the entire wire doesn't have to be dug up or pulled out. By creating a balanced bridge through various loop tests, the approximate location of the fault point can be determined. The Wheatstone bridge was developed by Samuel Christie. It is named after Charles Wheatstone because he described it in print in 1843 and often mentioned it in his lectures. See megger.

Wheler, Granville (1701–1770) An English cleric and experimenter who collaborated with Stephen Gray in discovering conductors and nonconductors, and demonstrated in the late 1720s that an electrical charge could be conducted through a thread of more than 600 feet in length. In fact, it was found it could be conducted simultaneously over multiple threads. See Gray, Stephen.

Whetstone When microprocessors were slow and rudimentary in design, and operating systems and software applications were limited, it was easier to run a few tests to evaluate and compare the relative speeds of various systems. Thus, a number of benchmark tests were devised to measure performance. The Whetstone test was developed in the 1970s by B. Wichmann with an Algol 60 compiler. It is named after the English town where it originated. The Whetstone monitored the number of floating point operations that could be carried out by a process in one second. Floating point operations (flops) are common in processor-intensive computing applications such as graphics and scientific work. See benchmark, Dhrystone, Rhealstone.

Whirlwind A historic, large-scale computing machine developed at MIT in the late 1940s and early 1950s. It is credited as being the first realtime-processing digital computer. It is also significant for incorporating random access memory in a matrix core memory.

White Alice Communications System WACS. A historic communications system installed by AT&T and Western Electric. About two dozen of the stations were installed across the state of Alaska, with the White Alice station located on Pillar Mountain in Kodiak. WACS operated from the mid 1950s until 1979 and was dismantled in 1997. It went through a number of hands during its history, including the U.S. Army and RCA. The system used 5-story-high curved rectangular vertical dishes supported with billboard-style scaffolding. Stations intercommunicated through tropospheric scattering

White Book 1. Any book in the set of 1992 ITU-T recommendations. 2. A document that specifies MPEG video and audio file structure, coding, and indexing. Up to 74 minutes of full-motion video can be recorded on a disc. White Book discs can be played on standalone video CD and DVD players and on suitably equipped computer systems. The format was introduced in the early 1990s by JVC, Philips, and other major vendors. See MPEG, Red Book.

white noise Human-audible signals that consist of a spectrum of frequencies more-or-less evenly distributed across the range so that no one tone predominates, as in background noise. White noise is sometimes used to create ambience in sound systems and it is used in audio experiments.

white pages 1. In most English-speaking countries, the directory portion of residential, or residential and business listings, in the local telephone directory. 2. On the Internet, directories of individuals' physical and electronic addresses usually accessible to the public through the World Wide Web. Web white pages can be searched through keywords and are more flexible and powerful than standard printed phone listings. You don't have to be using the Internet to be listed in these directories; many are compiled from telephone listings and other public sources of information. See ego surfing, yellow pages.

white paper *Colloq.* A technical document, usually describing research, experimental results, or details of a technology. White papers and Request for Comments documents are two of the primary communications venues used by developers of communications technologies to present and disseminate information about their theories, inventions, and operational observations.

White Paper The popular name for a paper titled *A Statement of Policy on the Management of Internet Names and Addresses* that was issued by the U.S. Department of Commerce's National Telecommuni-

W

cations and Information Administration (NTIA) in June 1998. This document supported a prior Green Paper call for the establishment of a not-for-profit entity to coordinate Internet domain name administration. This was an important step in the direction of privatizing the domain name registration system while still acknowledging the rights of national governments to coordinate their own country code TLDs. This led to the formation of the Internet Corporation for Assigned Names and Numbers (ICANN). See Green Paper, top level domain.

white room, clean room A controlled environment in which dust, smoke, bacteria, and moisture are eliminated or regulated in order to reduce interference with and contamination of the functioning of the environment. Controlled environments are used in component production, research labs, medical facilities, etc.

white transmission 1. In an amplitude-modulated transmission of an image, a black transmission means that the greatest divergence in amplitude in the signal represents the black tones, and the narrowest divergence represents the lightest tones (or no tone at all). In white transmission, the opposite is true. 2. In a frequency-modulated transmission, a black transmission means that the lowest frequency corresponds to black and the highest frequency corresponds to white, or no tone, and in a white transmission the opposite relationship is used. The concept applies to image scanners, facsimile machines, photocopiers, etc.

whiteboarding Communicating through means of text and graphics drawn on an erasable wall board or large sheets of paper. See whiteboarding, electronic.

whiteboarding, electronic 1. An electronic software application in which input from various keyboards and pointers is displayed or projected on a large white screen that resembles an erasable whiteboard. 2. A dedicated computer network display system that enables remote two-way communication of ideas through text and graphics. 3. The conceptual analog of a whiteboard communication, carried over a computer-based videoconferencing system. The author first saw this demonstrated by SGI with a videoconferenced paint program at a trade show around 1991, in which both participants, in different locations, contributed to the same illustration and text while another window simultaneously showed their faces as they talked to one another. As it was demonstrated, the whiteboarding concept was very broad and could conceivably include collaborative use of any type of software application by two or more participants. This is a powerful concept that goes several steps beyond mere conversation on a videoconferencing system and shows great promise. See audiographics, telecommuting, videoconferencing, whiteboarding.

Whittaker, Edmund Taylor (1873–1956) A British mathematician and professor, Whittaker made important contributions to mathematical theory and the instruction of mathematics. Whittaker served as the secretary of the Royal Astronomical Society from 1901 to 1906. In 1905 he was elected as a Fellow of the London Royal Society and, in 1906, he became the Astronomer Royal for Ireland and accepted a position as an astronomy professor at the University of Dublin. In 1912 he moved to Edinburgh and worked as a professor there until 1946. Whittaker is known for his work in numerical analysis and applied mathematics as it relates to astronomy. In the context of communications, Whittaker developed a mathematical framework for later sampling theories that led to the implementation of pulse code modulation in digital recording. The Sir Edmund Whittaker Memorial Prize is awarded for published works of high merit in mathematics or mathematical physics. See pulse code modulation, sampling theory.

whois An Internet username directory service application that responds to a name query by accessing a central database and returning a listing of users found to match the name or a portion of the name. Whois is based on the WhoIs Protocol (NICNAME) which is an elective proposed Draft Standard of the IETF. See finger, rwhois, RFC 812, RFC 954.

wicking The process of drawing a liquid out of a substance or along a path. Wicking occurs when solder runs along a wire or up underneath an insulating sheath. Diapers and hiking socks are designed to wick moisture away from the skin to prevent irritation. Oil lamps draw oil up through the wick as they burn.

Wide Area Augmentation System LAAS. A Global Positioning System (GPS) augmentation system intended to provide safer, more reliable satellite-based navigation services for aviation. WAAS provides geographically expansive en route and nonprecision approach navigation data which is further used in conjunction with the more precise Local Area Augmentation System (LAAS) in specific locations. WAAS meets Category I aviation requirements and encompasses most of North America through approximately 25 ground reference stations. Information is broadcast through GPS frequencies (1575.42 MHz) to onboard aircraft receivers, improving GPS accuracy to about 7 meters. Further details are available through the Federal Aviation Administration (FAA). See Local Area Augmentation System. http://gps.faa.gov/

wide area differential GPS WDGPS. An implementation of the Global Positioning System (GPS) that includes a network of reference stations that act as data collection sites to receive and preprocess GPS satellite signals. The information is forwarded to a central processing hub that creates correction vectors for each satellite, including clock corrections. WDGPS systems can provide more accurate local positioning information for a variety of industries, including surveying and navigation, particularly aviation, where precise positioning for takeoffs and landings is important. See differential GPS, local differential GPS, Global Positioning Service.

Wide Area Information Server WAIS. Developed in the early 1990s by Brewster Kahle, WAIS is a powerful search and retrieval system widely used on databases on the Internet. The WAIS URL scheme designates WAIS databases, searches, and individual documents available from a WAIS database. WAIS can be accessed on the Web through the following URL schemes:

WAIS database
   wais://<host>:<port>/<database>

Specific WAIS search
   wais://<host>:<port>/<database>?<search>

Specific WAIS document
   wais://<host>:<port>/<database>/<wtype>/<wpath>

There are freely distributable and commercial versions of WAIS available. With the burgeoning information on the Net, WAIS is a tool of some significance. See RFC 1625.

Wide Area Information Server gateway WAIS gateway. A computer used as a "go-between" between incompatible networks or applications to translate WAIS data.

wide area network WAN. Unlike LANs, which tend to be directly cabled and thus limited in scope, Wide Area Networks can connect users over broad geographical regions through the use of long-distance transmission technologies, such as telephone and satellite services. WANs are often used to connect LANs with a variety of architectures and protocols. A WAN accomplishes the transmission of various formats through *routers* or, alternately, through *bridges*, that are protocol independent in order to connect WANs and LANs. See bridge, Local Area Network, router.

Wide Area Network Management Center WANMC. There are a number of centers in different organizations operating under this general department title. This is a central facility that administrates, configures, and operates various aspects of wide area networks (WANs) depending upon the needs of the firm and the degree of centralization.

Wide Area Telephone Services See WATS.

wide band channel WBC. In FDDI-II isochronous networks, WBC is the circuit-switching capability. Any bandwidth not allocated to WBCs can be used for other data, such as statistical information about data traffic. See Fiber Distributed Data Interface.

wide characters Character codes consisting of two bytes (16 bits), rather than the traditional one byte (8 bits), in order to accommodate a much larger number of characters from different languages. See Unicode.

wide open receiver A receiver that is receiving a range of frequencies simultaneously. CB radios and various emergency systems are sometimes set to receive a range of transmissions at one time.

wideband 1. A band wider than that which is necessary for transmitting voice, sometimes called medium-capacity band, in the 64-Kbps to 1.5-Mbps range. 2. A range between narrowband and broadband, typically between about 1.5 Mbps and 45 Mbps. 3. A band with a broad range of frequencies, often multiplexed. See broadband, narrowband.

wideband digital cross-connect system W-DCS. A digital cross-connect system that accepts a variety of optical signals and is used to terminate SONET and DS-3 signals. In other systems, it may also cross-connect DS-3/DS-1. W-DCS can be used as a network management mechanism. Switching is carried out at the VT level. See broadband digital cross-connect system.

wideband modem A modem designed with a bandwidth (frequency spectrum) capability greater than that of common consumer modems designed to work over basic voice channels.

widescreen TV A home theater TV set that supports and can display a video transmission with a 16:9 horizontal to vertical picture ratio, as is found in movie theatres. These are the same proportions supported by letter-boxed laserdiscs and videos, those in which none of the original movie imagery is cut off of the sides when it is displayed. The 16:9 ratio is also supported by some of the better consumer camcorders.

Widrow, Bernard (ca. 1930–) An American electrical engineer, professor, and pioneer in the field of adaptive signal processing, systems that learn and adapt their behavior through interactions with their environments. Adaptive signal processing relates to neural networks and has practical applications in high speed networking over traditional wirelines, for example. Widrow was a historical early contributor to the theory of adaptive antennas. He coauthored "Adaptive switching circuits" in 1960. He was awarded the IEEE Alexander Graham Bell Medal in 1986, the IEEE Neural Networks Pioneer Medal in 1991, and the Benjamin Franklin Medal in Electrical Engineering in 2001.

Wiener, Norbert (1894–1964) An American mathematician who collaborated with Arturo Rosenblueth and a group of scientists from various disciplines in developing many fundamental concepts of artificial intelligence. He authored *Cybernetics: or, Control and Communication in the Animal and the Machine* in 1948 and updated it in 1961 to include ideas about self-reproducing machines and self-organizing systems. He also contributed to the fields of stochastic processes and quantum theory. See neural network.

WIG Wireless Interoperability Group. See WAP Forum.

wild virus See virus, wild.

wildcard A symbol which takes the place of and represents a series of characters, or an unknown quantity or any quantity, usually in a numeric or alphanumeric context. For example, the asterisk (*****) is a wildcard character frequently used in computing applications, especially file management, for representing unknown characters or any characters. Thus, the UNIX shell command rm myfile.* would *remove* any filename in the current directory beginning with "myfile." and ending in any extension (or no extension).

WildList Organization International WOI. An authoritative industry inventory of viruses that are a considered to be a threat to computers in general use, i.e., those "in the wild." To put it another way, it lists viruses that spread as a result of normal computer use by a wide general audience, as opposed to those that

W

are developed, tested, and used in labs. Like so many aspects of the computer industry, the WildList started as a freely distributable list developed by Joe Wells in his garage and has grown, with the help of volunteers, to be an important resource for research and testing of antivirus software. See virus.
http://www.wildlist.org/

WiLL See wireless local loop listing

Williams-Kilburn tube A historic cathode-ray tube-based system used to store information electronically for use with computing machines. The system was developed near the end of World War II by F. Williams and T. Kilburn who were associated with the Telecommunications Research Establishment (TRE) in Malvern. The cathode-ray beam was used to display dots and dashes on the surface of the tube representing the binary values. Unlike modern cathode-ray tubes with streamlined housings, the Williams-Kilburn tube (also known as the Williams tube) had a boxy industrial look, like a large metal toilet paper roll with a display surface at one end under which were a few small button-shaped dials.

The Williams-Kilburn tube became an important component in the subsequent design of the Small Scale Experimental Machine, a historic computer with randomly accessible program instructions and data stored in "tube memory." It was also incorporated into the later IBM 701. See Small Scale Experimental Machine.

Willis Graham Act of 1921 An act that not only recognized AT&T's monopoly in the phone industry, but legitimized it as well. However, 28 years later, the U.S. federal government filed antitrust proceedings against both AT&T and Western Electric. This long process was not settled until 1956 with a consent decree. See Modified Final Judgment.

WIM WAP Identity Module. See WAP Forum, Wireless Application Protocol.

WIN See Wireless Intelligent Network.

Win-OS/2 Microsoft and International Business Machines (IBM) were originally collaborating on an operating system for IBM's microcomputers which IBM released as OS/2, but at some point during the project, Microsoft stepped back and began concentrating on developing their own Windows products. The windowing component of OS/2 and a number of the basic concepts were incorporated into Microsoft's Windows product line.

WIN95 See Windows 95.

Winchester disk A random access hard disk drive marketed in 1980 by Shugart Associates, founded by Alan Shugart, formerly of International Business Machines (IBM). Considering that computer users were predominantly using cassette and reel-to-reel tapes to sequentially write and read data up until this time, the hard disk drive was an important improvement as a computer peripheral and is still one of the foremost storage media.

winding *n.* 1. A conductive path, coupled inductively to a magnetic core or cell, usually made of metal wire. Helical coil windings are used in simple armatures, inductors, and transformers. 2. A structure to enhance

the transmitting or receiving of electromagnetic waves. A type of compact antenna mechanism. Various types of windings can be found around small spool-like cores in old radio sets. These were used for frequency selection, employing different thicknesses of wire and winding patterns. Depending upon the purpose of the winding, the wire may be left open or may be sealed in paraffin, rubber, or some other protective material.

Winding can be somewhat tedious and exacting. For this reason, windings are sometimes done in sections and winding machines may be used to create the coil. Since the amount of wire that is wound is sometimes critical, the spool on which the wire is wound may be weighed before and after the winding, to check that the desired amount has been used.

Boiling the coil in linseed oil was one of the techniques used in early fabrication to drive out moisture and provide a tight insulating layer. Paraffin and rubber were sometimes also used. See basket winding, coil.

winding machine Any machine improvised or designed to facilitate wire windings around various cores for the creation of armatures, antennas, frequency tuners, spark coils, and other apparatus that utilize wire windings. Evenly wrapped, tight windings are important in many electronic applications and tedious to wind by hand. Thus, a spinning bobbin, lathe, or specialized winding machine is now typically used to increase speed and precision.

Basic Armature Winding Types

Type	Characteristics
ring	The two common types are spiral wound (single closed helix) and series-connected wave-wound (see the preceding diagram).
drum	The two common types are lap and wave, which are similar on bipolar machines, but different on multipolar machines.
wave	The winding passes along the conductor to the back of the armature once through each conductor.
multiplex	Two or more independent windings on a single armature. These are more commonly used on generators intended to supply large currents with small voltages.

window 1. An opening, entrance, time interval, or opportunity. 2. An opening or transparent material that permits light to penetrate, or permits a viewer to see beyond the structure in which the window is installed. 3. A graphical user interface structure developed at Xerox PARC in the 1970s and incorporated into the Alto computer. It first came into widespread use on Macintosh computers in 1984 (it was used also on the Apple Lisa in 1983, but the Lisa never caught on). It contains a group of related information or functions. Most computer interfaces now use similar conventions for sizing, scrolling, opening, closing, and iconizing a window. See graphical user interface. 4. In networking, an opportunity, space, or transmissions lull during which information can be sent, processed, or otherwise efficiently handled. System tuning, capacity monitoring, and flow control are all ways of taking advantage of windows.

windowing A description for a means of organizing and interacting with graphical user interface (GUI) structures called *windows*. A window is essentially a *portal* into a portion of the computer's data, visually represented within a bounded entity that can usually be moved around, sized, iconized, or placed in priority overlapping with other windows. There are text windows, graphics windows, sound "windows," and others. When they are moved about a computer screen in order to make them comprehensible and easily accessible, popped to the front, or pushed to the back to bring a relevant window to the front, it's called *windowing*.

Windows A Microsoft Corporation graphical operating environment that works in conjunction with Microsoft's text-oriented MS-DOS (Microsoft Disk Operating System), primarily on Intel platforms, although a number of Windows emulators for other systems are available and recent versions exist for handheld devices.

Windows is descended from user interface concepts developed in the 1970s at the Xerox PARC research lab that were first widely incorporated into microcomputers by Apple Computers, Inc. in the early 1980s. Apple shipped the Lisa computer with a graphical user interface in 1983, shortly before the announcement of Windows 1.0, which didn't actually ship until late in 1985. Current versions of Microsoft Windows use essentially the same basic concepts as their Xerox-inspired predecessors, including icon and gadget point-and-click interaction, sizable windows, visually accessed directory structures, and window overlap priority for the active window. The Windows Operating Systems chart describes some of the development of various versions.

Windows Application Binary Interface WABI. Windows Application Binary Interface. Software from Sun Microsystems that enables Microsoft Windows applications to run on the Solaris desktop system. Thus users can have access to the large library of software available for the Windows operating environment and can run them, along with Solaris applications, on computers installed with Solaris.

Windings – Various Types of Windings and Historic Armatures

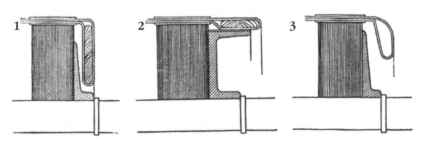

Three types of armature windings are shown here: (1) evolute-wound, (2) barrel-wound, and (3) bastard-wound (a variation on barrel-wound is also shown bottom right). [Cyclopedia of Applied Electricity, 1908.]

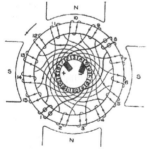

This schematic shows one type of ring armature, a wave-wound armature connected in series. The spacing of the wire windings is called 'pitch.' [Cyclopedia of Applied Electricity, Chicago American School of Correspondence, 1908.]

This is the armature from a Westinghouse generator. It is a bastard barrel winding, a type of winding whose end connections are inward and cylindrical, thus shortening the length of the armature parallel to the shaft compared with barrel winding.

Windows Application Binary Interface server
WABIserver. A Microsoft Windows application server providing integrated Windows/Solaris options to SPARCstation users running Sun Microsystems' Solaris, Solaris Intel, or an X terminal system.

Windows Internet Name Service WINS. A name resolution client/server application that resolves the internal names applied to networked Windows computers to corresponding Internet Protocol (IP) addresses.

Windows Open Services Architecture WOSA. A Microsoft Corporation distributed, semi-open, client/server-based architecture first announced in 1992. WOSA is built on a three-level model consisting of service providers, applications programming interfaces for each category of services, and the applications themselves. While the service provider interface (SPI) level is essentially open, the applications programming interfaces (APIs) and service provider interfaces (SPIs) are proprietary to Microsoft.

Windows Operating Systems

Name Description

Windows - Early Versions

Originally called Interface Manager, the first version of Windows was developed between 1981 and 1985. Windows 1.0 was officially announced in 1983, not long after the release of Apple's Lisa computer, although it was not commercially distributed until two years later, beginning late in 1985, around the same time the Atari ST and Amiga 1000 shipped with full graphical user interfaces (followed by the Apple IIGS). While these competitors never unseated MS-DOS/Windows as the prevalent operating system, they did give consumers and developers food for comparison and many of the ideas pioneered on the competitive machines were incorporated into later versions of Windows. In addition, the distribution of the Amiga in 1985 provoked a storm of discussion about whether operating systems should be multitasking. Despite the controversy, other platforms gradually began adapting task-switching as an early step toward multitasking. Thus, in 1987, when Microsoft released Windows 2.0, it incorporated some of these ideas, including the capability to open more than one application at a time and to overlap windows. Graphical elements were not yet significant, as on the Macintosh, Amiga, or Atari operating systems. Windows 2.0 was renamed Windows/286 when Windows/386 was released later the same year. The early versions Windows were limited to 640 Kbytes of address space and "16-color" palettes (which were actually eight colors plus the same eight displayed at half intensity).

Windows 3.0/3.1

Windows was greatly enhanced and overhauled between 1987 and 1990 to provide an improved user interface, the capability to support more than 16 colors (8 + 8 at half intensity), and memory addressing beyond 640 Kbytes. The result was released as Windows 3.0.

Two years later, in 1992, Windows 3.1 shipped with significant enhancements over 3.0, with scalable fonts, object linking and embedding, and better multimedia capabilities (Multimedia Windows was absorbed into this product). Windows 3.1 became a widely distributed version of the Microsoft graphical user operating environment, popular in the mid-1990s. Windows 3.11 was released as a free upgrade in 1994 to correct some of the networking functions of Windows 3.1. Windows 3.1 has, for the most part, been superseded by Windows 95, 98, 2000, and ME, although a sizeable number of corporations continued using it for many years after the release of Windows 95.

Windows for Pen Computing 3.1

Although not well known, this version of Windows, specifically designed for pen computers requiring handwriting recognition for the execution of commands, shipped in the spring of 1992.

Windows for Workgroups 3.1

This version of Windows, designed for integrated networking and sharing of resources, shipped in 1992, the same year as Windows 3.1. It should be noted that the Macintosh operating system provided file sharing network support off-the-shelf eight years earlier, so Workgroups was not a forerunner in microcomputer networking, but it was a welcome enhancement to corporate users running Windows systems and included network mail capabilities. Windows for Workgroups 3.11 was enhanced with a number of features, the most important being the addition of 32-bit file access.

Name	Description

Windows 95

Windows 95 was the highly promoted successor to Windows 3.1 and, to some extent, also succeeded Windows for Workgroups. Windows 95 first shipped in the summer of 1995. It did not have the robustness, multiprocessor support, or security features of Windows NT, but it was less expensive, and thus became a popular consumer and small business alternative to the more powerful Windows NT operating system.

Windows 95 differed from earlier versions in three important ways. It relegated DOS to secondary status, incorporated multitasking, and fully implemented graphics icons and menus. Thus, it followed the example of the Lisa and Macintosh line of computers (1983) and the Amiga and Atari computers (1985) in embracing the graphical interface ideas pioneered by Xerox PARC researchers in the 1970s. All major desktop computers now had graphical interfaces and most of them have multitasking or, at least, task-switching operating systems. Windows 95 was a strong commercial success and was still being used on many systems six years after its release.

Windows 98

The successor to Windows 95, released in June 1998. This version was significant in that it was the first to directly incorporate network browsing into its structure, an important trend and step toward integration with the Internet. The Net was growing as an important economic and communications force at the time 98 was released and Microsoft was seeking ways to facilitate Internet access through its operating system. Windows 98 shipped with Internet Explorer, one of the popular Web browsers competing with Netscape Navigator and OmniWeb. This integration of OS and browser software was a point of contention with other software manufacturers, as it was felt that this could lead to a stifling of competition through monopolistic practices. While Microsoft continued to ship Windows 98, the Justice Department investigated various allegations against the company, proceedings that continued for several years while Windows came out in new versions. The controversy became even more heated in 2002 when AOL/Time-Warner, who had acquired Netscape, challenged Microsoft's continued efforts to incorporate Internet Explorer into its OS.

Windows NT

Windows New Technology. A 32-bit multitasking, multithreaded, networking operating system first released in 1993, Windows NT shares many basic features and user interface concepts with Windows 95, but it is more polished, more reliable, more secure, and includes symmetric multiprocessor support. Bill Gates has publicly stated that Windows NT originated as "OS/2 3.0," with development work beginning in 1987 based upon the IBM/Microsoft collaboration to develop OS/2, with D. Cutler heading the evolution into Windows NT since 1989.

Windows NT is favored in development environments, server applications, and corporate networks. Windows NT was designed to run on processors other than just the Intel chips, providing portability to other platforms not well supported by earlier versions of Windows, except in the form of third-party emulators. It was offered in workstation and server versions.

Windows NT Advanced Server NTAS

An extended version of Windows NT specifically aimed at server applications.

Windows CE

Released in 1996, CE was aimed at the growing numbers of handheld computing devices and had a look and feel similar to the desktop Windows systems. Version 3 was released in 2000 along with applications development tools for programmers to build embedded device applications.

Windows 2000

Released early in 2000, this product was in many ways a bridge between Windows 98 and Windows NT. Support for small business networks, encryption, Internet access, and the many other features that users were now demanding were being incorporated into Microsoft operating systems by this time.

Windows ME

Windows Millenium Edition, released fall 2000. Windows ME was aimed as an economical option for home users. It includes some of the multimedia capabilities that were becoming

W

There are a variety of WOSA elements including:

Type	Notes
Messaging API (MAPI) – Provides electronic messaging access, including a variety of services such as voice mail, electronic mail, facsimile transmissions, etc. Address databases are supported by Extended MAPI.	
Speech API (SAPI) – Provides a means for a speech engine to run on a Windows client or server, thus enabling a telephony application, for example, to utilize speech capabilities.	
Telephony API (TAPI) – Telephony support for Windows NT systems to enable use of voice, data, and video over existing networks. The API provides generic means for connecting between multiple machines and accessing media streams being transmitted among them. TAPI supports standards-based H.323 conferencing and IP multicast conferencing.	

The Open Group Architectural Framework (TOGAF) has information and opinions on WOSA and WOSA-related open architectures at the following Web site: http://www.opengroup.org/public/arch/wosa.htm

WINF See Wireless Information Networks Forum, Inc.

wink In telephony, a handshaking signal indicating the transition between on-hook and off-hook states. A wink signal can indicate that the central office is ready to receive the dialed number. A wink is a fairly generic type of signal that can be used in many contexts and can be created in a number of ways, depending upon the type of communications protocol and whether the line is analog or digital. In an analog system, the wink can be indicated by a change or interruption in tone or a change in polarity. On a digital line, a wink may be indicated with signaling bits.

WINS See Windows Internet Name Service.

Winsock Windows sockets. Winsock provides a standardized applications programming interface (API) for Windows- and OS/2-equipped computers to interact with the Internet. It originated at an Interop conference discussion group in 1991. Winsock 2 is a further development effort of the Winsock Standard Group. In addition to the capabilities of the original version, Winsock 2 incorporates multiprotocol and multicast capabilities, quality of service (QoS), and a layered provider architecture. Microsoft's Windows Open Services Architecture (WOSA) incorporates Winsock into its APIs.

Winsocks were designed on the same concepts as UNIX-based Berkeley sockets, with extensions to accommodate Windows-specific functions; programming them is somewhat similar to programming Berkeley sockets.

Winsock functions by means of a dynamic link library file called *winsock.dll* that comes in 16-bit and 32-bit versions, and is loaded into memory. The .dll file format is compatible with Windows 3.x and newer versions. Winsock 32-bit drivers are included with Windows 95 and Windows NT. There are also variations in the .dll files to provide compatibility with various TCP/IP stacks. In a sense, the .dll provides a communications layer between the TCP/IP stack and the Winsock applications programs developed to communicate through the sockets.

Winsock Standard Group An organization of network software developers and vendors, established in the fall of 1991 that meets several times a year to enhance, refine, and extend the Winsock specifications. Winsock 2 specifications were developed and published by this group.

Wintel systems Windows and Intel. A term that refers to Intel hardware-based computers running Microsoft Windows operating software.

WIP See Women Inventors Project.

wipe pad An assembly and safety wipe that cleans surfaces to be bonded or cleans the bonding mechanism (popsicle stick, pipette, syringe, etc.). In fiber optics cable assembly, for example, wipes are used to clean the fiber filaments prior to fusing or bonding to connectors.

WIPO See World Intellectual Property Organization.

Wire Center A physical facility of a telecommunications provider, to which outside subscriber lines lead.

wire concentrator Any wire path or conduit that serves to bring together a number of wires running roughly through the same installation path.

wire pair Two separate conductors following the same transmissions path in close proximity to one another. It is called twisted pair when the wires are twisted around one another in order to reduce noise. Copper twisted-pair wire is very common in the telephone industry.

wire speed The raw speed at which data bits are transmitted over wire. This is not the same as the amount of information that can be transmitted. Various factors will hinder the transmission when all the devices and wires are interconnected, including noise, interference, connection devices, etc. Conversely, a variety of compression and multiplexing techniques can be applied to the data to increase the informational-carrying capacity of a wire.

wire stripper A tool that quickly removes the protective sheath from wires so they can be easily attached to mounting posts or to one another to make good electrical contact. Punchdown tools for installing telephone line often include a wire stripping blade.

wire tap See wiretap.

wireless A system that performs a function, or transmits information fully or predominantly, without wires. An older term for a radio set, since radio signals were intercepted as waves. Now the term is used more broadly for telecommunications technologies that work without wires, including cellular phones, intercoms, and video security systems that work over FM frequencies.

WIRELESS The largest annual wireless industry trade show, sponsored by the Cellular Telecommunications Industry Association (CTIA).

Wireless Access Controller WAC. A central point in a private branch or public switched network that provides a communications link to a host network or base station. Wireless systems require constant monitoring of signal strength and administration of user registration, hand-off, and roaming provided by the WAC so the user is not aware of the changing connections.

Wireless Application Protocol WAP. A wireless telephony network protocol specification developed by a group of wireless vendors, including Ericsson, Motorola, Nokia and Unwired Planet (Phone.com, now OpenWave.com), in 1998. The technology was designed to work with a variety of industry mobile standards and handsets worldwide to facilitate widespread commercial distribution. It is intended to be operable with CDMA, TDMA/D-AMPS, GSM, DECT, as well as a number of dedicated mobile standards, including CPDP, Mobitex, iDEN, and Data TAC. WAP further supports FLEX and ReFLEX paging standards. See WAP Forum.

Wireless Application Reader WAR. A Wireless Application Protocol (WAP) browser developed by Digital Mobility Limited, announced in March 2001. The browser is intended for use on Java-enabled phones and personal digital assistants (PDAs). It was developed for the J2ME Mobile Information Device Profile (MIDP) type of Java.

wireless cable Wireless cable doesn't utilize physical cables or, at least, is not predominantly based on wired connections, but is named that way to provide marketing familiarity for subscribers who are used to associating 'cable services' with television broadcast services. Thus, *wireless cable* refers to television broadcast services and, to some extent, interactive TV services delivered through the airwaves without physical cables (usually through microwave transmissions).

Wireless Certificate Profile WCP. One of the wireless security projects of the WAP Forum, also called WAP Certificate Profile, based on work done within the IETF's PKIX working group. Given the limited resources of most handheld wireless devices, WCP is a reduced-footprint, reduced-processing security mechanism for identifying the holder of a public key, providing a means to bind the key to the appropriate user. It is intended that the WCPs work interchangeably with other X.509 certificates for maximum compatibility with the current Internet infrastructure. In general, WCPs are based on the Internet Certificate Profile. See WAP Forum.

wireless communications protocol WCP. A transmission protocol specifically designed to accommodate wireless technologies.

Wireless Consortium WC. An association and long-term independent facility established in 1996 through the University of New Hampshire's InterOperability lab. It represents the cooperative interests of wireless products and services vendors wishing to develop and test IEEE 802-11-conformant products.

Wireless Data Forum WDF. Formerly the Cellular Digital Packet Data Forum, the WDF is a nonprofit organization established to promote the acceptance and development of wireless data products and services, particularly those employing the Commercial Mobile Radio Services (CMRS) spectrum assigned by the Federal Communications Commission (FCC). This consensus-building organization holds its meetings in a variety of locations around North America. http://www.wirelessdata.org/

Wireless Datagram Protocol WDP. A protocol of the WAP Forum associated with the Wireless Application Protocol. The WDP was specified as version 14 in May 1999.

WDP is one of two transport layer protocols defined as the Wireless Transaction Protocol (WTP) and the Wireless Datagram Protocol (WDP). WDP is a general datagram service. Hierarchically, the WDP layer operates above the data-capable bearer services, offering service to the upper layer protocol. It does not include an authentication mechanism. See Wireless Session Protocol.

wireless dongle In general, a dongle is a communications-influencing device that restricts access, provides security or copy protection, or otherwise limits the communications that pass through it. It is often attached to a serial or other communications port between the computer communications interface card and external devices. Thus, a wireless dongle is a wireless transmitting and/or receiving device that is attached to equipment that is either wired or wireless. In recent years, the term *dongle* has been more loosely used to describe a host that is interfaced with multiple wireless devices, thus mirroring the hierarchy of the interface device that it is being used to supplement. Thus, the dongle is described as an interface device to aid in adapting wired device protocols and hardware to wireless operations. However, it is preferable to call this type of device a *wireless adaptor* rather than a wireless dongle, or the specific security or limiting aspect that has long been ascribed to the term dongle becomes confused.

Wireless Information Networks Forum, Inc. WINF, WINForum. A trade association of manufacturers of unlicensed communication systems, including personal communications services (PCS). WINForum works with the Federal Communications Commission (FCC) in the allocation of spectrum with the Unlicensed National Information Infrastructure (U-NII). See band allocations.

Wireless Intelligent Network WIN. International Standard IS-41. Actual implementation of this standard is typically done in conjunction with time division multiple access (TDMA) technology. WIN is a network of realtime databases and transaction processing systems supporting the processing of the messages in a wireless transmission and manages roaming. See time division multiple access technology.

W

Wireless LAN Interoperability Forum WLI Forum. A trade organization of wireless product and service organizations established to promote and support the growth of wireless LAN. The WLI Forum has developed and published an open interface specification for interoperability not intended to conflict with IEEE 802.11. The specification is based upon the RangeLAN2 radio frequency wireless LAN technology developed by Proxim, Inc. in 1994. It uses a multichannel spread spectrum frequency hopping architecture in the 2.4 GHz range at a data rate of 1.6 Mbps per channel.

Wireless LAN Research Laboratory WLRL. A research laboratory established through the Center for Wireless Information Network Studies at the Worcester Polytechnic Institute (WPI) in Massachusetts. The lab serves as a center for the development, specification, and testing of hardware and software performance, compatibility, and interoperability.

wireless local area network WLAN. A data network connected through radio wave and/or infrared transmissions rather than through cables and wires. Interbuilding connections may be through line-of-site microwave connections.

The IEEE is working on standards for wireless LANs through the IEEE 802.11 Working Group. The basic structure of the standard specifies three physical layers (two for radio, one for infrared) and one Media Access Control (MAC) sublayer to provide authentication, association, integration, and distribution services. Two configurations for wireless stations have been generalized as independent (directly communicating stations) and infrastructure (communicating through inter-access points).

There are a number of campuses that have wireless LANs, and the technology is gradually being adapted by corporations. There are advantages to both wired and wireless communications. Signals can be kept separate more easily in wired communications, but wireless is sometimes the only practical means for individuals or groups on the move to keep in communication with the network. There will probably be many situations where hybrid systems serve the greatest needs. See air interface, HIPERLAN, Inter-Access Point Protocol, local area wireless network.

wireless local area network physical layer WLAN physical layer. There are three physical layer (PHY) specifications for the IEEE 802.11 wireless local area

network, two for radio and one for infrared transmissions. The radio PHYs specify spread spectrum frequency hopping and the direct sequence spread spectrum. The infrared PHY uses pulse position modulation with 16 positions.

wireless local loop 1. WLL. A local telecommunications services distribution system that uses low power radio waves instead of wires. This is particularly cost effective in rugged terrain or sparsely populated areas, where it is not practical to lay cables. Wireless services may also be easier to install in some areas from a legislative/jurisdictional point of view with regard to Federal Communications Commission (FCC) licensing to local carriers. 2. WiLL. A Motorola, Inc. fixed wireless system designed to provide an alternative to traditional landline systems. WiLL systems are provided in two configurations: as public switched telephone network direct connect (PSTN-based) or as mobile telephone switching office/mobile switching center (MTSO/MSC-based) networks. The WiLL System Controller (WiSC) provides performance monitoring and radio channel control functions.

Wireless Markup Language WML. A tag-based browsing language, similar to HTML, but specifically intended to be used on wireless handheld devices that have smaller screen sizes and more limited resources (memory, processing speeds, display characteristics, etc.) than desktop or laptop computer-based communications systems. WML evolved from HDML (Handheld Device Markup Language), which was developed in the mid-1990s by Unwired Planet (which became Phone.com and then Openwave). WML is XML-compliant and is now part of the WAP standard.

Because WML servers and devices do not directly use HTML and other standard Web protocols, it is necessary to implement specialized security features in conjunction with WML if security is required. See HTML, WAP Forum.

wireless packet switching A means of sending data over wireless networks similar in concept to packet switched wired networks. Like wired packet transmissions, a communication is subdivided into small data packages called *packets*. The packets are routed over the network, sometimes different portions of the message taking different routes, and reassembled at the receiving end. Routing will vary depending upon

Wireless Local Area Network Physical Layers

Type	Scheme	Modulation	Notes
radio	spread spectrum/ frequency hopping	Gaussian frequency shift keying	1 Mbps with 2-level GFSK 2 Mbps with 4-level GFSK
radio	spread spectrum/ direct sequence	differential binary phase shift differential quadrature phase	1 Mbps 2 Mbps
infrared	--	pulse position modulation	1 Mbps with 16 PPM 2 Mbps with 4 PPM

the current load on the routing stations, and the distance and number of 'hops' over which the message is transmitted. The message may even travel over wires for part of its journey. There are a variety of protocols for accomplishing this. See packet switching.

Wireless Priority Services WPS. Under the umbrella of the WPS, there have been a number of wireless communications priority services established for cellular, PCS, and enhanced satellite communications in order to support the needs of national security and emergency preparedness (NS/EP). In Executive Order 12472, responsibility was assigned for conducting technical studies and assessing research and development programs to identify resources and approaches to assist U.S. authorities in fulfilling security and emergency-related telecommunications needs. See National Communications System, Priority Access Service.

wireless public key infrastructure WPKI. A mechanism for adding security features to wireless communications through public key encryption and management techniques. WPKI is expected to enable more secure banking as well as more secure transmission of sensitive corporate and government data through wireless devices. There are many ways to implement WPKI through mobile devices but, in general, trusted registry agents are expected to generate and verify digital signatures while others will likely maintain digital signature directories to facilitate encounters and transactions. Secure servers operating over private or Internet links will handle and record the content and/or steps in a transaction to give it some of the legal force and protections of traditional transactions.

In March 2001, Baltimore Technologies, AU-System, and Gemplus announced an advanced security and authentication digital signature technology for General Packet Radio Services (GPRS) networks through Ericsson handheld devices. For this system, the digital certificates are issued by network operators and content providers and are stored on the mobile devices using Subscriber Wireless Identity Module (S/WIM), a type of chip card. See public key encryption, wearable public key infrastructure.

Wireless Server Certificate WSC. An administrative and operational mechanism for implementing server-side security for microbrowser-equipped wireless devices conforming with Wireless Transport Layer Security (WTLS). There are two defined formats for WTLS certificates:

- *X.509* optional, similar to certificates used in Secure Sockets Layer security
- *WTLS mini-certificates* mandatory, similar to X.509 but with a reduce-resource footprint adapted to the needs of small wireless devices (e.g., Personal Digital Assistants)

Service providers seeking to secure a WAP server must purchase a Server Certificate from a Certificate Authority (CA) (examples include VeriSign, Baltimore, MobileTrust), by submitting a Certificate Signing Request (CSR). The issued certificate will have an encryption/decryption key-pair. The registration cost is approximately $900 per year depending upon whether it is an initial or short-term certificate. New short-term certificates that change every day or two may be available, reducing the probably of security breaches. Some companies provide short-term trial certificates at no cost to help a provider become familiar with the process. The Certificate authenticates the WAP server-to-client wireless devices and then provides a transmission channel. Certificate issuers typically will provide third-party verification services associated with the Server Certificate. Note that the WAP microbrowser system works through proxy servers, so it is the gateway that handles the certificate. See WML Script Sign Text, Wireless Transport Layer Security.

wireless service provider An authorized wireless communications supplier, usually providing cellular phone and paging services.

Wireless Specialty Apparatus Company WSA. A pioneer radio parts manufacturer founded in 1906 by Greenleaf Whittier Pickard and Philip Farnsworth (who provided legal counsel to the company). By the mid-1940s, WSA had become one of the largest radio engineering manufacturers in the world, but it is chiefly remembered for its production of rugged, well-constructed radio wave crystal detectors in the early part of the century. Later, in conjunction with a number of other significant manufacturers (GE, Westinghouse, AT&T, etc.), WSA made an agreement for Radio Corporation of America (RCA) to be an exclusive distributor of radio receiving sets.

WSA equipment is now found in many radio collections around the world and a number of the WSA-related legal documents (including patents) are part of the *George H. Clark Radioana Collection* now in the Archives Center of the National Museum of American History.

wireless switching center WSC. A terminal center for switching wireless communications with other trunks or service. It's not uncommon for wireless centers to interface with public switched telephone networks (PSTNs) and various landline facilities in order to route a call that originated as a wireless call through traditional routes. This makes use of the best features of the various technologies, extends the range, and improves the transmissions clarity of wireless-originated calls.

Wireless Session Protocol WSP. A session layer protocol family within the WAP architecture, approved in July 2001 by the WAP Forum. WSP provides the upper-level application layer of WAP with a consistent interface for session services, including a connection-mode service and a connectionless service. WSP services are currently best suited for network browsing applications. They provide HTTP 1.1 functionality and support for data push functions. They are optimized for low-bandwidth bearer networks with relatively long latency characteristics. See WAP Forum.

Wireless Subscriber-Automatic Number Identification WS-ANI. A system for identifying the callback number (Mobile Directory Number) for wireless mobile phones. This is important in the context of 911 emergency calls from a mobile phone user, whose location will change. The callback number is used in the event of a 911 caller being cut off (or hanging up); it enables the operator to recontact the person in distress. Due to the importance of this service, fault-tolerant, redundant computer systems are provided to Public Safety Answering Points (PSAPs).

In the BellSouth WS-ANI system, for example, wireless carriers interface with the SCC Communications Corporation Automatic Location Identification (ALI) data centers that manage the BellSouth information and services for the E911 system. Through BellSouth interfaces, wireless carriers use out-of-band data connections between the switching network and the ALI database. Physical connectivity is through ports with synchronous DB-25 connections supporting either RS-232 or V.35 electrical interfaces. They are configured as Data Terminal Equipment (DTE) devices. See Automatic Number Identification, Pseudo-Automatic Number Identification.

Wireless Transaction Protocol WTP. A light weight transaction-oriented protocol within the WAP architecture that provides the services necessary for interactive request/response (browsing) applications. During a session, the client requests information from a fixed or mobile server and the server responds, completing the transaction. WTP is intended to reliably deliver transactions while balancing reliability levels required for the application. WTP runs on top of a datagram service and may run on top of a security service. WTP is suitable for 'thin' clients wireless datagram networks. See WAP Forum.

Wireless Transport Layer Security WTLS. A WAP Forum wireless protocol to enhance privacy, data integrity, and authentication for wireless transmissions. WTLS uses a client/server approach implemented at the security layer of the Wireless Application Protocol (WAP). It is similar to Transport Layer Security (TLS) and Secure Socket Layer (SSL) protocols with adjustments for the specialized characteristics of portable wireless devices. Unfortunately, these adjustments may make it vulnerable in some ways to data recovery, forgery, and datagram truncation attacks, issues that need to be resolved because it is expected WTLS may be incorporated into large numbers of consumer communications devices. See WAP Forum, Wireless Server Certificate.

wireless telephone service A variety of forms of wireless voice transmissions. The service now primarily means cellular telephone which became established in 1983 in the U.S. Almost half of wireless systems were purchased to be used in emergencies; the rest are divided among business and personal users.

wireline 1. Communications circuits that are comprised of physical wires and cables, usually copper twisted pairs as in traditional telegraph and telephone lines, or fiber optic. 2. In cellular communications, not all aspects of the transmission are necessarily carried through wireless radio waves. Sometimes the system interfaces with a wired system for part of the transmission. A wireline cellular license granted to the local telephone company from the Federal Communications Commission (FCC) is designated as a B Block. It provides permission to operate at FCC-specified frequencies.

wiretap *n.* A covert listening device or system of devices and procedures associated with a communications circuit (usually voice) for the purpose of eavesdropping and/or recording transmissions (conversations). The device need not necessarily be physically attached to the line being tapped; some devices can sense electrical emanations without directly touching the electrical conduit. Others can silently dial a phone and listen in to room noises even if the phone appears to be 'on hook.'

Other than high level encryption through a digital line, there are no surefire ways to prevent unauthorized wire taps, especially on wire-based analog connections. Strictly regulated, wire taps require legal authorization and are permitted only in specific, carefully evaluated circumstances, usually by law enforcement officials. Wiretapping is much more difficult over secured digital lines, especially if they are being communicated over fiber optic cables which cannot be physically 'tapped' from outside in the same way as wirelines without detectable light loss beyond the point of the tap. See Anti-Terrorism Act of 2001, bug, trace.

wiring closet In medium- and larger-sized wiring installations, whether for phone, computer networks, or other services, a wiring closet will often be designated on the premises as a central administrative and physical connections facility. This provides consolidated access to equipment and connections, and makes it possible to restrict access to unauthorized users. This is typically the location at which wires enter the premises from external sources or from the building distribution system, and may or may not be the demarcation point for the services, depending upon the type of service and the particular contract. Several wiring closets on a premises or campus may be cross connected.

wiring grid The wiring architecture of a department, section, or building. This may be illustrated with a diagram of locations, connections, and circuits. The wiring grid may be monitored on a computer system, with consoles for determining and configuring routes and pathways and for diagnostic testing.

WITS Wireless Interface Telephone System.

Wizard 1. An elite professional, as a programmer, who appears to perform miracles with his or her solutions to problems, or ability to troubleshoot or configure a system. 2. A shortcut computer software applications helper that may consist of hotkeys or other means to quickly perform often-repeated tasks. Some programming Wizards are user-scriptable. 3. A computer software applications informational helper that may pop up in a particular context or be requested by the user.

WLAN See local area wireless network and wireless local area network.

WLANA An organization established to promote

awareness and acceptance of wireless local area network technologies (also called local area wireless networks (LAWN)). WLANA provides education and support for the technology through standards committees and other venues.

WLI Forum See Wireless LAN Interoperability Forum.

WLL See Wireless Local Loop.

WLRL See Wireless LAN Research Laboratory.

WLS See wavelength-shifting.

WML 1. See Website Meta Language. 2. See Wireless Markup Language.

WML Script Sign Text A function within the client/server Wireless Transport Layer Security (WTLS) Certificate system that enables WTLS servers to verify digital signatures. These signatures are stored with Certificate Authorities (CAs) rather than on the reduced-resource wireless devices, thus not hindering the size of the certificate. Client certificates are issued in the same manner as server certificates, through a certificate-issuing service. See Wireless Transport Layer Security.

wobbled groove recording A recording technology for magneto-optical media. The region of the disc for user data storage has a series of irregular grooves, with the data stored in the groove. The wobbling of the groove is created based on a continuous sinusoidal deviation of the track from an average center line. Wobbled groove technology has the advantage of resistance to tiny defects. It is used in the production of DVD-RAM discs. See land/groove recording, wobbled land/groove recording.

wobbled land/groove recording A high-density recording technology for magneto-optical media described in the mid-1990s and introduced commercially by Hitachi in 1998. Wobbled land/groove recording is the chosen format for DVD-RAM rewritable storage media.

Wobbled land/groove recording takes advantage of the high-density recording features of land groove recording and the resistance to tiny defects of wobbled groove recording. Thus, data is recorded on both the upper surface (the land) and in the grooves enabling up to 2.6 GBytes per side. The address information is stored in the pit. Land and groove tracks alternate with each revolution of the disc to form a continuous spiral track. This enables large data streams (e.g., video animations) to be recorded in a long, continuous spiral.

To promote compatibility with existing DVD formats, the modulation and error-correction schemes for this recording format have been kept the same. The format has been supported by the DVD Forum for DVD-RAM discs. See land/groove recording, wobbled groove recording.

work In the human sense, expending energy in accomplishing or trying to accomplish something that is required or expected and which isn't necessarily fun. In the scientific sense, a transfer of energy through the application of force. When a phototube converts sunlight into electricity, it is performing work. When a circuit supplies power to a device, it is performing work. The basic unit for mathematically expressing and calculating work is the *joule* (J). See joule, power.

WOM write-only memory.

WOMBAT waste of money, brains, and time. One of the wittier of the acronyms seen in email and online public forums on the Internet.

Women Inventors Project WIP. A Canadian project that highlights women who have contributed to technology in a variety of industries. The results of a collaboration of this project with the Canadian National Museum of Science and Technology is available as an educational traveling exhibit for loan periods of 6 to 8 weeks.

Wood, Robert Williams (1868–1955) An American physicist and inventor who was known for improvising much of his own experimental equipment. Wood observed charged particle field emissions, in 1897. After joining Johns Hopkins University in 1901, Wood published "A suspected case of the electrical resonance of minute metal particles for light-waves. A new type of absorption" and "On a remarkable case of uneven distribution of light in a diffraction grating spectrum," in 1902. In 1905, he published *Physical Optics* which became a standard text in the field. In the early 1900s, Wood tested liquid mirror optics in the process of developing a telescope for astronomical research that was adapted by W. Warner. In the 1920s, he became a pioneer (and also an expert) in spectroscopy and researched applications of absorption screens in astronomical imaging.

In addition to his many inventions, Wood's observations led to new research in refraction and the development of diffraction gratings. Gratings are now essential components in many optical devices. See surface plasmon resonance, Wood-Rayleight anomaly.

Wood light Also called a "black light," this is a light that generates ultraviolet radiation (which makes reflective whites practically "glow in the dark"), named after R.W. Wood.

Wood-Rayleigh anomaly In 1902, R.W. Wood published an article on the uneven distribution of light in a diffraction grating spectrum. His observation was followed up by research on resonance in diffraction gratings conducted by J.W. Strutt (Lord Rayleigh). The Wood-Rayleigh anomaly occurs when higher order diffracted light reflects at the grazing angle along the grating surface.

Since the scientific understanding of diffraction in the early 1900s did not take into account all the possible variables, including selectivity of wavelength, Wood's observation was important in that it could not be explained with the body of existing theories. His paper, and later calculations by Lord Rayleigh that began including electromagnetic factors, led to an improved understanding of the phenomenon of diffraction and its subsequent prediction and modeling. See diffraction grating, Rayleigh expansion, Talbot effect.

Woodstock File Server A simple, shared file server with a limited command set that clients use to communicate over a network with the server. The system responds to page-level requests as though it were a

W

remote disc, as opposed to a connection. The client program is responsible for implementing the directory system, stream, etc. WFS utilizes connectionless protocols and atomic commands. It was developed in 1979 by Xerox Palo Alto Research Center (PARC).

WORA write once run anywhere. A slogan usually used in context with Java, the portable programming language from Sun Microsystems, but which can apply to any software designed to be portable, that is, platform- and/or application-independent. See Java.

word In computer programming, a contiguous group of a specified number of bits (basic information units) constituting an information entity transferred or processed as a unit. The length of a word varies from system to system, with smaller systems usually processing a word as 8 bits, and larger systems as 32 bits. Some supercomputing systems handle a word as 64 bits. Commonly, a word is a multiple of 8 bits, as a byte is 8 bits, and is a basic unit of information between a *byte* and a *longword*.

In networking, the term *octet* is often used to mean eight bits in order to prevent the ambiguity associated with *byte* and *word*.

word spotting In speech recognition, the selective recognition of a predetermined sound. This can be used to glean specific information or to filter extraneous noise or information. Word spotting is demonstrated in Star Trek episodes on the turbolift, where turbolift commands are recognized, but general conversation is ignored. The same concept applies to automated door and phone systems, and software programs configured to recognize only relevant material.

words per minute WPM. A description of the speed of a communication over time, as in typing, printing, code sending, and some data communications.

work order WO. A widely used administrative tool, especially in the trades, to request and record work to be done. Usually in the form of a document, a work order may be assigned a number and/or a priority, and provides a sequential record of the people, parts, and procedures involved in carrying out a requested job. Work orders help track customer requests, work and completion cycles, and sometimes include details as to who did the job and how. The work order may also include billing information or may be used as a reference for creating a separate invoice.

work station See workstation.

workflow The sequence and path of work in an organization, environment, or department. Workflow management is especially important in production lines, where actions must occur a particular order and location to keep the line running. In service-oriented environments, it refers to the efficient deployment of staff and resources, timed, allocated, and adjusted to process calls and requests as they come in. In computer technology, the term applies to the sequence and path of data transport and its management.

workgroup Individuals within an organization who share certain tools, software applications, or tasks, typically on a network. Priority and access levels within a specified workgroup are often similar. A person may belong to more than one workgroup, and workgroup access to resources may overlap from one group to another. Software companies have been targeting as workgroups people within an organization who share communications or applications needs without necessarily being in the same department or physical location. Workgroup applications include electronic mail, word processors, spreadsheets, scheduling programs, and others.

workgroup manager A supervisory individual who has authority to assign login, priority, and security parameters to individual computer network users organized into workgroups. The manager may use a software utility to set up the system, or may pass on information about the organizational structure to the system administrator, who then sets up the computer configuration.

workload 1. The maximum or effective capacity or capability of an employee or system to handle work tasks, usually within a given time period or with regard to deadlines. 2. In networks, workload may be measured by the number of transmissions or the quantity of information that can be processed within a specified interval.

workstation In telecommunications, a networked or networkable computer that fits somewhere between the low-end consumer market and the high-end computing market (scientific research, military, high-end medical imaging). Some people refer to all microcomputers as workstations; some refer only to microcomputers with enhanced processors, storage, and memory as workstations; and some refer to microcomputers with decent processing speed and robust operating systems suitable for professional work as workstations. In general, the least and most expensive computers are not called workstations.

World Administrative Radio Conference. WARC. A global space telecommunications conference that convened regularly. Founded in June 1971 by the ITU to review regulations relating to radio astronomy and extraterrestrial communication. At the 1979 general conference, significant access to bands between 1 and 10 GHz was provided for amateur radio satellite programs, bandwidths that the operators had previously been restricted from using.

World Administrative Telegraph and Telephone Conference WATTC. A body formed as a result of a resolution at the 1982 Plenipotentiary Conference of the International Telecommunication Union (ITU). It was scheduled to convene in 1988 to consider proposals for a new regulatory framework to handle new situations in the field of telecommunication services. IN 1984, the CCITT resolved to set up a Preparatory Committee (PC) in preparation for WATTC-88 to prepare the draft text of the new Regulations to be submitted to the CCITT Plenary Assembly.

The purpose and scope of the Regulations were to establish general principles for the provision, underlying transport, administration, and operation of international telecommunication services offered to the public (including legal and governmental bodies).

World Association of Community Radio Broadcasters AMARC. An international nongovernmental organization promoting global cooperation in community radio broadcasting in more than 100 countries. AMARC dates back to meetings in Canada and other countries in the early 1980s and the formation of the International Solidarity Network in the early 1990s. These efforts culminated in the official founding of AMARC at the 2nd World Assembly in 1996.

AMARC seeks to support diversity and friendship, to support women's efforts in the field of community broadcasting, and to promote solidarity, cooperation, cultural identity, and independence of all peoples through community programming. See Community Broadcasters Association, National Public Radio, Public Radio International.

World Congress on Insulator Technologies An annual trade conference on industrial and research aspects of insulators. Examples of some of the topics include performance assessment in various environments, design and reliability of different types of insulators, field handling, and materials used in new generation insulators and their operating performance.

World DAB Forum See Digital Audio Broadcasting Forum.

World Intellectual Property Organization WIPO. A Geneva-based global organization comprising over 160 member countries that promotes, supports, and educates about intellectual property protection through international treaties. Recently, WIPO has been actively involved in the debate over domain name assignments and jurisdictions. See Digital Millennium Copyright Act. http://www.wipo.org/ and http://www.wipo.int/

World Internetworking Alliance WIA. A coalition of parties working to develop an international, independent, autonomous Internet institution within which diverse bodies can work in a cooperative effort toward creating an open, robust, competitive marketplace. http://www.wia.org/

World Numbering Zone See World Zone.

World Radiocommunication Conference WRC. A month-long, international assembly administered by a number of committees and working groups to deal with important matters of spectrum allocations and technical matters related to global radio communications. The conference is convened every two or three years to officially assess and revise the international Radio Regulations. One of the important outcomes from WRC-2000 was support for global wireless systems (especially 3G) that facilitated commercial implementations of wireless technologies.

World Report on the Development of Telecommunications Information and economic reports on global telecommunications developments. This is a publication of the International Telecommunications Union (ITU).

World Telecommunication Policy Forum WTPF. An organization established as a result of the Kyoto Plenipotentiary Conference wherein members can discuss matters of telecommunication policy and regulation, and provide reports and opinions to the ITU. The first meeting was in 1996, with a discussion theme of *global mobile personal communications by satellite (GMPCS)*.

World Telecommunications Advisory Council WTAC. A global private and public sectors organization established in 1992 that studies telecommunications implementations and advises the Secretary-General of the ITU on matters of policy and strategy. Originally composed of top level managers, the organization now also includes consultants, government officials, and entrepreneurs. The WTAC has provided a number of publications and sponsorships, including a booklet entitled "Telecommunications: Visions of the Future," WorldTel sponsorship, and a global mobile personal communication systems (GMPCS) symposium.

The World Wide Web in Graphical Format

A portion of the CRC Press storefront on the World Wide Web, as viewed in Netscape. The Web address for this site is expressed in the browser as http://www.crcpress.com/. In this manner, the WWW acts as client/server resource topology that supports Web browsers as a graphical front-end to many millions of resources on the Internet. This makes it relatively easy for laypeople to surf the Web.

The Web is probably one of the most significant developments in commerce in the history of human transactions, in addition to its impact on communications and social evolution. Before the development of the Web, the Internet was used primarily by military and research organizations and a small percentage of technically inclined individuals. It now reaches more than 60 million people worldwide and is rapidly growing.

World Wide Web Web, WWW, W3. The W3 was first proposed in March 1989 by Tim Berners-Lee (CERN), who created the first Web software a few months later which he publicly introduced in 1991.

The Web is the single most significant implementation on the Internet and has spurred global access and acceptance not only by technical professionals, but by computer novices and laypersons. In a sense, it constitutes a simple graphical user environment for the Web (it is actually much more than that, but that is how 80% of new users first perceive it) and, as such, its use will probably greatly change and grow in future revisions of Web browsers, especially now that Netscape, the primary browser, is freely distributable with source code.

A browser actually gives a thin window on the Web, as it consists of more than just a browser front-end. It is built on a set of protocols and conventions that make access and transfer of information widely available. The first international WWW conference was held in 1993. See browser, HTML.

World Wide Web Conference Committee, International World Wide Web Conference Committee IW3C2. Founded in 1994 by NCSA and CERN, the IW3C2 was incorporated as a nonprofit Swiss Confederation association in 1996. IW3C2 organizes international telecommunications conferences focusing on the World Wide Web, and makes the information resulting from these conferences open to the widest possible audience. IW2C2 promotes and assists in the global evolution of the Web through a broad-based international body. Currently, the IW3C2 selects one academic institution per year to host the IW3C2 conferences. http://www.iw3c2.org/

World Wide Web Consortium WWWC, W3C. An industry consortium of more than 150 organizations, founded in 1994 to promote the development and use of common standards for the evolution of the World Wide Web (WWW). The W3C develops and maintains information repositories for programmers and users, reference code for implementation of recommended standards, prototypes, and sample applications. http://www.w3.org/

World Zone, World Numbering Zone Geographic divisions devised to facilitate the global linking of national telephone services, a system developed in the 1960s. Country codes of one to three digits in length were assigned to specified regions. As examples, World Zone 1 represents most of North America and some Caribbean countries, 2 is Africa, 3 and 4 represent parts of Europe. In an international long-distance call, the country code is the first part of the phone number dialed.

Worldwide Submarine Fiber Optic Systems Report WSFOSR. A commercial report on fiber optics published by Pioneer Consulting in 1999 in which investment in submarine fiber over the next five years was estimated at $31.8 billion.

Worldwide Submarine Fiber Optics 2001 WSFO2001. A followup to the *Worldwide Submarine Fiber Optics Systems Report* of 1999, published by Pioneer Consulting in 2001. The report identifies growth and flat patterns over about a five-year cycle. The report estimates expenditures of over $100 billion on submarine fiber optic cable between 1988 (when the first transatlantic cable was commissioned)

and 2007. Forecasts cover three major oceanic cable routing regions.

worm A computer program that replicates itself through host systems without the specific need to attach itself to a file. Since a virus is a self-replicating program that requires a host file, a worm program can be seen as a 'standalone' variation of a virus or can be considered a separate category. Here it is treated as a specific type of virus.

The worm may be self-replicating within the host system or may be controlled from a remote site or sites in such a way that it is not wholly dependent upon specific resources in the computer system it has infected. This is a less common and sometimes more sophisticated program than many of the computer viruses that are attached to downloaded files and email messages. A worm 'worms' its way through a network much as a biological worm wiggles its way through soil or wood, creating a trail of holes and tunnels that have been compared to the intermittent 'holes' that occur in file storage space or running processes on a system on which there is a worm program. A worm may also be combined with other types of invasive 'organisms' such as Trojan horses (e.g., a worm may be a conduit for spreading and implanting an accompanying Trojan horse program). See Computer Emergency Response Team, Trojan horse, worm. See virus for further details and specific examples.

WORM write once read many, write once read multiple. Although this commonly refers to optical storage media that are not as readily written as magnetic media, it generically means any storage medium that is typically written (easily) only once, and is usually read, or can be read, many times by the user. Most CD-ROMs fit into this category, although read/write optical systems are coming down in price to the point where they may become consumer items. Kodak PhotoCD discs demonstrate that consumer products on CDs can be written in several sessions. See compact disc.

WORM-ARQ WORM - Auto Repeat Request. ARQ is a means to achieve reliability in network transmissions by using an error detection scheme in conjunction with a block retransmission scheme for blocks found to have errors. WORM-ARQ is a technology developed by Nippon Telephone and Telegraph (NTT) DoCoMo for frame-based broadband network transmissions. Depending upon the transmission's quality, WORM-ARQ will switch, as needed, between Selective Repeat (SR) and Go-Back-N (GBN) systems to maintain stability at high efficiency levels.

WOSA See Windows Open Services Architecture.

WOTAN Wavelength-Agile Optical Transport and Access Network. An ACTS project for developing and providing a managed, end-to-end optical communications system for public communications networks. WOTAN investigates and tests a mix of optical and electronic combinations for switching and networking to determine an optimum mix for building an integrated, upwardly flexible system based on wavelength agility and distributed wavelength routing and switching. Project participants include a number of U.K. and continental European companies, universities, and labs. Test systems are established in cooperation with BT Laboratories and their East Anglia Network test bed. See BLISS, BROAD-BANDLOOP, and UPGRADE.

wow An undesirable undulating audio distortion from various causes including uneven rotation of a turntable or tape reel.

Wozniak, Stephen (Woz) (1950–) Steve Wozniak was a shy, inventive child, given to tinkering with electronics and entering projects in science fairs. When he first became friends with Steve Jobs, neither one realized they would soon be making history and millions of dollars.

Wozniak was constantly working on hardware projects in his parents' home and was naturally drawn to the hobbyists who began to meet and talk shop at the Homebrew Computer Club. Together with Bill Fernandez in 1971, Steve Wozniak built a simple computer from cobbled together parts called the Cream Soda Computer.

From 1973 to 1976, Wozniak worked in electronics at Hewlett-Packard. After meeting John Draper, Woz began to collaborate with Steve Jobs in developing blue boxes, phone devices that provided unauthorized access to long-distance services. After this ran its course, he was inspired by the homebrew members and a *Popular Electronics* feature of the Altair computer kit to develop the Apple I computer. This lead to formation of Apple Computer in 1976, a change of operations to Steve's home, and the subsequent creation of the historic Apple II computer.

Wozniak stayed with Apple Computer until 1985 when he left to form a wireless electronics company called CL-9 (Cloud Nine). Through all his ventures, he has participated in many educational and philanthropic activities. He is the founder and current president of Unuson Corporation in which he enjoys his time playing and teaching guitar.

WPG Wireless Protocols Group. See WAP Forum.

WPS See Wireless Priority Services.

WPKI 1. See wearable public key infrastructure. 2. See wireless public key infrastructure.

WRAM Windows Random Access Memory. A type of memory similar to VRAM (virtual RAM), but faster and less expensive; it is commonly used to enhance graphics performance. WRAM offers bisynchronous input/output. See VRAM.

wrap A redundancy mechanism used in International Business Machines (IBM) Token-Ring local area networks (LANs). Token-Ring cabling is set up with two trunks, one for normal use and one for a backup. If something happens to the main trunk, a disconnected TCU connection creates a signal that causes the path to 'wrap' onto the backup trunk in order to maintain the interconnections through the ring. See Fiber Distributed Data Interface.

wrap connector A diagnostic tool that interconnects cables or controllers to verify the circuit.

wrap plug A diagnostic device which, when plugged into a computer port, causes the data to be looped around back to the port.

wrap test A diagnostic test in which a signal is looped back through a device, usually to see if the input signal and the output signal match.

wraparound 1. In word processing, the automatic wrapping of text onto the next line, when it reaches the end of the line, which may be an indent, margin, window, or screen. This is similar to the line feed and carriage return on a typewriter. 2. A circumstance where visual or textual information goes off the edge of the display device, usually the bottom or right-hand side, and the information that would be clipped off by the edge is 'wrapped around' to the opposite side. For example, if a window drops off the bottom of a computer monitor, the bottom part may be displayed at the top of the monitor. Display wraparounds are usually undesired and are a result of programming bugs. However, there are circumstances where wraparound enhances the perception of a bigger screen, such as graphical scenes in which a landscape is intentionally wrapped, usually from side to side, to give the illusion of an infinite landscape.

WRC See World Radiocommunication Conference.

write To record data, as to a storage device, such as RAM, a hard drive, or a floppy drive. Many computer data storage media use a means of rearranging magnetic particles in order to form patterns that are subsequently recognized and interpreted by software. Since this can be done many times, these media can be reused. There are also media which can only be written once. See WORM, write protect.

write head A storage write mechanism that is designed to put data on the writable medium, which may be tape, a diskette, an optical disk, a vinyl record, etc. Write heads are sometimes separate from read heads, and are sometimes incorporated into combined read/write heads.

write protection A device or method for protecting a medium from accidentally being overwritten with other data or erased. Most computer diskettes and audio and video tapes have simple physical write protect mechanisms that usually involve covering up a small hole with a label or a plastic tab. This allows the drive or tape mechanism to recognize that it should not alter the diskette or tape if the user accidentally tries to save something to the diskette (or erase it) or record on the tape. With computer storage media, software write protect methods exist as well.

WRS Worldwide Reference System. A global indexing scheme designed for Landsat.

WS-ANI See Wireless Subscriber-Automatic Number Identification.

WSC See wireless switching center.

WSFOSR See Worldwide Submarine Fiber Optic Systems Report.

WSG Wireless Security Group. See WAP Forum.

WSP Wireless Service Provider.

WST See WAP Security Toolkit.

WTA Wireless Telephony Applications. See WAP Forum.

WTAC See World Telecommunications Advisory Council.

W

WTB Wireless Telecommunications Bureau. An organization of the Federal Communications Commission (FCC).

WTLS See Wireless Transport Layer Security.

WTO World Trade Organisation. An international organization established in 1994. As the Internet has become a significant vehicle for domestic and international commerce, the WTO has become involved in discussions about electronic commerce and in the debate regarding intellectual property and domain name registration. The functioning of the WTO has been scrutinized by critics who are concerned about ensuring that there is adequate accountability built into the function and makeup of the organization.

WTPF See World Telecommunication Policy Forum.

WWV A U.S. Government National Institute of Standards and Technology (NIST) radio facility providing time and frequency standard voice announcements at a variety of high frequency (HF) broadcast wavelengths. WWVH operates 24 hours a day, 7 days a week from a radio station in Colorado.

The time is announced by a male voice based upon a 24-hour clock at the Greenwich meridian.

WWYH broadcasts on 5, 10, and 15 MHz at 10,000 watts and on 2.5 MHz and 20 MHz at 2500 watts using individual transmitters for each frequency. WWV uses five half-wave vertical antennas. Modulation is double-sideband amplitude modulation (AM). See WWVB, WWYH.

WWVB A U.S. Government National Institute of Standards and Technology (NIST) radio facility providing time and frequency standards broadcast at 60 kHz with 50 kW of power from Fort Collins, Colorado. WWVH operates 24 hours a day, 7 days a week. Unlike WWV and WWYH, this is not a voice announcement service, but rather a reference based upon a stable radio frequency transmission that can be traced to the national standard.

At WWVB, a time code is synchronized with the carrier for continuous broadcast at a rate of 1 bps using pulse-width modulation. The time code bits are produced by reducing and restoring carrier power, thus creating a low/high binary system that is used to represent decimal numbers. In this manner, the time code conveys the current year, day, hour, minute, and second along with flags that indicate Daylight Savings Time (DST) and leap year/leap second status. WWVB uses two top-loaded dipole antennas utilizing four towers in a diamond configuration with capacitance cables suspended between the towers.

Commercially available clocks and wristwatches that derive their time from NIST-based WWVB broadcasts can be used within the continental U.S. There are also software clients that enable Internet-connected computer users to synchronize their computer time with NIST time. This type of capability will become increasingly important as auction sites, stock sites, and other types of time-sensitive public network-based electronic commerce services become more widely used. See binary coded decimal, WWV, WWYH.

WWVH A U.S. Government National Institute of Standards and Technology (NIST) radio facility providing time and frequency standard voice announcements at a variety of high frequency (HF) broadcast wavelengths. WWVH operates 24 hours a day, 7 days a week from a radio station in Hawaii. The service includes time announcements, standard time intervals, time frequencies, time corrections, BCD time code, and various weather and Global Positioning System (GPS) reports. The time is kept within 1 microsecond of Coordinated Universal Time (UTC), but will become delayed as it propagates out from the source. In other words, there is lag time (travel time) from the radio station to the receiver that is dependent upon distance and atmospheric conditions. There can be delays up to 10 or 40 milliseconds, depending upon whether the user is listening through a radio receiver or a telephone receiver.

The time is announced by a female voice ("At the tone the time will be ...") based upon a 24-hour clock at the Greenwich meridian. Since 1964, the female voice has been that of Jane Barbe, whose recorded messages are used on millions of telephone and time-related products.

WWYH broadcasts on 5, 10, and 15 MHz at 10,000 watts and on 2.5 MHz at 5000 watts using individual transmitters for each frequency. The 5-, 10-, and 15-MHz frequencies use phased-array vertical dipole antennas (driven 90° out of phase), whereas the 2.5-MHz frequency uses half-wave vertical antennas. Modulation is double-sideband amplitude modulation (AM). See Barbe, Jane; WWV; WWVB.

WWW See World Wide Web.

WWWC See World Wide Web Consortium.

WYGIWYD (*pron.* wiggy-wid) Abbreviation for "what you got is what you deserve." In other words, if you made a bad decision on your own behalf or one you communicated to others to direct their work, you should accept the responsibility and the consequences.

WYPFIWYG (*pron.* wip-fee-wig) Abbreviation for "what you pay for is what you get." The concept applies as much in fiber optics and computer technology as anywhere else. If you buy a cheap system, you may pay more later for adding extra memory, sound cards, graphics cards, hard drive controllers, CD-ROM drives, etc. The same applies to network server software.

WYSIWYG (*pron.* wiz-ee-wig) Abbreviation for "what you see is what you get." A desktop publishing term that refers to a computer display of a document or image that looks on the screen the way it would look on the intended output device, usually a printer. The term became prevalent when desktop publishing programs on the Macintosh could be printed on Adobe PostScript-capable printers, beginning in the mid-1980s. Up to this time, 9-pin dot matrix printers were prevalent and no matter how nice a document looked on the screen, it rarely looked that spiffy on the dot matrix printout.

WZ1 See World Zone.

X, x 1. *symb.* an unknown, situational, derivable, or arbitrary quantity. Commonly used in mathematics and software programming, written both lower- and uppercase, and often italicized as in $4 + x = 10$. The symbol is also often used in product identification to indicate a family of products, e.g., *x*DSL for the family of Digital Subscriber Line services. 2. *abbrev.* cross. 3. *abbrev.* exchange. 4. *abbrev.* external. 5. *abbrev.* trans- (prefix). 5. The USOC Federal Communications Commission (FCC) code for complex multiline or series jack. 6. *colloq.* See X Window System.

X axis, x axis A geometric convention for coordinate systems (e.g., Cartesian coordinates) designate the horizontal axis as the X axis. When graphing processes that may occur over time, the X axis is often used for the time variable.

X Consortium A group that continued the development and management of the X Window System, now part of the Open Group. See X Window System.

X Cut

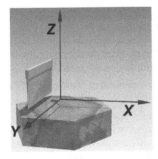

An X cut creates a crystal plate with the plane perpendicular to the crystal's X axis.

X cut A type of cut used with piezoelectric crystals. Crystals are used in radio wave detection and timing applications, and their piezoelectric properties are partly determined by their shape and size. See crystal, quartz, piezoelectric, X-ray goniometer, Y cut, Y bar.

X Protocol A low-level client/server standard communications protocol that handles window manipulation routines for the graphical user interface (GUI) X Window Systems. See X Window System.

X Series Recommendations A set of ITU-T recommended guidelines for interconnecting networks and network devices. These are available for purchase from the ITU-T and a few may be freely downloadable from the Net. This is a large category; some of the X Series Recommendations of particular interest are listed in the following charts, organized into three categories. See Appendix C for a general list of the different categories. See also individual listings under G, I, Q, and V Series Recommendations.

X Window System, The X Window System, X Windows, X, X.11 Hardware-independent foundation software for the development of graphical user interfaces (GUIs) based upon a client/server model. The X Window System is a nonproprietary, distributed, multitasking, network-transparent protocol that has been implemented on many different Unix-based systems. Originally used as a graphics display protocol for text-based UNIX platforms, developers are recognizing and exploiting its ability to enable popular OSs to run on a UNIX workstation or, conversely, to run UNIX applications on popular hardware platforms, and to run applications from within Web browsers. Development tools such as Motif facilitate the quick design of X GUIs.

The X Protocol is an X Windows System client/server protocol and the X server is a client/server process that controls a display device on the system.

X Windows code for noncommercial purposes is freely downloadable from The Open Group Web site. As of version X11R6.4, commercial users must pay a license fee to continue support for development efforts. See Athena project. http://www.opengroup.org/

X Window System 11 Release 6.x (X11R6.x) A substantial initiative by The Open Group to enable the X Window System to be used to create and access interactive World Wide Web applications through the X Window System and a downloadable plugin. Applications linked to the Web using X11R6.x can be found, accessed, and executed with the same Web browsing utilities used to access current static HTML documents. This may become a very significant means of networking through the Internet. See X Window System.

ITU-T X Series Recommendations	Recommendations of Particular Interest

Prevalent Formats

X.25 Definitions of the procedures for exchanging data between user devices (DTEs) and network nodes in a public switched packet data network (PSPDN) in order to provide a common interface across a variety of systems. X.25 is a layered packet transmissions protocol commonly used in wide area networks (WANs). A version of X.25 specifically designed for packet radio has been developed as AX.25.

X.400 An international ISO/ITU-T series of standards for electronic messaging architecture for the exchange of data between computer systems. X.400 was published by the ITU-T in 1984. The standard was jointly rewritten by ISO and ITU-T in 1988.

X.400 does not stipulate the formatting of data. It provides guidelines for internetworking various messaging systems, addressing individual messages, and describing message contents. Within X.400 there are also substandards and recommendations to X.400, some of which are: X.402 describes the overall architecture; X.420 describes email transfer; X.435 defines the electronic movement of Electronic Data Interchange (EDI); and X.440 describes voice messaging. See Electronic Data Interchange.

X.445 Asynchronous Protocol Specification (APS). A commercially promoted multiple media client/server extension of the X.400 standard that facilitates the exchange of digital data over public phone networks rather than X.25 standard leased lines.

X.500 A directory service protocol for building distributed global directories. It was developed in response to a need to design directories that would not experience the same problems and bottlenecks that were developing with many of the large databases being accessed by thousands or millions of users on the Internet. X.500 employs decentralized maintenance, searching capabilities for complex queries, homogenous global namespace, and a structured standards-based information framework.

X.1 *International user classes of service in, and categories of access to, public data networks and integrated services digital networks (ISDNs).*

Includes information on access to leased or switched circuits by data terminal equipment (DTEs) in various modes, access by facsimile terminals, and access to Frame Relay systems.

X.6 *Multicast service definition.*

Service definitions and capabilities of a multicast service providing a common model for the description of service elements. Interface specifications and protocol elements are not specified by X.6.

X.31 *Support of packet mode terminal equipment by an ISDN.*

Service and signaling procedures definitions operated at the S/T-reference point of an ISDN for subscribing packet mode terminal equipment and terminal adapter functionalities to support existing X.25 terminals at the R-reference point of the ISDN.

X.75 *Packet-switched signaling system between public networks providing data transmission services.*

A description of packet-switching signaling systems among public data networks.

X.76 *Network-to-network interface between public data networks providing the Frame Relay data transmission service.*

A description of interface interconnections between Frame Relay networks and public data networks. Layer, data transfer, and signaling information are provided.

X.77 *Internetworking between PSPDNs via B-ISDN.*

Definitions of procedures for internetworking that include reference configurations, protocol stacks, and signaling procedures.

X.121 *International number plan for public data networks.*

A description of the design, characteristics, and applications of the numbering plan for public data networks. The International Number Plan was developed to facilitate the linking of public data networks with the worldwide system. It describes country identification (assigned by the ITU-T), regional/local network identification, and a mechanism for integrating with other numbering plans. Guidance for efficient allocation of numbers is included, along with eligibility criteria and procedures. See Data Network Identification Codes.

X.122 *Numbering plan interworking for E.164 and X.121 number plans.*

Information on interworking with other numbering plans.

Brief Listing of Further X Series Examples

X.2 International data transmission services and optional user facilities in public data networks and ISDNs. X.3 defines a set of parameters used for regulating basic functions such as terminal characteristics, flow control, and data forwarding.

X.3 Packet assembly/disassembly facility (PAD) in a public data network.

X.4 General structure of signals of international alphabet No.5 code for character-oriented data transmission over public data networks.

X.5 Facsimile Packet Assembly/ Disassembly facility (FPAD) in a public data network.

X.7 Technical characteristics of data transmission services.

X.8 Multiaspect pad (MAP) framework and service definition.

X.10 Categories of access for data terminal equipment (DTE) to public data transmission services.

X.20 Interface between data terminal equipment (DTE) and data circuit-terminating equipment (DCE) for start-stop transmission services on public data networks.

X.20 bis Use on public data networks of data terminal equipment (DTE) designed for interfacing to asynchronous duplex V Series modems.

X.21 Interface between data terminal equipment and data circuit-terminating equipment for synchronous operation on public data networks. A digital signaling interface that includes specifications for physical interface elements, character alignment, and data transfer.

X.21 bis Use on public data networks of data terminal equipment (DTE) designed for interfacing to synchronous V-series modems.

X.22 Multiplex DTE/DCE interface for user classes 3-6.

X.24 List of definitions for interchange circuits between data terminal equipment (DTE) and data circuit-terminating equipment (DCE) on public data networks.

X.25 Interface between Data Terminal Equipment (DTE) and Data Circuit-terminating Equipment (DCE) for terminals operating in the packet mode and connected to public data networks by a dedicated circuit.

X.28 DTE/DCE interface for a start-stop mode data terminal equipment accessing the packet assembly/ disassembly facility (PAD) in a public data network situated in the same country.

X.29 Procedures for the exchange of control information and user data between a packet assembly/ disassembly (PAD) facility and a packet mode DTE or another PAD.

X.30 Support of X.21-, X.21 bis-, and X.20 bis-based data terminal equipment (DTEs) by an integrated services digital network (ISDN).

X.32 Interface between Data Terminal Equipment (DTE) and Data Circuit-terminating Equipment (DCE) for terminals operating in the packet mode and accessing a packet-switched public data network through a public network.

X.33 Access to packet-switched data transmission services via Frame Relaying data transmission services.

X.34 Access to packet-switched data transmission services via B-ISDN.

X.35 Interface between a PSPDN and a private PSDN based upon X.25 procedures and enhancements to define a gateway function provided in the PSPDN.

X.36 Interface between data terminal equipment (DTE) and data circuit-terminating equipment (DCE) for public data networks providing Frame Relay data transmission service by a dedicated circuit.

X.37 Encapsulating X.25 packets of various protocols including Frame Relay.

X

X Series Recommendations, cont.

Brief Listing of Further Examples, cont.

X.38	G3 facsimile equipment/DCE interface for G3 facsimile equipment accessing the Facsimile Packet Assembly/Disassembly facility (FPAD) in a public data network situated in the same country.
X.39	Procedures for the exchange of control information and user data between a Facsimile Packet Assembly/Disassembly (FPAD) facility and a packet mode Data Terminal Equipment (DTE) or another FPAD.
X.42	Procedures and methods for accessing a Public Data Network from a DTE operating under control of a generalized polling protocol.
X.45	Interface between Data Terminal Equipment (DTE) and Data Circuit-terminating Equipment (DCE) for terminals operating in the packet mode and connected to public data networks, designed for efficiency at higher speeds.
X.48	Procedures for the provision of a basic multicast service for Data Terminal Equipment (DTEs) using 25.
X.49	Procedures for the provision of an extended multicast service for Data Terminal Equipment (DTEs) using 25.
X.50	Fundamental parameters of a multiplexing scheme for the international interface between synchronous data networks.
X.50 bis	Fundamental parameters of a 48-kbps user data signaling rate transmission scheme for the international interface between synchronous data networks.
X.51	Fundamental parameters of a multiplexing scheme for the international interface between synchronous data networks using a 10-bit envelope structure.
X.51 bis	Fundamental parameters of a 48-kbps user data signaling rate transmission scheme for the international interface between synchronous data networks using a 10-bit envelope structure.
X.52	Method of encoding anisochronous signals into a synchronous user bearer.
X.53	Numbering of channels on international multiplex links at 64 kbps.
X.54	Allocation of channels on international multiplex links at 64 kbps.
X.55	Interface between synchronous data networks using a 6 + 2 envelope structure and single channel per carrier (SCPC) satellite channels.
X.56	Interface between synchronous data networks using an 8 + 2 envelope structure and single channel per carrier (SCPC) satellite channels.
X.57	Method of transmitting a single lower-speed data channel on a 64 kbps data stream.
X.58	Fundamental parameters of a multiplexing scheme for the international interface between synchronous nonswitched data networks using no envelope structure.
X.60	Common channel signaling for circuit switched data applications.
X.61	Signaling System No. 7 – data user part.
X.70	Terminal and transit control signaling system for start-stop services on international circuits between anisochronous data networks.
X.71	Decentralized terminal and transit control signaling system on international circuits between synchronous data networks.
X.80	Interworking of interexchange signaling systems for circuit-switched data services.
X.81	Interworking between an ISDN circuit-switched and a circuit-switched public data network (CSPDN).
X.82	Detailed arrangements for interworking between CSPDNs and PSPDNs based on Recommendation T.70.
X.92	Hypothetical reference connections for public synchronous data networks.
X.96	Call progress signals in public data networks.
X.110	International routing principles and routing plan for public data networks.
X.115	Definition of address translation capability in public data networks.
X.116	Address translation registration and resolution protocol.
X.123	Mapping between escape codes and TOA/NPI for E.164/X.121 numbering plan interworking during the transition period.
X.130	Call processing delays in public data networks when providing international synchronous circuit-switched data services.
X.131	Call blocking in public data networks when providing international synchronous circuit-switched data services.
X.134	Portion boundaries and packet-layer reference events: Basis for defining packet-switched performance parameters.

	Brief Listing of Further Examples, cont.
X.135	Speed of service (delay and throughput) performance values for public data networks when providing international packet-switched services.
X.136	Accuracy and dependability performance values for public data networks when providing international packet-switched services.
X.137	Availability performance values for public data networks when providing international packet-switched services.
X.138	Measurement of performance values for public data networks when providing international packet-switched services.
X.139	Echo, drop, generator, and test DTEs for measurement of performance values in public data networks when providing international packet-switched services.
X.140	General quality of service parameters for communication via public data networks.
X.141	General principles for the detection and correction of errors in public data networks.
X.144	User information transfer performance parameters for data networks providing international Frame Relay PVC service.
X.145	Performance for data networks providing international Frame Relay SVC service.
X.150	Principles of maintenance testing for public data networks using data terminal equipment (DTE) and data circuit-terminating equipment (DCE) test loops.
X.160	Architecture for customer network management service for public data networks.
X.161	Definition of customer network management services for public data networks.
X.162	Definition of management information for customer network management service for public data networks to be used with the CNMc interface.
X.163	Definition of management information for customer network management service for public data networks to be used with the CNMe interface.
X.180	Administrative arrangements for international closed user groups (CUGs).
X.181	Administrative arrangements for the provision of international permanent virtual circuits (PVCs).

X Window System history X was originally developed by Robert Scheifler and Ron Newman from the Massachusetts Institute of Technology (MIT) and Jim Gettys of Digital Equipment Corporation (DEC) to provide a user interface for the Athena Project. It has been further developed by The X Consortium and is now trademarked and managed by The Open Group. See Athena project, X Window System.

X9 standards An important set of standards defined by the American National Standards Institute (ANSI) for the financial industry that has many ramifications for secured transactions over computer networks. The X9 standards cover such aspects as personal identification number (PIN) management, electronic transfer of funds, transaction processing, security mechanisms. A brief summary of some of the X9 security-related standards includes

Title	Description
X9.9	U.S. wholesale banking standard for transaction authentication.
X9.17	Financial Institution Key Management standard for wholesale transactions. Defines protocols for the transfer of encryption keys using symmetric techniques.
X9.30	U.S. standards for digital signatures based upon the Digital Signature Algorithm (DSA) using the SHA-1 hash algorithm.
X9.31	U.S. standards for digital signatures based upon the RSA algorithm based upon the MDC-2 hash algorithm.
X9.57	Certificate management encryption schemes.
X9.42	A draft standard for key agreement based upon the Diffie-Hellman algorithm.
X9.44	A draft standard for key transport based upon the RSA algorithm.

See RSA Security Inc.

X.25 A widely implemented, significant, connection-oriented, packet-based communications protocol used in local and wide area networks. The protocol was developed in the mid-1970s, when analog networking over noisy copper connections was optimized for voice rather than data communications. It helped to fulfill a growing need for a common language to interconnect local area networks that often used proprietary network protocols and for error mechanisms that could overcome the problems associated with marginal connections. In 1976, the CCITT (now the ITU) recommended X.25 for international data exchange. It was approved and subsequently revised every 4 years or so.

In general, X.25 is a three-layer model that includes a physical level, a link level, and a packet level that are, in turn, associated with the lower physical, data,

X

and network layers of the hierarchical Open Systems Interconnect (OSI) model.

Basic Overview of X.25 Levels

Name	Level #	Description
Physical	1	The electromechanical, procedural, and functional interfaces between the data terminal equipment (DTE) and the data circuit-terminating equipment (DCE). This level is specified by ITU-T X and V Series Recommendations that apply to modems and interchange circuits (e.g., X.21).
Link	2	Also called the frame level, the link level ensures reliable data transfer between the DTE and the DCE in an efficient, timely manner. It synchronizes transmitter/receiver interactions, and detects and handles errors. A number of link access protocols may be implemented at the link level.
Packet	3	Also called the network level, the packet level creates the network data "chunks" with appropriate control/error information and user data payload. Various virtual circuit (VC) and datagram services are handled at this level.

Data is formatted into X-25 data packets by "packetizing" the data into smaller chunks and adding the appropriate protocol information into each packet header. Error mechanisms are also included to enable retransmission if packets are lost.

X-10 Protocol A protocol for sending radio frequencies (RF) signals over power line carriers (PLCs). In other words, a radio frequency-emitting device can be plugged into a wall socket and the protocol will enable device control signals to be sent through the wiring (over the alternating current power curve) to compatible devices set to the same settings as the controlling devices. Lately, wireless X-10-compatible transmitters are being added to commercial catalogs, with transmission ranges of up to about 100 feet. See CEBus, LONWORKS.

X-band An assigned spectrum in the microwave frequencies of approximately 8 to 12 GHz with wavelengths of just a few centimeters. X-band signals are not significantly hindered by the Earth's atmosphere unless a lot of moisture (e.g., rain) is present. X-band frequencies are used by military satellites and deep-space vehicles. See band allocations for a chart of assigned frequencies.

X-Bone A system designed to facilitate and automate the rapid deployment and management of multiple overlay networks. X-Bone is an overlay technology combined with teleconferencing-style coordination and management tools. X-Bone provides a virtual networking infrastructure that is configurable. While X-Bone is intended to be implemented with networks running more advanced systems, such as IPv6, some of the automatic tunneling services can be deployed to a limited extent on IPv4 systems. See 6bone, Mbone, overlay network, X-Bone xd.

X-Bone xd An X-Bone directory tool for performing a number of tasks including the coordination of resource sharing at the local site, the support of local daemons through authentication, configuration, and creation of IP-encapsulation tunnels between daemons, and the provision of a user interface and API for users or programs wishing to manually parameterize and override overlays. See X-Bone.

X-dimension of recorded spot In facsimile transmissions, a means of describing variation density in terms of the minimum density. The largest center-to-center space between recorded spots is measured in the direction of the recorded line. This can also be assessed perpendicular to the recorded line as the Y-dimension of a recorded spot. The same principles can be applied to assess the scanning spot.

X-ray A radiant energy within the spectrum of high energy, invisible, ionizing electromagnetic radiation that ranges about 0.08 nm in wavelength, between ultraviolet light and gamma rays. X-rays were somewhat naively and irresponsibly used in early radio signal and human anatomy experiments. These practices are now used with great care due to the damaging influence of X-rays on living cells. X-rays are used in many medical, industrial, and fabrication applications. See X-ray goniometer; Roentgen, Wilhelm Konrad; scintillator.

X-ray goniometer An instrument for determining the position of the axes in a quartz crystal. X-rays are aimed at the atomic planes of the crystal and the reflected rays are evaluated. Since crystals are physically manipulated to alter their oscillating properties and often cut in very thin slices, it is important to know the orientation of the crystalline structure before cutting. See quartz, X cut, Y cut.

X-ray spectrometer An instrument that is used, by means of reflected rays and evaluation of the resulting diffraction angles, to study the characteristics and composition of materials, including crystals. See X-ray goniometer.

X/Open A global, independent organization of computer manufacturers founded in 1984. X/Open seeks to promote an open, multivendor Common Applications Environment (CAE) to enhance application portability. This is a good concept, as it allows software developed by different vendors to run on a variety of platforms, leaving the choice of equipment up to the individual purchaser. See Common Applications Environment, Open Systems Interconnection.

X/Open Federated Naming A naming mechanism from the X/Open group for developers to access network naming services and to provide integration with industry-accepted naming services such as X.500, Domain Name Service (DNS), DCE, and others.

X/Open Portability Guide XPG. A guide documenting the X/Open common applications environment system.

XA extended architecture.

XA-SMDS See Exchange Access SMDS.

XAPIA X.400 Application Program Interface Association. See X Series Recommendations.

Xaw The Athena Widget set. A set of widgets distributed with the X Window System, which began as Project Athena. See Athena, X Window System.

Xbar *abbrev.* crossbar. See utility pole.

XBase, Xbase A generic designation for applications that read and/or write dBase-compatible files.

XC *abbrev.* cross connect.

XCA extended communication adapter.

Xcoral A multiwindow text editor for the X Window System, that can be used in conjunction with a mouse.

XCVR *abbrev.* transceiver.

xd See X-Bone xd.

XDL An object-oriented extension to the ITU-T-defined Specification and Design Language (SDL) for telecommunications systems. See Specification and Design Language.

XDMA Xing Distributed Media Architecture. A commercial streaming media architecture for delivery of live and on-demand audio-video from Xing Technology Corporation. It is built around standards such as TCP/IP and MPEG, and supports multicasting to multiple simultaneous users over local area networks (LANs) and wide area networks (WANs). XDMA can be implemented over ISDN networks for services such as news and distance learning.

XDMCP X Display Manager Control Protocol. A protocol used to communicate between X terminals and UNIX workstations.

XDR See External Data Representation.

xDSL Generic term for a variety of digital subscriber line technologies, which include ADSL, EDSL, and HDSL. See digital subscriber line and individual listings for further information.

XENIX A Unix implementation best known as being from the Santa Cruz Operation, Inc. (SCO), it was originally codeveloped by International Business Machines (IBM) and Microsoft as XENIX-11 for Intel machines. SCO is now marketing UnixWare 7.

xerographic printer A printer that uses the same basic electrostatic mechanisms and techniques as a xerographic copier. The information is imaged onto a drum with lasers, the printing medium is passed across the drum and picks up the dry transfer toner, which then fuses the toner to the printing medium.

Xerox Corporation One of the first companies to see the commercial benefits of new photocopying technology developed by Carlson in the early 1940s. When still a relatively new, small company, Xerox took a chance on the new photocopying invention that had been passed up by other companies. Xerox is now known throughout the world for its technology, especially in the replication industry, and many people refer to all photocopies as "xeroxes." See photocopy for further information and an illustration of Carlson's patent.

Xerox Network Services XNS. A multilayer, distributed file network architecture developed by the Xerox Corporation which is somewhat similar to TCP/IP. Unlike many networks from other vendors, XNS permits a user to use files and devices from a remote machine as if they were on a local machine. XNS functions compatibly with the third and fourth layers of the Open Systems Interconnection model (OSI).

Xerox PARC Xerox Corporation's Palo Alto Research Center. This research center provided enormous impetus to early computer companies and software developers (e.g., Apple Computer Inc., Microsoft Incorporated), especially those developing object-oriented systems and graphical user interfaces (GUIs). PARC researchers invented mice and various laser printing technologies, developed Smalltalk; they generated one good idea after another throughout the 1970s and early 1980s, yet surprisingly few of these were commercially implemented or marketed through Xerox. Charles Simonyi, one of the early founding members of PARC, was the demonstrator of the Alto, a pioneer desktop computer that inspired many of those fortunate enough to see it in the early days. Later, Simonyi was hired by Microsoft to move the company into graphical applications.

XFN See X/Open Federated Naming.

XFR *abbrev.* transfer.

XGA 1. See extended graphics adapter. 2. See extended graphics array.

Xi'an The capital of the Shaanxi province in northwest China; firms in this region are major suppliers of broadband and optical networking products, including ADSL and fiber optic cables and components.

XID exchange identification. In data networking, XIDs are request and response packets exchanged prior to communications between a router and a network host. XID is used for device discovery, address conflict, resolution, and sniffing. The XID packet includes the parameters of the serial device, and a connection can only be negotiated if this configuration is recognized by the host.

XIP *abbrev.* execute in place. A means to access memory and execute code on PCMCIA cards without having to load them into system memory first. See PCMCIA.

XIWT See Cross-Industry Working Team.

Xlib X Library. A program interface for the X Windows System.

Xmission, Xmit *abbrev.* transmission, transmit.

XML See Extensible Markup Language.

XModem A widely used error-correcting network file transfer data transmission protocol developed by Ward Christensen in the late 1970s. XModem utilizes 128-byte packets, so files of various lengths will be padded to adjust the packet length and may be longer than the original file. The filename is not sent with the transmission. YModem, a successor to XModem, with support for longer data packets and file attributes, was developed by Chuck Forsberg.

XModem is often used with computer modems to transfer files to and from bulletin board systems over traditional phone lines. XModem is not fast, but it has error correction and it's reasonably reliable. It's well supported, an important consideration since both ends of the connection have to agree on a protocol. Many service bureaus use XModem for file transfers, however, check if they have ZModem, which is faster and capable of restarting an interrupted transmission from where it left off. See Kermit, YModem, ZModem.

XModem-1K, XModem-K A variant of XModem that manages data in 1K (1024-byte) packets. See XModem.

XModem-CRC XModem with 16-bit cyclic redundancy checking (CRC) error detection mechanisms, instead of checksum. XModem-CRC can communicate with XModem versions that use checksum for error correction.

XMP X/Open Management Protocol.

XMS See Extended Memory Specification.

XMT *abbrev.* transmit.

XMTR *abbrev.* transmitter.

XNMS MICOM's commercial IBM-licensed Intel-based desktop computer packet data network (PDN) network management system software.

XNS See Xerox Network Services.

XO *abbrev.* crystal oscillator. See crystal detector, quartz.

XON/XOFF transmission on/transmission off. Common flow control signals used between two communicating devices or software programs, typically through modems. Since many transmissions media are inherently slow, there may be a delay between receiving a block of data and resuming transmissions. XON/XOFF signals allow the communicators to signal when to stop sending data and when to resume, in order to prevent loss or corruption.

XON/XOFF is also known as software flow control. In newer high-speed modems, flow control may be handled by hardware, often in conjunction with specific types of cables.

Flow control signals are not limited to modem communications. If a user is working on a terminal that understands XON/XOFF commands, usually signified by Ctrl-S (stop) and Ctrl-Q (resume), then it is possible to suspend a listing or other activity and resume when it is convenient.

Xover *abbrev.* crossover.

XPAD external packet assembler/disassembler.

XPG See X/Open Portability Guide.

Xponder transponder.

Xpress Transport Protocol XTP. A flexible, high-performance commercial multicast protocol to support a wide variety of applications from wide area networks (WANs) to multimedia and realtime embedded systems. It was developed by an international group of representatives from academia, industry, and the government from 1987 to the early 1990s and was described by Strayer et al. in the early 1990s. It has been adopted as part of MIL-STD-2204 (Survivable Adaptable Fiber Optic Embedded Network).

XTP brings together the functionality of UDP, TCP, and TP4, in addition to transport multicast, group management, Quality of Service (QoS) negotiation

capabilities, rate and burst control, and error and flow control mechanisms. It is designed to operate over any network or datalink layer or may be implemented directly over an ATM AAL. It can run in parallel with other transport protocols and thus can increase functionality without giving up interoperability. It is expressly intended to separate the communication paradigm from error control policies.

The protocol sparked interest in the development of a public domain version at Sandia National Laboratories. This protocol is distinct from the lightweight eXpress Transfer Protocol originally developed by Protocol Engines, Inc. See multicast, XTP Forum.

XRB transmit reference burst.

xref *abbrev.* cross reference.

XRF Extended Recovery Facility.

XSI X/Open System Interface Specification.

XSMP X Session Manager Protocol.

XT 1. *abbrev.* crosstalk. 2. IBM Personal Computer XT. See IBM.

Xtalk *abbrev.* crosstalk.

xterm A popular terminal emulator for the X Window System that has been ported to several other operating systems. Xterm lets you have more than one terminal window active at a time through a single modem, each with its own input/output process running independently of the others.

XTP 1. See eXpress Transfer Protocol. 2. See Xpress Transport Protocol.

XTP Forum A nonprofit group promoting the development and distribution of Xpress Transport Protocol (XTP). A number of significant vendors, including Apple Computer, Hughes Aircraft, Lockheed Martin, Northrop-Grumman, Philips Research, and Silicon Graphics are involved in this effort, in collaboration with XTP Forum Research Affiliates from universities and military labs around the world.

XWA The callsign issued to a historically significant Canadian radio station, it stood for X-perimental Wireless Apparatus. XWA was issued its first experimental license by the Department of Naval Service, in 1918, following about 3 years of discussion (and presumably some preliminary experiments).

The first general broadcast licence was issued in September 1919 to XWA as part of the Marconi Wireless Telegraph Company. Broadcasts apparently began in December 1919. The pioneer broadcasts were primarily musical and, since few consumer radios existed at the time, the listeners were amateur radio enthusiasts. In May 1920, XWA aired its first regularly scheduled broadcast and the radio industry evolved quickly, with XWA becoming station CFCF in November 1920. See CFCF, radio history.

XWindows, XWS See X Window System.

XXXX A designation for the last four digits in a telephone number, usually used when the numbers are not yet known (or assigned). It represents any number between 0 and 9.

XY cut A means of angle-cutting a piezoelectric crystal such as LGS so its electrical characteristics are between those of an X cut and a Y cut. See quartz, X-ray goniometer, X cut, Y cut.

y *symb.* yocto-. See yocto-.

Y 1. *symb.* admittance. The ease with which alternating current (AC) flows through a circuit, as opposed to impedance. See impedance. 2. Symbol for yttrium. See yttrium. 3. *symb.* yotta-. See yotta-. 4. A general purpose programming language distributed from the University of Arizona in the early 1980s that is semantically similar to C, but without C pointers and structures.

Y antenna A single-wire antenna with leads connected in a Y shape, with the top part of the Y corresponding to the transmission line. Since the top of the Y is closed, causing it to resemble the Greek "D" (*delta*), the Y antenna is sometimes also known as a *delta matched* antenna. This style of antenna is commonly used for very high frequency (VHF) and frequency modulated (FM) signals.

Y axis, y axis A reference baseline or vector within a coordinate system, most often associated with rectangular or Cartesian coordinates. The Y axis is oriented vertically by convention, perpendicular to a horizontal X axis in a two-dimensional system, and perpendicular to the Z and X axes in a three-dimensional system. See Cartesian coordinates, X axis, Z axis.

Y bar A type of cut used with piezoelectric crystals in which the plane of the long direction is parallel to the crystal's Y axis. See Y cut.

Y cable, Y connector A cable or cable connector that splits from a single line or bundle into two usually equivalent lines or bundles. Sometimes called a Y splitter. Y cables are frequently used in audio applications to split a mono signal into two jacks (not the same as real stereo) to connect systems with different inputs and outputs, or to combine a stereo signal onto one jack. Y cables are also used to split power sources, as when adding an extra drive to a computer system. Depending upon the application, the Y cable may or may not cause a degradation of the transmission once the signal is split. A Y cable may or may not be combined with other connectors or converters. A Y cable is usually functionally the same as a T cable, except that the "Y" angle of the split is narrower than 180 degrees. See converter.

Y cut A type of cut used with piezoelectric crystals. Crystals are used in radio wave detection and timing applications, and their piezoelectric properties are partly determined by their shape and size. See crystal, detector, quartz, piezoelectric, X cut, X-ray goniometer, Y bar.

Y Cut

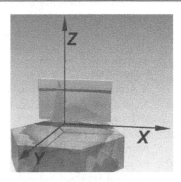

A Y cut creates a crystal plate with the plane perpendicular to the crystal's Y axis.

Y Series Recommendations A series of ITU-T recommended guidelines for global information insfrastructure and Internet protocol aspects. These guidelines are available as publications from the ITU-T for purchase on the Net. Since ITU-T specifications and recommendations are widely followed by vendors in the telecommunications industry, those wanting to maximize interoperability with other systems need to be aware of the information disseminated by the ITU-T. A full list of general categories is listed in Appendix C and specific series topics are listed under individual entries in this dictionary, e.g., G Series Recommendations. See Y Series Recommendations chart.

Y signal A monochrome (it's actually more descriptive to say grayscale) signal luminance transmission. When combined with a color signal, luminance provides brightness to the image. The relative absence of a luminance signal is used to represent black, while the highest level of power applied to the luminance signal is used to represent white, with shades of gray in between. See Y/C.

ITU-T Y Series Recommendations	
Recom.	**Description**
Y.100	General overview of the Global Information Infrastructure standards development
Y.101	GII Terminology - Terms and definitions
Y.110	Global Information Infrastructure principles and framework architecture
Y.120	Global Information Infrastructure scenario methodology
Y.130	Information communication architecture
Y.140	Global Information Infrastructure (GII): Reference points for interconnection framework
Y.801	Relationships among ISDN, Internet protocol, and GII performance recommendations
Y.1001	IP Framework - A framework for convergence of telecommunications network and IP network technologies
Y.1231	IP Access Network Architecture
Y.1241	Support of IP based Services Using IP Transfer Capabilities
Y.1301	Framework of Optical Transport Network Recommendations
Y.1302	Requirements for Automatic Switched Transport Networks (ASTN)
Y.1310	Transport of IP over ATM in public networks
Y.1311.1	Network-based IP VPN over MPLS architecture
Y.1321	IP over SDH using LAPS
Y.1322	Network node interface for the Synchronous Digital Hierarchy (SDH)
Y.1331	Interfaces for the optical transport network (OTN)
Y.1401	General requirements for interworking with Internet protocol (IP)-based networks
Y.1402	General arrangements for interworking between Public Data Networks and the Internet
Y.1501	Relationships among ISDN, Internet protocol, and GII performance recommendations
Y.1710	Requirements for OAM functionality for MPLS networks

Y-dimension of recorded spot In facsimile transmissions, the X-dimension of recorded spot is a means of describing variation density in terms of the minimum density. The largest center-to-center space between recorded spots is measured in the direction of the recorded line. When it is assessed perpendicular to the recorded line, it is the Y-dimension of recorded spot. The same principles can be applied to assess the scanning spot.

Y/C A color image information encoding scheme in which the chroma (color) signals (C) are separated from the luminance (brightness) signals (Y). A Y/C splitter cable enables the two signals to be handled separately.

Yablonovitch, Eli (ca. 1947–) A Bell research scientist and physics professor at Harvard and UCLA who pioneered concepts in photonic crystals, structures that have significant potential for selectively controlling the behavior and wavelength filtering of light which, in turn, is valuable in the development of dense wavelength fiber-based transmission systems and superfast computer systems. Yablonovitch described the structures in "Inhibited spontaneous emission in solid-state physics and electronics," in 1987, and by the early 1990s, had succeeded in producing a photonic crystal in a material with a relatively high refractive index (now called Yablonovite). In the mid-1990s, he collaborated with Sievenpiper et al. in devising means to adapt the technology to longer wave lengths using 3D circuit arrangements. Yablonovitch has been awarded numerous honors for his work with photonic crystals, including the Wood Prize of the Optical Society of America. See Kawakami, Sujiro; photonic crystal.

YAG See yttrium-aluminum-garnet.

Yagi, Hidetsugu (1886–1976) A Japanese researcher, lab director, and educator who worked with his lab engineer, Shintaro Uda, to develop and describe a new, more sensitive directional antenna structure. See Yagi-Uda antenna.

Yagi-Uda antenna A narrow bandwidth, linear, directional antenna array that resembles a driven dipole antenna with branches (passive directors), usually along one plane. The Yagi-Uda antenna improves gain with reflectors and directors (branches extending out from the main rods) and works in high frequency ranges (radar, television, etc.). Rooftop television aerials commonly use Yagi-Uda antennas.

This important antenna was designed and built ca. 1924-1926 by Shintaro Uda and possibly also by Hidetsugu Yagi, Uda's lab director at Tohoku University. The relative contributions of the two engineers are not clear. It was first described in a paper by Yagi and Uda in 1926, but did not come to international attention until it was described again by Yagi in an English language Institute of Radio Engineers (IRE) paper in 1928. Following this wider dissemination of information, many called it the Yagi antenna and, ironically, it came to be more widely used in Japan than abroad. Uda continued to refine the design and adapted it for television reception in the mid-1950s. An original Yagi-Uda antenna is housed in the

Japan Broadcasting Corporation's Broadcast Museum. In 1995, the antenna was awarded an IEEE Electrical Engineering Milestone.

Yahoo A significant, high-profile, extensive search and information site originally developed at Stanford University and established on the World Wide Web in the mid-1990s. Yahoo provides general search capabilities as well as a large database of sites organized according to popular categories and topics. Yahoo also sponsors Yahoo Groups (formerly E-Groups), discussion lists. http://www.yahoo.com/

Year 2000, Y2K A designation for the electronic changeover to the new century, a circumstance that was not anticipated and accounted for by all programmers when designing hardware, operating systems, spreadsheets, databases, backup software, schedulers, and the like. Many software programs and hardware clocks accommodated only up to the year 1999 and were not capable of rolling over to 2000. A significant number took into consideration only the last two or three digits of the year, and thus were unable to resolve a "00" that followed a "99," for example, causing potential file and data problems for backup systems as well as various application programs.

yellow alarm See yellow signal.

Yellow Book CD-ROM A CD-ROM authored and written according to *Yellow Book standards* in ISO 9660 format, a format very common in computer multimedia applications.

Yellow Book 1. Standard for the physical format of a CD-ROM or audio CD disk (as opposed to the logical format). DVD drives are able to read CD-ROMs created according to Yellow Book standards. The Yellow Book followed from the original Red Book digital audio (CD-DA) standards. Yellow Book made it possible to store computer data in addition to audio data and supported up to about 650 MBytes of data. There are two subcategories, Mode 1 and Mode 2, for data with and without logical error correction (LECC). Mode 2 can support both Mode 2 and Mode 1 on a disc. See CD-ROM, ISO 9660. 2. Standards for audits of U.S. government organizations, programs, and functions, and of government assistance paid to nonprofit organizations, and contractors.

Yellow Book, Jargon This is a common name for the illustrated printed publication *The New Hacker's Dictionary,* which is descended from the infamous Jargon File. See Jargon File.

Yellow Box An Apple Computer Inc. designation for an object-oriented, OS-independent developer platform that was initiated after NeXT was folded into Apple Computer. See Blue Box, Red Box, Rhapsody.

yellow pages 1. *colloq.* A common name for a number of business/advertising sections in various phone directories printed on yellow paper to distinguish the section from residential and government listings. Many online electronic business directories are colloquially called the yellow pages, although British Telecom has a trademark on the name. 2. A Sun Microsystems client/server protocol for system configuration data distribution now known as Network Information Service (NIS). See Network Information Service.

Yellow Pages A trademark of British Telecommunications.

yellow signal In telecommunications, an alarm condition, warning, or failure signal, usually in a network. The signal is often a yellow-colored light, though it may be an auditory or textual signal or a specific data value or sequence. On gauges, there may be a yellow band indicating a warning level, with a red band for more serious problems, such as danger levels or stop indicators.

Color-coded signals are industry and even product specific but, in general, red signals are used for more serious conditions and yellow signals for localized failures or less serious warnings (conventions similar to those used in railroad and traffic signals). Warnings may even be associated with a specific segment of a network (e.g., fiber optic transmission link).

In ATM networks alarm interfaces indicate whether alarm conditions are present. Data values are used to indicate a no alarm condition or that an incoming yellow signal, for example, has been received. Loss of frame failures are commonly signaled by alarms. Detection of a yellow or red signal alarm in a data network may trigger other signals, such as *unavailable signal states,* and timers to log the duration of an unavailable signal event.

In SONET networks, a yellow signal may remotely signal a failure or help in trunk conditioning. A far end receive failure is a more specific instance of a failure detection signal in the downstream direction. In DS3 network interfaces, a yellow signal may trigger a yellow alarm, described in ANSI T1.107-1989. A loss of signal, loss of frame, or alarm indication signal (AIS) may trigger a red alarm, which is cleared if there are no severely errored seconds (SESs). Commonly, in data networks, alarms are declared after a certain number of consecutive seconds have elapsed (e.g., 2 seconds) and cleared after a certain number of seconds have elapsed (e.g., 10 seconds).

yellow wire 1. A color designation used by IBM to indicate wires used to re-establish a broken connection in traces or flat cables (ribbon cables). See blue wire, purple wire, red wire. 2. A color commonly used for *ring* on the second phone in four-wire phone installations (two wires for each phone). The corresponding *tip* wire is usually black.

Yes A Novell product-compatibility certification program.

Yes, yes key A pushbutton shortcut key on some appliances or keyboards (e.g., teletype-style) to provide an affirmative response without typing "yes."

YIG See yttrium-iron-garnet.

YIG filter A wide bandwidth filter in which a YIG crystal is positioned within the field of a permanent magnet associated with a solenoid and tuned to the center of the frequency band.

YIQ A color model originating from hardware characteristics that is used in color television transmission and for some computer monitors. Y is luminosity; I and Q provide chroma signals separate from the luminosity in order to provide backward compatibility

Y

with black and white standards. On black and white (grayscale) televisions, only the Y component of the signal is displayed. See Y signal.

YModem A data transfer protocol commonly used with modems developed by Chuck Forsberg as a successor to XModem. Typically faster than XModem or Kermit, though not as well supported on BBSs and at services bureaus as XModem and ZModem. YModem comes in two flavors, batch and nonstop (YModem-G). The batch version allows multiple files to be sent in a transmission session and wildcard characters for filenames are supported. When errors in transmission are frequent, YModem is able to fall back automatically to smaller packets, thus accommodating a poor line connection, but is not compatible with the error correction of XModem. See XModem, ZModem.

YModem-G A nonstop streaming variation of the batch-oriented YModem network communications software developed by Chuck Forsberg. YModem-G does not wait for receiver acknowledgments and does not have built-in error correction, instead it relies on an error-correcting modem to supply the error logic. It's appropriate in situations where a good clean connection is available and speed is desired. If a problem with a bad block is encountered, the entire transfer is aborted. Unlike ZModem, it cannot resume a new transmission at the point at which errors and termination of the initial session occurred. See YModem, ZModem.

yocto- y. A prefix for an SI unit quantity of 10^{-24}, or .000 000 000 000 000 000 000 001. It's a mind-bogglingly small quantity. See yotta-, zepto-

yoke 1. A clamp or frame that unites or holds two parts or assemblies firmly together. 2. A coil assembly installed over the neck of a cathode-ray tube (CRT) to deflect the electron beam as currents pass through it. 3. A ferromagnetic assembly, without windings, that forms a permanent connection between two magnetic cores.

yotta- Y. A prefix for an SI unit quantity of 10^{24}, or 1,000,000,000,000,000,000,000,000 or 1000^8. It's a mind-bogglingly large quantity. See yocto-, zetta-.

Young, Thomas (1773–1829) A precocious English scientist and physician who was accepted into the Royal Society when just barely in his 20s. Young became the foreign secretary to the Society in 1802 and retained the position for the rest of his life. In 1800, he published *Experiments on Sound and Light* and described his influential research into the theory of optical interference. Young lived in a period of history when the wave and particle nature of light were only barely understood and still hotly debated. Fresnel was the first to really make use of the many important discoveries of Young. See interference, Newton's rings.

YSZ yttrium-stabilized zirconia. A material that is usually distributed in powder form for manufacture into a range of products that are corrosion-resistant and readily shaped with molds. It sells for about $400 – $700 per 100 grams of powder or suspension. The material is used in the production of oxygen sensors

and oxygen generation systems. See yttrium, zirconia.

yttrium A metallic element similar to the rare-earth metals with which it is typically found. Yttrium oxide is used with Europium to make red phosphors for cathode-ray tubes. See erbium, europium, gadolinium.

yttrium aluminate A substance with a low dielectric constant that is suitable for high-frequency applications.

yttrium-aluminum-garnet YAG. A crystalline substance used in laser electronics, YAG may be doped or undoped. YAG can be shaped into fibers, rods, single crystals and polycrystalline forms. It is useful in precision lasers, filter and bandgap components, and for reinforcing fiber in composite materials. When YAG is doped with cerium, it can be used to produce scintillation. A YAG scintillating crystal has a long lifespan and is useful in electron microscopy. See scintillator.

yttrium-aluminum-perovskite YAP. A crystalline substance used as a scintillating material in electron microscopy, YAP has high light-emitting properties and good resistance to high temperatures and radiation damage. YAP has a faster decay time than yttrium-aluminum-garnet (YAG). YAP is versatile in terms of emitted spectrum (may not require replacement of an existing photomultiplier) and long-lived, capable of outliving the lifespan of the instrument for which it is producing a scintillation material (e.g., electron microscope).

yttrium-iron-garnet YIG. A crystal used in the manufacture of amplifiers, filters, multiplexors, etc. that is used with a variable magnetic field to tune wideband microwave circuits. See YIG filter.

Yukawa, Hideki (1907–1981) A Japanese physicist who proposed a new nuclear force-field theory and a massive nuclear particle, the meson. The existence of the meson was verified by Cecil Powell a couple of decades later. Yukawa furthered the nuclear process of *K capture*, the absorption of an innermost encircling electron. This theory, too, was subsequently confirmed. Yukawa received the Nobel Prize in Physics in 1949 for his significant contributions to the understanding of quantum mechanics.

yuppie young urban/upwardly mobile professional. A takeoff on the term hippie from the sixties, a yuppie was generally someone born into a home with good financial resources or who was ambitious and was likely to enjoy a prosperous living.

yurt A sturdy, circular, temporary, mobile shelter of Asian origin. Yurts are used on work sites where shelter for equipment and workers is needed during construction or installation work. In the author's town, little yurts have been popping up around the various road construction sites under which new fiber optic cables are being installed.

YUV A color encoding scheme to accommodate human visual systems which are less sensitive to color variations than to intensity variations (particularly in individuals who are color blind). Thus, YUV uses full bandwidth to encode luminance (Y) and half bandwidth to encode chroma (UV). See Y/C.

yV Y-matrix of a vacuum tube.

Z 1. *symb.* impedance. See impedance. 2. *abbrev.* Zulu time. See Zulu time. 3. *abbrev.* Zebra time. See Zebra time. 4. *symb.* zetta-. See zetta. 5. The name of a formal specification language for describing and modeling computing systems, based on axiomatic set theory and predicate calculus. Z was developed at Oxford University in the early 1980s.

Z axis A reference baseline or vector within a coordinate system, most often associated by convention with rectangular or Cartesian coordinates. The Z axis is oriented perpendicular to the X and Y axes in a three-dimensional system. See Cartesian coordinates, X axis, Y axis.

Z axis modulation The varying of the intensity of an electron stream in a cathode-ray tube (CRT) by manipulating the cathode or control grid.

Z code In telegraphy, a system of shortcut codes related to short phrases to save transmissions time. Z codes were those prefixed with "Z," a rarely used letter, to reduce the chance of confusing them with the content of a message. For example, ZFB meant "Your signal is *Failing Badly*."

Z Fiber A commercial pure silica core optical fiber developed by Sumitomo Electric Industries, Ltd. (SEI) which held the record for low loss at attenuation of 0.154 dB/km from 1986 to 2002, when it was succeeded by a new fiber by the same company. Z Fiber is suitable for repeaterless submarine dense wavelength division multiplex (DWDM) transmissions applications. See Sumitomo Electric Industries, Z-PLUS Fiber.

Z force The pressure sensitivity of touch-activated devices, such as touchscreen monitors or touch-sensitive pads, as are often used in kiosks.

Z Series Recommendations A series of ITU-T recommended guidelines for programming languages and general software aspects of telecommunications systems. These guidelines are available for purchase from the ITU-T. Since ITU-T specifications and recommendations are widely followed by vendors in the telecommunications industry, those wanting to maximize interoperability with other systems, or conform to software conventions widely used, need to be aware of the information disseminated by the ITU-T. A full list of general categories is listed in Appendix C and specific series topics are listed under indi-

vidual entries in this dictionary, e.g., B Series Recommendations. See Z Series Recommendations chart.

Z-80 The Zilog Z-80 8-bit computer microprocessor was released in 1976 by the Zilog Corporation and was quickly incorporated into many control and robotics applications, and into a number of popular microcomputers such as the Tandy Radio Shack Model I and LNW-80 computers. While it was capable of clock rates up to 2.5 MHz, implementations of 1.4 to 2.4 were common.

The Z-80 evolved from the Intel 8080 with which it was more or less compatible; it was faster, with more instructions, not all of which were used to ensure compatibility with 8080 systems. The Z-80 has a simple register structure, including index registers and an accumulator, and is capable of 16-bit addressing through 8-bit double register pairs, something not found on most of the other microprocessing chips that were incorporated into 8-bit microcomputers at the time.

Many hobbyists acquired their first *machine language* and *assembly language* programming skills on the Z-80 chip. The original Z-80 was followed by faster versions, such as the Z-80A, Z-80B, and others, and has been used for over two decades in many control applications such as robotics and satellite telemetry, due to its simple efficiency, low cost, and practical instruction set. See TRS-80.

z-fold, zigzag fold, fanfold A term for a type of fold often used in continuous-feed forms and other computer printouts. The name is derived from the shape of three sheets of paper that alternately fold one way or the other along the perforations.

Z-marker See zone marker.

Z-PLUS Fiber A commercial pure silica core optical fiber with fluorine-doped silica cladding, developed by Sumitomo Electric Industries, Ltd. (SEI) which held the record for low loss at attenuation of 0.154 dB/km from 1986 to 2002, when it was succeeded by a new fiber by the same company. The Z-fibers are widely used in repeaterless submarine cable installations. See Sumitomo Electric Industries, Ltd., Z Fiber.

Z++ Just as C++ is seen as a more recent, object-oriented derivative of the C programming language, Z++

is an object-oriented extension of Z, a formal specification language for describing computing systems. See Z.

Z1 A historically important pioneer home-brewed binary relay computer developed in Germany in the mid-1930s by Konrad Zuse. While this was a significant pioneering computer construction, it was largely unknown outside of Germany due to the second World War and, as such, did not significantly influence the industry abroad. It is noteworthy that it was built in an apartment, rather than in a university research facility as were other pioneer computing platforms. See Zuse, Konrad.

Z3 A historic programmable, general purpose computer, developed by Konrad Zuse. The Z3 was released in 1941. See Z1, Zuse, Konrad.

zap *slang* To eradicate data, to burn out a circuit, or to apply charge to an object or environment. It may or may not be intentional, and can result from power fluctuations such as those caused by lightning (which is probably where the term originated). You can zap a file to kill specific data (or the whole file), you can zap food in a microwave, and you can zap or "fry" a circuit by accidentally shorting it. The term is occasionally used to indicate a quick change, such as the changing of a TV station with a remote, but this dilutes the meaning of the term and is better avoided (unless perhaps it's a lightning fast change) as zap is intended to describe an action or event that is potentially dangerous, lethal, or destructive, especially where electricity is involved. See kill.

zapper 1. A device for applying a sudden stimulus such as electricity or heat that affects an immediate change such as a burn or incision. Examples include shock devices such as a stun gun or a laser torch or scalpel. 2. A software utility that causes something to be instantly altered or removed, such as a file zapper that may be designed to seek out and permanently eradicate the entries of customers who have not purchased anything for more than a year.

ZBLAN A heavy metal fluoride glass used in the fabrication of fibers, ZBLAN is fluorine with zirconium, barium, lanthanum, aluminum, and sodium (Zr, Ba, La, Al, Na), sometimes called fluorozirconate. Fibers manufactured with ZBLAN have low attenuation and good transmission properties. Compared to silica-based glass, ZBLAN supports a broader spectrum of wavelengths, from ultraviolet to infrared. It is also possible to dope ZBLAN fibers for use as fiber amplifiers.

By 1998, D. Tucker at the NASA/Marshall Space Flight Center had reported that ZBLAN does not crystalize as readily, and is clearer, if constructed in a near-weightless environment away from Earth's gravity. This leads to a remarkably pure, clear fiber and ZBLAN's transmission properties approach that which is considered theoretically best. ZBLAN significantly outperforms silica glasses in the regions above which they "top off" at wavelengths of about 2.3μ. ZBLAN is capable of transmitting wavelengths of up to 4.5μ and higher. In actual use, ZBLAN is currently limited to short distances, but with improved fabrication processes to prevent or reduce crystallization, the potential of ZBLAN for fiber communications in a broad spectrum of wavelengths is excellent.

Scientists point out that the whole process doesn't have to take place at zero "g" to take advantage of space-based ZBLAN optical fiber. The fiber *preforms* could be created in space, with the filaments pulled back on Earth. See preform, vapor deposition process.

ITU-T Z Series Recommendations

Recom.	Description
Z.100	Specification and description language (SDL)
Z.105	SDL combined with ASN.1 modules (SDL/ASN.1)
Z.106	Common interchange format for SDL
Z.107	SDL with embedded ASN.1
Z.109	SDL combined with UML
Z.110	Criteria for use of formal description techniques by ITU-T
Z.120	Message sequence chart (MSC)
Z.130	ITU object definition language
Z.140	The tree and tabular combined notation version 3 (TTCN-3): Core language
Z.141	The Tree and Tabular Combined Notation version 3 (TTCN-3): Tabular presentation format
Z.200	CHILL — The ITU-T Programming Language
Z.301	Introduction to the CCITT man-machine language
Z.302	The meta-language for describing MML syntax and dialogue procedures
Z.311	Introduction to syntax and dialogue procedures
Z.312	Basic format layout
Z.314	The character set and basic elements
Z.315	Input (command) language syntax specification
Z.316	Output language syntax specification
Z.317	Man-machine dialogue procedures
Z.321	Introduction to the extended MML for visual display terminals
Z.322	Capabilities of visual display terminals
Z.323	Man-machine interaction
Z.331	Introduction to the specification of the man-machine interface
Z.332	Methodology for the specification of the man-machine interface - General

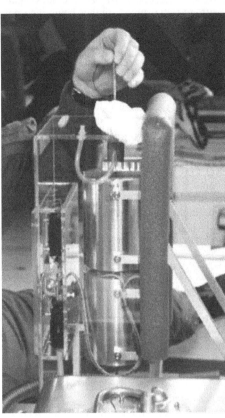

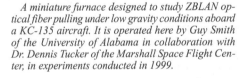

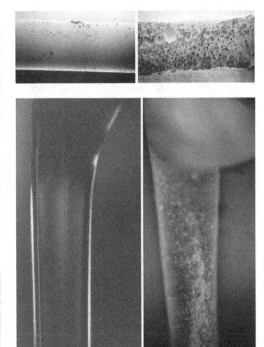

Comparison of ZBLAN fibers pulled at 0 G (left) and 1 G (right) aboard a low-gravity experiment conducted by the NASA/Marshall Space Flight Center. Bubbles and other defects in the example on the right will scatter optical signals and interfere with traansmission. The pure, uncrystallized optical fiber on the left has the potential to transmitt 100 times more data than traditional silica-based optical fibers. [NASA/Marshall photos.]

A miniature furnace designed to study ZBLAN optical fiber pulling under low gravity conditions aboard a KC-135 aircraft. It is operated here by Guy Smith of the University of Alabama in collaboration with Dr. Dennis Tucker of the Marshall Space Flight Center, in experiments conducted in 1999.

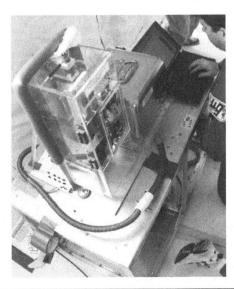

ZBLAN – Silica Comparison

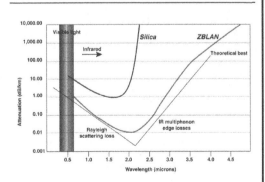

Optical fibers made from ZBLAN rather than silica exhibit a profile of efficiency far closer to the theoretical best. NASA has been conducting research into pulling ZBLAN fibers in space where conditions can be created for reducing crystallization. [NASA/Marshall Space Flight Center, 1998.]

Z

ZBTSI See zero byte time slot interchange.

Zebra time, Z time The same as Greenwich Mean Time. See Greenwich Mean Time, Z.

Zeeman effect An effect observed in 1896 on the structure of gas spectrum lines when subjected to the influence of a moderately strong magnetic field. It was observed by Pieter Zeeman that sharp spectral lines split into multiple closely spaced lines. This early puzzle gave rise to a number of lines of research in quantum physics. The number of lines is generally three, although anomalous effects have been observed and even stronger magnetic fields can cause some spectral lines to emerge.

The Zeeman effect is a useful tool in a number of areas of science including the study of laser light. It intrigues quantum physicists and is used by astrophysicists to map the magnetic field strengths of solar effects and to study other galactic magnetic fields. The Zeeman effect is often described along with the Stark effect (the influence of strong electrical fields on spectral lines). See Stark effect.

Zen An Asian Buddhist system of belief that encourages meditation and self-discipline, and the attainment of enlightenment through direct intuitive experience. See Zen mail.

Zen mail A tongue-in-cheek descriptive phrase for computer communications that erroneously (or deliberately) arrive with no content in the body of the message. See Zen.

zener current In an intense electric field, a current through an insulator sufficient to excite an electron from the valence band to the conduction band.

zener diode An electronic component that behaves like a rectifier below a certain voltage, but exhibits a sudden increase in current-carrying capacity above a specific voltage level (zener value) and a corresponding decrease in dynamic resistance resulting in a *reverse breakdown*. The dual nature (di-) makes it possible to represent conducting and nonconducting (or very low conducting) states as a binary system and the breakdown under reverse bias (functioning as an open circuit) is particularly useful as a reference circuit or as a shutoff circuit to respond to power surges above a certain voltage threshold.

Doping is a manufacturing process in which elements such as rare earth elements are introduced into a substance (e.g., fiber optic cable) to improve its qualities for a particular purpose such as the transmission of laser light. Doping is also used in semiconductors to influence their conducting qualities. Heavy doping in p-n junctions, for example, results in very thin depletion layers that hinder impact ionization but facilitate quantum tunneling and the flow of current. This, in turn, leads to zener breakdown which is somewhat related to avalanche breakdown from impact ionization, as may occur in lightly doped p-n junctions.

Zener diodes are used in voltage regulators and power supplies and help prevent electrostatic discharge in a wide variety of semiconductor technologies. They are becoming especially important in the mobile communications industry because small portable devices become vulnerable to voltage surges/static electricity when they are connected to other devices such as network cables, docks. Tiny zener diodes capable of absorbing static surges are thus being incorporated into mobile phones, pagers, etc.

zener effect A reverse-current breakdown effect that occurs at a semiconductor or insulator junction in the presence of a high electric field. See zener current, zener diode.

Zenith Corporation One of the early entrants to the radio industry, Zenith was founded in 1931 by a radio amateur and soon became a major manufacturer. Zenith now promotes consumer flat-screen, large-size television/media screens. Too bad the commercials aren't as classy as the products.

ZEONEX A commercial tradename for ZION Technology's implementation of cyclic olefin polymer. See cyclic olefin polymer.

zeppelin antenna A horizontally oriented antenna, which is a multiple of a half wavelength in length with a two-wire transmission coming into one end, which is also a multiple of a half wavelength.

zepto- A standard metric prefix of the Système Internationale (SI). The zepto unit is used in scientific measurement requiring very small numbers and represents 1000^{-7} in decimal. See zetta-.

zero balancing A telephone service accounting technique in which a specific dollar quantity is distributed over a large category of calls. The total base price of all the calls within the category is used to calculate an adjustment percentage so the full dollar amount produces a zero balance (rounding errors are not permitted).

zero beat In electronics and acoustics, a condition during which two combined frequencies match, and consequently do not create a *beat*.

zero beat reception, homodyne reception In radio transmission reception, a system that uses locally generated voltage at the receiving end of the transmission, which is the same frequency as the original carrier and combines it with the incoming signal. See beat reception, heterodyne.

zero bias 1. In a cathode-ray tube (CRT), the absence of any difference in potential between the cathode and its control grid. 2. In teletypewriter transmission circuits, zero bias is a state in which the length of the received signal matches the length of the transmitted signal.

zero bit insertion, bit stuffing In data communications, the process of inserting a zero bit after a series of one bits in order to specify a distinct break or change. Thus, the beginning or ending of a frame is not misconstrued. This is sometimes used in place of control signals.

zero compression A data compression technique in which nonsignificant leading zeros are removed.

zero fill A data manipulation technique in which zeros are inserted into a file or transmission without affecting the meaning of the data. See zero bit insertion.

zero dispersion In an optical waveguide, physical characteristics that result in the propagating phenomena traveling in step with one another from one end of the link to another (without dispersing). This situation is actually somewhat difficult to achieve, especially in long haul fiber optic links carrying multiple wavelengths.

Due to the different reflective lines of travel of different light beams, they tend to arrive at the endpoint at different times. Different frequencies in the same lightguide (e.g., in multimode fiber) also will have different travel speeds due to their different energy levels and characteristics. In addition, the wavelengths of the lightwaves are not usually coincident with acoustical vibrations that occur in the cable except at certain specific frequencies. Thus, various types of dispersion almost invariably occur and, in fact, tend to be detrimental to the composition of the signal at the endpoint (it's usually advantageous for related signals to arrive at the same time).

Thus, zero dispersion often a design goal in the engineering and construction of fiber optic communications systems. However, there are situations where radiant energy in a cable will travel without dispersion and may be in phase with energy that could cause noise. In very short cable lengths, there may not be sufficient distance/time for the different parts of a signal to shift from one another, especially if the lightguide is very straight and little internal reflection is necessary to channel the light. In a carefully engineered cable where graded index fiber is used to compensate for differences in reflectivity and travel time in different wavelengths, dispersion may be minimized and approach zero. Thus, there is a balance between trying to achieve zero dispersion and interference factors that may impinge upon the signal when it reaches that level. See dispersion, zero-dispersion wavelength.

zero-dispersion slope In a single-mode optical fiber, for example, the rate of change of dispersion at the zero-dispersion wavelength relative to the light beam's wavelength.

zero-dispersion wavelength (*symb.* $-\lambda_0$) In an optical waveguide, the wavelength(s) at which material disperson and waveguide dispersion cancel one another, in the sense that they are in phase. When unwanted wavelengths move in step with operational wavelengths, significant interference can occur. The λ_0 is sometimes called the "dispersion window."

In dispersion-unshifted silica-based optical fibers, the λ_0 occurs at about 1.3 µm, in dispersion-shifted fibers at about 1.55 µm. Dispersion-shifting may be accomplished by doping the fiber. It is important to understand the disperion wavelength phenomenon so as to avoid undesired nonlinear interference and modulation instability. It is also possible to achieve high data transmission rates near this frequency, so it helps to know just where that point lies.

Dispersion-shifted fiber (DSF) shifts the zero-dispersion point from 1310 to 1550 nm to move it out of the phase range. However, four-wave mixing may occur in these fibers, so they are more suitable for single-channel systems than others. To gain the advantages of dispersion-shifted fiber while avoiding the disadvantages of four-wave mixing, nonzero dispersion-shifted fiber was developed (e.g., OFS's TrueWave RS fiber).

Dispersion-compensating fiber (DCF) is another specialized type of fiber useful for endlink connections because it has a high negative dispersion value that reverses chromatic dispersion.

Zero-dispersion wavelength parameters have traditionally been difficult to measure precisely, hindering the development of a reference standard. Mechels et al. have described the development of a frequency-domain phase-shift system intended to reduce systematic λ_0 measurement errors. See dispersion.

zero insertion force ZIF. A type of socket used in integrated circuits that allows a chip to be inserted without undue pressure. A lever or screw is then pushed or turned to secure the component so that it is not dislodged due to bumping or transit.

This type of socket costs more than standard pressure sockets and tends to be used in specialized systems such as test systems or with specific chips that are larger or more expensive.

zero level A level established in order to have a reference from which to judge further states or activities of sounds and signals for observation, calibration, or testing. The definition of zero level is technology specific.

zero potential The potential of the Earth, used as a reference measure.

zero power peripheral A device that requires very little power and consequently can draw that power from the primary device to which it is attached, or from the circuit with which it is associated. Some modems and most telephones take their power from the phone line, unless they have extra features (e.g., speakerphone) requiring additional power.

Zero power peripherals are favored in constrained spaces and mobile communications as they are often less bulky than standard peripherals with power supplies, and easier to attach, since extra electrical outlets are not required.

zero punch A punch located specifically in the third row from the top of a punch card. See Hollerith card, punch card.

zero shift, zero drift A descriptive measure of the amount of shift or drift that has occurred from the original setting or calibration point, at a subsequent point in time. See zero stability.

zero stability The ability of an instrument to retain its original state or settings over time, that is, to withstand zero shift. Zero stability is generally considered a desirable characteristic. See zero shift.

zero stuffing See zero bit insertion.

zero suppression The elimination of zeros that are not meaningful. Zero suppression is often used to increase the readability of information with leading zeros, for formatting or transmission purposes, which would otherwise be distracting or confusing to a human reader. In these cases, the zeros are often replaced with a blank (a space character on printouts). Tables and columnar data (like financial statements) are usually printed with zero suppression.

Z

zero transmission level reference point For an arbitrarily selected point in a circuit, a level reading that is subsequently used against which to measure transmission levels in other points of the circuit or at the same point at another time. In telephone transmissions, the reference point is frequently selected at the location of the source of the transmission.

zero usage customer A listed subscriber who has not used the network to which he or she has access.

zero water peak fiber A type of optical communications fiber first commercially released by OFS in 2002. This fiber permanently eliminates a water peak that typically occurs at 1400 nm while preserving the 1300 nm wavelength band, thus substantially increasing capacity over traditional single-mode fibers carrying E-band coarse wavelength division (CWDM) multiplexed transmissions.

zerofill, zeroize To insert the zero character into unused storage locations. This is done for a variety of reasons: for formatting, for creating space savers, and sometimes as a delay mechanism to match up transmission speeds with output speeds of slower output devices (printers, facsimile machines, etc.).

zetta- Z. A standard metric prefix of the Système Internationale (SI). The zetta unit is used in scientific measurement requiring very large numbers and represents 10^{21} or 1,000,000,000,000,000,000,000 or 1000^7 in decimal. See yotta-, zepto-.

ZIF See zero insertion force.

Zimmermann, Philip R. A software engineer and cryptography specialist, best known for developing the Pretty Good Privacy (PGP) encryption scheme based upon the Blowfish technology, Zimmermann is the founder of PGP, Inc. He is a software engineer with a long history of experience in cryptography, data security, data communications, and realtime embedded systems development.

Zimmermann has been honored with numerous humanitarian awards due to his contribution to the safeguarding of personal privacy, including the 1996 Norbert Wiener Award, the 1995 Pioneer Award from the Electronic Frontier Foundation, and many more. Zimmermann is a member of the International Association of Cryptologic Research, the Association for Computing Machinery (ACM), and others. See Blowfish, PGP Inc., Pretty Good Privacy.

zinc (*symb.* – Zn) A malleable, light-bluish, active, metallic element (AN – 30) with a relatively high electropotential, useful in plating, dipping, and galvanizing, to prevent corrosion in other metals. It tends to be brittle and difficult to work with above and below 100 – 150° C. Combined with copper, it forms brass. It is used for piping, construction accessories, and household items.

zinc-fluoride (*symb.* – ZnF_2, F_2Zn_1) A white, crystalline zinc compound. Zinc-fluoride is one of a number of zinc compounds suitable for use in ZBLAN-type high-quality optical glass applications. See ZBLAN.

zinc-oxide (*symb.* – ZnO, O_1Zn_1) A white, crystalline zinc compound (familiar in commercial sun screens and antiseptive ointments) that is widely used in optics. Historically, zinc was used in early voltaic cells (the forerunners to modern batteries) such as the Daniell cell. See Daniell battery.

At some wavelengths, ZnO is transparent. It is used in deposition processes, photocopier photoconductive surfaces, light-emitting diode components, and blue lasers. The material behaves differently when it interacts with gas molecules, making it suitable for the detection of flammable gases. It may be used to improve the physical properties of rubber. When zinc-oxide is combined with aluminum (Al), fluoride (F), or indium (In), it can function as an electrically-conducting oxide in optoelectronic applications.

New fabrication applications for ZnO are being devised. In January 2002, Eagle-Picher Technologies announced the successful development of p-type thin film layers created from ZnO by a molecular beam epitaxy (MBE) process. This has potential for detectors and light-emitting diode and laser illuminators in the ultraviolet (UV) range. The addition of Magnesium (Mg) brings the spectral range even deeper into the UV spectrum.

zinc-selenide (*symb.* – ZnSe, Se_1Zn_1) An amber-to-reddish-colored crystalline zinc compound that has good uniformity and transmission properties, as well as low bulk absorption ratings, though it is more brittle and subject to cracking than glass. In commercial applications, it transmits in the ca. $0.5 - 15$ μm range and is popular for laser optics in the infrared spectrum.

Zinc-selenide is used in lenses for carbon-dioxide lasers and certain guidance systems and spectroscopes. It has also been used in experimental photoconductive semiconductor switching circuits. Compounds similar to zinc-selenide are sometimes used in vertical-cavity surface-emitting lasers (VCSELs).

zip *v.* To bundle up or compress a file or set of files, usually for transferring or archiving. The term is now commonly used for any such processing (zipping) of files, regardless of the archive utility used, but originally referred to programs called gzip and PKZIP. See arc, compress, gzip, PKZIP.

ZIP Zone Information Protocol.

zipped See zip, gzip, PKZIP.

zirconia A substance exhibiting different molecular structures at different temperature ranges. Oxides may be combined with zirconia to reduce structural changes that occur at different production temperatures. Once stabilized, the material becomes highly resistant to temperature-induced phase transitions. Zirconia is customarily made by injecting zirconium-dioxide powder into molds and grinding the molded components to industry standards.

Zirconia is a versatile material that is made into a ceramic for use in many industries, including the fabrication of joint prosthesis for surgical implantation. The material is favored for ferrules for fiber optic joint support due to its strength, resistance to environmental erosion, and low insertion loss qualities, though some companies are now creating plastic ferrules with good optical performance. See ferrule, zirconium oxide. See YSZ.

zirconium oxide A tan- or beige-colored substance with high refractory properties, low thermal conductivity, low friction coefficents, and high heat tolerance. Zirconium oxide is strong, versatile, and chemical and corrosion-resistant. It is created by stabilizing zirconia crystals to resist structural changes at different temperatures. The material is suitable for the production of ceramic ferrules for fiber optic connectors and, when combined with metal leads, can provide illumination for spectrometers. It is also used in oxygen-sensing components. See zirconia.

Ziv See Lempel-Ziv.

ZModem A fast, flexible, error-correcting, full-duplex file transfer data transmission protocol similar to XModem and YModem, but with updates and enhancements. YModem and ZModem were written by Chuck Forsberg.

ZModem is well supported by various telecommunications programs, and many BBSs and service bureaus use it.

ZModem includes fallback and dynamic adjustment of packet size, which is important if the connection is a line with fluctuating characteristics. One of ZModem's most valuable features is its ability to resume a file transfer that has been aborted. With almost all earlier desktop serial file transfer programs, if the line was interrupted or the file transfer aborted 99% of the way through, the program would start again from the beginning when reconnected, rather than resuming from where it left off. ZModem can detect a partial file, establish a new starting point at the end of the partial file, and continue transmitting until the transfer is completed or interrupted. If you've ever spent a couple of hours transferring a very large file on a slow link and lost it when the transmission was almost complete, you will appreciate ZModem's "resume" feature. See XModem, YModem.

ZOG A high-performance frame- and link-based hypertext system designed for local area networks (LANs) developed at Carnegie Mellon University. It has evolved into a commercial product called Knowledge Management System (KMS).

zone 1. An area, usually contiguous, which may or may not be self-contained; or has some common characteristics within itself; or which is distinguished as different, in some way, from the surrounding area; or which is assigned on the basis of size, population, or some other economic or practical characteristic (as in shipping zones). 2. A section of computer storage allocated for a particular purpose. 3. In telephone communications, a specified area outside the local exchange. 4. In cellular phone communications, a zone consisting of adjacent cells is called a cluster. 5. A region in which communications are banned or impeded, called a blackout zone. 6. A physical or virtual region of access within a network, sometimes delineated by physical structures such as a workstation or router. See firewall.

zone, optical recognition A region manually specified by the user, or automatically detected and enclosed by the optical recognition software, that cor-

responds to a particular type of information. For example, some optical document recognition systems, and some of the better optical character recognition systems, can distinguish columns and page numbers, images and formulas, and scan each separately from the others. Some software facilitates the preconfiguration of zones in a template. Order priority for the zones may be assigned. In this way a long document, such as a book in which most of the pages are identical in format, can be scanned without resetting the zones each time.

Zones in Optical Recognition

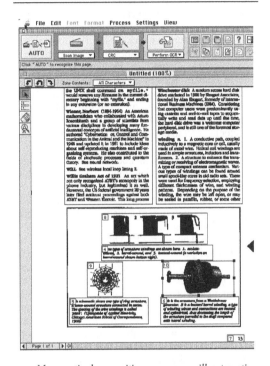

Many optical recognition programs will automatically determine zones or allow them to be manually configured. This allows the flow of text, and separation of graphics and text to be handled more efficiently by the software, as in this Caere Omnipage example.

zone, punch card One of three specific positions on the top of a punch card. See punch card, zone punch.

zone blanking Turning off a cathode-ray tube (CRT) at a point in the sweep of an antenna.

zone cabling A cabling architecture designed for open office systems in which various physical zones are designated and cabled so that office desks and equipment can be moved around without ever being too far from the various necessary outlets and connectors.

zone marker, Z-marker A vertically radiating beacon of light that signals a zone above a radio station transceiver.

zone method A wire installation ceiling distribution system in which the space above rooms is organized into sections or zones. Cables are centralized in each zone, with main arteries running between zones or to the central power source. See distribution frame.

zone of authority The set of names managed by, or under the authority of, a specific name server.

zone of silence, skip zone In radio transmissions, a geographic region that does not receive normal radio signals, frequently due to abrupt changes in terrain, (e.g., mountains).

zone paging The capability of an intercom or phone system to selectively page certain groups of speakers. See public address system.

zone punch In a punch card, a punch located in one of the upper three rows (the section that usually has less text displayed on the card). See Hollerith card, punch card. Contrast with digit punch.

zone time A system in which the Earth is divided longitudinally into 24 time zones of about 15 degrees each starting in Greenwich, England. It was developed in the late 1800s by a Canadian, S. Fleming, to establish a standard time. See Greenwich Mean Time.

zoning, stepping In microwave transmissions, displacement of portions of the surface of the microwave reflector in order to prevent changes in the phase front in the near field.

zoom To continuously reduce or enlarge an image, as on a monitor or in a viewfinder. Zooming capability is usually provided to improve the visibility of details (zoom in), or to provide a 'big picture, wide angle' view (zoom out) of a diagram, object, image, or scene. See zoom factor.

zoom factor The degree to which an image can be scaled, that is, decreased or enlarged. The X and Y axes may or may not be capable of sizing independent of one another. The enlargement zoom factor on consumer camcorders often ranges from 20 to 200 times (or more) the normal viewing factor. In some cases, the zoom factor on still cameras and camcorders may be digitally enhanced, that is, the zoom up to 20 times may be an optical zoom and, beyond that, it may be a digital zoom, which may show some pixelation at higher zoom factors. See zoom.

zoom lens An apparatus that provides the ability to reduce or enlarge the apparent size of an image in order to frame that image with the desired scope or to enhance detail. Commonly used on video and film cameras, and sometimes on telescopes and binoculars.

zsh, Z shell A Unix command interpreter shell, similar to ksh, developed by Paul Falstad. Zsh is said to be similar to a bash shell, but faster and with more features. Zsh is not a Posix-compliant implementation.

Zulu time Greenwich Mean Time, Coordinated Universal Time.

Zuse, Konrad (1910–1995) A German structural engineer and inventor who independently created pioneer general purpose calculating and computing technologies, coincident with the development of the Attanasoff-Berry computer in the U.S. and Alan Turing's computing theories in England. Zuse's accomplishment is remarkable, considering he didn't have the corporate and university facilities, funding, and personnel support that led to the development of other pioneer computing systems.

Zuse began building computers in his mid-20s from a studio he set up in his parents' apartment. He reports in his autobiography that he originated the concept in 1934, and constructed the V1, later called the Z1, between 1936 and 1938. His early notebooks describe important binary computing concepts. The Z1 included mechanical memory for data storage (for practical reasons of space and ease of construction) and used program instructions punched into film. Later, Zuse added the Z2, Z3 and Z4, with a friend, Helmut Schreyer, providing expertise on electronic relays and vacuum tubes. Zuse considered his Z3 relay-based version to be a true binary computer (the original Z3 was destroyed after World War II bombing raids). A reproduction has been installed in the Deutsches Museum. The Z4, which began development in 1942, was demonstrated in April 1945, and operated in Zürich from 1950 to 1955. From Zuse's agile mind came also a pioneer algorithmic programming language called PlanKalküel (plan calculus) that was developed in the mid-1940s. See Attanasoff-Berry computer; Hertzstark, Curt.

Zworykin, Vladimir Kosma (1889–1982) A Russian-born American physicist and electrical engineer who emigrated to America in 1919 and worked for Westinghouse Electric Company in the 1920s.

Zworykin developed an idea to control the passage of beams in an electron tube with magnets to devise a cathode-ray tube (CRT), which he patented in 1928. This historic CRT led to his development of the *iconoscope*, a practical television camera. In 1929 Zworykin demonstrated a cathode-ray-based display device, the same basic concept as current computer and television monitors. In the same year, Zworykin became the director of research for the Radio Corporation of America (RCA). In November 1935, he received a patent for his apparatus for producing images (U.S. #2,021,907).

Numerals

Network Technologies

4B/5B Fiber transmissions cable that is commonly used in asynchronous transfer mode (ATM) and Fiber Distributed Data Interface (FDDI) networks. This 4-byte/5-byte multifiber cable can support transmission speeds up to about 100 Mbps. See 8B/10B.

6bone An IETF-supported international collaboration testbed providing policies and procedures for the evolution of Internet Protocol (IP). The name 6bone is derived from *backbone* a major 'artery' of the Internet, IP version 6. 6bone is designed to be used in the development, deployment, and evolution of the Internet Protocol Version 6 (IPv6) which is intended to succeed the current Internet protocol IPv4.

This testbed and transition project is essential in that the Internet is not one machine and one agency running it, but a global collaboration of computing devices managed and owned by many different personal, commercial, and governmental entities. The 6bone provides not only a means to test the many features and concepts of the new systems, but also a means for developing and deploying a transition infrastructure.

The 6bone is a virtual network that is layered on portions of the current physical structure of the IPv4-based Internet. IPv4 routers are not designed to accommodate IPv6 packets. By layering IPv6 on the existing structure, the routing of IPv6 packets can be accomplished prior to the implementation of enhanced physical structures, particularly routers designed to take advantage of the features of IPv6.

To understand the 6bone virtual structure, imagine various workstation-class computers, such as those commonly used as servers in various communities and institutions. Provide these machines with operating system support for IPv6 so that they have direct support for the IPv6 packets. Now provide a means through the Internet for these machines to interconnect and communicate with one another through virtual point-to-point links called *tunnels*, thus managing the links on behalf of physical routers until IPv6 support is more widespread. Eventually, as the Internet is upgraded to IPv6, this interim system will be replaced by agreement with direct physical and virtual IPv6 support.

The 6bone Web site is sponsored by the Berkeley Lab Networking & Telecommunications Department. See IPv4, IPv6, MBone, X-Bone. See RFC 2546, RFC 2772. http://www.6bone.lbl.gov/6bone/

8B/10B The designation for a fiber transmission cable suitable for high speed networks. This 8-byte/10-byte multifiber cable can support transmission speeds up to about 149.76 Mbps. See 4B/5B.

10Base- After ratification of Ethernet as a standard (IEEE 802.3), a number of variations were defined to support twisted pair and fiber optics physical media and data formats transmitting at rates of 10 MBps. See Ethernet Standards chart.

10 Gbit Ethernet 10GbE. A telecommunications technology developed within the IEEE 802.3ae working committee (a subcommittee of the IEEE 802.3 Ethernet Working Group), based upon the evolution of Ethernet/Fast Ethernet technologies. The project began in spring 1999 to extend and update the capabilities of Ethernet and was approved by ballot in March 2001, a milestone toward standards ratification that is expected in 2002.

10GbE differs from earlier versions in a couple of physical interface aspects. It includes a long-haul optical transceiver capable of 40 kilometers or more, or physical medium-dependent (PMD) interface for single-mode fiber. This can be used with either a LAN or WAN physical layer to support metropolitan area networks (MANs). Secondly, 10GbE includes an optional WAN physical (PHY), enabling 10GbE to be transported transparently over SONET OC-192c or SDH VC-464c infrastructures.

In general, 10GbE multimode fiber connections are expected to transmit 65 or 300 meters or more and 9-micron diameter single-mode fiber connections are expected to transmit 10,000 to 40,000 meters or more. Initial implementations are expected to include high-speed local backbones in large-capacity installations (campuses, ISPs, etc.).

10 Gbit Ethernet Alliance A nonprofit trade alliance promoting standards-based 10 Gigabit Ethernet technology development, distribution, and utilization. The founding members include 3Com, Cisco Systems, Extreme Networks, Intel, Nortel Networks, Sun Microsystems, and World Wide Packets. http://www.10gea.org/

100Base-T A baseband signaling networking standard supporting Fast Ethernet with data transfer rates up to 100 Mbps. 100Base-T is intended to provide a faster option to Ethernet networks based originally upon 10Base-T carrier sense multiple access systems. 100Base-T is described in IEEE 802.3u. 100Base-T specifications can be supported over a variety of physical media configurations.

Type	Physical Medium/Notes
100Base-TX	*data-quality twisted pair* Requires at least Cat 5 cable. As in 10Base-T, the data is not split and may be used in full-duplex transmission modes.
100Base-T4	*4 pairs of regular twisted pair* Requires at least Cat 3 cable. The data stream is divided into three 33-Mbps streams with the 4th twisted pair used for error mechanisms. Half-duplex transmission.
100Base-FX	*dual-stranded fiber optics* Segments may be up to 412 meters.

100BaseVG-AnyLAN A commercial LAN from Hewlett-Packard that was later refined and standardized by the IEEE 802.12 committee and ratified in 1995 as 802.2. Like 100Base-T (Fast Ethernet), it provides data rates of 100 Mbps. It is similar to Fast Ethernet, and capable of carrying Ethernet and Token-Ring transmissions simultaneously. A VG network consists of nodes connected to hub ports in a star topology. Hubs are interconnected through a tree topology. This technology has not found the same widespread commercial acceptance as 100Base-T. See High Speed Token-Ring.

100BASE-SX TIA/EIA/ANSI-785. The Short Wavelength Fast Ethernet Standard developed as Project SP-4360 by the FO 2.2 committee, approved June 2001. This standard was developed to provide a clear, cost-effective means to upgrade local area networks (LANs) from 10-Mbps copper or fiber to 100-Mbps fiber with a cabling structure different from that of Gigabit Ethernet. Thus, 10BASE-FL can be upgraded to 100BASE-SX fairly readily and may be less expensive than Gigabit Ethernet deployed over Cat 5 or higher cabling. It uses 850-nm optical media for backward compatibility with 10BASE-FL and is similar to 100BASE-FX, with the same signal encoding structure. 100BASE-SX differs from 100BASE-FX in that it uses lower cost 850-nm optics, has a local distance of 300+ meters, and does

---

Ethernet Standards – Overview

Name	Description
10Base-2	10 Mbps baseband "thin" Ethernet 50-ohm coaxial cable as a network physical transmissions medium. Up to 30 stations can be supported per cable segment of up to 200 meters. This format is popular for small local area networks connecting personal computers. See RFC 1983.
10Base-5	Essentially Ethernet delivered over a standard or "thick" Ethernet 50-ohm cable at data rates of 10 Mbps over cable segments up to 500 meters.
10Base-F	A physical layer specification for 10-Mbps data rates over fiber optic cable.
10Base-FL	10-Mbps baseband fiber optic network links supporting transmission segments of up to two kilometers, with a maximum of two devices per segment (station and hub). Multiple stations may be connected through a hub in a star topology. There are commercial media converters available for connecting twisted pair (10Base-T) cables to 10Base-FL Ethernet-based networks to extend twisted pair cable lengths with optical connections. Converters may be used with regenerating repeaters.
10Base-T	A physical transmissions medium supporting up to 10 Mbps of baseband transmissions over twisted pair (T) that is commonly interconnected with RJ-45 cables (the ones that have connectors that look like fat RJ-11 phone connectors). Three or more stations can be interconnected in a star topology through a hub (and stars can be interconnected through a 'bus' topology). It's a convenient method of connection since the loss of a station in the star doesn't bring down the rest of the network as in a ring topology. The Manchester scheme of binary coding is typically used with 10Base-T.
10Broad36	A multiple-channel network broadband signaling system that can be implemented over single or dual coaxial cables. The bandwidth is subdivided into two or more channels for the simultaneous transmission of different types of signals as might be found with multimedia communications. Segments can support transmission distances of up to 3600 meters per segment.

not require a specific fiber optic connector, any of the standard connectors meeting minimum performance requirements may be used.

A new Physical Media Dependent (PMD) sublayer has been defined as part of the standard, as has an optional auto-negotiation mechanism. Above the physical layer, the standard remains the same as previous implementations and is intended to be used along with IEEE 802.3 (1998) that is prevalent in LANs. Thus, 100BASE-SX is intended to fit in well with established Ethernet/Fast Ethernet environments.

1000Base-T An IEEE standard approved in 1997, developed by the P802.3ab study group. This standard defines a full duplex Gigabit Ethernet signaling system for category 5 (Cat 5) network systems. Unlike 100Base-TX transceivers, which use only two pairs of wires, one in each direction, 1000Base-T transmits on all four pairs simultaneously from both directions of each pair. This creates a more complex system and a greater potential for crosstalk. See far end crosstalk, near end crosstalk.

802.3, 802.3u, 802.3z An IEEE-specified family of Ethernet standards ranging from 10 Mbps to Gigabit Ethernet that are commonly used on local area networks. The maximum distance depends in part upon the cabling. For example, for 802.3, a maximum segment length would be 500 meters for "thick" Ethernet (10Base-5) or 185 meters for "thin" Ethernet (10Base-2). See Ethernet Standards chart.

802.5 An IEEE-specified token-passing network system using differential Manchester coding for up to 250 workstations to a maximum distance of 101 meters at 1 or 4 Mbps.

General

1BL *abbrev.* 1. single bottom line. 2. one business line.

1EAX A GTE variant of Western Electric/AT&T's ESS1A telephone switching system. See 1A, Electronic Switching System 1. **214 License** A Section 214 license is issued by the Federal Communications Commission (FCC) to qualifying applicants pursuant to the Communications Act. It charges the FCC with the responsibility of determining whether applicants have demonstrated that their proposal will serve the public interest and need. Thus, activities such as corporate expansions and mergers require prior Section 214 FCC approval. In the mid-1990s, the FCC streamlined the applications process such that it is automatically granted if no one objects during a period set aside for public comment.

2D two-dimensional. Existing or described in two spatial dimensions or in terms of two selected realms of data. A Cartesian coordinate system is a graphical representation of data in two dimensions. The dimensions need not be spatial, they may be quantities, time, or other types of information that may be plotted. Two-dimensional systems representing spatial concepts commonly tend to be flat in the sense that they represent width and breadth but not depth. In call accounting, 2D representations may illustrate profits over time or calls over time. In network systems logs, connections over time or downtime over time are commonly represented as 2D graphs.

3D three-dimensional. Existing or described in three spatial dimensions or in terms of three selected realms of data or time. A Cartesian coordinate system can graphically represent three dimensions but often does so in two-dimensional spatial conventions that use an illusion of 'stretching into space' in the Z (third) dimension. A photograph is a representation of spatial relationships in two dimensions whereas a sculpture is a representation of spatial relationships in three dimensions. In graphical representations and statistical reports, time is often one of three dimensions represented. For example, an accounting log may track new versus established employees' sales over a period of three months. Thus, employee status is treated as one dimension, the number of sales as the second, and time as the third.

In telecommunications, one of the most significant developments is in the representation of three-dimensional space through graphical rendering or ray-tracing or through NCR-type controller systems that can translate three-dimensional data into physical structures. This opens up a new world of communications. It becomes feasible, given enough speed and resources, to model a sculpture remotely. One artist may supply the coordinates (or a model) in one location and another may 'render' the sculpture with, for example, a milling machine, in another location, perhaps thousands of miles away. Thus, a metalworking shop in the U.S. could conceivably craft a new tool in a remotely controlled milling machine in a rural region in Africa. A physician in Canada could remotely carry out a liver operation on a patient in Germany. An engineer could remotely control a repair robot on a space station or space probe, without leaving Earth. Once the 3D world can be represented by data that can be instantly transmitted over great distances, a whole new world of telecommunications applications becomes possible.

3G Wireless Third generation wireless systems for telecommunications services through radio technologies as described by global telecommunications trade associations and standards bodies, and the Federal Communications Commission (FCC). This category encompasses a wide variety of mobile systems that may be linked into terrestrial or satellite-based communications relays and implies a general overall compatibility with existing and emerging systems. For FCC administrative purposes these systems are considered to be capable of supporting high bit-rate circuit and packet data transmissions with GPS and roaming capabilities with a reasonable degree of interoperability and standardization.

In an October 2000 U.S. Presidential Memorandum the Secretary of Commerce was directed to work with the FCC to develop a plan for the administration of radio spectrum frequencies for third generation wireless systems so that spectrum frequencies could be allocated in 2001 with licenses auctioned by 2002. The 2500- to 2690- and 1755- to 2690-MHz frequency

Z

bands were pinpointed for feasibility studies. Regulations arising from these studies can impact existing services (cell phones, PCS, etc.) if reallocation of bandwidth is recommended. Final reports of discussions between the FCC and the NTIA came online in spring 2001. http://www.fcc.gov/3G/

Microprocessors

1 A processor used in AT&T electronic switching systems, developed by Western Electric in the 1960s. See 1A.

1A A processor used in AT&T electronic switching systems (ESSs), developed by Western Electric in the 1970s as a successor to the 1 processor.

In ESS1A (a.k.a. No.1AESS), a commonly-used electronic switching system, the 1A processor provides maintenance and administrative support and interfaces with the central control. For readouts of operating and maintenance data, the 1A also can be interfaced with operator terminals for receiving instructions and outputting status information. Control panels may be further used for 1A input/output.

The No.4ESS digital toll switch also uses this processor.

4004 An early 4-bit central processing unit (CPU) from Intel as part of the MCS-4 chipset released in 1969. See Intel, MCS-4.

68000, MC68000 A 32-bit central processing unit commonly used in Amiga, Macintosh, and other computers, the first in a series by Motorola. See Motorola.

8008 A historic 8-bit central processing unit (CPU) from Intel, released as a successor to the 4004 as part of the MCS-8 chipset in 1972. The historic Altair computer was based on this processor. See Altair, Intel.

8080 An 8-bit central processing unit (CPU) released by Intel in 1974. RAM addressing was limited to 64 kilobytes. It was incorporated into a number of early microcomputers including the first model released by International Business Machines (IBM) in 1980.

8086 A successor to the 8080, the 8086 was an Intel 16-bit central processing unit (CPU). It could address 1 Mbyte of RAM. This chip was quickly incorporated into new versions of the IBM computers and was also used by manufacturers licensing IBM technology in competition with IBM. See Intel and Motorola for charts of other numbered microprocessors.

See also listings under Intel and Motorola.

Publications

2001 Fiberoptic Undersea Systems Summary A market research summary and CD-ROM database published by KMI Corporation. It includes an overview of the submarine fiber optics industry, along with information on mergers, acquisitions, new system developments, installation cost analyses, maps and profiles of fiber optics systems, and regional activities. See KMI Corporation.

2001 Update to Worldwide Markets for Dense Wavelength-Division Multiplexing A market report on DWDM systems worldwide published by KMI Corporation, completed in November 2000. The report describes cyclic trends in the industry and predicts growth and flat periods based upon trends to date. See KMI Corporation.

Telephone Prefix Calling Numbers/Services

0345 A 'shared tariff' telephone prefix and service offered by British Telecom to enable callers to pay local call rates no matter where the physical location of the number may be within the area covered by the company offering the service, e.g., within the U.K. This can provide businesses with a way to give information to callers responding to a marketing campaign, for example, in such a way that the caller bears part of the cost and thus will probably not call out of idle curiosity alone. It is an alternative to a 0800 number where the callee bears the full cost of the call.

0500 A reverse charge telephone service introduced by Mercury in the U.K. in 1992 that is similar to the British Telecom (BT) 0800 service. See 0800.

0645 A 'shared tariff' telephone service introduced by Mercury in the U.K. that is similar to the British Telecom (BT) 0345 service. See 0345.

070 A European national telephone services providing a subscriber (typically a business) with a portable number at which they can be reached from any calling location in the country. The caller pays for the call, based upon National Call rates. This is especially useful when a business moves, as directory listings, stationery, and other business identifiers don't have to be changed to reflect a new number.

0800 A European 'FreePhone' service offered by WorldCom, Global Carrier Services, and others, similar to North American 800 service, that makes it possible for callers to call the 0800-prefixed number free of charge, with the company holding the 0800 number bearing the cost. It's essentially a 'collect' or 'reversed charge' call that is put through directly as a continuing service, without going through operators or authorization to accept the call. Golden 0800 numbers are those that are inherently easy to remember (e.g., 0800 555 0000) or that correspond to a mnemonic (e.g., 0800 callnow). Most phone companies charge a premium for golden numbers but sometimes you get lucky. See 0345, 800.

0990 A national pay telephone service number prefix and telephone service offered in the U.K. by British Telecom (BT). It is similar to U.S. 900-prefix services that are typically billed by the minute. 0990 numbers are used for dating services, psychic-style services, and sex services. Not all businesses using these numbers are ethical. They are required to play a recording to inform customers that they are being billed, but some 'neglect' this requirement. Some keep the caller on the line for a long time by asking personal questions or chatting, in order to increase the call duration and hence the charges. It is possible to make 0990 services accessible from Internet phone gateways (almost 1/3 of the call lines are accessed this way in the U.K.) and, once again, the vendor is required to inform the caller that they are being billed 0990 rates.

Due to problems with enforcing ethical and legal 0990

services, in 2000 Nippon Telephone and Telegraph (NTT) reduced the maximum charges a vendor could accumulate for each 0990 call.

155 In the U.K., the British Telecom (BT) code for connecting with an international operator.

555-1212 In the U.S., a dialing sequence for contacting the Directory Assistance service (formerly the Information service) for obtaining long-distance publicly listed telephone numbers. It is useful for obtaining new numbers that may not yet be listed, or numbers published elsewhere geographically. Prior to the mid-1960s, the dialing code for Information was 113. Most North American telephone exchanges use this number as a standard. See 555-1212.

611 A telephone dialing code for contacting telephone maintenance and repair services to request assistance with telephone services and equipment. Unlike services prior to the mid-1960s, 611 numbers were used through a transitional period during which the phone company's monopoly was challenged and subscribers began to select their own telephone equipment from third-party vendors and eventually were permitted to install or modify the line extending from the phone company's line attachment point into the premises (depending upon the service and region). Thus, repair services didn't just schedule the repair any more, they would query the 611 caller about the nature of the problem and try to guess whether it was being caused by the phone company's line or service or the subscriber's line and equipment. If it was suspected that the problem was with subscriber equipment, the phone company would warn the subscriber about minimum and hourly charges and then confirm whether they still wanted to call in a repair person. This was more complicated than previous procedures but not nearly as complicated as what happened a couple of decades later with further deregulation and burgeoning carriers and phone services.

800 A service in which calls are billed to the receiver. 800 numbers are widely used by businesses to encourage potential buyers to contact the company through a toll-free 800 number (or 888 number) without concern about long distances charges. These services are sometimes used internally, through an unpublicized 800 number for traveling employees to contact their main office or branch. When 800 began to be in short supply, 888 and 877 prefixes/services were added to extend the available numbers. See 0800.

800 Service Management System (SMS/800) Functions A Federal Communications Commission (FCC) Tariff document published by the Bell Operating Companies (BOCs) to describe the regulations, rates, and charges applicable to the provision of SMS/800 functions and support services through intelligent telephone networks. See SMS/800.

877/877/866/855/844/833/822 Toll-free calling service prefixes that were developed to extend the availability of 800 numbers when 800 numbers assignments were nearing capacity. These numbers are sequentially opened as needed. See 800.

900/976 A set of numbers billed to the caller through a rate determined by the callee. 900 services are used by information brokers, public opinion pollers, advice counsellors, astrologers, other prognosticators, and by vendors offering phone sex. Charges for 900 calls usually range from $1.95 to $4.95 per minute.

These calls are not only somewhat expensive, but some of the less scrupulous 900 vendors will keep the caller on the line longer by asking him or her many personal questions at the beginning of the call. It's not unusual for the average call to be around $25 and often they are much more. Because of the abuse or overuse of 900 numbers, subscribers demanded a way to disable 900 calling and this is now provided by the phone companies. This is mainly to curtail calls by children, 900-addicts, and 900 calls by unauthorized callers using a phone without permission. See 0900.

911 calling A telephony service designed to expedite connections to emergency services such as medical services, law enforcement, and fire departments. By dialing only three easy-to-remember digits, subscribers can more easily get help when needed. This concept was first introduced in the 1970s. The calls are connected quickly to a Public Safety Answering Point (PSAP) where a trained emergency dispatcher records the call, determines the origin and nature of the call, responds to the caller, and dispatches services as needed.

1571 In the U.K., the British Telecom (BT) standard telephone message retrieval number.

1660 In the U.K., the British Telecom (BT) Worldcom access number.

Telephone Quick-Dial Numbers

112 In Europe, the British Telecom (BT) standard emergency telephone number. The number 17099 is an alternate emergency code. See 911.

113 Historically, in the U.S., a telephone dialing sequence for contacting Information Services, a service that aided subscribers in getting publicly listed phone numbers that were not in the current directory (e.g., new numbers). In the mid-1960s, in North America this code was changed to 411. There are other countries where the use of 113 has been retained. For example, in Copenhagen, it is used for overseas directory inquiries. See 411.

114 Historically, in the U.S., a telephone dialing sequence for contacting telephone maintenance and repair services. If you had trouble with your line, you called 114. Since most people had only one phone in those days, you usually went next door to dial from your neighbor's phone while sitting down for a cup of coffee and a chat. The repair people would schedule a visit and repair the problem. It was a simpler process in those days. Not only was the telephone network far more homogenous than now, but the phone company owned the line and the equipment right up to the phone itself and thus could standardize hardware and procedures for making repairs. In the mid-1960s, the dial sequence was changed to 611 and now local phone companies have a variety of numbers to dial for repair services. See 611.

411 In telephony, a short dialing sequence for contacting the Directory Assistance service (formerly the

Z

Information service) for obtaining publicly listed telephone numbers that may not be listed in the local directory (or which the caller couldn't find in the local directory). It is useful for obtaining new numbers that may not yet be listed, or numbers published elsewhere geographically. Prior to the mid-1960s, the dialing code for Information was 113. Historically, in some areas the service was free, but it became a pay service charged by the call. More recently, some phone companies offer bundled services that permit up to a specified number of Directory Assistance calls per time period (e.g., per week or month). Most North American telephone exchanges use this number as a standard.

Until 2001, in most areas, a caller could request two numbers for about $.40 but deregulation resulted in increases of up to $1.99 for a single number, depending upon the service provider. (Some carriers still permit two or three calls free per month per subscriber.) As charges rise, it is likely that people will migrate to CD-ROM directories and Web-based directory services such as InfoSpace and Switchboard. See 555-1212.

Telegraph and Radio

1 A short telegraphic shortcut numeric code to express "Wait a moment," "Give me a second," or "Hold on." See 73 for the background to numeric codes.

"10-4" radio signal codes Numeric codes used by police departments to describe a situation in shorthand. For example, 11 might mean a burglary, while 34 could signal a suicide. These codes are regional and specific to their industries. In California, for example, a 10-4 *patrol code* indicates message received, while 10-15 signals a prisoner in custody, and 10-33 is an alarm or indication of an officer needing help.

Radio codes are also used by the rail industry for cross-country trains and subway trains. For example, in the New York City Transit system radio code signal system, 12-6 signals a derailment and 12-12 indicates disorderly passengers.

Codes are usually used for brevity and consistency, but may also be used to provide a small amount of security. See "Code 3" radio communications codes.

13 A shortcut telegraphic numeric code for "I don't understand." See 73.

30 A short telegraphic shortcut numeric code to express the end of a communication, thus "Done," "Finished." See 73.

73 A number in a telegraph numeric 'shortcut' code dating back to at least the mid-1800s. The number 73 was an abbreviated means of representing various sentimental, amorous, and fraternal greetings, depending upon the time, place, and operator. It was similar to the greeting at the end of a letter written by someone familiar with or fond of the addressee.

There is a reference in an 1857 issue of the *National Telegraphic Review and Operator's Guide* that lists "73" as a numeric shortkey code. Apparently, a committee was established at a convention in 1859 to assign meanings to the numbers from 1 to 92, so this may have been the original impetus for more broad use of standardized numeric codes.

Some have attributed the origin of 73 and other numeric shortcuts to Phillips shortcode, but Walter Polk Phillips didn't publish his code until 1879 and the Phillips code emphasized alphabetic rather than numeric relationships. Thus, while his contribution was substantial and influential, especially in the news industry, Phillips didn't create an entirely new code; his contribution was to expand, consolidate, and revise existing code practices. So, it appears 73 was likely in use before the Phillips code was developed.

Whatever their origin, certain of the numeric shortcodes became widely used and still retain their original meanings, while many have fallen into disuse and some have mutated in meaning. The railroad still uses some of the code numbers for train-related orders. See Q signal, Z code.

73 key A somewhat unique-looking, historic, palm-sized telegraph key intended to be portable and thus covered with a squarish metal housing with the user parts protruding from two sides. The number 73 was printed on the top of the covers, probably a tongue-in-cheek reference to the number 73 shortcode used in telegraphy to convey greetings or intimate best wishes. The key was distributed and labeled by the Ultimate Transmitter Company, Los Angeles.

73 Magazine A magazine of interest to telegraphers and amateur radio operators originated by Wayne Green and associates in 1960. Topics ranged from hobby projects, to in-depth looks at telegraph keys, to amateur radio enthusiasms, to the history of submarine cable communications, and there were even some parody issues. Probably not coincidentally, early issues sold for 37 cents each.

In the mid-1970s, controversy over the content of the magazine provoked Pacific Telephone & Telegraph Co. to file suit against 73, Inc., due to their concerns that the information provided might make it easy for readers to find ways to avoid phone charges. Given that this was right around the time that 'blue boxing' was beginning to spread, the concerns may have been based on actual phone service thefts. Wayne Green went on to found other magazines, notably *80 Microcomputing*, which attracted much of the same audience that had been interested in amateur radio and telegraphy prior to the development of personal computers.

"Code 3" radio communication codes A system of numeric codes used by police departments as a shorthand for describing a situation to dispatchers and other officers within radio contact. For example, Code 3 might indicate emergency lights and sirens. In Dallas, Code 5 is shorthand for officer en route to a scene while in California, it signifies a stakeout. In the Dallas Police Department, Code 10X is shorthand for a stolen vehicle. These are usually prefaced with the word "Code" to distinguish them from similar numeric radio signal codes. See "10-4" radio signal codes.

Appendices

A. Fiber Optics Timeline 1050

B. Asynchronous Transfer Mode (ATM) 1052

C. ITU-T Series Recommendations 1055

D. List of World Wide Web Search Engines 1056

E. List of Internet Domain Name Extensions 1057

F. Short List of Request for Comments (RFC) 1059

G. National Associations 1062

H. Dial Equivalents, Radio Alphabet, Morse Code,
   Metric Prefixes/Values 1066

I. ASCII Character and Control Codes................. 1067

Fiber Optics Timeline

Essential Concepts and Evolution of Fiber Optic Technologies

~1650 B.C.   Ahmose transcribes Egyptian mathematics, which include fractions.

~670 B.C.   Thales engages in abstract and deductive mathematics and investigates magnetism.

~500 B.C.   Pythagoras and the Pythagoreans make important contributions to mathematics and the study of sound frequency relationships.

~260 B.C.   Archimedes establishes many important, basic principles of physics.

~800   The concept of zero is used in mathematics in Asia.

~140   Ptolemy describes the interaction of light and matter, which was referred to in Arab texts.

1200s   L.P. Fibonacci authors *Liber Abaci* (*Book of the Abacus*) in which he promotes the use of Arabic numerals and positional notation.

15/1600s   Galileo Galilei makes important observations of the laws of physics, especially of gravity and bodies in motion.

late 1600s   Sir Isaac Newton makes important observations about basic physical laws, now widely known as Newtonian physics or classical physics. These highly significant discoveries form the basis of modern physics. He observes Newton's rings but doesn't fully appreciate their significance.

1676   Ole (Olaf) Røhmer calculates the velocity of light as a constant 227,000 kilometers/second.

1669   Rasmus Bartholin receives a piece of Iceland spar and describes double refraction.

mid-1700s   Benjamin Franklin conducts numerous experiments with electricity, inspires other scientists, and coins many terms associated with the emerging science.

1700s   Luigi Galvani studies electromagnetism in living tissue and galvanometer is developed.

1775   Alessandro Volta invents the electrophorus, the basis of subsequent electrical condensers, replacing the Leyden jar as an energy storage capacitor.

1795   F. Salvà i Campillo describes a system for an electric telegraph, which he was finally able to construct, in 1804, by incorporating Alessandro Volta's ca. 1800 invention of the voltaic pile.

1800   Alessandro Volta invents the voltaic pile, a pioneer wet cell, forerunner to capacitors.

1800s   F.W. Herschel discovers infrared light and J.W. Ritter discovers ultraviolet light.

1807   J.B. Fourier announces Fourier's theorem, which forms the basis for Fourier transforms, now widely used as analytical tools in mathematics and for analyzing/recomposing waves.

Humphry Davy uses battery power to separate out and discover potassium and, about a decade later, invents an arc lamp (arc lamps were also invented by others).

1819   H.C. Ørsted demonstrates the relationship of electricity and magnetism.

1820   A.M. Ampère studies the mathematical characteristics of electromagnetism and announces the right-hand rule.

1822   Charles Babbage develops important historic models for 'different engines.'

1826   J.N. Niepce develops a primitive type of photography, the forerunner of optical recording.

1834   Charles Babbage develops the concept of an 'analytical engine.' Technology has not yet developed to the point where his ideas can be fully carried out, but the design concepts are sound.

1841   D.-J. Colladon demonstrates that a curved jet of water can guide light.

1848   The laying of the first transatlantic telegraph cable. It only lasted a few days.

1850s   Thomas Young describes and demonstrates wave properties of light.

1854   John Tyndall duplicates Colladon's light-guiding principles.

1861   The first transcontinental telegraph line is built in a record four months. At the line's completion in October, the Pony Express ceased operations.

P. Reis demonstrates the transmission of tones through wire. See telephone history.

1866   The laying of the first successful installation of a transatlantic telegraph cable, thus revolutionizing communications. Previously, sea voyages of two or three months were necessary to 'transmit' overseas messages.

A. Loomis demonstrates essential basics of wireless radio wave transmissions.

1872   James Clerk-Maxwell publishes an important paper on electromagnetic wave theories.

1874   Elisha Gray submits a caveat to the U.S. patent office after Alexander Graham Bell submits a patent for the 'harmonic telegraph,' the forerunner to the telephone.

1876   A. Graham Bell reports having spoken intelligibly over wires to his assistant, Watson.

1878   Public telephones make their commercial debut in Connecticut.

The first telephone exchange is established in London.

1879   American Telephone & Telegraph is founded based on the technology in the Bell patents, later to be known as AT&T. The telephone infrastructure is gradually put in place.

1891   Almon B. Strowger invents the automatic telephone switching system so subscribers can dial the desired number, rather than depending upon a human operator to connect a call.

Essential Concepts and Evolution of Fiber Optic Technologies, cont.	
1880	Bell and Tainter invent the Photophone, communication through light.
1880s	Charles Vernon Boys uses quartz fibers to study values for the gravitational constant.
1900	Max Planck states quantum theory and Planck's constant is expressed.
1900s	Einstein builds upon the quantum theory foundation established by Planck and describes the particle nature of light and the photoelectric effect.
1904	Fleming releases the two-element Fleming tube, which leads to de Forest's development of the three-element tube, the Audion.
	Lee de Forest invents the Audion leading to the evolution of three or more element vacuum tubes that revolutionize the electronics industry. R. A. Fessenden broadcasts the first voice and music broadcasts, using an Alexanderson alternator to supply the power.
1905	Albert Einstein publishes a paper on the special theory of relativity.
1920s	Bell Laboratories begins transmitting picture phone images.
	Various types of facsimile transmissions are implemented by different inventors.
	John Logie Baird carries out pioneer television broadcasting experiments. Hansell works with Baird and investigates the use of fiber for transmitting broadcast images.
1927	Scheduled television broadcasts are begun by station WGY in New York state.
1928	Bell Laboratories develops pioneer color television technologies.
1930	V. Bush's idea for a difference analyzer is constructed at the Massachusetts Institute of Technology (MIT).
1930	H. Lamm creates an aligned bundle of optical fibers but distance was limited.
1936	Konrad Zuse applies for a patent for 'mechanical memory' in connection with his historic calculating computer devices.
1937	John Atanasoff conceives the Atanasoff-Berry Computer (ABC).
1937	Howard Aiken teams up with IBM to produce the Harvard Mark 1, which became operational six years later.
1947	Bell Laboratories scientists invent the transistor. The age of vacuum tube calculators and computing machines nears an end as the solid-state electronics age is born.
1949	Edmund C. Berkeley authors *Giant Brains or Machines That Think* describing the construction of a 'personal computer.' He designed and built Simon, GENIAC, and many robots.
1950s	Hansen, van Heel, Hopkins, Hirschowitz, Curtiss, Kapani, O'Brien, and Hicks, and others develop lightguides leading to more practical optical fibers.
1950s	Scientists Dicke, Townes, Schawlow, and others develop laser technology. See laser history.
1956	Bell Telephone demonstrates Picturephone technology to the Institute of Radio Engineers.
1957	The Russian Sputnik communications satellite is launched into orbit.
1960s	The capabilities of fibers and lasers come together and now the industry rocks and rolls with Werts, Kao, Roberts, Shaver, and others developing practical embodiments of fiber lightguides. The earliest desktop computers predating the Altair begin to emerge but don't sell well.
1967	The concept and design of the ARPANET are born.
1969	The ARPANET is put into operation.
1970s	Corning is at the forefront of many of the new fiber fabrication and commercial technologies. Bell Labs creates fiber processes and improves the speed and distance capabilities of fiber. Many private and university labs become involved in fiber optics R & D.
1972	The Canadian ANIK satellite becomes the first domestic television broadcast communications.
1974	The Altair becomes the first desktop computer to be commercially successful, launching the microcomputer age. IMSAI, Apple, TRS-80, and other computers follow in rapid succession.
1980s	The evolution of lasers, fiber lightguides and more powerful compact computers makes it possible to develop cost-effective light-based telephone and data communications networks.
	Fiber doping research and applications are developed to improve fiber conducting properties.
	Fiber-based transatlantic cables are installed and the phone system begins to evolve into fiber for backbone trunk lines.
1980	T. Berners-Lee develops Enquire hyptertext system.
1983	The ARPANET is split into Milnet and ARPANET, the precursor to the Internet.
1989	T. Berners-Lee develops a Web browser, the World Wide Web is born. Everyone gets online.
1990	The ARPANET is officially discontinued as the Internet is born.
1990s	Fiber begins to become an important aspect of computer networks, with protocols and faster/longer technologies developed. Speed and distance barriers are overcome almost monthly.
	Federico Capasso et al. make significant contributions to quantum well theory/applications.
1999	Lene Hau, a Harvard physicist, stops light and releases it again at full energy.
2000s	Fiber begins to reach the curb and local area networks begin deploying fiber. Fiber has arrived.

Asynchronous Transfer Mode (ATM) Information

asynchronous transfer mode ATM. ATM is a highly significant protocol due to its flexibility and widespread use for Internet connectivity. It is a high-speed, cell-based, connection-oriented, packet transmission protocol for handling data with varying burst and bit rates. ATM evolved from standardization efforts by the CCITT (now the ITU-T) for broadband ISDN (B-ISDN) in the mid-1980s. It was originally related to Synchronous Digital Hierarchy (SDH) standards.

ATM allows integration of local area network (LAN) and wide area network (WAN) environments under a single protocol, with reduced encapsulation. It does not require a specific physical transport, and thus can be integrated with current physical networks. It provides virtual connection (VC) switching and multiplexing for broadband ISDN, to enable the uniform transmission of voice, data, video and other multimedia communications.

Two methods for carrying multiprotocol connectionless traffic over ATM are routed and bridged Protocol Data Units (PDUs). Routed PDUs allow the multiplexing of multiple protocols over a single ATM virtual circuit through LLC Encapsulation. Bridged PDUs carry out implicit higher-layer protocol multiplexing through virtual circuits (VCs).

ATM employs fixed-length cells consisting of an information field and a header. The information field is transparent through the transmission. The U.S. and Japan proposed the use of 64-byte cells, and Europe proposed 32-byte cells. As a consequence of the discrepancy, 48-byte cells are favored by many as a compromise.

Charts and simplified diagrams on the following pages show an ATM system through user input and reception of a variety of media, including voice, video, and data. The data are inserted and extracted by the ATM adaptation layer (AAL) into a logical package called a payload which makes up part of the ATM cell. The ATM layer, in turn, adds or removes a five-byte header to this payload, and the physical layer converts the information into the appropriate format for transmission, which may extend over large areas and pass through other networks switches and routers. The physical layer is comprised of two sublayers, the physical medium (PM) sublayer and the transmission convergence (TC) sublayer. See dictionary entries for Ethernet, frame relay, HIPPI, TCP/IP.

ATM cell The ATM cell is the basic unit of information transmitted through an ATM network. An ATM cell has a fixed length of 53 bytes, consisting of a 48-byte payload (the information being transmitted) and a 5-byte header (addressing information). Interpretation of the signals from different types of media into a fixed length unit of data makes it possible to accommodate different types of transmissions over one type of network.

There are a number of important traffic flow control, congestion management, and error-related concepts related to ATM, including those listed in the ATM Cell Rate Concepts chart shown on page 1038.

ATM adaptation layer AAL. In ATM, a set of ITU-T-recommended, service-dependent layer types interface the user to the ATM layer. The AAL is the top of three layers in the ATM protocol reference model. Higher layer services are translated through one or more ATM cells. AAL0 to AAL5 perform a variety of connection, synchronization, segmentation, and assembly functions for adapting different classes of applications to ATM. Within the AAL, information is mapped between the PDUs and ATM cells. Upon creation of a virtual connection (VC), a specific AAL is associated with that connection. See the following diagrams for the relationships of the adaptation layers to the ATM format.

ATM Cell Header and Payload Format

```
|<------------------- Header ------------------------------->|<Payload>|
+---------+---------------+-----------------+-------------------+----/----+
|VCI Label | control | header checksum | option. adaptation | payload |
| 3 bytes | 1 byte  |    1 byte       |   layer 4 bytes    | 44 or 48|
+---------+---------------+-----------------+-------------------+----/----+
```

ATM Cell at the User Network Interface (UNI)

Item	Abbreviation	Bits	Notes
cell loss priority	CLP	1	Cell loss priority of '1' is subject to discard, without violating agreed upon quality of service (QoS). If CLP is '0,' resources are allocated.
generic flow control	GFC	4	Point-to-point and point-to-multipoint. The field appears at the user network interface.

Adaptation Layer	Description
AAL0	A layer implementation intended to provide a direct connection between the user and the ATM. It is limited in that it provides no service guarantee mechanisms. It is recent and rarely used, except in proprietary, standalone systems. Nevertheless, some standard commercial drivers support AAL0.
AAL1	A constant rate service level. It is useful for time-sensitive applications such as voice, video, and circuit emulation.
AAL2	A variable rate service. It is rarely used.
AAL3/4	A variable rate service. It is the most comprehensive of the adaptation layers, and was originally specified as separate AAL3 and AAL4 for connectionless and connection communications.
AAL5	A variable rate service similar to AAL3/4. It is sometimes called SEAL for Simple and Efficient Adaptation Layer. It is widely used, especially in TCP/IP implementations. This is a nonassured service, and retransmission must be accomplished by higher-level protocols. It specifies a packet with a maximum size of 64K minus 1 octets.
AAL6	A recent addition, designed to accommodate demand for some of the recent multimedia, high-bandwidth applications.

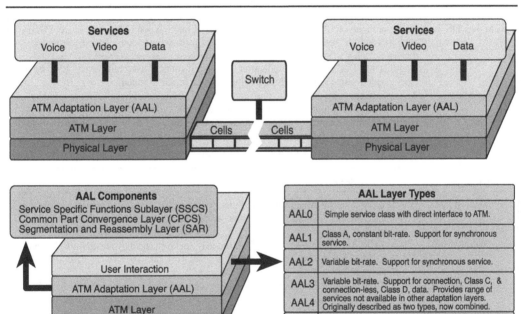

For further information related to ATM adaptation layers, see RFC 1483, RFC 1577, RFC 1626.

ATM models

Because of the great variety of needs in the networking community, many types and implementations of ATM networks have been developed. Information on some of the more common and emerging models is shown in the ATM Models chart. For further details on specific models, see dictionary entries under ATM Transition Model, Classical IP Model, Conventional Model, Integrated Model, Peer Model.

ATM Cell Rate Concepts

At its heart, ATM is concerned with moving and directing traffic; cells must be directed (and sometimes even discarded) such that signals, priority levels, and data are effectively transmitted and balanced with respect to the needs at hand. There is no single best way to 'tune' a network. The settings will vary, depending upon the system, the time of day, the quantity and priority levels of users, and many more subtle factors that are not necessarily known in advance.

Thus, there have been a number of basic cell rate concepts defined for system installers and administrators to assess and 'tweak' their systems for optimum information flow without loss. Some of the more important concepts are summarized in the ATM Cell Rates chart.

ATM Models and Test Systems of Interest

Model over ATM	Description
Classical IP	A model for enabling compatible, interoperable implementations for transmitting IP datagrams and ATM address Resolution Protocol (ATMARP) requests and replies over ATM adaptation layer 5 (AAL5). LLC/SNAP encapsulation of IP packets. IP address resolution to ATM addresses via an ATMARP service within the LIS. One IP subnet is used for many hosts and routers. Each virtual connection (VC) directly connects two IP members within the same LIST. TCP/IP applications. See RFC 1577.
IP Broadcast	An IP multicast service in development by the IP over ATM Working Group for supporting Internet Protocol (IP) broadcast transmissions as a special case of multicast. See RFC 2022, RFC 2226.
IP Multicast	Internet Protocol (IP) multicasting over Multicast Logical IP Subnetwork (MLIS) using ATM multicast routers. A model developed to work over the Mbone, an emerging multicasting internetwork. It is designed for compatibility with multicast routing protocols such as RFC 1112 and RFC 1075.
LANE	Local Area Network (LAN) Emulation. Protocol-independent applications aid in the transition from legacy internetworks to ATM.
Native ATM API	ATM-specific applications which take advantage of its quality of service (QoS) capabilities.
ATM over DS3	An experimental testbed network called XUNET II running at 45 Mbps to connect FDDI networks at eight sites across the continental U.S. from coast to coast. Internet Protocol routers at each site forward packets between connected local area networks (LANs) and long-distance DS3 links. The production version transmits IP datagrams over a PVC mesh fabric, with a single virtual circuit (VC) between each pair of routers.
Multicube	An experimental approach to the implementation of IP Multicast over ATM proposed by Schulzrinne et al. Multicube is a project to develop, test, and validate an ATM-based multipoint infrastructure for supporting CSCW applications. The majority of the multicast applications of endusers involved in the project are Internet Protocol (IP) based.

ATM Cell Rate Concepts

Cell Factor	Description
allowed cell rate	ACR. A traffic management parameter dynamically managed by congestion control mechanisms. ACR varies between the minimum cell rate (MCR) and the peak cell rate (PCR).
current cell rate	CCR. Aids in the calculation of ER and may not be changed by the network elements (NEs). CCR is set by the source to the available cell rate (ACR) when generating a forward RM-cell.
cutoff decrease factor	CDF. Controls the decrease in the allowed cell rate (ACR) associated with the cell rate margin (CRM).
cell interarrival variation	CIV. Changes in arrival times of cells nearing the receiver. If the cells are carrying information that must be synchronized, as in constant bit rate (CBR) traffic, then latency and other delays that cause interarrival variation can interfere with the output.
generic cell rate algorithm	GCRA. A conformance enforcing algorithm that evaluates arriving cells. See leaky bucket.
initial cell rate	ICR. A traffic flow available bit rate (ABR) service parameter. The ICR is the rate at which the source should be sending the data.
minimum cell rate	MCR. Available bit rate (ABR) service traffic descriptor. The MCR is the transmission rate in cells per second at which the source may always send.
peak cell rate	PCR. The PCR is the transmission rate in cells per second which may never be exceeded. It characterizes the constant bit rate (CBR).
rate decrease factor	RDF. An available bit rate (ABR) flow control service parameter that controls the decrease in the transmission rate of cells when it is needed. See cell rate.
sustainable cell rate	SCR. The upper measure of a computed average rate of cell transmission over time.
unspecified bit rate	UBR. An unguaranteed service type in which the network makes a best efforts attempt to meet bandwidth requirements.
variable bit rate	VBR. The type of irregular traffic generated by most non-voice media. Guaranteed sufficient bandwidth and QoS.

International Telegraph Union (ITU-T) Telecommunications Recommendations

Over the decades since its inception, the ITU has been developing international guidelines to promote compatibility and interoperability of communications systems, from the original telegraph to modern mobile communications systems.

These guidelines are available as publications from the ITU-T for purchase over the Internet and many in the A Series are downloadable without charge from the Web. Since ITU-T specifications and recommendations are widely followed by vendors in the telecommunications industry, those wanting to maximize interoperability with other systems need to be aware of the information disseminated by the ITU-T. The list below describes the general overall categories and specific series topics are listed under individual entries in this dictionary, e.g., B Series Recommendations. Note that some series topics include only a few documents, while others, such as the G Series Recommendations, include many hundreds of documents and thus some sections may be summarized, or described with examples. Note also that the author has taken time to categorize many of the documents, which can sometimes be difficult to locate in a numerical-only list, to aid the reader in finding the appropriate document and understanding the depth and breadth of the publications.

ITU-T Recommendations	
Categories	**Description**
Series A	Organization of the work of the ITU-T
Series B	Means of expression: definitions, symbols, classification
Series C	General telecommunications statistics
Series D	General tariff priniciples
Series E	Overall network operation, telephone service, service operation, and human factors
Series F	Telecommunication services other than telephone
Series G	Transmission systems and media, digital systems and networks
Series H	Audiovisual and multimedia systems
Series I	Integrated Services Digital Networks (ISDN)
Series J	Transmission of sound program and other multimedia signals
Series K	Protection against interference
Series L	Construction, installation, and protection of cables and other elements of outside plant
Series M	TMN and network maintenance: international transmission systems, telephone circuits, telegraphy, facsimile, and leased circuits
Series N	Maintenance: international sound program and television transmission circuits
Series O	Specifications of measuring equipment
Series P	Telephone transmission quality, telephone installations, local line networks
Series Q	Switching and signaling
Series R	Telgraph transmission
Series S	Telegraph services terminal equipment
Series T	Terminals for telematic services
Series U	Telegraph switching
Series V	Data communication over the telephone network
Series X	Data networks and open system communication
Series Y	Global information infrastructure and Internet protocol aspects
Series Z	Programming languages

World Wide Web
Major Search Engines

Name	URL (Web Address)	Notes
AltaVista	http://www.altavista.com/	Extensive searching, advanced search parameters, priority ranking. First introduced by Digital Equipment in 1995.
Ask Jeeves	http://www.ask.com/	Enables input of query sentences or phrases and provides intelligent natural language parsing of the query to provide a targeted list of hits and suggestions for related topics.
c|net Search	http://www.search.com/	Perhaps best known for its large repository of software updates, shareware, and public domain software, c|net also provides a Web search engine that displays a short list and enables users to look at further selections, if desired.
DejaNews	http://www.dejanews.com/	A huge archive of the posts to various USENET newsgroups. A remarkable record of public conversations online, searchable by keywords or author. This has now been acquired by Google.
DogPile	http://www.dogpile.com/	Looking for a short, targeted list of hits? Try this search engine. It also includes category searches and stores.
Excite	http://www.excite.com/	General search, weather, stocks.
Google	http://www.google.com/	Fast, extensive, with a lovely simple, uncluttered interface. Also includes Google News and lists USENET group postings as Google Groups (acquired from DejaNews).
i-Explorer	http://www.i-explorer.com/	Search in popular, general interest categories.
InfoSeek	http://guide.infoseek.com/	Web pages, newsgroups, and individuals.
InfoSpace	http://www.infospace.com/	Personal and business listings, maps, etc.
Inktomi	http://inktomi.berkeley.edu/	Fast distributed searchable database from the University of California at Berkeley.
LinkStar	http://www.linkstar.com/	Business directory search.
Lycos	http://www.lycos.com/	General searching, maps, and personal names from Carnegie Mellon University.
Magellan	http://www.mckinley.com/	Sites reviewed and rated by the McKinley Group, Inc.
Sleuth	http://www.isleuth.com/	The Internet Sleuth searches over 3,000 Internet databases. Selections can be found through general categories. (This may now be defunct.)
Starting Point	http://www.stpt.com/	Searches the Web and other Internet resources (selectable), includes advanced search capabilities.
Switchboard	http://www.switchboard.com/	Personal and business listings of names, addresses, and email addresses.
Webcrawler	http://www.webcrawler.com/	Quick, to-the-point listings.
Yahoo	http://www.yahoo.com/	An extensive service that includes a search engine and hundreds of topics organized under categories of interest.

For further information from the publisher: http://www.crcpress.com/
For further information from the author: http://www.4-sightmedia.com/

Internet Domain Name Extensions

North America and Generic International

.us	United States	U.S., not commonly used
.um	United States	Outlying islands
.gov	U.S. government	Local, state, and federal government agencies
.mil	U.S. military	Military agencies, bases
.arpa	ARPANET	Advanced Projects Research Agency
.ca	Canada	
.mx	United Mexican States	
.int	international	
.com	commercial	General business, services, suppliers
.biz	business	Retail business, malls, electronic storefronts
.pro	professional	Doctors, lawyers, consultants, home care nurses, realtors, vets, carpenters
.info	information	Noninstitutional educational, informational
.net	network	Net related.
.org	organization	Nonprofit, not-for-profit, charitable
.edu	education	Schools, colleges, universities, other educational facilities
.museum	museums	Public and private repositories in many disciplines
.aero	aeronautics	Airlines, aeronautical suppliers, contractors

Central and South America

.ag	Antigua, Barbuda
.ar	Argentine Republic
.aw	Aruba
.bb	Barbados
.bz	Belize
.bm	Bermuda
.bo	Bolivia
.br	Brazil
.cl	Chile
.co	Colombia
.cr	Costa Rica
.cu	Cuba
.dm	Dominica
.do	Dominican Republic
.ec	Ecuador
.sw	El Salvador
.fk	Falkland Islands (Malvinas)
.tf	French Southern Territories
.gd	Grenada
.gp	Guadeloupe
.gt	Guatemala
.gy	Guyana
.gf	Guyana (French)
.ht	Haiti
.hn	Honduras
.jm	Jamaica
.mq	Martinique
.ms	Montserrat
.ni	Nicaragua
.pa	Panama
.py	Paraguay
.pe	Peru
.pr	Puerto Rico
.kn	St. Kitts, Nevas
.lc	St. Lucia
.vc	St. Vincent, Grenadines
.gs	So. Georgia, So. Sandwich Islands
.pm	St. Pierre, Miquelon
.sr	Suriname
.uy	Uruguay
.ve	Venezuela

United Kingdom, Europe

.gb	Great Britain
.ie	Ireland
.im	Isle of Man
.uk	United Kingdom
.al	Albania
.ad	Andorra
.at	Austria
.by	Belarus
.be	Belgium
.ba	Bosnia, Herzegovina
.bg	Bulgaria
.hr	Croatia
.cy	Cyprus
.cz	Czech Republic
.dk	Denmark
.ee	Estonia
.fo	Faroe Islands
.fi	Finland
.de	Federal Republic of Germany
.fr	France
.fx	France
.de	Germany
.gi	Gibraltar
.gr	Greece
.gl	Greenland (Denmark)
.gg	Guernsey
.hu	Hungary
.is	Iceland
.it	Italian Republic
.je	Jersey
.lv	Latvia
.li	Liechtenstein, Principality of
.lt	Lithuania
.lu	Luxembourg, Grand Duchy of
.mk	Macedonia
.mt	Malta
.md	Moldova
.mc	Monaco
.nl	Netherlands
.no	Norway
.pt	Portuguese Republic
.ro	Romania
.sm	San Marino (Italy)
.sk	Slovakia
.si	Slovenia
.es	Spain
.se	Sweden
.ch	Switzerland
.ua	Ukraine
.va	Vatican City State
.yu	Yugoslavia

Middle East

.bh	Bahrain
.ir	Iran
.iq	Iraq
.il	Israel
.jo	Hashemite Kingdom of Jordon
.kw	Kuwait
.lb	Lebanon
.sa	Saudi Arabia
.tr	Turkey
.ae	United Arab Emirates

Internet Domain Name Extensions, cont.

Eastern Europe, Middle Asia

.af	Afghanistan
.al	Albania
.am	Armenia
.az	Azerbaijan
.bg	Bulgaria
.by	Bielorussia
.hr	Croatia
.ee	Estonia
.kz	Kazakhstan
.kg	Kirgistan
.lv	Latvia
.lt	Lithuania
.md	Moldavia
.pl	Poland
.ro	Romania
.ru	Russian Federation
.rw	Rwanda
.sk	Slovakia
.si	Slovenia
.tj	Tadzhikistan
.tm	Turkmenistan
.ua	Ukraine
.uz	Uzbekistan

Mediterranean, Caribbean

.an	Antilles (Netherlands)
.aw	Aruba
.cy	Cypress
.mt	Malta
.lc	Saint Lucia

Africa

.dz	Algeria
.ao	Angola
.bj	Benin
.bw	Botswana
.bv	Bouvet Island
.bf	Burkina Faso
.bi	Burundi
.cm	Cameroon
.cw	Cape Verde
.cf	Central African Republic
.td	Chad
.km	Comoros
.cd	Congo
.cg	Congo Republic
.ci	Côte d'Ivoire
.dj	Djibouti
.eg	Egypt, Arab Republic of
.gq	Equatorial Guinea
.er	Eritrea
.et	Ethiopia
.ga	Gabon
.gm	Gambia
.gh	Ghana
.gn	Guinea
.gw	Guinea Bissau

.ci	Ivory Coast
.ke	Kenya
.ls	Lesotho
.lr	Liberia
.ly	Libya
.mg	Madagascar
.mr	Mauritania
.mw	Malawi
.ml	Mali
.mu	Mauritius
.yt	Mayotte
.ma	Morocco
.mz	Mozambique
.na	Namibia
.ne	Niger
.ng	Nigeria
.re	Reunion
.sn	Senegal
.sc	Seychelles
.sl	Sierra Leone
.so	Somalia
.za	South Africa
.sh	St. Helena
.sd	Sudan
.sz	Swaziland
.tz	Tanzania
.tg	Togo
.tn	Tunisia
.ug	Uganda
.eh	Western Sahara
.zr	Zaire
.zm	Zambia
.zw	Zimbabwe

Asia, South Pacific, Antarctic

.bd	Bangladesh
.io	British Indian Ocean Territories
.in	India
.bt	Bhutan
.kh	Cambodia
.hk	Hong Kong (Xianggang)
.jp	Japan
.kp	Korea (North)
.kr	Korea (South)
.la	Laos
.mo	Macau
.my	Malaysia
.mn	Mongolia
.np	Nepal
.pk	Pakistan
.cn	People's Republic of China
.sg	Singapore
.kr	South Korea
.lk	Sri Lanka
.tw	Taiwan
.th	Thailand

.to	Tonga
.vn	Vietnam
.am	American Samoa
.au	Australia
.bn	Brunei
.ck	Cook Islands
.fj	Fiji
.gu	Guam
.id	Indonesia
.my	Malaysia
.mh	Marshall Islands
.fm	Micronesia
.nz	New Zealand
.nf	Norfolk Island
.pw	Palau
.pg	Papua (New Guinea)
.ph	Philippines
.pn	Pitcairn Islands
.pf	Polynesia (French)
.ws	Samoa
.sb	Solomon Islands
.tk	Tokelau
.tv	Tuvalu
.wf	Wallace, Futuna Islands
.pg	Antarctica

West Indies, Antilles

.ai	Anguilla
.ag	Antigua and Barbuda
.bb	Barbados
.bm	Bermuda
.bs	Bahamas
.ky	Cayman Islands
.cu	Cuba
.dm	Dominica
.do	Dominican Republic
.gd	Grenada
.gp	Guadeloupe (French)
.ht	Haiti
.jm	Jamaica
.mq	Martinique (French)
.pr	Puerto Rico (U.S.)
.lc	Saint Lucia
tc.	Turks, Caicos Islands
.tt	Trinidad and Tobago
.vg	Virgin Islands (British)
.vi	Virgin Islands (U.S.)

Miscellaneous

.aq	Antarctica
.nt	Neutral Zone

Request for Comments (RFC) Documents

Request for Comments (RFC) documents are an essential resource for understanding the implementation, structure, format, and evolution of the Internet. There are over 3,000 of these documents and, unfortunately, not sufficient space here to list abstracts or even the titles of all the RFCs. Nevertheless, the following quick lookup summarizes some of the most important general concepts and some of the RFCs with greater relevance to fiber optic technologies.

The reader is encouraged to consult the many excellent RFC repositories on the Internet archived in various formats including ASCII, editable PostScript, Adobe PDF, and HTML. The RFCs themselves include references to related documents of interest.

In addition to www.w3c.org, (for Web info) and www.rfc.net, some good Net archives of interest include

http://www.armware.dk/RFC	Searchable, nicely formatted, forward and backward references
http://www.faqs.org/rfcs/	Searchable and shows authors, dates, and references in search results
http://www.nexor.com/index-rfc.htm	Searchable from a selection of archives, includes Perl 5 expressions and Title/Author/Keyword searching
http://www.cis.ohio-state.edu/cs/Services/rfc/index.html	Categorizations and lists and links to RFCs of particular interest

General

RFC 1358	Charter of the Internet Architecture Board (IAB), August 1992.
RFC 1594	Answers to Commonly Asked "New Internet User" Questions, March 1994
RFC 1709	K–12 Internetworking Guidelines, November 1994.
RFC 1796	Not all RFCs are Standards, April 1995.
RFC 1920	INTERNET OFFICIAL PROTOCOL STANDARDS, March 1996.
RFC 1925	The Twelve Networking Truths, 1 April 1996.
RFC 1935	What is the Internet, Anyway?, April 1996.
RFC 1941	Frequently Asked Questions for Schools, May 1996.
RFC 1958	Architectural Principles of the Internet, June 1996.
RFC 1983	Internet User's Glossary, August 1996.
RFC 1999	Request for Comments Summary RFC Numbers 1900-1999,
RFC 2000	INTERNET OFFICIAL PROTOCOL STANDARDS, January 1997.
RFC 2014	IRTF Research Group Guidelines and Procedures, October 1996.
RFC 2026	The Internet Standards Process - Revision 3, October 1996.
RFC 2028	The Organizations Involved in the IETF Standards Process, October 1996.
RFC 2031	IETF-ISOC Relationship, October 1996.
RFC 2125	A Primer on Internet and TCP/IP Tools and Utilities, March 1997.
RFC 2223	Instructions to RFC Authors, October 1997.
RFC 2360	Guide for Internet Standards Writers, June 1998.
RFC 2799	Request for Comments Summary RFC Numbers 2700-2799, September 2000.
RFC 2900	Internet Official Protocol Standards, August 2001. Obsoletes 2800.
RFC 3000	INTERNET OFFICIAL PROTOCOL STANDARDS, November 2001.
RFC 3160	The Tao of IETF - A Novice's Guide to the Internet Engineering Task Force, August 2001.
RFC 3272	Overview and Principles of Internet Traffic Engineering, May 2002.

IP General

RFC 1919	Classical versus Transparent IP Proxies
RFC 1932	IP over ATM: A Framework Document
RFC 1954	Transmission of Flow Labeled IPv4 on ATM Data Links Ipsilon Version 1.0
RFC 2002	IP Mobility Support
RFC 2764	A Framework for IP-Based Virtual Private Networks
RFC 3168	The Addition of Explicit Congestion Notification (ECN) to IP

IPv6 (IPNG)

RFC 1902	Structure of Management Information for Version 2 of the Simple Network Management Protocol
RFC 1903	Textual Conventions for Version 2 of the Simple Network Management Protocol
RFC 1904	Conformance Statements for Version 2 of the Simple Network Management Protocol
RFC 1905	Protocol Operations for Version 2 of the Simple Network Management Protocol
RFC 1906	Transport Mappings for Version 2 of the Simple Network Management Protocol
RFC 1907	Management Information Base for Version 2 of the Simple Network Management Protocol
RFC 1924	A Compact Representation of IPv6 Addresses
RFC 1933	Transition Mechanisms for IPv6 Hosts and Routers
RFC 1955	New Scheme for Internet Routing and Addressing (ENCAPS) for IPNG
RFC 1970	Neighbor Discovery for IP Version 6 (IPv6)
RFC 1971	IPv6 Stateless Address Autoconfiguration
RFC 1972	A Method for the Transmission of IPv6 Packets over Ethernet Networks
RFC 1981	Path MTU Discovery for IP version 6SNMPv2

PPP

RFC 1172	Point-to-Point Protocol (PPP) initial configuration options
RFC 1332	The PPP Internet Protocol Control Protocol (IPCP)
RFC 1334	PPP Authentication Protocols
RFC 1661	The Point-to-Point Protocol (PPP)
RFC 1841	PPP Network Control Protocol for LAN Extension
RFC 1877	PPP Internet Protocol Control Protocol Extensions for Name Server Addresses
RFC 1915	Variance for The PPP Connection Control Protocol and The PPP Encryption Control Protocol
RFC 1962	The PPP Compression Control Protocol (CCP)
RFC 1963	PPP Serial Data Transport Protocol (SDTP)
RFC 1967	PPP LZS-DCP Compression Protocol (LZS-DCP)
RFC 1968	The PPP Encryption Control Protocol (ECP)
RFC 1969	The PPP DES Encryption Protocol (DESE)
RFC 1973	PPP in Frame Relay
RFC 1975	PPP Magnalink Variable Resource Compression
RFC 1976	PPP for Data Compression in Data Circuit-Terminating Equipment (DCE)
RFC 1979	PPP Deflate Protocol
RFC 1989	PPP Link Quality Monitoring
RFC 1990	The PPP Multilink Protocol [Obsoletes RFC 1717]
RFC 1993	PPP Gandalf FZA Compression Protocol
RFC 1994	PPP Challenge Handshake Authentication Protocol (CHAP)
RFC 2363	PPP over FUNI
RFC 2364	PPP over AAL5
RFC 2716	PPP EAP TLS Authentication Protocol
RFC 2823	PPP over Simple Data Link (SDL) Using SONET/SDH with ATM-Like Framing
RFC 2878	PPP Bridging Control Protocol (BCP)

ATM

RFC 1483	Multiprotocol Encapsulation over ATM Adaptation Layer 5
RFC 1680	IPng Support for ATM Services
RFC 1755	ATM Signaling Support for IP over ATM
RFC 1932	IP over ATM: A Framework Document
RFC 1946	Native ATM Support for ST2+
RFC 2022	Support for Multicast over UNI 3.0/3.1-based ATM Networks
RFC 2098	Toshiba's Router Architecture Extensions for ATM: Overview
RFC 2225	Classical IP and ARP over ATM
RFC 2226	IP Broadcast over ATM Networks
RFC 2331	ATM Signaling Support for IP over ATM - UNI Signaling 4.0 Update

ATM, cont.

RFC 2364	PPP over AAL5
RFC 2379	RSVP over ATM Implementation Guidelines
RFC 2380	RSVP over ATM Implementation Requirements
RFC 2492	IPv6 over ATM Networks
RFC 2512	Accounting Information for ATM Networks
RFC 2514	Definitions of Textual Conventions and OBJECT-IDENTITIES for ATM Management
RFC 2515	Definitions of Managed Objects for ATM Management
RFC 2684	Multiprotocol Encapsulation over ATM Adaptation Layer 5
RFC 2761	Terminology for ATM Benchmarking
RFC 2844	OSPF over ATM and Proxy-PAR

RFCs Related to Optical Networking

RFC 1044	Internet Protocol on Network Systems HYPERchannel Protocol Specification, February 1988.
RFC 1077	Critical Issues in High Bandwidth Networking, November 1988.
RFC 1152	Workshop Report Internet Research Steering Group Workshop on Very-High-Speed Networks, April 1990.
RFC 1259	Building the Open Road: The NREN as Test-Bed for the National Public Network, September 1991.
RFC 1323	A set of TCP extensions that help extend TCP into speeds that are characteristic of fiber optic networks, May 1992. Obsoletes RFC 1072 and RFC 1185.
RFC 1368	Definitions of Managed Objects for IEEE 802.3 Repeater Devices, October 1992.
RFC 1374	IP and ARP on HIPPI, October 1992.
RFC 1455	Physical Link Security Type of Service, May 1993.
RFC 1595	Definitions of Managed Objects for the SONET/SDH Interface Type, March 1994.
RFC 1619	PPP over SONET/SDH, May 1994.
RFC 1686	IPng Requirements: A Cable Television Industry Viewpoint, August 1994.
RFC 2067	IP over HIPPI, January 1997.
RFC 2171	MAPOS – Multiple Access Protocol over SONET/SDH Version 1, June 1997.
RFC 2558	Definitions of Managed Objects for the SONET/SDH Interface Type, March 1999.
RFC 2615	PPP over SONET/SDH, June 1999.
RFC 2625	IP and ARP over Fibre Channel, June 1999.
RFC 2816	A Framework for Integrated Services Over Shared and Switched IEEE 802 LAN Technologies, May 2000.
RFC 2823	PPP over Simple Data LInk (SDL) Using SONET/SDH with ATM-Like Framing, May 2000.
RFC 2834	ARP and IP Broadcast over HIPPI-800, May 2000.
RFC 2837	Definitions of Managed Objects for the Fabric Element in Fibre Channel Standard, May 2000.
RFC 2892	The Cisco SRP MAC Layer Protocol, August 2000.
RFC 3186	MAPOS/PPP Tunneling Mode, December 2001.
RFC 3255	Extending Point-to-Point Protocol (PPP) over Synchronous Optical NETwork/Synchronous Digital Hierarchy (SONET/SDH) with Virtual Concatenation, High Order and Low Order Payloads, April 2002.
RFC 3347	Small Computer Systems Interface Protocol over the Internet (iSCSI) Requirements and Design Considerations, July 2002.

Frame Relay

RFC 1586	Guidelines for Running OSPF over Frame Relay Networks
RFC 1973	PPP in Frame Relay
RFC 2115	Management Information Base for Frame Relay DTEs Using SMIv2
RFC 2427	Multiprotocol Interconnect over Frame Relay
RFC 2590	Transmission of IPv6 Packets over Frame Relay Networks Specification
RFC 2954	Definitions of Managed Objects for Frame Relay Service
RFC 3034	Use of Label Switching on Frame Relay Networks Specification

National Associations

National Association of Broadcasters NAB. A well-known American broadcast industry association providing support and education to its members through literature, standards activities, programming, conventions, and seminars. http:www.nab.org/

National Association of Radio and Telecommunications Engineers NARTE. An international professional association which provides support to members along with certification programs.

National Association of Regulatory Utility Commissioners NARUC. A Washington D.C.-based organization serving the needs of the various United States government utility commissioners. http://www.naruc.org/

National Association of State Telecommunications Directors (NASTD). An association of telecommunications professionals in state government engaged in the promotion and advancement of effective telecommunications policies and technology implementation to improve government operations. NASTD was founded in 1978 and is affiliated with the Council of State Governments (CSG). It includes representatives from the American states, territories, and the District of Columbia. http://www.nastd.org/

National Association of Telecommunications Officers and Advisors NATOA. A professional association which supports and services the telecommunications needs of local governments. NATOA provides education, information, and advocacy for their members. http://www.natoa.org/

National Bell Telephone Company A merger of the Bell Telephone Company and the New England Telephone Company, in a bid to achieve widespread national coverage of services. A court decree dissolved the company only 4 years later.

National Broadcasting Company NBC. A major broadcast company for many decades, formed in 1926 by David Sarnoff. NBC provides general television programming, entertainment, sports, news, local/interactive, programming transcripts, contests, games, and arts. See Sarnoff, David.

National Bureau Of Standards NBS. A bureau of the U.S. government which provides testing and standardization services. The NBS had an important role in the development of early computing devices in the 1940s when it undertook the construction of two large-scale computing machines for its internal needs, one to be installed on each coast. This resulted in the building of the Standards Eastern Automatic Computer (SEAC) and the Standards Western Automatic Computer (SWAC).

National Cable & Telecommunications Association NCTA. Formerly the National Cable Television Association, the NCTA is the primary trade association of the cable television industry in the U.S., founded in 1952. It provides members with industry information and a unified voice for advancing the technology and industry of cable telecommunications, serving more than 150 cable program networks. NCTA hosts an annual industry trade show. http://www.ncta.com/

National Cable Television Association NCTA. A trade association representing the cable broadcast industry founded in 1952. NCTA represents the interests of its members to public policy makers in the U.S. Congress, the judicial system, and the public. NCTA hosts a large annual trade show. NCTA is now the National Cable & Telecommunications Association.

National Committee for Information Technology Standardization NCITS. A U.S. organization for developing national information technology (IT) standards in cooperation with national and international standards bodies. http://www.ncits.org/

National Communications System NCS. A branch of the U.S. government formed in 1962 during the Cuban Missile Crisis. The recommendation of an interdepartmental committee reporting to President John F. Kennedy was to form a single communications system to serve the President, the Department of Defense (DoD), diplomatic and intelligence activities, and civilian leaders. The NCS was officially established in 1963 to link, improve, and extend the communication facilities and components of various federal agencies. It cooperates with various standards bodies, and develops emergency procedures for the American communications infrastructure. It also provides documents and CD-ROMs, including a Glossary of Telecommunications Terms (FS-1037C). The NCS is administered by the General Services Administration (GSA). See Glossary of Telecommunications Terms. http://www.ncs.gov/

National Continental Telephone, Telegraph & Cable Company of America Founded in 1899, this ambitious undertaking was an attempt to gain control of all the independent telephone companies, that is, all those not controlled by Bell. Despite backing by some of America's richest high-profile financiers, this project was unsuccessful.

National Coordinating Center NCC. A joint U.S. government-industry organization established by the National Communications System (NCS) to provide for the U.S. government's telecommunications service requirements. The NCC initiates, coordinates, and restores NS/EP telecommunications services. It is one of several divisions of the Office of the Manager of the National Communications System (OMNCS).

National Coordination Office for Computing, Information, and Communications NCO. A U.S. information and communications coordinating agency. http://www.hpcc.gov/

National Counterintelligence Information Center A U.S. government center within the National Security Council (NSC) for coordinating the identification and countering of foreign intelligence threats to the U.S. national/economic security.

National Digital Cartographic Data Base NDCDB. A U.S. database of digital cartographic/ geographic data files compiled by the U.S. Geological Survey (USGS). The database includes elevation, planimetric, landcover, and landuse data at various scales.

National Emergency Number Association NENA. A not-for-profit standard-setting organization that supports and promotes the development and availability of a universal emergency telephone number system (currently 911). NENA was founded in 1982 as a result of meetings of the National Telecommunications Information Administration (NTIA). NENA further supports research and education in the advancement and use of emergency number services. http://www.nena9-1-1.org/

National Exchange Carrier Association NECA. A nonprofit organization established by the Federal Communications Commission (FCC) in 1983 to administer issues related to service and access charges. The NECA serves the interests of incumbent local exchange carriers (LECs), and administers the universal service fund (USF), which subsidizes certain loop services. http://www.neca.org/

National Federation of Community Broadcasters NFCB. A national U.S. alliance of noncommercial community-oriented radio wave broadcast stations and producers. The NFCB was founded in the 1970s as a result of discussions at a meeting called the National Alternative Radio Konvention (NARK). It subsequently provided assistance and education to community broadcasters including the publishing of Audiocraft and The Public Radio Legal Handbook as standard references. In 1995, the head office was moved from Washington, D.C. to San Francisco.

NFCB member stations are committed to community support and participation. They are funded by listeners, grants, and the Corporation for Public Broadcasting (CPB). Almost half of them service rural communities, and a third are dedicated to minority radio broadcasting. The NFCB has become a lobbying voice at Congressional hearings and liaises with other national organizations. Lynn Chadwick has been credited with long-term leadership of the organization through some of its most important growth and development phases. See Community Broadcasters Association, Communications Policy Project. http://www.nfob.org/

National High Altitude Photography NHAP. Originally established in 1980 with satellites at 40,000 feet, the height was lowered to 20,000 feet, and the program renamed to National Aerial Photography Program in 1987. See National Aerial Photography Program.

National Communications System NCS. A governmental organization established in 1963 as a result of the Cuban missile crisis during which the U.S., NATO, and the U.S.S.R. experienced communications problems that could have had deadly repercussions. Following the crisis, President Kennedy ordered the investigation of national security communications and a committee recommended a single, unified system to serve the President, the Department of Defense (DoD), and other relevant bodies. The NCS focused on interconnectivity and survivability of an extended, interlinked governmental communication system. In 1984, the NCS's mandate was broadened to include national security and emergency preparedness. Support for the NCS within the Operations Division is provided by Training, Planning & Operational Support (TPOS). See National Coordinating Center for Telecommunications. http://www.ncs.gov/

National Coordinating Center for Telecommunications NCC. A joint government and industry organization for coordinating the initiation, restoration, and reconstitution of U.S. government national security and emergency preparedness telecommunications services in the U.S. and abroad. The National Communications System Operations Division (N3) provides guidance to NCC. See National Communications System. http://www.ncs.gov/ncc/

National Information and Communications Infrastructure – Africa NICI. A framework for developing policy, guidelines, regulations, and laws for directing and shaping Africa's communications infrastructure. In 2001, the number of Internet users in Africa was approaching 3 million, the majority being in the Sub-Saharan region, primarily South Africa.

National Institute of Optics NIO. The Canadian national organization located in Quebec city. This is now known as ONI.

National Institute of Standards and Technology NIST. This national standards organization is affiliated with the U.S. Department of Commerce. http://www.nist.gov/

National Associations, cont.

National Internet Services Provider NISP or NSP. An Internet Services Provider of national scope, usually with broader regional access and a variety of connection points. The Internet services provided by local ISPs and NSPs are usually similar. The main difference is that national providers often have dialups in major cities that the user can access with a local call when traveling, thus avoiding long-distance connect charges.

National ISDN Council See ISDN associations.

National Laboratory for Applied Network Research NLANR. An organization that researches leading-edge networks and supports the evolution of a U.S. national network infrastructure. Its main function is to provide technical, engineering, and traffic analysis support to National Science Foundation High Performance Connection sites. NLANR is divided into three main areas serving applications and users, engineers, and measurement and analysis professionals. It began as a collaborative project among NSF-supported supercomputer sites and was established in 1995.

National Oceanic and Atmospheric Administration NOAA. A U.S. government agency that sets strategic goals for environmental assessment, prediction, and stewardship and describes and predicts changes in the Earth's environment, to manage coastal and marine resources. http://www.noaa.gov/

National Public Safety Telecommunications Council NPSTC. A federation of associations representing and advocating telecommunications for public safety, founded in 1997. Recommendations from the Public Safety Wireless Advisory Committee (PSWAC) are followed up by NPSTC. The associations affiliated with NPSTC include the National Association of State Telecommunications Directors, the Association of Public-Safety Communications Officials – International (APCO), the National Association of State Emergency Medical Services Directors (NASEMSD), and others. http://npstc.du.edu/

National Research and Education Network NREN. A government-funded, gigabit-per-second, national research backbone proposed in the early 1990s after an initial proposal was presented in 1987 to the Congress by the Federal Coordinating Committee for Science Engineering and Technology (FCCSET). It was intended to support voice and video, and to become a significant means of finding and disseminating information. See National Science Foundation.

National Research Council NRC. A U.S. organization established in 1916 by the National Academy of Sciences to serve the needs of the science and technology community in advising the federal government. It is now the principal operating agency of the National Academy of Sciences and the National Academy of Engineering. The NRC provides services to the government, scientific and engineering communities, and the public. It is administered jointly by the Academies and the Institute of Medicine.

In 1989, the NRC expressed concern in a report about the vulnerability of a fully interconnected public switched network (PSN) and its implications for national security. See Network Reliability and Interoperability Council. http://www.nas.edu/nrc

National Rural Telecommunications Cooperative NRTC. A trade organization supporting over 1000 rural utilities organizations in 46 states in delivering telecommunications and information technology services. NRTC was founded in 1986 by the National Rural Electric Cooperative Association (NRECA) and the National Rural Utilities Cooperative Finance Corporation (CFC). In the early 1990s, in partnership with DIRECTV, Inc., NRTC made a significant investment toward launching the first U.S. high-power direct broadcast satellite (DBS) system, acquiring exclusive sales rights to 8% of DIRECTV subscribers. It is now the leading distributor of satellite broadcasting services and hardware to rural consumers. To promote legislative and distribution support for rural satellite technologies, NRTC also collaborated in the development of the Satellite Home Viewer Improvement Act in 1999. See LOCAL TV Act of 2000, Satellite Home Viewer Improvement Act. http://www.nrtc.org/

National Security Telecommunications Advisory Committee NSTAC. A committee of corporate leaders representing major telecommunications industries providing advice to the U.S. President on issues and vulnerabilities in national security and emergency preparedness telecommunications policies. See National Communications System. http://www.ncs.gov/

National Science Foundation NSF An independent U.S. government agency, established in 1950 to promote public welfare through science and engineering research and education projects through various types of educational and financial support. The NSF was established by the National Science Foundation Act of 1950, and provided with additional authority through the Science and Engineering Equal Opportunities Act. It is administered by the National Science Board appointed by the President with the advice and consent of the U.S. Senate. See NSFNET. http://www.nsf.gov/

National Space Development Agency NASDA. A Japanese national agency established to promote the development and peaceful use of space through the Japanese Space Development Program, NASDA was founded in October, 1969. The headquarters are located at the Tanegashima Space Center, where satellites are launched into orbit. In 1972, a further Tsukuba Space Center was established, followed six years later by an Earth Observation Center and, in 1980, by the Kakuda Propulsion Center. By the early 1990s, NASDA was involved in a number of national and collaborative space experiments. NASDA jointly supports the Geostationary Meteorological Satellite (GMS) system in conjunction with the Japan Meteorological Association. See Geostationary Meteorological Satellite. http://www.nasda.go.jp/

National Standards System NSS. A Canadian standards association which is managed by and works in conjunction with the Standards Council of Canada and a committee of volunteers to write standards, and to test and certify products and systems. See Canadian Standards Association, Standards Council of Canada.

National Storage Industry Consortium NSIC. A nonprofit consortium of more than 50 universities, research laboratories, and corporations dedicated to research and development in digital information storage systems. NSIC is headquartered in San Diego, California, and was incorporated in April 1991. Among other things, NSIC defines, organizes, and manages longer-range research projects such as the Extremely High Density Recording (EHDR) Project and the Network-Attached Storage Devices (NASD) Project. See NASD Project. http://www.nsic.org/

National Technical Information Service NTIS. An agency of the U.S. Department of Commerce, through the Technology Administration. NTIS is the official source for various types and formats of U.S. government-sponsored global scientific, technical, engineering, and business-related information, supplied by many U.S. government agencies. http://www.ntis.gov/

National Telecommunications and Information Administration NTIA. An Executive Branch agency of the U.S. Department of Commerce founded in 1978. NTIA is responsible for domestic and international telecommunications policy issues, and is a principal advisor to the President. NTIA works to promote efficient and effective uses of telecommunications information and resources in order to support U.S. competitiveness and job opportunities.
NTIA is descended from a reorganization of the Office of Telecommunications Policy (OTP) and the Office of Telecommunications (OT). It cooperates with the Federal Communications Commission (FCC) in managing broadcast spectrum administration and assignment.
Various endowment and grant programs have been transferred to the NTIA, including the Public Telecommunications Facilities Program from the Department of Health, Education, and Welfare. The NTIA has a laboratory for conducting applied research in telecommunications, located in Boulder, Colorado. See Federal Communications Commission, Institute for Telecommunication Sciences. http://www.ntia.doc.gov/

National Telecommunications Commission - Philippines NTC. An independent government agency created in the late 1970s, superseding the Board of Communications and the Telecommunications Control Bureau. The NTC performs regulatory and quasi-judicial functions for enforcing telecommunications in the Philippines. It is affiliated with the Department of Transportation and Communications; however, decisions are accountable directly to the Supreme Court. The NTC traces its origins to a regulatory office established when the Ship Radio Station Law was enacted in 1927.
In 1995, a Special Committee for Children was established to deal with the problem of sexual abuse and child prostitution using children as performers or models. Internet entrepreneurs were exploiting and promoting Filipino children as easily available for sexual tourism.
Due to the growing prevalence and concern over cable piracy, the NTC formed the Cable TV Piracy Task Force (CTP-TF) to cooperate with other agencies in educating the public about cable piracy with the goal of reducing cable broadcast signals theft.

National Telecommunications Damage Prevention Council NTDPC. Initially established in 1989 as the California Common Carrier Steering Committee, the group first addressed contractual issues related to railroad coastal cable ducts. Over time, the mandate of the group grew to support the protection of all below-surface telecommunications facilities. In the mid-1990s, the name was changed to The National Common Carriers Cable Hazard Prevention Committee and then changed to NTDPC to reflect the broader focus; a number of significant telecommunications carriers joined the organization. The NTDPC is a noncompetitive forum dedicated to protecting all telecommunications networks, regardless of ownership, with a focus on preventing damage to buried facilities forming part of the nation's communications infrastructure.

National Telephone Cooperative Association NTCA. A national trade association representing over 500 small and rural independent local exchange carriers (LECs), based in Arlington, Virginia. Small telephone cooperatives overseas are also becoming part of the organization, in addition to nonlocal exchange businesses (wireless, Internet, cable television, DBS). NTCA promotes the regulatory and educational needs of its members, and supports and coordinates a number of employee benefit programs. http://www.ntca.org/

Dialing Letter-Number Equivalents

Num.	Alphabetic Equivalent	
	U.S.	U.K.
1	---	---
2	ABC	ABC
3	DEF	DEF
4	GHI	GHI
5	JKL	JKL
6	MNO	MN
7	PRS	PRS
8	TUV	TUV
9	WXY	WXY
0	Operator	OQ

There is no Q in U.S. systems and no Z in U.S. or U.K. telephone systems. These letter-number equivalents are also found on ATMs.

Useful Units

Quantity	Units	Symbol
capacitance	farad	–
capacity	liter	L
current	ampere	A
force	newton, kilonewton	N, kN
frequency	cycle, hertz	Hz
magnetic flux	maxwell	–
length	meter, micron	m, μ
potential	volt	V
power flow	watt, kilowatt	W, kW
pressure	pascal, kilopascal	Pa, kPa
temperature	kelvin	K
velocity	meters/second	m/s, m/sec
volume	cubic meters	m^2
work, energy	joule, kilojoule	J, kJ

See also dictionary entries for illuminance, radiance, flux, candela, radian, and steradian.

Metric Prefixes/Values

Prefix	Symb.	Numerical Expression	Expon.
yotta-	Y	1,000,000,000,000,000,000,000,000	10^{24}
zetta-	Z	1,000,000,000,000,000,000,000	10^{21}
exa-	E	1,000,000,000,000,000,000	10^{18}
peta-	P	1,000,000,000,000,000	10^{15}
tera-	T	1,000,000,000,000	10^{12}
giga-	G	1,000,000,000	10^{9}
mega-	M	1,000,000	10^{6}
kilo-	k	1,000	10^{3}
hecto-	h	100	10^{2}
deca-	da	10	10^{1}
deci-	d	0.1	10^{-1}
centi-	c	0.01	10^{-2}
milli-	m	0.001	10^{-3}
micro-	m	0.000001	10^{-6}
nano-	n	0.000000001	10^{-9}
pico-	p	0.000000000001	10^{-12}
femto-	f	0.000000000000001	10^{-15}
atto-	a	0.000000000000000001	10^{-18}
zepto-	z	0.000000000000000000001	10^{-21}
yocto-	y	0.000000000000000000000001	10^{-24}

International Morse Code

Let.	Code	Let.	Code
A	•—	Á	•——•—
B	—•••	Ä	•—•—
C	—•—•	É	••—••
D	—••	Ñ	——•——
E	•	Ö	———•
F	••—•	Ü	••——
G	——•	1	•————
H	••••	2	••———
I	••	3	•••——
J	•———	4	••••—
K	—•—	5	•••••
L	•—••	6	—••••
M	——	7	——•••
N	—•	8	———••
O	———	9	————•
P	•——•	0	—————
Q	——•—	,	——••——
R	•—•	.	•—•—•—
S	•••	?	••——••
T	—	;	—•—•—
U	••—	:	———•••
V	•••—	/	—••—•
W	•——	'	•————•
X	—••—	!	—•—•——
Y	—•——	( )	—•——•—
Z	——••	—	••——••—

ASCII (Character and Control Codes)

Oct	Dec	Hex	Name		Oct	Dec	Hex	Name	
000	0	0x00	NUL		0100	64	0x40	@	commercial at sign
001	1	0x01	SOH		0101	65	0x41	A	
002	2	0x02	STX		0102	66	0x42	B	
003	3	0x03	ETX	Control-C	0103	67	0x43	C	
004	4	0x04	EOT		0104	68	0x44	D	
005	5	0x05	ENQ		0105	69	0x45	E	
006	6	0x06	ACK		0106	70	0x46	F	
007	7	0x07	BEL		0107	71	0x47	G	
010	8	0x08	BS	backspace	0110	72	0x48	H	
011	9	0x09	HT	tab	0111	73	0x49	I	
012	10	0x0a	LF	line feed, newline	0112	74	0x4a	J	
013	11	0x0b	VT		0113	75	0x4b	K	
014	12	0x0c	FF	form feed, NP	0114	76	0x4c	L	
015	13	0x0d	CR	carriage return	0115	77	0x4d	M	
016	14	0x0e	SO		0116	78	0x4e	N	
017	15	0x0f	SI		0117	79	0x4f	O	
020	16	0x10	DLE		0120	80	0x50	P	
021	17	0x11	DC1	XON, Control-Q	0121	81	0x51	Q	
022	18	0x12	DC2		0122	82	0x52	R	
023	19	0x13	DC3	XOFF, Control-S	0123	83	0x53	S	
024	20	0x14	DC4		0124	84	0x54	T	
025	21	0x15	NAK		0125	85	0x55	U	
026	22	0x16	SYN		0126	86	0x56	V	
027	23	0x17	ETB		0127	87	0x57	W	
030	24	0x18	CAN		0130	88	0x58	X	
031	25	0x19	EM		0131	89	0x59	Y	
032	26	0x1a	SUB		0132	90	0x5a	Z	
033	27	0x1b	ESC	escape	0133	91	0x5b	[	open square bracket
034	28	0x1c	F		0134	92	0x5c	\	backslash
035	29	0x1d	G		0135	93	0x5d	]	close square bracket
036	30	0x1e	RS		0136	94	0x5e	^	caret
037	31	0x1f	US		0137	95	0x5f	_	underscore
040	32	0x20	space	space	0140	96	0x60	`	back apostrophe
041	33	0x21	!	exclamation mark	0141	97	0x61	a	
042	34	0x22	"	double quote	0142	98	0x62	b	
043	35	0x23	#	number sign, hash	0143	99	0x63	c	
044	36	0x24	$	dollar	0144	100	0x64	d	
045	37	0x25	%	percent	0145	101	0x65	e	
046	38	0x26	&	ampersand	0146	102	0x66	f	
047	39	0x27	'	apostrophe/quote	0147	103	0x67	g	
050	40	0x28	(	open parenthesis	0150	104	0x68	h	
051	41	0x29	)	close parenthesis	0151	105	0x69	i	
052	42	0x2a	*	asterisk, star	0152	106	0x6a	j	
053	43	0x2b	+	plus	0153	107	0x6b	k	
054	44	0x2c	,	comma	0154	108	0x6c	l	
055	45	0x2d	-	minus	0155	109	0x6d	m	
056	46	0x2e	.	period, full stop	0156	110	0x6e	n	
057	47	0x2f	/	oblique stroke	0157	111	0x6f	o	
060	48	0x30	0	zero	0160	112	0x70	p	
061	49	0x31	1	one	0161	113	0x71	q	
062	50	0x32	2	two	0162	114	0x72	r	
063	51	0x33	3	three	0163	115	0x73	s	
064	52	0x34	4	four	0164	116	0x74	t	
065	53	0x35	5	five	0165	117	0x75	u	
066	54	0x36	6	six	0166	118	0x76	v	
067	55	0x37	7	seven	0167	119	0x77	w	
070	56	0x38	8	eight	0170	120	0x78	x	
071	57	0x39	9	nine	0171	121	0x79	y	
072	58	0x3a	:	colon	0172	122	0x7a	z	
073	59	0x3b	;	semicolon	0173	123	0x7b	{	open curly bracket
074	60	0x3c	<	less than	0174	124	0x7c	|	vertical bar, pipe
075	61	0x3d	=	equals	0175	125	0x7d	}	close curly bracket
076	62	0x3e	>	greater than	0176	126	0x7e	~	tilde
077	63	0x3f	?	question mark	0177	127	0x7f	delete	delete